PRINCIPLES OF PHYSIOLOGY,

GENERAL AND COMPARATIVE

PRINCIPLES OF PHYSIOLOGY,
GENERAL AND COMPARATIVE.

BY

WILLIAM B. CARPENTER

GYAN

Published by

Gyan Publishing House
5, Ansari Road
Daryaganj, New Delhi-110002
Phone: 011-47034999, 9811692060
E-mail: books@gyanbooks.com

Distribution Network
gyanbooks.com
India, USA, Canada, UK, Australia, France

ISBN: 978-93-6280-104-3 (PB)

First Published, 1839

2nd Impression 2023

Printed at: Gyan Press, Delhi.

PRINCIPLES OF PHYSIOLOGY, GENERAL AND COMPARATIVE
Author: WILLIAM B. CARPENTER

PRINCIPLES OF PHYSIOLOGY,

GENERAL AND COMPARATIVE.

PRINCIPLES OF PHYSIOLOGY,

GENERAL AND COMPARATIVE.

WITH THREE HUNDRED AND TWENTY-ONE WOOD-ENGRAVINGS.

BY

WILLIAM B. CARPENTER, M.D., F.R.S., F.G.S.,

EXAMINER IN PHYSIOLOGY AND COMPARATIVE ANATOMY
IN THE UNIVERSITY OF LONDON; PROFESSOR OF MEDICAL JURISPRUDENCE
IN UNIVERSITY COLLEGE, ETC. ETC.

MDCCCLI.

TO

SIR JOHN F. W. HERSCHEL, BART.,

K.H. F.R.S. L. AND E. ETC.,

THIS VOLUME IS

MOST RESPECTFULLY DEDICATED,

AS A TRIBUTE DUE ALIKE

TO

HIS HIGH SCIENTIFIC ATTAINMENTS,

AND MORAL WORTH,

AND AS AN EXPRESSION OF GRATITUDE FOR

THE BENEFIT DERIVED FROM HIS

"DISCOURSE ON THE STUDY OF NATURAL PHILOSOPHY,"

BY

THE AUTHOR.

PREFACE.

"SCIENCE IS THE KNOWLEDGE OF MANY, ORDERLY AND METHODICALLY DIGESTED AND ARRANGED, SO AS TO BECOME ATTAINABLE BY ONE."—*Sir John F. W. Herschel.*

THE Second Edition of his "Principles of General and Comparative Physiology" having been for some time out of print, the Author commenced, about two years ago, the task of preparing for the re-appearance of that work; in the expectation that, although much addition and modification would be requisite, in order to embody in it the most important of those discoveries in which the Physiological researches of recent times have been so fertile, yet that it would be found possible to interweave these with the original matter, in such a mode as to preserve the unity of the whole. On proceeding with his task, however, he soon perceived that it would be impossible to do it the justice he desired, without putting aside his former work altogether, and entirely reconstructing the Treatise; only employing the materials of the original, where they might chance to be appropriate to his purpose. And thus it happens that the present work may be considered as essentially a new one; since, out of the 1080 pages of which it consists, not above 150, or less than one-seventh, belong to the previous Edition. The Author hopes that this fact may be admitted as evidence of his desire to spare no pains, in order to render his Treatise a faithful exposition of the present state of Physiological Science; and that it will be deemed a sufficient justification of the slight change, by which he considers that the Title has been made to represent more appropriately the present character of the work.

It has been with considerable satisfaction, however, that, notwithstanding this almost complete renewal of the *materials* of his Treatise, the Author has found no important modification to be requisite in its *plan;* for, in consequence of his having originally selected the simplest Vegetable organism, consisting of a single cell, as the starting-point from which to trace out the essential character of each Physiological action, through its successive complications in the higher forms of Vegetable and Animal life, the 'cell-doctrine' of Schleiden and Schwann, perfected and extended by the labours of the numerous observers who have subsequently prosecuted the same line of research, has fallen (so to speak) into its appropriate place, and now serves as a complete confirmation of the validity of the principles which the Author had previously propounded.

In regard to certain points on which his opinions have undergone modification, the Author refers with satisfaction to the following passage in the Preface to his former edition :—" *Truth* is his only object; and, even if his own doctrines should be overthrown by more extended researches, he will rejoice in their demolition, as he would in that of any other error. The character of the true philosopher as described by Schiller,—*one who has always loved truth better than his system,*—will ever, he trusts, be the goal of his intellectual ambition."

In attempting to embody in a Systematic Treatise the general aspect of Physiology or any other Science of like comprehensiveness, it will be obvious that an Author, however extensive his own range of acquirement, must largely avail himself of the labours of others; and that the scientific character of such a treatise must depend, not so much on the amount of original matter it may contain, as on the degree in which "the knowledge of many" has been "orderly and methodically digested and arranged, so as to become attainable by one." It is by this standard that the Author desires his work to be tried; and he cheerfully leaves the verdict to the judgment of those, who are qualified by their own knowledge of the subject to pronounce it. He feels it due to himself, however, to state that he has devoted considerable time and attention to the verification of the statements of other observers, especially on points under dispute,—a kind of labour which is but little appreciated by those, who contemptuously designate works like the present as 'mere compilations ;' and that a large amount of materials, drawn from his own original enquiries, are scattered through the work. It would have been easy for him to bring these last into greater prominence, had he been so disposed; but as his constant aim has been, to work out his general plan harmoniously and methodically, rather than to force any one portion of it into undue prominence, he has generally preferred to allow his own contributions to pass undistinguished, rather than to be continually obtruding his originality upon the attention of his readers.—Since, however, he has sometimes found himself referred to, as the authority for views or statements whose merit or demerit legitimately belongs to others, he thinks it well here to specify the most important of the facts and doctrines which he regards as more particularly his own.

I. The *mutual connection of the Vital Forces, and their relation to the Physical* (CHAP. III., Sect. 1, and elsewhere): this doctrine is more fully developed in a paper on "The Mutual Relations of the Vital and Physical Forces," in the Philosophical Transactions for 1850.

II. The general doctrine that *the truly Vital operations of the Animal as well as the Vegetable organism are performed by the agency of untransformed cells* (§ 135); which was first developed in an Essay "On the Origin and Functions of Cells," published in the Brit. and For. Med. Rev., vol. xv., 1843.

III. The *Organic Structure of the Shells of Mollusca, Echinodermata, and Crustacea* (§§ 195-199); of which a full account is contained in the Reports of the British Association for 1844 and 1847.

IV. The application of Van Bär's Law of Development *from the General to the Special,* to the interpretation of the succession of Organic forms presented in Geological time (§ 345); here first brought forwards.

V. The relation between the two methods of Reproduction, that by *gemmation,* and that by *sexual union* (§§ 647, 648), with the application of this doctrine to the phenomena of the so-called 'Alternation of Generations;' first developed in the Brit. and For. Med.-Chir. Rev., Jan. 1848, and Oct. 1849.

VI. The relation between the different methods of *Sexual Reproduction* in Plants (§ 650); first developed in the Brit. and For. Med.-Chir. Rev., Oct. 1849.

VII. The application of the doctrine of *Reflex Action* to the *Nervous System of Invertebrata,* especially Articulated Animals (CHAP. XX.); first developed in the Author's "Prize Thesis," published in 1839.

VIII. The Functional relations of the *Sensory Ganglia* to the Spinal Cord on the one hand, and to the Cerebral Hemispheres on the other (CHAP. XX.); first sketched out in the Edinb. Med. and Surg. Journ., No. cxxxii. (1837), and more fully developed in the Brit. and For. Med. Rev., vol. xxii. (1846), and in the Brit. and For. Med.-Chir. Rev., Jan. 1850.

In the selection of his materials, the Author has endeavoured to avail himself of the best and most recent information he could procure upon each department of the subject; it is scarcely to be expected, however, that he should be equally well-informed upon every point; and those who have followed particular departments into detail, will doubtless find scope for criticism in what they may regard as deficiencies, or even as errors.* Here, again, the Author must beg that his work may be estimated by its *general* merits; and rather by what it *does,* than by what it *does not* contain. It would have been far easier to expand it by mere compilation to twice its present dimensions, than it has been found to compress the collected materials within the space they even now occupy.

It has been the Author's endeavour, wherever practicable, to draw the materials, both for his text and its illustrations, *direct* from original Treatises and Monographs; and thus to avoid the errors which too frequently arise from second-hand transmission. To have attempted, however, to assign each individual fact to its original discoverer, each doctrine to its

* As the printing of the work has been necessarily spread over a space of a year and a half, faults of omission, in regard to recent discoveries, may, it is probable, be especially noticed in the earlier portions; but the Author has endeavoured to correct, in the latter part, such as he deemed of most importance.

first enunciator, would have at least doubled the bulk of a volume which has already far outgrown the dimensions originally intended; and while most desirous to avoid taking credit for what is not his own, the Author has felt himself compelled to limit his references, for the most part, to those *new* facts and doctrines, which cannot be yet said to have become part of the common stock of Physiological Science. The Illustrations not his own are referred to their originals, in the list at the commencement of the volume; and this list, together with the occasional foot-notes, will serve to indicate the principal sources on which the Author has relied for his information, as to those points into which he has not personally examined. From the mode in which these materials are combined and arranged, he believes that he has frequently been able to develope relations between facts and phenomena, which, in the isolated form in which they were originally stated, had only that value which all statements of truth possess; and that he has thus been able to impart to them a new and unexpected significance.

The Author thinks it but due to himself to state, that as the almost constant occupation of his time in other indispensable avocations, leaves him much less opportunity than he would desire for the prosecution of those original researches in which he feels the greatest satisfaction, so he trusts that the same plea may in some degree avail him, in regard to any deficiencies which may be found in the present work. The whole of his disposable time has been devoted to it, for nearly two years, not merely without the anticipation of the slightest pecuniary reward, but under the certain loss involved in the relinquishment of other literary engagements of a remunerative character; and the sale of the entire edition will not do more than remunerate his liberal Publisher, for the very large outlay which has been incurred in the production of the work, and more especially for the beautiful series of illustrations with which it is embellished, most of them having been executed expressly for it by Mr. Bagg.

The Author cannot bring his task to a conclusion, without expressing his particular obligations to Prof. Owen, not merely for the readiness which he has at all times manifested to afford him such information as he might seek; but also for his kindness in revising those portions of Chapter VII., in which an outline view of the 'Vertebral Theory' has been embodied. This outline may, it is hoped, serve to convey an idea of this beautiful system to those who are not disposed to pursue it into detail; and may advantageously prepare the professional student for its application to the Osteology of Man.

REGENT'S PARK, LONDON,
May 1, 1851.

TABLE OF CONTENTS.

BOOK I.

GENERAL PHYSIOLOGY.

CHAPTER I.

CHAPTER II.

CHAPTER III.

CHAPTER IV.

CHAPTER V.

CHAPTER VI.

GENERAL VIEW OF THE VEGETABLE KINGDOM.

CHAPTER VII.

GENERAL VIEW OF THE ANIMAL KINGDOM.

CHAPTER VIII.

BOOK II.

SPECIAL AND COMPARATIVE PHYSIOLOGY.

CHAPTER IX.

CHAPTER X.

OF ALIMENT, ITS INGESTION AND PREPARATION.

CHAPTER XI.

OF THE ABSORPTION OF NUTRITIVE AND OTHER MATTERS.

CHAPTER XII.

OF THE CIRCULATION OF NUTRITIVE FLUID.

CHAPTER XIII.

OF RESPIRATION.

CHAPTER XIV.

OF THE EXHALATION OF AQUEOUS VAPOUR.

CHAPTER XV.

OF NUTRITION.

CHAPTER XVI.

OF SECRETION.

CHAPTER XVII.

OF THE EVOLUTION OF LIGHT, HEAT, AND ELECTRICITY.

CHAPTER XVIII.

OF THE REPRODUCTION OF ORGANISED BEINGS.

CHAPTER XIX.

CHAPTER XX.

OF THE FUNCTIONS OF THE NERVOUS SYSTEM.

CHAPTER XXI.

OF SENSATION AND THE ORGANS OF THE SENSES.

CHAPTER XXII.

LIST OF ILLUSTRATIONS.

TREATISES AND MEMOIRS REFERRED TO IN THE LIST OF ILLUSTRATIONS.

I. *Alder* and *Hancock*, Monograph of the British Nudibranchiate Mollusca. (Published by the Ray Society.)

II. *Bagge*, De Evolutione Strongyli et Ascaridis.

III. *Berres*, Anatomie der Mikroskopischen Gebilde des Menschlichen Körpers.

IV. *Bischoff*, Entwickelungs-geschichte des Hunde-Eies.

V. *Blanchard*, Observations sur le Système Nerveux des Mollusques Acéphales Testacés. (Ann. des Sci. Nat., Juin, 1845.)

VI. ——— Recherches sur l'Organisation des Vers. (Ann. des Sci. Nat., 1847-50 and XVII.)

VII. *Bojanus*, Anatome Testudinis Europeæ.

VIII. *Bowman*, On the Minute Structure and Movements of Voluntary Muscle. (Phil. Trans., 1840.)

IX. ——— On the Structure and Use of the Malpighian Bodies of the Kidney. (Phil. Trans., 1842.)

X. ——— (and Todd), Physiological Anatomy and Physiology of Man.

XI. *Brongniart*, Sur la Structure, etc. des Feuilles. (Ann. des Sci. Nat., 1830.)

XII. *Carter*, On the Freshwater Sponges of Bombay. (Ann. of Nat. Hist., Aug. 1849.)

XIII. *Chenu*, Leçons Elémentaires d'Histoire Naturelle. Paris, 1847.

XIV. *Coste*, Histoire Générale et Particulière du Développement des Corps Organisés.

XV. *Couch*, On the Metamorphosis of the Decapod Crustaceans. (Report of Cornwall Polytechnic Society for 1843.)

XVI. *Cuvier*, Mémoires sur les Mollusques.

XVII. ——— Règne Animal. (Edition accompagnée de Planches gravées, par une réunion de Disciples de Cuvier.)

XVIII. *Dalyell*, Rare and Remarkable Animals of Scotland.

XIX. *Dana*, Structure and Classification of Zoophytes.

XX. *Donné*, Cours de Microscopie (Atlas).

XXI. *Duméril* (et Bibron) Histoire Naturelle compléte des Reptiles.

XXII. *Ehrenberg*, Die Infusionsthierchen.

XXIII. ——— Des Leucthen des meeres. (Abhaldlungen der Königlichen Akademie der Wissenschaften zu Berlin, 1835.)

XXIV. ——— On the Animals of the Chalk formation still found in a living state. (Taylor's Scientific Memoirs, Vol. III.)

XXV. *Flourens*, Mémoires d'Anatomie et de Physiologie Comparées.

XXVI. *Forbes*, On Beröe pileus. (Ann. of Nat. Hist., Vol. II.)

XXVII. *Gerber*, Elements of the General and Minute Anatomy of Man and the Mammalia.

XXVIII. *Goodsir*, Anatomical and Pathological Observations.
XXIX. *Hartig*, in Mulder's Chemistry of Animal and Vegetable Physiology.
XXX. *Hassall*, History of the British Freshwater Algæ.
XXXI. *Henle*, Allgemeine Anatomie.
XXXII. *Henslow*, Principles of Descriptive and Physiological Botany.
XXXIII. *Hoffmeister*, Sur la Fécondation chez les Œnothérées. (Ann. des Sci. Nat., Févr., 1848.)
XXXIV. *Hollard*, Sur l'Organisation des Vélelles. (Ann. des Sci. Nat., Avril, 1845.)
XXXV. *Jussieu*, Cours Elémentaire de Botanique.
XXXVI. *Kiernan*, On the Structure of the Liver. (Phil. Trans., 1835.)
XXXVII. *Kölliker*, Sur le Développement des Tissus chez les Batraciens. (Ann. des Sci. Nat., Août, 1846.)
XXXVIII. *Kützing*, Phycologia generalis.
XXXIX. *Lebert*, Physiologie Pathologique.
XL. *Leszczyc-Luminski*, Zur Entwickelungs-Geschichte der Farrnkräuter.
XLI. *Leidy*, On the Comparative Structure of the Liver, and on the Intimate Structure and History of the Articular Cartilages. (Amer. Journ. of Med. Sci., Jan., 1848, and April, 1849).
XLII. *Lesson*, Histoire Naturelle des Acalèphes.
XLIII *Leuret*, Anatomie Comparée du Système Nerveux.
XLIV. *Mandl*, Manuel d'Anatomie Générale.
XLV. *Milne-Edwards*, Recherches sur les Polypes.
XLVI. ———— Observations sur les Ascidies Composées des Côtes de la Manche.
XLVII. ———— Observations sur la Circulation chez les Mollusques. (Ann. des Sci. Nat., Juin, 1845, et Juillet, 1847.)
XLVIII. ———— Observations sur la Circulation chez les Annélides. (Ann. des Sci. Nat., 2ième Sér. Tom. x.)
XLIX. ———— Observations sur le Développement des Annélides. (Ann. des Sci. Nat., Mars, 1845.)
L. ———— Elémens de Zoologie. (See also XVII.)
LI. *Mirbel*, Sur la Structure et la Développement de la Marchantia polymorpha. (Nouv. Ann. du Musée, Tom. III.)
LII. *Mohl*, Vermischte Schriften botanischen Inhalts.
LIII. *Müller*, De Glandularum secernentium Structurâ penitiori.
LIV. ——— Ueber die Larven und die Metamorphose der Echinodermen. (Abhaldlungen der Koniglichen Akademie der Wissenschaften zu Berlin, 1846 und 1848.)
LV. *Newport*, Memoirs on the Structure of Insects, Myriapoda, and Macrourous Arachnida (Phil. Trans. 1839—1843.)
LVI. *Owen*, Odontography.
LVII. ——— Lectures on Comparative Anatomy.
LVIII. ——— On the Homologies of the Vertebrate Skeleton.
LIX. *Quatrefages*, Mémoire sur l'Organisation des Pycnogonides (Ann. des Sci. Nat., Août, 1845.)
LX. ——— Mémoire sur quelques Planariées marines. (Ann. des Sci., Sept. 1845.)
LXI. *Ralfs*, Monograph of the British Desmideæ.
LXII. *Sars*, Fauna littoralis Norvegiæ.
LXIII. *Schleiden*, Principles of Scientific Botany.
LXIV. ——— The Plant, a Biography.
LXV. *Schwann*, Mikroskopische Untersuchungen über die Uebereinstimmung in der Structur und dem Wachsthum der Thiere und Pflanzen.
LXVI. *Sharpey*, Art. Cilia (Cyclop. of Anal. and Physiol.)

LXVII. *Slack*, On the Elementary Tissues of Plants, and on Vegetable Circulation. (Trans. of Soc. of Arts, Vol. XXXIX.)

LXVIII. *Solly*, The Human Brain, its Structure, Physiology, and Diseases.

LXIX. *Strauss-Durckheim*, Considérations Générales sur l'Anatomie comparée des Animaux Articulés.

LXX. *Teale*, Anatomy of Actinia Coriacea. (Trans. of Phil. and Lit. Soc. of Leeds, Vol. I.)

LXXI. *Thwaites*, On the Conjugation of the Diatomaceæ. (Ann. of Nat. Hist., 1847 and 1848.)

LXXII. *Tiedemann*, Anatomie der Rohrenholothurie, &c.

LXXIII. *Toynbee*, On the Organisation and Nutrition of the Non-vascular Animal Tissues. (Phil. Trans., 1841.)

LXXIV. *Trembley*, Mémoires pour servir à l'Histoire d'un genre de Polype d'Eau douce.

LXXV. *Unger*, Recherches sur l'Achlya prolifera. (Ann. des Sci. Nat., Juillet, 1844.)

LXXVI. *Van Beneden*, Mémoire sur les Campanulaires de la Côte d'Ostende. (Mém. de l'Acad. Roy. de Bruxelles, Tom. XVII.)

LXXVII. ———— Recherches sur les Bryozoaires de la Côte d'Ostende. (Mem. de l'Acad. Roy. de Bruxelles, Tom. XVIII.)

LXXVIII. *Vogt*, Embryologie des Salmones.

LXXIX. —— Recherches sur l'Embryologie des Gasteropodes. (Ann. des Sci. Nat., Juillet, 1846.)

LXXX. *Wagner*, Icones Physiologicæ.

LXXXI. *Willis*, On the organs of Voice. (Trans. of Cambridge Phil. Soc., Vol. IV.)

LXXXII. *Wilson*, Anatomist's Vade-Mecum.

BOOK I.

GENERAL PHYSIOLOGY.

CHAPTER I.

ON THE NATURE AND OBJECTS OF THE SCIENCE OF PHYSIOLOGY.

1. THE most superficial observer can scarcely fail to recognize, in the world around him, many evident distinctions between *living beings* and *inanimate bodies.* A scientific comparison is not required to discover such obvious differences in their external Form, as are usually sufficient for their ready discrimination. A very cursory examination of the internal Structure which they respectively present, adds many new indications of their fundamental diversity. And daily experience teaches, scarcely less satisfactorily than the refined analysis of the laboratory, that there is a contrast no less remarkable in their Chemical Composition.

2. But it is in their Mode of Existence, that we find this disagreement the most complete and the most obvious. In the kingdom of Inorganic matter, Permanence is the rule, change the exception. The particles of the Mineral mass remain coherent, and their primal constitution (whether elementary or compound) is retained, so long as no external forces overcome by their superior potency the existing attractions and affinities. The "everlasting hills" which rear their summits to the clouds, unaffected by the lapse of ages which have rolled by since their first elevation, bear testimony through countless generations to the vastness of the forces which first upheaved them—perennial monuments of Nature's power; and even the edifices erected by the hand of Man frequently convey to a remote posterity the attestation of the grandeur of a people, which may have subsequently relapsed into barbarism and insignificance, or which may have altogether disappeared from among the nations of the earth. But in the domain of Life it is altogether different. Here, Change is the rule, permanence the exception; and this change does not so much proceed from the agency of external forces, as from those inherent in and peculiar to the body itself. To be born—to grow—to arrive at maturity—to decline—to die—to decay, is the sum of the history of every being that lives, from the highest to the lowest, from the most transient to that of most protracted duration. "Our life," to use the expressive simile of the sacred writer, "is as a vapour, which appeareth for a little time, and then vanisheth away."

3. Yet amidst this constant change and succession of individuals, we observe the form and character first impressed upon each race by the Creator of all, uninterruptedly transmitted from parent to offspring; so that, as "one generation passeth away," "another cometh" similar to it in structure and actions, and, like it, destined to give place in due time to a successor. The lofty stem of the Date is even now hailed from afar

by the thirsty traveller in the Red Sea desert, as a welcome indication of the proximity of water, just as when the palm-trees and wells of Elim refreshed the weary Israelites on their route to Sinai. The Mistletoe flourishes on the oak of our own forests, not otherwise than when it was an object of superstitious veneration in the hallowed groves of our Druidical ancestors. The Bee still constructs her comb with the same wonderful regularity, and stores it with the same materials, as when her beautiful works attracted the notice of the poets and philosophers of classic ages. And Man, however various his degree of civilization, however great the development of his physical powers, however lofty the elevation of his spiritual nature, is yet born in the same dependent state, with the same dormant faculties of body and mind, as the first offspring of our original parents.

4. And when we examine more attentively into the succession of changes exhibited by each individual, we see that the history of the fabric as a whole is repeated in every portion of it. Each one of that assemblage of diverse components, of which the higher forms of living beings—whether Plants or Animals—are made up, has its own independent powers of growth and development, its own peculiar endowments when fully evolved, and its own limit to the term of their exercise ; and when this has been completed, it is either annihilated by death and decay, or it passes into the inactive condition of inorganic bodies, to which it becomes assimilated in structure and composition. Hence there is a constant change in the materials of which any living being is composed, even while its structure and constitution remain essentially the same ; and as the continuity of the race depends on the replacement of the departing generation by a young and vigorous progeny, so is the maintenance of the life of the individual dependent on the continual replacement of its decaying components by newly-developed tissue. Notwithstanding the constancy of this change, however, the form, structure, and endowments of the entire fabric are preserved unaltered, except in so far as they are modified in accordance with the progress of its own life ; and its individuality is maintained continuously, from its first origin to its final decay, although not a single particle may remain among its constituents during any considerable portion of its term of existence. There is nothing, perhaps, in the inorganic world, which affords so apt an illustration of this marvellous truth, as the continuity of the mighty cataract, without the least perceptible change in the form presented by its descending torrent, whilst every instant witnesses the renewal of its component particles,—the delivery of those which have performed their descent into the gulf below, and their replacement by those perpetually supplied from the source above.

5. Even a cursory glance, then, at the ever-varying conditions of the animated world, serves to impress us with the idea of regularity and arrangement ; and the more attentively we investigate the succession of changes which it exhibits, the more firm becomes our assurance, that all these changes take place according to a definite and stable *plan ;* or (to use the current phraseology of science) that they are governed by *laws* *

* The only legitimate use, in my apprehension, of the term *Law of Nature*, is as an expression of that uniformity observed by the philosopher in the phenomena of the Universe, which indicates the stability of the original plan. Thus when we speak of the "law of

as determinate as those which regulate the movements of the heavenly bodies, or control those molecular changes in the inorganic world of which the Physicist and the Chemist take cognizance. To arrive at the knowledge of these laws, and, having attained them, to trace their application to all the countless variety of phenomena presented by the myriads of living beings whose beautiful forms people this globe, is the object of the science of Physiology, using that term in its most extended sense, to which the designation *Biology* is perhaps rather applicable. That the most advantageous plan of studying it is that inductive method which has been successful in other sciences, will not perhaps now be disputed : yet the prevalence of a contrary system has long retarded its progress ; and it is only within a recent period that the ends to be attained have been generally understood, and the most satisfactory means of pursuing them fully determined.

6. In this, as in the Physical Sciences, the first object of the philosophic enquirer is to collect a body of facts, from the comparison of which the general truths common to all may be elicited. Now the facts which the observation of living beings brings under our notice, are obviously of two kinds ;—one class having reference to their *structure*, the other to their actions or *functions*. The investigation and comparison of the former class of particulars, is the object of the science of *Anatomy ;* whilst by the

gravitation," we simply refer to the general fact, that all masses of matter attract each other in a certain determinate ratio, from which we never witness any departure. But it must be remembered that all our so-called "laws" are but *human* expressions, based upon a very limited amount of observation, and liable to be modified, therefore, by the extension of our knowledge of the entire scheme of Creation; and to suppose that the Deity is in any way tied down to those "laws" which Man supposes that he has discovered, is equivalent to making Man rather than God the Legislator of the Universe.—The true idea appears to me to be this. The material universe was constructed in the first instance on a plan so perfect, that no departure from that plan has been required for the maintenance of its existence, and for the welfare of its sentient inhabitants. (To this statement, the occurrence of those apparent departures from the "order of nature" which we term Miracles, does not constitute a real exception ; for every believer in the Divine Omniscience must admit that they were as much foreseen and provided for in the original plan, in order to produce a special influence upon the mind of Man, as were those occurrences which impress him with the idea of uniformity and fixity, and which thus give rise to his notion of immutable laws.) But whilst we reject the idea of continual *interposition*, or change of plan, as inconsistent alike with our belief in the Divine perfections, and with the results of our observation of His agency, we cannot do otherwise, in my apprehension, than recognize his *constant sustaining action* as a necessary condition of the continued existence of the Universe. For if we are led to believe, by the indications of a Designing Mind, which are so abundantly presented to every one who seeks for them, that the *first existence* of matter was but an expression of an Intelligent Will, to what can its *continued existence* be attributed, but to the sustaining action of that same Will? Reason and Revelation here seem to me in perfect harmony. Further, I can conceive of no agency intermediate between an Infinite Deity and His works. Either all the phenomena of the material universe are the immediate results of His will, or they have no dependence upon it whatsoever. In the former case, the "laws of Nature" are simply expressions of what *we* know or imagine as to the mode of operation of that Will. In the latter they are nothing else than concise statements of comprehensive truths established by observation. In neither case can the "laws" be conceived to have any real force or agency in themselves ; such as is attributed to them by those who speak of the Deity as framing laws for the Universe, and then leaving them to their own independent operation. This mode of viewing the subject has probably originated from the supposed analogy of human legislatures ; but this analogy, when carefully examined, not only fails to support such a doctrine, but sustains that for which I have argued. For a human "law" is nothing else than an expression of the will of the governing power; and its action upon the community entirely depends upon the constant though silent operation of that will. Let the governing power be overthrown by political convulsion, and the laws no longer exert any controlling agency.

collection and generalisation of the latter, the science of *Physiology* (strictly so called) is built up.

7. The obstacles which interpose themselves to the prosecution of these sciences, result more from that difficulty in the ascertainment of facts and the observation of phenomena, which is occasioned by the peculiar conditions of living beings, than from any incapability on the part of these facts and phenomena to be comprehended within laws as stable and as definite as those of the physical sciences. Thus, although the structure of the human body has been carefully and minutely examined by so many thousands of Anatomists, many points are still uncertain, and much still remains to be discovered. Yet this structure is but one of those groups of *instances* (the Baconian term for *phenomena*) that must be collected from the many hundred-thousand species of Plants and Animals, which the Naturalist knows to exist on the present surface of the globe, or whose remains are disinterred from its depths by the Geologist, before we can form anything more than a limited and imperfect idea of the wondrous *plan*, on which it has been thus peopled with living inhabitants.

8. The difficulties that present themselves in the observation of the facts which it is the object of the science of Anatomy to ascertain and generalise, are as nothing to those which beset the path of the Physiological enquirer, who has to study the changes which all classes of living beings perform and undergo during the whole period of their existence. The sum of all the phenomena which occur during the Life of a single organised structure, and which result from the actions of that structure, is, in like manner, to be regarded as a collection of *facts*; of which each must be stated in a separate and concise form, before it can be made the subject of any general expression, founded upon the comparison of similar facts derived from the study of other living beings. Now the great difficulty in physiological investigation results from the complexity of the combinations in which vital phenomena present themselves, and from their dependence upon one another to a degree that almost entirely precludes their separate examination. Were we able to ascertain the changes which take place in the interior of the living body, with the same ease that the Astronomer watches the motions of a planet, or the Chemist observes the formation of a precipitate,—the very multiplicity of these changes, and the variety of the conditions under which they occur, would be of essential service in the determination of their laws, instead of being, as at present, sources of doubt and embarrassment. The Chemist, when desirous of establishing to which of the ingredients in a given mixture a particular effect is due, places each separately in the conditions required to produce the result: but the Physiologist finds that the attempt to insulate any one organ, and to reduce the changes performed by it to definite experimental investigation, necessarily destroys, or considerably alters, those very conditions under which alone its functions can be normally performed. Take away an important and essential part of a living being, and it ceases to exist as such; it no longer exhibits even a trace of those properties which it is our object to examine; and its elements remain subject only to the common laws of matter. We cannot, like the fabled Prometheus of old, breathe into the lifeless clay the animating fire; we cannot, by a judicious and skilful arrangement of those elements, combine them into new and artificial forms so as to produce new and unexpected phenomena; and

almost all our knowledge of the laws of Life must therefore be derived from *observation* only. *Experiment* can conduct us very little further in this enquiry, than the determination of the dependence of the functions upon one another, and upon the external agents, Heat, Light, &c., by the action of which upon the organism the phenomena of Life are produced. But a judicious and careful system of *Observation* will almost supply the place of experiment; for the ever-varying forms of organised beings by which we are surrounded, and the constantly-changing conditions in which they exist, present us with such numerous and different combinations of causes and effects, that it must be the fault of our mode of study if we do not arrive at some tolerably definite conclusions as to their mutual relations. To use the language of Cuvier with some extension, all the different forms of living beings may be regarded as "so many kinds of experiments ready prepared by Nature, who adds to or deducts from each of them different parts, just as we might wish to do in our laboratories, showing us herself at the same time their various results."

9. From such considerations as these, it will be evident that the laws of Life can only be searched for, with a probability of success, by investigating their operations *wherever* presented to us; and that the study of Physiology can only be scientifically prosecuted (the attainment of these laws being regarded as its ultimate object), by embracing within its range the examination of the phenomena exhibited by *all* classes of living beings. It is a great mistake to suppose that there is anything fundamentally different in the character of the vital operations, as performed in the Animal and in the Vegetable structures, or in the simpler and more complicated organisms of either kingdom. An enlarged view of their functions, not based upon the observation of their conditions in one or two instances only, but derived from an extended examination of their performance wherever manifested, will recognise an essential conformity throughout, wherever those which are really analogous are compared. There is an obvious advantage, therefore, in commencing the study of Physiology by an enquiry into the simplest manifestations of each of those functions, which in the higher organisms are so complicated in their nature. From no such enquiry should the consideration of the Vegetable Kingdom be excluded; for those vital functions which are performed by Plants in common with Animals, are presented by the former in a state of greater simplification than is ever exhibited by the latter; since all the changes necessary to the support of the individual and the continuance of the species, are performed without the influence or interference of those powers which are possessed in a greater or less degree by the whole Animal kingdom. Hence the Physiologist may advantageously resort to the study of Vegetable life, for the explanation of many of the proximate causes of those phenomena, which are complicated in the higher forms of organised beings by so great a variety of secondary influences. In the pursuit of his science on this plan, therefore, the student will learn what are the essential conditions of life; he will see the changes indispensable to its support manifested in their simplest circumstances; he will be able to ascertain what structures are necessary to their performance, and what additions and modifications these may undergo to suit the various purposes of their existence; and thus he will be saved the necessity of unlearning many erroneous notions, which he would almost unavoidably imbibe from the study of the complex

phenomena exhibited by the living human organism, if prosecuted without such guidance. Moreover, in those departments of Physiology which are capable of being elucidated by experiment, recourse may generally be had with advantage to the lower classes of Animals, and to Plants; since that bond of union which links together, in the higher Animals, all the changes concerned in the maintenance of the vitality of the system, is in them much less close and decided; so that particular organs may be insulated, and the study of the conditions of their actions prosecuted to a much greater extent.

10. The mass of facts which is gradually being accumulated relative to the structure of living beings, and to the vital phenomena they exhibit, must be classified and arranged, before they can become subservient to the purposes of science; and this object is accomplished in different ways, according to the nature of the laws of which the philosophical enquirer is in search. Thus the Anatomist and Physiologist, whose object it is to discover the peculiarities of organised structures, their adaptations to particular uses, and the conditions of the functional changes to which they are subservient, *analyse*, as it were, the group of facts which each living being exhibits (§ 7, 8): pursuing their enquiries through an extensive range of objects, they classify these individual results according to their similarity with each other, and their obvious tendency to the same end; or, to speak in less abstract language, they compare the *individual* organs and functions through all the forms of animated beings in which they are manifested: and their final aim is the attainment of some comprehensive formulæ, which shall express the essential characters of each of these groups of phenomena and of the sum of the whole. The object of the Naturalist, on the other hand, is to ascertain the plan according to which the *combination* of the separate organs into living fabrics, and their adaptation to different modes of existence have been effected: he, therefore, viewing each organism in its *totality*, arranges similarly formed beings into the same group, placing as the character common to the whole the points in which they agree, and leaving the subordinate differences to be added to this common character, in order to express the qualities of an individual. This classification is but a step towards the discovery of those general laws,—expressing the manner in which the organs are combined and adapted to each other, the relative development or simplicity of each, the modifications which the typical form of each group may undergo according to the circumstances in which the being is to be placed, and various other conditions of its formation,—which it is the ultimate object of the Naturalist to ascertain; and any mode or system of classification which he may adopt, is valuable in proportion as it keeps the determination of these laws in view, and facilitates the accumulation of the knowledge by which they may be reached. The connexion between these two branches of investigation is so intimate, that neither can be pursued with any probability of success, without a considerable knowledge of the data and principles upon which the other is founded; and he will evidently be the most likely to arrive at the discovery of important general truths in either, who includes the whole of the phenomena of Life in one extensive survey. The Physiologist refers to the Naturalist for instances in which a function is performed on the same general plan, but under a great variety of circumstances, as manifested by the adaptation of the structure of the organ to the medium

of existence, (*e. g.* the formation of the respiratory membrane into lungs or gills); whilst the Naturalist refers to the Physiologist to assist him, by the examination of the history of the development of an organ, in determining its real character, to which the consideration of its form and structure alone might not lead him. It has been by this conjoint kind of investigation that the most important of the recent advances have been effected, towards a more intimate acquaintance with the plan, on which it has pleased the Creator to construct and maintain the Organised World.

11. Although the special object of the present treatise is the exposition of purely Physiological principles, it seems desirable to preface these by such an outline of the general structure and arrangement of the organs on which the phenomena of Life are dependent, as may render subsequent details respecting their functions more intelligible. We shall first consider, therefore, what there is peculiar in the chemical composition and physical arrangement of the particles of which organised structures are composed, and in the forms which such fabrics present. The principal varieties of the *primary* or *elementary tissues*, of which the more complex organs of Plants and Animals are constructed, will next be described and compared with one another. And thirdly, the general characters of the principal groups in each of the animated kingdoms of Nature will be pointed out, the mode in which their individual organs are arranged and combined will be explained, and their relative positions assigned. Although such knowledge is readily accessible to the student of Natural History, the *embarras des richesses* may not be a little perplexing to such as seek only that extent of it, which will enable them to enter upon the study of Physiological Science, without being immediately checked by the want of this kind of information.

CHAPTER II.

OF THE GENERAL CHARACTERS OF ORGANISED STRUCTURES.

12. The changes which constitute the Life of every animated being, take place through an instrumentality which is peculiar to the bodies that exhibit such changes, and which is entirely different from anything we observe in the surrounding universe. These bodies are invariably composed of what is termed *organised structure*, a designation implying that it consists of separate parts or *organs*, each of which is adapted to perform some distinct part in the Vital economy. The whole organised structure of any living being is termed its *organism;* and the word *organisation* is used to imply that peculiar process by which the organism is constructed out of the materials supplied by the inorganic or mineral world, and sometimes also to indicate the state or condition of the matter upon which this process—one of the most remarkable of all the vital actions—has been effected. When we come to enquire into the nature of the functions which these organised structures perform, it will be seen that they all tend towards a common object—the maintenance of the integrity of the fabric. And it may be regarded therefore as the peculiarity of an organism, that its distinct parts or organs are destined thus to subserve, each in its own particular

way, some general purpose. This, indeed, is one of the peculiarities which distinguish organised structures from inorganic matter; for in a mineral, every particle possesses a separate individuality, and whatever changes the whole may undergo in obedience to physical agencies, such changes are the result of their separate and independent influence on each integer; whilst in a living being, the actions of all parts of the machine are so connected together, that whatever influences one single particle of the organism on which these actions depend, will more or less affect the entire system. Thus, we may suppose a mass of gold alloyed with a small quantity of silver, and immersed in nitric acid; this chemical agent will affect every particle of the silver as completely as if the mass consisted of nothing else, and will leave the gold in its previous condition, having of itself no power of dissolving it. On the other hand, a similar chemical agent applied to an organised structure, will not only destroy the integrity of the part itself, but will produce a disturbance of the general functions proportional to the importance of the organ which has been injured; and when we come to trace the influence of any of the external agents on which life is dependent (see CHAP. III.) we shall find that this is exercised not only on the parts with which they are in immediate relation, but through them on the entire organism.

13. But it may be said that this is no more than takes place in any engine of human construction, or in the complicated machine of the universe,—that in these, as well as in living bodies, there is an adaptation of parts to each other, and of their actions to some general purpose,—and that all forms of matter are possessed of properties, by whose mutual influence changes may be produced, as regular, as ceaseless, and as connected, as those which living beings exhibit. Thus, the uniform motions of the heavenly bodies, the alternation of the seasons, the continual alterations which the crust of the earth is undergoing, the earthquakes and volcanic eruptions which so remarkably excite our attention to those alterations, and which may be regarded as more prominent indications of the same latent causes, the successive evaporation and condensation of the watery covering of the globe, the perpetual variations in the force and direction of the wind, the occasional recurrence of those violent meteorological changes which spread terror and desolation wherever they occur, but which serve such important purposes in the economy of Nature,—all these phenomena, and many more which might be instanced, result, no less than the constant changes exhibited by the animated creation, from that adaptation of parts to a whole, which is so characteristic of Design in the universe at large. Hence some philosophers have gone so far as to embody these phenomena into a general expression—the Life of the World; and as far as the abstract meaning of the term is concerned, it would be difficult to show that a single piece of mechanism, or the entire universe, is not *organised* as completely as any animated structure. But a little consideration will show that, in the construction of a machine, man avails himself only of the ordinary or physical properties of matter; and that, in the most complicated arrangement of its parts, each single element possesses only the same capabilities as it would have if separated from the rest. Thus, the moving power of a clock is given by the gravitation of a weight,—that of a watch by the elasticity of a spring,—that of a steam-engine by the expansion and condensation of watery vapour; and all the

rest of the mechanism is contrived only to give effect to these agencies, by employing them in the manner most advantageous for the designed end. In the phenomena of the universe, again, we see nothing but the agency of the ordinary physical properties of matter. Thus the motions of the solar system are all dependent on that universal property of matter—gravitation—which, originally balanced against other forces, will continue to produce the same effects as long as may be consistent with the designs of the Creator. By these motions are produced the variations of climate and season in our own globe; and from the same property which causes them, results the constant movement of the waters of our ocean. By the heat of the central luminary, again, are probably occasioned, either directly or indirectly, most of the atmospheric changes which are of such consequence to the well-being of the plants and animals that people this earth; and the same agent has a most important and direct influence on the phenomena of their growth and development. But these changes are only exhibited, when different material bodies are brought into such a relation with one another, as shall call their respective properties into action; and so long as any mass of inorganic matter is placed out of the pale of their influence, it may remain perfectly unchanged for an indefinite period.

14. When, therefore, we compare the constant activity which we encounter in living organised beings, with the passive condition of inorganic matter, we are compelled to conclude, that to whatever extent the forces which control the latter contribute to the actions going on in the former, there must be additional forces resulting from the operation of properties, to which we know nothing analogous elsewhere. The degree in which these superadded forces harmonise or interfere with those common to other forms of matter, constitutes a fair and highly interesting branch of enquiry which will hereafter be pursued. (CHAP. III.) But it is at present sufficient to state that, since these properties are never exhibited by any forms of matter except those usually denominated 'organised,' our notion of an organised structure is founded not only upon the adaptation of its parts to one another, but upon the indisputable possession by each part, of independent properties, by which it is enabled to execute actions for which physical laws will by no means account. And the process of organisation implies, therefore, not only the conversion of the homogeneous materials into regular and complex structures, but the simultaneous development of vital properties in the organisms thus generated.

15. Although in most Animal and Vegetable fabrics there are many different kinds of organised tissue, differing from one another both in structure and properties; and although these present differences according to the class of beings in which the examination is made;—yet there are certain general peculiarities by which all are seen to be characterised, when contrasted with mineral or inorganic bodies; and these peculiarities are as manifest in the humblest and simplest member of the animated creation, as in the most elevated and complex. It has been a favourite attempt with some naturalists, to trace a regular gradation or scale through the whole material universe; and not only to prove that the line of separation is indistinct between the Animal and Vegetable kingdoms, but to show that there is not such a complete division between the Organised and Inorganic world as physiologists think themselves justified in erecting. It is doubtless true that the disco.eries of modern science are constantly

bringing to light connecting links, which were previously deficient in many parts of the chain ; and that, in particular, an increased acquaintance with the various races of Animals and Vegetables which formerly inhabited this earth, through many successive epochs, seems likely to fill up whatever chasms are left open between the groups at present existing. But it is no indication of a philosophical spirit to attempt to discover relations where none can by possibility exist. The simple cells which constitute the humblest forms of Vegetation (§ 36), as well as the most minute and apparently least complex Animalcule, exhibit, when carefully examined, all the characteristics of organised structure, as well as all that can be regarded as peculiar in vital actions. They grow from a germ, increase, reproduce their kind, die, and decay, as regularly as any of the higher members of their respective kingdoms ; and they present the same peculiar and definite arrangement of particles, the same combination of fluid and solid materials, the same mutual adaptation of organs, as the latter possess.

16. However close, therefore, may be the links by which the Animal and Vegetable kingdoms are connected together, the relation is confined to them ; and between Organised fabrics, and the products of Crystallisation, (or of any other mode of aggregation by which Inorganic matter is held together, in masses great or small,) there is a total want of resemblance, either in structure or in properties.—To show in what the state of Organisation essentially consists, it will be necessary to contrast, in more detail, the peculiarities common to all beings which exhibit it, with the condition of Inorganic or Mineral bodies.

17. Wherever a definite *form* is exhibited by Mineral substances, that form is bounded by straight lines and angles, and is the effect of the process termed *crystallisation.* This process results from the tendency which evidently exists in particles of matter, especially when passing gradually from the fluid to the solid state, to arrange themselves in a regular and conformable manner with regard to one another ; and there is, perhaps, no inorganic element which is not capable of assuming such a form, if placed in circumstances adapted to the manifestation of this tendency among its particles. The mineralogist is conversant with an immense variety in the forms of crystals ; these, however, may all be reduced to a few primary types, from which the mathematician can deduce the rest. But, on the other hand, if the particles be not placed in circumstances favourable to this kind of union, and the simple molecular attraction, or attraction of cohesion, is exercised in bringing them together without any general control over their direction, an indefinite and shapeless figure is assumed. —Neither of these conditions finds a parallel in the Organised creation. From the lowest to the highest of living beings, the shape is determinate for each species or race ; and this shape is such, that, instead of being circumscribed within plane surfaces, right lines, and angles, organised bodies are usually bounded by convex surfaces, and present curvilinear outlines. It is true that, in the Vegetable kingdom, and to a certain extent among Animals also, we find a considerable difference in the external configuration of individuals of the same species ; but this difference is restrained within certain limits, and may usually be referred to the influence of external causes. (§ 71.) It is generally of the lowest tribes of both kingdoms, that we find the typical form of each species least definite, and the ten-

dency to irregular departures from it the greatest; and there seems reason to believe that among these the same germ may assume a variety of distinct forms, according to the circumstances under which it is developed; and thus is presented an approach, on the one hand, to that entire indefiniteness which is characteristic of uncrystalline mineral masses; and, on the other, to that variety of crystalline forms which the same mineral body may present, according to the conditions under which its crystallisation takes place.

18. With regard to *size*, again, nearly the same remarks apply. The magnitude of Inorganic masses is entirely indeterminate; being altogether dependent upon the number of particles which can be brought together to constitute them. On the other hand, the size of Organised structures is restrained, like their form, within tolerably definite limits, which may nevertheless vary to a certain extent among the individuals of the same species. These limits are least obvious in Vegetables, and in the lower classes of Animals. A forest-tree may go on extending itself to an almost indefinite extent; certain species of sea-weed attain a length of many hundred feet, and their growth does not appear to undergo any check; and the same may be said of those enormous masses of coral, of which so many islands and reefs in the Polynesian Archipelago are composed, or of which the débris seem to have constituted a large proportion of the calcareous rocks of ancient formation. But in these cases the increase is produced, not so much by the consecutive development of the individual, as by the continued production of new individuals, which remain in connection with the original and with each other,—the older portion of the mass having not unfrequently perished, whilst the younger is in a state of active growth. Thus, each bud of a tree may be regarded as a distinct individual, because, if placed under favourable circumstances, it can maintain its life by itself, and perform all the actions proper to its species; and consequently the almost indefinite extension of a tree by the multiplication of buds is not so much an act of growth as of reproduction. The case is nearly the same with regard to the extension of a sea-weed; for although it does not present the same obvious separation into distinct elements as we notice in the tree, yet almost every part of its lengthened frond is but a repetition of other similar parts, and may continue to exist independently of them. In the coral-mass the resemblance to a tree is more complete; for its extension is effected by the multiplication of polypes by a process of budding from the original; and thus the entire structure is rather to be regarded as an assemblage of individuals, than as a single body. Hence the form and size of the aggregate mass may present very wide extremes of variation, while that of the individuals may be as constant as it is among the higher orders of organised beings.

19. It is, however, in the *internal arrangement* or *disposition* of the parts, of which Organised structures and Inorganic bodies are respectively composed, that we find the difference between the two most strongly marked. Every particle of a Mineral body (in which there is no mere *admixture* of ingredients) exhibits the same properties as those possessed by the whole; so that the Chemist, in experimenting with any substance, cares not, therefore, except as a matter of convenience merely, whether a grain or a ton be the subject of his researches. The minutest atom of carbonate of lime, for instance, has all the properties of a crystal

of this substance, were it as large as a mountain. Hence we are to regard a mineral body as made up of an indefinite number of constituent particles, similar to the whole and to each other in properties, and having no further relation among themselves than that which they derive from their juxtaposition. *Each particle* may be considered, therefore, as having a *separate individuality.* The Organised structure, on the other hand, whether of a Plant or Animal, is made up of a number of dissimilar parts or organs, each of which has a peculiar texture and consistence ; and it derives its character from the whole of these collectively. By their action with each other, and with external agents, the Life of the entire fabric is sustained ; and hence there is a relation between its elementary constituents much closer than that of proximity only, namely, that of mutual dependence ; so that, as no one part can continue to exist without the rest, it cannot be regarded as possessing that separate individuality which belongs to the whole system alone. Thus, in the perfect Plant which has roots, stem, and leaves, the entire destruction of any one set of organs is fatal to its life ; although, as many of these organs are but repetitions of each other, some of them may be removed without permanent injury to it, provided enough are left to maintain its present existence.

20. In the lower Animals, as in Vegetables, we find a marked tendency to the repetition of *similar* parts, which cannot themselves exist independently of the entire fabric, although a part of them may be removed without serious detriment to it. This is especially the case among the Radiated and the lower Articulated tribes. It is not uncommon, for example, to meet with specimens of the common five-rayed Star-fish, in which not only one or two, but even three or four, of the arms have been severed, without the destruction of the animal's life ; and this is the more remarkable, as the arms are not simply organs of locomotion or prehension, but contain prolongations of the stomach. And in the Tape-worm, a large number of the joints of the body, each containing a portion of the alimentary canal and of the generative apparatus, may be detached without any apparent injury to those which remain in connection with the head. In the bodies of the higher Animals, however, there are few or no such repetitions, the diversity of structure and function existing among their several organs being much greater ; hence the mutual dependence of their actions upon one another is much more intimate, and the loss of a single part is much more likely to endanger the existence of the whole. Such structures are said to be more *highly-organised* than those of the inferior tribes ; not because the whole number of parts is greater, for it is frequently much less ; but because the number of *dissimilar* parts is larger, and the consequent adaptation of the whole to a *variety* of purposes is more complete,—the principle of division of labour, in fact, being carried much further, a much larger range of purposes being kept in view, and a much greater elaborateness in the accomplishment of them being provided for.

21. Keeping in view, then, what has just been stated in regard to the divisibility of a Tree or a Zoophyte into a number of parts, each capable of maintaining its own existence, we may trace a certain gradation from the condition of the Mineral body to that of the highest Animal, in regard to the character in question. Thus, the individuality of a Mineral substance is sometimes said to reside in each molecule, for every one is an ex-

act repetition of the rest; that of a Plant or a Zoophyte in each member that possesses, or can develope, the organs essential to the maintenance of life; and that of one of the higher animals in the sum of all the dissimilar components. The distinction is much greater, however, between even the lowest Organised fabric and *any* Mineral body, than it is between the highest and the lowest organised structures; for the existence of even the least diversity of structure in the several parts, and of the slightest mutual dependence between their actions, is sufficient to indicate an essential difference between *organisation* and mere *aggregation;* whilst the difference between the highest Animal and the lowest Plant is essentially one only of *degree,* as will appear from the details hereafter to be given.

22. Between the very simplest Organised fabric and every form of Mineral matter, there is a marked difference in regard to *intimate structure* and *consistence.* Inorganic substances can scarcely be regarded as possessing a structure; since (if there be no admixture of components), they are uniform or *homogeneous* throughout, being composed of similar particles, which—whether simple or compound, solid, liquid, or gaseous,—are held together by attractions which affect all alike. Far different are the characters of Organised structures; for in the minutest parts of these may be detected a *heterogeneous* composition,—a combination of solid and liquid elements, which are so intimately blended as to impart peculiarities to the tissues, even in regard to their physical properties, of a kind which we never encounter amongst Mineral substances. In the latter, *solidity* or *hardness* may be looked upon as the typical condition; whilst of the former, *softness* (resulting from the large proportion of fluid components) may be considered the distinctive quality, being most obvious in parts that are most actively concerned in vital operations.

23. The greatest solidity exhibited by Organised fabrics is found where it is desired to impart to them the simple physical property of *resistance;* and this is attained by the deposition of solid particles, usually of a mineral character, in tissues that were originally soft and yielding. When mineral matter is thus introduced into the fabric, it appears in some instances to enter into actual combination with the organic base, as seems to be the case in bone; whilst in other instances it is deposited *per se* in the interstices of organised tissue, and is then sometimes amorphous, sometimes crystalline, in its aggregation,—the former being the case with the particles of carbonate of lime which are dispersed through the fibrous basis of the shell of the fowl's egg (§ 160), whilst the latter condition is seen in the solid columns of the same material which occupy the prismatic cells of the shells of many Mollusks (§ 197). Most of the higher Plants have a greater or less proportion of carbonate and oxalate of lime diffused through their tissues; the latter most commonly in the form of *raphides,* or long needle-shaped crystals, which are collected in bundles in the interior of particular cells; the former being generally dispersed in rhombohedral crystals in the interstices of the tissues.* A dense incrustation of carbonate of lime is found on the exterior of the tubular cells of many of the *Characeæ,* which hence derive their popular name of 'stone-

* The oxalate of lime is converted into the carbonate by calcination, as are also the compounds of lime with vegetable acids which are dissolved in the juices of the living plant; hence we are not to regard the amount of carbonate of lime in the ash as any measure of the quantity which existed in the solid form in the living fabric.

worts.' But the most remarkable example of consolidation of vegetable texture by this material is presented by the true *Corallines* and *Nullipores*, the greater part of whose fabric is rendered of stony hardness, by the large amount of carbonate of lime deposited in their tissues. In the *Grasses* and *Equisetaceæ*, a large quantity of silicious particles, retaining their crystalline form, is deposited beneath the epidermis, imparting great strength to the integument. In Animal structures, earthy depositions are usually more concentrated in particular spots, especially where the locomotive powers are considerable; since it is obviously essential to the exercise of those powers, that, whilst the frame-work which gives attachment to the organs of propulsion should be solid and unyielding, these organs themselves, as well as other parts of the fabric, should be capable of great freedom of action. In the higher animals, therefore, we find carbonate and phosphate of lime deposited in special situations, so as to give a firm basis for the attachment of softer structures; the former ingredient predominating where the skeleton is massive and external, serving chiefly for the protection of an inert body, as in the Mollusca in general; whilst the latter, of which certain mineral forms are among the hardest substances known, is chiefly met with where the skeleton is enclosed within the softer parts, as in Vertebrated animals, and where concentration of bulk without diminution of strength is therefore an important object. But there are some among the lowest, in which the adaptation for locomotive power is no object; and here we find the structure even more universally penetrated with calcareous matter, than that of vegetables. Thus, the masses of Coral, which were long supposed to be constructed as habitations by the Polypes found in connection with them, are now known to be produced by the deposition of calcareous matter in the soft jelly-like substance which constitutes the real flesh of the animal, and are therefore to be considered as properly belonging to the organised fabric itself (§ 194).

24. The parts in which depositions of this kind take place, however, are those which participate least in the operations which are essentially concerned in the maintenance of life. They may be from the first developed simply for the physical uses already adverted to; as is the case with the bones, shells, &c., of Animals. Or they may be tissues which have served for a time some more important purpose in the economy, but which, instead of then wasting by decay, are consolidated by interstitial deposit, and are thus rendered subservient to a mechanical purpose, at the very time that they cease to participate in the proper Life of the system. This is the case, for example, with the woody structure; which, when first formed, and in the condition of 'sap-wood,' assists in transmitting the nutritious fluids through the stem; but which, when condensed into 'heart-wood,' and acting as the skeleton of the fabric, can no longer serve that purpose. So the soft succulent tissue, which, when newly formed at the extremities of the rootlets, is known as the *spongiole*, and serves as the special organ for the absorption of fluid, gradually becomes condensed into the hard substance of the root, and henceforth merely takes a share in the mechanical office of fixing the stem, its original function being performed by newer tissue generated beyond it. This is the case, too, with the Corallines and Nullipores, the growing parts of which are always soft; and also with the Corals, whose soft fleshy matrix is continually

extending itself above, whilst it is progressively converted into an almost homogeneous stony mass below.—In contrast with these, we may advert to that texture of animals, whose functions are the most important, and the furthest removed from anything which we meet with in the domain of Physics, namely, the Nervous matter; this is the softest and most decomposable of all the tissues of the body, and is continually wasting by decay in proportion to the demands upon its functional activity, whilst in the healthy fabric it is regenerated no less constantly and rapidly.—As a general rule, then, it may be stated, that wherever organised fabrics approach mineral bodies in solidity and *homogeneity* of structure, they approach them also in the purely *physical* nature of their endowments, and in the want of those properties which are distinctive of living bodies. Whilst, on the other hand, these *vital* properties are most remarkably exhibited by textures and organs, which, in the *heterogeneousness* of their structure, present the greatest dissimilarity to mineral bodies.

25. Organised structures are further distinguished from Inorganic bodies, by the peculiarity of their *chemical constitution.* This peculiarity does not consist, however, in the presence of any elementary substances which are not to be found elsewhere; on the contrary, all the elements of which organised bodies are composed, exist abundantly in the world around. When the vast differences in the endowments of animated and of inert matter are considered, it appears at first sight surprising that the same elements should present themselves in both; but a little consideration will show, that this fact is a necessary result of the mode in which organised fabrics are generated. For the parent communicates to its offspring, not the structure itself, but the power of forming that structure out of the materials supplied to it from external sources,—these materials being derived by Plants directly from the Mineral world, whilst Animals obtain them through the intermediation of Vegetables. The Chemist enumerates *sixty-one* elementary or undecomposed bodies as existing in the solid crust of the globe, and as forming, by their union in different proportions, the various mineral substances of which it is composed. Of these, not more than *eighteen* or *nineteen* have been detected in organised structures; and several of them appear to be rather accidentally than necessarily present there. In fact, although the known Organic compounds are at least equal in number to those of the Inorganic world, and surpass them in variety of properties, yet the number of elementary substances which appear essential to their composition is not above *seven* or *eight.* Again, the bulk of the Mineral kingdom consists of the metals and their compounds, which form the earths, alkalies, and some of the acids; and these, when present in living bodies, rarely form part of their tissues, but are either held in solution in their juice for some purely chemical purpose, or are deposited (as already explained) in a completely unorganised condition to serve some mechanical object. On the other hand, the chief components of the organised tissues of Plants and Animals, and of the products of their activity, are the four non-metallic substances, Oxygen, Hydrogen, Nitrogen, and Carbon; and of these the first three in their uncombined state have a gaseous form, whilst the last is never met with in nature as a solid body, except under circumstances which lead us to regard it as derived from Vegetable or Animal substances. The two substances

next in importance as constituents of living bodies, are Sulphur and Phosphorus, also non-metallic. And Lime and Iron, which belong to the class of Metallic substances, may be regarded as the lowest in importance of all the components essential to organised structures; their presence being much less universal than that of the preceding, and the purposes to which they appear to be subservient having a close relation to physical and chemical agencies. Thus Lime, even when it enters into organic union with the animal basis of bone, is only destined to impart solidity to the structure; and Iron, whose chief seat is in the red corpuscles of the blood (§ 176) may be removed from the substance of which it is a component, without any other obvious change in its properties than chemistry can account for.*

26. In considering the distinctive Chemical characters of the substances, formed at the expense of the elements just enumerated, which are met with in the Vegetable and Animal fabric, it is proper to bear in mind the conditions under which they occur in the organism.—Thus in the circulating juices of the Plant, or stored up in the neighbourhood of its growing parts, we meet with certain *nutritive materials*, as *Gum*, *Sugar*, and *Starch;* substances which have a close chemical relation to each other, and which are all capable (as it would appear) of undergoing organisation, and of being converted into living tissue. So, again, in the blood of the Animal, and set apart in the egg, we find the substance termed *Albumen;* which appears to hold a similar relation to its organised fabric, being the pabulum at whose expense its tissues are formed. These must be regarded as purely Chemical compounds, in which no properties of a truly Vital nature are yet discernible; and they are, as such, proper subjects of analytical investigation.—But, secondly, the *living organised tissues* of Plants and Animals must be regarded in a very different light. So long as they retain their distinctive attributes, they do not appear subject to the laws of ordinary affinity; and it is only when these attributes have ceased to manifest themselves, that the Chemist can bring them within the sphere of his research. Their constitution as living tissues, therefore, cannot be expressed by formulæ, nor can their properties be accounted for by analysis; and their chemical relations can only be judged of by the nature of the materials at whose expense they grow, and by the products of their ultimate decomposition. In this manner it seems to be determined that the walls of the cells, vessels, and hollow fibres of Plants are composed of a substance, to which the name of *Cellulose* has been given, closely related in composition to gum, sugar, and starch. And in Animals a large proportion of the tissues formed upon the cellular type (§ 156) may be regarded as composed of *Fibrin*, which seems nothing else than organised albumen, and which presents itself in an incipient state in the blood; whilst the tissues formed upon the *simple fibrous* type (§ 161) are resolvable into the substance termed *Gelatin*, which, though not to be detected in the nutritious fluids, is undoubtedly

* It is a fact of some interest, that the Combining Equivalents of all these substances are, according to the latest and most exact researches, *exact multiples* of that of Hydrogen. Thus, Hydrogen being taken as 1, Oxygen is 8, Carbon 6, Nitrogen 14, Sulphur 16, Phosphorus 32, Calcium 20, and Iron 28. Out of the forty-eight remaining elementary substances whose combining equivalents are known, there are not above seven of which the same can be stated even with probability; these are, Arsenic, Cerium, Lanthanum, Mercury, Potassium, Silver, and Uranium.

formed at the expense of Albumen.—But from the original nutritive materials, in the third place, there is formed a class of bodies which are never destined to undergo organisation, but which are designed for various ulterior purposes in the economy of the being itself or of some other, and are deposited in the interior of the cells, or in the interspaces of the tissue. Such are the lignine, which gives solidity to the woody tissues of Plants; the oils, resins, alkaloids, acids, colouring-matters, &c., which they generate and set apart; as well as a large share of those albuminous compounds, which seem to be prepared by the Vegetable kingdom expressly for the maintenance of the Animal world, the proportion which answers any direct purpose in the economy of the plant being apparently very minute. So in the Animal body, we find oleaginous compounds laid up in certain cellular textures for special purposes; whilst other cells contain horny matter, in others are found colouring-particles, and so on. All these substances are formed at the expense of those of the first class; and there is no reason to consider them as having any other than purely chemical relations. They have not entered into the composition of the living tissues, and they do not appear capable of undergoing organisation; and the progress of research renders it highly probable, that the agencies by which they are elaborated are of a purely chemical nature, acting under that peculiarity of conditions which a living body can alone supply. Hence, like the substances of the first class, they rightly belong to the domain of the Chemist.—But, fourthly, we encounter even in the living structure, more especially in that of Animals, certain substances which cannot be looked upon in any other light than as the products of the continual decomposition of the nutritive fluids and of the solid textures. These, however, are never destined to be retained within the body, but are always met with on their way out of it. Among the most characteristic of such substances are urea, kreatine and kreatinine, ammonia, and carbonic acid. The last of these is the only one which appears to be regularly thrown off as a product of decomposition from Vegetable structures. Of the purely chemical character of all these, there can be no doubt whatever; the existence of all of them being the result of the tendency of the higher organic compounds to pass into simpler and more stable forms, when entirely freed from the influence of vitality, and subjected to the ordinary conditions of matter; and most of them being capable of synthetical production in the laboratory of the Chemist.

27. Of the substances belonging to the first three classes, it may be remarked that their atomic constitution is far more complex than that of Inorganic compounds generally; in regard alike to the number of the elements which they contain, and to the number of atoms or combining equivalents of these, which exist in a single combining equivalent of the compound. Thus all the substances included in the first and second classes, and all of the third also, except a few whose place might rather seem to be in the fourth, consist at the least of three elements, Oxygen, Hydrogen, and Carbon; another, Nitrogen, is present in a large proportion of them; and in addition, Sulphur and Phosphorus appear in many instances to be essential constituents. Again, in Sugar, which is one of the simplest of all organic bodies in its composition, we find 12 eq. of Carbon united with 10 of Oxygen and 10 of Hydrogen; a proportion

which is common also to gum, starch, and cellulose. In the nitrogenous compounds, the number of combining equivalents is much greater; thus the proportion of the four principal constituents of albumen is represented by Liebig and Mulder as follows;—

Liebig.... 49 Carbon, 36 Hydrogen, 6 Nitrogen, 14 Oxygen,
Mulder... 40 Carbon, 31 Hydrogen, 5 Nitrogen, 12 Oxygen:

and if the equivalents of sulphur and phosphorus, which are present in comparatively minute proportions, are to be represented by whole numbers, then it will be requisite to multiply the preceding formulæ by *twenty* throughout. So far as is at present known, these very complex combinations cannot be resolved into others of a simpler nature, without such a complete disruption as precludes that re-formation of the original product by the synthesis of its components, which can be always effected in the case of inorganic compounds. Thus, whatever may be the mode in which the atoms of Sulphur, Oxygen, and Potassium are disposed in the salt which we call Sulphate of Potash (concerning which Chemists are not at present in accordance), we may separate this salt into two substances of opposite characters, sulphuric acid and potass, by bringing which together, we reproduce the salt; and each of these substances may be again resolved into its component elements, the sulphuric acid into sulphur and oxygen, the potass into potassium and oxygen, and each may be regenerated by the synthesis of its elements. Nothing of this kind can be accomplished in regard to the saccharine, albuminous, or gelatinous compounds. Thus although sugar might be said to be made up of carbon united with the elements of water, no method of analysis with which we are acquainted could separate the carbon and the water, without the addition of elements which shall form new products whose sum does not represent the original; still less can the synthetical method enable us to obtain sugar, either by the direct combination of carbon with water, or by the union of its elements under any other form. So, again, the utmost that has been yet accomplished with regard to the nitrogenized compounds, has been to produce various new substances by decomposing them with powerful reagents; and no reunion of these substances, or of the elementary components into which they are resolved by ultimate analysis, can reproduce the original.

28. With regard, however, to the substances comprised in the third class, the Chemist has been more successful; many of them being artificially producible from those of the first class or from each other; and many being capable of separation into two or more components, from which the original may be reproduced by synthesis. Thus from Salicine, the alkaline base of willow-bark, may be obtained the fragrant essential oil of Spiræa ulmaria or "queen of the meadows," by treating it with bichromate of potass and sulphuric acid, whereby it receives two additional equivalents of oxygen, and is resolved into oil of Spiræa and grape sugar. The latter product, by a further oxidation, is converted into formic acid (naturally secreted by the red ant as a defensive fluid,) carbonic acid, and water.—So, again, valerianic acid, to which Valerian-root owes its pungent fœtid odour, may be obtained by the action of caustic potass on the hydrated oxyde of amyle, which is an analogue of alcohol, obtained during the distillation of spirit from fermented grain or potatoes.—The essential oil of Mustard may be converted into that of Garlic, by dropping into the former a piece of potassium, which subtracts from it two atoms of carbon,

one of nitrogen, and one of sulphur; so as itself to become sulphocyanide of potassium, whilst the residuary atoms undergo a new arrangement in virtue of which they form oil of garlic.—And the essential oil of the Gualtheria procumbens (a beautiful ericaceous plant of New Jersey), which has become an article of commerce on account of its beautiful perfume, combining the odours of hyacinth and cinnamon, may be obtained by distilling salicylic acid (which may be procured by heating either indigo, salicine, or the oil of spiræa, with caustic potass) with pyroxylic spirit or wood naphtha, and a little sulphuric acid.*—There appears strong reason to believe that even vegetable alkaloids may be artificially generated after a similar fashion. Thus when bran is treated with sulphuric acid, an oily substance termed Furfurol may be distilled over; and this, when brought into contact with ammonia, becomes transformed into a solid slightly crystalline substance, which is easily decomposable again into ammonia and furfurol. But when this substance is boiled with weak caustic potass, a rearrangement of its atoms appears to take place without any change in their relative number; and it now becomes a stable and powerful alkaline base, forming a regular series of salts with acids.—Again, it has been found that Piperine, the peculiar constituent of pepper, may be regarded as a neutral combination of a nitrogenous acid with the alkaline base Aniline which may be obtained from indigo, gas-tar, and a great variety of substances; piperine being producible synthetically by the union of the acid and the base. In like manner, Narcotine may be separated into a non-nitrogenous acid and a peculiar base. Little doubt is entertained by Chemists, that all the alkaloids having high equivalents will thus be separated into simpler compounds which may be common to many of them; and it is anticipated that some of the rarer and more valuable among them, such as quinine, morphine, &c., may be generated from others which are more abundant or less serviceable in themselves.—Still in regard to all such products, it must be borne in mind that they cannot be generated by the synthesis of their elements, in any mode with which the Chemist is acquainted, or by any means of which he can at present even conceive; it being requisite that he should have, as the basis of his operations, some compound which has been formed by the agency of a living body.

29. The substance of the fourth class stands in a much closer relation to inorganic compounds, than do any of the preceding; as regards alike the simplicity of their composition, the readiness with which they may be resolved into compounds of a yet simpler nature, and their capability of being formed by synthesis. Thus Urea, which is one of the products of the decomposition of the muscular tissue, and which is being continually excreted in large amount from the animal body, is found by ultimate analysis to be a compound of 2 eq. of Carbon, 2 Nitrogen, 4 Hydrogen, and 2 Oxygen. When strongly heated, however, it is resolved into cyanic acid and ammonia; and by the recombination of these two substances with an equivalent of water ($CyO + NH_3 + HO$), an artificial urea is produced. It has not yet been found possible thus to form kreatine or kreatinine (which also are products of decomposition, excreted in the urine) by the artificial union of their elements; but the comparative

* See Dr. Golding Bird's Lectures on Organic Chemistry in connection with Therapeutics, in Medical Gazette, 1848.

simplicity of their composition, their crystalline aggregation, and their behaviour with reagents, encourage the belief that they too may be found to be binary compounds, capable of being reproduced by synthesis. Such products of decomposition as Ammonia, Cyanogen, and Carbonic acid, stand on the border-line between Organic and Inorganic Chemistry, belonging alike to both domains. With regard to Cyanogen, however, it may be remarked that it has been only of late that chemists have succeeded in procuring it by the direct union of Carbon and Nitrogen ; in all the previous methods of generating it, the presence of a nitrogenous organic compound having been required. But the Carburets of Hydrogen, which are among the products of the decomposition of vegetable substances, have not yet been found capable of formation by synthesis ; although the apparent simplicity of their composition encourages the belief that this will ere long be effected.

30. There is an important difference amongst the chemical compounds peculiar to organised bodies, in regard to their tendency to spontaneous change or decomposition. As a general rule, those which include nitrogen are much more prone to such changes than those which are composed of oxygen, hydrogen, and carbon alone ; and we shall hereafter see that one very important end of the presence of azotized substances in the vegetable organism, is that they may, whilst themselves undergoing decomposition, effect *catalytic* changes in the ternary compounds (§ 39). But it is also a general rule, that the compounds belonging to the third and fourth of the classes just enumerated are for the most part much less prone to change than those of the first. Thus horny matters and urea, which are the products of the metamorphosis of albuminous compounds, may be preserved unchanged for any length of time ; whilst albumen can only be kept from speedy decomposition, either by completely drying it, or by entirely excluding it from air, or by subjecting it to a very low temperature,—all these conditions being the very opposites of those under which it exists in the living body. So, again, the lignine which gives to the heart-wood of trees its remarkable durability, oils, resins, and many other products of vegetable chemistry, are more durable in their nature than the saccharine or gummy fluids at whose expense they are elaborated. It is true that, when reduced to the solid form by drying, sugar and gum approach inorganic substances in their freedom from disposition to spontaneous change ; but this is not the condition under which they exist in the living plant, the juices of which are prone to fermentation at moderate temperatures, especially when any azotized matter is present. How far the actual components of living tissues (forming our second class of organic substances) are subject to decomposition, is a question which will be more fitly considered in the next chapter (§ 46).

CHAPTER III.

OF THE NATURE AND CONDITIONS OF VITAL PHENOMENA.

1. *Of Life, or Vital Activity.*

31. There is no department of Philosophy around which so much unnecessary mystery has been cast, as the investigation of the character and laws of the Phenomena of Life. And this veil of mystery can only be removed by the rigorous adoption of those methods of research, which have been proved to be most effectual in the study of Physics and Chemistry. To designate any of the actions of a living body as *vital*, or as effected by the *vital principle*, was long regarded as a sufficient account of them, and had the effect of placing a complete check upon further inquiry. The history of Physical Science shows, however, that *it* was once labouring under the same restraint; and that, until the true objects of investigation were understood, scarcely any advance was made. Thus, in past ages all the movements of the heavenly bodies were attributed to the operation of some vague "principle of motion," the laws of which it was considered impracticable to attain; and even after the empirical ingenuity of Kepler had led him to the discovery of some of these laws, it was still maintained that the powers concerned in the celestial motions could have nothing in common with terrestrial forces,—a doctrine which the genius of Newton subverted, by proving the universal applicability of the laws of motion, and the existence of the attractive force between all masses of matter. Again, the simple optical fact—that when the sun's light passes through a hole, the bright image, if formed at a considerable distance from it, is always round, instead of imitating the figure of the aperture,—was attributed by Aristotle to the "circular nature" of the sun's light; whilst the simple consideration that the rays of light travel in straight lines, would, if properly applied, have explained this phenomenon, not only in regard to the sun, but in the case of any other round luminous body placed at a sufficient distance. To attribute all the operations of Life, which cannot be referred to Physical or Chemical laws, to a "Vital Principle," is in reality to proceed just as unphilosophically as the ancients did in the cases just quoted; since a strict examination into their character will show that, although not *identical with* physical phenomena, they are analogous to them, in so far as they take place according to a regular *plan*, and present themselves under fixed *conditions*, a definite acquaintance with which would give to Physiological Science the same kind of precision and comprehensiveness as it is the aim of the Physical philosopher to attain in *his* department of study. The laws of Vital phenomena (§ 5), in fact, are as open to investigation as those which comprehend the phenomena of Gravity, Electricity, or Chemical Affinity; and although the intricacy of the combinations under which these phenomena are usually presented to our observation (§ 7, 8), renders their laws more difficult of attainment, the success which has attended the philosophical method of enquiry of late pursued by scientific Physiologists, is a most

satisfactory proof that they are not beyond the reach of persevering and well-directed research.

32. Few terms have been employed in a greater variety of significations, or more frequently without any definite meaning at all, than the word *Life*. The older philosophers regarded it as a distinct entity or substance residing in certain forms of matter, and as the cause both of their organisation and of the actions exhibited by them.* We have seen that the tendency to rest satisfied with vague hypotheses of this kind, operated in the retardation of Physical Science; and had it not been for the comparative prominence and simplicity of its phenomena, it is not improbable that, even at the present day, we should hear employed in *it* such terms as the 'Vital Principle,' 'Organic Agent,' or any other equally unphilosophical refuge of those physiologists, who neglect the substance to grasp at the shadow.—To the term 'principle' no very definite meaning can be attached. It has been remarked that "this word, characteristic of a less advanced state of science, has been generally employed (as the final letters of the alphabet are used by algebraists) to denote an unknown element, which, when thus expressed, is more conveniently analysed." Thus, it has been customary to speak of the 'principle' of Gravity, the 'principle' of Electricity, or of the 'principle' of Magnetism, as the unknown causes of certain phenomena as yet imperfectly comprehended. In so far, however, as the *laws* of these phenomena are understood, they are really nothing else than *generalised expressions of the conditions* under which they occur; our conviction of the uniformity of Nature (that is, of the unchangeableness of the Divine plan of government, § 5, *note*) leading us to expect the same result whenever the same conditions are presented. Thus, the "law of Gravitation" is nothing else than a statement of the fact, that all masses of matter attract one another in a certain ratio; the essential condition being, that two or more masses should exist; whilst the other conditions of relative bulk and distance determine the amount of force exerted by each mass upon the other. So the "law of Definite Proportions" in Chemistry expresses the universal fact, that when different substances combine with each other, they do so in one out of several definite ratios; the essential condition here being, that two or more substances capable of uniting should be brought into the requisite proximity with each other; whilst the particular ratio in which they will unite is determined by a variety of other considerations, of which the Chemist seeks to arrive at some precise and comprehensive expression. And as it is the whole aim of the scientific Physiologist to attain similar general and definite expressions of the conditions under which Vital phenomena occur, it seems obvious that nothing can be really gained by the use of the term 'Vital principle;' whilst the employment of it is really injurious

* Every sect had its own notion of the origin and nature of this entity; some regarding it as a kind of fire; others as a kind of air, or ether, or spirit; and others, again, as merely a kind of water. The fable of Prometheus embodies this doctrine in a mythological form, the artist being described as vivifying his clay statues by Fire stolen from the chariot of the Sun. And whatever was the idea entertained as to the character of this agent, all regarded it as universally pervading the World, and as actuating all its operations in the capacity of a Life or Soul; whilst a special division of it—*divinæ particula auræ*—regulated all the concerns of each individual organism. The vestiges of these ideas may be traced in the expressions 'vital spark,' 'vital spirit,' 'breath of life,' and others which are still prevalent.

in so far as it tends to limit the range of enquiry into the phenomena of Life.

33. Still more opposed to the progress of Physiological Science, is the doctrine propounded by one of the most distinguished philosophers of our age,* that Vital phenomena are to be attributed to the operation of "distinct intelligent agents, superior to, and possessing the power of directing and controlling, the common forces of matter." Of these "Organic agents" it is supposed that a series exists in every organised body; each agent "possessing more or less control over all the agents below itself, and having the power of appropriating their services, till at length in the combined operation of the whole series of agents at the top of the scale, we reach the perfection of organic existence." This hypothesis, in favour of which not a single fact can be advanced, may perhaps be best disposed of by carrying out the analogy of an army, which is suggested by the phraseology just cited. As an army is at present constituted, there is large room for the operation of intelligence in the subordinate officers or delegated powers; because it is upon them that the carrying-out of the will of the commander-in-chief entirely rests, his knowledge and skill being confined to the general plan; and they have not only to choose the best means of effecting this, but to modify and accommodate it to circumstances, sometimes even acting in opposition to it, should it prove dangerous or impracticable. But if the supreme commander had a perfect control over every component individual of a large army, and possessed the means of immediately operating upon the minds of every soldier so as to cause him to execute his will without any intervening command; and if he himself possessed the knowledge and skill requisite, not only to guide the general evolutions of the *masses* composing his army, but to direct the movements of every *man*, in the most advantageous manner;—would he trouble himself with the intermediate machinery of subordinate generals, colonels, captains, lieutenants, sergeants, and corporals, in order to convey his will and direct the operations of his troops? To imagine, then, that anything is gained by the interposition of 'agents,' intelligent or non-intelligent, between the Deity and the materials upon which He operates, is either to set limits to His knowledge and power, or to give to those agents an office purely nominal. For if they have any of that independence which is possessed to a certain extent by the officers of a human army, then the Almighty intelligence is not directing the operations of matter in their detail; whilst if that detailed direction has a real existence, the function attributed to them is a nonentity.

34. Any such idea of a delegated control over the actions of living beings is not only quite useless and unnecessary in the explanation of facts, but is totally unsupported by the analogies of Nature, and by what we know of the Divine Government in general. No reflecting mind has any doubt that this earth and its inhabitants form a system, of which every part is perfectly adapted to the rest, (so that we might almost call it an *organised* one, if the idea of a particular structure were not involved in the term, § 13,) and of which all the actions and changes, however in appearance independent or even contrary, have one common tendency, the ultimate happiness of the creatures of Infinite Benevolence. It cannot be regarded as an improbability that the other spheres and systems,—whose

* **Dr. Prout, in Bridgwater Treatise, 3rd ed. p. 396 *et seq.***

countless multitudes, revealed by the aid of science, impress our minds with the nearest conception of infinity, of which our finite comprehension is capable,—are peopled with beings, if not similar in structure with ourselves, a least equally worthy of the Creator's care. In the government of our own planet, itself but a point in the vast universe, we are able to recognise, to a small extent, the laws according to which its material operations are conducted; and we discern indications of those which affect the moral condition of sentient beings. So far as we can understand the mutual adaptation of these laws, we everywhere see them working to the same end; and we entertain the highest anticipations of that beauty and harmony which will be revealed to us, when our imperfect glimmerings of knowledge shall be extended and corrected by the light of Eternal Truth. Should we not consider it degrading to the dignity of Infinite Wisdom, to suppose that, at the creation of each world, He had found it necessary to delegate to a subordinate the control over its working,—instead of so devising the original plan, in all its details as well as in its general arrangements, that the various changes which it was His intention to produce, should be harmoniously effected by the uniform operation of the same all-sustaining energy? It cannot be said that the adaptation of means and ends is less evident in the scheme of the Universe at large, than it is in the structure and actions of a single Organised being. And if it be the aim of the Physical Philosopher to discover the simple but majestic plan, of which this wonderful adaptation is the result, why should not the Physiologist directly refer all the operations of the living body to the same primary agency, and seek, in like manner, to comprehend the scheme of its government?

35. If the application of the term 'Life' to some independent agent, which is assumed to be the immediate cause of the peculiar attributes of Plants and Animals, be found useless or injurious, it may be reasonably inquired what is to be understood by it. Even our most elementary notion of a *living* being, based upon the phenomena presented by the simplest organism that can maintain an independent existence, is so complex, as to render it impossible to express it in the form requisite for a concise definition; and although various attempts have been made to convey in words the essential 'idea' which it involves, none has been yet successful enough to satisfy the scientific Physiologist.* It has been already pointed out that there is a class of phenomena, to which we give the designation *Vital*, which is even more different in its character from those of Physics and Chemistry, than they are from each other; and that these 'vital' phenomena are only manifested by bodies of that peculiar structure which we term 'organised.' By the term 'Life,' then, we most appropriately designate the state or condition of a being that exhibits vital actions: and it is thus placed in opposition to the term

* One of the latest of these attempts is put forth in a posthumous work by S. T. Coleridge, entitled "Hints towards the Formation of a more Comprehensive Theory of Life." (London, 1848.) He considers Life as the 'principle of individuation,' that is to say, as a power which discloses itself from within, combining many properties into one individual thing. In his view, however, the term must be extended to Inorganic matter; each individual molecule being endowed with a low degree of this principle, the higher forms of which are manifested in Plants, and the highest in Animals. (See § 21.) Hence the term 'Life' as thus employed is no longer a distinctive one; and it would be requisite to invent some other to designate the peculiar activity of *Organised* bodies.

'Death,' which implies the state of a being in which those actions have altogether ceased, and whose structure is subject to no other forces than those of inorganic matter, which speedily effect its decomposition: whilst both are distinguished from the state which may be designated as that of 'Dormant Vitality,' in which the whole vital activity of the being has been temporarily suspended, without any such impairment of the structure or composition of the organism, as shall prevent the resumption of that activity when the requisite conditions are supplied.

36. The distinctive characters of Vital actions will be best understood, if we take a general view of the series of phenomena presented by a living being of the simplest kind, from its origin to its final decay. For this purpose, we must have recourse to the humblest forms of Cryptogamic Plants, which consist of mere aggregations of *cells*, every one of which may be regarded as a distinct individual, since it is perfectly independent of the rest, and performs *for* and *by* itself, all the functions of Growth and Reproduction. We shall find in the operations of the simple cell, an *epitome*, as it were, of those of the highest and most complex Plant; which again bear a close correspondence with those that are immediately concerned in the Nutrition and Reproduction of the Animal body. The functions peculiar to Animals will not enter into the present enquiry, not being common to all organised beings.—A *Cell*, in Physiological language, is a closed vesicle or minute bag, formed by a membrane in which no definite structure can be discerned, and having a cavity which may contain matters of variable consistence. Every such cell constitutes an entire organism in such simple plants as the *Protococcus nivalis* (Red Snow), or *Palmella cruenta* (Gory Dew); for although the patches of this kind of vegetation, which attract our notice, are made up of vast aggregations of such cells, yet they have no dependence upon one another, and the actions of each are an exact repetition of those of the rest. In such a cell, every organised fabric, however complex, originates. The vast *tree*, almost a forest in itself,—the *zoophyte* in which we discover the lowest indications of animality,—and the feeling, thinking, intelligent *man*—each springs from a germ, that differs in no obvious particular from the permanent condition of one of these lowly beings. But whilst the powers of this latter are restricted, as we shall see, to the continual multiplication of new and distinct individuals like itself, those of the former enable it to produce new cells which remain in closer connexion with each other; and these are gradually converted, by various transformations of their own, into the diversified elements of a complex fabric. The most highly-organised being, however, will be shown to consist in great part of cells that have undergone no such transformation, amongst which the different functions performed by the individual in the case just cited, are so distributed, that each cell has its particular object in the general economy, whilst the history of its own life is essentially the same as if it were maintaining a separate existence.

Fig. 1.

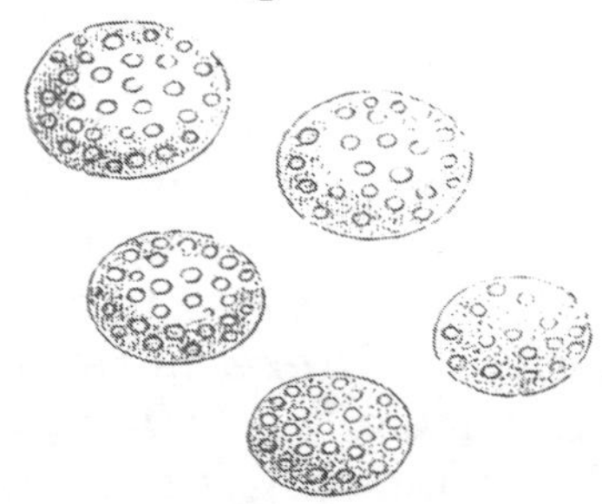

Simple isolated cells, containing reproductive molecules.

37. In the first place, then, the cell takes its origin from a *germ*, which may be a minute *molecule* not visible without a microscope of high

power.* This molecule appears, in its earliest condition, to be a simple homogeneous particle, of spherical form; but it gradually increases in size; and a distinction becomes apparent between its transparent exterior and its coloured interior. Thus we have the first indication of the *cell-wall*, and the *cavity*. As the enlargement proceeds, the distinction becomes more obvious; the cell-wall is seen to be of extreme tenuity and perfectly transparent, and to be homogeneous in its texture; whilst the contents of the cavity are distinguished by their colour, red or green according to the species. At first they, too, seem to be homogeneous; but a finely-granular appearance is then perceptible; and a change gradually takes place, which seems to consist in the aggregation of the minute granules into molecules of more distinguishable size and form. These molecules, which are the germs of new cells, seem to be at first attached to the wall of the parent-cell; afterwards they separate from it and move about in its cavity; and at a later period, the parent-cell bursts and sets them free. Now, this is the termination of the life of the parent-cell; but it is the commencement of a life of a new production; since every one of these germs may become developed into a cell, after precisely the foregoing manner, and will then in its turn propagate its kind by a similar process.

38. By reasoning upon the foregoing history, we may arrive at certain conclusions, which will be found equally applicable to all living beings.—In the first place, we witness the origin of the cell in the reproductive action of a pre-existing organism; a phenomenon peculiarly characteristic of Life. There is no sufficient reason to believe that there is any exception to this rule. So far as we at present know, every Plant and every Animal is the offspring of a parent, to which it bears a resemblance in all essential particulars; and the same may be said of the individual cells, of which the composite Animal and Vegetable fabrics are composed.—It does not always seem requisite, however, that a definite germ-particle should have been prepared by the parent-cell; for there appear to be cases in which new cells originate in the midst of a homogeneous 'protoplasma,' in which no distinct germs can be detected (§ 143). Still, that fluid must have been elaborated by cell-agency; the first stage of cell-formation consists in the appearance of aggregations of molecules, forming *nuclei* from which cells grow; and these new cells bear the same relation to those which prepared the 'protoplasma,' as if they had originated in distinct germs or reproductive molecules.

39. The first stage in the development of the germ or nucleus thus furnished, into a new cell, is the production of the requisite *pabulum* or material. This, however, is effected only in the Vegetable kingdom; for even the simplest Animal cell appears entirely dependent for its means of growth upon a ready-prepared supply of its pabulum, elaborated either directly or indirectly by the agency of Plants. The materials which the Plant requires are simply Water, Carbonic Acid, and a small quantity of Ammonia; for the first two supply it with Oxygen, Hydrogen, and Carbon, in the condition requisite for the production of saccharine or gummy matter; whilst the Ammonia furnishes azote for the albuminous matter, which (according to late researches) is required by every

* We here purposely restrict our history to one form of cell-development, that from *Zoospores*, seen in the *Confervæ* (§142); other modes will hereafter be noticed (§139-141).

Vegetable cell as an essential material for its development.—Now this combining process is probably of a purely Chemical nature (§ 26), the product being a mere chemical compound; and analogous cases are not wanting, amongst the phenomena of Inorganic Chemistry, in which one body exerts an influence upon other bodies, so as to occasion their union or their re-arrangement, without itself undergoing any change. Thus Platinum, in a finely-divided state, will cause the union of Oxygen and Hydrogen at ordinary temperatures; and finely-powdered Glass will do the same at the temperature of 572°. This kind of action is called Catalysis.—A closer resemblance, perhaps, is presented by the act of *fermentation;* in which a new arrangement of particles is effected in a certain compound, by the presence of another body which is itself undergoing change, but which does not communicate any of its elements to the new products. Thus, if a small portion of animal membrane, in a certain stage of decomposition, be placed in a solution of sugar, it will occasion a new arrangement of its elements, which will generate two new products, alcohol and carbonic acid. If the decomposition of the membrane have proceeded further, a different product will result; for, instead of alcohol, mannite and lactic acid will be generated.—There appears no improbability, then, in the idea, that the influence exerted by the germinal molecule is of an analogous nature; and that it operates upon the elements of the surrounding water and carbonic acid, according to purely Chemical laws, uniting the Carbon with the elements of Water, and setting free the Oxygen. This result of the nutritive operations of the simple cellular plants may be easily verified experimentally, by exposing the green scum, which floats upon ponds, ditches, &c., and which consists of the cells of a minute Cryptogamic Plant, to the influence of light and warmth beneath a receiver; it is found that oxygen is then liberated, by the decomposition of the carbonic acid contained in the water. We shall presently have to return to the consideration of the Chemical phenomena of living beings; and shall pass on, therefore, to consider those to which no such explanation applies.

40. The next stage in the process of cell-development consists in the appropriation of the new product thus generated to the enlargement of the living cell-structure;—a phenomenon obviously distinct from the preceding. It is well to observe, that this process, which constitutes the act of *organisation,* may be clearly made out in the higher Plants and Animals to consist of two stages; the first of these being the further preparation or elaboration of the fluid matter, by certain alterations whose nature is not yet clear, so as to render it *organisable,* or fit to undergo organisation; the second being the act of organisation itself, or the conversion of the organisable matter into the solid texture, and the development in it of the properties that distinguish that texture. Thus, for example, we do not find that a solution of Dextrin (or starch-gum) is capable of being at once applied to the development of Vegetable tissue, although it is identical in composition with Cellulose; for it must first pass through a stage, in which it possesses a peculiar glutinous character, and exhibits a tendency to spontaneous coagulation, that seems like an attempt at the production of organic forms. And in like manner, the Albumen of Animals is evidently not capable of being applied to the nutrition of the fabric, until it has first been converted into Fibrin; which also is distinguished by its

tenacious character, by its spontaneous coagulability, and by the fibrous structure of the clot. Now, in both these cases, there is probably some slight modification in Chemical composition, that is, in the proportions of the ultimate elements; but the *principal* alteration is evidently that, which is effected by the re-arrangement of the constituent particles; so that, without any considerable change in their proportions, a compound of a very different nature is generated.

41. Of the possibility of such changes, we have abundant illustrations in ordinary Chemical phenomena; for there is a large class of substances, termed *isomeric*, which, with an identical composition, possess chemical and physical properties of a most diverse character. It is by no means sufficient, however, to attribute the production of Fibrin from Albumen, the organisable from the unorganisable material, to the simple operation of the same agencies as those which determine the production of the different isomeric compounds; for the properties of Fibrin are much more *vitally* distinct from those of Albumen, than they are either *chemically* or *physically;* that is, we find in the one an incipient manifestation of *Life*, of which the other shows no indications. The spontaneous coagulation of fibrin, which takes place very soon after it has been withdrawn from the vessels of the living body, is a phenomenon to which nothing analogous can be found elsewhere; for it has been clearly shown not to be occasioned by any mere physical or chemical change in its constitution; and it takes place in a manner which indicates that a new arrangement of particles has been effected in it, preparatory to its conversion into a living solid. For this coagulation is not the mere homogeneous *setting*, which takes place in a solution of gelatine in cooling; nor is it the aggregation of particles in a mere granular state (closely resembling that of a chemical precipitate), which takes place in the coagulation of albumen: it is the actual production of a simple *fibrous tissue* by the union of the particles of fibrin in a determinate manner, bearing a close resemblance to the similar process in the living body (§ 158). We say, then, that the coagulation of Fibrin, and the production of a fibrous tissue, are the result of its *vital* properties, rather than of chemical or physical agencies; because no substance is known to perform any such actions, without having been subjected to the influence of a living body; and because the actions themselves are altogether different from any which we witness elsewhere. This production of an *organisable* or partially *vitalised* substance, from an unorganisable one, possessing none but chemical properties, and therefore as yet *inert*, so far as the living body is concerned, may be termed *Assimilation;* and it may be conceived, as we have seen, to consist of a new arrangement of the particles of the substance thus changed, *analogous* to that which occurs when one isomeric product is converted into another by some ordinary chemical agency,—Heat or Electricity for example; but *not identical* with it, because it cannot be produced by any other agency than that furnished by a *living* structure.

42. We now come to the act of *Organisation* itself: which seems to consist of a continuation of the same kind of change,—that is, a new arrangement of the particles, producing substances which differ both as to structure and properties from the materials employed, but which may be so closely allied to them in chemical composition, that the difference

cannot be detected. Thus, from the dextrin of plants is generated, in the process of cell-development, the membrane which constitutes the walls of the cells: chemically speaking, there seems to be no essential distinction between these two substances; and yet between the living, growing, reproducing cell, and the inert, unchanging, starch-grain, how wide the difference! So in the animal body, we find that the composition of the fibrin of muscular fibre scarcely differs, in regard to the proportion of its elements, from the fibrin, or even from the albumen, of the blood; and yet what an entire re-arrangement must take place in the particles of either, before a tissue so complex in structure, and so peculiar in properties, as muscular fibre, can be generated! Both in the Plant and the Animal, we find that tissues presenting great diversities both in structure and properties, may take their origin in the same organisable material; but in most cases (at least after the fabric has attained its full development) the new tissue of each kind is formed *in continuity* with that previously existing. Thus in the stem of a growing tree, from the very same glutinous sap or cambium, intervening between the wood and the bark, the wood generates, in contact with its last-formed layer, a new cylinder of wood; whilst the bark produces, in contact with *its* last-formed layer, a new cylinder of bark;—the woody cylinder being characterised by the predominance of ligneous fibre and ducts; and the cortical by the predominance of cellular tissue. In like manner we find that, in animals, muscle produces muscle, bone generates bone, nerve develops nerve, in continuity with it—all at the expense of the materials supplied by the very same blood.

43. The *Nutrition* of tissues, by the organisation of the materials contained in the nutrient fluid with which they are supplied, may be superficially compared, therefore, to the act of crystallization, when it takes place in a mixed solution of two or more salts. If in such a solution we place small crystals of the salts it contains, these crystals will progressively increase by their attraction for the other particles of the same kind, which were previously dissolved; each crystal attracting the particles of its own salt, and exerting no influence over the rest. And if one of the angles of a crystal be broken off, it will be reproduced by a new deposit, so as to preserve the typical form. But it must be borne in mind, that such a resemblance goes no farther than the surface; for the growth of a crystal cannot be really regarded as in the least analogous to that of a cell. The crystal progressively increases by the deposit of particles upon its exterior; the interior undergoes no change; and whatever may be the size it ultimately attains, its properties remain precisely the same as those of the original nucleus. On the other hand, the cell grows from its original germ by a process of *interstitial* deposit; the substance of which its wall is composed, extends itself in every part; and the new matter is completely incorporated with the old. Moreover, as the increase proceeds, we see an evident distinction between the cell-wall and its cavity; and we observe that the cavity is occupied by a peculiar matter, different from the substance of the cell-wall, though obviously introduced through it. Of the essential difference which may exist in composition, between the cell-wall and the contents of the cavity, we have a remarkable example in the case of the simple Cryptogamic plant, which constitutes Yeast, and which differs in no essential part of the history of

its growth from the species already referred to. The substance of its cell-walls is nearly identical with ordinary Cellulose; whilst the contents of the cells are closely allied in composition to the Albumen of Animal fluids. Again, in the fat-cells of Animals, the cell-wall is formed from an albuminous compound; whilst the oily contents agree, in the absence of nitrogen, and in their general chemical relations, with the materials of the tissues of Plants. It is evident, then, that one of the inherent powers of the cell, is that by which it not only combines the surrounding materials into a substance adapted for the extension of its *wall*, but that which enables it to exercise a similar combining power on other materials derived from the same source, and to form a compound,—of an entirely different character, it may be,—which occupies its *cavity*. Now this process is as essential to our idea of a living cell, as is the growth of its wall; and it must never be left out of view, when considering the history of its development.

44. Every kind of cell has its own specific endowments; and generates in its interior a compound peculiar to itself. The nature of this compound is less immediately dependent upon the nutrient materials which are supplied to the cell, than upon the original inherent powers of the cell itself, derived from its germ. Thus we find that the Red Snow and Gory Dew invariably form a peculiar red secretion; and that they will only grow where they can obtain, from the air and moisture around, the elements of that secretion. Again, the Yeast Plant invariably forms an albuminous secretion; and it will only grow in a fluid which supplies it with the materials of that substance. Hence the Red Snow would not grow in a fermentible saccharine fluid; nor would the Yeast Plant vegetate on damp cold surfaces. Yet there is little difference, if any, between their cell-walls, in regard to chemical composition.—So, also, we shall find hereafter, that one set of cells in the animal body will draw into themselves, during the process of growth, the elements of bile; another, the elements of milk; another, fatty matter; and so on:—the peculiar endowments of each being derived from their several germs, which seem to have an attraction for these substances respectively, and which thus draw them together; whilst the cell-wall appears to have a uniform composition in all instances.—The term *Secretion*, or setting-apart, is commonly applied to this operation, to distinguish it from Nutrition or growth; but it is obvious from what has now been stated, that the act of secretion is in reality the increase or growth of the cell-contents, just as the process of enlargement is the increase or growth of the cell-wall; and that the two together make up the whole process of Nutrition, which cannot be properly understood, unless both are taken into account. It is to be remembered, however, that the *contents* of the cell may not be destined to undergo organisation; indeed we shall find hereafter, that the main use of certain cells is to draw off from the circulating fluid such materials as are incapable of organisation; and the operation may be so far attributed, therefore, to the agency of Chemical forces. But we shall find that, in other instances, the cell-contents *are* destined to undergo organisation, and this either within the parent-cell, or after they leave it; here, then, we must recognise a distinct vitalising agency, as exerted by the cell upon its contents.

45. This organising or vitalising influence must be exerted upon a cer-

tain portion of the contents of every cell that is capable of reproducing itself; for it is in this manner that those germs are produced, in which all the wonderful properties are inherent, that are destined to manifest themselves, when they are set free from the parent-cell. This power of *Reproduction* is one of those, which most remarkably distinguishes the living being; and we shall find that, in the highest Animal, as in the humblest Plant, it essentially consists in the preparation of a cell-germ, which, when set free, gradually develops itself into a structure like that from which it sprang. The reproductive molecules or cell-germs are formed, like the tissue and the contents of the parent-cell, from the nutrient materials which it has the power of bringing together and combining; and in their turn they pass through a corresponding series of changes; and at length produce a new generation of similar molecules, by which the race is destined to be continued. Notwithstanding the mystery which has been supposed to attach itself to this process, it is obvious that there is nothing in reality more difficult to understand in the fact, that the parent-cell organises and vitalises the nutriment which it has elaborated, so that it should form the germ of a *new* individual possessing similar properties with itself; than in its incorporating the same material with *its own* structure, and causing it to take a share in its own actions.

46. Finally, the parent-cell having arrived at its full development, having passed through the whole series of changes which is characteristic of the species, and having prepared the germs by which the race is to be propagated, *dies* and *decays*;—that is, all those operations, which distinguish the living organised structure from inert matter, cease to be performed; and it is subjected to the influence of chemical forces only, which speedily occasion a separation of its elements, and cause them to return to their original forms, namely, water and carbonic acid. There is evidence, however, that *interstitial* death and decay are incessantly taking place during the whole life of the being; and that the maintenance of its healthy or normal condition depends upon the constant removal of the products of that decay, and upon their continual replacement. If, on the one hand, those products be retained, they act in the manner of poisons; being quite as injurious to the welfare of the body, as the most deleterious substances introduced from without. On the other hand, if they be duly carried off, but be not replaced, the conditions essential to vital action are not fulfilled, and the death of the whole must be the result.—Now it is to be observed that, as Plants obtain the materials of their *growth* from water and carbonic acid, taking the carbon from the latter and setting free the oxygen, so do they require, as the condition for their *decay*, the presence of oxygen, which may reunite with the carbon that is to be given back to the atmosphere. If secluded from this, the vegetable tissues may be preserved for a long time without decomposition. Generally speaking, indeed, they are not prone to rapid decay, except at a high temperature; and hence it is that we have so little evidence, in Plants, of the constant interstitial change, of which mention has just been made. Its existence, however (at least in all the softer portions of the structure), is made evident by the fact, that a continual extrication of carbonic acid takes place, to an amount which sometimes nearly equals that of the carbonic acid decomposed, and of the oxygen set free, in the act of Nutrition (§ 73).

The latter operation is only effected under the stimulus of sun-light; the former is constantly going on, by day and by night, in sunshine and in shade; and if it be impeded or prevented by want of a due supply of oxygen, the plant speedily becomes unhealthy. Now this extrication of the products of interstitial decay is termed *Excretion.* It is usually confined in Plants to the formation of carbonic acid and water, by the union of the particles of their tissues with the oxygen of the air,—a process identical with that, which occurs after the death of the entire structure. But in Animals it is much more complicated; owing to the larger number of constituents in their fabric, and to the much greater variety in the proportions in which these are combined; hence the products of interstitial decomposition are much more numerous and varied, and several distinct modes are devised for getting rid of them. Moreover, as the Animal tissues are much further removed than the Vegetable from the composition of Inorganic bodies, they are subject to much more rapid and constant decay; and we shall find that this decay is so considerable in amount, as to require on the one hand a very complex excretory apparatus to carry off the disintegrated matter, and on the other a large supply of nutrient material to replace it.

47. The preceding history may be thus summed up.—I. The Vegetable cell-germ draws to itself, and combines together, certain inorganic elements; and thereby produces a new and peculiar compound, as the *pabulum* for its further development. This compound, however, exhibits no properties that distinguish it from others, in which ordinary *Chemical* agencies have been concerned; and we may, therefore, regard the first act of the cell-germ as of a purely chemical nature.—II. The cell-germ then exerts an agency upon the *pabulum* thus prepared; by which a new arrangement of its particles is produced. This new arrangement is evidenced by the new and peculiar qualities of the fluid, which show that it is something more than a mere chemical compound, and that it is in the act of undergoing the process of organisation.—III. The Organisation of this elaborated pabulum then takes place; its materials are withdrawn from the fluid, and incorporated with the solid texture; and in thus becoming part of the organised fabric, they are caused to exhibit its own peculiar properties.—IV. At the same time, another portion of this pabulum is gradually prepared to serve as the germ of a new cell, or set of cells, by which the same properties are to be exhibited in another generation.—V. By a Chemical operation resembling that concerned in the first preparation of the pabulum, a certain secretion, more or less differing from it in character, but not destined to undergo organisation, is formed in the cavity of the cell.—VI. A decomposition or disintegration of the organised structure is continually going on, by the separation of its elements into simpler forms, under the influence of purely Chemical agencies; and the setting-free of these products by an act of excretion, is thus incessantly restoring to the inorganic world a portion of the elements that have been withdrawn from it.—VII. When the term of life of the parent-cell has expired, and its reproductive molecules are prepared to continue the race, the actions of nutrition cease; those of decomposition go on unchecked; and the death of the structure, or the loss of its distinguishing vital properties, is the result. By the decomposition which then takes place with increased rapidity, its elements are restored to the

inorganic world; presenting the very same properties as they did when first withdrawn from it; and becoming capable of being again employed, by any number of successive generations of living beings, to go through the same series of operations.

48. The germ, however, can in no wise appropriate the materials of its growth, nor develop itself into a cell, unless it be brought under the influence of certain Physical Agents. The first production of the pabulum out of water, carbonic acid, and ammonia, requires the agency of Light; this operation, however, is limited to the Vegetable kingdom, and (as already remarked, § 39) is probably rather of a *chemical* than of a *vital* nature.* But that a certain amount of Heat is necessary to all vital activity, is a fact of universal applicability; as is also the fact that, *ceteris paribus*, the amount of activity increases with the temperature, until this reaches an elevation which becomes injurious, either by over-stimulating the vital properties of the tissues, or by effecting some physical change. These truths will be more fully developed hereafter (Sect. 3); but it is necessary here to state them, in order that the relation which we shall presently attempt to demonstrate between Vital and Physical phenomena may be comprehended (§ 54).

49. The solitary cell which constitutes the entire organism of the humblest of living beings, is also the simplest organic element of the most elaborate and heterogeneous fabrics. We shall hereafter see (CHAP. IV.) that all the Vegetable tissues originate from similar cells, and that all the truly *vital* operations are performed, even in the fully-developed plant or tree, by tissues retaining their primitive cellular character; and notwithstanding that, at first sight, the Animal fabric appears to have very little in common with that of the Plant, we shall find that even in it the same general facts are demonstrable, both as to the origin of all the tissues (directly or indirectly) in cells, and as to the instrumentality of cells in the performance of all the truly vital actions of the organism. Hence we are justified in affirming that, as the simple cell presents us with the type of Organised structure (§ 15), so our idea of what is essential to Life should be founded upon those vital phenomena which every cell exhibits in the course of its history as an individual, and upon the distinctive attributes which we must consequently regard it as possessing. Putting aside, then, such portions of its series of actions as we appear justified in referring to Chemical agencies, we find that the incorporation of the *pabulum* into the substance of the cell itself, which is taking place not only during its first development but throughout the whole period of its vital activity, must be considered as its most important manifestation of life; on this all subsequent phenomena are dependent; and whatever may be the endowments which the cell is to possess in its perfect state, they are inseparably connected with this process of Organisation.

50. These endowments are extremely various; and their diversities

* The following experiment of Schleiden's confirms this view. Having allowed some grains of Oats to germinate in darkness until the buds were four inches long, he cut them down, washed them and dried them between blotting paper, and then placed them upon white paper to be dried in the sun. They were previously of a pale yellow colour; but after six hours' exposure to the solar rays, during which the plants were almost perfectly dried, they had all become of a perfectly green colour. It can scarcely be supposed that any peculiarly *vital* agency was concerned in the transformation under such circumstances.

constitute, in fact, the differences between the properties of the several tissues. Thus we find certain cells whose agency is exerted in developing a preliminary organisation or vitalisation in the liquid *pabulum* (§ 39); which, when thus prepared, is to be yielded up for appropriation by other cells in process of development. This process we have designated by the term *Assimilation*.—Again, there are other cells which are distinguished for their power of *Reproduction;* their function as individuals being apparently confined to the preparation of the germs of a new generation.—Others, again, are characterised by the change of form which they undergo, after having attained their development as cells; this change of form, rendering them subservient to some new purposes in the economy, originates entirely in their own vital activity, and must be considered as indicating a power of *Morphological Transformation*.—In many other instances, the endowments of the cells are manifested in their power of effecting *Chemical Transformations*, of a kind which we do not meet with under any other circumstances.—And both in Animals and Plants, but in the former more especially, we find a number of cells endowed with the power of giving rise to *Mechanical Motion;* either by a sudden and temporary change of form, as in the contraction of a muscle, and in most of the sensible movements of plants; or by the vibratile action of certain filaments attached to the exterior of the cells, and termed *cilia* (§ 183).—Other cells in the Animal body have it as their special endowment to generate that peculiar power which we call *Nervous Agency;* the operation of which is more nearly allied to that of Electricity than to that of any other power with which we are acquainted, though not identical with it (§ 246).

51. Now all these endowments may be regarded as related to each other through their common parentage; being all manifestations of the power existing in the living cell, and consequently of the power by which that cell was formed. We may best express the distinctive character of Vital Activity, therefore, by describing it as essentially manifested in *Cell-formation.** And it will be convenient to employ the term *Vital Force* as an expression of the agency by which this is accomplished.—That the various operations, enumerated in the preceding paragraph as effected by cell-agency, are to be regarded (however dissimilar in themselves) as so many manifestations of this Vital Force, acting through different instruments, appears probable from this consideration;—namely, that each endowment is so far inconsistent with the rest, that it would seem as if the expenditure of the vital force of the cell upon that one purpose unfitted it for any other.—Thus, the Assimilating cells do not exercise chemical transformation; they do not undergo change of form; they do not exert mechanical or nervous power; and it is doubtful whether they reproduce themselves. So, again, the cells which are specially endowed with the Reproductive power seem to possess no other; whatever may be the peculiar endowments of their progeny. And the cells which are operative in Chemical transformations are destitute of the power of

* The term *formation* is here intended to include all that we separately express by the terms *development*, *growth*, and *maintenance;*—the first expressing the progressive evolution of a new form; the second, its simple increase; and the third, its continued nutrition after the attainment of its complete development and full dimensions. See Mr. Paget's Lectures on Nutrition, &c., in Medical Gazette for 1847.

exciting mechanical movement, as they are also of reproductive force. And the same is true conversely of the cells concerned in Mechanical Movement, which are incapable of exercising chemical transformations, as they are of continuing their race. We shall hereafter meet with a variety of exemplifications of these general statements ; and shall at present only refer to the production of the secreting cells within a glandular follicle (§ 157), to that of the cells of the muscular fibrillæ within the tubular fibre (§ 226), to the difference of endowment between the ciliated and the secretory epithelium-cells (§ 183), and to the cessation of ciliary action in the *zoospores* of the Algæ when they begin to exercise chemical transformations (§ 185), as illustrations of the argument.

52. Another argument for the mutual relation here contended for, is furnished by the analogy of the Forces operating in the Inorganic world. Reasons have been lately given* for the belief, that Physical Forces of every kind,—namely, Heat, Light, Electricity, Magnetism, Chemical Affinity, and that which produces or resists Motion,—are all 'correlative,' or have a relation of mutual dependence ; each being capable of producing any one of the rest, either directly or through the medium of some other ; and each being consequently producible by any of the remainder. And we may observe that this conversion of one force into another is due to its passage through some material body. Thus the Motion of a body, retarded or resisted by friction, gives rise to Heat ; and conversely, heat applied to any form of matter produces its expansion, *i. e.* Motion. The friction of two dissimilar bodies produces not merely heat but Electricity ; and Heat itself, when made to act on certain combinations of metals, also produces Electricity ; whilst, on the other hand, the Electric current may produce Heat, Light, Magnetism, or Motion, according to the nature of the substances through which it is transmitted. Light, Heat, and Electricity, again, are closely related to Chemical Affinity, which is often specially excited by them (§ 68), and which can, in its turn, generate these Forces ; a material *substratum* being required in both cases. So in the beautiful experiments by which Faraday has brought Magnetism to act upon Light, the effect is only produced when the ray is being transmitted through certain material substances.

53. Now the relation which exists among the several endowments peculiar to Organised Structures, and therefore distinguished as *Vital*, appears to be most appropriately expressed in the very same form ;—they may be said to be mutually *correlated*, in such a manner that each can either directly or mediately produce the other.—The most universal and characteristic of these forces has been shown to be that from which 'Cell-formation' proceeds ; and out of this arise all the other agencies. Thus, the tissue of the Muscle is constructed solely with a view to its manifestation of contractile power ; and the development of Nervous matter has reference entirely to the peculiar agencies in which it is to be concerned. The contraction of a Muscle, and the production and transmission of a Nervous impression, are phenomena of a distinctly vital character ; and it will hereafter appear that, with the manifestation of the peculiar attributes of these tissues, they cease to maintain their existence as living structures,—as if the general Vital force which they previously possessed,

* See Prof. Grove on the "Correlation of Physical Forces," 1846.

had been expended by its conversion into the Muscular and Nervous powers. Here, too, we see that a certain material *substratum* is requisite for the conversion of one force into another; the vital force only being able to produce contractile action, by means of a certain variety of organised structure which we term Muscle; whilst it can only develope itself as nervous agency, through that peculiar form of tissue which is called Nerve. Again, we find that the nervous force can evoke the muscular, and this in a degree proportional to the intensity of its own excitement; and nervous agency would seem capable of influencing cell-formation, in such a manner as to justify the idea that it is actually re-converted into the form of vital force concerned in it.

54. If this view be considered to have a certain claim to reception as (to say the least) a probable hypothesis of the mutual relation of the various forms of Vital activity, we have next to consider whether any similar 'correlation' can be traced between the Vital and the Physical forces; since, if such can be shown to exist, the probability of the hypothesis will be greatly increased. Now it will be shown (Sect. 3) that all Vital activity is dependent upon the constant agency of one at least of the Physical Forces already enumerated—viz. Heat; and that Light and Electricity, operating through Chemical Affinity, are also concerned in sustaining particular forms of it. These agents, indeed, which have long been known under the term 'Vital Stimuli,' may perhaps be more justly considered as the real Forces concerned in the production of Vital phenomena, modified in their operation by the organised structure through which they are exerted; just as Heat becomes Electricity when it passes through a certain combination of metals. And this idea is borne out by the fact, that the Vital Forces appear capable of being in their turn reconverted into the Physical; as in those numerous cases already referred to (§ 50), in which the whole energy of one set of cells is directed to the performance of some *chemical* transformation, and that of another to the production of *mechanical* movement; as well as to other instances, in which it seems nearly certain that the liberation of Electricity, and perhaps also of Heat and Light, are due to Nervous power operating through a particular structure (CHAP. XVII.). Indeed, when we come to consider the manifestations of Nervous agency in more detail, we shall find that it is at least as closely correlated to Electricity, as Electricity is to Light or Heat (§ 246); and the intimacy of this relationship, existing between one of the most special or peculiar of all the Vital forces and one of an exclusively Physical nature, strongly confirms the idea just expressed in regard to the mutual connection of the remainder.

55. In the consideration of this or any other Theory of Life, it is very important to keep constantly in view, that it has relation only to the agencies concerned in the formation and conservation of the *material* organism. The phenomena of *mind* belong to a category altogether distinct, and cannot be considered as bearing any necessary relation to the former.—It seems also requisite to point out, that in speaking of *forces* as possessing an absolute existence, it is not intended on the one hand to imply that they are anything else than 'affections of matter;'* nor on the other to regard them in any other light than as the direct operation of the primal all-sustaining Cause. We can form no conception of matter,

* See Grove on the "Correlation of Physical Forces," p. 7.

except as possessing *properties,* which when in action give rise to *powers* or *forces;* whilst on the other hand, we cannot think of forces, except as operating through some form of matter, of whose properties they are the manifestation. The existence of matter, and the action of the forces to which material phenomena (whether physical or vital) are attributable, are alike the expressions of the Divine Will; and our aim must be limited to the discovery of the plan, according to which it has pleased the Creator to develop and maintain the existing condition of the Universe we inhabit. (See § 5, *note*).

2. *Of the Material Conditions of Life.—Food, Water, and Oxygen.*

56. Every organised structure requires to be placed in certain *conditions,* in order that its Vital properties may be manifested in action; and although this, as we have seen, is no more than is true of every form of Inorganic Matter, in regard to its Physical and Chemical properties, yet the conditions of vital activity are more numerous, and require to act in a more complex combination. It will simplify our consideration of them to arrange them into two groups; the first comprehending the *material substances,* of which a continual supply is required by all living beings; the second including certain *forces* brought to bear upon these substances, without the agency of which they could not be appropriated by the organism. There is an important difference in the relation of these two sets of conditions to Vital activity; for whilst that activity cannot be increased by any augmentation of the *material supplies* beyond the amount normally required, it can be promoted by a more energetic operation of these *forces,* as well as retarded by their diminution; and the demand for the former will be modified accordingly.—A simple analogy may render these distinctions obvious. For the development of the power of the Steam-engine, two things are essential; the material substance, Water; and the force of Heat, by which the particles of the water are made to repel each other, and a powerful mechanical agency thereby obtained. Now if the Steam-engine be supplied with as much water as is required to maintain the proper amount in the boiler, no increase of that supply can of itself produce the least augmentation of the power of the machine; but if increased heat be applied, the elastic force of the steam is augmented, and the power of the machine proportionably raised; whilst as the water is thereby more rapidly converted into steam, a larger supply of it is required for the maintenance of that power.

57. The material conditions essential to Life are three;—Food, or substances capable of being converted into the solid portion of the organised fabric; Water, for the maintenance of the due proportion of liquid in its constitution; and Oxygen, for promoting various changes in the assimilated matter, by which it is applied to a greater variety of purposes, and also for uniting with the particles destined for excretion, so that they may pass off in the form in which they can be most readily got rid of.

58. *Of Food.*—It has been shown that all Vital Action is dependent in some way or other upon Cell-formation (§ 51); and it is obvious that the supply of materials for that formation is essential to its continuance. These materials, in the case of Plants, are exclusively supplied by the

Inorganic world; for even when organic substances are used to promote vegetable growth (as in the case of manures), they act only by setting free the carbonic acid and ammonia, which, with water, constitute the staple food of Plants, and at the same time, it may be, by supplying certain mineral constituents of importance to their nutrition. It is only among the lowest tribes of the Vegetable kingdom, however, that every cell directly obtains its food from the Mineral world; for in all but the most homogeneous organisms, we find certain portions of the structure set apart for the purpose of effecting the requisite combination of Oxygen, Hydrogen, Carbon, and Nitrogen, and of thus preparing a *pabulum*, which shall serve not merely for their own growth and maintenance, but also for the formation of new parts elsewhere, which are destined for purposes of a different kind, and are not themselves capable of obtaining the materials of their growth from external sources. Thus we shall find that in the perfect Plant, the requisite water is taken in by the spongioles of the roots; whilst the carbonic acid and ammonia are in part imbibed in solution in that liquid, but are in part also derived (the former more especially) from the atmosphere, through the agency of the cells of the leaves and green surfaces; and the compounds formed in these cells serve as the pabulum for the growth of every other part of the organism,—the new leaf- and flower-buds being developed, and the increase of substance taking place in the solid framework of the stem, roots, and branches, entirely at its expense. The food of Animals entirely consists, as formerly stated, of compounds which have been already generated by the agency of Plants; but these compounds have first to be reduced to a condition in which they can be applied, by the various component tissues of the fabric, to their formative processes; and it is in the nutrient fluid thus prepared and circulated through the body, that each individual cell finds the materials of its growth.

59. The various sources of the demand for food in the living Plant and Animal, and the circumstances under which this demand varies from time to time, will hereafter have to be considered in detail (CHAP. X.); at present it will be sufficient to state the general principle—as one in which theory and experience alike coincide,—that the demand for Food is proportional to the Vital Activity; and this alike in the organism as a whole, and in each individual component of it. We may refer, by way of illustration, to the fact that cold-blooded animals require a much smaller amount of food, and are able to sustain a deprivation of it for a much longer period, at low than at moderately high temperatures; their vital activity being at its minimum in the former case, and at its maximum in the latter (§ 97). In this respect, also, the hybernating Mammal corresponds with the cold-blooded Reptile, to which it is for a time assimilated in its whole physiological condition (§ 95).

60. The following curious fact, well known to Naturalists, affords proof that the *kind* of food supplied may not only influence the ordinary processes of nutrition, but may produce an important modification in the development of the organism. In every hive of Bees, the majority of individuals consists of *neuters*, which have the organs of the female sex undeveloped, and are incapable of reproduction; that function being restricted to the *queen*, who is the only perfect female in the community. If by any accident the queen be destroyed, or if she be purposely

removed for the sake of experiment, the bees choose two or three from among the neuter eggs that have been deposited in their appropriate cells, which they have the power of converting into queens. The first operation is to change the cells in which they lie into *royal* cells, which differ from them considerably in form, and are of much larger dimensions; and when the eggs are hatched, the maggot is supplied with food of a very different nature from the farina or bee-bread which has been stored up for the nourishment of the workers, being of a jelly-like consistence and pungent stimulating character. After the usual transformations the grub becomes a perfect queen, differing from the neuter bee into which it would otherwise have changed, not only in the development of the reproductive system, but in the general form of the body, the proportionate length of the wings, the shape of the tongue, jaws, and sting, the absence of the hollows on the thighs in which the pollen is carried, and the loss of the power of secreting wax.

61. *Of Moisture.*—Independently of the utility of Water as an article of food, supplying two of the most important materials (oxygen and hydrogen) for the construction of the tissues, there can be no doubt that a certain supply of moisture is requisite as one of the conditions without which no vital actions can go on. It has been already remarked, indeed, that one of the distinguishing peculiarities of organised structures is the universal presence of solid and fluid parts (§ 22); and this in the minutest portion of the organism, as well as in the aggregate mass. And for all the *vital* as well as *chemical* actions in which these structures are concerned, the presence of liquid is essential. All nutrient materials must be reduced to the liquid form, before they can be assimilated by the solids; and, again, the solid matters which are destined to be carried off by excretion, must be reconverted to the fluid state, before they can be thus withdrawn from the body. In all these operations, water is the liquid employed; and no other could be substituted for it. Taken as a whole, the tissues of Animals contain a much larger proportion of water than those of Vegetables; but this is because a considerable part of the fabric of the latter is usually destined for a long-continued existence, and is removed by consolidation out of all participation in vital actions (§ 24). With such structures we might compare that of Coral among Animals; in which the proportion of organic matter is smaller than in any plants, and liquid can scarcely be said to exist; whilst, on the other hand, the tissues of many rapidly-growing Fungi are as soft, and contain as large a proportion of water, as any of those most actively concerned in the maintenance of Animal life. As a general rule, the amount of water contained in an organised tissue bears a proportion to its vital activity; but this rule is not without numerous exceptions. For although we never find any active vital operations carried on by organised structures in which the solid elements predominate, yet there are numerous cases in which a great preponderance of fluids is obvious, without any corresponding indication of vital energy. Thus many succulent Plants, such as the *Sedums* of our own country, the *Cacti* and *Euphorbiæ* of the tropics, are rather remarkable for their slow than for their rapid growth; the succulence of their tissues having relation to their scanty opportunities of obtaining moisture, and being designed to furnish a store for the wants of the organism, when external sources are deficient. So, again, among

Animals, we find an entire group, that of *Acalephæ* (§ 293), characterised by the extremely small proportion that the solid matter bears to the water of their tissues; nevertheless their vital activity is extremely low; and the purpose of the arrangement does not seem obvious, unless it be to endow them with a bulk altogether disproportionate to the amount of organic material consumed in its production.

62. When we take a general survey of the Vegetable kingdom, we find that there are some Plants which are capable of adapting themselves to a great variety of situations, differing widely as to the amount of Moisture which their inhabitants can derive from the soil and atmosphere; and we may generally notice a marked difference in the mode of growth, when we compare individuals that have sprung up under opposite circumstances. Thus a plant from a dry exposed situation shall be stunted and hairy, whilst another, of the same species, but developed in a damp sheltered situation, shall be rank and glabrous. But in general there is a certain quantity of moisture congenial to each species; and the excess or deficiency of this condition has, in consequence, as great an influence in determining the geographical distribution of Plants, as the amount of light and heat. Thus the Orchideæ and Tree Ferns of the tropics grow best in an atmosphere loaded with dampness; whilst the Cactus tribe, for the most part, flourishes best in dry situations. The former become stunted and inactive, if limited in their supply of aerial moisture; whilst the latter, if too copiously nourished, become dropsical and liable to rot. Among the plants of our own country, we find a similar limitation; a moist boggy situation being indispensable to the growth of some, whilst a dry exposed elevation is equally essential to the healthy development of others. There is a beautiful species of exotic Fern, the *Trichomanes speciosum,* the rearing of which has been frequently attempted in this country and elsewhere, without success; but which only requires for its healthy development an atmosphere saturated with dampness, being easily reared in one of Mr. Ward's closed glass-cases. In this, as in similar examples, it is only necessary to imitate as closely as possible the conditions under which the species naturally grows; and sometimes this can only be accomplished by surrounding the plant with small trees and shrubs, so as to give it a moister atmosphere than it could otherwise attain. Professor Royle mentions the growth under such circumstances, of a fine specimen of the *Xanthochymus dulcis,* one of the *Guttiferæ* or Gamboge-trees, in the garden of the king of Delhi; this tree is naturally found only in the southern parts of India; and the success of its cultivation in this northerly situation is entirely due to its being sheltered by the numerous buildings within the lofty palace wall, surrounded by almost a forest of trees, and receiving the benefit of perpetual irrigation from a branch of the canal which flows through the garden.

63. In regard to the influence of external moisture upon Animal life, there is much less to be said; since the mode in which fluid is received into the system is so entirely different. It may be remarked, however, that Animals habitually living beneath the water, like submerged Plants, are usually incapable of sustaining life for any length of time when removed from it, in consequence of the rapid loss of fluid which they undergo from their surface. It is, however, by the desiccation of the

respiratory surface, preventing the due aeration of the blood, that the fatal result is for the most part occasioned; since we find that when there is a special provision to prevent this, as in the case of certain Fishes and Crustacea, the animals can quit the water for a great length of time. The amount of Atmospheric Moisture is one of those conditions, which are collectively termed Climate, and which influence the geographical distribution of Animals, no less than that of Plants. But it is difficult to say how far the variations in moisture act alone. There can be no doubt, however, of their operation; for every one is conscious of the effect, upon his health and spirits, of such variations as take place in the climate he may inhabit. The two principal modes in which these will operate, will be by accelerating or checking the exhalation of fluid from the skin and from the pulmonary surface; for when the air is already loaded with dampness, the exhaled moisture cannot be carried off with the same readiness, as when it is in a condition of greater dryness; and it will consequently either remain within the system, or it will accumulate and form sensible perspiration (CHAP. XV).

64. Although, as already stated, no vital actions can go on without a reaction between the *solids* and *fluids* of the body, yet there may be an entire loss of the latter, in certain cases, without necessarily destroying life; the structure being reduced to a state of dormant vitality (§ 114), in which it may remain unchanged for an unlimited period; and yet being capable of renewing all its actions, when moisture is again supplied. Of this we find numerous examples among both the Vegetable and the Animal kingdoms. Thus the Mosses and Liverworts, which inhabit situations where they are liable to occasional drought, do not suffer from being, to all appearance, completely dried up; but revive and vegetate actively, as soon as they have been thoroughly moistened. Instances are recorded, in which Mosses that had been for many years dried up in a Herbarium, have been restored by moisture to active life. There is a Lycopodium inhabiting Peru, which, when dried up for want of moisture, folds its leaves and contracts into a ball; and in this state, apparently quite devoid of animation, it is blown hither and thither along the ground by the wind. As soon, however, as it reaches a moist situation, it sends down its roots into the soil, and unfolds to the atmosphere its leaves, which, from a dingy brown, speedily change to the bright green of active vegetation. The *Anastatica* (Rose of Jericho) is the subject of similar transformations; contracting into a ball, when dried up by the burning sun and parching air; being detached by the wind from the spot where its slender roots had fixed it, and rolled over the plains to indefinite distances; and then, when exposed to moisture, unfolding its leaves, and opening its rose-like flower, as if roused from sleep. And there is a blue Water-Lily, abounding in several of the canals at Alexandria, which at certain seasons become so dry, that their beds are burnt as hard as bricks by the action of the sun, so as to be fit for use as carriage roads; yet the plants do not thereby lose their vitality; for when the water is again admitted, they resume their growth with redoubled vigour.

65. Among the lower Animals, we find several of considerable complexity of structure, which are able to sustain the most complete desiccation. This is most remarkably the case in the common *Wheel-Animalcule*; which may be reduced to a state of most perfect dryness, and kept in

this condition for any length of time, and which will yet revive immediately on being moistened (§ 310). Experiments have been carried still further with the allied tribe of *Tardigrades;* individuals of which have been kept in a vacuum for thirty days, with sulphuric acid and chloride of calcium (thus suffering the most complete desiccation the Chemist can effect), and yet have not lost their vitality. It is singular that in this desiccated condition, they may be heated to a temperature of 250°, without the destruction of their vitality; although, when in full activity, they will not sustain a temperature of more than from 112° to 115°. Some of the minute Entomostracous Crustacea, which are nearly allied to the Rotifera, appear to partake with them in this curious faculty. Many instances are on record, in which Snails and other terrestrial Mollusks have revived, after what appeared to be complete desiccation; and the *eggs* of the Slug, when dried up by the sun or by artificial heat, and reduced to minute points only visible with the Microscope, are found not to have lost their fertility, when they are moistened by a shower of rain, or by immersion in water, which restores them to their former plumpness. Even after being treated eight times in this manner, the eggs have been hatched when placed in favourable circumstances; and even eggs in which the embryo was distinctly formed, have survived such treatment without damage.—That such capability should exist in the animals and eggs just mentioned, shows a remarkable adaptation to the circumstances in which they are destined to exist; since were it not for their power of surviving desiccation, the races of Wheel-Animalcules and Entomostraca must speedily become extinct, through the periodical drying-up of the small collections of water which they inhabit; and a season of prolonged drought must be equally fatal to the terrestrial Mollusca.

66. It would seem that many Cold-blooded animals are reduced, by a moderate deficiency of fluid, to a state of torpidity closely resembling that induced by cold; and hence it is, that during the hottest and driest part of the tropical year, there is almost as complete an inactivity, as in the winter of temperate regions. The common Snail, if put into a box without food, constructs a thin operculum or partition across the orifice of the shell, and attaches itself to the side of the box: in this state it may remain dormant for years, without being affected by any ordinary changes of temperature; but it will speedily revive if plunged in water. Even in their natural haunts, the terrestrial Mollusca of our own climates are often found in this state during the summer, when there is a continued drought; but with the first shower they revive and move about. In like manner it is observed that the rainy season, between the tropics, brings forth the hosts of insects, which the drought had caused to remain inactive in their hiding-places. Animals thus rendered torpid seem to have a tendency to bury themselves in the ground, like those which are driven to winter-quarters by cold. Mr. Darwin mentions that he observed with some surprise at Rio de Janeiro, that, a few days after some little depressions had been changed into pools of water by the rain, they were peopled by numerous full-grown shells and beetles.—The torpidity consequent upon drought is not confined to Invertebrated animals. There are several Fish, inhabiting fresh-water, which bury themselves in the mud when their streams or pools are dried up, and which remain there in a torpid condition until they are again moistened. This is the case, also, with the

curious *Lepidosiren* (§ 323), which is an inhabitant of the upper parts of the river Gambia, which are liable to be dried up during much more than half the year; and the whole of this period is spent by it in a hollow which it excavates for itself deep in the mud, where it lies coiled up in a completely torpid condition,—whence it is called by the natives the *sleeping-fish*. When the return of the rainy season causes the streams to be again filled, so that the water finds its way down to the hiding-place of the Lepidosiren, it comes forth again for its brief period of activity; and with the approach of drought, it again works its way down into the mud, which speedily hardens around it into a solid mass. In the same manner, the *Proteus* (§ 323), an inhabitant of certain lakes in the Tyrol, which are liable to be periodically dried up, retires at these periods to the underground passages that connect them, where it is believed to remain in a torpid condition; and it thence emerges into the lakes, as soon as they again become filled with water. The Lizards and Serpents, too, of tropical climates appear to be subject to the same kind of torpidity, in consequence of drought, as that which affects those of temperate regions during the cold of winter. Thus Humboldt has related the strange accident of a hovel having been built over a spot, where a young Crocodile lay buried, alive though torpid, in the hardened mud; and he mentions that the Indians often find enormous Boas in the same lethargic state; and that these revive when irritated or wetted with water.—All these examples show the necessity of a fixed amount of fluid, in the animal structure, for the maintenance of vital *activity*: whilst they also demonstrate, that the preservation of the vital *properties* of that structure is not always incompatible with the partial, or even the complete, abstraction of that fluid; the solid portions being then much less liable to decomposition by heat, or by other agencies, than they are in their ordinary condition.

67. *Of Oxygen.* All living beings have constant access to Oxygen, in a free or uncombined state, either as a constituent of the Atmosphere, or as diffused through the Water which they inhabit. It does not appear, however, that this free oxygen is ever appropriated by Plants, as one of the materials of the primary organic compounds which they generate from the Inorganic elements; for it seems to be an essential condition of the union of these elements in such peculiar modes, that they should be in a *nascent* state,—that is, in the act of being set free from a previous combination. When these compounds have been once formed, however, they may undergo further modifications within the living organism, whether Vegetable or Animal (§ 28); and for some of these modifications the addition of a further measure of oxygen seems needed, which is absorbed from the free oxygen of the atmosphere. Such a use of oxygen, however, properly falls under the head of Food; since this element, when thus introduced into the organism, becomes a constituent part of it.—The chief dependence of Vital Activity upon the access of Oxygen is quite of another character; for it results from the necessity that exists for the continual getting rid of the effete matter, which has served its purpose in the organism, and which would be injurious if retained there. By this action of free oxygen on the Vegetable tissues, carbonic acid and water are being continually produced, even during their state of greatest vital activity (§ 46); and thus the materials which they have drawn from the Inorganic world are returned to it again in the same form. The same excretions are constantly

set free, by the very same agency, from the Animal body; but in addition to these, other azotized compounds are thrown off, for the production of which free oxygen introduced from without is required.—This source of demand for Oxygen is common to all living beings, and its energy is proportioned to their vital activity. There are other purposes, however, of a less universal character, to which it is subservient; thus the manifestation of the Muscular and Nervous power of Animals appears to be essentially dependent upon the union of oxygen with the elements of their tissues (§§ 236 and 243); and the maintenance of the standard heat of Warm-blooded animals is dependent upon a process of combustion in the interior of their bodies, to which the presence of free Oxygen is as essential as it is to the burning of hydrogen or carbon elsewhere. This process, however, will require special consideration on a future occasion (CHAP. XVI. Sect. 3).

3. *Of the Dynamic Conditions of Life;—Light, Heat, and Electricity.*

68. No organised fabric, however perfect in its structure and composition, and however amply supplied with the materials requisite for its growth and maintenance, can perform a single Vital Action without the concurrence of certain other agencies, of the class commonly termed Physical Forces. These afford the *stimulus* or provocative to activity; and the germ could no more develope itself into a cell without the operation of one (at least) of them, than it could do so with their assistance but without the supply of the alimentary materials. Hence these agents,—Light, Heat, and Electricity,—when viewed with reference to their participation in the phenomena of Life, have been termed *Vital Stimuli.* Their operation on Organized structures is not altogether without analogy to the influence they possess over inorganic matter, which is exerted in the production of Chemical phenomena. Thus the union between two substances, which previously showed no tendency to combine, although in close contact with each other, may be frequently caused to take place, with various manifestations of energy, by the momentary influence of one of these agents. When Chlorine and Hydrogen are mingled together in a bottle, and this is placed in darkness, they may be kept for any length of time without change. But the agency of *Light* brings their previously-dormant affinities into a state of activity; for if they be subjected to the direct rays of the sun, an instantaneous explosion takes place, with production of Hydrochloric acid; whilst, if exposed for some time to diffused daylight, they unite more slowly. The same effect may be produced by *Heat;* for a union of the two gases occurs, with a violent explosion, when any incandescent substance is introduced into the mixture. And a like result is produced by *Electricity;* the transmission of a spark through the mixture producing an explosion, with a generation of hydrochloric acid. In neither of these cases can we consider that Heat was excluded; for although a high temperature was required to bring about the union when no other agent was in operation, yet neither Light nor Electricity could have effected the union, unless the elements had been in the gaseous condition which they present at all ordinary temperatures, in virtue of the latent heat they contain.

69. In like manner we shall find that, of these three agents, *Heat* is

the one which is most *universally* concerned in the production of Vital phenomena. The action of *Light* is indispensable for certain purposes, and we shall see that upon it essentially depends the first production of Organic compounds; but it would possess no efficacy whatever, without the simultaneous operation of Heat; and a very large proportion of the vital operations of Animals have but little direct dependence upon it. Of the operation of *Electricity*, we have little certain knowledge, although there is reason to believe that it occasionally exerts an important influence.* In regard to all these Vital Stimuli it may be observed, that the dependence of the vital actions of an Organism upon their constant influence varies with its degree of organisation (§ 20); so that beings of simple organisation are capable of enduring a deprivation of these stimuli, which would be fatal to those higher in the scale. This will be partly understood, when it is borne in mind that the higher the development of the living being, the more complete is the distribution of its different actions amongst separate organs,—the more close, therefore, is their mutual dependence,—and the more readily, in consequence, are they all brought to a close by the interruption of any one. But there is no doubt, that the actions of even the individual parts of the higher organisms require for their excitement a greater supply of these stimuli, than the similar actions of the corresponding parts in the lower: whilst if these stimuli be exerted upon the lower with the intensity that is required for the higher, they destroy the vital properties of the tissues altogether, by the excess of their action. This distinction is most obvious in regard to the relative influence of Heat upon warm-blooded and cold-blooded animals; of which examples will be given hereafter.

70. It may also be observed of the influence of these, as of that of other stimuli whose agency is less general, that it is rather *relative* than *absolute;* being frequently dependent upon the degree of *change,* rather than upon the actual measure of the *amount* of the stimulus. This constitutes a marked difference between the influence of these stimuli on mere chemical compounds, and their operation on bodies endowed with vitality. In the former case, their action is always uniform; thus the same amount of Heat, the same exposure to Light, the same charge of Electricity, would be required to produce a given Chemical effect, how often soever the action might be repeated. But this is not the case with living bodies; since an increase or diminution in the intensity of the vital stimuli, which, if made *suddenly,* would be scarcely compatible with the continuance of Life, may be so brought about as to produce no marked change in its phenomena,—the organism possessing a certain power of adapting itself to conditions which are habitual to it, and thus allowing great changes in these conditions to be *gradually* effected, without any serious disturbance.—Thus of two individuals of the same species, one may become torpid at a temperature of 60°, because it has been accustomed to a temperature of 70°; whilst another, habituated to a temperature of 60°, would require to be cooled down to

* Some allusion might here be expected to the influence of *Magnetism* on the living body; since the researches of Prof. Faraday have proved that even organized structures are subject to its agency. There is as yet, however, no proof whatever, nor even any indication, that the Magnetic force has any operation, either direct or indirect, in calling forth the ordinary phenomena of life.

50°, in order to induce torpidity;—the influence of temperature upon the vital condition being proportioned, more to the variation from the usual standard, than to the actual elevation of that standard. Yet the first of these individuals might be gradually habituated to live in the same temperature with the second; and to require the same amount of further depression to induce torpidity. (See § 99.)

71. It is a very curious fact that, whilst the lower classes of living beings are more capable than the higher of bearing the deprivation of these Vital stimuli, they are at the same time more liable to alterations in their own structure and development, in consequence of variations in the degree of the agency of these, or of other causes external to themselves. Thus the forms of the lower tribes of Plants and Animals are liable to be greatly affected by the conditions under which they grow; and these also modify their degree of development. It seems as if the formative power were less vigorous in the lower, than in the higher classes; so that the mode in which it manifests itself in the former is more dependent upon external influences; whilst in the latter it either predominates over them, causing the regular actions to be performed, or gives way altogether.—The same principle applies to the early condition of the higher organisms; their embryos, like those beings of permanently-low type which they resemble in degree of development, being liable to be affected by modifying causes, which the perfect beings of the same kind are able to resist. When the organism has so far advanced in its development, that its several parts have assumed different characters, the influence of the Vital Stimuli upon its development becomes more restricted. A remarkable example of this will be presently given in regard to Light (§ 78).

72. *Of Light.*—The importance of this agent, not only to the Vegetable but to the Animal World, is not in general sufficiently estimated. Under its influence alone can the first process be accomplished, by which inorganic matter is transformed into an organic compound, adapted by its nature and properties to form part of the organised fabric. The following is an example of the simplest phenomenon of this kind, which demonstrates the influence of Light the more clearly on account of that simplicity. "If we expose some spring-water to the sunshine, though it may have been clear and transparent at first, it presently begins to assume a greenish tint; and, after a while, flocks of green matter collect on the sides of the vessel in which it is contained. On these flocks, whenever the sun is shining, bubbles of gas may be seen, which, if collected, prove to be a mixture of oxygen and nitrogen, the proportion of the two being variable. Meanwhile the green matter rapidly grows; its new parts, as they are developed, being all day long covered with air-bells, which disappear as soon as the sun has set. If these observations be made upon a stream of water, the current of which runs slowly, it will be discovered that the green matter serves as food for thousands of aquatic Insects, which make their habitations in it. These insects are endowed with powers of rapid locomotion, and possess a highly-organised structure; in their turn they fall a prey to the Fishes which frequent such streams."* Such is the general succession of nutritive actions in the Organised Creation. The highest Animal is either directly dependent upon the Vegetable Kingdom for the materials of its fabric, or it is furnished with these by some other Animal, this again

* Prof. Draper, on the "Forces which produce the Organization of Plants," p. 15.

(it may be) by another, and so on; the last in the series being *always* necessitated to find its support in the Vegetable kingdom, since the Animal does not possess the power of causing the Inorganic elements to unite into even the simplest Organic compound. This power is possessed in a high degree by Plants; but it can only be exercised under the influence of *Light,* which is, therefore, the prime agent on which all the phenomena of Life depend.—We shall now examine more in detail the conditions of this influence, both in the instance just quoted, and in others drawn from the actions of the higher Vegetable organisms.

73. The "green matter of Priestley" (as it is commonly called), which makes its appearance when water of average purity is submitted to the action of the Sun's light, and which also presents itself on the surface of walls and rocks that are constantly kept damp, is now known by Botanists to consist of *cells* in various stages of development,—the early forms, it may be, of several different species of Confervæ. That these cells all originate from germs, and not from any direct combination of the inorganic elements, appears not only from general considerations, but also from the fact that, if measures be taken to free the water entirely from any possible infusion of organic matter, and to admit into contact with it such air alone as has undergone a similar purification, no green flocks make their appearance, under the prolonged influence of the strongest sunlight. We find, then, that the presence of a germ is one of the conditions indispensable to the new production in question.—It may be asked how it can be certainly ascertained that *light* and not *heat* is the essential condition of this process; seeing that the two agents are combined in the solar beam. To this it may be replied, that a certain moderate amount of heat is undoubtedly necessary; but that no degree of heat without light will be effectual in producing the change, as is easily proved by exposing the water to warmth in a dark place. Moreover, when a certain measure of light is afforded, variations in the amount of heat make very little difference; but we shall presently see that under the same degree of heat, the amount of the change is directly proportional to the intensity of the light. Although, therefore, heat furnishes an essential condition, it cannot be questioned that *light* is the chief stimulus to that process, by which the germ brings into union the elements to be employed in the development of its own fabric.—These elements are present in all water that has been exposed to the atmosphere, and has absorbed from it a certain amount of its constituent gases; their relative quantities, in a given measure of water, being proportional to the facility with which they are respectively absorbed by the liquid. Thus carbonic acid is most readily absorbable, oxygen next, and nitrogen least so. From the experiments of Prof. Draper it would appear, that notwithstanding the very small proportion of carbonic acid contained in the atmosphere (usually not more than 1–2000th part), this gas forms as much as 29 per cent. of the whole amount of air expelled from water by boiling. The absolute quantity of this water-gas contained in any measure of water, is subject to variation with the temperature; the quantity being diminished as the temperature rises.—Now when water thus impregnated with carbonic acid, oxygen, and nitrogen, and containing the germs of simple aquatic plants, is exposed to the sun's light, so that a development of vegetable organisms takes place, it will be found that the carbonic acid is diminishing in amount, and that oxygen is

being evolved. The growing mass increases in volume and weight; and after a time exhausts the whole carbonic acid originally contained in the water. If then it be prevented from receiving an additional supply, the process stops; but as conducted naturally, there is a free exposure to the atmosphere, through which carbonic acid is diffused; and hence, as fast as it is removed by decomposition, it is restored by absorption.

74. The process whose conditions we have thus examined, is carried on in the individual cells that compose the highest and most complex Plants, precisely as in those which constitute the entire forms of the lowest. Thus if a few garden-seeds of any kind be sown in a flower-pot, and be caused to germinate in a dark room, it will soon be perceived that although they can continue growing for a time without the influence of light, that time is limited; the *weight* of their solid contents diminishes, although their *bulk* may increase by the absorption of water; their young leaves, if any should be put forth, are of a yellow or gray-white colour, and they soon fade away and die. But if these plants are brought out sufficiently soon into the bright sun-light, they speedily begin to turn green, they unfold their leaves, and evolve their different parts in a natural way; and the proportion of their solid contents goes on increasing from day to day. If the fabric be then subjected to chemical analysis, it is found to contain oxygen, hydrogen, carbon, and azote; united in various proportions, so as to form compounds, many of which differ in the various species, though some are common to all. If plants be made to grow in closed glass vessels, under such circumstances that an examination can be accurately made as to the changes they are impressing on the atmosphere, it is discovered that they are constantly decomposing its carbonic acid,—appropriating its carbon, and setting free its oxygen,—so long as they are exposed to the influence of sunshine or bright daylight. They also appropriate a part of the minute quantity of ammonia which is diffused through the atmosphere; extracting its nitrogen to employ it in the production of their azotised compounds. It is capable of being demonstrated by experiment, that these changes are confined to the *green* surfaces of plants, and therefore to the leaves or leaf-like organs, the young shoots, and the stems of herbaceous plants or of those in which (as in the Cactus tribe) the leaves are wanting and the enlarged succulent stem supplies their place. When these surfaces cease to become green, the decomposing action also ceases; carbon is no longer fixed, and oxygen set free; but, on the contrary, carbonic acid is exhaled: this is the case when the leaves change colour, previously to their fall, in the autumn. The compounds which are thus generated in the green surfaces, are conveyed, by the circulation of the sap, to the remote parts of the fabric, and become the materials of their nutrition; and thus the green cells of the leaves have exactly the same function, in ministering to the growth of the largest tree, which the green cells of the humble Conferva perform in regard to themselves alone.

75. Various experiments have been recently made, with the view of determining more precisely the conditions under which Light acts, in producing the chemical changes that have been now discussed. These experiments for the most part agree in the very interesting result, that the amount of carbonic acid decomposed by plants subjected to the differently-coloured rays of the solar spectrum, but otherwise placed in similar circumstances, varies with the *illuminating* power of the rays, and not with

their *heating* or their *chemical* power. The method adopted by Prof. Draper (op. cit. p. 177), which seems altogether the most satisfactory, consisted in exposing leaves of grass, in tubes filled with water which had been saturated with carbonic acid (after the expulsion of the previously dissolved air by boiling), to the influence of the different rays of the solar spectrum, dispersed by a prism; these were kept motionless upon the tubes, for a sufficient length of time to produce an active decomposition of the gas, in the tubes which were most favourably influenced by the solar beams; and the relative quantities of the oxygen set free were then measured. It was then evident that the action had been almost entirely confined to two of the tubes, one of them being placed in the red and orange part of the spectrum, and the other in the yellow and green. The quantity of carbonic acid decomposed by the plant in the latter of these, was to that decomposed in the former, in the ratio of *nine* to *five;* the quantity found in the tube that had been placed in the green and blue portion of the spectrum, would not amount, in the same proportion, to *one;* and in the other tubes, it was either absolutely nothing, or extremely minute. Hence it is obvious that the yellow ray, verging into orange on one side, and into green on the other, is the situation of the greatest exciting power possessed by Light on this most important function of plants; and as this coincides with the seat of the greatest illuminating power of the spectrum, it can scarcely be doubted that *light* is the agent here concerned; more especially as the place of greatest Heat is in the *red* ray, and that of greatest Chemical power is in the *blue*, both of which rays were found to be nearly inert in the experiment just quoted. It must not be supposed from this experiment, however, that the yellow ray, and those immediately adjoining it, are the *only* sources of this power in the Solar spectrum; since it proves no more than that, when the leaves were exposed to a highly carbonated atmosphere, they could only decompose it under the influence of these rays. It is certain, from other experiments, that plants will grow, in an ordinary atmosphere, under rays of different colours; and it appears that the amount of carbon they severally fix, bears a constant proportion to the illuminating powers of the respective rays.

76. Although this fixation of carbon by the decomposition of carbonic acid, is the most essentially-dependent, of all the processes of the Vegetable economy, upon the influence of Light, yet it is not the only one, especially among the higher Plants, in which that agent becomes an important condition. The Exhalation of superfluous fluid by the stomata of the leaves (CHAP. XV.) is also very much affected by it; and through its influence upon these two important operations, Light affects, more or less directly, almost every process in the Vegetable economy. The effect of its complete and continued withdrawal from a growing plant, is to produce an *etiolation* or blanching of its green surfaces; a loss of weight of the solid parts, owing to the continued disengagement of carbon from its tissues, unbalanced by the fixation of that element from the atmosphere; a dropsical distension of the tissues, in consequence of the continued absorption of water, which is not got rid of by exhalation; a want of power to form its peculiar secretions, or even to generate new tissues, after the materials previously stored up have been exhausted; in fine, a cessation of all the operations most necessary to the preservation of the vitality of the structure, of which

cessation its death is the inevitable result. A partial withdrawal of the influence of light, however, is frequently used by the Cultivator, as a means of giving an esculent character to certain Plants, which would be otherwise altogether uneatable; for in this manner their tissues are rendered more succulent and less stringy, whilst their peculiar secretions are formed in diminished amount, and communicate an agreeable flavour instead of an unwholesome rankness of taste.

77. There is one period in the life of the Flowering Plant, however, in which the influence of Light is injurious instead of beneficial; this is during the first part of the process of germination of seeds, which is decidedly retarded by its agency. This forms no exception, however, to the general rule; since the decomposition of the carbonic acid of the atmosphere, and the fixation of carbon in the tissues, do not constitute a part of the process: on the contrary, the chemical changes taking place in that substance of the seed, which has been stored up for the nutrition of the embryo, involve the opposite change,—the extrication of carbon, which is converted into carbonic acid by uniting with the oxygen of the atmosphere. It is obvious, then, why light should not only be useless, but even prejudicial, to this process; since it tends to fix in the tissues the carbon which ought to be thrown off.—The non-luminous *chemical* rays, however, are stated by Mr. Hunt to favour this process; so that seeds will germinate in the soil beneath *blue* glass,—which allows these to pass freely, whilst keeping back a large proportion of the luminous rays,—even more readily than usual; whilst, on the other hand, germination is retarded by the use of *yellow* glass, which obstructs the chemical or 'actinic' rays, whilst it allows free passage to the luminous and calorific.

78. The influence of Light upon the direction of the growing parts of Plants, the opening and closing of flowers, &c., is probably due to its share in the operations already detailed. Thus the green parts of plants, or those which effect the decomposition of carbonic acid (such as the leaves and stems), have a tendency to grow towards the light; whilst the roots, through whose dark surfaces carbonic acid is thrown out by respiration, have an equal tendency to avoid it. That the first direction of the stems and roots of plants is very much influenced by this agent, appears from the fact, demonstrated by Schultze, that, by reflecting light upon germinating seeds, in such a manner that it shall only fall upon them from below, the stems are caused to direct themselves downwards, whilst the roots grow upwards.—The periodical movements of the flowers of certain plants appear to be partly under the control of light, although without doubt subject to other influences. Some flowers open in the morning and close at evening; others, on the contrary, expand at night, and fold together early in the day. When withdrawn for a long time from the stimulus of light, these movements cease and the flowers remain closed; but by artificial illumination, they may be made to recommence; and Decandolle found that, by preserving plants in a cellar, which was at night lighted with lamps, and by day kept dark, their natural times could be reversed. This, however, could only be effected when their original tendency had been subdued by the continued deprivation of the stimulus; and the experiment was not found to succeed with all plants.—Light has been found to have a still more direct influence on the development of particular

organs in certain Vegetables. Thus when the *gemmæ* * of the *Marchantia polymorpha* (§ 275) are in process of development, it has been shown, by the repeated experiments of Mirbel, that stomata are formed on the side exposed to the light, and that roots grow from the lower surface ; and that it is a matter of indifference *which* side of the little disk is at first turned upwards, since each has the power of developing stomata, or roots, according to the influence it receives. After the tendency to the formation of these organs has once been given, however, by the sufficiently prolonged influence of light upon one side, and of darkness and moisture upon the other, any attempt to alter it is found to be vain ; for if the surfaces be then inverted, they are soon restored to their original aspect by the twisting growth of the plant.

79. The same amount of this stimulus is not requisite or desirable for all Plants. Nature has provided in the constitution of each for the periodical changes, which the alternation of day and night, and the succession of the seasons, produce in the degree of illumination afforded to it. We find some adapted to exist only where they can be daily invigorated by the powerful rays of a tropical sun ; and others, whose energies, after remaining dormant during the tedious winter of the arctic regions, are aroused into a brief activity by the return of the luminary on whose cheering influence they depend. Neither race could flourish if transferred to the external conditions of the other ; for to each are the circumstances of its existence *natural*, its original constitution being framed to take advantage of them. Generally speaking, the succulent thick-leaved Plants require the largest amount ; their stomata are few in number ; and the full influence of light is requisite to induce sufficient activity in the exhaling process : accordingly we find them growing, for the most part, in exposed situations, where there is nothing to interfere with the full influence of the solar rays. On the other hand, plants with thinner and more delicate leaves, in which the exhaling process is easily excited to an excessive amount, evidently find a congenial home in more sheltered situations ; and there are some which can only develope themselves in full luxuriance, in the deep shades of a plantation or a forest. By a further adaptation of the same kind, some species of Plants are enabled to live and acquire their green colour, under an amount of deprivation which would be fatal to most others ; thus in the mines of Freyberg, in which the quantity of light admitted must be almost infinitesimally small, Humboldt met with Flowering Plants of various species ; and Mustard and Cress have been raised in the dark abysses of the collieries of this country. Generally speaking, however, the Cryptogamia would seem to be better adapted than Flowering Plants, to carry on their vegetating processes under a low or very moderate amount of this stimulus. Thus Humboldt found a species of Sea-weed near the Canaries, which possessed a bright grass-green hue, although it had grown at a depth of 190 feet in the sea, where, according to computation, it could have received only 1-1500th part of the solar rays that would have fallen upon it at the surface of the ocean. Many Ferns, Mosses, and Lichens seem as if they avoided the light, choosing the northern rather than the southern sides of hedges, buildings, &c.,

* These gemmæ are analogous to the *buds* of higher plants ; and they consist of little collections of cells, arranged in the form of flat disks ; which are at first attached by footstalks to the parent plant, but afterwards fall off, and are developed into new individuals.

for their residence; so that the former often present a luxuriant growth of Cryptogamic vegetation, whilst the latter are comparatively bare. It must not be supposed, however, that they avoid light altogether, but only what is to them an excessive degree of it. The avoidance of light seems to be much stronger in the Fungi, which grow most luxuriantly in very dark situations; and the reason of this is probably to be found in the fact, that, like the germinating seed, they form rather than decompose carbonic acid; their food being supplied to them from the decaying substances on which they grow; and the rapid changes in their tissues giving rise to a high amount of Respiration,—a change exactly the converse of that, on which, as we have seen, Light exerts such a remarkable stimulating power.

80. In regard to the influence of Light upon the functions of Animals, comparatively little is certainly known. It is evident that the influence it exerts on those chemical processes, which constitute the first stage of Vegetable nutrition, can have scarcely any place in Animals; because *they* do not perform any such acts of combination, but make use of the products already prepared for them by Plants. Hence we do not find that the *surface* of Animals undergoes that extension, for the purpose of being exposed to the solar rays, which is so characteristic a feature in the Vegetable fabric, and so important in its economy.—Still there can be no doubt, that the degree of exposure to light has a great influence upon the *colours* of the Animal surface; and here we seem to have a manifestation of Chemical agency, analogous to that which gives colour to the Vegetable surface. Thus it is a matter of familiar experience, that the influence of light upon the skin of many persons, causes it to become spotted with brown *freckles*; these freckles being aggregations of brown pigment-cells, which either owed their development to the stimulus of light, or were enabled by its agency to perform a chemical transformation which they could not otherwise effect. In like manner, the swarthy hue, which many persons acquire in warm climates, is due to a development of dark pigment-cells diffused through the epidermis; and an increased development of the same kind gives rise to the blackness of the Negro-skin. There can be no doubt that the prolonged influence of light upon one generation after another, tends to give a permanent character to this variety of hue; which will probably be more easily acquired, in proportion to the previously-existing tendency to that change. Thus it is well known that a colony of Portuguese Jews, which settled at Tranquebar about three centuries ago, and which has kept itself distinct from the surrounding tribes, cannot now be distinguished as to colour from the native Hindoos. But it is probable that a similar colony of fair-skinned Saxons would not, in the same time, have acquired anything like the same depth of colour in their skins.—There can be no doubt that the brilliancy of colour, which is characteristic of many tribes of animals in tropical climates, especially of Birds and Insects, is in great part dependent, like the brightness of the foliage and fruit of the same countries, upon the intensity of the light to which their surfaces are exposed. When birds of warm climates, distinguished by the splendour of their plumage, are reared under an artificial temperature in our own country, it is uniformly observed that they are much longer in acquiring the hues characteristic of the adult; and that these are never so bright, as when they have been produced by the influ-

ence of the tropical sun. And it has been also remarked, that if certain Insects (the Cockroach for example), which naturally inhabit dark places, be reared in an *entire* seclusion from light, they grow up almost as colourless as Plants that are made to vegetate under similar circumstances.

81. There is abundant proof, that Light exercises an important influence on the processes of *development* in Animals, no less than in Plants. Thus, the appearance of Animalcules in infusions of decaying organic matter is much retarded, if the vessel be altogether secluded from it. The rapidity with which the small Entomostracous Crustacea (Water-Fleas, &c.) of our pools, undergo their transformations, has been found to be much influenced by the amount of light to which they are exposed. Even when full-grown, they continue to exuviate their horny casing every two or three days; and this change is necessary for their welfare, since, if it does not take place, their bodies become clothed with minute plants, which attach themselves to their surface, impeding their motions through the water, and preventing them from breathing with freedom. If they be secluded from light, they are not able to renew their shells with the usual frequency; and cannot so often get rid, therefore, of their troublesome incumbrances. Further, it has been ascertained that, if equal numbers of Silk-worms' eggs be preserved in a dark room, and in one exposed to common daylight, a much larger proportion of larvæ are hatched from the latter than from the former.—The most striking proof of the influence of Light on Animal development, however, is afforded by the experiments of Dr. Edwards. He has shown that, if Tadpoles be nourished with proper food, and be exposed to the constantly-renewed contact of water (so that their respiration may be freely carried on, whilst they remain in their Fish-like condition), but be entirely deprived of light, their growth continues, but their metamorphosis into the condition of air-breathing animals is arrested, and they remain in the condition of large tadpoles. It is interesting to remark that the *Proteus anguineus* (§323) an animal which closely corresponds in its fully-developed condition with the transition-stage between the Tadpole and the Frog, finds a congenial abode in the dark lakes of the caverns of Styria and Carniola, and in the underground caverns that connect them; thus showing its adaptation to a condition, which keeps down to the same standard the development of an animal, that is empowered under other circumstances to advance beyond it.—Numerous facts, collected from different sources, lead to the belief that the healthy development of the Human body, and the rapidity of its recovery from disease, are greatly influenced by the amount of Light to which it has been exposed. Thus it was remarked by Dr. Edwards, that persons who live in caves or cellars, or in very dark and narrow streets, are apt to produce deformed children; and that men who work in mines are liable to disease and deformity, beyond what the simple closeness of the atmosphere would be likely to produce. Part of this difference is doubtless owing to the relative purity of the atmosphere in the former case, and the want of ventilation in the latter; but other instances might be quoted, in which a marked variation presented itself, under circumstances otherwise the same. Thus, it has been stated by Sir A. Wylie (who was long at the head of the medical staff in the Russian army), that the cases of disease on the dark side of an extensive barrack at St. Petersburgh, have been uniformly, for many years, in the proportion of three to one, to those on

the side exposed to strong light. And in one of the London Hospitals, with a long range of frontage looking nearly due north and south, it has been observed that residence in the north wards is much less conducive to the welfare of the patients, than in those on the south side of the building. On the other hand, the more the body is exposed to the influence of light, the more freedom do we find, *ceteris paribus,* from irregular action or conformation. Humboldt has remarked that, among several nations of South America who wear very little clothing, he never saw a single individual with a natural deformity; and Linnæus, in his account of his tour through Lapland, enumerates constant exposure to solar light as one of the causes, which render a summer journey through high northern latitudes so peculiarly healthful and invigorating.

82. These facts being kept in view, it is easy to perceive that there must be differences among the various species of Animals, as among those of Plants, in regard to the degree of light which is congenial to them. Among the lowest tribes, in which no special organs of vision exist, there is evidently a susceptibility to the influence of light, which appears scarcely to deserve the name of sensibility, but which seems rather analogous to that which is manifested by Plants; thus among those Polypes which are not fixed to particular spots, and amongst Animalcules, there are some species which seek the light, and others which shun it. And it appears from various observations upon the depths at which marine animals are found, especially from the extensive series of facts collected by Prof. E. Forbes,* that there is a series of *zones,* so to speak, to be met with in descending from the surface towards the bottom of the ocean, each of which is characterised by certain species of animals peculiar to itself, whilst other species have a range through two or more of the zones;—the extent of the range of depth, in each species, bearing a close correspondence with the extent of its geographical distribution. Now there can be no doubt, that the restriction of particular species to particular zones is due in great part to the degree of *pressure* of the surrounding medium; but there can be as little doubt, that the variation in the degree of light also exerts a most important influence, the solar rays in their passage through sea water being subject to a loss of one half for every seventeen feet. From the results of Prof. Forbes's researches, it appears that no species of Invertebrated animals habitually live at a greater depth than 300 fathoms; and although Fishes have been captured at a depth of from 500 to 600 fathoms, it is probable that they had strayed from their usual abodes. That an almost infinitesimally small amount of Light, however, is a sufficient stimulus to the vital activity of some animals, appears from the fact mentioned by Capt. Scoresby, that various species of Star-fish, of the most brilliant and beautiful markings, were brought up at the end of a line of 250 fathoms.

83. *Of Heat.*—The most perfectly organised body, supplied with all the other conditions requisite for its activity, must remain completely inert, if it do not receive sufficient stimulation from Heat. For each Vital action requires a certain amount of Caloric for its performance, and can only continue within a particular range of temperature; between the limits of which, it is excited by the additional application of the stimulus, and depressed by its abstraction. Different species of Animals and Vegetables

* Report on the Invertebrata of the Ægean Sea, in Transactions of British Association, 1843.

exhibit great varieties in the limitations of temperature which they require, and in their power of adaptation to extreme conditions; and hence in part arises their diversity of geographical distribution. As a general rule it may be stated, that the greater the amount and variety of vital action, the more immediate is the dependence of the individual upon the maintenance of its usual temperature. We shall hereafter see (CHAP. XVII.) that, whilst the higher Animals possess within themselves the means of generating the heat necessary to sustain their vital activity, the lower Animals, as well as Plants in general, are almost entirely dependent upon the medium they inhabit for the necessary supply of caloric; it being only during one or two periods of the existence of the more perfect kinds, that any sensible degree of heat is generated by them, and this only for local purposes. But, being thus dependent, their vital actions are so adjusted as to be carried on within wide extremes of heat and cold; and it is in them that we can most clearly trace the influence of Caloric upon vital activity. This is attested, on the large scale, by the striking contrast between the dreary barrenness of polar regions, and the luxuriant richness of Tropical countries, where almost every spot to which moisture is supplied teems with Animal and Vegetable life. And the alternation of Winter and Summer in temperate climates may be almost said to repeat under our own view the opposite conditions of these two extreme cases. The Vegetable world, in the depth of winter, is reduced to a state of death-like torpor; all the actions of growth are suspended below the freezing point, not merely in the tree completely bared of its foliage, but also in the 'evergreen' that presents more of the semblance of life; and this condition is prolonged until an elevation of temperature takes place. Even a low degree of warmth suffices to produce a renewal of vital activity; and this not merely in evergreens, but in trees unclothed with leaves, a movement of sap taking place in their stems and branches during every mild and sunny day in winter, and the leaf-buds being gradually prepared to start into full development with the steadier warmth of the returning spring. The influence of Heat upon Vegetation is easily made apparent by experiment; in fact, experimental illustrations of it, upon a large scale, are in daily progress. For the gardener, by artificial warmth, is not only enabled to rear with success the plants of tropical climates, whose constitution would not bear the chilling influence of our winter;—but he can also, in some degree, invert the order of the seasons, and produce both blossom and fruit from the plants of our own country, when all around seems dead.

84. To every species of Plant there is a temperature which is most congenial, from its producing the most favourable influence on the actions (whether physical or vital) concerned in the maintenance of its life; and by this admirable provision we find that each zone is endowed with the power of sustaining a vegetation of its own, which would not flourish elsewhere. Thus in the countries lying near the Equator, the vegetation consists in great part of dense forests of leafy evergreen trees, Palms, Bamboos, and Tree-ferns, bound together by clustering Orchideæ and strong creepers of various kinds. There are no verdant meadows, such as form the chief beauty of our temperate regions; and the lower orders of Vegetation are extremely rare. It is only in this torrid Zone, that Dates, Coffee, Cocao, Bread-fruit, Bananas, Cinnamon, Cloves, Nutmegs, Pepper, Myrrh, Indigo,

Ebony, Logwood, Teak, Sandal-wood, and many other of the vegetable products most highly valued for their flavour, their odour, their colour, or their density, come to full perfection.—As we recede from the Equator, we find the leafy Evergreens giving place to trees with deciduous leaves; rich meadows appear, abounding with tender herbs; the Orchideæ no longer find in the atmosphere, and on the surface of the trees over which they cluster, a sufficiency of moisture for their support, and the parasitic species are replaced by others which grow from fleshy roots implanted in the soil; but aged trunks are now clothed with Mosses; decayed vegetables are covered with parasitical Fungi; and the waters abound with Confervæ. In the warmer parts of the temperate regions, the Apricot, Citron, Orange, Lemon, Peach, Fig, Vine, Olive, and Pomegranate, the Myrtle, Cedar, Cypress, and Dwarf Palm, find their congenial abode.—These give place, as we pass northwards, to the Apple, the Plum, and the Cherry, the Chesnut, the Oak, the Elm, and the Beech. Going further still, we find that the fruit-trees are unable to flourish, but the timber-trees maintain their ground. Where these last fail, we meet with extensive forests of the various species of Firs; the Dwarf Birches and Willows replace the larger species of the same kind; and even near or within the arctic circle, we find wild flowers of great beauty, the Mezereon, the yellow and white Water-Lily, and the Globe-flower.

85. Where none of these can flourish, where trees wholly disappear, and scarcely any flowering plants are to be met with, an humbler Cryptogamic vegetation still raises its head, in proof that no part of the Globe is altogether unfit for the residence of living beings, and that the empire of Flora has no limit. It is for the most part, among the Cryptogamic tribes (the Ferns, Mosses, Liverworts, Fungi, Lichens, and Algæ) that the greatest power of growing under a low temperature exists; and we accordingly find that the proportion of these to the Phanerogamia, or Flowering Plants, increases as we proceed from the Equator towards the Poles. It has been estimated by Humboldt, that, in Tropical regions, the number of species of Cryptogamia is only about *one-tenth* that of the Flowering Plants; in the part of the Temperate zone which lies between Lat. 45° and 52°, the proportion rises to *one-half;* and the relative amount gradually increases as we proceed towards the Poles, until, between Lat. 67° and 70°, the number of species of Cryptogamia *equals* that of the Phanerogamia. Among the Flowering Plants, moreover, the greatest endurance of cold is to be found in those, which approach most nearly to the Cryptogamia in the low degree of their development; thus the Glumaceous group of Endogens, including the Grasses, Rushes, and Sedges, which forms about *one-eleventh* of the whole amount of Phanerogamic vegetation in the Tropics, constitutes *one-fourth* of it in the Temperate regions, and *one-third* in the Polar; and the ratio of the Gymnospermic group of Exogens, which chiefly consists of the Pine and Fir tribe, increases in like manner. Still the influence of a high temperature is evident even upon the Cryptogamia and their allies; for it is only under the influence of the light and warmth of tropical climes, that the Ferns,—the highest among the former,—can develope a woody stem, and assume the character of trees; and it is only there that the tall Sugar-Canes, and the gigantic Bamboos, which are but Grasses on a large scale, can flourish.

86. But distance from the Equator is by no means the only element,

in the determination of the mean temperature of a particular spot. Its height above the level of the sea is equally important; for this produces a variation in the amount of heat derived from the Sun, at least as great as that occasioned by difference of latitude. Thus it is not alone on the summits of Hecla, Mont Blanc, and other mountains of arctic or temperate regions, that we find a coating of perpetual snow; we find a similar covering on the lofty summits of the Himalayan chain, which extends to within a few degrees of the tropic of Cancer; and even on the higher peaks of that part of the ridge of the Andes, which lies immediately beneath the Equator. If, then, Temperature exert such an influence on Vegetation as has been stated, we ought to find on the sides of lofty mountains in tropical climates the same progressive alterations in the characters of the Plants that cover them, as we encounter in journeying from the equatorial towards the polar regions. This is actually the case. The proportion of Cryptogamia to Flowering Plants, for example, is no more than *one-fifteenth* on the *plains* of the Equatorial region; whilst it is as much as *one-fifth* on the *mountains*. In ascending the Peak of Teneriffe, Humboldt remarked as many as five distinct Zones, which were respectively marked by the products which characterize different climates. Thus at the base, the vegetation is altogether tropical; the Date-Palm, Plaintain, Sugar-Cane, Banyan, the succulent Euphorbia, the Dracœna, and other trees and plants of the torrid zone, there flourish. A little higher grow the Olive, the Vine, and other fruit-trees of Southern Europe; there Wheat flourishes; and there the ground is covered with grassy herbage. Above this is the woody region, in which are found the Oak, Laurel, Arbutus, and other beautiful hardy evergreens. Next above is the region of Pines; characterised by a vast forest of trees resembling the Scotch Fir, intermixed with Juniper. This gives place to a tract remarkable for the abundance of Broom; and at last the scenery is terminated by Scrofularia, Viola, a few Grasses, and Cryptogamic plants, which extend to the borders of the perpetual snow that caps the summit of the mountain.

87. There are some examples of the adaptation of particular forms of Vegetable life to extremes of temperature, which are interesting as showing the extent to which this adaptation may be carried. In hot springs near a river of Louisiana, of the temperature of from 122° to 145°, there have been seen to grow not merely Confervæ and herbaceous plants, but shrubs and trees; and a hot-spring in the Manilla islands, which raises the thermometer to 187°, has plants flourishing in it, and on its borders. A species of *Chara* has been found growing and reproducing itself in one of the hot-springs of Iceland, which boiled an egg in four minutes; various Confervæ, &c., have been observed in the boiling-springs of Arabia and the Cape of Good Hope; and at the island of New Amsterdam, there is a mud-spring, which, though hotter than boiling-water, gives birth to a species of Liverwort.—On the other hand, there are some forms of Vegetation, which seem to luxuriate in degrees of cold, that are fatal to most others. Thus the Lichen, which serves as the winter food of the Reindeer, spreads itself over the ground whilst thickly covered with snow; and the beautiful little *Protococcus nivalis*, or Red Snow, reddens extensive tracts in the arctic regions, where the perpetual frost of the surface scarcely yields to the influence of the solar rays at Midsummer.

88. Of the particular processes of Vegetation, those of Flowering and Germination appear to be most dependent upon a specific amount of heat. Thus we generally find that a degree of warmth which is barely sufficient for the development of leaves, will keep back that of flowers; whilst, on the other hand, flowers are rapidly withered by the scorching rays whose influence can be borne by leaves with impunity. And we occasionally find in Flowers a provision for generating heat independently of external sources.—This is still more remarkable in the germination of seeds, which can only take place within a comparatively limited range of temperature; a circumstance that greatly contributes to limit the geographical range of plants. We have a striking example of this in the case of the *Cerealia* (corn-plants), which civilized Man endeavours to cultivate wherever climate and soil admit of their growth. The limit, in regard to heat at least, is rather set by the necessity for a moderate soil-temperature in the stage of germination, than by the effect of the solar rays or of a heated atmosphere upon the growing plant. Thus it has been found by experiment that Wheat will not germinate in water at a higher temperature than 95°, that which is most suitable to the process being about 65°; and we at once see, therefore, why it cannot grow in the torrid zone, where the temperature of the soil is not unfrequently 120°, sometimes even rising to 140°. But Maize will germinate in water at 113°; and hence it will grow in situations too hot for wheat. On the other hand, the limit to the germination-temperature of Barley is lower than that of wheat, and it will consequently fail to grow in a soil whose heat can be borne by wheat; whilst, in virtue of the same property, it will thrive in climates too cold to produce wheat.

89. The effects of Temperature on Vegetation are not only seen in its influence upon the Geographical distribution of Plants, that is, in the limitation of particular species to particular climates; for they are also shown, perhaps even more remarkably, in the variation in the size of individuals of the same species, when that species possesses the power of adapting itself to widely-different conditions, which is the case with some. Thus the *Cerasus Virginiana* grows in the southern states of North America as a noble tree, attaining one hundred feet in height; in the sandy plains of the Saskatchawan, it does not exceed twenty feet; and at its northern limit, the Great Slave Lake, in Lat. 62°, it is reduced to a shrub of five feet.—Another curious effect of heat is shown in its influence on the sexes of certain Monœcious flowers; thus Mr. Knight mentions that Cucumber and Melon plants will produce none but male or staminiferous flowers, if their vegetation be accelerated by heat; and all female or pistilliferous, if its progress be retarded by cold.

90. The influence of excessive Heat in destroying life, can sometimes be traced through the direct *physical* or *chemical* changes which it occasions in the vegetable tissues and fluids. Thus it has been ascertained that grains of corn will vegetate, after exposure to water or vapour possessing a considerable degree of heat; provided that heat do not amount to 144° in the case of water, and 167° in that of vapour. At these temperatures, the structure of the seed undergoes a disorganising change, by the rupture of the vesicles of starch which form a large part of it; and the loss of its power of germinating is therefore readily accounted for. So, again, when a growing plant is killed by immersion in boiling water, it is easy to conceive that the loss of vitality is due to the coagulation of

these albuminous matters, which are actively concerned in the most important functions of the organism. The mode in which excessive heat usually operates, is to increase the aqueous evaporation, and thus to carry off the water that is requisite for the due performance of the various chemical and physical changes on which life depends. If the supply be adequate to the demand, the effect of heat will be to stimulate all the vital operations of the plant, and to cause them to be performed with increased energy; though this energy may be such as to occasion a premature exhaustion of its powers, by the excessive luxuriance which it occasions. But if the supply of water be deficient, the plant is burnt up by the continuance of heat in a dry atmosphere; and it either withers and dies, or its tissues become dense and contracted without losing their vitality. Thus it has been remarked, that shrubs growing among the sandy deserts of the East, have as stunted an appearance as those attempting to vegetate in the Arctic regions; their leaves being converted into prickles, and their leaf-buds prolonged into thorns instead of branches.

91. It scarcely admits of doubt, that the destructive influence of a very low temperature upon the Vitality of Plants, is immediately exerted through its chemical and physical effects upon the tissues and their contents. Thus it will produce congelation of their fluids; and the expansion which takes place in freezing will injure the walls of the containing cells,—distending, lacerating, or even bursting them. The same cause will probably occasion the expulsion of air from some parts which ought to contain it; and the introduction of it into other parts which ought to be filled with fluid. And a separation will take place, in the act of freezing, between the constituent parts of the vegetable juices; which will render them unfit for discharging their functions, when returning warmth would otherwise call them into activity. Hence we are enabled in some degree to account for the differences in the power of resisting cold, which the various species of Plants, and even the various parts of the same individual, are found to possess. For, other things being equal, the power of each plant, and of each part of a plant, to resist a low temperature, will be in the inverse ratio of the quantity of water contained in the tissue; thus, a succulent herbaceous plant suffers more than one with a hard woody stem and dense secretions; and young shoots are destroyed by a degree of cold, which does not affect old shoots and branches of the same shrub or tree. Again, the viscidity of the fluids of some plants is an obstacle to their congelation, and therefore enables them to resist cold; thus it is, that the resinous Pines are, of all trees, those which can endure the lowest temperature. The dimensions of the cells, too, of which the tissue is composed, appear to have an influence; the liability to freeze being diminished by a very minute subdivision of the fluids. And when the roots are implanted deep in the soil, where the temperature does not fall so low as that of the surface by many degrees, the fluidity of the sap may be maintained, in spite of an extremely cold state of the atmosphere. It is in Cryptogamic plants, that the greatest power of sustaining cold exists; as might be inferred from what has been already stated in regard to their geographical distribution. The little Fungus (*Torula Cerevisiæ*) which is one of the principal constituents of Yeast, does not lose its vitality by exposure to a temperature of 76° *below zero*, though it requires a somewhat elevated temperature for its active growth. It would appear

that *Seeds* are enabled to sustain a degree of cold, without the loss of their vitality, which would be fatal to growing plants of the same species; thus grains of corn, of various kinds, will germinate after being exposed for a quarter of an hour to a temperature equal to that of frozen mercury. It is not difficult to account for this, when the closeness of their texture, and the small quantity of fluid which it includes, are kept in view. The act of Germination, however, will only take place under a rather elevated temperature; and we find in the Chemical changes which it involves, a provision for maintaining this, when the process has once commenced.

92. The influence of Heat upon the vital activity of Animals, is quite as strongly marked as we have seen it to be in the case of Plants; but the mode in which it is exerted is in many instances very different. In those Animals which are endowed with great energy of muscular movement, and in which, for the maintenance of that energy, the nutritive functions are kept in constant activity, we find that a provision exists for the development of heat from within, so as to keep the temperature of the body at a certain uniform standard, whatever may be the climate in which they live. Their energy and activity are, in fact, so dependent upon the steady maintenance of a high temperature in their bodies, that, if this be not kept up nearly to its regular standard, a diminution or even a complete cessation of vital action takes place, and even a total loss of vitality may result. In these *warm-blooded* animals, as they are termed, we do not so evidently trace the effects of Heat, because they are constantly being exerted, and because *external* changes have but little influence upon them, unless these changes be of an extreme kind. But if those *internal* operations, on which the maintenance of the temperature is dependent, are from any cause retarded or suspended, the effect is immediately visible, in the depressed activity of the whole system. In the class of Birds, whose muscular energy, and whose general functional activity, are greater and more constant than those of any other animals, the temperature is pretty steadily maintained at from 108° to 112;° and we shall presently see, that a depression of the heat of the body to about 80° is fatal. Among Mammalia, the temperature is usually maintained at from 98° to 102°; and it seems that here too a depression of about thirty degrees is ordinarily fatal. In the different tribes of Birds and Mammals, we find a very diversified power of generating heat; and on this depends their adaptation to various climates. Where the usual temperature of the atmosphere is but little below the normal standard of the body, a small amount of the internal calorifying power is required; and accordingly we find that animals which naturally inhabit the torrid zone, cannot be kept alive elsewhere, except, like the Plants of the same regions, by external heat. On the other hand, the animals of the colder-temperate and frigid climes are endowed with a much greater internal calorifying power; and their covering is adapted to keep in the heat which they generate. Such animals (the Polar Bear for example) cannot be kept in health, in the summer of our own country, unless means be taken for their refrigeration. The constitution of Man seems to acquire, by habituation to a particular set of conditions through successive generations, an adaptation to differences of climate, of which that of few other animals is susceptible; and thus we find different races of human beings inhabiting countries, which are subject to the extremes of

heat and cold.—There is a small number of species, which, being endowed with a similar adaptiveness to temperature, can accompany Man in all his migrations.

93. The conditions on which the power of maintaining the heat of the body, in despite of external cold, is dependent, will become the subject of inquiry hereafter (CHAP. XVII.). It is sufficient here to state, that this power is in great part the result of numerous Chemical changes going on within the body; and especially of a process analogous to combustion, in which carbon and hydrogen taken in as food are made to unite with oxygen derived from the atmosphere. It is dependent, therefore, as to its amount, upon the due supply of the combustible material on the one hand, and of atmospheric air on the other. If the former be not furnished either by the food, or by the fatty matter of the body (which acts as a kind of reserved store laid up against the time of need), the heat cannot be maintained; and it is in part for want of power to digest and assimilate a sufficient amount of this kind of aliment, that animals of warm climates cannot maintain their temperature in colder regions. On the other hand, if the supply of oxygen be deficient, as it is when the respiration is impeded by diseased conditions of various kinds, there is a similar depression of temperature.—Now if, from either of these causes, the temperature of the body of a Bird or Mammal (except in the case of the *hybernating* species of the latter, to be presently noticed) be lowered to about 30° below its usual standard, not only is there a cessation of vital *activity*, but a total loss of vital *properties*; in other words, the *death* of the animal is a necessary result. This occurrence is preceded by a gradually-increasing torpidity; which shows the depressing influence of the cooling process upon the functions in general. The temperature of the superficial parts of the body is, of course, soonest affected; the circulation is at first retarded, causing lividity of the skin; but, as the temperature becomes lower, the blood is almost entirely expelled from the surface, by the contraction of the vessels, and paleness succeeds. At the same time, there is a gradually-increasing torpor of the nervous and muscular systems, which first manifests itself in an indisposition to exertion of any kind, and then in an almost irresistible tendency to sleep. At the same time, the respiratory movements become slower, from the want of the stimulus that should be given by the warm current of blood to the medulla oblongata, which is the centre of those movements; and the loss of heat goes on, therefore, with increased rapidity, until the temperature of the whole body is so depressed, that its vitality is altogether destroyed.—But when there is a deficiency of the proper *animal* heat, the vital activity of the system may be maintained by caloric supplied by external sources. This fact is of high scientific value, as giving the most complete demonstration of the *immediate* dependence of the vital functions of warm-blooded animals upon a sustained temperature; and its practical importance can scarcely be over-rated. It rests chiefly upon the recent experiments of Chossat, who had in view to determine the circumstances attending death by Inanition or starvation. He found that, when Pigeons were entirely deprived of food and water, their average temperature underwent a tolerably regular diminution from day to day; so that, after several days (the exact number varying with their previous condition), it was about 4½° lower than at first. Up to this time, it seems that the store of fat laid up

in the body supplies the requisite material for the combustive process; so that no very injurious depression of temperature occurs. But, as soon as this is exhausted, the temperature falls rapidly, from hour to hour; and as soon as the total depression has reached 29½° or 30°, death supervenes. Yet it was found by M. Chossat, that when animals thus reduced by starvation, whose death seemed impending (death actually taking place in many instances, whilst the preliminary processes of weighing, the application of the thermometer, &c., were being performed), were subjected to artificial heat, they were almost uniformly restored, from a state of insensibility and want of muscular power, to a condition of comparative activity. Their temperature rose, their muscular power returned, they took food when it was presented to them, and their secretions were renewed; and, if this artificial assistance was sufficiently prolonged, and they were supplied with food, they recovered. If the heat was withdrawn, however, before the time when the digested food was ready, in sufficient amount, to supply the combustive process, they still sank for want of it.

94. There is a certain group of warm-blooded animals, however, which offers a remarkable peculiarity in respect to heat;—their power of generating it being for a time greatly diminished or almost completely suspended; the temperature of their bodies following that of the air around, so that it may be brought down nearly to the freezing-point; their general vital actions being carried on with such feebleness as to be scarcely perceptible; and yet the vital properties of the tissues being retained, so that, when the temperature of the body is again raised, the usual activity returns. This state, which is called *hybernation*, appears to be as natural to certain animals, as sleep is to all; and it corresponds with sleep in its tendency to periodical return. No account can be given of the causes to which it is due: but the condition of the animals presenting it offers several points of much interest. There are some, as the *Lagomys*, in which it appears to differ but little from deep ordinary sleep; they retire into situations which favour the retention of their warmth; and they occasionally wake up, and apply themselves to some of the store of food, which they have provided in the autumn. In other cases, a great accumulation of fat takes place within the body in autumn, favoured by the oily nature of the seeds, nuts, &c., on which the animals then feed; and this serves the purpose of maintaining the temperature for a sufficient length of time, not indeed to the usual standard, but to one not far below it. The state of torpor in these animals is more profound than that of deep sleep, but it is not such as to prevent them from being easily aroused; and their respiratory movements, though diminished in frequency, are still performed without interruption. But in the *Marmot*, and in animals which, like it, hybernate completely, the temperature of the body (owing to the want of internal power to generate heat) and the general vital activity, are proportionably depressed; the respiratory movements fall from 500 to 14 per hour, and are performed without any considerable enlargement of chest; the pulse sinks from 150 to 15 beats per minute; the state of torpidity is so profound that the animal is with difficulty aroused from it; and the heat of the body is almost entirely dependent upon the temperature of the surrounding air, not being usually more than a degree or two above it. When the thermometer in the air is somewhat below the freezing-point, that placed within the body falls to

about 35° ; and at this point it may remain for some time without any apparent injury to the animal, which revives when subjected to a higher temperature. When, however, the body is exposed to a more intense degree of cold, the animal functions undergo a temporary renewal ; for the cold seems to act like any other stimulus in arousing them. The respiratory movements and the circulation increase in activity, so as to generate an increased amount of heat ; but this amount is insufficient to keep up the temperature of the body, which is at last depressed to a degree inconsistent with the maintenance of life ; and not only the suspension of activity, but the total loss of vital properties is the result.

95. Now the condition of a hybernating Mammal closely resembles that of a cold-blooded animal, in regard to the dependence of its bodily temperature upon external conditions. There is this important difference, however ;—that the reduction of the temperature of the former to 60° or 50° is incompatible with a state of activity, which is only exhibited when the heat of their bodies rises to nearly the usual Mammalian standard ; whilst a permanently low or moderate temperature is natural to the bodies of most cold-blooded animals, whose functions could not be well carried on under a higher temperature. Thus all the muscles of a Frog are thrown into a state of permanent and rigid contraction, by the immersion of its body in water no warmer than the blood which naturally bathes those of the Bird ; and we find, accordingly, that cold-blooded animals which cannot sustain a high temperature, are provided with a *frigorifying* rather than with a *calorifying* apparatus.—Although we are accustomed to rank all animals, save Birds and Mammals, under the general term *cold-blooded*, yet there exist among them considerable diversities as to the power of generating heat within themselves, and of thus rendering themselves independent of external variations. These diversities, which will be particularly considered, in their proper place (CHAP. XVII.), will usually be found to have reference to the nervous and muscular activity by which the species or tribe is characterized ; as is seen especially in the classes of Fishes and Insects.

96. But there is abundant evidence of the influence of Heat, not merely upon the movements of cold-blooded animals, but also upon the processes of nutrition and development. Thus the time of emersion of Insect-larvæ from their eggs,—or in other words, the rate at which the previous formative processes go on, is entirely dependent upon the temperature. In the case of the Bird we find that, if the temperature be not sufficient to develope the egg, chemical changes soon take place, which involve the loss of its vitality ; or if the temperature be reduced so low as to prevent the occurrence of those changes, the loss of heat is in itself destructive of life. But this is not the case in regard to the eggs of cold-blooded animals in general ; for, like the beings they are destined to produce, they may be reduced to a state of complete inaction by a depression of the external temperature ; whilst a slight elevation of this renews their vital operations, at a rate corresponding to the warmth supplied. Hence the production of larvæ from the eggs of Insects may be accelerated or retarded at pleasure ; and this is, in fact, practised in the rearing of Silk-worms, in order to adapt the time of their emersion from the egg to the supply of food which is ready for them. The same may be said in regard to the eggs of other cold-blooded animals ; those, for

example of the minute Entomostracous Crustacea which people our ditches and ponds. In many of these, the race is continued solely by the eggs which remain dormant through the winter; all the parents being destroyed by the cold. The common *Daphnia pulex* produces two kinds of eggs; from one, the young are very speedily hatched; but the others, which are produced in the autumn, and enveloped in a peculiar covering, do not give birth to the contained young until the succeeding spring. They may be at any time hatched, however, by artificial warmth.—Phenomena of an equally interesting and instructive character may be observed in the history of the Pupa-state of Insects; which, in those that undergo a complete metamorphosis, may be almost characterised as a re-entrance into the egg, for the attainment of a higher grade of development (§315). This process of development is remarkably influenced by external temperature; being accelerated by genial warmth, and retarded by cold. There are many Larvæ, which naturally pass into the Pupa state during the autumn, remain in it during the entire winter, and emerge as perfect Insects with the return of spring. It was found by Reaumur, that Pupæ, which would not naturally have been disclosed until May, might be caused to undergo their metamorphosis during the depth of winter, by the influence of artificial heat; whilst, on the other hand, their change might be delayed a whole year beyond its usual time, by the prolonged influence of a cold atmosphere. A very curious provision exists in the economy of the Social Bees, in order to hasten the development of their pupæ (CHAP. XVII. Sect. 3).—It has been remarked by Mr. Paget that the processes of *development* seem to require a higher degree of vital force than those of simple *growth;* and the facts just stated, when viewed in this light, afford an interesting confirmation to the hypothesis advanced in the early part of this chapter, as to the relation between the Vital Force and Heat (§54).

97. The influence of variations in the Heat of the body upon its vital activity, is further manifested by the very remarkable experiments of Dr. Edwards; who has shown that Cold-blooded animals *live much faster* (so to speak) at *high* temperatures, than at low; so that they die much sooner, when deprived of other vital stimuli. Thus when Frogs were confined in a limited quantity of water, and were not permitted to come to the surface to breathe, it was found that the duration of their lives was inversely proportional to the degree of heat of the fluid. When it was cooled down to the freezing-point, the frogs immersed in it lived during from 367 to 498 minutes. At the temperature of 50°, the duration of their lives was from 350 to 375 minutes; at 72°, it was from 90 to 35 minutes; at 90°, from 12 to 32 minutes; and at 108° death was almost instantaneous. The prolongation of life at the lower temperatures is not due to torpidity, for the animals perform the functions of voluntary motion, and enjoy the use of their senses; but it is occasioned by their diminished activity, which occasions a less demand for air. On the other hand, the elevation of temperature increases the demand for air, and causes speedier death when it is withheld, by increasing the general agility. The natural habits of these animals are in correspondence with these facts. During the winter, the influence of a sufficient amount of aerated water upon their exterior serves to maintain the required amount of respiration through the skin, so that they are not

obliged to come to the surface to take in air by the mouth. As the season advances, however, their activity increases, a larger amount of respiration is required, and the animals are obliged to come frequently to the surface to breathe. During summer, the yet higher temperature calls forth an increased energy and activity in all the vital functions; the respiration must be proportionably increased; the action of the air upon the cutaneous surface, as well as upon the lungs, is required; and if the animals are prevented from quitting the water to obtain this, they die, as soon as the warmth of the season becomes considerable.—The result of experiments on Fishes, in regard to the deprivation or limited supply of the air contained in the water in which they are immersed, is exactly similar; the duration of life being inversely as the temperature. And precisely the same has been ascertained with respect to hybernating Mammals; which, as already remarked, are for a time reduced, in all such respects, to the level of cold-blooded animals.

98. The energy of the *reparative* actions in higher animals appears to be much influenced by temperature; as might be inferred from what has been just said of their nutritive and developmental operations. Thus in Frogs, we find that wounds are healed much more rapidly in summer than in winter; and it has been recently stated that the *Triton* (water-newt) possesses the power of reproducing its lost limbs between the temperature of 58° and 75°, but cannot do so without the stimulus of that degree of warmth.*

99. We have seen that the animals termed *cold-blooded* are greatly influenced as to the temperature of their bodies, by the temperature of the surrounding medium; although many of them are endowed with the power of keeping themselves a certain number of degrees above it. Now the consequence of this is, that all of them which are subject to any considerable and prolonged amount of cold, pass into a state of more or less complete inactivity during its continuance; which state bears a close correspondence with the hybernation of certain Mammalia. Among the Reptiles of cold and temperate countries, this torpid state uniformly occupies a considerable part of the year; as it does also with Insects, terrestrial Mollusks, and other Invertebrated animals, which are subject to the influence of the cold. On the other hand, Fishes, Crustacea, and other marine animals, do not usually appear to pass into a state of torpidity; the temperature of the medium they inhabit never undergoing a degree of depression, nearly so great as that of the atmosphere. The amount of change necessary to produce this effect, or on the other hand to call the animals from a state of torpidity to one of active energy, differs for different species; and there is probably a considerable difference even among individuals of the same species, according to the temperature under which they habitually live. Thus one animal may remain torpid under a degree of warmth, which will be sufficient to arouse another of the same kind, accustomed to a somewhat colder climate; because the stimulus is *relatively* greater to the latter. It was observed by Mr. Darwin, that at Bahia Blanca in South America, the first appearance of activity in animal and vegetable life, a few days before the vernal equinox, presented itself under a mean temperature of 58°, the range of the thermometer in the middle of the day being between 60° and 70°. The plains were ornamented

* Mr. Higginbottom, in Proceedings of the Royal Society, March 18, 1847.

by the flowers of a pink wood-sorrel, wild peas, evening primroses, and geraniums; the birds began to lay their eggs; numerous beetles were crawling about; and lizards, the constant inhabitants of a sandy soil, were darting about in every direction. Yet a few days previously, it seemed as if nature had scarcely granted a living creature to this dry and arid country; and it was only by digging in the ground that their existence had been discovered,—several insects, large spiders, and lizards, having been found in a half torpid state. Now at Monte Video, four degrees nearer the Equator, the mean temperature had been above 58° for some time previously, and the thermometer rose occasionally during the middle of the day to 69° or 70°; yet with this elevated temperature, nearly equivalent to the full summer heat of our own country, almost every beetle, several genera of spiders, snails, and land-shells, toads and lizards, were still lying torpid beneath stones. We have seen that at Bahia Blanca, whose climate is but a little colder, this same temperature, with a rather less extreme heat, was sufficient to awake all orders of animated beings; —showing how nicely the required degree of stimulus is adapted to the general climate of the place, and how little it depends on absolute temperature.

100. We may learn much from the Geographical distribution of the different species of cold-blooded animals, in regard to the influence of temperature on Animal life. No general inferences of this kind can be founded upon the distribution of warm-blooded animals; since their own heat-evolving power makes them in great degree independent of external warmth. And it is probably from the distribution of the marine tribes, whose extension is least influenced by local peculiarities, that the most satisfactory deductions are to be drawn. In regard to the class of Crustacea, which is the one that has been most fully investigated in this respect, the following principles have been pointed out by M. Milne Edwards; and they are probably more or less applicable to most others.

I. *The varieties of form and organisation manifest themselves more, in proportion as we pass from the Polar Seas towards the Equator.*—Thus on the coast of Norway, where there is frequently a vast multiplication of *individuals* of the same species, the number of *species* is very small; but the latter increases rapidly as we go southwards. The number of species of Crustacea of the two highest orders, for example, known to exist on the coast of Norway and in the neighbouring seas, is only 16; but 82 are known to be inhabitants of the western shores of Britain, France, Spain, and Portugal; 114 are known in the Mediterranean Sea; and 202 in the Indian Ocean. A similar increase may be observed in following the coast of the New World, from Greenland to the Caribbean Sea.

II. *The differences of form and organisation are not only more numerous and more characteristic in the warm than in the cold regions of the globe; they are also more important.*—The number of natural groups which we find represented in the polar and temperate regions, is much smaller than that of which we find types or examples in the tropical seas. In fact, nearly all the principal forms which are met with in colder regions, also present themselves in warm; but a very large proportion of the latter have no representatives among the former. Of the three *primary* groups composing the Class, indeed, one is altogether wanting beyond the 44th degree of latitude; and in the other two there are whole Orders, as well as numerous subordinate divisions, which are as completely restricted to the warmer seas.

III. *Not only are those Crustacea, which are most elevated in the scale, deficient in the Polar regions; but their relative number increases rapidly as we pass from the Pole towards the Equator.*—Thus the Brachyoura, which must be considered as the most elevated of the whole series, are totally absent in some parts of the arctic region; and we find their place taken by the far less complete Edriophthalma, with a small number of Anomourous and Macrourous Decapods. In the Mediterranean, however, the Decapods surpass the Edriophthalma in regard to the number of species; and the Brachyourous division of the former predominates over the Macrourous, in the proportion of two to one. And in the East and West Indies, the species of Brachyoura are to those of Macroura, as three, four, or even five, to one. Again, the *Land Crabs*, which are probably to be regarded as taking the highest rank among the Brachyoura (and therefore in the entire class), are only to be met with between the tropics. Moreover, of the *fluviatile* Decapods (inhabiting rivers, brooks, and fresh-water lakes,) a large proportion belong, in tropical regions, to the elevated type of the Brachyoura; whilst all those found in the temperate and arctic zones (the River Cray-fish and its allies) belong to the Macrourous division.

IV. When we compare together the Crustacea of different parts of the world, we observe that *the average size of these animals is considerably greater in tropical regions, than in the temperate or frigid climes.*—The largest species of the arctic and antarctic seas are far smaller than those of the tropical ocean; and they bear a much smaller proportion to the whole number. Further, in almost every natural group, we find that the largest species belong to the equatorial regions; and that those which represent them (or take their place, as it were) in temperate regions, are of smaller dimensions.

V. It is where the species are most numerous and varied, and where they attain the greatest size,—in other words, *where the temperature is most elevated,—that the peculiarities of structure which characterise the several groups, are most strongly manifested.* Thus the transverse development of the cephalo-thorax, which is so remarkable in the Brachyourous Decapods (the *breadth* of the carapace or arched shell in the typical Crabs being much greater than its *length* from back to front), is carried to its greatest extent in certain Crustacea of the Equatorial region; and the same might be said of the characteristic peculiarities of most other natural groups. Further, it is in this region that we find the greatest number of those anomalous forms, which depart most widely from the general structure of the Class.

VI. Lastly, *there is a remarkable coincidence between the temperature of different regions, and the prevalence of certain forms of Crustacea.*—Thus there are few *genera* to be met with in the West Indian seas, which have not their representatives in the East Indian,—the *species*, however, being usually different. The same may be said of the genera inhabiting the temperate regions of the globe;—similar *generic* forms being usually met with in the corresponding parts of the Old and New World, and of the Northern and Southern Hemispheres, although the *specific* are almost invariably different.

101. Now although, as appears from the foregoing general statements, the number of *species* of Crustacea inhabiting the colder seas bears a very small proportion to that which is found within the tropics; and although the species formed to inhabit cold climates are so far inferior, both as to size, and as to perfection of development; yet it does not follow that the same proportion exists in regard to the relative amount of Crustacean life in the two regions, for this depends upon the multiplication of *individuals.* In fact it may be questioned whether there is any inferiority in this respect; so abundant are some of the smaller species in the Arctic and Antarctic, as well as in the Temperate seas. Thus we see that a low range of temperature is as well adapted to sustain *their* life, as a higher range is to call forth those larger and more fully-developed forms, which abound in the tropical ocean. There is an obvious reason why the *seas* of the frigid zones should be much more abundantly peopled than the *land;* the mean temperature of the former being much higher. And it would almost seem as if Nature had intended to compensate for the dreariness and desolation of the one, by the profuseness of life which she has fitted the other to support.

102. The influence of Temperature in modifying the size of individual Animals of the same species, is not so strongly marked as it is in the case of Plants; for this reason, perhaps, that an amount of continued depression or elevation, which might be sustained by a Plant, but which would exert a modifying influence upon its growth, would be fatal to an Animal formed to exist in the same climate. Instances are not wanting, however, in which such a modifying influence is evident; and these, as might be anticipated, are to be met with chiefly among the cold-blooded tribes. Thus the *Bulimus rosaceus*, a terrestrial Mollusk, is found on the mountains of Chili of a size so much less than that which it attains on the coast, as to have been described as a distinct species. And the *Littorina petræa* found on the *south* side of Plymouth Breakwater, acquires, from its superior exposure to light and heat (though perhaps also from the greater supply of nutriment which it obtains) twice the size common to individuals living on the north side within the harbour.—The following circumstance shows the favourable influence of an elevated temperature, in producing an unusual prolificness in Fish; which must be connected with general vital activity. Three pairs of Gold-fish were placed, some years since, in one of the engine-dams or ponds common in the manufacturing districts, into which the water from the engine is conveyed for the purpose of being cooled; the average temperature of such dams is about 80°. At the end of three years, the progeny of these Fish, which were accidentally poisoned by verdigris mixed with the refuse tallow from the engine, were taken out by wheelbarrows-full. It is not improbable that the unusual supply of aliment, furnished by the refuse grease that floats upon these ponds (which would impede the cooling of the water, if it were not consumed by the Fish), contributed with the high temperature to this unusual fecundity.

103. Although a very low temperature is positively inconsistent with the continuance of vital *activity*, in Animals as in Plants, yet we find that even very severe Cold is not necessarily destructive of the vital *properties* of organized tissues; so that, on a restoration of the proper amount of heat, their functions may continue as before. Of this, the case of frost-bitten limbs furnishes an example; but the fact is much more remarkable,

when considered in reference to the whole body of an animal, and the complete suspension of all its functions. Yet it is unquestionably true, not only of the lowest and simplest members of the Animal kingdom, but also of beings so high as Fishes and Reptiles.—In one of Captain Ross's Arctic Voyages, several Caterpillars of the *Laria Rossii*, having been exposed to a temperature of 40° *below zero*, froze so completely, that, when thrown into a tumbler, they chinked like lumps of ice. When thawed, they resumed their movements, took food, and underwent their transformation into the Chrysalis state. One of them, which had been frozen and thawed four times, subsequently became a Moth.—The eggs of the Slug have been exposed to a similar degree of cold, without the loss of their fertility.—It is not uncommon to meet, in the ice of rivers, lakes, and seas, with Fishes which have been completely frozen, so as to become quite brittle; and which yet revive when thawed.—The same thing has been observed in regard to Frogs, Newts, &c.; and the experiment of freezing and subsequently thawing them, has been frequently put in practice. Spallanzani kept Frogs and Snakes in an ice-house for three years; at the end of which period they revived on being subjected to warmth.—It does not appear, however, that the same capability exists, in regard to any warm-blooded animals; since if a *total* suspension* of vital activity take place in the body of a Bird or Mammal for any length of time, in consequence of the prolonged application of severe cold, recovery is found to be impossible. The power which exists in these animals, however, of generating a large amount of heat within their bodies, acts as a compensation for the want of the faculty possessed by the cold-blooded tribes; since they can resist for a great length of time (if in their healthy or normal condition) the influence of a temperature, sufficiently depressed to produce a complete suspension in the activity of the classes below them.

104. It only remains to say a few words regarding the degree of *Heat* which certain Animals can sustain without prejudice, and which even appears to be genial to them. Among the higher classes, this range *seems* to be capable of great extension. Thus many instances are on record, of a heat of from 250° to 280° being endured, in dry air, for a considerable length of time, without much inconvenience; and persons who have become habituated to this kind of exposure can (with proper precautions) sustain a temperature of from 350° to 500°. In all such cases, however, the real heat of the body undergoes very little elevation; for, by means of the copious evaporation from its surface, the external heat is prevented from acting upon it. But if this evaporation be prevented, either by an insufficiency in the supply of fluid from within, or by the saturation of the surrounding air with moisture, the temperature of the body begins to rise; and it is then found, that it cannot undergo an elevation of more than a few degrees, without fatal consequences. Thus in several experiments which have been tried on different species of warm-blooded animals, for the purpose of ascertaining the highest temperature to which the body could be raised without the destruction of life, it was found that as soon as the heat of the body had been increased, by continued immersion in a limited quantity of hot air (which would soon become charged with moisture), to from 9°—13° above the natural standard, the animals died. In

* **In the case of hybernating Mammals, the suspension is not total; and if it be rendered such, the same result follows as in other instances.**

general, Mammals die, when the temperature of their bodies is raised to about 111° ; the heat which is natural to the bodies of Birds. The latter are killed by an equal amount of elevation of bodily heat above *their* natural standard.

105. Hence we see that the actual range of temperature, within which vital activity can be maintained by warm-blooded animals, is extremely limited ; a temporary elevation of the bodily heat to 13° above the natural standard, or a depression to 30° below it, being positively inconsistent, not merely with the continuance of vital operations, but also with the preservation of vital properties : and a continued departure from that standard, to the extent of only a very few degrees above or below it, being decidedly injurious. The provisions with which these animals are endowed, for generating *heat* in their interior, so as to supply the external deficiency, and for generating *cold* (so to speak), when the external temperature is too high, are therefore in no respect superfluous : but are positively necessary for the maintenance of the life of such animals, in any climate, save one whose *mean* should be conformable to their standard, and whose *extremes* should never vary more than a very few degrees above or below it. Such a climate does not exist on the surface of the earth.

106. The range of *external* temperature, within which cold-blooded animals can sustain their activity, is much more limited, as well in regard to its highest as to its lowest point ; notwithstanding that the range of *bodily* heat, which is consistent with the maintenance of their life, is so much greater. In those which, like the Frog, have a soft moist skin that permits a copious evaporation from the surface, a considerable amount of heat may be resisted, provided the air be dry, and the supply of fluid from within be maintained. But immersion in water of the temperature of 108°, is almost immediately fatal. In many other cold-blooded animals, elevation of the temperature induces a state of torpidity, analogous to that which is produced by its depression. Thus the *Helix pomatia* (Edible Snail) has been found to become torpid and motionless in water at 112° ; but to recover its energy when placed in a colder situation. It would seem to be partly from this cause, but partly also from the deprivation of moisture, that the *hottest* part of the tropical year brings about a cessation of activity in many tribes of cold-blooded animals, as complete as that which takes place during the *winter* of temperate climates. The highest limit of temperature compatible with the life of Fishes has not been certainly ascertained : and it appears probable that there are considerable variations in this respect amongst different species. Thus it is certain that there are some which are killed by immersion in water at 104° ; whilst it is also certain that others can not only exist, but can find a congenial habitation, in water of 113°, or even of 120° ; and examples of the existence of Fishes in thermal springs of a much higher temperature than this, have been put on record. Various fresh-water Mollusca have been found in thermal springs, the heat of which is from 100° to 145°. Rotifera and other animalcules have been met with in water at 112°. Larvæ of Tipulæ have been found in hot springs of 205°; and small black beetles, which died when placed in cold water, in the hot sulphur baths of Albano. Entozoa inhabiting the bodies of Mammalia and of Birds must of course be adapted to a constant temperature of from

98° to 110°; and they become torpid when exposed to a cool atmosphere. These lowly-organized animals seem more capable of resisting the effects of extreme heat, than any others;—at least if we are to credit the statement, that the Entozoa inhabiting the intestines of the Carp have been found alive, when the Fish was brought to table after being boiled.—In all such cases, it is to be remembered, the heat of the animal body must correspond with that of the fluid in which it is immersed; and we have here, therefore, evident proof of the compatibility of vital activity, in certain cases, with a very elevated temperature. Additional and more exact observations, however, are much wanting on this subject.

107. *Of Electricity.*—Much less is certainly known as to the ordinary influence of this agent, than in regard to either of the two preceding; and yet there can be little doubt, from the effects we observe when it is powerfully applied, as well as from our knowledge of its connexion with all Chemical phenomena, that it is in constant though imperceptible operation. Electricity differs from both Light and Heat in this respect;—that no manifestation of it takes place so long as it is uniformly diffused, or is in a state of *equilibrium;* but in proportion as this equilibrium is disturbed, by a change in the electric condition of one body, which is prevented, by its partial or complete insulation, from communicating itself to others, in that proportion is a *force* produced, which exerts itself in various ways according to its degree. The *mechanical* effects of a powerful charge, when passed through a substance that is a bad conductor of Electricity, are well known; on the other hand, the *chemical* effects of even the feeblest current are equally certain. The agency of Electricity in producing Chemical change is the more powerful, in proportion as there is already a predisposition to that change; thus, as already remarked, the largest collection of oxygen and hydrogen gases, or of hydrogen and chlorine, mingled together, may be caused to unite by the minutest electric spark, which gives the required *stimulus* to the mutual affinities that were previously dormant. Hence it cannot but be inferred, that its agency in the Chemical phenomena of living bodies must be of an important character; but this may probably be exerted rather in the way of aiding decomposition, than of producing new organic combinations, to which (as we have seen) Light appears to be the most effectual stimulus. Thus it has been shown that pieces of meat, that have been electrified for some hours, pass much more rapidly into decomposition, than similar pieces placed under the same circumstances, but not electrified. And in like manner, the bodies of animals that have been killed by electric shocks, have been observed to putrefy much more readily than those of similar animals killed by an injury to the brain. It is well known, moroever, that in thundery weather, in which the electric state of the atmosphere is much disturbed, various fluids containing organic compounds, such as milk, broth, &c., are peculiarly disposed to turn sour; and that saccharine fluids, such as the wort of brewers, are extremely apt to pass into the acetous fermentation.

108. The actual amount of influence, however, which Electricity exerts over a growing Plant or Animal, can scarcely be estimated. It would, perhaps, be the most correct to say, that the state of Electric *equilibrium* is that which is generally most favourable; and we find that there is a provision in the structure of most living beings for maintaining such an equili-

brium,—not only between the different parts of their own bodies, but also between their own fabrics and the surrounding medium. Thus a charge given to any part of a Plant or Animal, is immediately diffused through its whole mass; and though Organized bodies are not sufficiently good conductors to transmit very powerful shocks without being themselves affected, yet a discharge of any moderate quantity may be effected through them, without any permanent injury,—and this more especially if it be made to take place slowly. Now the points on the surfaces of Plants appear particularly adapted to effect this transmission; thus it has been found that a Leyden jar might be discharged by holding a blade of grass near it, in one third of the time required to produce the same effect by means of a metallic point; and an Electroscope furnished with Vegetable points has been found to give more delicate indications of the electric state of the atmosphere, than any other. Plants designed for a rapid growth have generally a strong pubescence or downy covering; and it does not seem improbable that one purpose of this may be, to maintain that equilibrium between themselves and the atmosphere, which would otherwise be disturbed by the various operations of vegetation, and especially by the process of evaporation, which takes place with great activity from the surface of the leaves.

109. There appears to be sufficient evidence, that, during a highly electrical state of the atmosphere, the growth of the young shoots of certain Plants is increased in rapidity; but it would be wrong thence to infer that this excitement is useful to the process of Vegetation in general, or that the same kind of electric excitement universally operates to the benefit or injury of the Plant. From some experiments recently made by Messrs. Solly and Sidney, it would appear that potatoes, mustard and cress, cinerarias, fuchsias, and other plants, have their development, and, in some instances, their productiveness, increased by being made to grow between a copper and a zinc plate, connected by a conducting wire; while, on the other hand, geraniums and balsams are destroyed by the same influence. The transmission of a series of moderate sparks through plants has been found, in like manner, to accelerate the growth of some, and to be evidently injurious to others. It is not unreasonable to suppose, that, as a great variety of chemical processes are constantly taking place in the growing plant, an electric disturbance, which acts as a stimulus to some, may positively retard others; and that its good or evil results may thus depend upon the balance between these individual effects. This would seem the more likely from the circumstance, that, in the process of Germination, the chemical changes concerned in which are of a simpler character, Electricity seems to have a more decided and uniform influence. The conversion of the starch of the seed into sugar, which is an essential part of this change, involves the liberation of a large quantity of carbonic and of some acetic acid. Now as all acids are negative, and as like electricities repel each other, it may be inferred that the seed is at that time in an electro-negative condition; and it is accordingly found that the process of germination may be quickened, by connexion of the seed with the negative pole of a feeble galvanic apparatus, whilst it is retarded by a similar connexion with the positive pole. A similar acceleration may be produced by the contact of feeble alkaline solutions, which favour the liberation of the acids; whilst, on the same principle, a very small

admixture of acid in the fluid with which the seed is moistened, is found to produce a decided retardation.

110. It is well known that Trees and Plants may be easily killed by powerful electric shocks; and that, when the charge is strong enough (as in the case of a stroke of lightning), violent mechanical effects,—as the rending of trunks, or even the splitting and scattering of minute fragments,—are produced by it. But it has also been ascertained, that charges which produce no perceptible influence of this kind, may destroy the life of Plants; though the effect is not always immediate. In particular it has been noticed, that slips and grafts are prevented from taking root and budding. There can be little doubt that, in these instances, a change is effected in the chemical state of the solids or fluids; although no structural alteration is perceptible. The coagulation of the layer of albuminous matter which lines every growing cell, would be quite sufficient to account for the result.

111. In regard to the influence of Electricity upon the Organic functions of Animals, still less is certainly known; but there is evidence that it may act as a powerful stimulant in certain disordered states of them. Thus in Amenorrhea, a series of slight but rapidly-repeated electric shocks will often bring on the catamenial flow; and it is certain that chronic tumours have been dispersed, and dropsies relieved by the excitement of the absorbent process, through similar agency. Again, it is indubitable that a highly-electric state of the atmosphere produces very marked effects on the general state of many individuals; and brings on in some a degree of languor and depression, which cannot be accounted for in any other way. An instance is on record, in which the atmosphere was in such an extraordinary state of electric disturbance, that all pointed bodies within its influence exhibited a distinct luminosity; and it was noticed, that all the persons who were exposed to the agency of this highly-electrified air, experienced spasms in the limbs and an extreme state of lassitude.

112. Animals, like Plants, are liable to be killed by shocks of Electricity; even when these are not sufficiently powerful to occasion any *obvious* physical change in their structure. But there can be no doubt that such Mechanical changes may be produced in their delicate parts, as are quite sufficient to account for the destruction of their vitality, even though these may be only discernible with the Microscope. The production of changes in the Chemical arrangement of their elements, is, however, a much more palpable cause of death; since it may be fully anticipated beforehand, and can easily be rendered evident. To take one instance only;—it is well known, that *albumen* is made to coagulate, *i. e.* is changed from its soluble to its insoluble form, under the influence of an electric current; and it cannot be doubted that the production of this change in the fluids of the living body (almost every one of which contains albumen), even to a very limited extent, is quite a sufficient cause of death, even in animals that are otherwise most tenacious of life. "I once discharged a battery of considerable size," says Dr. Hodgkin, "through a common Earth-worm, which would in all probability have shown signs of life long after minute division. Its death was as sudden as the shock; and the semi-transparent substance of the animal was changed like Albumen which has been exposed to heat."—It has been noticed that the bodies of

animals killed by Electricity pass into decomposition with unusual rapidity; a fact which affords an additional indication that the chemical condition of the organic components of the fabric has been affected.

113. Electricity possesses, in a remarkable degree, the power of exciting the Contractility of Muscular fibre, and the Nervous force; but this series of phenomena will be more fitly described, when the properties of those tissues are under consideration (§§ 231 and 246).

4. *Of Dormant Vitality.*

114. There are many organized beings, whose vital activity is occasionally suspended, either wholly, or as regards some portion of their structure, without being altogether destroyed: and this may result either from the withdrawal of the stimuli necessary for the maintenance of the operations thus suspended, in which case the suspension may be brought about or terminated at any time; or it may proceed from some change in the organism itself, whereby its capability of responding to those stimuli is for a time diminished or lost, in which case we generally find it presenting a tendency to *periodical* recurrence.—This suspension of vital activity cannot long continue without the entire disintegration of the tissues, and consequent loss of their vital properties, through that breaking-up of their peculiar composition, which is continually taking place even when they are in the state of greatest vital activity (§ 46), and which would now go on unchecked and uncompensated; unless at the same time the conditions to which the organism is subjected, be such as are unfavourable to the play of chemical affinities. Thus if a man fall into the water, or receive a concussion of the brain, and be reduced to a state of death-like torpor without vitality being actually destroyed, he can only remain in that condition for a short period; for if means be not employed to restore the circulation, and thus to renew the other functions, death will speedily ensue, either from the lowering of the temperature below the point at which vitality departs (§ 93), or, if the bodily warmth be artificially maintained, from the decomposition which will then speedily ensue. But, on the other hand, if we subject a cold-blooded animal to a temperature of 32°, we reduce it to a state of torpidity in which it may remain for an unlimited period (§ 103); the reduction of its temperature not having in itself the power of destroying the vitality of the tissues, whilst it effectually prevents their decomposition; so that, when they are again acted on by warmth, the tissues are precisely in the condition in which they were when their activity was suspended, and are ready for the performance of their several functions.

115. Beings placed in this condition can scarcely be said to be *alive*, if Life be regarded as synonymous with *vital activity;* on the other hand, they cannot be rightly designated as *dead*, since this implies their incapability of ever resuming their previous state; and they may be most appropriately said to be in a state of *dormant vitality*, since they retain their peculiar attributes without any manifestation of them, just as a sleeping man possesses mental power although its operation is for a time suspended.—Some of the most remarkable examples of it are presented by the spores or reproductive germs of the Fungi, which appear to be diffused in incalculable multitudes through the air, ready to start into develop-

ment whenever the fitting conditions are afforded them (§ 271). We have no means, however, of ascertaining the length of time during which these almost infinitesimally minute particles can retain their vitality whilst floating about in the air; but with respect to the *seeds* of Flowering Plants, we have evidence that the term may be so prolonged as to be really unlimited.

116. One of the most interesting cases of this kind on record is thus related by Dr. Lindley. "I have now before me," he says, "three plants of Raspberries, which have been raised in the gardens of the Horticultural Society, from seeds taken from the stomach of a man, whose skeleton was found 30 feet below the surface of the earth, at the bottom of a barrow which was opened near Dorchester. He had been buried with some coins of the Emperor Hadrian, and it is probable, therefore, that the seeds were sixteen or seventeen hundred years old."—Instances are of very frequent occurrence, in which ground that has been turned up, spontaneously produces plants dissimilar to any in their neighbourhood. There is no doubt that, in some of these, the seed is conveyed by the wind, and becomes developed in particular spots which afford congenial soil; and this fact has a very interesting bearing upon the question of the production of Animalcules in infusions of decaying organic matter. Thus, it is commonly observed that clover is ready to spring up on soils, which have been rendered alkaline by the strewing of wood-ashes, or by the burning of weeds; and it is stated by Professor Graham that, after any hill-pasture in Scotland has been laid dry and limed, and the surface broken, white clover always makes its appearance. But there are many authentic facts, which can only be explained upon the supposition, that the seeds of the newly-appearing plants have lain for a long period imbedded in the soil, at such a distance from the surface as to prevent the access of air and moisture; and that, retaining their vitality under these circumstances, they have been excited to germination when at last exposed to the requisite conditions.—Several cases of this kind are related by Dr. Prichard,* on the authority of Professor Graham; amongst them is the following:—"To the westward of Stirling there is a large peat-bog, a great part of which has been flooded away by raising water from the river Teith, and discharging it into the Forth, the under-soil of clay being then cultivated. The clergyman of the parish standing by while the workmen were forming a ditch in this clay, which had been covered with 14 feet of peat-earth, saw some seeds in the clay which was thrown out of the ditch: he took some of them up and sowed them; they germinated and produced a crop of Chrysanthemum septum. What a period of years must have elapsed while the seeds were getting their covering of clay, and while this clay became buried under 14 feet of peat-earth!"

117. For the following very interesting case, not elsewhere published, the author is indebted to his valued friend, the late Dr. Tuckerman, of Boston, N. E. "About 25 or 30 years ago, Judge Thacher, then one of the Judges of the Supreme Court of Massachusetts, told me that he knew the fact, that in a town on the Penobscot river, in the state of Maine, and about 40 miles from the sea, some well-diggers, when sinking a well, struck, at the depth of about 20 feet, a stratum of sand, which strongly excited curiosity and interest, from the circumstance that no similar sand

* Researches on the Physical History of Mankind, vol. i., p. 39, &c.

was to be found anywhere in the neighbourhood, and that none like it was nearer than the sea-beach. As it was drawn up from the well, it was placed in a pile by itself; an unwillingness having been felt to mix it with the stones and gravel which were also drawn up. But when the work was about to be finished, and the pile of stones and gravel to be removed, it was found necessary to remove also the sand-heap. This, therefore, was scattered about the spot on which it had been formed, and was for some time scarcely remembered. In a year or two, however, it was perceived that a large number of small trees had sprung up from the ground over which the heap of sand had been strewn. These trees became, in their turn, objects of strong interest, and care was taken that no injury should come to them. At length it was ascertained that they were Beach-Plum trees; and they actually bore the Beach-Plum, which had never before been seen except immediately upon the sea shore. These trees had, therefore, sprung up from seeds which were in the stratum of sea-sand that had been pierced by the well-diggers. By what convulsion of the elements they had been thrown there, or how long they had quietly slept beneath the surface of the earth, must be determined by those who know very much more than I do."

118. In these and other similar instances, the complete seclusion of the seed from moisture and oxygen, and the maintenance of a uniform temperature, were sufficient to preserve its organic structure and composition entirely unchanged; so that after the lapse of thousands or even of tens of thousands of years, its vital properties were still found capable of being aroused into vigorous activity. If the seclusion from air and moisture be not complete, and the temperature be not sufficiently high for germination, a slow decay takes place in the seed, by which its germinating power is destroyed; so that we perceive that it is only by the withdrawal of all those conditions under which its decomposition can possibly take place, that its vitality can be preserved.—Various examples of the reduction of Plants and Animals to a state of dormant vitality, in consequence of the withdrawal of the Heat or Moisture requisite to maintain their activity, have already come before us (§ 103) and (§ 65); in all of which the same general fact holds good, that this state, if complete, can only be prolonged by such a reduction of temperature, desiccation of the tissues, or complete exclusion of oxygen, as must necessarily prevent the disintegration of the tissues by the play of Chemical Affinities. In the hybernating Mammal, however, this state is not complete; for the temperature of the body cannot be reduced below about 35° without the destruction of its vitality; and we find that a feeble amount of vital activity continues during the whole period, rather, it would seem, for the purpose of getting rid of the products of decomposition as fast as they are set free, than for the purpose of adding anything to the structure. The same is not improbably the case in regard to the torpidity of Reptiles, &c., in the hot and dry season (§ 66).

119. That this state of Dormant Vitality does not always result solely from changes in the external conditions on which Life is dependent, but is connected in part with alterations in the constitution of the animal itself, which make it a part of its regular history, is rendered particularly obvious by a consideration of the phenomena attending the *pupa* stage of Insects. This stage is one in which *formative* changes of a most important character take place, notwithstanding the appearance of complete torpor;

but the activity of these changes varies in different cases, as they are performed in some instances in a few days or even hours, whilst in others they are spread over almost as many months; and a considerable portion of the latter period must be passed in a state of complete inactivity. The influence of artificial heat or cold, in accelerating or retarding the transformation, has been already noticed (§ 96); but we have now to show that there are other causes, inherent in the animal itself, which tend to prolong the period of inactivity.—Thus in the *Papilio Machaon* there are two generations every year; for the butterfly that comes forth in the early summer lays eggs, which rapidly come to maturity and produce another set of eggs later in the season, whose larvæ pass into the pupa state before the winter. Now the pupa-stage of the first brood, which passes into it in July, lasts only *thirteen days;* but that of the second brood, which passes into it in September, lasts *nine or ten months,* the butterfly not appearing until the June following. The difference of temperature is obviously insufficient of itself to account for this remarkable diversity between the two periods.—The following, moreover, are cases in which a similar diversity exists, without any difference in external conditions. If a number of the pupæ of *Eriogaster lanestris,* a moth whose larvæ are common on the blackthorn in June, be selected at the same time, and placed under the same circumstances, the greater number of them will disclose the perfect insect in the February following; but some will not produce it until the February of the year ensuing; and the remainder not before the same month in the third year. The same has been observed of the *Arctia mendica,* of which thirty-six pupæ, grown from eggs laid by the same parent, produced twelve perfect insects in each of the three following seasons.*—The *purpose* of this curious adaptation is probably, as suggested by Mr. Kirby, to secure the race from being cut off by unfavourable seasons, or by some extraordinary increase of its natural enemies; but the *cause* of it can only be sought for in the state of the organism itself, since there is here no variation in the external conditions to which the several individuals of the three broods have been subjected.

120. That such a predisposition really exists, appears also from the *periodical* manifestation of the tendency to pass into a state of dormant vitality more or less complete, which is seen in a large proportion of Plants and Animals. Thus the 'winter sleep' (as it has been termed) of Plants is not referable solely to the depression of external temperature (§ 83); for it is easily shown that, in the trees of temperate climates, there is a tendency to its annual recurrence, arising out of their own succession of vital changes. Thus, the exuviation of the leaves is not the cause of their death, but its consequence; when the cells of which they are composed have accomplished their term of life, they cease to perform the function for which they are created, they begin to undergo decay, and the dead and useless members are thrown off; and this takes place whatever may be the temperature. Until new leaves shall have been developed, therefore, there must be a great reduction in the vital activity of the organism; and it is by its agency in hastening or retarding this development, that external temperature exerts so powerful an influence over the duration and intensity of the torpor. Thus in the trees of tropical climates, we frequently see a new generation of leaf-buds called forth, in the course of a few weeks

* Kirby and Spence's Entomology, vol. iii. p. 266.

after the tree has been completely bared by the fall of the preceding set of leaves; and two or even three generations may thus replace each other every year. In those of the coldest regions that can produce them, we see, on the other hand, a prolongation of the torpor until the year has far advanced; and then with the lengthening days of summer, into which the dreary winter seems suddenly transformed, every form of vegetation is rapidly excited to activity. The constitution of the tropical and of the arctic Plant is adapted to its own periodicity; so that neither can be charged by artificial means, without manifest injury.—So, again, in regard to Animals, it is certain that the tendency to *hybernation* (§ 94) results from a periodical alteration in the bodily constitution of the animal, which *corresponds* with the seasonal change, but is not *dependent* upon it. For although it is possible to reduce a Marmot or a Dormouse to this condition, even in summer, by the application of artificial cold, yet the external temperature must then be lowered to a point much below that, at which hybernation naturally commences on the approach of winter.

121. The diurnal *sleep* of Vegetables and Animals is a phenomenon of the same general character. In all the higher Plants, there is a tendency to diminished activity during a part of the cycle of twenty-four hours; which, though really affecting their entire series of nutritive operations, is chiefly manifested in the folding or drooping of the leaves and flowers. It appears from experiments made with a view to determine the conditions of this phenomenon, that it cannot be attributed solely to alterations in those external physical agents which most influence the processes of vegetation,—such as light, heat, and moisture; but that, although capable of being modified by them, its occurrence is essentially dependent on a periodically-recurring change in the vital condition of the plant itself.—The sleep of Animals differs essentially in its character from that of Plants; for it consists, not in a state of diminished energy of the *nutritive* functions, but in a cessation of the *sensorial* activity, dependent upon a suspension of the functional power of certain parts of the nervous system, during which there is reason to believe that the nutritive and reparative operations of those organs go on with even augmented rapidity. Still this complete suspension of the peculiar activity of the nervous substance is obviously a form of dormant vitality; and it is one in which we can trace, better than in most other cases, the influence of preceding changes in the organism itself in bringing it about. For the occasional suspension of sensorial activity is requisite for the reparation of the destructive effects of that activity (§ 243); so that, however unfavourable may be the external circumstances, sleep will supervene as a necessary result of exhaustion when this has been carried very far. Thus, sleep may come on amidst the roar of artillery, and the excitement of a battle; and it is related that Damien slept during his protracted tortures on the rack. The diurnal periodicity of sleep,—which is a fact common to most animals, although some are active during the repose of others,—is certainly favoured by the alternation of darkness with light in every twenty-four hours; but after we have subtracted all that can be fairly attributed to this, or to acquired habits, there still remains a periodical character about this condition, which, if not absolutely determinate in itself, accommodates itself to the period of the earth's rotation, more readily than to any other.*

* See Art. *Sleep*, in the Cyclopædia of Anatomy and Physiology.

5. *Of Death.*

122. The *limited duration* of the vital activity of organised structures has been several times referred to, as one of their most distinctive characters. We have seen that this activity may be suspended, even for lengthened periods, without the loss of the power of renewing it; but when the very organisms which appear thus tenacious of their vitality are called into active Life, they become subject, like all others, to that complete and final loss of their peculiar attributes, which we call *Death.* The duration of vital activity in different organised structures is extremely various; depending partly upon the peculiar attributes of the structures themselves, the arrangement of their chemical components, the proportion of their solid and watery constituents, &c.; and partly upon the activity of their Life—their duration being, *ceteris paribus,* inversely to the rate at which the various changes incident to their living condition take place in them. Hence that duration may be greatly affected by those external agencies which modify the vital activity; and on the whole it would be more correct to say, that the natural death of each organism takes place when it has accomplished the whole series of changes it was adapted to perform, whether these have been executed at a more or at a less rapid rate, than to attempt to assign any definite limit of time to its duration as a living being.

123. The state of *Death* differs from that of Dormant Vitality in this,—that the suspension of vital activity which is temporary in the latter condition, is permanent and complete in the former. Thus a seed which will not germinate under any treatment, must be regarded as dead; whilst another which does not differ from it in any perceptible character, but which retains its capability of development, must be considered as endowed with vitality. The idea of *decay* or *decomposition* is usually connected with that of death; but there is no essential relation between them. For we have seen that decomposition is continually taking place during life; whilst, on the other hand, decay does not necessarily supervene upon death. There are many structures both in the Plant and Animal, which, when once consolidated by their processes of growth, may resist decay for an unlimited period (§ 24); and this as well when they are detached from the living structure, as whilst they remain connected with it. Thus the heart-wood of a Tree may be preserved longer when used as timber, and placed in a dry situation, than when remaining in the trunk of which it forms the solid support, and subjected to various agencies that favour its decay (§ 125).—Even the softer azotized tissues of Animals, which much more readily pass into spontaneous decomposition under ordinary circumstances, may be kept from it by the complete exclusion of oxygen, as when meat is preserved in air-tight tin cases, or by the entire expulsion of their water, as in the desiccation of the Rotifera (§ 310).

124. The phenomena of Death, like those of Life, are displayed to us in the simplest form in those humble Plants in which every cell is really a distinct individual; since we have not there to make the distinction which becomes necessary in the case of the higher organisms, between the death of the individual parts and that of the fabric as a whole. And the case is nearly the same with those that consist of a homogeneous mass of

cells, which are developed so rapidly as to be almost of the same age, and which all perish together,—as is the case with the softer Fungi. Here we find that, as soon as vital activity has finally ceased, the whole structure decays; its components being resolved, by a new arrangement of their molecules in combination with oxygen derived from without, into carbonic acid, water, &c.—But with forms of vegetable structure somewhat higher in the scale, the case is different; for each of these consists of an aggregation of cells, every one of which has its own independent vitality, goes through its own series of changes, and dies at its own time, without reference to the rest; and yet by the development of new cells in continuity with those which are themselves about to undergo dissolution, the life of the mass may be prolonged almost indefinitely. Thus, as remarked by Schleiden, we might yet find floating about, on the great Fucus bank of Cervo and Flores, plants of Sargassum which were cut into strips by the bark of Columbus; and the masses of Fungous vegetation which form the progressively-extending fairy rings or magic circles of the grassy meadow, may be said to be several years of age, although, as fast as new tissue is generated on the exterior of the ring, that of the central side dies and decays, so that no individual part has more than a brief duration.

125. Passing upwards to the higher Plants, we find that with the increase of heterogeneousness in the structure and of variety in its functions, the bond of mutual dependence becomes more intimate; so that the death of one part has a tendency to destroy the vitality of the whole, by interrupting the actions which are requisite to maintain it. This is best seen in the herbaceous plants whose duration is limited to a single season; for to such the destruction of any entire set of organs, such as the leaves or roots, is necessarily and immediately fatal; and the whole fabric dies together, as soon as the seeds of a new generation have been matured. A perennial Tree, on the other hand, may be regarded as a composite structure, made up of a number of integral plants; and its continued existence depends upon the successive development of new individuals, as the old ones die and decay. This may take place to an unlimited extent. We have numerous examples of trees whose age is calculated at from one to five thousand years; and there is no proof that, under favourable circumstances, any such tree ever died from the 'weakness of old age.' In most cases, death takes place as a consequence of mechanical injuries. A branch is broken off by a storm, and the broken surface is exposed to the action of rain-water; this causes decay of the woody structure, and the decay gradually spreads inwards to the heart-wood of the trunk; finally, a new storm casts the whole tree to the ground, separates the leaves from the roots by the fracture of the trunk, and the death of the whole is the inevitable result.—If no such interference take place, however, the development of new leaf-buds, and the formation of new layers of wood and bark, may take place as long as the *cambium-layer* (§ 280) retains its activity; and as this is being continually regenerated, there does not seem any reason why the structure should perish from any causes inherent in itself. The periodical exuviation of the leaves is an obvious indication, that the life of the fabric as a whole is perfectly consistent with the death of certain of its constituent parts. On the other hand, the continued life of a bud or off-shoot, which we have detached from the stock and placed in circumstances favourable to its separate development, affords a clear proof

that the life of the individual segment does not depend upon its connection with the entire structure,—all that is requisite being that it should possess the organs essential to the maintenance of its own life, or should have the power of developing them, or should be able to make use of those of another plant. Thus a leaf of the *Bryophyllum* (one of the 'air-plants') will of itself develop roots, when detached and laid upon damp earth, and will in time grow into a new plant; on the other hand, a detached leaf-bud generally thrives best when it is enabled to obtain its nutriment through the roots and stem of another plant, on which it is fixed by 'grafting,' and with which it speedily forms an organic union.

126. The condition of Zoophytes and of the lower Mollusca and Articulata, which are of a similar composite nature with the higher Vegetable forms, corresponds so closely with that of the latter in regard to the topic now under consideration, that it is not requisite to notice them separately; and accordingly we shall pass on at once to those higher types of Animal existence, in which the heterogeneousness of structure and variety of function are the greatest, and in which, therefore, the bond of mutual dependence is the closest. We here still find that every part of the organism has its own limited term of vitality; and that there is a continual succession of new cell-life, in those parts more especially in which active vital changes are taking place. In fact, the periodical exuviation of the leaves of Plants is a phenomenon which is daily repeated in the body of every one of the higher Animals; the duration of life being extremely short in many parts of the fabric of the latter, and the production of new tissue taking place with wonderful rapidity. Even the most solid portions of the animal fabric have a limited term of existence as living structures; for if they become so condensed and mineralised as to possess the character of permanence (which is the case with Shells and Corals, § 24), they are then so completely cut off from participation in the Vital Activity of the organism, that they cannot be said to be alive; whilst, on the other hand, if they contain a larger proportion of organised tissue, and this remains capable of taking any share in the nutritive operations (as is the case with Bones, and to a less extent with Teeth), their duration in the living organism is limited,—although they may be preserved in their dead state, under favourable circumstances, for an indefinite period. Thus nearly the whole of the Animal fabric is undergoing continual change, by the death and decay of its integral parts; and this not merely during the period of growth, when it is requisite that the structure originally laid down should be progressively extended, by the removal of the parts first deposited, and the deposition of new material elsewhere; but even during the whole term of its existence. For there is no doubt that the exercise of the truly *Animal* powers,—which are the endowments of the Nervous and Muscular tissues,—involves the death and disintegration of those tissues, as its necessary condition; so that even to the end of corporeal life, every sensation, every movement, and probably every act of mind, takes place at the expense of the vital energy of a certain amount of organised structure, which thereby gives up its own Life and becomes subject to decay. Hence if it were not for the continual reconstruction of these organs by the Nutritive or Vegetative processes, the Animal fabric would speedily perish; and we accordingly find that after this has attained its full growth and development, the whole energy

of these processes is directed towards its maintenance; and, as we have just seen, even their performance involves the continual death and disintegration of certain parts of the fabric, in order that the materials may be prepared for the development of others of more permanent character.

127. The continual series of changes, however, to which the Animal fabric is subject in virtue of its essential nature, is harmonized and co-adapted with such perfection, that the *molecular death*, which is perpetually taking place in some or other of the component parts of the organism, shall contribute to the maintenance of the Life of the entire being. But this is only the case, so long as each important organ remains in its due functional activity; for if an interruption occurs in the vital actions of any one, those of the remainder are necessarily checked, in consequence of the relation of mutual dependence which exists among them. Thus, for example, the growth of every tissue in the body, and consequently its vital activity, is dependent upon a supply of blood; hence, if the circulation of the nutritive fluid be interrupted, every action of Life must speedily cease. The interruption is first seen in the Animal functions; since the Nervous tissue requires, as the condition of its activity, not so much nutritive material (for this is applied to its regeneration in the *intervals* of its action, § 236), as oxygen; and this can only be supplied to it by the *movement* of the blood, which conveys to it this element from the respiratory organs where it is introduced into the body. The vital activity of Muscle may continue for a longer period, especially in cold-blooded animals;—thus the heart of a Frog, when removed from the body, will pulsate for some hours; and it has been asserted that the heart of a Sturgeon, cut out and hung up to dry, continued alternately contracting and dilating, until the movement produced a crackling noise in consequence of the desiccation of the texture;—but its persistence in warm-blooded animals is comparatively brief, and it is necessarily extinguished after a time in all. The processes of Nutrition and Secretion may continue for a certain time after the suspension of the circulation, the requisite materials being supplied by the blood remaining in the vessels; thus Mr. T. Bell mentions * that in dissecting very carefully and minutely the poison-apparatus of a large Rattlesnake which had been dead for some hours, he found the yellow poison continually secreted by the gland, so fast as to require being occasionally dried off with a sponge. We find, again, that blood retains its vitality for some time, so as neither to coagulate nor to putrefy, when withdrawn from the current of the circulation, if it remain in contact with living tissues. Hence it is that parts which have been completely separated from the body may be reunited with it, if they were previously in a healthy state, and too much time has not elapsed; thus, there are many cases on record, in which fingers, toes, noses, or ears, that have been accidentally severed, have been made to adhere and grow as before, by bringing the cut surfaces into contact. It is evident, then, that the parts so separated cannot have lost their vitality; since no treatment could produce union between a dead mass and a living body. And we are fully justified in assuming that, in cases where attempts at such reunion have not been successful, the death of the separated part has resulted from the too-prolonged interruption of its regular nutritive operations, whereby chemical and physical changes have taken

* History of British Reptiles, p. 61.

place in it, which have destroyed the peculiar structure and composition of its several parts.

128. The cessation of the Circulation, which supplies the nutritive material for the various components of the fabric, and which also in warm-blooded animals sustains the temperature of the body, may be considered as constituting *Somatic Death*,* or the death of the organism as a whole. This must be followed at no distant interval by *Molecular Death*, or the death of every integral portion of the organism.—Somatic and Molecular death may, however, be coincident. Thus when an animal dies of pure 'old age,' there is a gradual diminution in the vital power of each part of the organism; so that, when the heart's action ceases, activity of every other component part of the body terminates at the same time. And the same thing happens when an animal is killed by a powerful electric shock, by concussion of the brain, by the introduction of certain poisons into the blood, or by any other cause which has an injurious action (direct or indirect) upon the whole frame alike.—But it more generally happens that Somatic death is the result of the suspension of some or other of the functional changes necessary to the Life of the organism as a whole; and this may be a consequence of the molecular death of a very small part of it. Thus if the muscular power of the heart should fail, the circulation of the blood is thereby brought to a stand, and death necessarily ensues, notwithstanding that every other part of the body may be in the most perfect condition. So it will be shown hereafter, that the failure in the vital activity of a very small portion of nervous matter (situated at the summit of the spinal cord, just within the skull) involves the cessation of the movements concerned in respiration; that the aeration of the blood will be thereby checked; that its movement in the capillaries of the lungs will be consequently stagnated; and that the whole current of the circulation will be thus brought to a stand. And again, if the liver or the kidneys fail to perform their office of removing from the blood the matters with which it is continually being impregnated by the decay of the tissues, these matters accumulate in the circulating fluid, and act as poisons, which more or less affect the powers of the whole organism, but those of the nervous system more especially.

129. It will be obvious, from the considerations already adduced, that the dependence of Somatic death upon the Molecular death of any individual organ, will depend upon the degree in which the actions of that organ form a necessary part of the chain or connected series of operations concerned in the maintenance of the Life of the entire organism. Thus even in the higher animals, the death or removal of the limbs, although they may constitute (as in Man) a large proportion of the fabric, is not necessarily fatal; because it does not involve any interruption in the nutritive operations of the viscera, or of the sensorial functions of the brain. On the other hand, the destruction of a certain minute portion of the nervous centres, or an injury of the heart which would be trivial elsewhere, will be the occasion of immediate death. It sometimes happens that life may be prolonged, after the cessation of the activity of an important organ, by the power which some other possesses of discharging its function; thus we find that in Man, thekidneys occasionally take upon themselves the elimination of bile from the blood; whilst in the Frog, the skin can perform

* See Art. *Death*, by Dr. Symonds, in the Cyclopædia of Anatomy and Physiology.

part of the office of the lungs, so as to effect to a certain extent the aeration of the blood. Of course the more this is the case, the greater is the power which animals possess of sustaining injuries of great severity without the loss of Life: and the power is further augmented by the capability which some of them possess, of re-forming the organs of which they have been deprived. The greatest amount of this appears to be exhibited by the Hydra (§ 289); almost every part of whose fabric can regenerate every other part; so that its life cannot be annihilated by any agencies, except such as effect the destruction of the entire organism.

130. It has been maintained, however, by those who consider Vitality as something superadded to an Organised Structure, essentially independent of it, and capable of being subtracted from it, that Death frequently takes place under circumstances which leave the organism as it was; so that "the dead body may have all the organisation it ever had whilst alive." But for such an assumption, there is not the least foundation. In nearly all cases in which death takes place as a result of disease, the connexion between changes of structure and composition, either in the tissues or in the blood, and such a loss of the vital properties of some part or organ as is sufficient to bring the Circulation to a stand, is so palpable as to require no proof; and in by far the greater number of cases in which it is not at once obvious, a more careful scrutiny will reveal it. It must be confessed on both sides, that our means of investigation, and our knowledge of the normal structure and composition of the tissues and the blood, are not yet sufficient to enable us to detect minute shades of alteration, nor to assert what extent of change is inconsistent with the continuance of life. But as no one has yet shown, by the careful and precise microscopical and chemical examination of the solids and fluids of a dead body, that it has the exact organisation and composition it had whilst alive, the assertion above quoted is totally unwarranted by experience, and is contradicted by all our positive knowledge of the matter.

131. But it has been urged, that Death may result from the sudden operation of some agency of an *immaterial* character, which leaves no trace behind it,—such as a powerful electric shock, or a violent mental emotion. Here, too, the argument entirely fails. It is scarcely possible that a powerful electric shock could be transmitted through a mass like the animal body, composed of elements in such a loose state of combination that they are continually undergoing decomposition, without producing important *chemical* changes in it (§ 112); and its imperfect conducting power renders it equally liable to *physical* disturbances. Thus it has been ascertained, that when Eggs in process of development have had their vitality destroyed by an Electric shock, the minute vessels of the *vascular area* (CHAP. XVIII.) have been ruptured.—Nor is it more difficult to explain the immediate cause of death, as a result of Mental emotion. In some cases, an obvious physical change has been produced, by the too violent action of the heart, the movements of which are stimulated by the emotion; thus, even in a healthy person, rupture of the heart or aorta has been known to take place,—an occurrence to which those affected by previous disease of that organ are much more liable. Where there is *any* disorder in the heart's action, resulting from thickened valves, narrowed orifices, &c., the physical influence of mental emotion can be easily

accounted for. But it must be admitted that cases have occurred, in which no such explanation can be offered; sudden death having taken place without any perceptible structural cause. We are not obliged, however, to have recourse, for an explanation of even these cases, to any hypothesis which is not borne out by ample analogy. For it is well known that mental emotions exert a powerful influence over the composition of the *fluids* of the body, and are capable of *instantaneously* altering these. Thus in many human beings, and still more in the lower animals, alarm or agitation will occasion the immediate disengagement of powerfully odorous secretions, which must have resulted from new combinations suddenly formed. And there can be no question that a fit of passion may immediately occasion such a change in the milk of a nurse, as renders it a rank poison to the infant. There is no reason to doubt, therefore, that the blood itself may undergo changes of analogous character from the same cause; and that it may become a violent poison to the individual himself, instead of being the source of wholesome nutriment, or the stimulus to vital activity.

132. To conclude, then;—we only know of *Life* or *Vital Activity* as exhibited by an *Organised Structure* under the influence of certain *Physical Forces.* And we only know of *Vitality,* the peculiar state or endowment of the being which exhibits that activity, as consequent upon that new arrangement of the molecules of matter entering into its composition, which we term *Organisation.* The conditions of that endowment, therefore, must be found in the properties of the original elements of the structure composing it, and in the forces by which they are brought into play. When we 'magnetise' a piece of steel, by the agency of another piece already magnetic, or by the transmission of an electric current across it, we are but calling forth the properties which were previously dormant or inactive in the metal. So when the cell-germ, under the influence of Light and Heat, effects the combination of the inorganic elements into new and peculiar compounds, and applies these to the formation of its own structure, it does nothing else than call forth the dormant properties of those forms of matter. For the existence of the properties which thus enable certain elements to become component parts of Organised structures and to perform Vital actions, we can assign no other cause than the Will of the Creator. They are ultimate facts in Physiology, just as Attractions and Repulsions constitute the ultimate facts in Physics, and as do Affinities in Chemistry. The constancy of the actions which result from them, when the conditions are the same,—that is, their conformity to a fixed plan, or (in the language commonly employed) their subordination to *Laws,*—indicates the constancy and unchangeableness of the Divine Will, as well as the Infinity of that Wisdom by which the plan was at first arranged with such perfection, as to require no departure from it in order to produce the most complete harmony in its results. (See § 5, *note.*)

CHAPTER IV.

OF THE COMPONENT STRUCTURES OF ORGANISED FABRICS.

1. *General Considerations.*

133. Having thus taken a general survey of the characters which are common to all Organised beings, and which distinguish them from Inorganic matter,—having also enquired into the conditions that are essential to the occurrence of all Vital phenomena, and enquired into the relations subsisting between the peculiar agencies to which we must attribute them, and those which are known as the Physical Forces,—we are prepared to examine, in more detail, into that combination of diverse elements, and that succession of mutually-related actions, which every living being presents to our observation. It is a fact now well-established by microscopic investigation, that just as the Chemist resolves the countless substances found in Nature or producible by Art into a comparatively small number of ultimate elements, each having its distinctive properties; so can the Anatomist resolve the fabrics of Plants and Animals, whatever may be their dimensions, into a limited number of 'Elementary Tissues,' each having a structure peculiar to itself; and in like manner, the Physiologist can reduce the most complicated series or combination of actions to its individual components, and can assign each of these to some form of the elementary tissues as its own special and peculiar instrument. Thus, then, by an examination into the structure and endowments of these 'elementary' or 'primary tissues,' we shall obtain that kind of preparation for the study of the more complex phenomena of Life, which the Chemical student seeks to acquire, when he investigates the properties of each elementary substance, before endeavouring to acquaint himself with the changes which this undergoes in its combination with others, or with the multitudinous variety of phenomena which arise out of such combinations.

134. It must be freely conceded that investigations into the structure and properties of the Elementary parts of which Organised structures are composed, are often attended with much difficulty and liability to error. The minuteness of the objects which are the subjects of our examination, and the production of changes in them by the preparation they are necessarily made to undergo before being submitted to Microscopic inspection, not to mention the deceptions arising from imperfection of the instrument itself or from ignorance of the right mode of employing it, have led to much discrepancy in the statements of different observers, which has brought the Microscope into undeserved disrepute. Too often the descriptions given have not been of what has been actually seen, but of what has been imagined; and thus, without any intention of falsifying them, they have been shaped according to the preconceived notions of the inquirer. Hence, an examination of the characters of the primary tissues, whether of Plants or Animals, requires not only considerable manual skill and dexterity in the use of the microscope, but an acquaintance with all the fallacies arising from the difference between the image presented to

the eye, and the object as it exists in its natural situation; besides—what is even more important—a perfect readiness to give up preconceived notions, when they are inconsistent with observation, and a determination to consider nothing as proved until every mode of investigation has been employed with the same result. The improvements which have been recently effected in the optical powers and in the mechanical construction of the Microscope, together with a more general diffusion of the knowledge 'how to observe,' have gone far to remove these discrepancies; and there is now so general an agreement among those who have studied the structure of the 'primary' or 'elementary' Tissues of Plants and Animals, in regard to their most important attributes, that no reasonable doubt can be any longer entertained as to the accuracy of their descriptions. There are, of course, many phenomena in the interpretation of which there is room for a difference of opinion; but this is the case, more or less, in all Sciences; and it is the almost invariable tendency of the progressive extension of knowledge, to clear up such obscurities.

135. There is no department of Anatomical or Physiological inquiry, in which more elucidation has been obtained from an extension of the survey to *all* forms of Organised structure, than that which concerns the Primary Tissues. Botanists had long been aware that the Cell is the type of Vegetable structure;—that the simplest Plants are composed of individual cells;—that those a little higher in the scale are made up of aggregations of Cells not very different from each other in form or endowments;—that where other structures are present, cells still constitute the essential part of the fabric, being the special instruments of its *vital* actions;—and that these superadded structures are still closely related to the cellular, and are formed in the first instance by metamorphoses of the cells, of which the original embryonic structure, and the tissue of all young parts, are exclusively composed. It was known also that certain tissues of Animals, especially the adipose or fatty, resemble those of Plants in their simple cellular character; and as it had been established by observation of *their* embryonic structure, that even the most complex fabric takes its origin in a mass of cells, it was anticipated that a more complete analysis of the Animal organism would show that many other parts of it are formed upon the same type. It is to the industrious and skilful researches of Schwann, however, that we owe the first establishment of the general fact, which has been since more fully developed by the researches of other Microscopists,*—that the Animal like the Vegetable tissues originate, directly or indirectly, in cells; many of them retaining their original cellular character through life; whilst in others a transformation takes place, by which this is more or less completely obscured. And it seems to have been further established, as the aggregate result of the labours of many observers, that, in Animals as in Plants, all the parts in which active vital changes are taking place essentially consist of cells, which may be regarded as the real instruments of these operations, the tissues with which they are blended having no other purpose than to supply the physical conditions requisite for them.† These two general

* Mikroscopische Untersuchungen über die Uebereinstimmung in der Structur und dem Wachsthum der Thiere und Phlanzen. Berlin, 1839.

† The Author believes that he may himself claim the credit of this second generalization; which he first put forth in regard to the *organic* functions, in his Report on the Origin and

facts lie at the foundation of the whole fabric of modern Physiological science; and it will therefore be requisite to go at some length into the exposition of them. As we proceed, it will become apparent how valuable is the light which the study of 'cell-formation' in one class of organised beings can throw upon its phenomena in others; and how comprehensive is the expression of these phenomena, which the Physiologist can now give; since, basing his account of the process upon that which he sees in the simplest form of Cryptogamic Vegetation, he finds it to hold good, as to all essential particulars, in regard alike to the original germ-cell of the most complex Plant and of the highest Animal, and to each of those individual cells which enter in such incalculable numbers into the composition of the fully-developed organisms of both.

2. *Of the Primary Tissues of Plants.*

136. *Formation of Cells.*—The general idea of the structure and history of a Cell, which has already been given (§ 47), has now to be more specially applied and more fully developed; and we have first to enquire, therefore, in what essential particulars the Vegetable cell differs from the Animal, and what is the mode in which the former comes into existence. Although we have hitherto spoken of the *cell-wall* as a simple membrane, yet it is now well known to be made up, in most if not in all instances, of two layers of very different composition and properties. The inner of these layers, which has received the name of 'primordial utricle,' appears to be the one first formed, and most essential to the existence of the cell; it is extremely thin and delicate, so that it escapes attention so long as it remains in contact with the external layer, and is only brought into view when circumstances occasion its separation from this (Fig. 2); it seems to consist of an azotised compound, probably of an albuminous nature; and it appears to participate actively in the vital operations of the cell. The external layer, on the other hand, though commonly regarded as the proper 'cell-wall,' seems to be generated on the external surface of the primordial utricle, after the latter has completely enclosed the cavity and its contents, so that it cannot be regarded as essential to the cell; it is usually thick and strong in comparison with the other, but it may possess various degrees of consolidation, from mere mucus to a firm and tenacious substance; it is composed of cellulose, a

FIG. 2.

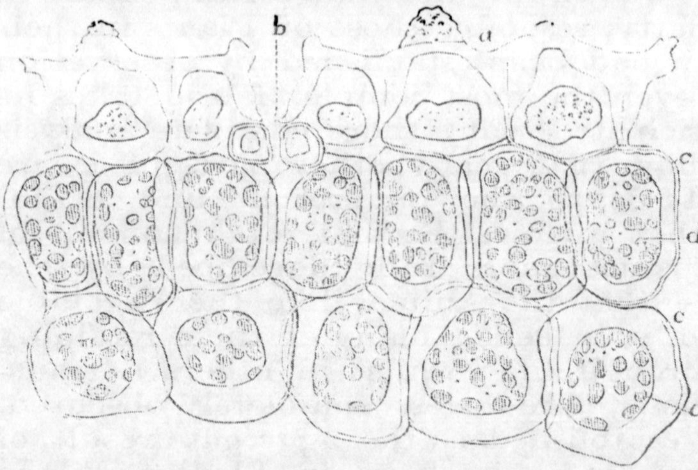

Section of leaf of *Agave*, treated with dilute nitric acid, showing the primordial utricle contracted in the interior of the cells: *a*, epidermic cells; *b*, boundary cells of the stoma; *c*, cells of parenchyma; *d*, their primordial utricles.

Functions of Cells in No. XXIX. of the British and Foreign Medical Review, Jan. 1843; and which, by the discovery of the cellular structure of the muscular fibrilla in 1846, he was enabled to extend to the *animal* functions.

substance nearly identical with starch ; and it does not appear to take any active share in the vital operations of the cell, its principal office being to protect, locate, and isolate the matter it contains.* This external layer may consist of many laminæ, the results of successive deposits from the surface of the primordial utricle (Fig. 9) ; but it still usually remains readily permeable to fluid, although no pores can be distinguished in it, even under the highest magnifying power.

137. The granular matter contained in the interior of the cell, being usually coloured, has received the name of *endochrome ;* it is this, with its containing sac or 'primordial utricle,' which essentially constitutes the cell ; and all the vital phenomena which the cell exhibits have their seat in it. Among these phenomena, there is one which, not being as yet certainly known to be common to all cells, has not been previously adverted to. This is the existence of a continuous motion in a certain part of the fluid contents of the cells, as marked by the revolution of the floating granules which are carried along in the stream. The motion appears to be confined to the somewhat viscid layer (probably of albuminous matter) which immediately lines the primordial utricle. In most plants of the aquatic families of *Characeæ, Nayadaceæ,* and *Hydrocharidaceæ,* and especially in the long tubular cells of the *Characeæ* (§ 269) we may observe a single broad current, sweeping up one side of the cell and down the other, and returning continuously into itself ; and this current is strong enough to carry along with it granular masses of starch, chlorophyll, and albuminous matter, of considerable size. The ascending and descending currents do not come into absolute contact, being separated by a clear space in which no motion can be detected. If the cell be carefully tied across, the current is in short time re-established in each segment, as if the cell had naturally subdivided itself. The vigour of the movement increases and diminishes in accordance with the general vital activity of the cell ; as appears, not merely from observation of its ordinary growth, but also from the exciting or depressing effects of physical agents. Although this

FIG. 3.

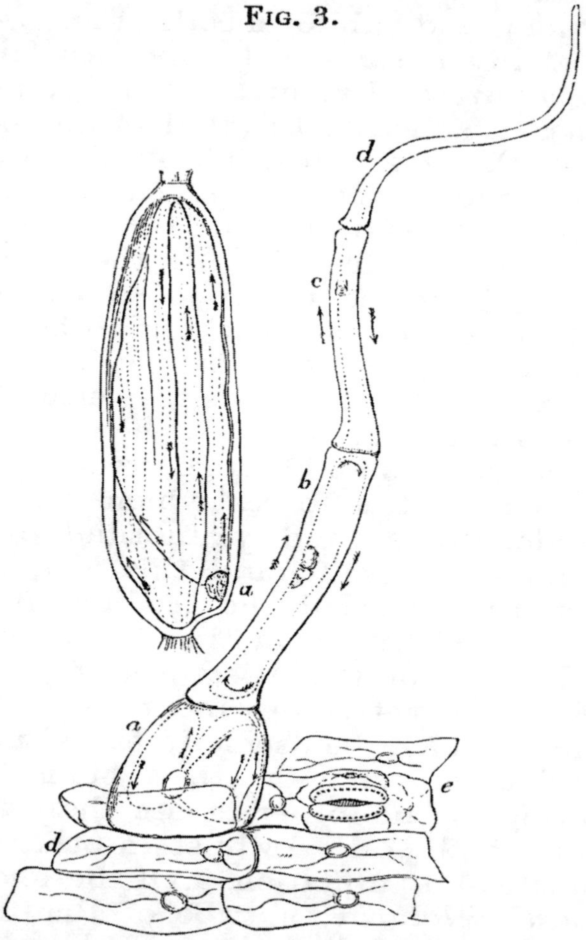

Circulation of fluid in hairs of *Tradescantia Virginica*:—A, portion of cuticle with hair attached; *a, b, c,* Successive cells of the hair; *d.* cells of the cuticle; *e.* stoma:—B, joint of a beaded hair, showing several currents; *a,* nucleus.

* See the admirable Memoir by Mr. Thwaites on the Cell-Membrane of Plants, in the Annals of Natural History, Vol. XVIII., to which the reader is referred for facts and arguments in support of the above position.

movement may be advantageously studied in the plants of the families just named, it is by no means confined to them; for it may be readily detected in the hairs of Flowering-plants, many of which afford very beautiful examples of it, and it has been observed also in the young cells of leaves, fruits, and other parts; so that it may with probability be regarded as existing in every vegetable cell at a certain stage of its development. The current is never so broad, however, as in the instances previously cited: and usually presents itself as a thread-like stream passing through a stratum of motionless matter. This stream may be single, passing up one side of the cell, and down the other, as is usually the case in long and tubiform cells (Fig. 3, *b*, *c*): but several distinct currents may exist in the same cell, and these are observed to have a common point of departure and return,—namely, a collection of granular matter in one mass attached to the wall of the cell, which is termed the *nucleus* (Fig. 3, A, B, *a*). Here, too, it would appear, that the currents are connected with the general activity of the life of the cell, and that their cessation indicates the cessation of its formative powers; and from the manner in which they are connected with the 'nucleus,' it would seem that this, where it exists, is to be regarded as the centre of the vital activity of the cell. Although such a nucleus is to be found, however, in a very large proportion of Vegetable cells, yet its absence in some instances is a matter of equal certainty with its presence in others; and as these do not manifest any inferiority of vital power, the nucleus is obviously not essential. Perhaps we should be correct in regarding it as concentrating in itself a portion of the forces which are elsewhere diffused through the entire 'endochrome' and 'primordial utricle,' rather than as having any powers peculiar to itself; and this idea is confirmed by the facts presently to be stated in regard to the production and multiplication of cells.

138. There is another phenomenon of occasional occurrence, which is indicative of the existence of properties in certain Vegetable cells, that have not yet come under our notice. This is the rapid change of form which may be sometimes witnessed in them, and which produces obvious movements; these being in some cases *spontaneous*, that is, originating in no other causes than those which are concerned in the formation of the cells; whilst in other cases they are *excited* by mechanical or chemical stimulation. Of the former, we have a very curious example in the *Oscillatoriæ*, a tribe of Plants of the simplest possible construction, consisting of filaments or elongated cells, which continue in a state of rhythmical vibration throughout the greater part if not the whole of their lives; many of these are remarkable as possessing no proper cell-wall, so that the cause of the movement must reside in the primordial utricle and endochrome. Of the latter we have examples in the Sensitive Plant, Dionæa, Berberry, &c., certain parts of which are put in motion by the contraction of tissues that are designated as 'irritable;' these contractions being due to a change of form in the component cells of those tissues, which takes place when certain stimuli are applied to them (CHAP. XIX.).

139. We have now to enquire into the origin and mode of multiplication of the Vegetable cell.—It may be stated as a general fact, that every cell owes its origin in some way or other to a pre-existing cell; but the mode in which new individuals are generated, is by no means constant; and there are some forms of the process which are not yet clearly understood.

That which is already known, however, leaves no room for doubt, that in every instance a portion of the endochrome of the parent cell is the starting point of its successor : and that, although the primordial utricle may be also concerned, the outer cell-wall is entirely passive in the matter. The most common method of multiplication in Plants is the subdivision of the original cell into two halves, such as is seen very characteristically in the *Hæmatococcus binalis* and others of the humblest forms of vegetation, in which every cell, being capable of existing by itself, may be regarded as a distinct individual. The cells of this little plant are originally of a globular form (Fig. 4, *a*) ; and the first step in the process of subdivision is their elongation into an oval shape, and the appearance of a slight constriction round them, as seen at *b*. This constriction indicates the tendency of the endochrome to separate into two halves, each included, it would appear, in an envelope of its own, so that two young cells are now included within the external wall of the parent ; and after this has been effected, the contiguous portions of the two 'primordial utricles' appear to develop or secrete a thick partition between them, so that the two young cells are now completely divided, as seen at *c, c*. An increase of a somewhat similar but less condensed secretion on their exterior forms a mass of mucus, in which the cells are imbedded ; and these are frequently carried, by the interposition of this new substance, to a considerable distance from each other, as is shown at *d*, where each of the first pair of 'twin' cells has itself undergone a similar subdivision.—This process may also be well studied in the *Confervæ*, which are filamentous aquatic plants, each filament composed of a single file of cells, adherent to each other, end to end ; and we shall derive from the examination of one of these a further insight into some stages of the process. The first step is here seen to be the subdivision of the endochrome, and the inflexion of the primordial utricle around it (Fig. 5, A, *a*) ; and thus there is gradually formed a sort of hour-glass contraction across the cavity of the parent-cell, by which it is divided into two equal halves (B). The two surfaces of the infolded utricle produce a double layer of permanent cell-membrane between them ; and the formation of this may be seen to commence even before the complete separation of the cavities of the twin-cells (D). This deposition is not confined, however, to the contiguous surfaces of the young cells, but takes place over the whole exterior of the primordial utricle ; so that the new septum is continuous with new layers that are formed throughout the interior of the original parent-cell (C).

FIG. 4.

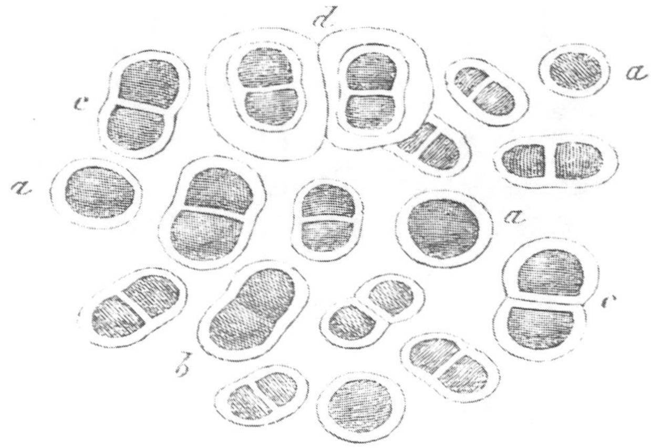

Various stages of development of *Hæmatococcus binalis* :— *a, a*, simple rounded cells ; *b*, elongated cells, the endochrome preparing to divide ; *c, c*, cells in which the division has taken place ; *d*, cluster of four cells formed by a repetition of the same process.

140. The foregoing is the method according to which the extension of Vegetable structures from cells already in existence most commonly takes place. Among the lower Algæ, for example, we find the single cell giving

rise to an amorphous cluster, to a prolonged filament, or to a flattened leaf-like expansion (Fig. 67), according to the mode in which the subdivision takes place, and to the degree in which the new cells remain attached to

FIG. 5.

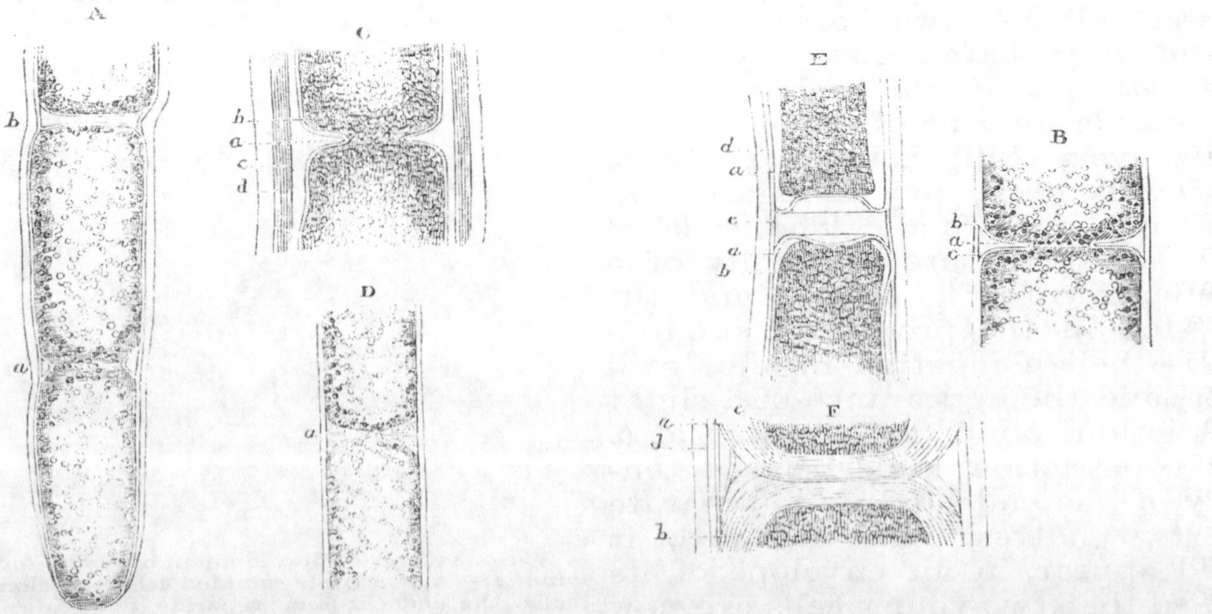

Process of cell-multiplication in *Conferva glomerata*. A, portion of filament within complete partition at *a* and complete partition at *b*. B, commencement of the formation of a partition; C, more advanced stage; D, the separation nearly complete; E, the two primordial utricles completely detached, and cell-membrane deposited between them; F, successive layers of cell-membrane, making up the thickness of the partition. In the last five figures, *a* indicates the primordial utricle; *b*, the endochrome; *c*, the cell-membrane; and *d*, the mucous investment.

each other. It is in this manner, again, that the cell-multiplication first takes place in that germinal mass, in which not merely the higher Plants, but the highest Animals also, have their origin (CHAP. XVIII.); and the process continues to take place on the same plan, until certain parts of that mass begin to undergo transformation into heterogeneous structures, in which case we commonly find it superseded by some other method. Of its persistence in certain individual parts, even of the highest Animal, we shall presently meet with a very characteristic example, in the multiplication of Cartilage-cells by division (§ 157).—It does not appear that in this process the 'nucleus' performs any essential part; for we find it taking place where no nucleus can be distinguished, in virtue, it may be surmised, of a sort of mutual repulsion between the two halves of the endochrome, which leads to their spontaneous separation. Where a nucleus is present, this also undergoes subdivision at the same time with the endochrome, so that half of it is appropriated by each of the 'twin-cells.'

141. But we sometimes observe that new cells originate in little bud-like prominences on the surface of the parent-cell, which are developed in continuity with that from which they arise. This is well seen in the *Conferva glomerata*, a common species, which increases not merely (like other Confervæ) by the repeated subdivision of the cells at the extremity of its filaments, but by the origination of new cells from every part of their surface. A certain portion of the primordial utricle seems to undergo

increased nutrition ; for it is seen to project, carrying the outer cell-wall before it, so as to form a protuberance, which sometimes attains considerable length before any separation of its cavity from that of the parent-cell begins to take place. This separation is gradually effected, however, by the infolding of the primordial utricle, just as in the preceding case ; and thus the endochrome of the young cell is completely severed from that of its parent, and its separate individuality may be said to begin from that time. We may consider this process of 'budding' as differing from that of 'subdivision' only in this,—that whilst in the latter case the individuality of the parent-cell is lost by the *equal* subdivision of its cavity into two similar parts, it is retained in the former through the *unequal* division of the cell, of which only a small portion is pinched off (so to speak) to form the new cell, whilst the greater part remains unaltered.* Of the extent to which the multiplication of cells by this budding process takes place among plants, we have as yet no certain knowledge. It is obviously the regular method of growth among the *Characeæ* (§ 269), in which the long tubiform cells that form the axis give off, at their points of junction with each other, circular rows of buds, from each of which is developed a whorl of lateral branches. And the same mode of increase is observable among the 'ferment-cells' (§ 271), which, whilst rapidly multiplying under favourable circumstances, shoot forth little buds from one or even both extremities, from each of which a new cell is developed. There is no reason to believe that in this process the 'nucleus' takes any direct share ; since the evolution of buds may take place from cells destitute of nuclei ; and even where nuclei exist in budding cells, they do not seem to be specially connected with the process, since the buds are not observed to originate in or near them.

142. In cases where a very rapid production of new cells is requisite, it would seem to be effected by the primary separation of the endochrome into numerous parts, each of which acquires for itself a covering of cell-membrane ; so that a whole brood of young cells may thus be at once generated in the cavity of the parent-cell, which subsequently bursts and sets them free. Of this plan we have the most characteristic examples in the formation of the 'zoospores' of the inferior Algæ (§ 267). Thus in *Achlya prolifera,* a plant composed of tubiform cells, which grows parasitically upon fish, the end of the filament dilates into a large cell, the cavity of which is cut off from the rest by the formation of a partition ; and within this dilated cell, an irregular circulation of granular particles can for a time be distinguished. (Fig. 6, C). Very speedily, however, it appears that the endochrome is being broken up into a large number of distinct masses, which are at first in close contact with each other and with the walls of the cell (A), but which gradually become more isolated, each seeming to acquire a proper cell-wall ; they then begin to move

* We have here, in fact, in their very simplest forms, the two methods of *fissiparous* and *gemmiparous* reproduction, which we shall hereafter find carried out not merely as regards the individual components, but in the entire fabric of many Plants and Animals. Thus the 'fissiparous' method of reproduction, in which the body is cleft (as it were) into two equal halves, each of which may be regarded as a new individual, is exhibited in the lower tribes of Infusorial Animalcules (§ 284) in a form almost as simple at that under which we have traced it in the humblest Algæ ; and the 'gemmiparous' plan is manifested in Zoophytes (§ 288) and in the lower Articulata and Mollusca, just as in almost the entire Vegetable Kingdom.

about within the parent-cell; and, when quite mature, they are set free by the rupture of its wall (B), to go forth and form new attachments, and to acquire for themselves the materials of development into tubiform cells resembling those from which they sprang. A similar process may be observed in the production of the 'zoospores' of Confervæ and Algæ in general; usually taking place, however, upon a smaller scale, and the number of new cells being generally less, than in the instance just quoted. And there appear to be some cases among the higher plants, in which parent-cells give origin to a new brood in their interior, by a process somewhat similar, without the successive duplication which is certainly the more usual method of the production of 'cells within cells,' or 'endogenous multiplication.' This seems the case, for example, in the 'embryo-sac' of Flowering-Plants (CHAP. XVIII.); which at one time contains only a mixture of albuminous and starchy matter, but which is afterwards filled up by a mass of cells, that have incorporated these materials into their own substance, forming the 'endosperm.' According to Nägeli, who has recently investigated this process with much care, the following are the essential points in its history. Minute globular particles of perfectly homogeneous matter, varying in diameter from 1 to 4-1000ths of a line, seem to be first formed in the midst of the mucilaginous contents of the embryo-sac; larger globular bodies are apparently produced by the aggregation of other (nitrogenous?) particles around these, forming the nuclei of the future cells. Each of these nuclei seems to attract around it a greater or smaller quantity of the contents of the parent-cell; and over this a membrane is subsequently generated.* Thus the history of such a formation is very nearly the same with that which we have traced in the inferior Cryptogamia; and it may be that even in the latter, the formation of nuclei, round which the contents of the parent-cell group themselves, may be really the first stage in the production of the mass of secondary cells.

FIG. 6.

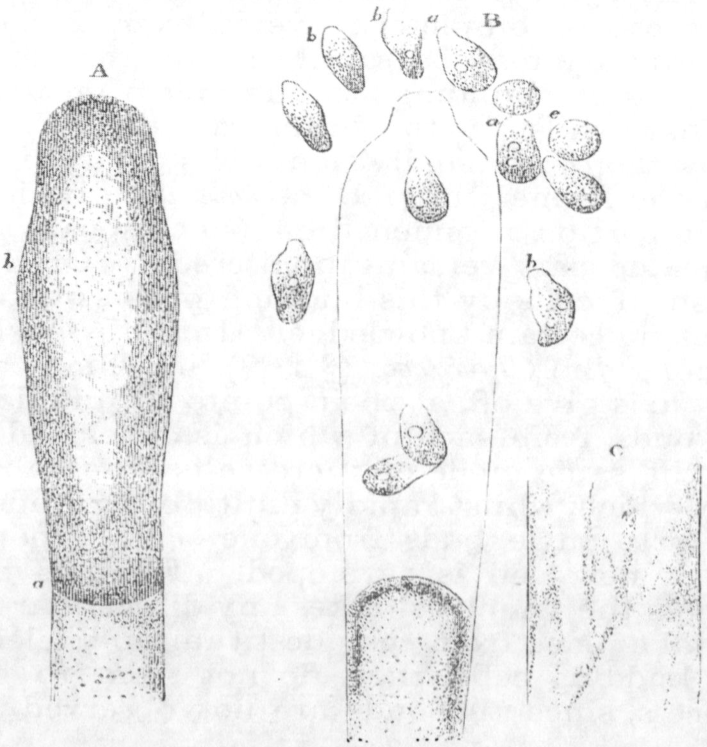

Development of *Achlya prolifera*:—A, dilated extremity of a filament *b*, separated from the rest by a partition *a*, and containing young cells in progress of formation;—B, conceptacle discharging itself, and setting free young cells, *a*, *b*, *c*;—C, portion of filament showing the course of the circulation of granular contents.

143. The first *visible* stages of the development of new cells, however,

* See Nägeli's Reports on the Formation of Vegetable Cells, in the Reports and Papers on Botany published by the Ray Society, 1845 and 1849.

do not always take place in the interior of a pre-existing generation ; for cells sometimes appear to originate *de novo*, in that mixture of starchy and albuminous fluids, which, being the appropriate pabulum for Vegetable cells, has been denominated *protoplasma*. This protoplasma, however, must have always been elaborated by cell-agency ; so that, even if the young cells appear to be developed quite independently in its substance, they must really be regarded as the offspring of the cells which formed it. In some instances which have been regarded in this light, it is probable that the protoplasma was contained in a cellular parenchyma which escaped observation through its extreme delicacy: whilst in other cases, there can be no doubt that definite cell-germs had been prepared and set free with the protoplasma, escaping observation on account of their minuteness.

144. There is evidence that the process of cell-production may take place with a rapidity almost inconceivable. Extensive tracts of snow, in alpine and arctic regions, have been seen to be suddenly reddened by the cells of the little *Protococcus nivalis;* and there are some minute blood-red Fungi, which occasionally make their appearance in almost equal multitudes upon the surface of every organic substance. Almost every one is familiar with the appearance of certain more elevated forms of Fungous vegetation, which shoot up in the course of a single night, and seem to melt away before the morning sun. A specimen of *Bovista giganteum*, a large Fungus of the *puff-ball* tribe, has been known to grow in a single night from the size of a mere point to that of a huge gourd ; and from a calculation of the average size of the cells, and of the probable number contained in the full-grown plant, it has been estimated that they must have been generated at the rate of *four thousand millions* per hour, or more than *sixty-six millions* per minute. In all such cases, the amount of solid matter present in the tissues bears a very small proportion to the fluids.—A very rapid growth of leaves may be occasionally noticed ; thus the leaf of *Urania speciosa* has been seen to lengthen at the rate of from $1\frac{1}{2}$ to $3\frac{1}{2}$ lines per hour, and even as much as from four to five inches per day. It is doubtful, however, how much of this result may be attributed to the formation of new cells, and how much is due to the enlargement of those of which the leaf was previously composed.

145. *Cellular Tissue.* — Such cells as those which we have been describing, present themselves not only as the constituents of those simplest Vegetable organisms, in which every cell may be accounted a distinct individual ; but also as the chief and frequently the sole components of fabrics of much higher rank, and of much greater complexity of structure. Thus we do not meet with any other forms of tissue than those referable to the simple Cellular type, in any of the Algæ, Lichens, Fungi, or Mosses : whilst in Ferns and Flowering Plants we find this tissue not merely in the soft substance of the leaves, flowers, and fruits, but also in the stem, branches, and roots ; uniformly making up the chief part of the organs most actively concerned in the vital operations, and being almost the sole component of young and growing structures. The simplest condition of this tissue is that in which the cells retain their primitive spheroidal form, and have simply membranous walls (Fig. 7, A). The rounded form is only exhibited when the cells are but loosely aggregated together, and it is then that the distinctness of their sides is most

evident. When the tissue is more solid, the sides of the vesicles are pressed against each other, so as to become flattened, and to be in close apposition (B); and sometimes they adhere in such a manner, that the

FIG. 7.

Various examples of Vegetable Cells:—A, spheroidal membranous cells; B, membranous cells with flattened walls; C, cells with internal spiral fibre; D, cells in which the fibre is obscured by its coalescence with the cell-wall; E, dotted cells exhibiting traces of spiral fibre; F, dotted cells without vestiges of spiral.

partition between two adjacent cells seems to be but a single instead of a double membrane. If the pressure to which the vesicle is subject be equal in all directions, the form it will assume is that which is mathematically termed a rhomboidal dodecahedron; that is to say, a twelve-sided solid with all its faces equal, and showing an hexagonal section when cut across. Each cell will thus be in contact with twelve others, which completely surround it without leaving interstices.

146. It is not very often, however, that this form is displayed with such extreme regularity; since there is usually, in the growing plant, a disposition to elongation in the direction of increase, and to compression in the transverse one, so that the cells are found to have rather a prolonged form; such cells are especially found in the lower tribes of plants, which have no other kind of tissue, and are destitute of vessels, the function of which is partly performed by them; and we also meet with them in the 'medullary rays' of Exogenous stems (§ 280), in which this elongation takes place horizontally, so as to favour the transmission of fluid, through their means, from the exterior to the interior of the trunk. Not unfrequently the cells are found to possess a cubical or prismatic shape; and their arrangement is occasionally such as to give them the appearance of bricks in a wall, in which case the tissue is said to be 'muriform.' One of the most curious varieties of form which Vegetable cells present, is that which is delineated in Fig. 8, and which constitutes the *stellate* cell. This kind of metamorphosis is of great importance, as illustrating the formation of some of the Animal tissues. It is obvious that, if the radiating prolongations of these cells were to coalesce at the points where they come into mutual contact, so as to throw together the cavities of the entire series, a network of vessels would be produced; this, however, does not seem to occur in Plants; in

FIG. 8.

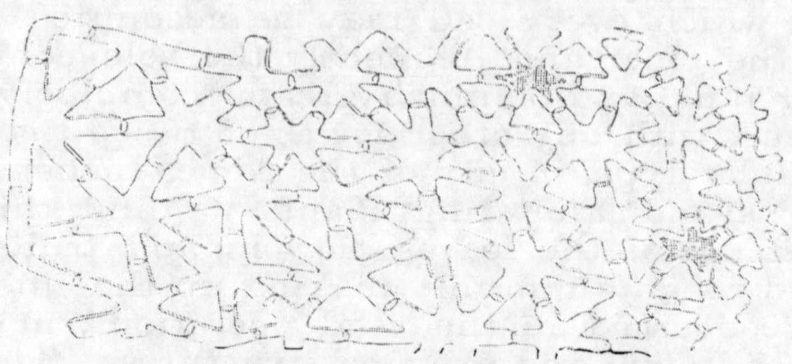

Spongiform Parenchyma composed of stellate cells.

whose fabric this stellate tissue appears to be generated for the purpose of forming a loose spongy parenchyma.—The dimensions of the component vesicles of Cellular Tissue are extremely variable ; they are usually from 1-300th to 1-500th of an inch in diameter, but they are occasionally as much as 1-30th of an inch across, whilst in other instances they do not measure the hundredth part of that amount. They seem to be held together by an *intercellular substance* which intervenes between them, and which is analogous to the mucus-layer that surrounds the isolated cells of the simple Algæ. This analogy becomes obvious, when we trace the gradual aggregation of isolated cells into such fabrics as we meet with in the higher Algæ ; between whose component cells a considerable thickness of this mucus is still to be found (§ 267, 268).

147. The walls of the vesicles of Cellular Tissue, however, do not always consist of simple membrane alone ; for these are occasionally strengthened by fibre that lies in contact with their internal surface ; or, as more frequently happens, are thickened by a deposit of hard matter secreted from the juices of the plant some time subsequently to the original formation of the cell. Of the first of these methods we have an example in Fig. 7, C, which shows cells with a fibre spirally coiled in their interior. This kind of Fibre, which may be termed *elementary,* is of extreme minuteness, its diameter frequently not exceeding 1-12,000th of an inch ; it is solid, and, when first formed, is usually very elastic. Of this elasticity we have a very interesting example in the fibre-cells which form the external coating of the seeds of *Collomia,* of *Salvia verbenaca* (Wild Clary), and of many other plants ; for the walls of these cells are so imperfectly consolidated, that they are readily softened by water ; and the spiral fibres suddenly elongate themselves, like springs which have been compressed, so soon as they are set at liberty by the rupture of the cell-wall which held them together. The *elaters* of the *Jungermanniaceæ* (§ 275) are constructed upon a like plan ; being fibrous cells, whose membrane gives way when mature and dry, so that they rupture with a jerk that disperses the spores among which they lie. There are a few plants which are almost entirely made up of fibrous cells ; this is the case, for example, with the *Sphagnum* (Bog-moss), in which we meet with this further peculiarity, that the cells occasionally communicate with one another by apertures large enough to allow of the passage of Wheel-Animalcules, which find a congenial abode in their interior.—The spire may wind in either direction, but towards the right is the most common ; sometimes we find two spires winding in the same direction in the same cell ; but there is no satisfactory example of two spires crossing each other in contrary directions.

148. Of the origin of this Spiral Fibre, we have at present no definite knowledge ; it would seem to result, however, from a general tendency to the arrangement of the components of the cell in a spiral direction ; for we see the endochrome presenting more or less of this disposition in many of the simpler plants, but more especially in the *Zygnema ;* and the cell-wall itself is sometimes found to tear most readily in a spiral direction.—Occasionally the fibre becomes so adherent to the cell-wall that it cannot be separated ; and the distinctness of its edges is lost, so that it almost forms an inner lining to the cell (Fig. 7, D). There would seem, in fact, to be a great tendency in any additional deposits which are

applied to the lining of the cell, to arrange themselves upon the spiral fibre; and thus it is increased in breadth and thickness, and the regularity of its spires disappears, in consequence of the coalescence of its several turns at various points. There are usually some parts of the cell-membrane, however, which are left uncovered by the secondary deposit; these can sometimes be clearly distinguished to be spaces intervening between the spiral fibres, as seen at E, Fig. 7; but more frequently all trace of the spire is obscured in these *porous cells* (Fig. 7, F), by the irregular mode in which the new matter is deposited. Such transition-forms as the preceding, however, serve to indicate their origin, which we shall presently see to be a point of some interest (§ 152). The secondary deposit often forms numerous layers, which present themselves as concentric rings when the cells containing them are cut through (Fig. 9); and these layers are sometimes so thick, as almost to obliterate the original cavity of the cell. In the long cells of Ligneous tissue, we usually find each layer continued over the whole interior of the cell; but in other cases we frequently meet with a continuation of the same arrangement as that which shows itself in the first layer of the dotted cell; each deposit being deficient at certain spots, which correspond in the successive layers, so that a series of passages is left, by which the cavity of the cell is extended at some points to its membranous wall. Where several cells of this kind are aggregated together, as happens in the 'stones' of fruit, the gritty tissue of the pear, &c., the points at which the deposit is wanting on the walls of two contiguous cells are coincident, so that the membranous partition is the only obstacle to the communication between their cavities.—Such deposits are not found in cells which are actively contributing to the vital operations of the plant; but are rather intended to give strength and durability to the tissues in which they occur; the radiation of the cell-cavity being chiefly found in those cases, in which it is necessary that the tissue thus consolidated should still have the power of conveying fluid. It would seem probable, although it cannot be stated with certainty, that the deposit is formed on the exterior of the 'primordial utricle,' which progressively contracts as it secretes new matter from its surface; so that at last, in such cells as those represented in Fig. 9, this utricle has a stellate form, resembling that depicted in Fig. 8. The secondary deposit, which thus gives firmness to the tissue which was originally tender and succulent, is sometimes termed *sclerogen*; it does not appear, however, that in its purest condition it is essentially different from Cellulose; but it is frequently intermingled with resinous and other matters.

FIG 9.

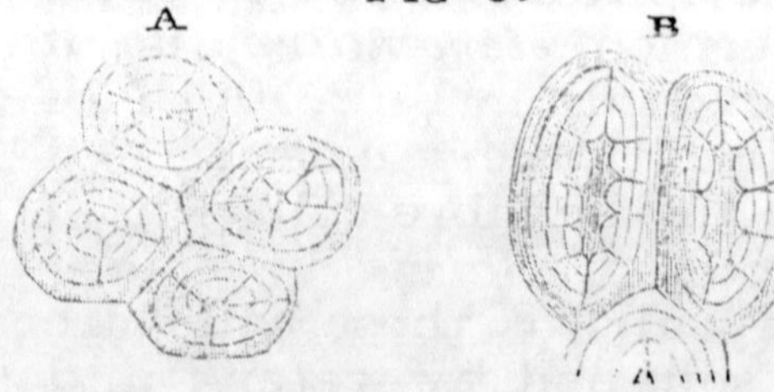

Cells from the gritty tissue of the Pear, divided transversely at A, longitudinally at B, showing concentric layers of internal deposit, with passages radiating from the central cavity.

149. *Woody Fibre.*—A large proportion of the denser fabric of the higher plants is made up of what is usually termed *Ligneous Tissue* or *Woody Fibre.* This, however, can only be regarded as a very simple variety of the Cellular; for it is composed of cells which have undergone elongation at one or both extremities, so as to become fusiform or spindle-shaped (Fig. 10, A, B), and which have a peculiar tendency to the production of an internal consolidating deposit (D). The elongation does not seem to

be the result of pressure, but of the unequal nutrition of different parts of the cell-wall. It is obvious that a tissue consisting of elongated cells, adherent together by their entire length, and strengthened by internal deposit, must possess much greater tenacity than any tissue in which the cells depart but little from the primitive spherical form; and we accordingly find this tissue introduced, wherever it is requisite that the fabric should possess not merely density, but the power of resistance to tension. In the higher classes of the Vegetable kingdom, it constitutes the chief part of the stem and branches, where these have a firm and durable character; and even in more temporary structures, such as the herbaceous stems of annual plants, and leaves and flowers of almost every tribe, this tissue forms a more or less important constituent, being especially found in the neighbourhood of the spiral vessels and ducts, to which it affords protection and support. Hence the bundles or fasciculi composed of these elements, which form the skeletons of leaves, and which give 'stringiness' to the tissues of various esculent Plants, are commonly known under the name of *fibro-vascular* tissue. In their young and unconsolidated state, the ligneous cells seem to conduct fluids with great facility in the direction of their length; and in the *Coniferous* tribe they afford the sole channel for the ascent of the sap, their stems and branches being destitute of vessels. But after their walls have become thickened by internal deposit, they are no longer subservient to this function; nor, indeed, do they then appear to fulfil any other purpose in the Vegetable economy, than that of affording mechanical support. It is this which constitutes the difference between the *alburnum* or 'sap-wood,' and the *duramen* or 'heart-wood,' of Exogenous stems (§ 280).

FIG. 10.

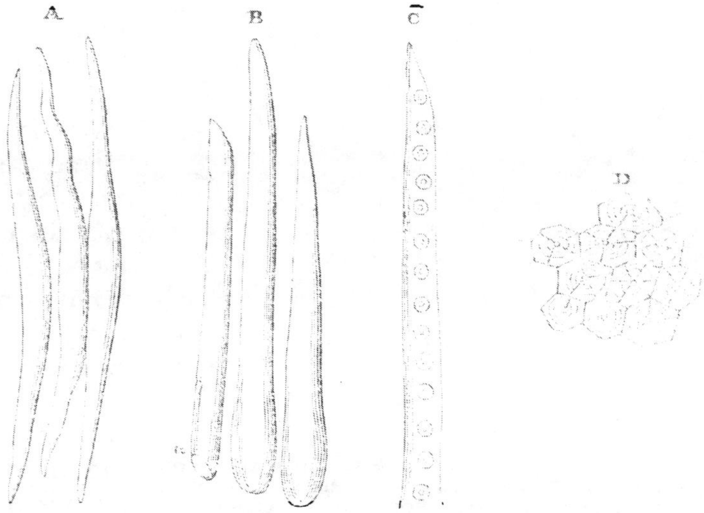

Elongated cells constituting Woody Fibre:—A, fusiform cells elongated at both ends; B, cells elongated in one direction only; C, dotted fibre from Pine; D, transverse section of a fasciculus of fibres, showing concentric internal deposits.

150. A peculiar set of markings, seen on the woody fibres of the *Coniferæ*, and of some other tribes, is represented in Fig. 10, C; in each of these spots, the inner circle appears to mark a deficiency of the lining deposit, as in the porous cells of other plants; whilst the *outer* circle indicates the boundary of a lenticular cavity, which intervenes between the adjacent cells at this point, and which contains a small globular body that may be sometimes detached. Of the purpose of these minute bodies interposed between the wood-cells, nothing is known; there can be no doubt, however, from the definiteness and constancy of their arrangement, that they fulfil some important object in the economy of the plants in which they occur; and there are varieties in this arrangement so characteristic of different tribes, that it is sometimes possible to determine, by the microscopic inspection of a minute fragment, even of a fossil wood, the tribe to which it belonged.

H 2

151. *Spiral Vessels.*—The spiral cells are sometimes elongated in a similar manner, and then become what are commonly known as *Spiral Vessels* * (Fig. 11, C, D). These, like the cells out of which they are transformed, are characterized by the presence of a spiral fibre winding from end to end, remaining distinct from the cell-wall, and retaining its elasticity; this fibre may be single, double, or even quadruple,—this last character presenting itself in the very large elongated fibre-cells of the *Nepenthes* (Chinese pitcher-plant). These cells in their perfect state contain air only; they are especially found in the delicate membrane surrounding the pith of Exogens, and in the midst of the woody bundles occurring in the stem of Endogens; thence they proceed in each case to the leaf-stalks, through which they are distributed to the leaves. By careful dissection under the microscope, they may be separated entire; but their structure may be more easily displayed by cutting *round*, but not *through* the leaf-stalk of the strawberry, geranium, &c., and then drawing the parts asunder. The membrane composing the tubes of the vessels will thus be broken across; but the fibres within, being elastic, will be drawn out and unrolled, as seen in Fig. 10, D.

FIG. 11.

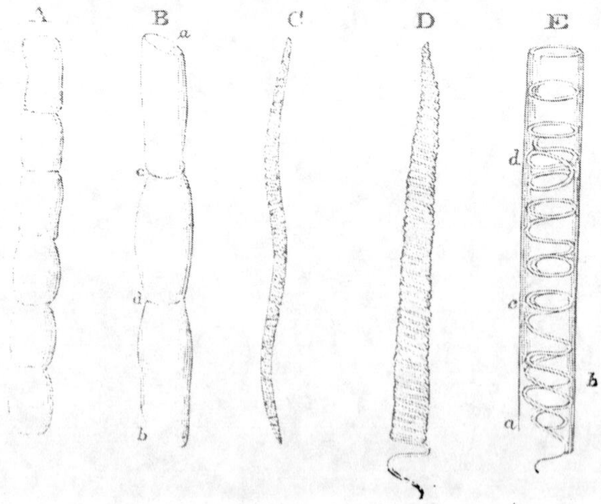

A and B, rows of cells coalescing into ducts, the partitions being still indicated at *a, b, c, d*; C, elongated spiral cell, or spiral vessel; D, portion of the same enlarged; E, fibrous duct, spiral at *a, b*, annular at *c*, spiral again at *d*.

152. *Tubular Tissues.*—Although fluid generally finds its way with tolerable facility through the various forms of Cellular tissue, especially in the direction of the greatest length of its cells, a more direct means of connection between distant parts is required for an active circulation (CHAP. XII.). This is afforded by what has been termed *Vasiform tissue*, which consists merely of cells laid end to end, the partitions between them being more or less obliterated, so that a continuous *Duct* is formed. The origin of these ducts in cells is occasionally very evident, both in the contraction of their calibre at regular intervals, and in the persistence of the remains of their partitions (Fig. 11, A, B); but in most cases it can only be ascertained by studying the history of their development, neither of these indications being traceable. The component cells appear to have been sometimes simply membranous, but more commonly to have possessed the fibrous type. Some of the ducts formed from the latter are so like continuous spiral vessels, as to be scarcely distinguishable from them, save in the want of elasticity in the spiral fibre, which causes it to break when the attempt is made to draw it out (Fig. 12, 2). This would seem to have taken place, in some instances, from the natural elongation of the cells by growth; the fibre being broken up into rings, which sometimes lie close together, but more commonly at considerable intervals; such a duct is said to be *annular* (Fig. 12, 1). Intermediate forms between the

* So long, however, as they retain their original cellular character, and do not coalesce with each other, these fusiform spiral cells cannot be regarded as having any more claim to the designation of *vessels*, than have the elongated cells of the ligneous tissue.

spiral and annular ducts, which show the derivation of the latter from the former, are very frequently to be met with (Fig. 11, E). The spires are sometimes broken up still more completely, and the fragments of the fibre extend in various directions, so as to meet and form an irregular network lining the duct, which is then said to be *reticulated.* The continuation of the deposit, however, gradually contracts the meshes, and leaves the walls of the duct in the condition of those of the porous cells ; and canals upon this plan, commonly designated as *Dotted* ducts, are among the most common forms of vasiform tissue, especially in parts of most solid structure and least rapid growth (Fig. 12, 3). The ducts of Ferns are for the most part of the spiral type ; but spiral ducts are frequently to be met with also in the rapidly-growing leaf-stalks of Flowering-Plants, such as the Rhubarb. Not unfrequently, however, we find all forms of ducts in the same bundle, as seen in Fig. 12. The size of these ducts is not unfrequently so great, as to enable their openings to be distinguished by the unaided eye. They are usually largest in stems whose size is small in proportion to the surface of leaves which they support, such as the common Cane, or the Vine ; and generally speaking they are larger in woods of dense texture, such as Oak or Mahogany, than in those of which the fibres, being softer, can themselves be subservient to the conveyance of fluid. They are entirely absent in the Coniferæ.

FIG. 12.

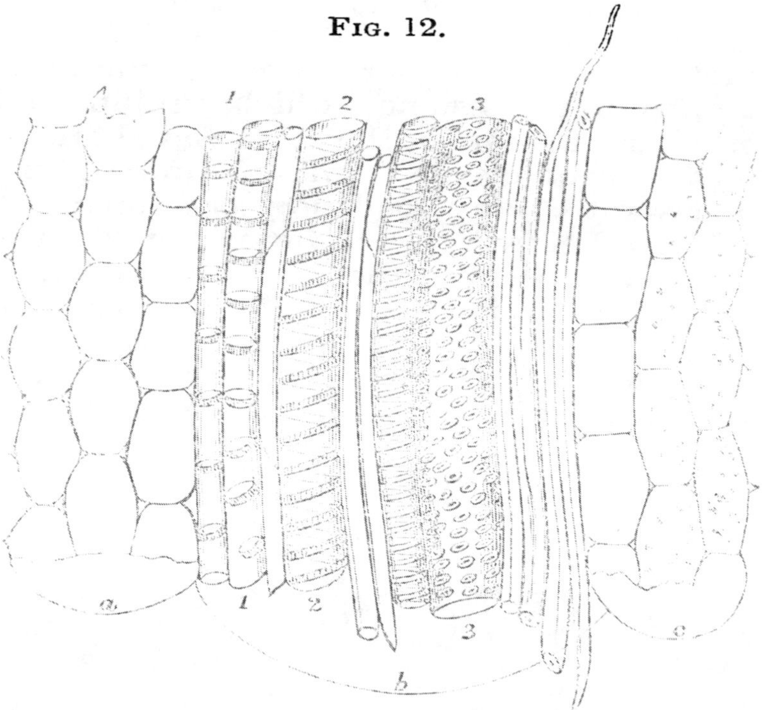

Longitudinal section of stem of Italian Reed :—*a*, cells of the pith ; *b*, fibro-vascular bundle, containing, 1. annular duct ; 2. spiral duct ; 3. dotted duct with woody fibre ; *c*, cells of the integument.

153. Some curious analogies to the several forms of vessels and ducts just described, are presented in the Animal fabric.—Thus, the *tracheæ*, or air-tubes, of Insects, which ramify by minute subdivisions through the whole of their bodies (CHAP. XIII.) are formed, like the Spiral vessels of plants, of an external membrane distended by spiral fibre, which is coiled with the most beautiful regularity ; the principal difference in these two structures being, that the air-tubes of Plants are closed vessels, and that their gaseous contents find their way by permeation through the delicate membrane which composes their walls ; while the tracheal system of Insects exhibits the most beautiful and minute ramifications, formed by the subdivision of its principal trunks, which communicate directly with the atmosphere.—Again, we are reminded by the Annular duct of the structure of the *trachea* or windpipe of air-breathing Vertebrata, which is composed of a membranous tube kept open by cartilaginous rings regularly arranged at near intervals. In some Birds, however, traces of

a spiral are to be met with in the cartilage; and this appears to be the regular structure of the trachea of the *Dugong* (an aquatic Pachyderm), in which the cartilage forms two or three turns of a spire ending in a ring, and then recommences in a spire, precisely like the fibre represented in Fig. 10, E. A very similar structure is seen in the trachea of the abdomen of the female *Termes* (White Ant), when distended with eggs to many times its original size; the fibre does not seem to keep pace with the rapid growth of the tube, and breaks irregularly into rings, which are here and there connected by a spire.—The bronchial ramifications of the trachea, in the higher animals, exhibit a 'patchy' arrangement of the cartilaginous substance, which reminds us of the irregular extension of the secondary deposit in the Reticulated and Dotted ducts of plants.

154. The several forms of Ducts just described are distinguished by their want of any tendency to branch or anastomose; they run straight and parallel with each other; and little or no lateral communication takes place between them. The purpose of this arrangement will become evident, when it is seen that the office of these Ducts is simply to convey fluid in the most direct manner possible from the extremities of the roots to those of the leaves (CHAP. XII. Sect. 2).—There is another class of vessels, however, existing in most of the higher plants; the structure and purpose of which are very different. The vessels of this class consist of branching tubes, which anastomose with each other so as to form a network (Fig. 13, B); their walls are usually very thin and transparent in the young plant, so that they may be easily overlooked (whence it has happened that these anastomosing canals have been represented by some as mere intercellular passages); but they are gradually increased in thickness by secondary deposit, which occasionally presents a fibrous character. These vessels contain the peculiar milky juice, known as the *latex*, of such plants as possess it; hence they are spoken of as *laticiferous* vessels. The resemblance which they bear to the capillary vessels of Animals (§ 221) is very striking; and this resemblance is not confined to their peculiar arrangement, but extends to their origin,—the network in both instances being formed by the coalescence of cells, as is shown with regard to the laticiferous vessels in Fig. 13, A.—Besides the vasiform and laticiferous tissues, we find another provision in Plants for the conveyance of fluid; namely, a series of passages left in the parenchyma by the want of junction

FIG. 13.

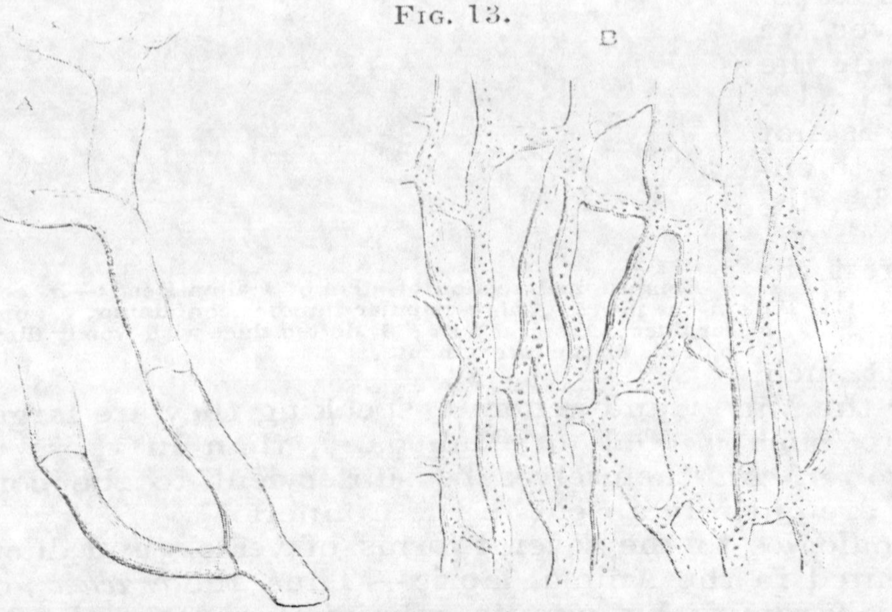

Laticiferous vessels:—A, their formation from cells; B, network of milk-vessels from the stipule of *Ficus elastica*.

between its component elements. These passages, having no other boundary-walls than those of the tissues among which they are excavated, are termed *intercellular spaces.* In some instances they contain air, which is the case with the large and freely-communicating passages that traverse the loose parenchyma of the under surfaces of leaves. In other situations, however, they serve for the conveyance of fluid; and this, which is certainly the case with the intercellular passages that are found in the interior of the stems of many Flowering Plants, is probably true also of those which frequently present themselves in Cryptogamia. If the circulation of a nutritious juice or elaborated sap, equivalent to the milky juices of those plants which possess them, be a general occurrence (CHAP. XII. Sect. 2), it may take place through the intercellular passages of the leaves and bark of such plants as do not possess laticiferous vessels with distinct walls.

155. Such are the principal forms of the Primary Tissues of Plants, which can scarcely be studied with attention, without impressing the inquirer with a feeling of the wonderful Unity of that Design, by means of which such a countless variety of results has been wrought out from the simplest elements. For we have seen that in the analysis of the most complex Vegetable fabrics, we never meet with any structures that have so far lost their original cellular type, as to prevent our easy recognition of their origin; the greatest departure from that type consisting in the formation of tubes, for the simple physical purpose of the transmission of fluid. And when we come to inquire in detail into the structure of the organs which minister to functions of a strictly *vital* nature, we shall find that cells are still their essential components (whatever tissues are added being merely for physical purposes), and that the cells most actively concerned in these operations do not depart widely from their simple primitive form and character. In fact, even the 'secondary deposits' have a purely physical purpose, that of imparting density and resisting power to the tissue; and the elongation of the cells in woody fibre is destined to impart to them the purely physical endowment of tenacity.

3. *Of the Primary Tissues of Animals.*

156. In applying to the Animal fabric those methods of investigation which have been attended with so much success in the analysis of Vegetable structure, we at once perceive that the former is much more *heterogeneous* than the latter; that is to say, the number of dissimilar tissues is much greater, and the differences between them are far more decided;—and this in proportion as we ascend the Animal scale. In fact there are many animal tissues, whose relation to the primitive *cellular* type is so obscure, that it can only be established by attentively watching the history of their development. This cannot be deemed surprising, when it is considered that the endowments peculiar to Animals must necessarily be connected with 'organs' or instrumental structures, of a kind which the Vegetable does not possess; and that the possession of these endowments involves numerous important modifications in the apparatus that ministers to the functions of Vegetative life. Still we shall find the general rule to hold good, that all the Animal tissues are

developed in the first instance through the medium of cell-life; and that in the organs subservient to the strictly *vital* operations, cells remain the essential instruments.—It is doubtful whether there are any beings unquestionably entitled to be regarded as Animals, whose organisation is so simple that they consist in their perfect state, like the humblest Plants (§ 139), each of a single isolated cell. But this is the earliest condition of all Animals, as of all Plants; and the history of the Animal cell, whether thus temporarily ranking as a distinct individual, or occurring as a component of an aggregate fabric, is essentially the same as that of the Vegetable cells into whose history we have examined, except as regards its dependence on external supplies for that pabulum which the latter can generate for themselves. In so far as is yet known, the composition of the *cell-wall* is everywhere the same, being that of the albuminous substances; and in this respect the cell-wall agrees, therefore, with the 'primordial utricle' of Plants, of which it is probably in other respects to be regarded as the representative, rather than of that external deposit of cellulose which usually ranks as the wall of the Vegetable cell. It is in the nature of the *contents* of the cells, that the greatest diversity exists; and we shall find that the purposes to which the several groups of cells are to be subservient in the Animal economy, depend upon the nature of the materials which they select for their development, and upon the mode in which these products are subsequently disposed of.

157. *Formation of Cells.*—In Animals, as in Plants, there are two principal modes in which cells may be developed;—namely, within the cavity of a previously-existing cell, in which case the process is said to be *endogenous;* or in the midst of a plastic fluid or *blastema*, probably containing cell-germs, which has been prepared or elaborated by cells of a

Fig. 14.

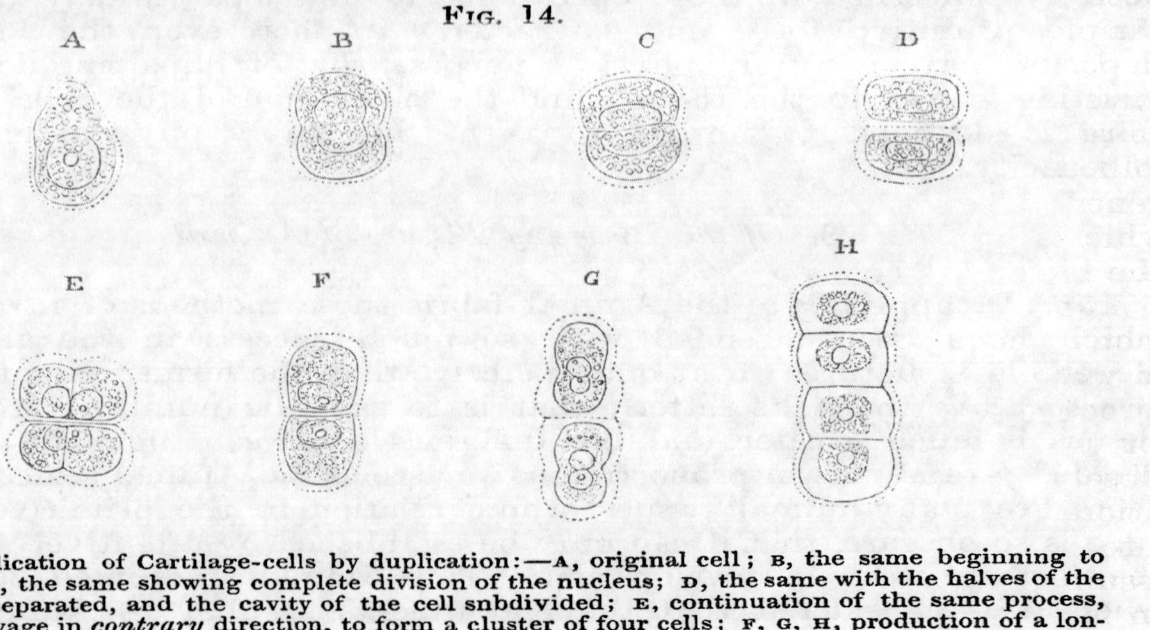

Multiplication of Cartilage-cells by duplication:—A, original cell; B, the same beginning to divide; C, the same showing complete division of the nucleus; D, the same with the halves of the nucleus separated, and the cavity of the cell snbdivided; E, continuation of the same process, with cleavage in *contrary* direction, to form a cluster of four cells; F, G, H, production of a longitudinal series of cells, by continuation of cleavage in the *same* direction.

previous generation (of whose life it may be said to partake), but which has been set free by their rupture. In both these processes, as well as in other acts of cell-life, we usually find the 'nucleus' performing a more important part than it seems generally to take in Plants. Thus in the multiplication of cells by 'endogenous' growth, we find that, where a nucleus exists, it begins to undergo subdivision as soon as the first inflection can be traced in the walls of the cell; and this subdivision is frequently completed, each fragment of the nucleus drawing around itself a portion of the contents of the parent-cell, before any division of the cavity of the latter becomes apparent. We have characteristic examples of this process in the multiplication of the cells of the germinal mass (CHAP. XVIII.), and in that of Cartilage-cells, of which the successive steps are exhibited in Fig. 14. In these instances, the nucleus and the contents of each parent-cell undergo division into *two* parts; so that the number of cells is successively *doubled.* If the division continue to take place in the same direction, a *filament* is the result, as seen at H; but if the second cleavage take place in a direction transverse to the first, a *cluster* will be produced, as at E.—In other cases, however, the nucleus appears to break up into several fragments, each of which may draw around it a portion of the contents of the parent-cell, whose cavity it may thus fill with a new brood of cells, without any such progressive subdivision as the preceding plan would involve. Of this process we frequently have examples in morbid growths, in which the multiplication of cells often takes place with great rapidity (Fig. 15). Generally speaking, the former method seems to prevail in structures which have a comparatively *permanent* destination; whilst the latter is adopted in cases where the life of the cells thus generated is but transitory. Thus the follicles of Glands (§ 185) are but parent-cells, in whose wall an opening has been formed for the liberation of the cells of the new generation (which are the real agents in the secreting process) as fast as they are formed; and from the nuclei of these parent-cells, which occupy the blind extremity of the follicles, successive crops of young cells are generated, at the expense of the fresh materials which the nuclei are continually drawing from the blood.

FIG. 15.

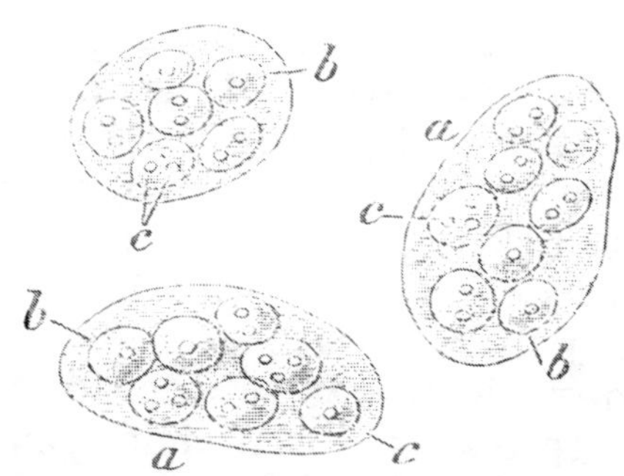

Parent-cells, *a*, *a*, of cancerous structure, containing secondary cells, *b*, *b*, each having one, two, or three nuclei, *c*, *c*.

158. Such are the principal modes in which 'endogenous' cell-development appears to take place in Animals; we have now to consider those in which new cells originate in plastic or formative material, without any direct intervention of pre-existing cells. The material which is required for this purpose is one containing *Fibrine;* a substance which, although closely resembling Albumen, if not identical with it, in chemical composition, differs from it in possessing the tendency to pass spontaneously into a solid condition when withdrawn from the living vessels, and to assume, in thus coagulating, a more or less distinct fibrous arrangement. In fact we may regard it as Albumen undergoing the first stage of Organisation and Vitalisation (§ 41). The degree to which this process has already taken place, affects the formative power of the fluid *blastema;* thus we find that the 'coagulable lymph' thrown out for the reparation of injuries, possesses it in a higher degree than does the ordinary fibrinous plasma of the blood, or 'liquor sanguinis'; so that, whilst the latter is

only ready for appropriation by tissues already existing, the former possesses in itself the power of originating new tissue, when placed in favourable circumstances to do so. This tissue may consist of little else than simple fibres, or it may contain fully-developed cells intermingled with these, or it may be composed almost entirely of cells, according to the circumstances under which it is developed.* In the first instance the effused *blastema* or plastic fluid is apparently homogeneous; but as it solidifies, it becomes dimly shaded by minute dots; and as it is acquiring further consistence, some of these dots seem to aggregate so as to form little round or oval clusters, bearing a strong resemblance to cell-nuclei. These bodies, which have been termed 'cytoblasts' or cell-formers, appear to be actively concerned in the further changes which take place in the blastema; for if it be about to undergo development into a *fibrous* tissue, they seem to be the centres from which the fibres proceed; whilst if a *cellular* structure is generated, it is clearly from them that the cells take their origin, although the mode in which the cell-membrane is first produced in connection with them has not been completely made out.

159. This development of cells in the midst of a blastema which originally contained no ostensible germs, appears to be the mode in which the *epidermic* and *epithelial* cells (§ 177) are constantly being developed; for the layers of the Epidermis which are in closest contact with the surface of the true Skin, are found to consist of a plasmatic fluid containing molecules and nuclei in various stages of development into cells; and the material for this production can be nothing else than a formative liquid, capable of transuding through the apparently-impervious membrane which intervenes between the vessels of the Skin and the Epidermic layer. There are probably many other cases, in which a similar production takes place among the higher animals, as a part of the regular formative processes.—Although this might seem at first sight to afford an exception to the general rule of the origin of every new cell in a parental generation, yet it does not really do so. For it is pretty certain that the blastema is itself the product of the formative agency of certain cells expressly provided for its elaboration; and it does not seem improbable that these cells, in bursting and setting free the plastic fluid which they have prepared, should diffuse through it their own nuclear or germinal particles in a state of solution or extremely minute division; and that these, attracting each other in the act of solidification, should act as new centres of cell-growth, just as if they were still contained within the parent-cell. The connecting link, in fact, between the two modes, is afforded by cases in which the parent-cell sets free by its rupture the evident germs of new cells, which it does not retain long enough for their ulterior development to take place in its interior, but which it throws at once upon their own resources.—Thus we seem entitled to say, that every Animal cell has its origin in a pre-existing cell; either arising directly from its nucleus, and being developed in its interior; or springing out of the formative liquid, which it has prepared and impregnated with its own 'germinal' or developmental power.

160. That *all* the Animal tissues are immediately developed from Cells, was the doctrine originally put forth by Schwann, who first attempted to

* See Mr. Paget's Lectures on Repair and Reproduction after Injuries, in Medical Gazette, 1849.

generalise upon this subject. By subsequent research, however, it has been found that this statement was too hasty ; and that, although many tissues retain their original cellular type through the whole of life, and many more are evidently generated from cells and are subsequently metamorphosed, there are some in which no other cell-agency can be traced, than that concerned in the preparation of the plastic material.—This would appear to be the case with regard to certain forms of the very delicate structureless lamella of membrane, now known under the name of *Basement* or *primary membrane*, which is found beneath the epidermis or epithelium on all the free surfaces of the body, internal as well as external. In its very simplest form, the basement-membrane is a pellicle of such extreme delicacy, that its thickness scarcely admits of being measured ; it is to all appearance perfectly homogeneous, and presents not the slightest trace of structure under the highest powers of the microscope, appearing rather to resemble the membrane of which the walls of the cells are themselves constituted. Such a membranous lamella is easily obtained by subjecting a portion of the inner layer of the shell of almost any bivalve Mollusk to the action of dilute acid, so as to dissolve away the calcareous matter ; its presence there being evidently the result of its exuviation from the surface of the true skin. In other instances, however, the membrane has a somewhat granular appearance, as if the blastema by whose consolidation it was produced had contained nuclei ; such a membrane is found lining the blood-vessels of higher animals, and it may be occasionally obtained by the decalcification of the nacreous substance of shell. And thus we are led towards other forms of basement membrane, in which it can be distinctly seen to consist of flattened polygonal cells with granular nuclei.—Not only the simpler forms of basement-membrane, however, but also *fibres* may originate in the formative fluid without any direct cell-agency ; simply in virtue, as it would seem, of the tendency of fibrine to assume a fibrous arrangement in the act of coagulation (§ 158). Such an arrangement is seen, with greater or less distinctness, in every coagulum of which fibrine has been a sufficiently large component ; a very beautiful example of it is presented in the animal basis of the common egg-shell, a very thin pellicle of which exhibits the structure shown in Fig. 16.—It seems probable that on a process of this kind is chiefly dependent the formation of those simple Fibrous tissues of Animals, with the description of which, as the *least organised* part of their structure, we shall first proceed.

Fig. 16.

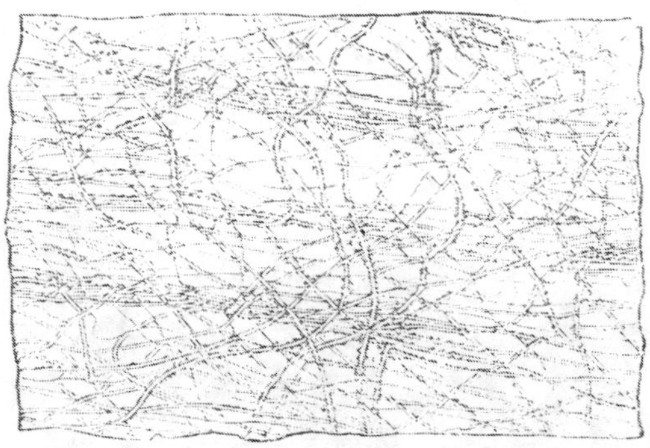

Fibrous membrane from Egg-shell.

161. *Simple Fibrous Tissues.*—A large part of the Animal fabric, especially among the higher classes in which the parts have the greatest amount of motion upon one another, is composed of tissues, which seem as if they consisted of nothing else than homogeneous *fibres*, woven together in various ways, according to the purposes they are destined to serve. These fibres are altogether different from those hereafter to be described as constituting the Muscular and Nervous tissues, and must not be confounded with them.

The former are *solid*, and possess none but *physical* properties; the latter are *tubular*, and are distinguished by their peculiar *vital* endowments, which seem chiefly, if not entirely, to reside in the *contents* of the tubular fibre. The simple fibrous tissues, of which we have now to treat, appear to have it for their sole office in the animal body, to bind together the other elementary parts into one whole, without uniting them so closely as to render them immoveable; and we find the same elements arranged in very different modes, according to the purposes they are destined to fulfil. Thus in the *Tendons*, by which the Muscles are connected with the Bones and impart motion to them, the only property required is that of resisting strain or tension in *one* direction; and in these we find the fibres disposed in a parallel arrangement, passing continuously in straight lines between the points of attachment. In the *Ligaments* which connect the bones together, and which also have for their purpose to afford resistance to strain, but which are liable to tension in a greater variety of directions, we find bundles of fibres crossing each other according to these directions; and in some instances we find the ligaments endowed also with a certain degree of elasticity. The structure of the strong *Fibrous Membranes*, which form the envelopes to different organs, and bind together the contained parts, is very similar; each of these membranes being composed of several layers of a dense network, formed by the interweaving of bundles of fibres in different directions. In the *Fibro-Cartilages*, we find a mixture of the characteristic structure of Ligament with that of Cartilage; bundles of fibres, similar to those which constitute the former, being disposed among the cells, which are the chief organised constituents of the latter. In certain Fibro-Cartilages, however, these fibres are endowed with a high degree of elasticity.

162. These two qualities,—that of resistance to tension without any yielding,—and that of resistance combined with elasticity,—are characteristic of two distinct forms of Fibrous tissue, the *White* and the *Yellow*.—The *White Fibrous* tissue presents itself under various forms; being sometimes composed of fibres so minute as to be scarcely distinguishable, but more commonly presenting the aspect of bands, usually of a flattened form, and attaining the breadth of 1-500th of an inch. These bands are marked by numerous longitudinal streaks, but they can seldom be torn up into minute fibres of determinate size; hence they must be regarded as made up of an aggregation of the same elements as those which may become developed into separate fibres. The fibres and bands are occasionally somewhat wavy in their direction; and they have a peculiar tendency to fall into undulations, when it is attempted to tear them apart from each other (Fig. 17). This tissue, which is perfectly inelastic, is easily distinguished from the other by the effect of Acetic acid, which swells it up and renders it transparent, at the same time bringing into view certain oval nuclear corpuscles.—The *Yellow Fibrous* tissue exists in the form of long, single, elastic,

Fig. 17.

White Fibrous Tissue, from Ligament.

branching filaments, with a dark decided border; which are disposed to curl when not put on the stretch (Fig. 18). They are for the most part between 1-5000th and 1-10,000th of an inch in diameter; but they are often met with both larger and smaller. They frequently anastomose, so as to form a network. This tissue does not undergo any change, when heated with acetic acid. It exists alone (that is, without any mixture of the white) in parts which require a peculiar elasticity, such as the middle coat of the Arteries, the Vocal Cords, the Ligamentum Nuchæ of Quadrupeds, the elastic Ligament which holds together the valves of a Bivalve shell (§ 301), and that by which the claws of the Feline tribe are retracted when not in use; it enters largely into the composition of certain parts, which are commonly regarded as Cartilaginous, such as the external ear; and it is also a principal component of other tissues to be presently described.

Fig. 18.

Yellow Fibrous Tissue from Ligamentum Nuchæ of Calf.

163. These tissues are very different in Chemical composition. Those which are composed of the *White* fibrous element,—namely, Tendons, Ligaments, &c.—are almost entirely resolved by long boiling into the substance termed *Gelatine* * or Glue; and this is also largely obtained from

* The composition of Gelatine is much simpler than that of the Albuminous compounds; so far, at least, as regards the number of atoms of its several elements; for it consists (according to Mülder) of 13 Carbon, 10 Hydrogen, 2 Nitrogen, 5 Oxygen. The distinctive characters of Gelatine are its solubility in warm water, its coagulation on cooling into a uniform jelly, and its formation of a peculiar insoluble compound with Tannic acid. Its power of forming a jelly on cooling is such, that a solution of one part in 100 of water will become a consistent solid. And its reaction with Tannic acid is so distinct, that the presence of one part of Gelatine in 5000 of water is at once detected by infusion of Galls.—There can be no doubt that Gelatine does not exist *exactly as such* in the Fibrous tissues; since none can be dissolved out of them by the continued action of cold water, and it usually requires the *prolonged* action of hot water, to occasion their complete conversion. There are some substances, however, in which this is not requisite; and from which the gelatine may be more readily extracted. This is the case, for example, with the air-bladder of the Cod and other fish; which, when cut into shreds and dried, is known as Isinglass. It is the case also with the substance of bones, from which the calcareous matter has been removed. In both these instances, it would seem that the state of organization is very imperfect; scarcely any traces of the fibrous structure being perceptible. When the fibrous arrangement is more complete, the solubility of the tissue is much diminished. Hence it is obvious that the particles have a different arrangement in the tissues, from that which they have in the product obtained by boiling. Their ultimate composition, however, is the same; for when any serous membrane, or other tissue principally composed of the white fibrous element, is analysed by combustion, the elements are found to have the same proportion to each other as in Gelatine, allowance being made for the small admixture of other substances. The action of Tannic acid, too, is the same on the organized tissue, as it is on the gelatine extracted from it; and hence results its utility in producing an insoluble compound, not liable to undergo decomposition, in the substance of the skin, converting it into leather.—It is not yet known how Gelatine is produced in the Animal body. There can be no doubt that it may be elaborated from Albumen; since we find a very large amount of Gelatine in the tissues of young animals, which are entirely formed from albuminous matter; and also in the tissues of herbivorous animals, which cannot receive it in their food, as Plants yield no substance resembling gelatine.

the Skin, from Mucous and Serous Membranes, and from the animal basis of Bones, into all of which, as we shall presently see, that element enters largely.—The composition of the *Yellow* fibrous tissue appears to be altogether dissimilar. It scarcely undergoes any change by prolonged boiling; it is unaffected also by the weaker acids; and it preserves its elasticity, if kept moist, for an almost unlimited period. According to Scherer it consists of 48 Carbon, 38 Hydrogen, 6 Nitrogen, and 16 Oxygen; and he considers it to be composed of an atom of Proteine with two atoms of water.—The simple Fibrous tissues appear to be very little susceptible of change in the living body; and we find them very sparingly supplied with blood-vessels. In the solid Tendons, the bundles of straight parallel fibres are a little separated from each other by the intervention of the Areolar tissue to be presently noticed; and this permits the sparing access of vessels to their interior. In the Fibrous Membranes and Ligaments, this is found in somewhat larger amount; and the vascularity of these tissues is rather greater.

164. The great use of the foregoing Tissues appears to be, to afford a firm resistance to tension; with which, however, a considerable amount of elasticity may be combined. But we have now to notice a tissue, in which a very different arrangement of the same elements presents itself; and the object of this is, to bind together the elements of the different fabrics of the Animal body, and at the same time to endow them with a greater or less degree of freedom of movement upon one another. This tissue, which is called the *Areolar*, consists of a network of minute fibres and bands, which are interwoven in every direction, so as to leave innumerable *areolæ* or little spaces, which communicate freely with one another. Of these fibres, some are of the Yellow or elastic kind; but the majority are composed of the White fibrous tissue; and, as in that form of elementary structure, they frequently present the form of broad flattened bands, or membranous shreds, in which no distinct fibrous arrangement is visible. The proportion of the two forms varies, according to the amount of elasticity, or of simple resisting power, which the endowments of the part require. The interstices or *areolæ* are filled during life with a fluid, which resembles very dilute serum of the blood; consisting chiefly of water, but containing a sensible quantity of common salt and albumen. It is the undue accumulation of this fluid, which constitutes dropsical effusion; the influence of gravity upon the seat of which, shows the free communication that exists among the interstices. This freedom of communication is still more shown, however, by the fact, that either air or water may be made to pass, by a moderate continued pressure, into almost every part of the body containing Areolar tissue, although introduced at only a single point. In this manner it is the habit of butchers to *inflate* veal; and impostors have thus *blown-up* the scalps and faces of their children, in order to excite commiseration. The whole body has been thus spontaneously distended with air by emphysema in the lung; the air having escaped from the air-cells into the surrounding areolar tissue, and thence, by the continuity of this tissue with that of the body in general at the root or apex of the lungs, into the entire fabric.

165. As already mentioned, we find this tissue in almost every part of the bodies of higher Animals; thus it binds together the ultimate fibres of the Muscles into minute fasciculi, unites these fasciculi into larger

ones, these again into larger ones which are obvious to the eye, and these into the entire muscle ; and also forms the membranous divisions between distinct muscles. In like manner it unites the elements of nerves, glands, &c. ; binds together the fat-cells into minute masses, these into larger ones, and so on ; and in this manner penetrates and forms a considerable part of all the softer tissues of the body. But it is a great mistake to assert, as it was formerly common to do, that it penetrates the harder organs, such as bones, teeth, cartilage, &c. Its purpose obviously is, to allow a certain degree of movement of the parts which it unites ; and hence we find it entering much more largely into the composition of the Mammary gland (which, from its attachment to the great pectoral muscle, must have its parts capable of being shifted upon one another), than into that of the Liver, Kidneys, &c. It also serves as the *bed*, in which blood-vessels, nerves, and lymphatics may be carried into the substance of the different organs ; but these are not largely distributed to its own substance. It cannot be certainly affirmed that Areolar tissue, any more than the elementary forms of which it is made up, possesses any distinctly vital endowments. A slight degree of contractility has been said to be displayed by it upon the application of stimuli ; but this may be due to the muscular tissue of the vessels which traverse it.—Among the lower animals, however, it would appear that a semi-fluid gelatinous substance, containing granules in its areolæ, but without distinct fibres, and strongly resembling the incipient form of Areolar tissue in the higher (§ 158), possesses a considerable amount of contractility, and is in fact the chief instrument of their motions.

166. Two different views of the mode of development of the Fibrous tissues have been taken by those who have studied it. By some it has been asserted that the White fibres are first developed as cells, which progressively become elongated and solidified, their nuclei at the same time disappearing until brought into view by acetic acid (Fig. 19) ; and the Yellow fibres have been supposed to have a similar origin. By others it has been considered that the White fibres are produced by the direct *fibrillation* of a *nucleated blastema* (§ 158) ; and that from the nuclei proceed the Yellow fibres ; no development of cells being requisite for the production of either. The recent enquiries of Mr. Paget * tend to show that *both* these methods are adopted in the formation of the fibrous tissue that is produced for the repair of injuries in the adult body ; the former being seen in the reparation of external wounds to which air has access ; the latter in the organisation of ' coagulable lymph ' effused into internal cavities altogether secluded from it.

FIG. 19.

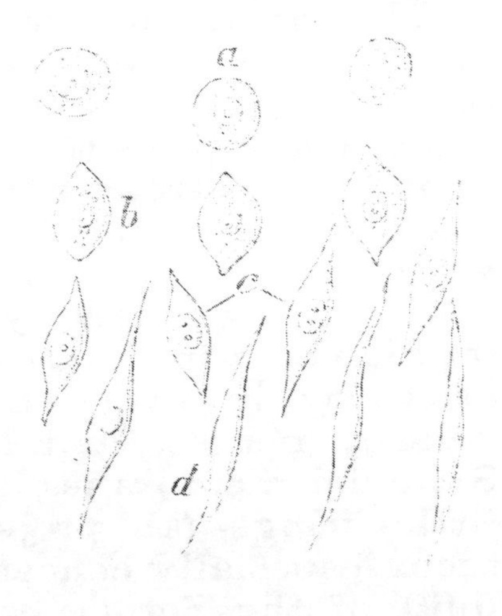

Development of fibres from cells:—*a*, circular or oval nucleated cells ; *b*, the same becoming pointed ; *c*, the same become fusiform, the nuclei still being apparent ; *d*, the same elongating into fibres, the nuclei having disappeared.

167. The structure of the *Serous* and *Synovial* Membranes is essentially the same with that of Areolar tissue. It is the peculiar character of these membranes to form *closed* bags or sacs, having a very smooth and glisten-

* Medical Gazette, June 22, 1849.

ing inner surface, and containing a fluid more or less closely allied in composition to the serum of the blood. Their arrangement is often very complicated, and will perhaps be best understood by studying one of the simpler forms of it, such as is found in the *Synovial* membranes lining the joints; the disposition of which is explained in the accompanying diagram. At *a, a,* are seen the extremities of the two articulated bones; at *b, b,* the layers of cartilage which cover them: and the dotted line shows the course of the synovial membrane, which covers the ends of the bones as far as *c, c,* and is then reflected from one to the other, so as to form a continuous bag, which is completely closed.* The arrangement of the *Serous* membranes is usually much more complicated. These line the three great cavities of the body,—the head, chest, and abdomen,—together with their subdivisions; enveloping the viscera which these contain, so as to afford them an external coat over every part save that by which they are suspended; and being then reflected over the interior of the cavity, so as to form a shut sac intervening between its outer walls and its contents.

FIG. 20.

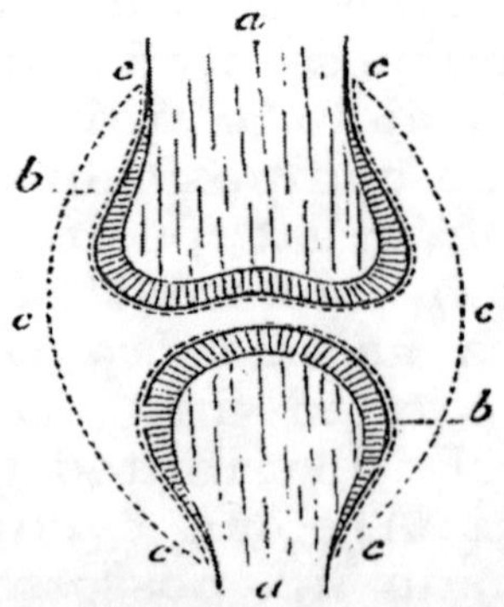

Ideal Section of a Joint.

The purpose of this appears to be, to facilitate the movements of the contained organs, by forming smooth surfaces which shall freely glide over each other; this is evidently of great importance, where such constantly-moving organs as the heart and lungs are concerned.—The free or unattached surface of these membranes is covered with a layer of cells; but these constitute a distinct tissue, the *Epithelium,* of which an account will be given hereafter. The epithelium lies upon a continuous sheet of *basement* or *primary-membrane,* which thus completely isolates it from the tissues beneath. Subjacent to this is a layer of condensed Areolar tissue, which constitutes the chief thickness of the serous membrane, and confers upon it its strength and elasticity; this gradually passes into that laxer variety, by which the membrane is attached to the parts it lines, and which is commonly known as the *sub-serous* tissue. The yellow fibrous element enters largely into the composition of the membrane itself; and its filaments interlace in a beautiful network, which confers upon it equal elasticity in every direction. The membrane is traversed by blood-vessels, nerves, and lymphatics, in varying proportions; some of the Synovial membranes, especially that of the knee-joint, are furnished with little fringe-like projections, which are extremely vascular, and which seem especially concerned in the secretion of the synovial fluid. The fluid of the Serous cavities is so nearly the same as the serum of the blood, that the simple act of transudation is sufficient to account for its presence in their sacs; on the other hand, that of the Synovial capsules, and of the Bursæ Mucosæ which resemble them, may be considered as serum with from 6 to 8 per cent. of additional albumen.

168. The elements of Areolar tissue enter largely also into two other textures, which perform a most important share in both the Organic and

* The continuity of the synovial capsule over the Articular cartilages has been denied by many anatomists; the truth, however, appears to be, that the arrangement above described always prevails at the first formation of the joint, although, when it has come into use, the synovial membrane becomes obliterated at the parts where it is subject to friction.

the Animal functions ;—namely, the *Mucous Membranes* and the *Skin.* These textures are continuous with each other; and may, in fact, be considered as one and the same, modified in its different parts according to the function it is destined to perform. The *Mucous* membranes, unlike the serous, line *open* cavities of the body; thus, one commences at the mouth (being there continuous with the skin), communicates with that which lines the nostrils and covers the spongy bones (upon which the olfactory nerve is minutely distributed), and then divides into two branches ; one of these passes down the air-passages, and is continuous over the whole interior of the lungs ; the other lines the alimentary tube through its whole extent, communicating again with the skin at its farther extremity, and sending prolongations along the ducts of the glands which pour their secretions into it,—these prolongations ramifying and subdividing in such a manner, as to be, in fact, the essential constituents of the glands themselves. Another mucous membrane covers the eye, and lines the eyelids, sending one prolongation which forms the lachrymal gland, and another which lines the lachrymal duct and thus communicates with the membrane of the nose. Another lines the urinary passages, and forms the tubuli of the kidney; and another has a similar connection with the tubes and cavities of the generative system. All these, it is obvious, are continuous with the Skin at some point or other ; and anatomical examination of the cutaneous tissue shows, that it is not itself organically different from the mucous membrane, consisting, like it, of simple fibres interwoven together in all directions, and having almost identically the same chemical composition,—gelatine predominating in both. Although, in the higher animals, the functions which the external and internal portions of this membrane (for so they may be regarded) have to perform, are so different as to lead to such modifications in their structure as render them incapable of altogether fulfilling each other's offices, they would seem, in some of the lowest, to be mutually convertible ; the lining of the stomach in the Hydra (§ 289) being capable of becoming the skin by inversion, and what was previously the skin serving equally as well as the other to line the digestive cavity. In the higher classes, the Skin is the principal organ of common sensibility, the nerves of touch being minutely distributed upon it ; and it also furnishes the means for dissipating a large proportion of the superfluous fluid of the system by exhalation. On the other hand, the Mucous Membrane lining the alimentary canal is specially modified for absorption ; that of the lungs, for the interchange of gaseous ingredients between the blood and the air ; that of the various glands, for the separation or elaboration of their products from the blood,—and so on.

169. The Mucous Membranes are among the parts of the fabric most actively concerned in the *organic* or constructive functions ; their especial office being to maintain the communication between the nutrient system and the external world, by obtaining from the latter the materials requisite for the supply of the body, and by returning to it the superfluous or effete portions. Instead of being smooth and glistening, like the serous membranes, they have a soft unpolished surface, which sometimes has somewhat the appearance of the 'pile' of velvet, or the rind of a ripe peach. This is due to the elevation of the membrane into minute folds, or its prolongation into rows of filaments (or *villi*), especially where

it is to be subservient to absorption (Figs. 21, and 32, *a*); and to its depression into little pits or follicles (Fig. 32, *b*), where some special secretion is to be formed, or where it is necessary that a large amount of *mucus* should be thrown out for its protection (Fig. 22). This peculiar

FIG. 21.

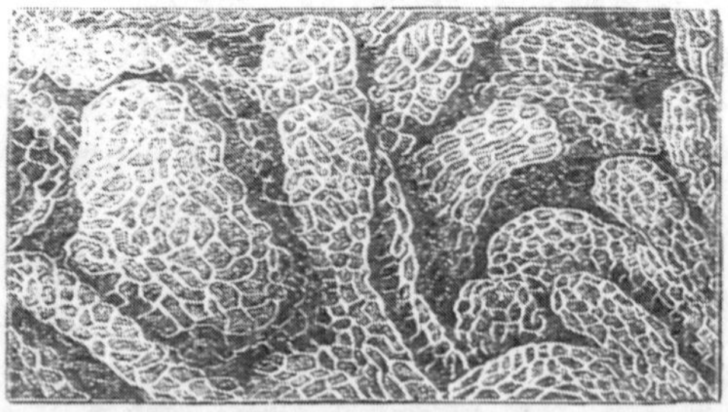

Portion of Mucous Membrane of Intestine, showing villous surface; the blood-vessels injected.

FIG. 22.

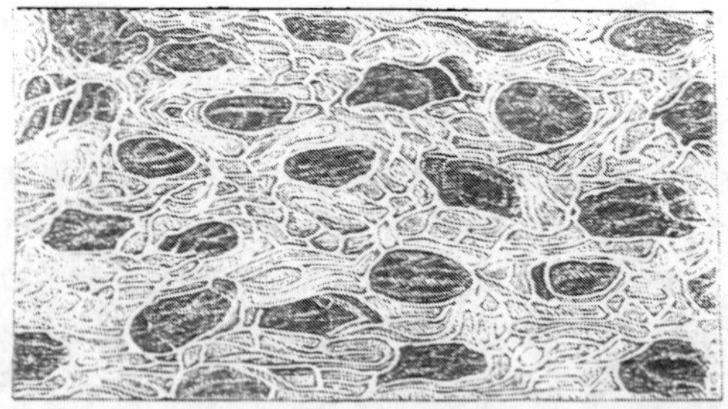

Portion of Mucous Membrane, showing orifices of secreting follicles, and arrangement of vessels around them.

substance appears to be the product of certain secreting cells which form an epithelial layer on Mucous membrane (§ 182), and may therefore be produced from any part of its surface. Its presence is necessary to shield the membrane from the irritation, which the contact of the various solids and fluids, and even of air, would otherwise produce.

170. The *Skin* and *Mucous Membrane* may be said, like the Serous, to consist of three chief parts;—the Epidermis or Epithelium covering its free surface; the subjacent Basement-membrane; and the Areolar tissue, which, with blood-vessels, nerves, and lymphatics, forms the thickness of the membrane, and connects it with the surrounding parts. The Epidermis which forms the outer layer of the skin, and the Epithelium which occupies the same position on the Mucous Membrane, alike consist of cells; but the function of the former (which consists of several layers, of which the outer are dry and horny) is simply protection to the delicate organs beneath; whilst that of the latter is essentially connected with the process of Secretion, as will be shown hereafter. The Basement-membrane resembles that of the serous membranes; but its separate existence is unusually evident in some parts where it exists alone, as in the tubuli uriniferi of the kidney; whilst it can with difficulty be demonstrated in others, as in the skin. The Areolar tissue of Mucous membranes usually makes up the greatest part of their thickness; and it is so distinct from that of the layers beneath, constituting the sub-mucous tissue, as to be readily separable from them. It differs not in any important particular, however, from the same tissue elsewhere; and the white and fibrous elements may be detected in it in varying proportions, in different parts,—the latter being especially abundant in the skin and lungs, which owe to it their peculiar elasticity. Hence the Mucous membranes yield Gelatine in abundance, on being boiled. The Skin also appears to contain some of the non-striated Muscular fibre (§ 225), in varying proportions in its different parts.—The relative amount of Blood-vessels, Nerves, and Absorbents, is subject to great variation, according to the part of the system examined. The first, however, are the most constantly-abundant, being required in the Skin for sensation (Fig. 23), and in the Mucous membranes for absorption and secretion. In fact we might say of many of the mucous

membranes, especially those of the glands, that their whole purpose is to give support to the secreting cells, and to convey blood-vessels into their immediate neighbourhood, whence these cells may obtain materials for their development. The Skin is almost the only part of the whole system, which is largely supplied with Nerves ; hence the sensibility of the internal mucous membrane is usually low, although its importance in the organic functions is so great. The Skin is copiously supplied with Lymphatics ; and the first part of the alimentary canal with Lacteals ; some of the glandular organs are also largely supplied with Lymphatics.

FIG. 23.

Distribution of Capillary blood-vessels in Skin of Finger.

171. The White fibrous tissue appears to have a peculiar power of combining with calcareous matter, so as to become solidified into a texture of great firmness. This conversion may take place in the substance of the skin, and thereby are formed the shells of Echinodermata (§ 195). It also occurs in the *periosteum,* or fibrous membrane covering bones, and thus forms successive additions to their surface (§ 210). And it not unfrequently takes place abnormally, forming irregular deposits of imperfect bone in unusual situations. Such transformations, occurring in Man as a result of disordered action, not unfrequently tend to establish an analogy with the usual condition of the part affected in some other animal. Thus it is not uncommon to find bony plates existing in the fibrous membrane (*dura mater*) which surrounds the brain of Man, and especially in those projections of it which divide and support the different parts of that organ ; and these projections exist in a state of more or less complete ossification in many quadrupeds, especially among the Carnivora. Again, the ligamentous substance which connects the muscular fibres at the base of the heart, is not unfrequently ossified by disease in Man ; whilst in the Ox and other Ruminating quadrupeds, bone naturally exists there.

172. *Simple Isolated Cells.*—The very simplest and most independent condition of the Animal Cell, is probably to be found in the Blood, the Chyle, and the Lymph ; in all of which liquids we meet with floating cells, which are as completely isolated from one another, and which are consequently just as independent, as the vesicles of the Red Snow or other simple cellular Plants. Indeed in the nature of their *habitat,* we may compare them with the Yeast-Plant (§ 271) ; for as this will only vegetate in a saccharine fluid containing vegetable albumen, so do we find that these floating cells will only grow and multiply in the albuminous fluids of animals. Those of which we are first to speak are found in the nutritious fluids of *all* animals in which a proper Circulation exists ; being nearly the only corpuscles in the *blood* of Invertebrata, and in the *chyle* and *lymph* of Vertebrata ; and being present also in the blood of the latter, in which they are known as the *white* or *colourless* corpuscles. They are usually nearly spherical ; and their diameter ranges between 1-2000th and 1-3000th of an inch. They sometimes contain a number of separate molecules, scattered through their cavity, and these are occasionally seen to be in active movement. More frequently, however, the molecules are

aggregated into a soft irregular nucleus, which sometimes appears to occupy nearly the whole of the cavity.—From these cells, the *red* corpuscles of the blood are now generally supposed to take their origin; but as these are nearly peculiar to Vertebrata, it is almost certain that the colourless corpuscles must have some other more general function; and there seems reason to think that this consists in the transformation of Albumen into Fibrine,—that is to say, the elaboration of the spontaneously-coagulating and fibrillating substance, from the mere chemical compound which forms the raw material of the Animal tissues. For we find them in every situation in which we know this transformation to be going on; and we observe their number to bear a close relation with the amount of fibrine produced in the fluid. Thus in the Inflammatory process, the quantity of fibrine in the blood is very greatly augmented; and the number of white corpuscles found in that fluid, when it is drawn from the body, is very largely increased. Moreover they are observed to accumulate in great numbers in the vessels of inflamed parts; and not only in these, but in all parts where processes of growth and reparation are going on, which require a large supply of highly-elaborated fibrine.—Thus we now see how the separate life of the individual cells is made to contribute to the general life of the entire organism, and is at the same time dependent upon it. If the nutrient material were not prepared by other processes, these cells could not exist; on the other hand, if this nutrient material were not further elaborated by *their* action, no subsequent processes of growth could take place.

Fig. 24.

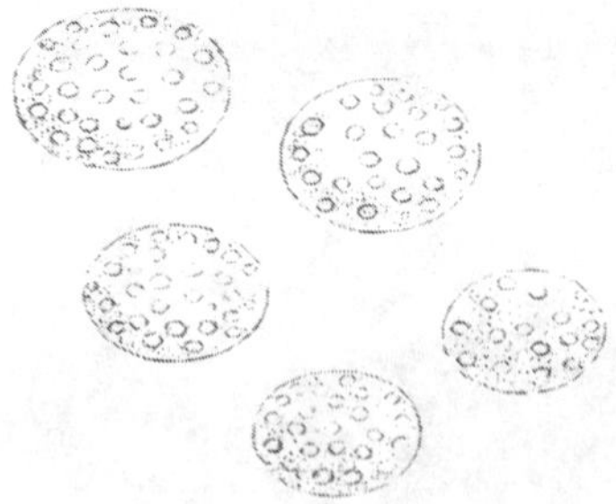

Colourless Corpuscles of Blood.

173. Besides the foregoing, the blood of Vertebrated animals also contains other cells, which are distinguished by their *red* colour and flattened form. These are equally isolated, and lead an independent life; undergoing all their changes whilst floating in the rapidly-circulating current. These Red Corpuscles are found but very sparingly in the blood of Invertebrated animals; and only in that of the higher classes. Their proportion in the blood of Vertebrata varies considerably in the several groups of that sub-kingdom; and seems to be closely connected with the relative activity of respiration in each case. They present, in every instance, the form of a flattened disk, which is circular in Man and in most Mammalia (Fig. 26), but which is oval in Birds, Reptiles, and Fishes, and in a few Mammals (Fig. 25): in both instances this disk is a flattened cell, whose walls are pellucid and colourless, but whose contents are coloured. They may be caused to swell up and burst, however, by the imbibition of water; and the perfect transparency and the homogeneous character of their walls then become evident. The Red corpuscles in the blood of Oviparous Vertebrata are distinguished by the presence of a distinct central spot or *nucleus*, which appears to be composed of an aggregation of minute granules; this is most distinctly brought into view by treating the blood-discs with acetic acid, which renders the remainder of the particle extremely transparent, while it increases the opacity of the nucleus (Fig. 25, *d*). It is remarkable, however, that the red corpuscles of the blood of Mammals should possess no obvious nucleus; the dark spot which is seen in their centre (Fig. 26, *b*), being merely an effect of

refraction, in consequence of the double-concave form of the disk. When the corpuscles are treated with water, so that their form becomes first flat, and then double-convex, the dark spot disappears; whilst, on the other hand, it is made more evident when the concavity is increased by the partial emptying of the cell, which may be accomplished by treating the blood-corpuscles with fluids of *greater* density than their own contents.

FIG. 25.

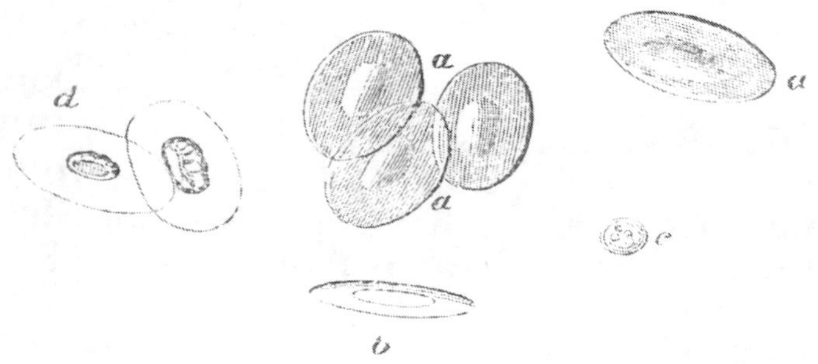

Red Corpuscles of Frog's Blood:—*a*, *a*, their flattened face; *b*, particle turned nearly edgeways; *c*, colourless corpuscle; *d*, red corpuscles altered by dilute acetic acid.

FIG. 26.

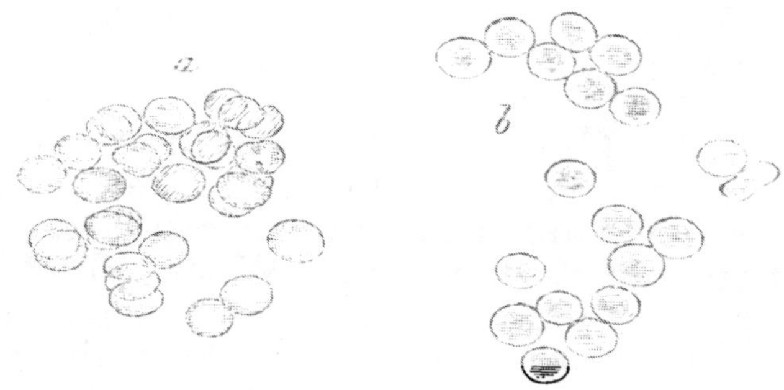

Red Corpuscles of Human Blood; represented at *a*, as they are seen when rather beyond the focus of the microscope, and at *b*, as they appear when within the focus.

According to the statements of Mr. Paget (§ 175), the absence of a nucleus in these corpuscles,—which is observable not merely in the circular discs of Mammalia in general, but also in the oval discs peculiar to the Camelidæ, —is to be regarded as indicative of a more advanced stage of their development than presents itself in Oviparous Vertebrata.

174. The size of the Red Corpuscles is not altogether uniform in the same blood; thus it varies in that of Man, from about the 1-4000th to the 1-2800th of an inch. But we generally find that there is an average size, which is pretty constantly maintained among the different individuals of the same species; that of Man may be stated at about 1-3200th of an inch. The following Table* exhibits the average dimensions of some of the most interesting examples of the Red blood-corpuscles, in the four classes of Vertebrated Animals, expressed in fractions of an inch. Where two measurements are given, they are the long and the short diameters of the same corpuscle.

MAMMALS.

Man	1-3200	Camel	1-3254, 1-5921
Dog	1-3542	Llama	1-3361, 1-6294
Whale	1-3099	Java Musk-Deer	1-12325
Elephant	1-2745	Caucasian Goat	1-7045
Mouse	1-3814	Two-toed Sloth	1-2865

BIRDS.

Golden Eagle	1-1812, 1-3832	Ostrich	1-1649, 1-3000
Owl	1-1830, 1-3400	Cassowary	1-1455, 1-2800
Crow	1-1961, 1-4000	Heron	1-1913, 1-3491
Blue-Tit	1-2313, 1-4128	Fowl	1-2102, 1-3466
Parrot	1-1898, 1-4000	Gull	1-2097, 1-4000

* These measurements are chiefly selected from those given by Mr. Gulliver in his edition of Hewson's Works, p. 236, *et seq*.

REPTILES.

Turtle	1-1231, 1-1882	Frog	1-1108, 1-1821
Crocodile	1-1231, 1-2286	Water-Newt	1-814, 1-1246
Green Lizard	1-1555, 1-2743	Siren	1-420, 1-760
Slow-worm	1-1178, 1-2666	Proteus	1-337
Viper	1-1274, 1-1800	Lepidosiren	1-570, 1-941

FISHES.

Perch	1-2099, 1-2824	Pike	1-2000, 1-3555
Carp	1-2142, 1-3429	Eel	1-1745, 1-2842
Gold-Fish	1-1777, 1-2824	Gymnotus	1-1745, 1-2599

175. There can be little doubt that the Red Corpuscles go through the same history as other cells; and there is evidence that they are rapidly regenerated, under favourable circumstances, when a large number of them have been lost. When much blood has been drawn from the body, the proportion of red corpuscles in the remaining fluid is at first considerably lowered: since the fluid portion of the blood is replaced almost immediately, whilst these floating cells require time for their reproduction. Their amount progressively increases, however, until it has reached its proper standard, provided that a due supply of their materials be afforded, one of the most important of these being Iron (§ 176). It appears probable from observations made by Dr. G. O. Rees, that the Red Corpuscles occasionally multiply by self-division after the manner of the simplest Algæ (Fig. 4); but their reproduction would seem chiefly to depend upon the metamorphosis of the Colourless corpuscles, of which a constantly new supply is furnished by the chyle and lymph.* According to the description given by Mr. Paget,† the white corpuscles, at first tuberculated, containing many granules, and darkly shaded, become smoother, paler, less granular, and more dimly shaded or nebulous; the corpuscles then acquire a pale tinge of blood-colour, and the granules progressively vanish; and the colour gradually deepens, the corpuscles at the same time exchanging their spherical for a flattened form. In the oviparous Vertebrata, some of the granules remain aggregated in the centre to form the nucleus. The Red corpuscles make their first appearance in the embryo, however, long before the formation of chyle and lymph commences; and they appear to be formed by the metamorphosis of some of the cells which constitute the inner layer of the germinal membrane (CHAP. XVIII.). These cells are at first nearly spherical, and are full of particles of a yellowish substance like fatty matter; in the midst of which, though somewhat obscured by them, a central nucleus may be seen. The development of these embryo-cells into the complete form of

* Although this is the doctrine now taught by most Physiologists, yet there appear to the author to be certain difficulties connected with it, which have not yet been explained away. These difficulties are chiefly connected with the entire absence of any fixed proportion between the size of the colourless and that of the red corpuscles. In Man and the Mammalia generally, there is no considerable difference between their diameters, that of the colourless corpuscles being the greatest. In Reptiles, on the other hand, the colourless corpuscles are about the size of the nuclei of the red. The Proteus, whose red corpuscles are of such extraordinary dimensions, have colourless corpuscles little larger than usual; whilst, on the other hand, the Musk-deer, whose red-corpuscles are so small that nearly *a thousand* of them would be covered by *one* of those of the Proteus, has lymph-globules nearly of the ordinary diameter.

† Kirkes's Handbook of Physiology, p. 68.

the corpuscles, is stated by Mr. Paget * to be effected by the gradual clearing-up, as if by division and liquefaction, of the contained particles, the acquirement of blood-colour and of the elliptical form, the flattening of the cell, and the more prominent appearance of the nucleus. This first set of blood-disks is nucleated in Mammalia, as well as in Oviparous Vertebrata ; and they occasionally present indications of being in course of multiplication by subdivision. They gradually disappear from the blood, however, when the lymph- and chyle-corpuscles first present themselves in it ; and thenceforth the Red corpuscles seem to be formed at the expense of the latter alone. It is curious that this change should usually coincide with the time at which the external branchiæ of the tadpole disappear, and, in warm-blooded animals, with the period at which the branchial fissures are closed in the neck, and the course of the circulation altered (CHAP. XII.).

176. The Chemical composition of the walls and nuclei of the Red corpuscles is very different from that of their contents. The substance of the former has been termed *Globuline ;* but it does not seem to differ in any essential character from other substances resulting from the organization of the albuminous compounds. The compound which forms the contents of the red corpuscles, however, and gives them their characteristic hue, is altogether peculiar, and has received the name of *Hæmatine.* Its composition is notably different from that of the albuminous compounds ; the proportion of Carbon to the other ingredients being very much greater, and a definite quantity of iron being an essential part of it. Its formula is 44 Carbon, 22 Hydrogen, 3 Nitrogen, 6 Oxygen, and 1 Iron. The iron may be separated from the hæmatine by strong reagents which combine with the former, yet the latter still possesses its characteristic colour ; its hue cannot be dependent, therefore, on the presence of iron in the state of peroxide, as some have supposed. Regarding the nature of this compound, and the changes which it undergoes in respiration, there is still much to be learned ; and until these points have been more fully elucidated, the precise uses of the Red corpuscles in the animal economy cannot be understood. There is evidence, however, that the production of Hæmatine is (like the production of the red colouring matter of the Protococcus nivalis, § 44), a result of chemical action taking place in the cells themselves ; for no substance resembling Hæmatine can be found in the liquid in which these cells float, and scarcely a trace of iron can be detected in it ; whilst, on the other hand, the fluid portion of the chyle holds a large quantity of iron in solution, which seems to be drawn into the red corpuscles, and united with the other constituents of hæmatine, as soon as ever it is delivered into the circulating current.—The colouring matter appears to exist in two states, the precise chemical difference between which has not yet been ascertained. In arterial blood it is a florid scarlet ; whilst in venous blood it is of a purpler hue. By circulating through the capillaries of the system, the arterial or bright hæmatine becomes converted into dark or venous hæmatine ; and the converse change takes place in the capillaries of the lungs, the original florid hue being recovered. Now it is certain that the blood, in its change from the arterial to the venous condition, loses oxygen, and becomes charged with an increased amount of carbonic acid, although its

* Kirkes's Handbook of Physiology, p. 65.

precise mode of combination is not known; on the other hand, in its return from the venous to the arterial state, the blood gives off this additional charge of carbonic acid, and imbibes oxygen (CHAP. XIII.). These changes in the condition of the contents of the Red corpuscles,—taken in connexion with the fact that these bodies are almost completely restricted to the blood of Vertebrata (whose respiration is much more energetic than that of any Invertebrated animals, save Insects, which have a special provision of a different character), and that their proportion to the whole mass of the blood corresponds with the activity of the respiratory function,—leave little doubt that they are actively (but not exclusively) concerned as *carriers* of Oxygen from the lungs to the tissues, and of Carbonic acid from the tissues to the lungs; and that they have little other direct concern in the functions of Nutrition, than the fulfilment of this duty. Their complete absence in the lower Invertebrated animals, in the earliest condition of the higher, and in newly-forming parts until these are penetrated by blood-vessels, seems to indicate that they have no immediate connexion with even the most energetic operations of growth and development; whilst, on the other hand, there is abundant evidence that the normal activity of the *animal* functions, together with the power of generating heat, for both of which a copious supply of oxygen is requisite (§ 67), are mainly dependent upon their presence in the blood in due proportion.

177. *Epidermic and Epithelial Cells.*—Next in independence to the cells or corpuscles floating in the animal fluids, are those which cover the free membranous surfaces of the body, and form the *Epidermis* and *Epithelium.* Between these two structures there is no more real difference, than there is between the Skin and the Mucous membranes. The one is continuous with the other; they are both formed of the same elements; they are cast off and renewed in the same manner; the history of the life of the individual cells of each is nearly identical; but there is an important difference in the purposes, which they respectively serve in the general economy.—The *Epidermis* or Cuticle covers the exterior surfaces of the body, as a thin semitransparent pellicle, which is apparently homogeneous in its texture, is not traversed by vessels or nerves, and was formerly supposed to be an inorganic exudation from the surface of the true Skin, designed for its protection. It is now known, however, to consist of a series of layers of cells, which are continually wearing-off at the external surface, and are being renewed at the surface of the true skin; so that the newest and deepest layers gradually become the oldest and most superficial, and are at last thrown off by slow desquamation. In their progress from the internal to the external surface of the Epidermis, the cells undergo a series of well-marked changes. When we examine the innermost layer, we find it soft and granular; consisting of *nuclei,* in various stages of development into cells, held together by a tenacious semi-fluid substance. This was formerly considered as a distinct tissue, and was supposed to be the peculiar seat of the colour of the skin; it received the designation of *rete mucosum.* Passing outwards, we find the cells more completely formed; at first nearly spherical in shape; but becoming polygonal where they are flattened one against another. As we proceed further towards the surface, we perceive that the cells are gradually more and more flattened, until they become mere horny scales,

their cavity being obliterated ; their origin is indicated, however, by the nucleus in the centre of each. This flattening appears to result from the gradual desiccation or drying-up of the contents of the cells, which results from their exposure to the air. Thus each cell of the Epidermis is developed from the nucleus on the surface of the basement-membrane (§ 160), and is gradually brought to the surface by the development of new cells beneath, and by the removal of the superficial layers ; whilst at the same time it is progressively changed in form, until it is converted into a flattened scale (Fig. 27). This change in form is accompanied by a change in the Chemical composition of the tissue, which seems to be due to the metamorphosis of the contents of the cells into a product identical with that of which hair, horn, nails, hoofs, &c., are composed.—The Epidermis appears solely destined for the *protection* of the true Skin ; both from the mechanical injury and the pain which the slightest abrasion would produce, and from the irritating effects of exposure to the external air, and of changes of temperature. We perceive the value of this protection, when the Epidermis has been accidentally removed. It is very speedily replaced, however ; the increased determination of blood to the Skin, which is the consequence of the irritation, being favourable to the rapid production of Epidermic cells on its surface.

FIG. 27.

Oblique section of Epidermis, showing the progressive development of its component cells ,—*a*, nuclei, resting upon the surface of the cutis vera *f* ; these nuclei are seen to be gradually developed into cells, at *b*, *c*, and *d* ; and the cells are flattened into lamellæ, forming the exterior portion of the epidermis at *e*.

178. Mingled with the Epidermic cells, we find others which secrete colouring matter instead of horn ; these are termed *Pigment-cells.* The most remarkable development of Pigment-cells in the higher animals is on the inner surface of the Choroid coat of the eye, where they have a very regular arrangement, and form several layers, known as the *Pigmentum nigrum.* When examined separately, they are found to have a polygonal form (Fig. 28, *a*), and to have a distinct nucleus (*b*) in their interior. The black colour is given by the accumulation, within the cell, of a number of flat rounded or oval granules, of extreme minutenes, which exhibit an active movement when set free from the cell, and even whilst enclosed within it. The Chemical nature of this black pigment has not yet been made evident ; it has been shown, however, to have a close relation with that of the Cuttle-fish ink or Sepia, which derives its colour from the pigment-cells lining the ink-bag ; and to include a larger proportion of Carbon than most other organic substances,—every 100 parts containing $58\frac{1}{2}$ of this element. The pigment-cells are not always of a simple rounded or polygonal form ; they sometimes present remarkable stellate prolongations, under which form they are well seen in the skin of

FIG. 28.

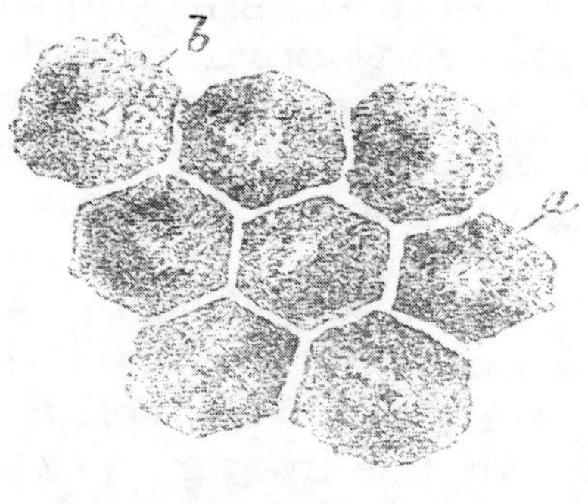

Cells from Pigmentum Nigrum :—*a*, pigmentary granules concealing the nucleus ; *b*, the nucleus distinct.

the Frog (Fig. 52, *c*, *c*.) The gradual formation of these prolongations may be traced in the pigment-cells of the Tadpole during its metamorphosis (Fig. 29). Similar varieties of form are to be met with in the pigmentary cells of Fishes and some Crustacea, which also present a great variety of hues. On the surface of the shell of the common Crab, there is a pigmentary layer composed of regular polygonal cells, resembling those of the choroid (Fig. 28), but containing red colouring matter; and similar cells are found in the margin of the mantle of many shell-forming Mollusca, communicating various hues to the shell, by setting free their coloured secretion to be incorporated with its substance (§ 197). That the development of the Pigment-cells, or at least the formation of their peculiar secretion, is in some degree due to the influence of Light, seems evident from the facts already mentioned (§ 80).

FIG. 29.

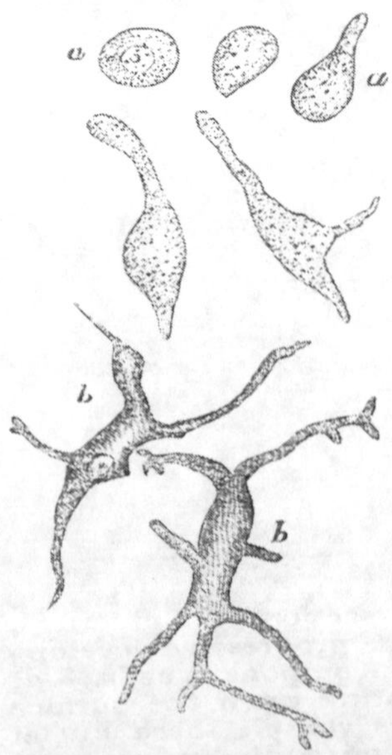

Pigment-cells from tail of *Tadpole*:—*a*, *a*, simple forms of recent origin; *b*, *b*, more complex forms subsequently assumed.

179. The *Epidermic Appendages*, as they are commonly termed,—namely, hoofs, nails, claws, scales, &c.,—seem to partake of the same character. They were formerly regarded, like the epidermis itself, in the light of an inorganic exudation from the true skin,—a sort of hardened glue—which could undergo no subsequent change. It has now been ascertained, however, that they are produced in the first instance by the growth of cells, the contents of which gradually evaporate, so that the walls are left to approximate with each other. In the most recently-formed portion of the *hoof*, the separate cells of which it is composed may be distinctly recognised; they are somewhat flattened against each other, but retain their general rounded form. At a subsequent period, however, there seems to be a deposit of horny matter on the interior of the cell, by which the membranous walls, originally thin and smooth, are thickened as well as roughened. The cells afterwards change their form in a considerable degree, and their walls approximate to each other so closely, that their several boundaries can be scarcely distinguished. The *nails* of the new-born infant have been found to exhibit a similar structure. The thin horizontal laminæ into which they may be split, consist of cells resembling those of the epithelium, which afterwards become much extended and flattened, whilst horny matter is at the same time deposited within them, just as *sclerogen* is in the cells and woody fibres of Plants (§ 148). When this is effected, no further change appears to take place in the tissue; so that it may be regarded as a dead and almost inorganic substance. The cellular character may be traced with perfect distinctness in the spongy portion, or medullary centre, of the stem of the *feather*, which closely resembles the parenchyma of plants; it is also very evident in the soft interior of the Porcupine's quill. In the outer hard envelope, or cortical portion, however, an indistinctly fibrous structure may be seen. The Epidermis and all its appendages are completely *extra-vascular*; that is, they are not traversed by nutritious and absorbent vessels. They have little tendency to spontaneous decomposition; and

therefore they do not require that constant *interstitial* change, which is so characteristic of the tissues that are actively employed in the performance of the vital functions (§ 46). In this respect, then, they closely resemble the heart-wood of the tree, and the horny sheaths or solid stony axes of the Polypifera ; which are originally formed by a consolidation of living tissue, but which, when once fully developed, undergo little or no subsequent change.

180. Under the same head it is proper to consider *Hair*, which is truly an 'epidermic appendage,' although not developed upon the external surface, but in the interior of a follicle formed by a depression of the true skin. This follicle is lined by a continuation of the Epidermis, the cells of which are developed in peculiar abundance from a spot at its deepest portion ; the dense exterior of the cluster thus formed being known as the *bulb* of the hair, while the softer interior is termed its *pulp*.—Although the Hairs of different animals vary considerably in the appearances they present, we may generally distinguish in them two elementary parts, corresponding with those which we meet with in the stem of a feather ; namely, a *cortical* or investing substance, of a dense horny texture ; and a *medullary*, or pith-like substance, usually of a much softer character, occupying their interior. The fullest development of both substances is seen in the spiny hairs of the Hedgehog, and in the quills of the Porcupine, which are but hairs on a magnified scale ; they are very characteristically seen, however, in some of the hairs of the Sable (Fig. 30, B), whose medulla is composed of large rounded cells, whilst the cortical substance is obviously composed of flattened epidermic scales, arranged in an imbricated manner. In some instances, almost the entire hair seems made up of cortical substance ; whilst in others the medullary predominates, as in those of the Musk-deer (Fig. 30, A). In the hair of the Mouse and other small Rodents, the cortical substance forms a tube, which we see crossed at intervals by partitions that are sometimes complete, sometimes only partial ; these are the walls of the single or double line of cells, of which the medullary substance is made up. In the Human hair, we do not find these structures so characteristically developed ; they are still readily distinguishable, however, in a transverse section ; the medullary matter presenting a granular arrangement and often containing pigmentary molecules ; to which the cortical sheath forms a thin but dense and apparently homogeneous envelope, its origin in epidermic scales being still indicated by wavy lines crossing its surface irregularly. The hair of Man has been commonly reputed to be tubular ; but this is proved by microscopical examination to be seldom or never the case. The mistake seems to have arisen from a misinterpretation of the appearance of a dark band in the interior of the hair, when viewed by

FIG. 30.

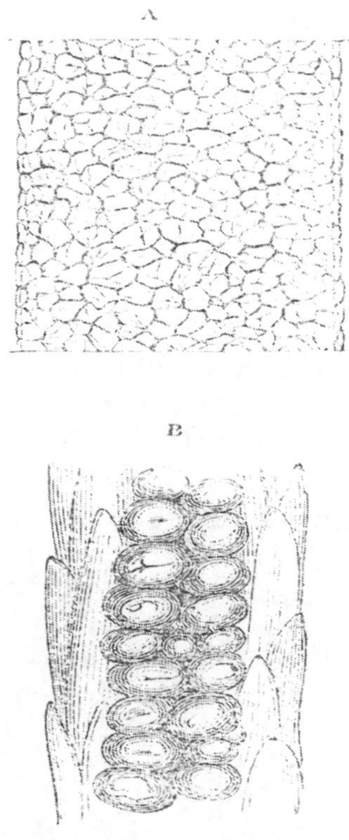

Structure of Hair:—A, hair of Musk-deer, consisting almost entirely of polygonal cells; B, hair of Sable, showing large rounded cells in its interior, covered by imbricated scales or flattened cells.

transmitted light; this band being really due in part to the refraction of light by the cylindrical surface, and in part to the presence of pigmentary matter in the central portion of the shaft.—The cortical envelope of hairs is a continuation of the outer scaly layers of the epidermis that lines the follicle; whilst the medullary is derived from the deeper stratum, whose cells are produced in unusual abundance at its cœcal extremity: and it is by the constant development of new cells at this point, that the continual growth of the hair is kept up.

181. The *Shells* of Mollusca, also, are to be regarded in the light of 'epidermic appendages;' and were formerly supposed to be formed, like the Epidermis itself, by an *exudation* from the surface of the true skin. It has been demonstrated, however, by microscopic inquiry, that their structure is as much organised as that of the Epidermis is now known to be; and that the substance of the shell is formed by the consolidation of epidermic cells by calcareous matter which the cells imbibe into their interior.—It will be more convenient, however, to consider this structure in connection with that of other forms of Shell, and of Bone (§ 197).

182. The *Epithelium* may be designated as a delicate cuticle, covering all the free *internal* surfaces of the body; and apparently designed, in some instances, simply for their protection; whilst in other cases, as we shall presently find, it serves purposes of far greater importance. The Epithelial cells, being always in contact with fluids, do not dry up into scales like those of the Epidermis; and they differ from them also in regard to the nature of the matter which they secrete in their interior. In this respect, however, the Epithelial cells of different parts are unlike one another, fully as much as any of them are unlike the cells of the Epidermis; for we shall find that *all* the secretions of the body are the product of the elaboration of Epithelium-cells; and consequently there are as many varieties of endowment in these important bodies, as there are varieties in the result of their action.—The Epithelium covering the Serous and Synovial membranes, and forming the lining of the blood-vessels, is composed of flattened polygonal cells (resembling those shown in Fig. 31), lying in apposition with each other, so as to form a kind of pavement; hence this form is termed *pavement-* or *tesselated*-Epithelium. There is no reason to believe that it possesses any active endowments in these situations: since it does not appear to be concerned in the elaboration of any peculiar secretion (§ 167). The Epithelium of the Mucous membranes and their prolongations are found under two principal forms, the *tesselated*, and the *cylindrical*. An example of the Tesselated form is shown in Fig. 31, which shows the flattened epithelium-cells of the mucous membrane of the mouth, as they are frequently met with in saliva. The Cylinder-epithelium is very differently constituted. Its component cells are cylinders, which are arranged side by side; one extremity of each cylinder resting upon the basement-membrane, whilst the other forms part of the free surface. The perfect cylindrical form is only shown, when the surface on which the cylinders rest is flat or nearly so. When it is *convex*, the lower ends or basements of the cells are of

FIG. 31.

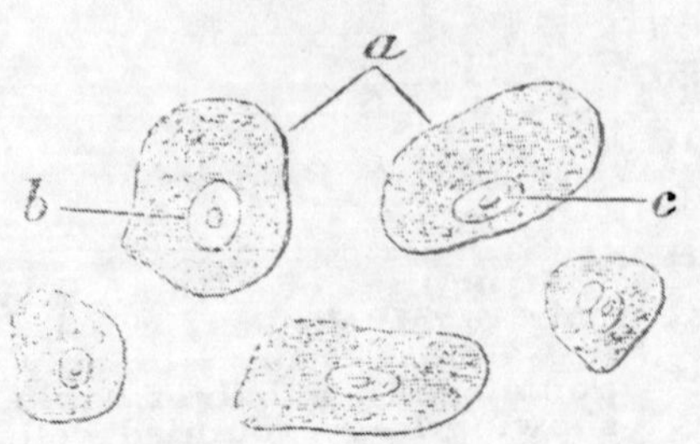

Separated Epithelium-cells, *a*, with nuclei, *b*, and nucleoli, *c*, from mucous membrane of mouth.

much smaller diameter than the upper or free extremities ; and thus each has the form of a truncated cone rather than of a cylinder, as is well seen in the cells which cover the *villi* of the intestinal canal (Fig. 32, *a*). The two forms of Epithelium pass into one another at various points, and various transitional forms are then seen.

FIG. 32.

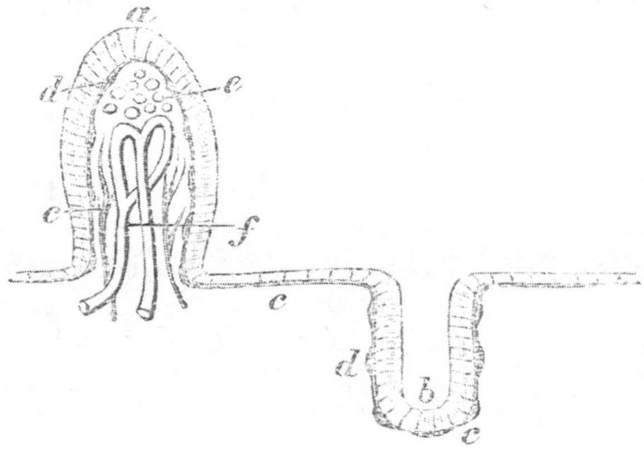

Diagram of mucous membrane of jejunum, when Absorption is not going on ; *a*, protective epithelium of a villus ; *b*, secreting epithelium of a follicle ; *c*, *c*, *c*, primary membrane, with its germinal spots or nuclei, *d*, *d* ; *e*, germs of absorbent vesicles ; *f*, vessels and lacteals of villus.

183. Both these principal forms of Epithelial cells are frequently observed to be fringed at their free margins with delicate filaments, which are termed *Cilia ;* and these, although of extreme minuteness, are organs of great importance in the animal economy, through the extraordinary motor powers with which they are endowed. The form of the ciliary filaments is usually a little flattened, and tapering gradually from the base to the point. Their size is extremely variable ; the largest that have been observed being about 1-500th of an inch in length, and the smallest about 1-13,000th. When in motion, each filament appears to bend from its root to its point, returning again to its original state, like the stalks of corn when depressed by the wind ; and when a number are affected in succession with this motion, the appearance of progressive waves following one another is produced, as when a cornfield is agitated by successive gusts. When the ciliary motion is taking place in full activity, however, nothing whatever can be distinguished but the whirl of particles in the surrounding fluid ; and it is only when the rate of movement slackens, that the shape and size of the cilia, and the manner in which their stroke is made, can be clearly seen. The motion of the cilia is not only quite independent (in all the higher animals at least) of the will of the animal, but is also independent even of the life of the rest of the body ; being seen after the death of the animal, and proceeding with perfect regularity in parts separated from the rest. Thus, isolated epithelium-cells have been seen to swim about actively in water, by the agency of their cilia, for some hours after they have been detached from the mucous surface of the nose ; and the ciliary movement has been seen fifteen days after death in the body of a Tortoise, in which putrefaction was far advanced. In the gills of the River-Mussel, which are among the best objects for the study of it, the movement endures with similar pertinacity.

FIG. 33.

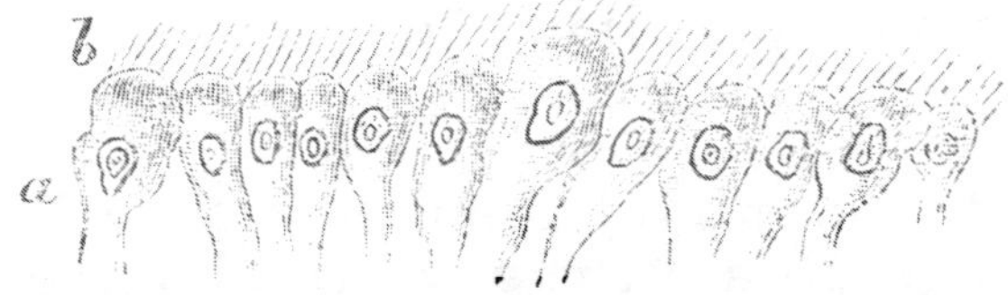

Ciliated Epithelium ;—*a*, nucleated cells, resting on their smaller extremities ; *b*, cilia.

184. The purpose of this Ciliary movement is obviously to propel fluids over the surface on which it takes place ; and it is consequently limited in the higher animals to the internal surfaces of the body, and

always takes place in the direction of the outlets, towards which it aids in propelling the various products of secretion. The case is different, however, among animals of the lower classes, especially those inhabiting the water. Thus the external surface of the gills of Mollusks, Tadpoles, &c., is furnished with cilia ; the continual movement of which renews the water in contact with them, and thus promotes the aeration of the blood. In the lower Mollusca, and in many Zoophytes, which pass their lives rooted to one spot, the motion of the cilia serves not merely to produce currents for respiration, but likewise to draw into the mouth the minute particles that serve as food. And in the free-moving Animalcules of various kinds, the cilia are the sole instruments which they possess, not merely for producing those currents in the water which may bring them the requisite supply of air and food, but also for propelling their own bodies through the liquid. This is the case, too, with many larger animals of the class Acalepha (Jelly-fish), which move through the water, sometimes with great activity, by the combined action of the vast numbers of cilia that clothe the margins of their external surfaces. In these latter cases it would seem as if the ciliary movement were more under the control of the will of the animal, than it is where concerned only in the organic functions. In what way the will can influence it, however, it does not seem easy to say ; since the ciliated epithelium-cells appear to be perfectly disconnected from the surface on which they lie, and cannot, therefore, receive any direct influence from their nerves.—Of the cause of the movement of the Cilia themselves, no account can be given ; they are usually far too small to contain even the minutest fibrillæ of muscle ; and we must regard them as being, like those fibrillæ, organs *sui generis,* having their own peculiar endowment,—which is, in the higher animals at least, that of continuing in ceaseless vibration, during the whole term of the life of the cells to which they are attached. The length of time during which the ciliary movement continues after the general death of the body, is much less in the warm-blooded than in the cold-blooded animals ; and in this respect it corresponds with the degree of persistence of muscular irritability, and of other vital endowments.

185. The simplest office which the Epithelium-cells of Mucous membranes perform, appears to be that of elaborating their protective *Mucus* (§ 169); but where these membranes are prolonged into the follicles or tubuli of which *glandular* structures are essentially composed (CHAP. XVI.), we find that the epithelial cells of these follicles (Fig. 32, *b*) are the real agents in the process of Secretion which the glands carry on ; drawing from the blood, as the materials of their own growth, the substances which the glands are destined to separate ; and setting free these products, by their own death and decay, at the appropriate time and place. The epithelial cells of different glandular surfaces differ widely from one another, in regard to the kind of matter which they appropriate and assemble in their cavities, although the nature of their walls is probably the same throughout. Thus we find biliary matter and oil, easily recognisable by their colour and refracting power, in the cells of the liver ; milk in the cells of the mammary gland ; sebaceous or fatty matter in the cells of the sebaceous follices of the skin ; and so on. All these substances are derived from the blood ; being either contained in it previously, or being elaborated from its constituents by a simple process of transformation,—as, for example,

that which converts the albumen of the blood into the caseine of milk. Hence they may be considered as the peculiar aliments of the several groups of cells ; whose acts of *nutrition* are the means of drawing them off, or *secreting* them, from the general circulating fluid. When they have attained their full growth, and accomplished their term of life, their walls either burst or dissolve away, and thus the contents of the cells are delivered into the cavity, or upon the surface, at which they are required. Now as all the canals of the glands open either directly outwards upon the surface, or into cavities which communicate with the exterior, it is evident that the various products of the action of these epithelial cells must be destined to be cast forth from the body. This we shall find to be the case ; some of them, as the bile and urine, being excretions, of which it is necessary to get rid by the most direct channel ; whilst others, like the tears, the saliva, the gastric fluid, the milk, &c., are separated from the blood, not so much for *its* purification, but because *they* are required to answer certain purposes in the economy.—We shall find, too, that the cells in which the *Spermatozoa* are developed for the fertilisation of the ova (CHAP. XVIII.), are the offspring of the epithelial cells lining the tubes or follicles of the Spermatic gland.

186. *Aggregated Cells, forming Permanent Tissues.*—We now pass on to consider those Cells, which enter as component elements into the solid and permanent fabric of the body, and which do not take so active a part in its vital operations. These we shall find to be usually more or less closely connected together, either by a general enveloping membrane, or by an *intercellular substance*, which is interposed between their walls, and holds them together by its adhesive properties. This generally presents no distinct traces of organisation ; and usually consists of Gelatine, or of a substance allied to it in composition. The proportion which this substance bears to the cells may vary in different cases ; and very different characters may thus be presented by tissues made up of the same elements. Thus the subjoined figure represents a portion of one of the animal layers included

FIG. 34.

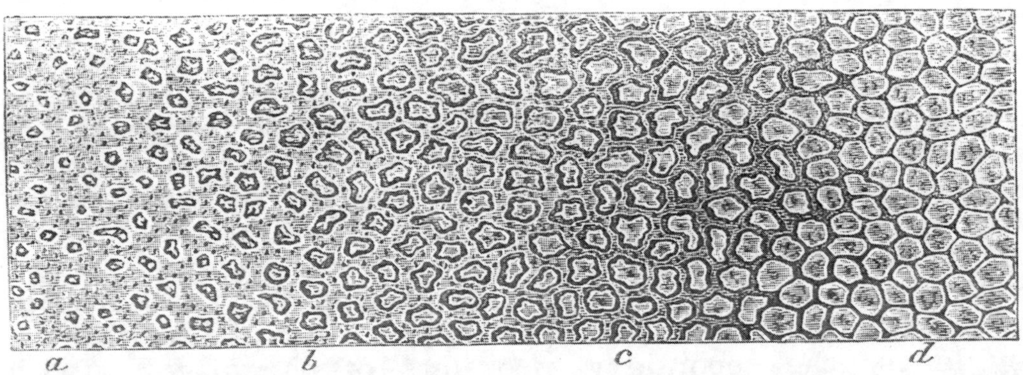

Portion of shell-membrane, showing the origin of cells in the midst of horny intercellular substance ; *a*, nuclei ; *b*, incipient cells ; *c*, the same further advanced, but separated by intercellular substance; *d*, the cells become polygonal by mutual pressure.

between the calcareous laminæ of a bivalve shell ; in which we see on the one side (*a*) a number of nuclei or incipient cells, scattered through a bed of homogeneous intercellular substance, and bearing but a very small proportion to it ; while the opposite end (*d*) exhibits a set of polygonal cells, in close contact with each other, the intercellular substance being only represented by the thick dark lines, which mark the boundaries of the cells, and which are rather thicker at the angles of the latter. Between these two extremes, we observe every stage of transition.—The presence of a very large amount of intercellular substance, through which minute cells are scattered at considerable intervals (Fig. 34 *a*), is characteristic of various forms of Cartilage; and more particularly of that soft semi-cartilaginous structure, of which the Jelly-fish are for the most part composed. In other forms of cartilage, we find the cells more developed, and in closer proximity to each other, the proportion of the intercellular substance being at the same time diminished (as seen at *b* and *c*, Fig. 34); but it is not often, save in embryonic structures, that we find the cells in such close proximity, and the intercellular substance so nearly wanting, as at *d*. Such examples do occasionally present themselves, however, even in the soft tissues. Thus the *chorda dorsalis*, which replaces the vertebral column in the lowest Fishes, and of which the analogue is found in the embryos of the higher Vertebrata, is made up of a structure of this kind. The true Skin, in the Short Sun-fish, is replaced by a similar layer of cellular tissue, which extends over the whole body, varying in thickness from one-fourth of an inch to six inches. And in the *Amphioxus* (§ 321), a considerable portion of the fabric is made up of a similar parenchyma.

187. Now we shall find that one method, by which the requisite firmness and solidity are given to the animal fabric, consists in the deposition of earthy substances in the interior of such cells, by a peculiar secreting action of their own. Thus in Shell, we find them completely filled up with carbonate of lime; and in the enamel of Teeth, chiefly with phosphate of lime. When this is the case, there is often a tendency to a *coalescence* of the cells, by the removal of the whole intercellular substance from between them, and then by the obliteration of the partitions themselves; so that the solid mass appears homogeneous, and its cellular origin can only be detected by watching its development, and by observing the various stages of transition. This is the case, for example, in regard to the shells of many Mollusca, especially those of a *porcellanous* character; in which the amount of animal matter is very small.

188. We often find, however, a coalescence between the *cavities* of cells, as in Plants (§ 152), for the production of straight or anastomosing tubes. Thus the smaller Blood-vessels of Animals seem to originate in rows of cells, the cavities of which have run together by the obliteration of the transverse partitions; and the persistent nuclei of such cells may be generally brought into view in the walls of the capillaries. And the same appears to be the origin of the tubular fibres of Muscular and Nervous tissue, which contain the elements characteristic of those tissues; these elements,—the fibrillæ of muscle, and the granular pith of the nerve-tube,—being evidently the secondary products of parent-cells, which seem to remain as their investing tubuli, the original nuclei being often to be seen in their walls (§ 226 and 238).—Besides these changes, the original cells may often undergo marked alterations of *form;* and this quite independently of any pressure to which they may be subject. Thus the pigment-cells, as already mentioned (§ 178), frequently exhibit a curious *stellate* form; arising from the development of radiating prolongations, which are put forth from the original spheroid. And we

sometimes find cells undergoing such an elongation, that they first become *fusiform* or spindle-shaped, and are at last converted into solid fibres (§ 166).

189. We now proceed with the description of the various tissues in the Human body, which are composed of cells united or transformed in the foregoing manner ; and we shall commence with *Adipose* or *Fatty* tissue, which may be considered as a sort of link, connecting the permanent tissues with those which are more actively concerned in the processes of Nutrition, Secretion, &c. The Adipose tissue is composed of isolated cells, which have the power of appropriating fatty matter from the blood, precisely in the same manner as the secreting cells appropriate the elements of bile, milk, &c. These cells are sometimes dispersed in the interspaces of the Areolar tissue ; whilst in other cases they are aggregated in distinct masses,—constituting the proper Adipose tissue. The individual fat-cells always present a nearly spherical or spheroidal form ; sometimes, however, when they are closely pressed together, they become somewhat polyhedral, from the flattening of their walls against each other. Their intervals are traversed by a minute net-work of blood-vessels, from which they derive their secretion ; and it is probably by the constant moistening of their walls with a *watery* fluid, that their contents are retained without the least transudation, although they are quite fluid at the temperature of the living body. If the watery fluid of the cell-walls of a mass of Fat be allowed to dry up, and it be kept at a temperature of 100°, the escape of the contained oily matter is soon perceptible. By this provision, the fatty matter is altogether prevented from escaping from the cells of the living tissues, by gravitation or pressure ; and as it is not itself liable to undergo change when secluded from the air, it may remain stored up, apparently unaltered, for a period only limited by the demand for it in the system.—The consistency, as well as the Chemical constitution, of the fatty matter contained in the Adipose cells, varies in different animals, according to the relative proportions of three component substances, which may be distinguished in it,—Stearine, Margarine, and Oleine. The two former are solid when isolated, and the latter is fluid ; but at the ordinary temperature of the warm-blooded animal, they are dissolved in it. These fatty matters are abundantly supplied by the Vegetable kingdom, and are found to exist largely in substances which were not previously supposed to contain them ; and it is not requisite, therefore, to suppose that Animals usually elaborate them by any transforming process from the elements of their food. The portion of fatty matters separated from the circulating fluid to form the Adipose tissue, is only that which can be spared from the other purposes to which they have to be applied ; and hence the production of this tissue depends in part upon the amount of Fatty matter taken in as food. This is not entirely the case, however, as some have maintained ; for there is sufficient evidence that animals *may* produce fatty matter by a process of chemical transformation, from

FIG. 35.

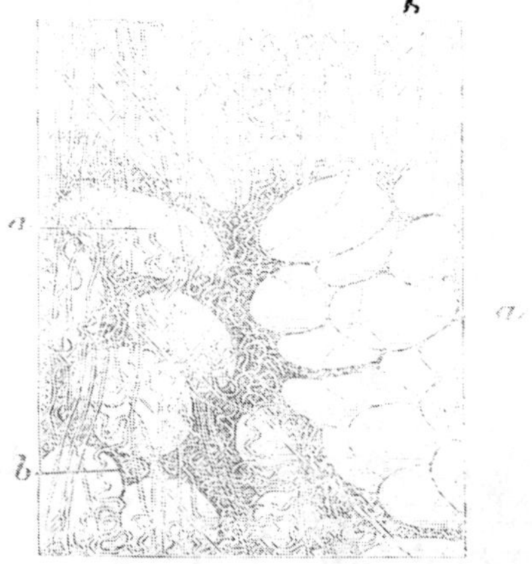

Areolar and Adipose tissue ; *a, a,* fat-cells ; *b, b,* fibres of areolar tissue.

the starch or sugar of their food, when there is an unusual deficiency of it in their aliment (CHAP. X.).

190. The development of Adipose tissue in the animal body appears to answer several distinct purposes. It fills up interstices, and forms a kind of pad or cushion for the support of moveable parts; and so necessary does it seem for this purpose, that, even in cases of great emaciation, some fat is nearly always found to remain. It also assists in the retention of the animal temperature by its non-conducting power; and we accordingly find a thick layer of it, in those warm-blooded mammals that inhabit the seas, either immediately beneath their skin, or incorporated with its substance. And it also serves as a reservoir of combustible matter, at the expense of which the respiration may be maintained when other materials are deficient: thus we find that the respiration of hybernating animals is kept up, during the period when they cease taking food (§ 94), by the consumption of the store of fat which was laid up in their bodies, previously to their passing into that state; and it is also to be noticed that herbivorous animals, whose food is scanty during the winter, usually exhibit a strong tendency to such an accumulation, during the latter part of the summer, when their food is most rich and abundant, in order to supply the increased demand created by the low external temperature of the winter season. Other circumstances being the same, it appears that the length of time during which a warm-blooded animal can live without food, depends upon the quantity of fat in its body; for the rapid lowering of its temperature, which is the immediate cause of its death (§ 93), takes place as soon as the whole of this store has been exhausted. It has been shown by the experiments of Matteucci, that fatty matter will pass by Endosmose through animal membrane, when a slightly alkaline liquid is on the other side; and it would not seem improbable that it is in this mode that the contents of the fat-cells are progressively taken back into the blood, in proportion as the quantity of fatty matter which it usually contains is expended.

191. In the simpler forms of *Cartilage*, we have an example of a tissue of remarkable permanence, composed entirely of cells scattered through an inter-cellular substance. This substance has a close resemblance to Gelatine in composition and properties, but is not identical with it; and it has received the distinguishing appellation of *Chondrine*, which marks it as the solidifying ingredient of Cartilage. It requires longer boiling than Gelatine for its solution in water; but the solution fixes into a jelly in cooling, and dries by evaporation into a glue that cannot be distinguished from that of gelatine. When a pure Cellular Cartilage is examined microscopically, its cells are seen to lie, sometimes singly, and sometimes in clusters of two, three, or four, in cavities excavated in the intercellular substance; and these occur at very variable distances. From the various appearances which may be observed in the same cartilage, at different stages of its growth, it would appear that the component cells multiply by the *doubling* process already described (§ 157); that they then separate from one another, each of them drawing towards itself (as it were) an envelope of intercellular substance; and that, by the repetition of the same process, the number of cells in the cartilage may be indefinitely multiplied. Various stages of this history are shown in the accompanying figure (Fig. 36), which is taken from a section of the cartilaginous

branchial ray of the tadpole of the *Rana esculenta,* or Edible Frog. This primitive *cellular* organisation is retained in some Cartilages through the whole duration of their existence ; as, for example, in most of the *articular* cartilages which cover the ends of bones where they enter into joints.—

FIG. 36.

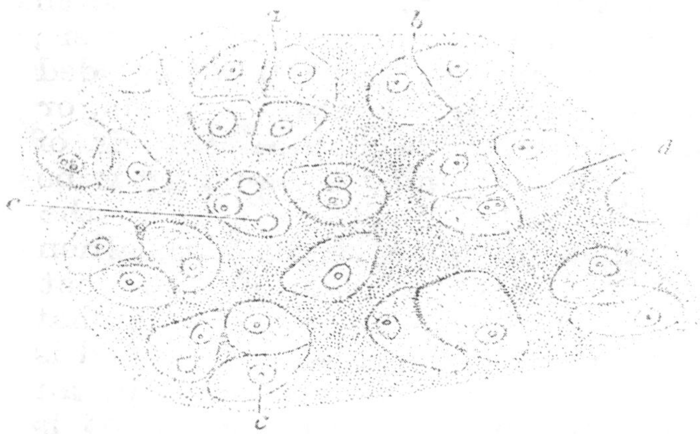

Section of the Branchial cartilage of Tadpole; *a*, group of four cells, separating from each other; *b*, pair of cells in apposition ; *c*, *c*, nuclei of cartilage cells; *d*, cavity containing three cells.

FIG. 37.

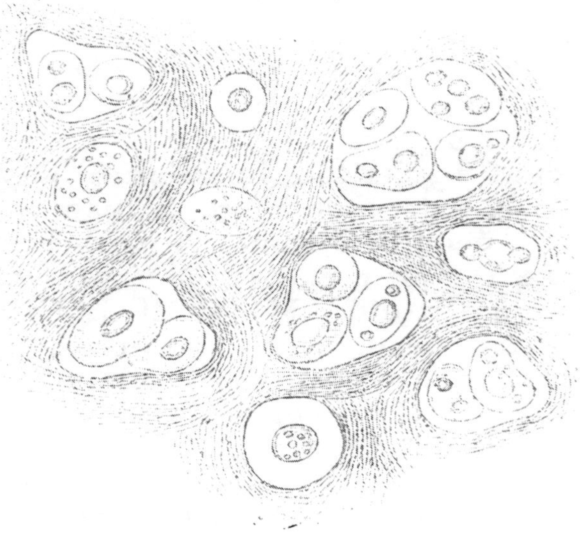

Section of Fibro-Cartilage ; showing disposition of Cartilage-cells in areolæ of fibrous tissue.

But the intercellular substance, instead of being homogeneous, may have a fibrous character ; the tissue called *Fibro-Cartilage* is thus produced, and this may be either elastic or non-elastic, according as the yellow or the white form of fibrous structure prevails (Fig. 37). In some instances, the fibrous structure is so predominant over the cellular, that the tissue has rather the character of a ligament than of a cartilage. The white fibrous structure is seen in all those cartilages, which are destined not merely to sustain pressure, but also to resist tension ; this is the case especially in the substances, which intervene between the vertebræ, and which connect the bones of the pelvis. The yellow-fibrous tissue is introduced where more elasticity is required, as in the epiglottis, and in the concha of the ear.

192. It is certain that the substance of the permanent cellular Cartilages is *not* permeated, in a state of health, even by the minutest nutrient vessels ; none such being brought into view under the highest magnifying power. They are, however, surrounded by vessels, which form large dilatations at their edges, or spread over their surfaces (Fig. 38) ; and it is by the fluid which is drawn from them by the Cartilage-cells, that the latter are nourished. The nutrition of a mass of Cartilage thus seems to bear a strong resemblance to that of the thick fleshy Sea-weeds, which are in like manner composed entirely

FIG. 38.

Dilated loops of capillary vessels between Articular Cartilage and attached Synovial membrane.

of cells, with intercellular substance disposed between them in greater or smaller amount (Figs. 69, 70). The cells in nearest proximity to the nutrient fluid, draw from it the requisite materials, and transmit these to the cells in the interior of the mass, receiving a fresh supply in their turn from the source in their own neighbourhood.—The Cartilaginous tissue, however, appears to be more removed than almost any other in the body, from the general tide of nutritive action. Its properties are simply of a physical character; and they are not impaired for a long time after the death of the tissue, its tendency to decomposition being very slight, so long as it is exposed to ordinary temperatures. It is protected by its toughness and elasticity from those mechanical injuries, to which softer or more brittle tissues are liable; and consequently it has little need of any active power of reparation.

193. With the subject of Cartilage, that of *Bone* seems naturally to connect itself. We shall find this, however, to be a fabric of great complexity, peculiar to Vertebrated animals; being especially adapted, in its mode of growth, to the changing dimensions of their *internal* skeletons. And it will be convenient to consider, in the first instance, the principal modes in which the tissues of Invertebrated animals are condensed by deposits of solid matter, in order that they may become subservient to those purposes of mechanical support and protection, which are required in various degrees by the different tribes of that division of the Animal Kingdom.

194. Commencing with ZOOPHYTES, we find that in some of their more delicate forms the requisite support and protection are derived from the consolidation of the *external* or tegumentary membrane by a deposit of horny matter; this is the case, for example, more or less, with most of the compound *Hydroida* (§ 290). In other instances, we find this horny deposit taking place in the *centre* of the stem, as in the common *Gorgonia* (Sea-Fan), *Antipathes* (Black Coral), &c. There are many Zoophytes of the *Alcyonian* group (§ 292) which have a spongy structure; and in these, as in true *Sponges*, we find the tissue strengthened by the presence of a number of detached crystalline *spicules*, sometimes composed of carbonate of lime, sometimes of siliceous matter, which remind us of the raphides of Plants (§ 23). These spicules present a great variety of forms, and are interspersed without any obvious arrangement among the elements of the softer tissues; thus we even find them in the soft flesh which clothes the horny stems of Gorgonia and its allies. The stony *Corals* are formed, however, by the consolidation of the soft animal tissue with a deposit of carbonate of lime; and this, again, may be either external or internal, the two opposite conditions presenting each other in animals which are otherwise closely allied. Thus in the group of Alcyonian Zoophytes (§ 292), we find the Organ-pipe Coral (*Tubipora musica*) to be composed of a series of calcareous tubes formed by the deposition of carbonate of lime in the integument of the polypes; whilst the common Red Coral, is formed by a similar consolidation of the central axis which connects them. So, again, in the Flustra, Eschara, and others of the *Bryozoa* (§ 298), we find the coral chiefly formed by the calcification of the whole external integument; whilst in the massive corals formed by Actiniform Polypes (§ 291), it is only in the soft tissue at their base, and in that which intervenes between the polypes, that the consolidation takes

place. Thus we see that among these lowest members of the Animal kingdom, neither the material nor the position of the skeleton is by any means constant; and we may consider both the external skeletons of the Articulata and Mollusca, and the internal skeletons of Vertebrated animals, as here pre-figured or sketched out. These last seem especially foreshadowed by the *Isis hippuris,* in which the coral stem acquires a jointed character from the alternation of calcareous and of horny deposit in the soft tissue which forms its basis.

195. In the class of ECHINODERMATA, the skeleton is generally *dermal;* being formed by the union of the fibrous tissue of the true skin with calcareous matter. Although usually ranked as an *external* skeleton, it is not altogether so; for it is covered with a contractile membrane, by the agency of which the movements of its different parts (where they are capable of motion), are effected. Thus in the *Echinus* (§ 295) the body is enclosed in a shell of somewhat globular form, having an opening at each pole, one for the mouth and the other for the anus; this shell is made up of a number of plates more or less regularly hexagonal (Fig. 39, *a*), on the outer surface of which are little tubercles for the articulation of the *spines,* which serve as instruments of locomotion and defence. The substance of the shell and of the spines is exactly alike, being a sort of *calcified areolar tissue,* formed by the union of carbonate of lime with gelatinous fibres (§ 164), which are so disposed as to leave multitudes of *areolæ* or minute interspaces freely communicating with each other (Fig. 39, *b*). The arrangement of this calcareous network in the spines, is most varied and elaborate; and causes thin sections of them to be among the most beautiful of all microscopic objects. The external and internal surface of each plate, in the shell of the living Echinus, is covered with a membrane from which its nutrition is derived; this membrane, also, dips down into the spaces between the adjacent plates. A similar membrane covers and encircles the spines; and it also connects these with the shell, being continuous with the membrane that envelops the latter. The substance of which the shell is composed, although of extreme delicacy, is of great permanence, and does not exhibit the slightest tendency to decay, however long it is preserved; so that, when once consolidated, it appears to undergo no further change in the living animal. The growth of the animal, however, requires a corresponding enlargement of its enveloping shell; and this is provided for by the simple process of adding to the surface, through the subdivision of the whole shell into component plates. For by the addition of new matter at the edge of each plate, through the consoli-

FIG. 39.

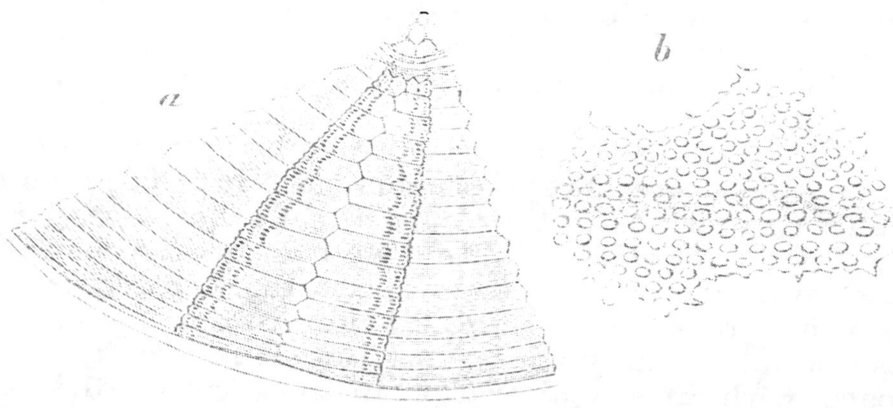

Portion of the shell of the Echinus, showing at *a* the constituent plates; with the calcified areolar tissue, of which they are composed, at *b*.

dation of a portion of the soft membrane that intervenes between the adjacent plates, the whole shell is enlarged, without losing its globular form; and at the same time it is strengthened in a corresponding degree, by the consolidation of the soft tissue covering the *surface* of each plate. In like manner, the spines are enlarged and lengthened by the progressive formation of new layers, each on the exterior of the preceding; so that a transverse section exhibits a number of concentric rings, like those of an Exogenous tree (§ 280).—Thus even in the growth of this complex and elaborate structure, we recognize the principle of *superficial deposit*, which we shall find to be universal amongst the hard parts of the Invertebrata; notwithstanding that, at first sight, it would have appeared impossible to provide on this plan for the gradual enlargement of a globular shell, completely enclosing the animal, and therefore required to keep pace with the latter in its rate of increase.

196. Among the MOLLUSCA we find the body sometimes altogether destitute of solid organs of support, protection, or locomotion,—as is the case, for example, in the Slug: and the movements are feeble and the habits inert, the muscles having no fixed points for their attachment, and acting without any of the advantages of leverage. In other cases, we find the body more or less completely protected by a Shell; which is sufficiently large in some instances to cover the soft parts completely; whilst in others, it affords them only a partial investment. The plan on which this shell is formed, however, is very different from that which has just been described, being much less complex. The *Univalve* shells, or those formed in one piece, are always of a conical form: the axis of the cone being sometimes straight, as in the Limpet; in other cases being spirally coiled, as in the Snail: and through the open base of this cone the animal can project its moveable parts. When its increasing size requires additional accommodation, it is obvious that an addition to the large end of the cone will increase its diameter and its length at the same time; so as to afford the required space, without any alteration in the form or dimensions of the older and smaller portions of the cone. This last, indeed, is frequently quitted by the animal, and remains empty; being sometimes separated from the later portions, by one or more partitions thrown across by the animal,—as is seen especially in the Nautilus and other chambered shells. Besides the new matter added to the mouth of the shell, a thin layer is usually formed over its whole interior surface; so that the lining of the new part is continuous with that of the old.—In the Bivalve shells, we trace this mode of increase without any difficulty; especially in such shells as that of the Oyster, in which the successive laminæ remain distinct. Each lamina is *interior* to the preceding, being formed on the living surface of the animal; but it also projects beyond it, so as to enlarge the capacity of the shell; and as the separation of the valves affords free exit to those parts of the animal, which are capable of being projected beyond the shell, there is obviously no need of any other provision to maintain the shell in its natural form.—Thus in the shells of the Mollusca, increase takes place at the surfaces and edges only.

197. The proportion of organic and calcareous matter in Shell differs considerably in the various tribes. The former is sometimes present in such small amount, that it can scarcely be detected; and the condition of the calcareous matter then obviously approaches that of a crystalline

deposit. But in other instances, the animal basis is very obvious; remaining as a thick consistent membrane, after all the calcareous matter has been dissolved away by an acid. This membrane is formed of an aggregation of cells arranged with great regularity (Fig. 40, *a*) ; the cavities of which are filled with carbonate of lime in a crystalline state. The form of the cells approaches the hexagonal ; their diameter varies in different shells from 1-100th to 1-2800th of an inch ; their thickness also is extremely

FIG. 40.

a 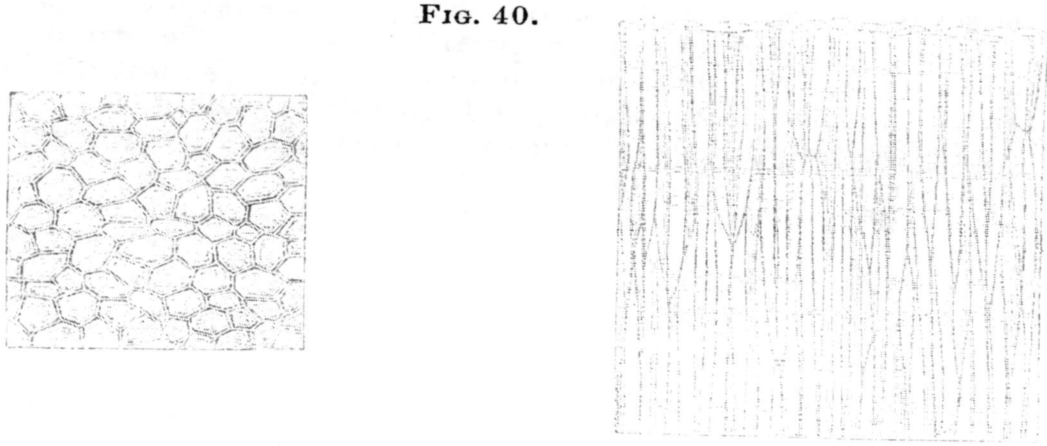 b

Prismatic cellular structure of shell of Pinna ; *a*, surface of lamina; *b*, vertical section.

variable, even in different parts of the same shell. Thus we sometimes meet with a lamina of such tenuity, as not to measure a hundredth of an inch in thickness ; whilst in other instances, a single layer may have a thickness of half an inch, or even (in certain large fossil species) of an inch or more. In this case, the cells, instead of being thin flat scales, like the pavement-epithelium (§ 182), are long prisms, with their walls flattened against each other ; and the appearance which is then presented by a vertical section of them, is represented in Fig 40, *b*. The long prismatic cells are there seen to be marked by delicate transverse striæ ; and these, taken in connection with other indications, appear to show, that every such prism is in reality formed by the coalescence of a pile of flat cells, resembling those which are seen in the very thin laminæ just described; so that the thickness of the layer depends upon the number of the cellular laminæ, which have coalesced to form its component prisms. This character is of interest, as representing on a magnified scale a corresponding appearance in the enamel of human Tooth, which we shall presently find to be formed upon the very same plan (§ 214).—We are to regard this kind of shell-substance, therefore, as formed by the secreting action of the epithelial cells covering the *mantle* of the animal ; which membrane, though it answers in position to the skin, has the soft, spongy, glandular character of a mucous membrane. These draw calcareous matter into their cavities, as a part of their own process of growth ; this matter being supplied from the fluids of the vascular surface beneath. Now when these calcigerous cells are separated by intercellular substance (Fig. 34), they remain distinct through the whole of their lives, and they form by their cohesion a tenacious membrane, that retains its consistency after the removal of the calcareous matter. But this is only the case in certain groups of shells, chiefly belonging to the bivalve division. When the intercellular substance is wanting, and the cells come into close contact, their partitions become indistinct, on account of their extreme tenuity; and not unfrequently a *fusion* of the whole substance appears to take place, by the dissolution of the original cell-walls, so that it becomes more or less homogeneous,—traces of the original cellular structure being here and there distinguishable (§ 187).

198. Sometimes where this fusion has taken place, so as to obliterate the original cell-structure, we find the almost homogeneous substance traversed by a series of tubuli; these not being arranged, however, on any very definite plan, but forming an irregular net-work. The tubes vary in size from 1-2000th to 1-20,000th of an inch; but their general diameter, in the shells in which they most abound, is 1-4500th of an inch. Although it might be supposed that this structure is destined to convey nutrient fluid into the substance of the shell, yet there is no evidence that such is the fact; and, on the contrary, there is ample evidence, that, even in shells most copiously traversed by these tubuli, no processes of interstitial increase or removal take place. The *permanent* character of the substance of all Shells, when once it is fully formed, is as remarkable as that of Coral; and as the adaptation of their size to that of the animals to which they belong, is entirely effected by additions to their surfaces and edges, no interstitial deposit can have a share in producing it. Still there is a difference, well known to the Conchologist, between a *dead* shell and one which remains in connection with the living animal; so that it is probable that the organic basis of shell has its composition preserved by communication with the nutrient fluid of the mantle that lines its inner surface, so long as the latter can supply it.—Such are some of the principal forms of Shell-tissue; there are, however, very numerous varieties in the structure of the shells of different groups of Mollusca, and these varieties are occasionally so characteristic as to allow of the determination of the tribe to which the shell belonged from the microscopic examination of a minute fragment.*

FIG. 41.

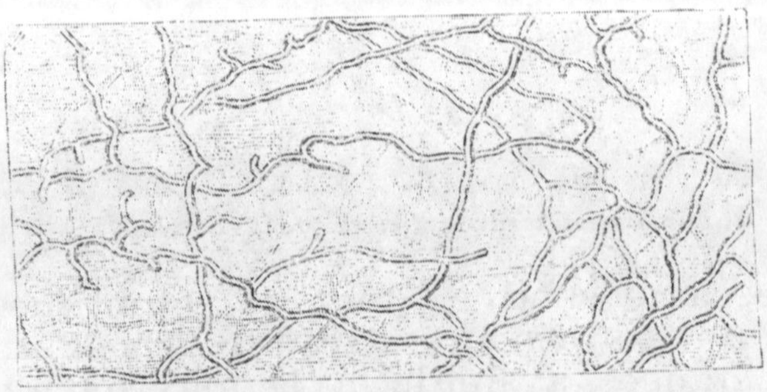

Tubular Shell-structure, from Anomia.

199. Among the ARTICULATED classes, we still find that the skeleton is altogether external, and belongs therefore to the cutaneous system; but it is formed upon a very different plan from the shells of the Mollusca, being closely fitted to the body, and enveloping every part of it; consequently it must increase in capacity, with the advancing growth of the contained structures. Moreover it is destined not merely to afford support and protection to these, but to serve for the attachment of the muscles by which the body and limbs are moved; and the hard envelopes of the latter serve, like the bones of the Vertebrata, as levers by which the motor powers of the muscles are more advantageously employed. Again,

* See the Author's Researches on the Microscopic Structure of Shells, in "Reports of the British Association" for 1844 and 1847.

the hard envelopes of the body and limbs are not formed of distinct plates, like those of the Echinus-shell ; but are only divided by sutures at the joints, for the purpose of permitting the requisite freedom of motion. It might have been thought that here, if anywhere, a process of interstitial growth would have existed, to adapt the capacity of the envelopes to the dimensions of the contained parts, as the latter increase with the growth of the animal ; but, true to the general principle, that epidermic structures are not only extra-vascular, but that they undergo no change when they are once fully formed, we find that the hard envelopes of Articulated animals are thrown off, or exuviated, when the contained parts require an increase of room ; and that a new covering is formed from their surface, adapted to their enlarged dimensions. This is well known to occur at certain intervals in Crabs, Lobsters, and other Crustacea ; which thus exuviate not merely the outer shell, with the continuation of the epidermis over the eyes, but also its internal reflexion, which forms the lining of the œsophagus and stomach, and the tendinous plates by which the muscles are attached to the lining of the shell. A similar moulting may be observed to occur in some of the minute Entomostracous Crustacea of our pools, every two or three days, even after the animals seem to be full grown. During the early growth of Insects, Spiders, Centipedes, &c., a similar moult is frequently repeated at short intervals ; but after these animals have attained their full size, which is the case with Insects at their last change, no further moulting takes place, the necessity for it having ceased. —This moulting is precisely analogous to the exfoliation and new formation of the Epidermis, in Man and most other Vertebrata ; differing from it only in this, that the latter is constantly taking place to a small extent, whilst the former is completely effected at certain intervals, and then ceases. We have examples of a periodical complete moult in Vertebrata, however, among Serpents and Frogs.—The structure of the hard envelopes of Articulated animals corresponds with that of the Epidermis and its appendages in Man. The firm casings of Beetles, for example, are formed of layers of epidermic cells, united together, and having their cavities filled by a horny secretion. The densest structure is found in the calcareous shells of the Crustacea ; which consist of a substance precisely analogous to the *Dentine* of Teeth (§ 212) ; covered on the exterior with a layer of pigment-cells. The calcareous matter consists chiefly of carbonate of lime ; but traces of the phosphate are also found. The animal basis has a firm consistent structure, resembling that of teeth.

200. Now the condition of the osseous skeleton of VERTEBRATED animals is altogether different. It forms a part of the internal substance of their bodies ; and as these grow in *every* part, and not merely by addition to this or that portion, so must the Bones also, in order to keep pace with the rest of the structure. Hence we find them so formed, that the processes of *interstitial* deposition may be continually going on in their fabric, as in that of the softer tissues ; and the changes in their substance do not cease, even when they have acquired their full size. The continuance of these changes appears destined, not so much to repair any *waste* occasioned by decomposition,—for this must be very trifling in a tissue of such solidity,—as to keep the fabric in a condition, in which it may repair the injuries in its substance occasioned by accident or disease. The degree of this reparative power is proportional to the activity of the normal changes

which are continually taking place in the bone; and is thus much greater in youth than in middle life, and greater in the vigour of manhood than in old age.—The presence of Bone is not solely confined, however, to the internal skeleton of Vertebrata, which, being especially destined for the protection of their Nervous System, is sometimes termed the *neural* skeleton; for there are many instances in which true osseous structure is found in the dermal *scales,* which form a more or less complete external skeleton. This is more especially the case with Ganoid and Placoid Fishes (Fig. 43), whose neural skeleton is generally but imperfectly ossified; but true bone is also met with in the scaly armour of some Reptiles and Mammals.

201. In describing the structure of *Bone,* it will be desirable to commence with the simplest form of osseous tissue, such as is presented in a very thin natural plate of bone,—the scapula (blade-bone) of a Mouse, for example;—since the real essentials of the tissue are thus better comprehended. Such a lamella is found by microscopic research to be composed of a substance apparently almost homogeneous, in the midst of which a number of dark spots are to be observed, the form of which is very peculiar. In their general outline they are usually somewhat oval; but they send forth numerous radiating prolongations of extreme minuteness, which may be frequently traced to a considerable distance (Fig. 42). These spots, now known as the *bone-cells* or *lacunæ* (formerly termed the *Purkinjean corpuscles,* after the name of their discoverer), are highly characteristic of the true osseous structure; being never deficient in the minutest parts of the bones of the higher animals, although those of Fishes are occasionally destitute of them. They were formerly supposed, from their dark appearance, to be opaque, and to consist of aggregations of calcareous matter which would not transmit the light: but it is now quite certain that they are open spaces, and that the radiating prolongations from them, which are far smaller than the minutest capillary vessel, are *canaliculi* or delicate tubes. Of these canaliculi, some may be seen to interlace freely with each other, whilst others proceed towards the surface of the bony lamella; and thus a system of passages, not by any means wide enough to admit the blood-corpuscles, but capable of transmitting the fluid elements of the blood, or matters selected from them, is established through the whole substance of the lamella.—Thus the nutrition of the ultimate osseous texture seems to be carried on upon the same plan with that of Cartilage; being effected by the imbibition of nutrient matter from the surface, through the agency of cells. But it differs in this;—that there is a provision in Bone for the ready transmission of nutrient matter through its texture, by means of minute channels, which does not exist in Cartilage;—a difference obviously required by the greater solidity of the substance of the former, which does not allow of the diffused imbibition that is permitted by the softer and moister nature of the latter. We shall presently find that these channels are only formed

FIG. 42.

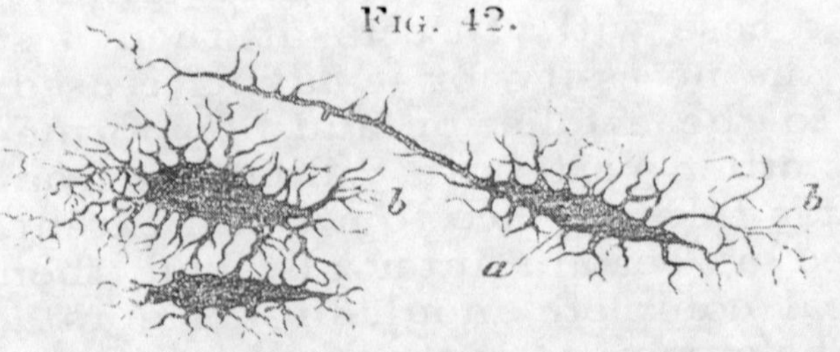

Lacunæ of Osseous substance, magnified 500 diameters;—*a*, central cavity; *b*, its ramifications.

at a late stage of the development of bone, when the remaining tissue is acquiring its completest consolidation.

202. The size and form of the lacunæ differ considerably in the several Classes of Vertebrata, and even in some instances in the Orders; so as to allow of the determination of the tribe to which a bone belonged, by the microscopic examination of even a minute fragment of it. The following are the average dimensions of the lacunæ, in characteristic examples drawn from the four principal classes, expressed in fractions of an inch.

	Long Diameter.	Short Diameter.
Man	1-1440 to 1-2400	1-4000 to 1-8000
Ostrich	1-1333 to 1-2250	1-5425 to 1-9650
Turtle	1-375 to 1-1150	1-4500 to 1-5840
Conger-eel	1-550 to 1-1135	1-4500 to 1-8000

The lacunæ of *Birds* are thus distinguished from those of *Mammals* by their somewhat greater length and smaller breadth; but they differ still more in the remarkable tortuosity of the canaliculi, which wind backwards and forwards in a very irregular manner. There is an extraordinary increase in length in the lacunæ of *Reptiles*, without a corresponding increase in breadth; and this is also seen in some *Fishes*, though in general the lacunæ of the latter are remarkable for their angularity of form, and the fewness of their radiations,—as shown in Fig. 43, which represents the lacunæ and canaliculi in the bony scale of the *Lepidosteus* (one of the few existing representatives of the once numerous tribe of Sauroid fish), with which the bones of the internal skeleton perfectly agree in structure. The dimensions of the lacunæ in any bone do not bear any relation to the size of the animal to which it belonged; thus there is little or no perceptible difference between their size in the enormous extinct Iguanodon, and in the smallest Lizard now inhabiting the earth. But they bear a close relation to the size of the blood-corpuscles in the several classes; and this relation is particularly obvious in the perennibranchiate Batrachia (§ 323), the extraordinary size of whose blood-corpuscles has been already noticed (§ 174).

FIG. 43.

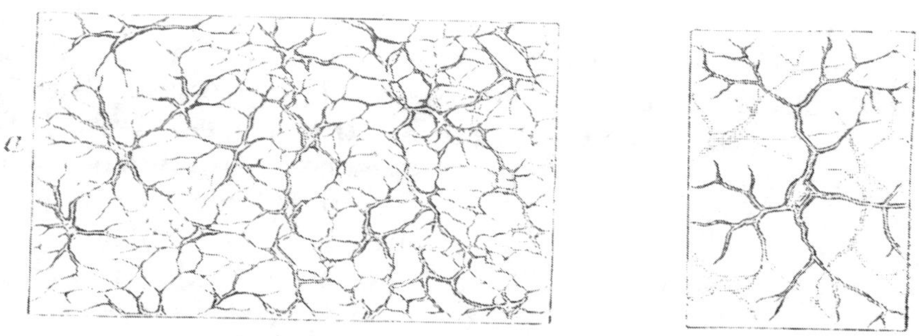

Section of the bony scale of Lepidosteus;—*a*, showing the regular distribution of the lacunæ and of the connecting canaliculi; *b*, small portion more highly magnified.

	Long Diameter.	Short Diameter.
Proteus	1-570 to 1-980	1-885 to 1-1200
Siren	1-290 to 1-480	1-540 to 1-975
Menopoma	1-450 to 1-700	1-1300 to 1-2100
Lepidosiren	1-375 to 1-494	1-980 to 1-2200
Pterodactyle	1-445 to 1-1185	1-4000 to 1-5225

It will be seen from the preceding table, that the lacunæ of the *Proteus*, *Siren*, and *Menopoma* differ from those of ordinary Reptiles in their great breadth; and that in this respect the lacunæ of the *Lepidosiren* agree with them, as do the dimensions of their blood-discs. The lacunæ of the *Pterodactyle* agree in dimensions with those of ordinary Reptiles, and may be at once distinguished from those of Birds, to which class fragments of their wing-bones have been occasionally referred.*

203. When the calcareous matter has been dissolved away from a very thin lamella of bone, and the remaining substance is carefully examined, it is found to consist, not (as commonly stated) of Cartilage, but of a substance made up of indistinct fibres interwoven with each other. These fibres correspond in appearance and composition with those of the White Fibrous tissue (§ 162); and it seems probable that the solid mass of fully-formed bone is formed by the *calcification* of this tissue. The presence of lacunæ and canaliculi, however, is essential to perfect Bone; and these appear to have their origin in *cells*, interspersed in the midst of the fibrous basis, which shoot out radiating prolongations (analogous to those of pigment cells, § 178), insinuating themselves through the fibrous substance at the time it is undergoing consolidation, and thus forming the canaliculi.†

204. Although the Osseous texture is not itself penetrated by blood-vessels, yet it would seem essential to its nutrition that no part of it should be far removed from a vascular surface, from which its canaliculi may absorb the fluid they are destined to convey. Hence we find a special arrangement requisite, for carrying blood-vessels into the neighbourhood of every portion of the substance of the larger bones. The solid osseous texture, which forms the cylindrical shafts of the long bones, and of the thick external plates of the denser flat bones, is penetrated by a series of large canals, termed the *Haversian* (after Clopton Havers, their discoverer), which form a net-work in its interior (Fig. 44), and which serve for the transmission of blood-vessels into its substance. These canals, in the long bones, run for the most part in a direction parallel to the central cavity; and they communicate with this, with the external surface, and with each other, by frequent transverse branches; so that the whole system forms an irregular network, pervading every part of the solid texture, and adapted for the establishment of vascular communications throughout. The diameter of the Haversian canals varies from 1-2500th to 1-200th of an inch, or more; their average diameter may be stated at about 1-500th of an inch. They are lined by a membrane, which is continuous with that of the external surface, and which carries this inwards (so to speak) to form the lining membrane of the central cavity, and of the cancelli. On this membrane, a plexus of blood-vessels is distributed,

FIG. 44.

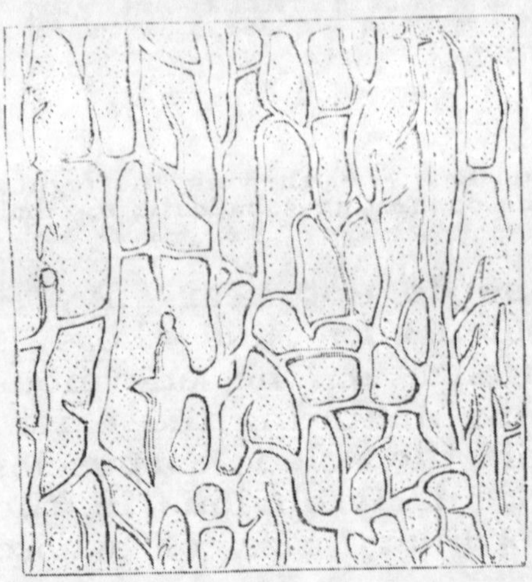

Vertical section of tibia, showing the net-work of Haversian canals.

* See Mr. J. Quekett's Researches, in Trans. of Microsc. Soc. vol. ii.

† This view of the origin of the lacunæ and canaliculi, which the author's own observations had led him to entertain, has been recently advocated by an observer of distinguished ability, Dr. Leidy of Philadelphia. See American Journal of Medical Sciences, for April, 1849, p. 536.

where the size of the canal admits it ;—otherwise, the tube encloses a single twig of an artery or vein. Thus we may consider the whole Osseous texture as inclosed in a membranous bag ; on which blood-vessels are minutely distributed ; and which is so carried into the bone, by involutions and prolongations, that no part of the latter is ever far removed from a vascular surface.

205. The arrangement of the elementary parts of the osseous substance around the Haversian canals, is very interesting and beautiful. When a transverse section of a long bone is made, the open orifices of the longitudinal canals present themselves at intervals, sometimes connected by a transverse canal, where the section happens to traverse this. Around these orifices, we see the osseous matter arranged in the form of cylinders, which appear to be marked by concentric circles (Fig. 45). Now when one of these circles is minutely examined, it is found to be made up of a series of lacunæ, analogous to those already described ; these, however, are seldom or never so continuous as to form a complete circle. The long sides of the lacunæ are directed, the one towards the Haversian canal in the centre, the other towards the circular row next beyond it. And when the course of the canaliculi is traced, it is found that they converge on the inner side towards the central canal, inosculating with those of the series next within, whilst those of the outer side pass outwards in a radiating or diverging direction, to inosculate with those of the series next external. Thus a complete communication is formed, by means of this system of radiating canaliculi and intervening lacunæ, between the central canal, and the outermost cylindrical lamella of bony matter ; and each of these lamellæ derives its nourishment from the vessels of the central canal, through the lamellæ which intervene between itself and the vascular membrane lining that tube.—Thus every one of the Haversian canals is the centre of a cylindrical *ossicle*, which is complete in itself as far as its elementary structure is concerned, and which has no dependence on, or connexion with, other similar ossicles. These are arranged, however, side by side, like sticks in a faggot ; they are bound together by a thin cylinder of bone, on the exterior of all, which derives its nourishment from the periosteum or enveloping membrane ; in like manner, the hollow bundle is lined by a similar cylinder, which surrounds the great medullary cavity, and is nourished by its vascular membrane ; and the spaces that here and there intervene between the ossicles, are filled up

FIG. 45.

Minute structure of Bone, as seen in transverse section ; magnified 300 diameters. 1, an ossicle surrounding an Haversian canal, 3, showing the concentric arrangement of the lamellæ; 2, the same, with the lacunæ and canaliculi ; 4, portions of the lamellæ parallel with the external surface.

with laminæ, (Fig. 45, 4), which are parallel to those of the external and internal cylinders, and which seem to derive their nutriment through them. In this manner, the whole structure acquires great density and solidity.—The structure of the outer and inner tables of the skull, and of other thick solid layers of bone, is precisely similar; except that the Haversian canals have no such definite direction, and form an irregular network.—In the extremities of the long bones, and between the solid layers of the flat bones, we find what is termed a *cancellated* texture; that is, a sort of spongy substance composed of osseous lamellæ and fibres interwoven together (like those of areolar tissue, on a larger scale) so as to form a multitude of minute chambers, or *cancelli*, freely communicating with each other and with the cavity of the shaft; the whole being enclosed in a thin layer of solid bone. These cancelli, like the Haversian canals, are lined by prolongations of the investing membrane, accompanied by blood-vessels. The Haversian canals and cancelli are repetitions of each other in all essential particulars, their form alone being different.

206. In the long bones of Man and of most *Mammalia*, we find a central cavity, which may be considered as an enlarged Haversian canal, and which is filled with the fatty matter known as *marrow*. This cavity does not exist in the bone in its early state; but is formed by the removal of a portion of the cancellated osseous tissue first developed in its interior; and the primitive condition is retained in the bones of the Sloths and their great extinct congeners, the Mylodon and Megatherium (which, of all Mammalia, present the nearest approach to Reptiles in the inertness of their habits); and also in the bones of the Cetacea and Seals, which may be regarded as the representatives of Fishes. In some of the bones of the head, even in Man, but still more in the Giraffe, Mylodon, and Elephant, we find the cancelli of the bones of the skull enlarged and occupied with air instead of marrow; but these have no direct communication with the lungs.—Among *Birds*, however, we meet with a very different condition; the central cavity being usually unoccupied by marrow, and communicating with the lungs, so that the membrane lining it becomes an auxiliary organ of respiration (CHAP. XIII.), whilst the lightening of the bones thus produced diminishes their specific gravity. In the Swift, and other birds of peculiarly rapid flight, this condition extends throughout the entire skeleton; even the ordinary cancellated structure being permeated with air. But in birds of inferior powers of flight, the proportion of bones receiving air is diminished; and in the completely wingless Apteryx and aquatic Penguin, not a single bone receives air, though a medullary cavity still exists in all bones except those of the last-named bird.—In *Reptiles* we meet with considerable diversity as regards this character. The Saurians and most of the Batrachians usually have their long bones hollowed in the centre, the cavity being occupied by marrow; the same is the case with the ribs of the larger Serpents; and in the bones of the extremities of the gigantic Iguanodon and Megalosaurus (the highest genera among Reptiles, so far as can be ascertained from their osseous conformation) the medullary cavities are as capacious as those of any Mammal. On the other hand, in the aquatic Crocodiles the medullary cavities are small; and in the Turtle tribe they are reduced to the condition of cancelli.—In *Fishes* we usually find that the osseous structure presents a still less advance upon the original

condition ; the outer portion of the bones, in many instances, being the only parts that have undergone consolidation, and the interior remaining filled with the original cartilage (so that in dried specimens there may be the semblance of a medullary cavity) ; and no regular system of Haversian canals and cancelli being established,—the large irregular lacunæ and canaliculi apparently serving to convey the plasmatic fluid, through distances which it could not otherwise traverse. The bones of some Fishes, however, closely approximate to those of Reptiles in their general structure.*

207. The principal features of the Chemical constitution of Bone are easily made evident. After all the accessory parts have been removed, and nothing remains but the real Osseous texture, this may be separated, by simple processes, into its two grand constituents,—the animal basis, and the calcareous matter. The latter may be entirely dissolved away, by maceration of the bone in dilute Muriatic or Nitric acid ; and a substance of cartilaginous appearance is then left, which, when submitted to the action of boiling water for a short time, is almost entirely taken up by it, and the solution forms a dense jelly on cooling. This same substance, Gelatine, may be obtained by long boiling under pressure, from previously unaltered bone ; and the calcareous matter is then left in a friable condition.—By submitting a bone to a heat sufficient to decompose the animal matter, without dissipating any of the earthy particles, we may obtain the whole calcareous matter *in situ ;* but the slightest violence is sufficient to disintegrate it. The bones of persons long buried are often found in this condition ; their form and position being retained, until they are exposed to the air, or are a little shaken ; when they crumble to dust. —The proportion of the earthy matter of Bones to the animal basis may be differently stated ; according as we include in our estimate of the latter, the contents of the medullary cavity, the Haversian canals, and the cancelli ; or confine ourselves to that portion only of the animal matter, which is united with the calcareous element in the proper osseous tissue. According to the experiments of Dr. Stark,† the relative amount of the two elements, in the latter estimate, is subject to very little variation, either in the different classes of animals, or in the same species at different ages ; the animal matter composing about one-third, or $33\frac{1}{3}$ per cent. ; and the mineral matter two-thirds, or $66\frac{2}{3}$ per cent. The degree of hardness of bone does not altogether depend, therefore, on the proportion of mineral deposit they may contain ; for the flexible, semi-transparent, easily-divided bones of Fish contain as large a proportion of earthy matter, as the ivory-like leg-bones of the deer or sheep. It seems to be influenced in part by the proportion of phosphate to carbonate of lime, which is greater in the higher animals ; in part, by the quantity of water that may be united with the osseous matter ; and in part, by the degree in which the Haversian canals and cancelli become filled up with bony substance. It is from this last circumstance, that the bones of old people have a greater density than those of young or middle-aged persons ; for after the full dimensions of the bones have been attained, they are progressively consolidated during the whole remainder of life, by the formation of new

* See Prof. Owen's Lectures on Vertebrate Animals, vol. i., p. 33, *et seq.*

† Edinburgh Medical and Surgical Journal, April, 1845.

layers of osseous substance within the preceding, so as to contract the calibre of the Haversian canals.

208. The first development of Bone is commonly preceded by the formation of a Cartilaginous structure, which occupies the place the bone is afterwards to take; and it has been commonly considered that the bone is formed by the ossification of cartilage. This, however, does not appear to be the case; for none of the peculiar substance of the cartilage—*Chondrine* (§ 191)—can be found in perfect bone; and microscopic examination of the process shows, that although a calcareous deposit does in the first instance take place in the intercellular substance of cartilage, the structure thus formed has not the character of true bone, and is destined but for a temporary purpose.—The process of bone-formation always commences in the immediate neighbourhood of blood-vessels, which pass down into canals excavated in the substance of the cartilage and lined by a continuation of its investing membrane; hence the spots where these vascular canals are especially developed are termed *centres of ossification.* We usually find one of these in the centre of the shaft of a long bone, and one at each end, with an additional one for any considerable projection or 'process;' in the flat bones there is generally one in the middle of the surface, and one in each of the principal 'processes.' Up to the period when the bone attains its full dimensions, the parts which contain distinct centres are not connected by osseous union, but only by cartilage, so that they fall apart when this decomposes; the purpose of this is, to allow of an increase in the size of the bone, by the growth of cartilage between its detached portions, which cartilage may give place to bony structure when there is no further need of increase. There is a remarkable difference in this respect, however, between the warm-blooded and the cold-blooded Vertebrata; for the bony pieces which arise from separate centres of ossification have a much greater tendency to *remain* distinct in the latter than in the former; so that the adult skeleton is composed of a much larger number of bones. On the whole, a close correspondence exists between the number of ossific centres, in the early condition of the skeleton of all Vertebrated animals; and it is on the basis of this conformity, that the Philosophical Anatomist forms his idea of the fundamental plan of the Vertebrate skeleton (§ 320); but this conformity is often obscured by the multiplication or suppression of ossific centres, in order to meet certain peculiar requirements in the economy of one or another class. Thus, contrary to the general rule, we find the femur (thigh-bone) of Man and the Mammalia developed from *four* distinct centres,—one for the shaft, one for each of the extremities, and one for the great trochanter;—whilst in Birds and most Reptiles, it is developed from a single centre. This peculiarity seems to have reference to the use which is to be made of these bones in the Mammalia, the young of which soon begin to run and leap upon the ground, in a manner that would produce a jar or shock to their yet delicate fabric, especially to their nervous centres, if it were not for the interposition of layers of cartilage between the shaft of the bone and its extremities. And this view is confirmed by the fact, that in the Frog, which, unlike other Reptiles, moves by leaps, the extremities of the thigh-bones are developed from distinct centres as in Mammalia.* On the other hand, we not

* See Prof. Owen's Lectures on Vertebrate Animals, vol. i., p. 36, *et seq.*

unfrequently meet with a reduction in the number of ossific centres below their typical amount : thus the regular number of distinct bones in the carpus (wrist) and tarsus (instep), as deduced from an extensive comparison, is undoubtedly *ten;* but these are reduced in the Human carpus to *eight*, and in the tarsus to *seven*, each having but a single centre ; and the reduction is still greater in the hoofed Mammalia.*

209. The structure of the *temporary* Cartilage, which occupies the place to be subsequently taken by Bone, is essentially the same with that of the *permanent* Cartilage already described (§ 191). If we make a section of it, however, in a direction vertical to the plane of ossification, we find the cartilage-cells undergoing multiplication longitudinally (§157), so as to form cylindrical piles or columns, which are separated from each other by intercellular substance (Fig. 46). It is in this substance that the first calcareous deposit takes place ; this deposit forming a sort of network, in the meshes of which the lower ends of the columns are imbedded. The further stages of the process have not been precisely made out ; but it appears that the cartilage-cells speedily degenerate and die, leaving the spaces void which they previously occupied ; and that of the first calcified network (which, as it contains no lacunæ or canaliculi, has scarcely a title to be designated true bone) the greater portion soon breaks down, leaving larger open areolæ of no very definite form. These spaces become filled, from the neighbouring vessels, with a formative blastema containing cells ; and it is probable that the production of all the succeeding layers of bone, which line these spaces, and gradually contract them into the definite form of Haversian canals and cancelli, are generated by the consolidation of this blastema into a fibrous tissue, which subsequently becomes calcified, whilst the interspersed cells are metamorphosed into lacunæ and canaliculi (§ 201).

Fig. 46.

Vertical Section of Cartilage at the seat of ossification, showing the arrangement of the cartilage-cells in columns ; at the lower part, ossific deposit has begun to take place between these, whence arises the opacity.

210. Although, in a large proportion of the Skeleton, the formation of Bone is thus preceded by that of Cartilage, yet this is by no means necessarily the case ; for it is certain that true bone may be developed in the substance of Fibrous Membrane. This may be seen in the original production of the tabular bones forming the roof of the skull ; these having usually only a centre of cartilage, beyond which the ossifying process extends in membrane only. But it is equally well displayed in the successive additions which are made to the surfaces of all other bones, subsequently to their first development ; these additions having their origin in the progressive calcification of the inner layers of the *periosteum* (or fibrous membrane covering the bones), whilst new fibrous tissue is

* See Prof. Owen's Discourse on the Nature of Limbs, p. 24, *et seq.*

formed upon their exterior. It is not difficult to trace the consolidation of these fibres, on the growing surface of a long bone; and the gradual extension of the canaliculi from the originally spherical cells included in their meshes, which become the lacunæ. The first rude outline of the Haversian canals is produced by the development of the bony fabric in longitudinal ridges; and these gradually rise up and arch over, so as to meet and include irregular lengthened spaces, which are gradually contracted by the subsequent formation of new laminæ in their interior, whereby the new osseous layer is progressively consolidated.*

211. The growth of Bones, like that of Shells, is thus in reality effected by the addition of new tissue to the part already formed; but this addition may take place in three modes, namely,—by the development of new bone in the cartilage intervening between the parts yet remaining separated, somewhat as in the enlargement of the shell of the Echinodermata (§ 195), — by the development of new bone in the membrane covering the surface, as in the addition of new layers to the shell of Mollusca,—and by the *interstitial* formation of new layers within the Haversian canals of the osseous fabric already developed, which is a process peculiar to Vertebrata. But it is not by addition only, that we find that 'moulding' of the bones effected, which is requisite for the maintenance of their form whilst their dimensions are undergoing increase. For this purpose, a *removal* of osseous tissue already formed is also necessary; and of such a removal we have a very remarkable example in the first formation of the central cavity in the long bones of the higher Vertebrata (§ 206), and in its continued enlargement during their growth. —The perfect *reparation* of Bone, after severe injüries, is one of the most remarkable features in its history; and it seems to be for the effectual perforance of this, in the event of its being demanded, that the complex nutritive apparatus which has been described is kept up after the bone has attained its full dimensions; since but very little interstitial change can be requisite, in a tissue so free from tendency to spontaneous decay. The reparation of bone is effected by the development of new osseous tissue in the substance of membrane or cartilage formed in the seat of injury; and the new fabriç gradually becomes perfectly continuous with the old, progressively acquiring the form and solidity of the original bone. The reparative power is even more remarkably displayed, when the original bone has been destroyed by disease, than when it has been the seat of mechanical injury.

212. We next proceed to the *Teeth*, which are organs of mechanical attrition, developed in the first part of the alimentary canal, for the purpose of comminuting the food conveyed into it. Their place of origin is altogether different from that of Bone, as they commence in little papillary elevations of the mucous membrane covering the jaw; but they are very analogous to Bone in their structure and composition. It is in the inferior Vertebrata that this resemblance is the strongest; and it is there, too, that

* For a fuller account of the structure and development of Bone, the Author would more especially refer his readers to the admirable researches of Professor Sharpey, as detailed in his Introduction to the fifth edition of Dr. Quain's Elements of Anatomy. His own researches have led him to concur with Professor Sharpey on every point save the origin of the lacunæ and canaliculi, of which he thinks that Professor Sharpey has given an imperfect account in regarding them as mere interspaces in the fibrous tissue.

the history of the development of Teeth may be most conveniently studied. Thus, in the fœtal Shark, the first appearance of the tooth is in the form of a minute papilla on the mucous membrane covering the jaws ; the substance of this papilla is composed of spherical cells, which are imbedded in a kind of gelatinous substance resembling that of incipient cartilage ; whilst its exterior is composed of a dense, structureless, pellucid membrane. The cellular mass is not at first permeated by vessels ; but a small arterial branch is distributed to each papilla, and spreads out into a tuft of capillaries at its base. The papilla gradually enlarges, by the formation of new cells at the part immediately adjacent to the blood-vessels, which supply the material requisite for their development ; and when it has acquired its full size, the process of *calcification* takes place, by which it is converted into *Dentine* or tooth-substance. In the Shark tribe, as in many other Fishes, the general structure of the Dentine is extremely analogous to that of Bone ; the tooth being traversed by numerous medullary canals, which are continuous with the Haversian canals of the subjacent bone, and receive blood-vessels from them (Fig. 47, A) ;

FIG. 47.

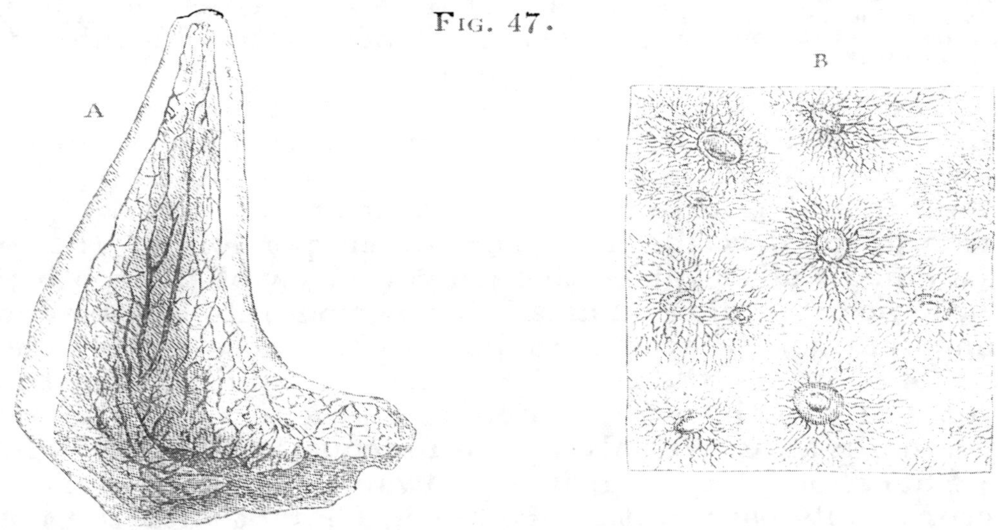

Structure of the Teeth of Fish :—A, perpendicular section of tooth of *Lamna* moderately enlarged, showing network of medullary canals ; B, transverse section of portion of tooth of *Pristis*, more highly magnified, showing orifices of medullary canals, with systems of radiating and inosculating tubuli.

and each of these canals being surrounded by a system of tubuli (B), which radiate into the surrounding dense and apparently-homogeneous substance. These tubuli, however, do not enter lacunæ, nor is there any concentric lamellar arrangement around the medullary canals ; but each system of tubuli is continued onwards through its own division of the tooth, the individual tubes sometimes giving off lateral branches, whilst in other instances their trunks bifurcate.—In the teeth of the higher Vertebrata, however, we usually find the centre excavated into a single cavity (Fig. 49), and the remainder destitute of vascular canals ; but there are intermediate cases (as in the teeth of the great fossil Sloths) in which the inner portion of the dentine is traversed by prolongations of this cavity, conveying blood-vessels, which do not pass into the exterior layers. The tubuli of the 'non-vascular' dentine, which exists by itself in the teeth of nearly all Mammalia, and which in the Elephant is known as Ivory, all radiate from the central cavity, and pass towards the surface of the tooth in a nearly parallel course (Fig. 48). Their diameter at their largest part, averages 1-10,000th of an inch; their smallest branches are immeasurably fine. It is impossible that even the largest of them can receive blood, as their diameter is far less than that of the blood-discs; but it is probable that, like the canaliculi of bone, they may absorb nutrient matter from the vascular surface upon which their inner extremities open.—There is a greater predominance of mineral matter in Dentine than in bone; but still the former has a definite animal basis, which retains its form when the calcareous matter has been removed by maceration in acid. In every 100 parts, the animal matter forms about 28; and of the mineral portion, phosphate of lime constitutes about $64\frac{1}{3}$ parts, carbonate of lime $5\frac{1}{3}$ parts, and phosphates of magnesia and soda, with chloride of sodium, about $2\frac{1}{3}$ parts.

Fig. 48.

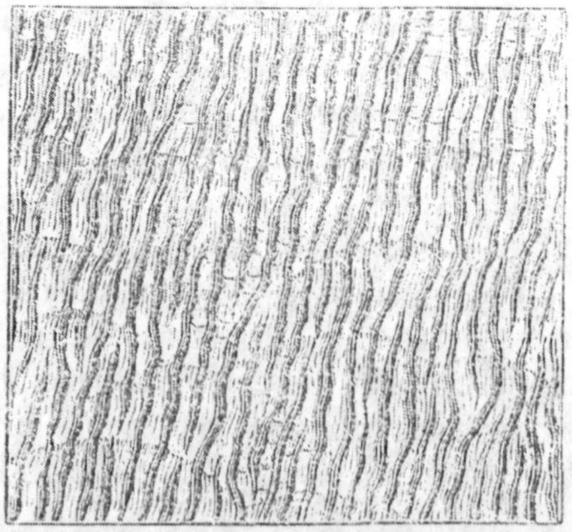

Oblique section of Dentine of human tooth, highly magnified, showing the calcigerous tubuli, and the outlines of the original cells (?).

213. That the Dentine has its origin in the calcification of the original cellular pulp, and is not an excretion from its surface as formerly supposed, is now generally admitted. Observers are not agreed, however, upon the mode in which this conversion takes place, and we shall therefore not attempt to describe it. The inner portion of the pulp still remains uncalcified, and occupies the central cavity of the tooth; but this sometimes becomes consolidated at a later period into a substance resembling vascular dentine, or even into bone.—In those simple teeth which consist solely of Dentine, the consolidation of a papilla upon the mucous membrane of the mouth is all which is requisite; and the tooth thus formed may remain attached only to the mucous membrane, which is the case in the Shark, or it may grow downwards, by the addition of new dental structure at its base, until it comes in contact with the bone of the jaw. Where it is only attached to the mucous membrane, it is very liable to be torn away; but a new tooth, formed from a distinct papilla, is ready to replace it; and this process is continually repeated, the development of new papillæ being apparently unlimited. On the other hand, where the root of the tooth comes in contact with the jaw, it may completely coalesce with it, which is the case in many Fishes, the Haversian canals of the bone being continued as medullary canals into the dentine; or it may send long spreading roots into the bone, which are united to it at their extremities.

214. It is obvious that there is no provision, in the simple calcification of the dental papilla, for any variations of density, other than those which may result from the different degrees of hardness in the substance of the Dentine itself. Now in the teeth of Man and most other Mammals, and in those of many Reptiles and some Fishes, we find two other substances, one of them harder, and the other softer, than Dentine; the former is termed *Enamel;* and the latter *Cementum* or *Crusta Petrosa.* For the development of these, a peculiar modification of the apparatus is requisite. —The *Enamel* is composed of long prismatic cells, exactly resembling

those of the prismatic shell-substance formerly described (§ 197), but on a far more minute scale; the diameter of the cells not being more, in Man, than 1-5600th of an inch. The length of the prisms corresponds with the thickness of the layer of enamel; and the two surfaces of this layer present the ends of the prisms, which are usually more or less regularly hexagonal. The quantity of animal matter in the enamel of the adult tooth is extremely minute,—not above 2 parts in 100; and it is only at a very early age, that the true character of the organic structure can be distinctly seen. The course of the prismatic cells is more or less wavy; and they are marked by numerous transverse striæ, resembling those of the prismatic shell-substance, and probably originating in the same cause,—the coalescence of a series of shorter cells, to form the lengthened prism.—Of the 98 parts of mineral matter in the enamel, 88½ consist (according to Berzelius) of phosphate of lime, 8 of carbonate of lime, and 1½ of phosphate of magnesia. In density and resisting power, the Enamel far surpasses any other organised tissue, and approaches some of the hardest of mineral substances. In Man, and in Carnivorous animals, it covers the crown of the tooth only, with a simple cap or superficial layer of tolerably uniform thickness (Fig. 49, *a*), which follows the surface of the dentine in all its inequalities; and its component prisms are directed at right angles to that surface, their inner extremities resting in slight but regular depressions on the exterior of the dentine. In the teeth of many Herbivorous animals, however, the Enamel forms (with the Cementum) a series of vertical plates, which dip down into the substance of the dentine, and present their edges alternately with it, at the grinding surface of the tooth; and there is in such teeth no continuous layer of enamel over the crown. The purpose of this arrangement is evidently to provide, by the unequal *wear* of these three substances,—of which the Enamel is the hardest, and the Cementum the softest,—for the constant maintenance of a rough surface, adapted to triturate the tough vegetable substances on which these animals feed. —The Enamel is the least constant of the Dental tissues. It is more frequently absent than present in the teeth of the class of Fishes; it is wanting in the entire order of Ophidia (Serpents) among existing Reptiles; and it forms no part of the teeth of the Edentata (Sloths, &c.) and Cetacea (Whales) amongst Mammals.

FIG. 49.

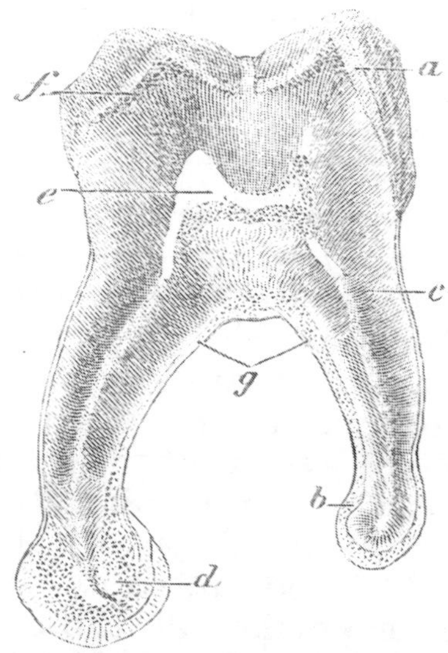

Vertical section of human molar tooth;—*a*, enamel; *b*, cementum or crusta petrosa; *c*, dentine or ivory; *d*, osseous excrescence, arising from hypertrophy of cementum; *e*, pulp-cavity; *f*, osseous cells at outer part of dentine.

215. The *Cementum*, or *Crusta Petrosa*, has the characters of true Bone; possessing its distinctive stellate lacunæ and radiating canaliculi. Where it exists in small amount, we do not find it traversed by medullary canals; but, like Dentine, it is occasionally furnished with them, and thus resembles bone in every particular. These medullary canals enter its substance from the exterior of the tooth, and consequently pass towards those which radiate from the central cavity towards the surface of the dentine, where this possesses a similar vascularity,—as was remarkably the

case in the teeth of the extinct Megatherium. In the Human tooth, however, the Cementum has no such vascularity; but forms a thin layer, which envelops the root of the tooth, commencing near the termination of the capping of Enamel (Fig. 49, *b*). This layer is very subject to have its thickness increased, especially at the extremity of the fangs, by hypertrophy, resulting from inflammation; and sometimes large exostoses are thus formed (Fig. 49, *d*), which very much increase the difficulty of extracting the tooth. When the tooth is first developed, the Cementum envelops its crown, as well as its body and root; but the layer is very thin where it covers the Enamel, and being soft, it is soon worn away by use. In the teeth of many Herbivorous Mammals, it dips down with the Enamel to form the vertical plates of the interior of the tooth; and in the teeth of the Edentata, as well as of many Reptiles and Fishes, it forms a thick continuous envelope over the whole of the surface, until worn away at the crown.

216. The development of these additional structures is provided for by the enclosure of the primitive papilla, from which the Dentine is formed, within a Capsule, which, at one period, completely covers it in: between the inner surface of the capsule and the outer surface of the dentinal papilla, a sort of epithelium is developed, by the calcification of which the Enamel is formed; and the Cementum is generated by the conversion of the fibrous capsule itself into a bony substance. The Capsule is at first formed by the sinking-in of the Mucous Membrane around the papilla, so that this lies at the bottom of an open *follicle,* analogous to those which are common in mucous membranes elsewhere (§ 169), or to those in which Hair originates (§ 180). After a time, however, the edges of the follicles are lengthened into little valve-like processes or *opercula*, which are destined to meet and form covers to the follicles, so as completely to enclose the pulps within. Before the calcification of these pulps commences, a provision is made for the production of a subsequent set of teeth; for a bud or offset springs from the upper part of each capsule; and this, although at first in the condition of an open follicle communicating with the primitive capsule, is gradually detached from it and closed in, so as to form a new and independent capsule.—From the inner surface of the capsule, which is in a villous and vascular condition, a copious formation of epithelial cells takes place, so as to occupy the space left between the surface of the dentinal papilla and the lining of the capsule; this is termed the enamel-pulp; and its conversion into enamel-prisms probably takes place in the same manner as the formation of the prismatic shell-substance (§ 197). The ossification of the capsule itself, so as to form the investment of Cementum, does not take place until the rest is completed.—The layers of Enamel and Cementum, which are interposed between the plates of Dentine in the molar teeth of the Elephant and of Herbivorous animals generally, are formed from prolongations of the opercula of the capsules, which dip down and are received into depressions of the dentinal pulp before its calcification begins.

217. The history of the first development of the higher forms of Teeth may thus be divided into three stages, the *papillary,* the *follicular*, and the *saccular*.—The papillary corresponds precisely with the complete mode of dental development in the Shark and some other Fishes, as already mentioned.—The follicular, which commences with the enclosure of the

papillæ in open follicles, and terminates when the papillæ are completely hidden by the closure of the mouths of those follicles, has also its permanent representation in the development of the teeth of many Reptiles and Fishes ; the primitive papillæ of which, though enclosed in follicles, are never covered-in at the summit, and thus free themselves from their envelopes by simply growing upwards through their open mouths. But in Man, and in all other animals which agree with him in going on to the saccular stage, there must also be an *eruptive* stage, which consists in the bursting-forth of the tooth from the enclosing capsule ; the summit of the tooth being carried against the lid of the sac, by the growth of its root. By the continuance of the same growth, the teeth are caused to penetrate the gum, and are gradually raised above its surface.—The duration of the first-formed teeth is limited in all Animals. They die and are cast forth ; and are then replaced by others, which originate either in buds derived from the primitive capsules, or in newly-formed papillæ. The teeth of the second set, in Man and the Mammalia, are comparatively permanent, being destined to last through the whole of life ; and we do not usually find that new capsules are derived from them by gemmation, except in the case of the posterior molars, which were not represented in the first set. Thus in Man, behind the last temporary molars, which are replaced by the permanent bicuspids, three permanent molars are to be developed, on either side of each jaw. The *first* of these is formed, like the milk-teeth, from a primitive capsule ; but is not completed until the seventh year. The capsule of the *second* is formed at a later period from that of the first, by a process of budding exactly analogous to that by which the other permanent capsules are formed from the corresponding temporary ; and at a still later period, the capsule of the *third* permanent molar is formed as a bud from that of the second. The evolution of this molar does not usually take place until the system has acquired its full development ; and the process of budding then ceases in Man, being limited to a single act of reproduction in the case of the ordinary Milk-teeth, and to a double one in that of the first permanent Molar. In those Fishes and Reptiles, however, whose dentition goes on to the follicular stage, this reproductive process continues through the whole of life without any limit ; the newly-formed teeth usually taking the places of those of the previous set, instead of coming up at their sides like the second and third permanent molars of Man. The analogy between the continued succession of teeth in the lower Vertebrata, by the gemmiparous reproduction of their capsules, and the development of the capsules of the permanent teeth of Man from those of the temporary set, is made further evident by the fact, that a *third* set occasionally makes its appearance in persons advanced in life ; the development of which would not be intelligible, if we could not refer it to the continuance of the same process in the other capsules, as that which regularly takes place to a limited extent in the permanent molars in Man, and which goes on without limit through the whole lives of the lower Vertebrata.—In those Fishes, however, in which the dental development stops at the papillary stage, the successive teeth are formed from new and independent papillæ.

218. There are certain cases, however, in which the teeth are not changed more than once, but continue to grow through the whole remainder of life from *persistent* pulps ; as for instance, the tusks of the

Elephant, the front teeth of the Rodentia, and the grinders of the Edentata. In such teeth, the base of the pulp remains uncalcified, and a new development of cells is continually taking place in that situation; these new cells are in their turn converted into dentine, in continuity with that previously formed; and thus the tooth or tusk is continually lengthening at its base, in a degree which compensates for its usual wear at its summit. If anything should prevent that wear,—as when the opposite tooth has been broken off,—there is an absolute increase in the length of the tooth, from the continued growth at its base; which may become a source of great inconvenience to the animal. There is nothing, in the human subject, at all analogous to this mode of development from persistent pulps; the process being checked by the closure of the root around the base of the pulp, which obstructs the supply of blood it receives.

219. We have seen that the Teeth are formed, in the first instance, upon the *surface* of the Mucous membrane of the mouth; and consequently they really form a part of the *external* or dermo-skeleton, and not of the *internal* or osseous skeleton. They correspond, therefore, with the external skeletons of the Invertebrata; and thus the analogy which has been pointed out, between the Enamel of teeth and the prismatic cellular substance of the shells of Mollusca, and between the Dentine and the shells of the higher Crustacea, holds good also in regard to the situation of these bodies. Great diversities exist amongst the teeth of different tribes of Vertebrata, not only in regard to the general arrangement of their component tissues, but also with respect to the minute characters which these severally present; and a knowledge of these diversities enables the Microscopist, in a great number of instances, to determine from a minute fragment of a tooth, not merely the class and order, but sometimes even the genus and species, to which the owner of it belonged.*

220. *Simple Tubular Tissues.*—In all but the very simplest forms of Animal structure, we find a system of canals for the conveyance of fluid from one part of the fabric to another. These canals frequently seem to be nothing else than channels excavated in the midst of the other tissues, like the 'intercellular passages' of Plants (§ 154); such appears especially to be their character in the lower tribes of animals, and in the early embryonic condition of the higher. In the more advanced condition of the latter, however, we find that all the canals for the conveyance of the nutritive fluids,—whether blood, chyle, or lymph—have definite walls; which are present equally in the largest and in the minutest vessels. The vessels of Animals do not, like the ducts of Plants (§ 152), pursue a straight and independent course; for the plan of the Animal circulation requires that they should be distributed on a ramifying arrangement;—those which convey blood towards any part being progressively subdivided and diverging from each other, like the branches of a tree from its trunk; whilst those which collect and return the blood which has traversed the part, converge and reunite into trunks, like the roots of a tree or the sources of a river. Between these two sets of vessels, of which the former

* See the "Odontography" of Prof. Owen, and the "Recherches sur les Poissons Fossiles" of Prof. Agassiz.

are termed *Arteries*, and the latter *Veins*, intervenes a network of extremely minute tubes, usually so narrow as only to allow the blood-corpuscles to pass in single file; and it is when traversing these, which are termed *Capillaries*, that the blood undergoes the various changes to which it is subservient in the processes of Nutrition. The form of the meshes of the Capillary network is adapted to that of the elementary parts of the tissue between which they pass; thus in Fig. 50 is seen the

FIG. 50.

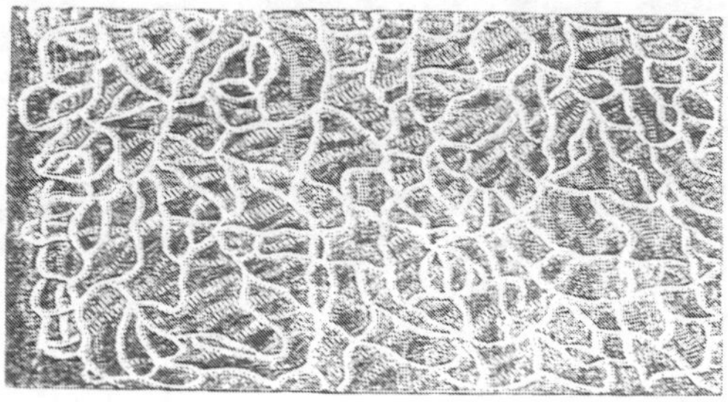

Capillary network around Fat-cells.

FIG. 51.

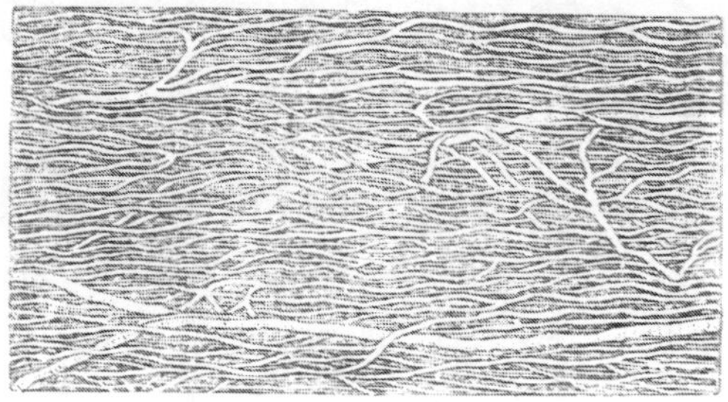

Capillary network of Muscle.

capillary network around the Fat-cells; and in Fig. 51 the arrangement of these vessels in a fasciculus of Muscular fibres. The disposition of the

FIG. 52.

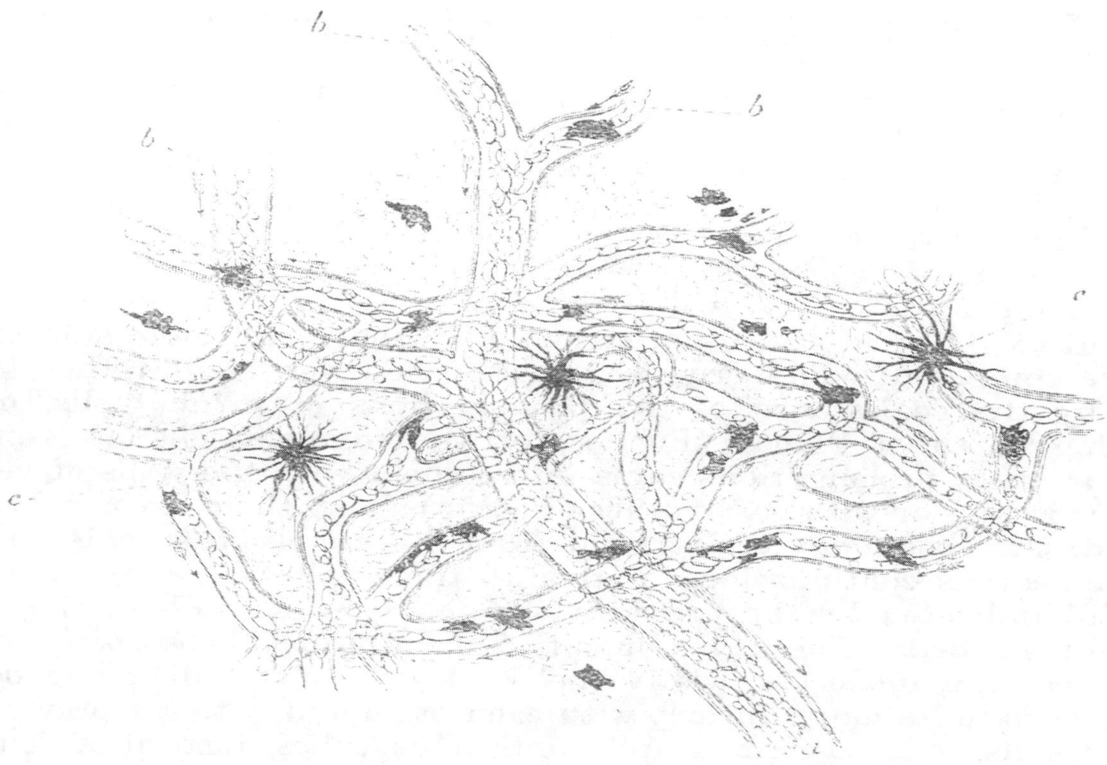

Capillary circulation in a portion of the web of a *Frog's* foot, magnified 110 diameters; *a*, trunk of vein; *b*, *b*, its branches; *c*, *c*, pigment cells.

Capillaries, with the blood-corpuscles in their interior, as seen in the web of the Frog's foot, is shown on a larger scale in Fig. 52. Into this network the blood is conveyed by the ramifications of the arteries on one side; and it is collected and returned by the rootlets of the veins on the other; as is shown in Fig. 53.

FIG. 53.

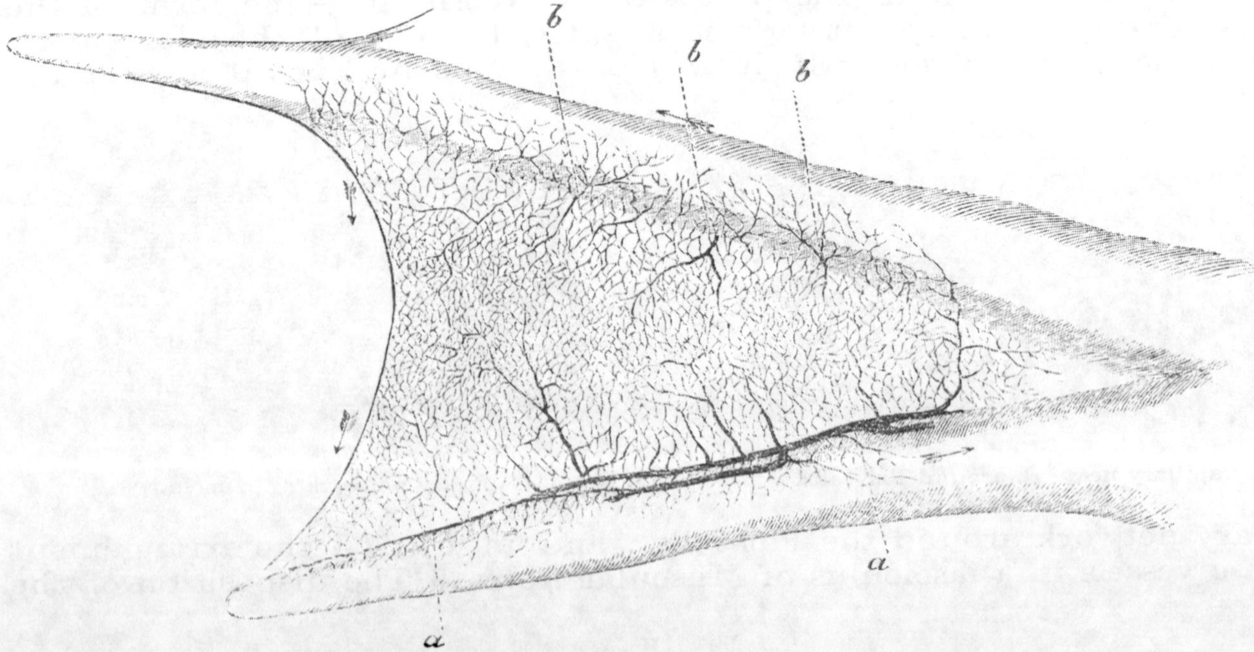

Web of *Frog's* foot, stretching between two toes, magnified 3 diameters; showing the blood-vessels, and their anastomoses: *a*, *a*, veins; *b*, *b*, *b*, arteries.

221. The trunks and branches of the blood-vessels appear to be formed, like the ducts of Plants, by the coalescence of cells arranged in linear series; those of moderate size taking their origin in single or double files of such cells, whose coalesced walls form the primitive simple membranous tubes of these vessels; whilst the principal trunks, like the heart, are formed out of aggregations of cells, of which those in the interior liquefy to form the cavity, whilst those on the periphery are metamorphosed into the fibrous and other tissues of which their more substantial walls are composed. The true Capillaries, however, seem rather to be formed by the coalescence of the prolongations of *stellate* cells, much resembling those of Plants (Fig. 8), or the pigment-cells of Animals (Fig. 29); these prolongations come into contact with the walls of vessels already existing, or with each other (Fig. 54); and their cavities having coalesced, they begin to receive fluid from the vessels, then enlarge, and finally appear as continuations of them,—their cellular origin, however, being still indicated by the presence of nuclei in the walls of the capillaries thus formed. The development of the minutest Absorbent vessels takes place in a mode essentially the same; the chief difference being, that the cells unite more linearly with each other and with the previously-formed vessels, so as to form simple continuous tubes, instead of sending off prolongations to form a network.*

* See the observations of Kölliker in the "Annales des Sciences Naturelles, Zool., Aout, 1846;" and a summary of them, with confirmatory observations by Mr. Paget, in the "Supplement to Müller's Physiology," p. 101.

222. *Compound Tubular Tissues.*—We now come to two tissues of the highest importance in the Animal fabric ; the presence of which is, indeed, its distinguishing characteristic. These are the *Muscular*, and the *Nervous* tissues. The former is the one, by which all the sensible *movements* of the body are effected ; and the latter serves as the instrument, by which *sensations* are received, and by which the *will* excites the muscles to action, —besides serving as a medium for other operations, in which motion is produced, without the intervention of either sensation or will. These tissues, with the apparatus of bones and joints on which the muscles act, constitute the purely *animal* portion of the fabric ; and if a being could be constructed, in which they should be capable of continued activity without any other assistance, it would be in all essential particulars an Animal. But, as we shall presently see, the plan on which these tissues are formed, in fact the very conditions of their existence and activity, are such, that they require constant *nutrition* and *re-formation ;* so that the Animal cannot exist, without an apparatus for preparing, circulating, and maintaining in constant purity, a fluid by which nutrient operations may be effected, and which shall also be the means of carrying off the products of the *waste* consequent upon the action of those tissues. This apparatus constitutes the Vegetative portion of the frame ; the elementary parts concerned in which have been already noticed.—Now it might not have been unreasonable to expect, that tissues altogether so peculiar in their endowments, and in the purposes to which they are destined, should have had a structure departing widely from the type of the simple Cell. Such, however, is not the case. We shall find the *Muscular Fibre* (as it is commonly called) to be a *tube* formed by a coalescence of cells, and distinguished by the nature of its contents, which, in the most perfect form of the tissue, are composed of numerous linear series of very minute secondary cells. And in like manner, the *Nervous Fibre* is a tube having a similar origin in cells; whilst that portion of the Nervous matter by which its most active functional changes are effected, retains its original cellular character without any change except in form.

FIG. 54.

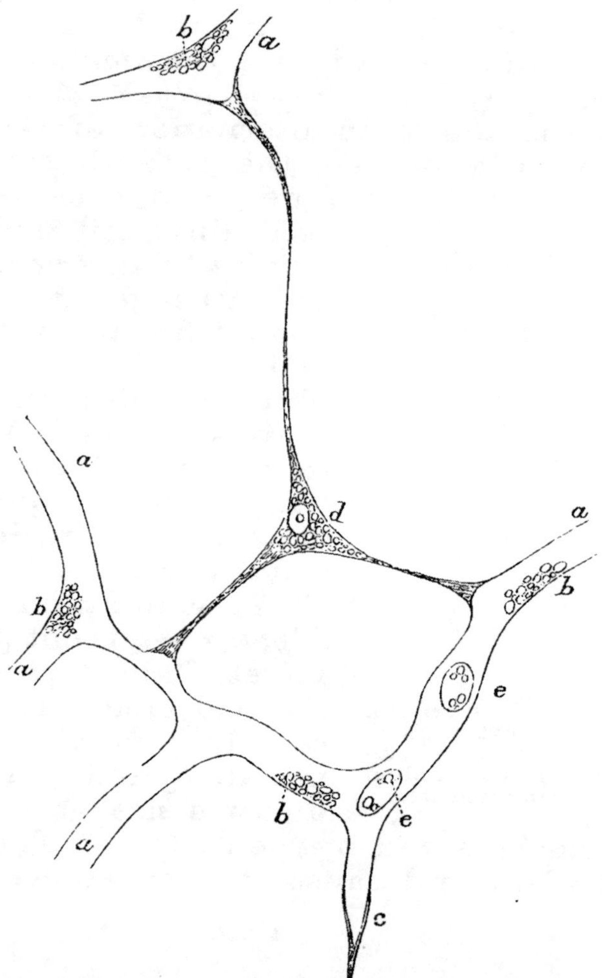

Formation of Capillaries in tail of Tadpole :— *a, a,* **capillaries permeable to blood ;** *b, b,* **fat granules attached to the walls of the vessels, and concealing the nuclei ;** *c,* **hollow prolongation of a capillary ending in a point;** *d,* **a branching cell, with nucleus and fat-granules, communicating by three branches with capillaries already formed ;** *e,* **blood-corpuscles, still containing granules of fat.**

223. When we examine an ordinary Muscle with the naked eye, we

observe that it is made up of a number of *fasciculi* or bundles of fibres; which are arranged side by side with great regularity, in the direction in which the muscle is to act; and which are united by areolar tissue. These fasciculi may be separated into smaller parts, which appear like simple fibres; but when these are examined by the microscope, they are found to be themselves fasciculi, composed of minuter fibres bound together by delicate filaments of areolar tissue. By carefully separating these, we may obtain the ultimate *Muscular Fibre.* This fibre exists under two forms, the *striated* and the *non-striated.* The former is chiefly distinguished by the *transversely-striated* appearance which it presents, and which is due to an alternation of light and dark spaces along its whole extent (Fig. 55, *a*); the breadth and distance of these striæ vary, however, in different fibres, and even in different parts of the same fibre, according to its state of contraction or relaxation. The *longitudinal striæ,* which are also frequently visible (*b*), are rather due to a partial separation between the component fibrillæ, into which the fibre may be broken up. Neither class of striæ shows itself in the *non-striated* fibre (Fig. 60) which more resembles that of the simple fibrous tissues.—When a fibre of the former kind is more closely examined, it is seen to consist of a delicate tubular sheath, quite distinct on the one hand from the areolar tissue which binds the fibres into fasciculi, and equally distinct from the internal substance of the fibre. This membranous tube, which has been termed the *Myolemma,* is not perforated either by nerves or capillary vessels; and forms, in fact, a complete barrier between the real elements of Muscular structure and the surrounding parts. These elements appear to be very minute cylindrical particles with flattened faces of nearly uniform size, and adherent to each other both by their flat surfaces and by their edges. The former adhesion is usually the most powerful; and causes the substance of the fibre, when it is broken up, to present itself in the form of delicate *fibrillæ,* each of which is composed of a single *row* of the primitive particles (Fig. 56). On the other hand, the lateral adhesion is sometimes the stronger; and causes the fibre to break across into *disks,* each of which is composed of a *layer* of the primitive particles (Fig. 57). That the fibre is a solid collection of these elementary parts, and not hollow in the centre, as some have supposed, is shown by making a thin transverse section of a fasciculus (Fig. 58); by which also it is seen, that the fibre is rather polygonal than cylindrical. The size of the fibres is subject to great variation, not merely in different classes of animals, but in different species, in the two sexes of the same species, and even in different parts of the same muscle. Thus

FIG. 55.

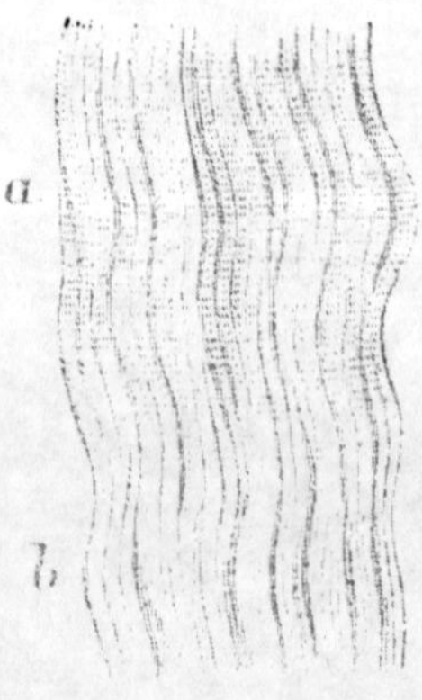

Fasciculus of striated Muscular Fibre, showing at *a* the transverse striæ, and at *b*, the longitudinal striæ, more distinctly.

FIG. 56.

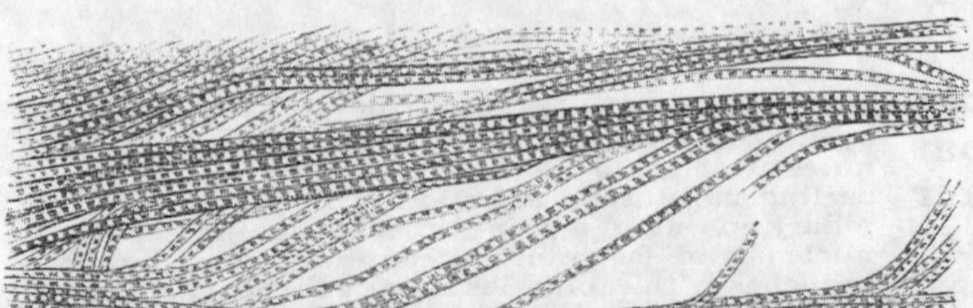

Striated Muscular fibre separating into fibrillæ.

Mr. Bowman estimates the *average* diameter of the fibres in the Human *male* at 1-352nd of an inch ; the *largest* being 1-192nd, and the *smallest* 1-507th. In the *female*, he found the *average* to be 1-454th of an inch ; whilst the largest was 1-384th, and the smallest 1-615th. The average

Fig. 57.

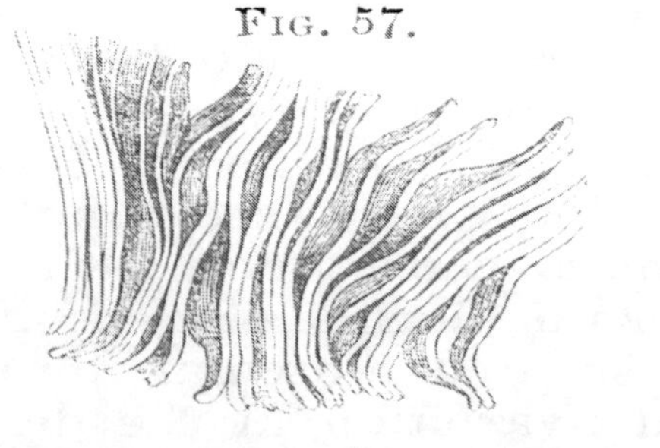

An ultimate fibre, in which the transverse splitting into discs, in the direction of the striation of the ultimate fibrils, is seen.

Fig. 58.

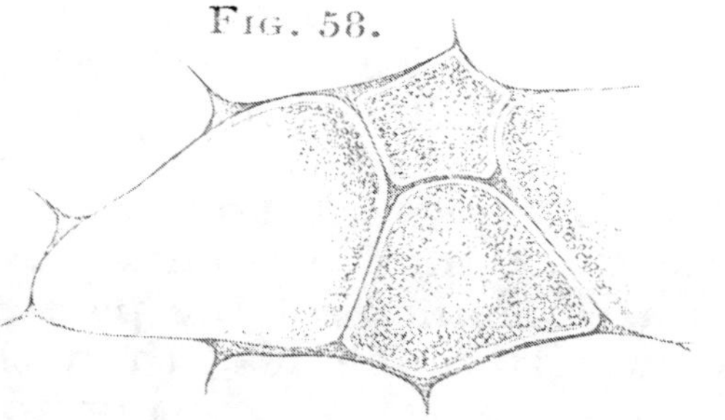

Transverse section of ultimate fibres of the biceps. In this figure the polygonal form of the fibres is seen, and their composition of ultimate fibrils.

size of the Muscular fibre is greater among Reptiles and Fishes, than in other Vertebrata ; but on the other hand the extremes are much wider. Thus its dimensions vary in the Frog from 1-100th to 1-1000th of an inch ; and in the Skate from 1-65th to 1-300th. In Insects the fibres often present the form of flattened bands, on which the transverse striæ are very beautifully marked ; and in some minute Worms and Zoophytes they seem to consist of a very small number of ultimate fibrillæ.

224. When the fibrillæ are separately examined, under a high magnifying power, they are seen to present a cylindrical or slightly-beaded form, and their linearly-aggregated particles then appear to be *cells*. We observe the same alternation of light and dark spaces, as when the fibrillæ are united into fibres or into small bundles ; but it may be distinctly seen, that each light space is divided by a transverse line ; and that there is a pellucid border at the *sides* of the dark spaces, as well as between their contiguous extremities (Fig. 59). This pellucid border seems to be the cell-wall ; the dark space enclosed by it (which is usually bright in the centre) being the cavity of the cell, which is filled with a highly-refracting substance. When the fibril is in a state of relaxation, as seen at *a*, the diameter of the cells is greatest in the longitudinal direction : but when it is contracted, the fibril increases in diameter as it diminishes in length ; so that the transverse diameter of each cell becomes equal to the longitudinal diameter, as seen at *b* ; or even exceeds it.—Thus the act of Muscular contraction seems to consist in a change of form in the cells of the ultimate fibrillæ, consequent upon an attraction between the walls of their two extremities ; and it is interesting to observe, how very closely it thus corresponds with the contraction of certain Vegetable tissues, of which the component cells (§ 138) are capable of producing movements, when they are irritated, by means of a similar change of form. The essential difference, therefore, between the Muscular tissue

Fig. 59.

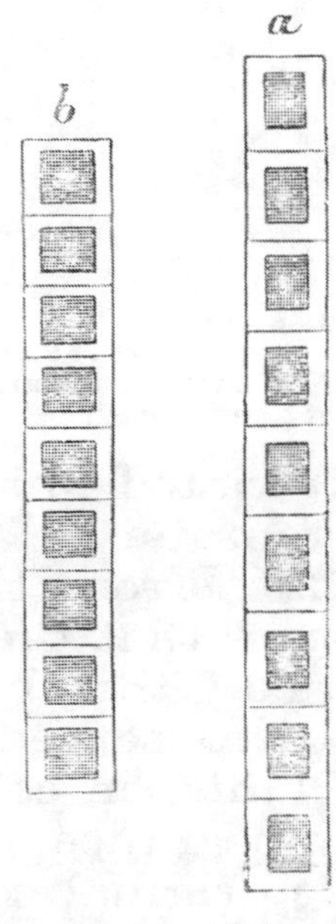

Structure of the ultimate fibrillæ of striated muscular fibre:—*a*, a fibril in a state of ordinary relaxation; *b*, a fibril in a state of partial contraction.

of Animals, and the contractile tissues of Plants, consists in the subjection of the former to nervous influence (§ 232).—The diameter of the ultimate fibrillæ will of course be subject to variations, in accordance with their contracted or relaxed condition; but seems to be otherwise tolerably uniform in different animals, being for the most part about 1-10,000th of an inch. It has been observed, however, as high as 1-5000th of an inch, and as low as 1-20,000th, even when not put upon the stretch.

225. The *non-striated* Muscular fibre consists of a series of filaments, which do *not* present transverse markings; but which are tubular like the preceding, their contents having a granular consistence, without any definite arrangement of the particles into disks or fibrillæ (Fig. 60). Their size is usually much less than that of the striated muscular fibre; but owing to the extreme variation in the degree of flattening which they undergo, it is difficult to make even an average estimate of their dimensions. They generally present enlargements at intervals, caused by the persistence of their nuclei. These fibres are, like those of the other muscles, arranged in a parallel manner into bands or fasciculi; but these fasciculi are generally interwoven into a network, without having any fixed points of attachment. It is of this kind of structure, in Vertebrated animals, that the proper muscular coat of the alimentary canal and of the bladder, and the substance of the pregnant uterus are composed; and it is found in no inconsiderable amount in the trachea and bronchial tubes. Such organs as these make up, as we shall hereafter see, a large part of the body of the Mollusca in general (§ 297); and we find the non-striated fibre so predominant in them, as to be even for the most part employed in their locomotive apparatus, the actions of which are usually slow. On the other hand, in Articulated animals, which are generally remarkable for the high development of their motor powers (§ 308), the striated fibre prevails; as it does also in the locomotive apparatus of Vertebrata.

FIG. 60.

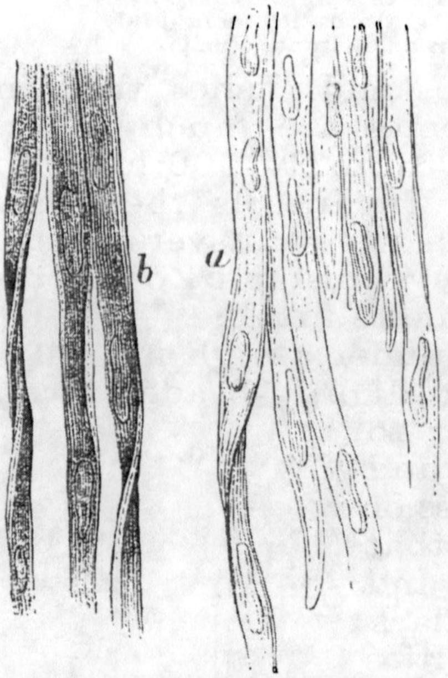

Non-striated Muscular Fibre, at *b*, in its natural state; at *a*, showing the nuclei after the action of acetic acid.

226. From the study of the early development of Muscular Fibre, it appears that the Myolemma, or external transparent tube, is the part first formed; this being distinctly visible, long before any traces of fibrillæ can be observed in it. This tube takes its origin, like the straight ducts of Plants, in cells laid end to end; the cavities of which coalesce, by the disappearance of the partitions, at a subsequent period. The nuclei of these original cells may be distinctly seen, for some time after the appearance of the transverse striæ, which indicate the formation of the fibrillæ in their interior; and they project considerably from the sides of the fibres. In the fully-formed striated fibre, however, they are not perceptible, except when it is treated with weak acid; the effect of which is to render the nuclei more opaque, whilst the surrounding structure becomes more transparent (Fig. 61). They are usually numerous in proportion to the size of

the fibre. There is every probability that these nuclei continue to act, like the "germinal spots" of the glandular follicles, as centres of nutrition ; from which the minute cells that compose the fibrillæ are developed as they are required. The diameter of the Muscular Fibre of the fœtus is not above one-third of that which it possesses in the adult ; and as the *size* of the ultimate particles is the same in both cases, their *number* must be greatly multiplied during the growth of the structure. But we shall find reason to believe, that the decay of these particles is constantly taking place, with a rapidity proportional to the functional activity of the Muscle ; and their generation, which occurs as continually, when the nutrient operations proceed in their regular course, is probably accomplished by a development from these centres, at the expense of the blood with which the Muscle is copiously supplied.—From the preceding history it appears, that there is no difference, at an early stage of development, between the striated and non-striated forms of muscular fibre. Both are simple tubes, containing a granular matter in which no definite arrangement can be traced, and presenting enlargements occasioned by the presence of the nuclei. But whilst the striated fibre goes on in its development, until the cells of the fibrillæ are fully produced, the non-striated fibre retains throughout life its original embryonic condition, the contents of the tube remaining granular.

FIG. 61.

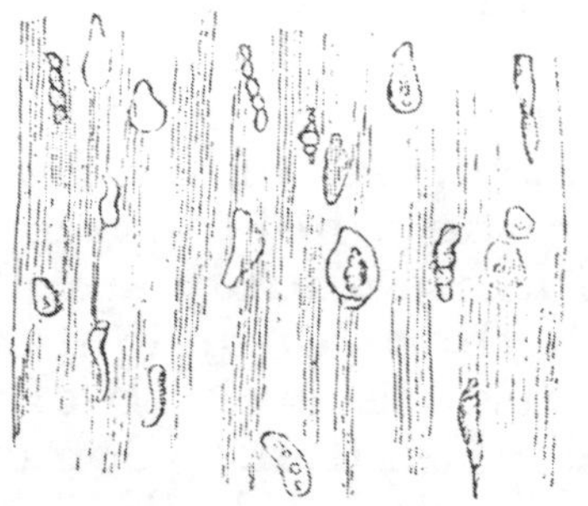

Mass of ultimate fibres from the pectoralis major of the human fœtus, at nine months. These fibres have been immersed in a solution of tartaric acid ; and their numerous corpuscles, turned in various directions, some presenting nucleoli, are shown.

227. We have seen that the Muscular tissue, properly so called, is as extra-vascular as cartilage or dentine ; for its fibres are not penetrated by vessels ; and the nutriment required for the growth of its contained matter is drawn by absorption through the myolemma. But the substance of Muscle is extremely vascular ; the capillary vessels being distributed in nearly parallel lines, in the minute interspaces between the fibres (Fig. 51) ; so that it is probable that there is no fibre, which is not in close relation with a capillary. Hence there is every provision for the active nutrition of this tissue ; the arterial circulation bringing the materials for its growth and renovation ; whilst the venous conveys away the products of the waste or disintegration, which is consequent upon its active exercise. The supply of blood is not merely requisite for the *nutrition* of the muscular tissue ; but it also affords a condition which is requisite for its *action*. This condition is Oxygen. It is not enough that blood should circulate through the muscles ; for that blood, to exercise any beneficial influence, must be arterialized. Consequently the muscles of warm-blooded animals soon lose their contractile power, after the supply of arterial blood has been suspended, either by the cessation of the circulation, or by the want of aeration of the blood ; but those of cold-blooded animals preserve their properties for a much longer period, in accordance with the general principle formerly stated,—that, the lower the usual amount of vital energy, the longer is its persistence, after the withdrawal of the conditions on which it is dependent.—The Muscles composed of striated fibre are, of all the tissues except the Skin, the most copiously supplied with Nerves.

These, like the blood-vessels, appear to lie on the *outside* of the Myolemma of each fibre, and their influence must consequently be exerted through it; their arrangement is shown in Fig. 64.

228. Every Muscular Fibre, of the striated kind at least, is attached at its extremities to fibrous tissue; through the medium of which it exerts its contractile power on the bone or other substance, which it is destined to move. The tendon appears to unite itself abruptly with the myolemma at the truncated extremity of the muscular fibre; but from the recent observations of Dr. Leidy, it appears that the tendinous fibres are prolonged over the whole myolemma, crossing it diagonally in both directions, so as to form a doubly-spiral extensible sheath, some filaments of which interlace with those of neighbouring fibres. Thus the whole muscle is penetrated by minute fasciculi of tendinous fibres; and these collect at its extremities into a tendon. Sometimes the muscular fibres are attached obliquely to the tendon, which forms a broad band that does not subdivide; this is seen in the legs of Insects and Crustacea, in which the muscular fibres have what is called a *penniform* arrangement, being inserted into the tendon, on either side, like the laminæ of a feather into its stem.—The *forms* which different muscles present, have reference purely to the mechanical purposes, which they have respectively to accomplish. The elements are the same in all, both as regards structure and properties.

229. It is probable that the pure Muscular Fibre is identical in Chemical composition,—or nearly so,—with the Fibrine of the blood. It is, however, impossible to separate it completely from the areolar tissue, nerves, blood-vessels, fatty matter, &c., which enter into the substance of the muscle; so that it cannot be precisely analysed. In ordinary muscle, the solid matter forms about 23 parts in 100; the remainder consisting of water.—The solid matter contains about 7½ per cent of fixed salts.

230. We now come to investigate the remarkable property, which is the distinguishing characteristic of Muscular tissue; the Contractility called into action by a stimulus. Some approaches to this property are manifested, as we have seen (§ 138), by certain Vegetable structures; and the simplest form under which it manifests itself in the Animal body, bears a close relationship with that which is displayed in Plants. For among some of the lowest organisms included in the Animal Kingdom, we find the movements of the body effected only by the change of form in the cells of which the entire fabric is composed (§ 286), or by the contractions of the simple fibrous tissue, which is the principal constituent of some others (§ 289); and it is only as we ascend the scale, and find the various actions of the body distributed among structures more expressly adapted to perform them, that we meet with a distinct Muscular tissue. Again, the non-striated fibre of the alimentary canal, which is subservient to the functions of Vegetative life alone, is called into action much more readily by a stimulus directly applied to itself, than it is in any other mode. Such is not the case, however, with the striated fibre, of which the muscles of Animal life are composed; this being much more readily called into action by a peculiar stimulus conveyed through the nerves supplying those muscles, than by any other more directly applied to them.

231. That the Contractility of Muscles is a property inherent in them, and in this respect analogous to the peculiar vital endowments of any other forms of tissue, cannot be any longer a matter of doubt;—though

many Physiologists have sought to show, that it is in some way derived from the nerves. Not only may an entire Muscle be made to contract, by the application of a proper stimulus, long after the division of the nervous trunks supplying it, or after they have been rendered incapable of conducting the influence of stimuli, by the application of narcotic agents ; but even a single fibre, completely isolated from all its nervous connexions, may be seen to contract under the Microscope. Moreover, in the non-striated muscular fibre, it is often difficult to excite contractions through the nerves at all, when a stimulus directly applied to itself will immediately produce sensible and vigorous movements.—The energy of the contractile power depends in great part upon the state of nutrition of the muscle ; and this, again, is influenced by the degree in which it is exercised. Now as the Muscles of Animal Life are all excited to action, in the usual state of things, through the medium of their nerves, it follows that if the nerves be paralysed, the muscles will be seldom or never called into use. When disused, they will receive very little nourishment ; the disintegrating changes will not be counterbalanced by reparative processes ; and in consequence, the muscular structure will be gradually so far impaired as to lose its peculiar properties, and will even, in time, almost totally disappear. Yet even after the almost complete departure of muscular contractility, through the metamorphosis of the structure consequent upon disuse, it may be again recovered, if the muscles be called into exercise ; but the recovery of the power is very slow, and proceeds *pari passu* with the improvement in the nutrition of the part, being more tedious in proportion to the length of the previous disuse.—Various attempts have been made to show that the contraction of Muscle is an electrical phenomenon ; but no proof has been given that such is the case ; and every probability seems to be in favour of its being one of the manifestations of the Vital force, which is itself dependent upon the preceding nutritive operations (§ 51). But still we may regard it as 'correlated' alike to the Nervous agency, to Electricity, and to other Physical Forces. For we find that whilst it is especially capable of being called into action by the Nerve-force, it is in its turn capable of exciting the Nerve-force, either directly, or through the medium of Electricity. For if the nerve of a galvanoscopic frog be laid across a muscle, so as to touch it at two points, and this muscle be excited to contraction, similar contraction will be thereby induced in the leg of the galvanoscopic frog.*—But it is not only through the medium of Nervous agency, that the Muscular force may be developed ; for all Muscle which retains its vitality may be made to contract by a Physical stimulus applied *directly* to itself ; and this stimulus may be of different kinds. The simplest is the contact of a solid substance ; thus we may excite muscular contractions by simply touching the fibre, just as we cause contraction in the tissue of the Dionæa or Sensitive plant. Most substances of strong Chemical action, such as acids and alkalies, will call forth the contractility of muscular fibre, when applied directly to it ; and the same result is produced by Heat, Cold, and Electricity,—the last-named agent being the most powerful of all.

* For this remarkable discovery we are indebted to Prof. Matteucci ; who, in his latest reference to it (Dr. Pereira's Edition of Matteucci's "Lectures on the Physical Phenomena of Living Beings") expresses himself as unable to say with certainty, whether or not the medium of this induction is an Electric current generated during Muscular contraction.

Conversely, every exertion of Muscular force involves Chemical action (§ 236); and there are indications that it may also generate Heat, Electricity, and even Light (CHAP. XVII.). Thus it may be said to be as closely 'correlated' to these Physical Forces, as these are to each other, although not identical with any one of them.

232. The effect of the application of any of these stimuli varies considerably, according to the kind of Muscle on which it is exerted. If we irritate a portion of any ordinary muscle, composed of striated fibre, the fasciculus of fibres which is touched will immediately contract, and that one only; and the contracted fasciculus will soon relax, without communicating its movement to any other, unless a powerful irritation be rapidly repeated, as when successive shocks are given by an electro-magnetic machine.—If we irritate a portion of non-striated fibre, however, as that of the Alimentary canal, the fasciculus which is stimulated will contract less suddenly, but ultimately to a greater amount; its relaxation will be less speedy; and before it takes place, other fasciculi in the neighbourhood begin to contract; their contraction propagates itself to others; and so on. In this manner, successive contractions and relaxations may be produced through a considerable part of the canal, by a single touch; a sort of wave of contraction being transmitted in the direction of its length, and this being followed by relaxation.—In the Heart, again, the muscular structure of a large part of the organ is thrown into rapid and energetic contraction, by a stimulus applied at any one point; and this contraction is speedily followed by relaxation.—On the other hand, when the stimuli which excite muscular contraction are applied to the Nerve, which supplies a muscle composed of striated fibre, they produce a simultaneous contraction in the whole muscle; the effect of the stimulus being at once exerted upon every part of it. In the ordinary action of such muscles, the nervous system is always the channel through which they are called into play, whether to carry into effect the determinations of the mind, or to perform some office necessary to the continuance of life, such as the "automatic" movements concerned in Respiration (§ 244). The nerves of the striated fibre are all derived at once from the cerebro-spinal centres, or from the ganglia that represent these organs in Invertebrated animals. The ordinary actions of the non-striated fibre, on the contrary, are executed (in the higher animals at least) in respondence to stimuli applied directly to themselves. It is so difficult to excite contractions in it through the medium of its nerves, that many Physiologists have denied the possibility of doing so: and the nerves lose their power of conveying the influence of stimuli very soon after death, although the contractility of the muscles may remain for a considerable time. The nerves of the non-striated fibre are chiefly those belonging to the Sympathetic system; but, as will be shown hereafter (CHAP. XX.), those which excite motion are probably derived in reality from the Cerebro-spinal system, through the communicating branches which unite the two.

233. When a Muscle is thrown into contraction, its bulk does not appear to be at all affected. Its extremities approach, so that it is shortened in the direction of its fibres; but its diameter enlarges in the same proportion. It was formerly supposed that the ultimate fibres, in the act of contraction, throw themselves into ziz-zag folds; but this is now well ascertained not to be the case. The fibre, like the entire muscle,

preserves its straight direction in shortening, and increases in diameter. The fibrillæ themselves, as already mentioned (§ 224), exhibit an evident change, in regard to the distances of their successive light and dark portions ; and the fibre, which is made up of these, displays, in its contracted state, a very close approximation of the transverse striæ ; to such an extent that they become two, three, or even four times as numerous in a given length, as they are in a similar length of a non-contracted fibre. It appears that, even when considerable force of contraction is being exerted, the *whole* fibre is seldom or never in contraction at once ; but that a continual interchange is taking place amongst its different parts,—some of them passing from the contracted to the relaxed state, as shown by the separation of the transverse striæ,—whilst others are taking up the duty, and passing from the relaxed to the contracted condition, as shown by the approximation of the striæ. But it is not only among the different parts of the individual fibres, that this interchange seems to take place. There is good reason to believe that, when a muscle is kept in a contracted state, by an effort of the will, for any length of time, only a part of its fibres are in contraction at any one time ; but that a constant interchange of condition takes place amongst them, some contracting whilst others are relaxing, so that the entire muscle remains contracted, whilst the state of every individual fibre may have undergone a succession of alterations. When the ear is applied to a muscle in vigorous action, an exceedingly rapid faint silvery vibration is heard, which seems to be attributable to this constant movement in its substance.—Hence it appears that the prolongation of the contraction of a muscle, through any length of time, is not opposed to the fact that, in the individual fibres, relaxation speedily follows contraction ; but is only a peculiar manifestation of it. The ordinary movements of the Heart exhibit a different manifestation ; its fibres contracting simultaneously, and relaxing together, instead of alternating amongst themselves like those of a voluntary muscle.

234. It has been already pointed out, how close is the dependance of the property of Contractility upon the due *nutrition* of the tissue ; but it cannot be long exercised except under another condition, which is consequently of almost equal importance,—the circulation of oxygenated blood through the substance of the muscle. The length of time during which the contractility remains, after the circulation has ceased, has been shown by Dr. M. Hall to vary inversely to the activity of the respiration of the animal. Thus in cold-blooded animals, the standard of whose respiration is low, the contractility remains for many hours after death, even in the voluntary muscles ; and the muscles of organic life retain it with great tenacity ; the heart of a Frog, for example, will go on pulsating for many hours after its removal from the body. An exceedingly feeble Galvanic current is sufficient to excite the muscles of these animals to contraction ; so that Matteucci, in his experiments upon Animal Electricity, has been accustomed to use the prepared hind-leg of a Frog as the best indicator of the passage of an electric current. Among warm-blooded animals, the same rule holds good, in regard to the inverse proportion of the duration of irritability and the amount of respiration ; for the muscles of Birds lose this property at an earlier period after the cessation of the circulation, than do those of Mammals.

235. That the circulation of arterial or oxygenated blood through the

muscles, is the essential condition of the continuance of their irritability, appears from this,—that after the general death of the system, and even after the removal of the brain and spinal cord, the muscles will preserve their irritability, and the action of the heart itself will continue for a long time, provided that the circulation be kept up through the lungs by artificial respiration. But if, whilst the general circulation continues, the circulation through a particular muscular part be interrupted, that organ will lose its contractility earlier than usual. Further, if blood charged with carbonic acid, instead of with oxygen, circulate through the muscles, their irritability is speedily impaired, and is even destroyed. This is best seen, when animals are killed by being caused to breathe an atmosphere highly charged with carbonic acid; for the irritability of their muscles departs as soon as they are dead. In fact, the destruction of the irritability of the heart, by the circulation of venous blood through its substance, is one of the immediate causes of death. A similar effect is produced by the respiration of other gases, which are either poisonous in themselves, or which prevent that interchange of carbonic acid and oxygen which ought to take place in the lungs. On the other hand, when animals have been made to respire oxygen, and their blood has been consequently highly arterialized, the contractility of their muscles is retained for a longer time than usual.—Hence we may conclude that the presence of Oxygen in the blood is one of the conditions of muscular contraction; although a much smaller amount of it suffices in the case of cold-blooded, than in that of warm-blooded animals. It is interesting to remark, that the muscles of *hybernating* warm-blooded Mammals (§ 94) are reduced for a time to the level of those of cold-blooded animals, their contractility being retained almost as long as that of the latter;—thus confirming the general principle already stated, as to the relation between the amount of respiration, and the duration of the irritability.

236. The Muscles, as we have seen, are largely supplied with blood; and the flow of blood into them increases with the use that is made of them. The demand for nutrition is obviously augmented, in proportion to the activity of the exercise of the Muscular system; for the slightest observation suffices to show, that a much smaller amount of nourishment is sufficient to sustain the body in its normal condition, when the muscular system is not actively exercised, than when it is in energetic operation. The quantity of food which is ample for an individual leading an inactive life, is far too little for the same person in the full exercise of his muscular powers. Again, there is evidence derived from the relative amount of urea and other azotized matters excreted from the body under different circumstances, that a *waste* or *disintegration* of the muscular tissue takes place, whenever it is being actively employed; and this in a degree strictly proportional to the amount of force which it is called upon to exert. In fact, it would appear that this waste is a necessary consequence of the exercise of the muscle;—every act of contraction involving the death and decomposition of a certain amount of tissue. And as the presence of oxygen is always necessary for the decomposition of organic substances, so do we find that the penetration of the muscular tissue by oxygenated blood is essential to the manifestation of its contractile power. Every act of contraction, then, may be said to involve the Death of a certain amount of muscular tissue; and the products of decomposition,

which consist of the elements of muscular fibre united with the oxygen of the arterial blood, are carried off by the venous current. On the other hand, the muscular substance is repaired by an act of nutrition, at the expense of the fibrine supplied to it by the arterial circulation. The *rest* of muscles whose action involves *fatigue,* is essential to the recovery of their powers ; and this recovery is due to the restoration of their original condition by means of the nutritive operations, which then take place unchecked, and which repair the losses previously sustained. There are certain muscles, however,—as the heart, and the muscles of respiration,—whose action is necessarily constant and does not involve fatigue ; the reparation of these must take place as unceasingly as their waste. The permanently-increased flow of blood to a muscle, which takes place when it is continually being called into vigorous action, is thus on the one hand occasioned by the demand for oxygenated blood created by its use, whilst on the other hand it tends to increase the power of the muscle, by an augmentation of its nutrition. Hence it is, that, the more a muscle is exercised, the more vigorous and more bulky does it become, as we see in the arms of the smith and in the legs of the opera-dancer ; and this is equally the case, whether the exercise of the muscle be voluntary or not.

237. We have lastly to treat of the *Nervous Tissue,* which, taken as a whole, is the instrument of those operations that peculiarly distinguish the Animal from the Plant ; whilst it serves many additional purposes, connected with the Organic or Vegetative functions, which the peculiar arrangements of the Animal body involve. Wherever a distinct Nervous System can be made out (which has not yet been found possible in the lowest Animals), it consists of two very different forms of structure ; the presence of both of which, therefore, is essential to our idea of it as a whole. We observe, in the first place, that it is formed of *trunks,* which are distributed to the different parts of the body, especially to the muscles and to the sensory surfaces ; and of *ganglia,* which sometimes appear merely as knots or enlargements on those trunks, but which, in other cases, have rather the character of central masses, from which the trunks proceed. Now it is easily established by experiment, that the *active powers* of the nervous system reside in the *ganglia;* and that the *trunks* serve merely as *conductors* of the influence which is to be propagated towards or from them. For if a trunk be divided in any part of its course, all the parts to which the portion thus cut off from the ganglion is distributed, are completely paralysed ; that is, no impression made upon them is felt as a sensation, and no motion can be excited in them by any act of the mind. Or if the substance of the ganglion be destroyed, all the parts, which are exclusively supplied by nervous trunks proceeding from it, are in like manner paralysed. But if, when a trunk is divided, the portion still connected with the ganglion be pinched, or otherwise irritated, sensations are felt, which are referred to the points supplied by the separated portion of the trunk ; which shows that the part remaining in connexion with the ganglion is still capable of conveying impressions, and that the ganglion itself receives these impressions and makes them felt as sensations. On the other hand, if the separated portion of the trunk be irritated, motions are excited in the muscles which it supplies ; showing that it is still capable of conveying the motor influence, though cut off from the usual source of that influence.

238. When we minutely examine the trunks of the nerves, we find that they are composed, in the first place, of a *Neurilemma* or nerve-sheath, consisting of white fibrous tissue; the office of which is evidently that of protecting the nerve-tubes, and of isolating them from the surrounding structures, at the same time that it allows blood-vessels to pass into the interior of the trunk. From the interior of the neurilemma, thin layers of areolar tissue pass into the midst of the enclosed bundle of nervous fibres; separating it into numerous smaller fasciculi, which are thus bound together and supplied with blood-vessels. When the neurilemma has been removed, and the trunk has been separated into its component fasciculi, we may still further subdivide the fasciculi themselves, by careful dissection, until we arrive at the ultimate Nervous Fibre, which is the essential element of the structure.—Two forms of this fibre exist in the nerves of higher animals, bearing a considerable analogy to the two forms of the Muscular fibre; one being termed the *tubular*, and the other the *gelatinous*. The former, which seems to be in the condition of more complete development than the latter, is composed externally of a very delicate transparent membrane, which is apparently quite homogeneous; this is obviously analogous to the myolemma of the Muscular fibre, and serves, like it, to isolate the contained substance most completely from surrounding structures. This membranous tube is not penetrated by blood-vessels, nor does it branch or anastomose with others; and there is reason to believe it to be continuous from the origin to the termination of the nervous trunk. Within the tube is a hollow cylinder of a material known as the *White substance of Schwann*, which differs in composition and refracting power from the matter that occupies the centre of the tube, and of which the outer and inner boundaries are marked out by two distinct lines. And the centre or axis of the tube is occupied by a transparent substance, which is termed the *axis-cylinder*. There is reason to believe that this last is the essential component of the nervous fibre; and that the hollow cylinder which surrounds it, serves, like the external investment, chiefly for its complete isolation. The whole of the matter contained in the tubular sheath is extremely soft; yielding to very slight pressure. The tubular sheath itself varies in density in different parts; being stronger in the nervous trunks, than in the substance of the brain and spinal cord. The diameter of the tubuli is usually between 1-2000th and 1-4000th of an inch. Sometimes, however, it is as much as 1-1500th; and occasionally as little as 1-14000th. They are larger in the nerve-trunks than in the brain; and they diminish in the latter as they approach the cortical substance. The fibres of the nerves of special sense are smaller than the average, in every part of their course.—The *gelatinous* fibres cannot be shown to consist of the same variety of parts as the preceding; no tubular envelope can be distinguished; and the white substance of Schwann seems wanting. They are flattened, soft, and homogeneous in their appearance, bearing a resemblance in these respects to the unstriped Muscular fibres; and, like them, they contain numerous cell-nuclei, which are arranged with tolerable regularity. These nuclei are brought into view by acetic acid, which dissolves the rest of the fibre, leaving them unchanged. These fibres are usually of smaller size than the tubular, their diameter averaging between the 1-6000th and the 1-4000th of an inch; and they sometimes show a

disposition to split into very delicate fibrillæ. Being of a yellowish-grey colour, they have been sometimes distinguished as the *grey* fibres.—These two classes of fibres have been supposed to be essentially distinct in character and function; the 'tubular' having been regarded as ministering to the Animal functions of Sensation and Motion; and the 'gelatinous' as connected with the Organic or Nutritive. The facts which will be presently stated regarding their origin, however, as well as their joint existence in almost every nerve, are decidedly adverse to this view; and we shall find reason to consider them as differing chiefly in grade of development. Both classes of fibres appear to run continuously, from one extremity of the nervous cord to the other, without anything like union or anastomosis; each ultimate fibre probably having its distinct office, which it cannot share with another. The fasciculi, or bundles of fibres, however, occasionally intermix and exchange fibres with each other; and this interchange may take place among either the fasciculi of the same trunk, or among those of different trunks. Its object is evidently to diffuse among the different branches the endowments of a particular set of fibres.

239. The second primary element of the Nervous System, without which the fibrous portion would seem to be totally inoperative, is composed of nucleated cells, containing a finely granular substance, and lying somewhat loosely in the midst of a minute plexus of blood-vessels. Their original form may be regarded as globular (hence they have been termed *nerve-* or *ganglion-globules*); but they often present an extension into one or more long processes, which gives them a *caudate* or a *stellate* aspect (Fig. 62). These processes are composed of a finely-granular substance, resembling that of the interior of the vesicle, with which they seem to be distinctly continuous; and if traced to a distance, they are frequently found to become continuous with the axis-cylinders of the nerve-tubes. Besides the finely-granular substance just mentioned, these cells usually contain a collection of pigment-granules, which give them a reddish or yellowish-brown colour. This, however, is frequently absent, especially among the lower animals.—The size of the vesicles is liable to great variation; the globular ones are usually between 1-300th and 1-1250th of an inch in diameter. The substance, which is made up of these peculiar cells, of the plexus of blood-vessels in which they lie, and of the granular matter that is disposed amongst them, is

Fig. 62.

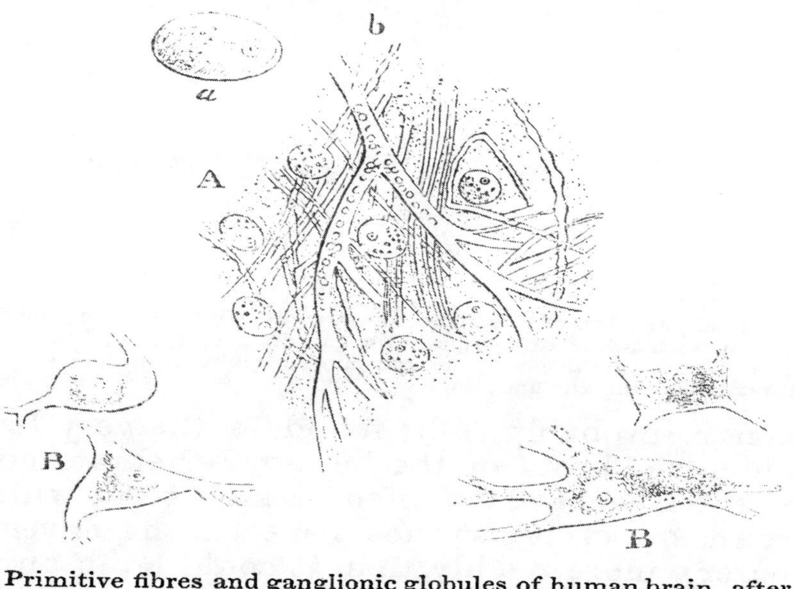

Primitive fibres and ganglionic globules of human brain, after Purkinje. A, ganglionic globules lying amongst varicose nerve-tubes, and blood-vessels, in substance of optic thalamus; *a*, globule more enlarged; *b*, small vascular trunk. B, B, globules with variously-formed prolongations, from dark portion of crus cerebri.

altogether commonly known as the *cineritious* or *grey* substance; being distinguished by its colour, in Man and the higher animals at least, from the *white* substance (composed of nerve-tubes) of which the trunks of the nerves, as well as a large part of the brain and spinal cord, are made up. But this distinction is by no means constant; for the grey colour, which is partly due to the pigment-granules of the cells, and partly to the redness of the blood in the vessels, is wanting in the Invertebrata generally, and is not characteristically seen in the classes of Fishes and Reptiles. Moreover, when the ganglionic substance exists in small amount, even in Man, its colour is not sufficiently intense to serve to distinguish it; and, as we have already seen, there are nerve-fibres which possess a greyish hue. The real distinction evidently lies in the *form* of the ultimate structure, which is *fibrous* in the one case, and *cellular* or *vesicular* in the other; and these terms will be henceforth used to characterize the two kinds of Nervous tissue, which have been now described.

240. A *ganglion*, then, essentially consists of a collection of nerve-vesicles or ganglion-globules, interspersed among nerve-fibres (Fig. 63). When a nerve enters a ganglion, its component fibres separate and pass through the ganglion in different directions, so as to be variously distributed among the branches which pass out of it. The only exception to the general fact, that the vesicular matter occupies the centre of the ganglia, occurs in the Cerebrum of Vertebrata, in which it is chiefly disposed on the exterior, forming the *cortical* envelope. The reason for this variation is probably to be found in the very large amount of this substance, which the brain of the higher Vertebrata contains; and in the necessity of the free access of blood-vessels to it, which is provided for by a great extension of its surface beneath the investing vascular membrane (pia mater), more readily than it could be in any other mode.

FIG. 63.

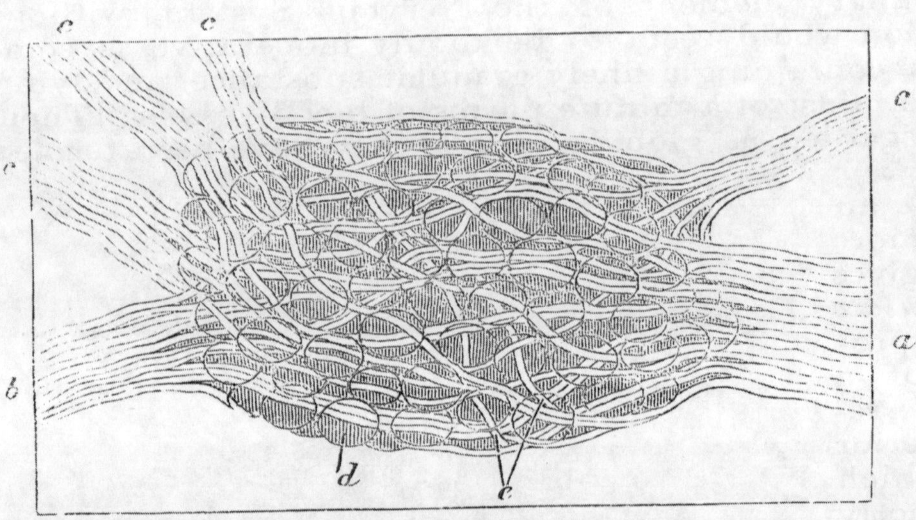

Dorsal ganglion of Sympathetic nerve of Mouse;—*a*, *b*, cords of connexion with adjacent sympathetic ganglia; *c*, *c*, *c*, *c*, branches to the viscera and spinal nerve; *d*, ganglionic globules or cells; *e*, nervous fibres traversing the ganglion.

241. The manner in which the nerve-fibres are connected with the ganglionic vesicles, appears not to be always the same. That some of the fibres are continuous with the prolongations of the vesicles, is now well established; and even some of the globular vesicles seem to be included within dilatations of tubular fibres, the membranous wall of which is expanded over them. But there is also good reason to believe, that many of the nerve-tubes simply come into contact with the globular vesicles, passing between them, or forming loops around them. That some functional diversity is connected with this difference of structural arrange-

ment, can scarcely be doubted ; but of the relative offices of these two kinds of nerve-cells we have no certain knowledge. It may be surmised, however, that the cells from which the nerve-fibres seem to spring, are those by which they are formed ; whilst the globular cells among which they pass, are rather the instruments of their functional changes. This idea derives confirmation from the researches of Kölliker on the peripheral origin of the nerve-fibres; for he has found that, in the tail of the tadpole, the nervous plexuses are formed after the same fashion as the capillary network (§ 221), namely, by the inosculation of the prolongations of radiating cells, whose centres are at a considerable distance from each other. The nerve-fibres of the trunks with which the plexus becomes connected, appear to originate in cells which become fusiform (§ 166) by elongation, and which then coalesce at their extremities ; and these seem to increase, after the first formation of the trunks, by the longitudinal subdivision of fusiform cells which had not previously undergone complete metamorphosis into fibres, or by the development of cells *de novo.** Such a new production of nerve-fibres may continue during the whole period of growth ; and there is ample evidence, in the phenomena of *regeneration* of nervous tissue, that it may take place with great rapidity. The fact of this regeneration is satisfactorily proved, by the return of motor and sensory endowments in a part which had been deprived of them by the division of its nerves, and by the rapidity with which a new and very sensitive skin is often formed in the healing of a wound ; since all our knowledge of the functions of the nervous system leads to the belief, that *perfect continuity* of the nerve-tubes is requisite for the conduction of an impression of any kind. — The nerve-fibre in its incipient state seems always to have the 'gelatinous' character ; and it is not at all uncommon to find that fibres which are tubular in other parts of their course, are gelatinous near their central or peripheral terminations, where they are about to become continuous with a caudate or stellate cell, or to enter the ultimate plexus or network. Hence we may consider the gelalatinous fibre as the early stage of development of the tubular fibre ; but it is not improbable that, as in the case of the Muscular tissue (§ 226), there are many gelatinous fibres which never go on to the higher stage of development ; and these seem to be especially abundant in the Sympathetic System of Vertebrated Animals (CHAP. XX.).†

242. The distribution of the ultimate nerve-fibres in the various organs which their trunks connect with the nervous centres, also appears to differ too much to allow any general statement to be made on the subject. In Muscle, whose nerves may be said to proceed to it from the nervous centres, the trunks subdivide into branches which seem to return as loops either into themselves or into some other trunks (Fig. 64). But from the recent observations of Wagner and others, it appears that these looping fibres do not constitute the ultimate distribution of the nerve-tubes, but that minuter fibres pass off from these, which lose themselves amongst the fibres of the muscle. In various Sensory organs, again, from

* Annales des Sciences Naturelles, Zool., Aout, 1846.

† The results of late inquiries into the relations of these two classes of fibres are admirably summed up by Prof. Sharpey, in his Introduction to the Fifth Edition of Dr. Quain's Anatomy, p. cxc. *et seq.*

which the nerve fibres may be said to pass towards the nervous centres, similar loops have been seen; but it is probable that here, too, there is a more minute distribution of the essential component of the fibres, whose central axes appear to split into fibrillæ, which form a plexus resembling the capillary network; and in this the tubular envelope seems altogether wanting, the fibrillæ apparently coalescing with each other. And in the retina and some other sensory surfaces, the fibres come into relation with vesicular substance, which appears to resemble, in all essential particulars, that of the nervous centres (§ 245).

FIG. 64.

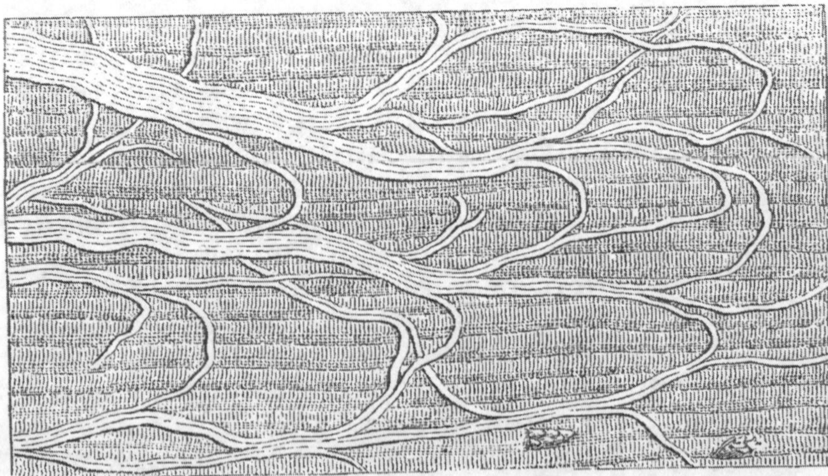

Portion of Muscle, showing the arrangement of the motor nerves supplying it.

243. We have now to speak briefly of the Chemical Composition of the Nervous matter;—a consideration which will be presently shown to be of much importance. As formerly remarked (§ 24), the vital activity of a tissue is usually greater, as the proportion of its solid to its fluid contents is less; and this rule holds good most strikingly in regard to the Nervous substance, the vital activity of which is far greater than that of any other tissue, and the solid matter of which usually constitutes no more than a fourth, and occasionally does not exceed an eighth, of its entire weight. Of the solid matter of the brain, about a third consists of an albuminous substance; which is probably the material of the membrane of the tubuli, as well as of the tissue that connects them. It is chiefly with the Fatty matter, which constitutes about a third of the solid substance, that the attention of Chemists has been occupied. This is stated by M. Fremy (one of the most recent analysts) to contain, besides the ordinary fatty matters, and Cholesterine or biliary fat, two peculiar fatty acids, termed the Cerebric and the Oleo-phosphoric, both of which contain Phosphorus, the latter in large amount. The proportion of phosphorus in the brain is considerable; being from 8 to 18 parts in 1000 of the whole mass, or from 1-20th to 1-30th of the whole solid matter.—Various circumstances lead to the belief, that the Nervous tissue, during the whole period of active life, is continually undergoing changes in its substance, by decay and renewal. We know that, after death, it is one of the first of all the animal tissues to exhibit signs of decomposition; and there is no reason to suppose that this tendency is absent during life. Hence, for the simple maintenance of its normal character, a considerable amount of nutritive change must be required. But many circumstances further lead to the conclusion, that, like all other tissues actively concerned in the vital operations, Nervous matter is subject to a *waste* or *disintegration*, which bears an exact proportion to the activity of its operations; or, in other words, that every act of the Nervous system involves the death and decay of a certain amount of Nervous matter, the replacement of which will be

requisite in order to maintain the system in a state fit for action. Now as the amount of Muscular tissue that has undergone disintegration, is represented (other things being equal) by the quantity of urea in the urine, so do we find that an unusual waste of the *Nervous* matter is indicated by an increase in the amount of *alkaline* phosphates present in the urine, these having their source in the union of the phosphoric acid generated in the oxidation of nervous matter, with the alkaline bases of the blood.*—As the disintegration of the Nervous System is thus proportional to its exercise, so must its reparation make a corresponding demand upon the nutritive processes. And accordingly we find, that it is very copiously supplied with blood-vessels ; and that the amount of food required for its maintenance in an active condition, is very considerable.

244. The functions of the Nervous System,—thus composed of ganglionic cells, and of fibres connected with them,—are two-fold :—first, to bring the conscious mind (using that term in its most extended sense, to denote the psychical endowments of animals in general,) into relation with the external world ; by informing it, through the medium of the organs of sensation, of the changes which material objects undergo ; and by enabling it to re-act upon these, through its motor apparatus ;—and also, to connect and harmonise different actions in the same individual, without necessarily exciting any mental operation. These two sets of purposes, however, are fulfilled by a mechanism of the same kind. An 'impression' made upon some part of the general surface of the body, or upon a special organ of sense, is received by the nerve-fibres which originate in it, and is propagated by them to some part of the central ganglionic apparatus. If the impression reach the portion termed the *sensorium* (which is always seated in the *head*, where this can be distinguished), it then affects the *consciousness* of the animal, and becomes a *sensation*, provided the sensorium be in a condition of activity. The sensation thus produced may give rise to ideas, and these to reasoning processes, which may terminate in an act of the will ; and this, playing as it were upon the nervous apparatus, produces a change in the condition of that portion of the nervous centres from which the motor nerves arise, whence it is propagated by these nerves to certain muscles, and excites them to contraction (§ 232). But the sensation may in itself more directly excite a motor action, without any reasoning process or voluntary effort ; and the movement is then said to be 'automatic.' Further, even if the impression do not reach the sensorium, it may still excite a motor action through some other nervous centres ; and the movement thus produced is 'automatic' like the preceding, from which it differs in not being prompted by a sensation, but in resulting from the conveyance of the simple 'impression' to a portion of the nervous centres capable of originating a respondent action. We shall hereafter see that every living Animal executes a large number of automatic movements, some of them in respondence to sensations, others without any excitement of consciousness ; and that these are, for the most part, intimately connected with the maintenance of the actions necessary to its production, development, and growth, which have

* The amount of *earthy* phosphates present in the urine has been shown by Dr. Bence Jones to depend upon the quantity taken in as food. A deposit of alkaline phosphates may arise from other causes than excessive waste of nervous matter ; but this does not invalidate the above statement.

no direct dependence upon the exercise of the reasoning powers or the exertion of the will (CHAP. XX.).

245. All our knowledge of the functions of the Nervous System leads to the belief, that the office of its Fibrous component is simply *conductive;* its purpose being to receive impressions, and to transmit them from one part of the system to another. On the other hand, it is only through the Vesicular matter, that these impressions can produce any effect upon the conscious mind; and it is from changes which take place in the same substance, that all motor impulses originate. In fact, the vesicular matter of the ganglionic centres, and the nerve-fibres leading from it to the muscles, may be considered as standing in the same kind of relation to each other as the galvanic trough and its conducting wires. How far the origination of the impressions which travel from the periphery towards the centres, is to be attributed to anything like the same nerve-cell influence, is a point still doubtful; but analogy seems in favour of it. For in the retina of the eye, which receives the impressions of luminous rays, and transmits them through the optic nerve to the sensorium, we have a structure closely resembling that of the nervous centres; and it seems probable that the plexuses which appear to constitute the ultimate distribution of the sensory nerves in the skin, are really composed of nerve-cells which have sent out very slender prolongations to inosculate with each other, like those in which capillaries originate (Fig. 54).

246. Of the mode by which the effects of changes in one part of the Nervous System are thus instantaneously transmitted to another, nothing whatever is known. There is evidently a strong *analogy* between this phenomenon, and the instantaneous transmission of the Electric power along good conductors; and there is this farther analogy between the Nervous and Electric agencies, that the latter will produce many of the effects of the former. Thus, a very feeble galvanic current transmitted along a motor nerve, serves to excite contractions in the muscles supplied by it; and in like manner, a galvanic current transmitted along one of the sensory nerves, gives rise to a sensation of the kind to which that nerve ministers. Moreover we shall hereafter see, that certain animals are capable of generating Electric power in a very remarkable manner (CHAP. XVII.); and that the nervous system is in some way essentially concerned in this operation. But, on the other hand, it seems quite certain that the influence transmitted along the nerves of the living body is *not* ordinary electricity; for all attempts to procure manifestations of electric changes in the state of nerves, that are acting most energetically on muscles, have completely failed; and a nerve remains capable of conveying the influence of electricity, when it has been rendered unable to transmit the influence of its ganglion, by tying a ligature round it, or by tightly compressing it between the forceps, whereby no interruption is given to the one agency, whilst the other is completely checked.—Notwithstanding, then, the strong *analogy* which exists between these two powers, we are *not* warranted in regarding them as *identical;* and we should probably best express their true affinity, by saying that they are so closely 'correlated,' that each may be the means of exciting the other. It will perhaps assist us in understanding the correlation between other Vital and Physical phenomena (§ 54), if we examine that of the Nervous and Electrical agencies a little more in detail.—In the first place, then, if a

current of Electricity be made to traverse the trunk of a *motor* nerve for a short distance only, it will produce contraction of the muscles which are supplied from its branches. This cannot be the direct result of the Electric current, which does not itself pass on to the muscles; but must be attributed to the excitement of the Nervous force by its means. On the other hand, if the current be passed for a short distance along a *sensory* nerve, it will excite in the *sensorium* the peculiar sensations which impressions conveyed by that nerve are adapted to produce; that is to say, not merely the ordinary tactile impressions, but those of a special kind also, according as the current is transmitted along the nerves of common sensation, the optic, auditory, olfactive, or gustative; so that, by proper management, we may be made conscious at one and the same time of pricking sensations, of flashes of light, of distinct sounds, of a phosphoric odour, and of a peculiar taste, all excited by the same cause, the transmission of an electric current along sensory nerves. But as these effects are produced when the current merely traverses the nerves for a short distance, we must here too regard them as directly due to the excitement of the Nervous force by the Electrical.—But that, conversely, the Nervous force will excite Electricity, appears from the phenomena displayed by the Electric Fishes, the operation of whose electric organs has been fully proved to depend upon their connexion with the nervous centres, and to vary in intensity according to the amount of that connection (CHAP. XVII. Sect. 3). And if it should be proved that disturbance of electric equilibrium during the contraction of a muscle is the result of nervous agency, rather than of the molecular changes taking place in the muscle itself, this, so far from proving (as some have imagined) that Electricity and Nervous agency are identical, merely confirms the idea of their 'correlation,' by affording another instance of the power of the Nervous force to excite Electricity.

247. But Electricity is by no means the only agent through which the Nervous force may be excited; for *Heat*, when applied to motor nerves, will produce muscular contractions, and, when applied to sensory nerves, will occasion sensations both common and special; whilst, conversely, there are phenomena which have not yet been explained upon the purely chemical doctrine of calorification, and which seem to indicate that heat may sometimes be generated more directly by the expenditure of nervous force (CHAP. XVII. Sect. 2).—Precisely the same may be said of Chemical Affinity; for the application of various reagents may be made to excite both motor and sensory changes in the nerve-trunks; and we shall find strong reason to believe, that the nervous force may produce or modify chemical actions in the parts to which it is conveyed.—The power of Light to excite the Nervous Force is clearly indicated by the influence of this agent on the Optic nerve; whilst, on the other hand, there are certain curious phenomena of luminosity among the *Annelida*, which seem altogether different from the persistent luminosity of the *Medusæ*, &c., and which appear to be rather dependent upon a direct exertion of Nervous power (CHAP. XVII. Sect. 1).—Any similar relations that may exist between Magnetism and the Nervous Force have yet to be developed.—The relation between Nervous Force and Motion, however, is too striking to be overlooked. For if we pinch or prick a living nerve, we call its vital endowments into activity; that is, if it be a motor nerve which is thus treated,

muscular contractions are excited; if it be a nerve of common sensation, pain is produced; whilst if it be a nerve of special sensation, and the mechanical irritation be applied through the organ in which it originates, it will call forth the special sensations to which these nerves minister. Thus, pressure may be so made upon the ball of the eye, as to produce sensations of light and colours in utter darkness; pressure against the meatus of the ear, will, in like manner, generate a ringing sound; and by quickly but lightly striking the surface of the tongue, near its tip, with the finger, a distinct taste, sometimes acid, sometimes saline, is occasioned. Conversely, the Nervous force will produce Motion through the medium of the muscular apparatus, just as it excites Electricity through the instrumentality of the electric organs of Fishes.

248. We have next to inquire into the conditions under which the Nervous power is manifested in the living body; and these are different as regards the two forms of nervous substance and their respective endowments. Thus the conducting power of the nerve-fibres appears to remain with little decrease for some time after death, especially in cold-blooded animals; for we can, by pinching, pricking, or otherwise stimulating the motor trunks, give rise to contractions in the muscles supplied by them, exactly as during life. This power is much lessened by the influence of narcotics; so that if a nervous trunk be soaked in a solution of opium, belladonna, or other powerful narcotic, it ceases to be able to convey the effects of stimuli to the muscles, some time before the muscles themselves lose their contractile power. On the other hand, it seems to be exalted by various irritating influences; so that, when the nervous trunk has been treated with strychnia, or when it has been subjected to undue excitement in other ways, a very slight change is magnified (as it were) during its transmission, and produces effects of unusual intensity.—The peculiar powers of the vesicular substance, however, in virtue of which it *originates* the changes which the former transmits, are *only* manifested, when blood is moving through its capillaries. Thus, if the circulation through the nervous centres cease but for a moment, total insensibility, and loss of the power of motion, immediately supervene; as in the state called *Syncope*, or fainting, which consists in a failure of the heart's action, brought on by some cause which acts first upon that organ, and through it upon the system at large. The due activity of the vesicular nervous matter is not only dependent upon a sufficient supply of blood, but it requires that this blood should be in a state of extreme purity; for there is no tissue in the body, whose functions are so readily deranged, by any departure from the regular standard in the circulating fluid,—whether this consist in the alteration of the proportions of its normal ingredients, or in the introduction of other substances which have no proper place in it. Such changes may have the effect of depressing the activity of the vesicular substance of the nervous centres, and consequently producing torpidity, not merely in regard to the reception of impressions, and the performance of voluntary motions, but also in the mental operations generally; as exemplified in the narcotic effects of opium, chloroform, &c. On the other hand, various conditions of the blood, especially those depending on the presence of certain external agents, produce an undue energy in the functions of the nervous centres; which energy, however, is almost invariably accompanied by irregularity, or want of balance among

the different actions. Of this we have examples in the operation of alcohol and nitrous oxide on the brain, and of strychnine on the spinal cord.

249. The influence of these conditions may be traced on the endowments of the peripheral expansions of the sensory nerves, as well as upon those of their ganglionic centres ;—a fact which tends to confirm the idea already suggested (§ 245) as to the operation of the same agency at the two extremities of the conducting fibres. For it is easily shown that the circulation of the blood through the parts in which sensory nerves originate, is just as necessary for the original reception of the impressions, as is the circulation through the brain to their conversion into sensations, or to the origination of motor impulses by an act of the will. We find that anything which retards the circulation through a part supplied by sensory nerves, diminishes its sensibility ; and that if the flow of blood be completely stagnated, entire insensibility supervenes. A familiar example of this is seen in the effects of prolonged cold ; which, by diminishing, and then entirely checking, the flow of blood through the skin, produces first numbness, and then complete insensibility of the part. This result however, may be partly due to the direct influence of the cold upon the peripheral nerve-vesicles themselves ; depressing their peculiar vital powers. The same effect is produced, however, when the supply of blood is checked in any other way ; as, for example, by pressure on the artery, or by obstruction in its interior. Thus when the main artery of a limb is tied, numbness of the extremities is immediately perceived ; and this continues even though the limb be kept warm, until the circulation is re-established by the collateral branches, when the usual amount of sensibility is restored. On the other hand, increased circulation of blood through a part produces exaltation of its sensibility ; that is, the ordinary impressions produce changes of unusual energy in its sensory nerves. This is particularly evident in the increased sensibility of the genital organs of animals during the period of heat ; and in those of man, when in a state of venereal excitement.

250. To sum up, then, we may compare the vesicular substance, wherever it exists, to a galvanic combination : the former being capable of generating nervous influence, and transmitting it along the fibrous structure to the part on which it is to operate ; in the same manner as the latter generates electric power, and transmits it along the conducting wires, to the point at which it is to effect a decomposition or any other change. In one of the most perfect forms of the galvanic battery (that invented by Mr. Smee), although the metals remain inserted in the acid solution, and are consequently always ready for action, no electricity is generated until the circuit is complete ; and the *waste* of the zinc, produced by its solution in the acid, is therefore exactly proportional to the electric effects to which it gives rise. The condition of the nervous system, in the healthy and waking state, bears a close analogy to this ; for it is in a state constantly ready for action, but waits to be excited ; and its waste is proportional to the activity of its function. The vesicular matter, diffused over the surface of the body, is inactive, until an impression is made upon it by some external agent ; but a change then takes place in its condition (of which we know no more, than that the presence of arterial blood and a certain amount of warmth are necessary for it), which is transmitted to the central organs by the sensory trunks. It would appear that the

excitement of this change has a tendency to increase the afflux of blood to the part; thus when a lozenge or some similar substance is allowed to lie for a time in contact with the tongue or with the side of the mouth, a roughness is produced, which is due to the erection of the sensory papillæ, by the distension of their blood-vessels. On the other hand, the change in the vesicular matter of the central organs, by which motion is produced in the distant muscles, may be excited either by the stimulus conveyed by the sensory nerves, or by an act of *Mind*. Of the mode in which the mind thus acts upon the nervous system, we know nothing whatever, and probably never shall be informed in our present state of being. But it is sufficient for us to be aware of the physiological *fact* of the peculiar connection between the mind and the brain; a connection so intimate, as to enable the mind to receive through the body a knowledge of the condition of the Universe around it, and to impress on the body the results of its own determinations; and of such a nature, that the regularity of the working of the mind itself is dependent upon the complete organisation of the brain in the first instance, upon the constant supply of pure and well-elaborated blood, and upon all those influences which favour the due performance of the nutritive operations in general.

4. *Transformation of Tissues.*

251. There exists, to a certain extent, a capability on the part of the several tissues now described, to assume each others' characteristic forms and properties. This Transformation of Tissues, however, evidently takes place according to a determinate plan; and we shall find the simplest manifestations of this plan in the Vegetable kingdom, where, as we have seen, the variety of the tissues, as also of the purposes to which they are subservient, is the smallest. In the very simplest Plants, as there has been frequent occasion to repeat, every cell may be regarded as a separate individual; and even in those of larger dimensions, and of more apparent complexity of structure, among the inferior Cryptogamia, we still find the whole fabric composed of aggregations of cells, none of which depart in any considerable degree from the simplest type, whilst all of them seem to retain the power of increase by giving origin to new cells. Such is precisely the condition of the early embryo of the Flowering Plant; for this takes its origin in a single cell, which gradually forms a cluster of similar cells, by the ordinary process of multiplication; and the embryo attains a considerable degree of development, the outline of its principal organs being (as it were) sketched out, before there is the least indication of the presence of any other form of tissue. The same is the case, also, in regard to the formation of buds, which always take their origin in some part of the cellular tissue, being formed from it by the ordinary processes of cell-multiplication; and which do not contain any other kind of tissue, until their development has considerably advanced.—In the higher Cryptogamia we occasionally meet with Spiral cells, and also with Ducts of various kinds for the conveyance of fluid; and in the Ferns, which most nearly approximate to Flowering Plants, we also find true Wood-cells. In like manner, in the germinating Phanerogamic plant, or in the growing bud, we observe the gradual evolution of these higher forms of tissue; the Ducts usually presenting themselves first, whilst the

production of true Woody-fibre seems to take place only when the most complete vegetative activity has been attained. These tissues are formed, as already explained, by the metamorphosis of cells; and the occurrence of such a metamorphosis appears to be inconsistent with the retention of the reproductive power; since we never find new ducts or woody-fibre arising out of those already formed, but always directly originating in cellular structures. Further, after this metamorphosis has once taken place, the character of the tissue remains fixed during the whole remainder of its existence; the ducts never transforming themselves into woody structure, nor the fusiform cells of wood into ducts. Again, in parts which are imperfectly developed, or which have undergone 'morphological transformation' (as when a carpel or stamen is produced from the elements which would have otherwise gone to form a leaf, § 278), we find that the ducts and woody-fibre are replaced by cellular tissue. And it is of cellular tissue, or of its simplest transformations, that all those adventitious growths are composed, which cannot be regarded as originating in the metamorphosis of normal organs, but which must be considered as strictly *morbid* products,—such as those originating in the puncture of insects, or in the application of irritating substances.

252. Although the variety of tissues in the Animal Kingdom, and of the purposes to which they are subservient, is so very much greater than in the Vegetable, yet we can trace them all back to the same origin, the simple cell; and we may observe somewhat of the same relation between the successive evolution of these tissues in the ascending series of Animals, and their progressive development in the embryo of the highest, that we have traced among Plants; although the requirements of the several tribes of Animals more frequently necessitate an apparent departure from this plan for some special purpose.—Many of the simpler forms of (so-called) Animalcules must be probably regarded as nothing else than simple cells, which are endowed by their ciliary appendages with the power of free locomotion, but which contain no variety of organs, and multiply themselves (like other independent cells) by spontaneous fission (§ 284). Now such beings cannot be properly compared to any of the higher Animals, in the fully-developed state of the latter; but they do present a very close correspondence to their earliest embryonic condition. For the life of the highest commences in a single cell, in which not the least vestige of its future organs or tissues can be traced; and this cell speedily multiplies itself by a process of binary subdivision, until a homogeneous mass of cells is produced, bearing a very close resemblance to some of those composite clusters of Animalcules, which are produced in like manner by the repeated bi-partition of a single individual (§ 285).—Ascending to those forms of Animal existence, in which distinct organs begin to show themselves, as in Zoophytes (§ 288) and Acalephæ (§ 293), we still find the tissues composing the mass of the body almost homogeneous; consisting of a nearly structureless 'blastema,' with interspersed fibres and cells, more or less completely developed; the cells, in some instances, the fibres in others, being the elements most fully formed. It is difficult to distinguish anything like Muscular or Nervous structure in such animals, nor do we meet with vessels having distinct walls for the conveyance of fluid: and the dense skeletons which many of them possess, are formed upon the simplest possible plan (§ 194). In the most important of these

particulars, they correspond with the more advanced embryo of higher animals; in which "morphological transformation," or the evolution of distinct organs, proceeds to a considerable extent, before "histological transformation," or the production of the more special forms of tissue, distinctly commences; a sort of mould or model of the organ being generally formed of cells, before it presents any appearance of the structure which is ultimately to characterise it. The principal difference is one which has reference entirely to the conditions of existence of the two classes of beings, and is therefore purely *teleological;* namely this, that the embryo, not being adapted for active movement, does not require those fibrous tissues which enter largely into the composition of animals that have to live by the exercise of their own active powers, and is composed more exclusively of cells; and further, that the embryo, growing in security within the ovum, does not require, until a much later period of its existence, the support or protection of those hard parts which are developed even in the humblest Animals whose mode of life necessitates their possession of them.—The same general facts may be observed in the development of new structures at any period of life; whether occurring as a part of the normal history of the animal, as in the evolution of the limbs of the Batrachia (§ 323); or taking place for the purpose of repairing losses of substance. The new material is at first invariably found to exhibit the simplest histological condition; namely, a "blastema," or organisable material, interspersed with nuclear particles, from which a cellular or a fibrous tissue, or a mixture of both, may be developed, according to circumstances. This tissue, like the early embryonic mass, at first contains no blood vessels; but these gradually extend into it from the neighbouring capillaries; and when the supply of blood is fully established, a further series of transformations may take place, by which a variety of tissues may be evolved.

253. Where the transformation of the primitive elements of any particular organ, into some perfect form of tissue, involves a succession of distinct processes, the same gradations are often traceable in the permanent conditions of that organ, which we encounter in ascending the animal series. Of this we have a very beautiful illustration in the development of the vertebral column of Vertebrated animals. This commences as a delicate tube, termed the *chorda dorsalis*, which encloses the spinal cord, and which is composed of thin-walled polygonal cells, in close apposition with each other; and this is precisely the condition which is presented by the rudiment of the vertebral column, in the Lamprey and in others of the lowest group of Cartilaginous Fishes. After a time, the bodies and arches of the vertebræ begin to be marked out in cartilage, around the 'chorda dorsalis' of the embryo; and this is the condition of the vertebral column in some of the higher Cartilaginous Fishes. The ossification of these cartilaginous rudiments commences in several distinct centres, and it is not until a later stage that the different elements of the vertebræ are united by osseous tissue; this detached condition remains permanent in the partly-ossified vertebræ of the highest Cartilaginous Fishes. In the formation of new bone, again, for the reparation of a fracture, the primitive blastema is usually developed at first either into cartilage or into a fibrous tissue;* and the development of bone subsequently takes

* Paget's Lectures on the Processes of Repair, &c., Med. Gaz. 1849.

place by a process of ossification precisely corresponding with that by which bone is normally developed in the first instance. Further, where particular bones remain undeveloped, we usually find their places marked out either by cartilage or by fibrous tissue; thus in the Mammalia generally, the ribs are connected to the sternum (breast-bone) by cartilages, which represent the osseous sternal ribs of Birds and Reptiles; the white fibrous bands that cross the 'recti' muscles of the abdomen represent the abdominal ribs of Reptiles; and in those Mammals which do not possess a clavicle, its place is usually indicated by a ligamentous band. There is a special tendency in cartilaginous or fibrous textures to undergo ossification abnormally, where their ossification normally takes place in some other part of the animal series; thus we find the cartilages of the ribs in old people to be so frequently converted into bone, that we might almost regard their ossification as an ordinary occurrence in advanced life, whilst fractures of these cartilages are generally united by osseous substance; so, again, it is very common to find the lumbar vertebræ of old people anchylosed together, like those of Birds, by the ossification of their intervening fibro-cartilages; and we have already noticed a correspondence between the abnormal development of bony substance in certain fibrous structures in Man, and its normal presence in the same parts of other animals (§ 171).—The simple fibrous tissues, again, may be developed into any of those more highly organised structures, serous, mucous, or cutaneous membranes, which are formed upon the fibrous type; thus a new serous or synovial membrane is produced to obviate friction, where a new joint results from an unreduced dislocation or an ununited fracture; mucous membrane is generated, where an artificial passage or opening remains in connection with any part of the open cavities of the body (as where a fistulous orifice has been established, leading directly from the stomach or intestine to the surface of the abdomen); and a cutaneous membrane is produced, where protection from the external air is necessary. A mucous membrane constantly exposed to the air gradually acquires a resemblance to skin; on the other hand, skin when turned inwards so as to form part of an internal cavity, and kept moist, gradually assumes the characters of mucous membrane. These, as we have seen, are a part of one and the same system of tissues; and we shall find that in the *Hydra*, the external integument and the lining of the digestive cavity are mutually convertible (§ 289). But we do not find that serous membranes are disposed to acquire the characters either of skin or mucous membranes. Nor do we find that any tissues formed upon the cellular type are disposed to undergo transformation into one of the fibrous type, or *vice versâ;* thus, for example, cartilage and mucous membrane, adipose and ligamentous tissues, are not in any degree mutually interchangeable.

254. When tissues have fulfilled their proper offices, or are not called upon to perform them, they undergo *degeneration;* and this consists in the gradual disappearance of the normal elements of the tissue, so that nothing is left but the framework of areolar texture, by which they were held together. Not unfrequently, however, the place of those elements is occupied by adipose matter, which has a remarkable tendency to occupy the places left by the degeneration of muscular fibres, gland-cells, &c. Sometimes this degeneration takes place as a part of that regular succession

of changes, which marks the advance of life; as when the urachus, ductus arteriosus, &c., of the fœtus, shrivel up into ligamentous cords; or when, at a later period of infancy, the thymus gland ceases to increase, and its peculiar substance gradually disappears. But such degenerations not unfrequently occur, as the result of the forced or accidental disuse of parts which ought normally to continue in activity, or as the consequence of a disordered condition of the nutritive actions; of the former we have a characteristic example in the degeneration of muscles consequent upon paralysis (§ 231); whilst of the latter one of the most remarkable forms is that known as "fatty degeneration," in which it would seem that the proper elements of the tissue are themselves converted into fat. Thus it has been remarked by Mr. Paget,* that when the muscles are the subjects of this change, the ultimate fibres are still to be distinguished, but that the sarcolemma is occupied by fat-particles arranged in the same manner as the proper constituents of the fibrillæ. So, again, a fatty degeneration may take place in bones, giving rise to one form of the disease termed *mollities ossium;* which has been attributed to a deficiency in the earthy components of the osseous tissue, but which seems to consist, in many cases at least, in the conversion of their organised basis into fatty matter. —It has been found by experiment that when fibrine undergoes decomposition in the open air, butyric acid (the fatty acid of butter) is formed as one of its products; and there is still stronger evidence, derived from the conversion of the entire bodies of men and animals into adipocire, that the albuminous compounds may undergo spontaneous transformation into adipose matter, without the production of proper adipose tissue; this being, in fact, only developed for the purpose of storing it up in special situations or for special objects. There is no difficulty, then, in understanding that this fatty degeneration is connected with imperfect or insufficient nutrition, and that it forms no exception to the general rule just now stated, as to the non-metamorphosis of any fully-formed tissue into another constructed upon a different type.

255. We not unfrequently meet with new and abnormal growths, as the result of an extraordinary development of some of the ordinary tissues of the body, especially the simple-fibrous, the adipose, and the cartilaginous; and in such tumors we occasionally find spicules or plates of bone, formed by the ossification of the fibrous or cartilaginous elements, apart from all connection with the normal osseous textures; but these fragments seldom possess the structure of perfect bony tissue.—Morbid growths not unfrequently present themselves, however, the structure of which is more or less distinct from that of the normal tissues of the body, but which are yet regularly organised, and grow and develope themselves according to the laws of cell-formation. When these have the power of rapid increase, and tend to destroy the normal tissues in the midst of which they may be developed; and especially when there is a disposition to the development of similar growths in other parts of the body, after the removal of the original diseased mass; the growth is said to be *cancerous* or *malignant.* In those which extend themselves most rapidly, we usually find the tissue to contain a large number of nucleated cells, which are developing new cells in their interior (Fig. 15); but in other forms of slower growth, we meet with dense bands and laminæ composed of fusiform cells (Fig. 19, *d*).

* Lectures on Nutrition, Hypertrophy, and Atrophy, Med. Gaz. 1847.

The whole mass may be considered as developed upon a very low and simple type, which admits of rapid increase, but predisposes to early degeneration. In this respect Cancerous growths possess a remarkable analogy with the parasitic Fungi, which develope themselves in the interior of Vegetable and even of Animal structures (§ 271, 272); and the supposition long ago entertained, that Cancer might be regarded as an independent growth of corresponding nature, does not now appear so extravagant as it was at one time considered. There can be little doubt that a cancerous tumor of any size may be developed from a single cell; and it is probable that the origin of such growths in parts distant from their primary centre, is to be traced to the conveyance of cancer-cells or of their germs by the circulating current; so that it seems very difficult to draw the line, which shall separate such independent growths, on the one hand from the ordinary tissues of the body, and on the other from structures really parasitic. It is interesting to remark, that blood-vessels cannot be traced in these productions at an early period of their formation, but that they make their appearance, as in the normal development of the tissues, at a later date.

CHAPTER V.

OF THE DISTINCTIVE CHARACTERISTICS OF THE VEGETABLE AND ANIMAL KINGDOMS.

256. From the considerations which have been set forth, in preceding Chapters, as to the attributes of Organised Structure in general, and the distinctive nature of Vital Action; and from the more detailed survey which has just been taken of the peculiarities in structure and properties, by which the principal elementary forms of organised structure, or Primary Tissues, are respectively characterized; we naturally pass on to a more special examination of those diversities which are presented by entire Living Beings, in respect both to their conformation and to their mode of existence; diversities which (as heretofore remarked, § 8) present to the Physiological inquirer that same variety of combinations, which the Chemist and the Physicist aim at reproducing in their field of investigation by experiments devised for the purpose, and which thus afford him the means of determining what are the real and essential conditions of vital phenomena, and what are to be considered as accessory and modifying circumstances. On the threshold of this inquiry we are met by those obvious and striking distinctions, by which the whole assemblage of Organised beings naturally arranges itself into the two great series known as the *Vegetable* and the *Animal* kingdoms; and we are thus led, in the first instance, to examine into their respective peculiarities of structure and conditions of existence. The precise determination of these, however, is much more difficult than it may at first sight appear; for whilst the prevalent notions of the distinctive attributes of Plants and Animals are founded chiefly upon the prominent features which are exhibited in the conformation of the higher tribes of both kingdoms respectively, we find, as we descend the scale, that these features are progressively softened

down (so to speak), until the distinctions on which we have at first relied become completely obliterated. Hence we are led to believe that such features cannot be looked upon in any other light, than as indications or exponents of some more essential though less apparent differences; and it is thus a most important object with the Physiologist to determine what these differences really are. The enquiry cannot be even entered upon, however, without our attaching some definite ideas to the terms "Plant" and "Animal;" and if we found such ideas rather upon a comprehensive survey of both kingdoms, than upon the characters presented by individuals selected from each, we shall find that they may be expressed somewhat as follows.—A Plant is an organised being whose vital powers are directed solely to the performance of *formative* operations, by which its fabric is not merely built up in the first instance, but is continually receiving additions during the term of its existence; and any movements which it may exhibit, are destined solely for the furtherance of these operations, and must be regarded as originating in *physical* or *vital* forces. On the other hand, an Animal is an organised being whose vital powers are not merely directed to the construction and maintenance of its corporeal fabric, but are also subservient to the operations of the conscious mind, which involve a continual *disintegration* of the structures that minister to them, on the repair of which, rather than on the extension of the fabric, after it has attained its full development, the formative energy is chiefly expended; and of the movements which it may exhibit, though a part are still to be regarded as directly dependent (like those of Plants) on causes inherent in its material organisation, there is another part, small though it may often be, in which the consciousness and spontaneity of the individual are necessarily concerned, and which must therefore be distinguished as originating in *psychical* causes.

257. Thus, then, the existence of consciousness and the power of spontaneous motion must be regarded as the chief attributes, which must be superadded in our 'idea' of an Animal to those which, being involved also in our 'idea' of a Plant, are, therefore, the characteristics of organised beings in general; and if we always possessed the means of determining where consciousness and spontaneity *do* and *do not* exist, we should have comparatively little difficulty in drawing a definite line of demarcation between the two. But we have no other means of judging of the presence or absence of these psychical endowments, than by studying the *movements* of the beings whose character we seek to determine; and it is frequently impossible to say, with any thing like certainty, from what source these movements proceed. This difficulty is increased by the fact, that the most remarkable movements exhibited by undoubted Plants are seen in a group whose lowest forms present the nearest approximation to the simplest types of the Animal kingdom (§ 267); and hence it has happened that beings have been invested with the attributes of Animality, whose subsequent history has clearly indicated their Vegetable nature. On the other hand, there are beings to which, from the absence of sensible movements, we should be disposed to refuse the endowment of consciousness; although the general characters of their structure would lead us to suppose their place to be rather in the Animal than in the Vegetable Kingdom. Such beings, however, present a strong resemblance to the embryonic or transitory forms of higher types of organisation; and

as the (presumed) absence of consciousness in the early embryo does not lead us to exclude it from the Animal Kingdom, so it should not warrant the exclusion of a class of beings, the *tendency* of whose development is the same with that of undoubted animals, although the higher stages of that development are never attained. Here, as elsewhere, it happens that in the separation and circumscription of really 'natural' groups, we are obliged to leave a margin around each, for the reception of forms to which the definitions that we frame to express their most characteristic features are not applicable; and in the assignment of such 'aberrant' forms to one or the other of these groups, we must be rather guided by their general resemblance to those which display its typical characters, than by the degree in which they may themselves possess them. Thus, when we meet with beings possessing the general structure of Plants, and corresponding with them in the mode of their development, we are justified in referring them to the Vegetable Kingdom, although their movements may bear a strong resemblance to those of Animals. On the other hand, when we encounter beings possessing the general structure of Animals, and corresponding with them in the mode of their development, we seem justified in referring them to the Animal Kingdom, even if we never witness any such movements in them, as can be fairly considered indicative of consciousness.—We have now to enquire, therefore, what other distinctive characters can be found, for the separation of the Vegetable and Animal Kingdoms; and the best starting-point in this investigation presents itself in a comprehensive survey of the relations which they respectively bear to the Inorganic World.

258. It has been already pointed out (§ 25) that, so far as regards the number of Elementary substances entering into their composition, Organised bodies present a remarkable simplicity; since only *four*, out of the *sixty-one* elementary substances that exist in the Mineral World, appear essential to the production of living organised tissues, although several more are employed for special purposes in most Vegetable and Animal fabrics. These four elements, oxygen, hydrogen, carbon, and nitrogen (or azote) exist alike in the Vegetable and the Animal Kingdoms; but the fourth is present in much larger amount in the animal than in the plant; since, as we have seen, the great bulk of the tissue of the latter is made up of compounds into which it does not enter; whilst in the former there is no part of the solid fabric, of which it is not an essential ingredient. This difference in composition, however, is not by any means so important as that which exists in the mode in which Plants and Animals respectively obtain the materials of their growth and development; for whilst Animals can only appropriate, as food, organic compounds which have been already prepared for their use, Plants have the power of generating these, by effecting new combinations of their inorganic elements. These elements are all drawn by the Vegetable Kingdom, directly or indirectly, from the Atmosphere; which contains not merely oxygen and nitrogen, but also a minute proportion of Carbonic Acid (about 4 parts in 10,000), and a still more minute proportion of Ammonia, together with a variable quantity of watery vapour. It does not appear, however, that either the oxygen or the nitrogen of the atmosphere is directly appropriated by Plants; for there is reason to believe that they can only make use of these and the other elements, when they are already in the state of binary compounds.

Thus their oxygen and hydrogen are derived from water, their carbon from carbonic acid, and their nitrogen from ammonia ; and although any or all of these compounds may be taken up by their roots, yet directly or indirectly the Atmosphere is the real source of them. This is demonstrated by the history of the mode in which the barren unproductive rock, upheaved by submarine agency from the depths of the ocean, becomes gradually clothed with a succession of Vegetable forms; until at last it is capable of sustaining those, which are most elevated in their type of structure, and most beautiful and majestic in their aspect. No soil at first exists, from which nutriment may be extracted ; the disintegrating surface of a granitic or a calcareous rock can afford no other pabulum than the liquefied vapour which has descended upon it from the clouds ; and every particle that is contained in the fabric of the apparently insignificant Lichens and Mosses which first invade its sterile uniformity, must have been drawn from the air. And when, by the death and decay of successive generations of such humble Plants, a more productive soil has been created, wherein the higher Vegetable forms can take root and flourish, that soil must itself be regarded as condensed or consolidated air ; and it can only yield nutriment to the vegetation it supports, by undergoing a decomposition, whereby its components are again restored to the condition of carbonic acid, water, and ammonia ; which products, when absorbed by the roots, are subservient to precisely the same purposes in the economy of the Plant, as if they had been taken in by the surfaces more directly exposed to the atmosphere. Now this history, there is every reason to believe, precisely repeats that by which all the dry land that has been upheaved from time to time above the oceanic waters, dating back to the remotest periods in the history of our globe, has been rendered capable of sustaining the various forms of Vegetable life which have successively appeared on its surface ; and thus the atmosphere of the primeval Earth must have contained not merely its present amount of carbonic acid, ammonia, and watery vapour, but also the whole of that which would be generated by the reconversion, into those forms, of the entire mass of the existing Vegetation of the globe, as well as of those remains of past generations, which are imbedded in the Earth's crust under the forms of coal, lignite, peat, bitumen, naphtha, &c. That such is the real relation of the Vegetable Kingdom to the Atmosphere, is further apparent from the fact experimentally ascertained, that peas and other plants will grow, if their seeds be caused to germinate in sand, and they be watered with distilled water alone ; all the carbon which is consolidated in their fabric, when it is fully evolved, having then been obviously derived from the air, since no other source can have afforded it. And if it be further considered that most trees add much more solid matter to the soil around them, by the annual exuviation of their leaves, than they remove by absorption through their roots ; and that in the solid trunk of an oak of a century's growth, there are many thousand times as much solid carbon as the soil itself can ever have supplied; it will be evident that, even where the soil appears to contribute most to the support of the vegetation which it aids in developing, the atmosphere is in reality the great storehouse from which it is drawn.

259. From the elementary materials which the Plant thus obtains, it generates that wonderful variety of organic compounds, which the fabric of even a single individual includes. From oxygen, hydrogen, and carbon

are produced Gum, Starch, Sugar, and Oleaginous matters; and at the expense of these, the greater part of its solid tissue is built up. Its combining powers, however, are not limited to the production of the compounds which it requires for its own immediate use. In all the growing Vegetable tissues, as we have seen, there is a certain amount of azotised matter, which seems to be essentially and actively concerned in the vital operations of the part; but in most Plants there is a much larger production of azotised compounds than can be thus accounted for; and in some species (whose peculiar endowments in this respect have been experientially discovered by Man) there is a remarkable tendency to generate them, and to store them up in particular parts of the fabric. Such a production, then, would seem altogether purposeless as regards the Plant itself; and hence, if the earth were tenanted only by the Vegetable creation, the process would be without a meaning. Its intention in the economy of Nature, however, becomes obvious as soon as we enquire into the mode in which Animals obtain their supplies of aliment; for we then learn that they are entirely destitute of that power of *forming* organic compounds which is so remarkable an attribute of Plants, and that they are entirely dependent, therefore, upon the materials for organisation supplied to them by the latter. Further, these materials do not so much consist of these *ternary* compounds, which are the chief materials of the Vegetable fabric,—such compounds being but little employed in the formative processes of the Animal,—as of the *quaternary* or azotised compounds, whose use in the Vegetable economy we have seen to be so limited; and thus we perceive that, in what may be considered the highest operations of their vital chemistry, Plants are not labouring for themselves, but are the unconscious instruments whereby the atmosphere is rendered capable of maintaining a population of sentient beings, whose entire existence is thus bound up (so to speak) in theirs. But the materials thus elaborated by Plants are not solely, or even chiefly, employed in the first construction or building-up of the Animal fabric; for we shall find that the greater part of them are required for that incessant reparation, which the peculiar conditions of Animal existence cause it to require. Every exertion of the strictly *Animal* powers (as already pointed out, § 236 and 243) would seem to involve the death and decay of a proportional amount of the tissue by which those powers are developed; and it is only by the regeneration of this, that the vigour of the body can be sustained. Hence we see that whilst, in Plants, the extension of the fabric is continually taking place during the whole period of their existence, and is the chief source of their demand for aliment, it ceases to take place in Animals as soon as the evolution of their organs has been fully completed; and thenceforth the demand for aliment is chiefly regulated by the amount of waste or disintegration of tissue which has taken place. We also find, however, that food is required for a special or accessory purpose in some Animals; namely, for the development of Heat by a process analogous to combustion, whereby a fixed temperature may be maintained, this temperature being requisite for the higher manifestations of vital activity (§ 92); and the pabulum made use of by animals for this chemical operation, is that which it cannot employ as the material for more than a small proportion of its own fabric.

260. Now by the disintegration or waste of its own tissues, consequent

upon the exercise of its peculiar powers, and by the combustive process which is necessary to the highest exertion of those powers, the Animal is restoring to the atmosphere, during its whole life, those very compounds of which the Plant is as constantly depriving it; and thus we find the chemical effects of Animal life upon the Inorganic world to be directly antagonistic to those of Vegetation. For whilst the Plant *decomposes* the carbonic acid of the atmosphere, fixes its carbon, and gives back its oxygen; the Animal *reproduces* carbonic acid, setting free carbon from its body, and combining it with oxygen which it takes from the air. Whilst the Plant exhales *less* water than it takes in, fixing some of it in its tissues by the separation of its elements and their reunion with other elements in organic compounds; the Animal exhales *more* water than it takes in, the surplus being generated by the decomposition of its organic constituents. And whilst the Plant in like manner *fixes* ammonia, and uses its elements for the production of organic compounds; the Animal continually *sets free* azotised substances, as the products of the waste or decomposition of its tissues, which speedily resolve themselves into ammoniacal compounds after their exit from the body.* Thus from the time when the creation of the first Plants and Animals took place upon our planet, there has been an uninterrupted cycle of mutually-balanced operations, which in their totality have concurred to develope the existing Flora and Fauna of our globe. The Vegetable kingdom is the great elaboratory of organic compounds; producing not merely such as are required for its own existence, but also those which are needed for the development of the fabrics of Animals; all being alike formed at the expense of the water, carbonic acid, and ammonia of the atmosphere. From Vegetables, these substances pass ready-formed into the bodies of Herbivorous animals; which, while consuming one portion of them, and giving it back to the air, convert another into the various forms of animal tissue. And the bodies of these animals, in their turn, being devoured by the Carnivorous orders, supply to them the very same materials, which are by them appropriated in the same manner; part of

* The principal of these substances is *Urea*, whose constitution has been already noticed (§ 29). A remarkable adaptation has been pointed out by M. Dumas, between the purposes of the urinary excretion, and the chemical form of its components. It might have been anticipated, he observes, that as Animals restore oxygen, hydrogen, and carbon to the atmosphere, in the very forms (water and carbonic acid) in which they were originally appropriated by Plants, they would do the same in regard to their nitrogen, and would set it free in the form of ammonia. The contact of so caustic a substance as ammonia, however, or even of carbonate of ammonia, would have been incompatible with the vital endowments of the delicate tissues concerned in the elaboration of the secretion; in place, therefore, of carbonate of ammonia, the animal has been made to produce urea, which is carbonate of ammonia *minus* two proportionals of water, and which, through its inert and unirritating properties, can pass through the delicate tissues of the kidneys, the ureters, and the bladder, without producing any injurious effect upon them. But when brought into contact with the air, urea speedily undergoes a true fermentation, which restores to it the two equivalents of water that are requisite to convert it into carbonate of ammonia; and thus an extremely volatile substance is generated, which is readily exhaled into the atmosphere, and which, being easily soluble in water, is caught up by the atmospheric moisture and precipitated in dews and rains, to become the pabulum for new generations of plants and animals. The small amount of albuminous matter contained in the urine, which is derived from the mucus of the bladder, aids in this conversion by acting as a 'ferment;' and thus it is that, if the urine be long retained within the bladder, or if the mucous secretion be excessive, the urine becomes ammoniacal before it is discharged. (See MM. Dumas and Boussingault "On the Chemical and Physiological Balance of Organic Nature," 3rd ed., p. 41.)

them being continually restored to the air, by the disintegration of their tissues, and by the combustive process, during the whole life of the animal; and part remaining solidified in their bodies, to be given back to the atmosphere by their decomposition after death, if they are not appropriated as the food of some other animal. In either case, the entire amount of carbonic acid, water, and ammonia, first drawn forth from the atmosphere by the plant, is in the end restored to it, in some way or other; unless it should happen that the usual processes of decay are prevented from taking place, and that portions of the solid animal or vegetable fabrics are preserved without change; thus keeping abstracted from the atmosphere, so to speak, some of its original constituents, which will be restored to it whenever the processes of decomposition or combustion may occur in these substances. Thus, the vast mass of carbonaceous matter, which has remained locked up in the Earth's crust ever since it was first solidified from the atmosphere by the luxuriant Flora of the Carboniferous epoch, is now being restored at the rate of tens of millions of tons per annum, by the reconversion of that carbon into carbonic acid, in the various combustive operations to which *coal* is being subjected, to supply the diversified needs of civilized Man.

261. It is not only in their relations to the Inorganic world, on which they are dependent for the *materials* of their respective fabrics, that Plants and Animals occupy an antagonistic position; for nearly the same antagonism presents itself in their relation to those Physical *forces*, whose agency may be regarded as the immediate source of their vital powers (§ 54). For, in the first place, all the processes of Vegetable growth, save the act of germination, are so entirely dependent upon *Light*, that if its operation were withdrawn but for a few days, the whole existing Flora must perish; and the Animal Kingdom, dependent as we have seen it to be upon the Vegetable, must speedily be involved in the same destruction. With the luminous beams of the sun, Plants receive a chemical force (to which the name of *actinism* has been given), that enables them to decompose carbonic acid, water, and ammonia; and they may thus be regarded, to use the language of Dumas, as "embodiments of a reducing power, of greater virtue than any other that is known, for no other will decompose carbonic acid in the cold." The amount of vegetative activity exhibited by any Plant, appears to be so strictly commensurate with the measure of solar Light and Heat which it has received, that, according to Boussingault, the same annual plant, in arriving at its full development, *everywhere receives the same amount of this influence*, whether it be grown at the equator or in the temperate zone. Thus we are led to look upon Light (with its accompanying Heat) as the prime agent in the development of Vegetable, and therefore of Animal, life; and to perceive that the intellectual deductions of the philosopher are here in the fullest harmony with the records of Revelation as to the order of Creation.*—

* "Organisation, sensation, voluntary motion, life," said Lavoisier, "only exist on the surface of the earth and in places exposed to light. It might be said indeed, that the fable of Prometheus was the expression of a philosophical truth which had not escaped the penetration of the ancients. Without Light, nature were without life and without soul: a beneficent God, in shedding light over creation, strewed the surface of the earth with organisation, with sensation, and with thought."—"These words," it is justly remarked by Dumas (On the Balance of Organic Nature, p. 8), "are as true as they are eloquent. If sensation and thought,—if the noblest faculties of the soul and the understanding, require

Although, as we have seen, (§ § 80-82) Light exerts an important influence on some of the organic processes of Animals, besides being necessary to the employment of that visual sense which is so important a means of bringing their consciousness into relation with the external world, the existence of the Animal kingdom cannot be said to be directly or essentially dependent upon its agency; indeed we may consider that, on the whole, more light is produced or set free by Animals, than is absorbed or embodied by them. The chief mode, however, in which the dynamic agency exerted by Light upon Plants, is restored by Animals to the Universe around them, is by its conversion into sensible Motion; for the decomposition of those peculiar compounds which are formed under its influence in the Vegetable fabric, seems to be the immediate source of the mechanical power which Animals can exert; thus presenting a remarkable example of the metamorphosis of one kind of physical force into another, through the intermediation of an organised body (§ 54). In their relation to Heat, again, there is a corresponding antagonism between Plants and Animals; for although all organised beings, as we have seen, are dependent upon Heat for the maintenance of their vital activity, yet much of the heat supplied to Plants may be said to be absorbed and rendered latent in its operation on their formative processes, whilst a continual generation of it is taking place in Animals by the oxidation of the materials of their fabric; and although it is only in warm-blooded animals that a sufficient amount is set free at any one time to become sensible and capable of measurement, yet the chemist well knows that during the comparatively slow reproduction of carbonic acid, water, and ammonia, which is effected by cold-blooded animals, the same amount of Heat must be really set free, as if these products were generated by a more energetic process of combustion. The same is true of the decomposition of Animal bodies after their death; for in this process, also, there is a disengagement of heat, exactly equal to that which would take place if the same materials were more rapidly consumed. And thus we may say that, during the whole life of Animals, and after their death, there is a continual restoration to the surrounding Universe, of that Heat which was abstracted from it by the Vegetative operations; so that, by the time that the organic compounds generated by the latter have been restored to the mineral world in the form of water, carbonic acid, and ammonia, the whole dynamic agency exerted by Heat in the first production of those compounds, together with that which has directly operated upon the Animal body itself, has been regenerated, either in its original form, or in that of Motion.

262. In comparing the organised structures of Plants and Animals respectively, we find the most obvious and striking differences between them to consist in the two following particulars:—1st. The presence of a Stomach or digestive cavity in Animals, and its absence in Plants:—2nd. The presence of a Nervo-muscular apparatus in Animals, and its absence in Plants. Now of the Stomach it may be remarked, that although its presence among Animals is so constant,

a material vesture for their manifestation, vegetables are the labourers charged with the task of building it up, from elements which they derive from the air and elaborate under the influence of the light, which the sun, its inexhaustible fountain, pours in ceaseless floods upon the earth."

that some Physiologists have regarded it as the most distinctive feature in their organisation, yet a little consideration will show that it is an *adaptive* rather than an *essential* character. For, on the one hand, it is required by the nature of their food; since this, not being always in immediate relation with the Animal (as that of Plants is with them), and being only occasionally obtainable, must be stored up within its body during the intervals; and, being, moreover, for the most part in a solid form, it must be reduced to the liquid state before it can be absorbed, and thus really introduced into the system. Again, the possession of a stomach is requisite to confer upon Animals the power of active locomotion; for if they were rooted in the ground, like Plants, and drew from it a part of their nutriment, they could not exercise this power even if they possessed it; whilst, on the other hand, if altogether deprived of this power, Animals could not obtain the food which they require. It is quite conceivable, however, that an Animal might be placed in the midst of a medium adapted for its nutrition, and might be so organised as to be enabled to imbibe all that it requires for its support through its external surface alone, like a Plant; in which case a stomach might be dispensed with. This appears to be the case with some of the simplest of the Entozoa (§ 309), which live upon the juices of the animals whose cavities they infest; and it is the case, also, with the embryo of every animal in an early stage of its development, the mass of cells of which it is composed being surrounded by alimentary matter, in a state already prepared for appropriation by it. But it is only in this earliest period of the life of the Animal embryo, that the stomach is wanting; for the first manifestation of a difference between the Animal and Vegetable germ consists in the extension of the former around the 'yolk,' so as to enclose it within a membranous expansion, which is, in fact, a temporary stomach, and which in some animals remains as the permanent digestive cavity; whilst in the latter, the germ continues to lie in the midst of the 'albumen,' which, instead of being enclosed by it, remains external to it, until absorbed into its own substance. Thus we see that, although the presence of a stomach is very intimately connected with the other attributes of Animals, it cannot be justly regarded as an *essential* feature of their structure; and hence, although its existence may be held to indicate the Animal nature of a being whose place in the scale is doubtful, yet its absence must not be considered as equally decisive of its Vegetable character.—The Nervous apparatus of Animals, however, must be regarded as more intimately connected with their peculiar endowments, and therefore as a more essential part of their fabric, than is their digestive cavity. In all animals in which a nervous system can be detected, we find it to be the instrument by which the conscious mind is brought into relation with the external world, receiving impressions from it through the organs of sensation, and reacting upon it through the instruments of motion; and the ganglionic masses to which the *sensory* nerves proceed, and from which *motor* nerves issue, appear to be the organs in which that consciousness is centred, and to which it is restricted. All our knowledge of such animals, therefore, would lead us to infer, that although the existence of ganglia and nerves does not necessarily indicate the existence of a conscious mind (since, as we have seen, many of the movements excited through the instrumentality of the nervous apparatus do not involve consciousness at all, § 244), yet that wherever a 'nervous system' can be detected, a conscious mind must be connected with some part of it; whilst, on the other hand, we seem equally justified in the inference that a conscious mind, even its very lowest form, cannot be connected with an Animal body, without the instrumentality of a Nervous structure, which, in its very simplest form, must consist of a ganglionic centre, and of cords of communication with the sensory and motor organs. Such a structure, it must be admitted, has not yet been certainly detected by the closest scrutiny, in the bodies of many beings (zoophytes for instance), to which we can scarcely refuse the attributes of animality; but as there are adequate reasons for believing that it would form a very small part of their fabric, and as the ganglionic vesicles and nerve-fibres may very probably not be isolated from their other tissues as distinctly as they are in the higher animals, we do not seem justified in regarding these cases as presenting a valid exception to the statement just made.* There is strong reason to believe, however, that the beings to which the term *Protozoa* will be applied (§ 283) remain in a condition analogous to the early embryonic state of higher animals, during their whole period of existence; their development never advancing so far as to the evolution of a nervous system. If this be really the case, we cannot regard them as possessed of consciousness; and their deficiency in the most characteristic attribute of Animality, leaves their place in the series to be decided by other considerations.—To sum up, then, we may always regard the presence of a Nervous system as affording positive evidence of the animality of the

* Some Physiologists have supposed that, in such Animals, the nervous matter is present in 'a diffused form,' that is to say, incorporated with their other tissues; but it is difficult to understand how the essential purposes of a nervous system can be thus answered; or to comprehend what attributes can be possessed by nervous matter in any such condition. It is quite true that Muscular power may be said to be thus 'diffused;' for the entire substance of many of the lower animals seems to possess some measure of that contractility, which, in the higher, is the attribute of the Muscular tissue only. In like manner, there is an absence, in many such beings, of anything which can be considered as a special apparatus for receiving Sensory impressions; movements being excitable by stimuli applied to any portion of the surface. But this cannot be justly regarded as warranting the assumption that the Nervous system is 'diffused' in like manner; on the contrary, it leads us to perceive that to such animals a nervous system would be of little use, except in the form of a central sensorium which may serve as an instrument of consciousness. For as the essential office of a nervous system is that which was felicitously termed by John Hunter *internuncial*, it is obvious that when all parts of the structure possess the same vital endowments, and these all concur (as in the Plant) to the maintenance and welfare of the general system, there can be no occasion for that mutual communication which is required where the *impressional* and the *motor* portions of the apparatus are distinct from each other. Hence the chief if not the only purpose of a nervous system in such animals, must be to serve as the instrument of the consciousness; and this purpose cannot be answered (according to anything which we learn from the experience of higher animals on this point) except by an organ of distinct structure and peculiar endowments. An argument in favour of this 'diffusion' of the Nervous system has been also erected upon the fact, that certain of the lower animals, such as the Hydra (§ 219) can be multiplied by the subdivision of their bodies. But this, too, is entirely destitute of validity. For that each separated fragment of any animal is capable of re-producing the whole, does not by any means prove that it possesses the endowments of the whole at the time of its separation; and there is no more reason to suppose that such a fragment contains a sensorium, than that it contains a stomach or any other *single* organ. Like the single cell, from which the being is at first evolved, it possesses the capability of *developing* a fabric, which shall possess, when complete, all the endowments of that from which it was detached; and whenever the nervous system has been generated in its body (as in that of the embryo of higher animals), then, and not till then, we may regard it as becoming possessed of consciousness.

being in which it exists. But its apparent absence must not be considered as an equally valid proof to the contrary; since it may be accounted for in some instances by our want of power to distinguish it; whilst in other cases it is due to the low type of development of the being, whose chief structural claim to the Animal character rests upon its resemblance to the transitory embryonic states of higher Animals, rather than to the permanent forms of the humbler Plants.

263. In any doubtful case, then, we shall do right to take the following considerations into account.—*First*, the Chemical composition of the fabric. If its cell-walls and fibres be found to consist solely of albuminous or gelatinous matter, the presumption is strongly in favour of its Animal nature. If, on the other hand, the tissues should be chiefly composed of cellulose, the presumption is equally strong in favour of its Vegetable character. In the application of such a test, however, great discrimination is required; for, as already pointed out, albuminous matter is not merely stored up in the cell-cavities of Plants, but even forms the most essential part of their cell-walls; whilst, on the other hand, it is quite certain that cellulose (contrary to what was formerly maintained) may be an integral part of the bodies of certain animals. A distinction is to be drawn between the cases in which, as in the *Corallines* and *Nullipores*, it forms part of the proper cell-walls; and those in which it is merely deposited within or around the cells, as happens in certain Compound Tunicata (§ 299) whose food consists largely of starchy matters, the cellulose appearing to be stored up in certain parts of their fabric, just as fatty matter is in other instances. Thus, although the mere presence of starchy substances in the fabric is not of itself a sufficient proof of the Vegetable nature of a doubtful being, it may be held to be so when there is no distinct proof to the contrary, and when the starch is found as an integral part of the solid fabric, and is generated by the being itself, instead of being a mere deposit derived from external sources.—This leads us to consider, in the *second* place, the inferences derivable from the mode in which the nutritive materials are taken in. For if the aliment of a being whose other characters leave its place doubtful, be *formed* by it at the expense of carbonic acid, water, and ammonia,—as is indicated especially by its decomposition of carbonic acid, and its liberation of oxygen under the influence of light,—there is a strong presumption of its Vegetable nature:* whilst, on the other hand, if it be dependent for its support upon organic compounds previously formed by other beings, and especially if these be received into internal cavities for the purpose of digestion, the presumption is equally strong in favour of its Animal character. Here, too, discrimination is needed; for Plants may seem to live at the expense of matters previously organised (as is the case with the whole group of Fungi, §§ 271–3), when they really only thrive by the decay of these, taking in the carbonic acid and ammonia which are thus set free; whilst, on the other hand, there are Animals which appear to subsist on inorganic matters, such as sand or earth, although they are really supported by the small quantity of organic substances which these may include.—*Thirdly*, although movements which seem to indicate sen-

* It has been asserted that some true *Animalcules* possess this power; but in the Author's opinion without any adequate foundation. Beings possessed of active powers of motion undoubtedly decompose carbonic acid under the influence of light; but there is no proof, save the completely fallacious one of *movement*, of their Animal nature.

sibility, or which cannot be accounted for otherwise than as originating in a mental cause, must be held to justify the Naturalist in attributing an Animal character to the beings that exhibit them, even where no nervous system can be detected, yet it is extremely difficult to say with certainty, in particular instances, whether such movements are or are not to be considered in this light; and it is in great part from this cause, that so much doubt at present exists with regard to the true character of many of the beings occupying the border-ground between the two kingdoms. When *all* movements that appeared to be spontaneous were held to indicate the existence of consciousness, the 'zoospores' of many of the inferior Algæ, propelled hither and thither by ciliary action, were considered as Animalcules; and were thought to yield up their Animal life, and to return to the Vegetable kingdom, so soon as they cease to move and begin to evolve themselves into filaments. But such an explanation of this phenomenon cannot any longer be accepted; and the Physiologist is now compelled unhesitatingly to admit that regular movements of a *spontaneous* character (that is, originating in the vital endowments of the being itself, and not in any direct operation of external agencies) may be exhibited by organisms indubitably belonging to the Vegetable Kingdom. Such movements are of two kinds; being either *ciliary*, as in the case of the 'zoospores' of certain Algæ (§ 267); or being produced by the contraction of the cell-wall itself, as in the *Oscillatoriæ* (§ 138). In each of these cases, the constancy of the movements and their rhythmical character give to the observer the impression that they are not under any *psychical* guidance or control, but that they simply result from the peculiar vital endowments of the bodies that exhibit them; which endowments may be manifested in mechanical motion, as well as in the generation of tissues or in the production of chemical changes (§ 50). Now the existence of such movements among beings of undoubted Vegetable character, should obviously lead the Naturalist to question the usually-received idea, that the movements of the so-called 'Polygastric Animalcules' (§ 284) are indicative of consciousness, and are under psychical control; more especially as, in these beings, no traces of a nervous system can be discovered. Indeed it may be stated with certainty that no nervous system *can* exist in them; the transparency of their bodies being such that it must be readily discerned if it were really present; and the general simplicity of their structure being equally opposed to the notion that this, the highest form of Animal tissue, could be present in organisms so little advanced in development. Moreover, the movements in question, like those of the inferior Algæ, always depend either upon *ciliary* action, or upon a change of form of the cells of which their whole body is composed; and are never traceable to anything like muscular structure. Now we have seen that ciliary action takes place, not only without any exertion of the will, but even independently of consciousness; and as this is known to be its condition alike in the lowest forms of Vegetable and in the highest types of Animal life, we have no adequate reason to suppose that it anywhere indicates the existence of the least amount of sensibility, much less the influence of a guiding or controlling will. The same may be said of those changes of form in the component cells of a part or the whole of a fabric, which depend, like ciliary action, upon causes inherent in themselves, instead of taking place in respondence to nervous stimulation; for it is only when they are excited in the latter mode, that they can be certainly regarded as indications of animality. We shall find (CHAP. XVIII.) that movements very similar to those of the Animalcules in question are exhibited by the embryos of many among the higher animals in an early stage of their development, when as yet the body is homogeneous or nearly so, and when the first traces of a nervous system have not made their appearance; and as they seem to take place under very similar conditions in both cases, we have no right to assume the existence of a consciousness in the former class of beings, which is certainly wanting in the latter. Where the motions are performed, however, by the agency of *muscular* structure, a strong presumption exists in favour of the Animal nature of the being which exhibits them; and this presumption is raised to certainty, when the muscles are seen to be called into contraction by nerves. Further, where (as in the *Hydra*, § 289) the movements are evidently of the same kind with those which are elsewhere performed under the direction of a nervous system, a strong presumption must be felt in favour of the animality of their character, even though no nervous system can be detected. Thus, then, it is obvious that the performance of movements cannot in itself be regarded as any sufficient criterion between the Animal and Vegetable kingdoms; but becomes so, when these movements clearly indicate a dependence on *consciousness* for guidance or control.—*Fourthly*, important assistance may often be derived, in doubtful cases, from a comparison of the general affinities of structure; or from the mode in which particular functions are performed. Thus, it is a valuable indication of the animality of the *Sponge*, that certain *Alcyonian* zoophytes (§ 292), in the progress of their development, pass through a condition nearly allied to its own; whilst the Vegetable nature of the *Diatomaceæ*, as well as of the *Desmideæ* (both of which groups were ranked by Ehrenberg among Animalcules), may be predicated with great probability, from the occurrence among them of that process of *conjugation*, which seems to be especially characteristic of the inferior Algæ (§ 267).—In conclusion, then, we must look for guidance in the determination of the Vegetable or Animal nature of a doubtful organism, (1) from the predominance of starchy compounds on the one hand, or of albuminous on the other, in its solid tissues; (2) from its power of decomposing carbonic acid under the influence of light, or from its reception of solid alimentary matters, previously elaborated by some other organism, into a digestive cavity; (3) from the absence, or the presence, of distinct manifestations of consciousness; and (4) from the general resemblance which its structure or its mode of existence may bear, to that of beings undoubtedly belonging to one or the other kingdom.

CHAPTER VI.

GENERAL VIEW OF THE VEGETABLE KINGDOM.

264. The scientific *Biologist,* who desires to acquire some comprehension of the plan according to which the countless varieties of organised fabrics are developed, and to frame some general statements, or 'laws,' expressive of the conditions under which the different manifestations of vital activity occur, may pursue two very different modes of investigation, according as the first or the second of these objects is the one to which he more especially devotes himself. For he may study each individual fabric, whether Plant or Animal, as one whole; may compare it with others bearing more or less resemblance to itself; and may so classify and arrange the entire assemblage of such forms into groups of greater or smaller comprehensiveness, that every single one shall be found in the midst of those to which it is most nearly allied in its general plan of structure, and at the greatest distance from those to which it is most dissimilar. Of this process, which is termed *Classification,* it is the ultimate aim to construct a *Natural System,* which shall present the nearest approximation that the limited capacity of the Human Mind will admit, to the plan on which the Divine Author of the Universe has formed the Organised portion of his Creation. Those only, however, can devote themselves to the work of systematic arrangement with a fair prospect of ultimate success, who have a competent knowledge, not merely of the *external configuration,* but also of the *internal structure,* of the beings whose relations to each other they aim at determining; since but a very small part of these relations can be discerned from the most elaborate study of outward form alone. It is now coming to be generally admitted, also, that a knowledge of the *history of their development* is required in addition; and this on three accounts more especially. For (1) it is only by such knowledge, that the Naturalist can be preserved from the grievous error of assigning distinct places in his arrangement to forms which are really those of similar individuals in different stages of their development. Again (2), it is only by watching the evolution of the several parts of the fabric, that he can ascertain, in a great variety of cases, what are those which essentially correspond with others, so as to admit of just comparison, in different beings; that original or fundamental conformity among organs, to which the term *homology* has been applied, being frequently obscured, in their subsequent development, by their adaptation to some special purposes in the economy of the beings of which they form part. Further (3) as the changes just alluded to frequently tend to mask or obscure the real affinities of the beings in which they occur, it follows that the study of their development will afford direct and important assistance in Classification: the place of a particular group in the series being often determinable much more positively by its embryonic characters, than by those which it presents in its perfect state. This is peculiarly the case, for example, with the class of *Cirrhopoda* (§ 311). In the construction of his system of Classification,

then, the philosophic Naturalist must have continual recourse for assistance to the Anatomist and Physiologist.*—But the Physiologist, on the other hand, whose special aim it is to ascertain the Conditions of Vital Activity, cannot expect to arrive at any really comprehensive expressions of these, unless he prosecute his researches through the entire series of living beings, every one of which presents a different combination of them (§ 8); and he can make very little progress in his survey, unless he avail himself of the systematic arrangements which the Naturalist is at the same time endeavouring to construct; these, even in their imperfect state, being capable of affording him the most important assistance. Hence, although the purpose of the present treatise is purely Physiological, it will be of advantage to its reader that he should be led, before proceeding further, to take a general survey of the principal types of structure presented by the Vegetable and Animal kingdoms.

265. It is computed that above 120,000 distinct 'species' (or permanently-dissimilar forms) of Plants are at present known to Botanists;† and although it is probable that many of these are not real 'species,' being either *transitional forms* presented by the same plant in different stages of its development, or *varieties* produced by the influence of soil, climate, &c., and always capable of reproducing the original type, yet there can be no question that however much the list might be reduced by a fuller acquaintance with the history of the plants of which it is composed, it is destined to receive a far larger augmentation from more extended research: since a great part of the surface of the globe has been as yet but very imperfectly, or not at all, examined by the Botanist; whilst new discoveries are continually being made, among the humbler and especially the microscopic forms of vegetation, even in the countries which (like our own) have been the most thoroughly explored. We should almost

* It is a most encouraging sign of real progress in Natural History, that these truths are now coming to be generally admitted and acted upon. The Zoologist and the Botanist have, until recently, devoted by far too large a proportion of their time and attention to the mere business of collecting species, of defining their external characters, and of framing systematic arrangements based almost exclusively upon these. The Anatomist and Physiologist, who have looked more deeply into the mysteries of organisation, who have investigated the internal structure and the living actions of these beings, and who have urged the necessity of basing all classification upon the points of agreement or difference in fundamental characters thus brought into view, have been too often looked upon as well-meaning enthusiasts, who put themselves and others to a great deal of unnecessary trouble, and whose labours tend to confuse what is already plain and simple, to disturb what has long since been established. Thus a degree of isolation, and even of opposition, has grown up between the votaries of different departments of Natural History, which has greatly tended to retard the progress of the science. What God has joined, Man has foolishly striven to put asunder. The mere systematist has rejected, as long as he could possibly do, the friendly overtures of the physiologist; and the physiologist has repaid this distrust, by a not undisguised contempt of the mere species-collector. This state of things, however, is rapidly giving way; and it is almost universally acknowledged, that the most valuable contributions to Natural History, of late years, have been from those who have combined both methods of investigation.

† The result of a recent enumeration of the plants contained in the Muséum d'Histoire Naturelle, at Paris, gives from 115,000 to 120,000 species as the number collected in that one Herbarium; and it is certain that other Herbaria must contain many species which it does not include. The total number of species scientifically described by Botanists now exceeds 100,000.—The estimate given above of the total number of Plants at present existing, is certainly rather *within* than *beyond* the mark: for Baron Humboldt, than whom there is perhaps no greater authority on such a subject, considers that as many as 160,000 species of Flowering Plants alone are either described or preserved in herbaria; and that these do *not* constitute *half* the number of those at present covering the surface of our globe.

certainly be within the mark, therefore, in saying that at least *two hundred thousand* species of Plants are in existence at the present epoch. These present us with every gradation of complexity of structure; from those simple tribes in which every cell is similar to the rest, and may, from its independence of them, be regarded as a distinct individual; to those aggregations of diverse tissues, which constitute the highest types of plant-formation, and which present us, in their roots, stems, branches, leaves, and flowers, as even in the subordinate parts of these, with a great variety of organs, whose combined action is essential to the well-being of the whole, and whose sum, therefore, must be regarded as making up a single individual. Now all Botanists are agreed that this vast assemblage of species may be arranged under a small number of primary groups; and yet the principles of such an arrangement have not yet been satisfactorily determined. Thus Linnæus divided the Vegetable kingdom, in the first instance, into PHANEROGAMIA, or 'flowering-plants,' in which the sexuality of the reproductive organs is apparent; and the CRYPTOGAMIA, or 'flowerless plants,' of which the sexuality is obscure, or is altogether wanting. Now this division is acknowledged by every Botanist to be quite *natural*, since it brings together those plants which have most agreement in their general structure, whilst it separates those which are most dissimilar. But the greatest diversity of opinion exists with regard to the character on which it is based; for whilst some maintain that there is no sexuality in the Cryptogamia analogous to that of the Phanerogamia, others assert that the Cryptogamia possess all the essentials of the sexual apparatus of the flowering-plant, and others again affirm that the Phanerogamia have not, any more than the Cryptogamia, a sexual apparatus at all comparable to that of animals. So, in regard to the subordinate divisions of the Cryptogamia, we find all Botanists agreeing to group together certain plants under the titles of Ferns, Mosses, Fungi, Lichens, and Algæ, although no two of them shall give precisely the same definition of any one of these assemblages; and the principles on which the classification of the Phanerogamia shall be founded, are equally far from being settled.—As the purpose of this work, however, is not to present a systematic classification of Plants and Animals, but to guide the Physiologist in the search for the essential conditions of vital activity, it will be best answered by an outline-view of the plan, on which the principal groups of each kingdom seem to be constructed, and of the relations which they appear to bear to each other.

266. Commencing our survey with the lowest forms of vegetation, we find three groups, those of ALGÆ, LICHENS, and FUNGI, possessing an equal simplicity of structure, and having the same starting-point in the isolated cell; but developed under very different conditions, and presenting in their higher grades an obvious diversity in the plan of their conformation; so that, however difficult it may be to draw a definite line of demarcation between their lowest and most similar forms, no such difficulty exists in regard to the determination of their more elevated types. Thus, no one is in any danger of confounding a Mushroom with a Sea-weed, or either of these with the Lichen that encrusts the trunk of an aged tree; yet the *Protococcus nivalis*, a little plant that sometimes reddens extensive tracts of snow in arctic and alpine regions, has been referred in turn to all three groups; and it is not even yet settled, in the

opinion of some Botanists, whether the *Torula cerevisii,* or yeast-plant, is to be ranked among the Algæ or the Fungi. All these plants agree in the extreme simplicity of their structure ; the entire organism, even of their most complex forms, consisting of nothing but 'cells,' without a trace of woody fibres, or of vessels of any kind. Hence they are sometimes distinguished as *Cellular* plants. Again, we find among them no complete distinction between stem, roots, and leaves ; the primordial cell being developed into an expanded 'thallus,' whose form has but little definiteness, and whose structure is nearly homogeneous throughout ; whence they have been designated as *Thallogens.* And as presenting the simplest types of vegetable structure, which bear a considerable resemblance to the transitory forms through which the higher tribes pass in the progress of their development, they have not been unaptly designated *Protophyta.*—The three subdivisions of this group may be most readily characterised by the conditions under which they are respectively developed, and by the *nisus* or direction of evolution which they severally manifest. Thus, the ALGÆ vegetate exclusively in water (though not all, as implied by their vernacular synonym 'sea-weed,' in *salt* water), or in damp situations ; they require no nutriment but such as is supplied by water and the air dissolved in it, which they absorb equally by every part of their surface ; and they show a great tendency to the extension of the 'thallus' by the multiplication of cells in continuity with the existing fabric, which frequently takes place with great rapidity, and to which the evolution of the reproductive apparatus appears generally subordinate. On the other hand, LICHENS grow upon living vegetables, stones, or hard earth, in situations where they are not abundantly supplied with moisture, and are fully exposed to light and air ; they derive their food from the atmosphere and from the water which it conveys to them ; and their 'nisus' or tendency of development is to form a hard dry crust-like thallus, of slow growth and no great dimensions, but of very durable character, whose upper side, or the one most exposed, contains the fructification, whilst it is through the lower and softer surface that the absorption of nutriment chiefly takes place. The group of FUNGI differs from both these, in requiring, as the most favourable condition for the development of its members, the presence of dead or decaying organised matter, which shall afford by its decomposition a larger supply of carbonic acid and ammonia than the atmosphere and its moisture would alone furnish ; their growth is favoured by darkness rather than by light ; their 'nisus' is manifested in the predominant evolution of the reproductive apparatus, to which the development of the thallus seems quite subordinate (the latter being in fact very often overlooked altogether, so that the plant is erroneously imagined to consist entirely of a mass of 'spores' or reproductive cells with their envelopes) ; and the formative operations are carried on with extraordinary rapidity under favourable circumstances, the tissues thus generated, however, seldom possessing much permanence of character, but decaying as soon as they have passed through all the stages of their evolution.—Of all these Protophytes it may be remarked, that the conditions under which they are developed produce a considerable modification in their mode of growth, and may even effect such a change as to obscure their characteristic 'nisus.' Thus if Lichens be removed from the influence of the light, and be over-supplied with moisture, they show a tendency to the extension of the vegetative or foliaceous portion of the thallus, and to the non-development of the fructification; and there are certain tribes of Lichens inhabiting the sea-shore, where they are liable to be occasionally submerged, which bear so strong a resemblance to Algæ, that they have been ranked under that group, although closely allied to true Lichens in structure and fructification. So, again, when the simpler forms of Fungi develope themselves in liquids, they show an unusual tendency to the extension of the vegetative thallus, and have sometimes so much of the characteristic appearance and mode of growth of Algæ, that their true nature becomes apparent only when the fructification is evolved.* It would seem, indeed, as if the vital force of these simple organisms is not sufficient to give a complete determination to their mode of development; this being much more influenced by external agencies, than is that of the higher types of organised structure.

267. The group of ALGÆ is one of peculiar interest to the Physiologist, since it presents him with an almost continuous succession of forms, which connect the very simplest members of the Vegetable kingdom with plants of a considerable degree of complexity, in whose fabrics we may observe a shadowing-forth of the diverse organs of the higher plants. We must confine ourselves, however, to a notice of some of the principal gradations

FIG. 65.

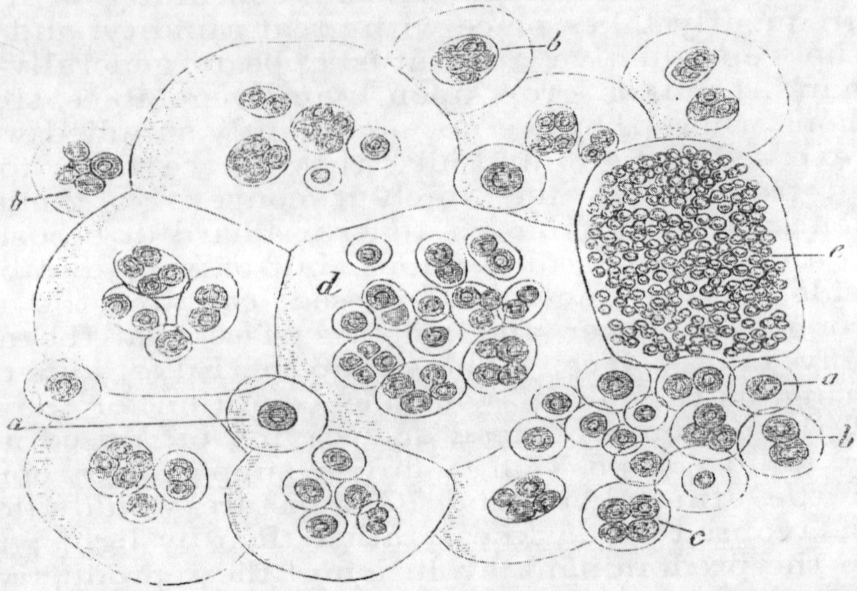

Hæmatococcus sanguineus, in various stages of development;—*a*, single cells, enclosed in their mucous envelope; *b*, *c*, clusters formed by subdivision of parent-cell; *d*, more numerous cluster, its component cells in various stages of division; *e*, large mass of young cells, formed by the continuance of the same process, and enclosed within a common mucous envelope.

of structure which the group exhibits. Its lowest division includes those tribes (already so often referred to), in which every cell may be regarded as constituting a distinct individual, living and growing for itself, though generally aggregated with others to form composite masses; of such plants, constituting the greenish or reddish slime often seen on damp surfaces, we have already had an example in the *Hæmatococcus binalis*

* See especially a description of a Confervoid state of *Mucor clavatus*, by the Rev. M. J. Berkeley, in the Magazine of Zoology and Botany, vol. ii. p. 340.

(Fig. 4), which displayed to us the method in which a single cell may become multiplied by subdivision to an indefinite extent. It is a peculiarity of the greater number of the Algæ, that the cells have the tendency to form on their exterior, apparently by secretion from within, a mucous investment, which seems almost like an imperfectly consolidated cell-wall; it is this which gives to the ordinary 'Sea-weeds' their peculiarly slimy surface; and the same substance connects the isolated cells of these simpler tribes, and holds them together in masses, so as to form a rude sort of thallus, which is then termed *gelatinous.* As an illustration of this kind of aggregation, we may refer to the preceding figure of *Hæmatococcus sanguineus;* which shows the component cells clustered together, and surrounded by their mucous envelope, in different stages of multiplication. It is evident that there is here no tendency whatever to the assumption of any definite form or dimensions; but that, so long as the process of cell-multiplication continues, so long may the thallus extend itself in all directions. Sometimes, however, the mode of development is such, that a tolerably regular globose form is preserved in the aggregate mass and in its component parts (Fig. 66).—Many of these simple cellular plants have their cell-wall strengthened by a coating of siliceous matter, which frequently presents very beautiful markings; this is the case with various forms of *Desmideæ* and *Diatomaceæ,* two tribes which have been assigned by Ehrenberg and several other Naturalists to the Animal kingdom, but of whose Vegetable character we can scarcely any longer entertain a doubt, notwithstanding that several of them exhibit very curious and regular movements, strongly resembling those of Animalcules. In these tribes, moreover, we observe a tendency to the production of *filaments,* by the repeated subdivision of cells in the same direction, the new cells remaining adherent to each other in a linear series. This, however, is the special characteristic of the tribe of *Confervæ;* every one of whose long, green threads is made up of a longitudinal succession of cells, gradually produced by the process of subdivision from a single individual (Fig. 69, A, *b*); whilst the occurrence of the same process in the contrary direction will cause such a filament to be developed into an expanded thallus. These various gradations of structure are displayed in Fig. 67. In the filamentous *Desmideæ,* each individual cell continues to multiply by subdivision, so that the number of cells in the filaments may be doubled by one act repeated along its whole length. In the *Confervæ,* however, we more commonly find the extension of the filament effected by the repeated subdivision of its terminal cell alone, the portion already formed undergoing no further increase; but when a confervoid filament is in process of development into an expanded thallus, we find each

FIG. 66.

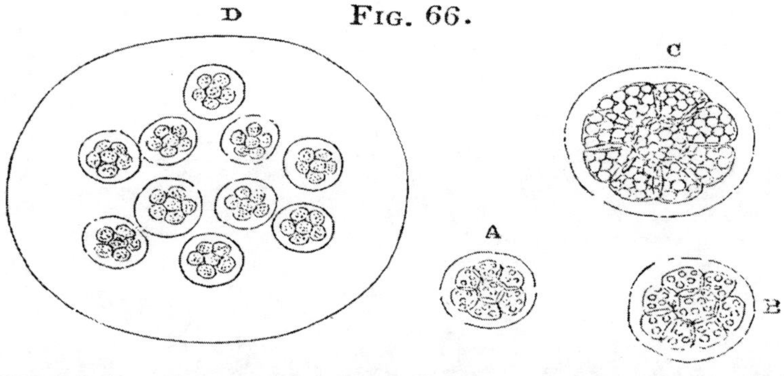

Successive stages of multiplication of *Botryocystis morum;*—A, parent-cell with incipient secondary cells; B, the same, more advanced; C, the same at a later stage, the secondary cells now filled with cells of a third generation; D, globular mass, formed by the separation of the secondary cells, of which each has now acquired the form and character of the parent-cell A, but which remain invested by the mucous envelope common to the whole.

individual cell undergoing multiplication, as is shown at the two ends of the filament E, Fig. 67, in whose central part the cells still form but a linear series. The flat leafy expansion of the *Ulvæ* may be looked upon as nothing else than an assemblage of confervoid filaments remaining adherent to each other; whilst the simple confervoid filament is itself but a linear series of cells united end to end. In fact, every Ulva, in the course of its evolution from its primordial cell, goes through this series of forms; the cell first giving rise to a small cluster; the cluster elongating itself, by multiplication in one direction, into a filament; and this subsequently extending itself by subdivision in the opposite direction, so as to form the leaf-like expansion characteristic of that tribe of sea-weeds.—Now in these stages of development, we trace a gradual progress towards that *composite individuality*, which is characteristic of higher types of structure. In such aggregate masses as those of *Hæmatococcus sanguineus* (Fig. 66), there is no approach to definiteness of form or to limitation of size, any more than to a relation of mutual dependence among the component cells; but in the *Confervæ* a definite form, though of the simplest possible kind, begins to manifest itself; and we no longer find all the cells possessing precisely the same endowments, the filament being attached at one extremity to some fixed object, whilst it is at the other extremity alone that it continues to grow: and in the *Ulvæ*, the limitation of form and size are yet more decided, so that we can scarcely refuse to speak of an entire plant as one individual, although it is made up of an assemblage of parts, each of which is almost exactly identical with the rest.

FIG. 67.

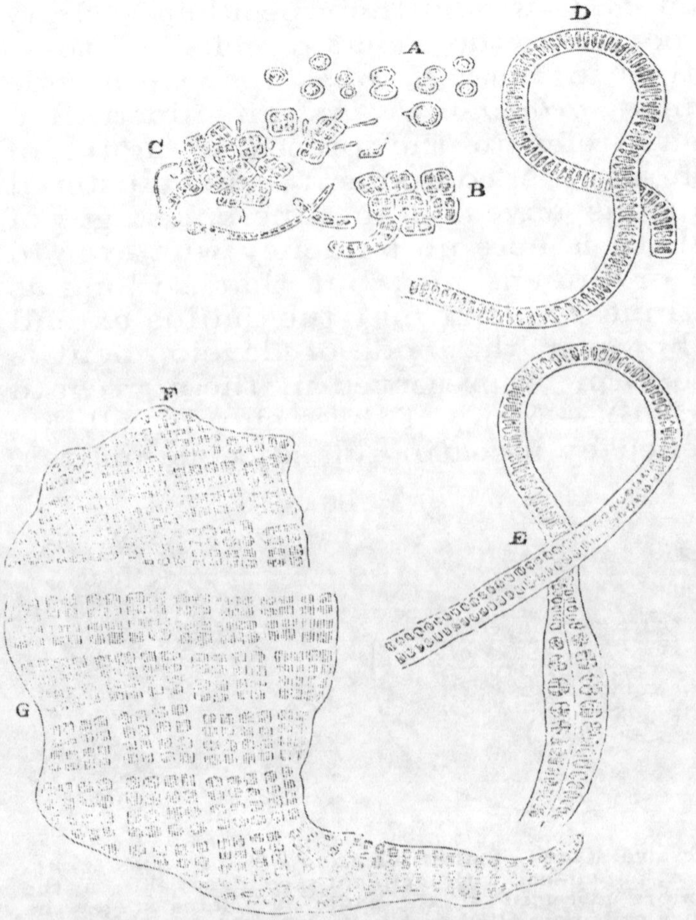

Successive stages of development of simpler Algæ:—A, individual cells of *Protococcus viridis*; B, C, clusters formed by their multiplication; D, filament of *Schizogonium murale*; E, a similar filament, subdividing laterally, which constitutes the early form of the *Ulvaceæ*; F, G, portions of the expanded thallus of *Ulva furfuracea*, formed by the continuance of the same process of transverse subdivision.

267*a*. The propagation of these simpler Algæ appears to take place in two principal modes; of which one may be considered as extending the original structure, after the manner of the 'budding' of the higher plants; whilst the other originates a really 'new generation' by a process which is termed 'conjugation,' and which is of the very opposite character to that of the multiplication of cells by subdivision. The first of these processes consists in the separation of the endochrome of the individual cells into a number of small particles, each of which acquires a proper cell-wall of its own, and thus becomes an incipient cell. The young cells thus formed in the interior of their progenitor, may be seen for a time to move about within its cavity, and to strike against its walls; an orifice then forms, by which they are set free; and they then swim forth with an active movement, and disperse themselves through the surrounding liquid. Their motion seems to depend, in some cases, upon the presence of *cilia* (§ 183) on the exterior of the cell; whilst in other instances it appears due to the undulation of one or two filaments, of larger size, attached to the cell. During their state of activity, which is not of long duration, the 'zoospores' (as they are termed) may be easily mistaken for Animalcules. The history of their development, in the case of *Achlya prolifera*, has been already detailed (§ 142); where it has been shown that the formation of these 'zoospores' is to be regarded as merely a variation of the ordinary multiplication by binary subdivision; the principal difference consisting in this, that the young cells are detached from the parent fabric, and are endowed with locomotive power, whereby they may be dispersed to a distance so as to originate new colonies, instead of remaining to develope themselves in connection with it. This propagation by 'zoospores' has been hitherto noticed chiefly in the Confervoid Algæ; their development taking place in some instances within every one of the cells of which the plant is composed; whilst in other cases they are formed within specially-enlarged cells, developed either at the sides or the ends of the filaments (Fig. 68).—An entirely distinct method of propagation, however, has been observed in certain forms of the inferior Algæ; and recent observations appear to indicate that it is not confined, as was formerly supposed, to a single genus, but occurs throughout the whole series, at some period or other in the existence of every tribe. One of the most characteristic illustrations of the process is furnished by the *Zygnema*, a common confervoid plant, inhabiting fresh waters, which derives its name from the 'conjugation' of the filaments of which it is composed. The filaments are developed separately, like those of ordinary Confervæ (Fig. 69, A), increasing in length by the subdivision of their component cells after the usual manner; as they approach to maturity, the endochrome generally assumes a spiral arrangement, as shown at B, and the process of 'conjugation' then takes place. This consists in the approximation of two filaments, and the formation of protuberances from the corresponding cells of each, which meet and adhere; the partitions which still separate the cavities of the cells thus joined, disappear after a time, so that the endochromes of the two may freely intermingle; and by

FIG. 68.

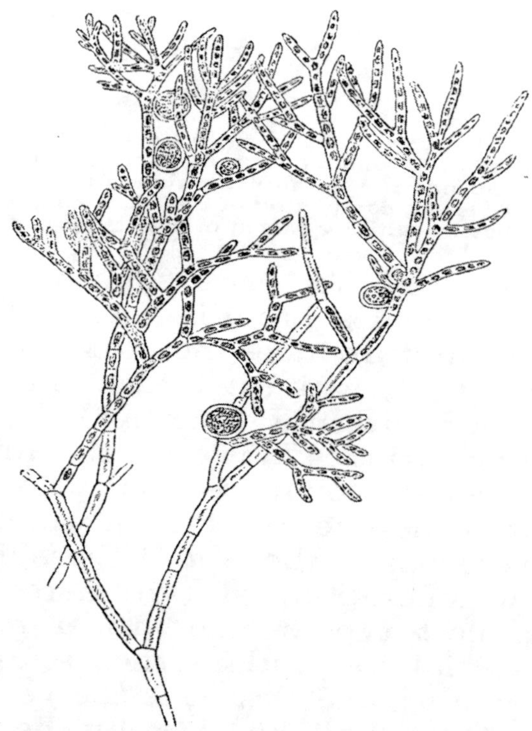

Chætophora tuberculosa, a branching confervoid plant, each filament consisting of a linear series of cells, and having its zoospores developed within capsules or dilated cells springing from the sides of the filaments.

their junction a new mass is formed, which is called the *sporangium.* In the *Zygnema,* the endochromes of all the cells of one filament (Fig. 69, C *a*) pass over into the corresponding cells of the other filament (*b*), and the sporangia are consequently formed in the cavities of the latter; but

FIG. 69.

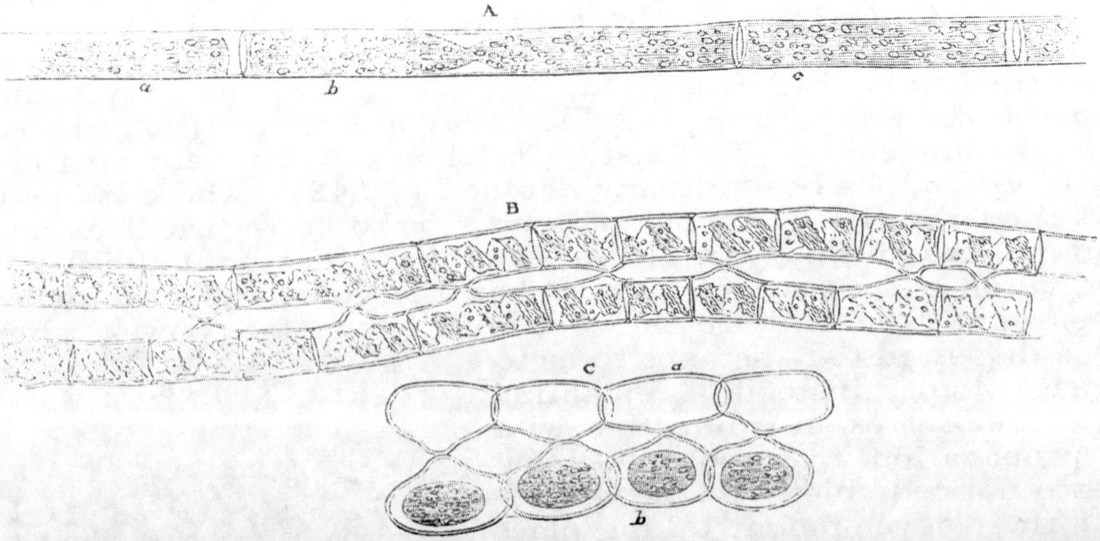

Various stages of the history of *Zygnema quininum*:—A, three cells *a*, *b*, *c*, of a young filament, of which *b* is undergoing subdivision; B, two filaments in the first stage of conjugation, showing the spiral disposition of their endochromes, and the protuberances from the conjugating cells; C, completion of the act of conjugation, the endochromes of the cells of the filament *a* having entirely passed over to those of filament *b*, in which the sporangia are formed.

there are other cases (some of them to be noticed hereafter), in which both endochromes are discharged from the conjugating cells, so that their intermixture does not take place within the cavity of either; and there are others, again, in which the endochromes meet, and the sporangium is formed, within a dilatation of the passage that unites them. In these and other varieties, however, the fundamental nature of the process remains the same; for it essentially consists in the *reunion* of the endochromes of two cells, the product of which is a new and peculiar body, apparently answering to the *seed* of the higher plants, and giving origin to an entirely new generation. It will hereafter be shown (CHAP. XVIII.) that this is the simplest type of the proper *generative* process; that it represents in all essential particulars the 'sexual' propagation of the higher plants and animals; and that it is the very converse of propagation by *gemmation*,—whether this take place by the continuous extension of the original fabric, as by the formation of new branches in the *Chætophora* (Fig. 68),—or by the detachment and dispersion of Zoospores, as in the multiplication of *Achlya prolifera* (Fig. 6).

268. In the higher tribes of this group, we meet with forms which strongly remind us of those of the more perfect plants; the thallus being frequently evolved in the likeness of a distinct stem, with roots, leaves, and fruit (Figs. 70, A, and 71, A). This resemblance, however, is only superficial; for there is none of that difference in internal structure, which is characteristic of these parts in the Phanerogamia, the whole substance, even of the highest Algæ, being made up of cells which nowhere depart widely from the typical form (Fig. 70, B, D, Fig. 71, C, D);

whilst, again, there is but little difference in function, each part of the plant for the most part living for and by itself, and being but little dependent upon the rest. The chief advance is in the more complete separation of the reproductive from the nutritive apparatus; and in the higher type which the latter presents.—The substance of which the higher Algæ are composed is often remarkably tenacious, having in some instances the toughness of leather, in others the firmness of wood; and there is one remarkable tribe, that of *Corallines* and *Nullipores* (long ranked in the Animal kingdom, on account of the resemblance of its members to various forms of Zoophytes), in which the tissue is so completely consolidated by carbonate of lime as to possess a stony hardness, although, when this is removed by the action of an acid, it has all the characters of that of the ordinary Algæ. The component cells, even in the highest Algæ, are generally separated from each other by a large quantity of mucilaginous intercellular substance; and a layer of this covers their surface also, giving it that peculiar 'sliminess' through which these plants are generally at once distinguished by the touch, from the Zoophytes to which many of them bear a close resemblance in form and aspect. The greatest heterogeneousness of structure which is anywhere to be met with in this group, is seen in some of the larger *Fucoids;* in which a section of any division of the thallus shows four distinct portions concentrically arranged, foreshadowing in some degree the concentric arrangement of the pith, wood, and bark, of Exogenous stems (§ 280). The central portion, which occupies fully a third of the diameter of the branch, is composed of densely-packed, longitudinal, parallel fibres, or strings of cellules, firmly cohering into one compact mass. Outside this is a much less dense layer, of a paler colour, composed of cells elongated into the semblance of fibres, branching out and anastomosing with each other, sò as to form a network. This, again, is surrounded by a third and much denser layer, composed of elongated cells, lying parallel to the surface, and very closely packed together. Outside this portion, and forming the external coating of the thallus, is a very thin layer of small cells, which is often but loosely attached, and separates much in the manner of an epidermis. In other cases, however, the whole central portion is composed of large and loosely-aggregated cells, whilst the outer portion is very dense (Fig. 70, D), thus prefiguring the Endogenous type of stem-structure in Phanerogamia (§ 279); whilst in certain genera, as the *Laminaria,* we find the stem hollow, foreshadowing that of the Grasses. Most of the higher Sea-weeds have a root-like organ at the base of the stem; this sometimes consists of a simple disk, whilst in other cases it is formed of thick clasping fibres, which sometimes extend for a considerable distance upon or around the body to which they attach themselves. There can be no question that this root chiefly serves to anchor the plant in its place, and that it cannot be regarded as a special organ for the absorption of nutriment, like the root of the higher Plants. Some species, indeed, as the *Sargassum bacciferum* or Gulf-weed (which forms vast fields floating in the Gulf-stream), do not seem ever to possess roots; whilst others are attached by them in their young state, but are afterwards carried about freely by the waves. These floating Sea-weeds are almost universally provided with air-vesicles (Fig. 71, A, *d*); and these are found also in many of the rooted species, such as the common *Fucus vesiculosus,* and serve to keep the upper portion of the plant near the surface of the water. Many of the smaller and more tender kinds of Algæ are of rapid growth and quickly perish; appearing, for instance, early in the spring, and passing away before the summer is far advanced. Others live through the greater part of the year; and there are many which partly die down on the approach of winter, but send out fresh growths from the old stumps with the return of spring. In others, again, the entire plant may endure for several years, and may continue to extend itself by additions to its extremity, or by sending out new branches, according to its type of growth; and thus, as the older portion dies away at one end, whilst new tissue is formed at the other, the life of the aggregate fabric seems capable of almost indefinite prolongation, notwithstanding the change continually taking place in its individual components (§ 124).* In the *Laminaria*, however, the new growth begins at the base, and pushes the older portion before it; as is most remarkably seen in the *L. digitata*, in which the growths of successive years can be distinguished by their diversity of form. This development of the individual is sometimes carried on to an enormous extent; thus plants of *Macrocystis pyrifera* of the South Seas are found to range from 500 to 1500 feet in length, though their stems are not thicker than the finger, whilst the upper branches (which are buoyed up with air-vesicles) are as slender as common pack-thread; and the *M. luxurians* is said by Dr. Hooker to range from 200 to 700 feet in its horizontal growth at the surface of the ocean, lining the beach at the Falkland Islands for miles with entangled cables, much thicker than the human body. In some instances, the diverging root-fibres send up new stems at a distance from the principal trunk; whilst in other cases, portions of the branches spontaneously detach themselves, and become separate plants.

268 *a*. Besides these methods of propagation, however, we find in most of the higher Algæ, as in the inferior tribes, two more special provisions for the multiplication and renovation of the race; the one consisting in the simple production of cells, by a slight variation of the ordinary process of 'subdivision,' which are destined to be set free from the parent structure, and to originate new plants; whilst the essential feature of the other is the 'reunion' of the contents of two cells, to form a new and peculiar body, the *sporangium*, with the production of which an entirely new generation must be considered as originating. The reproductive bodies of the first class, which are analogous to the 'zoospores' of the inferior Algæ, are known in the higher division of the class under the title of 'tetraspores,' in consequence of their occurring in groups of four (Fig. 70, *c*, *b*); they are sometimes found imbedded in the general substance of the thallus, whilst in other instances they are developed only at the extremities of the branches, or in other cases, again, are produced in special receptacles or *thecæ*. They are set free by the rupture of the parent-cell,

* It may be said that in this way the identity of the individual is lost; since, after the lapse of years, no part of the original structure remains. But it must be borne in mind that (as will be shown hereafter) this is equally true of the most highly organised animals, such as Man; the materials of whose fabric are so constantly being changed by decay and replacement, that probably no single particle of the body of the infant remains in the child after its second dentition; and nothing which the child possesses, save the teeth, enters into the composition of the body in adult age; yet the individuality of the whole is universally admitted to be maintained throughout this continual change of component parts.

by the subdivision of which they were generated; but when they have made their escape, they have no power of spontaneous movement, their dispersion being effected by the action of the waves and currents. These 'tetraspores,' like the 'zoospores' of the inferior tribes, must be regarded in the light of detached *gemmæ* or buds; and we shall find that this method of propagation is by no means restricted to Algæ, but that it occurs also throughout the Cryptogamic series, and is very frequent among Zoophytes.—The proper fructification in the higher Algæ is usually restricted to some particular portion of the plant; being either developed at the extremities of the ordinary branches, or in branches expressly modified to evolve it (Fig. 71, A *c*, B, C), or in special capsules still more completely separated from the vegetative portion of the thallus. From the recent researches of MM. Decaisne and Thuret* it appears that the fructification is not simple, as was formerly supposed, but that it consists of two classes of cells, which differ essentially from each other, but the conjunction of whose contents is necessary for the production of a fertile embryo. Of these cells, those from which the new plant is subsequently to proceed may be termed the 'germ cells;'† they are commonly found in the middle of a

FIG. 70.

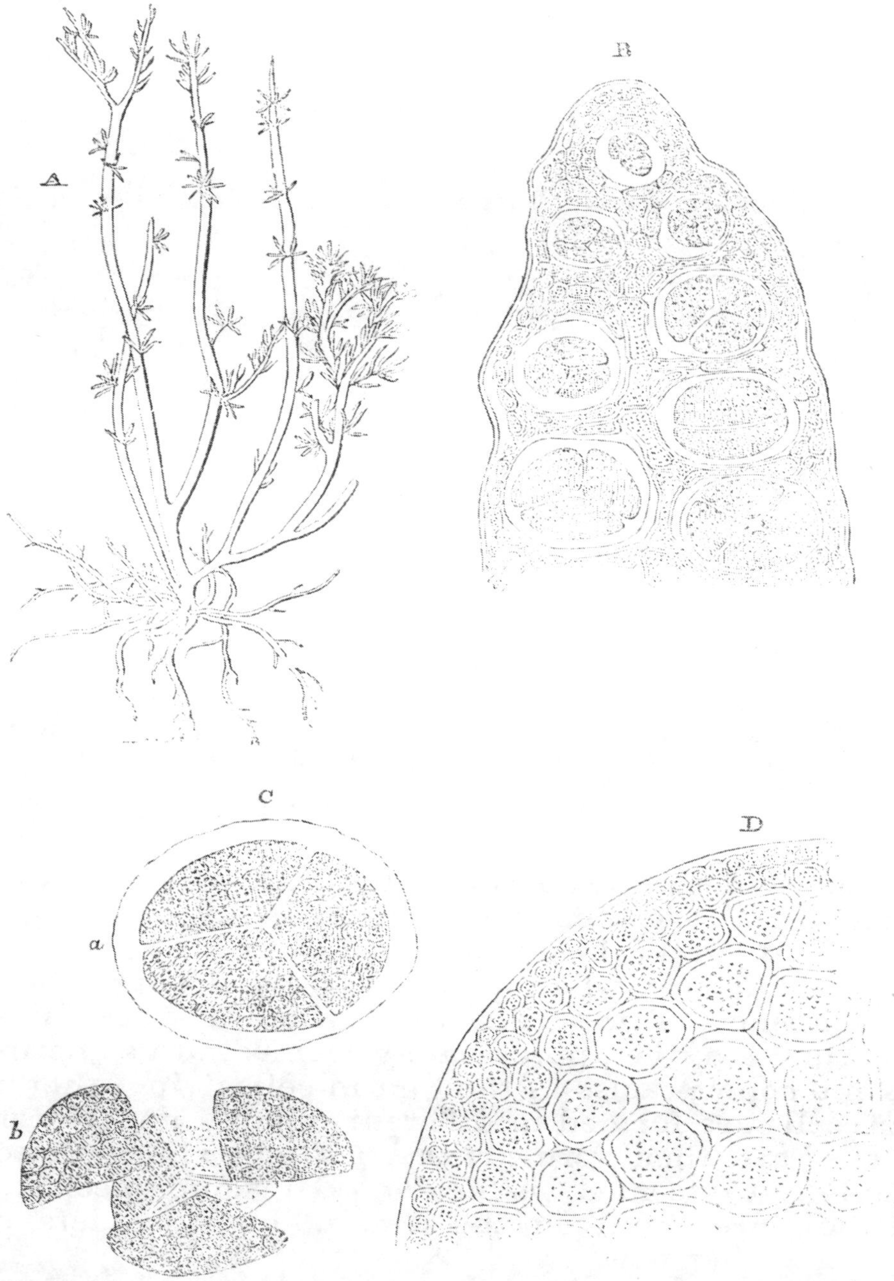

Structure of *Carpocaulon mediterraneum*:—A, entire plant; B, longitudinal section of branch, with tetraspores; C, tetraspores, *a*, within parent-cell, *b* separated; D, transverse section of stem.

* Annales des Sciences Naturelles, Botan., tom. iii. 1845.

† The terms 'germ-cell' and 'sperm-cell,' which will be frequently employed in this treatise, are adopted from Prof. Owen (Lectures on Parthenogenesis, 1849).

cavity lined with filamentous cells, as seen in Fig. 71 C, in which are shown the successive stages of development of the germ-cells in the fruit-bearing branch of *Sargassum.* The contents of the filamentous cells (*a*) seem to be gradually absorbed into the germ-cells, so that the former are emptied of their endochrome (*b*), and may at last disappear altogether (*c*, *d*). But

FIG. 71.

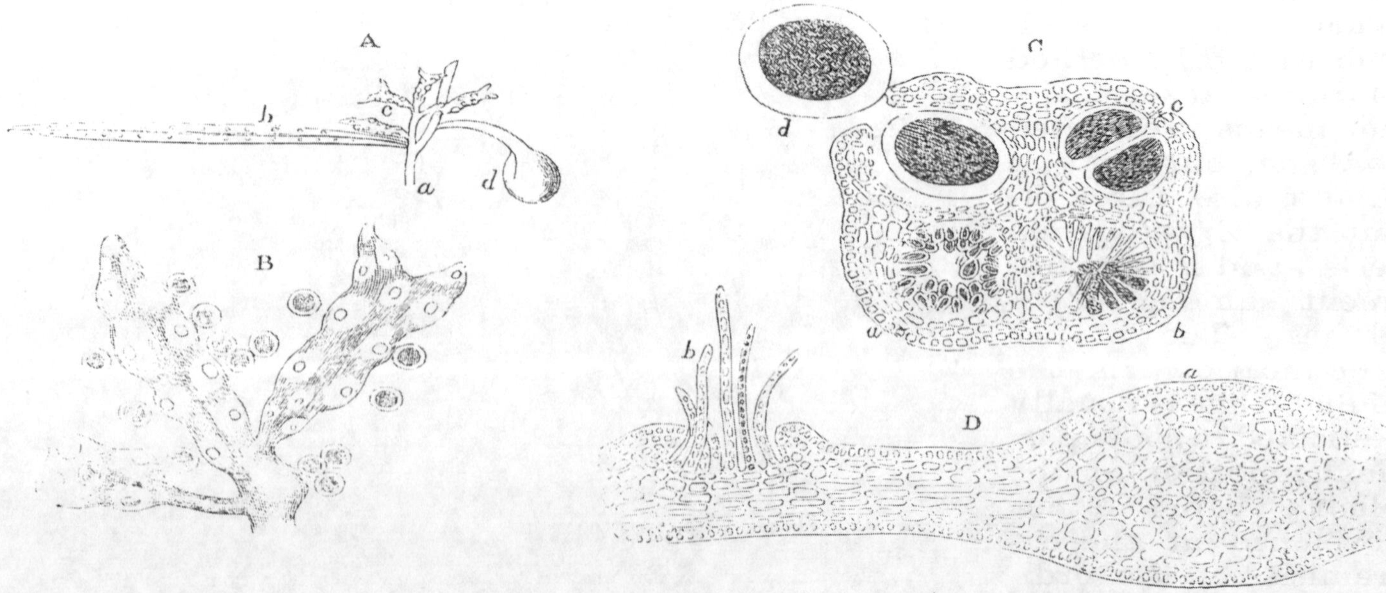

Structure of *Sargassum coarctatum*:—A, portion of the plant, showing *a* stem, *b* leafy portion, *c* a fruit-bearing branch, *d* an air-vesicle; B, fruit-bearing branch, with sporangia, enlarged; C, section of fruit-bearing branch, showing successive stages, *a*, *b*, *c*, *d*, of the evolution of the sporangia; D, section of leafy portion, passing through its central part at *a*, and showing one of the tufts of hairs at *b*.

besides these sterile cells, there are others whose contents are altogether peculiar; their endochrome arranging itself into the form of a spiral filament, which lies coiled up within the cavity of the cell, until set free by its rupture, when it moves freely through the waters for some time, by the agency of the cilia with which it is furnished. These 'sperm-cells,' as they may be appropriately termed, are sometimes contained within the same conceptacles as the 'germ-cells:' in other instances the two classes of cells are formed in different organs, and those containing the 'sperm-cells' are then termed *antheridia*,* whilst those containing the 'germ-cells' are termed *pistillidia*. In some species the antheridia and pistillidia are found on the same plant; in others, on distinct individuals;†

* The term *antheridia* is applied by MM. Decaisne and Thuret (loc. cit.) to the 'sperm-cells' of the Fuci; but it appears to the Author much more appropriate to employ it, here as elsewhere, to designate collections of sperm-cells within a common receptacle.

† It is very curious to find such a marked variation in this respect, between species that are otherwise closely allied. Thus in the *Fucus canaliculatus*, the sperm-cells and the germ-cells are mingled with each other throughout the interior of the same conceptacles; in the *F. tuberculatus* the upper portion is occupied by the sperm-cells, the lower by the germ-cells; in the *F. vesiculosus*, the sperm-cells and germ-cells are sometimes mingled in the same conceptacles, but are more commonly developed in distinct antheridia and pistillidia, which are seated upon different individuals; in the *F. nodosus*, the two organs are always separate, but are sometimes developed on the same, sometimes on distinct individuals; whilst in the

and thus we have, even in this lowest class of Plants, the 'monœcious' and 'diœcious,' as well as the 'hermaphrodite' arrangements of the reproductive apparatus, which are more distinctly seen in the Phanerogamia. It is from the 'germ-cell,' when fertilized by the agency of the 'sperm-cell,' that the new generation originates. The process of 'fertilization' appears to consist in the passage of certain of the contents of the sperm-cell into the 'germ-cell, being thus essentially the same with that of the 'conjugation' of the inferior Algæ;* and the product, though commonly termed a 'spore,' would be more correctly designated a *sporangium*, since it is homologous with the product of conjugation in Zygnema, &c., whilst it is essentially different, not only from the 'zoospores' and 'tetraspores' of the Algæ, but also (as will hereafter appear) from the 'spores' of the higher Cryptogamia.

269. We may now rank with the ALGÆ a very interesting little group of plants, termed the *Characeæ*, in which we have a vegetative apparatus as simple as that of the Confervæ, whilst the reproductive organs are more highly developed than those of the Fucaceæ. Each plant is composed of an assemblage of long tubiform cells, placed end to end; with a distinct central axis, around which the branches are disposed at intervals with great regularity (Fig. 72, A). In one of the genera, *Nitella*, the stem and branches are simple cells, which sometimes attain the length of several inches; whilst in the true *Chara*, each central tube is surrounded by an envelope of smaller ones. Some species have the power of secreting carbonate of lime from the water in which they grow, if this be at all impregnated with calcareous matter; and by the deposition of it beneath their teguments, they have gained their popular name of 'Stone-worts.' These humble plants have attracted much attention, in consequence of the facility with which the 'rotation,' or movement of fluid in the interior of the individual cells, may be seen in them. Each cell, in the healthy state, is lined by a layer of green oval granules, which cover every part except two longitudinal lines that remain nearly colourless (Fig. 72, B); and a constant stream of semi-fluid matter, containing numerous jelly-like globules, is seen to flow over this green layer, the current passing up one side, changing its direction at the extremity, and flowing down the other side, the ascending and descending spaces being bounded by the transparent lines just mentioned. That the currents are in some way directed by the layer of granules, appears from the fact recently noticed by Mr. Varley,† that if accident damages or removes them near the boundary between the ascending and descending currents, a portion of the fluids of the two currents will intermingle by passing the boundary; whilst if the injury be repaired by the development of new granules on the part from which they had been detached, the circulation resumes its regularity, no part of either current passing the boundary. In the young cells, however, the circulation may be seen, before their granu-

F. serratus, the antheridia and pistillidia seem to be always found on separate individuals. The antheridia of the Fucoids may be generally distinguished by their yellow colour, whilst the pistillidia are of an olive-green.

* The actual process of fertilization has not yet been observed in the Algæ; but there can be little doubt, from the great development of their sperm-cells, and from the analogical evidence hereafter to be stated (CHAP. XVIII.), that it takes place as above stated.

† Transactions of the Microscopical Society, vol. ii. p. 99.

lar lining is formed. The rate of the circulation is affected by anything which influences the vital activity of the plant; thus, the movement is accelerated by moderate warmth, whilst it is retarded by cold, and may be at once checked by a slight electric discharge through the plant. The moving globules, which consist of starchy matter, are of various sizes; being sometimes very small, and of definite figure, whilst in other instances

FIG. 72.

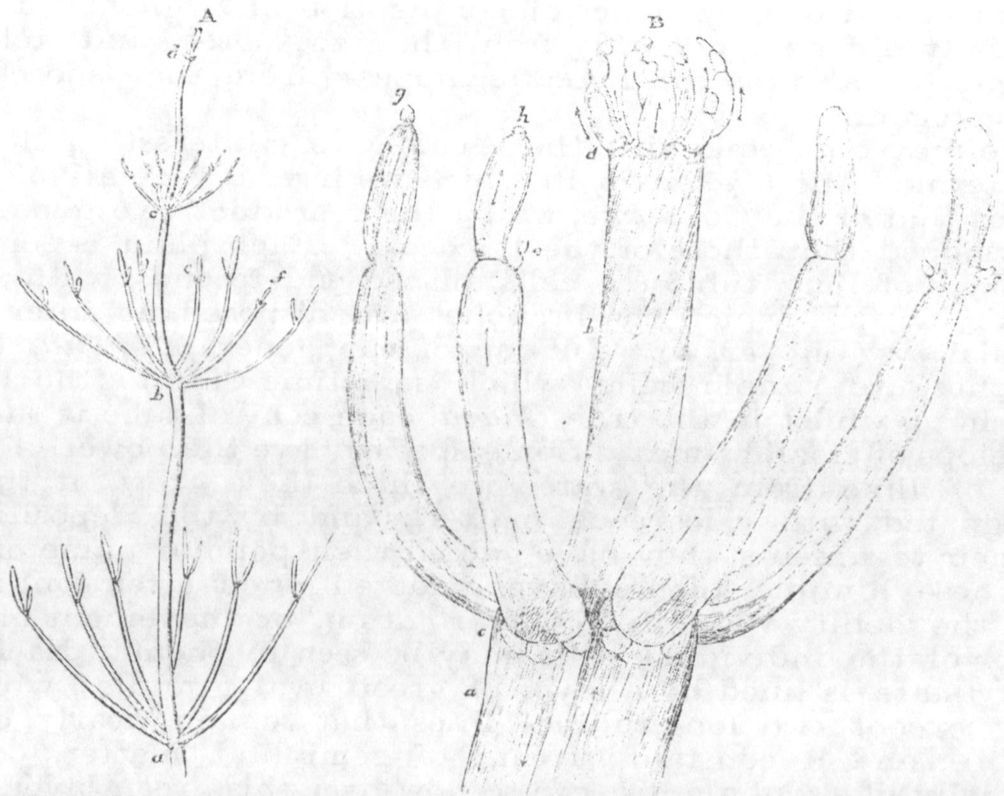

Nitella flexilis:—A, stem and branches of the natural size; *a, b, c, d*, four verticils of branches issuing from the stem; *e, f*, subdivision of the branches;—B, portion of the stem and branches enlarged; *a, b*, joints of stem; *c, d*, verticils; *e, f*, new cells sprouting from the sides of the branches; *g, h*, new cells sprouting at the extremities of the branches.

they are seen as large irregular masses, which appear to be formed by the aggregation of the smaller particles. The production of new cells, for the extension of the stem or branches, or the origination of new whorls, is not here accomplished by the subdivision of the parent-cell, but takes place by the method of out-growth (Fig. 72, B, *e, f, g, h*) which, as already shown (§ 141), is nothing but a modification of the usual process of cell-multiplication. In this manner, the extension of the individual plant takes place with considerable rapidity; but no provision seems to exist for the multiplication of individuals by the spontaneous detachment either of buds or of spores, although parts of the plant artificially separated will continue to vegetate under favourable circumstances, and will develope themselves into the typical form.—The true process of 'generation' in the Characeæ appears to be effected by the conjoint action of two different bodies, the 'globules,' and the 'nucules,' both of which grow from the bases of the branches. The peculiar structure of each of these will be hereafter explained more in detail (CHAP. XVIII.); and at present it will be sufficient to state that the 'globules' are *antheridia*, each containing vast numbers of 'sperm-cells;' whilst the 'nucules' are *pistillidia*, each containing but a single 'germ-cell,' from which, when fertilized by the sperm-cells, the young plant springs up.

270. The contrast between the conditions under which the typical Lichens and Algæ are respectively developed, is not more remarkable than that which exists between their general aspect and mode of growth; and yet we shall find that both the lower and the higher forms of the two groups are closely connected by intermediate links.—In the simpler LICHENS, the primordial cell gives origin, by the usual process of subdivision, to a single layer of cells, which usually spreads itself over the surface to which it is attached, in a more or less circular form; and one or more additional layers being afterwards developed upon its free surface, a *thallus* is formed, which has no very defined limit, and which, in consequence of the very slight adhesion of its component cells, is said to be 'pulverulent.' There are some tribes, however, in which gelatinous matter is interposed among the cells, and which thus connect these simpler Lichens with the inferior Algæ; such is the case with the genus *Collema*, in which we find a 'membranous' thallus essentially consisting of green spherical cells, arranged in bead-like filaments, and imbedded in a firm gelatine; whilst these are tied together (as it were) by cylindrical threads formed of elongated cells, which are closely interwoven at the surface, so as to form a sort of cuticle.* Passing on towards the higher tribes of Lichens, we find these interlaced filaments forming a tough envelope to both surfaces; whilst from the under side of the thallus, which is then said to be 'crustaceous,' thread-like fibres are put forth, which serve for the attachment of the plants, and probably also for the absorption of nutriment. One of the highest forms of thallus seen among the Lichens, is that which is termed the 'foliaceous' or leaf-like; since there is here the greatest approach to regularity and definiteness of form (Fig. 73, C) as well as the greatest complexity of internal structure (A, B). Three principal layers are here easily distinguishable in a vertical section; namely, the 'medullary' layer (B, *c m*), and the upper and under 'cortical' layers (*c c*). The former consists of long, thin, dry, and usually branching cells, loosely interwoven together; and the latter of shorter cells, more closely packed together, and firmly connected by intercellular substance. On the confines between the medullary and the upper cortical layers, is found a layer of rounded green cells, sometimes nearly continuous, in other cases more interrupted, the cells being aggregated in little masses of variable size, which are termed *gonidia*. These green cells probably constitute the true vegetative portion of the plant, being to the medullary and cortical tissues of the foliaceous Lichens what the bead-like rows of green cells in *Collema* are to the gelatine and filamentous substance in which they are enveloped.† These green cells sometimes increase in particular spots, and make their way through the cortical layer, so as to appear as little masses of dust upon the surface, which are termed *soredia*; and when dispersed and separated, they are capable of originating new plants.—A more special reproductive apparatus exists, however, in the higher Lichens; consisting

* See Thwaites in Annals of Natural History, March, 1849. † Thwaites *loc. cit.*

of a series of *asci* or spore-cases (Fig. 73, B, *t t*), each containing a small number of spores (from four to sixteen, but always a multiple of two), and congregated in groups, sometimes immediately upon the surface of the thallus,, sometimes upon an intermediate substance. The groups of asci are often confined to certain projecting portions of the thallus; and these have received various names according to their form, being termed 'shields' or *apothecia* in the Parmeliæ (C, *s*). The asci are imbedded in a mass of straight elongated cells or filaments, which are termed *paraphyses* (B, *p*). The fructification appears to originate in the medullary layer of the thallus (probably in its green cells), and to burst through the upper cortical layer, in order to make its way to the surface, and expand into the shield.—Nothing is yet known as to the existence of any process, among the Lichens, analogous to the 'conjugation' of the lower Algæ, or to the fertilisation of the 'germ-cell' by the 'sperm-cell' in the higher; but analogical considerations afford a strong probability that such a conjugation does take place between certain pairs of the green cells, and that the *asci* with their contained spores are its product. In this case, we must look upon the *soredia* as the representatives of the 'zoospores' and 'tetraspores' of the Algæ; whilst the *asci* are the homologues of the sporangia, which frequently give origin in the Algæ to four or more reproductive cells, of which every one is capable of developing itself into a perfect plant.—It is by their fructification, that *Lichina* and certain other genera inhabiting the sea-shore between the tide-marks, are determined to belong to this group; although both in their general appearance, and in the gelatinous character of their thallus, they so closely resemble the higher Algæ, that they have been frequently ranked among them.—The tribe of Lichens is of peculiar interest, as the one by whose agency vegetation is spread over the bare surfaces of newly-formed rocks. Pulverulent Lichens are first observed, as minute specks upon the pumice or lava of the volcano; these are succeeded by the crustaceous forms; these again by the foliaceous kinds; and to these follow Liverworts and Mosses. Many Lichens form oxalic acid as one of the products of their vital chemistry; and this acts upon the substance of the rock (especially if it be calcareous), forming a hollow which retains the decaying particles of the plant, when its term of

FIG. 73.

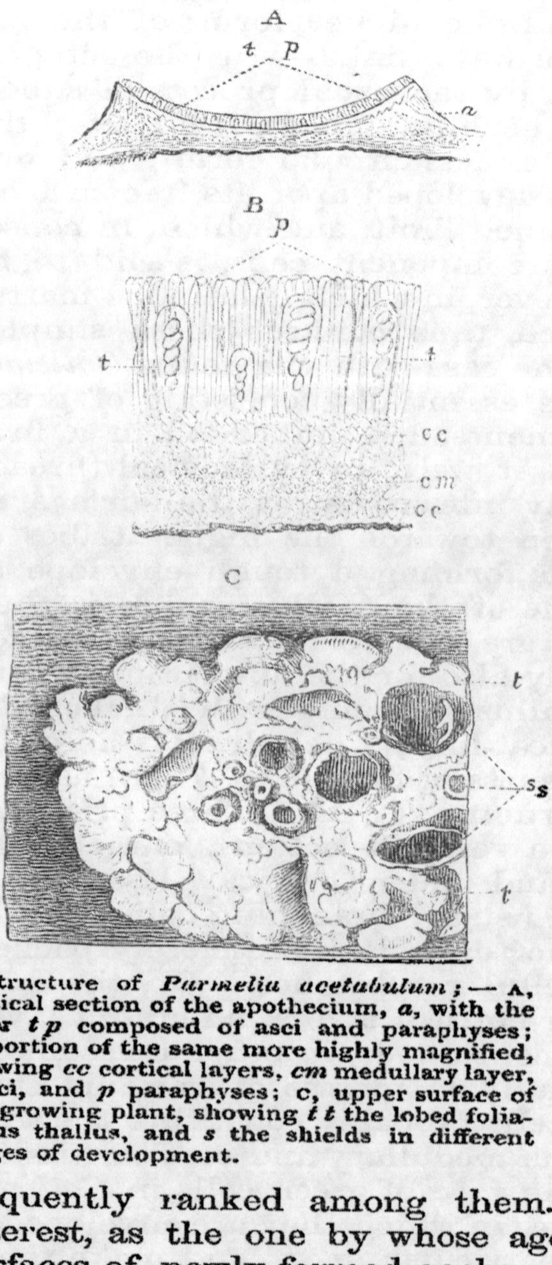

Structure of *Parmelia acetabulum*; — A, vertical section of the apothecium, *a*, with the layer *t p* composed of asci and paraphyses; B, portion of the same more highly magnified, showing *cc* cortical layers, *cm* medullary layer, *t* asci, and *p* paraphyses; C, upper surface of the growing plant, showing *t t* the lobed foliaceous thallus, and *s* the shields in different stages of development.

life is concluded; this hollow is deepened by the growth of successive generations, which leave their débris behind them; and the moisture which is caught in them, finding its way into the crevices of the rocks, assists in their disintegration (especially when expanded by freezing), and thus continually adds to the forming soil. This is no sooner adequate to the support of higher forms of vegetation, than we find such developing themselves upon it, their seeds being conveyed by the winds, by the agency of birds, or by other means; and by the growth and decay of successive generations of these, a sufficient thickness of soil is produced for the nourishment of the luxuriant herbage, and the support of the lofty forest-tree. And thus, by the labours of these apparently insignificant plants, the naked and desolate rock is gradually fitted to afford sustenance to various tribes of animals, and at last to become the dwelling-place of Man.

271. In the simplest forms of FUNGI, we return to the lowest type of Vegetable existence, namely, the single cell; whilst in the higher divisions of the group, we meet with fabrics of much greater complexity, in which there is a considerable degree of heterogeneousness of structure and specialization of function. The most distinctive physiological character of the entire series of Fungi, as already remarked, is their dependence upon dead or decomposing organic matter as the material of their growth. They seem to be for the most part, if not entirely, deficient in the power of decomposing carbonic acid under the influence of light; accordingly, they generally seem to grow in dark situations by preference, their *vegetative* portion being frequently buried beneath the ground, even when their *reproductive* apparatus raises itself in the air, as in the common Agarics, which rank among the highest members of the group; and they give out carbonic acid, derived (no doubt) from the decomposing matters on which they grow, and thus increase the amount of that component of the atmosphere, instead of diminishing it after the manner of other Plants. Further, their cells usually contain a large proportion of azotised substances, identical in composition, or nearly so, with the albuminous compounds of animals; and thus they approximate more closely in composition to the animal tissues, than do any other entire Plants. The character of their texture is usually soft; it is frequently very rapid in its growth (§ 144), but it is also prone to pass with nearly equal rapidity into a state of decay; so that there are probably scarcely any plants, whose entire term of existence is so brief as that of many Fungi. There are instances, however, in which the texture possesses an almost woody hardness; and it is then comparatively persistent and of slow growth. Nothing like real woody structure, however, exists in this group; the whole fabric being composed of cells, whose forms are often strangely modified by the outgrowth of processes in different directions, although they still preserve their independence, no coalescence into tubes being as yet observed.—Of one of the simplest forms of Fungi, we have a good example in the *Torula cerevisii,* which so abounds in Yeast that this substance may be said to be almost entirely made up of it. When a small quantity of yeast is placed under the Microscope, it is found to be full of globules, which are clearly *cells;* and these cells vegetate, when placed in a fermentable fluid containing some form of albuminous matter in addition to sugar, in the manner represented in the accompanying figure. Each cell puts forth one or two projections, which seem to be

young cells budding off from their parents (§ 141); these, in the course of a short time, become complete cells, and again perform the same process; and in this way the single cells of yeast develope themselves, in

FIG. 74.

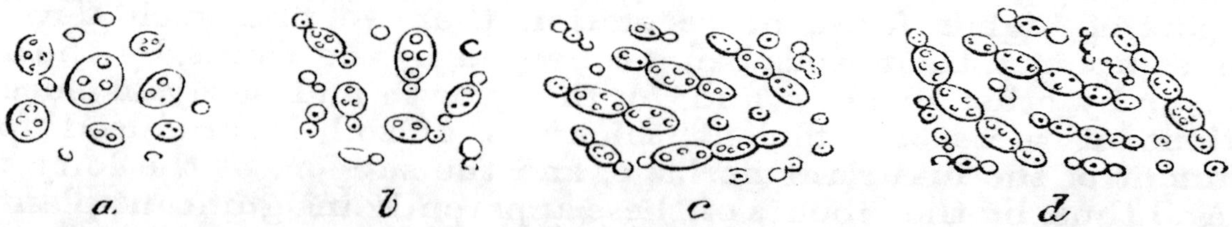

Different stages of the vegetation of *Torula Cerevisii*, or Yeast Plant;—*a*, single cells of which it at first consists; *b*, cells with buds; *c*, the same more advanced; *d*, rows of cells produced by continuance of the budding process.

the course of a few hours, into rows of four, five, or six, which remain in continuity with each other whilst the plant is still growing, but which separate if the fermenting process be checked, and return to the condition of those which originally constituted the yeast. Thus it is that the quantity of yeast first introduced into the fermentable fluid is multiplied six times or more, during the changes to which it contributes.* This little plant is very tenacious of its vitality; its power of growth not being destroyed by exposure either to the heat of boiling water, or to a cold of 76° below zero, or by being completely dried up. It is as yet uncertain whether the processes just described constitute its entire growth, or whether they are anything else than the early stages of its development; for there is reason to believe that if its vegetation be not checked, it will go on to produce a continuous filament of some length, bearing a regular fructification analogous to that of the plants next to be described.—According to Schleiden's observations, many of the simpler Fungi (such as the 'moulds' which appear on cheese, fruits, preserves, &c.), consist of single cells, which, instead of increasing in number by subdivision or gemmation, augment in size, and send out ramifying prolongations, which give to the plant a complex form, whilst its structure remains in the simplest possible condition.† This seems to be the case, for example, with regard to the *Botrytis*, a little plant of which some species are parasitic upon leaves (Fig. 75, A), whilst another developes itself in the bodies of silk-worms. In the accompanying figure, we see the lower part of the tubular cell subdividing into root-like prolongations, which insinuate themselves among

* Much discussion has taken place, as to whether the vegetation of the Yeast-plant or ferment-cells is to be regarded as the essential cause of fermentation. Some explain this process altogether on chemical principles, and deny that the vegetation has any connection with it; whilst others assert that fermentation is entirely caused by the decomposition effected by the Fungus, and cannot take place without it. The truth appears to lie between the two extremes. The elements of the saccharine fluid are arranged in such a manner, that they may be easily induced to enter into new combinations; and these may be formed under a great variety of conditions, several of which are examples of the *catalytic* action already noticed (§ 39). The introduction of the germs of the Yeast-plant appears to be the most favourable of all conditions; the fermentation which is occasioned by its vegetation being more active and complete than that which can be produced in any other way. It appears to act upon the fluid in the same way as common Mould does upon a sweet preserve; and this action is not very dissimilar to that of a germinating seed upon the starch laid up for it by its parent. See on this subject the *Comptes Rendus* for August 20, 1838; Meyen's Report on Vegetable Physiology for 1837, p. 83—87; and Liebig's Organic Chemistry.

† Principles of Scientific Botany, p. 153.

the cells of the leaf on which it is parasitic; the stem, which is a part of the same tubular cell, rises through the stomate, or cuticular orifice, of the leaf; and it then divides into branches, each of them an extension of the same individual, and each bearing a 'spore' or reproductive cell at its termination. This spore is at first developed within the extremity of the branch; as it gradually increases in size, it distends its envelope without rupturing it; and when mature, it detaches itself from the branch with the integument it has thus derived from the parent cell. The stages of this process are well seen in the fructification of the common *Mucor* or 'mould;' in which the spores are formed on a spherical enlargement at the extremity of the stem (Fig. 75, B, *a*), which is furnished with minute pear-shaped processes, projecting in all directions (*b*). Within the extremity of each of these processes, a 'spore' or reproductive cell originates, as seen at *c*; and this, when mature, is seen to be attached to the lower part of the process (whose extremity is expanded over it) by a narrow neck, the rupture of which ultimately sets it free. Each spore in its development, seems to expand into similar irregularly branching cells; and in its turn developes a multitude of new spores after the same method.—Thus we see that even in these simplest forms of Fungous vegetation, as in the highest, the *nisus* displays itself especially in the development of the reproductive apparatus; and the knowledge of this fact, together with the excessive minuteness of the spores, renders it not difficult to account for the apparently universal diffusion of these bodies, and the readiness with which various forms of Fungi spring up, wherever the proper soil exists for their development.

Fig. 75.

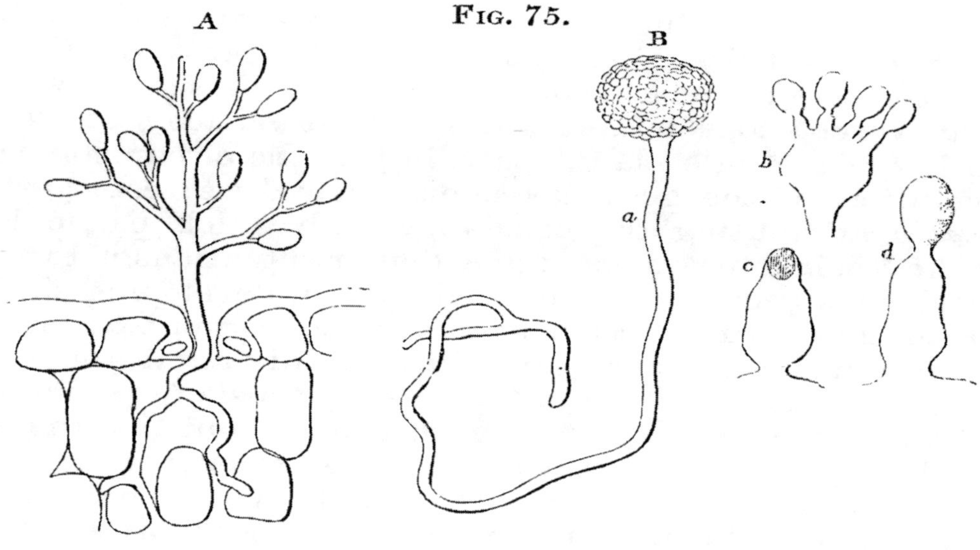

Lower forms of Fungi:—A, a *Botrytis* growing out of the stoma of a leaf of *Allium*, which is shown in section;—B, a *Mucor*, showing *a* the whole plant; *b*, head of the spore-case with a few of its spore-bearing protuberances; *c*, earlier condition of such a protuberance with the spore originating from it; *d*, protuberance with mature spore.

272. The conditions under which that large variety of forms is generated, which presents itself among the inferior Fungi, constitute a subject of inquiry of great interest alike to the Physiologist and to the Systematic Botanist; since it is the business of the former to determine how far the diversified influence of external circumstances may modify the characters of plants produced from similar germs; whilst the latter has to ascertain, on the basis of the data thus obtained, what among these forms are to be regarded as real *species*, and what must be considered as mere *varieties*. That much is yet to be learned on these subjects, and that we are only in the infancy of our knowledge as to the influence of external conditions

upon the development of these low forms of vegetation, may be judged of from the following facts, in addition to that already mentioned (§ 266) as to their occasional *confervoid* aspect. It has been shown by Professor Henslow* that the 'rust' of corn (*Uredo rubigo*) is only an earlier form of the 'mildew' (*Puccinia graminis*); the one form being capable of development into the other, and the fructification characteristic of the two supposed genera having been produced from the same individual. And it is asserted by Fries, that out of a single *species* of *Thelephora*, more than eight *genera* have been constructed by various authors. Hence it is evident that a very considerable variety of forms may be developed as the offspring of a single plant, according to the circumstances under which the evolution of the germ takes place; and we seem justified, therefore, in *partly* attributing the constancy with which particular forms of Fungi make their appearance in particular situations,† to the influence of the conditions peculiar to each situation in developing similar germs into one or another form. Still there is ample reason to consider that this mode of accounting for such diversities is only applicable in a limited degree; and that an immense number of germs of species really distinct and inconvertible are constantly floating about in the atmosphere, ready to develope themselves wherever and whenever the fitting conditions of food and temperature are afforded them. The difficulty which may be felt in regard to such an admission, ceases to present itself with any force, when it is borne in mind how vast is the development of reproductive particles in this group of Plants. Of this, any one may convince himself by examining a puff-ball in a state of maturity. It has been well remarked by Fries on this point, that the sporules are in such vast numbers (in a single individual of *Reticularia maxima* he has estimated that ten millions must be present), are so subtle (being invisible to the naked eye except when collected in masses, and appearing as a thin smoke when diffused in vast multitudes through the air), are so light (being raised, perhaps, by evaporation into the atmosphere), and are dispersed in so many ways (by winds, insects, elasticity, &c.), that it is difficult to conceive a place from which they can be excluded. There can be no reasonable doubt, then, that notwithstanding the rapidity and constancy with which particular forms of Fungi make their appearance under particular combinations of circumstances, they all arise from germs produced by a pre-existing generation; and that, although there may be a considerable degree of latitude as to the precise form into which any particular germ may evolve itself, so that the new plant may not be an exact copy either of its parent or of any one of its fellows of the same generation,

* Annals of Nat. Hist. vol. vi. p. 379.

† Thus, no *Puccinia* but the *P. rosæ* is found upon rose-bushes, and this is seen nowhere else; *Onygena exigua* is said to be never seen but upon the hoof of a dead horse; and *Isaria felina* has only been observed upon the dung of cats deposited in humid and obscure situations. It is affirmed by Dutrochet, that whilst a solution of albumen in distilled water may be kept for a year without giving rise to any fungoid production (portions of such productions, introduced from without, not developing themselves in it), this same solution, if acidulated, is speedily found to contain a species of *Monilia;* whilst, if a small quantity of alkali be added, a kind of *Botrytis* developes itself in it. On the other hand, a solution of fibrin with alkali produced *Monilia;* and water distilled over lettuce and acidulated with phosphoric acid gave rise to *Botrytis*: so that the form which the fungus assumed was not due to the *direct* influence of acids or alkalies respectively.

still there is a certain fundamental type for each assemblage of forms, the amount of departure from which is limited. Now as this is true of *every* species of Plant or Animal, whether of low or high organisation,—there being probably no one which is not capable of being in some degree modified in form and manner of development by external conditions,—the latitude in this respect which seems to be characteristic of the lower Fungi is only a peculiarity in *degree,* and does not by any means place them on a different footing from other Plants. An ingenious attempt has been made to show, that certain reputed Fungi which develope themselves in the substance (usually near the surface) of the higher plants, are not independent structures of extraneous origin, but are the result of morbid alterations in the tissues of the plant in which they are found, analogous to the eruptive diseases of animals. But the latest investigations have shown, that although morbid growths do occasionally present themselves in Plants,* which present a certain superficial resemblance to fungous vegetation, yet that in the case of *Uredo, Puccinia, Æcidium,* and other genera whose appearance it has been attempted to explain in this manner, no such idea is tenable, since they have a distinct fructification of the regular fungous type, and their whole development may be traced from the spore in which they originate. Still it would appear probable, from the obvious influence exerted upon their multiplication by the condition of the plant upon which they vegetate, that when the higher forms of vegetation are attacked by 'rust,' 'smut,' 'mildew,' or other like forms of 'blight,' an unhealthy condition of the plant itself is usually the first and most necessary condition of the growth of the fungi; their spores, which are probably everywhere diffused, not developing themselves, except where they meet with their appropriate *pabulum* in the decomposing products afforded by the juices of an unhealthy plant.†

272*a.* It is not only in the substance or in the cavities of Plants, that several of the inferior Fungi are disposed to become parasitic; for there are many instances in which they develope themselves in the interior of the Animal body. Thus the Silk-worm breeders of Italy and the South of France, especially in particular districts, have been subject to considerable loss by a disease termed *Muscardine,* which sometimes attacks the caterpillars in large numbers, just when about to enter the chrysalis state. This disease has been ascertained to be due to the growth of a minute Fungus, nearly resembling the common *mould,* within their bodies. It is capable of being communicated to any individual from one already affected, by the introduction, beneath the skin of the former, of some particles of the diseased portion of the latter; and it then spreads in the fatty mass beneath the skin, occasioning the destruction of this tissue,

* Such, for example, as the 'oak-spangles' and other minute 'galls,' which owe their origin to the puncture of various species of *Cynips.*

† The Author has been recently informed by Mr. John Marshall, that in the course of some investigations made to determine the frequency of the presence of the *Uredo segetis* in wheaten bread, he has detected a few sporules near the apex of *every grain* in very fine samples of wheat; thus fully confirming the idea of the universal presence of the germs of such fungi, and of their dependence for the conditions of their development upon an unhealthy state of the plant on which they are lodged. The spores, although thus deposited in the ears, seem to remain dormant, unless a cold and wet season, by inducing a morbid state of the growing wheat-grain, favours their evolution at the expense of the food supplied by its decomposing juices.

which is very important as a reservoir of nutriment to the animal, when it is about to pass into a state of complete inactivity. The Fungus spreads by the extension of its own stems and branches; and also by the production of sporules, which are taken up by the circulating blood, and carried to distant parts of the body. The disease invariably occasions the death of the Silk-worm; but it seldom shows itself externally until afterwards, when it rapidly shoots forth from beneath the skin. The Caterpillar, Chrysalis, and Moth, are all susceptible of having the disease communicated to them by the kind of inoculation just described; but it is only the first which usually receives it spontaneously. By a careful investigation of the circumstances which favour its propagation, the breeders of Silk-worms have been able greatly to diminish the mortality.*—Again, it is not at all uncommon in the West Indies to see individuals of a species of *Polistes* (a wasp-like insect) flying about with plants of their own length projecting from some part of their surface, the germs of which have been probably introduced through the breathing-pores at their sides, and have taken root in their substance, so as to produce a luxuriant vegetation. In time, however, this fungous growth spreads through the body and destroys the life of the insect; and it then seems to grow more rapidly, the decomposing tissue of the dead body being still better adapted than the living structure to afford it nutriment. A similar growth of different species of fungi of the genus *Sphæria* takes place in the bodies of certain caterpillars in New Zealand, Australia, and China; the Chinese species is valued as a medicinal drug. Even Man is not free from the invasion of these parasites; for simple forms of fungi have been found growing in the stomach and in the lungs; in certain diseased conditions of the cutaneous surface, named *Porrigo favosa* and *Sycosis menti*, a considerable development of fungous vegetation takes place; and the same has been recently discovered to be true of these white patches (*Aphthæ*) on the lining membrane of the mouth in children, which are known as *thrush*, these being found to contain filamentous forms of fungi, with distinct fructification. In all these cases, however, it seems most probable that a certain morbid condition of the animal fluids must exist, in order that the germs of the fungus may develope themselves; so that this condition, rather than the presence of the fungus, must be looked upon as the essence of the disease.

273. The principal advance which is seen in the tribe of Fungi towards the higher types of vegetation, consists in the separation of the nutritive from the reproductive apparatus. In many of the lower forms of the group, even when the entire plant is made up of an aggregation of distinct cells, instead of consisting of a single one (as in the instances just described, Fig. 75), we still find each cell capable of originating reproductive cells or spores; but in other cases, the fertile or spore-producing cells are collected at the extremities of the stalks or branches. In higher forms we find the sporiferous cells clustered together in a mass of tolerably determinate figure, and enclosed in a distinct envelope, as in the *Lycoperdon* or puff-ball; but this mass of fructification, termed the *sporangium*, arises from a filamentous expansion, which is the real vegetating portion of the plant, and is termed the *mycelium*. Even in the highest

* See Annales des Sci. Nat. 2me Série. Zool. tom. viii.

forms of Fungi, this mycelium—which is, in fact, the real plant, the part of the fungus that appears above ground being only its reproductive apparatus—still retains its simple condition; being composed of a sort of flocculent mass of elongated cells, loosely interlaced with each other, and sending out prolongations which extend themselves through the soil; it is this which constitutes the 'mushroom-spawn,' from which the edible mushrooms are usually grown. The highest form of reproductive apparatus among the Fungi is that of the *Hymenomycetous* group, of which the *Agarics* may be taken as the type. In these, when the fructification is fully developed, we see a dome-shaped body, termed the *pileus* (Fig. 76, A, *a*), surmounted upon a *stipes* or stem (*c*), which rises from the *mycelium* (*d*) that constitutes the nutritive portion of the fabric. The pileus is composed of two membranes; of which the upper and outer is simple and sterile like the cortical layer of Lichens; whilst the inner and lower, which is termed the *hymenium*, bears the spores. The surface of this membrane is usually extended by duplication or involution; thus in the *Agarics* it forms vertical plates termed *lamellæ* or gills, which radiate from the stipes towards the circumference of the pileus (Fig. 76, A, *b*, and B, *a*); in *Boletus* and *Polyporus* it lines a mass of vertical tubes arranged like the cells of a honeycomb; and in *Hydnum* it covers the exterior of a similarly arranged series of solid columns. In all these cases, however, the mode in which the fructification is produced on the surface of the hymenium is essentially the same; the spores being developed from projecting bodies termed *basidia*, each of them usually having four prolongations, in every one of which a spore is developed in precisely the same manner as in the lower forms of Fungi (Fig. 76, B, *b*, *c*, *d*, *e*). Among the basidia are found a number of other vesicular bodies of large size, which do not bear spores; these, which have been termed *cystidia*, have been supposed to have the nature of *antheridia*; but this is very doubtful, more especially as they are not universally present; and they may be more probably considered as analogous to the paraphyses or sterile filaments which lie among the asci of Lichens (§ 270). If anything like a conjugation takes place in this group, of which no evidence has hitherto been discovered, it probably occurs between the cells of the part of the mycelium from which the

Fig. 76.

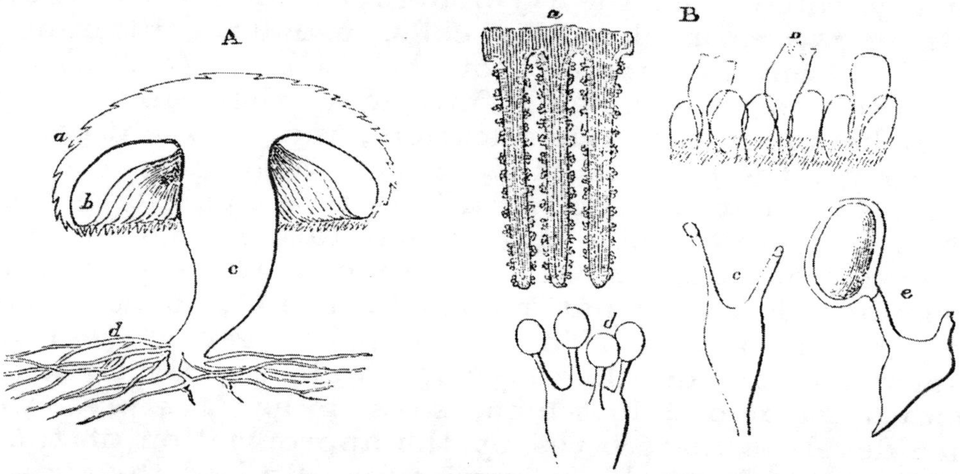

Structure of *Agaricus campestris*;—A, longitudinal section of the entire plant; showing *a*, the pileus; *b*, the lamellæ covered by the hymenium; *c*, the stipe; *d*, the mycelium:—B, portions of the fructification enlarged; *a*, section through the lamellæ, showing the investing hymenium, and the spores clustered upon it; *b*, portion of hymenium with basidia in four different stages of formation; *c*, basidium somewhat more developed, one of the processes showing the spore in its first stage, the other more advanced; *d*, basidium with four processes and as many half-developed spores; *e*, the upper part of a basidium with a fully-developed spore on one of its processes.

apparatus of fructification is evolved. The first stage of its evolution which has been yet traced, consists in the development of a small round hollow body called the *volva,* within which a spherical body is found, attached to the base by a short stalk. The fructification is at first developed in the interior of this body, as in the lower forms of Fungi; but whilst its upper wall increases in thickness, so as to form the solid disk of the pileus, the lower part of it gives way, so as no longer to enclose the hymenium, and allows the pileus to expand; its central portion sometimes remaining traceable, however, as a membranous fringe round the base of the stipes. Whilst these parts are being developed, they break through the volva in which they were at first enclosed, and this usually decays and disappears very speedily.—There are certain forms of Fungi usually ranked in the Hymenomycetous group, whose fructification consists of *thecæ,* or elongated cells, usually containing eight free spores, closely resembling the *asci* of the Lichens (§ 270); these obviously connect the two groups, being fungoid in their mode of growth, whilst they are lichenoid in their fructification; and we ought perhaps (with Schleiden) to transfer these tribes of *thecaspori* to the group of Lichens, on the same principle that this is now considered to include the genus Lichina and its allies, in spite of their marine habitation.—One of the great purposes of the group of Fungi in the economy of Nature, appears to be the removal of refuse and decaying matter, which, if left to decompose, would prove injurious to animal life. Their vapour-like germs seem to float about in the atmosphere, in countless myriads, waiting only for a soil adapted to afford them nutriment; in which, so soon as it is presented to them, they at once develope themselves, by the appropriation of the products of decomposition which would otherwise be diffused through the air. The more rapid the decomposition, and the greater the quantity of noxious exhalations which would thus be given off, the greater is the tendency to luxuriant growth and rapid multiplication in these humble plants, which may not unappropriately be designated by the title more expressly appropriated to the class of Insects, that of "Scavengers of Nature."

274. Having thus studied the principal characters of the three important groups which constitute the inferior subdivision of the Cryptogamic series, we proceed to those more elevated tribes, which present a manifest approach in their general conformation to the highest types of Vegetable structure. Their comparative elevation is most obviously displayed in the distinction which now shows itself, more or less completely, between stem, root, and leaf; that is, in the separation and specialisation of the several parts of the nutritive apparatus. The separation of the leaf from the stem is that which first occurs, being commenced in the *Hepaticæ,* and fully carried out in the *Mosses,* in neither of which groups are true roots present; indeed it is only in the *Ferns* and their allies, that we meet with that downward prolongation of the stem which corresponds with the true roots of Phanerogamia, the radical fibres of the inferior tribes being merely elongated cells, and rather comparable to the hairs with which the surfaces of leaves are frequently beset. The growth of the stem is always limited to its extremity, the part already formed being subject to very little increase, if any; and from this character it has been proposed (by Dr. Lindley) to designate this higher division of the Cryptogamic series by the title *Acrogens,* significant of growth at

their points.—The advance in complexity of outward form is accompanied by a corresponding elevation in the character of the internal organisation. Throughout the whole series, we find a cuticle or enveloping membrane, very distinct from the subjacent tissue; and this is perforated by *stomata* or breathing pores, which are frequently of very complex structure, and which often communicate with air-chambers of very regular form, that are excavated in the parenchyma. There is a tendency, also, to a higher development of the individual cells, as manifested in the first instance by the formation of spiral fibres in their interior (§ 147); such spiral cells are for the most part confined to the reproductive apparatus of the Hepaticæ, where they answer a special purpose; but we find them occasionally constituting nearly the entire plant in the Mosses. In neither of these tribes, however, do we meet with true woody fibre or spiral vessels; the existence of these tissues being confined to the Ferns and their allies, which are the first plants in the ascending series in which we meet with a true woody stem composed of ligneous and vascular bundles regularly arranged. And it is in the same group that we first meet with continuous ducts for the conveyance of fluid; their introduction into the vegetable fabric being obviously related to the separation between the absorbent and the foliaceous organs which here takes place, the former being carried downwards by the extension of the root, whilst the latter are borne upwards by the elongation of the stem. The absence of woody tissue from the leaves of Mosses forms a marked distinction between them and the leaves of the Ferns, which have a regularly 'veined' skeleton of fibro-vascular tissue, like that of the leaves of higher plants; in the distributions of the 'veins,' however, there is this marked distinction, that those of Fern-leaves do not unite again after their subdivision, so that none of that reticulation occurs, which is generally so remarkable in the leaves of Phanerogamic, and especially in those of Dicotyledonous plants.—In the reproductive apparatus of Acrogens, we see but a very remote approximation to the type which distinguishes the Phanerogamia. Although it sometimes appears to be (as in Mosses) very distinct from the nutritive apparatus, the spores being enclosed in an elaborately constructed 'urn' which is borne aloft upon a long stalk, yet we shall find reason to believe that the *essential* part of the generative operation is performed, in the whole of this series, whilst the organs concerned in it are as yet undistinguished from the simply vegetative portion of the plant, instead of in organs which are already obviously set apart for it, like the flowers of Phanerogamia.

275. The group of HEPATICÆ, or Liverworts, forms a very natural and complete transition between the Lichens and the Mosses; its lowest subdivision, the *Ricciaceæ* or Crystal-worts, being nearly allied to the former; while its highest, the *Jungermanniaceæ* or Scale-mosses, approximates equally closely to the latter.—It is in the *Marchantiaceæ*, or ordinary Liverworts, that we see the characters of the group in their most distinctive form. These are humble plants, growing on the earth or on the surface of trees, where they can obtain an adequate supply of moisture; they generally exhibit a definite axis of growth, but the distinction between leaf and stem is not clearly manifested (Fig. 77); and from the whole of the under surface of the frond (or foliaceous expansion) radical fibres are put forth. This frond is composed of a soft green

parenchyma, which contrasts strongly with the hard dry crusts of the Lichens. In the common *Marchantia polymorpha* we find the surface divided into minute diamond-shaped areolæ (Fig. 78, A, *a*), which mark the boundaries of chambers excavated in the substance of the frond. These chambers are enclosed by very regular walls, and are covered-in by the cuticle, which is perforated by *stomata* of remarkable complexity of structure, hereafter to be described (CHAP. XIII.). The development of this frond takes place by the continuous multiplication of cells, originating in the spore or reproductive cell detached from the parent, very much as in the case of the *Ulvæ* (§ 267). The 'spores' or reproductive cells are produced within *thecæ* or spore-cases, which are clustered together in the substance of an umbrella-like disk (termed the *pelta* or shield) that is mounted upon a long stalk, which rises from the centre of the frond (Fig. 77). With the spores, we find a number of curious bodies termed *elaters*, each of which consists of a double spiral thread resembling that of a spiral vessel, but more elastic, and destitute of an enveloping cell-membrane. The elaters are developed, however, as ovoidal spiral cells, lying in the midst of the parent-cells within which the spores are generated; when the cell has attained its maturity, the cell-membrane gives way, and the spires no longer confined by it have a tendency to elongate themselves: this they cannot do, however, so long as they are packed with the spores in the interior of the spore-case; but, when both are set free by the opening of the thecæ (which usually takes place by irregular fissures), the elaters spring out suddenly, and jerk forth the spores which may be adherent to their coils, thus assisting in their dispersion. Besides these organs, the Marchantiaceæ produce *antheridia*, or clusters of sperm-cells containing peculiar self-moving filaments (closely resembling those of the *Characeæ*, § 269), included within special envelopes; these also are generally collected in disk-like bodies, which sometimes are but little elevated above the frond, whilst in other cases they are borne on stalks like those of the peltæ. In some instances the antheridia are borne on the same shields with the spore-cases; thus in *Marchantia androgyna*, we find the upper surface of one half of the pelta developing antheridia, whilst the under surface of the other half bears thecæ. The contents of the antheridia are mature before those of the spore-cases; and it seems probable, from considerations hereafter to be stated, that the entire theca with its contents is to be regarded as the product of an act of fertilisation (consisting in the union of the filament of the 'sperm-cell' with a special 'germ-cell'), which takes place previously to its development.—There is, however, another very curious provision for the multiplication of these plants; consisting in the production of little disk-like bodies, which may be distinguished as 'bulbels,' within basket-like conceptacles; these may commonly be seen in different stages of growth, on various parts of the surface of the frond (Fig. 78, A, B). These 'bulbels' must be regarded as *gemmæ* or buds, having the power of independent existence; they are at first developed upon little footstalks which rise from the base of the receptacle (C, D); and becoming spontaneously detached from these, they

FIG. 77.

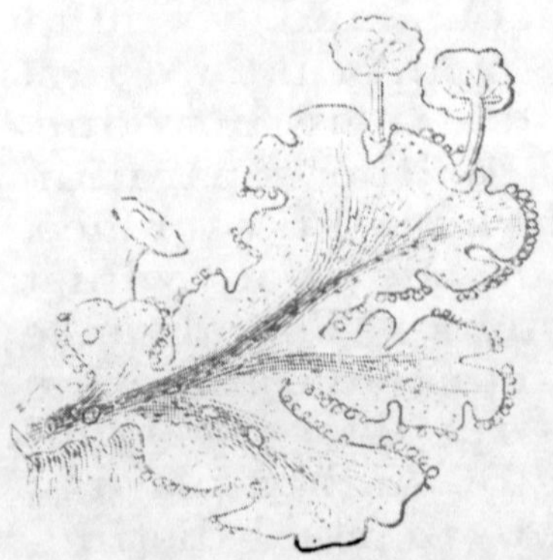

Portion of a frond of *Marchantia polymorpha* in fructification.

are commonly washed out by rain and carried to different parts of the neighbouring surface, on which they grow very rapidly when well supplied with moisture; sometimes, however, they may be found growing whilst still contained in their receptacle, and seen to graft themselves (as it were) on the parent plant.—The principal departure from the type of the Marchantiaceæ presented by the *Ricciaceæ,* which consist of irregular

FIG. 78.

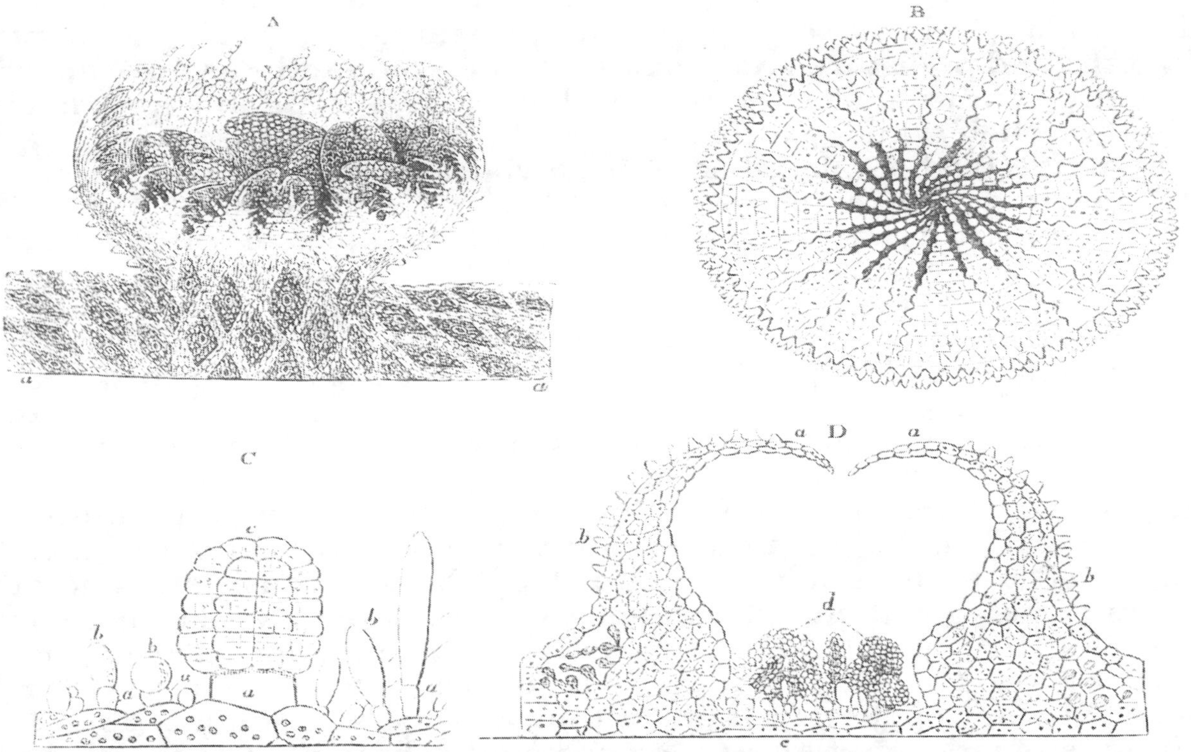

Gemmiparous conceptacles of *Marchantia polymorpha*:—A, conceptacle fully expanded, rising from the surface of the frond *a a*, and containing disks already detached:—B, first appearance of the conceptacle on the surface of the frond, showing the manner in which its fringe is formed by the splitting of the cuticle:—D, section of a conceptacle in progress of expansion, showing *a, a,* its thin fringed edges, *b, b,* the thicker walls of its lower portion, and *c* its base whence spring *d* the bulbels:—C, portion of the base more enlarged, showing the bulbels in various stages of development; *a*, pedicel, *b, b,* bulbels in their earliest condition, *c*, bulbel more advanced.

foliaceous expansions of diminutive size that usually float on water or mud, is shown in the want of elevation of the fructification above the surface of the frond, and in the absence of elaters; both of which points indicate an approximation to the Lichens. Here, too, we find distinct antheridia, which are sometimes imbedded in the substance of the frond which bears the thecæ, but are sometimes produced by a different individual.—The *Jungermanniaceæ,* on the other hand, present many points of approximation to the true Mosses, for which they might be readily mistaken. They usually possess a distinct axis of growth; and although the leaves and stem are not disjoined in some instances, yet in other cases they are evolved separately; the leaves being very symmetrical in form, with the indication of a central midrib; and being very regularly arranged in an imbricated manner upon the stem, which still, however, trails upon the ground. The distinctive character of this group is found in the structure of its spore-cases, which are borne separately at the extremities of long stalks, instead of being clustered together as in the Marchantiaceæ, and which open regularly by four valves. In some genera, each theca contains a central *columella,* resembling that of the Mosses (§ 276) ; but although the approximation of the Jungermanniaceæ to that group is thus rendered very close, yet we never find their thecæ possessing the characteristic *operculum* of the Mosses ; and they are further distinguished by the presence of elaters among their spores, these being never found in the thecæ of the true Mosses.—The Hepaticæ are found abundantly in nearly all climates, seeming especially to luxuriate in warmth and moisture ; the extremes of cold and dryness appear to be the only conditions under which they cannot flourish ; and many of them retain their vitality after being subjected to these, so as again to vegetate actively when placed in more favourable conditions.

276. In the general structure of the MOSSES, we find but a very trifling advance upon the type presented to us in the more elevated Hepaticæ. We can always recognise a distinct axis of growth, commonly more or less erect, on which the minute and delicately-formed leaves are arranged with great regularity. The stem shows some indication of the separation of a *cortical* or bark-like portion from the *medullary* or pith-like, by the intervention of a closed circle of bundles of elongated cells ; some of them bearing a resemblance to woody fibre in the smallness of their diameter and the thickness of their walls, whilst others seem in like manner to prefigure the ducts of higher plants. From this rudiment of a fibro-vascular circle, we find prolongations passing into the leaves, in each of which they form a sort of midrib. The leaf consists of one or two flat layers of polygonal cells, adherent to each other ; when there is but one, the midrib separates the two lateral halves from each other ; but when there are two, it is interposed between them. The leaves rarely possess a distinct epidermic layer, although this is commonly found (perforated by stomata) on the *setæ* or footstalks bearing the fructification ; sometimes, however, we find an epidermic layer above and below the midrib of the leaf, even where no such layer exists on its lateral expansion. Root-fibres are still developed from every part of the lower extremity of the axis, and even from the under surfaces of the leaves.—The fructification of Mosses, when mature, consists of a capsule or 'urn,' (Fig. 79, A, B, *u*, *u*) borne at the top of a long bristle-like footstalk termed the *seta* (*s*), which springs from the centre of a cluster of leaves. When examined in its earliest state, however, before its stalk is developed, this urn is found to be one of several small bottle-shaped bodies, clustered together, and enveloped in a membrane that tapers upwards to a point. The greater number of these bodies, which are termed *pistillidia*, are abortive ; but one of them, becoming developed, bursts its membranous envelope, and carries the principal part of it upwards upon its summit, so as to form the *calyptra* or hood (A, B, *c*). The 'urn' or *theca* of Mosses is closed by an *operculum* or lid (*o*), which falls off when the fructification is mature, to allow of the escape of the contained sporules. The mouth of the urn, when the operculum has fallen off, is seen to be surrounded by a fringe termed the *peristome* (B, *p*) ; this is originally double, one portion springing from the outer, and the other from the inner, of two layers of cells which may be distinguished in the immature capsule (C, *p*) ; but

frequently, at the time of maturity, one or other of these is wanting, and sometimes both are obliterated. The peristome usually has its border divided into minute 'teeth,' the number of which is always a multiple of four, varying from 4 to 64; sometimes the teeth are prolonged into straight or twisted hairs. The 'spores' are contained in the upper part of the capsule, where they are clustered round a central pillar, which is termed the columella (C, *c*). Besides the urns, Mosses produce *antheridia*, usually in the form of cylindrical or club-shaped or sometimes bottle-shaped bodies, attached at their base to the axis, and opening by an irregular perforation at their free extremity; and these antheridia are filled with 'sperm-cells,' each of them containing its self-moving filament. The antheridia and pistillidia are frequently produced on separate individuals; and it has been asserted that no true fructification is developed in such Mosses, unless these individuals be in each other's neighbourhood. Many circumstances, hereafter to be stated (CHAP. XVIII.), tend to favour the idea, that a real act of fertilization is effected by penetration of the spermatic filaments of the antheridia into the pistillidia, long before the evolution of the capsules; and that each capsule, with its contained mass of sporules, is in fact the product of that fertilization.* The development of the spore or reproductive cell into the perfect plant commences by the rupture of its outer wall, and the protrusion of its inner coat; and from the projecting extremity new cells are put forth, which at first grow into a sort of confervoid structure, made up of filaments composed of cells linearly arranged. At certain points of this structure are formed clusters of rounded cells, from every one of which an independent plant may arise; so that several individuals may be evolved from a single spore. After the stem and leaves have been developed from these clusters of cells, the confervoid filaments for the most part remain as their radical fibres. As to the mode in which the capsule or urn is developed from

FIG. 79.

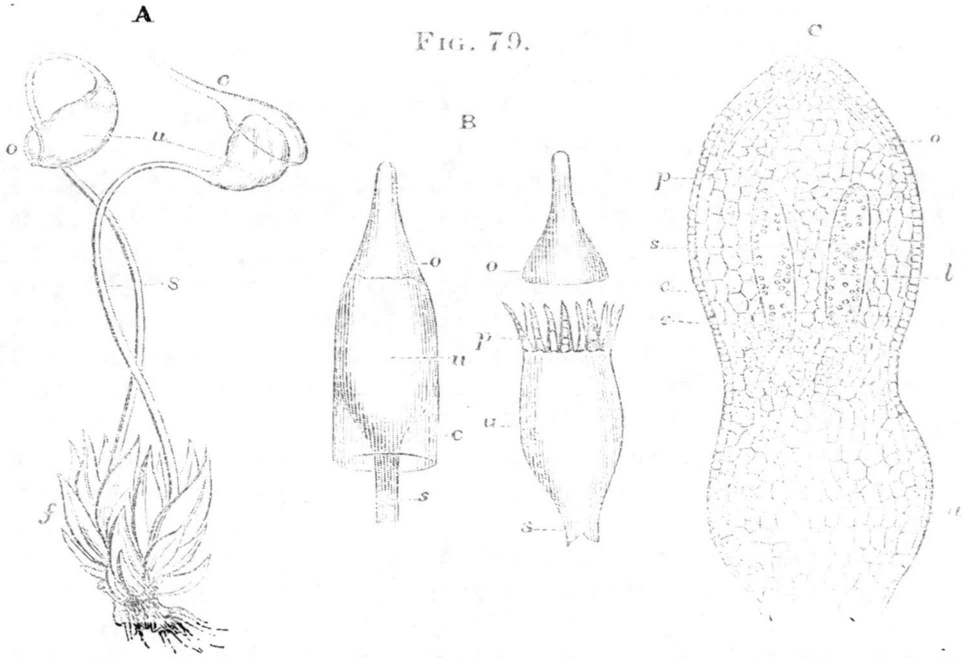

Structure of Mosses:—A, Plant of *Funaria hygrometrica*, showing *f* the leaves, *u* the urns supported upon the setæ or footstalks *s*, closed by the operculum *o*, and covered by the calyptra *c*:—B, urns of *Encalyptra vulgaris*, one of them closed and covered with the calyptra, the other open; *u*, *u*, the urns; *o*, *o*, the opercula; *c*, calyptra; *p*, peristome; *s*, *s*, setæ:—C, longitudinal section of very young urn of *Splachnum*; *a*, solid tissue forming the lower part of the capsule; *c*, columella; *l*, loculus or space around it for the development of the spores; *e*, epidermic layer of cells, thickened at the top to form the operculum *o*; *p*, two intermediate layers, from which the peristome will be formed; *s*, inner layer of cells forming the wall of the loculus.

* See Thwaites in Annals of Natural History for Nov. 1848.

the pistillidium, there is still much obscurity; this much, however, is certain, that it is at first solid throughout, consisting of a mass of cells in which no distinction of parts can be perceived. Its structure at a later period, however, is shown in Fig. 79, C; in which we see that the *loculus* or cavity for the spores (*l*), which surrounds the columella, is still very small in proportion to the entire mass. This space was previously occupied by the parent cells, within which the spores are developed; these, when mature, disappear, and the spores are left free in the cavity, ready to be dispersed as soon as the falling-off of the operculum sets them free. In this dispersion, the elasticity of the peristome seems in some degree to supply the want of elaters.—Independently of the regular fructification, there are several instances among the Mosses, in which buds or *gemmæ* develope themselves, sometimes from the stem, sometimes from the leaves, and sometimes from the confervoid root-fibres, and then spontaneously detach themselves, so as to multiply the individual plant to a considerable extent. It is well known to Botanists, that some species of Mosses rarely if ever occur with true fructification in certain localities; and such must owe the chief part of their multiplication and dispersion to the agency of this *gemmiparous* process. Like the Hepaticæ, Mosses flourish best in moist situations, but can also endure a long-continued drought, though their vegetation is entirely checked whilst it lasts. In common with the Liverworts and Lichens, this group seems to labour to produce vegetation on newly-formed surfaces, where soil can scarcely be said to be: thus we see Mosses clinging to the faces of bare rocks, and gradually accumulating particles of soil about their roots; the first green crust upon the cinders with which the surface of Ascension Island was covered by a volcanic eruption, consisted of minute Mosses; and the black and lifeless surface of New South Shetland, one of the islands nearest to the South Pole, is studded with specks of Mosses struggling for existence.

277. In the group of FERNS, we have in many respects a near approximation to Phanerogamia; but this is shown rather in their general structure and mode of growth, than in the condition of their reproductive organs; for these not only retain the peculiar type which is characteristic of the Cryptogamia, but continue in more intimate connection with the nutritive apparatus than is ever seen in Flowering Plants. The Ferns of temperate climates for the most part present themselves as herbaceous plants, consisting of fronds (or foliaceous expansions bearing fructification) usually supported upon strong stalks, which shoot up from a *rhizoma* or horizontal stem, that creeps upon or below the surface of the ground (Fig. 81, A). Between the tropics, however, and sometimes beyond that limit (as in Australia and New Zealand), we find certain species of Ferns presenting a completely arborescent character, which is in striking contrast to the ordinary types of Cryptogamic vegetation; their humble creeping stems being replaced by tall woody trunks of from 40 to 80 feet in height, and these being crowned at their summit with a most graceful circle of fronds, the length of each of which is sometimes 10 or 12 feet. The stem is for the most part composed of cellular parenchyma; but this is separated into a cortical and medullary portion, by the interposition of a circular series of fibro-vascular bundles, which contain true woody tissue and ducts, so that we have here a distinct foreshadowing of the *pith, wood,* and *bark* of Exogenous stems. We do not find, however, any such arrangement of layer

around layer of these fibro-vascular bundles, formed at successive epochs of growth, as is characteristic of Exogens; the addition to the substance of the stem being almost entirely confined to its extremity, and being produced by the agency of the successive circles or whorls of fronds, of which every one as it falls is replaced by a new frond above it. The scars left by the detachment of the leaf-stalks remain visible on the surface of the stem; and the wider separation between them, which is seen on the lower part of the trunk, together with a slight excess in its diameter over that of the upper portion, seem to indicate that a certain amount of increase does take place in it subsequently to its first formation; though this may perhaps be attributed with more probability to the expansion of the original tissue, than to the formation of new substance. The fibro-vascular bundles in the stem form a kind of irregular net-work, from which prolongations are given off, that pass into the leaf-stalks, and thence into the mid-rib and its lateral branches. The fronds are usually constructed upon the same type with the leaves of Phanerogamia; being composed of at least two layers of cellular parenchyma, supported upon the fibro-vascular skeleton which is derived from the leaf-stalk, and enclosed by a regular epidermis, which on the under surface is perforated by stomata. They are frequently subdivided into *pinnæ* or leaflets, which may be again subdivided more than once (Fig. 80); and as these subdivisions are effected with extreme regularity, there are no leaves which are more beautiful than the fronds of some of the commonest Ferns. The mode in which the fronds of Ferns are developed, is very characteristic of the group; each leaflet being at first rolled upon itself in a spiral towards its base, and the whole leaf being in like manner rolled upon itself towards the stem; as is shown in the young fronds which are unfolding themselves at the summit of the Tree-fern (Fig. 80), and also in those of the herbaceous ferns of our own country (Fig. 81, A, f' f''). The root of Ferns is formed upon the same plan with the stem; being a regular *descending axis* (that is, having the same tendency to grow downwards, which the stem has to grow upwards), as is seen in its first development; and containing fibro-vascular bundles, instead of consisting of a mere extension of the cellular parenchyma.

FIG. 80.

Tree-Fern.

277*a*. What is usually considered as the 'fructification' of the Ferns, is most commonly found on the under surface of the fronds, in the form of spots or bands of various kinds, which are termed *sori* (Fig. 81, A, f'''', and B, *s*). Each of these sori, when closely examined, is found to be made up of a multitude of capsules or *thecæ* (C, *c*, and D), which are sometimes *sessile* (or closely attached to the surface of the leaf), but which more commonly spring from it by a pedicel or foot-stalk (D, *p*). The wall of the capsule is composed of flattened cells, whose edges are applied to each other; but there is generally one row of these, thicker and larger than the rest, which springs from the pedicel, and is continued over the summit of the capsule, so as to form a projecting ring (D, *a*) which is known as the *annulus*. This ring appears to have an elasticity superior to that of the rest of the capsular wall; causing it to split across, when mature, so that the spores (*s*) may escape; and in many instances carrying the two halves of the capsule widely apart from each other, when the fissure extends so deep as to separate them completely. There are some tribes of Ferns, however, in which this ring is imperfect, or is altogether wanting; as is the case in the *Osmunda* and its allies. The sori are sometimes found naked on the under surface of the fronds; but they are frequently covered with a delicate membrane, termed the *indusium* (C, *i*), which is a prolonged fold of the cuticle.—Although this fructification is usually borne indifferently upon all the fronds, which thus combine the functions of nutrition and reproduction, yet in certain Ferns (as in the *Osmunda regalis*, commonly but improperly termed the 'flowering fern') we see a distinction between the 'fertile' or spore-bearing fronds, and the 'sterile' or leafy fronds. The former are usually distinguished by their shrunken aspect, the development of the fructification being accompanied by a diminished growth of the true leafy part; and not unfrequently the green parenchyma disappears almost entirely, so that there is nothing left but the fibro-vascular skeleton formed by the mid-rib and its nerves, from which the capsules spring. The transition from one form to the other is frequently seen in specimens of *Osmunda* growing in a richer soil than usual; for the

FIG. 81.

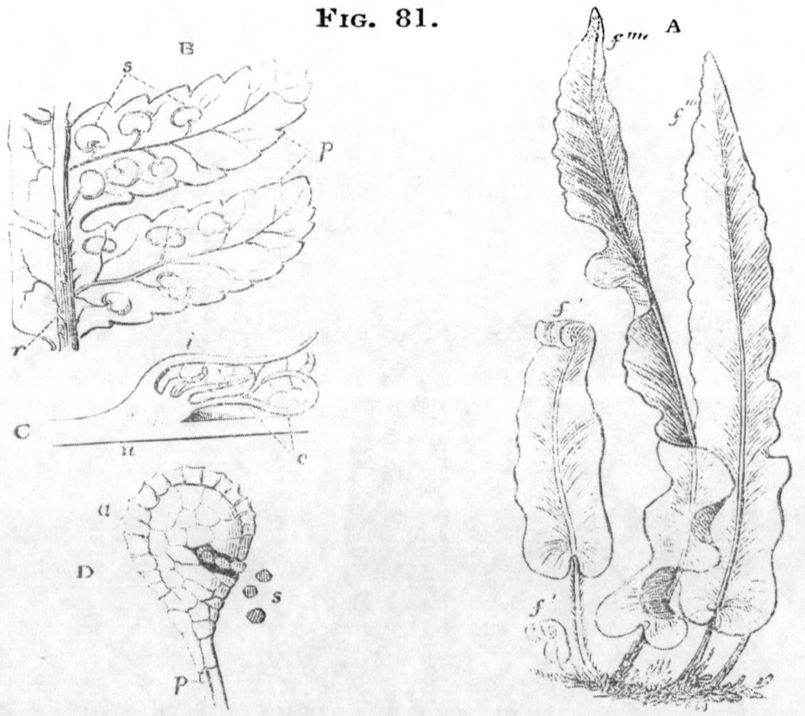

Fructification of Ferns.—A, plant of *Scolopendrium officinale* (Hart's tongue) with four fronds f', f'', f''', f'''', in different stages of development; on the lower side of f'''', the sori are seen as dark oblique lines:—B, portion of the under side of a frond of *Nephrodium angulare*, showing the sori *s*, upon two pinnules *p*, supported by the rachis *r*:—C, vertical section of a sorus; *n*, the nerve which bears it; *i*, indusium; *c*, capsules or thecæ:—D, one of the capsules at the moment of rupture; *a*, annulus or cellular ring; *p*, pedicel; *s*, spores escaping.

fertile fronds are then often sterile or leafy towards their base, bearing capsules only towards their extremity; and the same change is frequently seen in the individual pinnæ. The greatest departure from the ordinary type of fructification among the Ferns is seen in the *Ophioglossum*, or 'adder's tongue,' and its allies; in which the fertile frond is but a spike or stalk, whilst the spore-cases attached to this (which are probably not analogous to the thecæ of other Ferns, but seem to be formed by a metamorphosis of the leaf itself) are entirely destitute of the elastic ring, and open by two distinct valves; in these characters approaching the Lycopodiaceæ.

277b. It would appear, however, from recent investigations, that these spore-producing organs do not constitute the true generative apparatus of the Ferns, which has only been discovered by the attentive study of the history of their evolution. The development of the spore, when this is set free from the capsule and falls upon a damp surface, commences with the rupture of its outer coat, and the protrusion of the inner or proper cell-wall; this first extends itself into a long tube, at the extremity of which new cells are generated; at first in a linear series, forming a conferva-like filament; but afterwards in several such series, so as to form a sort of thallus, which so strongly resembles that of the *Hepaticæ* as to be readily mistaken for it. A few of the cells of this thallus extend themselves downwards, so as to form radical fibres exactly resembling those of the *Marchantia*, by which it clings to the subjacent surface and absorbs nutriment from it. The young Fern was formerly supposed to arise by continuous growth from the central portion of this thallus; but it is now known that it is the product of a true generative process which takes place in it. When the under surface of the thallus is examined with a sufficient magnifying power, a large part of it is seen to be studded with numerous *antheridia*, or collections of 'sperm-cells' (each containing its spermatic filament), which project slightly from the general surface; whilst a small number of *pistillidia* may be distinguished, usually collected in a distinct portion of it. The details of the structure and actions of these organs (the recently-acquired knowledge of which has greatly assisted in the formation of a more correct idea of the reproductive process in Cryptogamia) will be given hereafter (CHAP. XVIII.); at present it will be sufficient to state that the 'germ-cell' contained in the pistillidium, being fertilised by the contents of one of the 'sperm-cells,' becomes a true embryo, from which the young Fern is gradually evolved. In its early condition it is supplied with nutriment prepared for it by the thallus, or Marchantia-like expansion; but as soon as the first of its own fronds has been developed and unfolded, and its root has extended itself into the soil, the thallus or *pro-embryo* decays and disappears. Thus it is obvious that we cannot regard the 'spore' as having any true relation to the 'seed' of Flowering Plants; for the true representative of the latter is to be found in the fertilised germ-cell imbedded in the pro-embryo; whilst the former must be considered in the same light with the 'zoospore' of the Algæ, namely as a sort of detached *gemma*, whose function is rather to multiply the individual than to regenerate the race. And it is further apparent that although the thallus is functionally analogous to the cotyledon of flowering plants, in so far as it furnishes the embryo with nutriment until its own organs are sufficiently developed to enable it to obtain this for itself, it cannot be regarded as bearing any structural resemblance to it, since its origin is altogether different; for, instead of being a portion of the embryo itself, which is developed early for a temporary purpose, it serves, as we have seen, to produce the embryo, being itself derived from the parent-plant by a process of gemmation. In fact, the thallus of the Ferns, viewed in relation to Phanerogamia, would rather have to be regarded as a *capitulum* or head of florets (like that of the Daisy), developed independently from a detached flower-bud, the spore.—By Linnæus, the group of Ferns was poetically designated by the term *Novacolæ*, or new settlers, which very happily expresses their habits, and their importance in the economy of Nature; for they are among the first colonists of barren tracts, which will not as yet bear a more luxuriant vegetation, but which they gradually prepare for the sustenance of higher plants; and it would seem to have been with a view to this office, that the propagation of even the largest species of the tribe is provided for by the production of fine dust-like spores, which can be readily transported by the wind for unlimited distances, like those of Mosses and Fungi; instead of by larger and weightier seeds, which are less easily wafted from place to place.*

* It is probable that with the Ferns we are to associate the group of *Lycopodiaceæ* or 'Club-Mosses,' which appears in some degree to connect the Ferns with Mosses on the one hand, and with the Coniferous Phanerogamia on the other. The organisation of the existing species is somewhat above that of Mosses, though strongly resembling it. They have a regular stem, in which, as in the Ferns, there are bundles of true fibro-vascular tissue in the midst of the cellular parenchyma; these are usually disposed in the form of a hollow cylinder; separating the central or medullary from the peripheral or cortical portion of the stem; but sometimes they are clustered together in the centre of the axis. This stem, which usually creeps along the ground, grows only at its extremity, and dies off from below upward. It developes roots along its whole length; but each of these seems to be a true descending axis as in the Ferns, instead of being a mere radical fibre as in Mosses. The leaves are very simple in their form, resembling those of Mosses, or even of Liverworts; they have distinct midribs, however, composed of fibro-vascular tissue, which is connected with the vascular bundles of the stem; and they also have a distinct epidermis, perforated by stomata, which, however, are confined to the neighbourhood of the midrib. Their fructification consists of capsules; which are sometimes found at the base of the leaves, along the whole length of the stem; whilst in other instances they are clustered together in club-shaped masses at the extremities of the branches, whence the tribe has received its name. Each of these contains a mass of fine powdery granules, which obviously correspond with the 'spores' of Ferns and Mosses. Some species of Lycopodiaceæ, however, produce two-valved capsules of another kind, within which are found only three or four large fleshy bodies; and from these, as well as from the spores, it appears that new individuals may be produced. The history of the development of this group has not yet been studied with sufficient minuteness to ascertain how far it corresponds with that of the Ferns, and whether a pro-embryo is formed from the spore, in which the true generative process takes place. It would not seem improbable that the larger bodies just described are peculiar *gemmæ*, resembling the 'bulbels' of Marchantia. The Lycopodiaceæ at present existing are probably the mere remnant of a much more important group which flourished in an earlier epoch of the earth's history; the *Lepidodendra* and other arborescent vegetables of the coal formation appearing to have had the fructification of the Lycopodiaceæ, with woody stems in some respects approaching those of the Coniferæ, though never presenting the complete type of Exogenous structure.—Near to the Ferns, or to the Mosses, also, we should place the *Equisetaceæ* or 'Horse-tails,' not unfrequently called 'jointed-ferns.' Like the herbaceous ferns, they have a creeping stem that runs along the surface of the ground, sending up at intervals its curious branches, which are commonly regarded as stems. These branches are cylindrical, hollow, and jointed like a bamboo-cane, having a complete partition at each node; and in most instances there springs from every node a whorl of short bristly processes, which may be considered either as secondary branches, or as the representatives of leaves. In the substance of the hollow cylinder itself, other longitudinal channels (intercellular passages) exist, which appear to be

278. The distinctive characteristic of the PHANEROGAMIA, or 'Flowering-Plants,' is *not* the possession of what are commonly designated as 'flowers,' since these may sometimes be reduced to a condition in which they are scarcely distinguishable from the fructification of the Cryptogamia. This is the case, for example, with the whole *Coniferous* tribe, in which, however, the essential parts of the flower can be easily distinguished, its leafy investments or 'floral envelopes' being alone deficient: but it is still more remarkably the case with the *Marsileaceæ*, a little group in which the real character of the reproductive organs is so obscured by their peculiar conformation, as to have led to the inclusion of the tribe, until recently, among the higher Cryptogamia. The essential distinctions between the Phanerogamia and the Cryptogamia will be best understood from the following sketch of the typical structure and mode of development of the former; a more particular account of their generative apparatus being reserved for a future opportunity (CHAP. XVIII.).—If we analyse the fabric of any common Flowering-plant, we find that it consists essentially of an *axis* and *appendages:* the former being made up of an ascending portion or *stem* (Fig. 82, a^{i} to a^{vi}), and of a descending portion or *root* (a b), with their respective ramifications; the latter of *leaves* and *flowers* (c to c^{vii}), which will be shown to be modifications of the same fundamental type. The axis is in part composed of cellular parenchyma, and in part of bundles of fibro-vascular tissue; these last may be dispersed through its whole substance, the spaces between them being everywhere filled up with cellular tissue; or they may be arranged in the form of a hollow cylinder, including one portion of the cellular substance within it, which is then termed the *pith*, whilst it is itself included in an envelope of the same substance, which is known as the *bark*. These two distinct

air-cavities. The tissue is for the most part cellular; but the stem contains from six to ten fibro-vascular bundles, which are disposed alternately with the air-passages, so as to form a cylinder beneath the cortical envelope of cells. From these vascular bundles are given off prolongations, which pass into the leaves and lateral branches. The leaves are formed very much upon the plan of those of Mosses. The fructification forms a sort of cone at the extremity of certain of the branches, and consists of a cluster of pellate or shield-like disks, each of which carries a circle of thecæ or spore-cases, that open by longitudinal slits to set free the spores. Each spore has attached to it a pair of elastic filaments, which are caused to move by the slightest application of moisture; these are formed as spiral fibres on the interior of the wall of the parent-cell, within which each spore is separately generated, and are set free by its rupture; and they appear to be analogous in function to the 'elaters' of the Hepaticæ. No other organs of fructification have been detected in this tribe; but as the spore first developes itself into a pro-embryo, from one point of which the stem is seen to grow upwards and the root downwards, there seems every probability that, as in the Ferns, a true process of fertilisation takes place at that period in the history of the plant. Like the Club-Mosses, it seems probable that the Equisetaceæ constituted a much more important part of the Flora of the ancient earth, than they do of the vegetation of the modern epoch; the existing forms being humble plants growing in or near rivers, brooks, and ditches; whilst they were formerly developed into arborescent forms of considerable size. The tribe of *Calamites*, of which abundant remains are preserved to us in the Coal formation, is considered by most Botanists to have been Equisetaceous in its nature; whilst by others it has been regarded as belonging to the Phanerogamia, on account of the Exogenous character of the stem. From the most recent investigations of M. Ad. Brongniart, however, it would appear that under the general term 'Calamites' two distinct types of structure have been associated; one of which is unquestionably allied to the true Equisetaceæ (whose remains are abundant in the same formation); whilst the other exhibits a transition to the *Gymnospermæ*, if, indeed, it should not be actually transferred to that group (Dict. Univ. d'Hist. Nat. tom. xiii., p. 97.

types of stem-formation will be more particularly described in subsequent paragraphs. From the axis, fibro-vascular bundles pass off into the *leaves* (c^{i}, c^{ii}, c^{iii}), of which they form the midrib and its ramifications; the space between them is filled up with cellular parenchyma, of which several layers usually intervene between the two surfaces of the leaf; and the whole is inclosed in an epidermis, which is perforated (especially on the lower side of the leaf) with 'stomata' or breathing-pores. The *flowers* consist externally of an assemblage of leafy organs, which are known as the 'floral-envelopes;' these are usually arranged in two series, of which the exterior is termed the *calyx*, and its component parts the *sepals* (c^{iv}); whilst the interior is termed the *corolla*, and its component parts the *petals* (c^{v}). The sepals are usually green, their leafy character being then sufficiently apparent; whilst the leaves which are in nearest approximation to the flower, not unfrequently approach the sepals in their form; so that it becomes obvious that no real distinction exists between the two. The petals, also, possess a leafy structure, and their chief point of difference from the sepals lies in their greater variety of hue: although they are seldom green, in the properly developed flower, they frequently revert to the true leafy condition, in what are termed 'monstrosities;' and it is hence obvious that they, too, are but modifications of the same fundamental type. Within the floral envelopes, or *perianth*, we find the essential part of the flower; namely, the *stamens* (c^{vi}), which bear the 'sperm-cells,' and the *carpels* (c^{vii}) which include the 'germ-cells.' The stamens consist of two parts, the *anthers*, within which the sperm-cells or *pollen-grains* are generated, and which burst to set them free, and the *filaments* or foot-stalks on which these are supported; they depart more widely than the sepals and petals from the ordinary condition of the leaf; but it is quite certain, from the history of their development, and from their occasional reversion (either wholly or in part) to the form of the petal or of the sepal, or even to that of the ordinary leaf, that they too belong

FIG. 82.

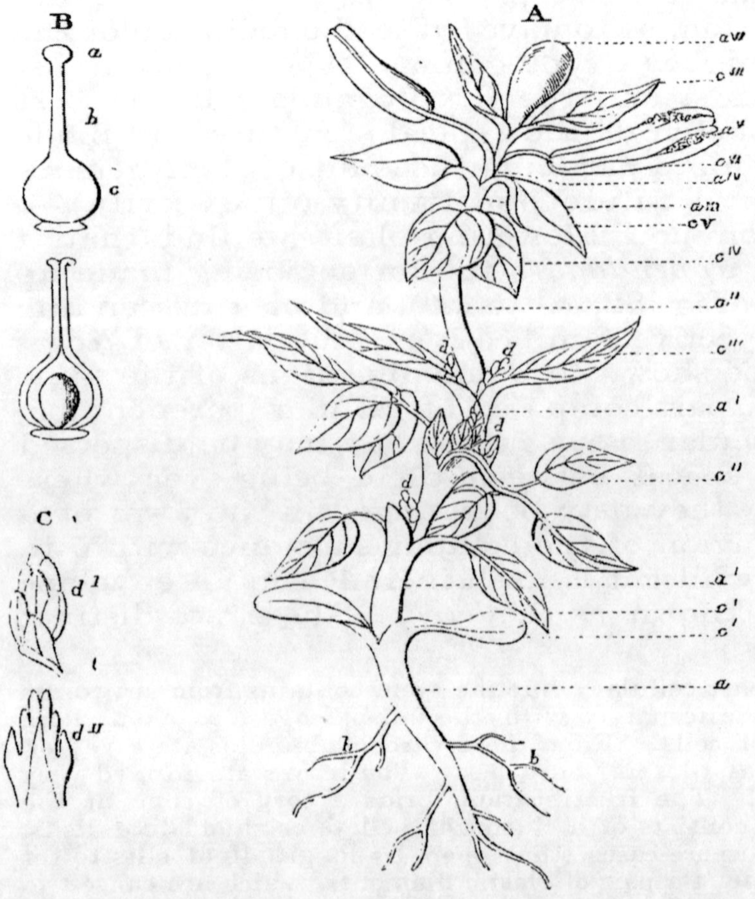

A, Ideal Plant, after Schleiden; *a* to a^{vi}, the axis, *a* being the root, a^{i}, a^{ii}, a^{iii}, a^{iv}, and a^{v} the successive internodes of the stem, and a^{vi} the terminal development of the axis into an ovule; *b*, rootlets; *c* to c^{vii} the successive foliaceous appendages to the axis, *c* being the cotyledons, c^{i}, c^{ii}, and c^{iii} the ordinary leaves, c^{iv} the outer floral leaves or sepals, c^{v} the inner floral leaves or petals, c^{vi} the stamens, and c^{vii} the carpellary leaves; *d*, leaf-buds:—B, carpel enclosing an ovule, seen externally and in section, showing *a* the stigma, *b* the style, *c* the ovary:—C, leaf-buds, as seen externally at *d'*, and in section at *d''*.

to the same type of structure with it.* The carpels which occupy the centre of the flower are sometimes developed separately (as in the common *Ranunculaceæ*), whilst in other instances they are grouped together into a single body; the lower part, which contains the *ovules* or young seeds, is termed the *ovary* (Fig. 82, B, *c*), whilst the column which rises from this is known as the *style* (*b*), and its dilated extremity as the *stigma* (*a*), the whole together forming the *pistil*. When each carpel is separately developed, it has its own style and stigma; and sometimes where the carpels are adherent at their lower portion, so as to form but a single ovarium, their styles and stigmas remain distinct. Each carpel, enclosing one or more ovules, may be considered as a leaf folded together at its edges; this is indicated by its frequent retention of the leafy character, even in the normally-developed flower (as seen in the pod of a Pea, when the adherent edges of the two halves are separated, and they are turned back so as to form one leafy surface, of which the prolongation of the foot-stalk serves as midrib); but still more by its occasional more or less complete reversion to the ordinary form of the leaf, in 'monstrous' blossoms. The ovules spring from a thickened fleshy portion of the carpel, which is termed the *placenta*; this is usually regarded as a part of the carpellary leaf itself; but such cogent reasons have been urged by Schleiden for regarding it as a prolongation from the axis, that we can scarcely do otherwise than admit that it may have this origin in some instances, if not universally. (Fig. 82, A, c^{vii}.)

278*a*. The *ovule* is composed of a germ-cell imbedded in a mass of nutritious matter; the whole being included within two or more envelopes, the 'seed-coats.' Its fertilisation is accomplished in a manner very different from that which we have seen in the Cryptogamia; for the pollen-grain, when cast upon the surface of the stigma, instead of rupturing and emitting a spermatic filament, merely projects its inner coat as a long tube, which insinuates itself down the lax tissue of the style, and makes its way to the ovary; and the end of this tube, entering an aperture left in the seed-coats, comes into contact with the 'germ-cell' or *embryonic vesicle*, and thus conveys to it a portion of the contents of the cell of which it is a prolongation. The process of fertilization thus effected has much in common with the 'conjugation' of the simplest Algæ; and the immediate result of it is the same, namely, the production of a germ, capable of developing itself into an entirely new structure resembling that of its parent. This germ at first makes its appearance as a single cell within the embryonic vesicle; and the early development of this cell in the Flowering-plant is quite comparable to that of the germ of the simplest forms of vegetation, except that it takes place at the expense of the store of nutriment previously prepared by the parent, and whilst the ovule is still deriving support and protection from it. In this manner is

* In the common White Water-Lily (*Nymphæa alba*) every gradation may be seen from the stamen to the petal; the lobes of the anther becoming separated and the filament expanded in the outer rows of stamens, so as to present the appearance of a petal with its edges thickened by the development of pollen; and the expansion increasing, whilst the thickening of the edges disappears, as we trace the series from within outwards, until we reach the perfect form of the petal. In the common double Pæony, again, we may trace every gradation between the crimson petal, and the green sepal of equally simple form; and also between this last, and the deeply-cut ordinary leaves which present themselves lower down on the stem.

produced the 'seed;' within which, at the time of its detachment from the parent, the embryonic rudiments of the stem and roots are already formed, and a temporary leaf-like expansion (the single or double cotyledon) is also prepared to evolve itself; whilst a further supply of nutriment is stored up within it, either forming a separate albumen external to the embryo, or absorbed into its cotyledons, which are in that case thick and fleshy. The 'germination' of the seed consists in the further development of these parts, and in their escape from the interior of the seed; the stem raises itself into the air, carrying with it the cotyledons, which for a time perform the functions of leaves; the root buries itself in the ground, and soon begins to absorb liquid nutriment from it; and by the time that the whole of the previously-assimilated store has been exhausted, the first true leaves are evolved, and are capable of obtaining carbon from the atmosphere and of combining it with the water supplied by the roots, so as to form the materials required for the continued growth of the plant; after which the cotyledons or 'seed-leaves' decay away.

278*b*. The increase of the individual plant is accomplished by the continued extension of its axis; which from time to time puts forth leaves at certain points termed *nodes*, the spaces between the nodes being termed the *internodes* (Fig. 82, a^{i} to a^{vi}): there are certain tribes of plants, however, in which no leaves are developed (as is the case with the *Cacteæ*), and the stem is then soft and fleshy, and its surface not merely has a leafy aspect, but performs the functions of foliaceous organs. But the growth of a Phanerogamic plant does not merely consist in the extension of the original structure; for it involves in most cases the development of bodies as offsets from the axis, which are miniature representations of itself in every respect save the want of roots; and which, being usually capable of developing roots when separated from the parent-axis and placed under favourable circumstances, are generally accounted distinct individuals, even whilst still remaining in connection with the parent fabric. These bodies, which are termed *leaf-buds* (Fig. 82, A, *d*, and C, d^{i}, d^{ii}), consist of prolongations of the central axis, around which are found rudimentary leaves, all of which seem to arise from the same node; this appearance, however, is simply due to the want of development of the intervening internodes; and when these have been evolved by subsequent growth, the nodes are separated from each other by the elongation of their axis, and the bud becomes a leaf-bearing branch. This extension or multiplication of the individual, however, is simply a more complicated form of that gemmiparous production, whose simplest manifestation is the multiplication of cells by subdivision; and the sexual re-union, whereby a new generation is to be originated, is effected by the agency of the floral organs, which are the parts of the plant last developed. In the flower, we find the extension of the axis almost entirely checked; so that, by the non-development of the internodes, a large number of nodes are brought together, the successive verticils of floral organs thus apparently arising from the same part of the axis; it occasionally happens, however, that the internodes are developed, and the parts of the flower are then separated from each other in the same manner as the ordinary leaves. The axis usually terminates in the carpels, no extension taking place beyond the point from which they spring, unless it be to give off the ovules; in certain cases of 'monstrosity,' however, we find the axis continued through the centre of the mass of carpels, and developing another set of floral organs at a higher point.—With the development of the first set of appendages to the axis, and the maturation of their seeds, many plants terminate their existence; when this takes place in one season, the plant is said to be *annual;* whilst if its life be spread over two seasons, in the first of which the leaves are put forth, and in the second the flowers, the plant is termed *biennial.* But in other cases, although the leaves and flowers all fall off at the end of the season, yet the axis retains its vitality; and a preparation is evident for the development of new appendages in the ensuing season, leaf-buds being found at the extremities of the branches (when these have not borne blossoms), and also in the *axillæ* of the former leaves, or the points at which their leaf-stalks join the stem. These in their turn are developed into branches bearing leaves; and the latter, after fulfilling their term of active life, die and decay; but not before preparation has been made, through their agency, for the evolution of a series that shall succeed them. Thus, in a perennial plant, it is the axis only which remains permanent; and this is continually receiving additions, not merely to its length (as in Acrogens), but also to the amount of substance contained in the parts already formed. These additions are effected, however, in two very different modes, according to the plan on which the fibro-vascular bundles are arranged in the stem; as will be explained in the succeeding paragraphs.

279. In those Phanerogamia which are distinguished as *Monocotyledonous,* having only a single 'cotyledon,' the stem presents the following type of construction. When it is cut transversely, as seen at A, Fig. 83, its section exhibits a number of fibro-vascular bundles, disposed without any regularity in the midst of the mass of cellular tissue (*a a*), which forms as it were the matrix or basis of the fabric. Each bundle contains

FIG. 83.

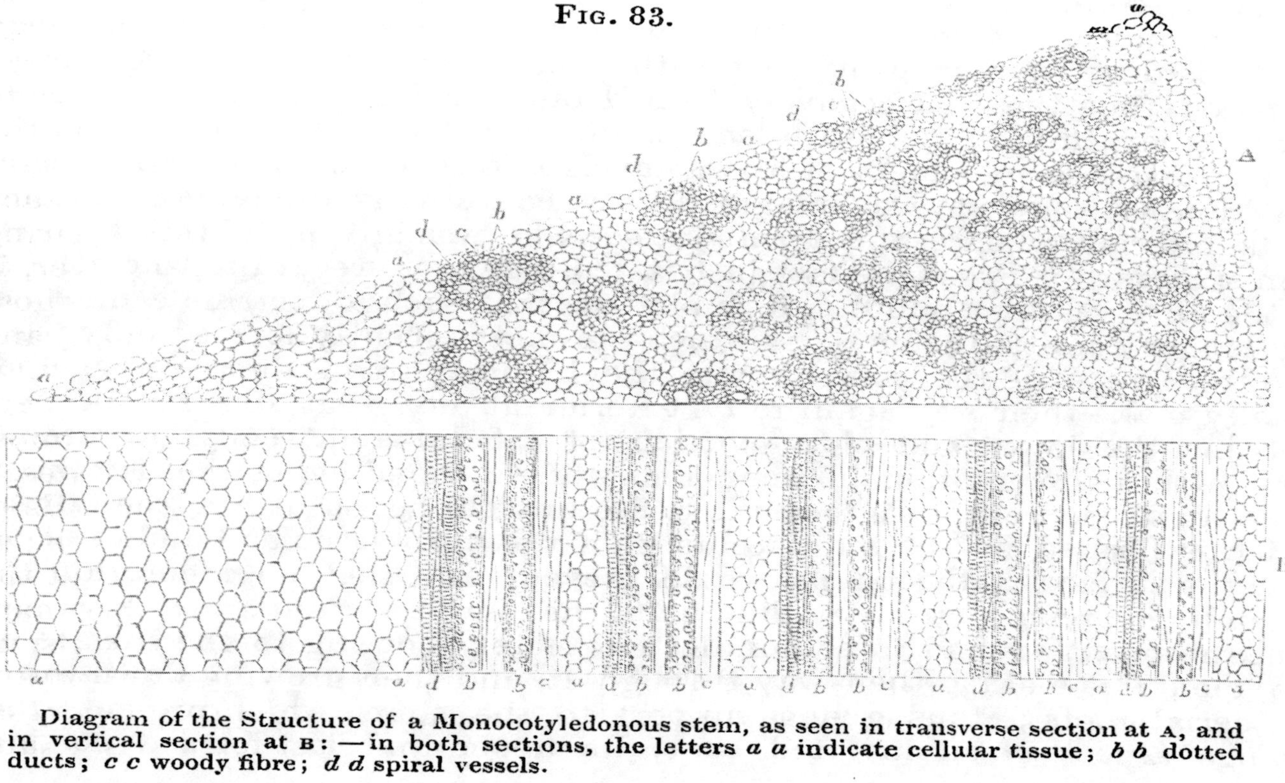

Diagram of the Structure of a Monocotyledonous stem, as seen in transverse section at A, and in vertical section at B:—in both sections, the letters *a a* indicate cellular tissue; *b b* dotted ducts; *c c* woody fibre; *d d* spiral vessels.

large ducts (*b b*), woody fibre (*c c*), and spiral vessels (*d d*), usually combined with cellular tissue. The bundles are least numerous in the centre of the stem, and become gradually closer towards the circumference; but it frequently happens that the portion of the area in which they are most compactly arranged is not absolutely at its exterior, this portion being itself surrounded by an investment composed of cellular tissue only; and sometimes we find the central portion, also, completely destitute of fibro-vascular bundles; so that a sort of indication of the distinction between pith, wood, and bark is here presented. This distinction, however, is very imperfect; for we do not find either the central or peripheral portions ever separable, like pith and bark, from the intermediate woody layer. In its young state, the centre of the stem is always filled up with cells; but these not unfrequently disappear after a time, except at the nodes, leaving the stem hollow, as we see in the whole tribe of Grasses. When a vertical section of a woody stem (as that of a Palm) is made of sufficient length to trace the whole extent of the fibro-vascular bundles, it is found that whilst they pass at their upper extremity into the leaves, they pass at the lower end, also, towards the surface of the stem, and assist, by their interlacement with the outer bundles in forming that extremely tough investment, which the lower ends of these stems present. The fibro-vascular bundles once formed receive no further additions; and the augmentation of the stem in diameter depends upon the development of new woody bundles in continuity with the leaves which are successively evolved at its summit. The stems of Palms (which are the chief arborescent Monocotyledons now existing) grow for the most part from a terminal bud only, like the Tree Ferns; one whorl of leaves being cast off after another, and new leaves being developed from the part of the axis that is extending itself above them. But whilst, in Acrogens, the stem increases merely by addition to its extremity, its lower portion receiving no augmentation of tissue, we find that in these stems a new fibro-vascular bundle is developed in continuity with every new leaf. It was formerly supposed that these successively-formed bundles descend in the interior of the stem through its entire length, until they reach the roots; and as the successive development of leaves involves a successive development of new bundles, the stem was imagined to be continually receiving additions to its interior, whence the term *Endogenous* was given to this type of stem-structure. From the fact just stated, however, regarding the course of the fibro-vascular bundles, it is obvious that such a doctrine cannot be any longer admitted; for those which are most recently formed only pass into the centre of the stem during the higher part of their course, and usually make their way again to its exterior at no great distance below; and thus the lower and older portions of a Palm-stem really do receive very little augmentation in diameter, while a rapid elongation is taking place at its summit. In fact, the dense unyielding nature of the fabric, which is formed by the interlacement of the fibro-vascular bundles at or near the surface of the trunk, would prevent any such augmentation by expanding pressure from within. In the curious *Pandanus* or Screw Pine, however, we find these bundles not merely coming to the surface of the stem, but making their way through it, and then growing downwards as 'aerial roots;' thus giving support to the stem, which presents the strange aspect of an inverted cone, its upper and last-formed portion being of larger diameter than its lower and older.—Monocotyledons are generally distinguished, not merely by the presence of a single cotyledon, and by the structure and mode of growth of their stems, but also by the arrangement of the 'veins' in their leaves, which mostly run *parallel* to each other with but little connection or reticulation; and by the development of the parts of the flower in *threes* or in multiples of three. These characters, however, are by no means constant.

280. In the stems of *dicotyledonous* Phanerogamia, we find a method of arrangement of the several parts, which must be regarded as the highest form of development of the axis. When the woody stem of any tree or shrub of temperate climates (of which all belong to this division of Flowering Plants) is cut across, we always find in the section a distinct division between the three concentric areæ of the *pith*, the *wood*, and the *bark;* the first being central, the last peripheral, and the wood being interposed between them (Fig. 84).—The Pith (*a*) is almost invariably composed of cellular tissue only; so also is the bark, except in its inner layer which adjoins the wood; but the woody portion is made up of woody fibres (*c c*), usually with the addition of ducts of various kinds (*b b*),—these, however, being absent in one large group, the *Coniferæ* or 'Fir-tribe,' in which the woody fibres are of unusual diameter, and are 'dotted' (§ 150). The pith is immediately surrounded by a delicate membrane, consisting almost entirely of spiral vessels (*d d*), which is called the 'medullary sheath.'—In a stem or branch of more than one year's growth, the Woody structure presents a more or less distinct appearance of division into concentric

FIG. 84.

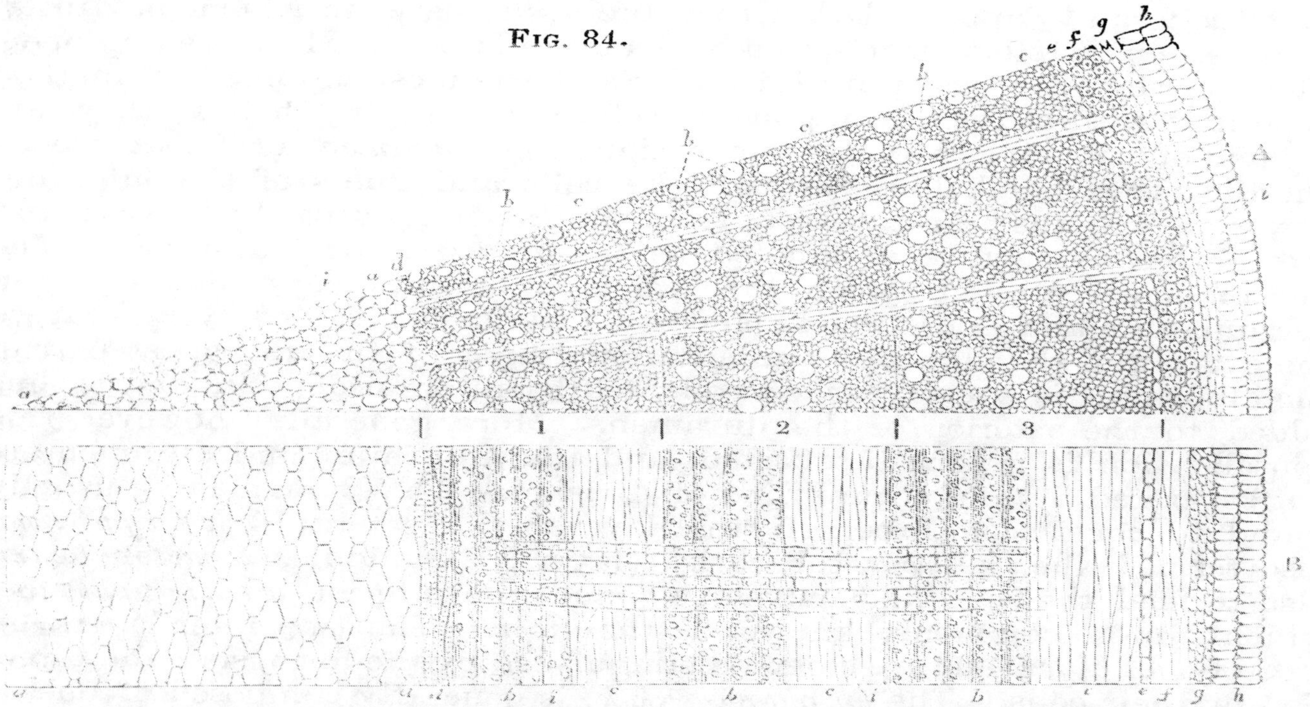

Diagram of the Structure of an Exogenous stem, of three years' growth, as seen in transverse section at A, and in vertical section at B:—in both sections, *a* represents the cellular tissue of the pith; *b b* the dotted ducts, and *c c* the woody fibre, of the successive annual layers; *d d* the spiral vessels of the medullary sheath; *e e* cambium layer; *f f* liber; *g g* cellular envelope; *h h* corky layer; *i i* medullary rays. (N.B. In the vertical section, the medullary ray is shown in only part of its length; since the continuity of the medullary rays from the pith to the bark, owing to the slight flexure which almost always occurs in them, is rarely or never shown by such a section, however large a part of one of them may be traversed by it.)

rings, the number of which varies with the age of the tree. The composition of the several rings, which are the sections of so many cylindrical layers, is uniformly the same, although their thickness may vary; the arrangement of the two principal elements, however,—namely, the woody fibre and the ducts,—varies in different species, the ducts being sometimes almost uniformly diffused through the whole layer, whilst in other instances they are confined to its inner part; but the general fact is, that the ducts predominate towards the inner side of the ring (which is the part of it first formed), and that the outer portion of each layer is almost exclusively composed of woody tissue. This alternation of ducts and woody fibre frequently serves to mark the succession of layers, when, as is not uncommon, there is no very distinct line of separation between them. The number of layers is usually considered to correspond with that of the years during which the stem or branch has been growing; and this is, no doubt, generally true in regard to the trees of temperate climates. There appears strong reason to believe, however, that such is not the universal rule; and that we should be more correct in stating that each layer indicates an 'epoch of vegetation;' which, in temperate climates, is usually (but not invariably) a year, but which is commonly much less in the case of trees flourishing in tropical regions. This question, however, will be more fully considered hereafter (CHAP. XIV.). The woody rings are seen in the transverse section to be traversed by bands or lines (*i i*) radiating from the centre to the circumference; these, which are termed the *medullary rays*, are thin plates of cellular tissue, connecting the cellular substance of the pith with that of the bark, but are usually compressed so closely as to appear darker than the tissue between which they intervene, except in stems whose woody structure is peculiarly loose and open. The inner layers of wood are the oldest and the most solidified by matters deposited within their component cells and vessels; hence they are spoken of collectively under the designation *duramen* or 'heart-wood.' On the other hand, it is through the cells and ducts of the outer and newer layers, that the sap rises from the roots towards the leaves; and these are consequently designated as the *alburnum* or 'sap-wood.' The line of demarcation between the two is sometimes very distinct, as in Lignum-vitæ and Cocos wood; and as a new layer is added every year to the exterior of the alburnum, an additional layer of the innermost part of the alburnum is every year consolidated by internal deposit, and is thus added to the exterior of the duramen. More generally, however, this consolidation is gradually effected, and the alburnum and duramen are not separated by any abrupt line of division.—The Bark may be usually found to consist of three principal layers: the external, or *epiphlœum*, also termed the *suberous* (or corky) layer; the middle, or *mesophlœum*, also termed the 'cellular envelope;' and the interior, or *endophlœum*, which is more commonly known as the *liber*. The two outer layers are entirely cellular; and are chiefly distinguished by the form, size, and direction of their cells. The *epiphlœum* (*k k*) is generally composed of one or more layers of colourless or brownish cells, which usually present a cubical or tabular form, and are arranged with their long diameters in the horizontal direction: it is this which, when developed to an unusual thickness, forms *cork*, a substance which is by no means the product of one kind of tree exclusively, but which exists in greater or less abundance in the bark of every exogenous stem. The *mesophlœum* (*g g*) consists of cells, usually of green colour, prismatic in their form, and disposed with their long diameters parallel to the axis; it is more loosely arranged than the preceding, and contains intercellular passages and laticiferous vessels; and although usually less developed than the suberous layer, it sometimes constitutes the chief thickness of the bark. The *liber* or inner bark (*f f*), on the other hand, usually contains woody fibre in addition to the cellular tissue and laticiferous vessels of the preceding; and thus approaches more nearly in its character to the woody layers, with which it is in close proximity on its inner surface. The liber may generally be found to be made up of a succession of thin layers, equalling in number those of the wood, the innermost being the last formed; but no such succession can be distinctly traced in the cellular envelope, or in the suberous layer; although it is certain that they too augment in thickness by additions to their interior, whilst their external portions are frequently thrown off in the form of thickish plates, or detach themselves in smaller and thinner flakes. The bark is always separated from the wood by the *cambium-layer* (*e e*), which is the part in which all new growth takes place; this seems to consist of mucilaginous semifluid matter; but it is really made up of cells of a very delicate texture, which gradually undergo transformations, whereby they are for the most part converted into woody tissue, ducts, and laticiferous vessels. These materials are so arranged, as to augment the fibro-vascular bundles of the wood on their external surface; thus forming a new layer of alburnum, which *encloses* all those that preceded it; whilst they also form a new layer of liber, on the *interior* of all those which preceded it: they also extend the medullary rays, which still maintain a continuous connection between the pith and the bark; and a portion remains unconverted, so as always to keep apart the liber and alburnum.—This kind of stem-structure is termed *exogenous;* a designation which applies very correctly to the mode of increase of the woody layers, although (as we have just seen) the liber is formed upon a truly endogenous plan.

280 *a.* In its first-developed state, the Exogenous stem consists, like the so-called endogenous, of cellular tissue only; but after the leaves have been actively performing their functions for a short time, we find a circle of fibro-vascular bundles (*d d*, Fig. 55) interposed between the central or medullary portion (*a*), and the peripheral or cortical portion (*b b*) of the cellular matrix; these fibro-vascular bundles being themselves separated from each other by plates of cellular tissue (*c c*), which still remain to connect the central and the peripheral portions of the matrix. This first stage in the formation of the Exogenous axis, in which its principal parts—the pith, wood, bark, and medullary rays — are marked out, is seen even in the stems of herbaceous plants, which are destined to die down at the end of the season; but in such, the difference between the

Fig. 85.

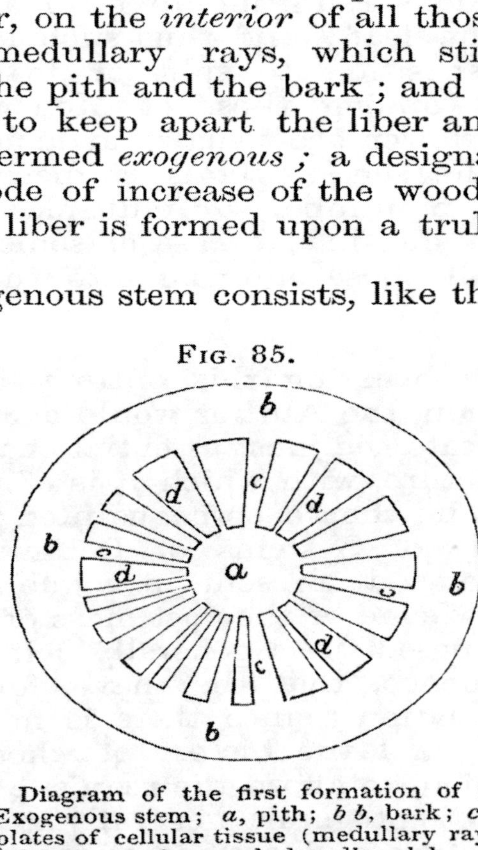

Diagram of the first formation of an Exogenous stem; *a*, pith; *b b*, bark; *c c*, plates of cellular tissue (medullary rays) left between the woody bundles *d d*.

endogenous and exogenous types is manifested in little else than the disposition of the fibro-vascular layers, which are scattered through nearly the whole of the cellular matrix (although most abundant towards its exterior) in the former case, but are limited to a circle within the peripheral portion of the cellular tissue in the latter. It is in the further development which takes place during succeeding years in the woody stems of perennial Exogens, that those characters are displayed, which separate them most completely from the Ferns and other Cryptogamia whose stems contain a cylindrical layer of fibro-vascular bundles, as well as from (so called) Endogens. For whilst the fibro-vascular layers of the latter, when once formed, undergo no further increase, those of Exogenous stems are progressively augmented by the metamorphosis of the cambium-layer; so that each of the bundles which once lay as a mere series of parallel cords beneath the cellular investment of a first year's stem, may become in time the inner boundary or edge of a wedge-shaped mass of wood, extending continuously from the centre to the exterior of a trunk of several feet in diameter, and becoming progressively thicker as it passes outwards. The fibro-vascular bundles of Exogens are therefore spoken of as 'indefinite;' whilst those of Exogens and Acrogens are said to be 'definite' or 'closed.' From the fibro-vascular bundles of the stem, prolongations pass off, accompanied by spiral vessels derived from the medullary sheath, through the leaf-stalks into the leaves, of which they form the mid-rib and veins. The veins of the leaves of Dicotyledonous plants have for the most part a *reticulated* arrangement, their subdivisions reuniting to form a network throughout the entire substance of the leaf; and thus such leaves possess a degree of firmness, and an indisposition to split or tear in any direction, which is seldom met with among those of Monocotyledonous plants. Finally, the parts of the flower are usually arranged in a *quaternary* or *quinary* plan; that is, the sepals, petals, stamens, and carpels are usually either four or five in number, or multiples of these numbers; but a departure from this rule, in the case of some particular set of organs, or with respect to all those entering into the composition of the flower, is extremely common.

In bringing this outline-view of the Vegetable Kingdom to a conclusion, the Author would again point out that his aim has been solely to indicate the most essential characters of those primary types of vegetable structure, with which it is of the greatest importance to the Physiologist that he should be acquainted; and not even to lay a foundation for the study of Systems of Botanical Classification. However valuable the systems at present in vogue may be, as helps to the attainment of a knowledge of the vast mass of forms of which the Vegetable Kingdom is composed, it is generally felt by Botanists of the highest philosophic eminence, that they have, for the most part, but a very unsatisfactory foundation; since there is no one tribe of Plants, whose internal structure, and the history of whose development, have yet been sufficiently studied, to afford the requisite data for a truly scientific arrangement. Of the Cryptogamia generally it may most assuredly be said, that until the structure and actions of their reproductive apparatus have been clearly made out, any attempt at classification can have but a temporary value, like the 'artificial' system of Linnæus. And with regard to Phanerogamia, although there is less difficulty in associating together plants of similar conformation and habits into 'Natural Orders,' yet for the grouping together of these natural orders into primary subdivisions of the Monocotyledonous and Dicotyledonous divisions, no fixed principles have yet been established in general estimation, simply because no adequate search has yet been made for the data which the history of development would furnish. The Botanist, therefore, may do a service fully as important to science, by devoting himself to observing and recording the phenomena presented by the most familiar plants, in the progress of their evolution; as he can by collecting a number of species previously unknown, from some of those regions which still afford a fertile harvest to the enterprising collector. The Author is far from wishing to undervalue the energy and skill displayed in such labours; but he is desirous of urging upon his readers, that the same energy and skill may find ample scope in researches, which may be prosecuted as a relaxation in the intervals of the toilsome avocations of daily life, and which will afford a no less abundant harvest of novelties, of at least equal scientific value, however inferior in external attractiveness.

CHAPTER VII.

GENERAL VIEW OF THE ANIMAL KINGDOM.

281. Of the number of distinct specific forms of Animals at present existing upon the Earth's surface, it is scarcely possible to form even an approximative estimate: since, although we may be probably not far wrong in our calculation of the number of existing species, in certain classes which have been especially studied (such as those of Mammals and Birds), and of which by far the greater part are certainly known to us; it is at least equally probable that our present acquaintance is limited (from various causes) to a very small proportion of other classes, whose total amount, therefore, we can do little more than guess at. The number of species of *Mammals* known to Naturalists is about 1700; and it is probable that scarcely 300 more remain to be discovered. Of *Birds*, about 8000 species are certainly known; and to these we may perhaps add 4000 for those not yet discovered or not yet clearly distinguished. Of *Reptiles*, about 1200 species are known; but it is probable that the proportion not yet discovered is larger, and that for this we should add at least 800 species. Of *Fishes*, about 8000 species are known; and to these, also, numerous additions may be expected, probably at least 4000 species. Thus of Vertebrated Animals alone, nearly 19,000 species are known, and 9000 more are probably in existence. The number of *Mollusks* has been hitherto reckoned chiefly by that of the *shells* contained in collections, no account being taken of any but the testaceous species. Of these about 15,000 species have been collected; and probably at least as many more are yet unknown to the Conchologist. But the number of 'naked' or shell-less Mollusks is undoubtedly extremely large; and of these it is

probable that only a small proportion are yet known.* The class of *Insects* far outnumbers all the preceding, both as to the number of species already known, and still more as to the number of whose existence we have presumptive evidence. It is certain that at least 150,000 species are at present to be found in collections; and that these do not by any means include the total number existing even in the countries whose entomology has been best explored. So little, in fact, is this the case, that if anything like the same proportion holds good elsewhere between Flowering Plants and Insects, as obtains in our own country (namely at least *ten* species of Insects to every species of Flowering Plant), we should have to estimate the total number of existing species of Insects at little less than *two millions*. In regard to none of the inferior classes, have we at present adequate means of forming any estimate whatever.

282. This vast assemblage of species was formerly divided into two primary groups; the VERTEBRATA, possessing a jointed spinal column, within which the chief ganglionic centres of the nervous system are inclosed; and the INVERTEBRATA, which are destitute of any such structure. The first division included only Mammalia, Birds, Reptiles, and Fishes; the second comprehended all the Insect and Vermiform tribes, the Mollusca or Shell-fish, together with Zoophytes (using that term in its most extended signification). But it is now generally acknowledged that this method is by no means a natural one; since the Invertebrata contain at least three and perhaps four groups, differing as much from each other as that of Vertebrata does from either of them, and therefore entitled to hold the same rank with the latter. The primary groups or sub-kingdoms are, therefore, to be regarded as consisting of—I. VERTEBRATA, which are characterised by the possession of an internal bony column, composed of jointed pieces or *vertebræ*, within which, or their modifications, the central organs of the nervous system are inclosed; to this column all the other bones in the body are more or less directly attached; and these are covered by soft flesh, which chiefly consists of the muscles whereby they are moved. Of all animals, their structure is most complicated; they all possess the power of active locomotion, which is usually effected by members or limbs, supported by bony projections of the vertebral column, and never exceeding four in number; and nearly all enjoy the senses of taste, smell, sight, and hearing, as well as that of touch. In some the blood is warm, in others cold; but in all it is of a red colour. These characteristic peculiarities undergo various modifications among the lower forms of this group, by which the affinities to the other types are indicated.—II. ANNULOSA or ARTICULATA, animals in which the hard parts composing the skeleton are external, and formed into jointed rings. This is the character of a large number of classes included in this division, which present, with much difference in complexity, a very general conformity of structure. Thus, from the soft and simple Vermiform tribes, such as the Leech or Earthworm, we pass by almost insensible gradations to the Centipede, and from this, to the highly organised Insects and Crustacea. Although some of the animals con-

* Thus, of the tribe of Nudibranchiate Gasteropods, only about a dozen species were formerly known as inhabiting the coast of Britain; but in the beautiful Monograph of Messrs. Alder and Hancock (in course of publication by the Ray Society) more than a hundred British species will be described.

tained in this division border upon the lowest of the whole kingdom, yet others are inferior only to the Vertebrata in the complexity of their organisation. A distinct head and eyes are almost universally present. The muscles that execute the movements of the body, are attached to the interior of the hard envelope, which, where distinct members are developed, incloses them as well as the trunk. In some of the Annulosa the blood is red, in others it is nearly colourless; and among the Insect tribes there is a power of generating heat, almost as great as among any of the Vertebrata, although less constantly exercised. The locomotive powers are usually very considerable; and the general structure of the body is peculiarly adapted to the predominance of this faculty. When distinct members are developed, they are never less than six in number, and are usually much more numerous. There is one group, however, that approaches the Mollusca, in which this tendency is reduced to a subordinate condition, in conformity with other peculiarities of its organisation (§ 311).—III. MOLLUSCA, animals whose body is soft throughout, being sometimes enclosed in a shell, but being frequently destitute of any such protective habitation. The shell, where it exists, is not constructed upon any very definite type, nor does it serve to give points of attachment to the muscles concerned in the movements of the body; its sole purpose (apparently) being to afford a defensive covering to animals capable neither of active resistance, nor of seeking safety in flight. This group also includes many animals of high organisation, such as the Cuttle-fish, which approach the Vertebrata very closely in several points of their structure; as well as many whose conformation is very simple. Instead of long jointed bodies equally developed on the two sides, they almost always present an irregular rounded form, with no distinct locomotive organs; or when such are developed, they are but fleshy tentacula (as the *arms* of the Cuttle-fish), or muscular disks (as the *foot* of the Snail), or tongue-like protuberances (as the *foot* of the Cockle), quite different from the complex jointed limbs of Vertebrata or Insects. In none are the locomotive powers developed to a high degree; many remain affixed during nearly their whole lives to other substances; and in most, the nutritive system appears to predominate above the animal functions. Some have a distinct head, with eyes, ears, and mouth; whilst others are destitute of any organs of special sensation, and the entrance to their alimentary canal is not indicated by its situation on any prominent part of the body.—IV. RADIATA, or radiated animals, a group which is usually considered to include a vast quantity of materials apparently most heterogeneous; uniting the comparatively symmetrical and complex Star-fish and Echini, with those simple beings which form the transition to the Vegetable Kingdom. The name properly applies, however, to those animals only, which have their organs arranged in a radiated or star-like form around the orifice to the stomach; such are the Star-fish and Sea-urchin tribe, the Medusæ or jelly-fish, and all true Polypes. Some of these are protected by firmly-jointed external skeletons, like those of the Articulata; whilst others deposit calcareous matter in the interior of their soft fleshy structure, as if sketching out the internal skeleton of the Vertebrata. Their locomotive powers are usually inconsiderable, many of them remaining fixed to one spot during all but the embryonic period of life; their organs of sensation are indistinct, though rudiments of eyes are supposed to exist in the higher tribes; but the nutritive processes appear to be performed with great activity, and some of the softest and most delicate of these animals are known to seize upon and digest the hard bodies of others much higher in organisation. —There are certain beings, however, which cannot be included in the preceding groups, but which must be considered on various accounts as entitled to a place in the Animal kingdom, although their manifestations of 'animality' are very obscure (§ 262). In these, no definite type of form presents itself, and their structure is the simplest possible; hence they have received the designation *Amorphozoa;* but from the analogy they bear to those lowest types of Vegetation which have been characterised as *Protophyta* (§ 266), as well as to the early embryonic condition of higher animals, they may be more characteristically designated as PROTOZOA.

282 *a.*—Now it is found that every one of these groups may be characterised by the conformation of its Nervous System; and as this has an obvious relation with all the functions, both animal and nutritive, it is probably the best single character which could be adopted (see CHAP. XX.). —Thus, Vertebrated animals have a nervous cord, consisting of a series of coalesced ganglia, inclosed in the spinal column which supports the back; and this is dilated, by the addition of new ganglionic masses, into the *brain,* which is contained within the cavity of the head, where also it is connected with the organs of special sensation. Hence, they may be termed *Spini-cerebrata.*—Annulose or Articulated animals, again, present, as the typical form of their nervous system, a double cord, studded at intervals with ganglionic enlargements; this runs along the lower or abdominal surface of the body, protected by its general envelope, and sends its principal branches to the legs; and it is connected with similar ganglia within the head, the size of which is proportional to the development of the organs of special sensation, but which never correspond entirely with the brain of Vertebrata. In those species in which the locomotive apparatus is most connected with one part of the body, as in Insects, the ganglia are concentrated in its neighbourhood, and no longer present their regular disposition through the whole trunk, so that their peculiar arrangement is less evident; but in these, at an earlier period of life, the typical conformation is observable, to express which, the term *Diplo-neura* has been applied to this sub-kingdom.—In the Mollusca, the nervous system is principally concentrated around the entrance to the alimentary canal, forming a circle, through which the œsophagus passes, and which is connected with other ganglia, disposed without symmetry among the viscera, or in the neighbourhood of the organs of locomotion, if such should be specially evolved. In some of the highest of this division, the nervous system approaches very closely in its arrangement to the form it presents in the lowest Vertebrata, and receives a corresponding protection by a rudimentary internal skeleton; but, in general, it is more connected with the immediate supply of the nutritive functions, and wants that symmetrical arrangement and close connection with the locomotive organs, which may be regarded as characters of elevation in the nervous system of the Articulata. From the irregular distribution of their ganglia, Mollusca have been termed *Hetero-gangliata.*—The Radiata present such a form of nervous system as might be expected, when the

peculiarity of the arrangement of their organs is considered. It is composed of a filamentous ring, which surrounds the mouth, and sends off branches to the different divisions of the body; a slight ganglionic enlargement being usually perceptible where these fibres are given off. Hence these animals have been termed *Cyclo-neura*. In the lowest of the Radiated tribes, however, no nervous system has yet been certainly discovered; still we seem to have some right to presume upon its existence, since the movements of these animals are such as we do not witness in beings that are destitute of this apparatus, and it has not been proved by sufficiently careful examination of their tissues, that no nervous structure exists among them.—In the Protozoa, on the other hand, while the closest scrutiny has failed to detect the slightest traces of a nervous system (which the extreme transparency of their tissues would allow to be easily discovered, if they were really present), the character of the movements furnishes no sufficient reason for referring them to any such agency, since they bear on the whole a much stronger resemblance to those of Plants, than to those of the higher Animals.

283. The beings which are here grouped together under the designation PROTOZOA, have a very near relationship to those which have been already characterised as *Protophyta* (§ 266), and considered as the simplest forms of the Vegetable kingdom. And it is in the separation of these two groups, that we have most occasion to employ those means of discrimination, on which we have already dwelt at length (CHAP. V.). A large proportion of the Protozoa may be considered, with much probability, either as single cells, or as aggregations of cells, which, being similar to each other, are therefore independent. If, therefore, the doctrines already laid down respecting the essentials of a nervous system (§ 262) be correct, we have a right to conclude that, notwithstanding the active movements which many of them perform, *no consciousness* exists among them; and in this conclusion we are justified by the fact (on which we shall presently have to enlarge), that these movements are performed by the same agency as that which effects the motions of Plants, and which also operates in those movements in the higher Animals that are not attended with consciousness. And it is a remarkable proof of the futility of any attempt to establish a distinction upon the mere character of the movements of these beings, that many which were formerly ranked as undoubted Animalcules, are now known to be the 'zoospores' of Algæ, the movements of which have frequently just as much appearance of *spontaneity*, as have those of Animalcules.—As we are unable, then, to draw any inferences from the movements of these beings, as to their Animal or Vegetable nature, we fall back upon the other characters already enumerated (§ 263); namely, the presence or absence of a digestive cavity, the chemical composition of their bodies, the mode in which their nourishment is obtained, and the general affinities which they may present to undoubted Animals or Plants. Now the presence of a digestive cavity is by no means universal among them; for we shall find it to be wanting in some of those beings, which, from their other characters, seem most clearly entitled to a place on the Animal side of the line. In regard to chemical composition, it may be said with certainty that Sponges and other composite Protozoa, whose masses are large enough to allow of analysis, are rather allied to Animals; whilst, on the other hand, the Vegetable nature

of certain other tribes formerly accounted Animalcules has been suspected, from the predominance of starchy compounds in their cells. Where no starch or cellulose can be detected by the iodine-test, the Animal character of the tissue may be presumed. But it is in the mode in which they obtain their food, that we seem to find the most marked and constant difference between the two kingdoms; and we shall be probably justified in ranking among *Protophyta*, all such as decompose carbonic acid, and form a green or red endochrome by the direct combination of the inorganic elements; and in considering as *Protozoa* such as derive their nutriment from organic compounds already elaborated by some other living being.

283*a*. The best idea of the essential characters of the Protozoa may perhaps be formed from an examination of the structure and actions of the *Amœba* (commonly known as the *Proteus*,* on account of its capability of assuming an unlimited variety of forms), an inhabitant of fresh and stagnant waters, vegetable infusions, &c. At first sight, this creature seems to be nothing else than a minute particle of gelatinous matter; but its spontaneous changes of shape soon give evidence of its possession of vitality; and it becomes evident, on further examination, that it may be considered as a *simple cell*, of unusual dimensions (being sometimes, in the largest species, as much as 1-70th of an inch in diameter), analogous to that which constitutes each individual of the simplest Protophytes. Like these, it lives for and by itself, and it imbibes its nutriment through its cell-wall, having no oral orifice; but it possesses a motor power with which they are not endowed; and this power is obviously related to the character of the food on which it is to be supported. For, whilst *they* can obtain, wherever they may be situated, the materials required for their development, this Animal cell, being dependent upon matter previously assimilated, and not endowed with any means of drawing such matter towards itself, must go (as it were) in search of it. This movement is accomplished by the continual changes of form that take place in its body, the typical figure of which may be considered globular, but which may assume almost any shape whatever (Fig. 86). The change of form seems due rather to actions taking place in the interior of the cell, than to any irritability of the cell-wall; for if the movements of an Amœba be attentively watched, the extension of the gelatinous body in any particular direction (so as to form one of the digitate prolongations) will be seen to be preceded by the setting of a current of the moving molecules within the cell, in that direction; to which current the protrusion of the cell-wall is really due. A continuation of the same current distends the prolongation, and the whole mass of the body is gradually carried onwards (so to speak) into it, so that its place in the field of the microscope is slowly changed. After a short time, the particles in the interior of the cell are again seen to move in a definite course; a new prolongation is thrown out, in the same or some other direction; and the body is again absorbed into it. Thus its locomotion is effected by the agency of the currents which traverse the interior of the cell, and which are obviously connected (as in the cells of *Chara* and other plants, §§ 137 and 269) with the organic processes concerned in its growth and maintenance; and it must

* The name *Proteus*, although peculiarly appropriate to this being, is inadmissible, being in use as the designation of a much higher animal (§ 323).

consequently be regarded as entirely independent of any spontaneous control on the part of the individual, and cannot even be allowed to afford the least indication of consciousness. The nutrition of the Amœba is derived from animal and vegetable substances; but these are not received into an internal cavity (as affirmed by Ehrenberg), their enclosure within

Fig. 86.

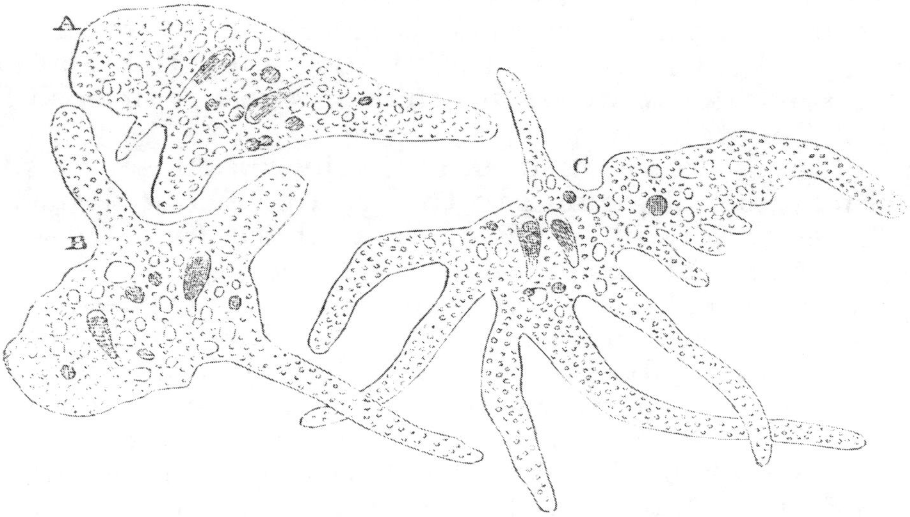

Amœba princeps, in different forms, A, B, C.

the gelatinous body being rather apparent than real. When the creature, in the course of its progress, meets with a particle capable of affording it nutriment, its body spreads itself over or around this, so as even to invest it completely; and the particle thus seems to be included in a digestive cavity, while it is really in contact with the external surface only, through which its nutritive elements are absorbed into the cavity of the proteus-cell. It is interesting to see such a creature thus manifesting the peculiar *nisus* of animal development; *making*, as it were, a stomach for itself, by wrapping its body round the alimentary matter, which it is not able to receive into its interior.*—Thus we have, in the *Amœba* and its allies, just such a manifestation of animal *tendencies*, rather than of actual *powers*, as might be expected in beings that constitute the transition-group between the two kingdoms. In the higher forms of the Protozoa, we meet with more distinct manifestations of the same tendencies. Thus in the (so-called) *Polygastric Animalcules*, we seem to find the highest development which animal cells can attain as individuals; their nourishment being directly received into their interior by an orifice in the cell-wall; and the movements of the cilia with which they are beset, being adapted not merely for the propulsion of the body through the water, but also for the introduction of food through this orifice. But this is a type of structure, which does not seem capable of any further development; for although we meet with forms, among these Animalcules, which seem like sketches of fabrics much more elevated in the scale, we do not meet with any

* The foregoing account of the structure and actions of the *Amœba* is given from the author's own observations; which he is glad to find in precise harmony with those of Mr. Carter on the proteiform sponge-cell (§ 286).

which present a decided approximation towards higher groups. So in the *Rhizopods,* we have composite fabrics presenting a considerable regularity of form, which seem to have their origin in the gemmiparous multiplication of beings closely allied to the Amœba in their general structure and mode of life, each individual still living for and by itself. In the *Sponges* and their allies, on the other hand, we meet with vast aggregations of proteiform cells, united into composite fabrics; of which the several parts, though almost precisely similar to each other, have yet a certain degree of interdependence; and through these we are led towards the true *Zoophytes,* in which the individuality of the component cells and fibres is merged (as it were) in that of those composite structures which are termed Polypes.

284. To the class of POLYGASTRICA are referred those simpler forms of *Animalcules,* which do not belong to the group of *Rotifera* (§ 310). Its name was conferred upon it by Prof. Ehrenberg (who was the first to point out the essential distinctions between these groups), under the idea that the beings composing it are distinguished by the possession of numerous digestive sacs or stomachs; but upon this point there is much doubt, many competent observers being far from coinciding with the views of Prof. Ehrenberg in regard to it.* There can be no doubt that he has been in some instances mistaken in his account of the organisation of the beings which he has grouped together under the designation *Polygastrica,* and to which he has assigned a complicated digestive apparatus, organs of sense,† a generative system, &c.; since many of them have been proved by further research to belong unquestionably to the Vegetable kingdom; whilst many others will probably be assigned to it, when the history of their development shall have been more fully ascertained. Until, however, the limits of the group shall have been more precisely determined, by the removal from it of all the beings which can have a more appropriate place assigned to them elsewhere, it will be premature to attempt to define the characters of the class, or to substitute any other designation for it; and all that will be here attempted, will be to give a general view of its leading peculiarities.—The *forms* presented by the Polygastrica are extremely various. In some, as among the *Amœbæ* already described, there cannot be said to be any definite shape; their bodies being capable of projection into any figure. In others, there is a considerable variety in the forms assumed by the same individual under different circumstances; but still a prevailing shape can be recognised. In others, again, the body, although still unprotected by any firm envelope, appears to undergo but little change in figure, except when affected by temporary pressure. But there are some species which cannot even be influenced by this, their soft bodies being inclosed in a delicate but firm integument, occasionally strengthened by a deposit of calcareous or siliceous matter; such are termed *loricated* Animalcules. Among these, there are certain composite forms, which most remarkably prefigure the

* See CHAP. X., for a fuller notice of this question.

† The red spot, which has been asserted by Prof. Ehrenberg, to be the rudiment of an eye (for which doctrine there is no other foundation than the fact, that the rudimentary eyes of animals, in which such organs *may* be presumed to exist, are commonly red spots), has been observed in certain moving cells, which are undoubtedly of a Vegetable nature; its existence among Animalcules, therefore, cannot be admitted as any proof of their conscious perception of light.

arborescent Zoophytes, especially those of the *Hydraform* group (§ 290); each Animalcule having a bell-shaped horny cell, open at one end like the polype-cells of the *Campanularia* (Fig. 95), and formed by the condensation of its integument; and the development of these, one from another, taking place by gemmation upon the same plan as in that class. Of those forms, however, which were ranked by Ehrenberg under the loricated series, and whose envelope is composed of *silex*, a large proportion (included in the families *Diatomaceæ* and *Desmideæ*) are now generally ranked among Vegetables; whilst of those which possess calcareous envelopes, the greater number seem more properly to belong to the group of *Rhizopoda* to be presently described.—The true Polygastrica all appear to have a distinct *mouth*, or entrance to the cavity of the body; and this is usually surrounded by vibratile cilia (Fig. 87, A, B, C, &c., *a*), by

FIG. 87.

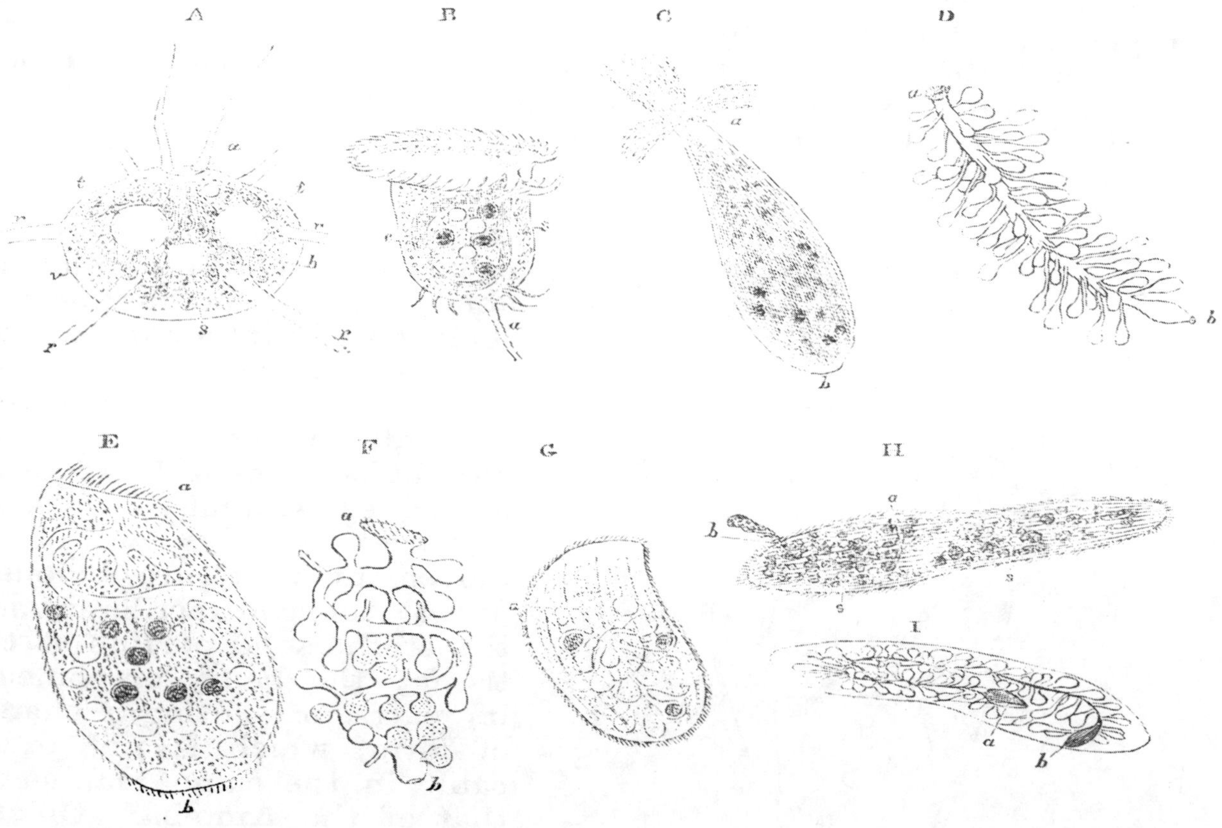

Polygastric Animalcules:—A, individual Monad of *Volvox globator*, showing *a*, the mouth; *b*, *b*, gastric vesicles (?); *s*, *t*, *t*, contractile vesicles; *r*, *r*, *r*, cords of communication;—B, individual of *Vorticella citrina*; *a*, the stalk; *s*, contractile vesicles; *t*, intestinal tube;—C, external aspect of *Enchelis pupa* in the act of taking food, showing *a* the mouth, and *b* the anal orifice,—D, supposed digestive apparatus of the same animalcule;—E, F, similar views of *Leucophrys patula*; G, *Chilodon cucullulus*; showing *a* the dental apparatus; H, I, *Paramecium aurelia*, showing *a* the mouth, *b* the anus, *s* contractile vesicles.

whose agency the alimentary particles are introduced through it into the interior. There is usually, but not invariably, a distinct anal orifice (C, D, E, F, H, I, *b*), through which refuse matter is expelled. When these Animalcules have been allowed to remain a short time in water, in which finely-divided particles of colouring matter are suspended, the whole of the transparent body is seen to be studded with coloured globules of tolerably uniform size, each of them composed of an aggregation of minute particles (B, C, E, G, H): and it is chiefly on this fact that Prof. Ehrenberg's doctrine rests, that a number of distinct globular cavities exist in the substance of the body, into which food is received; all these cavities being supposed, in some instances, to communicate directly with the mouth (A, G); while, in other cases, they are represented as disposed round an intestinal tube, which may proceed either directly or sinuously to the opposite extremity of the body (D, F), or which may return upon itself so as to terminate near the mouth (B, I).—It is to the action of the vibratile *cilia*, with which some part of the body appears always to be furnished in the true Polygastrica, that the great variety of movements exhibited by these beings is almost entirely due. Some propel themselves directly forwards, with a velocity which appears (when thus highly magnified) like that of an arrow, so that the eye can scarcely follow them; whilst others drag their bodies slowly along like the leech. Some make a fixed point of some portion of the body, and revolve around it with great rapidity; others move by undulations, leaps, or successive gyrations; in short there is no kind of animal motion, or method of progression, that is not exhibited by these Animalcules. As already shown, however (§ 283), there is no sufficient reason to believe, that such motions indicate consciousness on the part of the beings which perform them; and we can only regard them as indicative of a wonderful adaptation, on the part of these simple ciliated cells, to a kind of life which enables them to go in quest of their own nutriment, and to introduce it when obtained into the interior of their bodies, without any of that complex mechanism which is required in higher Animals. In some instances, however, we notice movements which are obviously due to the contractile properties of the solid portion of the tissue; the body, or some of its parts, presenting a change of form, which is not explicable in the same manner as that of the Amœba. Of this we have a very good example in the *Vorticella*, a bell-shaped animalcule, whose small extremity is prolonged into a stalk, by which it commonly attaches itself to a solid body; whilst this stalk, in its extended condition, is straight or nearly so (Fig. 88, A), it will suddenly contract into a spiral (B), from which, after a little time, it will usually return to the extended state. This contraction

FIG. 88.

Group of *Vorticella nebulifera*, showing A the ordinary form, B the same with the stalk contracted, C the same with the bell closed, D, E, F, successive stages of fissiparous multiplication.

often takes place when the Animalcule, by its ciliary action, has drawn towards itself floating particles too large to be received into its mouth; it also occurs when the animalcule is touched by some other in the course of its movements; and it may be produced at any time by smartly tapping the stage of the microscope.

284 *a*. The reproduction of the Polygastric Animalcules is effected in nearly every instance, so far as is at present known, by the processes of *fission* and *gemmation*, which, as formerly explained (§ 141), are but variations of the same fundamental plan. In many species, the multiplication is effected by fission alone; the body subdividing, sometimes longitudinally, sometimes transversely, into two equal parts, each of which soon becomes a perfect individual (Fig. 88, D, E, F). This operation takes place with such rapidity under favourable circumstances, that it has been calculated by Professor Ehrenberg that no fewer than 268 *millions* might be produced in a month from a single *Paramecium*. The process of gemmation consists in the formation of little buds or offsets, like the young cells of Chara (Fig. 72), which gradually increase nearly to the dimensions of the parent, and then become detached; this is chiefly seen in the *Vorticellinæ*, and some other Polygastrica, which also multiply by fission. In the *Kolpodinæ*, however, a method of reproduction exists, which seems more like the propagation of the Algæ by Zoospores; for the whole body dissolves, as it were, into granular particles, each of which seems capable of becoming a new Animalcule. Nothing analogous to the 'conjugation' of the inferior Algæ, or to the sexual congress of higher animals, has yet been observed in them; and although certain minute contractile vesicles, observable in most Animalcules (Fig. 87, A *s t t*, B *s*, H *s*), have been considered by Professor Ehrenberg to be generative organs, there is no sufficient foundation for such an hypothesis. There is strong reason to believe that the process of cell-division cannot be continued indefinitely, and that a sexual reunion must consequently take place at some time and in some mode or other; and this should lead to protracted observation of the reproductive process in the Polygastrica, by which, perhaps, the mystery may be solved.—There are many cases, however, in which the individuals thus produced by fission or gemmation do not become completely detached from each other; but remain in more or less intimate connection. This is the case, for example, with some of the *Vorticellinæ*, in which an arborescent structure is produced by the partial subdivision of the stalks, with which the newly-developed individuals remain in connection, instead of separating themselves and forming new and distinct stalks. The *Ophrydium*, a genus nearly allied to the Vorticellinæ, exists under the form of a motionless jelly-like mass, of a dull, green colour, some specimens of it attaining a diameter of four or five inches; but this mass is made up of millions of distinct and similar individuals, imbedded in a gelatinous connecting substance, so as to remind us on the one hand of the gelatinous thallus of the lowest Algæ (§ 267), whilst, on the other, it seems to prefigure the analogous composite masses of the *Alcyonian* Zoophytes (§ 292) and of the compound Tunicata (§ 299). All the individuals of such an aggregation seem to have been produced by repeated fission from a single individual; and to the extension of the entire mass, there seems no limit, either as regards form or size. In the group of *Monadinæ*, again, we find several composite clusters, having considerable locomotive

power, whose production is entirely due to the continued attachment of the individuals that have been multiplied by fission; one of the most interesting of these, from the resemblance which it bears to the embryonic cell-clusters of higher animals, is the *Uvella*, or 'grape-monad,' which strongly resembles a transparent mulberry rolling itself across the field of view by the ciliary action of its component individuals. The most interesting examples of this kind of conformation, however, are those which are included in the family *Volvocinæ;* of which the common *Volvox* may be taken as the type. The nature of these beings was formerly altogether misunderstood; the bodies which are now known to be aggregate clusters, having been regarded as single individuals. The *Volvox globator* (Fig. 89), as its name imports, has a spherical form; and its diameter being often as much as 1-30th of an inch, it can be seen with the naked eye, under favourable circumstances, moving through the water which it inhabits. Its onward motion is usually of a rolling kind, but it sometimes rotates, like a top, without changing its position, whilst at other times it glides smoothly along, without turning on its axis. When sufficiently magnified, the *Volvox* is seen to be formed of a hollow sphere, composed of a very pellucid material, which is studded at regular intervals with minute green spots; and when these spots are examined,

FIG. 89.

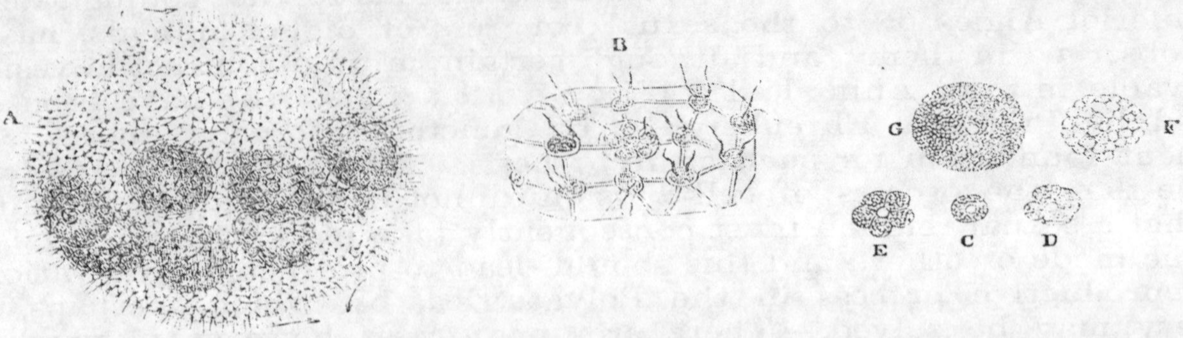

Volvox Globator:—A, the entire composite body; B, a portion enlarged; C, D, E, F, G, successive stages in the development of the embryonic (?) mass.

with a sufficient magnifying power, it is found that each of them is a monad-like Animalcule, giving origin to two filaments that project from the surface, and connected with its fellows by threads which proceed from it along the pellucid sphere (Figs. 87 A, and 89 B). Thus, the whole surface of the sphere is beset by vibratile filaments, to whose combined action its movements are due. Within the parent-sphere may generally be seen other globes of darker colour, varying in number from two to twenty. These are young formations of a like compound nature, originating in the subdivision of one of the monad-like individuals of the parent-sphere (as seen in the centre of Fig. 89, B), to the inner wall of which they remain for some time attached. After their detachment, they still remain for some time in the interior of the parent-sphere, and may be observed revolving in its cavity, by the action of their own ciliary filaments; and during this period, the number of individuals of which they are composed undergoes a considerable augmentation (Fig.

89, C, D, E, F, G). After a time, the parent-sphere bursts, and the young ones swim forth, and speedily become developed into its likeness. The dark colour of the immature Volvox is owing to the close proximity of its individual monads; the transparent portion of the sphere, which forms the interval between these, not showing itself until after its emersion from the cavity of the parent, and its development occasioning a great increase of size. Not unfrequently, a second generation may be observed in the interior of the spheres which are still included within the parent-volvox; and a third generation has even been distinguished within these. The family of *Volvocinæ* contains several other cluster-animalcules of the same general character, but presenting a great variety of forms; and it is in them, perhaps, that we find the highest examples any where to be met with among Polygastric Animalcules, of the union of a number of distinct elementary parts into *one whole*, possessing a definite form, and exhibiting actions peculiar to itself. Yet the individuality of the component monads obviously predominates over that of the composite cluster; and they have no relation of mutual dependence among themselves. The movements of the Volvox as a whole are due to the vibration of some hundreds or thousands of filaments, appertaining to separate Monads; and the perfect consentaneousness of these is an obvious indication of their purely *automatic* character.*

284*b*. The beings of this class are remarkable for their universal diffusion. Wherever any decaying organic matter exists in a liquid state, and is exposed to air and sufficient warmth, it will speedily be found peopled with minute inhabitants, of the most varied forms and diversified movements; whose numbers (as they are far too great to admit of being counted) can only be estimated by the data furnished by their size. Thus the *Monas crepusculum*, which presents itself in infusions of putrid flesh, is not more than 1-2000th of a line in diameter; and it frequently occurs in such dense crowds, that there is scarcely any space whatever among the individual animalcules. In such a case, a drop of water equivalent to a *cubic line* will contain *eight thousand millions* of these animalcules; whilst in a cubic inch there would be nearly *fourteen billions*. There can be no doubt, then, that they are by far the most numerous of all the existing forms of living beings; and the question naturally arises, how they come to present themselves in such vast numbers, in situations where none existed but a short time previously. Some have had recourse, for the solution of the difficulty, to the supposition that their germs form part, in a latent state, of the living tissues of Animal and Vegetable fabrics, by the decomposition of which they are called into activity; whilst others have even supposed them to originate in new and spontaneous combination of inorganic elements. There does not seem, however, to be any necessity for either of these suppositions; and they are both of them opposed to the fact, that unless ordinary atmospheric air be

* It is not by any means certain that the *Volvocinæ* are not to be transferred to the Vegetable kingdom. Their green colour leads to the suspicion that they decompose carbonic acid; and the stomachs described by Ehrenberg in the component Monads, are not more distinct than the stomachs which he has represented as existing in several other beings, whose Vegetable nature is now generally admitted. It is considered by Braun, who has paid much attention to the development of the inferior Algæ, that the *Volvocinæ* are of the same type with certain 'Zoospores,' which become composite by fissiparous multiplication. See Valentin, in Canstatt's Jahresbericht for 1848.

freely admitted to the decomposing infusion, no animalcules make their appearance in it. This has been ascertained by the experiments of Prof. Schultz, who kept infusions of decaying animal and vegetable matter in air which had been filtered (so to speak) by passing through a red-hot tube, or through strong sulphuric acid, by either of which processes any organic germs which it might contain must have been destroyed; and who found that no animalcules made their appearance under these circumstances, even after the lapse of some weeks; although they were seen in abundance after the free exposure of the same infusions to the atmosphere for a few hours only. It would hence appear that their dried bodies, or their germs, must be everywhere floating in the air, ready to develope themselves wherever the appropriate conditions are presented; and as a great number of kinds usually make their appearance successively, in different stages of the decomposition of the same infusion, it would seem as if each set of germs underwent development only when a particular combination of circumstances presents itself. As yet, however, somewhat of the same obscurity hangs over the origin of these Animalcules, as attends the question of the propagation of the inferior Fungi (§ 272); since it is by no means certain that the same germ may not develope itself into a variety of different forms, according to the conditions under which it is evolved. The idea of the diffusion of the germs of the Polygastrica through the atmosphere, seems incompatible with the fact that their bodies do *not*, like those of the more highly-organized Rotifera (§ 310) bear desiccation without the loss of their vitality; but if (as has been already stated to be probable on other grounds) they have another method of reproduction, of a proper sexual character, by which *ova* are produced, it is not in the least degree improbable that *these* should be taken up and wafted about in a perfectly dry state, since we find that the eggs even of the higher Mollusca will bear desiccation without the loss of their vitality.—It must not be supposed, however, that these Animalcules, as the name 'infusory' sometimes given to them would seem to imply, are confined to infusions of organic matter; for although they most abound in such, there is no collection of water, fresh, salt, or stagnant, in which they are not present; and they constitute the principal means of support to many of the larger aquatic animals, such as Zoophytes, the inferior Mollusca and Crustacea, and even (it would appear) to certain Fish. Thus their function in the economy of Nature is to appropriate organic matters that would otherwise pass into decomposition, and to bring them back into a state in which they may afford nutrition to higher animals.

285. Under the title of RHIZOPODA* is now designated a group of animals, which, though at present inconspicuous, has had a much more remarkable degree of development in former periods of the Earth's history, as we know from the vast accumulation of remains of them in its crust; the designation being given to them on account of the root-like character of the filamentous prolongations which constitute their sole locomotive organs (Fig. 91). In their simplest forms, we trace a close approximation to the *Amœba* (§ 283 *a*); their chief peculiarity being the consolidation of

* This term, first proposed by M. Dujardin, has been adopted as being much more appropriate than those of *Polythalamia*, or *Foraminifera*, which are only applicable to a part of the class.

the exterior of the body, usually by calcareous deposit, so that a kind of testa or shell is formed, within which the soft parts can be entirely withdrawn, whilst through one or more apertures these soft parts may put forth long digitate extensions, of no regular form or size, to which the term *pseudopodia* has been appropriately given (Fig. 90). As far as is yet known, there is neither mouth nor digestive cavity in these beings; but the entire body is either a single cell, or an aggregation of cells, deriving its nutriment by absorption through the external surface; and the extended pseudopodia seem rather to serve as rootlets whereby nutriment may be imbibed, than as instruments of locomotive action; these animals being usually very inert in their habits. The *Difflugia*, *Arcella*, and

FIG. 90.

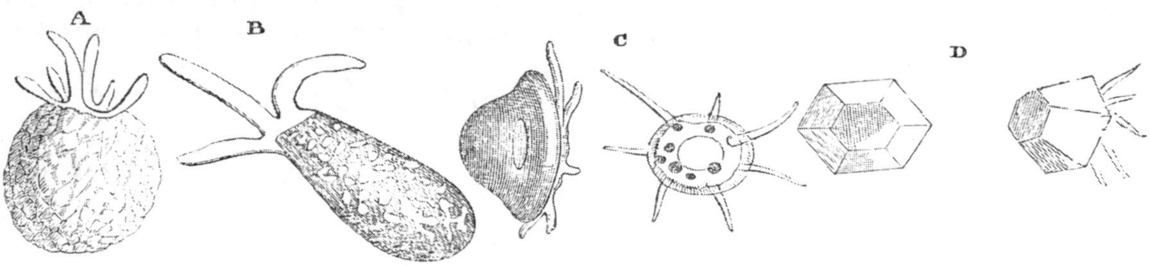

A, *Difflugia proteiformis*: B, *Difflugia oblonga*: C, *Arcella acuminata*: D, *Arcella dentata*.

Cyphidium of fresh waters, bear so close a resemblance to the *Amœba*, that the existence of a shell constitutes almost the only difference; but in the vast multitudes of marine forms which have been distinguished as *Foraminifera*, we find the pseudopodia more numerous, more slender, longer, and more subdivided. The greater number of the so-called Foraminifera are composite structures, formed in the manner to be presently described; but single-bodied species do exist among them; and of these, some are inhabitants of fresh water, and are obviously allied so closely to the amœbiform tribe (which was ranked by Ehrenberg among the Polygastrica), that little hesitation can be felt as to the propriety of bringing them together under one group. Whether any process analogous to sexual generation ever takes place among them, is as yet entirely unknown; their ordinary method of multiplication being obviously, like that of the Polygastrica, by spontaneous fission or gemmation. The new individuals which thus originate, remain, in a large proportion of cases, in connection with those from which they sprang; and thus composite structures are produced, the form of which has usually a considerable degree of regularity, being dependent on the mode in which the gemmation takes place. In most instances, the original form of the body, with its investing shell, seems to be globular or nearly so; but from some part of this, a bud springs forth, which in time becomes clothed in a testaceous investment of its own; from this bud another proceeds at a later time, and is in its turn, covered with a shell of its own. As the buds remain in connection with each other and with the parent, so do their newly-formed shelly coverings; and thus a composite shell is produced, which is 'polythalamous' or many-chambered, each chamber or cell giving lodgment to a distinct individual. The chambers may be arranged in a straight line, in which case the shell bears a strong resemblance to that of the *Orthoceratite;* or they may form a continuous spiral (Fig. 91), reminding us of the shell of the *Nautilus.* In fact, the strong resemblance of these minute shells to the chambered shells of existing and extinct Cephalopods (§ 306), for a long time led to the supposition that the animals which formed them must have belonged to that class; and it has only been recently, that the extreme simplicity of their structure has been ascertained. From the polythalamous shells of Cephalopods, however, those of Rhizopods may be distinguished by the absence of that single definite *siphuncular* aperture, by which a continuous passage is established from the mouth of the shell to the innermost chamber; and in place of it, we usually find only a few minute perforations in the septa (on whose presence it has been proposed to distinguish the entire group as that of Foraminifera), or, where a larger passage exists (as in *Nummulina*), it is manifestly formed by the confluence of minute apertures, and is not such as to allow of free communication between one chamber and another.* It has been supposed by some, that these apertures serve for the passage of an intestinal tube, and that the relation of the soft parts lining the chambers is that rather of segments of one individual body, than of a succession of distinct individuals. No such intestine, however, has ever been demonstrated; and it is certain that the apertures are in many instances far too small to give it passage, their size merely admitting the extension of minute thread-like processes of the gelatinous body, from which, as from the 'stolons' or creeping stems of Plants or Zoophytes, new gemmæ may originate. Moreover, whilst the individuals contained within the successive chambers of these composite shells have no such ready communication with that which inhabits the outer chamber, as would enable them to draw their nourishment through it, they are provided with a much

FIG. 91.

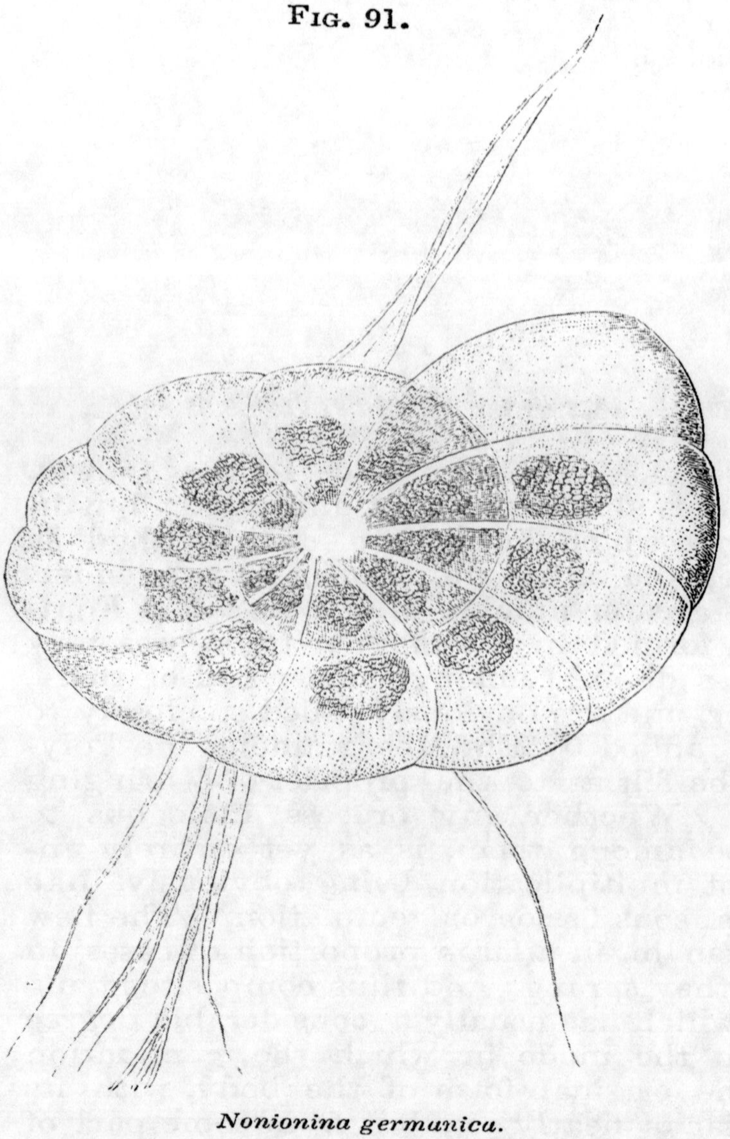

Nonionina germanica.

* See a Paper by the Author, on the Genera *Nummulina*, *Orbitolites*, and *Orbitoides*, in the Quarterly Journal of the Geological Society, Feb. 1850.

more direct means of imbibing that which they require for their support; for each chamber has a set of apertures in its side-walls, through which the soft pseudopodia can be protruded; and there is evidence that even where, as in *Nummulina*, each successive whorl of the shell completely invests those which preceded it, the pseudopodia of the oldest and innermost chambers can pass directly to the exterior, so that their soft parts can still be nourished by direct absorption from without. In the *Nummulites*, which were comparatively gigantic forms of this group, especially numerous at the commencement of the Tertiary period, the shell itself was everywhere minutely porous, apparently to permit the free imbibition of water. And the same appears to have been the case with that of the remarkable genus *Orbitoides*, which has been confounded on the one hand with true Nummulites, and on the other with *Orbitolites*, coralline structures probably formed by Bryozoa (§ 298).—The species of this group at present in existence are of very minute size, the largest of them being only just visible to the naked eye. Of such, however, a considerable part of the Chalk formation appears to be composed; and a formation of the same kind is at present taking place on the sea-bottom of the Levant. Vast beds of tertiary limestone, chiefly made up of Nummulites and Orbitoides, and of their comminuted fragments, are found extending across the south of Europe, from the Pyrenees, through Asia Minor, to the East Indies; and similar beds present themselves also in North America. And even in the Palæozoic strata, extensive beds of limestone are met with, which are entirely composed of a Rhizopod shell, termed *Alveolina*. The importance of this group in the economy of Nature, therefore, must not be estimated by the apparent insignificance of its existing forms.

286. Under the designation PORIFERA are grouped together the various forms of the *Sponge* tribe; this designation being founded upon the only obvious character that is common to them all,—namely, the perforation of their substance by minute channels. These pores lead to a set of intercommunicating passages and excavations, more or less regular in form; and these are usually connected also with a set of larger canals, which ramify through the mass, and terminate in a set of larger orifices, the *oscula* or vents; in some Sponges, however, no regular vents or ramifying canals can be found. The fabric of Sponges is made up of two distinct parts; namely, the fibrous skeleton, and the animal flesh which clothes it. The skeleton is composed of an irregular reticulation of fibres, the structure and composition of which vary considerably in the different sections of the group. In the genus *Spongia*, of which the Sponge of commerce may be taken as a type, the fibre is solid, and is for the most part *Keratose* or horny; minute siliceous spicules, however, being imbedded in its substance, as was first shown by Mr. Bowerbank. In *Verongia*, to which genus many fossil as well as some recent sponges belong, the horny fibre is tubular. In *Halichondria*, a genus which includes a large proportion of the sponges of our own shores, the horny portion of the fibre is much less tenacious, whilst the siliceous spicules are much increased in number, being aggregated together in bundles, and forming a considerable portion of the skeleton. In the genus *Duseideia*, we find, instead of spicules regularly formed by the organic processes, grains of sand agglutinated together, and imbedded in the substance of

the fibre. And in the curious and beautiful *Dictyochalix pumiceus* of Barbadoes, the whole fibrous skeleton is composed of silex, which is so transparent, that it has the appearance of being formed of spun glass, worked into a reticulation resembling that of the ordinary sponge. The skeleton, of whatever materials it is composed, has the same texture in every part of the body; and its fibres are so arranged as to give support to the soft gelatinous matter, which constitutes its living formative tissue; whilst they also form the boundaries of the various pores, channels, and vents, which are excavated in its substance. The spicules of the fibrous skeleton are usually more or less fusiform or needle-shaped; sometimes, however, having a kind of head like that of a pin at one end, whilst they are pointed at the other. That they are not, like the raphides of Plants (§ 23), formed by simple crystalline action, seems proved not merely by the peculiarity of their form (Fig. 89, H, I), but by their possession of an internal cavity, which has been shown by Mr. Bowerbank to contain organic matter. Hence it is probable that each spicule was originally an elongated cell, which has secreted siliceous matter either on the outer surface or on the lining of its wall. Other spicules, however, are found in the soft animal tissue, and especially in the integument, which is frequently of denser texture than the internal substance; these spicules are usually *stellate* in their form, sometimes having three radii, but often more; and like the preceding, they contain a central cavity, which is continuous through all the radii, as if they had been formed by the silification of a stellate cell (§ 146). The spicules, though usually siliceous, are calcareous in the genus *Grantia;* in which the usual fibro-siliceous skeleton would seem to be deficient, or very imperfectly developed; the pores and canals being simply excavations in the gelatinous substance, instead of having regular boundaries.

286 *a*. The fleshy parenchyma which clothes the skeleton is usually so extremely soft, as to drain away when the mass is removed from the water, like the white of an egg, or the vitreous humor of the eye; in *Verongia*, however, it is of a firmer texture, not unlike that of the substance of the liver. From the observations of MM. Dujardin and Gervais, but more especially from those recently made public by Mr. Carter,* it appears that this parenchyma may be really regarded as made up of an aggregation of cells, of about the 1000th of an inch in diameter, each of which has the characters of a distinct Animalcule, having a certain power of spontaneous motion, obtaining and assimilating its own food, and altogether living *by* and *for* itself, except so far as it may contribute materials for the formation of the fibro-siliceous skeleton. The form of this sponge-cell, as shown in the accompanying delineations (Fig. 92) is continually changing. When first separated from the mass, it is usually globular (A *a*); but it soon puts forth extensions of its wall, in the form of obtuse or globular projections, or digital or tentacular prolongations (B, G); and by the attachment of the extremities of these extensions, whilst they are gradually taken back again (so to speak) into the body of the cell, the latter is caused to move across the field. If in this progression it meets another cell, the two combine; and if more are in the immediate neighbourhood, they unite together into one common globular mass. The

* Annals of Natural History, August, 1849.

'protean' sometimes spreads its cell-wall over bodies with which it comes into contact, such as a portion of a spicule, the body of an animalcule, or even a sluggish or dead one of its own kind (C, D, E, F); and this expansion may take place to such an extent, that the sponge-cell will completely enwrap the object, which would thus appear (if the method of

FIG. 92.

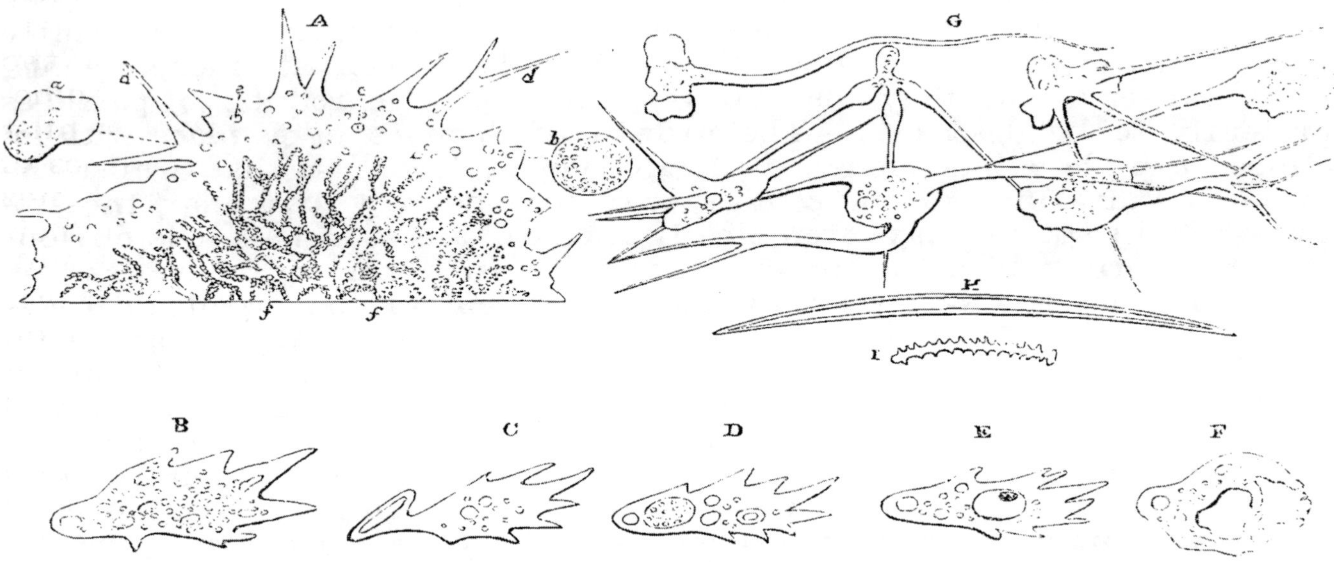

Structure and development of *Spongilla*, or Fresh-water Sponge:—A, magnified view of a portion grown in water from a seed-like germ, showing *a* a separated sponge-cell highly magnified, with its granules and hyaline vesicles, *d*, *d* its thin irregular edge, containing *e*, *e* hyaline contracting vesicles; *f*, *f*, sponge-cells *in situ*:—B, denticulated proteus in progression, showing its granules and hyaline vesicles;—C, magnified view of a denticulated proteus, with a portion of a spiculum in a fold of its cell-wall;—D, another proteus with a loricated animalcule and a sponge-germ(?) similarly included;—E, another proteus with a transparent cavity, to the side of which a bud seems attached;—F, a proteus in the act of surrounding a foreign body;—G, remarkable forms of proteans developed from the seed-like bodies; H, general form of large spiculum; I, peculiar spiniferous spiculum.

enclosure had not been watched) to be embedded in its interior. The individual sponge-cells, when *in situ*, are connected by a gelatinous intercellular substance, which seems to possess in itself the property of growth and extension into various forms (A, *d*), being apparently a kind of 'nucleated blastema,' such as constitutes the most rudimentary form of tissue among the higher animals (§ 158). Certain phenomena are exhibited by the entire mass, however, which show that it is not a *mere* aggregation of individual cells, but that the separate vital actions of these are made to concur in the performance of operations which are for the welfare of the whole. These will be now described.

286 *b*. In their living state, Sponges appear to be continually imbibing fluid by their minute pores; this passes through the communicating cavities of the interior of the body, until it reaches some one or other of the ramifying canals; and having entered this, it is carried back towards the surface, and discharged by one of the vents. It has been hitherto found impossible to assign a cause for this movement; no cilia having been discovered in any part of the surfaces, internal or external; and no such contractions or dilatations being visible, as would in the least account for it on simply mechanical principles. It must be probably regarded as one of those cases, hereafter to be considered (CHAP. XII.), in

which a movement of fluid is kept up by forces arising out of its relations to the living tissues over which it passes. It would appear to be from the minute organic particles contained in the surrounding water, and thus carried through its substance by the imbibition of that liquid, that most Sponges derive the materials of their growth; but there is some evidence that the *Spongilla fluviatilis*, or River-sponge, acquires a green colour, and decomposes carbonic acid, under the influence of light, after the manner of Plants. The currents ejected from the vents are frequently seen to carry out delicate flocculi, which probably consist of excrementitious matter; and they also convey forth the reproductive particles presently to be described.—The growth of Sponges takes place with an almost complete independence of any typical form; so that, whilst the forms of individuals of the same species may be very dissimilar, those of two different species may often closely approximate. In fact, it is obvious that the form which the mass may take is usually more determined by external circumstances, than by causes inherent in the organism itself; and no reliance can be placed on it, therefore, as furnishing characters for the classification of these bodies, which must be sought in the structure of their minute components. Still there is a certain mode of growth which is sufficiently characteristic of most of the species, to enable such as have become familiar with them to distinguish them by their external aspect; one of the chief differences among them being the consistence of the tegumentary investment, which seems in some degree to determine the mode in which the vents open upon it. Thus if the texture be loose and fibrous, it yields easily to the outward current, and the oscula are level with the surface, or nearly so; if it be more compact, the integument is raised into a papillary eminence, on the centre of which the vent is found, like the crater on the summit of a volcano; whilst if it be too dense to yield to the pressure from behind, the oscula are not elevated. But even this character is liable to be modified by external agencies; "for the littoral sponge, which, in a sheltered hollow or fringed pool, will throw up craters and cones from its surface, may only be perforated with level oscula, when it is swept over and rubbed down by the waves at every tide." No distinct indications of sensibility have been obtained, either by watching the ordinary actions of the Sponge, or by experiments made to test it. The mass may be torn, cut, burned, or otherwise irritated, without performing respondent movements; but slow changes of form and size have been occasionally observed to take place in the oscula, both of marine and fresh-water sponges, when these have been measured at intervals of some minutes. These changes, however, evidently result from causes (such as removal from the water for a short period) which affect the whole mass alike; and do not in the least indicate such a communication between the different parts, as is necessarily involved in the idea of a particular *sensorium* in which such impressions can be *felt*.

286*c*. The multiplication of Sponges seems to be provided for in two ways; in the first place by the production of *gemmules*, which seem like detached portions of the gelatinous flesh lining the canals, and which, being furnished with cilia, issue forth from the vents, and transport themselves to distant spots where they may lay the foundation of new sponges; and, secondly by *capsules*, which are bodies of larger size, frequently having a very peculiar investment strengthened by siliceous particles, and containing numerous globular bodies that are set free by the rupture of their case, and may develope themselves each into a new individual. The former are produced during the period of most active growth of the sponge, and must be considered, like the *zoospores* of the simpler Algæ (§ 267 *a*), to which they are closely analogous, in the light of *gemmæ* or buds. The latter, on the other hand, are developed (like the ova of the Hydra, § 290) towards winter, when the nutritive operations are failing, and when in many cases the whole parent-structure is about to die, so that on them alone depends the continuance of the race; they must be pretty certainly regarded as the analogues of the true *ova* of higher Animals; but of the manner in which they are produced, nothing can yet be stated from actual observation.—The claim of Sponges to a place in the Animal kingdom has been much canvassed. They certainly seem to be destitute of that which is generally regarded as the essential characteristic of Animals, namely a consciousness of external impressions, and the power of executing respondent movements. But, on the other hand, they certainly appear to live at the expense of matter previously organised; and even if it should be proved that the River-sponge possesses some traces of the combining power which is characteristic of Plants, this would merely serve to indicate that it is more closely related to the Vegetable kingdom, than are the other members of the group. The general character of the fabric is decidedly more Animal than Vegetable, both in point of chemical composition, and structural arrangement. For, not to speak of the fleshy parenchyma (whose composition has not yet been ascertained), it is certain that nothing like the horny fibre of the 'keratose' Sponges is met with among Plants, whilst similar horny matter is a common constituent of Animals. Furthermore, we have in the system of communicating excavations and ramified canals, through which the nutrient material is made to pass, the first sketch of a digestive apparatus, of which the similar parts are indefinitely multiplied; and we shall find that traces of the same arrangement are discernible in the complex and extensive ramifications of the digestive cavity in some higher Animals; whilst no arrangement of the kind exists in Plants. There is nothing that is characteristic of either kingdom in the reproductive process.—We may trace in the Sponge the first manifestation of that tendency to the individualization of a composite mass, which is exhibited by the simpler Algæ amongst Plants (§ 266); for whilst each component sponge-cell must be regarded as a repetition of the rest, and the different parts of the aggregate mass are altogether similar and enjoy a separate vitality, still a Sponge is *one whole*, and is not a *mere* aggregation of independent elements, as is proved by the connection of the life of the individual parts with the continuance of the currents which circulate through the mass.

287. Excluding the two preceding groups, the sub-kingdom RADIATA is composed of three classes, in all of which the *radiated* type of conformation predominates. This arrangement is not always to be looked for, however, in those entire fabrics, the continuity of whose parts might lead us to regard them as single individuals; for in by far the larger proportion of the group of *Polypifera*, or true 'Zoophytes,' such fabrics are *composite*, being made up by the aggregation of parts which are all exactly similar to each other, and which are capable of maintaining an independent existence; and it is in each of these component parts that the radial

symmetry exists, the form of the aggregation being extremely variable. In this respect, then, there is an obvious analogy to the higher forms of the Vegetable Kingdom; the relation of the several polypes to each other, and to the 'polypidom' to which they belong, being (as we shall presently see) very much the same as that of the leaves and flowers of a Phanerogamic plant to each other, and to its central axis. In each case, the composite structure is produced by a process of 'gemmation,' or budding, from a single individual; this may take place to an indefinite extent; and the mode in which it occurs is the chief determining cause of the particular plan or type of growth, which is traceable in each species, but which is liable to such great variation from the influence of external causes, as to produce great difficulty in the determination of what are really identical and what dissimilar. This tendency to the production of composite structures by gemmation is almost entirely absent in the higher classes of *Acalephæ* and *Echinodermata;* for although instances are not wanting of the production of gemmæ in the former class, yet these gemmæ are usually detached when mature, and thenceforth rank as completely distinct individuals; whilst in the latter class, the production of new individuals by gemmation is not known to occur at all. In them, consequently, we may always look for radial symmetry in the entire fabric; and although it is not always perfectly exhibited, yet there are very few cases (among the animals undoubtedly belonging to the group) in which it is not predominant. The chief departures from the radial type are presented among certain tribes which are usually ranked among the Acalephæ, but which are considered by some distinguished Zoologists as being rather degraded forms of Mollusca; and among some of the higher Echinodermata, which present, in their tendency to *bi-lateral* symmetry and to elongation of the body, a manifest approximation to the Annulose series (§ 296).—The *radial* symmetry must be regarded as a *vegetative* character, for it corresponds with that which is seen in the disposition of the appendages around the axis in the leaf-buds and flower-buds of Plants (§ 278); and it is intimately connected with another vegetative character, *the repetition of similar parts,* which, although observable also in the lower Articulata, is especially characteristic of this group. Thus, in all true *radiated* animals, we find the parts arranged in one or more circles around the mouth to be exact repetitions of each other; whether these be merely tentacular appendages, as those of the *Polypes* (Figs. 93, 97, 101), of the *Medusæ* (Fig. 106), or of the *Holothuria* (Fig. 112); or be divisions of the body itself, each ray containing a portion of the vital organs, as in the *Star-fish* (Fig. 110). In the class *Acalephæ,* we find the number *four* and its multiples to predominate, and among the *Echinodermata,* the number *five;* whilst among *Zoophytes* there seems to be no fixed standard, the smallest number (seen in some of the *Asteroid* polypes) being *six,* in others of the same group (Fig. 102) *eight,* in the *Hydraform* polypes (Figs. 93, 95) from *six* to *twenty* in a single whorl, but the whorl being sometimes double, and in the *Actiniform* polypes (Fig. 97) the number in each whorl being much greater, and the whorls more multiplied. We shall have occasion to notice the same tendency to repetition in the internal organs.

287 *a.* The Radiated series presents us with a remarkable diversity, among the different types which it includes, both as to general elaborateness of structure, and as to the development of particular tissues and

organs. This will become at once apparent on the comparison of the simple *Hydra* (Fig. 93), which consists merely of a stomach and appendages composed of the most primitive form of tissue, with the complicated *Holothuria* (Fig. 113), in which a great variety of organs and tissues presents itself to the most superficial observer. It will be useful to follow this comparison somewhat more into detail.—The digestive apparatus among Zoophytes consists of a cavity having but a single entrance, that serves alike for the introduction of food and for the expulsion of undigested or excretory matters; this cavity is simple in the solitary species (as in the *Hydra*, Fig. 93, and in the *Actinia*, Fig. 98); but in many of the composite forms, of which the greater part of the group is composed, the bottom of the stomach of each polype is extended into the general mass or 'polypidom,' and comes into connection with similar prolongations of the stomachs of other polypes, so that the whole forms one complex cavity (as in the *Campanularia*, Fig. 95, and the *Alcyonidium*, Fig. 102). In the *Acalephæ* there is a great variety in this respect; for in some the stomach has but a single orifice either for ingestion or egestion (as in the *Velella*, Fig. 108); in others it has an anal outlet distinct from the oral orifice (as in the *Beroe*, Fig. 107); in others, again, the oral orifices are very numerous, whilst no distinct anal orifice can be found (as in the *Rhizostoma*, and probably also in the *Physalia*, Fig. 109); and finally there are some, and these among the most characteristic forms of the entire group, in which the oral orifice is single, whilst the anal outlets are multiplied (as in the *Cyanœa*, Fig. 106). In most of these, the digestive cavity is little else than an excavation in the gelatinous tissues of the animal. In all the *Echinodermata*, however, we find a digestive sac possessing distinct membranous walls of its own, and suspended in the midst of the general cavity of the body: in the *Star-fish* (Fig. 110), however, it has but a single orifice, and sends prolongations into the lobes of the body (which are commonly, but erroneously, termed the arms); whilst in the *Echinus* (Fig. 111) and in the *Holothuria* (Fig. 112) there is a distinct anal orifice, situated at the opposite extremity of the body from the oral.—Of the proper circulating apparatus, no distinct traces are found in *Zoophytes*, the transmission of liquids from one part of the fabric to another being accomplished, in the solitary species, simply by *lacunæ*, or irregular interspaces, burrowed out in the tissues; whilst the connection of the several parts of the composite fabric is accomplished, either by similar lacunæ in the connecting tissue (as would seem to be the case in the compound *Helianthoida*), or by the extension of the digestive cavity into a system of canals (as in the compound *Hydroida* and *Asteroida*). It is by the prolongations of the digestive cavity alone, that nutritious fluid seems to be carried into the remoter parts of the fabric of the *Acalephæ;* no distinct system of vessels having been yet proved to exist in that class. In all the *Echinodermata* whose anatomy is known, however, such vessels exist; and they are frequently found to have a very complex and extensive distribution (as in the *Holothuria*, Fig. 113), even though no distinct heart is yet developed.—So, again, a distinct respiratory apparatus is altogether wanting in *Zoophytes* and *Acalephæ*, and even in the lower *Echinodermata;* but in the higher forms of the latter class (as the *Holothuria*) we find an elaborate organ set apart for the aëration of the circulating fluid.—The generative apparatus exists under its simplest possible condition in the *Hydra*, the sperm-cells and germ-cells being developed in the midst of the general substance of the body; but in the *Actinia*, we not only find special organs set apart for their production, but also a provision for the nurture of the embryo for some time after its emersion from the egg; the sexes, however, are still united in the same individual, and it is not until we pass to the *Acalephæ* and the *Echinodermata*, that we find them distinct.—Finally, of a nervous system, special organs of sense, and true muscular tissue, no traces whatever have yet been detected in *Zoophytes;* the indications of them in the *Acalephæ*, are obscure, and far from constant; but no doubt of their existence in the *Echinodermata* can be entertained, although their actions probably form but a very small part of the entire operations of the complex fabric of these animals.

288. No doubt can any longer exist as to the Animal character of the beings composing the class of POLYPIFERA; although the Plant-like aspect which many of them present, deceived the Naturalists of former days, as it does the uninformed observer at the present time, into the belief of their Vegetable nature. In the works of the older Botanists, the *Zoophytes*,* whether hard and stony, or flexible and horny, were arranged and described with Sea-weeds and Mosses, without any misgivings as to the propriety of doing so. Even the Sea-Anemone was described as a veritable flower; and the analogous beings found on living Corals and Madrepores were spoken of as the *blossoms* of these stony plants. This resemblance arises out of two peculiarities in their mode of growth; in the first place, from that *radial symmetry* which is so exact in the arrangement of the tentacula surrounding the mouths of the individual polypes (Figs. 97, 102), and which corresponds so closely with that of the component parts of a 'regular' flower; and secondly, from the extension of the fabric by the process of *gemmation*, the gemmæ retaining their connection with their parent, just as the buds of a growing tree remain in continuity with the parent stem, a more or less ramifying structure being thus produced in both instances.—Although the Animal character of Zoophytes has been almost universally admitted among Naturalists for more than a century, and the structure and mode of life of particular species have been carefully investigated, it has been only of late that the general character of the class has been understood. A large proportion of those compound structures, which are known as *Coral, Madrepore*, &c., come under the inspection of Naturalists only in their dry state, stripped of all that characterises the living animal. Some of these are massive and stony (*Lithophytes*); others of more delicate conformation and of horny consistence (*Keratophytes*). Some of them, again, serve as a central axis or stem, which is clothed with the living flesh; whilst others form a tube, which encloses the softer tissues;—this variety being found both among the *lithophytes* and the *keratophytes*. Moreover, there are some polypes which are never developed into associated groups; either producing gemmæ which become detached when they have attained the power of independent existence, as is the case with those of the *Hydra* (§ 289); or not propagating by gemmation at all, which is

* This term has been applied to the whole of the Radiata, but may be advantageously restricted to this class.

the case with the ordinary Actiniæ (§ 291). In the first attempts to classify the group, these latter forms, together with a few composite Zoophytes which are not furnished with a firm skeleton of any description, were arranged in an order by themselves; and the remainder were divided and subdivided according to the stony or horny texture of their axis, and its internal or external position. By this method, however,—as by any which depends upon a single set of characters, and which is therefore *artificial* instead of *natural*,—beings the most dissimilar were associated together; and those which were really allied most closely, as regards the general structure of the individual polypes, were separated, on account of a dissimilarity in the material by which the aggregate structures are consolidated, and in the mode in which their buds and branches are put forth. It is now universally admitted, however, that the structure of the individual Polypes, and their mode of connection with each other, are the characters of greatest importance; and that all those derived from the structure and composition of the axis are but secondary. Putting aside the *Bryozoa*, which have been ranked until very recently as Zoophytes, but which have now been satisfactorily proved to be truly Molluscan in their character (§ 298), the class of *Polypifera* or *Zoophytes* may be divided into three Orders, each of which will need to be separately considered, in order that the character and relations of the entire group may be properly understood. Of the first of these, that of *Hydroida* or 'Hydraform' Polypes, the common *Hydra*, or 'Fresh-water Polype', may be taken as the type; the composite fabrics which it includes consisting of repetitions (so to speak) of polypes, having the same general conformation with it. In like manner, the order *Helianthoida* or 'Actiniform' Polypes consists of such Zoophytes, whether single or composite, as are constructed upon the same general plan with the common *Actinia* or Sea-Anemone. And lastly, the order *Asteroida*, or 'Alcyonian' Zoophytes (which contains no solitary forms), is made up of those which, in the structure and mode of connection of their component polypes, correspond with the common *Alcyonium*. Of these we may regard the *Helianthoida* as the 'type' of the class; not merely on account of the large proportion of the known species which it includes, but also because it presents all the peculiar characteristics of the class in the most complete development, and is the furthest removed from any other group. On the other hand, we shall find that by the *Hydroida* we are led towards the class of *Acalephæ;* whilst by the *Asteroida* we are carried back to the *Porifera*. As the *Hydra*, however, presents us with the *polypoid* structure in its greatest simplicity, it will be advantageous to consider it in the first instance.

289. Of the *Hydra*, or 'Fresh-water Polype,' two species are commonly known, the *H. viridis* or green polype, and the *H. fusca* or brown polype; these differ not merely in their colour (this being liable to some variation according to the nature of the food on which they have been subsisting), but still more in the length of their tentacula, which in the former are scarcely as long as the body, whilst in the latter they are, when fully extended, many times longer (Fig. 93). The body of the Hydra consists of a simple bag or sac, which may be regarded as a stomach, and which is capable of varying its shape and dimensions in a very remarkable degree; sometimes extending itself in a straight line, so as form a long narrow

cylinder, at other times being seen (when empty) as a minute contracted globe, whilst, if distended with food, it may present the form of an inverted flask or bottle, or even of a button. At the upper end of this sac is a central opening, which is commonly termed the 'mouth;' and this is surrounded by a circle of tentacula or 'arms,' usually from six to ten in number, which are arranged with great regularity around the orifice, diverging from it at their bases like the spokes of a wheel; so that the orifice with its appendages, when seen from above, exhibits a truly radiated arrangement. Strictly speaking, however, the so-called 'mouth' should rather be likened to the entrance to the stomach (the 'cardiac orifice') of higher animals; and the tentacula to the muscles of the pharynx and œsophagus, which grasp the food presented to them, and convey it downwards to the digestive cavity. The distinction is not without importance, as will be seen when we come to inquire into the import of the movements which this animal exhibits. — The body is prolonged at its lower end into a narrow base, which is furnished with a suctorial disk; and by means of this, the Hydra usually attaches itself to some fixed object, most commonly to the stems and leaves of aquatic plants, whilst it allows its tendril-like tentacula to float freely in the water, like so many fishing-lines.—The wall of the stomach has been stated to be composed of several layers of cells;* but the more correct account of it is that which describes it as a kind of 'nucleated blastema,' consisting of an amorphous, homogeneous, semi-transparent substance, through which granular particles, apparently cells, and cell-nuclei, are dispersed, and having its continuity interrupted by cavities, which inosculate so as to form an irregular network, wherein fluids are seen to move. This substance may be divided into three layers, which do not differ from each other, however, in any essential particular, and the reticulated channels of which are continuous from

FIG. 93.

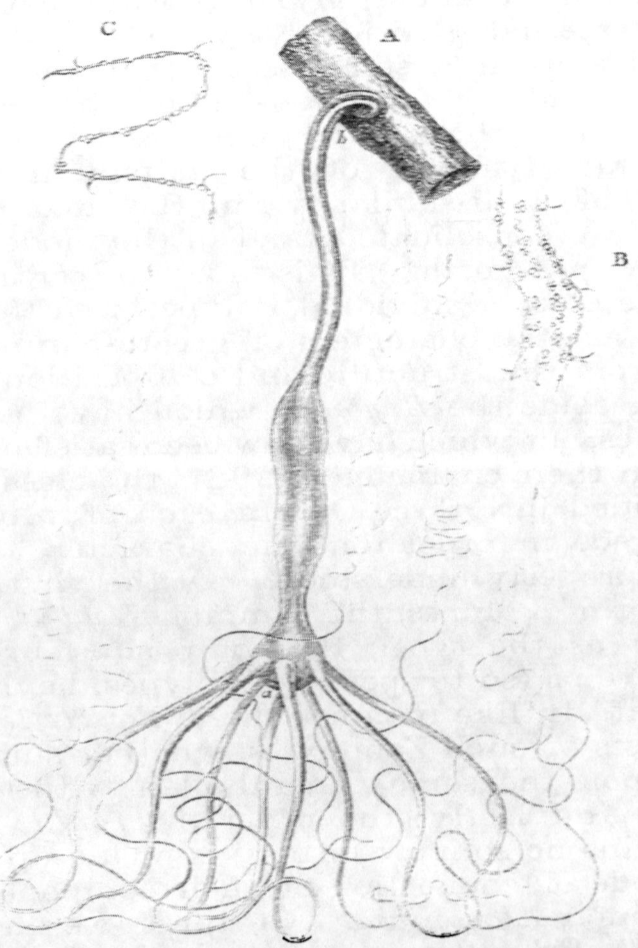

A, *Hydra fusca*, or Brown Fresh-water Polype, attached to a piece of stick, with its arms extended, as in search of prey; *a*, the mouth surrounded by tentacula: *b*, foot or base, with its suctorial disk: at B is seen a portion of one of the arms near its origin, and at C another portion near its termination, more highly magnified.

* See Corda on the structure of the *Hydra fusca*, in the Acta Acad. Naturæ curios. Vol. xviii. The Author is satisfied from his own observations, however, and from the concurring testimony of other British Microscopists, that on this and some other particulars of Corda's description, no reliance is to be placed.

one to another.* In the arms, however, we find a more complicated organisation, which has many features of interest. Their internal structure seems to be nearly the same with that of the parietes of the stomach; and, like the latter, they contain irregular cavities in which fluids are seen to move. Their surface, however, is covered with a set of wart-like protuberances; in the centre of every one of which is seen a long and sharp calcareous spiculum, that is capable of being protruded or retracted by an apparatus at its base; whilst around this is a cluster of little vesicles, each of them furnished with a very minute filament, so that the whole protuberance has a bristly appearance (Fig. 93, B, C). This apparatus, repeated many times on each tentacle, is doubtless intended to give to the organ a great prehensile power; the minute filaments forming a rough surface adapted to prevent the object from readily slipping out of the grasp of the arm; whilst the central spiculum is projected into its substance, and probably conveys into it a poisonous fluid secreted by a vesicle at the base of the dart. The latter inference is founded upon the oft-repeated observation, that if the living prey seized by the Hydra have a body destitute of hard integument, as is the case with the minute aquatic worms which constitute a large part of its aliment, it speedily dies, even if, instead of being swallowed, it escapes from their grasp; this result having even been observed when the worm has merely touched one of the tentacula, whilst itself freely swimming through the water. On the other hand, minute Entomostracous Crustacea, Insects, and other animals with hard envelopes, may escape without injury, even after having been detained for some time in the grasp of the polype. The contractility of the tentacula is very remarkable, especially in the *Hydra fusca;* whose arms, when extended in search of prey, are not less than seven or eight inches in length; whilst they are sometimes so contracted, when the stomach is filled with food, as to appear only like little tubercles around its entrance. By means of these instruments, the Hydra is enabled to derive its subsistence from animals, whose activity, as compared with its own slight powers of locomotion, might have been supposed to remove them altogether from its reach; for when, in its movements through the water, a minute worm or a water-flea happens to touch one of the tentacula of the polype, spread out as these are in readiness for prey, it is immediately seized by this, other arms are soon coiled around it, and the unfortunate victim is speedily conveyed to the stomach, within which it may frequently be seen to continue moving for some little time. Soon, however, its struggles cease, and its outline is obscured by a turbid film, which gradually thickens, so that at last its form is wholly lost. The soft parts are soon completely dissolved, and the harder indigestible portions are rejected through the mouth. A second orifice has been observed at the lower extremity of the stomach; but this would not seem to be properly regarded as anal, since it is not used for the discharge of such exuviæ; it is probably rather to be considered as representing, in the Hydra, the entrance to that ramifying cavity, which, in the compound Hydroida, brings into connection the lower extremities of the stomachs of all the individual polypes (Fig. 95).—A striking proof of the simplicity of the structure of the *Hydra*, is the fact that it may be turned inside-out like a

* See Prof. Ecker, in Ann. des Sci. Nat. Zool. Dec. 1848.

glove; that which was before its external tegument becoming the lining of its stomach, and *vice versâ*. Another very curious endowment seems to depend on the same condition,—the extraordinary power which one portion possesses of reproducing the rest. Into whatever number of parts a *Hydra* may be divided, each will retain its vitality, and give origin to a new and entire fabric; so that thirty or forty individuals may be formed by the section of one. The ordinary mode of reproduction in the animal bears some analogy to this. Little bud-like processes are developed from its external surface, which are soon observed to resemble the parent in character, possessing a digestive sac, mouth, and tentacula; for a long time, however, their cavity is connected with that of the parent, but at last the communication is cut off, and the young Polype quits its attachment, and goes in quest of its own maintenance. A second generation of buds is sometimes observed in the young Polype, before quitting its parent; and as many as nineteen young Hydræ, in different stages of development, have been seen thus connected with one central Polype (Fig. 94).—This process is obviously analogous to the propagation of certain Plants by *gemmæ*, which become detached as soon as they have acquired the power of maintaining an independent existence (§ 275); and its occurrence in this animal will afford us the clue to the explanation of certain other phenomena hereafter to be described.—The *Hydra* also propagates itself, however, by a truly sexual process; the fecundating apparatus, or vesicle producing 'sperm-cells,' and the ovum (containing the 'germ-cell,' imbedded in a store of nutriment adapted for its early development,) being evolved in the substance of the parietes of the stomach, the former just beneath the arms, the latter nearer to the lower end of the body. It would appear that sometimes one individual Hydra developes only the male cysts or sperm-cells, while another developes only the female cysts or ovisacs; but the general rule seems to be, that the same individual forms both organs. The fertilisation of the ova, however, cannot take place until after the rupture of the spermatic cyst and that of the ovisac also; so that the parent has no further participation in it, than has the *Fucus* in the analogous fertilisation of its germ-cells after their discharge (§ 268). Although the production, from such an egg, of a new Hydra, similar in all respects to its parent, has not yet been witnessed, there seems no reason to doubt the fact. It would seem that this alternation in the method of reproduction, between

FIG. 94.

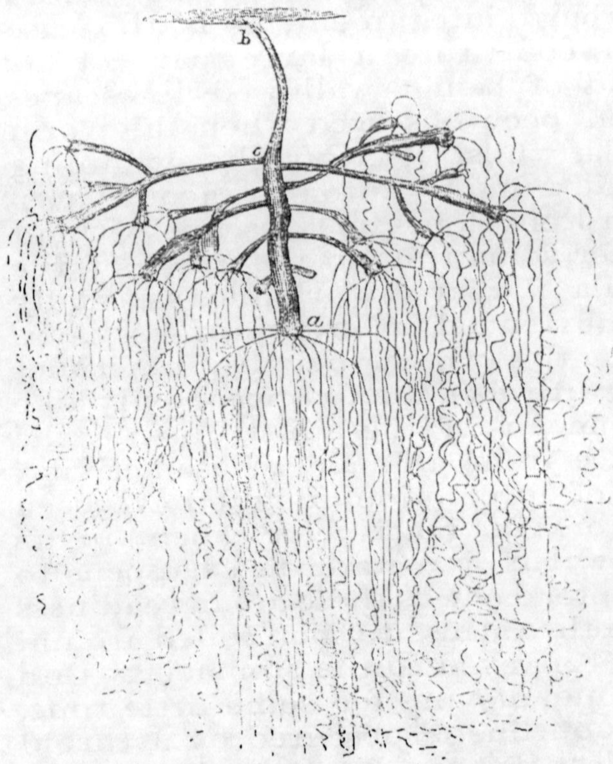

Hydra fusca in gemmation; *a*, mouth; *b*, base; *c*, origin of one of the buds.

the gemmiparous and the sexual, is greatly influenced by external temperature; the eggs being produced at the approach of winter, and serving to regenerate the species in the spring, the parents not being able to survive the cold season; whilst the budding process naturally takes place only during the warmer part of the year, but may be made to continue through the whole winter, by keeping the water inhabited by the polypes at a sufficiently high temperature.—The Hydra possesses the power of free locomotion, being able to remove from the spot to which it has attached itself, to any other that may be more suitable to its wants; and this it is able to effect, as will be hereafter explained (CHAP. XIX.), in various modes. Its changes of place, however, seem rather to be performed under the influence of *light,* towards which the Hydra seeks to move itself, than with reference to the search after food; there is no reason, however, to suppose that the Hydra is conscious of its presence *as* light, since no rudiments of organs of vision can be detected in it; and it will be shown hereafter, that in these as well as in its other movements, there is much more of a vegetable than of a proper animal character.

290. Some of the simpler forms of the composite *Hydroida* may be likened to a *Hydra,* whose gemmæ, instead of becoming detached, remain permanently connected with the parent (Fig. 94); and as these in their turn may develope gemmæ from their own bodies, a structure of more or less arborescent character may be produced (Fig. 95, A), the exterior of which has usually a horny consistence. The form which this will present, and the relation of the component polypes to each other, will depend upon the mode in which the gemmation takes place; in all instances, however, the entire cluster is produced by continuous growth from a single individual; and the stomachs of the several polypes are united by tubes, which proceed from the base of each, along the stalk and branches, to communicate with the cavity of the central stem. Where a horny *polypidom* (or rather polype-skeleton) presents itself in this order, it is invariably formed by the consolidation of the external integument; and this frequently expands into bell-shaped 'cells' (Fig. 95, *c*), the interior of each of which is occupied, in the living Zoophyte, by a hydraform polype (*d*). It is, however, from the soft matter (*f*) contained within the stem and branches, rather than from the polypes themselves, that the new polype-buds (*b*) take their origin; and it is from this, also, that the bodies which have been erroneously designated as 'ovarian capsules' (*e*), are put forth. The independent vitality of the stem and branches of some of the compound Hydroida is remarkably shown by the fact, that their polypes periodically die, and are cast off, like the leaves of a tree. The capture of prey and the digestion of food would seem to be performed by these polypes in the same manner as by the Hydra; the digested product, however, is not applied merely to the nutrition of the polypes themselves, but to that also of the entire structure. This is brought about by the passage of the nutrient material of the several polypes (through an orifice guarded by a sphincter muscle), from the stomachs, into the tubular cavity of the stem and branches; and in this it is seen to undergo a kind of circulation strongly resembling that of the fluid contained within the elongated cells of *Chara* (§ 269).—The extension of the parent-structure is effected in the compound Hydroida, as just shown, by gemmation, not from the bodies of the polypes, but

from some part of the stem or branches ; which fabric, however, may be strictly considered as the composite base of the whole, uniting in itself (so to speak) the prolonged footstalks of the several polypes which it bears, its external horny tube being continuous with the horny cell, which is really the external wall of the polype-body, whilst its soft internal membrane is a prolongation of that which lines the stomach of each individual polype. But for the origination of new structures, at least two different plans are observed in operation ; one of which is of a kind altogether different from anything which we might have *à priori* expected. The individual polypes do not, like the Hydra, put forth gemmæ from their sides ; nor does it appear that these Zoophytes themselves develope either true ova or spermatic vesicles : but gemmæ capable of development into *Polypes* are put forth from certain parts of the stem or branches, being detached, however, long before they have attained their polypoid form ; whilst gemmæ of another kind are put forth from similar parts, which, instead of developing themselves into polypes, become *Medusæ* (§ 293) : these, being provided with proper sexual organs (the sperm-cells and ova being developed, however, in distinct individuals) produce fertile ova ; and each ovum produces a single polype, from which an extensive zoophytic structure, sometimes numbering many hundred thousand polypes, may be gradually evolved by continuous gemmation.—This curious process has a very close analogy with the history of the development of the Flowering Plant (§ 278).

FIG. 95.

Campanularia gelatinosa; A, upper part of the stem and branches, of the natural size: B, a small portion enlarged, showing the structure of the animal; *a*, terminal branch bearing polypes; *b*, polype-bud partially developed; *c*, horny cell, containing the expanded polype *d; e*, ovarian capsule, containing medusiform gemmæ in various stages of development; *f*, fleshy substance extending through the stem and branches, and connecting the different polype-cells and ovarian capsules; *g*, annular constrictions at the base of the branches.

The seed and the egg, which are their respective starting-points, are *homologous* (or essentially similar) bodies ; from it spring, in the one case, a stem and leaves, in the other, a stem and polypes ; these may extend by continuous gemmation to any degree, producing new *leaf-buds* or new *polypes,* and occasionally producing also 'bulbels' or free gemmæ,* which shall spontaneously detach themselves and evolve themselves into similar fabrics elsewhere ; but after a time, a different set of organs appears, the *flower-buds* and the *Medusæ*, containing distinct sexual organs, by means of which a true process of *generation* takes place, and seeds and ova are again produced. The only difference that even *seems* essential, lies in the *detachment* of the Medusa-buds, whilst flowers ordinarily remain *in continuity* with the plant ; but this is only that the former may possess locomotive powers which shall carry them to a distance, in order that the ova may be widely scattered through the ocean ; and an example is not wanting (the *Vallisneria spiralis*) in which the male flowers separate from the plant that produced them, before the process of fertilisation takes place.—In

FIG. 96.

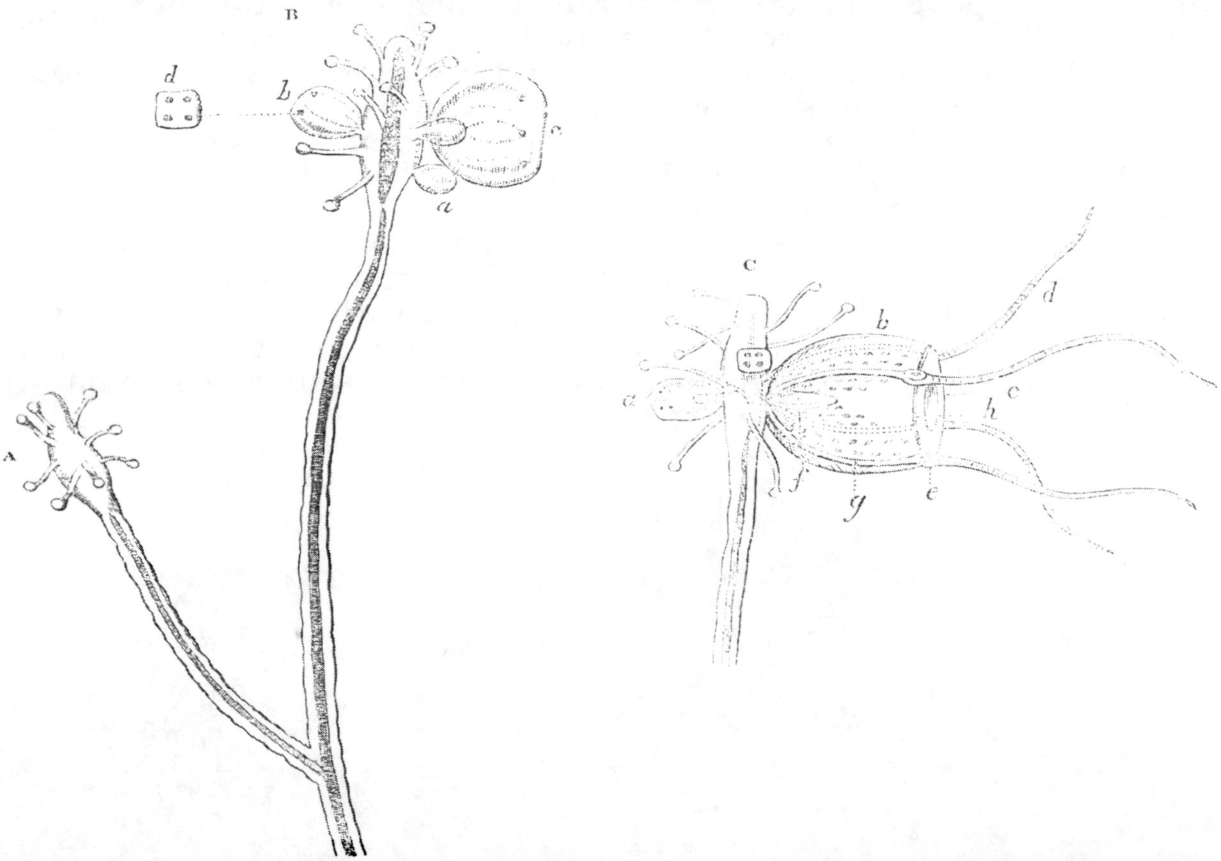

Development of Medusa-buds in *Syncoryna Sarsii*.—A, an ordinary polype, with its club-shaped body covered with tentacula:—B, a polype putting forth medusan gemmæ ; *a*, a very young bud ; *b*, a bud more advanced, the quadrangular form of which, with the four nuclei whence the cirrhi afterwards spring, is shown at *d* ; *c*, a bud still more advanced :—C, a polype with a young medusa-bud *a*, and one much more developed *b* ; in the latter are seen at *c*, the annular membrane partly closing the mouth of the bell ; *d*, the marginal cirrhi ; *e*, the ocelli at their base ; *f*, the stomachal proboscis : *g*, the four radiating canals ; *h*, the marginal canal.

some of the simpler forms of the compound Hydroida (namely, the families *Corynidæ* and *Tubularidæ*), we find both classes of gemmæ budding forth from the upper part of the stem, just beneath the stomach

* The following list of Phanerogamic plants propagating by bulbels has been furnished by M. Decaisne; the first two reproducing themselves by this method exclusively;—*Lunularia vulgaris*, *Lemna gibba*, *Dentaria bulbifera*, *Dioscoræa*, *Globba amarantina*, *Gagea villosa*, *Ornithogalum umbellatum*, and *Lilium bulbiferum*.

or body of the polype, without any investment whatever; the production of Medusa-buds in this manner is shown in Fig. 96. In the more complex forms, however, in which the external integument is more consolidated (as in the families *Campanularidæ* and *Sertularidæ*), there is a distinct apparatus for the purpose. This consists of a large cell or capsule (Fig. 95, *e*), which has been shown by Prof. E. Forbes * to be a metamorphosed branch, having exactly the same kind of relation to an ordinary branch bearing polype-cells, that a flower has to a branch bearing leaves (§ 278). This capsule was formerly termed the "ovarian vesicle," from the idea that the bodies produced within it are true ova. This is now known, however, not to be the case; these bodies being really gemmæ, sprouting from the axis, like those of a Coryne or Tubularia; some of them going forth in a condition nearly allied to that of the 'gemmules' of Sponges (§ 286), to develope themselves into new polype-structures; whilst others are Medusa-buds, which acquire a somewhat higher development within the capsule, and then go forth to propagate the species by the deposition of ova at a distance.†

291. The order *Helianthoida* derives its designation from the resemblance which the polypes of which it consists, bear to a Sun-flower or other composite blossom. The common *Actinia*, or Sea-Anemone (Fig. 97), may be taken as its type; the individual polypes of all the composite structures included in the group being constructed upon the same model. The body of this animal is broad and flat, compared with that of the Hydra-form polypes; the mouth is situated in the centre of its upper surface, and is surrounded by concentric rows of short tubular tentacula, the

Fig. 97.

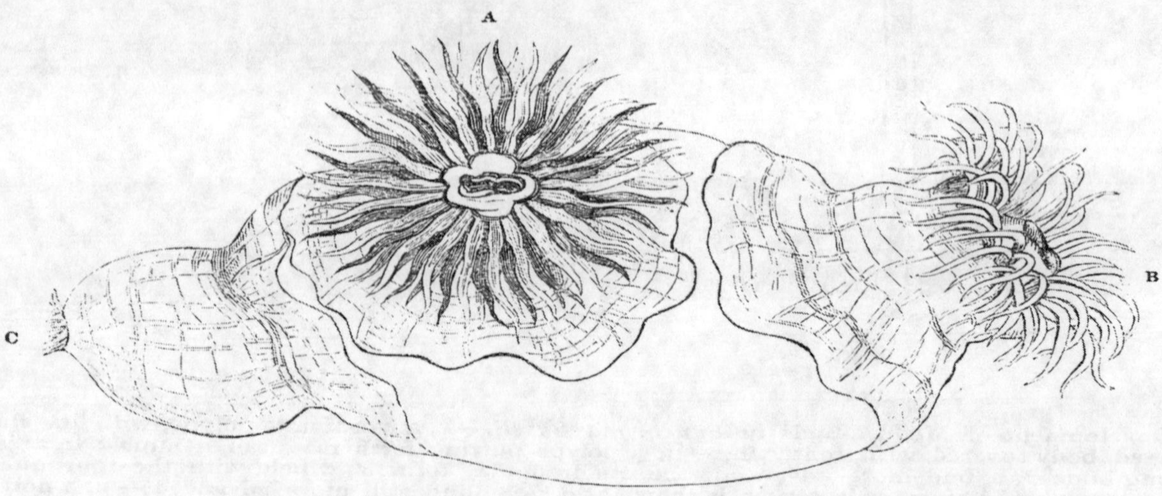

Actiniæ, attached to a pebble, (A, B) in open, and (C) in closed condition.

number of which is extremely variable, increasing with the age of the animal; and the body spreads out at its base into a fleshy disk, which takes a very firm attachment to the solid surface to which it is applied.

* Annals of Natural History, vol. xiv.

† This appears to the Author to be the most feasible mode of accounting for the variety in the results of the observations upon the reproduction of the Compound Hydroida, which have been made by different Naturalists. For a fuller examination of the question, and for a comparison of their statements, see the Brit. and For. Med. Chir. Rev. vol. i. p. 208, *et seq.*

The skin is firm, and in some species even leathery in its consistence; but it is very irritable, like that of the Hydra, and great changes are produced in the form of the animal by its contraction. The commonest is that represented in Fig. 97, C, which is seen when the animals are left dry by the recession of the tide, the tentacula being drawn in, and the skin of the upper part of the body contracted over them, so as almost or completely to conceal them, and to give to the whole body somewhat of a conical form. The essential difference between the *Actinia* and the *Hydra* consists in this,—that the stomach of the former does not, like that of the latter, occupy the whole cavity enclosed by the integument, but that it is suspended, as it were, in the centre of that cavity (Fig. 98, *e*), by a set of vertical partitions (*g*, *g*, *k*, *k*), which pass in a radiating direction from the walls of the stomach to the external integument, so as to divide the intervening space into numerous chambers. Several of these radiating partitions, however, do not extend inwards as far as the parietes of the stomach. The interseptal chambers communicate with each other by openings (*g'*, *g'*) in the lamellæ; and there is also a free passage (*h*) from them into the hollow tentacula, which are provided with orifices at their extremities, allowing the introduction or expulsion of water through them. The stomach is closed at the bottom, alike in the solitary and in the compound Helianthoida; its walls usually present a series of plaits, extending vertically from the mouth downwards, by the unfolding of which its capacity may be greatly increased. The food, drawn within the mouth by the agency of the tentacula, is gradually subjected to the digestive process; and its exuviæ, after some time, are cast forth by the mouth. No true vessels exist in the walls of the stomach, by which the nutrient matter may be taken up from its cavity; but it seems to be at once absorbed into the substance of the membrane, as in the Hydra, and to be transmitted by lacunæ in its fibrous texture (apparently a more advanced form of that of the Hydra) from the coats of the stomach to the external portions of the body.—The movements of the *Actiniæ* are scarcely so active as those of the Hydra and its allies; consisting in little else than the protrusion and retraction of the tentacula, in the introduction of food into the mouth and the rejection of the exuviæ, and in the introduction of water through the tentacular and other orifices into the ovarial chambers, and its ejection from them through the same pores; all of these operations being performed under circumstances, which lead to the conclusion that

FIG. 98.

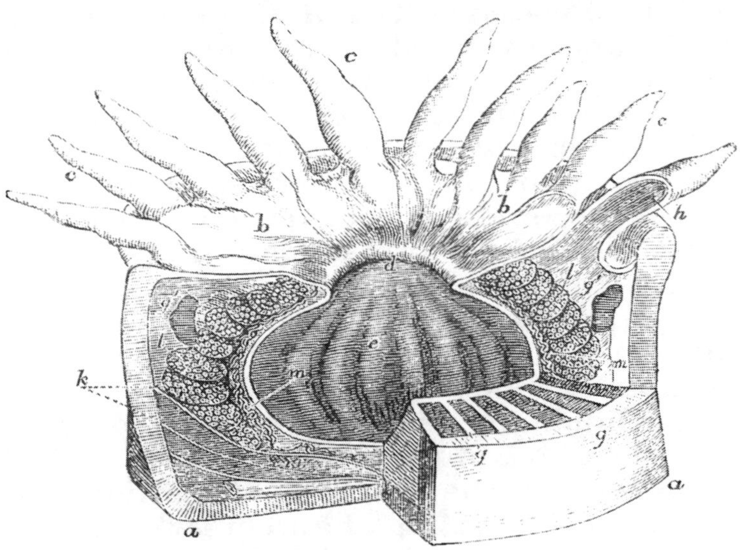

Diagrammatic section of *Actinia*, showing its internal structure; *a*, *a*, base or foot; *b*, *b*, oral disk; *c*, *c*, tentacula; *d*, mouth; *e*, stomach; *g*, *g*, *k*,*k*, vertical partitions cut across in different directions; *g'*, *g'*, apertures in these; *h*, passages opening into the tentacula; *l*, *l*, ovaria; *m*, *m*, testes.

they can be but little regarded as indications of consciousness, still less of voluntary power. The animals do not seem to be in the least aware of the proximity of food, unless it be absolutely in contact with their tentacula; nor do they make any effort to move towards water, although they may have been long kept out of it and may be close to its surface. They are influenced, however, by light, and apparently also by atmospheric conditions; being observed to expand themselves much more fully in a bright clear sunshine, than in dull hazy weather. They occasionally travel slowly along the surface to which they have attached themselves, by the alternate contraction and expansion of their disk; and sometimes they completely detach themselves, and allow their bodies to be carried hither and thither by the waves. No nervous system has yet been certainly detected in them; nor have the least indications of organs of special sense been discovered.—The *Actiniæ* possess great power of reproducing lost parts; and even when a specimen is divided into two or more pieces, each may in time develope itself into a new individual. They do not, however, usually propagate by gemmation; though this mode of increase has been observed (by Sir J. G. Dalyell) in one species, from the expanded base of which small portions occasionally detach themselves, which subsequently become perfect Actiniæ. In the *Zoanthus*, however, which agrees with the Actinia in its general organisation, and which is, like it, unprovided with a calcareous skeleton, several individuals are found springing from a common base, which is sometimes broad and flat, but more commonly a creeping stem; and the individuals thus connected have all originated by gemmation from the one first formed. In the compound Helianthoida, as will be presently pointed out, the process of gemmation takes place to a much greater extent. All these Polypes, however, reproduce themselves by the true sexual method; the generative organs being developed in the radiating chambers which surround the stomach. The 'sperm-cells' are formed within the long convoluted tubuli (*m*) which these chambers contain in great abundance; whilst the ova are produced in large plaited masses (*l*, *l*), enclosed in a common membrane, and attached along the inner border of such vertical leaflets as do not extend as far as the stomach. The ova escape from these ovarial masses, apparently by the rupture of their investing membrane, into the interseptal chambers; there they are probably fertilised by the contents of the "sperm-cells;" and in these, or in the cavities of the tentacula, they may remain during a larger or smaller proportion of their early period of development, so as even to acquire mouths and tentacula of their own. The young Actiniæ are discharged, however, not by the tentacular orifices, which are too minute to give them passage, but by the mouth; although the manner in which they pass from the ovarial chambers into the stomach, is yet an unsolved mystery. They are sometimes discharged in that earlier stage of development, in which they are found living in the tentacula; that, namely, of ciliated globular bodies, presenting no appearance of either mouth or tentacula, but having the power of free locomotion, which is lost when these organs are evolved.

291 *a*. In the greater number of Helianthoid Zoophytes, the bases of the polypes and the lower part of their radiating lamellæ are consolidated by calcareous deposit; and the Corals thus formed are distinguished as *lamelliform*. As the lower and older part of the animal tissue is thus condensed, it is withdrawn from any connection with the vital operations,

and is converted into a mere organ of support; still the coral-mass cannot be regarded as *dead*, for it remains in complete continuity with the softer tissue above (§ 194), and when this dies, it loses much of its firmness and tenacity. As the soft tissue is in a state of progressive upward growth, it is progressively consolidated below; just as the heart-wood of an Exogenous tree is increased in diameter by the consolidation of the internal layers of alburnum, in accordance with the additions which the alburnum receives on its external surface (§ 280); and thus the stony skeleton gradually rises from its base. It is quite erroneous, therefore, to speak of Coral as being 'built up' by the agency of the polypes; when it really 'grows' as a part of their own substance. If the stony mass be the skeleton of a single polype, as is the case with the well-known *Fungia* of tropical seas (known also as the 'Mushroom coral' from its resemblance to the under side of the pileus of the Agarics, § 273), or with the *Caryophyllia* (of which one species is found on British shores), it is marked on its upper surface with a single series of radiating lamellæ; but if it have been the axis of a compound mass, the radiating lamellæ will be seen in every one of the individual polype-cells (Fig. 99), which are frequently very numerous and minute, especially in the family *Madreporidæ.* The cells are not by any means regularly circular; but still the laminated plates project inwards from their circumference towards a common centre. Sometimes a number of cells coalesce (as it were) into a prolonged groove or furrow; this is the case, for example, in the hemispherical masses of *Meandrina,* commonly known as 'brain-stone coral,' from the resemblance of its convoluted furrows to those of the surface of the brain. The individual polypes, in these compound lithophytes, are connected together by a kind of fleshy crust which invests the whole stony mass; this may be considered as an expansion of the disk or base, which is common to all of them; but it does not contain prolongations of their stomachs, nor have any vessels been detected in it. Still it must receive nourishment from them, like the medulla contained within the stem and branches of the Hydroida; for it has in itself the power of adding to the stony deposit; this being sometimes formed, not only *upon* the cells, as it is in the arborescent corals (Fig. 99) which grow by additions to the extremity of the stems and branches, but also *between* the cells; the latter being the distinctive peculiarity of the *Madreporidæ.* This gelatinous crust, also, seems to have the power of giving origin to new polypes, when it is detached from the axis, in this respect also resembling the medulla of the compound Hydroida; and it is probably from it that new polypes are developed in the Madreporidæ. In the *Madrephyllidæ,* however, whose

FIG. 99.

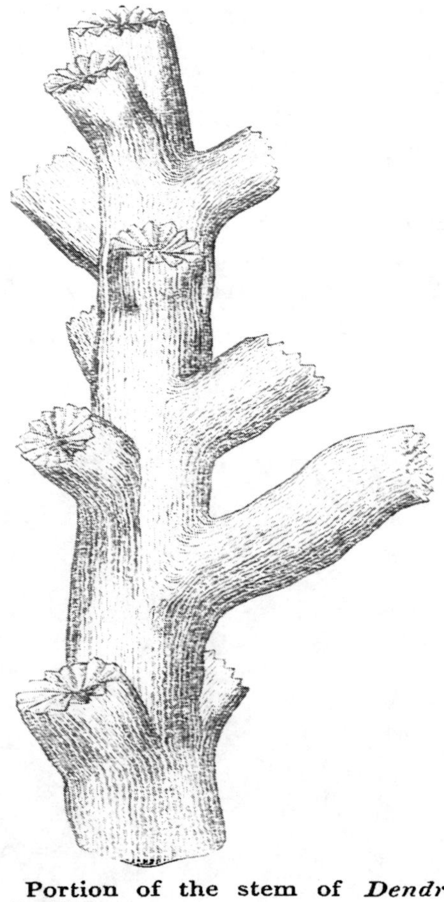

Portion of the stem of *Dendrophyllia.*

polype-cells are terminal, the increase in their number seems to take place by the subdivision of the polypes themselves; the progressive stages of which are shown in Fig. 100, which represents the process as taking place in the individual polypes of an *Astræa.* It is in this mode that a single vertical column subdivides into two, or that a lateral branch is put forth from the vertical axis; and it is in this mode, too,

FIG. 100.

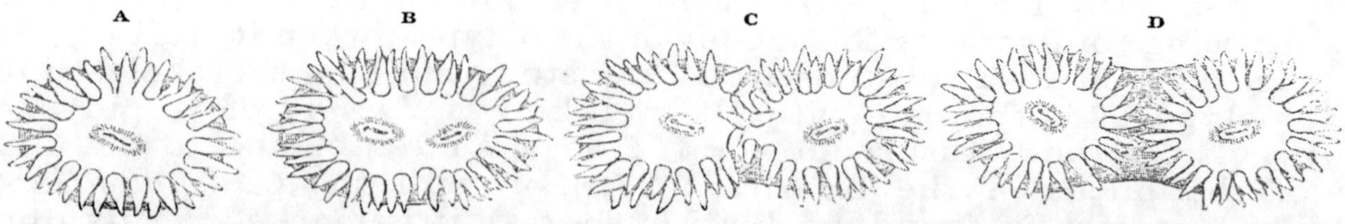

Multiplication of polypes of *Astræa* by subdivision; A, original simple disk; B, the same elongated, with a second polype mouth; C, subdivision commencing; D, subdivision completed, two distinct polypes being formed.

that the number of individuals in the furrows of the *Meandrina* is increased, in accordance with the progressive increase in their length by the augmentation of the mass in diameter.—It is to this group, and especially to the family *Madreporidæ,* that nearly all the massive Corals belong, to which the formation of coral reefs and islands is due; and its operation in thus adding to the amount of dry land on the earth's surface, would seem to have been yet more extensive in former epochs of its history, than it is at present.

292. The order *Asteroida* is so-named from the 'star-like' appearance presented by its polypes when expanded, in consequence of the very regular arrangement of the six or eight broad short tentacula which radiate from the mouth of each (Fig. 102, *a*). This order contains no solitary species; and the essential character by which it is most distinguished from the Helianthoida, is the intimate connection which exists between the individuals of the same composite mass or polypidom. This is well seen in the com-

FIG. 101.

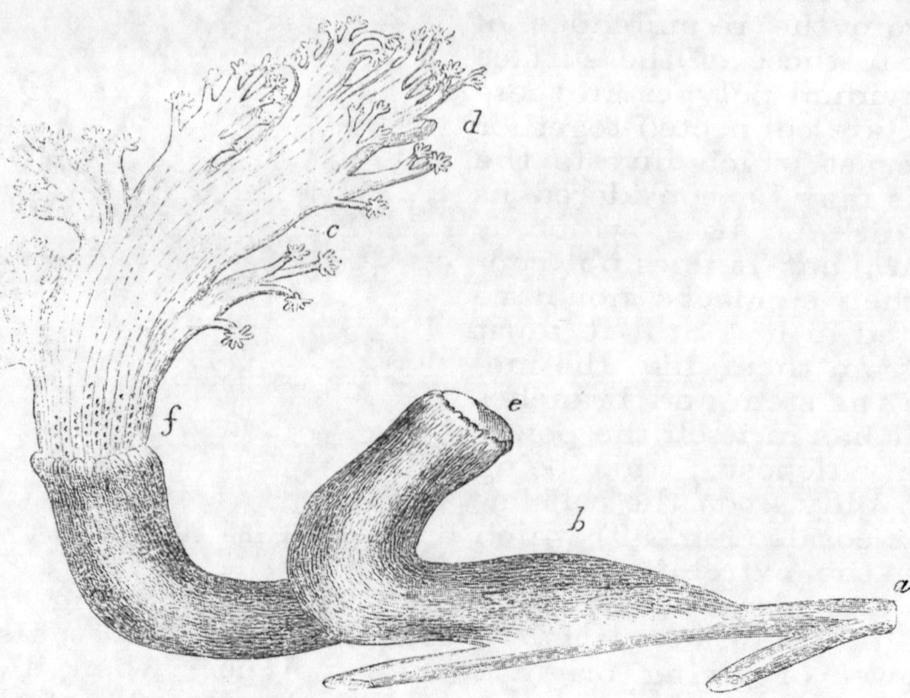

Portion of *Alcyonidium elegans* of the natural size, attached to the stem (*a*) of a marine plant; *b,* basal portion or stem of the polypidom; *c,* softer or membranous portion of the polypidom; *d,* polypiferous branches; *e,* stem with the soft parts retracted; *f,* yellowish spots occasioned by the presence of ova in the interior.

mon *Alcyonium* of our coasts (commonly known by the name of 'dead man's toes,' which is very expressive of its appearance), whose polypidom has a kind of spongy texture, and is traversed by a network of canals, at the outlets of which the polypes are situated. The mouth of each polype leads to a stomach (*c*, Figs. 102, 103), which (as in the Helianthoida) is suspended in the midst of the general cavity of the body by partitions (Fig. 103, *f*) radiating from its walls; the number of these partitions, and consequently that of the chambers surrounding the stomach, being the same with that of the tentacula. Instead of being closed at its lower extremity, however, like that of the Actiniform polypes, the stomach of the Alcyonians opens into the canals that ramify through

Fig. 102.*

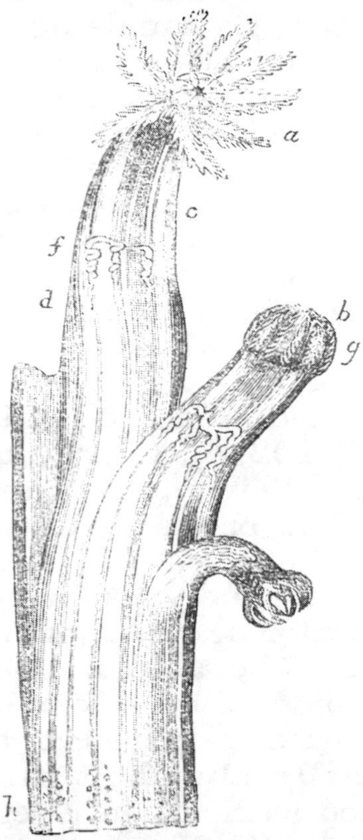

Fig. 103.†

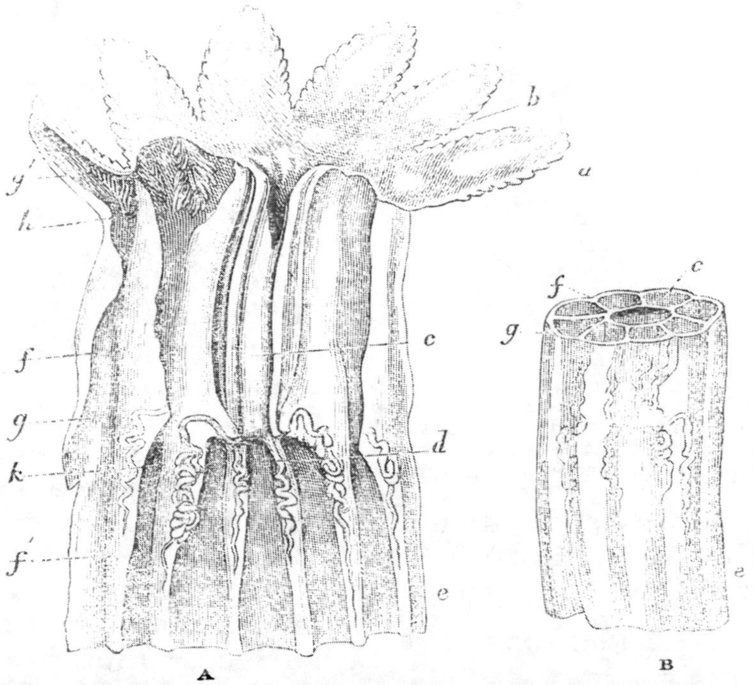

* Portion of a branch of *Alcyonidium elegans* considerably enlarged; *a*, polype expanded; *b*, polype contracted; *c*, stomach, seen through the transparent parietes of the body; *d*, portion of the body enclosing the superior part of the general cavity; *f*, filiform appendages (testes) seen through the transparent walls; *g*, lines projecting on the base of the tentacles, and formed by brownish spicules of cartilaginous substance; *h*, ova, seen through the parietes of the body.

† Terminal portion of one of the polypes, considerably enlarged, and laid open (A) longitudinally, (B) transversely, to show the interior structure; *a*, tentacula; *b*, mouth; *c*, stomach; *d*, its inferior opening into the general cavity, of which the upper part is seen at *e*; *f*, longitudinal partitions traversing the space between the stomach and the external parietes; *f'*, prolongation of these, as folds in the wall of the general cavity; *g*, canals surrounding the stomach, and communicating with the cavity of the tentacula; *g'*, one of tentacula laid open; *h*, groups of spicules situated at the base of the tentacula; *k*, filiform appendices, probably testes.

the fleshy mass in which the polypes are imbedded; the orifice (Fig. 103, *d*) being surrounded by a circular muscle or sphincter, by the action of which it may be expanded or entirely closed. The chambers which surround the stomach communicate above with the cavity of the tentacula, each of which has a small orifice at its extremity; whilst below they are continuous with the ramifying canals of the polypidom; and the membranous septa which support the stomach do not cease at its lower extremity, but are prolonged downwards as plaits or folds of the lining of these canals (Fig. 103, *f′*), until they gradually disappear. Thus the stomachs of the individual polypes, and the chambers which surround them, are so intimately connected through the polypidom, that the entire mass might almost be regarded as one composite individual; and it is obvious that, as in the Compound Hydroida, the lives of the polypes are subordinate to that of the polypidom, since it is from the latter and not the former that the extension of the fabric by gemmation takes place. When a bud is put forth from the common *Alcyonium*, this bud has all the essential characters of *Sponge*, except that its canals do not open upon the surface (Fig. 104); and the texture of the polypidom is almost completed, by nourishment derived from the parent mass, before the polypes begin to make their appearance. — A true sexual generation takes place in this group. The sperm-cells are probably formed within the convoluted filaments (Fig. 102, *f*, Fig. 103, *k*) which hang down from the base of the stomach; whilst the ova are developed in the substance of the membranous plaits that are continued along the lining of the canals (Figs. 101, *f*, 102, *h*, 103, *f′*), and after being fertilised make their way outwards through the mouth, in what precise grade of development, however, is not known. — Although the structure of the polypes in this group is remarkably uniform, that of the polypidoms presents an unusual amount of variation. The animal tissue of the Alcyonium is strengthened, like that of the Sponge, by calcareous spicules, which form a sort of loose skeleton that extends through the entire mass, being especially abundant in its surface, which is furnished with a firm integument. In the *Tubipora musica* (or 'organ-pipe coral') the external integument of each polype is completely consolidated into a calcareous tube, reminding us of the tubular horny skeletons of the Hydroida. On the other hand, in the *Corallium rubrum* (or 'red coral') it is the centre which is thus hardened; a dense stony axis being formed, on the smooth surface of which not a vestige of polype-cells can be detected; whilst the flesh that clothes this is channelled out by canals, which connect the stomachs of the polypes imbedded in it. In the *Gorgonia* (Sea-Fan) and *Antipathes* (Black Coral), the axis, though very firm, is altogether composed of horny matter; but in its investing flesh a large number of spicules present

FIG. 104.

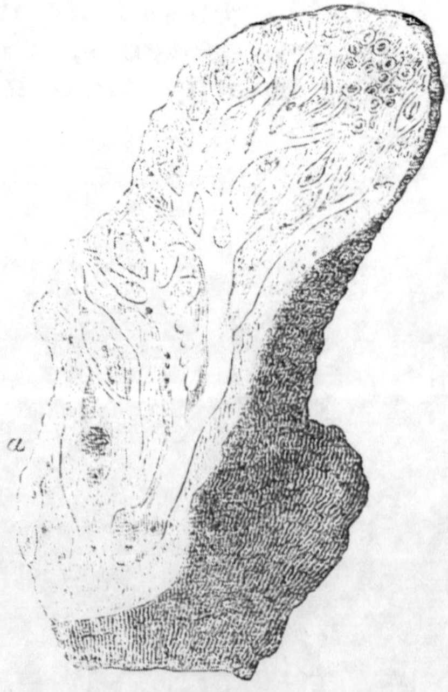

Portion of a young branch of the polypidom of *Alcyonium stellatum*, divided longitudinally to show the spongiform character of its structure; *a*, incipient state of young polype.

FIG. 105.

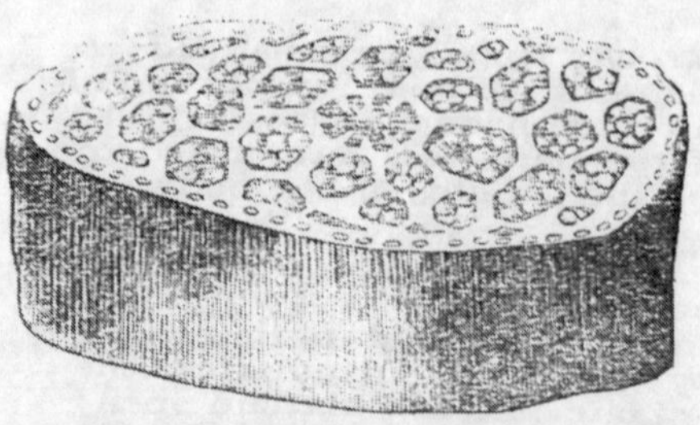

Transverse section of the basal portion of the polypidom of *Alcyonidium elegans*, showing the manner in which it is traversed by canals continued from the stomachs of the polypes, with the longitudinal folds in their walls, and the ova projecting from these.

themselves; and these form a thick friable crust around the axis, when the animal tissue has been allowed to dry upon it. In some few cases, instead of being attached by roots to fixed objects, the polypidoms are free to be carried about by the action of the waves and currents of the ocean; this is the case, for example, with the *Pennatula* (or 'sea-pen'), and a few other genera, in none of which, however, does any power of spontaneous motion appear to exist.

293. The animals composing the class ACALEPHÆ, or 'jelly-fish,' differ so widely in external form and internal structure, that it is difficult to assign any character which shall be applicable to them all. Much, indeed, remains to be known in regard to the internal structure of many of these beings, with whose external forms we are familiar; and when to this it is added, that the history of their development has not yet been studied with any attention, save in the case of one tribe only, and that the results of the observations made upon this are such as to give an entirely new view of its position in the series, it will be apparent that no such general account can at present be given of the entire class, as shall satisfactorily represent its essential peculiarities. The name by which the class is scientifically designated, has reference to the stinging power possessed by a large proportion of the species it includes; and some of the vernacular names by which these animals are known, such as 'sea-nettles' or 'stang-fishes,' have the same source. This stinging power appears due to an acrid secretion from the surface of their bodies, which remains after the death of the animals, and may be communicated to other substances placed in contact with them. Some of the commonest of the Acalephæ are also frequently denominated 'sea-blubber,' 'jelly-fish,' &c.; the former term apparently having reference to their shapeless and uncouth aspect when cast upon the shore; while the latter indicates their transparent gelatinous appearance, which is due in great part to the large quantity of water that enters into the composition of their tissues. This proportion is often so great, that the body of a *Medusa* (for example), weighing fifty pounds when first taken out of the water, is reduced by drying to little more than as many grains. In the greater number of Acalephæ, there is no hard support whatever, so that the animals when taken out of the water, lose their form completely; in the *Velella* (Fig. 108) and *Porpita*, however, the upper surface of the body is covered by a thin cartilaginous plate, which retains its shape when dry. Another property which seems common to a considerable part of the class, is that of emitting light from the general surface or from particular parts of it; and it is chiefly to the presence of Acalephæ, that the beautiful phenomenon known as the "phosphorescence of the sea" is attributable (CHAP. XVII. Sect. 1). All the animals of this class are marine; and they are more frequently met with in the open sea, than in the neighbourhood of land. The voyager frequently encounters vast fleets of them, extending as far as the eye can reach, basking in the sunshine that illumines the surface of the ocean, and reflecting its rays with all the gorgeousness of the most brilliant iridescence; and after passing through one such fleet, he will frequently within no long period encounter another, composed of individuals of another species.—In place of dwelling further, however, upon the general characters of the class, it will be desirable to pass in brief review the principal types of structure which it contains; and these

(so far as at present known) are capable of being reduced to four, which are characteristic of as many orders, named after their respective methods of locomotion.

294. The order *Pulmograda,* or Discophoræ, including all the ordinary *Medusæ,* is characterised by the regular discoidal or circular form of the animals composing it; and by the mode in which they are propelled through the water by the rhythmical movements of the disk. The body is of very soft jelly-like texture, being chiefly composed of a sort of cartilaginous tissue (that is, of cells dispersed through a structureless intercellular substance) which contains an enormous proportion of water; and it possesses no skeleton whatever, either internal or external. The stomach (Fig. 106 *d d*) occupies the centre of the disk, and usually opens by a single central mouth of quadrangular form (*a*); its cavity is frequently divided into four lobes, which extend beyond the ovaries and in great part surround them. From the outer margin of the stomach, a series of canals passes off towards the margin of the disk; some of which open there by anal pores (A, *e e*, &c.). The mouth is usually furnished, either with a sort of proboscis formed by the prolongation of the lips, or with a set of four tentacula (B, *g g*); each instrument serving for the reception of the kind of food most adapted to the wants of the species. The ovarial chambers (which may contain either ovaries or testes) are four in number (A, *b b b b*), and are disposed around the entrance to the stomach; they open externally by separate orifices (*c c c c*), which are seen near the mouth. At the margin of the disk, intermediate between the anal pores, are found a set of red spots (*f f*), which are believed (with much probability) to be rudimentary eyes. The margin is usually furnished, also, with a set of *cirrhi* or tendril-like appendages, which are sometimes of considerable length; but these are not extensible and contractile like the oral tentacula, which serve to lay hold of animals of considerable activity (such as small Crustacea, or even Fishes), and to bring them to the mouth. The body moves slowly through the water by a sort of flapping movement of the disk, which is furnished with muscular fibres; this movement bears much more resemblance, in its rhythmical character, to the pulsations of the heart of the higher animals, than to their ordinary locomotive actions; and it is a confirmation of this view of their character, that the nervous system, if it exist at all, must be extremely limited in its distribution and operations, since, notwithstanding the transparency of the bodies of these animals, none has yet been detected in them.—The study of their mode of development has been attended with most unexpected results. A true sexual process of generation takes place in them; ova being produced within the ovarial chambers of some individuals, and spermatic cells in those of others. The ova, when discharged and fertilised, are carried for some time upon the tentacula (Fig. 106 B, *g g*), on the surface of which, little marsupial sacs are formed for their reception. The embryo first issues forth in the condition of a ciliated Animalcule, and for a time swims freely through the water; it then fixes itself, however, by one extremity, and is developed into a hydraform Polype, in which condition it has been observed and described, without any suspicion of its peculiar nature. In this state it may continue for a considerable time, conforming to the condition of the Hydra, not merely in its general habits of life, but also in its

propagation by gemmation; but it then gives off (in the manner hereafter to be described, CHAP. XVIII.) a number of Medusa-buds, each of which is developed into the likeness of its parent; the original polypoid body, however, still remaining, and subsequently developing a new set of gemmæ of both descriptions. Even when the Medusan form has been attained, the propagation is still occasionally effected by gemmæ, which

FIG. 106.

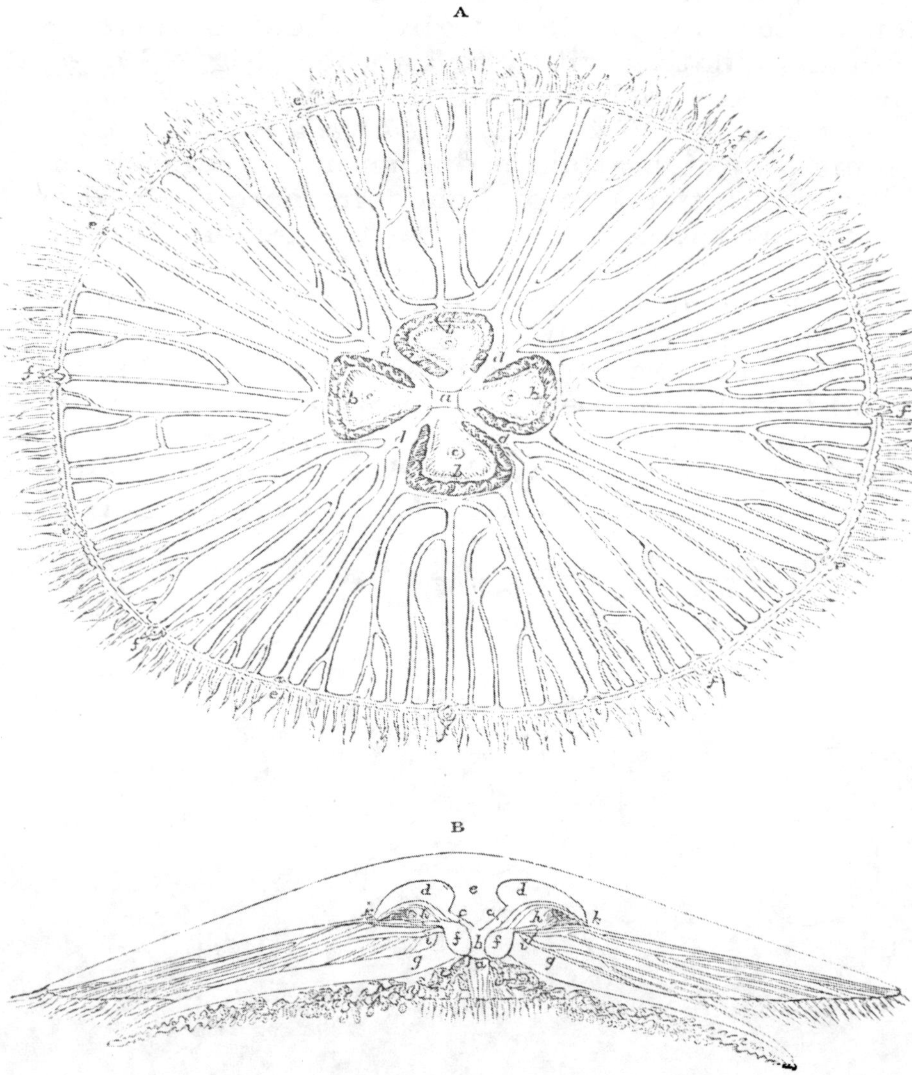

Structure of *Cyanæa aurita*.—A, disk seen from above, showing the quadrilateral mouth *a*, the four ovaries *b b b b*, the four orifices of the ovarian chamber *c c c c*, the stomach *d d d d*, and its radiating prolongations, the eight anal orifices *e e*, &c., and the eight ocelli *f f*, &c. :—B, vertical section, showing *a* the entrance to the mouth *b*, which leads by four canals, two of which *c c* are laid open by the section, to the four-lobed stomach, two divisions of which *d d*, separated by the partition *e*, are laid open by the section; at *f f* are seen the vertical plates, from which hang the fleshy tentacula *g g*; at *h h* two of the ovaries; at *i i* the situation of two of the ovarial orifices; and at *k k* the commencement of the radiating canals.

may grow from various parts of the body, as will be shown hereafter.—The order Pulmograda is again subdivided by Prof. E. Forbes* into two sub-orders, the *Steganophthalmata*, or 'hooded-eyed,' and the

* "Monograph of the British Naked-Eyed Medusæ," published by the Ray Society, 1848.

Gymnophthalmata, or 'naked-eyed;' the former consisting of those which have the *ocelli* or rudimentary eyes at their margin protected by membranous hoods or coverings more or less complicated, while the latter have the ocelli unprotected. This character, though apparently trivial, is one of great value, from its correspondence with diversities in their internal structure, less capable of being readily distinguished. Thus, whilst the former of these orders possesses a much-ramified and anastomosing series of tubular prolongations of the stomach, extending over the surface of the disk, the second has a very simple gastric apparatus, the canals often proceeding to the margin without branching, or, if they do subdivide, not inosculating with one another (Fig. 96 C, *g*, *h*). In the first of these families are included all the larger *Medusæ*, such as those belonging to the genera *Aurelia*, *Pelagia*, *Chrysaora*, *Rhizostoma*, *Cassiopea*, and *Cyanæa*, some of which attain a diameter of four feet and a weight of 60 lbs; whilst the latter comprehends numerous smaller and more delicate forms, such as those belonging to the genera *Oceania*, *Æquorea*, *Geryonia*, and *Thaumantias*, several of which, however, will probably be found to be the medusoid states of certain Hydraform polypes already well known to Naturalists (§ 290).

294*a*. The *Ciliograda* are distinguished from the preceding by the variety of form which they present, but more especially by their mode of progression; this not being accomplished by any undulating or pulsatile movement of the body itself, but by the vibration of the cilia with which certain parts of the surface are clothed. In *Cydippe* (Fig. 107, A) the body

FIG. 107.

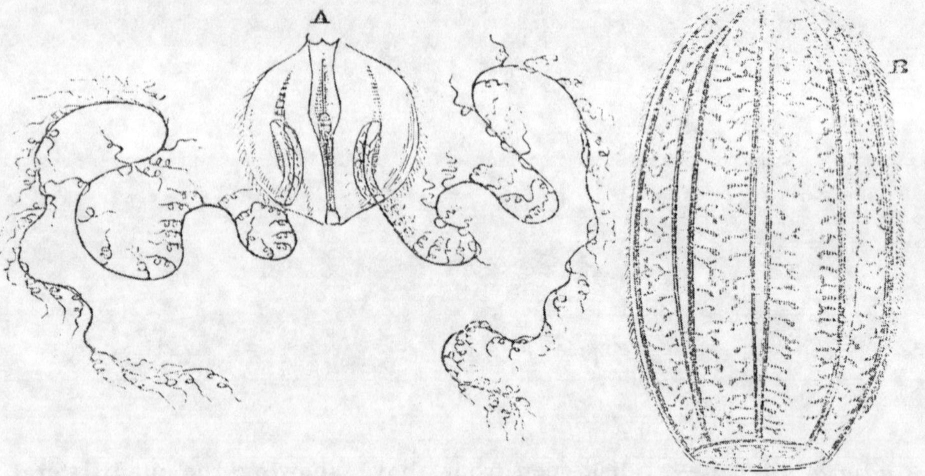

A. *Cydippe pileus*. B. *Beroe Forskalii*, showing the tubular prolongations of the stomach.

is nearly globular, and the cilia are disposed in eight bands, which extend like meridian lines from pole to pole. In *Cestum Veneris*, the body has the form of a long flat riband; and both edges of it are fringed with cilia. Between these two forms there would not seem to be much relation; but they are connected by a succession of intermediate links. Thus, in *Callianira*, the globular body is extended laterally, so as to form a wing-like appendage on either side; in other genera, these appendages are still more extended, and the central globe is lost in them; until at last the flat riband-like form of Cestum Veneris is attained. The position of the alimentary canal, which has here two distinct orifices, is the same throughout this series; for whilst in *Cydippe* it runs from pole to pole of the globe (Fig. 107 A), in *Cestum Veneris* it is equally short and straight, running *across* the body at the middle of its length. From this alimentary canal, however, as from the stomach of the Medusæ, a system of gastric canals extends throughout the gelatinous body (Fig. 107 B), conveying the nutritious portion of its contents to the parts of the structure most remote from the digestive cavity, and also exposing them to aëration. In no animal of this order is there any proper skeleton, either external or internal; but in Cydippe, the bands upon which the cilia are seated are of firmer texture than the rest. Many of the Ciliograda are very active in their habits, contrasting strongly with the sluggish Pulmograda. The *Cydippe pileus* (a species very abundant on many parts of the British coast) is peculiarly energetic. It is provided with two long tendril-like filaments, each furnished with lateral branches, arising from the bottom of a pair of cavities in the posterior part of the body (Fig. 107, A); the filaments can be entirely doubled up within these cavities, so that they are not visible externally; and when they are put forth, the main filaments are first ejected, apparently by the contraction of the cavity, the lateral tendrils uncoiling subsequently. Of the propagation of the Ciliograda, little is yet known, but there is reason to believe, from the observed presence of ovaries, that it is truly sexual; it does not appear, however, that, in the case of *Cydippe* at least, any such metamorphosis takes place as occurs among the Pulmograda.

294*b*. The order *Cirrhigrada* is a small group, distinguished by the presence of a kind of skeleton of greater density than is elsewhere found in the class; but more especially by the possession of numerous contractile *cirrhi*, which seem in some degree concerned in the propulsion of the animal through the water. In *Porpita*, the body is a circular disk; and the skeleton is a flat cartilaginous plate, lying close to its upper surface. In *Velella* (Fig. 108), the body is more oval, or rather quadrangular; and the cartilaginous disk has a vertical plate rising from it (A), which acts as a sail, whereby the animal is propelled by the wind along the surface of the water. The mouth or entrance to the stomach is situated at the extremity of a sort of flask-shaped proboscis (B, C, *a*); and the base of this opens into a cavity of a somewhat cylindrical form (C, *b b*), which lies in the direction of the length of the body, and subdivides at each end into two prolongations, from which a system of canals appears to pass off into other parts of the body, analogous to those of the Medusæ. Between the stomach and the cartilaginous plate, is a glandular organ (C, *c*) which is probably a liver. The cirrhi nearest to the mouth are tubular, and are furnished with suckers at their extremities, with an orifice at the centre of each; their canals all open into a cavity distinct from the stomach, in which nothing but water has ever been found. At the base of these cirrhi, and attached to them like grapes upon a stalk, are cœcal appendages, in which ova are produced; and these would seem to pass forth through the canals and orifices of the cirrhi. No spermatic cells, however, have yet been detected; and the history of the production, fertilization, and development of the ova is as yet entirely unknown.—In the position of the mouth in the midst of a set of tubular tentacula, and in the connection of these at their base with the ovaries and with a cavity to which water

FIG. 108.

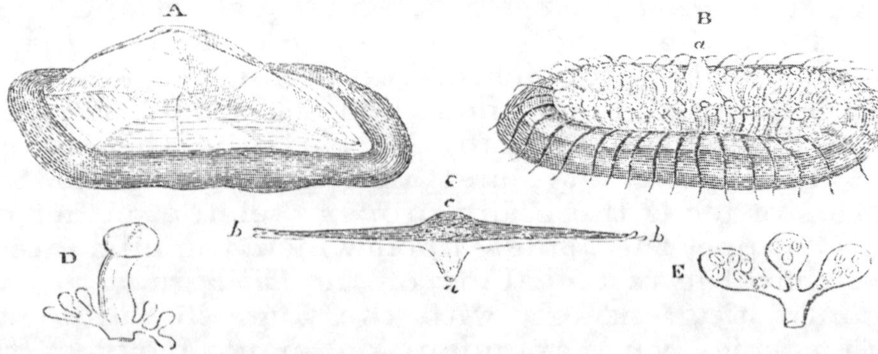

Velella limbosa: A, in its usual position: B, the under surface, showing *a* the stomach, surrounded by the tubular cirrhi and tentacula; C, side view of the digestive apparatus, showing *a* the mouth, *b b* the bifurcating extremities of the gastric cavity, *c* the liver: D, one of the cirrhi, with the ovarian cœca attached: E, ovarian cœca enlarged.

is admitted, the Velella and its allies present an obvious relationship to *Actinia* (§ 291).

294*c*. The order *Physograda* seems nearly allied to the preceding, both in the conformation of the digestive apparatus, and in the mode of progression. It chiefly consists of the *Physalia* (Portuguese man-of-war) and its allies, in which we altogether lose the *radiated* form, but find a *lateral* symmetry; the two halves of the body, divided by a plane passing from one end to the other, being equal and similar. The principal part of the bulk of the *Physalia* (Fig. 109), is composed of a membranous sac or bladder which is filled with air, and floats partly above the surface of the water; this air-bag is prolonged at one end into a conical beak, perforated by an aperture through which air can be expelled; and it is surmounted by a vertical ridge or crest, which, like the cartilaginous plate of the Velella, is acted upon by the wind in the manner of a sail. Beneath the air-bag is the digestive cavity, which is entirely distinct from it; this has no single mouth, but receives its supplies of aliment through a number of

FIG. 109.

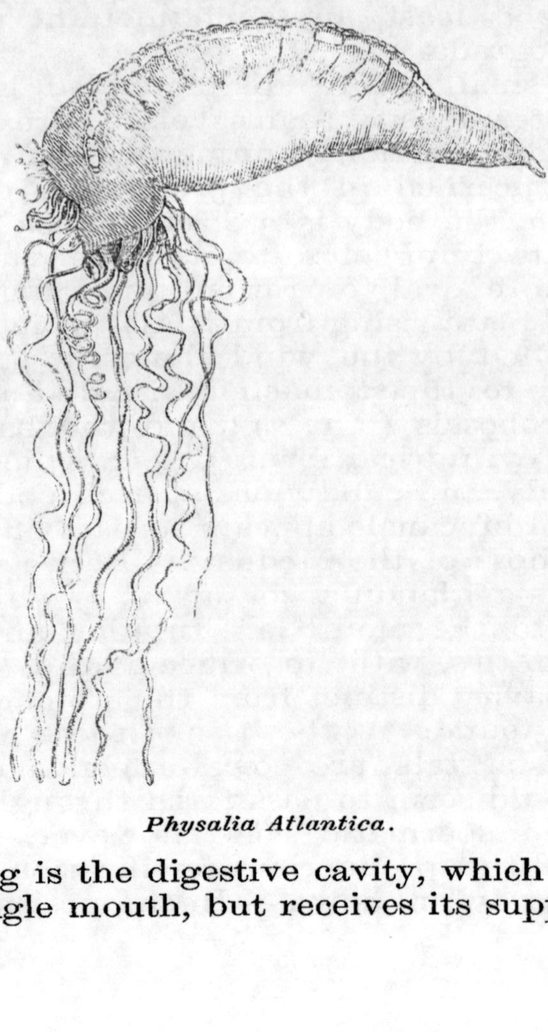

Physalia Atlantica.

flask-shaped cirrhi which hang down beneath, each having an orifice at its extremity, surrounded by a sucker. The suctorial cirrhi are entirely distinct from the long contractile tentacula, which are employed for grasping prey, at the same time paralysing it by their peculiar stinging power. These tentacula can be drawn up to within half an inch of the air-bladder; and may then be suddenly shot forth to a length of eighteen or twenty feet; and by means of these instruments, which are furnished with glandules secreting a highly irritating fluid, the Physaliæ can attack small fishes, &c. at a considerable distance, and can then draw them within reach of their suctorial appendages. Besides the suctorial cirrhi and the prehensile tentacula, there are other appendages, covered with vibratile cilia, which have been supposed to be respiratory organs. A glandular apparatus exists in the immediate neighbourhood of the digestive cavity; and this may probably be regarded as a liver. No generative organs have been hitherto found in these curious beings; and nothing is yet known of the history of their development. There can be little doubt, however, that their early form, whatever it may be, must be very different from that of the animal whose peculiar structure has been just described. And the same remark may be made with respect to certain other forms still more aberrant, which are referred to the Acalephæ chiefly because they have no claim to admission into any other class.

295. The class of ECHINODERMATA (or 'prickle-skinned' animals) derives its name from the spiny integument with which some of the best known species belonging to it—such as the Star-fish and Sea-Urchin—are covered. The term, however, is far from being generally appropriate, for the spiny surface is only found in three out of the six orders of which the class is composed. A more universal character is the presence of cirrhi or tendril-like suckers, capable of being put forth from different parts of the surface of the body, and subservient to progression rather than to the prehension of food. But a still more constant character is the existence of a calcareous skeleton, which, however different in the general arrangement of its parts, is composed of the peculiar 'calcified areolar tissue' formerly described (§ 195), which is found in no other tribe of animals. Of this material there is formed in the *Encrinite* a branching stem, bearing a more or less massive body, which it does not entirely enclose; in the *Star-fish*, a regular frame-work of detached plates, lying just beneath the skin; and in the *Echinus*, a complete globular shell, armed on its exterior with spines, which are sometimes mere prickles, but which sometimes attain considerable length and massiveness. On the other hand, in the *Holothuria*, the skeleton is merely represented by a few scattered plates, which still, however, present its characteristic structure; whilst in the aberrant *Sipunculus*, which forms the transition to the Annulosa, it is altogether wanting. The class is made up of six groups, which are so closely connected, one with another, that there can be no doubt as to the propriety of their being thus associated; and yet, as in the preceding case, it is difficult to find any general characters which shall include them all under one definition. On the whole, however, they are much more elevated in the scale than any of the forms of Animal life which have been yet brought under notice. A very distinct nervous system exists in them all; and some of them exhibit a high

degree of complexity and elaborateness in the arrangement of their muscular structure, and in the manner in which the muscles are made to give motion to the hard parts. The most remarkable example of this is found in the dental apparatus of the Echinus. It is only in the Star-fish, however, that any organs of special sense have been yet detected.—The digestive apparatus here no longer presents the obscurity which characterized it in some of the preceding class; but consists of a definite stomachal cavity, with distinct membranous walls, opening by a single mouth: this mouth, in some instances, also serves as the anal orifice; whilst in other cases the alimentary canal is continued as an intestinal tube, which terminates by a separate aperture, usually at the opposite extremity of the body. The stomach is in some instances extended into the remoter parts of the body, as in most of the Acalephæ; but it is more commonly restricted to the central portion; and we here for the first time meet with a proper vascular apparatus, consisting of a system of anastomosing tubes which are *not* continuous with the digestive cavity, but which take up by absorption the nutritious fluid prepared by the digestive process, and carry it by a regular circulation through the system. The circulation, however, is not carried on by the agency of any muscular organ of impulsion, or heart; the only approach to such being a pulsatile enlargement upon some of the vascular trunks. Distinct respiratory and glandular organs make their appearance in this class, as will be presently seen. In the higher Echinodermata, we entirely lose the capability of gemmiparous or fissiparous propagation; some approach to it being shown in the remarkable power of regenerating *parts* of the body, which is possessed by the Star-fish, Holothuria, &c. The reproductive process, therefore, is entirely performed by a sexual apparatus; and so far as is yet known, the male and female organs are possessed by distinct individuals. In the development of the embryo of some of the orders, several curious phases are presented; the most remarkable of which will be hereafter described (CHAP. XVIII.). We shall now take a rapid survey of the principal features of each of the six orders, into which the existing forms of this class have been distributed by Prof. E. Forbes,* and which he has characterised according to the structure of their locomotive organs; and shall also notice a seventh, now altogether extinct, which is of peculiar interest to the Palæontologist.

296. Of the first order, *Crinoidea*, the existing forms are few; but it was extremely abundant in former periods of the earth's history; and its remains form no inconsiderable portion of the solid crust of the globe. Its characteristic structure is best known from the *Comatula;* an animal which, at first sight, does not depart very widely from the type of the Star-fish, but which differs from this in several important particulars. It consists of a central disk with arms, which is attached during the early period of its life to solid bodies, by means of a jointed stalk proceeding from the under side of the disk; this stalk, in the ordinary Crinoidea, is permanent; and in the freedom of its adult life, the Comatula may be regarded as forming a transition from them towards the unattached forms of the class. The digestive apparatus is confined to the central disk, and consists of a stomach opening by a mouth in the centre

* "History of British Star-fishes."

of the disk, with an intestinal tube which passes off from its side, winds round the body, and opens externally by a projecting vent, situated between the mouth and one side of the body. The arms arising from this disk are solid, being composed of the calcareous framework already described, covered with a thick, soft, contractile integument; they are five in number, but they speedily subdivide, each usually separating into four. To the central stem of each arm, jointed lateral appendages of a similar structure are attached; and these also are clothed with the fleshy integument, which extends on either side in a sort of fin-like expansion. By the simultaneous movement of the arms, and the stroke of these numerous *pinnæ,* or lateral appendages, against the water, the Comatula swims through the ocean very much after the manner of a Medusa; hence the order is distinguished by the name *Pinnigrada.** Sometimes, however, the Comatula attaches itself to sea-weeds or other floating bodies, and employs its long arms in bringing food to its mouth. The position of the ovaries constitutes one of the most curious features in the structure of this animal; for they are dispersed through the integument of the arms, in which they form many thousand distinct spots. The Comatula, in its young and attached state, has been described as the *Pentacrinus Europæus,* and has all the characters of a true crinoid. The stem is made up of the same kind of calcareous framework as the rest of the skeleton, and is covered with the same irritable integument, by the contractions of which the head may be made to turn in any direction; the attachment of its base to solid bodies is effected by means of a suctorial disk. The *pinnæ* are not developed, until near the close of the period of attachment—A much larger *Pentacrinus* (*P. Caput-Medusæ*) has been found, however, in the West Indian Seas; which probably passes its whole life in the attached condition, and is thus a truer representative of the vast assemblage of extinct Crinoidea. The disk and arms are formed like those of the Comatula; the latter are very numerous, and are thickly set with jointed pinnæ. The stem is more than a foot long, and is composed of a large number of pieces similar to those of the arms; from this stem there arise, at regular intervals, several verticils of secondary arms, which do not subdivide, and are destitute of lateral appendages. The ovaries are not so dispersed as in the Comatula; but they are still *external* to the central disk, being situated on the arms near their base. In some of the fossil species of Pentacrinus, which are especially abundant in the Lias formation, the subdivision and ramification of the arms is carried to a much greater extent than in either of the existing forms; and the number of pieces in the skeleton thus becomes very large. Thus in the *P. Briareus* it has been calculated that at least 100,000 separate joints exist, exclusive of those of the lateral appendages, which are probably more than 50,000 additional. The base of the stem of the recent *P. Caput-Medusæ* has not yet been obtained; so that its mode of attachment to solid bodies cannot be positively stated; but from the circumstances under which fossil remains of other species are sometimes met with, there is strong reason to believe that the animals of *this* genus were not permanently adherent to solid bodies, but had the power of occasionally detaching themselves, and

* This designation, however, is open to the objection that it is applicable to only a very small part of the order; the character of which, *as a whole,* is its *attached* condition.

moving freely from place to place in search of a new and more appropriate basis.—The bulk of the order, however, is made up of the very numerous tribe of *Encrinites,* of which we have no existing representatives, but which seems to have been in many respects of lower organization, connecting the free Echinodermata with true Zoophytes. The body and jointed stem exhibit a rounded instead of a pentagonal form ; the latter is usually destitute of secondary arms ; and the principal arms do not ramify with the same minuteness as in the Pentacrinus. The stalk seems to have been attached by a sort of spreading root, resembling that of many Corals ; and we must believe this tribe of Crinoidea, therefore, to have been entirely fixed during the whole of life. It contains a numerous series of forms, which vary considerably in the degree in which the body is enclosed by the calcareous skeleton. In some instances it is nearly shut in, on its upper as well as its under side, by calcareous plates, so that it almost resembles a stalked Echinus ; whilst in other cases, the under side of the body alone seems to have been supported by the solid disk, which is formed in the shape of a cup, more or less deep, from the margin of which the arms spring, its upper portion or opening having been apparently enclosed by soft integuments alone, as is the case with the existing species of this order.

296 *a.* The second order, *Ophiurida,* receives its name from the long serpentiform arms, which are appended to the round depressed urchin-like bodies. These animals, too, are commonly associated with the Star-fish, but are constructed upon a very different type. As in the *Crinoidea,* the viscera are entirely confined to the central disk, and the arms are solid, being composed of a jointed calcareous framework, covered with a contractile integument, by which they are moved with great freedom. On the other hand, they differ from the Crinoidea, and agree with the *Asteriada* (or true Star-fish), in having but a single orifice to the digestive cavity, which is nearly as simple in its structure as that of the Actinia (§ 291). In the position of their ovaries, the Ophiurida differ from both these orders ; for these bodies are situated within the disk, but discharge their contents by separate orifices near the base of the arms. The disk is never attached, except in the early embryonic state of these animals, which in their adult condition are among the most active of the whole class ; their arms being very efficient instruments of locomotion. These are sometimes simple and undivided from their base to their free extremity, gradually tapering to a point, as in the ordinary *Ophiuræ ;* whilst in *Euryale* they ramify minutely, dividing regularly into branches, which again subdivide, so as to form a complex series of appendages. Their surface is usually beset with spines, or covered with rough scales ; and these appear to be of great use in the movements of the animals, giving to the arms the power of *resting* against points of the surface over which they are moving, from which they can propel themselves in any direction. Hence, as the *Ophiuridæ* mainly depend upon their spines for locomotion, they have been termed *Spinigrada.* The body and arms in this order are furnished with *cirrhi ;* but these are not sufficiently developed to assist in locomotion ; although those near the mouth are enlarged into tentacula, which seem to draw the food towards that orifice. Most of the animals of this order are remarkable for their brittleness, and their tendency to separate into a number of pieces when laid hold of. It

is certain that the body can reproduce the arms ; and it has been affirmed that if a portion of the disk remain attached to a separated arm, the entire individual may be reproduced. Whether in this manner two or more individuals can be developed by the subdivision of a single one, or whether this process of fission ever takes place spontaneously, has not yet been ascertained. The ordinary mode of propagation is undoubtedly by sexual generation ; but the young animal, on its emersion from the egg, bears a form which differs so completely from that of the adult, that it would not be recognized as having any relation with it. The larval or embryonic structure, which has received the distinctive name of *Pluteus*, and from which the Ophiura is subsequently to be developed, is a transparent gelatinous body, conical above, and prolonged into eight branches below, thus having somewhat the form of an easel with eight supports instead of three ; it is beset with rows of cilia, by whose agency it is freely propelled through the water, after the manner of the ciliograde Acalephæ (§ 294*a*), for one of which it might readily be mistaken, were it not for the presence of calcareous rods and spicula in the body and branches. A more particular description of this curious larva, and of the mode in which the Ophiura is developed from it, will be given hereafter (CHAP. XVIII.).

296*b*. We now come to the *Asteriada*, or true 'Star-Fishes,' in which the real *arms* disappear ; what are commonly known as such, but more properly termed *rays*, being lobes or divisions of the body itself. In some instances there is scarcely any central disk, the body being almost entirely divided into rays ; whilst in other cases there is scarcely any appearance of rays, the central disk being but slightly indented at its edges. The skeleton of these animals is composed of calcareous plates, covered with a dense integument ; and this is very commonly raised into prickles, which are sometimes horny, but are sometimes calcareous in their interior, like rudimentary spines. The calcareous plates have usually no great regularity on the dorsal surface of the body and rays ; but on the lower surface of the rays, they are arranged with the most exact symmetry, in such a manner as to afford protection and support to the tubular cirrhi, which now become the principal instruments of progression (Fig. 110 *e f*). These cirrhi are tubes composed of a contractile membrane, and furnished with suckers at their extremities ; every one of them is connected internally with a vesicle filled with water, by the contraction of which the tube can be projected through a passage that is left between the calcareous plates ; whilst, on the other hand, by the general contraction of the membrane of the tube itself, it can be shortened and drawn in, or, by its contraction on one side more than on the other, it may be turned in any direction. The vesicles are themselves connected with a vascular canal which surrounds the mouth. The Asteriada possess a considerable amount of locomotive power, but it is for the most part sluggishly exerted. Their progression is accomplished entirely by means of the tubular cirrhi ; which, when extended, take an attachment to some fixed object, and when retracted, draw the body towards it ; and hence they are appropriately termed *Cirrhigrada*. The rays, too, in many instances, are very flexible, so that they can be drawn up towards the mouth, or moved from side to side towards each other ; and in this manner food can be brought to the entrance of the stomach, and the body can be insinuated through narrow apertures. The mouth of the Asteriada is situated in the centre of the disk, and opens at once into the stomach, which occupies a considerable part of the cavity. The surface on which the mouth opens, and which is that whence the cirrhi are put forth, is designated as the *lower* surface; the Star-fish and its allies habitually living in what would be (as compared with other animals) an inverted position. From the central stomach (Fig. 110, *a*), which is thin

FIG. 110.

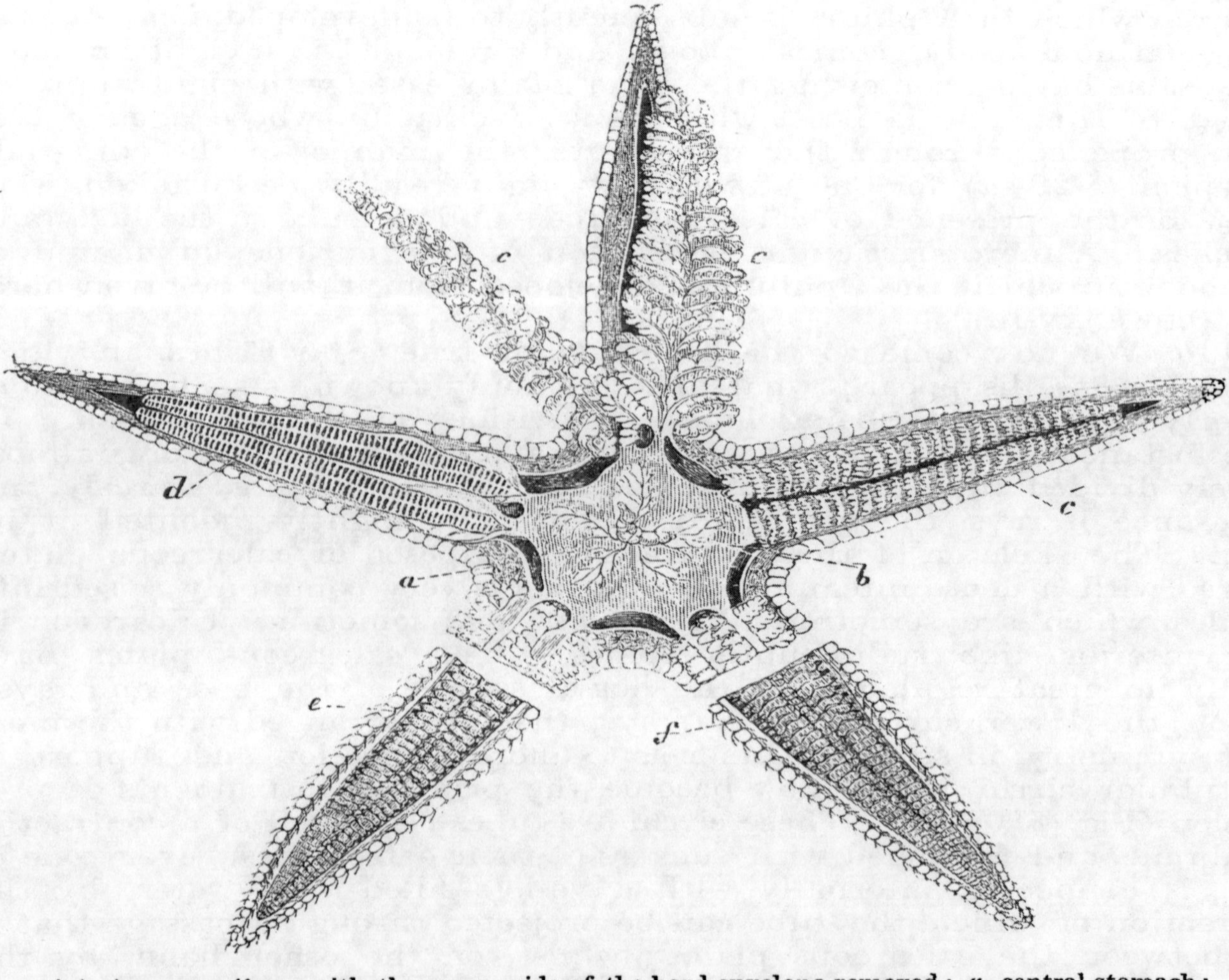

Asterias aurantiaca, with the upper side of the hard envelope removed: *a*, central stomach; *b*, cœca upon its upper surface (salivary glands?); *c c*, cœcal prolongations of the stomach into rays; *c'*, the same empty; *d*, the same laid open; *e*, the under surface, seen from within after the removal of the cœca, showing the vesicles of the tubular cirrhi; *f*, the same in a contracted state, showing the skeleton between them.

and membranous, a pair of cœcal prolongations passes out into each ray (*c, c*); these are considerably subdivided and ramified, and their walls are beset with numerous follicles, which are not improbably to be regarded as a rudimentary form of a liver. Above the central stomach, in some species, are seen five other cœca (*b, b*), which appear to be secreting organs, and may be the analogues of the salivary or pancreatic glands of higher animals. No separate anal orifice exists; the indigestible matters being voided by the mouth. A regular system of blood-vessels, for the absorption of the products of the digestive process, and for the conveyance of them to the different parts of the fabric, exists in the Asteriada; its arrangement being conformable to the very regular radial symmetry which exists in the body generally. No special respiratory apparatus, however, yet exists; but the surrounding water is freely admitted into the general cavity of the body, and bathes the surfaces of the viscera, which are covered with vibratile cilia. A distinct nervous system is here found, arranged upon the most regular radiated type (§ 282 *a*); and many species possess *ocelli* at the extremities of the rays.—These animals possess a remarkable capability of reproducing lost parts; but the preservation of the disk seems essential to their power of doing so; and hence, although three or four of the rays may be removed without the destruction of the disk, which gradually regenerates them, the separated rays cannot form a new disk; so that no multiplication of individuals can ever take place by this process. The true generative apparatus is very obvious in them at the breeding season; consisting of a set of ramified tubes, which are arranged in pairs in each ray, and which seem to open by minute pores in the circle of calcareous plates that surrounds the mouth. These tubes, in the male, are filled with sperm-cells, and in the female with ova; the sexes being thus distinct. The ova, when matured and fertilized, are retained (in some species at least) in a kind of marsupium formed around the mouth of the parent by a peculiar change in its form; and there the embryo undergoes the early phases of its development, attached at its base by a sort of footstalk divided into three lobes, which contracts as the disk is developed, and is at last entirely withdrawn into the body of the animal, remaining only traceable as the 'madreporiform tubercle' which is found on one of the plates of the dorsal surface of the body.

296*c*. Of the order *Echinida*, the type is the well-known *Echinus* or 'Sea-Urchin,' which seems to have received its name from the resemblance presented by its prickly surface to that of the hedgehog or 'urchin.' This animal has a shell of nearly globular form, composed of polygonal plates (Fig. 39) accurately fitted together, so as to constitute a very complete investment to the body. The surface of the shell is beset with tubercles, on which are implanted a vast number of spines, having regular sockets at their bases, enabling them to be readily moved in any direction by the contractile integument which invests the shell, and which is attached to the bases of the spines, round which it is considerably thickened. These tubercles, however, are not equally present on *all* the component plates of the shell; for we find rows of smaller plates, passing from pole to pole, the tubercles of which are comparatively few and small, whilst they are themselves perforated with apertures for the tubular cirrhi. The structure of these organs is precisely the same as in the Asteriada; but their development is greater, for they are always capable of being projected further than the spines (which are frequently several inches in length), and of taking an attachment to fixed objects beyond. By their agency, the shell can be drawn onwards in any direction; the movement being effected by the contraction of the tubes, whilst the body is supported upon the spines; and from this compound mode of progression, the Echinida may be appropriately termed *Cirrhi-Spinigrada*. In all the Echinida, the alimentary canal possesses two distinct orifices; which, in the *Echinus* and its nearest allies, are both of them central, being situated at the two poles of the globular shell. In the *Clypeaster* and *Scutella*, however, we have an approach towards the Asteriada; the shell being more or less flattened and divided at its margin, so as to resemble the body of a Star-fish; whilst the anus leaves its central position on the upper surface, and approaches the mouth, which still retains its central position below. In the *Spatangus* and its allies, on the other hand, the radiated form is considerably departed from; the shell being oval instead of globular, and the mouth and anus being neither of them central; in fact, the radiating arrangement shows a tendency to give place to a bi-lateral symmetry; and in this and some other particulars, the Spatangaceæ may be considered as leading towards the next order.—Notwithstanding these variations, however, the general plan of structure in the Echinida is nearly the same. The mouth (Fig. 111, *a*), situated at the lower pole of the globe, and surrounded in the Echinus and its allies by a complicated apparatus of teeth set in a calcareous framework (*a,c*), leads by an œsophagus (*d*) to the most dilated portion of the alimentary canal (*e*) which may be considered as the stomach. From this passes off the intestinal tube (*g, h*), which makes two turns round the shell, and terminates at the anal orifice (*i*). The alimentary canal is attached to the interior of the shell by a fold (*f*) of the membrane which lines the cavity; and this 'mesenteric' membrane contains distinct blood-vessels, by which (as in the Asteriada) nutriment is absorbed from the digestive cavity, and distributed to other parts of the body. Respiration is chiefly provided for, as in them, by the admission of the surrounding water into the cavity of the shell; the membrane lining which is clothed with cilia, whose action occasions a constant renewal of the liquid. The water is introduced into the cavity by ten respiratory tubes, which are placed around the mouth, and which bear a strong resemblance to the tubular tentacula of the Actinia. In the neighbourhood of the mouth, also, are glandular structures, which would seem to be salivary organs. No distinct traces of a liver have yet been detected. The nervous system is formed upon the same plan as that of the Asterias; but no organs of special sense have yet been detected.—No trace of the fissiparous or gemmiparous modes of reproduction seems to remain among the Echinida; and it is not yet known what extent of reparative power they possess. Their proper generative apparatus, however, is highly developed; and consists

FIG. 111.

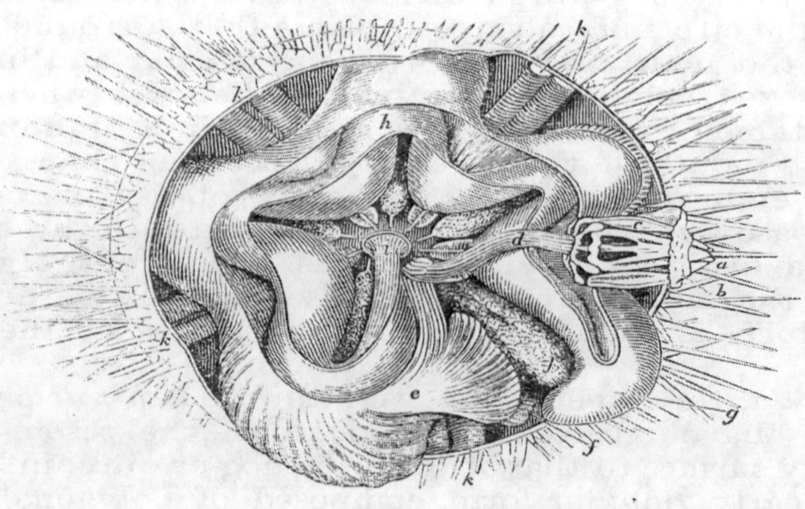

Anatomy of *Echinus lividus*, laid open from the under side; *a*, buccal apparatus turned to one side; *b*, portion of tegumentary membrane surrounding the mouth; *c*, calcareous jaws; *d*, œsophagus; *e*, commencement of intestine; *f*, mesenteric membrane; *g*, duplicature of the intestine, which forms a second convolution, *h*, along the course of the first, and terminates at the anal orifice *i*, around which are seen the five oviducts; *j*, ovaria; *k k*, ambulacral vesicles.

of five large sacculi, which open by as many separate orifices in the plates that surround the anus. These sacculi contain spermatic cells in the male, and ova in the female. No provision here exists for the protection of the embryo during its immature state; but it comes forth from the egg in the condition of a free-moving 'pluteus,' and it is only by a long process of development, during which the animal is supported upon nutriment obtained and assimilated by itself, that the typical form of the Echinus is evolved.

296*d*. The recent discovery of an entirely new series of forms of Echinodermata, which abounded in the early ages of the earth's history, but which seems to have become entirely extinct before the Pentacrinites were called into existence, has rendered it necessary to institute a new order, that of *Cystidea*, the place of which seems to be intermediate between the *Crinoidea*, the *Ophiurida*, the *Asteriada*, and the *Echinida;* for it combines within itself, in the most remarkable manner, some of the distinctive characters of each of these groups. "The Cystidea are more or less spherical bodies covered with polygonal plates, varying in number according to the genus, closely fitting together so as to invest the entire surface with a compact coat of mail, except at four points; viz. inferiorly, where the body unites with a stem; centrally, or above the centre on one side, where there is an opening closed by valves, supposed with good reason to be the orifice of the reproductive system; and superiorly, where the mouth is found, usually if not always with a small perforation, supposed to be a vent, alongside of it. These parts, viz. the plates investing the body, the three orifices (for the fourth perforation, that of the base, is continuous with the canal of the stem where the latter is well developed), and probably the stem, are common to all Cystidea. There are certain other parts, apparently of great consequence in the organisation of the animal, which are common only to certain members of the order. These are, the brachial appendages (arms and tentacula), and certain curious organs or appendages connected with the plates, to which the name of 'pectinated rhombs' may be appropriately given." *—Thus in the attachment of the body by a stem, the *Cystidea* resemble the *Crinoidea,* and some of the aberrant forms of the two orders come into very close approximation with each other: their types differ essentially, however, in the relative position of the ovaries. In the complete enclosure of the body by polygonal plates, they correspond with the *Echinida;* but differ from them in the attachment of the body by a stem, in the single outlet of the reproductive apparatus, and in the possession (in some instances) of arms. In the division of the body of certain genera into lobes, again, they approach the *Asteriada;* but differ from them in the characters just enumerated. The arms, where they are present, are more allied in structure and origin to those of *Ophiurida,* than to those of Crinoidea; and it is through this group that the Cystidea show the nearest approach to the free-moving forms of Echinodermata; certain genera having five ovarial orifices, instead of a single one, and these orifices being situated near the base of the arms, as in the Ophiura and its allies. In the *singleness* of the generative apparatus, however, which is the ordinary character of the group, it may even be thought to manifest a tendency towards the *Holothuriada.*

* See Prof. E. Forbes on the British Cystidea, in "Memoirs of the Geological Survey of Great Britain," vol. ii.

296*e.* In the order *Holothuriada,* we find united in a very curious manner the radiated arrangement of the exterior, and especially of the oral appendages, which is characteristic of the Echinodermata in general, with that elongation of the body and bi-lateral symmetry of the internal

FIG. 112.

organs, which are marked features in the structure of Articulated animals; and some of them, moreover, exhibit obscure indications of a transverse division of the body into segments. We here take leave almost entirely of that calcareous skeleton, which has been so remarkable a feature in the structure of all the preceding orders; for we find the body inclosed within a soft distensible and contractile integument, unsupported, save (in some species) at the base of the tentacula and in a few other situations, by calcareous plates; and these disposed in such a manner as to allow the utmost freedom of motion, so that both the size and form of the body can undergo great variation. The tubular cirrhi, however, still exist (Fig. 112); and are sometimes found in distinct rows, extending from one pole of the body to the other, whilst in other cases they are scattered irregularly over the surface, or are limited to one side of it on which the animal usually crawls. The progressive movement of the body is partly effected by the cirrhi, but partly also by the extension and contraction of the body itself, as in the vermiform tribes: hence this order has been designated by the term *Cirrhi-vermigrada.* The general form of the body is not unaptly expressed by the vernacular name of 'sea-cucumber,' which is given to some of the commonest Holothuriæ of British coasts; but in some species it is more globular, approaching that of the Echinus, whilst in others it is more prolonged.—In their internal structure, the *Holothuriada* present us with the highest type anywhere encountered in this order. The mouth and anus are always placed at the opposite extremities of the body, and never show any tendency to approximate. The mouth (Fig. 113, *b*) is surrounded by ramifying tentacula (Fig. 112), which may be regarded as highly developed cirrhi; these can be completely drawn back into the mouth, and are sometimes not put forth for days together. Within the mouth is a dental apparatus, strongly resembling that of the Echinus; to which is appended a set of glandular cœca (*t'*), which are probably salivary organs. The œsophagus passes through this, and opens into a stomach, the walls of which are sometimes distinctly muscular, but which is generally not clearly distinguished from the intestinal canal. From the stomach proceeds the intestine (Fig. 113, *i*), supported by its mesentery *m*, to terminate in the funnel-shaped cloaca *c*. In the mesentery is seen a copious distribution of blood-vessels; and this forms part of a complex vascular system, which conveys the blood through all portions of the body. Several parts of its main trunks seem contractile; but the impulsive power appears essentially concentrated in one organ, a contractile vesicle (*p*), which is the rudiment of a heart. A distinct and elaborate respiratory apparatus here exists, consisting of a pair of arborescent cœca (*r r*), which originate in the cloaca, and extend and ramify throughout the body; and water appears to be alternately introduced into these organs, and expelled from them, by the action of the powerful muscles which

FIG. 113.

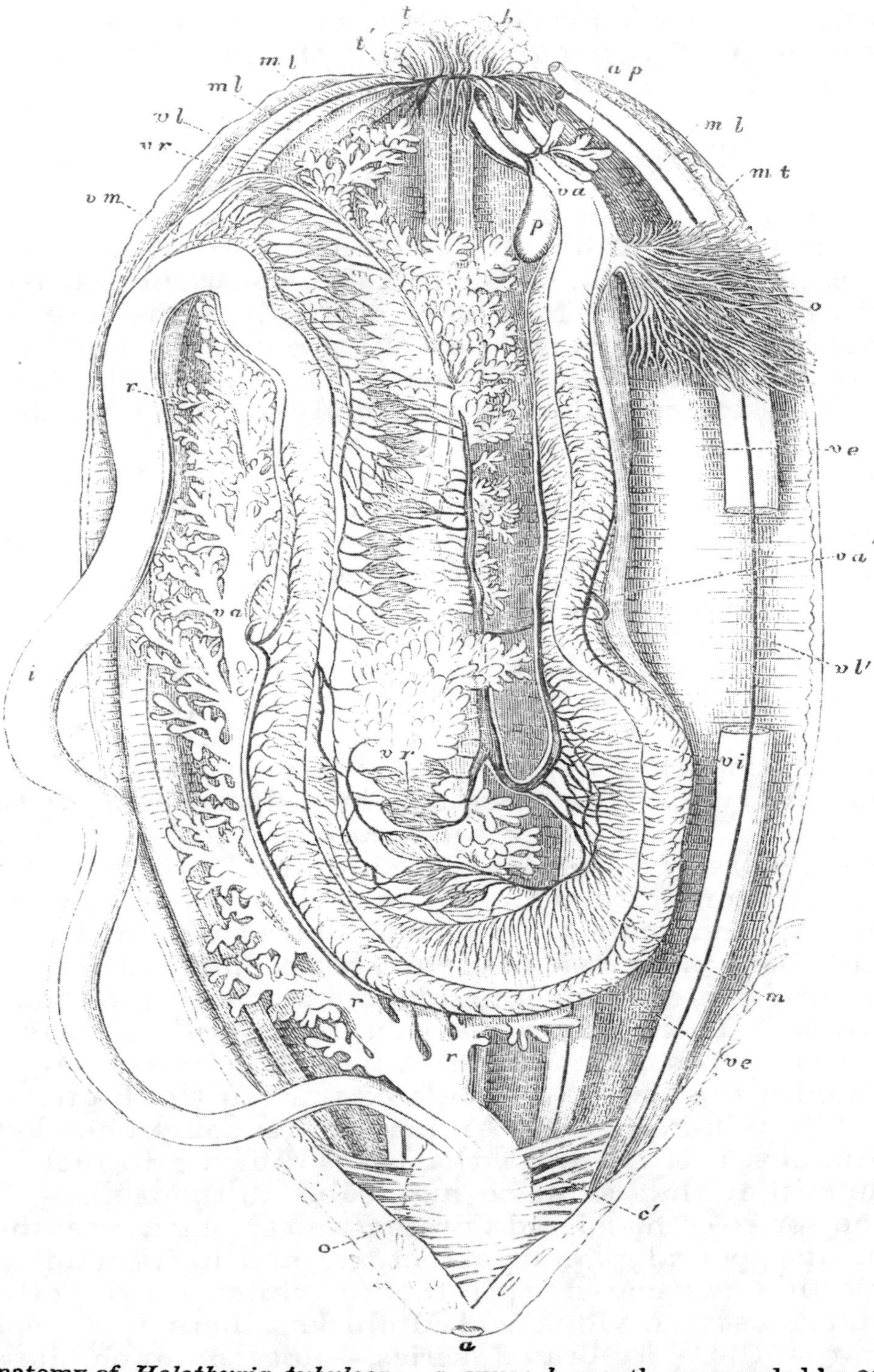

Anatomy of *Holothuria tubulosa*;—*a*, anus; *b*, mouth, surrounded by 20 tentacula; *c*, cloaca, surrounded by muscular dilators *c′*; *i*, intestinal tube; *m*, mesentery; *ml*, *ml*, longitudinal muscles; *mt*, transverse muscles lining the entire inner surface of the integument; *o*, ovary; '*ap*, cœcal appendages, probably seminiferous; *p*, contractile vesicle, probably a heart; *r*, *r*, respiratory apparatus, originating in the cloaca; *t*, oral tentacula; *t′*, cœcal reservoirs; *va*, annular vessel surrounding the mouth and supplying the tentacula; *ve*, external intestinal vessel, giving off a large anastomotic branch *va′* which enters another part of the same trunk; *vi*, internal intestinal vessel, with contractile dilatations; *vl*, longitudinal tegumentary vessel, giving off transverse branches *vl′*, seen by removing the longitudinal muscles; *vm*, mesenteric vessels, connecting the branches of the external intestinal vessel with those of the respiratory system of vessels, *vr*.

form the walls of the visceral cavity. These muscles, which are disposed with great regularity (*ml*, *mt*), appear to be put in action by a nervous system, which is formed upon the general plan of the group, consisting essentially of a gangliated ring that surrounds the mouth. Among the most curious facts in the economy of these animals, is their tendency to eject the entire mass of viscera by a kind of convulsive action ; this they are usually said to do in alarm when captured ; but Professor E. Forbes states that he has never seen such an occurrence. He has met with specimens of many species, however, in which there was not a trace left of any of the viscera, although the creature was taken alive and seemed healthy. According to Sir J. G. Dalyell, the viscera are all regenerated in the course of a few months. It is affirmed by the same observer, that some species of this order divide spontaneously through the middle into two or more parts, each becoming ultimately perfect by the development of new organs. This self-division would seem especially to occur in the most vermiform tribe of the group (that of *Synaptæ*) ; and it is to be regarded rather as a character of approximation to the lower Annulosa, in some of which we shall find it remarkably manifested, than as representing the gemmiparous multiplication of the lower Radiata. The ordinary mode of reproduction in the Holothuriada is by a true generative apparatus, which differs from that of all the preceding orders now existing, but agrees with what seems to have been the plan of conformation among the Cystidea, in being a *single* organ, instead of being repeated five or more times round a common centre. This organ (Fig. 113 *c*) is a cluster of cœca, which opens, by a canal of its own, near the mouth ; and it contains spermatic cells, or ova, according to the sex. Of the development of the embryo in this order nothing whatever is yet known.

296*f*. In the order *Sipunculida*, the radiated arrangement still more completely gives place to the annular. In their external appearance, the animals composing it are Worms; they have no calcareous plates, nor cirrhi ; the oral tentacula are wanting, or are irregular in number and disposition ; and their progression is entirely accomplished, like that of worms, by the contraction of their integuments ; whence they may be designated *Vermigrada*. Their internal structure is still imperfectly understood ; but it appears probable, from the latest enquiries, that their true place will be found rather in the Articulated than in the Radiated series.*

297. The range of Animal forms comprehended in the Sub-kingdom MOLLUSCA is so great, that it is difficult to include them in any positive definition that shall be applicable to them all. The lowest group which the series comprehends, presents so strong a resemblance to *Zoophytes*, both in its general plan of structure, and in its habit of life, that it was until recently associated with them ; whilst, on the other hand, in the highest and most individualized Mollusks, there is a decided approximation towards the Vertebrated series,—not so much, however, in their general type of conformation, as in the structure of their separate organs. The zoophytic Mollusks for the most part exhibit, more or less distinctly, that *circular* arrangement of the parts about the mouth, which is the most obvious character of the Radiated classes ; but it is comparatively seldom that we meet with that exact *bilateral* symmetry of external form, which is nearly an invariable characteristic of Articulated and Vertebrated animals.

* See Blanchard, in Ann. des Sci. Nat. Juillet, 1849.

With the exception of the *Chiton* (Fig. 125) and its allies, which form a connecting link between the Molluscan and Articulated series, we do not find any trace of that division of the *external* skeleton into segments, which is the distinctive feature of Annulose animals ; and even in those highest Mollusks, which present the rudiment of an *internal* skeleton for the support and protection of the nervous centres, we never perceive the least indication of that segmental division of this skeleton, which is the characteristic feature in the structure of the Vertebrated series.—Taking them as a whole, the Mollusca are characterized rather by the *absence,* than by the possession, of any definite form ; and there is a corresponding absence of any regular organs of support, by which such a form could be maintained. The *shell,* where it exists, is to be regarded rather in the light of an appendage, designed for the mere protection of the body, and deriving its shape from it, than as a skeleton giving attachment to muscles, and regulating the form of the whole structure, which is the case with the skeletons both of Articulated and of Vertebrated animals. It is in no instance a fixed point for the muscles of locomotion ; and it is only, indeed, where the body is uncovered by a shell, or where the locomotive organs can be projected beyond it, that any active movements can be executed. In a considerable proportion of the series, no shell exists : and it is a clear indication of the non-essential character of this appendage, that we find it absent in the one, and present in the other, of two groups which are formed in every other respect upon the same general plan, the Slugs and the Snails. Leaving the shell, therefore, out of consideration, and attending merely to the general characters of the body, we notice, in the first place, that this, as the designation of the group implies, is altogether *soft,* and that even where locomotive appendages exist, they are as destitute of firm supports, either internal or external, as is the body itself. The whole fabric is enclosed in a thick fleshy glandular skin, which is called the *mantle ;* it is on the surface of this envelope that the shell is formed, by the development and subsequent calcification of epithelial cells, as already explained (§ 197) ; and it is by the muscular fibre which the mantle contains, that the movements of the animal are chiefly effected. The muscular structure is frequently concentrated, as it were, in some particular part or parts of the mantle, so as to constitute a single or multiple locomotive organ or *foot ;* thus in the Gasteropoda (§ 302), it forms a sort of flattened disk, by the alternate contraction and extension of different parts of which, the animal can slowly crawl onwards ; in the free moving Conchifera (§ 301) it is a tongue-like organ, capable of being projected from between the valves of the shell, and serving to propel the body by more sudden movements ; whilst in the Cephalopoda (§ 306), we find the prolonged and contractile tentacula (or oral appendages) to be the chief instruments of locomotion. The disposition of the nervous centres is chiefly remarkable for its want of constancy (§ 282 *a*), and for the absence of either circular or longitudinal repetition. In the lowest forms of this group, we are presented with the simplest known type of a nervous system, namely, a single ganglion, with afferent and efferent fibres ; and this is adequate to call the muscular mantle into contraction upon the application of stimuli, which is the only movement that such animals have occasion to perform. But with the advance in complexity of structure, and more especially with the development of sensory organs and of special locomotive appendages, we find the number of ganglia increasing; still, however, without any approach to a definite arrangement, save that they are for the most part disposed in the neighbourhood of the mouth.—From this absence of any uniform plan in the arrangement of the nervo-muscular apparatus, and from the want of any solid framework which might afford fixed points of attachment to the muscles and enable them to act at a mechanical advantage, it is obvious that the development of the proper *animal* portion of the fabric of the Mollusca is quite subordinated to that of the organic or *vegetative*, which we shall find to present a remarkable elaborateness and complexity.

297 *a*. The alimentary canal of the Mollusca invariably possesses two distinct orifices, an oral and an anal; and it usually presents an obvious separation between the œsophagus or gullet, the stomach or digestive cavity, and the intestinal tube. The mouth, or entrance to the œsophagus, is not situated, in the lower Mollusca, on a prominent part of the body, nor is it surrounded by organs of special sense; in some instances, indeed, it is buried deeply amidst other organs; but in the higher classes, it is situated on a *head* distinct from the trunk, which is also furnished with well-developed eyes, and with rudimentary organs of smell and hearing. In the lower Mollusca, again, the mouth leads directly to a funnel-shaped œsophagus, down which a current of water is continually driven by ciliary action; and it is from the animalcules and other minute organic particles which the water may contain, that these beings derive their support. But in the higher members of the group, the mouth is furnished with instruments for the mechanical reduction of solid food, which are put in action by a complex muscular apparatus; and they have locomotive power enough to go in search of that food, instead of being dependent upon such as the currents in the water they inhabit may chance to bring them. The alimentary canal occupies in the Bryozoa, as in Polypes, a large part of the general cavity of the body; but in all the higher Mollusca it is more limited in its dimensions, and is completely isolated from other organs. In the curious group of Nudibranchiate Gasteropods, however, it is prolonged into a multitude of tubular appendages, which radiate into almost every part of the body (§ 303 *d*), in a manner that reminds us of the extension of the gastric cœca into the rays of the Star-fish (§ 296 *b*). Throughout the whole series, the liver is very distinctly developed. The form in which it presents itself in the Bryozoa is quite rudimentary; but even in the higher Acephala, such as the Mussel or Cockle, it constitutes a large part of the mass of the body; and in the Cephalopoda it is constructed upon a plan nearly as elevated as that of Fishes. Other secreting structures, also, present themselves in these classes, in a more or less distinct form; such as salivary glands, pancreas, and urinary organs. In all but the very lowest Mollusca, we find a set of vessels adapted to absorb from the walls of the alimentary canal the nutrient fluid which has been prepared by the digestive operation; and to convey this into the parts of the body remote from the visceral cavity. In the class in which these vessels first present themselves, we find that, as in the Holothuria (§ 296 *e*), the heart is little more than a pulsating dilatation of the principal vascular trunk; but as we ascend the series, we find it becoming more and more distinct from the vessels with which it is connected, presenting a division into two or

more cavities, and acquiring distinct muscular walls. The blood is invariably colourless, and contains but few floating corpuscles. In all but the very lowest, again, we find a distinct respiratory apparatus, to which the blood is transmitted for aeration. With the exception of the terrestrial Gasteropods, which constitute but a small proportion of the entire series, we find this apparatus uniformly constructed with a view to aquatic respiration; and in all but the Cephalopods, in which streams of water are propelled over it (as in Fishes) by muscular action, we find that the renewal of the water in contact with the surface of the gills is chiefly due to ciliary movement. The propagation of the lower Mollusca is effected, like that of Zoophytes, by gemmation as well as by the true generative process; the gemmæ being sometimes detached, so as to make their way freely through the water; but more frequently remaining connected with the parent structure and with each other, so as to form aggregate or composite fabrics, strongly resembling those of Zoophytes, for which some of them have been mistaken. In all instances, however, a true generative apparatus is very distinctly developed. In the lowest classes, the two sets of organs are united in the same individual, in such a manner that its 'sperm-cells' may impregnate its 'germ-cells' and produce fertile ova. In certain higher forms, although both sets of organs are present in each individual, yet the congress of two is requisite, the germ-cells of each being fertilized by the sperm-cells of the other. But in a large proportion of the higher classes of the Molluscous series, the sexes are completely distinct; and the power of propagation by gemmation is altogether wanting.—Thus it appears that the predominant character of the group consists in the high development and functional activity of the apparatus of organic or vegetative life, in comparison with that of the locomotive and sensorial organs. A considerable number of the lower Mollusca are fixed to one spot during all but the earliest period of their lives; the movements of one part of the body upon another are of the simplest possible kind; and their food is obtained by an action of the same kind as that which serves for the alimentation of Animalcules. And thus we find beings possessed of an elaborate digestive apparatus, of a complex system of blood-vessels with a distinct contractile heart, and of special respiratory organs of great extent and completeness, furnished with no organs of sense but such as Zoophytes seem to possess in common with themselves, and leading a life almost as inactive (save in regard to the movements of the cilia which clothe their internal surfaces) as that of Plants. Even among the Gasteropod Mollusca, which may be regarded as the types of the whole group, although the body is free to move, yet the term 'sluggish,' derived from one of the best-known forms of the class, is applicable to the locomotion of the whole of it; and there is every reason to believe that the sensations are as obtuse as the movements are slow, the use of both these kinds of animal power being limited to the search for food, which, when found, is voraciously devoured. It is only in the Cephalopod Mollusks, which constitute an aberrant group, leading us in many respects towards the Vertebrated series, that we meet with very active powers of locomotion, or with evidence of acuteness of sensation.—The distinctive characters of the principal classes into which this Sub-kingdom is divided, will be now more particularly described.

298. In the class of BRYOZOA, we have a remarkable adaptation of the Molluscan type of structure to the Zoophytic mode of life; and it is not surprising, therefore, that, so long as the former was ill understood, and the latter received almost exclusive attention, the group should have been associated with that which contains the true Polypes. That its true place, however, is in the Molluscan series, is now generally admitted by those most qualified to appreciate its affinities, and will become sufficiently obvious when the close alliance between this class and the next shall have been shown by the comparison of the essential features of both.—The animals of the Bryozoa, in consequence of their universal tendency to multiplication by gemmation, are seldom or never found solitary, but form clusters of various kinds; these are sometimes connected by a stolon or creeping stem, as in the *Laguncula repens* (Fig. 114), but sometimes spread out in every direction over a plane surface, as in the ordinary *Flustræ, Lepraliæ,* &c., and not unfrequently form (by their arborescent mode of growth, and by the consolidation of the connecting tissue) a coral-like mass, such as that of *Eschara cervicornis* (Fig. 115). Each individual is composed externally of a sort of sac, of which the outer or tegumentary layer is either simply membranous, or is horny, or in some instances calcified; and this is always continuous with that of other individuals, either by direct contiguity, or by the intermediation of the stem, or of a connecting tissue which may be either spongy or solid. This tegumentary sac is lined by a more delicate membrane, which closes its orifice, and which then becomes continuous with the wall of the alimentary canal; this lies freely in the visceral sac, floating (as it were) in the liquid which it contains. The further details of the anatomy will be best understood from the examination of a characteristic example, such as the *Laguncula repens;* which is shown in the state of expansion at A, Fig. 114, and in the state of contraction at B and C. The mouth is surrounded by a circle of tentacula, which are clothed with vibratile cilia; these tentacula, in the species we are considering, vary from ten to twelve in number; but in some other instances they are more numerous. By the ciliary investment of their tentacula, the Bryozoa are at once distinguishable from those Hydraform polypes to which they bear a superficial resemblance, and with which they were at one time confounded; and accordingly, whilst still ranked among the Zoophytes, they were characterized as *Ciliobrachiata.* The mouth, situated in the centre of this circle of tentacula, leads to a funnel-shaped cavity, or pharynx, *b,* which is separated from the œsophagus *d* by a valve at *c;* and this œsophagus opens into the stomach *e,* which occupies a considerable part of the visceral cavity. In the *Bowerbankia,* and some other Bryozoa, a muscular stomach or gizzard, for the trituration of the food, intervenes between the œsophagus and the true digestive stomach. The walls of the stomach *h* have considerable thickness; and they are beset with minute follicles, which seem to have the character of a rudimentary liver. This, however, is more obvious in some other members of the group. The stomach is lined, especially at its upper part, with vibratile cilia, as seen at C *g;* and by the action of these, the food is kept in a state of constant agitation during the digestive process. From the upper part of the stomach, which is (as it were) doubled upon itself, the intestine *i* opens, by a pyloric orifice *f,* which is furnished with a regular valve; within the intestine are

seen at k particles of excrementitious matter; which are discharged by the anal orifice at l. No circulating apparatus here exists; but it is probable that the liquid which fills the visceral cavity is not mere water, but contains the nutritive matter which has been prepared by the digestive operation, and which has transuded through the parietes of the alimentary canal; a few corpuscles of irregular size are seen to float in it. We shall hereafter see that this sac, in several other instances, may be regarded as a sort of diffused form of the circulating apparatus, and even constitutes a part of it when it is more specialized. The visceral sacs of the different individuals put forth from the same stem, appear to communicate with each other. No other respiratory organs exist than the tentacula, which are hollow, their cavities being connected with that of the visceral sac; and into these it is probable that the nutritive fluid is sent for aeration. — The sperm-cells are developed in a glandular body, the testicle *m*, which lies beneath the base of the stomach; when mature, they rupture, and set free the spermatozoa *q q*, which swim freely in the liquid of the visceral cavity. The ova, on the other hand, are formed in an ovarium *n*, which is lodged in the membrane lining the tegumentary sheath, near its outlet; the ova, having escaped from this into the visceral cavity, as at *o*, are fertilized by the spermatozoa which they there meet with; and

FIG. 114.

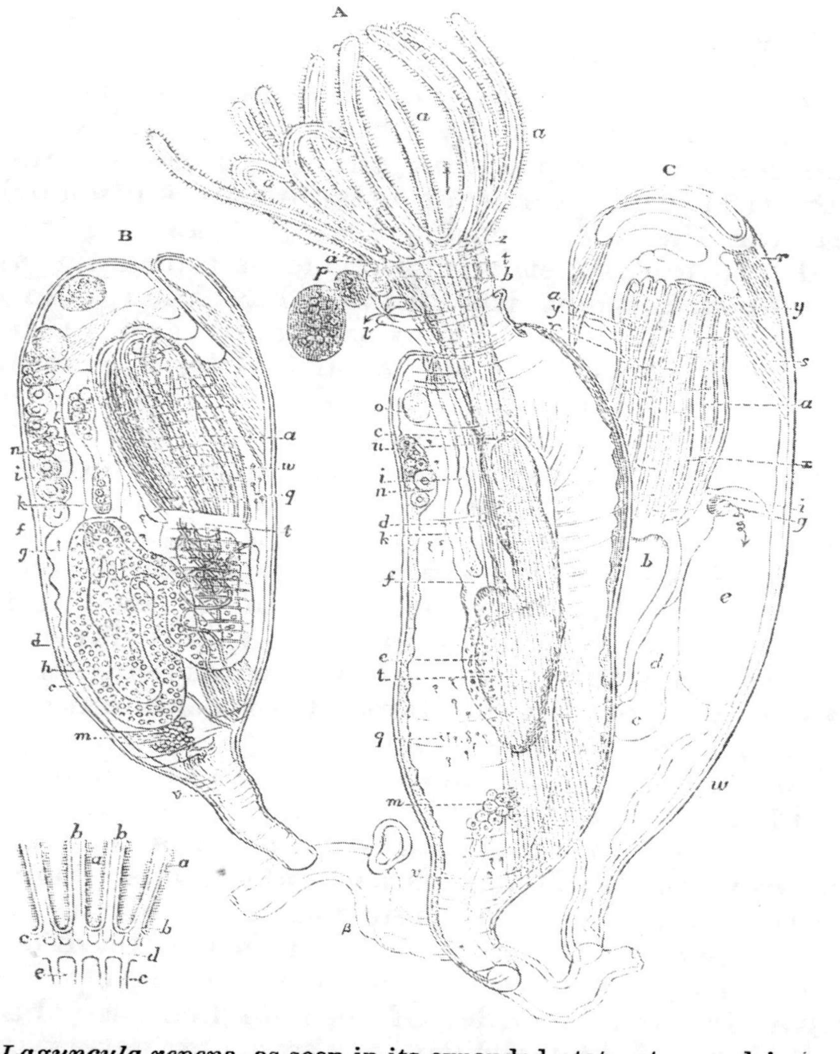

Laguncula repens, as seen in its expanded state at A, and in its contracted state, in two different aspects, at B and C. The same references answer for each figure:—*a a*, tentacula clothed with vibratile cilia; *b*, pharyngeal cavity; *c*, valve separating this cavity from *d* the œsophagus; *e*, the stomach, with *f* its pyloric valve, and *g* the circle of cilia surrounding that orifice; *h*, wall of the stomach with biliary follicles; *i*, the intestine, containing *k* excrementitious matter, and terminating at *l* the anus; *m*, the testicle; *n*, the ovary; *o*, an ovum set free from the ovary; *p*, openings for the escape of the ova; *q*, spermatozoa freely moving in the cavity that surrounds the viscera; *r*, retractor muscle of the angle of the aperture of the sheath; *s*, retractor of the sheath; *t*, retractor of the tentacular circle; *u*, retractor of the œsophagus; *v*, retractor of the stomach; *w*, principal extensor muscle; *x*, transverse wrinkles of the sheath; *y*, fibres of the sheath, themselves probably muscular; *z*, muscles of the tentacula; *α* (at the base of the tentacular circle in A) nervous or œsophageal ganglion; *β*, stem.—D, a portion of the tentacular circle shown separately on a larger scale: *a a*, the tentacula clothed with cilia; *b b*, their internal canals; *c*, muscles of the tentacula; *d*, transverse muscles forming a ring at the base of the tentacula; *e*, muscles of the tentacular circle.

are finally discharged by an outlet at *p*, beneath the tentacular circle. Besides this mode of propagation, however, we find a provision for the multiplication of individuals by gemmæ; these are not here put forth from the bodies of the individuals previously completed, but from the stem. In those cases, however, in which the tegumentary sheaths are in close contiguity, instead of being united by a stem or by other connecting tissue, the gemmæ are put forth from the individuals already formed, as in Flustra and its allies.—These beings possess a considerable number of muscles, by which their bodies may be projected from their sheaths or drawn within them; of these muscles, *r, s, t, u, v, w, x,* the direction and points of attachment sufficiently indicate the uses; they are for the most part *retractors,* serving to draw and double up the body, to fold together the circle of tentacula, and to close the aperture of the sheath, when the animal has been completely withdrawn into its interior. The projection and expansion of the animal, on the contrary, appears to be chiefly accomplished by a general pressure upon the sheath, which will tend to force out all that can be expelled from it. The tentacula themselves are furnished with distinct muscular fibres, by which their separate movements seem to be governed; the arrangement of these is seen at D. At the base of the tentacular circle, just above the anal orifice, is a small body (seen at A, *a*), which is probably a nervous ganglion; this being the situation of the single ganglion in the Tunicata. As yet no branches have been distinctly seen to be connected with it in this species; but its character is less doubtful in some other Bryozoa.—The general structure of the animals whose sheaths are imbedded in a coralline fabric (formed by the deposit of carbonate of lime in their connecting tissue), is essentially the same; as shown in Fig. 115. When this fabric is decalcified by the action of dilute acid, it leaves a much more definite residuum than that which is obtainable from the ordinary corals; and it is a curious circumstance, which shows that it possesses a power of growth inherent in itself, that, as the individuals lodged in its hollows become less active, the mouths of their sheaths gradually contract around them; and at last, when the minute inhabitants have ceased to live, the cavities in which they were imbedded are completely closed by the growth of new tissue from the border of their orifices, so that no trace of these at last remains.—If we scrutinize these characters, we shall find that the most important of them are Molluscan, rather than Zoophytic. In the first place, all true Polypes use their tentacula to grasp their food and convey it to the mouth; and these tentacula are destitute of cilia: whilst, on the other hand, in all the Acephalous Mollusca, the nutritive matter is drawn in by a ciliary current, which also serves to aerate the fluids. Now, the latter, as we have just seen, is the case with the Bryozoa; and thus, although their arms very commonly present a circular disposition around the mouth, they may be considered as representing, in their relation to the economy of the animal, the ciliated branchial sac of the Ascidians (§ 299). But they do not by any means constantly present this radial symmetry: thus, in the *Plumatella,* a beautiful fresh-water genus of Bryozoa, the ciliated arms are set upon two lobes or projections, one on either side of the mouth. The structure of the alimentary canal, again, removes the Bryozoa from the zoophytic series. In no true polype is there a separate intestine and anal orifice; nor does the whole apparatus hang freely in the visceral cavity; and the existence of a gizzard-like organ (foreshadowing that of Cephalopods, § 307), and of a rudimentary liver (closely resembling that found in the lowest Tunicata), are also characters of elevation. The most important of all the single characters furnished by the anatomy of these animals, is their nervous system; which, as already pointed out, is distinctly Molluscan in its type. The absence of a heart and a distinct circulating system is, it is true, a Zoophytic character; but we shall presently find that even in the Tunicata, which are true Mollusks, the character of the circulating apparatus is extremely degraded. The propagation by gemmation, although formerly

Fig. 115.

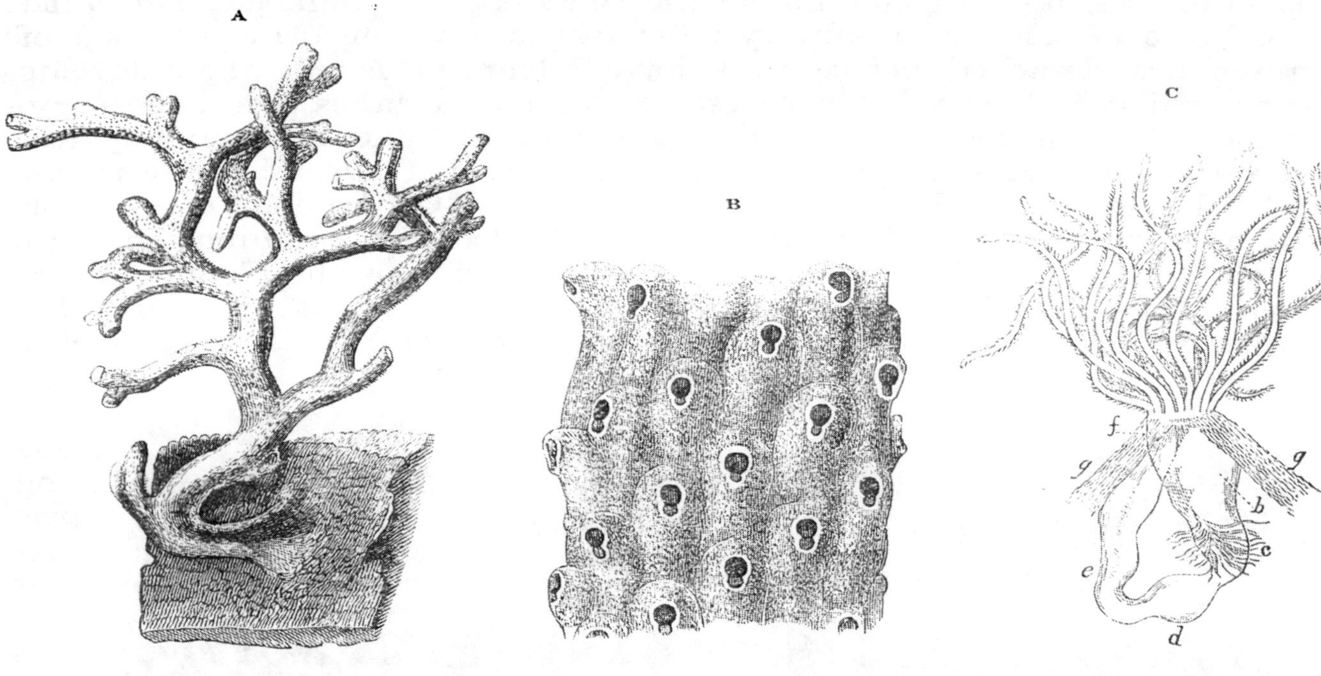

Eschara cervicornis; A, portion of its coralline fabric, of the natural size; B, portion of a young branch enlarged, to show the arrangement of the superficial apertures; C, an individual withdrawn from its cell, and highly magnified; *a*, mouth; *b*, pharyngeal cavity; *c*, secreting tubuli (?); *d*, digestive stomach; *e*, intestinal canal; *f*, anus; *g*, retractor muscles.

supposed to be a character exclusively Zoophytic, is now known to belong also to the greater part of the Tunicated Mollusks; and from this, therefore, no argument can be drawn in favour of the Zoophytic nature of the Bryozoa. And although many of their composite fabrics have a stony density, and closely resemble the solid polypidoms of the Helianthoid and Asteroid polypes, yet in others, especially amongst the fresh-water species, we find a very close resemblance to the gelatinous bed or leathery crust in which the Compound Ascidians are lodged; and if we imagine calcareous matter to be deposited in this bed or crust, we should have a fabric closely resembling that of many Bryozoa. In their power of projecting their bodies from their sacs, the Bryozoa must be admitted to resemble Polypes rather than Tunicata; but this is a character of no particular importance, and some approaches to it are seen among the compound Ascidians.

299. The class of Tunicata contains a series of forms which are closely allied to each other in the most important features of their structure, and

which constitute a chain of connection between the Zoophytic *Bryozoa* on the one hand, and the more truly Molluscous *Conchifera* on the other. Its lower tribes are allied to Bryozoa, not merely in the general plan of their conformation, but also in their tendency to produce composite structures by gemmation; they are distinguished, however, by the absence of the ciliated tentacula which form so conspicuous a feature in the external aspect of the Bryozoa, by the presence of a distinct circulating apparatus, of which no traces have as yet been found in that class, and also by the special modification of the pharyngeal cavity for respiratory action.* In their habits, too, they are more inactive; exhibiting scarcely anything comparable to those rapid movements of expansion and retraction, which it is so interesting to watch among the Bryozoa. The highest forms of the class, however, are quite solitary; for in those which form gemmæ, the gemmæ are detached before they have advanced far in their development; whilst in many other cases, no gemmation takes place, the reproduction being effected only by the sexual process: and it is among these that we find the closest approximation, in general structure, to the type of the Mollusca contained within 'bivalve' shells.—Of the characters of the class, as a whole, we shall derive our best idea from an examination of the structure of one of the 'solitary' Ascidians, such as the *Cynthia microcosmus* (Figs. 116, 117). Its body, like that of Tunicata in general, is completely enclosed within a general integument or 'tunic' (Fig. 116, A,

FIG. 116.

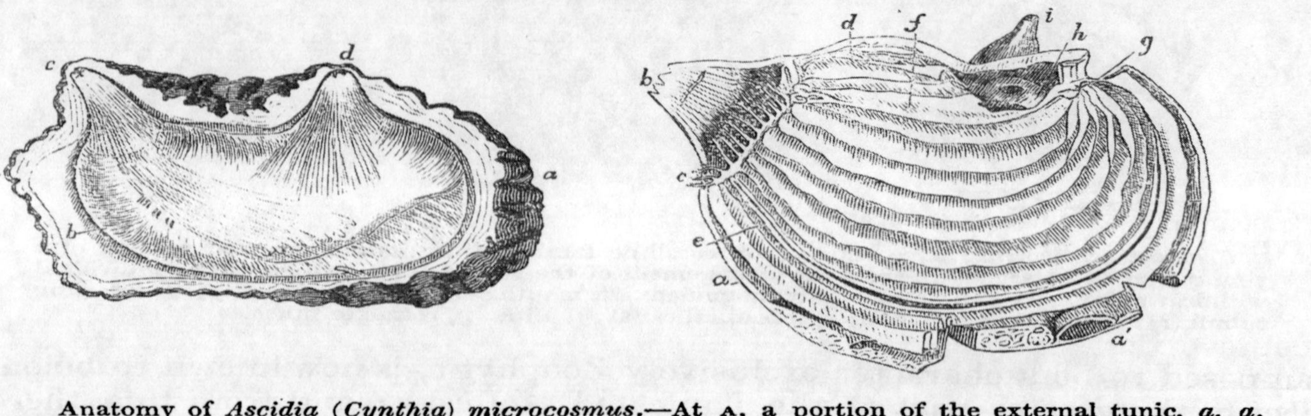

Anatomy of *Ascidia* (*Cynthia*) *microcosmus*.—At A, a portion of the external tunic, *a*, *a*, having been removed, the muscular tunic *b* is shown; *c*, the oral orifice; *d*, the anal orifice.—At B, the muscular tunic *a* having been divided, the cardiac cavity *a* is shown, partly excavated in its substance; *b*, the oral orifice; *c*, oral tentacles surrounding the entrance to the branchial sac; *d*, branchial membrane; *e*, ventral furrow, corresponding to the great thoracic sinus; *f*, dorsal fold, corresponding to the dorsal sinus; *g*, œsophageal orifice; *h*, extremity of the intestine, situated in the cloaca; *i*, anal orifice.

a a) possessing two orifices, a mouth (*c*), and a vent (*d*); this tunic, in the Ascidians, usually possesses some degree of toughness, frequently being leathery or even cartilaginous in its texture; and it is occasionally rendered still more resisting by the agglutination of grains of sand, gravel,

* It may be thought that these are not characters sufficient for the separation of the Bryozoa from the Tunicata, and the erection of the former into a distinct class; but the Author would again request his readers to bear in mind, that he is only endeavouring to make them acquainted with the characters of the groups which are most *Physiologically* distinct from each other, without aiming to determine their relative *Zoological* value. The rank of the Bryozoa will not improbably prove in the end to be that of the lowest order of Tunicata.

&c., to its external surface, so that it almost takes the place of a shell. Within the tunic or 'test' we find the muscular 'mantle' (A *b*, B *a*), which is adherent to the tunic only near the orifices, and which is able by its contraction to produce a general compression of the cavity it encloses, or a constriction of either of the apertures. A large part of the cavity is occupied by the branchial or respiratory sac (B *d*), which may be considered as a dilated pharynx, and which fills the whole of the space not taken up by the nutritive and reproductive organs. This branchial sac is lined with cilia; and by the agency of these, a continual current is drawn in through the oral aperture, part of it traversing the respiratory surface, whilst another part is driven through the alimentary canal, the whole being finally ejected through the anal orifice. The digestive organs of the Tunicata present but a slight advance in their plan of structure upon those of the Bryozoa. In all instances, the alimentary canal has two orifices; these may be situated near to each other, as in most of the *Ascidians* (Fig. 116 A, *e*); or they may open at the opposite extremities of the body, as in the *Salpæ* (Fig. 124). The oral orifice is often furnished with short tentacula (Fig. 116 B, *c*), which appear to be sensory organs; their purpose apparently being to indicate the presence of any such particles, in the stream that enters the orifice, as may be unfit for admission into the branchial cavity; and to call forth a reflex contraction of the muscular tunic, which serves not only to prevent the entrance of the offending particle, but also to eject a gush of water that shall convey it to a distance.* The stomach and intestinal tube are formed, in the inferior Tunicata, upon the plan of those of the Bryozoa, the lower part of the pharyngeal (branchial) sac narrowing to form the entrance to the stomach; and the liver still consists of a set of isolated follicles imbedded in the walls of the latter. In the higher forms, however, we find a short œsophagus commencing by a distinct orifice (Fig. 117 A, *c*) at the bottom of the pharyngeal chamber; this œsophagus leads to the stomach (*d*), which receives the terminations of a number of separate biliary canals, that proceed from a liver (B, *k*) altogether distinct from the stomach, but composed of numerous lobules that cluster around it; and from the stomach, a short wide intestine (A, *e*) passes off, which soon reaches (at *f*) the cloaca, or common receptacle of the various matters that are to be ejected through the anal outlet.—All the Tunicata possess a distinct circulating apparatus; and in all of them, too, the movement of blood seems to take place upon a very peculiar plan. The heart is very simple in its structure, being sometimes a mere contractile dilatation of the principal trunk (Fig. 119, *o*), and sometimes appearing as an excavation in the muscular tunic itself (Fig. 116, B, *a*); in neither case, however, is there any division into an *auricle* or receiving cavity and a *ventricle* or impelling cavity. The course of the blood is not uniform; for after the heart has received it from one set of vessels for a short time, and propelled it into another, its contractions become slower and the current feebler, and after a while they cease; recommencing, after a short pause, in the opposite direction. After continuing for a further period in this *reversed* direction, the current returns again, in the same manner, to its original course. This remarkable phenomenon has been witnessed in species from each of the principal divi-

* This action is exactly analogous to that of *coughing* in higher animals.

sions of the class; and may, therefore, be considered as a peculiarity common to the whole of it.—The Tunicata all possess distinct generative organs, which are usually united in the same individual; and from the ova thus produced and fertilized, are developed embryoes, which, in their early state, are extremely unlike the form they are ultimately to attain, and which possess a power of free locomotion, even in the species most completely destitute of it in the adult condition. These embryoes, after swimming through the water for a time, so as to disperse themselves to wide distances from the parent and from each other, become stationary, attach themselves, and go through a series of phases of development, which end in their assumption of the perfect form. The varieties in the methods according to which the extension by gemmation takes place in

FIG. 117.

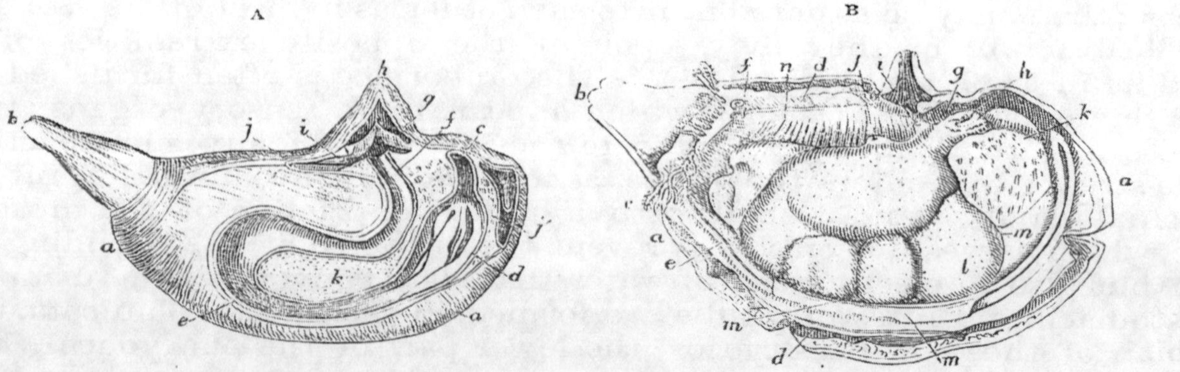

Internal anatomy of *Ascidia (Cynthia) microcosmus.*—At A, the external tunic has been removed, and the internal opened in such a manner as to display the alimentary canal; *a*, muscular tunic; *b*, oral orifice; *c*, œsophageal orifice at the bottom of the branchial chamber; *d*, stomach and hepatic orifices; *e*, intestine; *f*, its termination in the cloaca; *g*, valves dividing the cloaca; *h*, anal orifice; *i*, orifice of the generative canal; *j*, glandular organ supposed to be a testicle; *j*, branchial membrane; *k*, liver.—At B, the branchial membrane has been removed to show the arrangement of the thoracic chamber and its contained organs; *a*, external integument; *b*, oral orifice; *c*, oral tentacula; *d*, *d*, portions of the branchial membrane turned back; *e*, ventral furrow; *f*, dorsal furrow, with gangliform tubercles; *g*, cloaca; *h*, œsophageal orifice; *i*, termination of intestine; *j*, orifice of generative canal; *k*, liver; *l*, portion of the thoracic chamber lodging the organs of generation; *m*, *m*, channels for the passage of water to the cloaca; *n*, portion of cloaca on which are seen the fibres of the muscular tunic.

the different subdivisions of the class, will be presently more particularly explained.—The nervo-muscular apparatus in this class seems to present its simplest possible form; the nervous portion of it consisting of but a single or double ganglion (Fig. 117, B, *f*), connected by cords of communication with the sensory tentacula that guard the entrance of the pharyngeal sac, and with the muscular tunic which envelopes it; whilst the muscular system consists only of the general envelope, composed of fibres crossing in various directions, and of the sphincters formed by the circular disposition of these fibres around the orifices (Fig. 117, A *a*). Its functions appear to be limited to the forcible ejection of the contents of the branchial sac through one or both orifices; the contraction of the muscular tunic being excited either by the stimulus of the contact of some hard body with the external integument, or by an impression made upon the oral tentacula, or by distension of the respiratory sac. It is only by this ejection of water, that the nervo-muscular apparatus takes a share in the general locomotion of the body of the free floating Salpæ; and the movement, being simply respiratory, cannot be regarded as distinctly indicative even of consciousness, still less as an evidence of volition.

FIG. 118.

Compound mass of *Amaroucium proliferum*.

FIG. 119.

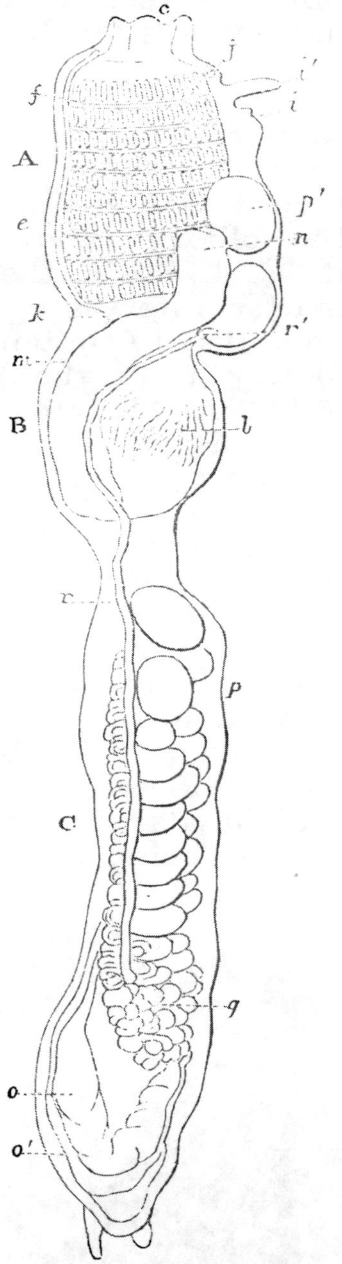

Anatomy of *Amaroucium proliferum*:—A, thorax; B, abdomen; C, post-abdomen:—*c*, oral orifice; *e*, branchial sac: *f*, thoracic sinus; *i*, anal orifice; *i'*, projection overhanging it; *j*, nervous ganglion; *k*, œsophagus; *l*, stomach surrounded by biliary tubuli; *m*, intestine; *n*, termination of intestine in cloaca; *o*, heart; *o'*, pericardium; *p*, ovarium; *p'*, egg ready to escape; *q*, testis; *r*, spermatic canal; *r'*, termination of this canal in the cloaca.

299 *a*. The Tunicata most nearly allied to Bryozoa, are those which are designated Compound Ascidians or *Botryllidæ;* being constructed upon nearly the same plan with the 'solitary' Ascidians, of which some account has already been given; but presenting themselves in composite masses, each made up of numerous animals which appear to have been developed from a single individual by the process of gemmation. Of these we have a characteristic example in *Amaroucium proliferum;* of which the form of the composite mass is shown in Fig. 118, whilst the anatomy of a single individual is displayed in Fig. 119. The clusters of these animals present themselves on rocks, sea-weeds, &c., very commonly between the tide-marks. They appear almost completely inanimate, exhibiting no very obvious movements when irritated; but if they be placed when fresh in sea-water, a slight pouting of the orifices will soon be perceptible, and a constant and energetic series of currents will be found to enter by one set and to be ejected by the other, indicating that all the machinery of active life is going on within these apathetic bodies. In the tribe of *Polyclinians,* to which this genus belongs, the body is elongated, and may be divided into three regions, the thorax (Fig. 119, A) which is chiefly occupied by the respiratory sac, the abdomen (B) which contains the digestive apparatus, and the post-abdomen (C) in which the heart and generative organs are lodged. At the summit of the thorax is seen the oral orifice *c*, which leads to the branchial sac *e;* this is perforated by an immense number of slits, which allow part of the water to pass into the space between the branchial sac and the muscular mantle, where it is especially collected in the thoracic sinus *f*. At *k* is seen the œsophagus, which is continuous with the lower part of the pharyngeal cavity; this leads to the stomach *l*, which is surrounded by biliary tubuli; and from this passes off the intestine *m*, which terminates at *n* in the cloaca. The long post-abdomen is principally occupied by the large ovarium *p*, which contains ova in various stages of development. These, when matured and set free, find

their way into the cloaca; where two large ova are seen (one marked p', and the other immediately below it), waiting for expulsion. In this position they receive the fertilizing influence from the testis q, which discharges its products by the long spermatic canal r, which opens into the cloaca at r'. At the very bottom of the post-abdomen, we find the heart o, enclosed in its pericardium o'.—In the tribe we are now considering, a number of such animals are embedded together in a sort of gelatinous mass, and covered with an integument common to them all. We are probably to regard this common integument as representing the 'tunic' or 'test' of the solitary Ascidians, since the individuals of the cluster are destitute of it; and both in the compound and solitary species, we find the gelatinous matter intervening between the test and the muscular tunic or proper 'mantle,' to contain *cellulose*, which is, however, probably derived immediately from the food of these animals, and not generated by them (§ 263). The mode in which new individuals are developed in this mass, is by the extension of 'stolons' or creeping stems from the bases of those previously existing; and from each of these stolons several buds may be put forth (as in Fig. 121 A, *t t*), every one of which may be evolved into the likeness of the stock from which it proceeded, and may in its turn increase and multiply after the same fashion. A communication between the circulating systems of the different individuals is kept up, through their connecting stems, during the whole of life; and thus their relationship to each other is somewhat like that of the several polypes on the polypidom of a *Campanularia* (§ 290).—In the family of *Didemnians*, the post-abdomen is absent, the heart and generative apparatus being placed by the side of the intestine in the abdominal portion of the body. The individuals are frequently arranged in star-shaped clusters, the anal orifices being all directed towards a common vent which occupies the centre.—This is still more remarkably the case, however, with the family of *Botryllians*, whose beautiful stellate gelatinous incrustations are extremely common upon sea-weeds and submerged rocks (Fig. 120). The anatomy of these animals, as will be seen from the accompanying figures (Fig. 121, A, B), is very similar to that of the *Amaroucium* already described; with this exception, that the body exhibits no distinction of cavities, all the organs being brought together in one, which must be considered as thoracic. In this respect, there is an evident approximation towards the solitary species.—The approximation is still closer, however, in the 'social' Ascidians, or *Clavelinidæ*; in which the general plan of structure is nearly the same, but the individuals are simply connected by their stolons, instead of being included in a common

Fig. 120.

A B

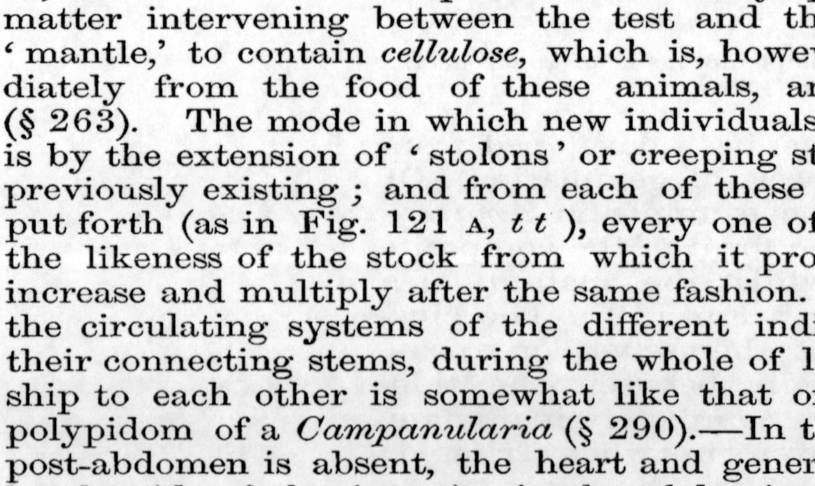

Botryllus violaceus:—A, cluster on the surface of a Fucus:—B, portion of the same enlarged.

investment, so that their relation to each other is very nearly the same as that of the individuals of a *Laguncula* (Fig. 114), the chief difference being that a regular circulation takes place through the stolon in the former case, such as has no existence in the latter.—Between the foregoing and the 'solitary' Ascidians, or *Ascidiadæ*, there are no very important structural differences; save that in the latter, the two sets of sexual organs are always in distinct individuals, instead of being combined in the same. In these, however, we entirely lose sight of the power of gemmation; the only means of propagation amongst them being through the true generative process, except in one or two instances (as *Cynthia aggregata*), in which the power of putting forth buds from the root-fibres appears still to be retained, although the general plan of conformation is that of the solitary species. One of the chief structural differences exhibited by these is in the branchial or pharyngeal sac, which is often folded into longitudinal plaits (Fig. 116 B), so as to increase its extent of surface, thus in some degree prefiguring the branchial lamellæ of Conchifera. The muscular tunic, too, seems to be more employed in the ordinary acts of respiration, than it is in the 'compound' Ascidians; for whilst in the latter the respiratory currents are maintained by ciliary action alone, there appears in some of the solitary species to be a regular alternate movement of expiration and inspiration; the former, which is sudden and energetic, being produced by the contraction of the muscular sac; whilst the latter, which is more gradual, is apparently dependent in part upon the elasticity of the external tunic, which tends to return to the dilated state, and in part upon the distending influence of the ciliary current. In both cases, however, a part of the water that is introduced through the oral orifice escapes through the slits in the branchial membrane into the surrounding space, from which it is conducted by special channels (Fig. 117 B, *m m*) to the cloaca; whilst another portion is driven through the

Fig. 121.

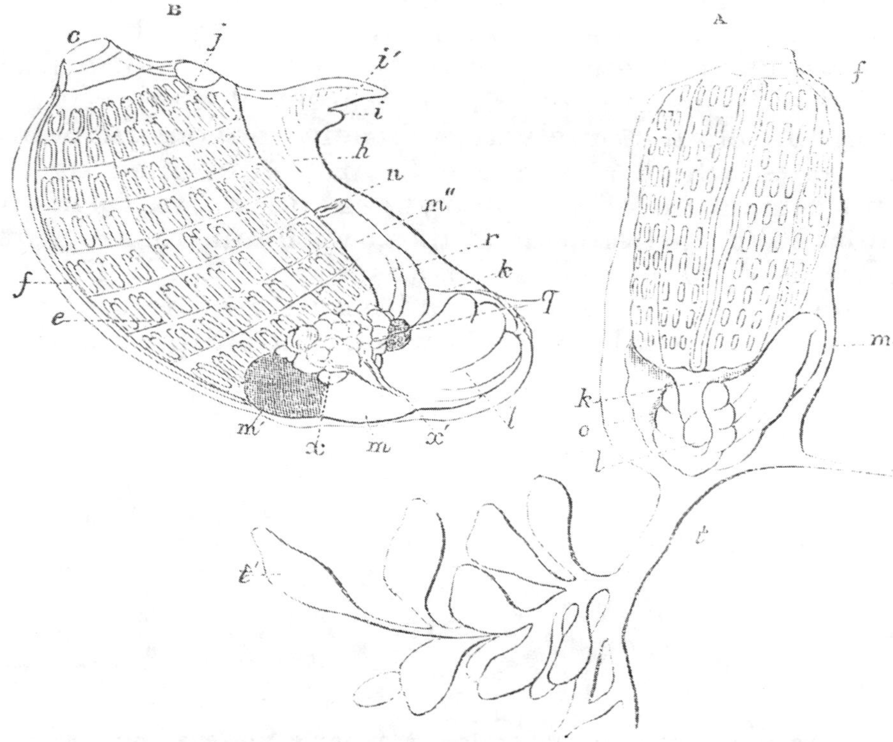

Anatomy of *Botrylloides rotifera*; A, front view; B, side view:—*c*, oral orifice; *e*, branchial sac; *f*, thoracic sinus; *h*, cloaca; *i*, anal orifice; *i′*, projection overhanging it; *k*, œsophagus; *l*, stomach; *m*, first portion of the intestine, or duodenum; *m′*, second portion of the intestine, or chylific ventricle; *m″*, third portion of the intestine, or rectum; *n*, termination of the intestine in the cloaca; *o*, heart; *q*, testis; *r*, spermatic canal; *t*, root-like appendage; *t′*, reproductive buds; *x*, glandular mass (liver ?); *x′*, its excretory duct.

alimentary canal by the cilia with which it is lined, and reaches the cloaca through the anal orifice of the intestinal tube. From observation of some of the microscopic species, it would appear that the ciliary action is specially arranged in such a manner as to drive down all solid particles towards the œsophageal orifice, allowing liquids only to pass through the branchial fissures.

299 *b*. The animals composing the other principal division of the Tunicata, that of *Salpidæ*, differ from the Ascidians as well in their habits as in their structure. They are not attached to solid bodies, but habitually swim through the waters of the ocean, sometimes singly, sometimes in chains or clusters; their movement being due to the respiratory currents, which are in great part maintained by the alternate contraction and relaxation of the pharyngeal sac. In form (Fig. 122), they resemble short but rather wide tubes, with an opening at each end; the tube is composed of a delicate transparent tunic, lined by the muscular coat, which here forms bands adherent to the tunic and running across the body. The oral orifice is wide, and unprovided with tentacula, but it is

FIG. 122.

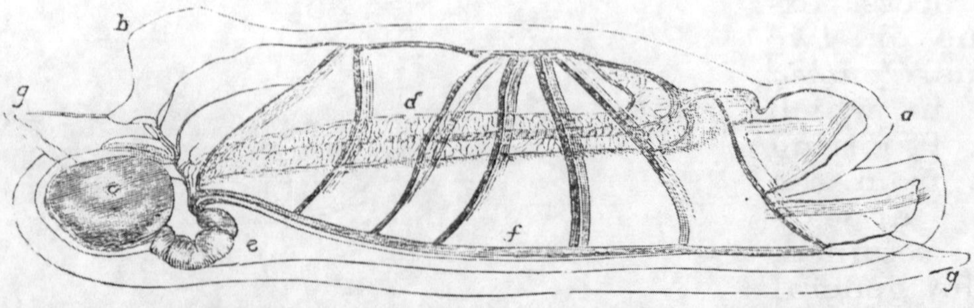

Salpa maxima; *a*, oral orifice; *b*, vent; *c*, nucleus, composed of the stomach, liver, &c.; *d*, branchial lamina; *e*, the heart, from which proceeds the longitudinal trunk *f*, sending transverse branches across the body; *g g*, projecting parts of the external tunic, serving to unite the different individuals into a chain.

furnished with a valve that permits the free ingress of water whilst it prevents its return; the vent (*b*) is narrower and somewhat conical, and the stream that issues from it impinges against the surrounding water, with sufficient force to propel the body in the opposite direction. The pharyngeal sac, as a whole, does not appear to be subservient to the respiratory process; but a more special organ is here developed, consisting of a double riband-like fold of its wall, minutely furnished with blood-vessels, and extending through the entire length of the body (*d*). This distinctly prefigures the branchial apparatus of the Lamellibranchiata (§ 301). The viscera are collected into a single mass at the posterior extremity of the body (*c*); and this mass or *nucleus* is usually very conspicuous, owing to the brilliant orange, brown, or reddish hue of the liver, of which the external part of it is composed, and to the transparency of the investing membranes.—The *Salpæ* are met with in two states; the 'solitary,' and the 'aggregated.' The latter are simply adherent to each other by little suckers, instead of being organically united like the Compound Ascidians; this adhesion is so strong in some species, that it is easier to tear the bodies of the animals, than to separate them from each other; in other species, however, it is less powerful, so that when a mass is placed in a vessel of water, the sides of which are smartly struck, the individuals fall asunder. The 'aggregate' and the 'solitary' Salpæ, as appears from the researches of Chamisso and Krohn, are related to each other in a most remarkable manner; for each solitary Salpa may give origin to an aggregate chain or cluster, not by the ordinary process of external gemmation (as in the Compound Ascidians), but by the formation of an *internal stolon,* which seems to take the place of the true ovarium, and to produce chains or clusters of buds; these become detached from the stolon, but remain adherent to each other; and in this condition they swim forth from the cavity of the parent, and maintain an independent existence in the aggregate state. Each of these 'aggregate' Salpæ possesses true sexual organs; the ovaria and testes, however, are not united in the same individuals, and are not even present in the different individuals of the same chain; but the ova produced by one set are fertilized by the spermatic fluid of another, and from each of these fertile ova a 'solitary' Salpa is produced, which in its turn developes another chain of aggregate individuals. The 'solitary' and 'aggregate' forms thus alternate in continuous succession; although they differ so much from each other, as to have been regarded as distinct species until their true relationship was known. There are many forms, however, of both; but it would appear that every 'solitary' species has a corresponding 'aggregate' form, and *vice versa.** The differences in conformation between the aggregate and the solitary individuals of the same species, have reference chiefly to their adhesion to one another.—Among the 'aggregate' Salpæ, there is one curious group, that of *Pyrosomidæ,* which seems to unite this tribe with the *Botryllidæ.* Each composite mass is a hollow cylinder, closed at one end; and this cylinder may be regarded as consisting of a pile of star-shaped clusters, resembling those of *Botrylli* (Fig. 120, B); the oral orifice of each individual being on the outer margin, while its vent opens into the central channel of the cylinder. As this is closed at one end, the water forced into it by the respiratory currents of the component individuals is obliged to issue from the other; and the continual stream which is thus ejected causes the entire mass to move in the opposite direction.—It is a curious fact, that *Salpæ* are sometimes found making their way through the sea by the rhythmical contractions of their muscular sac, after having been deprived of their nuclei by birds or fishes; but whether, like the *Holothuriada* (§ 296 *e*), they can in time regenerate the important organs they have thus lost (those of digestion, circulation, and reproduction), has not yet been ascertained.

299 *c.* In addition to the connection which is so obviously established by the *Bryozoa* between Compound Ascidians and Zoophytes, we find the

* This method of propagation is one of the most characteristic examples of what has been termed the "alternation of generations;" in which, (it is affirmed) generation A (*e.g.* a 'solitary' Salpa) produces generation B, a being of a different form (*e.g.* an 'aggregate' Salpa); and the offspring of B resembles generation A; so that each individual is unlike its own immediate parents and offspring, but resembles its grandparents and grandchildren. In this mode of stating the doctrine, however, the fact is too much lost sight of, that the two *modes* of reproduction alternate, as well as the two *forms* produced. In fact, the real nature of the process is essentially the same as that by which the clusters of compound Ascidians are produced; the only difference being that the 'stolon' projects into the cavity of the body, instead of being prolonged from its exterior (very much as if the finger of a glove were turned inwards so as to project into the cavity of the hand), and that the gemmæ formed from it spontaneously detach themselves, instead of remaining in connection with the parent. This subject, however, will be more fully considered hereafter (CHAP. XVIII.).

Tunicata related to the Echinodermata through the medium of the very curious genus *Pelonaia ;* which in its general form, and in some points of its structure, exhibits an obvious approximation to the *Sipunculida* (§ 296 *f*) ; whilst in other respects there is (as in that group) a decided approximation to the Annulose division, more especially in the bi-lateral symmetry of the organs, and the transverse wrinkling of the external tunic.*

300. From the Tunicata we pass on to the Mollusca inhabiting *bivalve* shells, which are sometimes distinguished as CONCHIFERA. These, like the preceding, are *acephalous*, or headless ; that is, the mouth is not situated upon a prominent part of the body, nor is it assisted in its choice of food by organs of special sensation in its neighbourhood. In their general organisation, the Conchifera present a certain degree of advance upon the Tunicata ; but this is not very strongly marked ; and the principal differences in structure which we shall encounter, have reference more to the general plan of conformation, than to the comparative elevation of the former in the scale. This plan presents itself under two very dissimilar aspects, the Conchiferous Mollusca being capable of arrangement, according to well-marked peculiarities of structure, into two subdivisions, which are perhaps as dissimilar to each other as either is to the Tunicata, and which should rank, therefore, as distinct classes.—The one which will be first noticed, as being on the whole of inferior organisation to the other, is that of BRACHIOPODA ; which receives its name from the two long arms, fringed with filaments, that spring from either side of the mouth in most of the animals composing the class, and occupy a considerable part of the cavity of the shell (Fig. 123, A). These arms, however, do not appear to be instruments of prehension ; for in most of the existing Brachiopoda

FIG. 123.

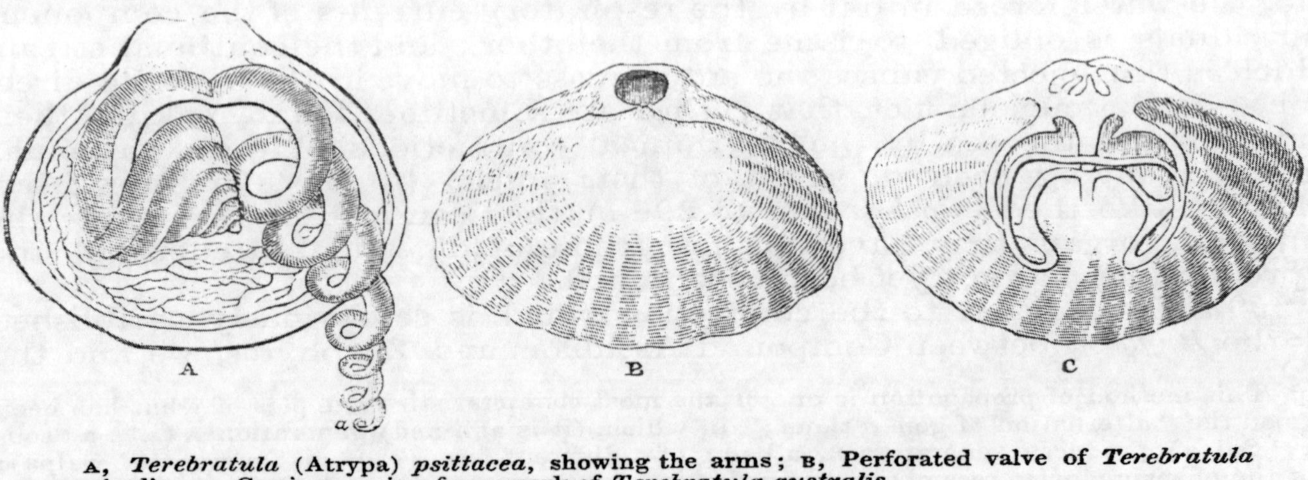

A, *Terebratula* (Atrypa) *psittacea*, showing the arms ; B, Perforated valve of *Terebratula australis ;* C, Carriage-spring framework of *Terebratula australis.*

they are incapable of being extended, being attached along their entire length to a very curious and delicate framework, composed of shelly substance, which is attached to one of the valves (C). Nor do they appear to be respiratory organs; for a careful examination of their structure shows that the blood-vessels are not distributed to them in a degree that would

* See Forbes and Goodsir, in Edinb. New Philos. Journ. vol. xxxi. p. 29.

warrant our attributing to them any direct participation in this function. It is probable that their special purpose is to create currents of water, by the agency of the cilia that clothe their surfaces, which shall convey food to the mouth, and shall at the same time aerate the blood, which is transmitted for this purpose to the extended surface of the mantle. Each valve of the shell is lined by half of the mantle, or general investment of the body; and the two halves are completely unattached, except near the hinge at which the two valves meet, so that, by this splitting-open of their containing sac, the viscera that lie between them are freely exposed to the contact of the water, which enters the cavity of the shell. Of this cavity, however, the viscera occupy a very small proportion; being limited to the central part of it, so as only to fill (in the Terebratulæ) the space surrounded by the 'carriage-spring' framework of the arms, which is closed-in by a firm membrane that fills up the space between its loops. The entrance to the stomach thus lies buried deeply between the lobes of the mantle; the digestive apparatus is itself constructed nearly upon the plan of that of the Ascidia (§ 299); and, as in that animal, the short intestine terminates not far from the mouth. The principal distribution of blood-vessels is upon the lobes of the mantle, which apparently perform the functions of a respiratory or aerating organ; and the heart or organ of impulsion is here double, consisting of a dilatation of the trunk that connects the visceral vessels with those of the mantle, on each side of the visceral mass. In this, as in the conformation of the shell, arms, and various other parts, there is a remarkable manifestation of bi-lateral symmetry.—We lose, in this group, all trace of the power of gemmation; and the continuance and multiplication of the race is altogether effected by proper sexual process. The ovaries in the female, and the spermatic organs in the male, are spread out in an arborescent form over the surface of the mantle, accompanying its great veins.—All the Brachiopoda are fixed to one spot, except during the earliest period of their development; but the mode in which they are attached presents a considerable diversity. In the *Terebratula*, the shell, which is chiefly composed of carbonate of lime,* is affixed to solid bodies by means of a tendinous peduncle, which is attached to the valve that bears the framework, and passes through a hole in the beak-shaped prolongation of the other valve (Fig. 123, B). These valves fit together with somewhat of a hinge-like arrangement, and are further connected by a complex muscular apparatus, which draws them closely to one another. In many species, the 'carriage-spring' framework is so constructed, that the valves cannot be completely closed together without somewhat compressing it; to this compression its elasticity allows it to yield without breaking; but in this state it will tend to force the valves again apart; and one of its purposes would thus seem to be, to keep the shell usually somewhat open. This, in the living animal, must be its ordinary condition; currents of water, created by ciliary action, being thus continually drawn through the mouth, and transmitted

* It is remarkable that in all the true Terebratulæ now existing, the shell is perforated by a multitude of minute canals which extend through its thickness, these canals giving lodgment to glandular cœca, the use of which is at present unknown: in many of the extinct species, however, such perforations did not exist; but these species constitute a group, which is distinguished from the perforated species in several other particulars. See the Author's "Reports" to the British Association for 1844 and 1847.

over the surface of the mantle. But when the animal is alarmed, the valves are probably drawn together, as in other Bivalves; this action being due to an impression which is made through its sensory organs upon its nervous system, and thence reflected upon the muscles. A distinct nervous system has been found, in the form of a collar round the œsophagus, with slight ganglionic enlargements: but no special organs of sense have been yet detected.—In *Lingula*, the shell is of horny texture, and neither of the valves is perforated; but the peduncle, which is longer, is attached to both alike, and serves to hold them together. There is no framework for the arms, which are unattached except at their bases, and may be extended beyond the shell; and the animal further departs from the ordinary type, in having the mantle raised into fringes, which may be considered as rudiments of the special branchiæ of the Lamellibranchiata. It is probable, from the circumstances under which the Lingulæ are found, that they possess more locomotive power than the Terebratulæ; the former live near the surface, whilst the latter are inhabitants of very deep water.—The *Orbicula*, like the Lingula, has a horny shell; this is attached, however, not by a long peduncle, but by a short muscle which is affixed to one valve and passes through a fissure in the other.—These are the principal forms of existing Brachiopoda; and the variety of conformation which they present is not considerable. There is ample evidence, however, that in former periods of the Earth's history, the class contained a much larger number of more diversified types than it does now; and that these constituted a far larger proportion of the Bivalve Mollusca, than do the Brachiopoda of the existing epoch.

301. The class of LAMELLIBRANCHIATA is chiefly distinguished from the preceding by the presence of special respiratory organs, or *branchiæ*, which, in nearly every case, form four riband-shaped lamellæ, two of them attached to each lobe of the mantle, and lying within the marginal portion of the shell. The variety of forms which we encounter in this class is very considerable; and the modes of life of the different animals which it includes are no less dissimilar. Some of them, as the *Oyster*, are fixed by the adhesion of their valves during all but the earliest period of life; others, as the *Pinna* and the *Mussel*, are attached to solid bodies by a tendinous cord or *byssus;* others, as the *Pecten*, float freely through the water, and are said to propel themselves by a flapping movement of their valves; others, as the *Cockle*, can move by leaps over hard surfaces, by means of their fleshy foot; others, as the *Mya, Solen, &c.*, bore deeply into sand or mud, and live in the excavations they have formed; and others, as the *Teredo* and *Pholas*, make their way into wood, or even into stone. In all but the small and peculiar group which may be considered to connect this class with the Tunicata, we find the bivalve shell forming the chief external feature of the structure. The valves are connected by a more or less perfect *hinge*, which is often beset with 'teeth' that lock into one another; and this joint is sometimes very perfect, and so peculiar in its character, that, even when the shells are dry, it allows free motion of the valves without permitting them to be separated. In general, however, the retention of the valves in apposition to each other, is due to a *ligament* attached to the hinge in such a manner, that its elasticity keeps the valves somewhat apart, unless counteracted by the action of the muscle in the interior of the shell, which draws them together. This is a very beautiful

provision for the performance of the animal functions without difficulty or effort; for when the creature is undisturbed, the ligament keeps the valves open; but when danger is apprehended, or circumstances require it, the 'adductor' muscle contracts, overcomes the resistance of the hinge, and shuts the valves close, until they may be opened with safety. One of the earliest signs of the loss of vitality in Conchifera, is the unusually wide gaping of the shell; this arises from the continuance of the elasticity of the ligament (which does not depart as long as its structure is undecomposed), unbalanced by the vital contractility of the muscle. The valves are formed and increased by the calcification of the epithelial layer of the mantle, by which the shelly substance is generated (§ 197). There are usually two distinct layers in the shell; the inner one being often more or less pearly or 'nacreous' in its character; whilst the outer is usually duller in aspect, and contains a larger proportion of animal matter, in which the original cellular constitution is more distinctly recognised. The augmentation of the shell, in accordance with the growth of the contained animal, is effected by additions made to the previously-existing valves; but this is done in a different manner for the two valves; for whilst the *outer* layer is simply extended by a new growth proceeding from its margin, an entirely new *inner* layer is formed in the interior of the old one and projecting beyond it. This is easily seen in shells in which the successive laminæ are not closely adherent, as in the common *Oyster;* but it is not less true of those, in which the boundaries of the successive additions cannot be detected by an ordinary scrutiny.* It is not only for the augmentation of the shell, that new layers are thus formed; for it would seem as if the animal were impelled by a necessity of its nature to *contract* its cavity, when this is too large for its body. Thus an Oyster kept without food will frequently expend its last energies in secreting a new pearly layer, at a distance from the old internal surface of the concave valve; so as to reduce the cavity to the dimensions to which its body has shrunk during its fast. In certain species of *Spondylus*, the new layers are always formed at a considerable interval from the old ones; so that each valve, when cut across, presents a tolerably regular chambered structure. In the *Pholas*, the valves have 'accessory pieces,' constituting what is called a multivalve shell; these accessory pieces may probably be regarded as a rudiment of the shelly tube of the Teredo. In the *Teredo* and its allies, the respiratory siphons, as well as the proper mantle, form shell on their surface; and thus, in addition to the two valves, we find a long shelly tube, lining the whole burrow of the animal. In certain other animals formed on the same plan, the valves are scarcely discoverable, being imbedded (as it were) in the substance of this tube, which thus seems itself to constitute the shell.—The position of the animal in the shell, as determined by the place of the mouth and anus, is not the same in the Lamellibranchiata, as in the Brachiopoda. In the latter, the two valves may be regarded as forming the back and belly of the animal (the perforated valve of the Terebratula being the ventral, and the non-perforated the dorsal); whilst the two halves of each valve correspond to the two *sides* of the animal, which, as already remarked, are usually symmetrical. But in the former, the two valves are the two sides of the animal, being some-

* See the Author's "Report" to the British Association for 1847.

times symmetrical and sometimes not; whilst each valve has an *anterior* extremity, towards which the *mouth* is directed (Fig. 119, *o*), and a *posterior*, in the neighbourhood of which is the anus (*a*); and it thus has also a *dorsal* margin, of which the hinge forms part, and a *ventral* margin, along the side opposite the hinge.

301 *a.* Each valve, in the ordinary Lamellibranchiata, is lined by one of the lobes or divisions of the mantle; and these, in the *Oyster* and other genera which approach most nearly to Brachiopoda, are completely disunited except near the hinge, so that the water has free access to the internal surfaces of the mantle, to the gills, and to all the viscera included between them. In the *Mussel*, however, the two lobes are adherent along the posterior margins of the valves, except where a slit is provided for the egress of the water, which is drawn in through the fissure that separates its lobes elsewhere. In the *Cockle*, again, the lobes of the mantle are adherent around the greater part of the margin of the shell; a large fissure being left, however, for the passage of the fleshy foot, at the part opposite to the hinge; and two openings being left, one for the ingress and the other for the egress of water, at the posterior margin. These last, in many other Bivalves, are prolonged into tubes or siphons (Fig. 122, *y*, *z*), which are sometimes of considerable length, especially when the animal makes for itself a deep excavation in sand, mud, wood, &c.; that through which the water enters (*y*) is termed the *oral* siphon, whilst that through which it escapes (*z*) is termed the *anal* siphon. In the *Solen* we find the body elongated, and the mantle entirely closed except at the two extremities of the shell; the opening for the foot, at the anterior extremity, being only just large enough to allow it to pass; and no water being admitted to the interior, or passing out from it, except by the oral and anal siphons prolonged from the posterior extremity. The inclosure of the animal is still more complete in some of the Cuvierian family of *Inclusa;* and we are thus conducted back into close proximity with the higher Tunicata. For if the membranous integument of a *Salpa* (whose branchial leaflets approach those of the Lamellibranchiata, § 299 *b*) were to form a layer of shell, and the two orifices of its mantle were in proximity (as in the Ascidians), we should have a structure resembling that of the Inclusa in the most important particulars; the presence of a *foot*, in the latter, which is the strongest point of difference, not being essential, since many Lamellibranchiata are destitute of it.—The general organisation of the Lamellibranchiate bivalves will be understood from the accompanying representation of the structure of the animal of *Mactra* (Fig. 124). The mouth, or rather the entrance to the œsophagus (*o*), is buried deep between the two lobes of the mantle, and looks towards the anterior extremity of the shell; it is furnished with four long sensory tentacula, of which one is seen at *p*. The short œsophagus leads to the stomach (*h*), which has a long cœcal appendage *v*. From the stomach passes off a convoluted intestine (*t*), which, after making several turns, runs towards the dorsal margin of the shell (*e*), and is continued along it to the posterior extremity, where it terminates in the anal orifice (*a*), which discharges itself into the anal siphon (*z*). The stomach and intestine are partly invested by the liver (*k*), which now assumes a very distinct form, and attains a considerable size. The rudiment of a kidney, also, is discoverable in the higher Bivalves. The heart (*i*) lies along the dorsal margin of the shell, and is frequently perforated

by the intestine ; a distinction now becomes apparent between the *auricle* or receiving cavity, to which the blood is transmitted from the branchial lamellæ (*w*), and the *ventricle* or propelling cavity, from which it is sent forth to the general system ; and the auricle is not unfrequently double, a distinct cavity appertaining to the branchiæ of each lobe of the mantle. Between the heart and the posterior extremity is seen the reproductive

FIG. 124.

Dorsal Margin.

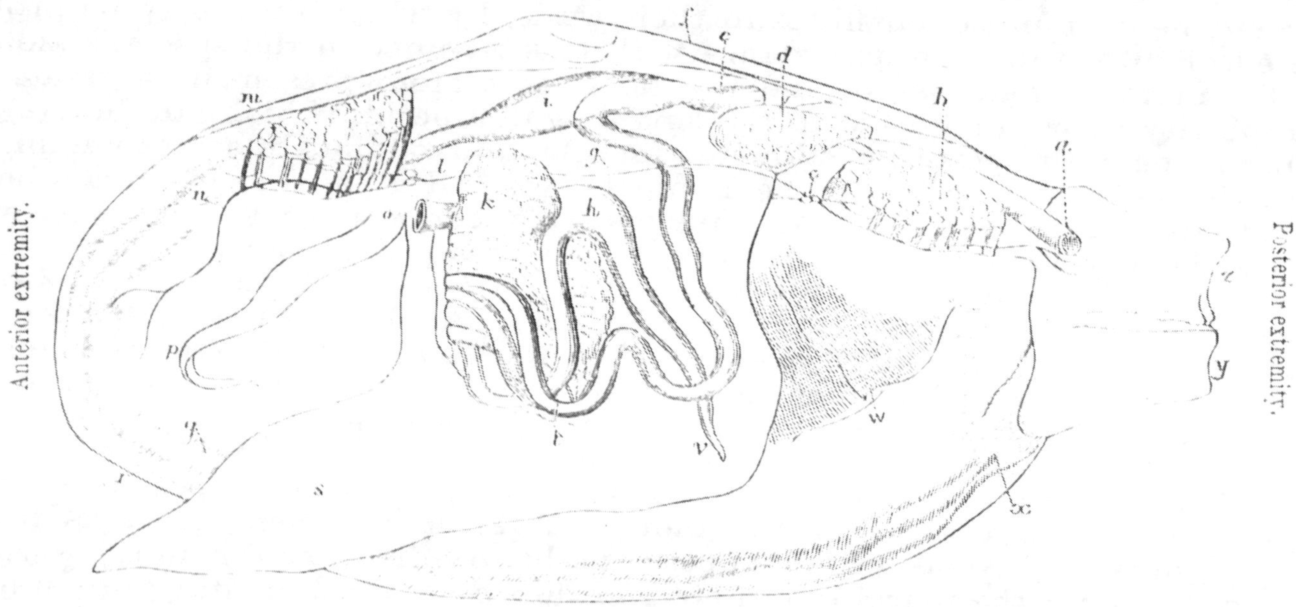

Ventral Margin.

Anatomy of *Mactra* ;—*a*, anus ; *b*, posterior muscle ; *c*, branchial ganglion ; *d*, ovary ; *e*, *t*, intestine ; *f*, shell ; *g*, nervous cord, connecting œsophageal and branchial ganglia ; *h*, stomach ; *i*, heart ; *k*, liver ; *l*, anterior or œsophageal ganglia ; *m*, anterior adductor muscle ; *n*, nervous filaments ; *o*, mouth ; *p*, one of the oral tentacula ; *q x*, mantle ; *r*, margin of the shell ; *s*, foot ; *w*, branchial lamellæ ; *y*, oral siphon ; *z*, anal siphon.

organ (*d*), which is an ovary in the female, and a spermatic gland in the male. In the *Cyclas*, the two sets of sexual organs are combined ; but in other bivalves they are set apart in distinct individuals, so that the ova developed by one are fertilized by the spermatozoa set free from another. This fertilization must be usually effected, however, before the ova are sent forth from the body of the female ; for the early development of the embryo takes place in some instances, and probably in all, within the shell of the parent.—The muscular apparatus of the Lamellibranchiata is very simple. The muscle most constantly present is the *adductor*, by which the valves are drawn together ; this is sometimes single, but is more commonly double, one portion of it (*m*) being situated near the anterior extremity, and the other (*d*) near the posterior extremity of the shell, on both valves of which its attachments are plainly distinguishable. But the greater number of this class have a fleshy tongue-like organ (*s*), termed the foot, which is capable of being put forth from between the valves, and of being applied to a variety of purposes. In some of those in which it is least developed (as the Mussel), its principal office appears to be to form the *byssus* or cord of horny filaments by which the animal attaches itself to rocks, &c. But in others it is an instrument of active locomotion ; thus, the common Cockle can take considerable leaps, by suddenly extending this organ, which was previously bent at an acute angle; and the first specimen of *Trigonia* discovered by Mr. S. Stutchbury on the coast of New South Wales, having been placed in the stern of the boat from which he was dredging, leapt over the gunwale, a height of about six inches, into the sea. And in others, again, it is the instrument for excavating and perforating: although upon the mode in which it is used for these purposes, there is still some difference of opinion. Besides these muscles, most of the Bivalves furnished with long respiratory siphons possess an arrangement of muscular fibres in their walls, by which they may be projected or drawn in. In accordance with this advance in the specialisation of the motor apparatus, we find the nervous system presenting a greater complexity; for in addition to the single or double branchial ganglion (*c*), situated near the posterior adductor muscle, and obviously corresponding with the solitary ganglion of the Tunicata, we find a pair of small ganglia (*l*) situated near the œsophagus, and specially connected with the sensory tentacula that surround its entrance; from this proceed branches (*n*) to the anterior portion of the body, whilst a cord of communication (*g*) passes backwards to the posterior ganglion. In those Bivalves which have a foot, a third ganglionic centre is always to be found at or near its base, connected with the œsophageal ganglia. And where the respiratory siphons are muscular, these muscles are supplied from a small ganglion of their own, situated at the point of junction of the tubes. The tentacula prolonged from the mouth are the most constant organs of sense possessed by these animals; but some of the more active among them, such as the *Pecten*, have red spots which appear to be ocelli or rudimentary eyes, disposed around the margin of their mantle; and a rudimentary organ of hearing has also been detected in the substance of the anterior ganglion of *Cyclas*.—We thus find the Lamellibranchiata advancing, in several particulars, towards the higher type of the Mollusca; but the predominance of the organic over the animal life is still very strongly marked. Some of them acquire a very considerable size and weight; thus the *Pinna* sometimes attains a length of four feet (and some of the extinct species appear to have been much larger), whilst the *Tridacne* (or Giant Clamp-shell) has been known to weigh 600 lbs. Those which are most bulky are the least active, their lives being passed almost entirely in the same situation, and scarcely any movement (save the ciliary action which supplies food to the stomach and water to the gills) being exhibited by them, except the closure of their valves which takes place when the animal is in any way irritated.*

302. The class of GASTEROPODA, which receives its name from the situation of the muscular disk on which the animal creeps, may be regarded as the type of the whole Molluscous series; for it presents in their most

* It is by taking advantage of this action, that the natives of the East Indian Archipelago obtain the shells of the *Tridacne*. In their young state, the animal is anchored by a *byssus;* but as it approaches adult age, and the shell acquires considerably increased weight, the byssus disappears, and the groove in the shell is filled up by a solid growth. When it is thus lying free at a depth of ten or fifteen feet in the water, a long pole is pushed down between the opening of its valves, which then close upon it (just as the valves of an Oyster do upon the knife inserted between them) with so strong a hold, as to allow the massive shell to be lifted by the pole out of the water.

distinct and complete form all its characteristic features, and is further removed than any other class from either of the remaining sub-kingdoms. Moreover, it contains by far the greatest number of species; and these present a remarkable variety in their plan of conformation, whilst the grade of development is nearly the same in all. A large proportion of them possess a shell, within which the body can be entirely retracted; and this shell, being formed in one piece, is said to be univalve. Its typical form may be considered to be a cone with a broad open base, as we see in the common Limpet (*Patella*); in the *Pileopsis*, the point of the cone is prolonged and somewhat turned to one side, so as to resemble a 'fool's cap,' thus presenting the rudiment of a convolution. The increase of this tendency to deflexion in the axis of the cone produces a regular *spiral* shell, such as that of the *Planorbis*, in which all the whorls or convolutions are on the same plane. But if these whorls do not revolve in the same plane, the axis of the cone then forms a *helix* instead of a simple spire, as is the case in the common *Snail*, and in the greater proportion of the shells of this class. From this form we may return to the straight cone by the *Scalaria*, in which the coils of the spire touch each other only by their ribs; and by the *Vermetus* and *Magilus*, in which the commencement only of the shell presents a helical form, the remainder being prolonged into a tube which is but slightly curved, thus approaching the *Dentalium*, the shell of which is an elongated cone, whose axis is almost a straight line. Where the whorls revolve around a vertical line, instead of remaining in the same plane, a sort of central pillar is formed, which is termed the *columella*; this is frequently grooved at its lower part, for the passage of water to the respiratory organs.—The substance of univalve shells generally contains less animal matter than that of bivalves; and we frequently find it extremely hard or *porcellanous*, exhibiting little indication of its organic origin. There can be no doubt, however, that it is formed as a calcified epithelium from the surface of the mantle, as in other cases. The size of the shell is augmented, in accordance with the growth of the animal, by the extension of the margin or opening of the cone; this, of course, at the same time enlarges its diameter, and increases its length; so that as, in a helical shell, the number of whorls increases with age, so does also the size of the successive whorls. As in the Bivalves, however, with every such addition, a new shelly layer is formed within the whole interior of the previously-existing shell; but this layer is so thin, except near the mouth, that it cannot be detected except by microscopic examination.—When the full growth of the individual has been attained, a thickened rim or lip is formed around the aperture, which serves to strengthen the whole fabric; but in several marine species, which attain to large dimensions, such a lip is formed after each extension; so that the shell has a ribbed appearance, each rib having been at one time the boundary of the aperture. This margin is sometimes fringed with spines formed from prolongations of the mantle, as in the *Murex*; and the dissimilar number of these has led to the establishment of many distinct species, which, when the habits of the animal came to be better known, have proved to be but different forms of the same. For it now appears that the animal has not only the power of forming new spines, but of removing old ones, especially such as would interfere with the continued growth of the shell. How the absorption of shelly matter from their base, which causes them to drop off, is effected, is still unexplained; various analogous phenomena may be witnessed among other species, portions of the shell first formed being wholly or partly removed. Sometimes the walls of the older portion are thinned for the purpose of lightening the shell, and the same object seems to be attained by other inhabitants of helical shells in a different manner; these withdraw their bodies from the highest part of the cone, throw a partition across the cavity, and then allow the point (which, not being internally supported, is brittle, and appears to have been purposely thinned) to be broken off, leaving the shell *decollated* as it is termed.* It must be borne in mind, however, that the changes thus effected in the shell are not the consequence of any *interstitial* absorption, such as takes place in the osseous structures of Vertebrata,—but result from the same kind of power of *superficial* absorption, as appears to be exercised by many Gasteropoda upon calcareous rocks, which they perforate for their habitations, as well as upon the shells of other species upon which they feed.—Several of the aquatic species of this class form not merely a spiral shell, but an accurately fitted cover to its mouth, so attached to the body, that, when the latter is entirely withdrawn, the *operculum*, as it is called, completely shuts it in. Sometimes this is horny, but not unfrequently calcareous; and occasionally it bears so large a proportion to the shell, as almost to appear like a second valve, such as is characteristic of the Conchifera. Some of the land species also possess an operculum; but in general they are destitute of it, and form during hybernation a temporary closure to the mouth of the shell by a viscid secretion, which becomes hard and includes a bubble of air; behind this a second and even a third similar partition are occasionally found, as in the common Snail—The presence of a shell, however, is by no means universal in this group; as many Gasteropods are 'naked' or shell-less, and many more have shells too small to allow their bodies to be completely withdrawn into them. Of these we find some, especially among the terrestrial species, that are so closely allied to the proper testaceous Gasteropods, that it is evident that the presence or absence of the shell is a feature by no means essential; thus between the Slug and the Snail, the former a 'naked,' and the latter a 'testaceous' Mollusk, there is a very close alliance; and the two extremes are connected by a very complete series of intermediate forms, many Slugs having small shells upon their tails, and many Snails having shells far too small to include their bodies. There is one marine order, however, that of *Nudibranchiata* (§ 304 *c*), in which the absence of the shell is a characteristic feature, being connected with the extension of the respiratory apparatus over so large a part of the external surface, that the investment of the body by a shell would obstruct its functions. It is a very remarkable fact, however, that in their *young* state the Nudibranchiate, as well as (it is believed) all other marine Gasteropod Mollusca, possess a shell.—The most remarkable departure from the general type of conformation in the shell of this group, is presented by the *Chiton;* the animal of which is more closely allied to the Limpet than to any other Gasteropod; while the shell, instead of being a simple cone, is composed of a number of pieces jointed to each other in such a manner as to present a remarkable resemblance to the covering of

* Mr. S. Stutchbury has informed the Author, that he has seen the *Bulimus* forcibly strike the apex of its shell against a stone, as if for the purpose of decollating itself.

an Articulated animal (Fig. 125, B). This resemblance is not confined to the shell; for its different portions are connected together by a complex muscular apparatus, which enables the animal to move them one upon another in such a manner as to roll itself up; and there are other points in the internal structure of the animal, such as the symmetrical arrangement of the generative apparatus, and the resemblance of the heart to the 'dorsal vessel' of an Annelide rather than to the heart of an ordinary Gasteropod, which induce the Philosophic Naturalist to regard the Chiton as a connecting link between the Molluscous and the Articulated series.* It is worthy of note that in the *Chitonellus* (Fig. 125, A) we find the shell not sufficiently developed to cover the animal, yet its several rudimentary pieces are arranged at regular intervals along the back, precisely as in the Chiton, thus strongly marking the tendency to segmentation in the body.

FIG. 125.

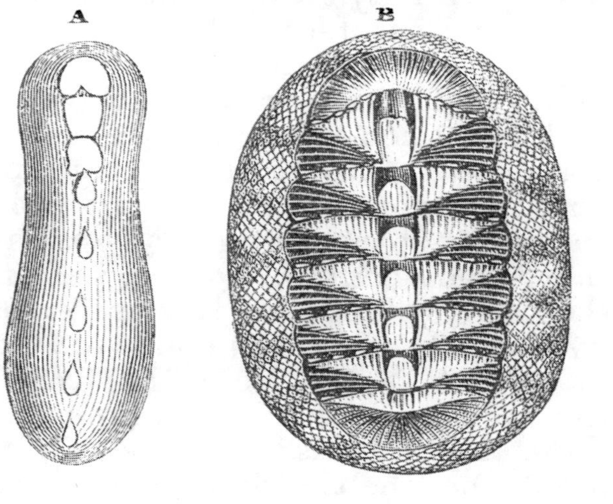

A, *Chitonellus.*—B, *Chiton.*

303. Excluding the *Chitons,* for the reasons just stated, the class of Gasteropoda may be characterised as a whole by the presence of a *head,* on which is situated the mouth, with organs of sense, more or less numerous; by the possession of a locomotive disk, formed by a fleshy expansion from the posterior part of the head; by the want of symmetry in the viscera generally, and especially in the generative apparatus; and by the tendency to a *spiral* disposition of the organs, which almost invariably manifests itself in the embryo, and frequently continues through the whole period of life. The class thus defined must be separated, however, into two sub-classes; namely, the *True Gasteropoda,* in which the fleshy disk is broad and flattened horizontally, lying beneath the visceral mass, and serving to enable the animal to creep over solid surfaces, and in which the abdominal region of the body is fully developed; and the *Heteropoda,* in which the locomotive organ is flattened into a vertical fin (Fig. 129), whereby the animal swims freely through the water but cannot crawl, and in which the abdominal portion of the body is rudimentary, whilst the cephalo-thoracic is enormously developed. The latter must be considered as an aberrant group; and the species which it contains are few in comparison with those of the true Gasteropods.—Putting these on one side for the present, the sub-class of *True Gasteropoda* may next be divided into the *Branchiferous* and *Pulmonated* sections; the former including all that breathe by gills and inhabit the water; whilst the latter comprehends those which breathe air by means of a pulmonary sac, and which are for the most part inhabitants of land. There are other important differences between these two sections, which will be noticed hereafter.—In the Branchiferous Gas-

* See the admirable Memoir on the Classification of the Gasteropoda, by Prof. Milne Edwards, in the Ann. des Sci. Nat., Ser. III., Zool. vol. ix. p. 102.

teropods, moreover, we find two dissimilar types of structure, which are designated by M. Milne-Edwards,* in accordance with the position of the gills, as *Prosobranchiata,* and *Opistobranchiata.* In the former, of which we may take *Paludina vivipara* (Fig. 128) as an example, the body is always protected by a shell of sufficient dimensions to include the whole of it; the mantle is prolonged forwards, and forms behind the head an arched chamber, in which are found the orifices of the excretory canals, and in which the gills are usually lodged. The gills (Fig. 128, *g*) are composed of a series of straight and simple lamellæ, inserted along a vascular stem, and having a pectinated arrangement (that is, resembling the teeth of a comb); they are usually situated in front of the heart, so that the blood flows *backwards* from them towards the heart; and even when they are prolonged to the hinder part of the body, the course of the vessels by which the blood is conveyed from them to the heart is such that it flows into that organ from the front. Further, these animals are diœcious, the male and female organs of generation not being possessed by the same individuals. In the Opistobranchiata (Fig. 127), on the other hand, the shell, although existing in the larva, becomes rudimentary, or even disappears altogether, in the adult; the gills are usually arborescent in their form (Fig. 126), and are never included within any special cavity, being always more or less obviously displayed on the back or hinder part of the animal; the blood flows from them *forwards* to the heart; and the reproductive apparatus is hermaphrodite, the male and female organs being combined in the same individuals.—The general structure of the Pulmonated Gasteropods most nearly corresponds with that of the Prosobranchiata (Fig. 128), with the exception that the blood is not transmitted for aëration into a pectinated branchial apparatus immersed in water, but is distributed over the walls of an internal cavity to which air is admitted. The presence or absence of the shell in this group does not seem to be a matter of as much importance as it is among the branchiferous Gasteropods; as we find the naked Pulmonata corresponding in all essential respects with the testaceous.

FIG. 126.

Doris Johnstoni, showing the tuft of external gills.

304. We shall now take a general survey of the structure and arrangement of the principal organs of the body, in these different groups, commencing with the *Gasteropoda* proper. The mouth is almost invariably furnished with instruments for cutting or rasping solid substances; these animals not being dependent, like the Acephalous tribes, upon the minute particles brought by their ciliary currents, but being able to go in search of their own food. This they separate, in many cases, by means of a pair of horny jaws, fitted for the division of vegetable or animal substances, and worked by a muscular apparatus of some complexity (Fig. 128, *a*). In the *Patella* (Limpet), however, the food is obtained by means of a tongue, which, when extended, exceeds the body in length,

* Loc. cit.

and which is furnished with sharp, recurved, horny teeth, that render it a very efficient instrument for rasping the sea-weeds which constitute the nourishment of this animal. In the *Buccinum* (Whelk) and other

FIG. 127.

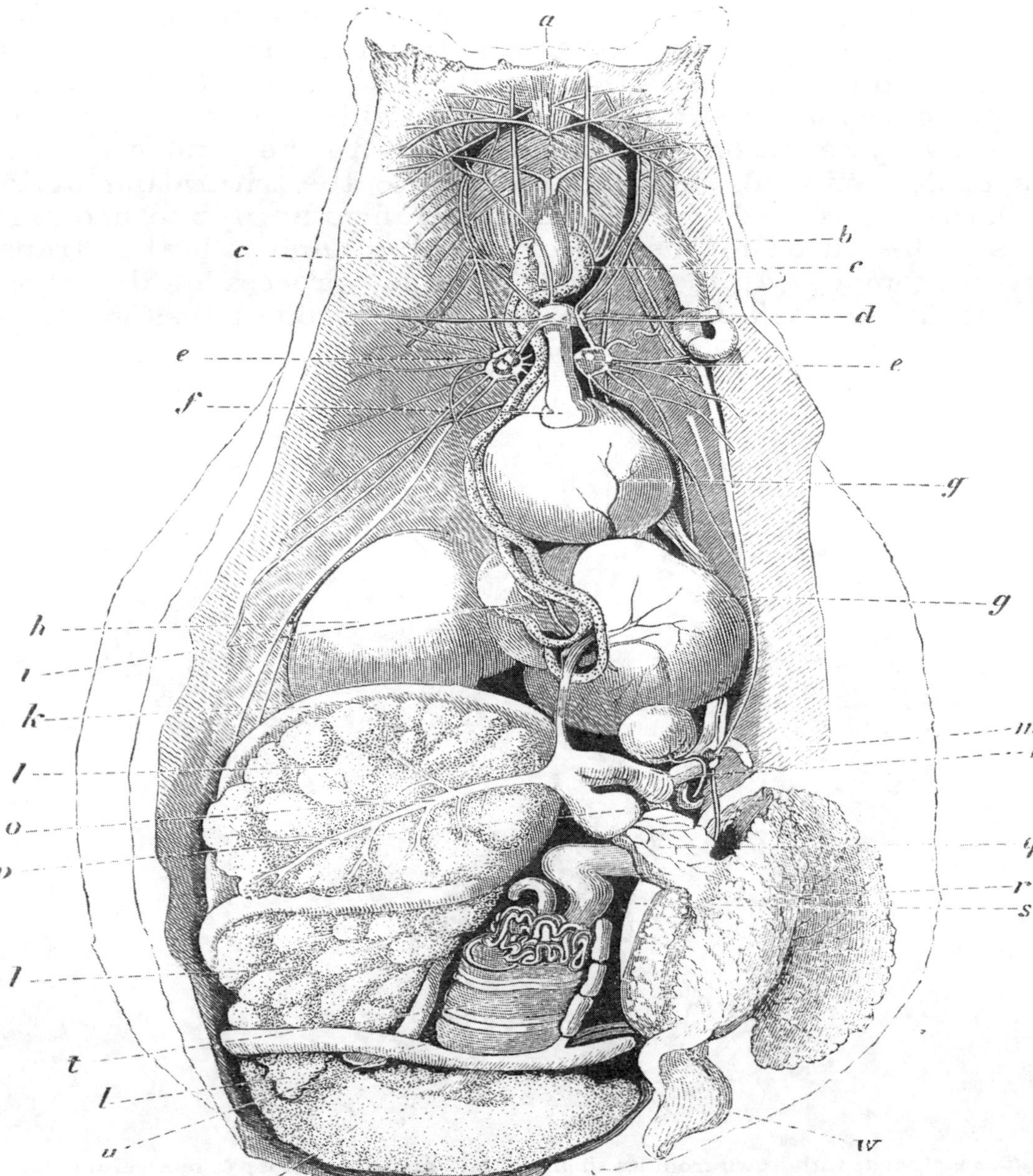

Aplysia **laid open, to show the arrangement of the viscera:—*a*, upper part of the œsophagus; *b*, penis; *c, c*, salivary glands; *d*, superior or cephalic ganglion; *e, e*, inferior or subœsophageal ganglia; *f*, entrance of the œsophagus into *g, g*, the first stomach or crop; *h*, the third or true digestive stomach; *i*, the second stomach or gizzard; *k*, intestine; *l, l, l*, liver; *m*, posterior or branchial ganglion; *n*, aorta; *o*, hepatic artery; *p*, ventricle of heart; *q*, auricle; *r, s*, branchiæ; *t*, testis; *u*, lower part of intestine; *v*, ovary; *w*, anus.**

carnivorous species, again, a tongue of a very similar description, but shorter, is inclosed within a proboscis-like organ (Fig. 128, *p*), which is capable of considerable elongation, or of being entirely retracted within the cavity of the body, and which is vertically cleft at the extremity into two lips, the inner surface of each being covered with recurved spines; by the actions of this apparatus, the Buccinum can perforate the shells of other Mollusks, and extract the soft parts of the animals which they contain.—Behind the mouth there is usually found a pair of salivary glands (Fig. 127, *c c*, Fig. 128, *v v*), between which passes down the œsophagus (Fig. 127, *f*, Fig. 128, *q*) towards the stomach. In the *Aplysia* and some other Gasteropods, the œsophagus first dilates into a sort of *crop* or *ingluvies* (Fig. 127, *g g*), in which the food is macerated with the salivary secretion before it passes into the true stomach; from this it is passed onwards, to be subjected to the triturating action of a gizzard (*i*), the interior of which is beset with firm horny teeth, or with hard calcareous plates, fitted for rasping or crushing; and it is then transmitted to the true stomach (*h*), in which the actual process of digestion takes place. But in the *Paludina*, as in the carnivorous tribes generally, this

Fig. 128.

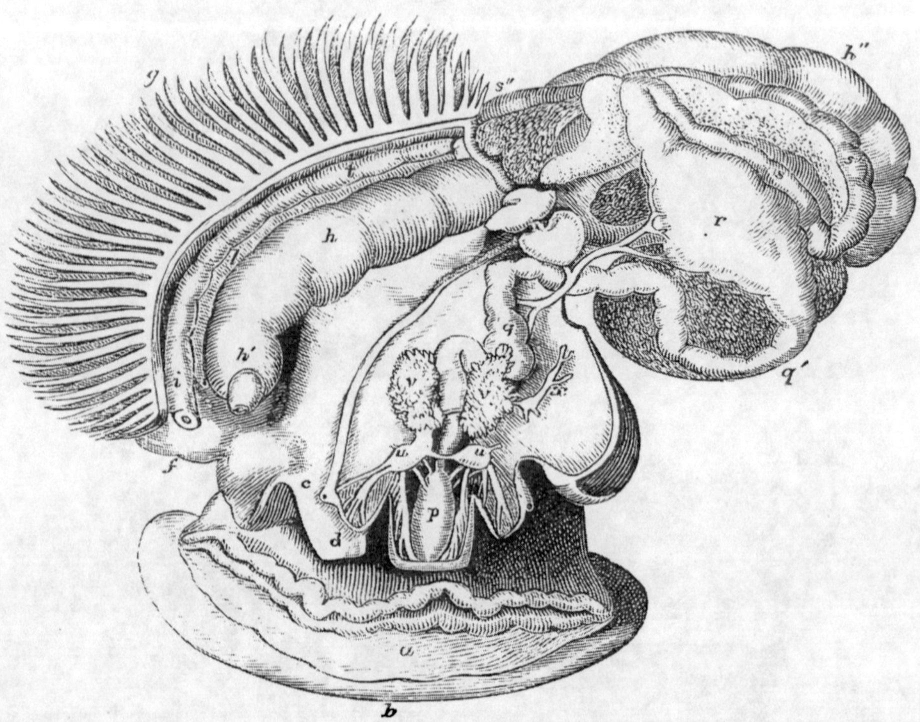

Paludina vivipara, withdrawn from its shell and laid open;—*a*, foot; *b*, operculum; *c*, one of the tentacula with its ocellus; *d*, siphon; *f*, border of the mantle; *g*, pectinated branchiæ; *h*, *h'*, the oviduct dilated for the retention of the ova; *h''*, portion of it situated within the spire of the shell; *i*, termination of the intestine; *l*, canal for the urinary (?) secretion; *n*, heart; *o*, liver; *p*, proboscis; *q*, *q'*, œsophagus; *r*, stomach; *s*, *s'*, *s''*, intestine, lying at *t* within the branchial cavity; *u*, *u*, cephalic ganglia; *v*, *v*, salivary glands; *x*, the principal muscular nerve.

complex apparatus is wanting; and the stomach dilates at once into a stomach of simple form (Fig. 128, *r*), which suffices to perform the digestive operation of the whole. The intestinal tube (Fig. 128, *s*, *s'*, *s''*, Fig. 127, *k*, *u*), which is most prolonged in the herbivorous species, usually makes several convolutions in the substance of the liver (which organ here attains a high degree of development, Fig. 127, *l*, *l*, *l*, Fig. 128 *o*); and at last generally turns upon itself, so as to open behind the head (Fig. 128, *i*), although in some of the naked Gasteropods, it terminates near the posterior part of the body (Fig. 127, *w*). The heart (Fig. 128, *n*) is more muscular than in the Acephalous Mollusca, and is always distinctly divided into an auricle and a ventricle (Fig. 127, *n*, *p*); the auricle, which receives the blood from the respiratory organs, being sometimes double. From the ventricle proceeds the aorta (Fig. 127, *o*), which transmits the blood to the system in general; the fluid, in its return thence, however, is not confined within distinct vessels, but meanders through sinuses or passages excavated in the tissues; and through these it reaches the branchial apparatus (Fig. 127, *r*, *s*, Fig. 128, *g*) or the pulmonic cavity, whence it is transmitted back to the heart through the branchio-cardiac canals. Besides the liver and the salivary glands, we find other glandular organs in this class. Thus the rudiment of a pancreas distinctly presents itself, in some instances, in connection with the stomach; and a follicular gland is frequently to be found discharging a mucous secretion, which contains uric acid, and is therefore to be considered as urine, through a duct that opens near the anus (Fig. 128, *l*). Secretions of colouring matter, as well as of other special products, are effected by distinct follicular glands in particular cases. Thus the whole apparatus of organic life displays a considerable advance upon the type presented in the Acephala.

304*a*. A similar advance is manifested in the apparatus of Animal life; the Gasteropods being generally possessed of various organs of special sense, and of considerable locomotive power, although their actions are slow and apparently feeble. The number of ganglionic centres is increased, in accordance with the number of organs to whose functions they minister; and there is a greater tendency to a concentration of them in the neighbourhood of the mouth, so as to form a sort of collar around the œsophagus. Of this collar, the upper and lateral part, which sometimes forms a single ganglion (Fig. 127, *d*) but is more commonly double (Fig. 128, *u*, *u*), is composed of the ganglionic centres of the organs of special and general sense, which are situated in the head; whilst the lower (Fig. 127, *e*, *e*) is usually formed by the ganglionic centres of the mantle and foot, which are sometimes combined, sometimes distinct. Besides these, there is a separate ganglionic centre for the respiratory apparatus (Fig. 127, *m*), which is always found in its neighbourhood, and whose position, therefore, varies with that of the gills; and there is also generally one that is connected with the muscles of mastication and deglutition. Nearly all the Gasteropoda possess one or two pairs of sensory tentacula; and where, as in the *Snail*, two pairs are present, one of them (the smaller) usually bears the eyes, which in other instances are situated at the base of the principal pair. These eyes never exceed two in number, and they are generally well developed; but in some species they are minute, and in a few are entirely wanting. A rudimentary organ of hearing exists, imbedded in the posterior part of the œsophageal nervous collar; and there is reason to believe that the terrestrial Gasteropods (to say the least) enjoy the sense of smell, although no distinct organ for this purpose can be detected. The tongue, where it exists, is a mere mechanical instrument, and does not appear to minister to the sense of taste; but this may have its seat in the soft membranes of the mouth and pharynx. The softness of the general surface of the body might naturally lead to the supposition that it is a delicate organ of touch ; but the fact that the terrestrial Gasteropods show no signs of pain when their integument is being devoured by their carnivorous enemies, would seem to prove that its sensibility must be extremely low. It seems probable, however, both from the copious supply of nerves and blood-vessels to the foot, as well from the consideration of the purposes to which that organ is applied, that it possesses more tactile sensibility, as well as more muscular power, than the remainder of the mantle, of one part of which it may be regarded as an augmented development. It is interesting to observe that a symmetrical arrangement prevails in the organs of sense, and in other parts which are specially concerned in the functions of *animal* life ; that symmetry being usually deficient in the posterior part of the body, which simply encloses the apparatus of organic life, but extending also to it (so far as external form is concerned) when the locomotive powers are greatest, as is seen especially in the *Nudibranchiata* (Fig. 126).

304*b*. Of the power of reproduction by gemmation, there is no known example in the Gasteropoda ; although we observe its traces in their capability of repairing injuries, which extends to the regeneration of the head, tentacula, and eyes, when the cephalic ganglia have been left uninjured. The proper generative apparatus is usually developed largely. The male and female organs are combined in the same individuals in the Pulmonated Gasteropods and in the Opistobranchiata ; but the sexes are distinct in the Prosobranchiata. It is uncertain whether any of the hermaphrodite Gasteropods are self-impregnating, like most of the inferior Mollusks ; but it is certain that the pulmonated tribes are not, the congress of two individuals being requisite, each to fertilize the ova of the other. We find in this class a more elaborate provision for the accomplishment of this act, than we have hitherto seen ; namely, an intromittent organ, or penis (Fig. 127, *b*), frequently of considerable dimensions, which conveys the spermatic fluid of the male organs into the immediate neighbourhood of the ova which it is to fecundate ; instead of this fluid being simply diffused through the surrounding water, and left to find its way to the ova by that medium. The ovarium usually occupies a considerable space in the posterior part of the body (Fig. 127, *v*) ; but the orifice of its oviduct is generally behind the head. This oviduct, in the *Paludina* (Fig. 128, *h*, *h′*, *h″*) and some other Gasteropods, is greatly dilated at its termination, and serves to retain the eggs until they are hatched, so that the young are produced alive.—There is an important difference between the condition of the embryos of the Branchiferous, and that of the Pulmonated Gasteropods. All the former, whether naked or testaceous in their adult state, are provided, at their emersion from the egg, with a minute spirally-convoluted and transparent shell, from the mouth of which is protruded a rudimental head, in which the eyes are distinctly perceptible ; but instead of the proper disk-like foot which is characteristic of the class, we find a bilobed expansion fringed with cilia, by the agency of which the embryo is propelled through the water it inhabits ; and the viscera which should occupy the interior of the shell are as yet scarcely distinguishable, that space being chiefly filled up by the mass of yolk, which is not yet converted into the fabric of the embryo. In the latter, the shell is not formed in the embryo, except where it is to be retained in the adult, and even then it is not formed until a period which is comparatively much later; and the ciliated lobes are altogether wanting, although ciliary movement is seen in the embryo before its emersion from the egg, especially on the portion of the surface which may be considered as the rudiment of the pulmonic organ.

304 *c.* A remarkable departure from the ordinary Gasteropodous type is presented by a portion of the *Nudibranchiate* section of Opistobranchiata, consisting of the *Eolis* and its allies, which some have considered as constituting a distinct order, in virtue of the peculiar conformation of their digestive and respiratory organs. The stomach of these animals seems to be extended, by a series of ramifying prolongations, throughout the entire body; these prolongations terminating in numerous papillæ, which arise from the dorsal surface, and which are themselves sometimes arborescent in their form (as in *Dendronotus*). But a careful examination of these terminations shows that they are surrounded by hepatic secreting cells; so that the whole of this ramifying apparatus must be regarded, not as a prolongation of the digestive cavity, but as an unfolding and extension (so to speak) of the liver.* The external surface of these papillæ is covered with cilia; and from this circumstance, as well as from the manner in which the blood is transmitted to it, there seems little doubt that it is specially adapted for the respiratory function, there being no other branchial apparatus. In some of the most degraded forms of this group, however, these papillæ are absent, notwithstanding that the same ramification of the biliary apparatus through the body still presents itself; and the general integument here constitutes the only respiratory surface. Through such Mollusks as these, we may trace a relationship with the *Planaria* and its allies, which are usually ranked among the Entozoa (§ 309 *d*).—The generative apparatus of these Mollusca is enormously developed, the ovaries occupying a considerable part of the body; and it would seem as if this enlargement of one set of organs necessitated a different distribution of the other, on whose place it encroaches.

304 *d.* The *Heteropoda*, although separated from the Gasteropoda by some Naturalists, seem properly to belong to the same class, of which, however, they constitute an aberrant section; corresponding with the true Gasteropods in their grade of development, but differing from them in the arrangement of the organs, and in the proportional size of the several parts of the body. Thus the expanded discoidal foot is converted into a vertical fin-like expansion (Fig. 129, *f*), the edge of which in some species bears a sort of sucker (*g*), the only rudiment of the flattened disk of the remainder of the class. The body has an elongated form, and is symmetrical through its whole length, and the principal part of it is composed of a pellucid gelatinous substance, included in a thin muscular layer, which is itself sufficiently transparent to allow the organs contained within it to be distinctly seen. The mouth, which is situated at the anterior extremity, is furnished with a muscular mass (*k*), and with a rasplike tongue;

* By M. de Quatrefages, and some other French Naturalists, it has been supposed that the circulating apparatus of these Mollusks is very imperfect, and that, in fact, this extension of the alimentary canal answers the purpose of a venous system; and from this supposed combination of the venous and intestinal cavities, the term *Phlebenterata* has been applied to the group. The researches of Messrs. Alder and Hancock and other British Zoologists, however, have shown that the circulating system is as complete in itself as is that of other Gasteropoda; and have established the hepatic nature of the ramifying prolongations of the alimentary canal.

and the head (l) bears two tentacula (i), at the base of which are usually found distinct eyes (h). The alimentary canal passes directly backwards from the mouth towards the posterior and superior part of the body, where it leaves the trunk to enter the visceral mass, which is borne upon the back, as if detached from it; in its course this canal undergoes a dilatation, which must be considered as the stomach; and thus the anterior portion constitutes the œsophagus, whilst the posterior forms the

Fig. 129.

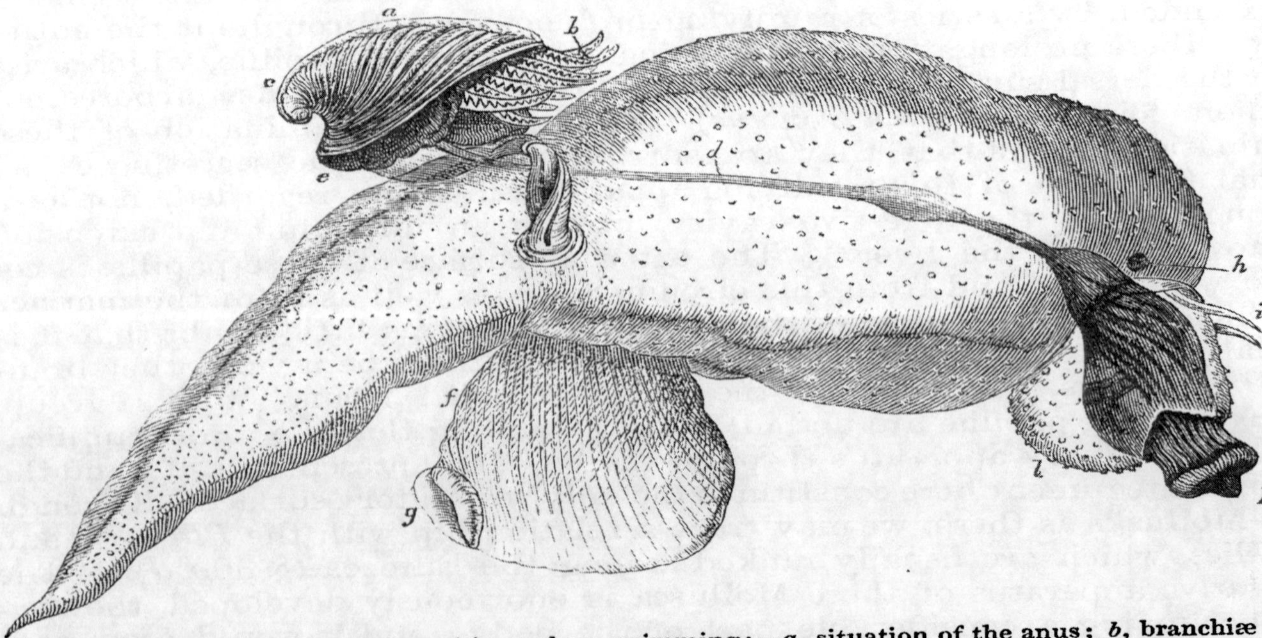

Carinaria, in its natural position when swimming;—*a*, situation of the anus; *b*, branchiæ; *c*, shell; *e*, liver; *f*, fin-like foot; *g*, sucker; *h*, eye; *i*, tentacula; *k*, proboscis; *l*, head.

intestine. The body (which, as already remarked, must be regarded as formed by an extraordinary development of the thoracic portion of the animal) does not contain any other parts of the nutritive apparatus; the heart, the branchiæ (b), the liver (e), and the generative organs, being aggregated together in a small mass, which reminds us of the *nucleus* of the Salpæ (§ 299 b), except that it is external to the remainder of the body, instead of being imbedded within it. This nucleus, in the Carinaria, is protected by a delicate shell (c), strongly resembling that of the Argonaut; but in other Heteropods the shell is wanting. The sexes are believed to be disjoined in some forms of this group, but united in others. The nervous system is formed upon the type of that of the Gasteropoda in general; consisting of a pair of cephalic ganglia, at the sides of the œsophagus, which receive the nerves of the organs of sense, and supply the muscular apparatus of the mouth; with a large four-lobed ganglion near the foot, which may be considered as combining within itself the pedal and palleal ganglia. These animals swim freely through the water, by the strokes of their fin-like foot, and of the vertically-flattened tail; those which possess the sucker at its margin occasionally anchoring themselves (it is presumed) to solid bodies, either fixed or floating.

305. Another small group which should perhaps be rather regarded as an aberrant subdivision of the Gasteropod class, than as ranking as a class of itself, is that of Pteropoda, which is so named from the presence of a

wing-like membranous expansion on each side of the neck (Fig. 130), by the agency of which the animals swim through the water with greater activity than do most of the tribes to which they seem most nearly allied. Their zoological position is a matter of considerable interest. By some eminent Zoologists they have been regarded as ranking *above* the Gasteropoda, and as connecting them with Cephalopods. But there can be no doubt that on the whole they are inferior in organisation to most Gasteropods, and that in many particulars, and especially in that which is their peculiar characteristic, they bear a striking resemblance to the embryonic condition of that class. Some of them, as the *Hyalœa* and *Criseis* (Fig. 130, A and B) have the hinder part of the body encased in a delicate shell; whilst others, as the *Clio* C, are destitute of any such protection. The shell, where it exists, may present many different forms, one of the most remarkable is that of the *Hyalœa*, which is like a bivalve shell cemented together at the hinge, and having a narrow fissure in front and at the sides, from which the anterior part of the body projects; that of the *Criseis*, again, is a straight cone; and that of *Limacina* is a convoluted cone. In all instances, however, the shell, like the body, is perfectly symmetrical laterally; whilst it usually presents some indication of a division into a dorsal and ventral valve; thus resembling, as respects its relation to the contained animal, the shell of the Brachiopoda (§ 300), whose long ciliated arms may be considered to be represented by the short but expanded lobes which occupy a somewhat similar position in the Pteropoda.* Thus the Pteropods may perhaps be considered as bearing the same relation to the Brachiopoda, that the Gasteropods do to the Lamellibranchiata; that is, as carrying on the same plan of organisation into a more advanced grade of development. They have always a more or less distinct head, this being usually most prominent in the naked species; but although this is usually furnished with a complex apparatus for the prehension and mastication of food, the organs of sense are rudimentary. In *Hyalœa* and many other Pteropods, there are no cephalic tentacula whatever. In *Clio* there are six tentacula, which seem to prefigure the arms of Cephalopods, being

FIG. 130.

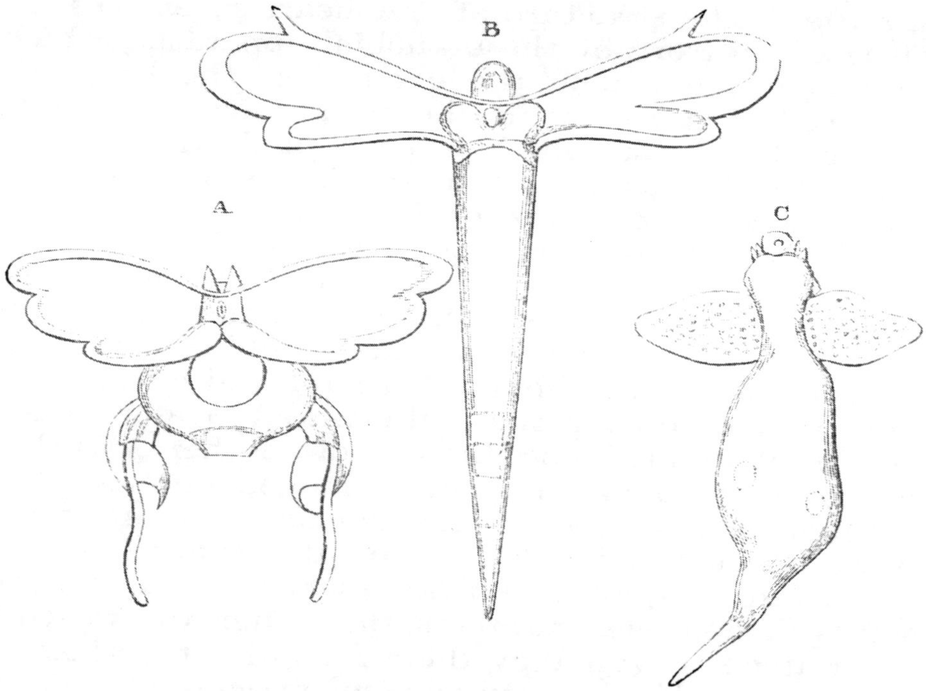

A, *Hyalœa*; B, *Criseis*; C, *Clio*.

* See Owen, Lectures on the Comparative Anatomy and Physiology of the Invertebrata, p. 292.

covered with minute suckers, the whole number of which is estimated at 360,000; and there are also two slender tentacula, which appear to be instruments of the tactile sense. In *Cymbulia*, there are two small sensory tentacula, with a rudimentary eye at the base of each. A rudimentary organ of hearing has also been detected. The digestive apparatus is formed upon the general plan of that of Gasteropods. The mouth is usually furnished either with horny jaws or with a rasp-like tongue; in *Clio*, both are present. The stomach varies in its degree of complexity, some genera being provided with curiously-armed gizzards, whilst others are destitute of them, and have merely a simple globular cavity; the intestine usually forms three or four coils in the midst of the voluminous liver, and terminates at the right side towards the anterior part of the mantle. The structure of the heart resembles that of Gasteropoda; it is situated in front of the branchial apparatus, which is a network of blood-vessels, developed on the exterior of the body in the naked species, but enclosed in a cavity in the testaceous.—The muscular apparatus, independently of that which is connected with the mouth, chiefly consists of the fibres that give motion to the lateral fin-like expansions, by means of which these animals swim with considerable activity. The cephalic ganglia are relatively smaller than those of Gasteropods, as might be expected from the inferior development of the organs of sense; the pedal and palleal ganglia are both double, and are symmetrically disposed beneath the œsophagus. The sexual organs are combined in the same individuals; but it is probable that, although hermaphrodite, they are not self-impregnating, but that, as among the Pulmonated Gasteropods, the spermatic fluid of one individual fertilizes the ova of another. Of the history of the development of these animals, little is yet known.—Although the number of distinct forms contained in this class is comparatively small, yet the multiplication of individuals is frequently enormous. Thus the *Clio borealis* in the Arctic seas, and the *Clio australis* in the Antarctic, often present themselves in such vast numbers, that, when the weather is calm, the surface of the water is covered by them for miles together; and these little animals are among the most important articles of food to the Whale, which is enabled, by the curious sieve-like apparatus in its mouth, to strain them (as it were) out of the water which it gulps-in.

305 *a*. The group of *Pteropods* may be regarded as representing among the Mollusca the class of *Insects* in the Articulated series, and that of *Birds* in the Vertebrated; all these being characterised by the highest degree of locomotive power that is to be found in their respective sub-kingdoms, by the exact bilateral symmetry of their bodies, and by the possession of lateral appendages, put in action by muscles, as instruments of progression. The highest manifestation of the same plan of construction in the Radiated division, is perhaps to be found in the *Ciliograde Acalephæ* (§ 294 *a*); in which, as we have seen, there is an obvious tendency to bilateral symmetry, and to the formation of wing-like expansions, whilst their movements are more rapid and energetic than those of any other Radiata. It is interesting to remark, further, that whilst the class of Insects, which carries into fullest development what is the most characteristic tendency of the whole Articulated series, is by far the *largest* class, not only in it, but in the whole Animal kingdom;—and whilst the class of Birds, which is the exponent (as it were) of a similar tendency, extremely

well marked, if not absolutely the most characteristic, in the Vertebrated series, is also very numerous ;—the group of Pteropods, which, in its manifestation of this tendency, runs counter to the general 'idea' of Molluscous organisation, is the *smallest* class (if class it is to be considered) in the whole sub-kingdom ;—whilst the Ciliograde Acalephæ, whose organisation and habits are still more alien to the plan of structure and to the mode of life of Radiated animals generally, constitute but a subordinate and still more restricted subdivision of that series.

306. The class of CEPHALOPODA, so named from the attachment of the feet or principal locomotive organs to the head, includes a series of Molluscous animals, of which the lowest are not far removed above the Gasteropoda, whilst the highest present, as regards the development of their individual organs, a certain degree of approximation to Vertebrated animals. As regards their general conformation, however, we do not find any considerable departure from the ordinary Molluscous type ; and the only indication of an approach to the plan of structure peculiar to Vertebrata, is given by the presence of cartilaginous plates within the body, disposed in such a manner as to support and protect the nervous centres. The feet, which are disposed in a radiating manner around the mouth, so as to remind us of the arms of the Polypes (§ 289), must be regarded as highly-developed

FIG. 131.

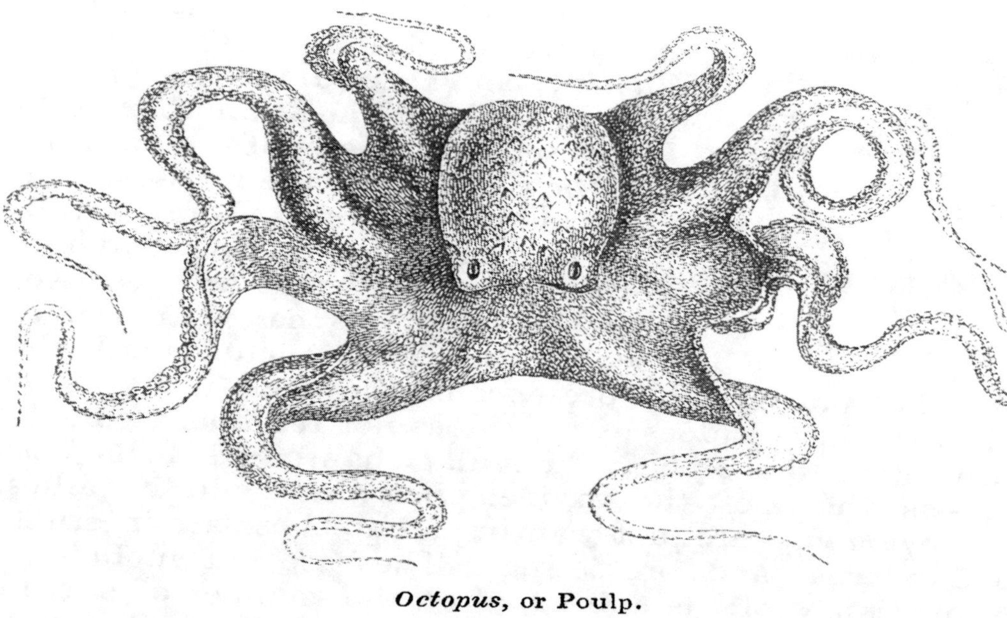

Octopus, or Poulp.

tentacula, and have nothing in common with the locomotive organs of any other animal. They are sometimes connected at their bases by a circular fin ; and by the alternate expansion and contraction of this, the animals which possess it can swim through the water with considerable rapidity. In other instances, however, the body is more elongated, so as to possess somewhat of the form of a Fish ; and it is provided with lateral fin-like expansions (Fig. 134), by the strokes of which the animal can propel itself through the water with considerable velocity. We do not find any trace of the muscular disk or foot of the Gasteropods, except in the *Nautilus*, which approaches more nearly to that class in its general structure, than does any other living Cephalopod. This animal is the only existing type of a section of the class, which was once extremely numerous, and which appears to have contained a great number of diversified forms; all of them, however, being now extinct except this genus, of which only three species remain, out of the many scores which are known to the Palæontologist. In the Nautilus and its allies (Fig. 133), the body is protected by a firm shell, the cavity of which is divided by transverse septa or partitions, into chambers, their number depending upon the age of the individual; it is in the last or outer chamber alone, that the soft parts are contained; but a membranous tube, termed the 'siphon,' runs backwards from this, through the successive septa, even to the most remote chamber. These 'chambered' shells, like the simple shells of Gasteropods, may be regarded as formed upon the type of a cone, which is prolonged, and at the same time widened, by successive additions to its base; when these additions are made in such a manner that the axis remains a straight line, the shell is a simple cone, the cavity of which is divided by transverse septa, as in the *Orthoceratite* and *Baculite;* but if the axis of the cone wind round a centre, the convolutions still remaining in the same plane, a spiral chambered shell is produced, as that of the *Nautilus* or *Ammonite* (Fig. 132); whilst if the axis be obliquely coiled, so that the successive convolutions do not lie in the same plane, a helical shell is formed, as that of the *Turrilite*. Between these, as in the Gasteropoda, there are intermediate forms; some shells, as the *Lituite* and *Hamite,* being spiral at one period of their growth, and straight at another; others, as the *Toxoceras,* having a gentle curvature, like that of a bow, through their whole length; and others, as the *Crioceras,* being spirally curved, like the Ammonite, but not having the turns of the spire in contact with each other.—On the other hand, in that section of the Cephalopod class, which includes all the existing forms except the Nautilus, the presence of the shell is by no means constant; and even where it does exist, it does not serve as a complete investment or protection to the body. Regular chambered shells, both straight and spiral,—as those of the extinct *Belemnite* and its allies, and of the existing *Spirula,*—are not wanting; but these are internal and supporting, not external and protecting. They are not included, however, within the soft tissues of the animal, after the manner of a true internal skeleton; but are simply enfolded by prolongations of the mantle. This is the case, too, with regard to the so-called 'bone' of the Cuttle-fish, which is a rudimentary chambered shell, with its last or outer chamber largely developed on one side only; and it is the case, also, with what seems an obvious exception to the general rule, namely, the delicate shell of the *Argonauta,* commonly but erroneously termed the Paper Nautilus. This shell, in the ordinary state of the animal, is nearly or entirely covered over by the two expanded arms; the animal is not attached to its interior by any muscular connection; and it seems to be wanting in the male, so that it is probably to be regarded less as a proper shell, than as a receptacle for the eggs. In many of the Cuttle-fish tribe, the internal

FIG. 132.

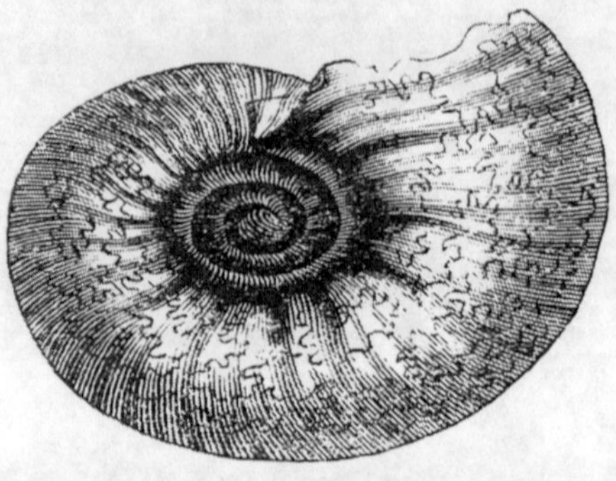

Ammonite.

shell is reduced to a mere horny rudiment, whilst in others it is absent altogether; and it is in the latter that we find the greatest flexibility of the body, the greatest general activity of movement, and the highest condition of the nervous system, as well as of that rudiment of an internal skeleton which is developed for its special support and protection. The naked species, whether possessed of an internal shell or not, are furnished with an 'ink-bag,' which can discharge a black pigment, that renders the surrounding water quite opaque; and this is an important means of passive defence to the animals that are provided with it, since, under cover of the inky cloud which thus hides them from their enemies, they can make their escape. The ink-bag appears to be invariably present when the external chambered shell is wanting, and to be deficient when the testaceous covering exists; and these two conditions are respectively characteristic of the two principal sections of the class, which have been designated by Prof. Owen as *Dibranchiata* and *Tetrabranchiata*, according to the number of the gills, which are two in the one case, and four in the other.

307. The order *Tetrabranchiata*, being the one whose conformation approaches most nearly to the Gasteropod type, is that which we should first consider. Although the order contains only a single recent genus, the true *Nautilus*, yet we seem justified in referring to it all those numerous extinct forms of siphoniferous chambered shells, whether straight, curved, or spirally or helically convoluted, in which the size of the outer chamber evidently adapted it for the reception of the body of the animal, in which the external surface does not present any indications that it was once invested by soft vascular membranes, and in connection with which no traces of an ink-bag have ever been discovered; and there is a strong presumption that the animal inhabitants of such shells did not depart very widely from the type of conformation which we shall find in that genus. The body in *Nautilus* is an oblong mass, adapted in its general form to the shape of the cavity in which it lies (*a a*, Fig. 133), but presenting some indication of a division into an anterior portion, which includes the organs of sense and motion, and a posterior, which encloses the viscera. The posterior segment, or abdomen, is completely enveloped by the mantle, which is thin and membranous; and lies in close contact with the testaceous walls of the cavity; and from the deepest portion of this cavity, a tubular prolongation of the mantle, the 'siphon,' passes off through an aperture in the septum, into the next chamber, and thence through the whole succession of chambers, penetrating their successive septa, which not only give it passage, but also afford it a shelly covering for about a quarter of the distance between one septum and another.*

* In some of the fossil chambered shells belonging to this group, the membranous tube appears to have been enclosed in a complete shelly covering throughout its entire length; and this fact disproves the idea that the distention or contraction of the siphon could have had any connection with that alteration of the specific gravity of the animal, by which it raises or sinks itself in the water. The most probable use of the siphon, in the author's estimation, is that assigned to it by Mr. Searles Wood; who considers that its office is to maintain the vitality of the portion of the shell most removed from the soft parts. A small artery and vein have been found by Professor Owen to pass inwards with the membranous tube; and he has also pointed out that the septa and walls of each chamber continue to be lined by a delicate membrane. The vitality of this membrane can only be kept up by vascular communication with the animal, of which it forms a remote part; and that of the shell must be dependent upon that of the animal membrane in contact with it. There are many cases, in which we find an erosion of shelly structures to take place, when they are far

Towards the front of the abdominal portion, however, the mantle increases in thickness, becomes more muscular, and extends freely outwards, forming a wide fold (*e*) which is reflected over part of the involuted whorl; whilst it elsewhere presents a sinuous margin, thickened by horny matter, so as to form a ring of adhesion to the internal surface of the shell, with which it is further connected by a pair of muscles that take a firm attachment to the sides of the latter. Towards the lower part, we find the mantle perforated for the passage of the muscular expiratory and excretory tube (*i*), which is termed the 'funnel.' The anterior portion, which may be considered as the head, though not

FIG. 133.

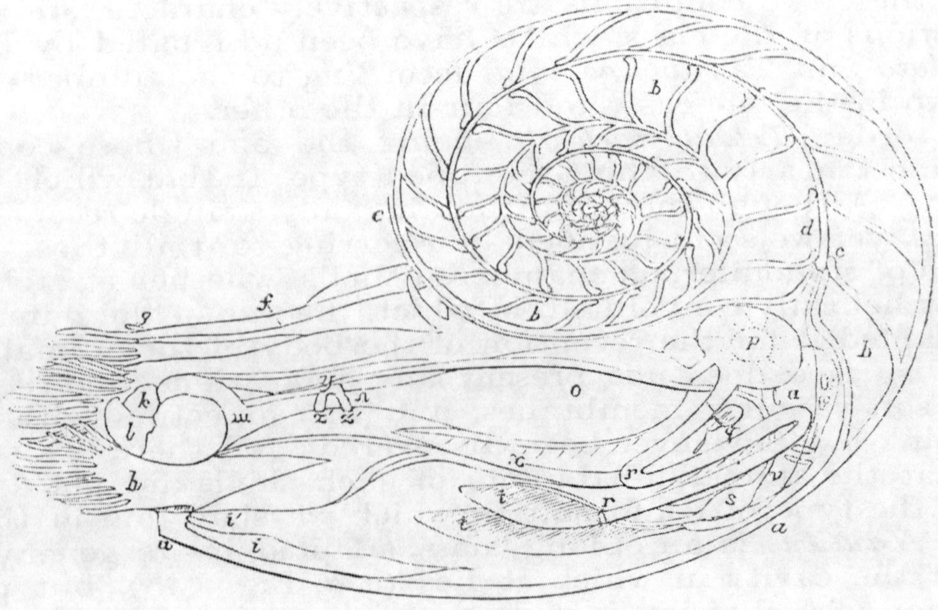

Diagram of the structure of *Nautilus pompilius* ;—*a*, *a*, outer chamber of the shell ; *b*, *b*, *b*, its internal chambers ; *c*, calcareous siphon perforating the septa ; *d*, membranous siphon, passing back through the whole series ; *e*, fold of the mantle, reflected over the convexity of the involuted whorl ; *f*, muscular hood ; *g*, digital tentacula ; *h*, labial tentacula ; *i*, mantle forming the funnel *i'* ; *k*, *l*, horny mandibles ; *m*, mouth ; *n*, œsophagus ; *o*, crop ; *p*, gizzard ; *q*, globular cœcal appendage (rudimentary pancreas) ; *r*, *r*, intestine ; *s*, heart ; *t*, *t*, branchiæ ; *u*, ovary ; *v*, oviduct ; *x x*, liver ; *y*, aorta ; *z z*, nervous collar around œsophagus.

apparently capable of being projected beyond the shell, is covered at its back or upper side with a fleshy expansion of the mantle (*f*), which is termed the 'hood ;' this organ appears to serve in part for the closure of the entrance to the shell ; but it probably serves also, like the foot of the Gasteropods, to enable the animal to creep over a solid surface. The head is furnished with a large number of tentacula, amounting in all to about ninety ; of these, about thirty-eight arise directly from its muscular envelope, and are named by Prof. Owen *digital* tentacula ; whilst about forty-eight arise from the circular fleshy lip that surrounds the mouth, and are hence designated as 'labial ;' and four others, again, which spring

removed from a living surface ; and there can be no question that a dead shell is much more brittle than a living one. There seems a peculiar necessity for the preservation of the shell uninjured in these Cephalopods ; since the fracture or loss of any part of it would probably destroy the hydrostatic balance of the animal ; and to such fracture or loss they must be peculiarly liable, from their habit of swimming backwards. See Mr. Edwards's Monograph on the Cephalopoda of the London Clay, published by the Palæontographical Society, p. 12.

from the immediate neighbourhood of the eyes, are termed 'ophthalmic.' These numbers, however, are not constant. None of these tentacula are furnished with suckers, as are the fewer but more powerful arms of the Dibranchiata; but they seem extremely well adapted both for prehension and sensation. This multiplication of similar parts in the tentacula, as well as in the gills, is a character which of itself indicates the inferiority of the Tetrabranchiate to the Dibranchiate Cephalopods. This inferiority, however, is not strikingly apparent in the apparatus of organic life. We find the mouth, in the Nautilus, furnished with a pair of horny mandibles (*k*, *l*), resembling those of certain Fishes, Reptiles, and Birds; and, as in those Vertebrata, opening vertically instead of laterally, which is not the case in any other Invertebrata furnished with analogous instruments. These jaws are moved by powerful muscles, and inclose a fleshy tongue, of which part is clothed with horny spines, whilst another part is covered with sensory papillæ. The œsophagus (*n*) dilates into a capacious crop (*o*), and this again contracts into a short canal which leads to the muscular gizzard (*p*). The intestine (*r r*) commences near the cardiac orifice of the stomach; and it soon communicates with a small globular pouch (*q*) through which the biliary secretion arrives at the intestine; this pouch itself contains secreting cells, and is probably to be regarded as a rudimentary pancreas. The liver (*x x*) is a bulky gland, extending on each side along the crop, as low down as the gizzard. The intestinal tube, after making two abrupt bends, terminates in the branchial cavity, close to the funnel. The heart (*s*) is situated at the posterior and inferior part of the body, near the base of the branchiæ (*t*, *t*) of which there are two pairs; it has no distinct auricle; but the blood returning from the gills passes at once to the ventricle (*s*), and from this it is transmitted forwards by the aorta (which is seen running towards *y*), and by other arteries. From the general system it returns by venous sinuses and channels in different parts of the body, of which the peritoneal lining of the visceral cavity itself forms a part; by these it finds its way to the gills, at the base of which we find certain peculiar organs of a glandular character, of which the use is unknown; where also, at its entrance into the proper branchial arteries, it may possibly receive a fresh impulse from the contractile fibres of what appear to be rudiments of the branchial hearts in the Dibranchiate Cephalopods. The blood which is distributed by the branchial arteries to the leaflets of which the gills are composed, obtains its aëration from the water admitted into the branchial cavity through a wide fissure at the base of the head; and this water, after passing over the branchiæ, is ejected through the funnel (*i*, *i'*), along with the matters voided by the intestine, and with the products of the generative organs, which open into the same cavity.

307, *a*. The sexes are distinct in all Cephalopods; and the male and female organs of the Nautilus probably occupy a corresponding position at the back of the abdomen. The ovary is shown at *u*, lying in the immediate neighbourhood of the gizzard, but in a separate division of the peritoneal cavity; its products are discharged by the wide oviduct *v*, which terminates very near the anus. In the immediate neighbourhood of this orifice is found a peculiar glandular apparatus; the office of which is probably to afford a 'nidamentum' or protective envelope to the ova when discharged, similar to that which is found in many Gasteropods and in the higher Cephalopods. Of the development of the embryo, nothing is yet known.—The cephalic portion of the nervous system is considerably larger, relatively to the remainder of the body, than it is in the inferior Mollusca; but the increase in its dimensions is seen rather in the inferior ganglia of the œsophageal ring (*z′ z′*), than in the superior portion (*y*), which is still little else than a connecting band or commissure; and its chief departure from the type of the Gasteropoda is seen in the distinctness of the optic ganglia, which now present themselves as separate masses appended to the sides of the supra-œsophageal ring, in accordance with the higher development of the eyes, which present a near approach to those of Vertebrata. The œsophageal ring is double at its lower side; the anterior portion supplying the tentacula, whilst the posterior sends its nerves backwards to the viscera. Besides the great cephalic centres, we find other smaller ganglia supplying certain parts of the tentacular and pharyngeal apparatus; and others again in connection with the branchiæ. The cephalic ganglia are supported by a true internal cartilage; and in this is found a cavity on either side, which has been supposed to be an organ of hearing, but this is doubtful. In front of the mouth is a curious laminated structure, copiously supplied with nerves; which is regarded by Prof. Owen, with great probability, as an organ of smell. The conformation of the tongue and of the interior of the mouth are such as to indicate that the Nautilus possesses the sense of taste; and the numerous tentacula, especially those proceeding from the circular lip, are doubtless instruments of very acute touch. The movements of this animal through the water, however, can scarcely be supposed to be very active; since it appears to be destitute of all other means of progression than that supplied by the respiratory current proceeding from the funnel, the constant ejection of which will tend to give it a backward motion. The air included in the chambers of its shell, however, gives it such a degree of buoyancy, that the whole, with the animal, is almost exactly equal in weight to its own bulk of water; and thus the animal will be made to rise or fall by a very slight change in its specific gravity. This change it certainly has the power of making; as the living Nautilus has frequently been seen floating on the surface of the ocean, and the difficulty of its capture arises from the suddenness with which it sinks when alarmed.

FIG. 134.

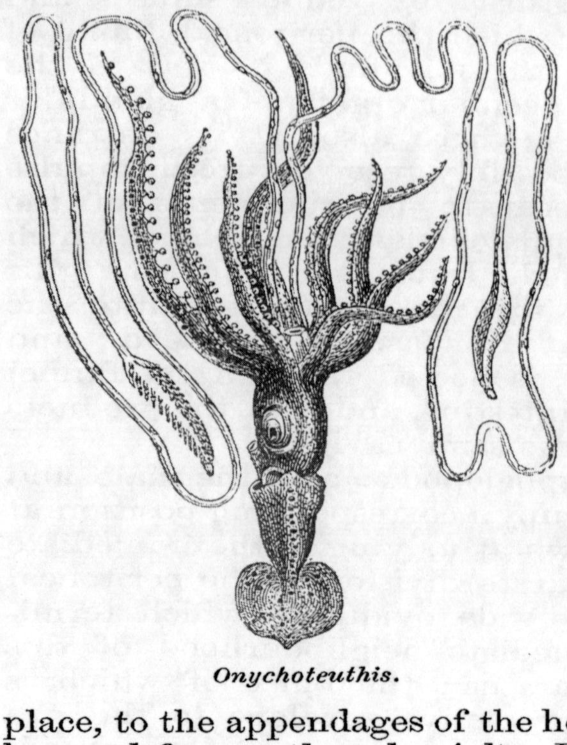

Onychoteuthis.

307 *b.* In describing the *Dibranchiate* Cephalopods, it will be sufficient to indicate the chief points of difference which exists between them and the preceding order. These have reference, in the first place, to the appendages of the head, which are never more than *ten* in number, and frequently only *eight*. Where there are ten, one pair is frequently

prolonged to a considerable extent beyond the rest, as is the case with the *Onychoteuthis* (Fig. 134), and in a less degree with the common Sepia (Cuttle-fish). In the *Octopod* group, we frequently find the bases of the arms connected by a circular fin (Fig. 131) by the alternate expansion and contraction of which, these animals are enabled to swim backwards through the water with considerable rapidity. In the *Argonauta,* which belongs to this division, two of the arms are extended into broad flattened expansions ; these were formerly supposed to act as sails which the animal might spread to the wind ; but it is now known that they are never employed as such, but that they are applied to the exterior of the delicate shell (which, not being chambered, and having no muscular attachment to the body, has been consequently supposed by many Naturalists to belong to some other animal, and to be only parasitically tenanted by the Argonaut,) and that they serve not merely to hold it in its place, but also to extend it, or to repair it when injured, by a new formation from their epithelial surface. In all instances, the arms are covered on their inner sides with suckers, which enable them to take a very firm grasp of the object which they seize ; and in the *Onychoteuthis* we find the two long arms furnished also with horny hooks, by which the grasp is rendered still more powerful.—The mouth is furnished, as in Nautilus, with a pair of strong horny parrot-like jaws, enclosed in a mass of muscles (Fig. 133, *f*) that serves to put them in action ; and it receives the secretion of two pairs of salivary glands (*g, h*). The œsophagus (*i*) dilates (in the Octopods, but not in the Decapods) into a crop (*j*), and then enters the gizzard-like stomach (*k*), from which passes off the intestine (*m, m′*) ; this is connected, near its commencement, with a spirally-convoluted appendage (*l*), which is obviously the rudiment of a pancreas, more highly developed than that of Nautilus. Into the same part is discharged the biliary secretion, formed by a large liver, which (though removed in the figure, in order to show the other parts) occupies a similar position to that which it has in the preceding order. The intestine terminates in the branchial chamber, near the base of the funnel (*e*), as in the Tetrabranchiata; and lying by the side of its latter portion, and opening in the same situation, we find the ink-bag. In the structure of the heart, we see a decided advance upon that of Nautilus. The systemic heart (*p*), placed at the base of the aorta (*q*), is still the principal impelling cavity ; but the blood returning from the body at large by the venous canals (of which one is seen at *r*), then enters a pulmonic heart or impelling cavity (*s*) which is situated at the base of each gill ; and by its agency, the branchial circulation is rendered more energetic. Although the branchiæ (*o, o*) are single instead of double, on each side, yet the single gill, from the extent of surface which it presents, and the more active movement of blood through it, is a more efficient apparatus for aëration than the double gill of the Nautilus. At the base of the pulmonic hearts, we find the same glandiform appendages (*r′*) as in Nautilus. The blood impelled through the branchial arteries (*s′*) by the pulmonic hearts, returns by the branchial veins (*t*) ; and these, in their passage towards the heart, present dilatations (*u, u*) which might be almost regarded as auricles. In this division of the impelling power (so to speak) between the commencement of the systemic and that of the respiratory circulation, we see a transition between the plan of the circulating apparatus characteristic of the Mollusca in general, which have the heart entirely *systemic*, and that hereafter to be described in the class of Fishes, which have the heart entirely *pulmonic*.—The nervous system of the Dibranchiate Cephalopods presents a considerable advance upon that of the Tetrabranchiate, especially in the development of the superior portion of the œsophageal nervous collar, and in the still larger dimensions of the ganglia of sense, although in other respects formed upon the same general plan. The cephalic portion is completely enclosed in a thick cartilage; and in the basal portion of this is an excavation on each side, which is evidently a rudimentary organ of hearing. The eyes are of considerable size, and present some remarkable peculiarities of structure hereafter to be described (CHAP. XXI). No distinct traces of a special organ of smell can be discovered in the Dibranchiata; but there is reason to believe that the olfactive sense is possessed by them; and the external lips may probably be its seat. The mouth, as in Nautilus, contains a highly organised tongue; which is probably in part subservient to the sense of taste. Of the sense of touch, the arms are doubtless the peculiar seat, especially towards their slender terminations; but these are rather instruments of prehension; and the adhesive power of the suckers does not require sensibility for its exercise, as is shown by the fact that it will continue in operation when the arms are entirely detached from

FIG. 135.

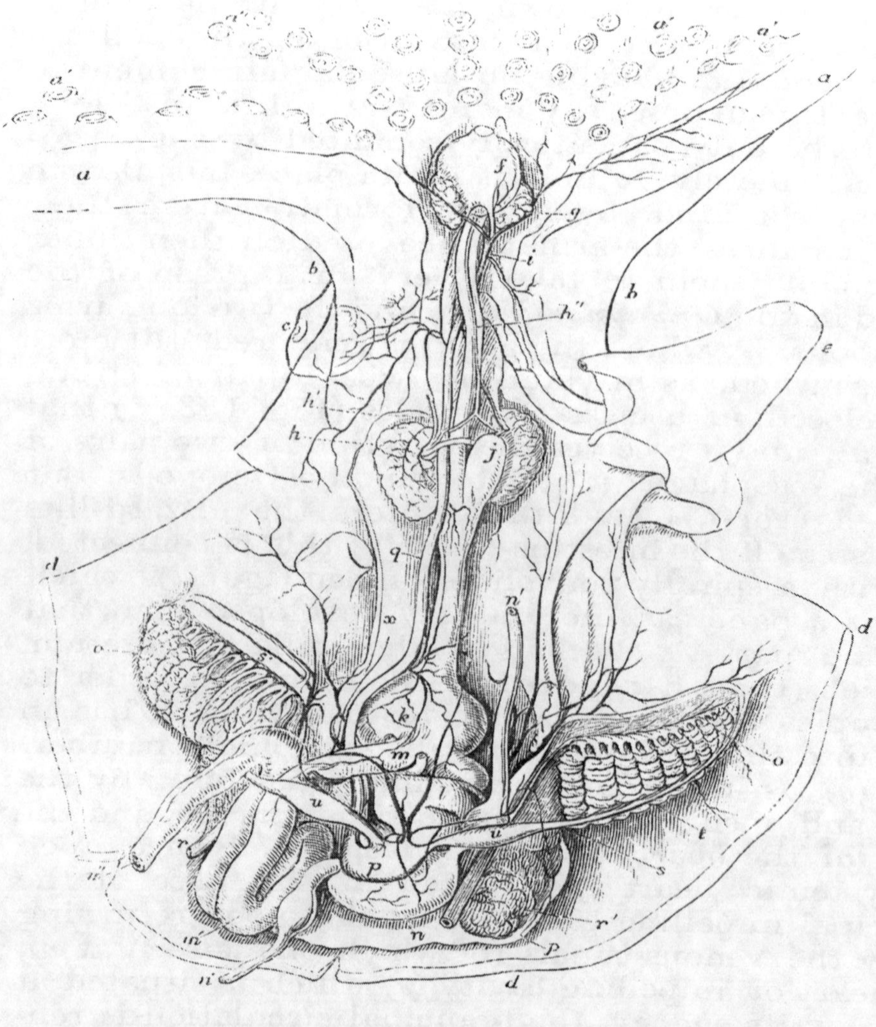

Anatomy of *Octopus*, the animal being laid open on the ventral side, and the inferior wall of the abdominal cavity, with the liver, having been removed; *a*, *a*, base of the tentacula, with the suckers *a′ a′* on their inner surface; *b*, head; *c*, eye; *d*, *d*, mantle turned back; *e*, funnel; *f*, fleshy mass surrounding the mouth; *g*, salivary glands of the first pair; *h*, salivary glands of the second pair, with their suspensory ligaments *h′*, and their excretory canal *h″*; *i*, œsophagus; *j*, crop; *k*, stomach; *l*, spirally convoluted cœcal appendage (rudimentary pancreas?); *m*, commencement of intestinal tube, with biliary canal on each side; *m′*, intestinal convolution; *m″*, anal extremity, turned downwards and to one side; *n*, ovarium; *n′*, *n′*, oviducts, of which one is in its natural position, and the other turned downwards; *o*, *o*, branchiæ; *p*, heart; *q*, ascending aorta; *r*, venous trunk passing towards the pulmonary heart; *r′*, its glandiform appendage; *s*, pulmonary heart; *s′*, branchial artery; *t*, branchial vein; *u*, *u*, bulbous dilatations of branchio-cardiac veins.

the body. The general surface is usually beset with papillæ, which appear to be tactile; and these are especially developed in some species about the head and eyes, taking the place in some degree of the numerous tentacula of the Nautilus. The movements of the Dibranchiata are generally very active. Their principal instruments of locomotion are their powerful tentacula, by the agency of which they can swim backwards; and this retrogression will be aided by the respiratory current, which is constantly issuing forth from the funnel. The shorter-bodied Octopi can also crawl over solid surfaces, with the body upwards, and the mouth downwards, in the position represented in Fig. 131; whilst, on the other hand, the long slender-bodied Calamaries seem to propel themselves through the water, like Fish, by the fin-like expansions at the posterior extremity of the body, and are even said to raise themselves above the surface, and to dart through the air for a short distance, like Flying-fish, by the force of the strokes they can make with these organs. In all their movements, the Dibranchiate Cephalopods seem guided by their very acute sense of vision: being enabled, by the degree in which their eyes project from the sides of their head, to see in every direction.—In the structure of their generative apparatus, the Dibranchiata do not present any essential departure from the type of the Nautilus. The ovary of the female (*n*) is single, and lies at the posterior part of the body; but in the Octopus the oviduct (*n′ n′*) is double, though it remains single in the Sepia; in both cases, however, it passes through nidamental glands, from which the ova receive an additional investment, before terminating in the branchial chamber. The testis or spermatic organ of the male lies in the same situation as the ovary of the female; and its efferent canal passes through a glandular organ, in which the spermatozoa become enclosed in peculiar cases, having a power of spontaneous movement, which are termed the 'filaments' or 'animalcules of Needham,' after their first discoverer. The spermatic fluid containing these is discharged through a penis: it does not appear, however, to be conveyed by this into the body of the female; but seems to be rather applied to the masses of ova, at the time when they are being deposited by the female. The ova, enclosed in their nidamental capsules, are aggregated together in large masses, almost every species having its own peculiar arrangement; and it would appear to be for the purpose of penetrating these clusters, that the filaments of Needham are endowed with their very curious motor powers; as they burst and set free their contained spermatozoa, within a short time after they have themselves come into contact with water. The young of the Dibranchiate Cephalopods do not undergo any metamorphosis after coming forth from the egg, but already possess at their emersion the general form of the adult, whilst all their organs are sufficiently developed to enable them to obtain their food and to escape from their enemies.

307 *c*. The differences which have been shown to separate the ordinary Dibranchiate Cephalopods from the Tetrabranchiate order, are not less characteristically displayed by those which possess chambered shells, notwithstanding the very close resemblance which these appendages bear to the shells of Nautilus, and its extinct allies. And it is a remarkable proof of the justice of the grounds on which the separation of these orders was proposed by Professor Owen, that the recently-acquired knowledge of the animals of *Spirula* and *Belemnite*, of which the former possesses a

spirally-convoluted, and the latter a straight conical chambered shell, has shown them to be, in all essential points, what the prediction founded upon a small number of data had asserted that they would prove.—The shell of the *Spirula* is enclosed between two lobes of the mantle, at the posterior end of the body of a Decapod; the animal has eight short arms, beset with numerous minute suckers promiscuously arranged, and two long tentacula terminating in club-shaped enlargements; it possesses an ink-bag; its gills are single on each side, and have pulmonic hearts at their base; and it has a distinct auditory apparatus, excavated in the cephalic cartilage; in all these characters conforming to the regular Dibranchiate type, as had been predicted from the conformation of its cephalic appendages.*—So in regard to the *Belemnite*, the fact that it is an internal shell having been guessed from the markings upon its external surface, and from its possession of an ink-bag (the fossilized remains of which are not unfrequently preserved in its last chamber), it followed upon the same train of reasoning, that the animal must have been of the Dibranchiate order, rather than allied to that of Nautilus, and that it must have possessed not more than ten arms; whilst the elongation of the chambered portion of the shell in a straight line (Fig. 136, *b*), and still more the elongation of the solid calcareous cone (*c*) that includes it (which, from its crystalline texture, has been termed the 'spathose guard'), naturally led to the inference that the form of the animal must also have been elongated, like that of the existing Calamaries. The recent discovery of specimens of Belemnites with the soft parts sufficiently preserved to allow their form to be clearly made out, has fully justified this prediction; as will be seen from Fig. 136, which exhibits Professor Owen's restoration of the entire animal from the materials thus obtained.† The arms, instead of being furnished with suckers, were provided with hooks, like those of the long tentacles of the Onychoteuthis; and the body had a pair of small lateral fins (*f, f*) at about the middle of its length. From the weight of its dense internal shell, the Belemnite may be supposed to have commonly maintained a vertical position; but, like the Nautilus, it probably had the power of ascending and descending in the water with facility. It would rise swiftly and stealthily, to fix its claws in the belly of a fish swimming near the surface above; and would then descend, perhaps as swiftly, and devour its prey at the bottom. We cannot doubt that, like the hooked

FIG. 136.

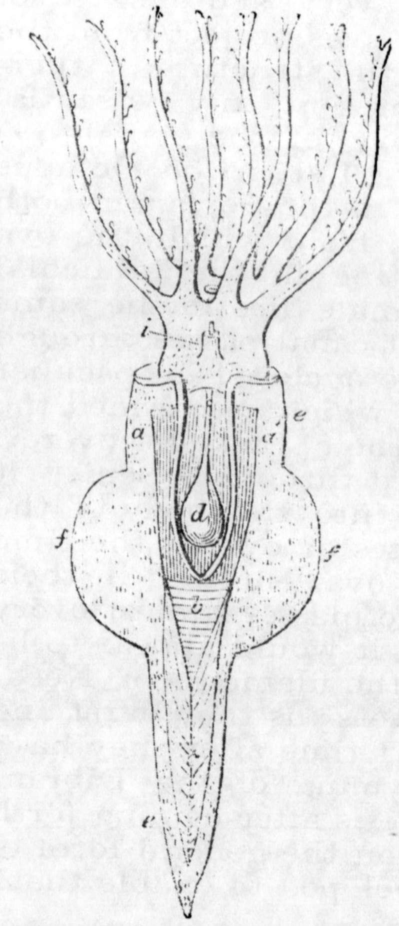

Belemnite restored; *a, a*, outer chamber of conical shell; *b*, chambered cone; *c*, spathose guard; *d*, ink-bag; *e, e*, mantle; *f, f*, fins; *g, g*, hooked arms; *i*, funnel.

* See Prof. Owen's description of the Anatomy of *Spirula*, in the Zoology of the Voyage of the *Samarang*:—Mollusca, p. 6.

† See Prof. Owen's Memoir on the Belemnite, Phil. Trans. 1844.

Calamaries of the present seas, the ancient Belemnites were the most formidable and predaceous of their class.

307 *d.* In now quitting the Molluscous type, we may notice that notwithstanding the high development of their organs of nutrition, there are yet strongly-marked peculiarities in the arrangement of these organs in that series; and thus, notwithstanding their individual resemblance to the corresponding organs of the lower Vertebrata, there is still an entire absence of any *general* approximation to the latter type. Thus we find in all Cephalopods, as in Mollusca generally, the intestinal tube doubling back upon itself, so as to terminate near the mouth; instead of passing, as it invariably does in Vertebrated animals, to the posterior extremity of the trunk. And we find in Cephalopods, as in most Gasteropods, the branchial apparatus to be in connection with the anal termination of the alimentary canal; instead of being, as in all Vertebrata, in relation with its oral commencement. So again, although we find the assemblage of cephalic ganglia, which are the centres of the organs of sense, approaching the lower forms of the brain in Fishes,—although we find these included in a cartilage which may be likened to a rudimentary cranium,—and although we find in the instruments of sight, hearing, smell, and taste, a similar approximation to those of Vertebrata;—yet no such resemblance exists between the ganglia connected with the locomotive apparatus of Cephalopods and the spinal cord of even the lowest Vertebrata, the locomotive apparatus itself, as we have seen, being constructed upon an entirely different plan in the two series. The muscles which move the various parts of the body and arms of the Cephalopod may be considered as merely a development of particular parts of the general muscular envelope; they have no fixed points of attachment, and have no levers to act upon; and they do not present the least approach to that *segmental* arrangement, which is characteristic of the locomotive apparatus of Vertebrata, in common with that of Articulata. Consequently, as there is no succession of segments in the body, each endowed with a muscular apparatus of its own, there is no trace of that corresponding series of ganglionic centres which forms the gangliated cord in the Articulata, and which is consolidated into the spinal cord of Vertebrata. Hence, whilst the Cephalopods show no rudiment of that segmental division of the integument, which we have seen in the Chiton (Fig. 125), they are equally far from possessing any traces of the 'vertebral column' which is the distinguishing characteristic of the Vertebrata, and which, in its full development, affords support and protection not merely to the principal nervous centres, but also to the viscera in general, together with fixed points for the attachment of the muscles, and levers by which their power may be most advantageously applied. These differences, however, will come before us hereafter; and they are only here alluded to, for the sake of bringing into clear view the points on which the Cephalopod Mollusca *do*, and those on which they *do not*, approximate to the Vertebrata.—It is particularly instructive to see how the higher development of the organs of animal life, and the more active locomotive powers of the Cephalopoda, involve a much more complete lateral symmetry than we meet with (except in a few comparatively rare exceptions, and these among the inferior forms of the class) in the Gasteropoda. This is seen, not merely in the general contour of the exterior, but also in the conformation of the apparatus of organic life; thus the respiratory organs in the

Cephalopods are always double and symmetrical; and the ovaries show a tendency to become so. There is also an almost exact bi-lateral symmetry in the nervous centres.—On the whole we may say that the group of Cephalopoda presents us with as close an approximation to the Vertebrated sub-kingdom, as it could well do without a departure from the general Molluscan type, and consistently with the plan of structure peculiar to itself.

308. The characters of the sub-kingdom ARTICULATA are much more definite than those of the Molluscous series, and are for the most part in remarkable contrast with them. One prevailing plan of construction runs through all the classes associated under this designation; and this is generally easy of recognition, although it is subject in particular cases to considerable modification, and is sometimes partially obscured. This plan, the idea of which is conveyed by the terms 'articulated' and 'annulose,' may be defined as follows. The body is made up of an assemblage of *segments*, which succeed each other *longitudinally;* it is completely inclosed in an integument, consisting of a series of rings of firm texture, jointed or 'articulated' one to another, and held together by a flexible membrane, which allows considerable freedom of motion; and there is a very complete bilateral symmetry, the two sides of the body being exactly similar, not merely in their external form, but for the most part also in their internal structure. In the lowest grades of this type, the successive segments are nearly exact repetitions of each other, so that each can maintain an independent existence and can reproduce the rest, thus almost deserving to be regarded as distinct individuals. Ascending higher, we still find the segments maintaining their equality, and presenting a close resemblance to each other externally (Fig. 137); but the organs they contain are fused together (so to speak) into an apparatus that is subservient to the life of the whole, and no part can maintain a separate existence. In the highest forms of the series, such as Insects, Crustacea, and Arachnida, we no longer meet with this equality of the segments; some being developed in excess whilst others remain in an almost rudimentary condition, and their endowments being alike dissimilar; so that the fabric (according to the general principle of elevation in the empire of Life, § 20) loses its homogeneity, and acquires more special powers.—So, again, in the lower forms of the group, we find the integument altogether so soft, that the intervals of the articulations are scarcely distinguishable from the rings themselves; the whole body is thus extremely flexible; and its progression is effected by the movement of its parts one upon another. In certain parts of the highest, on the other hand, the hard segments of the skeleton are often so closely united together, that the line of separation between them is not apparent; hence these segments have no power of motion one upon another; and the propulsion of the body is delegated, more or less completely, to members developed for this especial purpose. These members, in the lowest tribes in which they make their appearance, are all similar and equal one to

FIG. 137.

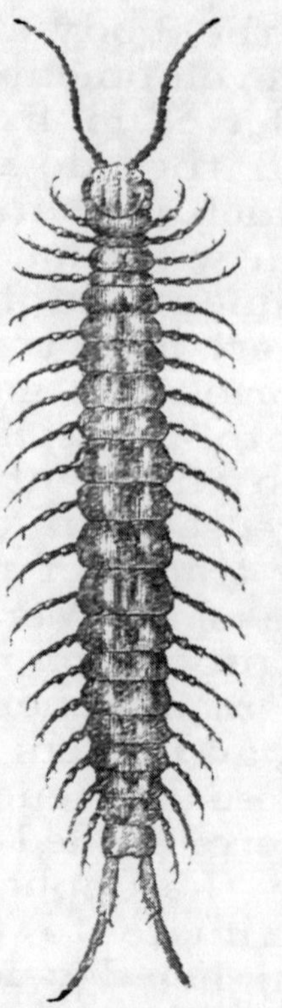

Centipede.

another, and are developed alike from every segment of the body (Fig. 137); but in the higher Articulata, we find them restricted to certain parts of the trunk, especially to those which are most firmly united together; this very union, it would seem, being subservient to the power of the limbs, by giving a firmer attachment to the muscles that move them. These members, like the body itself, are enclosed in an articulated envelope, or external skeleton; and this, so far as regards the mechanical action of the muscular apparatus, answers essentially the same purpose as the internal skeleton of Vertebrata, giving to the muscles fixed points of attachment, and furnishing a system of inflexible levers for the advantageous application of their contractile power.—A large proportion of the fabric of the Articulata is made up of the Muscles by which the several segments and their various appendages are put in action (Fig. 138); and the development of the organs of nutrition would seem to be for the most part subservient to that of the locomotive apparatus, their function being chiefly to supply the nerves and muscles with the aliment necessary to maintain their vigour, and with the oxygen required for their functional activity. The power of these muscles is so great in proportion to the size of the animals, that in energy and rapidity of movement, some of the Articulated tribes surpass all others. Their locomotive actions are directed by organs of sensation, which, though developed on a very different plan from those of Mollusca, are evidently very acute in their powers. It is curious to observe that even in these we have the principle of repetition strikingly manifested; the eyes of all, save the highest Articulata, being 'compound,' that is, being made up of an assemblage of *ocelli*, sometimes to the number of several thousand. There are very few instances in which Articulated animals are in any way restrained as regards freedom of motion; and these occur, for the most part, in a single group, the *Cirrhipoda* or Barnacle tribe (§ 311), which connects this sub-kingdom with the preceding. In general they roam freely abroad, whether in search of prey, or for sexual purposes; and in their various actions we trace an obvious, and frequently a most remarkable adaptation to particular purposes. This adaptation, however, seems rather to be the result of the *instinctive* propensities of the animals, than of *intelligent will*; these animals apparently acting like machines, adapted by the Designing Mind which constructed them to execute a certain set of operations; and doing this blindly, without any choice or purposive intention of their own, yet producing results which Man, by the highest efforts of his reason, could not surpass in the perfection of their adaptiveness to the required end.

308 *a*. All the Articulata, save a few of the very lowest species, possess a distinct head at one end of the body, furnished with organs of special sensation, and also (except in those which live by suction) with lateral jaws for the prehension and reduction of food. The alimentary canal is almost uniformly provided with a second orifice, which is situated towards the posterior extremity of the body; and it frequently passes towards this in a perfectly straight line, exhibiting an exact bilateral symmetry both in its own structure and in that of its glandular appendages. In most of the higher Articulata, however, the intestine is more or less convoluted, so that this symmetry is in some degree lost; but it still manifests itself in the early development of the apparatus, and in the glandular appendages connected with it (Figs. 153, 154). In

regard to the development of these, which for the most part represent the liver, we have a remarkable variety, that seems to have a relation to the medium which the animals are destined to inhabit. In the aquatic Articulata, the hepatic apparatus, at first consisting merely of secreting cells lodged in follicles of the stomach, comes at last (in the Crustacea) to form a large glandular mass, which occupies, as in the Mollusca, a considerable part of the abdominal cavity (Fig. 156, *f*). In Myriapods, Insects, and Spiders, on the other hand, all of which breathe air, the liver consists of but a small number of tubuli, containing secreting cells; and although these are sometimes considerably elongated, yet they never occupy any large proportion of the visceral cavity.—In all save the very lowest Articulata, we find a distinct provision for the Circulation of nutritive fluid. This consists of a set of blood-vessels, which, in the inferior tribes, are nothing else than channels or interspaces among the tissues, receiving the nutritious fluid which transudes from the walls of the alimentary canal; but which we find, as we pass towards the higher, to assume the characters of regular vessels, of which a part are distributed upon the alimentary canal, and act as absorbents, whilst another division conveys the absorbed fluid to the system at large, and to the respiratory organs. In the lower forms, moreover, there is no central organ of impulsion; but the blood meanders through its canals, partly under the influence of forces developed during its circulation (CHAP. XII.), and partly by the agency of the continual restless movements of the body itself, one part upon another. As we pass upwards, we frequently meet with a numerous succession of organs of impulsion; the tendency obviously being to the development of one or more for each segment. In the higher classes, however, these separate centres are fused together (so to speak), in such a manner as to form a continuous *dorsal vessel*, which occupies the median line, and propels the blood from behind forwards; this vessel still presenting indications of its composite origin, in the valvular partitions which correspond to the successive segments. It is only in the Spiders and Crabs, which, in the shortness of their bodies and concentration of their organs, present the greatest departure from the ordinary Articulated type, that we find any approach to the compact heart of the Mollusca (Fig. 156, *c*). The blood is usually colourless, as in other Invertebrata; but it contains a larger number of corpuscles than are seen in that of most of the Mollusca.

308 *b*. The Respiratory apparatus of Articulated animals everywhere presents a similar tendency to repetition of parts. In the lowest forms of this type, as in the inferior Radiata, we have no distinct provision for the aëration of the blood; the demand for oxygen not being great, and the integument being so soft as to allow the free transmission of gases through every part of its substance, so that all the blood sent to it is subjected to the respiratory process. As we pass upwards, however, we meet with distinct organs for this purpose, which are in the first instance developed equally in every segment: in the aquatic tribes they consist of leaf-like or arborescent gills, which are developed from the exterior of the body, so as to expose the blood transmitted to them to the influence of the air contained in the water which bathes them; whilst in the air-breathing classes, they are so constructed as to introduce atmospheric air into the interior, the action of each organ, however, being

limited to its own segment. Even in the highest Articulata, we still find much of this repetition of the respiratory organs; but the portions belonging to the different segments are brought into more combined action; and the whole apparatus attains such an extension in Insects, as to give it an amount of functional power that is nowhere surpassed. In all instances, the respiratory organs are perfectly symmetrical.—In regard to the development of other Secreting organs, such as salivary glands, pancreas, and kidneys, there is little to be said. In the lower classes of the series, no rudiment of them can be traced; and even in the highest, there is by no means the same approximation to the more elevated type of these organs, that we have seen in the Mollusca. In fact, although salivary glands, in the form of long cœcal tubes, are very commonly found in Insects and Crustacea that masticate their food, it is not by any means certain that even the rudiment of a pancreas exists; and the urinary organs even in Insects, Spiders, and Crustacea, are never more developed than are the biliary tubuli of Insects, which they strongly resemble, except in position.—The Reproductive process, in some of the lowest Articulata, as in the lower Radiata and Mollusca, takes place after a double method; that is, by the multiplication of the individual, as well as by true sexual generation. The former operation, however, is not so frequently accomplished by a process of 'gemmation,' or the formation of an offshoot from the parent fabric (which we have seen to be the usual method of multiplication in Zoophytes and the lower Mollusca), as by the subdivision of the latter into two or more parts, of which every one is capable of reproducing the entire structure. This subdivision takes place spontaneously in some species; in others we only find the multiplication effected, when it has been artificially performed; and in others, again, although considerable power of reproducing new parts is retained, it is limited to only one of the divisions of the body, so that no multiplication can be thus effected, all but that one of the severed pieces ceasing to exist. This power of reproducing lost parts is retained to a considerable extent even in the highest Articulata. In the lowest members of the Articulated series, and especially in those which have comparatively little freedom of locomotion (such as the Entozoa), we find the two sets of sexual organs (which are almost invariably symmetrical) united in the same individual, which has the power of self-fertilization, and these organs are sometimes repeated in every segment; in all save these, however, the sexes are distinct, and a sexual congress is required, in order that the ova may be fertilized by the spermatic fluid of the male, whilst yet within the body of the female. In a few instances, as in certain Mollusca, the eggs are hatched within the maternal oviduct, so that the young are born alive; but generally speaking, the embryos come forth in a very immature state, and have to pass through a long series of changes of form and condition, before they attain the type of their parents. The nature of these changes, however, differs considerably in the several classes; and no statement can be made respecting them that shall be even generally applicable.

308 *c.* Reference has already been made, on more than one occasion, to the high development of the Muscular apparatus of the Articulata generally. In the lowest forms of the series, however, no distinct muscular fibres can be detected; and the movements of the body appear

to be produced by the contractility of an almost homogeneous tissue resembling that of the Hydra (§ 289). Passing on to the higher forms, we find that where the segments of the body are similar to one another in form and general endowments, the muscular apparatus of each is, in like manner, a repetition of that of the rest, as is the case in the Centipede ; but where, as in the Insect, the locomotive appendages are restricted to only one portion of the body, and the segments are dissimilar, the muscular apparatus is modified accordingly, its greatest development being of course in those segments to which the members are attached (Fig. 138).—The same general principle applies also to the

FIG. 138.

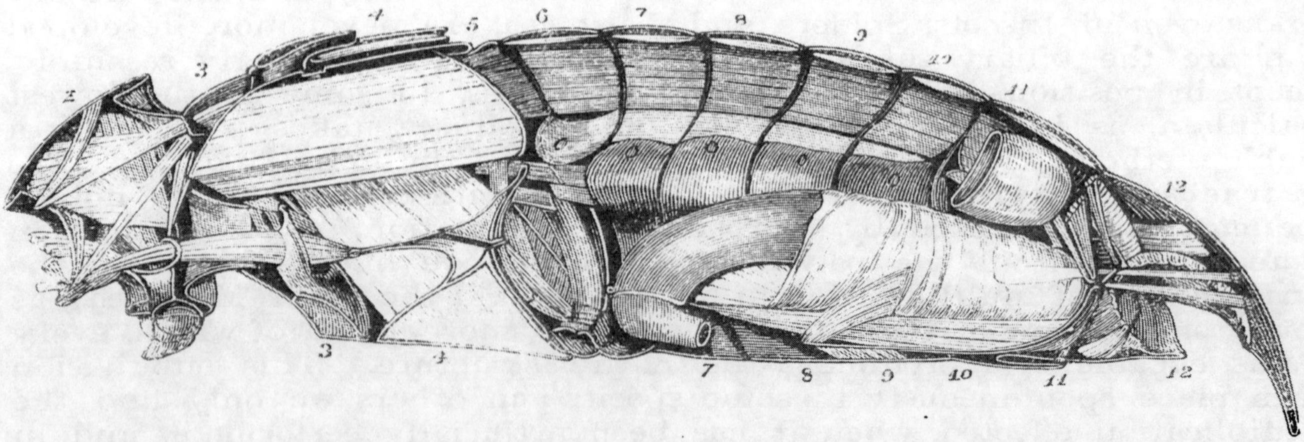

Section of the trunk of *Melolontha vulgaris* (Cockchafer), showing the complexity of the muscular system. The first segment of the thorax (2) is chiefly occupied by the muscles of the head, and by those of the first pair of legs. The second and third segments (3 and 4) contain the very large muscles of the wings, and those of the two other pairs of legs. The chief muscles of the abdomen are the long dorsal and abdominal recti, which move the several segments one upon another.

Nervous system. It is not distinguishable in some of those beings, which, as presenting the Articulated type in its most degraded form, must be placed in this series, although but little elevated above the Protozoa. Where it first presents itself, it is little more than a double thread, running from end to end of the body, with scarcely any ganglionic enlargement. As the segmental division of the body, however, becomes more distinct, and the locomotive apparatus more developed, so do the ganglionic enlargements on the ventral cord increase ; and wherever the locomotive apparatus is uniformly repeated in the successive segments, there do we find its ganglionic centres (which may be regarded as so many repetitions of the *pedal* ganglion of the higher Mollusca) repeated likewise. But with the concentration of the locomotive apparatus in a particular division of the body, we find the ganglia of the ventral cord undergoing a corresponding enlargement in that portion, and sometimes even coalescing into one mass (Fig. 161). The size and relative importance of the cephalic ganglia bear a constant relation to the development of those organs of sense of which the head is the seat. These variations are traceable, not merely as we pass from one group to another, but as we follow the history of development in any one of the higher Articulata, such as an Insect (Figs. 152—154): for it is uniformly to be observed that equality of the segments, and of all they contain, is the first condition; and that the differences in the endowments of these, arising from variations in the degree of development of particular organs, are of subsequent origin. The respiratory organs of the higher Articulata usually possess, as in the Mollusca generally, distinct ganglionic centres of their own; and where these organs are repeated in successive segments, the ganglia are repeated also.

308 *d*. Thus we perceive that the Articulated series, in its comprehensive acceptation, includes an extensive range of forms; the lowest of which are not more highly organized than the inferior Radiata, and differ from them chiefly in the *longitudinal* in place of the *circular* repetition of their parts; whilst the highest approximate the Vertebrated series in the development of their nervo-muscular apparatus, and in the disposition of their viscera, without losing the essential peculiarity of the group, namely, the articulated external skeleton. The peculiar confirmation of this skeleton involves a special mode of adapting it to the increasing dimensions of the animal during the period of growth, such as we have not yet elsewhere met with. Although this skeleton is made up of a number of pieces, yet it could only be increased *longitudinally* by additions to the edges of these; for, as each ring completely encircles the body, an augmentation in its *diameter* would be impossible, without an interstitial growth, of which these skeletons are not capable. The requirement is met by the exuviation or 'moulting' of the entire tegumentary skeleton, which takes place during the period of increase of the body, as often as this becomes too large for its investment; preparation being made for the formation of a new skin or shell, before the older one is cast off. It is only when these animals have attained their full growth (as in the case of perfect Insects), that this exuviation ceases to be performed; and in many tribes of Crustaceans it appears to take place occasionally through the whole of life.

309. The lowest forms of the Articulated type are found in the class which has received the designation of Entozoa, from the circumstance that most of the animals which it includes are parasites of other animals; their *habitat* being in some instances the intestinal canal, whilst in others it is the substance of the muscles, glands, &c.; and their nourishment being derived by suction from the juices of the bodies which they thus infest. In this respect, therefore, they bear a striking analogy to the parasitic Fungi (§ 271, 272). But no such peculiarity of habit would of itself justify the association of the animals displaying it into a distinct zoological group, which must be founded on peculiarities of structure common to all: and taking these characters as our proper guide, we find that some of the parasitic suctorial Articulata are really aberrant forms of Crustacea; whilst with the remainder, which present a general accordance in structure and mode of development, we must associate other forms which are *not* parasitic, but which agree with the true Entozoa in their general plan and grade of organisation. It is very difficult, however, to determine what are are to be regarded as the true distinctive characters of the class; for, like other groups which occupy a low place in their respective series, it contains a much wider variety of forms than we find in those of more elevated grade; and these forms are such as

connect it closely with all the principal types of the Animal Kingdom, except the Vertebrated series. We shall find in it a manifest approximation to the Protozoa, to the Radiata, and to the Mollusca, as well as to the higher Articulata; and in fact, it is only because the most numerous and characteristic forms of Entozoa possess a distinctly *articulated* structure, that the class as a whole is best referred to the last-named series. It has been proposed to break it up, and to distribute it amongst the other types to which it is related; but such an arrangement would have the effect of widely separating from each other a number of beings, which, notwithstanding the variety of structure exhibited by their most dissimilar forms, are brought into mutual approximation by such a succession of intermediate grades, as manifestly links them into a continuous series. We may, on the whole, most philosophically regard the class as occupying an intermediate position between the Protozoa, Radiata, Mollusca, and Articulata; its closest affinity, however, being with the last-named series. And the best idea of its distinctive peculiarities will be acquired from the examination of a succession of forms, which will manifest these relations, and which, at the same time, will display the typical characters of its principal groups. Any attempt at the systematic arrangement of the Entozoa (and many such attempts have been made), must be regarded as merely provisional; since there is doubtless much still to be learned, not merely with regard to their anatomy, but also with respect to the history of their development; and what is already known on this last point leads to the suspicion, that many forms which are so dissimilar as to have been ranked in distinct orders, are really but different stages of evolution of one and the same being.

309 *a*. The simplest form of Entozoa at present known, is that which has received the name of *Gregarina*, from the aggregation of large numbers of individuals in the same situation. It infests the alimentary canal of insects, worms, &c.; and a number of species have been described, differing from each other so slightly in their structure, as to suggest the doubt whether they may not proceed from similar germs, their differences being due (as in the inferior Fungi, § 271) to slight variations in the circumstances under which their evolution takes place. Each individual essentially consists of *a single cell*, the membranous wall of which is very pellucid, whilst its contents are milk-white and minutely granular. This cell is usually more or less ovate in form: and is sometimes considerably elongated; a sort of beak or proboscis frequently projects from one extremity; and in some instances this is furnished with a circular row of hooklets, closely resembling that which is seen on the head of the *Tænia* (Fig. 140, A). Within the body is generally seen a pellucid cell-like nucleus, which usually contains one or more nucleoli. By the division of the nucleus, and the formation of a partition across the cavity, precisely after the manner of the cartilage-cell (Fig. 14), the single Gregarina-cell becomes double; and the two cells thus divided appear to separate when fully developed, and to exist thenceforth as distinct individuals. But it would seem that in some instances the partition-wall is not completely formed, and no separation takes place; the body then exhibiting an indication of segmental division. This, however, seems preparatory to the development, within each of the cells, of a cluster of minute bodies, having a close resemblance in form to the *Navicella* (a genus of Dia-

tomaceæ, § 267), but destitute of its siliceous envelope; and these are set free by the bursting of the parent-cell.* It is not known, however, whether they undergo development into Gregarinæ, or whether they are evolved into some other form; and the latter would seem most probable, from the fact that the production of the Navicella-like bodies is rather analogous to the formation of 'zoospores' or 'free gemmæ,' than to that of true ova, as well as from the resemblance of these bodies to the cercariform stage of the *trematoid* Entozoa (§ 309 *d*); so that it may be inferred that some intervening stage exists, in which a true generative process occurs, of which the Gregarinæ are the immediate products.—Much discussion has taken place as to the claim of these beings to a place in the Animal Kingdom, since they have neither mouth nor stomach, and no trace of nervous or muscular tissues; the movements they exhibit being due simply to the irritability of the cell-wall. This question is capable of easy decision, however, if the views formerly advanced (CHAP. VI.), respecting the essential distinctions between the two kingdoms, be admitted. For the *Gregarina* is an animal in its chemical constitution, its cell-wall being soluble in acetic acid; and still more decidedly, in its dependence for nutriment upon matter previously elaborated; and further, it presents a striking resemblance to the embryonic forms of other beings whose animal nature is unquestionable. If we consider its relations to other groups, we shall see that its organisation is on a level with that of the Protozoa; for it consists, like the *Amœba* (Fig. 86), of a single proteiform cell, multiplying by spontaneous fission; and it is removed from that group rather by the peculiarity of its *habitat*, than by any essential difference in its structure. The circular arrangement of the hooks on its proboscis, however, is a point of resemblance to the Radiata, which, though trifling in itself, becomes important when it is considered that this is the only indication of any definite plan exhibited in the entire structure. In the usual form of the body, we trace a sort of sketching-out of that of the *Trematoda* (§ 309 *d*); whilst in the occasional tendency to elongation, and to the formation of segments by transverse fission, we have an indication of affinity to the *Cestoidea* (§ 309 *c*). The complete affinities, however, of this interesting protozoic being, can only be understood, when we shall have become acquainted with the entire history of its development.

309 *b*. In the group of Entozoa which has received the appellation of *Cystica*, we start from a point which is at least as low as the preceding; for the simplest forms of these animals can be regarded in no other light than as independent cells, adapted to develope themselves, and to multiply their kind, in particular situations; whilst in their higher forms, they manifest more of the radiated than of the articulated type. The Cystic Entozoa are never found in the alimentary canal of the animals they infest; but always in 'cysts' or membranous sacs, imbedded in the substance of their various organs, especially their glands. These cysts are formed as exudation-membranes by the infested tissues, and partake of their vascularity; and it is by absorption from their inner surface, that the contained parasites are nourished. The simplest of all forms of these parasites is that which is known as the *Hydatid* or *Acephalocyst*.

* See Kölliker's researches on the Gregarina, in Siebold and Kölliker's Zeitschrift, Part I. 1848.

This consists of a vesicle filled with fluid, and apparently possessing no further organisation. A careful examination of the wall of this vesicle, however, shows that, besides the enveloping cyst, it consists of two or more coats; and that in the substance of the inner one are rudiments of new cells in various stages of growth.* These, as they advance in development, project more and more into the cavity of the parent-cell, and at last become detached from its wall, and lie loosely within it: shortly before this separation, however, the young Hydatid is seen to contain smaller cells, which increase in size along with it. This increase continues, until the new brood thus formed entirely fills the cavity of the parent, and a further increase causes the rupture of its sac and the escape of its progeny; and these in their turn undergo the same evolution, becoming parent-hydatids in distinct cysts, and setting free their contained cells as a subsequent generation. The Acephalocyst appears, however, to require for its proper development, that it should be completely and immediately enveloped in the tissues of the infested being; and consequently, if the original Hydatid be buried too deep in these tissues, or the surrounding cyst be of too firm a texture, to allow of its giving exit to its young, the latter do not undergo their proper evolution, the process of multiplication ceases after having been two or three times performed, and the whole mass degenerates, sometimes disappearing so completely as to leave nothing else than a firm cicatrix, but in other cases being converted into a fatty cretaceous mass. In one species of Acephalocyst, the young are found to exhibit, at the time of their attachment to the germinal membrane of the parent, a circlet of rudimentary hooks, resembling those of the higher Cystica; but these disappear as soon as the young are cast off and lie freely within the cavity of the parent-cell.—The *Echinococcus* presents us with a further development of this last character, still preserving the general type of the Acephalocyst; for in this genus, the young has not merely a circle of hooklets, but four suckers, projecting from a certain point of its sac; and by these it seems to hold on to the lining of the parent-cell, and to imbibe its nourishment from it, after its organic connection with it has ceased. Great numbers of Echinococci have been thus found, both free and attached, in the interior of vesicles that could not externally have been distinguished from ordinary Acephalocysts; and hence it has been supposed, though erroneously, that the Hydatid is never anything else than the nidus of this Entozoon.—In *Cœnurus*, on the contrary, the gemmation from the parent vesicle takes place externally, and the circle of the hooklets is borne at the end of a pedicle, so as to bear a strong resemblance to the head and neck of the *Tænia* (§ 309 *c*); and the gemmæ thus formed do not detach themselves from it, so that from a single vesicle a considerable number may be found to project.—In the *Cysticercus*, on the other hand, the parent-vesicle has never more than one pedicle; which is often considerably elongated, and which so closely corresponds to the (so-called) head and neck of the Tænia, that no positive distinction can be established between them. In the substance of this pedicle are formed peculiar cells, which seem to be the germs of a subsequent brood; these find their way into the vesicle, and there they are developed into the form of the parent, from which they probably escape at a

* See the Memoir of Mr. Harry D. S. Goodsir, on the Cystic Entozoa, in his "Anatomical and Pathological Observations."

subsequent time by its rupture. Some forms of this genus present a close approximation to the Tænia, the body being much elongated, whilst the terminal vesicle is of comparatively small size; and it is affirmed by M. Em. Blanchard,* that so close a resemblance exists in their internal structure, that no difference exists between them, save in regard to the development of the proper generative apparatus. This, which forms a large proportion of the body in the *Cestoid* order, seems altogether wanting in the *Cystic;* the only mode of multiplication at present known to occur in the latter, being by the process of gemmation. From this circumstance it may be inferred, that we are not yet acquainted with the complete history of their evolution; and it would seem probable that they go through some stage, in which proper sexual organs are evolved, and true ova developed and fertilized. By M. Blanchard it is surmised that the Cystic Entozoa are nothing else than Cestoidea whose development has been checked by the unfitness of their position; and that the eggs which, in the intestinal canal, would be evolved into Tæniæ, are abortively developed in the substance of the muscles, glands, &c., into Cysticerci. If this opinion should prove to be well founded, it will, of course, be necessary to merge this order into the succeeding.—In whatever light, however, we view the Cystic Entozoa, it is manifest that the first and most essential manifestation of symmetry in them is *radial;* the circle of hooklets possessed by the greater number of them, and the suckers which in their higher forms are borne at equal distances around the so-called head, being their principal distinctive features; whilst the annular pedicle on which these are borne, is fully developed only in such as approach the succeeding group.

309 *c.* Of the order *Cestoidea,* the members of which inhabit the intestinal canal, the common *Tænia,* or 'tape worm,' may be selected as a typical example. The elongated form and segmental division of this animal (Fig. 139) indicate its alliance to the Articulated series; and this

FIG. 139.

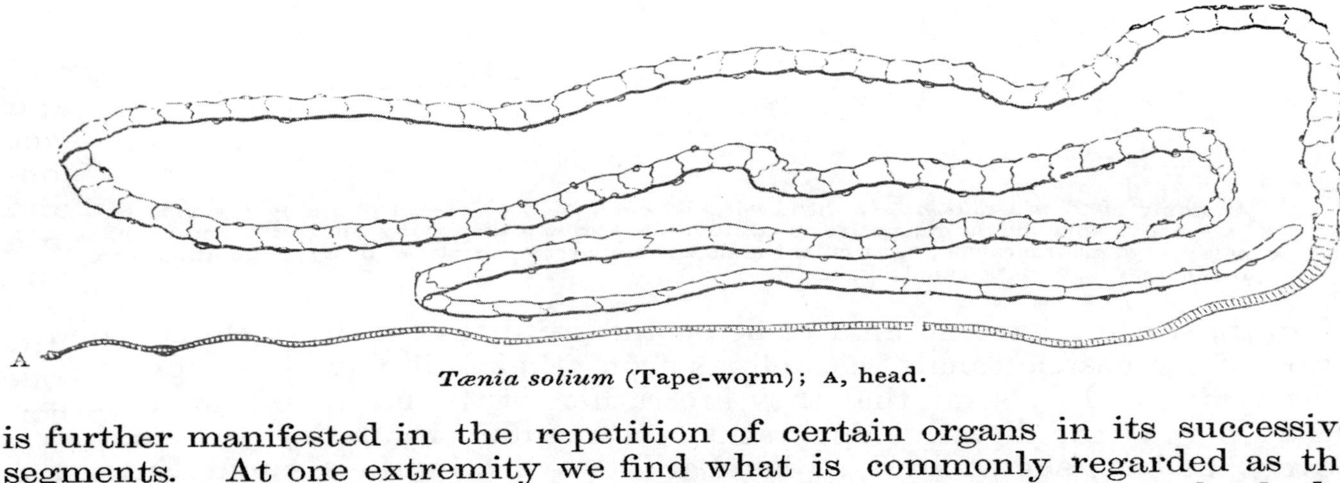

Tænia solium (Tape-worm); A, head.

is further manifested in the repetition of certain organs in its successive segments. At one extremity we find what is commonly regarded as the 'head' (Figs. 139, 140, A); this is very small in proportion to the body, and is nearly spherical. It is furnished with a circular row of hooklets at its most anterior part, surrounding a sort of proboscis; and behind

* Ann. des Sci. Nat. 3me Série, tom. x., p. 348.

this are four suctorial disks, arranged at equal distances from it and from each other. No oral orifice can be found, either in the centre of the circle of hooks, or in the suckers;* but the juices imbibed through the latter appear to pass by transudation of tissue into a gastric cavity which is situated in the interior. The anterior part of the body is very narrow, and the segmental division is only indicated by transverse wrinkles; but it widens posteriorly, and becomes flatter; and the segmental divisions are more strongly marked. The gastric cavity is prolonged through the entire body, in the form of a pair of canals which originate from it, and then pass backwards along the margins of the segments, being connected by transverse branches at the two extremities of each (Fig. 140, B, C, *a a*); these gastric canals nowhere exhibit any dilatations or contractions, and their structure in each segment is precisely the same as in the rest. They possess distinct membranous walls having considerable power of resistance, and are surrounded by muscular fibres; instead of being mere excavations in soft gelatinous tissue, as some have supposed. As these animals have no

FIG. 140.

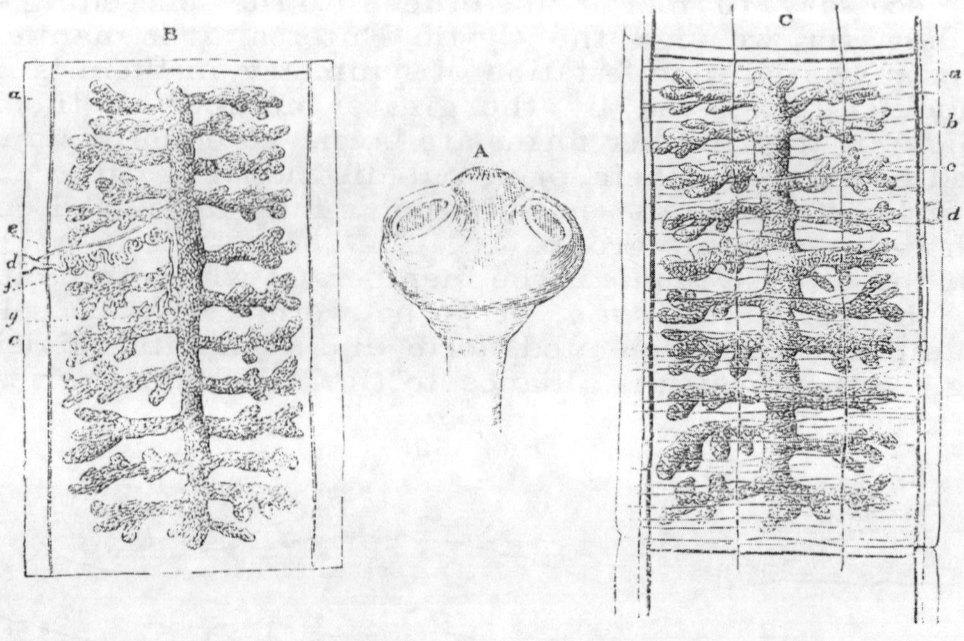

Anatomy of *Tænia solium*:—A, head with its circle of hooklets surrounding the mouth;—B, segment displaying the generative apparatus;—C, segment displaying the gastric and vascular canals;—*a a*, gastric canals; *b*, vascular canals and transverse vessels; *c*, ovary; *d*, genital orifice; *e*, oviduct; *f*, spermatic canal; *g*, testis.

mouth, so they have no anal orifice; and from this entirely-closed condition of the gastric canals, as well as from their tubular character, it has been supposed by some that they are rather of the nature of blood-vessels than part of the digestive apparatus. A regular circulating system, however, has been shown by M. Blanchard to exist in the Cestoid Entozoa; this is composed of four very minute trunks, which run, like the gastric tubuli, from one end of the body to the other, two of them alongside of

* This statement is made upon the authority of M. Emile Blanchard; from whose admirable researches upon the Entozoa (as detailed in the "Annales des Sciences Naturelles" for 1848 and 1849) most of the facts stated in the text are derived.

the gastric tubuli but externally to them, and two others nearer to the median line; and of a multitude of transverse branches which unite them in each segment, some of these being straight and direct, whilst others are more sinuous and ramifying (Fig. 140, C). No rudiments of special respiratory organs can be detected; nor of any other secreting organs than those connected with the generative function. These animals are not destitute, however, as some have maintained them to be, of a distinct nervous system; for M. Blanchard describes the Tænia as possessing a pair of ganglia united by a transverse commissure, immediately behind the hooked proboscis; from this he has traced nerve-fibres proceeding to the different parts of the head, and communicating with a minute ganglion at the base of each sucker, from which the muscular fibres of the sucker are supplied; whilst slender filaments pass backwards from the principal ganglia, accompanying the gastric canals.—A large proportion of each segment is occupied by the generative apparatus; both male and female organs being thus repeated through the entire length of the body. The testis or spermatic organ (Fig. 140, B, *g*) is of comparatively small size, and occupies the centre of the segment; its product is conveyed by the convoluted spermatic duct (*f*) towards the genital pore (*d*), which is seen on the middle of one of the margins, usually alternating from one margin to the other in successive segments; and thence it seems to find its way to the ova, through the oviduct (*e*), whilst they are as yet within the ovarium. This last organ (*c*, *c*) is of enormous dimensions; consisting of a central canal which is nearly as long as the segment itself, and of large ramifying branches which nearly extend to its margins. Thus each segment contains within itself the means of producing fertile ova; and its gastric and vascular canals may be regarded as destined chiefly to minister to the reproductive function. The embryo, when first distinctly visible within the egg, seems to possess nothing but the so-called *head* of the Tænia; no vestige of an articulated *body* can then be seen; but this is formed, after the emersion of the embryo from the egg, by the processes of gemmation and fission. The first segment is budded off from the back part of the head, and this soon undergoes subdivision into two; each of these again subdivides; and thus the total number continues for a time to augment with great rapidity. At a certain distance from the head, however, this mode of increase ceases; and the whole nutritive power of the segments is then devoted, not to their multiplication, but to their individual growth, and especially to the development of their generative apparatus. Hence it is that the anterior part of the body, in which the increase in length is still continuing, is made up of the smallest segments; and that the generative apparatus is most advanced in the terminal segments. These are detached from time to time, as the ova are approaching maturity; a provision which seems to have reference simply to their dissemination. The separated segments thus loaded with fertile eggs, may be compared to mature fruits or seed-vessels which have fallen off the plant that produced them; they have no power of reproducing the head, or of multiplying themselves into new segments; and their vitality is only retained, so long as is requisite for the preservation of their contents. As the terminal segments are cast off, others approach to maturity, and are cast off in their turn; while new segments are at the same time being developed in the part of the body nearest to the head, these to be in their turn the subjects of the same enlargement, and to be cast off when perfected. Thus there appears to be no limit to the development of these segments, save the continued vitality of the so-called 'head,' without which the remainder of the structure cannot maintain its existence; and this must hence be regarded as the essential part of the composite fabric,—as, indeed, we might infer it to be, from the fact that it is the part first developed. Now if we examine into the characters of this organ, we shall see that it may be legitimately considered as combining with the head the essential part of the real 'body'; for it contains the proper gastric cavity, of which the canals prolonged into the segments are merely extensions; while these segments are nothing else than deciduous egg-sacs appended to it, not in any way subservient to locomotion, or possessed of ganglionic centres of their own. The arrangement of the external parts of the 'head' is distinctly radiated; and its interior structure displays much of the same plan; so that, notwithstanding the articulated character of the body of the Tænia and its allies, we may consider them as still retaining somewhat of that alliance to the Radiata, which is shown so decidedly in the Cystic Entozoa.

309 *d.* From the Cestoid order, which may perhaps be regarded as the type of the class, we may naturally pass to that of *Trematoda;* the members of which are for the most part parasites of the interior of animals,—living in their cavities, however, rather than in the substance of their tissues; but of which one important section (as well as the young of most of the others) inhabits the water, though deriving its nutriment, like the rest, from the suction of the juices of animals. These Entozoa show no indication whatever of an Articulated arrangement; but each individual may be regarded as analogous to a single segment of the Tænia with the 'head' fused into it; for in a flattened ovate body, from which the head is seldom distinctly separated, we find the gastric, circulating, and generative apparatus, with the nervous system and organs of sense, all formed upon a plan very much resembling that of the Tænia, save that there is no longitudinal repetition of parts. In the general form of the body, and in the disposition of the organs, of many species of this order, and especially of the free-moving *Planariæ,* we may trace such a resemblance to the Eolidan family of Nudibranchiate Mollusca (§ 304 *c*), that they may be almost considered as degraded forms of that group; whilst, on the other hand, in certain other tribes we find an approximation to the lower Annelida, especially in the nervous system. The general structure of the former tribe will be for the most part understood from an examination of the accompanying figure (Fig. 141) of one of the marine Planariæ. The mouth (*a*), which is situated at a considerable distance from the anterior extremity of the body, is surrounded by a circular sucker that is applied to the living surface from which the animal draws its nutriment; in some species the mouth and sucker are prolonged into a sort of proboscis; but in this, a short œsophagus (*c*) leads at once to the cavity of the stomach, which occupies a large part of the anterior portion of the body. This cavity does not give origin to any intestinal tube, nor is it provided with any second orifice; but a large number of ramifying canals (*e*) are prolonged from it, which carry its contents into every part of the body. Notwithstanding this extensive distribution of the gastric apparatus, a distinct vascular system is also present; this essentially

consists of a pair of principal trunks, which pass backwards at some distance from the median line, and which subdivide into ramifications that again inosculate, so as to form a minute network through the body; and these terminate anteriorly in a pair of pulsatile cavities, united on the median line, which surround the cephalic ganglia (*f*), and give off smaller vessels to the anterior part of the body. No special respiratory apparatus exists in these animals, any more than in the parasitic Trematoda; their soft integument being sufficiently pervious to allow of the penetration of air to the fluid which is circulating immediately beneath it. The generative apparatus occupies a considerable part of the body; and we here find, as in Tænia, the two sets of sexual organs in close relation to each other. The testes, or spermatic organs (*g*, *g*), lie one on each side of the stomach; their secretion is poured into the central receptacle, or vesicula seminalis (*h*); and from this proceeds the efferent canal, which terminates in a sort of penis (*i*). The ovaries would seem to be much more extensively diffused; in fact, ova may be seen in nearly all parts of the body not occupied by other organs; and it would hence seem probable that the ovary sends ramifications, commencing in the dilated oviducts (*k*, *k*), into almost every part of the body. The oviducts unite in a dilated cavity (*l*), and from this there is an outlet by a short wide canal (*m*). These animals, although combining in themselves both kinds of sexual organs, do not seem to be capable of self-impregnation; a congress of two individuals being apparently necessary, in which the spermatic fluid of each may exert its fertilizing power on the ova of the other. It is not yet known whether their embryoes undergo any metamorphosis in their progress to maturity. The Planariæ possess a distinct nervous system, which combines in some degree the characters of that of the Mollusca with that of the Articulata. Its principal centre is a pair of ganglia (*f*), situated in the anterior part of the body, from which nervous filaments are seen to proceed to the parts in front and at their sides; whilst the posterior part of the body is supplied from two large trunks, which run backwards at some distance from the median line on either side. This arrangement closely resembles that which is seen in most of the naked Gasteropods; but we observe a tendency to departure

FIG. 141.

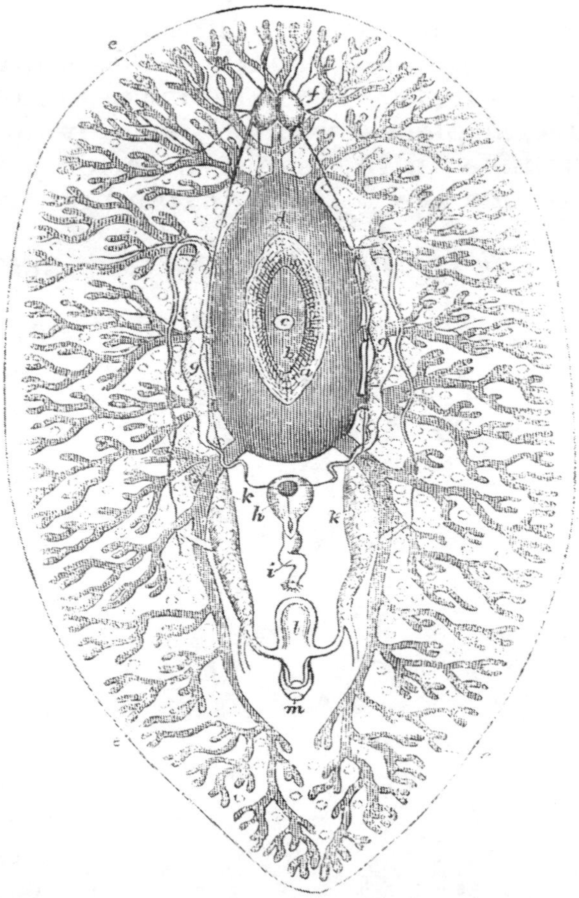

Structure of *Polycelis levigatus* (one of the Planariæ):—*a*, mouth, surrounded by its circular sucker; *b*, buccal cavity; *c*, œsophageal orifice; *d*, stomach; *e*, ramifications of gastric canals; *f*, cephalic ganglia and their nervous filaments; *g*, *g*, testes; *h*, vesicula seminalis; *i*, male genital canal; *k*, *k*, oviducts; *l*, dilatation at their point of junction; *m*, female genital orifice.

from it, and to an approximation to the Articulated type, in certain Planariæ, in which the longitudinal trunks are more closely approximated to each other, and are dilated at intervals into minute ganglionic enlargements. These animals have a distinct set of muscular fibres, running longitudinally and transversely, after the manner of those of a leech; and by means of the alternate contraction and extension of the body, they crawl with considerable activity. Their movements appear to be in some degree guided by rudimentary eyes, which are found in the neighbourhood of the cephalic ganglia, their number varying in the different species from two to forty. It is remarkable, however, that if the body be divided into several pieces, whilst it is moving forwards, each segment will continue to advance in the same direction; thus showing how purely automatic its motions are. Notwithstanding their comparatively high organisation, the *Planariæ* possess a power of regenerating lost parts, which is scarcely inferior to that of the *Hydra* (§ 289); for not merely do we find that any portion of the body which is removed is speedily replaced, but that even the separated parts can reproduce the entire body; so that several individuals can be thus generated by the subdivision of one. Such a method of multiplication seems natural to these animals; the fission taking place spontaneously in a transverse direction.—The organisation of the parasitic Trematoda is for the most part essentially the same with that of the Planariæ. The mouth, however, is usually at the anterior part of the body (Fig. 142, *a*); and besides the suctorial disk with which it is furnished, we find one, two, or more suckers on different parts of the body, which serve for its adhesion merely. One of the commonest of these Entozoa is the *Fasciola hepatica*, or fluke, which inhabits the bile-ducts of the sheep; and the arrangement of its gastric and vascular apparatus is shown in the accompanying figure, the former being represented by the portions most deeply shaded. The stomach is situated at a short distance behind the head; and from it proceed anterior, lateral, and posterior prolongations, the latter being the largest, and giving off lateral branches which extend themselves into every part of the body. Notwithstanding this, a distinct vascular system also exists, the centre of which is a dilated trunk that is seen on the median line between the two longitudinal gastric cœca, in the situation of the 'dorsal vessel' of the higher Articulata. The nervous system is constructed upon the plan of that of the Planariæ; but in the adult condition of these animals there are no

FIG. 142.

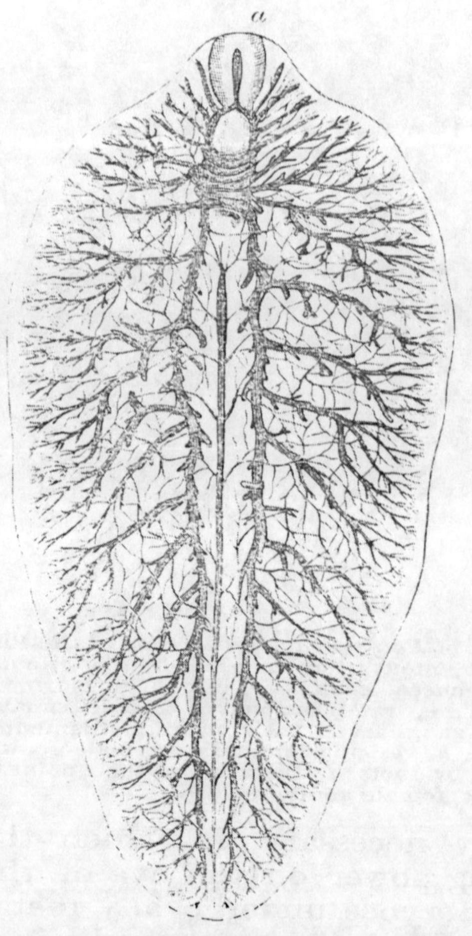

Anatomy of *Fasciola hepatica* (Distoma hepaticum) enlarged, showing the ramifications of the digestive cavity through the whole body of the animal, and the vascular network connected with a median trunk; *a*, the mouth.

eyes, and the cephalic centres are relatively smaller; and the ganglionic enlargements on the longitudinal trunks are more distinct. Thus in the terminal position of the mouth, and in the conformation of the vascular and nervous systems, the Fasciola presents a closer approximation to the Articulated series, than we can see in the Planariæ.—In some of this division of the order, a very curious series of distinct forms is passed through, between the production of one true sexual generation and another. Thus the ova of the *Distoma* which inhabits the *Lymnæus stagnalis* (a fresh-water snail) are not developed into the likeness of their parent, but into little worm-like bodies, in which no complete organs are evolved, but which seem to consist of the cells of the germinal mass, inclosed in a contractile integument. These cells are in their turn developed into independent beings, which escape from the parent cyst, and from the animal which contains it, in the condition of free ciliated animalcules; in this condition they remain for some time, and then imbed themselves in the mucus that covers the tail of the Mollusk, in which they undergo a gradual development into true *Distomata;* and having thus acquired their perfect form, they penetrate the soft integument, and take up their habitation in the interior of the body. Thus a considerable number of Distomata may be produced from a single ovum, by a process of gemmation in an early stage of development, closely resembling the production of the navicellar bodies within the Gregarina (§ 309, *a*). In some instances, the free ciliated larva possesses distinct eyes, although they are wanting in the fully-developed Distoma, the peculiar *habitat* of which would render them useless.—Among the curious forms contained in this order, the *Diplozoon paradoxum* is especially worthy of notice; as it presents the curious phenomenon of a body almost completely divided along the median line, each half containing in itself all the organs of nutrition, circulation, reproduction, and locomotion, and the two being united merely by a narrow band, through which, however, the digestive cavities on the two sides intercommunicate.—The genus *Malacobdella* is regarded by M. Blanchard as belonging to the Trematode Entozoa, although it has also a close relationship to the Leech-tribe; so that it unites this order with the Suctorial Annelida (§ 311 *e*).

309 *e.* Passing over some small intermediate groups which connect the lower and the higher orders of Entozoa, we come to the order *Nematoidea,* in which the Articulated character is unmistakable. The body of these animals is greatly elongated, and is usually cylindrical, like that of the ordinary Worms; it does not, however, present externally any true annulations, though a transverse wrinkling of the integument is generally perceptible. The integument is rendered firm by layers of muscular fibre, which present in some of these animals a considerable degree of development, and give them the power of active movement. The oral orifice is situated at one extremity of the body (Fig. 143, *a*); and this leads to the alimentary canal (*c, c, c*), which runs through the axis of the cylinder to the anal aperture (*d*) at its opposite extremity, presenting in its course but little flexure or undulation, and rarely exhibiting any distinction of stomach and intestine, though the first part (*b*), being rather narrower than the rest, may be considered as an œsophagus. The walls of this canal are not in immediate contact externally with the soft tissues of the body, as they are

in the preceding orders, but lie in a cavity, to the parietes of which they are attached by transverse muscular filaments; hence the Nematoid Entozoa have been designated by Prof. Owen as *Cœlelmintha* (or *cavitary* worms) in contradistinction to the preceding orders, which are ranked together in his group of *Sterelmintha* (or *solid* worms). No gastric cœca, nor appendages of any kind, are connected with this simple alimentary canal. The vascular system essentially consists of a pair of longitudinal trunks on each side, enclosed in a spongy envelope common to the two vessels; one of these trunks passes from one extremity of the body to the other; but the other quits the envelope, at a point not far behind the mouth, and passes over to join the corresponding trunk on the other side, a small contractile vesicle being developed upon the arch thus formed. The trunks which lie nearest the median line, communicate with each other freely by transverse branches; but there are few other ramifications. These internal trunks are considered by M. Blanchard to be the representatives of arteries, and to distribute the blood impelled by the contractile sac; whilst the posterior are supposed to collect it again, and thus to perform the functions of veins. No rudiments of a respiratory apparatus exist here, any more than in the inferior Entozoa.—The cephalic centres of the nervous system consist of a pair of very minute ganglia at each side of the œsophagus, which are connected by two commissural bands that pass the one above and the other below the œsophagus, so as to form the same kind of collar as we see in the higher Articulata; from this collar proceed two longitudinal trunks, which traverse the length of the body; but these do not present any visible minute gangliform enlargements; nor do any organs of special sense appear to be connected with the cephalic ganglia. The very small size of the nervous centres of these animals, as compared with their well-developed muscular apparatus, is a striking feature in their organisation; and seems to indicate that the movements of these worms, which often seem to possess a rhythmical character, are in great degree independent of nervous influence. The generative apparatus

FIG. 143.

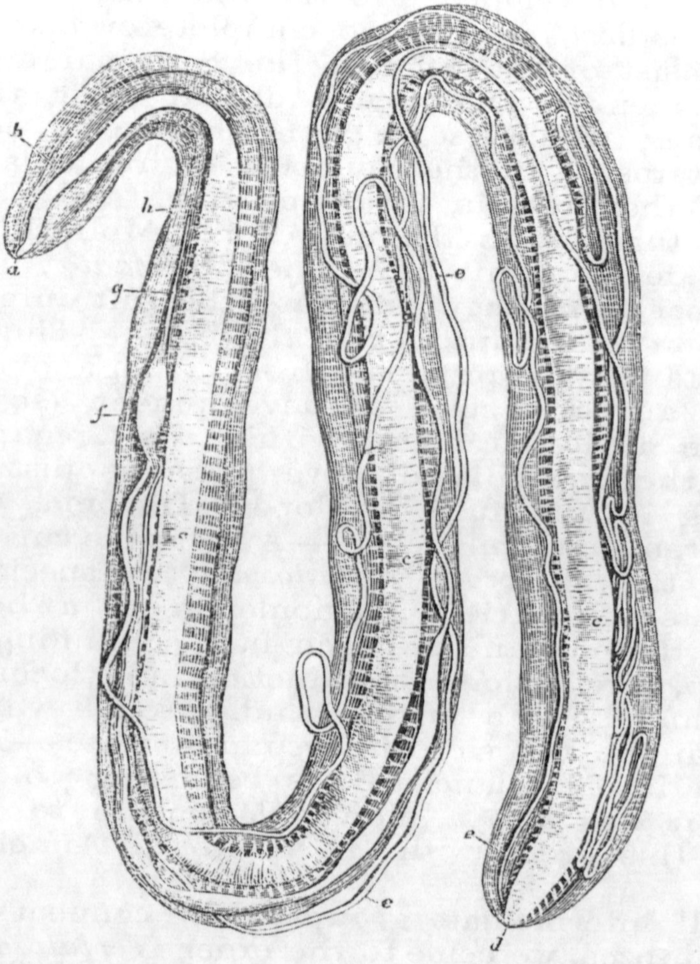

Strongylus gigas (female) laid open to show its internal structure; — *a*, mouth; *b*, œsophagus; *c*, *c*, *c*, intestinal canal; *d*, anus; *e*, *e*, *e*, ovary; *f*, uterine dilatation; *g*, narrow oviduct; *h*, its orifice.

in all the Nematoid worms is formed upon the diœcious type; that is, the male and female organs are restricted to distinct individuals, instead of being united in the same, as in the inferior Entozoa. The male genital organ usually consists of a single long tube, enlarged at its lower part, and opening near the anus. The female ovary, on the other hand, sometimes consists of a similar tube very much prolonged (Fig. 143, *e*, *e*, *e*), whilst in other instances it is multiple; in either case, however, it opens into a wide oviduct, which has frequently a dilatation (*f*) near its termination, and which opens externally at a considerable distance from the anus, and frequently in the middle or even the anterior portion of the body (*h*). The ova of the Nematoidea are fertilized by sexual congress, the anal extremity of the male attaching itself to the genital orifice of the female; and as this is usually situated at a distance from either end of the body, and as the male is almost always smaller than the female (sometimes very much so), he has the appearance of a branch or young individual sent off by gemmation, and attached at an acute angle to her body. The number of ova developed by these Entozoa is scarcely inferior to that produced by the Cestoidea; no fewer than *sixty-four millions* have been calculated to be contained in a single female *Ascaris lumbricoides*, one of the commonest parasites of the human intestine. In some of this order, the ova are retained within the body of the parent until they are hatched, so that the young come forth alive; in other instances, the eggs are hatched subsequently to their deposition. In either case, however, the young worm at its emersion from the egg has the aspect and structure of its parent; not undergoing a metamorphosis, like the Trematoda; nor subsequently evolving the greater part of its structure by a gradual development, like the Cestoidea.—To this order belong a large number of simple worms, many of which are not parasitic, although they live in situations in which they can obtain fluid nutriment, being unpossessed of organs for the prehension and reduction of solid food. There is a considerable variety among them in regard to complexity of organisation. That which has just been described is exhibited by the higher forms of the order, such as the *Ascaris*, the *Filaria*, and other parasites of the intestinal canal, urinary passages, and other cavities in Man and the higher animals. But we find that certain minute worms, developed in the substance of Animal and even Vegetable tissues, have a much inferior organisation. A remarkable example of the lower type of Nematoid structure is presented by the *Trichina spiralis*, which not unfrequently infests the substance of the human muscles, in enormous numbers. It is enveloped in a double cyst, of about 1-50th of an inch in length; of which the outer wall is evidently formed by condensation of the surrounding tissue; whilst the inner (which is completely isolated from it, and is apparently structureless) seems to appertain to the animal itself. Within this cyst, the worm itself, which is about 1-30th of an inch long, lies coiled up; the remaining space being filled with a minutely granular fluid. The mouth, anus, and alimentary canal are very indistinct; and no generative organs have yet been detected in it. Altogether, it bears a strong resemblance to the early embryonic condition of the higher Nematoidea; and it may perhaps be imagined to proceed from the ovum of one of these, which, through being developed in an unfavourable situation, has become abortive, the embryo having acquired the worm-like form, but having never ruptured its egg-sac, within which (forming the inner cyst) it remains inclosed. The so-called 'eels' of vinegar and sour paste are little worms of the same grade of organisation; and so also is the *Vibrio tritici*, which is parasitic in the ears of unhealthy wheat, and the presence of which constitutes the morbid condition known as 'cockle.'—It is a remarkable feature in the history of the Nematoid worms, that they are able to bear extremes of temperature, which are fatal to animals whose organisation is apparently less complex; thus many of them have been found to resist prolonged immersion in boiling water, whilst on the other hand they will endure being frozen, without the loss of their vitality; and the *Vibrio tritici* has been revived by moisture, after being completely dried up for many years.

309*f*. Such are the most characteristic features of the principal divisions of the class of Entozoa; in the survey of which we have had the opportunity of seeing how much variation there may be in external form, and even in general plan of organization, whilst the most discrepant examples are connected by intermediate links, and a certain correspondence in grade of development is traceable throughout; so that no reasonable doubt can be entertained of the *unity* of the group thus constituted. We have seen that in each of its three principal divisions,—the *Cestoidea*, *Trematoda*, and *Nematoidea*,—there is an obvious tendency towards the Articulate type, even where that type is not clearly displayed; and thus, whilst the first of these divisions connects it with the Radiata, and the second with the Mollusca, the place of the whole must be considered to be at the foot of the series of Articulated animals. The class of Entozoa presents a vast field for attentive study, more especially as regards the history of the development of the animals composing it, which, when more fully known, will doubtless throw much light on the relations of its different forms to each other.

310. The position of the class of ROTIFERA, or *Wheel Animalcules*, and their allies, would seem to be in the Articulated sub-kingdom: for although the Annulose character is by no means clearly marked in it, yet it is sufficiently indicated to show that these beings are neither Radiata nor Mollusca, whilst their complexity of structure raises them far above the condition of the Protozoic Infusoria, with which they were formerly confounded. We shall, in fact, see reason to regard this group as an osculant one; presenting, amidst the variety of forms which it includes, links of connection between the lower part of the Articulated series on the one hand, and the Polygastrica and Bryozoa on the other. That the complexity of structure, by which the Rotifera are for the most part distinguished, was for a long time overlooked, is owing to their minute size; scarcely any of them exceeding a line in length, while many are less than 1-500th of an inch. All of them are more or less exclusively aquatic in their habits. Some live only in salt water; others frequent stagnant ponds; while others seem confined to vegetable infusions, where they generally succeed Animalcules of inferior organisation. But the *Rotifer vulgaris* is sometimes found in the interior of the cells of *Sphagnum* (§ 147), and even lives in those parts of the plant which are not immersed in water; whilst other species can maintain their existence in the midst of damp earth.—The name of the class is derived from the appearance of revolving wheels, which is presented, in some of the most abundant forms

of it, such as the *Rotifer vulgaris* (Fig. 144), by the vibration of the two circular rows of cilia (B, *c*), which are situated one on either side of the head. Such an arrangement of these cilia, however, is by no means constant, it being only in a comparatively small number of species, that they form circles returning into themselves; in some of the aberrant forms of the group (Fig. 145), we have a close approximation to the arrangement of the cilia on the tentacula of the Bryozoa (Fig. 114); whilst in others there is but a single row of cilia, sometimes arranged in a circular mode around the funnel-shaped anterior portion of the body, almost as in the *Vorticella* (Fig. 88), and sometimes (where the body is more flattened) extending in a sinuous line from one side to the other (Fig. 146). The great transparency of the Rotifera permits their general structure to be easily recognized. They have usually an elongated form, similar on the two sides; but this rarely exhibits any traces of segmental division. The body is covered with a double envelope; both layers of which are extremely thin and flexible in some species; whilst in others the outer one seems to possess a horny consistence. In the former case, the whole integument is drawn together in a wrinkled manner when the body is shortened; in some of the latter the sheath has the form of a polype-cell, and the body lies loosely in it, the inner layer of the integument being separated from the outer by a considerable space (Fig. 145); whilst in others, the envelope or *lorica* is tightly fitted to the body, and strongly resembles the horny casing of an insect or the shell of a crab, except that it is not jointed, and does not extend over the head and tail, which can be projected from the openings at its extremities, or completely drawn within it for protection (Fig. 146). In those Rotifera in which the flexibility of the body is not interfered with by the consolidation of the external integument, we usually find it capable of great variation in shape, the elongated form being occasionally exchanged for an almost globular one, as is seen especially when the animals are suffering from deficiency of water; whilst by alternative movements of contraction and extension, they can make their way over solid surfaces, after the manner of a worm or a leech, with considerable activity,—some of even the loricated species being rendered capable of this kind of progression, by the contractility of the head and tail. All these, too, can swim readily through the water, by the action of their cilia; and there are some species (those most allied to the *Vorticellinæ*) which are limited to the latter mode of progression. The greater number have an organ of attachment at the posterior extremity of the body, which is usually prolonged into a tail, by which they can affix themselves to any solid object; and this is their ordinary position when keeping their wheels in action for a supply of food or of water. They have no difficulty, however, in letting go their hold, and moving through the water in search of a new attachment, and may therefore be considered as perfectly free; the polypoid species, however, remain attached by the posterior extremity to the spot on which they have at first fixed themselves, and their cilia are consequently employed for no other purpose than that of creating currents in the surrounding water.

310*a*. In considering the internal structure of Rotifera, we shall take as its type the arrangement which it presents in one of the commonest and most characteristic forms of the class, the *Rotifer vulgaris* (Fig. 144);

and specify the principal variations exhibited by others. The body of this animal, when fully extended, possesses greater length in proportion to its diameter than that of most others of the class; and the tail is composed of three joints or segments, which are capable of being drawn up, one within another, like the sliding-tubes of the telescope, each having a pair of prongs or points at its extremity. Within the external integument of the body, are seen a set of longitudinal muscular bands (*h*), which serve to draw the two extremities towards each other; and these are crossed by a set of transverse annular bands, which also are probably muscular, and serve to diminish the diameter of the body, and thus to increase its length. Between the wheels is a prominence bearing two red spots (*b*) supposed to be rudimentary eyes, and having the mouth (*a*) at its extremity; this prominence may be considered, therefore, as a true head, notwithstanding that it is not clearly distinguishable from the body. This head also bears upon its under surface a projecting tubular organ (*d*), which is thought by Professor Ehrenberg to be a siphon for the admission of water to the cavity of the body for the purpose of respiration; but this, for reasons hereafter to be stated, must be considered very doubtful. The œsophagus, which is narrow in the *Rotifer*, but which is dilated into a crop in *Stephanoceros* (Fig. 145) and in some other genera, leads to the masticating apparatus, which in these animals is placed far behind the mouth, and in close proximity to the stomach. It consists of a pair of semicircular jaws, each having from one to five teeth (in *Rotifer*, two), which appear to contain mineral matter and to be of harder texture than the rest of the fabric; these jaws are put in action by powerful muscles, and are so moved that all the food which passes into the stomach is subject to be divided and torn by these teeth; this movement is sometimes seen to occur, when no fresh supply of food is being ingested by the agency of the wheels. The jaws in some species may even be projected beyond the mouth, and may be employed as instruments of prehension.* The form of the alimentary canal varies; it being sometimes a simple tube, passing without enlargement or constriction from the masticating apparatus to the anal orifice at the posterior

FIG. 144.

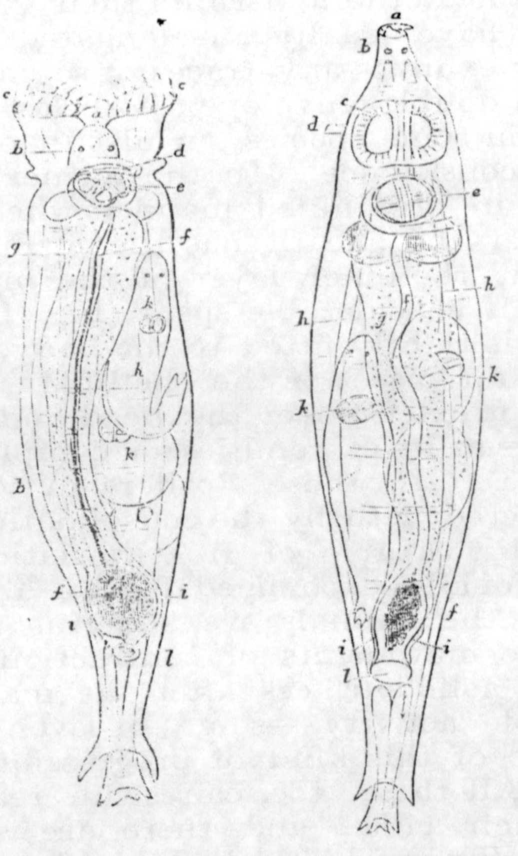

Rotifer vulgaris, as seen at A with the wheels drawn in, and at B with the wheels expanded:—*a*, mouth; *b*, eye-spots; *c*, wheels; *d*, respiratory siphon; *e*. jaws and teeth; *f*, alimentary canal; *g*, glandular (?) mass inclosing it; *h*, longitudinal muscles; *i*, *i*, seminal (?) vessels; *k*, young animal; *l*, cloaca.

* See Lankester in Cyclop. of Anat. and Phys. vol. iv. p. 412.

part of the body; whilst in other instances there is a marked distinction between the stomach and intestinal tube, the former being a large globular dilatation immediately below the jaws, whilst the latter is cylindrical and comparatively small. The alimentary canal of *Rotifer* most resembles the first of these types; but it presents a dilatation (*l*) close to the anal orifice, which may be considered as a cloaca. Connected with the alimentary canal are various glandular appendages, more or less developed; sometimes clustering round its walls as a mass of separate follicles, which seems to be the condition of the glandular investment (*g*) of the alimentary canal in *Rotifer;* in other cases having the form of cœcal tubuli. Some of these open into the stomach close to the termination of the œsophagus, and have been supposed to be salivary or pancreatic in their character; whilst others, which discharge their secretion into the intestinal tube, have been regarded, and probably with correctness, as the rudiment of a liver. In a curious animalcule of this class, recently minutely described by Mr. Dalrymple,* although the mouth, masticating apparatus, and stomach are constructed upon the regular type of the genus *Notommata*, to which it seems nearly allied, yet there is neither intestine nor anal orifice, and the indigestible matters are rejected through the mouth. This, so far as is yet known, is a solitary example of the existence of this character of degradation in the class Rotifera. —It is as yet uncertain whether any special circulating or respiratory apparatus exists in these animals; and considering the thinness of their tissues, their minute size, and the consequent facility with which the nutritive matter may find its way from the alimentary canal to any part of the body (as it seems to do in the Bryozoa, § 298); considering also that the action of their ciliary apparatus will not only bring food to their mouths, but currents of aërated water to their surface; we need not be surprised if further examination should disprove the idea entertained by Prof. Ehrenberg, that true vessels and respiratory organs are to be found amongst them. The fluid which is contained in the general cavity of the body, between the exterior of the alimentary canal and the inner tegumentary membrane, is probably to be regarded as nutritive in its character; and its movement appears to be provided for by the continual agitation of a number of ciliated 'tags' (supposed by Ehrenberg to be respiratory organs), which are disposed symmetrically along a filament on either side of the cavity.† Such a movement of fluids within the general cavity of the body, we shall hereafter find to be the condition of the circulation in certain animals of higher organisation (§ 316 *c*).—There is some uncertainty, also, with regard to the structures which Prof. Ehrenberg has described as ganglia and nerves; although probability seems here decidedly in favour of the admission of his view of their nature. The single, double, or multiple red spots seen upon the head of the Rotifera, correspond precisely in situation with those which, in the higher Articulata, are unquestionably eyes; and the principal ganglionic mass described by Prof. Ehrenberg, is in immediate connection with them. From this cephalic ganglion, a pair of principal filaments passes backwards along the ventral surface of the body; and upon these, in some species, minute ganglionic enlargements may be observed at intervals, so

* Phil. Trans. 1849, p. 331.

† Dalrymple, *loc. cit.*

as to give to the nervous system a decidedly articulated type. Besides the principal cephalic ganglion, a pair of lateral ganglia is observable in some species, in connection with the muscles that work the ciliary apparatus; this being especially the case with such as have this apparatus most widely extended laterally.

310*b*. The reproduction of the Rotifera has not yet been completely elucidated. There is no known instance, in this group, in which multiplication by gemmation or spontaneous fission is certainly known to take place; but the occurrence of clusters formed by the aggregation of a number of individuals of *Conochilus*, adherent by their tails, and enclosed within a common *lorica*, would seem to indicate that these clusters, like the aggregations of Polygastrica, Bryozoa, and Tunicata, must have been formed by continuous growth from a single individual; and this seems the more probable, as the genus is one of those that approaches most nearly to the *Vorticellinæ*, the clusters of which are formed by a similar method (Fig. 97). The ordinary method of multiplication, however, is by a proper generative act, for which two sets of sexual organs are provided, which are usually combined in the same individual. The male organs in the *Rotifer* are believed to be a pair of tubes (*i*, *i*) which open into the cloaca, but as no spermatozoa have yet been observed in them, their function must be considered doubtful; the female organ consists but of a single ovarian sac, which frequently occupies a large part of the cavity of the body, and which opens at its lower end by a narrow orifice into the cloaca; and these animalcules are probably self-fertilizing. In the remarkable genus, however, described by Mr. Dalrymple (*loc. cit.*), the sexes are separate, and the ova are impregnated by an act of copulation. The condition of the male of this genus is a most extraordinary one; for it possesses no mandibles, pharynx, œsophagus, stomach, nor hepatic glands; having, in fact, no other organs fully developed, than those of generation. It would appear, therefore, quite unfit to obtain aliment for itself; and its existence is probably a very brief one, being continued only so long as the store of nutriment supplied by the egg remains unexhausted.—Although the number of eggs in these animals is so small, yet the rapidity with which the whole process of their development and maturation is accomplished renders the multiplication of the race very rapid. The egg of the *Hydatina* is extruded from the cloaca within a few hours after the first rudiment of it is visible; and within twelve hours more the shell bursts, and the young animal comes forth. In the *Rotifer* and several other genera, the development of the embryo takes place whilst the egg is yet retained within the body of the parent, and the young are extruded alive; whilst in some other instances, the eggs, after their extrusion, remain attached to the posterior extremity of the body, until the young are set free. In general it would seem that, whether the rupture of the egg-membrane takes place before or after the egg has left the body, the germinal mass within it is developed at once into the form of the young animal, which resembles that of its parent; no preliminary metamorphosis being gone through, nor any parts developed which are not to be permanent. The transparency of the egg-membrane, and also of the tissues, of the parent Rotifer, allows the process of development to be watched, even when the egg is retained within the body; and it is curious to observe, at a very early period, not merely the red eye-spot of the embryo, but also a distinct ciliary movement. The multiplication of *Hydatina* (in which genus three or four eggs are deposited at once, and their development completed out of the body) takes place so rapidly, that, according to the estimate of Prof. Ehrenberg, nearly *seventeen millions* may be produced within twenty-four days from a single individual. This extraordinary rate of increase, taken in connection with the power which many species possess of sustaining the complete desiccation of their bodies without loss of their vitality (§ 65), removes all difficulty in accounting for the extent of the diffusion of these animals, and for their occurrence in incalculable numbers in situations where, a few days previously, none were known to exist. For their entire bodies may be wafted in a dry state by the atmosphere, from place to place; and their return to a state of active life, after a desiccation of unlimited duration, may take place whenever they meet with the requisite conditions,—moisture, warmth, and food. It is probable that the ova are capable of sustaining treatment even more severe than the fully-developed animals can bear; and that the race is frequently continued by them, when the latter have perished. Even in those species which usually hatch their eggs within their bodies, it appears that a different set of ova is occasionally developed, which are furnished with a thick glutinous investment: these, which are extruded entire, and are laid one upon another, so as at last to form masses of considerable size in proportion to the bulk of the animals, seem not to be destined to come so early to maturity, but very probably remain dormant during the whole winter season, so as to produce a new brood in the spring.—Notwithstanding that, in most of the Rotifera, the embryo at its emersion from the egg possesses all the most important characteristics of the parent, yet according to the observations of Dr. Mantell,* the embryo of the *Stephanoceros* first comes forth from the egg in a form and condition much resembling that of the *Stentor* (one of the Vorticellinæ), and only gradually attains the characteristic form and structure of its parent. Should any such resemblance present itself in other cases, it will naturally lead to the suspicion of a closer relationship between these groups, than is at present known to exist.

310 *c.* We shall now briefly notice some of those *aberrant* forms, by which the class of Rotifera is connected with other groups. As already remarked, there are certain Wheel-Animalcules, which, in the general simplicity of their structure, and the form of their rotatory organs, present a decided approximation to the *Vorticellinæ*; and in the genus *Chætonotus*, the cilia are no longer arranged so as to form a distinct rotatory organ, but cover the whole body, whilst at the same time the internal structure is very indistinct, so that it stands on the confines of the two classes of Rotifera and Polygastrica.—Again, in *Stephanoceros* (Fig. 145) and other Rotifera which have the external wall separated from the body so as to form a cell, there is a manifest approximation to the *Polypifera* and *Bryozoa*; and the resemblance to the latter group is increased in Stephanoceros, by the conversion of the rotatory organ into a set of ciliated tentacula. However, the internal structure of these animals still remains conformable to the ordinary rotiferous type, in every respect save the

* Thoughts on Animalcules, pp. 68, 69.

dilatation of the pharynx into a sort of crop, which may remind us of the dilated pharynx of the Bryozoa. In *Melicerta,* the exterior of the cell is strengthened by the agglutination of foreign particles, which are arranged with great regularity; reminding us of the sheaths of some of the Tubicolous Annelida (§ 311 *b*).—The passage between the Rotifera and the Annelida, however, is more directly established by a very curious group, that of *Tardigrada,* which, with the general organisation of the former, possesses the segmental division of the body, the setigerous appendages,

FIG. 145.

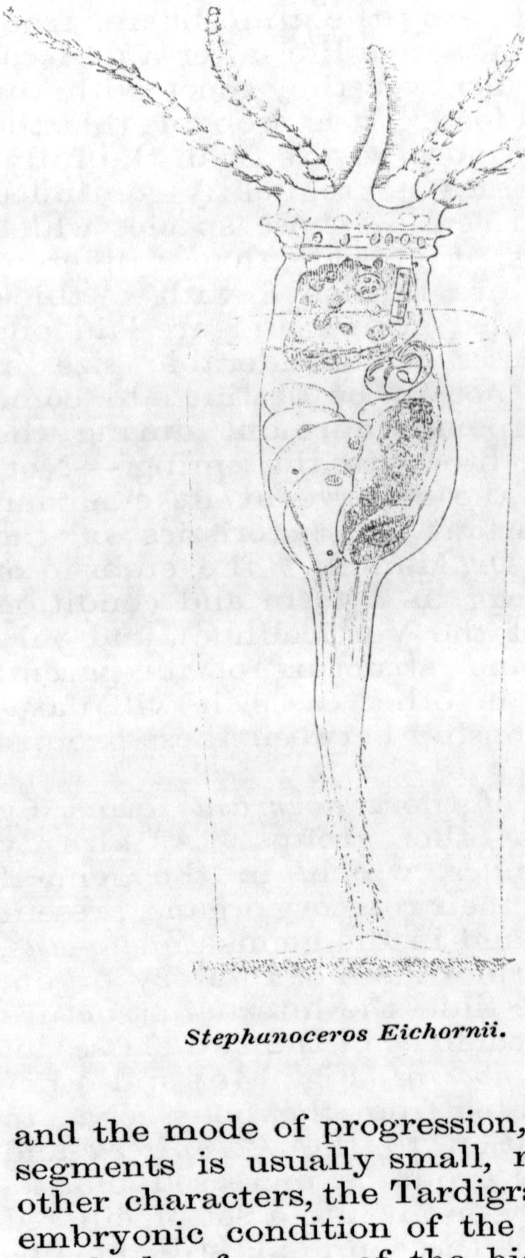

Stephanoceros Eichornii.

FIG. 146.

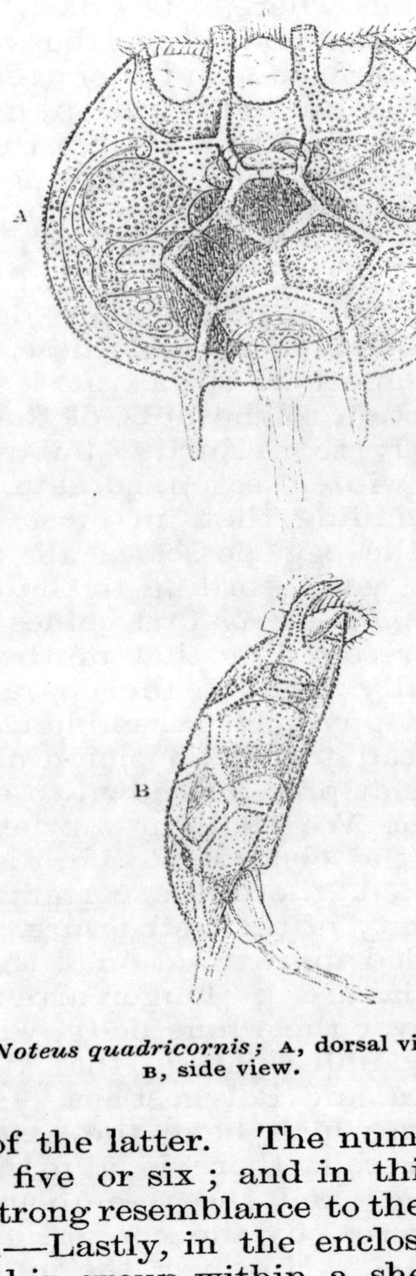

Noteus quadricornis; A, dorsal view; B, side view.

and the mode of progression, characteristic of the latter. The number of segments is usually small, rarely exceeding five or six; and in this and other characters, the Tardigrades present a strong resemblance to the early embryonic condition of the true Annelides.—Lastly, in the enclosure of the body of some of the higher forms of this group within a shell-like envelope (Fig. 146), and in some points of their internal organisation, there appears to be an approximation towards the simpler forms of *Entomostracous Crustacea* (§ 317).

311. The class ANNELIDA includes the higher types of the *Vermiform* subdivision of the Articulated series,—that in which the body is more or less worm-like in aspect, and is destitute of articulated members. We find in it, however, a foreshadowing of the presence of such members; most of the segments of the body, in the greater part of the class, being furnished with projecting *setæ*, or bristles, which serve to assist in progression. The body in this class is usually very long, and nearly always presents a well-marked segmental division; the segments are for the most part equal and similar to each other, it being the head alone that is different. In the lower forms, however, such as the Leech and its allies, the segmental division is very indistinctly seen, on account of the general softness of the integument; and in these we scarcely find the cephalic segment ostensibly different from the rest. In their general organisation, they present a considerable advance upon the higher Trematoid and Nematoid Entozoa, to which they seem most nearly allied; the head being provided with special organs of sense, and the mouth having an apparatus for the reduction and prehension of food; the intestinal canal, though sometimes simple, being frequently furnished with complicated glandular appendages; the circulating system being much increased in extent and complexity, and a special respiratory apparatus being almost invariably developed; and the nervous system being formed upon a higher plan, the two ventral cords being closely approximated to each other on the median line, and being studded with a succession of distinct ganglia which correspond with the segments. In the general structure of the different orders of this class, however, there are some important differences, which require that they should be separately noticed.

311 *a*. The order which may probably be considered as the type of the group, is that of *Dorsibranchiata;* so named from the attachment of the respiratory appendages to the dorsal surface of the body, along the greater part or the whole of its length (Fig. 147). The animals which it includes are all aquatic, worm-like in their appearance, and active in their habits, crawling with facility and swimming rapidly; and from this last circumstance they are sometimes designated as *Errantia*, in order to distinguish them from the succeeding group, in which an opposite mode of life prevails. The head of these worms is always quite distinct from the trunk, and is furnished with one or two pairs of rudimentary eyes, which are seen as black or reddish specks upon its upper side, and also with one or more pairs of tactile appendages, resembling the antennæ of Insects, &c., except in being soft and unjointed. The body is frequently of very considerable length, some of the tropical Nereids being not less than four or five feet long, and possessing four or five hundred segments. These segments are externally similar to each other except in size; and their appendages, also, are almost precisely repeated from one extremity of the body to the other. These appendages vary in the different genera; but we usually observe that there are the rudiments at least of two fleshy tubercles on either side of each segment (thus making four for the entire segment), from which these appendages proceed; and these are termed the 'dorsal oars' (D, *A′*) or the 'ventral oars' (*B′*), according as they project from the upper or the lower half of the segment. Each oar usually possesses a long soft cylindrical appendage, or *cirrhus* (D, *a*, *e*), analogous to the antenniform appendages of the cephalic segment, which proceeds

from the base of the tubercle; whilst its summit bears a tuft of *setæ*, or bristles, which serve as instruments of locomotion when the animal is crawling over solid surfaces. In the ordinary Nereids, each oar has also a membranous lobe (that of the dorsal oar is shown at *b*), which is the instrument of propulsion in water. All these probably serve as organs of respiration, exposing to the air diffused through the surrounding water the blood transmitted to them; but there is a more special respiratory organ (*d*) developed from the under side of each dorsal oar, consisting either of a branching tuft, or of a flattened vesicle, to which the blood is minutely distributed; this, however, is not always continued through the entire length of the animal.—The mouth is usually situated at the extremity, or rather on the under surface, of the head; it is armed in some species with one, two, or even three pairs of jaws, opening sideways; whilst others have a proboscis (Fig. 147, B, C) which is capable of being projected from the head, and everted in such a manner as to bring into play a set of horny teeth which are at other times withdrawn within it. In most of the Nereids, which constitute the principal part of this order, there is no distinction between stomach and intestine, but the alimentary canal proceeds with little alteration in size from the mouth to the posterior extremity; it is, however, frequently furnished with cœcal appendages, apparently of a glandular character. In all of them there is a complex vascular apparatus; and we find this to be provided, not merely with a contractile impelling cavity in the front of the body, but with a multitude of pulsatile organs, which are situated at the bases of the branchiæ, and which propel the blood through these organs. The blood is red in these as in the greater number of Annelida; but this is not a character of much importance, since the hue of the fluid does not depend upon the presence of red corpuscles (§ 173) floating in it, but is derived from colouring matter diffused through the plasma.—The nervous system

FIG. 147.

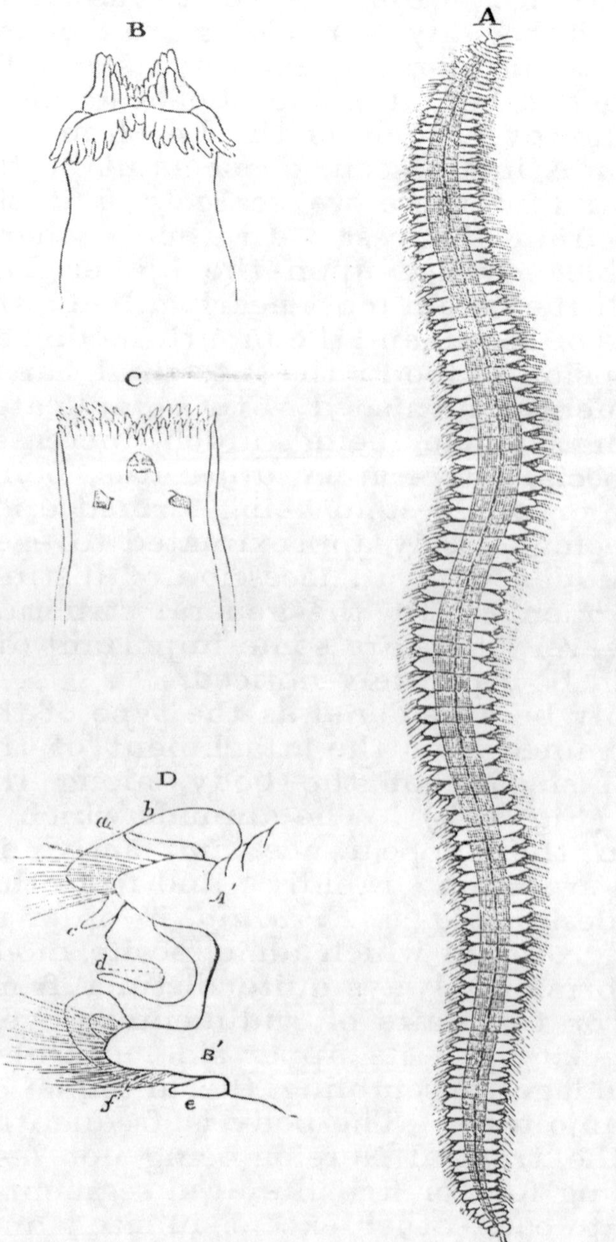

A, *Nephthys Hombergii*; B, its proboscis; C, the same laid open, to show the horny teeth it contains; D, one of the feet, showing A′, dorsal oar; B′, ventral oar; *a*, dorsal cirrhus; *b*, membranous lobe of dorsal oar; *c*, tentaculiform appendage; *d*, branchial appendage; *e*, ventral cirrhus; *f*, membranous lobe of ventral oar.

of the Dorsibranchiata presents the highest type which it anywhere attains in this class; the cephalic ganglia being much superior in size to the ganglia of the ventral cord, in virtue of the development of organs of special sense with which they are connected; and indications being observable of a set of nerves and ganglia appertaining to the digestive and respiratory organs, distinct from the proper ventral gangliated cord which is subservient to motion alone. Moreover, the two halves of the latter are closely approximated on the median line; but their ganglia are still usually bilobed. The multiplication of these ganglia is conformable to that of the segments; so that as many as five hundred double ganglia, repeating each other precisely in function and connections, are sometimes met with. The complex muscular apparatus of each segment is repeated in like manner.—Notwithstanding this complex organisation, we find even in the highest Annelida a remarkable power of regenerating portions of the body which have been cut off; and in certain species, which undergo spontaneous fission, the whole body is reproduced from the parts that are detached. This power is obviously related to the original mode of development of these animals, which, like the Tæniæ, come forth from the egg without the least rudiment of a body; the formation of this being subsequently commenced by a process of gemmation from the posterior part of the cephalic segment (which at first almost resembles a ciliated animalcule), and being continued by the subdivision of the last segment that has been developed. This process would seem to go on, in some species at least, without any precise limitation; and in some of the Nereids we find the same detachment of egg-bearing segments from the posterior part of the body, which we have seen in the Tænia (§ 309 *c*); the chief difference consisting in this, that whilst in the Tænia these segments are detached singly, and are unprovided with organs of locomotion, those now alluded to are detached in groups, each of them having a head and eyes, with the ordinary apparatus of motion, but their digestive apparatus being so imperfect, that these moving egg-sacs do not seem able to assimilate food for themselves. The resemblance of these Annelida to the Cestoid Worms, as regards their reproductive operation, is further borne out by the universal diffusion of their true generative apparatus; the sexes not being distinct, and both male and female organs being repeated in every segment except the cephalic, and each discharging its products by an orifice of its own.

311 *b*. The order *Tubicolæ* consists of Annelida which closely resemble the preceding in general structure, but which have the respiratory organs developed from the first or cephalic segment alone. The name of the order is derived from the fact, that the animals composing it are for the most part inhabitants of tubes; and to this habit their structure is most beautifully adapted, since it is only by such a transposition of the branchial tufts, that the blood of animals thus enclosed could be effectually aërated. The materials of these tubes, however, vary considerably. In the *Serpula* (Fig. 148), a true *shell* is formed, agreeing in all essential respects with that of Mollusca, and hardly in fact to be distinguished (when the animal has been removed) from that of the Gasteropod *Vermetus*. The tubes of Serpulæ are generally found clustering in masses, and attached to the surface of stones, shells, or other bodies which have been immersed for any length of time in the sea; they

are usually more or less contorted in form, varying in this respect according to the position in which they grow; but they are always closed at one end, which tapers to a point, whilst the wide open mouth gives exit to the head of the inhabitant with its branchial appendages. The *Sabella* appears to have some power of exuding calcareous matter from its surface, but strengthens its shell by the agglutination of particles of fine sand, clay, or mud to its exterior. The *Terebella* forms its tube entirely by the agglutination of grains of sand, of pieces of shell, &c.; and this tube is frequently branched near its open end, to form sheaths for the tentacula and for the gills, which last do not come off from the head itself, but from the segments immediately behind it. Some species of Terebella live in groups, so that their clustering together forms solid masses, which may go on increasing to a considerable size, so as even to constitute reefs across a shallow channel. The *Pectinaria* forms a tube of grains of sand, cemented together with the most beautiful regularity; but this tube is open at both ends, and is not attached; and the contained animal has the roaming propensities of the Errantia. It is perhaps a consequence of its locomotive powers, which give it a greater facility of *selection*, that the Pectinaria constructs its tube with so much more exactness than the other sand-worms. — All these animals are constructed upon the general plan of the preceding group, except as to the disposition of the respiratory organs. The segmentation of the body is less marked; but the segments are provided with setæ. The head is less distinctly separated from the body; and the eyes seem to be wanting. In the *Serpula*, the head bears, besides the branchial tufts, only one pair of filamentous appendages; and one of these is prolonged and dilated into a flat disc, which fits to the mouth of the shell, and serves to close it when the animal is withdrawn into the tube (Fig. 148); whilst the other remains of small dimensions. In the *Terebella*, on the other hand, the head is furnished with numerous tentacular appendages, which are capable of wide extension, and which probably serve for the prehension of food. The mouth of these animals seldom possesses the

Fig. 148.

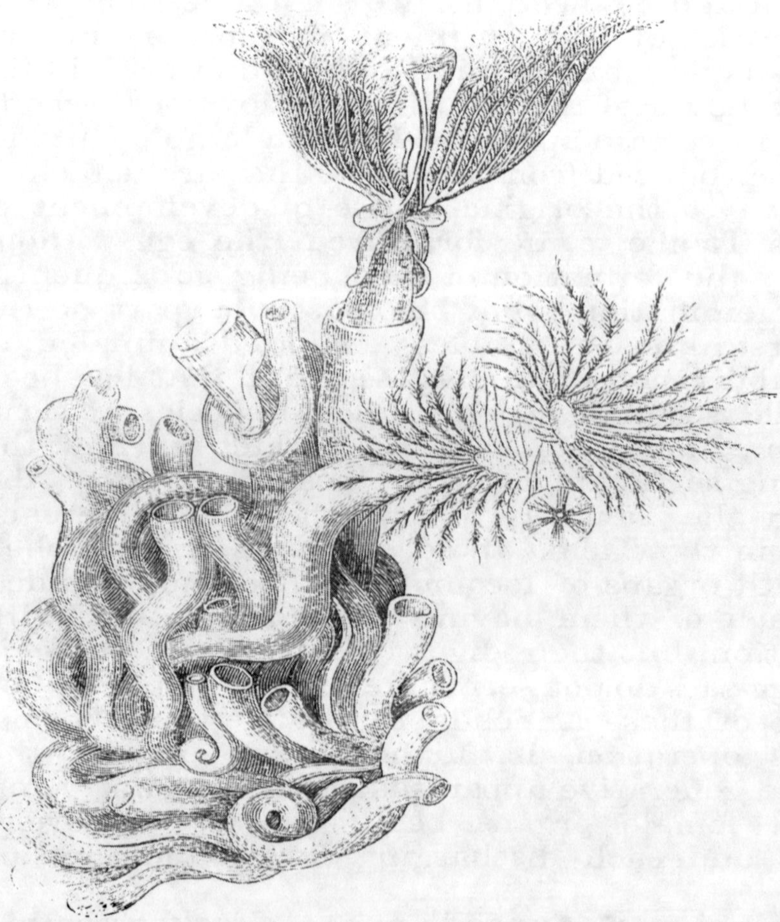

Cluster of shells of *Serpula contortuplicata*, with the animals of two of them expanded, showing their branchial tufts.

complex armature of that of the Nereids; the character of the alimentary canal, however, is for the most part the same as in that group, except that we usually find the secreting cœca replaced by simpler sacculi. The changed position of the respiratory organs involves, of course, a corresponding alteration in the arrangement of the circulating apparatus; and we find that the pulsating trunk at the anterior and dorsal portion of the body is here no longer destined to propel the blood to the system alone, leaving it for the multiple branchial hearts to urge it through the respiratory organs; but that the 'dorsal vessel' takes upon itself this function, and carries on the circulation through the tentacula and the branchiæ with great energy.—The close alliance between this group and the preceding is shown by the fact, that the early embryoes of the *Tubicolæ* go through exactly the same phases of development as those of the *Errantia*, and cannot be distinguished from them either in structure or mode of existence; those modifications which adapt the *Tubicolæ* to their peculiar habit of life, manifesting themselves only when their embryoes are approaching their adult dimensions, and consisting chiefly in the unequal development of certain segments and of their appendages, of which some are almost abortive, whilst others are evolved to a most disproportionate degree.*

311*c*. The foregoing orders are considered by M. Milne-Edwards as properly constituting the true Annelida; those which remain, namely the *Terricolæ* or Earth-worms, and the *Suctoria* or Leeches, being aberrant forms which serve to connect the Annelidan type with the Nematoid worms (§ 309 *e*) on the one hand, and with the Trematoda (§ 309 *d*) on the other. —The *Terricolæ*, like the Nematoidea, have a cylindrical body, tapering to a point at both ends, and without well-marked distinct segmentation; the head is not distinct from the trunk; and there are neither eyes, antennæ, mandibles, cirrhi, nor tubercular feet. The order includes two principal groups; one of which is terrestrial, while the other is aquatic. Of the former, the common *Lumbricus* or 'earth-worm' may be taken as the type; whilst of the latter, the *Arenicola* or 'lob-worm' is an example.† The Lumbricus has no external appendages, save a few setæ or bristles on each segment, which are scarcely visible to the naked eye, but which may be distinguished by the resistance they make when the finger is passed along the body from behind forwards; their points are directed backwards, in order to give the animal a firm hold of the earth through which it is boring. The mouth of the earth-worm has a short proboscis, but this is unprovided with teeth; its food, which consists of the decaying animal and vegetable particles mingled with soil, is conveyed by a short wide œsophagus to a sort of muscular gizzard; from which the wide intestine, constricted at intervals by the transverse septa of the common cavity of the body, so as to present somewhat of a sacculated appearance, passes direct to the posterior extremity. It contains within it, attached to its inner surface, a peculiar organ called the *typhlosole;* this is a long and slender tube, destitute of outlet, which appears to be formed by a duplicature of the intestinal wall; and its function is probably to receive the nutrient matter, which is filtered off into it (so to speak) from

* See Milne-Edwards, in Ann. des Sci. Nat. 3ième Ser. tom. iii.

† This animal has been commonly placed among the Dorsibranchiata on account of its external gills; but in its general plan of structure it agrees with the Terricolæ.

the coarse contents of the intestinal canal. The circulating system is complex in its distribution; but it is less provided with distinct organs of impulsion than that of the true Annelida. It essentially consists of a 'dorsal vessel' in which the blood moves forwards, and of a ventral trunk in which it moves backwards; but passing from one to the other, transversely to the body, are a set of beaded contractile vessels, one pair for each segment, each of which seems, like the multiple branchial hearts of the Nereids, to communicate to the blood a separate impulse, and thus maintains the circulation conjointly with the rest. The respiratory organs may be regarded as consisting of a series of sacculi opening by pores on the surface of the body, one pair being contained in each segment; these, at least, are certainly the rudiments of the internal respiratory apparatus in the higher air-breathing Articulata; but in the Earth-worm their function appears to be rather the secretion of mucus than the aëration of the blood, the latter operation being probably accomplished sufficiently by the general integument.—The nervous system of the Terricolæ is less developed than that of the true Annelida; and in particular the cephalic ganglia are much inferior in size (in accordance with the absence of sensory organs in the head), and are situated rather at the two sides of the œsophagus than above it. The muscular apparatus, however, is powerful; enabling the animal to contract its body in length, or to elongate it by contracting it in diameter, with considerable force.—Of the aquatic worms of this group, the *Arenicola* (Fig. 149) and its allies possess external gills (i^1 to i^{13}), and present other points of relationship to the Errantia; but they are formed in most respects upon the general plan of the Lumbricus, and agree with it in the history of its development, which (as we shall presently see) constitutes a marked distinction between the true Annelida and these aberrant orders. The Arenicola burrows in sand, as the Lumbricus burrows in earth, filling its digestive cavity with that substance, for the sake of the organic particles it may contain; and in the conformation of its proboscis (*a*), its dilated œsophagus (*d*), and the remainder of its alimentary canal (*f*, *g*) with its biliary cœca (*m*), it closely corresponds with the earth-worm. Its circulating apparatus, too, is constructed upon the same general type; the chief difference being that instead of a repetition of the beaded contractile vessels, passing from the dorsal to the ventral trunk in every segment, we have here one principal impelling cavity on each side at the commencement of the stomach (*n*, *n*), by the contractions of which the blood is sent backward along the ventral trunk, and through the branchial arteries (*r*) into the respiratory tufts, from which it returns by the branchial veins (*s*) into the dorsal vessel, and is by it carried forwards again towards the head and anterior part of the body. Here, too, there is a set of cœca opening on each side of the body, a pair for every segment; and these are obviously, as in the earth-worm, rather secretory than aërating organs, although homologous with the respiratory apparatus of the higher Articulata. In the *Nais* and its allies, there appears to be an absence of all indication of special respiratory organs; so that its blood must be aërated, as in the Nematoid worms to which it approximates very closely, by the integument alone. —The regenerative power in the Terricolæ is even greater than in the true Annelida. In several species of *Lumbricus* and *Nais,* as in the

Hydra and Planaria, when the body has been divided into a number of parts, each will in time develope itself into a complete worm; and in all it would appear that if the head or tail be cut off, the body will reproduce the lost organs. In the *Nais*, a spontaneous fission takes place; not merely for the separation and dispersion of the egg-masses (as in some of the Nereids), but for the production of perfect individuals. There is always, however, a highly-developed generative apparatus in these animals; but this is more localized than in the true Annelida, being confined to a small number of segments, and each set of organs having but a single external orifice. Although both male and female organs are present in each individual, they are not self-impregnating; but the congress of two individuals is requisite, the spermatic fluid of each fertilizing the ova of the other. This fertilization is effected by the passage of the male spermatic fluid, along the female generative canal, to the ova contained in the ovarium; these ova are not discharged, however, by this canal, but are set free into the cavity of the body by the bursting of the ovary, and are gradually propelled by the muscular contractions of the integument, to a receptacle near the anus. Here they usually undergo their early phases of development, and sometimes emerge from the egg-membrane before they escape from their parent-envelope. The eggs of the Arenicola are set free from the ovaria after a similar method. In whatever condition, however, the embryo is liberated from the parent, the course of its development is the same; resembling that of the Nematoidea, and differing from that of the true Annelida, in this important particular,—that the young, at the time of their emersion from the egg, have the worm-like form of their parents, instead of coming forth in the condition of ciliated Animalcules, and acquiring a longitudinal development and segmental division by progressive gemmation.

311 *d.* The order *Suctoria* must be considered as the lowest of the Annelidan class in its general organisation, which is adapted for a mode of life resembling that of the Entozoa. The group derives its name from the suckers which terminate the two extremities of the body, and which constitute its chief instruments of locomotion, these animals possessing neither bristles nor tuberculated feet; but the name is also appropriate as

FIG. 149.

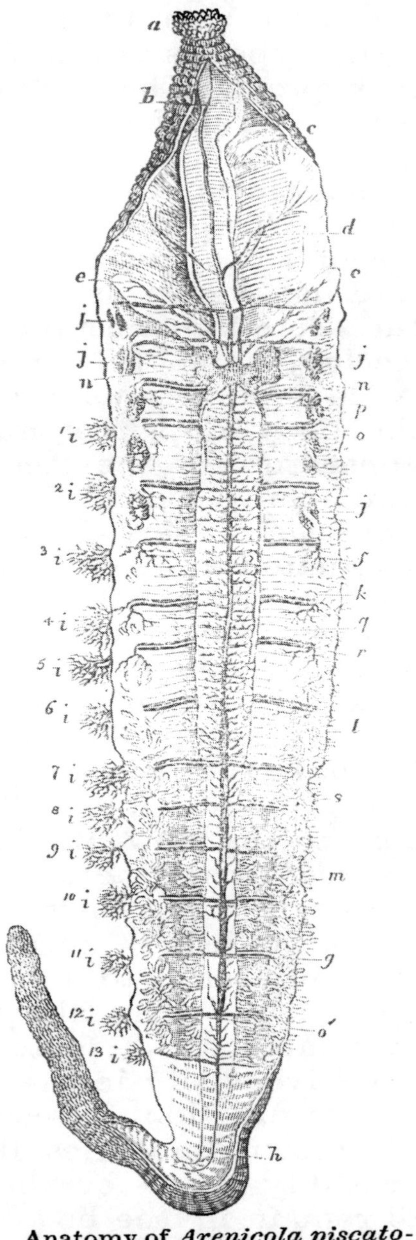

Anatomy of *Arenicola piscatorum* laid open and viewed from above:—*a*, proboscis; *b*, pharynx; *c*, retractor muscles; *d*, dilated œsophagus, or crop; *e*, cœcal appendages; *f*, stomach; *g*, intestine; *h*, muscular partitions surrounding the abdominal portion of the digestive tube; i^1 to i^{13}, thirteen pairs of branchiæ; *j*, organs of generation; *k*, setiferous tubercles, and their muscles; *l*, secreting cœca of the yellow matter exuded from the skin; *m*, secreting cœca (biliary?) surrounding the intestine; *n, n*, heart; *o*, dorsal vessel; *o'*, abdominal portion of the vessel; *p*, lateral intestinal vessels; *q*, subcutaneous vascular network; *r*, branchial arteries and veins; *s*, branchial veins returning to the dorsal vessel.

an indication of the method in which they obtain their food, which is for the most part by the suction of the juices of other animals, although certain species are enabled to swallow the entire bodies of those on which they prey. The body of the Leech is even less distinctly marked by segmental divisions than that of the earth-worm ; in fact, the indications of this division are scarcely more decided than they are in the higher Nematoidea, being chiefly seen in the transverse wrinkles of the integument. The number of these wrinkles, however, bears no constant correspondence to the marks of segmental repetition exhibited by the internal organs, as the alimentary canal, the ventral nervous column, and the secreting sacculi ; being five times as numerous in the fresh-water leeches, three times as numerous in some of the marine species, and twice as numerous in others. The head of the Leech is not distinct from the body, notwithstanding that it bears numerous ocelli and a complex apparatus for the laceration of the tissues from which it is to draw its nutriment ; indeed the two extremities of the body cannot be distinguished from each other without a minute inspection. Round the mouth, which is situated in the middle of the suctorial disk, are disposed, in a triradiate manner three little cartilaginous plates, usually called *teeth*, but more properly *jaws*. Each of these has two rows of minute teeth at its edge, so that it resembles a small semicircular saw ; and it is imbedded at its base in a bed of muscle, by the action of which it is worked in such a manner, as to make its teeth cut into the skin, a sawing movement being given to each piece separately, and a triradiate form being thus given to the whole incision. The lacerated character of the wound is very favourable to the flow of blood, and this is further promoted by the vacuum created by the action of the sucker. This apparatus of jaws and teeth, however, is not constant throughout the group. The sucking surface of the oral disk is furnished with eight or ten rudimentary eyes, which probably serve to indicate the proximity of a surface adapted to furnish nutriment. The alimentary canal is straight and wide, but is deeply sacculated at its posterior part, reminding us of the ramification of the digestive cavity in the Trematoda ; these sacculi would not seem destined to the function of secretion, since a distinct biliary apparatus, consisting of clustered follicles, lies upon the exterior of the intestine ; and they probably serve for the retention of the ingested blood, which appears to remain in the body of the Leech for weeks or months, the progress of digestion being extremely slow, notwithstanding the rapidity with which the canal is filled. The circulating apparatus consists of a dorsal vessel in which the blood moves forwards, and a ventral trunk in which it passes backwards ; but in addition to these there are two lateral vessels the course of the circulation in which has not been ascertained. None of these trunks present any distinct contractile enlargement for the propulsion of the blood ; and in the absence of such special organs, as well as in the multiplication of the principal trunks, the circulating system of the leech bears a close resemblance to that of the Entozoa. The secreting sacculi form a row on either side, as in the Earth-worm ; but there is no reason to regard them as having any participation in the respiratory function, which seems to be sufficiently performed by the soft integument. The greater number of this tribe are aquatic, though all of them can live for some time in air ; one small species, however, habitually lives on land, and inhabits the woods of Ceylon, attacking men and horses that are passing through them, with such voracity as to become a most troublesome pest; and there are other species which frequent moist situations near water, although seldom resorting to it. Some of the marine species have rudimentary locomotive appendages, like those of the true Annelida.—The nervous system of the Suctoria is formed upon the same general plan with that of the Terricolæ; the cephalic ganglia, however, meet above the œsophagus, and are larger in proportion, being the centres of the numerous ocelli; and the first subœsophageal ganglion, also, is of greater size than the rest, as from it the muscular apparatus of the jaws is supplied with motor power. The muscular fibres of the integument are disposed in three layers; the outer one being transverse; the middle one diagonal, and the innermost longitudinal. By the variety of action of which these are capable, Leeches are endowed with considerable locomotive power, being able not merely to crawl and swim like other Worms, but also to advance after a method of their own, which is accomplished as follows:—Having fixed its anterior extremity by means of the oral sucker, the animal doubles its body together, so as to bring the anal sucker into immediate proximity with the oral; and then, holding fast by the anal, it quits hold at the anterior extremity, extends its body to its utmost limit, and then takes a fresh attachment with the oral sucker at a point considerably in advance, to which the anal sucker is forthwith brought up; all this being accomplished with great rapidity.—It has not yet been determined how far the *Suctoria* correspond with the Terricolæ and Planariæ in their regenerative powers. Their sexual apparatus, however, is highly developed. The spermatic organs are found in pairs in several distinct segments; but their products are collected by two canals, which meet in a single orifice. The ovaria, however, are not thus multiplied, only one pair being developed; their two oviducts meet in a common receptacle, or uterus, of considerable size; and here the ova are retained, after being fertilised, so as to be extruded in one mass. This mass is enveloped in a sort of 'cocoon,' furnished by the mucous secretion from the glandulæ of the parent; and within this the eggs remain, until the embryo is ready to escape. The embryo, as in the previous order, acquires the form of the parent before its emersion, and subsequently undergoes no considerable change except in size.

311 *e.* Between the Suctorial Annelida and the Entozoa may be placed the very curious genus *Malacobdella*, which unites many of the characters of both groups. Its body is elongated, and cylindrical anteriorly like that of the Leech, but flattened posteriorly like that of a Planaria. There is a sucker at each extremity as in the Leech, and the mouth is situated in the centre of the oral sucker; but there are neither jaws nor eyes. The body is smooth, unmarked even by tranverse wrinkles; and there is not the least rudiment of appendages of any kind. The alimentary canal is comparatively simple in its structure, passing directly from one end of the body to the other, without the sacculi of the Leech, or the ramifying extensions of the Trematoda; and it has a distinct anal outlet. There are no rudiments of respiratory organs, or secreting sacculi. The widest departure, however, from the tribe of the Suctoria, is shown in the nervous system, which approaches that of the Trematoda. For instead of a single ganglion, formed by the approximation of the two cephalic

centres on the median line, the Malacobdella has the cephalic ganglia at the sides of the œsophagus widely separated from each other; whilst, instead of a single gangliated column, formed by the apposition of the two ventral cords on the median line, a slender cord, studded with minute ganglionic enlargements, passes backwards from each cephalic ganglion, these two cords being nearer to the sides of the body than to each other.* The genus may be thus regarded as the most degraded of the Suctoria, or as the most leech-like of the Trematoda, according as we attach most importance to the characters of the digestive apparatus, or to those of the nervous system.

312. The class MYRIAPODA occupies an intermediate position between the lower and the higher divisions of the Articulated series. It agrees with the Vermiform tribes in the longitudinal extension of the trunk, in the similarity of its segments from one extremity to the other, and also (as regards the lower forms of the class) in the cylindrical form of the body. On the other hand, in the presence of distinctly articulated members, in the possession of more complete eyes, in the adaptation of the respiratory apparatus for the introduction of air into the interior of their bodies, and in the general elaborateness of their structure, the Myriapoda agree with Insects, with which some Naturalists consider that they ought to be ranked. They differ from true Insects, however, not merely in the absence of wings, but in the great multiplication of the segments and their appendages, and in the absence of any division of the trunk into thoracic and abdominal regions; and they differ still more, as will hereafter appear, in the history of their development. They bear, however, a considerable resemblance to certain Crustacea, which may almost be regarded as Myriapoda adapted for aquatic respiration.—The covering of the bodies of these animals is firm and of a somewhat horny character, resembling that of many Insects. The division into segments is very distinct, a flexible membrane being interposed between each pair of firm rings or plates. The legs and other appendages are enclosed in the same kind of integument, and their joints are formed in the same manner as those of the body. We find in this class, however, two distinct types of conformation, of which one approximates most nearly to the Vermiform tribes, and the other to that of the higher Articulata. In the former, of which the *Iulus* (Gally-worm) may be taken as an example (Fig. 150), the body is generally cylindrical, or nearly so, the number of segments is considerable, and most of them bear two pairs of thread-like legs, so that the number of these members sometimes amounts to 160 pairs. The legs are very imperfectly developed, being scarcely large or strong enough to sustain the weight of the body, and their articulations being indistinct; and the animal seems rather to glide or crawl with their assistance, like a serpent or a worm, than to use them as its proper instruments of locomotion. This kind of movement is facilitated in some species by the incomplete enclosure of the body in the consolidated integument; for this merely forms plates above and below, which are connected at the sides by soft membrane; so that the trunk can be easily flexed in any direction. When at rest, the body is rolled up in a spiral form; so that the legs, concealed in the concavity of the spire, are protected from

* See Blanchard, in Ann. des Sci. Nat. 3me Ser. tom. iv.

injury. These animals do not move with rapidity, and they chiefly feed upon decomposing organic matter. In the higher division, on the other hand, of which the *Scolopendra* (Centipede) may be taken as the type (Fig. 137), the body is flattened, and each segment is completely enclosed in its horny envelope; the number of segments is not very great, never exceeding twenty-two, and being sometimes as low as twelve; and each

FIG. 150.

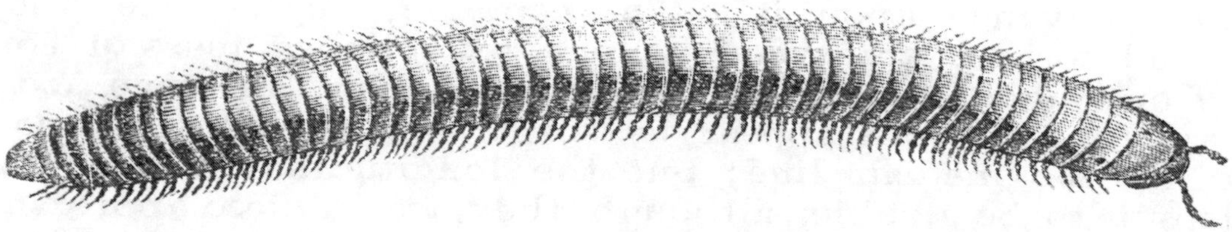

Iulus.

segment bears a single pair of well-developed legs, on which these animals can run with considerable rapidity. Still their bodies are possessed of considerable flexibility; and they are thus enabled to wind their way with facility through very narrow and tortuous passages, in search of the Insects, &c., which constitute their food.—In both orders, the first segment, or head, is furnished with numerous eyes on each side, and also with a pair of jointed antennæ; the mouth is adapted for mastication, being furnished with a pair of powerful cutting jaws; and it is also provided, in the Centipede and its allies, with a pair of appendages, formed by a metamorphosis of the legs of the first segment of the body, which are adapted not merely to hold and to tear the prey, but to convey poison into the wounds thus made, this poison being ejected through a minute aperture near their points.

312 *a*. The alimentary canal of the Myriapods for the most part exhibits a division into œsophagus, stomach, and intestine; the first of these is long and narrow in the carnivorous species, but is broader, and even dilated into a sort of crop, in those which devour decaying vegetable matter. The stomach usually possesses distinct muscular walls, and is sometimes lined by a plaited horny membrane, so as to constitute a gizzard; the intestine is straight and wide, but its surface is increased by longitudinal foldings, and it is somewhat sacculated by transverse constrictions. In connection with various parts of the digestive apparatus, we find a small number of long tubular cœca; of these the upper ones appear to be salivary glandulæ; the middle ones, which open near the stomach, to be hepatic; and the inferior, which discharge their product near the anal outlet, to be urinary. The circulation is chiefly carried on by means of a 'dorsal vessel,' which now exhibits a distinct segmental division and contractile power; each division may be considered as the heart or impelling cavity of the segment to which it belongs; but the entire series acts in succession, propelling a current of blood from behind forwards, which is then distributed to the body and respiratory organs, whence it returns to the posterior extremity of the multiple heart. The respiratory organs of the higher Myriapods consist of air-vessels or *tracheæ*, resembling those of Insects; these ramify through the body, commencing from the *stigmata* or breathing pores, of which a pair is usually found in

each segment, at the sides or on the under surface of the body; the tracheæ do not form one entire system, however, as in Insects, those of the different segments having little communication with each other. The worm-like *Iulus*, on the other hand, has a series of air-sacs resembling the sacculi of the Leech and Earth-worm, and possessing, like these, a secretory function; an offensive fluid being exhaled from them, apparently for the protection of the animal.—The nervous system of these animals is constructed upon the same type as that of the higher Annelida, but exhibits a considerable advance upon it. The cephalic ganglia meet above the œsophagus, and form a transversely-elongated bilobed mass of considerable size, from which nerves proceed to the eyes, antennæ, &c. The ganglia of the ventral cord appear as single masses, by their complete coalescence on the median line; but the longitudinal strands may be easily made out to be double, although they are in close approximation. As in the Annelida, these ganglia are of very uniform size from one extremity of the body to the other; except that the first sub-œsophageal ganglion, from which the masticating apparatus is supplied with motor power, is larger than the rest. The system of respiratory ganglia and nerves, formed by the repetition of the respiratory ganglion of the Mollusca (as the principal ventral cord is formed by a repetition of their pedal or locomotive ganglion), here becomes very distinct; and the separate *stomato-gastric* system, of which traces were found in the Annelida, is more fully developed. The muscular apparatus is highly elaborated; no longer consisting of mere layers of fibres, but being made up of distinct muscles; each having its separate office: and the movements are more obviously under the direction and control of sensations, especially the visual. The eyes in the *Scolopendra*, as in the Nereids, are distinct from each other; but they are more numerous, of higher organisation, and clustered in closer proximity to each other. In the *Iulus* they coalesce into a composite mass on each side, closely resembling the composite eye of Insects and Crustacea. The antennæ are, without doubt, instruments of tactile exploration; and they may possibly be subservient also to the reception of olfactive or auditory impressions.—In regard to the structure of the generative apparatus, a remarkable difference exists between the lower and the higher tribes of Myriapods; for whilst in the former it is arranged upon the plan of that of Annelida, in the latter it more resembles that of Insects. The sexes, however, are always distinct, the two sets of organs never being united in the same individual. In the *Iulus*, both the male and female organs extend through a great part of the body; but instead of being separately repeated in each segment, they coalesce into continuous elongated sacs. In each sex, however, the generative opening is near the *anterior* part of the body. In the *Scolopendra*, on the other hand, the generative apparatus is entirely restricted to the posterior part of the body, where it occupies seven segments; it is remarkable, however, that in the structure of the spermatic organs there is more of segmental distinctness than in the *Iulus*. The history of the development of the embryo of Myriapoda presents a remarkable resemblance to that of the true Annelida; for the embryo, at the time of its emersion from the egg, possesses but a small number of segments; and these continue to increase, by the repeated subdivision of the penultimate segment, until the number characteristic of the species has been attained. The original number of segments rarely exceeds nine ; but it is sometimes multiplied, during the period of growth, to eight times that amount. In the *Iulus*, the larva at its emersion from the egg seems entirely destitute of legs, these members not making their appearance until after the first moult; they may be found, however, beneath the outer skin when this is stripped off. Whilst the progress of development is still going on, a considerable power of regenerating lost parts is exhibited by these animals, the antennæ and legs being reproduced after being cut off; but it does not appear that any portion of the body which has been separated can be renewed, or that it can itself regenerate the other parts ; and the reparative power appears to be lost, as soon as the animals have attained their full development.

313. In the highest division of the Articulated series, which is chiefly made up of the three principal groups of *Insecta*, *Crustacea*, and *Arachnida*, we find the general tendency, of which we have seen an indication in the superior Myriapoda,—that, namely, to concentration and specialization,—more completely carried out. The number of segments in the body is considerably reduced ; they are more intimately united one with another ; they no longer present the same repetition of similar parts, but differ considerably in their respective endowments ; the motion of the body is almost entirely delegated to the appendages ; the number of which is reduced, whilst a greater variety of conformation is observable among them ; and the internal organs exhibit a more or less complete restriction to particular divisions of the body, instead of being uniformly diffused through the whole series. It is peculiar to the history of the development of the animals included in the three groups just named, that at the time of the emersion of the embryo from the egg, it has nearly or completely attained the number of segments characteristic of its perfect state ; and that in the changes which it may subsequently undergo, therefore, there is little or no multiplication of segments, but rather a higher elaboration of the structure of some or of the whole of them. Although presenting this general accordance, however, in regard to their *nisus* of development, these classes are constructed upon very different types, and each of them includes an extensive range of forms ; so that it is difficult to say which should be considered as the most elevated, and which as the lowest, in the series. The animals of two of these classes, namely, Insects and Spiders, are atmospheric in their respiration ; breathing by means of air introduced into passages or cavities in the interior of their bodies. But between these two there is a considerable difference. The body of the Insect, in its perfect form, manifests a division into three distinct parts, head, thorax, and abdomen ; in the Spider, on the other hand, the head is united to the thorax, forming a large 'cephalothorax.' The Insect in its perfect state never possesses more than six legs ; whilst the animals of the Spider tribe have never fewer than eight. The Insect, in its typical form, possesses also two pairs of wings ; and of these not a rudiment exists among the Spiders or their allies. The Insect usually comes forth from the egg in the condition of an Annelide, and only attains its characteristic form and structure after a series of metamorphoses, in which its original constitution is greatly modified ; in the greater part of the Spider tribe, the young is completely formed at the time of its emersion, and in none does it undergo any real metamorphosis.—On the other hand, the class of

Crustacea includes a more varied series of forms; some of which may be regarded as the aquatic representatives of Insects and Spiders, and agree with them in the general elevation of their plan of structure; others would seem, in like manner, to represent the Myriapods; whilst in others we find a very inferior grade of organisation and a much wider departure from the characters of the class, so that the true affinities of these animals are chiefly to be determined by the history of their development. Thus it has been ascertained that certain reputed *Entozoa*,* and a group which has received the designation of *Epizoa*, must be considered as aberrant Crustacea; whilst even the *Cirrhipoda*, whose claim to a place in the Articulated series was long overlooked, must be regarded as finding their true place on the borders, if not actually within the limits, of the same class. —In many parts of their organisation, the highest Crustacea and Arachnida approach the Vertebrated series much more closely than Insects do; especially in the concentration of their nervous centres, of their circulating apparatus, and of their respiratory organs, and (in the Crustacea) in the high development of the liver. But, on the other hand, we find in Insects the highest perfection of the Articulated type of structure; the characters which are most peculiar to the series, taken as a whole, being most completely developed in that class. Thus it is in Insects that we find the greatest activity of locomotion, the largest predominance of the nervo-muscular apparatus over the nutritive, the most acute sensations, and the most extraordinary display of instinctive power, unmodified by intelligence. And it is in Insects, as already stated (§ 281), that we find the greatest multiplication of distinct forms; the number of species in this class being probably at least twenty times that of all the other classes of Animals put together.

314. In considering the general characters of the class of INSECTS, we find, that, putting aside these differences in the conformation of the mouth, which serve as the chief foundation for the subdivision of the class into orders, there is on the whole a remarkable conformity to one general type. The tegumentary skeleton of Insects, which in some tribes (especially the *Coleoptera* or beetles) possesses a remarkable thickness and resisting power, is composed of epidermic cells, consolidated partly by an albuminous deposit, but still more by a peculiar horny substance, to which the distinctive appellation of *chitine* has been given. This differs from ordinary horn in not being soluble in caustic potass, and in not being softened by heat; so that parts composed of it retain their form even after being incinerated. With this substance a small quantity of phosphate of lime appears to be usually blended. The regular number of segments is thirteen; the first being the head; the thorax being constituted of the three following; and the remaining nine forming the abdomen. Any departure from this number is on the side of diminution rather than of excess, the last segment, or even two segments, being sometimes imperfectly developed, and not distinctly separated from the preceding; but where this is the case, the last ganglion of the ventral cord is of unusual size, and is sometimes bilobed, as if formed by the coalescence of two distinct centres.—The cephalic segment bears a pair of jointed appendages, the *antennæ*, which are

* The *Acanthotheca* have been recently proved by M. Van Beneden to be Crustaceans allied to *Lernæa*. See Ann. des Sciences Nat., Août, 1849, p. 67.

frequently of considerable length, and which present a great variety of form.—From the thoracic region are developed the proper locomotive appendages; each of its three component segments having its pair of legs, and the second and third segments each bearing a pair of wings in addition. The legs consist of several pieces, which, from the analogy they have been suppósed to present to the limbs of Vertebrated animals, have received the names of *coxa* or 'hip', *femur* or 'thigh', *tibia* or 'shank', and *tarsus* or 'foot',—terms, however, which are by no means strictly applicable. The wings have not the least analogy, except in

FIG. 151.

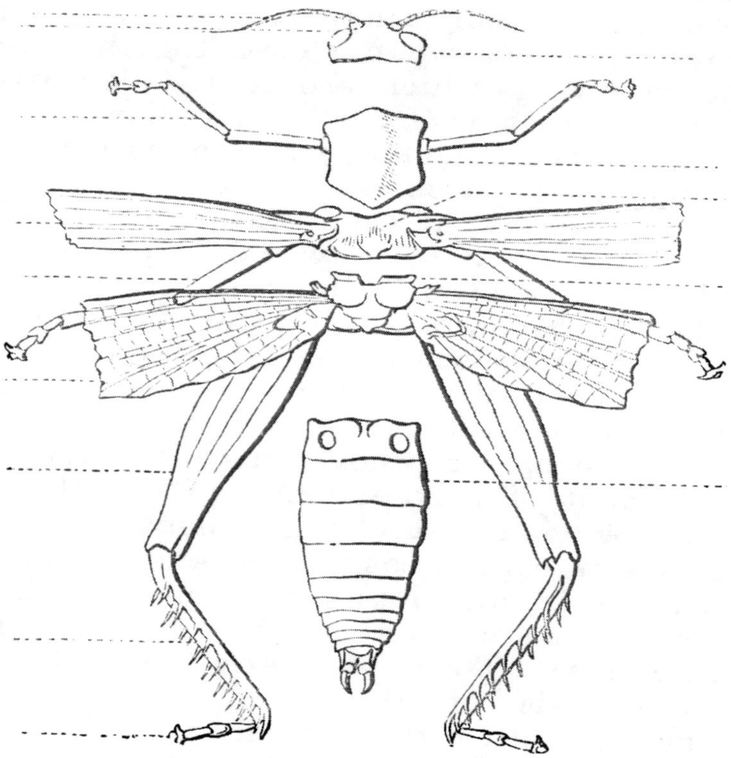

External skeleton of an Insect.

function, to the wings of Vertebrata,—these last (when adapted for actual flight) being formed by a modification of the anterior pair of limbs for the support of an expanded surface,—but are, in fact, extensions of the superficial tegumentary membrane (which is usually very thin and transparent) over a framework formed by an extension of the denser portion of the integument; and this framework is penetrated, throughout its ramifications, by 'tracheæ' or air-tubes, communicating with those of the interior of the body. The essential structure of the wings, in fact, corresponds closely with that of the external respiratory organs with which the aquatic larvæ of certain Insects (as the *Libellula*, or 'dragon-fly') are provided: and they have hence been not inappropriately designated by Oken as 'aërial gills,' a term of which the applicability becomes more apparent, when the method in which the first expansion of the wings is accomplished (§ 315 *g*) is understood. The texture of the wings varies in different orders; but the variation depends chiefly upon the degree in which the denser layer of the integument of the body is continued into the wings, and in which their surface is covered by epidermic scales. Thus in the *Neuroptera* and *Hymenoptera*, the wings are composed of two layers of an extremely thin membrane, extended over a light framework of the horny tissue with which the body is enclosed; and their surface is nearly smooth, bearing only a few scattered hairs. But in the *Lepidoptera*, the wings, themselves of similar texture, are covered with a layer of epidermic cells or 'scales,' arranged in an imbricated manner; and it is to these scales that the colours of the wings of butterflies, moths, &c., are entirely due. In the *Coleoptera*, the wings of the posterior pair are simply membranous; but those of the anterior are entirely consolidated by the amount of horny tissue they contain, and are converted into *elytra* or 'wing-covers', not being employed in flight, but being only of sufficient size to cover the back when closed together, and thus serving as protectors of the true wings which are folded beneath them. In the orders *Orthoptera*, *Hemiptera*, and *Homoptera*, we have various intermediate gradations between these two extremes; one or both pairs of wings being sometimes of a coriaceous or parchmenty texture throughout; whilst in other cases, the wings are dense at their base and membranous towards their extremities. In the *Diptera*, only one pair (the anterior) of these organs is developed; the posterior being represented by a pair of small club-shaped appendages, which are termed the 'balancers.' In nearly all the orders of Insects, certain species are to be met with, in which one or both sexes are *apterous* or destitute of wings; but there is a considerable group, made up chiefly of parasitic insects (such as the Flea, Louse, &c.) in which the absence of wings is a constant and important character.—The relative development of the segments of the thorax bears a close relation to that of the wings; and it is frequently so unequal, that the regular segmental division cannot easily be discovered, this being further obscured by the partial or complete union of one piece with another, whereby a more secure attachment is given to the powerful muscles that move the legs and wings. Thus, in many cases the first segment is extended so far backwards, on the dorsal surface of the thorax, as completely to cover-in the other two; so that these can only be recognised by examining the ventral surface, where they may be distinguished by observing the intervals between the origins of the three pairs of legs. —The segmentation of the abdomen, however, is usually very regular; the divisions of the tegumentary skeleton being simple rings, which differ in little else than in size. The chief exception is in regard to the last segment, which is frequently modified in its development for particular purposes. These abdominal segments, in the perfect Insect, are never furnished with the least rudiments of locomotive appendages.

315. There is an immense variety in the class of Insects, as to the mode in which food is taken into the mouth; but we may take the oral apparatus of the *Coleoptera* and the *Lepidoptera* as presenting us with its two principal types, which are designated as the *mandibulate* and the *haustellate* respectively. In the Coleoptera and other 'mandibulate' insects, we find an apparatus adapted for the division and prehension of solid substances; and this consists of two pairs of jaws opening laterally, the one above and the other below the entrance to the cavity of the mouth, the former being designated as the *mandibles*, and the latter as the *maxillæ;* with an upper lip or *labrum*, and a lower lip or *labium*.

This labium is elsewhere seen to be composed of a pair of elongated pieces which constitute the *tongue*, and of a central piece on which these are supported, which is termed the *mentum* or chin. The maxillæ usually bear one or two pairs of small jointed appendages, termed *palpi*; and another pair of palpi is borne by the labium. These palpi appear to be sensory organs for taking cognizance of the food when it is near the mouth; and they also serve to hold the food between the mandibles, whilst these divide it. The maxillæ are sometimes enormously developed, forming a large pair of pincers on the top of the head, as in the Stag-Beetles.—The 'haustellate' Insects, on the other hand, are adapted to imbibe their food by suction; and where, as is the case with the Lepidoptera, there is no occasion for any apparatus for the division of the tissues whose fluids are to be drawn up (these insects imbibing the juices of flowers which are already poured forth), we find the most complete deficiency of all the parts of the mouth, save those which enter into the formation of the *haustellium* or 'proboscis.' This organ is in reality composed of the two maxillæ, which are excessively prolonged, and united on the median line; and each maxilla being channelled on its internal margin, the union of the edges of the two channels forms a complete tube. At the base of the proboscis is seen a small membranous piece, which is the representative of the labrum; and on each side a small tubercle, which is the only vestige of the mandible. Rudiments of maxillary palpi are also discoverable; whilst behind the proboscis is a small triangular labium, having two very large labial palpi, composed of three joints, and nearly always clothed with scales. In the *Hymenoptera* (or Bee and Wasp tribe) we find the mouth presenting a curious adaptation of the mandibulate type to the suctorial mode of obtaining the food. The labrum and the mandibles are constructed as in mandibulate insects; and the mandibles serve as the instruments with which the remarkable fabrics of these animals are constructed, or with which the insects, &c., are killed, whose juices serve for their food. The maxillæ and the tongue, however, are greatly elongated; the former being so shaped as to inclose the latter and form a tube when they are brought together, but not being adherent along the line of junction; and the tongue having a pair of lateral appendages of its own. The labial palpi, also, are of unusual size. In the suctorial *Hemiptera* (Bugs, Cicadæ, &c.), the mandibles and maxillæ are converted into lancet-shaped organs, which serve to puncture the vegetable or animal tissues whence these insects imbibe the juices on which they live; whilst it is through the grooved labium, with which the labial palpi seem to have coalesced, that these juices are drawn into the mouth.—Such are some of the principal methods in which Insects obtain their food; the more detailed scrutiny of them would still further demonstrate the wonderful variety, which is produced by different modes of development and combination of the same elementary organs.

315 *a*. The structure of the alimentary canal presents, as might be expected, a similar variety; its most complex forms, however, are found in the mandibulate insects, especially such as live on coarse vegetable food; whilst its simplest condition is seen in those which live upon animal juices. In the former, the mouth is furnished with salivary glands, whose secretion assists in the act of mastication; the œsophagus is generally wide, and is frequently dilated into a 'crop;' from this, the

food passes into a muscular stomach or gizzard, the walls of which are frequently armed with horny plates or teeth, which assist in the comminution of the contents; and below this is the true digestive stomach (Fig. 152, *d*), whose texture is soft and delicate, and which often has a number of minute follicles (for the elaboration of the gastric secretion) projecting from its walls. From the stomach proceeds the intestinal tube (*e*), which is usually of much smaller diameter than the preceding portions of the canal, and which is frequently much convoluted; this receives, close to its commencement, the terminations of the secreting

Fig. 152.

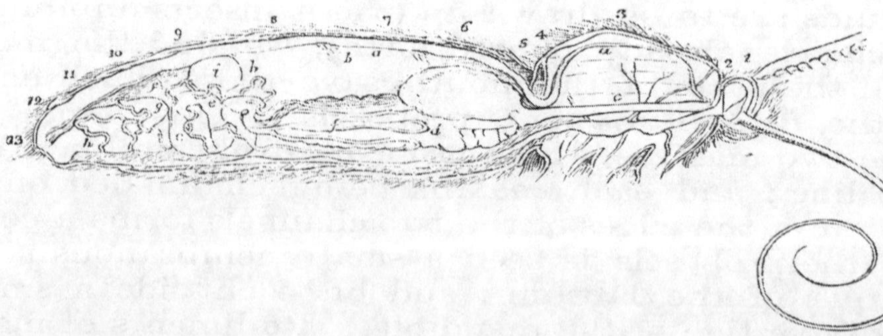

Ideal section of *Sphinx ligustri* (Privet Hawk-moth), to show the relative position of its organs:—1-13, the successive segments; *a, a*, dorsal vessel; *b, b*, ligamentous bands holding it in its place; *c*, œsophagus; *d*, stomach; *e*, intestinal canal; *f*, biliary tubuli; *g*, cœcum; *h*, cloaca; *i*, ovary;—the gangliated nervous cord is seen to commence in the large cephalic ganglia beneath 1, and to pass backwards along the floor of the thorax and abdomen.

tubuli (*f*), which constitute the only rudiment of a liver possessed by these animals; whilst, lower down, it sometimes receives another set of tubuli, which seem to be urinary in their character. Not unfrequently the intestine dilates in its course into a cœcum (*g*); its anal orifice is always at the posterior extremity of the body, where we find a sort of cloaca (*h*), which receives the termination of the intestine, and is also the outlet of the generative organs.—The circulation of the blood in Insects is carried on in part by means of distinct vessels, and in part by channels excavated in the tissues. Its central organ is the 'dorsal vessel' (*a, b*), which, as in Myriapods, is segmentally divided; the compartments are separated by valves, which do not allow the blood to pass in any other direction than from behind forward. This segmental division, however, in the perfect insect, does not extend into the thorax; the dorsal vessel in that region being converted into an aortic trunk, which carries the blood onwards, that it may be distributed to the head and thoracic appendages. From these it returns backwards along the limbs and body, to re-enter the dorsal vessel, either by veins which open into its several chambers, or by larger vessels that collect the whole to convey it into the posterior chamber. In its course, however, it is brought into very close relation with the air that is conveyed through the whole interior of the body by the complex tracheal apparatus; for it appears from recent observations * that the blood not only bathes the exterior of the air-tubes, but moves through that space between

* Blanchard, in Ann. des Sci. Nat., Juin, 1848.

the outer and inner membranes, in which a spiral filament winds (as in the spiral vessels of Plants, § 151) to keep them from being closed by lateral pressure.—These air-tubes form a complex system, which is distributed with the most elaborate minuteness throughout the body, commencing from lateral *stigmata* or breathing-pores, of which each segment normally contains a pair, though some of them are frequently closed up, so that the number is considerably reduced; between all the parts of this system there is the freest communication; and in some parts of it, especially in Insects of most rapid flight, we find the air-tubes dilated into large air-sacs, which both serve as reservoirs of air, and contribute to diminish the specific gravity of the body. By this extra-ordinary development of the respiratory apparatus, the apparent imperfection of the circulating system is compensated; since the chief demand for a very rapid movement of the blood, in animals which (like Birds) put forth a vast amount of muscular energy and activity, arises not so much out of the demand for nutrition, as from the necessity for a constant supply of oxygen to the tissues, which is here provided for by the penetration of the air itself into their substance.

315 *b*. The nervous system consists, as in other Articulata, of a series of ganglionic centres corresponding to the successive segments, united by a pair of fibrous cords, and disposed along the ventral surface of the body; but we no longer meet with the same uniformity which we have seen in the lower classes of this sub-kingdom. The cephalic centres here possess a much greater proportionate size, in accordance with the high development of the eyes and other sensory organs; they are placed entirely above the œsophagus (Fig. 152); and a cord of communication passes down on either side, to unite in the first infra-œsophageal ganglion, from which the ventral cord is continued backwards. The thoracic centres, from which the nerves of the legs and wings proceed, are much larger than those of the abdomen, which do not supply any locomotive organ; and the former are sometimes found to have coalesced into a single mass of large dimensions, whilst some of the latter seem altogether wanting. The respiratory and stomato-gastric systems of nerves, also, attain a higher development; the ganglia of the former being distinctly traceable along the ventral cord, intermediate between the locomotive centres; whilst those of the latter are found in front of and behind the great cephalic ganglia, and also among the viscera to which the nerves are distributed. There are also indications of the existence of a set of nerves comparable to the 'sympathetic system' of Vertebrated animals.—The development of the organs of sense, and especially of the visual apparatus, attains a much higher elevation in Insects than in any of the inferior classes. The clustered eyes of the Myriapods are here aggregated into large compound masses, usually of nearly hemispherical form, which are so large as to occupy a considerable part of the sides of the head. The structure of each individual eye (which is of nearly cylindrical form) seems most perfectly adapted to bring to a focus the rays which impinge upon it in the direction of its own axis; and by the mode in which the single eyes of each hemispheric mass are disposed, the range of vision is extended in every direction, although the eyes themselves are perfectly motionless. This multiplication of cylindrical eyes precisely similar to each other, to gain an end which is

answered in Vertebrated animals by a single globular eye, endowed with the power of motion, on either side, is in remarkable accordance with the general plan of structure characteristic of the Articulata (§ 308), and is sometimes carried to a most wonderful extent; the number of single eyes in the common House-Fly being 4000, in the Cabbage-Butterfly 17,000, in the Dragon-Fly 24,000, and in the Mordella Beetle 25,000. Besides the great compound eyes, most insects have a few minute simple eyes, disposed upon the top of the head, in the narrow space between the aggregate masses; and these appear to be of considerable use in directing their upward flight.—There can be no doubt that Insects are possessed of the sense of hearing; for although the precise organ which is subservient to it has not been determined, there is ample evidence that they are guided and influenced by sounds, one of the most striking indications of which is, that the males of many Insects (such as *Cicadæ*, *Crickets*, &c.) possess the power of emitting peculiar sounds, which attract the females to them. These sounds, however, are produced by purely mechanical means, and cannot be regarded as *vocal*. It would seem probable, from various considerations, that some part of the base of each antenna is the auditory organ.—There is similar evidence that Insects possess the sense of smell; and there is a strong probability that this is exercised by the soft surfaces of some of the appendages to the mouth. Thus, in some cases, the palpi are swollen at their extremities into club-shaped dilatations, composed of a soft membrane which shrivels when dry; and the insects may be seen to apply these to particles of food and liquid, as if taking cognizance of their odorous properties, before taking them into the mouth. In the common house-fly, the same use is made of the sheath of the proboscis. It is certain that other actions of insects are guided by the sense of smell; thus the flesh-fly deposits its eggs in the thick fleshy petals of the *Stapelia* (carrion-flower), deceived by its odour which resembles that of tainted meat.—The nicety of selection shown by many Insects in their choice of food, seems further to indicate that they are possessed of the sense of taste; how far the tongue is the instrument of this sense, however, remains to be ascertained.—As the enclosure of the body of Insects in a dense integument must prevent any considerable exercise of the sense of touch by the general surface, we find particular parts of it specially modified for the exercise of the tactile function. That such is the principal purpose of the antennæ, notwithstanding that they are themselves usually covered by an integument as firm as that of the rest of the structure, cannot for a moment be doubted by any one who watches the exploratory actions of these organs; the extremities of these antennæ are frequently furnished with plumes, tufts, or delicate laminated expansions, which must peculiarly fit them for such purposes; and they are always copiously supplied with nerves. In many cases, too, we find that the extremities of the feet are furnished with peculiar organs, such as soft cushions, or delicate expanded suckers, that are apparently adapted to receive tactile impressions from the surfaces to which they may be applied.

315 *c*. The muscular system of Insects (as already remarked, § 308 *c*) is highly developed; being entirely made up of the *striated* muscular fibre in its most perfect form (§ 223); and consisting not only of muscles for the contraction and elongation of the trunk by the approximation or

separation of its segments; but also of numerous muscles which give motion to the legs and wings (Fig. 138). In insects of rapid and powerful flight, these last are so highly developed as almost to fill the cavity of the thorax. The joints are for the most part so constructed, however, as to admit but of two kinds of movement, namely, flexion and extension; and the muscular apparatus has consequently not that variety of action which is seen where the 'ball-and-socket joint,' which permits movements of circumduction, takes the place of the simple 'hinge-joint.' Nevertheless there are no animals which surpass Insects in command over the organs of flight. Even the swallow, as Prof. Owen remarks, cannot match the dragon-fly; "this insect has been seen to outstrip and elude its swift pursuer of the feathered class: nay, it can do more in the air than any bird, it can fly backwards and sidelong, to right or left, as well as forwards, and alter its course on the instant without turning." When, moreover, we compare the actual space traversed by an Insect in a given time, with the dimensions of its body, we find that it vastly exceeds the similar ratio in the Bird; and thus we perceive that the locomotive powers of Insects are far higher than those of any other animals whatever. This power is most remarkably developed in the orders *Neuroptera* (Dragon-flies, Termites, &c.), and *Hymenoptera* (Bees, Ants, &c.); and it is remarkable that these are the very orders in which we find the most extraordinary manifestation of those *instinctive* tendencies, the high development of which, with an almost complete absence of intelligence, is a striking characteristic of the Articulated series in general (§ 308), and of the class of Insects in particular. These tendencies, as will be hereafter shown (CHAP. XX.), may be considered as dependent upon an association between sensory impressions and muscular movements, which arises from the original constitution of the nervous systems of these animals; and they may thus be regarded as necessities of their nature, not in the least indicative of intelligence, design, or voluntary choice on their own parts, but rather indicating the wise adaptation, by which they have been constructed to work out plans of most admirable elaborateness, with the most wonderful perfection. Now these and all other instinctive actions have for their object the maintenance of *animal* life; as distinguished, on the one hand, from the mere *organic* life of Plants; and, on the other, from the *mental* or *psychical* life of higher beings. And thus if we consider the Animal kingdom, as holding an intermediate position between the Vegetable world on the one side, and the domain of Mind on the other, we should be led to regard the class of INSECTS, and especially the orders Neuroptera and Hymenoptera, as its type.

315 *d*. Insects are endowed with extraordinary powers of multiplication; but this is accomplished, with only one exception, by means of the sexual process of generation. The exception referred to is that of the *Aphis*, which is capable of propagation by a process that appears to be analogous to the gemmation of the *Salpæ* (§ 299 *b*); the new individuals being budded off, so to speak, from internal stolons, instead of being developed from ova provided by the female and fertilized by the male. This method of propagation may be several times repeated; the individuals thus generated being all apparently of the female sex, and generating others like themselves. At the end of the season, however, perfect winged males and females are developed; and these concur to

produce ova, which retain their vitality through the winter, and give birth to a new generation in the spring, long after the parents have perished. Each viviparous Aphis is capable of producing a hundred repetitions of itself; and as the process may be repeated ten times in the season, it has been calculated that, if no destructive agencies were at work, the last brood would number 1,000,000,000,000,000,000 individuals. The conditions under which this curious phenomenon takes place will be more particularly examined hereafter (CHAP. XVIII.); at present we must be satisfied with noticing it as a remarkable fact, that such an extraordinary power should be confined (so far as is yet known) to a single genus among Insects, no traces of it having been elsewhere observed; and that a method of propagation which is restricted to the lower classes in the Radiated and Molluscous sub-kingdoms, should be met with at so high a place in the Articulated. The sexes are distinct in all Insects; and the males and females are frequently distinguished by diversities in size, configuration, or colour, as well as by the difference in their generative organs. The spermatic glands of the male present numerous varieties in arrangement, in the different tribes of Insects; being sometimes united together on the median line, and forming but a single globular mass; sometimes elongated and double; and sometimes multiplied on each side. Whether they be single or multiple, however, their product is at last poured into a single seminiferous duct, from which it is ejected, through a penis or intromittent organ (which is a modification of the one or two last segments of the abdomen), into the female organs. The ovaries of the female present as many varieties in their conformation, as do the testes of the male; in general, however, they consist of one or more cæcal tubuli on each side, in which the eggs are found lying in regular rows, those nearest the outlet of the tube being in a state of most advanced development. These tubuli are often extremely elongated, and are laid in coils which are closely approximated to each other. In many instances, the masses of eggs reach an enormous size, which is to be especially seen where the sexual organs are only developed in a few perfect females, the others remaining sterile, and acting as 'workers,' which is the case with Bees, Wasps, Ants, &c.; the most remarkable example of this is probably the female *Termes* (white-ant), whose abdomen, at the time she commences laying, is from 1500 to 2000 times larger than the rest of the body, the eggs being subsequently deposited at the rate of 80,000 a day, for several successive weeks. The oviducts of the two sides unite into one canal on the median line; and on this we find, near its outlet, a vesicular dilatation for the reception and retention of the spermatic fluid, which is thus brought into contact with the eggs as they are successively extruded. The last segment of the body of the female is usually modified to perform the functions of an ovipositor; and this, in some tribes of insects (especially the *Ichneumonidæ*), is prolonged to a length considerably exceeding that of the animal's body, thus enabling it to lay its eggs in situations to which it could not itself gain access. The ova are usually deposited whilst the embryo is as yet very immature; and frequently they remain in this condition during a period of many weeks, months, or even years (§§ 96 and 116). In a few instances, however, the eggs are retained until their included embryoes are ready to escape, so that these are sent forth alive into the world, either in the *larval* stage of development, as is the case with the common *Musca vomitoria* (flesh-fly), or in the condition of *pupæ*, as is seen in the *Hippobosca* (forest-fly).

315 *e*. No Insects come forth from the egg in their perfect condition; and their state in many cases at the time of emersion is quite embryonic; so that it is usually not until a series of very considerable changes have taken place in external configuration and internal structure, together constituting what is known as the 'metamorphosis,' that the complete development of the specific type is attained. The amount of this metamorphosis, and the mode in which it is accomplished, vary considerably in the different orders of insects; but three stages are usually marked out, more or less distinctly, in the life of each individual. The term *Larva*, in the ordinary language of Entomology, is applied to the insect from the date of its emersion from the egg, up to the time when the wings begin to appear; the term *Pupa* is in like manner employed to mark the period during which it is acquiring wings; and from the time when these and other organs characteristic of its perfect state are completed, it is spoken of as the *Imago*. The grade of development, however, at which the insect comes forth from the egg, is very different in the several orders and families; and it is consequently very unphilosophical to associate under the same designation beings which are in conditions essentially diverse. In all cases the embryonic mass within the egg is first converted into a footless worm, resembling the higher Entozoa or the inferior Annelida in its general organisation, but possessing the number of segments—thirteen—which is typical of the class of Insects. Such, in the *Diptera* and *Hymenoptera*, and in some of the *Coleoptera*, is the condition of the larva at the time of its emersion from the egg; and it is remarkable that many of the larvæ of the first of these groups resemble Entozoa in their parasitic habits. The head, in larvæ of this kind (which are familiarly known as 'maggots'), differs but little from the segments of the body; the eyes in many instances not being developed, and the mouth being furnished with a mere suctorial disk. In the *Lepidoptera* and most of the *Coleoptera*, however, the larva at the time of its emersion possesses the rudiments of the three pairs of thoracic legs, although they are little else than simple claws (Fig. 153), save in the carnivorous Beetles; whilst in addition to these, several of the abdominal segments are furnished with fleshy tubercles or pro-legs (generally to the number of four or five pairs), which are peculiar to the larva-state. In such larvæ (which are commonly designated as 'caterpillars') we observe a remarkable equality in the different segments, both as to size, form, and plan of construction, which strongly reminds us of the Annelida. The alimentary canal occupies nearly the whole of the cavity of the body, and passes without flexure from one end of it to the other. The compartments of the dorsal vessel, the respiratory organs, the nervous centres, and the muscular bands, are repeated with great regularity; and there is as yet no distinction between the thoracic and abdominal portions of the trunk (Fig. 153). The head, however, is usually protected by a horny covering, and is provided with simple or clustered eyes like those of the higher Annelida and Myriapoda; and the mouth is furnished with powerful cutting jaws for the division of the food, which is usually vegetable in its nature. In the Orthopterous and Hemipterous orders, however, these

stages of development are passed through within the egg; and as the young Insect does not emerge thence until it has attained a higher grade, in which it presents a close resemblance to its parents in almost every particular save the want of wings, it cannot be regarded as having the characteristics of a real *larva*. This is the case, too, with some of the *Coleoptera*, in which order we find a considerable variety as regards the stage of development at which the embryo quits the ovum.—In the true Larva condition, the whole energy seems concentrated upon the nutritive functions; the quantity of food devoured is enormous; and the increase in the bulk of the body is very rapid. According to Lyonnet, the comparative weight of the full-grown caterpillar of the Goat-Moth, and that of the young one on its emersion from the egg, is as 70,000 to 1; and the maggot of the Flesh-fly is said to increase in weight 200 times in the course of twenty-four hours. During this rapid growth, the caterpillar throws off and renews its epidermis several times; but the larvæ of the Hymenoptera and Diptera do not undergo this exuviation until they pass into the pupa state, their integument being soft enough to yield to the distension from within. The sexual organs are but little developed during the larval life; but their rudiments may be detected. The activity of *growth*, however, seems to supersede in the larva the progress of *development*; for the tissues remain in an embryonic state, and the organs retain their original condition with little or no essential change except in size, until a sufficient store of nutriment has been taken into the system, to serve as the *pabulum* for all the subsequent developmental operations, by which the fabric of the perfect Insect is to be completed.

FIG. 153.

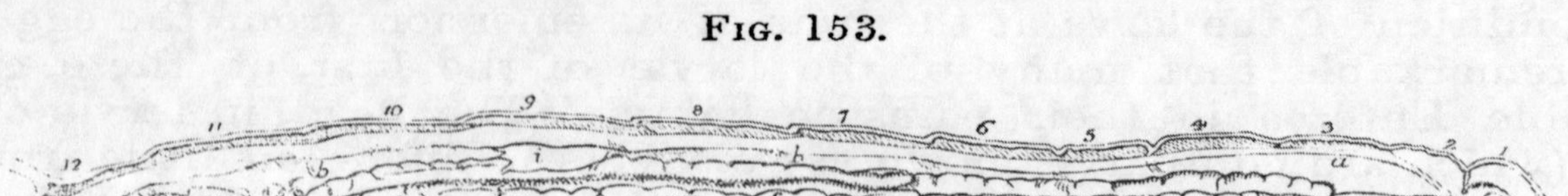

FIG. 154.

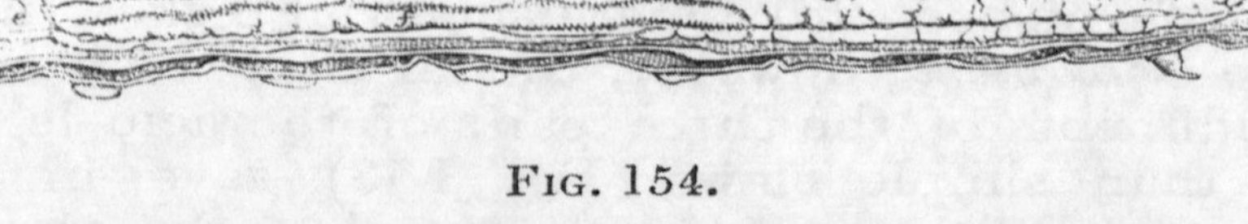

Ideal sections of Larva and Pupa of *Sphinx Ligustri* (Fig. 152), showing the relative state and position of their organs:—1-13, the successive segments; *a*, *a*, dorsal vessel; *b*, *b*, ligamentous bands holding it in its place; *c*, œsophagus; *d*, stomach; *e*, intestinal canal; *f*, biliary tubuli; *g*, cæcum; *h*, cloaca; *i*, sexual organ; *k*, cephalic ganglion;—the ventral gangliated cord is seen passing backwards from this along the floor of the thorax and abdomen.

315 *f*. Of the mode in which the Insect enters the *Pupa* state, and of its condition in that state, no general statement can be made; since they correspond for the most part with the grade of development which the larva has previously attained. Where it already possesses the general form and structure of the Imago, and little else is required for its completion than the development of the wings and of the sexual organs,

this is usually effected without any cessation of its activity; and after the first moult, which is regarded as marking the commencement of the pupa state, it continues to move about and to take food as usual, whilst its wings are sprouting and gradually becoming elongated beneath the next skin. This, again, is exuviated, and the wings as yet unexpanded are seen on the exterior of the body. After a third moult, the wings which were previously short, thick, and soft, are caused to expand, probably by the injection of air into their tracheæ; and no change subsequently takes place. This is the case with the *Orthoptera*, *Hemiptera*, and some *Neuroptera*. It is curious to observe that in the viviparous broods of Aphides (§ 315 *d*), neither metamorphosis nor moulting takes place; but that the process of gemmation is performed in a condition which corresponds with that of the larvæ of the perfect insects that are developed at the end of the season; as if the vital force which would otherwise be directed to the complete evolution of the individual, is here employed in the multiplication of the race.—In the *Coleoptera*, *Lepidoptera*, *Hymenoptera*, *Diptera*, and some *Neuroptera*, however, the pupa-state is one of complete inactivity as regards all the manifestations of *animal* life; although the formative processes are carried on with extraordinary energy. The imperfect larvæ of these orders, as we have already seen, are truly embryonic in their condition; and the processes of development which were commenced within the egg, and which were then only carried far enough to enable the larvæ to come forth and obtain their own nutriment, are now continued at the expense of the food which they have collected and stored up within their bodies; so that the passage into the pupa-state, in such cases, may be fairly likened to a re-entrance into the egg. The pupa is enclosed in the last skin exuviated by the larva, which, instead of being thrown off, dries up and remains to encase the proper skin of the pupa that is formed beneath it; and in addition to this, it is frequently protected by a silken 'cocoon,' the construction of which was the last act of larval life. The duration of the pupa condition, and the rate at which the developmental changes take place, vary considerably in different cases; some Insects remaining in this state for years, whilst others pass through it in a few days or even in a few hours; in both cases, however, we perceive that an important influence is exerted by external temperature (§ 96). As the state of the Pupa is one of rapid transition, it cannot be said to have any characteristic organisation; the intermediate condition of its structure, however, between that of the Larva on the one hand, and that of the Imago on the other, is shown in Fig. 154; and the nature of the changes which its principal organs undergo, will be hereafter explained in more detail (CHAP. XVIII).

315 *g*. The assumption of the *Imago* or perfect type of Insect life, is always marked by an exuviation of the integument which covered the pupa; and with this are cast off all the vestiges of the organs peculiar to the larva-state, while the wings, the true legs, the compound eyes, the antennæ, the complete masticating or suctorial apparatus, and many other organs, are now revealed for the first time in all those whose pupa condition was inactive. The wings, however, are seldom ready for use at the time of the Insect's emersion from the pupa-case; being usually soft

and moist, hanging loosely at the sides of the body, and having none of that rigidity which is requisite to give them the power of serving as organs of impulsion in air. It is not until the Insect has forcibly injected their tracheæ with air,—by taking several full inspirations, and then making an expiratory effort whilst the spiracles are closed,—that the wings are expanded; they then soon become dried up by exposure to the air, and by the cessation of the circulation which previously took place within them; and from that time they are the chief instruments of locomotion in all the Insects in which they are fully developed. The nutritive apparatus of the Imago is far less developed relatively to the muscular, nervous, and sexual organs, than it is in the preceding conditions; and its subordination to the offices of these is shown by the fact, that many Insects take no food whatever after their last change, the sole purpose of their existence, in their perfect state, being the propagation of the race by the generative process. In many instances, the duration of the Imago state is very brief, even where that of the preparatory periods has been very long; as in the case of the *Ephemera* (day-fly), which usually dies within a few hours after its last change, although the term of its previous life as a larva and an active pupa has not been less than two or three years. And even where the length of the life of the perfect insect is much greater, as in Bees, Wasps, &c., it seems to have a special relation to the nurture of the offspring, which are tended and supplied with food during the whole of their larva state. This duty, in the 'social' Insects, is performed by the 'neuters' or 'workers,' which are females whose sexual organs have not been evolved; their sexuality being proved (in the case of the Hive Bee at least) by the fact that they may be developed into perfect females by a different treatment during the larva state (§ 60). In the *Ant* tribe, the neuters do not acquire wings; and some of them, which are two or three times the size of the rest, and are somewhat differently formed, are characterized as 'soldiers,' their special office being the defence of the nest rather than the nurture of the young. Among the *Termites* (white-ants), however, the 'soldiers' appear to be *pupæ* arrested in their development; whilst the 'workers' have the characters of permanent *larvæ*.—In the *Apterous* orders of Insects, we find some tribes undergoing a regular metamorphosis, which is complete in every respect save the non-development of the wings. Thus the larvæ of the *Pulex* (flea) are footless worms, which afterwards pass into the pupa state, spinning for themselves a silken cocoon; in this they remain inactive for about twelve days, after which the Imago comes forth, having the rudiments of wings attached to the second and third segments of the body, though without any proper distinction of thorax and abdomen. The *Pediculus* (louse), *Podura* (spring-tail), and some other Aptera, however, undergo no metamorphosis; coming forth from the egg in the condition in which they remain all their lives; and this being far from the type of the perfect Insect.

316. The class of CRUSTACEA includes all those Articulated animals which are provided with articulated limbs, but which are furnished with organs fitted for aquatic instead of for atmospheric respiration. Thus it may be considered as containing within itself the aquatic representatives of all those Annulose animals, which are characterized, like it, by the possession of articulated members, but which are adapted for living in the

air: and we find such a representation to be really indicated by the diversity in form and structure presented by the members of this class; some of which (the *Isopoda*), in the equality of their segments and in their general organization, form a group that may be regarded as the parallel of that of Myriapods; whilst the Lobster may be considered as a gigantic Marine Insect; and the Crab in many points of its organization may be looked upon as a Spider adapted for a predaceous life along the shore, instead of for pursuing or entrapping its prey upon land. The classes of Crustacea and Arachnida, indeed, are very closely connected by a curious little group, that of *Pycnogonidæ* (§ 316 *e*), in which no special organs of respiration, either atmospheric or aquatic, are to be found, and in which the distinctive features of neither class present themselves in any marked degree; so that by some Zoologists it has been placed in the former, and by others in the latter. With the lower Articulata, also, this class has some very interesting relations; for some of its aberrant forms are so peculiarly modified, in accordance with the important purposes they are destined to perform in the economy of Nature, as to approach the higher Entozoa (§ 309 *e*); whilst others bear no inconsiderable resemblance to the higher Rotifera (§ 310). Deferring the notice of these for the present, we shall address ourselves to the description of the most important peculiarities of that higher division or sub-class of the Crustacea, which includes all its best-known forms, such as the Crab, Lobster, Cray-Fish, Shrimp, &c.—All these animals possess a dermo-skeleton of considerable firmness; its animal basis being usually consolidated by calcareous matter. In its most perfect form, such as that which it presents in the common Crab, it is made up of three layers; the outer one a horny epidermic membrane (easily detached from the shell by maceration for a short time in dilute acid), which presents no distinct trace of structure; next, a thin cellular layer, to which the colouring matter is exclusively confined; and beneath this, the thick calcareous shell, composed of a tubular structure closely resembling *dentine* (§ 212), which is peculiarly dense in its texture (so as closely to resemble ivory) in the black extremities of the claws, where its greatest strength is required for the powerful prehensile actions of the animal. The relative proportion of these layers is different, however, in other cases; for the cellular and calcareo-tubular layers are sometimes almost or entirely deficient, the shell being composed entirely of the horny layer; whilst the cellular layer is sometimes the thickest of the three.* The composition of the shell, when it retains its horny character, is very nearly allied to that of the dermo-skeleton of Insects, *chitine* (§ 314) being its principal ingredient; and this is found, with an albuminous constituent, in the dense shell of the Crab, though forming but a relatively small proportion of it.—As the shell of Crustacea completely envelopes their body and limbs, and is of an unyielding texture, it has no power of adapting itself to the increasing dimensions of the body during the period of growth; and we accordingly find that it is continually 'exuviated,' or cast off, and speedily replaced by a new formation. The *exuvium*, or cast shell, consists not only of the entire external tegument, including even the facetted membrane which forms the front of the compound eyes; but also carries with it the lining membrane of

* See the Author's "Report on the Microscopic Structure of Shells," &c., in the Report of the British Association for 1847, p. 128.

the stomach, and the calcareo-tendinous plates to which (in the principal limbs of the larger species) the muscles are attached. The surface of the animal remains for a few days in a soft condition ; but the formation and consolidation of a new shell take place with great rapidity, the calcareous material being supplied from two reservoirs, commonly known as 'crabs-eyes,' situated at the sides of the stomach. This exuviation takes place very frequently during the process of growth,—as often, in fact, as the body becomes too bulky for the shell ; it is still occasionally performed, however, when the animal attains its adult age, and in the Crab and Lobster is observed to take place at least once a year, at the reproductive period.

316 *a*. The crustaceous envelope, in its simplest condition, may be considered as a series of rings, like those which invest the body of a Myriapod ; each ring being furnished with a pair of members, the form of which varies according to the uses to which it is to be subservient. These rings may be articulated together, so as to allow a considerable freedom of motion ; or they may be adherent to one another, so as to be only distinguishable by furrows at their line of juncture ; or they may be so completely fused together, that the distinctness of the segments they enclose can only be recognised by the members they bear, or by observation of their separate development. The typical number of these segments in the higher Crustacea is twenty-one ; each of the three regions—the *head, thorax,* and *abdomen,*—consisting of seven. It is not often possible, however, to trace out all these segments distinctly ; for one or more of them are frequently obscured, either by coalescence with others, or by the entire change of form which they undergo, or by unusual development of others which conceal them. Not unfrequently, too, one or more of the segments may be deficient ; and these are usually such as form the posterior termination of the body, as is readily understood when it is known that the segments of Crustacea, like those of *Myriapoda* (§ 312 *a*) and many *Annelida* (§ 311 *a*), are developed from before backwards, so that a premature arrest of development will cause the number of terminal segments to be smaller than usual. In this manner it happens that in the order *Lœmodipoda* (to which the *Cyamus* or 'whale-louse' belongs), the whole abdominal region is deficient.—It is in the head that the difficulty of recognising the segmental division is most frequently met with, especially in the higher forms of the class, as we shall hereafter see to be the case in Vertebrated animals (§ 320 *i, k*); and the number of segments is usually to be best made out, by that of the appendages which they respectively bear. Thus we have ordinarily a pair of pedunculated eyes, marking the *first* segment ; the *second* bears the first pair of antennæ (Fig. 155, *a*), which are here of small size, whilst the *third* bears the second or larger pair of antennæ (*b*), the presence of *two* pairs of these instruments being very characteristic of the class ; the *fourth* bears the mandibles, which are the lateral jaws that meet above the mouth ; whilst the *fifth, sixth,* and *seventh* bear three pairs of 'feet-jaws,' which form the lower or posterior part of the complex buccal apparatus of these animals. These 'feet-jaws' usually present a gradation in form, from before backwards ; the anterior pair most resembling the mandibles, and the posterior being more conformed to the type of the thoracic appendages. The separation of the segments of the head is most complete in the order

Stomapoda, especially in the *Scilla,* sometimes termed 'sea-mantis' from the strong resemblance which it bears to the insect of that name; in many other orders, however, the rings are fused together, as it were, into a single piece. The head is sometimes moveable, and distinct from the thorax; but is sometimes so closely united with it, that the whole forms but one mass, analogous to the 'cephalo-thorax' of Arachnida.—The distinctness of the segments of the thorax may nearly always be clearly made out; for although, in many of the higher Crustacea, its upper surface is covered by a *carapace,* formed by an extraordinary backward extension of one of the cephalic rings, which does not present the least trace of division, the transverse lines of separation are readily discernible on the under side of the body, between the origins of the successive pairs of members (Fig. 155). In the order *Decapoda* (ten-footed), to which the Cray-fish belongs, with the Lobster, Crab, Prawn, &c., the number of 'legs' or locomotive appendages is reduced to five pairs, by the appropriation of the first two pairs of thoracic members as additional 'feet-jaws,' the second of them (*e*) being larger than the rest, and covering-in the whole series; whilst the first pair of 'legs' (*f*) is frequently developed rather for prehension than for locomotion, constituting the formidable *chelæ* or 'nippers' of these animals. But in the orders *Amphipoda* (of which the common 'sandhopper' is an example) and *Isopoda* (the 'wood-louse' and its allies), the number of legs is raised to the normal amount of fourteen, by the development of all the thoracic members as instruments of locomotion; and in the *Stomapoda,* the last pair of cephalic appendages is sometimes developed in nearly the same form, making the number of legs sixteen. The legs of the higher Crustacea have no fewer than six joints; and are furnished at their base with one or two curious appendages, named the *palp* and *flagellum,* which are sometimes as much developed as the principal member, or even more so. Thus in the *Mysis* or 'opossum-shrimp,' the 'palp' has nearly the semblance of the ordinary leg, so that the animal seems to have double the usual number of members; on the other hand, in those which have the limbs converted into respiratory organs, this is usually accomplished by the extension of the 'flagellum' as a flattened leaf-like plate. The general conformation

FIG. 155.

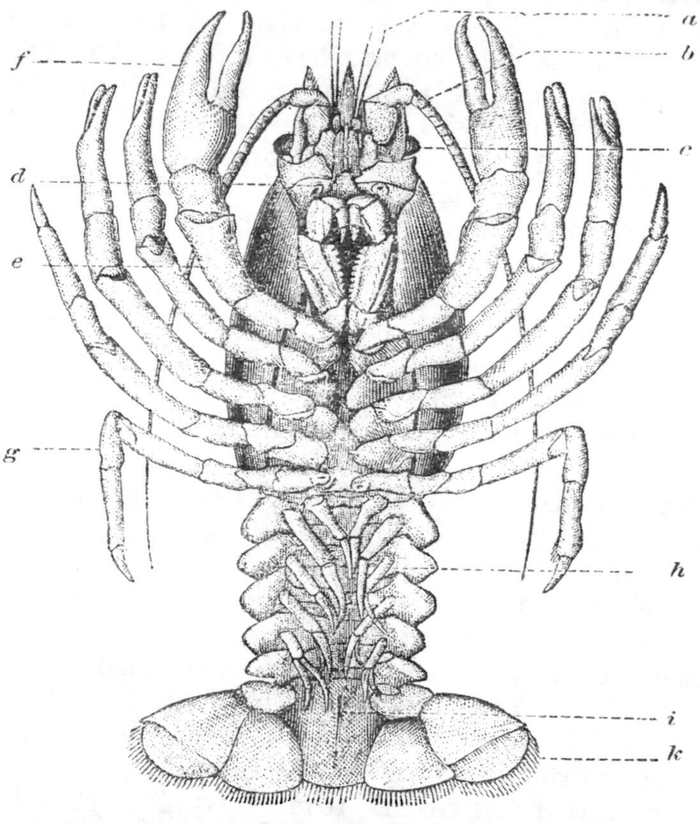

Under side of *Astacus fluviatilis* (Cray-fish): — *a,* antennæ of the first pair; *b,* antennæ of the second pair; *c,* eyes; *d,* auditory tubercle; *e,* external feet-jaws; *f,* first pair of thoracic appendages; *g,* fifth pair of thoracic appendages; *h,* abdominal false legs; *i,* anus; *k,* tail-fin.

of these extremities varies considerably, according to the purposes to which they are to be applied; thus, in some cases, they are used chiefly for walking, and are then simple, jointed, nearly cylindrical limbs (Fig. 155); in other cases they are also employed for digging, and are expanded at their extremities into a set of little spades; they may be also adapted as instruments of prehension, by the development of a moveable finger which can be opposed to the fixed one, as in the three anterior legs of *Astacus* (Fig. 155), or by the last joint being made to fold back against the preceding, as in *Squilla;* or they may be chiefly adapted for swimming,—an adaptation, however, which is more remarkably exhibited in the inferior division of the group. The greatest *specialization* of function displayed in the members of the more highly-organised Crustaceans, is where the first pair alone is adapted for prehension, as we see in certain Crabs, whose *chelæ* are frequently enormously developed in comparison with the ordinary limbs, and are capable of taking a most powerful grasp, that of one side having a sharp edge for cutting, and that of the other being blunt and adapted for crushing; whilst the ordinary legs are destitute of the second finger, and are adapted for locomotion only. On the other hand, in the curious *Limulus* or 'king-crab,' which departs most widely from the higher type, connecting the more perfect Crustacea with the Entomostraca (§ 317), we find the greatest degree of community of function; for the ordinary cephalic instruments of mastication are wanting, and the thoracic limbs are so arranged around the mouth as to be subservient to this function; the basal joint of each being converted into a jaw, and working against that of the opposite side, whilst the remainder of the limb is conformed after the ordinary type of the didactyle legs of the Astacus, and serves the same purposes. The *abdomen,* as already mentioned, may be altogether deficient; and even when present, it may constitute a very small part of the body, although none of its segments may be wanting. This we see in the Crab, in which the viscera are entirely contained in the thorax, whilst the abdomen, whose seven segments are still readily distinguishable, is reduced to a mere rudiment, resembling a tail, and is concealed beneath the thorax. In the swimming Decapods, on the other hand, such as the Cray-fish, Lobster, Prawn, &c., the abdomen often forms by far the largest proportion of the body; its segments are very moveable one upon another; and the shell of its terminal segment is expanded horizontally into a flattened tail-fin (Fig. 155, *k*), which is the principal instrument of locomotion in water. The abdominal appendages are never developed as true walking 'legs', but they are often modified for swimming, and are then known as 'fin-feet;' whilst in the *Talitrus* or 'sandhopper' and its allies, the posterior pairs are united into a sort of bundle, by the sudden jerking-back of which these animals are enabled to leap high into the air. Very frequently they participate, in some mode or other, in the respiratory function; thus, in the group of *Amphipoda* (to which the Talitrus belongs), the three first pairs of abdominal appendages are employed to produce a constant renewal of the respiratory current over the gills which are attached to the thoracic members; whilst in *Squilla* the abdominal members themselves bear the gills as feathery tufts developed from them; and in the *Isopods* we usually meet with a further development of scale-like appendages, by which the gills are enclosed and protected.

In most of the *Decapods*, they are nearly rudimentary, as in *Astacus* (Fig. 150, *h*), and their chief purpose appears to be that of supporting the eggs, which are carried about by the parent after their extrusion; and in the *Mysis* or 'opossum-shrimp' (which belongs with Squilla to the order *Stomapoda*), we find them rendered still more subservient to the generative function, being so constructed as to form, by their folding-together on the under side of the abdomen, a complete pouch or 'marsupium,' within which the young undergo their early development. In the Isopods, whose abdominal members are entirely subservient to respiration, the support and protection of the eggs is performed by the thoracic limbs, which have large plates attached to their bases, forming a similar cavity by their mutual overlapping.—Thus in the division of the functions of locomotion, mastication, respiration, ovi-protection, &c., among these appendages, we have a remarkable want of constancy, which is in strong contrast with the uniformity that characterises (with few exceptions) the functions of the articulated members in the whole class of Insects; and we can scarcely help being reminded of a similar want of constancy in the Molluscan series (§ 297), which the Crustacean class may be regarded as in some degree representing, including, as it does, those members of the higher part of the Articulated series, which are specially adapted for an aquatic life.

316 *b*. As an example of the general structure and arrangement of the internal organs, we shall take the common *Crab*, in which the apparatus of organic life may be regarded as attaining the highest perfection, and the apparatus of animal life the greatest concentration, which it any where presents in this group.—In this, as in all the higher Crustacea, the mouth is adapted for mastication, by the complex buccal apparatus already described; the edges of the mandibles and of the maxillary plates being usually sharpened for cutting, and sometimes provided with a row of teeth. The alimentary canal is short and simple; the stomach (Fig. 156, *g*) being a capacious sac, situated almost close to the mouth (in *Limulus*, anteriorly to it), and the intestine passing directly backwards from it to the anus, without any convolutions. In the Crab, Lobster, and most of the higher Crustacea, the walls of the stomach are connected with a complicated frame-work, which bears hard calcareous teeth that project into its cavity; these gastric teeth are worked by powerful muscles (*a*), and give important assistance in the reduction of the food. The liver (*f*) is a large and important organ in all Crustacea, and is developed symmetrically, occupying a considerable part of the visceral cavity on each side; in the general plan of its structure it closely corresponds with the liver of Mollusca, and is strikingly different from that of Insects.—The Circulating system partly consists of distinct vessels, and partly of uninclosed sinuses or passages which are left between the several organs and tissues of the body. Its centre, however, is no longer the elongated and compound dorsal vessel which is so characteristic of the Articulata generally, but is a simple cavity (*h*), furnished with firm muscular walls; in some of the lower tribes (as in the *Limulus*), however, it is considerably lengthened, but does not present any other segmental division than is indicated by the mode in which its vessels issue from it. The circulation of blood in this group may be considered as taking place essentially upon the plan of the Gasteropod Mollusks (§ 304), but with a modification that seems like a foreshadowing

of the peculiar condition of this function in Reptiles (§ 324*k*). Of the blood sent forth from the heart by the arteries (Fig. 156, *d*, *e*), only a part is transmitted by the venous sinuses, after its passage through the system, to the respiratory organs, thence to return to the heart in an aerated condition by the branchio-cardiac canals; the remainder is directly returned to the heart, by other venous sinuses, without passing through the respiratory organs at all; and thus the blood which enters that cavity,

FIG. 156.

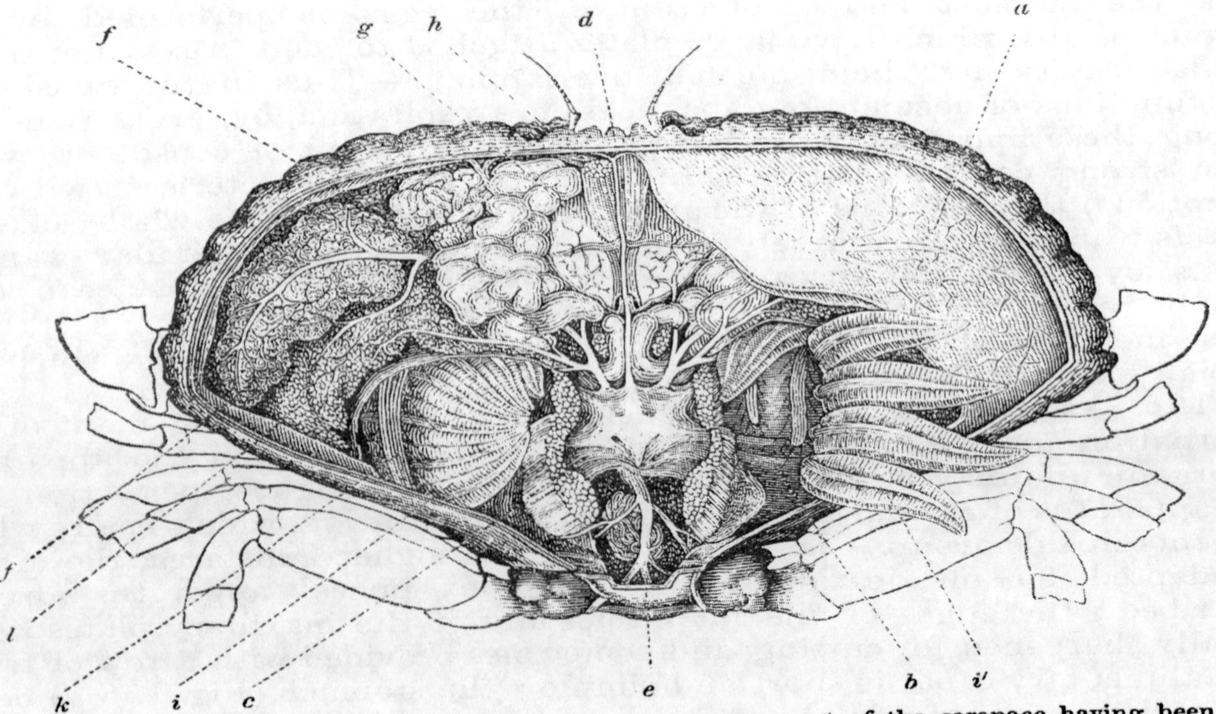

Anatomy of *Cancer pagurus* (Common Crab), the greater part of the carapace having been removed;—*a*, portion of the membrane lining the carapace; *b*, vault of the flanks; *c*, heart; *d*, ophthalmic artery; *e*, abdominal artery; *f*, liver; *g*, stomach; *h*, muscles of the stomach; *i*, branchiæ in their natural position; *i'*, branchiæ everted; *k*, flabella; *l*, testis.

and is again sent forth from it, is (like that which is propelled by the ventricle of the Reptilian heart) of a mixed quality, partly aerated or arterial, partly non-aerated or venous. The condition of the sanguiferous canals of the systemic circulation is very imperfect; for only the arteries possess proper walls, and the blood is conveyed by them into a system of *lacunæ* or capillary passages through the tissues; and from these it is collected by venous sinuses of very irregular form, through which, however, its course is determinate, partly towards the heart and partly to the gills. At the base of the gills we sometimes find them provided with muscular walls, and serving by their pulsation as accessory hearts, whereby the blood is propelled through the respiratory circulation, as in many Annelida.—The Respiratory organs differ greatly, as we have seen, in their position; being generally attached externally, either to the abdominal or to the thoracic members; whilst in the Decapod Crustacea they are enclosed within the general envelope, and seem disconnected with the locomotive appendages. They differ also in their form; being sometimes membranous vesicles, as in *Læmodipoda;* sometimes simple expanded plates, as in *Amphipoda;* sometimes plates dividing into filaments, so as in some degree to resemble a feather, as in *Stomapoda;* whilst in the *Decapoda* they have the form of long slender quadrangular pyramids (from 9 to 22 on either side), each of them consisting of a central stem, to which are attached a multitude of thin plates or cylindrical filaments (Fig. 156, *i, i'*). Further, the interchange of the current of water in contact with these organs is dependent, in the lower tribes, upon the general movement of the limbs; in the *Amphipoda* we have seen that certain pairs of the abdominal appendages are specially adapted to maintain this current by their repeated strokes upon the water; but in the *Decapoda* we find what seems to be a peculiar apparatus, which serves not merely to maintain a current of water through the branchial chamber, but also to keep the branchial filaments or lamellæ apart from each other, so as to facilitate the action of the water upon them. A careful examination of the development of the respiratory organs, however, even in this their highest condition, shows that they are here (as elsewhere) to be regarded as appendages of the legs; that the chamber which encloses them is formed by the gradual folding-in (so to speak) of the carapace around them; and that the parts by whose action the stream of water is maintained and brought into contact with every individual filament, are nothing else than parts of the regular organs, specially adapted by a peculiarity of conformation for this purpose. (See CHAP. XIII.). Neither in the terrestrial Isopods (such as the 'wood-louse') nor in the terrestrial Decapods (such as the 'land-crab'), is the type of the respiratory apparatus changed; for all that is necessary to adapt it for atmospheric respiration, is to keep the gill-filaments constantly moist, for which these animals have a special provision.—The only apparatus known to exist in Crustacea, which can at all be regarded as a urinary organ, consists of a set of simple, long, unbranched, slender tubes, which open into the intestinal canal of some of the higher tribes; but it has not yet been shown that the contents of these have a urinous character; and as their place of discharge into the intestine is high up, near the termination of the biliary ducts, it would seem by no means impossible that they are rather to be regarded as a rudimentary pancreas.

316*c*. The condition of the *Nervous System* in Crustacea presents a regular series of gradations, between the type on which it is constructed in the Annelida and Myriapoda, and one of higher concentration than exists in any Insect. The former is seen in those whose bodies display the greatest equality of segmental division, and in which there is the greatest similarity in the endowments of the several members; for in such we find the ganglia of the ventral cord corresponding in number with the segments of the body, nearly equal to each other in size, and placed at uniform distances from each other; and in *Talitrus* we find the distinctness of the two lateral halves unusually obvious, the two strands of the ventral cord being separate along their whole length, each having its own series of ganglia, and the ganglia which are thus arranged in pairs in the successive segments being connected transversely by commissural bands. In the genus *Astacus* (lobster and cray-fish), on the other hand, we find (as in Insects) the thoracic ganglia developed to a greater size than those of the abdomen, that the locomotive members may receive their due supply of nervous influence; but as the tail is a powerful swimming organ in these animals, the difference is not so great as it would have been if the thoracic members had been the sole instruments of locomotion; and we find the last ganglion, situated above the anus, and radiating nerves to the swimming plates of the tail, particularly large, being apparently made up by the coalescence of the sixth and seventh abdominal ganglia. In *Palemon* (prawn) and *Palinurus* (rock-lobster) we find a coalescence of the thoracic ganglia into a long elliptical perforated nervous mass; but this coalescence reaches its greatest degree in the Crab, in which all the ganglia of the ventral cord are blended into one large oval ganglion, with a perforation in its centre, which is situated near the middle of the under surface of the body, and from which nerves radiate to all parts of the trunk, to the legs, and to the short tail. The relative development of the cephalic ganglia varies with that of the organs of sense and motion which are situated in the head. They usually appear to be made up of those of the four anterior segments of the head, from which proceed nerves that connect them with the eyes, the two pairs of antennæ (the second containing the organ of hearing), and the mandibles; whilst those of the two posterior segments ordinarily coalesce with the first ganglion of the ventral cord, to form the great sub-œsophageal ganglion, whence proceed the nerves to the feet-jaws. There is not, as in Insects, a distinct system of respiratory nerves, these being blended with those of the general sensori-motor apparatus; but the 'stomato-gastric' system is here well developed, and its relations to the visceral system of Vertebrata become apparent.—In regard to the development of the sensorial apparatus in this class, it may be stated generally that, with the exception of the visual organs, no part of it attains a high standard. The investment of the entire body in a dense crustaceous envelope must necessarily almost completely destroy its tactile sensibility; and the *antennæ*, being themselves enclosed in a similar casing, cannot be delicate instruments of touch, although they may readily convey impressions produced by the mere contact of hard bodies, to the nerves that supply them, just as an inflexible rod will be the means of conveying such impressions to the hand that holds it. The *palpi*, which are appended to the mandibles and feet-jaws, probably serve (as in Insects) to take cognizance of the qualities of the substances which are being employed as food, and may possibly minister to the sense of smell. For taste there is no special organ; and whatever be the degree of this sense which these animals may possess, it must depend upon the endowments of the membrane lining the mouth and œsophagus. For hearing there appears to be a special organisation, in the highest Crustacea at least; the first joint of the second or larger pair of antennæ being excavated into a cavity with a round orifice closed by a membrane, and this cavity containing a vesicle filled with fluid, upon which the antennal nerve expands. It is in the Crabs that this apparatus presents its most complex condition.—The visual organs, in all the higher Crustacea, are elaborately constructed; and it is chiefly by the indications which they afford, that these animals are guided in their movements. These organs are formed upon the same plan as in Insects; each of the large 'compound eyes' being made up by the aggregation of a number of *ocelli* or single eyes, every one of which is a separate instrument of vision, perfect and complete in itself. In the highest division of the group, the compound eyes are mounted upon jointed peduncles, by the flexure of which they can be turned in any direction; but still, except in a few species, they are protected by the over-arching of the carapace, which forms a sort of orbital cavity for their lodgment. It is interesting to remark, that the compound structure of the eyes is one of the most unequivocal indications of the Crustacean nature of the extinct *Trilobites*, which might otherwise have been associated with the Chitons (§ 302); this structure being so beautifully preserved, notwithstanding the very early period at which these animals became fossilized, that it even serves to indicate the particular division of the Crustacean class to which they should be referred.—The muscular apparatus attains a great development in this class, forming masses whose size renders the larger Crustacea available articles of food. They are attached to the shell by means of strong tendinous plates, which are continuous with the animal substance of the latter, and which are thrown off with its exuviation. Still there is no great *variety* in their actions; those which move the trunk and the limbs being alike limited to the simple movements of flexion and extension of the joints, which are usually so constructed as to allow of motion in only one plane.

316 *d*. Although no propagation by spontaneous fission or gemmation takes place in the Crustacea, yet they possess a power of regenerating lost parts which is truly astonishing, the complexity of their organisation being considered. The mode in which this takes place will be hereafter explained in more detail (CHAP. XVIII.); at present it will be sufficient to state that it extends to the regeneration of all the limbs, and that it seems essentially connected with the 'moulting' process, the new limb making its first appearance when the shell has been cast off. The sexes are distinct throughout the entire class; and the generative apparatus of each sex presents this remarkable difference from that of Insects, that the two lateral halves of it do not unite on the median line so as to have a common outlet, but that the orifices of the seminal organs and of the ovaria are separate on the two sides. A progressive complication in their structure presents itself, as we ascend from the lower to the higher forms of this division of the Crustacea; for in the inferior tribes the testes consist of a small number of vesicles opening into a common tube on either side, whilst in the higher they are made up of a mass of minute convoluted tubes (Fig. 156, *l*), opening into a common duct, which is often dilated into a receptacle for the seminal secretion; and the ovaria, also, are a pair of simple sacculi in the inferior tribes, but in the superior show a division into the proper ovaries (which are long branching cæca communicating with each other on the two sides), the oviducts, and the *spermothecæ* or 'copulatory pouches' within which the spermatic fluid is stored up. The orifices of the male organs are usually stated to be in the first joints of the last pair of thoracic limbs; and it has been supposed that the first and second pairs of abdominal appendages, or false legs, in the male serve the purpose of exciting organs. From recent observations, however, it appears that the spermatic duct of each side passes on as a membranous tube from the base of the thoracic limb to the first or second abdominal appendage, and runs to the extremity of this, which thus becomes a true penis or intromittent organ.* The spermatic fluid, received and stored up in the spermothecæ, may remain there for a considerable period, and may fertilize ova which are far from maturity at the

* See Mr. Spence Bate's "Notes on Crustacea," in the 'Annals of Natural History' for August, 1850.

time of copulation; and it even appears that, in many instances, the spermatic fluid itself is far from being ready for its office at the time that it is deposited in the female organs, the final development of the spermatozoa within the spermatic cells having yet to take place,*—a striking example of the independent life of cells, which can be thus transferred from one organism to another, without any interruption to their development. The separation and independence of the lateral halves of the sexual apparatus in the Crustacea, appear to have some relation with the frequency of what is termed 'lateral hermaphrodism' in this class; monstrosities being of no unfrequent occurrence, in which the sexual organs present the male character on one side, and the female on the other.—The ova, after their extrusion from the oviduct, are usually carried about by the female parent, supported either by the thoracic or the abdominal appendages, until the embryo is nearly ready to come forth from its envelope; and special provisions for their protection, made by the peculiar development of these appendages, are found in a considerable number of Crustacea, as well as in the *Mysis* (§ 316 *a*). The relation which the young animal, at the time of its emersion from the egg, bears to the adult, seems to differ very greatly in different tribes, and even in genera which are otherwise closely allied. Thus in the *Astacus fluviatilis* (cray-fish), according to the observations of Rathké, the embryo passes through what may be considered as its successive stages of metamorphosis, whilst yet within the egg, and comes forth in a form which corresponds in all essential particulars with that of its parent. The same would appear to be the case (from the observations of Mr. Westwood) with the young of the *Gecarcinus* (land-crab). On the other hand, several competent observers have confirmed the discovery originally announced by Mr. V. Thompson† as to the metamorphosis of the common *Crab*; which comes forth from the egg in a condition so unlike the adult, that in its early or larval condition it was accounted a distinct genus, *Zoea*, and placed among the inferior Crustaceans; and which only acquires the characters of the adult by a series of progressive modifications, which present themselves at the successive exuviations of the shell. Similar metamorphoses have been observed in *Astacus marinus* (lobster), *Palinurus* (rock-lobster), *Palemon* (prawn), *Crangon* (shrimp), and many other species among the higher Crustacea; so that the absence of any such change would seem to be the exception rather than the rule. Generally speaking, a strong resemblance exists among the young of all the species which undergo a metamorphosis; the head and thorax being included within a large carapace, while the abdominal portion is so much developed as to make up the chief part of the length of the body; and the thoracic segments being furnished with 'fin-feet', whilst the abdominal segments are destitute of appendages, the last, however, being flattened out into a large tail-fin. In the *Macrourous* (long-tailed) species (as the lobster, prawn, &c.), metamorphosis chiefly consists in the replacement of the thoracic 'fin-feet' by true locomotive appendages and by a special apparatus of respiration, and in the development of abdominal appendages; but in the *Brachyourous* (short-tailed) Crabs and their allies, a much more complete change of

* See Mr. H. D. S. Goodsir's paper on "The Testis and its Secretion in the Decapodous Crustaceans," in "Anatomical and Pathological Observations."

† Zoological Researches, vol. i.

form takes place, the thorax being gradually developed at the expense of the abdomen, which becomes rudimentary. The following description of this remarkable series of changes in the young of *Carcinus mœnas* (small edible crab), is condensed from that given by Mr. R. Couch.*— The young, when it *first* escapes from the egg, is of very small size, rarely exceeding half a line in length. The most bulky portion of its body is formed by the almost hemispherical dorsal shield, which covers the head and thorax, and has a long spine projecting from the middle of its upper part (Fig. 157, A); the tail (formed by the six abdominal segments) is long, rounded, and slender, and its extremity is forked, the lateral portions being attached to the last ring by joints. The eyes are large, and situated under the front of the dorsal shield, but are not mounted on peduncles; two pairs of flattened appendages are seen in their neighbourhood, behind which are three pairs of claws, each composed of three joints, and terminating in four long slender hair-like appendages; whilst at the posterior part of the shield are two pairs of 'fin-feet.' These little creatures are extremely active in the first stage of their lives, swimming freely by continual flexions and extensions of their tails, and repeated beating motions of their claws; but as the shell becomes more solid, they retire to the bottom of the water, and become more stationary. The progressive solidification of the shell is curiously seen in the dorsal spine, which is quite soft within the egg, being doubled back upon the shield; for some little time after the emersion of the animal from the egg, the movement of blood can be seen to proceed as far as its apex; but as the consolidation advances, the circulation becomes more and more limited in its extent, and is finally confined to the base. When the first shell has been exuviated, the animal presents itself under a *second* form (B); the dorsal

FIG. 157.

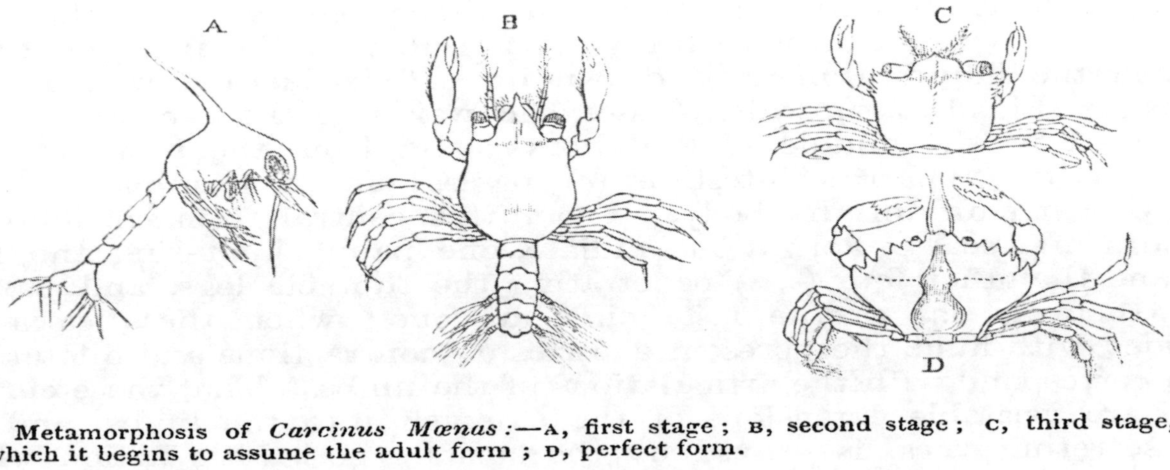

Metamorphosis of *Carcinus Mœnas*:—A, first stage; B, second stage; C, third stage, in which it begins to assume the adult form; D, perfect form.

spine has disappeared; the shield is more depressed; the eyes are mounted on footstalks, and the appendages in their neighbourhood are converted into antennæ; the first pair of claws becomes stouter than the rest, and is armed with a pair of nippers on each side; the other claws resemble those of the lobster, except that the posterior pair are branched near the base, and one of the branches ends in a brush-like tuft. The

* Report of the Cornwall Polytechnic Society for 1843.

tail is greatly diminished in its relative size and proportions, and is sometimes partially bent under the body, but is more commonly extended. In this state, the active swimming habits of the earlier larva are still preserved. When this shell has been cast, the animal comes forth under a *third* form (C), which presents a nearer approximation to that of the adult. The cephalo-thoracic shield is now flattened, and somewhat squarer in its form ; the tail also is flattened, and is provided with false feet ; the first pair of claws is still more developed in relation to the rest, approaching more nearly to the *chelæ* of the adult ; and the remainder of the thoracic limbs are well developed for walking. In this condition, the swimming habits of the larva are but little preserved ; the tail is usually bent under the body ; and progression is chiefly accomplished by the thoracic members. After the next moult, the animal comes into the world under a *fourth* form (D), which is that of the adult, although its size is still very minute.

316 *e.* In the very curious tribe of *Pycnogonidæ,* we have the higher Crustacean type reduced to its extreme of simplification. The general form of the body (Fig. 158) reminds us of that of the Crab ; the abdomen being entirely deficient or represented only by a small tubercle ; the head also being very small, and generally fused (as it were) into the thorax ; so that the last-named region, with its appendages, constitutes almost the whole bulk of the animal. In the complete obliteration of the abdomen, a resemblance is presented to the *Cyamus* and other Læmodipoda (§ 316 *a*); but the departure from that type is manifested in every organ of the body. The mouth is formed for suction, and the parts of which it is elsewhere composed appear to be either rudimentary or fused together ; the only appendages to the head, in several genera, being a single pair of antennæ and one pair of feet-jaws. The thorax bears four pairs of legs, each composed of several joints, and terminated by a hooked claw ; and these legs all radiate, as it were, from one centre. The mouth opens by a narrow orifice at the extremity of a trunk-like prolongation of the head, and it seems to be furnished with vibratile cilia, by the instrumentality of which the liquid aliment is drawn in. It leads to a very narrow œsophagus (Fig. 158, *a*), which passes backwards to the central stomach (*b*) situated in the middle of the thorax ; and from the posterior part of this cavity, a narrow intestine (*c*) passes off, to terminate at the posterior extremity of the body. From this central stomach, however, five pairs of cæcal prolongations radiate, one pair (*d*) entering the feet-jaws, and the other four (*e, e*) penetrating the thoracic legs, and passing along them as far as the last joint but one ; when these cæca are distended with fluid, they present a series of contractions and dilatations, which correspond with the articulations of the limbs. That these curious organs are veritable extensions of the stomach into the limbs, and not mere secreting cæca, is proved by the curious peristaltic action which the transparency of the body permits to be observed in them ; for they all contract and dilate alternately, so as to produce a continual flow of the contents of this curious digestive cavity from one part into another. Their membranous walls are covered, however, with a layer of brownish-yellow granulations ; and these, which are scarcely perceptible on the wall of the central stomach, and not at all upon that of the intestine, may probably be regarded as a diffused and rudimentary condition of the liver. In the fluid which is transmitted from one portion of the digestive cavity

to another, minute granular particles can be seen, which are probably of a fæcal nature. The digestive sac is freely suspended in the general solid cavity of the body, by occasional fibrous bands; and nothing intervenes between its external surface and the muscular lining of the crustaceous envelope. Thus a wide *lacuna* is left around it, occupying the greater part of the thoracic cavity, and also the hollow of the legs; and this lacuna is filled with a transparent liquid, in which are seen floating a number of minute transparent corpuscles of irregular size. This fluid is kept in continual motion, not only by the general movements of the body and limbs, but by the more special and constant action of the different parts of the digestive sac; for when the cæcum in any one of the legs undergoes dilatation, a part of the circumambient liquid will be pressed out from the cavity of the limb, either into the thorax or into some other limb whose cæcum is contracting; and when, in its turn, the first cæcum contracts, the space around it will be again filled with liquid forced into it by the dilatation of the central stomach or of some other cæcum. There is no other circulating apparatus whatever; and it cannot be doubted that the liquid which fills the great cavity of the body must be regarded as the nutrient material derived from the digestive organs, that is, blood; and that its movement from one part of the body or limbs to another is to be regarded as a great lacunar circulation, no part of it being as yet circumscribed within distinct vessels, and no special organ of impulsion being developed. There is no special organ of respiration; but the aeration of the circulating fluid must be accomplished by exposure to the surrounding medium through the crustaceous envelope, which does not possess great density in any of these animals.— The nervous system consists of a supra-œsophageal or cephalic ganglion, and of four ganglia in the thoracic region; the latter are adherent to each other, so as to form one continuous mass; and this is connected with the cephalic ganglion in the usual mode, by a cord that passes round the œsophagus on either side. The front of the cephalic ganglion expands into a sort of tubercle, on which are set four eyes at a little distance from each other, resembling in their fewness and in their

FIG. 158.

Ammothea pycnogonoides;—*a*, narrow œsophagus; *b*, stomach; *c*, intestine; *d*, digestive cæca of the feet jaws; *e e*, digestive cæca of the legs.

separate condition the eyes of the Annelida. No other organs of sense have been observed. Each of the thoracic ganglia sends off a pair of nerves to one of the pairs of legs. The movements of these animals are very sluggish. They have no swimming power, but crawl over the surfaces of Sea-weeds, &c., by means of their long flexible legs, taking an attachment to some fixed point by means of the hooked claw, and then drawing themselves towards it. The reproductive apparatus of the *Pycnogonidæ* is as yet unknown.*

317. Besides the Crustacea to which the descriptions given in the preceding paragraphs are more or less completely applicable, the class includes a great variety of simpler forms, all of them purely aquatic, but many of them inhabiting fresh water; some of these having appendages fitted for swimming, and having their mouths furnished with a masticating apparatus; whilst others attach themselves as parasites to the bodies of different aquatic animals, especially Fishes, and live by the imbibition of their juices through a mouth adapted for suction. The former of these groups, all the existing representatives of which are animals of minute size, is commonly known under the designation of *Entomostraca;* from the circumstance that the body is in many instances completely enclosed within a horny casing, which sometimes closely resembles a bivalve shell in form and in the mode of attachment of its pieces to each other, whilst it sometimes forms but a single piece like the hard casing of certain Rotifera (Fig. 146). There is no limitation in the number of segments of the body in this group; and it almost always exceeds that which is characteristic of the higher Crustacea, being sometimes increased to an extraordinary extent. Thus in *Branchipus stagnalis* (a very common little Entomostracan of stagnant waters), the number of thoracic segments is eleven, and that of the abdominal, nine; the segmentation of the head, however, being altogether undistinguishable. In *Apus*, the thorax and abdomen together contain thirty segments, and the appendages to these present a remarkable multiplication in the number of their parts; thus the first pair of legs gives off a set of branches, whose articulations altogether amount to about 150, and the caudal segment gives off two long cylindrical filaments, each of which has no fewer than 480 joints; and each segment of the body bears two pairs of legs, so that the number of these members is sixty pairs,—the total number of articulations in this creature being (according to Latreille) not less than two millions. In the greater part of these animals, the number of proper locomotive appendages is small, the anterior pairs of thoracic appendages alone being specially modified for this function, and the remainder (if present at all) being converted into 'fin feet', which are subservient rather to the aeration of the blood than to locomotion; this is well seen in *Limnadia* (Fig. 159), and other *Branchiopodous* (or gill-footed) Crustacea. Sometimes the 'fin-feet' constitute the principal organs of progression; and the movement is then of an easy gliding character, the animals frequently swimming on their backs (as is the case with the *Branchipus*), or moving with equal facility on the back,

* The above account of this very curious group is given from the Author's own observations, which were made previously to the publication of the excellent paper of M. de Quatrefages, "Sur l'Organization des Pycnogonides," in the Ann. des Sci. Nat. for 1845. He has found M. de Quatrefages' observations accordant with his own, in every particular.

belly, or sides, as is done by the *Artemia salina*, or 'brine-shrimp.' The action of the fin-feet also serves to bring a current of water to the mouth, and thus to supply the animal with food. But where the anterior thoracic limbs and their appendages are chiefly used for locomotion, as in *Apus, Limnadia, Daphnia* (the common 'water-flea') and many others, the animal moves in a series of jerks; and this is the case also with the *Cypris* and other Entomostraca of the order *Ostrapoda*, in which there are no proper respiratory appendages whatever, and only a small number of pairs of legs.—In the general formation of their nutritive apparatus, the Entomostraca present a type of structure much simpler than that which characterises the higher division of the class. The mouth usually has a pair of mandibles, with only one or two pairs of feet-jaws; and the alimentary canal presents very little dilatation from one end to the other, nor do the biliary organs rise above the condition of simple cæca. The heart is elongated into a dorsal vessel; and this, in some species, may be observed to be partially divided into a series of distinct chambers (as in Insects), no fewer than nineteen having been counted in *Branchipus*. The movement of the circulating fluid does not seem to take place, in any great degree, within distinct vessels, but rather through sinuses or lacunæ excavated among the various tissues and organs; and there is no special respiratory circulation, the blood being aerated in its course through the ordinary appendages of the body. In regard to the structure and arrangement of the nervous centres, little is certainly known; but that of the visual organs presents a variety of interesting conditions, all gradations being traceable, from the moveable single eye made up of a small number of ocelli aggregated together, to the large compound eyes which very closely approximate to those of the higher orders of the class. In some instances, too, single eyes are met with, in addition to the compound masses.

Fig. 159.

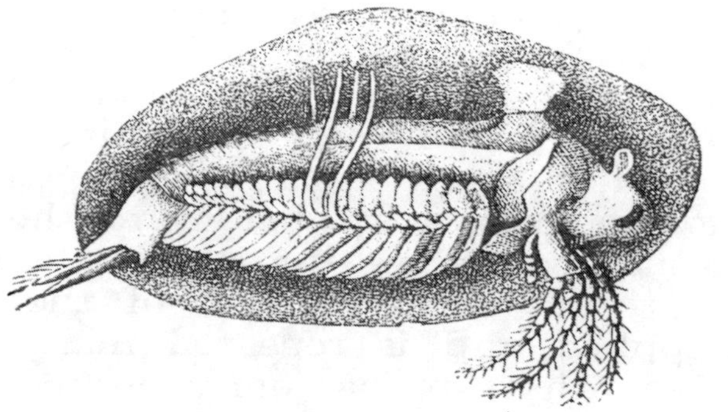

Limnadia, one of the valves of the carapace having been removed.

317 *a*. Some of the most interesting points in the natural history of this group, lie in the peculiar modes in which the generative function is performed, and in their tenacity of life when desiccated, in which last respect they correspond with many Rotifera (§ 310 *b*). This provision is obviously intended to prevent them from being completely exterminated, as they might otherwise soon be, by the drying up of the pools, ditches, and other small collections of water which constitute their usual 'habitats.' It does not appear, however, that the adult animals can bear a *complete* desiccation, although they will preserve their vitality in mud that holds the smallest quantity of moisture; but their eggs are more tenacious of life, and there is ample evidence that they will become fertile on being moistened, after having continued for a long time in the condition of fine dust. Most Entomostraca, too, are killed by severe cold, and thus the whole race of adults perishes every winter; but their eggs seem unaffected by the lowest temperature, and thus continue the species which would otherwise be exterminated. It is a curious fact that the *Artemia salina*,

or 'brine-shrimp,' should never make its appearance in large numbers, except in water that is highly charged with salt, as in the 'salt pans' in which sea-water is undergoing concentration by evaporation; the proportion of salt which is found to be most favourable to its presence, being as much as 4 oz. to a pint of water.—Again, we frequently meet in this group with that reproduction by 'gemmation,' of which so remarkable an example has been brought under notice in the class of Insects (§ 315 *d*). In many species there is a double mode of multiplication, the sexual and the non-sexual. The former takes place at certain seasons only, the males (which are often so different in conformation from the females, that they would not be supposed to belong to the same species, if they were not seen in actual congress) disappearing entirely at other times; whilst the latter continues at all periods of the year, so long as warmth and food are supplied, and is repeated many times (as among the Aphides), so as to give origin to as many successive 'broods.' Further, a single act of impregnation serves to fertilize not merely the ova which are then mature or nearly so, but all those subsequently produced by the same female, which are deposited at considerable intervals. In these two modes, the multiplication of these little creatures takes place with great rapidity, the young animal speedily coming to maturity and beginning to propagate; so that, according to the computation of Jurine, founded upon data ascertained by actual observation, a single fertilized female of the common *Cyclops quadricornis* may be the progenitor in one year of 4,442,189,120 young. The eggs of some Entomostraca are deposited freely in the water, or are carefully attached in clusters to aquatic plants; but they are more frequently carried for some time by the parent in special receptacles developed from the posterior part of the body; and in many cases they are retained there until the young are ready to come forth, so that these animals may be said to be ovo-viviparous. In the common *Daphniæ*, the eggs are received into a large cavity between the back of the animal and its shell, and there the young undergo almost their whole development, so as to come forth in a form nearly resembling that of their parent. Soon after their birth, a moult or exuviation of the shell takes place; and the egg-coverings are cast off with it. In a very short time afterwards, another brood of eggs is seen in the cavity, and the same process is repeated, the shell being again exuviated after the young have been brought to maturity. At certain times, however, the *Daphnia* may be seen with a dark opaque substance within the back of the shell, which has been called the *ephippium* from its resemblance to a saddle. This, when carefully examined, is found to be of dense texture, and to be composed of a mass of hexagonal cells; and it contains two oval bodies, each consisting of an ovum covered with a horny casing, enveloped in a capsule which opens like a bivalve shell. The first traces of the 'ephippium' are seen after the third moult, as a green matter in the ovaries, which differs both in colour and appearance from that of the eggs; after the fourth moult, this green matter passes from the ovaries into the matrix or open space on the back, and there becomes developed into the ephippium; and at the fifth moult this is thrown off, and the ephippium, with the two eggs inclosed, floats on the water until the next spring, when the young are hatched with the returning warmth of the season. This curious provision is obviously destined to afford protection to the eggs which are to endure the severity of winter cold; and some approach to it may be seen in the remarkable firmness of the envelopes of the 'winter eggs' of some of the Rotifera (§ 310 *b*). It has been ascertained by Dr. Baird, that the young produced from the ephippial eggs have the same power of continuing the race by non-sexual reproduction, as the young developed under ordinary circumstances. In most Entomostraca, the young at the time of their emersion from the egg differ considerably from the parent, especially in having only the thoracic portion of the body as yet evolved, and in possessing but a small number of locomotive appendages; the visual organs, too, being frequently wanting at first. (See Fig. 160, B, C.) The process of development, however, takes place with great rapidity; the animal at each successive moult (which process is very commonly repeated at intervals of a day or two) presenting some new parts, and becoming more and more like its parent, which it very early resembles in its power of multiplication, the female laying eggs before she has attained her own full size. Even when the Entomostraca have attained their full growth, they continue to exuviate their shell at short intervals during the whole of life; and the purpose which seems to be answered by this repeated moulting, is the preventing the animal from being injured, or its movements obstructed, by the over-growth of parasitic Animalcules and Confervæ; weak and sickly individuals being frequently seen to be so covered with such parasites, that their motion and life are soon arrested, apparently because they have not strength to cast off and renew their envelopes. The process of development appears to depend in some degree upon the influence of light, being retarded when the animals are secluded from it; but its rate is still more influenced by heat; and this appears also to be the chief agent that regulates the time which elapses between the moultings of the adult, which, in the *Daphnia*, take place at intervals of two days in warm summer weather, whilst several days intervene between them when the weather is colder. The cast shell carries with it the sheaths of the limbs and plumes, and of the most delicate hairs and setæ which are attached to them. If the animal have previously sustained the loss of a member, it is generally renewed at the next moult, as in higher Crustacea.*

317 *b*. All the Entomostraca at present existing are animals of minute size, many of them being scarcely discernible by the unassisted eye, and the largest of them (such as the *Branchipus*) scarcely exceeding an inch in length. The exuviæ of even the smaller species, however, may accumulate to such an extent, as to form masses of no inconsiderable size. Thus in certain fresh-water strata of the Secondary and Tertiary formations, we find layers, sometimes of great extent and thickness, which are almost entirely composed of the fossilized shells of a species of *Cypris*, none of them much exceeding a line in length; and even in the marine Cretaceous deposit, considerable masses are sometimes to be found, composed of the remains of bivalve Entomostracous shells, probably belonging to the genus *Cytherina*. It was in the earlier ages of the earth's history, however, that the Entomostraca seem to have attained their highest development.

* The statements of experimenters upon this subject do not agree with each other, some asserting that little or no reparative power of this kind exists among the Entomostraca. But it is probable that, as in other instances, this power may be greater in the young animal than in one which is nearer the close of its life, and that it may be partly dependent upon the temperature (§ 98).

During the whole of the Palæozoic series of formations, we meet with the fossils termed *Trilobites* (from the division of the body longitudinally into three lobes), which seem to have been essentially Entomostracous, although resembling the Isopod order among the higher Crustacea in certain points of their conformation. The absence of all vestiges of appendages of any kind to the under side of the body, makes it certain that, whatever was their form, these must have been so soft as to be readily decomposable; since other parts, even to the facetted structure of the compound eyes, are found in most perfect preservation. It seems probable, therefore, that they must have belonged to the group of *Branchiopoda;* to some existing members of which they seem most nearly allied in the structure of their eyes also, as well as in other points of their organisation. The cephalic shield of the Trilobites did not extend far back over the body; but the segments of the thorax and abdomen were enclosed in a succession of distinct rings, of which the dorsal portion (which is alone preserved) was of firm consistence; and the last segment also frequently bore an expanded caudal shield. The segments of the thorax were from ten to fifteen in number, and those of the abdomen, in some species at least, eight; so that in this essential particular the Trilobites corresponded with the lower rather than with the higher division of the Crustacean class. Generally speaking, the segments of both thorax and abdomen were nearly similar to each other in form, and gradually diminished in size from before backwards; sometimes the abdomen narrowed into a tail, which was bent under the body like that of Crabs. The flexibility of the body, in a large proportion of the group at least, has been made evident by the fact that the remains of many species have been found coiled up like 'wood-lice.' Evidence has recently been obtained that these animals came into the world in a condition very unlike that of their parents; and that they only attained this after a long succession of exuviations, some progress being apparent at every change.*—It is interesting to remark that the *Limulus* (§ 316 *a*), which may be regarded in some respects as a gigantic Entomostracan, seems to have been introduced at nearly as early a period as the Trilobites, its remains being coeval with theirs in the Carboniferous strata; and that the earliest forms of this genus present characters of approximation to the Trilobites, which are also shown in the immature states of the existing Limuli.

317 *c*. The most complete and remarkable departure from the ordinary Crustacean type is presented by the sub-class of *Suctorial* Crustacea, in which the mouth, instead of being furnished with mandibles and maxillæ adapted for the division of solid aliment, is prolonged into a kind of proboscis adapted only for the reception of liquid substances. As in Insects (§ 315),

* M. Barrande of Prague, who is preparing a work on the Silurian system of Bohemia, in studying the numerous Trilobites which he has collected in that country, has traced for the first time the development of a Trilobite from its embryonic state to its adult condition; and has observed twenty successive stages, during which this one species undergoes very remarkable changes of organisation, passing from a simple disk-like body to a fully formed trilobite with seventeen free thoracic segments and two caudal joints. This discovery diminishes the number of the so-called species; by showing that as many as ten genera, including eighteen species, have been made out of the successive forms of one and the same trilobite (Athenæum, July 7, 1849). There can be no doubt that a similar reduction in the number of species will take place in many other cases, alike amongst existing and extinct groups, when the *whole history* of each species shall have been made known.

however, the parts which compose the mandibulate mouth of the preceding groups may be recognized in that which is modified for suction; the proboscis itself being formed by a prolongation of the upper and lower lips, and being usually accompanied by a pair of long appendages that represent the mandibles and serve to make the punctures in which the proboscis is inserted; whilst other appendages, representing the feet-jaws, are furnished with hooks by which these parasites attach themselves to the animals from whose juices they derive their nutriment. Many of the Suctorial Crustacea bear a strong resemblance, even in their adult condition, to certain Entomostraca; but more commonly it is between the immature forms of the two groups that the resemblance is the closest, most of the *Suctoria* undergoing such extraordinary changes in their progress towards the adult condition, that if their complete forms were alone attended to, they might be excluded from the class altogether, as has (in fact) been done by many Zoologists. As an example of this curious transformation, we may take the *Lernœa*, a parasite which infests the gills and eyes of Fish. The form of the adult female is represented in Fig. 160, A; where we see its long suctorial proboscis at *a*, its short thorax at *b*, bearing the single pair of legs, *c*, by which the animal is attached, the large abdomen, *d*, and the two egg-capsules, *e*; the legs are united to each other at their extremities, and bear a kind of sucker, *f*, which is applied to the surface of the animal attacked. The integument of the whole body and of the limbs is soft, and scarcely displays a trace of segmentation; the animal has hardly any locomotive power, its whole anterior portion being usually buried in the substance of the part from whose juices its nutriment is derived. But however dissimilar the form and characters of such a creature may appear to those of the ordinary Crustacea, its connection with them is established in two ways. In the first place, there are numerous parasitic species which depart much less from the ordinary type, forming a gradual transition from the Entomostraca to the most aberrant Suctoria; so that, if the latter are to be excluded, it would be difficult to say where the line is to be drawn. And, again, the early forms of *Lernœa* and its allies, so strongly resemble those of the *Cyclops* and other Entomostraca, that it would be impossible to point out any obvious distinction between them. It is interesting to observe how the activity of the young animal, in most or all of these parasitic Crustacea, compensates for the comparative immobility of the adult, which shows scarcely any other sign of life than taking food and producing eggs; thus effecting the dispersion of the progeny, which, if they were all to attach themselves to one individual, would soon destroy it, and would then become themselves extinct from the want of food. The two large egg-sacs, which, when filled with ova, give such a peculiar aspect to these animals, are only a higher development of the receptacles which are

FIG. 160.

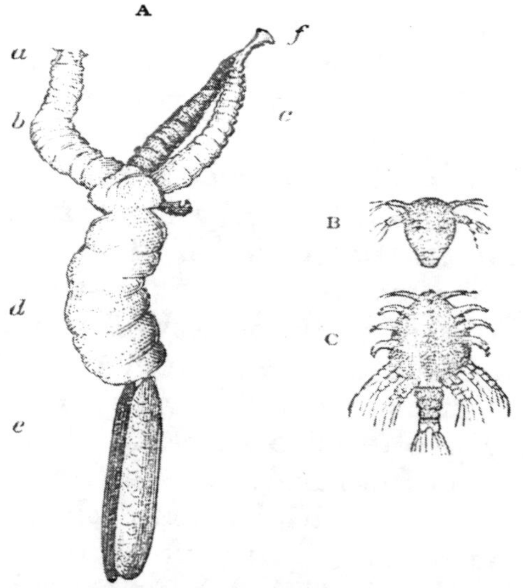

Lernœa in adult state (A), and in larva conditions (B and C):—*a*, suctorial proboscis; *b*, thorax; *c*, single pair of legs, serving for attachment by the sucker *f*; *d*, abdomen; *e*, egg-capsules.

possessed by many of the Entomostraca, and of which we trace the rudiments in some Rotifera. Within these egg-sacs the embryoes very commonly undergo their whole development, up to the point at which they are ready to come forth from the egg, and to enter upon active life. At this period they are of an oval shape, having a large eye situated in the centre of the anterior and upper part of the body, and two pairs of swimming feet, with a pair of jointed antennæ (Fig. 160, B). After the first moulting, the body is seen to manifest a division into two parts, the anterior of which is furnished with three pairs of hooked feet, and the posterior with two pairs of swimming feet (C). Not only the greater number of the feet, but also the eye, subsequently disappear, when the animal becomes attached; but the precise grade of development at which this takes place has not yet been accurately determined. As in the Entomostraca generally, the adult male differs from the adult female in many important characters, although the larval state of the two seems to be precisely the same. Instead of the single pair of united feet, provided with a suctorial disk, he has two pairs of strong hooked feet; and the relative proportions of the several parts of his body are very different; the whole being much smaller than that of the female, sometimes not above a fifth part of her size. It is curious, too, that the males bear a strong resemblance to each other, even in different genera, of which the females are very dissimilar.—Among those suctorial Crustacea which present a nearer approach to the ordinary Entomostracous type, may be specially mentioned the *Argulus foliaceus*, which attaches itself to the surface of the bodies of fish, and is commonly known under the name of the 'fish louse.' This animal has its body covered with a large firm oval shield, which does not extend, however, over the posterior part of the abdomen. The mouth is armed with a pair of styliform mandibles; and on each side of the proboscis there is a large short cylindrical appendage, terminated by a curious sort of sucking disk, with another pair of longer jointed members, terminated by prehensile hooks. These two pairs of appendages, which are probably to be considered as representing the feet-jaws, are followed by four pairs of legs, which, like those of the Branchiopoda, are chiefly adapted for swimming; and the tail, also, is a kind of swimmeret. This little animal can leave the fish upon which it feeds, and then swims freely in the water, usually in a straight line, but frequently and suddenly changing its direction, and sometimes turning over and over several times in succession. The stomach is remarkable for the large cæcal prolongations which it sends out on either side, immediately beneath the shell; for these subdivide and ramify in such a manner as to be distributed almost as minutely as the cæcal prolongations of the stomach of the *Planariæ* (Fig. 141). The proper alimentary canal, however, is continued backwards from the central cavity of the stomach, as an intestinal tube, which terminates in an anal orifice at the extremity of the abdomen.

318. From the parasitic suctorial Crustacea, the transition is not really so abrupt as it might at first sight appear, to the class of CIRRHIPODA, consisting of the *Barnacles* and their allies, which, like many of the Suctoria, are fixed to one spot during the adult portion of their lives, but come into the world in a condition that bears a strong resemblance to the early state of many of the true Crustacea. The departure from the ordinary Crustacean type in the adult, is, in fact, so great, that the

animals belonging to this group were formerly ranked among the Mollusca, to which they bear many points of resemblance. Thus, the body and its appendages are themselves quite soft; and the skin has the loose, spongy, muscular character which is characteristic of the 'mantle' of the Mollusks (§ 297). The shell, which is formed upon its surface, corresponds closely with that of Mollusks in its general characters; and although it is 'multivalve,' or composed of several pieces, yet these pieces are not arranged in such a manner as to bear any resemblance to the dermal skeleton of an articulated animal,—such a resemblance, in fact, being much more strongly presented by the shell of the *Chitons* (Fig. 125) whose general structure is essentially Molluscous. Further, the shells of the Cirrhipoda are either themselves firmly united at their bases to rocks or other solid masses (not unfrequently to the bodies of living animals); or they are attached by a long peduncle or foot-stalk (Fig. 161, A); no one species being free in its movements in any but the early period of its life: so that the conditions in which these animals exist, closely resemble

FIG. 161.

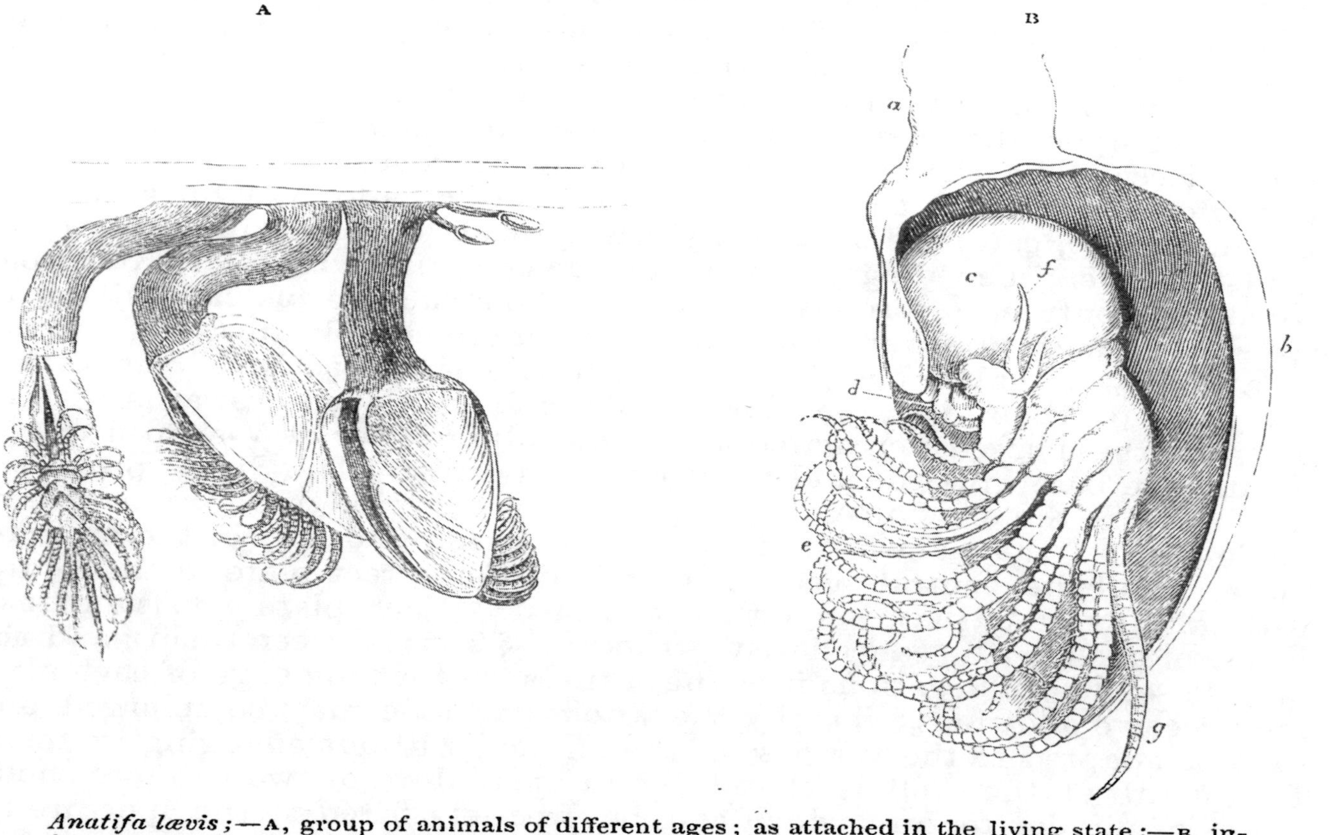

Anatifa lævis;—A, group of animals of different ages; as attached in the living state:—B, interior structure, enlarged, as shown by the removal of the valves of one side; *a*, peduncle; *b*, mantle; *c*, cephalic portion of the body; *d*, mouth; *e*, articulated members; *f*, flabelliform appendages (or branchiæ); *g*, abdominal appendage.

those to which we find the Mollusca especially adapted.—On the other hand, when we examine the animals themselves, we find them presenting a perfectly symmetrical form, with indications of a segmental division of the body into six parts (B); whilst from each of these segments there arises a short pair of feet, obscurely divided into three joints, and bearing a pair of long, slender, many-jointed *cirrhi*, or tendril-like appendages (*e*) fringed with delicate filaments, and curled towards the mouth. These appendages, from which the class derives its name, are furnished with cilia, the continual vibration of which produces currents in the water, that bring food to the mouth, and also aerate the blood transmitted to the respiratory organs; hence they may be likened, on the one hand, to the ciliated legs of many Entomostraca (§ 317), whilst, on the other, they bear no inconsiderable resemblance to the spirally-coiled 'cirrhi' of the Brachiopoda, which seem to perform a similar function (§ 300).—The anterior extremity (c) of the body is so modified as to form the organ of attachment of the animal; being expanded into a broad disk or basis of adhesion in the *sessile* tribes, such as the *Balanus* or 'acorn-shell' and its allies, which attach themselves directly to the body that affords them support, by the entire base of their somewhat conical shells; but being produced into a long and flexible stalk (a) in the *pedunculate* forms of this class, as the common *Anatifa* or 'barnacle,' which float more freely in the water, being generally attached to some object that has motion through it, such as a log of timber or a ship's bottom. The general organisation is very nearly the same, however, in both these tribes, notwithstanding that they differ widely, not merely in the mode of their attachment, but also in the structure of their shells. The shell of the 'pedunculate' species (in some of which this protection is very imperfect) is formed upon the plan of that of bivalve Mollusca, with certain accessory pieces; and the two valves can be drawn together, so as completely to enclose and protect the animal, by means of an adductor muscle. The peduncle, also, can be shortened by the contraction of its muscular coat; whilst it is lengthened again by the elasticity of its external fibrous coat, when the contractile force is removed. The structure of the shell in the 'sessile' species, however, is usually much more complicated. It consists of a nearly circular base, from which rise six Λ-shaped plates with the base of their triangle downwards and the apex upwards; whilst between these are intercalated a corresponding series of V-shaped plates having the base upwards and the apex downwards. The dorsal piece consists of one of the erect or Λ-shaped plates; on each side there are two pieces made up by the adhesion of one of the erect, and one of the inverted plates; and the ventral piece is composed of an erect plate in the centre, with an inverted plate on either side of it. Each plate consists of two layers of shell, with 'cancellated' structure (§ 205) between them; and the mantle sends prolongations into the sutures, so that the edge of each plate may receive additions, whereby the whole envelope may be enlarged, and thus be adapted to the increase of the animal, without changing its form. The mouth of the shell is closed by an operculum of two or more shelly plates of similar texture, which can be precisely fitted to the aperture by means of the muscles that act upon it, so as to afford a complete protection to the animal within. The shell of one of the sessile Cirrhipods, *Tubicinella*, which is parasitic upon the whale, is developed as a nearly cylindrical tube of considerable length; whilst, on the other hand, in the *Coronula*, which is also parasitic upon the whale, the shell is a flattened cone, its base being extended by the enormous development of the cancellated structure which intervenes between its inner and outer wall, and its cavity being small in comparison with its external diameter. Scarcely a trace of this cancellated tissue is to be found in the shells of Mollusks;

and its presence in very large amount in the curious extinct group of *Rudistes* (including the *Hippurite*, *Sphærulite*, and their allies), taken in connection with other characters presented by them, is an indication that this group had very intimate relations with the Cirrhipods, which it probably served to connect with the Bivalve Mollusks.

318*a*. Turning now to the internal structure of these animals, we find, on the under side of the cephalic portion of the body (*c*), the mouth (*d*), which is provided with a broad upper lip, with two palpi or feelers, and with three pairs of jaws, toothed at their edges, and opening laterally as in all Articulata. The marine animalcules, brought by the currents that are kept up by the action of the cirrhi, after passing through the mouth, are transmitted by a short œsophagus to a dilated stomach lying beneath the cephalic hood; into this stomach the ducts of two long glandular follicles, probably salivary, discharge themselves; and from its walls are developed numerous hepatic cæca, which constitute a rudimentary form of liver. From the stomach there passes off an intestinal tube, which follows a slightly sinuous course, and terminates by an anal orifice at the base of the long, jointed, caudal appendage (*g*), which is seen at the opposite extremity of the body. The circulating system has not been thoroughly made out; but its organ of impulsion is a 'dorsal vessel,' which transmits the blood through the body, and to the respiratory organs. Although its circulation through the cirrhi, in which a double current can be seen, probably contributes to the aeration of the blood, yet there seem to be also special respiratory organs, though of a different character in the two groups respectively. In the 'pedunculate' Cirrhipods they resemble those of the Crustacea, in being attached, like flabelliform appendages (*f*) to the base of the maxillary feet, and to that of some of the cirrhigerous legs; whilst in the 'sessile' tribes, the branchiæ are formed by duplicatures of the mantle, as in the Mollusca generally.—The nervous system of this group has the truly articulated character. The cephalic or supra-œsophageal ganglia are but little developed, in accordance with the almost entire deficiency of organs of special sense; but a pair of cords passes backwards from them, diverging to surround the œsophagus, at the two sides of which we find the ganglia of the first segment of the body, meeting again below in the coalesced ganglia of the second segment, and then proceeding along the ventral surface of the body to its caudal extremity, with a pair of ganglia for every segment. No traces of eyes or of auditory organs can be found in the adult; and the only indication of sensation which these animals exhibit, consists in the retraction of their cirrhi and the closure of their shells, at the sound, or more probably at the vibration, of an approaching footstep, or even, in some of the sessile species, at the atmospheric movement caused by the approach of the hand within some inches' distance—The nature of the generative organs in these animals has not yet been clearly elucidated. All the Cirrhipods with which we are at present acquainted possess ovaria and deposit fertile ova; and these must consequently either possess male in addition to female organs, or the male must exist separately under some entirely different form, which conceals its relations to the class. A large glandular organ is found in the pedunculated species, covering the viscera immediately beneath the muscular tunic of the body, and extending from the mouth to the anus; its numerous ducts unite into three or four principal trunks, which terminate in a lateral receptacle at the side of the intestine; and from this a single or double canal is prolonged to the very extremity of the extensile tail (Fig. 161, *g*), where it terminates. This organ is by some supposed to be an ovarium, whilst by others it is considered as a testis. Certain it is, that ova are found in the hollow peduncle; and those who regard the organ just described as a testis, believe that the ova are developed in that situation; whilst those who consider that organ as an ovarium, believe that the ova are conveyed by the long tail or ovipositor to the peduncle, whose hollow serves as a receptacle in which their early development may take place. In whatever way the ova are fertilized (and the idea of a distinct, smaller, and differently formed male, is the one conformable to the analogy of the Entomostracous and Suctorial Crustacea to which these animals are nearly allied), it is certain that the early history of the young Cirrhipods presents phenomena of the most interesting character. So strange, indeed, are the metamorphoses which these animals undergo, that when they were first made known by Mr. V. Thompson,* his account of them was scarcely credited by Zoologists, who were not then aware, as now, in how large a number of instances, phenomena of the same order present themselves among the Invertebrated classes. —The young of *Balanus*, at the time of its emersion from the egg, is enclosed in a bivalve shell, of which the valves are united by a hinge along the back, and open along the lower margin, for the protrusion of a large and strong anterior pair of prehensile limbs, provided with an adhesive sucker and hooks, and of six pairs of posterior legs, adapted for swimming. The form and general condition of the animal thus bears a strong resemblance to that of the Daphnia (common 'water-flea'); and its movements, too, are similar, consisting of a succession of jerks. A pair of compound eyes, supported on a peduncle, is seen on the anterior part of the body. Nearly all these organs are thrown off together; and the animal assumes an entirely different form and condition. It first attaches itself to a neighbouring surface by a layer of calcareous matter, which is subsequently to become the base of its new shell; the valves that surround its body are gradually developed on its envelope or mantle; the new ciliated arms begin to grow within; and the eyes disappear, the animal being henceforth destitute of visual organs.—The form of the young *Barnacle*, however, is different, having more resemblance to that of *Cyclops*. The animal is not enclosed in a bivalve shell, but its back is covered by an expanded shield like that of *Argulus* (§ 317, *c*); it has a single eye, not pedunculated; and three pairs of members, of which the anterior is simple, while the posterior are bifid.—Hence if their early state be alone considered, the Cirrhipods might be regarded as an aberrant order of Crustacea; but the complete change, not merely of form, but of condition, which they undergo in their passage to their adult state, and the peculiar transition which this presents towards the Molluscan type, seem to remove them to a far greater distance from the true Crustaceans, than that at which the Lernæa and its allies may be placed; and may be held to justify the erection of this group into a distinct class,—unless, indeed, it should prove that separate males exist, retaining more of the ordinary Crustacean conformation.

* Zoological Researches, Fourth Memoir, 1830; and Philosophical Transactions, 1835.

319. Returning now to the air-breathing Articulata, we find in the class of ARACHNIDA, which includes *Spiders*, *Scorpions*, and *Mites*, much of that concentration of structure which has presented itself to our notice in the highest members of the Crustacean class, and which, whilst it is a departure from the typical condition of Articulata, may be considered as in some degree indicating an approximation to the Vertebrated series,—an approximation, however, not so much in *form*, as in *character*. The Arachnida are nearly related to Insects in many points of their organisation; but are usually distinguishable from them at the first glance by the general form of their bodies, and by the number of their limbs. Instead of the three regions which may always be distinguished in perfect Insects, —the head, thorax, and abdomen,—the Arachnida have only two, the head and thorax being (as it were) fused together into one mass, which is termed the *cephalo-thorax:* whilst instead of the *six* legs which characterise the perfect Insect, the Arachnida have always *eight*, all of them attached to the cephalo-thorax. Of *wings* there is not the least trace. The *antennæ*, of which Insects possess *one* pair, and the higher Crustaceans *two*, are here entirely wanting; but the *palpi*, which are appendages to the maxillæ (or lower pair of jaws), are usually developed to a large size in the Arachnida, so as often to resemble the *chelæ* or pincers of Crabs and Lobsters. The tegumentary skeleton of Arachnida varies considerably in its degree of firmness, in the different members of the class; and the segmentation of the body is least apparent in those which have the softest integument. Thus in Spiders and Acari, the integument is generally soft, especially in the abdomen, which is a soft spheroidal mass without any external trace of segmental division; whilst in the 'cephalo-thorax' and limbs it is a little harder, and some degree of articulation may be perceived. In the Scorpions, on the other hand, it is almost as dense as among Beetles, and (as in that tribe) is hardened by *chitine* (§ 314); and not only is the segmentation of the 'cephalo-thorax' and of the limbs there quite distinct, but the abdomen also, the posterior part of which is very slender and much prolonged, is obviously divided into several rings.—This class may be distributed into two principal groups, distinguished by several important differences in their internal structure, and therefore requiring separate consideration. The first of them, which is that of *Pulmonaria*, containing the Spiders and Scorpions, is especially characterised by the condition of their respiratory apparatus, which consists of pulmonary cavities instead of tracheæ (as in Insects), and by the completeness of their circulating system; they are also distinguished from some of the second group, which greatly resemble them in external appearance, by the number of their eyes, which is six, eight, or even more. The two great tribes of *Araneidæ* (Spiders) and *Scorpionidæ* (Scorpions), which make up this group, differ rather in the form of their bodies, and in certain special modifications of particular organs, than in their general plan of conformation; thus we find in Spiders a soft abdomen of more or less spheroidal form, provided with 'spinnerets' at its extremity, and the palpi are comparatively small: whilst Scorpions have a prolonged and slender abdomen, which is not furnished with spinnerets, but usually bears a poison-apparatus in their place; and the palpi are large, equalling or exceeding the legs in size, extended like arms, and furnished with pincers. These two groups are connected by the family

Thelyphonidæ (inhabiting tropical regions only), in which the general form of the body is that of the Spider, but the spinnerets are wanting, whilst the palpi are developed into large members as in the Scorpion.—The second principal subdivision of the Arachnida is composed of the *Trachearia*, so named from the mode of respiration, which is carried on by tracheæ (as in Insects), instead of by pulmonary sacs. In this group the circulating apparatus is very incomplete, and in some members of it the digestive cavity is much extended, as if to compensate for the want of vessels. Altogether the plan of conformation is much lower; and although the *form* of some of the members of this group frequently presents a close resemblance to that of the Spiders and Scorpions, they may be at once distinguished by the absence or by the fewness of the eyes, the number of which never exceeds four. The typical family of this group is that of *Acaridæ* (Mites), which presents many characters of approximation to the wingless parasitic Insects.—The Arachnida in general differ from Insects, not merely in the structure of their respective adult forms, but in the mode in which those forms are attained; for the higher members of this class undergo no metamorphosis, but come forth from the egg in a form and condition resembling, in all essential particulars, that of their parents, to the production of which form, the whole course of evolution within the egg manifestly tends; whilst in Insects, as we have seen, the first process of evolution tends to produce a worm-like being, which only attains the higher type by the subsequent development of the organs that characterise it. The *Acaridæ*, however, undergo a slight change of form, coming forth from the egg with only six legs, and not acquiring the additional pair until after the first moult; and in certain of the aquatic and parasitic species of this family, a greater modification takes place, reminding us of that which is seen among the Suctorial Crustacea (§ 317*c*).

319 *a*. The mouth, in all the higher Arachnida, is formed upon the *Mandibulate* type; for although many of these animals live exclusively upon the juices of others, they do not obtain these by imbibing them as parasites from living bodies, but by first destroying their victims and then availing themselves of the liquids which their dead bodies afford. The mandibles, or anterior pair of jaws, situated at the front of the head (Fig. 165, *a*), are long in Spiders, and are furnished with a moveable hook, which has a small orifice near its extremity; and from this orifice a poisonous liquid is poured out, which is secreted by a gland situated in the preceding joint, and which almost immediately proves fatal to the small animals that serve as their usual prey, although too weak to be injurious to Man. In Scorpions, on the other hand, the mandibles are shorter and are terminated by a pair of pincers; and the poison-apparatus is removed to the opposite end of the abdomen. The maxillæ, or second pair of jaws (Fig. 164, *b*, and Fig. 165, *f*), are in themselves usually simple; and the maxillary palpi, which are developed to such an extraordinary extent as instruments of prehension in the Scorpions, are comparatively short in the Spiders, in the males of which, however, they take a remarkable share in the generative function, being (in fact) the intromittent organs. No labial palpi present themselves in ordinary Arachnida; but in the genus *Galeodes* (one of the family of *Pseudo-Scorpionidæ,* which connects the Scorpionidæ and the Acaridæ) we find the head

unusually distinct from the thorax, and it supports the first pair of legs, which nearly resemble the maxillary palpi in form, being short and slender, and destitute of the terminal hooks ; whence it appears that the first pair of legs in the ordinary Arachnida really represent the labial palpi of Insects. In this genus, rudimental antennæ are present, thus presenting another interesting approximation to the Insect type ; and as they are found attached to the mandibles, it is probable that they are to be regarded as ordinarily confluent with them, which the disposition of the nerves of the latter organs would seem to indicate.—The Alimentary canal of the *Scorpionidæ* is very simple in its character, extending in a direct line, without any dilatation into a stomach, or convolution of the intestinal portion of the tube, from the mouth to the anus. In the thoracic region, however, it sends off five short and straight cæcal pouches, which are probably to be regarded as constituting the liver ; whilst at the beginning of that portion of the tube which is continued into the tail-like abdomen, it receives two delicate secreting tubuli on each side, which may perhaps be considered as renal organs. In the *Araneidæ* we meet with a conformation of the digestive apparatus, which reminds us of that of the *Pycnogonidæ* (§ 317 *e*). The aperture of the mouth (Fig. 165, *g*) is very narrow ; since it only needs to receive the juices which are pressed from the body of the victim by means of the maxillary plates (*f*) ; and in several instances the cavity of the mouth communicates with the œsophagus by two or more narrow pharyngeal apertures (Fig. 162, *b*). The œsophagus passes backwards through the nervous collar, and then enters the stomach, which is situated in the middle of the cephalo-thorax, and sends off cæcal prolongations (*c*), which extend into the legs as far as the first joint, and also into the maxillary palpi. These do not seem, however, to be connected with any secreting structure, but appear destined to increase the capacity of the gastric cavity by their dilatation. From the stomach a straight intestine passes backwards, which is narrow at first where it enters the abdomen (at *d*), but then dilates (*e*) to receive four ducts on either side, proceeding from a large biliary apparatus that occupies a considerable part of the abdominal cavity (Fig. 164) ; below these it again contracts (*f*), and forms two short convolutions, after which it dilates into the cæcal enlargement (*g*), receiving two long secreting tubuli (*h*, *h*), which are probably uriniferous. Thus it appears probable that the cæca of the intestinal tube of the Scorpion are the representatives, not of the gastric but of the hepatic cæca of the Spider ; their place in the cephalo-thorax being determined merely by the extreme contraction of the post-abdomen, which throws forward (so to speak) the appendages to the alimentary canal.—The centre of the Circulation is formed upon the plan of the dorsal vessel of the lower Articulata, being a prolonged trunk,

Fig. 162.

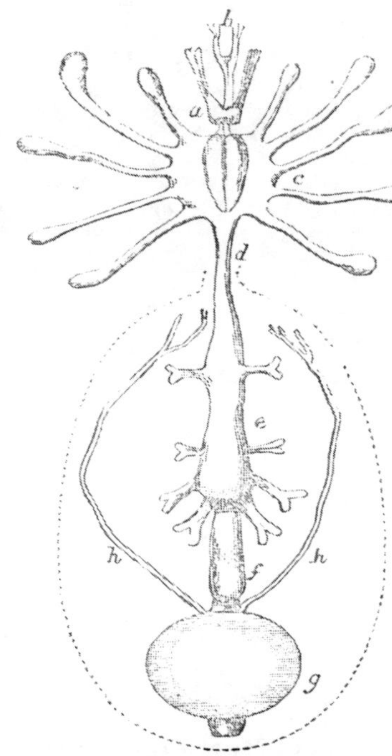

Digestive apparatus of *Mygale* :—*a*, cephalic ganglion ; *b*, bifurcated origin of œsophagus ; *c*, stomach with lateral cæca passing into the limbs and palpi ; *d*, portion which traverses the abdominal peduncle ; *e*, dilated intestine, receiving biliary vessels ; *f*, small intestine ; *g*, cæcal enlargement, receiving the urinary vessels *h*, *h*.

divided by valvular folds into a succession of chambers; but its walls are much thicker and more muscular than they are in Insects; the multiplication of its chambers is not so great; and (in the *Scorpionidæ* at least) the valvular partitions are not complete. In the *Araneidæ,* we find the anterior portion of the dorsal vessel, which lies in the cephalo-thorax, contracted into an arterial trunk; and the pulsatile chambers exist only in its abdominal portion (Fig. 163), where they are reduced in number to four or five. In the *Scorpionidæ,* on the other hand, the pulsatile portion of the dorsal vessel occupies nearly the whole of the body (that is, of the cephalo-thorax, and anterior portion of the abdomen), and contains eight chambers; whilst in the post-abdomen or tail, it is contracted into an arterial trunk. Each chamber, on its dilatation, receives aerated blood from the respiratory organs, by the large venous trunks (Fig. 163, *b, c*); and from each chamber also proceed systemic arteries, through which a part of this blood is impelled by its contraction to the various organs of the corresponding segment. The remainder is propelled forwards, in the *Spider* tribe, by the successive contractions of the pulsatile cavities, through the great dorsal trunk *a,* to be distributed to the cephalo-thorax and its members; whilst in the *Scorpions,* whose segmental heart lies much more in front, the propulsion seems to be partly forward and partly backward, the post-abdomen requiring a supply, as well as the head and thoracic members. The arterial system supplying the body of the higher Arachnida appears to be very complete and minute in its distribution, as will be more particularly explained hereafter (CHAP. XII.); but it seems probable that the canals by which the blood is returned, have rather the character of *sinuses* or *lacunæ* channelled out among the tissues, than of true vessels. By this system of canals, the blood is conveyed to the Respiratory organs, which, from their peculiar conformation, have been termed 'pulmonary branchiæ,' —combining, as they do, the characters of pulmonic cavities and of branchial expansions, especially when the latter are enclosed in cavities of their own, as is the case in the higher Crustacea (§ 316 *d*). These organs are usually four in number on each side, and they open by separate orifices or stigmata, although communicating freely with each other by a connecting tube, so that uniformity in the respiratory action throughout the whole series is secured. Each cavity, covered in by an opercular plate (Fig. 164, *d, d'*), contains a number of lamellæ (*e, e'*) formed by folds of its lining membrane; and over these, the blood which is brought by the respiratory system of vessels is distributed for aeration, to be then returned to the dorsal heart. In certain genera of Spiders, however, this pulmono-branchial apparatus is less complete; and respiration is partly carried on, as in Acaridans and Insects, by means of *tracheæ,* which pass from the pulmonic cavities into the body and limbs.

FIG. 163.

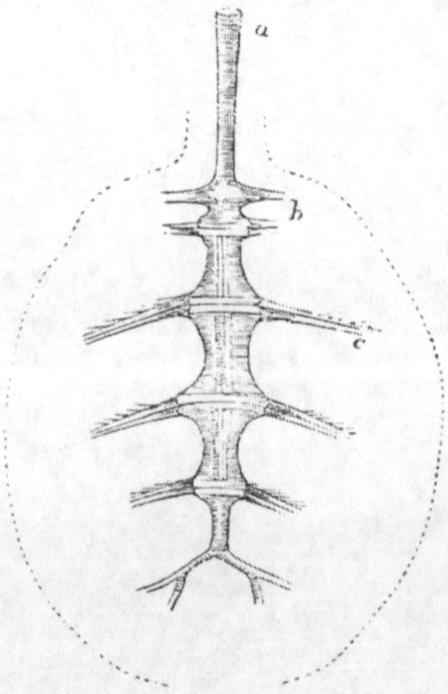

Heart of *Mygale*;—*a,* arterial trunk proceeding to cephalo-thorax; *b,* vessels of the anterior, and *c,* vessels of posterior pulmonic apparatus.

—The character and position of the biliary and urinary organs having been already noticed, it remains to speak of the special organs of Secretion, of which either one or two are found in all the higher Arachnida. The poison-apparatus of the Scorpions is situated at the extremity of the post-abdomen or tail, the slenderness and flexibility of which seems to have special reference to the wielding of this weapon; it consists of a sharp recurved sting, with a minute slit near the point, into which open the ducts of two secreting vesicles which are situated in the last dilated joint of the tail. In the Spiders, on the other hand, the perforated sting forms the second joint of the mandible; and the elongated ovoid vesicle which forms the poison-gland, is lodged in the basal joint, sometimes extending into the cephalothorax. This apparatus corresponds in position with the salivary organs in Insects, which in the Silk-worm are modified to produce the silken thread for the cocoon. The spinning apparatus of the Araneidæ, on the other hand, consisting of the 'spinnerets,' and of the glandular vesicles in which the fluid is elaborated, takes the place of the poison-apparatus of the Scorpions. The usual number of the spinnerets (Fig. 164, *h*) is six, two of them being longer than the rest, and somewhat differently constructed; they are all, however, mammillary projections, of which the apex is perforated, not by one simple aperture, but by a multitude of microscopic pores; and through these the glutinous secretion is forced out in a liquid state, drying into a tenacious thread, however, immediately on coming into contact with air. Thus the fibre of each spinneret is composed of a bundle of as many minute fibrils as there are pores in the spinneret, and the fibrils of all six spinnerets may unite to form a single thread; so that each thread is composed of many thousand fibrils adherent to each other. The size and complexity of the secreting glandulæ vary with the demand for their product occasioned by the habits of the species; thus in the Spiders which are most remarkable for the large size and the regularity of their webs, they occupy, when in full activity, as much as one-fourth of the abdominal cavity, and are composed of slender branching tubes, whose length is increased by numerous convolutions; whilst in those which have only occasional use for their threads, the secreting organs are smaller, and are either short and simple follicles, or undivided tubes of moderate length. There is some reason to think that the character of

Fig. 164.

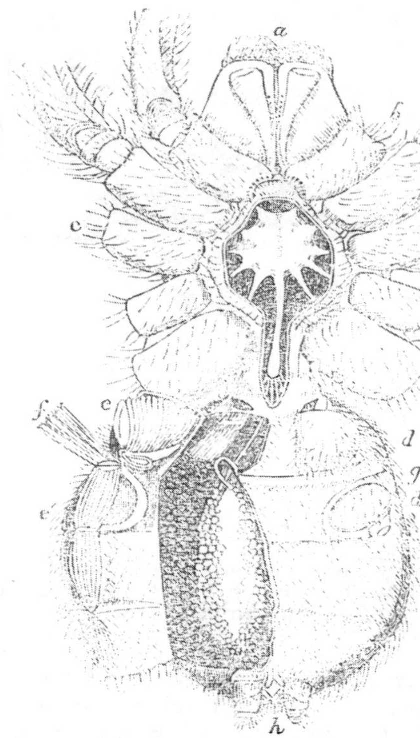

Interior surface of *Mygale* (maconæ), showing the thoracic and abdominal cavities laid open;—*a*, mandibles; *b*, maxillæ, bearing palpi; *c*, first joints of legs, issuing from the thorax, within which is seen the large thoracic ganglion; *d*, *d'*, opercular plates, covering-in the pulmonic organs; *e*, anterior pulmonic cavity laid open; *e'*, posterior pulmonic cavity, partly covered by the longitudinal muscle *f*; *g*, transverse furrow, in the middle of which is the slit forming the entrance to the vulva; from the left-hand corner of this proceeds the oviduct, leading to the ovarium which fills a large part of the abdominal cavity; the chief part of the remainder is occupied by the liver, and by the gland for the secretion of the silk, seen just above the spinnerets *h*.

the secretion formed by these several glandulæ is not precisely the same in all; for differences are perceptible between the circular and the radiating threads in the common Spider's web, which make it probable that they issue from different spinnerets, and that their material may be provided by different glandulæ.*

319 *b*. The Nervous system of the higher Arachnida exhibits a high degree of concentration, and is very analogous (especially in Spiders) to that of Crabs (§ 316 *c*); the cephalic ganglia, however, being in yet closer relation to the subœsophageal mass. The former are fused together into a single bilobed ganglion (Fig. 165, *j*), lying above the œsophagus (*h*), a little behind the mouth; this is connected anteriorly with the eyes (*c*) by as many nerves (*k*) as there are ocelli; and it sends off two great nerves (*k*) to the mandibles; whilst beneath, it gives off two peduncles, which closely embrace the œsophagus (*h*), and

FIG. 165.

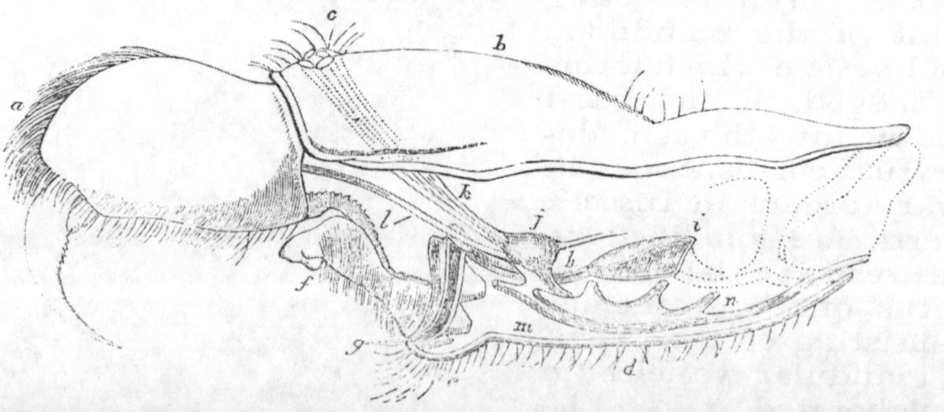

Cephalo-thorax of *Mygale*, viewed from the left side and considerably enlarged;—*a*, mandible; *b*, carapace; *c*, eyes; *d*, section of plastron; *e*, section of lip; *f*, camerostoma; *g*, entrance to the mouth; *h*, œsophagus; *i*, commencement of stomach; *j*, cephalic ganglion, receiving the optic nerves *k*, and sending off the mandibular nerve *l*; *m*, great thoracic ganglion; *n*, its terminal cord.

* The purposes to which these curious threads are applied, vary greatly in the different families of the Spider tribe. Some (as the *Clubiones*) construct for themselves silken tubes, with which they line a cylindrical or conical retreat, formed sometimes of a coiled-up leaf; and from this tube, which has an outlet at both extremities, they issue forth in search of prey, which they take means to entrap by sending forth loose glutinous threads, like so many fishing lines. Others (as the *Segestriæ*) form similar tubes, of five or six inches in length, in the fissures of walls, or other corresponding situations. The *Mygale cementarea*, or mason-spider, again, burrows deeply into the ground, forming a long tortuous gallery, which it lines with a silken tube; and also constructing with the same fabric a very curious trap-door, which effectually closes and conceals the entrance of the burrow. The *Argyroneta aquatica*, or diving-spider, constructs a nest for itself beneath the water; spinning a close hemispherical web, of which the threads adhere to each other so completely that the fabric is impervious to air or water; this web, which remains open at the bottom, is gradually filled with air by the successive descents of the little spider, which carries down a few minute bubbles at each descent, entangled in the hairs beneath its body; and when this operation is complete, this Spider (which, notwithstanding its atmospheric respiration, is aquatic in its habits) takes up its abode in this diving-bell, rears its young there, and passes the winter within it in a state of torpidity. The *Clotho*, again, fabricates a kind of tent, composed of several layers of the most delicate silken web, beneath which the female lives and rears her young.—The Spiders which make such large use of their silken threads in the construction of their habitations are not those which employ it the most for the purpose of entrapping their prey. The webs spun by the latter vary greatly in their degree of regularity and complication; some being little else than long threads of silk rudely matted together; others being close in their texture and without any definite arrangement; whilst the highest type is seen in those beautiful webs, of an open and regular texture, which are widely spread out for the entanglement of flying insects; these spiders either clinging to the centre of their webs, or lying in wait in the neighbourhood, until warned that their net has caught a victim, by the agitation of a thread passing from it to their hiding-place.—But besides these purposes, which have reference to the well-being of the individual, Spiders employ their silken webs for another object, the protection of the eggs (§ 319 *c*).

connect it with the great stellate mass (*m*), which is formed by the fusion of all the thoracic ganglia. (The form of this mass, as seen from below, is shown in Fig. 164). From this mass, five pairs of nerves are given off on each side ; the first to the maxillæ and their great palpi, and the remainder to the four pairs of legs ; whilst posteriorly, a double cord (*n*) is sent backwards towards the abdomen, where it soon subdivides (as seen in Fig. 164) into a bundle of nerves, which radiate to the several parts of the abdominal mass. In *Mygale*, an additional small ganglion is found upon the cord anteriorly to its subdivision ; but this seems to be wanting in other Spiders.—In *Scorpions*, however, the nervous system is less concentrated, as might be anticipated from the prolongation and more complete segmentation of their bodies ; and it bears somewhat the same relation to that of Spiders, that the nervous system of the Macrourous Decapods bears to that of the Brachyourous (§ 316 *c*). The ganglia of the thoracic and of part of the abdominal region coalesce into one stellate mass (Fig. 166, *e*), as in Spiders ; and this supplies the thorax and its members, and the anterior portion of the abdomen, including the pulmonary branchiæ ; the ventral cord of the posterior part of the abdomen, however, has ganglia of its own (Fig. 166, 10–16), which are very small in its caudal prolongation. The general distribution of the nerves proceeding from these ganglia will be seen in the accompanying figure.—The Eyes of the higher Arachnida exhibit a very interesting intermediate condition between the articulate and the vertebrate type. We nowhere meet with the large compound eyes, composed of multitudes of clustered ocelli, which are so characteristic

FIG. 166.

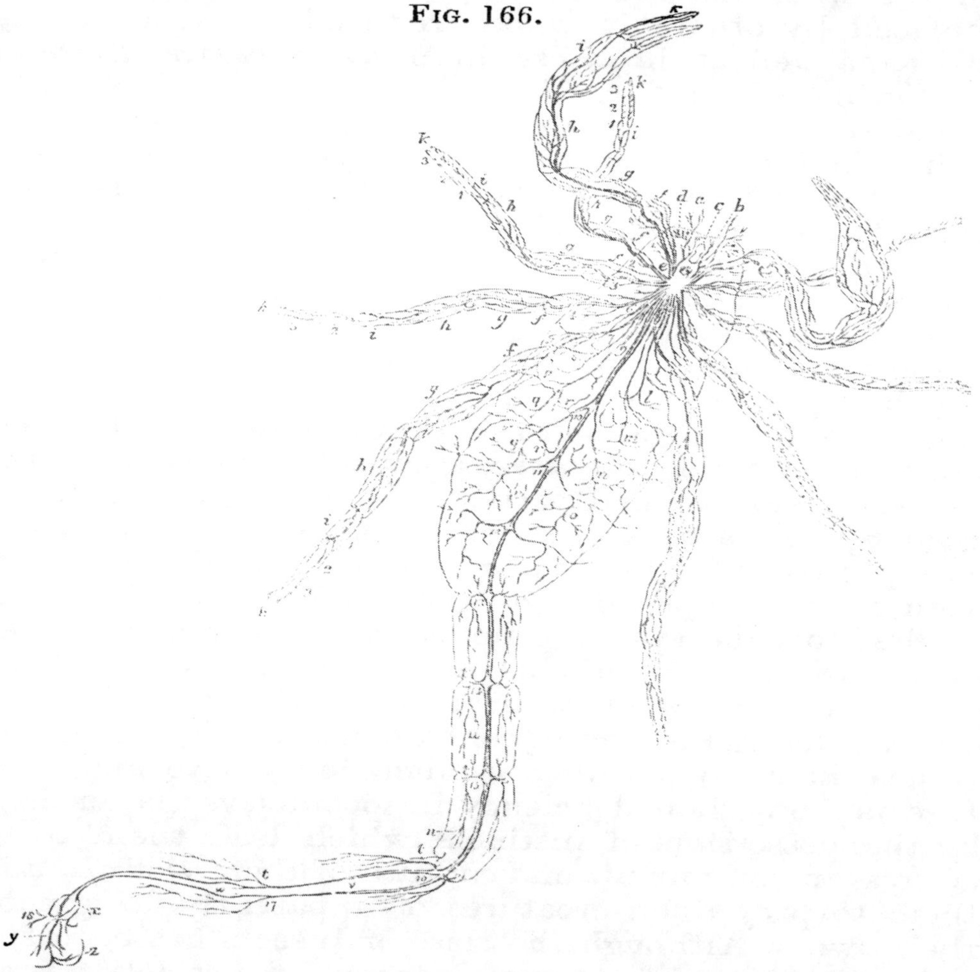

Nervous system of *Androctonus* (Scorpion):—*a*, antennal nerves; *b*, cephalic ganglia; *c*, principal optic nerves; *d*, lateral ocelli and nerves; *e*, subœsophageal ganglion; *f*, coxa; *g*, femur; *h*, tibia; *i*, basal joint of tarsus; 2, 3, second and third joints; *k*, terminal nerves to double claw; *l*, *m*, *n*, *o*, nerves to abdominal branchiæ; *p*, *q*, *r*, *s*, nerves to the segments; *t*, *u*, nerves of the caudal ganglia; *v*, terminal nerves; *w*, fifth joint of tail; *x*, anal collar; *y*, *z*, termination of the cord in the extremity of the sting; 1 to 9, nerve-trunks from the great subœsophageal ganglion; 10 to 16, ganglia of the ventral cord.

of Insects and Crustaceans; but in their stead we find six or eight simple eyes, sometimes all clustered together in one group at the anterior part of the cephalo-thorax (Fig. 165, *c*), sometimes divided in such a manner that some of them are on the median line, while the remainder form two lateral and symmetrical groups. In their internal structure, these eyes have at least as much resemblance to those of Vertebrated animals, as to the single ocelli which make up the large composite eyes of Insects; but it is interesting to remark that the simple eyes, placed at the top of the head in many members of the latter class, represent the more perfect organs of the same kind in the Arachnida. Thus on the borders of the sub-kingdom Articulata, we find the peculiar visual organs, which are on the whole so characteristic of the group, entirely disappearing, and replaced by others in whose reduced number and isolation from each other, as well as in whose internal structure, there is an obvious foreshadowing of the Vertebrated type. There can be no doubt that the visual sense is very acute in these animals, especially in such as do not spin webs for the capture of their prey, but lie in wait for it, and pounce upon it as soon as their sight informs them of its sufficient proximity.—It is obvious from the effect of sounds upon them, that Spiders have the sense of Hearing; but no special organ of this sense has yet been discovered. The same may be said of the sense of Smell. It is possible that the membrane lining the mouth and pharynx may have sufficient gustative sensibility to influence these animals in their selection of food; but their predaceous operations appear to be principally directed by the sense of sight. The softness of the general integument of Spiders probably allows them to possess a considerable share of the Tactile sense over the whole surface, and particularly in that of the anterior limbs, which in many Spiders are used rather for palpation than for locomotion; and this may probably be the explanation of the absence of the *antennæ*, which seem to be the special organs of touch in the higher Articulata generally.—The Muscular system is not so greatly developed in Arachnida, relatively to the size of these animals, as it is in Insects; and this might be anticipated from the want of firm points of support and attachment for the contractile organs, consequent upon the softness of their dermo-skeleton. Hence their movements are much less active and vigorous than those of Insects; but their deficiency in locomotive power is amply compensated by the possession of instincts, which lead them to employ the greatest address in the pursuit and capture of their victims, and which thus enable them to prey upon creatures of apparently far greater capabilities than their own. Although the class of Insects has been spoken of (§ 315 *c*) as presenting the highest manifestation of instinctive tendencies not complicated by the operation of intelligence, yet it can scarcely be said that the Arachnida are inferior to them in the former respect; for whether we look at the construction of their habitations, the various devices to which they have recourse in order to secure their prey, or the care and protection which they afford to their eggs, it is difficult to point to any tribe of Insects that surpasses them,—the economy of the *social* Hymenoptera and Neuroptera being wonderful rather as regards the *combination* of the operations of multitudes of individuals, than as regards the actions themselves, which are perhaps equalled, in the elaborateness of their adaptation to particular ends, by those of many Spiders. There is every reason to believe, that, neither in the one case nor in the other, is this adaptation *designed* by the animal; both the Insect and the Spider appear, from the extraordinary uniformity of their operations, to be working out (so to speak) the law of their respective natures; but there are indications, in the Spider tribe, of the occasional operation of a degree of *intelligence*, which Insects do not possess. This is especially manifested in their *educability;* many instances being on record, in which Spiders have learned to recognize the individuals who cherished them, and have been trained by them to actions foreign to their nature. In this respect, then, as in several others that have been mentioned, do the higher Arachnida depart from the general characters of the Articulate series, and exhibit a certain degree of approximation towards the Vertebrata.

319*c*. Like the other classes which form the higher group of Articulata, the Arachnida possess only one method of propagation; that, namely, which is accomplished by sexual union; and there are no known exceptions to this rule, such as present themselves among Insects (§ 315 *d*) and Entomostracous Crustacea (§ 317). They retain, however, some vestiges of the method of multiplication by gemmation, in the power of reproduction of lost parts, which they seem to possess in about an equal degree with the higher Crustacea (§ 316 *d*). If either of their limbs be broken off, it is replaced by a new growth; but this must take place from the articulation nearest the body, and the remainder of the limb is cast off at that joint. The new limb is not seen externally until after the next moult of the integument, and it is then of small size; after the succeeding moult, however, it presents its usual dimensions.—The condition of the sexual apparatus in the higher Arachnida presents certain very remarkable peculiarities. The sexes are always distinct, and the male (among the Araneidæ at least) may generally be distinguished from the female by his smaller size, longer limbs, and brighter colours. The spermatic organs of the male Spider are two long worm-like tubes, which commence at the posterior end of the abdomen, either by a simple cæcal termination, or by an oblong vesicle; and which terminate, either by a single or double orifice, on a slit in the integument of the under side of the abdomen, near its anterior extremity, in a position corresponding to that of the vulva of the female (Fig. 164). The palpi, however, contain the organs by which the spermatic fluid is introduced; each of these being furnished with a tubular projection, terminated by a horny appendage, which, when not in use, is retracted within the last joint of the palp. No connection whatever can be traced between the organs which prepare the spermatic fluid and these intromittent instruments; nevertheless it has been ascertained by repeated observations, that no closer sexual congress takes place, than the introduction of these appendages of the palpi within the vulva of the female; and it would seem probable that the male himself applies these appendages to his abdominal aperture, and charges them with the fertilizing fluid, before the act of intromission. The ovarium of the female Spider is a simple elongated vesicle, closed at one extremity, and communicating at the other with a slender oviduct, that terminates at the corresponding side of the transverse sexual fissure, which, as in the male, is situated between the anterior pulmonic cavities; when this ovarium is dilated with ova, it occupies a considerable part of the abdominal cavity (Fig. 164). The peculiar mode of sexual congress just described seems to have relation to the remarkable instinct which prompts the female spider to attack and devour the male, as soon as the fertilization of her ova is accomplished; which her superior size and strength enable her to do, if he remain within her reach. It is obvious that if the orifice of his sexual canal were applied to hers, his danger would be augmented with this closer approximation. It is curious that, with such an instinct towards the opposite sex, the attachment displayed by the female towards her offspring should be of most extraordinary strength. The ova are generally enveloped in a soft and warm silken cocoon, which she guards with the most jealous care. Some species carry this about with them; others attach it to trees, or hide it in empty snail-shells. Within this the young remain until their development is completed, and they then make their way out. The mother generally foregoes all nourishment during her watch; and if the young are prevented by the coldness of the weather from coming forth at the accustomed time, she will die of hunger, rather than quit her post.—In the *Scorpionidæ,* the testes and ovaria are formed upon the same general plan as in the Araneidæ, but are more complex in their structure; the outlet by which they terminate is situated at the middle of the under side of the last segment of the thorax; and as the palpi are destitute of the peculiar appendages which they possess in the Spiders, it is probable that they do not take any share in the sexual operation. The ova are retained in a dilatation of the oviduct, until the young are mature; so that the egg-case is ruptured within the body of the parent, and the young are born alive.—In both of these tribes, the embryo attains the form and the organisation of the parent, previously to its emersion from the egg; and it is remarkable that not even in its course of intra-oval development does it pass through stages which can be said to repeat, with any degree of closeness, those which are permanent in the lower Articulata; the whole tendency of its development being at once to evolve the Arachnoid type, as will be hereafter shown (CHAP. XVIII.).

319*d*. The class of Arachnida is not limited, however, to animals of the high organisation first described; for, like that of Crustacea, it includes beings of a structure so simple as to connect it with the inferior division of the Articulate series. Of the *tracheary* Arachnida (§ 319) there are some that approach the Spiders and Scorpions in their general organisation; but in the tribe of *Acaridæ* (mites), and especially in those which are parasitic, we meet with a general degradation of structure which leads us towards the Entozoa, whilst yet there is a remarkable character of elevation in the *type* of their respiratory apparatus. In the bodies of these animals, there is a more complete absence of segmentation, and a greater fusion of the principal divisions, one into another, than even in the Araneidæ; the head, thorax, and abdomen forming but one mass, of which the cephalo-thoracic portion is but very little developed in relation to the abdominal; and the integument being extremely soft throughout, so that the segmentation is only marked by the points of origin of the limbs. The integument of the members, however, is generally more firm and horny, and their joints more distinct; though in the curious parasitic *Demodex,* which infests the sebaceous sacs and hair-follicles of the human skin, the legs are of the simplest and most rudimental kind, bearing a strong resemblance to the pro-legs of a caterpillar. In most of the true Acaridæ, they are terminated either by adhesive suckers, or by bristle-like filaments, instead of by curved claws. The structure of the mouth varies in the different members of the group. In some, the mandibles are forcipated like the claws of Scorpions; in others they are provided with recurved hooks, as in the Spiders; whilst in the parasitic Acari, again, they are usually prolonged into piercers, for puncturing the integument of the animals whose juices they imbibe. When formed after the pattern of those of the Spider, the mandibles of the Acari are provided with what appears to be a poison-apparatus. The maxillæ present a corresponding series of transitional states, leading to the production of a prolonged channel for the imbibition of the juices, not unlike that of some of the parasitic Insects; and the mouth of those Acaridæ which are formed for suction, has its cavity surrounded by a distinct layer of muscular fibres, the contraction of which will augment the capacity of the pharynx. Behind this pharynx, however, it has hitherto been found impossible to trace any distinct alimentary tube; the digestive cavity being to all appearance a mere excavation amongst the soft tissues of the body, everywhere surrounded by a mass of cells, which, from their brownish colour, appear to form the biliary secretion. This cavity occupies a considerable part of the interior of the body, and penetrates to nearly the extremities of the limbs, as in the *Pycnogonidæ* (Fig. 150). As in that group, the unusual extension of the alimentary cavity renders any special circulating apparatus unnecessary; whether any movement of fluid takes place, however, in spaces around that cavity, has not been ascertained.—In the lowest parasitic Acaridæ (such as the *Sarcoptes*, or 'itch-insect') there is no respiratory apparatus; so that the aeration of the fluids must take place through the general surface. In the higher members of the group, however, there is a regular tracheal system, nearly resembling that of Insects; and between these two extremes there are several gradations, of which that which is seen in *Trombidion* and in *Hydrachna* (water-mite) and its allies, is the most curious. The whole tracheal system (the branches of which are destitute of the usual spiral filament) diverges from two large trunks, which pass forwards to an orifice situated at the base of the mandibles; it appears, however, that this merely serves for *expiration*, the introduction of air into the body being accomplished by a different means. A network of passages, excavated in a transparent homogeneous substance, is found beneath the skin; and the integument is covered, in *Trombidion*, by plumose hairs, by which its surface is increased; whilst in *Hydrachna* it is pierced with a number of orifices (resembling the *stomata* of the leaves of plants) each of them covered by a very thin membrane, and in communication with the lacunæ beneath, whilst by its side is constantly found a hair, the movements of which seem to contribute to the renewal of the water in contact with the surface of the body, and thus promote the introduction of air into the passages beneath. The air, thus taken in by the general surface, appears to be distributed through the aeriferous canals of the body, and then to be expelled by the tracheal apparatus through the single expiratory orifice, the position of which presents us with the nearest approach anywhere seen in Invertebrated animals, to that connection of the respiratory passages with the alimentary mouth, which is so characteristic of the Vertebrated series.—The condition of the nervous system in the group of Acaridæ is as singular as the other parts of their structure already described; for the only ganglionic centre which it appears to possess is a single large globular ganglion, which receives the optic nerves, and gives off trunks both anteriorly and posteriorly, without forming a collar round the œsophagus; as if the cephalic ganglia of the Spider were fused with its thoracic mass. The eyes are usually four in number; but sometimes they are reduced to two, or even to one; and in some of the parasitic Acari they are entirely absent.—There seems reason to believe that in some, at least, of the *Acaridæ*, the sexes are united; and in the larger proportion of this group, there appears to be no definite ovarium, the ova being lodged in the general substance of the tissues, as in many Entozoa (Fig. 141). In the genus *Oribates*, the young are born alive; but in other Acaridæ the ova are deposited and left to themselves, the young coming forth in due time, in a form resembling that of their parents, except that they have only six legs instead of eight, the deficient pair being supplied at the first moult.*

320. We now arrive at the sub-kingdom VERTEBRATA, which unquestionably ranks as the highest division of the Animal Kingdom, since it contains those classes which display the greatest perfection of organic structure, as manifested in the special adaptation of each part of their organism to some different purpose, and in the consequent number and variety of their faculties. It must not, however, be left out of view, that there are members of the Vertebrated series, which must be accounted lower in their general organization than some of the higher Invertebrata; but these are aberrant forms, in which (as we shall hereafter see) the Vertebrated characters are but very imperfectly developed, and which correspond in many respects to the embryonic conditions of the proper types of the group.—It is in the high grade presented by the *Nervo-Muscular system*, and in the obvious subordination of the whole structure to its purposes, that we see the most characteristic difference between the Vertebrata and the sub-kingdoms already passed in review. In the latter, the functions of this system are obviously restricted to the maintenance of the nutritive and reproductive operations, and to the guidance of the individual, by means of its sensory endowments, in its search for food, or for an individual of the opposite sex; the locomotive portion of the apparatus being predominant in the Articulata, whilst the organs of sense attain a higher relative development in the Mollusca. In the Vertebrata, on the other hand, we find both these elements in a state of much greater concentration and completeness; and we meet with superadded organs obviously ministering to the *Intelligence*, which seems almost exclusively restricted to this group of animals, and which is, perhaps, its most characteristic endowment. The nervo-muscular apparatus thus becomes no longer the mere instrument of sustaining animal life; it now ministers to higher endowments; and the intelligence, called into activity by impressions received through the organs of sense, and executing its voluntary determinations by the apparatus of motion, has an existence independent of either. It is in Man that we see the most complete subordination of the entire organism to the purposes of *psychical* existence; but through the whole Vertebrated series, this direction of development may be easily traced; and it is in no

* Most of the facts stated in this paragraph are derived from the Memoir of M. Dujardin, '*Sur les Acariens*,' in the Annales des Sciences Naturelles, Troisième Serie, Zool., tom. iii. p. 5.

particular more obvious, than in the peculiarities which the *vertebrated skeleton* presents, in contrast with the skeletons of the various classes of Invertebrata.

320 *a.* Wherever a proper *skeleton* exists among Invertebrata, that skeleton—with very few and unimportant exceptions—is tegumentary or *dermal:* and like the epidermis and its appendages, it is non-vascular, and can only be increased by additions to its edges (§§ 193–199). This hard insensible armour serves alike to prevent the animal from being cognisant of hurtful agencies, and to protect it from their injurious operation; and whilst it affords fixed points of attachment to the muscles that move the body and limbs, and may even serve as a framework of levers on which they may act with mechanical advantage, these muscles are always attached to its interior, and are enclosed by it, as are the viscera in general. In scarcely any instance among the Invertebrata, do we find that the skeleton has any special relation to the nervous centres; and these are far from presenting such a fixity in their arrangement, as could serve to govern its development. Thus the number and situation of the nervous ganglia vary in the Mollusca with the development of the locomotive organs and of different portions of the muscular layer of the mantle, and with the position of the branchiæ; all of which conditions are subject to so much diversity, that scarcely any two genera are alike in the arrangement of their nervous centres. And in Insects, Crustacea, and other Articulata, although the ventral chain of ganglia is usually disposed on a more uniform plan, their number varies with that of the segments of the body, and their relative size with that of the development of the parts of the muscular apparatus with which they are respectively connected, whilst their degree of concentration is most diverse, and has not yet been shown to follow any fixed plan whatever. The nervous system, in the Invertebrata generally, receives only the same amount of support and protection from the skeleton, as the other tissues possess;* and its comparatively low importance in the life of these animals is shown by the severe injuries which it may sustain, without fatal results necessarily ensuing.

320 *b.* On the other hand, in the Vertebrated series, the Skeleton is for the most part *internal;* and is composed of true bone, whose relations with the vascular system are such, that its form and dimensions may be changed not merely by deposition but by removal; so that its configuration can be adapted to that of the organs which it encloses and supports, not only during the period of growth and increase, but during the whole of life (§§ 200–211). This internal osseous skeleton is essentially connected with the Nervous system, which is formed upon one uniform plan throughout the entire sub-kingdom, its scattered centres being collected (with the exception of those of the sympathetic system) into a linear series, the *cerebro-spinal axis,* in which they always occupy the

* It has been imagined that certain internal projections in the dermo-skeleton of the Insect and Crustacean were specially adapted to the protection of the nervous cord; but, as Prof. Owen remarks, there is no conformity between these and the number of ganglia, and they are so situated as to inclose the communicating cords rather than the ganglia themselves; so that their relation to the nerve-trunks seems merely accidental, their special purpose being to afford attachment to the muscles. It is only in the Cephalopoda, especially the *dibranchiate* order (§ 307 *b*), that we meet with a true rudiment of an internal skeleton, specially designed to support and protect the nervous centres.

same order, and differ only in their relative development; and this axis is included in a *vertebral column*, which constitutes the fundamental portion of the skeleton, and the number of whose pieces is in constant relation with the number of pairs of nerves given off from the nervous axis. The vertebral column is expanded at one end into the *cranium*, which affords protection to the cephalic portion of the nervous axis, and to the organs of sense connected with this; and although in the higher animals the relation of the cranial bones to the vertebræ of the spinal column is obscured by the high development of the portion that arches over the brain, and by the adaptation of other parts for special purposes, this relation becomes obvious, when the comparison is made in those forms of Vertebrata in which the cephalic ganglia are not largely developed in proportion to those of the spinal cord. The vertebral column, at its opposite extremity, is usually contracted instead of dilated, forming a *tail*, or the rudiment of one, from which the nervous centres are entirely withdrawn; and the alteration which the vertebræ there undergo, by the loss of one element after another, is scarcely less in degree (though more easily traced) than that to which they are subjected in the cranium by the excess of development of some of these elements. But besides the canal for the lodgment of the nervous or 'neural' axis, which lies on the upper side of the central osseous column, there is another (less constantly present, however), for the lodgment of the great vascular trunk, which may be called the 'hæmal' canal. This is dilated, in the trunk of the body, to an enormous size; being made to include the great centres of the circulation, the digestive and respiratory organs, and in fact the whole visceral apparatus concerned in the preparation and purification of the blood. The walls of this cavity are chiefly formed by the ribs and sternum, which, though commonly regarded as distinct bones, are really to be considered, in a philosophical sense, as parts of the vertebral segments. In addition to the bones of the head, trunk, and tail, the Vertebral skeleton commonly contains a set of bones that give support to the two pairs of locomotive organs, or 'extremities,' which are developed with greater or less completeness in by far the larger part of the series. The bones proper to each pair of limbs are connected with the vertebral column by a peculiar arch, of which the anterior is termed the 'scapular', and the posterior the 'pelvic.' Even these bones are reducible (as will be presently shown) to the general type of the vertebra; the scapular and pelvic arches being modifications of the hæmal arches which elsewhere serve to protect the visceral cavity; whilst the bones of the limbs themselves may be regarded as originating in an extraordinary development of certain 'diverging appendages' of those hæmal arches, which are elsewhere adapted to a very different purpose. When it is added that the 'hyoid' bone, which is detached from the remainder of the skeleton (lying between the summit of the larynx and the root of the tongue), is also a 'hæmal arch' of one of the cranial vertebræ, all the parts of the proper vertebral skeleton will have been (so to speak) accounted for.

320 *c*. In many Vertebrated animals, however, a very distinct *dermal* skeleton is present, notwithstanding that the internal skeleton is fully developed; and although this dermo-skeleton is sometimes only horny in its texture, yet there are several instances in which it is composed of true

bone. This is the case, for example, in the *Lepidosteus* and *Ostracion* among existing Fishes, as well as in the whole of the great group of *Ganoids* now almost extinct ; in the *Crocodiles*, many of the dermal plates present almost the same complete ossification ; and even among Mammalia, the tesselated armour of the existing *Armadilloes* and gigantic extinct *Glyptodons* affords a striking example of the co-existence of a well-developed bony envelope with a complete osseous endo-skeleton. Even where no such complete osseous envelope exists, bones are occasionally found, especially among Fishes and Reptiles, which, though intercalated amongst those of the vertebral skeleton, do not really belong to it, but are derived from the dermal system. Such are, in Fishes, the 'rays' upon which the dorsal fin is supported (Fig. 173); whilst among Reptiles, the Turtle presents a remarkable example of this derivation, in the bones forming the margin of the carapace (Fig. 176), as well as in part of both the dorsal and ventral shields themselves (§ 324 *f*). Besides the osseous elements of the vertebral and dermal skeletons, certain other bones occasionally present themselves in Vertebrated animals, which seem to fall naturally into a distinct category ; being developed to afford a special support and protection to particular parts of the visceral apparatus. We find some rudiment of such structures, even in the Invertebrata ; thus in the *Lobster*, the walls of the stomach contain a calcified framework, supporting the gastric teeth, and giving attachment to the muscles that work them ; and in the *Bulla*, there is a pair of calcareous plates imbedded in the walls of the muscular stomach or gizzard, which adds considerably to the crushing and triturating powers of the organ. Now in most air-breathing Vertebrata, we find the larynx, trachea, and bronchial tubes more or less supported by a cartilaginous framework, which sometimes becomes ossified ; in Fishes and in the tadpole state of Batrachians, the gills are supported upon a cartilaginous or osseous framework developed independently of the vertebral skeleton ; and in many Mammalia, the heart contains a bone that serves as a support to its muscular and ligamentous fibres. These, with some other isolated bones whose presence is only occasional, may be regarded as constituting the 'splanchno-skeleton ;' and to this we are probably to refer the osseous cups or plates which are found in the outer coat of the eye of many Birds and Reptiles,—the hard bony envelope which surrounds the internal ear, and which becomes incorporated with the neural skeleton in most Vertebrata, forming what is commonly known as the 'petrous portion of the temporal bone',—and the turbinate or spongy bones of the nose, with the cartilages which enclose some parts of the nasal cavity. It is quite plain, from the history of their development, that these 'sense-capsules' are originally independent of the vertebral skeleton, however closely they may be subsequently incorporated with it. The *teeth* of Fishes, Reptiles, and Mammalia, which originate in papillæ developed from the surface of the mucous membrane lining the mouth (§ 212), must be regarded as belonging to the splanchno-skeleton ; and in their well-known mode of implanting themselves in the jaws, we have a good example of the process by which parts of the dermo- and splanchno-skeletons elsewhere form an intimate connection with the vertebral.

320 *d*. Thus the skeleton of Vertebrata has to be regarded as consisting of the *endo-skeleton*, osseous or cartilaginous ; which is composed of a

series of vertebræ, and their appendages, developed in different modes and degrees, but always especially related to the cerebro-spinal axis;—of an *exo-skeleton,* more or less complete, which may even envelope the entire body in a bony casing, or may be reduced to the condition of hairs, quills, or horny scales, or may be entirely wanting;—and of a *splanchno-skeleton,* the various parts of which have no other relation to each other, than that of being developed with express reference to the visceral system. It is not always easy to discriminate the class to which a particular bone should be referred; indeed, a knowledge of its form and connections in a great variety of animals, and of the history of its development, is frequently necessary to enable this classification to be satisfactorily carried out. "As in all arrangements of natural objects," observes Prof. Owen, "where nature is followed in selecting their characters, so in classifying the parts of the general skeleton of Vertebrata, the primary groups blend into one another at their extremes, and make it difficult to draw a well-defined boundary line between them. But we must not on that account abandon the advantage of arrangement and classification, in acquiring an intelligible and tenable knowledge of a complex system of organs, when typical characters clearly indicate the general primary groups." This analysis has been pursued by Prof. Owen far more satisfactorily than by any other Philosophical Anatomist; and in the more particular description which will be now given of the vertebral skeleton, and of its modifications in the several classes, his views of its constitution will be adopted and explained so far as space will permit.*

320*e.* A complete 'typical vertebra,'—"one of those segments of the endo-skeleton which constitute the axis of the body and the protecting canals of the nervous and vascular trunks,"—consists, according to Prof. Owen, of the parts represented in the accompanying figures, of which A is a mere ideal diagram, whilst B shows one of the modes in which they are actually seen united in nature. In the former, the parts which are fully shaded are those which are usually developed from distinct and independent centres, and which are termed by Prof. Owen 'autogenous' elements; those, on the other hand, which are left in outline, shoot out as 'processes' or projections from some of the others, and are termed 'exogenous.' These elements may be united with each other in various modes and degrees, sometimes remaining almost entirely disjoined even in the adult animal, whilst in other cases they are united into a single piece, so that their real distinctness is only to be recognized by tracing the history of their development. In a large proportion of instances, some one or more of these elements may be entirely deficient; whilst in other cases, one element or set of elements is developed in so much larger proportion to the rest, as to throw them (so to speak) into the shade. Thus in the Mammalian cervical vertebra (Fig. 169) we have a mere trace of the hæmal arch; whilst in the thoracic vertebra (Fig. 167, B) we find it expanded to

* These views are set forth in detail in Prof. Owen's admirable works on the "Homologies of the Vertebrated Skeleton," and on the "Nature of Limbs;" a general account of them, with their special application to the class of Fishes, being also contained in his "Lectures on Comparative Anatomy," vol. ii. These works are examples of that rare combination of logical appreciation of *facts,* with originality in the conception of *ideas,* which distinguishes the true philosopher from the rash speculator on the one hand, and from the mere plodding observer on the other.

many times the size of the neural arch, in order that it may include and protect the great centre of the circulating system, with the respiratory organs. It is in the cranial segments, however, that the vertebral ele-

FIG. 167.

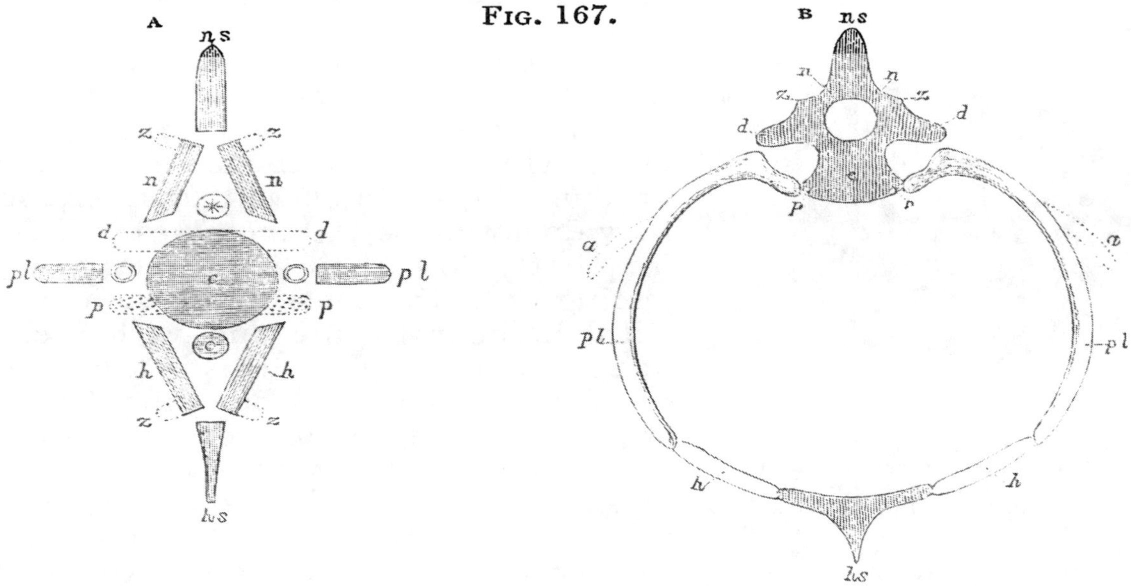

Elements of a *Vertebra*, according to Prof. Owen:—A, ideal typical vertebra:—B, actual thoracic vertebra of a Bird:—*c*, centrum, giving off *d*, *d*. the diapophyses, and *p*, *p*, the parapophyses; the neural arch, enclosing the spinal cord*, is formed by *n*, *n*, the neurapophyses, and *ns*, the neural spine; the hæmal arch, enclosing the great centres of the circulation, is formed by *h*, *h*, the hæmapophyses, and *hs*, the hæmal spine. From both the neurapophyses and hæmapophyses may be given off the zygapophyses, *z*, *z*. The lateral arches, which may enclose the vertebral arteries o o, are completed by the pleurapophyses, *pl*; these in B are bent downwards, so as to form part of the hæmal arch, and give off the diverging appendages *a*, *a*.

ments undergo their most remarkable transformations; and it is only by tracing them through their simplest to their most complicated forms and arrangements, that the true nature of the latter can be elucidated.

320*f*. The *centrum* (*c*, Figs. 167, and 170) forms the axis of the vertebral column, and most commonly serves as the central bond of union between the peripheral elements of the vertebra. In some instances it retains its primitive soft state, when some of the other elements have been completely ossified; and it is remarkable that such should have been the condition of the vertebral column of all the Fishes, whose fossil remains have been preserved to us in strata earlier than the Permian epoch, whilst among existing Fishes it is comparatively rare. Towards the extremity of the caudal prolongation of the vertebral column, the segments are often found reduced to the body or centrum alone. Though it may present a great variety of shapes, its normal form may be regarded as cylindrical; the two extremities of the cylinder being either flat, more or less concave, or convex, or one concave and the other convex. The individuality of the bodies of the vertebræ is frequently lost by their coalescence with each other; of this the 'sacrum' of Man (Fig. 168) and of Mammals and Birds generally, is one of the most common examples. The centrum supports the 'neural axis' or spinal cord (Fig. 167, *) on its upper side, the vertebral column being supposed to be in a horizontal position); whilst on its under side it is in relation with the 'hæmal axis' (*o*) or principal trunk of the vascular system.—The *neurapophyses,* which constitute what are commonly known as the 'laminæ' or 'arches' of the vertebræ (Fig. 169, 2) are constantly found protecting the neural axis wherever this passes through the vertebral segment, and are even more constant as osseous or cartilaginous elements than the centrum itself; it being only near the extremity of the tail, along which neither the neural axis nor its prolongations into nerve-trunks extend, that they do not present themselves. They very frequently lose their individuality by coalescing with the centrum at their base, and with the neural spine and with each other at their apex. When this is the case, they form a complete osseous ring (as seen in any one of the human vertebræ, Fig. 169), through which the neural axis passes. Between every pair of vertebral segments, a pair of nerve-trunks makes its exit from the central axis, sometimes through the interspaces between the contiguous neurapophyses, sometimes through apertures formed by the junction of notches in the margin of each (Fig. 169, 3). It occasionally happens that the neurapophyses, like the bodies, of several successive segments coalesce, so as to form a continuous bony canal; but their original distinctness is always marked by the apertures for the exit of the nerve-trunks (Fig. 168, 2, 2).—The *neural-spine* commonly retains, in the trunk of the body, the long-pointed form which its name implies, and constitutes the 'spinous process' of the ordinary vertebra (Fig. 169, 4) which is specially adapted for the attachment of muscles; occasionally, however, it is flattened and greatly increased in breadth and solidity, so as even to coalesce with the pleurapophyses. It is seldom altogether absent, but it frequently loses its individuality by coalescing with the neurapophyses. —The foregoing elements constitute, when united, the 'neural arch,' which is the part of the vertebra not only the most constant in its presence, but also the least disposed to vary widely from a common type. From some part of this arch, we generally find developed the 'oblique' or 'articulating' processes (Fig. 169, 7, 8), which serve to connect the vertebræ together; these, which are termed by Prof. Owen the 'zygapophyses,' are never developed as distinct elements, but shoot out from the neurapophyses.

FIG. 168.

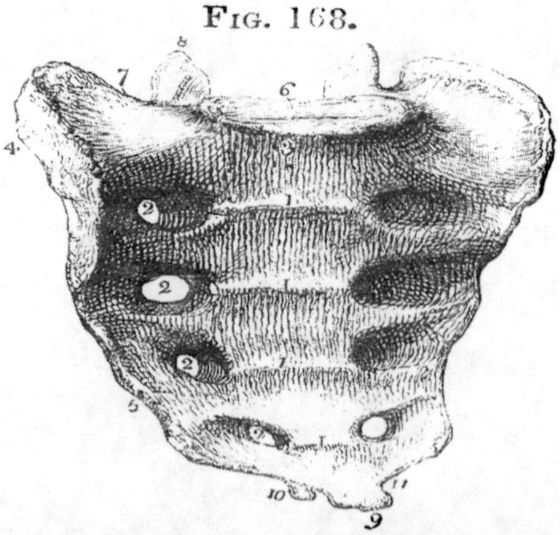

Human Sacrum, seen upon its anterior surface;—1, 1, the transverse lines marking the original constitution of the bone of four pieces; 2, 2, the anterior sacral foramina; 3, the promontory of the sacrum; 4, the ear-shaped surface which articulates with the ilium; 5, the sharp edge to which the sacro-ischiatic ligaments are attached; 6, the vertebral articular surface; 7, the broad triangular surface which supports the psoas muscle and lumbo-sacral nerve; 8, the articular process of the right side; 9, the inferior extremity, or apex, of the sacrum; 10, one of the sacral cornua; 11, the notch which is converted into a foramen by the coccyx.

FIG. 169.

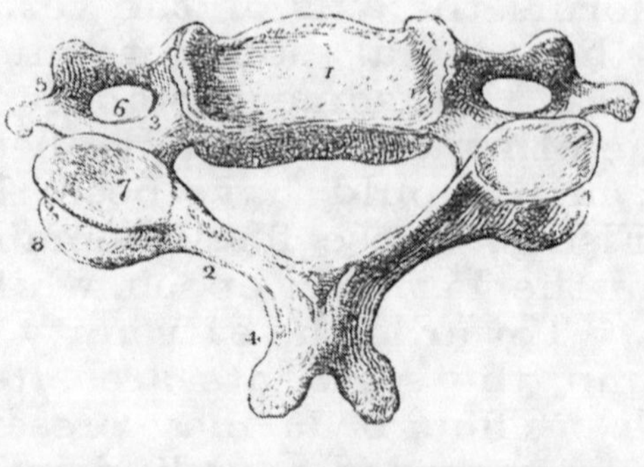

Human Cervical Vertebra;—1, the body; 2, one of the laminæ; 3, intervertebral notch; 4, spinous process; 5, transverse process; 6, vertebral foramen; 7, superior articular process; 8, inferior articular process.—N.B. This figure must be considered as *inverted,* with reference to Fig. 167, the neural arch being here *below* the centrum.

320 *g.* Projecting from the centrum, we find a pair of processes on either side, between which is a canal that serves for the passage of a blood-vessel. The 'diapophyses,' which project above this canal, are nearly always 'exogenous' elements, shooting forth from the centrum or from the neurapophyses; the 'parapophyses,' which form the lower boundary of the canal, are 'autogenous' or independent elements in Fishes, but are developed from the centrum in all the higher Vertebrata. These elements never attain any high development in Mammals, Birds, or Reptiles; in which classes they serve merely to enclose the vascular canal, and to support the 'pleurapophyses.' Thus in the Human cervical vertebra, they form the two roots of the 'tranverse process,' surrounding the foramen for the passage of the vertebral artery (Fig. 169, [6]); whilst in the thoracic vertebra (Fig. 167, B), the diapophyses (*d, d*) form the 'transverse processes,' and the parapophyses (*p, p*), reduced to mere rudiments, constitute the articulating surfaces with which the 'heads' of the ribs come into contact. In Fishes, however, they are more highly developed, and in the Cod-tribe are even longer and broader than the 'pleurapophyses' or true ribs. The ordinary function of these lateral processes is to give support to the pleurapophyses, to afford attachment to muscles, and to protect the lateral vascular trunks.—Usually connected with the foregoing, where they exist, but not unfrequently absent, and sometimes directly connected with the centrum, are the 'pleurapophyses' (*pl, pl*), whose development varies extremely in different parts of the same trunk, as well as in different animals. Thus in the cervical vertebræ of Man, they form the short bifid 'transverse processes' (Fig. 169, [5]), which are anchylosed at their base to the two 'roots,' the diapophyses and the parapophyses, which surround the vertebral canal. In the thoracic segments, on the other hand, they are developed separately and constitute the 'ribs,' which form the greater part of the boundary of the hæmal arch. The ribs of Birds usually possess certain peculiar 'diverging appendages' (*a, a*), each of which projects backwards over the next rib; their special function appears to be, to give increased solidity to the thoracic framework, by thus connecting the ribs with each other (§ 325 *a*). These appendages to the ribs of birds, however, will be hereafter shown to be the rudiments of the elements, which, in a state of higher development, constitute the 'extremities' or 'limbs;' these bearing the same relation to the pleurapophyses of the scapular and pelvic arches, as do the connecting pieces of the ribs of birds to the ribs themselves (§ 320 *k*). Similar diverging appendages, of much larger relative size, are found proceeding from the ribs of many Fishes.

320 *h.* On the under side of the centrum, we find, in a complete vertebra (Fig. 167, A), the 'hæmal arch,' formed of two 'hæmapophyses' (*h*), and of a hæmal spine (*hs*), and thus answering to the neural arch. The entire hæmal arch, however, is very frequently wanting, as in the cervical and lumbar vertebræ of Man and Mammalia; but in the tail of some Mammals and of Reptiles, a hæmal arch exists, and presents a remarkable similarity, both in form, size and composition, to the neural arch. It is made up of two bony laminæ, the 'hæmapophyses,' which form by their junction with each other at their apex a set of V-shaped bones, but which retain their primitive independency at their bases, where they simply articulate with the centrum; these hæmal arches protect the caudal

artery and vein. In tracing the tail from its attached towards its free extremity, we find the hæmal arch first disappearing, and then the neural, the centrum alone being left. It is in the thoracic region of Mammalia, Birds, and Reptiles, that we find the greatest expansion of the hæmal arch, which is made, by the articulation of the hæmapophyses with the extremities of the ribs, instead of with the centrum, to surround the entire visceral cavity (Fig. 167, B). In Man and the Mammalia, the hæmapophyses, remaining unossified, are known as the 'cartilages of the ribs;' but they are ossified in Birds and Reptiles, and are there known as the 'sternal ribs.' The 'hæmal spine' presents a much greater variety in form than the neural, and is much less constantly present. In the thorax of Mammalia it constitutes the flattened 'sternum;' which is originally formed (even in Man) of as many distinct segments as there are costal cartilages connected with it; but which may be at last made, by the coalescence of these, to appear as a single bone. The sternum of Birds, on the other hand, usually presents a prominent keel on the median line, so that its transverse section more resembles that of the neural spine (Fig. 167, B). In most Reptiles, again, the hæmal spine (sternum) is flattened out laterally as in the Mammalia; and in the Turtles this lateral extension takes place to an unusual extent (§ 324 *h*); thus offering an interesting analogy to the expansion and flattening of the neural spine in the cranium of large-brained animals. The hæmapophyses and hæmal spines (sternum and sternal ribs), which are wanting in the abdominal region of Mammalia and Birds, are usually continued backwards in the Lizard tribe; although, for want of the pleurapophyses, the hæmal arch is not completed. In Serpents, on the other hand, the hæmal arches are wanting along the whole trunk, the ends of the ribs being everywhere free. In Fishes, also, the hæmapophyses and hæmal spines are altogether absent or unossified; but a hæmal canal is frequently formed by the inclination of the parapophyses towards each other; and in the herring a sort of sternum is formed of dermal plates, articulated to the ends of the ribs.

320 *i*. The number of the segments entering into the skull has been a subject of much discussion among those who admit the 'vertebral theory' of its composition. By Oken, who first (in 1807) distinctly enunciated the idea, the number was fixed at *three*, although he recognized the rudiment of a *fourth* anteriorly, which he has since adopted into his system; Spix and Boganus also made out *four*; but Geoffroy St. Hilaire maintained the existence of *seven*; whilst Carus would find *six*. Prof. Owen (in whose determinations on this point the author has entire confidence) returns to the original number, *four*, as corresponding with that of the four primary divisions, succeeding each other in a linear series, which are distinctly marked in the early development of the Brain; as well as with that of the nerves of special sense, which issue from this part of the neural axis with the same regularity that the ordinary sensori-motor nerves do elsewhere. These segments are named (proceeding from behind forwards) the *Epencephalic*, the *Mesencephalic*, the *Prosencephalic*, and the *Rhinencephalic*, according to the divisions of the brain to which they respectively correspond; and from the Epencephalic segment proceed the *auditory* nerves; from the Mesencephalic, the lingual or *gustative;* from the Prosencephalic, the *optic;* and from the Rhinencephalic, the *olfactive*. In the skull

of the Fish, it is not at all difficult to make out the vertebral elements of the neural arches of these vertebræ, the relative position of which is nearly the same as it is represented in the ideal "Archetype Skeleton" (Fig. 170), although the forms are much modified. Thus in the 'epencephalic' segment, we have the 'basi-occipital' (1), representing the centrum; the 'par-occipital' (4) on each side, representing the parapophysis; the 'exoccipital' (2) on each side, representing the neurapophysis; and the 'supra-occipital' (3) on the summit, representing the neural spine. All these bones coalesce in the single 'occipital bone' of Man, in which, however, most of these elements are distinct from each other at an early period of development; and this segment is thus conveniently known as the *occipital* vertebra.—Of the 'mesencephalic' segment, the centrum is formed by the 'basi-sphenoid' (5); the parapophyses are represented by the 'mastoid' bones (8); the neurapophyses by the 'ali-sphenoids' (6); and the neural spine by the junction of the parietals (7). In the adult human cranium,

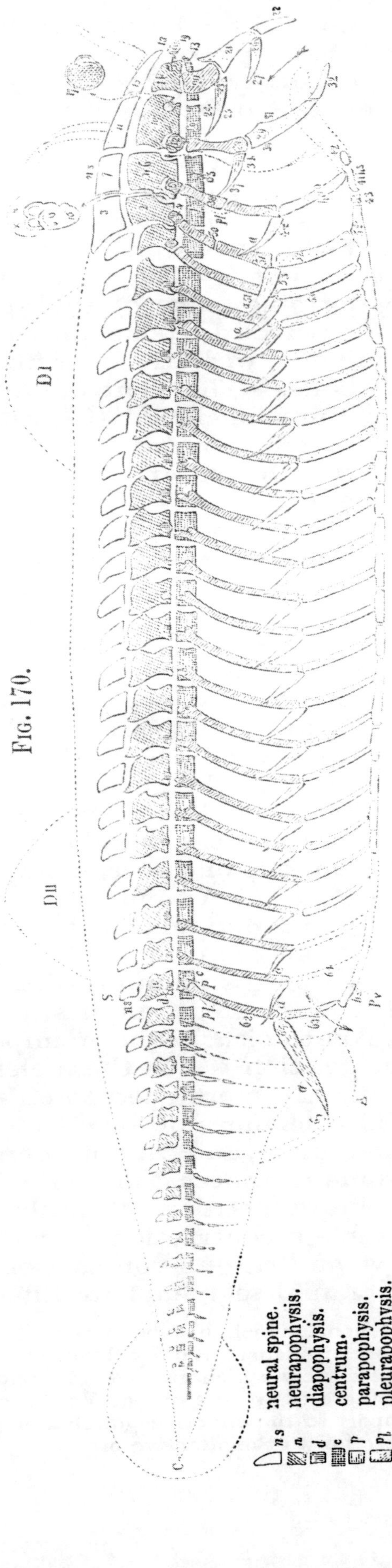

Fig. 170.

ns neural spine.
n neurapophysis.
d diapophysis.
c centrum.
p parapophysis.
pl pleurapophysis.
h hæmapophysis.
hs hæmal spine.
a diverging appendage.

Diagram of *Archetype Vertebral Skeleton*, according to Prof. Owen, composed of the following parts:—1, basioccipital, 2, exoccipital, 3, supraoccipital, 4, paroccipital, constituting the neural arch of the first or occipital vertebra of the head;—5, basisphenoid, 6, alisphenoid, 7, parietal, 8, mastoid, constituting the neural arch of the second or parietal vertebra;—9, presphenoid, 10, orbitosphenoid, 11, frontal, 12, postfrontal, constituting the neural arch of the third or frontal vertebra;—13, vomer, 14, prefrontals, 15, nasal, constituting the neural arch of the fourth or nasal vertebra;—16, petrosal, or acoustic sense-capsule; 17, sclerotic, or ophthalmic sense-capsule; 18, 19, ethmo-turbinal, or olfactive sense-capsule;—20, palatine, 21, maxillary, 22, premaxillary, 23, 24, 25, pteregoid, 26, malar, 27, squamous, constituting the hæmal arch of the nasal vertebra with its diverging appendages;—28, tympanic, 29-32, mandibular or lower jaw, 34-37, operculars, constituting the hæmal arch of the frontal vertebra, with its diverging appendages;—38, stylohyal, 39, epihyal, 40, ceratohyal, 41, basihyal, 42, glossohyal, 43, urohyal, 44, branchiostegal, constituting the hæmal arch of the parietal vertebra, with its diverging appendages;—50, suprascapula, 51, scapula, 52, coracoid, 52′, episternum, 53, anterior member, constituting the hæmal arch of the occipital vertebra, with its diverging appendages; 58, clavicle, the hæmapophysis of the post-occipital vertebra;—62, ilium, 63, ischium, 65-69, posterior extremity, constituting the hæmal arch of the sacral vertebra S, with its diverging appendages; 64, pubis, the hæmapophysis of one of the lumbar (?) vertebræ. The chief osseous developments of the dermo-skeleton are added in outline, as the median horn supported by the nasal spine, 15, in the Rhinoceros; the pair of lateral horns developed from the frontal spine, 11, in most Ruminants; the median dorsal fins, D I, D II, in Fishes and Cetaceans the caudal fin, C, and the anal fin, A, in Fishes.

the 'parietals' attain an enormous relative development, in accordance with that of the cerebral hemispheres over which they arch, so that this vertebra is most appropriately termed the *parietal;* and the basi-sphenoid and ali-sphenoids are united (with other elements) into one bone, the 'sphenoid,' of which the former constitutes the 'body,' whilst the latter are known as the 'great wings.' The acoustic cavity is always formed by an excavation in the parietal and occipital vertebræ, which are disposed around it in the same manner as the elements of the frontal vertebra are arranged to form the orbit of the eye, or those of the ethmoidal to form the nasal cavity. The organ of hearing has usually a bony capsule of its own, forming part of the splanchno-skeleton; and this, which is called the 'petrosal' bone on account of its hardness, coalesces, in the higher animals, with the 'mastoid' and other elements, to form the 'temporal' bone.—The centrum of the 'prosencephalic' segment is formed by the 'pre-sphenoid' (9); which, though developed from the same centre of ossification with the basi-sphenoid, may be regarded as distinct from it; the parapophyses are represented by the 'post-frontals' (12); the neurapophyses by the 'orbito-sphenoids' (10); whilst the 'frontal' (11) obviously constitutes its neural spine. The 'pre-sphenoid' and the 'orbito-sphenoid' enter into the composition of the 'sphenoid' bone of man and the higher animals; the former constituting the anterior part of its 'body,' and the latter its orbitar portion. The 'frontal' and 'post-frontals' coalesce to form the single 'frontal bone,' which, like the parietals, is enormously developed, and from which this vertebra is appropriately termed the *frontal.* It is chiefly by the elements of this segment, that the 'orbit' of the eye is formed. The capsular bones of the eye itself, where they exist, are always distinct from the vertebral elements which form the general protective envelope to the organ.—Of the 'rhinencephalic' or *nasal* vertebra, the centrum is formed by the 'vomer' (13), which in fishes is a broad flat bone; the parapophyses are not present as distinct elements in this segment; the neurapophyses are represented by the 'pre-frontals' (14); and the neural spine, by the single or double 'nasal' bone (15). In Man and the higher animals, the vomer is a thin vertical plate very unlike the centrum of a vertebra; and the pre-frontals form the orbital portion (os planum) of the 'ethmoid' bone.*

320 *k.* In tracing out the 'hæmal' arches of the four cranial vertebræ, the Philosophical Anatomist is more impeded by the extraordinary changes of form and position which these elements undergo, and his determinations are consequently more open to objection, than in the case of the neural arches. These changes of form and position have reference to the special wants of the animal, and cannot, therefore, be reduced to any general expression. In some instances, there is not only an extraordinary enlargement, but an absolute repetition of similar parts; whilst in other cases, there is as remarkable a contraction or complete deficiency. It is only from a most extensive and careful comparison of the facial and other bones connected with the cranial segments in different animals, that Prof.

* The body of the 'ethmoid' bone, which has been regarded by many anatomists as the centrum of the anterior vertebra, is considered by Prof. Owen in the light of a 'sense-capsule' to the olfactive organ; a view which its position and condition in Fishes appears fully to sanction. It does not there (as in the higher Vertebrata) form part of the floor of the cranial cavity, and give support to the olfactive ganglia, or 'rhinencephalon;' but forms its anterior wall, and gives protection to the olfactive organ.

Owen has felt justified in the following determinations.—The hæmal arch of the 'occipital' segment is in reality the *scapular* arch, to which the anterior extremities are attached; its detachment from the cranium, in the fully developed condition of the higher animals, masking its original connection with it, which is obvious enough in the class of Fishes, where the scapular arch is as distinctly articulated with the cranium at its hinder part, as the lower jaw is in front. The several parts of which this arch consists (Fig. 170, 50, 51, 52, 53), will be more particularly described in the next paragraph.—The hæmal arch of the 'parietal' segment is the *hyoidean,* which is suspended from the mastoid bones, and consists, when complete, of a considerable number of independent pieces. The pleurapophyses are represented by two bones on each side; the 'stylo-hyals' (38), and the 'epi-hyals' (39); the hæmapophyses by the 'cerato-hyals' (40); and the hæmal spine by the 'basi-hyal' (41), to which the 'glosso-hyal' (42) is added in front, and the 'uro-hyal' (43) behind. The 'diverging appendages' of this hæmal arch present themselves in Fishes as a set of simple elongated rays, projecting from the outer margins of the hæmapophyses, and supporting the membrane which closes the gill-chamber externally, whence they are termed 'branchiostegal' rays; their number and size vary considerably, their most common number being seven, as in the Cod. The hyoid arch is reduced to a very small relative size in Man and the higher Vertebrata, and is comparatively incomplete (§ 326 *l*, Fig. 191).—The hæmal arch of the 'frontal' segment is the *tympano-mandibular,* consisting of the lower jaw, with the tympanic pedicle which it has in all oviparous vertebrata, connecting it with the centrum of that segment. The 'tympanic' bone (38), which is subdivided in most Fishes into four bones articulated end to end, constitutes the pleurapophysis on either side; the articular portion of the lower jaw (29), which is subdivided into numerous pieces in Reptiles, constitutes the hæmapophysis; whilst the body of the jaw (32), on which the teeth are implanted, is the hæmal spine completing the arch. A remarkable set of bones is developed in Fishes, on either side of the head, as the diverging appendage of the pleurapophysis of this arch; those, namely, which form the osseous portion of the gill-cover, and are known as *operculars.**—The most anterior of the hæmal arches, that belonging to the 'nasal' vertebra, is termed the *palato-maxillary;* it forms the palate and upper jaw, and is always much more developed in point of size than the neural arch of the same segment. The pleurapophyses are represented by the 'palatine' bones (20), which are suspended from the vomer; the hæmapophyses by the 'maxillary' (21); and the hæmal spine by the 'premaxillary' or 'intermaxillary' bone (22). This arch frequently has diverging appendages developed from its pleurapophysis, constituting the 'pterygoid' (24), and ento-pterygoid (23); whilst the hæmapophysis also has an articulating appendage which brings it into connection with the posterior segment.

* Philosophical Anatomists have long since repudiated as utterly untenable the strange doctrine propounded by Geoffroy St. Hilaire, that the opercular bones are the homologues of the minute *ossicula auditûs* (bones of the ear) of higher animals,—a doctrine which was an apt illustration of what genuine philosophical anatomy is *not.* By some, as once by Prof. Owen himself, these bones are regarded as belonging to the dermal skeleton; but his present determination seems fully borne out by the size, importance, and tendency to multiplication of parts by subdivision, which the 'diverging appendages' elsewhere present in the class of Fishes.

In Man and the higher Vertebrata, the pterygoid bones coalesce with the sphenoid; and the articulating appendage is developed into the malar bone and the zygomatic and squamous portion of the temporal. The premaxillary bone in Man usually coalesces with the maxillary. On the other hand, in Fishes, the bones entering this arch are multiplied, and are so disposed as to appear like three parallel and independent arches. This arrangement seems to have reference to the peculiar mobility which the lips of fish, in the absence of other prehensile organs, require to possess; and is one of the most striking examples yet brought under notice, of a departure from the regular type of conformation, for the sake of answering a specific purpose.

(320 *l*). We have now to enquire into the nature of those appendages to the trunk, which are commonly known as 'members' or 'limbs,'—to consider how far they can be regarded as offsets from the vertebral column, or as peculiar developments of any of its elements,—and, if such be their nature, to ascertain from what segments they are put forth. In this enquiry, it is obviously necessary to take as the guide those simplest or most elementary forms, in which there is the least departure from the primitive fundamental type; and a most beautiful example of this kind is furnished by the *Lepidosiren* (Fig. 171, A, C), in which the extremities are severally formed of a single series of bones, articulated to each other end to end. Now in this animal, the *scapular* arch, with which the anterior pair of extremities is connected, departs in no considerable degree from the hæmal arch of the typical vertebra. We trace in it (Fig. 171, A) the rib-like pleurapophyses (*pl*, 51) which answer to the 'scapula' of higher animals; and from the extremities of these hang down the hæmapophyses (*h*, 52), which represent the 'coracoids,' meeting on the median line; the hæmal spine is alone wanting to complete the parallel, and this we know to be frequently deficient. The single jointed ray, which constitutes the rudiment of the bones of the extremities (*a*, 53-57), is obviously the 'diverging appendage' of this arch; its connection with the hæmapophysis, instead of with the pleurapophysis, being but a trivial displacement. Now, this hæmal arch hangs down from the *occipital vertebra* in the Lepidosiren; and is obviously required to complete the segment, whose neural arch is formed by the bones that support and cover-in the epencephalon. Consequently, the scapular arch of the Lepodisiren must be regarded as the hæmal arch of its occipital vertebra; and the jointed rod-like anterior extremities as its diverging appendages. But the Lepidosiren is not an exceptional instance of this connection and relation. The pectoral fins of all osseous and of some cartilaginous Fishes, whatever their form and size, are supported by an inverted bony arch, which is as closely connected with the centrum and neural arch of the occipital vertebra, as it is in the Lepidosiren; and which must consequently be regarded (notwithstanding its somewhat wider departure from the original simple type) as the hæmal arch of the first cranial segment, being even present when the pectoral fin itself is not developed. And the pectoral fin is formed by a mere multiplication of parts resembling the single elements of the spiny appendage in the Lepidosiren; just as the branchiostegal rays are the multiplied or subdivided diverging appendages of the frontal vertebra.—When we go to the lower Batrachian Reptiles, as the *Amphiuma*, we find the scapular arch disconnected from the occipital vertebra; but

when these parts are again brought together (as in Fig. 171, B), the original relation between them is no less evident than in the preceding case. The bones of the extremities now begin to exhibit the arrangement characteristic of terrestrial animals; the distinction of humerus (53), ulna and radius (54, 55), carpus (56), and digits (57) being now apparent; but the latter are as yet only two in number. The humerus (or origin of

FIG. 171.

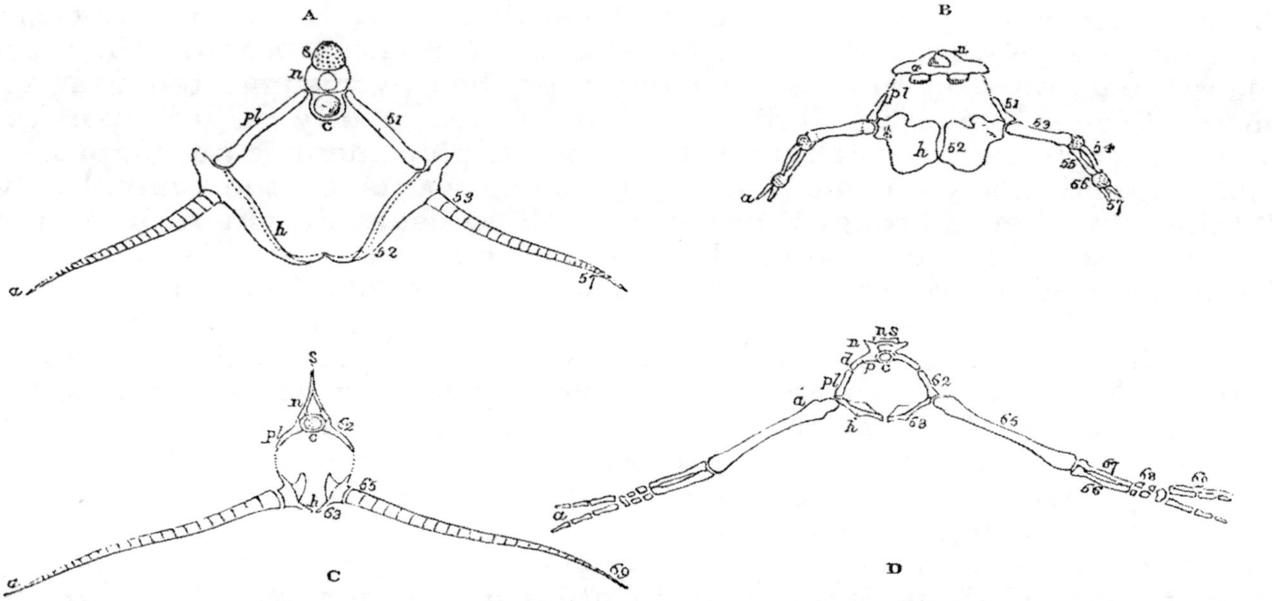

Diagram illustrating the *Nature of Limbs*:—A, posterior view of the occipital vertebra of *Lepidosiren annectens*:—B, posterior view of the occipital vertebra of *Amphiuma didactylum*:—C, posterior view of the pelvic vertebra of *Lepidosiren*:—D, posterior view of the pelvic vertebra of *Proteus anguinus*. In the several diagrams, the following references indicate corresponding parts; *c*, centrum; *n*, neurapophyses; *s*, neural spine; *pl*, 51, pleurapophysis of the occipital vertebra, or scapula; *h*, 52, hæmapophysis of the occipital vertebra, or coracoid bone; *a*, 53-57, diverging appendages of the occipital vertebra, or anterior limbs; *pl*, 62, pleurapophysis of the pelvic vertebra, or iliac bone; *h*, 63, hæmapophysis of the pelvic vertebra, or ischiac bone; *a*, 65-69, diverging appendages of the pelvic vertebra, or posterior limbs.

the diverging appendage) is here articulated with the hæmal arch at the junction of the pleurapophysis and hæmapophysis, both the scapula and coracoid entering into the composition of the shoulder-joint; and this is the arrangement which we meet with in all oviparous vertebrata, in which the clavicle is a subordinate bone. In the Mammalia, however (except among the non-placental orders), the coracoid bone is reduced to a mere rudiment, being known as a 'process' of the scapula; and its function (that of keeping the shoulders apart) is taken-on by the 'clavicle;' which is regarded by Prof. Owen as the hæmapophysis of the 'atlas' or first cervical vertebra (Fig. 170, 58), the portion of the 'sternum' (properly the episternum) to which it is attached, being the hæmal spine of the same arch.—Now although to such as have not studied Anatomy with a view to the determination of the original or primitive relations of the component parts of the structure, it may be difficult to conceive of the scapular arch in Man and the air-breathing Vertebrata, as being thus detached from its original segment and carried backwards, so as to be implanted on the trunk *below* the neck instead of at its summit; yet a more careful survey of the phenomena of Nature shows that the possibility or even probability of such a displacement must be freely admitted. For, not to mention other instances in which it is less striking, we find that in Fishes the *abdominal* fins, which obviously answer to the posterior extremities of higher Vertebrata, may be carried forwards to any part of the trunk, and may even occupy a position *anterior* to that of the pectoral fins. There would be as much reason, then, for regarding these organs as no longer the same with the normal abdominal fins, when they are thus displaced forwards, as there would be for contending that the scapular arch and anterior extremities of air-breathing Vertebrata are not the representatives of the occipital appendages of Fishes, because they are displaced backwards; and as no Anatomist has ever suggested a doubt upon the former point, no difficulty can be reasonably urged upon the latter. In each case, it is obvious that the displacement is an adaptation of the original plan, to meet the particular wants of the animal; the abdominal fins being brought forwards in those fishes in which the power of readily ascending and descending in the water is specially required; and the pectoral extremities being carried backwards from the front to the hinder end of the neck, in order to give advantageous support to the trunk of those animals which are to be modified for progression on land.

320 *m*. That the *pelvic* arch, however greatly it may be modified, is the representative of the scapular,—the 'ilium' answering to the scapula, the 'ischium' to the coracoid, and the 'pubis' to the clavicle,—is not doubted by Anatomists who have studied their relations in different animals. Such being the case, the pelvic arch must be regarded as the hæmal arch of one or more of the pelvic vertebræ; and such it plainly appears to be, when it is studied in its simpler forms. Thus in the *Lepidosiren* (Fig. 171, C), we find the posterior jointed spines attached to a pair of bones (63), which hang beneath the pleurapophyses (62) of the pelvic vertebra, and which are obviously its hæmapophyses. In common with the whole class of Fishes, the hæmal arch is here divided, the pleurapophyses being articulated with the centrum, whilst the hæmapophyses are detached from them; and this arrangement obviously has reference to the mobility of the arch in this class. In the lower Batrachian Reptiles, however, we find, with a very little wider departure from the primitive type, that arrangement of the pelvic arch which generally prevails in air-breathing animals, in which the situation of the posterior extremities is constant. Thus in the *Proteus* (Fig. 171, D) we see the rib-like 'ilium' (62), articulated as a pleurapophysis with the parapophysis of the centrum; whilst the arch is completed below by the 'ischium' (63), which is evidently the hæmapophysis. The posterior extremity, like the anterior, is here connected with the pleurapophysis and the hæmapophysis at their junction; both these entering into the composition of the hip-joint. The extremities themselves, in the Proteus, show a slight advance upon their condition in the Amphiuma; for the tarsal bones (68) are now distinct, and the digits (still only two in number) are divided into metatarsus and phalanges (69). The 'pubis,' which does not exist in Fishes, makes its appearance in Reptiles as an arch altogether distinct from the ischiatic; and this condition it retains in Birds, where it evidently holds the same relation to the ischium in the pelvic arch, that the clavicle does to the coracoid in the scapular, and would thus appear to be the hæmapophysis of another vertebral segment. In Man and the

Mammalia, the ischia do not meet on the central line, the arch being completed by the pubis.—Thus it appears that the position of the *posterior* extremities is constant and normal in the air-breathing Vertebrata, whilst it is exceptional in Fishes. On the other hand, the position of the *anterior* extremities is constant and normal in Fishes alone, being exceptional (that is, departing from the fundamental type) in Reptiles, Birds, and Mammalia. If the former exception be allowed, there need be no difficulty about the latter, since it is obviously due to the operation of the very same principle,—that of special adaptation to a particular mode of existence, involving a departure from the 'archetype' or primitive design.

320 *n.* The condition of the vertebral skeleton, and its degree of development, vary greatly in different parts of the series. In the lowest Fishes, as we shall presently see, it retains through life its embryonic condition; no bone, nor even true cartilage, being formed around the neural axis; the indications of segmental division being very slight; and no members or appendages being developed; the progression of these animals being accomplished, like that of Worms, by the flexion of the body itself. In proportion, however, as distinct members are developed, and the power of locomotion is delegated to them, we find the firmness of the spinal column increasing, and its flexibility diminishing; of this we have the most remarkable example in the class of Birds, in which (as in Insects) it is requisite that the muscles of the organs of flight should have very firm points of support, and in which the internal skeleton of the trunk (like the tegumentary skeleton of the thorax of Insects) is so closely knit together as to form an almost unyielding framework. As a general rule, then, the firmness and consolidation of the vertebral column, and the locomotive power of the extremities, vary in a like proportion. The number of these extremities never exceeds *four;* and two of them are absent in certain members of each Vertebrate class; whilst in the entire group of Serpents, and in some worm-like Fishes, both pairs are wanting. We find them adapted, by variations in their mode of development, to a great variety of purposes; thus, in Fishes, aquatic Reptiles, and Mammals, they are modified for the mere propulsion of the body through water, and not for giving it support; in the terrestrial quadrupeds, whether Reptilian or Mammalian, they are adapted to support the body on the solid ground, as well as to give it motion over its surface; whilst in Birds, as also in certain members of each of the other classes, the anterior pair of extremities is developed into wings, by whose agency the body is (more or less completely) supported and propelled through the air. Notwithstanding the *variety* of locomotive power possessed by Vertebrated animals, however, its *energy* must be regarded as inferior to that which is seen in the Articulata; the swiftest Bird or Fish not traversing nearly so many times its own length in a given period, as a large proportion of the Insect tribes pass through with ease. It is far greater, however, than is enjoyed by the Mollusca, among which only a few of the Cephalopoda can rival Fishes in the rapidity and energy of their movements; and there is no species that is fixed to one spot, without the power of changing its place. In the segmental division of the body, the articulation of the limbs, and the general arrangement of the locomotive apparatus of the Vertebrata, we trace a strong resemblance to the plan which we have seen to be pursued in the Articulata; with the important

and typical difference in the position of the skeleton, on which we have already so fully dwelt. This resemblance is especially apparent in the *bilateral symmetry*, which, with very few exceptions,* is as constantly exhibited in the entire external configuration of Vertebrated, as in that of Articulated animals; and also in the repetition of the ganglionic centres of the nerves of the body and limbs, though these unite in Vertebrata into a continuous tract—the spinal cord,—instead of remaining disjoined as they are in the ventral cord of Articulata generally. Consequently, in regard to the nervo-muscular apparatus of locomotion, we may consider the Vertebrata as carrying upwards the type of the Articulata. It is interesting to remark that the muscular fibre of the whole apparatus of locomotion in the Vertebrata is of that higher type, presenting transverse striæ (§ 223), which is found throughout nearly the whole of the Articulated series; whilst that of the apparatus of organic life resembles that generally found even in the locomotive organs of the Mollusca, in being of the lower or non-striated kind (§ 225).

320 *o*. It is upon the type presented by the highest Mollusca, however, that we find the development of the organs of special sense, and of the ganglionic centres with which they are connected, proceeding in Vertebrated animals. Thus the eyes of Fishes bear a much closer resemblance to those of the *Octopus* (Fig. 131), than they do to the compound eyes of Articulata; and there is never, in any Vertebrated animal, more than a single pair of these organs,—the only vestige of that extraordinary multiplication of them which is such a characteristic feature of the Articulata (§ 308) being seen in the curious *Anableps*, in which the cornea and pupil are double on each side, though the whole of the posterior part of the eye is single. So, again, the rudimentary apparatus of hearing in Gasteropods and Cephalopods bears a very close resemblance to that of even the highest Vertebrata at an early period of its development. All the four special senses—sight, hearing, smell, and taste—appear to be possessed by most Vertebrated animals; but it is not uncommon for the visual organs to remain undeveloped, when the circumstances in which the animal is adapted to exist are such as would render eyes of no use to it. There are Fishes, Reptiles, and even Mammals, which cannot possess the power of sight, even if they have the perception of light; but there are no known Birds that are destitute of this sense. The organ of hearing is more universally present, not being known to be deficient in any vertebrated animal save the *Amphioxus* (§ 321). It is difficult to determine in some cases how far the senses of smell and taste are present; but it is probable that they are never entirely deficient. The sense of touch is more generally

* The upper and under surfaces of the *Pleuronectidæ* (Turbot, Sole, &c.), although differently coloured, are in reality the two *sides* of their bodies; and the external conformity between what are popularly regarded as their right and left halves, is produced by the similarity in form, and in the osseous framework, of the dorsal and ventral regions. The principal departure from the bilateral symmetry of the skeleton, is seen in the configuration of the head, in which both the eyes are brought to one side.—In the Dolphin and Physeter, the right nasal orifice is much the larger, and the right nasal orifice is pushed to one side, the nasal bones lying rather behind than by the side of each other.—In the Narwhal, only one of the two long tusks is usually developed, and the half of the upper jaw which bears it is longer and larger than that on the other side, the same want of symmetry extending itself also to the lower jaw.—The position of the anal orifice is unsymmetrical in *Amphioxus* (§ 321) and *Lepidosiren* (§ 325 *a*).

diffused over the bodies of most Vertebrated animals, owing to the absence or low development of their dermal skeleton, than it is in the greater number of Invertebrata; and we find it rendered subservient in the former, by its remarkable perfection in particular organs, to the attainment of a much more complete knowledge of the conditions of external objects, than it can afford to the latter.

320 *p.* The fundamental elements of the Nervous system of the Vertebrata, then, may be looked upon as consisting of a set of *locomotive* ganglia, arranged in a continuous longitudinal series (as in the Articulata), forming the Spinal Cord; and of a set of *sensory* ganglia, more resembling those of the Mollusca, which may be regarded as constituting the essential part of the Encephalic centres, being the parts most universally present, and being those first developed at the commencement of embryonic life. Interposed between these, however, in the portion of the neural axis termed the Medulla Oblongata, we find the ganglionic centres which answer to the respiratory and pharyngeal ganglia of Invertebrated animals, and which in them are usually disunited from the general sensori-motor apparatus, being connected with the cephalic ganglia alone by communicating cords. This incorporation of so many distinct centres into one system, would seem destined in part to afford to them all the protection of the vertebral column, and in part to secure that consentaneousness of action, and to afford the ready means of that mutual influence, which are peculiarly requisite in animals in whom the activity of the nervous system is so predominant.—Although the sensory ganglia may be regarded as fundamental elements of the brain of Vertebrata, yet they do not, except in the lowest Fishes, constitute the whole of it; for it is one of the most constant and remarkable of the general characters of the animals composing this sub-kingdom, that they possess an additional nervous mass, the *Cerebrum*, superimposed (as it were) upon the sensory ganglia; not itself receiving the central terminations of the nerves of sense, nor giving off the central origins of those of motion; but being connected with both, through the medium of their proper ganglionic centres. The development of the Cerebrum in the Vertebrated series bears so close a relation to that of the *intelligence*, which has been already mentioned as their special characteristic (§ 320), that little doubt can exist of its being the instrument of those peculiar psychical endowments, which may be conveniently grouped together under this designation. It is this organ, more than any other, which is especially characteristic of the Vertebrated sub-kingdom; for if any vestige of it exist among the inferior classes (which is by no means certain), it is only found to be in such as present some degree of general approximation to the Vertebrated type;* and whilst it is a mere rudiment in those lowest Fishes which present but little that is distinctive of the Vertebrata in their general conformation, it manifests a regular progressive development in the ascending series, until it reaches its culminating point in Man, in whom (more than in any other animal) we find that highest grade of development of the corporeal organism, which is exhibited in *the specialisation of each part*—even of such as most closely resemble one another—*for distinct uses.* Besides the Cerebrum, we find in Vertebrata another superadded nervous centre, which is equally distinct from the

* By this is meant *not* the vertebral conformation itself, so much as the general condition of the group.

ganglionic centres of the nerve-trunks. Of this organ, the *Cerebellum*, the function is less certainly known. Its development is less uniformly progressive in the different classes; and it seems to bear the closest relation to the number and variety of muscular actions, which it is necessary to combine in the movements of the animal, as if it were the instrument for the automatic selection and combination of the various single operations, of which a considerable number is involved in almost every movement that is executed by a determination of the will. The Cerebro-spinal axis, formed by the union of these several centres, is wholly lodged in the neural canal of the vertebral column, and in its cranial expansion; and it lies entirely above the alimentary canal, the cords which connect the cephalic ganglia with those of the trunk (the fibrous strands of the *medulla oblongata*) not being separated from each other, as they are both in Articulata and Mollusca, for the passage of the œsophagus between them.—But besides the cerebro-spinal nervous system, which is the instrument of the *animal* life of Vertebrata, we find another set of ganglionic centres and nerve-trunks especially connected with the viscera, and constituting what is known as the Sympathetic system, or nervous system of *organic* life. Very distinct approaches to this system are presented by some of the higher Invertebrata; but it is not so completely isolated in them, as it is in Vertebrated animals, from the nervous system of animal life.

320 *q*. The principal modifications presented by the various parts of the apparatus of Organic life in Vertebrated animals, all seem to bear more or less directly upon the exercise of their psychical endowments; and cannot, therefore, be regarded as so characteristic of the class, as are those peculiarities of the apparatus of animal life, to which we have given the first and fullest consideration.—The Digestive cavity invariably possesses two orifices, a buccal and an anal: of these the former is usually situated at the most anterior part of the head, and is a *transverse* fissure provided with a pair of jaws opening vertically (the mouth of the Amphioxus and of the Cyclostome Fishes being the only exception to this general statement); whilst the latter is placed at the posterior extremity of the trunk, on the median line, beneath the origin of the tail where this organ exists. The alimentary canal usually exhibits a distinct division into œsophagus, stomach, and intestinal tube; and it is for the most part disposed, with its glandular appendages, in the posterior or abdominal portion of the cavity of the trunk. In all Vertebrata, the liver is a very important organ, and presents not only a large size, but a high grade of development; it has also a special circulation of its own, a part of the venous blood on its return from the system being transmitted through its capillaries, before it reaches the heart. This organ is usually non-symmetrical, from the unequal development of its two halves, so that it occupies more of the right side in Mammals, and of the left side in Fishes; but in an early stage of its development, its symmetry is exact in all Vertebrata; and this character is retained in Birds, in which the internal symmetry is as great as that of the Articulata. The pancreas, which made its first appearance in the Cephalopods, is obviously an organ of considerable importance in the higher Vertebrata; and its function seems to be specially related, by the peculiar property of the secretion which it forms, to the assimilation of those fatty constituents of the food, which are the chief sources of their

heat-producing power. It is in Vertebrata, again, that we first meet with the peculiar organ termed the spleen ; though some approaches to it may be seen in the higher Mollusca. The most important peculiarity, however, in the apparatus of 'sanguification,' exhibited in the Vertebrata, consists in the mode in which the crude alimentary materials are taken into the current of the circulation. We no longer find this effected by the agency of the blood-vessels alone ; for, although these are still concerned in the process, a special set of *absorbent* vessels is found, commencing, like so many rootlets, in the walls of the intestine, and uniting into one or more main trunks, which discharge into the systemic veins the fluid they have taken up, there to mingle with the circulating blood. The fluid of the 'lacteals,' known as 'chyle,' is especially characterized by the presence of fatty matter in a state of very minute subdivision, which seems to perform an important part in the subsequent processes of organisation. The 'absorbent system' is not confined, however, to the walls of the intestinal tube, but is distributed throughout the entire body ; the 'lymphatics,' as these interstitial absorbent vessels are termed, being especially adapted to carry back into the current of the circulation such nutritive material as has escaped from it without being appropriated, or such as, having once formed a part of the living structure, has been set free by its death, without having been rendered unfit for re-assimilation by a change of composition. The current of blood is always put in motion, save in the anomalous *Amphioxus* (§ 321), by a single compact muscular organ, the heart ; which is situated in the anterior or thoracic part of the cavity of the trunk. The bloodvessels form a complete *closed* system ; the blood being returned by veins, within which it is entirely confined, instead of escaping from the vessels into the great cavities of the body or the interstices of the tissues, as it does in most Invertebrata. The blood (save in the Amphioxus) is reddened by the presence of 'red corpuscles' (§ 173), which, being exclusively present in Vertebrated animals, impart to their blood a distinctive character. The colourless blood in the Invertebrata may, in fact, be regarded as resembling the *chyle* of Vertebrated animals more than it resembles their *blood*, which is a fluid of much higher elaboration.

320*r*. The apparatus for the depuration of the blood presents a very high development in the Vertebrated classes. The Respiratory organs, whether formed for aquatic or atmospheric respiration, are always placed in immediate connection with the heart ; and of the venous blood, which has been returned to it by the systemic veins, either the whole or a part is sent directly to the aerating surface, before being again propelled through the systemic arteries to the body at large. The respiratory circulation is therefore carried on at a much greater advantage, than when it is a mere offset from the venous portion of the general or systemic circulation, as it is in most of the Invertebrata. The efficiency of the respiration is further provided for by this,—that the apparatus is everywhere connected with a set of muscles, which, by an automatic action of the nervous system, keep up a regular inspiratory and expiratory movement ; whereby the aerating medium, whether air or water, in contact with the respiratory surface, is continually renewed. The respiratory organs are always so connected with the mouth, that the *in*spiration takes place through it ; the *ex*piration takes place in Fishes, and in the perennibranchiate Batrachia (§ 323), through one or more fissures on each side of the neck (which fissures present themselves, up to a certain period of embryonic development, in all the higher classes); but in all the air-breathing Vertebrata, the expiratory as well as the inspiratory current passes through the mouth.—In no organs is the principle of *concentration* more remarkably manifested, than it is in the Lungs; which possess a much greater functional activity in the Mammalia than they do in Reptiles (notwithstanding their much larger relative size in the latter), in virtue of their very minute subdivision, and the energetic working of those subsidiary arrangements, whereby the blood and the air which they contain are continually renewed.—The Kidneys, of which scarcely a vestige can be traced in the greater number of Invertebrata, are here constantly present; and from their size, as well as their high grade of glandular development, may be considered as organs of essential importance in the depuration of the blood,—a conclusion which we shall hereafter find to be confirmed by physiological evidence. In the cold-blooded Vertebrata, they partake, with the liver, in the special circulation just described as diverting a portion of the venous current on its return to the heart; but in Birds and Mammals, they are supplied exclusively with arterial blood from the general circulation.—In the mucous membrane of the alimentary canal, and in the skin, we find an immense number of minute glandulæ, apparently destined for the elimination of other effete matters from the blood; the constant maintenance of whose purity is especially necessary, in animals that are destined to support a life of continual activity.

320*s*. It is in the higher Vertebrata alone, that we meet with a provision for the constant maintenance of an elevated temperature. The calorific power is possessed to a certain degree by particular tribes of Insects (CHAP. XVII., SECT. 2); but it depends for its exercise upon the state of activity of the individual, the temperature of the animal when in repose being scarcely above that of the surrounding medium. In Birds and Mammals, however, we find a provision for the maintenance of a constantly-elevated temperature, which is scarcely at all dependent upon the general activity of the individual, being sustained up to the regular standard during repose, though somewhat augmented by continued muscular exertion. This provision is connected with the energy of the respiration of these animals, and with the large number of red corpuscles in their blood; it serves to render them almost entirely independent of moderate changes in external temperature, enabling them to keep up the same amount of activity, both in the organic and nervo-muscular functions, through all variations of the seasons; and it answers another important purpose, also, in promoting the development of the embryo.

320*t*. In the Vertebrate series, we find the power of reproduction by 'gemmation' limited to the reparation of lost or injured parts, and not extending to the multiplication of individuals; and in no known instance does this multiplication take place in any other way, than by true sexual generation. This function is always performed by the congress of two individuals; the sexes being invariably separate. In both sexes the outlet of the generative organs is situated near the posterior extremity of the body, in close proximity with the anal and renal orifices; and where the organs are of moderate size, they are restricted to the posterior part of the abdominal cavity, though they may extend far forwards in certain cases. The ova are not fertilized by the contact of the spermatic fluid until after they have escaped from the body of the female, in a considerable proportion of the class of Fishes, and in the lower Reptiles ; but in the higher Fishes and Reptiles, and in all Birds and Mammals, the ova are fertilized whilst yet within the oviducts, and are retained for a shorter or longer period afterwards, so that the embryo undergoes the early processes of development before the deposition of the ovum. In the higher Cartilaginous and in some Osseous Fishes, and in a few Reptiles, the ova are thus retained even until they are hatched, and the young are produced alive ; in the greater number of Reptiles, however, as in Birds, they are deposited before their development has made any great advance ; but in the latter class the process of evolution is completed under the influence of heat supplied by the body of the parent. In the Mammalia, the embryo is retained within the body of the mother, and continues to receive from her not only warmth but nutriment, its development taking place in a dilated portion of the oviduct, known as the *uterus.*—In nearly all Vertebrated animals, the young animal, when it first comes forth into the world, has the characters of the *class* to which it belongs ; and those of its order, family, genus, and species, if not at once distinguishable, speedily become so ; consequently there is not anything like true 'metamorphosis,' but only a gradual evolution of a special type from one more general. In one remarkable group, however, a true metamorphosis takes place ; the members of the group of *Amphibia,* or Batrachian Reptiles, coming forth from the egg in the form and condition of Fishes, and gradually assuming that of Reptiles, by a series of changes in which every part of their organism is concerned (§ 323). Still, as we shall find hereafter, these changes are not more remarkable than those which take place in the embryoes of *all* Vertebrata during their development within the ovum ; and the chief peculiarity of the Batrachia consists in their emergence from the ovum at a comparatively early period, and in their adaptation to maintain their own existence at a grade of development, at which other embryoes are dependent upon the supply of nutriment afforded to them, directly or indirectly, by their parents. In this particular, there is an exact correspondence between the metamorphosis of Batrachia and that of Insects (§ 315 *e*).

321. In the class of FISHES, we have, on the one hand, the lowest development of the Vertebrated type of structure, not only in the skeleton, but in the entire fabric ; and, on the other hand, a variety of special modifications, whereby a Vertebrated animal is enabled, not merely to inhabit the water and to move through it with rapidity, but also to use that medium for the purpose of respiration, extracting from it the oxygen which it contains. Before proceeding, however, to our general survey of the characteristic structure of this important class, it will be desirable to bestow some attention on the anatomy of an animal, which serves to connect together the Vertebrated and Invertebrated series in a most remarkable manner ; linking the lower tribes of Fishes, not with the highest, but with some of the lowest forms both of Mollusca and Articulata. This animal is the *Amphioxus** or Lancelet ; of which it has been justly

* The history of our knowledge of this extraordinary creature is worth a short notice in this place. The first mention of it in any scientific treatise appears to have been that of Pallas, the eminent Russian Naturalist; who, in his *Spicilegia Zoologica*, described a single

remarked, that "it presents a combination of characters which must excite the wonder and interest equally of the Physiologist and the scientific Naturalist." Its usual length is about 2 inches; its greatest breadth from the dorsal to the ventral margin is about 1–5th of an inch; and its thickness, from one side to the other, about 3–20ths of an inch. It tapers gradually, however, towards the two extremities, the head being somewhat more rounded than the tail; and it also thins away towards its dorsal and ventral margins, the former of which is bordered along its whole length by a membranous fin-like expansion, whilst on the under side of the body there are two strong lateral folds, extending from the mouth nearly to the abdominal pore, but giving place behind this to the median fin which extends to the end of the tail.* On the head, which is scarcely distinguishable from the body, there are no external indications of the existence of organs of sense; and the mouth, situated on its under surface, is a *longitudinal* slit, fringed on each side by a row of *cirrhi* or tendril-like filaments (Fig. 172, *ff*), which can close-in towards each other and clasp together, so as to protect the oral orifice. These cirrhi are attached at their base to a cartilaginous framework (*h*), formed of separate pieces, of which each sends a prolongation into one of the filaments.† About half way between the posterior extremity of the mouth and the tail, is the abdominal pore (*o d*), which opens at once into the abdominal cavity, and serves as an outlet for the water that has passed over the branchial surface, as also for the genital products. At some little distance behind this is the anus (*a s*), which is not situated precisely on the median line, but on the left side of the caudal fin. The body is marked externally by a succession of oblique striæ, which are the indications of the lateral muscular bundles, seen through the translucent skin.—Such are the external characters of this curious little animal, in which there is an almost entire want of those peculiar to Vertebrated animals, whilst some of them (particularly the unsymmetrical position of the anus) remind us of Mollusca, and others (especially the *longitudinal* direction of the oral fissure, with the

specimen, which he had received from Cornwall during the latter part of the last century, under the designation of *Limax lanceolatus;* Pallas having believed it to be a naked Mollusk, though he remarked in his description of it, on the resemblance of some of its characters to those of a fish. It seems to have been lost sight of for more than half a century; when, in 1834, it was rediscovered by Costa on the Neapolitan shores, and described by him as a Fish, under the name of *Branchiostoma.* Although this name has the priority to that of *Amphioxus*, conferred by Mr. Yarrell in 1836, yet as the former designation is founded on an erroneous conception of the office of the filamentous appendages with which the mouth is guarded, the author thinks with M. De Quatrefages (from whose admirable Memoir on the anatomy of this animal, in the Ann. des Sci. Nat. for 1845, his account of it is chiefly condensed), that it would be better to abandon it for *Amphioxus.* In 1836, Mr. Yarrell, without being acquainted with Costa's account of the animal, described it in his 'British Fishes,' from a specimen obtained by Mr. Couch on the coast of Cornwall; and about the same time it was taken by several naturalists on the coast of Sweden. It has been subsequently found in considerable numbers on the west coast of Scotland, as well as in several other localities; and has been the subject of many admirable Anatomical Memoirs. The Author is informed by Prof. Owen that it is regularly brought to market by certain fishermen on the Neapolitan coast, to whom it has been known from time immemorial, being used, as we employ shrimps, in *sauce* to larger fish; and it is not a little remarkable that an article in such common use should have so long escaped the attention of the Italian Naturalists.

* These lateral folds, being mistaken by Pallas for the margins of a ventral disk, were among the characters which led him to rank it as a Gasteropodous Mollusk.

† These filaments, which are obviously analogous to the cirrhi of Cyclostome fishes, were erroneously supposed by Costa to be branchiæ.

lateral movement of the cartilaginous framework that bounds it) indicate a relation to the Articulata. We shall see that, in the internal structure, the Vertebrated type is most distinctly manifested, although in a most rudimentary condition; whilst approximations to the Mollusca on the one hand, and to the Articulata on the other, are also very clearly exhibited.

FIG. 172.

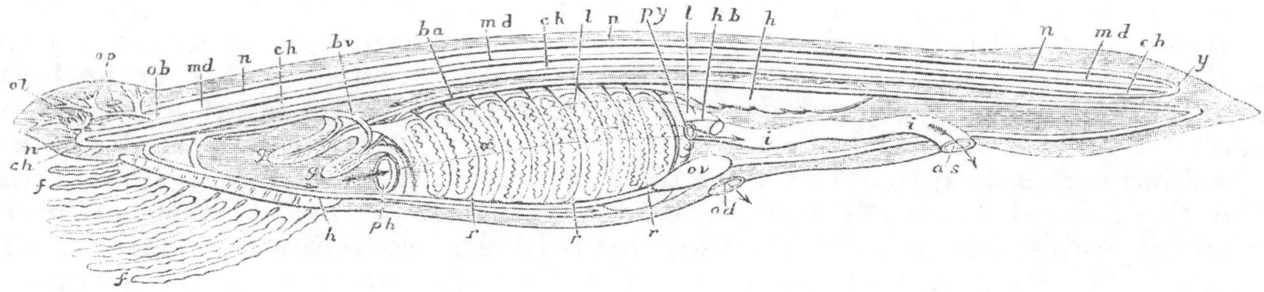

Diagram of the Anatomy of *Amphioxus*, or *Branchiostoma* (Lancelet);—*ch, ch,* chorda dorsalis; *n, n,* fibro-membranous wall of neural canal; *ol,* olfactory capsule; *op,* optic nerve; *ob,* trigeminal nerves; *md, md,* neural axis; *bv,* branchial vein; *ba,* branchial artery; *l, l,* hepatic cæcum, opening at *hb*; *py,* cardiac orifice; *h,* renal organ; *y,* caudal extremity of neural axis; *as,* anus; *i, i,* intestine; *od,* abdominal pore; *ov,* cardiac sinus; *r, r, r,* dilated œsophagus, with branchial fissures; *ph,* pharyngeal orifice; *g, g,* vascular intra-buccal processes; *h,* cartilaginous arch, supporting *f, f,* the ciliated labial tentacula.

321 *a.* Almost the only trace of the vertebral column in the *Amphioxus*, is to be found in the fibrous and cellular structures surrounding and supporting the neural axis. Passing along almost the entire length of the animal, with but little change in its dimensions from one extremity to the other, is a nearly cylindrical tube (*n, n*), formed of a delicate fibrous membrane; this encloses the neural axis with the 'chorda dorsalis', and sends off lateral fibrous prolongations, which pass between the muscular bundles, and constitute the only representatives of ribs. The neural axis is nearly uniform in size from one extremity to the other; consisting of a series of ganglia, obviously distinct from each other, although in close approximation; and not presenting any such enlargement at its cephalic termination, as could be denominated a brain. The only vestiges of organs of special sense consist of rudimentary eyes, which, though possessed of a distinct crystalline lens with pigmentary matter, are lodged in the substance of the fibrous investment of the neural axis; and of another peculiar body resembling the rudimentary organ of smell in the embryoes of other fishes: no trace has yet been detected of an organ of hearing. The neural axis rests upon the *chorda dorsalis* (*ch, ch*), a cylinder composed of a membranous envelope, filled with a soft cellular substance which resembles cartilage in its earliest grade of development. In this cylinder, which represents the column formed by the 'bodies' of the vertebræ of higher animals, there is but the most indistinct indication of subdivision into segments. The arrangement of the lateral muscles, which intervene between the costal prolongations of the fibrous axis, is precisely that of the intercostal muscles of fishes; besides these, there is a set of longitudinal muscles running along the under side of the abdominal cavity (which, unlike the intercostals, are composed of non-striated fibre); and there are also special groups of muscles connected with the oral ring and with the branchial apparatus. The movements of the animal are some-

what snake-like, and are rapid and powerful; but it seems to pass a large part of its time either buried in sand, or lying flat upon its surface.—In all the foregoing particulars, the structure is obviously that of a Vertebrated animal; resembling that of higher Vertebrata at a very early period of development, and being (so to speak) a further degradation of the already low type which is presented in the Myxinoid fishes. The want of an osseous or even a cartilaginous skeleton is a simple result of 'arrest of development;' and the same may be said of the rudimentary condition of the brain and organs of sense. In the distinctness of the ganglionic segments of the neural axis, we are reminded of the Articulata; but this, also, may be regarded as a mere result of the want of that coalescence which ordinarily takes place in higher Vertebrata; and it does not indicate a real approximation to the Articulated sub-kingdom, from which the neural axis differs in its dorsal position (lying *above* the alimentary canal along its entire course) and in its relations to the chorda dorsalis. The arrangement of the lateral muscles more resembles that of the Fish than that of the Worm; and their attachment is to the rudiment of an *internal*, not to any trace of an *external* skeleton.

321 *b*. It is in the arrangement of the organs of nutrition, however, that the greatest approximation to Invertebrate types is displayed. The oral cavity is lined with cilia; and these are especially disposed upon a set of digitate prolongations of the mucous membrane (*g*, *g*), which have been denominated 'rotatory organs,' from their resemblance to the ciliated lobes of the Rotifera (§ 310). It opens at the back, by the orifice *ph*, into a wide pharyngeal cavity (*r*, *r*), the walls of which are strengthened by a sort of cage, formed of delicate, transparent, hair-like cartilaginous arches, seventy or eighty in number; whilst between these are a series of clefts, the edges of which are fringed with cilia. This appears to serve as the principal respiratory apparatus; a considerable part of the water which is taken in by the action of the oral ciliated lobes, and which enters the pharyngeal orifice (*ph*), being driven through these slits into the abdominal cavity, from which it makes its exit by the abdominal pore (*od*). The remainder of the water passes backwards through the cardiac orifice (*py*) into the digestive cavity, which consists of a short and nearly straight intestine (*i*), having no distinct dilatation into a gastric cavity at its anterior part, but being there connected with a peculiar cæcum (*l*, *l*) which extends forwards, and which is probably to be regarded as a rudimentary liver, being lined by a layer of greenish cells, which extends, however, along the intestine also. The intestinal tube, which terminates in the anal orifice (*as*), is lined throughout by vibratile cilia; and these appear, as in the lower Invertebrata, to be the sole instruments by which the alimentary as well as the respiratory current is sustained. The blood, which contains no red corpuscles, and only a few colourless corpuscles of irregular size and form, moves for the most part through a regular system of closed vessels, though these appear to give place in some parts to lacunæ; its movement, however, is not due to the action of any single propelling organ, or heart, but is accomplished by the agency of several pulsating dilatations, which are developed on different parts of the vascular system. As in Articulata, we find two principal trunks, a dorsal and a ventral; these are connected together by arches at the two extremities of the body, and also by transverse vessels which pass along the

walls of the branchial sac (*r, r, r*), between its fissures. In these vessels, however, the blood moves in a direction contrary to that in which it flows in the Articulata; for it passes backwards along the *dorsal* trunk (*ba*), receiving in its course the blood which has been aerated on the branchial surface; it is then distributed to different parts of the body, whence it is collected by veins which carry it forwards towards one of the principal contractile dilatations (*ov*) situated close to the under surface of the branchial sac; and by this it is propelled, partly through the branchial vessels which deliver it into the dorsal trunk, and partly through the anterior continuation of the ventral trunk. The latter portion finds its way through the aortic arches at the anterior extremity of the body, which are themselves pulsatile; and by their agency it is distributed to the head, and especially through the vessel *bv* to the ciliated appendages *g, g* (which are considered by Prof. Owen as the chief respiratory organs), whence it is collected by the dorsal vessel (*ba*), and transmitted backwards, mingling with that which has crossed the pharyngeal sac. At the base of each of the branchial arteries given off from the principal heart (*ov*), is a small pulsating dilatation; there is another on the large vein (the vena portæ) which collects the blood from the intestinal tube and conveys it to the hepatic cæcum; and one more upon the principal venous trunk, the vena cava. These several pulsating organs do not contract together, but alternate with each other, one filling as another empties itself; and in this manner a constant flow is maintained, which is, however, far from rapid, the interval between the successive pulsations of the principal branchial heart (*ov*) being as much as a minute.—It only remains, for completing the description of this curious animal, to mention that the upper part of the abdominal cavity contains a peculiar organ (*h*) apparently glandular in its character, and probably performing the office of a kidney; whilst the ovaries or testes lie along its floor, and discharge their products into its cavity, whence they make their exit by the abdominal pore.

321 *c*. In looking at the structure and relations of the digestive and respiratory organs of the *Amphioxus*, we are at once struck with their resemblance to those of the Ascidian Mollusks (§ 299). In these, as in the Amphioxus, the respiration is carried on through a pharyngeal sac whose conformation is almost identically the same in both cases, that of the Amphioxus being strengthened by a splanchno-skeleton of delicate cartilaginous arches; in both alike is the water drawn into the oral orifice by ciliary action, and thus propelled partly along the alimentary canal, and partly through the pharyngeal slits into the visceral cavity; in both alike does the expiratory current pass out through the same orifice with the genital products; and the only important difference lies in the circumstance, that the intestinal tube terminates in a separate external orifice, instead of opening into a common cloaca with the preceding. On the other hand, the plan of the circulating apparatus is distinctly Annelidan (§ 311); and in the peculiar disposition of the separate pulsating cavities, it bears a very close resemblance to that which will be hereafter described in the *Eunice* (CHAP. XII.).—When we examine, on the other hand, into the relations of these peculiarities to the structure characteristic of the respiratory and circulating organs in ordinary Fishes, we shall see that the departure is not so great as might be at first supposed. For if we suppose the pharynx of Amphioxus to be considerably shortened, the

number of its arches to be greatly reduced, the slits to be fringed with filaments suspended to the cartilaginous arches, and the surrounding space to be shut off from the abdominal cavity, and to communicate externally by an orifice or set of orifices of its own, we have the ordinary branchial apparatus of Fishes; which further differs, however, in this—that the passage of the water over the branchial filaments is accomplished in them by muscular contraction, instead of by the vibration of cilia. So, again, we may regard the dorsal trunk as the representative of the aorta of Fishes, which receives its blood from the branchial arteries and transmits it backwards along the body; the posterior part of the ventral trunk in like manner corresponds with the vena cava; the principal pulsating cavity, situated at the origin of the branchial arteries, with their proper heart; whilst even the anterior arches, which establish a communication between this heart and the aorta, independently of that which is formed by the branchial circulation, find their representatives in certain Fishes whose respiratory organs depart in some degree from the regular type.—Thus we have, in this most interesting little creature, the following anomalous combination:—an internal skeleton, formed on the Vertebrated type, but in so rudimentary a condition as not even to show a trace of separation into vertebræ, although its segmental division is marked in the rib-like intermuscular fibrous bands; a nervous system formed on the Spini-Cerebrated type, but destitute of any distinguishable brain, and having the spinal cord composed of a succession of ganglia; an oral orifice which is rather that of an Articulated than of a Vertebrated animal; a ciliated alimentary canal, such as most closely resembles that of the lower Mollusca; an anal orifice on one side of the median plane, as in the higher Mollusca; the circulating system of an Annelide; the colourless blood of Invertebrata generally; and the respiratory apparatus of an Ascidian. There can be no question that it is to be regarded as *essentially* a Fish; and whilst some of its peculiarities are those by which embryonic Fishes are distinguished from those whose development is complete, others obviously link this class to the lower members of the two great parallel series of Mollusca and Articulata. It can scarcely be thought improbable by the Philosophic Naturalist, that this curious being is "the relic of some great order of Fishes, which in their organisation brought down the vertebrated series to a parallel with the lower forms of Mollusca, and which became extinct in the earlier epoch of the world's geological history, and from the unpreservable nature of their bodies, and the absence of hard parts, left 'not a wreck behind.'"

322. The class of FISHES has been usually divided, according to the material of the skeleton, into two principal groups, the *Osseous* and the *Cartilaginous*. Of the latter division, however, some are inferior to the former in their general organisation, whilst others are much superior; and it cannot, therefore, be regarded as a natural assemblage. In our general account of the skeleton of Fishes, however, it will be convenient to adopt it.—Among the inferior members of the Cartilaginous group, constituting the order *Cyclostomi* (or circular-mouthed), the characteristic structure of Vertebrata is but little more developed than in the Amphioxus. Thus in the *Myxine*, or Hag, the neural axis still rests upon a 'chorda dorsalis', which is a soft and flexible cylinder, semifluid towards its centre, and not presenting the least indication of segmental division.

In the *Lamprey,* a series of proper cartilaginous rings is developed upon the chorda dorsalis at intervals ; these are the rudiments of the bodies of the vertebræ, the neural arches of which begin to make their appearance in the form of cartilaginous laminæ, in the substance of the fibrous envelope of the neural axis. There are no corresponding rudiments, however, of hæmal arches ; the cartilaginous framework, which supports the seven rows of gills, not being composed of true ribs, but being a part of the splanchno-skeleton, like that of the branchial arches both in the Osseous Fishes and in the larvæ of Batrachia ; and no members or appendages of any kind being developed. In the head of the *Myxine* are a set of cartilages, protecting the cephalic ganglia, and closely resembling, in their degree of development, the first rudiments of the cranium in the embryo of higher fishes ; whilst with these are connected a set of rudimentary hæmal arches, specially modified to give support to the suctorial lips and to the rasp-like tongue. This animal presents several interesting approximations to the Amphioxus ; being, like the latter, destitute of eyes ; its mouth being guarded by cirrhi, which (being eight in number, and arranged in a circle) strongly remind us of the oral tentacula of the Cephalopods ; and the branchial orifices, through which the respiratory current passes out, being situated far back, nearly in the situation of the abdominal pore, though double (one on each side) instead of single.* In the head of the Lamprey, the cranial cartilages are more developed ; as are also those supporting the buccal apparatus, which consists of a flat circular disk formed by the expanded lips, and covered with horny tooth-like processes that gradually increase in size towards the centre of the mouth.

322*a.* In the higher Cartilaginous Fishes, we might trace (did space permit) the progressive steps of the development of the vertebral column, from the first indication of segmentation in the chorda dorsalis, to its complete replacement by a pile of vertebral 'centra ;' and from the first appearance of neurapophyses, to the complete development of the neural and hæmal arches. A very general account, however, must suffice. In the *Sturgeon,* we meet with firm cartilaginous rings, formed in the outer part of the 'chorda dorsalis,' increased in thickness, and evidently forming the bodies of a series of vertebræ, with which the peripheral elements are articulated ; these elements, though themselves also cartilaginous and disconnected from each other, form complete neural arches, with the addition of hæmal arches in some parts. Amongst the *Sharks,* the progression may be traced from this point, up to the complete development of the bodies of the vertebræ, as a succession of cylindrical pieces, partly osseous and partly cartilaginous, having deep conical excavations at each end, so that, when fitted together, bi-conical spaces are left between them. These spaces, occupied by the unconsolidated portions of the 'chorda dorsalis,' communicate with each other, in the inferior Sharks, by a portion of the unconsolidated 'chorda' which still remains in the middle of each vertebral 'centrum ;' but as we pass through the ascending series, we find this canal gradually diminished in size by the development of the vertebral body towards its centre ; until, in great *Selache* (basking-shark), nothing remains in its place but a small impervious cord, the intervertebral spaces

* This current, however, does not pass into the visceral cavity, but makes its exit by a special canal leading from the branchial chambers.

being now occupied by a series of disconnected elastic capsules, filled with gelatinous fluid.* The neural arches are here rendered firmer by the partial ossification of their elements; the pleurapophyses of the dorsal vertebræ, or ribs, are also partly ossified; whilst in the posterior part of the vertebral column, the hæmal arches are nearly as complete as the neural. In these Fishes we find several *fins*, or locomotive appendages, some of them upon the median line, others in pairs on the two sides. The former, which are known (according to their position) as the *dorsal*, *anal*, and *caudal* (Fig. 173, *d*, *a*, *c*), are supported upon a set of cartilaginous or osseous rays, belonging to the dermo-skeleton, but intercalated amongst the neural or hæmal spines, in the manner shown. Of the latter, the *pectoral* fins, or anterior members (Fig. 173, *p*), are not, as in Osseous Fishes, connected with the occipital vertebra, but are suspended to a firm scapular arch, which encircles the body behind the branchial apparatus, and is connected with the vertebral column. In the *Rays*, these pectoral fins are immensely expanded, and are supported upon a remarkable number of digits; these sometimes amounting to a hundred, every one of which bifurcates towards its extremity. It is interesting to remark, that, in this group, the vertebræ of the anterior part of the spinal column are immovably fixed together by an incrustation of earthy matter, that forms a kind of tube or sheath in which they are encased; thus giving a much firmer support to the muscles of the great wing-like fins, in accordance with the general principle already laid down (§320 *n*), that the consolidation of the spinal column takes place *pari passu* with the delegation of the locomotive power to the extremities. The posterior or *ventral* fins are seldom highly developed as swimming organs; but portions of them in the male are converted into 'claspers', to be used in the sexual congress of these Fishes; the pelvic arch is less perfect than the scapular, not being connected with the vertebral column. The tail, in all these Cartilaginous fishes, is formed upon a different plan from that of most Osseous; for whilst, in the latter, the framework of the caudal fin consists of dermal rays, intercalated amongst the neural and hæmal spines of a number of vertebræ whose bodies have coalesced together (Fig. 173), the caudal fin of the former is supported upon a prolongation of the vertebral column itself, which passes on to the termination of the tail, as in all other Vertebrata. The first of these methods of conformation is termed *homocercal*, from the equality between the upper and lower lobes of the caudal fin; whilst the latter is termed *heterocercal*, the vertebral column being continued into the upper lobe of the tail, which is consequently the largest.†

* "There are few examples in the animal economy," says Prof. Owen (Lectures on Comparative Anatomy, Fishes, p. 55), "in which the smallest possible quantity of earthy matter is arranged according to such beautiful and clearly-manifested mechanical principles, for affording the greatest amount of strength, and that degree of resistance which the necessarily light semi-ossified vertebra of a gigantic Shark, maintaining itself near the surface by muscular exertion, without help from a swim-bladder, must have to sustain during the vigorous inflexions of the vertebral column, producing the violent compressions of their interposed elastic balls."

† It is very interesting to remark that the Heterocercal conformation is found in *all* Fishes at an early period of their development; giving place to the Homocercal, in the Osseous Fishes, in proportion as their other peculiarities manifest themselves. We may consequently regard the heterocercal tail as the *most general* type, whilst the homocercal is a *special* modification of this, peculiar to those of the class of Fishes which may be regarded as its more typical members. It is further to be mentioned, that all the earlier Fishes whose remains are

In their retention of the heterocercal tail, as well as in many other parts of their conformation, the higher Cartilaginous Fishes show a decided approximation to the class of Reptiles; the transition to which is also established by the *Lepidosteus* (or gar-pike) an Osseous Fish having many Reptilian characters; by certain extinct Saurian Reptiles, which had many important Ichthyic characters; as well as by the *Lepidosiren** (Fig. 175), a Batrachian Reptile, which retains in very many points of its organisation the low grade of development that is generally characteristic of Fishes.—Of the skull of the higher Cartilaginous Fishes it must here suffice to remark, that it is formed on the same general plan with that of the Osseous; but as the cartilage of which it is composed is consolidated rather by a general granular deposition of osseous matter, than by regular ossification from distinct centres, there is no well-marked division into separate pieces; so that, although the situation of the different bones may be pointed out under the guidance of the skull of an Osseous Fish, to define their limits is impossible. Not only are the component parts of the brain-case thus consolidated into one mass, but the bones of the face are united with this, —the great maxillary ring, however, composed of the upper and lower jaw-bones articulated together, being but loosely connected with the rest.

322 *b*. In describing the skeleton of the Osseous Fishes, we shall for the most part confine ourselves to a notice of its chief departures from the 'Archetype' already surveyed; as in this mode will the special modifications of the common vertebrate type to the requirements of the life of the Fish, be best brought into view.—The bodies of the vertebræ, with few exceptions, retain their cupped extremities; and these work as sockets over a set of doubly-convex intervertebral capsules (the remnants of the 'chorda dorsalis', and the representatives of the 'intervertebral cartilages' of higher animals), in such a manner as to give great freedom of motion to the entire column. The number of these vertebræ varies greatly in the different families of the class, being greatest in the long serpent-like Fishes, and least in those whose spinal column possesses comparatively little extension longitudinally. Thus in the *Conger-Eel* the number of vertebræ is 162, and in the *Gymnotus* 236; whilst in the *Mackerel* there are but 31, and in the *Diodon* no more than 17. The distinction into regions, which is so well marked in most of the higher Vertebrata, cannot be made in Fishes; for as the pectoral fins are attached to the occipital segment of the skull, and as the formation of the expanded hæmal arches for the protection of the visceral cavity commences immediately behind it, there is no neck or cervical region, but the dorsal portion of the vertebral column must be considered as commencing with the very first vertebra; again, as the pelvic arch has no bony attachment to the spinal column, so that there is no sacrum, and as its position is inconstant, it does not mark out the separation between the dorsal and

preserved to us, possessed the heterocercal tail; these Fishes having either belonged to the *Placoid* order (Agassiz) which comprehends all the Sharks and Rays, or to the *Ganoid*, a large series, now nearly extinct, some members of which carried onwards the same general type of conformation into closer connection with the Reptilian class.

* In ranking the *Lepidosiren* amongst Batrachia, rather than in the class of Fishes, the Author ventures to dissent from his friend Prof. Owen, on whose authority he generally places the firmest reliance; and to express his accordance with the views of several other eminent Anatomists, who regard its general characters as indicating an affinity to the former group, notwithstanding the ichthyic (or fish-like) condition of its skeleton.

caudal portions of the spine, which is usually determined merely by the extent of the visceral cavity, those vertebræ whose hæmal arches are expanded for its protection being regarded as *dorsal,* and those in which they are again narrowed, so as merely to enclose the great vascular trunk, being reckoned as *caudal;* there is, therefore, no proper *lumbar* region in Fishes. In most Fishes, but especially in those of the Eel kind, the caudal portion forms a considerable part of the whole length of the body; and it is in this, that the principal multiplication in the number of vertebræ presents itself. A mere glance at the vertebral column suffices to show that the successive segments resemble one another far more closely than they usually do in the higher animals; or, to use Prof. Owen's phrase, that the principle of 'vegetative repetition' is more obviously manifested in Fishes than in higher Vertebrata. In nearly all cases, the neural spines are strongly developed, from the head to the end of the tail; and in the caudal region, the hæmal arches are similarly prolonged in a vertical direction. In this manner, the movement of the spinal column is limited in a vertical direction, whilst its lateral flexion is not interfered with; and the lateral surface of the body is so augmented, that its stroke against the water (which is the chief means whereby the fish is propelled through the liquid) is rendered far more effectual. We generally meet with a further augmentation of the surface with which the water is struck, in the various median fins which are implanted upon the summits of the neural and hæmal arches. These are named the *dorsal* (Fig. 173, d^1, d^2) when they are erected upon the back; the

Fig. 173.

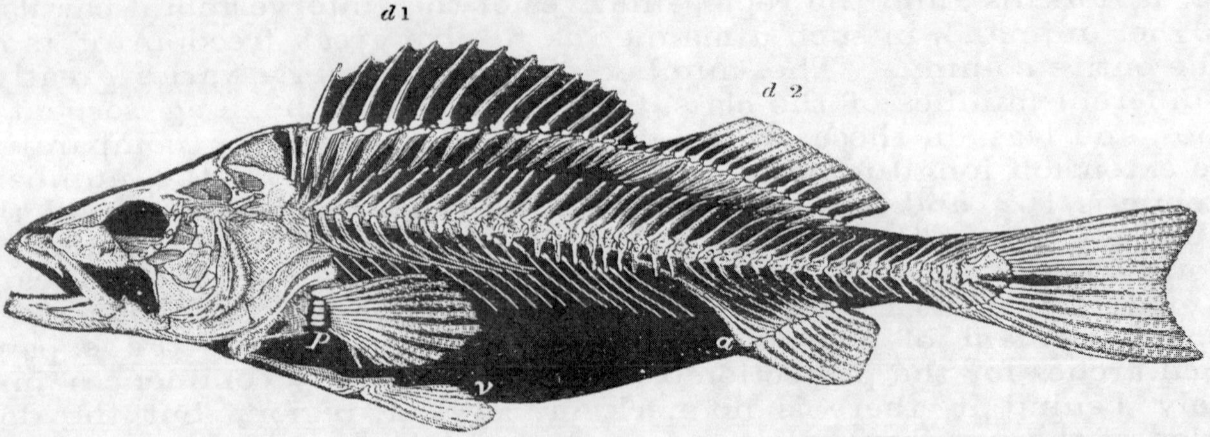

Skeleton of *Perch;*—*d* 1, *d* 2, dorsal fins; *c*, caudal fin; *a*, anal fin; *v*, ventral fin; *p*, pectoral fin.

caudal (*c*) when they are supported upon the caudal vertebræ; and the *anal* (*a*) when they project from the under side of the body in the immediate neighbourhood of the anus. As already remarked, the rays (sometimes osseous, sometimes cartilaginous) upon which these median fins are supported, belong to the dermo-skeleton; and the intercalary bones upon which they rest are probably to be regarded in the same light. In the dorsal region, we usually find each vertebra made up of its centrum or body, with a complete neural arch: whilst the hæmal arch is only partially formed by the expansion of the pleurapophyses (or ribs), which are articulated to the parapophyses, over the upper part of the visceral cavity; its lower part being unprotected by those hæmapophyses and hæmal spines, which we find constituting the sternal ribs and sternum of higher Vertebrata. The pleurapophyses often bear 'diverging appendages,' in the form of simple undivided bony rays, sometimes of considerable length, analogous to those of the ribs of Birds. The hæmal arches of the caudal region are completed in a manner that constitutes a very marked departure from the archetype, and distinguishes Fishes from higher Vertebrata; for they are formed by the bending-down of the 'parapophyses' until they meet on the median line, the proper hæmapophyses and hæmal spines being here also deficient, and the pleurapophyses being either absent altogether, or so articulated to the parapophyses as to take no part in the formation of the arch. At the extremity of the caudal region, the bodies of a number of the vertebræ are usually found to have coalesced; but their neural and hæmal arches and spines remain distinct, and give support to the caudal fin, which is equally developed above and below, or 'homocercal,' in nearly all the Osseous Fishes of the present epoch.

322 *c.* The looseness of the framing of the vertebral column of Fishes, which so greatly favours their free movement through the water, obviously incapacitates the body from deriving support from its lateral appendages; but no such support is required, since, when immersed in a fluid of equal specific gravity with itself, every part of the body is borne up by it. The functions of these appendages, therefore, in regard to locomotion, are limited to the balancing of the body, and to the change in the direction of its movement, especially from the horizontal to the vertical. In the *pectoral* fins (*p*), which are always situated immediately behind the head, and connected with the occipital segment, where their power can be most advantageously exerted, we notice that the arm and fore-arm are scarcely developed, and are hidden within the trunk, from which the hand alone projects, its movement being confined to the wrist-joint; and the extent of surface required to strike the water is given by the elongation and multiplication of the fin-rays, which represent the carpus and digits. The same is observed in the ventral fins (*v*), which are generally of less size and importance, are not connected with the vertebral axis, and are very variable in situation; being usually placed near the anus in the long-bodied and small-headed fishes, where their chief use is to serve as accessory balancers; whilst in most of the shorter and larger-headed fishes which enjoy a considerable range of depth, they are brought forwards (as in the Perch) to assist in raising the head, being sometimes placed beneath the throat, even *anteriorly* to the pectorals. In the 'abdominal' fishes, in which the pelvic arch may be considered as occupying its normal position, this is connected with the pleurapophysis of a vertebra at the extremity of the dorsal region, by a fibrous band on either side; whilst in the 'jugular' fishes, in which it is moved forwards to the neighbourhood of the scapular arch, it is sometimes connected in like manner with this. Both pectoral and ventral fins are frequently modified for special purposes; thus the former are occasionally furnished with tactile appendages, are sometimes developed into wing-like organs, capable of supporting the fish in the air, and may even be adapted for an imperfect progression over the surface of the earth; whilst the latter are often reduced to the condition of mere filamentary 'feelers,' and may even be modified in such a manner as to form a suctorial disk or organ of adhesion.

322 *d.* In the conformation of the skull, we find several remarkable departures from its ordinary structure in higher Vertebrata, which adapt it to the particular conditions of ichthyic existence, independently of those which mark its low grade of development. The latter is especially shown in the want of consolidation of the vertebral elements, which remain for the most part as distinct bones, whose edges overlap each other, without being united by 'sutures;' but the former are manifested in the extraordinary size and figure which some of these elements exhibit. That which first strikes the observer, on looking at the head of most Osseous Fishes, is its large size, and the absence of neck; the head being, as it were, the anterior rounded termination of the large extremity of the body, which gradually tapers off behind towards the tail. These peculiarities are favourable to free and rapid movement through the water; for where the propulsive power comes from the opposite extremity, it is mechanically advantageous that the parts in advance should be the largest, and able to bear the most pressure. The large external size of the skull does not involve a large cranial cavity; it being the face and not the brain-case which is thus expanded, and the greater part of this expansion being due to the extraordinary development of the whole buccal apparatus. The limbs of Fishes have no prehensile power; the mouth may be propelled and guided by them to the food, but the act of prehension must be performed entirely by the jaws; and these are frequently endowed with the power of protrusion and retraction, as well as of opening and shutting to a very wide extent. This peculiar adaptation for the prehension of food, is further displayed in the disposition of the *Teeth* in the interior of the mouth. These appendages are usually extremely numerous, and are seldom confined (except in the Sharks) to the upper and lower jaws. There is, in fact, scarcely one of the bones which forms part of the walls of the cavity, that is not, in some tribe or other, beset with teeth; and in some instances, we find teeth disposed on nearly all these at once, so as to line the mouth as completely as possible. The intermaxillary and premandibular bones, the palatines, the vomer, the lingual bones, the branchial arches, and sometimes also (though more rarely) the pteregoid and the sphenoid bones, are found carrying teeth, very commonly of different shapes, and adapted for different purposes (§ 322, *f, g*).—But further, the peculiar mode in which the act of Respiration is performed in Fishes, necessitates an extraordinary adaptation of the osseous framework for this purpose; and parts which are elsewhere rudimentary, are here developed in their most complicated form. This is the case with the diverging appendages of the tympano-mandibular arch, here developed into the broad flat 'opercular bones' which support the valvular flaps that cover-in the external orifices of the gill-chamber; and with the whole of the hyoidean arch, which is the chief point of suspension of the branchial arches (a part of the splanchno-skeleton) that support the gills; whilst its diverging appendages, the 'branchiostegal rays,' give support to the branchiostegal membranes that regulate the course and exit of the respiratory currents. The last pair of branchial arches is reduced to the capacity of the pharynx which it surrounds, and is converted into an accessory pair of jaws, which, by means of the teeth implanted in them, aid in reducing the food in its passage from the mouth towards the stomach. In no other Vertebrate animals is the mouth thus

provided with maxillary instruments at its posterior as well as at its anterior aperture; and it is very interesting to see how essentially this peculiarity is connected with the special modification of its bony skeleton for aquatic respiration.

322 *e.* The development of the dermo-skeleton varies greatly in this class, and bears no constant relation, either direct or inverse, to that of the endo-skeleton. Thus in the *Amphioxus* and the *Cyclostome* fishes, in which the vertebral skeleton is rudimentary, the dermo-skeleton is rudimentary also; the surface being entirely unprotected by scales, and the median fins not being supported by dermal bones. In the *Lepidosteus*, again, the body is completely enclosed in a casing of bony scales, which are developed into large plates on the head; whilst the internal skeleton is highly developed and completely ossified. In the *Sturgeon*, on the other hand, the dermo-skeleton is developed in the form of large osseous plates, completely enclosing the body and head, whilst the internal skeleton is for the most part cartilaginous; and such appears to have been the condition of a large proportion of the fishes which lived in the earlier epochs of the Earth's history,—their dermo-skeletons, consisting of bony scales or plates, having been perfectly preserved to us, whilst scarcely any trace exists of their vertebral column. In the *Sharks* and *Rays*, again, whose neuro-skeleton is but slightly ossified, the dermo-skeleton is also very imperfect; and of the vast numbers of this tribe which must have existed in the ancient seas, the greater part are known to us only by their teeth and by osseous fin-rays belonging to the dermal skeleton, every portion of the neural skeleton, and every other part of the dermal and visceral, having disappeared by decay. In most Fishes, the dermal skeleton affords (as we have seen) the chief support to the median fins; in the *Herring* and its allies, it supplies a sort of sternum with sternal ribs, which have been supposed (but erroneously) to be parts of the neuro-skeleton; and in many other cases it undergoes an unusual development for some special purpose.—It is, however, in its protective relation, as a part of the general integument, that we have now more especially to view it. With the exception of the Amphioxus and the Cyclostomi, all the existing Fishes have the skin strengthened by the development of cartilaginous or osseous 'scales' in its substance;* these scales being of various shapes and sizes, in some instances merely studding the skin at intervals, sometimes coming into close contact with each other at their edges, sometimes partially covering each other in an imbricated manner, and sometimes being developed into large plates firmly united into a sort of cuirass. Each of these scales is developed within a distinct capsule, and grows by its own inherent powers of nutrition; and where, as in the *ctenoid* or comb-like scales (that of the *Sole* for example), it has a set of spiny projections from its edge or surface, each of these has a separate capsule, and attaches itself to the scale in the progress of its development, like a tooth to the jaw which supports it. The structure of the cartilaginous scales is best seen in those cases in which (as in the *Eel*) they are very thin, being composed of only a single layer of cells held

* The scales of Fishes are not, like those of Reptiles, *epidermic* formations of horny composition; but are always inclosed within the substance of the true skin, and are covered by its outer layer, as well as by the epidermis; and though sometimes containing horny matter, are essentially constituted of cartilage.

together by intercellular substance; when the scales are thicker, the cells are arranged in several layers, and their under part has more of a fibrous than of a cellular arrangement. They are usually marked by a set of radiating and of concentric lines; the former are the indications of furrows or canals, which probably lodge nutrient vessels; the latter are due to the peculiar transformation undergone by some of the concentric rows of cells on the outer surface of the scale, these assuming an elongated form so as to touch each other at their ends, and becoming filled with horny matter.* Nearly all the existing Osseous Fishes have cartilaginous scales; and according to the form of these scales, they are grouped by Prof. Agassiz under the two heads of *Cycloid* and *Ctenoid*,—the former comprehending those in which the scales have a more or less regular circular form with smooth edges, as in the *Carp;* whilst the latter includes those, such as the *Perch,* whose scales have the posterior margin serrated, or furnished with comb-like teeth.† The simplest condition of the *bony* scales is seen in the *Shark* tribe, in which the skin is beset with a number of little hard points or tubercles, giving to it the roughness which is well known in 'shagreen;' these, when examined microscopically, are found to consist of miniature teeth, developed in distinct capsules near the surface of the skin, and differing in no essential particular, save in size, from the teeth which are developed in the mucous membrane covering the jaws (§ 212). In some cases, these small tooth-like scales are replaced by much larger bony plates, which are not, however, covered by enamel, and do not come into close contact with each other. The Fishes whose scales present either of these conditions, agree amongst themselves in their general structure (constituting the *Plagiostome* tribe of Cartilaginous Fishes in Cuvier's arrangement), and are designated by Prof. Agassiz as *Placoid.*—The most perfect condition of the dermo-skeleton of Fishes, is that in which the bony scales or plates come into close contact at their edges, and are covered externally with a layer of a peculiar lustrous substance which has received (though improperly) the designation of 'enamel.' The Fishes thus protected were ranked by Prof. Agassiz under the single order *Ganoid;* but the few existing species which have an envelope of ganoid scales or plates, are too diverse in their internal structure to be thus associated, and require to be grouped in at least three orders. To that of *True Ganoids,* however, of which the *Sturgeon* and the *Lepidosteus* are the principal existing representatives, the fossil remains of all those Fishes of the Palæozoic period and of nearly the whole of the Secondary, which were not Placoid, appear to be referable,—the Cycloid and Ctenoid orders making their first appearance at the Cretaceous epoch. The structure of the Ganoid scales, though commonly spoken of as osseous, and corresponding essentially with that of true bone (Fig. 43), presents several remarkable peculiarities, which

* For additional information on this subject, see the Memoir of Dr. Mandl on the Scales of Fishes, in his *Anatomie Microscopique*, or in the Ann. des Sci. Nat. 2ième Serie, tom. xi.

† Trivial though such a character may seem, it is found to correspond very closely with the distinctive peculiarities of the internal structure of these groups, the Cycloid Fishes being nearly identical with the *Malacopterygii* of the Cuvierian arrangement, and the Ctenoid with the *Acanthopterygii;* and thus it is of essential service in the determination of the fossil remains of Fishes, whose scales are often better preserved than the internal skeleton.

are characteristic of particular families and genera.* It is interesting to remark that the Placoid fishes of the earlier epochs, whose prey consisted of Ganoids invested by a dense bony armour, had teeth of very different conformation from those of the existing Sharks, which are destined to restrain the undue multiplication of the soft-scaled Ctenoids and Cycloids; for whilst the latter are generally thin and lancet-shaped, the former had broad flat surfaces, fitted for crushing and grinding, like those of the existing *Cestracion Philippi*, or Port Jackson Shark.

322 *f*. The *Teeth* may be considered (as already remarked) either as portions of the dermo-skeleton, or as constituting part of the splanchno-skeleton. To the former they would seem to belong in virtue of their origin from papillæ developed on the surface of the mucous membrane of the mouth, which is the internal continuation of the dermis; being thus analogous to hair, feathers, &c. With the latter they may be associated in consideration of their instrumentality as parts of the digestive apparatus. Although there are a few Fishes in which the teeth are entirely wanting, and a few more in which they are reduced to a very small number (the *Myxinoids*, for instance, having only a single pointed tooth on the roof of the mouth, and two serrated dental plates on the tongue,—and the *Tench* having but a single dental plate in each jaw-bone, with two small denticles on the nasal bone), yet as a general rule the teeth are extremely numerous in this class; being sometimes so small, and so closely crowded together over nearly the whole internal surface of the mouth, as to be almost countless. In such cases they repeat each other almost precisely in size and form, thus presenting another example of the principle of vegetative repetition which so extensively prevails in this group. They are found to present great diversities, however, in the several tribes constituting the class, according to the purposes for which they are adapted. Thus in some instances they are nothing more than minute, slender, sharp pointed filaments, closely set together like the 'pile' of velvet (the 'dents en velours' of Cuvier); sometimes equally fine and numerous, but longer (the 'ciliiform' teeth of Prof. Owen); sometimes long and slender, but a little stronger, like the bristles of a brush (the 'dents en brosse'); sometimes of larger size, and either straight or recurved cones (the 'dents en rape,' or 'en cardes'); and sometimes depressed into broad flat plates, with the surfaces roughened, for crushing and grinding.—The structure of these teeth, the history of their development and succession, and their mode of connection with the bones on which they are set, offer several interesting peculiarities that are very characteristic of the class. They are almost entirely composed of various modifications of 'dentine;' that form of it which is penetrated (like bone) by vascular canals, and hence called 'vascular dentine' (§ 212), constituting a far larger proportion of the dental structures in this class, than in any other. In fact we might say that the typical structure of a Fish's tooth is for its central portion to consist of vascular-dentine, and its peripheral or superficial of hard or non-vascular dentine. In some instances the former predominates (as in the tooth of *Lamna*, Fig. 47, A), and the latter is reduced to a thin layer of almost enamel-like hardness, which covers the exterior of the tooth. On the other hand, the non-vascular dentine occasionally forms the greater

* See the admirable memoir on this subject by Mr. W. C. Williamson in the Philosophical Transactions for 1849.

part, or even the whole, of the dentinal portion of the tooth; and this is covered externally by a layer of 'cementum,' the crown being even occasionally protected by a thin coating of a substance resembling 'enamel.' The most complex dental organisation in any existing Fish is that presented by the *Scarus* (parrot-fish), which is adapted to browse upon the branches of the stony Corals, deriving its food from the gelatinous bodies of the soft polypes to which these corals serve as the skeleton (§ 291 *a*); for in its pharyngeal teeth there is a very remarkable resemblance, both in their structure and in their mode of successional development, to the molars of the Elephant (§ 326 *p*). On the whole, however, the teeth of Fishes may be considered as bearing a much closer resemblance to bone in their structure, than they do in higher animals; and this resemblance is peculiarly remarkable in certain cases, in which the entire tooth is made up of a bundle of 'denticles,' each having its own medullary cavity or canal, and its own system of radiating tubuli (as in the *Pristis*, or saw-fish, Fig. 47, B), and is thus nearly allied to an ordinary long bone made up of its bundle of 'ossicles' (§ 205, Fig. 45). In fact, when teeth become connected with the jaws by continuous ossification, which is not uncommon in Fishes, the vessels of the Haversian canals of the bone pass on into the medullary canals of the tooth; and sometimes these medullary canals are surrounded by concentric lamellæ, as in bone; thus forming the peculiar modification of tooth-substance, which has been termed by Prof. Owen 'osteo-dentine.'

322 *g*. The development of the teeth in Fishes takes place in a mode that is strictly comparable with the early part of the process by which the teeth of higher animals are developed (§ 216, 217); and it is very interesting to trace its arrestment, in different members of the class, at different stages of the process. Thus in the *Shark* tribe, the developmental process is stopped at the 'papillary' stage; and it is in these, as already pointed out, that the largest proportion of the tooth is made up of the 'vascular dentine' which constitutes the lowest form of dental tissue. As Prof. Owen remarks, "the simple crescentic maxillary plate, with the open groove behind containing the germinal papillæ of the teeth, offers in the Shark a magnified representation of the earliest condition of the jaws and teeth in the human embryo." In the *Pike*, and many other Fishes, the process goes on to the 'follicular' stage; the follicle does not become invested, however, by the bony socket of the jaw, but the tooth completes its growth within the substance of the vascular gum. In the *Balistes* (file-fish), *Scarus* (parrot-fish), and some others, in which the peculiarly dense nature of the food necessitates a higher elaboration of the dental structure, the process of development goes through all the stages which characterise it in the higher Vertebrata; for the orifice of the follicle into which the papilla sinks, becomes completely closed so as to form a 'capsule,' which, again, is included in the 'alveolus' or socket formed by the growth of the jaw around it; in those which are to possess enamel, a distinct 'enamel-pulp' (§ 216) is developed: and the eruptive process takes place just as in Mammalia.—In all Fishes, the greater number of the teeth are shed and renewed, not once only (as in Mammals), but frequently, during the whole course of their lives; the exceptions to this rule being extremely few. In by far the larger proportion of the class, this successional development takes place from new and independent

papillæ ; the germs of the replacing teeth being produced from the free surface of the buccal membrane, and advancing in their regular course of development in proportion as the teeth previously formed are 'exuviated' (that is, cast off as parts that have completed their term of life), or are torn away or worn down in the predatory operations of the animal. In those, however, in which the teeth are developed within capsules, the new or replacing teeth have their origin in germs which are budded-off (so to speak) from the capsules of those which preceded them, just as in higher Vertebrata (§ 217) ; sometimes coming up side by side with their predecessors, sometimes displacing them in a vertical direction, as in Man. In this endless succession and decadence of the teeth of Fishes, we have another example of the principle of 'vegetative repetition,' as it manifests itself in the introduction of new organs in the ascending series of the animal kingdom.—The most common mode of attachment of the fully-formed teeth of Fishes, is by a continuous ossification between their expanded bases and the jaw on which they rest. Sometimes, however, the connection remains in the ligamentous condition in which it is before this ossification of the intervening substance has taken place. Even when the tooth is received into a distinct socket, it is almost invariably anchylosed to some part of the surrounding bony wall, rarely exhibiting that complete freedom which we see in the teeth of Mammals. And although, as in the latter group, the teeth of Fishes may be thus implanted by distinct 'fangs,' yet these fangs never diverge from each other. In the powerful teeth of *Dendrodus*, an extinct Sauroid fish, the broad base subdivides into many root-like processes, which are implanted in the coarse osseous substance of the jaw ; but these become so completely continuous with the bony tissue, that the line of separation cannot be traced.*

322 *h*. The digestive apparatus of Fishes is generally formed upon a very simple plan, their food being usually animal, and very easy of digestion. They do not masticate their food, and there is consequently no need of salivary glands ; but it passes at once, with little or no division, through the short wide œsophagus, into a capacious stomach (Fig. 173, 2), the walls of which are generally thin, although sometimes so thick and muscular as to give to the organ the character of a gizzard. The intestinal tube (4, 5) is usually short, and makes but a small number of convolutions in the visceral cavity ; the real extent of the internal canal, however, is greatly augmented in the higher Cartilaginous Fishes, by a spiral fold of membrane which winds from one extremity of it to the other. The liver (6) is a large and compact gland, usually rich in oil ; and it pours its secretion by a single duct into the intestinal tube near its commencement. The pancreas is commonly represented in Osseous Fishes by a set of large cæca (3), which seem to be prolongations of the stomach ; in the Cartilaginous Fishes, however, it is composed of a mass of smaller cæca, clustered together into a glandular mass, and conveying its secretion into the intestine by a single duct,—some approach to which structure is seen in the pancreas of the Cod (Fig. 231). A spleen is found, under some form or other, in nearly all Fishes ; it usually presents itself as a flattened highly vascular body, lying between the folds of the mesentery ; and the large quantity of blood which it receives, is transmitted, after circulating through it, to the vena portæ. The 'absorbent system' of

* For further information on this subject, see Prof. Owen's 'Odontography.'

vessels in Fishes is abundantly distributed through the body and on the walls of the intestines; and it discharges its contents into the general venous system at several points. No lymphatic or mesenteric glands present themselves upon its trunks.—The heart of the Fish is composed of a single auricle and ventricle, like that of the Gasteropod Mollusks (§ 394); but it is placed at the commencement of the branchial circulation, instead of at the root of the great systemic artery, and the blood is consequently propelled through the respiratory organs with its fullest energy. The auricle (Fig. 174, *a*) receives the blood which has been returned by the great systemic veins (*b′b″*), and collected in the great venous

FIG. 174.

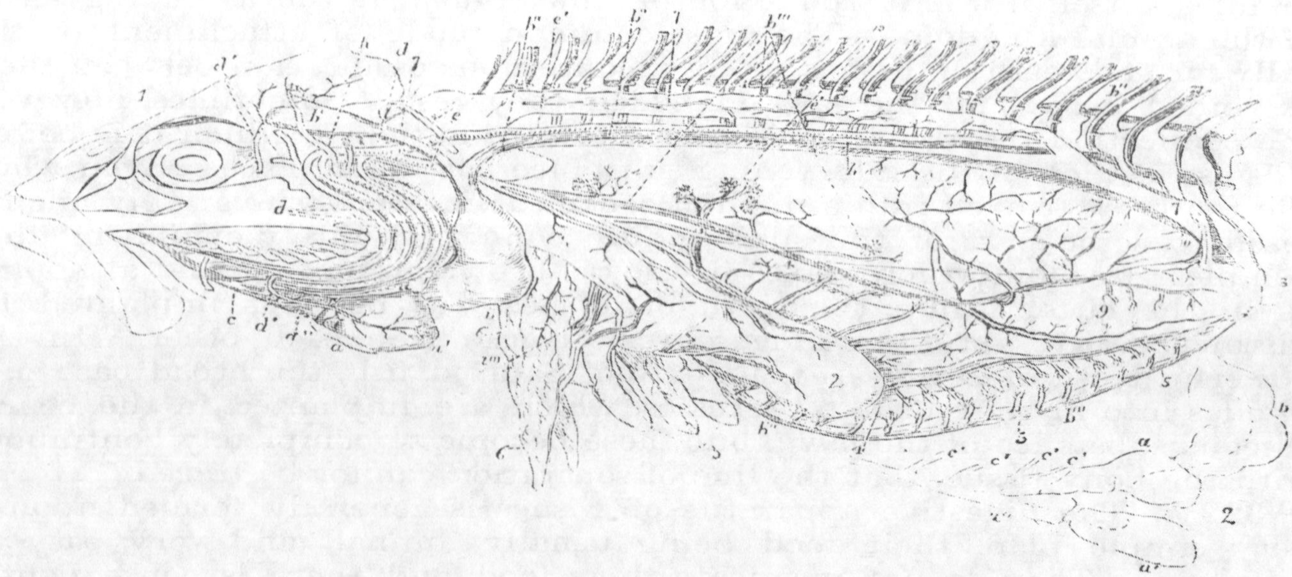

Anatomy of the Circulating apparatus, and other viscera, of a Fish;—*a*, auricle; *a′*, ventricle; *a″*, bulbous dilatation of the branchial trunk *c*; *b*, sinus venosus; *b′*, trunk and sinus of the cephalic veins; *b″ b″*, great venous trunks from the locomotive organs; *b‴*, venous trunks from the digestive organs, the liver, kidneys, generative apparatus, and air-bladder; *c*, branchial artery, giving off a branch to each branchial arch; *d*, branchial veins, whose union forms the aortic trunk that supplies the body with arterial blood, the head and heart being supplied by arteries, *d′ d′*, which originate immediately in the branchial veins; *e*, visceral branch of the aorta; *e′*, spinal trunk, supplying the apparatus of locomotion; 1, œsophagus; 2, stomach; 3, pancreatic cæca; 4, small intestine; 5, large intestine and rectum; 6, liver; 7, renal organs; 8, urinary bladder; 9, organs of generation; 10, swimming bladder.
A. Diagram of the heart, to show the course of the blood through it; the references as above.

sinuses (*bb*); and from this the blood is transmitted into the ventricle (*a′*), from which springs the great branchial trunk (*c*) having a bulbous dilatation—the *bulbus arteriosus* (*a″*)—at its base. This trunk proceeds forwards, and gives off an arterial trunk on either side to each branchial arch; which trunk subdivides into minute branches, that distribute the blood through the whole series of respiratory filaments attached to the arch. From these it is collected by the branchial veins (*d*), the greater part of which coalesce to form the great systemic trunk or 'aorta,' which carries the now-oxygenated blood into the different parts of the body. The head is supplied by arteries (*d′ d′*) which arise independently from the anterior branchial veins; and the heart also is supplied from the same source. Of the blood which has been distributed by the systemic artery to the body at large, a part returns directly to the heart through the systemic veins and sinuses ; but of that which has been transmitted to the viscera, the greater part (if not the whole), when collected by the great visceral venous trunks $b''' b'''$, is carried by the portal circulation into the liver and kidneys, through which it ramifies before returning to the heart ; and the portal circulation sometimes receives, also, a portion of the blood which has been brought back from the locomotive apparatus of the posterior part of the body and tail.—Besides the branchial apparatus, which is the true respiratory organ of Fishes, we find in many members of the class a rudiment of the pulmonary organ, constituting the 'air-bladder' (10) ; this does not appear, save in a few exceptional cases, to take any part in the aeration of the blood ; but it would seem to be destined simply as an appendage to the locomotive apparatus, serving to assist the Fish in rising or sinking in the water, by an alteration of its specific gravity.—The renal organs (7, 7, 7) for the separation of the urinary secretion, attain a considerable size in many Fishes, extending along the whole length of the visceral cavity, and even advancing as far forwards as the back of the skull, where their anterior termination lies above the branchial apparatus ; they are not, however, to be regarded as representing the true kidneys of higher Vertebrata, but rather as corresponding with the Corpora Wolffiana which are developed at an early period of their embryonic life. The secretion which they form is collected by a couple of ducts, which unite into a sort of bladder-like dilatation that opens externally behind the anus. — One of the most remarkable secretions in this class, is the mucus with which their surface is covered ; this is formed, either in a number of small follicles imbedded in the substance of the skin, and discharging their products all over the surface ; or by a set of long tubular organs seated beneath it, which pour forth their secretion through ducts which either pierce or groove the scales along what is termed the *lateral line.*

322 *i*. The chief peculiarity of the Nervous System of Fishes consists in the very low development of the Cerebrum and Cerebellum in relation to the Spinal Cord and Sensory Ganglia. The true Cerebrum, in the Osseous Fishes, seldom attains a size equal to that of the optic ganglia alone, and is frequently much smaller ; in the higher Cartilaginous fishes, on the other hand, it is evidently an organ of considerably greater importance. The Cerebellum is quite rudimentary in many Fishes ; but attains a larger size in the Sharks and Rays. That the Spinal Cord is to be considered as an assemblage of locomotive or pedal ganglia, appears from the circumstance that its relative size in different parts of the body is always proportionate to the amount of muscular development of those segments or their appendages ; thus it is quite uniform throughout in the Eel and other snake-like fishes, but is greatly enlarged at its anterior part in those which, like the Ray and the Flying-Fish, have pectoral fins of unusual power.—The Sympathetic or visceral system of ganglia and nerves presents itself as distinct from the Cerebro-spinal, in all but the Cyclostome fishes. It is much more closely connected, however, than in higher animals, with the spinal nerves, especially with the pneumogastric, which takes its place along the intestinal canal of the Lamprey, &c.—The organs of Vision in Fishes are usually well developed (except in a few species formed to inhabit situations in which they are cut off from the access of light), and frequently attain a very large size. The eye of Fishes is chiefly remark-

able for the great density and nearly spherical form of the crystalline lens, which enable it to exert the necessary refracting influence upon rays which come to the organ through water; and also for the presence of the body termed the 'choroid gland,' a vascular erectile organ at the back of the eye-ball, which appears (like the 'pecten' in the eye of the Bird, § 325 *i*) to be concerned in the adaptation of the eye for distinct vision at different distances. The organs of Hearing present a decided advance upon the auditory apparatus of Invertebrata; the principal cavity being nearly always prolonged into three 'semicircular canals' excavated in bone or cartilage, and having some part of its wall membranous, so that the fluid which it contains can be readily thrown into vibration by external impulses; these vibrations are strengthened by the calcareous bodies or *otolithes* which are suspended in the cavity, but there is as yet no vestige of the tympanic apparatus of higher Vertebrata. An organ of Smell is almost invariably present, in the form of a pair of cavities in the front of the snout, to which the surrounding water is admitted by nasal orifices; their lining membrane, on which the olfactive nerve is distributed, is commonly thrown into folds whereby the surface is increased; the 'posterior nares,' however, which in air-breathing animals establish a communication between the nasal cavities and the back of the mouth, are here wanting. There is reason to believe that the olfactive sense is very acute in many Fishes, serving to guide them to their food. How far they possess the sense of Taste cannot be determined, but it is probably low; for although the tongue is frequently large, it is furnished with prehensile teeth, rather than with sensitive papillæ; and the food seems to be swallowed as soon as it has been taken into the mouth. The sense of Touch cannot be generally acute in Fishes that are covered by a scaly armour; but it would seem to be specially possessed by the parts about the mouth, and by the pectoral fins which are used by many Fishes for the purpose of exploration. In the *Gurnards* and some other Fishes having a hard covering on their cheeks, certain digitate appendages are developed from the pectoral fins, which appear to be specially modified as tactile organs.—The Muscular apparatus of Fishes, which constitutes a large proportion of the bulk of the body, is for the most part disposed in two great lateral masses running from the head to the tail; the arrangement of whose elements partakes of the simplicity and uniformity which characterise the vertebral skeleton, the muscles of the different segments repeating each other with great precision. These either bend the tail from side to side, giving impulses whereby the body is propelled directly onwards through the water or even raised above it; or, the trunk being made a fixed point, they act upon the head, moving it rapidly from side to side, so that it may deal powerful blows with the weapons with which it may be armed, such as the serrated pectoral spines in the *Siluroids,* and the opercular spines in the *Perch* and the *Cottus* (Bull-head). But there are also special sets of muscles connected with the anterior and posterior members, and with the respiratory apparatus; and those belonging to the head have a complexity of arrangement, which corresponds with that of the parts of the cranium itself.

322 *k*. The Generative apparatus of Fishes is developed upon a very low type, although it usually attains an enormous size at the period of its functional activity; these animals being by far the most prolific of

Vertebrata, and bearing comparison in this respect with Insects. In the Osseous fishes, the ovaries of the female and the testes of the male occupy a corresponding position at the posterior part of the visceral cavity, extending far forwards, however, when distended with their mature contents. The ovary is a capacious membranous bag, the interior of which is thrown into folds or festoons; and the ova, at first developed in the substance of these, fall into the central cavity, and are expelled from it (in most Osseous Fishes) by an oviduct which terminates externally behind the anus and in front of the urinary orifice. In like manner, the seminal fluid, formed within the reticulated tubuli of the testes, is carried out of the body by an excretory duct which terminates in the same situation. In the greater number of Osseous Fishes, there is no sexual congress; but the ova deposited by the female are fertilized by the seminal fluid which the male diffuses through the surrounding water. In a few cases, however, the ova are fecundated whilst yet within the body of the female, the male being provided with an external prolongation of the orifice of the seminal duct, by which its products are conveyed into the oviduct of the other sex; and in some of these instances, the ova are retained within the female passages until the embryonic development is complete, and the young then come forth alive. In a few Osseous Fishes, as in Cartilaginous Fishes generally, we find an approach towards the mode in which the ova are set free in the higher Vertebrata. In the Eel and the Lamprey, the ovarium, instead of having an internal cavity and an excretory duct, is a solid mass, composed of membranous folds attached beneath the spine; the ova developed within the substance of these, escape into the abdominal cavity, within which they are loosely dispersed; and they make their exit through a pair of simple orifices, placed one on either side of the anal opening. The seminal fluid of the males of these Fishes is discharged in precisely the same manner. In the higher Cartilaginous Fishes, however, we find the external orifices prolonged inwards, so that each forms a tube with a funnel-shaped dilatation at its extremity; into this the ova are received when they are detached from the ovarium, and by it they are conveyed towards the outlet. A regular sexual congress takes place in these Fishes, the males of which are provided with 'claspers' developed from the ventral fins, for the purpose of holding the female; the seminal fluid is introduced within the female oviduct, so as to fertilize the ova soon after they have quitted the ovary; and these ova (which are far less numerous than those of most Osseous Fishes) undergo their early embryonic development before they are discharged, or are retained until the embryoes are mature;—the ova, in the former case, being enclosed in peculiar protective capsules; whilst in the latter, a continuous supply of nutriment is in some instances afforded to the young, by means of an apparatus specially developed for the purpose, foreshadowing the 'placenta' of Mammalia. Thus in their Nervous and Generative systems,—the two which most characteristically mark the position of any group or individual,—as well as in many other particulars, we find the Sharks and their allies presenting a decided superiority to Fishes in general. The development of the embryo Fish within the egg takes place in such a manner, that at no time does it present the characters of any inferior animal; and whether the egg be hatched before or after its emersion from the oviduct, the young fish is already so far advanced as distinctly to manifest the characters of its class, although its subordinate distinctions generally make their appearance at a later period. For some time, however, after it comes forth into the water, it lives upon the store provided for it in the yolk-bag, which still hangs beneath the abdomen; its tissues and organs present quite an embryonic condition; and its usual state is by no means one of activity, until it begins to seek food for itself.

323. The transition from the class of Fishes to that of Reptiles, which is the lowest of the air-breathing Vertebrata, is accomplished by a peculiarly interesting group, which is ranked by some Naturalists as the *Batrachian* order of Reptiles, by others as a distinct and independent class under the name of AMPHIBIA. The peculiar character of this group is derived from the 'metamorphosis' which its members undergo; all of them coming forth from the egg in the condition of Fishes, having the respiration and circulation, the digestion and nutrition, the locomotion and sensation, of that group; and gradually advancing towards the state of true Reptiles, which some attain much more completely than others. This metamorphosis is most complete in the *Frog* and its allies, which constitute in their adult state the *anourous* or tail-less family of Batrachia. Its tadpole, or larva, which is an inhabitant of the water exclusively, has the large head and vertically-expanded tail of a Fish; it is entirely destitute of locomotive members, but is urged through the water by the lateral undulations of its body. As its growth proceeds, two pairs of limbs are developed, and the tail is proportionally atrophied; and at last, the power of progression is entirely committed to the members, which are formed for movement upon the solid ground. Whilst this change is taking place, the whole organic apparatus is undergoing important modifications. The branchial fringes of the Fish give place to the pulmonic cavities of the Reptile, and the aquatic respiration is exchanged for the aerial; the heart acquires an additional auricle, and the whole course of the circulation is changed; the intestinal canal undergoes considerable modification, in accordance with the change in the alimentation of the animal; the skeleton is raised from the type of that of the Fish to that of the Reptile by various essential changes, such as the ossification of the intervertebral capsules, and the union of these with the bodies of the vertebræ, so that each is convex at one end and concave at the other, and is connected with those before and behind it by a ball-and-socket joint at each end; and the nervous and muscular systems undergo a corresponding advance. The true Reptilian condition is also attained by the Batrachians of the *Salamander* tribe; in which, however, the limbs are not developed to the same degree, and the tail is retained, the habits of these animals for the most part continuing aquatic. And there is another curious tribe, that of the *Cæcilia* and its allies, which attains the true Reptilian state by a metamorphosis, and represents the Serpents in form. These are all distinguished from ordinary Reptiles, however, by the simple external character of the smoothness of their skin and the absence of scales; and this, as we shall presently see, corresponds with certain peculiarities in internal structure, by which (as well as by their metamorphosis) the Batrachian Reptiles are distinguished from Lizards, Serpents, Turtles, &c.

323*a*. Besides those animals, however, which attain the true Reptilian condition by a complete metamorphosis, this group contains several whose development is arrested, as it were, in an intermediate or transition state;

their adult form presenting a remarkable mixture of the characters of the two classes which they thus connect; and their respiration being partly aquatic and partly atmospheric,—lungs being developed, without the gills ceasing to act. Of this group (which is known as that of the *perennibranchiate* Batrachia, this name indicating the retention of the gills) may be especially mentioned the *Proteus*, which inhabits the waters of the underground lakes of the Tyrol, and has a long serpent-like body, with four short and slender limbs on which it crawls; the *Axolotl*, an inhabitant of the lake of Mexico, which is nearer in form to ordinary Fishes, and whose limbs are more adapted for swimming; the *Siren*, which is found in the marshy grounds of Carolina, and has no posterior extremities, the anterior pair being short feeble rudiments; and finally the curious *Lepidosiren* (Fig. 175), of which one species inhabits the Gambia and other rivers of Western Africa, whilst another is found in some of the South American rivers, and which presents such a remarkable combination of the characters of the Fish and the Reptile, that some of the most eminent Zoologists are at issue (as already remarked) with regard to its proper place in the series. In the general form and aspect of its body, it bears a considerable resemblance to some of the 'perennibranchiate' Batrachia; but its skin is beset with cycloid scales, and its dorsal and caudal fins are supported by soft elastic rays, which rest upon the extremities of the neural and hæmal spines of the vertebral column, as in Fishes. Its peculiar rudimentary extremities, which are single slender-jointed rods, bear more resemblance to the limbs of the Proteus reduced to a state of further simplification (Fig. 171), than they do to the simplest fins of Fishes, which always present a great multiplication of digits, and they are used by the animal to crawl along the bottom of the water it inhabits; but the anterior pair is connected with the occipital segment, as in Fishes (§ 322 *c*); whilst in the pelvic arch we find the same absence of completion by attachment to the vertebral column, as in that class. It is curious to observe in this animal, as in the Amphioxus, a want of symmetry in the position of the anal orifice, which is situated a little to one side of the median plane. The vertebral column presents an organisation which is found in no existing true Reptile; for the central elements of the vertebræ are not more consolidated than they are in the Sturgeon, whilst the neural arches are completely ossified. In the condition of the cranium, the Lepidosiren is stated by Prof. Owen to bear a resemblance to the higher Cartilaginous Fishes, as regards the confluence of its elements; but in this respect the Siren also resembles them. It has, however, a pre-opercular bone, which is not found in any of the Perennibranchiates; and the composition of the lower jaw corresponds rather with that of the Osseous Fishes, than with that of the Amphibia. In its respiratory and circulating apparatus, on the other hand, it certainly bears a closer resemblance

FIG. 175.

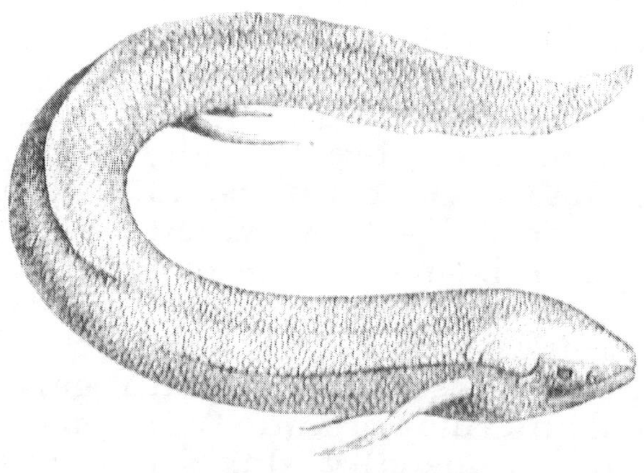

Lepidosiren paradoxa.

to the Perennibranchiata than to any existing Fishes; its lungs being as much developed, and as much used for breathing, as those of the Siren; and the general arrangement of the bloodvessels being the same, although the additional auricle of the heart is wanting. The brain presents characters of resemblance to both groups, and it would be difficult to say which predominates. As in Fishes, the nasal sacs, though possessing a double orifice on either side, have no connection with the back of the mouth by 'posterior nares;'* on the other hand, the Reptilian alliance of the Lepidosiren is indicated by the large size of its blood-corpuscles, which nearly approach in dimensions and appearance the remarkable blood-disks of the Proteus and Siren (§ 174). From recent enquiries into the anatomical relations of the Lepidosiren, not only to the Perennibranchiata, but also to the tadpoles of the Frog tribe (especially the gigantic tadpole of the *Rana paradoxa*) in their transition state, its correspondence with these appears to be much closer than it is with any Fishes.—Whatever be the view taken, however, in regard to its Zoological position, its interest to the Physiologist remains the same, as an animal in which the characters of Fish and Reptile are so blended, that it completely bridges over (so to speak) the hiatus between the two classes.

324. The class of REPTILES is oviparous and cold-blooded like that of Fishes; but the animals belonging to it are formed to breathe air, and to inhabit the surface of the earth, along which they for the most part creep or crawl, rather than run or leap; the few which are adapted to make the water their dwelling, being obliged to come to the surface at intervals to breathe. Although they breathe air, however, their respiration is much less energetic than that of Fishes, and their blood is for the most part less perfectly aerated,—that which is transmitted along the systemic arteries being a mixture of the oxygenized blood just returned from the lungs, with venous blood brought back by the systemic veins and sent again into the circulation without aeration. Reptiles present a remarkable contrast both to Fishes and Birds, in regard to the general inactivity both of their animal and their organic functions. Their perceptions are obtuse, and their movements usually sluggish. Their demand for food is not frequently repeated, and they can sustain a long privation of it; their demand for oxygen, also, is by no means urgent, and they can exist in a much less pure atmosphere than that which higher air-breathing animals require. Altogether they may be said to live very slowly; but their rate of life and their degree of general activity (of which the amount of oxygen consumed by them may be regarded as the exponent) vary remarkably with the temperature to which they are subjected, their whole vital activity being increased by warmth—within certain limits—and diminished by cold (§ 97). Not only in all the foregoing particulars, but in the grade of development of the several parts of the vertebral skeleton, and in the condition of the nervous centres and organs of sense, there is a very close agreement between the four most diverse groups of Reptiles,—*Batrachia* (or the Frog tribe), *Ophidia* (Serpents), *Sauria* (Lizards); and *Chelonia*

* The possession of a second orifice to each nasal sac (of which the posterior was at first overlooked by Prof. Owen, through the partial decomposition of the specimen examined by him,) has been advanced as a Reptilian character; but both the orifices open within the lip, instead of externally to it, whilst *neither* of them communicates with the proper cavity of the mouth, still less with the respiratory passages; so that this character must be regarded as essentially Ichthyic.

(Turtles); notwithstanding the strongly-marked differences which exist between them, both in regard to external form and general configuration, and with respect to their adaptation for entirely different habits of life. These differences, however, are not so great as those which present themselves, when we come to include not merely the existing but the extinct forms of Reptilian life; for among the remains which represent the latter, we find some which combine many of the characters of *Fish* with a general organisation essentially Saurian, whilst others exhibit to us the Saurian organisation adapted to the mode of life of *Birds*, and others again show a nearer approximation to the *Mammalian* type than is manifested by any existing species. In order to include these, it is necessary to add three orders to those which have been already enumerated; and it has been also found advantageous to separate the *Crocodiles* and their allies from the Saurian order, and erect them into a distinct group intermediate between the Sauria and Chelonia.—The distinctive peculiarities of the Reptilian skeleton, and the chief modifications which it undergoes in the different orders, will be now briefly noticed; and for this purpose it will be convenient to begin with that of the Crocodile.

324*a*. In the skeleton of the *Crocodile*, and of *Lizards* in general, the distinction between the neck, trunk, and tail is well marked, as it is in higher animals; the anterior extremities being now removed backwards from the head, and the intervening space being considered as the neck; whilst the posterior have now a fixed position at the back of the visceral cavity, the pelvic arch being attached to the spinal column, so as to divide the tail from the trunk posteriorly, as the neck is marked off anteriorly. The total number of vertebræ is frequently very great; but the multiplication is chiefly in the caudal region. Thus in the neck we never find more than 7; those of the trunk do not exceed 25, except in the curious group which forms the transition to the Ophidian order; whilst those of the tail are 36 in the Crocodile, 72 in the Iguana, and 115 in the Monitor.—The neural arches of the spinal column are everywhere completed (save towards the extremity of the tail, where they are deficient), by the adhesion of the neurapophyses to the bodies of the vertebræ; but the connection is established by 'suture' only, so that they remain as really distinct bones, instead of by continuous ossification as in the higher Vertebrata. The vertebræ of the neck are furnished in the Crocodile with short pleurapophyses (cervical ribs) which are articulated with the transverse processes; these lie over those of the adjoining vertebræ in such a manner as to limit the movements of the spinal column in a lateral direction, giving that strength to the head and neck which is required by the predacious habits of the Crocodile, but preventing the animal from easily turning to one side. These cervical ribs, however, are wanting in many Lizards, or are only developed as continuations of the transverse processes. No portion of the hæmal arch, strictly so called, is developed in the neck; but behind the scapular arch we find the pleurapophyses (ribs) greatly elongated, and coming into connection at their extremities with the hæmapophyses (sternal ribs) and hæmal spines (sternum); the whole being so highly developed, as to enclose a capacious visceral cavity. The pleurapophyses are articulated to the transverse processes merely, having no direct connection with the bodies of the vertebræ as in Mammalia; but in the Crocodile we see the double articulation foreshadowed (as it were) by the presence of two articulating surfaces between the rib and the transverse process, one of them corresponding to the 'head' and the other to the 'tubercle' of the rib. Between each ossified spinal and sternal rib, a cartilaginous piece intervenes, which may be regarded as an unossified portion either of the pleurapophysis or of the hæmapophysis,—probably of the former, as it bears a 'diverging appendage' in the shape of a somewhat expanded plate. The sternum varies considerably, among different Sauria, in its degree of ossification. The anterior portion of it is generally T-shaped, having a central median portion and two lateral bars; these last probably represent the distinct 'episternal' pieces which are so remarkable in the scapular arch of Enaliosaurians (§ 324 *h*) and Monotremata (§ 326 *c*). Posteriorly the sternum is cartilaginous in some Sauria, osseous in others, being generally compounded of several distinct pieces which do not unite; and these extend, with their sternal ribs, far back over the abdominal cavity, even where the spinal ribs are not sufficiently developed to meet them and thus to complete the hæmal arch. The presence or absence of distinct pleurapophyses determines the distinction between 'dorsal' and 'lumbar' vertebræ, as in Man. The two vertebræ to which the pelvic arch is attached, are usually immoveably united together to form a 'sacrum,' which is attached to the ilium by long transverse processes; the pelvic arch itself, as already explained (§ 320 *m*), is the hæmal arch of one of these vertebræ, and probably includes that of another. In the vertebræ of the tail, the hæmal arches are carried backwards in nearly the same form as the neural, the hæmapophyses being articulated with the bodies of the vertebræ instead of with the ends of the pleurapophyses (as they are in the trunk of the body), and uniting below into a hæmal spine; the presence of these arches, which, when detached from their bodies, are known as the V-shaped or 'chevron bones,' constitutes a marked distinctive feature between the caudal vertebræ of a Saurian and those of nearly all Mammals having the same general form (§ 326 *c*). Both neural and hæmal arches gradually become rudimentary as we approach the extremity of the tail; the bodies only of the vertebræ being left. In the existing Sauria, as in Reptiles in general, one surface of each centrum is concave and the other convex; the gelatinous capsules that intervene between the vertebræ of the Fish having become replaced by bone, which has united itself in some instances with the anterior extremity of the body, so that all the convexities look forwards, but more commonly with the posterior, so that the convexities look in the opposite direction. This, with several other differences between the skeleton of the Saurian Reptile and that of the Fish, obviously tends to the greater consolidation of the osseous framework, which is now constructed to bear up the body on four points of support, instead of to allow every part of it free play in a liquid element whose pressure is equably distributed over its whole surface.—The bones of the extremities now present very nearly the same general aspect as those of higher Quadrupeds, and are developed in proportion to the weight which they have to sustain, and to the use that is to be made of them in the act of progression. We have in the anterior extremity an arm with its 'humerus,' a fore-arm with its 'radius' and 'ulna,' a 'carpus' composed of several distinct bones, and five digits, each with its 'metacarpal' and three 'phalangeal' bones; and in the posterior extremity a thigh with its 'femur,' a leg with its 'tibia' and 'fibula,' a 'tarsus' composed of several distinct bones, and five digits with their 'metatarsal' and 'phalangeal' bones. This number of digits is not exceeded in any existing Reptile. The limbs, in certain of the Lizards which approach most nearly to Serpents (as in the genus *Seps*), are but very feebly developed ; and one of the pairs of extremities may even be entirely deficient ; the anterior alone being present in *Bimanus*, and the posterior in *Bipes*.—Notwithstanding the general similarity between the bones of the extremities of Lizards and those of Mammalian Quadrupeds, the composition of the scapular and pelvic arches presents a marked difference ; for in the first of these we find the connection of the humeral articulation with the sternum established by the large 'coracoid' bones, the true clavicles being either deficient or but slightly developed, abutting on the median line against each other, instead of upon the sternum, and not entering into the composition of the shoulder-joint ; whilst in the second we have the proper pelvic arch completed below by the 'ischia,' the pubic bones forming a separate arch in front of this, and not entering into the composition of the 'acetabulum' or articular cavity of the femur.

324 *b*. It is in the Cranium, perhaps, more than in any other part of the skeleton, that the inferior condition of the Reptile as compared with the Mammal is most strongly manifested ; for the vertebral elements, although not as distinct from each other as in the Fish, undergo much less coalescence than in warm-blooded animals ; and the number of separate bones is therefore very considerable. Thus in the Crocodile the 'occipital' bone, or neural arch of the occipital segment, is made up of four pieces, which remain permanently distinct ; although they all coalesce to form the single 'occipital bone' of Man. The 'sphenoid bone' of Man, again, is represented by six distinct bones ; its body by the 'anterior' and 'posterior' sphenoids, whilst its great wings remain separate as the 'alisphenoids,' as do also the 'pteregoids,' which form part of its hæmal arch. In the 'frontal' we find three pieces ; namely, the principal 'frontal' on the central line, and the post-frontals. And in the 'temporal' on each side we have four pieces ; the 'petrosal' which forms the auditory capsule, the 'tympanic' which not merely gives support to the membrana tympani but forms a pedicle for the attachment of the lower jaw, the 'mastoid' which enters into the composition of the second or parietal neural arch, and the 'squamous' bone which is considered by Prof. Owen as part of the diverging appendage from the hæmal arch of the next vertebra. Among the bones constituting the upper part of the face, we find no greater complexity than in Man ; but in the front of the upper jaw, we find the 'intermaxillary' remaining permanently separate on each side of the median line ; and the lower jaw presents us with a peculiarity of composition, to which we elsewhere find no parallel. Each lateral half of it is subdivided into at least *five* and generally *six* distinct pieces, which are united together by suture ; the purpose of this arrangement being apparently to diminish the risk of fracture, which would otherwise attend the snapping-together of these long jaws.* Whilst in the bones of the face and lower jaw, we have usually a high development of the maxillary

* See Dr. Buckland's Bridgwater Treatise, p. 176.—It is very interesting to see how the necessity for a particular adaptation, to meet a special purpose, has here prevailed over the tendency to adhesion to the 'archetype' or fundamental plan. For the lower jaw does not contain any other vertebral elements than the hæmapophyses and the hæmal spine of the

and mandibular arches, which belong to the frontal and nasal vertebræ, the hyoidean arch, which belongs to the parietal vertebra, and attains to such an extraordinary development in Fishes, is here greatly reduced in size and completeness, but comes into connection with the laryngeal apparatus, the support of which, as well as of the tongue, is its chief office in air-breathing Vertebrata. Of the scapular arch, which belongs to the occipital vertebra, mention has been already made.—In the connection of the skull with the spinal column, we find an arrangement that is peculiar to the *scaly* Reptiles (Saurians, Chelonians, and Ophidians); for the articulation is formed by a single condyle projecting from the basi-occipital bone, which is received into a hollow in the body of the 'atlas' (or first vertebra of the neck), just as the convexity at the back of each ordinary vertebra is received into a concavity of the vertebra behind it. A portion of the body of the atlas is developed out of connection with the rest, and coalesces with the body of the second to form the 'odontoid process,' round which the first vertebra and the head can be made to rotate.

324 *c.* Such are the chief peculiarities in the vertebral skeleton of the Saurian order; and we have now to notice briefly the condition of their *dermal* skeleton. This is represented, in the ordinary Lizards, simply by a covering of horny scales, which are truly 'epidermic appendages;' but we find in some species of the genus *Scincus*, beneath the external scaly layer, a firm armour of bony scutes developed in the substance of the true skin. In the *Crocodiles* and their allies, the presence of this defensive armour is a very important and distinctive feature. The dermal bones along the midde line of the back serve not merely for protection, but also to support the crest which rises from the neural spines of the body and tail, and acts as a sort of fin in swimming; and on either side of this ridge are several rows of thick flat bony scutes, which possess great resisting power. Hence these animals have received the distinctive appellation of *Loricata;* and, as we shall presently see (§ 324 *h*), the high development of this part of their dermo-skeleton is a character in which they present a decided approximation to the Chelonian order.—All existing Saurians are furnished with numerous teeth, implanted on the margins of the jaws. These, in the Crocodile, are fixed in distinct bony 'alveoli' or sockets; but in many tribes of Lizards, they have a much less complete mode of attachment. Some of their most interesting peculiarities will be described hereafter (§ 324 *o, p*).

324 *d.* When we turn from the skeleton of a Lizard to that of a *Frog*, the contrast is very striking. Whilst the former is remarkable for the great multiplication of vertebræ, and for the small size of the extremities in proportion to the trunk, the latter presents us with a remarkable diminution in the number of vertebræ, whilst the extremities (especially the hinder pair) are immensely increased in dimensions, and take upon themselves the whole locomotive function. In the common Frog there are but nine moveable segments in the spinal column, and this number is reduced in the genus *Pipa* to seven; behind these is the single sacral vertebra to which the pelvic arch is attached; and to the posterior extremity of this

mandibular arch; yet these two parts are multiplied or subdivided into six, and this not merely by simple repetitions of each other (like those which we see in the digital bones of a Fish's fin), but in such a manner as to form a very complex arrangement.

is fixed a long sword-shaped bone, which may be regarded either as a second sacral or as a caudal vertebra. The bodies of the vertebræ are concavo-convex; their neural arches are everywhere complete, and their articular processes (zygapophyses) are well developed; but their pleurapophyses are rudimentary, the ribs being reduced to the condition of transverse processes; and of the hæmal arches of the trunk, no parts are present save those that go to make up the sternum, which bone is largely developed, and serves to give extensive attachment to the muscles of the abdomen. In the scapular arch we find the scapula made up of two distinct portions, the proper 'scapula' and the 'acromial bone;' and the latter unites with both the coracoid and the clavicle to form the glenoid cavity of the shoulder joint. The clavicle is more developed than in the Saurians, but is still inferior in size to the coracoid, which is a broad expanded bone, meeting its fellow on the median line, besides being articulated to the sternum. In the anterior limbs we have the usual number of bones; but those of the fore-arm are united together.* The posterior extremities are enormously developed, and are connected with the trunk by a pelvic arch of very unusual dimensions, the chief part of which, however, is formed by the extension of the iliac bones. Like their representatives in the scapular arch, the pubic bones take a share in the formation of the articulating cavity for the reception of the head of the femur. The two bones of the leg, like those of the fore-arm, coalesce together, and their original distinctness is more completely lost. The hind foot of the Frog presents a peculiarity which especially fits it for leaping; not merely the digital bones, but those forming the first row of the tarsus, being extremely elongated, as in the foot of the Kangaroo. —The head of the Frog retains in many respects more of the fish-like character, than we see in the higher Reptiles; and it is remarkable for the great lateral extension of the bones of the face, especially of those which form the jaws. The occipital segment is connected with the atlas by a double articulating surface or 'condyle;' the Batrachia differing in this respect from other Reptiles. The lower jaw is not furnished with teeth in the Frog, although they are present in it, as well as in the upper jaw, in other Batrachia; but, as if to make up for the deficiency, the vomers have each a transverse row of teeth, as in some Fishes.

324*e*. The conformation of the skeleton in the inferior Batrachia, however, and especially in the Perennibranchiate group, by no means corresponds with that which it presents in the Frog; some of the differences consisting in diversities in the proportions of its component parts, adapting them to different purposes; whilst others result from the inferior advance in development. Thus if we look at the proportion between the spinal column and the limbs in the *Salamandridæ* (water newts) we shall find that it is much the same as in Lizards; they have 15 or 16 vertebræ in the trunk, and from 25 to 30 in the tail, which becomes their chief instrument of progression; they possess short ribs, although the sternum is but very imperfectly developed; and the extremities are only so far developed

* An interesting departure from the usual type presents itself in the mode in which the head of the humerus of the Frog is articulated with its glenoid cavity; for it is provided with the ligamentous connecting cord, which is elsewhere confined to the hip-joint. The use of this provision becomes apparent, when it is remembered that although the Frog makes its leap with its hind-legs only, it alights upon its fore-limbs, which might be dislocated by the shock, if it had not been for this unusual attachment.

as to aid their movements in the water, and to enable them to crawl upon land. Their general form and proportions thus bear so close a resemblance to those of Lizards, that it is not surprising that they should be popularly associated with them; they are readily distinguishable, however, on the slightest inspection, by the want of scales upon their skins.—In the *Cæcilia*, again, the Batrachian type is so modified as to present an extraordinary resemblance to the group of Serpents, in which this genus was long placed, notwithstanding the nakedness of its integument. The vertebræ are greatly multiplied, their number in one species reaching even 200; nearly all of them possess short ribs; and there are no traces either of sternum or of extremities. Yet the structure of the skull is distinctly Batrachian; and this animal has been found to possess gills during the early part of its life, and to undergo subsequently the characteristic metamorphosis of the Batrachia.—It is in the *Perennibranchiate* division, more especially, that we meet with those variations from the structure of the Frog, which depend upon general arrest of development. Among these is one which most obviously marks their approximation to Fishes; namely, the persistence of the bi-concave form of the bodies of the vertebræ, the intervertebral capsule remaining unossified. And in the head we find, not merely the large proportional development of the hyoid arch in connection with the branchial apparatus, but the ready separability of the bones of the skull, and many other characters, which show a yet nearer approximation to the type of Fishes in the skulls of these animals, than in those of such as have advanced, by the completion of their metamorphosis, to the true Reptilian condition. As already remarked (§ 323 *a*), the *Lepidosiren* completes the transition between the two classes; its osteology presenting a still closer conformity with that of Fishes.*

324 *f*. Turning now to the order *Ophidia* or Serpent tribe, we meet with another extraordinary modification of the Reptilian skeleton, effected by the multiplication of some parts, and the non-development of others. The vertebral column here presents us with the greatest number of segments which it attains in any vertebrated animal, the number of vertebræ being no less, in the *Python* (or Boa of the Old World) than 422; whilst, on the other hand, there is no external appearance either of anterior or of posterior extremities. About six-sevenths of the vertebræ possess ribs, which are articulated to their bodies by a kind of ball-and-socket joint that allows them free play; whilst their opposite extremities are nearly unattached, not the least vestige of a sternum being present, and the sternal ribs only existing as short cartilages tipping the ends of the spinal ribs, which are connected with the abdominal scuta of the integument. In virtue of their unusual mobility, the ribs can be employed

* It is scarcely possible, in surveying this order, not to be struck with its remarkable *representation* of the other existing orders of Reptiles. The *Cæcilia* is so like a *Serpent*, that it was long regarded as an Ophidian, even by scientific Naturalists. The *Salamandridæ*, in like manner, represent the *Saurians*. And although it might appear difficult to find any analogy between *Frogs* and *Turtles*, yet in the delegation of the locomotive power to the extremities, in the relative shortness of the vertebral column, and in the conformation of the scapular and pelvic arches, there are many points of resemblance. In the genus *Pipa*, this resemblance is also seen in the conformation of the cranium, and in the condensation of the skin, which is beset with horny tubercles. And in another South-American Batrachian, this condensation proceeds so far in the skin of the back, as to protect the body in a kind of carapace. (Ann. des Sci. Nat. Deuxième Série, tome iii. p. 318.)

in some degree as instruments of progression, their extremities moving backwards and forwards beneath the skin, like the slender legs of the Iulus (Fig. 150); but the remarkable locomotive power which these animals enjoy, is chiefly due to the extraordinary flexibility of the spinal column, and to the power of the muscles by which movement is given to it. "By the endless combinations and inflections of their long spine," remarks Prof. Owen, "the absence of locomotive extremities is so compensated, that the degraded and mutilated Serpent can overreach and overcome animals of far higher organisation than itself; it can outswim the fish, outrun the rat, outclimb the monkey, and outwrestle the tiger; crushing the carcase of the great carnivore in the embrace of its redoubled coils, and proving the simple vertebral column to be more effectual in the struggle than the most strongly developed fore-limbs with all their exquisite rotatory mechanism for the effective and varied application of the heavy and formidably armed paws."* Although no true limbs can be seen externally in the typical Serpents, yet in the *Python* we have a trace of a pelvic arch, with 'diverging appendages,' terminated by horny hooks which appear externally and serve as 'claspers.' In some members of the order, however, which present slight deviations from its ordinary characters (such as the genera *Amphisbæna, Tortrix,* and *Typhlops*) there is a rudiment of a scapular as well as of a pelvic arch. And in the *Anguis fragilis* (commonly known as the 'blind-worm' or 'slow-worm'), whose form is that of a Snake, but which approaches more nearly to the Lizards in some other important characters, we find not merely a distinct though minute scapular and pelvic arch, but also some traces of limbs as well as of a sternum. Through this animal, the order is connected with those serpent-like Lizards, in which the body is extremely elongated, and one of the pairs of extremities deficient; or in which they are both present, but almost in a rudimentary condition.

324 *g*. The head of Serpents is chiefly remarkable for the extraordinary dilatability of the entrance to the mouth. This results in part from the want of union between the two halves of the lower jaw on the median line, so that they can be opened out from one another; and in part from the peculiarity of the method by which it is connected with the cranium,—the 'tympanic pedicle' to which it is articulated being not only moveable in itself, but having a moveable base; for the 'mastoid' bone is here itself lengthened out into a pedicle, and united to the rest of the cranium only by ligaments and muscles. The bones of the upper jaw, too, are less closely approximated to each other than usual; and being only connected by ligaments, possess some degree of mobility upon each other. In the non-venomous Serpents, the palate-bones as well as the upper and lower jaws have a row of teeth of simple conical form, pointing backwards; these are not strong enough, however, to serve any other purpose than that of prehension of the prey. In the venomous Serpents, the 'poison-fangs' with which their deadly wounds are inflicted, are two large teeth situated in front of the superior maxillary bone, which does not bear any other teeth at its margin; these teeth are either simply grooved for the passage of the poison-duct, or the groove is converted into a complete canal (especially near the point of the tooth) by the arching-over and meeting of its walls. The poison-fangs of some venomous serpents are

* Homologies of the Vertebrate Skeleton, p. 182.

immoveably fixed in the jaw; but in those which are most dreaded on account of the fatality of their bite (as the Rattlesnake, or the Cobra di Capello), the poison-fangs when not in use lie flat upon the roof of the mouth, and are erected by special muscles when the animal makes its attack, the poison-gland being at the same time compressed by the action of the muscle that closes the jaw, so as to force out some of its secretion through the duct which deposits it in the wounds made by the teeth.—The dermo-skeleton of Serpents consists of epidermic horny scales, which are sometimes of small size and arranged in an imbricated manner over the whole surface, but which, in the typical Snakes, present themselves along the ventral surface as broad plates arranged transversely to the axis of the body. The purely epidermic character of this formation is remarkably shown in its frequent *exuviation;* for the whole of the scaly tegument of Snakes is thrown off at least once a year, sometimes more frequently; in some instances coming away piecemeal; but in other cases being detached entire, and even carrying with it (like the *exuviæ* of the Crustacea, § 316) the outer layer of the transparent cornea of the eye.*

324 *h.* The most remarkable modification of the archetype vertebral skeleton presented to us in the class of Reptiles, is that which we meet with in the order *Chelonia,* which includes the Tortoises and Turtles. We here find the trunk of the body enclosed within a bony casing, into which the head, tail, and the four extremities can be withdrawn in most species. This casing is composed of two shields, covered with horny plates; the upper one, which is more or less highly arched, is termed the *carapace;* whilst the lower, which is usually nearly flat, is called the *plastron.* When we look at the carapace on the exterior, it is seen to be composed of a longitudinal series of 'central' plates united together by sutures; on each side of these is a corresponding series of 'lateral' plates, which form the principal part of the expanded arch; and the edges are bounded by a series of 'marginal' plates, articulated with each other and with the edges of the lateral plates. On looking at the under side of the carapace (Fig. 176), these plates are seen to be intimately connected with the vertebral skeleton; for the 'median' plates are supported by the neural arches of the dorsal vertebræ (*e*) with which they have coalesced; and the lateral plates (*f*), in like manner, appear like expansions of the ribs. It is not surprising, then, that Comparative Anatomists of the greatest eminence (such as Cuvier) should have regarded each 'median plate' as nothing else than a modification of the 'spinous process' (neural spine) of the dorsal vertebra on which it rests, and each 'lateral plate' as an expansion of the rib (pleurapophysis); and further, that the 'marginal plates' (*g*) should have been likened to the sternal ribs (hæmapophyses) of other Reptiles. This mode of viewing the composition of the carapace has been generally adopted, notwithstanding the suggestion of Carus that the carapace is partly formed by the coalescence of dermal bones with those of the vertebral skeleton; but Prof. Owen, having investigated the history of its

* Of the exuviation of the common Ringed Snake of this country, Mr. Bell says:—"It is a mistake to assign a particular period to this process; some have stated it to occur once, some twice in the summer; but I have found it to depend upon the temperature of the atmosphere, and on the state of health, and the more or less frequent feeding, of the animal. I have known the skin shed four or five times during the year. (British Reptiles, p. 52.)

development in the existing Chelonia, and taken a comprehensive survey of its various modifications, not merely in existing but in extinct species, has recently demonstrated that the greater part of the carapace is really *dermal*,—the 'median plates' being dermal bones, representing the median dermal scutes of the Crocodile's back-shield, but developed in connection with the neural spines of the vertebræ which bear them; the 'lateral plates' in like manner representing the medio-lateral rows of dermal scutes in the Crocodile, but never being distinct from the ribs, which become (so to speak) more or less completely buried in them; whilst the 'marginal plates' are wholly dermal, the representatives of the sternal ribs being to be found in the plastron.*—The *plastron*, or floor of the thoracico-abdominal chamber, consists in all Chelonia of nine pieces, of which one (Fig. 177, *a*) is median, while the others (*bb*, *cc*, *dd*, *ee*) are arranged in pairs. Their development varies considerably in different sections of the group. Thus in the Land-Tortoises (A), the plastron covers the ventral surface as completely as the carapace protects the dorsal; being rendered entire and solid through its whole length by the extension of its component plates into contact with each other, and by their adhesion to each other at their edges; whilst it is connected laterally by cartilaginous union with the marginal pieces of the carapace, leaving only an anterior fissure for the passage of the head and fore-limbs, and a posterior fissure for the tail and hind-limbs. And in those which are known as 'box-tortoises,' the plastron is divided by transverse articulations into three portions, of which the central one alone is united immoveably with the carapace, whilst the anterior and posterior move upon it as upon a hinge, in such a manner

FIG. 176.

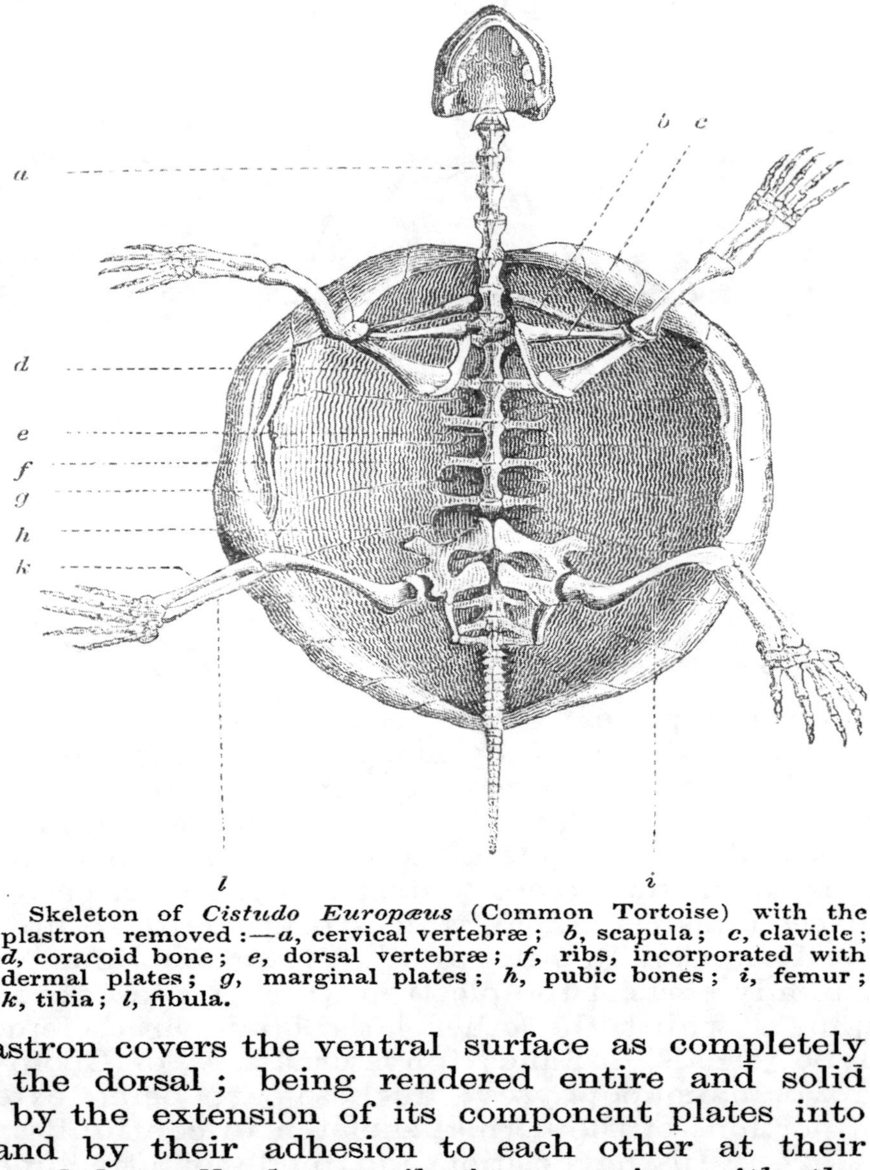

Skeleton of *Cistudo Europæus* (Common Tortoise) with the plastron removed:—*a*, cervical vertebræ; *b*, scapula; *c*, clavicle; *d*, coracoid bone; *e*, dorsal vertebræ; *f*, ribs, incorporated with dermal plates; *g*, marginal plates; *h*, pubic bones; *i*, femur; *k*, tibia; *l*, fibula.

* See Prof. Owen's masterly paper on the "Development and Homologies of the Carapace and Plastron of the Chelonian Reptiles," in the Philosophical Transactions for 1849; and his Memoir on the "Chelonia of the London Clay," published by the Palæontographical Society in 1849.

that they can completely close the fissures, after the head, tail, and limbs have been drawn within their secure protection. On the other hand, in the Marine Turtles (B), these nine pieces do not by any means form a continuous shield, but only afford an incomplete osseous protection to the under surface, which is uncovered both in the centre and in parts of the sides, save by the external envelope supplied by the skin and by its horny

FIG. 177.

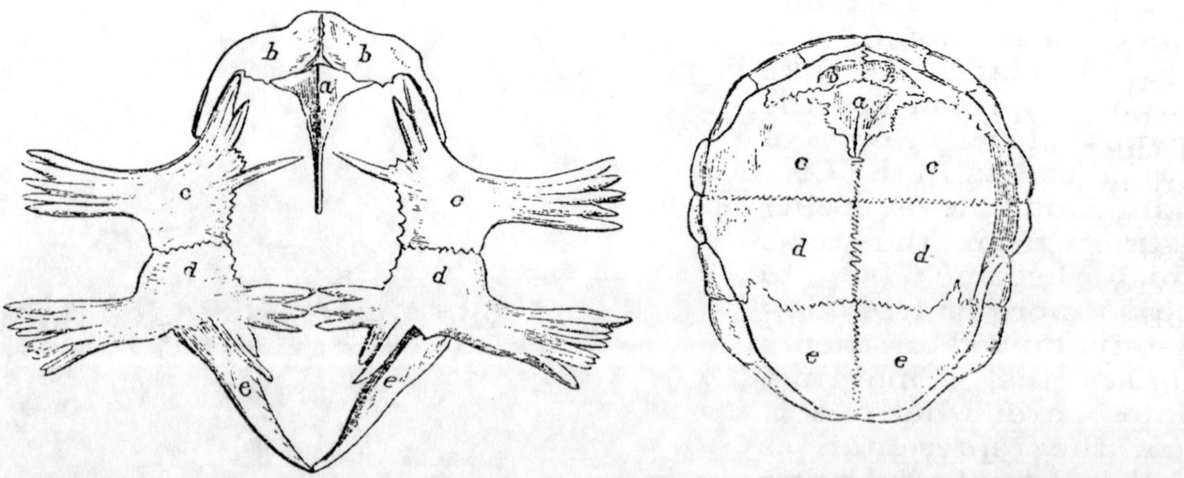

Plastron of Chelonia; A, *Cistudo;* B, *Chelone:—a*, entosternal, or median piece, representing the true sternum; *b b*, episternals; *c c*, the so-called hyosternals, *d d*, hyposternals, and, *e e*, xiphysternals, of Geoffroy.

epidermic plates. The plastron was considered by Geoffroy St. Hilaire (and, following him, by many other Comparative Anatomists) as composed of the sternum, immensely extended laterally; others have considered it as altogether furnished by the dermal skeleton; whilst others, again, have regarded its elements as derived from both sources. By Prof. Owen, however, the median piece (entosternal, *a*) with perhaps the first pair of lateral pieces (episternal, *bb*), which together make up the T-shaped sternum in the Lacertian Reptiles and in Monotrematous Mammalia, are regarded as the only parts of the plastron which represent the true sternum or 'hæmal spine;' whilst the other lateral or parial elements (the hyosternals, *cc*, hyposternals, *dd*, and xiphysternals, *ee*, of Geoffroy) represent the sternal ribs or hæmapophyses; all these parts being extended by the development of portions of the dermal skeleton in continuity with them.

324 *i.* In the position and mode of attachment of the extremities of Chelonian Reptiles, we have peculiarities which at first sight appear quite exceptional. Whilst in all other Vertebrated animals the scapular arch is *outside* the walls of the visceral cavity, we here find it *within;* the scapula lying on the *under* side of the pleurapophyses, instead of on their upper. This displacement could be scarcely accounted for, did we not bear in mind that this arch is really part of the occipital segment (§ 320 *l*), and that in being (as it were) drawn backwards, it may be carried *under* the dorsal pleurapophyses, without a much greater departure from the archetype, than if it were carried *over* them as it usually is. The pelvic arch, also, is included between the carapace and plastron; but this is simply effected by the great extension of the bony plates of which these are composed,

whereby they are made to cover-in many segments of the vertebral column posterior to those from which they are developed.—As in the structure of the trunk of the Chelonia we have the most complete consolidation of which its vertebral segments are capable (presenting in this respect the condition most strikingly opposed to that which has been just described in Serpents), the locomotive power is delegated to the extremities in a greater degree than in most other Reptiles; and we accordingly find, at least in the aquatic members of the group, a greater development, not merely of the bones of which they are themselves composed, but also of the arches by which they are connected with the vertebral column. The 'scapula' (Fig. 176 *b*), which is suspended by ligaments from the second vertebra of the carapace, is a narrow slender bone; but on the under side of the arch we find not merely a very broad and strong coracoid bone (*d*), but a well developed clavicle (*c*), which is not separated, however, from the scapula, and was regarded by Cuvier as an elongated 'acromion process.' In like manner, the pelvic arch, which is attached by somewhat narrow ilia to two sacral vertebræ, is completed below by the very broad pubic arch (*h*), as well as by the union of the ischiatic bones. In the Marine Turtles, the bones of the hand and foot are greatly elongated, but the digits are not separated; the whole of each extremity being enclosed in a continuous skin, which is strengthened with a covering of horny plates; and constituting a very efficient 'oar,' or instrument of aquatic propulsion, which presents the nearest approach that is shown by any existing animal to the 'paddles' of the Enaliosauria. In the proper Land Tortoises, on the other hand, the bones of the extremities are very much reduced in length, the digits of the anterior member being almost rudimentary, whilst those of the posterior are somewhat longer, and are furnished with claws that serve for digging in the ground. The Common Tortoise belongs to the group of *Emydæ* or Mud-Tortoises, the conformation of whose members presents a condition intermediate between these two extremes (Fig. 176).—The skull of the Chelonia does not present us with any such considerable departure from the general Reptilian type, as to require special description; except in the replacement of the teeth by a horny beak, and in the peculiar osseous covering which is furnished to the temporal fossa in certain members of the group. The horny beak, which invests the margin of the jaws along their whole extent, is extremely firm in its texture; in the carnivorous species it is raised into a sharp cutting edge, that of the lower jaw shutting within that of the upper, so that the two act against one another like the blades of a pair of shears; whilst in the vegetable-feeders, its working surface is variously sculptured, and is adapted for bruising as well as cutting. According to Prof. Owen, the development of this horny sheath commences "from a series of distinct papillæ, which sink into alveolar cavities, regularly arranged (in *Trionyx*) along the margins of the upper and lower jaw-bones: these alveoli are indicated by the persistence of the vascular canals, long after the originally separate tooth-like cones have become confluent, and the horny sheath completed."* Thus it appears that the horny beak is really the representative of the two rows of teeth which most other Reptiles possess; and it is a confirmation of this view to find that a similar

* Cyclopædia of Anatomy, Art. *Teeth*, vol. iii. p. 882.

beak is possessed by the larvæ of Batrachia, in most of which it is replaced by teeth when they have assumed the Reptilian form; the casting-off of the beak being analogous to the shedding of the deciduous or milk-teeth in Mammalia. In the Marine Turtles, we find the temporal fossa (in which is lodged the temporal muscle) covered-in externally by a complete arch, which is prolonged on either side from the cranial vault; this arch is formed by the combination of unusual expansions of the parietal, posterior frontal, malar, squamosal, and mastoid bones; and its purpose appears to be, to furnish a peculiarly firm attachment to the temporal muscle, which is lodged in the temporal fossa, and is the principal elevator of the lower jaw. It is but little developed in the Emydæ, and still less in the Land Tortoises.—Although the Chelonia appear so completely isolated from other Reptiles by the peculiarities of their organisation, yet the transition to the Loricated Saurians is by no means abrupt. For in the *Emydosaurus* of North America (known as the 'alligator tortoise' or 'snapping turtle'), the shell is so small in proportion to the long neck, tail, and limbs, that these cannot be retracted into it; whilst in its general habits, the animal has at least as much of the Saurian as of the Chelonian. And in the *Trionyx* (or 'soft tortoise') not merely the plastron but the carapace is partly destitute of bony support, the ribs not reaching to its edges, and its marginal pieces being formed of flexible cartilage. Thus we pass, without any very wide hiatus, to the Loricated Saurians (Crocodiles, Alligators, &c.) in which the dermal portion of the Chelonian carapace is represented by the rows of dermal bones which protect the back.—The epidermic portion of the dermo-skeleton of Chelonia is generally remarkable for the large size and thickness of the plates which cover the carapace and plastron, and which either fit together at their margins, as in ordinary Tortoises and Turtles, or have an imbricated arrangement, as in the *Chelonia imbricata* (hawksbill turtle) which yields what is known as 'tortoiseshell.' The neck, tail, and limbs are usually covered with scales not unlike those of Lizards and Serpents. In the *Sphargis* (leathery turtle), however, and in the *Trionyx*, the surface is everywhere destitute of scales. In the latter genus, the skin is prolonged at the sides, beyond the margins of the carapace, and serves as a sort of lateral fin, which helps these animals to shuffle along beneath the sand or mud at the bottom of the rivers in which they reside.

324 *k*. This general account of the principal varieties in the conformation of the skeleton in the principal groups of Reptiles would be by no means complete, without a notice of three extinct orders, which seem to have occupied a very conspicuous place among the Vertebrated animals of former epochs.—Through the whole of the 'Secondary Period,' commencing with the Triassic Series, and ending with the Chalk, we find the remains of a remarkable tribe in which the characters of Fishes and Whales are most curiously blended with those of Lizards, these animals having been obviously adapted to an exclusively aquatic life, and dwellers in the ocean rather than in rivers or lakes. Hence this group has received the name of *Enaliosauria* or Marine Lizards. The greater part of the species it includes belong to the genera *Ichthyosaurus* and *Plesiosaurus;* the most obvious distinction between which lies in the size of the head, and in its mode of connection with the trunk. In the *Ichthyosaurus* (or fish-lizard) the head is very large, and is joined-on (as it were) to the

trunk, without any intermediate neck, as in Fishes and Cetacea. In the *Plesiosaurus,* on the other hand, the head is comparatively small in proportion to the size of the body, and is carried at the end of a long swan-like neck. The large size of the head of the Ichthyosaurus is due rather to the elongation of the jaws, and to the extension of the bones to which their muscles would be attached, than to any other unusual development; and it is here that we best see the obvious purpose of that peculiar subdivision of the bones of the jaws, which has been noticed as characteristic of Reptiles (§ 324 *b*),—the shock occasioned by the sudden closure of a pair of jaws 7 feet long (for to such a length do we find them occasionally extending) being quite sufficient to snap asunder bones that might be put together with less regard to elasticity and tenacity. These jaws are furnished, along their whole extent of margin, with a set of conical teeth, the peculiarity of whose implantation will be noticed hereafter (§ 324 *p*). The eye of the Ichthyosaurus was extremely large, the diameter of the orbit being sometimes as much as 18 inches; and the eyeball was protected by a complete osseous capsule, formed of several plates articulated together. The nostrils were placed, as in the Cetacea, nearly at the top of the head; and thus the animal could breathe whilst the entire body and nearly all the head were immersed.—The spinal column was composed of numerous vertebræ, usually from 100 to 120. Each vertebra was biconcave, as in Fish, and must have moved over a soft and elastic intervertebral capsule; and the freedom of movement must have been further increased by that want of ossific union between the bodies of the vertebræ and the neural arches, which is characteristic (as we have seen) of the highest Cartilaginous Fishes. From these, however, they differ in the completeness of the ossification of their bones, which has caused them to be preserved to us with the most remarkable perfection. The cervical vertebræ of the Ichthyosaurus are thin, and closely united together, as in the Whales; on the other hand, the tail is greatly elongated and produced to a point, as in the Saurians. Its vertebræ, however, do not bear any vestige of either neural or hæmal arches for any great distance behind the posterior paddles; but from a slight lateral compression which they exhibit, and from the circumstance that the tail is broken off and twisted round at nearly the same part in most of the specimens in which it is preserved, it has been inferred by Prof. Owen that it was furnished with a caudal fin, resembling that of the Cetacea in having no bony support, but corresponding with that of Fishes in being vertical instead of horizontal. The ribs are very numerous, commencing at the second vertebra of the neck, and extending backwards as transverse processes along even a portion of the tail. Those of the anterior portion of the trunk are not united to a sternum, but are connected to each other by a set of 'costo-sternal arches,' each of which contains five pieces; of these pieces the central one obviously represents the 'hæmal spine,' whilst one of the lateral pairs is equivalent to the sternal ribs or hæmapophyses, and the other is an intercalated bone resembling that which intervenes between the sternal and spinal ribs in Crocodiles.

324 *l*. In the form of its paddles, the Ichthyosaurus seems to have resembled the Cetacea; but in the number of their component bones, which sometimes amount to two hundred in each member, and in the multiplication of the rows in which they are disposed, it presents a much

nearer approach to Fishes. The mode of attachment of the anterior extremities to the trunk, however, by the scapular arch, is essentially Reptilian, whilst it presents peculiarities found in no other Reptiles. The broadest part of the arch is formed by the two expanded coracoids, coming into contact with each other on the median line, as well as with the sternum, which is here the mere keystone of the scapular arch, not entering at all into the formation of the thoracic cavity; the clavicles form an arch of their own, uniting together on the central line, so as to form a single bone resembling the 'furcula' of Birds; but besides these, the upper portion of the sternum sends off a lateral prolongation on either side (corresponding with the episternal pieces of the plastron of Chelonia, Fig. 177, *b b*) which extends itself along the curve of the clavicular arch, and comes into close proximity with the margin of the scapula. To this curious arrangement, the nearest approach presented among existing animals is that which is shown in the scapular arch of the *Ornithorhyncus* (§ 326 *d*), which in many features of its organisation reminds us of the Reptilian type. The posterior extremities, on the other hand, which are generally smaller than the anterior, are not attached to the trunk by a pelvic arch of corresponding firmness; the iliac bones not being connected with sacral vertebræ, but being detached from the spinal column, like the arch which supports the ventral fins of Fishes. The skin seems to have been for the most part naked, like that of existing Cetaceans; although, from the regularity of certain markings on the paddles, the integument of those organs appears to have been divided into a number of shield-like compartments, analogous to those of the paddle of the Turtle.—The *Plesiosaurus* (as its name imports) approaches more nearly than the Ichthyosaurus to the ordinary Saurian type; and this is especially apparent in the conformation of the head, and in the reception of the teeth into distinct sockets, as in the Crocodile. The chief elongation of the vertebral column here exists in the neck, which in some species contains as many as thirty segments,—a considerably larger number than that which we find in the neck of the Swan. The surfaces of the vertebræ of the trunk are nearly flat, as in Mammals and some land Reptiles, and are locked together by the articulating processes, in such a manner as to increase the strength of the spine, whilst diminishing its flexibility; but still the neural arches are unconnected with the centra. The costosternal arcs are each composed of seven pieces, as if to allow of a great dilatation of the cavity of the trunk. The scapular arch is more conformable to the ordinary Reptilian type, but is still of most extraordinary strength; the pelvic arch is more closely connected with the spine, and the members it supports are usually of the same size with the anterior, sometimes even larger; the tail is shorter, and was obviously less used to propel the body, but probably served rather to steer it, the locomotive power being chiefly delegated to the extremities. The Enaliosauria were still more closely connected with Crocodilian Reptiles (the most aquatic in their habits of all existing Sauria) by the extinct genus *Pliosaurus*, which combined in some degree the characters of both the genera already mentioned, with those of the higher orders.—From the number of remains of distinct species of Enaliosauria, preserved in the several formations of the middle and later Secondary Period, and from the frequency with which these remains present themselves in strata whose mode of deposit was favourable to their entombment and preservation, it may be inferred without improbability that they were then among the most numerous of the larger tenants of the seas; as their great size, predaceous habits, and great activity rendered them among the most formidable. It is impossible to contemplate their skeletons, without perceiving that they were altogether framed for rapid motion through the water; and that these animals must have rather participated in the speed and energy so characteristic of Fishes, than in the comparatively slow and feeble progression which is ordinarily seen amongst Reptiles. It is not improbable that the peculiar mode in which the hæmal arches are completed, allowing great distensibility of the cavity inclosed by them, may have been destined to afford space for unusually capacious lungs. Another purpose, however, may perhaps be assigned to it,—that of allowing the trunk to participate in the mobility of the spinal column, and to take upon itself the form most favourable for rapid progression through the water, when the animal was urged to its highest speed.*

324 *m.* It need not much surprise us to meet, in the group just considered, with Saurian Reptiles, in many respects degraded (so to speak) to the type of Fishes; because we have been led, by the succession of forms presented to us in the Batrachia, to view the classes of Fishes and Reptiles as very nearly related to each other. But Reptiles and Birds may be regarded almost in the light of *physiological antipodes;* and an animal which should in any degree unite their seemingly incompatible qualities, must necessarily present a peculiar interest for the Zoologist and Physiologist. Such a combination is presented by the extinct genus *Pterodactylus,* which is a Saurian Reptile so modified as to enjoy the powers and to lead the life of a Bird. The Reptilian character of this remarkable animal is at once demonstrated by the structure of its skull, which approaches, not only in its general proportions, but even in the details of its different parts, that of the Crocodile. Thus we find the lower jaw composed of six bones on each side; both jaws are furnished with teeth, and these are implanted in distinct sockets. The orbits, as in the Ichthyosaurus, are very large; but there is no trace of any bony capsule for the eye. The neck only contains the usual number of cervical vertebræ, 7; but these are of great length, and also of unusual size in relation to the body. The head was partly supported by stout tendinous cords running along the sides of the neck; and these are even found partially ossified in some species. The number of vertebræ in the back is much greater than in Birds; the ribs are narrow and almost thread-like, and are destitute of the peculiar 'diverging appendage' possessed by the pleurapophyses of Birds; the tail is short and rudimentary, as in Birds.—It is in the structure of the organs of flight, however, that the great peculiarity of

* The author learns from Mr. J. Scott Russell, well known for his admirable researches on the forms of Waves, and upon the best construction of vessels in regard to speed, that if a boat be built of slight planks, without any supports to determine a particular form, and be then drawn rapidly through the water, she will be moulded (as it were), by the pressure of the medium through which she moves, into that which he designates as the 'wave form,'—that is, the form of least resistance, adapted to move through the water with the slightest possible propelling force. It is well known that certain ships will sail best when their framing is loose or slack, apparently because they can then accommodate themselves in some degree to the 'wave form.' Is not this looseness in the framing of the *Enaliosauria* a provision of the same nature?

the *Pterodactylus* consists, from which it derives its name. The bones of the arm and forearm, especially the latter, are of considerable length; but it was not upon these bones, which give the chief support to the wings of Birds (Fig. 185), nor upon the outstretched and lengthened hand that supports the membranous wings of Bats (Fig. 190), that the Pterodactyle was borne through the air; for its wings were extended by a single finger, developed out of all proportion to the rest (Fig. 178). The remaining digits of the anterior members seem to have been left perfectly free; and may have either served for walking, when the wings were folded, or, more probably, were used for climbing, or for suspending the animal in the mode that bats hang by the hooked nail of the thumb. Of these Flying Lizards a number of species have been found, sufficient to show that they were once probably the most numerous vertebrated inhabitants of the air; if indeed they were not, at the time when they most flourished, the only vertebrated animals capable of flight. And if it should prove that certain wing-bones found in the Chalk formation, whose microscopic structure seems decidedly Reptilian, were those of a Pterodactyle, we shall have to extend our notion of the spread of wing of the largest species of this genus, from four or five feet to twelve or fourteen.—The only existing Reptile that presents the least approximation to the Pterodactyle in the development of organs of flight, is the little *Draco volans*

FIG. 178.

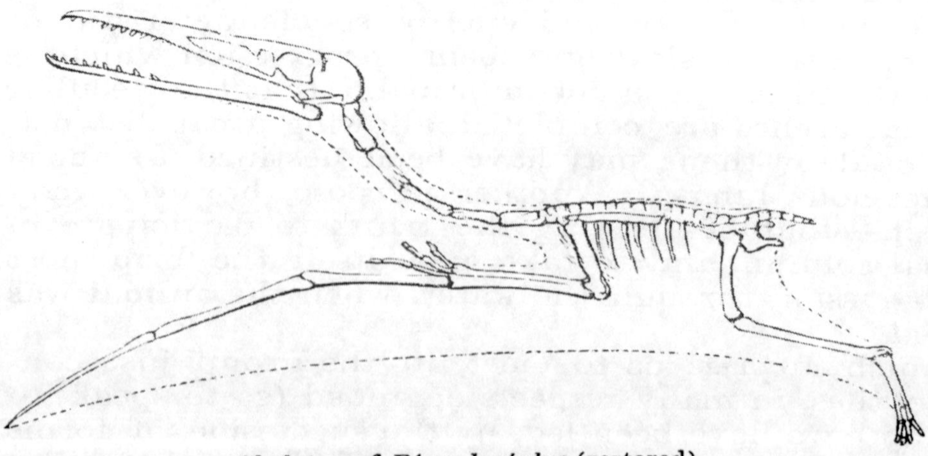

Skeleton of *Pterodactylus* (restored).

FIG. 179.

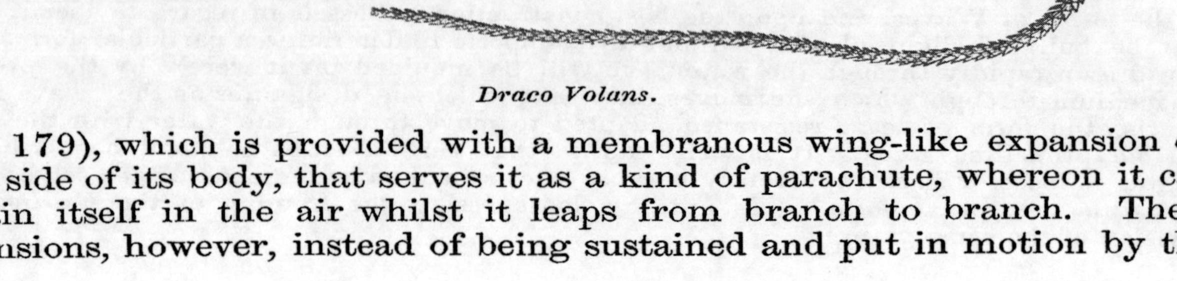

Draco Volans.

(Fig. 179), which is provided with a membranous wing-like expansion on each side of its body, that serves it as a kind of parachute, whereon it can sustain itself in the air whilst it leaps from branch to branch. These expansions, however, instead of being sustained and put in motion by the anterior members, are altogether independent of them, being supported by the first six false ribs, extended horizontally in a straight line; and they cannot be moved with sufficient force to sustain the animal in the air for any length of time, far less to propel it in the manner of true wings.

324 *n.* We have lastly to notice an order of Reptiles that has been formed by Prof. Owen for the reception of a group of fossil Saurians, which present, with a mixture of the Lacertian and Crocodilian types, a closer approximation than either of them exhibit towards the class of Mammalia. Of this order, which has been denominated *Dinosauria* from the gigantic size of the Lizard-like Reptiles which it includes, the principal genera are the *Megalosaurus,* the *Iguanodon,* and the *Hylæosaurus;* of which the first was undoubtedly carnivorous, and the second herbivorous. The principal points in which these Saurians departed from the ordinary Reptilian type, and approached the Mammalian, were the much greater size of the extremities (especially the posterior) in proportion to the trunk, so that their heavy bodies were raised above the ground at least as much as those of a Rhinoceros; the connection of their ribs with the spinal column by a double articulation, as in Mammalia (§ 326 *b*), instead of by a single one, as in ordinary Reptiles (§ 324 *a*); the consolidation of at least *five* vertebræ into a sacrum as in many Mammalia, instead of *two* as in Reptiles, in order to give firmer support to the pelvic arch; the flatness of the surface of most of the bodies of the vertebræ; and the hollowness of the bones. In the structure of the teeth of the Iguanodon, moreover, are peculiarities adapting them for crushing and grinding vegetable food, which are found in no other Reptiles, and to which the nearest approach is presented in the gigantic extinct Sloths of South America; and it is curious that a further resemblance to these animals should be presented by the Iguanodon, in the absence of teeth from the front of the lower jaw, as well as in the massive proportions of the body, and the great development of the posterior extremities; so that it would not seem improbable that their general mode of life was the same, and that the Iguanodon performed the same part in restraining the inordinate extension of arboreal vegetation during the later Secondary period, as the Mylodon and Megatherium did in part of the Tertiary (§ 326 *h*).*

324 *o.* Having thus sketched the chief peculiarities in the neuro- and dermo-skeletons of Reptiles, it remains for us to speak of their *Teeth;* from a brief survey of which we shall naturally pass to the consideration of the Digestive apparatus.—The proportion of the Reptilian class in which there is an entire absence of teeth, is much greater than that of Fishes; for the whole order Chelonia is destitute of them; in addition to which they are absent in the *Bufonidæ,* or Toad family of the Batrachian order; and although they are found in all existing Sauria, yet certain extinct genera, which present a remarkable combination of Chelonian characters with a predominant Saurian organisation (*e. g.* the

* For a vast body of information respecting the Fossil Reptiles, and for the grounds upon which the order *Dinosauria* has been constituted, the reader may be especially referred to Prof. Owen's Report on the British Fossil Reptiles in the "Reports of the British Association" for 1841, and to the various Memoirs contributed by him to the "Transactions of the Geological Society;" also, to the Memoirs contributed by Dr. Mantell to the successive volumes of the "Philosophical Transactions."

Rhynchosaurus), were destitute of them. The larvæ of the Batrachia in general are without teeth, and their jaws are covered with a continuous horny sheath like that of the Chelonia; this is shed, during the metamorphosis, in all these animals save the *Siren*, by which it is retained even though true teeth are subsequently developed. Dissimilar as the horny sheath of the jaws of Turtles and Birds may appear to the proper dental organisation, yet it would really seem, as already pointed out (§ 324 *i*), to be the representative of this.—The teeth are situated upon the jaws alone, in most Saurians; and it is in that group that they attain their highest development, and show the closest approximation to Mammalian teeth. In some Lizards, however, they are found upon the pteregoid bones also; in most Serpents, they are implanted on the palatine as well as the pteregoid bones; and in the Batrachia, they are commonly met with on the vomer in addition. In no existing Reptile furnished with teeth, is the number less than 30. The extinct *Dicynodon*, however, which, though a Saurian in its general organisation, had only two long tusks in front of the upper jaw, growing through life from persistent pulps, presents a remarkable departure from the ordinary type; and it is probable that the teeth which are wanting at the sides of the jaws, were replaced by a horny sheath. The greatest number of teeth is found in the Batrachian order, where the small size and great multiplication of the teeth remind us of Fishes, though the teeth are nowhere so numerous as in some of the last-named class. It is rarely that the number of the teeth is so fixed and determinate in any Reptile, as to be characteristic of the genus or species. The *form* of the teeth, with few exceptions, is that of a simple cone, more or less acute at its point, but sometimes compressed, so as to form sharp edges. They are adapted rather for seizing and holding prey, than for dividing and masticating their food. The most remarkable departure from the typical form is presented by the herbivorous *Iguana*, and by the extinct *Iguanodon* which resembled it in regimen; in the teeth of these animals, the crown is expanded, both in length and breadth, and the margins are notched or dentated; and those of the Iguanodon are further marked by longitudinal ridges, presenting the most complicated external form anywhere met with in this class. A trenchant edge is formed, on the outer side, by a layer of enamel; and the dentine is softer as it recedes from the enamelled edge; so that the crown of the tooth is worn away in a regular oblique plane.

324 *p*. The varieties of dental *structure* are few in Reptiles, as compared either with Fishes or with Mammals. The teeth are most commonly composed of 'ordinary' or 'non-vascular dentine,' 'enamel,' and 'cementum;' and, in addition to these, we occasionally meet with a substance closely resembling bone, in the interior of certain teeth, where it is formed by the calcification of the remains of the pulp. The inner portion of the dentine of the Iguanodon is traversed by medullary canals extending from the pulp-cavity; but these are not, as in the 'vascular dentine' of Fishes, centres of radiation for their own set of tubuli. This, however, is the nearest approach to that texture, which the class of Reptiles presents. The most constant of the foregoing elements are the dentine and the cementum; these constitute the entire substance of the tooth in *Ophidia*, the body of dentine being invested with a thin layer of cementum; in some *Batrachia* and in most *Sauria*, the enamel usually forms a

thin coat over the crown; and it is in the latter order, that we occasionally find a deposit of coarse bone in the interior. The peculiar conformation of the poison-fang of the venomous Serpents has been already noticed (§ 324 *g*); and the only other important variety of structure is one which is presented by certain of the lower Reptiles in common with some among the higher Fishes; namely, an implication or doubling-in of the dentinal cylinder that surrounds the pulp cavity, so as to form a series of vertical folds, which are usually marked by 'flutings' upon the surface, as in the lower part of the teeth of the *Ichthyosaurus.**—The *development* of the teeth of Reptiles is intermediate, as to the grade to which it proceeds, between that of the teeth of ordinary Fishes, and that of the same organs in Mammalia. In all cases, the process passes beyond the 'papillary' stage to the 'follicular' and 'capsular;' and in some Reptiles the capsule becomes more or less surrounded by true bone; but it is never so far invested by it, that the emergence of the tooth can be considered as a true 'eruptive' stage. The teeth of Reptiles, like those of Fishes, are 'successional;' that is, new teeth are constantly in progress of development, whilst the old are as continually undergoing exuviation. The new teeth are not here developed from independent germs; but each formative pulp is derived as an offset from the follicle of the preceding tooth, and its development takes place in a cavity at the side of its base. As the tooth acquires hardness and size, it presses against the base of the contiguous attached tooth, causes a progressive absorption of that part, and finally undermines, displaces, and replaces its predecessor. The rapid succession of tooth-germs, which stamps the impress of decay upon their predecessors, often before the growth of these is completed, is most strikingly manifested in the Crocodiles; in which three, and sometimes even four generations of teeth, sheathed one within the other, are contained in the same socket. The process of replacement is subject to certain variations, in accordance with the diversities in the mode of attachment of the teeth which are characteristic of particular tribes of Reptiles; but it is performed in every case in a manner essentially the same, the extinct *Dicynodon* constituting the only known exception.—The simplest mode of *attachment* of the teeth of Reptiles to the jaws that bear them, is that in which the tooth becomes anchylosed by its base to the margin of the jaw, or to a conical process of bone rising above it; this, which is the case in the *Mososaurus,* corresponds with the condition of the human jaw at that early period of embryonic development, when as yet not even the 'primitive dental groove' has been formed. In the greater number of ordinary *Lizards* and in *Frogs,* the teeth are partly imbedded in a groove formed by the elevation of an alveolar ridge on the external margin of the jaw, and to this ridge they adhere by anchylosis at the outer side of their base; to this stage of the formation of the primitive dental groove, we have the analogue in the human embryo at about the seventh week. In the *Ichthyosaurus,* as in the slightly more advanced human embryo, the dental groove is completed by the elevation of a ridge on the inner margin of

* According to Prof. Owen, the wonderfully-complicated structure of the tooth of the extinct *Labyrinthodon* (one of the earliest known Reptiles in order of geological time, which was essentially a Batrachian with some Saurian affinities,) is referable to this type; by Mr. Tomes, however, it has been recently explained as resulting from the aggregation of a series of denticles,—the only known example of such a composite structure in the class of Reptiles.

the jaw, and indications of a transverse subdivision into sockets may be perceived; here the tooth does not become anchylosed to its socket, but remains free. In most *Ophidia* and some *Batrachia* the tooth is surrounded by a shallow socket, to which its base becomes adherent; and in the extinct *Megalosaurus* and *Thecodon*, this socket was moderately deep, and the tooth was adherent to its walls. In the *Plesiosaurus* and the whole *Crocodilian* group, the teeth are enclosed in sockets, and remain free; a condition which approaches towards that of Mammalian teeth,—the difference being that in Reptiles the base of the tooth is seldom closed and contracted to a 'fang,' and there is never any implantation of Reptilian teeth by diverging fangs.*

324 *q*. The structure and arrangement of the several parts of the Nutritive apparatus in Reptiles do not offer nearly the same important diversities, as we have seen to exist in the Osteology of these animals; for although the *form* of the organs may vary, in accordance with that of the body, yet their essential characters remain unchanged. Thus in Serpents, we have an extraordinary elongation of most of the viscera (Fig. 180), the general arrangement of which strongly reminds us of the disposition of parts in the Nematoid Entozoa (Fig. 143). Although the food of Reptiles is generally seized by their jaws, armed as these usually are with prehensile teeth, or furnished with a sharp cutting horny ridge, yet the tongue is the chief instrument by which it is obtained in some of the smaller Reptiles of insectivorous habits, such as the Chameleon, Frog, and Toad; among which we find a provision for its being suddenly darted out to a considerable distance, and either attaching the victim to its point by the glutinous saliva with which it is furnished, or grasping it between the points of its bifid extremity. Among very few Reptiles does it seem that anything like a proper mastication takes place; yet a considerable salivary secretion is formed by a glandular apparatus, which is sometimes disposed in the tongue itself, sometimes along the margin of the jaws; and the purpose of this secretion would seem to be chiefly to lubricate the prey, which has frequently to be forced down the œsophagus with considerable effort. The œsophagus is usually capacious, and opens into a stomach, which is sometimes a scarcely perceptible enlargement of the alimentary canal (as in Serpents, Fig. 180, *i*), but which in other instances presents itself as a large sac of variable form; being in both cases susceptible of very great dilatation. The greatest departure from the usual type is shown in Crocodiles, in which the stomach has strong muscular walls, the fibres of which radiate from a central tendon, in a manner that reminds us of the conformation of the gizzard of Birds. The pancreas now presents more of the character of the ordinary glands, and its secretion, with that of the liver, is poured into the upper part of the intestinal tube; this is usually short, but is divided into two portions, representing the small and the large intestines respectively, by a prominent valve analogous to the ileo-cæcal valve in man; but it is very interesting to remark, that the intestine of the Enaliosauria must have had the spiral fold, which augments the extent of the intestinal tube in the Cartilaginous Fishes (§ 322 *h*), as is shown by the preservation of its markings on the 'coprolites' or

* For full information on this subject, see Prof. Owen's 'Odontography,' from which the preceding statements are derived.

fossilized fæces, found abundantly in the strata in which the remains of these animals are entombed.—The absorbent system appears to be very highly developed in Reptiles, the trunks being unusually large and numerous, forming extended plexuses, and sometimes being furnished

Fig. 180.

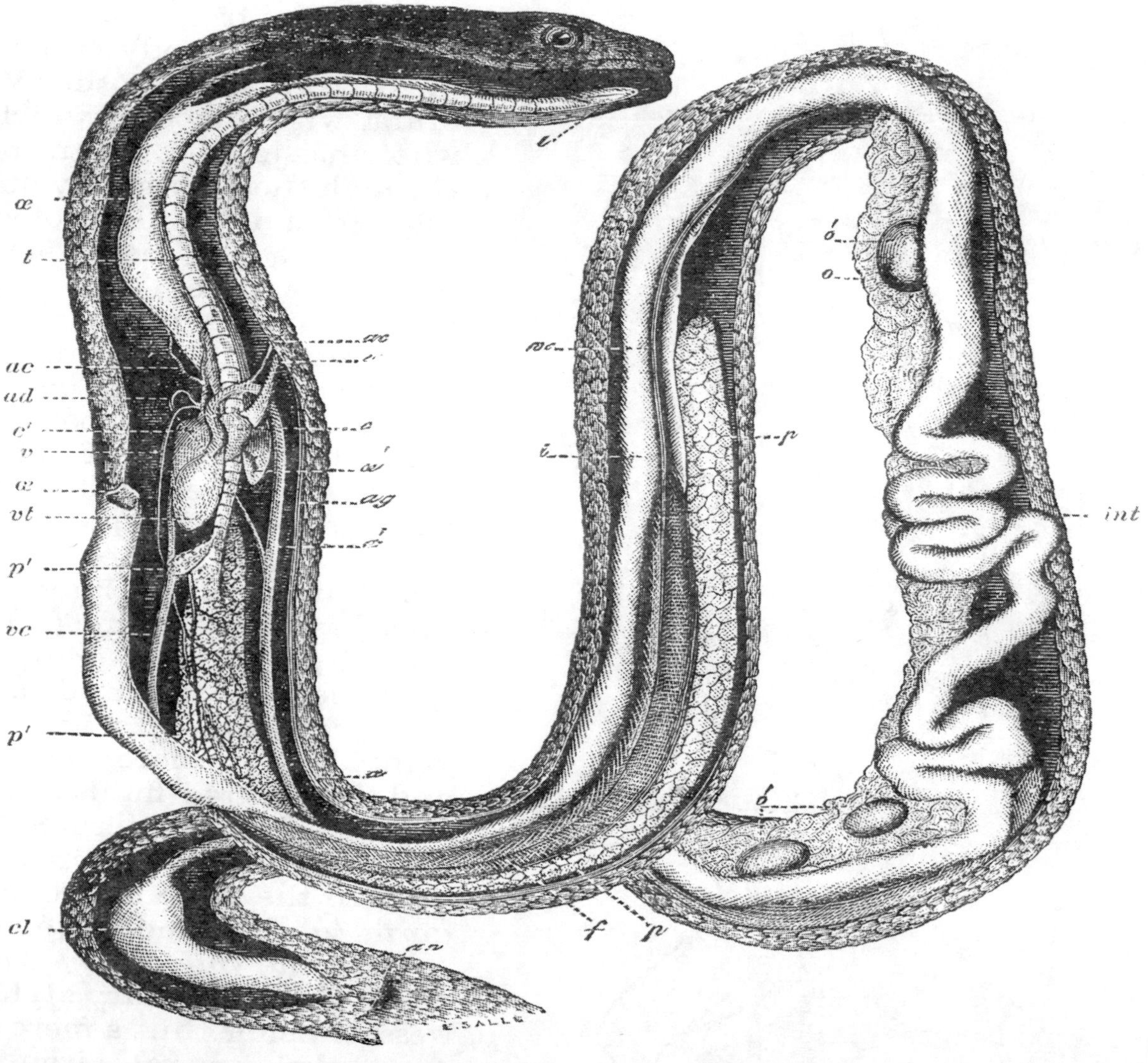

Anatomy of *Coluber*;—*l*, tongue and glottis; *œ*, *œ*, œsophagus, divided at *œ′* to show the heart, &c.; *t*, trachea; *ac*, *ac*, carotid arteries; *ad*, right aortic arch; *ag*, left aortic arch; *c*, left auricle; *c′*, right auricle, *vt*, ventricle; *v*, vena cava superior; *vc*, *vc*, vena cava inferior; *p*, *p*, principal lung; *p′*, rudimentary lung; *f*, liver; *i*, stomach; *int*, intestine; *cl*, cloaca; *an*, anus; *o*, ovary; *o′*, ova.

with pulsating dilatations or lymphatic hearts; but they are still destitute of those glandulæ or rather 'ganglia' (knots), which, in the higher Vertebrata, concentrate a great length of tube within a small space; and they are also unpossessed of valves. Moreover, they terminate by numerous communications with the venous system in different parts of the body.

324 *r*. It is, however, by the peculiar arrangement of their Circulating and Respiratory apparatus, that Reptiles are most remarkably characterized. Like Birds and Mammals, they breathe air; but like Fishes,

they are cold-blooded; that is, their respiration is not sufficiently active to enable them to sustain a fixed standard of temperature, above that of the surrounding medium. This arises from the very slight degree of subdivision that exists in their lungs, whereby the aerating surface is but little extended; and from the arrangement of the *pulmonary* with reference to the *general* circulation, from which it happens that the body is not supplied, as it is in all other Vertebrata, with blood of which the whole has just been transmitted through the respiratory organs, but with a mixed fluid, of which part alone has been aerated, while the other part consists of the venous blood that has just returned from the systemic circulation. The heart, in all save the Crocodiles and their allies, consists of but three cavities, one ventricle and two auricles (Fig. 181); of these auricles, one (*b*) receives the blood from the *vena cava* or great systemic vein (*f*), the other from the *pulmonary veins* (*p*); both of them discharge their contents into the single ventricle (*s*), where the venous and the aerated fluids are mingled; and this half-aerated blood is sent forth by its contraction, both to the system through the *aorta* (*a, d, i*), and to the lungs (*o*), by the *pulmonary arteries* (*e,e*); the pulmonary system of vessels being thus a mere offset from the general circulation. The venous blood which has passed through the capillaries of the stomach (*n*) and of the intestines (*k*), is not at once discharged into the systemic veins: but is collected by the *vena portæ* (*l*), which conveys it for distribution to the kidneys (*h*), and to the liver (*g*); and it is not until after it has passed through the capillary system of these organs, in which it is subjected to their secreting action, that it returns to the heart through the vena cava.—The lungs in most Reptiles are

FIG. 181.

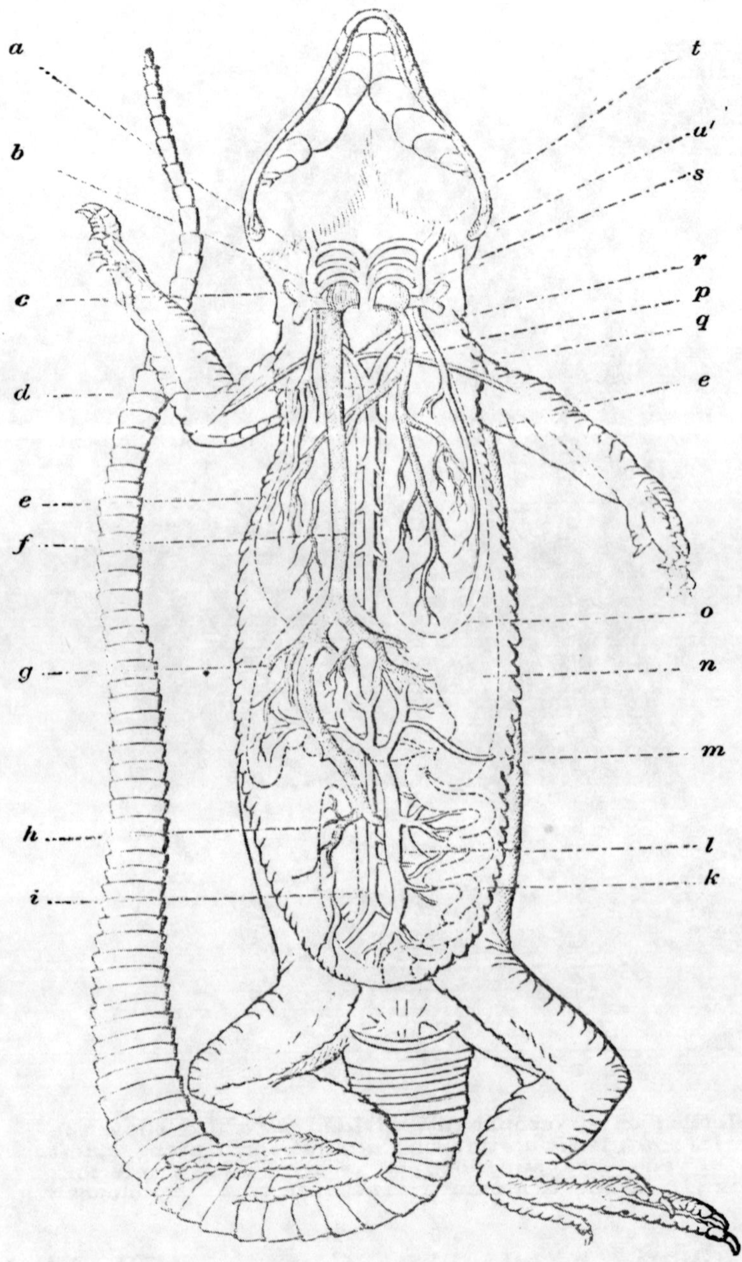

Circulating Apparatus of *Lizard*;—*a, a'*, arches of the aorta; *b*, right auricle; *c*, vena cava superior; *d*, ventral aorta; *e, e*, pulmonary arteries; *f*, vena cava inferior; *g*, liver and hepatic vein; *h*, kidneys and renal vessels; *i*, ventral aorta; *k*, intestines; *l*, vena portæ; *m*, gastric vein; *n*, stomach; *o*, lung; *p*, pulmonary vein; *q*, brachial artery; *r*, single ventricle; *s*, left auricle; *t*, carotid artery.

very capacious; but in their internal structure they present, among the inferior forms of the class at least, but little advance above the type which is exhibited in the air-sacs of some of the higher Fishes. There is but little subdivision of the cavity by those partitions, which, in the higher Vertebrata, separate it into air-cells; and consequently the blood is only exposed to the influence of the air by being distributed upon the walls of the cavity, and upon the extension of the surface which is gained by the duplications or inflexions of these; so that the capillary network is in relation with the air on one side only. In the lungs of the *Chelonia*, however, and in the upper part of those of some of the higher Sauria, we find a degree of subdivision which leads us towards the more complicated arrangement presented by the lungs of warm-blooded Vertebrata. The relative inferiority of the respiratory apparatus of Reptiles is further shown in the absence of those means for effecting a continual interchange in the gaseous contents of the lungs, which we see in Birds, and still more in Mammals. In Batrachia and Amphibia, the lungs can only be filled with air by an action that resembles swallowing; and in Serpents and Lizards, although the movements of the ribs can in some degree effect such a dilatation of the pulmonary cavity as shall occasion the ingress of air, yet this is a slow and imperfect process compared with the descent of the diaphragm in the Mammalia. It is a curious circumstance that in Serpents the lung is single (Fig. 182, C), that of one side (*c*) being developed to a great length, whilst that of the other side (*c′*) remains in a rudimentary condition. And it is further remarkable that the lungs of those Saurians, which approach the Ophidian type in the great elongation of their bodies and the imperfect development of their limbs (§ 324 *a*), present an approach to the same inequality (Fig. 182, A, B).—In regard to the other excreting organs, it will be sufficient to mention that the urinary apparatus now manifests an advance in its condition, which is really as considerable as that of the respiratory organs; the permanent Corpora Wolffiana of Fishes being replaced by true

FIG. 182.

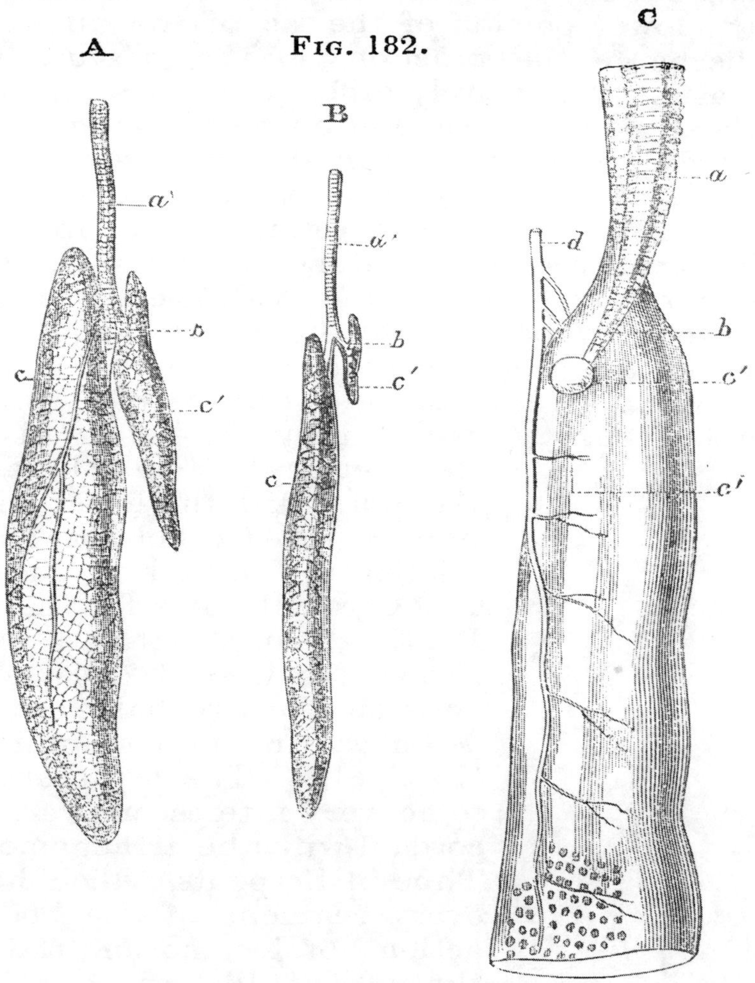

Lungs of the two sides, in A, *Bipes lepidopus*; in B, *Bimanus canaliculatus*; in C, *Coluber natrix*;—*a*, trachea: *b*, bronchial tubes; *c*, right lung; *c′*, left lung; *d*, pulmonary artery.

Kidneys. These organs are generally placed far back in the abdominal cavity, and are sometimes found even within the cavity of the pelvis. In Serpents, however, as in Fishes, the urinary organs are greatly elongated, and advance towards the head; and it is curious to observe that the want of symmetry, which prevails in the pulmonary apparatus of these animals, shows itself also in their urinary organs, the right kidney of Serpents being placed in advance of the left, as if for the convenience of packing. The two ureters, by which the secretion is collected and conveyed away, open into the *cloaca* (Fig. 180, *cl*), which is a common vent for the discharge of the fæces, of the urinary secretion, and of the genital products. A sac is found in some Reptiles in connection with the cloaca, which has received the name of urinary bladder; but although its mode of development is essentially the same with that of the urinary bladder of Mammalia (being the lower portion of the sac of the allantois), yet it does not appear to discharge the functions of that viscus, but ministers (in Batrachia at least) to an entirely different purpose, being a reservoir of fluid to supply the continual evaporation which takes places from their surface.—The hard dermal or epidermal envelopes with which the body is covered in the scaly Reptiles, prevent their skin from being rendered in any considerable degree an excreting organ; in the naked soft-skinned Batrachia, however, the tegumentary surface takes a very important share in the respiratory function, and it also sets free a large quantity of fluid in the state of vapour. This fluid, however, would seem to be little else than water, and not to contain other excreted matters in any large amount.

324 *s*. In the nervous centres of Reptiles, we find a decided advance upon the type which they present in Fishes, in the much larger size of the Cerebral Hemispheres in relation to the Sensory Ganglia; still the latter make up a considerable proportion of the Encephalon; and the entire brain is small in relation to the bulk of the body and to the Spinal Cord. The Cerebellum is for the most part of very small size in Reptiles, being relatively less developed than in the higher Fishes; so that it does not always cover-in that fissure between the two halves of the Medulla Oblongata, which is known in higher animals as the 'fourth ventricle.' (Fig. 183).—The structure of the Spinal Cord presents some very interesting variations, which are precisely accordant with the arrangement of the locomotive apparatus. Thus in Serpents, which have no members, and in which every segment of the body takes a similar part in the actions of locomotion, the spinal cord is of uniform size throughout, like the ganglionic cord of the Annelida and Myriapoda, only differing from these in the continuity of the ganglionic tract, which is not broken up into distinct centres, although its segmentation is marked by the uniformity of the exit of the successive pairs of spinal nerves, which come off between the vertebræ. On the other hand, in Frogs and Turtles, in which the locomotive power is chiefly delegated to the extremities, there are distinct ganglionic enlargements in the spinal cord, at the parts from which their nerves issue; and this is particularly the case

FIG. 183.

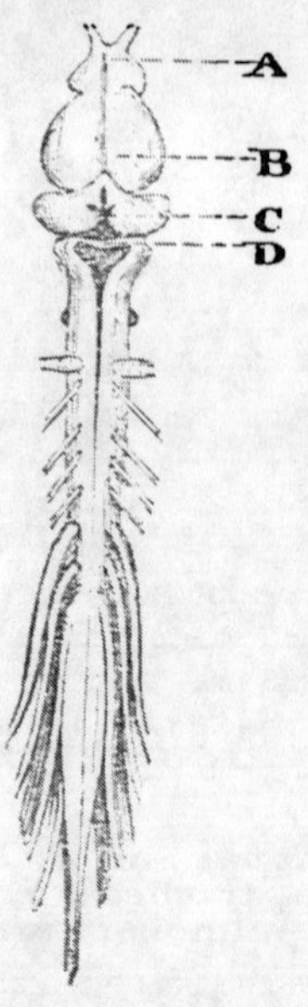

Nervous Centres in *Frog*;—A, olfactive ganglia; B, cerebral hemispheres; C, optic ganglia; D, cerebellum, not covering-in the fourth ventricle which is seen below it.

in the Frog in the *posterior* segment of the spinal cord, in accordance with the greater use which this animal makes of the posterior extremities. In regard to the development of the Organs of Sense, we see, as in the case of the respiratory apparatus, an advance in type, without a corresponding advance (indeed it may even be with a positive diminution) in functional power. Thus the nasal cavity, in all Reptiles, as in higher Vertebrata, has a posterior communication with the mouth, and through it with the air passages; so that, of the air drawn in by the inspiratory movement, a part, at least, shall pass through the Olfactive organ. But in this organ we find no provision for extending the surface on which the nerve is distributed; and there is every reason to believe that the sense of smell is very obtuse.—The Eye is always present in Reptiles, except in the *Proteus,* in which it is a mere rudiment, quite unfitted for visual use. It presents no remarkable peculiarity, except in the rudimental existence, among certain Chelonia and Sauria, of that osseous capsule which is so largely developed in the extinct Enaliosauria, and which performs a most important office in the eyes of Birds. We here first meet, however, with a special arrangement of the integuments of the face for the protection of this delicate organ; for whilst in Serpents the skin of the head passes continuously in front of the eyes, merely becoming transparent where it covers the cornea, it is doubled in most other Reptiles into two folds, constituting the upper and lower eyelids, which can be drawn together by a sphincter muscle; and we also find a rudiment of a third eyelid, formed by an additional fold of membrane at the inner angle, which is so completely developed in Crocodiles as to form a 'nictitating membrane,' that can be drawn completely across the eye, as in Birds, by a muscle specially adapted for the purpose. In the Batrachia, although the upper and lower eyelids are nearly motionless, the eye can be covered by the third eyelid, as in the Crocodile.—The organ of Hearing presents a great advance in the higher members of this class, a tympanic apparatus and chain of bones being interposed between the entrance of the vestibule and the outer air; the purpose of which is obviously to cause the sonorous vibrations to act more effectually upon the fluid in the auditory cavity. This tympanic apparatus is very imperfectly formed, however, in Serpents and Frogs; and it is entirely wanting in the Perennibranchiate Batrachia, the condition of whose organ of hearing presents but very little advance upon that of Fishes, save that the membranous labyrinth is now entirely imbedded in bone, by the complete ossification of the capsule by which it is surrounded. It is only in the Crocodile and its allies, that the rudiment of a cochlea first makes its appearance; and throughout the class, the auditory capsule contains the 'otolithes' which are so characteristic of the inferior grades of this organ.—The tongue in Reptiles would seem to be less an organ of Taste than of Touch; being also modified, in certain cases, for some special purpose, as the capture of prey, or the formation of the salivary secretion (§ 324 *m*). The firmness of the dermal skeleton, in the scaly Reptiles, obviously unfits their general surface for the exercise of any acute tactile power; and although in the naked species there is no mechanical obstacle to the sensibility of the skin, yet the general want of true sensory papillæ seems to indicate that its feeling cannot be acute.

324 *t*. Notwithstanding the general advance in the development of the Nervous system of Reptiles, its functional activity would seem to be kept

down by the imperfect oxygenation of their blood; their movements being usually slow, and their perceptions obtuse. And even when, as in the case of Serpents, we see unusual manifestations of nervo-muscular energy and of acuteness of sensation in the pursuit and capture of prey, such manifestations alternate with long periods of almost torpid repose, during which these animals seem as if they were recovering from the fatigue of such unwonted exertions. From their want of power to sustain an independent temperature, and from their exposure to a degree of winter cold which is incompatible with the maintenance of their activity, the Reptiles of temperate climates pass a considerable part of the year in a state of profound torpor (§ 99); and even those of tropical regions are not exempt from this periodical inactivity, the immediate cause of which, however, appears in their case to be the deficiency of moisture occasioned by extreme and continued heat (§ 66).

324*u*. Reptiles are uniformly oviparous; and, being also cold blooded, can afford but little of that assistance in the development of the ova, which is so remarkably imparted by Birds. Nevertheless we observe a decided advance in the character of the generative function, in this respect, that a sexual congress of some kind always takes place; so that the fertilisation of the ova is not left (as in most Osseous Fishes) to the chance-contact of the seminal fluid that may have been dispersed through the neighbouring water, but is either accomplished before they have been expelled from the female passages, by the introduction of the seminal fluid within these, or in the very act of expulsion. The latter is the case in the Frog and its allies; the former, in most of the higher Reptiles, many of which possess a penis or intromittent organ. The testes or spermatic organs of Frogs show an approach to the structure which is characteristic of higher Vertebrata, being composed of an aggregation of numerous short cæca, the representatives of the tubuli seminiferi; but they discharge their secretion by several excretory ducts, which open into the ureters, instead of uniting into a distinct spermatic canal or 'vas deferens.' The ovaries of the female are composed of duplicatures of a vascular membrane in which the ova are developed, and these lie in the posterior or pelvic portion of the visceral cavity; but, on the other hand, the oviducts are prolonged forwards nearly as far as the heart, and there terminate by open fimbriated extremities; the ova, therefore, set free in the abdominal cavity by dehiscence from the surface of the ovary, must travel from the posterior towards the anterior extremity of the visceral cavity, before they can enter the dilated orifice of the oviduct which is to convey them forth.—In the higher Reptiles, the testes exhibit a decided advance in development, in the greater length and convolution, and in the diminished multiplication, of the tubuli seminiferi; and the products of each testis are collected into a single excretory duct or 'vas deferens,' which does not open into the ureter, but either discharges the spermatic fluid into the cloaca, or is continued as a groove along the penis, and conveys the fluid to its extremity. The internal orifices of the oviducts in the female approach more closely to the ovaries; and these organs present a more compact structure. The oviducts invariably open externally into the cloaca, and the ova are usually discharged soon after fertilization, if not before. In Batrachia, Ophidia, and Sauria, however, there are certain species which retain their ova in a sort of uterine cavity formed by a dilatation of the oviduct near its extremity, until the development of their contained embryo is so far advanced that the enveloping membrane ruptures in the very act of the expulsion of the ovum, or even previously, so that the young are born alive. There is not, however, in any Reptile, the least trace of that provision for affording a continued supply of nutriment to the embryo during its development, which is characteristic of true viviparous animals, and to which we have seen a slight approach in the higher Cartilaginous Fishes; and the proper designation of this method of generation is *ovo-viviparous.*—In all Reptiles, save in the Batrachia (the phenomena of whose metamorphosis have been already considered, § 323), the young come forth from the egg in a condition which does not essentially differ from that of their parents, and are so far advanced towards their complete evolution, that on their entrance into the world they are at once qualified to obtain their own support, and are but little, or not at all, dependent on assistance from their parents. With this higher development, the greater relative size of the eggs in Reptiles seems to have an important connection; for their increased bulk, as compared with that of the eggs of Fishes, is given by the larger provision of nutrient material which they contain; and we accordingly find that whilst, in the higher Reptiles, the *number* of eggs is comparatively small, the Batrachia alone, whose young come forth at an earlier grade of development than those of other orders, and whose eggs do not contain yolk enough to serve for the evolution of the animal on the perfect Reptilian type, deposit ova at all comparable numerically with those of Fishes.

325. From Reptiles to Birds, the transition would seem rather abrupt, since the latter class is in its general characters the almost precise opposite of the former. But we have seen in the class of Reptiles a special modification of the Saurian type to the life and habits of a Bird; and the class of Birds presents us with certain forms, which are deficient in some of the distinctive attributes of the group, and which show an obvious degradation towards the Reptilian type. Taken as a whole, however, the members of this class exhibit a remarkable conformity to one general type, presenting in this respect a marked contrast to the diversity which we have seen to prevail amongst Reptiles; and this conformity is both structural and physiological, being shown alike in the plan of organisation, and in the working of the organs.—Birds have been denominated, and not inappropriately, "the Insects of the Vertebrated series." As in the animals of that class, we find the whole structure peculiarly adapted for motion, not in water, nor upon solid ground, but in the elastic and yielding air. It is impossible to conceive any more beautiful series of adaptations of structure to conditions of existence, than that which is presented to us in the conformation of the Bird, with reference to its intended mode of life. In order to adapt the Vertebrated animal to its aerial residence, its body must be rendered of as low a specific gravity as possible. It is further necessary that the surface should be capable of being greatly extended; and this by some kind of appendage which should be extremely light, and should at the same time possess considerable resisting power. The degree of nervo-muscular energy required for support and propulsion through the air, involves the necessity of a very high amount of respiration (§ 67); and as the general activity of the vital processes depends greatly upon the high temperature (sometimes reaching 112°) which this

energetic respiration contributes to maintain, a provision is required for keeping in this heat, and not allowing it to be carried away by the atmosphere, through which the Bird is rapidly moving. Now of these four objects, the lightening of the body and the extension of the respiratory surface are beautifully attained, in a mode which will be found to correspond closely with the plan adopted for the same purpose in Insects (§ 315 *a*). The air which enters the body is not restricted to a single pair of lungs placed in the upper part of the visceral cavity, but is transmitted from these organs to a series of large air-sacs, disposed in the abdomen and in various other parts of the body ; and even the interior of the long bones is made subservient to the same purpose, their shafts being hollowed out, and lined with a delicate membrane over which the blood-vessels are minutely distributed. In this manner the respiratory surface is greatly extended ; whilst by the large quantity of air introduced into the mass, its specific gravity is diminished.* Again, the extension of the external surface, and the retention of the heat within the body, are accomplished by the same beautiful and refined contrivance, the covering of feathers. Like hair or scales, feathers are to be regarded as epidermic appendages to the cutis. The stem is formed by an apparatus which may be likened to a 'hair-follicle' on a very large scale (§ 180) ; but there are some additional parts for the production of the laminæ that form the vane of the feather, which are joined to the stem during its development. These laminæ, when perfectly formed, are connected by minute barbs at their edges, which hook into one another, and thus give to the entire vane the necessary continuity of resisting surface. The substance of which feathers consist, is a very bad conductor of heat ; and when they are lying one over the other, small quantities of air are included, which still further obstruct its transmission by their non-conducting power. In the two most aberrant forms of the class, however, one of which leads us back towards Reptiles, whilst the other presents certain approximations to the Mammalia as well as to Reptiles, we meet with remarkable modifications in the typical structure of feathers. In the *Penguin,* those which cover the wings have a strong resemblance to scales ; and the wings are not employed to raise this bird in the air, but only to propel it through water (as fins would do) by their action on the liquid. On the other hand, in the *Ostrich* tribe, which consists of Birds adapted simply for running on the ground like Mammals, the laminæ of the feathers are so separated from each other, as no longer to form a continuous surface ; in the *Emeu* they resemble mere branching hairs, of which two arise from each root, one of them being the 'accessory plume' of other feathers developed to the same size with the principal shaft ; and in the *Cassowary* some of them are developed as shafts without any vane, so as almost to resemble Porcupine's quills.

* The degree in which the bones of Birds are thus connected with the respiratory function, and their interior occupied with air instead of with marrow, bears a very exact proportion to their powers of flight (§ 206). In the Ostrich and its allies, the cavity of the shaft of the long bones is partly occupied by medullary cancelli; and it was from finding this cavity, in the fragment of a femur of the extinct *Dinornis* first submitted to him, still more divided into cancelli than that of the corresponding bone in the Ostrich, that Prof. Owen ventured to predict, with true philosophic sagacity, that the Struthious Bird to which that bone belonged must have been still more *wingless* than the Ostrich itself,—a prediction which has been amply verified by the result, since remains of many hundreds of individuals have now been collected, without the least trace being discovered of even such rudimentary wing-bones as the Ostrich possesses.

325 *a.* The bony framework constituting the Bird's vertebral skeleton, presents many remarkable adaptations to the same purposes. In the first place it is to be remarked, that the faculty of locomotion is here most completely taken from the spinal column and delegated to the extremities; so that the skeleton of the trunk must be consolidated, in proportion to the power with which the members are to be endowed, as we have seen in numerous other cases. The relative length of the neck to that of the body varies considerably, and seems to follow no constant rule; thus we find it very great in the long-legged Waders and Struthious birds, as if to enable their heads to reach the ground without flexure of their limbs; but it is equally great in the Swan and in some other short-legged aquatic birds, in which its purpose is obviously to allow the bill to be plunged deep into the water in search of food, whilst the body is floating upon the surface. The tail, on the other hand, is always very short, or even almost rudimentary; and it serves no other purpose than to support the tail-feathers, by whose agency the birds of active flight are steered (as by a rudder) in their rapid course through the air.—In each of the vertebræ of the neck of Birds, whose number ranges from 9 to 23, we find the rudiments of all the principal elements of the neural, hæmal, and lateral arches, consolidated into a single bone, and including four canals, — the superior, or neural, for the passage of the spinal cord; the inferior, or hæmal, for the passage of the two carotid arteries, here brought together on the median line; and the two lateral for the passage of the vertebral arteries, whilst they also enclose the trunks of the sympathetic nerve. The articulating surfaces of these vertebræ are connected, like those of other joints, by regular synovial capsules, the bodies of the vertebræ not being held together by intervertebral fibro-cartilage; so that great mobility is provided for, whilst the requisite tenacity is given by the numerous muscles and ligaments which connect the vertebræ with each other. In the posterior part of the cervical region, we find the pleurapophyses elongating, and becoming detached from the 'transverse processes,' so as to constitute 'cervical ribs'; and in some Birds one or two pairs of such ribs attain considerable length. The proper *dorsal* portion of the spinal column, which is considered to commence with the first vertebra whose ribs are attached to the sternum, presents a remarkable contrast to the cervical in regard to

FIG. 184.

Skeleton of *Vulture*; — *a*, cervical vertebræ; *b*, clavicle: *c*, carpus; *d*, phalanges; *e*, sternum; *f*, tibia; *g*, tarsus; *h*, bones of fore-arm; *i*, humerus; *k*, sacral vertebræ; *l*, femur; *m*, caudal vertebræ.

the mode of connection of its component segments; for the dorsal vertebræ, which never exceed 11, and are commonly no more than 7 or 8 in number, are so united as to provide for the greatest fixity which is compatible with the degree of mobility that the osseous framework is required to possess for respiratory and other purposes. This is especially seen in the Birds of most rapid and powerful flight, in which the bodies of a part of the dorsal vertebræ are 'anchylosed' (or united by continuous ossification) to each other; some one or more of the articulations, however, being usually left free, so that the body can be inflected from side to side, which is especially necessary in such as seize their prey whilst on the wing. But the union of the bodies of the vertebræ is not the only mode of consolidation adopted; for the spinous processes (neural spines), also, are frequently anchylosed to each other; whilst the transverse processes (diapophyses) send off splint-like prolongations, which overlap those of the contiguous vertebræ. In the Struthious birds and Penguins, on the other hand, which cannot fly, all the dorsal vertebræ are moveable one upon another. The anterior cervical ribs are connected only with the diapophyses (Fig. 167, *d*), by an articulation corresponding with that of the 'tubercle' of the human ribs; but as we pass backwards towards the thoracic region, each rib is found to become furnished with a distinct 'head,' which is articulated with a rudimentary parapophysis (Fig. 167, *p*) on the body of the vertebra. The spinal ribs, or pleurapophyses, are connected with the sternum, not by cartilaginous ribs as in Mammalia, but by true osseous sternal ribs; which are regularly articulated at the one end with the spinal ribs, and at the other with the sternum, by synovial capsules and smooth inter-articular cartilages, so as to form a perfect joint, allowing the free dilatation and contraction of the thorax. The costal portion of the thoracic framework is further strengthened by a peculiar disposition of the 'diverging appendages' (Fig. 167 B, *a*) which the pleurapophyses here possess; for by these splint-like bones, a connection is established between the successive pairs of true ribs (seen most perfectly in the Raptorial birds), the diverging appendage of each rib projecting from its posterior margin, and overlapping the rib behind it, to which it is attached by fibrous ligaments. These 'diverging appendages' are not mere 'processes' shooting forth from the ribs, but are developed from independent centres; and retain their individuality through life in some, as the Apteryx and Penguin, in which they are even provided with proper muscles for their elevation and depression; so that their relation to the limbs, through such rudimental forms of the latter as are presented by the Lepidosiren, Proteus, &c. (§ 320 *l*), becomes apparent. The sternum (Fig. 183, *e*) is more remarkably developed in Birds than in any other class of Vertebrated animals,* in accordance with the enormous size of the muscles for the depression of the wings, to which it furnishes the chief attachment. Its peculiar character is given to it, not merely by such an enormous increase in its length and breadth that it covers the ventral surface of a great part of the abdominal as well as of the thoracic cavity (extending in some cases even to the pubis), but also by the projection of

* We have seen that of the 'plastron' of the Chelonian Reptiles, which is commonly described as a sternum whose elements are extended laterally, only a small part is constituted by the true sternum, the remainder being made up of sternal ribs and dermal bones (§ 324 *h*).

its central portion as a keel or ridge, the degree of prominence of which bears a regular proportion to the strength of the pectoral muscles, and consequently to the power of flight. This is well seen in those exceptional forms, whose wings are merely rudimentary; for the Ostrich, Emeu, &c., have a comparatively small sternum, entirely destitute of keel. The ribs are developed (except in some aquatic birds in which the last pair does not present itself) from all the vertebræ intervening between the scapular and pelvic arches; although the last pair is seldom continuously connected with the sternal ribs and sternum. Hence all these vertebræ rank as 'dorsal,' and the usual 'lumbar' region would seem to be wanting in Birds: but it has been pointed out by Prof. Owen, that we are probably to regard as *lumbar* vertebræ those that form the anterior part of the sacrum, and are connected at their sides with the ilia; which bones are prolonged very far forwards, and are connected with a far larger number of segments than we elsewhere find giving support to them. The vertebræ which are anchylosed together to form the *sacrum* of Birds (Fig. 183, *k*) are scarcely ever fewer than 10 in number, and are sometimes as many as 19; the greatest number presenting itself, as might be expected, in Birds whose posterior extremities are those chiefly concerned in the propulsion and support of the body.—The *caudal* portion of the spinal column (Fig. 183, *m*) is made up of from 7 to 9 vertebræ, gradually diminishing in size from before backwards, but terminating in a singularly formed bone, like the body of a vertebra in size, but compressed laterally so as to form a sharp edge above and sometimes below. It is with this bone that the tail feathers are more particularly connected, whose function is the important one of guiding the bird in its movements through the air.

325 *b*. The *scapular* arch is still formed upon the type which we have seen to prevail through the Oviparous Vertebrata in general; for the scapula, a long narrow bone lying upon the thoracic ribs, is chiefly connected with the sternum by the strong coracoid bone, which is firmly articulated to the scapula where it helps to form the glenoid cavity, and with the sternum at its opposite extremity: whilst the clavicles, united with each other on the central line so as to form a single bone (known as the *furcula* or 'merry-thought'), constitute an independent arch, which is much less closely connected with the shoulder-joint, and of which the centre or keystone is connected either by ligament or by osseous junction with the summit of the sternal keel. It is interesting to remark that, in the Ostrich tribe, the two clavicles are separated from one another at their median extremities, but are anchylosed to the scapula; a condition which exists in the Turtles, and thus indicates a certain relationship, however slight, between these two groups. The clavicles are sometimes entirely wanting, or are merely rudimentary, especially in certain birds which live much upon the ground, or are climbers among trees, having little power of active flight; whilst they are strongest, and their arch most open, in birds of great force of wing, in which their function is to oppose the forces that would tend to drag the shoulder-joint towards the median plane, in the downward stroke of the wing. Notwithstanding the modification which the anterior extremity has externally undergone, in order to convert it into an instrument of flight, it is not difficult to recognise its osseous elements as corresponding to those of other Vertebrata. The

humerus (Fig. 185, *a*) varies considerably in its relative length, even in birds of great powers of flight, being short in the Swifts and Humming-birds, whilst it is two-thirds the length of the body in the Albatross and other sea-birds. At the elbow-joint (I), we sometimes find a detached 'sesamoid bone,' analogous to the 'patella' of the lower extremity; this in Mammalia is usually anchylosed to the ulna, of which it forms the 'olecranon process.' The two bones of the fore-arm, the ulna and radius, are sometimes nearly equal in size, but where there is a marked difference the ulna (*b*) is always the strongest; they are so connected with each other and with the humerus, that scarcely any movement of rotation is permitted in the fore-arm, the firm hinge-like motion of the elbow-joint being what is alone required in the act of flight. The bones of the hand are extended in length, but drawn together (as it were) laterally; and the movements of the wrist-joint (II) are limited to the hinge-like action requisite for the folding and extension of the wing. The carpal bones are only two in number, and these support two principal metacarpal bones (*c*) anchylosed together at both extremities, with another separate bone (*o*) which is to be regarded as a rudimentary thumb. Of the digits, only one (1) is considerably developed, possessing two or even three phalanges, which are united together and to the metacarpal bones by strong ligaments, so as to be nearly inflexible; but rudiments of another digit, or even of more, are generally discoverable. In the *Struthious* birds, the bones of the anterior extremity are reduced to their smallest dimensions; still affording support in most cases to wing-feathers, and being moved by muscles which enable these rudimentary wings to give some assistance to the animals in their rapid course over the ground; whilst in the *Apteryx* they do not appear to give any assistance in the act of locomotion, but serve only to support a horny spur which it uses as a defensive weapon.—The *pelvic* arch is not merely remarkable for the great number of sacral vertebræ which are anchylosed together to afford it support; but it contrasts very strongly with the scapular arch in this respect,—that whereas in the latter the greatest strength is found at its centre or keystone, where resisting power is most required, in the former there is an incompleteness at that very part, neither the ischial nor the pubic elements of the two halves of the pelvic arch meeting on the median line. This incompleteness seems to have reference to the relatively larger size of the eggs of Birds; since to allow their passage through a complete bony ring, would have involved an

FIG. 185.

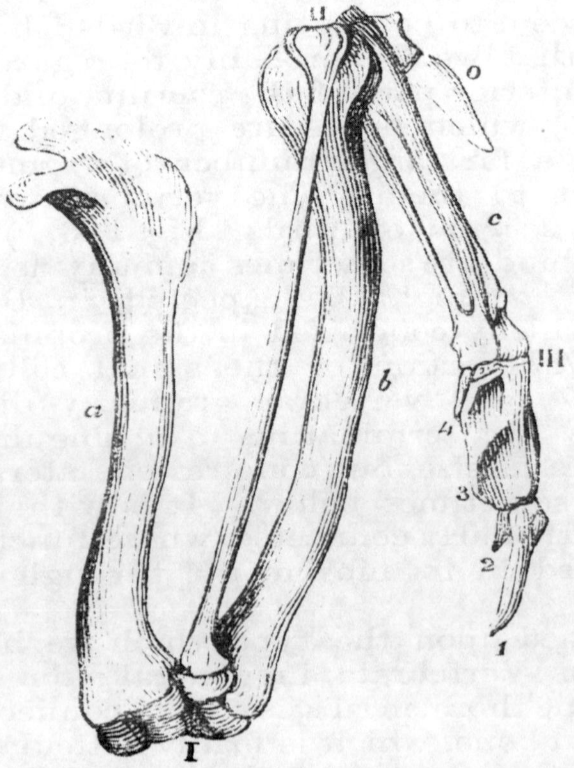

Bones of the wing of *Falco gyrfalco* (Jerfalcon);—*a*, humerus; I, elbow-joint; *b*, ulna; II, wrist-joint; *c*, metacarpus; *o*, rudimentary thumb; III, knuckle-joint; 1, 2, 3, 4, rudimentary phalanges of fingers.

increase in its dimensions, which would have been unsuitable to the general conformation of the posterior part of the trunk in birds of flight. And it would seem as if the elongation of the sacrum was intended, by giving a firmer attachment to the corresponding elongated ilium, to afford the requisite fixation to the hip-joint, in default of the support which it usually receives from below. The ilium is not merely very much lengthened, but it is even anchylosed to the sacral vertebræ; and as the ischial and pubic bones are anchylosed to it (forming an 'os innominatum' as in Man), the entire pelvis presents a remarkable degree of firmness, notwithstanding the incompleteness of its descending arch. This consolidation of the pelvis takes place to the most extraordinary extent in the *Struthionidæ*, in which the iliac bones arch over and meet each other on the dorsal surface of the sacrum, whilst in some instances the ischiac bones do the same on its ventral surface, so that the sacrum is enclosed between a pair of bony walls formed by the ossa innominata. In the Ostrich, too, we find the under side of the pelvic arch completed by the slender pubic bones, which arch downwards towards one another, and meet on the median line; this junction of the pubic bones, whilst the ischial remain separate, is a decidedly Mammalian character. In the conformation of the lower extremities, we have chiefly to notice that the femur (Fig. 184, *l*) remains short, even when the general elongation of the limb is the greatest; this elongation being given by the extraordinary development of the tibia and metatarsus. The fibula is rarely present as a separate bone, being usually anchylosed with the tibia (*f*), and sometimes being almost undistinguishable. The tarsal bones are anchylosed with the metatarsal at an early period in the life of most birds, though rudiments of them may often be subsequently detected. The metatarsus is represented in most Birds by a single bone (*g*) in each leg; but this bone is usually grooved, in such a manner as to show that it is made up of three bones fused together; and these bones actually remain separate in the Penguin through a considerable part of their length. The bones of the foot are usually developed considerably, so as to afford to the bird the extent of support necessary for the maintenance of its equilibrium; in those which inhabit marshy places, the toes are not only elongated, but radiate widely from each other, so as to render these birds less likely to sink into the oozy ground on which they tread; whilst in the aquatic birds, they support the web by whose agency the body is propelled along the surface of the water. The typical number of true digits is four; but in Gallinaceous Birds we find the metatarsal bone bearing a 'spur' on its inner side, which is perhaps to be regarded as the rudiment of an additional toe; whilst in the Ostrich, on the other hand, the number of toes is reduced to two. The number of phalanges in these digits increases regularly from within outwards; being *two* in the innermost of the ordinary toes, which ranks as the first; *three* in the second; *four* in the third; and *five* in the fourth or outermost. Of the four moveable toes possessed by all save the Struthious Birds, the innermost is turned backwards, while the other three are usually directed forwards; the prehensile power of the foot is increased, however, in the *Scansores* (Parrot and Woodpecker tribe) by the opposition of two of the toes to the other two.

325 *c*. In the structure of the *Cranium* of Birds, we trace, more than in any other part of their Osteology, the striking contrast between this

group and the preceding; for whilst the skull of Reptiles is characterized by the number of pieces of which it remains permanently composed, that of Birds is remarkable for the closeness of the union which is early established between the different parts; in this respect differing as much from Mammalia, as Mammalia differ from Reptiles. If the skull of a Bird be examined at a sufficiently early period, the several elements of the neural arches of its cranial vertebræ may be readily distinguished; but the ossifying process advances so rapidly, that the sutures soon become obliterated, and the whole brain-case seems formed of but one piece,—the tympanic portion of the temporal bone, which forms a part of the maxillary arch, being the only element that remains detached. The bones of the face, however, do not undergo the same consolidation; for they are always so connected with those of the cranium, as to possess considerable mobility; and their union amongst each other is seldom complete. In their number and relative position, they correspond with those of Mammalia; but they differ greatly in their forms and proportions. The principal part of the elongation of the upper jaw, constituting the *palato-maxillary* arch, is due to the great development of the intermaxillary bone (hæmal spine), of which the two halves usually coalesce at an early period of development; with this, the slender maxillaries and the palatine bones are connected by bony union; and the three pairs together constitute the most anterior of the hæmal arches (the palato-maxillary), which is in great part detached from the other bones of the face, and is so connected by the malar or zygomatic bone with the tympanic or 'os quadratum,' that the movements of the latter are in some degree transferred to it. This tympanic bone, as in all the Vertebrata already considered, is interposed between the cranium and the lower jaw; being, in fact, the pleurapophysis of the frontal vertebra, with which the proper hæmal arch that constitutes the jaw is articulated. In Mammalia it is always anchylosed with the temporal bone; and thus its presence as a separate element is an invariable mark of the Oviparous cranium. It is curious to observe, in the Rodent Mammalia, among several other points of degradation towards Ovipara, that the tympanic bone long remains detached from the temporal, although it does not enter into its articulation with the lower jaw; whilst, on the other hand, in some of the Struthionidæ, the tympanic bone is so fixed that it can add very little by its own movements to the mobility of the jaws, thus presenting an approximation to the Mammalian type. The lower jaw, which, with the tympanic bone, constitutes the *temporo-mandibular* arch, is originally composed, like that of Reptiles, of six pieces on each side, making twelve in all; but the number of these is soon diminished by their mutual coalescence. This begins in the two pieces, one on each side of the median line, which form the body of the jaw; for these, like the intermaxillaries of the upper jaw which they represent in their own segment (being, like them, hæmal spines) coalesce completely at the symphysis; and an union of other portions follows, until there are only two pieces left on either side of the jaw, and these are very firmly connected by suture with each other and with the central or symphyseal element. The traces of the original separation of these parts remain longest in those aquatic birds which show the closest approximation to the Reptilian type; whilst, on the other hand, the sutures themselves are partly obliterated by complete osseous union in the Struthionidæ. The *hyoidean* arch, which was found to be so highly developed among Fishes in connection with their branchial apparatus, and which is so nearly rudimentary in Reptiles, again rises into importance in this class, in which it is subservient to the actions of the complicated laryngeal apparatus. It is considered by Prof. Owen, however, that a large part of what is usually reckoned as the os hyoides of Birds really belongs to the visceral skeleton; for we find, in addition to the small cerato-hyals (Fig. 170, 40) and basi-hyal (41), which form the hæmapophyses and hæmal spine, with the glosso-hyal (42), projecting forwards, and the uro-hyal (43) passing backwards, a pair of long branches, analogous to the 'great cornua' of the human hyoid (Fig. 191, 2), but very much more elongated, especially in the Humming-birds and Woodpeckers, in which they pass round the back and top of the head, to be inserted into a pair of canals on the sides of the upper mandible (§ 325 *i*). These are stated by Prof. Owen to consist of retained portions of the branchial arches, which appear at an early period in the development of the Bird, and of which every other portion speedily vanishes. On the other hand, the stylo-hyal and epi-hyal elements (together making up the pleurapophyses of this segment) are not developed into bone, but remain ligamentous. The skull of Birds is articulated to the spinal column by a single tubercle, which projects from the occiput, and is received into a hollow, on the body of the 'atlas' or first cervical vertebra. This articulation allows much greater extent and freedom of motion to the head, than exists in the Mammalia. In the Penguin, however, part of the lateral or condyloid portions of the occipital bone are included with the basilar in the articulating tubercle, as in the Chelonian Reptiles; whilst in the Ostrich, there is a vertical fissure in the tubercle, which indicates an approach to that separation of the articulating surface unto two halves, which is characteristic of the Mammalia.

325*d*. The covering of feathers, which so remarkably distinguishes Birds from all other Vertebrated animals, constitutes their sole dermal skeleton; with the exception of the horny scales that invest the lower part of the legs and feet, and the horny sheath which encases the bony jaws, and forms the 'bill.' This last is developed into various forms, in accordance with the uses to which it is to be applied by the Bird in the search for, or the prehension of, its appropriate food; and accordingly its shape is so far an index of the general habits of the species, that it furnishes characters of great importance to systematic Ornithologists. These, however, need not here detain us; it being sufficient for our present purpose to know that the horny bill serves the purpose of teeth, which it replaces in this group (as in Chelonian Reptiles, § 324 *i*, *o*), in seizing, cutting and tearing the food; but the act of trituration or mastication it is entirely unable to perform. In the Parrot tribes, the strong hooked beak is also used as an instrument for locomotion; the bird frequently raising itself in climbing, by employing it like a third foot.

325 *e*. As might be expected from the analogy of Birds with Insects, the development of their organs of nutrition (excepting that of the respiratory organs) is much less striking in point of size, than is that of the locomotive apparatus; and we find the abdominal portion of the visceral cavity small in size in comparison with the bulk of the body. But what is wanting in size is made up in activity of function; and the demand for

food is more active in Birds than in any other animals. As their food is not subjected to any mechanical reducing process within the cavity of the mouth, the salivary secretion (except in some special cases) is only required for its lubrication previously to being swallowed. Accordingly, the salivary glands are but little developed; and it is interesting to observe that in certain Birds the secreting follicles constitute part of the tongue itself, as in some of the Reptilian class. In nearly all the members of this class, the food is delayed in a receptacle formed by the dilatation of some part of the mouth, pharynx, or œsophagus, before it passes into the stomach; and in this receptacle it sometimes undergoes a sort of preliminary digestion. In the Pelican we find it in the form of a wide pouch suspended from the lower jaw, the two halves of which bone are not united at the symphysis, so that the pouch is enormously dilatable; in the Swift, and other birds that catch insects on the wing, a similar distensibility exists in the membranous wall of the 'fauces,' or upper part of the throat; in the Cormorant and other fishing birds, it is the wide œsophagus which serves as the receptacle; but in the granivorous birds, as well as in many others which take in a large quantity of food at once, we find a pouch, termed the 'crop' (Fig. 186, *b*), developed from the side of the gullet, into which the food passes when it is first swallowed, and in which it lies for some time; during which it is moistened by a secretion copiously poured out from glandular follicles in its walls, and is thus prepared for the further stages of the digestive process. Below the crop, the œsophagus is usually again dilated, before its termination in the gizzard, into another cavity, known as the 'proventriculus' (*c*), from the walls of which, the true gastric fluid is poured out upon the food that is delayed in it; this may be considered as the first or cardiac portion of the true gastric cavity; whilst the 'gizzard' (*d*) or muscular sac which succeeds it, corresponds with its second or pyloric portion. The gizzard is usually a somewhat lengthened sac, having its intestinal orifice, as well as that by which it communicates with the proventriculus, at its upper part. In those whose food requires the aid of mechanical reduction for its solution, the walls of the gizzard are very thick and muscular, and its fibres radiate from two central tendons situated on opposite sides of the cavity, which is lined by a horny epithelium of remarkable toughness. The triturating power of the gizzard is frequently much increased by the presence of hard angular stones, which the bird occasionally swallows, and which are retained in its cavity. In Birds, however, whose food consists of flesh, fish, succulent fruits, or other substances easily reduced, the muscular structure is so little developed, that the ordinary character of the 'gizzard' is entirely wanting. The intestinal tube varies considerably in length, and also in the degree of definiteness of the division between the small and the large intestines. At no great distance from the stomach, it receives the secretions of the liver (*e*) and of the pancreas (*h*); the former gland, which is of moderate size, possesses a gall-bladder (*f*), in which the bile can be stored up when not required for use, and yet has a second excretory duct (*g*) by which the bile can be poured direct into the intestinal tube; the latter, which is now much increased in size, and has attained a higher grade of glandular development, still pours forth its products, sometimes by two ducts, in other cases by three. After a larger or smaller number of convolutions, the intestinal tube terminates in the cloaca (*o*); first, however, receiving the products secreted from one or two cæcal appendages (*k*), whose opening into the canal is considered to mark the boundary between the small and the large intestines (*l* and *l′*). — The *Absorbent* system of Birds, although not so extensive in appearance as that of Reptiles, is really developed upon a higher type, since we find its vessels not merely provided with valves, but in some parts possessed of the so-called 'glandulæ,' which may be more compared to knots formed by the convolution of the tube, and in which (as will be shown hereafter, CHAP. XI.) it is probable that the elaborating function of this apparatus is performed with peculiar activity. The greater part of the fluid taken up by the Absorbents is collected into two dorsal trunks, or 'thoracic ducts,' which pour their secretion into the veins of the neck, as in Man; but we do not yet lose either the 'lymphatic hearts' of Reptiles, or the communications of the absorbent system with the venous in the posterior part of the body. The Spleen is of very small size.

FIG. 186.

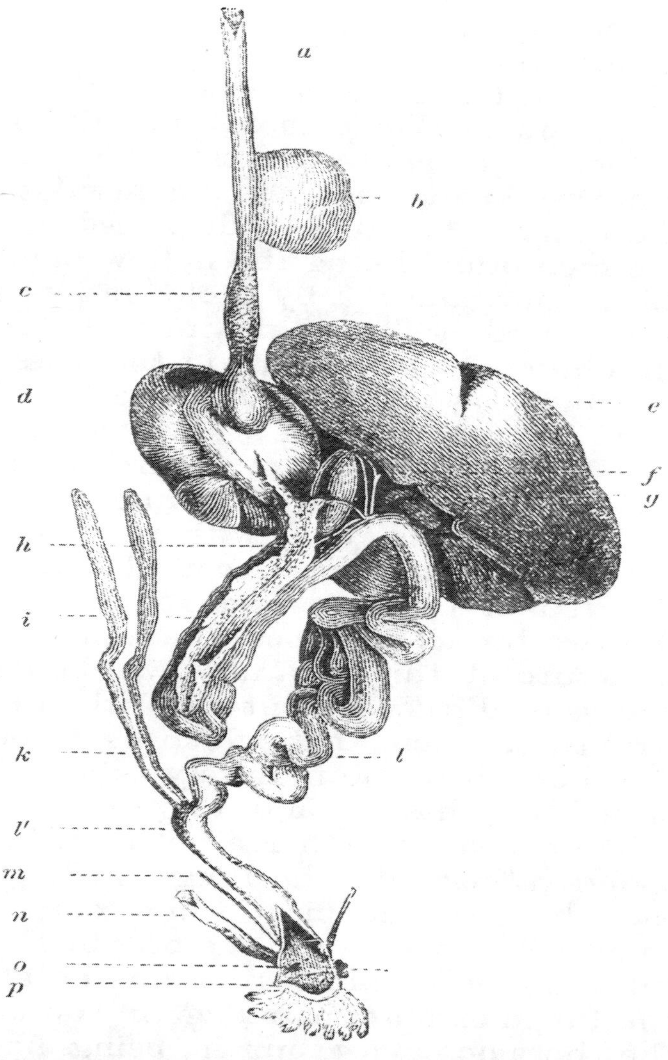

Digestive apparatus of *Fowl*:—*a*, œsophagus; *b*, crop; *c*, ventriculus succenturiatus; *d*, gizzard; *e*, liver; *f*, gall-bladder; *g*, hepatic ducts; *h*, pancreas; *i*, duodenum; *k*, cæcum; *l*, small intestine; *l′*, large intestine; *m*, ureter; *n*, oviduct; *o*, cloaca; *p*, anus.

325*f*. The Circulating apparatus of Birds is formed upon what is called the 'complete double' type; the pulmonary and systemic circulations being entirely distinct from each other, each having an organ of impulsion at its commencement; and the relation of the two being such, that every part of the blood which has passed through the system must be transmitted through the lungs, before being again sent into the body. Thus we have the *systemic* heart of the Mollusca superadded (so to speak), to the *respiratory* heart of Fishes; a superaddition of which we saw some traces in the Cephalopoda, whose systemic and respiratory hearts, however, were detached from each other (§ 307*b*). In the warm-blooded Vertebrata, the pulmonic and systemic hearts are brought together, in accordance with that principle of concentration which elsewhere prevails, so as to form but one organ; and this contains two auricles or receiving cavities, and two ventricles or propelling cavities (Fig. 192). The blood returning from the system is discharged by the vena cava into the auricle of what may be termed the *pulmonic* *heart,* which is situated on the right side; from this it is emptied into the ventricle, and by its contraction is propelled, through the pulmonary artery, to the capillaries of the lungs. After undergoing aeration in those organs, it is returned by the pulmonary veins to the auricle of the left or *systemic* heart; and being discharged by it into the ventricle, it is propelled by the latter along the great systemic trunk or aorta, by the ramifications of which it is carried into the capillary network of the whole body, to be returned again by the veins. Now it is peculiar to Birds among Vertebrated animals, that the exposure of the blood to the influence of the atmosphere takes place, not merely during the pulmonary, but also during the systemic circulation; for the vessels which ramify on the walls of the air-sacs dispersed over different parts of the body, and on the membrane lining the hollow bones, are supplied from the neighbouring trunks, and not by prolongations of the pulmonary system. Thus we are carried back to that intimate union of the systemic and respiratory circulations, which exists in the class of Insects (§ 315 *a*).—The *trachea* or windpipe, by which the air receives admission to the lungs, is furnished at its upper part in Birds with a 'larynx,' which is obviously the equivalent, both in position and structure, of that of Mammalia; this larynx, however, seems rather to be connected with the regulation of the supply of air to the lungs, than with the production of vocal tones; for the organ of voice in Birds is a second larynx, which is peculiar to the class, and is situated at the lower end of the trachea, just above its bifurcation into the two bronchial tubes which proceed to the right and left lungs. The structure of this apparatus will be described hereafter (CHAP. XXII). The *Lungs* of Birds present us with an immense advance in their type of structure, upon that which is to be found even amongst the highest Reptiles; being so minutely subdivided, as entirely to lose the character of hollow air-sacs, and to present instead a somewhat spongy texture. Still we shall find hereafter, that, in their minute structure, they correspond rather with the lungs of Reptiles than with those of Mammals; each lung of the bird being composed, as it were, of an aggregation of small 'lunglets,' every one of which is furnished with its own bronchial tube, and is a miniature representative of the lung of the Frog. From the lungs of Mammalia, again, those of Birds differ in not lying free in the thoracic cavity, but in being bound down on their dorsal aspect to its walls, as in many Reptiles; and the thoracic cavity itself differs from that of Mammals in this essential point, that it is not closed-in below by a complete diaphragm, but is only separated from the abdominal cavity by the serous sacs which respectively line the two, and by the partial extension of the tendinous expansion of the imperfect diaphragm over the bases of the lungs, on which it loses itself. The pleural sac which surrounds the lungs is itself made part of the respiratory surface, the air being admitted into its cavity by bronchial tubes (Fig. 187, *b*), whose open orifices (*d*) may be seen upon the surface of the lungs; and from the pleural sac it passes into the peritoneal, through perforations in their walls which also pierce the diaphragm. It is by reflected prolongations of these membranes, that the 'air-cells' are formed, which occupy, when distended, a large part of the visceral cavity. These communicate freely with each other, with the cavity of the long bones, and with a set of tegumentary air-cells, which seem to take the place of the 'bursæ mucosæ' that are elsewhere found to be interposed among the muscles and tendons, for the purpose of preventing them from pressing unduly upon one another. Several mechanical advantages seem to be gained by the peculiar arrangement of these air-cells, which is adapted in each case to the particular wants and habits of the species; but such uses are all subservient to the essential purpose of these cavities, which is to contain a supply of air that may at once diminish the specific gravity of the body and contribute to the aeration of the blood. The mechanism by which the interchange of the air in the lungs and visceral air-cells is effected, is as peculiar as are the provisions we have already noticed for the energetic performance of this function. From the elasticity of the bony framework by which the thoracico-abdominal cavity is surrounded, the state of fulness is natural to it, and that of emptiness is forced. When the lungs and air-cells are filled with air, they are partly emptied by the agency of muscles which draw the sternum nearer to the spinal column, and which consequently diminish the capacity of the visceral cavity; but when these muscles are no longer caused to contract, the sternum springs outwards again, the cavity is restored to its original dimensions, and the air rushes in to fill the lungs and air-cells. The distension of the lungs is further aided by the contraction of the diaphragm, which, by its attachment to their bases, draws these downwards in the act of inspiration. This muscle is more developed in the *Struthionidæ,* than in any other birds; and it is most interesting to see how closely, as regards the respiratory functions, some of this tribe approximate to Mammals. The small sternum of these birds cannot take the same share in the process of filling and emptying their pulmonic cavities, that it performs in the rest of the class; and some other provision is requisite. This we find most complete in the *Apteryx;* in which the abdominal cavity is entirely separated from the thoracic by a musculo-tendinous diaphragm, capable by its contraction of enlarging the capacity of the latter, as in Mammalia; and although the air passes into the pleural cavity, it does not proceed elsewhere, neither into the abdominal cavity, nor into the bones, which are occupied by marrow; so that, although the *amount* of respiration in this bird is doubtless much less than is usual in the class, the function is performed upon a plan which shows a decided advance towards the Mammalia.—It

FIG. 187.

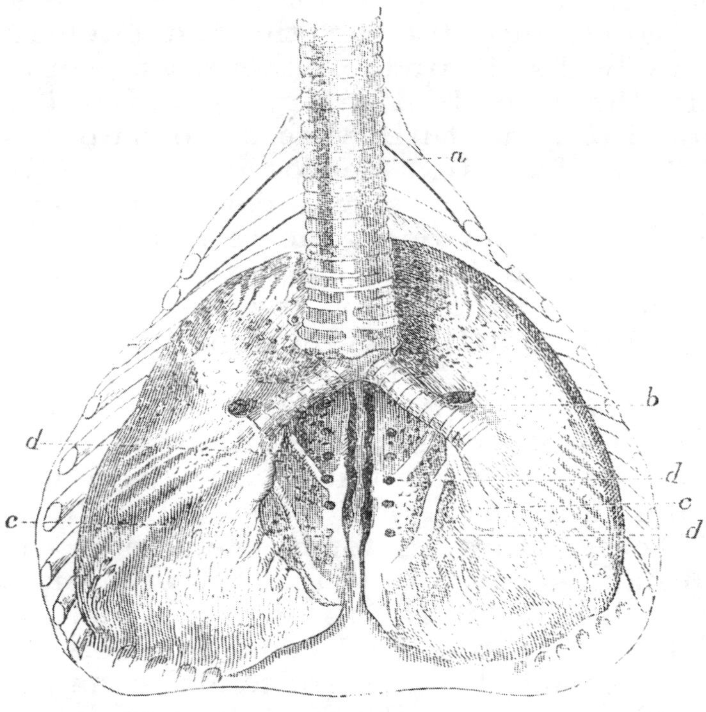

Pulmonary apparatus of a *Pigeon*, as seen on removing the anterior wall of the thorax;—*a*, trachea; *b*, bronchi; *c*, lungs; *d*, apertures of communication with air-cells.

is interesting to observe that a more exact symmetry prevails in the arrangement of the circulating and respiratory apparatus of Birds, than we find in the Mammalia; and this is evidently connected with the necessity involved in the peculiar mode of locomotion of the former, of an exact balance between the weight of the two sides of the body, and of a perfect equality in their functional power. The lungs and air-sacs are precisely similar in size and situation on the two sides; consequently the heart is placed on the median plane; and the mode of origin, from the aortic arch, of the trunks supplying the head and upper extremities, is alike on the two sides,—there being on each an 'arteria innominata,' that subdivides into the carotid and subclavian arteries, the latter of which is chiefly distributed on the great pectoral muscles, the vessel that represents the 'brachial artery' of Man being quite a subordinate branch instead of constituting the main trunk.

325 *g.* The great nervo-muscular activity of Birds, which involves the necessity for a large and constant supply of food, for a rapid circulation, and for an energetic respiration, also requires that ample provision shall be made for carrying off that part of the *waste* of the system, which it is not the function of the liver and lungs to eliminate; and we accordingly find that the Kidneys of birds are very large, and their functional activity great. A large part of these organs lies within the pelvic cavity, where they occupy deep depressions on the internal surface of the prolonged ilia; but they extend anteriorly as far as the lungs. The urinary secretion, which contains very little water, but is loaded with urate of ammonia, is discharged from each kidney through several excretory ducts; but these unite to form the ureter on each side, and the two ureters terminate separately (Fig. 186, *m*), very near to the orifices of the genital ducts, in a portion of the cloaca which has hence received the designation of the 'urethro-sexual canal.' Between the termination of the large intestine and this urethro-sexual canal, the cloaca generally exhibits more or less of a pouch-like dilatation, which is bounded, both above and below, by valvular folds; and this, which is obviously the rudiment of a urinary bladder, attains its greatest development in the Ostrich and its allies, thus presenting another character of approximation to the Mammalia.—The skin would not appear to contain, in this class, any large amount of excreting glandulæ. It is obvious that there will be little need of cutaneous exhalation, as a means of keeping down the temperature of the body, where every means is adopted to maintain this at a high fixed standard; whilst, again, as Birds drink so little that even their urinary secretion is almost solid, there is but little superfluous water in their bodies to be thus eliminated. One superficial gland, however, known as the *Glandula Uropygii,* which nearly all Birds possess, deserves special mention; this is situated above the last caudal vertebra, and its office is to secrete an oily fluid, which is applied to the feathers by the bill, and thus preserves them from being wetted by water. However short may be the neck of Birds, it is always long enough to enable the bill to reach this gland, and to carry its product to every part of the surface; and, as might be expected, the gland is the largest, and its secretion most abundant, in the aquatic Birds, whose plumage is thus enabled to keep the surrounding medium from contact with the skin, which would otherwise be chilled by it.

325 *h.* The condition of the *Nervous System* in Birds shows a decided advance upon the Reptilian type, in the larger proportion which the Encephalic centres bear to the Spinal, and in the greater development of the Cerebrum and Cerebellum in relation to that of the Sensory Ganglia. The optic ganglia, however, still form a very conspicuous part of the brain, the cerebral hemispheres not extending sufficiently far back to cover them completely when viewed from above; and we find that the *instinctive* propensities still take the lead in prompting and guiding the actions of the animal, although they are often modified (in a manner and degree which strikingly contrast with the *fixity* of the methods followed by Insects) by the operations of their superior *intelligence,* in which we can indubitably trace an adaptation of means to ends which springs from *their own* design or purpose, not being provided for in the ordinary instinctive habits of the species. The proportionate development of this intelligence, and of the instinctive propensities, varies considerably in the different orders and families of Birds; and it is interesting to remark that whilst the latter predominate among the *Insessores,* which may be regarded as the *typical* Birds—possessing in the highest degree the distinctive attributes of the class,—the former is most strikingly manifested in the *Parrot* tribe, which are not only (as every one knows) the most *educable* of all Birds, but which are pre-eminently distinguished for the *hand-like* use which they make of the inferior extremity (§ 325 *b*). This variety in the relative development of the intelligence in Birds, bears a close correspondence with the relative development of the *Cerebrum.* On the other hand, the development of the *Cerebellum* seems to bear a like conformity to the variety and complexity of the movements habitual to different tribes of Birds. The relative development of the anterior and posterior ganglionic enlargements of the Spinal Cord, again, bears a constant proportion to the relative use that is made of them in the act of locomotion. In Birds whose power of wing is very great, and the use made of the legs and feet very small, except for support in 'perching,' the anterior enlargement equals the posterior in size; and would probably exceed it, if it were not that the posterior enlargement supplies not only the legs but the tail, which is an important accessory organ of flight. In the Struthious birds, on the contrary, whose progression is almost entirely performed by the legs, the posterior enlargement is by far the greatest; and an approach to this extreme inequality is presented by the wading and aquatic birds, in which these members take a large share in the general movements.—The decided inferiority in the nervous system of Birds to that of Mammals, is marked, not only by the relatively low development of the Cerebrum and Cerebellum, but by the deficiency of those *commissural* bands, which unite the lateral halves of the nervous centres on the median line. Thus in the Cerebrum we have no trace of the *corpus callosum,* whilst in the Cerebellum we find no *pons varolii;* and at the lower extremity of the Spinal Cord, there is a wide divergence between its columns, leaving a space (termed the *sinus rhomboidalis*) that resembles the fourth ventricle in the medulla oblongata (§ 324 *o*).—The Sympathetic system of nerves in Birds bears a close resemblance to that of Mammalia; its chief peculiarity consisting in the lodgment of its cervical chain of ganglia, along with the vertebral arteries, in the canals formed by the lateral arches of the cervical vertebræ (§ 325 *a*).

325 *i.* The class of Birds is as much distinguished by the acuteness of its sensations, as that of Reptiles is by the opposite condition. This is especially to be noticed in regard to the sense of *Sight,* which is that whereby the active locomotion of these animals must be especially guided. Accordingly we find the visual organs never absent, and generally of remarkably large size. The tough fibrous investment of the globe of the eye is very commonly strengthened in front by a circle of bony plates, usually from twelve to fifteen in number; and among other purposes which these plates appear to serve, is that of giving a fixed point of action to a peculiar muscular apparatus, by which the adaptation of the eye to distances is effected. The eye of Birds is also remarkable for the presence of a peculiar body, termed the 'pecten' or 'marsupium,' which projects forwards from the fissure of entrance of the optic nerve, into the vitreous humour; it is chiefly composed of a layer of blood-vessels, constituting an 'erectile tissue,' folded in numerous plications, the size, form, and number of which vary considerably in different Birds, even such as are otherwise closely allied; and the whole is covered by a continuation of the black pigment which lines the choroid. The uses of this organ are not certainly known; it may, perhaps, serve more than one purpose, but its action seems not improbably to be connected with the adaptation of the eye to distances, the relative positions of the crystalline lens and of the retina being altered when its vessels are distended with blood. The *Apteryx* is the only bird, in whose eye the marsupium is known to be absent. In Birds, as in the higher Reptiles, the front of the eye is protected by three eyelids; of these, the third, or 'nictitating membrane,' is thin and semitransparent, and is employed not so much for the exclusion of light, as for sweeping off impurities from the surface of the eye, being continually drawn rapidly across it by a special muscular apparatus; whilst of the two which move vertically, the under one is the largest, and is alone furnished with a tarsal cartilage.—The organ of *Hearing* is not yet marked, except in a few of the Owl kind, by any trace of those external appendages, which, in most of the Mammalia, constitute such important features in the general appearance of the head; in other respects, however, it presents a decided advance upon the Reptilian type, the tympanic apparatus being more complete, and the rudiment of a cochlea being considerably developed, the otolithes being almost entirely absent, and the surface upon which the auditory nerve is distributed considerably augmented.—The Nasal cavity of Birds is of considerable size, and its lining membrane is greatly extended by prolongation over the 'turbinated bones,' which project into its cavity. The degree of this extension depends, however, upon the amount of convolution of the turbinated bones, which varies considerably in different tribes of Birds; but it never seems to approach that which is presented in the corresponding part of those Mammals which are distinguished for their acute scent; and it seems probable, from experiments made for the purpose, that many of the actions commonly regarded as indicative of great acuteness in the olfactive sense (in *Vultures* more especially) are really to be attributed to the guidance of sight. In Birds, as in the preceding classes, the nostrils are mere perforations leading to the nasal cavity; and are not furnished with moveable cartilages and muscles for dilating and contracting their apertures, such as are possessed by most Mammalia.—

The sense of *Taste* appears to be but little developed in Birds. The tongue, as in the preceding class, seems to be organised rather for the capture and prehension of food, than for taking cognizance of its gustative qualities; and the full examination of these is further prevented by the circumstance, that the food is not retained in the mouth for mastication, but is swallowed as soon as it is received into the cavity. The tongue of most Birds is nearly or entirely destitute of true gustative papillæ, and is invested in a horny sheath; and it is by the prolongation of this sheath into filamentous processes, that the brush-like appendage is formed at its extremity, in the Lorikeets, which serves to take up the nectar of flowers upon which they feed. In the Humming-birds, the tongue presents a most peculiar conformation, which strongly brings to mind the *haustellium* of Lepidopterous Insects (§ 315); for it is composed of two musculo-fibrous tubular cylinders, which are adherent to each other (like the two barrels of a double-barrelled gun) until near their extremity, where they diverge, and open out into a pair of spoon-like terminations, well fitted to lick up honeyed exudations, which are transmitted from front to back of this tubular tongue by the contraction of its muscular walls, as well as to seize upon the small insects which form part of the food of these birds. In the Humming-birds, Woodpeckers, and some other birds, the central part of the hyoidal apparatus, upon which the tongue is fixed, is rendered very moveable by the extraordinary prolongation and flexibility of its lateral cornua (§ 325 *c*); and by a muscular arrangement specially adapted for the purpose, the tongue can be instantaneously projected out of the mouth to nearly the length of the body, and as suddenly drawn in.—The exercise of the sense of *Touch* must of course be limited in Birds, by the entire want of digital subdivision in the anterior extremities (which are elsewhere its chief instruments), and by the character of the investments of the skin, which must greatly diminish its power of receiving external impressions. The only part which would seem to possess any special tactile sensibility, is the soft skin with which the bill is covered in the Duck tribe, and in other birds which, like these, obtain their food by digging into soft mud; this skin is copiously supplied with branches of the fifth pair of nerves, and has a regular papillary structure.

325 *k*. The disposition of the *Muscular* system in Birds may be, for the most part, easily anticipated from what has been already said of their mode of locomotion, and of the conformation of their bony skeleton. In all Birds distinguished by their powers of flight, the pectoral muscles—attached at one end to the keel and expanded surface of the sternum, and at the other to the humerus, so as to draw the latter downwards by their contraction,—are those whose development chiefly attracts attention; and there are some instances in which the 'pectoralis major' alone weighs as much as all the rest of the muscles of the body put together. On the other hand, in proportion as the anterior extremities are reduced in size, and the posterior take a larger share in locomotion, do we find the pectoral muscles dwindling down, and those of the legs presenting the chief development; and this, of course, is most remarkably seen in the Struthious tribe. The muscles of those Birds which habitually rest on 'perches,' rather than on the ground, are so disposed as to cause the foot to grasp the perch during sleep, and thus to keep the bird in its position, without

any muscular effort; for besides those which ordinarily connect the thigh with the body, and put in motion the lower part of the member, we find one of the thigh-muscles (whether the representative of the *rectus femoris* or of the *gracilis* in Man is not yet determined) which arises from the superior part of the pubis, passing continuously over the convexity of the knee-joint, and then over the projection of the heel, into the foot, where it becomes connected with a flexor of the toes, in such a manner that when the knee and ankle-joints give way to the weight of the body, the toes must be drawn firmly round any object within their grasp.—The most remarkable of the special modifications of the muscular apparatus of Birds, consists in the development of a vast number of small muscular fasciculi in the substance of the skin, in such connection with the feathers as to be able to move them in different directions. These are best seen in the larger birds, such as the Goose, in which as many as four or five of these slips can be traced to each feather; and as this bird has about 3000 quill-feathers on its whole body, the number of its quill-muscles must be at least 12,000.

325 *l.* Notwithstanding the higher grade of embryonic development which Birds present, as compared with Reptiles, at the time of their entrance into the world, this is not dependent upon any decided advance in the conformation of the genital organs, which retain the type characteristic of the preceding class, with very little essential modification. The 'testes' or spermatic organs of the male, are compact bodies of a globular or elongated form, situated close to the upper part of the kidneys; they are composed, as in the Mammalia, of long convoluted tubuli; and they undergo a remarkable periodical development, their size at the period of sexual activity being from twenty to fifty as great as it is at other times. The left testis is generally the largest; but we never find it alone developed, and the right testis sometimes equals it in size. The two seminal ducts discharge themselves into the cloaca by two distinct orifices; and a pair of papillary elevations in which these terminate, constitute, in most Birds, the sole rudiment of a penis or intromittent organ. In the Ostrich and some other Birds, however, a distinct penis, furnished with erectile tissue, is present; and the seminal ducts are made to open into a groove upon its upper surface, which, though not completed into a canal, serves by the apposition of its lips to convey the fluid to its extremity, and thus to deposit it within the female passages. When not in a state of erection, the organ is entirely concealed within the cloaca.—The genital apparatus of the female is usually remarkable for its unsymmetrical development; the right ovary and oviduct remaining undeveloped (Fig. 188, *f*), like the left lung of Serpents. Up to a certain period of embryonic life, however, these organs are developed with perfect symmetry on the two sides; and in some of the Raptorial birds, they present an equal development even in adult age. The ovarium, which is attached to the superior or anterior extremity of the right kidney, presents an approach to the structure of this organ in Mammalia; being composed of a 'stroma' or bed of compact fibrous tissue, in the midst of which the ova are developed. These, as they increase in size, begin to protrude from its surface; and at last, before they are set free, hang merely by foot-stalks, composed of prolongations of the investing membranes of the ovary. At the time of greatest functional activity, especially

in the domestic Fowls, in which the number of eggs is not limited, and their stages of development different, the surface of the ovary presents an appearance like that of a cluster of fruits of various sizes (Fig. 188, *a*). When set free, by the bursting of their membranous capsules, the eggs are received into the dilated extremity (*c*) of the oviduct (*b*) which conveys them towards the cloaca. In their way, however, they receive an important addition; for each ovum quits the ovarium in the condition of a 'yolk-bag' only, and this is surrounded, as it passes along the oviduct, with layer over layer of albumen poured out from its walls and forming the 'white;' and over all is formed the peculiar fibrous membrane (Fig. 16), of which the outer layer is consolidated, by the deposition of chalky particles in its meshes, into the calcareous shell, whilst the inner remains as its lining membrane. The last of these formations takes place in a dilatation at the lower part of the oviduct (opposite to *g*, Fig. 188), which is called the 'calcifying segment.' The oviduct opens at its lower extremity into the cloaca; which, in the females of the species whose males possess a penis, contains a 'clitoris,' or organ of sexual excitement. — The ovum of Birds chiefly differs from that of the lower Vertebrata in its great size; and this is due to the presence of a very large store of yolk within the yolk-bag, and to the subsequent addition of a large amount of albumen, which answers the same purpose, being taken into the yolk-bag as the contents of the latter are exhausted. Hence it is that the number of eggs laid by Birds, in their natural state, is but small in proportion to that which is deposited by Fishes and Insects, and does not even approach the number deposited by the higher Reptiles. The peculiar deficiency in the pelvic arch already noticed (§ 325 *b*) would seem to have reference to the large space required for the exit of the egg; and the non-development of the ovarium on the right side may very probably be intended to prevent injury to the eggs from mutual pressure, during their development and their passage down the oviduct. The special provision which we find in the parental instinct of the Bird, for the care and maturation of the ova, and for the subsequent nurture of their young, has no parallel in any other class of oviparous animals, save in Insects. The eggs are deposited

FIG. 188.

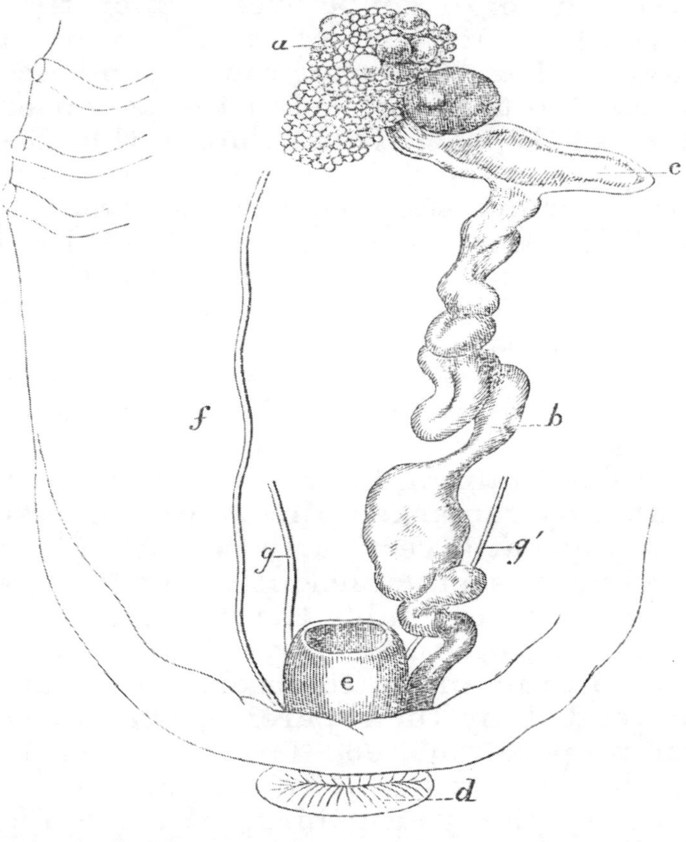

Generative apparatus of the common *Fowl*;—*a*, single ovary, situated nearly on the median line; *b*, left oviduct; *c*, its funnel-shaped opening; *d*, cloaca; *e*, rectum; *f*, right oviduct, atrophied; *g*, *g'*, ureters.

in an artificially-constructed nest, which frequently displays the most ingenious and elaborate workmanship; and the heat necessary for the development of the embryo is imparted to it by the body of the parent. Enabled by the continued agency of this developmental force (§ 96) to appropriate the large store of nutriment laid up in the egg, the embryo of the Bird does not stop short in the process of evolution at the phase which marks the reptilian type; but advances to a far higher grade before quitting the shell. It comes forth into the world with all the attributes of a Bird; and although many of its distinctive characters are at first but slightly manifested,* yet these are very speedily evolved, under the influence of the high temperature which the body possesses, this being partly the result of its own calorific power, but still in part derived from the parents. The degree of the dependence of the young upon the warmth imparted to them by their parents, varies greatly, however, in the different orders of Birds. Thus among the *Raptores* and *Insessores*, the young usually come forth from the egg in a blind and naked state; and it is only by the warmth supplied by the parents, and by the obstacles presented to the dissipation of their own heat by the shelter of the nest and the mutual contact of their own bodies, that their temperature can be kept up to its standard, even during the heat of summer. On the other hand, the young of most birds that live on the ground or on the water, belonging to the orders *Rasores, Grallatores,* and *Natatores,* come forth from the egg in a state which is relatively much more advanced; being capable of running or swimming almost as soon as they quit the shell, and having the power of sustaining their own heat with but little assistance from the parent. As already remarked, the complete ossification of the skeleton of Birds takes place at a very early period of their lives; and the development of the wings is soon sufficiently advanced, even in the Raptorial and Insessorial orders, to enable the young birds to sustain themselves upon them for short flights in the air. During the period which elapses before these are able to maintain themselves by their own exertions, they are nurtured and tended by their parents, who in some instances not only select but even prepare the food for their offspring, by macerating it in the crop, and then returning it to the mouth, from which it is transferred to the mouth of the young bird. In the Pigeon tribe, the 'crop,' during the breeding season, secretes a milky fluid, which is mingled with the macerating grains, and is returned with them to the mouth, to be imparted to the young. This constitutes the nearest approach seen among Birds to the lacteal secretion of the Mammalia. The young of those inferior orders, however, which live upon the ground or on the waters, are far less dependent upon their parents for their early sustenance; being able to provide for themselves almost from the time of quitting the egg.—Here, as in many other instances, we notice that the prolongation of the period during which the early development is aided by parental nurture, is closely related to the elevation in the grade of development which the being is destined ultimately to attain.

* It is in the downy character of its plumage, and the consequent imperfect development of the feathery expansions which constitute the acting surface of its wings and tail, that the young Bird chiefly differs from its parent. These were the very characteristics of the now extinct *Dodo*, of which (as remarked by Mr. Strickland) we may form a very good idea, by imagining a young duck or gosling enlarged to the dimensions of a swan.

326. The MAMMALIA are universally regarded as the highest group in the Animal kingdom; not only as being that to which Man belongs, but also as possessing the most complex and heterogeneous organisation,—that is, having a structure made up of the greatest number of dissimilar parts, which are severally and specially adapted for the performance of distinct operations. It is in this group also, even putting Man out of view as belonging to it, that we have the greatest manifestation of *intelligence* which we any where meet with in the Animal series. This is intimately connected with the high development of the Cerebrum, which presents a rapidly increasing complexity of structure, as well as augmentation of relative size, as we trace it from the lower to the higher Mammalia. And this, again, is closely related to the greatly-prolonged connection between the parent and the offspring, which we find to be a characteristic of this class; for it is a general rule, to which reference has already been several times made, that the higher the development an organized being is ultimately to attain, the longer is it dependent, during the early stages of its evolution, upon the fostering aid of its parent. Now it is the essential characteristic of the Mammalia, considered as a class, that the young are brought into the world alive; and this not merely, as in certain Reptiles and Fishes, by the retention of the egg within the oviduct, until the development of the embryo has so far advanced that it is ready for hatching before the deposition of the ovum; but by the formation of a new connection between the embryo and its parent, while the former lies within the maternal cavities, so that the means of its continued development is provided, not (as in Birds) by the accumulation of a large store of nutriment in the ovum itself, but by a supply of freshly elaborated material continually furnished by the maternal blood. The ovum of Mammals, owing to the very small quantity of its yolk, is of extremely minute size; and its contents only serve to carry on the very earliest portion of the developmental process, which is especially directed towards the formation of the temporary nutrient apparatus. This consists, in the first instance, of a set of root-like tufts, which are prolonged from the surface of the ovum, and insinuate themselves among the maternal vessels, without, however uniting with them; and which absorb from the maternal fluid the ingredients necessary for the support of the embryo, whilst they convey back to it the effete particles of the growing structure, to be eliminated by her secreting operations. Up to this stage, the connection of the mother and fœtus is essentially the same in all Mammalia; but in the lower subdivision of the class (which includes the orders *Marsupialia* and *Monotremata*) the connection of the ovum with the parent ceases before it becomes more intimate, and the young are brought into the world in a condition which (as compared with the young of other Mammalia) is one of extreme immaturity. But in the higher division which constitutes the great bulk of the class, the ovum is delayed for a longer period, within a cavity formed by the dilated union of the two oviducts, termed the *uterus*, which only exists in a rudimentary condition in the two orders just named; and here the embryo forms a new and more direct connection with its parent. The tufts proceeding from the chorion become especially developed at one point, and root-like prolongations of the fœtal vessels insinuate themselves into them; these vascular tufts dip down (as it were) into a chamber formed by an extension of the inner lining of the

uterine vessels; and thus the fœtal blood is brought into such intimate relation with the maternal—without any intercommunication by open orifices—that a ready transudation of elements can take place from one liquid to the other. The peculiar organ thus formed is termed the *placenta;* and its fœtal portion may be considered as bearing a resemblance alike to the roots of plants and to the branchiæ of aquatic animals, absorbing from the maternal blood the materials required for the nourishment of the embryo, and aerating the blood of the fœtus by exposing it to the influence of that of the parent. The two sub-classes into which the Mammalian group is divided, are thus designated as 'placental' and 'implacental,' according as this organ is or is not developed; and we shall presently see that they differ in many remarkable particulars of their organisation, the *Implacentalia* occupying an intermediate position between the Oviparous Vertebrata and the *Placental* Mammals. In both alike, however, do we find that a provision exists for nourishing the animal, during the early period of its extra-uterine life, by a lacteal secretion formed in external mammary glands; and as this is found in no other class, it is the character by which, in a zoological arrangement, the entire group is most positively defined.

326 *a*. Taken as a whole, this class cannot be said to be so much characterized by the possession of any one particular faculty, like the power of flight which is so typical of Birds, as by the combination of different powers which we find in the group generally, and also (for the most part) in each individual. Thus whilst, in the greater number of Mammalia, the body is supported upon all four extremities, and is adapted for progression along the surface of the earth, this typical structure undergoes such a modification in the Bat tribe, as to adapt these animals to the mode of life characteristic of Birds. So again in the Whale tribe, we find the ordinary Mammalian type so modified as to adapt these animals to the life of Fishes; the locomotive power being almost completely taken from the extremities, and given back to the trunk; and the form of the body being such, as to enable it to move rapidly through an aquatic medium. In neither case, however, do we find that these modifications involve any essential departure from the mode of organisation characteristic of the class, or any important variation from its usual physiological conditions. The wing of the Bat does not represent that of the Bird in any other respect than its analogous use; but is formed upon a plan of its own, the extended skin being chiefly stretched over the elongated fingers of the anterior member (Fig. 190). And the expanded tail-fin of the Whale is not extended vertically upon the neural and hæmal spines of its caudal vertebræ; but is flattened out horizontally, and supported by cartilages that do not form a part of the vertebral skeleton. Turning, again, to the endowments of the individual members of the class, we find that whilst there are none that can rival Birds in rapidity of locomotion, there are few that cannot perform several different kinds of progression; and whilst there are probably none that can compete with Birds in acuteness of sight, there are few that do not possess a greater range of special sensibility.—The inferior energy of muscular movement in the Mammalia is accompanied by an inferior amount of respiration; but we shall find the type of the respiratory apparatus to be decidedly higher than that of Birds. The diminished amount of respiration, again, involves the pro-

duction of a lower degree of animal heat (the temperature of this class seldom rising above 104°); and there is consequently the less need of means for effectually confining their caloric; especially, too, as their greater average size causes their radiating surface to be much less, in proportion to their bulk, than it is in Birds. Accordingly we find them provided only with a covering of hair or fur; and even this is sometimes very scanty, or is almost entirely wanting. In the Cetacea, the same end is answered by another provision, the interposition of an immense quantity of oleaginous matter in the meshes of their enormously-thickened skin,—thus forming the 'blubber,' which not only constitutes an admirable badly-conducting septum between the warm body of the animal it incloses and the cold water of the surrounding ocean, but serves, by its peculiar resisting power, to antagonize the enormous pressure to which the surface is subjected when the animal sinks into the depths of the sea.—The inferior nervo-muscular energy of Mammals as compared with Birds, involves not merely a less active respiration and an inferior calorifying power, but renders it unnecessary that the nutritive functions in general should be carried on with that extraordinary activity which characterizes the former class. Accordingly we find that the demand for food is less constant, the digestive process less rapidly accomplished, and the circulation slower, than in Birds; whilst, on the other hand, the 'waste' of the body, as indicated by the amount of the *excreta*, is less considerable.

326 *b*. In the skeleton of the ordinary four-footed Mammalia, we find a return to the same general plan as that on which the skeleton of the Saurian Reptiles is constructed, the method of locomotion being essentially the same in both cases; but we shall observe in many points a considerable advance in the grade of osseous development, not manifested by any conspicuous change of external form. This is remarkably the case with regard to the intimate structure of the bones themselves, which retain, in many Reptiles, the characters which they present in the higher Mammalia during only the early period of their development (§ 206); and it is the case also with regard to the degree of consolidation which the skeleton undergoes, by the union of parts originally developed from distinct centres, of which several examples have already been brought under notice. Now although Mammalia might be considered as inferior to Birds in both these respects,—the bones being filled with oil like those of the Reptile, instead of being penetrated by air, and the separate osseous elements not being consolidated to nearly the same extent as in the preceding class,—yet this inferiority is only apparent; for it is obvious that the peculiarities of the Bird's skeleton must be considered as special modifications having reference to its mode of life, and not as naturally presenting themselves in the general progress of development, as intermediate conditions between Reptiles and Mammalia. In fact, through the lower or non-placental group of Mammalia, and in some degree also through the gigantic extinct Sloths (which seem to have approximated the herbivorous Dinosauria of a still older epoch in many points of their organisation), the connection of the Mammalian and Reptilian groups is better established without the intermediation of Birds; which must be considered as on the whole more separated from both of them, than they are from each other. In our general view of the osteology of Mammals, therefore, we shall take the Saurian skeleton as our standard of comparison, and make little

reference to that of the Bird.—In regard to the conformation of the vertebral column we have first to remark, that the number of vertebræ in the different regions is more constant than we elsewhere find it ; seldom departing much from that which Man possesses, excepting in the caudal region. The typical number of *cervical* vertebræ (Fig. 189, *vc*) is 7 ; and however long the neck may be (as in the Giraffe), or short (as in the Whale), this number is almost uniformly observed ; the only known exceptions on the side of diminution being two of the whale-like Pachyderms (*Manatus* and *Rytina*) which have but 6 ; whilst an excess has only been met with in the *Ai* or three-toed Sloth, which has 9. In cases where little or no more power of moving the head upon the trunk is required, the cervical vertebræ are very thin, and are more or less closely united by the anchylosis of their bodies and arches ; this we especially see in the Cetacea. The cervical vertebræ are composed of bodies and complete

Fig. 189.

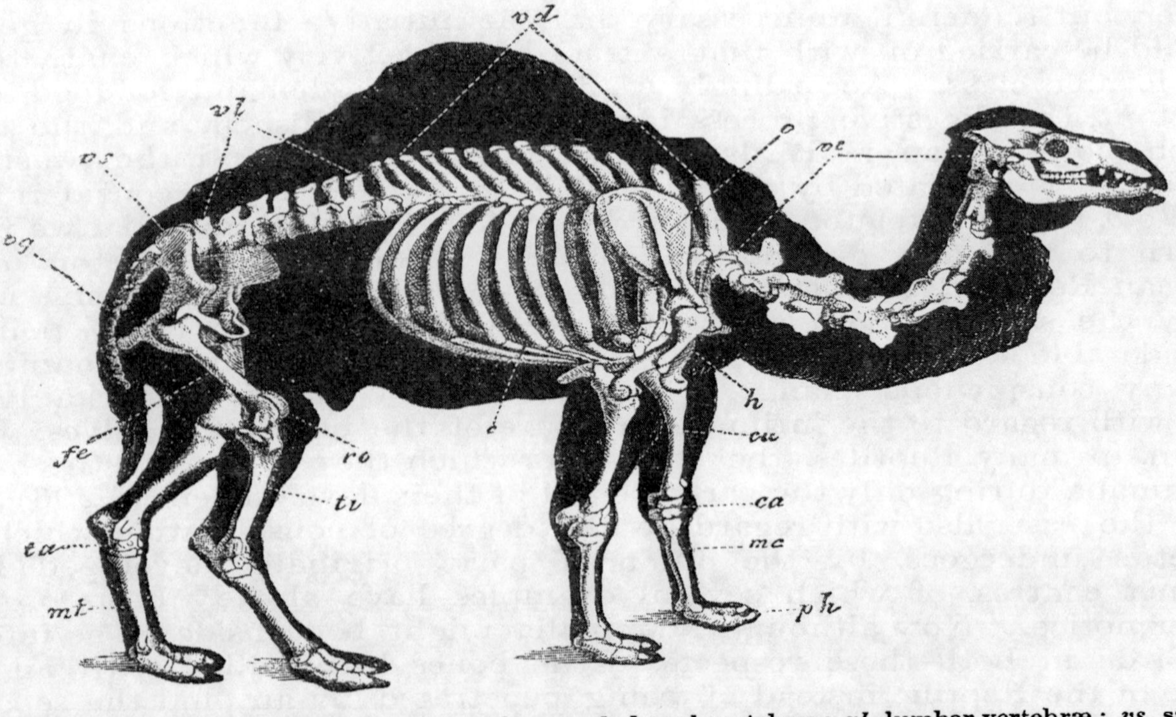

Skeleton of *Camel*;—*vc*, cervical vertebræ; *vd*, dorsal vertebræ; *vl*, lumbar vertebræ; *vs*, sacral vertebræ; *vq*, caudal vertebræ; *o*, scapula; *h*, humerus; *cu*, radius; *ca*, carpus; *mc*, metacarpus; *ph*, phalanges; *c*, ribs; *fe*, femur; *ro*, patella; *ti*, tibia; *ta*, tarsus; *mt*, metatarsus.

neural arches, with 'transverse processes' which are formed by the union of the diapophyses and parapophyses with the rudiments of pleurapophyses, so as to form the vertebral canals (Fig. 169). These pleurapophyses are very seldom developed as distinct elements; we find them, however, in the two lower cervical vertebræ of the *Ai*, where they constitute true 'cervical ribs,' differing only from those of the dorsal region in not being sufficiently developed to reach the sternum; aud they are occasionally seen as a 'monstrosity' in Man. No other trace of the hæmal arch, save the clavicles (§ 326 *d*), exists in the cervical vertebræ of Mammalia. The number of *dorsal* vertebræ, from which distinct ribs are developed (Fig. 189, *v, d*), is usually from 11 to 20 ; the greatest number known is in the two-toed Sloth, which has 23. In the upper part of this region, the hæmal arch is completed by the elongation of the pleurapophyses (ribs), and by the development of the hæmapophyses and hæmal spine. Excepting in the *Monotremata*, the ribs of Mammals always have a double articulation with the spinal column, one by their 'heads' with the bodies of the vertebræ, and the other by their 'tubercles' with the transverse processes; and it is curious that the greater number of the ribs should thus articulate in part with the bodies of other vertebræ than those to which they properly belong, each head being received into a cavity partly formed in the centrum of its own vertebra, and partly in the one anterior to it. The hæmapophyses generally remain cartilaginous, and hence are not commonly known by their proper appellation of 'sternal ribs,' but are designated as 'costal cartilages;' they show a marked tendency to ossification, however, in advancing life; and when fractured, they are commonly repaired by bone. The hæmal spines of the successive segments, which are always distinct in the fœtus, continue so in some Mammalia, constituting so many subdivisions of the 'sternum;' they frequently coalesce, however, to such a degree as to form only three distinct bones (which process takes place during the period of adolescence in Man); and even these may be at last united into one, as we occasionally find in advanced age. The central portion of the sternum forms a projecting keel, as in Birds, in certain cases in which it is necessary that the anterior members should be endowed with unusual strength; and we meet with this conformation, not merely in Bats (which, as the representatives of Birds, might be most expected to present it), but also in the Moles and Armadilloes, whose fore-feet are used as spades in burrowing. The lower portion of the sternum, which usually remains cartilaginous in Man until a late period of life (being known as the 'ensiform cartilage'), is more developed in the Edentata, and is prolonged nearly to the pubis; thus reminding us of the abdominal sternum of Reptiles. In other Mammalia, however, this portion of the sternum is only represented by the fibrous aponeurosis which stretches from the sternum to the pubis, intervening between the abdominal muscles on the median line; and the sternal ribs, by the transverse aponeurotic bands dividing the 'recti' muscles into distinct segments. In some of the posterior segments of the dorsal region, the hæmal arch remains incomplete, by the non-development of the hæmal spine and of part of the hæmapophyses; so that the ribs are either connected to those in front of them by the adhesion of their cartilages, or they are altogether free at their sternal extremities. The proportion of these 'false' ribs is very great in the Cetacea, which have only one or two true ribs; in this respect presenting a very curious reversion to the conformation of Fishes, which have no sternum, and in which all the ribs are consequently free (§ 322 *b*).—In the *lumbar* vertebræ (Fig. 189, *vl*), we again lose the pleurapophyses as distinct elements; and the rudiments of the abdominal sternum and sternal ribs, just alluded to, usually constitute the only representatives of the hæmal arches of the lumbar segments. The number of these vertebræ varies from 2 to 9; but it is commonly either 4 or 5. In some of the larger Quadrupeds, the lower lumbar vertebræ are joined to each other and to the sacrum by bony union of the transverse processes; a condition which sometimes presents itself in advanced age in Man.

326 *c.* The *sacral* vertebræ (Fig. 189, *vs*), which unite to form the 'sacrum' (Fig. 168) and to support the pelvic arch, are by no means constant in number; and we perceive in the conformation of this part of the Mammalian skeleton, a variety which becomes very striking when we consider its character and importance. In Man the number is 5, and in the Mole, 6; but in other placental Mammalia it is diminished to 4, and even to 3; whilst in Marsupialia and Monotremata it is reduced, as in Reptiles, to 2; and in the *Ornithorhyncus* these two remain permanently separated from each other. In Cetacea, as in Fishes, there cannot be said to be any sacrum; the rudimentary pelvis being but loosely connected with a single vertebra, which is not united with those before or behind it.—The number of *caudal* vertebræ (Fig. 189, *vq*) varies more widely than that of the segments of any other region; for whilst in Man and the higher Apes they are no more than 4 or 5, these being abortive and becoming anchylosed with each other (forming the 'os coccygis') or even with the sacrum, the number rises in other Mammalia to 27 (Whale), 31 (Long-tailed Monkey), 36 (Long-tailed Opossum), and even 40 (Long-tailed Ant-Eater). It is interesting to remark that the highest numbers occur in members of those orders,—the *Marsupialia* and the *Edentata*,—which present the greatest number of other approximations to the Reptilian type of structure; but it must be recollected that the tail is a part of the vertebral column, whose development or non-development seems to vary according to the special use to be made of it, rather than according to any general plan. Thus we find it in the Kangaroo serving as a third leg of the tripod, on which this animal rears itself erect; and there can be no doubt, from the conformation of their skeletons, that the gigantic Sloths of former ages were enabled to support their massive bodies in the same manner. In the American Monkeys, on the other hand, as well in some of the Opossums inhabiting the same Continent, and in a few other animals, the tail is prehensile; being capable of coiling round branches of trees with such a firm grasp, as to support the whole weight of the body without difficulty. In the Cetacea, again, as in Fishes, the tail is the chief instrument in the propulsion of the body through the water; and in some other Mammals (as the Squirrel) it seems to steer the animal in its leaps through the air, just as it steers Birds in their flight, being converted moreover into a kind of parachute by the wide lateral expansion of its bushy hair. The tail of the Beaver, again, which is flattened horizontally and covered with scales instead of with hair, is the principal instrument of propulsion in water; it has also been said, but erroneously, to serve as a trowel for laying on the mud with which the habitation is plastered.—The form of the caudal vertebræ, and the number of elementary parts discoverable in them, vary in accordance with the purposes to which the member is to be applied. In some instances, the pleurapophyses may be traced in the anterior vertebræ; but they coalesce with the diapophyses to form the elongated transverse processes, and do not bend down to enter into the hæmal arch. In the Kangaroo, and some other Marsupials, however, some of the caudal vertebræ have proper hæmal arches, formed (as in Reptiles) of the hæmapophyses and hæmal spine, which are anchylosed to the under side of the bodies of the vertebræ, leaving a hæmal canal through which the caudal artery passes. The terminal vertebræ are reduced to the 'centra' alone.—In the conformation of the tail of Mam-

malia, we have thus a very remarkable instance of that *specialization* which has been pointed out as a distinguishing feature of the structure of this class; the same instrument being modified in various ways, so as to be adapted to a number of purposes very distinct from each other, including all those for which it is employed elsewhere, with others which are peculiar to the Mammalian group.

326 *d*. This specialization is carried to a still greater extent in the conformation of the extremities; in which a very remarkable degree of variety is presented. The *scapular arch*, in nearly all Mammalia, differs from that of Oviparous Vertebrata in this; that its coracoid element is not sufficiently developed to reach the sternum, or to meet its fellow on the median line; but that where the scapula has any osseous connection with the sternum, that connection is established by the 'clavicle,' which is probably to be regarded as the hæmapophysis of another segment (possibly that of which the 'atlas' constitutes the centre with its lateral and neural arches). The clavicle itself, however, is very frequently wanting; and we find it absent in some Mammals, whilst present in others very nearly allied to them (as in the group of Rodentia). It is obvious that, taking the whole Vertebrated series together, the completion of the scapular arch by coracoid is the regular or typical plan; whilst the introduction of the clavicle always seems to have a view to some special purpose. As a general rule it is present in those Mammalia in which the anterior extremities are drawn inwards (towards the median line), as well as downwards or backwards, in the act of locomotion; and it serves the double purpose of giving attachment to the muscles by which this movement is accomplished, and of keeping the shoulder-joints from being drawn together when they are put in action. It is also present when the anterior members are unguiculated, and are used for prehension; as in Man, the Quadrumana, most Carnivora, and in part of the Rodentia. On the other hand, it is usually absent in the 'hoofed' Quadrupeds, in which the anterior extremities are simply concerned in supporting the weight of the body, and in propelling it forwards by their own backward movement. The coracoid bone is for the most part reduced to a mere rudiment, and is anchylosed with the scapula, of which it ranks as a 'process.' In the *Monotremata*, however, with many other Reptilian affinities, this most remarkable one is presented; that the scapula is connected with the sternum by a fully-developed coracoid, as well as by a clavicle; that, in addition, the clavicles do not (as in Birds) unite on the median line, but articulate with a bone which is placed upon the summit of the ordinary sternum; and that the lateral prolongations of this T-shaped bone (termed the 'episternum') pass outwards parallel with the clavicles, just as in the *Ichthyosaurus* (§ 324 *k*), and thus give additonal assistance in the fixation of the shoulder-joint.—The structure of the *pelvic* arch presents a much greater constancy in the majority of Mammalia, although its form is liable to undergo great modification. The three bones of which it is composed on each side,—the ilium, ischium, and pubis,—generally coalesce together, as in Man, at an early period of life; but in the *Monotremata*, as in Reptiles, they remain separate much longer. And just as we have found that it is generally characteristic of the scapular arch of Mammalia, to be closed-in by the clavicle instead of by the coracoid, so is it characteristic of their pelvic arch to be completed by the pubic instead of the ischiac bones.

In most Mammals, the symphysis, or inferior median junction of the two lateral halves of the pelvis, is entirely formed by the rami of the pubis; but in the Implacental sub-class, the ischium also has a share in the junction, thus carrying us back towards the Reptilian type. It is very interesting to see how strongly the conformation of the pelvis in the Bats and Whales resembles that of the groups which they respectively represent; for in the former it is greatly elongated, and neither pubic nor ischiac bones unite to form a symphysis, so that the lower front of the arch remains open as in Birds; whilst, on the other hand, in the latter the pelvis is reduced almost to a minimum, consisting merely of a simple hæmal arch (not unlike the hyoid bone of Man), formed by the convergence of the hæmapophyses of the single sacral vertebra, and by their junction with each other, either directly, or through the mediation of a hæmal spine. The pelvis of the Implacental sub-class presents this further peculiarity; that from the pubic symphysis two additional bones project forwards, being articulated at their bases with the pubis, whilst their free extremities are connected with the tendons and aponeuroses of the abdominal muscles. These additional bones have been termed 'marsupial,' from their supposed function in supporting the 'marsupium' or pouch, whose presence characterises the principal of the two orders that make up the group. As they are present, however, in both sexes of Marsupials, and also in animals of the order *Monotremata* which do not possess the pouch, it seems likely that they must be subservient to other purposes. They are not, it would appear, to be considered as part of the vertebral skeleton; but are rather adventitious bones, developed in the substance of the abdominal tendons, as the 'sesamoid bones' (of which the patella is a gigantic example) are in some of the powerful tendons of the limbs.

326 *e*. The anterior extremities are never deficient in Mammalia, although they present very diversified modes of conformation in the several orders of the class, according to the purposes for which they are designed. Where they are adapted merely as instruments of support and locomotion on land, as in the *Ungulata* or 'hoofed' Quadrupeds, the extremity is reduced to its simplest condition; only a pair of digits, or even a single one, being developed in each member, and these being entirely enveloped in hard horny casings; the bones of the fore-arm being consolidated into one, so as entirely to prevent any rotation of the hand; and all the articulations, even that of the shoulder-joint, being so formed, as to limit the movements of the limb to that one plane (backwards and forwards) in which its actions are required for the onward propulsion of the body. The opposite extreme is where (as in Man) a thin nail covers only one side of the extremity, leaving the other possessed of all its tactile delicacy; where several such fingers exist, capable of being used independently, one of them being opposed to the rest, so as to render prehension more perfect, and to adapt the member for a great variety of actions; where the plane of the whole hand can be turned in any position, by the rotation of which the fore-arm is capable; and where the form of the humeral articulation, and the arrangement of its muscles, confer upon the arm, as a whole, the power of free and extensive motion. Between these two extremes, however, there are many intermediate gradations; and some of the more important of these will be now considered. —Starting with the anterior extremity of *Man*, as presenting us with its

highest type of conformation as a prehensile organ, we find it composed of a humerus, articulated at its upper extremity by a ball-and-socket joint with the scapula, and at its lower with the radius and ulna; of which the latter alone enters into the proper 'hinge' of the elbow-joint, the radius being kept out of that articulation in order that it may freely rotate (carrying the hand with it) around the axis of the ulna. The radius alone forms the wrist-joint, with the carpal bones; and this joint has more of the 'ball-and-socket' than of the 'hinge' character; so as to allow great freedom of movement to the hand. The 'carpal' bones in Man are eight in number, forming two rows; but the typical number is probably *ten* (as seen in the Turtle), its reduction being the result of the coalescence of four bones into two, the 'unciform' and 'scaphoid,' the latter of which is double in many Quadrupeds. The carpal bones support the five 'metacarpal,' which are included in an undivided integument; that of the thumb, however, being free to move independently of the rest. The metacarpal bones support the 'phalanges,' of which each digit possesses three, the thumb alone having two; and these are all perfectly independent, so that either phalanx of each digit may be flexed or extended without any movement of the others. The chief peculiarity of the hand of Man, as distinguished from that of those higher Quadrumana which are like him possessed of an opposable thumb, consists in the much greater length of this thumb, and in the power which Man alone possesses of bringing it into forcible opposition to the extremities of each one of the fingers.—Passing from the hand of Man to that of the *Quadrumana*, we see a much nearer approximation, than in him, to a similarity between the anterior and posterior extremities. The possession of 'four hands' is not, as might be supposed from the name, a character which raises them in the least degree above Man; for none of these four hands are adapted to the same variety of actions of which his are capable; and all of them are in some degree required for support. In this respect their character approaches much nearer to that of the extremities of the inferior Mammalia; and there are several among them, in which, the thumb being rudimentary, and its opposable power deficient, there is no very marked distinction between the so-called 'hand,' and the 'foot' of some Carnivora. The arm of the Ape has as wide a range of motion as that of Man, so far as its articulations are concerned; but its freedom of movement can only be fully exercised when the animal is in the erect position, which is unnatural to it.—The next degree of degradation is seen in the *Carnivora;* in which the anterior extremity is used, almost as much as the posterior, for support and progression; but in which it is still employed as an instrument of prehension, though in a comparatively limited degree. Here we find the fore-arm still composed of two distinct bones; but these have little capability of rotation; and the carpal articulation does not give the hand any freedom of movement in more than one plane. The fingers are inclosed in a common integument, as far as their last phalanx, and cannot be used independently of each other; and the first or innermost digit, representing the thumb of Man, is but very little developed, or is altogether wanting. The horny appendage to the extremities of the fingers is no longer a mere 'nail,' covering only one side of their last phalanx; but is a strong 'claw' embracing its whole termination. In the Insectivora and 'plantigrade' Carnivora, we find a gradual transition in the conformation of the anterior extremity, from the type which it presents in the lower Quadrumana, to that which characterizes the 'digitigrade' portion of the order, of which the Felines are the type; the carpal and metacarpal portion of the hand being elongated at the expense of the phalanges, and the latter alone resting upon the ground; an arrangement which is in beautiful conformity with the 'springing' habits of the order, but one that diminishes the prehensile power of the extremities. The clavicles also disappear; and in the shortening of the humerus, its inclusion within the integuments of the trunk, and the greater limitation of the movement of the shoulder-joint to the antero-posterior direction, the Feline shows the complete reduction of this member from the quadru*manous* to the quadru*pedal* type.—The same reduction is carried still further in the group of *Rodentia;* in some of which we find the bones of the fore-arm, though still distinct from each other, incapable of rotation; whilst in others they are partially fused together. And among the *Edentata* (Sloths, &c.), we find a reduction in the number of digits to three (as in the *Ai*) or even to two (as in the *Unau*); whilst in the gigantic extinct forms of this order we have a transition to the hoofed quadrupeds, which they resembled in their general proportions,—two of the five digits of the fore-foot of the *Mylodon* having been furnished with flattened callous coverings to their stunted terminations, so as evidently to serve as points of support, and the other three having been provided with long sharp claws adapted for digging.—In the *Ungulated* Quadrupeds, we meet with a considerable variety in the degree of consolidation of the anterior extremity. Thus in the Elephant, the five digits are present, but their terminations are all included together in a callous skin, which takes the place of a hoof: in the Hippopotamus there are only four digits (the first being deficient), but these are inclosed in separate hoofs; in the Hog there are also four, but the two middle ones are longer than the others, and it is on these alone that the body rests: in the Rhinoceros and Tapir, the number of digits is reduced to three, by the non-development of the fifth or outer one: in the entire *Ruminant* group, the third and fourth digits are enormously developed, and rudiments only of the second and fifth are present (Fig. 189); whilst in the Horse and other *Solidungula*, the third digit alone is fully developed and furnished with a hoof, the rudiments of the second and fourth being also present, forming what are known as the 'splint-bones.' * This progressive reduction in the number of digits, and the increased development of some of those which remain, may be conveniently represented by the following table, in which the figures 1, 2, 3, 4, 5, used for Man and the Elephant, indicate the presence of these respective digits in a condition of nearly equal development, the smaller figures expressing the rudimentary condition of the digits they represent, and the largest indicating an extraordinary excess in their relative size.

* This statement is made on the authority of Prof. Owen, who opposes the ordinary notion that the great single digit of the Horse is formed by the coalescence of the two digits of the Ruminants; affirming that the extremity is rather constructed upon the plan of that of the Rhinoceros, the central digit of its tridactyle foot becoming enormously developed, whilst the others are reduced to the rudimentary condition. The extinct *Palæotherium* presents the transitional link between the tridactyle and the apparently monodactyle extremities. See his "Discourse on the Nature of Limbs," pp. 31, 32.

Man, Elephant	1	2	3	4	5
Spider-Monkey	1	2	3	4	5
Hippopotamus, Pig		2	3	4	5
Bandicoot (Marsupial)		2	3	4	5
Ox (Ruminant)		2	3	4	5
Rhinoceros		2	3	4	
Horse		2	3	4	

With this reduction in the number of the digital extremities, and the limitation of their functions to those of mere support and progression, we find a corresponding simplification in the other portions of the member. The bones of the carpus are reduced in number, only six presenting themselves in the Horse. The two bones of the fore-arm, where distinct from each other, have no power of rotation; and both of them enter into the composition of the elbow and wrist-joints; but they are generally connected with each other by continuous ossification, especially at their carpal extremity; and in the Ruminants and Solipeds, the radius is much more developed than the ulna, and forms the greater part of the elbow-joint, as well as the carpal articulation. The humerus is shortened in proportion to the lower portion of the extremity, and its articulation with the scapula has almost the character of a hinge-joint, allowing it only to play in the antero-posterior direction; whilst the clavicles are altogether deficient, whereby considerable mobility is allowed to the scapulæ. All this modification seems destined to contribute towards that fleetness in running, which is the especial attribute of the Ruminant and Solipede Mammalia, and which can only be maintained by the complete sacrifice of the prehensile power which the anterior extremity elsewhere possesses.

326 *f*. In the posterior or pelvic extremities of all the Mammalia alluded to in the preceding paragraph, there is a very close resemblance to the anterior. The difference is greater in Man, than it is in any other case; in consequence of the special adaptation of these extremities for the support of his body in the erect posture. Thus we find the two bones of his leg, which represent those of the fore-arm, arranged more after the fashion of the bones of the fore-arm in some Herbivora: the weight of the body being almost entirely thrown upon one bone, the 'tibia,' which corresponds with the radius; whilst the 'fibula,' which represents the ulna, takes no share whatever in the knee-joint, but only guards the exterior of the ankle-joint; and there is consequently no power of rotating the foot upon the leg. The 'os calcis' is extraordinarily developed, so as to project far backwards, and to serve for the attachment of the powerful 'extensor' muscles which keep the leg erect upon the foot; and it is regarded by Prof. Owen as representing *two* bones of the carpus (the 'cuneiforme' and 'pisiforme'), thus accounting for the inequality in number between the carpal and tarsal bones.* The

* There are other differences, however, in the relative positions of the bones constituting the first row of the carpus and tarsus, which make it impossible to understand their true relations without the aid of Comparative Anatomy. Indeed it is only in this way, that the mutual correspondence of the other bones of the anterior and posterior extremities in the Human subject, can be satisfactorily determined. Thus Vicq D'Azyr, who was the first anatomist who attempted to work out these 'serial homologies' (or 'homotypes,' as they are

tarsal and metatarsal bones make up the principal part of the length of the foot, the phalanges being comparatively short; and the great toe is not separated from the rest (though still distinguished by its possession of only two phalanges), and is not capable of being opposed to them. —Now in the Quadrumana, this dissimilarity is greatly diminished; the posterior and anterior extremities approximating much more closely (as already mentioned) to a common type; and thus conducting us towards the ordinary Quadrupeds, in which the hind-foot undergoes a series of modifications, that repeat, with scarcely any considerable departure, the various conditions already described as presented by their anterior extremities. The reduction in the number of digits by the disappearance of some, and the excessive development of others, go on for the most part *pari passu* in both pairs of extremities; whilst the coalescence of the tibia and the fibula almost always takes place to about the same degree as that of the radius and ulna. It is curious to observe among many of these, that the fibula is much larger in proportion to the tibia than it is in Man, and that its analogy to the ulna thus becomes more obvious. This analogy, however, is most completely presented among the Implacental Mammalia; for there are many species in that group, in which the hind-foot possesses a rotatory movement in virtue of the rotation of the tibia round the fibula; and the latter bone has a prominence at its upper extremity, which distinctly represents the olecranon. In the Opossums and Phalangers, there is even an opposable thumb on the *hind* foot; whilst in the fore-foot the first digit is parallel with the rest, as in the foot of Man; hence these animals have been termed *Pedimana* (foot-handed), from their reversal of the ordinary relations between the anterior and posterior extremities.

326 *g*. We have now to consider some of the most remarkable modifications which the extremities undergo, for special purposes not comprehended in the foregoing general description. These modifications present themselves for the most part in the *anterior* extremity; which, in the Bat, is made to answer the purpose of the wing of the Bird; in the Whale it is modified into the likeness of the fin of the Fish; whilst in

now termed by Prof. Owen), though correctly interpreting the correspondence between the scapula and the ilium, the humerus and femur, the bones of the hand and those of the foot, was misled by the different proportions of the ulna and fibula in the human skeleton, to the adoption of the belief that the tibia is the representative of the ulna, and the fibula of the radius, the olecranon being supposed to be represented by the patella; so that each anterior extremity is not paralleled or repeated by the posterior limb of the same side, but by that of the opposite side. This notion was adopted by Cuvier. A still more absurd hypothesis was put forth by Cruveilhier; who made the upper end of the tibia the representative of the upper half of the ulna, and its lower end the representative of the lower end of the radius; whilst the fibula in like manner represents the radius above and the ulna below. "Nature, however (as Prof. Owen justly remarks), when rightly interrogated and propitiated by due observant service, extricates us from these complex involutions and alternations of serial homology, and makes the simple truth plain." The true view, put forth by Dr. Barclay and M. Flourens, has been still further elucidated by Prof. Owen; who has removed the difficulty that had been felt in regard to the olecranon, by showing that it is not represented by the patella, which is developed even where (as in the Wombat) the fibula presents the broad, flat, high process which obviously corresponds with the olecranon of the ulna; whilst in the elbow-joint of the Bat, the true representative of the patella presents itself, not (as has been supposed) in its detached olecranon, but (according to Prof. Owen) in a sesamoid bone contained in the tendon of the biceps. See Prof. Owen's "Discourse on the Nature of Limbs," p. 19.

the Mole and other burrowing animals, it is metamorphosed into a most efficient instrument for the excavation of earth.—The skeleton of the *Bat* is remarkable, in the first place, for the high development of all the bones connected with the movement of the anterior extremity; for the breadth of the scapula, the strength of the clavicles, and the projecting keel of the sternum. The humerus (Fig. 190, *h*) is not of disproportionate length; the bones of the fore-arm, however, are unusually long, and we find that, as no movement of rotation is required, the ulna (*cu*) is for the most part in a state of coalescence with the radius (*r*), as in the Horse. The bones of the carpus are small, but they serve to support the metacarpal bones, which are of extraordinary length, and are capable of a wide divergence from each other. These metacarpal bones support the phalangeal or true

FIG. 190.

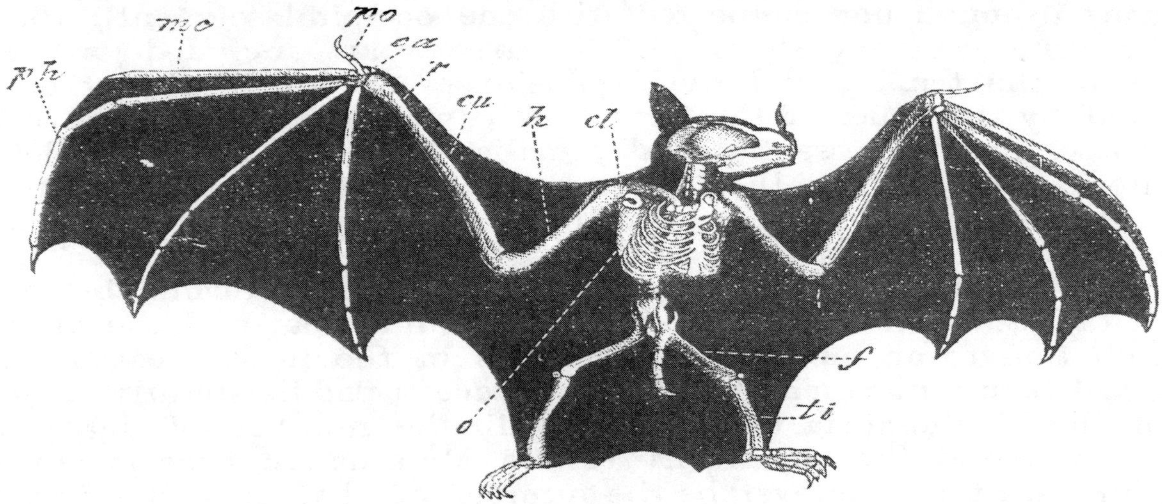

Skeleton of *Bat*;—*o*, scapula; *cl*, clavicle; *h*, humerus; *cu*, ulna; *r*, radius; *ca*, carpus; *po*, thumb; *mc*, metacarpus; *ph*, phalanges; *f*, femur; *ti*, tibia.

finger-bones, the number of which is the same in Man; and it is over these, with the assistance of the hind-legs and tail, that the membrane of the wings is extended. The thumb, however, does not partake of this extension in length, nor does it assist in the support of the wing-membrane; but it is short and free, and is terminated by a hooked claw, with which the animal can suspend itself when at rest.—It is marvellous to see how the same elements are adapted to such widely different purposes, as the support and propulsion of the body through the yielding air, and the excavation of a passage through the solid earth; yet in the burrowing trowel of the Mole we have the same elements as those which present themselves in the expanded wing of the Bat, developed (as it were) in precisely the opposite direction; all that is elongated and attenuated in the bat being shortened and thickened in the mole, with the exception of the scapula, which is here a long and narrow but very strong bone. The clavicle has almost a cubical form; the humerus is nearly as broad as it is long, its extraordinary expansion being not merely for strength, but also for giving an extended surface for muscular attachment; the radius and ulna are likewise short and very thick bones, yet are so constructed as to be capable of admitting the rotation of the fore-arm; the carpal bones form the usual series, with the addition of a large sabre-shaped ossicle, which strengthens the digging edge of the broad palm; and these support the usual number of metacarpal and phalangeal bones, nearly all of them as broad and thick as they are long, and all five digits being buried, as far as the strong digging claws, in a sheath of tough skin. With such rapidity does the Mole accomplish its work of excavation, when the soil is light, that it has been figuratively said to 'swim through the earth.'—In the Cetacean group, on the other hand, we find the anterior extremities converted into paddles for assisting in the propulsion of the body through the water; and in the general arrangement of their parts, we trace a close analogy to the pectoral fins of Fishes, of which they are the representatives. Thus, in consequence of the thinness of the cervical vertebræ (§ 326, *b*), and the consequent rudimentary condition of the neck, they are placed immediately behind the head, although not connected with the occipital segment; the arm and fore-arm are very short, and in great part included within the muscles of the trunk; and the expansion of the fin is almost entirely supported by the bones of the hand. Still we trace in the conformation of all these parts the essentially Mammalian type, so that the skeleton of the anterior extremity of the Whale bears far more resemblance to that of the arm of Man, than it does to that of the pectoral fin of the Fish; the two bones of the fore-arm, although partly anchylosed together, have their normal relations; the carpal bones are all determinable as the representatives of those of the terrestrial Mammalia; and the only real approximation to an inferior type consists in the multiplication of the phalangeal bones, in which respect the Cetacean paddle slightly resembles that of the Enaliosauria (§ 324, *l*). Still the number of digits never exceeds five; and the conformity to the Mammalian type is curiously shown in this, that whatever be the number of phalanges, the first digit (representing the thumb) has a smaller number than the rest.

326 *h*. With regard to the posterior extremities, there is comparatively little to remark. In the *Seals*, which are ordinary Carnivora *partially* modified for the aquatic life of Fishes, we find the posterior extremities carried backwards in a horizontal direction, and supporting two broad expanded fins, by which they swim with great facility and strength; their motions on land, which are chiefly effected by the anterior extremities, being proportionally awkward and shuffling. In the *Cetacea*, which are adapted to an exclusively aquatic habitation, the posterior extremities are altogether deficient, only a rudiment of the pelvic arch being discoverable (§ 326 *d*); but their place is taken by the horizontally-expanded tail, which still more effectually answers the purpose of a propelling instrument; and the power of movement on land is so completely lost, that a Whale which is 'stranded' on the shore cannot get itself off. On the other hand, there are several Mammalia in which the posterior extremities are, as in Frogs, the chief instruments of locomotion; and this not, as in Man, by their alternate movement in walking or running, but by their combined action in leaping. When this is the case, they are developed to a much greater size than the anterior members; and the strength of the body is chiefly shown in the pelvic arch which supports them, and in the muscles by which they are put in action. We see an approach to this type of conformation in many of the ordinary *Rodentia*, such as the

Hare and Rabbit, which, though four-footed in their ordinary progression, can move by leaps in which the posterior extremities are the chief agents; in the Beaver and Squirrel, the disproportion is still more strongly marked, the body being very commonly supported upon the posterior extremities, whilst the anterior are employed for prehension by opposition with each other; and in the Jerboa, the difference is still more striking, the power of progression being almost entirely committed to the posterior members. It is in the Kangaroo, however, that this peculiar type of conformation is most strikingly exhibited; for, although this animal partly supports itself upon its fore-paws when browsing upon the herbage, its movements are entirely accomplished by its hind-limbs and tail, this last member, which is of enormous thickness and strength, affording important assistance in them. The extraordinary length of the feet chiefly depends upon the elongation of the metatarsal bones, especially that of the fourth toe; the first or innermost toe being generally absent altogether, the second and third being present only as rudiments, and the fifth being of large dimensions but inferior in size to the fourth. The under surface of the foot is covered with a callous sole along its whole length; whilst the fourth toe is furnished with an elongated hoof, which is sharp-pointed and three-sided like a bayonet, and serves as a powerful weapon of defence. The anterior extremities, though comparatively small, are highly developed as prehensile instruments; approaching in this respect to those of some other Marsupials, in which we almost see the Quadrumanous type repeated.—Somewhat of the same remarkable contrast, having reference, however, to a very different purpose, is presented between the anterior and posterior extremities of the great extinct Sloths (*Mylodon*, *Megatherium*, &c). For whilst the former are constructed (as already noticed) partly as excavating instruments, and partly for support, the whole conformation of the posterior extremities, of the pelvic arch which supports them, and also that of the tail, points to these members as having possessed the capability (as in the Kangaroo) of supporting the principal weight of these enormous animals in the semi-erect position; and thus leads, with other data derived from the structure of their teeth and the general conformation of their skeleton, to the almost irresistible conclusion, that whilst these giants of a period long gone by must have been supported upon food of precisely the same nature as their pigmy representatives of the present time,—namely, the leaves and soft shoots of forest-trees,—they obtained that food, not, like the existing Sloths, by climbing the stems and clinging to their branches, which would not have sustained their weight, but by excavating the soil from around and beneath their roots with their scraper-like fore-feet, and then, by rearing themselves upon their hind-legs and tail, placing their fore-legs against the trunk, and applying to it the whole force of their monstrous bodies, laying it prostrate upon the earth, to browse upon it at their leisure.*

326 *i.* The *skull* of Mammalia, so far as regards the number and arrangement of its individual bones, presents a general conformity to one type, which is highly characteristic of the class; and the greatest departure

* See Prof. Owen's 'Memoir on the Mylodon;' a masterpiece of philosophical reasoning, built upon the most careful and exact comparison of the peculiarities of conformation of this singular group of extinct animals, with those of all existing forms to which it has any alliance.

from this presents itself (as might be anticipated) in the lowest order of the Implacental sub-class, the cranial conformation of the Monotremata presenting many curious points of resemblance to that of Birds and Reptiles; whilst the next degree of modification is perhaps exhibited by the Cetacea, in which the skull has certain characters that remind us of Fishes. Excluding these, however, we find that the ordinary Placental Mammalia do not present any wider departure from the common type in one direction, than Man exhibits in the other; and it will be convenient to take his cranium as the standard of comparison, in pointing out the chief peculiarities which distinguish the Mammalian skull from that of other Vertebrata, and the principal varieties which present themselves in the conformation of the several orders of the group. The most important distinctive peculiarities of the skull of Mammalia are the following:—the cranial cavity, which is of a rounded form, is shut in by a set of bones connected with each other by sutures, which usually remain visible during the whole of life, although they are sometimes obliterated with age; the number of these bones is much less than in Reptiles (the difference being at once seen in the single 'occipital' bone, which represents the four bones forming the neural arch of the occipital vertebra in the Crocodile); the articulation of the cranium with the spinal column is now no longer effected by the 'centrum' of the occipital vertebra, but is formed by two prominent 'condyles,' one on either side of the 'foramen magnum' or neural canal, which represent the 'zygapophyses' or articulating processes of ordinary vertebræ, and are received into corresponding depressions in the broad zygapophyses of the 'atlas' or first vertebra; the bones of the face are all (with the exception of the lower jaw) immoveably articulated to each other and to those of the cranium; and the lower jaw articulates directly with the cranium, having a single convex 'condyle' on either side, which is received into a 'glenoid cavity' at the base of the skull, instead of being connected with it (as in all oviparous Vertebrata) by the intermediation of a moveable 'os quadratum' or tympanic bone.—The principal variations, on the other hand, which we encounter in the crania of different orders of Mammalia, have reference chiefly to the relative development of the cranial and facial bones, the adaptation of the jaws to different kinds of food, and the support of organs specially developed in particular groups, such as the horns, tusks, proboscis, &c. The development of the cranial bones, and the capacity of the cranial cavity, attain their highest proportion in Man, and next to him in the Quadrumana; and this development, with the greater or less tendency to the erect posture which accompanies it, coincides with an important change in the position of the 'foramen magnum' and of the 'occipital condyles,' which in the lower Mammalia are at the back of the skull and usually in a nearly perpendicular plane, whilst in Man they are nearly in the middle of the under side of the skull and have a horizontal direction, and in the Quadrumana present a regular transitional gradation from the former to the latter condition. On the other hand, the facial portion attains its greatest relative size in the Cetacea, in which the jaws are immensely elongated; whilst among ordinary terrestrial Mammals, the elongation of the jaws is greater in the Herbivorous series than in the Carnivorous, in order to adapt them for that triturating action which the food of the former requires. The contrast between the facial and the cranial portions of the skull of the Horse and of Man respectively, is peculiarly striking; the facial being four times larger than the cranial in the former; whilst precisely the reverse proportion obtains in the latter.

326 *k*. In examining into the composition of the cranial vertebræ, we find that the disproportion just alluded to entirely depends upon the relative development of their *neural* and *hæmal* arches. The former attain their greatest size in Man; and it is curious to observe that the greatest increase is to be found in the 'neural spines,' rather than in the neurapophyses,— this mode of development being an example of the general law, of which the most remarkable examples are to be found in the extremities, that any excessive development, either in size or in multiplication of parts, takes place rather in the most peripheral elements of the vertebra, than in those nearer to the centrum. In giving a general description of its conformation upon the 'vertebral theory,' it will be most instructive to analyse each bone of the Human cranium, and reduce it to its proper elements; rather than to go over again, in nearly the same form, what has been already stated (§ 320 *i*, *k*).—Commencing, then, with the *Occipital* bone, we find this representing in itself the entire neural arch of the occipital or Epencephalic vertebra. This is completely proved by the history of its development; for its ossification commences from four centres, so that for a considerable time before and after birth this bone consists of four pieces, as in the adult cranium of the Crocodile; namely, a 'basi-occipital,' representing the centrum of the vertebra, two 'lateral' or 'condyloid portions, representing the 'neurapophyses' and 'parapophyses' combined (the latter being especially indicated in Man by the scabrous ridge overhanging the transverse processes of the atlas, and giving attachment to the 'rectus lateralis' muscle, but in some of the lower Mammalia by a process of considerable size, the 'paramastoid'), and the 'proper occipital,' which is the 'neural spine' enormously developed for the protection of the cerebellum and the posterior lobes of the cerebrum. In the human species, these four portions usually coalesce into one piece by about the sixth year of childhood; but in the Monotremata they remain separate up to the period of full growth; and if they coalesce at all, it is at an advanced period of life.—The *Parietal* bones together constitute the 'neural spine' of the Mesencephalic vertebra, enormously developed for the protection of the cerebral mass. Each half of this spine is developed as one bone from a single ossific centre; the two sometimes coalesce along the median line, so as to obliterate the 'sagittal suture;' and this coalescence is normal in many of the lower Mammalia, the vault of their cranial cavity being covered-in by a single 'parietal bone.'—That the composition of the *Temporal* bone of Mammalia is very complex, is indicated by the number of distinct centres in which it originates. These are at least five; namely, one for the 'zygoma and its squamous expansion,' one for the 'tympanic,' one for the 'petrosal,' one for the 'mastoid' (which is sometimes ossified continuously from the petrosal in Man), and one for the 'styloid process.' The light of comparative anatomy makes it evident that the mastoid portion is the 'parapophysis' of the Parietal vertebra, and the styloid process its 'pleurapophysis;' that the petrosal portion is not part of the vertebral skeleton at all, but is an ossified 'sense-capsule' which coalesces at an early period with the mastoid; that the tympanic, here reduced to its subordinate

function of supporting the membrana tympani, is really the pleurapophysis of the Frontal vertebra; and that the squamosal and zygomatic portions are really diverging appendages of the hæmal arches of the Nasal vertebra, carried backwards (so to speak) from the malar, and intercalated in this neural arch, the one to complete the bony enclosure of the cranium, and the other to strengthen the upper jaw by articulating it with the solid head-piece through the zygomatic arch.—In the *Sphenoid* bone, again, the complexity of composition is in some degree indicated by its peculiar form and connections; but more decidedly by the history of its development. Ossification commences from independent centres in the 'body,' in the 'great wings,' and in the 'pterygoid processes,' which together constitute the posterior or 'spheno-temporal' part of the bone; and in the 'body,' and in the lateral portions or 'small wings,' of the anterior or 'spheno-orbital' part. The body or 'basi-sphenoid,' is the 'centrum' of the Parietal or mesencephalic vertebra; the great wings or 'ali-sphenoids,' on which the parietals are reared, are its 'neurapophyses;' and the pterygoid portions represent the 'diverging appendages' of the hæmal arch of the Nasal vertebra, which pass backwards from the 'palatine' bones to form part of the posterior osseous boundary of the cavities of the mouth and nose. In like manner, the spheno-orbital part, representing the 'pre-sphenoid' and 'orbito-sphenoids,' constitutes the 'centrum' and 'neurapophyses' of the Frontal or prosencephalic vertebra. The body and wings of the orbito-sphenoid portion coalesce long before the end of fœtal life; and a partial coalescence in the other portions of this composite bone takes place before birth, the great wing and the pterygoid processes having then united on each side, and the basi-sphenoid (or body of the posterior division) having coalesced with the pre-sphenoid (or body of the anterior division). During the first year of infancy, the great wings unite with the body; and at a later period, the body of the sphenoid becomes anchylosed with the basilar portion of the occipital, so that some anatomists have described the whole as one bone.—The structure of the *Frontal* bone, on the other hand, is very simple. Though usually a single bone in the adult, it is developed in two symmetrical pieces, each of which has a single ossific centre; these, at birth, are separated by a continuation of the sagittal suture that passes down to the root of the nose; but in the course of a few years, this suture is usually obliterated in the Human subject, though it sometimes remains persistent during the whole of life, as it normally does in many of the lower Mammalia. The frontal bone is the 'neural spine' of the Frontal or prosencephalic segment; highly developed, like the parietal bones and the upper part of the occipital, for the protection of the cerebral mass beneath. We do not, in the Mammalia, find the 'parapophyses' of this segment developed as distinct elements; they are represented by the 'external angular processes' of the frontal bone, which shoot forth by exogenous growth from the proper frontal, instead of constituting independent elements like the 'postfrontals' of Fishes.—Of the *Ethmoid* bone, which completes the category of bones usually ranked as forming the cranium proper, only the 'os planum' (according to Professor Owen) forms part of the vertebral skeleton; the remainder being a true 'sense-capsule,' though developed in the higher animals in an unusual degree, and forming most intimate relations with the bones among which it is intercalated

(§ 320 *i*, note). The 'ossa plana,' supported upon the vomer, represent the 'neurapophyses' of the Nasal or rhinencephalic vertebra. In their turn they support the *Nasal* bones, which together represent the 'neural spine' of the same vertebra; these, each developed from a single centre, coalesce into a single bone in certain Mammals, like the frontals and parietals. The 'centrum' of this vertebra is formed by the *Vomer*, whose form and relations in Man would seem to give it no title to such a distinction, but whose conformation in Fishes is altogether different, and marks its true character (§ 320 *i*). When we consider the modifications which the terminal caudal vertebræ undergo, by the suppression of some of their elements, and by the peculiar conformation or arrangement of others, it is not surprising that this, the segment which forms the anterior termination of the body, should be subject to similar irregularities. Neither in the Human subject, nor in the lower Vertebrata, do we find any distinct representative of its parapophyses.

326 *l*. Turning now to the bones which represent the hæmal arches of the cranial vertebræ, it will be convenient to proceed first with those which constitute the face; and in this manner we shall be carried from before backwards, as in the enumeration of those of the cranium we have passed from behind forwards.—The *Superior Maxillary* bone is more simple in its composition, than its form in Man would at first seem to indicate. It appears to be developed from several distinct centres on either side; but most of the parts thus formed coalesce at a very early period; and each bone consists essentially of two portions, the true 'maxillary,' and the 'intermaxillary,' of which the former represents the 'hæmapophysis,' and the latter a ramus of the 'hæmal spine.' In the Human subject, the intermaxillary bone usually coalesces with the maxillary long before birth; but it remains distinct in most of the lower Mammalia, and attains an enormous development in the Rodentia and also in the Elephant, in which it bears the peculiar 'tusks' growing from persistent pulps, which are characteristic of these animals; and it is sometimes found to continue separate even in Man.—The *Palatine* bones, which, notwithstanding the complexity of their form and connections, are developed from single centres, are the 'pleurapophyses' of this segment; and it is very interesting to see how completely this view of their character harmonizes with, and even (so to speak) explains, their very peculiar connections. Thus we see them coming into connection above, not merely with the 'centrum' (vomer) but with the 'neurapophyses' (ossa plana of the ethmoid), and also with the 'sense-capsule' (cellulæ ethmoideæ) of this segment; whilst below they articulate with the 'hæmapophyses' (maxillary bones) and with the 'diverging appendages' (pterygoid bones) which belong to them. —The *Malar* bone is another 'diverging appendage' of this segment, passing backwards (like the branchiostegal rays of Fishes) towards the temporal region, and there expanding, as it were, into the squamous portion of the temporal bone. Thus we have, in this anterior segment, the highest development of the hæmal arch, with the lowest condition of the neural.—The *Inferior Maxilla*, or lower jaw, of Man, and of the Mammalia generally, presents a striking contrast, in its simplicity of conformation, with the extraordinary complexity which the tympano-mandibular arch exhibits in the lower Vertebrata. It is always developed in two distinct halves; and these remain separate during life in a large proportion

of the inferior Mammalia, being only united at the symphysis by ligamentous connections. Each half is developed from at least two nuclei in the human subject, one producing the 'articular' portion that forms the joint, the other forming the 'dentarium' which contains the teeth; and of these, the former constitutes the 'hæmapophysis,' and the latter a ramus of the 'hæmal spine,' of the Frontal vertebra. The 'pleurapophyses' of this segment enter into the composition of the temporal bone, forming its tympanic element, and no longer participating in the articulation of the lower jaw. This tympanic portion is far more largely developed in some orders, than it is in Man; thus in the Cetacea it forms an almost solid mass of ivory, of considerable size, which is scarcely at all connected by bony union with the mastoid.—Thus we have accounted for all the component bones of the Human skull, save the 'lachrymals,' the 'turbinate bones' of the nose, the 'bones of the internal ear,' and the 'teeth.' The lachrymals do not in reality belong to the neuro-skeleton, but are portions taken in from the dermal skeleton, and connected with the proper vertebral elements; just in the same way that the marginal and other bones of the carapace of Turtles are introduced and blended with the spinal column and ribs (§ 324 *h*). The turbinate bones constitute part of the 'sense-capsule' of the olfactive organ, being analogous to the petrosal element of the temporal bone, which is the 'sense-capsule' of the organ of hearing; there are not in Mammalia any bony plates in the fibrous sclerotic, forming a sense-capsule to the eye. The bones of the internal ear, and the teeth, belong, like the sense-capsules, to the 'splanchno-skeleton.'—We have as yet noticed, however, the constituents of the hæmal arches of only the two anterior cranial vertebræ. That of the third, of which the greater part is known as the *Hyoid* bone, is yet more reduced than the second; yet can we clearly trace its principal elements. In the Hyoid bone, simple as its form is in Man, there are no fewer than five centres of ossification, one for the 'body,' one for each of the 'great cornua,' and one for each of the 'lesser cornua.' The body, or 'basi-hyal,' is obviously the 'hæmal spine' of this arch; the lesser cornua, or 'cerato-hyals,' which are connected by ligaments to the styloid processes of the temporal bone, are its 'hæmapophyses;' the greater cornua, or 'thyro-hyals,' developed as appendages to the larynx, are its 'diverging appendages;' and the styloid processes of the temporal bone, as already mentioned, are its 'pleurapophyses.' This styloid process is the last to coalesce with the other elements which unite to form the temporal bone. In several of the lower Mammalia, the cerato-hyals are considerably longer than the thyro-hyals, instead of being far inferior to them in size, as they are in Man; and in some cases they form a continuous bony arch with the styloid bones to which they are articulated.

FIG. 191.

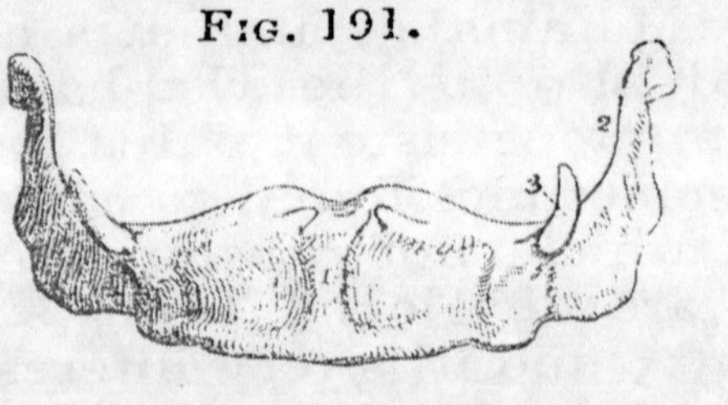

Front view of the *os hyoides*;—1, the body; 2, the great cornu, and 3, the lesser cornu of the left side.

326 *m*. The dermo-skeleton, in the greater number of Mammalia, is reduced for the most part to the condition of *Hair* (§ 180); a form of tegumentary appendage which is peculiarly characteristic of this class, being rarely or never entirely absent, and not presenting itself in the

same form elsewhere.* An exuviation or moulting of the hair, which falls off and is replaced by a new growth, takes place to a considerable extent in most Mammalia provided with this covering; sometimes with, and sometimes without, an alteration of colour. This change usually takes place either in spring or autumn, or at both periods. The degree in which the hair covers the body, varies greatly in the several orders and genera of the class; and it is interesting to observe that its deficiency is most marked in those which in other respects represent the *squamigerous* (scale-bearing) Ovipara. Thus in the *Cetacea* (Whales), which are (so to speak) Mammalian Fishes, we have an almost entire absence of hair, without, however, any approach to a scaly covering in its place; one of the great purposes which the hairy investment usually serves,—that of keeping in the heat of the body,—being here provided for by the thickening of the skin itself, and the deposition of an enormous quantity of fatty matter in its areolar interspaces, forming the 'blubber.' On the other hand, in many of the *Edentata* (the group which, of all placental Mammals, approaches most nearly to Reptiles), we have the hairy covering in great part replaced by horny scales or by osseous plates. The former is the case in the *Manis*, or 'scaly ant-eater,' whose body is almost entirely covered with dense horny scales which have an imbricated arrangement; these scales, however, seem as if composed of a congeries of hairs; and in certain parts on which they are deficient (generally speaking, the under surface of the head, trunk, and tail, and the inner surface of the limbs), the skin is furnished with scattered hairs in their stead. When alarmed, the Manis rolls itself into a ball, wraps its tail over its head, and raises all its pointed and sharp-edged scales in such array, as to defy the attack of almost any possible enemy. In the remarkable *Chlamyphorus*, the upper side of the body is covered by a tesselated cuirass, composed of a series of plates, in apposition with each other at their edges, but each enclosed in a membranous sac of its own; so that these are probably (like the scales of Fishes) *dermal* rather than epidermic; the under side of the body of this animal is covered by soft silky fur, resembling that of a Mole. In the *Armadillo*, the body is much more completely inclosed by its dermal cuirass, which invests the same parts as those covered by scales in the Manis, with the addition of the under side of the tail, which member is completely enclosed in a succession of bony rings. The bony armour consists, first, of a triangular or oval plate on the top of the head, which projects backwards over the neck; secondly, of a large cuirass or buckler which covers the shoulders; thirdly, of a series of transverse bands, somewhat moveable one upon another, which nearly encircle the body, their number varying with the species; fourthly, of another cuirass which covers the haunches; and fifthly, of the series of bony rings investing the tail. Each of these distinct segments consists of a number of small pieces united to each other, like the separate portions of a mosaic, or of a tesselated pavement; and this arrangement allows of a certain degree of yielding at every part of this coat of mail, although it is only at the joints between the segments that any proper movement of one part upon another takes place. The limbs, which are short and thick, are almost

* The so called 'hairs' of Insects and Annelida are not properly comparable with those of Mammals; being, in fact, only elongated epidermic scales, and not being developed from cutaneous follicles.

entirely concealed by the edges of this projecting shield; and these animals, when alarmed, can so roll themselves together, and draw the head and limbs within the protection of the cuirass, as to present nothing to their enemies but a bony coat of mail enveloping them on every side,—an arrangement which strongly reminds us of that which exists in the 'box-tortoises' (§ 324 *h*). The skin of the parts unprotected by the bony armour is tough, and is beset with long scattered hairs, of which some are seen to issue forth between its joints. In the gigantic extinct *Glyptodon*, which was formed upon the plan of the existing Armadillo, the bony cuirass was of still greater density, and formed a vast undivided dorsal shield entirely composed of tesselated plates united together at their edges, without any joint through its entire extent, and overhanging the sides of the body, without, however, covering any part of its under surface. The tail was completely enclosed in a similar envelope. It is obvious that this animal could not have possessed the means of passive defence which the puny Armadilloes of the present time enjoy, whilst its vast dimensions were in themselves a sufficient protection; for if it crouched, by the flexure of its short limbs, in such a manner as to allow the edges of the shield to rest upon the ground, and then withdrew its head beneath the front of the cuirass, none but a beast of far greater power than any Carnivora of the present day could have overthrown it, so as to get at the unprotected parts beneath.

326 *n*. The other modifications of hair, which depart less from its ordinary character, such as the spiny bristles of the Hedgehog, and the quills of the Porcupine, have been already noticed (§ 180); and the structure of the horny substance that forms the nails, claws, and hoofs, which are organs of great functional importance in the Mammalia, constituting part of their dermo-skeleton, has also been described (§ 179). It remains, however, to speak of the *horns*, the possession of which is a remarkable feature of the *Ruminant* group, though some approaches to it are seen elsewhere. With the exception of the Camels and Musk Deer, the males (and sometimes the females) of all the species of this order bear upon their frontal bones one and occasionally two pairs of bony prominences; which are sometimes simple cones, more or less elongated and twisted; but which in other cases branch and spread out to a very great distance from their base. The bony protuberances may be covered by the skin of the forehead, which does not differ at this part from that of the rest of the body, and which does not undergo any change; such is the character of the horn in the Giraffe, in which the bony core is articulated with the frontal bone, and not an outgrowth from it. In other instances, the bony core has more the character of an out-growth from the frontal bones, and is covered with a horny sheath, which grows with it during life and never falls off; this is the case with that division of the group which includes the Oxen, Sheep, Goats, and Antelopes. In the *Deer* tribe, on the other hand, the bony portion of the horns is at first covered with a soft bony skin, termed the 'velvet,' and never becomes invested with horn; this skin dries up and wears away, so that the bone becomes bare; and at the same time, the vascular canals of the interior of the bone become gradually blocked up by the progressive consolidation of its substance. At last the nutrition of the bone ceases completely; it dies and falls off, like a dead leaf; and is very speedily renewed by the growth of a new horn,

which is usually larger, and possesses more branches, than that which it replaced. The horns of the Deer, being thus purely *bony*, are properly called 'antlers;' and their exuviation and renewal usually take place in the spring of every year. It is only in the Reindeer, that the female, as well as the male, possesses antlers.—To describe these bony 'antlers' as a part of the dermo-skeleton, rather than of the endo-skeleton, may appear inconsistent with the history of their development, which seems to take place by exogenous growth from the frontal eminences. But their annual *exuviation* is a very marked indication of their dermal character, no such exuviation taking place in any portion of the neural skeleton, whilst it is a very ordinary occurrence in the dermal; and they seem naturally to fall into the same category with the dermal portions of the carapace and plastron of the *Chelonia* (§ 324 *h*), which are developed in continuity with the parts of the vertebral skeleton that support them.—The single or double horn which the Rhinoceros bears upon its snout, is a purely epidermal formation, and may be compared to a bundle of hairs adherent together. In this animal, as in many other Pachydermata, the epidermis covering the general surface of the body is of extraordinary thickness and of almost horny density; whilst the hairy covering is very scanty. A similar local development of a horny epidermis in unusual amount, forms the callous pads on the under side of the feet of most Unguiculated Quadrupeds, the callosities on the buttocks of the Quadrumana, &c.

326 *o*. The *Teeth* of Mammalia constitute so characteristic a feature in their organisation, and afford such important assistance in classification, as to require a much more special mention than was necessary in the case of Reptiles and Fishes. There are few animals belonging to this class, in which these organs are altogether absent; and these animals belong to the groups which have been already alluded to as the most aberrant. No rudiments of teeth are found either in the true *Ant-eaters*, which belong to the *Edentate* order, or in the *Echidna* (or 'spiny ant-eater' of New Holland), one of the members of the *Monotrematous* order; and in the *Ornithorhyncus*, which is the only other existing Monotreme, the teeth are only represented by horny projections on the surface of the bill-like sheath with which its jaws are covered, the purpose of these projections, as of those which are found in many piscivorous Birds, being simply to increase the prehensile power of the bill. In the Whalebone-Whales, no bony teeth ever appear above the gum, although their rudiments are found already calcified in the jaws of the embryo. These are the only exceptions to the general rule of the presence of teeth in this class.—The *number* of teeth in Mammalia is usually much more restricted than in the lower Vertebrata. Through a considerable proportion of the class, it is the same as in Man, namely 32; but the typical number is considered by Prof. Owen as 44. Now it is curious that here, again, the greatest departure, both in the way of reduction and excess, should be found among the Fish-like Cetacea. In two genera of this order (*Hyperoodon* and *Ziphius*) only *two* teeth are developed; in both cases from the lower jaw. In the male *Narwhal*, although the rudiments of two teeth are formed and sometimes developed, one of them is usually abortive, whilst the other grows to an enormous size, this unequal development being accompanied with a want of symmetry in the jaw-bones themselves (§ 320 *n*); hence this genus received from Linnæus the name of *Monodon* (single-toothed), which is

not in reality much more appropriate than its common name of 'Sea-Unicorn.' On the other hand, the *Cachalot* (Spermaceti Whale) has upwards of 60 teeth, although most of them are confined to the lower jaw; the Porpoise has between 80 and 90 teeth; and the Dolphin as many as 190. The only other similar example of great excess is presented in one of the Reptile-like Armadilloes, which has 98 teeth. In all these cases of unusual multiplication, the teeth are, like those of most Reptiles, comparatively small, and nearly similar to each other in size and shape. This similarity extends also to the teeth of other members of the Edentate order, in which these organs present themselves in their simplest and least specialized condition; such, as the ordinary Armadilloes, the Orycterope, and the three-toed Sloth. In the two-toed Sloth, we perceive the first approach to specialization, in the increased size of the two anterior teeth of the upper jaw, which are adapted for cutting and tearing, whilst the remainder are only fit for crushing substances of no great hardness. In nearly all the other Mammalia, we find the principle of specialization remarkably carried out in the teeth; diversified forms being given to them, whereby they are adapted to different uses. The front teeth usually terminate in a thin cutting edge; and being intended simply to effect the first division of the food introduced into the mouth, are termed *incisors.* On the other hand, the back teeth, which are adapted for the further division or comminution of the aliment, are termed *molars,* or grinding teeth; a term, however, which is by no means appropriate to the molars of the Carnivora, whose action is simply cutting. Between these, we very commonly find certain teeth of conical form, projecting beyond the rest, which are not adapted either for cutting or grinding the food, but for taking a firm hold of it, whereby the animal may tear it asunder; these, from being well developed in the Dog, are called *canine* teeth, and are the especial attribute of the Carnivorous order, though developed in some Pachydermata (as the Boar) and in many Quadrumana, as weapons of attack and defence. The first and the last of these kinds of teeth may be deficient; the molars, however, are never wanting.

326 *p.* In the mode in which the teeth of Mammalia are implanted in the jaw, we find a marked distinctive character of the class. In all save those which grow from persistent pulps (§ 218), we find the dental cavity closed-in at its lower part, and the base of the tooth prolonged into a 'fang,' which is implanted into a proper 'socket' formed by a projection of the bony substance of the jaw that grows up to invest it. Each tooth has its particular socket, and the exterior of the fang, which is formed of the ossified capsule of the tooth (§ 215), adheres firmly to its periosteal lining; there is never, however, such a continuous ossification of the tooth to the jaw, as constitutes its mode of attachment in many Fishes (§ 213). The most imperfect form of the regular mode of connection between the teeth and jaws of Mammalia, shows itself in some of the Cetacea, in which the sockets, at the posterior part of the jaw, are wide and shallow, and the teeth adhere more strongly to the gum than to their periosteal lining; so that, when the gum has been stripped away, in the Cachalot, it has brought all the teeth away with it. The simplest form of the fang, in the higher Mammalia, is that which is presented by the incisors and canines, in which it is for the most part a simple cone, sunk in a corresponding conical socket. In the molar teeth, however, the fang

is usually subdivided into two, three, or even four portions, which diverge more or less from one another; and as the sockets are moulded (as it were) upon these diverging processes, it is obvious that this method of implantation gives to the jaw a most secure hold of the teeth. Although some approach to the implantation of the teeth in distinct sockets, by simple fangs, is seen in Fishes and in Crocodilian Reptiles, yet there is no known Fish or Reptile, existing or extinct, in which teeth are thus implanted by bifid fangs; and this character, therefore, appears sufficient to determine the Mammalian nature of a jaw, or even a fragment of a jaw, in which it occurs, such as that of the *Amphitherium* and *Phascolotherium* of the Stonesfield Slate, whose Reptilian nature was advocated by many zoologists, but which are undoubtedly to be ranked as extinct Mammals.—True teeth implanted in sockets are found nowhere else in the Mammalian class, than in the margins of the upper and lower jaws; they may be confined to either one or the other of these, as we see in the Cetacea; they may be wanting in the intermaxillary bone, as in the Ruminantia, which have incisor teeth developed only in the lower jaw, although their germs may be detected in the upper; or they may be wanting in the front of both jaws, as in the Sloths. Generally speaking, they include the three kinds of tissue which have been described as making up the complete tooth; namely, the *Dentine*, which is usually non-vascular, the *Enamel*, and the *Cementum* (§ 212). The first of these is the most constant element, and makes up the bulk of the teeth; the second is absent in the Sloths (both extinct and fossil) and in some of the Cetacea, thus presenting another character of degradation in the dental apparatus of these orders; the third is present in all instances around the lower part of the tooth, and originally forms a thin layer over the crown (which, however, is soon worn away) in Man, Quadrumana, and Carnivora; but it forms a large part of the substance of the molar teeth of many Herbivorous Quadrupeds, in which it is imbedded, with vertical plates of enamel, in the dentine of the crown during its development (§ 216), in such a manner as to form an uneven surface for trituration, the roughness of which is constantly preserved, however much the tooth may be worn down, by the unequal density of these tissues. In the molar teeth of the Sloths, the only provision for the maintenance of a grinding surface consists (the enamel being absent) in the difference between the wear of the hard or non-vascular dentine which forms the outer part of each tooth, and the softer or vascular dentine which occupies its centre (§ 212). The teeth of the Orycterope present us with a very interesting reversion to the inferior type, being of a composite nature, like that of the teeth of many Fishes, which are made up by the coalescence of several little 'denticles,' each having its own medullary canal (the representative of the pulp-cavity) and its system of radiating tubuli.

326 *q*. The Mammalia further present us with certain marked peculiarities in the *succession* of their teeth. We have seen that in Fishes and Reptiles the teeth have not a permanent character; but that, like most epidermic formations, they are continually being cast off and renewed; the new development taking place either from new and independent papillæ, or from offsets from the follicles of the previous teeth. In Mammalia, on the other hand, this process is restricted in a much less degree; the shedding and renewal not taking place, except in some peculiar

cases, more than once during the whole of life. In the *Cetacea*, the inferiority of whose dental organisation has been already noticed, the follicles of the teeth first formed never give off gemmæ for the production of successors; so that when once completed, and worn down or torn away, these teeth are not replaced. In those *Edentata*, too, which are provided with teeth (Sloths and Armadilloes), this germinal power seems to be deficient; but its want is here compensated by the persistence of the matrix (or tooth-pulp), and by the uninterrupted growth of the teeth. In most other Mammalia, we find the entire series of teeth first developed (which are commonly known as the 'deciduous' or 'milk-teeth') replaced at an early period of life by a second or 'permanent' set; the latter having a size and form as suitable to the jaws of the adult, as the former were to those of the young animal. Of these permanent teeth, all that occupy the places of the milk-teeth are developed from offsets detached from their follicles (§ 217); but the number of the 'permanent' set is greater than that of the 'deciduous,' and all the additional teeth are *molars*, which come up behind the *premolars* (known in Man under the name of 'bicuspids') that replace the molars of the milk set. The foremost of these true molars originates (as in Man) in a new and independent matrix, and is therefore like a milk-tooth retarded in its development, and endowed with unusual permanence of character. It further resembles the milk-teeth in its power of producing another by gemmation; this, however, does not come up in place of it, but is developed posteriorly to it, and forms the second true molar; and from this, again, another follicle is budded off, which gives origin to the third true molar that comes up at a later period of life behind the second. In most of the Placental Mammalia, the process of gemmation stops short at this point, so that only three true molars are ever developed; but in many of the Marsupials, it advances one more stage, and produces four true molars. It is remarkable that, in the Kangaroo, the first of these molars should be shed before the last is developed; and it is by a still more complete change of the same kind, that the curious succession of molars takes place in the jaw of the Elephant, which was long regarded as an altogether exceptional phenomenon. Each of these molars is a large composite tooth, occupying a considerable part of the lateral segment of each jaw; and it is formed of a succession of vertically-alternating plates of enamel, dentine, and cementum. The chief wear of these teeth takes place in front; and a reproduction of new teeth is continually going on behind; so that each tooth, as it wears down, is pushed forward by the new tooth behind it, which comes to occupy its place. The jaw almost always contains two such teeth on either side; and sometimes three may be met with,—one that has been nearly worn away, another in full maturity, and a third in an early stage of advancement. The molars are thus changed six or eight times; each tooth is of greater dimensions, and contains a larger number of vertical plates, than the one which preceded it; and the last formed, which is the largest of all, remains in its place for the rest of the animal's life. In the 'tusks' of the same animal, and of its extinct allies, we have a gigantic example of a very different mode of development; namely, the continued growth of a persistent tooth, whose formative pulp remains uninclosed (through the absence of root) during the whole of life, and is continually undergoing conversion; these tusks are known, by their implantation in the intermaxillary bones, to be really 'incisors.' In the extinct genus *Mastodon*, the lower as well as the upper jaw possessed tusks, in the young state, of dimensions inferior to those of the others ; and one or both of this second pair were retained by the males in adult age, although both were shed by the females without being renewed. The mode of development of the molar teeth, also, was strikingly different in these two genera ; for the first and second deciduous molars in each jaw of the Mastodon were replaced by permanent molars, which pressed upon them not from the *back* but from the *base* of the socket, as in ordinary Mammals. In the gigantic *Dinotherium*, the tusks were developed from the lower jaw alone. In several other *Pachydermata*, as the Hog, Babiroussa, and Hippopotamus, the canines are developed as 'tusks,' continuing to grow from persistent pulps ; this is the case also in the Musk-deer, which in some degree connect this order with the Ruminants ; and we meet with it again in the Walrus among the Carnivora. This peculiar arrangement, in the foregoing cases, would seem rather to have reference to the special habits of the animal, than to follow any regular plan. But in the *Rodentia*, the entire order is characterised by the growth of the long incisors from persistent pulps, whereby the teeth are pushed up by additions to their base as fast as they are ordinarily worn away at their cutting edges ; and in some of this order, as also in the *Edentata*, the molars also are developed from persistent pulps, so that in proportion as their surfaces are worn away by the trituration of the tough vegetable substances that constitute their food, they are kept up to their level by new growth at the base.—It is from the great variety of such special modifications as these, of the ordinary type of dental structure, in accordance with peculiarities in the food of the several animals that present them, and in the mode in which they are organized to obtain it ; and from the constancy in the form, number, position, structure, and succession of the teeth, in the greater number of families of Mammalia ; that these organs furnish characters of such value to the Zoologist for the construction of his systematic arrangements. They cannot be taken, however, as the sole or even as the principal basis of such arrangements, without serious infractions of the natural affinities of the several groups ; as happened when Cuvier associated the Monotrematous *Echidna* and *Ornithorhyncus* with the ordinary *Ant-eaters*, in the order Edentata. No arrangement can be really valuable, which is not based upon the whole collection of the characters of each tribe, and which is not in conformity with the phenomena of development ; and the value of the characters furnished by the teeth, therefore, entirely depends upon the degree in which they are exponents of the general type of organisation. This is especially the case in the *typical Carnivora*, and in the Ruminant Quadrupeds which may be regarded as the *typical Herbivora*. On the other hand, the characters furnished by the teeth especially fail in the aberrant *Edentata*, in which we have to associate the Sloths, Armadilloes, and Ant-eaters, on account of marked features of accordance in their general structure and physiological condition, although they differ widely in the nature of their food and in the mode in which they obtain it ;* and

* It is very interesting to note the curious *representations* of the Reptilian class that occur in this order, which, of all ordinary Mammalia, presents the closest approximation to it. The *Manis*, or Scaly Ant-eater, in its general form and scaly covering reminds us of the *Sauria* ;

also in the *Cetacea,* which, in their dental organisation so closely resemble Fishes.*

326 *r.* The *Digestive* apparatus acquires its greatest completeness in this group ; its several parts here attaining their highest development, and the combinations which they present being more elaborate than those elsewhere met with. It is evident, from the increased size and fleshiness of the tongue, and from the large supply of nerves sent to its papillary surface, that the sense of Taste is more acute than in the inferior classes, and that it becomes an important guide in the selection of the food. Its introduction into the mouth is usually accomplished by a pair of fleshy lips, the *Ornithorhyncus* being the only Mammal whose lips are horny and inflexible, and even *its* lips being soft at an early age, when it draws its support from the mammary glands of its parent. For the prehension of the food, however, whereby it may be brought within the reach of the lips, we find a variety of provisions, some of which are special to this group, whilst others carry into a higher development the plan which has already been shadowed forth in other classes. Thus of the former kind are the 'trunk' of the Elephant, formed by a prolongation of its nose ; the elongated proboscis of the Tapir and Desman ; the prehensile tongue of the Giraffe ; and the prolonged upper lip of the Ruminants generally, the pressure of which against the incisor teeth of the lower-jaw enables them to crop the herbage on which they feed. But the highest form of prehensile instrument is undoubtedly that which is constituted by the special modification of the anterior extremities for this purpose. This modification is most complete (as we have seen) in Man, in whom the anterior members are not used for support or progression ; but we observe that these limbs are constantly employed in the prehension of food by the Quadrumana, and in a less degree by the Carnivora and Rodentia.—The food, when received into the mouth, is almost always submitted, in greater or less degree, to the masticating process ; which in Carnivorous animals, however, consists in little else than the division of the food by the sharp-edged molar teeth ; whilst in Herbivora, it subjects the food to a crushing and triturating action, either before it is swallowed, or, in the Ruminant order, on its return to the mouth after being macerated in the fluids of the first stomach. In all Mammals to which the masticating process is important, the salivary glands are highly developed ; and their secretion appears not merely to moisten the food, but to accomplish a change in certain elements of it, which is the first act of the digestive process. The cavity of the mouth is sometimes dilated in a very remarkable manner into 'cheek-pouches' for the storing-up of food ; these are met with in most of the *Simiadæ,* or Monkeys of the Old World, and also in some *Cheiroptera ;* but they attain their highest development in certain *Rodentia,* as the

the *Armadillo* presents us with the closest Mammalian analogy to the *Chelonia ;* and the great extinct Sloths (*Gravigrada* of Prof. Owen) had much in common with the herbivorous *Dinosauria.*

* Ample details upon the Teeth of Mammalia will be found in the elaborate 'Odontography' of Prof. Owen ; to which the Author has been chiefly indebted for the materials of the preceding sketch of their dentition, as for the account previously given of the teeth of Reptiles and Fishes. Prof. Owen's latest views on the *succession* of Teeth in Mammalia, as the guide to the determination of their homologies in different groups, will be found in his article *Teeth* in the 'Cyclopædia of Anatomy and Physiology;' and in his Memoir on the Teeth of the Wart-Hogs in the Philosophical Transactions for 1850.

Hamster, in which they extend deeply down the neck, and are compressed by peculiar tegumentary muscles.

326 *s.* The œsophagus is usually a long and narrow canal, passing down through the thoracic cavity until it reaches the Stomach. The form of the latter organ varies remarkably in the different orders of Mammals; being, for the most part, a simple saccular dilatation of the alimentary tube, in those which are most purely carnivorous in their diet; but being rendered more or less complex, in these which feed on vegetable substances of various kinds, by the subdivision of its cavity into two or more chambers, in the first of which (as in the proventriculus of Birds) the food undergoes a preparatory maceration, whilst it is in the last that the true process of digestion takes place. It is in the *Ruminant* tribe, as will be described hereafter (CHAP. X.), that the most complex organisation of the stomach is met with. The length of the Intestinal canal, again, is subject to great variations; and between these, also, and the food of the animal, some relationship may be generally traced. Thus in the purely Carnivorous tribes, the intestine is not above 3 or 4 times the length of the body; whilst in the frugivorous Monkeys, it is 8 or 10 times; in the Ruminants it is very commonly 20 times, and in the sheep it is as much as 28 times, the length of the body. The Liver and Pancreas attain a very high development in this class. The secretion of the former is poured into the intestinal tube, at a short distance from the stomach, usually by a single duct, which is generally furnished with a dilated receptacle, the gall-bladder, for the storing-up of the secretion in the intervals of the digestive process. The secretion of the latter, also, is conveyed by a single duct, which frequently discharges itself into the intestine at the same point with the hepatic duct. The respective offices of these secretions in the digestive process will be considered hereafter (CHAP. X.)—The second portion of the intestinal canal is usually much more dilated than the first, and is known as the 'large,' the first part being distinguished as the 'small' intestine. At the termination of the small intestine, there is frequently a very considerable dilatation, answering to the *cæcum* of Man; this is very conspicuous in the *Ruminants*, and still more in the Horse; but it is in the *Rodentia* that it attains its greatest development, being usually, in that order, many times larger than the stomach. In this part of the alimentary canal, the food appears to undergo a sort of secondary digestion; the secretion formed in its walls being of an acid character, like the gastric juice. The walls of the small intestine usually contain a large number of peculiar glandulæ, known as the *Peyerian* from the name of their discoverer; these, which are sometimes solitary, sometimes arranged in large clusters, probably serve more as *ex*creting than as *se*creting organs, the matters which they separate from the blood being rather destined to be directly carried out of the system, than to perform (like the biliary and pancreatic secretions) any purpose subservient to the digestive function; and there is reason to believe that their office is the elimination of the peculiar putrescent matter which characterises the fæces, and which is evidently not derived from the undigested residue of the food. This putrescent excretion is found more abundantly in the fæces of Mammalia, than in those of other classes. The anal termination of the intestinal canal is quite distinct from the outlets of the urinary organs, in all Mammalia, save in the *Monotremata*,

in which, as in Birds, the intestinal and urinary passages open into a common *cloaca*.— The *Absorbent* system of vessels attains its highest development in the Mammalia ; for although the length and apparent convolution of the tubes are not nearly so great as in Reptiles, yet the glandulæ which are developed in their course present the same elements in a form of much greater concentration. The contents, both of the lacteals and lymphatics, are for the most part collected into a single tube, the *thoracic duct*, and are discharged by it into the venous current, at the angle between the jugular and subclavian veins on the left side ; a smaller duct, which terminates at a corresponding point on the right side of the venous system, receives and discharges the contents of the lymphatics of the right side of the head and upper part of the trunk.

326 *t*. The Circulating apparatus of Mammals is constructed upon the same plan as that of Birds (§ 325*f*) ; the heart possessing four cavities (Fig. 192) ; and the whole of the blood returned from the system being transmitted for aeration to the lungs, before being again propelled through the systemic arteries. The chief peculiarities of this apparatus in Mammals consist in the diversified arrangements for the distribution of the blood, and more particularly in the development of enormous extensions both of the arterial and venous systems in the Cetacea and other diving animals. The most important of these modifications will be described hereafter (CHAP. XII.).— The Respiratory organs, however, present a very different arrangement. The lungs are more minutely subdivided into air-cells, so that they must be considered as attaining a higher development than those of Birds, although, in the activity of the respiratory function, Mammals are on the whole inferior. The thoracic cavity, in which they lie, is separated from the abdominal by a complete muscular diaphragm; by the action of which, concurrently with that of other muscles which act upon the ribs, a continual exchange of the air contained in the lungs is provided for. There is no extension of the respiratory surface beyond the cavity of the chest; but the large supply of oxygen required by the most active and energetic Mammals, and the free removal of the carbonic acid which they abundantly generate, are amply provided for by this arrangement. For, supposing the expiratory movement to have been just performed, the cavity of the chest, reduced to its smallest dimensions by the ascent of the diaphragm, the descent of the ribs, and the falling-back of the sternum, is enlarged in all its dimensions by the contraction of the diaphragm, which diminishes the convexity of its arch, and by the elevation of the ribs, which brings them more nearly into a straight line with their cartilages, and thus pushes forwards the sternum, whilst it separates the middle points of the opposite ribs from each other laterally, so as to increase the breadth of the thoracic cavity. A vacuum is thus created,

FIG. 192.

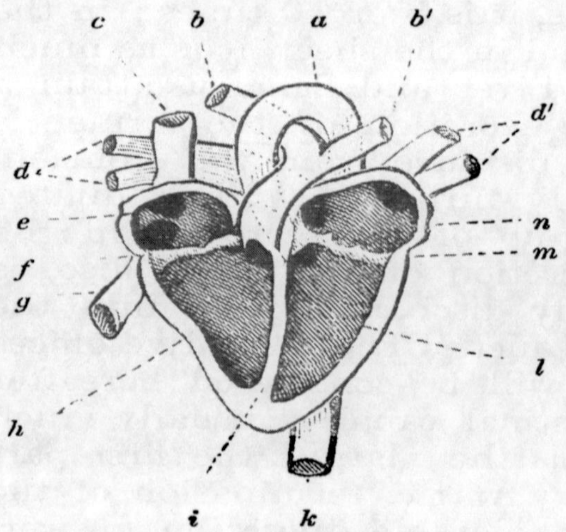

Ideal Section of Mammalian Heart:— *a*, arch of aorta ; *b*, *b′*, pulmonary arteries; *c*, superior vena cava; *d*, *d′*, pulmonary veins; *e*, right auricle; *f*, tricuspid valves; *g*, inferior vena cava ; *h*, right ventricle; *i*, septum ventriculorum; *k*, descending aorta ; *l*, left ventricle ; *m*, mitral valve ; *n*, left auricle.

which can only be filled by the expansion of the lungs; and they consequently become distended with atmospheric air, which rushes down the trachea, and makes its way through the ramifications of the bronchial tubes, to the ultimate air-cells. When the inspiratory movement has been completed, the diaphragm relaxes, and is pushed up by the abdominal viscera, under the pressure of the abdominal muscles which are now called into contraction; whilst the ribs, no longer sustained by the contraction of their elevators, are drawn down again, partly by the contraction of another set of muscles, and partly by the elasticity of their own cartilages; so that the thoracic cavity is reduced in all its dimensions, and a portion of the air contained in the lungs is expelled from them. There is usually an inequality in the size of the lungs in Mammalia, the heart being situated rather to the left side of the median line, so that a larger space remains for the right lung than for the left; in the Musk-deer this inequality is very considerable, the right lung being double the size of the left. The Larynx, which is situated at the upper extremity of the trachea, serves the double purpose of regulating the ingress and egress of air, and of producing vocal tones; two functions which appear to be separately performed in Birds by the upper and the lower larynx. The laryngeal cavity is increased in many of the Quadrumana, by the development of large sacculi in connection with it; and in the "howling monkeys" these sacculi are enclosed by an extraordinary expansion of the laryngeal cartilages and even of the hyoid bone, which concur to form a resonant apparatus, by which the power and volume of the voice are most remarkably increased.

326 *u*. The Kidneys of Mammalia are for the most part of moderate size, but possess a very compact structure, and are copiously supplied with arterial blood,—not being connected, as in oviparous Vertebrata, with the portal circulation, which is here restricted to the liver only. In some Mammalia they possess a lobulated structure, which is observable in all of them in their fœtal condition. What they want in size appears to be made up by their functional activity, to which the large quantity of water taken in by Mammals obviously contributes. The urinary excretion of Birds and Reptiles is nearly solid; but that of Mammals consists chiefly of water, in which the greater part of the solid excretory matters are held in solution, so that their transudation appears comparatively easy. All Mammals are furnished with a urinary bladder, in which the excretion may accumulate so as only to require being discharged at intervals. The urethra, or outlet of this receptacle, usually coalesces with the genital canal, both in the male and in the female; being continued in the former into the penis, and receiving the spermatic ducts in its course; whilst in the latter, it opens into the vagina. The *Monotremata*, as will be hereafter explained (§ 326 *z*) present a curious exception to this general statement, in which we see an approximation towards the oviparous Vertebrata; and an exception of an opposite kind is met with in a few cases in which the three orifices are distinct in the female, the urethra being continued onwards through the clitoris, instead of discharging itself into the vagina.—The skin, in most of the Mammalia, is an important excreting organ, being usually furnished with a multitude of glandulæ, which discharge their products upon its surface. The products of these glandulæ differ very considerably in different animals, and in different parts of the same animal. The ordinary perspiratory glands draw off watery fluid

from the blood, with a small quantity of excretory matter closely resembling that which is eliminated by the kidneys; and their action, which is specially augmented by external heat, serves to resist its undue influence, the vaporization of the perspiratory fluid abstracting warmth from the surface, and thus keeping it cool. Besides these, however, many of the cutaneous glandulæ secrete a sebaceous or fatty matter; which serves to lubricate the skin, and to prevent it from being dried up by exposure to the sun and air; these are generally restricted, however, to particular situations. In most Mammalia, too, an odoriferous secretion, characteristic of the particular species, is formed by glandulæ which pour out their product on some part of the surface, their situation, however, being extremely variable in the different tribes; this secretion is usually formed most abundantly at the period of sexual excitement, and appears to be concerned in the attraction of the sexes towards each other. A peculiar gland is found in the male *Ornithorhyncus* and *Echidna* on the outer side of the leg, and an excretory duct proceeds from it to the extremity of the peculiar horny 'spur' which these animals possess; it does not seem, however, that the spur is used as a weapon of offence or defence, so that the gland connected with it is probably not to be considered as a 'poison-gland,' but rather as an accessory genital organ.

326 *v.* It is in the development of their Nervous Centres, that the chief superiority of Mammals to all other Vertebrata is shown; yet even here, we find that the transition from the type of the Ovipara is by no means abrupt,—the Encephalic ganglia of the Implacental Mammals, and even of the Rodents, not presenting any great advance in development, as regards either size or complexity of arrangement, upon that of Birds. The *Cerebrum* of the lower Mammalia, the size of which does not bear any large proportion to that of the sensory ganglia, cannot be considered as representing the whole cerebrum of Man on a diminished scale, but only the anterior portion of it. The *corpus callosum*, or great commissure between the cerebral hemispheres of the two sides, is but very little developed in the Rodentia, and scarcely at all in the Marsupialia; its place is in some degree supplied, however, by the higher development of the *anterior* commissure, whose connection, however, is not so much with the *cerebral hemispheres*, as with the *corpora striata*. As we ascend the series, however, we find the Cerebrum increasing in size, and becoming more and more elongated posteriorly, by the development of the middle lobes; coincidently with which the corpus callosum is developed. No vestige of the posterior lobes is seen, however, until we ascend as far as the Carnivora; and even in that group they are present in merely a rudimentary form. This rudiment is somewhat enlarged in the Quadrumana; but it is only in Man that the posterior lobes attain such a development, as to cause the Cerebrum to cover-in, not merely the Sensory Ganglia, but also the Cerebellum. It will be hereafter shown that the evolution of the Cerebrum in the Human fœtus presents a series of phases, which are fairly comparable with the foregoing.—The inferior development of the Cerebrum in the lower orders of Mammalia, is marked also by the deficiency of those 'convolutions,' which, in the higher tribes, greatly increase the surface of grey matter, and also allow the blood-vessels to come into contact with it over a much larger extent. These are entirely absent in the brains of the Implacental

Mammals; the furrows present themselves only as a few shallow depressions in the Rodentia; their number increases in the Carnivora, being greater in the Dog than in the Feline tribe; among the Herbivorous Mammals, it is most considerable in the Horse and the Elephant; whilst it is in the higher Quadrumana, that the convoluted arrangement of the surface presents the closest approximation to that of Man.—The *Sensory Ganglia* are thrown (so to speak) into the shade, by the enormous relative development of the Cerebrum; but they are not really smaller in proportion to the bulk of the body, than they are in the inferior classes; and if (as is probable) we are to regard the *thalami optici* as the ganglionic centres of the sense of Touch, their relative bulk would really on the whole be greater than it is even in Birds.—The dimensions of the *Cerebellum,* also, and its complexity of structure, present a very marked increase, as we pass from the lower to the higher Mammalia; the increase being especially seen in the augmented size of its hemispheres as compared with its central portion, and in the amount of grey matter which it contains, as well as in the peculiar lamellated arrangement whereby the extent of communication between the grey and the white matter is increased. It is in the semi-erect Apes, that the condition of this organ approximates most nearly, both in form, size, and structure, to that which it presents in Man.—On the whole it may be affirmed that a very close correspondence may be traced between the relative development of the *Cerebrum* (as marked not merely by its size, but by its complexity of structure, the depth of its convolutions, &c.), and that of the Intelligence, in the different orders of Mammalia; whilst the development of the *Cerebellum* appears to be related rather to that power of combining a diversity of muscular movements, not only in acts of progression, but more especially in the maintenance of the equilibrium of the body, which is required in animals that are endowed with such a variety of powers as we have seen to be the special characteristic of the Mammalian class, and more especially in those which habitually sustain themselves without any assistance from the anterior extremities, in the erect or semi-erect position.—The *Spinal Cord,* as in the other classes, presents enlargements that correspond with the issue of nerves to the extremities; the relative size of these enlargements bearing a constant correspondence to the development of the limbs, and to the use which is made of them in the act of progression. In the Cetacea, as might be expected, the posterior enlargement is entirely deficient. Generally speaking, the spinal cord extends as far as the sacrum; beyond this point, the spinal canal contains only a bundle of nerves (cauda equina) which is prolonged from it; and this is the case even in the Cetacea, in which the tail is the chief organ of propulsion.

326 *w.* The organs of Sense in the Mammalia are remarkable for their generally high development; although perhaps that of Vision may be considered as inferior in power to that of Birds, its function in directing the movements of Mammals being partly taken-on by the organs of Smell and Hearing. The eye, except in the Ornithorhyncus, is not protected by a circlet of bony plates, but only by a tough fibrous membrane; and there is reason to believe that the range of distance to which it may be accommodated, is not so great as in Birds. In its general plan of structure, it departs little from the eye of Man, though there are several

minute variations in the different orders; thus in the Cetacea, the crystalline lens is nearly globular, as in Fishes; and in many Mammalia, a portion of the choroid is lined with a pigment-layer of metallic lustre, silvery in the Carnivora, blue-green in the Herbivora, which is called the *tapetum.* In addition to the six muscles by which the eye is moved in Man, this organ, in most of the lower Mammalia (except the Quadrumana), has an additional muscle, the 'suspensor' or 'retractor,' which is of conical form, embracing the back of the eye and the optic nerve, and serving to draw it back into the orbit. Most Mammalia possess the third eyelid or 'nictitating membrane,' of which the semi-lunar membranous fold at the internal angle of the eye of Man is a rudiment; this covering, however, is not moved across the eye by the special muscular apparatus which it possesses in Birds, being merely projected by the pressure of the globe, drawn back by the retractor muscle, against the cartilage which it contains.—The organ of Hearing, in Mammalia generally, corresponds closely in its structure with that of Man; and its chief advance in conformation, beyond the type which it presents in Birds, may be considered to lie in the more complete development of the *cochlea,* and in the great augmentation in the size of the external ear, as well as of the cancellated structure contained in certain bones which surround the auditory apparatus, whereby its resonance appears to be increased. The external ear is least developed in the aquatic Mammals, in most of which it is entirely wanting; it attains its highest development in point of size, in the nocturnal Cheiroptera, whose movements are in great degree regulated by impressions received through the hearing, and through their delicate apparatus of touch; whilst in many Herbivora, as the Horse, it is remarkable for its extreme mobility, and for the number of muscles by which it may be acted upon, so as to be directed to any point whence sound proceeds. It seems as if the want of external ears is compensated, in the Cetacea, by the extraordinary development of the extensions of the tympanic cavity, which penetrate into the cranial bones, but are partly covered with membrane only. It is remarkable, however, that in this order the semicircular canals should be of very small size; not being so large in the Whale, the largest of Mammalia, as in the Field-mouse, which is the smallest. The *cochlea* always forms a spiral, except in the Ornithorhyncus, whose cochlea resembles that of a Bird; the number of turns of this spiral varies from $1\frac{1}{2}$ to 5, the fewest being found in the Cetacea (in which order, however, this organ is very large in proportion to the semicircular canals), and the most numerous occurring in the Rodentia.—The organ of Smell is for the most part highly developed, the nasal cavity being large, and the surface over which the olfactive nerve is distributed being augmented by the convolution of the turbinated bones; of these, the lowest is the one that is usually most developed; and in the Carnivora, to which animal this sense is of the greatest importance in guiding them to their prey, this bone is divided into a number of irregular lateral laminæ, so that a transverse section of it has an arborescent form. The nasal cavity is usually more or less extended by sinuses, which communicate with large cancelli in the surrounding bones; the frontal, maxillary, sphenoidal, temporal, and even the occipital thus contribute to the formation of the olfactive organs, which are obviously much more acute in most of the lower Mammalia than they are in Man,—serving not merely to guide the Carnivora to their prey, but to warn the timid Herbivora of the approach of their enemies, and being probably also one of the chief means whereby the opposite sexes are made aware of each other's proximity. Of the extension of the nose into a tactile and prehensile organ in the Elephant, and (in a less degree) in some other Pachydermata, mention has already been made; but another curious metamorphosis of the olfactive organ has now to be mentioned, which is presented by the *Cetacea*. In this order, the sense of smell, which could not be efficiently exercised in the water, is sacrificed to two objects more particularly connected with their general organisation and habits;—namely, the reception of air into the lungs, whilst the head and body are immersed in water; and the ejection of water which has been received into the mouth, but which the animal does not require to swallow. The former object is provided for by the location of the nostrils upon the highest part of the head, and by the communication which can be established between the posterior nares and the larynx, so as to allow air to enter the trachea freely without ever passing into the cavity of the mouth. For the latter purpose a very curious valvular apparatus is provided, whereby the water can be forced from the back of the mouth into a pair of capacious cavities placed above the posterior nares, instead of into the pharynx; and by a muscular envelope surrounding these cavities, the water can be spouted forth through the 'blow-holes' or nostrils, at the will of the animal, the valves by which it was admitted to the contractile sacs being then closed. The turbinated bones are altogether wanting, and neither olfactive ganglia nor nerves are developed; so that it is probable that these animals are entirely destitute of the sense of smell. The apparatus for 'blowing' is the chief means, in many species, by which the food is obtained; thus the common Whale gulps in an enormous quantity of water, which strains through the sieve-like apparatus with which the interior of its mouth is provided; and the small animals which it may contain being thus filtered out, the superfluous liquid is ejected through the blow-holes.

326 *x*. The organ of Taste is more developed in Mammalia than in any other animals; for we find them almost universally furnished with a thick fleshy tongue, some part of whose surface is beset with sensory papillæ. The relative development of these papillæ, however, differs greatly in the several orders. Thus in the *Cetacea*, the gustative papillæ are scarcely to be found; and the tongue is comparatively small and but slightly moveable. In the *Carnivora*, the tongue is usually beset in front with horny recurved spines (the rudiments of which exist in Man as the 'filiform' papillæ), and the gustative papillæ are sparingly scattered amongst these, being more numerous, however, at the back of the organ. It is probable that, as the food of these animals is very limited in its kind, and as they are guided to it by other senses than that of taste, the development of the gustative apparatus is in them rendered subordinate to that of an instrument for the prehension and reduction of the food; for the use of the rasp-like surface of their tongue is obviously to remove from the bones of their prey the small particles of flesh that may still remain adherent to them, a single stroke of the tongue of a lion being said to be capable of abrading the whole thickness of the human skin. In the Ant-eaters, the tongue is made subservient to the acquirement of

food; this organ being used by them very much in the same mode as that of the insectivorous Reptiles (§ 324 *q*). In the Giraffe, again, as already mentioned, it is endowed with prehensile power, and is used to lay hold of the young branches upon whose leaves this animal is chiefly supported, and to draw them into the mouth. These, however, are peculiar modifications of this organ, which, in the Mammalia generally, may be regarded as the chief though not the only seat of the sense of Taste, and as specially modified for this purpose. It is in the Herbivorous, and still more in the Omnivorous Mammals (such as the Quadrumana and Man), that this sense appears, from the predominance of the gustative papillæ, to be most developed, and to have the most influence in the selection of food.—The sense of Touch, also, appears to be more developed in Mammalia than in any other class; attaining in them its greatest acuteness and discriminating power. Although the whole integument is probably endowed with sensibility, except when enclosed in a dense corneous or osseous envelope, yet it is undoubtedly in those portions of it which are furnished with vascular papillæ supplied with nerve-fibres, that the tactile power is chiefly located. Thus in the Quadrumana generally, both pairs of extremities are thickly set with papillæ; and in those which have a prehensile tail, the under surface of this organ possesses them in abundance. But in the Carnivorous and Herbivorous Mammals, whose extremities are furnished with claws or encased in hoofs, we find the lips and the parts surrounding the nostrils to be the chief seats of tactile sensibility, these being copiously furnished with papillæ; this is especially the case with those which have the lips or nostrils prolonged into a snout or proboscis, as the Pig, the Rhinoceros, the Tapir, and the Elephant; and it is particularly remarkable also in the Mole, in which animal the tactile sensibility of the snout appears to be the means whereby its movements are immediately guided, their general direction, in the search for food, or in the pursuit of the opposite sex, being given by the extraordinary development of the organ of smell. Besides the ordinary papillary apparatus, however, we occasionally find the bulbs of certain hairs to be copiously supplied with nerves, in such a manner as to become tactile instruments of great acuteness. Such is the case, for example, with the long stiff hairs which are known as the 'whiskers' of the Feline tribe, and which are so particularly large in the Seal; these are also highly developed in many of the Rodentia, such as the Hare and Rabbit; and it has been proved by experiment that if they be cut off, the animal loses in great degree its power of guiding its movements in the dark. The nocturnal Bats, on the other hand, seem to derive their chief guidance in locomotion from another very curious modification of the organ of Touch; for in them the whole membranous expansion of the wing is endowed with exquisite sensibility to the impressions derived from undulations of *air;* so that even after they have been deprived of sight, smell, and hearing, these animals will fly about a room across which numerous threads have been stretched, without ever touching one of them, or striking against the walls. It is probable that the additional expansions of the cutaneous surface, which constitute the large external ears and the 'nose-leaf' of some of the most lucifugous of these bats, contribute to this power; the impressions derived from the movement of the air struck by the wings, which will vary according to the propinquity of any solid body, being apparently their chief means of guidance in the dark penetralia of caverns and other similar haunts frequented by them, to which no appreciable amount of light ever gains access.

326 *y*. The Muscular apparatus of Mammalia presents a series of most remarkable variations, which are in accordance with the form of the skeleton, the relative development of the trunk and members, the use of the several organs in the act of locomotion, the habitual posture of the animal, and the development of special instruments for particular purposes. Of these variations, a few characteristic examples only can be here given.—In the muscular apparatus of the *Cetacea*, we are almost carried back to the simple type of conformation which prevails in that of Fishes (§ 322 *i*). The movements of these animals through the water are chiefly accomplished by means of the muscles of the trunk and tail; but as the latter organ is expanded horizontally, instead of vertically, in order that, by its vertical impulses on the water, they may be brought up to the surface for the respiratory act, the muscles which move it are chiefly disposed above and below the vertebral column, instead of on its sides as in Fishes. So, again, as their anterior extremities are converted into fins, the chief purpose of which seems to be to sustain the equilibrium of the body, and as there is no external separation of digits, or movement of the several parts of the member one upon another, the muscles which in other Mammalia are destined for the motion of the fore-arm, hand, and fingers, are altogether wanting, whilst those of the scapula and humerus are nearly all present, but under peculiar modifications. The muscles of the posterior extremities are of course entirely absent, these extremities themselves being deficient. Again, as the head is nearly immoveable upon the trunk, owing to the thinness and anchylosis of the cervical vertebræ, the muscles of the neck are but little developed.—In the *Cheiroptera*, on the other hand, whose locomotion closely resembles that of Birds, the muscles of the anterior extremities, and especially the pectorals which are the chief depressors of the wings, are extraordinarily developed, as in that class. In the *Rodentia*, on the contrary, and still more in the Kangaroo, in which the posterior extremities are the chief instruments of locomotion, their muscles are enormously developed, and those of the anterior extremities are insignificant in proportion. In the Kangaroo, again, the muscles of the tail attain a relative size which has no parallel among existing Mammalia, although it might have probably been paralleled by that of some of the extinct *Gravigrada* or gigantic Sloths (§ 326 *c*). The muscular apparatus of the tail is also highly developed in the American Monkeys, in which that organ is enabled by its prehensile power to serve as a fifth hand.—The greatest uniformity in the muscular apparatus of the limbs is to be found in the *Ungulated* quadrupeds, in which both pairs are used for progression simply, in a mode and degree almost exactly alike. In many of these, we find the muscular apparatus of the neck most extraordinarily developed, in order to sustain the weight of the large head. In their limbs, as in the legs of Birds, we find that the muscles are situated at the upper part; the simple movements of flexion and extension, which alone they are needed to effect, being communicated by means of long tendons that pass between the muscles and the bones on which they act. In the ordinary *Carnivora*, there is the same general correspondence between the muscles of the

anterior and those of the posterior extremities ; but in proportion as these are endowed with prehensile power, do we find a greater variety and complexity in the arrangement of the muscular apparatus. In this order we still find the muscles of the neck very strong ; not so much to support the weight of their own heads, which are seldom as bulky as those of the Herbivora, as to enable them to drag or carry the bodies of the animals on which they feed. That which is most remarkable in them, however, is the enormous power of the muscles (the temporal and masseter) which elevate the lower jaw. In the *Quadrumana* we have a gradual transition from the ordinary Quadrupedal type to one which approximates the Human conformation ; and this is shown in the arrangement of the muscular apparatus, as well as in that of other parts. The prehensile muscles, as well as those which serve for locomotion, are developed in all the limbs ; the phalanges, more especially, being endowed with a power of flexion and extension, which they do not elsewhere possess ; and the digits being more or less capable of separation from, and of approximation towards, each other. In these and other respects, however, there is a much closer conformity between the muscles of the anterior and posterior extremities, even in the most anthropoid Apes, than exists in Man ; in whom the *arm* and *hand* are modified more exclusively for prehension, and the *leg* and *foot* for support. It is in him that we find the most complete *specialization* of parts, every one having its own distinct function, and in consequence not being a repetition of any other. This is peculiarly obvious in the power which he possesses of flexing or extending each finger separately, and of bringing the thumb into opposition to every one of them ; which is not possessed even by the most anthropoid Apes. He is further distinguished by the great number of distinct facial muscles which he possesses ; whence he derives that power of *expression*, which is so much greater in his countenance than in that of any other animal. The tegumentary muscles, however, are much less developed in Man, than in most of the lower Mammalia ; for whilst in him they are very thin, and are limited to particular situations, chiefly about the head and neck, they frequently present themselves elsewhere as thick muscular layers, spreading over the whole anterior part of the body ; and in those mammals which can roll themselves up into a ball, such as the Hedgehog and the Porcupine, they form a sort of fleshy cap, which in this contracted state covers the whole back, sides, and part of the extremities, and is the chief means whereby this peculiar position is assumed and maintained. Another special development of muscular structure, of most remarkable elaborateness, is that which is found in the proboscis of the Elephant ; this consists of numerous longitudinally-arranged muscular fasciculi, with tendinous contracted portions, which act as flexors of the proboscis ; whilst their antagonists are transverse or oblique fasciculi, imbedded in a net-work of adipose tissue. In all there are estimated to be from 30,000 to 40,000 of these fasciculi.

326 *z*. The condition of the Generative organs, more especially those of the female sex, presents a remarkable diversity in the several orders of this class ; for we find almost every gradation in them, from a type which presents but a very slight advance upon that of Birds, to that which is found in Man, and which may be considered as the highest example of the Mammalian conformation, being destined to afford support and nur-

ture to the offspring during the longest relative period.—The spermatic organs or *testes* of the male usually correspond pretty closely in structure with those of Man; being composed of long tortuous *tubuli seminiferi*, in whose interior the spermatic cells are formed, and the fluid part of the secretion elaborated. They are always developed within the posterior part of the abdominal or in the pelvic cavity, occupying nearly the same situation as the ovaria in the female; and this situation they retain in many of the lower Mammalia, such as the *Cetacea* and *Monotremata*, as also some Pachydermata and Rodentia. But in most of the other orders, they pass out into a peculiar prolongation of the abdominal cavity, invested by an extension of its muscles and integuments, which is called the *scrotum;* the cavity of the scrotum frequently communicates freely, however, with that of the abdomen, by an open inguinal canal; and it is only in Man, the Quadrumana, and a comparatively small proportion of the lower orders, that the passage between the two cavities is closed by the contraction of the canal, so as to prevent that descent of the intestines into the scrotum which might take place in the erect posture. The spermatic ducts often communicate with *vesiculæ seminales*, which appear to serve both as receptacles for the seminal fluid, and as accessory glandular organs; they are, however, frequently wanting. In all instances, the spermatic canals have their final outlet at the extremity of the penis, being carried onwards to it even where the urinary canal is not continued to that point, which is the case in the *Monotremata*. In this order, the seminal ducts unite with the urethra to form a 'uro-genital canal,' which opens into the cloaca; the penis, however, is perforated by a duct, that subdivides at its bifid extremity, first into two branches, and then into further ramifications which terminate in the papillary elevations of its surface; and it would seem that during coitus, the commencement of the duct of the penis is applied to the orifice of the uro-genital canal, in such a manner that the fluid ejected from this shall be transmitted along the intromittent organ, although the urine discharged from the same orifice at other times passes into the cloaca. Thus, as Prof. Owen remarks, "if the canal of the penis were slit open along its under part, and thus converted into a groove, the male organs of the Ornithorhyncus would be then essentially like those of a Tortoise. The adhesion to the Mammalian type is manifested in a highly interesting manner by the completeness of the urethral canal; whilst the complete separation of the uro-urethral from the semino-urethral passages beautifully illustrates the fact, that the existence of the penis is essentially and subordinately related to the sexual organs and not to the renal." *—In the *Marsupialia*, the seminal ducts discharge themselves into the urethra, and this is continued onwards to the extremity of the penis; in some members of this order, the penis is bifid, and each half receives a branch of the urethral canal, as in the Monotremata; whilst in others it has a single termination, as in all higher Mammalia.—Various accessory glandular organs are found in the males of most Mammalia, discharging their contents into the urethra, and apparently connected with the generative function; their use, however, is not known.

326 *aa.* The *ovaries* of the Mammalian female are always found near

* Cyclopædia of Anatomy and Physiology, vol. iii. p. 392.

the posterior part of the abdomino-pelvic cavity; and they consist of a dense fibrous *stroma*, in the substance of which the ova are developed. These, when mature, are set free by the thinning-away of their envelopes; and are received into the trumpet-shaped dilatations of the oviducts, which canals are known in this class as the 'Fallopian tubes.' The oviducts remain quite distinct from each other in the *Monotremata*, and terminate separately in the uro-genital canal, one on either side of the orifice of the bladder; each of them having first undergone dilatation into a uterine cavity, so that these animals have two completely distinct uteri. The uro-genital canal opens below into the *cloaca* or common vestibule, its orifice being guarded by a sphincter muscle; and this vestibule also receives the rectum, as in Birds; so that the female generative apparatus of the Monotremata differs but little from that of the highest Ovipara, in which even the uterine dilatation of the oviduct has its representative (Fig. 188). There is not, however, as in Birds, any tendency to the unequal development of the ovaries and their appendages on the two sides.—In the *Marsupialia*, there is a closer approximation of the two lateral sets of organs on the median line; for the oviducts converge towards one another, and meet (without coalescing) on the median line; so that their uterine dilatations are in contact with each other, forming a true 'double uterus.' Each uterus, at its lower extremity, opens by an *os tincæ* into a separate vaginal canal; and the two canals diverge again, so as to terminate separately in the uro-genital canal, which receives both their orifices and that of the urinary passage, but terminates externally without coalescing with the rectum in a common vestibule. The vaginal canals are frequently furnished, in this order, with peculiar sinuses or dilatations for the reception of the fœtus after it has left the uterus, in which it has undergone a part of its development.—As we ascend the series of placental Mammals, we find the lateral coalescence becoming gradually more and more complete. It shows itself first in the vagina, which is everywhere single, although a trace of separation into two lateral halves is seen in the Mare, Ass, Cow, Pig, and Sloth, in which animals it is traversed, in the virgin state, by a narrow vertical partition. In many of the *Rodentia*, the uterus still remains completely divided into two lateral halves; whilst in others, these coalesce at their lower portion, forming a rudiment of the true 'body' of the humerus in the Human subject. This part increases at the expense of the lateral 'cornua,' in the higher Herbivora and Carnivora; but even in the lower Quadrumana, the uterus is somewhat cleft at its summit, and the 'angles,' into which the oviducts enter, form a considerable part of the whole organ. As we ascend through the Quadrumanous series towards Man, we find the 'body' increasing, and the 'angles' diminishing in proportion; until the original division is completely lost sight of, except in the slight dilatation of the cavity at the points at which the Fallopian tubes enter it. In most Mammalia the female possesses a clitoris, which resembles a rudimentary penis, and which in some instances attains a considerable size. The urethral canal generally terminates near the base of this organ; but in some of the Marsupialia, this canal seems to be continued as a groove along its under side; and in the Lemming, the Mole, and some of the lower Quadrumana, it is completed into a tube, which passes along the whole length of the clitoris, just as the urethra of the male traverses the penis. The

Marsupialia are remarkable for the peculiar addition to their generative apparatus, which consists of a sac or pouch upon the front of the pelvis, formed by a pair of folds of the integument, and covering-in the mammary glands and nipples ; and the function of this 'marsupium' is obviously to afford protection to the young animal, in that state of extreme immaturity which characterizes it at the date of its emersion from the uterus and for some time afterwards ; during which period it is clinging to the nipple, at first apparently almost unpossessed either of sensibility or of power of movement, but gradually acquiring the ability to sustain an independent existence.—The possession of *mammary glands,* as already remarked, is the exclusive attribute of the class Mammalia ; and they are universally present throughout the series, being even found in the Ornithorhyncus. In the greater part of the class, they are found in close connection with the generative apparatus, their orifices in the Cetacea being very near the outlets of the rectum and vagina, whilst in many other groups (as the Ruminants) they are only a little removed from these ; but as we ascend towards Man, we find them occupying a more and more advanced position on the trunk, until they are limited, as in him, to the pectoral region. In complexity of structure, also, the advance is alike gradual : for the Mammary glands of the *Ornithorhyncus* are of the most simple conformation, consisting of a mere cluster of isolated follicles without any nipple (Fig. 230) ; those of the *Cetacea* are but little above them in structure, and the nipples are buried in a cleft of the integument ; whilst in the higher Mammalia, these glands are formed upon a very elaborate type, and the nipples are prominently developed, so as to be received into the mouth of the offspring, which draws forth the secretion by suction. The most remarkable development of the nipple, however, is found in the *Marsupialia,* in which it serves for the attachment of the 'marsupial fœtus,' and for the conveyance into its œsophagus of the secreted fluid, which is here expelled from the mammary gland by the action of a compressor muscle. A similar expulsor action seems to be exercised by the abdominal muscles of the *Ornithorhyncus ;* for although (as Prof. Owen has shown) the young possesses flexible lips instead of a horny bill, and a tongue adapted for suction, these can scarcely act effectually without a nipple, and the nutriment is probably imparted and received by both actions conjointly.

326 *bb.* Although the class of Mammalia, as a whole, is essentially *viviparous,* yet the condition in which the young are brought into the world varies greatly in its different orders ; and the subsequent attainment of that of the adult does not always depend upon growth or increase alone, but usually involves a certain alteration in form, as well as the development of certain organs which are only rudimentarily present at the time of birth. The young Mammal, however, never presents the least approximation to any inferior class, in those special modifications of the general plan of Vertebrate organisation by which they are respectively characterized ; and, as will be shown hereafter, the whole course of its development, before as well as after birth, is destined to the evolution of the Mammalian type, and of no other, out of that which, at a certain grade of the process, is common to this class with Birds, Reptiles, and Fishes.—The condition of the young of the *Monotremata* at the time of their birth, has not yet been ascertained ; but at the earliest subsequent

period at which they have been met with, they are entirely destitute of hair, the body is doubled up so that the head and tail approximate each other, the limbs are but very little developed, and the eyes are completely covered by the integument; so that it would appear that for some time after their entrance into the world, they can scarcely possess more locomotive power than that which belongs to the human fœtus at about half its term of uterine gestation. The young *Marsupial*, at the time of its exit from the maternal vagina, is extremely small in proportion to the size of its parent, the new-born offspring of a Kangaroo that measures 7 feet in height (when standing erect on the hind feet and tail), not being more than an inch in length. Like the young Ornithorhyncus, it is blind and destitute of hair, its body is bent forwards, its hind-legs (which are so enormously developed in the adult) are smaller than the fore-legs, and the short tail is tucked in between them. It is probably conveyed to the *marsupium* by the mouth of the mother; and there it attaches itself to the very elongated nipple, its mouth being so formed as not merely to receive this, but to hold firmly to it, whilst its larynx is continued upwards into such close proximity with the posterior nares, that the air-passage is kept distinct from the alimentary tube, and the milk ejected by the mammary gland of the mother passes into the œsophagus, without interfering with the act of respiration. In this condition it remains for some months, gradually increasing in size and advancing in development; it then no longer suspends itself to the nipple, but lies freely in the pouch, occasionally protruding its head from its aperture, and cropping the grass at the same time that the mother is browsing; and after a further period of marsupial gestation, it quits the pouch, still occasionally returning to it for shelter and supply of milk, until it has attained a weight of ten pounds.—In no other Mammalia do the young quit the uterus at so early a grade of development; their relative size at birth being far greater, and their organisation in every respect more advanced. Still we find that the young of the *Carnivora* and of most *Rodentia* are brought forth in a very feeble condition, their eyelids being adherent, their movements being feeble, and their calorifying power being insufficient for the independent maintenance of their heat, for some time after birth. On the other hand, in the *Ruminants* and several other orders, the young possess the sense of vision from their first entrance into the world; they speedily acquire considerable locomotive power; and they are from the first capable of sustaining their normal temperature, in an atmosphere not too much below this, without the calorifying assistance of the parent. In all cases, however, the young are for some time entirely or chiefly dependent upon the maternal parent for their supply of food; and the teeth, by whose instrumentality they are afterwards to prepare it for themselves, are not developed until near the close of the period of lactation. The generative organs are the last to undergo their full development, and to come into functional activity; this not taking place until the dimensions of the adult have been nearly attained. The period required for this purpose varies considerably, among the different orders of Mammalia, in its proportion to the whole term of life; being usually shortest in the lower orders, such as the Rodentia, in which the evolution of the nervous system is checked at a very early period, whilst it is longest in Man and in those that approach him.

CHAPTER VIII.

ON THE GENERAL PLAN OF ORGANIC STRUCTURE AND DEVELOPMENT.

327. There are few things more interesting to those who feel pleasure in watching the extraordinary advancement of almost every department of knowledge at the present time, than the rapid progress of philosophical views, in sciences which have hitherto been too much confined to mere observation. The laws of *Life* were long considered beyond the reach of human investigation; and the mind shrunk from attempting to analyse the complex and varied phenomena, which, though constantly under observation, must be reduced to their simplest form, before any inductive reasoning can be founded upon them. It is recorded, however, of Newton, that, whilst contemplating the simplicity and harmony of the plan according to which the Universe is governed, as manifested in the relations which his gigantic mind developed between the distant and apparently unconnected masses of the solar system, his thoughts glanced towards the organised creation; and reflecting that the wonderful structure and arrangement which it exhibits, presents in no less a degree the indications of the order and perfection which can result from Omnipotence alone, he remarked, "I cannot doubt that the structure of animals is governed by principles of similar uniformity." ("Idemque dici possit de uniformitate illâ, quæ est in corporibus animalium.") "Why," asks Cuvier in his eloquent discourse on the revolutions of the globe, "should not Natural History some day have its Newton?"—Although the labours of the Naturalist and Comparative Anatomist have not yet unveiled more than a small part of that general plan, the complete discovery of which may perhaps be reserved for another Newton, many subordinate principles have been based on a solid foundation, and many more, which were at first doubtful, are daily receiving fresh confirmation. Several of these laws are alike important from their extensive range, and interesting from the unexpected nature of the results to which they frequently lead; and though their application may sometimes appear forced, and inconsistent with the usual simplicity of Nature, further investigation will generally show that the difficulty is more apparent than real,—frequently arising solely from our own prejudices, and diminishing in proportion as we fix our attention upon that *combination of unity of plan with variety of purpose*, by which is produced the endless diversity united with harmony of forms, so remarkable in the animated world.

328. In comparing phenomena of any kind, for the purpose of arriving at a principle common to them all, it is necessary to feel certain that they are of a *similar character*. Indeed the sagacity of the philosopher is often more displayed in his discovery of that relation amongst his facts, which allows of their being compared together, than in the inferences to which such comparison leads him. The brilliancy of Newton's genius was shown in the perception that the fall of a stone to the earth, and the

motion of the moon around it, were *analogous* phenomena, subject to the same law; not in the mere deduction of the numerical law from the ratios supplied by those facts. In the sciences which have Life for their subject, the dissimilarity of the facts which are made the object of comparison, often prevents the true relation between them from being readily detected. Here it is that the mental training which the previous cultivation of Physical science affords, becomes peculiarly valuable to the Physiologist. "The most important part of the process of Induction," says Professor Powell,* "consists in seizing upon the probable connecting relation, by which we can extend what we observe in a few cases to all. In proportion to the justness of this assumption, and the correctness of our judgment in tracing and adopting it, will the induction be successful. The analogies to be pursued must be those suggested from already-ascertained laws and relations. Thus, in proportion to the extent of the inquirer's previous knowledge of such relations subsisting in other parts of Nature, will be his means of guidance to a correct train of inference in that before him. And he who has, even to a limited extent, been led to observe the connection between one class of physical truths and another, will almost unconsciously acquire a tendency to perceive such relations among the facts continually presented to him. And the more extensive his acquaintance with Nature, the more firmly is he impressed with the belief that some such relation must subsist in all cases, however limited a portion of it he may be able actually to trace. And it is by the exercise of unusual skill in this way, that the greatest philosophers have been able to achieve their triumphs in the reduction of facts under the dominion of general laws."—If, as was formerly stated (CHAP. I.), the true objects of Physiological investigation are only now beginning to be understood, it is no less certain that the true mode of pursuing them has not long been followed. From the time of Aristotle downwards to the commencement of the present century, Anatomists have been in the habit of regarding similarity of external form, and of evident purpose, as indicating the analogies between different parts. Now this mode of estimating the character of organs is perfectly correct when they are considered as *instrumental structures*; that is, when we are inquiring into the conditions of the function performed by them. But it totally fails, when we are examining into the laws according to which the development of the organs has taken place; since it is frequently found that two structures, which are not unlike in external form, and have corresponding functions in the system, originate from elements entirely different, and are therefore fundamentally dissimilar.

329. If, for example, we take a cursory glance at the organs of support or motion in the air with which different Animals are furnished, we shall observe a community of function, and a general similarity of external form, concealing a total diversity of internal structure and of essential character. Amongst all the classes which are adapted for atmospheric respiration, we encounter groups of greater or less extent, in which the resistance of this element becomes the principal means of progression; and even among aquatic animals, there are many instances in which the function of locomotion is partly dependent upon the same agency.

* Connection of Natural and Divine Truth, p. 33.

Wherever *true wings* exist among the Vertebrata, some modification of the anterior member serves as their basis; but there is considerable variety in the mode in which the apparatus is constructed. Thus, in the *Bat* (§ 326 *e*) the required area for the surface of the wing is formed by an extension of the skin over a system of bones, of which those of the hand form a very large part; and this membrane is extended also from the posterior extremity, and is attached to the whole length of the trunk, as well as to the tail where one exists. In the *Bird* (§ 325 *b*), on the contrary, the wing is formed by the skin and its appendages attached to the anterior member alone; and here the bones of the hand are developed in a comparatively slight degree, those of the arm and fore-arm being the principal support of the expansion. From what is preserved of the *Pterodactylus* (§ 324 *m*), it seems that the wing of this extraordinary animal was extended, not over the whole member, as in the Bird,—nor over the hand, as in the Bat,—but over one of the fingers only, which was immensely elongated in proportion to the rest. In the *Flying-fish*, again, the pectoral fins may be regarded as, in some sort, its wings; though it does not appear that the animal has the power of raising itself by means of their action on the air, the impulse being given at the moment of quitting the water. These fins evidently represent the anterior members of higher Vertebrata; but the bones of the arm and fore-arm are scarcely developed, while the hand is expanded, and joined immediately, as it were, to the trunk.

330. A very different structure prevails among those imperfect wings, which serve rather to *support* the animals which possess them, in their movements through the air, than to *propel* them in that medium. Thus, in the *Flying Squirrels*, *Flying Lemurs*, and *Phalangers* or 'flying opossums,' there is an extension of the skin between the fore and hind legs (Fig. 193), which, by acting as a parachute, enables these animals to descend with safety from considerable heights. In the *Draco Volans* (Fig. 179), on the other hand, the wings are affixed to the sides of the back, being supported by prolongations of the ribs, and are quite independent of the extremities. Here we have still the same function and general form; but it would evidently be absurd to say that the organs are of the same structural character.—Among the Invertebrated classes, there is still greater variety in the construction of the organs, which make use of the resistance or impelling power of the air as a means of locomotion. Details on this subject have already been given in various sections of the preceding chapter; and it will suffice to refer to what was there stated in regard to one of the most remarkable kinds of propelling organs met with in these classes,—the wings of Insects (§ 314),—namely, that there now seems adequate reason to consider them as essentially appendages to the respiratory system.*

FIG. 193.

Galæopithecus, or Flying Lemur.

* That they bear no real analogy to the wings of Vertebrata, would appear almost self-evident, when their structure is compared; and yet there are Entomologists who have

331. These instances will show the caution which must be exercised in deciding upon the relations of the organs of different Animals, from correspondence in external form and function merely. Many similar examples might readily be adduced from the Animal kingdom; but the Vegetable world affords them in even greater abundance. To take a very simple case;—the *tendril* is an organ developed to serve a particular purpose, that of supporting the plant by twining round some neighbouring prop; but this varies much in its real character, being in the Vine a transformation of the peduncle or flower-stalk, in the Pea a prolongation of the petiole or leaf-stalk, in Gloriosa the point of the leaf itself, whilst in the singular genus Strophanthus, it is actually the point of the petal which becomes a tendril and twines round other parts.

332. In all these instances we might with perfect propriety found any inquiries regarding the *function* of the organs respectively compared, upon their external analogy. For example, we might estimate the respective rates at which Birds, Bats, and Insects could be propelled through the air, by ascertaining the superficial extent of their wings, and the force and rapidity with which these are moved; although, as just shown, the wings are really formed upon a different plan in each case. But to maintain that the same laws of *formation* can apply to them, is evidently absurd. The organs are not related in *this* respect; and no such comparison can be erected between them, as could help to the determination of the fundamental principles of which we are in search. Although in each case the extent of surface is given by an expansion of the general integument, that expansion is supported in the Bat and Bird by one part of the osseous system; in the Draco Volans by another; and in the Insect by prolongations of the respiratory apparatus. The Philosophical Anatomist, who seeks to determine the *structural relation* of these organs, first considers their internal conformation, and examines into the *elements* of which they are composed. In the cases just alluded to, he would find these elements nearly the same in the wings of the Bat and Bird, consisting of the bones, muscles, &c. of the superior extremity; whilst he would discover nothing corresponding with these in the Insect, the 'nerves' of whose wings are composed of tracheæ and vessels.—He will then trace their *connections;* and he will thus gain a clearer insight into the nature of the respective parts. It is a principle of very extensive application, that similar parts are connected with similar parts, in different Animals of the same type. Thus, we never find a hand or foot springing directly from the spinal column of Vertebrata; but always through the medium of other bones, which, however different their size and shape, are never wholly wanting. This principle, as we shall presently see, may be applied pretty rigorously within the limits of each principal type of structure; and the application of it often assists the Anatomist in ascertaining the real character of an organ. He will thus see that the principal bones supporting the wing of the Bat are those of the hand; whilst in the Bird they are those of the arm.—But the most certain of all the means of discovering the structural relations of organs, is the study of their *development;* and this, if carefully pursued, will scarcely ever fail to clear

maintained that the wing of an Insect is a modification of its leg. A very little attention to the relative positions of these parts, and to the history of their development, will disprove this doctrine; and the true homology of the wings of Insects must be sought in the lower Articulata.

up whatever doubts may be left by other modes of investigation. It is in this manner, as already shown, that the true solution has been at last attained, of many of the most difficult and controverted questions in the Osteology of Vertebrated Animals; such as the nature of the spiny rays supporting the median fins of Fishes (§ 322 *b*), and the composition of the bony carapace and plastron of the Chelonia (§ 324 *h*).

333. We may have recourse to the Respiratory system for another instance, which will bring the difference between *functional* similarity, or *analogy*, and *structural* correspondence, or *homology*,* into clear view. An uninstructed observer would scarcely perceive any resemblance between the gills of a Fish and the lungs of a Quadruped, or between the elegant tufts on the body of a Sand-worm, and the air-tubes ramifying through the structure of an Insect; and those who are in the habit of forming exclusive notions upon a hasty survey, might be led to deny that any real analogy could exist. When the character of the function is investigated, however, with the structure which it requires for its performance, it becomes evident that, in order to bring the circulating fluid into the due relation with the atmosphere, all that is needed is a membrane which shall be in contact with the air on one side, and with the fluid on the other. And this key, applied to the examination of all the forms of respiratory apparatus which exist in the Animal kingdom, shows that they all possess the same essential character as instrumental structures, and that their modifications in particular instances (which will hereafter be specially described, CHAP. XIII.) are only to adapt them to the conditions of the organism at large. There is therefore, *functionally* considered, a relationship of *analogy* amongst all these organs; although they are not really the *homologues* of one another. Thus the gills of the Fish, and the branchial tufts of the Sand-worm, are external prolongations of the tegumentary surface; whilst the tracheæ of the Insect, and the lungs of air-breathing Vertebrata are internal reflexions of that surface; and farther, the two former sets of organs, as the two latter, differ from each other in regard to the *part* of the surface from which the prolongation or inversion takes place. In the *Perennibranchiate Batrachia* (§ 323), both kinds of organs are present; and their essential difference of character is most apparent, whilst their correspondence as instruments of the same functional operations is equally evident. Now in the air-bladder of the Fish, we have an apparently anomalous organ; the only known use of which is to assist in locomotion. What is its real character? It is certain, from a comparison of its most developed forms with the simplest pulmonary sacs of Amphibia, that it is to be regarded as a rudimentary lung (§ 322 *h*); and the study of its development leads to the same conclusion.—Here, then, we have a structural analogy, affording a very interesting addition to the

* In the former editions of this work, the terms *functional* and *structural analogy* were used to express the mutual relations of parts, on the one hand as *instrumental structures*, on the other as *fundamentally* or *organically correspondent*. It will be found convenient to limit (as Prof. Owen has done) the use of the term *Analogy* to functional resemblance, and to employ *Homology* as indicative of structural correspondence. Thus by Analogue we now understand "a part or organ in one animal, which has the same function as another part or organ in a different animal:" whilst by Homologue is implied "the same organ in different animals under every variety of form and function." (Prof. Owen's Hunterian Lectures, Vol. I. Glossary.) Thus, for example, the wings of an Insect are the analogues of those of a Bat or Bird, but not the homologues; whilst the latter are homologues with the arms of Man, the fore-legs of a Quadruped, and the pectoral fins of a Fish.

collection of *instances* upon which the laws of formation are based, and nevertheless totally valueless as regards the laws of the respiratory function. It would be easy to adduce parallel examples from the Vegetable kingdom; where organs which correspond in structure, connections, and development, and which are therefore *homologous*, are observed to assume the most varied forms, and to perform the most different functions; whilst the same functions are often performed, as just shown (§ 331), by organs which are really different, though *analogous* as regards the form they assume and the uses to which they are applied.—These illustrations, however, must at present suffice.

334. Much controversy has taken place among Physiologists, in regard to the general laws founded on the comparison of different plans of organisation. This has been chiefly due to a neglect of the difference just pointed out, between *functional* and *structural* correspondence, that is, between *analogy* and *homology*. The important general law, designated as that of *Unity of Composition*, is applicable to both, as will presently appear. By some, however, who have passed by functional analogy altogether, this law has been stated in a form which facts by no means warrant; these Anatomists maintaining that the same elementary parts exist in all Animals, and that the only difference between the various classes is in the respective development of these parts. This is by no means true as a general statement; but in a more restricted form it may be correctly upheld, and its application is then very striking. If we separately consider the four great divisions of the Animal Kingdom, we observe that the classes forming each agree in certain leading characters; and that their differences among each other are not so much produced by alterations in the general plan of structure, as by variations in the proportionate development of parts which are present under some form in all, at some period of their existence.—This conformity to a common plan is the closest in *Vertebrated* animals; in which (as we have seen) the neuro-skeleton, which is the most distinctive feature in their structure, may be referred, throughout the whole series, to a common 'archetype,' so that the skeleton of a Fish or a Reptile may be shown to be formed of essentially the same parts with that of a Bird or a Quadruped, though the form of each individual bone may be totally dissimilar in the four cases. It is very easy to trace a similar conformity in the structure of their nervous system, with which, as formerly remarked (§ 320 *b*) the vertebral skeleton has a peculiarly close relation; and the same may be said of the organs of sense and the muscular apparatus. The apparatus of vegetative life, again, is evidently constructed upon an uniform plan; the only apparent departure from this being in the case of the respiratory organs; and even this anomaly ceasing to be such when carefully looked into, since we not only find the lung of air-breathing Vertebrata existing in a rudimentary state in Fishes, but also discover (by the study of their development) that the rudiments of a branchial apparatus are found in the embryoes of all the higher classes (CHAP. XIII.).—Among the several classes of the sub-kingdom *Articulata*, moreover, the same general correspondence may be traced, alike in their dermal skeleton, in their sensori-motor apparatus, and in their organs of nutrition; but their general plan of structure is so different from that which prevails in Vertebrata, that no comparison can be instituted between the classes of

the two sub-kingdoms, except in the way of *analogy;* the adaptation of Birds and Insects (for example) to the same general mode of life, being effected by modifications in their respective organisms, which present (as we have seen) a certain degree of correspondence. In both the foregoing cases, it is in the *bi-lateral symmetry,* which is so peculiarly characteristic of locomotive power, and in the development of the sensori-motor apparatus, and of the organs in immediate connection with it, that we find the most striking manifestation of a general plan.—In the *Mollusca,* on the other hand, the sensori-motor apparatus is altogether subordinate to that of vegetative life; and as the plan on which the latter is constructed does not involve any such conformity to a definite figure as we see among the Radiata, the evidence of a "Unity of Composition" throughout the entire series is much less obvious than in the preceding groups. Still it is not really less complete; for, as has been shown in the preceding Chapter, although the *general form* of the body is so indeterminate, there is a remarkably close conformity to one general type in the structure and connections of the apparatus of vegetative life; and notwithstanding that the locomotive apparatus is constructed upon a different plan in every one of the classes included in this series, it may still be regarded, in each case, as an increased development of some portion of that muscular integument, which, in the lowest group (the Tunicata), is the sole instrument of motion. The respiratory organs are, perhaps, the portion of the apparatus of nutrition which is the least constant in position and connections; but this want of constancy seems to be related to those diversities in the arrangement of the muscular system, and in the general conformation of the body, on which we have just remarked.—In the *Radiata* we observe, for the most part, a very regular *form,* resulting from that regularity in the disposition of the component parts round a common centre, which prevails not merely in the organs of nutrition, but in the sensori-motor apparatus also, where this has a distinct existence. By this *radial symmetry,* we are conducted towards the Vegetable kingdom, and it is interesting to observe that where there is a longitudinal axis of growth, as well as a plane of radial extension, the addition of new parts really takes place, as in the higher Plants, in a *spiral* direction,—such being the mode in which, according to Prof. Agassiz, the new plates are successively added in the shell of the *Echinus* (§ 296 *c*), and such also appearing to be the plan on which the progressive multiplication of tentacula takes place in the *Helianthoid* Polypes (§ 291). In the general condition, however, of the organisms included in the Radiated sub-kingdom, there is, perhaps, a greater amount of diversity than prevails within the limits of the Molluscous, Articulated, and Vertebrated sub-kingdoms respectively.

334*a*. Even in this diversity, however, some indications of conformity to a general plan may be perceived, which are strengthened by facts that may be noted elsewhere. Thus, one of the leading characteristics of the class *Echinodermata,* is the possession of an elaborate external skeleton, which gives the general form to the body, and serves for the attachment of its muscles; in many members of this class, moreover, a bilateral symmetry is apparent; and in the Holothuridæ there is an evident approximation to the Vermiform tribes. Here, then, we perceive a sort of sketching-out of the leading features of the *Articulated* series. On the other hand, in the softness of the bodies of the *Acalephæ*, the absence of any definite skeleton, the very low development of their muscular apparatus in comparison with that of their nutritive, and the consequently small amount of any other movement than such as depends on ciliary action, we seem to have a foreshadowing of *Molluscan* characters. Even in the one order of *Asteroid* Zoophytes (§ 292), we find the skeleton exhibiting the two opposite plans which are characteristic of the Articulated and Vertebrated series respectively ; for whilst in the *Tubipora* the soft parts are ensheathed within a stony tube that is formed by the consolidation of the integument, as in the Articulated sub-kingdom, the stony support is internal in the *Red Coral*, being invested by the soft flesh, and in the *Isis Hippuris* we are still more reminded of the Vertebrated structure by the jointed character of this central axis, produced by the alternate deposition of earthy and calcareous matter in the soft tissue first developed.—It may, perhaps, be stated as a general fact, that a similar diversity of conformation presents itself in the lower grades of almost every considerable group ; and that the various modifications exhibited by these *represent* (to a certain extent) the most important characters of the tribes above them. Of this we have already pointed out two remarkable examples, in the combination presented by the *Cystidea* of the characters of the other orders of the class Echinodermata (§ 296 *d*), and in the assemblage of diverse forms included in the order *Batrachia* (§ 324 *e*). Another very marked instance may be found in the *Implacental* Mammalia, among which may be found representatives of the Quadrumanous, Insectivorous, Carnivorous, Ruminant, Rodent, and Edentate orders of the Placental sub-class. And in like manner, among those aquatic Birds which agree together in structural and physiological peculiarities, to such an extent as to be justly ranked together in the order *Natatores*, it would not be difficult to point out the representatives of the Raptorial, Insessorial, Rasorial, and Grallatorial types.*

335. On turning to the Vegetable Kingdom, we find ample evidence, among the higher classes more especially, of the prevalence of common types of structure. In many of the simpler *Protophyta*, it is difficult to recognize, in the general indefiniteness of form and homogeneity of composition, any well-marked symmetrical characters ; and the growth of each individual seems to be so much governed by external conditions, that it is frequently very difficult to determine what are the distinctive characters of the species. But among those higher Cryptogamia, in which the distinction of parts becomes apparent, we perceive at the same time an obvious tendency to a definite plan of arrangement. Thus in the more perfect *Fungi*, the disposition of the apparatus of fructification is as distinctly 'radiate' as in Radiated animals. In *Mosses*, we find abundant evidence of a spiral arrangement of the leaves around the axis ; and in the arborescent *Ferns*, this disposition of the organs is a remarkable and characteristic feature. The spiral arrangement of the appendages to the axis, may be considered as the regular law of growth in the *Phanerogamia*, although it is frequently obscured by special modifications. Where the internodes between the successive leaves are developed, the leaves have

* This principle of *representation* has been carried out by Mr. Swainson, in combination with Mr. Macleay's *quinarian* system, to a most extravagant extent ; as above stated, however, the author believes that it will be found to be only a generalized expression of facts.

what is termed an *alternate* arrangement; the second leaf not arising exactly above the first, but from a point more or less to one side of the vertical line; the position of the third leaf being the same with respect to the second; and so on. Thus a line carried through the points of origin of the successive leaves, will not only ascend the stem, but will gradually turn round it, and will at last pass through a point directly above the origin of the first leaf. The leaves whose origin has been intersected by this line, whilst it makes one turn round the stem, are said to form a *cycle;* and the number of leaves which this cycle contains, is subject to great variations. Thus in *Dicotyledonous* plants generally (§ 280), it may be said to be *five;* that is, the sixth leaf will be directly above the first, the eleventh directly above the sixth, and so on. In *Monocotyledons* (§ 279), however, the typical number is *three;* the fourth leaf being above the first, the seventh above the fourth, and so on. There are cases in which the cycle seems to consist of only *two* leaves; each leaf springing from the side of the stem precisely opposite to that from which the leaf below it, as well as the one above it, arises.—The most common departures from the spiral type, shown in the disposition of the leaves, are those which are known as the *opposite* and the *verticillate* arrangements. These may be explained in three modes, each of which has some evidence to recommend it; and perhaps the deviation really does not always take place in the same way. Thus, 'opposite' leaves would be produced in a plant whose 'cycle' consisted only of two, by the non-development of every alternate internode, so that each leaf and its successor on the opposite side come to be developed from the same part of the stem, whilst separated by an interval from the next pair. But this explanation does not suit those cases, in which the successive pairs of leaves are arranged on the stem at right angles to each other; and this arrangement may either be attributed to the development of two opposite leaves from each node, the successive pairs being then arranged in a cycle of four; or to the existence of two spirals proceeding up the stem simultaneously. In like manner, a 'verticil' of five leaves originating from the same point of the stem, may be conceived to result from the non-development of the internodes between five successive nodes; and it sometimes happens that leaves which have a verticillate arrangement at one part of the stem, are alternate at another, being separated by the development of the intermediate internodes. But this does not account for the fact, that the successive whorls themselves usually alternate with each other; each leaf of the verticil being over the spaces between the leaves of the verticil beneath it. And here again it would seem necessary, either to imagine that all the leaves of one verticil may originate from a single internode, or to suppose several spirals to be passing round the stem. In either way, however, this very common arrangement is reconcilable with the general theory of spiral development; and this theory is thus carried into application as regards the disposition of the parts of the Flower. Each of these, whether bract, sepal, petal, stamen, or carpel, may be regarded as a foliaceous appendage to the axis (§ 280), the elements from which it is developed being capable of evolution as true leaves; and the law of their arrangement is the same. For in a perfectly regular flower, the bracts form the first or outermost whorl; the sepals a second, alternating with the first; the petals a third, alternating with the second, and opposite to the first; the stamens a fourth, opposite to the second; and the carpels a fifth, opposite to the first and third. In the Dicotyledonous division of the Phanerogamia, the typical number of the elements of each whorl, as that of the cycle of ordinary leaves, is *five;* whilst in the Monocotyledonous, it is *three.* The regularity of a flower may be interfered with by the suppression or by the multiplication of whorls; but the greatest departures from the common type are those which result from the unequal development of different parts of the same whorl, some being altogether suppressed or very imperfectly evolved, whilst others are extraordinarily augmented in size, and strangely altered in figure and character. The scientific Botanist, however, has no difficulty in solving these apparent difficulties; and in showing, partly by the aid of intermediate forms, and partly by the study of their development, that even those which seem to present the greatest departures from the 'archetype' are really reducible to it (§ 339).

336. Thus we are led by the study of *Morphology* (that is, by the recognition of 'homologous' organs, under whatever forms they may present), to the perception of that great general truth, which is, perhaps, the highest yet attained in the science of Organisation, and which is even yet far from being fully developed; that in the several tribes of organised beings, we have *not* a mere aggregation of individuals, each formed upon an independent model, and presenting a type of structure peculiar to itself;—but that we may trace throughout each assemblage *a conformity to a general plan,* of which every modification has reference either to the peculiar conditions under which the race is to exist, or to its relations to other beings. Of these special modifications, again, the most important themselves present a conformity to a plan of less generality; those next in order, to a plan of still more limited extent; and so on, until we reach those which are peculiar to the individual itself. This, in fact, is the philosophic expression of the whole science of Classification. For, to take the Vertebrate series as our illustration, we have found that Fishes, Reptiles, Birds, and Mammals agree in certain leading features of their structure, which constitute them *vertebrated* animals; but this structure is displayed under diversified aspects, in these *classes* respectively, which constitute their distinctive attributes. Thus, of the *general* vertebrated type, the Fish presents one set of *special* modifications, adapting it to its peculiar mode of life; the Reptile, another; the Bird, a third; the Mammal, a fourth. So, again, in each of these classes, we find its general type presenting subordinate modifications in the respective *orders;* thus, for example, the Reptilian type exhibits itself under the diverse aspects of the Frog, the Snake, the Lizard, and the Turtle; the Mammalian, under those of the Whale and the Bat, the Sloth and the Deer, the Elephant and the Tiger, the Kangaroo and the Monkey, the Ornithorhyncus and Man. Each order, again, is subdivided into *families,* in accordance with the subordinate or more special modification which the type of the order presents; every one of these families displaying the type of the class and order, with distinctive variations of its own. Each family consists of *genera,* in every one of which the family type is presented under a somewhat diversified aspect. Each genus is made up of an aggregation of *species,* which exhibit the generic character under a variety of modifications; these, however slight, being uniformly repeated through successive

generations. Lastly, each species is composed of an assemblage of *individuals*, every one of which repeats the type of its kingdom, sub-kingdom, class, order, family, genus, and species, through its whole line of descent. —Thus, in assigning to any particular being its place in the Organised Creation, we have to proceed *from the general to the special.* We will suppose an unknown body to be brought for our determination; the first business is to determine whether it be an Organised fabric, or a mass of Inorganic matter. This is soon discriminated, in the majority of cases, by an appeal to those *most general* characters which distinguish *all* organised structures; and the next question is to determine its Animal or Vegetable nature, by the aid of those characters which are not common to both, but are distinctive of each respectively. We will suppose the Animal nature of our unknown body to have been ascertained; the question next arises, to which of the four sub-kingdoms shall it be referred; and this, again, has to be ascertained by an appeal to characters which are less general than the preceding, not being common to all organised structures, nor yet to all animals, but being restricted to each of the four sub-kingdoms respectively. Then having ascertained that it is a Vertebrated, Molluscous, Articulated, or Radiated animal, as the case might be, the Naturalist would determine its order by characters of still less generality, which are peculiar to that order; its family, by features which are still more limited; its genus, by those modifications of family character which are presented by the several genera it includes; and lastly, its species, by characters which are the most special of all, that is, which are limited to that race alone.

337. Now if our classification were perfect, it would be comparatively easy to determine the 'archetype' or ideal model of each group; because we should have all the forms before us, by the comparison of which the Philosophical Zoologist seeks to educe what is common to the whole. But in practice it has often been found extremely difficult to determine what shall be considered as characters of classes, what of orders, and so on; since their respective values are not always evident on a simple comparison, even to the instructed Naturalist, much less to those ignorant of the principles of Classification. Thus in popular ideas, a Bat ranks as a Bird, because it flies by wings through the air; whilst the Whale ranks as a Fish, on account of its fish-like form, habitation, and mode of progression. But the scientific Naturalist has no hesitation in placing the Bat amongst the Mammalia, because it presents all the characters which are essential to that class, and which distinguish it from that of Birds; namely, its viviparous and placental generation, its subsequent nurture of its young by lactation, its covering of hair, the dental armature of its mouth, its diaphragmatic respiration, its highly-developed cerebrum, and many other peculiarities of conformation; whilst its apparent resemblance to the class of Birds merely results from the adaptation of the Mammalian type to an aerial life. So, again, notwithstanding its fish-like habits, and the peculiarities of structure which adapt it to these, the Whale is a Mammal in all that is essentially characteristic of the class, and distinguishes it from that of Fishes,—namely, its atmospheric respiration, its complete double circulation, its warm blood, its viviparous generation and subsequent lactation, its well-developed cerebrum, its osteological and many other peculiarities.—Here, then, the determination is easy to those

who possess but a smattering of Zoological knowledge. But we will take another case, in which a fundamental error was committed even by a great Master in modern science, owing to his misapprehension of the value of characters. Following too closely the indications afforded by the teeth (which are valuable in so far only as they serve as a key to the general plan of conformation), Cuvier placed the *Marsupial Mammalia* (§ 326) in the first instance as a subdivision of his order *Carnaria;* and even when he subsequently raised them to the rank of a distinct order, he gave them a position intermediate between the Carnaria and the Rodentia. Likewise, on account of the absence of teeth, he associated the *Monotremata* with the Sloths and Ant-eaters, in his order Edentata; satisfying himself with indicating that they presented a certain degree of affinity to the Marsupiata. Now the mutual resemblance of these two orders, as we have seen, is extremely close; and their unlikeness to all other Mammalia, in the structure of their cerebrum, and in the mode in which the genital function is performed,—two characters of fundamental importance,—is such as unquestionably to require their detachment as a distinct sub-class.—So, again, it is now coming to be perceived that the adaptation of the Mammalian structure to a fish-like habit of life, is not of itself sufficient to assemble all the animals which present it into a distinct order; for whilst the greater part of those which agree in possessing the Cetacean form, agree also in structure and in carnivorous habit, there are certain genera (the *Dugong*, *Manatee*, and *Stellerine*) which have been until recently ranked with them, but which are found rather to correspond in the more essential peculiarities of their organisation with the great series of *herbivorous* Mammals, and are connected with that series by forms now extinct.—Many other examples might be cited, illustrative of this difficulty, which is one that especially presents itself among the lower classes of animals, with whose structure and physiology the acquaintance of the Naturalist is as yet so imperfect. It is one, however, which is continually lessening with the progress of research; and whilst, therefore, we should avoid placing too much confidence in existing systems of classification, and in existing ideas of what really constitute natural groups, we may look forwards with hope, if not with absolute confidence, to the gradual accumulation of those materials, which shall enable the Philosophic Naturalist to do that for each group, which has been already effected, in great measure, for the Vertebrated series. In the determination of the relative importance of characters, it is certain that great assistance may be expected from the study of Embryology (§ 341); although we may not perhaps go the full length with those who maintain, that 'development' should constitute the groundwork of all classification.

338. With the general law of *Unity of Composition,* as here stated, some very interesting corollaries are connected. For, in the first place, if it be true that such a unity exists, we should expect to find that when any new function, or great modification of function, has to be performed, there would not be an entirely new structure evolved as its instrument, but that it would be effected by a corresponding modification in some organ, which is already present under some other form in neighbouring groups. This we find to be actually the case. Thus, the proboscis of the Elephant is but an extended nose; and an approach to the same

extension is presented by the Tapirs among existing Mammals, as well as (judging by the conformation of the cranium) by various extinct Pachydermata. The wing of the Bat, as we have seen, is not an additional member, but is stretched upon an extended hand; that of the Pterodactylus, upon a single finger. The neck of the Giraffe contains no additional vertebræ; but is adapted to its offices by peculiar modifications in the structure of those already existing. So, the protective shield of the Turtles is not so complete an *addition* as it would at first appear, to the ordinary skeleton of the Reptile; nor in the horny mandibles which cover its jaws, have we an organ which is altogether new to the group, since (as formerly shown, § 324 *i*) these are formed by a different development of the same elements as those from which teeth are elsewhere produced. So, turning to the Vegetable kingdom, we find that special organs, such as tendrils, pitchers, fly-traps, &c., are evolved out of the more general type of the leaf, and are not introduced as additions to the ordinary fabric.—Again, it might be expected that if the plan of structure in a particular tribe involves the *non*-development of some organ which is possessed by neighbouring groups, its conformity to the archetype or general model should be manifested in the presence of that organ in a *rudimentary* or undeveloped condition; and this, as a general rule, we find to be the case. Thus, we find some rudiment of the lung in most Fishes, even where it is not sufficiently developed to serve as an 'air-bladder' in regulating the specific gravity of the body. In the abdominal muscles of Mammals, again, we find the abdominal sternum and ribs of Saurian Reptiles indicated by white fibrous bands; and in those Mammals which do not possess a clavicle, that bone is usually represented by a ligament, just as the stylo-hyoid ligament in Man represents a portion of the hyoidean arch which is elsewhere completely ossified. Such rudimentary structures, however, often display themselves only at an early period of development, and are subsequently lost sight of. Thus the rudiments of teeth, which are never developed, and which at a later period cannot be detected, are found in the embryo of the Whale, both in the upper and under jaws; and Prof. Goodsir has ascertained that the rudiments of canine teeth, and of the incisors of the upper jaw, which are not subsequently developed, exist in the embryoes of Ruminating Mammals.* The most remarkable example of this kind, however, is the existence of *branchial arches*, resembling those of the Fish, in the early embryo of *all* air-breathing Mammalia, as will be hereafter explained (CHAP. XIII.).—The same is true, as a general rule, in the Vegetable kingdom; thus when a whorl or part of a whorl in a flower is suppressed, the deficiency is manifested, either by the presence of the undeveloped organs in a rudimentary form, or by the leaving of a space for them (so to speak) in the arrangement of the parts which are present. Thus in the *Primrose* tribe, we commonly find a single row of stamens *opposite* to the petals, instead of alternating with them according to the regular plan of floral development (§ 335); and hence the Botanist would conclude that a whorl has been here suppressed, which ought to intervene between the petals and the stamens. This is found to be the case in the genus *Samolus*, whose flower, formed in other respects upon the same type with

* Report of the British Association for 1839, p. 82.

the Primrose, possesses the rudiments of the intermediate row, in the form of a whorl of little scales, not developed into stamens. In the common *Sage*, again, we find only two stamens, where the general plan of the flower would lead us to expect five; but upon looking attentively at the interior of the corolla, two little scales are often to be seen growing in the place where two of the deficient stamens should have been; these two scales are frequently developed as perfect stamens, in flowers which are otherwise constructed exactly like the Sage; and even the fifth makes its appearance in some instances, exactly where it should be regularly found. Sometimes, again, the conformity to a common type is manifested by the full development, under cultivation, of organs which are not usually evolved; thus there are plants in which one set of flowers is purely 'staminiferous,' from the non-development of the carpellary whorl, whilst another set is 'pistilline' only, from the non-development of the stamens; and in which the effect of increased nutriment is to develope the deficient carpels in one set, and the deficient stamens in the other, so as to render both of them complete and 'hermaphrodite.'

339. The same tendency to conformity to an ideal 'archetype' is frequently shown, in a most remarkable manner, by the occurrence of *monstrosities;* which, though once regarded by men of science with feelings very little higher than those with which they are looked on by the vulgar, may now be considered as among the most interesting and suggestive of all the illustrations of 'Unity of Design;' since of these malformations, a considerable proportion are such in virtue of their closer conformity to the *general* model, those modifications of it which are characteristic of the *special* form not having been evolved. Some of the most curious examples of this are furnished by the Vegetable kingdom. Thus, the families *Lamiaceæ* and *Scrophularineæ* are distinguished by that peculiar form of corolla which is denominated *labiate*, from the two large lips bounding its mouth; and the stamens, instead of being five (like the sepals of the calyx), are only four in number, and are 'didynamous,' that is, two are longer than the other two.—Now in these points, there is a departure from that symmetrical arrangement, and that equality of parts, which are characteristic of the *regular* flower; this departure being a modification special to the order, superinduced upon the more general type. A further modification is seen in certain genera of the *Scrophularineæ*, such as the common *Antirrhinum* (Snapdragon), in which a long spur is developed from *one* only of the petals of the corolla, whilst the upper lip is developed in such a manner as to form an arch, against which the lower lip closes completely, forming what is termed a *ringent* corolla. Now in cultivated specimens of this plant, it is not uncommon to meet with a reversion to the regular type; the petals being all equal and similar, so as to produce a circularly-symmetrical corolla, and each of them having a spur developed from its under side; while the stamens are augmented to five in number, and are all of equal length.—So, again, in the common *Tropæolum* (Nasturtium), which in its normal state possesses one spurred petal, the tendency to regularity is exhibited, sometimes in the disappearance of the spur, sometimes in the development of the same appendage from other petals. The breaking-up of the whorls of a flower by the development of the internodes, so that its parts are found to be disposed in a

regular spiral round the axis (which may be occasionally seen in the double Tulip and in some of the Euphorbiacæ), is an example of reversion to a still more general plan, of which the flower itself is a special modification; and the same may be said of that reversion of each of the parts of the flower to the foliaceous type, which is sometimes witnessed in a single whorl, sometimes in the whole flower at once (§ 278).—In the Animal kingdom it is not difficult to trace the same tendency; the departures from the normal type, however, being for the most part greater in internal structure than in external conformation. Of the former kind, some of the most interesting are those which will be hereafter described as presenting themselves in the Circulating apparatus (CHAP. XII.); of the latter, the most frequent (as among Plants) are those which occur in the generative system. For example, among the higher Mammals, in which the *specialization* of the sexual apparatus is the greatest,—that is, in which the differences of the male and female organs are most strongly marked,—it is not at all uncommon to meet with instances of 'spurious hermaphrodism,' which are really nothing else than approximations to the community of type that is seen in the lower tribes, whose external generative organs are nearly alike in the two sexes. Thus in an animal whose general characters are those of the male, we may find the penis imperfectly developed, and not perforated by the urethra, which terminates in a uro-genital fissure that opens posteriorly; and the two halves of the scrotum, separated by this fissure, may resemble the labia of the female, the testes not having descended to occupy them. On the other hand, in an animal in which the characters of the female on the whole predominate, the upper part of the vaginal canal may be closed, so that it cannot be distinguished from the uro-genital fissure found in an imperfect male; the clitoris is sometimes developed to a very large size, and the urethra may be continued to its extremity, either as a complete canal or as a groove; and the ovaries, descending into the labia, may give them the character of a divided scrotum.

340. Great assistance in the study of the homologies between organs, and in the determination of the relations between natural groups, is afforded by the examination of *transitional* or *intermediate* forms. Thus it has been by the regular progression exhibited in the structure of the pulmonic apparatus, from the simple closed undivided air-sac of most Fishes, through the higher forms which this organ presents even in that class, and through the various phases of development exhibited in the Perennibranchiate Batrachia (§ 323 *a*), that the true relation of the air-bladder of the Fish to the lung of air-breathing Vertebrata has been established. So, again, in the curious *Amphioxus,* we may trace the relation of the gills of the Fish to the branchial apparatus of the Tunicated Mollusks (§ 321 *c*). The identity of composition between the jaws and true legs of the Crustacea, moreover, is shown by the transitional gradations presented by the feet-jaws (§ 316 *a*). And, turning to the Vegetable kingdom, we find the homologous relations of the parts of the flower to be indicated by those cases, in which there is a gradual instead of a sudden transition from one form to another (§ 278 *note*). These allusions are sufficient to show how important to the Philosophic Anatomist must be the study of *all* the gradations of structure presented to his notice; and how great would be his liability to be misled, if he were

to fix his attention only on those 'types,' which are most strongly marked out by the special characters that place them at the greatest distance from each other.—In like manner, the Philosophic Naturalist, whether his attention be directed to the Animal or the Vegetable kingdom, derives the most important guidance from such forms of transition, as present him with a mixture of the characters of two or more natural groups. He may be perplexed in assigning a precise place to these in his systematic arrangement (as we have seen to be the case in regard to the *Lepidosiren*, § 323 *a*); and yet from this very perplexity he derives increased knowledge of the mutual relations of the two groups thus connected, and of the fundamental harmony of the plans on which they are respectively constructed. And it can scarcely be doubted that, in proportion as our knowledge of internal structure becomes more complete, and external form is less recognised as the guide in classification, it will be found that a large number of those variations from the archetype, which constitute the characters of classes, orders, families, genera, and species, will themselves be referable to general rules, being, in fact, so arranged as to form a series mutually connected in every direction. Such connections have been continually adverted to, in the outline view of the Vegetable and Animal kingdoms, presented in the two preceding Chapters; and it will be sufficient here to state it as a general fact, that in every one of the great subdivisions of the Organised world, we meet with subordinate groups, in which not only are the distinctive features of the class shaded-off (so to speak), in such a manner as to make it a less characteristic or typical specimen of its kind, but there is an absolute blending of these with characters of neighbouring groups, which are in like manner modified into accordance with those of the first. We cannot have a better example of this nature, than is afforded by the *Amphioxus*; in which, as we have seen, the vertebral skeleton and the cerebro-spinal axis are only traceable in their most rudimentary state; whilst its circulation is on the whole most like that of the lower Articulata, and its respiration like that of the lower Mollusca. And the same animal furnishes an apt illustration of another important fact, that it is by the *lowest*, rather than by the highest forms, of two natural groups, that they are brought into closest relation. Thus, although both the classes *Cephalopoda* and *Arachnida* present certain approximations to the Vertebrated type of structure, yet these are not such as to modify in any considerable degree the Molluscous and Articulated types which they respectively present. Nor do either the Amphioxus, or any other of the lowest Fishes, present anything more than a slight relationship to the Cephalopoda; whilst none at all is apparent to the Arachnida. Yet we find that with certain groups, which are themselves far from presenting the Molluscous and Articulated characters in a well-marked degree, the Vertebrated sub-kingdom is closely linked, through *Amphioxus*, in a manner which would not formerly have been thought possible.

341. The most important and satisfactory evidence of the general principle of 'Unity of Design,' however, is derived from the study of the *Development* of living beings. The perfect organism of any one among the higher Plants or Animals is not more dissimilar in form and condition to that of the lowest and simplest member of either kingdom, than it is to the germ of its own kind in the earliest periods of its evolution; and

in fact, when we go back to the very commencement of the process, we observe that the *most general* type of organised structure,—the simple cell,—is that in which every living being commences. The evolution of the germ commences in the multiplication of this cell, precisely after the fashion of the multiplication of the simplest Protophyta and Protozoa; and it is not until this has proceeded to a considerable extent, that it could be stated with certainty, from the examination of the germ alone, whether it is that of a Plant or of an Animal. At the time, again, when the distinctive characters of *animality* first present themselves, it could not be predicated whether the germ is that of a Radiated, Molluscous, Articulated, or Vertebrated animal; the special characters of these sub-kingdoms not being evolved until a later period. These, however, are the next to appear, but still the distinctive peculiarities of the class are wanting; the germ of a Vertebrated animal (for example) being at first destitute of anything that can mark it out as a Fish, Reptile, Bird, or Mammal. When the distinctive characters of the class have been made manifest by the further progress of development, those of the order still remain indeterminate; these are evolved in *their* turn; and then those of the family, genus, species, sex, and individual, in succession.* This, at least, is the result of observations made in a considerable number of cases; and where such an accordance does not exist, the want of it is probably due to imperfections in the system of classification with which the comparison is made.—Thus we see that in watching the history of the development of any one of the higher forms of organised structure, we find the realization of that ideal evolution of *the more special characters from the more general*, which it is the object of the Philosophic Naturalist to bring into view by the methods of proceeding already pointed out.

342. It is maintained, indeed, by many distinguished Naturalists, that the information derivable from the history of Development, in regard to the relative value of characters and the affinity of groups, is so much more certain and satisfactory than that of any other kind, that it ought to furnish the fundamental data for a truly scientific classification; those tribes being considered as most nearly related to each other, whose embryonic development advances furthest along the same course without divergence; whilst those are considered as most fundamentally dissimilar, whose directions of development are distinct from the earliest period. This principle may be admitted as one which deserves to be fully taken into account, in any attempt at a systematic arrangement on philosophical principles; but to adopt it to the exclusion of all comparison of forms in their state of perfect and complete evolution, would be to deprive the changes which may occur at a comparatively late period of develop-

* Although certain approximations had been previously made to this general truth, yet the first definite and complete statement of it, with its application to Classification, will be found (the Author believes) in two papers contributed by Dr. Martin Barry to the "Edinburgh Philosophical Journal" for 1837. It has been subsequently developed in a very admirable manner by Prof. Milne-Edwards in a Memoir on the Principles of the Natural Classification of Animals ("Annales des Sciences Naturelles," 1844), which bears evidence of having been written without the knowledge of what Dr. Barry had put forth; the principle, in fact, having been advanced in a more limited form by Prof. Milne-Edwards himself, in a Memoir on the Changes of Form exhibited by various Crustacea during their development, read by him to the French Academy in 1833, and published in the "Annales des Sciences Naturelles," Ser. I., tom. XXX., Ser. II., tom. iii.

ment, of what appears to be their due claim to consideration. An illustrative example will perhaps make this apparent.—In the class of *Crustacea*, as long since observed by Prof. Milne-Edwards, the young of such as come forth from the egg in an early period of development, and have many changes to undergo, resemble one another very closely. As they increase in size, however, the peculiarities of the respective tribes to which they may belong gradually manifest themselves, partly through an alteration in the rate of development of different parts, and partly by the evolution of new and special organs. Thus in one case, it is the thorax which grows more rapidly than the abdomen, and greatly preponderates; in another, it is the abdomen which presents the greatest increase in its dimensions; in other instances, again, an extraordinary development is seen in certain extremities, or even in certain articulations of these extremities. So, again, it is not uncommon for certain parts which were possessed by the embryo Crustacean, to become atrophied and to disappear; thus still further tending to *specialize* the particular form, in its progress towards its complete development. In these and other modes, the larva, which, at the time of its emersion from the egg, may have presented no characters that serve to distinguish it from the larvæ of numerous other very dissimilar forms, gradually comes to present in succession the characters of its tribe, genus, species, and sex.*—If we turn, however, to the *Cirrhipods*, we find that whilst they agree with Crustacea in their larval condition, and must, if the principle of development be followed as the sole guide, be placed at no great distance from the Entomostracous sub-class, if not as actual members of it, they undergo such extraordinary metamorphoses at a later stage, that their character is changed in a degree sufficient to exclude them from any definition, however comprehensive, which may be framed for that class; whilst they are brought into much closer approximation with the Mollusca, than is exhibited by any other group of the Articulated series (§ 318).

343. But although the application of this general principle to the purposes of classification may require more limitation in particular cases, than some of its advocates are willing to admit, yet there can be no question that it is itself an expression of fact; and that in the Development of all living beings, to use the words of Von Bär, who first enunciated the doctrine,†—"*A heterogeneous or special structure arises out of one more homogeneous or general, and this by a gradual change.*" In the subsequent portion of this treatise, there will be found such ample illustrations of this law, in the progressive complication of structure, witnessed alike in the ascending scale of Vegetable and Animal life, and in the advance from the earliest embryonic condition towards the perfect form of the higher types, that it need not be more fully dwelt on at present.

344. It is peculiarly interesting to remark, that the same general plan appears to have been followed, in the organisation of the vast series of living forms which have successively appeared on the face of this globe. The entombment of the remains of many of these, in the strata in progress of formation at the time of their existence, has enabled the Palæontologist to reconstruct, to a certain extent, the Fauna and Flora of each

* See the Memoir of Prof. Milne-Edwards already cited; also his 'Histoire des Crustacés,' and his article *Crustacea* in the 'Cyclopædia of Anatomy and Physiology,' vol. i.

† Uber Entwickelungsgeschichte der Thiere, Königsberg, 1828–37.

of those great epochs in the Earth's history, which are distinctly marked out in Geological time, both by extensive disturbances in the earth's crust, and by striking changes in the structure and distribution of the living beings which dwelt upon it. Each of these epochs was characterized by some peculiar forms, or combinations of forms, of Animal and Vegetable life, which existed in it alone; and the farther we go back from the existing period, the wider are the diversities which we meet with, both in that general aspect of these kingdoms of nature which depends upon the relative proportions of their different subordinate groups, and in the features and structure of the beings composing these groups. The attempt has been made to prove, that these changes might be reduced to a law of 'progressive development;' meaning by this, that the lowest forms of Vegetable and Animal life were first introduced, that those of the least degree of elevation next presented themselves, and so on consecutively, until we reach Man, who, as the highest in the series, was the last to make his appearance upon the globe. Further, it has even been surmised that an actual 'transmutation' of the lower forms into the higher took place in the course of geological time; so that, from the germs first introduced, or from others which have since originated in combinations of inorganic matter, the whole succession of organic forms, from the simplest Protophyte up to the Oak or Palm, from the Protozoon up to Man, has been gradually evolved; not, however, in a single series, but from several distinct *stirpes*, whose development has taken different directions.* The facts of Geological science, however, do not seem to bear out the first of these doctrines; and the facts of Physiology lend no real support to the second. For it is easily capable of being shown, that although the doctrine of 'progressive development' as just stated *may be* true in some of its main features,—Radiata, Mollusca, and Articulata having *perhaps* existed before any Vertebrated animals left traces of their existence, Fishes having been abundant before we have any evidence from the remains of Reptiles that the latter had been introduced, and Reptiles having been for a time the sole air-breathing Vertebrata, and having occupied the place (so to speak) of Birds and Mammals, when as yet these had been either very scantily produced or were altogether wanting,—yet that when we come to apply it more closely, it altogether fails. And even if the doctrine of *progressive development,* in its usual form, were true in every particular, it would afford no ground whatever for the doctrine of *transmutation,* which is not only opposed to all our experience, but which fails to account for the intimate *nexus* that so commonly unites together, not merely the higher and the lower forms of each series, but the members of different series with each other. The question how far any real support is afforded to this hypothesis by the physiological facts advanced in its behalf, and which have been supposed to prove the *possibility* of such transmutation, will be more advantageously considered hereafter (CHAP. XVIII).

* See the "Vestiges of Creation," in which this hypothesis is put forth and sustained with great ingenuity. The writer of that remarkable work has drawn some of his materials from the former edition of this treatise; but the Author does not regard them as legitimately bearing the construction which has been put upon them, and dissents altogether from the conclusions above stated. He does not think it desirable, however, to enter upon a lengthened discussion of them in this work; which is not designed to examine and criticise error, but to develope what the Author believes to be established or probable Truth.

345. A more satisfactory account of the succession of Organic Life on the surface of the globe, may probably be found in the general *plan* which has been shown to prevail in the development of the existing forms of organic structure; namely, *the passage from the more general to the more special.* Many indications present themselves, that the types of each principal group first introduced were *not* the lowest, but that they presented *in combination* those characters which are found to be *separately* distributed, and more distinctly manifested, among groups that subsequently made their appearance. One of the most curious exemplifications of this principle in the *Radiated* division of the Animal kingdom, is to be found in the history of the class *Echinodermata;* for the group which seems to have attained a high development at the earliest period, is not that of *Crinoidea* (§ 296), by which the class in question is most closely connected with Zoophytes, but that of *Cystidea* (§ 296 *d*), which (there is reason to believe) was much superior to this in general organisation. Now this order seems to have presented, as formerly pointed out, a most extraordinary combination of the distinctive characters of the remaining groups; of which some appear not to have existed, and the rest to have presented a very limited range of forms, at the time when it was predominant. Thus, the *Crinoidea* of the Palæozoic period, though very numerous, exhibit but little variety of type; and in the complete enclosure of the body by polygonal plates, they present a closer approximation to the Cystidea, than do the Crinoidea of the Secondary period, in which the variety of forms is much greater. So, again, the *Asteriada* and *Ophiurida* of the Palæozoic period appear to have represented only a small part of the forms which those groups have since included. It is probable that the true *Echinida* did not exist at all in the Palæozoic period;* and although we are unfortunately not likely ever to obtain proof or disproof of the existence of *Holothuriada,* it cannot but be thought probable that they, too, were as yet absent. In the Secondary period, on the other hand, when the *Cystidea* had ceased to exist, we have evidence that they were replaced by all the orders just named (save the Holothuriada, the softness of whose bodies would be likely to prevent their preservation); and these soon came to present a very high degree of development, dividing among them (so to speak) the characters possessed by the Cystidea, and carrying these out separately as the distinctive peculiarities of their respective structures.—So among the higher *Mollusca,* we find that a prominent place in the earlier formations was occupied by that group (§ 307) which presents the least development of the distinctive characters of the Cephalopod class, and which has much in common with the testaceous Gasteropods. Now there is no evidence of the existence of the *higher* order of 'Dibranchiate' Cephalopods, at that early date in the Palæozoic period at which the *tetrabranchiate* order had acquired an extraordinary multiplication and variety of forms; and so far it might seem that we have a progression from the lower to the higher. But the paucity of remains of typical Gasteropods, at the same period, is almost as remarkable; and some of those forms which are most abundant (*e. g.* Euomphalus and Bellerophon) present indications of close proximity to Cephalopoda. So that it would seem as if the *Nautiloid*

* The genera *Palæchinus* and *Palæocidaris,* which have been usually referred to this group, are considered by Prof. E. Forbes as connecting links between the *Cystidea* and true *Echinida,* approximating most nearly to the former.

type is really to be regarded as having occupied the place at that period, not merely of the order above, but also (in part) of the class below; its decline and almost complete disappearance, during the Secondary epoch, being coincident with the multiplication of forms of the more typical Gasteropods, and of the higher Cephalopods.—Again, among the Fishes which were the earliest of the *Vertebrated* inhabitants of the globe, we find a remarkable assemblage of characters; some of them presenting, in the extraordinary development of the dermo-skeleton, and in the softness and probably rudimentary condition of their vertebral skeleton, an evident leaning towards the Invertebrated series; whilst others seem to have foreshadowed the class of Reptiles, an approach to which is presented, not merely by the Sharks and their allies, but by the *Sauroid* tribe of Osseous fishes, which was extremely abundant towards the end of the Palæozoic period. The Ctenoid and Cycloid orders, which (on a review of the whole class) may be undoubtedly considered as comprehending the most *typical* Fishes, did not make their appearance (so far as can be determined from the evidence of their fossil remains) until the Cretaceous period. The universal possession of the *homocercal* tail (§ 322 *a*) by the earlier Fishes, is to be considered on this view, *not* as an essentially embryonic character, but as the *most general* character of the class, which is presented in every member of it at an early stage of development, but which subsequently gives place in certain cases to a special modification.—Turning to the air-breathing Vertebrata, again, we find that during the Secondary period, this series was chiefly represented by the class of Reptiles, which then attained its greatest importance, and included groups which represented Fishes, Birds, and Mammals respectively; thus having a more *general* character than the class at present exhibits. These groups subsequently gave place to the more *special* forms, which carry out most exclusively the Reptilian type. And when we look at the earliest forms of Reptilian life, of which we have any cognizance, we find them to present very remarkable combinations of the characters which are now distributed among different groups. Thus, the *Labyrinthodon* (§ 324 *p*) of the Triassic formation, appears to have been essentially *Batrachian* in its structure, but to have possessed some characters of the *Crocodilian* order. And the same formation contains remains of Reptiles, which, while essentially *Saurian* in their general structure, had the horny mandibles, and probably many other characters, of the *Chelonia*.—In the early history of the class Mammalia, so far as known to us, the same general plan may be traced. The only order that is distinctly recognizable by the remains preserved in the Secondary strata, is that of *Marsupialia*, which has much in common with the Oviparous Vertebrata. Near the commencement of the Tertiary epoch, remains of *Pachydermata* are abundant; but these were for the most part different from those of the present epoch, containing combinations of characters which are now distributed among several distinct families, and presenting also a closer approximation to the Herbivorous Cetaceans on the one hand, and to the Ruminants on the other, than is exhibited by any existing species of the order. So among the early *Edentata*, we meet with a group now entirely extinct, which connected that order with the massive Pachyderms.

346. In regard to the Geological history of the Vegetable kingdom, it must be admitted that our knowledge is still very imperfect; in con-

sequence of the small number of cases in which the internal structure and fructification of the earlier plants have been preserved, in a condition that allows of the exact determination of their characters and affinities. So far as our present information extends, however, it is fully in harmony with the above doctrine; the characteristic Flora of the Coal-formation appearing to have been chiefly composed of *Coniferæ*, which constitute a connecting link between the Phanerogamia and Cryptogamia; and of these Coniferæ, while some may have been nearly allied to existing forms, the great majority (*Sigillariæ*, *Lepidodendra*, *Calamites*, &c) appear to have presented such a combination of the characters of the Coniferæ with those of the higher Cryptogamia, as no existing group exhibits (§ 277 *b*).

347. So far as at present known, therefore, the general facts of Palæontology appear to sanction the belief, that *the same plan* may be traced out in what may be called the general life of the globe, as in the individual life of every one of the forms of organised being which now people it; and that in the successive introduction of the several groups composing the Animal and Vegetable kingdoms respectively, the progression was not so much from the lower to the higher forms, as from the more general to the more special,—from those which were in closest relationship to each other, to those that are most isolated as types of their respective groups. And thus it has happened, that, as every Palæontologist must be ready to admit, a large proportion of the extinct forms of Animals and Vegetables must rank, in any philosophical system of classification, as *osculant* or *transitional* forms; connecting together the groups which seem naturally to assemble round existing types, and seldom standing as centres round which existing forms should be arranged.—It would be premature and presumptuous to assert that such *was* the plan, on which the progressive evolution of the great scheme of Organic Creation has proceeded; but the foregoing indications may be thought sufficient to justify the assertion that such *may have* been the plan; and the *à priori* evidence in its behalf, as a manifestation of that great Unity of Design which we have been attempting to trace in the history of the living beings that now exist, may be urged as entitling it to candid consideration. If it have a foundation in truth, the full development of the principle in all its completeness must be left for the time, when Palæontology shall possess, as the result of the accumulated labours of many generations (it may be) of industrious explorers, a collection of information respecting the *past* distribution of Animal and Vegetable life upon our globe, in some degree comparable to that to which the Natural History of the present time is rapidly attaining.

348. The general principle of Von Bär affords the real explanation of those resemblances which are sometimes discernible, between the transitory forms exhibited by the embryoes of the higher beings, and the permanent conditions of the lower. When these resemblances were first observed in the study of Animal Embryology, an attempt was made to generalize them in the assertion, that the higher animals, in the progress of their development, pass through a series of forms corresponding with those encountered in ascending the animal scale. But this statement was hasty and unphilosophical; and it is only now referred to, for the sake of showing what amount of real truth there is in it.—No animal *as a whole* passes through any such series of changes, except where it comes forth

from the egg at an early period of development, but in a condition that enables it to sustain its own existence, and to lead the life of a class below, from which it is afterwards raised by metamorphosis. This is the case, for example, with such Insects as resemble Annelida in their larva condition, and with Batrachian Reptiles, which are essentially Fish during the early period of their lives. But in neither of these instances, is the condition of the Larva that of the *perfect* animal which it represents in form and grade of organisation; for the condition of its tissues and organs is altogether embryonic; so that the Caterpillar bears a much closer accordance with the embryonic than with the adult Annelide, while the Tadpole is more nearly related to the embryonic than to the perfected Fish. These and other cases of the same kind must be regarded as special modifications of the general plan to meet a particular purpose; and while they present nothing discordant with that plan, they cannot be taken as examples of the usual mode in which it is followed out. On studying the development of any one of the higher animals, which remains within the ovum until it has attained the form characteristic of its class, we find that its entire structure does *not* present at any time such a resemblance to either of the classes beneath, as would justify the slightest analogy; thus, the Human embryo is never comparable with a Fish, a Reptile, or a Bird, much less with an Insect or with a Mollusk. In its very earliest grade, indeed, it might be likened to the cells, or clusters of cells, of which the *Protozoa* are constituted; but so soon as the multiplication and conversion of these has proceeded to such an extent, as to give it a form and structure in which a resemblance can be traced to *any* higher animal, it is to the Vertebrated type that we should at once assign it. Now, whilst it is passing through this condition, a close correspondence may be traced between the several parts of *its* structure and those of *any other* vertebrated embryo at a similar grade of development;—there is, for example, no essential difference between the vertebral column of the early embryo of Man, and that of an embryo Fish; the evolution of the nervous centres begins in both upon the same plan; so also does that of the circulating apparatus. And as the progress of development is arrested in the lower tribes, at the stages thus indicated in the transitional conditions of the higher, a mutual resemblance in the condition of particular organs may most assuredly hence arise. Thus, in the Cyclostome and higher Cartilaginous Fishes, we find permanently represented the various stages in the development of the vertebral column, which may be detected in the embryo of higher Vertebrata. But each of these animals in its adult condition presents a special adaptation of the general plan to its own organism; and this special modification is *not* represented in the human embryo. Thus, whilst the cranium of the human embryo is developed from a great number of distinct centres, which represent the bones that remain permanently separate in the skull of the Fish,—so that there is a correspondence in the condition or grade of development, as presented in the two cases respectively,—yet it never exhibits those peculiar characters, which distinguish the skull of the Fish from that of all other Vertebrata. Or, to take an illustration from another source, the circulation is carried on, at an early period in the development of all vertebrated animals, by a system of blood-vessels distributed upon the same plan as that which is met with in the adult Fish. It is not, however, correct to affirm, that the circulating apparatus of Man ever passes through the condition of that of a Fish; for although the 'branchial arches' are developed in *all* Vertebrated animals, so that their presence may be considered as the *most general* fact in the history of their arterial system, yet the twigs which they give off in the adult Fish for the supply of the branchial filaments are never developed, except in animals that are to be adapted for aquatic respiration; so that the blood flows onwards continuously through the branchial arches, and is delivered by them into the aorta, instead of being distributed amongst the gills, to be returned from them by a distinct set of vessels, the branchial veins. Hence, however close may be the resemblance between the *embryo* of Man and the *embryo* Fish, there is no real correspondence between the *embryo* of Man and the *completed* Fish; since every departure from the general plan or 'archetype,' which gives to the embryo Fish the special characteristics of its class, does in reality diminish the resemblance borne to it by the embryo of either of the higher classes. Thus, whilst there is at an early period a very close correspondence between the embryoes of all classes of Vertebrata, in virtue of the general principle of Von Bär, each one of these, as it proceeds in its course of development, takes a direction that separates it from the rest; and the mutual divergence consequently becomes greater and greater, in proportion as the perfected form and condition, that are characteristic of each class respectively, are approximated.

349. Now although the life of all Organised beings commences in the simplest and most general type of organic structure, so that there is no perceptible distinction between their germs, yet we see that each germ must have a certain capacity of development peculiar to itself; since it is a general law of Organic Development that *like produces like.* However varied may be the series of forms through which the parent passes, the offspring repeats these with the greatest exactness;* and the whole scheme of development may be described, as one in which the primordial cell is tending towards the attainment of the perfect form and condition of its parent. In proportion to the mutual resemblance of the parents, will be the conformity of the processes by which their respective forms are attained; in proportion to the dissimilarity of their adult conditions, will be the divergence of their directions of development: thus the development of the heart of the Bird and of the Mammal proceeds upon a method essentially the same, the single ventricle being divided first, and the single auricle subsequently (the septum remaining imperfect in the Mammal until birth); but in the Reptile, the auricle is first divided, its circulation being carried on upon a plan, to which the embryo Bird and Mammal never present anything comparable. And in accordance with the degree of proximity of each complete form to the general model or 'archetype' of the entire series, will be the degree in which it will be represented in the transitional states of the higher forms: thus the vertebral skeleton of the Fish as a whole departs much less from the archetype, than does that of the Bird; and consequently, that of the embryo Mammal is much more nearly related to the former, than it ever is to the latter.—These examples will serve, it is hoped, to show the distinction between the fundamental principle of development, first enun-

* It will be shown hereafter (CHAP. XVIII.) that the phenomena ranked under the term 'Alternation of Generations' do not constitute a real exception to this rule.

ciated by Von Bär, which is applicable (as the author believes) to all the facts hitherto ascertained, and that crude and illogical generalization which has brought discredit upon Philosophical Biology, and has led to a host of erroneous inferences.

350. We have hitherto spoken of organisms and their development, in relation rather to their *structural* composition, than to their *functional* character as instruments for particular ends. Under the latter aspect it is perhaps more difficult to trace a fundamental Unity, concealed as this is by that extraordinary variety, which is the chief source of the diversity of forms presented by living beings. It cannot be doubted that this variety is itself the expression of a systematic plan; but at present we have but very faint glimpses of such a plan, and can only trace its leading outlines. In place of looking at the origin and connections of the parts which we compare functionally, we must now regard them only in reference to the acts to which they are subservient; and must endeavour to seize upon the essential or fundamental character of that relation. If in this manner we compare the organs which are common to all living beings, Plants as well as Animals, we shall perceive amongst them a more general analogy, than any that we can trace by the guidance of evident structure only. The simplest Plant, for example, differs from the most complex, principally in this,—that the whole external surface of the former participates equally in all the operations which connect it with the external world, as those of Absorption, Exhalation, and Respiration,—whilst in the latter we find that these functions are respectively confined to certain portions of the surface. Although, therefore, the leaves and roots of a Vascular plant are distinct organs, they both have a functional analogy with the same simple membrane of the lowest species of Protophyta. So, again, in the highest Animals, we find that the organs specially adapted to the functions of Absorption, Exhalation, Respiration, Secretion, and Reproduction, are all essentially composed of a membrane which is a prolongation of the general surface, as may be proved, not only by its continuity with this, but by its development from it during the evolution of the embryo; and this general surface is the sole instrument for the performance of these functions in the lowest Animals, and shows no special adaptation for one or for another of them.—Where the greatest specialization of function presents itself, the particular arrangement of the organs will depend upon the general plan of conformation, and on the circumstances in which the being is destined to exist. Thus, whilst the absorbing organs of Plants are prolonged externally into the soil, they are distributed in Animals upon the walls of a cavity fitted to retain and prepare the food, in order that there may be no impediment to locomotion. Still, the same fundamental unity exists; and the spongiole of the vascular Plant, and the origin of the absorbent vessel in the Animal, have precisely the same essential character with the membrane which constitutes the general surface of the Sea-weed or Red Snow. It may, then, be enunciated as a general truth, that, "*throughout the whole animated Creation, the functional character of the organs which all possess in common, remains the same; whilst the mode in which that character is manifested varies with the general plan upon which the being is constructed.*" The latter part of this law may be rendered more intelligible by another illustration. The respiratory surface of Plants is always prolonged externally;

as they have no means of introducing air into cavities, and effecting that constant change in it which is necessary for the function. The same is found to be the case in nearly all aquatic Animals, the gills of which are evidently analogous to the leaves of Plants. In terrestrial Animals, on the contrary, the respiratory membrane is prolonged internally, so as to form tubes or cells exposing a large amount of surface. The different cavities which we find adapted to this office, are all lined by a membrane which is either derived immediately from the external surface, as in Insects, terrestrial Mollusks, &c., or from that inversion of it which forms the digestive cavity, as in Vertebrata; and this membrane is everywhere functionally or instrumentally the same, although the organs of which it forms a part are not homologous, those of one kind often existing in a rudimentary state, where another is fully developed.—Throughout the second division of the work, illustrations will be found of this essential unity in the functional character of the different organs common to all; and it need not, therefore, be further dwelt upon in this place.

351. Now the great principle of *progression from the more general to the more special*, appears to hold good as well in regard to the functional character of organs, as with respect to their structural and developmental conformity; as may be seen in proceeding from the lower to the higher forms of organised being, and in following the successive stages of development of any one of the higher organisms. If we compare the forms which the same instrumental structure presents in different parts of the series, we shall always observe that it exists in its most general or diffused form in the lowest classes, and in its most special and restricted in the highest; and that the transition from one form to the other is a gradual one. The function, therefore, which was at first most general, and so combined with others performed by the same surface as scarcely to be distinguishable from them, is afterwards found to be confined to a single organ, or to be *specialized* by separation from the rest; these also, by a similar change, having been rendered dependent on distinct organs. Thus, to refer again to the provisions existing in different Plants for the *Absorption* of fluid; we find that this action is performed by the entire surface of the lower tribes, which may be thus said to be *all root;* as we ascend through the series of Cryptogamia, it becomes more and more limited to certain parts of that surface; until in Flowering plants it is chiefly performed by its special organs, the 'spongioles,' or succulent extremities of the root-fibres, which draw-in fluid far more energetically than any portion of the general surface can do. And so in the early development of Phanerogamia, the germ absorbs by its whole surface the nutriment in which it lies embedded; it continues still to absorb by a part of that surface, even after the expansion of its plumula and the extension of its radicle into the soil; and it is not until the store laid up by the parent has been nearly exhausted, that the proper root-fibres are evolved, which are henceforth to be its special instruments of imbibition. Abundant illustrations of this principle will be found in the subsequent portions of this treatise.—But it would seem as if, even in the most complex beings, the specialization of function seldom proceeds so far, as to incapacitate the more general instrument for taking some part in it; for observation of the functions of the more complex forms of animated beings leads to the knowledge of another law, which is simply an expression of this fact;

namely that, *in cases where the different functions are highly specialized, the general structure retains, more or less, the primitive community of function which originally characterized it.** As this principle, also, will be copiously illustrated in subsequent chapters, it is unnecessary here to do more than point out its mode of application; and we shall again refer to the Absorbent system for this purpose. As, in the simplest or most homogeneous beings, the entire surface participates equally in the act of imbibition, so, in the most heterogeneous, every part of the surface retains some connection with it; since, even in the highest Plants and Animals, the common external integument admits of the passage of fluid into the interior of the system, especially when the supply afforded by the usual channels is deficient. In the same manner we find, that whilst, in the lowest Animals, the functions of Excretion are equally performed by the whole surface, there is in the highest a complicated apparatus of Glandular organs, to each of which some special division of that function is assigned; but as all these glands have the same elementary structure, and differ only in the peculiar adaptation of each to separate a particular constituent of the blood, it is a necessary result of the law just stated, that either the general surface of the skin, or some of the special secreting organs, should be able to take on, in some degree, the function of any gland whose duty is suspended; and observation and experiment fully bear out this result, as will hereafter appear (CHAP. XVI.).

352. In comparing the different forms of Plants and Animals with each other and with the 'archetype,' we observe that many of their differences depend upon diversities in the proportional development of the organs which all possess in common. The attempt has been made to bring these diversities under a general expression,—the *balancing of organs,*—which is nothing else than that which is alluded to by Paley and other authors as the "principle of compensation." Like many other generalisations, this has been carried too far by many writers who have dwelt upon it. Thus, it has been stated in the following most objectionable form;—that the extraordinary development of one organ *occasions* a corresponding deficiency in another, and *vice versâ.* It is perfectly true that, in a great majority of cases, *the extraordinary development of one organ is accompanied by a corresponding deficiency of development in another;* but the development and the deficiency are both parts of one general plan, and neither can be regarded as the cause or the effect of the other. Thus, in the human cranium, the elements which form the covering or protection of the brain are very largely developed, whilst those which constitute the face are comparatively small. In the long-snouted herbivorous Mammals, Reptiles, and Fishes, on the other hand, the great development of the bones of the face is coincident with a very small capacity of the cerebral cavity. In the Bat, we find the anterior extremity widely extended, so as to afford to the animal the means of rising in the air; whilst the posterior is very much lightened, so as not to impede its flight. In the Kangaroo, on the other hand, the posterior members are very large and powerful, enabling the animal to take long leaps; whilst the fore paws are proportionally small. The Mole, again, requires for its underground burrows the power of excavating with its

* See Edinburgh Philosophical Journal, July, 1837.

fore-feet, whilst the hind legs are used for propulsion only; and the relative development of these members follows the same proportion as in the Bat, although the *plan* in the two cases is widely different. Moreover it is obvious that, from the peculiar habits of this animal, eyes would be of little or no use to it; and accordingly we find them merely rudimentary, and no cavity in the skull for their reception; whilst to compensate for the want of them, the organ of smell, and its capsule, the ethmoid bone, are amazingly developed. In other classes of animals, similar illustrations abound; thus, the Birds of most active and energetic flight usually have the smallest and feeblest legs; and the Struthious birds, in which the legs are enormously developed, have only rudimentary wings. So, again, among Reptiles, we find the vertebral column most lengthened, and the tail especially developed, in those whose limbs are feeblest, or altogether deficient, as among Serpents, and serpent-like Sauria and Batrachia; whilst, if the limbs are the principal instruments of locomotion, as in Frogs and Turtles, the vertebral column is shortened and the tail contracted. And in Fishes the same general rule holds good. In no class, however, is this rule without its exceptions; and it must be taken rather as an expression of facts, possessing a certain *empirical* value, than as entitled to the character of a 'law' of development which some would claim for it.

353. Another principle, propounded by Cuvier, and supported by those who have adopted the 'functional' rather than the 'homological' relations of organs as their guide, is that of the *harmony of forms*, or the *coexistence of elements*. It implies that there is a necessity, arising out of the conditions of organic existence, for the combination of organs according to their several actions; that there is a constant harmony between organs which are functionally connected; and that the altered form of one is invariably attended with a corresponding alteration in the others.—That this statement is true as far as it goes, no one can deny; and the researches which have been based upon it have been most successful in repeopling the globe, as it were, with the forms of animals which have long been extinct, but which can be certainly predicted even from minute fragments of them. A general comparison of the skeleton of the Carnivorous with that of an Herbivorous quadruped, will show the manner in which this enquiry is pursued. The Tiger, for example, is furnished with a cranial cavity of considerable dimensions, in order that the size of the brain may correspond with the degree of intellect which the habits of the animal require. The face is short, so that the power of the muscles which move the head may be advantageously applied. The canine teeth are large and pointed; whilst the molars have sharp edges, adapted only for cutting, to which purpose they are most effectively applied by the scissors-like action of the jaw. The lower jaw is short, and the cavity in which its condyle works is deep and narrow, allowing no motion but that of opening and shutting; the fossa in which the temporal muscle is imbedded, is very large; and the muscle itself is attached to the jaw in such a manner as to apply the power most advantageously to the resistance. The spinous processes of the vertebræ of the back and neck are very strong and prominent, giving attachment to powerful muscles for raising the head, so as to enable the animal to carry off his prey. The bones of the extremities are disposed in such a manner, as to allow the union of strength with free-

dom of motion ; the head of the humerus is round, and the articular surfaces on the fore-arm indicate that it possesses the power of pronation and supination. The toes are separate, and armed with claws, which are retracted when not in use by a special apparatus that leaves its mark upon the bones.—On the other hand, in the conformation of the Herbivorous quadruped, we are at first struck with the diminished capacity of the cranium, and the increased size of the bones of the face. The jaws are long, and have a great degree of lateral motion, the glenoid cavity being broad and shallow : and whilst the pterygoid fossa, in which the muscles that *rotate* it are lodged, is of large size, the temporal fossa is comparatively small, no powerful *biting* motions being required by the nature of the food or the mode of obtaining it. The front teeth are fewer and smaller ; but the surfaces of the grinding teeth are extended, and are kept constantly rough by the alternation of dentine and enamel. The extremities are more solidly formed, and have but little freedom of motion, the shoulder being scarcely more than a hinge-joint ; the extremities are encased in hoofs, which are double if the animal ruminates, and either single or multiple if it does not. The whole body is heavier in proportion, the nutritive system being more complicated ; and the muscles which enable the tiger to lift considerable weights in his mouth, are here necessary to support the weight of the head itself.

354. A little consideration, however, will show that the existence of this adaptation of parts is nothing more than a manifestation of the general plan of development, and gives us no information of the nature of that plan. It is evident that, if it were deficient, the race must speedily become extinct, the conditions of its existence being no longer fulfilled ; and that whatever be the laws of development, they must operate to this end, in order that the world may be peopled with life. An animal with the carnivorous propensity of the Tiger, for instance, and the teeth or hoofs of a Horse, could not remain alive, from the want of power to obtain and prepare its aliment ; nor would a horse be the better for the long canine teeth of the tiger, which would prevent the grinding motion of his jaws required for the trituration of the food. The statement above given cannot, therefore, be regarded as a *law ;* since it is nothing more than the expression, in an altered form, of the fact, that, as the life of an organised being consists in the performance of a series of actions, which are dependent upon one another, and are all directed to the same end, whatever seriously interferes with any of these actions must be incompatible with the maintenance of its existence. The splendid discoveries of Cuvier and other anatomists, who have succeeded in determining, from minute fragments of bones, the characters of so many extraordinary species of remote epochs, have resulted from the sagacious appreciation of this truth, and from the use made of it in the laborious comparison of these remains with the similar parts of animals now existing. Until that comprehensive Plan shall have been discovered, of which these are so many individual results, no briefer process can be adopted.—In seeking to discover the plan of *Organic Structure,* then, the Philosophic Anatomist does not regard the ostensible object or function of a particular organ as a sufficient account of its existence, but follows it through all its varieties of form and condition, that he may learn what are its real elements, and its fundamental relations to other parts. Of this kind of comparison we have traced out

the general method, in the exposition of the views of Prof. Owen as to the homologies of the Vertebrate Skeleton.—On the other hand, in seeking to determine the laws of *Vital Action,* the Philosophic Physiologist must in like manner disregard the 'homological' relations of organs, and must look at them simply in their 'teleological' or purposive capacity ; and it will be to this that the reader's attention will be directed in the remainder of the present treatise.

BOOK II.

SPECIAL AND COMPARATIVE PHYSIOLOGY

CHAPTER IX.

GENERAL VIEW OF THE FUNCTIONS OF ANIMATED BEINGS, AND OF THEIR MUTUAL RELATIONS.

355. It has been stated (§§ 5-9) to be the ultimate object of the Physiogist, to ascertain the laws in accordance with which the Vital Actions of Organised beings take place; and it has been pointed out that in order to arrive at any certain general conclusions respecting them, he must collect and compare all the facts of similar character, with which the study of the animated creation furnishes him. The changes which occur during the Life of any one being, are of themselves inadequate to furnish the required information: since this presents us only with a group of dissimilar phenomena, incapable of comparison with each other, or permitting it but to a low degree. Were we to derive all our notions of Physiology from the history of one of the simple Cellular Plants, we should obtain but very vague ideas as to the character of its different nutritive processes; since we cannot separate these from one another, and investigate them apart. And, on the other hand, we should be apt to form very erroneous conceptions of the essential conditions of these processes, were we to study them only in that specialized condition, which they present in the most complete Animal, and reason thence as to their dependence upon particular kinds of structure. It is only, then, from a comprehensive survey of the whole organised creation,—embracing the unobtrusive manifestations of life which Nature presents at one extremity of the scale (as if to show the simplicity of her operations), as well as those obvious changes which she every moment displays to us in her most elaborate works, —that any laws possessing a claim to general application can be deduced.

356. In every living Organism of a complex nature, we can readily distinguish a great variety of actions, resulting from the exercise of the different powers of its several component parts; and these actions are said to be the *functions* of the structures by whose instrumentality they are performed. Thus it is the function of a Muscle, to contract on the application of a stimulus; and that of a Nerve, to receive and convey sensory and motor impressions. When we look at such an organism, however, as a whole, we perceive that these changes may be associated into groups, in accordance with their mutual relations to each other; each group consisting of an assemblage of actions, which, though differing among themselves, concur in effecting some determinate and important purpose. To these groups of actions, also, the name of *Functions* is applied, but not in the same sense as before. Thus when we speak of the "Function of Respiration," we imply that assemblage of actions which concur in effecting the

aeration of the nutritious fluid, by exposing it to the atmosphere, or to the gases diffused through water, so as to effect a certain change in its composition. Simple as this operation appears, many provisions are required in the higher organisms for its effectual performance. In the first place, there must be an aerating surface, consisting of a thin membrane permeable to gases; on one side of which the blood may be spread out, while the air is in contact with the other. Secondly, there must be a provision for continually renewing the fluid which is brought to this surface, in order that the whole mass of it may be equally benefited by the process. And thirdly, the stratum of air must also be renewed, as frequently as its constituents have undergone any essential change. In each of these subdivisions, a great many diverse actions are involved; yet all of them are included under the same general term, since they concur to the same fundamental purpose.—Now if we analyse that series of phenomena, which constitute the "Function of Respiration" in one of those higher Animals (Man, for example) in which it presents its greatest complexity, we shall find that whilst some of these are indispensable to the continuance of life, and can only be performed under the conditions supplied by the organised system, others are merely superadded for the purpose of facilitating them; and these, if from any cause not performed by the mechanism contrived for their production, may be artificially imitated, with a degree of success exactly proportional to the perfection of the imitation. The essential part of that function being the aeration of the blood, all the other changes which are associated together as partaking in it must be regarded as non-essential, sharing in it only by contributing to this—the real constituent of it. Thus the alterations in the capacity of the chest, which are effected by the actions of the diaphragm, and have for their object the renewal of the quantity of air in contact with the aerating surface, are really a part of the functions of the Muscular and Nervous systems; and are only associated under that of Respiration, on account of their obvious tendency towards its essential purpose. They have no share in the production of the aeration of the blood, except by supplying its conditions; and if these conditions can be supplied independently of them, the essential part of the function will be performed as when they were concerned in it.*

357. By an analysis of this kind applied to the other functions, similar conclusions might be arrived at respecting their essential character; for it will appear in every one of them, that some of the changes which are thus grouped together are *essential*, and others *superadded*. But these conclusions do not possess the same certainty as if they were founded upon a broader basis; nor are they so easily attained. For, to revert to the instance just quoted, *observation* alone of the vital phenomena of the lower

* Thus in *Asphyxia*, the deficient supply of arterialised blood to the brain soon paralyses its functions; and the nervous stimulus required for the respiratory movements being withheld, those movements cease. But, if the chest be artificially inflated, and emptied again by pressure, and these alternate movements be sufficiently prolonged to re-excite the circulation of the blood through the lungs, by aerating that which had been stagnated there, the whole train of vital actions may be again set in motion. Or, if the cessation of the respiratory movements result from a cause primarily affecting the nervous system—as when narcotism is induced by poisoning with opium—and the blood be, in consequence, stagnated in the lungs by the want of aeration, this change, so essential to the continuance of vitality, may be prolonged by artificial respiration, until the narcotism subsides, unless the dose have been too powerful.

animals, will reveal what could only be determined in the higher by *experiment.* Until an experiment (the insufflation of the chest) had been found successful in continuing the aeration of the blood, it could not be certainly known that the respiratory movements had not some further share in the function, than that of mechanically renewing the air in apposition with the circulating fluid. But when the conditions of the function are examined in the lower animals, it is found that these are varied (the essential part being everywhere the same) to suit the respective circumstances of their existence. Thus, many Reptiles, having no diaphragm, are obliged to fill the lungs with air by a process which resembles swallowing. In Fishes and other aquatic animals, to have introduced the necessary amount of the dense element they inhabit into the interior of the system, would have occasioned an immense expenditure of muscular power; and the required purpose is answered by sending the blood to meet the water, which is in apposition with the external surface. And in those simple creatures, in which the fluids appear equally diffused through the whole system, their required aeration is effected by the mere contact of the water with the general surface: the stratum in immediate apposition with it being renewed, either by their own change of place; or, if they are fixed to a particular spot, through the means they possess of creating currents, by which their supply of food also is brought to them. And, going still further, we find the essential part of the function of Respiration performed in Plants without any movement whatever; the wide extension of the surface in contact with the atmosphere affording all the requisite facility for the aeration of the circulating fluid.—It is by such a mode of analysis, then, that we are most certainly enabled to distinguish the essential conditions of vital phenomena, from those which are superadded or accidental; and it is this which will form the subject of the remainder of the present Treatise.

358. When we examine and compare the several Functions, or assemblages of Vital Actions grouped together according to the principle just set forth, we find that they are themselves capable of some degree of classification. Indeed the distinction between the groups into which they may be arranged, is one of fundamental importance in Physiology. If we contemplate the history of the life of a Plant, we perceive that it grows from a minute germ to a fabric of sometimes gigantic size, generates a large quantity of organic compounds which it appropriates as the materials of its own structure, and multiplies its species by the production of germs similar to that from which it originated; but that it performs all these complex operations without (so far as we can perceive) either feeling or thinking, without consciousness or the exertion of will. All its vital operations, therefore, are grouped together under the general designation of Functions of *Organic* or *Vegetative* life; and they are subdivided into those concerned in the maintenance of the structure of the *individual,* which are termed functions of *Nutrition;* and those to which the *Reproduction* of the *species* is due.—The great feature of the Nutritive operations in the Plant, is their *constructive* character. They seem as if destined merely for the building-up and extension of the fabric; and to this extension there seems in some cases to be no determinate limit. It is important to remark, however, that the growth of the more *permanent* parts of the fabric is only provided for by the successive development,

decay, and renewal of parts whose existence is *temporary*; the growth of the dense, durable, but almost inert woody structure being dependent upon the continual production of new leaves, composed of a soft, transitory, but active cellular parenchyma.—Sooner or later, however, the life of the individual must come to an end; and the race itself must become extinct, were it not for the special provision which is made for its continuance, in the Generative function. This consists in the evolution of germs, which, becoming detached from the parent, are able to support an independent existence, usually at the expense, however, in the first instance, of nutriment in some way provided by the being that gave them origin; they gradually become developed into its likeness, perform all the vital operations characteristic of it, and in their turn originate a new generation by a similar process.

359. Now, it may be observed, before proceeding further, that there is a certain degree of antagonism between the Nutritive and Reproductive functions, the one being executed at the expense of the other. The reproductive apparatus derives the materials of its operations through the nutritive system, and is entirely dependent upon it for the continuance of its function. If, therefore, it be in a state of excessive activity, it will necessarily draw off from the individual fabric some portion of the aliment destined for its maintenance. It may be universally observed that, where the nutritive functions are particularly active in supporting the *individual*, the reproductive system is in a corresponding degree undeveloped,—and *vice versâ*. Thus, it has been stated that in the Algæ, the dimensions attained by single plants exceed those exhibited by any other organised being; and in this class the fructifying system is often obscure, and sometimes even undiscoverable. In the Fungi, on the other hand, the whole plant seems made up of reproductive organs; and as soon as these have brought their germs to maturity, it ceases to exist. In the Flowering-Plant, moreover, it is well known that an over-supply of nutriment will cause an evolution of leaves at the expense of the flowers, so that what actually would have been flower-buds, are converted into leaf-buds; or the parts of the flower essentially concerned in reproduction, namely, the stamens and pistil, are converted into foliacious expansions, as in the production of 'double' flowers from 'single' ones by cultivation; or the fertile florets of the 'disk' in composite species, such as the Dahlia, are converted into the barren but expanded florets of the 'ray.' And the gardener who wishes to render a tree more productive of fruit, is obliged to restrain its luxuriance by pruning, or to limit its supply of food by trenching round the roots.—The same antagonism may be witnessed in the Animal Kingdom; but as a third element (the sensori-motor apparatus) here comes into operation, it is not always so apparent. It appears to be a universal principle, however, that during the period of rapid growth, when all the energies of the system are concentrated upon the perfection of its individual structure, the reproductive system remains dormant, and is not aroused until the comparative inactivity of the nutritive functions allows it to be exercised without injury to them. Thus, in the Larva condition of the Insect, the assimilation of food and the increase of its bulk seem the sole objects of its existence (§ 315 *e*); its locomotive powers are only adapted to obtain nourishment that is within easy reach, to which it is directed by the position of its egg, and by an unerring

instinct that seems to have no other end. The same is the case, more or less, with all young animals; although there are few in which voracity is so predominant a characteristic. In the Imago, or perfect Insect, on the other hand, the fulfilment of the purposes of its reproductive system appears to be the chief and often the only end of its being. The increased locomotive powers which are conferred upon it, are evidently designed to enable it to seek its mate; its instinct appears to direct it to this object, as before to the acquisition of food; it now shuns the aliment it previously devoured with avidity, and frequently dies as soon as the foundation is laid for a new generation, without having taken any nutriment from the period of its first metamorphosis. In the adult condition of the higher Animals, again, it is always found that, as in Plants, an excessive activity of the nutritive functions indisposes the system to the performance of the reproductive; a moderately-fed population multiplying (cæteris paribus) more rapidly than one habituated to a plethoric condition.

360. On analysing the operations which take place in the Animal body, we find that a large number of them are of essentially the same character with the foregoing, and differ only in the conditions under which they are performed; so that we may, in fact, readily separate the *Organic* functions, which are directly concerned in the development and maintenance of the fabric, from the *Animal* functions, which render the individual conscious of external impressions, and capable of executing spontaneous movements. The relative development of the organs destined to these two purposes, differs considerably in the several groups of Animals, as we have already in part seen (CHAP. VIII.). The life of a Zoophyte is upon the whole much more 'vegetative' than 'animal;' and we perceive in it, not merely the very feeble development of those powers which are peculiar to the Animal kingdom, but also that tendency to indefinite extension which is characteristic of the Plant. In the perfect Insect, we have the opposite extreme; the most active powers of motion, and sensations of which some (at least) are very acute, being combined with a low development of the organs of nutrition. In Man and his allies, we have less active powers of locomotion, but a much greater variety of Animal faculties; and the instruments of the organic or nutritive operations attain their highest development, and their greatest degree of mutual dependence. We see in the fabric of all beings, in which the Animal powers are much developed, an almost entire want of that tendency to indefinite extension, which is so characteristic of the Plant; and when the large amount of food consumed by them is considered, the question naturally arises, to what purpose this food is applied, and what is the necessity for the continued activity of the Organic functions, when once the fabric has attained the limit of its development.—The answer to this question lies in the fact, that the exercise of the Animal functions is essentially *destructive* of their instruments; every operation of the Nervous and Muscular systems requiring, as its necessary condition, a disintegration of a certain part of their tissues, probably by their elements being caused to unite with oxygen. The duration of the existence of those tissues (as formerly stated, § 236 and 243) varies inversely to the use that is made of them; being less, as their functional activity is greater. Hence, when an Animal is very inactive, it requires but little nutrition; if it be in moderate activity, there is a moderate demand for food; but if its Nervous and Muscular energy be frequently

and powerfully aroused, the supply must be increased, in order to maintain the vigour of its system. In like manner, the amount of certain products of excretion, which result from the disintegration of the Nervous and Muscular tissues, increases with their activity, and diminishes in proportion to their freedom from exertion.*

361. In the Animal fabric, then, among the higher classes at least, the function or purpose of the organs of Vegetative life is not so much the extension of the fabric, for this has certain definite limits; as the maintenance of its integrity, by the reparation of the destructive effects of the exercise of the purely Animal powers. Thus, by the operations of Digestion, Assimilation, and Circulation, the nutritious materials are prepared and conveyed to the points where they are required; the Circulation of Blood also serves to convey oxygen, which is introduced by the Respiratory process; and it has further for its office to convey away the products of the decomposition of the Muscular and Nervous tissues, that results from their functional activity,—these products being destined to be separated by the Respiratory and other Excreting operations. In the performance of the Organic functions of Animals, as in those of Plants, there is a continual new production, decay, exuviation, and renewal, of the cells, by whose instrumentality they are effected; which altogether effect a change not less complete than that of the leaves in Plants. But it takes place in the penetralia of the system, in such a manner as to elude observation, except that of the most scrutinizing kind; and it has been in bringing this into view, that the Microscope has rendered most essential service in Physiology.

362. The regular maintenance of the functions of Animal life is thus entirely dependent upon the due performance of the Nutritive operations. But there also exists a connection of an entirely reverse kind, between the Organic and Animal functions; for the conditions of Animal existence render the former in a great degree dependent on the latter. In the acquisition of food, for example, the Animal has to make use of its senses, its psychical faculties, and its power of locomotion, to procure that, which the Plant, from the different provision made for its support, can obtain without any such assistance. Moreover, the propulsion of the food along the alimentary canal of the former, requires a series of operations, in which the Nervous and Muscular systems are together involved at the two extremes, simple muscular contractility being alone employed through the greater part of the intestinal canal; and thus we see that the change in the conditions required for the ingestion of food by Animals, has rendered necessary the introduction of additional elements into the apparatus, to which nothing comparable was to be found in Plants. The same may be observed, as already pointed out, in the operations of the Respiratory apparatus. And it may be stated as a general rule, that the more exalted is the *animality* of any particular being (or, in other words, the more complete the manifestation of characters peculiarly animal), the more closely are the organic functions brought into relation with it.†

* This doctrine, though propounded in general terms by previous writers, was first pointedly stated by Prof. Liebig in his Treatise on Animal Chemistry.

† A simple illustration will render this evident. In certain of the lower tribes of animals, whose locomotive powers are feeble and general habits inactive, the circulation of nutritive fluid is carried on nearly in the same manner as in plants; there is no central organ for

363. From what has been said, then, it appears that all the functions of the body, among the higher Animals, are so completely bound up together, that none can be suspended without the cessation of the rest. The properties of all the tissues and organs are dependent upon their regular Nutrition, by a due supply of perfectly-elaborated blood; this cannot be effected, unless the functions of Circulation, Respiration, and Secretion, be performed with regularity,—the first being necessary to convey the supply of nutritious fluid, and the two latter to separate it from its impurities. The Respiration cannot be maintained, without the integrity of a certain part of the nervous system: and the due action of this, again, is dependent upon its regular nutrition. The materials necessary for the replacement of those, which are continually being separated from the blood, can only be supplied by the Absorption of ingested aliment; and this cannot be accomplished, without the preliminary process of Digestion. The introduction of food into the stomach, again, is dependent, like the actions of Respiration, upon the operations of the muscular apparatus and of a part of the nervous centres; and the previous acquirement of food necessarily involves the purely Animal powers. On the other hand, the functions of Animal life are even more closely dependent upon the Nutritive actions, than are those of organic life in general; for many tissues will retain their several properties, and their power of growth and extension, for a much longer period after a general interruption of the circulation, than will the Nervous structure; the action of which is instantaneously affected by a cessation of the due supply of blood, or by a depravation of its quality.

364. Yet, however intimate may be the bond of union between the Organic and Animal functions, the former are never *immediately* dependent upon the latter; although, as already shown, they generally depend upon them for the conditions of their maintenance. There is no good reason to believe that "nervous agency" is *essential* to the processes of Nutrition and Secretion in Animals, any more than to the corresponding processes in Plants. This is a question which may be more certainly determined by *observation*, than by any possible *experiment*. That these processes are very readily influenced by changes in the condition of the nervous system, is universally admitted; and it is the intimacy of this connection, which has given rise to the idea of a relation of *dependence*, and which prevents that idea from being experimentally disproved. In order to cut off *all* nervous communication from any portion of the organism—a gland for example,—so violent an operation is required (involving no less than the complete division of the blood-vessels, on which a plexus of ganglionic nerves is minutely distributed), that it is impossible to say that the disturbance of the function may not be owing to the shock produced on the general system. Observation, however, shows us that these processes are performed in the most complex and elaborate manner by Vegetables, in which all the attempts that have been made to prove the existence of a nervous system have signally failed (these attempts seeming to have been

propelling it through the vessels, and ensuring its regular and equable distribution; and its motion appears dependent upon the forces created in the individual parts themselves. In the higher classes, on the other hand, the comparative activity of all the functions, and the peculiar dependence of those of animal life upon a constant supply of the vital fluid, require a much more elaborate apparatus, and especially a central power, by which the movements of that fluid through the individual parts may be harmonised, directed, and controlled.

only excited by an indisposition to admit the possibility of any vital actions being independent of "nervous influence");—that the lowest Animals appear equally destitute of a nervous apparatus destined to influence them;—that in the higher classes there are many tissues into which nerves cannot be traced, and which yet exhibit as much vital activity as those which are in most intimate relation with nerves;—and that in their early embryonic condition, the formative operations are performed with their greatest energy, at a period when it is quite certain that no nerves have yet been developed. The intimacy of the connection which seems to be established between processes essentially independent of each other, by the instrumentality of the Nervous system, and especially by that part of it known as the 'Sympathetic,' seems to have relation to the necessity which arises, from the complication and specialisation of the Organic functions, for their being harmonised and kept in sympathy with each other, and with the conditions of the Animal system, by some mode of communication more certain and direct than that afforded by the circulating apparatus, which is their only bond of union in Plants.* If the views already offered, respecting the 'correlation of the vital forces' (§ 53), be accepted, they offer a definite expression of the relation of 'nervous influence' to 'cell-growth' and other organic functions; for just as an electric current, set in motion by certain chemical reactions, is capable of influencing the chemical reaction of substances submitted to its agency, so may it be fairly anticipated that the nervous force, itself the result of a certain class of nutritive operations (of which it may be considered the highest product), should be able to modify the course of those operations elsewhere.

365. The *Absorption* of alimentary materials is the first in the train of strictly Vital operations, and is common to Plants and Animals, although performed under somewhat different conditions in the two Kingdoms. The *Plant* derives its support immediately from the surrounding elements; it is fixed in the spot where its germ was cast; and it neither possesses a will to move in search of food, nor any locomotive organs for so doing. By the peculiar properties of its roots, however, it is endowed with some power of obtaining aliment not immediately within its reach (§ 421). The *Animal* possesses a recipient cavity, in which its food, consisting of matter previously organised, undergoes a certain preparation or *Digestion*, before it is taken up by the absorbent vessels which are distributed on its sides. The introduction of food into this cavity, or its *ingestion*, seems more and more dependent upon the animal functions, in proportion as we ascend the scale. The ciliary movements of the lower classes of animals, which produce rapid currents in the water that surrounds them, and thus bring a supply of food to the entrance of the digestive cavity, are probably to be regarded as of the same involuntary character as those which exist in the higher (§ 184). In animals of more complex structure, the process of obtaining food requires a much greater variety of movements, which are evidently dependent on the Muscular and Nervous systems; but these

* This, it may be safely affirmed, is all that has yet been *proved* of the functions of the Sympathetic or Ganglionic system of nerves in Vertebrata, and any hypothesis which presumes further, must be regarded as unphilosophical, because unnecessary to explain facts. The *onus probandi* certainly rests with those who maintain what is contrary to the important analogy just adduced, and not with those who frame their opinions in harmony with it.

may still be regarded as not involving changes of a psychical character. Rising still higher, however, we find the psychical endowments of the animal evidently concerned in procuring its support; and in Man, in whom these exist in their highest perfection, the reliance upon them is necessarily the greatest, his bodily organisation not being adapted for the supply of his physical wants, except under the direction of his Intelligence.

366. The alimentary materials taken up by the absorbent vessels, are carried by the *Circulation* into all parts of the fabric. This movement, so important in the highest classes of Plants and Animals, becomes less necessary in the lower, where the absorbent surface is in more immediate relation with the parts to be supplied with nourishment (§ 266). Besides affording a continued supply of nutrient material for the maintenance of the formative operations, the circulating system of Animals is usually the means of conveying to their nervo-muscular apparatus the oxygen whose presence is a necessary condition of its vital activity. It also serves to take up the effete matters which are set free by the 'waste' of the system, and to convey these to the organs provided for their elimination. The Circulation in Animals, as in Plants, is entirely independent of the will, cannot be in the least degree controlled by it, and, in its usual condition, is even unaccompanied with consciousness. The Muscular apparatus is concerned in it, only to give to it the energy and regularity which the conditions of animal existence require (§ 362 *note*); and Nervous agency merely brings it into sympathy with other operations of the corporeal and mental systems.—In the higher Animals, we find that the absorption of crude alimentary materials is not only effected by the blood-vessels, but also by a special set of *Lacteal Absorbents* spread over the walls of the digestive cavity; these discharge the matters they have taken up, into the current of the circulation; and it is probable that in this 'absorbent system' that function of Assimilation commences, by which the crude material is prepared for taking part in the formative processes. There is another division of this 'absorbent system,' which extends itself through the body, and which seems destined to collect the superfluous nutritive material which may have escaped from the blood-vessels, together (perhaps) with such as may have served its purpose in the system, and has died without decomposing, so that it may be again employed as an alimentary substance; and this *Lymphatic* system of vessels, also, would seem to partake in the Assimilative operation, of which we have next to speak.

367. The alimentary materials first taken up by the absorbent process, must undergo various changes by *Assimilation*, before they can become part of the organised fabric. There is much difficulty, however, in tracing these with precision, either in the Animal or the Vegetable system. The first step which is perceptible in the latter, is the formation of *organic compounds* by a new combination of the elements supplied by the food, which appears to commence as soon as these elements are absorbed. In the former, these organic compounds are directly supplied by the food; and this preliminary operation is consequently not required. From whichever source they are derived, however, these organic compounds appear to be subjected, whilst yet circulating in the liquid form, to a vital influence, exerted either by the living tissues through which they flow, or by cells that are floating in the stream, or perhaps by both; for we find them exhibiting peculiar properties, without any detectible change in ultimate composition, which

mark the transition from the crude form in which they are received into the body, towards the condition of living tissue. The 'protoplasma' of the Plant, and the 'liquor sanguinis' of the Animal, are very different from mere admixtures of gum, albumen, water, and other organic compounds, but show a capacity for becoming organised, which these do not possess; hence their peculiar ingredients are designated as *organisable* or *plastic* substances. From these materials the individual tissues of the fabric are developed and renewed, by the process of *Nutrition;* the elements of each tissue deriving from the plastic fluid that portion which their composition requires. This process is influenced in the Animal, through the Nervous system, by conditions of the mind or of the general fabric; but it does not seem to be maintained by any such influence (§ 364). It is of course dependent, however, upon the continued supply of blood, and cannot long continue if the Circulation be brought to a stand; there is evidence, however, that it may go on for a time, at the expense of the blood contained in the vessels of a part, after the current has ceased to flow; and a retardation of the current is often seen, where the formative processes are most actively going on.

368. In order to preserve the circulating fluid in the state required for the due performance of its important functions, means are provided for separating and carrying out of the system whatever may be superfluous or injurious in its constituent parts; as well as for elaborating from it certain fluids having a destined use in the economy. These changes may be comprehended in the general term *Secretion*, the former constituting the special function of *Excretion*. This is one no less important to the welfare of the system, than the absorption of aliment; and in proportion to the complexity of the structure, and to the diversity of its actions, do we find a multiplication of the excreting organs, as well as a variety in their products. The elimination of fluid by *Exhalation*, and of carbonic acid by *Respiration*, are constant, however, in all living beings. These changes seem to have no more immediate dependence upon the Nervous system, than have those of nutrition; and they will take place, to a certain extent, after the final extinction of the animal powers. Wherever a proper Circulation exists, however, they are most intimately dependent upon its maintenance, and soon come to an end if it ceases; but it is probable that, in particular cases, they are kept up by the *capillary* circulation, after the general propulsive force is extinct.

369. The essential difference between the function of *Reproduction* and that of *Nutrition* consists in this,—that in the latter case the alimentary materials are appropriated in renovating the structures of the *individual*,—whilst in the former they are applied to the production of a new structure, which, although for some time a part of the parent being, subsequently becomes a *new* individual. In the lowest classes of Plants and Animals, we often find these two functions completely blended together; thus, in the Protophyta and the Protozoa, each new cell may be regarded either as a part of the parent structure, or as a distinct individual, being capable of living, growing, and multiplying by itself. Even in higher members of both kingdoms, we have seen that a multiplication of individuals is effected by the development of *gemmæ*, by a process which corresponds in every essential particular with that of ordinary Nutrition, and which is no more dependent than *it* is upon the activity of the Animal functions. In

the true *generative* function, the concurrent action of two sets of organs, the 'sperm-cell' and the 'germ-cell,' is necessary; but this we see taking place in Plants without any interference of will, or excitement of consciousness, on the part of the individual; the two organs being sometimes united in the same being (as in hermaphrodite flowers), or, if separated (as in monœcious species), their functions being made to concur by external influences, as when the pollen of one flower is conveyed to the stigma of another at some distance, by the agency of the wind, insects, &c. There are some Animals, in which the two sets of organs are united in the same individual; and the actions necessary to bring them into relation are probably of a purely instinctive character, like those which are designed to bring food to the digestive organs. But in the higher classes, where the organs exist in separate individuals, the will, excited by a powerfully-stimulating propensity, is evidently the instrument by which they are brought into relation with one another; and in Man, where this propensity is connected with a nobler and purer passion, not only the will, but the highest powers of the intellect, are put in action to gratify it. But even here, the *essential* part of the function, which consists in the fertilization of the 'germ-cell' by the contents of the 'sperm-cell,' is as completely independent of mental influence, as it is in the plant or in the simplest animal.

370. The function of *Muscular Contraction*, to which nearly all the sensible motions of the higher Animals are due, is one which has an important connection with almost every one of their vital operations; although, as already explained, this connection is mostly of an indirect character. The property of Contractility on the application of a stimulus is not, however, confined to animals; since it is possessed by many of the Vegetable tissues, and has an important relation with their nutritive processes. Nor, even in Animals, is it confined to the muscular tissue. For in the lowest tribes it seems generally diffused through the system, and appears to be for the most part excited by external stimuli. But in the higher classes, it is concentrated in a special texture, and is called into operation by a new stimulus, the Nervous power, which originates in the individual itself. By this means it is brought under subordination to the will, and is made the instrument of changing the relations between the living organism and the external world.

371. The functions of the *Nervous System* are twofold. First, to bring the conscious Mind (using that term in its most extended sense, to denote the psychical endowments of animals in general) into relation with the external world; by informing it, through the medium of the organs of *sensation*, of the changes which the material universe undergoes; and by enabling it to react upon them through the organs of motion. And secondly, to connect and harmonise different actions in the same individual, without necessarily exciting any mental operation. But, in the words of a profound writer on this subject, "mental-acts, and bodily changes connected with them, are not merely *superadded* to the organic life of animals, but are intimately connected or interwoven with it; forming in the adult state of all but the very lowest animals, part of the conditions necessary to the maintenance of the quantity, and of the vital qualities, of the nourishing fluid on which all the organic life is dependent."* In proportion

* Prof. Alison's Outlines of Human Physiology, p. 13.

to the complexity and extent of the Psychical endowments of each species of Animals, may their influence over the conformation of the Organic structure be perceived; so that it becomes more and more removed from that which is presented by Vegetables, the chief end of whose existence appears to be the elaboration of an organised fabric from the elements furnished by the inorganic world. In Man, the being that possesses the largest share of these capabilities, the whole apparatus of Organic life would seem destined but to serve for the maintenance of the Animal functions. The processes of Nutrition are here chiefly directed towards perfecting the Nervous and Muscular systems, and bringing their functions into most advantageous operation; whilst in many of the lower animals this would seem quite a subordinate object, the extension of the organs of vegetative life being the direction taken by their development, and their organic functions being performed under conditions nearly the same (except as regards the nature of the food) with those which prevail in Plants.

CHAPTER X.

OF ALIMENT, ITS INGESTION AND PREPARATION.

1. *Sources of the Demand for Aliment.*

372. WE have seen that all *Vital Action* involves a *change* in the condition of the Organised Structure which is its instrument.—The vital activity of the Plant is chiefly manifested in its *increase, development,* and *reproduction;* in the multiplication of its component cells, in the metamorphoses which these cells undergo, and in the formation of germs which are destined to be cast off by it, and to originate new organisms elsewhere. These operations can only be performed, however, when the plant is supplied with such alimentary substances as can be converted by it into the proximate materials of its own tissues; which materials are then appropriated by these, as the *pabulum* at whose expense their growth takes place. In those simple Cellular Plants whose structure is nearly homogeneous throughout (§ 266), every act of cell-multiplication conduces directly to the growth of the entire mass; since the new cells thus produced remain as constituent parts of it, so long as the organism holds together. In the higher tribes of Plants, however, this is not the case; for we find, as we have seen, that certain organs are periodically developed, the term of whose existence is comparatively brief, so that they only form constituent parts of the structure for a short time, being cast off as soon as their term of life is over, to be replaced by another set possessing attributes of precisely the same kind. It is, however, through the agency of these temporary organs,—namely, the leaves and the flowers,—that the means of increase and reproduction are provided for the more permanent parts of the organism, which could neither grow nor regenerate its kind without their agency. Thus it would seem to be a general rule, that wherever true *woody* tissue (the production of which seems to be the highest exertion of the vital force of Plants) is produced, the Plant shall be furnished with a set of organs whose peculiar function it is to prepare

and elaborate its materials, and which accomplish this with such vigour and activity, that their own vital energy is soon exhausted, so that they die, and are cast off in a state of decay ; the most *permanent* portion of the highest vegetable fabrics being thus built up through the instrumentality of the most *transient* part of their organisation, and the growth of the wood-cells being entirely dependent upon the vital activity of the leaf-cells. So again, in the higher Plants, we find that instead of that simple liberation of germs from the interior of certain cells, more or less distinctly set apart from the general structure, which is the common method of reproduction in the lowest, a complex apparatus is provided for their evolution ; and that this apparatus,—consisting in fact, of elements which might be developed into leaves, and which, though metamorphosed into the parts of the flower, still present an accordance with the general laws of leaf-growth,—is itself of yet more transient duration than the leaves, being cast off as soon as the germs which it has brought into existence are capable of being separated from the parent structure. Further we have to note, that it is one part of the office of the apparatus of fructification, to prepare and store up a supply of nutriment for the early development of the germ ; and this store, which makes up the chief bulk of the *seed,* is derived from the materials provided by the roots and leaves of the parent plant.

373. Thus, then, if we look at the sources of demand for Aliment in the Vegetable Organism, we shall see that they may be reduced to the following heads.—I. The extension of the individual fabric, by the multiplication and development of its component parts ; which parts may themselves possess the same degree of duration as the rest of the fabric, and may in their turn be the subject of further multiplication ; or may be destined only for a temporary activity, replacing for a time those that have already expended their powers and passed through their term of life, and themselves going through the same phases of existence, which have as their ultimate object the extension of the more permanent parts of the structure.—II. The production of germs for the continuance of the race ; which not only incorporate in their own substance a certain amount of elaborated aliment, but which, in the higher plants, carry with them a further supply, at the expense of which their early development takes place.—It is, then, by the activity of the functions of growth and reproduction, that the demand for food is alone determined in the Plant ; and we shall find that, in their turn, the activity of these functions is dependent upon, and in some degree determined by, the supply of nutriment afforded to it. Thus it is when the plant is most abundantly supplied with appropriate materials by its roots and leaves, that it will most tend to throw out new leaf-buds, to form new wood, and thus to grow as an individual ; whilst, on the other hand, it is when more sparingly supplied with food, that it will produce the greatest number of flower-buds, and develope the most numerous progeny. In the life of the simpler plants, there may thus be said to be no 'waste' whatever ; for so long as their tissues are growing and multiplying, so long do they resist the influences which tend to their decomposition. But in the life of the higher plants, a source of 'waste' arises from the temporary nature of some of their organs ; but even this is connected, as we have seen, with the *constructive,* not with the *destructive* class of operations, the successional development and death of the leaves

being subservient to the building-up of the individual fabric, and to the regeneration of the race. Thus a very large proportion of the aliment taken into the Vegetable system is directly appropriated to these purposes; the amount of carbonic acid and of other excretory matters given off during the period of growth, being very small in proportion to that which is taken in; and the fall of the leaves, which restores to the condition of inorganic substances a certain quantity of matter that has undergone the organising process, not taking place, until by their instrumentality a considerable addition has been made to the solid fabric of the tree. To these processes of extension and reproduction, there would not seem to be,—at least in those more complex Vegetable structures, in which one part after another may die and decay, without destroying the life of the organism in its totality,—any very definite limit (§ 124, 125).

374. The case is very different, however, in the Animal. In it, also, the activity of the functions of growth and reproduction becomes a source of demand for food. But the period of increase is limited. The full size of the body is usually attained, and all the organs acquire their complete evolution, at a comparatively early period. The continued supply of food is not then requisite for the extension of the structure, but simply for its *maintenance;* and the source of this demand lies in the constant 'waste,' to which, during its period of activity, it is subjected. We have seen that every action of the Nervous and Muscular systems involves the death and decay of a certain amount of the living tissue, as indicated by the appearance of the products of that decay in the Excretions; and a large part of the demand for food will be consequently occasioned by the necessity for making good the loss thus sustained. Hence we find that the demand for food bears a close relation to the activity of the 'animal' or *destructive* functions; and thus the Birds of most active flight, and the Mammals which are required to put forth the greatest efforts to obtain their food, need the largest and most constant supplies of nutriment; whilst even the least active of these classes stand in remarkable contrast with the inert Reptiles, whose slow and feeble movements are attended with so little waste, that they can sustain life for weeks and even months, with little or no diminution of their usual activity, without a fresh supply of food.—This waste and decay, however, do not affect the muscular and nervous tissues alone; for as in the Plant we have found that the higher parts of the structure are developed by the instrumentality of the vital activity of the lower, so do we find in the Animal, that the exercise of those *constructive* operations, by which the materials for the first growth and the subsequent maintenance of the fabric are prepared and kept in a state of the requisite purity, involves the agency of a set of organs, which may be said to be entirely 'vegetative' in their character, and in which, as in the higher Plants, a continual renewal of the cells that constitute their essential structure, seems necessary for their functional activity. Thus all the glandular and mucous surfaces are continually forming and throwing off epithelial cells, whose formation requires a regular supply of nutriment; and only a part of this nutriment (that which occupies the *cavity* of the cells) consists of matter, that is destined to serve some other purpose in the system, or that has already answered it; the remainder (that of which the solid walls are composed) being furnished by the nutritive materials of the blood, and being henceforth altogether lost to

it.—Thus every act of Nutrition involves a waste or decay of Organised tissue.

375. We may observe a marked difference, however, between the amount of aliment required, and the amount of waste occasioned, by the simple exercise of the *nutritive* or *vegetative* functions in the building-up and maintenance of the animal body, and that which results from the exercise of the *animal* functions. The former are carried on, with scarcely any intermixture of the latter, during fœtal life. The aliment, in a state of preparation, is introduced into the fœtal vessels; and is conveyed by them into the various parts of the structure, which are developed at its expense. The amount of *waste* is then very trifling, as we may judge by the small amount of excretory matter, the product of the action of the liver and kidneys, which has accumulated at the time of birth; although these organs have attained a sufficient development to act with energy when called upon to do so. But as soon as the *movements* of the body begin to take place with activity, the waste increases greatly; and we even observe this immediately after birth, when a large part of the time is still passed in sleep, but when the actions of respiration involve a constant employment of muscular power.—In the state of profound *sleep*, at subsequent periods of life, the vegetative functions are performed, with no other exercise of the animal powers, than is requisite to sustain them; and we observe that the waste, and the demand for food, are then diminished to a very low point. This is well seen in many animals, which lead a life of great activity during the warmer parts of the year, but which pass the winter in a state of profound sleep, without, however, any considerable reduction of temperature; the demand for food, instead of being frequent, is only felt at long intervals, and the excretions are much reduced in amount. And those animals which become completely inert, either by the influence of cold, or by the drying-up of their tissues, do not suffer from the prolonged deprivation of food; because not only are their animal functions suspended, but their nutritive operations also are in complete abeyance; and the continual decomposition of their tissues, which would otherwise be taking place, is checked by the cold or by the desiccation to which they are subjected; so that the whole series of changes which goes on in their active condition, is completely at a stand.

376. But there is another most important source of demand for food, amongst the higher Animals, which does not exist either amongst the lower Animals, or in the Vegetable kingdom. We have seen (§ 92) that Mammals and Birds, and to a certain extent Insects also, are able to sustain the heat of their bodies at a fixed standard, and thus to be independent of variations in external temperature. This they are enabled to do, as will be explained hereafter (CHAP. XVII.), by a process analogous to ordinary combustion; the carbon and hydrogen which are directly supplied by their food, or which have been employed for a time in the composition of their living tissues and are then set free, being made to unite with oxygen introduced by the respiratory process, and thus giving off as much heat as if the same materials were burned in a furnace. And it has been experimentally proved, that the immediate cause of death in a warm-blooded animal from which food has been entirely withheld, is the inability any longer to sustain that temperature, which is requisite for the performance of its vital operations (§ 93). Hence we see the necessity for

a constant supply of aliment, in the case of warm-blooded animals, for this purpose alone; and the demand will be chiefly regulated by the difference between the external temperature and that of the animal's body. When the heat is rapidly carried off from the surface, by the chilling influence of the surrounding air, a much greater amount of carbon and hydrogen must be consumed within the body, to maintain its proper heat, than when the air is nearly as warm as the body itself; so that a diet which is appropriate to the former circumstances, is superfluous and injurious in the latter; and the food which is amply sufficient in a warm climate, is utterly destitute of power to enable the animal to resist the influence of severe cold. Again, the Bird, whose natural temperature is 110° or 112°, and whose body exposes a large surface in proportion to its bulk, must consume a greater quantity of combustible material for the maintenance of its normal heat, than is required by a Mammal, whose natural temperature is 100°, and whose body, being of much greater bulk, exposes a much smaller proportional surface to the cooling influence of the surrounding medium.

377. Thus we find that, in the Animal body, aliment is ordinarily required for four different purposes; the first two of which are common to it and to the Plant, whilst the others are peculiar to it.—I. The first construction or building-up of the organism, by the multiplication and development of its component parts. II. The production of germs for the continuance of the race; and, in addition, in the female, the provision of the store of aliment required by these germs during their early development. III. The maintenance of the organism, both during its period of growth, and after its attainment of its full size, notwithstanding the 'waste' occasioned by the active exercise of the nervous and muscular systems. IV. The supply of the materials for the heat-producing process, by which the temperature of the body is kept up.—The amount required for these several purposes will vary, therefore, not only with the general activity of the nutritive processes, but in accordance with the conditions of the body, as regards exercise or repose, and external heat or cold. It is also subject to great variation with difference of *age*. During the period of growth, as might be anticipated, a much larger supply of food is required in proportion to the bulk of the body, than when the full stature has been attained; but this results, not so much from the appropriation of a part of the food to the augmentation of the fabric (the proportion of its whole amount which is thus employed being extremely small), as from the much greater rapidity of *change* in the constituents of the body of the young animal, than in that of the adult, which is evidenced by the large proportional amount of the excretions of the former, by the rapidity with which the effects of insufficiency of aliment manifest themselves in the diminution of the bulk and firmness of the body, by the short duration of life when food is altogether withheld, and by the readiness with which losses of substance by disease or injury are repaired, when the nutritive processes are restored to their full activity. The converse of all this holds good in the state of advanced age. The excretions diminish in amount, the want of food may be sustained for a longer period, losses of substance are but slowly repaired, and everything indicates that the interstitial changes are performed with comparative slowness; and, accordingly, the demand for food is then far less in proportion to the bulk of the body, than

it is in the adult. This contrast is most remarkably shown in the Insect tribes, which, as we have seen, are far more voracious in the *larva* than in the *imago* state ; many species, indeed, taking no food whatever, after their last metamorphosis ; and most others taking very little, except such as they may be preparing to apply to the sustenance of their progeny.

378. The influence of the supply of food upon the size of the individual, is very evident in the Vegetable kingdom ; and it is most strikingly manifested, when a plant naturally growing in a poor dry soil is transferred to a rich damp one, or when we contrast two or more individuals of the same species, growing in localities of opposite characters. Thus, says Mr. Ward,* "I have gathered, on the chalky borders of a wood in Kent, perfect specimens in full flower of *Erythræa Centaurium* (Common Centaury), not more than half an inch in height ; consisting of one or two pairs of most minute leaves, with one solitary flower : these were growing on the bare chalk. By tracing the plant towards and in the wood, I found it gradually increasing in size, until its full development was attained in the open parts of the wood, where it became a glorious plant, four or five feet in elevation, and covered with hundreds of flowers." On the other hand, by *starvation*, naturally or artificially induced, Plants may be dwarfed, or reduced in stature : thus the Dahlia has been diminished from six feet to two ; the Spruce Fir from a lofty tree to a pigmy bush ; and many of the trees of plains become more and more dwarfish as they ascend mountains, till at length they exist as mere underwood. Part of this effect, however, is doubtless to be attributed to diminished temperature ; which, as formerly remarked (§ 89), concurs with deficiency of food in producing inferiority of size.

379. Variations in the supply of food would not appear to be effectual in producing a corresponding variety of size in the Animal kingdom : this is not, however, because Animals are in any degree less dependent than Plants upon a due supply of food ; but because such a limitation of the supply, as would *dwarf* a Plant to any considerable extent, would be fatal to the life of an Animal. On the other hand, an excess of food, which (under favourable circumstances) would produce great increase in the size of the Plant, would have no corresponding influence on the Animal ; for its size appears to be restrained within much narrower limits,—its period of growth being restricted to the early part of its life, and the dimensions proper to the species being rarely exceeded in any great degree. Even in the case of *giant* individuals, it does not appear that the excess of size is produced by an over-supply of food ; but that the larger supply of food taken-in is called for by the unusual wants of the system,—those wants being the result of an extraordinary activity in the processes of growth, and being traceable rather to the properties inherent in the system, than to any external agencies. The influence of a diminished supply of food, in producing a marked inferiority in the size of Animals, is most effectually exerted during those early periods of growth, in which the condition of the system is most purely 'vegetative.' Thus it is well known to Entomologists, that, whilst it is rare to find Insects departing widely from the average size on the side of excess, dwarf-individuals, possessing only half the usual dimensions, or even less, are not uncommon ; and there can be

* "On the Growth of Plants in closely glazed Cases," p. 7.

little doubt that these have suffered from a diminished supply of nutriment during their larva state. This variation is most apt to present itself in the very large species of Beetles, which pass several years in the larva state; and such dwarf-specimens have even been ranked as sub-species. Abstinence has been observed to produce the effect, upon some Caterpillars, of diminishing the number of moults and accelerating the transformation; in such cases, the Chrysalis is more delicate, and the size of the perfect Insect much below the average. A most characteristic example of the influence of particular kinds of food in modifying the processes of development, seen in the economy of the Hive-Bee, has been already cited (§ 60). That insufficiency of wholesome food, continued through successive generations, may produce a marked effect, not merely upon the stature, but upon the form and condition of the body, even in the Human race, appears from many cases, in which such influence has operated on an extensive scale. Of these cases, some of the most remarkable are those of the Bushmen of Southern Africa, and the aborigines of New Holland; whose low physical condition appears to be in great part due to imperfect nutrition.

2. *Nature of the Alimentary Materials.*

380. Amongst the general differences between the Animal and Vegetable kingdoms, none are more striking, as already pointed out (CHAP. V.), than those existing between the aliments whereon they are respectively supported, and the mode of their *ingestion* or introduction into the system. The essential nutriment of Plants appears to be supplied by the inorganic world, and to consist chiefly of the elements of *water*, with certain saline impregnations, *carbon*, and *nitrogen*. The water is partly derived from the fluid which percolates the soil and is absorbed by the roots; and partly from the moisture of the atmosphere, which is imbibed by the leaves. The carbon is principally obtained from the carbonic acid of the air; but most plants are assisted in their growth by its introduction through the roots also. In all soils of moderate richness, there exists a large quantity of the remains of organised structures, the upper layer of which is constantly undergoing some degree of decomposition by contact with the atmosphere, so that carbonic acid is formed in it. The water which traverses such a soil, therefore, will become charged with this gas; and this state of solution appears to be that in which carbon may be most advantageously introduced into the vegetable system. It may be regarded as certain, that the organic matter which rich soils contain is not itself applied to the nutrition of the plant without this previous decomposition; for it is found that those soils which afford the most steady and equable supply of carbonic acid, are the most favourable to vegetable growth; and that this end may be answered, not merely by an admixture of decomposing organic matter, but by the introduction of substances, such as gypsum or powdered charcoal, which have the property of condensing carbonic acid from the atmosphere. The opinion that air and water furnish the essential food of plants, is confirmed by the fact, that not only will the simpler forms of Lichens appear on barren rocks in the midst of the ocean, increasing by absorption from the atmosphere alone, and preparing by their decomposition a nidus for the reception of the

germs of higher orders of vegetation; but that many, even of the more highly organised species, will grow in circumstances where no other kind of nutriment is accessible to them. The small amount of earthy or saline matter contained in the tissues of such plants, must be derived from the atmosphere, which is known to hold such particles in suspension. A constant supply of the mineral ingredients naturally found in the tissue of each species, is more important than is generally supposed; and the fertility of a soil, and the efficiency of a manure, will often depend as much upon this, as upon any other cause. It is only within a recent period, that the dependence of all Vegetable growth upon a due supply of *nitrogen* has been ascertained; but it is now known that, although usually existing in only a small proportion, its presence in the vegetable tissues is peculiarly important at the time of their greatest formative activity; the 'primordial utricle,' which is the seat of the most active vital operations (§ 136), being composed of albuminous matter, in which nitrogen is an essential ingredient. The small quantity of nitrogen which the usual rate of growth of ordinary Plants causes them to require, appears to be derived from the minute proportion of ammonia existing in the atmosphere; this being condensed from it in rain or dew, or absorbed in the gaseous state by porous soils, so as in either case to find its way to the roots in the liquid which they imbibe. But the growth of most plants is powerfully stimulated by an additional supply of ammonia, such as they derive from the introduction of decaying animal substances into the soil, as manures; and the efficacy of these is peculiarly manifested in the large increase of the amount of *azotized* compounds, then generated by such Plants as naturally produce them in considerable proportion,—such, for example, as the corn-grains. To the fertility of a soil, then, it is essential that it yield a sufficient and regular supply of moisture, carbonic acid, and ammonia; the two latter being either attracted from the atmosphere, or evolved by its own decomposition.—The only class of Plants which even seems to be *dependent* for its support upon matters already organised, is that of *Fungi* (§ 271—273); but it is probable that this dependence only arises from the peculiarly large and constant supply of carbonic acid and ammonia, which they require as the condition of their growth; as well as (perhaps) from their being only able to appropriate these compounds in the 'nascent' state. There is no reason to believe that they can make use of organic compounds in any other than a state of decomposition; and hence it is that their great utility in the economy of Nature arises, the products of decay, which might otherwise have poisoned the atmosphere, being converted into living and growing tissues. Fungi present us with two curious analogies to the Animal kingdom; both resulting, no doubt, from the mode in which they receive their aliment. The large quantity of carbonic acid with which their absorbent apparatus furnishes them, prevents the necessity of their drawing any additional supply of it from the atmosphere; but on the contrary, like animals, they have only to get rid of what is superfluous. And again, the proportion of azotized matter contained in their tissues is much greater than in any other vegetable; so that their substance, if capable of being digested, is almost as nutritious as animal flesh.

381. It is a general law of vitality, that the materials of nutrition can only be introduced into the living system in the fluid state; and although

the ingestion of solid aliment by the higher Animals might seem to contradict such a principle, a little examination into the character of their nutritive organs will show that they are framed in conformity with it. In addition to the absorbing organs with which Plants are furnished, and which form the medium of communication between the organic life and the external world, nearly all Animals are provided with cavities for the reception of their food, and for its reduction to a state fit to enter the vessels. The necessity for these cavities arises out of the nature of the aliment required by Animals, which usually pre-exists in a form more or less solid; and also from the occurrence of intervals between the periods at which it is obtained. Whilst the roots of Vegetables are fixed in the soil, and ramify through it in pursuit of their nutriment, Animals, whose locomotive powers are necessary for the search after the food they require, may be said to carry their soil about with them; for their absorbents are distributed on the walls of the digestive cavity, just as those of Plants are externally prolonged into the earth. This cavity is in all instances formed by a *reflexion* of the external surface, of which the *Hydra* (§ 289) may be regarded as presenting us with the simplest example. It is merely a bag with one opening (A), which may be regarded as *all stomach.* A higher form is that in which the cavity has two orifices, and thus becomes a canal (B); and all the complicated intestinal apparatus of the higher animals may be considered as a more extended development of this simple type. That the presence of the stomach, however, is not an *essential* character of the Animal (as taught by some Physiologists), but is rather a special adaptation of their organism to the peculiarity of their food, which may be dispensed with under peculiar circumstances, has been already pointed out (§ 262).—The food which is introduced into this cavity, is acted upon mechanically by the motion of the walls, and chemically by the secretions poured from their surface; so that the nutritious parts of it are separated from those which may be rejected, and are reduced to a fluid form.—That the process of Digestion in Animals is really of no higher a character than this, and that it has nothing to do with 'organising' or 'vitalizing' the materials submitted to it, appears from *a priori* considerations, and from experiment also. For the substances contained in the alimentary canal, and in contact with that reflexion of the external integument which constitutes its lining, are really as much external to the living body, as if they were placed in contact with the skin; we cannot regard them as introduced into it, until they have been absorbed; and up to that period, they hold precisely the same relation to the absorbent vessels, as the fluid diffused through the soil bears to the roots of Plants that ramify upon the surface of that Earth, which has been expressively said to be their 'common stomach.' All the experiments which have been performed upon artificial digestion, have precisely the same bearing; since it appears from them, that if the food be subjected to the action of the same solvent fluids, with the same assistance from heat and from mechanical movement, the result is the same out of the stomach as in it.

A

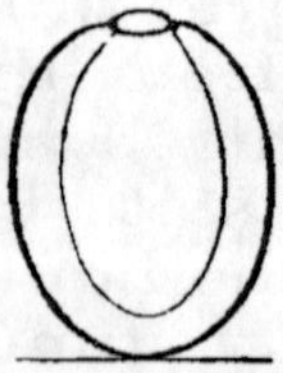

B

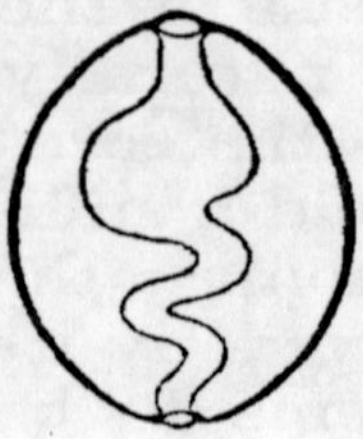

382. The particular articles which constitute the food of the different races of Animals, are as various as the races themselves. Some appear to

draw their nutriment from the Inorganic world; but this is not the case in reality. Thus the *Spatangus* (§ 296 *c*) and *Arenicola* (§ 311 *c*) fill their stomachs with sand, but really derive their nutriment from the minute animals which it contains. The Earth-worm and some kinds of Beetles are known to swallow earth; but they only derive from it the particles of organised matter which it includes, and reject the rest.* Some tribes in almost every division of this kingdom are maintained solely by Vegetable food; and wherever Plants exist, we find Animals adapted to make use of the nutritious products which they furnish, and to restrain their luxuriance within due limits. Thus, the Dugong browses upon the submarine herbage of the tropics; whilst the Hippopotamus roots up with his tusks the plants growing in the beds of the African rivers; the Giraffe is enabled by his enormous height to feed upon the tender shoots which are above the reach of ordinary quadrupeds; the Rein-deer subsists during a large part of the year upon a lichen buried beneath the snow; and the Chamois finds a sufficient supply in the scanty vegetation of Alpine heights. Many species of Animals, especially among the Insect tribes, are restricted to particular Plants; and, if these fail, the race may for a time disappear. But there is probably not a species of Plants which does not furnish nutriment for one or more tribes of Insects, either in their larva state or perfect condition, by which it is prevented from multiplying to the exclusion of others. Thus, on the Oak not less than 200 kinds of Caterpillars have been estimated to feed; and the Nettle, which scarcely any beast will touch, supports fifty different species of Insects, but for which check it would soon annihilate all the plants in its neighbourhood. The habits and economy of the different races existing on the same plant, are as various as their structure. Some feed only upon the outside of the leaves; some upon the internal tissue; others upon the flower or on the fruit; a few will eat nothing but the bark; while many derive their nourishment only from the woody substance of the trunk. It is very curious to observe that many plants injurious to Man afford wholesome nutriment to other animals; thus, Henbane, Nightshade, Water-hemlock, and other species of a highly poisonous character, are eaten greedily by different races of quadrupeds. Some cattle, again, will reject particular plants upon which others feed with impunity.

383. Every class of the Animal kingdom has its Carnivorous tribes, also, adapted to restrain the too rapid increase of the vegetable-feeders, by which a scarcity of their food would soon be created,—or to remove from the earth the decomposing bodies, which might otherwise be a source of disease or annoyance. The necessity of this limitation becomes evident, if we consider the rapid multiplication which the prolific tendency of the Herbivorous races would speedily create, until checked by the famine that would necessarily result from their inordinate increase. Thus, the myriads

* Among the human race, some savage nations are in the habit of introducing large quantities of earthy matter with their food; and this sometimes through ignorant prejudice, but more frequently to give bulkiness to the aliment, so that the stomach may be distended,—as among the Kamschatdales, who mix saw-dust or earth with their train-oil. It has been until recently supposed that the siliceous earth, which has been employed in Lapland in times of scarcity, admixed with flour and the bark of trees, merely answered this purpose; but recent microscopic examination has shown, that it consists of the *exuviæ* of Infusoria (§ 284), and contains a large portion of animal matter. If the latter be dissipated by incineration, the earth loses about 20 per cent. of its weight.

of Insects which find their subsistence on our forest trees, if allowed to increase without restraint, would soon destroy the life that supports them, and must then all perish together; but another tribe (that of the insectivorous Birds, as the Woodpecker), is adapted to derive its subsistence from them, and thus to keep within salutary bounds the number of these voracious little beings. Sometimes, however, they increase to an enormous extent. The pine-forests of the Hartz mountains have been several times almost destroyed by the ravages of a single species of Beetle, the *Bostrichus typographus*, which is less than a quarter of an inch in length. The eggs are deposited beneath the bark; and the larvæ, when hatched, devour the alburnum and inner bark in their neighbourhood. It was estimated that, in the year 1783, a million and a half of pine-trees were destroyed by this insect in the Hartz alone; and other forests in Germany were suffering at the same time. The wonder is increased when it is stated, that 80,000 larvæ are sometimes found on a single tree. Their multiplication is aided by their tenacity of life. It is found that, even if the trees infested by these larvæ be cut down, floated in water, kept for a length of time immersed either in water or snow, or even placed upon ice, they remain alive and unhurt. In the pupa state, however, they are more susceptible; and vast numbers perish in this condition from the influence of unfavourable seasons, which operate as the principal check to their multiplication.—A very curious instance of the nature of the checks and counter-checks, by which the "balance of power" is maintained amongst the different races, is mentioned by Wilcke, a Swedish naturalist. A particular species of Moth, the *Phalœna strobilella*, has the fir-cone assigned to it for the deposition of its eggs; the young caterpillars, coming out of the shell, consume the cone and superfluous seed; but, lest the destruction should be too great, the *Ichneumon strobilella* lays its eggs in the caterpillar, inserting its long tail in the openings of the cone until it touches the included insect, for its body is too large to enter. Thus it fixes upon the caterpillar its minute egg, which when hatched destroys it.*

384. The alimentary value of the various substances used as food by the several races of Animals, is not so different as, from the diversity of the sources whence it is drawn, we might be led to suppose. It depends, in the first place, upon the quantity of solid matter they respectively contain; being of course the greater, as the solids form the larger proportion of the entire weight. Many esculent vegetables contain so large a quantity of water, that the nutriment they afford is very slight in proportion to their bulk.—Next, it depends upon the proportion of digestible matter which the solid parts include; for it is not every substance containing the requisite ingredients, that is capable of being reduced to a state which enables it to be absorbed. Thus, woody fibre is composed of the same elements as starch-gum; but it passes out of the intestinal canal of the higher animals unchanged, and therefore affords them no nutriment; yet there are many tribes of Insects, which seem to draw their supply of nutriment exclusively from wood, and this even in its driest condition.

* The Chapters on the "Economy of Nutritive Matter," in Dr. Roget's Bridgwater Treatise, and those of Mr. Lyell's Principles of Geology, on the "Equilibrium of Species," may be referred to for a more extended view of this interesting subject, than the limits of the present work permit.

So, again, the horny tissues of animals, though nearly allied to albumen in their composition, are completely destitute of nutritive value to Man and the higher animals, because not capable of being reduced by their digestive process; though certain Insects appear capable of living exclusively upon them.—But when the watery and indigestible parts of the food are put out of consideration, and our attention is directed only to the soluble solids, we find most important relations in the chemical composition of the several alimentary materials, whether furnished by the Animal or the Vegetable kingdom, which render them more or less appropriate to the different purposes that have to be answered in the nutrition of the body. It has been already pointed out (§ 259), that Vegetables possess the power of combining the elements furnished by the Inorganic world into two classes of compounds,—the ternary, consisting of oxygen, hydrogen, and carbon,—and the quaternary, which consist of these elements, with the addition of azote or nitrogen. These two classes are hence termed the *non-azotized*, and the *azotized*.

385. Now the *azotized* compounds which are formed by Plants, are essentially the same with those *Albuminous* substances which are furnished by the flesh and by the nutritious fluids of Animals; and are equally adapted with the latter for the reparation of the waste of the muscular tissue, and for the general nutrition of the body. The quantity of these, however, which Plants yield, is usually in small proportion to that of the non-azotized; being considerable only in the Corn-grains, and in the seeds of Leguminous plants, which the universal experience of ages has demonstrated to be the most nutritious of Vegetable substances. But, unless the food contain a sufficient proportion of these compounds, the body must be insufficiently nourished, and the strength must diminish, even though other elements of the food be in superabundance; and consequently, if the food be of a kind which contains but a small proportion of albuminous matter, a very large amount of it must be ingested, to afford the requisite supply of the essential ingredients. We see a provision for this requirement, in the capacity of the alimentary canal of Vegetable-feeding Animals; which is almost invariably far greater than that of the carnivorous members of the same groups.—There is another azotized compound, *Gelatine*, that is furnished by Animals, to which nothing analogous exists in Plants; and this, although it cannot sustain life by itself, and is not an essential article of food, is a valuable adjunct to the albuminous compounds. For as the gelatinous tissues (§ 163) suffer waste, in common with the others, it is evident that if the gelatine be supplied already-prepared, it may be at once applied to their nutrition; and thus the proportion of albuminous material, which they would otherwise require, is not demanded, and the labour of transformation is also saved. But there is no evidence that Gelatine can ever be transformed into an albuminous compound, and can thus be applied to the nutrition of the muscular and other fibrinous tissues; and the presumption, derived from the results of various experiments, is very strong the other way.

386. The *non-azotized* compounds supplied by Plants, exist under various forms; of which the principal are starch, sugar, and oil. The two former may be regarded as belonging to one class, the *Saccharine* or *Farinaceous;* because we know that starch, and the substances allied to it, may be converted into sugar by simple chemical processes, and that this trans-

formation takes place readily both in the Vegetable and Animal economy. On the other hand, the *Oily* matters are usually ranked as a distinct group of alimentary substances; and it has been maintained that, under no circumstances, has the Animal the power of elaborating fatty matter from starchy or saccharine compounds. But this is now known to be an unfounded limitation; since the transformation of a saccharine into a fatty compound takes place in the case of Bees, which form wax when fed upon pure sugar; and it has been recently shown that it may be effected in the laboratory of the Chemist, butyric acid (the characteristic fatty acid of butter) being one of the products of the 'lactic fermentation' of sugar, excited by animal substances. Thus, then, whether derived from Vegetable or from Animal bodies, the non-azotized substances available as food are essentially the same. The former kingdom supplies them chiefly in the saccharine or farinaceous form, the latter chiefly in the oleaginous; but a considerable quantity of oil is furnished by certain Plants; and there are Animals which incorporate the cellulose of plants with their own substance (299 *a*). The only Animal tissue to which the non-azotized compounds apparently serve as the appropriate *pabulum*, is the Adipose; the peculiar character and offices of which have been already explained (§§ 189, 190). There is reason to believe, however, that oleaginous matter performs a most important part in the incipient stages of Animal nutrition; and that its presence is not less essential to the formation of cells, than is that of the albuminous matter which forms their chief component. If such be the case, it is not surprising that oil in some form, or substances capable of conversion into it, should be such universal constituents of the food of Animals.—Besides serving these purposes, however, the non-azotized compounds have a most important use in warm-blooded Animals; that of supporting the respiratory process, and thus maintaining the temperature of the body. We have seen that, in the compounds of the *Saccharine* group (in which Starch is included), the amount of oxygen is no more than sufficient to form water with the hydrogen of the substance (§ 27); so that the carbon is free to combine with the oxygen taken in by the lungs, and thus becomes a source of calorifying power. Again, in the *Oily* matters taken in as food, the proportion of oxygen is far smaller; so that they contain a large quantity of surplus hydrogen, as well as of carbon, ready to be burned-off in the system, and thus to supply the heat required. The extraordinary power of oleaginous substances to impart heat to the system by the combustive process, is indicated by the experience of the Human inhabitants of frigid zones, who feed upon whales, seals, and other animals loaded with fat, and who devour this fat with avidity, as if instinctively guided to its use. It is by the enormous quantity of this substance taken in by them, that they are enabled to pass a large part of the year in a temperature below that of our coldest winter, spending a great portion of their time in the open air; as well as to sustain the extremes of cold, to which they are occasionally subjected. And in consequence of its being more slowly introduced into the system than most other substances, a larger quantity may be ingested at one time, without palling the appetite; whilst its bland and non-irritating character favours its being retained until it is all absorbed. In this manner, the Esquimaux and Greenlanders are enabled to take in 20 or 30 pounds of blubber at a meal; and, when thus supplied, to pass several days without food.—On the other hand, among the inhabitants of warm climates, there is comparatively little disposition to the use of oily matter as food; and the quantity of it contained in most articles of their diet is comparatively small.

387. The greatest economy in the use of Aliment is therefore exercised, when the diet contains a sufficient proportion of azotized substances to repair the 'waste' of the albuminous and gelatinous tissues; and a sufficient amount of non-azotized compounds, to develope (with the aid of other processes) the requisite amount of heat by combination with oxygen. Now in the Milk, which is the sole nutriment of young Mammalia during the period immediately succeeding their birth, we usually find an admixture of albuminous, saccharine, and oleaginous substances; which seems to indicate the intention of the Creator, that all these should be employed as components of the ordinary diet. The Caseine, or cheesy matter, is an albuminous compound; the Butyrine of butter is but a slight modification of the ordinary fats; and the Sugar differs from that in common use, only by its larger proportion of water. The relative amount of these ingredients in the milk of different animals, is subject, as we shall hereafter see, to considerable variation; but they are constantly present in the milk of the Herbivorous Mammalia, and of those which, like Man, subsist upon a mixed diet. It has been recently found, however, that the milk of the purely Carnivorous animals is destitute of Sugar, consisting, like their food, of albuminous compounds and fatty matter only; though even *their* milk is found to contain sugar, when saccharine or farinaceous compounds have formed part of their diet.—No fact in Dietetics is better established, than the impossibility of long sustaining health, or even life, upon any single alimentary principle. Neither pure albumen or fibrine, gelatine or gum, sugar or starch, oil or fat, taken alone for any length of time, can serve for the due nutrition of the body. This is partly due, so far as the non-azotized compounds are concerned, to their incapability of supplying the waste of the albuminous tissues. This reason does not apply, however, to the azotized compounds; which can not only serve for the reparation of the body, but can also afford the carbon and hydrogen requisite for the sustenance of its temperature. The real cause is to be found (partly at least) in the fact, that the continued use of *single* alimentary substances excites such a feeling of disgust, that the animals experimented on seem at last to prefer the endurance of starvation, to the ingestion of them. Consequently it is quite impossible to ascertain, by such experiments, the nutritive power of the different alimentary principles; no animal being capable of sustaining life upon less than two of them. The same disgust is experienced by Man, when too long confined to any article of diet which is very simple in its composition; and a craving for change is then experienced, which the strongest will is scarcely able to resist. The *natural combinations* in which the alimentary substances present themselves, appear to be those which are best adapted for the healthful nutrition of the animal body.

388. The Organic Compounds, which have been enumerated as supplying the various wants of the system, would be totally useless without the admixture of certain Inorganic substances, which also form a constituent part of the bodily frame, and which are constantly being voided in the excretions, especially in the urine. These substances have various

uses in the system. Thus common *Salt*, or the Chloride of Sodium, appears to afford, by its decomposition, the muriatic acid which is concerned in the digestive process, and the soda which is an important constituent of the bile. Its presence in the serum of the blood, also, and in the various animal fluids which are derived from this, probably aids in preventing the decomposition of the organic constituents of these fluids.—*Phosphorus* seems to be chiefly requisite as one of the materials of the Nervous tissue (§ 243); and also, when acidified by oxygen, to unite with Lime in forming the bone-earth by which bone is consolidated.—*Sulphur* exists in small quantities in several animal tissues; but its part appears to be by no means so important as that performed by Phosphorus.—*Iron* is an essential constituent of Hæmatine (§ 176); and is consequently required for the production of the red corpuscles of the blood in Vertebrated animals.—*Lime* is required for the consolidation of the bones of Vertebrata, and for the shells and other hard parts that form the skeletons of the Invertebrata; and it exists in the animal body in combination either with carbonic or with phosphoric acid. The Carbonate would seem principally destined to mechanical uses only; and we find it predominating, or existing as the sole mineral ingredient, in those non-vascular tissues of the Invertebrated animals, which give support and protection to their soft parts (§§ 194–199). The degree of development of these tissues depends in great part upon the supply of carbonate of lime which the animals receive. Thus the Mollusca which inhabit the sea, find in its waters the proportion of that substance which they require; but those which dwell in streams and fresh-water lakes, that contain but a small quantity of lime, form very thin shells: whilst the very same species inhabiting lakes, which, from peculiar local causes, contain a large impregnation of calcareous matter, form shells of remarkable thickness. The curious provision by which the Crustacea, which periodically throw off their calcareous envelope, are enable to renew it with rapidity, has been already noticed (§ 316). The large amount of carbonate of lime which is required by the laying Hen, is derived from chalk, mortar, or other substances containing it, which she is impelled by her instinct to eat; and if the supply of these be withheld, the eggs which she deposits are soft on their exterior,—not being destitute of shell, as commonly supposed, but having the fibrous element of it unconsolidated by the intervening deposit of calcareous particles (§ 160).

389. These substances are contained, more or less abundantly, in most of the articles generally used as food; and where they are deficient, the animal suffers in consequence, if they be not supplied in any other way. — Common *Salt* exists, in no inconsiderable amount, in the flesh and fluids of animals, in the milk, and in the substance of the egg; it is not so abundant, however, in Plants; and the deficiency is usually supplied to herbivorous animals in some other way. Thus, salt is purposely mingled with the food of domesticated animals; and in most parts of the world inhabited by wild cattle, there are spots where it exists in the soil, and to which they resort to obtain it. Such are the "buffalo-licks" of North America.—*Phosphorus* exists also, in combination with albuminous compounds, in all animal substances composed of these; and in the state of phosphate, combined with lime, magnesia, and soda, it exists largely in many vegetable substances ordinarily used as food. The phosphate of

lime is particularly abundant in the seeds of the grasses; and it also exists largely, in combination with caseine, in Milk.—*Sulphur* is derived alike from vegetable and from animal substances. It exists, in union with albuminous compounds, in flesh, eggs, and milk; also in several vegetable substances; and, in the form of sulphate of lime, in most of the river and spring water used as drink.—*Iron*, also, is very generally diffused, in small quantity, through the tissues of plants; but it exists in much larger proportion in the flesh, and more particularly in the blood, of animals; and it appears from comparative analyses of the blood of Carnivorous and Herbivorous Mammals, that the proportion of red corpuscles, and consequently of iron, is greatest in the former; which circumstance seems partly attributable to the nature of their diet.—*Lime* is one of the most universally diffused of all mineral bodies; for there are very few Animal or Vegetable substances, in which it does not exist. The principal forms in which it is an element of Animal nutrition, are the carbonate and phosphate. Both these are found in the ashes of the grasses, and of other plants used as food; the phosphate of lime being particularly abundant in the corn-grains. Phosphate of lime unites readily with albumen, caseine, &c.; and a large quantity of it is found in the milk of the Mammal, and in the egg of the Bird, for the promotion of the ossifying process in the bones of the young animals which they are respectively to nourish.

390. The dependence of Animal life upon a constant supply of aliment, is more close in some cases than in others. As a general rule it is the most immediate, when the vital processes, particularly those of Nutrition, are being most actively performed. Thus, we find that young animals are never able to bear the deprivation of food to the same extent with older ones of the same species; and that the warm-blooded Vertebrata—viz. Mammals and Birds,—are usually less capable of abstinence than Reptiles and Fishes. Even of the first of these classes, however, many species pass several months without eating, during the state of hybernation (§ 94); whilst among many cold-blooded animals, the period of abstinence from food may be indefinitely prolonged, under the influence of those agencies which keep them in a state of complete torpidity or 'dormant vitality' (§ 118).—When we carry our enquiries further, however, it becomes difficult to give any general explanation of the varieties which we meet with in this respect, among the different species of animals. Thus, it has been observed by Flourens and Dugès,* that the Mole perishes, when in a state of confinement, if not fed every day, or even more than once a day; whilst the Dog has lived without food for 36 days, the Antelope and the Wild Cat for 20, and the Eagle for five weeks. It is in Reptiles, that the power of abstinence appears to exist to the greatest extent among Vertebrata. Putting aside those cases in which the natural period of torpidity has been artificially extended, we find numerous instances in which these beings have performed all the functions of life for many months together, without the ingestion of food. Tortoises, Lizards, Serpents, and Batrachians seem to agree in this respect. It is to be borne in mind, however, that a large supply of food is frequently ingested at once by these animals; and that, owing to the slowness of their digestive process, the introduction of the Aliment into the system is protracted over a very long period,—as is

* "Physiologie Comparée," tom. ii. p. 288.

seen, for example, in the case of the *Boa constrictor*, which occupies a month in the digestion of a single meal. Little is known regarding the powers of abstinence possessed by Fish; but it has been stated that some of this class, such as the Perch, naturally take food but once a fortnight.* —It is perhaps among the Insect tribes, that we find the power of sustaining a deprivation of aliment the most remarkably evidenced. The Scorpion has been known to endure an abstinence of three months, the Spider of twelve months, and the *Scarabæus* beetle of three years, without inconvenience or loss of activity; the *Melasoma*, also one of the beetle tribe, has lived for seven months pinned down to a board. We notice in the class of Insects a very striking illustration of the general fact already stated, respecting the difference between the old and the young animal of the same species. The Larva is not only extremely voracious, but is usually incapable of sustaining a long abstinence; whilst in many tribes the Imago never eats, but dies as soon as its share in the propagation of the race is accomplished.—From what is known regarding the power of abstinence in the Mollusca, it may be stated generally, that they are not capable of maintaining their activity if not frequently supplied with food, in this respect corresponding with the larvæ of Insects; but that, when reduced to a state of torpidity, whether by cold or by the deprivation of food, they may continue without any aliment for a very protracted period.

391. Some have attempted to show that Herbivorous animals are more dependent upon a constant supply of food, than Carnivorous species; and that domesticated animals less easily sustain a deprivation of it, than wild ones. But these statements, though generally true, are found to be wanting in accuracy when applied to individual cases. It would probably be more correct to state, that, in proportion to the facility with which each species usually obtains its food, will be the directness of its dependence upon this, and its inability to sustain a protracted abstinence. In accordance with this principle, we observe that, though vegetable-feeders have in general their food within reach, and are very dependent upon its constant supply, some, as the Camel, are enabled to sustain abstinence better than many carnivorous species; and that some carnivorous animals, being enabled from the nature of their food to obtain it with little difficulty, are comparatively unable to bear the want of it. On the same principle it is evident that domestication may induce a change in the character of the animal, in this respect, as in others, by causing it to become accustomed to frequent or constant supplies of food. The same adaptation may be found among the Larvæ of Insects. Those which feed upon vegetables or upon dead animal matter, speedily die if out of the reach of their aliment; whilst those that lie in wait for living prey, the supply of which is uncertain, are able to endure a protracted abstinence, even to the extent of ten weeks, without injury.†

392. The general facts which appear to have been substantiated, in regard to the mode in which the Animal body is nourished by Food, may be thus summed up.

* Gold-fish have been known to live and thrive in small vessels of water, without any perceptible nutriment, for two or three years. But, if they be not fed in any other way, it is requisite that the water they inhabit should be frequently changed; and the minute quantity of organic matter which it holds in solution is probably the source of their aliment.

† Lacordaire, "Entomologie," tom. ii. p. 152.

I. The 'waste' of the tissues of which *Albumen* is the basis, must be supplied by Albuminous compounds, whether these be derived from the Animal or from the Vegetable kingdom; the amount of this 'waste,' and consequently the demand for albuminous aliment, depends essentially upon the degree of vital activity which has been put forth, and especially upon the exercise of the nervo-muscular apparatus; and therefore, *cæteris paribus*, it is greater in cold-blooded animals in proportion to the *elevation* of the external temperature.

II. The 'waste' of the tissues of which *Gelatine* is the basis, may be supplied either by Albuminous or by Gelatinous compounds, since there is no doubt that albumen may be converted into gelatine, although the reverse process cannot be performed. As Gelatine does not exist in Plants, it must be formed in Herbivorous animals at the expense of the albuminous elements of their food; whilst in Carnivorous animals, it is probably derived immediately from the gelatinous components of the bodies on which they prey.

III. The materials of the *Adipose* tissue, and the oleaginous particles which seem requisite in the formative operations of the system generally, are derived, in the Carnivorous races, from the fatty substances which the bodies of their victims may contain; whilst the Herbivorous not only find them in the oleaginous state in their food, but have the power of producing them by the conversion of farinaceous and saccharine matters.

IV. The foregoing statements are applicable to all tribes of Animals, 'cold-blooded' as well as 'warm-blooded;' we have now to consider the special case of the latter.—In the *Carnivorous* tribes, the 'waste' of the tissues is so great, in consequence of the restless activity which is habitual to them, that it appears to furnish a large proportion of the combustible material required for the maintenance of their proper temperature. The remainder is made up by the fat of the animals upon which they feed; and it is to be observed that the amount of this is much greater in the bodies of animals inhabiting the colder regions of the globe, than in the inhabitants of tropical climates.—In the *Herbivorous* tribes, the case is different. They are, for the most part, much less active; and the 'waste' of their tissues consequently takes place in a less rapid manner, and is far from supplying an adequate amount of combustible material, especially in cold climates. Their heat is in great part sustained by the combustion of the saccharine and oleaginous elements of their food, which are appropriated to this purpose without having ever formed part of the living tissues; and the demand for these will be larger in proportion to the *depression* of the external temperature, a greater generation of caloric being then required to keep up the heat of the body to its proper standard.

V. Hence 'cold-blooded' animals can usually sustain the privation of food longer than warm-blooded; and this more especially when they are kept *cool*, so that they are made to *live slowly;* and death, when at last it does ensue, is consequent upon the general deficiency of nutrition. On the other hand, 'warm-blooded' animals, whose temperature is uniformly high, must always *live fast;* and deprivation of food is fatal to them, not only by preventing the due renovation of their tissues, but also by destroying their power of sustaining their heat (§ 93). The duration of life under these circumstances depends upon the amount of fat previously stored up in the body, and upon the retardation of its expenditure by external *warmth*, or by the enclosure of the body in non-conducting substances; and there is evidence that, if this be duly provided for, and all unnecessary waste by nervo-muscular activity be prevented, the life even of a warm-blooded animal may sometimes be prolonged for many weeks without food.

3. *Ingestion and Preparation of Aliment in Plants.*

393. Although, as already explained, the Vegetable world as a whole is supported by the introduction of the alimentary materials, derived from the earth and air, into the organism, without any preliminary alteration, yet there are particular cases in which adaptations of structure are met with, that appear to be subservient to the reception and preparation of nutritive materials; and some of these it would not be easy to exclude from any definition we might frame of a 'stomach.' Concavities in different parts of the surface, fitted for the collection of the moisture caught from rain or condensed from dew, may frequently be observed; and these vary in the completeness of their structure, from the simple hollow formed in the leaf of the *Tillandsia* (wild pine of the tropics), or of the *Dipsacus* (teasel), to the extraordinary *ascidia* of the 'Pitcher-plants.' The exact method in which the fluid thus obtained is applied to the nutrition of the plant, is not always evident. Sometimes the channelled leaves seem to convey it to the roots, by which it is absorbed in the usual manner. The function of the pitcher of the *Nepenthes* or 'Chinese pitcher-plants' (Fig. 194, *b*), has not been certainly determined; it is difficult to ascertain, how much of the fluid which it contains is collected from the atmosphere by the downy hairs that line its interior, and how much is secreted by the plant itself; and it is certain that the young pitcher contains fluid, secreted from an organ within it, before its lid first opens. The object of the pitchers of the *Dischidia* is, however, less doubtful, and their structure far more complicated. This curious plant grows by a long creeping stalk, which is bare of leaves until near its summit; and as, in a dry tropical atmosphere, the buds at the top would have great difficulty in obtaining moisture through the stem, a sufficient supply is provided by the pitchers, which store up the fluid collected from the occasional rains. "The cavity of the bag," says Dr. Wallich,* "is narrow, and always contains a dense tuft

FIG. 194.

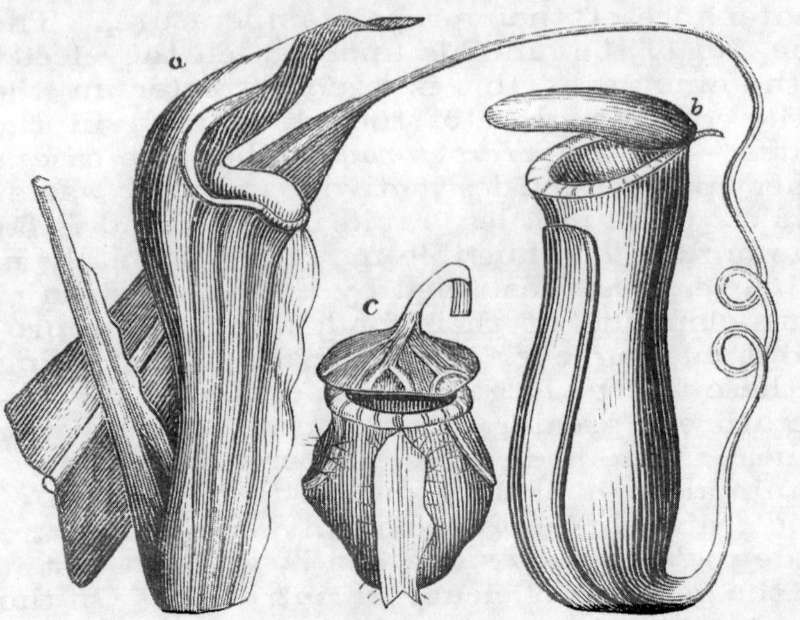

Different forms of *Ascidia* (pitchers);—*a*, pitcher of *Sarracenia*; *b*, of *Nepenthes*; *c*, of *Cephalotus*.

* "Plantæ Asiaticæ rariores," vol. ii. p. 35.

of radicles, which are produced from the nearest part of the branch, or even from the stalk on which the bag is suspended, and which enter through the inlet by one or two common bundles. The bags generally contain a great quantity of small and harmless black ants, most of which find a watery grave in the turbid fluid which frequently half fills the cavity, and which seems to be entirely derived from without." Thus it would seem as if the failure of the ordinary means of support in this curious Plant, has been compensated by the addition of an organ, which, like the stomach of Animals, serves as a receptacle for the supplies it may occasionally obtain. According to Mr. Burnett,* in the pitcher of the *Sarracenia* (Fig. 194, *a*), a process still more like that of animal digestion goes on; for it appears that the fluid it contains is very attractive to insects, which, having reached its surface, are prevented from returning by the direction of the long bristles that line the cavity. The bodies of those which are drowned seem, in decaying, to afford a supply of nutriment that is favourable to the growth of the plant; like the similar process on the leaves of the well-known *Dionœa muscipula* (Venus' fly trap), to the health of which a supply of animal food appears to be essential. —Although such instances as these may seem to contradict the general statement, that Plants derive the materials of their nutrition from the inorganic world, yet they probably do so more in appearance than in reality. In all cases where previously-organised matter influences their growth, it is only whilst in a decomposing state, during which it is separated into its ultimate elements or very simple combinations of them. In Animal digestion, on the contrary, the proximate principles contained in the food appear to be immediately subservient to the formation of others of a higher order; and whatever tendency to disunion their elements might have previously manifested, this is immediately checked by the antiseptic qualities of the gastric fluid.

4. *Ingestion and Preparation of Aliment in Animals.*

394. In considering the various organs which Animals possess for the *in*gestion and *di*gestion of their food, it is right to take notice, in the first instance, of those cases in which there appears to be an absence of that provision for its reception, which is on the whole so characteristic of the beings included within the limits of that kingdom. There are several examples, even amongst animals of high organisation, in which, during the last stage of their evolution, there is an entire absence of any power of receiving nourishment into the system. These are principally met with in the class of *Insects;* among which there are many (as already stated, § 315, *g*) which take no food in their perfect or *Imago* state, the duration of this being very short, and serving merely for the performance of the reproductive act. In some of these cases, the mouth appears to be actually closed; although the digestive apparatus remains, but in an atrophied condition. In all such examples, however, the appropriation of food has been actively performed during a previous stage of the animal's existence; but we have now to notice a peculiar case, in which an animal seems to take-in no nutriment, from the time of its quitting the egg, until its death. The animal referred to is the *male* of the curious *Notommata* described by

* "Brande's Journal," vol. vi.

Mr. Dalrymple (§ 310 *a*), which does not possess either oral apparatus or digestive cavity, and must be entirely incapable of deriving support from the nutriment upon which the female subsists.* Its organism is built up at the expense of the store prepared for it in the egg; and it is probable that the duration of its life is extremely limited, and that it ceases to exist soon after it has performed its part in the reproductive function. Such instances, then, are not really so exceptional as they at first sight appear; we have now to consider those cases, in which the growth and development of beings included in the Animal kingdom takes place at the expense of aliment appropriated by themselves; and in which there is, nevertheless, no digestive cavity for its reception.

395. The group of *Protozoa* includes many forms which can scarcely but be reckoned among Animals; when we consider, on the one hand, their close alliance to tribes which unquestionably belong to that kingdom, and with which they are connected by a continuous series of links; and when we bear in mind, on the other, the probable evidence we possess, that they derive their nourishment from organic compounds which have previously been prepared by the Vegetable kingdom, and not, like Plants, from those which they have themselves elaborated from inorganic elements. It is true that some eminent Naturalists, who still hold to the doctrine that *a stomach* is a necessary constituent of an Animal, are ready to affirm the presence of this organ in the very lowest members of the series; but the Author regards it as much more philosophical to believe, that a stomach *may be* absent in the simplest forms of Animal life, as in the earliest period of the embryonic life of the highest; and when the facts of the case are carefully examined with an unprejudiced eye, he believes that they will be found in perfect accordance with this view. For although the simple cell of the *Amœba* (§ 283 *a*), or the *Sponge-proteus* (§ 286 *a*), may *appear* to receive into its interior the solid particles from which it derives its nutriment, yet there is strong reason to believe that they are really external to it, and that the fluid which they yield is imbibed through the cell-wall; notwithstanding that, by folding around them, it may form a sort of extemporaneous stomach. And the most careful examination of the anatomy of many of the *Rhizopoda* (§ 285) has failed to discover in them a mouth, stomach, or intestinal canal; and seems to justify the conclusion, that their *pseudopodia* are really to be regarded as 'radical filaments' like those of Plants, and that the masses of cells of which their bodies are composed, are nourished by imbibition, partly through these special prolongations, and partly by their general surface, to which the surrounding liquid gains access by means of the porosity of the shell. So again, the *Gregarina* (§ 309 *a*), which seems to be the simplest known form of Entozoon, appears undoubtedly to be nourished by absorption through its external surface alone; and even the *Cystic* and *Cestoid* Entozoa (§ 309, *b*, *c*) seem to possess no mouth, though an alimentary canal may be distinctly made out in them.

396. In the group of *Sponges* (§ 286) we seem to have the transition from this condition to that of higher forms of Animal life; for whilst, on the one hand, we may regard each mass as an aggregation of Amœba-like cells, held together by a common integument, and supported by a frame-

* "Philosophical Transactions," 1850, p. 342.

work which all participate in forming; we may, on the other hand, trace in the higher types of the group a manifest approximation to the *Alcyonian* Zoophytes (§ 292). And thus whilst each of the component cells, like the elements of the tissues of higher animals, lives for and by itself, they all derive their nutriment from the fluïd which is imbibed through the pores of the surface, and which circulates through its canals. These canals, if considered with reference to the entire body, must be regarded in the light of an alimentary cavity, which is so diffused through the mass, as to supply to every part of it the materials of its growth, without any special circulating apparatus.* We shall presently find that in many Animals of far more complex organisation, the digestive cavity is extended through the body in a manner essentially the same as this.

397. Before proceeding to the description of the principal forms under which the Digestive apparatus presents itself, in those organisms of which it forms a definite and limited part, something must be said of its condition in the (so called) *Polygastrica,* in the true members of which group there is undoubtedly a reception of food into the interior through an oral orifice, although it is very questionable whether it passes into any special receptacle, or into the general cavity of the body. The doctrine of Prof. Ehrenberg, that each of these Animalcules contains a number of distinct sacculi, sometimes opening externally by distinct orifices at the same part of the body, sometimes communicating with an alimentary canal that is common to them all, has been already stated (§ 284); and some of the principal forms of the apparatus, described by him, have been represented (Fig. 87). Other observers, however, have been far from confirming the correctness of these statements; and the Author's own investigations have led him to place little reliance upon them. The belief in the existence of a number of distinct sacculi, opening from a common intestinal tube, is founded upon the appearance of Animalcules which have been fed with coloured particles, and which exhibit numerous coloured globules in their substance (Fig. 87, B, C, E, G, H). These globules have been regarded as consisting of aggregations of the coloured particles within the spherical sacs, into which they have been received through the intestinal tube. Not only, however, has no such tube been ever seen, but no separate particles have been traced in their passage from one part of the body to the other along such a canal; and, what appears a still more fatal objection, the globules themselves have been seen to perform a kind of circulation through the body, changing their relative positions to each other, and sometimes escaping by the anal orifice. Moreover, many Animalcules have been seen to receive as prey, into the cavity of their bodies, other animalcules of nearly their own size; which could scarcely be, if the intestinal tube were of the nature described by Ehrenberg.—An account of the nature of the digestive system of these animals, which appears much more consistent with observed facts, has been given by Meyen. He states that behind the oral orifice is a small globular cavity; and that the

* Since the account of this group (§ 286) was printed, ciliary movement has been discovered by Mr. Bowerbank in the canals of *Grantia compressa;* whence it seems probable that the motion of fluid in other Porifera is due to the same agency. The extreme minuteness of the cilia, and the difficulty of making observations upon these bodies in their living state with the high magnifying power requisite for their discernment, furnish the explanation of the cilia having so long remained undetected.

particles of the food which are drawn into the mouth by ciliary action, are here aggregated into a globular mass, and pressed into a degree of consistence, sufficient to cause them to hold together. When a ball has thus been formed, it is expelled into the general cavity of the body; and the formation of a second globular mass commences. The general cavity is undivided by any membranous partitions, and is chiefly occupied by a semifluid gelatinous substance, in which no distinct organisation can be discovered. Into the midst of this, the globule is propelled by the contraction of the first cavity; and, as one after another is thus forced in, each, as it is introduced, pushes on the rest, and a kind of circulation of the globules is occasioned, as already described,—those first formed making their escape (after yielding their nutritive materials) by the second orifice.* Besides the globular masses of colouring particles, the cavity of the body is often seen to contain peculiar transparent vesicles; and as these are much more obvious when no colouring matter whatever has been introduced, it has been supposed by Prof. Ehrenberg that they are unfilled stomachs. It is scarcely questionable, however, that these vesicles are closed and unattached; for if the body of any Polygastric Animalcule in which they are distinctly visible, be caused to rupture by pressure, they are forced out entire, and are seen to be destitute of either aperture or peduncle; and this result may also be obtained by the application of liquor potassæ,† which causes the vesicles to undergo a perceptible enlargement whilst still within the general investment. The vesicles thus discharged are seen to contain numerous minute molecules in active motion; and it would seem not improbable that they are 'assimilating cells,' whose function it is to select from the materials introduced into the general cavity, and to prepare for the nutrition of the body at large, such constituents as may be appropriate to that purpose.

398. The food is usually introduced into the digestive cavity of higher animals, through a single orifice; the only known exception to this statement being in the case of the curious family of *Rhizostome* (root-mouthed) Acalephæ (§ 294). In these animals, the single oral orifice of the ordinary *Medusæ* is replaced by a multitude of pores at the sides and extremities of their tentacula (Fig. 195); from each of these pores a narrow tube commences, which terminates in a canal running upwards through the centre of the tentacle; and the canals of the eight tentacles unite in pairs, so as to form four large trunks, which then discharge themselves into the cavity of the stomach. It has been shown by the experiments of Prof. Milne-Edwards, that colouring particles are absorbed through the pores; but they are too small to allow any save very minute Animalcules to enter.—The relative size and form of the single oral orifice that is common to all other animals, differ very greatly, according to the nature of the food upon which the species is destined to live. The Invertebrate division contains many groups, that are supported by *suction* of the juices of higher animals or of plants; and in them we usually find the oral orifice narrowed and sometimes immensely prolonged. As such a means of obtaining food usually involves the necessity of locomotive powers, whereby the animal may go in search of it, we find, as might be expected, that it is most common in the Articulated sub-kingdom; all the

* See "Annals of Natural History," vol. iii. pp. 100 and 170.

† See Dr. Addison's "Experimental Researches on the Functions of Cells," p. 44.

principal classes of which,—Entozoa, Annelida, Insects, Crustaceans, and Spiders,—contain groups of greater or less extent, which are especially adapted for obtaining their food by suction. The *Pycnogonidæ* (Fig. 158) present a characteristic example of the most simple form of suctorial mouth, in which the aperture is merely narrowed and somewhat prolonged;* but the most remarkable modifications for this purpose are to be met with in the class of Insects (§ 315), especially in the *Lepidopterous* and *Dipterous* orders. Of the first of these groups, a considerable proportion feed upon the juices of flowers, which the large expanse of their wings prevents them from entering; and their long *haustellium*, which is coiled up beneath the head when not in use, can be so extended as to suck up the honey from the bottom of a deep blossom, while the insect rests upon its outer edge. Of the latter, however, a considerable proportion feed upon the juices of animals; but there are some that have recourse to the honeyed exudations of flowers; and among these the *Nemestrina longirostris* (a Dipterous insect of the Cape of Good Hope) deserves especial mention, the length of its proboscis being about 3 *inches*, whilst that of its body is only about 8 *lines*; so that it is enabled to feed upon the juices of a flower whose long tubular corolla equals that of its trunk. Many animals whose mouths are adapted for suction, possess some means of attaching themselves to the spots in which they can best obtain the nutriment they require; thus we find the *Cystic* and *Cestoid Entozoa* very commonly provided with a set of hooklets on their heads (Fig. 140, A); the *Trematoda* usually have the mouth surrounded by a circular sucker (Fig. 141, *a*); the *Suctorial Annelida* (§ 311 *d*) have not merely a sucker for attachment, but an incising apparatus for making incisions into the blood-vessels of the animals whose juices they suck; the *Suctorial Crustacea* usually have a sucker formed by the peculiar development of one of the pairs of members (Fig. 160, *f*), whilst the mouth is elongated into a proboscis armed with penetrating instruments; and the

FIG. 195.

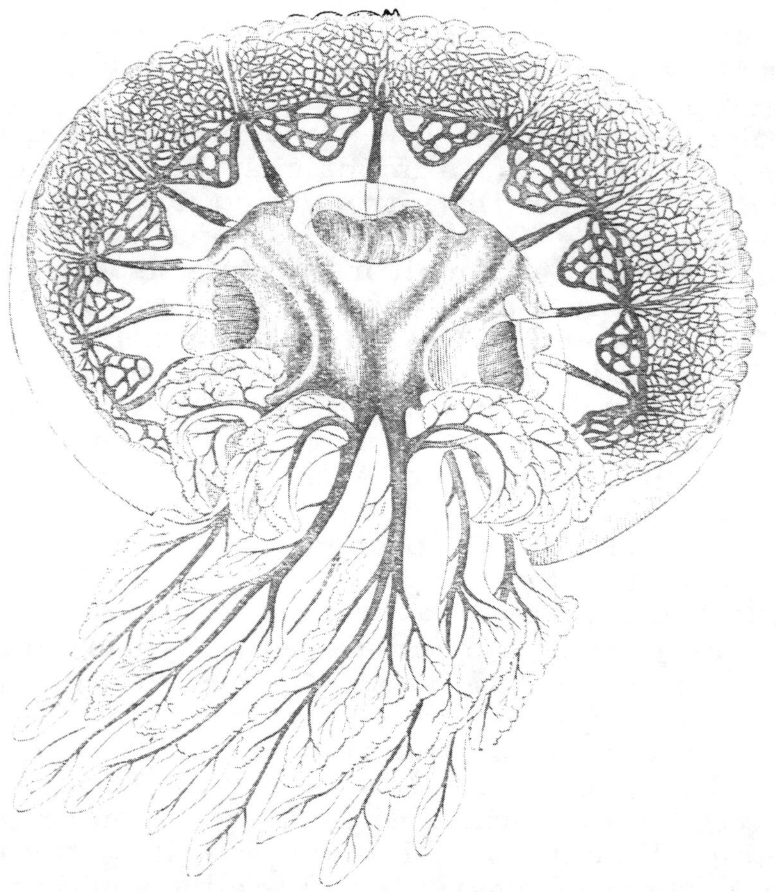

Rhizostoma injected with coloured fluid, to show the central digestive cavity, and the canals branching off from it to ramify in the arms and in the disk.

* From the situations in which these creatures are commonly found, the Author would infer that they draw their nourishment from the mucus that covers the surface of sea-weeds.

Suctorial Insects, which seldom attach themselves permanently in any one situation, but make a fresh puncture whenever their appetite incites them to feed, are for the most part destitute of any special organs of adhesion, but have a most elaborate set of instruments for incising the skin of the animals they attack (§ 315).—The *Cyclostome* Fishes (§ 322) are the only Vertebrated animals, which present any similar adaptation to a suctorial mode of nutrition.

399. When the food consists of solid matters, which is the case in by far the larger proportion of the Animal kingdom, we find the entrance to the digestive cavity of much greater proportionate size ; and it is seldom prolonged forwards, but far more frequently has somewhat of a funnel shape, so that the aliment may be more readily drawn within it. The means by which the introduction of food is accomplished, are extremely various. The simplest plan is one which carries us back to the type of the Infusoria, although it presents itself in a much more elaborate form. Among many aquatic animals, whose food consists of minute particles diffused through the water, we find that the action of cilia situated around and within the entrance of the alimentary canal, or upon organs in immediate connection with it, whereby a current of water may be driven into the stomach, is the sole means by which aliment is introduced. The most remarkable examples of entire dependence upon this agency are furnished by the Acephalous Mollusks ; for in the *Bryozoa* (§ 298) the polype-like arms are unfitted for prehension, and effect the introduction of alimentary matter solely by the ciliary currents they produce ; and the *Tunicata* (§ 299), *Brachiopoda* (§ 300), and *Lamellibranchiata* (§ 301), are all supplied with aliment in a manner essentially the same. Among Radiated animals, recourse seems less frequently had to this plan, which is obviously one that is unsuitable to those forms of the digestive cavity that have but a single orifice, since the requisite current could not be effectually maintained through such ; but among the *Ciliograde Acalephæ* (§ 294 *a*), which possess an anal orifice, the ciliary action appears to be the means of introducing aliment, as well as of propelling the body through the ocean. In the Articulated classes, too, this ciliary action is seldom the means of introducing aliment into the stomach ; most of them feeding upon solid bodies of comparatively large dimensions, and being enabled by their locomotive powers to go in search of their food. But in those which resemble the Molluscous tribes already alluded to, in their habits of life, we find the same method of obtaining nutriment ; thus it appears to be by the currents set in motion by the cilia clothing the respiratory tufts which are seated on their heads, that the *Serpulæ* (Fig. 148) and their allies draw in their food ; and in the *Rotifera* (§ 310) and the *Cirrhipoda* (§ 318) we find special ciliated organs developed, which appear subservient both to the purposes of respiration and to the ingestion of aliment. It is probable that the *Amphioxus* (§ 321 *b*) is nourished in a similar manner ; but that curious animal affords the only known example among the Vertebrata, of the employment of ciliary action for this purpose. It is curious to observe, however, in the Greenland Whale, a mode of ingestion which is essentially the same, though accomplished by a different instrumentality ; for this huge animal gulps enormous volumes of water into its capacious mouth, and then ejects these through its blow-holes, straining out, through its whalebone-sieve, the small Medusæ, Pteropods, Crustacea,

Fishes, or other free-swimming marine animals which the water may contain ; it being upon such alone that it is adapted to subsist.—The foregoing plan is one that seems intermediate, in regard to the kind of nutriment for which it is adapted, between the suctorial method, which is fitted only for the introduction of liquid aliment, and the one we have next to consider, which is appropriate to the ingestion of solid masses.

400. The organs with which the digestive apparatus is provided, for the introduction of *solid* food into its cavity, vary greatly in complexity in the different tribes of animals which derive their subsistence from aliment of this kind. Thus we shall find that in the lowest and simplest types, these organs are in immediate connection with the stomach ; but that they are partly superseded in the higher by the addition of an oral apparatus, with prehensile appendages for laying hold of the food, which, after passing through the oral cavity, is drawn downwards into the stomach ; and that in the highest of all, the introduction of food into the mouth is itself aided by accessory organs, which do not form part of its own organization. Of the first of these types, we have a characteristic example in the Hydra (§ 289), the entrance to whose stomach is surrounded by 'tentacula,' which, as formerly remarked, appear homologous with the œsophageal muscles of higher animals. The tentacula of the *Actinia* (Fig. 97) correspond with these in character and mode of action ; and it may be said that the same type prevails through the whole class of Polypifera. The *Medusæ* (Fig. 106) and their allies are formed upon a plan which must be considered as essentially similar, the tentacula being there, also, placed around the entrance to the stomach ; and it is by the adhesion of the edges of these tentacula, that the 'proboscis' is formed in those of the Pulmograde Acalephæ which possess it (§ 294). It is interesting to observe in the class of *Echinodermata*, a gradual transition towards a more elevated type. Thus in the *Asteriada* (§ 296 *b*), which have no oral apparatus, we find the food brought to the stomach, not by tentacula, but by the flexion of the lobes of the body, commonly known as their 'arms.' "Starfishes," says Prof. E. Forbes,* "are not unfrequently found feeding on shell-fish; in such cases they enfold their prey within their arms, and seem to suck it out of its shell with their mouths pouting out the lobes of the stomach. They can project the central parts of their stomachs in the manner of a proboscis." Whether the true 'arms' of the *Crinoidea* (§ 296), and *Ophiurida* (§ 296 *a*) serve to bring food to the stomach, is not certainly known ; but their position in the former of these orders, at least, would lead to the inference that they are subservient to this purpose. In the *Echinida* (§ 296 *c*) and *Holothuriada* (§ 296 *e*), however, we find that the entrance to the alimentary canal is no longer the entrance to the stomach ; but that the former becomes a true 'mouth,' being furnished with a prehensile apparatus of its own, and being separated from the stomach by the intervention of the œsophagus. The mouth is provided, in the typical Echinida, with a dental apparatus (Fig. 111, *a*), which is capable of being projected beyond the oral orifice ; whilst in the Holothuriada it is surrounded by prehensile tentacula, which are here to be regarded as *labial*, being extensions of the lips, rather than a divided pharynx.—Turning to the Articulated series, we find that in a considerable proportion of the

* "History of British Echinodermata," p. 86.

Annelida (§ 311 *a*), the mouth is furnished with powerful jaws, sometimes to the number of three pairs, opening laterally; or with a proboscis which is capable of being everted, and which then displays an armature of teeth on its exterior. In the *Mandibulate Insects* (§ 315), and in the greater part of the *Crustacean* class (§ 316 *a*), we find the jaws highly developed, and taking the place of any other kind of apparatus; the former group always possessing two pairs of these organs, which open laterally, one above and the other below the oral orifice; whilst in the latter, besides the regular mandibles and maxillæ, there is a variable number of feet-jaws.—In the higher part of the Molluscous series, again, we find the mouth usually provided with jaws; these animals being adapted to go in search of their food, and to select and divide it for themselves. In many of the *Gasteropoda,* we find a single pair of horny jaws, opening laterally; but there are several whose buccal apparatus rather consists of a proboscis-like organ, somewhat like that of the Annelida, which can be everted, and which then displays a set of powerful teeth, adapted to abrade substances of even shelly hardness. The *Cephalopoda,* on the other hand, possess a pair of horny jaws opening vertically, like those of Vertebrata (§ 306); and the oral orifice is surrounded with 'arms,' which, though resembling those of the Polypes in uses, and apparently also in position, are not really homologous with them, being the representatives of the *labial* tentacula, which are possessed by many of the inferior Mollusks, as well as by several Fishes. The difference between two sets of organs, which are ostensibly so like one another, will become apparent when it is remembered that the 'arms' of the Polype surround the entrance to the stomach; whilst those of the Cephalopod surround the oral orifice, which is separated from the stomach by the jaws, buccal cavity, pharynx, and œsophagus. The latter would seem to be more truly prefigured by the oral tentacula of the Holothuriada.—Throughout the Vertebrated series, it is almost invariably by the action of the jaws and lips, that food is introduced into the mouth; the suctorial *Cyclostomi* already referred to, and the *Syngnathidæ* or 'pipe-fishes,' which have the jaws prolonged and adherent together, and which seem to draw up worms, small mollusks, and crustaceans, and the ova of other fishes, through the tube thus formed, being the only cases of even partial reversion to the lower type.

401. But among the more elevated forms of the Articulated and Vertebrated sub-kingdoms, we find a special provision for the *prehension* of food, by organs which have no immediate connection with the digestive apparatus; the aliment being thus brought within reach of the mouth and of its proper appendages. Some approaches to such an arrangement we have seen even in the lower Echinodermata, in which the 'arms' and lobes of the body are made to bring the food to the entrance of the stomach; and in the *Echinida* it seems not improbable that the spines and 'cirrhi,' especially the latter, may be occasionally used in subservience to the ingestion of food, as well as in locomotion.—In most of those higher *Articulata* which possess numerous well-developed members, we find that those nearest the mouth are used to a greater or less extent in the prehension of food; and that in many instances one pair is specially developed for this purpose. The transition from the general to the special form of the prehensile apparatus is well seen among the *Decapod Crustacea*

(§ 316 *a*) ; in some of which (as the Cray-fish) we find all the true legs of nearly equal size, and most of them terminated by forcipated appendages ; whilst in others (as the Crab) the first pair are enormously developed as prehensile organs, and are furnished with powerful pincers, whilst the remainder are restricted to locomotive action.—In the *Vertebrated* series, it would appear as if the restricted number of members, and the necessity for employing these in locomotion, placed a greater limitation upon the use of them as prehensile organs. Among Fishes and Reptiles, the prehension of food is solely accomplished by the jaws, save where the tongue (as in the Chameleon) is subservient to this function. In Birds, it is only in a few cases that the foot is employed for this purpose, the anterior members being never modified into subserviency to it, and the tongue (as in Reptiles) being the only instrument that is ever specially developed as a prehensile organ. Among Mammals, as has been fully pointed out already (§ 326 *e*), the special modification of the anterior extremities for prehension is limited to those few groups in which they are comparatively little needed for locomotion : thus, among most of the Ungulated quadrupeds which have the power of standing erect upon the hind legs, as the Rodentia, Plantigrade Carnivora, and many Marsupialia, we find the anterior extremities employed to clasp objects between them ; in the Quadrumana, the prehensile power is exercised by a single extremity, which can hold an object of moderate size in the grasp of the digits ; but it is in Man that this prehensile power is carried to its fullest extent, by the peculiarity of conformation of the thumb, which brings its extremity into opposition with those of each of the fingers.—Some of the Mammalia whose extremities are not adapted for the prehension of food, are furnished with other organs for the purpose, which are special modifications of those elsewhere existing under the ordinary type ; it is sufficient to refer to the proboscis of the Elephant, the prolonged snout of the Tapir, the prehensile tongue of the Giraffe, and the very extensible tongue of the Ant-eater.

402. Thus, between the simple stomach furnished with prehensile tentacles, of which the *Hydra* consists, and the complex apparatus for the prehension of the food which is placed in front (so to speak) of the cardiac orifice of the stomach in *Man*, a regular gradation may be traced. The tentacula are brought together to form a pharyngeal tube ; a buccal cavity is placed at the entrance of this ; jaws and lips are developed at the orifice of the buccal cavity ; and external members are developed or modified, for the purpose of bringing the food within their reach. Yet the development of these more special organs does not supersede the necessity of those more generally diffused through the Animal kingdom. Although the hand of Man brings his food *to* his mouth, yet it is by his lips and jaws that his food is taken *into* its cavity, as in the Herbivorous Mammalia ; and notwithstanding that the muscular apparatus of the mouth propels the food backwards into the pharynx, it only thereby carries it within reach of the pharyngeal constrictors, which then lay hold of it and draw it into the œsophagus, by the muscles of which it is carried down into the stomach, precisely as by the tentacula in the *Hydra*. And it is curious to observe, that just as the tentacula of the Hydra no longer make any attempts to grasp an object that touches them, when the stomach is already gorged with food, so does Man experience a difficulty

in the act of swallowing at the end of a full meal, which seems to result from a like want of readiness to contract on the part of the pharyngeal constrictors.—In the progress of the food along the alimentary canal of higher Animals, we may observe the gradual removal of it from connection with the functions of *animal* life. To procure it in the first instance, is one important office of these functions; and the highest exercise of the locomotive, sensorial, and intellectual powers is often required for this purpose. Its introduction into the mouth is an act of pure volition in Man; whilst the masticatory movements to which it is there subjected, may be regarded as essentially automatic (§ 244) in their character (though capable of being controlled and directed by the will), and as closely corresponding with the instinctive actions of the lower animals. The act of 'deglutition,' or swallowing, is of a purely reflex nature (CHAP. XX.), being the result of a nervous influence in which neither the will nor sensation is concerned: for when the solid or fluid contents of the mouth are brought in contact with the surface of the pharynx, the impression made upon the nerve distributed on it is transmitted to the upper part of the spinal cord; and an automatic impulse is propagated along the motor fibrils, by which the muscular movements requisite for the action of swallowing are produced. A similar action assists the propulsion of food down the œsophagus, and the movements of the stomach seem to be in part excited in the same manner; but the proper fibres of the œsophagus itself are *not dependent* for their power of action upon nervous influence, nor are those of the stomach. Beyond the stomach, the connection of the motions of the alimentary canal with the nervous system almost wholly ceases, the ordinary peristaltic movements of the intestines appearing to depend upon the stimulus directly applied to their muscular coat by the contact of food; although they may be in some degree controlled by a system of muscles disposed around the outlet of the canal, which are, like those at its entrance, partly involuntary, and partly under the direction and restraint of the will.—So, in descending the Animal scale, we find that the introduction of food into the digestive cavity is first removed from the agency of the intellect and will, and provided for by the instincts alone; this is probably the case in even the highest Invertebrata, and likewise in the lower Vertebrata, as it undoubtedly is in the infantile condition of Man. When we descend to Zoophytes, we find strong reason to believe that the movements of the tentacula, by which the food is grasped and drawn into the stomach, are not merely involuntary, but are not even dependent upon sensation (CHAP. XX.); and there is still stronger reason to believe that such is the case, in those inferior Mollusks in which the supply of food is obtained by ciliary currents. And thus we may perceive a regular gradation between those beings, in which the supply of food has been made (for a wise purpose) dependent upon the highest exercise of the functions of *Animal* life, and those in which the operation is as purely *Organic* as it is in Plants.

403. One more class of accessory organs has to be noticed, before we proceed to the account of the essential part of the Digestive apparatus, and of the operations to which it is subservient. In order that solid food may be more easily dissolved in the stomach, it is frequently submitted in the first instance to a process of mechanical reduction by a triturating apparatus; and either at the same time, or at some other, it is incorpo-

rated with a Salivary fluid, the admixture of which not only renders the subsequent process of solution more easy, but also appears to effect a change in the alimentary matters themselves. It is in animals that are destined to feed upon vegetable substances, and especially upon such as are of tough consistence, that we find this reducing apparatus most powerful and efficient; for as the flesh of animals is of comparatively easy solubility, there is less occasion for any such preliminary preparation. This reduction is not by any means constantly performed, however, in the mouth; for almost any part of the first division of the alimentary canal may be the seat of the apparatus. Among the Radiata, there are none, save the typical *Echinida*, which possess such a reducing apparatus; but it is remarkable that it should attain in that group an extraordinary complexity, and should occupy a situation corresponding to that which it possesses in the highest Vertebrata. The 'lantern of Aristotle' (Fig. 111, *a*), as it is sometimes termed, is a five-sided conical mass, composed of five 'jaws' in apposition with one another, each of them having the form of a triangular pyramid; two of their sides, which are flattened, look towards those of the neighbouring jaws, and have their surfaces roughened by grooves like those of a file; whilst the third, which is somewhat convex, is external, and serves for the attachment of the muscles that fix the apparatus to the interior of the shell, which is furnished with projections round the oral orifice developed for this purpose. Each jaw contains a long pointed tooth of remarkable hardness, which projects to a considerable length from its socket, and this appears (like the 'tusks' of Mammalia) to be continually renewed by new growth at its base, as it is worn away at its apex. The whole is moved by a most complex set of muscles, which can bring the whole 'lantern' nearer to the oral orifice, and can cause the points of the teeth to project beyond it; which can separate the jaws, and consequently the points of the teeth, from one another, and then draw them together, so as to enable the latter to grasp and divide any hard substance adapted for food; and which can subject this to trituration, not merely between the points of the teeth, but also between the flattened file-like surfaces of the jaws, which can be made to work against each other.—Among the Acephalous Mollusks, no reducing apparatus seems usually to be required; their food being in a state of very fine division at the time it is introduced; nevertheless in the *Bowerbankia* and some other *Bryozoa* (§ 298), a 'gizzard' or muscular stomach, apparently subservient to this purpose, intervenes between the œsophagus and the true digestive stomach. A gizzard of considerable power, closely resembling that of Birds, is found in many Gasteropods, especially in those whose mouth is unfurnished with a reducing apparatus; this is the case, for example, with *Aplysia* (Fig. 127), whose gizzard (*i*), situated between the crop (*g*, *g*) and the true digestive stomach (*h*), is lined with firm horny teeth; and in some instances, these teeth are replaced by firm calcareous plates, adapted to crush the shells of the smaller Mollusks that are devoured by these animals, which is the case in *Bulla*.* But in the greater number of instances, the reduction of the food is accomplished by some part of the buccal apparatus; either by the cutting jaws with which some Gasteropods are furnished; or, as in *Buccinum*, by the teeth which form a

* The Author once met with an entire shell of *Pandora rostrata* in the gizzard of a Bulla.

sort of palatal lining to the proboscis, and which are carried outwards by its eversion; or, as in *Patella* and some other phytophagous Gasteropods, by the long rasp-like tongue. In the *Cephalopoda*, notwithstanding the presence of jaws for the division of the food, its reduction seems to be chiefly accomplished by a muscular gizzard closely resembling that of Birds (Fig. 133, *p*, and 135, *k*).—In most of these instances, we find salivary glands pouring their secretion into the mouth or pharynx; and where a gizzard exists, there is usually also a 'crop,' or 'first stomach' (which is rather to be regarded as a dilatation of the lower part of the œsophagus), into which the food is received in the first instance, and in which it is probably macerated in the salivary fluid, before it is subjected to the trituration of the gizzard.

403 *a*. In the lower Articulata, there is seldom any special reducing apparatus; in the *Rotifera*, however, which obtain their food (like the Bryozoa) by ciliary action, we find, in place of a gizzard, a curious pair of jaws (Fig. 144, *e*), furnished with sharp hard teeth, and worked by a complex muscular apparatus; these are situated in the pharynx, and serve to divide the larger particles as they pass downwards to the stomach. Owing to the transparency of these animals, the movements of their jaws can be distinctly seen, when the 'wheels' are in action, and are driving downwards the particles which have entered the mouth.—In the *Mandibulate Insects*, the reduction of the food, as well as its division and ingestion, is partly performed by the jaws, and the mouth is usually furnished with salivary glands; but many insects are also provided with a gizzard (Fig. 199, *d*), which, as in the cases already alluded to, does not act upon the food until it has been received into the crop (*c*). In many of the higher *Crustacea*, notwithstanding their complex buccal apparatus, the stomach is furnished with a powerful set of teeth, affixed to a complicated framework which is worked by powerful muscles, so as to constitute very efficient reducing instruments. Among the mandibulate Articulata, salivary glands are almost invariably found opening either into the mouth, pharynx, or œsophagus; but sometimes the stomach itself is clothed with cæca (Fig. 199, *e*), that appear to have this character.—Of that dental reducing apparatus, which is so especially characteristic of the Vertebrata, a full account has already been given in the description of the respective classes of that sub-kingdom (Fishes, § 322 *f*, Reptiles, § 324 *o*, Mammals, § 326 *o*); and it is unnecessary to do more here than refer to the substitute for it in the class of *Birds*, which is altogether destitute of the means of reducing the food in the mouth, but carries us back, in the modification of the stomach for the purpose, to the plan so common among the Invertebrata. There is this curious distinction, however, that the gizzard succeeds, instead of preceding, the proper gastric cavity; the food being not only macerated in the salivary and other secretions furnished by the crop, but also impregnated with the proper gastric fluid, before it is subjected to the triturating action of the gizzard. Although in the *Mammalia* the reducing apparatus is limited to the mouth, yet the stomach frequently presents a complex arrangement, of which the purpose seems to be to favour the mechanical reduction of the food, and its impregnation with fluid, before it is subjected to the true digestive action. The most remarkable example of this kind is presented in the group of *Ruminantia*, in which the stomach is subdivided into four distinct cavities; the first

two of these, however, being rather dilatations of the œsophagus, than parts of the stomach itself. The food, on its first passage down the œsophagus in a crude unmasticated state, enters the large cavity termed the *ingluvies* or 'paunch' (Fig. 196, *b*), which, like the crop of Birds, serves as a temporary receptacle for it, and moistens it with the fluid secreted from its walls; the liquid swallowed, on the other hand, seems to be specially directed into the second cavity (*c*), which is termed the *reticulum* or 'honeycomb-stomach' from the reticulated appearance of its interior, occasioned by the irregular folding of its internal membrane. It is here that the peculiar provision of 'water-cells' is found, for which the Camel has long been so celebrated, but which exists in a greater or less degree in all the Ruminants. These cells, which are the spaces between the deepest reticulations, are bounded by muscular fasciculi, by the contraction of one set of which their orifices may be closed and their contents retained; whilst by that of another set, the fluid they contain may be expelled into the general cavity of the stomach. After remaining there until it is sufficiently impregnated with fluid, the solid matters which have passed into the first and second stomachs are returned at intervals, in the form of little balls or pellets, to the mouth; where they undergo a thorough mastication, and are completely incorporated with salivary fluid. When finally swallowed, the food is directed, in the manner to be presently described, into the third stomach (*d*), the *omasum*, commonly termed the 'many-plies' from the peculiar manner in which its lining membrane is disposed; this presents a number of folds, lying nearly close to each other like the leaves of a book, but all directed by their free edges to the centre of the tube, a narrow fold intervening between each pair of broad ones. The food, now reduced to a pulpy state, has therefore to pass over a large surface, before it can reach the outlet of that cavity; which leads to the *abomasum* or fourth stomach (*e*), commonly called the *reed*. This is the seat of the true process of digestion, the gastric fluid being secreted from it alone; and it is from this part of the calf's stomach, that the 'rennet' is taken, which derives its extraordinary power of coagulating milk from the organic acid it contains. In the suckling animal, the milk which forms its nourishment passes directly into this stomach; the aperture leading to the first and second being closed, and the folds of the third adhering together so as to form a narrow undivided tube. The direction of the food into one or another of these cavities appears to be effected without any voluntary effort on the part of the animal itself, but to result simply from the very peculiar endowments of the lower part of the œsophagus. This does not entirely terminate at its opening into the first

Fig. 196.

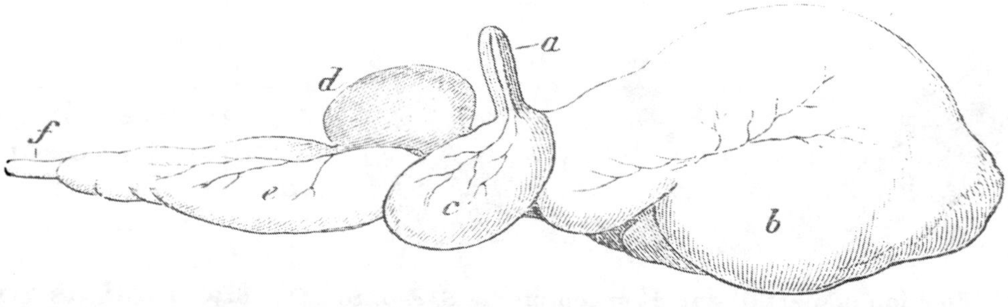

Stomach of *Sheep*;—*a*, œsophagus; *b*, paunch; *c*, second, or honeycomb-stomach; *d*, third stomach, or many-plies; *e*, fourth stomach or reed; *f*, pylorus.

stomach or paunch, but it is continued onwards as a deep groove with two lips (Fig. 197): by the closure of these lips it is made to form a tube, which serves to convey the food onwards into the third stomach; but when they separate, the food is allowed to pass either into the first or the second stomach. When the food is first swallowed, it has undergone but very little mastication; it is consequently firm in its consistence, and is brought down to the termination of the œsophagus in dry bulky masses. These separate the lips of the groove or demi-canal, and pass into the first and second stomachs. After they have been macerated in the fluids of these cavities, they are returned to the mouth by a reverse peristaltic action of the œsophagus; this return takes place in a very regular manner, the food being shaped into globular pellets by compression within a sort of mould formed by the ends of the demi-canal, drawn together, and these being conveyed to the mouth at regular intervals, apparently by a rhythmical movement of the œsophagus. After its second mastication, it is again swallowed in a pulpy semi-fluid state; and it now passes along the groove which forms the continuation of the œsophagus, without opening its lips; and is thus conveyed into the third stomach, whence it passes to the fourth. Now, that the condition of the food as to bulk and solidity is the circumstance which determines the opening or closure of the lips of the groove, and which consequently regulates its passage into the first and second stomachs, or into the third and fourth, appears from the experiments of Flourens; who found that when the food, the first time of being swallowed, was artificially reduced to a soft and pulpy condition, it passed for the most part along the demi-canal into the third stomach, as if it had been 'ruminated,'—only a small portion finding its way into the first and second stomachs.

FIG. 197.

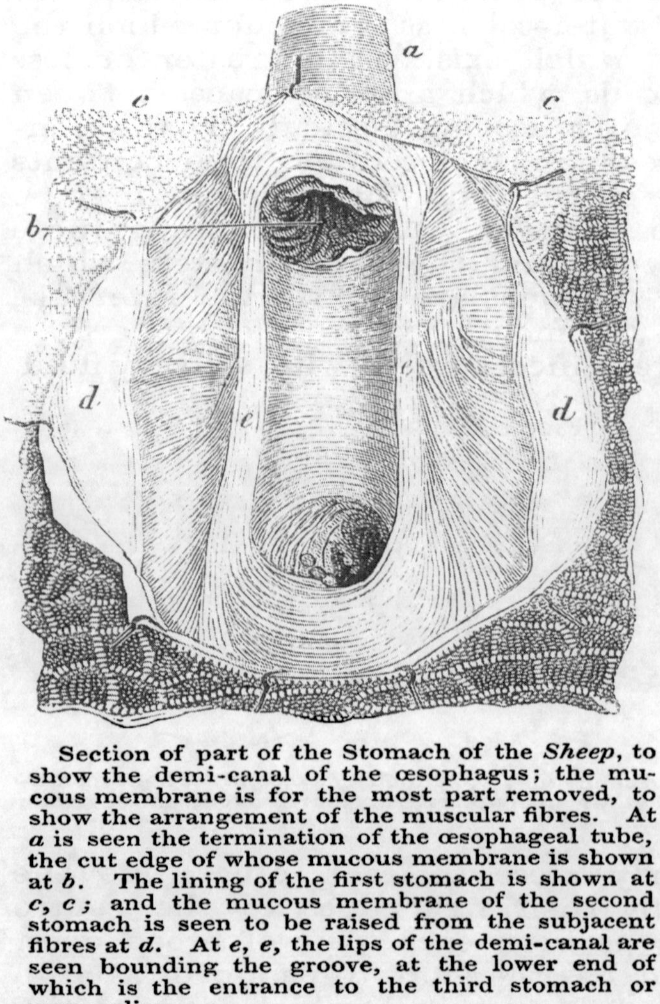

Section of part of the Stomach of the *Sheep*, to show the demi-canal of the œsophagus; the mucous membrane is for the most part removed, to show the arrangement of the muscular fibres. At *a* is seen the termination of the œsophageal tube, the cut edge of whose mucous membrane is shown at *b*. The lining of the first stomach is shown at *c, c*; and the mucous membrane of the second stomach is seen to be raised from the subjacent fibres at *d*. At *e, e*, the lips of the demi-canal are seen bounding the groove, at the lower end of which is the entrance to the third stomach or many-plies.

404. We have now to trace out the principal forms, under which the proper Digestive apparatus presents itself in the different classes of animals; and the mode in which it operates. Putting aside that of the *Protozoa*, which has been already considered, the simplest of all its conditions is that in which we find it in the *Hydra*, *Actinia*, and other solitary Zoophytes. Thus in the *Hydra* (§ 289), it will be remembered, the stomach occupies the whole of the cavity enclosed by the outer walls; and the membrane with which it is lined is so completely identical with

that which forms the external integument, that the one may be made to take the place of the other without any injury to the animal. Yet it is obvious that a powerfully-solvent fluid is secreted from the walls of the gastric cavity; for the soft parts of the food which is drawn into it (usually in a living state) are gradually dissolved, and this without the assistance of any mechanical trituration; whilst the hard parts are ejected again by the orifice through which they were introduced. The product of this operation appears to be absorbed from the lining of the digestive sac, by the whole of the tissue which forms it; and to be conveyed into the tentacula (which are the only parts not in immediate relation with the digestive cavity) by interstitial channels left for that purpose.—The same general condition presents itself in the *Actinia;* the only difference being, that the digestive sac does not occupy the whole of the cavity of the body, but is surrounded by the ovarian chambers (§ 291).—In the compound *Hydroida*, on the other hand, the lower part of the digestive sac of each polype is prolonged into a tube, which communicates with similar prolongations from other polypes, so that the stem and branches of the entire arborescent structure contain a continuation of the gastric cavities of the several polypes which they bear (Fig. 95); and the fluid that has been prepared by the digestive operations of the latter, passes through the aperture at the lower extremity of each (which is guarded by a sphincter muscle), and is received into the system of ramifying tubes, which may be considered as an extension of the digestive cavity throughout the general structure, for the purpose of conveying to every portion of it, and especially to such parts as are undergoing increase by the formation of new branches or polype-cells, the materials prepared for nutrition. Through this system of canals, the contained fluid has been observed to move with considerable regularity, and in a manner that cannot be accounted for by any mechanical agency communicated by the contraction of the stomachs of the polypes. Thus, when the stem and branches of a living *Sertularia* are examined with a sufficiently high magnifying power, a current of granular particles is seen moving along the axis; which, after continuing one or two minutes in the same direction, changes and sets in the opposite one, in which it continues about as long, and then resumes the first; thus alternately flowing down the stem to the extremities of the branches, and back again. The change of direction is sometimes immediate; but at other times the particles are quiet for a while, or exhibit a confused whirling motion for a few seconds, before it takes place. The current extends into the stomachs of the polypes; in which, as well as in their orifices, a continual agitation of particles is perceptible. When these particles are allowed to escape from a cut branch, they exhibit for a time an apparently spontaneous motion. No contraction of the tube, any more than of the stomach, seems concerned in the production of the currents; and their rapidity and constancy appear intimately connected with the activity of the nutritive processes taking place in the parts towards which they are directed, the 'set' of the current being especially *towards* any portion which is undergoing development, and *away from* any part which exhibits a diminution or loss of vitality. In the *Tubularia,* in which the polype-bearing stem ramifies but little, or not at all, and is sometimes divided by nodes or partitions somewhat resembling those of the *Chara* (§ 269), the movement of fluid very strongly resembles that which is seen in that plant; for the current passes down one side, crosses the septum, and then ascends the other side, following a somewhat spiral line of particles deposited on the walls of the tube.—The same prolongation of the gastric cavity through the whole of the mass on which the polypes are seated, is characteristic of the *Asteroid* Zoophytes (§ 292); and here, also, this extension of the digestive sac through the entire body, serves to convey to every part the materials of its growth, which it can at once absorb for itself without the mediation of a special circulating apparatus.

405. In the lowest members of the Articulated series, we find the same general type of conformation of the digestive apparatus still retained, though its *form* is very different. The gastric cavity, in the lower *Entozoa*, is excavated (as it were) in the soft tissues of the body, so that, although provided with walls of its own, these are blended externally with the surrounding parts; and further, the cavity is extended through the entire body, being repeated in every segment of it, so that there is but little occasion for any separate system of vessels for the conveyance of nutritive fluid. We have seen this to be the case in the *Tænia* (Fig. 140, c), in which, with its allies, the gastric cavity appears to be filled by imbibition through its walls, instead of by suction through an oral orifice (§ 309 *b*, *c*); and it is also the case in the *Planariæ* (Fig. 141), *Fasciolæ* (Fig. 142), and the other *Trematoda*, in which the aliment is received through a suctorial orifice, but in which no anal opening exists. —The same type is repeated under a higher form and more complete development, in some of those more elevated members of the Radiated series, which constitute the class *Echinodermata*. In the *Ophiurida* (§ 296 *a*), and *Asteriada* (§ 296 *b*), the gastric cavity is still a single sac, occupying the centre of the body, and furnished with but one orifice, through which food is drawn in, and undigestible or fæcal matter is ejected. But the sac is now suspended freely in the general cavity of the body, its walls being quite distinct from the surrounding tissues. In the cæcal prolongations of the central stomach of the Asterias (Fig. 110), and in the abundance of the cells which line them, we find the first manifestation of special glandular organs for the elaboration of the fluids required in the digestive processes, though the precise nature of these is as yet unknown. The proper digestive apparatus appears to be limited in the Star-fish, as in the Ophiura, to the *central* cavity; and being removed by a considerable interval from other parts of the body, it does not here directly convey the nutritious fluid to the tissues which make use of it; and the intervention of a set of vessels becomes requisite, for its absorption and transmission to distant parts.—On nearly the same grade with the preceding, as regards the conformation of its digestive apparatus, we may place the curious *Rotiferous* Animalcule formerly noticed (§ 310 *a*) as possessing no intestine nor anal orifice, and as ejecting the indigestible particles of food through its mouth. This is the only known instance among the Articulata, save in the class of Entozoa, in which the digestive cavity has but one orifice. No conformation of a similar kind is ever met with in Molluscous and Vertebrated series.

406. We have now to glance at those higher forms of the digestive apparatus, in which a second orifice or 'anus' is present, for the discharge of indigestible or fæcal matters; and this, as we shall see, is by far the

most general plan of conformation. Although this orifice may externally be situated in the neighbourhood of the mouth, yet it always communicates with the part of the digestive cavity that is most remote from it; this cavity being usually more or less prolonged into a tubular form. The lowest of the animals which present this great advance in the type of conformation, are the *Acalephæ;* in most of which we find the digestive cavity provided either with a single or a multiple outlet, the former being seen in the *Beroe* (Fig. 107, A), the latter in the *Medusæ* (Fig. 106). This class is further remarkable for the absence of true blood-vessels; their place being supplied by the tubular prolongations of the gastric cavity, which extend themselves into the mass of the tissues, and which convey to them the materials for their growth and renovation (Figs. 106 and 195). In the Beroe we see the first indication of that division between the 'stomach' and 'intestinal canal,' which becomes so much more obvious and important in the higher classes. This division is by no means well marked, however, even in the highest Radiata; the alimentary canal, which commences with the mouth and terminates at the anus, being of nearly uniform size throughout, and apparently of similar endowments; as we see in the *Echinus* (Fig. 111), and the *Holothuria* (Fig. 113). In the *Nematoid* Entozoa, again, the same uniformity presents itself (Fig. 143); and it is curious to observe this almost exactly repeated in the *Serpent* tribe, whose organisation is so much higher; but the prolongation and uniformity of whose body seems to necessitate a somewhat similar conformation of the alimentary canal. Even in the *Annelida,* the same general plan is continued with but little modification; the alimentary canal passing in a straight line from one extremity of the body to the other, without distinction of stomach or intestine (Fig. 198); but from its sides we find cæcal prolongations extending, sometimes as a single pair of prolonged tubes (B, *c,* C, *b*), sometimes as mere sacculi in the walls of the stomach (B, *b*), and sometimes as a multiplied and extended series of appendages (A, *d*), which seem, like the radiating cæca of the

FIG. 198.

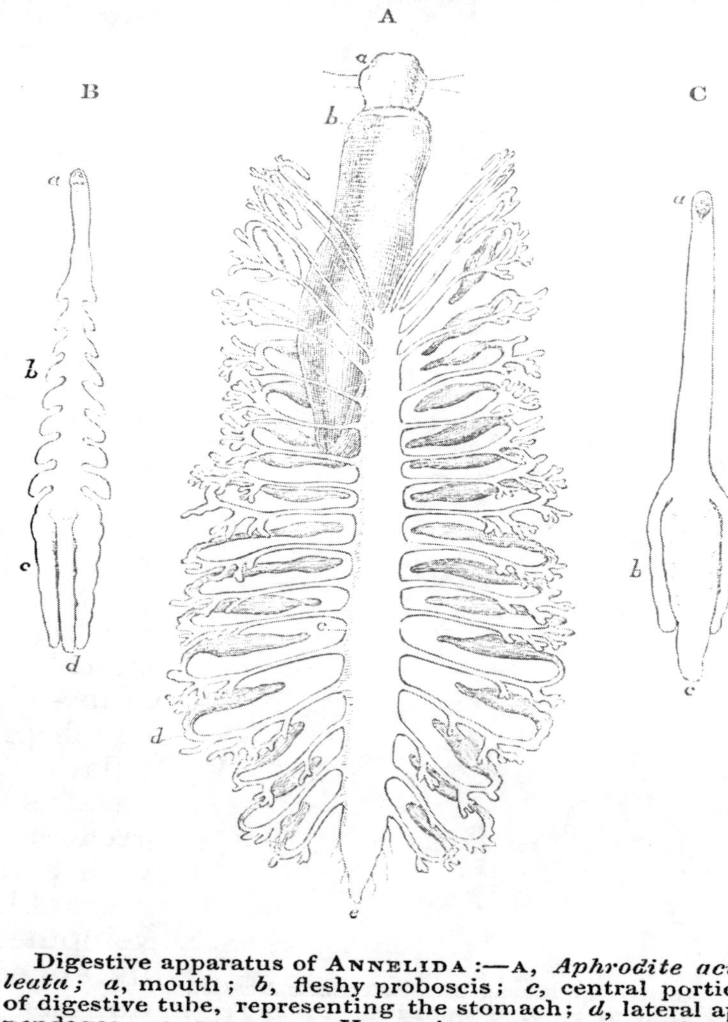

Digestive apparatus of ANNELIDA:—A, *Aphrodite aculeata; a,* mouth; *b,* fleshy proboscis; *c,* central portion of digestive tube, representing the stomach; *d,* lateral appendages; *e,* anus:—B, *Hæmopis vorax; a,* mouth; *b,* lateral sacculi of the stomach; *c,* two large terminal cæca; *d,* intestine:—C, *Aulostoma nigrescens; a,* mouth; *b,* cæca; *c,* intestine.

Asterias, to be rather glandular in their character, than destined to admit the passage of food into them.

407. From these forms we might gradually ascend towards the more complex digestive apparatus of Insects and Crustacea ; but we shall first revert to that of the lowest Mollusca, as the simplest case in which the division of the canal into stomach and intestine is clearly marked out. This is well seen in the *Bryozoa* (§ 298), in which, as in the Hydra, the whole process of digestion may be watched, owing to the transparency of the tissues of the animal. The digestive stomach is here separated from the comparatively narrow intestinal tube, by a valvular orifice, or true 'pylorus ;' and whilst the whole process of solution takes place in the former division of the canal, whose walls are beset with secreting follicles that pour forth a bile-like fluid, the latter seems to have little to do, save to convey to the anal orifice at the mouth of the polype-cell the excrementitious particles which are to be ejected from it. The same essential type is retained through the whole of the Molluscous series: the principal advance being shown in the higher development of the liver, by the withdrawal of its follicles from the parietes of the stomach, and by their aggregation into distinct glands which discharge their secretion into the upper part of the intestinal tube ; in the development of a rudimentary pancreas (in the Cephalopoda) ; and in the increased length given to the intestinal portion of the tube, which usually makes several 'convolutions' before it finally terminates at the anus.—The same is the mode of advance which presents itself in the higher Articulata ; among which the digestive apparatus of the Crustacea presents the greatest resemblance to that of Mollusca, whilst that of Insects (Fig. 199) is remarkable for the very low grade of development of the liver, which only makes its appearance in the form of a small number of cæcal tubuli (*f*) discharging their contents into the intestine. The numerous short cæcal follicles which cluster round the chylific stomach (*e*), are perhaps to be regarded as pancreatic. As yet no distinction between the 'small' and the 'large' intestine presents itself ; this being only manifested fully in the higher Vertebrata.—It is interesting to observe that, among the higher tribes of both the Molluscous and Articulated sub-kingdoms, we should find

FIG. 199.

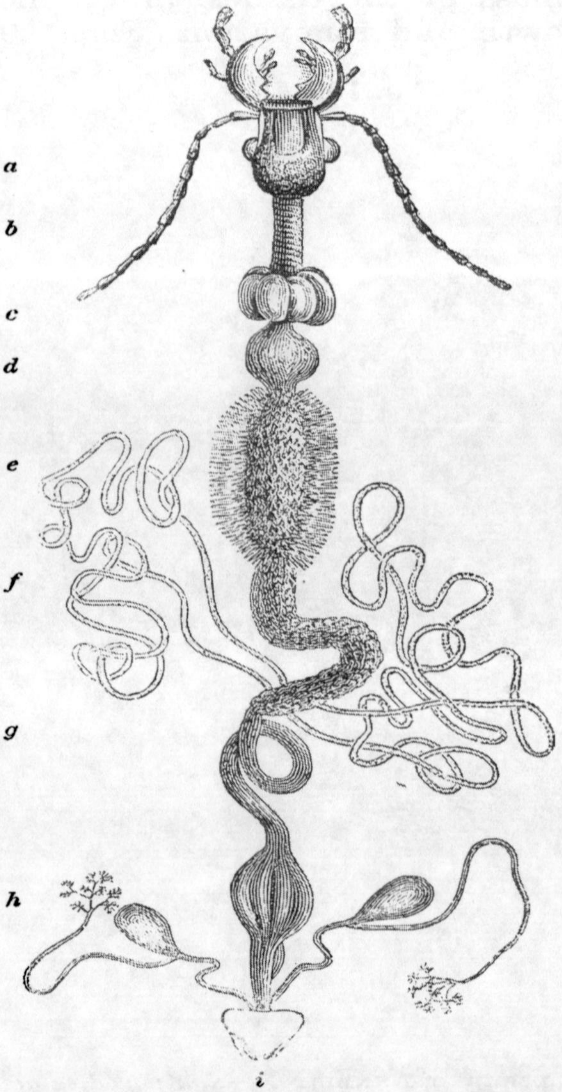

Digestive apparatus of *Carnivorous Beetle* ;—*a*, pharynx ; *b*, œsophagus ; *c*, crop ; *d*, gizzard ; *e*, chylific stomach ; *f*, biliary vessels ; *g*, small intestine ; *h*, special secreting organs ; *i*, anus.

examples of reversion to that lower type of conformation of the digestive apparatus, which is manifested in its extension into cæcal prolongations that radiate into parts of the body remote from the proper digestive sac, and in the diffusion of the glandular apparatus over the whole or the greater part of their surface. This is the case with the *Nudibranchiate* Gasteropods (§ 304 *c*), in which the alimentary canal sends off as many cæca as that of the *Planariæ* (Fig. 141), each one of them, however, being surrounded at its termination by a group of hepatic cells, the liver being thus subdivided (so to speak) amongst them all. So among the Crustacea, we find the curious group of *Pycnogonidæ* (§ 316 *e*) especially characterized by the extension of the digestive cavity into the articulated members, almost to their extremities (Fig. 158), and by the spreading-out of the hepatic cells over the entire surface of these cæca; and it is peculiarly worthy of note, that whilst the alimentary matter passes into these cæca, and is conveyed by them (as microscopic observation shows) into the immediate neighbourhood of every portion of the body, the special circulating apparatus presents its very lowest grade of development. An approximation to the same structure is seen in the cæcal prolongations of the digestive cavity, which pass into the thoracic members of the Arachnida (§ 319 *a*). These, in the *Araneidæ* (Fig. 162), would not seem to possess any special secreting function, true biliary vessels being found in the abdomen; in the *Scorpionidæ*, however, they seem to be the only hepatic organs; whilst in certain *Acaridæ* (§ 319 *d*) we return to a condition even lower than that of the Pycnogonidæ, the parietes of the cæcal prolongations of the digestive cavity not being distinctly separable from the tissues around, and not even the rudiment of a proper circulating apparatus being distinguishable.

408. The digestive apparatus of Vertebrated animals may be considered as carrying onwards the general plan which has been seen to prevail in the Molluscous sub-kingdom. Thus in *Fishes*, there is usually a well-marked distinction between the stomach and intestinal tube; the liver attains a high development, and is completely withdrawn from the parietes of the alimentary canal; and the pancreas, also, begins to appear as a distinct gland, instead of being a mere collection of cæca clustering around the digestive cavity. In the *Amphioxus*, however, we see a reversion to a very inferior type (§ 321 *b*); the digestive portion of the alimentary canal being a short tube of nearly uniform size throughout, which commences behind the pharyngeal respiratory apparatus, and runs backwards almost in a straight line to the anal orifice, receiving in its passage the fluid discharged by the simple large cæcum, which is the only rudiment of the liver. In the *Cyclostomi* the stomach is merely a dilated portion of the tube, which lies in a straight line between the œsophagus and the intestine; but in nearly all the higher Fishes, both Osseous and Cartilaginous, it forms a large cavity which lies out of the course of the alimentary canal, and is adapted to retain the food while the process of solution is being carried on. It is usually separated from the intestine by a narrow 'pyloric' orifice; and it is also generally divided from the œsophagus by a 'cardiac' valve, the aperture of which, however, is very wide, so as to admit the food which is swallowed in an undivided state, and does not completely prevent its regurgitation. Indeed it seems common for Fishes to disgorge the shells and other indigestible parts of

their food, through the mouth, like Polypes; and there are some (especially of the *Carp* tribe) in which a sort of *rumination* takes place (§ 403), the food being sent back to the pharynx to be masticated by the pharyngeal teeth, and then returning to the stomach to undergo its final digestion. The intestinal tube is usually short and almost destitute of convolutions, as well as of nearly uniform diameter throughout, so that the distinction between the small and the large intestine is only marked (and this not constantly) by the ileo-colic valve. The surface of the mucous membrane lining the canal is considerably extended, however, by being thrown into *rugæ* or wrinkled folds, more or less projecting; these in the Osseous Fishes have seldom any great regularity of arrangement; but by the great projection and spiral continuity of one of these folds, the real length of the intestinal tube is greatly increased in the higher Cartilaginous Fishes, just as a spiral staircase is much longer than the cylindrical cavity within which it winds.—In the omnivorous tadpoles of *Batrachian Reptiles,* the alimentary canal is of great length, and forms numerous convolutions in the abdominal cavity; but the stomach is narrow and elongated, the intestinal tube is of nearly equal size throughout, and the boundary between the small and the large intestine is not distinctly marked. This condition is retained in many of the *Perennibranchiata* (§ 323 *a*); but in the adult condition of the Frog and its allies, a nearer approach is exhibited to the character presented by the alimentary canal in the higher *Reptiles,* most of which are carnivorous. In these we usually find the gastric dilatation larger, and more completely distinguished from the intestinal tube; although the intestine is relatively much shorter, yet the extent of its mucous surface is augmented by *rugæ* and *villi* (§ 169); and the small intestine is now more constantly separated from the colon, by a valvular constriction. It is in the vegetable-feeding Turtles, that we find the intestinal tube possessing the greatest length, and presenting the greatest difference of diameter in its two divisions; and in most of these, as well as in some other Reptiles, the large intestine has a cæcal dilatation at its commencement. In *Birds,* the length of the alimentary canal varies greatly in the different families, in accordance with their habitual diet, being greatest in the granivorous and frugivorous tribes, and least in the predaceous; the intestinal tube is always distinctly separated from the stomach by a pyloric valve; and the large intestine is somewhat more dilated relatively to the small, and is frequently furnished at its commencement with a pair of long cæcal appendages (Fig. 186). The mucous membrane is much more regularly plicated, and its villi are more highly developed, than in Reptiles. It is in the Ostrich and its allies, that the intestinal canal presents the closest resemblance to that of Mammals.

408 *a.* The length and complexity of the alimentary canal in the *Mammalia,* as in Birds, vary in accordance with the diet on which the several tribes are adapted to exist; and no part exhibits a greater diversity of conformation than does the stomach,—even putting aside that peculiar apparatus which has been already described (§ 403 *a*) as concerned in the preparation of the aliment in the Ruminantia. Thus in the carnivorous and insectivorous Mammals, whose food is easy of solution, and consequently requires to be but little delayed for the digestive operation, the stomach is usually simple in its form, being a mere dilatation of the

alimentary canal, and lying nearly in the direction of its course; but in proportion as the composition of the food departs more widely from that of the body itself, and it is less capable of reduction by the gastric fluid, do we find (in most cases at least) the dimensions of the stomach increasing, and its form undergoing alteration, so that it has more and more the character of a *diverticulum*, into which the aliment is turned aside, and in which it is retained during the time required for its subjection to the solvent power of the gastric juice. The difference in form is principally given by the increased development of the large or left-hand extremity of the stomach, which thus forms a sort of cæcal dilatation off the line that connects the cardiac and pyloric orifices; and not unfrequently this portion of the stomach is separated from the other, by an incomplete partition. In addition to this, numerous smaller cæca are frequently developed in connection with the principal cavity; whilst in many *Cetacea* we find a succession of constrictions intervening between the principal gastric cæcum and the pyloric orifice, which partially divide the entire stomach into numerous cavities, all of them to be traversed by the food during its passage from the œsophagus to the intestine. The small intestine is chiefly remarkable for the great length to which it extends in some of the herbivorous Mammalia, and for the increase of its mucous surface by 'valvulæ conniventes,' which are deep folds of membrane, succeeding one another at tolerably regular intervals, but not embracing above two-thirds of the circumference of the tube; towards the lower extremity of the smaller intestine, also, we usually meet with the 'glands of Peyer,' which are either altogether absent, or are but slightly developed, in the lower Vertebrata. In nearly all Mammals, save in the *Cetacea* (which exhibit a return to a fish-like inferiority in this respect), a well-marked distinction between the small and the large intestine is manifested, not only by the presence of the ileo-cæcal valve, but also by that difference of diameter from which these two divisions of the canal derive their names. The large intestine is here not only larger in proportion to the small, but is longer, than in the lower Vertebrata; and in many Mammals, as in Man, a sacculated character is given to the colon by the peculiar arrangement of its muscular bands, so that the extent of its lining membrane is greatly increased. The cæcal enlargement, in which the ileum terminates and the colon commences, is frequently of great size, especially in herbivorous animals; being sometimes larger than the stomach, as is the case in the Horse, or even many times larger, as may be seen in some of the *Rodentia*. The *villi*, which in all the lower Vertebrata are dispersed over the surface of the large as well as of the small intestine, are now limited to the latter; whilst from the great number of glandulæ contained in the former, and from the change in the character of the contents of the alimentary canal which takes place when they arrive there, it seems probable that whilst the small intestine is the part of the canal in which *absorption* specially takes place, the large is particularly destined for *excretion*.

409. Thus then, we see, that although the development of the Digestive apparatus is subjected, perhaps more than that of any other portion of the apparatus of organic life, to variations which are immediately connected with the purposes to which it is to be subservient, a gradual *specialization* can be most distinctly traced, when we take a general comparative survey

of the succession of forms presented by the principal groups of animals. For, in its lowest condition, we have seen it to consist of a simple cavity excavated in the tissues, for the reception of alimentary substances, which, when reduced to the state of solution, are taken up from every part of its walls, and are carried, without any special system of canals, into the parts of the body most remote from it: whilst the indigestible matters are ejected by the same orifice as that by which the food was taken in. Next, we find the digestive cavity furnished with parietes distinct from the surrounding tissues; and the dissolved nutriment is taken up from its walls by a set of vessels, which carry it to the most distant organs. Next, we find the cavity changed into a canal, and provided with a second orifice, through which the indigestible and fæcal matters are ejected. Next, we find the stomach more or less distinctly separated from the intestinal tube; and the former becomes possessed of special endowments which are not shared by the latter, the gastric fluid being secreted by its walls alone, and the secretion of the liver being either poured into it, or into the first portion of the intestine. Next, we find the mucous surface of the intestinal canal considerably extended by folds and duplicatures; and these are thickly set with *villi*, each of which not merely contains a copious net-work of blood-vessels (Fig. 21), but also includes the commencement of one of the special absorbents or 'lacteals' peculiar to Vertebrated animals. Next, we find the distinction between the small and the large intestine gradually becoming more distinctly marked, and each part becoming characterised by its own peculiarities; whilst at the same time, the concentration of structure is still further indicated by the greater extension of the mucous surface within the canal, by the larger number of villi in the small intestine, and by the increase of the glandulæ and follicles in the large.—Now in its general outlines, the history of the embryonic development of the digestive cavity in the higher animals closely corresponds with this; but the entirely different mode in which the alimentation of the embryo is provided for, involves a considerable variation in the particular method adopted. The embryonic cell, when it is first distinguishable in the ovum, lies in the midst of a store of nutriment (the yolk) provided for it by the parent; and this it begins to draw in, like the simple cells of the Protozoa which it resembles in grade of development, by imbibition through its cell-wall. The successive generations produced by fissiparous multiplication, are at first nourished after the same fashion; but these soon tend to spread themselves in the form of a membranous expansion around the yolk, and thus to include it completely within the embryonic cavity. This cavity may be likened to the stomach of the Hydra, in every respect save that it has no orifice; but this is not needed, since it is already filled with the requisite supply of aliment. The further introduction of this into the embryonic tissues, is accomplished at first by the cellular layers of the germinal membrane; but blood-vessels are soon developed, which thenceforth are the special channels for its reception and conveyance. This cavity, however, serves but a temporary purpose,—in this respect corresponding with the cotyledons, by which nutriment is received into the embryo of the higher plants; and the permanent digestive apparatus commences in the more advanced embryo, as a small portion of the temporary vitelline sac, that is gradually detached from it, like the stomach of the budding *Hydra* from that of its parent (§ 289). The form which this detached portion at first exhibits, is that of a simple tube, closed at both extremities, whilst its middle portion remains connected with the yolk-bag. An oral and an anal orifice are then formed; and the tube thus comes to present the characters of the alimentary canal of the lower Articulata. The next grade in development is the evolution of the stomach, which first shows itself as a projection of the tube towards the left side; and the accessory glands, the liver and pancreas, soon make their appearance. The short straight tube gradually increases in length, and is thrown into convolutions; that part being most increased in length, which remains of the smallest diameter, and thus arises the difference between the small and the large intestine. The folds of the mucous membrane, the villi, and the glandular apparatus, which are the parts of the apparatus most restricted to higher animals, are the last to appear in the course of their development.

410. We have now to consider the proper function of Digestion, which may be said to commence with the introduction of the food into that part of the alimentary canal, in the walls of which is secreted the fluid destined for its solution. In the higher animals, this secretion is restricted to the gastric cavity or stomach; and hence it is named the 'gastric juice.' The only chemical change which the food appears to have undergone, before being submitted to this, is that which is effected by the admixture of saliva in the mouth; a peculiar animal principle contained in this fluid having the power, like the *diastase* in Plants, of converting starch into sugar. The process thus commenced in the mouth, is checked in the stomach, the acid character of the gastric fluid being unfavourable to it; it is recommenced, however, in the intestinal canal, after that acid has been neutralized by the alkali of the bile and pancreatic fluid. The great purpose of the *gastric* digestion appears to be, to dissolve the *albuminous* and *gelatinous* constituents of the food; and, by the withdrawal of these, the remainder are usually reduced to a state of fine division, in which they are afterwards more easily acted upon. In regard to the constitution of the gastric fluid, there is at present much diversity of opinion; and it does not seem improbable that it may vary in different animals. It appears essentially to consist, however, of a free acid (either the hydrochloric, acetic, or lactic) which is the real solvent; and of an animal principle in a state of change, which acts as a *ferment*, and disposes the organic compounds to solution. Water slightly acidulated with these acids, is capable of dissolving albuminous compounds with the aid of a high temperature; but if a solution of 'pepsin' (which is the animal matter obtained by macerating the stomach of a pig in cold water, after it has been repeatedly washed) be added, the acid solvent will then be able to act efficaciously at the ordinary temperature of the body. The solvent action of the gastric fluid is aided by the movements of the walls of the stomach, which are produced by the successive contractions and relaxations of their muscular fibres. The purpose of this motion is obviously to keep the contents of the stomach in that state of constant agitation, which is most favourable to their chemical solution; and particularly to bring every portion of the alimentary matter into contact with the lining of the stomach, so as to subject it to the action of the solvent fluid which is poured forth from it. Thus, little by little, the reduction of the alimentary

materials to the homogeneous pulpy mass termed *chyme* is accomplished; this escapes into the intestine, as fast as it is formed, through the pyloric orifice, which closes itself to solids, but allows liquids to pass; whilst the solid residue is continually subjected to the same action, until its solution has been effected. There is no doubt, however, that a portion of the nutritious matter dissolved by the gastric fluid is at once absorbed into the blood-vessels of the stomach, and never passes into the intestinal tube, nor into the special lacteal system of vessels.—That the action of the gastric fluid is in all respects one of a purely chemical nature, there can be no longer any question. When drawn direct from the stomach of Man or of the lower Mammalia, it is found to possess the power of dissolving various kinds of alimentary substances, provided that these are submitted to its action at a temperature equal to that of the body, and are frequently agitated. The solution appears to be in all respects as perfect as that which naturally takes place in the stomach; but a longer time is required to make it,—a difference which is easily accounted for, when the impossibility of fulfilling all the conditions under which gastric digestion takes place, is borne in mind. The quantity of food which a given amount of gastric fluid can dissolve, is limited; precisely as in the case of the acidulous solution of 'pepsin,' or 'artificial gastric juice,' whose solvent power is chiefly regulated by the quantity of acids which it contains, the same quantity of pepsin being capable of disposing the solution of many successive amounts of the substance to be acted on, provided that acid be added as required.

411. The *Chyme* which passes into the intestinal tube, is commonly a greyish, semifluid, and homogeneous substance, possessing a slightly acid taste, but being otherwise insipid. When the food has been of a rich oily character, the chyme possesses a creamy aspect; but when it has contained a large proportion of farinaceous matter, the chyme has rather the appearance of gruel. The state in which the various alimentary principles exist in it, has not yet been accurately determined; the following, however, may be near the truth. The albuminous compounds, whether derived from the Animal or from the Vegetable kingdom, whether previously possessing the form of fibrin, of casein, of gluten, &c., are all reduced by solution to the condition of Albumen; but a part of these compounds may still remain undissolved in the chyme of the intestine. Gelatine will be dissolved, or not, according to the previous condition of the substance containing it; for if this be a tissue which does not readily yield gelatine to hot water, the gastric fluid will have little influence upon it. The gummy matters of plants are dissolved, when they exist in a soluble form; but starch is not dissolved unless it has been previously acted upon by the saliva, save in the case of the Granivorous Birds and Ruminant Mammals, in which the provision for the mechanical reduction of this element of the food is greater than in other tribes. Sugar, whether introduced as such, or formed by the transformation of starch, is undoubtedly reduced to a state of complete solution, and is probably taken up by the blood-vessels as fast as it is dissolved. Oily matters, whether of animal or vegetable origin, are reduced to minute particles, and these are dispersed through the other constituents of the chyme. Most other substances, as the woody fibres, all the firmer cell-walls, and the resinous matters, of Plants, the horny matter, yellow fibrous tissue,

epidermic and other thick-walled cells, of Animals, pass unchanged from the stomach, and undergo no subsequent alteration in the intestinal canal, and form part of the fæcal matter which is discharged from it.—On passing into the small intestines, the Chyme soon becomes mingled with the *biliary* and *pancreatic* secretions, which appear to effect important changes in its character, though the nature of their respective agencies has not yet been clearly made out. It may, however, be pretty certainly stated, that by the excess of alkali which they contain, they neutralise the acid of the gastric juice, so that thenceforth there is no further solution of albuminous compounds, but the conversion of the starch into sugar, which was interrupted in the stomach, now recommences, and is probably carried on while the food is passing through the small intestines; and further, that either by their separate or their combined actions, the fatty matter of the chyme is reduced to that state of fine division which is known as an 'emulsion,' and that it is thus rendered capable of being received into the absorbents. Thus, then, during the continued passage of the food along the upper part of the alimentary canal, the process of reduction and solution are still carried on; those operations being gradually performed in the intestine, which would have interfered with the digestion of albuminous matters if performed in the stomach; and the prolongation of this process being also the means of preventing the too rapid entrance of those saccharine and oily matters into the blood, which are only taken into it with a view to being speedily eliminated by the respiratory process, for the maintenance of the animal temperature. It is chiefly in warm-blooded animals, that we find the farinaceous or starchy substances, supplied by plants, entering largely into the composition of the food; and notwithstanding the evidence we possess that they are actually received into the vessels, yet there is a difficulty in detecting them under any form in the blood or chyle, in the healthy condition of the system. It is obvious, then, that their removal from the intestinal canal by the process of absorption must take place slowly; and thus we seem to have the explanation of the prolongation of the intestinal tube in herbivorous animals. It has been supposed that the contents of the intestinal canal are subjected to a sort of second digestion in the cæcum, for the purpose of dissolving any albuminous matters which have not been previously reduced, especially in those animals in which the cæcum is of large size in proportion to the stomach; and this idea seems to derive confirmation from the fact, that the secretion of the cæcum has been found to possess an acid reaction. It cannot, however, be at present regarded as more than a probable hypothesis.—The residuary undigested portions of the food, mingled with the excrementitious portion of the biliary and pancreatic secretions, together form the principal part of the fæcal matters which are discharged from the large intestine; but there is strong reason to believe, that the putrescent matter, which, in the Mammalia especially, gives the peculiar odour and appearance to the excrement, is not derived from either of these sources, but is a real excretion, separated from the blood by the glandulæ that are so thickly set in the intestinal walls, and especially at the lower part of the small intestine.

CHAPTER XI.

OF THE ABSORPTION OF NUTRITIVE AND OTHER MATTERS.

1. *General Considerations.*

412. THE process by which the alimentary materials, whether in their original fluid form (as in the case of Plants), or after they have been reduced to that form by the digestive process (as in Animals), are introduced into the living body, is termed *Absorption.* Before considering the particular conditions under which this operation is performed in the different classes of Organised beings, it will be right to consider what is its essential character, and how far Physical principles can be applied to its elucidation.—It was formerly the general opinion that Absorption is always effected by vessels, the open mouths of which, being in contact with the fluid, might imbibe it by capillary attraction, suction, or other like agencies; and hypothetical 'absorbent vessels' were inferred to exist in all beings, even in the simplest Cellular Plants, as the channels whereby fluids are received at the surface and conveyed into the interior. But it has been shown by minute Anatomical inquiry, that in no one instance are absorbent vessels thus brought into immediate relation with the fluid to be received by them, but that the transmission always takes place in the first instance through some tissue of a membranous character; and further, that in a great number of cases, vessels are not concerned in the process at all, the fluid being at once received into the tissues which it is destined to nourish. Thus, we shall find that neither in the roots of Plants, nor in the walls of the alimentary canal of Animals, do the absorbent vessels commence in open mouths; but that all the fluid which enters them, must first traverse the membrane which covers their extremities; whilst, in the lowest members of both kingdoms, and in the early embryonic condition of the higher, the external investment, or that reflexion of it which lines the digestive cavity (§ 381), has an equal power of imbibition throughout its entire surface, and communicates a portion of that which it has taken up to the tissues in contact with it, whence it is progressively transmitted to the more remote parts,—and this without the aid of any system of vessels, either for the Absorption or for the Circulation of nutritive fluid. And even in the most elaborately-constructed organisms, we find that a considerable portion of the tissues is nourished by the same kind of imbibition,—not, however, from the external surface, or from the walls of the digestive cavity, but from nearer sources of supply. Thus, as we have seen, the interior of a mass of proper cellular *Cartilage* is entirely nourished by imbibition from the blood-vessels which bring the nutriment to its surface and edges (§ 192); the dense tissue of *Bones* and *Teeth* is in like manner transversed by nutritious fluid, which is drawn from the vessels of the nearest vascular canal or cavity (§ 204); and the *Epidermic* and *Epithelial* tissues must derive their nourishment from vessels ramifying on the opposite side of the basement-membrane

whereon they rest (§ 177).—Thus it appears that the first introduction of fluid into the organism by *Absorption,* is but a particular case of that *Imbibition,* which takes a most important part in its subsequent diffusion through the system, even when the most complete vascular apparatus exists; and it will be right, therefore, to consider in the first instance what are the physical conditions of Imbibition.

413. When any porous substance (not already saturated) is brought in contact with a liquid which has such a molecular attraction for its particles as to be capable of 'wetting' it, the liquid is imbibed by it, and, if the force of imbibition be strong enough, is speedily diffused through the whole mass. This force depends in part upon *the degree of attraction subsisting between the particles of the solid and those of the fluid;* and in part upon the size of the capillary pores or canals. Thus it was found by Professor Matteucci,* that when glass tubes of about $\frac{3}{4}$ of an inch in diameter were filled with fine sand, previously dried, and introduced without pressure, and were immersed at their lower ends into the following liquids, the action of imbibition (which took place at first rapidly, then more slowly, and then ceased after about ten hours) raised the liquids in the tubes to the following heights respectively:—

Solution of carbonate of potash, . .	85 millimètres.
Solution of sulphate of copper, ...	75 ,,
Serum of blood, ...	70 ,,
Solution of carbonate of ammonia, ...	62 ,,
Distilled water, ...	60 ,,
Solution of common salt, ...	58 ,,
Milk, ...	55 ,,
White of egg diluted with its own volume of water,	35 ,,

When thick solutions of gum or starch, or fixed oils, were employed, scarcely any imbibition took place; and it was but little more, when strong saline solutions were used. — The degree in which imbibition is affected by the peculiar attractions subsisting between the solids and the liquids employed, is further illustrated by the following experiment. Three tubes, respectively filled with sand, pounded glass, and sawdust, were immersed at their lower extremities in water, and three similar tubes in alcohol; the following were the comparative results:—

	Sand.	Pounded Glass.	Sawdust.
Alcohol . .	85 millim.	175 millim.	125 millim.
Water . .	175 ,,	182 ,,	60 ,,

Thus we see that, whilst the imbibition of the two liquids took place into the pounded glass in nearly an equal degree, the quantity of water drawn up by the sand was more than double that of the alcohol, whilst precisely the reverse effect obtained in the case of the sawdust, the quantity of alcohol imbibed being more than double that of the water.—That the force of imbibition is dependent in part upon *the size of the capillary channels,* was proved by the fact, that when the experiment was tried with two tubes, both holding pounded glass, but one of them containing twice as much as the other (the powder being finer, and the capillary channels being consequently more minute), water was found to rise in the fullest

* "Lectures on the Physical Phenomena of Living Beings," translated by Dr. Pereira, pp. 21, et seq.

tube to 170 millim., in the same time which it occupied to rise to 107 millim. in the other.—*Temperature,* also, was found to have a remarkable influence upon the result; for two tubes, similarly prepared with sand, having had their extremities immersed in water, and having been kept, the one at 131° Fahr., and the other at 59° Fahr., the water rose in eleven minutes to 175 millim. in the former, whilst in the same period it only rose to 12 millim. in the latter. Hence we see that an elevation of temperature has not only a direct influence in augmenting vital activity, but that it assists in supplying an essential condition of that activity, by promoting the transmission of nutritive fluid through the textures.

414. But further, when two liquids capable of mixing together freely, without chemical decomposition, are allowed to do so, there is a tendency in each to a uniform diffusion through the other. This phenomenon of the 'Diffusion of Liquids' has been recently studied by Prof. Graham (the discoverer of the law of the 'diffusion of gases'); and the following are some of the results obtained by him, which bear most directly upon the present subject.*—The phenomenon was studied by means of a simple apparatus, consisting of an open phial to contain the liquid to be diffused, which was immersed in a large jar of pure water. The diffusion was stopped, generally after seven or eight days, by closing the mouth of the phial with a plate of glass, and then raising it out of the water-jar; and the quantity of the liquid which had found its way into the water-jar (or the 'diffusion-product') was then determined by evaporating it to dryness. The characters of 'liquid diffusion' were first examined in detail in reference to common salt; and it was found that when the amount of diffusion from solutions containing 1, 2, 3, and 4 per cent., was compared, it varied nearly in the same proportion, being in each case about one-eighth of the whole amount, after a period of eight days. But here, too, temperature has a remarkable influence; for an elevation of 80° Fahr. was found to double the quantity of salt diffused in a given time.—Different substances possess this property of diffusibility in very varying degrees; thus when solutions of the following substances were employed, of the strength of 20 parts to 100 of water, the relative quantities diffused in a given time were as follows:—

Chloride of sodium,	58·68	Crystallized cane-sugar,	26·74
Sulphate of magnesia,	27·42	Starch-sugar (glucose),	26·94
Nitrate of soda, .	51·56	Gum-arabic, .	13·24
Sulphate of water,	69·32	Albumen, . .	3·08

The low diffusibility of albumen is very remarkable; and it is further to be noticed that if common salt, sugar, or urea (which is as highly diffusible as chloride of sodium) be added to the albumen under diffusion, they diffuse away from this as readily as from their aqueous solutions, leaving the albumen behind in the phial. So, when solutions of two salts are mixed in the phial, they diffuse out into the water-atmosphere separately and independently of each other, according to their respective individual diffusibilities. In comparing the diffusibility of different salts, it was found that equal weights of 'isomorphous' compounds, dissolved in ten times their weight of water, diffused themselves to the same amount. And further, one salt, such as nitrate of potash, will diffuse into a solution of another salt, such

* "Philosophical Transactions," 1850, p. 1, et seq.

as nitrate of ammonia, as rapidly as into pure water; the solutions appearing to have that mutual diffusibility, which gases are known to possess (§ 495).—The properties of different substances in regard to their relative diffusibility, cannot but have an important influence on the course of the vital operations. Thus the low diffusibility of albumen obviously tends to the retention of the serous fluids within the tissues; whilst the high diffusibility of urea will favour its escape from them.

415. Both of these agencies,—the imbibition of fluids by porous solids, and the mutual diffusion of miscible liquids,—seem to be concerned in the production of the curious process, to which the term *Endosmose* was given by its discoverer, Dutrochet; and this phenomenon bears so close a resemblance to a vast number of the operations continually taking place in the living body, that it is scarcely possible to doubt that the same causes are in operation in both cases. The following is a general account of the process in question.—If into a tube, closed at one end with a piece of bladder or other membrane, be put a solution of gum or sugar, and the closed end be immersed in water, a passage of fluid will take place from the exterior to the interior of the tube, through the membranous septum; so that the quantity of the contained solution will be greatly increased, its strength being proportionably diminished. At the same time, there will be a counter-current in the opposite direction; a portion of the gummy or saccharine solution passing through the membrane to mingle with the exterior fluid, but in much less quantity. The first current is termed *Endosmose*, and the counter-current *Exosmose*. The increase on either side will of course be due to the relative velocity of the currents; and the changes will continue, until the densities of the two fluids are so nearly alike, as to be incapable of maintaining it. The greater the original difference (provided that the denser fluid be not actually viscid, but be capable of mixing with the other), the more rapidly and powerfully will the process be performed. The best means of experimenting upon these phenomena is afforded by a tube narrow above, but widely dilated below, so as to afford a large surface to the membrane, compared with that of the superincumbent column, which will then increase in height with great rapidity. By bending this tube into the form of a syphon, and introducing into its curve a quantity of mercury, the force as well as the rapidity of the endosmose between different fluids may be estimated with precision. In this way it was ascertained by Dutrochet, in some of his experiments, that fluid might be raised against a pressure of no less than 4½ atmospheres, or nearly 70 pounds to the square inch.*—Although it is not universally true that the activity of the process depends upon the difference in density of the two fluids (for in one or two cases the stronger current passes from the denser to the lighter), it seems to be so with regard to particular solutions, as those of gummy or saccharine matter. No endosmose takes place between fluids which will not mingle, such as oil and water; and very little between such as act chemically on each other. Although an organic membrane forms the best septum, yet it has been found that thin laminæ of baked

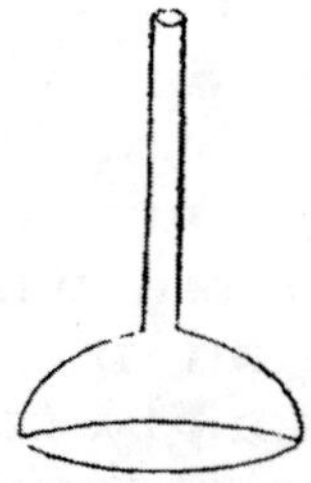

* For further information on this curious subject, see the Article *Endosmosis* in the "Cyclopædia of Anatomy and Physiology," Dutrochet's "Memoires Anatomiques et Physiologiques," tom. i., and Matteucci's "Lectures on the Physical Phenomena of Living Beings."

pipe-clay will suffice for the evident production of the phenomenon; and that porous limestones possess the same property in an inferior degree.

416. Although it may not as yet be possible to explain all the phenomena of Endosmose upon physical principles, yet these will go so far towards it, that the general conditions of the process may be considered as well understood. Supposing that two mutually diffusible liquids are on the opposite sides of a porous septum, which is not equally penetrable by them, then the one which is most readily imbibed will tend to occupy the capillary passages of the septum, and will thus be brought into contact with the liquid on the opposite side. This contact will permit the diffusion of that which has passed through the pores of the septum; and as fast as that which occupies these pores is removed by diffusion, so fast will it be renewed from the other side,—just as oil continues to ascend through the capillary channels in the wick of a lamp, so long as it is being dissipated by the combustive process at its summit. In this way, then, an Endosmotic current is produced, the force of which will depend upon the diffusion-powers of the two liquids, and upon the difference of the attractive power which the capillary tubes of the septum have for the two respectively. Thus, when a solution of sugar or gum is on one side of the septum, and water on the other, the water is the most readily imbibed; and consequently, the chief mixture and diffusion of the liquids, the one through the other, takes place at the surface of the septum in contact with the more viscid liquid. But at the same time, this liquid is tending to diffuse itself through the water which occupies the capillary channels of the septum; and, as it is not repelled by the septum, but is only attracted by it in a less degree than the water, a portion of it finds its way in a direction opposed to the principal current, and diffuses itself through the water on the other side, thus constituting Exosmose.—Thus it happens that the *direction* of the principal current, or Endosmose, will be determined by the attractive power of the septum for one or other of the liquids; though the diffusion-power of the liquids through each other will help to determine its *force*. When alcohol and water, for example, are separated by a septum composed of animal membrane, the endosmotic current will be from the water towards the alcohol, because the former liquid most readily 'wets' the membrane, and consequently tends most strongly to occupy its capillary passages: but, on the other hand, when the separation is made by a thin lamina of caoutchouc, the endosmotic current is from the alcohol towards the water, because the former is most readily imbibed by the septum.

417. It has further been ascertained by the experiments of Matteucci, that, when an organic membrane is employed as a septum, the rapidity of transmission is considerably affected by the direction in which the endosmotic current traverses the membrane. Thus when the skin of the Torpedo was employed, with a solution of sugar on one side of it, and water on the other, although there was always an endosmotic current from the water to the sugar, yet this current was strong enough to raise the interior liquid to 80 degrees, when the water was in contact with the *internal* surface of the membrane, in the same time that was occupied by its rise to 20 degrees, when the *external* surface of the membrane was turned towards the water. Again, when the mucous membrane of the stomach of a dog was used as the septum, and its *external* (or muscular) surface

was placed in contact with alcohol, the passage of water from the other side took place with such rapidity as to raise the liquid in the tube to 130 degrees; whilst if the *internal* (or mucous) surface of the membrane was placed in contact with the alcohol, and the muscular surface with water, the current was only sufficient to raise the liquid 6 degrees in the same time; so that it is evident that the transudation of water takes place much more readily from the mucous lining of the stomach towards the outer side of the viscus, than in an opposite direction, in virtue simply of the physical properties of the membrane. In fact, according to Prof. Matteucci, the cases are very rare in which, with fresh membranes, endosmose takes place with equal readiness, whichever of their two sides is exposed to the water. The direction which is most favourable to endosmose through skins, is usually from the internal to the external surface, with the exception of the skin of the frog, in which the endosmotic current, in the single case of water and alcohol, takes place most readily from the external to the internal surface. But when stomachs and urinary bladders are employed, the direction varies much more, according to the nature of the liquids employed. This variation appears to have some relation to the physiological conditions in which these membranes are placed in the living animal; thus, the direction most favourable to endosmose between water and a saccharine solution, is not the same for the stomach of a Ruminant as for that of a Carnivorous animal: as yet, however, no positive statement can be made on this subject. When membranes are employed, that have been dried or altered by putrefaction, we either do not observe the usual difference arising from the position of the surfaces, or endosmose no longer takes place; thus affording another indication, that it is to the physical condition of the perfectly-organised membrane that we are to look, for many of the peculiarities which are noticeable in the transudation of fluids through them.—The Exosmotic current does not bear any constant relation to the endosmotic, as may be easily comprehended from the preceding explanation; for if the liquids have a strong tendency to mutual diffusion, and the difference in the attractive power which the septum has for them respectively is not great, each may find its way towards the other, and a considerable exosmose may ensue, with very little change of level. The amount of the *ex*osmotic as of the *end*osmotic current, varies with the direction in which it traverses the membrane; thus when sugar, albumen, or gum was employed in solution, its transudation towards water took place most readily from the *internal* towards the *external* surface of all the skins examined by Matteucci;—a fact which is not without its significance, when it is remembered that it is in this direction that the secretion of mucus takes place on the skins of fishes, frogs, &c.

418. Applying these considerations to the phenomena of imbibition of liquids into the tissues and canals of the living body, we shall have to enquire how far they are capable of being accounted for on the physical principles which have been now brought forwards. It has been maintained by some that Absorption is a purely vital operation, because it does not occur save during the continuance of life. But this is not true; since imbibition will take place into dead tissues, though more slowly than into the same parts when living; and the difference of rate seems to be fully accounted for, by the difference of the condition between a mass of tissue all whose fluids are stagnant, and another in which an active circulation

is taking place. Thus, as Matteucci has shown, if the hind-legs of a frog recently killed be immersed for some hours in a solution of ferrocyanide of potassium, it will be found that every part of the viscera is so penetrated with the salt, that by touching it with a glass rod moistened with a solution of the chloride of iron, a more or less deep blue stain is the result. Now the same effect is produced much more speedily in a living frog; and it is easily proved that the imbibition takes place, in the latter case, into the blood-vessels, and that the salt is conveyed to the remoter parts of the body by the circulation, instead of having slowly to make its way by transudation through the tissues, as in the dead animal.—But further, not only does the movement of blood in the vessels promote the diffusion of liquid which has been already absorbed; it also increases the rapidity of the absorption itself, in a very extraordinary degree. Thus, if a membranous tube, such as a piece of the small intestine or of a large vein of an animal, be fixed by one extremity to an opening at the bottom of a vessel filled with water, and have a stopcock attached at the other extremity, and be then immersed in water acidulated with sulphuric or hydrochloric acid, it will be some time before the acid will penetrate to the interior of the tube, which is distended with water; but if the stopcock be opened, and the water be allowed to discharge itself, the presence of the acid will be immediately discovered (by tincture of litmus) in the liquid which flows out, showing that the acid has been assisted in its penetration of the walls of the tube, by the current traversing its interior. Thus the continuance of the Circulation is obviously one of the most potent of all the conditions of Absorption; and the difference in the rate of the process in dead and living organisms placed under the same circumstances, may be accounted for in great part, if not entirely, by the stoppage of the circulation in the former. All the circumstances which are laid down by Physiologists as favouring Absorption, are in strict accordance with the Physical principles which have been now explained. These circumstances are :—1. The ready miscibility of the liquids to be absorbed with the juices of the body. 2. The penetrability of the tissue through which the absorption takes place. 3. The absence of previous distension in the tissues or canals towards which the flow takes place. 4. The elevation of the temperature, within certain limits. 5. The vascularity of the tissue, and the rate of movement of the blood through the vessels. —And the results of experiments upon recently-dead membranes, which retain almost exactly the same physical conditions as those which they possessed during life, but have entirely lost their vital properties, seem most decidedly to indicate, that the relative facility with which different substances are absorbed, and the direction most favourable to their passage through the tissues, are determined in great part by the physical relations of those tissues (and of the vessels which traverse them) to the liquid which is seeking to enter them. In this way, then, many of the phenomena of *selective* absorption are probably to be explained, especially in Plants and the lower Animals; we shall hereafter see, however, that the special absorbent vessels of Vertebrata seem to possess properties which can scarcely be thus accounted for.

2. *Absorption in Vegetables.*

419. In the lowest orders of Plants, we find this function performed under its most simple conditions. Their substance is composed of vesicles more or less firmly united to each other, and but slightly altered from their original spheroidal form; and the envelope which surrounds them can seldom be regarded as a distinct structure, as it generally differs but little from the remainder of the cellular tissue. In all the *Algæ* (§§ 267–269), the whole surface appears to be endowed with the power of absorption to nearly an equal degree; and although the semblance of a stem and roots presents itself in the higher orders, yet these seem to have no other function than to give the means of attachment to the frondose or leaf-like expansion, which performs not only the nutritive but the reproductive function on all parts of its surface. The preference of particular species of Algæ for particular rocks, cannot be fairly considered as indicating that any absorption of the mineral particles takes place through their roots; it is much more probably due to the fact, that the materials of these rocks are in some degree diffused by solution through the neighbouring water; and something may be also attributable to the mechanical qualities of the rocks, as affording advantageous surfaces for attachment.—The difference in the situation inhabited by the *Lichens* (§ 270) appears to involve a separate appropriation of portions of their surface to the nutritive and reproductive functions, and a certain specialization of the absorbing organs. The upper surface of these plants, being exposed to the sun and air, becomes hard and dry, a condition which seems to favour the evolution of the fruit; whilst it is mostly by the lower surface, which is usually soft and pale, that the nutriment is introduced into the system. The latter is not unfrequently furnished with hair-like prolongations, which not only serve to fix the plant, but appear to be much concerned in the absorption of its aliment; being so much developed in some Lichens, which are located upon the ground, as almost to resemble roots.—In the *Fungi* we find the same evolution of the more special organ from the more general type. The lower forms of this group (§ 271) seem to imbibe their aliment by their whole surface; but in the more complex structures, in which the reproductive system is separated from the nutritive by the intervention of a stalk (as in the Mushroom, § 273), the *mycelium* at its base is prolonged into filaments, whereby the decaying matter, that constitutes the food of this remarkable group of plants, is introduced into the system. In some species, too, the whole surface is covered with hair, which may assist their very rapid development by absorption of fluid from the atmosphere.—In the *Mosses* and their allies (§ 276), we find a somewhat higher form of the same structure. From the base of the stem there usually proceed slender radical filaments, which sometimes ramify through the soil to a considerable extent; and other similar filaments are frequently developed from the sides of the stalk, and from the lower surface of the leaves. In Mosses that exist on rocks, however, these filaments are but little developed, and appear to serve rather for mechanical support, than for absorption of nourishment, which must in such circumstances be derived from the atmosphere through the leaves. These, as is well known, are very permeable to fluid, so that the application of moisture

will cause Mosses to recover the appearance of life after being long dried; and the same property enables these beautiful little plants to vegetate rapidly during a moist season, whilst their tenacity of life enables them to withstand a subsequent drought.—In ascending through these tribes of Cryptogamia, then, we may trace a gradual development of separate absorbent organs, and may observe the *specialization* of the function, by its restriction to one particular part of the surface, instead of being diffused over the whole. The absorbent filaments, however, are very inferior in their structure to true 'roots,' being little else than elongated cells, resembling the hairs with which other parts of the surface are covered; and they seem to absorb fluids equally throughout their entire length. Further, we find that, when these special organs are not developed, or are insufficiently supplied with nutriment, the general surface can take on its original function, and thus supply the deficiency.

420. It is in the *Ferns* (§ 277) that we first meet with a regular 'descending axis' of growth, from which the absorbent fibres are given off; and this is evolved in its completest form in the *Phanerogamia*. In these, moreover, it seems to be through the newly-formed succulent extremities alone, that fluid is admitted; and the function is, of course, more actively performed by them, in proportion to the diminution in the amount of surface they expose. The 'root' presents a great variety of forms in different plants; there are, however, some parts which are essential, and others that are merely accessory. The simplest form, as well as the most essential part, consists of single fibres; these occasionally exist alone (as at the base of 'bulbs'), but more often proceed from ramifying branches of woody texture (as in most trees and shrubs), or from 'tubers' (as that of the turnip). Each radical fibre is of a structure far more complex than the absorbent filaments just mentioned as existing in Cellular plants; for it is a cylinder, in whose axis lies a bundle of fibro-vascular tissue, whilst its exterior is composed of firm cellular parenchyma; at its free termination, however, the extremities of the vessels are covered with loosely-formed cellular tissue, through which the fluid passes into them. The *spongiole*, as this point has been termed, is sometimes spoken of as a distinct organ; but it is nothing more than the growing point of the root, which, with a few exceptions, lengthens only by additions to its extremity. The soft lax texture of the newly-formed part causes it to possess, in an eminent degree, the power of absorption; but as the fibre continues to grow, and additional tissue is formed at its extremity, that which was formerly the spongiole becomes consolidated into the general structure of the root, and loses almost entirely its peculiar properties. That it is to the spongioles that the principal absorbing power of the root is due, was fully proved by the experiment of Senebier. Having fixed two roots in such a manner, that the extremity of one was in contact with water, whilst of the other every part was immersed except the extremity, he found that the first root absorbed nearly as much as usual, whilst the second scarcely took up a sensible quantity. It is not improbable that the relative absorbent power of the spongioles and of the general surface of the root, may vary in different plants, according to the character of the texture of each, and the situation in which it grows; but it appears to be a general fact, that, in Vascular plants, the spongioles are the organs *specially* destined for introducing the fluid nutriment into the system.

421. There are evident limits to the supply of alimentary materials to the roots of Plants, as long as they remain in the same spot ; and some change must take place to ensure its continuance. As the Plant cannot remove itself to a new situation, its wants are provided for by the simple elongation of its radical fibres ; and their extension takes place, not by increase throughout their whole length, but by addition of fresh tissue to their points. This addition, being made in the direction of least resistance, enables the fibrils to insinuate themselves into the firmest soil, and even to overcome the obstacle presented by solid masonry ; for however narrow the crevice may be into which the filament enters, the subsequent expansion of the tissue by the infiltration of fluid is so great, as to enlarge the opening considerably, and even to rupture masses of stone. This tendency to increase in the direction of least resistance, will also evidently cause the root to grow towards a moist situation ; and by keeping this in view, many of the facts regarding the so-called *instinct* of plants, which at first sight appear so remarkable, may be satisfactorily explained. Thus, it was noticed, when the water of the New River was conveyed through wooden pipes, that if these pipes were carried within thirty yards of trees, they were very likely to be in time obstructed by their roots ; which 'found' the joints, and then spread out in 'foxtails' of fibres, two or three feet long. It is well known to the agriculturist, that the course of large drains, even at a considerable depth in the ground, is liable to be interrupted by the extension of roots, not only from trees, but also from apparently insignificant plants. Thus at Saucethorpe, in Lincolnshire, a drain nine feet deep was filled up by the roots of an elm-tree, which was growing at upwards of fifty yards from the drain : a deep drain outside the garden-wall at Welbeck was entirely stopped by the roots of some horse-radish plants, which grew seven feet into the ground : and at Thoresby Park, a drain fourteen feet deep was entirely stopped by the roots of gorse growing at a distance of six feet from it.*—In other cases, we must attribute the result to the dispersion of vapour through the atmosphere in a particular direction. Thus, in a case which fell under the Author's cognizance, a lime-tree, which grew at the distance of about fifteen feet from the shaft of a well, sent a single long root through the soil, in a direct line towards a point of the shaft at which there was a small aperture left by the deficiency of a brick ; this aperture was at a height of eleven feet above the usual level of the water in the well ; and having passed through it, the root divided into a brush-like mass of fibres, which descended into the water, and formed a large mass in the lower part of the well. Again, in a peculiar case known to the Author, in which one tree grew upon the trunk of another, having originated from a seed deposited at about twelve feet above the ground, one of the large roots which it sent down, subdivided about two feet above the surface of the ground, instead of proceeding directly down to it, as did all the rest. Now this subdivision took place above a large stone, on the centre of which the root would have impinged if it had continued to grow directly downwards ; and it would appear as if its division, half proceeding to one side, and half to the other, was due to the direction given to its growth by the ascent of vapour from the soil beneath. On the same principle

* "Journal of the Royal Agricultural Society," vol. i. p. 364.

we are probably to explain the following case. "Near the waterfall at the head of the river Leven, in the Western Highlands, is the trunk of a decayed oak, rotten within, but alive on some parts of the outside. From one of these, a shoot grows out, about fifteen feet from the ground; and this shoot has protruded from its lower part a root, which, after having reached the ground (a bare rock), runs along the rock in a horizontal position, about thirty feet further, till it reaches a bank of earth in which it has embedded itself." *

422. The absorbent power of the Spongioles appears limited by the size of their pores; for if the roots be immersed in coloured solutions, they take up the most finely divided particles, leaving behind the larger molecules, which are only absorbed when the spongioles have been damaged. The pores are liable to be blocked up by fluids which are of too viscid or glutinous a consistence to pass readily through them; and if the roots be immersed in a thin solution of gum or sugar or neutral salts, the watery particles are absorbed in the greatest degree, so that the portion which is left contains a larger proportion of the ingredient in solution. The power of *selection*, however, would seem to extend beyond this: since of two substances equally dissolved, some plants will take one, and some the other; whilst some neutral salts are rejected altogether. It does not appear that the selecting power is employed to prevent matter, which is capable of exerting a deleterious influence upon the plant, from being introduced into its tissue; for many substances are taken up by the roots, which speedily put a stop to vital action, if opportunity be not afforded for their excretion. From the little that is at present known on the subject, it seems a reasonable inference, that the rejection of any particular ingredient of the fluid in contact with the roots, results either from an organic change effected by it on their delicate tissue, such as is proved by the experiments of M. Payen † to occur when tannin enters into the solution, even in very minute proportion; or from the want of molecular attraction between its particles and the substance of the spongioles. That some kind of relation between the living tissue and the crystalline character of the salt, is concerned in the selection, appears from the interesting results of the experiments of Dr. Daubeny on the absorption of mineral substances by Plants. He has found, that, if a Plant naturally absorbs the compounds of any particular base, it will also take up those of another base which are *isomorphous* with them: for instance, most vegetables will absorb the salts of Lime and Magnesia with equal readiness; but salts, however soluble, which have a crystalline arrangement different from theirs (such as those of Strontia), are not absorbed.—The *quantity* of fluid absorbed, and the force with which it is propelled upwards in the stem, vary not only in different species and individuals, but in the same plant at different periods of the year, and even of the day. The former seems intimately connected with the activity with which the other processes of vegetation are being carried on, and especially to depend upon the quantity of vapour transpired from the leaves (CHAP. XIV.); all the causes which increase exhalation, may therefore be considered as stimulants to absorption also. The force of the roots in the propulsion of the sap is sufficiently proved by the celebrated experiments of Hales on

* "Gardeners' Magazine," Oct. 1, 1837.

† "Annales des Sciences Naturelles," Deuxième Série, Botan., tom. iii. p. 5, &c.

the vine. By gages affixed to the stem during the "bleeding season," when the sap rises rapidly, he found that a column of mercury 26 inches high, equal to a column of water of nearly 31 feet, might be supported by the propellent force of the absorbent organs ; but if the upper part of the plant was cut off, this power soon diminished, and after a time ceased altogether.

423. There would seem much reason to believe, that the mere act of Absorption, in this and other cases, is due to the physical property already referred to as possessed by many organised tissues,—viz. the capability of producing Endosmose (§ 415). The succulent extremities of the spongioles serve as the medium required for this process ; but it may be reasonably enquired whence the other condition is furnished, namely, that difference in density of the fluids on the opposite sides of the septum, which is necessary for the commencement and continuance of the action. This is supplied, in the first instance, by the store of nutritious matter obtained by the embryo from its parent, and contained within its tissues ; and, possibly, at a later period, when the plant is supporting an independent existence, by the admixture of a portion of the dense elaborated sap, with the crude and watery ascending fluid. If this be the true explanation of the phenomenon, a counter-current ought to exist, and an *exosmose* of the fluids within the system should take place into the surrounding medium. That this is actually the case, would appear from the fact, that an excretion of the peculiar products of the species may be often detected around the roots of the plant ; a fact of very important practical application. The cessation of this action of admixture (a change evidently depending upon other vital actions) at the death of a plant, fully accounts for the non-continuance of endosmose, which is also checked if the superincumbent column of fluid be not drawn off by the leaves. It has been very justly remarked by Professor Henslow, that, "if we suppose the plant capable of removing the imbibed fluid as fast as it is absorbed by the spongioles, then we may imagine the possibility of a supply being kept up by the mere hygroscopic property of the tissue ; much in the same way as the capillary action of the wick in a candle maintains a constant supply of wax to the flame by which it is consumed." *

424. It is an axiom in Vegetable Physiology, which has been laid down by De Candolle, "that when a particular function cannot, according to a given system of structure, be sufficiently carried into effect by the organ which is ordinarily appropriated to it, it is performed wholly or in part by another." This is a single case of the general principle which has been already laid down (§ 351) ; and the reason that it is more evident in the Vegetable than in the Animal kingdom, is simply, that in the former the specialization of function is nowhere carried so far as in the latter ; so that any part of the general surface of a plant can perform in a considerable degree the functions of all the rest. We might then *a priori* expect, that whilst the roots are, in the usual condition of the perfect plant, the organs by which its fluid nutriment is absorbed, and the leaves its organs of transpiration and respiration, some traces of the primitive community of function enjoyed by the general surface of the simpler tribes, would be found in the capacity of each of these organs to perform in a certain

* "Cabinet Cyclopædia: Botany." p. 177.

degree, if required, the function of the other. Thus, it is evident that when the roots are either absent or imperfect, or are implanted in an arid or barren soil, serving merely to fix the stem (as happens with many *Orchideæ* and the generality of aerial parasites), the plant must derive its chief supply of nutriment through the absorption performed by the leaves, or, in leafless plants (as the *Cacteæ*), through the general surface. And it must be obvious to all who have observed the manner in which plants, faded by the intense action of light and heat, are refreshed by the natural or artificial application of moisture, that absorption takes place, in these instances also, by the general surface, as well as by the roots.—Various experiments have been devised, with the view of determining the relative extent to which the plant is supplied by these two channels; but the proportion appears to depend upon the circumstances of its growth. Thus, Bonnet took some specimens of *Mercurialis,* and immersing the roots of part of them in water, he placed others so that only their leaves touched the fluid. A small shoot of each plant was kept from contact with water; and after the experiment had proceeded for five or six weeks, those which had derived all their nutriment through the leaves were nearly as vigorous as those which had imbibed it by the roots. It is by the under surface of the leaf, where the cuticle and cellular tissue beneath are least compactly arranged, that absorption is performed with the greatest rapidity; and the downy hairs with which some plants are plentifully furnished, seem to contribute to this function, acting like so many rootlets. These prolongations of the surface are usually wanting in such plants as grow in damp shady situations, where moisture already exists in abundance; but in hot, dry, exposed localities, where it is necessary that the plant should avail itself of every means of collecting its food, we find the leaves thickly set with them; and this diversity may be observed in different individuals of the same species of plant, according to the soil and climate in which they exist, and even in the same individual if transplanted.

425. In tracing the gradual evolution of the special Absorbent apparatus of the more perfect Plants, we may observe many interesting relations between the progressive stages of its development, and the permanent forms of the system in the lower orders. Thus, the embryo at its first appearance within the ovule (CHAP. XVIII.) is nothing but a single cell, like that of the *Protococcus,* in the midst of the store of semifluid nutriment prepared by its parent, which it gradually absorbs by its whole surface, just as do the simplest cellular plants. At the time of the ripening of the seed, we find the rudiment of the future root, which is developed during germination; but in the early stages of this process, the radicle simply prolongs itself into the ground, and appears to be equally capable of imbibing moisture through its whole length, like that of the Liverworts or Mosses. It is not until the true leaves are evolved, that the root begins to extend itself by ramification; then first protruding perfect fibrils, composed of woody fibre and vessels, and terminated by spongioles. —Thus, then, in the development of the Absorbent system of Vegetables, the first which we have been called upon to study in detail, we have perceived the application of the laws which have been already enunciated (CHAP. VIII.); for it has been found that, whether we trace its various forms through the ascending scale of the different tribes of Plants, or watch the progress of its evolution in the more perfect orders, it is con-

stantly to be observed that the *special* structure and function arise by a gradual change out of one more *general ;* and that, even where the *special* organ is most highly developed, the *general* structure retains, in some degree, the primitive community of function which originally characterised it.

3. *Absorption in Animals.*

426. It has been shown in the preceding chapter, that the conditions under which the function of Absorption is performed in Animals, are so far different from those which obtain in Plants, that a preparatory process of *Digestion* becomes necessary in the former, for the reduction of the food to the fluid form required for its entrance into the system. This process is effected in cavities of the body, which are bounded by a continuation of its external surface, modified, by its secreting power, to supply the means necessary for the solution of the aliment, and, by its absorbent faculty, for the selection of the part of it capable of contributing to the nutrition of the fabric. But, so long as this aliment remains unabsorbed, it cannot be regarded as introduced into the system ; since it merely holds the same relation with the absorbent vessels, as the nutritious fluid in which the roots of plants may be immersed, bears to the ducts which they enclose. This is brought into clear view by the remarkable fact, that the poison of the most venomous Serpents, and the Woorara-poison of the South American Indians, of either of which a very small quantity will produce death when it is introduced by the minutest puncture or scratch into the current of the circulating fluid, are perfectly innocuous when taken into the stomach ; the mucous membrane of the alimentary canal apparently having a peculiar inaptitude for allowing them to penetrate by imbibition, either into the blood-vessels or the absorbents. Some experiments recently made upon the latter substance, by MM. Bernard and Pelouze, are peculiarly interesting, as confirming the statement already made (§ 422), that the absorption of particular substances, however favourable their condition may appear, may be prevented by the simply-physical conditions of the membrane which they have to traverse.*—It

* That the absence of poisonous effects from the *Woorara* poison, when it is simply introduced into the stomach, does not arise from any modification effected in its properties by the agency of the gastric juice, is shown by the fact that the poison, after digestion in that fluid for 24 or 48 hours, remains as virulent as ever ; whilst the gastric fluid to which Woorara has been added, loses none of its solvent power. The various secretions which make up the intestinal juices, have been experimented on with the same results. Hence it appears that the cause of the innocuousness of the poison, under these circumstances, must be looked for in the gastro-intestinal mucous membrane, which will not give passage to the active principle of the poison, soluble as this is. Experiment proves this to be the case. If the gastric mucous membrane of a recently-killed animal be adapted to an endosmometer, so that the mucous surface looks outward, and the endosmometer containing sugared water is then placed in a watery solution of woorara, endosmose will have been found to have taken place in three or four hours, for the liquid will have risen in the tube, and yet this will contain no trace of woorara, as may be ascertained by inoculating with it. If the experiment were allowed to go on for a much longer time, the endosmosis of the poison might occur ; but we should then find that the mucous membrane had undergone modification, the mucus and epithelia covering it being altered, so that imbibition and endosmosis of the woorara becomes possible ; and if, in place of taking a quite fresh mucous membrane, we take one that has undergone some change, the endosmosis of the poisonous fluid occurs instantly.

As it was interesting to ascertain whether other mucous membranes possessed this resisting power, those of the bladder, nasal fossæ, and eyes were tried, and constantly with the same results. An injection was retained in the bladder without inconvenience, for from six

has further been shown that the introduction of the nutritive material into the system, is effected in the lowest animals by simple imbibition into the tissues that surround the digestive cavity, and by percolation through them towards the more remote parts; and where such is the case, the digestive cavity either itself occupies a very large part of the body, as in the Polypifera, or prolongations of it, in the form of canals, extend to the parts remote from the principal cavity, as in many of the Acalephæ. In the higher Radiata, however, as well as in all the Articulata (save the lowest Entozoa) and Mollusca, we find that the conveyance of nutriment to the organism at large is specially accomplished by the agency of the circulating system, some part of which is in direct relation with the alimentary canal. Thus, in the *Holothuria* (Fig. 113), one of the principal vascular trunks runs parallel with the intestinal canal for a considerable part of its length, and receives from it the liquid imbibed by the venous ramifications which are copiously spread out upon its surface; whilst on the other hand, in the *Pycnogonidæ*, nearly the whole alimentary canal is included in what may be considered as a great venous sinus, into which the nutritive material transudes from its cavity, and by which this is conveyed throughout the body (Fig. 158). In most Invertebrated animals, these two plans seem to exist in combination, one or the other being predominant in different groups; but in all cases, the nutritive fluid, when withdrawn from the alimentary canal, is introduced at once into the general current of the circulation; and this introduction would appear to be effected simply by endosmotic imbibition.

427. In the Vertebrata, however, it is by *vessels* alone, that the nutritive fluid is removed from the alimentary canal, the walls of which do not allow it to transude into the surrounding cavity. And we here find provided, in addition to the blood-vessels that are copiously distributed upon the coats of the gastro-intestinal canal, a special set of *Absorbent* vessels, which seem to be set apart for the introduction of *nutritive* materials into the system, and for submitting these to a certain degree of preparation, before they are admitted into the current of the circulation. The 'Absorbent System' of vessels consists of two principal divisions, which may be compared to two sets of roots proceeding from a common trunk; one of these commences upon the walls of the intestines, and is termed the 'lacteal' system, from the milky character of the 'chyle' which it contains; whilst the other takes its origin in various parts of the substance of the organism at large, especially in the skin and subcutaneous textures, and is known as the 'lymphatic' system, from the transparent watery aspect of the liquid it conveys.—Although the walls of the whole gastro-intestinal canal are furnished with Absorbent vessels, in common with other musculo-membranous parts, yet it is on those of the small intestine, below the point at which the liver and pancreas discharge their secretion, that the *Lacteals* especially abound; and they seem, in the higher Vertebrata at least, most commonly to originate in the interior of the villi (§ 169), where they are surrounded by the plexus of capillary blood-vessels that lies immediately beneath the external surface of these fila-

to eight hours, by a dog; but the urine it passed after this time had all the toxical properties of woorara. One mucous membrane alone is excepted from this immunity, the pulmonary; for the poison, when applied to it, produces the same effects as when introduced into the subcutaneous cellular tissue.—" L'Union Médicale," 1850, No. 125.

mentous processes (Fig. 21). In Fishes, however, the villi are few or altogether absent, and the lacteal trunks receive their supplies through a coarse plexus situated in the walls of the intestinal canal. Such a plexus is seen also in Reptiles, in which villi are developed ; and it seems probable that it exists even in higher Vertebrata, in which the mucous membrane is much more thickly set with villi, and in which it appears to be chiefly through the lacteals contained in these, that the chyle gains admission into the larger trunks. The precise mode in which the lacteals commence near the free extremities of the villi, cannot be stated with certainty ; but it is probable that they form loops by anastomosis with each other (Fig. 32), so that there is no proper free extremity in any case. It is beyond all doubt, however, that the lacteals never commence by orifices upon the internal surface of the intestine, as was formerly imagined. When these vessels are turgid with chyle, the extremity of each appears to be imbedded in a collection of globules, presenting an opalescent appearance, which give to the end of the villus a mulberry-like form ; and it is supposed by Prof. Goodsir, by whom this circumstance was first observed, that these globules are cells, which are developed during the act of absorption, from granular germs that may be seen in the same situation during the intervals of the process (Fig. 32) ; and that these cells, drawing to themselves during their growth certain of the nutritive materials contained in the intestines, are thus the real agents in the *selection* of the substances that are to be introduced into the lacteals, delivering them to these, by the rupture or deliquescence of their walls, as soon as their term of life is completed.* It is further held by Prof. Goodsir, that the epithelial cells which cover the extremities of the villi (Fig. 32) fall off during the process of absorption, so as to leave the villi more free to imbibe the fluids in contact with their surface ; and thus that a new set of absorbent cells is developed with every recurrence of the act of absorption, and a new set of protective epithelial cells in the subsequent interval. These views have been partly confirmed by other observers ; but there is still considerable doubt whether the supposed 'absorbent cells' are anything else than oil-globules : and it is maintained by Prof. E. Weber and others, that, so far from the epithelium falling off as useless or obstructive during the act of absorption, it is by these epithelial cells that the selection is first made, the chyle which *they* have absorbed being afterwards transferred to the cells within the villi, or at once to the lacteals. Some, indeed, have argued that the liquid contained in the lacteals has been first absorbed by the blood-vessels, and has been eliminated from them by a process resembling glandular secretion ; and this view would seem to derive confirmation from the fact, that if the circulation be stopped in the blood-vessels of any part of the intestinal canal, the introduction of chyle into the lacteals is checked also. But if a process of cell-development be required for the selection of the materials to be received by the lacteals, a sufficient reason is furnished for the dependence of the absorption of chyle upon the continuance of the circulation ; since the growth of the cells within the villi could scarcely take place, without a continual supply of elaborated blood to form their *walls* and maintain their vital activity, although their *contents* are drawn from the intestine.—The whole question of the precise

* "Anatomical and Pathological Observations," pp. 5—10.

mode in which the contents of the lacteals find their way into them, must be regarded as at present *sub judice.*

428. The *Lymphatic* vessels are distributed in the greater number of tissues and organs which possess vessels for the conveyance of blood; but to this general statement, there are some remarkable exceptions; for they appear to be entirely wanting in the substance of the brain and spinal cord, although they are found in their investing membranes; they have not been detected in the interior of the eye, nor in the placenta and umbilical cord; and it is doubtful whether they accompany blood-vessels in the Haversian canals and cancelli of bone. That they are wanting in cartilage, the epidermic appendages, and other non-vascular parts, follows naturally from the general principle just stated. It appears to be in the skin and subcutaneous textures, at least in Man, that the lymphatics are most plentifully distributed; and they seem there to originate, like the lacteals of the intestinal walls, in plexuses of which the meshes are very close. According to Prof. Kölliker, the lymphatics in the tail of the Tadpole do not form a network, but branch out like rootlets, their ultimate extremities, or rather their commencing radicles, having free but closed ends, running out into fine points;* it may be doubted, however, whether this is not the result of a want of completeness in their development, and whether these ramifications would not meet and inosculate in the fully-developed Frog. Like the capillary blood-vessels (§ 221), they take their origin in stellate cells, which send forth long projections; but these branches do not inosculate with each other in any other way than to form continuous tubes; and if they subsequently constitute a plexus, it must be by the development of connecting arches at a later period.—It has been maintained that the minute lymphatics communicate with the capillary vessels in their neighbourhood; and these communications have been supposed by some to allow the direct transmission of the lymph into the sanguiferous system; whilst by others it has been inferred that the *liquor sanguinis,* or fluid portion of the blood, (which, when diluted, closely resembles the contents of the lymphatics in its composition) finds its way into the absorbents. It is nearly certain, however, that no such apertures exist; and that when any direct passage of fluid does take place from one set of vessels into the other (as is often the case in artificial injections, and was noticed by Prof. Kölliker in watching the circulation in the tail of a Tadpole which had been injured), it is through an abnormal opening.

429. The walls of the Absorbent vessels (whether lacteals or lymphatics) are extremely thin, so that the character of their contained fluid can be readily discerned through them. Those which form the ultimate ramifications of the system appear to be limited only by a very delicate homogeneous membrane; and it is not certain that even this is universally present in the absorbents of Fishes. A similar membrane, covered with a layer of pavement-epithelium upon its inner or free surface, constitutes the lining of the mid-sized and larger trunks; but these possess, in addition, a middle or fibrous layer, in which non-striated muscular fibres may be distinguished, and an external sheath of areolar tissue. In the higher Vertebrata, the Absorbents are furnished with valves, which allow the passage of their contents in only one direction, namely, from their origin towards their termination in the sanguiferous system; these valves,

* "Annales des Sciences Naturelles," Zool. (1846), tom. vi., p. 98.

however, seem to be wanting in the 'plexuses of origin,' which are filled by mercury injected into any part of them. In Fishes and Reptiles, however, in which this system is obviously developed upon an inferior type, the valves are few or are altogether wanting.—In the higher Vertebrata, again, certain small solid bodies are found in the course of the lacteals and lymphatics, which are termed 'absorbent glands.' The structure of these, however, does not correspond with that of ordinary *glands*; and they have been more appropriately named 'ganglia,' for they are essentially composed of plexuses of absorbent vessels, convoluted (so to speak) into knots, and dilated into larger cavities, amongst which capillary blood-vessels are minutely distributed; the whole being bound together by areolar tissue, and invested in a capsule of the same. These blood-vessels have no direct communication with the interior of the lacteals, but are separated from them by the membranous walls of both sets of tubes; so that whatever passage of fluid normally takes place from one set of vessels to the other, must be accomplished by transudation through these.* According to the observations of Prof. Goodsir, the absorbents, when they enter a gland, lay aside all but their internal coat and epithelium; and the latter, in place of retaining its pavement-like character, presents itself as an irregular layer of spherical nucleated corpuscles, measuring about 1-5000th of an inch in diameter (Fig. 200, B); which layer is thickest in the dilated lymphatics which form the 'cells' of the centre of the gland, and

FIG. 200.

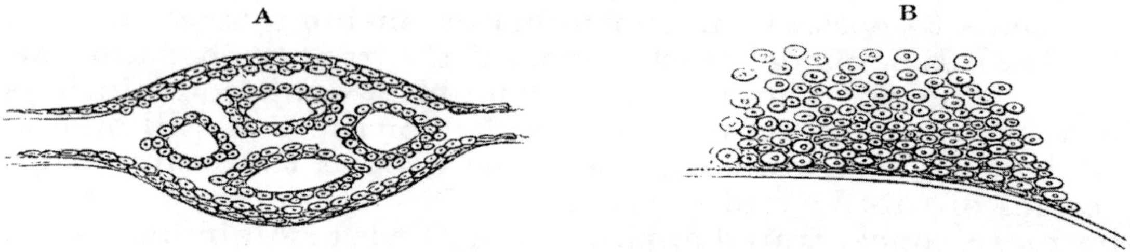

A, Diagram of a lymphatic gland, showing the intra-glandular network, and the transition from the scale-like epithelia of the extra-glandular lymphatics, to the nucleated cells of the intra-glandular.—B, Portion of intra-glandular lymphatic, showing along the lower edge the thickness of the basement-membrane, and upon it, the thick layer of glandular epithelial cells.

becomes gradually thinner towards the periphery, where it is continuous with the epithelium of the afferent and efferent vessels (A). The inner layers of the central epithelium appear to have no tenacity; so that the component cells may be readily detached from one another, and carried off in the fluids which traverse the cavity.—Having thus considered the general structure of the Absorbent system, we shall proceed to notice its more special peculiarities in the different classes of Vertebrata.

430. The proper Absorbent system is exhibited in its simplest and most diffused form in *Fishes*, the lowest class in which its existence has been demonstrated. Where it consists of distinct vessels, their walls are very thin and distensible. The Lacteals commence in a somewhat coarse network of canals, that seem channelled-out (as it were) beneath the mucous

* It has been asserted by many anatomists, that free apertures exist, by which the contents of one set of vessels can pass directly into the other; but these statements are founded on the results of injections, which can be very easily *forced to make* such apertures; and the most careful examination has failed to detect them, when no such procedure has been employed.

lining of the intestinal canal, and cannot be shown to possess definite walls; from these, however, the chyle is conveyed away by proper vessels, which form capacious plexuses in the mesentery, along the course of the alimentary canal. The lymphatics are distributed extensively through both the superficial and deep-seated parts of the body; and they, also, by the convolutions and anastomoses of their trunks, form numerous plexuses in different parts of the body, especially around the veins, which may be regarded as the first indications of the so-called 'glands' that are presented in the higher classes. Some of these trunks, moreover, dilate into sinuses, which appear to be contractile, and pre-figure the 'lymphatic hearts' of Reptiles. Such a sinus may be seen in the tail of the Eel; and another is found on each side of the cranial cavity of many Fishes, externally to the jugular veins. Although a considerable proportion of the lymphatic trunks unite with the lacteal vessels, to form principal canals (corresponding with the thoracic duct in higher animals), which empty their contents into the systemic veins near the heart, there are many other communications between the two systems, as Fohmann appears to have satisfactorily demonstrated. Thus, the caudal sinus of the Eel discharges its contents into the caudal vein, the orifice being provided with a valve.

431. The conformation of the Absorbent system presents several interesting peculiarities in the class of *Reptiles*. As in Fishes, the vessels are generally destitute of valves, though these may occasionally be observed in the larger trunks; but they everywhere seem to possess distinct walls. When compared with that of Birds and Mammals, the absorbent system of Reptiles seems to possess an enormous extension; large and capacious lymphatic plexuses being developed around the great veins, and the length of the trunks being often augmented by doublings and convolutions. But this extension is rather apparent than real; for there is still an absence of the 'glands,' which seem to concentrate, as it were, the assimilating power of a long series of tubes; and the relation of the Absorbent system of Reptiles to the more concentrated apparatus of Birds or Mammals, thus comes to resemble that which the extended tracheal system of the Insect bears to the lungs of the higher Vertebrata.—The *lacteals* in Reptiles, as in the classes above them, partly commence in the 'villi' of the intestinal canal; but, as in Fishes, there is also a very coarse plexus of absorbents beneath its mucous lining. The fluid which they absorb is collected into a *receptaculum chyli*, situated at the root of the mesentery; and from this it passes by two or more ducts, which also receive many of the lymphatic trunks, into the great systemic veins. The *Lymphatic* portion of the system is furnished, in most Reptiles, with certain pulsating dilatations, or *lymphatic hearts*, which aid in the propulsion of the contents of the vessels; the walls of these contractile cavities are formed of striated muscular fibres. In the *Frog* there are two pairs of these; one situated just under the skin, through which its pulsations are readily seen in the living animal, immediately behind the hip joint; while the other pair is more deeply seated at the upper part of the chest. The former receive lymph from the posterior part of the body, and pour it into the veins proceeding from the same part, by orifices furnished with valves; the latter collect that which is transmitted from the anterior part of the body and head, and empty their contents in like manner into the jugular vein. Their pulsations are totally independent of the heart and of the acts of

respiration, since they continue after the removal of the former, and for an hour or two after somatic death and the complete dismemberment of the animal. Neither are they synchronous with each other on the two sides of the body, nor always performed in the same space of time; for the pulsations are not only generally irregular, but sometimes exhibit long and frequent intermissions; when in constant action, they occur about sixty times in a minute. From the observations of Volkman, however, it appears that these movements are dependent upon the connection of the lymphatic hearts with the corresponding segments of the spinal cord; since they cease when these are destroyed, although all other parts may be left uninjured; and continue so long as this connection remains perfect, although all other parts of the nervous centres be destroyed.*—A pair of vesicles similar to the posterior pair in the Frog, has been detected in *Salamanders, Lizards*, and *Crocodiles*, in which they are situated near the root of the tail, and are connected, in like manner, with the veins of the lower extremity; they have also been discovered in *Serpents*, where they lie under the last rib. It was for some time believed that the *Chelonia* formed an exception to the general fact of the existence of these pulsating receptacles; but it has been shown by Müller that they are particularly large in that group, in which they lie behind the superior extremity of the iliac bones, receiving the lymph by capacious trunks from the posterior extremities, and pouring it into veins that discharge themselves into the reno-portal system.†

432. In *Birds*, we find the Absorbent system existing in a more perfect form; its trunks being everywhere provided with valves, and the diffused plexuses being partly replaced by 'glands' or 'ganglia,' which may be probably considered as performing the same function by an organisation of more concentrated character. The *lacteals*, which are not furnished with these glands, all converge towards a *receptaculum chyli*, from which proceed two thoracic ducts, one on either side, to terminate in the angles formed by the junction of the jugular and subclavian veins. These ducts receive, also, most of the lymphatic trunks; but the lymphatic system has two other communications, as in Reptiles, with the veins of the lower extremity. These are connected with two large dilatations of the lymphatic trunks, which are evidently analogous to the lymphatic hearts of Reptiles, but which do not seem to have any power of spontaneous movement. In the Goose, they are about the shape and size of a kidney-bean, and are situated in the angle between the tail and the thigh. They were supposed by Panizza to possess an automatic power of alternate contraction and dilatation; but this motion has been shown by Müller to be due to the respiratory actions,—being synchronous with them, and ceasing when they are interrupted.

433. In the Absorbent system of the *Mammalia*, we witness its most concentrated and highly-developed form; the vessels are copiously provided with valves; and their parietes are firmer than in the lower classes. Instead of the extensive plexuses of Fish, we find small dense 'glands' disposed in different parts of the system; these are more numerous than in Birds, and present themselves on the lacteals as well as on the lymphatics, being known in the one case as the 'mesenteric,' and in the other as 'lymphatic' glands. In some Mammalia, especially of the order

* "Müller's Archiv.," 1844. † Ibid. 1840.

Carnivora, the mesenteric glands cluster together into a single mass, named the *pancreas Asellii,* which lies at the root of the mesentery. The lacteals all discharge their contents into the *receptaculum chyli,* which is situated in the lumbar region, close to the spine; into the same receptacle many lymphatic trunks pour the fluid which they have collected from the posterior part of the trunk and extremities; and from it arises the thoracic duct of the left side, which passes forwards along the spine, receiving other lymphatic trunks in its course, and terminates at the junction of the left jugular and subclavian veins. A smaller trunk on the right side receives the lymphatics of the right side of the head and upper extremity, with those of the right lung and right side of the liver; and this terminates, in like manner, at the junction of the right subclavian and jugular veins. It is a beautiful instance of mechanical adaptation, that as the angle formed by the convergence of two veins is a point of much less resistance than any other part of the walls of the vessels, the easiest possible entrance is thus provided for the fluid discharged from the thoracic ducts into the current of the circulation. Although these are the only two canals by which the Absorbents usually communicate with the veins in Man, their number is greater in many species of Mammalia; they all terminate, however, in the same part of the venous system. Thus, the left thoracic duct often resembles rather a plexus of vessels than a single tube; branches proceeding from it and then reuniting, and at last terminating in the veins by several apertures. Sometimes it consists throughout of two tubes, which anastomose with each other and with the duct on the right side, and terminate separately in the veins; and in the Pig a branch of communication is sent off to the vena azygos, which is a small trunk running in proximity with it along the spinal column. All these modes of distribution occur as irregularities of conformation in the human subject, the former not being uncommon; the last, however, is rare.

434. We have now to enquire into the cause of the onward movement of the contents of the Absorbent vessels in the higher Vertebrata, which do not possess 'lymphatic hearts;' and also of the flow of fluid towards these pulsating cavities in the animals which possess them, situated (as they are) close to the points where the fluid of the absorbents is discharged into the veins. A part of the movement may be attributed to the *vis a tergo,* produced by the continual entrance of fresh fluid into the rootlets (so to speak) of the vascular tree; and although it may be thought, from the extreme distensibility of the walls of the absorbents, that this force will be rather expended in dilating them, than in pushing onwards the column of liquid which they contain, yet it must be remembered that they are surrounded by tissues, whose tonicity, during the living state, gives to them much more resisting power than they possess after death. Further, in all the moveable parts of the body, assistance is doubtless afforded by the occasional pressure which will be exercised upon the absorbents by the surrounding tissues; for while this pressure is operating, it will tend to empty them of their contents, which are only permitted by their valves to pass in one direction; and when the pressure is relaxed, they will be re-filled from behind. But it seems probable that the regular propulsion of the fluid mainly depends upon an alternate contraction and dilatation of successive portions of the vessels, slowly repeated at intervals; such alternations having been witnessed by Prof. Kölliker in the tail of the

tadpole; and it being apparently by such contraction, without subsequent dilatation and re-filling, that the absorbents are emptied after death, and this with considerable rapidity.

435. We have now to enquire into the relative parts which are performed in the function of *Absorption*, by the proper Absorbents, and by the Blood-vessels; and although these cannot yet be said to be precisely definable, yet there can be little doubt that we are in possession of the general truth with regard to them.—From the time when the Lacteal vessels were first discovered, down to a comparatively recent period, it was supposed that they constitute the channel through which *all* fresh nutritive material is taken into the body; and this idea seemed to derive confirmation from the great uniformity in the composition of the chyle, which, though different as a whole from that of blood, appeared adequate to supply those substances to the latter, which are most constantly being eliminated from it by the nutritive and respiratory operations. Various considerations, however, would lead to the conclusion, that the nutritive materials do not enter through the lacteals alone; but that the most soluble portions of them, together with other substances not nutritious, find their way directly into the blood-vessels.—In the Invertebrata, it will be recollected, no special Absorbent system exists, and in all those which possess blood-vessels, it is by them alone that alimentary matters are introduced into the system; it would not seem likely, therefore, that this most general method of performing the function should be entirely superseded in Vertebrata by the more special one. But further, when the extraordinary vascularity of the whole gastro-intestinal membrane is considered, together with the peculiarity of the special distribution of the capillaries in the villi; and when it is remembered also that the rapid movement of blood through these, creates the condition most especially favourable to the passage of liquids into them from the outside (§ 418), it might be almost certainly affirmed that endosmose *must* take place between the contents of the alimentary canal and the blood in the vessels.—This conclusion has been confirmed by numerous experiments. Thus it was ascertained by MM. Tiedemann and Gmelin, that when various substances were mingled with the food, which might be easily detected by their colour, odour, or chemical properties,—such as gamboge, madder, camphor, musk, asafœtida, and various saline substances,—they were seldom found in the chyle, though many of them were detected in the blood, and some had even passed into the urine. So, again, it was found that if any of these substances be introduced into a portion of the intestine separated by ligatures from the remainder, and all the vessels of that portion be divided or tied save its artery and vein, the substance may speedily be detected in the blood, and if it be poisonous its effects are manifested nearly as soon as usual; whilst, if the blood-vessels be tied, the lacteals being left entire and uninterrupted, a long period elapses before there is any evidence of absorption. It cannot be doubted, then, that alimentary substances in a state of solution, such as albumen, gelatine, or sugar, may pass into the blood-vessels by simple endosmose, when the relative densities of the blood and of the intestinal liquids are such as to favour the inward current. Conversely, it might be inferred that if the liquid in the intestines be of a nature to determine the endosmotic current in the contrary direction, some of the constituents of the blood would be

drawn from them into the alimentary canal. Now it has been shown by the experiments of Poisseuille, that an endosmotic current takes place through animal membranes *from* the serum of the blood towards certain saline solutions; and thus it happens that when these are taken into the alimentary canal, they produce a copious exudation of fluid from its walls, which fluid contains a considerable quantity of albumen.—It appears, then, that an interchange between the contents of the blood-vessels and those of the alimentary canal takes place, in either direction, in a manner which is in all respects conformable to physical principles; so that Absorption must be effected through the blood-vessels of the higher animals, as of the lower.

436. On the other hand, the Lacteals, as already pointed out, would appear to receive only substances of a particular class, more especially fatty matters in a state of more or less fine division, with which albuminous compounds are intimately mixed. These, whether received at once into the lacteals by the instrumentality of epithelial and other cells at the extremity of the villi, or first imbibed by the blood-vessels which enclose the lacteals, and then transferred to the latter, are doubtless obtained almost directly from the food; for the quality of the Chyle depends upon that of the aliment last digested, it being of an opaque white if that food contained much oily or fatty matter, and of a more transparent aspect if such matters were deficient. The experiments of Sandras and Bouchardat have shown, that particular kinds of fatty substances with which animals may have been fed, are recognisable in the chyle; and on the whole it may be considered, that the special function of the lacteals is to take up the oleaginous portion of the food, and to bring it into that intimate relation with albumen which seems to be requisite for its subsequent assimilation (§ 560). Still it appears by no means certain that the blood-vessels also may not absorb oleaginous substances, as they do in Invertebrated animals; particularly as it has been shown by the experiments of Prof. Matteucci,* that if an emulsion be made by shaking a few drops of oil in water to which a little alkali has been added, and an endosmometer filled with a weak alkaline solution be then immersed in this emulsion, the oil penetrates in a short time through the membranous partition, and makes its appearance in the interior of the endosmometer. Now as the blood is slightly alkaline, and as its density favours the imbibition of fluid, there seems no reason why fatty matters should not thus find their way into the blood-vessels, when they have been reduced to a state of fine division in the alimentary canal, and have been rendered alkaline by the admixture of the biliary and pancreatic fluids.

437. With regard to the relative share taken by the Lymphatics and the blood-vessels, in the absorption of fluids by the external surface or from the closed cavities of the body, and in that "interstitial" absorption which removes the solid particles of the fabric when they no longer retain their vital endowments, there has been a yet greater amount of misapprehension. It was long imagined (and the doctrine was strongly sustained by John Hunter and his immediate followers), that the office of the Lymphatic system is to take up and remove all the *effete* matter which is to be cast out of the body, as no longer adapted to form part of it, and as inconvertible into any other useful product. For such an idea,

* "Lectures on the Physical Phenomena of Life," Dr. Pereira's Translation, p. 111.

however, there is not the least adequate foundation. The liquid contained in the lymphatics has all the characters of dilute 'liquor sanguinis' (the liquid portion of the blood in which the corpuscles float); and it differs from that of the lacteals chiefly in the absence of fat. The lacteals, indeed, when the alimentary canal is empty, seem to perform the function of lymphatics; being found to contain a fluid which resembles lymph in every respect, and is probably derived from the same source. Again, as the lymphatics do not discharge their contents into any of the outlets through which they might be carried off, or convey them to the excretory glands by which they might be eliminated under some other form, but pour them into the same receptacle with the nutrient materials newly imbibed from the food, whence both are propelled together into the general current of the circulation, it seems almost certain that their contents, in whatever mode obtained, are destined to be employed in the formation of the tissues, and not to be forthwith eliminated from the system.—With regard to the source of the lymph, and the manner in which it is imbibed, there is at present a deficiency of accurate knowledge. It is very probable, however, that it partly consists of the residual fluid, which, having escaped from the blood-vessels into the tissues, has furnished the latter with the materials of their nutrition, and is now to be returned to the current of the circulation. But it also seems not unlikely, that it may partly be derived from those particles of the solid framework, which have lost their vital powers, and are therefore unfit to be retained as components of the living system, but which have not undergone such a degree of decay, as to prevent them from serving, like the aliment derived from the dead bodies of *other* animals, as a material for reconstruction, when it has been again subjected to the assimilating process. In what way the *selective* power is exercised, by which extraneous substances are usually prevented from entering the lymphatic system, when they are quickly received into the blood-vessels, cannot at present be even guessed at: since there is no reason to believe that the lymphatic 'plexuses of origin' are in relation with cells (like those of the villi) to which such a function might be attributed. Although the rule is not so constant for the lymphatics as for the lacteals, yet it does not appear that these absorbents readily take up liquids in contact with the skin, unless these be of an alimentary character, assimilating in composition to lymph. Thus it has been found that saline compounds, applied in solution to the surface, are sooner to be detected in the blood than in the lymph, provided the circulation be not checked; although, if the blood be not in motion, they will pass by imbibition through the thin walls of the cutaneous lymphatics, perhaps even sooner than into the blood-vessels. The peculiar aptitude of the lymphatics for the absorption of milk is shown by the experiments of Schreger, who found that the lymphatics of a limb long immersed in it became turgid with this fluid, whilst none of it could be detected in blood drawn from the veins. The fact that the lymphatics in the neighbourhood of collections of peculiar animal fluids, sometimes become filled with those fluids,—as bile from an over-distended gall-bladder, or pus from an abscess,—need not be regarded as invalidating the general statement already made with regard to their probable function; for there can be no doubt that under such circumstances a direct entrance might be readily gained into the absorbents, either through rupture or ulceration of their walls; and in this way alone could particles so large as pus-globules find their way into these vessels. The case is different, however, with regard to substances introduced through the skin by friction; for these often seem to find their way quickly into the lymphatics, as is shown by the circumstance that if they be of an irritating character, red streaks appear along the course of the absorbents, and the neighbouring glands are swollen. This is probably to be explained by the peculiarly abundant distribution of the lymphatics in the skin, and the ready access which liquids can obtain to their walls.

438. There can be no reasonable doubt that it is by the Blood-vessels, rather than by the Lymphatics, that all that 'interstitial' and 'superficial' absorption is performed, which is not of a directly nutritive character; and in particular that those effete matters are carried away from the tissues, which are destined to immediate elimination. Thus, the most rapid and constant in its formation of all the excretory products, and the most injurious if retained, is carbonic acid; and this is conveyed by the venous system to the Lungs, in which it is forthwith removed from the blood. So, again, it is by a special arrangement of a part of the venous system, that the Liver is supplied with venous blood for the elaboration of bile; indicating that it is in such blood that the elements of the bile most abound. And we have seen this to be the case with regard to the Kidneys also, in the lower Vertebrata; although in the higher, it is from the arterial system that they receive their supply of blood for secretion as well as for nutrition. Further, as no lymphatics exist in the whole Invertebrated series, it is obvious that nothing but the sanguiferous system can perform the entire function of interstitial absorption; and the same must be the case in those tissues of Vertebrata, which are destitute of special absorbents. It is not a little remarkable, that organs in which such rapid changes of composition take place, as in the brain and spinal cord, should be destitute of lymphatics; and the fact is confirmatory of the view already given of the special function of these vessels; since the composition of the nervous matter is such, that, when once dead, it would be speedily oxidized and carried off in the form of carbonic acid, water, and phosphoric acid, no element remaining that would be fit for again entering into the composition of the body.—Experiments demonstrating the exercise of the absorbent power by the blood-vessels, upon substances placed in contact with them, are most numerous and convincing; it will be sufficient to mention two. Mayer, having injected a solution of prussiate of potash into the lungs, detected it in the left cavities of the heart sooner than in the right, whence it is obvious that it must have been absorbed by the pulmonary blood-vessels more speedily than by the lymphatics of the lungs. And when the jugular vein of a young dog was laid bare by Magendie, and a solution of nux vomica was applied to its external surface, the symptoms of poisoning manifested themselves in four minutes.

439. It may be stated, then, as a general fact, that the Blood-vessels are the principal channels, through which water and substances dissolved in it are introduced into the body, either from the walls of the alimentary canal, or from the general surface; and by which that interstitial absorption is effected, whereby the particles that have served their purpose in the solid fabric are removed from it:—But that the Absorbent

system, possessed by the higher animals, is the special channel for the introduction of oleaginous matters, suspended in an albuminous fluid, from the intestinal tube; and for the return to the sanguineous circulation of such matters as may be yielded by the tissues (whether from the superfluity imparted to them by the blood, or as products of their own disintegration), in a state to be again employed for the purposes of nutrition. We shall hereafter find reason to believe (CHAP. XV.), that during its slow movement through the absorbents, and especially during its passage through their glandulæ, the chyle and lymph undergo changes by which it is brought into a nearer likeness to the blood, more especially in regard to the vital properties of that fluid.

440. Notwithstanding that it is through the walls of the alimentary canal, in most of the higher animals, that the introduction of nutriment from without is chiefly effected, yet it must not be lost sight of that the *general surface* of Animals, as of Plants, is still to a certain degree capable of taking this function upon itself; and this not merely when the regular channels have been closed, but even in some cases as the regular functional duty of that part. Thus *Frogs* seldom or never drink, but they habitually live in a moist atmosphere; and when they have lost fluid by exposure to hot dry air, they will regain their weight by being left for a time upon moist sand, and the bladder (which serves as a reservoir of water for cutaneous exhalation) though previously emptied, will be refilled. It is probable that the same may occur in all animals with a soft naked skin. The experiments of Dr. Madden * show, that a positive increase usually takes place in the weight of a man immersed in a warm-bath, even though there is at the same time a loss of weight by pulmonary exhalation and by transudation through the skin; and he found that when this loss was taken into account, the quantity of water absorbed was about an ounce and a half in half an hour. The absorption will be more rapid, when the fluids of the body have been previously diminished by unusual exhalation; thus Dr. S. Smith mentions that a man who had lost nearly three pounds by perspiration during an hour and a quarter's labour in a very hot dry atmosphere, regained *eight ounces* by immersion in a warm bath for half an hour. And a patient into whose stomach no aliment of any kind could be introduced, has been kept alive for some time chiefly by cutaneous absorption, the body having been immersed night and morning in a bath of milk and water, and imbibing from 24 to 36 ounces per diem.

CHAPTER XII.

CIRCULATION OF NUTRITIVE FLUID.

1. *General Considerations.*

441. In beings of the most simple organisation, whether belonging to the Animal or to the Vegetable kingdom, we have seen that every part of the surface is equally capable of absorbing the liquid aliment brought into contact with it; and that the materials of the tissues are supplied by the

* "Prize Essay on Cutaneous Absorption," pp. 59—63.

continual imbibition of the nutriment thus immediately derived from external sources. In such, therefore, it might be inferred that no transmission of fluid from one portion to another would be required for the purposes of the economy; and we find no evidence of its existence, either in a structure specially adapted to it, or in any visible motion of such fluid. But as, in more complex organisms, a small part only of the surface is particularly appropriated to the function of Absorption, it becomes evidently necessary that means should exist, of conveying to distant parts the nutriment they require. This is effected by the *Circulation* of the fluid taken up by the absorbents, through vessels or passages adapted to this purpose; and it may be regarded as a general statement of the condition of this *vascular* system in all classes of living beings, that *its development is proportional to the degree of limitation of the power of absorption,* by which the parts imbibing aliment are removed from those requiring supplies. But the conveyance of nutrient fluid to the remote parts of the system, is not the only object to be fulfilled by the circulating apparatus; since the crude aliment must be exposed to the influence of the air before it becomes fit for its ultimate purpose, and that which has once passed through the tissues of Animals must undergo a similar process to restore it to its proper condition. This process, which constitutes the function of Respiration (CHAP. XIII.), requires that the circulating fluid should pass through certain organs adapted for its performance; and hence the arrangement of the vascular system is modified, not only for conveying the alimentary materials from the part of the system where they are introduced to that where they are required; but also for causing it to be brought, during some part of its transit, into relation with the atmosphere. It is very evident, therefore, that the uninterrupted performance of this function is necessary to the continuance of life; since not only does the nutrition of the tissues, or 'vegetative life,' wholly depend upon the materials thus supplied; but the presence of oxygen conveyed by the vital fluid is necessary for the active performance of the 'animal functions' by the nervo-muscular apparatus (§ 67).

442. In the study of the Circulation, we shall have reason to see the peculiar advantage to be derived from the investigation of the simplest conditions under which it may be performed. It has been from the confinement of their attention to this function, as it exists in the higher Animals only, that many Physiologists have adopted incorrect and narrow views as to the powers by which it is maintained;—views which are incapable of extension to the whole Animal kingdom, far less to the Vegetable creation, and which must therefore be fundamentally erroneous. We shall endeavour to show that principles of higher comprehensiveness may be attained, by comparing the principal facts relative to the circulation of nutrient fluid, derived from all the classes of living beings in which it presents itself.

2. *Circulation in Vegetables.*

443. The tissues of the lower tribes of *Cryptogamia,* being almost entirely *cellular* in their structure, do not seem to be adapted for any very regular or definite transmission of fluid. It has been already stated (§ 266) that the *Algæ* absorb by their whole surface; and there appears to be so little communication between different parts of the same indivi-

dual, that, if one portion be suspended out of the water, it will dry up and die, whilst that which remains immersed will preserve its freshness. No trace of vessels is discoverable in this order; the cells present a rounded form in almost every part; and the only deviation from this arrangement occurs in the veins which strengthen the foliaceous expansions of some of the higher species, in which we find the cells somewhat elongated, and presenting an approach in form to woody fibre (§268).—Amongst the *Lichens,* a similar uniformity of structure prevails; no appearance of vessels is perceptible; but wherever the form of a stem is assumed, the cells, which are rounded in the foliaceous expansions, possess more or less of elongation. As in this tribe the power of absorption is usually restricted to the side least exposed to light, some capability of diffusing the nutrient fluid is required; and it appears that, when the absorbent surface is placed in water, the liquid is transmitted in the course of the elongated cells, to the whole plant.—In the higher *Fungi,* we may trace a further development of this simple form of the Circulating apparatus. In those species whose reproductive apparatus is elevated on a *stipes* (as in the Mushroom tribe), the nutriment, which is entirely received by the mycelium at its base, is transmitted by its elongated cells, and probably through certain hollows left by the separation of the tissue (termed *intercellular spaces*), to the expansion on its summit, where it is diffused in every direction.—It may be regarded, therefore, as a general expression of the function in these Cellular plants, that, when there is no tendency to prolongation in a particular direction, and the cells retain their rounded form, they transmit fluid with equal readiness towards all sides; but that, when any separation of the different parts takes place, by the restriction of the function of Absorption to one portion of the surface, there is a tendency to the evolution of an axis formed of prolonged cells, in the direction of which the fluid is conveyed most readily to the other parts of the system.

444. In the higher group of *Cryptogamia,* consisting of the *Mosses* and *Ferns* with their allies, we find a much more evident approach to the vascular structure and general circulation of the Phanerogamia. Still, however, its lower tribes (such as the *Hepaticæ,* § 275) are so closely connected with the more perfect forms of the preceding group, that what has been said of them will be equally applicable to these.—Among the *Mosses,* strictly so called, we find several species in which a complete stem is developed, furnished with radical fibres at its base, and bearing a number of veined leaves regularly arranged upon it. In these, the cellular tissue of the stem and of the veins of the leaves becomes considerably elongated, so as almost to resemble woody and vascular structure; and there can be little doubt that the transmission of the fluid absorbed by the roots is actively performed by this channel, especially as the Mosses are partly furnished with that special *exhalant* apparatus for the transpiration of fluid, which is fully developed in the more perfect plants (CHAP. XIV.).—In the *Ferns* the evolution of a true woody stem proceeds to a much greater extent; and in it is found a vascular structure scarcely differing from that of the Phanerogamia. Although little has been observed as to the circulation of sap in this order of plants, it can scarcely be doubted that the fluid absorbed by the roots ascends, through the fibrovascular tissue of the stem, to the leaves, as in Flowering plants; since we

find in them vessels of precisely the same character with those which are known to be the peculiar channels of this fluid in the latter, namely, the dotted and reticulated ducts (§ 152).

445. We shall therefore pass on at once to describe the circulation in the *Phanerogamia*, in which it has been more fully investigated; and we have first to speak of the *ascending* current, which is produced by the movement of the fluid that is absorbed at one extremity of the axis, towards the leaves by which a large proportion of it is exhaled at the other.—Each annual layer that composes the wood of the stem of *Exogens* (§ 280), consists of ducts and woody fibre, intermixed with more or less of cellular tissue; the vessels being usually situated at the inner part of the ring, and the fibrous tissue, which is not formed until later in the year, lying externally to them. The vessels have usually the greatest diameter in long slender stems, belonging to plants of active vegetation, in which the sap has to be conveyed with rapidity to a considerable distance, as in the Vine and the Clematis; and they are usually larger, also, where the stem is dense, as in the Oak, Elm, Mahogany, &c., than where its softness and laxity allow the whole texture to convey fluid more readily, as in the Pine tribe, which is destitute of any distinct sap-vessels, or in herbaceous plants, in which they are usually small in proportion, or in the young shoots of woody branches in which the intercellular passages abound. The deposition of the products of secretion, which gives strength and firmness to the *duramen*, destroys or greatly diminishes its power of transmitting fluid; and it is consequently through the external layers, which constitute the *alburnum* or sap-wood, that movement of fluid chiefly takes place.—Of the precise course of the ascending sap in *Endogens*, we have no certain knowledge; there can be little doubt, however, that it is conveyed through all the tissues of the stem which are not consolidated by interstitial deposit, but more especially by the ducts when these are peculiarly large and open, as is especially the case in long, firm, slender stems, whose leaves are borne at a considerable distance from the roots. Of this arrangement, the various species of *Calamus* or 'reed-palm' (one of which furnishes the well-known 'rattan-cane') furnishes a most characteristic example; the ducts being of great size and freely pervious through considerable lengths, whilst the remaining tissues of their wiry stems are so dense as to be quite unfit for the conveyance of fluid.*

* It has been affirmed by Prof. Schleiden, that the idea of the ascent of the sap through vessels is altogether a fiction, created by the imagination of those who have desired to find in Plants the analogues of the Animal functions. But the Author cannot help believing that the desire of that distinguished botanist to establish the entire *absence* of analogy between the two kingdoms, has much to do with his somewhat dogmatic denial of a movement of fluid through vessels in Plants. And whilst he is far from denying that the cells, woody fibres, and intercellular passages of the stem, *all* assist in the transmission of fluid from the roots towards the leaves, he cannot but think that the ducts, when fully developed, are the *special* channels by which this transmission is effected. How else could the ascending sap find its way through the wiry stems of the *Calamus rudentum*, or 'cable-palm,' which are sometimes 500 feet long?—The experiments of Honninger ('Botanische Zeitung,' 1843) upon the absorption of ferro-cyanide of potassium, the course of which upwards through the stem was afterwards tested with sulphate of iron, led him to the conclusion (which has been adopted by Link and other distinguished Botanists) that it is *only* through the tubular tissues of Vascular plants, that fluid ascends. His experiments, however, were chiefly made upon species in which, for the reasons above stated, the vessels afford a much readier channel for the sap than do the other tissues. On the other hand, Mr. Rainey concluded from experi-

446. The cause of the *ascent* of the sap in the stem has long been a disputed question amongst physiologists; some attributing it altogether to mechanical influences, and some regarding it as a purely vital (and therefore completely inexplicable) phenomenon. A very simple experiment will show that two sets of causes must be in constant operation. If the top of a young tree be cut off in the spring, and the divided extremity be immersed in water, it will absorb a sufficient quantity of fluid for the temporary supply of the leaves; whilst, on the other hand, the portion of the stem left in the ground will continue for a time to discharge the fluid drawn up by the roots. It is then evident, that the propulsive power of the roots, for which we have already endeavoured to account (§ 423), is a partial but not the entire cause of the ascent of the sap in the stem; since the latter will continue by simple imbibition, when the open extremities of the vessels are placed in fluid, provided that the functions of the leaves are sufficiently active to occasion a demand for it. Moreover, there would seem no reason why the spongioles should not be as capable of absorbing fluid in the winter as in summer; and if the ascent of the sap depended entirely upon them, we should expect that it would be continued. That they are thus capable has been frequently shown, by grafting a shoot of an evergreen upon a stock whose leaves are deciduous; and it is found that the uninterrupted continuance of the demand meets with a corresponding supply. A still more striking experiment is to train a shoot of an out-door vine, or other plant, into a hot-house during the winter; the unusual warmth will cause an immediate development of the buds, for which a supply of nutriment is required; and this is derived from the roots, whose usual torpidity at this season is thus remarkably interrupted. Careful examination of the first movement of the sap in spring, also leads to the same result; for it is now ascertained that the upward flow begins near the buds, and that it may be progressively observed in the branches, trunk, and roots,—the latter not commencing their action until the superincumbent column has been removed. It can scarcely, then, admit of a doubt, that the demand for fluid, occasioned by the vital processes which take place in the leaves, is the essential cause of the motion of the sap in the higher parts of the tree; and that the propulsive power of the roots is principally expended in raising it to the sphere of that influence. It is evident that the quantity of fluid absorbed by the roots will be proportioned to the rapidity of its removal by the leaves above; just as the continued rise of oil in the wick, by simple capillary attraction, is regulated by the combustion at its apex.

447. It has been commonly supposed that the 'crude sap' of the ascending current is entirely unfit to nourish the growing tissues; and that they derive the materials of their support from a *descending* current of 'elaborated sap,' which is prepared in the leaves, and is thence returned by a distinct set of vessels to the axis, through which it is transmitted even as far as the extremities of the roots. This doctrine, however, can by no means be admitted in the form here stated; although it probably

ments of a very similar kind, made with bichloride of mercury, that the ascent of sap chiefly takes place in the intercellular passages ("Experimental Inquiry into the Cause of the Ascent and Descent of the Sap," 1847).—The Author cannot but believe that the discrepancies of these and numerous other observations are chiefly due to differences in the structure of the stems of the respective species upon which they have been made.

contains a certain amount of truth. The ascending current does not contain those elements *only* which are absorbed from the soil, namely, water, carbonic acid, ammonia, and some mineral ingredients; for, as Prof. Schleiden remarks, "from whatever part and at whatever time we examine the sap of a plant, we find that it contains organic principles which cannot come from the soil, because they do not exist there; such are sugar, gum, albumen, malic, citric, and tartaric acids, &c." The presence of these substances in the ascending current is accounted for by some physiologists, on the idea that they were previously contained in the tissues, and have been taken up by it in its progress through them; whilst by Schleiden and others it is affirmed that they must have been generated by the assimilating action of these tissues upon the materials drawn in by the roots.* Whichever of these two views is the correct one (and it is possible that both may be partly true), it is certain that the ascending current of sap must be capable of affording nutriment to the growing tissues, in so far as it holds gum, sugar, and albumen in solution; and that there is no sufficient reason for looking to a supply of 'elaborated sap,' afforded by a hypothetical descending current, as their sole pabulum. But, on the other hand, there appears sufficient evidence that the leaves are the chief agents in the introduction of carbon into the system, by the decomposition of the carbonic acid of the atmosphere (§ 500); and that a much greater quantity of the organic compounds, at the expense of which the new tissues of the plant are formed, is prepared by their instrumentality, than by any other means. These organic compounds must be dispersed through the axis; and there appears strong reason to believe that this dispersion is especially effected by the *cellular* portion of it, and especially, in Exogens, by the bark (with which the leaf-stalks are in very intimate connection) and by the medullary rays. By the bark, these compounds are especially conveyed to the interspace between the liber and the alburnum, in which the formation of new wood takes place; whilst by the medullary rays they are carried in towards the centre of the stem, and afford the means of consolidation to the duramen. It is in accordance with this view, that if a ring of bark be removed from a stem, the parts above the ring undergo an unusual amount of increase, whilst those below the ring are comparatively atrophied. There is no reason to suppose, however, that this dispersion is effected by anything like a *current;* still less, that any special system of vessels is provided for conveying back the 'elaborated sap' from the leaves to the stem. Its transmission appears to be simply a process of imbibition, taking place between contiguous cells, whereby each communicates to the rest a portion of the nutritive materials with which it is charged; every one making that use of them which is in accordance with its own endowments. Additional evidence in favour of the view here advocated, will be given hereafter (CHAP. XV.).

448. It appears from microscopic observation, however, that a special circulation of peculiar juices does take place in certain parts of plants, through that system of anastomosing vessels, which has been termed

* It is difficult to understand, however, how this process can be effected by the cells of the interior of the stem and roots, secluded as they are from the direct influence of light; without which (we have every reason to believe) no direct production of organic compounds from inorganic elements can take place.

laticiferous (§ 154). This has been especially observed in plants with 'milky' juices, that is, in those in which the 'latex' is rendered opaque by the presence of floating particles of resin, caoutchouc, or other substances, which enable the movement of the fluid to be distinguished. There is reason to doubt, however, whether it is really confined to such; and whether it might not be discerned in many plants with transparent juices, if their movement could be in any way made evident. When the laticiferous vessels are cut or broken across, a flow of fluid takes place from the wounded part; and if a piece of the bark or leaf of a 'milky' plant be cut out and placed under the microscope, the fluid contained in the vessels will be seen in rapid movement throughout. This, however, has no relation to the true circulation, or *cyclosis,* which was first described by Prof. Schultz as taking place in the laticiferous vessels of an uninjured part.* The movement seems to take place in all directions, the currents often running contrariwise in contiguous vessels. Sometimes one of these currents may be observed to stop, its cessation being preceded by a temporary oscillation; it afterwards recommences, or a new current is established in a contrary direction. The rate of movement is greatest in parts which are in progress of development, other things remaining the same; it is also accelerated, within certain limits, by warmth; and is retarded or entirely brought to a stop by cold, recommencing on the renewed application of warmth. A strong electric shock puts an end to it immediately.

449. The resemblance between this *cyclosis* in Plants and the *capillary circulation* in Animals, makes it a point of peculiar interest and importance to determine the nature and source of the forces by which the former is sustained. It is quite obvious that the movement cannot be due to any *vis a tergo;* both because it is far from being constant in its direction in particular vessels, and because there is no organ to supply a propelling force,

* The statements of Schultz have been called in question by many distinguished Botanists, amongst others by Prof. Schleiden, who have regarded the movements described by him as the result of injury to the vessels, permitting the discharge of their contents, and consequently establishing a current towards the point of exit. So many competent observers, however, have satisfied themselves that such is not a sufficient explanation, that the existence of a *cyclosis* in certain plants must, in the Author's opinion, be regarded as an established fact, whatever be its degree of generality. The latest recorded observations on this point are those of Prof. Balfour; which are to the following effect.—"From observations made last summer, I am disposed to agree with Schultz's statements. It is true, as Mohl remarks, that any injury done to the part examined causes peculiar oscillatory movements, which speedily cease. Thus if the young expanded sepal of the *Celandine* is removed from the plant and put under the microscope, or if the inner lining of the young stipule of *Ficus elastica* be treated in a similar manner, very obvious motion is seen in the granular contents of the vessels, and this motion is affected by pricking the vessels or by pressure. In order to avoid fallacy, however, I applied the microscope to the stipules of *Ficus elastica,* while still attached to the plant, and uninjured; and I remarked that, while pressure with any blunt object on the stipule caused a marked oscillation in the vessels, showing their continuity, there could, nevertheless, be observed a regular movement from the apex towards the base, independent of external influences, when the stipule was simply allowed to lie on the field of the microscope without any pressure or injury whatever. This movement continued for at least twenty minutes during one of the experiments, and I have no doubt might have been observed longer. It is of importance to distinguish between those molecular movements which are caused by injury and pressure, and those which depend on processes going on in the interior of the living plant. My experiments are by no means complete; but they lead at present to the adoption of Schultz's opinion relative to the existence of the cyclosis."—"Manual of Botany," 1849.

which could extend itself through such a complex system of anastomosing canals. Nor, again, can we attribute it to any *vis a fronte*, like that which takes part in producing the ascending current from the roots towards the leaves. It is certain, too, that no such contraction of the laticiferous vessels themselves takes place, as could be effectual in propelling the fluid through them. Further, the movement continues for some time in parts that have been completely detached from the rest, and on which neither *vis a tergo* nor *vis a fronte* can have any influence. On the other hand, the facts stated in the preceding paragraph all indicate that, like the 'rotation' within the individual cells of Plants (§§ 137, 269), the movement of fluid within the laticiferous vessels (whatever may be its purpose in the vegetable economy) is intimately connected with the formative operations of the part, and is dependent upon forces which arise out of these. The manner in which they become so, is the next object of our enquiry; and on this subject, some views have been put forth by Prof. Draper,* which seem to help towards an explanation of the phenomena.—It is capable of being shown, by experiments on inorganic bodies, that, if two liquids communicate with each other through a capillary tube, for the walls of which they both have an affinity, but this affinity is stronger in the one liquid than in the other, a movement will ensue; the liquid which has the greatest affinity being absorbed most energetically into the tube, and driving the other before it. The same result occurs when the fluid is drawn, not into a single tube, but into a network of tubes, permeating a solid structure; for if this porous structure be previously saturated with the fluid for which it has the less degree of attraction, this will be driven out and replaced by that for which it has the greater affinity, when it is permitted to absorb this. Now if, in its passage through the porous solid, the liquid undergo such a change that its affinity be diminished, it is obvious that, according to the principle just explained, it must be driven out by a fresh supply of the original liquid, and that thus a continual movement in the same direction would be produced.—Now this is precisely what seems to take place in an organised tissue which is permeated by a fluid, between whose particles, and those of the tissue which it penetrates, affinities exist, which are concerned in the formative changes that take place during its circulation. Hence these affinities are continually being newly developed by acts of growth, as fast as those which previously existed are satisfied or neutralised by the changes that have already occurred; and thus in the circulation of the nutritive fluid, there is a constant attraction of its particles towards the walls of the vessels, and a continual series of changes produced in the fluid as the result of that attraction. The fluid, which has given up to a certain tissue some of its materials, no longer has the same attraction for that tissue; and it is consequently driven from it by the superior attraction then possessed by the tissue for another portion of the fluid, which is ready to undergo the same changes, to be in its turn rejected for a fresh supply. Thus in a growing part, there must be a constantly-renewed attraction for that portion of the nutritive fluid which has not yet traversed it; whilst, on the other hand, there is a diminished attraction

* "On the Forces which produce the Organisation of Plants," by John William Draper, M.D., New York, 1844; pp. 29 *et seq.*

for that which has yielded up the nutritive materials required by the particular tissues of the part; and thus the former is continually driving the latter before it. But the fluid which is thus repelled from one part, may still be attracted towards another, because that portion of its contents, which the latter requires, may not yet have been removed from it; and in this manner the current may be maintained through the whole capillary network, until the liquid has been entirely taken up by the tissues which it permeates. The source of the movement being thus attributable to the formative actions to which it is subservient, it is obvious that it must be affected by any external agencies which quicken or retard these; and it is thus that the influence of heat, cold, and electricity upon the rate of the flow seem most readily explicable.—These principles will be hereafter shown to have a most definite application to the phenomena of the 'capillary circulation' in Animals (§ 481).

450. The *development* of the Circulating system during the growth of Vascular Plants, has not yet been made an object of special attention; the general facts with which we are acquainted, however, correspond exactly with the principles which have been previously stated. As the absorption of nutriment by the embryo within the ovule (§ 278 *a*) appears to take place through the whole surface, there is no transmission of fluid from one portion to another; nor do we find, even at the period of the ripening of the seed, any distinct vascular structure. As far as its circulating system is concerned, therefore, the young plant, at the commencement of germination, is on a level with the simpler cellular tribes. During the rapid longitudinal development, however, which then takes place in the stem and root, there is of course a peculiar transmission of fluids in those directions; and this appears to be at first performed, as in the stem of the Fungi, by elongated cells and intercellular passages. It is not until the true leaves are expanded, that we find a distinct formation of woody or vascular structure; and it is very interesting to remark, that the ducts of young plants often present the appearance which is characteristic of the Ferns, having the spiral fibre more or less regularly disposed within them (Figs. 11, E, 12, 1, 2); whilst, after the stem has ceased to increase rapidly in length, these canals are converted into dotted ducts (Fig. 12, 3) by the process already described (§ 152). The anastomosing vessels of the latex in like manner originate from cells of less regular form, which open into one another at several points, and not, as in the formation of ducts, by their extremities alone (§ 154). This change is represented in progress in Fig. 13, A.

3. *Circulation in Animals.*

451. In tracing the evolution of the Circulating system through the Animal scale, it will be easy to discover its conformity to the same general plan, as that which has just been traced out in the Vegetable kingdom. In proportion as the power of absorbing aliment is restricted to one part of the surface, whether external or internal, does it become necessary that means should be provided for conveying the nutritious fluid to distant organs,—not merely that it may furnish the supplies, which they are constantly requiring for the maintenance of their respective structures, and for the manifestation of their vital properties, but also that it may itself

undergo certain changes, which are essential to the continuance of its characteristic qualities. Not only does the circulation of fluid through the system enable the new materials to be deposited in their appropriate situations, but it also takes up and removes the particles, which, having manifested a tendency to disorganisation, are no longer fit for the offices which they previously contributed to perform; so that, by the various processes of secretion, these may be separated from the general mass, and be either appropriated to some other purpose in the economy, or altogether carried out of the structure. The excretion of carbonic acid by the Respiratory apparatus is one of the most considerable and important of these processes; and it will be found that the distribution of the circulating apparatus has always an express relation to the conditions under which it is performed. In fact, so peculiar is this adaptation in the higher Animals, that many have considered the vascular system under two heads,—that belonging to the *general* circulation of nutritious fluid through the system,—and that which performs the *respiratory* circulation, conveying the blood, which has been rendered impure by the changes it has previously undergone, to the organs where its physical and vital properties are to be renewed by contact with the air. Respiration differs not in *kind*, however, from the other functions of Excretion, but only in its relative importance; and although in warm-blooded Vertebrata, whose nervo-muscular energy can only be maintained in full vigour by a constant supply of oxygenated blood, its cessation even for a short time is fatal, there are many amongst the lower classes, in which it can be suspended for a considerable period with impunity, and in which the increased amount of other secretions appears to counterbalance the diminution in its products. We find too, even in the highest Vertebrata, peculiar modifications of the circulating apparatus in connection with other secreting organs, as the liver in Mammalia, and the kidneys in Birds: so that it should rather be stated as a general fact, that, in proportion to the variety of the organs, and the importance of the functions they perform, is the complexity of the circulating apparatus which supplies them,—than that it undergoes modification according to the conditions of the respiratory system alone, as Cuvier maintained. In proportion as the function of Absorption is restricted to one part of the surface, that of Respiration will be limited to another; and the processes of Nutrition, and the formation of Secretions, will go on in parts of the structure distant from both; and all these must be brought into harmony by vascular communication, the arrangement of which will evidently vary from the most simple to the most complicated form, according to the number and variety of the actions to which it is subservient, and the vigour with which these are performed. Still, in most of those animals whose nervo-muscular energy is the greatest, the arrangement of the Circulating apparatus, and the rate of the movement of blood through it, appear to have special reference to the demand for oxygen created in the discharge of the Animal functions, and to the necessity for the removal of the carbonic acid generated in the same processes; whilst there is evidence that the Organic operations of growth and development might be carried on by means of a much less rapid flow, and by a less perfectly oxygenated blood. A remarkable confirmation of this view is presented by the condition of the Circulation in the class of Insects, in which the activity of the animal functions is relatively greater than in any other group; for we find the movement of their nutritive fluid, which is subservient in them to nutrition alone, to be comparatively slow and feeble; the very active aeration of their nervo-muscular apparatus being accomplished, not so much by the medium of the blood, as by the penetration of air-tubes into the tissues themselves (§ 520).

452. The Circulating apparatus of Animals, among the higher tribes at least, differs in one important particular from that of Plants. We have seen that, in the latter, the sap which has been elaborated in the leaves is dispersed through the fabric, giving up its nutritive constituents to the parts to which it finds its way; and that if any of it mixes with the ascending current, and circulates a second time through the system, the amount of this is comparatively small (§ 447). In the higher Animals, on the contrary, we observe that the same fluid is repeatedly transmitted through the body; and that the alterations which are effected in one part of its course by certain processes of Nutrition, are counterbalanced in others by different nutritive operations, as likewise by those of Respiration and Secretion, and by the continual admixture of new alimentary materials. These, as we have seen, are wholly taken up, in the Invertebrata, by the blood-vessels which ramify on the sides of the intestine; whilst in Vertebrated animals they are partly absorbed by a special set of vessels, which after a time empty them into the general circulating current. The variety of changes, which the blood undergoes in different parts of the body, requires the admixture of the portions returned from each, that a uniform fluid may be again sent forth into the system. This is provided for, by the mode in which the apparatus is constructed. In all but the simplest forms of it, we find a central reservoir, the *Heart*, into which the whole fluid returned from the body is poured, and on whose contractile force the maintenance of the current chiefly depends. From this it passes out by one or more large trunks, which convey it to the several organs and tissues of the body; these are called *Arteries*. The arteries gradually subdivide into ramifying vessels, which, repeatedly undergoing the same change (Fig. 53), terminate in a complex system of *anastomosing* (intercommunicating) tubes, of nearly uniform size, which are termed *Capillaries* (§ 220). It is in these only, that the blood comes into sufficiently intimate relation with the tissues which it supports, or by which secretions are elaborated from it, for the performance of chemical or vital reactions between them; so that we may consider the function of the arteries to be the simple conveyance of the nutritive fluid from the general reservoir to this network of capillaries, which exists in most of the living tissues of the body, and in near proximity to the remainder. Even the walls of these trunks are furnished with a distinct set of branches (the *vasa vasorum*), proceeding from neighbouring vessels, for their own nutrition. After traversing the capillaries, the blood is received into another series of vessels, formed by their reunion, which are termed *Veins*; and these, gradually coalescing into larger trunks, return it to the central reservoir.—The walls of the Arteries and Veins are formed of several layers of tissue, which, however, constitute three principal 'coats.' The *inner* coat is a pellucid structureless membrane, continuous both with that which lines the heart, and with that which forms the sole boundary of the capillaries; and this is covered on its free or internal surface by a layer of epithelial cells. The *outer* coat is simply protective, and is formed of condensed areolar

tissue. The *middle* or 'fibrous' coat, however, which is much thicker in the arteries than in the veins, is partly composed of yellow elastic tissue; and partly of non-striated muscular fibre. By the agency of the former, which is especially abundant in the larger arteries that receive the blood direct from the heart, the intermitting jets in which the fluid is at first propelled by its contractions, are gradually converted into a continuous stream. The latter, on the other hand, is more abundant in the smaller arteries, and appears to supply a propulsive force in some degree complementary to that of the heart; but its main purpose seems to be, to regulate the diameter of the vessels, in doing which it is probably influenced in some degree by the Sympathetic system of nerves that is minutely distributed upon it. Of a rapid alteration in diameter, which, being emotional in its source, we can only regard the nervous system as the instrumental agent in producing, we have a characteristic example in the act of 'blushing.'

453. The description above given, however, is by no means applicable to the entire Animal kingdom; for in many of the simpler tribes, there is no circulation of a proper nutritive fluid; and in many others, the vascular system is far from possessing the completeness of organisation which it exhibits in the higher groups. Thus we shall see that in many of the lower Articulata, there is no central cavity for the reception and propulsion of the blood; but the principal vessels are of nearly uniform diameter throughout, and are everywhere in some degree contractile; whilst the fluid within them does not appear to move in a determinate direction, but seems to be transmitted irregularly from part to part by undulating contractions, extending themselves through the tubes. In many Articulata, again, and still more in the Mollusca, we find the blood, for at least a part of its course, not contained within distinct vessels, but dispersed through the *lacunæ*, which, like the 'intercellular passages' of Plants, intervene between the tissues; and even the visceral cavities (such as that of the pericardium and peritoneum) sometimes form part of the spaces which it traverses, from the time it leaves the heart until its return to it. Indeed, there are some instances in which it seems to be in the general visceral cavity only, that any movement of nutritious fluid, distinct from the contents of the alimentary canal, takes place; this has been shown to be the case, for example, in the *Bryozoa* (§ 298), probably also in the *Rotifera* (§ 310), and also in the small tribe of *Pycnogonidæ* (§ 316 *e*); and it would appear to be in great part true also of the *larvæ* of many Insects, in which, however, the elongated visceral cavity seems connected at its two extremities with a contractile 'dorsal vessel,' whereby a constant movement of its contents in a determinate *direction* is secured. In the Vertebrated classes, the only trace that seems to remain of this *lacunar* circulation, is the commencement of the lymphatic system of Absorbents, which, as already mentioned (§ 437), seems to take up, for return into the circulating current, the nutritious fluid that has escaped from the capillaries into the interstices of the tissues.*

454. In the *Protozoa*, and in the lower *Radiata*, as already shown, the extension of the digestive cavity into all parts of the fabric does away with the

* See the admirable Memoir of Prof. Milne-Edwards on the "Circulation of Nutritive Fluid," in the 'Ann. des Sci. Nat.,' Mai, 1845.

necessity for a special circulation of nutritive fluid; and as the materials escaping from this cavity at once penetrate the tissues by which it is surrounded, there is no opportunity for that dispersion of it through the general visceral sac, which is so remarkable in the lower Mollusca and Articulata.—In the class of *Echinodermata*, however, which presents us with the highest development of the Radiated type, a true circulation of nutritive fluid, that has been absorbed through the walls of the alimentary canal, unquestionably exists; and this takes place, through the greater part of its course at least, in vessels having distinct parietes. At some part of the vascular system, there is usually found a pulsating dilatation which seems the first rudiment of a heart; but this is neither capacious nor powerful enough to possess much influence on the movement of the blood through the long series of trunks, branches, and capillary plexuses, which it must traverse after leaving this organ, before returning to it again. The arrangement of the vascular system differs in the several orders of Echinodermata, with the general diversity in their type of conformation; it has been especially studied in the *Asterias*, *Echinus*, and *Holothuria*.—In the *Asterias*, in which the digestive cavity is prolonged into the 'rays' or lobes of the body (Fig. 110), a vascular trunk is found lying on the surface of each of the radial cæca, and receiving minute branches from its subdivisions; and the several trunks, converging from the rays to the central disk, unite with other branches from the stomach, to form a circle or vascular ring round the upper part of the disk. This is connected with a similar ring surrounding the entrance to the stomach on the lower surface, by means of a vertical descending vessel, which Tiedemann found to possess muscular irritability, and regards as the rudiment of a heart; whilst, from the lower ring, other vessels proceed which are distributed through the disk and rays. The vessels first mentioned probably act as absorbent veins, and convey the nutritious fluid to the central rings, which are furnished with numerous dilatations; whence it is propelled through the second set of vessels, which may be regarded as arterial trunks, to the system at large. No communication has yet been detected, however, between the terminations of the second set of vessels and the commencement of the first; and it seems probable that this is only established by *lacunæ*. Moreover, the supposed course of the vital fluid has not been verified by observation, being merely conjectured from the distribution of the vessels. As no special respiratory organs appear to exist in the Asterias, so is there no special respiratory circulation.—In the *Echinus*, the vessels which arise from the sides of the alimentary canal unite into a trunk (apparently analogous to the mesenteric vein of higher animals); and this conveys it to a vascular ring, which surrounds the anal orifice, and which may be likened to the upper ring of the Asterias. From this circle, which also receives the venous blood of the ovaries, five trunks (branchial arteries) pass off, distributing the blood over the mesenteric membrane that lines the visceral cavity, to which water is admitted for aerating the blood, and also to the vesicles at the base of the tubular cirrhi, which are probably in part subservient to the same purpose. Here, therefore, we find an express modification of the circulating apparatus for the purpose of carrying into effect the second of its principal objects; and it is interesting to trace this modification so low down in the scale, before a heart is distinctly evolved, and whilst the motion of fluid in the vessels seems dependent upon the changes which it undergoes in them. After ramifying on the membrane which lines the shell, and being there submitted to the respiratory process, the circulating fluid is carried back by a set of branchial veins to another ring formed round the anal outlet of the alimentary canal; from this the ovaries are supplied with arterialized blood, whilst the remainder of the aerated fluid enters a canal, which proceeds directly to the œsophageal extremity of the alimentary tube, and there dilates into a short oval sac, which, possessing muscular parietes, and exhibiting during life slow but distinct contractions, is regarded as a heart. From this canal and its pulsating dilatation, which seems to represent the vertical contractile trunk of the Asterias, arterial trunks pass forth, which form a ring round the mouth, and distribute the blood to the several parts of the complex buccal apparatus; and the venous blood, returned from these to a venous ring round the mouth, thence proceeds to the intestinal vein first described, mingling with the newly-absorbed alimentary materials, and passing on with them to the respiratory system of vessels.—Such, at least, appears the most probable account of the course of the blood through the vessels; but the information hitherto obtained is not sufficient to enable it to be affirmed with certainty that its direction is not precisely the reverse.

455. In the *Holothuria*, as already stated (§ 296 *e*), a transition may be perceived from the Radiated to the Annulose form; and this is in some degree manifested in the arrangement of its circulating apparatus. The vessel which seems best to answer to the mesenteric vein of the Echinus, is the trunk which runs along the *external* border of the intestinal canal (Fig. 113, *ve*), receiving the venous branches which have collected the blood distributed over it, together with the fresh nutritive materials which they have absorbed. From this trunk, the blood is again sent forth, to be distributed by means of the 'mesenteric,' but rather 'branchial,' vessels (*vm*), upon the respiratory apparatus. This, as formerly shown, differs entirely from that of the Echinus, in being formed much more upon the plan of that of Insects, though destined for aquatic respiration; and the blood is distributed in a beautiful capillary network over the walls of the curious 'respiratory trees' (*rr*), into which water is introduced from the cloaca. After being aerated by this means, the circulating fluid is collected by branchial veins, which unite into a trunk (*vi*) that runs along the *internal* border of the intestine; this trunk presents numerous dilatations, and these appear to be contractile, thus foreshadowing the dorsal vessel of Insects. From it the blood is again transmitted to the mesentery (*m*) by numerous branches, and thence over the intestinal surface, to find its way through the capillary network into the intestinal vein (*ve*). Besides this system of vessels, however, there is another, consisting of an annular trunk (*va*), which surrounds the mouth (as in Echinus), and sends off branches to the buccal apparatus and the tentacula, as also to a set of tegumentary vessels (*vl*, *vl'*), which supply the general surface. In connection with the œsophageal collar is found a saccular dilatation (*p*), which seems to represent the pulsatile vessel of the Echinus; but whether it possesses a similar contractile power, is not known. The connection of this set of vessels with the preceding has not yet been made out.—It is scarcely possible to conceive that the impulse which the blood receives, in any of the three preceding cases, from the central propelling organs, can be

sufficiently strong to convey it through all the complex ramifications that have been described, and through a double or even triple system of capillary tubes, back to the centre from which it was originally distributed; especially since the pulsations which have been witnessed in those organs are slow and feeble. That the passage of the blood through the respiratory apparatus is independent of it, there seems good reason to believe; and the very complete analogy, which exists between this circulation and that of the higher plants, affords a striking confirmation of this view. As in plants, the nutriment taken up by the absorbent vessels enters at once into the general circulating system, by which it is conveyed to the respiratory organs; and, after leaving these, it is distributed to the fabric at large, to supply to it the materials for its nutritive operations. We have seen that in plants, the motion of the fluid in the laticiferous canals appears to be entirely independent, not only of the propulsion of the roots, but of every direct mechanical impetus (§ 449); and there seems no difficulty, but, on the contrary, the highest probability (if consistent with other facts, as it will hereafter appear to be), in supposing that the same causes, whatever may be their nature, are in operation in this instance also.

456. The *Articulated* classes are usually regarded as inferior to the Mollusca in the evolution of their circulating apparatus; and it certainly never presents the same concentrated condition in the former group, as in the latter. That which is chiefly remarkable, however, in the vascular system of the Articulata, is the multiplication of propelling cavities, in accordance with the segmentation of their bodies; it being only in the very highest tribes of the series, that we find all the blood transmitted to a single receptacle, and distributed by branches proceeding from it alone. It may also be remarked, that a very exact bi-lateral symmetry almost uniformly prevails in the conformation of the central organs, and in the distribution of the vessels; and that the plan of circulation which may be regarded as typical of the entire group, is for the blood to pass along the body *from behind forwards* through a *dorsal* trunk in which the propelling power is concentrated, and then to return *from before backwards* through *lateral* or *ventral* trunks. The principal variations from this plan have reference to the arrangement of the Respiratory organs, the great functional importance of which, in the Articulated series, almost invariably necessitates some special provision for the distribution of blood to them; and thus, with every change in their position and character, a new modification of the circulating apparatus is presented.

457. Putting aside the cases already referred to (§ 453), in which there is no proper *vascular* circulation, the distribution of the nutritious fluid being altogether *lacunar*, we find that the simplest condition of the circulating apparatus is presented to us in the lower Entozoa. Thus in the *Tænia* (§ 309 *c*) we have seen that four minute trunks pass from end to end of the body, which are of nearly uniform size throughout, and communicate by transverse branches; and in neither of these would any special propulsive power seem to exist. In the *Fasciola*, on the other hand, the capillary network that traverses the body communicates with a single trunk (Fig. 142) which passes along the dorsal region on the median line, and which is evidently analogous to the 'dorsal vessel' of higher Articulata. In the *Planaria*, again, the capillary network is chiefly formed by the ramifications of a pair of trunks which traverse the body longitu-

dinally at some distance from the median line; and these possess, at their anterior extremities, a pair of pulsatile cavities, which unite on the median line, and surround the cephalic ganglia, giving off smaller vessels to the anterior part of the body (§ 309 *d*). And lastly, in the *Strongylus* and other Nematoidea, we seem to have a combination of this plan with that which is presented to us in the *Tœnia;* for a pair of trunks run along the length of the body on either side, of which those that are nearest to the median line communicate with each other freely by transverse branches, whilst at the anterior extremity they arch towards each other, and a small vesicle is developed on the median line at their point of junction. This vesicle is supposed to perform the functions of a heart, and to send the blood along the internal trunks, which may perform the functions of arteries, whilst those which accompany them may possibly act as veins. But on this subject nothing is certainly known; and it may be doubted whether there is any determinate sanguineous current in these comparatively simple beings. None of them have any distinct respiratory apparatus; and consequently we do not find any special arrangement of the blood-vessels for the aeration of the blood.

458. The arrangement of the Circulating apparatus among the different tribes of the class *Annelida* is found to present many remarkable varieties: some of these have relation to the form and situation of the respiratory organs, which are no less variable; but for the purpose of others it seems difficult to account. This class has long been noted for the redness of the blood of the animals composing it,—a character by which they are distinguished from others of the Articulated series, as well as from the Mollusca; and which, combined with the transparency of their bodies, allows the circulation to be distinctly seen. This shade does not, however, exist throughout the class; for in a few species, the blood is nearly colourless; and in one, at least, it is of a green tint, approaching olive. It is not dependent upon the presence of coloured corpuscles, but is proper to the plasma or liquid element of the blood. The condition of the vascular system in the *Leech* and its allies appears to approach more nearly to that which has been just pointed out as characteristic of the Entozoa, than does that of any other Annelida; as might be expected from the fact that the order *Suctoria* forms the connecting link between the two classes (§ 311 *d*). It essentially consists of four longitudinal vessels, a dorsal, ventral, and two lateral trunks, together with a multitude of communicating branches. The *dorsal* vessel receives the blood posteriorly from the intestinal canal, and from the dorsal portion of the integument, and carries this forwards towards the head; arrived near the mouth, it subdivides into a pair of arches which pass round the œsophagus, uniting below to form the *ventral* trunk; and this runs backwards towards the posterior sucker, giving off branches in its course, which are chiefly distributed on the alimentary canal. The *lateral vessels* pass along the two sides of the body in a wavy course; they communicate with each other at both extremities, and likewise at intervals along their whole length, by transverse branches; and their ramifications are principally distributed to the integuments and to the so-called 'pulmonary vesicles,' which appear, however, to be merely receptacles for the mucous secretion. The branches of all four sets of vessels freely inosculate with each other. The walls of the main trunks are furnished with muscular fibres disposed

circularly, and it appears to be chiefly by the successive or peristaltic contractions of these, that the blood is kept in motion ; no determinate course, however, can be assigned to it ; as it seems to have an undulating movement, sometimes in one direction, sometimes in another,—reminding us of the want of constancy in direction which is so remarkable in the Tunicated Mollusks (§ 466). The lateral trunks, from their connection with the tegumentary system, would seem to be the part of the apparatus that is chiefly subservient to the respiratory function.

459. In the *typical* Annelida (§ 311 *a, b*), the most constant parts of this apparatus are the dorsal and ventral trunks ; the lateral vessels, however, often exist, although they are seldom of the same relative importance as in the Suctoria ; and the dorsal and ventral trunks are themselves sometimes double along a part or the whole of their course, running on the two sides of the median line, or at a little distance from it, and thus reminding us of the Entozoa. Moreover, we generally find special contractile dilatations on some part of the vascular system ; sometimes only a single one exists ; but more commonly they are greatly multiplied. Our acquaintance with the circulation in this group, has chiefly resulted from the skilful observations of M. Milne-Edwards.*—In the *Terebella*, whose gills are filamentous, and situated round the head (Fig. 201, *k,k*), there lies in the anterior part of the body, on the dorsal aspect, a large trunk (*l*), which receives at its posterior extremity, from a venous sinus (*n*) surrounding the œsophagus, the contents of an extensive vascular plexus, that has ramified on the walls of the intestine, and on the muscles, integuments, &c. This trunk, or dorsal vessel, propels forwards the blood, which it receives from behind, by irregular contractions. At its anterior extremity it subdivides into numerous branches, of which the principal enter the respiratory organs (*k, k*), whilst others pass to the head and tentacula (*b, b*) ; so that a large proportion of the blood is aerated, before it is again circulated through the system. The vessels that return it from the gills reunite into a trunk (*o, o*), which passes down the ventral surface of the body, giving off a pair of transverse vessels to each segment, and then returning along the

FIG. 201.

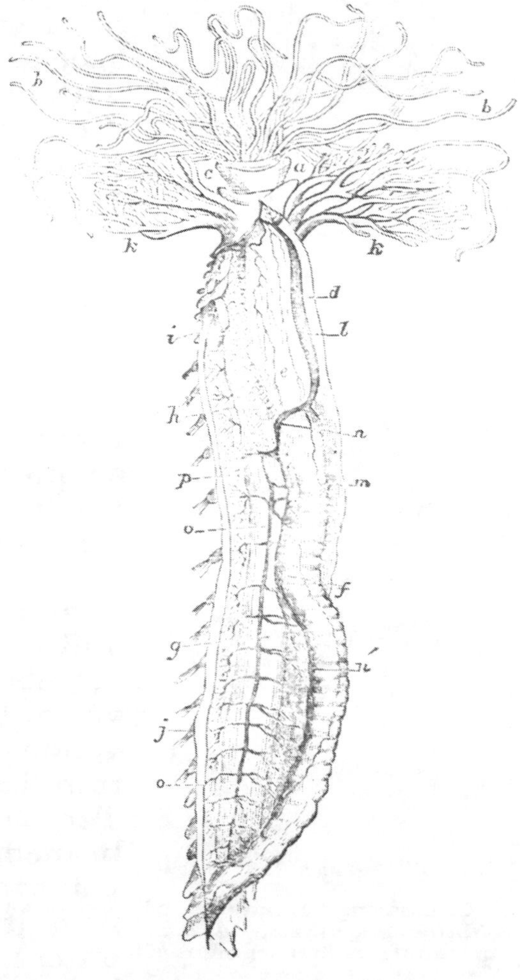

Circulating apparatus of *Terebella conchilega*:—*a*, labial ring; *b, b*, tentacula; *c*, first segment of the trunk ; *d*, skin of the back ; *e*, pharynx; *f*, intestine ; *g*, longitudinal muscles of the inferior surface of the body ; *h*, glandular organ (liver?) ; *i*, organs of generation ; *j*, feet; *k, k*, branchiæ ; *l*, dorsal vessel acting as a respiratory heart ; *m*, dorso-intestinal vessel ; *n*, venous sinus surrounding œsophagus ; *n'*, inferior intestinal vessel ; *o, o*, ventral trunk ; *p*, lateral vascular branches.

* See his elaborate Memoir in the " Ann. des Sci. Nat." 2ième Série, Zool. tom. x.

inferior side of the intestine (n') ; this trunk is of course to be regarded in the light of an artery. The blood is conveyed back into the venous sinus, both from the intestine and from the parietes of the body, by the dorso-intestinal vessels (m), which must be considered as veins. The propulsion of blood into the gills seems principally due to the contractions of the dorsal vessel, which may here be regarded as a sort of respiratory heart ; but its motion through the arterial trunk would appear to be partly owing to contractions of the gills themselves, which are occasionally seen to take place. The irregularity of these, however, requires some supplemental force, such as that which has been already described in the inferior tribes, for the maintenance of a steady current ; and there are many allied species, in which the blood circulates as energetically, without any such evident propelling agents.—In the *Eunice* (Fig. 202) we find the same general distribution of vessels, but an important change in the position in the respiratory organs, which involves a complete alteration in the character of the different parts of the system. The branchial organs (u) are not concentrated round the head, but are disposed in tufts along the whole body. The dorsal trunk (l') receives from the dorso-intestinal vessels (l), as in the former case, the blood which has ramified on the intestines ; but this fluid, as will presently appear, is as much arterial as venous. The contractions of this trunk are not so regular and powerful as in the Terebella, and seem to be but little concerned in maintaining the circulation. The vessels into which it divides anteriorly, are distributed only to the head and neighbouring parts ; and, by the reunion of the vessels which return the blood so distributed, the ventral trunk (q) is formed, which here possesses a venous character. The general distribution of its ramifications is very similar to that described in the Terebella, except that it transmits blood to the branchiæ as well as to the general system ; but each transverse branch (t) presents a dilatation or bulb (t') near its origin, which seems to propel the blood that enters it, by regular contractions, partly through the branchial tufts, and partly upon the intestines, cutaneous surface, muscles, &c., after permeating which it re-enters the dorsal vessel. This multiplication of propelling organs is very interesting, when viewed in reference to the general tendency to repetition of parts manifested in this class. It is found that, in the Annelida which possess it, the vitality of portions of the body is preserved during a very long time after the subdivision of the animal. In higher tribes, however, this multiplication is restricted to a particular division of the body, as will be presently seen in the Earthworm.

FIG. 202.

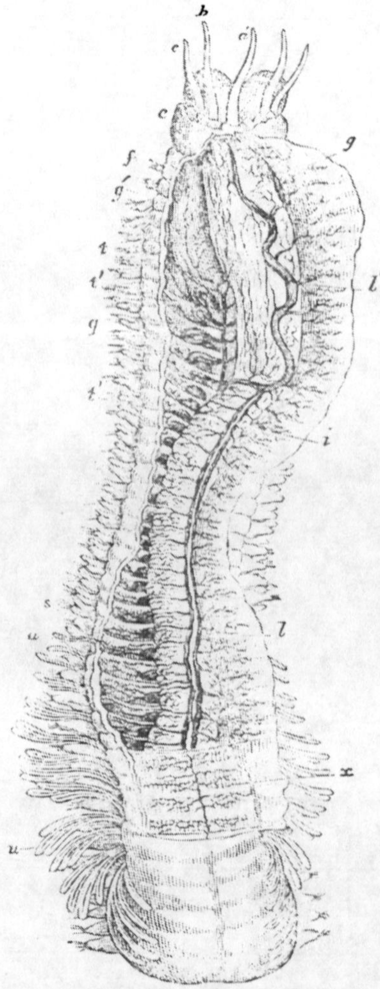

Circulating apparatus of *Eunice sanguinea*;—*a. b, c*, antennæ ; *e*, first segment of the body; *f*, feet; *g*, pharynx; *g'*, mandibular muscles; *i*, intestine; *l'*, dorsal vessel ; *l*, superior intestinal vessels ; *s*, their lateral branches; *q*, ventral vessel ; *t*, its lateral branches; *t'*, contractile bulbs on these branches ; *u*, branchiæ ; *x*, subcutaneous vessels of the back.

460. In the *Arenicola* (Fig. 203) we observe another interesting variety in the arrangement of the vascular system, which partly resembles the forms already noticed, and partly conducts us to others which would at first sight appear entirely different. The dorsal vessel (*o,o*) traverses almost the entire length of the body posteriorly, and it receives, as before, the blood which has circulated on the intestine and external surface, as well as some directly transmitted by the branchiæ. It terminates, however, at about the anterior fourth of the body, in a kind of contractile ventricle (*n*), which answers the purpose of a heart; but it first sends forward branches (*x*) to the head, the vessels returning from which enter the ventral trunk (*t*) that passes backwards from the propelling cavity. The branches of this trunk are almost entirely distributed to the gills; and the blood which is returned from them is partly transmitted to the intestine by the lateral intestinal vessels (*p*), partly to the integuments, and partly to the dorsal vessel direct. The branchiæ here, as in the Terebella, seem to have a direct propelling power on the blood which passes through them.—In the *Lumbricus* or 'Earth-worm,' again, we find the dorsal vessel communicating with the ventral trunk, not by *one* contractile cavity at its anterior extremity, but by *several* loop-like dilated canals, which seem to exercise a similar propelling agency. The waves of blood can be distinctly seen, if the animal be kept without food for a time, until it has discharged the black earth which usually fills the intestinal canal. The blood which is thus forcibly propelled into the ventral trunk, is conveyed backwards along the body, and distributed to its different organs, especially to the secreting follicles which appear to represent the respiratory sacs of higher air-breathing Articulata (§ 311 *c*); and from these it is returned to the dorsal vessel, its aeration having been effected through the medium of the general surface.—The great extent and importance of the capillary system in the Annelida, compared with the feebleness of the central propelling powers, is an interesting feature in the character of their vascular apparatus, and shows that we have not yet arrived at a condition of the circulating system very far removed from that which it presents in Plants.

FIG. 203.

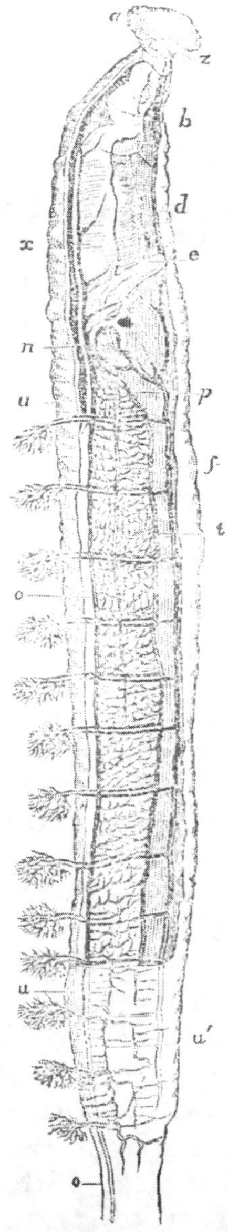

Circulating apparatus of *Arenicola piscatorum*, as seen from one side:—*a*, proboscis; *b*, pharynx; *d*, dilated œsophagus; *e*, cæcal appendage; *f*, stomach; *n*, heart; *o, o*, dorsal vessel; *p*, lateral intestinal vessels; *t*, ventral trunk; *u, u'*, cutaneous vessels coming off from the branchial; *x*, lateral pharyngeal vessels; *z*, labial vascular ring. (See also Fig. 149.)

461. In the Circulating apparatus of the *Myriapoda*, as in other parts of their structure, we meet with the condition which may be regarded as *typical* of the Articulated series; for this class presents the full evolution of that multiple heart, which is developed (so to speak) out of the principal dorsal trunk usually found in the lower Articulata; whilst it also exhibits that uniformity in its structure, and in the distribution of the vessels proceeding from it,

which are obscured in the higher Articulata by the unequal development of the successive segments, and by the specialization of particular organs. This change of type, however, is by no means abrupt. It appears from the researches of Mr. Newport,* to whom we owe the greater part of our knowledge of the vascular system of this class, that in the *Iulidæ* (§ 312), which form the connecting link between the Worms and the higher Annelida, the walls of the dorsal vessel are very thin, and the valvular constrictions between its successive segments, which are formed by reduplications of the muscular tunics, are by no means complete; further, the number of these chambers, which corresponds with that of the moveable segments of the body, is very large, being no less in one genus than seventy-five. A greater multiplication even than this is seen at an early period in the life of the Iulidæ; for each of their moveable segments, to which two pairs of legs are attached, is formed by the coalescence of two original segments; and every one of the latter at first possesses its own subdivision of the dorsal trunk, the indications of which are seen, even in the adult animal, in the duplication of the cardiac muscles and of the arterial trunks on each side, whilst the cavities have completely coalesced.—In some of the *Geophilidæ*, the number of distinct chambers is even greater than in the Iulidæ, being no fewer than a hundred and sixty in one species; but the whole apparatus presents a higher type of structure.—In the *Scolopendridæ*, the number of chambers is reduced, in accordance with that of the segments of the body, being never more than twenty-one, and sometimes as small as fifteen; the muscular portion of their walls is much more developed; and the valvular partitions which isolate the successive segments from each other, as well as the valves which guard the orifices of the vessels, are much more complete than in the lower tribes. The anterior and posterior portions of this *dorsal vessel*, and the parts in immediate connection with it, are shown in Fig. 204; in which the figures 1, 2, indicate its first and second chambers, and 17, 18, 19, 20, 21, those of the corresponding segments at the opposite extremity. The walls are formed of two layers of muscular fibres, some of them annular and others longitudinal; and similar fibres may be traced upon the principal systemic arteries. The purpose of these fibres is obviously to produce contraction of the chambers; their dilatation being accomplished by the bands of alar fibres, *f*, *f*, which extend to a considerable distance from either side of the dorsal vessel, and are inserted into the dermo-skeleton of their own segment. Each chamber has a pair of apertures guarded by valves, which is probably for the entrance of venous blood; but the source from which they receive it has not been clearly made out; it is probably, however, the great sinus formed by the pericardial sac which surrounds the dorsal vessel. From each chamber, also, a pair of systemic arteries, *h*, *h*, is given off, which are especially distributed to the organs of the upper side of their own segment, but inosculate with those of other segments. From the most anterior chamber (1) is given off a pair of large arches, *pr*, *pr*, which encircle the œsophagus, to meet again upon its under side in the ventral trunk, *xy*; but it also sends forwards a median trunk, *q*, that gives off two smaller pairs or arches, *u*, *v*, which in like

* "On the Nervous and Circulatory Systems of Myriapoda and Macrourous Arachnida." Philos. Transact., 1843.

manner meet in the anterior continuation of the ventral trunk, *w*. From the median trunk are also given off the vessels which supply the cephalic segment and its organs of sensation; the lateral branches, *s*, *t*, which supply the mouth and its appendages, being furnished by a trunk that comes off from the principal aortic arch on each side. The distribution of the arterial branches of the dorsal vessel is extremely minute; even its own parietes being furnished with distinct nutrient branches, *z*. The *ventral* trunk, which lies upon the ganglionic cord, and is principally formed by the union of the arches that surround the œsophagus, may be regarded as representing the 'aorta' of Vertebrated animals; it is of large diameter in the anterior portion of the body, but gives off a pair of arterial branches in each segment; and at its posterior extremity is reduced to a comparatively small size, there subdividing into two principal branches, which supply the last pair of legs. The branches successively given off from this aortic trunk chiefly proceed to the muscles of the legs; but one set of them is distributed to the tracheæ; and it is worthy of note that the tracheæ and blood-vessels are mixed up together in a remarkably intricate manner in the peritoneal membrane.—The *venous* system has not been made out by Mr. Newport; but there is a strong probability from analogy that it is altogether *lacunar*, and that the blood which has circulated through the system, together with that which has been sent to the tracheæ for respiration, finds its way to the great pericardiac sinus, and thence re-enters the dorsal vessel through the apertures in its successive chambers. A portion of this mixed fluid will be at once sent outwards by the contraction of each chamber, into the arterial trunks proceeding from it; but the principal portion will be transmitted from behind forwards by the peristaltic action of the chambers, those of the anterior segments dilating whilst those of the posterior contract; and thus the current will be directed through the aortic arches into the ventral trunk, and also through the arterial branches supplying the head. Besides the dorsal and ventral trunks, we find a pair of *lateral* trunks, *o*, which are specially connected with the hepatic organs; the branches of these inosculate with those of the ventral trunk; but in what way the blood is propelled through them is unknown.—In the family of *Scutigeridæ*, an interesting transition is presented to the structure which the dorsal vessel possesses in Insects; for

Fig. 204.

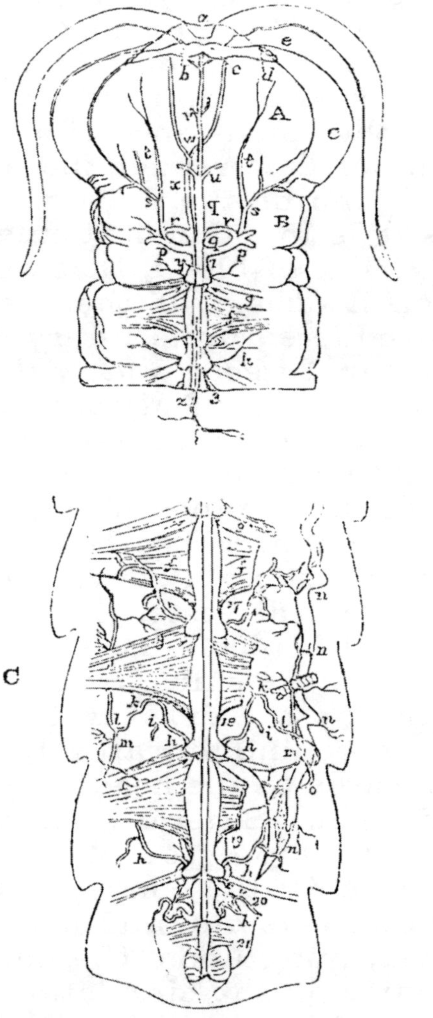

Circulating apparatus of *Scolopendra*:—A, cephalic segment; B, basilar segment; C, foot-jaws; 1, 2, anterior cardiac chambers; 17, 18, 19, 20, 21, cardiac chambers of the posterior extremity of the body; *a*, antennæ; *b*, *c*, cephalic ganglia; *d*, eyes; *e*, extremity of foot-jaws; *f*, *f*, alar muscles; *g*, *g*, valvular partitions; *h*, *h*, systemic arteries; *i*, *k*, *l*, *m*, *n*, their subdivisions, some of them inosculating with the hepatic vessels *o*; *p*, *p*, aortic arches; *q*, median trunk continued from the dorsal vessel; *r*, *r*, aortic arches reuniting, to form the ventral trunk; *s*, mandibular artery; *t*, cephalic artery; *u*, *v*, secondary arches; *w*, anterior continuation of the ventral trunk *x*, *y*; *z*, nutrient arteries of the dorsal vessel.

every alternate chamber is much smaller and shorter than the one before and behind it, and receives very little blood from its auricular orifices, so that, the total number of chambers being sixteen, the number of the principal chambers is no more than eight; whilst at the same time, the whole organ is shorter and of more compact form.

462. It is not a little remarkable that *Insects* should have been long regarded as unpossessed of a proper circulation; the peculiar provision for the conveyance of air through the interior of their bodies having been supposed to render the movement of blood unnecessary. Such, however, is by no means the case; for the circulation in Insects is at least as active as it is in other Articulata; and it is only the extraordinary rapidity in the flow of blood, that would have been required by animals of such wonderful energy, if their respiration had been localised in one organ, which is rendered unnecessary by the universal diffusion of that function.—We find in Insects a *dorsal vessel* (Figs. 152–4, *a a*), formed upon the same plan as that of Myriapoda; but the division into chambers is restricted to the abdominal portion of the body, the vessel being continued through the thorax to the head, as a simple contractile trunk. Thus it happens that the number of distinct chambers never exceeds eight; and it appears that in many Insects it is even less. The muscular walls of these chambers are considerably developed, and their valvular apparatus is very complete. The dorsal vessel is surrounded, as in Myriapoda, with a pericardial sac, in which blood has been seen to move; this may, therefore, be regarded as a venous sinus, from which the blood is received into the several chambers, by a pair of valvular orifices within each. The fluid is propelled from behind forwards, through the successive chambers; and is then driven onwards to the head, through the trunk which is the anterior continuation of them. Several branches have been detected, into which this subdivides for the supply of blood to the parts of the head; but no pair of aortic arches for the conveyance of blood to the under side of the body, has yet been made out, although a ventral trunk has been discovered by Mr. Newport, lying upon the gangliated nervous column, as in Myriapoda. The course of the circulation has been chiefly watched in transparent Larvæ, and in Pupæ during their development; and it has appeared as if, when the currents had once passed out of the dorsal vessel, they made their way rather through lacunæ among the tissues, than through distinct vessels. The principal stream of blood from before backwards does not seem to flow along the ventral trunk, but through two lateral passages which lie near the ventral surface; and it is from these that the secondary currents diverge, which pass into the wings and legs, and then return back to the main stream. These currents, too, seem to be rather 'lacunar,' than restrained by distinct parietes; but it is probable that the walls of the passages may be more complete in the perfect Insect. The mode in which the aeration of the blood is provided for in Insects, has not until recently been made out; but from the recent enquiries of M. Blanchard * it appears that the tracheæ are ensheathed, even to their minutest ramifications, by prolongations of the sanguiferous canals, and

* "Annales des Sciences Naturelles," June, 1848.—The statements of M. Blanchard have been called in question by several Anatomists, more especially by M. Joly ("Ann. des Sci. Nat.," Nov. 1849); he has replied to them, however, in his Memoir on the Circulation in the Arachnida, to be presently referred to.

that the blood is therefore aerated wherever the tracheæ penetrate. This view is in perfect accordance with the fact long since observed by Mr. Bowerbank* and others, that the current of blood in the 'nerves' of the wings moves in a space which completely surrounds the tracheæ.—So far, then, as the course of the circulation in Insects is yet known, it may be probably considered to be as follows. The blood impelled forwards by the dorsal vessel, is transmitted to different parts of the body, either by distinct vessels, or by lacunæ; after passing through the tissues, it finds its way by imbibition into the outer sheaths of the minuter tracheæ, which thus serve to collect it for aeration; and from these it is transmitted, whilst undergoing exposure to the air contained in the tracheæ, towards their external orifices, where it is collected from their sheaths by a system of canals which convey it back into the great pericardiac sinus, whence it enters the dorsal vessel.†—The dorsal vessel of the *Larva* is far less perfect in its structure, than is that of the Imago; being sometimes almost as destitute of valvular partitions, as is that of the Annelida; and very commonly presenting no higher a development, than that of the lower Myriapoda. But in its advance towards the perfect state, a gradual thickening of its walls, and a completion of its valvular partitions, may be noticed; and at the same time the whole organ becomes contracted in length, and presents a more concentrated condition. In many aquatic larvæ, especially of the order *Neuroptera*, there are leaf-like appendages affixed to the tail, in which the circulation may be distinctly seen, the streams passing off in loops from the main trunks, and entering them again, so as to be conducted to the posterior extremity of the dorsal vessel. Previously to the metamorphosis, the currents cease in these organs; and this is for the most part true also of the currents in the wings, which may be uniformly observed in these organs in the *Pupa* state, but which are very seldom to be seen for any length of time after the last metamorphosis, although they may be frequently seen in individuals that have recently emerged. This cessation of the circulation in the wings, is obviously the cause of their complete deficiency in reparative power; no losses of substance in them being ever made good, so that old bees may always be distinguished from young ones by the chipped indented edges of these organs, resulting from the accidental injuries to which they have been subjected.—The completely lacunar condition of the circulation in certain aquatic larvæ, especially of the *Gnat* tribe, has been already noticed (§ 453). The venous sinus surrounding the dorsal vessel seems to fill the entire visceral cavity; and the movement of blood, which can be most distinctly seen, on account of the multitude of corpuscles which it contains and the transparency of the bodies of these larvæ, appears to take place from behind forwards in the dorsal vessel, and from before backwards in the great venous sinus, without any diverging currents; not only the parietes of the body, but all their contents, being in such immediate relation with the circulating fluid, that no further provision is necessary.

463. The Circulating apparatus of the *Arachnida* presents us, at least in the higher forms of the class, with a much greater completeness than

* "Entomological Magazine," April, 1833, and October, 1836.

† Such, at least, appears to the Author to be the most probable interpretation of the facts ascertained by M. Blanchard's injections, taken in connection with those observed in the living animal.

that of Insects, as might be anticipated from what has already been said of the higher development of most of the organs in this class (§ 319); and the greater localisation of the respiratory organs involves the addition of a set of vessels, which shall be specially subservient to the aeration of the blood. Nevertheless there is no essential departure from the plan of structure already described as prevailing in Myriapods and Insects; and between the higher members of the former class, and the Macrourous Arachnida, such as the *Scorpion*, the conformity in the arrangement of the vascular system is extremely close. The following are the most important of the facts ascertained by Mr. Newport, in his minute investigations into the Circulating apparatus of the *Scorpionidæ*.*—The *dorsal vessel* runs continuously from the posterior extremity of the tail as far as the cephalo-thorax, and may be described as consisting of a *cardiac* portion (Fig. 205, *h, h*, 1–7) which occupies the abdomen, a *dorsal artery*, *a*, which is the anterior continuation of this, and a *caudal artery*, *q, q*, which is its posterior portion. The structure of the cardiac portion, which contains eight chambers (the posterior one, however, being very imperfect), is similar in most respects to that of the dorsal vessel of the higher Myriapods and perfect Insects; but it differs in this, that the valvular partitions between them are much less complete; a difference which probably has reference to the fact, that the cardiac portion has to send a great arterial trunk backwards as well as forwards. The blood enters the cardiac chambers from the surrounding sinus, through the auricular orifices *u, u*; and a portion of it is sent forth again, without being propelled into the adjacent chambers, through the systemic arteries *p, p*. The anterior continuation, or dorsal trunk, *a*, passes through the septum *x*, which divides the cephalo-thorax from the abdomen; and soon afterwards gives off a pair of large branches, which pass round the œsophagus, like the aortic arches of Myriapoda, to reunite below it into the ventral trunk *i, i*. It then gives off large branches, 1, 2, to the two posterior pairs of legs, and a smaller one, 2*, to the thorax; after which it separates into three principal trunks, of which one, *f*, is median, and runs above the œsophagus, giving off branches (6–14) to the cephalic ganglia and organs of sense, whilst the others, *g*, pass forwards at the sides and below the œsophagus, to supply the two remaining pairs of legs (3, 4) and the great prehensile claws (5). In this arrangement, there is a very marked conformity to the type of the Scolopendra (§ 461). Besides these vessels, the dorsal trunk gives off proper visceral branches, *z*, which proceed backwards along the anterior portion of the alimentary canal, inosculating with branches from the systemic arteries *p, p*. The posterior continuation of the multiple heart, constituting the caudal artery, *g, g*, passes backwards to the extremity of the tail, giving off in its course the systemic arteries *l, l*, and also the lateral branches *r, r*, which communicate with the portal trunk *s, s*.—The *ventral trunk*, *i, i*, which lies upon the upper surface of the gangliated nervous cord, extends backwards from its commencement in the aortic arches, nearly to the termination of the tail; gradually diminishing in diameter, as it gives off minute vessels for the nutrition of the cord, and successive pairs of branches, *k, k*, which pass downwards to communicate with the portal trunk; its terminal branches in the last caudal segment being distributed

* "Philosophical Transactions," 1843.

with the terminal nerves proceeding from the ganglionic cord.—Beneath the ganglionic column we find another trunk, designated by Mr. Newport as the *portal;* the purpose of which appears to be, to collect the blood from the systemic vessels, and to transmit it for aeration to the pulmonary branchiæ, C, C. This trunk is formed by the coalescence of branches from various sources; but especially, in the abdominal region, from the ventral trunk, and in the tail, from the caudal artery. Its branches are almost entirely distributed upon the respiratory organs. — Such is the general distribution of the proper 'vascular' portion of the circulating apparatus, which seems to be altogether arterial in its character; the venous circulation in the body at large would seem to be altogether 'lacunar;' whilst the return of the blood from the pulmonary branchiæ to the heart probably takes place by definite canals.— The course of the blood, therefore, would seem to be as follows. Of that which is returned from the respiratory organs to the chambers of the multiple heart, one part is sent forth by the systemic trunks proceeding from each chamber which it has just entered, another part is transmitted forwards from chamber to chamber into the dorsal trunk, whilst a third portion seems to be propelled backwards into the caudal trunk. The dorsal trunk distributes the blood to the cephalo-thorax and its organs of sensation and motion; whilst a portion of it is carried backwards through the aortic arches and ventral trunk, partly for the nutrition of the ganglionic cord, and partly for transmission into the portal system. So, again, the caudal trunk conveys the

FIG. 205.

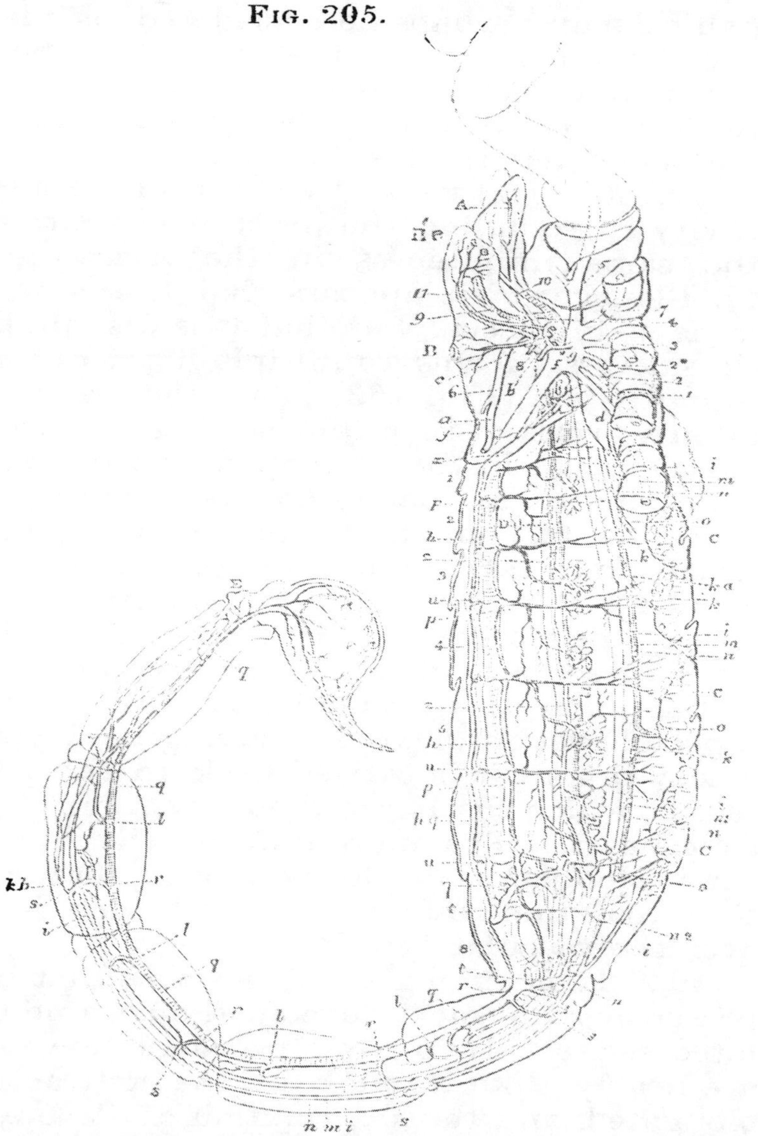

Diagram of the Circulatory and other organs in *Buthus*: —A, antennal claws; B, eyes; C, C, pulmono-branchiæ; D, alimentary canal; E, anal orifice; F, poison-glands; *a*, anterior continuation of the dorsal trunk, giving off 1, 2, 2*, branches to the posterior legs, besides the two great arches which form the ventral trunk; *b*, cephalic ganglia; *c*, optic nerve; *d*, great subœsophageal ganglion; *e*, ocelli; *f*, supraœsophageal artery, giving off 6—14, branches to the cephalic ganglia and organs of sense; *g*, lateral trunks of subœsophageal arteries, giving off 3, 4, 5, branches to the anterior members; *h*, *h*, chambers of the heart, 1—7; *i*, *i*, ventral artery; *k*, *k*, branches given off in each segment; *l*, *l*, branches of caudal artery; *m*, *m*, ganglionic cord; *n*, *n*, portal system of vessels; *o*, *o*, its branches; *p*, *p*, systemic arteries given off from the cardiac chambers; *q*, *q*, great caudal artery; *r*, *r*, its branches of communication with the portal trunk *s*, *s*; *t*, termination of cardiac portion in anterior dorsal trunk; *u*, *u*, auricular openings into cardiac chambers; *x*, diaphragm dividing cephalo-thorax from abdomen; *y*, anterior pair of systemic arteries distributed on this; *z*, visceral arteries.

blood backwards to the extremity of the tail, giving off nutrient branches to the various parts of that organ, and also transmitting a part of its contents directly into the portal system. The great portal trunk, (which may perhaps be considered as made up by the coalescence of the two lateral canals that exist in Insects and many Articulata) receiving blood from these sources, and collecting that which has been distributed through the tissues by the systemic branches of the dorsal and ventral trunks, transmits this fluid for aeration to the pulmonary branchiæ, from which it is returned by a set of 'branchio-cardiac canals' to the great cavity surrounding the heart.—The circulation has not been studied with the same minuteness in the *Araneidæ;* but from the researches of M. Blanchard* it appears that the course of the blood is essentially the same as in Insects, but that it is distributed by more perfect vessels. The dorsal vessel forms a multiple heart of four or five chambers in the abdominal region (Fig. 163); but the partitions between these chambers are often scarcely perceptible, so that the cardiac cavity is really single but elongated, as in the Stomapod Crustacea (§ 464). The blood is sent forth from this organ, partly by systemic vessels directly proceeding from its segments, but chiefly by the dorsal trunk which forms its anterior continuation; and from this, on its entrance into the cephalo-thorax, numerous branches are given off to the various organs of sensation and motion, and also to the viscera contained in that division of the body. The blood thus distributed through the system is stated by M. Blanchard to find its way to the pulmonary branchiæ by lacunar passages, neither ventral trunk nor a portal system of vessels having been yet made out; and from the respiratory organs it is carried back to the heart by distinct branchio-cardiac canals, which discharge themselves into the several chambers of the multiple heart. The more diffused the respiratory function is rendered, by the prolongation of the pulmonary branchiæ into tracheæ, the more does the circulation resemble that of Insects in the predominance of the lacunar over the vascular type.

464. It is among the *Crustacea,* that we find the sanguiferous system presenting the most developed form under which it exists in the Articulated series. In the lower orders, however, the segments of whose bodies are nearly alike throughout, the contractile portion of the dorsal vessel is elongated, and the distribution of its branches is nearly uniform in each segment; but the advance towards a higher type is shown in the order *Stomapoda,* the members of which have usually an elongated fusiform heart, developed as a muscular dilatation of a part of the dorsal vessel, which, towards its anterior and posterior extremities, possesses the character of a sanguiferous trunk. The same is the case also in the *Limulus.* But in the order *Decapoda,* which includes the most elevated forms of this class, we find the heart contracted into a short fleshy sac, possessed of considerable muscular power, and concentrating in itself the propellent force which is diffused in the lower tribes through a large part or the whole length of the dorsal trunk. This organ in the Crab and Lobster is situated on the median line, at the posterior part of the cephalo-thorax (Fig. 156, *c*); from its anterior part is given off a large *cephalic* trunk (*d*) which passes forwards, and soon subdivides into branches for the supply of the eyes,

* "Annales des Sciences Naturelles;" November, 1849.

antennæ, stomach, and neighbouring organs; whilst from its posterior extremity is given off the abdominal or *caudal* artery (*e*), which is, of course, of largest dimensions in the 'macrourous' Decapods. These two trunks, with the heart, obviously represent the entire dorsal vessel of the lower Articulata; the transitional form being shown in the Scorpion. Besides these trunks, a pair of large *hepatic* arteries is given off from the sides of the heart, which are exclusively distributed to the liver; whilst beneath the caudal artery, a large *sternal* trunk originates, which bends down towards the ventral aspect of the body, where it divides into an anterior and a posterior branch, the former of which supplies the thoracic members, whilst the latter distributes blood to the under surface of the abdomen. This may probably be regarded as homologous with the *ventral* trunk of the Scorpion, though it has not yet been shown to have any connection with the cephalic by aortic arches. The blood conveyed to the body by these arteries, appears to return through the lacunæ of the tissues into two sets of large venous sinuses; of which one consists of a series of flattened cavities, freely communicating with each other, which lies immediately beneath the shell of the back, covering the upper surface of the heart and dorsal trunks, and obviously representing the pericardiac sinus which incloses the dorsal vessel in other Articulata; whilst the other series lies at the bases of the branchiæ, on each side of the inferior surface of the thorax. The former collects the venous blood from the dorsal and caudal portions of the body, and carries it back to the heart, which it enters by two pairs of orifices, guarded by semilunar valves. The latter (which apparently corresponds to the 'portal' trunk of the Scorpion) collects the venous blood from the maxillæ and legs, and distributes it to the branchiæ for aeration. From these organs it is again collected by veins, which all coalesce in a pair of large trunks, the *branchio-cardiac* canals, that convey the aerated blood back to the heart. Hence, in that central organ, the venous blood received from a portion of the body is mingled with the arterial blood that is transmitted direct from the gills; and the fluid which is propelled through the systemic arteries is hence of a mixed character, as in Reptiles,—a class to which the Crustacea have many points of analogy. Thus the localisation of the respiratory organs in the higher Crustacea involves the existence of a special respiratory circulation. In the lower orders, however, the blood is aerated in its progress through the general system, as in Insects; and in many of them, the arterial as well as the venous portion of the system appears to be altogether 'lacunar.' In one of the most degraded forms of the class, we revert to the simplest possible type of the Circulating apparatus; even the dorsal vessel, which is so characteristic of the Articulata, being apparently deficient in the *Pycnogonidæ* (§ 316 *e*).

465. From the view which has thus been taken of the Circulating apparatus in the *Articulated* series, it will be seen that, notwithstanding the diversity of detail, there is a very general conformity to a definite plan; and that the tendency, as we ascend the series towards the higher forms of this apparatus, is to a concentration or specialisation in a single organ, of that impulsive power, which, in the lower tribes, was diffused through various parts of the system. Now when we follow a similar course with regard to the *Mollusca*, it will be seen that almost in the lowest forms of that series, the central organ is as powerful, and the circulation as much carried

on through distinct vessels, as in the highest Crustacea. The explanation of this general inferiority of the circulation in the Articulated series is partly to be found, as we have already seen, in the general diffusion of the respiratory apparatus; but it seems to be in part connected with the mechanical conditions of the system, the movements of whose several parts furnish an additional means of propulsion to the blood. Even in the highest animals, the alternate contractions and relaxations of the muscular apparatus assist in maintaining the flow of blood through the veins, the position of which renders them peculiarly subject to this kind of influence: for in every instance in which muscular contraction takes place, a portion of the veins of the part must undergo compression; and as the blood is prevented by the valves from being forced backwards, it is driven onwards towards the heart; whilst, as each set of muscles is relaxed, the veins that were compressed by it fill out again, to be again compressed by a renewal of the force. Thus it is that the general muscular movements, even in Man, have an important influence in promoting the venous circulation; so that continued exercise, involving the alternate contraction and relaxation of several groups of muscles, sends the blood more rapidly towards the heart, and thus increases the rapidity of its pulsations; whilst the sudden and simultaneous action of a large number of muscles after repose (as in rising up from the sitting or lying, to the standing posture), drives the blood to the centre of the circulation with a powerful impetus. Applying these facts to the case of Articulated animals, it becomes obvious that the general muscular movements, which are usually so active in them, may well be an important auxiliary to the circulation of their nutritive fluid, or may even be the principal means of maintaining it.*—In most of the instances in which we have hitherto found an organ of propulsion materially affecting the current of the circulation, it has transmitted the blood which it has received from the venous sinuses or canals, into the principal 'systemic arteries,' and may therefore be designated as the *systemic ventricle*. Among the Annelida, however, the impelling cavities are frequently situated at the commencement of the branchial circulation, and are to be considered as representing the *pulmonary ventricle* of higher animals. In the *Mollusca* generally, we find superadded to the ventricle, an *auricle* or contractile cavity, adapted to receive the blood transmitted to the heart, and to propel it into the ventricle; and the existence of these two cavities constitutes the

* To use the language of Dr. Grant (Lectures on Comparative Anatomy), "It is the restless activity of the Worm and of the Insect, that makes every fibre of their body as it were a heart to propel their blood and circulate their fluids. They require no complicated apparatus to accelerate the ever-active current of their blood, and hence the imperfect development of the great centre of their vascular system. Indeed it has been shown by Ehrenberg and by Nordnann, that, in the simplest of these animals, the trematoid Entozoa, the blood flows through the system by the mere motions of the body, without the least motion or impulse from the vessels which contain it. How differently circumstanced are the Mollusca! The inert Tunicata, the lowest of the Molluscous classes, fixed like plants upon the sea-beaten cliffs, and in which we can scarcely discover a trace of life, enclose in their motionless carcase a heart as highly developed as that of the Crustacea, the highest of the Articulated classes; and if they did not, their blood would soon stagnate in the complicated labyrinth of vessels and organs through which it has to pass. The slow-creeping snail that feeds upon the turf, has a heart as complicated as that of the red-blooded vertebrated fish which bounds with such velocity through the deep. It is because the fish is muscular and active in every point, that it requires no more heart than a snail to keep up the necessary movements of its blood."

typical character of the heart through nearly the whole of that sub-kingdom, although many variations are presented in their form and situation. Notwithstanding this higher development of the heart, the rest of the circulatory apparatus remains in a state of relative incompleteness; for in the lower classes of this series, the distribution of blood to the general system is almost entirely effected by lacunæ, of which the visceral cavity forms part, it being only in the respiratory organs that it moves through distinct vessels; and even in the higher tribes, in which the arterial part of the systemic circulation is generally truly vascular throughout, the venous portion of it is still in some degree lacunar.

466. Putting aside the *Bryozoa*,—in which there is no proper Circulation, the nutritive fluid that has transuded through the walls of the alimentary canal, being diffused through the general visceral cavity and passing into the tentacula (§ 298),—we commence our survey of the Circulating apparatus of the Mollusca with an examination of the condition which it presents in the class *Tunicata*. We here find a distinct heart, whose parietes although very thin are muscular, usually of a somewhat elongated form, and generally situated in the neighbourhood of the ovarium; but it is not yet divided into auricular and ventricular cavities; and the passages with which it communicates, are rather sinuses than proper vessels. In the long-bodied *Polyclinians*, the heart is situated at the extremity of the post-abdomen (Fig. 119, *o*); but in the *Botryllians*, which have no proper abdominal cavity, it is brought into much closer approximation with the pharyngeal sac (Fig. 121, *o*); and in the *Salpæ* it holds nearly the same relative position (Fig. 206, A, *e*). In the solitary *Ascidians*, however, it usually lies between the pharyngeal sac and the ventral surface of the body, partly excavated in the muscular tunic (Fig. 116, B, *a*). In the curious osculant genus *Pelonaia*, there seems to be no heart, although the vascular portion of the circulating apparatus is more complete than usual; but it is not improbable, that, as in many Articulata, to which that genus is related, the dorsal trunk is contractile throughout, and supplies the place of this organ.—The most curious feature in the action of the Circulating apparatus in this group, is the *alternation* which presents itself in the direction of the flow of blood; and this renders it impossible to designate either set of vessels or passages as arterial or venous, since both are arterial, and both venous, in their turns. The contraction of the heart, in all the species which are sufficiently translucent to allow it to be observed during life, is somewhat peristaltic in its character, proceeding from one extremity of its cavity towards the other; after the pulsations have continued in either direction for a minute or two, a short intermission takes place, during which the current of blood comes to a stand; and the peristaltic contraction then recommences in the opposite direction, the flow of blood being now first directed towards those parts from which it had last returned. If the course of the circulation be watched in an individual of the *Amaroucium* (Fig. 119) separated from the common envelope, it will be seen that the blood is not propelled towards the thorax in distinct vessels, but that it passes through the space left between the viscera and the lining membrane of the cavity; chiefly, however, along either its dorsal or its ventral aspect. When the ascending current takes the latter direction (which corresponds to its course in the higher Acephala), it enters a large vertical canal, which lies

in front of the branchial sac, and which is designated as the great thoracic or ventral sinus. From this sinus, a great number of vessels pass off transversely around the branchial sac; and these are connected together by numerous communicating twigs, which pass between the branchial fissures: so that a minute network is thus formed, in which the blood is exposed for aeration to the water that enters the sac. From this network the aerated blood is collected into another sinus lying on the dorsal aspect of the body; and this sinus also receives blood which has been transmitted to it, direct from the thoracic sinus, by means of a vessel that passes round the branchial orifice, so that the fluid which it contains is of a mixed character. From the dorsal sinus the blood returns to the heart, bathing the viscera in its course. When the reversal of the circulation takes place, the blood is first transmitted along the dorsal side of the body to the dorsal sinus; thence it is distributed over the branchial sac; and it finally returns to the heart by the ventral sinus.—The circulation in all the Tunicata is performed upon a plan essentially the same; but its peculiarity in certain compound *Clavellinæ* deserves a special notice. In the *Perophora* (as first observed by Mr. Lister),* the several individuals are not included in a common 'test,' but grow by footstalks from a common 'stolon.' Each of these stalks contains a double channel for the passage of blood, and these channels communicate in the common stem with those proceeding from other individuals; so that a vascular communication exists among them all, and this extends also to the undeveloped buds which sprout forth from the stolons. One of the channels in each peduncle enters the heart of the animal which it supports; and of the blood, when thus received into the propelling cavity, a great part is transmitted along the ventral canal over the branchial surface, whilst the remainder is distributed to the visceral apparatus and the mantle. Both these currents reunite in the dorsal sinus, which then conducts the blood, not back to the heart again, but into the peduncle, in which it may be seen to flow towards the principal stem. As in other cases, however, a reversal takes place about every two minutes; the blood then flowing *towards* the body in the peduncular channel that communicates with the dorsal sinus, and returning *from* it in that which is continuous with the heart. When one of the animals was separated from its peduncle, the circulation was at first disturbed, but soon regained its usual regularity; a new communication being apparently formed, by which a free passage of blood could take place between the dorsal sinus and the heart.—In the *Salpidæ,* which, as formerly remarked (§ 299 *b*), present the nearest approximation to the bivalve Acephala, we find a much more complete system of vessels, adapted to diffuse the blood over the general surface of the mantle, as well as upon the special respiratory organ. From the heart (Fig. 206, A, *e*) proceeds the ventral trunk (*f*), which passes towards the anterior extremity of the body, giving off numerous branches at right angles on either side, from which arise numerous smaller branches, the whole forming a minute net-work over the entire body (B). A separate set of vessels distributes blood to the branchial membrane, where it undergoes aeration. From all these vessels it is collected into the dorsal sinus, which conveys it back to the heart. Its movement among the viscera, however, is entirely lacunar. The same alternation may be witnessed in the direction

* "Philosophical Transactions," 1834.

of the flow, in the Salpæ, as in other Tunicata; but here, as elsewhere, it has been noticed that what may be considered as the *direct* current,—*i.e.* the forward movement of the blood from the heart, along the ventral trunk,—is of much longer duration than the *reverse*, in which the blood is propelled from the heart through the dorsal trunk; the number of pulsations being usually about twice as great for the former as for the latter.

FIG. 206.

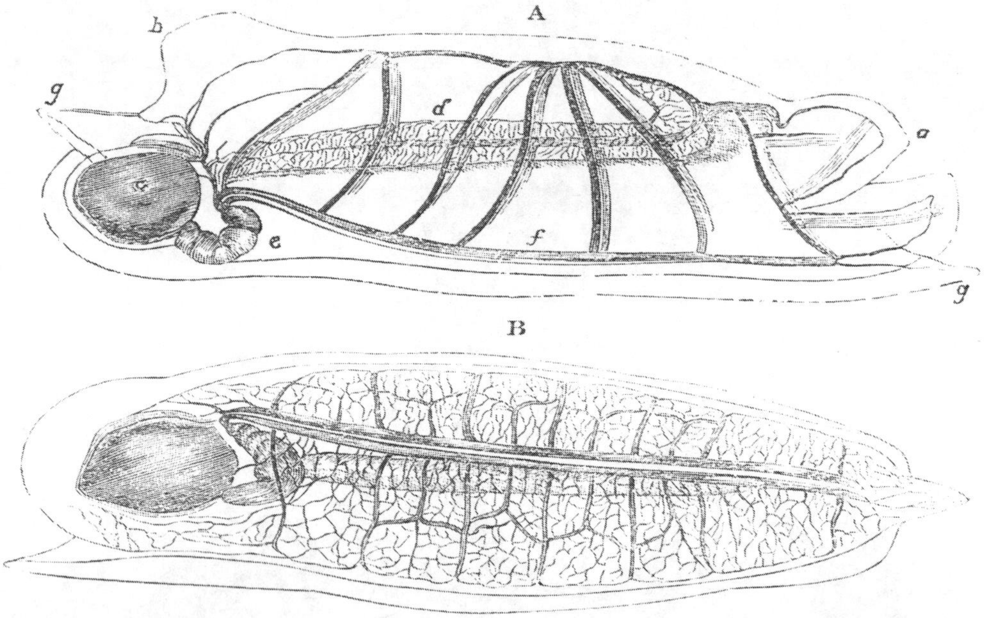

Circulating apparatus of *Salpa maxima*, as seen from the side at A, and from the ventral surface at B;—*a*, oral orifice; *b*, vent; *c*, nucleus, composed of the stomach, liver, &c.; *d*, branchial lamina; *e*, heart, from which proceeds the longitudinal trunk *f*, sending tranverse branches across the body; *g*, *g*, projecting parts of the external tunic, serving to unite the different individuals into a chain.

467. In the *Conchifera* we find the circulation carried on upon the same general plan, but the central organs are more highly developed, and the systemic vessels are more complete. The ventricle or impelling cavity of the heart (Fig. 207, *a*) is a distinct sac, of which the walls are formed by muscular fibres interlacing in every direction, and even projecting into the interior. From this centre, which is situated between the adductor muscle and the rectum, the blood is sent by two principal arterial trunks, an anterior (*d*) and a posterior (*e*) to the system at large; and thence it is returned, not by distinct veins, but by a system of lacunæ in the substance of the foot (B), of the labial tentacula (E), of the adductor muscle (H), of the glandular organs, and of the viscera generally; the contents of which are collected, in great part by a furrow (*i*) along the free margin of the mantle, into the afferent vessels (*h*, *h*), by which it is conveyed to the branchiæ. After being exposed in these organs to the action of the air contained in the surrounding water, it repasses to the heart by two large branchio-cardiac canals, which do not enter the ventricle, but terminate in auricles (*b*), of which one is usually placed on each side. The branchio-cardiac canals also receive the blood brought back by the venous sinuses from the general surface of the mantle, which still continues to act as a respiratory organ, although the proper branchiæ of the *Lamellibranchiate* bivalves are undoubtedly the special instruments of this function. In the

Brachiopoda, in which these special instruments are not developed, there are of course no proper branchial vessels; and the blood is aerated by the

FIG. 207.

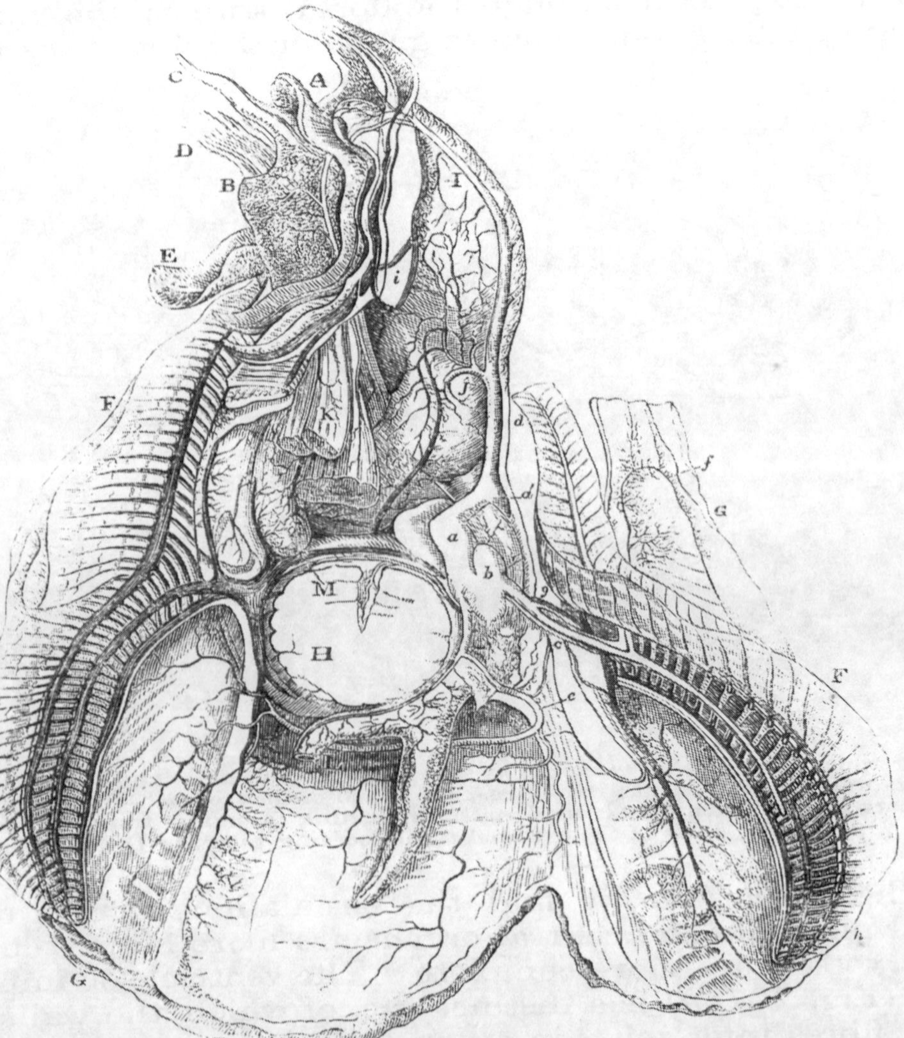

Pinna marina laid open, to show the arrangement of the Circulating apparatus:—A, mouth; B, foot; C, digitiform appendage to the foot; D, byssus; E, labial tentacula; F, F, branchiæ, those of the right side being in place, those of the left being divided near their anterior extremity, and turned back, so as to expose their anterior surface; G, G, the mantle, of which the left lobe has been detached and folded back; H, posterior adductor muscle; I, first stomach, covered by the liver; K, retractor muscles of the foot; L, anus; M, glandular organ, probably urinary; between this organ and the branchiæ of the right side is seen the second stomach; *a*, aortic ventricle; *a′*, anterior portion of this ventricle, passing round the rectum; *b*, one of the auricles turned back, the other being seen in its natural position on the opposite side of the ventricle; *c*, one of the branchio-cardiac canals; the other is seen in front of the adductor muscle H; *d*, anterior aortic trunk; *e*, posterior aortic trunk; *f, f*, palleal veins, proceeding to empty themselves into the branchio-cardiac canals at *g*; *h, h*, afferent vessels of the branchiæ; *i*, canal of communication between these last and the general lacunar system of the abdomen.

medium of the palleal vessels and sinuses alone.—On returning from these, it spreads itself through a large sinus that surrounds the visceral mass; and from this it is received into the auricle on either side.*—

* See Prof. Owen's account of the Circulating Apparatus of Brachiopoda in the "Annales des Sciences Naturelles," Mai, 1845. It is from the admirable series of Memoirs on the Circulation in Mollusca, contributed by Prof. Milne-Edwards to the same Journal, that the sketch in the text is for the most part condensed.

Although the auricular cavity is double in most Bivalve Mollusca, it is not to be regarded as representing the two auricles of warm-blooded animals, of which one receives the blood from the system at large, and the other that which is transmitted from the lungs; since here the two auricles have the same function, being both respiratory, and being merely repetitions of one another, separated for the sake of convenience. In the *Oyster*, in fact, they are united into a single cavity; whilst in another tribe, the *Archidæ*, the ventricle is divided like the auricle, in conformity with the breadth of the back of the animal, and the consequent separation of the gills from one another; and in the *Brachiopoda*, also, there are two ventricles at a considerable distance from each other.

468. Among the *Gasteropoda*, we find the same general arrangement of the circulating system; but the situation of its centre, and the distribution of its vessels, present great variations in different orders (§ 303). The heart is composed of a ventricular and of an auricular sac (Fig. 127, *p*, *q*); the former of which is usually provided with walls of considerable firmness, having muscular bands interwoven with its coats and projecting slightly into its cavity. In some Gasteropoda, as in Conchifera in general, the auricle is double; and in a few genera, even the ventricle is partly divided by the intestine which passes through it. The blood which is propelled from the ventricle by one principal systemic artery, or aorta (*u*), is distributed by its branches to the various organs of the body; in these it is dispersed through a capillary network possessing distinct parietes; and from this network it passes, not into a proper venous system, but into a set of intercommunicating lacunæ, which even pour it into the visceral cavity, so that it bathes the external surface of the alimentary canal. From a part of this lacunar system, it is collected into the afferent trunks which distribute it to the respiratory organs, whether branchial or pulmonary; and after having been aerated in these, it is returned to the auricle by the branchio-cardiac canals. Into these, however, is also poured the venous blood that is returned by a portion of the lacunar system, without passing through the respiratory organs; so that the general plan of the circulation strongly resembles that which we have seen in the Crustacea (§ 464), only a part of the blood which is sent forth from the heart, being subjected to the aerating process before returning to it again. The proportion which thus escapes aeration, however, seems to differ considerably in the several orders and genera of the class.—Many curious varieties in the arrangement of the vascular system of Gasteropoda might indeed be enumerated; but it must here suffice to mention that they principally have reference to variations in the condition and position of the respiratory organs, and do not present any essential departures from the type just described. Thus in *Haliotis* and *Patella*, certain parts of the arterial system present the lacunar condition; and in *Thethys*, the branchio-cardiac canals unite into an immense venous sinus, which occupies the dorsal region, and which is only separated from the visceral cavity by a thin membrane.*—In Gasteropod Mollusks, as in the higher

* The attempt has been made by M. de Quatrefages to show that the *Nudibranchiate* Mollusca (§ 304 *c*) differ from other Gasteropoda in the deficiency of a proper venous system; but this has been negatived, on the one hand, by the discovery that in the Gasteropoda in general the systemic venous circulation is lacunar; whilst, on the other hand, the Nudibranchiata have been proved to possess, like other Gasteropoda, a proper branchio-cardiac canal.

Crustacea, we find the liver very largely developed, and supplied with blood by a large arterial trunk (Fig. 127, *o*). We shall subsequently find that the function of the liver is in part supplementary to that of the respiratory organs; and that the former organ is in general the largest, when the changes for which the latter are adapted, are least actively performed.

469. Hitherto we have usually found the respiratory organs, whether branchial or pulmonary, interposed between the capillaries of the system and the central propelling cavity; the canals which collect the blood from the different organs of the body, uniting only to distribute it again, without any fresh impulse being given to their contents. The only instance of the interposition of any thing like an impelling cavity, between the systemic and the respiratory vessels, was seen in certain *Annelida*, which are provided with *branchial hearts* (§ 459). Now in the class of Fishes, the heart is entirely respiratory, the arterial trunk which proceeds from it being distributed at once to the gills, and the blood which has been aerated in them being returned into a systemic artery or aorta, whence it proceeds to the body at large; and in the class of *Cephalopoda* we meet with a condition of the circulating apparatus, which manifestly establishes the transition between that of the Mollusca in general, and that which is peculiar to Fishes. The *systemic heart* of the *Octopus* consists of only one cavity or ventricle (Fig. 135, *p*), which is usually of a nearly globular form, tolerably strong and muscular, and exhibiting on its internal surface bundles of fibres (*carneæ columnæ*) interlacing with one another, as well as distinct valves protecting the orifices by which the blood enters it. The aorta (*q*), with other arteries which proceed from it, distributes arterial blood to the general system; and this is returned by means of a regular system of veins possessing distinct parietes, of which one is seen at *r*. Of these, however, one pair opens directly into the visceral cavity, which thus, as in the inferior Mollusca, is made to take part in the venous circulation. By the union of the systemic veins on either side, the blood is conveyed, not immediately to the gills, as in the other Mollusca, but to two superadded impelling cavities, or *branchial hearts* (*s*), one of which is situated in connection with each row of gills. These branchial ventricles are less powerful than that which propels the blood through the system; but they are still sufficiently muscular and contractile to accelerate the circulation through the respiratory organs, and thus to prepare the blood for the maintenance of the increased muscular exertions which are required for the superior locomotive powers of these animals, as well as for the general activity of the functions of their highly-organised bodies. The blood that has been thus impelled through the branchial arteries (*s'*), is returned in an aerated condition by the branchial veins (*t*), which unite into a single branchio-cardiac canal on each side. These canals, before entering the ventricle, present dilatations (*u, u*), which have been usually regarded as mere sinuses; but they have been observed to be distinctly contractile, and must therefore be considered as representing the double auricles of the lower Mollusca. Here, therefore, we have, sketched-out as it were, the complicated form of the vascular system in warm-blooded animals possessed of a complete double circulation; the trunks which convey the blood to the respiratory organs being furnished, like that which distributes it to the system at large, with an impelling cavity, by which a constant

and regular current is maintained. This structure, however, is peculiar to that order of Cephalopoda, which, in the general superiority of its organisation, its deficiency of external shell, and its possession of a rudimentary internal skeleton, exhibits many points of resemblance to Fishes (§ 307 *b*). In the *Nautilus* tribe (§ 307), on the other hand, whose general structure is more analogous to that of the inferior Mollusks, the vascular system presents nearly the same arrangement as in the Gasteropoda; for the veins that return the blood from the system enter a common sinus, which has not, however, a distinctly muscular character, and does not seem to possess contractile powers; and from this proceed the four branchial trunks which distribute the blood to the two pairs of gills, whence it is conveyed back to the heart or systemic ventricle by branchio-cardiac canals.—From various parts of the systemic venous trunks, both in the Tetrabranchiate and Dibranchiate orders, a curious series of follicles or little sacs is seen to proceed, forming spongy masses, sometimes of considerable size (Fig. 135, *r'*). The use of these is not certainly known. Their glandular aspect, and the distribution of arterial branches over their surface, joined with the peculiar character of the fluid found in them, has caused them to be regarded as secreting organs, destined to purify the circulating fluid; and it has been thought probable that they are really Kidneys.

FIG. 208.

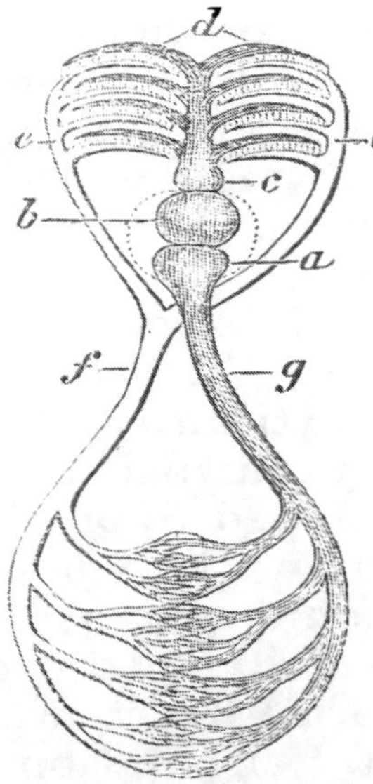

Diagram of the Circulating apparatus of *Fishes*;—*a*, the auricle; *b*, the ventricle; *c*, the trunk supplying the branchial arteries, *d*; the aerated blood returning from the gills is conveyed by *e*, *e*, the branchial veins, to *f*, the aorta, which distributes it to the system; thence it is collected, and returned to the auricle, by the veins which unite in the vena cava, *g*.

470. Although in *Fishes* we find the same simple conformation of the heart as that which exists in the Molluscous classes, the alteration in its position, relatively to the other parts of the vascular system, occasions its influence on the function of the circulation to be greatly modified. The blood expelled from the single ventricle (Fig. 174, *a'*), the cavity of which possesses firm muscular parietes, is carried at once to the gills through an arterial trunk, *c*, which presents a bulbous enlargement, *a''*, at its origin; this *bulbus arteriosus*, as it is termed, will be hereafter shown to exist at an early period of development in the higher Vertebrata, and to conduce towards the formation of the two principal trunks, which are subsequently to arise from the heart (§ 488). This trunk subdivides into four or five branches on each side (*c'*, *c'*, *c'*, *c'*, Fig. 174, A), which run along the branchial arches, sending ramifications to every filament. After being thus aerated, the blood is collected by the branchial veins, *d*, into the great systemic artery or *aorta*, which then distributes it to the different organs of the body; and thence it is returned to the auricle by the systemic veins (*b'*, *b''*, *b'''*), which, before entering it, unite in a large dilatation, the *sinus venosus* (*b*).—The circle just described, of which a plan is given in Fig. 208, appears simple in character; but it possesses one peculiarity which is worth notice, as foreshadowing more important modifications in higher classes. Two or three small arteries are usually seen passing off from the branchial arches, so as to convey the pure aerated blood directly

to the head, instead of transmitting it to the general systemic trunk. It will be hereafter shown that a similar provision exists in the Crocodile, and has a very important purpose in its economy; and that the same condition is manifested up to the termination of the embryo state of the higher Vertebrata, including the Human species.—Of the blood which is being returned by the veins from the systemic capillaries, a part is diverted into another channel before reaching the heart; for the veins of the digestive and generative organs, together with a part of those proceeding from the posterior part of the body and tail, reunite into trunks which convey the blood to the liver and kidneys; and it is minutely distributed through these organs, in order that it may undergo purification by the elimination of their respective secretions. After this process has taken place, the blood is conveyed to the heart by the hepatic and renal veins, which enter the vena cava, or proceed direct to the sinus venosus. Thus the *portal* circulation, as it is termed, holds precisely the same relation to the general circulation in Fishes, as did the respiratory circulation in Crustacea; being interposed, for the purification of the blood which has circulated through the system, between a part of its capillaries and the heart.

471. The foregoing is the general plan upon which the Circulating apparatus of Fishes is constructed; but there are some remarkable departures from it, which require special notice. The most important of these is that which is presented in the *Amphioxus* (§ 321 *a*), the condition of whose sanguiferous system almost carries us back to the type of *Eunice* (§ 459); for the impelling power is not here concentrated in a single organ, but is distributed among a number of separate pulsatile dilatations developed upon the vascular trunks. Thus we have not only a principal branchial heart (Fig. 172, *ov*), which corresponds in its position to the heart of the higher Fishes; but there is also a minute contractile bulb at the origin of each branchial artery, so that there are from twenty-five to fifty of these bulbs on either side. The arterial arches, also, at the anterior extremity of the body, appear to be contractile. The venous system, too, is furnished with its own pulsatile dilatations; for a venous heart is developed upon the vena cava or great dorsal vein, and another upon the trunk of the vena portæ, which runs along the ventral side of the intestine.—Some traces of a similar arrangement may be seen in other Fishes, especially in those of the Cartilaginous group. Thus in *Myxine* there is a portal heart, which contracts regularly, and assists in maintaining the portal circulation; in *Chimæra*, which has no bulbus arteriosus, each of the pair of large arteries given off from the aorta to the pectoral fins, has a contractile bulb at its origin; and a pulsating dilatation is found at the extremity of the tail of the *Eel*, where it receives the blood from the delicate veins of the end of the caudal fin, and propels it into the caudal vein.—There is another set of modifications, however, in the Circulating apparatus of Fishes, which conducts us towards the Reptilian type. For there are numerous instances in which the filaments on one or more of the branchial arches remain undeveloped; so that the artery of that arch, instead of subdividing into capillaries, carries on the blood at once into the aorta; and thus the fluid transmitted through that trunk to the system has only in part undergone aeration, the head, however, being always supplied with pure arterial blood from the branchial veins. Concurrently with this change, we find some provision for *atmospheric* respira-

tion; either in the advanced development of the air-bladder into a rudimentary lung, to which air gains access through a tracheal canal, as in the *Lepidosteus* and *Polypterus*, the two existing representatives of the great *Sauroid* family so abundant in the earlier periods of the Earth's history; or in the development of peculiar organs, analogous to these in function, but not homologous in structure, being saccular prolongations of the lining membrane of one of the gill-chambers, such as are found in the *Cuchia*, an eel-like fish of the Ganges, and in some others of the same tribe (§ 524). The blood is sent to these pulmonary organs by prolongations of the branchial arteries, and is returned from them in an aerated state into the aorta. In some of the Fishes whose Reptilian affinities are the greatest, there is a slight indication of that subdivision of the bulbus arteriosus into two distinct trunks, which takes place during the development of higher animals (§ 488).

472. Quitting now those classes which are modified for existing in water, and passing on to the air-breathing Vertebrata, we find that very important modifications of the Circulating system are necessary, to adapt the animal to the conditions of atmospheric respiration. It is evident that the blood will be aerated much more rapidly when exposed to the air itself, than when merely submitted to the small quantity which is diffused through the watery element. If, therefore, the whole amount of the circulating fluid be thus exposed, the changes which it undergoes will be performed with such increased energy, that, if the other vital processes be made to conform to them, a "warm-blooded" animal is produced at once. But as *Reptiles* are intended to lead a life of comparative inertness, and to exist in circumstances which would be fatal to animals of higher organization, the respiratory process is reduced in amount, by the peculiar arrangement of the sanguiferous system now to be described. The ventricle of the heart is either single (which is the case in *Batrachia*) or it is divided by an imperfect septum (as in most of the higher orders), so that the blood which is poured into it from its two sides can mingle more or less freely in its cavity. From this ventricle, a single *truncus arteriosus* is given off in Batrachia, which distributes the blood, by a series of arches very closely resembling the branchial arches of Fish,—partly to the system, through the cephalic arteries (Fig. 212, *o*, *t t*) which come off from the first and second branchial arches (*br*, 1, 2), and through the aorta (*av*) which is formed by the union of the second pair with branches from the third,—and partly to the lungs through the pulmonary arteries (*ap*). The distribution of the blood in the higher Reptiles is very nearly the same in effect; but the pulmonary and systemic trunks are kept distinct at their origin, by the division of the 'truncus arteriosus;' the former arising from the right, and the latter from the left side, of the partially divided ventricular cavity. Still the general appearance of branchial arches is preserved, as is shown in Fig. 181; and a part of the blood expelled by each contraction of the ventricle, is sent to supply the requirements of the nutritive system, and a part is separated for aeration. The pure arterialised blood which returns from the lung is conveyed to the left auricle, whilst the venous blood which is transmitted by the systemic veins enters the right; these two auricles are hence not repetitions of one another, but have distinct functions. Both empty themselves into the ventricle, in which the blood derived from these different sources is mixed, and from

which one part is again sent to the body, and another transmitted to the lungs, according to the plan represented in Fig. 209. The portal system of Reptiles corresponds with that of Fishes, in the circumstance that the kidneys are supplied by it with venous blood, as well as the liver; and also in deriving part of its supply from the veins of the tail, posterior extremities, and genital organs, instead of (as in higher animals) from the veins of the digestive organs alone. The degree in which the renal, hepatic, and portal circulations are united, however, and in which they are supplied from the systemic veins, differs considerably in the several orders; the closest approximation to Fishes being presented (as might be anticipated) in the order Batrachia.

FIG. 209.

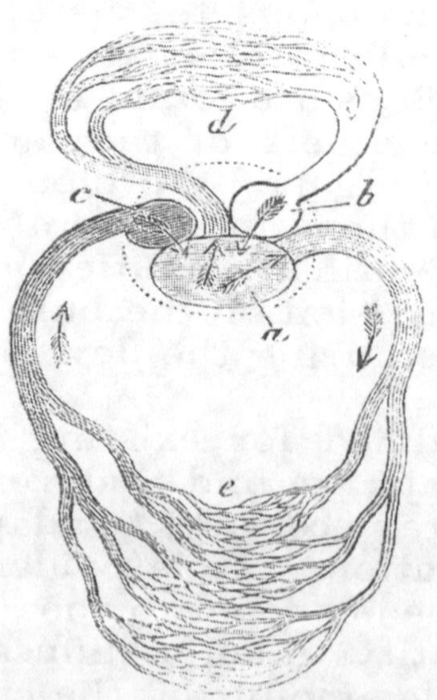

Diagram of the Circulation in *Reptiles*:—*a*, single ventricle, receiving the aerated blood from *b*, the pulmonary auricle, and venous blood from *c*, the systemic auricle; and propelling part of this mixed fluid to the pulmonary capillaries *d*, and part to the systemic capillaries, *e*.

473. Although the foregoing may be regarded as the *general* type of the Circulating apparatus in the class of Reptiles, yet there are some very curious modifications of it, which connect it very closely with that of Fishes on the one hand, and with that of Birds and Mammals on the other. These are shown in the several genera of *Perennibranchiate Amphibia*, which, as formerly explained (§ 323 *a*), present us with a very complete series of transitional forms connecting the two classes: and also in the progress of the metamorphosis of the *Batrachia*, which in their tadpole or larva condition must be regarded as Fishes in every essential particular of their organisation. Their circulation is for a time performed exactly upon the same plan as in that class; the blood being transmitted from the simple bilocular heart to the branchial arches, then being propelled through the branchial filaments, and after aeration being circulated through the system. The mode in which the transition is effected from this condition of the vascular organs, to that which they present in the perfect Reptile state of the animal, described in the preceding paragraph, will be rendered intelligible by the accompanying figures. In Fig. 210 is seen the arrangement of the parts before the metamorphosis has commenced. Three branchial trunks (*ab*, *ab*) pass off on each side of the heart, terminating in a minute capillary network which is contained in the branchial arches (*br*, 1, 2, 3), and by which the blood is aerated during the aquatic existence of the animal; from this network the returning vessels take their origin, which unite into trunks (*vb*), one for each gill; and of these the first gives off the main arteries (*t t*) to the head, while the second and third join to form the great systemic artery (*av*), as in Fishes. But, besides these vessels, there are some small undeveloped branches (1, 2, 3), which establish a communication between each branchial artery and the returning trunk that corresponds with it. There is also a fourth small trunk (*ap*) given off from the heart, which unites with another small branch from the aorta, to be distributed upon the (as yet) rudimentary lungs. After the metamorphosis has begun, however, by which the animal from a Fish has to be converted into a Reptile, the branches that connect the arteries of

FIG. 210.

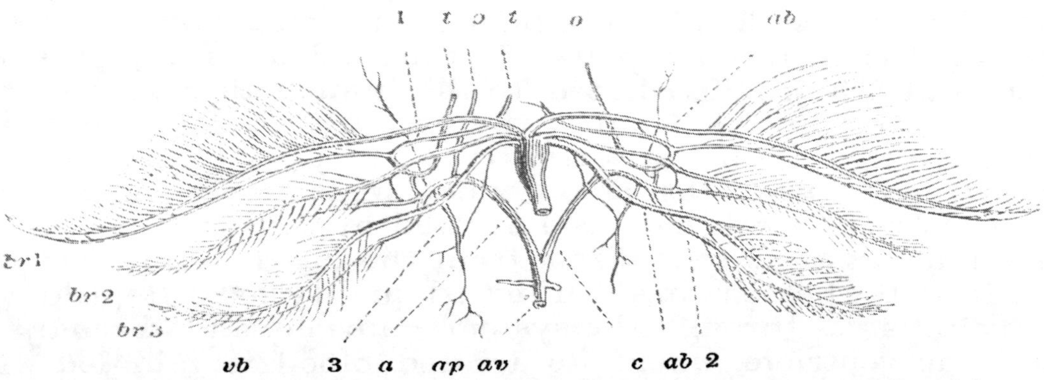

Respiratory circulation of the *Tadpole*, in its first state:—*a*, arterial bulb, giving off three pairs of branchial arteries, *ab*, to supply the gills *br* 1, *br* 2, *br* 3, from which return the branchial veins *vb*; *av*, aortic trunk, formed by the union of two arches *c*, each of which receives the veins of the second and third pairs of branchiæ; *t*, *t*, arterial trunks proceeding to the head, derived from the first pair of branchial veins; 1, 2, 3, communicating branches, between the branchial arteries and veins, as yet but little developed; *ap*, pulmonary arteries, almost rudimentary; *o*, orbital artery.

the gills with their returning trunks are much increased in size, so that a large part of the blood flows continuously through them without being sent to the gills at all, and the branchial vessels are themselves relatively diminished (Fig. 211); at the same time, the pulmonary trunk, which was before the smallest, becomes the largest, so that an increased proportion of blood is sent to the lungs. By a continuance of these changes, the branchial vessels gradually become obliterated, and the communicating branches, which were at first like secondary or irregular channels, now form part of the continuous line of the circulation (Fig. 212); the upper one sending off the cerebral vessels, the second and third uniting to supply the trunk, and the fourth passing as before to the lungs.—Turning from these to the *Perennibranchiata*, we find in the *Lepidosiren* a plan of circulation but little elevated above that which has been just described as existing in certain Fishes that present approxi-

FIG. 211.

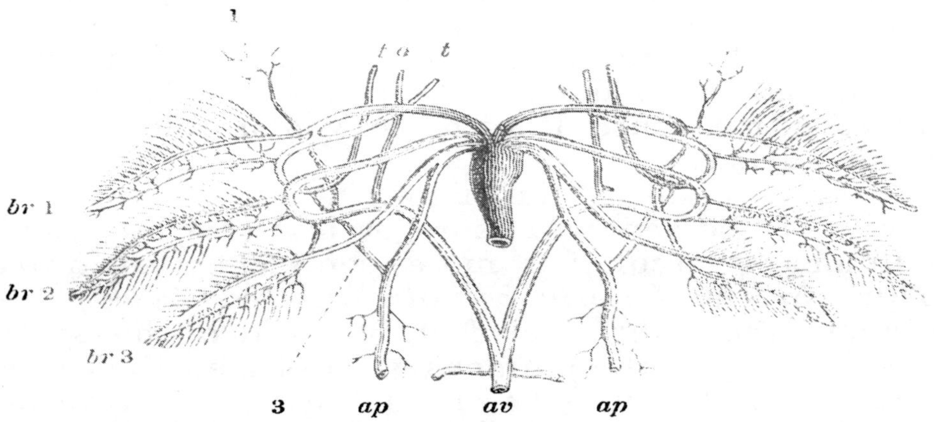

The same in a more advanced state; the communicating branches and the pulmonary arteries being now greatly enlarged in proportion to the branchial vessels.

FIG. 212.

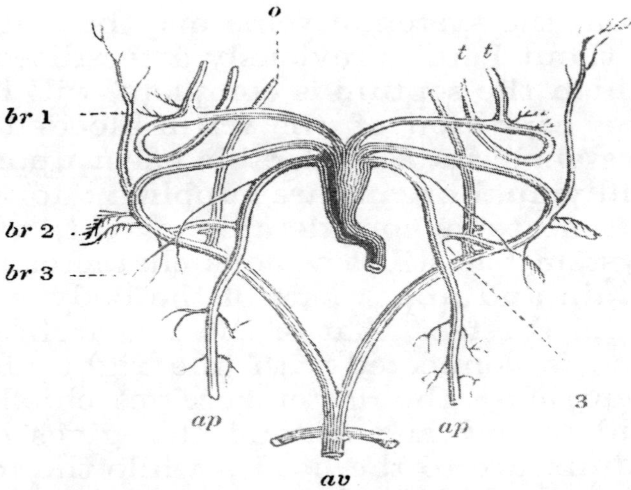

The same in the perfect *Frog*; the pulmonary arteries and the communicating branches being now fully developed, and the branchial vessels atrophied.

mations to the Reptilian class. The ventricular cavity is single, and gives off but one primary trunk, which is furnished with a 'bulbus arteriosus.' This trunk subdivides into six pairs of branchial arteries; of which the 1st, 4th, 5th, and 6th are distributed to the branchial fringes; whilst the 2nd and 3rd are not thus distributed, their arches having no gills attached to them, but reunite to form the aortic trunk, first giving off, however, the pulmonary arteries. Thus the blood which finds its way into the aorta, consists partly of that which has been aerated in the branchiæ, and partly of that which has passed to it direct from the heart. But that which is transmitted from the heart is itself of mixed quality, as in Reptiles; for the pulmonary vein passes through the systemic auricle, and discharges itself directly into the ventricle, where its aerated blood is mingled with that returned by the systemic veins.—In the *Proteus*, the arrangement of the vascular system permanently resembles that which has been represented as intermediate between the larva and perfect condition of the Frog. This animal is provided with lungs slightly developed, as well as with permanent gills; and the blood which is expelled from the ventricle is partly transmitted through the gills, partly finds its way directly into the aorta by means of the communicating branches, and a small quantity is transmitted to the lungs; the latter is returned perfectly arterialised to the auricle here developed upon the pulmonary vein, and is afterwards mixed in the ventricle with the venous blood transmitted from the systemic auricle.

474. In many of the higher Reptiles, on the other hand, we find the cavity of the ventricle more or less perfectly divided into two; and the pulmonary circulation more completely separated from the systemic. Thus, in the *Lacerta ocellata* (spotted lizard), whose ventricle is partly divided, the right side of it, into which the systemic auricle discharges itself, gives off the pulmonary trunks, so that a large proportion of the venous blood returned from the system is transmitted to the lungs for aeration; and this being returned to the pulmonic auricle, is discharged into the left side of the ventricle, from which the systemic arteries proceed. As long as there is any direct communication, however, between the two sides of the heart, it is obvious, that a part of the blood returned from the systemic veins may be sent immediately into the aortic trunks, without being previously arterialised; and in proportion to the degree in which the septum is complete, will be the approach of the animal towards the condition of the warm-blooded Vertebrata. The distribution of the vessels has a considerable effect upon the character of the fluid with which individual organs are supplied; for in Reptiles which manifest this separation to a considerable extent, a part of the blood transmitted to the system has still a venous character, whilst that which is furnished to the brain and upper part of the body is purely arterial. This difference arises from the fact, that of the two arches which unite to form the aortic trunk, one is connected with the right and the other with the left side of the ventricle; the latter receives chiefly the arterial blood from the left or pulmonary auricle, and this gives off branches which convey it without admixture to the head; while the main trunk passes on to unite with the second aortic arch that arises from the right side of the heart, and consequently is supplied with blood almost entirely venous, which has been discharged from the system into the right auricle. This second arch,

before its union with the first, however, gives off a large branch, which is distributed to the intestines and other viscera, and which, therefore, contains venous blood with little admixture of arterial ; and the common aortic trunk, formed by the union of the two arches, conveys mixed arterial and venous blood to the remainder of the trunk and members. It is beautiful to observe how by these simple contrivances the greatest economy of material is obtained, whilst each organ is supplied with blood sufficiently oxygenated to maintain its functions.—The *Crocodile* presents us with a condition of the vascular system still more allied to that of warm-blooded Vertebrata ; the ventricular septum being complete, and the circulation, as far as the heart is concerned, being truly double. Still, however, whilst the principal aortic trunk arises from the left ventricle, which contains nothing but arterialised blood, a second arch arises from the right (or venous side) along with the pulmonary artery, of which it might almost be considered a branch ; and this, after giving off its intestinal branches, enters the first trunk, which has already furnished the cerebral arteries with pure arterial blood, and transmits the mixed fluid to the rest of the system. There is another communication between the trunks arising from the two sides of the heart, by means of an aperture which passes through their adjoining walls just after their origin ; so that although the blood in the heart is entirely venous on one side and arterial on the other, it undergoes admixture in the vessels, according to the character of the functions to which it is to minister. We shall presently see a remarkable analogy to this distribution of the vascular system, exhibited in the fœtal condition of Birds and Mammalia (§ 489).

475. In the highest form of the Circulating system, that possessed by the 'warm-blooded' Vertebrata, *Birds* and *Mammals*, there is a complete double circulation of the blood ; each portion of it, which has passed through the capillaries of the system, being aerated in the lungs, before being again distributed to the body. This is effected by a form of the vascular apparatus, of which a sketch was presented in the *Cephalopoda* (§ 469), and to which a near approach is exhibited by the higher Reptiles. The heart consists of four cavities, two auricles and two ventricles ; those of the right or venous side having no direct communication with those of the left or arterial side ; and the vessels proceeding from them being entirely distinct, and having no connection whatever except at their capillary terminations. The blood transmitted by the great veins of the system to the *right* auricle or receiving cavity, passes into the ventricle or propelling cavity, and is transmitted by it through the pulmonary arteries to the lungs of the two sides. After being there arterialised by exposure to the atmosphere, it is brought back to the *left* auricle ; and having been poured by it into the corresponding ventricle, is transmitted through the great systemic artery or aorta to the most distant parts of the body (Fig. 213). The heart is therefore completely duplex in structure, and, so far as its functions are concerned, might be regarded as consisting of two distinct portions ; for economy of material, however, these are united, the partition of the ventricles serving as the wall to each. In the *Dugong* (one of the aquatic *Pachydermata*), however, the heart is bifid at its apex, and thus presents a partial division into two separate organs, not only functionally but structurally.—The portal circulation is limited in Mammalia to the liver, the kidneys being supplied with arterial blood only. In Birds, how-

ever, we find a trace of that arrangement of this peculiar off-set from the general circulation, which has been pointed out as existing in Reptiles and Fishes; for the great portal trunk receives its blood, not only from the veins of the digestive apparatus, as in Mammalia, but also by branches from those of the pelvis and posterior extremities; and it still communicates with the renal circulation, although this connecting branch seems rather destined to convey blood *from* than *towards* the kidney.

FIG. 213.

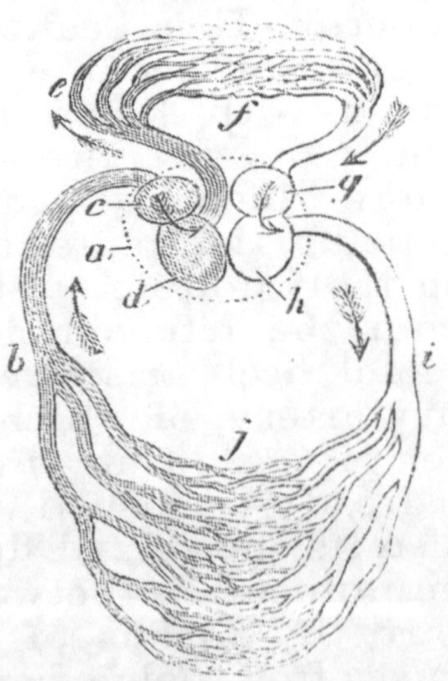

Diagram of the Circulating apparatus in *Mammals* and *Birds*:—*a*, the heart, containing four cavities; *b*, vena cava, delivering venous blood into *c*, the right auricle; *d*, the right ventricle propelling venous blood through *e*, the pulmonary artery, to *f*, the capillaries of the lungs; *g*, the left auricle, receiving the aerated blood from the pulmonary vein, and delivering it to the left ventricle, *h*, which propels it, through the aorta, *i*, to the systemic capillaries, *j*, whence it is collected by the veins, and carried back to the heart through the vena cava, *b*.

476. Various peculiarities in the distribution of the vascular system, which are presented by different orders of Birds and Mammals, would be worthy of notice if our limits permitted. Of these one of the most remarkable is the modification both of the venous and arterial trunks, existing in the *Cetacea* and other diving animals, which are occasionally prevented from respiring for some time, and in which, therefore, the arterialisation of the blood is checked. Various arteries of the trunk are here found to assume a ramified and convoluted form, so that a large quantity of blood may be retained in the reservoirs formed by these plexuses; whilst the venous trunks exhibit similar dilatations, capable of being distended with the blood which has been transmitted through the system, so as to prevent the heart being loaded with the impure fluid, whilst the lungs have not the power of arterialising it. In some diving animals, this object is effected, not so much by a number of venous plexuses, as by a single great dilatation of the vena cava before it enters the heart, resembling the 'sinus venosus' of Fishes.—In other instances, the force with which the blood is sent to particular organs seems to be purposely diminished, by the division of the trunk that conveys it into a number of smaller vessels, which, after a tortuous course, unite again and are distributed in the usual manner. A structure of this kind is found in the arteries of the long-necked grazing animals, to which the blood would be transmitted with too great an impetus, owing to the additional influence of gravitation, were it not retarded by such a contrivance. A similar distribution of the arteries is found in the trunks supplying the limbs of the Sloths, and of other animals which resemble them in tardiness of movement. In other cases, the arterial canals are specially protected from compression by surrounding organs, in order that there may be no obstruction to the passage of blood through them, and that they may be guarded from injury; thus, in the fore leg of the Lion, where all possible force and energy is to be attained, the main artery is made to pass through a perforation in the bone, as if to secure it from the pressure of the rigid muscles, which, when in a state of contraction, might otherwise altogether check the current through it. In most Mammals, as in Man, the right anterior extremity is

more directly supplied with blood from the aorta than the left; so that the superior strength and activity of this limb would seem to be not altogether the result of habit and education, as some have supposed; in Birds, however, where any inequality in the powers of the two wings would have prevented the necessary regularity in the actions of flight, the aorta gives off its branches to the two sides with perfect equality. Some further peculiarities in the distribution of the arterial system will be hereafter noticed (§ 492).

477. Having now traced the Sanguiferous system to its highest form, it is proper to inquire how far this differs functionally from that simple condition which it presents in the lowest forms in which it exists. There can be no doubt that, in the higher animals possessed of a distinct muscular heart, *this* is the chief agent in keeping up, by its successive contractions and dilatations, the motion of the blood through the vessels. But a careful survey of all the phenomena of the circulation would seem to lead to the conclusion, that the impulse of the heart is not the *only* means by which the motion of the blood is sustained; but that an additional impulse is given by the contraction of the muscular walls of the arteries upon the jets of blood successively impelled into them by the heart; and that the changes which this fluid undergoes in the capillaries have some share in its production, and have at any rate a very considerable modifying effect upon the quantity transmitted through the individual organs. We have seen that in Vegetables the laticiferous circulation is entirely capillary; that in the Radiata and lower Articulata it is chiefly so, the pulsatile action of some arterial trunk being the only force which can extend itself through the system, and this being too feeble to have much influence; whilst even in the higher Articulata and in all but the Cephalopod Mollusca, so large a part of the systemic circulation is 'lacunar,' that it seems impossible to imagine that the action of the heart can impel the blood through the branchial vessels which succeed. Some more 'diffused' forces, therefore, would seem to be in operation; and the following are among the facts which appear to support the conclusion, that, even in the highest animals, these most general forces are not obliterated, but are merely superseded by the energy of the special organ, which is developed as the centre of the whole circulation, and is endowed with an amount of power sufficient to govern and harmonise the numerous actions going on in different parts of the system.

478. In many warm-blooded Vertebrata, and still more in the cold-blooded Reptiles (amongst which the vitality of individual parts much longer survives injury to the general system), motion of blood in the capillaries has been seen to continue for some time after the heart has ceased to act, or has been removed, or after the great vessels have been tied; and this motion may be immediately checked by certain applications to the parts themselves. After most kinds of slow natural death, the arterial trunks and left side of the heart are found to be almost or even completely empty, and the venous cavities full of blood. This effect has been ascribed to the contraction of the arterial tubes after the heart has ceased to beat; but it seems impossible that it can be entirely due to that cause, since their calibre is not found to have diminished in a proportional degree; it must be partly attributed to a continuance of the capillary movement, after the general systemic circulation has ceased.

The continuance of various processes of secretion and even of nutrition, subsequently to general or *somatic* death, affords an excellent proof of this lingering vitality; and it is scarcely possible that these can be maintained without some degree of capillary circulation. There are some kinds of sudden death, however, in which the vitality of the whole system appears to be simultaneously destroyed, and the blood remains in the vessels as it was at the moment of decease.—Further, a careful examination of the circulation in the living animal discloses many irregularities in the rate of the capillary currents, which it is impossible to attribute to an influence derived from the heart or from the vessels that supply them; and such variations may present themselves, either in the capillary network of a part, or in a portion of it; the circulation taking place with diminished rapidity in one part, and with increased energy in another, though both are supplied by the same trunk. The change sometimes extends to a complete reversal of the direction of the movement, in certain of the transverse or communicating branches; this movement taking place, of course, from the stronger towards the weaker current: and not unfrequently an entire stagnation, of longer or shorter duration, precedes the alteration of the direction. Irregularities of this kind are most frequent, when the heart's action is enfeebled or partially interrupted; and it would thus appear, that the local influences by which they are produced, are overcome by the propelling power of the central organ, when this is acting with its full vigour. When the whole current has nearly stagnated, and a fresh impulse from the heart renews it, the movement is seldom uniform through the entire plexus supplied by one trunk; but is much greater in some of the tubes than in others,—the variation being in no degree connected with their size, and being very different in its amount at short intervals.

479. Amongst the most remarkable proofs of the influence of forces generated in the capillary circulation, on the general distribution of the blood, is one derived from the observation of organs which undergo changes in activity that are quite independent of alterations in the heart's action. Thus, when the uterus begins to develope itself during pregnancy, the unusual activity of the capillary circulation occasions an increased demand for blood, which is supplied by an increase in the diameter of the trunks that transmit fluid to the organ; and this is entirely independent of any increased energy in the heart's action, which would have affected the whole system alike. The same may be said of the occasional development of the mammæ for the secretion of milk; of the rush of blood through these organs during the act of suckling; and of similar changes in other parts, of which the activity is not constant or uniform. In certain diseased states, also, of particular portions of the system, which do not occasion any appreciable alteration in the heart's action, the quantity of blood sent to the part is much increased, and the pulsation of the arterial trunk leading to it is evidently stronger than that of the corresponding vessels on the outside of the body. These phenomena, and many others which might be mentioned, are evidently analogous to one formerly stated as having been ascertained by experiments on Plants (§ 446); and, when taken in connection, they seem to indicate without much doubt, that the quantity of blood sent to individual organs, and the force with which it is transmitted through them, are augmented with any increase of

energy in the vital processes taking place in them, the *vis a tergo* derived from the impulsive power of the heart remaining the same.—Additional evidence of the influence of the forces generated in the capillaries on the general circulation, is derived from cases in which the normal changes to which the capillary circulation ministers are suspended, and in which it then appears that the heart's impulse is not alone sufficient to maintain the current of blood. One of the most conclusive of these proofs is drawn from the phenomena of *Asphyxia*, or suffocation; since it now seems distinctly ascertained, that the check given to the circulation, and thence to all the other functions, arises from the stagnation of the blood in the capillaries of the lungs, consequent upon the cessation of the reaction between that fluid and the air.* So, again, cases of spontaneous gangrene of the lower extremities are by no means of unfrequent occurrence, in which the local stagnation of the circulation has been clearly dependent upon the cessation of the nutrient actions to which it was subservient; it being found, by examination of the limb after its removal, that both the larger tubes and the capillaries were pervious throughout, so that no mechanical impediment existed, to prevent the propulsive power of the heart from transmitting the blood through them. The influence of the prolonged application of Cold to a part, may be referred to in support of the same general proposition; for although the calibre of the vessels is diminished by this agent, yet their contraction is not sufficient to account for that complete cessation of the flow of blood through them, which precedes the entire loss of their vitality.—A periodical retardation or suspension of the circulation in particular portions of the body, unaccompanied by any other ostensible change, and not dependent upon any failure of the heart's power, is by no means an uncommon phenomenon. It frequently presents itself, for example, in one of the fingers; and a curious case is recorded by Dr. Graves,† in which the whole of one leg was thus affected, with remarkable periodicity, for about twelve hours out of the twenty-four; whilst in the intervals the circulation in the limb was unusually active, the action of the heart being quite natural throughout, and the circulation in the rest of the body not being in the least affected.

480. In the development of the embryo of the higher Vertebrated animals, moreover, there is a period at which a distinct movement of red blood is seen, before any pulsating vessel can be detected to possess an influence over it (§ 484). Further, instances not very unfrequently occur, of fœtuses having attained nearly their full development, which have been unpossessed of a heart, and in which the circulation has been, as it were, entirely capillary; and although in most, if not all, of these cases, the monster has been accompanied by a perfect child, the heart of which may have been suspected to have influenced its own circulation, yet, in one of those most recently examined, the occurrence of this has been disproved. From a careful examination of the vascular system, it appeared impossible that the heart of the twin fœtus could have caused

* For a fuller discussion of this part of the subject than the limits of this Treatise permit, see the Author's "Human Physiology," and his Article on *Asphyxia* in the "Library of Practical Medicine." See also the very conclusive experiments of his late valued friend Dr. John Reid, in the "Edinb. Med. and Surg. Journal" for April 1841; and Dr. Reid's "Physiological, Pathological, and Anatomical Researches," CHAP. II.

† "Lectures on Clinical Medicine," second edition, vol. i. p. 73.

the movement of blood in the imperfect one; and this must, therefore, have been maintained by forces arising out of the nutritive changes occurring in the capillaries.*

481. All these circumstances indicate that the movement of blood through the Capillaries is very much *influenced* by local forces; although these forces are not sufficiently powerful, in the fully developed state of the higher animals, to *maintain* it by themselves. And from other facts it appears, that the conditions necessary for the energetic flow of blood through these vessels, are nothing else than the active performance of the nutritive and other operations, to which its movement is subservient.—The principle already noticed (§ 449) as having been developed by Prof. Draper, seems fully adequate to explain these phenomena. It will be convenient to take the Respiratory circulation as an example of its application; since the changes to which this is subservient are more simple than those which take place elsewhere. The venous blood transmitted to the lungs, and the oxygen in the pulmonary cells, have a mutual attraction, which is satisfied by the exchange of oxygen and carbonic acid that takes place through the walls of the capillaries; but when the blood has become arterialised, it no longer has any such attraction for the air. The venous blood, therefore, will drive the arterial before it, in the pulmonary capillaries, whilst respiration is properly going on: but if the supply of oxygen be interrupted, so that the blood is no longer aerated, no change in the affinities takes place whilst it traverses the capillary network; the blood, continuing venous, still retains its need of a change and its attraction for the walls of the capillaries; and its egress into the pulmonary veins is thus resisted, rather than aided, by the force generated in the lungs. In the Systemic circulation, the changes are of a much more complex nature, every distinct organ attracting to itself the peculiar substances which it requires as the materials of its own nutrition; and the nature of the affinities thus generated will be consequently different in each case. But the same law holds good in all instances. Thus, the blood conveyed to the liver by the portal vein, contains the materials at the expense of which the bile-secreting cells are developed; consequently the tissue of the liver, which is principally made up of these cells, possesses a certain degree of affinity or attraction for blood containing these materials; and this is diminished, so soon as they have been drawn from it into the cells around. Consequently the blood of the portal vein will drive before it, into the hepatic vein, the blood which has already traversed the capillaries of the portal system, and which has given up, in doing so, the elements of bile to the solid tissues of the liver.

482. We are now prepared, therefore, to understand the general principle, that the rapidity of the local circulation of a part will depend in great measure upon the activity of the functional changes taking place in that part,—the heart's action, and the state of the *general* circulation,

* For the details of this interesting case, which was communicated by Dr. Houston of Dublin to the British Association in 1836, see the "British and Foreign Medical Review," vol. ii. p. 596, and the "Dublin Medical Journal" for 1837. An attempt was made by Dr. Marshall Hall ("Edinb. Monthly Journal," 1843) to disprove Dr. Houston's inferences; but a most satisfactory reply was made by Dr. H. in the "Dublin Journal" for Jan. 1844. See also the "Edinb. Med. and Surg. Journal," July, 1844.—A similar case is recorded by Dr. Jackson, of Boston (N.E.) in the "American Journal of the Medical Sciences," Feb. 1838.

remaining the same. When, by the heightened vitality, or the unusual exercise, of any organ, the changes which the blood naturally undergoes in it are increased in amount, the affinities which draw the arterial blood into the capillaries are stronger, and are more speedily satisfied, and the venous blood is therefore driven out with increased energy. Thus a larger quantity of blood will pass through the capillaries of the part in a given time, without any enlargement of their calibre, and even though it be somewhat diminished; but the size of the arteries by which it is conveyed soon undergoes an increase, which adapts it to supply the increased demand. Any circumstance, then, which increases the functional energy of a part, or stimulates it to increased nutrition, will occasion an increase in the supply of blood, altogether irrespectively of any change in the heart's action. This principle has long been known, and has been expressed in the concise adage "*Ubi stimulus, ibi fluxus*;" which those Physiologists, who maintain that the Circulation is maintained and governed by the heart alone, cast into unmerited neglect.

483. The development of that Circulating system which has been described as peculiar to the higher classes of Vertebrated animals, is not completed until the moment of birth; and the progressive changes which the heart and vascular apparatus undergo, in the evolution of the fœtus of Birds and Mammals, afford a most beautiful illustration of the principles already laid down (CHAP. VIII.), respecting the amount of correspondence between the transitory stages of each system in the higher animals, and the forms permanently exhibited by the lower. It has been seen that in the organs of Circulation, as well as in all others, the tendency, as we rise from their lowest to their highest condition, is one of specialization. In the simplest Animals, as in Plants, whatever motion of fluid takes place is effected by each individual part by and for itself; whilst in the complex and highly-developed structures that occupy the other extremity of the scale, the development of a powerful organ of impulsion, the influence of which extends over the whole system, has superseded the diffused agency by which the circulation was previously maintained. This progress from a more general to a more special type is equally manifested in the vascular system of the embryo; and the analogy which thus arises between the forms it presents at different epochs of its development, and those presented by the lower tribes of animals, is not superficial only, but extends even to minute particulars. The egg of the Bird affords the best opportunity for studying the early changes which it undergoes, and these have been described with great minuteness; but such a sketch of them only can here be given, as will serve to illustrate the principles alluded to. The preliminary stages of the process will be described in their proper place.

484. At an early period of incubation, the yelk is found to be enveloped by a "germinal membrane," composed of distinct cells, which is divisible into three layers; and a thickened portion of this is easily distinguishable, at which the embryo will be subsequently evolved. The middle layer gives origin to the Circulating system, and is therefore termed the *vascular* layer. The thickened portion of this that surrounds the germ, soon becomes studded with numerous irregular points and marks of a dark yellow colour; and as incubation proceeds, these points become more apparent, and are gradually elongated into small lines, which are united together, first in small groups, and then into one network, so as to form what is

called the *Vascular area* (Fig. 214). A large dark spot of a similar kind is seen in the situation to be subsequently occupied by the heart. These dark points and lines are formed by collections of blood-corpuscles, which originate in the transformation of the cells of the embryo and of the germinal membrane; and the rows and masses of blood-disks seem at first to lie in mere channels, the walls of the heart and blood-vessels that subsequently enclose them, being of later formation. From the first, however, a definite plan is perceptible; the network of capillaries that is formed over the vascular area, being supplied with blood by the ramifications of a pair of arterial trunks, *f, f*; whilst the blood is collected from them by the circular venous sinus which bounds the area, and is returned to the embryo by the venous trunks *g, g*. In the blood-vessels which are first observed in the body of the embryo, as well as in the vascular area, no difference is at first perceived between the characters of the arteries and those of the veins; and these are only to be distinguished by the direction of the currents of blood circulating through them. But at about the fourth or fifth day of incubation, the coats of the arteries begin to appear thicker than those of the veins, and the distinction between them soon becomes evident.

FIG. 214.

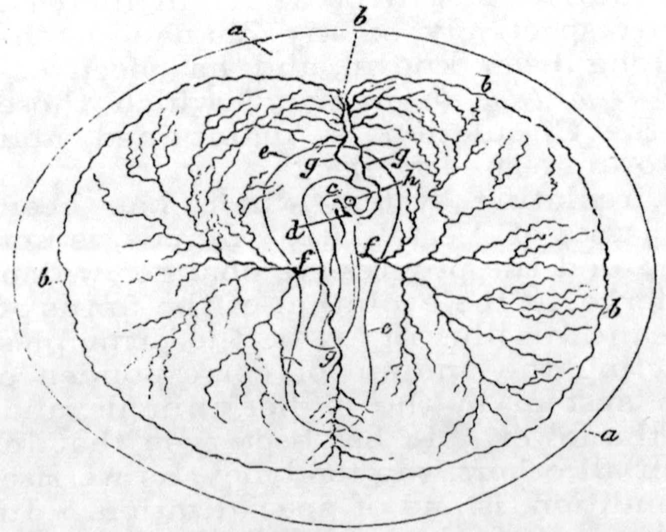

Vascular Area of *Fowl's* egg, at the beginning of the third day of incubation;—*a, a*, yelk; *b, b, b, b*, venous sinus bounding the area; *c*, aorta; *d*, punctum saliens, or incipient heart; *e, e*, area pellucida; *f, f*, arteries of the vascular area; *g, g*, veins; *h*, eye.

After the principal vessels are formed, the development of new ones appears to take place in two modes, according as they are to occupy the interspaces existing among those previously formed, or are to extend themselves into out-growing parts. In the first of these cases, the new capillaries appear to be formed, like the original ones, from stellate cells, whose prolongations meet the vessels in which blood is already circulating, coalesce with them, and thus receive the current into their own cavities, to transmit it to some other vessel (§ 221, Fig. 54). But in the second, the new vessels are formed entirely by extension from those already existing. This takes place in the following mode. "Suppose a line or arch of capillary vessel passing below the edge or surface of a part to which new material has been superadded. The vessel will at first present a slight dilatation in one, and coincidently, or shortly after, in another point, as if its wall yielded a little, near the edge or surface. The slight pouches thus formed gradually extend, as blind canals or diverticula, from the original vessels, still directing their course towards the edge or surface of the new material, and crowded with blood-corpuscles, which are pushed into them from the main stream. Still extending, they converge, they meet; the partition-wall, that is at first formed by the meeting of their closed ends, clears away; and a perfect arched tube is formed, through which the blood, diverging from the main or former stream, and then rejoining it, may be continuously propelled."*—This last process may be seen in the growing parts of the tail of the Tadpole, in the development of the filamentous gills and legs of the Water-Newt, in the first evolution of the extremities of the embryoes of higher animals, and in the formation of new structures in the fully developed organism, either for the repair of injuries, or as the result of morbid processes. In some instances it would appear that the wall of the newly-forming vessel gives way, and that the blood-corpuscles escape from it into the parenchyma, at first collecting in an undefined mass, but soon manifesting a definite direction, and coming into connection with another portion of the arch or with some adjacent vessel. Thus, then, a *channel* and not a *vessel* is formed; and it is probably in this way that those passages are excavated, which take the place of distinct vessels in many of the lower tribes of animals, and also, according to Mr. Paget, in some of the softer and least organised morbid growths in Man.†

485. The first rudiment of the heart appears about the 27th hour, and is a mass of cells, of which the innermost soon break down, so as to form a tubular cavity; for some time it is simple and undivided, extending, however, through nearly the whole length of the embryo; but the posterior part may be regarded as corresponding with the future auricle, since prolongations may be perceived extending from that part into the transparent area, which indicate the place where the veins subsequently enter. Although the development has proceeded thus far at about the 35th hour, no motion of fluid is seen in the heart or vessels until the 38th or 40th hour. When the heart, which may be considered as analogous at this period to the dorsal vessel of the Annelida, first begins to pulsate, it contains only colourless fluid mixed with a few globules. A movement of the dark blood in the circumference of the vascular area is at the same time perceived; but this is independent of the contractions of the heart, and it is not until a subsequent period that such a communication is established between the heart and the distant vessels, that the dark fluid contained in them arrives at the central cavity, and is propelled by its pulsations. This fact, which we have just seen to possess a very important bearing on the theory of the circulation, and which has been denied by some observers, appears to have been positively established by the latest researches of Von Baer.‡—The contraction of this dorsal vessel (for so it may be termed) begins, as in the Annelida, at its posterior extremity, and gradually extends itself to the anterior; but, between the 40th and 50th hours, a separation in its parts may be observed, which is effected by a constriction round the middle of the tube; and the dilatation of the posterior portion becomes an auricular sac, and that of the anterior a ventricular cavity. Between the

* See Mr. Paget's "Lectures on Reproduction and Repair," Medical Gazette, 1849.

† Op. cit.

‡ He says that there is no doubt of the blood being formed before the vessels. The formation of the blood goes on in every part of the body; and when formed, it is put in motion by some unknown cause that impels it in the proper direction, until at length it reaches the central formation of blood, around which is developed a tubular canal afterwards to be further modified and changed into a heart. The first motions of the blood are *towards* the heart, and consequently the first vessels formed are *veins;* a fact of itself sufficient to disprove the hypothesis that the motive power which presides over the circulation resides exclusively in the ventricles of the heart. "Uber Entwickelungsgeschicte der Thiere," &c. Königsberg, 1837, part II. p. 126.

50th and 60th hours, the circulation of the blood in the vascular area becomes more vigorous, and the action of the ventricle is no longer continuous with that of the auricle, but seems to succeed it at a separate period. At the same time the tube of the heart becomes more and more bent together until it is doubled; so that this organ now becomes much shorter relatively to the dimensions of the body, and is more confined to the portion of the trunk to which it is subsequently restricted. The convex side of the curve which the tube presents, is that which subsequently becomes the apex or point of the heart; and, between the 60th and 70th hours, this is seen to project forward from the breast of the embryo, much in the situation it subsequently occupies. About the same time, the texture of the auricle differs considerably from that of the ventricle; the auricle retaining the thin and membranous walls which it at first possessed, while the ventricle has become stronger and thicker, both its internal and external surfaces being marked by the interlacement of muscular fibres, as in the higher Mollusca. About the 65th hour, the grade of development of the heart may be regarded as corresponding with that of the Fish, the auricle and ventricle being perfectly distinct; but their cavities are as yet quite single.—The heart of the Dog at the 21st day bears a great resemblance to that of the chick at the 55th or 60th hour; it consists of a membranous tube twisted on itself and partially divided into two principal cavities, besides the bulb or dilatation which at this period is found at the commencement of the aorta, and which corresponds with the bulbus arteriosus of Fishes.

486. Having thus traced the evolution of the heart of the Chick up to the grade which it presents in Fishes, we may now enquire what is the condition of the other parts of the vascular system at the same time. At the end of the second day, the primitive arterial trunk (Fig. 215, *a*) is seen to have divided into two canals (1, 1′,) which separate from one another to enclose the pharynx, and then unite again to form the aortic trunk, A, which passes down the spine. During the first half of the third day (about the 60th hour), a second pair of arches, 2, 2′, is formed, which encompasses the pharynx in the same manner; and towards the end of the third day, two other pairs of vascular arches, 3, 3′ and 4, 4′, are formed; so that the pharynx is now encompassed by four pairs of vessels, which unite again to supply the general circulation. These evidently correspond with the branchial arteries of Fishes, although no respiratory apparatus is connected with them; and in fact the distribution of the vascular system of

FIG. 215.

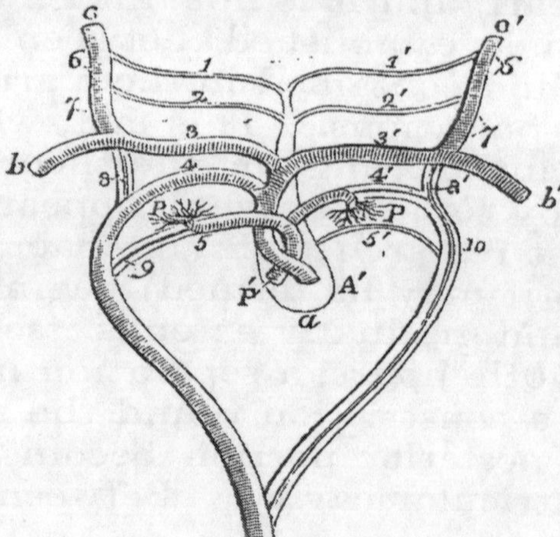

Diagram of the formation of the great Arterial trunks in the *Chick*:—A, descending aorta; A′, ascending aorta; P, P, pulmonary arteries; P′, their common trunk; *a*, truncus arteriosus; *b*, *b*′, subclavian arteries; *c*, *c*′, carotid arteries; 1, 1′, 2, 2′, 3, 3′, 4, 4′, 5, 5′, five pairs of branchial arches; 6, 6′, 7, 7′, communicating branches which become the trunks of the carotids; 8, 8′, communicating branches afterwards obliterated; 9, ductus arteriosus; 10, aortic arch of the left side.

the Bird, on the fourth and fifth days, exactly resembles that presented by many Cartilaginous Fishes, as well as by the tadpoles of the Batrachia. The first pair of arches is obliterated about the end of the fourth day; but a pair of vessels, which is sent from it to the head and neighbouring parts, and which afterwards remains as the carotid arteries, *c*, *c'*, continues to be supplied through a communicating vessel, 6, 6', from the second arch. While the first pair is being obliterated, a fifth, 5, 5', is formed behind the four which had previously existed; and proceeds, in the same manner as the fourth, from the ascending to the descending aorta. On the fourth day, the second arch also becomes less, and on the fifth day is wholly obliterated; whilst the third and fourth become stronger. From the third arch, now the most anterior of those remaining, the arteries are given off which supply the upper extremities, *b*, *b'*; and the vessels of the head, *c*, *c'*, are now brought into connection with it, by means of the communicating branches, 7, 7', which previously joined the third with the second arch. When these vessels are fully developed, the branches, 8, 8', by which these arches formerly sent their blood into the aorta, shrink and gradually disappear; so that, by about the thirteenth or fourteenth day, the whole of the blood sent through the two anterior arches (the second and third) is carried to the head and upper extremities, instead of being transmitted to the descending aorta as before. There now only remain the fourth and fifth pair of branchial arches, the development of which into the aorta and pulmonary arteries will be described in connection with the changes, which are at the same time going on in the heart.

487. During the fourth day, the cavities of the Heart begin to be divided, for the separation of the right and left auricles and ventricles. About the 80th hour, the commencement of the division of the auricle is indicated externally, by the appearance of a dark line on the upper part of its wall; and this, after a few hours, is perceived to be due to a contraction, which, increasing downwards across the cavity, divides it into two nearly spherical sacs. Of these the right is at first much the larger, and receives the great systemic veins; the left has then the aspect of a mere appendage to the right, but it subsequently receives the veins from the lungs, when these organs are developed, and attains an increased size. The septum between the auricles is by no means completed at once; a large aperture (which subsequently becomes the *foramen ovale*) exists for some time at its lower part, so that the ventricle continues to communicate freely with both auricles. This passage is afterwards closed by the prolongation of a valvular fold, which meets it in the opposite direction; it remains pervious, however, until the animal begins to respire by the lungs, and sometimes is not completely obliterated even then. The division of the ventricle commences *some time before* that of the auricle, and is effected by a sort of duplicature of its wall, forming a fissure on its exterior and a projection on its interior; and thus a septum is gradually developed within the cavity, which progressively acquires firmness, and rises higher up, until it reaches the entrance to the bulb of the aorta, where some communication exists for a day or two longer. At last, however, the division is complete, and the inter-ventricular septum becomes continuous with the inter-auricular, so that the heart may be regarded as completely a double organ. The progressive stages presented in the development of

this septum are evidently analogous to its permanent conditions in the various species of Reptiles (§ 473–4); but it must not be lost sight of that in all Reptiles the inter-auricular septum is first developed, and that it is completely formed in many instances in which the inter-ventricular septum is absent or imperfect.—The changes which occur in the heart of the Mammalia are of a precisely similar character; and, as they take place more slowly, they may be watched with greater precision. Soon after the septum of the ventricles begins to be formed in the interior, a corresponding notch appears on the exterior, which, as it gradually deepens, renders the apex of the heart double. This notch between the right and left ventricles continues to become deeper until about the eighth week in the Human embryo, when the two ventricles are quite separated from one another, except at their bases; this fact is very interesting, from its relation with the similar permanent form presented by the heart of the Dugong (§ 475). At this period, the internal septum is still imperfect, so that the ventricular cavities communicate with each other, as in the chick on the fourth day. After the eighth week, however, the septum is complete, so that the cavities are entirely insulated; whilst at the same time their external walls become more connected towards their bases, and the notch between them is diminished; and at the end of the third month the ventricles are very little separated from one another, though the place where the notch previously existed is still strongly marked.

488. Returning again to the distribution of the arterial trunks, we are now prepared to follow their final modification, by which they are adapted to the existence which the individual is soon to commence as an air-breathing animal.—The first, second, and third branchial arches have been shown to be replaced by the brachial and carotid arteries, and to have lost all communication with the primitive arterial trunk (Fig. 215, *a*) except at its commencement, where the third pair of arches arises with the other trunks from its dilated bulb. This remains as a single cavity even after the ventricles have been separated; but towards the end of the fifth or the beginning of the sixth day, in the Chick, the bulb becomes flattened, and the opposite sides adhere together, so as to divide it into two tubes running side by side. Of these, one communicates with the left, and the other with the right ventricle. The former, which subsequently becomes the ascending aorta (A′), is continuous with the fourth branchial arch (4) on the right side only; but from this the carotid and brachial arteries arise by two principal trunks. This arch becomes gradually larger, so as to form the freest mode of communication between the heart and the descending aorta; it subsequently becomes, in fact, the arch of the aorta. The trunk (P′), which is connected with the right ventricle, on the other hand, and which subsequently becomes the pulmonary artery, transmits its blood through the fourth arch of the left side, 4′, (the two primary tubes twisting round each other) and the fifth arch, 5, of the right; but the fifth arch on the left side, 5′, now ceases to convey blood. From the two trunks, 4′, 5, which still discharge their blood into the descending aorta, the pulmonary vessels, P, P, branch off as the lungs are developed; and the prolongation, 10, of the former of these, which previously constituted one of the arches of the descending aorta, soon afterwards becomes impervious. The original prolongation, 9, of the latter trunk, which meets the descending aorta, still remains; so that a portion of the blood

sent from the right ventricle is transmitted through this communicating branch directly into the descending aorta, just as in the adult Crocodile. After the first inspiration, however, the whole of the blood transmitted through the pulmonary artery passes into the lungs, and does not enter the aorta until it has been returned to the heart; and this communicating vessel, which is termed the *ductus arteriosus*, soon shrinks and becomes impervious. Thus the *third* pair of branchial arches becomes converted into the two *arteriæ innominatæ*, or common trunks, from each of which, in Birds and in some Mammals, the carotid and subclavian arteries of one side originate. The *fourth* branchial arch of the *right* side becomes the *arch of the aorta*. And the *fifth* branchial arch of the *right* side, with the *fourth* of the *left*, become the right and left pulmonary arteries.—The general plan of the changes which occur in the vascular system of the Mammalia, is the same as that which has been described in Birds; the differences being only in detail,—as for instance, that the aortic arch is formed, not from the right, but from the left branchial arch.

489. Up to the period of the hatching of the egg in Birds, and the separation of the fœtus from the parent in the Mammalia, the circulation retains some peculiarities, characteristic of the inferior type which is permanent in the Reptile tribes. Of the blood which is brought by the venous trunks to the right auricle, part has been purified by transmission to the respiratory organ (the *allantois* in Birds, and the *placenta* in Mammals), whilst a part has been vitiated by circulation through the system. The former is brought from the abdomen by the ascending vena cava, mixed with the blood which has circulated through the lower extremities; whilst the descending cava brings back that which has passed through the capillaries of the head and upper extremities, and which, having received no admixture of arterial blood, is not fit to be again transmitted in the same condition. It will be recollected that a communication still exists between the two auricles, the 'foramen ovale' yet remaining pervious; and by a fold of the lining membrane of the right auricle, forming the Eustachian valve, the ascending and descending currents are so directed, that the former (consisting of the most highly-arterialised blood) passes at once into the left auricle, whilst the latter flows into the right ventricle. From the left auricle, the arterial blood is propelled into the left ventricle, and thence through the arch of the aorta to the vessels of the head and upper extremities, a comparatively small part finding its way into the descending aorta. The venous current is propelled through the pulmonary artery; but the lungs not yet being expanded, little of it is transmitted to these organs, and the greater part finds its way through the ductus arteriosus into the descending aorta, where it mixes with the remainder of the first-mentioned portion. This trunk not only supplies the viscera and lower extremities (which are thus seen to receive, as in Reptiles, blood of which only a portion has been oxygenated), but sends a large proportion of its contents to the umbilical vessels, by which it is conveyed to the oxygenating organ, and returned again to the venous trunk of the abdomen. The peculiar course taken by the blood through the heart, which was suspected from anatomical investigation, has been actually demonstrated by means of coloured injections by Dr. J. Reid.*—Another peculiarity in the fœtal

* "Edinb. Med. and Surg. Journ.," vol. xliii. pp. 11 and 308.

circulation is the mode in which the blood passes through the liver. The vena portæ in the fœtus receives not only the venous blood of its abdominal viscera, but the arterial blood sent from the respiratory surface; and as it would not be desirable that the whole of this should pass through the liver before being transmitted to the heart, an immediate passage into the vena cava is provided for a part of it, in the *ductus venosus*, which, not being required after birth, shrivels into a ligament. A similar communication exists permanently in Fishes, and to a less degree in other oviparous Vertebrata: and it seems there intended to transmit directly to the heart whatever proportion of the blood, brought to the vena portæ, may be at the time superfluous as regards the function of the liver.

490. Thus we have traced, in the development of the Circulating apparatus of the higher Vertebrata, the same progressive advance from a more *general* to a more *special* condition, as that which we have witnessed in ascending the Animal series; and when considered *analogically* rather than *homologically* (§ 333), the correspondence is extremely close. For in the state of the circulating system in the early embryo, when the heart is as yet but a pulsating enlargement of one of the principal trunks, and the walls of the vessels are far from being complete, we have the representation of its condition in the higher Radiata, and in the lower Articulata and Mollusca. In the subsequent division of the cardiac cavity into an auricle and a ventricle, an advance is made corresponding to that which we encounter in passing from the Tunicata to the higher Mollusca. And when the branchial arches are formed, which enclose the pharynx and meet in the aorta, the type of the Fish is obviously attained. But it will be observed that, notwithstanding this similarity, the Vertebrated embryo never presents any of those features of the circulating apparatus, which are characteristic of the other sub-kingdoms respectively; thus, it does not exhibit that radiated distribution of the vascular trunks, which is seen in the *Echinodermata* (§ 454); nor does the heart, even when most like a 'dorsal vessel,' ever present the least approach to that transverse division into successive segments, which is typical of the *Articulata* (§ 461); and in its position, and connections, being situated in the immediate neighbourhood of the pharynx and sending its primitive trunks around it, the heart of every Vertebrated animal differs from that of the *Mollusca*, whose relations are with the opposite extremity of the alimentary canal (§ 466). In the subsequent progress of the Circulating apparatus, from the grade of the Fish, through that of the Reptile, to that of the Bird or Mammal, we have a characteristic illustration of the principles formerly laid down (§ 348); for although the branchial arches are formed in *all* Vertebrated animals, yet it is only in Fishes and Batrachian Reptiles that they give origin to branchial tufts; and although at a subsequent period the condition of the heart and great vessels presents a strong general resemblance to that of the typical Reptiles, yet that resemblance is wanting in the essential feature of the complete separation of the auricles, and the mixture of arterial and venous blood in the single ventricle. It is obvious that this want of conformity has reference to the difference in the seat of the respiratory process; the pulmonary vessels in the embryo being developed for future use, but the actual aeration of the blood being performed elsewhere.

491. The knowledge of these different stages of the development of the

Circulating apparatus, enables us to explain many of the malformations which it occasionally presents in Man. One of the most common of these gives rise to the malady termed *Cyanosis;* for this results from the *foramen ovale,* which establishes a communication between the auricles, remaining open after pulmonary respiration has been established; so that a considerable portion of the blood transmitted to the right cavity passes into the left, without having been previously arterialised by passage through the lungs. Persons thus affected have always a livid aspect, from the quantity of venous blood circulated through the arteries; they are deficient in muscular energy and in power of generating heat, and they are seldom long-lived. A consequence partly similar would probably have resulted from a curious malformation mentioned by Kilian, had the infant remained alive: in this case, the aortic arch had not been developed, so that the primary aortic trunk gave off only the vessels to the head and upper extremities; whilst the communicating branch between the pulmonary artery and descending aorta, which usually is of a secondary character, constituting the ductus arteriosus, was here the only means by which the blood could be transmitted to the latter; so that the circulation through the lower part of the trunk and extremities would have been entirely venous. A malformation of this kind in a diminished degree has not been found incompatible with the continuance of life; several cases being on record, in which the ductus arteriosus has remained pervious, and has brought part of the blood from the pulmonary artery to the descending aorta. Cyanosis is of course, as in the former instance, the result of this imperfect arterialisation; and the individual is reduced, as far as his vascular system is concerned, to the condition of the Crocodile. An arrest of development at an earlier period may cause still greater imperfections in the formation of the heart. Thus, the septum of the ventricles is sometimes found incomplete, the communication between the cavities usually occurring in the part which is last formed, and which in most Reptiles remains open. In other cases it has been altogether wanting, although the aorta and pulmonary artery were both present, and arose side by side from the common cavity; and this form of the circulating apparatus is evidently analogous to that presented by Reptiles in general. A still greater degradation in its character has been occasionally evinced; for several cases are now on record, in which the heart has presented but two cavities, an auricle and a ventricle, thus corresponding with that of the Fish; and in one of these instances the child had lived for seven days, and its functions had been apparently but little disturbed. The occasional entire absence of the heart has already been noticed; and coexistent with this, there is always great deficiency in the other organs, the brain, and sometimes the liver and stomach, being undeveloped. The bifid character of the apex, which presents itself at an early period of the development of the heart, and is permanent in the Dugong, sometimes occurs as a malformation in the adult human subject; evidently resulting, like the others which have been mentioned, from an arrest of development. On similar principles, some occasional peculiarities noticed in the distribution of the vessels may be accounted for, of which a striking example will be presently given. The Vena Cava is occasionally observed to consist of two parallel trunks, which are sometimes partially united, and then separate again; a similar condition is permanent in some Cartilaginous Fishes, and the explanation of it is to be sought for in the history of the development of the venous system in general. We have seen that in many of the lower animals, such as the Crustacea, where the arteries are perfect canals having distinct coats, the veins seem to be merely channels through the tissues, having no definite walls: in like manner, at an early period of the fœtal development of the higher animals, several small vessels are found where one vein subsequently exists; and, if the coalescence of these has been from any cause checked, they will remain permanently separated to a greater or less extent.

492. Several interesting varieties have been detected in the arrangement of the principal trunks given off from the Aorta: and though we cannot account for them on the principles already mentioned, it is not a little curious, that nearly all of these irregular forms possess analogues in the arrangements, which are peculiar to some or other of the Mammalia. The mode in which the cephalic and brachial vessels usually arise in the Human subject, is shown in the subjoined figure, A, where *a b* is the arch

Fig. 216.

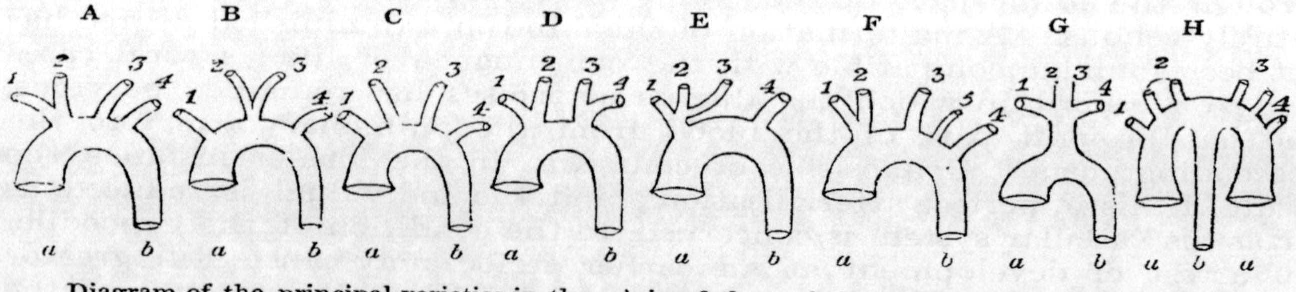

Diagram of the principal varieties in the origin of the cephalic and brachial trunks from the arch of the aorta; A, Man; B, Elephant; C, some Cetacea; D, Bat, &c.; E, Carnivora, &c.; F, Seal; G, Ruminants; H, Reptiles:—1, right subclavian; 2, right carotid; 3, left carotid; 4, left subclavian; 5, vertebral; *a*, ascending aorta; *b*, descending aorta.

of the aorta, 1 and 2 the trunks of the right carotid (which supplies the head) and of the right subclavian (which is distributed to the upper extremity), arising by a common trunk—the arteria innominata; while the left carotid, 3, and the left subclavian, 4, arise separately. At B is seen a distribution which is rare in the human subject, the two carotids arising by a common trunk, and the right as well as the left subclavian being given off separately; this is the regular arrangement of branches in the Elephant. It is not so unusual for all the branches to arise from single trunks, as at C; and this appears to be the regular type in some of the Cetacea. Sometimes, again, there is an arteria innominata on each side; subsequently dividing into the carotid and subclavian, as at D; and on this plan the branches are distributed in the Bat tribe, and also in the Porpoise. A not unfrequent variety in the human subject, is for both carotids to arise with the right subclavian from a single trunk, as at E, the left subclavian coming off by itself; this is observable as the regular form among many animals, being common among the Monkey tribe, the Carnivora, the Rodentia, &c. Another variety which is not unfrequent is shown at F, the vertebral artery on the left side, 5, which usually arises from the subclavian, springing directly from the aorta; it is on this plan that the branches are given off in the Seal. A form which is very uncommon in man is that represented at G; here the aorta divides at once into an ascending vessel, from which the two subclavian and two carotid arteries arise, and a descending trunk; this is the regular distribution of the vessels in Ruminating animals, and appears to be most general in Mammalia possessing a long neck. Lastly, at H, is seen a form which evidently results from an arrest of the usual changes in the arterial trunks described in §§ 486, 488; the aorta continuing to possess a double arch, from the ascending part of which the subclavian, external carotid, and internal carotid arteries are given off on each side, the single descending trunk being formed by the union of the two original branches. This, it will be recollected, is the normal type of formation in Reptiles.*

CHAPTER XIII.

OF RESPIRATION.

1. *General Considerations.*

493. THE function of Respiration essentially consists in the *evolution of carbonic acid* from the fluids of Organised beings, and the *absorption of oxygen* from the surrounding medium, usually in a nearly equivalent proportion. This process is performed by Plants as well as by Animals; and it may be regarded as arising out of the same *general* requirements in both kingdoms, although it answers some *special* purposes in the latter, which render it more immediately essential to the maintenance of their vital activity, than it seems to be in the former. For we shall hereafter find that the imperious necessity for the continual introduction of oxygen and exhalation of carbonic acid, which requires a most active performance of the respiratory function, and causes even a brief suspension of it to be fatal, in the higher Animals, is consequent upon the energetic exertion of their peculiarly *animal* powers, and upon the performance of that combustive operation by which their high temperature is maintained; whilst, on the other hand, when we pass to those tribes which are most remarkable for the inertness of their habits, and for their entire want of power to sustain an independent temperature, the demand for oxygen is greatly diminished, and the exhalation of carbonic acid may be checked for a time without injury. The amount of Respiration, then, which is required for the performance of the *organic* or constructive functions of Animals, is comparatively small; and it is not surprising that the existence of this function should have been long overlooked in Plants, in which its effects on the atmosphere are masked by a change of an entirely opposite nature, that is subservient to the introduction of alimentary material into the system,—namely, the decomposition of the carbonic acid of the air, under the influence of light, the fixation of its carbon in the vegetable tissues, and the consequent liberation of its

* In the foregoing account of the development of the Vascular system, the Author has availed himself freely of the valuable papers of Dr. Allen Thomson, in the "Edinb. Philos. Journal," vols. ix. and x.; in the sketch of the malformations of the heart, he has made use of the paper of Dr. Paget in the "Edinb. Med. and Surg. Journal," vol. xxxvi.; and the last paragraph, with the accompanying figures, has been entirely derived from the magnificent work of Tiedemann on the Arteries.

oxygen. To this last process the term Respiration has been commonly applied; and the Respiration of Plants is ordinarily spoken of as antagonistic to that of Animals. This statement is perfectly true, if under the term Respiration be included the sum-total of the changes produced in the air by the growth of a Plant; but it will be presently shown, that whilst Animal life gives rise to but one set of changes in the atmosphere (namely, the removal of a portion of its oxygen, and a replacement of this by carbonic acid), Vegetable life produces two sets of changes, which ought to be kept quite distinct from each other in a scientific description of them, their nature and their sources being alike different; and that it is only on account of the excess of one set of these changes beyond that which it antagonises, that *it* alone has received general attention, and has been commonly regarded as the proper respiration of Plants.

494. Restricting the meaning of the term *Respiration*, then, to the removal of carbonic acid from the living system in a gaseous form, and the introduction of oxygen into it, we have to enquire what are those most general sources of demand for this action in the vital economy, which are common to Plants and Animals. These appear to be two-fold; one arising out of the *disintegrating* changes which are always going on in the living system; the other being consequent upon some of those chemical operations, which necessarily participate in the *constructive* functions. The former seem to be the most general; the latter are rather of a special character, and manifest themselves most strongly, as we shall see hereafter, at certain periods in Vegetable life (§ 503).—All organised bodies, as already explained, are liable to continual disintegration, even whilst they are most actively engaged in performing the actions of life; in fact, a succession of organs whose individual duration is short, but whose functional energy is great, seems necessary for the maintenance of the life of the more permanent parts of the organism (CHAP. X., Sect. 1). The necessary result of this disintegration is decay; and one of the chief products of that decay is carbonic acid. A large quantity of this gas is set free, during the decomposition of almost every kind of organised matter, the carbon of the substance being united with oxygen supplied by the air. Hence we find that the formation and liberation of carbonic acid go on with great rapidity after death, both in the Plant and in the Animal; and that it is then but the continuation (so to speak) of that which has been taking place during life. Thus in Plants, as soon as they become unhealthy, the extrication of carbon in the form of carbonic acid takes place in greater amount than its fixation from the carbonic acid of the atmosphere; and the same change normally takes place during the period that precedes the exuviation of the leaves, their tissue being no longer able to perform its proper functions, and its incipient decay giving rise to a large increase in the quantity of carbonic acid set free. In some of these cases it would seem that the carbon of the decomposing tissue unites with the oxygen contained in the fluids of the system, and that carbonic acid is thus generated, the extrication of which contributes to the introduction of fresh oxygen (§ 495): in other instances, however, the oxygen may be more directly derived from the atmosphere.—The other source of demand for Respiration, which is common alike to Plants and Animals, arises out of the chemical transformations which are always going on in their systems,

as a part of their nutrient operations. These are as yet but very little understood; but enough is known to justify the belief, that in many of them the presence of oxygen is essential, and that carbonic acid is among their products. Some examples of such transformations, drawn from the Vegetable kingdom, have been already cited (§ 28); but in addition to these it may be remarked that the conversion of starch into sugar, a change that takes place in the neighbourhood of many growing parts, involves the combination of carbon with oxygen to a considerable amount; and that, in general, the production of the vast multitude of organic compounds yielded by Plants, from the gum, starch, and sugar which are first generated by them at the expense of the inorganic elements, requires a series of chemical changes, in many of which oxygen is taken in and carbonic acid given forth. Although the number of organic compounds generated by Animals is much less than that which we find in Plants, yet there can be no doubt, from a comparison of their atomic constitution, that oxygen must be taken into combination, and carbonic acid given off, in many of the chemical transformations which take place in the living body; and among these, the metamorphosis of Albumen into Gelatine has been pointed out by Dr. Prout as one of the most constant and important.

495. Besides the evolution of Carbonic acid and the absorption of Oxygen, it would appear that the exposure of the circulating fluid to the air is the means of keeping the *Nitrogen* of the system at its proper standard; and the whole series of reactions which take place between the living body and the air that surrounds it, or that is contained in the water in which it lives, may be conveniently included under the general term *Aeration.* This aeration would appear to be, like absorption, a change dependent on physical agencies, and taking place in conformity with their laws, when the requisite conditions are supplied by the structures of an organised being, and by the functional alterations which the living state involves.—All gases of different densities, which are not disposed to unite chemically with one another, have a strong tendency to mutual admixture. Thus, if a vessel be partly filled with hydrogen, and partly with carbonic acid, the latter, which is 22 times heavier than the former, will not remain at the bottom, but the two gases will be found in a short time to have uniformly and equably mixed; and it is on this principle, that the constitution of the atmosphere is everywhere the same, although the gases which compose it are of very different specific gravities. So strong is this tendency to admixture on the part of different gases, that it will take place when a membrane or other porous medium is interposed between them. This interchange, therefore, evidently resembles the endosmose and exosmose of fluids (§ 415); and although the tendency to admixture of the two gases is the fundamental cause of their movement, the nature of the septum has so much influence over the phenomenon, as sometimes to reverse the results. When plaster-of-paris is employed as the medium of diffusion, the exchange will take place with simple relation to the relative densities of the gases; and a general law has been ascertained by Prof. Graham, which applies to all instances,—that the 'replacing' or 'mutual-diffusion volumes' of different gases vary inversely as the square-roots of their densities. Thus, if a tube, closed at one end with a plug of plaster-of-paris, be filled with hydrogen, the gas will soon be entirely

removed, and will be replaced by something more than one-fourth of its bulk of atmospheric air; the density of hydrogen being about 1-14th that of the atmosphere. But when organic membranes are employed, the result is much influenced by the relative facility with which each gas permeates the septum. Thus, carbonic acid passes through moist bladder much more readily than hydrogen; and, in consequence, when a bladder of hydrogen is placed in an atmosphere of carbonic acid, a certain quantity of hydrogen will pass out; but a much larger proportion of carbonic acid will enter, so as to distend the bladder even to bursting. Further, it is found that, if a fluid be charged with any gas which it readily absorbs (as, for example, water with carbonic acid), it will speedily part with it when exposed to the attracting influence of another gas, such as atmospheric air; and the more different the densities of the two gases, the more rapidly, and with more force, will this take place. As in the former instance, this attraction will go on with little interruption through a porous membrane; and part of the exterior gas will be absorbed by the fluid (if of a nature to be so imbibed), in place of that which has been removed.

496. These simple phenomena will be found a key to the explanation of the changes which take place in the aeration of the circulating fluid by exposure to air; for it seems a universal fact, that *carbonic acid* existing in that fluid is exhaled, and is replaced by absorbed *oxygen;* and that an exhalation and absorption of *nitrogen* take place in animals, and perhaps also in plants.

2. *Respiration in Plants.*

497. Under the above designation have been associated two distinct changes, both nearly constant throughout the Vegetable kingdom. The atmosphere being the chief source whence Carbon is supplied to the living plant, the introduction of that element has been confounded with the contrary change, which is also necessary for the continued health of the structure, and which corresponds exactly with the respiration of Animals. The introduction of carbon is effected by the power which the green surfaces of Plants possess, of decomposing, under the stimulus of Light, the carbonic acid contained in the air or in the liquids supplied to them; and of retaining or fixing the carbon, whilst they set free the oxygen (§§ 72–75). In the *Phanerogamia*, the green surfaces of the leaves, and other appendages to the axis, are those by which this fixation of carbon, which may be considered as a process of alimentation, is chiefly, if not entirely effected; and where, as in the Cactus tribe, the leaves are deficient, but the stems are succulent and their surfaces green, it is obvious that these perform the same function. In the *Ferns, Mosses,* &c. there is the same separation of parts as in the Flowering plants; and the process is here also, without doubt, performed by the green parts of the surface. Of the inferior *Cryptogamia*, however, we know very little. The *Fungi* would not seem to depend upon the atmosphere for any part of their supply of carbon, which is altogether furnished by their peculiar aliment (§ 380); and these plants scarcely ever present any green surface, and flourish most in situations to which light has but little access. The same may be said of the *Cuscuta* (dodder) and other leafless parasitic plants of

more complex structure, that live upon the prepared juices they derive from the plant to which they attach themselves. There can be no doubt that *Lichens* ordinarily obtain the carbon which enters into their structure, entirely from the atmosphere ; and, that the *Algæ* are supported, in like manner, by the carbonic acid contained in the circumambient water ; but experiments are yet wanting to ascertain the precise conditions under which its assimilation is effected. Few Lichens have any green surfaces ; and although many of the Algæ are very brilliantly coloured, yet we find them occasionally existing at such depths, as make it difficult to believe that light is the only stimulus under which they can attain this appearance. The simpler forms of Algæ, especially the *Confervæ,* which inhabit fresh water, appear to exercise an important influence in maintaining it in a state fit for the support of animal life ; since it seems probable that they absorb the products of the decomposition of that foul matter by which all ponds and streams are constantly being polluted, and at the same time yield a supply of oxygen to the water. It is a notorious fact that Fishes are never so healthy in reservoirs destitute of aquatic plants, as in ponds and streams in which these abound.

498. The entire mass of Vegetation upon the surface of the globe is thus mainly dependent upon the minute proportion of carbonic acid contained in the atmosphere, which is not above 4 parts in 10,000. This is probably as much as Plants in general, under the feeble illumination which those of them are liable to receive whose 'habitat' is in variable climates, could advantageously make use of ; and a larger proportion would probably have been injurious to them, as well as to Animals. But it has been ascertained by direct experiment, that Plants will thrive in an atmosphere containing six or eight per cent. of carbonic acid, or even more, so long as they are exposed to strong sun-light ; and it would appear that in climates where the solar light is less obscured by clouds than it is in our own, the growth of plants may be favoured by an unusual supply of this alimentary substance. Thus the floating islands which are constantly being formed on the lake Solfatara in Italy, exhibit a striking example of the luxuriance of cryptogamic vegetation in an atmosphere impregnated with carbonic acid. These islands consist chiefly of Confervæ and other simple cellular plants, which are copiously supplied with nutriment by the carbonic acid that is constantly escaping from the bottom of the lake, with a violence which gives to the water an appearance of ebullition.* Dr. Schleiden mentions that the vegetation around the springs in the valley of Gottingen, which abound in carbonic acid, is very rich and luxuriant ; appearing several weeks earlier in spring, and continuing much later in autumn, than at other spots in the same district.†—A very ingenious hypothesis has been raised by M. Brongniart upon the fact, that an increased quantity of carbon may, under particular circumstances, be assimilated by Vegetables. He supposes that, at the epoch of the growth of those enormous primeval forests which supplied the materials of the coal-formation, the atmosphere was highly charged with carbonic acid, as well as with humidity ; and that from this source, the Ferns, Lycopodiaceæ, and Coniferæ of that era were enabled to attain their gigantic development. He imagines that they not only thus con-

* Sir H. Davy's "Consolations in Travel," 3rd ed., p. 116.
† "Wiegman's Archiv." iii. 1838.

verted into organised products an immense amount of carbonic acid, which had been previously liberated by some changes in the mineral world, but that, by removing it from the atmosphere, they prepared the earth for the residence of the higher classes of Animals. The hypothesis is a very interesting one, and well deserves consideration. It may be regarded as an almost absolute certainty, that the whole of the carbon now solidified in the coal-deposits of various ages, must have previously existed in the atmosphere; and if we were acquainted with the extent of these, it would be a simple matter of computation to determine, whether, if all this carbon were reconverted into carbonic acid, it would sensibly affect the proportion of that ingredient in the atmosphere.—The recent experiments of Dr. Daubeny * to a certain extent justify the hypothesis of M. Brongniart, by showing that the existing Plants and Animals most allied to those of the Carboniferous period can exist without injury in an atmosphere containing nearly 5 per cent. of carbonic acid; and if such a difference of climate then prevailed (either in consequence of a different distribution of land and water, or as a result of the internal heat of the globe), as would enable the solar rays to act with more constancy and power than they do in Britain at the present time, there seems no reason for asserting that such *might* not have been the case.†

499. The change which, strictly speaking, constitutes the *Respiration* of Vegetables is not, like that we have been describing, an occasional one; but is constantly taking place during the whole life of the plant, and appears to be more immediately necessary to its healthy existence. This consists in the disengagement of the superfluous carbon of the system, either by combination with the oxygen of the air, or (which is more likely) by replacing with carbonic acid the oxygen that has been absorbed from it; and it does not cease by day, by night, in sunshine, or in shade. If the function be checked, the plant soon dies,—as when placed in an atmosphere with a large proportion of carbonic acid, and without the stimulus of light which enables it to decompose the deleterious gas. Plants which are being 'etiolated' by the want of light (§ 76), absolutely diminish in the *weight* of their solid contents, owing to the continued excretion of carbon by the respiratory process, although their *bulk* may be much increased by the absorption of water; and if the proportion of carbonic acid in the surrounding air be augmented by its accumulation, they become sickly and die, from the impediment to their respiration.

* "Report of the British Association" for 1849; p. 56.

† The Author would suggest it as a point worthy of further consideration, whether there may not have been a special relation between the luxuriant growth of the plants that furnish those carbonaceous deposits, which are found especially to be intercalated amongst the Carboniferous, Oolitic, Wealden, and Cretaceous formations; and the deposit of those vast calcareous beds with which they are so remarkably associated. The latter, there is strong reason to believe, are almost entirely of *animal* origin; the carbonate of lime having been drawn by Zoophytes, Echinoderms, and Mollusks, from the waters of the ocean, just as the carbon of the Vegetation of those periods was drawn from the carbonic acid of the atmosphere. Now if we imagine that during the progress of these deposits, there was an unusual discharge of carbonate of lime, held in solution by free carbonic acid, by submarine springs issuing from the interior of the earth (like the Solfatara and other calcareous springs, which furnish the 'travertine' deposits, and at the same time promote the growth of plants, at the present day), we seem to have a probable account of the extraordinary abundance of carbonate of lime in the ocean-waters, and of carbonic acid in the atmosphere, at those periods; since the greater part of the latter would have escaped into the air, as soon as it became free to do so by the removal of the pressure which previously restrained it.

The parallel, therefore, between Plants and Animals appears to be complete, as regards the influence of carbon upon their growth ; for to both it is deleterious when breathed, and to both it is invigorating to the digestive system when absorbed as food ; and whenever Plants, or parts of Plants, derive their nutriment, like Animals, from organic compounds already prepared for them, there do we find the true respiratory process taking place, without the counterbalancing fixation of carbon as aliment. This is the case, for example, with Fungi in general ; it is the case, too, with the leafless Phanerogamic parasites, which draw their materials from the elaborated juices of other plants, instead of preparing them for themselves (§ 558) ; it is the case also with the growing embryo, whose food is derived from the store laid up in the seed, and whose action upon the air is contrary to that of the developed plant, until it has exhausted this store, and has unfolded its leaves to the light so as to take in fresh carbon from the atmosphere ; it is the case, again, with all the growing parts which derive their nutriment from the leaves (not being themselves able to decompose carbonic acid), and especially with the flowers ; and it is the case, too, even with the leaves themselves, when their functional activity is diminishing, and the decomposing changes in their tissue are commencing.

500. It becomes a question of much interest to determine the relative amount of carbon thus absorbed and excreted by Vegetables. Since a large part of the solid material of their tissues is derived from the atmosphere, it is evident that the whole quantity of carbonic acid in the air must be diminished by their growth ; but as a certain proportion of that carbonic acid is taken in by the roots, which are supplied with it through the absorbent agency of the soil (§ 380), the agency of the leaves requires to be tested by experiment. Such experiments have been repeatedly made by very careful and experienced observers ; and the general result of them is, that, so long as the leaves continue in healthy action, and are exposed to the influence of light, they are actively engaged in taking in carbon from the atmosphere ;* and that, when entire plants, consisting not only of leaves, but of stems and other parts, are confined in the same portion of air, day and night, and are duly supplied with carbonic acid gas during sunshine, they will go on adding to the proportion of oxygen present, so long as they continue healthy ; the slight diminution of oxygen and increase of carbonic acid which take place during the night, bearing no considerable ratio to the degree in which the opposite effect occurs by day.† The balance of nutrition, therefore, between the Animal and Vegetable kingdoms, is thus maintained in a very perfect and interesting manner (CHAP. V.).

501. With regard to the changes effected by vegetation upon the principal constituent of the atmosphere—Nitrogen—no very certain or definite statement can be made. We have seen that this element enters largely into the composition of Plants ; but there is reason to believe that all which they require is derived by them, not directly from the atmosphere, but by the decomposition of the ammonia absorbed by the soil,

* See the experiments of Mr. Pepys in the "Philosophical Transactions" for 1843, p. 329.

† See Dr. Daubeny's letter to Prof. Lindley, in his "Introduction to Botany," vol. ii. p. 297.

and taken up in solution by the roots (§ 380); indeed proof is wanting that the free nitrogen of the air has any concern with Vegetable Respiration; for the few experiments which have been performed with express view to this subject, lead to the belief that azote is more frequently exhaled than absorbed.

502. In the *Fungi* among *Cryptogamia*, however, and in the 'leafless parasites,' such as the tribe of *Orobancheæ*, among *Phanerogamia*, we have examples of the performance of the true respiratory process, without the antagonizing action of the alimentative; the only change which the growth of these plants induces in the surrounding air, being the replacement of its oxygen by carbonic acid. Thus from the experiments of Marcet upon Fungi, it appeared that growing Agarics absorbed from the air a large quantity of oxygen; a portion of which appears to combine with the carbon of the plant, and thus to form the carbonic acid which replaces it; whilst the remainder seems to be retained in its structure. The recent experiments of M. Lory upon the respiration of the Orobancheæ * are peculiarly interesting. He found that in every stage of their vegetation, all the parts of these plants, whether they are exposed to solar light, or whether they are in the dark, absorb oxygen and give out carbonic acid; the volume of the carbonic acid generated being nearly equal to that of the oxygen which disappears. But that the production of carbonic acid is not a result of a mere union of carbon excreted by the plant, with atmospheric oxygen, but takes place in the interior of the plant as a part of the changes in which its growth consists, appears from the fact, that the disengagement continues for a time when the plants are surrounded by an atmosphere of pure hydrogen. (A parallel fact will hereafter be cited in regard to the respiration of Animals, § 534.) The amount of carbonic acid exhaled is augmented by warmth, which increases the activity of the nutrient operations; and, as in other plants, it is peculiarly great during the period of flowering. The contrast between the action of one of these leafless parasites upon the atmosphere, and that of the plants from which they draw their nutriment, was curiously displayed by placing in two receivers of the same capacity, a portion of the stem of an *Orobanche*, and a portion of the leafy stem of the *Teucrium* on which it grew, each piece being of the same weight; the atmosphere surrounding them was composed of six volumes of common air mingled with one of carbonic acid; and they were exposed to light from 9 A.M. until 3 P.M. of the succeeding day. At the end of this time, the atmosphere of the jar containing the Teucrium did not present a trace of carbonic acid, whilst that in which the Orobanche had been immersed exhibited such an augmentation of carbonic acid, that whilst its original proportion to the oxygen was as 20 to 25, it was then as 36 to 9.—The explanation of this difference is doubtless to be found in the mode in which parasitic plants obtain their nourishment. Being supported, like Animals, upon organic compounds that have been already elaborated by the agency of other Plants, they are entirely destitute of the power of decomposing the atmospheric carbonic acid for the purpose of alimentation; and the sole change which they produce in the surrounding air is of the same kind with the respiration of animals. As already remarked,

* "Annales des Sciences Naturelles," Botan., Sept., 1847.

there is every reason to believe that the same change takes place in *all other* Plants, and that carbonic acid is continually given off from their *interior*, whilst oxygen is absorbed; although this is masked by the opposite change, which is effected by their *green surfaces* during the influence of light upon them. For so soon as the light ceases to act upon the latter, the respiratory change makes itself manifest; and it has been shown by the experiments of Saussure, that through the *dark portions* of plants, this change is continually taking place.

503. Further, there are certain processes in the life of all Phanerogamia, in which the function of Respiration seems to go on with remarkable activity, and in which its manifestation is not concealed by the converse operation. One of these is *Germination*, or the development of the young plant from seed (§ 278 *a*), which requires that the starch laid up by the parent for the support of the embryo should be converted into sugar, the latter being the form in which it is applied to the purposes of nutrition. This conversion involves the liberation of a quantity of carbon, which is disengaged by means of its combination with the oxygen of the surrounding atmosphere; and the young plant may then be regarded as living under the same conditions as the parasitic tribes just referred to, its nutriment being supplied to it without the necessity for the alimentative operation which it will be afterwards required to perform. Germination takes place most readily in the dark, since the most essential part of the change—the extrication of carbon—would be antagonised by the influence of light. The young plant is, therefore, much in the condition of one which is being etiolated; and it is accordingly found that, during the early period of germination, the weight of the solid contents of the seed diminishes considerably, though its bulk increases by the absorption of moisture. This is its state until the *cotyledons*, or seed-leaves, have arrived at the surface, and temporarily perform the functions of leaves. It is an interesting fact that, after many trials, germination has been found to take place most readily in an atmosphere consisting of 1 part oxygen and 3 parts nitrogen, which is nearly the proportion of the air we breathe. If the quantity of oxygen be much increased, the carbon of the ovule is abstracted too rapidly, and the young plant is feeble; if the proportion be too small, carbon is not lost in sufficient quantity, and the young plant is scarcely capable of being roused into life.—The changes which take place during *Flowering*, are very similar to those occurring in germination. A large quantity of oxygen is converted into carbonic acid by the action of the flower; and it is believed that the starch, previously contained in the *disk* or receptacle, is changed by this process into saccharine matter adapted for the nutrition of the pollen and young ovules, the superfluous portion flowing off in the form of honey. It is remarkable that this analogy between germination and flowering holds good, not only in their products, but in the conditions essential to their development. Neither will commence except in a moderately warm temperature; both require moisture, for flowers will not open unless well supplied with ascending sap; and the presence of oxygen is in each case necessary. It has been well ascertained that the carbonisation of the air bears a direct relation to the development of the glandular disk, and that it is principally effected by the essential parts of the flower, or organs of fructification. Thus, Saussure found that the *Arum Italicum*, whilst in

bud, consumed in twenty-four hours 5 or 6 times its own volume of oxygen; during the expansion of the flower, 30 times; and during its withering, 5 times. When the floral envelopes were removed, the quantity of oxygen consumed by the remaining parts was much greater in proportion to their volume. In one instance, the sexual apparatus of the Arum Italicum consumed in twenty-four hours 132 times its bulk of oxygen. Saussure also observed that double flowers, in which petals replace sexual organs, vitiate the air much less than single flowers in which the sexual organs are perfect. (See also § 616.)—The same is the case, also, during the development of leaf-buds from parts in which (as in the tuber of the Potato) a supply of starchy matter has been laid up as the material for their evolution, until they are so far expanded that they can obtain carbon from the atmosphere. Here, too, the growing bud is in the condition of a parasite, being supported upon materials provided for it by other agencies than its own; and here, again, we find that the conversion of the starch into sugar is the process to which the production of carbonic acid appears to be chiefly due.

504. Besides the means of aeration which the transmission of the nutritive fluid to the external surface affords, the more highly organised Plants seem to have the power of admitting air into cavities existing in the leaves (especially beneath their inferior cuticle, Fig. 224), through their *stomata*; and in this manner a much larger extent of membrane is exposed to its influence. The peculiar organisation which is probably subservient to this purpose will be hereafter described (§§ 540, 541) under the head of *exhalation*, for which function it appears more particularly designed. But, superadded to this, we find in the Phanerogamia a system of tubes apparently intended to connect the interior of the structure with the external air. These are the *spiral vessels* (§ 151), which, in their perfect form, are never found to contain any but gaseous fluids. In Exogens they usually exist in only one part of the stem, being confined to the 'medullary sheath' (§ 280); in Endogens they are more universally distributed through the stem, forming part of every bundle of fibro-vascular tissue (§ 279). In each case, however, they traverse the stem for the purpose of entering the leaves; and they seem to communicate with the intercellular passages, and, through their medium, if not more directly (as some have supposed), with the external air. We have already noticed the curious analogy between these respiratory tubes and the *tracheæ* of Insects (§ 153); and although their exact office is not fully ascertained, there can be little doubt that they contribute in some way to the aeration of the internal fluids. It has been found that they contain a larger quantity of oxygen, by 7 or 8 per cent., than that which exists in the atmosphere.—In a great number of the aquatic tribes, both among the simpler and the more highly organised plants, we find cavities expressly adapted for the inclusion of air; but these would seem designed rather to give buoyancy to the structure, than to take any share in the Respiratory function. The air which they contain, however, is seldom identical in composition with that of the atmosphere.

505. Regarding the progressive evolution of the Respiratory system in Plants, much might here be said, which will perhaps be more advantageously deferred to the account of their general development (CHAP. XVIII.).

It may be remarked, however, that the early form of the embryo of the Flowering plants resembles, in its want of special organs, the simple vegetation of the cellular Cryptogamia, although it differs in regard to the mode in which nutriment is supplied; the latter deriving it by their unassisted powers from the surrounding elements, whilst the former is provided with it by the parent. At the first period of the germination of the seed, a close analogy exists, as we have seen, between the growing embryo and the tribe of Fungi. Both are supplied with nutriment previously organised, the one by its parent, and the other by the decay of animal or vegetable matter; both are developed most rapidly when supplied with warmth and moisture, and in the absence of light; and both liberate carbon to a large amount, without assimilating any from the atmosphere. By the time, however, that the cotyledons have risen to the surface and acquired a green colour, the plant has advanced a stage in its growth, and the respiratory system has now arrived on the level of the Marchantia (§ 275), possessing, like it, stomata and intercellular spaces, but being destitute of spiral vessels. These do not appear until true leaves are evolved; and by the time that this last stage in the development has taken place, the cotyledons, which may be regarded as temporary respiratory organs, decay away.—When we have traced the evolution of the respiratory system of Animals in a similar manner, we shall observe a most interesting correspondence between the consecutive phenomena, as they occur in the two kingdoms.

3. *Respiration in Animals.*

506. The dependence of the life of Animals upon the constant performance of the Respiratory function, is more immediate than that of Plants; and this arises, not merely from the circumstance that there are sources of demand for oxygen and of production of carbonic acid, in the former, which do not exist in the latter; but also from the fact, that from those which are common to both (§ 494), the amount of carbonic acid generated, as well as of oxygen required, is far larger in the Animal than in the Plant. The more active are the organic functions, and the softer and more prone to decomposition are the tissues, the more considerable will be that constant decay to which all organised fabrics are exposed, even during life; and thus in 'warm-blooded' animals, the high temperature of the body, which favours the vital activity of its component organs, and causes them to live fast, will accelerate their decay, and thus give rise to a more rapid production of carbonic acid, and a greater demand for oxygen; whilst in 'cold-blooded' animals, so long as the temperature of their bodies is low, the 'waste' of the tissues from this source is kept down. But when the temperature of the Reptile is raised by external heat to the level of that of the Mammal, its need for respiration increases, owing to the augmented waste of its tissues. When, on the other hand, the warm-blooded Mammal is reduced, in the state of hybernation, to the level of the cold-blooded Reptile, the waste of its tissues diminishes to such an extent, as to require but a very small exertion of the respiratory process to get rid of the carbonic acid which is one of its chief products. And in those animals which are capable of retaining their vitality when frozen (§ 103), or when their tissues are completely dried up (§ 65), the decom-

position is for the time entirely suspended, and consequently there is no carbonic acid to be set free.

507. But another source of Carbonic acid to be set free by the Respiratory process, and one which is peculiar to Animals, consists in the rapid changes which take place in the Muscular and Nervous tissues, during the period of their activity. Every development of muscular force is accompanied by a change in the condition of a certain amount of tissue, to which change the presence of Oxygen is essential; and one of the products of the union of oxygen with the elements of Muscular fibre, is carbonic acid. The same may probably be said of the Nervous tissue. Hence it may be stated as a general principle, that the peculiar waste of the Muscular and Nervous substances, which is a condition of their functional activity, and which is altogether distinct from the general slow decay that is common to these tissues with others, is another source of the generation of carbonic acid, and of the demand for oxygen in the animal body; and that the amount of the one gas produced, and of the other gas required, will consequently depend upon the degree in which these tissues are exercised. In animals which are chiefly made up of the organs of vegetative life, in whose bodies the nervous and muscular tissues form but a very small part, and in whose tranquil plant-like existence there is but very little demand upon the exercise of these structures, the quantity of carbonic acid thus liberated will be extremely small, and the dependence upon a supply of oxygen by no means close. On the other hand, in animals whose bodies are chiefly composed of muscle, and whose life is an almost ceaseless round of exertion, the quantity of carbonic acid thus liberated is very considerable, and the demand for oxygen is incessant; so that vital activity is speedily suspended, if the respiratory function be not performed. We are enabled to trace the connection between the amount of muscular exertion, and the energetic performance of the act of respiration, in the class of Insects, better than in any other. They have no fixed temperature to maintain; and they are consequently not in the condition of warm-blooded animals, in which the quantity of carbonic acid set free is kept up to a more regular standard by the provision to be presently noticed. On the other hand, they are pre-eminent among all animals, in the energy of their muscular power as related to the bulk of their bodies, and the waste of muscular tissue during their state of activity must therefore be very great; and we shall hereafter find that the amount of carbonic acid generated in a given time bears a close correspondence with this.

508. Besides these sources of demand for Respiration, which are common to all Animals, there is another, which appears to be peculiar to the two highest classes, Birds and Mammals. These are capable of maintaining a constantly-elevated temperature, so long as they are supplied with a proper amount of appropriate food; and their power of doing so appears to depend upon the *direct* combination of certain elements of the food, with the oxygen of the air, by a process analogous to combustion (§ 376). The quantity of carbonic acid that is thus generated, seems to vary considerably in different animals, and in different states of the same individual; but the principal source of difference lies in variations of external temperature; for the energy of the Respiration increases with its diminution, as more heat must then be generated; and diminishes with its increase.—In

all cases, if a sufficient supply of food be not furnished, and the store of fat be exhausted, the animal dies of cold (§ 93).

509. To recapitulate, then; the sources of the production of carbonic acid, and of the demand for oxygen, in the Animal body, are fourfold.— 1. The continual decay of the tissues; which is common to all organised bodies; which is diminished by cold and dryness, and increased by warmth and moisture; which takes place with increased rapidity at the approach of death, whether this affect the body at large, or only an individual part; and which goes on unchecked, when the actions of nutrition have ceased altogether. — 2. The various changes in composition that take place in the tissues in the progress of their organic construction, many of which involve a liberation of carbon and a higher oxygenation. —3. The metamorphosis that is peculiar to the Nervous and Muscular tissues; which is the very condition of their activity; and which therefore bears a direct relation to the degree in which they are exerted.—4. The direct conversion of the carbonaceous materials of the food into carbonic acid; which is peculiar to warm-blooded animals; and which seems to vary in quantity, in accordance with the amount of heat to be generated.

510. The organs appropriated to the performance of the function of Respiration, in the various classes of the Animal Kingdom, appear at first sight so very different, that a superficial observer would hardly trace any relation between them (§§ 333, 350). A little reflection, however, will show, that all their forms are reducible to the simple element of which the respiratory organs are constructed in the Vegetable kingdom;—an extension of the external surface, peculiarly adapted, by its permeability to gases, for the interchange of ingredients, between the circulating fluid brought in contact with one side of it, and the atmosphere which it touches on the other. Considered, therefore, under this instrumental character, there is a complete 'unity' amongst them all; but when considered with reference to the general plan of structure, we find them to be 'homologically' diverse. Thus, the extension may take place from very different parts of the surface, so that its connections with other organs may be altogether dissimilar in the several tribes of Animals. Again, it usually takes place internally or externally, according as the animal is to be an inhabitant of the air or of the waters. In animals modified for atmospheric respiration, the air enters the system to meet the blood: a peculiar set of movements, more or less complicated, being appointed for its constant renewal by successive inhalation and expulsion. In those adapted to an aquatic residence, a different plan is required. The small quantity of air contained in the water, is all that the respiratory system employs; and it would have been a useless expenditure of muscular exertion, to have provided means for the constant inspiration and expiration of a large amount of so dense a fluid. In most aquatic animals, therefore, the aerating surface is extended outwardly, instead of being prolonged inwards; and the blood is propelled through it so as to come in relation with the surrounding medium; the portion of which in apposition with it is continually being renewed, either by the natural movements of the animal, or by others more expressly contrived for the purpose. The relation between these organs will, perhaps, be made more apparent by a simple diagram. Let A B represent the general external

surface of the body; then at *a* is shown the character of a simple *outward* extension of it, forming a foliaceous gill, such as is seen in the lower Crustacea; and in like manner, *b* may represent a simple *internal* prolongation or reflexion, such as that which forms the pulmonary sac of

FIG. 217.

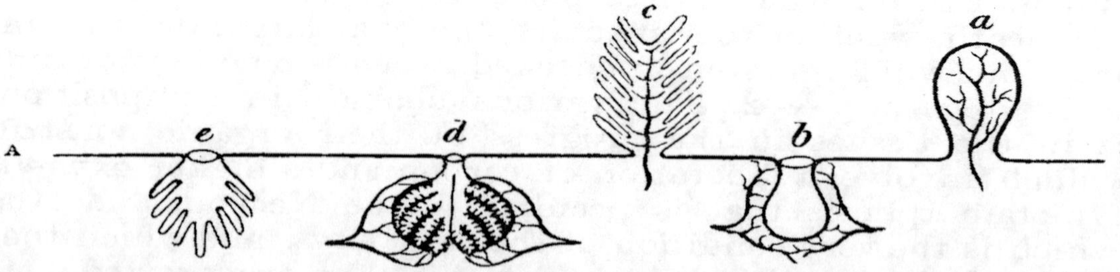

Diagram illustrating different forms of Respiratory apparatus:—*a*, simple leaf-like gill; *b*, simple respiratory sac; *c*, divided gill; *d*, divided sac; *e*, pulmonary branchia of Spider.

the air-breathing Gasteropods. A higher form of the branchial apparatus is shown at *c*, the respiratory surface being extended by the subdivision of the gill into minute folds or filaments, as we see in Fishes; and a more elevated form of the pulmonary apparatus is seen at *d*, the membranous surface being extended by subdivision of the internal cavity, as we find to be the case especially in Birds and Mammals. Lastly, at *e* is shown a plan of one of the 'pulmonary branchiæ' of the Arachnida, which forms a kind of transition between the two sets of organs; the extent of surface being given by gill-like plications of the membrane lining the interior of a pulmonic cavity.—Putting aside such modifications, however, as are destined to suit the particular conditions under which the function is to be performed, and looking simply at the essential characters of the Respiratory organs, we shall observe, on tracing them upwards through the principal classes of animals, the same gradual specialisation which has been noticed in the other systems; for, beginning with the lowest, it will be seen that the general surface is the organ of respiration as well as of other functions; whilst, in the highest, the aeration of the blood is almost entirely effected in one central apparatus adapted to it alone, although the general surface is not altogether destitute of participation in it.

511. In the simpler forms of Animal structure, we do not find any special provision for the performance of this function; the aeration of their fluids being accomplished solely by their exposure to the surrounding medium, through the thin membrane which forms their external integument. Such is the case, for example, with all the *Protozoa*, and also with the Zoophytes among the *Radiated* tribes, and with the Entozoa, the lower Annelida, and the Rotifera, among the *Articulated*. These animals are all aquatic; and the only atmospheric air with which they come into relation, is that which is diffused through the liquids they inhabit. Many of them are provided with the means of renewing the stratum of liquids in immediate contact with their bodies, by ciliary action; the same mechanism thus serving for the acquirement of food (§ 399), and for the aeration of their fluids. Passing to the higher classes of the Articulated series, we find that among the more inert and

inferiorly-organized Crustacea, such as the *Pycnogonidæ* (§ 316 *e*), and even in some of the lowest Arachnida, such as the simplest *Acaridæ* (§ 319 *d*), there are no special respiratory organs; but it is a curious mark of the superiority of the *Mollusca* in regard to the development of their apparatus of organic life, that in scarcely any instance is there such a complete deficiency of special respiratory organs in the higher parts of that series, the only case of the kind being among the most degraded forms of the *Nudibranchiate* order (§ 304 *c*). In the *Vertebrated* classes they are never wanting.

511 *a*. The first indication of a special provision for the aeration of the fluids, presents itself among the *Radiata*, in animals which have no proper circulating apparatus. Thus in many of the *Pulmograde* Acalephæ, we find the digestive cavity sending out prolongations, which form by their inosculation a close network along the margin of the disk (Fig. 195); and as this margin is furnished with cilia, a constant renewal of the water in contact with its external surface is provided for. In the *Ciliograde* members of the same class, we find a provision of a similar nature; the canals prolonged from the digestive cavity being extended beneath the rows of cilia which give movement to the body (Fig. 107, B); and thus imparting the materials of nutrition to the tissues which they penetrate, whilst they expose the fluids which they contain to the influence of the aerated water that bathes their surface.—In the class *Echinodermata*, however, we find a more special provision for the performance of the respiratory function; which is required, not merely by the increased nervo-muscular energy of these animals, and by the development of a special circulating system, but also by the condensation of their external tegument, which is no longer capable of serving, as in the classes we have been considering, for the aeration of the fluids contained within the tissues it invests. Contrary, however, to the general principle which has been just stated,—that in aquatic animals the circulating system is prolonged outwardly, bringing the blood to meet the air contained in the dense element,—we find that the respiratory apparatus for the most part consists of the membrane lining the cavity which intervenes between the viscera and the interior of the shell; into this cavity, the water from without is freely admitted, probably in all the members of this class, save the Holothuriada; and a network of blood-vessels is minutely distributed upon its walls. The water seems to be introduced into its interior, in the *Asteriada*, by means of an immense number of short, conical, membranous tubes, which pass between the pieces of the shelly framework, and project externally in little tufts; each of these tubes has a minute orifice at its extremity, corresponding with that existing at the extremity of each of the tentacula of the Actinia (§ 291), which they much resemble; and through this multiplied series of apertures, the water appears to be drawn in and expelled, by means of the cilia that clothe the respiratory surface. In the *Echinida*, the pieces of whose shell are in such close apposition as not to allow the passage of such tubes, we find that they are all collected into ten bundles, which pass through the flexible membrane surrounding the mouth; and it is possible that water may also find its way into their visceral cavity, through the interspaces of the teeth. The respiratory tubes may themselves act as external branchiæ; and it does not seem improbable that the admission of water

into the tubular cirrhi or 'feet' (§ 296, *b*, *c*) may also be subservient to this process, since we find the blood-vessels distributed in a minute net-work on the vesicles at their base. The visceral cavity of the Echinodermata is obviously homologous with the peritoneal cavity of Vertebrata; and it is remarkable that in many Fishes, and even in some aquatic Reptiles (as the Crocodile), a provision should exist for admitting water to its interior, apparently for the purpose of aiding in the aeration of the blood.—In the *Holothuriada* we find a very remarkable development of a *special* respiratory apparatus; for instead of the admission of water into the visceral cavity, we find that it is drawn into a long membranous sac, which opens externally in the cloaca, and extends nearly as far as the mouth. This sac is sometimes single and undivided; but in all the higher forms of the order it is double, and presents a series of beautiful and minute ramifications, forming what are known as the 'respiratory trees' (Fig. 113, *r*, *r*). Both the stem and branches of this organ contain distinct circular and longitudinal muscular fibres, which contract on being irritated, and which expel the water it contains at tolerably regular intervals (such as once, twice, or three times in a minute), its re-introduction being apparently accomplished by ciliary action. The contraction of the muscular fibres of the external integument may probably assist in the expulsor action; but it is not required, since the respiratory movement continues even after the sac has been laid open. The nearest approach to this curious form of respiratory apparatus, that is presented by any other class of animals, is that which we meet with in the tracheal system of Insects (§ 520); but although there is an obvious relation of *analogy*, yet they are not really *homologous*, for the 'respiratory tree' of the Holothuriada is an offset from the cloacal termination of the intestinal tube, and might thus perhaps be likened to the *allantois* (CHAP. XVIII.) more correctly than to any other organ of higher animals.

512. In the various classes of the *Molluscous* sub-kingdom, we find the respiration provided for by the adaptation of special organs for the purpose; but there is a remarkable difference in the mode in which this adaptation is brought about in the several groups. With the exception of the *Pulmonated Gasteropods*, however, they are all aquatic; and the aeration of the blood is therefore provided for, by bringing it into relation with the surrounding water; a continual interchange in the particles of the circulating fluid being secured by its regular movement through the vessels of the respiratory surface, and a like interchange in the aerating medium being effected by the action of the cilia with which it is clothed, as well as, in certain cases, by the muscular contraction of the sac into which it may be received.—In the *Bryozoa*, however, we find a deficiency of special respiratory organs; but this deficiency is easily understood, when it is remembered that they have no distinct circulation. Still, the aeration of the nutritious fluid that occupies the visceral cavity, seems to be provided for by its transmission along channels in the interior of the ciliated tentacula (Fig. 114, D, *b*, *b*); and it has also been observed that the wide pharynx is sometimes dilated with water, which is expelled again without passing into the stomach, as if it had been taken in for the aeration of the fluids of the body. This pharyngeal cavity, in the *Tunicata*, constitutes the special respiratory organ, and is now known as the 'branchial sac;' in all those which are formed upon the Ascidian type it is divided

by an immense number of fissures (Figs. 119 and 121), through which a part of the water received into it by the oral orifice passes to the interspace between its outer side and the muscular mantle; in the Salpians, however, there are no such apertures, and the respiratory function seems to be chiefly performed by the longitudinal fold of the branchial sac (Fig. 206, *d*), to which the blood-vessels are minutely distributed. In the Ascidians, again, the renewal of the aerating fluid is chiefly provided for by the action of the cilia lining the branchial sac; but the cavity is occasionally emptied by the sudden contraction of its muscular walls, which ejects the contained water with considerable force. In the Salpians, on the other hand, the respiratory current is chiefly sustained by the rhythmical contractions and relaxations of the muscular sac, which also effect the propulsion of their bodies through the sea (§ 299 *b*).—Turning from these to the *Brachiopoda*, we find that the mantle which lines their bivalve shells is the principal organ of respiration; the surrounding water being freely admitted to its internal surface by the fissure which passes along the whole margin of the valves, and thus coming into the closest possible relation with the blood contained in the vessels which are minutely distributed upon it. But as the water also bathes the external surface of the viscera, as well as the ciliated arms, the blood of these parts also must be acted on by the air which it contains; so that they may be regarded as accessory organs of respiration.—In the *Lamellibranchiata*, or ordinary 'Bivalves,' the internal surface of the mantle lining the valves, is doubled (as it were) into four riband-like folds, which are delicately fringed at their edges, and which have, in fact, the same essential structure as the gills of higher animals (Fig. 207). To these the blood is transmitted, when it has been rendered venous by traversing the tissues of the body generally; and in these it is exposed, through a surface which is greatly extended by the minute division of the fringes, to the action of water introduced from without, and constantly renewed by ciliary action. In many of these animals, as in the common Oyster, the two lobes of the mantle are so completely separated, that the water can still enter freely between the valves; but in general, they are more or less united, so that the cavity in which the gills lie is partially closed. There is always a provision, however, for the free access of water from without, by means of two apertures, one for its entrance and the other for its ejection; and in certain species which burrow deeply in sand or mud, these apertures are furnished with long tubes, or *siphons* (Fig. 124, *y*, *z*), which convey the water from nearer the entrance of the burrow, and carry it thither again. In these also, a continual flow of water over the respiratory surfaces is maintained by the vibration of the cilia, with which they are clothed.

513. The position of the gills in the *Branchiferous Gasteropoda* is extremely variable. In the *Nudibranchiata* they are disposed, without any protection, upon various parts of the general surface of the body; sometimes forming tufts of delicate leaf-like or arborescent appendages (Fig. 218), which are collected into one cluster (Fig. 126); sometimes, on the other hand, presenting themselves as papillæ which are scattered over the whole external surface; whilst in a few of the inferior species of this tribe, these special organs are undeveloped, and the function is performed by the general surface alone. In all these cases, the renewal of the water in contact with the external surface is effected by ciliary movement

only. In most of the other branchiferous Gasteropods, the gills are more or less covered by the shell; and even where this is but little developed, as in the Aplysia, we find it specially devoted to their protection. The order *Pectinibranchiata*, which is the highest and most numerous subdivision of the class, receives its name from the peculiar pectinated or comb-like arrangement of its gills (Fig. 128), which are lodged within a cavity formed by the arching of the mantle behind the head; and into this cavity the water is admitted by a special channel or siphon.—The highest development of the branchial apparatus presented by the Mollusca, is that which we find in the *Cephalopoda*. In these animals, the gills are very large, especially in the Dibranchiate order (Fig. 135, *o*, *o*); and they are lodged in a cavity formed by the folding-over of the mantle, to which water is admitted through a wide fissure at the base of the head; whilst the current, after passing over them, is ejected through the funnel (*e*). The respiratory surface is not here covered with vibratile cilia; but the respiratory current is sustained entirely by the alternate contractions and dilatations of the muscular parietes of the cavity; whilst in the presence of special branchial hearts for the propulsion of the blood through the gills (§ 469), we have another provision for the increased vigour of the respiratory function, required by the active habits of these animals, which present a remarkable contrast to the sluggish inert character of the Mollusca in general. — In these classes, taken as a whole, the respiration is low in its amount. The blood contains no *red* corpuscles, excepting perhaps in the highest class; and the change in its composition, which is effected by the air, is confined, therefore, to the fluid plasma, or liquor sanguinis. And as it is not exposed directly to the air, except in a few species, but to the air contained in the water inhabited by the animals, this change cannot be very energetically performed. But as the life of these animals is chiefly vegetative, as their movements, except in the highest classes, are few and feeble, and as they maintain no independent heat, there is comparatively little need of the interchange which it is the object of the Respiratory process to effect; and they can sustain the complete suspension of it for a long time.

FIG. 218.

One of the arborescent processes, forming the gills of *Doris Johnstoni*, separated and enlarged.

514. Among many of the *Articulated* tribes whose habitation is aquatic, the respiration is carried on upon a similar plan. It is in the *Annelida* that we first meet with a special provision for this function; the blood being transmitted, in the course of its circulation, to a series of gill-tufts, which are composed of a delicate membrane prolonged from the external surface of the body, and which sometimes have the form of branching trees, and sometimes of delicate brushes made up of a bundle of separate filaments. In either case, the filaments are traversed by blood-vessels, and are adapted to bring the blood into close relation with the surrounding water; and the continual interchange of the latter is provided for by the restless movements of the body, as well as by the

action of the cilia covering the respiratory organs. The tufts are sometimes arranged along every segment of the body, and their multiplication prevents them from individually attaining any considerable size, this is the case in the Nereidans and other *Dorsibranchiata* (Fig. 147). In other instances, they are disposed at intervals, and are then larger, being less numerous, as in the *Arenicola* (Fig. 149). Their most beautiful development is where they are present on the head only, the rest of the body being enclosed in a shelly or sandy tube, as in the *Serpulæ* and *Terebellæ* (Figs. 148, 201). The gill-tufts then frequently present the appearance of a flower, endowed with the most brilliant and delicate hues.* In many animals of this group, as we have seen in the preceding chapter, there are special provisions for the propulsion of the blood through the gills, for which the feeble action of the dorsal vessel would not furnish sufficient power.—The higher Articulated classes are, for the most part, adapted to atmospheric respiration, on the plan to be presently explained; but there is one class, that of *Crustacea*, whose respiration is still carried on through the medium of water. In the lowest forms of this group, there is no special respiratory apparatus; the general surface being soft enough to admit of the required aeration of the fluids through its own substance, and the animal functions being performed with so little activity, that a very small amount of interchange is required. In the higher orders, however, whose bodies are encased within a harder envelope, a special respiratory organ is almost invariably found. In some of these (as the *Branchiopoda*) the last joints of all or of the greater number of the legs are flattened out into a surface which is soft and vascular, and which, by its continual strokes upon the water, appears calculated to facilitate the influence of the air upon the nutritious fluid. Proceeding higher (to the order *Amphipoda*) we find a particular portion only of the extremity, the flabellar appendage, devoted to respiration; but this is developed to an increased extent, and the water in contact with its surface is incessantly renovated by currents set in motion by the abdominal members. The next stage in the specialisation of this function, is the restriction of the branchial apparatus (as in the order *Isopoda*) to the abdominal members, which are entirely devoted to it, and cease to have other uses. In a still higher order (*Stomapoda*), the gills have assumed more of the character which they present in Fishes and some Mollusca; the laminated or leaf-like form which they at first possessed, having given place to one in which the surface is greatly extended by minute subdivision into delicate filaments. The most developed form of respiratory apparatus possessed by Crustaceans, is that which exists in Crabs, Lobsters, and other *Decapods*. In this order, not only is the function thrown upon particular organs entirely set apart for the purpose, but these organs are lodged and protected within a special cavity; and the renewal of the water necessary to their operation is secured by

* There are few sights more striking to the observer of nature in tropical regions, than the unexpected view of a bed of coral in shallow water, having its surface scattered with the brilliant tufts of the Serpulæ which have formed their habitations in it; the glowing and variegated tints of which, when lighted up by the mid-day sun, and contrasted with the sombre hues of the surrounding rocks, present an appearance compared to which the most beautiful garden of carnations (which flower the animals much resemble in form) sinks into insignificance.

the motion of distinct appendages. This cavity is formed by a reduplication of the external tegument, and is provided with two orifices, one for the introduction and the other for the expulsion of the fluid. Through these orifices a constantly-renewed supply of water is made to pass, by the agency of a large valve-like organ, placed in the efferent canal, which, by its movements, drives a continual current from behind forwards, and thus occasions a constant ingress through the afferent opening; this organ is nothing else than the flabelliform appendage of the second pair of feet-jaws, specially developed to answer this purpose. The perfect contact of the water with the respiratory surface is further provided for by the actions of the flabelliform appendages of other maxillary or ambulatory members, which, in most Decapods, penetrate into the branchial cavity, and incessantly sweep the surface of the branchiæ, apparently for the purpose of combing out their filaments (so to speak) by means of the stiff hairs with which they are furnished. In those Crustacea which are adapted to live for a time on land, the orifices of the branchial cavity are very small, so that but a trifling amount of evaporation can take place from them; and it appears that, in all the species, the gills can be subservient to aerial as well as to aquatic respiration, provided their surface be kept moist,—the asphyxia of the animals in a dry atmosphere being due to the desiccation of the membrane, and its consequent unfitness for the performance of its functions. There are other species, known as Land-Crabs, which not only live habitually out of water, but are infallibly drowned if kept long immersed in that fluid. The membrane lining their branchial cavities is sometimes disposed in folds capable of serving as reservoirs for a considerable quantity of water; and sometimes presents a spongy texture equally well adapted for storing up the fluid, which is necessary to keep the organs of respiration in the state of humidity required for the performance of their functions. Land-crabs are never known to remove far from damp situations; and this humidity may be either derived from the atmosphere, or may be secreted, as in higher animals, from the circulating fluid. It can scarcely be doubted that the spongy lining of the branchial cavity in these Crustacea is peculiarly subservient to aerial respiration; and it appears to be owing to the check given to its activity, that the land-crabs are drowned when plunged under water.

515. The stages in the development of the branchial apparatus of the *Astacus fluviatilis* (Cray-fish) have been so beautifully traced by M. Milne-Edwards, in connection with the various forms of the same in adult species of different tribes, that it seems advantageous to notice them here for the sake of ready comparison, rather than to defer the account of them to the general description of the progressive evolution of the system in the embryo of higher animals.—In the earlier periods of embryonic life, no trace of branchiæ can be discovered; so that the embryo may then be considered as on a level with those inferior Crustaceans, whose respiration is entirely carried on by the general surface. When these organs are first evolved, during the process of incubation, they consist of simple laminated expansions, representing the flabelliform appendages of the three pairs of maxillary members, and thus corresponding with the branchial feet of the Branchiopoda. These soon subdivide, and one part assumes a cylindrical form, and seems no longer

to belong to the apparatus, whilst branchial filaments begin to appear on the other, which are subsequently prolonged into complete gills; during this interval the thoracic extremities have made their appearance, and they also become furnished with branchial appendages; and thus in possessing respiratory organs which are quite distinct from the instruments of locomotion, but are altogether external, the embryo Decapod corresponds with the higher Stomapoda. At a subsequent period, a narrow groove or furrow is seen along the under edges of the thorax, the margins of which, in no long time, are prolonged so as to meet each other and enclose the gills; openings being left for the entrance and exit of water, which are at first large, but subsequently become contracted to the proper size. It is thus evident that the lining membrane of the cavity, as well as that which covers the filaments of the branchiæ, is but a prolongation of the external tegument.

516. Although the respiratory apparatus in *Fishes* retains the type which characterised it in the inferior aquatic classes, it undergoes great increase both in extent and importance. In order to keep up with the rapid advance in the development of the other systems, the respiration requires to be conducted, though by means of an aquatic element, with great velocity and effect. For this purpose, it is not sufficient that Fishes should have merely filamentous tufts hanging loosely at the sides of the neck; but it is requisite that they should possess the means of rapidly and constantly propelling large streams of water over their surface, and of forcing the whole blood of the system through the respiratory apparatus, to be submitted to the action of the air that is contained so scantily in the water. The former of these ends is effected by the connection of the gills with the cavity of the mouth, the muscles of which send a rapid current of water through the branchial passages; and the latter, by the alteration in the position of the heart, which is placed so as to propel the blood through the respiratory organs *before* it proceeds to the system at large (§ 470). The gills in most Fishes are disposed in fringed laminæ, the lamellæ of which are set close together like the barbs of a feather, and are attached on each side of the throat, in double rows, to the convex margins of four or five long bony or cartilaginous arches, which are very similar to the ribs (Fig. 219, A). The extent of surface exposed by these gills is very great, and the network of capillaries spread over them is extremely minute (B). In the *Osseous* fishes, the gills are lodged in a capacious chamber, which communicates internally with the pharynx by a distinct orifice for each interspace between the branchial arches; whilst externally it opens by a single large orifice on each side, which is furnished with a valvular bony *operculum*, that allows free passage to the currents ejected from the mouth, but serves for the protection of the delicate organs beneath it. In the higher *Cartilaginous* fishes, the gill-chamber is more completely subdivided; the branchial lamellæ being attached, not merely at their bases to the branchial arches, but by their opposite extremities to the membrane lining the chamber; and each branchial interspace has its own external orifice for the discharge of the water. We shall hereafter find that the branchial fissures in the neck, which thus form a direct communication between the pharynx and the external surface, may be seen in all the higher Vertebrata at a certain stage of development. The curious variation from the ordinary type of the respiratory apparatus of

Fishes, which presents itself in *Amphioxus,* and which carries us back to the type of the Tunicated Mollusca, has been already pointed out (§ 321 *b*). Another variation, which in some degree reminds us of the rudimental respiratory apparatus of the Leech and Earth-worm, is presented in the *Cyclostomi,* in which we find six or seven pairs of gills on each side, and these not attached to cartilaginous arches, but developed as folds from

FIG. 219.

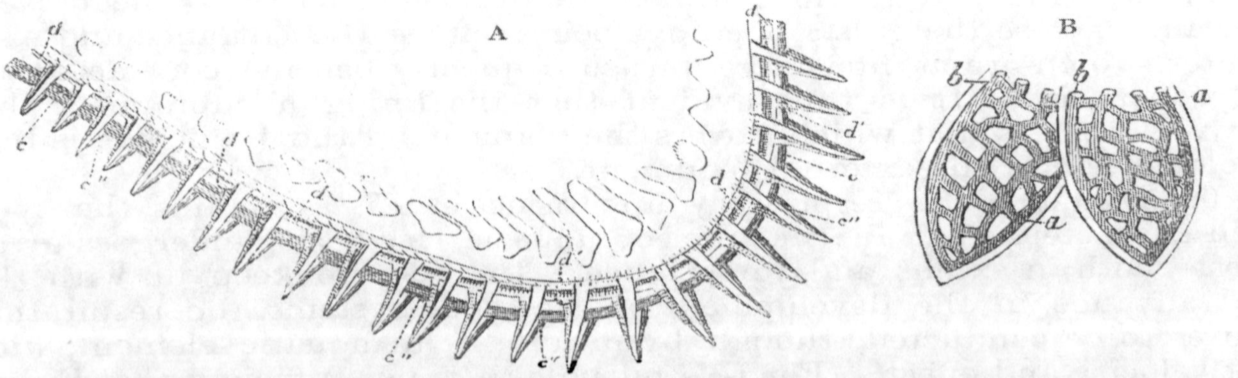

A, Branchial arch, with its boodvessels; *c c'*, branchial artery, diminishing in size from *c* to *c'*, in proportion as it furnishes the twigs *c'' c''* to the branchial lamella: *d d*, the branchial vein, which augments in the same proportion, as it receives the venous twigs *d' d'*.—B, Capillary network of a pair of leaflets of the gills of the Eel:—*a, a*, branches of the branchial artery conveying venous blood; *b, b*, branches of branchial vein, returning aerated blood. The disappearance of the dark shading in the net-work, as it traverses the gill, is designed to indicate the change in the character of the blood, as it passes from one side to the other.

the lining membrane of as many distinct sacculi; and to understand the relation between these and the gills of ordinary fishes, we must suppose (as Prof. Owen has remarked*) each compressed sac of the *Myxine* to be split through its plane, and each half to be glued by its outer smooth side to an intermediate septum, which would then support the opposite halves of two distinct sacs; if then these vascular surfaces be prolonged into lamellæ, tufts, or filaments, and an intermediate basis of support be developed, we have the branchial arch of one of the higher cartilaginous Fishes, which is the homologue, not of a single gill-sac, but of the contiguous halves of two distinct gill-sacs, in the Myxine. The first part of this change is seen in the Lamprey; in which we also find the branchial sacculi communicating with the pharynx, not directly by a separate fissure for each, but through the medium of a tube which comes off from the pharynx on either side, and conveys the water by a distinct branch to each gill-sac. In most of the Cyclostomi, as in the Lamprey, each sac opens externally at the side of the neck by a separate orifice; but in the Myxine the efferent current is conveyed by a special tube, which collects it from all the sacculi of one side, and passes backwards to discharge itself near the middle line of the ventral surface,—thus presenting a connecting link between the provision for the discharge of the respiratory current in Amphioxus, and that which prevails in ordinary Fishes. Some even of the Osseous fishes present a departure from the ordinary type in the prolongation and contraction of the efferent canal; and this is particularly remarkable in the *Eel* tribe, in which we find the two orifices approxi-

* "Hunterian Lectures," vol. ii. p. 259.

mating each other on the under side of the neck, until, in *Synbranchus*, they actually meet, so that the respiratory current is discharged from both sides by a single pore on the median line.—During the embryo condition of both of the principal divisions of Fishes, the gills may be seen hanging loosely from the back part of the neck; for, in Osseous fishes, they have attained considerable development before the prolongation of the integument has been formed into the valve which covers them; and in the Cartilaginous fishes, the branchial openings are at first large, and the filaments of the gills are prolonged much beyond them,—other filaments also, which subsequently disappear altogether, being produced from their edges.

517. The mechanism of respiration is very complex in Fishes; and is evidently adapted to produce the most effectual aeration possible. The mouth is first distended with water; and its muscles are then thrown into contraction, in such a manner as to expel the fluid, through the apertures on either side of the pharynx, into the gill-cavity. At the same time, the bony arches are lifted and separated from each other, by the action of muscles especially adapted to this purpose; so that the gill-fringes may hang freely, and may present no obstacle to the flow of the water between them. When they have been thus bathed with the aerating liquid, and their blood has undergone the necessary change, the water is expelled through the external apertures by muscular pressure, its return to the pharynx being prevented by the valves with which the orifices are furnished. It is well known, that most Fishes speedily die when removed from the water; and it can be easily shown, that the deficient aeration of the blood is the immediate cause of their death. But as it might have been expected that the atmosphere would exert a much more energetic influence upon the blood contained in the gills, than that which is exercised by the small quantity of air contained in the water, the question naturally arises, how this deficient aeration comes to pass. It is chiefly due to the two following causes;—the drying-up of the membrane of the gills themselves, where it is exposed to the air, so that the aeration of the blood is impeded;—and the flapping-together of the filaments of the gills, which no longer hang loosely and apart, but adhere in such a manner as to prevent the exposure of the greater portion of their surface to the air. Those among ordinary fishes can live longest out of water, in which the external gill-openings are very small, so that the gill-cavity may be kept full of fluid: but there is a particular family, that of *Labyrinthibranchii*, in which the anterior branchial arches give origin to a curious lamellated apparatus, in whose interspaces water may be retained for a considerable length of time, so as to keep the gills moist; and by this provision, analogous to that which exists in the Land-Crab (§ 514), such fishes are enabled to remain for a considerable time out of water, performing long migrations over land in search of food, and even (it is said) ascending trees.—The respiration of Fishes is much more energetic than that of any of the aquatic Invertebrata; and this is partly due to the great extension of the surface of the gills, partly to the provision just explained for maintaining a constant flow of fresh water over their surface, and partly to the position of the heart at the base of the main trunk that conveys the blood to the gills, by which the energetic propulsion of that fluid through these organs is secured. Their blood,

too, is furnished with red corpuscles, which seem to give important aid in conveying oxygen from the gills to the remote tissues of the body, and in returning the carbonic acid to be excreted. The proportion of these, however, varies considerably in the different species of the class; being very small in those that approach most nearly to the Invertebrata, whilst they are present in large numbers in the blood of certain Fishes, which have great muscular activity, and can maintain a high independent temperature.

518. The branchial apparatus of the *larvæ* of *Batrachia* and of the *Perennibranchiate* division of the same order, is constructed in all essential particulars upon the plan of that of Fishes. In the first instance the branchiæ are external, hanging like tufts from the sides of the neck; and these external gills, like the branchiæ of Invertebrated animals, are covered with cilia, whose actions are of considerable importance in renewing the stratum of water in contact with them. This state continues in most of the Perennibranchiata through the whole of life; but in the Amphiuma and Menobranchus, the external gills disappear, without being replaced by internal organs of a like kind, and nothing remains of their branchial apparatus but the fissures leading from the pharynx and the surface of the neck. In those Batrachia, however, whose development proceeds further (as Frogs and Water Newts), the branchiæ are subsequently more or less enclosed by a fold of the skin, which forms a membranous valve, analogous to the bony operculum of Fishes; and in the tadpole of the Frog, the branchial cavity thus formed is closed completely on the right side, the water which has passed through it being ejected through the opening that remains on the left.

519. Having thus traced the organs of *aquatic* Respiration, from their simplest and most general, to their most elaborate and most specialised forms, we have to follow the same course with those which are provided for *atmospheric* respiration. None such are met with among the *Radiated* classes, which are all aquatic; and the Pulmonated Gasteropods (§ 303) are the only *Mollusca* which are adapted for breathing air. This group includes not merely the Snails, Slugs, and other terrestrial Gasteropods; but also several, which, although habitually living beneath the water, receive air (taken-in at the surface) into a pulmonic cavity, instead of respiring through the medium of branchiæ. This pulmonic cavity is a simple undivided sac, on the walls of which the blood-vessels are minutely distributed; it lies beneath the dorsal portion of the mantle, and communicates with the external air by a single orifice on the right side of the neck, provided with a sphincter muscle by which it can be occasionally closed.—In a large proportion of the *Articulated* series, however, we find the apparatus for atmospheric respiration presenting a high degree of development; but, like the branchial organs of the lower members of that series, it exhibits a tendency to repetition in successive segments, which is in striking contrast with the conformation just described. The air-breathing *Annelida*, such as the Earth-worm, may be regarded as presenting us with the lowest type of this apparatus, each segment of the body being provided with a pair of sacculi, which open outwards by orifices corresponding in position to those of the stigmata of Myriapods and Insects; but it does not seem that either in the Earth-worm or in the Leech (which has a like series of vesicles), these organs are subservient to

respiration, as they are always found to contain mucus ; and it would appear that the aeration of the blood must be accomplished through the medium of the general surface of the body. In *Myriapoda*, however, we find a regular transition presented by the respiratory apparatus of the different families, from that just described, to that which is characteristic of Insects ; for whilst in the lower *Chilognatha* the stigmata lead only to a double series of sacculi presenting no ramifications, the higher members of that group have a set of ramifying tracheæ issuing from each sacculus, and proceeding to the various organs of its own segment ; and in the *Chilopoda*, the tracheal systems of different segments come into connection with each other by mutual inosculation, some even possessing the two longitudinal canals connecting together the whole series of stigmata and the main trunks of the tracheæ, as in Insects.

520. In studying the Respiratory system of *Insects*, we shall have occasion to observe several peculiar modifications which it undergoes for particular purposes, whilst its essential character remains unaltered ; and we shall have also an opportunity of noticing the varieties of form and function which the same apparatus may present at different periods of life, and under changes in external conditions. The muscular energy required for the locomotive powers of the perfect Insect, and the general activity of its organic processes, necessarily involve a large amount of communication between the nutritious fluid and the atmosphere ; but, on the other hand, the low development of the circulating system would prevent the aeration from being accomplished with sufficient rapidity, by the transmission of the blood through one particular organ. The difficulty is obviated by the introduction of the vivifying agent into every part of the body, by means of a complex and minutely-distributed system of tubes ; which ramify through even the smallest and most delicate organs, and thus bring the air into immediate relation with all their tissues, whilst they also dilate in certain parts into sacculi, which sometimes attain a considerable size (Fig. 220). The proportional size of these sacculi

Fig. 220.

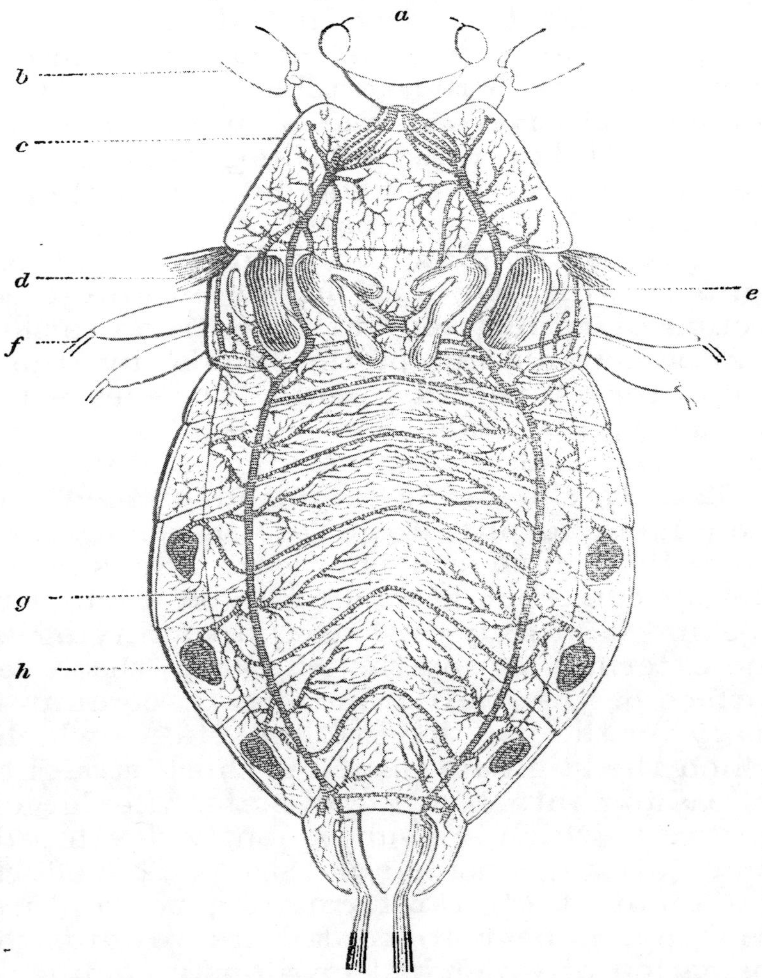

Tracheal system of *Nepa* (Water-Scorpion) ;—*a*, head ; *b*, 1st pair of legs ; *c*, 1st pair of wings ; *d*, 2nd pair of wings ; *e*, air-sac ; *f*, 2nd pair of legs ; *g*, tracheal trunk ; *h*, one of the stigmata.

varies greatly, however, in different families. They are usually most developed in those that sustain the longest and most powerful flight (such as the Bee tribe), which are generally those whose larva condition has been most imperfect, and in which there has been originally no appearance of these enlargements; on the other hand, they are almost entirely absent in the Insects destined to live upon the ground, or in them are little larger than the slight expansions found in the early conditions of such as undergo no complete metamorphosis. There can be little doubt that one use of these cavities is to diminish the specific gravity of the Insect, and thus to render it more buoyant in the atmosphere; but it would not seem improbable that they are intended to contain a store of air for its use while on the wing, as a part of the spiracles are at that time closed, so that less can enter from without.—The extent of respiratory surface thus created is such, that the amount of the aerating changes performed by an Insect in a state of activity, is not less in proportion to its bulk, than that effected by the most energetic of the Vertebrata. It is impossible to view this subject philosophically, without being struck by the fact, that this very high degree of respiratory power is given, not by a sudden advance to a more complicated and perfect system of organs, such as exists in the Vertebrated classes of animals, but by an extension of the comparatively simple plan, of which we observed the first traces in the Annelida; thus affording a beautiful example of the great law of regular progression in the development of organs, which has few apparent and perhaps no real exceptions. Nor would it be easy for any reflecting mind to contemplate the manner in which the air is thus brought into contact with the blood in the minutest textures of the body, without a feeling of admiration at the contrivance shown in the compensation of the limited circulation of the fluids by the extensive distribution of the respiratory apparatus; and at the means by which the necessary lightness, elasticity, buoyancy, and muscular energy are imparted to the bodies of these beautiful and interesting inhabitants of the air.

521. In the *Larva* condition of such aerial Insects as undergo a complete metamorphosis, and are therefore most different in their early state from their ultimate character, the structure of the respiratory system closely corresponds with the type it had attained in the higher Myriapoda. We find it entirely consisting of ramifying *tracheæ* (§ 153), connected with the external air by the *stigmata* that open on either side of the ventral surface of the body; and freely communicating with each other, especially by the two longitudinal tubes which traverse its length, and into which the stigmata open by short straight passages. Just as the *Larva* is passing into the *Pupa* state, the larger tracheæ exhibit dilatations at intervals, which are subsequently developed into expanded sacs that sometimes attain considerable size. The efforts which the animal makes at the moment of transformation, to rupture its skin by the distension of its body, appear to contribute towards the expansion of these sacs, the formation of which had previously commenced.* One remarkable portion of the tracheal system, also, the incipient evolution of which may be detected in the Larva state, now shows an increased tendency to prolongation;—that, namely, which forms the wings. It may be regarded as

* Newport on the Respiration of Insects, "Philosophical Transactions," 1836.

absurd to maintain that the wings of Insects are a part of the respiratory apparatus ; but that such is their real homology is shown by their perfect structure (§ 314), and by the history of their development (§ 315 *f, g*). During the first metamorphosis of the *Sphinx ligustri*, as observed by Mr. Newport, the wings, which, at the moment of slipping off the larva skin were scarcely as large as hemp-seeds, have their tracheæ distended with air ; and, at each inspiration of the insect, are gradually prolonged over the trunk by the propulsion of the circulating fluid into them. The enlargement of the tracheæ may also be observed in the *antennæ*, which just before the change were coiled up within the sides of the head, but are now extended along the sides and abdomen. The complete development of the respiratory apparatus only takes place, however, at the time of the last metamorphosis ; when the wings become fully distended with air, and prepared for flight by the active respiratory movements of the body ; and the expansion of the pulmonary sacs proceeds to a greater extent. It may frequently be noticed that, for some hours or even days after the perfect Insect has emerged from the pupa state, it makes no effort to fly, but remains in almost the same torpid condition with that it has quitted ; when stimulated to move, however, it makes a few deep inspirations, its wings rapidly become fully expanded, and it soon trusts itself in the element which was intended for its habitation.

522. The various provisions which are made for the respiration of such Insects as inhabit the water, are of a nature too interesting to be passed by. In those aquatic Larvæ which breathe air, we often find the last segment of the abdomen prolonged into a tube, the mouth of which remains at the surface while the body is immersed. The larva of the gnat may often be seen breathing in this manner, which calls to mind the elevation of the trunk of the elephant, when crossing rivers that entirely conceal his head and body. Sometimes this air-tube, which is to be regarded as a prolonged spiracle, is several inches in length, and its mouth is furnished with a fringe of *setæ* (or bristles), which entangle bubbles of air sufficient to maintain respiration when the animal descends entirely to the bottom ; and the large tracheæ proceeding from this tube convey the air through the body in the usual way. Most aquatic Larvæ which are unpossessed of such an air-tube, have their spiracles situated only at the posterior extremity of the body, and may be seen apparently hanging from the surface, whilst taking in the necessary supply. There are some Larvæ, however, more particularly adapted to aquatic respiration, by the development of the tracheal system externally into branchial plates or tufts ; the object of which is not so much to carry the circulating fluid into contact with the water, as to absorb from that element the air which it contains, and which is then carried into the internal respiratory apparatus. Sometimes the membrane which covers the branchiæ, and which is a prolongation of the external surface, is continuous, so that the gills have a foliaceous appearance like that of the wings ; but in other cases, it is divided, so that the branchiæ more resemble the filamentous tufts of the Nereis. Their position is constantly varying ; sometimes they are attached to the thorax, sometimes to the abdomen, sometimes even situated within the intestine ; but in every case they have an important relation with the movements of the animal, and are frequently the sole organs of progression with which it is furnished.

Thus, the sudden darting motion of the larva of the *Libellula* (dragon-fly) is caused by the violent ejection from the intestine, of the water which has been taken in for the supply of the gills it contains, whence air is imbibed into the tracheal system. A very little examination into the structure of the *wings* will show, that it is essentially the same as that of the expanded gills of aquatic larvæ; each consisting of a prolongation of the superficial covering of the body over a system of ramifying nerves or ribs, which are principally composed of tracheæ in connection with those of the interior of the fabric. Hence the designation of *aerial gills*, which was given by Oken to the wings of Insects, however ridiculed by succeeding writers on Entomology, and however absurd it may seem to those who look at the organs in their functional capacity alone, is really supported by the strictest analogies in structure, situation, and development. It is only by taking an extensive view of comparative organisation, that we can have any hope of arriving at accurate results; and great care is necessary to dismiss from the mind all prejudice in favour of a particular structure as the standard or type of the rest (§ 330, *note*). All perfect Insects (with the exception of a Neuropterous insect, *Pteronarcys*, described by Mr. Newport as retaining its larval branchiæ in the imago state) being adapted for tracheal respiration only, many curious contrivances may be witnessed among such as inhabit the water, for carrying down a sufficient supply of oxygen to aerate their blood whilst under the surface. Some, for example, enclose a large bubble beneath the *elytra* (wing-cases), which, not being closely fitted to the exterior of the body, leave a cavity into which the spiracles open; whilst others have the whole under surface of the body covered with down, which entangles minute bubbles of air, in such large quantity, as to render the insect quite buoyant, and to oblige it to descend by creeping along the stem of a plant, or by a strong muscular effort.

523. In the lower tribes of the *Arachnida*, such as the *Acaridæ*, the respiratory apparatus is usually constructed on the plan which prevails among Insects, being composed of a system of tracheæ, ramifying through the body, and opening externally by stigmata; this apparatus, however, would seem to be in some cases very imperfectly developed (§ 319 *d*). In the higher Arachnida, however, such as the *Spider*, *Scorpion*, &c., the circulating system is more complete, and there is no longer occasion for such universal aeration of the individual parts, that of the nutrient fluid being sufficient. Accordingly the respiratory apparatus exists in a more concentrated state, which approximates nearly to that which has been described as possessed by the higher Crustacea; but, being adapted for aerial respiration only, it must be regarded as belonging rather to the pulmonary than to the branchial system. The stigmata in these animals, usually four on each side, instead of opening into a prolonged set of ramifying and anastomosing tubes, enter at once into distinct sacs, disposed along the sides of the abdomen, to which the air has therefore ready access. The interior of these cavities is not smooth, however, like that of the pulmonary sacs of Insects, but is prolonged into a number of duplicatures or folds; these lie close to each other like the laminæ of gills, and may be regarded either as analogous to them, or as rudiments of the partition of the cavity into minute cells, like those of the lungs of higher animals. Each of these laminæ is described by Mr. Newport as consisting

of an exceedingly delicate and apparently structureless double membrane, which includes within it a parenchymatous tissue formed of single cells aggregated together, their arrangement having a degree of regularity in some parts, but being quite irregular in others. The blood is brought to this organ by a vessel which runs round its convex margin; but it must be distributed over its surface by a lacunar circulation between the component vesicles of the parenchyma.*—Before quitting the Invertebrated classes, whose Respiratory apparatus, both atmospheric and branchial, has now been described, it is worthy of note that in no instance do their respiratory organs communicate with the mouth, which is an organ solely appropriated to the reception and subdivision of the food. It may also be remarked, that it is only in the higher tribes that any muscular movements are performed, with express reference to the maintenance of the respiratory process by the renewal of the air or water which is required for it; this renewal, in the lower tribes, being accomplished by ciliary action, or by the general locomotive actions of the body.

524. We now pass on to the consideration of the pulmonary apparatus of Vertebrated animals; a rudiment of which is found in most members of the Osseous division of the class of *Fishes,* notwithstanding that the aeration of their blood is otherwise provided for. This rudiment is commonly known as the 'air-bladder;' which in many Fishes, as in the embryo of Mammalia, is a simple sac, placed along the middle of the back (Fig. 174, 10); whilst in others, its cavity is divided by one or more membranous partitions; and this division proceeds to such an extent in some instances, as to give to the organ the character of the lung of a Reptile. The air-bladder, in a large proportion of the Fishes which possess it, is closed on every side, its cavity having no connection with any other organ;† this is the case, for example, with that of the Perch, Cod, Mackerel, and other *Acanthopterygii* of Cuvier. But in most of the *Malacopterygii,* as the Herring, Salmon, Pike, Carp, Eel, &c., the air-bladder communicates with some part of the alimentary canal near the stomach, by means of a short wide canal termed the *ductus pneumaticus;* and in the *Cyprinidæ* (Carp tribe) this communication is formed with the œsophagus. The true relations of this organ are most remarkably shown in the *Lepidosteus* or 'bony-pike' of the North American lakes. This curious fish, which presents many Reptilian affinities, has the air-bladder divided into two sacs that possess a cellular structure; the trachea which proceeds from it opening high up in the throat, and being surmounted with a glottis. Another Fish may be mentioned as presenting an apparatus adapted for atmospheric respiration, which is rather a peculiar development of a portion of the branchial apparatus, than the rudiment of the lung of air-breathing Vertebrata; this is the *Cuchia,* the peculiarity of whose circulating system has been already noticed (§ 471). Here, as in the Synbranchus, there is but a single branchial orifice, which is situated under the throat; and this leads by a passage on each side to the gills, of which there are only two rows, and these but slightly developed. The principal organs of respiration are two vascular sacculi, prolonged

* "Philosophical Transactions," 1843, p. 296.

† According to Von Bär, the air-bladder is developed as a *process* or *diverticulum* from the upper part of the alimentary canal; so that, when it forms a closed sac, the original communication must have been obliterated. See *Note* to § 531.

from the branchial chamber, and placed on either side of the head; these communicate with the cavity of the mouth by two orifices, each of which is provided with a sort of constrictor muscle that serves to contract or entirely close it, and which thus resembles a glottis; and into these air is received for the oxygenation of the blood which is distributed upon their walls. From what has been said of the anatomical structure of this curious animal, it is obvious that it possesses the circulation of Reptiles, and the respiration partly of that class and partly of Fishes. Its blood will, therefore, be less oxygenated than in the regular types of either class; since the respiratory organs are less adapted for its aeration than those of Reptiles, and only a part of the blood is sent to them, instead of the whole as in Fishes. To this deficiency we may attribute the obtuseness of its senses, and sluggishness of its movements, which form a striking contrast to the vivacity of the Eel. It is generally found lurking in holes and crevices, on the muddy banks of marshes or slow-moving rivers. The power which the animal possesses of distending the respiratory sacs with air, while on land, and the necessity it is under of rising to the surface of the water for the same purpose, prove beyond a doubt that they perform the function of lungs; and lead us to the conclusion, therefore, that the Cuchia is amphibious in the strict sense of the word—forming a connecting link between the Ophidian Reptiles and the Synbranchus among Fishes.*—In some other Fishes, especially such as naturally inhabit small collections of fresh water whose temperature is liable to be considerably raised during the heat of summer, the mucous lining of the alimentary canal appears to serve as an additional organ of respiration; for such fishes are frequently seen to rise to the surface, and to *swallow* air, which is subsequently discharged by the anus with a large quantity of carbonic acid substituted for its oxygen. This is the case, for example, with the *Cobitis* (loach); and it would seem as if, under these circumstances, some such supplemental means is required for carrying on the respiratory process with unusual activity (§ 538).

525. It does not seem improbable that in those Fishes which possess a short and wide *ductus pneumaticus*, the air-bladder may serve as an accessory organ of respiration; atmospheric air being taken in, and carbonic acid ejected, through the alimentary canal, in the manner just described. But in those whose air-bladder is a closed sac, it seems evident that it cannot in any way conduce to the aeration of the blood; and this seems one of the instances, of which many might be pointed out both in the Vegetable and Animal kingdoms, where the rudimentary form of an organ, that attains its full development in other classes, is adapted to discharge some office quite different from that to which it is destined in its perfect state. The gas which the air-bladder contains, is composed of the same elements as atmospheric air, namely oxygen, nitrogen, and carbonic acid; but these are mixed in proportions that are very liable to variation. It has been said that oxygen is deficient in the contents of the air-bladder of fresh-water fishes, and is predominant in that of fishes which inhabit considerable depths in the sea. This organ is altogether absent in fishes accustomed to remain at the bottom, and, whose movements are slow, such as the *Pleuronectidæ*, or 'flat-fish;' whilst it is of large size in

* Taylor, in "Brewster's Journal," 1831.

those remarkable for vehement and prolonged movements, especially in Flying-fish of various species. It is generally supposed that the fish is enabled by means of the air-bladder to alter its specific gravity, by compressing the bag or permitting its distension ; but experiment shows that, after the organ has been removed, a fish may still retain the power of raising or lowering itself in the water.—In many Fishes, especially such as have either no ductus pneumaticus, or a narrow one, the blood-vessels of the air-bladder are developed into vascular tufts, which are sometimes spread over its interior, but are more commonly aggregated into one mass, which is spoken of as a 'vascular gland' or 'vaso-ganglion ;' the use of this organ is altogether unknown.

526. The members of the family of *Perennibranchiata* (which is the only true *amphibious* group) all possess lungs more or less developed ; those of the *Proteus* being very similar to the air-bags of Fishes, whilst those of the *Lepidosiren* and of the *Siren* exhibit saccular depressions upon their internal surface, which are the first indications of a subdivision of the cavity into air-cells. The tube by which they open into the mouth bears a greater resemblance to the 'ductus pneumaticus' of Fishes, than to the trachea of higher animals, being simply membranous without an appearance of rings ; and the glottis in which it terminates is a mere slit in the throat. Thus, the transition from the simple closed sac of Fishes, to the more complex subdivided lung of the Frogs, is perceived to be very gradual ; whilst, at the same time, the point of connection between the respiratory cavity and the alimentary tube may be observed to ascend, by similar gradations, from the stomach or some neighbouring part, to the œsophagus, and at last to the pharynx. Although all the animals which retain their gills, at the same time that they acquire lungs, are more or less adapted both to aerial and aquatic respiration, the relative degree of the two varies with the comparative development of their organs. Thus, in the *Siren,* the pulmonic respiration is more extensive and important than the branchial ; but the reverse is the case in the *Proteus.* Even the Siren, however, dies, if its branchial respiration be prevented by dryness of the gills. In the *Amphiuma* and *Menopoma* (whose early condition is not known) the lungs are more developed ; and the animals seem to breathe by them alone, no branchial filaments being present.

527. The lungs of the several orders of *Reptiles* are for the most part formed upon one type, being capacious sacs, the extent of whose vascular surface is but little augmented by sacculi developed in their walls ; in the *Loricata* and *Chelonia,* however, there is an incipient subdivision of the principal cavity, which foreshadows, so to speak, that which is presented by the lungs of Birds and Mammals. The lungs of the *Frog* (Fig. 221) present a good illustration of the inferior type of pulmonated structure in the class of Reptiles. On laying them open, each is found to consist of a capacious cavity, in which the bronchus of its own side terminates, and on the walls of which the pulmonary vessels are distributed ; these walls at the lower part are thin and membranous, but at the upper part they are strengthened by prolongations of the cartilaginous framework of the trachea, which forms a ring (*b*) at the root of the lungs ; and these prolongations produce by their inosculation a cartilaginous network, in the interspaces of which are minute sacculi, whose

lining membrane is crowded with blood-vessels. In this manner a set of 'air-cells' is formed in the wall of the upper part of the lung, which communicates with the general cavity, and very much increases the extent of surface by which the blood comes into relation with the air introduced into the pulmonary organs; but each air-cell has its own capillary network on its walls, and consequently the blood is only exposed to the air on one side of that network, instead of on both as in Mammals (§ 529).—In *Serpents*, we usually find the pulmonary apparatus to consist of a single long cylindrical sac (Fig. 180), simply membranous at its lower part, but furnished at its upper extremity with a cartilaginous reticulation projecting into its cavity, and inclosing in its interstices several layers of 'air-cells' with minutely vascular walls, which communicate through each other with the general cavity. The lung of the left side is generally undeveloped. From the great capacity of the respiratory sac, the mobility of their ribs, and the power of their intercostal muscles, Serpents are capable of rapidly inspiring and expiring a large quantity of air, by which the want of an extensive surface is compensated, and energy is imparted to their muscular exertions. It is the prolonged expulsion of the air after the lung has been fully inflated, that gives rise to the continued hissing sound by which these animals sometimes alarm their prey. In the aquatic Serpents, the large volume of air contained in the body serves to render it buoyant, and at the same time supplies the wants of the animal during a prolonged immersion.—In the *Saurian* Reptiles, we still find a very imperfect subdivision of the pulmonary sacs; but they are equally developed on both sides of the body, excepting in those genera which approach the preceding order (§ 324 *r*). In the lower forms of these organs, there is scarcely any appearance of cells, and they are usually much prolonged, frequently extending through the whole trunk, as in the Chameleon, as well as in many other Lizards; and their fulness or emptiness of air gives rise to the plump or lean appearance, either of which these animals have the power of assuming by the simple processes of inspiration or expiration. But when we have advanced upwards to the Crocodile, we find the lungs, though externally small, subdivided to a great degree of minuteness by internal partitions; and we also find the lungs more restricted to the thoracic region, with some indications even of a diaphragm, which is entirely wanting in all the inferior genera. It is not a little curious that in the Crocodile two openings are found near the cloaca, leading from the external surface to the interior cavity of the abdomen, which is lined by the peritoneum. This structure is evidently similar in character to that which has been described in the Echinus (§ 511 *a*); whether it is adapted to the same purpose, is not yet fully ascertained. It has been supposed by Geoff. St.

Fig. 221.

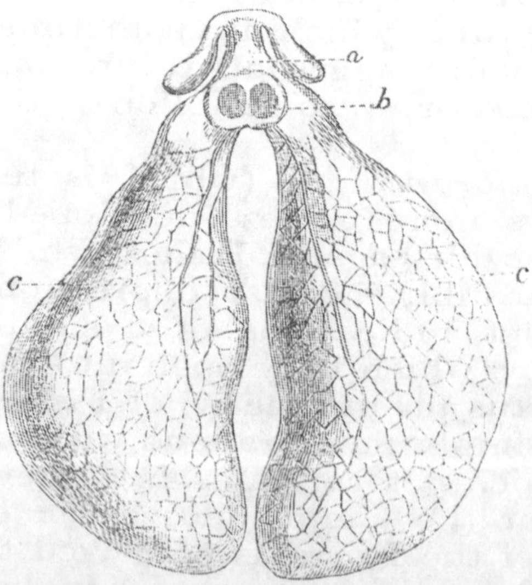

Respiratory organs of a *Frog*, as seen on their anterior surface;—*a*, hyoidean apparatus; *b*, cartilaginous ring at the root of the lungs; *c*, pulmonary sacs, covered with vascular ramifications.

Hilaire, that the superior energy of the Crocodile when immersed in water is due to the penetration of that fluid into the abdominal cavity, and the consequent conversion of the peritoneum into an additional respiratory surface.—The structure of the lungs in Turtles and other *Chelonia* is very similar to that exhibited by the Crocodile tribe; the sacs have their cavities subdivided by incomplete partitions (Fig. 222); but they are still very capacious, and materially assist, by the quantity of air they contain, in buoying up the heavy trunk of these animals when sailing on the surface of the water.—The very imperfect character of the mechanism by which air is drawn into the lungs in Reptiles, has been already noticed (§ 324 *r*); and it will here suffice to add, that, taken as a whole, this class is remarkable for the feebleness of its respiratory actions, and for the length of time during which the process can be suspended without injury. The demand for aeration is very much regulated, however, by the temperature to which the animal is subjected (§ 97).

Fig. 222.

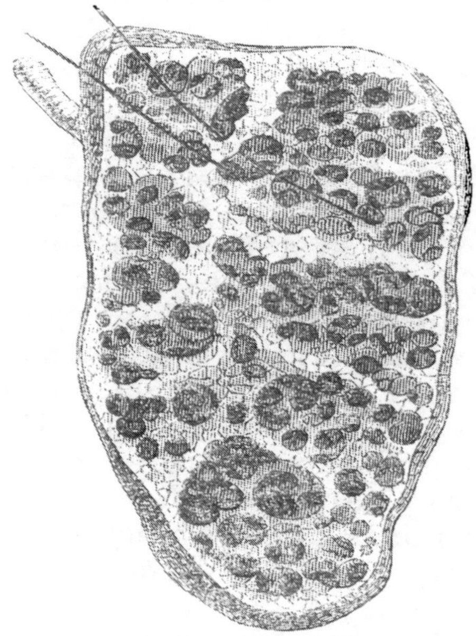

Section of the Lung of the *Turtle*.

528. The respiratory apparatus of *Birds,* notwithstanding its extraordinary extension, is intermediate, in its grade of development, between that of Reptiles and that of Mammalia. In this class, as in Insects, it extends through a great part of the body; large sacs connected with the lungs being contained in the abdomen, and even continued beyond the cavity of the trunk, as under the skin of the neck and extremities (§ 325 *f*).* Even the bones are made subservient to this function; for though at an early period they possess a spongy texture, like those of Reptiles, and are filled with thin marrow, they subsequently become hollow, and their cavities communicate with the lungs; in the aquatic

* Various surmises have been formed as to the particular uses of these air-sacs in the economy of the Bird; and it does not seem improbable that, besides contributing to the function of Respiration by the extension of surface they afford, they have some subsidiary purposes. One of the most evident is that of rendering the body specifically lighter, as in Insects; and this will be obviously assisted by the great heat of the system, which rarifies the contained air. Again, the distension of the air-cells assists in keeping the wings outstretched; as is shown by the fact that inflation of those situated in the neighbourhood of their muscles is followed by their expansion; this must be a most important economy of muscular action, in birds which hover long in the air. Their evident analogy to the pulmonary sacs of Insects (§ 520) is confirmed by their relatively larger dimensions in Birds of long-continued and rapid motion, than in the slow-moving tribes which are almost confined to the earth or waters. It has been remarked in addition, that "the same air which exerts its renovating influence upon the blood, supports all the more delicate structures which it reaches and surrounds, as a cushion of the most perfect softness and elasticity; so that by the most rapid motion, and the most violent twitches which the body receives in the changes and turnings of that motion, there can be no concussion of the parts more immediately necessary for the life of the birds." It would scarcely seem improbable that the large air-cells, which are found extending beneath the integument of the whole surface, especially the under side, of the Pelican and Gannet, serve to deaden the concussion which the body must experience, when the bird, after raising itself to considerable height in the air, lets itself suddenly fall upon the water in pursuit of its finny prey.

species, however, the original condition is retained through life. In those Birds of which the bones are thus permeated by air, the trachea may be tied, and the animal will yet continue to respire by an opening made in the humerus or even in the femur. The general structure of the Respiratory apparatus of Birds, and the mechanism of its action, having been already sufficiently explained (§ 324*f*), it is unnecessary here to do more than describe the minute structure of the lungs, which presents features of great interest. The entire mass of each lung may be considered as subdivided into an immense number of 'lobules,' or 'lunglets,' each of which has its own bronchial tube (or subdivision of the windpipe), and its own system of vessels, which have but little communication with those of other lobules. Every lobule has a central cavity, which closely resembles that of the Frog in miniature, its walls being strengthened by a network of cartilage derived from the bronchial tube, in the interstices of which are openings leading to sacculi in their substance. But these sacculi are not, as in Reptiles, bounded by a distinct membrane, prolonged from that of the general cavity; for, with the exception of the part nearest the latter, the whole thickness of the wall may be considered as made up of a very close plexus of blood-vessels, between the meshes of which the air penetrates freely without any limitation; and thus every capillary is in immediate relation with air on all sides,*—a provision which is obviously very favourable to the complete and rapid aeration of the blood which it contains. Of all animals, Birds are most dependent upon a constant renewal of the air in their lungs, and upon the purity of that with which they are supplied. Most Birds will die in air which has been but slightly charged with carbonic acid, and which can be respired by Mammals without immediate injury; whilst a greater degree of impurity, which is at once fatal to Mammals, can be sustained for a long period by Reptiles.—It is beautiful to observe that in Birds, as in Insects, the great extension of the respiratory surface is given by a simple increase in the capacity and prolongation of the sacs, and not by that concentration of it into a small bulk, which is effected by the minute partitioning of their cavity, and which indicates the highest form of the respiratory organs. Another analogy presented by their respiratory system to that of Insects, is this: in the latter the whole of the aeration is effected, by bringing the air in contact with the blood actually circulating through the system; whilst, in the higher air-breathing animals, possessed of a more centralised apparatus (whether consisting of lungs or gills), the blood is transmitted through it by a special adaptation of the vascular system, in the intervals of its circulation through the body. In Birds we find a curious adaptation of the latter more elevated type to the Insect-like conditions of their existence; for, whilst the air introduced into the lungs acts upon the blood transmitted by the pulmonary vessels, that which fills the air-cells and cavities of the bones comes into relation (as in Insects) with the capillaries of the system at large.

529. The respiration of *Mammalia* is not, like that of Birds, extended through the system, but is restricted to the lungs; and as a perfect diaphragm is now developed, which completely separates the thoracic from the abdominal cavity, these organs are confined to the former.

* See Mr. Rainey's description of the 'Minute Anatomy of the Lung of the Bird,' in the "Medico-Chirurgical Transactions" for 1849.

Although their bulk is proportionally so much smaller than that of the pulmonary sacs of Reptiles, the actual amount of surface over which the blood is exposed to atmospheric influence, is beyond comparison larger, owing to their very minute subdivision into cells. The want of capacity, too, is compensated by the active movements of inspiration and expiration, which constantly and most effectually renew their contents (§ 326 *t*). The 'lungs are greatly developed in all the more powerful Mammalia, especially in the Carnivorous species ; but they are comparatively smaller in their extent of surface, in the feeble and inferiorly-organised Herbivora ; and the red corpuscles are much more numerous in the blood of the former, than in that of the latter. The varieties in the conformation of the lungs presented by the different order of Mammals, relate chiefly to their exterior divisions, and to their greater or less capacity ; the plan of structure being nearly the same in all. The whole interior of the lungs of Man, and of Mammals generally, is divided into minute 'air-cells,' which freely communicate with each other, and with the ultimate ramifications of the bronchial tubes. The partitions between these cavities, the diameter of which in Man varies from 1-200th to 1-70th of an inch, are formed by double folds of basement-membrane, between which is a capillary plexus arranged in a single layer (Fig. 223); so that the blood in these capillaries is exposed to the air contained in the cells on *both* sides of it, but is not, as in Birds, brought into relation with the atmosphere without the intervention of a basement-membrane.* It has been calculated that the number of these air-cells grouped round the termination of each bronchial tube,—which cluster represents a 'lobule' of the lung of the Bird, and the entire lung of a Frog,—is not less than 18,000 ; and that the total number in the Human lungs is not less than *six hundred millions.* Some idea may be formed from this estimate, of the vast extent of surface which is thus provided for bringing the blood into relation with the air.

FIG. 223.

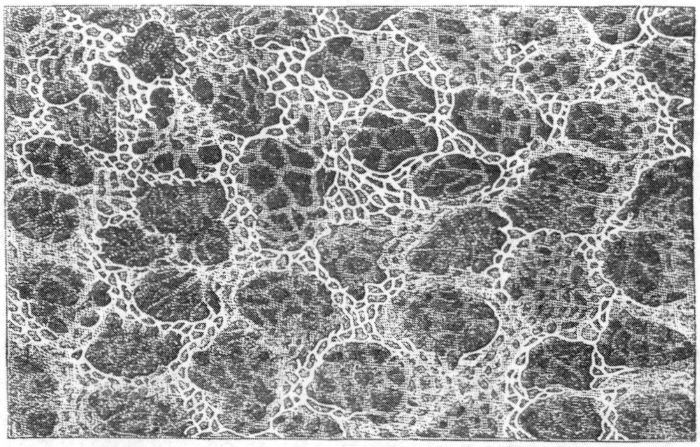

* Arrangement of the Capillaries of the air-cells of the *Human Lung.*

530. From the preceding sketch of the progressive evolution of the Respiratory system in the Animal scale, it will have been seen that the fundamental character of the respiratory organs is everywhere the same, however different their external form ; and that it is only the disposition of their parts that is varied, in accordance with the circumstances in which their function is to be performed. The progressive specialisation of the function has been traced in ascending the series, by marking the

* It is stated by Mr. Rainey (loc. cit.), that the lung of the Kangaroo, and even that of some Rodentia, presents a condition intermediate between that of Birds and that of the higher Mammalia ; the air-cells, in the parts most remote from the surface, being very small and being but imperfectly bounded by a basement-membrane.

evolution of a particular apparatus for its exercise, and the restriction of it to that apparatus; in no instance has any sudden change in character been witnessed; but, in the classes adjoining those in which a new organ was to be introduced, has been found some adumbration of it; yet even where the function is most highly specialised, the general surface is found to retain in some degree its participation in it. The function of Respiration is not confined to the lungs, even in animals which possess them in their most developed form. The blood which circulates through the capillaries of the skin is aerated by communication with the atmosphere, wherever there is no impediment offered by the density of the tegumentary covering, In *Batrachia*, especially Frogs, the cutaneous respiration is of such importance to the animal, that, if impeded by covering the skin with oil or other unctuous substance, death will take place almost as soon as if the lungs were removed; and the animal may be supported for a considerable time by it alone, if the temperature be not too high (§ 97). In such circumstances, it is found that carbonic acid is generated in an atmosphere of hydrogen, as by pulmonary respiration. In like manner, if Birds or Mammalia be enclosed in vessels out of which their heads protrude, carbonic acid will be found to replace a portion of the oxygen; and the same result has been obtained by the similar enclosure of a limb of the Human body.

531. We shall now briefly trace the evolution of the Respiratory apparatus in the embryo of the higher Vertebrata; reserving, as before, the account of the earliest changes in the ovum to a future period (CHAP. XVIII.), and leaving until then the description of the organs which are peculiar to the fœtal condition, and which serve only to assist in the conversion of the nutriment that is supplied from the parent system, as during the germination of seeds.*—At about the third day of the development of the Chick, four pairs of clefts or tranverse slits are observable behind the mouth, in the situation of the branchial apertures of Fishes; and at the same time, the branchial vessels are developed from the aorta, as already described (§ 486). One the apertures is intermediate between each pair of vascular arches, just as in the gills of Fishes and Tadpoles. Nothing like branchial tufts, however, are developed; and the appearance described is very transitory, the vessels changing their direction and condition within two days. The development of perfect gills would have been useless, as the animal is not destined to be for a time an inhabitant of water like the tadpole, but has the aeration of its blood provided for, until the time of the perfect evolution of its respiratory system, by an apparatus specially evolved for the purpose. The lung is developed, like the air-bladder of Fishes, as a *diverticulum* or process from the upper part of the alimentary canal. Soon after the middle of the third day, two minute wart-like projections are seen upon the tube, which are found to be hollow, and to communicate with its

* It may here be remarked, however, by way of completing the series of analogies, that the germinal membrane of the ovum, which serves as its first respiratory organ, may be regarded with much probability as corresponding with the 'mantle' of the Mollusca; and that the allantois seems to be the representative of the 'respiratory tree' of the Holothuria. The gills of many of the Gasteropods are developed immediately around the anus (Fig. 126); and they may thus be considered as an external prolongation of that part of the surface, which in the Holothuria is extended inwards.

cavity.* These gradually increase in size; and the channels of communication become elongated into tubes. A little later, the tubes partly coalesce into one, and enter the pharynx by a single aperture. This is what we observe in the Proteus, and, as in that animal, the sacs are still simple undivided bags; after a little time, however, they send out prolongations in various parts, which again put forth others, so that the cavity becomes gradually more complex. The larynx and glottis are not perfectly formed until a late period.—The history of the evolution of these organs in Mammalia is precisely analogous. It is usually at about the sixth of the entire period of uterine gestation, that the rudiments of the branchial apparatus are seen, as marked by the shortness and thickness of the neck, the penetration of the sides of the pharynx by the branchial clefts, and the division of the aorta into vessels corresponding in number and distribution with the branchial arteries of fishes. These general features have been observed in the embryoes of most orders of Mammalia, not excepting Man himself; and they are probably common to all. A few days after the appearance of the fifth arch, which is the last developed, the neck begins to elongate, the apertures are closed gradually on the outside, while the vascular arches undergo those changes by which the permanent arterial trunks arising from the heart are formed. The lungs in Mammalia are developed much in the same manner as in Birds. They are not discernible before the period when the branchial apertures begin to close; a single mass is first perceived, which is soon divided into the rudiments of a right and left lung by a longitudinal groove; and the trachea and bronchi are subsequently developed, as in Birds.

532. Scarely a more beautiful illustration of the Unity of Design, manifested in the creation of different classes of animals, could be adduced, than this hidden but not obscured correspondence. Nor is the analogy confined to Animals alone; for it is impossible to compare the stages of the evolution of the perfect respiratory apparatus in the higher forms of the two kingdoms, without being struck with their essential correspondence. In the Flowering Plant we have seen a temporary respiratory organ, the *cotyledon*, first developed, like the branchiæ of a tadpole; and disappearing altogether, when the evolution of the permanent aerating apparatus renders it unnecessary. And just as the system which is the permanent one of the lower tribes of animals, is transiently indicated in the early development of the higher, so will it hereafter be shown (CHAP. XVIII.) that the foliaceous expansions of the inferior stemless Cryptogamia are to be regarded as the homologues of the cotyledons of Flowering plants, which continue, in the inferior tribes, to perform their functions during the whole of life, like the gills of aquatic animals. That which has been said of the correspondence of the essential

* The account in the text is given on the authority of Von Bär. Many subsequent observers, however, agree in stating, that the bud-like process in which the lungs originate, is not hollow but solid; being produced by a multiplication of cells of the external layer of the alimentary canal, into which its external tunic is not prolonged. According to this view, the downward extension of the cellular mass in which the lungs originate, is the consequence of the progressive multiplication of cells in that direction; and the production of the trachea and bronchial tubes is the consequence of the *fusion* of the cells in particular lines, like the development of the great vessels. Viewed in this aspect, the absence of the 'ductus pneumaticus' in many Fishes is a result rather of non-development than of subsequent obliteration; and such would certainly seem to be by far the most probable account of it.

structure of the Respiratory apparatus, through all its varieties of external form, will apply with equal truth to its function also; for, in whatever tribe of Animals the changes composing it have been investigated, they are found to be of a very uniform character. The object of these changes appears to be, in all instances, the liberation of carbonic acid from the blood, the replacement of it by oxygen, and the exchange of nitrogen on one side or the other. It will be more convenient to enquire into the particular character of these changes, in the distinct form in which they are presented to us in the higher Animals, before proceeding to investigate their more obscure manifestations in the inferior tribes. These changes may be examined, either in the circulating blood, or in the air to which it has been exposed.

533. The most obvious difference between the Blood brought to the lungs for aeration after passing through the capillaries of the system, and that which has undergone the process—or, in short, between 'venous' and 'arterial' blood—is its colour, which is dark purple (sometimes called black) in the former, and bright red in the latter. The alteration in colour may be produced by agitating venous blood with oxygen, or even by exposing it for a time to the atmosphere; in the latter case, however, only the surface acquires the arterial tint. The bright scarlet colour may also be given by the admixture of neutral salts; whilst the addition of acids renders it still darker and prevents the change. It has been supposed until recently, that these effects are due to a chemical change produced by these agents in the hæmatine of the red corpuscles. But such would not appear to be the case; for when the hæmatine has been separated and diffused through water, it is neither darkened by carbonic acid, nor brightened by oxygen, unless some corpuscles be floating in the solution. Moreover, it is found that the action even of distilled water will darken an arterial clot. Taking all these circumstances in connection with the facts already stated (§ 173), in regard to the changes of form to which the red corpuscles are liable from the influence of reagents, it appears probable that the immediate effect of oxygen, like that of saline solutions, is to contract the corpuscles and thicken their walls, and thus, by altering their mode of reflecting light, to make them appear bright red; whilst carbonic acid, like water, may be seen to occasion a dilatation of the corpuscle, and a thinning of its walls (which are at last dissolved by it), in a degree that may be well supposed to produce the darkening of the mass formed by their aggregation.*—In what state of combination these gases exist in the blood, or whether they are present in a state of simple solution, has not yet been clearly determined. When venous blood is placed under the vacuum of an air-pump, a small quantity of carbonic acid gas is given out; but a larger amount, sometimes one-sixth of the whole volume, is evolved when the blood is agitated with atmospheric air, hydrogen, or nitrogen. Gas may be extracted also from arterial blood, by means of an air-pump capable of producing a very complete vacuum; and this is found to consist of a larger proportion of oxygen. From the experiments of Magnus, the latest and most satisfactory on the subject, it appears that the Oxygen in arterial blood

* For a full discussion of this subject, see Scherer's Reports in the recent volumes of "Canstatt's Jahresberichte," and the works and memoirs to which he refers.

amounts to about one-half of the quantity of Carbonic acid which it contains, whilst in venous blood it is only about one-fifth ; for whilst about 10 per cent. of Oxygen, and 20 per cent. of Carbonic acid, may be extracted from *arterial* blood, the quantity of Oxygen removable from *venous* blood is diminished to 5, whilst that of Carbonic acid is increased to 25. The relative quantity of Nitrogen is extremely variable.

534. The changes in the Air which has been respired, are capable of being examined with greater accuracy. They may be considered under four heads :—1. The disappearance of Oxygen which is absorbed. 2. The presence of Carbonic acid, which has been exhaled. 3. The absorption of Nitrogen. 4. The exhalation of Nitrogen.—It was formerly supposed that the Oxygen which disappears, is the precise equivalent of the Carbonic acid which is generated, the latter gas containing its own bulk of the former. But it is now known that the amount of Oxygen that disappears, is usually more than that which is contained in the Carbonic acid expelled, so that the surplus must be actually absorbed into the system. The amount of this surplus varies in such proportion, that it sometimes exceeds the third part of the carbonic acid formed, and is sometimes so small that it may be disregarded,—the difference depending, not only on the constitution of the species, but on the comparative degree of development, and more especially on the nature of the diet. This last fact has been established by the very careful experiments of MM. Regnault and Reiset,* who have shown that if the same animals be fed at one time upon flesh, and at another upon farinaceous food, the quantity of oxygen which is absorbed into the system is much greater in the former case, than in the latter. Thus it would seem evident, that the demand for oxygen in the system is partly connected with the metamorphoses which the alimentary materials undergo, in their passage through the body. Of the nature of these metamorphoses, our information is still very imperfect.—There is now quite sufficient evidence to prove that the generation of carbonic acid is not due, as was formerly supposed, to the union of carbonaceous matter brought by the blood, with the atmospheric oxygen introduced into the lungs ; since carbonic acid is not only found to exist in venous blood, but in the products of the respiration of gases entirely free from admixture with oxygen. Such an experiment can only be performed on animals, which can sustain for a time the absence of the stimulus of oxygen. That Snails confined in hydrogen will generate carbonic acid, was long ago shown by Spallanzani ; but the later experiments of Edwards, Müller, &c. upon Frogs are more satisfactory, both from their superior accuracy, and from their freedom from the objection which might be raised against the others, on the ground of the low place of their subjects in the Animal scale. It appears that, when confined in hydrogen, frogs will give out carbonic acid, for a time at least, as rapidly as in atmospheric air ; and that the quantity generated in nitrogen is not much inferior. These results are evidently conformable with the principles formerly stated as regulating the mutual diffusion of gases. Owing to its energetic reaction with carbonic acid (occasioned by its great difference in specific gravity), hydrogen removes it from the blood with greater force than any other gas ; so that venous blood will give off

* "Annales de Chimie," 1849.

carbonic acid when exposed to an atmosphere of hydrogen, even after it has been submitted to the exhausting power of a vacuum. It is obvious, however, that, for the continued generation of carbonic acid, oxygen must be supplied from without, as there is no superfluity of it in the system.—The following, therefore, appears to be the history of the changes which the blood undergoes in its passage through the body. In the capillaries of the lungs it becomes charged with oxygen, which it carries into those of the system; in the course of the actions which there occur between the nutritious fluid and the textures it supports and stimulates, part of the oxygen disappears, and carbonic acid takes its place; the venous blood, therefore, returns to the lungs, holding this in solution, together with the unabsorbed oxygen; and, in the capillaries of the lungs, the former gas is removed by the atmosphere, and replaced again by oxygen,—the interchange being entirely in accordance with the physical principles already stated.*

535. With regard to the absorption and exhalation of Nitrogen, it appears probable that both these processes are constantly going on, but that their relative activity varies in different species, at different parts of the year, and under different circumstances. It appeared from the experiments of Dr. Edwards, that an increase in the volume of nitrogen in the respired air took place in most young animals, and during the summer months; but that, in the autumn and winter, there is a considerable absorption when adult animals are employed. According to the recent experiments of MM. Regnault and Reiset (op. cit.), the exhalation of nitrogen is more common than its absorption; since warm-blooded animals in general, when subjected to their ordinary regimen, increase the amount of nitrogen in the atmosphere, in the proportion of from 1 to 2 per cent. of the oxygen consumed. But when food is withheld, or animals are fed upon a diet to which they are unaccustomed, an absorption of nitrogen takes place; and this was found to take place to a considerable extent in hybernating Mammals, whose production of carbonic acid does not go on at more than 1-50th of the ordinary rate; so that their absorption of oxygen and nitrogen even exceeds the loss of carbon and that sustained by perspiration, and the animal actually increases in weight, without taking food, until it voids its excretions.

536. Animals whose respiration is aquatic do not decompose the water they breathe, but merely abstract the oxygen from the air contained in it; for if one of this class be placed in a limited quantity of water, from which it speedily exhausts the air, or in water from which the air has been expelled by boiling, it dies almost as soon as would an animal whose respiration is aerial, when placed in a vacuum. If, however, the surface

* This view of the function of Respiration was given in a paper which the Author published in the "West of England Journal" in 1835, as that which best accorded with the facts then known. It has been fully confirmed by subsequent experiments, especially those of Magnus; and it is the one now generally received. Notwithstanding that the proportions between the oxygen absorbed and the carbonic acid exhaled are so inconstant, that they cannot be reconciled with the numerical ratio of 1174 of the former to 1000 of the latter, which would constantly prevail if the force of 'mutual diffusion' were the only one that regulates the exchange, yet there seems to the Author no adequate reason for doubting that the process is mainly effected by the agency of that force; the minuter variations being determined by other conditions,—in part, not improbably, by the relative amount of the gases existing in the blood, and the tenacity with which it may hold them.

of the water be in contact with the atmosphere, it will absorb air from it; and the life of the animal will be longer, the more fully the quantity thus obtained compensates for that which is consumed. When a Fish, in a limited quantity of aerated water, has reduced the proportion of air until its respiration has become difficult, it rises to the surface, and takes in air from the atmosphere; and, if prevented from doing so, it dies much sooner. The air thus taken in probably acts upon the lining membrane of the intestines; for, after being expelled, it is found to contain a large proportion of carbonic acid.—The respiration of some of the inferior aquatic tribes, such as Crustacea, Mollusca, and Annelida, has been examined with similar results. According to the researches of Humboldt and Gay Lussac, the air contained in water is richer in oxygen than that of the atmosphere; the proportion being 32 per cent. in the former, and but 21 in the latter.

537. The respiration of Insects has been made the subject of accurate research by Mr. Newport;* and the results which he has obtained correspond in a remarkable manner, with those of Dr. Edwards's experiments on Vertebrated animals under different conditions. In those tribes which undergo a complete metamorphosis, the proportion of air consumed by the *larva* is much smaller than that which the perfect Insect requires, when their relative bulk is allowed for, and their condition is the same as to rest or activity. If a larva of the common Butterfly, for instance, has arrived at its full size at the time of making the observation, it appears to respire in a given time more than the perfect insect; but the result is liable to this fallacy—that the former is at least two-thirds larger than the latter, and is almost always in a state of activity, whilst the latter is frequently in a state of quiescence. This fact is evidently analogous to one ascertained by Dr. Edwards, that in the higher animals, a greater quantity of oxygen is required in the adult state, in proportion to the size of the respiratory apparatus, than in the infant condition. Again, many larvæ can support a degree of privation of oxygen which would be fatal to the perfect Insect; thus, there are some which inhabit the bodies of other insects, or are buried deeply in the soil, or seek their subsistence in noxious and unaerated places, all of which situations would be soon destructive to life in their advanced condition. This, too, finds its parallel in the history of the Vertebrated classes: for Dr. Edwards found that puppies, soon after birth, will recover after submersion in water for 54 minutes, thus bearing the privation of oxygen much better than the adult animal. The amount of respiration in the perfect Insect depends chiefly upon its state of activity or excitement. When its movements are rapid and forcible, the aeration of the tissues must be performed to a greater extent than when it is at rest; and the difference is manifested, as well by the respiratory motions, as by the amount of oxygen consumed. Thus, the number of respirations in an Humble Bee (*Bombus terrestris*), while in a state of excitement soon after its capture, was from 110 to 120 in a minute; after the lapse of an hour they had sunk to 58, and subsequently to 46. Moreover a specimen of the same insect, confined in a limited quantity of air, produced in one hour after its capture, whilst still in a state of great activity, about 1-3rd

* "Philosophical Transactions," 1837.

of a cubic inch of carbonic acid; yet during the whole twenty-four hours of the succeeding day, the animal evolved a quantity absolutely less. The amount of respiration in the Pupa state is much less than in any other condition of the insect, which will readily be understood when its complete inactivity is remembered; the state of the animal at that time being comparable (as far as its respiration at least is concerned) to that of the hybernation of warm-blooded Vertebrata.

538. In *cold-blooded* animals in general, the activity of respiration is increased with elevation of the temperature of the surrounding medium. This has been shown in a very striking degree with regard to the Amphibia, by the researches of Dr. Edwards (§ 97). The influence of temperature is seen also on the existence of Fishes in limited quantities of water; and the degree of heat which obliges frogs to increase their respiration by quitting the water entirely, causes fishes to take-in air from the surface, as may be frequently witnessed during the summer, especially in small collections of water (§ 524). The amount of carbonic acid generated by Insects, also, increases with the surrounding temperature, other things being equal.—The activity of respiration in *warm-blooded* animals, on the other hand, is increased by a depression of the external temperature. Thus it was found by Letellier, that when small Birds and Mammals were enclosed in limited quantities of air, the quantity of carbonic acid produced in a given time was in close accordance with the difference between the temperature of the air and that of their bodies. The following were the amounts generated in an hour, by the respective animals named, at the temperature of 32°, at one near to that of their own bodies, and at an intermediate grade:—

	Temp. 86°—106°. Grammes.	Temp. 59°—68°. Grammes.	Temp. about 32°. Grammes.
A Canary . .	0·129	0·250	0·325
A Turtle-dove . .	0·366	0·684	0·974
Two Mice . .	0·268	0·498	0·531
A Guinea-pig . .	1·453	2·030	3·006

Thus it would appear that the quantity of carbonic acid exhaled by Birds between 86° and 106° is not much more than *half* of that which is exhaled between 59° and 68°; and is only about *two-fifths* of that which is given off at 32°. In Mammals, the quantity exhaled at 86°—106° is about half that which is given off at 32°. It is easy to understand the meaning of this difference in the effect of external temperature upon warm and cold-blooded animals, when viewed in its relations to their economy. For in the latter, the increase of respiration consequent upon the elevation of temperature is simply an exponent (so to speak) of the general vital activity of the animal, which heat tends to augment. But in the former, an elevation of external temperature decreases, whilst a depression increases, the demand for that internal combustive process, by which the standard heat of the body is maintained (CHAP. XVII., Sect. 3).

CHAPTER XIV.

EXHALATION OF AQUEOUS VAPOUR.

1. *General Considerations.*

539. As all the alimentary materials taken into living bodies, for the nutrition of their solid tissues, are in a fluid form, being either dissolved or suspended in water, it is evident that a large quantity of that liquid must be superfluous, and that means must be provided for carrying it out of the system. This is partly accomplished, in Animals more especially, by its combination with various other ingredients,—which have either been introduced in greater quantity than the processes of nutrition require, or have already served their purpose in the vital economy,—into the *fluid excretions,* for the separation and deportation of which various arrangements are provided (CHAP. XVI.). But besides the means thus afforded for the diminution of the superfluous fluid of the system, we find that the external surface has this function peculiarly imposed upon it, and that the disengagement of nearly pure aqueous vapour, though really the effect of simple evaporation, is principally dependent upon a special arrangement, by which its liberation from the circulating fluid is facilitated. This is most evident in Plants, where the quantity of fluid absorbed bears a much larger proportion to the amount of the solid matter contained in it, than it does in Animals ; and where, from the little opportunity which there is for the introduction of superfluous nutriment, and the slight tendency to decomposition in the solid structures, the necessity for a constant excretion of other ingredients unfit to be retained is much less.

2. *Exhalation in Plants.*

540. The soft and succulent tissues of Vegetables, if freely exposed to the atmosphere, would soon lose so much of their fluid as to be incapable of performing their functions ; and in all plants, therefore, which are subject to its influence, we find a provision for restraining such injurious effects. In the *Algæ,* however, and other tribes constantly immersed in water, or in a very moist atmosphere, no such loss can take place in their natural condition, and no means are required to prevent it. The outer layer of cells composing their integument, differs but little from those which it holds together, except in density (Figs. 70, 71) ; and it is accordingly found that such plants, when exposed to a dry air, speedily desiccate. All but the simplest Cellular Plants, whose natural residence is the air, on the other hand, are covered with a membrane of peculiar character, which is termed the *cuticle* * (Fig. 224, *a*, *b*). This is composed of cellular

* Notwithstanding the usage of many distinguished Botanists, who apply the designation *epidermis* to this membrane, the Author has preferred retaining the term 'cuticle,' not only as being the one in common use in this country, but likewise as being the more appropriate. For by the term *epidermis* is implied, a membrane lying *upon* the dermis or skin ; and therefore it can only be appropriately employed to designate the external pellicle presently to be described.

tissue, the vesicles of which are flattened, and arranged with great regularity, in close contact with each other; and they differ from those of the parenchyma beneath, not only in form, but also in the nature of their contents. The form of these vesicles is different in almost every tribe of plants; thus in the cuticle of the *Iris* and of many other Monocotyledons, they are elongated, and possess straight walls and regular angles; whilst in that of the *Apple* and *Lily* (Fig. 225) their boundaries have a sinuous character. The cells of the cuticle are filled with a colourless fluid; and their walls are usually thickened by secondary deposit, especially on the side nearest the atmosphere; this deposit is usually of a waxy nature, and consequently renders the membrane very impermeable to liquids. In most European plants, the cuticle contains but a single row of cells, which are moreover thin-sided; whilst in the generality of tropical species, there exist two, three, or even four layers of thick-sided cells, as in the *Oleander*, the cuticle of which, when separated, has an almost leathery toughness. This difference in conformation is obviously adapted to the respective conditions of growth; since the cuticle of a plant indigenous to temperate climates would not afford a sufficient protection to the interior structure, against the rays of a tropical sun; whilst the diminished heat of this country would scarcely overcome the resistance afforded by the dense and non-conducting tegument of a species formed to exist in warmer latitudes.—Externally to this membrane, there usually exists a very delicate transparent pellicle, without any decided traces of organisation, though occasionally somewhat granular in appearance, and marked by lines which seem to be the impressions of the junction of the cells with which it was in contact. This membrane is obviously formed by the agency of the cells of the cuticle; and is either a secretion from their exterior (as maintained by Schleiden, Payen, and others), analogous to that which forms a thick gelatinous layer on the surface of the Algæ, and which elsewhere constitutes the 'intercellular substance;' or itself consists of the external layers of the thickened walls of the cuticular cells, which have coalesced with each other, and which become detached from the subjacent layers by maceration (as affirmed by Mohl and Henfrey).

FIG. 224.

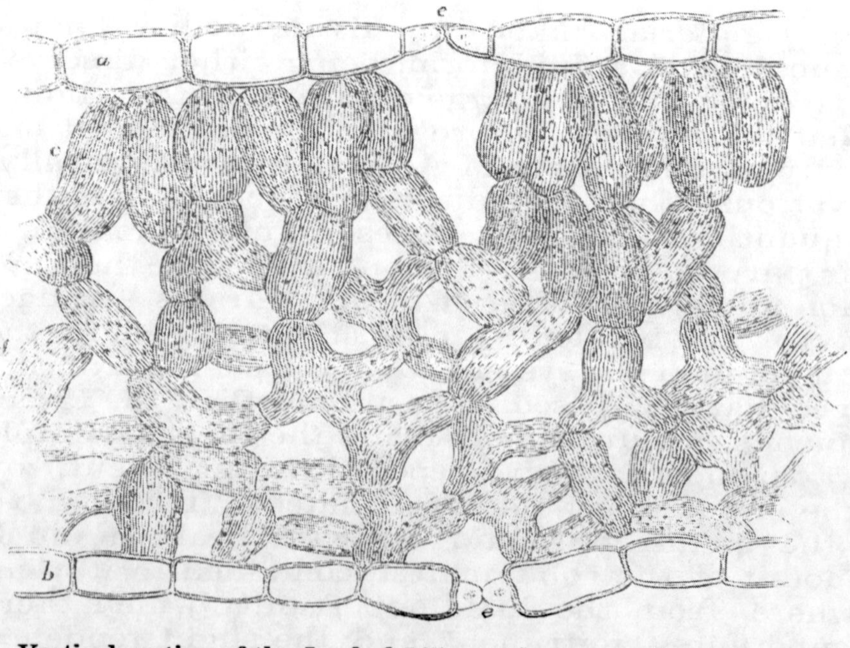

Vertical section of the Leaf of *Lilium album*; *a*, cells of the upper cuticle; *b*, cells of the under cuticle; *c*, *d*, cells of the parenchyma; *e e*, stomata.

541. In the Cuticle of most Plants which possess such a structure distinctly formed, there exist minute openings termed *Stomata* (Fig. 224, *e*, *e*)

which are bordered by cellules of a peculiar form, distinct from those of the cuticle, and more resembling in character those of the tissue beneath. These boundary cells are usually somewhat kidney-shaped (Figs. 225, *b*, *b*), and the opening between them oval; but, by an alteration in their form, the opening may be contracted or completely closed. They are sometimes more numerous, however, and the opening angular; and in the curious *Marchantia polymorpha*, their structure is extremely complicated. The openings in the cuticle of this plant (Fig. 226, A, *b*, *b*) are surrounded by five or six rings placed one below the other, so as to form a kind of funnel or chimney, each ring being composed of four or five cellules (B, *h*). The lowest of these rings (*i*) appears to regulate the aperture, by the contraction or expansion of the cells which compose it. Wherever stomata exist in the cuticle, they are always found to open into cavities in the tissue beneath, which are thus brought into immediate relation with the external air. In the Marchantia these chambers are very large, and are surrounded by regular walls (*c*, *c*); whilst in the leaves of higher plants they exist simply as intercellular spaces, left by the deficiency of the tissue. Stomata do not exist where there is no regular cuticle; and

FIG. 225.

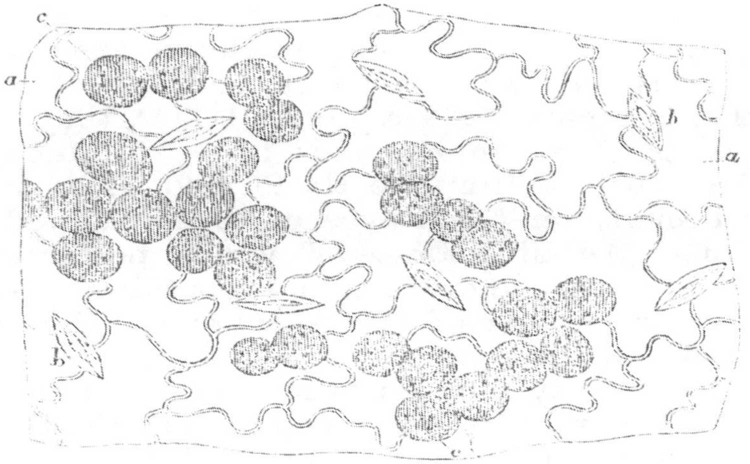

Under surface of the leaf of *Lilium album*; *a a*, cells of the cuticle; *b b*, stomata; *c c*, cells of the parenchyma in contact with the cuticle.

FIG. 226.

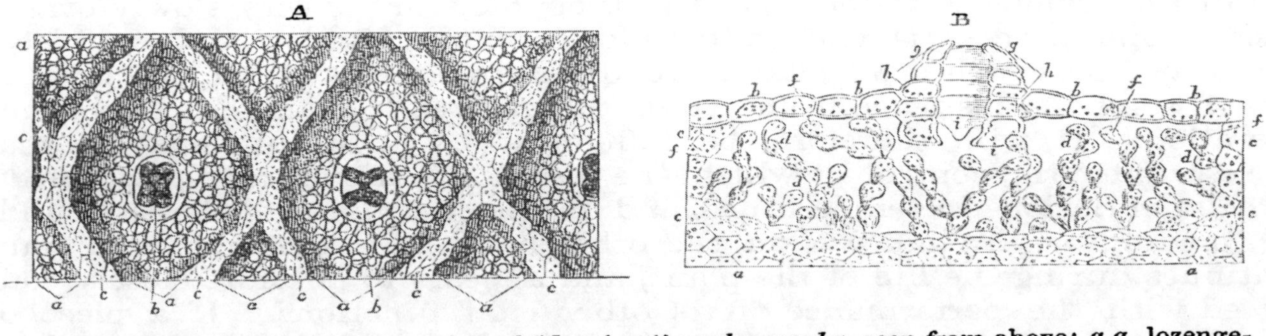

A, Portion of foliaceous expansion of *Marchantia polymorpha*, seen from above; *a a*, lozenge-shaped divisions; *b b*, stomata situated in the centre of the lozenges; *c c*, greenish bands separating the lozenges;—B, vertical section of the foliaceous expansion, showing *a a*, parenchymatous cellular tissue, forming the floor of the cavity; *b b*, superficial layer, covering-in the cavity; *c c*, cellular tissue forming the walls of the cavity; *d*, air-cavity, partly occupied by loose cells *f f*; *g*, stoma divided perpendicularly; *h*, layers of cells forming its wall; *i*, cells forming the obturator ring.

they are consequently not found upon the lower cellular plants, and but very rarely on Mosses, in which group they seem to be restricted to the capsules or 'urns,' and to their 'setæ' or footstalks, on which parts the cuticle is more distinct from the subjacent tissue than it is elsewhere. They are not formed upon the cuticle of any plants growing in darkness, nor upon the roots, nor the ribs of leaves; but they exist in general on all foliaceous expansions, and on herbaceous stems, especially on those of which the surface performs the functions of leaves, as in the *Cacteæ*. They are most abundant on the under surface of leaves, except when these float on water, and then they are found on the upper side alone; but they exist equally on both surfaces of erect leaves, as in the Lily tribe and Grasses. It has been estimated that no fewer than 160,000 are contained in every square inch of the *under* surface of the leaves of *Hydrangea*, and of many other plants; and the greatest number seems to present itself in species, in which the *upper* surface of the leaf is absolutely destitute of these organs. As a general fact, they are least abundant on succulent plants whose moisture is to be retained in the system; and they are frequently so imperfectly formed, as not to have any tendency to open, especially on the leaves of those adapted to exist in hot and dry situations. In all instances in which perforations exist, the tissue beneath is very loosely arranged, and contains many intercellular spaces; in the greater number of leaves, therefore, the most closely-packed cells will be found on the upper side (Fig. 224, *c*), whilst the parenchyma of the lower part (*d*) only comes into contact with the cuticle at intervals (Fig. 225, *c*, *c*); and it is to this arrangement that the darker hue generally possessed by the superior surface is principally due. If a leaf be placed in water, and the pressure of the air above be taken off, a number of minute globules will be seen to escape from these cavities, and to stud its exterior with brilliant points.

542. The loss of fluid from the surface of Plants may take place, as has been said, by simple *Evaporation*, or by proper Exhalation. The quantity of the former will be regulated by the degree of moisture in the tissue exposed to the atmosphere, and by the compactness of its arrangement. Thus, although the simpler terrestrial cellular plants, such as Lichens, have no true cuticle distinct from the subjacent tissue, their external layer of cells is generally of so dense a consistence, as to be almost impervious to water; so that their moisture is very slowly evaporated. The process is one quite independent of vitality, and is, indeed, the means by which dead plants are dried up, and by which the gradual loss of weight takes place from fruits, tubers, &c., that undergo no other alteration. It will, therefore, be influenced by those obvious external causes, under the control of which the process is universally performed; namely, variations in temperature, and in the humidity of the surrounding medium. — *Exhalation*, on the other hand, is a change which only continues during the life of the plant, and appears to be more closely connected with the performance of its other vital functions. If a piece of glass be held near the upper surface of a vine-leaf in full growth in a hot-house, it is scarcely dimmed after some time; but if in proximity with the lower surface of the same leaf, it is speedily bedewed with moisture, which accumulates in a short time so as to form drops. This rapid transpiration of fluid appears to take place through the stomata, as it is now satisfactorily proved that it bears a strict relation (other things being equal) with the number of stomata in the plant, or on the particular part of it made the subject of examination. Various experiments have been made at different times, with the view of ascertaining the quantity of water thus transpired from different plants, and the circumstances most

favourable to the process. With regard to *quantity*, the results obtained by Dr. Woodward * are among the most worthy of attention, although probably the earliest on record. Four plants of Spearmint were placed with their roots in water, and in a situation fully accessible to light, during 56 days (from June 2nd to July 28th); and the following table exhibits the quantity of water which each plant absorbed (proper allowance being made for the evaporation from the surface of the fluid), and its increase in weight at the end of the experiment. The *difference* must of course be the quantity exhaled, and would scarcely express the whole amount of it, as part of the increase in weight would be due to the fixation of carbon from the atmosphere.

	Original Weight.	Gain.	Water expended.	Difference.	
No. 1.	127 grs.	128 grs.	14,190 grs.	14,062 grs.	Nos. 3 and 4 were immersed in water with a little earth at the bottom.
No. 2.	110 grs.	139 grs.	13,140 grs.	13,001 grs.	
No. 3.	74 grs.	168 grs.	10,731 grs.	10,563 grs.	
No. 4.	92 grs.	284 grs.	14,950 grs.	14,666 grs.	
Total	403 grs.	719 grs.	53,011 grs.	52,292 grs.	

These experiments afford satisfactory evidence of the very large proportion of the absorbed fluid which is given out again by transpiration; and, joined with others by the same individual, they show that the activity of this function is much greater in summer than in the autumn.—A valuable series of experiments, communicated by Guettard to the Académie Royale in 1740, confirms this conclusion. He stated that transpiration is so much less active during the winter than at other parts of the year, even in evergreens, that a Laurel parts with as much fluid in two days of summer, as during two months in winter. He also maintained that transpiration goes on much more rapidly under the influence of light and a moderate degree of heat, than at a high temperature and without light.—The experiments related by Hales in his essays on 'Vegetable Statics' will ever remain, like those which he performed on the Animal Circulation, a monument of his skill and perseverance. The results which he obtained from the accurate observation of a specimen of *Helianthus annuus* (Sunflower) during 15 days, are those most frequently quoted by succeeding authors; but there are many others scarcely less interesting. This plant was 3½ feet high, weighed 3 lbs., and the surface of its leaves was estimated at 5616 square inches. The mean perspiration during the whole period was found to be 20 oz. per day; but on one warm dry day it was as much as 30 oz. During a dry warm night it lost 3 oz.; when the dew was sensible though slight, it neither lost nor gained; and by heavy rain or dew it gained 2 or 3 oz.—The following table shows the results of similar experiments on other plants.

Subject.	Surface.	Mean Transpiration.	Greatest Transpiration.	Depth.
Cabbage	2736 sq. in.	19 oz.	25 oz.	$\frac{1}{80}$
Vine	1820 sq. in.	5½ oz.	6½ oz.	$\frac{1}{191}$
Apple	1589 sq. in.	9 oz.	15 oz.	$\frac{1}{102}$
Lemon	2557 sq. in.	6 oz.	8 oz.	$\frac{1}{248}$
Plantain	2024 sq. in.	5 oz.	11½ oz.	$\frac{1}{112}$

The last column shows the mean quantity of water transpired from equal

* "Philosophical Transactions," 1699.

areas in the different plants (its depth being stated in parts of an inch) for the sake of ready comparison. That of the Sun-flower would be 1-165th; and is shown, therefore, to be less than half that of the Cabbage. The Lemon may be remarked to have exhaled far less than any of the others; and the same observation seems true with regard to evergreens in general. An experiment performed by Bishop Watson will assist in giving an idea of the extraordinary amount of change effected by this function in Plants. He placed an inverted glass vessel, of the capacity of 20 cubic inches, on grass which had been cut during a very intense heat of the sun, and after many weeks had passed without rain; in two minutes it was filled with vapour, which trickled in drops down its sides. He collected these drops on a piece of muslin, which he carefully weighed; and, repeating the experiment for several days between twelve and three o'clock, he estimated as the result of these inquiries, that an acre of grass land transpires in 24 hours not less than 6400 quarts of water. This is probably, however, an exaggerated statement; as the amount transpired during the period of the day in which the experiment was tried, is far greater than at any other.

543. The water exhaled is very nearly pure, so that what is furnished by different species varies but little in taste or odour. Duhamel remarked, however, that fluid thus obtained sooner becomes foul than ordinary water. Senebier analysed the liquid which he had collected from the exhalation of a vine at the commencement of the summer, and found that 40 oz. contained scarcely 2 grains of solid matter; and in a similar experiment on fluid collected at the end of the summer, 105 oz. gave but little more than 2 grains, or about 1-25000 part of solid matter. This minute impregnation does not indicate that any vital process of secretion is concerned in the process; since it is well known that the vapour of all water carries with it a small proportion of whatever substance the liquid may have held in solution.

544. All experiments point to the conclusion, that *Light* has a most important influence on the function of Exhalation. Thus it was shown by Senebier that if Plants, in which the process is being vigorously performed, are carried into a darkened room, the exhalation is immediately stopped; and that the absorption by the roots is checked almost as completely, as if the plant had been stripped of its leaves. Again, from the experiments of Dr. Daubeny,* it appears that exhalation is stimulated by the coloured rays of the solar spectrum in proportion to their *illuminating* not to their *heating* power, these two being separated by the prism; and Dr. D. further states that exhalation is not promoted by the most intense degree of artificial light, in which he contradicts the opinion expressed by Decandolle. Still, it is certain that heat also, especially when combined with dryness of the atmosphere, has a greater effect upon the loss of fluid than light alone. It would not seem improbable, then, that the effect of Light is confined to the opening of the stomata, which it is known to perform; and that the large quantity of fluid discharged from them, may be due to the effect of simple evaporation, from the extensive surface of succulent and delicate tissue, which is thus brought into relation with the air, and to the constant supply of fluid from within

* "Philosophical Magazine," May, 1836.

by which it is maintained in a moist condition.—If Plants are exposed to a light of too great intensity, especially if they are not at the same time well supplied with water, their tissue becomes dried up by the increased exhalation which then takes place, and which is not sufficiently counter-balanced by absorption, so that their vegetation is materially checked;—a fact of which we see abundant evidence in dry sandy soils and exposed situations. If, on the contrary, the leaves are shaded, and the roots take up much moisture, the growth of the plant is active and luxuriant, but its tissue is soft;—an effect partly owing to the retention of fluid, and partly to the diminution of the quantity of carbon fixed from the atmosphere. If a plant be kept for some time in total darkness, so that it becomes *etiolated* (§ 76), its texture is soft and succulent, and its tissue is distended with the moisture it has absorbed, and with which it cannot part; and if this state be allowed to continue too long, the leaves disarticulate and drop off, and the plant dies of dropsy. Succulent plants naturally require most light to secure for them a regular discharge of moisture; and there are some of this character which possess so few stomata, that they may be preserved out of the ground for many days and even weeks, without perishing from want of moisture. The thickness of their cuticle and their deficiency of stomata render it very difficult to dry them, even with the aid of heat; and it sometimes happens that *Sedums* and other such plants push considerable shoots, when placed under pressure whilst being prepared for the Herbarium.

545. The quantity of fluid lost by Exhalation, though ultimately dependent upon the degree of moisture supplied to the roots, does not appear to be increased by the propellent force of the sap; and this, observes the sagacious Hales, "holds true in animals, for the perspiration in them is not always greatest in the greatest force of the blood; but then often least of all, as in fevers." On the contrary, it is chiefly by the activity of Exhalation, that the amount of fluid absorbed is determined; just as the rate of ascent of oil in the wick of a lamp is dependent on the rapidity of combustion at its summit. Plants which exhale largely, therefore, must also absorb largely; and thus they are enabled to take in and to appropriate, in considerable quantity, whatever nutrient materials the soil may contain. These plants, moreover, have usually a great extent of green surface, by which they can obtain a large additional supply of carbon from the atmosphere; and thus their vegetative functions are performed with great activity, and they rapidly add to their growing parts. On the other hand, the plants which exhale slowly absorb but little from the soil; and their fixation of carbon from the atmosphere is usually performed by a comparatively small surface, and at a comparatively slow rate; so that their whole vital activity is very inferior, and the amount of solid tissue added to the fabric in a given time is far less. This is the case especially with succulent plants, whose soft and pulpy tissues are enclosed in a cuticle of remarkable density, that prevents the evaporation of their fluids; and whose life may be said to be remarkably slow. It is the case, also, to a certain degree, with most Evergreens; whose exhalation is slow, although their leaf-surface may be large, so that they fix a great quantity of carbon from the atmosphere; and it seems to be from this peculiarity, that their leaves possess a remarkable degree of firmness of texture, whilst they have a much longer

duration of life than those of deciduous trees.—That by regulating the rate of Exhalation, through the influence of light and heat, we can affect the *rate of vital activity*, and consequently the *duration of life* (§ 122), appears from the well-known fact, that the freshness of a *bouquet* of flowers may be preserved for many days longer than usual, by keeping them in the dark.

3. *Exhalation in Animals.*

FIG. 227.

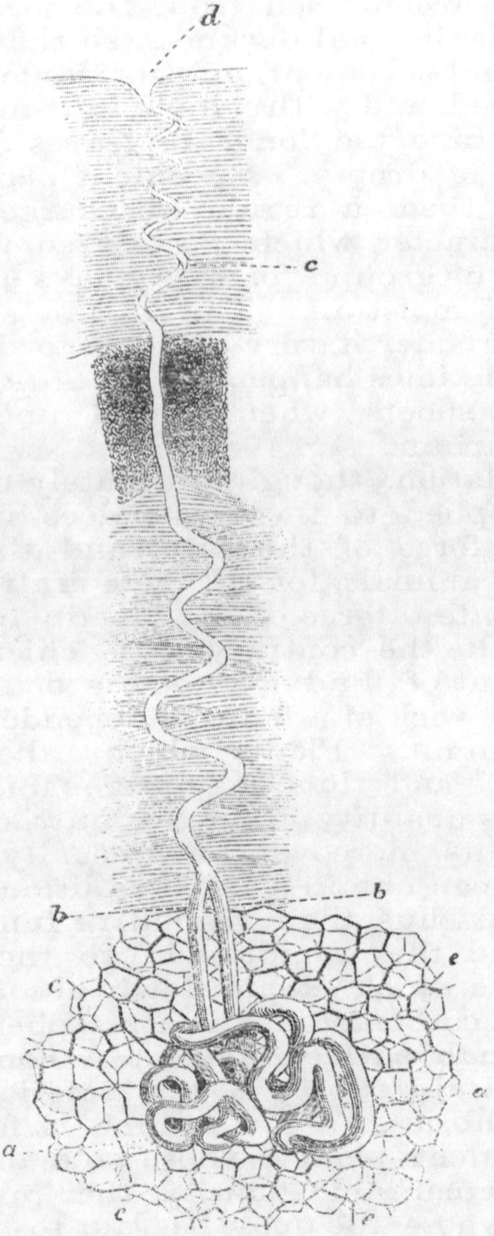

Sudoriferous gland from the palm of the *Human hand*, magnified 40 diam.; *a*, *a*, convoluted tubes, composing the gland, and forming two excretory ducts, *b*, *b*, which unite into one spiral canal, that perforates the epidermis at *c*, and opens on its surface at *d*; the gland is imbedded in fat-vesicles, which are seen at *e*, *e*.

546. The loss of fluid which is constantly taking place from the surface of all Animals inhabiting the air, or at least from some part of it, appears due, like the exhalation of plants, partly to its physical, and partly to its vital conditions. There can be no doubt that from all soft moist surfaces, *evaporation* will take place in a warm and dry atmosphere; and the quantity of fluid lost in this manner will be in strict relation with the temperature of the surrounding medium, and the rapidity with which this medium passes over the evaporating surface. The process will of course be impeded by a humid state of the atmosphere, and entirely checked by contact of water, whether warm or cold, with the part which previously effected it.—But there is another process by which fluid is exhaled from the surface, and which is more closely connected with the vital actions of the economy; this is the *transudation* from the blood of a watery fluid, usually containing a small quantity of saline and animal matter in solution, through the medium of a set of minute glands imbedded in the substance of the cutis or true skin. Each of these little bodies consists of a convoluted tube, in the neighbourhood of which the blood-vessels ramify minutely; this tube is continued to the surface of the skin as an excretory duct (Fig. 227), traversing the remaining thickness of the cutis and epidermis in a spiral manner, and opening by a very minute pore on the exterior of the latter, passing through it so obliquely that a kind of valve is formed by the membrane over its orifice. When the transudation of the sweat or sensible perspiration is observed with a glass, as it occurs on the palms of the hands or the tips of the fingers, the first drop from each pore will be seen to be preceded by an elevation of this little

valve. The ducts are visible in the form of delicate fibres passing from the cutis to the epidermis, when the latter is torn off.

547. The special object of the disengagement of fluid in the form of vapour, from the surface of the bodies of Animals, appears to be the reduction of the temperature. Animals which inhabit the water have no need of any special provision for keeping down the temperature of their bodies within a certain limit; since the rapidly-conducting power of the medium is sufficient to reduce any superfluous amount of caloric which may be generated. The tenants of the deep, indeed, have very little power of maintaining a temperature above it, unless they are provided, like the Whale tribe, with a layer of non-conducting fat, or, like diving Birds, with a downy covering possessed of a similar property (§ 325). Moreover, the vicissitudes of temperature in large collections of water are never great, so that there is no demand from this source for a means of regulating the temperature of the individual inhabitants. But an Animal living upon the surface of the earth, exposed to constant and extensive atmospheric changes, and deprived of the power of rapidly parting with its heat, when superfluous, by mere contact with a conducting medium, has need of some special means, not only of generating caloric, but also of getting quit of it. The former will be hereafter described in detail (CHAP. XVII.); the latter is simply effected by the watery transudation from the surface, which, being poured out of the perspiratory ducts in a fluid form, and carried off as a vapour by the atmosphere, necessarily renders latent a large quantity of caloric, and thus diminishes the sensible heat of the exhaling body.

548. There is no evidence that either evaporation or transpiration takes place in *aquatic* animals; for whilst simple evaporation from their surface is of course prevented by the contact of water, the secretions which are formed in their integument would seem to be purely protective. When exposed to the air, all those which are formed of soft tissue, unprotected by a hard envelope, are rapidly desiccated, and usually perish; and it is evident that such Animals are, when exposed to the atmosphere, in the same condition with the Algæ among plants, which lose weight very rapidly owing to the softness of their tissues and the want of a cuticle. Even amongst those which are provided with a hard envelope, there is always a peculiar tendency to evaporation from some part of the surface; thus, a very rapid evaporation of fluid takes place from the gills of the *Crustacea,* which would speedily offer a fatal impediment to the performance of their functions, if a special provision were not made for preserving their membrane in a humid condition (§ 514). From the experiments of Dr. Edwards on *Fishes,* it appears that the loss of fluid by evaporation from the general surface of the body and from the gills, when the animal is exposed to the air, is so great as to be one of the chief causes of its death. Sometimes the impediment to respiration, which is produced by desiccation of the gills, is the immediate cause of death; but where this is prevented, and the action of these organs continues during life, the surface parts with so much fluid by evaporation, that the body becomes stiff and dry, and previously to death loses from 1-14th to 1-15th part of its weight. Further, if the lower part only of the body be immersed in water, no absolute diminution in the weight of the whole takes place, and life is prolonged, although death at last results, seemingly from the un-

favourable influence of dry air upon the branchial apparatus ; but if, on the other hand, the head and gills be immersed and the trunk suspended in air, life may be almost indefinitely prolonged, although the drying of the surface of the part of the trunk exposed to the air is as marked, as in the case where these animals are entirely exposed to the atmosphere, and where they die after a considerable diminution in weight.

549. It is among *terrestrial* Animals, that the process of exhalation assumes a higher rank amongst the vital functions ; and, even in the lowest classes, we find it exercising a very important influence on the condition of the system. Thus, in *Insects,* it has been ascertained by Mr. Newport, that the transpiration of fluid takes place to a considerable extent ; and this not only in the species which have a soft external tegument, but among those which have the body encased in a dense horny envelope, such as the Beetle tribe. It is of course difficult to ascertain what proportion of the loss of fluid takes place in each case from the external surface, and from the prolongation of it that lines the air-passages, which in this class are so extensive and minutely ramified ; probably it is from the respiratory membrane, as in the Crustacea, that the principal liberation of it occurs. These observations show that Insects have the power of reducing their temperature, by this means, when it has been excessively raised by a continuance of rapid movements, or when the heat of the surrounding medium is too great.—It is among the *Batrachia,* however, that the exhalation of fluid from the surface is carried on to the most remarkable degree, and seems to answer the most important purpose in the economy ; and it is here, therefore, that its conditions may be most advantageously studied. Of the numerous experiments performed by Dr. Edwards on this subject, the following are the general results. He found that, when a Frog was placed in a dry calm atmosphere, the loss of weight during different succeeding hours varied considerably, but with a marked tendency to progressive diminution : that is to say, the more fluid the animal had lost, the less actively did exhalation go on. The actual quantity lost was influenced by various external agents, such as the rest or movement of the air, its temperature, and degree of humidity. Thus, frogs, hung in the draft of an open window, lost double, triple, or quadruple the amount exhaled by others placed at a closed window in the same room. The influence of the humidity of the air was tested by placing animals of the same kind in a glass vessel inverted over water ; and it was ascertained that exhalation, if not then entirely prevented, was reduced to its minimum. On the other hand, when the dryness of the air was maintained by quicklime during the progress of the experiment, the diminution of weight was found to be increased, the perspiration being from five to ten times greater in dry air than in extreme humidity, according to the duration of the experiment. The influence of temperature is shown principally in increasing the *transudation,* or Secretion from the skin ; since the amount of fluid lost in a heated atmosphere differs but little whether the medium be humid or dry, and increases in much more rapid proportion than mere evaporation would do. When Frogs were placed in an atmosphere saturated with humidity, by which mere evaporation would be almost or entirely suppressed, the loss by transudation between 32° and 50° was very slight, as also between 50° and 68° ; but between 68° and 104° it was so great, that at the last-named degree its

amount was 55 times that at 32°. The secretion is not even altogether suppressed by immersion in water. When Frogs are exhausted by excessive transpiration, and are placed in water, they speedily repair the loss, by absorption from the surrounding fluid (§ 440); and the quantity thus gained sometimes amounts to *one-third* of their entire weight.

550. From his experiments on warm-blooded Animals, Dr. Edwards obtained results of a similar kind; but the influence of changes in external conditions was not quite so marked. The distinction between the simple *evaporation* which takes place in accordance with physical laws, and the *transudation* which is the result of a secreting process, must be kept in view in order to account for their effects under different circumstances. It might, at first sight, appear to correspond with that between *insensible* or vaporous, and *sensible* or liquid transpiration; but this is not altogether true, since the *secretion* of the skin, if not very abundant, may pass off in the same form with the vapour which arises from its surface. The degree of *evaporation* from the skin of warm-blooded Vertebrata is modified, as in the Batrachia and other cold-blooded animals, simply by the temperature, degree of humidity, movement, or pressure of the surrounding medium. Wholly to suppress it, the air must not only be of extreme humidity, but also at a temperature not inferior to that of the animal; since, if the air be colder, it will be warmed by contact with the body, and thus be capable of holding an additional quantity of aqueous vapour in solution. Although cold, therefore, diminishes or even altogether suppresses transudation, evaporation will continue to a certain extent. In Man, as in the Batrachia, it seems probable that *heat* alone stimulates the function of *secretion* from the skin; so that, at moderate temperatures and in ordinary states of the atmosphere, the quantity transuded is not more than one-sixth of that which is evaporated: whilst at an elevated temperature, especially if the air be already humid, the amount of secretion will much surpass that lost by evaporation; but if the air be dry and sufficiently agitated, evaporation may increase nearly in the same ratio.—Of the quantity of fluid which may be set free by exhalation within a short time, an idea may be formed from the observations of Dr. Southwood Smith upon labourers at the Phœnix Gas Works.* These men are employed twice a day in emptying and charging the retorts and making up the fires, for about an hour on each occasion; the labour is performed in the open air, but is attended with much exposure to heat. On a foggy and calm day at the end of November, when the temperature of the external air was 39°, and the men continued at their work for an hour and a quarter, the greatest loss observed was 2 lb. 15 oz., and the average of eight men was 2 lb. 1 oz. On a bright clear day in the middle of the same month, when the temperature of the surrounding air was 60°, with much wind, the greatest loss was 4 lb. 3 oz., the average being 3 lb. 6 oz. And on a very bright and clear day in June, when the temperature of the external air was 60°, without much wind, the greatest loss (occurring in a man who had worked in a very hot place) was 5 lb. 2 oz., and the average of all was 2 lb. 8 oz.—From the experiments of Lavoisier and Seguin it appears, that the average amount of aqueous exhalation passing off insen-

* "Philosophy of Health," vol. ii. p. 322.

sibly from the human body is about $3\frac{1}{2}$ lbs. daily; the maximum being 5 lb., and the minimum $1\frac{3}{4}$ lb. Of this, however, a considerable proportion is set free from the pulmonary surface; the pulmonary being to the cutaneous exhalation, according to the estimate of Seguin, as 7 to 11. There is no reason to believe that the pulmonary exhalation is liberated in any other way than by *evaporation*, under the peculiarly favourable circumstances afforded by the delicacy and permeability of the respiratory membrane, its constant supply of fluid blood, and the frequent renewal of the air in contact with it. It is obvious that changes in the external conditions will have much less influence upon *its* amount, than upon the quantity evaporated from the skin; since the temperature of the air in the pulmonary cells will be nearly uniform under all circumstances (in the healthy state at least), and its movements are uninfluenced by the variations of the atmosphere. If, however, the external air be saturated with moisture, and be of the same temperature with the body (so as to be unable to acquire by its heat an increased capacity for vapour), it is obvious that the evaporation from the lungs, as well as that from the skin, will be entirely checked. This, indeed, appears to be the explanation of the peculiar feeling of oppression, which is consequent upon exposure to an atmosphere that is not only hot but moist. Under such circumstances, the temperature of the body is no longer kept down to its proper standard by vaporisation from its surface; and at the same time the capillary circulation of the skin and lungs is probably disturbed, by the obstruction to the elimination of fluid which should normally take place through these organs.

551. From the foregoing facts it would appear that we may look upon the process of Exhalation as essentially *physical* in its character. That portion of it which consists in simple evaporation from the cutaneous surface, is obviously so; and as to that which is eliminated by the agency of the sudoriparous glandulæ, it seems probable that this also may arise from the mere exudation of the watery parts of the blood through their thin-walled capillaries. We shall hereafter see that the escape of the superfluous water of the blood through the Kidneys, may be looked upon in precisely the same light (§ 600); and as we know that any medicinal agent which specially determines the blood towards those organs, will produce an augmentation in the watery portion of the urine, so does it seem probable that the stimulus of external Heat, which specially determines the blood to the cutaneous vessels, should occasion an increased pressure from within, and consequently an augmented passage of fluid through the walls of the capillaries. That the watery portion alone should pass, whilst the albuminous and saline matters are kept back, can be easily comprehended, from the facts already cited (§ 414) in regard to the very low tendency to 'diffusion' exhibited by albumen, and the special power of retaining it possessed by animal membranes.

CHAPTER XV.

OF NUTRITION.

1. *General Considerations.*

552. The function of *Nutrition* may be regarded as consisting of that series of changes in the Alimentary materials, by which they are converted into Organised tissues, and are caused to manifest Vital properties. The new tissues thus generated may either constitute an augmentation of the previous fabric, as is the case with Plants even to the end of their lives, with the Zoophytic and other forms of Animals whose increase by gemmation is indeterminate, and with all Animals during their period of immaturity ; or they may take the place of tissues of an inferior order, without any necessary increase of bulk ; or they may simply replace, by a *similar* production, the loss by disintegration which the fabric at large, or any particular portion of it, may have undergone from the various causes formerly enumerated (§§ 373, 374), as in Animals which, having attained their full dimensions and complete organisation, exhibit no change in size, form, or intimate structure during long periods of time. In the first case, the Nutritive operation manifests itself in *Growth ;* in the second, in *Development ;* in the third, in *Maintenance :* in all cases, however, it appears to be essentially the same, and may be considered as the end and aim of all the other Organic functions, in so far as they are concerned in maintaining the life of the individual. For it is dependent for its continuance, in the first place, upon the *Absorption* of alimentary materials, and hence (in Animals), upon the preliminary process of Digestion ; secondly, it requires the maintenance of the *Circulation,* for the transmission of the absorbed fluid from the points where it is taken in, to those at which it is to be applied ; thirdly, it cannot long be effected without the continuance of *Respiration,* the vital properties of the fluids being only maintainable by free communication with the atmosphere ; fourthly, it requires the separation of whatever is superfluous or injurious to the process, and thus becomes dependent on the due performance of the process of *Exhalation* in Plants, and on that of *Excretion* in Animals.

553. It has been already shown that every component part of each organism has an individual Life of its own, whilst contributing to support the Life of the entire being, which last is dependent upon the due performance of the vital actions of all the subordinate parts. But the degree of this dependence differs greatly in the different ranks of Organised beings ; and thus the conditions under which the Nutritive function is performed, present a very remarkable diversity, which influences the apparent character of the result, notwithstanding that the essential nature and tendency of the operation is the same in every case. Thus in the lowest forms of Organised beings, the entire fabric is made up of repetitions of the same simple elements, and every part can perform its functions independently of the rest. In such, every act of nutrition is in fact an act

of *growth;* for the tissues never advance beyond the simple cellular character which is common to every part of these homogeneous bodies; and it consists merely in the multiplication of cells by the act of subdivision (§ 139), each cell thus produced living for itself alone, and going through all its operations without influence or assistance from the rest. Precisely the same is the case with the early embryo of all the higher forms of organisation; for this, too, is composed of a homogeneous mass of cells, increasing in number by subdivision, and having no dependence upon one another for the conditions of their vital activity. But as we ascend the series, we find that the simple cellular tissues give place in greater or less degree to others, which are evolved out of them by a process of 'histological transformation;' thus, from the simple cellular Plants we ascend to those in which woody fibre and various forms of ducts are combined with cells; and in the Animal series, we meet, very low in the scale, with fibrous and membranous, muscular and nervous, horny and calcified tissues, which depart more or less from the original cellular type. Here, then, the act of nutrition does not consist in growth alone, but involves *development,* not merely in the first evolution of the organism from the embryonic state, but also in every extension which it subsequently undergoes. And it is in proportion to the heterogeneousness of this development, as already shown, that the several parts of the organism become more and more dissimilar to each other in structure and function, and consequently more dependent upon one another for the conditions of their vital activity; so that at last the integrity of the whole structure, and the complete performance of its entire series of actions, comes to be necessary for the continued maintenance of any one. But notwithstanding that the life of the individual parts seems to be thus merged (so to speak) into that of the body at large, it is so more in appearance than in reality; for although a bond of mutual dependence is created by the 'division of labour' which has been established in the organism, yet the integrity of each individual part remains as much dependent upon the due discharge of its own vital activity, as it is in those simpler forms of organic existence, in which every cell lives *for* and *by* itself alone.

554. As every component part of the most complex Organised fabric has an individual *life* of its own, so must each have a limited *duration* of its own, quite irrespective of the condition of the fabric at large, except in so far as this may tend to increase or to diminish its functional activity. We find, however, that this duration varies greatly in the different kinds of organised tissue; and they may be classified in relation to this attribute under the following heads.

I. Cells may be generated, which have a very transient existence, and which die and disappear again without undergoing any higher development. This is the case, for example, in the course of the formation of the pollen-grains, and at a certain stage of that of the ovule, of Flowering-Plants; in which the production of temporary cells, that afterwards liquefy again, appears to take place as a preparation for the formation of higher and more permanent structures. In Animals we have very numerous examples of the same general fact. The 'germinal vesicle' of the ovum seems to become filled with cells before its disappearance, and no trace of these cells can be afterwards detected; so that it would appear as if they too had deliquesced, leaving their component particles to perform some further part in the process of development. The special 'absorbent cells,' if such there really be (§ 427), the 'assimilating cells' of which an account will be given in the present chapter, the greater part of the 'secreting cells' (§ 576), and also the red blood-corpuscles (§ 173), appear, in like manner, to have a *very* transient existence, in the warm-blooded animal at least; their allotted functions being all performed within a few hours or days, and their term of life being brought to a close as soon as this is the case.

II. There are certain component parts of both Vegetable and Animal organisms, whose duration is not less determinate, although it is more extended. This is the case, for example, with the reproductive organs of a large number of Plants (as the flowers of Phanerogamia, the urns of Mosses, and the pileus of Agarics), and with the capsules which seem to represent them in the Hydroid Zoophytes; for these parts are developed only to serve a particular purpose, which is speedily accomplished, and they then die and are cast off. So, again, the leaves of Plants, and the polypes of the compound Hydroida, which are the organs most actively concerned in preparing nutritive material for the permanent fabric, have a limited though longer duration; and in the higher Animals, we find a considerable number of structures which undergo a periodical exuviation, these being for the most part epithelial or epidermic in their character. Thus, in a considerable proportion of the Articulated tribes, the external integument is thrown off many times during life, and is replaced by a new covering; a similar repeated moulting of the whole epidermis at once takes place in Frogs, Serpents, and other Reptiles; in Birds, the feathers are thus periodically cast off and renewed; in Mammalia, generally, the hair is regularly shed at certain parts of the year, whilst in Man there is a continual exuviation of the outer layers of the epidermis, and in the Deer tribe even the massive antlers are cast off and renewed every year. Even the Teeth, as we have seen, are limited in their duration among Reptiles and Fishes, being continually cast out to be renewed again; and a similar limit exists in the case of the deciduous or milk-teeth of Mammalia, which are shed at a determinate period of life. Now in most of these cases, it is capable of being clearly proved that the exuviation is consequent upon the death of the component elements of the part thrown off, not the death upon the exuviation. For with respect to the leaves of Plants, we have seen (§ 494) that they cease to perform their peculiar vital operations, and that the changes characteristic of decomposition commence, whilst they are still connected with the axis. So it has been pointed out by Mr. Paget,* that the death of the hair-bulb precedes the falling-off of the hair; and it is a familiar observation that the absorption of the fang of the deciduous tooth, which could not take place without a previous degeneration, very commonly occurs before it is pressed upon by the permanent tooth which is rising up to occupy its place. So, also, it may be shown by acetic acid or potash, that the outer layers of epidermis are in a state of degeneration; for these reagents produce very little effect upon them, whilst they render the younger cells of the interior layers transparent, as they do other tissues in a state of active growth. The ova, too,

* 'Lectures on Nutrition,' &c. "Medical Gazette," 1847.

as Mr. Paget remarks, exhibit a remarkable fixity in their term of existence. "These attain their maturity in fixed successive periods of days; they are separated (as some of the materials of several other secretions are) while yet living, and with a marvellous capacity of development, if only they be impregnated during the few days of life that remain to them after separation; but if these days pass, and impregnation is not effected, they die, and are cast out, as impotent as the merest epithelium-cell."

III. In striking contrast with the limited duration of such parts, is the condition of those tissues, whose function, instead of being transient, is to be indefinitely prolonged; but such prolongation is seen only when the function, instead of being *vital*, is simply *physical*,—as in the case of parts that afford mechanical *support*, or resist *tension*, or supply *elasticity*. Thus we have seen that, among Plants, mechanical support is afforded by the deposit of hard matter in the interior of the cells, tubes, &c., of which the part may be composed, as in the 'heart-wood' of the axis, the 'stones' of fruit, &c.; and that the tissues so consolidated are cut off from the general current of vital action. Among Animals, again, we meet with the same method of consolidation, by deposit within *cells*, in the shells of Mollusca (§ 197) and in the epidermic tissues (§ 179) of Vertebrata; whilst the shells of Echinodermata (§ 195) and the bones of Vertebrata appear to be formed rather by the chemical union of calcareous matter with a fibrous basis. When this consolidation has been once effected, the hard tissues, if not subjected to disintegrating agencies external to themselves,* may undergo little or no change for an almost indefinite period. Thus, the heart-wood of trees, the bones and shells of animals, and still more their hair, hoofs, horns, &c., may remain unaltered through an unlimited series of years. Of some of these parts it can scarcely be said that they are less alive, when removed from the organism to which they belonged, than when included in it. In the heart-wood of a Plant, for example, no vital change takes place, from the time that the woody tubes and fibres have been consolidated by internal deposit; it may decay, whilst still forming part of the stem, without the nutritive operations of the fabric being thereby interfered with; and if we could possibly remove it entirely, without doing injury by the operation to the rest of the structure, its absence would be productive of no other evil consequences, than such as would result from the withdrawal of the mechanical support afforded by it. The same may be said of the stony 'corals' formed in the soft bodies of Zoophytes and Bryozoa, and of the external skeletons of Echinodermata and Mollusca; which remain unchanged, except by addition, from the time of their first formation, not only during the whole life of the animal to which they belong, but for an indefinite period after its termination. It is probable that the same would hold good, also, of the Osseous skeleton of Vertebrata, if it were not for the necessity which exists for continually re-moulding this in accordance with the growth of the body, and for providing for its reparation when it has been injured after the attainment of its full growth (§ 211).—The same indisposition to spontaneous change shows itself in the simple Fibrous tissues, which, after their first formation, seem to require but little maintenance, their chemical

* Thus the shells of many Mollusca are altered in form, not merely by the addition, but by the removal, of parts (§ 302); this, however, seems to be effected by some *superficial* rather than by *interstitial* action, like that by which many Mollusca bore into rocks.

composition being such as indisposes them to spontaneous decay, and their functions in the economy being purely physical (§§ 161-165). Hence, when these tissues are formed by the transformation of cells (§ 166), it seems as if these cells, in becoming converted into fibres, had almost entirely parted with the distinguishing attributes of vitality, and had thus passed into a condition in which no necessary limitation is imposed on their continued existence. And when developed simply by the fibrillation of a blastema, or by the coalescence or extension of nuclear particles, it seems equally true that the vital attributes of the matter in which they originate, cease with the conversion of that matter into the form of an organised tissue.

IV. The duration of the component tissues of the Nervo-Muscular apparatus seems to depend mainly upon the degree in which their vital energies are called forth. When they are left in a state of inaction, they appear to partake, with other soft tissues, in the general attribute of limited duration; for we find that they undergo a gradual degeneration, and slowly waste away, the rate at which they do so being principally determined by the temperature. But when they are called into functional activity, it would seem as if the expenditure of their vital force put an *immediate* termination to their lives; so that their elements are at once resolved into inorganic compounds, by the oxygenating power of the blood. Hence we see that no rule can be laid down, as to the duration of the muscular and nervous tissues; since the exertion of their vital powers for no more than a few minutes, will occasion a larger amount of disintegration, than would otherwise occur during many weeks in warm-blooded animals, or during as many months or even years in the prolonged torpidity of cold-blooded animals or even of hybernating Mammalia.

555. It may be stated, then, as a general rule, applicable to all the foregoing cases, that the duration of an organised structure is very closely related to the activity of its vital manifestations; and that this, again, is related, on the one hand, to the character and attributes of the tissue, and on the other, to the conditions in which it is placed. Thus there are certain tissues (such as the Nervous substance of Animals), which would seem disposed to undergo very rapid change, in virtue alike of their chemical composition and of their vital endowments; whilst there are others (such as the leaf-cells of Plants) whose component elements have not the same inherent tendency to separation, and the discharge of whose functions does not seem to involve the same immediate loss of vital power. But in a cold-blooded Animal in a state of torpidity, the duration of the nervous tissue would seem to be protracted almost indefinitely by cold and inactivity; whilst that of the leaf-cells of a Plant may be reduced by unusual intensity of heat and light to a much shorter period than usual; and it has been ascertained by Mr. Paget, that "if the general development of the Tadpole be retarded by keeping it in a cold dark place, and if hereby the function of the blood-corpuscles be slowly and imperfectly discharged, they will maintain their embryonic state for even several weeks longer than usual, the development of the second set of corpuscles will be proportionally postponed, and the individual life of the first will be, by the same time, prolonged." * We see, then, that

* "Kirkes' and Paget's Hand-book of Physiology," p. 284.

external agencies have a most important influence upon the length of the life, either of the entire organism, or of its constituent parts. The greater the intensity of vital activity which they excite, the less is its duration; for, so soon as the series of vital operations, which each cell or other integral element is adapted to perform, has come to a close, the very same agencies that excited and maintained them, hasten its decomposition and decay. Of this we have a characteristic example in the case of annual Plants, whose term of life is inversely proportional to the amount of light, heat, &c., by which their vital activity has been called forth. And we see it also still more remarkably in the case of their seeds, which may retain their vital endowments for an unlimited period, provided the vegetative actions are prevented by the deficiency of warmth, moisture, and oxygen; but which, when called into germination, develope themselves into structures whose duration is extremely transient.—It does not seem an unfair inference, then, from these and similar facts, that the *Vital force* takes the place, in an organised structure, of the *Chemical Affinity*, which holds together the component particles of Inorganic bodies; and if, as formerly suggested (§ 54), this Vital force be nothing else than the *modus operandi* in Organised bodies, of the very same forces which are termed Physical when acting through Inorganic matter, the relation between the two agencies seems to justify such a conclusion. For, to take the case of the growing plant, so long as it is acted upon by heat and light, and supplied with nutritive materials, it continues to act upon these and to extend its own organism by the appropriation of them, giving origin to new tissue, and thus developing an additional amount of vital force; during the period of its most active growth, it shows little or no tendency to decay; but no sooner does the series of vital operations which it is fitted to perform, approach its termination, than the signs of degeneration show themselves, and the influence of Physical agencies is shown in promoting Chemical rather than Vital changes.

556. The operations in which the act of Nutrition appears essentially to consist, may be conveniently divided into two principal stages; the first of which, *Assimilation*, comprehends the gradual preparation for organisation which the crude alimentary materials undergo, whilst they are yet in a fluid state; the second, *Formation*, on the other hand, consists in the application of the assimilated or organisable particles to the generation of new tissues, or to the replacement of those which have become effete. This distinction cannot be recognised in the lower forms of either Vegetable or Animal existence, in which every component part executes both these operations alike, assimilating the nutritive matter, and either incorporating it with its own substance, or applying it to the production of new parts developed by its own vital endowments. It is in the higher Plants and Animals that we find distinct evidence, that the act of Assimilation may be performed by parts of the organism, which do not themselves apply the organisable material to the formation of tissue; and that the act of Formation may take place in parts, which have had no participation in the previous preparative changes. Such a modification is quite in accordance with the general rule of 'division of labour' which has been so frequently alluded to as a characteristic of the higher organisms.—Both these operations are purely *vital*. They have

nothing in common with physical or chemical changes; and cannot be produced (so far as our present knowledge extends) except through the instrumentality of an organised structure. It would seem as if, in the act of Assimilation, the vital force is imparted by the solid tissues already formed, to the nutritive fluids in contact with them; for there can be no doubt that the fluid becomes possessed of properties which are themselves purely vital, and that its manifestation of these is most complete when it is in immediate relation with the living solids. Thus, the coagulation of the 'plasma' of the blood, and the formation of a fibrous tissue which is the result of that act when it is most perfectly performed (§ 41), can obviously be attributed only to the peculiar endowments of which the material has become possessed, subsequently to its introduction into the animal body; it takes place most perfectly, when the plastic material is poured out upon the surface, or effused into the interstices, of living tissues; and it may be altogether prevented by such a violent 'shock,' as suddenly and completely destroys the vitality of the body. But we shall see reason to believe that there are certain organs, alike in the Animal and in the Vegetable, which are *specially* destined to effect the elaborating process; without superseding, however, the more general agency first alluded to.—In the act of Formation, a still higher measure of Vital Force seems to be communicated to the same elements; since they are then enabled, not merely to assume a more perfectly-organised form, and to manifest endowments of a much higher character, but also to develope, in their turn, similar vital endowments in other nutritive materials. We may trace at least four distinct modes in which it takes place.—In the *first* place, the material may be applied solely to the *interstitial augmentation* of the existing elements of the structure; its component cells or fibres increasing in size without undergoing multiplication, and the whole organ being enlarged without the generation of any new integral part. This is probably the case with the leaves of Plants in general, which continue to enlarge after the number of cells they contain appears to have ceased increasing, and it may also happen in some other instances; but in general, the enlargement of the cells, &c., already developed, takes place in conjunction with the *next* form of the nutritive process.—This, *secondly*, consists in the *multiplication* of the cells or other component parts, by the subdivision of those previously existing, or by outgrowth from them. Such appears to be the mode in which the most rapid increase of living structures takes place; and it is seen alike in the Plant and in the Animal, in the most complex as well as in the simplest forms of both. The process has already been fully described as it takes place in the lower Cellular Plants (§§ 137—142); and this seems to be the type according to which it occurs even among the highest Animals. In the latter, however, it is especially during embryonic life, and in the development of entirely new parts at subsequent periods, that we meet with examples of it; still there are certain tissues, as Cartilage (§ 157), in which it continues to take place throughout the whole of life.—In the *third* mode of Formation, the new tissues seem to arise from certain 'germinal centres,' which appear to have the character of cell-nuclei; sometimes, perhaps, not having been themselves fully developed into cells, and sometimes being residual components of cells which have undergone transformation, but which leave to their nuclei the developmental power. Of this mode of formation, which seems peculiar to Animals, we have examples in the development of gland-cells from the nucleus at the cæcal extremity of the follicle or tubule from which they are to be cast forth (§ 185), and in the production of the component cells of muscular fibrillæ from the nuclei contained within their enclosing tube (§ 226). Whether the epidermic and epithelial tissues generally take their origin in germinal centres contained in the 'basement membrane,' as maintained by Prof. Goodsir, or are developed in the fourth mode, has not been yet certainly determined. In the adult condition of Animals, in which no new parts are formed, but maintenance of those already existing is alone required, this mode of formation is undoubtedly more common than any other. For such 'germinal centres' or 'cytoblasts' are scattered through a large part of the fabric, their relative abundance in different organs being nearly proportional to their respective activity of growth, and their disappearance being (as Mr. Paget has justly remarked) a sure accompaniment and sign of degeneration. — But, *fourthly*, new tissue may be formed at once by the coagulation of the plastic fluid, without the direct intermediation of pre-existing cells or germinal centres (§ 158); but it usually possesses a very low degree of vitality, consisting, if it possess the fibrous type, of little else than an aggregation of similar particles into homogeneous fibres; whilst, if it be developed into cells, these show a peculiar proneness to degeneration. This mode of formation is rare in Plants, and is most commonly seen in Animals in the reparation of injuries.

557. Under whichever of these methods the Formation of tissues may be performed, the same general conditions are required.—In the *first* place, it is requisite that a due supply of the material be afforded, in a state in which it can be appropriated. The degree in which the formation of each tissue is dependent upon the previous preparation of its nutritive materials, seems to vary greatly. Thus, as we have seen, among the lower tribes of Plants, every cell can effect the preparative operations, as well as appropriate the material so prepared; whilst in the higher, a large part of the organism is dependent for its pabulum upon the materials supplied elsewhere. But we shall presently see (§ 560), that out of the very same organic compounds, the different cells of the higher Plants can elaborate a great variety of new products, differing widely in their chemical character; so that they must retain a certain degree of converting power. Other circumstances being the same, the activity of growth is proportional to the abundance of the nutritive material, provided that the formative power be not wanting. This is especially seen in Plants, in which there are fewer counteracting agencies; and it is, in fact, the foundation of a large part of the art of Cultivation, which aims especially to accelerate the growth of the entire plant, or to augment the growth of some particular part, by increasing the supply of the aliment which it may require. But we have also very characteristic examples of it in Animals; such as the increase in the Adipose tissue, when the food has been abundant in quantity and rich in quality; the increase in the size of glands, when the blood contains an unusual quantity of the special product to be eliminated by their agency; and (according to many eminent Pathologists), the development of abnormal growths (such as Cancer, Tubercle, Lepra, &c., &c.), which remove particular morbid

matters from the current of the circulation, with a rapidity proportional to the supply of the peculiar pabulum which their cells respectively require. It seems probable, indeed, that the Animal tissues have less *converting* power than those of Plants ; and that not only each tissue, but each part of the same tissue, *selects* some different material from the blood. For there are certain morbid matters, whose presence in the blood is manifested by the perversion of the nutritive process in certain spots only of the body, these spots being similar in size and situation on the two sides ; so that it would seem that the only parts of any tissue which are really identical in composition, are those which occupy symmetrical positions on the opposite sides of the body.* Now in the healthy state, as in those diseased conditions which afford more striking exemplifications of this principle, every part of the body, by taking from the blood the peculiar substances which it needs for its own nutrition, thereby removes from it a certain part of its constituents, which would interfere with the nutrition of the rest of the body if retained in the circulation. And this, as Mr. Paget has well remarked, seems to be the purpose answered by the development of many structures that perform no ostensibly useful part in the economy ; such as rudimental organs of various kinds, the *lanugo* or downy covering of the human fœtus which falls off some time before birth, and the coat of hair which is formed in many of the lower Mammals during fœtal life, and which gives place before birth to a more complete coat of a different colour. The development of these and other structures, for which no other purpose can be assigned, may be fairly regarded, on the principle which has been now laid down and illustrated, as resulting from the presence of their peculiar materials in the blood, and as leaving it fitter by the removal of them, for the nutrition of other parts, or as adjusting the balance which might else be disturbed by the formation of some other part. "Thus," to use Mr. Paget's own words, "they minister to the self-interest of the individual, while, as if for the sake of wonder, beauty, and perfect order, they are conformed to the great law of the unity of organic types, and concur with the universal plan observed in the construction of organised beings." When the due supply of nutritive material is *not* afforded, imperfect or deficient formation must be the result ; and this is probably the explanation of the atrophy which normally occurs in the course of the life of higher animals, in many organs, which at earlier periods of life were of considerable size and importance, such as the thymus gland, the ductus arteriosus and venosus, the mammary glands, &c.

557 *a.* In the *second* place, the formative process is mainly dependent upon a due supply of heat. This agent has been commonly regarded as the mere *stimulus* to its activity (§ 68) ; but reasons have been already given for the belief, that Heat, acting through a substance previously organised, *itself becomes* the formative power (§ 54). This much, however, is certain ; that the activity of growth bears a very close relation to the temperature of the fabric, so that the formative processes may be artificially retarded by cold or accelerated by warmth ; and it further appears that *development* requires a higher temperature (as it certainly

* See Dr. W. Budd's admirable paper on 'Symmetrical Diseases,' in the "Medico-Chirurgical Transactions," vol. xxv. ; and Mr. Paget's 'Lectures on Nutrition, &c.' in the "Medical Gazette" for 1847.

seems to depend on a higher measure of vital force) than simple *growth* (§§ 96, 98). Hence we see that the very same agent which exerts so remarkable an influence on the duration of the life of the tissues, exerts a corresponding influence on those processes by which their disintegration is compensated; and the perfect balance between waste and repair is thus maintained. In Plants, indeed, we see this beautiful adjustment extending still further back; for the heat which promotes the formative processes is accompanied (when not artificially supplied) by a corresponding amount of Light; and these two agents, in proportion to the intensity of their operation, accelerate the absorption of fluid by the roots (§ 544), and also produce a corresponding increase in the supply of carbon fixed by the leaves.

2. *Nutrition in Vegetables.*

558. The phenomena of Nutrition in the simplest *Cellular* Plants having been already sufficiently considered (§§ 39—47), it is not requisite here to detail them again; and it need only be mentioned, that the same facts hold good as to *all* Plants, in which the functions of Absorption and Aeration are not *specialised* by restriction to distinct organs. In the greater number of *Vascular* Plants, however, there is no doubt that the greatest proportion of the fluid imbibed into the system is derived from the soil surrounding the roots; and that it holds in solution carbonic acid and ammonia, from the combination of whose elements are produced the proximate principles, gum, sugar, albumen, &c., at the expense of which the tissues are generated. It would seem that the fluid thus absorbed is in all plants nearly the same, under corresponding circumstances, except as regards the mineral ingredients of the soil (§ 422); and that, provided it contain an adequate supply of the above-named compounds, the presence of organic matter is not favourable to growth. Even in the roots, however, the fluid that is passing upwards through the axis is found to contain gum and sugar; * and during its upward ascent, its specific gravity still further increases, and the quantity of sugar and gum contained in it becomes sensibly greater. If the results of experiments and observations, however, on the functions of the leaves (§§ 497—500), be duly considered, it seems difficult to avoid the conclusion, that the greatest addition to the materials for the formation of the solid tissues of plants is made through *their* agency (or by that of other leaf-like surfaces); and that the greater part of the process of conversion of the oxygen, hydrogen, carbon, and nitrogen, obtained by the plant from the water, carbonic acid, and ammonia, which it imbibes, into organic compounds, is effected by their instrumentality. We have seen that of the water taken in by the roots, a large proportion must necessarily be exhaled by the leaves; since, in order to obtain the requisite supply of the matters which are sparingly dissolved in this liquid, the plant must absorb far more of it than it can apply to its own use. Hence the crude sap brought to the leaves undergoes a double change; a large proportion of its water being got rid of, whilst a great addition is made to its carbon;

* It is inferred by Prof. Schleiden and his followers, that these are generated by the tissues among which the liquid is diffused, which begin to exert an assimilating power upon the crude sap, as soon as it is brought within their reach. There are cogent objections, however, to such an hypothesis; and the fact may be as well explained in another way (§ 447).

and thus a great increase is presented in the proportion of organic compounds which the 'proper juices' of the leaves contain. These proper juices appear to furnish the chief part of the *pabulum,* at the expense of which the development of new parts takes place in every part of the organism; for although it may not be distributed, as some have supposed, by any regular 'descending current' (§ 447), yet it is impossible to account for the large quantity of additional carbon taken in by the leaves, unless it is thus applied, since it is certain that these organs themselves do not undergo any proportional increase, during the period when they are most actively performing the function of aeration. That a marked difference exists between the crude ascending sap whose materials are supplied by the roots, and the 'proper juices' whose formation thus seems chiefly to depend upon the leaves, is shown in a variety of ways. Thus, the inhabitants of the Canary islands tap the trunk of the *Euphorbia canariensis,* and obtain a refreshing beverage from the ascending current, although the proper juice of the plant is of a very acrid character. The phenomena of *vegetable parasitism* have a peculiar interest when viewed with reference to this subject. The Phanerogamic parasites may be arranged under two groups; those provided with leaves, belonging for the most part to the order *Loranthaceæ* or mistletoe tribe; and those which are destitute of leaves, such as the *Cuscuta* (dodder), *Orobanche* (broom-rape), and the *Lathræa squamaria.* Now the plants of the first group may be regarded as natural grafts, organically uniting themselves with their stock; for the wood and bark of the Mistletoe grow in continuity with the wood and bark of the tree to which it has attached itself, although the line of demarcation between the two may be seen when a vertical section is carried down through the part where they meet. It is obvious that it must receive its supply, like the branches of the stock itself, from the current of sap ascending in the latter; and that this must be converted, by the elaborating action of its leaves, into the materials of its growth. On the other hand, the leafless parasites attach themselves to the *bark* alone, by means of suckers which apply themselves to its surface, or of fibres which penetrate into its substance; and it seems obvious that, as they are not able to elaborate sap for themselves, they are dependent for their support, not upon the crude ascending current, but upon the 'proper juices' of the plants on which they live. And this view derives confirmation from the fact, that whilst the *leafy* parasites will grow almost indifferently upon a great variety of trees, their ascending sap differing but little in quality, the *leafless* parasites are much more limited in their range, each kind being usually restricted to a few species whose 'proper juices' are suitable for its support, and those of other plants not being adapted to its nutriment.

559. The leaves of the higher Plants may be regarded, then, as the chief *assimilating* organs, by which the materials are prepared for the *formative* processes that take place in different parts of the fabric. That which seems to be the essential pabulum of the vegetable tissues, is the *protoplasma* (§ 143), containing saccharine, gummy, and albuminous matters in a peculiar state of combination, which is found in all rapidly-growing parts; and if this be supplied, it is probable that each component cell of these tissues can elaborate for itself the peculiar compounds which it is destined to contain. And thus it happens that the growth of

a 'stock' is not changed from its natural method, by the somewhat different quality of the proper juice that may be supplied by the leaves of a 'graft'; nor does the mistletoe impart any of its peculiarities to the branch of the tree with which it has united itself. From the very same pabulum, the wood of the mistletoe and that (*e.g.*) of the apple grow after their respective fashions; just as in the different parts of the mistletoe itself, the wood grows after one pattern, the bark after a second, the leaves after a third, the flowers after a fourth, and the fruit after a fifth. Every growing part, in fact, turns the 'pabulum' which it receives to its own account, and forms it into its own peculiar tissues. This 'pabulum' appears to be especially diffused through all the *cellular* portion of the fabric; and it is in this, as formerly explained, that all new formation originates. Thus we find new leaf-buds in *Dicotyledonous* stems (§ 280) developing themselves as extensions of the medullary rays; whilst in the *Monocotyledonous* (§ 279) they spring from the general cellular mass of the axis; and in the cells of the parts from which they spring, we usually find a much larger accumulation of starchy and other nutritious materials, than could be derived from the quantity of ascending sap attracted towards them, before exhalation is actively established by the expansion of the buds.—The addition to the central axis, again, is effected in Dicotyledons by the continual growth and transformation of the *cambium-layer*, which intervenes between the last-formed layers of wood and bark. The cambium was formerly supposed to be a mere glutinous sap, and various notions were entertained in regard to the mode of its conversion into the new wood; but it is now quite ascertained that the cambium is really a mass of young cells turged with protoplasma, and that as the inner layers of these are converted into fibro-vascular bundles, they continue to increase by the multiplication of cells on their outer side; and thus there is an absolute continuity of growth between the inner and outer parts of the axis, although, from the vegetative functions being periodically suspended in trees all whose leaves are cast off at once, lines of demarcation are more or less distinctly left between the portions formed at successive epochs. Thus are produced the so-called 'annual layers' in these stems; which are not, however, to be regarded as by any means *uniformly* indicating the number of years during which a stem or branch has been growing. For there are many trees in tropical climates, whose leaves are thrown off and renewed twice or even thrice in every year, or five times in two years; and as the whole series of nutritive operations then receives a check, it cannot be doubted that a new line of demarcation will be left by every exuviation. Even in temperate climates the same thing may be occasionally observed; thus the author has known a long continuance of heat and dryness in the early part of the summer, to be followed by the entire exuviation of the leaves of trees growing in exposed situations; a new covering of leaves making its appearance within a few weeks afterwards. And temporary checks to the vegetative processes may arise from other sources; thus the growth of the larch, which naturally thrives in cold and moist situations, is stopped by heat and dryness; and the author has been assured by a competent authority, that he has ascertained by actual observation that two thin layers of wood have been formed by larches, instead of one thick one, when the nature of the season had been such as to occasion a prolonged interruption. On the other hand, in 'evergreen' trees, in which the leaves are not all cast off at once, the lines of demarcation are usually much less distinct, since the growth of the cambium-layer is not at any time completely stopped, except by intense cold. As the additional wood is always developed on the exterior of that previously formed, it is obvious that the term *Exogenous* may be appropriately applied to this mode of growth ; it has been also designated as growth by *indefinite* fibro-vascular bundles.—On the other hand, in the stems of Monocotyledons, the fibro-vascular bundles once formed are not susceptible of increase, and they are therefore said to be *closed.* Hence the new bundles are not developed in continuity with the old, but take their origin in the midst of that part of the cellular mass of the stem which is in closest connection with the new leaves ; and instead of extending through the entire axis, so as to augment its diameter from the extremities of the branches to those of the roots, they pass towards its exterior at no great distance from the summit, and there terminate, so that the lower part of the stem undergoes no augmentation. To these stems, therefore, the term *Endogenous* cannot be justly applied ; and it should be dropped altogether.

559 *a.* But besides ministering to the development of new tissue, the elaborated sap of Plants supplies the materials for the production of that immense variety of organic compounds, in which the Vegetable World is so rich ; compounds whose use in the economy of the Plant is frequently by no means apparent, but whose value to the Animal Creation and to Man would frequently appear to be the sole end of their preparation. These compounds are frequently designated as Vegetable *Secretions ;* but this term cannot be applied to them in the sense in which it is used in Animal physiology. For all these products are contained and stored up in cells, which continue to form part of the organized fabric, instead of being cast forth from it (§ 573) ; and they might rather be likened to the fat of animals, which is in like manner separated from the blood by the development of adipose cells, that constitute permanent components of the organism. Now of the immense variety of these products that presents itself in different plants, and even in different parts of the same plant,—the fixed and volatile oils, resins, gums, colouring matters, alkaloids, acids, &c., &c.,—no other account can be given, than that each component cell generates its own peculiar products, at the expense of the nutrient materials supplied to all alike, just as, among the simplest Cellular Plants, each species may form a distinct organic compound. Thus, to take a simple case, in the petal of a *Heartsease* or any similar parti-coloured flower, one cell forms purple colouring-matter, while another in close proximity with it forms red colouring-matter ; just as *Protococci* and *Hæmatococci,* growing under the same circumstances, and at the expense of the same inorganic compounds, generate green, and blood-red endochrome respectively. —To enter into the chemical characters of the Vegetable secretions, would be foreign to the purpose of this work ; but there is one which is too obviously related to the vital operations of the economy, to be passed without notice. This is *Starch,* a substance very universally diffused through the Vegetable kingdom. When removed from the plant, Starch exists in the form of minute granules, presenting great diversities of figure and dimension ; but having, for the most part, a limit of size, and a characteristic aspect, in each tribe of plants, by which its source may frequently be determined. Each granule, when examined by the microscope, is seen to

be marked by numerous lines, having more or less of a concentric arrangement; these, however, are not striations on the surface, as might be at first supposed, but seem to result from the existence, in each granule, of a number of layers, one investing another. The composition of these layers appears to be uniformly the same; but the outer ones contain less water than those which they inclose, and have even an almost membranous firmness. Several of these granules, of different sizes, are usually found in one cell; and it seems obvious that they are formed by successive additions to their substance, though whether the outer layers are formed over the inner ones, or the latter are formed by imbibition through the former, has not yet been determined. When exposed to the heat of about 160°, the starch-grain bursts, and the inner layers are readily dissolved by water; and this is the explanation of the fact, that Starch once dissolved in hot water can never be restored to its original form. In composition, Starch is very closely related to gum and cellulose; and it may, in fact, be regarded, as a store of nutritive matter which has undergone such an alteration that it can be kept apart from the surrounding juices, and can thus be reserved for some special purpose. Thus, we find it stored up in the *seeds* of most species, either forming a separate albumen, as in the Grasses, or taken into the structure of the embryo and constituting the mass of the fleshy cotyledons, as in the Leguminosæ, &c. (§ 278 *a*); in each of these cases it serves as a magazine of food for the nutrition of the embryo, previously to the development of those organs which enable it to maintain an independent existence. Similar reservoirs are occasionally formed by the enlargement of the *stem* into tubers, for the nutrition of the buds to be developed from them,—as in the Potato, Arrowroot plant, &c.; or by the accumulation of the same material in fleshy *roots, bulbs*, &c., from which stems rapidly grow up. Starch is also found abundantly in the soft interior (improperly called pith) of the stem of the Sago Palm and other Monocotyledons, where it seems destined to assist the evolution of the young leaves; and in the fleshy expansions of the flower-stalk (termed *receptacles*), on which, in many orders, the flower is situated, and in which it seems to answer a corresponding purpose.—In all these cases, the immediate end of the accumulation of Fecula is, that it may be ready for the nutrition of the young germ before it is capable of obtaining food for itself; and it may be observed that the deposit continues to increase as long as the plant is in active vegetation,—arrives at its maximum,—and then, remaining stationary during the winter, begins to decrease in the spring. The deposition of fecula fulfils, therefore, an obvious purpose in the Vegetable economy; but we cannot doubt the wise and benevolent intention of the Creator, in thus providing a store of nutritious and palatable food for Man, in situations whence he can so easily obtain it; and it is interesting to remark that, as it almost always exists in an insulated form, it may be obtained in a state of purity from many vegetables which would otherwise be very poisonous. Before it can be applied to the nutrition of the plant, however, its condition must be changed. Thus, in the germination of seeds, it is converted into sugar; the same change takes place in the tuber of the potato; and, from the researches of Dunal (§ 503), it seems probable that the Starch deposited in the receptacle is converted, during the period of flowering, into Sugar. This conversion is a process which the chemist can imitate; for if the fecula be first heated, so that its vesicles are ruptured, and it be then treated with dilute sulphuric acid, it is converted into sugar; and the salivary matter of animals possesses the same converting power in a remarkable degree (§ 410). This change is effected in the Vegetable economy by the operation of a secretion called *diastase;* which seems to be formed for the express purpose, and which may be obtained in a separate state most readily from the neighbourhood of the *eyes* or *buds* of the potato. It is stored up in that situation, for the purpose of being conveyed, by the vessels connected with the bud, into the substance of the tuber, when the demand for nutrition is occasioned by the development of the shoot; and, in the laboratory of the experimenter, it produces exactly the same effects as in the Vegetable economy. It is probable that the secretion of diastase takes place in every instance, in which fecula previously deposited is to be reabsorbed.

3. *Nutrition in Animals.*

560. In tracing the gradual incorporation of the alimentary materials ingested by Animals, into their organized fabric, it is desirable, as in the study of Plants, to give our chief attention to the cases in which the different steps of the process are most distinctly separated from each other. For we see among the Protozoa nothing more than the fact, that each cell obtains from the fluids which it inhabits the materials of its growth, absorbs these into its own cavity, and applies them to its own enlargement, or to its multiplication by subdivision. Whilst at the opposite extremity of the series, in which a Circulating system is interposed between the points at which nutriment is absorbed and those at which it is appropriated, we are able to trace also a gradual metamorphosis in the components of the nutritive fluids, whereby it is *assimilated* in nature to the tissues, whose *formative* operations are performed at its expense. This metamorphosis partly consists in the conversion of Albumen (the form to which fibrine, caseine, and other substances of like composition, as well as albumen itself, are reduced in the digestive process) into *Fibrine;* a change which is not manifested by any appreciable difference in ultimate composition, but which consists in the acquirement of properties that are altogether different from those exhibited by any mere chemical compound, and are indicative of an incipient vitalization (§ 41). A further preparation, however, seems to be effected, by the intimate incorporation of *Fatty* matter with albuminous substances; which there is strong reason for regarding as one of the most important parts of the Assimilating operation in Animals. For oily matter, as we have seen, is a constituent of the food of nearly all animals; and where it does not exist as such in the food, it is generated in the system at the expense of farinaceous compounds; and since this takes place in cold- as well as warm-blooded animals, it is obvious that the oleaginous particles must have some special purpose in the system, altogether irrespective of the maintenance of the combustive process. Further, we shall see that one of the chief peculiarities of the chyle consists in the peculiar state in which these fatty particles are found; and this state corresponds so closely with that in which we find them dispersed as 'molecules' through the solids and fluids of the body, that we can scarcely hesitate in attributing to it

some peculiar relation to the nutritive operations. Of the nature of this relation we obtain some further indication from the fact, that oleaginous particles may be constantly distinguished in the nuclei or cytoblasts of growing cells or fibres, as well as in the nuclear bodies scattered through an 'organisable blastema;' so that it may be affirmed with much probability, that the presence of fatty matter is not less essential to the formative operations, than is that of the albuminous compounds themselves.* The successive stages of the assimilating process, so far as they are at present known, will be now described.

561. The nutritive materials prepared by the digestive process, are taken into the circulation of Vertebrated animals (as already shown, CHAP. XI.) through two distinct channels, the blood-vessels and the absorbents. Now all the veins which return the blood from the capillaries of the gastro-intestinal canal, converge into the portal trunk, which distributes the blood, thus charged with crude materials, to the secreting apparatus of the Liver. It was formerly supposed that the agency of that gland was limited to the elimination, from the blood subjected to its influence, of the materials of the biliary secretion; but there is now evidence that the blood itself is changed by its means, in a manner that indicates an *assimilating* as well as a *depurating* action. For it has been observed by Prof. E. H. Weber, that during the last three days of incubation of the chick, the liver is made bright yellow by the absorption of the yolk, which fills and clogs all the minute branches of the portal veins; and that in time, the materials of the yolk disappear, part being developed into *blood-corpuscles* and other constituents of blood, which enter the circulation, and the rest forming bile, and being discharged into the intestines.† There is no evidence that blood-corpuscles are thus generated in the liver during later life; but recent researches have shown that other materials of the blood here undergo an important assimilating action. For the blood which comes to the liver during digestion, contains a large quantity of albuminous matter and very little *fibrine;* whilst in that which has passed through the organ, there is a large increase in the amount of fibrine. Again, it appears that *fatty matters* are elaborated in the liver, from saccharine or some other constituents of the blood; so that even when no fat can be detected in the blood of the vena portæ, that of the hepatic vein contains a considerable amount of it. A portion of this fat may be destined to immediate elimination in the lungs; but if the supply that should be introduced by the lacteals be deficient, it would doubtless be made subservient to the formative processes. So, again, it appears certain that the liver elaborates from some other constituents of the blood a *saccharine* compound (diabetic sugar), which is destined for immediate elimination by the lungs, and which, being much more readily

* This view derives strong confirmation from the very remarkable effect of Cod-liver oil in improving the nutrition of Tubercular subjects; and from the curious prophylactic influence of the oleaginous diet of the Icelanders, a people whose habits are such as would peculiarly favour the development of Scrofula, but who are most remarkably free from any form of it. The importance of oleaginous matter to the process of textural nutrition, and the probable *rationale* of the beneficial effects of Cod-liver oil (which may be regarded as by far the most important among the recent improvements in Medical practice), were first developed, the Author believes, by his friend Prof. J. H. Bennett, in his work on Cod-Liver Oil, published in 1841.

† Henle and Pfeufer's "Zeitschrift;" 1846.

carried off by the respiratory process than either grape-sugar or cane-sugar, may be regarded as its most appropriate pabulum.*—Thus we may look upon the liver as an assimilating organ of very comprehensive endowments; being the seat (under certain circumstances at least) of the production of red-corpuscles, of the elaboration of fibrine, and of the generation of fatty matter, which are the three most important elements in the preparative stages of the nutrient process; and being further capable of preparing the blood most advantageously for the function next in importance, the generation of heat.

562. We have now to enquire whether the Chyle absorbed into the Lacteal system undergoes any corresponding changes, during its progress from the walls of the intestines to the thoracic duct. The fluid drawn from the lacteals that traverse the intestinal walls, has no power of spontaneous coagulation; whence we may infer that it contains little or no Fibrine. It contains Albumen in a state of complete solution, as we may ascertain by the influence of heat or acids in producing coagulation. And it includes a quantity of Fatty matter, which is not dissolved, but suspended in the form of globules of variable size. The quantity of this evidently varies with the character of the food; it is more abundant, for instance, in the chyle of Man and of the Carnivora, than in that of the Herbivora. It was formerly supposed that the milky colour of the chyle is owing entirely to its oil-globules; but Mr. Gulliver has pointed out that it is really due to an immense multitude of far more minute particles, which he has described under the name of the *molecular base* of the chyle. These molecules are most abundant in rich, milky, opaque chyle; whilst in poorer chyle, which is semi-transparent, the particles float separately, and often exhibit the vivid motions common to the finely-divided molecules of various substances. Such is their minuteness, that, even with the best instruments, it is impossible to determine either their form or their dimensions with exactness; they seem, however, to be generally spherical; and their diameter may be estimated at between 1-36,000th and 1-24,000th of an inch. As they are readily soluble in ether, there would seem to be no doubt of their oleaginous nature; but it appears probable that each of them is surrounded with that thin membranoid film, which, as first pointed out by Ascherson, is formed whenever oily and albuminous matters are brought into contact.—No other particles than these are observable in the chyle drawn from the lacteals near the villi; but after the fluid has passed through the Mesenteric glands, it exhibits very marked changes. The presence of fibrine (which must have been formed at the expense of the albumen) now begins to declare itself, by the spontaneous coagulation of the fluid; the quantity of molecules and larger oily particles diminishes, perhaps by their passage into the blood when the two fluids are brought into close relation in the mesenteric glands; and the peculiar floating cells or 'chyle-corpuscles' formerly adverted to (§ 172) now for the first time make their appearance. The average diameter of these is about 1-4600th of an inch; but they vary from about 1-7000th to 1-2600th,—that is, from a diameter about half that of the human blood-corpuscles, to a size about a third larger. This variation probably depends in great part upon the period of their growth. They are usually minutely

* See the Lectures of M. Cl. Bernard on the 'Functions of the Liver,' delivered before the Collége de France, and published in "L'Union Médicale" for 1850.

granulated on the surface, seldom exhibiting any regular nuclei, even when treated with acetic acid ; but three or four central particles may sometimes be distinguished in the larger ones. These corpuscles are particularly abundant in the chyle obtained by puncturing the mesenteric glands themselves ; and it would seem not improbable, that they are identical with the spherical nucleated particles, which are so copiously developed within the dilated lacteal tubes in their course through those bodies (§ 429). The glandular character of these cells, and their continued presence in the circulating fluid, seem to indicate that they have an important concern in the process of Assimilation. It is only in the Chyle which is drawn from the lacteals intervening between the mesenteric glands and the receptaculum chyli, that the spontaneous coagulability of the fluid is so complete, as to produce a perfect separation into *clot* and *serum*. The former is a consistent mass, which, when examined with the microscope, is found to include many of the chyle-corpuscles, each of them being surrounded with a delicate film of oil ; the latter bears a close resemblance to the serum of the blood, but has some of the chyle-corpuscles suspended in it. Considerable differences present themselves, however, both in the perfection of the coagulation, and in its duration. Sometimes the chyle sets into a jelly-like mass ; which, without any separation into coagulum and serum, liquefies again at the end of half an hour, and remains in this state. The coagulation is usually most complete in the fluid drawn from the receptaculum chyli and thoracic duct ; and here the resemblance between the floating cells, and the white or colourless corpuscles of the blood, becomes very striking.

563. The Lymph, or fluid of the Lymphatics, sensibly differs from the Chyle in its comparative transparency : its want of the opacity or opalescence which is characteristic of the latter, being due to the absence, not merely of oil-globules, but also of the 'molecular base.' It contains floating cells, which bear a close resemblance to those of the Chyle on the one hand, and to the colourless corpuscles of the Blood on the other ; and these, as in the preceding case, are most numerous in the fluid which is drawn from the lymphatics that have passed through the glands, and in that obtained from the glands themselves. Lymph coagulates like chyle ; a colourless clot being formed, which encloses the greater part of the corpuscles. The chief chemical difference between the Chyle and the Lymph, consists in the much smaller proportion of solid matter in the latter, and in the almost entire absence of fat, which is an important constituent of the former. This is well shown in the following comparative analyses, performed by Dr. G. O. Rees, of the fluids obtained from the lacteal and lymphatic vessels of a donkey, previously to their entrance into the thoracic duct ; the animal having had a full meal seven hours before its death.

	Chyle.	Lymph.
Water	90·237	96·536
Albuminous matter (coagulable by heat) .	3·516	1·200
Fibrinous matter (spontaneously coagulable) .	0·370	0·120
Animal extractive matter, soluble in water and alcohol	0·332	0·240
Animal extractive matter, soluble in water only	1·233	1·319
Fatty matter	3·601	a trace.
Salts ;—Alkaline chloride, sulphate, and carbonate, with traces of alkaline phosphate, oxide of iron .	0·711	0·585
	100·000	100·000

The Lacteals may be regarded as the Lymphatics of the intestinal walls and mesentery ; for, during the intervals of digestion, they contain a fluid, which is in all respects conformable to the lymph of the lymphatic trunks.

564. Thus by the admixture of the aliment newly-introduced from without, with the matter which has been taken up in the various parts of the system, and by the elaboration which these undergo in their course towards the thoracic duct, a fluid is prepared, which bears a strong resemblance to blood in every particular, save the presence of *red* corpuscles. Even these may sometimes be found in the contents of the thoracic duct, in sufficient amount to communicate to them a perceptible red tinge ; but there can be little doubt that they have found their way thither accidentally,—some of the lymphatic or lacteal trunks, which have been divided in the dissection necessary to expose the duct, having taken up blood by their open mouths, and rapidly transmitted it into the general receptacle. The fluid of the thoracic duct may be compared to the blood of Invertebrated animals ; from which the red corpuscles are almost or altogether absent, but which contains white or colourless corpuscles ; and whose coagulating power is comparatively slight, in consequence of its small proportion of fibrine. And we hence see, why these animals should require no *special* absorbent system ; since the blood-vessels convey a fluid, which is itself so analogous to the chyle and lymph to be absorbed, that the latter may be at once introduced into it, without injuring its qualities. The elaboration of this fluid in the Vertebrata would seem most probably due to the assimilating agency of the cells contained within the absorbent glandulæ, and of those which float in it during its passage through the vessels ; and we may, in fact, regard the whole of the special Absorbent system in the light of an assimilating apparatus, fitted not merely to take up, but also to prepare for admission into the circulating fluid, such appropriate materials as may be presented to it in the requisite condition.

565. The same function may with great probability be assigned to certain bodies, known as 'Vascular Glands' or 'Glands without ducts,' which are found in Vertebrated animals, in intimate relation with the lymphatic system. These are, the *Spleen*, the *Thymus Gland*, the *Thyroid Gland*, and the *Supra-renal Capsules*. Many points in the structure of these organs, especially of the one first-named, are still far from being satisfactorily made out ; and of their function it would be rash to speak too confidently. Nevertheless it may be said of them all, that when in most active operation they are largely supplied with blood, and that they are made up for the most part of a cellular parenchyma, which is contained either in isolated vesicles (as in the spleen, thyroid gland, and supra-renal capsules), or in one large branching cavity (as in the thymus). Thus, in every essential particular but the want of direct outlets, they correspond with the ordinary glands ; and it can scarcely be questioned that their function is to separate certain materials from the blood, and, after submitting them to an elaborating or assimilating action, to return them again to the current of the circulation, either directly or through the medium of their lymphatics, which are commonly very large. It is a strong confirmation of this view, that the greatest activity of the Thymus and Thyroid Glands, and of the Supra-renal capsules, is during fœtal life

and early infancy, when the formative processes are being performed with extraordinary energy, and are making a large demand upon the assimilative powers; and it has been further shown by Prof. Goodsir, that these three organs may be regarded as involuted portions of the 'germinal membrane' which is the first assimilating organ possessed by the fœtus (CHAP. XVIII.), and are in absolute continuity with each other at an early period of fœtal life. But there is no improbability that they may severally have some subsidiary or supplementary function to perform, varying according to their respective structure, position, and connections. Thus it has been observed by Mr. Simon, that the Thymus of hybernating Mammalia, instead of dwindling away (as in other instances) when full growth has been attained, greatly enlarges and becomes laden with fat; and that this accumulation takes place especially at the approach of winter; so that the organ, after it has ceased to perform its original function, is then used as a sort of storehouse of combustive material.—The Spleen, which has a different origin from the other three, would appear to serve, through the extraordinary distensibility of its vessels, as a kind of *diverticulum*, to relieve the vessels of the digestive viscera, when they are compressed by undue accumulation of the contents of their cavities, or when they are congested by obstruction to the flow of blood through the liver or the heart; but no such mechanical arrangement can be fairly supposed to be the chief purpose of such a complicated organ in the economy. The spleen has been repeatedly removed, without any obviously injurious consequences; whence it appears, either that its function is not of vital importance, or (which is more likely) that it is discharged by some other organ in its stead. In some of the instances in which animals have been allowed to survive longest after removal of the Spleen, the *lymphatic glands* of the neighbourhood have been found greatly enlarged and clustered together, so as nearly to equal the original spleen in volume; and hence it appears to be a fair inference, that the elaborating function of the spleen corresponds closely to that of the lymphatic glands.*

566. Having thus traced the steps by which the Blood is elaborated and prepared for circulation through the body, and having in an earlier part of this treatise (CHAP. IV.) examined into the characters of its chief constituents, we have now to consider the fluid as a whole, to study the usual proportions of these constituents, and the properties which they impart to it.—The Blood, whilst circulating in the living vessels, may be seen to consist of a transparent, nearly colourless fluid, termed *Liquor Sanguinis;* in which the *Red Corpuscles,* to which the Blood of Vertebrata owes its peculiar hue, as well as the *White* or *Colourless* corpuscles, are freely suspended and carried along by the current.—On the other hand, when the blood has been drawn from the body, and is allowed to remain at rest, a spontaneous coagulation takes place, separating it into clot and serum. The clot is composed of a network of *Fibrine,* in the meshes of which the *Corpuscles,* both red and colourless, are involved;

* In his article *Spleen* in the "Cyclopædia of Anatomy and Physiology," Prof. Kölliker endeavours to show that one function of the Spleen is to dissolve and melt down the effete blood-corpuscles, in preparation for the secreting action of the liver. He has since, however, considerably modified this view, having met with similar collections of disintegrating blood-corpuscles elsewhere. See Kölliker's and Siebold's "Zeitschrift" for 1849.

and the serum is nothing else than the liquor sanguinis deprived of its Fibrine. When the Serum is heated, it coagulates, showing the presence of *Albumen*. And if it be exposed to a high temperature, sufficient to decompose the animal matter, a considerable amount of earthy and alkaline *salts* remains.—Thus we have four principal components in the Blood;—namely, Fibrine, Albumen, Corpuscles, and Saline matter. In the *circulating* Blood they are thus combined :*—

Fibrine Albumen Salts	In solution, forming Liquor Sanguinis.
Red Corpuscles,—Suspended in Liquor Sanguinis.	

But in *coagulated* blood they are thus combined :—

Fibrine Red Corpuscles	Crassamentum or Clot
Albumen Salts	Remaining in solution, forming Serum.

The solid matter of the blood also contains various *Fatty* substances, which may be removed from it by ether. Some of these appear to correspond with the constituents of ordinary Fat (§ 189); whilst another contains phosphorus, and seems allied to the peculiar fatty acids of Neurine (§ 243); and another has some of the properties of Cholesterine, the fatty matter of the Bile (§ 590).—Besides these, there are certain substances known under the name of *Extractive;* one group of which is soluble in water, and another in Alcohol. Of the precise nature of these, little is known. They have been aptly termed 'ill-defined' animal principles; and it is probable that they may include various substances in a state of change or disintegration, which are being eliminated from the Blood by the processes of Excretion.

567. The proportion of these components varies greatly in the different classes and orders of Animals. Thus in Man, in a state of health, we may reckon the whole solid matter of the blood at about 212 parts in 1000; the proportion of the several components averaging nearly as follows :—

	Fibrine	2
	Corpuscles	130
Solids of Serum	Albumen	70
	Extractive Matters and Salts	8
	Fatty Matters	2

In Carnivorous Mammalia, the average proportion of corpuscles is rather greater, being nearly 150; whilst that of the solids of the serum is somewhat less, the fibrine remaining the same. On the other hand, in Herbivorous Mammalia, the proportion of corpuscles is considerably less, being reduced to about 95; that of the solids of the serum is about the same, but that of the fibrine seems to be above the human average. In Birds, the total amount of solid matter is greater than in the average of Mammals; the excess being especially in the Red Corpuscles. In the cold-blooded Vertebrata, on the other hand, the proportion of solid matter, and especially of the Corpuscles, is greatly reduced; and the blood is

* The corpuscles of Frog's blood may be separated from the liquor sanguinis by filtration; but this experiment cannot be performed with Human blood, because its corpuscles are small enough to pass through the pores of any filter that allows the liquor sanguinis to permeate it.

manifestly paler and thinner.—In the blood of Man and the Mammalia in general, the Colourless Corpuscles usually bear a small proportion to the Red; and being nearly of the same character, they have until recently attracted but little notice. In Reptiles and Fishes, however, they differ so much in form, size, and general appearance, that they force themselves on the attention, even whilst the blood is moving through the capillaries; and they are the more easily watched, owing to the comparatively small number of Red corpuscles, to which they are found to bear an increasing proportion as we descend through the lower orders of the class of Fishes, until we come at last to the *Amphioxus*, in whose blood the red corpuscles are altogether wanting (§ 321). Among the Invertebrata, there are no decidedly *coloured* corpuscles, although, in their form and grade of development, some of the colourless corpuscles which their blood contains, present an approximation to the early condition of the red corpuscle in the Oviparous Vertebrata. According to Mr. Wharton Jones,* who has attentively studied the characters of the Blood-corpuscles in different groups of animals, their progressive stages of development are as follows:—

1. Coarse granule-cell.
2. Fine granule-cell.
3. Colourless nucleated cell.
4. Coloured nucleated cell.
5. Free cellæform nucleus.

The first and second of these stages are seen in the blood of Invertebrata generally, and in the lymph and blood of Vertebrata. The third presents itself in the blood of the higher Invertebrata; the number of such cells being greater, the closer the approximation to the Vertebrated series; and it is seen also in the blood of the oviparous Vertebrata, as a transition-stage between the lymph-globule and the red corpuscle. The fourth is the highest form which the red corpuscle attains in the Oviparous Vertebrata; and this may be occasionally seen in the blood of adult Mammalia, as a transition-stage between the colourless nucleated cell and their peculiar red corpuscles. These transition-stages may be best seen, however, in the embryo. The non-nucleated red corpuscles of Mammalia are regarded by Mr. Wharton Jones as the escaped nuclei of the preceding, which have undergone development into cells; but the account of them already given (§ 175) seems much the more probable one.—Of the relative amount of Fibrine in the blood of different animals, no other estimate has yet been formed, than that which rests upon the coagulating power of the liquids. This coagulating power is much weaker in cold-blooded than in warm-blooded Vertebrata; and it is still weaker in Invertebrata, in many of which the clot that separates itself from the serum is rather due to the aggregation of the corpuscles, than to the fibrillation of the plasma.—Of the relative proportions of the other constituents of the blood, little is certainly known.

568. It cannot be doubted that the due admixture of *all* the components of the Blood, is necessary to the regular performance of its actions. But the *vital* properties of the fluid are still more immediately dependent upon the *Fibrine* it contains; since, as we have seen reason to

* "Philosophical Transactions," 1846.

believe, it is the material which is most completely prepared for organisation, and which supplies what is requisite for the nutrition of the larger proportion of the solid tissues of the body. It is, therefore, continually being withdrawn from the blood by the nutritive operations; and the demand appears to be supplied, in part by the influx of that which has been prepared in the Absorbent system, and in part by the continued transformation of albumen which takes place during the circulation of the blood. — The *Albumen* of the Blood is the raw material, at the expense of which not only the Fibrine, but many other substances, are generated during the nutritive process. All the Albuminoid compounds of the Secretions, the Horny matter of the Epidermic tissues, the Gelatine of the simple Fibrous tissues, and the Hæmatine of the Red Corpuscles, may be regarded as almost certainly produced by the transformation of the Albumen of the Blood; and a continual supply of this from the food is therefore requisite, to preserve its due proportion in the circulating fluid.—The *Red Corpuscles* in their fully developed state appear to be more connected with the function of Respiration than with that of Nutrition (§176); and the stimulating action of Arterial blood, especially upon the Nervous and Muscular tissues, seems to depend upon their presence. It appears from the experiments of Dieffenbach on transfusion, that the Red Corpuscles are more effectual as stimuli to the Heart's action, than is any other constituent of the blood. The rapidity with which they may be decomposed and reconstituted, is made remarkably evident by the experiments of Magendie; who found that, when the blood of one animal was injected into the veins of another having discs of very different size and form (care being taken to prevent the coagulation of the Fibrine during the operation), the Red particles so introduced soon disappeared, and were replaced by those characteristic of the species in whose veins the fluid was circulating. — The use of the *Saline* matter is evidently in part to supply the mineral materials, requisite for the generation of the tissues, and for the production of the various secretions. Further, it is by the Saline and Albuminous matters in conjunction, that the specific gravity of the Liquor Sanguinis is kept up to the point, at which it is equivalent to that of the contents of the Red corpuscles; and it is only in this condition, that the latter present their proper characters (§ 173).—The *Fatty* matters of the Blood are evidently derived from the food, either directly, or by the transformation of its farinaceous ingredients; and while they are employed, as we have seen, in the ordinary operations of cell-formation, a special demand for them must be created in the nutrition of the Nervous tissue; and they are also continually drawn upon, in warm-blooded animals, for the maintenance of the combustive process. The quantity of them contained in the blood is subject to great variation; for after a full meal including a considerable proportion of oleaginous matter (especially if preceded by a long fast), a great increase takes place, and the serum is often rendered opaque by the introduction of the 'molecular base' of the chyle; but after a few hours this excess is brought down again, through the elimination of fatty matters by the nutritive, secretory and combustive processes.

569. The new arrangement of the elements of the blood, which takes place when it is withdrawn from the living body and left to itself, or even within the body when its vitality is lost, consists chiefly in the passage of

the Fibrine from the state of solution to that of a fibrous network, in the meshes of which the Corpuscles are included. The rapidity of this change, and the completeness of its result, vary considerably; and usually in an inverse ratio to each other. When the fibrine is imperfectly elaborated, so that the coagulum is deficient in firmness, the process usually takes place rapidly, and is soon completed; but when it has undergone a higher degree of assimilation, so that it transforms itself into a definite fibrous tissue, the process is comparatively slow. The most perfect fibrillation is seen, not in blood itself, but in those effusions of modified liquor sanguinis, known as 'plastic' or 'organizable lymph,' which are thrown out in the inflammatory process, or are effused in the first stage of the reparative operation consequent upon an injury. This may be due in part to the higher elaboration of the fibrine itself; and in part to the influence of vital force imparted from the living parts around (§ 556). If the influence of this living surface be continued, and vessels shoot from it into the primitive tissue thus formed, it becomes part of the organised fabric; and thus we see that coagulation is not an indication of the *death* of the plastic fluid, but is a stage in its metamorphosis into a living solid. The tissue thus formed, however, is of a low order, and very prone to degenerate. When the Blood has been withdrawn from the body, so that the coagulation takes place without any influence from a living surface, the fibrillated mass soon passes into decomposition; so that the process may then be considered as the last act of the life of the vital fluid. Even under the most favourable circumstances, it does not seem that the plastic force of the blood, or of the plasma effused from it, is able to develope any form of tissue higher than cells or simple fibres; for these are the sole kinds of organised structure, which ever directly result from the development of the 'nucleated blastema' thrown out for the repair of injuries. And under less favourable circumstances, the same material resolves itself into a substance, *pus*, of far inferior character, in which no further organisation can ever take place, and which is only fit, therefore, to be cast out of the system.

570. In the foregoing account of the formation, composition, and properties of the Blood, it has been shown that the albuminous materials obtained from the aliment, or received back from the body itself, are gradually prepared for organisation, whilst yet remaining in the fluid state, until they have reached the condition of Fibrine; and that in this condition, they have been so far modified by the vitalising influences to which they are subjected, that they are capable of spontaneously passing into a state of incipient organisation. The production of all the higher forms of organised tissue, however, is manifestly dependent, not upon the properties of the plasma alone, but upon the Formative powers of the tissues themselves; each tissue, from the time when it first presents its characteristic form in the embryo, continuing to grow, develope, and maintain itself, at the expense of the materials which it draws from the blood. To the statements already made (Sect. 1) in regard to the general conditions under which this function is performed, it will be only requisite here to add certain facts which have reference solely to the Animal kingdom. It has been shown (§ 557) that in the Nutrition of the parts concerned in the maintenance of Organic life alone, the quantity of new tissue produced (the requisite amount of formative power being supplied) depends chiefly

upon the quantity of material specially adapted for its generation. But in the case of the tissues which minister to the Animal functions, a different rule seems to hold good; for in these we find that the activity of reparation is dependent upon the degree of previous waste or disintegration, caused by the performance of their peculiar operations; so that, if they be kept in complete inactivity, no nutritive changes can take place in them, notwithstanding the greatest abundance of their respective *pabula*, and degeneration immediately commences; whilst, if their functions be performed with unusual energy, so that an increased amount of 'waste' takes place, this waste is more than repaired by the nutritive activity of the part (provided that a sufficient amount of duly prepared material be supplied), and a positive *increase* of its substance, with a like increase of functional power, is the result. It is a very remarkable circumstance that the nutrition of *bones*, at least in those situations in which they serve to afford fixed points of attachment to muscles, rather than to support and protect the softer parts, is closely related to the development of the muscles; so that when this is augmented by continued activity, the bones become stronger, and their prominences and ridges more decided; whilst if it be checked by disuse, the nutrition of the bone also is impaired, and its bulk speedily diminishes.*

570 *a.* Thus the condition of nutritive activity in the tissues which are the instruments of the Animal functions, is as strikingly opposed to that which prevails in the part of the organism appropriated to the Vegetative functions, as are the conditions under which those functions are respectively performed. For, as we have seen, *Vegetative* activity of every kind is entirely *constructive*, and its conditions are merely (1) the supply of organisable matter, and (2) the requisite organising force. On the other hand, *Animal* activity is in its essential nature *destructive*, and its conditions involve the re-conversion of organised tissue into the inorganic state. But, in some mode not yet understood, the performance of this *de*structive operation excites a corresponding exertion of the *con*structive; and the tissues are renewed or even augmented. The explanation may perhaps lie in the fact, that the exercise of the endowments of the Nervous and Muscular tissues, which depends upon conditions external to themselves,† itself becomes a means of determining an increased afflux of

* Thus Dr. John Reid found that the relative weights of the bones and muscles of the two hind-legs of a rabbit, from which a portion of the sciatic nerve on one side had been removed seven weeks before, so as entirely to paralyze the limb, were as follows.

	Sound limb.	Paralyzed limb.
Muscles . .	327 grains	170 grains
Tibia and Fibula .	89 grains	81 grains.

There was an obvious difference, also, in the ultimate structure of the muscles of the two limbs; for the fibres of the paralyzed muscles were considerably smaller than those of the other side, exhibited the longitudinal and transverse striæ much less distinctly, and had a shrivelled appearance. ("Physiological, Pathological, and Anatomical Researches," p. 10.)

† The activity of Muscle must be called forth by the stimulus of innervation, or by other excitants immediately applied to itself. That of the *afferent* portion of the Nervous substance is entirely dependent upon the reception of impressions from without, so that if these be withheld it remains in a state of torpor, and its nutrition is proportionably affected,—as we see in the case of the eye, when the access of light to the retina is completely prevented by opacity of the cornea. In like manner, the activity of the central organs is dependent upon the excitement which they receive through the afferent nerves, or from mental operations;

blood to them on the principles already laid down (§ 481); and it is on the increase of vegetative activity which thus takes place, that the augmented nutrition seems to depend. On the other hand, when these tissues are not called into action, the circulation through them becomes extremely languid, and their nutrition suffers accordingly.

571. In addition to the foregoing peculiarities, it is to be remarked that the activity of nutrition among higher Animals varies greatly in proportion to the age of the individual. Among Plants, one part of the fabric may die, whilst another is vigorously growing and extending itself, the two having no relation of mutual dependence; and the same may happen among Zoophytes. But among Animals, we find, as we ascend the series, that as the individual parts become more and more intimately dependent upon one another, the nutritive activity of each is more and more closely related to that of the remainder, and consequently to that of the organism at large. Moreover, we find that *growth* consists less in mere addition to the parts already formed (as it does for the most part in Plants), but more and more in interstitial enlargement; and this involves a continual remodelling of the entire fabric. Thus all the tissues, even those most consolidated, are undergoing continual changes in the young animal; and it would seem as if the duration of their component parts is much less than in more advanced life, so that disintegration and renewal more rapidly succeed each other (§ 377). Here, as elsewhere, we find that *duration* and *functional activity* are in the inverse ratio to each other. The dull perceptions, and slow and feeble movements of the aged individual, are in no less striking contrast with the acute sensibility, and the rapid and vigorous muscular actions, of the child, than is the sluggishness of the interstitial changes in the former, as compared with their energy and activity in the latter. Hence it may be stated as a general fact, that the vital activity of the component parts of the organism diminishes with the prolongation of the general life of the whole.

571 *a*. That the function of Animal Nutrition is capable of being considerably *modified* by the influence of the Nervous system, cannot be doubted by any one who duly considers the facts which have been brought to light by experiment and observation. But still there appears to be no sufficient reason for regarding the ordinary Nutritive operations as in themselves necessarily *dependent* upon Nervous agency (§ 364). In fact, the results of injury to the nerves of a part seem to be exerted rather in a disturbance or perversion, than in an actual suspension of them. And seeing as we do, that active nutrition takes place in the embryo, long before nerves have been developed, and that it goes on in parts (*e. g.* cartilage) into which it may be said with certainty that no nerves enter, we seem to have full confirmation of the inference, which rests upon the essential conformity between the processes of Animal and Vegetable Nutri-

and thus we find that a continuance of intellectual exertion, if not too severe, tends to augment the nutrition of the brain. And the activity of the *efferent* portion of the nervous system is entirely dependent upon that of the central organs, so that the nutrition of any part of it ceases, as soon as it is separated from them. But it is a remarkable circumstance, first clearly ascertained by Dr. Waller, that if the connection of the sensory nerves with the central organs be interrupted, so that they cannot propagate onwards the impression they have received, their nutrition is immediately impaired. ("Philosophical Transactions," 1850, Part II.)

tion,—that in the one case, as in the other, the operation is effected by the formative powers of the part itself, exerted upon the plastic material supplied to it ; but that these, in the Animal, are modified and directed by Nervous agency, in such a manner as to harmonize them with each other, and with the general requirements of its organism.

CHAPTER XVI.

OF SECRETION.

1. *General Considerations.*

572. THE parallel between the Vegetable and Animal kingdoms, which has hitherto been so close, fails us when we arrive at this department of the enquiry ; for the conditions of Vegetable existence do not require that provision for the maintenance of the purity of the nutritive fluid, by the removal of effete matters, which becomes the most constant and urgent necessity of Animal life. It would seem as if all the nutritive material assimilated by the leaves of Plants, is applied to their growth and development, or is stored up as a provision for future operations of the same kind ; and the supply is so exactly proportioned to the demand (§ *557 a*), that there is seldom any unappropriated residue of assimilated matter left to become injurious by its superfluity and by its tendency to decomposition. Moreover, the parts which have fulfilled their purposes in the system, and whose term of life is expired (such as the leaves and flowers), are cast off *en masse ;* their ultimate decay, in which they return to the state of simple binary compounds, takes place after their separation ; and no such interstitial death and decomposition occur in the regular progress of Vegetable life, as we have seen to be necessary conditions of Animal activity. There is nothing, then, save Carbonic acid and Water, which needs to be cast out of the Vegetable organism ; and for the elimination of these we have found a special provision to be made. —In the Animal, on the other hand, there are numerous modes in which the Circulating fluid may be rendered impure. In the first place, a larger amount of nutriment may be introduced into the blood, than the Nutritive operations can appropriate ; for we have seen that the formative energy of those Nervo-muscular tissues, which make up by far the largest proportion of the bodies of the higher animals, has no relation whatever to the supply of food, but depends upon the exercise to which they have been subjected ; so that unless the amount of food ingested be proportionably reduced, there must remain a superfluity which soon becomes injurious by decomposition, its highly-azotized nature rendering it peculiarly liable to undergo such a change, at the elevated temperature of warm-blooded animals. Further, it seems scarcely possible that the varied demands which are made upon the nutritive fluid of higher animals, for the supply of the requisite components of the several tissues, can be so accurately adjusted under all circumstances, as not to leave some residue incapable of further use, and therefore unsuitable to be retained. But the great and constant occasion for the excretory opera-

tions, undoubtedly arises from the disintegration, to which, as so frequently mentioned, the whole apparatus of Animal life is subject as the very condition of its vital activity; and therefore the demand for their performance will vary with the degree of that activity. All that has been already said on this point under another head, is here equally applicable (§§ 506, 507). So urgent is the necessity for this process, that, as Dr. M. Hall has justly remarked, the functions of *egestion* are more immediately necessary to the maintenance of Animal life, than those of *ingestion*. This will more fully appear hereafter.

573. But besides that separation of effete matters from the blood, for the purpose of maintaining its purity, which is usually distinguished as *Excretion*, we find that certain products are elaborated from it for special purposes in the economy; and it is to the process by which this is accomplished, that the term *Secretion* in its more restricted sense is applicable. But even this has scarcely any parallel in the Vegetable kingdom. For although there is a large class of substances which are commonly designated as 'Vegetable Secretions,' yet these are not poured out upon the surface nor into the cavities of the Plant, but are stored up in its constituent cells, of which they form the characteristic contents; and thus they bear the same relation to it, as the oleaginous contents of the fat-cells, or the calcareous deposit in the cells of shells, &c., bear to the Animal organism (§ 559 *a*). We shall presently see, however, that the production of the true Secretions and even of the Excretions of Animals, is a process which is referable to the same general category as Nutrition; the separation of the several components of each, from the blood, being effected by the development of cells, to which they form the appropriate pabulum (§ 576).

2. *Secretion in Vegetables.*

574. It might seem almost superfluous to add anything under this head, to what has been just said, as to the absence of any true secreting or excreting process in Plants. But there are certain doctrines current upon this subject, to which it seems necessary to make more particular allusion.—The nearest approach to a true *Excretion* seems to be presented by those cases, in which peculiar organic compounds are exuded from the surface; as honey in flowers, wax upon leaves and fruits, gums and gum-resins upon bark, &c. In these and other similar instances, however, the exudation appears chiefly to result from the superabundance of the product in the subjacent tissues, and is not to be considered in any other light than as relieving their distention. It may be subservient at the same time, however, to some secondary purpose; thus the honey of the flowers is attractive to insects, whose services in the act of fertilization are frequently important and sometimes absolutely essential. And this, with the acrid fluids secreted in the so-called 'glands,' or agglomerations of cells, at the base of the hair of the Nettle, may be regarded as the nearest approximations exhibited by Plants to the true *Secretions* of Animals.

575. It has been imagined that Plants have the power of throwing off from their roots, substances injurious to their economy; and the matters thus separated have been regarded as real Excretions. Thus it was shown by Bonnet, that if a rapidly-growing plant have its roots divided into two bundles, and one of these be introduced into a solution of some saline

substance which it will absorb, whilst the other is immersed in pure water, the latter fluid will be found after a few days to have received a certain impregnation from the former; showing that the acetate of lead must have been taken up through one set of roots, and cast out again through the other. Again, if a *Leguminous* plant be placed in distilled water, the fluid will be found in a few days strongly impregnated with mucilaginous matter, which it has derived from the roots immersed in it. So the *Cichoraceæ* and *Papaveraceæ* exude a large quantity of brownish bitter matter analogous to opium; *Euphorbiaceæ*, an acrid gummy-resinous matter; and so on,—the substances exuded being in all instances those which are characteristic of the proper juices of the plants which furnish them. Hence it seems fair to conclude, that this exudation is a simple act of *exosmose*, necessarily connected with the endosmose by which water is taken up into the organism (§ 423). The existence of such exudations, however, occasionally comes to be a matter of much practical importance. It has probably something to do with the fact, that the growth of particular species of plants favours the subsequent development of other kinds; whilst there are other species, whose growth seems to leave the soil in a much worse condition than before. Now there can be little doubt that much of the latter result is to be attributed to the removal of particular ingredients from the soil, which render it less fitting to sustain the continued growth of its own kind, whilst it may still be capable of affording the requisite support to such as need some different materials; and it is probably by this principle that we are to explain the well known fact in forestry, that where a wood principally composed of one species of timber-trees has been cleared (by fire or otherwise), the trees which spring up spontaneously and supply the place of the former growth are for the most part of a different species; and the success of the 'rotation of crops' in agriculture seems chiefly attributable to the same cause. But this is not all; for it is well known that the soil is more benefited by the growth of a crop of certain Leguminous plants, such as Trefoil, than it would be by lying absolutely fallow; and this can scarcely be due to anything else, than its impregnation with gummy exudations from the roots of these plants. On the other hand, there are exudations which are positively injurious; this has been proved to be the case with tannin, which is given off by the roots of the oak, and which, even in very minute proportion, destroys the vitality of the growing tissues of the spongioles of other plants; thus affording an explanation of the well-known fact, that trees transplanted into a soil in which oaks have previously grown, seldom flourish, and generally die. And it seems not unlikely, too, that Poppies and other 'rank' weeds, do more damage by communicating narcotic exudations to the soil, which deaden the vital powers of other plants growing in it, than by any exhaustive influence exerted by them.

3. *Secretion in Animals.*

576. We have seen that, in the process of Animal Nutrition, the circulating current not only *deposits* the materials which are required for the renovation of the solid tissues; but also *takes back* the substances which are produced by the decay of these, and which are destined to be thrown off from the body. We have also seen that it supplies the materials of

certain fluids, which are separated from it to effect various purposes in the economy;—such as the Salivary and Gastric fluids, which have for their office to assist in the reduction of the food. Now the process, by which the fluids of the latter kind are separated from the Blood, is precisely the same in character as that, by which the products of decay are eliminated from it; and the structure of the organs concerned in the two is essentially the same. Hence both processes are commonly included under the general term *Secretion;* which, considered in its most general point of view, may be applied to *every* act, by which substances of any kind are *separated* from the blood; but which is usually restricted to those cases, in which the substances withdrawn are not destined either to be restored again to the circulating current (as in Assimilation), or to form part of the textures of the living fabric (as in Nutrition); and in which the separated product has a liquid or more rarely a solid form, and not a gaseous (as in Respiration). Viewed under this limitation, the essential character of the true *Secreting* operation seems to consist,—not in the nature of the action itself, for this is identical with those of Assimilation and Nutrition, being a process of cell-growth,—but in the position in which the cells are developed, and the mode in which the products of their action are afterwards disposed of. Thus the cells at the extremities of the Intestinal villi (if such there be, § 427), the cells of which the Adipose tissue is made up (§ 189), and the cells of which the greater part of the substance of the Liver is formed (§ 587), all have an attraction for fatty matter; and draw it from the neighbouring fluids, at the expense of which they are developed, to store it up in their own cavities. But the cells of the first kind, when they have come to maturity, set free their contents, which are delivered to the absorbent vessels, to be introduced into the circulating current:—those of the second kind seem more permanent in their character, and retain their contents, so as to form part of the ordinary tissues of the body, until they are required to give them up;—whilst the cells of the third class cast forth *their* contents into the hepatic ducts, by which they are carried into the intestinal canal, whence a portion of them at least is directly conveyed out of the body. It is, then, in the *position* of the Secreting cells,—which causes the product of their action to be delivered upon a *free surface*, communicating, more or less directly, with an external outlet,—that their distinctive character depends. All the proper *Secretions* are thus either poured out upon the exterior of the body, or into cavities provided with orifices that lead to it. Thus we have seen that a very large quantity of fluid, containing a considerable quantity of solid matter, of which it is desirable that the system should be freed, is carried off from the Cutaneous surface. Another most important secretion, containing a large quantity of solid matter, and serving also to regulate the quantity of fluid in the body,—namely, the Urinary,—is set free by a channel expressly adapted to convey it directly out of the body. The same may be said of the Mammary secretion; which is separated from the blood, not to preserve *its* purity, nor to answer any purpose in the economy of the individual, but to afford nutriment to another being. And of the matters secreted by the very numerous glandulæ situated in the walls of the Intestinal canal, a great part are obviously poured into it for no other purpose, than that they may be carried out of the body by the readiest channel.

577. The cells covering the simple membranes that form the free surfaces of the body, whether external or internal, are all entitled to be regarded as *secreting* cells; since they separate various products from the blood, which are not again to be returned to it. But the secreting action of some of these seems to have for its object the *protection* of the surface; thus the Epidermic cells secrete a horny matter, by which density and firmness are imparted to the cuticle (§ 177); whilst by the epithelial cells of the Mucous membranes of the alimentary canal and of other parts (§ 168), their protective mucus seems to be elaborated. But in general we find that special organs, termed *Glands,* are set apart for the production of the chief secretions; and we have now to consider the essential structure of these organs, and the mode of their operation.—A true Gland may be said to consist of a closely-packed collection of follicles, all of which open into a common channel, by which the product of the glandular action is collected and delivered. The follicles contain the secreting cells in their cavities; whilst their exterior is in contact with a network of blood-vessels, from which the cells draw the materials of their growth and development (Fig. 228). In any one of the higher animals, we may trace out a series of progressive stages of complexity, in the various glands included within their fabric; and in following any one of the glands that attain the highest degree of development (such as the Liver or Kidney), through the ascending scale of the Animal series, we shall be able to trace a very similar gradation from the simplest to the most complex form. Hence we see that there can be nothing in the *form* or *disposition* of the components of the glandular structure, which can have any influence upon the character of the secretion it elaborates; since the very same product,—*i. e.* the Bile, or the Urine,—is found to issue from nearly every variety of secreting structure, as we trace it through the different groups of the Animal kingdom. The peculiar power, by which one organ separates from the blood the elements of the Bile, and another the elements of the Urine, whilst a third merely seems to draw off a certain amount of its albuminous and saline constituents, is obviously the attribute of the ultimate secreting cells, which are the real agents in the secreting process. *Why* one set of cells should secrete bile, another urea, and so on, we do not know; but we are equally ignorant of the reason, for which one set of cells converts itself into Bone, another into Muscle, and so on. This variety in the endowments of the cells, by whose aggregation and conversion the fabric of the higher Animals is made up, is a fact which we cannot explain, and which must be regarded (for the present, at least,) as one of the "ultimate facts" of Physiological Science.

FIG. 228.

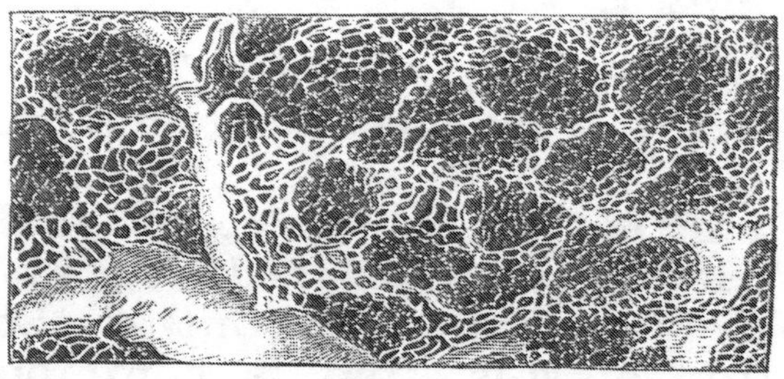

Capillary network around the follicles of Parotid Gland.

578. Passing by the extended membranous surfaces, and the protective or secreting cells with which they are covered, we find that the simplest form of a secreting organ is composed of an *inversion* of that surface into follicles, which discharge their contents upon it by separate orifices. Of this we have a characteristic example in the *gastric* follicles,

even in the higher animals; the apparatus for the secretion of the Gastric fluid never attaining any higher condition, than that of a series of distinct follicles, lodged in the walls of the stomach, and pouring their products into its cavity by separate apertures. In Fig. 229, A, is

FIG. 229.

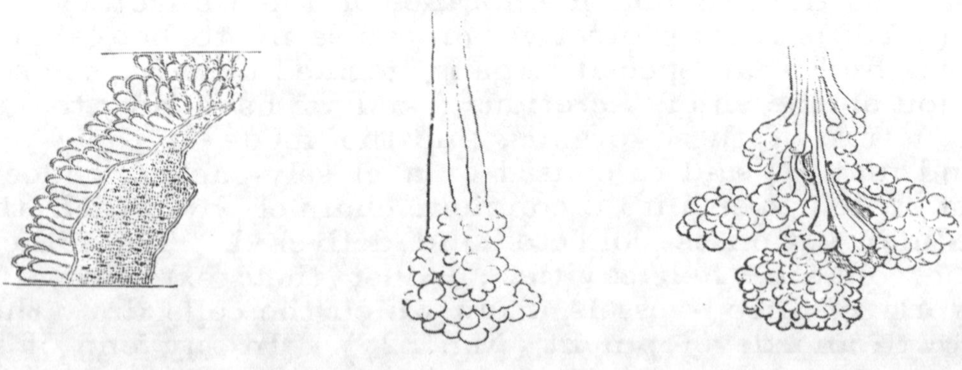

A, Glandular follicles in *Ventriculus Succenturiatus* of *Falcon*:—B, Gastric gland from the middle of the *Human Stomach*;—C, a more complicated form in the neighbourhood of the pylorus.

represented a portion of the 'ventriculus succenturiatus' of a Falcon, in which the simplest form of such follicles is seen. A somewhat more complex condition is seen in some of the Gastric follicles of the Human stomach (B, C); the surface of each follicle being further extended by a sort of doubling upon itself, so as to form the commencement of secondary follicles, which open out of the cavity of the primary one.—Now a condition of this kind is common to *all* glands, in the first stage of their evolution; and in this stage we meet with them, either by examining them in the lowest animals in which they present themselves, or by looking to an early period of their embryonic development in the highest. Thus, for example, the *Liver* consists, in certain Polypes and in the lowest Mollusca, of a series of isolated follicles, lodged in the walls of the stomach, and pouring their product into its cavity by separate orifices; these follicles being recognized as constituting a biliary apparatus, by the colour of their secretion. And in the Chick, at an early period of incubation, the condition of the Liver is essentially the same with the preceding; for it consists of a cluster of isolated follicles, not lodged in walls of the intestine, but clustered round a sort of bud or diverticulum of the intestinal tube, which is the first condition of the hepatic duct, and into which they discharge themselves (Fig. 239). So, again, the *Mammary Gland* presents an equally simple structure in that lowest type of the Mammalian class, the *Ornithorhyncus*; for it consists of nothing else than a cluster of cæcal follicles, discharging their contents by separate orifices upon the surface of the mammary areola. And in like manner, the *Pancreas* makes its appearance in many Fishes in the condition of a

FIG. 230.

Mammary Gland of *Ornithorhyncus*.

group of more prolonged cæcal follicles, clustered round the commencement of the intestinal canal, and partly coalescing with each other before they enter it (Fig. 231). A still simpler condition of the pancreas is found in the Cephalopoda (§ 307 *b*), and in the Cyclostome Fishes. Thus, it may be stated as a general rule, that every gland presents itself under this most elementary state, alike in the lowest grade of the animal series in which it is first traceable, and at its earliest appearance in the embryo of the higher.—The next grade of complexity is seen, where a cluster of the ultimate follicles open into one common duct, which discharges their product by a single outlet; a single gland often containing a number of such clusters, and having, therefore, several excretory ducts. A good example of such a condition, in which the clusters remain isolated from one another, is seen in the Meibomian glands of the eyelid; each of which consists of a double row of follicles, set upon a long straight duct, that receives the products of their secreting action, and pours them out upon the edge of the eyelid. And of the more complex form, in which a number of such clusters are bound together in one glandular mass, we have an illustration in the accessory glands of the genital apparatus in several animals, which discharge their secretion into the urethra by numerous outlets; or in the Mammary glands of Mammalia in general, the ultimate follicles of which are clustered upon ducts that coalesce to a considerable extent, though continuing to form several distinct trunks even to their termination. Such glands may be subdivided, therefore, into *glandulæ* or *lobules*, that remain entirely distinct from each other.—In the highest form of Gland, however, all the ducts unite, so as to form a single canal, which conveys away the products of the secreting action of the entire mass. This is the condition in which we find the Liver to exist in most of the higher animals; also the Pancreas, the Parotid Gland, and many others. In some of these cases, we may still separate the gland into numerous distinct lobules, which are clustered upon the excretory duct and its branches, like grapes upon a stalk (Fig. 232); in others, however, the branches of the excretory duct do not confine themselves to *ramifying*, but *inosculate* so as to form a network, which passes through the whole substance of the gland, and which connects together its different parts, so as to render the division into lobules less distinct. This seems to be the case in regard to the Liver of the higher Vertebrata.

FIG. 231.

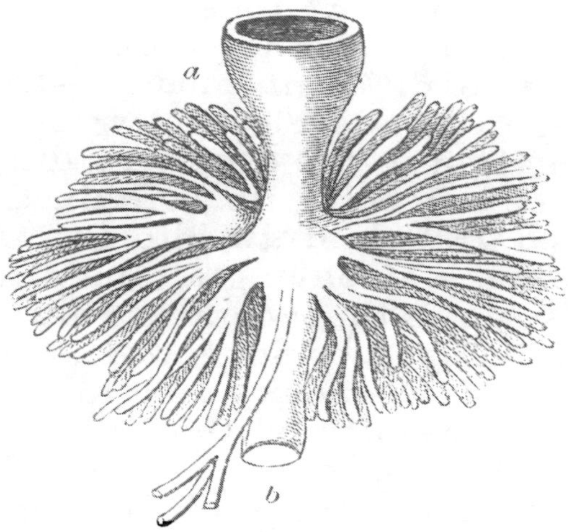

Rudimentary Pancreas from *Cod*:—*a*, pyloric extremity of stomach; *b*, intestine.

FIG. 232.

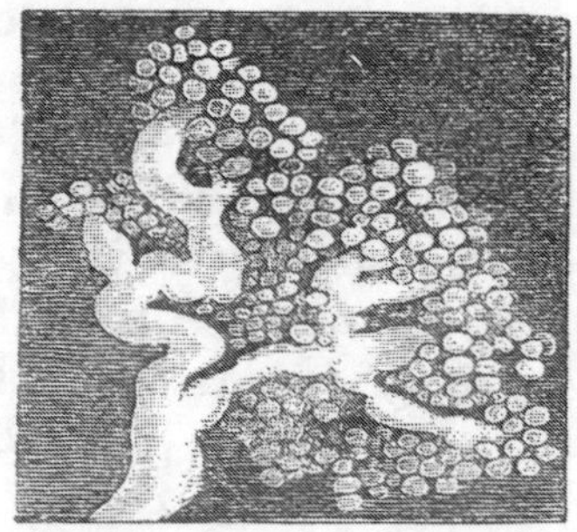

Lobule of Parotid Gland of *Human Infant*, filled with mercury; magnified 50 diameters.

579. Whatever degree of complexity, however, prevails in the general arrangement of the elements of the Glands in higher animals, these ele-

ments are themselves everywhere the same; consisting of *follicles* or *tubuli* that enclose the real secreting cells (Figs. 233 and 234). Now from the history of the development of Glands in general, it appears that the *follicles* may be considered as *parent-cells;* and that the *secreting* cells in their interior constitute a *second* generation, developed from the nuclei or germinal spots on the walls of the first. It has been pointed out by Prof. Goodsir,* that the continued development and decay of the glandular structure, in other words, the elaboration of its secretion, may take place in two different modes.—In one class of Glands, the parent-cell, having begun to develope new cells in its interior, gives way at one point, and bursts into the excretory duct, so as to become an open follicle, instead of a closed cell: its contained or secondary cells, in the progress of their own growth, draw into themselves the matter to be eliminated from the blood, and, having attained their full term of life, burst or liquefy, so as to discharge their contents into the cavity of the follicle, whence they pass by its open orifice into the excretory duct: and a continual new production of secondary cells takes place from the germinal spot, or nucleus, at the extremity of the follicle, which is here a permanent structure. In this form of gland, we may frequently observe the secreting cells existing in various stages of development within a single follicle; their size increasing, and the character of their contents becoming more distinct, in proportion to their distance from the germinal spot (which is at the blind termination of the follicle), and their consequent proximity to the outlet (Fig. 234). In some varieties of such glands, however,—as in the greatly-prolonged follicles, or tubuli uriniferi, of the kidney,—the production of new cells does not take place from a single germinal spot at the extremity of the follicle, but from a number of points scattered through its entire length.—In the second type of Glandular structures, the parent-cell does not remain as a permanent follicle; but, having come to maturity and formed a connection with the excretory duct, it discharges its entire contents into the latter, and then shrivels up and disappears, to be replaced by newly-developed follicles. In each parent-cell of a gland formed upon this type, we shall find all its secondary or secreting cells at nearly the same grade of development; but the different parent-cells, of which the parenchyma of the gland is composed, are in very different stages of growth at any one period, some having discharged their contents and being in progress of disappearance, whilst others are just arriving at maturity and connecting themselves with the excretory duct; others exhibiting an earlier degree of development of the secondary cells; others presenting the latter in their incipient condition; whilst others are themselves just starting into existence, and as yet exhibit no traces of a second generation.—The former seems to be the usual type of the ordinary Glands; the latter is chiefly, if not entirely, to be met with in the Spermatic glands.

580. The purposes answered by the function of *Secretion* we have seen to be two-fold; namely, the separation of some material from the circulating fluid, which would be injurious to the welfare of the system if retained in it; and the elaboration of a product which is destined to some other use in the economy. Now it is probable that in almost every

* "Anatomical and Pathological Observations," chap. xv.

act of secretion, both these purposes are in some degree served; the blood being freed from some ingredient whose accumulation would be superfluous, if not more positively injurious; and the matter separated having some secondary purpose to answer. Thus, whilst biliary matter becomes a positive poison if it be retained in the blood, it contributes, when poured into the duodenum, to complete the digestive process, and to prepare the nutrient contents of the intestinal canal for absorption. So, again, the cutaneous exhalation not only removes the superfluous water of the blood, but is one of the chief means of regulating the temperature of the body. And even the urine, which seems to be eliminated merely for the removal of the products of the disintegration of the tissues from the circulating current, is sometimes made to serve an additional purpose; its acridity, or its peculiarly offensive odour (increased under the influence of terror), frequently rendering it an effectual means of defence. On the other hand, the secretions which are separated from the blood for the purpose of discharging some important office in the economy, usually, if not always, contain some substances whose retention in the blood would be injurious, and which are therefore advantageously got rid of through this channel. Thus the Salivary, the Gastric, and the Pancreatic fluids, all contain an animal principle nearly allied to albumen; but this principle seems to be in a state of change, or of incipient decomposition; and it would not seem improbable, that whilst this very condition renders the albuminous matter useful in promoting the reduction of the aliment, it renders it unfit to be retained in the circulating current. So, again, the albuminous matter of Milk appears to have undergone a change, whilst yet within the sanguiferous system, which renders it unfit to be retained within the maternal blood; for we find that when the Mammary secretion is checked, the peculiar substance termed *Kiestine*, which is nearly related to caseine, and which was eliminated by the kidneys during pregnancy, again passes off in the urinary excretion.—It is impossible, therefore, to divide the secreted products strictly, as some Physiologists have attempted to do, into the *excrementitious* and the *recrementitious*; that is, into those which are purely excretory in their character, and those which are subservient to further uses in the economy; since most, if not all of them, partake more or less of both characters. Still it is convenient to group them, for practical purposes, according to the *predominance* of one or other of the objects first mentioned; those being regarded as *excretions*, in which the depuration of the blood is manifestly the chief end, any other purpose being rendered subservient to this; and those being ranked as *secretions*, in which the ulterior purpose of the separated fluid would seem to be the principal occasion of its production.

581. Another classification has been proposed, of which the foundation is the degree of resemblance which the secreted products bear to the normal constituents of the blood: those being associated into the first group, in which the characteristic ingredients are altogether unlike those of the nutritive fluid; whilst a second group is formed of those, whose elements seem nearly allied to the components of the blood. This classification is, in fact, almost identical with the preceding; for, as a general rule, all the cases in which the secreted products are very unlike the constituents of the blood, are those in which they are most directly and speedily removed from the body; whilst those in which they serve some

ulterior purpose, are for the most part also those, whose elements differ least from the components of the blood.—The products of the first class may be said to have their immediate origin in the disintegration of the tissues; and we find the amount in which they are generated, to be in close relation to the operation of the various causes, which tend to increase or diminish that disintegration (§§ 374–377). They act all of them as *poisons*, if retained within the system and allowed to accumulate even for a short time; and after their excretion, they speedily resolve themselves into simple combinations of inorganic elements. This is the case, for example, with urea, biliary matter, and the putrescent portion of the fæces; all of which may be regarded as the products of the metamorphosis of the tissues, on their way towards their restoration to the inorganic universe (§ 260); and it would seem to be *because* they have undergone this complete metamorphosis, that their presence in the circulating current is not only useless but injurious. We may consider that portion of the carbonic acid thrown off by Respiration, which is the result of the disintegration of the tissues (§ 507), as the most important of these excretory products, being the one whose briefest accumulation gives rise to the most pernicious results; but the accumulation of biliary matter or of urea, caused by the *complete* stoppage of the hepatic and renal excretions, usually proves fatal in from one to two days. Of the products of the first class it may be said, further, that many of them may be detected in the circulating blood, on account of their marked dissimilarity in chemical properties to its other constituents; but their quantity is normally extremely minute, since they are eliminated almost immediately that they are generated. But if any such secretory operation be checked, then the product which it is destined to remove speedily makes its presence apparent in the blood; being alike detectible by chemical analysis, and recognisable by the symptoms of poisoning to which it gives rise. — On the other hand, the products of the second class seem to be supplied rather from the materials of the blood itself, than from the disintegration of the tissues; and usually bear such a close relation to the normal constituents of the circulating fluid, as not to be distinguishable from them. Such are the elements of the Salivary, Gastric, Pancreatic, Lachrymal, and Mammary secretions. Hence, when these secretions are checked, the consequences are more apparent in the suspension of the function to which they are especially subservient, than in that general disorder of the system which indicates a contamination of the blood.

582. As it would be quite foreign to the purpose of this Treatise, to enter into a comparative examination of all the Secreting structures which present themselves in the different parts of the Animal series, it will be advantageous to restrict ourselves for the most part to two sets of organs,—the Biliary and Urinary—which are most generally distributed, and which present themselves under the greatest variety of forms and conditions. In the examination of these, we shall find ample illustration of the great principles of *specialisation* and *concentration* already so frequently referred to; and shall also meet with some remarkable examples of that interchange of function, which occasionally takes place even in the most complex organisms.

583. *The Liver and the secretion of Bile.*—There are few animals possessed

of a distinct digestive cavity, in which some traces of a biliary apparatus (recognizable by the colour of the secretion) cannot be distinguished. Thus in the *Hydra*, cells containing a yellowish-brown matter may occasionally be seen in the lining of the stomach, into the cavity of which they probably discharge their contents, by the rupture or solution of their own walls. Secreting cells of a similar kind are more distinctly seen in furrows, formed by duplicatures of the lining membrane of the stomach of the *Actinia* (Fig. 98). In the *Laguncula* and other *Bryozoa* (Fig. 114), very distinct spots may be seen in the parietes of the stomach, which seem to be composed of clusters of biliary cells contained within follicles; and during digestion, the contents of the stomach are seen to be tinged with a rich yellow-brown hue, derived from the matter discharged from these follicles. In the *Earthworm*, again, the large annulated alimentary canal is completely encased in a flocculent external coating, which consists of a mass of minute flask-shaped follicles, filled with cells; and several of these coalesce to discharge their contents by a common orifice into the digestive cavity.—Passing a little higher, as well in the Radiated, as in the Articulated and Molluscous groups, we find the hepatic cells clustering, not immediately around the digestive stomach, but around cæcal prolongations of this, which thus present us with the first approximation to the condition of the separate glandular mass, which we meet with in the higher animals. Thus in the *Asterias* (Fig. 110), the radiating extensions of the stomach have their walls dilated into numerous culs-de-sac; and these are lined with large glandular cells, whose colour and aspect indicate their hepatic character. So in the *Leech* and many other *Annelida*, whose digestive cavity is more or less sacculated (Fig. 198), the walls of these sacculi are covered with biliary cells, as are also, although in a less degree, those of the central canal. In the *Pycnogonidæ*, again, the only trace of hepatic organs is to be found in the biliary cells, which are dispersed through the walls of the cæca prolonged into their limbs (Fig. 158); and it is apparently in the same diffused condition, that the biliary apparatus exists in the *Acaridæ* (§ 319 *d*). The most remarkable example of this type of structure in the Molluscous series, is presented by certain of the *Nudibranchiate* Gasteropods (§ 304 *c*), in which, notwithstanding the high development which the liver attains in most other orders of the class, it is reduced to the condition of clusters of simple cells arranged round cæcal prolongations of the stomach. In the *Eolis* and its allies, the stomach gives off on either side a number of branches, which usually re-divide, and then give off smaller tubes, which are continued into the numerous papillæ that clothe the dorsal surface of these animals, and are subservient to the respiratory function. Each of these papillæ contains a central canal, which is sometimes a mere dilated tube, but which in most species is more or less deeply sacculated; and the inner surface of these simple or complex cæca is lined with hepatic cells of irregular form.

584. In the greater number of *Mollusca*, however, as well as in most *Myriapoda*, *Insecta*, *Crustacea*, and *Arachnida*, we find the biliary organs considerably more detached from the digestive apparatus; having the form either of prolonged tubes, or of clusters of shorter follicles, which coalesce with each other in such a manner, as to discharge their contents into the alimentary canal by a small number of orifices; and being only

connected with the alimentary canal, by the ducts through which their secretion is conveyed. The most simple type of this kind of structure is that which is met with in *Insects;* in which class the biliary organs are usually regarded as consisting of a number of distinct filiform tubes (Fig. 199, *f*), lying in close apposition with the sides of the alimentary canal, but quite disconnected from it, except at the points where they enter it. Where their number is small, they are usually greatly elongated, being sometimes three or four times the length of the alimentary canal, and are tortuous and convoluted; and they usually open separately into the intestinal tube. But when numerous, they are proportionally short; and two or three of them frequently coalesce to form a common trunk, before entering the intestine. Their number varies considerably; in some *Coleoptera*, there are but two; in *Diptera*, there are generally four; in *Lepidoptera*, six; whilst in other orders, there are many more. In many larvæ the biliary tubes are themselves furnished with lateral cæca; but these almost always disappear as the insect approaches the imago state; a less active biliary secretion being then apparently sufficient. In the *Melolontha* (Cockchaffer), however, which has but two biliary vessels, these are of great length, and retain their lateral cæca. When carefully examined, these tubes are found to consist of a delicate tube of transparent membrane, the inner surface of which is covered with secreting cells; from the thinness of the tube, the cells often project, so as to give it a granulated appearance when viewed with the naked eye; and generally at a distance from their outlets, the sides of the tubes are so irregular, that they appear as if folded upon the secreting cells to keep them together (Fig. 233). From the difference in size and degree of development between the cells of each tubulus, it seems probable that they originate at its upper extremity, that they are gradually being pushed on towards its outlet by the growth of new generations behind them; and that, as they thus advance, they acquire an increase in size by their own inherent powers of development, at the same time drawing into themselves the peculiar matters which they are destined to eliminate from the circulating fluids. The cells, having attained their full growth, and completed their term of life, give up their contents by the rupture or deliquescence of their walls; and these pass down the central cavity of the tube, to be discharged into the alimentary canal. It is by no means certain, however, that this constitutes the whole of the biliary apparatus of Insects. The hepatic tubuli, when traced as far as possible from their

FIG. 233.

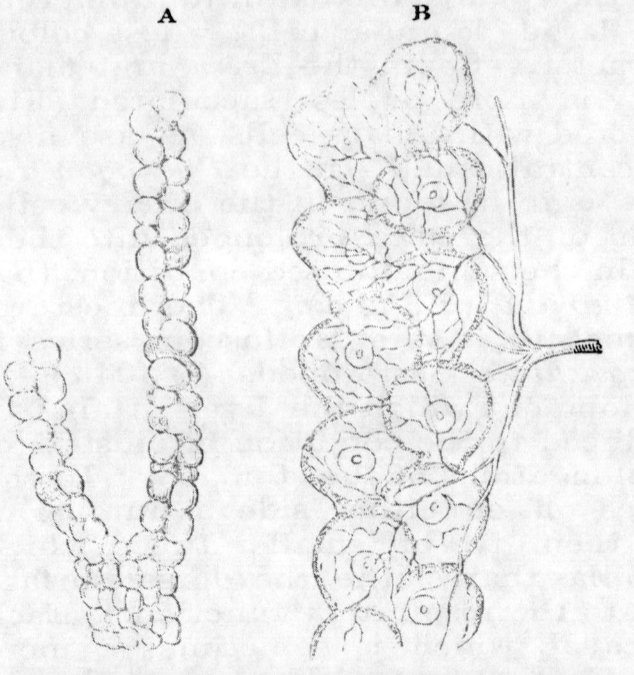

Structure of the biliary tubuli of *Musca carnaria* (flesh-fly):—A, portion of trunk and two branches, viewed by reflected light, and under a low magnifying power:—B, portion of a biliary tube more highly magnified, exhibiting the arrangement of the secreting cells, and the mode of distribution of the tracheæ.

intestinal connections, do not seem to terminate in free cæcal extremities but lose themselves in a mass which has been usually regarded as simply adipose, but which has been shown by Dr. T. Williams and Dr. C. H. Jones to be composed of cells in various stages of development, closely resembling those found within the hepatic tubuli; these cells being usually inclosed in vesicles, which are sometimes free, but are sometimes connected with a network of tubuli. This has been termed by Dr. C. H. Jones the *parenchymatous* portion of the liver; and we shall hereafter see that it appears to combine the functions of the liver and of the adipose tissue of higher animals (§ 594).

585. The biliary apparatus of the *Myriapoda* closely resembles that of Insects; but in that of the *Crustacea* we have a much closer approximation to the Molluscan type, in which the biliary cells are inclosed within a multitude of follicles with distinct cæcal terminations (Fig. 234), aggregated together into a lobulated glandular mass. On a careful examination of these follicles, and a comparison of the size and contents of the cells at the bottom and towards the outlet, it becomes evident that they originate in the former situation, and gradually increase in size as they advance towards the latter. It is also to be observed, that the cells which lie deepest in the cæcum (*a*, *b*) contain for the most part the yellow granular matter, which may be regarded as the proper biliary secretion: but as they increase in size, they also increase in the quantity of oil-globules which they contain (*c*); until beyond the middle of the tube, where they are found full of oil, so as to have the appearance of ordinary fat-cells (*d*, *e*). From this circumstance it happens, that when a cæcum is examined with the microscope, its lower half appears filled with a finely granular matter, intermingled with nucleated particles; and the upper half with a mass of fat cells, whose nuclei are obscured by the oily particles. These follicles are clustered into lobules (Fig. 235), and discharge the products of their secreting action into a cavity in the centre of each; and from these it is collected by ducts that coalesce with each other to form a small number on either side, by which the product of the whole mass (Fig. 156, *f*, *f*) is discharged into the alimentary canal. Here, as in other Articulata, the hepatic organs present a perfect bi-lateral symmetry. In some of the Entomostracous Crustacea, however, as the common *Daphnia pulex*, the biliary apparatus is reduced to that simple condition, which it has been already described as presenting in the Bryozoa and lower Annelida; namely, a set of cells lodged in the walls of the intestine itself.—The structure of the biliary organs of the *Arachnida* has not been clearly made out; but they would

FIG. 234.

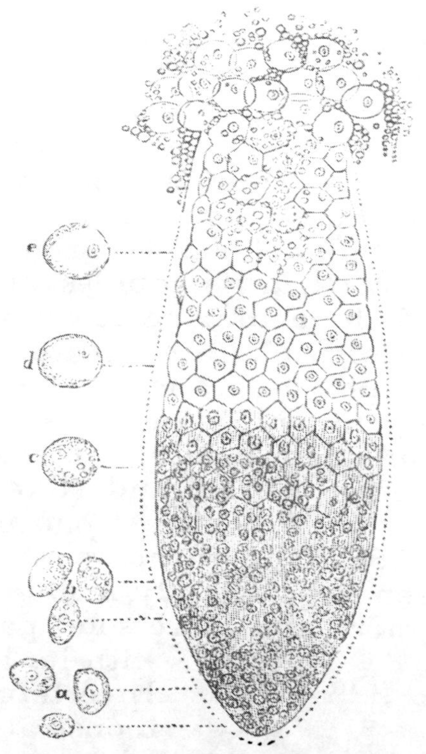

One of the hepatic cæca of *Astacus affinis* (Cray-fish), highly magnified, showing the progress of development of the secreting cells from the blind extremity to the mouth of the follicle; specimens of these, in their successive stages, are shown separately at *a*, *b*, *c*, *d*, *e*.

seem to partake rather of the character of those of the Crustacea, than of those of Insects. From the short, straight, alimentary canal, there pass off on either side, in the abdominal region of the *Araneidæ*, but in the thoracic region of the *Scorpionidæ*, several pairs of cæca (Fig. 162, *e*), which subdivide, and then lose themselves in a fatty mass of considerable size; this fatty mass has been usually regarded as nothing else than adipose tissue, but its hepatic character appears from recent observations to be unquestionable; and this circumstance adds weight to the opinion of those, who consider the great adipose mass in the body of Insects as biliary in its nature.*

FIG. 235.

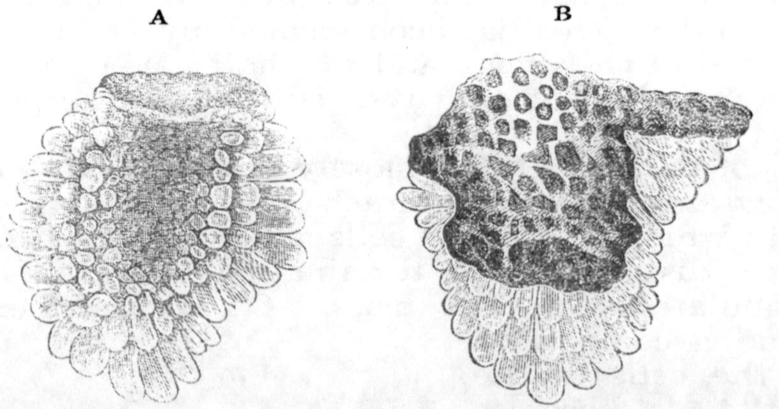

Lobule of Liver of *Squilla Mantis*; A, exterior; B, cut open.

586. Throughout the *Molluscous* series, with the exceptions already named, the liver presents the same type of structure as in the Crustacea; and the chief advance which we see in ascending the series, consists in the increasing compactness of the glandular mass, and its isolation from the walls of the digestive canal. For whilst, in the higher *Tunicata* and *Conchifera*, its lobules cluster round the pyloric portion of the stomach and the commencement of the intestine, and discharge their secretion by a multitude of separate orifices, these cluster in the higher *Gasteropoda* around the branches of a single or multiple common duct (Fig. 127, *l*, *l*, *l*), very much as in the Crustacea; and in *Cephalopoda*, the lobules are so connected together as to form a compact mass, nearly resembling the liver of Fishes, and (like it) removed to a distance from the intestinal canal, into which it pours its secretion by a single long duct.

587. The intimate structure of the Liver in Vertebrata has not yet been satisfactorily made out. The organ presents, as we ascend the series, a more and more solid parenchymatous texture, which strikingly contrasts with its loosely-lobulated appearance in all but the highest Invertebrata; and there is not the least difficulty in demonstrating that this parenchyma is composed of cells, which are obviously the instruments by which the biliary matter is eliminated. These biliary cells, in the liver of Man and of higher Vertebrata, are of a flattened spheroidal form (Fig. 236), their diameter being usually from 1-1500th to 1-2000th of an inch; and they commonly lie in piles, their faces adhering to one another. Each of them presents a distinct nucleus; and the cavity of the cell is occupied by yellow amorphous biliary matter, usually having one or two large adipose globules, or five or six small ones, intermingled with it. It is interesting to remark that, in the *Amphioxus* (§ 321), the liver

FIG. 236.

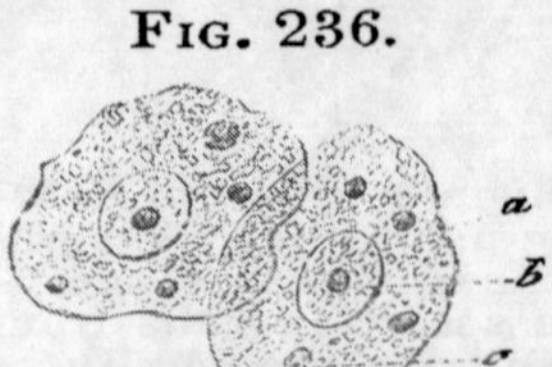

Glandular cells of Liver:—*a*, nucleus; *b*, nucleolus; *c*, adipose particles.

* See, on this part of the enquiry, the paper of Dr. C. H. Jones on the Structure and Development of the Liver, in the "Philosophical Transactions" for 1849.

reverts to the type which it presents in the lower Articulata; for it is nothing else than a cæcal prolongation from the alimentary canal (Fig. 172, *l, l*), surrounded by hepatic cells, which are found also in the walls of the intestinal canal itself, as in the Earth-worm. But in the higher Fishes, as in the classes above them, it has been hitherto found impossible to obtain an entirely satisfactory demonstration of the relation of the cells of the hepatic parenchyma to the biliary ducts, by which the products of their secreting action are carried off.—The following are the general results of the observations which have been made upon the structure of the Liver in Man and the Mammalia, on which attention has been chiefly bestowed. The entire mass is made up of a vast number of minute *lobules* of irregular form, but about the average size of a millet seed; and each of them contains the elements of which the entire organ is composed, namely, a plexus of biliary ducts connected with their main trunks, and a mass of biliary cells; and each is connected, in like manner, with the three blood-vessels which minister to the circulation in this organ, namely, the *hepatic artery* which brings blood for its nutrition, the *vena portæ* which brings blood for the secreting operation, and the *hepatic vein* which carries back the blood received from both sources. On endeavouring to trace the main excretory canal, or 'hepatic duct,' from its termination in the intestine to its sources in the liver, it is easy to make out that it subdivides and ramifies among the lobules of the whole mass; and in general it would appear that the ducts connected with neighbouring lobules inosculate with each other, so as to form a plexus that is continuous through the entire organ. This plexiform arrangement was inferred by Mr. Kiernan to be continued into the interior of the lobules (Fig. 237); and from various recent observations, it seems probable that this is really the case, and that the cells of the parenchyma are actually included within a limitary membrane, which is a continuation of that lining the biliary plexus, so that they discharge their contents immediately into the ducts.

FIG. 237.

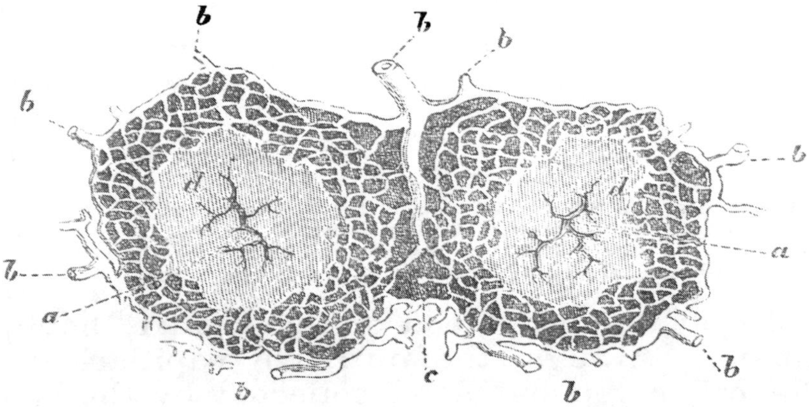

Horizontal section of two superficial lobules of the *Human Liver*, showing the probable arrangement of the biliary ducts; *a, a,* intralobular branches of the hepatic veins; *b, b,* trunks of biliary ducts, proceeding from the plexus which is supposed to traverse the lobules; *c,* interlobular tissue; *d,* parenchyma of the lobules.

588. The blood which is brought to the liver for the elimination of the biliary secretion, is distributed by the branches of the vena portæ to a similar *inter-lobular* plexus; and from these, a set of converging twigs proceed towards the centre of each lobule (Fig. 238), which supply a capillary network in its substance. In the centre of the lobule a radicle of the hepatic vein takes its origin, collecting the blood from the capillary network, and uniting with other radicles to form the trunk by which it is carried back to the vena cava. Owing to the peculiar position of the

branches of the hepatic vein in the centre of each lobule, the lobules are appended (as it were) to these branches, like fruits upon a stalk. The branches of the hepatic artery are principally distributed upon the walls of the hepatic ducts, and upon the trunks and branches of the portal and hepatic veins, supplying them with their *vasa vasorum*; and also upon their fibrous sheaths (formed by an inversion of the investing membrane of the gland, and termed 'Glisson's capsule'), from which prolongations extend into the substance of the liver, to hold together its elements. The precise relation of the capillaries of the hepatic artery to those of the portal and venous systems, has not yet been ascertained beyond all doubt; but there seems reason to believe, with Mr. Kiernan, that the arterial capillaries discharge themselves into the ultimate ramifications of the portal vein; and that thus the blood of the former, having become venous by transmission through the nutritive capillaries of the liver, mingles with the other venous blood collected by the vena portæ, to supply the materials of the secretory function, which are eliminated from it during its passage into the hepatic vein. Thus the parenchyma of each lobule will exercise its secretory action upon the blood, during the passage of the fluid from its periphery towards its centre; and the bile thus eliminated will be discharged, by the rupture or dissolution of the hepatic cells, into the plexus of ducts by whose limitary membrane they are invested.—It is only in the warm-blooded Vertebrata, that the cavity of these cells is chiefly occupied by biliary matter. In Fishes and Reptiles, the parenchyma is as fatty as in the inferior classes; and the oily matter fills the cells almost as completely as in the Crustacea (Fig. 234), so that the livers of the Cod and other large Fishes yield a considerable supply of it. In Birds, on the other hand, the hepatic cells are even more free from fat globules than are those of Mammalia; and they are almost entirely filled with amorphous biliary particles. These differences will be seen to have an important signification, when taken in connection with the probable action of the organ in these different groups (§ 594).*

FIG. 238.

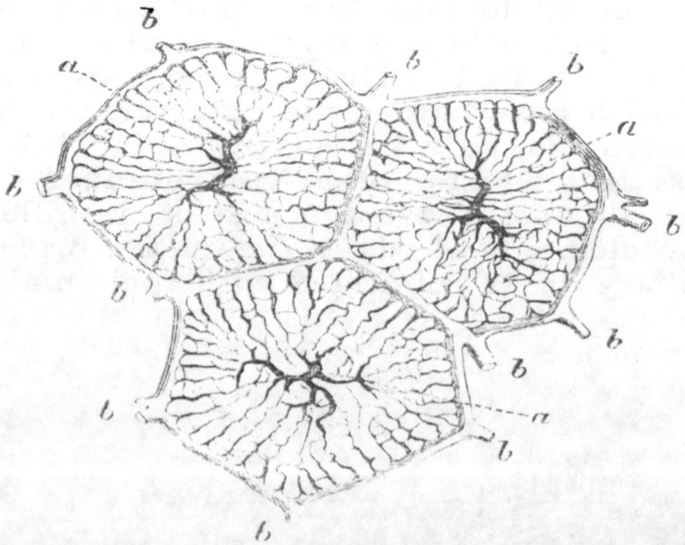

Horizontal section of three superficial lobules, showing the two principal systems of blood-vessels; *a, a, intra-lobular* veins, proceeding from the Hepatic veins; *b, b, inter-lobular* plexus, formed by branches of the Portal veins.

* Much valuable information has lately been contributed by microscopic examination of the minute structure of the Liver in different classes of animals. The Author has especially availed himself, in the foregoing account, of the "Anatomical and Pathological Observations" of Prof. Goodsir (1845); Dr. T. Williams's Essay in the "Guy's Hospital Reports" for 1846; Dr. Leidy's valuable communication in the "American Journal of the Medical Sciences" for Jan. 1848; a Report on the researches of Prof. Retzius in "Müller's Archiv.," 1849; and Dr. C. Handfield Jones's paper in the "Philosophical Transactions" for 1849. Mr. Kiernan's paper in the "Phil. Trans." for 1833 will always remain of standard value.

589. The history of the *development* of the Liver in the higher animals, presents many points of most interesting analogy to the permanent conditions which the organ has been thus shown to possess in the lower. The first rudiment of the gland is formed by the thickening of the cells in the wall of the alimentary canal, at the spot in which the hepatic duct is subsequently to discharge itself. This thickening increases, so as to form a projection upon the exterior of the canal; and soon afterwards the lining membrane of the intestine dips down into it, so that a kind of cæcum is formed, surrounded by a mass of cells, as shown in Fig. 239. Now here we have the obvious representation of the hepatic organs of the lower Invertebrata; which, in their simplest forms, are nothing else than isolated cells, situated in the walls of the alimentary canal itself; whilst in their next grade, they are clustered round a cæcal diverticulum from its cavity. The condition of the Liver in *Amphioxus* is almost precisely represented by that of the embryo Fowl at the fourth or fifth day. The increase of the organ seems to take place by a continual new budding-forth of cells from its peripheral portion; and a considerable mass is thus formed, before the cæcum in its interior undergoes any extension by ramifications into it. Gradually, however, the cells of the exterior become metamorphosed into fibrous tissue for the investment of the organ; those of the interior break down into ducts which are developed in continuity with the intestinal cæcum, and which are lined by muscular and fibrous tissues developed from the primitive cellular blastema; whilst those which occupy the intervening space, and which form the bulk of the gland, give origin to the proper secreting cells, which are now to come into active operation. As this is going on, the hepatic mass is gradually removed to a distance from the wall of the alimentary canal; and the cæcum is narrowed and lengthened, so as to become a mere connecting pedicle, forming, in fact, the main trunk of the hepatic duct.—Such is believed to be the general history of the process; and it applies equally to the other glands developed from the parietes of the alimentary canal, such as the salivary glands and the pancreas; but many of the minuter questions which it involves, still remain for elucidation.

FIG. 239.

Origin of the Liver from the intestinal wall, in the embryo of the *Fowl*, on the fourth day of incubation:—*a*, heart; *b*, intestine; *c*, everted portion giving origin to liver; *d*, liver; *e*, portion of yolk-bag.

590. We have now to enquire into the characters of the *Bile*, and the purposes to which it ministers in the Animal economy.—Of the solid matter, which forms nearly a tenth part of the whole secretion, about an eighth part is composed of alkaline and earthy salts, corresponding with those found in the blood, but not bearing the same proportions to each other; soda being present (apparently in a state of actual combination with biliary matter) in such amount as to be peculiarly characteristic of this product. The organic constituents, which make up the remainder of the solid matter of the bile, are acted upon by reagents with peculiar facility, and are thus liable to be changed by the processes which are

merely intended to separate them ; and hence the accounts of them given by different chemists are far from being conformable with each other. For our present purpose it will be sufficient to state that the proper *Biliary matter*, or *Biline* (of Berzelius), whether a single or a composite substance, is allied to resinous and fatty matters in its composition and properties, and appears to unite (like the fatty acids) with soda or other bases ; and its ultimate composition is represented by Prof. Liebig to be 76 Carbon, 66 Hydrogen, 22 Oxygen, 2 Nitrogen, with a considerable proportion of sulphur.—The fatty matter of the Bile of Mammalia (which has been the principal subject of chemical research) for the most part consists of the peculiar spermaceti-like substance termed *Cholesterine ;* the proportion of this, however, in healthy bile, is extremely small ; but in many disordered states it accumulates in large amount, and it usually forms the principal if not the sole ingredient in biliary concretions. It is a white crystallisable fatty matter, somewhat resembling spermaceti, free from taste and odour, and composed almost entirely of carbon and hydrogen, its formula being 36 Carbon, 32 Hydrogen, and 1 Oxygen. The ordinary fatty acids (§ 189) may occasionally be detected in the bile, but seldom in large proportion.—The peculiar *Colouring Matter* of bile is distinct from both the preceding substances ; its chemical characters have not yet been clearly made out ; but there is much probability that it is related to the colouring-matter of the blood. For the addition of a mineral acid to the colouring-matter of bile produces remarkable changes in its hue, converting the yellowish matter successively into green, blue, violet, red, and brown ; in fact, precisely into those colours which are exhibited by the colouring-matter of the blood, as left in an ‘ ecchymosis ’ in process of departure.

591. There is every probability that these substances exist in the blood, in a condition not very dissimilar to that in which they are found in the product secreted from it. Thus Cholesterine may be obtained from blood-serum by an analytical process of no great complexity ; and its presence there is manifested by its occasional deposit, as a result of diseased action, in other parts of the body, especially in the fluids of local dropsies. The Colouring-matter appears to be derived, directly or indirectly, from the hæmatine of the blood. Of the Biline, however, it can only be said that its indefinite nature and reactions have hitherto prevented its detection in the blood by chemical tests ; and much yet remains to be learned regarding its history. From the considerations to be presently adduced, it would seem probable that biliary matter does not exist *as such* in the blood, previously to the formation of the secretion ; but that its elements, derived from the disintegration of the tissues, are present in the circulating fluid under some more pernicious form, and are transformed into biline, by the agency of the liver, in order that they may be re-absorbed in a less noxious form, to be finally eliminated by the respiratory process. It is certain that the effects of the *re-absorption* of bile into the blood, as seen in ordinary cases of jaundice dependent upon obstruction of the biliary ducts, are not nearly so injurious as are those of the *retention* of the elements of the secretion, consequent upon deficiency of the secreting power of the liver ; for whilst in the latter case, death speedily supervenes, if no other outlet be found for the excrementitious matter, in the former, no severe injury necessarily arises from the accumulation of biliary matter, to even such an

extent that the tissues in general are tinged by it. And, as will presently appear, there is strong reason to believe that the re-absorption of biliary matter into the circulating current, is the means by which it is finally carried out of the system.

592. The special purposes answered by the secretion of Bile are still involved in considerable obscurity; but more definite and probably more correct ideas in regard to them are gradually being evolved; and the following may, perhaps, be regarded as tolerably well established truths. In the first place, the bile must be considered as an *excrementitious* substance, derived from the disintegration of the tissues, or from the decomposition of the elements of the blood; and it serves especially to remove from the blood the *hydro-carbonaceous* portion of the effete matters, the *nitrogenous* being eliminated in the urine. It has been pointed out by Prof. Liebig, that if we add to half the formula representing the ultimate composition of *bile*, the formula of *urate of ammonia* (which is the characteristic element of the urine of all animals save Mammalia), the sum gives the proportionals of the ultimate components of dried *blood* or of *flesh*, with the addition of 1 eq. of oxygen and 1 eq. of water. For ;—

Half an equivalent of Biliary matter	.	= 38 C, 33 H, N, 11 O
One equivalent of Urate of Ammonia	.	= 10 C, 7 H, 5 N, 6 O
The sum of which	.	= 48 C, 40 H, 6 N, 17 O

And, in like manner ;—

Formula of Blood	. . .	= 48 C, 39 H, 6 N, 15 O
1 Eq. of Water + 1 Eq. of Oxygen	.	= H 2 O
		48 C, 40 H, 6 N, 17 O

Now although it must be admitted that, by a dextrous management of formulæ, almost any kind of transformation may be effected *on paper*, yet this coincidence, without any management at all, is so close that it cannot be regarded as accidental; and we seem fairly entitled to look upon the principal materials of the animal fabric as partly resolving themselves, in their disintegration, into the characteristic components of the two principal excretions, the urinary and the biliary. Of course this resolution only expresses *a part* of the metamorphoses which the tissues undergo; for the kreatine and kreatinine of the urine, and the fæcal matters of the alimentary canal, must be regarded in the same light; and there are doubtless other excrementitious matters (included in the general term 'extractive'), of which we know still less, that must be attributed to the same origin. But it seems satisfactorily to account for the components of the biliary matter; since they form the *complement* of the urinary; so that the formation of each excretion seems to involve that of the other.—Now, that the bile is in its essence an excrementitious product, and that the assistance it may afford in the digestive process is *not* the principal purpose of its secretion, appears further from this; that it is eliminated during the latter part of embryonic life, and that it then accumulates in the alimentary canal, forming a large part of the meconium which is discharged soon after birth. But from the time that respiration commences, and during the whole subsequent life, it appears from chemical analysis of the fæces, that not much of the bile, save its colouring matter, is evacuated in the state of health; *so that a large part of that which is poured into the alimentary canal, must be re-absorbed through the blood-vessels and lymphatics of its walls.* And there can be little doubt that, when thus re-absorbed, and taken into the current of the circulation in a less noxious form than that in which its elements previously existed, the biliary matter is chiefly eliminated by the respiratory process; its sulphur, however, being oxidised, that it may be carried off by the urine, and its soda being eliminated by the same channel. In this point of view, the secretion of bile may be regarded as the intermediate stage between the disintegration or 'waste' of the tissues, and the final elimination of the combustible products of that waste by the instrumentality of the lungs.

593. But that the bile performs some subsidiary function in the Digestive process, would also seem beyond doubt. It is not, however, a sufficient indication of this, that we should find the outlet of the hepatic organs, through the entire Animal Kingdom, to be in the part where digestion is most actively going on; for it may be, that the discharge of the bile into the upper part of the alimentary canal, has reference merely to its re-absorption. But it appears from the results of numerous experiments, that when the bile-duct is divided, and its extremity is drawn out of the body, in such a manner that it can freely discharge the hepatic secretion, but this is prevented from passing into the alimentary canal, the animals, although they do not at first appear to suffer much in health, gradually become emaciated, and at last die of inanition, unless they are allowed to receive back the bile by licking the wound, in which case they survive. And it seems proved, by the recent investigations of Lehmann and Frerichs, that although the pancreatic fluid certainly possesses considerable power of 'emulsifying' the oleaginous matters in the alimentary canal, yet that it answers this purpose much more effectually when mingled with bile, which possesses an emulsifying power of its own. It also appears from the experiments of H. Meckel, that bile has the power of effecting the transformation of sugar into oleaginous matter; thus effecting that for a portion of the contents of the intestine, which the passage of the portal blood through the liver effects for a portion of the saccharine matter which it contains (§ 561). And lastly, from the experiments of Gorup-Besanez it appears that bile has a special solvent action upon caseine.—Hence it may be stated that the function of the bile in digestion is chiefly the preparation of the combustive material for the respiratory process; since by its means the saccharine matters are partly transformed into fat, and the absorption of the fatty matters is assisted by its admixture with the pancreatic fluid. And thus we see why animals should die of inanition when the bile is wholly discharged out of the body; since they thus lose, not only the combustive materials supplied by the bile itself, but also the greater part of those contained in the food, which they are prevented from assimilating.

594. Thus, then, if we bring together all the facts at present in our possession, with reference to the offices of the Hepatic organs, they seem to lead to these conclusions.—1. That they are essentially organs of excretion, designed to remove from the circulating fluid that portion of the products of disintegration, of which the principal component of the

urinary excretion is the 'complement.'—2. That in doing this, they convert the greater part of the excrementitious matters into Biline; a substance which can be reabsorbed with less injury, and which, after performing its part in the digestive process, is taken back into the blood, to be eliminated by oxidation through the lungs.*—3. That the temporary presence of bile in the alimentary canal is subservient to the digestion and absorption of the non-azotized compounds, and perhaps also in some degree to that of the albuminous.—4. That not only by the separation of biliary matters from the blood, and by the operation of this upon the alimentary substances, but also by the change which it effects in the constituents of the blood itself, the liver aids in preparing oleaginous materials for the combustive process.—5. It also aids (if we are to rely upon the statements of M. Bernard, § 561) in the assimilation of albumen, taking a share in effecting its metamorphosis into fibrine.—Now if we are to regard the production and separation of fatty matter as one of the great purposes answered by the Liver, it need not surprise us to find that this organ should frequently serve, not only to prepare this, but also to store it up; and this we might expect to be the fact especially in those classes of animals, in which its final elimination by the respiratory actions is slow. Such we have seen to be the case; the great bulk of the liver in the Crustacea, Mollusca, and cold-blooded Vertebrata, having reference apparently, not to a large production of bile, but to an accumulation of fat: whilst in Mammalia the fat is in very small amount, unless the respiration be impeded; and in Birds, whose respiration is pre-eminently active, scarcely any traces of fat are to be found. The fat thus stored up in the liver will probably be taken into the current of the circulation, as it is wanted, by the blood which passes through the organ; and may never be discharged as an excretion into the alimentary canal.†

595. *Of the Kidneys and the Urinary Excretion.*—The Kidneys are to be regarded as purely *excreting* organs; their sole function being to separate from the blood certain matters which would be injurious to it if retained; and these matters being destined to immediate and complete removal from the system. These glands almost always present a tubular structure; the required extent of surface being given, not by the multiplication of separate follicles, but by a great prolongation of the individual cæca.—No distinct traces of urinary organs can be made out among the lower Invertebrata. In *Insects* and *Arachnida*, however, they present themselves as a group of cæcal tubuli, discharging themselves into the cloacal termination of the alimentary canal (Fig. 162, *h, h*); and similar tubuli are found in some of the higher Crustacea.—In the Molluscous series, however, the urinary organs, where they exist, possess rather a follicular character. It is probable that a glandular mass found in some *Conchifera* near the termination of the intestine (which has been regarded by some as a calcifying gland) is really a rudimentary kidney; and this

* When an unusually large quantity of bile is poured into the intestinal canal, as after the action of a mercurial purgative, or in 'bilious diarrhœa,' the greater portion of it escapes reabsorption, and is ejected per anum.

† The doctrine that the agency of the Liver is preparatory to the Respiratory function, was first propounded by Prof. Liebig ("Animal Chemistry," 1842); and although it has had to contend against many important objections, it has gradually made its way into Physiological Science:—in the form in which it is above stated, the Author believes that it will be found consistent with all the facts at present known.

may be said with more certainty of a conglomerate mass of small follicles, with a single efferent duct or ureter which opens externally near the anus, in many *Gasteropoda,* this body having been found to contain uric acid. It is remarkable that in animals so highly organised as the *Cephalopoda,* the presence of a proper renal organ is as yet undetermined. Some have attributed this character to the 'ink-bag;' but this only exists in a part of the class; and it is thought more probable by Prof. Mayer, that the masses of follicles connected with the great veins (Fig. 135, *r*) are to be regarded as a urinary apparatus.

596. When we pass to the Vertebrated series, however, we usually find the renal organs attaining a large size in the class of *Fishes;* but their type of conformation is low. They very commonly extend through the whole length of the abdomen; and consist of tufts of uniform-sized tubules, which shoot forth transversely at intervals from the long ureter, and are connected together by a loose web of areolar tissue, that supports the network of vessels distributed upon their walls. This condition of the urinary apparatus is very analogous to that of the Corpora Wolffiana, or temporary renal organs of the embryo of the higher Vertebrata, which are afterwards superseded by the permanent kidneys (§ 599). In the lowest members of the class, however, the structure is yet simpler; for in the *Cyclostomi,* each long duct sends off at intervals, instead of a bundle of tubuli, a short wide tube which communicates with a single cæcum; and at the bottom of this is a small 'vaso-ganglion,' or convoluted plexus of blood-vessels, which reminds us of the Corpora Malpighiana in the kidneys of higher Vertebrata. In the *Amphioxus,* the only rudiment of a kidney is the slightly opaque, slender, elongated glandular-looking body (Fig. 172, *h*), which is considered by Prof. Owen to possess this character, though its structure has not yet been fully elucidated.

597. The size of the Kidneys is usually considerable in *Reptiles;* but their form differs greatly in the several orders of the class, being narrow and much elongated in Batrachia and Ophidia, but broader and shorter in Sauria and Chelonia. Their essential structure, however, is nearly the same throughout; for the ureter gives off a large number of transverse cæca, which are short and nearly straight in the lower Reptiles and in the early condition of the higher (Fig. 240), but which in the more developed conditions of the organ become long and convoluted (Fig. 241), each group forming a lobule, which receives branches from the portal trunk that supplies the organ with blood. In the Crocodile the distinction between the *cortical* and the *medullary* portion of the kidney becomes evident; the former being the part in which the blood-vessels are most copiously distributed, and in which the

FIG. 240.

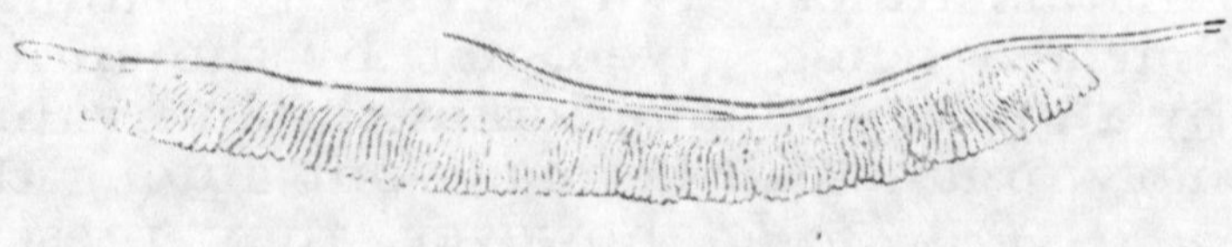

Kidney of fœtal *Boa:*—the urinary tubes as yet short and straight.

FIG. 241.

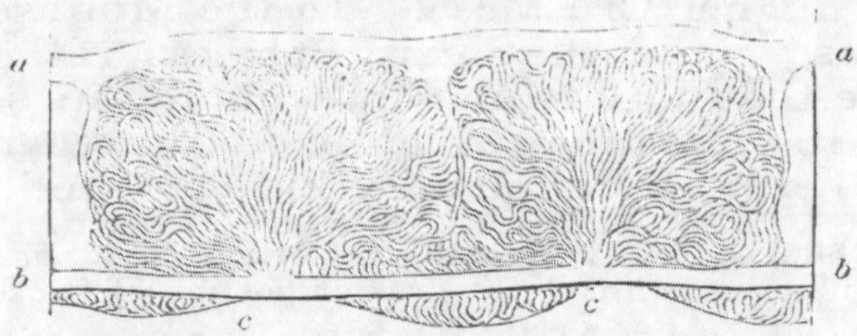

Portion of Kidney from *Coluber:*—*a, a,* vascular trunk; *b, b,* ureter; *c, c,* converging fasciculi of tubuli uriniferi.

tubuli have the most convoluted arrangement; whilst the tubuli in the latter are straighter, and converge more directly to the points at which they discharge themselves into the ureter. The *Corpora Malpighiana* (§ 598), where they exist in this class, are scattered through the entire substance of the kidney; not being restricted, as in the higher animals, to its cortical portion.—In *Birds*, too, the kidneys are of large size; and they present also a greater compactness of texture. The lobules of which they are composed, can often not be distinguished externally; but they are connected with separate branches of the ureter, and each consists of a converging bundle of tubuli uriniferi, forming its 'medullary' portion (Fig. 242), and of the dichotomous ramifications of these in the outer part of the lobule, of which they constitute the 'cortical' portion.—It is in *Mammalia*, however, as in Man, that the organ becomes most compact, and the distinction between its cortical and medullary portions most clearly marked. The separate clusters of tubes do not now open into distinct branches of the ureter; but discharge their contents into a capacious cavity formed by the dilatation of the ureter in the interior of the kidney (Fig. 243), which looks (in section) as if it were doubled together around it. The lobules composed of these separate clusters are usually blended together so closely, that they cannot be distinguished externally, and cannot be separated from each other anatomically. In the Human fœtus, however, when the kidney is first being developed, its lobular character is very apparent; and this type is permanently retained in many of the lower Mammalia.

FIG. 242.

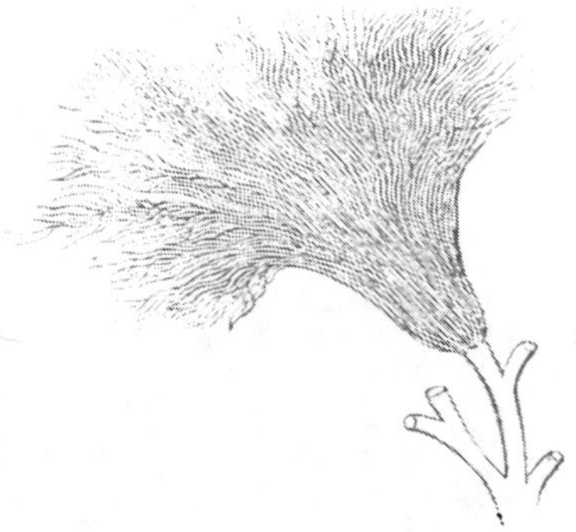

Pyramidal fasciculi of tubuli uriniferi of *Bird*, terminating in one of the branches of the ureter.

FIG. 243.

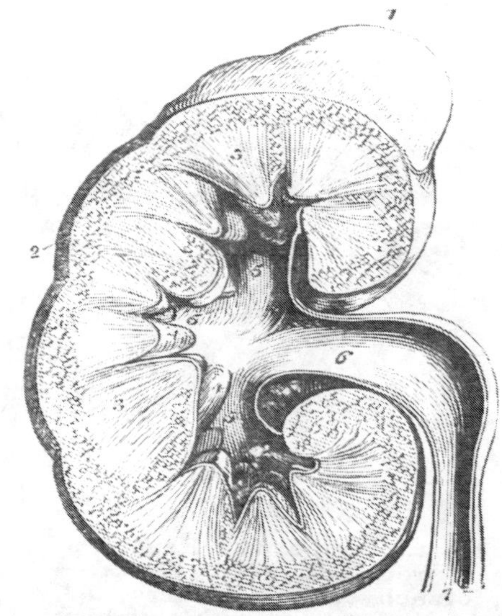

A section of the Kidney, surmounted by the supra-renal capsule; the swellings upon the surface mark the original constitution of the organ, as made up of distinct lobes. 1. The supra-renal capsule. 2. The vascular portion of the kidney. 3, 3. Its tubular portion, consisting of cones. 4, 4. Two of the papillæ projecting into their corresponding calices. 5, 5, 5. The three infundibula; the middle 5 is situated in the mouth of a calyx. 6. The pelvis. 7. The ureter.

598. The proper secreting apparatus consists of the epithelial cells lining the tubuli uriniferi (Fig. 244), which draw the peculiar elements of the urinary excretion from the vascular plexus which surrounds the exterior of the tubes, and then deliver them up to be carried off through the interior of these, to their terminations in the ureter. It does not appear that this process necessarily involves the continual exuviation and renewal of the secreting cells; for it would seem as if they could give up, by their free surface, the matters which they have taken in, when these are in a state of such perfect solution as to be able to transude the cell-wall, and thus to escape without rupturing it. For the liberation of the constituents of the almost

solid urine of Reptiles and Birds, however, it is probable that the exuviation of the secreting cells is necessary.—But the Kidney contains another apparatus, of a very peculiar description, which appears specially destined for the elimination of the superfluous *fluid* of the blood, by a process of simple transudation. When a section of the cortical substance is slightly magnified, the cut surface is seen to be studded with a number of little dark points, each one of which, when examined under a higher magnifying power, is found to consist of a knot of minute blood-vessels, formed by the convolutions of thin-walled capillaries (Fig. 245, *m*). These bodies, termed *Corpora Malpighiana* after their first discoverer, are included in little flask-shaped capsules, the necks of which are continuous with the tubuli uriniferi; so that these may be considered as sending off little diverticula to invest the Corpora Malpighiana, which appear to lie freely within them.* Each of these little vascular knots is directly supplied by a branch of the renal artery (Fig. 245, *a*, *b*), which, when it reaches the capsule, subdivides into a group of capillaries; and these, after forming the convoluted tuft (*m*), coalesce into a single efferent trunk (*ef*), which may be considered as representing a branch of the vena portæ. For the efferent trunks of the Malpighian bodies discharge their blood into the capillary plexus, which surrounds the tubuli uriniferi, and from which the solid matter of the urinary secretion is elaborated; just as the vena portæ supplies the capillary plexus, from which the biliary secretion is elaborated in the liver. Thus the whole Malpighian system of vessels may be considered (as Mr. Bowman has pointed out) in the light of a *portal* system within the kidney; the efferent vessels of the Corpora Malpighiana being collectively the representatives of the Vena Portæ, since they convey the blood from the first (or systemic) to the second (or secreting) set of capillaries. In Reptiles (in which, as in Fishes, the Kidney is partly supplied from the Hepatic Portal

FIG. 244.

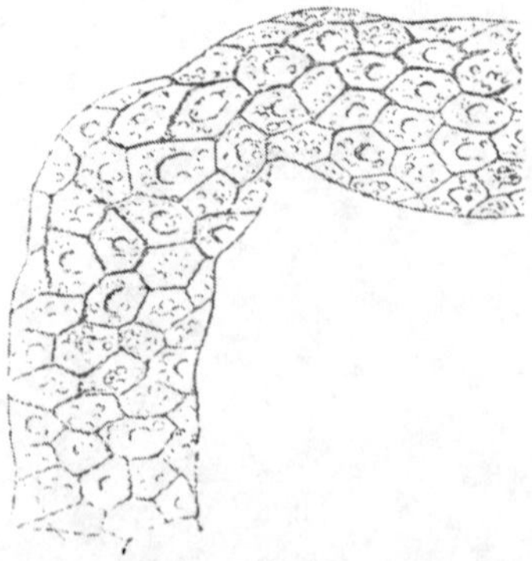

Portion of tubulus uriniferus, with its tesselated epithelium.

FIG. 245.

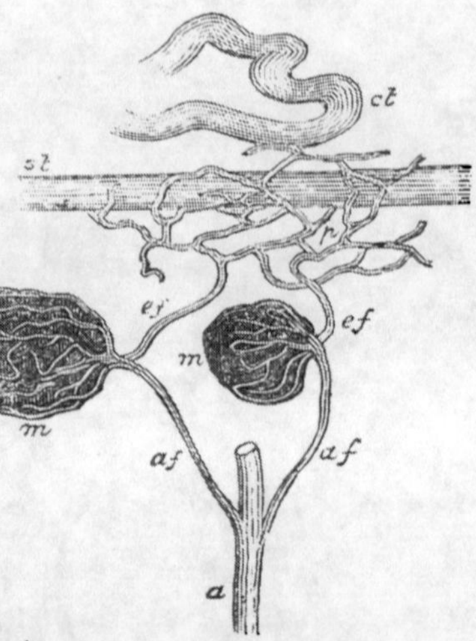

Distribution of the Renal vessels; from Kidney of *Horse*:—*a*, branch of Renal artery; *af*, afferent vessel; *m*, *m*, Malpighian tufts; *ef*, *ef*, efferent vessels; *p*, vascular plexus surrounding the tubes; *st*, straight tube; *ct*, convoluted tube. Magnified about 30 diameters.

* By Mr. Bowman, the first discoverer of this curious relation, it was supposed that the flask-shaped dilatation was formed by the expansion of the *extremity* only of the tubulus uriniferus; and that the membranous wall of the capsule is perforated by the afferent and efferent vessels, so that the capillary knot is really in its inside, and has no investment of its own. ("Phil. Trans." 1842). Subsequent investigations, however, have proved that the Corpora Malpighiana are connected with the sides, as well as with the extremities of the tubuli, so that each tubulus is in relation with several of them; and have rendered it probable that the flask-shaped capsule does *not* enclose the Corpus Malpighianum, but is reflected over it like the pericardium over the heart, so that the capillary knot really lies on its exterior, although embraced by it in such a manner that fluid transuding from the vessels at once enters the cavity of the capsule.

system, § 472), the efferent vessels of the Malpighian bodies, which receive their blood (as elsewhere) from the renal artery, unite with the branches of the Portal vein to form the secreting plexus around the tubuli uriniferi; so that this plexus, like the secreting plexus of the Liver (§ 588), has a double source, the vessels which supply it receiving their blood in part from the capillaries of the organ itself, and in part from those of other viscera. In Mammalia, however, the secreting plexus is entirely supplied by the efferent vessels of the Corpora Malpighiana; though in Birds the oviparous type of distribution seems still to prevail to a certain extent (§ 475).

599. In the embryological *development* of the Kidneys of the higher Vertebrated animals, we have a very interesting example of the evolution of an organ, which is to serve only a temporary purpose in them, but which remains as the permanent instrument of the function in the lowest class, although superseded in the higher by an organ formed upon a more elaborate type. The first appearance of anything like a Urinary apparatus in the Chick, is seen on the second half of the third day; and the form then presented by it is that of a long canal, extending on each side of the spinal column, from the region of the heart towards the allantois: and the sides of this present a series of elevations and depressions, indicative of the incipient development of cæca. On the fourth day, the *Corpora Wolffiana,* as they are then termed, may be distinctly recognised as composed of a series of cæcal appendages, which are attached along the whole course of the first-mentioned canal, opening into its outer side; and thus closely corresponding with the condition of the (so called) kidneys of Fishes (§ 596). On the fifth day, these appendages are convoluted, and the body which they form acquires increased breadth and thickness; they evidently then possess a secreting function, and the fluid which they separate is conveyed by the duct of each side into the *allantois,* a sac which, though employed as a temporary respiratory organ (CHAP. XVIII.), is also used as a urinary bladder. Vestiges of Corpora Malpighiana may even be detected in connection with the secreting cæca.—The development of the true *Kidneys* commences in the Chick at about the fifth day. They are seen on the sixth as lobulated grayish masses, which sprout from the outer edges of the Wolffian bodies; and they gradually increase in size, the temporary organs diminishing in the same proportion. The sexual organs also originate as offsets from the Wolffian bodies; and it seems to be in connection with them, that the last traces of these temporary organs are found. The Kidneys, in the Human embryo, soon after their first development in the manner just described, consist of seven or eight lobes, the future pyramids; their excretory ducts still terminate in the same canal, as that which receives those of the Wolffian bodies and of the sexual organs; and this opens, with the rectum, into a cloaca, analogous to that which remains permanent in the Oviparous Vertebrata. The lobulated appearance

FIG. 246.

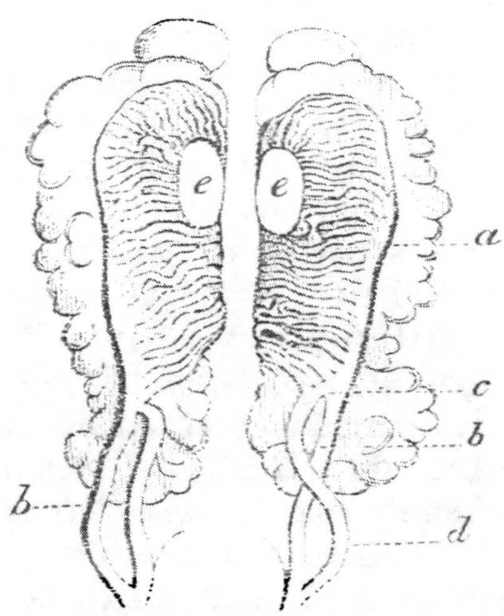

Corpora Wolffiana, with incipient Kidneys and Testes, from *Chick*:—*a*, kidney; *b*, *b*, ureters; *c*, corpus Wolffianum; *d*, its excretory duct; *e*, *e*, testes.

of the kidney gradually disappears, partly in consequence of the condensation of the areolar tissue which connects the different parts, and partly through the development of additional tubuli in the interstices; it is persistent, however, in the kidney of many of the lower Mammalia, as the Bear.—Thus we have, in the development of the Urinary apparatus, the same kind of progress from the more general to the more special type, as we have seen in the Respiratory; and it is not a little curious that the more general form of both should be retained in the same class, namely that of Fishes. There is this difference, however, between the two cases; that whilst the branchial arches of higher animals are not ever developed so far as to be instrumental in the respiratory function, their Corpora Wolffiana appear to be true temporary kidneys, eliminating a real urinary product (§ 601).

600. The Urine of Man and of *Mammalia* generally, is characterised by the large proportion of *water* which it contains, in comparison with its solid constituents; the latter being seldom above 5 parts in 100, and being very commonly less. The two, in fact, bear no constant relation to each other; for the amount of liquid in the secretion depends mainly on the degree of fulness of the blood-vessels; while that of the solid matter is governed by that of the previous 'waste' of the tissues. It would seem as if a much larger quantity of water is habitually taken-in, than is needed in the system; in order to provide for the reduction of the temperature of the body by cutaneous Exhalation, when it might otherwise be unduly elevated (§ 547); but if the usual quantity of water be not thus drawn off, in consequence of the depression of the external temperature, or the saturation of the atmosphere with dampness, or if an unduly large amount have been absorbed, the Kidneys afford the channel for its elimination. This appears to be the special function of the Malpighian bodies; whose thin-walled capillaries allow the transudation of water to take place, under a certain pressure, into the tubuli uriniferi; and thus act the part of regulating valves, permitting the passage of whatever is superfluous, while they retain the liquid that is needed in the system. In *Birds*, on the other hand, it would seem as if there is much less occasion for any provision to reduce the temperature, which is habitually kept up at a standard higher than that of any other animal; and they accordingly drink very little, so that the proportion of water in their urine is only sufficient to give it a semifluid consistence. The urinary excretion of *Reptiles* appears to be in general yet more solid; for these animals usually ingest but little water, and a part of this is given off by cutaneous exhalation when the external temperature is high. The condition of that of *Fishes* and *Invertebrata* appears to be generally the same; but from this statement the 'bombardier' beetles must be excepted, which emit their urine, as a means of defence, in little puffs of vapour, having a very acrid character, and believed to contain nitric acid.—The *solid matter* of the Urine partly consists of organic compounds formed within the body, as the result of the disintegration of the tissues, or of the decomposition of substances taken in as food; and partly of inorganic salts, such as normally exist in the serum of the blood, their proportion being liable to an increase, however, under circumstances to be presently alluded to.

601. The Organic Compounds are not the same in all animals; but yet they are nearly related to each other, and agree in the very large

proportion of nitrogen which they contain. The most characteristic of them, when completely isolated, present a *crystalline form,* which seems to be completely incompatible with the possession of plastic or organisable properties, and marks their affinity to inorganic substances. In the Urine of Man, the most characteristic ingredient is *Urea,* a neutral substance, isomeric with cyanate of ammonia (§ 29), which is very soluble in water, and may be crystallised-out in transparent, colourless, four-sided prisms. When pure, it has very little tendency to decomposition; but when associated with other substances which act as 'ferments,' it takes to itself the oxygen and hydrogen of 2 equiv. of water, and resolves itself into carbonate of ammonia. The uses of this arrangement in the economy of Nature have already been remarked upon (§ 260 *note*). The Urine of Man also contains a small quantity of *Uric Acid,* a substance which is not readily soluble in pure water, but which is more easily dissolved by water holding phosphate of soda in solution, especially when warm; and this seems to be its condition in human urine. It is also more readily dissolved, when in combination with ammonia; and in this condition it forms a large part of the almost solid urine of Serpents, which also contains, however, like the urine of Birds, a large quantity of undissolved uric acid. When separated and purified, uric acid forms a glistening snow-white powder, apparently amorphous, but shown by the microscope to consist of minute but regular crystals. It is tasteless and inodorous; and its acid reaction is very feeble. Uric Acid is replaced in the Urine of Herbivorous Mammals by *Hippuric Acid;* which is much more soluble in cold water, and dissolves readily in warm. When pure, it crystallizes in long transparent four-sided prisms, and has a strong acid reaction, with a bitterish taste. When dissolved in a liquid containing putrescent albuminous compounds, hippuric acid is converted into benzoic acid, ammonia being at the same time given off. In the fluid of the Allantois of the fœtal Calf (and probably also of other animals), which may be regarded as a temporary urinary bladder, receiving the product of the secreting action of the Corpora Wolffiana or temporary kidneys, is found another substance, termed *Allantoin,* which may be artificially obtained from uric acid by boiling it with peroxide of lead. This is a neutral substance, forming small but most brilliant prismatic crystals, which are destitute of taste, and moderately soluble in cold water; when decomposed by strong acids it is resolved into ammonia, carbonic acid, and carbonic oxide; and when acted on by alkalies, it is resolved into ammonia and oxalic acid. Two other substances, *Kreatine* and *Kreatinine,* have recently been discovered in the urine of Man and the Mammalia, which seem intermediate in character between the albuminous compounds and the characteristic components of the urinary excretion. *Kreatine,* which may be obtained from the juice of raw flesh, is a neutral substance, having the form of colourless, prismatic crystals, sparingly soluble in cold water, but dissolving readily in warm. By the action of strong acids, kreatine is converted into *kreatinine,* which only differs from it in composition by containing two proportionals less of the elements of water, but is a substance of very different chemical relations, having a strong alkaline reaction, and serving as a powerful organic base to acids. It pre-exists in the juice of flesh to a small extent; and is found, in conjunction with kreatine, in the urine. When long boiled with caustic baryta, kreatinine is gradually resolved into urea.—The composition of these substances, in relation to each other, and to that of Albuminous compounds, is shown in the following table; which gives for each the number of combining equivalents of its individual components, and their respective atomic weights, so as to show the proportion which the nitrogen bears to the whole.

	Carbon.	Hydrogen.	Nitrogen.	Oxygen.	Total.
Albumen, Liebig	49 (294)	36 (36)	6 (84)	14 (112)	526
Albumen, Mulder	40 (240)	31 (31)	5 (70)	12 (96)	437
Urea . .	2 (12)	4 (4)	2 (28)	2 (16)	60
Uric Acid . .	10 (60)	4 (4)	4 (56)	6 (48)	168
Hippuric Acid .	18 (108)	8 (8)	1 (14)	5 (40)	170
Allantoin . .	4 (24)	3 (3)	3 (42)	3 (24)	93
Kreatine . .	8 (48)	9 (9)	3 (42)	4 (32)	131
Kreatinine .	8 (48)	7 (7)	3 (42)	2 (16)	113

Thus we see that the proportion of Nitrogen in the whole, by weight, is in *albumen* as 1 : 6·24; in *urea* as 1 : 2·14; in *allantoin* as 1 : 2·21; in *kreatinine* as 1 : 2·69; in *uric acid* as 1 : 3·00; in *kreatine* as 1 : 3·12; and in *hippuric acid* as 1 : 12·14. Hence the proportion of nitrogen in the compounds of urine ranges from double to triple that which exists in the albuminous constituents of the living fabric; the only exception being in the case of hippuric acid, whose proportion of nitrogen is only half of that which exists in albumen.*—Besides the foregoing substances, the Urine contains others whose nature has not yet been clearly determined; these are at present included under the general designation *Extractive matters*, and appear to consist in part of non-azotized compounds in a state of change.—The Inorganic Compounds which are found in the urine, partly consist of salts which are taken-in as such in the food; and partly of salts which are formed in the economy, the acids being furnished by the oxygenation of bases contained in the aliment, and an ammoniacal base being supplied by the decomposition of the albuminous compounds. To the former class belongs *chloride of sodium* (common salt), of which the urine always contains a large amount, obviously derived directly from the serum of the blood; and also the *phosphates of lime and magnesia*, the proportion of which in the urine appears entirely to depend upon the amount ingested in the food. To the latter class belong the *alkaline sulphates and phosphates*; whose acids appear to be chiefly formed by the oxygenation of the sulphur and phosphorus which are constituents of all the albuminous compounds used as food; while their alkaline bases, when not ammoniacal, are supplied by the potass and soda that were ingested in combination with citric, tartaric, oxalic, and other organic acids, these acids being decomposed in the system, and carried off by the respiratory process. Such weakly-combined bases abound in the food of Herbivorous animals, but they are for the most part wanting in that of the purely Carnivorous; and the fixed alkalies are there replaced in greater proportion by ammonia.

602. Hence we may say, that the Urinary excretion is specially des-

* It would seem fair to surmise, therefore, that as Hippuric acid is restricted (so far is known) to the urine of animals of whose food non-azotized substances form a large part, it must have some other source than the metamorphosis of the organised tissues, and must be formed by the union of the products of this operation with some of the farinaceous or other superfluous components of the food.

tined (1), for the elimination of those products of the disintegration of the tissues, and of the metamorphoses taking place in the living body, which are of a highly azotized nature, or which, being in the condition of soluble salts, readily find their way by transudation through membranous cell-walls that hold back the albuminous element of the serum of the blood (§ 598); it is also obvious (2) that the kidneys are destined to remove, in the same form, whatever components of the food are superfluous, and are undergoing decomposition from not being applied to the purposes of nutrition; and it is further their office (3) to draw off any soluble saline matters taken into the system, which are either useless or injurious to it. —Although the relations of the amount of the organic compounds in the urine, to food, exercise, &c., have been as yet studied almost entirely in the Human subject, there can be no reasonable doubt that the same general rules will be found to hold good elsewhere. The proportion of *urea* which is voided in a given time, is proportional, *ceteris paribus*, to the amount of *muscular exertion* that has been put forth; showing that its presence depends in part upon disintegration of the Muscular tissue. But this is not its sole source. For it is greatly augmented, also, by an excess of azotized compounds in the food; these compounds, as already shown (§ 570), not being applied to the nutrition of the muscular substance, unless a demand for augmented formation has been created by previous functional activity. Thus, the average proportion of Urea in the Human urine, under ordinary circumstances as to food and exercise, appears to be from about 20 to 35 parts in 1000; but it may be raised to 45 parts by violent exercise, and to 53 parts by an exclusively animal diet; whilst it may fall as low as to 15 or even 12 parts, when the diet is deficient in azotized matter. The average daily amount excreted by adult *males*, is about 430 grains; by adult *females*, about 300 grains; in *children* of eight years old, it is nearly *half* what it is in adults; whilst in *very old persons*, the quantity sinks to *one-third* or even less; showing that the proportion is greatly influenced by the rapidity of interstitial change at different periods of life (§ 571). There can be no doubt that *kreatine* and *kreatinine* have the same origin and character, since they are actually found in the juice of flesh, as well as in the urine.—So the proportion of the *alkaline phosphates* in the urine is found to bear such a close relation to the previous energy of the *nervous system*, that there can be little doubt that, *ceteris paribus*, their amount may be taken as a measure of its disintegration by functional activity. It has been pointed out that, for the maintenance of this activity, a constant supply of arterialised blood is a necessary condition; and whilst the other elements of the nervous tissue (whose composition is almost entirely adipose, § 243) will be carried off by oxygenation in the form of carbonic acid and water, the phosphorus which largely enters into it will be oxygenated, and taken back into the blood in the form of phosphoric acid, uniting there with alkaline bases, as already pointed out.—The proportion of *extractive matters* appears chiefly to depend upon the nature of the food; being greatly augmented by an exclusively vegetable regimen, and greatly diminished by an exclusively animal diet.—The importance of the urinary excretion in removing superfluous or injurious saline compounds from the system (the introduction of which into it has taken place by endosmotic action, § 435), is further shown by the increase in the secretion which most of these substances produce, this increase being the result of an augmented determination of blood to the kidneys, and a consequently increased transudation of its watery portion carrying these substances with it. And further, it has been shown by Dr. Letheby, that poisonous substances (such as arsenious acid), whose rate of elimination through this channel is not in general sufficiently great to prevent them from exerting their injurious effects upon the system, may be carried out of the body with such rapidity as to render them innoxious, if diuretics (or medicines that augment the urinary secretions) be given at the same time.

603. *Cutaneous and Intestinal Excretions.*—The exhalation of superfluous water is by no means the only function performed by the Cutaneous glandulæ (§ 546). For the perspiratory fluid contains a considerable amount of solid matter; the proportion of this being sometimes as much as 12½ parts in 1000. The greatest part of this consists of animal matter, which is apparently an albuminous compound in a state of incipient decomposition, being not improbably composed in great part of the epithelium-cells cast out from the tubes of the glandulæ; but in addition to this, urea has been recently detected in the perspiratory fluid in no inconsiderable quantity; so that the skin may be considered as supplementary to the kidney in its excretory action. Besides this animal matter, the perspiratory fluid also contains saline substances, which are for the most part those existing in the blood. The compounds of lactic and acetic acid, however, seem to be specially determined to this surface, and the perspiration thus occasionally possesses a very sour odour and an acid reaction.—The Mucous surface of the Alimentary canal, besides containing glandular organs for the elimination of the Gastric and other secretions concerned in the Digestive process,* and follicles for the separation of its protective mucus, contains (in the higher animals at least) a large number of glandulæ, known as those of *Peyer*, after the name of their discoverer. These, which are confined to the large intestines and to the lower part of the small intestines, consist of closed vesicles, lying immediately beneath the mucous membrane, which is elevated over them into a circular raised spot. Their interior is found to contain cells in various stages of development; and thus each of them may be regarded in the light of a parent-cell, corresponding with the follicles of which the ordinary glands are composed, in every respect but this,—that the latter are situated at the extremities of the ramifying ducts, by which their secreted product is collected and conveyed away; whilst the former remain closed, until they open by a kind of dehiscence, discharging their contents into the alimentary canal; after which it is probable that they shrink up and disappear, their function being then taken-on by other similar vesicles newly developed for the purpose. There is strong evidence that the office of these glandulæ is to eliminate from the blood those *putrescent* matters, which would otherwise accumulate in it to its injury, whether as one of the results of the normal *waste* of the system, or as the products of the action of substances introduced into it which operate as *ferments*. It has been already mentioned (§ 411), that the peculiar putrescent matter which is characteristic of the *fœces*, is not directly derived from the decomposition of the indigestible

* The glandulæ of *Brunner* strongly resemble the Salivary glands or pancreas in miniature; and as they are restricted to the duodenum, it is probable that their secretion takes some share, with that of the liver and pancreas, in the act of chylification.

residue of the food, but is a product of the metamorphosis of the fluids and solids of the body itself; which seems necessarily to follow from this consideration among others,—that fæcal matter is still discharged in considerable quantity, long after the intestinal tube has been completely emptied of its alimentary contents. It has been shown by Prof. Liebig, that a substance having the characteristic odour of fæces may be artificially obtained by the imperfect oxidation of albumen, fibrine, caseine, and gelatine.

604. The foregoing are the Secreting organs, whose function seems most directly subservient to the depuration of the blood; but besides these, a vast number of glandular bodies are met with in the different classes and orders of Animals, which eliminate products that have special uses in the economy, but are not in themselves excrementitious. Some of these have a very extensive diffusion; others a more limited one. Under the former head may be ranked the secreting organs which minister to the Digestive operation; for example, the *Gastric follicles*, the *Salivary glands*, and the *Pancreas*. The last of these is the most restricted of the three; but it is met with (as we have seen) under a very simple form, in the highest Mollusca, and presents itself throughout the whole of the Vertebrated series, gradually advancing to a higher type of structure as we ascend the scale. The *Lachrymal* and *Mammary glands*, on the other hand, are more limited in their distribution; for the former are confined to the three higher classes of Vertebrata, and the latter to Mammalia only. Various glands for eliminating *odoriferous* matters, such as musk and castor,— or *poisonous* substances, as those connected with the 'stings' of Hymenopterous insects, or contained in the mandibles of Spiders, or placed at the end of the tail of Scorpions,—or a *glutinous* matter which hardens into a thread when expressed through a narrow orifice, as that which supplies the spinnerets at the end of the abdomen of the Spider, and furnishes the material for the 'cocoon' spun by the mouth of the larvæ of many Insects,—are of very limited diffusion; being confined to smaller groups, and even in some instances to a particular genus or species. —The Skin of many animals, again, is abundantly furnished with *Mucous* and *Sebaceous* follicles; whose secreting action is obviously rather protective as regards the integument, than depurative as regards the blood. The Mucous secretion is generally found in aquatic animals; and prevents the water from coming into direct contact with the skin. The follicles by whose cells it is eliminated from the blood, are usually very simple in their character, resembling those of ordinary Mucous membranes (Fig. 32, *b*); but sometimes they are more complex, especially in Fishes (§ 322 *h*). So the Sebaceous follicles are more commonly found in the skin of animals which live on land; and the office of their secretion appears to be, to prevent its surface from being dried up and cracked by the action of the sun and air. It is especially abundant in those tribes which are formed to inhabit warm climates. The sebaceous glandulæ present a degree of variety in complexity, which is similar to that which exists among the mucous glandules; some of them being simple follicles lodged in the substance of the skin; whilst others are composed of similar follicles, more or less branched, elongated, or convoluted; and others, again, seem to consist of little else than clusters of fat-cells, out of which an excretory duct arises. In the Mammalia, they very commonly open into the hair-canal;

and in Birds, into the socket of the stem of the feather. Various peculiar glands, moreover, whose uses are but little known, are connected with the genital apparatus. The Testes, or proper spermatic glands, will be described in the next chapter.—In all these cases, the general plan of structure is the same; and the difference in the products can be attributed to nothing else, than to a peculiarity in the endowments of the epithelial cells which are the real instruments of the secretion.

605. *Metastasis of Secretion.*—Although the number and variety of the secretions become greater, in proportion to the increased complexity of the nutritive processes in the higher classes, and although each appears as if it could be formed by its own organ alone, yet we may observe, even in the highest animals, some traces of that community of function which characterises the secreting apparatus of the lowest. It has been shown that, although the products of secretion are so diversified, the elementary structure of all glands is the same; that wherever there is a free secreting surface, it may be regarded as an extension of the general envelope of the body, or of that reflexion of it which lines the digestive cavity; that its epithelium is continuous with the epidermis of the integument, or with the epithelium of the mucous membrane from which it is prolonged; and that the peculiar principles of the secreted products preexist in the blood, in a form which is at least closely allied to that which they assume after their separation. If, then, the general law formerly stated (§ 351) be correct, we should find that, when the function of any particular gland is suspended, or when it is not performed with sufficient activity to separate all the excretory products from the blood, other secreting organs, or even the general surface, should be able to perform it in some degree. That this is actually the case, pathological observation is continually showing; and so striking are the 'metastases of secretion' which are thus exhibited, that it was even asserted by Haller that almost all secretions may, under the influence of disease, be formed by each and every secreting organ. This statement, however, needs to be received with some limitation; and it would probably be safest to restrict it to the *excretions*, whose elements pre-exist in the blood, and accumulate there when the elimination of them by their natural channel is suspended. Thus it seems to be established by a great mass of observations, that *Urine*, or a fluid presenting its essential characters, may pass off by the mucous membrane of the intestinal canal, by the salivary, lachrymal, and mammary glands, by the testes, by the ears, nose, and umbilicus, by parts of the ordinary cutaneous surface, and even by serous membranes, such as the arachnoid lining the ventricles of the brain, the pleura, and the peritoneum; and such a metastasis has not only taken place in cases in which the normal excretion was checked or impeded by disease, but has been induced experimentally by extirpating the kidneys, or by tying the renal artery. — So, again, if the elimination of the *Bile* be checked by disease of the liver, or by the application of a ligature to the vena portæ, or if its passage out of the system be prevented by the application of a ligature to the biliary duct, some (at least) of its elements are discharged through other channels; the urine, the pancreatic fluid, the milk, the cutaneous transpiration, and even the sputa derived from the respiratory passages, being more or less deeply tinged with the yellowish-brown colouring-matter of bile, and possessing its characteristic taste; and the

same matters being also found in the fluids of the serous cavities, and passing even into the solid tissues.—The secretion of *Milk*, also, has been thus transferred to different parts of the skin, to the gastro-intestinal mucous membrane, to the mucous membrane lining the bronchial tubes, and even to the surface of an ulcer.—Thus we see that those products of decomposition, at least, which accumulate in the blood when their usual exit-pipe is no longer open, may find their way through other channels; a provision which is obviously intended to diminish the injurious results of a suspension of the excretory functions (§ 581), and which is at the same time in complete and beautiful harmony with the general principle, that the specialisation of a function does not involve the complete extinction of its original generality (§ 351).*

CHAPTER XVII.

EVOLUTION OF LIGHT, HEAT, AND ELECTRICITY.

1. *General Considerations.*

606. In considering, on a former occasion (CHAP. V.), the relations of the Animal and Vegetable kingdoms to each other, and to the Inorganic universe, it was pointed out that whilst the existence of the *Vegetable* world depends upon the constant agency of certain Physical Forces, by which the germ is enabled to draw-in and to appropriate the inorganic elements, which it combines into organic compounds, and incorporates with itself into an organised fabric,—that of the *Animal* kingdom is rather dependent upon the supply of food which it derives from the vegetable world, by means of which its higher forms at least are rendered comparatively independent of external agencies, requiring no Light to enable them to appropriate their aliment, and being able to generate within themselves the Heat which is necessary to sustain their vital activity. In the production of *Heat*, then, we have one of those cases in which Animals restore to the Physical Universe the forces which it has imparted to Plants (§ 261); and we shall find it to be effected by the very act of restoring to the condition of inorganic matter, those elements which the agency of Light and Heat upon the vegetable germ had enabled it to withdraw. Although the production of Heat is most considerable and regular in those higher animals which are termed warm-blooded, yet it takes place in an inferior degree probably in all. It is also a phenomenon of occasional occurrence in Plants, but only under the same conditions as in Animals, *viz.* when organic compounds are being partially or completely restored to the condition of organic matter; and it would seem as if it was in them rather a necessary result of transformations which are being effected for other purposes, than a purpose for which such transformations are to be made.—The same may be said of the production of *Light*. It is by no means an ordinary phenomenon in the Animal kingdom; but where it does occur, it appears to have some

* For a more detailed examination of this interesting topic, see the Author's article on *Secretion* in the "Cyclopædia of Anatomy and Physiology," vol. iii.

special purpose; and although the processes by which it is maintained are not clearly understood, yet there can be little doubt that it too is dependent upon a slow combustion, in which the carbon and hydrogen of the living system are given back to the atmosphere as carbonic acid and water; the oxidation of other substances, also, perhaps contributing to the effect. On the other hand, its appearance in Plants is a much rarer occurrence, and seems to be (so to speak) accidental. — Of the generation of *Electricity*, we know comparatively little. There is strong evidence that its production must be going on, in every action of Organic as well as of Inorganic Chemistry; and that a disturbance of electric equilibrium must be continually taking place in each molecule of the living Plant and Animal. But it would seem as if, in general, the generation of electricity is simply a result of changes which are directed to other ends; and that, so far from any use being made of it in the economy, there is usually a set of provisions for the speediest possible restoration of the disturbed equilibrium. In certain Animals, however, the case is very different; for we find them endowed with an apparatus whose special purpose is obviously the generation of Electricity, in considerable amount and intensity; and although we may not be acquainted with all the objects which this curious organisation may answer, yet some of its more obvious uses can be clearly made out.

2. *Evolution of Light.*

So little is known of the causes or purposes of that evolution of Light, which is of no unfrequent occurrence amongst organised beings, of the lower classes especially, that it would be useless to speculate upon them. It is well, however, to bring together the principal facts relating to the phenomenon itself, and to the conditions of its occurrence.

607. *Evolution of Light in Vegetables.*—It has been asserted that many flowers, especially those of an orange colour, such as the *Tropæolum majus* (Nasturtium), *Calendula officinalis* (Marigold), *Helianthus annuus* (Sun-flower), &c., disengage light in serene and warm summer evenings, sometimes in the form of sparks, sometimes in a more feeble and uniform manner; but many physiologists are disposed to question these assertions, from their not having been themselves able to witness the phenomena. There is no doubt, however, that light is emitted by many *Fungi*, whilst actively vegetating, and in some instances to a very considerable extent.*

* The following is one of the most recent and authentic instances yet recorded.—"One dark night, about the beginning of December, while passing along the streets of the Villa de Natividada, I observed some boys amusing themselves with some luminous object, which I at first supposed to be a kind of large fire-fly; but on making inquiry, I found it to be a beautiful phosphorescent Fungus, belonging to the genus Agaricus; and was told that it grew abundantly in the neighbourhood on the decaying leaves of a dwarf palm. Next day I obtained a great many specimens, and found them to vary from one to two and a half inches across. The whole plant gives out at night a bright phosphorescent light, of a pale greenish hue, similar to that emitted by the larger fire-flies, or by those curious soft-bodied marine animals, the Pyrosomæ. From this circumstance, and from growing on a palm, it is called by the inhabitants 'Flor do Coco.' The light given out by a few of these Fungi in a dark room was sufficient to read by. I was not aware at the time I discovered this fungus, that any other species of the same genus exhibited a similar phenomenon; such, however, is the case in the *Ag. olearius* of De Candolle; and Mr. Drummond of Swan River Colony in Australia has given an account of a very large phosphorescent species occasionally found there."—Gardner's Travels in Brazil, 2nd Ed. p. 264.

The light is perceived in all parts of the plant, but chiefly in the young white shoots; and it is more vivid in young than in old plants. The phosphorescence is stronger in such as grow in the moist and warm localities of mines, than in those inhabiting dry and cold situations. It ceases if the plant be placed in vacuo, or in any atmosphere which does not contain oxygen; but reappears when it is restored to the air, even after remaining for some hours in vacuo or in azote. No phosphorescence is perceived after the death of the plant. Some *Algæ*, also, have been observed to be luminous when in a growing state.—On the other hand, luminosity is sometimes observed under circumstances that forbid our regarding it as in any degree a vital phenomenon. Thus it is stated by Martius, that the juice of the *Euphorbia phosphorea,* a Brazilian plant, emits light, especially when heated. An evolution of light has frequently been observed to take place from dead and decaying wood of various kinds, particularly that of roots; it seems connected with the conversion of oxygen into carbonic acid, but is not increased when the substance is placed in pure oxygen. Decomposing Fungi, also, frequently exhibit luminosity; but this is very different from that displayed by some of the same tribe during their living state.—Considering that in all the circumstances mentioned, the combination of carbon and oxygen is taking place to some amount, it seems difficult to believe that there is not some connection between the phenomena; but no speculation can yet be raised on the subject, with any prospect of stability, from the want of sufficient facts as its basis.

608. *Evolution of Light in Animals.*—A large proportion of the lower classes of aquatic Animals possess the property of luminosity in a greater or less degree. The phosphorescence of the sea which has been observed in every zone, but more remarkably between the tropics, is due to this cause. When a vessel ploughs the ocean during the night, the waves,—especially those in her wake, or those which have beaten against her sides,—exhibit a diffused lustre, interspersed here and there by stars or ribands of more intense brilliance. The uniform diffused light is partly emitted by innumerable minute Animals, which abound in the waters of the surface; and these, if taken up into a glass vessel, continue to exhibit it, especially when the fluid is agitated. Most of these Animals belong to the class of *Acalephæ,* a large proportion of whose members appear to be more or less phosphorescent, those of tropical seas being the most so. The *Noctiluca miliaris,* a minute jelly-like animal with an elongated proboscis, barely perceivable by the naked eye, is the most common source of the luminosity in the British seas. The light is emitted, particularly round the tentacula, and from the ciliated surfaces, during the movements of the animal: and it seems to proceed from a mucus secreted from the integument, which may continue to exhibit the same property for a time when removed from it. This mucus, which has a very acrid character when applied to the human skin, communicates to it a phosphorescent property; and, when mixed with water or milk, it renders these fluids luminous for some hours, particularly when they are warmed and agitated. From this source it is probable that the diffused phosphorescence of the sea is partly derived; whilst the brilliant stars and ribands, with which the surface is bespangled, indicate the presence of the larger tenants of the deep. The effects of physical and chemical agencies upon the luminosity of the

Noctiluca have been recently studied by Dr. Pring.* He found that when the water containing them was subjected to a simple galvanic current, no very perceptible effect could be observed; but when an electro-magnetic current was employed, after a time a steady and continued flow of light was given out from the whole of the water, the surface of which appeared as if spangled with numberless minute but persistent points of light. The light ceased after a quarter of an hour, and could not be reproduced, evidently in consequence of the death of the animalcules. When a portion of the luminous sea-water was placed in a bottle filled with oxygen gas, the phosphorescence of the animals was increased whenever the water was agitated with the oxygen; the animals lived in this state for more than a week. Sulphuretted hydrogen instantly destroyed all the luminosity, being at once fatal to the animals. With carbonic acid, the luminous property of the water was strongly and continuously brought out for about fifteen minutes, the light being bright enough to enable the hands of a watch to be seen in a dark room; but at the expiration of that time the light gradually became fainter, and in five or ten minutes more had totally ceased. When sulphuric acid was dropped upon the water, it emitted for a minute or two a bright light, which then disappeared; nitric acid had a similar effect; and with hydrochloric acid the increased luminosity was much less conspicuous, and the darkness ensued almost instantaneously. A few drops of ether let fall into the sea-water in the dark, appeared instantly to deprive it of its luminous property. On substituting chloroform for ether, in a second experiment, a very bright and persistent phosphorescence was given out for a few minutes, after which the water speedily became dark, the animals being evidently destroyed.—Thus, then, all the agents which produced a marked increase in the light, with the exception of oxygen, were speedily fatal to the life of these animals; and hence it would seem as if the augmented luminosity were the result of the disordered actions called forth by these injurious influences.

609. The Acalephæ, however, are not the only Radiated animals which exhibit luminosity. In the class of *Zoophytes*, the phenomenon seems to be very common among the members of the order *Hydroida*; being most distinctly seen when the animals are in a state of vigour, and are subjected to some shock or irritation. It is most remarkably exhibited, however, by the several species of the family *Pennatulidæ*, belonging to the order *Asteroida*; but their phosphorescence seems to be only displayed under the influence of some mechanical or chemical irritation. It has been observed by Prof. E. Forbes, that when any portion of the stem or branches of a Pennatula is touched, the luminosity first shows itself there, and then spreads itself, in a wave-like manner, towards the polypiferous extremities; whilst, if any of these extremities be touched, the luminosity does not spread backwards from the point of contact, but remains confined to the part irritated. When plunged into fresh water, the Pennatula scatters sparks about in all directions, and then ceases to be luminous; but when plunged into spirits, it does not do so, but remains phosphorescent for some minutes, the light dying gradually away, and vanishing last of all from the uppermost polypes.† Certain *Echinoder-*

* "Philosophical Magazine," Dec. 1849.

† See Dr. Johnston's "British Zoophytes," pp. 25–27, 150–155.

mata, also, have exhibited luminosity ; but it seems to be restricted to the orders *Asteriada* and *Ophiurida.*—Each class of the Molluscous series, also, contains phosphorescent animals ; but the phenomenon is especially frequent in the class of Tunicata, in which, however, it appears to be chiefly or entirely restricted to the families *Salpidæ* and *Pyrosomidæ,* which float freely on the waters of the ocean, abounding especially in the warmer seas. Of these, too, it may be observed that the luminosity is augmented by agitation or by friction. Among the Conchifera, this phenomenon has been chiefly witnessed in the *Pholas* and *Lithodomus.* No marine Gasteropod has yet been observed to possess luminosity ; but phosphorescence has been observed in a species of Slug (*Phosphorax noctilucus*) inhabiting the higher mountains of Teneriffe. Among Pteropods the *Cleodora,* and among Cephalopods the *Octopus,* have been seen to exhibit phosphorescence.—In all these, the general phenomena are analogous ; the luminous matter appearing to be a secretion from the surface of the animals, which communicates its peculiar property to water or to solid substances that come in contact with it. The light disappears in vacuo, but reappears in air ; it is increased by moderate heat, and by gently stimulating fluids ; whilst a cold or boiling temperature, or strong stimulants, soon extinguish it. It continues for some days after death, but ceases at the commencement of putrefaction.—The same power has been attributed to *Fishes ;* but it is not improbable that, with regard to these, there has been a partial deception, arising from the excitement given by their movements to the sources of phosphorescence in the surrounding water. Late observations, however, lead to the belief that, in some species of Fish, there is an inherent luminosity : a species of *Scopelus,* three inches long, has been seen to emit a brilliant phosphorescent light, in stars or spangles, from various parts of the scaly covering of the body and head ; and this continued to be displayed at intervals during the life of the animal, in a glass of sea-water,—ceasing entirely with its death.

610. Among the Articulated classes, the evolution of light is by no means an uncommon phenomenon ; but in some of these it appears due to a different agency from that on which the luminosity of Zoophytes and Mollusks is dependent. The luminosity which is observable in many of the marine *Annelida* is not a steady glow, but a series of vivid scintillations (strongly resembling those produced by an electric discharge through a tube spotted with tin-foil) that pass along a considerable number of segments, lasting for an instant only, but capable of being repeatedly excited by any irritation applied to the body of the animal. The peculiar character of this emission of light seems to remove it altogether from the category of ordinary 'phosphorescence', and leads to the supposition that it is dependent upon a direct conversion of Nervous Agency into Light (§ 54).* A very similar kind of luminosity is observable in many minute Crustacea, which emit light in brilliant jets ; and it is a curious fact, mentioned by Prof. E. Forbes, that the cavity of Salpæ which have

* This was the conclusion at which the Author had arrived from his own observations upon the luminosity of the Annelida, made at Tenby in the year 1843. About the same time, M. de Quatrefages arrived at similar conclusions from his observations on the Annelida of La Manche. See the Report of Prof. Milne Edwards, in the "Annales des Sciences Naturelles" for Jan. 1844, p. 23.

been deprived of their visceral 'nuclei' (§299 *b*), often contains multitudes of minute Crustacea, which give out such a succession of phosphorescent flashes, as possibly to deceive the observer into the belief that it is the mollusk itself which is luminous.

611. Among *Insects*, however, we find numerous examples of a luminosity which is obviously of a different character, being clearly traceable to a combustive process; and this is restricted to particular portions of the body, sometimes even to minute points. The luminous Insects are most numerous among the order *Coleoptera* (Beetle tribe); and are nearly restricted to two families, the *Elateridæ*, and the *Lampyridæ*.* The former contains about 30 luminous species, which are all natives of the warmer parts of the New World. The light of these 'Fire-flies' proceeds from two minute but brilliant points, which are situated one on each side of the front of the thorax; and there is another beneath the hinder part of the thorax, which is only seen during flight. The light proceeding from these points is sufficiently intense to allow small print to be read in the profoundest darkness, if the insect be held in the fingers and moved along the lines. In all the luminous Insects of this family, the two sexes are equally phosphorescent.—The family *Lampyridæ* contains about 200 species known to be luminous, the greater part of which are natives of America, whilst others are widely diffused through the Old World. These are known as 'Glow-worms' (*Lampyris noctiluca and Lam. splendidula*); their light issues from the under surface of the three last abdominal rings; it is most brilliant in the female, and exists in a feeble degree in the egg, larva, and chrysalis. The luminous matter consists of little granules, and is contained in minute sacs, covered with a transparent horny lid. These sacs are mostly composed of a close network of finely-divided tracheæ; which also ramify through every part of the granular substance. The lid exhibits a number of flattened surfaces, so contrived as to diffuse the light in the most advantageous manner. The phosphorescence appears to be occasioned by the slow combustion of a peculiar organic compound, the production of which is dependent for its continuance upon the life and health of the animal; the activity of this combustion is stimulated by anything which excites the vital functions of the individual, and it is particularly influenced by the energy of the respiratory process. If the opening of the trachea which supplies the luminous sac, be closed, so as to check the access of air to its contents, the light ceases; but if the sac be lifted from its place, without injuring the trachea, the light is not interrupted. In all active movements of the body, in which the respiration is energetic, the light is proportionably increased in brilliancy. If the luminous segments be separated from the rest of the body, they continue phosphorescent for some time; and if they be crushed between the fingers, long streaks of light are perceived to issue from the yellowish matter which they contain. By careful experiments upon the luminous

* Of the reputed luminous power of the *Fulgora*—a very remarkable genus of the order *Homoptera*, of which one species inhabits Guiana, whilst another is a native of China,—there is, to say the least, very considerable doubt. The authority on which it has been asserted is a very questionable one; and naturalists who have themselves carefully observed these insects, have seen no traces of it. There may, however, be some ground for the statement; particularly if, as it has been suggested, the luminosity be exhibited by one sex only, and during only a portion of the year.

product thus separated from the body, Prof. Matteucci has been able to prove that the emission of light is dependent upon a combustive process, in which carbonic acid is rapidly generated at the expense of the surrounding oxygen; and he has also ascertained by analysis, that the luminous matter does not contain any appreciable quantity of phosphorus.*—Phosphorescence is a rare phenomenon among aerial animals of the higher classes. An emission of light has been seen from the egg of the grey Lizard; and it has been stated that a species of Frog or Toad inhabiting Surinam is luminous, especially in the interior of its mouth. (See § 613.)

612. Of the particular objects of this provision in the animal economy, little is known, and much has been conjectured. It is generally imagined, that it is destined to enable the sexes of the nocturnal animals (especially Insects) to seek each other for the perpetuation of the race; and this hypothesis would seem to derive support from the fact, that the light is generally most brilliant at the season of the exercise of the reproductive functions, and at that period exists in some species (such as the Earth-worm) which do not manifest it at any other. Moreover it is well known that the male Glow-worm, which ranges the air (whilst the female, being destitute of wings, is confined to the earth), is attracted by any luminous object; so that the poetical language of Dumeril, who regards the phosphorescence of the female as "the lamp of love — the pharos — the telegraph of the night, which scintillates and marks in the silence of darkness the spot appointed for the lovers' rendezvous," would not seem so incorrect as the ideas of Poets on subjects of Natural History usually are. It may be objected on the other hand, that there are many moths and beetles, which have a similar tendency to fly towards the light, and among which no phosphorescence is exhibited. Some of these, however, are faintly luminous; and it would not seem improbable that the insects which are attracted by flame, and thus show that they are seeking for objects which emit light, may be cognisant of more feeble degrees of its emission, than our eyes can appreciate.† Still it must be remembered that certain animals (as Zoophytes) are phosphorescent, which have no occasion to seek each other with this object; and it does not seem impossible that the property may be conferred upon them (like the stinging power possessed by some) as a means of self-defence, in the deficiency of active powers of locomotion, or of dense external covering. It may serve, too, for the illumination (however faintly) of those depths of the ocean, which are known to

* "Lectures on the Physical Phenomena of Living Beings," p. 181.

† It has been objected that, as the male is luminous, and also the larva and pupa, the meeting of the sexes can scarcely be the object of the provision. But this difficulty is easily surmounted. Mr. Kirby justly remarks that "as the light proceeds from a peculiarly-organised substance, which probably must be in part elaborated in the larva and pupa states, there seems nothing inconsistent in the fact of *some* light being then emitted, with the supposition of its being destined solely for use in the perfect state. And the circumstance of the male having the same luminous property no more proves that the superior brilliancy of the female is not intended for conducting him to her, than the existence of nipples and sometimes of milk in man proves that the breast of woman is not meant for the support of her offspring." The luminosity of the Insect in all these states may have the more remote purpose, also, of making its presence known to the nocturnal birds, &c. which are destined, in the economy of nature, to feed upon it. The larva of the *Lampyris occidentalis* has been observed, when alarmed, to feign death and extinguish its light.

be tenanted by Fishes and other marine tribes, but which receive no appreciable portion of solar light.

613. An evolution of Light during the incipient decay of dead animal matter, is by no means of uncommon occurrence. It has been most frequently observed to proceed from the bodies of Fishes, Mollusca, Medusæ, and other marine tribes; but it has been seen also to be evolved from the surface of terrestrial animals, and even of Man. This phosphorescence ceases immediately on the commencement of fœtid putrefaction; and it would appear to proceed from the formation of luminous matter during an early stage of decomposition, by some of those primary changes in the combination of the organic elements, which immediately succeed dissolution, or which may even precede it for a brief period. Such would seem to have been the case in certain well authenticated instances of the evolution of light in the living human subject.* In most of these cases, the individuals exhibiting the luminosity had suffered from some wasting disease, and were near death. One instance is recorded, in which a large cancerous sore of the breast emitted light enough to enable the hands of a watch-dial to be distinctly seen, when it was held within a few inches of the ulcer; here, too, decomposition was obviously going on, and the phosphorescent matter produced by it was exposed to the oxygenating action of the atmosphere.†

3. *Evolution of Heat.*

614. As it is a part of the peculiar character of living organised beings, that their vital activity can only be sustained under the constant influence of Heat (CHAP. III. Sect. 3), it is obvious that those only can be rendered independent of variations in the temperature of the surrounding medium, so as to maintain that which is most favourable to the performance of their various actions, which have the power of *generating* Heat when it is not sufficiently imparted to them from external sources, and of *resisting* its influence when it is excessive. Having already considered the means by which the temperature of the living body is kept *down* to its proper standard (CHAP. XIV.), we have now to enquire into the sources of that generation of Heat, within the living body, which keeps *up* its temperature in 'warm-blooded' animals to a certain fixed point, and which assists in enabling even the 'cold-blooded' to resist the effects of extreme depres-

* "An Account of several cases of the Evolution of Light in the Living Human Subject;" by Sir Henry Marsh, M.D., M.R.I.A., &c.

† Such facts appear to give support to the idea, that a *preternatural combustibility* may sometimes exist in the body, owing to the retention of phosphorus-compounds, which should normally be excreted from it by the urine after undergoing oxidation. It has been observed that the *breath* of drunkards has sometimes exhibited luminosity, as if it contained the vapour of phosphorus or of some of its compounds; and it has been found by experiments upon dogs, that if phosphorus be mixed with oil and injected into the blood-vessels, it escapes unburned from the lungs. ("Casper's Wochenschrift," 1849, No. 15.)—The Author has seen a remarkable case, drawn up under the hand of the subject of it (a highly respectable clergyman), and shown him by his friend Dr. M. Barry, in which a troublesome sore, occasioned by the combustion of phosphorus on the hand, twice at distant intervals emitted a flame which burned the surrounding parts. It was particularly stated that ignition could not have been effected by any neighbouring flame, and that the combustion could not be due to any particles of phosphorus remaining in the wound; and it does not seem improbable that, in the peculiar condition of the sore, an unusual amount of phosphorus-compounds had been deposited in it, so as even to become spontaneously inflammable on the contact of oxygen.

sion of the external temperature.—It is well known that almost all Chemical changes are attended with some disturbance of the temperature of the agents concerned; and it may not unreasonably be surmised that, of those which are so constantly occurring in the living system, some may be connected with the disengagement of the Heat peculiar to it. Much uncertainty still prevails on this subject; but there can be little doubt that a large proportion of the caloric liberated by organised beings, is generated by the combination of atmospheric oxygen with the carbon and hydrogen furnished by them, to form the carbonic acid and water which they are constantly excreting; since we find these two changes everywhere bearing a close relation with each other. Several other changes of composition are going on, however, in the living body, to which a part of the effect must be attributed; and there are some residual phenomena, which seem to indicate that Heat may occasionally be a direct product of the metamorphosis of the Nervous Force.

615. *Evolution of Heat in Vegetables.*—Much dispute has occurred at different periods, as to whether Plants could be considered as having a *proper heat* or not; and this has resulted from the limited view which has been taken of the processes of the Vegetable Economy. Although the excretion of carbonic acid is constantly going on, under the conditions formerly described (§ 499), it usually takes place so slowly, and from a surface so openly exposed to the atmosphere, that it could scarcely be expected that there should be any sensible elevation of the temperature of the part from this source,—especially when it is considered that a constant loss of heat is taking place by evaporation: and as there is no provision for the conveyance of caloric set free in one part to distant portions of the system, a *general* maintenance of vital warmth would be still less anticipated. In plants of small or moderate size, accordingly, the temperature is found to vary with that of the atmosphere; but the interior of large trunks seems to maintain a more uniform degree, being colder than the atmosphere in summer, and warmer in winter. This fact may be accounted for on two different grounds. The slow conducting power of the wood, which is much less transversely to the direction of its fibre, than with it, would prevent the interior of a large trunk from being rapidly affected by changes in the heat of the external air; and accordingly, it is found that, the larger the trunk on which the observation is made, the greater is the difference. Again, some motion of the sap takes place even in winter; and as the earth, at a few feet below the surface, preserves a very uniform temperature, it is not improbable that the transmission of fluid derived from it through the stem, may have an influence on the state of the latter; a supposition which is countenanced by the fact, that the temperature of the interior of a large trunk, and that of the soil four feet below the surface (which may be regarded as the medium depth of roots), bear a very close correspondence. It is reasonable to suppose that both these causes may be in operation.—By experiments, however, made with instruments of great susceptibility to changes of temperature, Dutrochet ascertained that Plants do possess some power of generating heat, in the parts in which the most active changes are taking place. In order to obtain unexceptionable evidence to this effect, it was necessary to exclude the influence of evaporation in depressing the temperature. This was effected by making the comparison, not between the temperature of the

plant and that of the surrounding air, but between similar parts in a living plant, and in one recently killed by immersion in hot water, which would be (after cooling) equally susceptible with the former of the diminution of temperature which evaporation causes. In some instances, this source of error was still further guarded against, by the immersion of both plants in an atmosphere saturated with aqueous vapour. The temperature of the leaves and young shoots was ascertained, in preference to that of the stem; both in order to avoid the source of fallacy already mentioned, and because in them the greatest *proper heat* might be expected. With these precautions, the result was constantly the same. An elevation of temperature, sometimes to the amount of nearly a degree (Fahr.), was observed in the herbaceous parts of actively-growing plants; differing with the species, the energy of vegetation, and the time of the day. The highest temperature is observed about noon; it increases previously, and afterwards diminishes. This diurnal change is partly influenced by that of the light to which the plant is exposed.*

616. It is, however, when the processes of vegetation give rise to an extraordinary liberation of Carbonic acid (§ 503), that the evolution of heat becomes manifest. This is the case during Germination, when the elevation of temperature, scarcely manifested by a single seed, becomes evident if a number are brought together, as in the process of malting, in which the thermometer has been seen to rise to 110°.—The same may be said of the other period of vegetable growth, in which the function of respiration is carried on to a remarkable extent; that of Flowering. From the large surface exposed, it is evident that, in by far the greater number of instances, the heat will be carried off by the atmosphere the instant it is developed; nevertheless the flowers of a Cistus showed a temperature of 79° whilst the air was at 76°, and those of a Geranium 87° when the air was at 81°. It is in plants of the Arum tribe, however, where flowers are collected in great numbers within cases which act as non-conductors, that the elevation of temperature becomes most appreciable; and it bears a definite relation with the quantity of oxygen converted into carbonic acid. Thus, a thermometer placed in the centre of five spadixes of the *Arum Cordifolium* has been seen to rise to 111°, and in the centre of twelve to 121°, while the temperature of the external air was only 66°; but the production of heat was wholly checked, by preventing the spadix from coming in contact with the air.—The truth of statements of this sort, which has been questioned by many physiologists, has been placed beyond all doubt by the observations of Adolphe Brongniart. He found that, at the first opening of the spathe of *Colocasia odora*, the temperature of the spadix was 8·1° above that of the surrounding air; that this increased during the next day to 18°; and, during the emission of the pollen on the three succeeding days, to 20°; after which it began to diminish with the fading of the flower. More recently these observations have been confirmed by MM. Vrolik and Vriese; who have added to them some important facts. The rise of the temperature was found to be more rapid and considerable, in a spadix placed in oxygen, than in one at a corresponding stage surrounded by common air; and a larger proportion of carbonic acid gas was evolved. On the other

* "Ann. des Sci. Nat.," Deuxième Série, tom. xii.

hand, when a spadix, of which the flowers had already begun to expand, and the temperature to rise, was placed in nitrogen, the temperature sank, and exhibited no elevation during the emission of the pollen; nor was any carbonic acid evolved.*

617. *Evolution of heat in Animals.*—Although we find many instances in the Animal kingdom, in which the capability of maintaining an elevated and uniform temperature is exhibited in a degree to which nothing comparable exists in plants, yet this is by no means a *constant* function of animal any more than of vegetable organisms. Among the lower tribes of Animals, in which the power of locomotion is but feeble, and the supply of the wants of the system not immediately dependent upon it, very little more heat is generated than in Plants. But wherever a high degree of muscular energy is required, in connection with a general activity of the functions of the nervous system, the evolution of caloric to a remarkable extent is provided for in the nutritive processes. We may regard it, therefore, as in *its degree* essentially connected with the development of the Animal powers relatively to the system of Organic life; although really *dependent*, as it would appear, upon the changes occurring in the latter.—It is worthy of notice that, although the temperature of the various parts of the Animal body is usually much more uniform than that of the different organs in Vegetables (owing to the comparative rapidity with which the general circulation of the former diffuses the heat evolved in any one part, and thus tends to equalise the whole), wherever processes are going on which call the nutritive functions into extraordinary activity, *there* a corresponding elevation of temperature occurs. Thus, a slightly increased evolution of heat from the stomach has been observed during the determination of blood to its capillaries, which takes place during digestion; the same is observable in the reproductive organs of those animals, in which the aptitude for the function is periodic only; and the temperature of a muscle (as ascertained by MM. Becquerel and Breschet) rises a degree or more during its contraction.

618. Our knowledge of the heat evolved by the lower Invertebrata is very limited. The *Infusoria* have been observed to possess a certain degree of power of resisting cold. When the water containing them is frozen, they are not at once destroyed; but each lives for a time in a small uncongealed space, where the fluid seems to be kept from freezing by the caloric liberated from the animalcule. What is known in regard to other classes is principally derived from the experiments of John Hunter. He found that a thermometer, introduced in the midst of several *Earthworms*, stood at $58\frac{1}{2}°$, when the temperature of the external air was 57°; and in another instance, when the atmosphere was at 55°, the worms were at 57°. The amount of heat manifested by *Leeches* appeared to be nearly the same, viz. from one to two degrees above that of the atmosphere. Of the Mollusca, nearly the same may be said. Hunter found that the black slug (*Limax ater*) exhibited a temperature of $55\frac{1}{4}°$, when that of the atmosphere was 54°; and the garden snail (*Helix pomatia*) has been observed by others to evolve about the same amount of heat. Further experiments, however, are desirable, for the purpose of ascertaining whether the power of generating caloric varies in such

* Op. cit. tom. xi.

animals with different degrees of external temperature; or whether the heat of their bodies always bears the same close relation with that of the medium in which they exist. The experiments of Hunter furnish the only information on this subject which we possess. He put several leeches into a bottle which was immersed in a freezing mixture, and, the ball of the thermometer being placed in the midst of them, the quicksilver sunk to 31°; by continuing the immersion for a sufficient length of time to destroy life, the quicksilver rose to 32°, and then the leeches froze. A similar result was obtained with a snail. It would appear, therefore, that these animals have the power of resisting, *for a time*, the physical effects of cold; but how far this resistance is due to the power of generating heat, or to the causes arising from their structure (as in Vegetables § 615), cannot be determined without further inquiries. The simple maintenance of a temperature equal to that of the atmosphere, by an animal whose body has a soft surface exposed to the air, implies a certain degree of power of generating caloric; since that surface (as in plants) is constantly being cooled by the evaporation of its moisture.

619. In many Vertebrated animals, the heat of the body is almost equally dependent upon that of the surrounding medium. Thus, *Fishes* in general do not seem capable of maintaining a temperature more than two or three degrees higher than that of the water in which they live. There are, however, some remarkable exceptions; for Dr. J. Davy found that certain marine fishes, as the Bonito and Thunny, whose gills are supplied with nerves of unusual magnitude, and which have also a very powerful heart, and a quantity of red blood sufficient to give the muscles a dark red colour, maintain a temperature much higher than that of the white fishes of fresh water on which Hunter experimented. Thus, Dr. D. observed in the bonito a temperature of 99°, whilst that of the sea was but 80½°. Although the conditions of existence in Vertebrata, in which the animal powers are developed to their greatest extent, might have seemed to require a greater power of generating heat than Fishes usually possess, it is to be remembered that this class is less liable to suffer from alternations of temperature connected with the seasons, than those which inhabit the air. In climates subject to great atmospheric changes, the heat of the sea is comparatively uniform through the year, and that of deep lakes and rivers is but little altered. Many have the power of migrating from situations where they might otherwise suffer from cold, into deep waters; and it is an unquestionable fact, that the species which are confined to shallow lakes and ponds, and which are thus liable to be frozen during the winter, are frequently endowed with tenacity of life, sufficient to enable them to recover after a process which is fatal to animals much lower in the scale. Fishes are occasionally found imbedded in the ice of Arctic Seas; and some of these have been known to revive when thawed. (See § 103.)

620. In *Reptiles* the power of generating caloric is somewhat greater. In all cases, however, the temperature of their bodies is greatly dependent upon that of the medium which they inhabit; but in proportion to the depression of the latter, do they seem endowed with the power of maintaining their own above it. Thus, when the air was at 68°, a *Proteus* manifested the same degree of heat; but when the air was lowered to 55°, the temperatnre of the animal was 65°. In the same manner, it appeared

that the edible frog (*Rana esculenta*) possessed a temperature of 72½°, when examined in an atmosphere of 68°; and that in ice of 21°, the animal maintained a heat of 37½°. The *Chelonia* do not seem endowed with the power of evolving heat to the same degree with the *Saurian* and *Ophidian* reptiles. In some of the more agile of the Lizard tribes, the high temperature of 86° has been noticed, when that of the external air was but 71°.—In all experiments on the influence of *change* of temperature on such animals, it is necessary to guard against the fallacy arising from the slowness (resulting from their non-conducting power) with which their bodies acquire the altered heat of the medium, whether it be increased or diminished. By attending to this precaution, it has been shown that many of the statements which have been made regarding their power of modifying their temperature are liable to exception; but it cannot be questioned that Reptiles have some capability of generating heat, which is called into action in resisting the depressing influence of cold. This is unequivocally proved by the fact, that frogs will remain alive in water which is frozen around them (even when the thermometer has fallen to 9°), the water in contact with the body remaining fluid, and the temperature of the body being 33°.

621. The classes of animals, which are especially endowed with the power of producing and maintaining heat, are Insects, Birds and Mammalia. The temperature of *Insects* has been very ably investigated by Mr. Newport.*—In the *Larva* condition, the temperature of the animal corresponds much more closely with that of the atmosphere, than in the perfect state; thus, the larva of the higher species of *Hymenoptera* (Humble-bees, &c.) is usually from 2° to 4° above the surrounding medium, whilst the perfect Insect has a range of from 3° to 10°, or even more; and the Caterpillar of the *Lepidoptera* is seldom more than from ½° to 2° warmer than the atmosphere (the amount varying in close relation with the activity of the individual); whilst the perfect insect is, when much excited, 5° or 9° above it. It is probable that in those tribes, in which no complete metamorphosis exists, but in which the difference between the development of the larva and that of the perfect insect is but trifling, there is not the same variation with regard to the production of heat.—The *Pupa* state being, in all Insects which undergo a complete metamorphosis, a condition of absolute rest, the temperature of the individual is in general lower than at any previous or subsequent period of its existence; and it is only equal to, or at most very little above, that of the surrounding medium. But in those species, which, not undergoing a complete metamorphosis, continue active during the whole of life, this diminution of the power of maintaining heat probably does not occur. Within a short period after the first change, however, the Pupa often retains some of the characteristics of the larva state, and exhibits a temperature somewhat elevated; and if it be at any time excited to motion, a slight degree of heat is manifested. The pupa appears to follow variations in atmospheric temperature more rapidly than the larva; and as an elevation of temperature becomes necessary towards the epoch when the final metamorphosis is to take place, means are provided for it. In the Lepidoptera, the Chrysalis has itself the power of generating heat, at the

* "Philosophical Transactions," 1837.

period when its energies are aroused, and it is about to burst forth from its silky envelope; whilst in the Hymenoptera, it is most curious to observe an artificial warmth communicated to the pupæ, by an increased evolution of heat from the bodies of the perfect insects which crowd over their cells (§ 623).

622. The increase in the power of generating heat which is characteristic of the *Imago* or perfect Insect, is not manifested immediately on its emersion from the pupa state; in fact at that period, when the body is soft and delicate, and the unexpanded wings hang uselessly from its sides, it parts with its heat with great rapidity. It is not until its active respiratory movements have commenced, and the whole system has been stimulated by the exercise of its locomotive powers, that the evolution of heat takes place to any remarkable extent; and whether these processes be delayed or hastened by the influence of external circumstances, the elevation of the temperature of the individual is still proportional to them. Thus, a specimen of the *Sphinx ligustri* which had only left the pupa state about an hour and a quarter, had a temperature of but 4° above the atmosphere; whilst, at the expiration of two hours and a quarter, when it had become strong and had just taken its first flight, it had a temperature of 5·2°; and another specimen, which had been longer exerting itself in rapid flight, was as much as 9° warmer than the surrounding air. In the states of abstinence, inactivity, sleep, and hybernation, the evolution of heat is checked; and the temperature of the perfect insect may fall very nearly to that of the atmosphere. By inordinate excitement, on the other hand, a very rapid evolution of heat may be produced. Thus, a single individual of *Bombus terrestris* (Humble-bee) enclosed in a phial of the capacity of three cubic inches, had its temperature gradually raised, by violent excitement, from that of rest (2° or 3° above that of the atmosphere) to 9° above that of the external air, and had communicated to the air within the phial as much as 4° of heat within five minutes. In an experiment upon another species, *Bombus Jonella,* the temperature of the air within the phial was raised by the motion of the insect, during six or eight minutes, as much as 5·8° above that of the atmosphere; but when the bulb was held near enough to the insect to touch the tips of its wings, the mercury sunk 2·2°. This observation, which was repeated several times with the same results, shows that the vibration of the wings tends to cool the body of the insect during its flight.—In regard to the relative amount of heat evolved by different tribes of perfect Insects, Mr. Newport has ascertained that the volant insects in their perfect state have the highest temperature, while those species which have the lowest temperature are located on the earth. Among the volant insects, those *Hymenopterous* and *Lepidopterous* species have the highest temperature, which pass nearly the whole of the daytime on the wing; of these the Hive-bee, with its long train of near and distant affinities, and the elegant and sportive Butterflies, have the highest. Next to these are probably their predatory enemies the Hornets and Wasps, and others of the same order; and lastly, the Ants, the temperature of whose dwelling has been found to be considerably above that of the atmosphere. Next below the Diurnal insects, are the Crepuscular, the highest of which are the Sphinges and Moths, and almost equal with them are the Chaffers. In some of the *Coleoptera* (Beetle tribe) the

amount of heat is found to approach very nearly to that in Hymenoptera; in both of these tribes the organs of respiration are of large extent, and the quantity and activity of aeration considerable. On the other hand, the inferior temperature of crepuscular Insects to that of diurnal species of the same orders is associated with a lower degree of respiration. Nearly all the Hymenoptera are diurnal, and bear the privation of atmospheric air with greater difficulty than many other tribes. Further it would appear that some of the volant Coleoptera have, even in a quiescent state, a higher temperature than some of the terrestrial Coleoptera in a state of moderate activity, the difference being much increased in the active condition of the former.

623. It is among the Insects which live in societies, however (nearly all of them belonging to the order of *Hymenoptera*), that the greatest evolution of heat is manifested. Mr. Newport's observations were made principally upon the *Bombus terrestris* (Humble-bee) and *Apis mellifica* (Hive-bee).—A single individual of the former species has frequently, when moderately excited, a temperature 9° above that of the atmosphere; but that of the nest, examined in its natural situation, was from 14° to 16° above that of the atmosphere, and from 17° to 19° above that of the chalk bank in which it was formed. But the generation of heat is increased to a most extraordinary degree, at the period when the last change is about to take place in the inclosed pupæ, which require an elevated temperature for the completion of their developmental processes (§ 96). This is furnished by the individuals denominated by Huber *Nurse-bees*, of which Mr. Newport gives the following interesting account:—"These individuals are chiefly young female bees; and, at the period of hatching of nymphs, they seem to be occupied almost solely in increasing the heat of the nest, and communicating warmth to the cells by crowding upon them and clinging to them very closely, during which time they respire very rapidly, and evidently are much excited. These bees begin to crowd upon the cells of the nymphs, about ten or twelve hours before the nymph makes its appearance as a perfect bee. The incubation during this period is very assiduously persevered in by the nurse-bee, who scarcely leaves the cell for a single minute; when one bee has left, another in general takes its place: previously to this period the incubation on the cell is performed only occasionally, but becomes more constantly attended to nearer the hour of the development. The manner in which the nurse-bee performs its office, is by fixing itself upon the cell of the nymph, and beginning to respire very gradually; in a short time its respiration becomes more and more frequent, until it sometimes respires at the rate of 130 or 140 per minute." In one instance, the thermometer introduced among seven nursing-bees stood at 92½°, whilst the temperature of the external air was but 70°. The greatest amount of heat is generated by the nurse-bees just before the young bees are liberated from the combs, at which period they require the highest temperature. It is just after its emersion that the young insect is most susceptible of cold; it is then exceedingly sleek, soft, and covered with moisture; it perspires profusely, and is highly sensitive of the slightest current of air. It crowds eagerly among the combs and among the other bees, and everywhere that warmth is to be obtained. It is not until after some hours, that it becomes independent of external warmth. It is

interesting to remark that these bees do not incubate on cells that contain only *larvæ;* the temperature of the atmosphere of the nest being sufficiently high for the young in that condition, as well as to perfect their change into the pupa state.—Similar observations have been made by Mr. Newport upon the temperature of the *Hive-bees;* and he has shown that the fallacy of the statements of other experimenters, as to the degree of heat maintained by them during the winter, is caused by the rapidity with which, when aroused, they can generate caloric. The temperature of individual bees in a state of moderate excitement, is usually from 10° to 15° above that of the atmosphere; but it is greatly increased about the swarming season, when incubation of the pupæ is going on, and also when clusters are formed round the entrance of the hive. At such times Mr. N. has seen the thermometer raised as high as 96° or 98°, when the range of atmospheric temperature was only between 56° and 58°. The *mean* temperature of a hive during May was 90°, that of the atmosphere being 60°; whilst, in September, the mean of the atmosphere being also 60°, that of the hive was only 66½°. During the winter, it appears that bees, like other insects, exist in a state of hybernation; though their torpidity is never so profound, as to prevent their being aroused by moderate excitement. The temperature of the hive is usually from 5° to 20° above that of the atmosphere; but it is sometimes depressed even below the freezing point. It is when artificially excited in a low temperature, that their power of generating caloric becomes most evident. Mr. N. mentions one instance in which the temperature of a hive, of which the inmates were aroused by tapping on its exterior, was raised to 102°; whilst a thermometer in the air stood at 34½°, and the temperature of a similar hive which had not been disturbed was only 48½°.

624. In regard to the degree of Heat which Insects are capable of generating, therefore, it appears that they may be ranked between 'cold-' and 'warm-blooded' animals. Like the former, they are much influenced by external temperature; although the higher species are, when in a state of moderate exercise, relatively warmer than the least cold-blooded among the Reptiles. The degree of heat they are *occasionally* capable of evolving, is nearly equal to that generated by Mammalia; but this is only required for the performance of particular functions, and, if constantly maintained in Insects, would have occasioned an unnecessary activity in the processes on which it is immediately dependent, and, by consequence, in the whole of the nutritive system. In Birds and Mammalia, however,—where, from the high development of the animal powers, the *constant* maintenance of an elevated temperature is necessary,—all the functions are adapted to its support; and in them we no longer find any dependence upon the state of the external medium, the calorific and frigorific processes being so delicately adjusted, as to render the heat of the system extremely uniform.

625. The temperature of *Birds* is, almost without exception, higher than that of the Mammalia, varying from 100° to 111¼°. The first is that of the *Gull,* the last that of the *Swallow.* In general, the same statement may be applied to Birds, as has been made with respect to Insects,—that the temperature is greater in the species of most rapid and powerful flight, and less in those which principally inhabit the earth, as the Fowl tribe; but we find the lowest temperature of the class in the

aquatic birds (which most closely approximate Reptiles in their general organisation), notwithstanding that they possess, in the thick and soft down with which they are clothed, and which is rendered impervious to fluid by the oily secretion applied with the bill, a special provision for retaining that heat within their bodies, which would otherwise be too rapidly conducted away. It is to be remembered that, from the comparatively small size of most of the members of this class, and the larger surface which they consequently expose in proportion to their bulk, the cooling action of the surrounding medium will have a greater relative effect upon them, than upon larger animals; and the amount of heat they must generate to maintain the same internal temperature, will be greater. —The embryo of Birds requires for its development a heat nearly equal to that of the body of the parent; and this is afforded by the process of *incubation*. The contents of the egg, when lying under the body of its parent, are so situated, that the germ-spot (§ 685) is brought into closest proximity with the source of warmth. Eggs may, however, be artificially incubated,—a practice which is carried to a great extent in Egypt; and in tropical climates the heat of the sun is in some instances sufficient. Thus, the Ostrich is said to leave her eggs to be hatched by the sun's rays alone, when she breeds in the neighbourhood of the Equator; and to sit upon them, if inhabiting a more variable climate. It was observed by Mr. Knight, that a Fly-catcher, which built for several successive years in one of his stoves, quitted its eggs whenever the thermometer was above 71° or 72°, and resumed her place upon the nest when the thermometer sunk again. The young of many kinds of Birds are deficient, for some time after their emergence from the egg, in the power of maintaining an independent temperature. Thus, Dr. Edwards found that young Sparrows, a week after they are hatched, have, while in the nest, a temperature of from 95° to 97°; but when they are taken from the nest, their temperature falls in one hour to $66\frac{1}{2}$°, the temperature of the atmosphere being at the same time $62\frac{1}{2}$°; and this rapid cooling was shown by parallel experiments not to be owing to the want of feathers. This fact, however, is not common to all Birds (§ 325 *l*).

626. The temperature of *Mammalia* seems usually to range from about 96° to 104°; but more accurate observations are still required for the sake of comparison. It is remarkable that a still higher point should be attained by the animals inhabiting the coldest regions; the Arctic Fox having been found by Capt. Lyon to possess a temperature of nearly 107°, when that of the atmosphere was 14°. That of the *Cetacea* (Whale tribe) does not seem to be inferior to that of other orders; and, to retain it within the body, the skin is enormously thickened and penetrated with oil (so as to form the substance known as *blubber*), by which the conducting power of the medium they inhabit is prevented from operating too energetically and injuriously. As far as is yet known, the temperature of the *Cheiroptera* (Bat tribe) seems more variable than that of any other order; for it has been found by Mr. Paget* that the amount of heat evolved by a Noctule under his observation, varied (like that of Insects) with its degree of activity, its body being but a few degrees warmer than the atmosphere after a period of prolonged repose, but its temperature rising

* 'Lectures on Inflammation,' in "Medical Gazette," June 7, 1850.

to 99° when it had been making active exertions. The heat of different parts of the body varies a good deal according to the degree of surface exposed; and it seems greater among the viscera, than in any situation ever exposed to the air. Thus, the temperature of the Human body is usually stated at 98° or 99°, from the height of thermometers placed in the mouth, axilla, &c.; but that of the stomach, according to Dr. Beaumont, is generally 100°; and that of the blood from $100\frac{1}{2}$° to $101\frac{1}{2}$°.—In the young Mammalia also, there is usually a considerable deficiency of calorifying power; but the degree of this varies considerably in the different orders of the class (§ 326 *bb*).

627. The phenomena of *Hybernation,* a state in which certain warm-blooded animals are reduced to the condition of those of least independent temperature, having been already described (§ 94), it need only here be added, that, like many Insects, hybernating Mammals still preserve the *capability* of evolving heat, when a stimulus of any kind arouses the animal functions, and gives a temporary excitement to those of organic life. This effect may be produced by mechanical irritation, which arouses the animal from its torpidity, and accelerates its respiratory movements. Extreme cold will produce the same effect; but it does not last long; and a more profound torpidity then comes on, which speedily ends in death, if the cold continue.—Hybernating animals, however, are by no means the only ones which exhibit a *periodical* change in the power of generating heat. It appears probable that all species of animals inhabiting climates in which the seasons are subject to much variation, vary in this respect at different parts of the year. Their power of evolving heat is greatest in the winter; so that, if exposed to severe cold during summer, their bodies are speedily cooled. The change in the colour of the fur, from dark to white, which many animals exhibit at the approach of winter, has the evident purpose (besides other objects) of diminishing the radiation of heat from its surface. That this change is occasioned by the cold of the air around, has been proved by an experiment of Capt. Ross upon a Lemming, which he had kept in his cabin until Feb. 1, and which retained its summer fur. It was then exposed on the deck to a temperature of 30° below zero; and on the following day the change of the fur to white commenced on the cheeks and shoulders, from which it gradually extended itself over the body. The Common Stoat of this country turns completely white in Scotland and the North of Continental Europe (whence it is obtained as the Ermine); in the North of England this change is occasional only; and in the midland and southern counties it is of rare occurrence. It is interesting to perceive the depressing cause thus counteracting itself, by means of that provision in the structure of the animal, which occasions the change of the colour of its fur.

628. We have now to enquire what are the *conditions* of the evolution of Heat in the animal economy. That many of the nutritive processes are subservient to it, can scarcely be doubted; but it seems peculiarly to depend upon those changes in which the function of Respiration is concerned—viz. the union of oxygen derived from the atmosphere, with compounds of hydrogen and carbon existing in the living system. Wherever the aeration of the blood is extensively and actively carried on, there is a proportionate elevation of temperature. And, on the other hand, wherever the respiration is naturally feeble, or the aeration of the blood

is checked by disease or accidental obstruction, the temperature of the body falls. Thus, in spasmodic Asthma, the temperature of the Human body during a paroxysm has been found as low as 82° ; in the Asiatic Cholera, a thermometer placed in the mouth has indicated but 77° ; and in *Cyanosis* (or 'blue disease,' arising from malformation of the heart impeding perfect arterialisation, § 491), the same low temperature has been observed. Again, whenever the temperature of an animal is, by any extraordinary stimulus, quickly raised above that which it was previously maintaining, it is always in connection with increased activity of the respiratory movements, and increased consumption of oxygen. Thus, during the incubation of Bees, the insect, by accelerating its respiration, causes the evolution of heat and the consumption of oxygen to take place at least *twenty* times as rapidly as when in a state of repose.—Now it has been seen that arterial blood contains a larger proportion of oxygen, than exists in venous blood ; whilst, on the other hand, the latter contains a larger proportion of carbonic acid than exists in the former ; and it seems obvious, therefore, that during the passage of blood through the capillaries, a part of its oxygen has been exchanged for carbonic acid. The source of this product is obviously the union of atmospheric oxygen with Carbon, furnished by disintegration of the tissues, or (more directly) by the elements of the food (§ 392) ; and by this union, caloric must be generated, precisely as by the more rapid union of the same materials in the ordinary process of combustion. Further, since these materials yield Hydrogen as well as carbon, and since more oxygen almost always disappears from the air which has been respired, than is contained in its carbonic acid (§ 534), it seems probable (although this cannot be demonstrated) that this element also is subjected to the combustive process, with a further generation of caloric ; and that, of the water which is exhaled from the lungs, a part has been thus produced. Further, there can be no doubt that other combustive processes take place in the system, although these may be the chief ; for example, of the sulphur and phosphorus which the albuminous elements of the food contain, the greater proportion is thrown off by the urinary excretion (§ 601) in the condition of sulphuric and phosphoric acids, having been combined with oxygen in the system. When all these actions are taken into account,—when the amount of heat that should be generated by the production of the quantity of carbonic acid found in the air expired during a given time, is carefully estimated, and the oxygen which has disappeared is considered as having been similarly employed in other combustive processes, especially in the formation of water,—it is found that the total so closely corresponds with the amount of heat *actually* generated by the animal during the same time, that it can scarcely be doubted that this process is the main source of calorification.*

629. Notwithstanding, however, that the Chemical theory of Animal Heat may be considered as accounting for the ordinary maintenance of a fixed temperature in the body at large, yet there are some 'residual phe-

* From the experiments of Dulong and Despretz it appeared that the whole amount of caloric generated by an animal in a given time could not thus be accounted for ; but it has been more recently shown by Prof. Liebig, that the discrepancy nearly vanishes, when the calculation is based on the more correct data since determined, as to the amount of caloric generated in the combustion of carbon and hydrogen.

nomena' to which it scarcely appears applicable. Of this kind are the sudden elevation of temperature that occurs under the influence of nervous excitement, which may be either general or local; the equally sudden diminution which marks the influence of the depressing passions; and the rapid cooling of bodies in which the nervous centres have been destroyed, notwithstanding that the respiration is artificially maintained, and the circulation continues. So, again, when the spinal cord of a warm-blooded animal is divided, there is a temporary elevation of temperature in the lower extremities; but this soon subsides, and the temperature of the paralysed parts then remains permanently below the natural standard. This last fact, which corresponds with the constant deficiency of warmth observable in paralysed limbs in the Human subject, may be explained by attributing the imperfect calorification of the part to the torpor of the nutrient operations, consequent upon the deficiency of nervous energy and the want of functional activity. But this will scarcely apply to the cases first cited; and it seems not unreasonable to regard them as indications of that direct conversion of Nervous Power into Heat, of which we have strong evidence in regard to Light and Electricity. (See §§ 54 and 639).

4. *Evolution of Electricity.*

630. Electricity is a force which may be made to act through any form of matter, and which may be produced by any other of the physical forces, acting under certain conditions. Thus *Motion* will generate Electricity, as in the ordinary mode of obtaining it by the friction of two dissimilar substances; and it is scarcely possible for such friction to occur, without some degree of electric disturbance. So, *Heat* will generate Electricity, when applied to two dissimilar metals in contact; and the electric equilibrium is disturbed, even when two parts of the same bar are unequally heated. *Magnetism*, again, may be made to develope Electricity; as in the action of the ordinary Magneto-Electric machine. But the most frequent and powerful source of Electric disturbance, is *Chemical Action*; there being probably no instance of chemical union or decomposition, in which the electric condition of the bodies is not altered, although it may not be always easy to demonstrate the change. And, by means of the chemical actions which it produces, *Light* also may become the generator of the Electric force.—Now as many of these forces are operating in the living body, under circumstances which would appear to be peculiarly favourable to the excitement of Electricity, there can be no difficulty in accounting for its production in the organic processes of Nutrition and Secretion; and the wonder perhaps is, that the manifestations of it should be ordinarily so trivial. Thus, when the change of form of a body from solid to liquid, or from liquid to gaseous, is attended with the least chemical decomposition (as when water containing a small quantity of saline matter in solution is caused to evaporate and to leave it behind), a very decided electric disturbance is produced; and if it were not for the provisions which everywhere exist for the neutralization of the effects of such actions, by the conducting power of the parts in which they occur and of the bodies around them, such electric disturbances could scarcely fail to exert an important influence on the organic functions. (See § 108).

631. *Electricity in Vegetables.*—That the ordinary processes of Vegetable growth are attended with a disturbance of electric equilibrium, which is manifested when the bodies in which it takes place are effectually insulated, seems to have been proved by the experiments of Pouillet. Several pots filled with earth, and containing different seeds, were placed on an insulated stand in a chamber, the air of which was kept dry by quick-lime; and the stand was placed in connection with a condensing electrometer. During germination, no electric disturbance was manifested: but the seeds had scarcely sprouted, when signs of it were evident; and when the young plants were in a complete state of growth, they separated the gold leaves of the electrometer half an inch from each other. It was calculated by him that a vegetating surface of 100 metres square in extent produces in a day more electricity than would be sufficient to charge the strongest battery; and he not unreasonably considers that the growth of plants may be one of the most constant and powerful sources of atmospheric electricity. The disengagement of vapour from the surface of the leaves would alone be sufficient to produce such a disturbance, as the fluid from which it is given off is always charged with saline and other ingredients; and the gaseous changes which are effected by the leaves, upon the oxygen and carbonic acid of the atmosphere, may be regarded as additional sources of its development. During the various processes of decomposition and recomposition which take place in the assimilation of the Vegetable juices, we should expect that Electric equilibrium would be sometimes disturbed, sometimes restored. Of this, the following facts, amongst others, appear to be sufficient evidence. If a wire be placed in apposition with the bark of a growing plant, and another be passed into the pith, contrary electrical states are indicated, when they are applied to an electrometer. If platinum wires be passed into the two extremities of a fruit, they also will be found to present opposite conditions. In some fruits, as the apple or pear, the stalk is negative, the eye positive; whilst in such as the peach or apricot, a contrary state exists. If a prune be divided equatorially, and the juice be squeezed from its two halves into separate vessels, its portions will in like manner indicate opposite electrical states, although no difference can be perceived in their chemical qualities.*

632. *Electricity in Animals*—All that has been said of the effects of vegetation in producing a disturbance of Electric equilibrium, will manifestly apply to the nutritive processes of Animals also; and there is no deficiency of indications that such is the case. Thus, Donné found that the skin and most of the internal membranes are in opposite electrical states; and Matteucci has seen a deviation of the needle amounting to 15° or 20°, when the liver and stomach of a rabbit were connected with the platinum ends of the wires of a delicate galvanometer. It may be questioned whether or not the differences in the secretions of these parts were the causes or the effects of their electric conditions. According to Matteucci, it could not be by their chemical action on the wires that the manifestation was produced, since it became very feeble or entirely ceased on the death of the animal. The more recent experiments of Mr. Baxter,†

* "Annales de Chimie," tom. lvii.

† "Philosophical Transactions," 1848, p. 243.

which were directed to the determination of the relative electrical condition of secreting surfaces, and of the blood in the veins returning from them, seem to confirm the belief that an electric disturbance takes place in the very act of secretion. He found that when one of the electrodes was placed on the intestinal surface, and the other inserted into the branch of the mesenteric vein proceeding from it, a deflection of the needle to the extent of 4° or 5° was produced, indicating a positive condition of the blood; no effect, on the other hand, was produced, when the second electrode was inserted into the artery of the part. These effects cease soon after the death of the animal, which is not the case with those which proceed from simple chemical differences between the blood and the secreted product.—The foregoing experiments may seem confirmatory of Dr. Wollaston's theory respecting Secretion. Observing the connection between electricity and chemical action, he was led to think that all the secretions in the body are the effect of electrical agency acting in various modes, and that the qualities of each secretion point out what species of electricity preponderated in the organ which forms it: thus, the existence of free acid in the urine and gastric juice, and of free alkali in the bile and saliva, mark the prevalence of positive electricity in the kidneys and stomach, whilst an excess of negative electricity is indicated in the liver and salivary glands. But it is more likely that the disturbance of Electricity is the *result* of chemical changes, whose source is the Vital force, than that it is itself the cause of those changes.

633. From experiments on the Human subject, it would appear that the living body would be never in perfect equilibrium with those around it, were this not constantly maintained by free contact with them; thus, if two persons, both insulated, join hands, sufficient electricity is developed to affect the electrometer. Some electric disturbance is manifested by almost every individual, if it be carefully sought for. In men it is most frequently positive; and irritable men of sanguine temperament have more free electricity than those of phlegmatic character; whilst the electricity of women is more frequently negative than that of men. Some individuals exhibit these phenomena much more frequently and powerfully than others. There are persons, for instance, who scarcely ever pull off articles of dress which have been worn next the skin, without sparks and a crackling noise being produced, especially in dry weather; this may, however, be partly due to the friction of these materials on the surface and with each other, as it has been proved to be greatly influenced by their nature. The most remarkable case of the generation of electricity in the human subject at present on record, is one related a few years since in America.* The subject of it, a lady, was for many months in an electric state so different from that of surrounding bodies, that, whenever she was but slightly insulated by a carpet or other feebly-conducting medium, sparks passed between her person and any object which she approached. From the pain which accompanied the passage of the sparks, her condition was a source of much discomfort to her; when most favourably circumstanced, four sparks per minute would pass from her finger to the brass ball of the stove at a distance of 1½ inch. The circumstances which appeared most favourable to the generation of elec-

* "American Journal of Medical Science," January, 1838.

tricity, were an atmosphere of about 80°, tranquillity of mind, and social enjoyment; while a low temperature and depressing emotions diminished it in a corresponding degree. The phenomenon was first noticed during the occurrence of a vivid Aurora Borealis; and though its first appearance was sudden, its departure was gradual. Various experiments were made, with the view of ascertaining if the electricity was generated by the friction of articles of dress; but no change in these seemed to modify its intensity.

634. Electricity may be generated, however, not merely by the Organic functions, but by changes more peculiarly connected with the exercise of the Animal powers; and this through both the Muscles and the Nerves.—It has been shown by Matteucci,* that if an incision be made into a muscle of a living animal, and the nerve of a 'galvanoscopic frog'† be introduced into the wound, in such a manner that its extremity is applied to the deepest part, and another portion to the lips, or to the surface of the muscle, the leg of the frog is thrown into contraction; thus showing that a difference exists in the electric condition of the deeper and more superficial parts of the animal. This phenomenon is exhibited even when the muscle is separated from the body; but when the experiment is many times repeated, the contractions are observed to become feebler and feebler, and at last cease altogether, their duration being greater when a muscle of a cold-blooded animal, than when that of a warm-blooded animal is employed. A galvanic pile may even be formed of pieces of fresh muscle, provided they be so arranged that the internal substance of each is in contact with the external surface of the next; and thus it may be shown, by the galvanometer, that a current is continually proceeding from the interior to the surface of every muscle. This current exists when the muscle is in a completely passive state; and its intensity seems then to depend upon the activity of the nutrient changes taking place in it. But its intensity is greatly increased when the muscle is thrown, by a stimulus applied to its own nerve, into a state of energetic contraction. There can be little doubt that the source of this electric disturbance lies in the molecular changes taking place in the muscle, either during the ordinary process of nutrition, or in the alterations connected with its contractile action; and the explanation of its constant direction from the interior to the exterior of the muscle, appears to lie in the difference in the rate of these changes in these two situations; since they will go on more energetically in its interior, than they do nearer its surface, where the proper muscular fibres are mingled with a larger proportion of areolar and tendinous substance.—The existence of an Electrical current in the Frog, passing during its whole life from its extremities towards its head, has been known since the time of Galvani; but it has been usually supposed to be peculiar to this animal. Even Matteucci, for some time after his

* "Lectures upon the Physical Phenomena of Living Beings;" LECT. IX.

† The 'galvanoscopic frog,' which has been continually employed by Prof. Matteucci to test minute electric disturbances which are scarcely appreciable by a galvanometer, is simply the leg of a recently-killed frog, with the crural nerve, dissected out of the body, remaining in connection with it; the leg being enclosed in a glass tube covered with an insulating varnish, and the nerve being allowed to hang freely from its open end, when two points of the nerve are brought in contact with any two substances in a different electrical state, the muscles which it supplies are thrown into contraction.

discovery of the 'muscular current,' regarded the 'proper current of the frog' as something essentially different. More recently, however, he has shown that the latter is probably but a special case of the former. For he has found that in every living muscle, of which the tendinous extremities are not equally disposed (one, for example, forming a cord, whilst the other is broad and ribbon-like), a current circulates in the muscle, from the tendinous extremity to the muscular surface ; so that a galvanic pile may be formed by a succession of such muscles, the tendinous extremity of one being placed in contact with the middle of the surface of the next. And the arrangement of the muscles and tendons in the limbs and body of the frog, is such as peculiarly to favour the development of such a current, in the direction in which it is actually discovered. The community of origin of the 'muscular current' and of the 'proper current of the frog,' is further indicated by the circumstance that they are both modified in the same manner by agents which affect the vitality of the muscle ; and it is particularly remarkable that poisoning with sulphuretted hydrogen should almost immediately put an end to both, although narcotic poisons have very little influence. The 'proper current of the frog' bears this curious analogy to the electric discharges of Fishes presently to be described ;—that it is not manifested if the connection be made between corresponding points of the opposite sides ; but that it shows itself when the communication is made between points higher or lower than the body, whether on the same or opposite sides.

635. There are certain animals which possess the power of accumulating Electric force within their bodies, and of discharging it at will in a violent form ; and with the exception of some Insects and Mollusca, which have been said (though this is doubtful) to communicate sensible shocks, these animals are all included in the class of *Fishes.* About seven species of this class, belonging to five genera, are known to possess electric properties ; and it is curious that these genera belong to tribes very dissimilar from one another, and that, though each has a limited geographical range, one species or other is found in almost every part of the world. Thus, the three species of *Torpedo,* belonging to the Ray tribe, are found on most of the coasts of the Atlantic and Mediterranean, and sometimes so abundantly as to be a staple article of food. The *Gymnotus,* or electric Eel, is confined to the rivers of South America. The *Silurus* (more correctly the Malapterurus), which approaches more nearly to the Salmon tribe, occurs in the Niger, the Senegal, and the Nile. The *Trichiurus,** or Indian Sword-fish, is an inhabitant of the Indian Seas ; and the *Tetraodon* (one of a genus allied to the Diodon or globe-fish) has only been met with on the coral banks of Johanna, one of the Comoro Islands.—These Fishes have not all been examined with the same degree of attention ; but it seems probable that the phenomena which they exhibit, and the structural peculiarities with which these are connected, are essentially the same throughout. The peculiar characteristic of all is the power of giving, to any living body which touches them, a shock resembling in its effects that produced by the discharge of a Leyden jar. This is of very variable intensity in different species and individuals, and at different times. The *Gymnotus*

* There is some doubt, however, as to the real character of the fish to which this name has been given.

will attack and paralyse horses, as well as kill small animals; and the discharges of large fish (which are 20 feet long) sometimes prove sufficient to deprive men of sense and motion. The effects of the contact of the *Torpedo* are less severe, and soon pass off; but the shock is attended with considerable pain when the fish is vigorous. The electrical organs appear to be charged and discharged to a certain extent at the will of the animals. Their power is generally exerted by the approach of some other animal, or by some external irritation; but it is not always possible to call it into action, even in vigorous individuals. It usually diminishes with the general feebleness of the system, though sometimes a dying fish exerts considerable power. All electrical fishes have their energy exhausted by a continued series of discharges; hence it is a common practice with convoys in South America, to collect a number of wild horses and drive them into the rivers, in order to save themselves, when they pass, from being injured by the fish. If excessively exhausted, the animals may even die; but they usually recover their electrical energy after a few hours' rest.

636. The *Torpedo*, from its proximity to European shores, has been most frequently made the subject of observation and experiment; and the following are the most important results of the investigations which have been made upon it by various enquirers.—That the shock perceived by the organs of sensation in Man is really the result of an Electric discharge, has now been fully established. Although no one has ever seen a spark emitted from the body of one of the fish, it may be easily manifested, by causing the Torpedo or Gymnotus to send its discharge through a slightly interrupted circuit. The galvanometer is influenced by the discharge of the Torpedo, and chemical decomposition may be effected by it, as well as magnetic properties communicated to needles. It seems essential to the proper reception of the shock, that two parts of the body should be touched at the same time, and that these two should be in different electrical states. The most energetic discharge is procured from the Torpedo, by touching the back and belly simultaneously, the electricity of the dorsal surface being positive, and that of the ventral negative; and by this means the galvanometer may be strongly affected, every part of the back being positive with respect to every part of the opposite surface. When the two wires of the galvanometer are applied to the corresponding parts of the two sides of the same surface, no influence is manifested; but, if the two points do not correspond in situation, whether they be both on the back or both on the belly, the index of the galvanometer is made to deviate. The degree of proximity to the electric organ appears to be the source of the difference in the relative state of different parts of the body; those which are near to it being always positive in respect to those more distant. Dr. Davy found that, however much Torpedos were irritated through a single point, no discharge took place; and he states that, when one surface only is touched and irritated, the fish themselves appear to make an effort to bring the border of the other surface, by muscular contraction, into contact with the offending body; and that this is even done by fœtal fish. If a fish be placed between two plates of metal, the edges of which are in contact, no shock is perceived by the hands placed upon them, since the metal is a better conductor than the human body; but, if the plates be separated, and, while still in contact with the opposite sides of

the body, the hands be applied to them, the discharge is at once rendered perceptible, and it may be passed through a line formed by the moistened hands of two or more persons, the extremities being brought into relation with the opposite plates.—The electrical phenomena of the *Gymnotus* are essentially the same with those of the Torpedo; but the opposite electrical states are found to exist, not between the dorsal and ventral surfaces, but between the head and tail; so that the shock is most powerful, when the connection is formed between these two extreme points.

637. It has been ascertained by experiment, that the manifestation of this peculiar power depends upon the integrity of the connection between the Nervous centres and certain organs peculiar to Electrical fishes. In the Torpedo the Electric Organs are of flattened shape (Fig. 247, *d*), and occupy the front and sides of the body, forming two large masses, which extend backwards and outwards from each side of the head. They are composed of two layers of membrane, between which is a whitish soft pulp, divided into columns by processes of the membrane sent off so as to form partitions like the cells of a honeycomb; the ends of these columns being directed towards the two surfaces of the body. The columns are again subdivided horizontally by more delicate partitions, which form each into a number of distinct cells; the partitions are extremely vascular, and are profusely supplied with nerves, of which the fibres seem to break up into minuter fibrillæ to form plexuses upon these membranes. The fluid contained in the electrical organs forms so large a proportion of them, that the specific gravity of the mass is only 1026, whilst that of the body in general is about 1060; and, from a chemical examination of its constituents, it seems to be little else than water, holding one-tenth part of albumen in solution, with a little chloride of sodium.—The electrical organs of the *Gymnotus* are essentially the same in structure, though differing in shape in accordance with the conformation of the animal; they occupy one-third of its whole bulk, and run along nearly its entire length; there are, however, two distinct pairs, one much larger than the other. The prisms are here less numerous, but are much longer; for they run in the direction of the length of the body, a difference which is

FIG. 247.

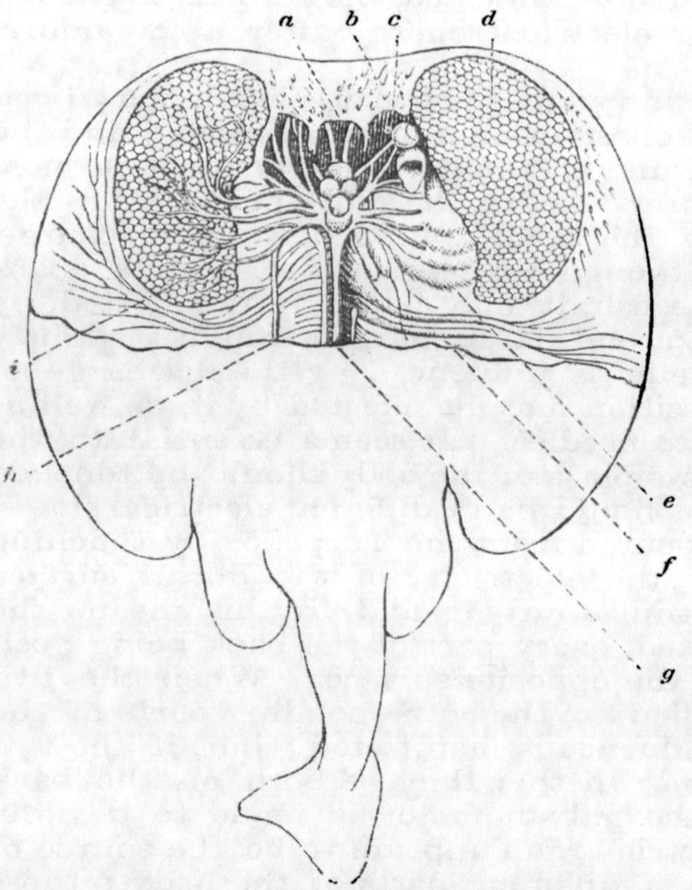

Electrical Organs of the *Torpedo*:—*a*, brain; *b*, cutaneous glandulæ; *c*, eye: *d*, electrical apparatus; *e*, branchiæ; *f*, spinal nerves supplying the pectoral fins; *g*, spinal cord; *h*, lateral nerves to the body; *i*, pneumogastric nerves to the electric organ.

productive of a considerable modification of the character of the discharge. —In the *Silurus* there is not any electrical organ so definite as those just described; but the thick layer of dense cellular tissue, which completely surrounds the body, appears to be subservient to this function; it is composed of tendinous fibres interwoven together, and of an albuminous substance contained in their interstices, so as to bear a close analogy with the cellular partitions in the special organs of the Torpedo and Gymnotus. —The organs of the other known electrical fishes have not yet come under the notice of any anatomist.

638. In all these instances, the Electrical Organs are supplied with Nerves of very great size, larger than any others in the same animals, and larger than any nerve in other animals of like bulk. They all arise in the *Torpedo* from a ganglionic mass situated behind the Cerebellum, and connected with the Medulla Oblongata, to which the name of 'electric lobe' has been given; the first two of them issue from the cranium in close proximity with the 5th pair, and have been regarded as belonging to it, although their real origin is different; whilst, from the distribution of the third electrical nerve to the stomach, after sending its principal portion to the electrical organ, it would seem analogous to the 8th pair or pneumogastric.—The electrical nerves in the *Gymnotus* are believed to arise from the spinal marrow alone; and those of the *Silurus* are partly intercostals and partly belong to the 5th pair.—The integrity of the nerves is essential to the full action of the electrical organs. If all the trunks be cut on one side, the power of that organ will be destroyed, but that of the other may remain uninjured. If the nerves be partially destroyed on either or both sides, the power is retained by the portion of the organs still in connection with the centres. The same effects are produced by tying the nerves, as by cutting them. Even slices of the organ entirely separated from the body, except by a nervous fibre, may exhibit electrical properties. Discharges may be excited by irritation of the brain when the nerves are entire, or of the part of the divided trunk distributed on the organ; but on destroying the 'electric lobe' of the brain, the electric power of the animal ceases entirely, although all the other ganglionic centres may be removed without impairing it. It is remarkable however that, after the section of the electrical nerves, Torpedos appear more lively than before the operation, and actually live longer than others not so injured, which are excited to discharge frequently.—Poisons which act violently on the nervous system, have a striking effect upon the electrical manifestations of these fish; thus, two grains of muriate of Morphia were found by Matteucci to produce death after about ten minutes, during which time the discharges were very numerous and powerful; and Strychnia also excited powerful discharges at first, succeeded by weaker ones, the animals dying in violent convulsions. When the animals were under the influence of strychnia, it was observed that the slightest irritation occasioned discharges; a blow given to the table on which the animal was placed, being sufficient to produce this effect. If the spinal cord were divided, however, no irritation of the parts situated below the section called forth a shock.—It has also been ascertained by Matteucci, that the electric power is suspended when the Torpedo is plunged into water at 32°, and is recovered again when it is immersed in water of a temperature from 58° to 68°; and that this alternation may be repeated

several times upon the same fish. But if the temperature be raised to 86°, the Torpedo soon ceases to live, and dies while giving a great number of violent discharges.*

639. From all these facts it seems an almost unavoidable inference, that the Electric force is developed in the electric organ, by a disturbance of its equilibrium consequent upon Nervous Agency. Such a disturbance may be conceived to take place in every one of those minute cells, into which the prism is divided by transverse partitions; the two electricities (to use the current phraseology) being separated by nervous agency, as they are in the tourmaline by heat. Now by the multiplication of such cells in each prism, a 'pile' would be produced, at the two extremes of which the greatest differences in electric condition would be found; and the *intensity* of the discharge would depend upon the number of elements in the pile, while its quantity would be proportional to the multiplication of the separate prisms. Now this is precisely what holds good in nature; for the electric discharge of the Gymnotus is far more intense than that of the Torpedo, as might be expected from the multiplication of its cells; so that, according to Prof. Faraday, a single medium discharge from this animal gives a shock equal to that of a battery of fifteen Leyden jars, containing 3500 square inches, charged to its highest degree.—Further evidence that the force which enables the Electric Fishes to give sensible manifestations of Electricity, is the same as that which excites Contraction when transmitted to the Muscles, is derived from the remarkable conformity between the conditions under which the two phenomena respectively occur. The connection of the organs specially appropriated to each of these actions with the nervous system, the dependence of their functions upon the integrity of this connection and upon the will of the animal, the influence of stimulation applied to the nervous centres or trunks, the effect of ligature or section of the nerve, and the results of poisonous agents, are all so remarkably analogous in the two cases, that it seems scarcely possible to refuse assent to the proposition, that the Nervous power is the agent which is instrumental in producing both sets of phenomena. Still, however, no proof whatever can be derived from this source, of the *identity* of nervous influence with any form of Electricity; since all that can be inferred from it is, that, as by the influence of the Nervous system on one class of organs, sensible Contraction is produced, so by its influence on another class of organs, Electricity is generated. On the other hand, all the experiments of Prof. Matteucci confirm the conclusion formerly stated (§ 246), that the two forces are *not* identical, but that they are 'correlated,' each being able to generate the other.†

* See Matteucci's "Lectures," p. 216.

† A remarkable experiment confirmatory of this view, is recorded in Prof. Matteucci's most recent paper ("Philosophical Transactions," 1850, p. 295). When an electric current is transmitted through the muscles of the thighs of a living animal, the positive pole being placed above and the negative pole below, so that the current passes in the direction of the efferent or motor nerves, there is not only a strong contraction of the muscle traversed, but also of the muscles of the leg below; indicating that an influence is transmitted through the nerves which proceed to them. On the other hand, when the current is transmitted in the inverse direction, the positive pole being below, so that it is directed along the course of the afferent nerves, or towards the nervous centres, *pain* is produced, without any muscular contraction save in the muscle traversed; showing that the influence is now transmitted towards the nervous centres. Now it has been clearly established by other experiments,

640. Regarding the uses of the Electrical organs to the animals possessing them, no very certain information can be given. It is doubtful to what extent their power is subservient to the prehension of food, which was once supposed to have been their principal object; since it is known that the Gymnotus eats very few of the fishes which it kills by its discharge; and young Torpedos kept by Dr. Davy for five months ate nothing, though supplied with small fishes both dead and alive, but nevertheless increased in strength and in electrical energy. The electric power of young Torpedos is much less exhaustible than that of the adults; this is readily accounted for by the fact of the greater energy of the vital processes in young than in old animals. Dr. D. experienced shocks from fœtal fish, which he was removing from the abdomen of the parent. He believes that the electric action may assist the function of respiration, by decomposing the water in the neighbourhood of the gills, when the animal, being buried in sand or mud, might be unable to obtain the requisite supply of oxygen in the ordinary way; but its chief use he considers to be to guard the fish from its enemies. Another function, however, may not improbably be influenced by it,—that of digestion; and this in two ways. It is well known that the vital properties of living tissues are so completely destroyed by a violent electric discharge, that they are disposed to pass more readily than in other cases into decomposition, the incipient stage of which is favourable to digestion; the shortness of the intestinal canal of the Torpedo would seem to render some assistance of this kind peculiarly necessary. The process of digestion may also be aided by the continued action of electricity through the nerves of the stomach, which have been mentioned to be peculiarly connected in the Torpedo with those of the Electrical organs; a supposition which derives some confirmation from the fact mentioned by Dr. Davy, that digestion appeared to be arrested in an individual which had been exhausted by frequent excitement to discharge itself: it should be kept in view, however, that the general depression of the system may have been the cause of the check put to the process. It is not unlikely, as Dr. Roget has suggested, that the electrical organs may communicate to the fish perceptions of electrical states and changes in the surrounding bodies (very different from any that we can feel), in the same way as other organs of sense convey perceptions with regard to light and sound. Such perceptions may be conceived to be very useful and pleasurable to animals living in the dark abysses of the waters.

that an electric current traversing a muscle, never quits its substance to travel along the filaments of the nerves distributed through it; so that it is obvious that the influence which in the one case excites muscular contraction, and in the other case occasions pain, is *not* Electricity, but Nervous Force generated by the passage of the electric current through the muscle, in the nerves which are distributed to it.

CHAPTER XVIII.

REPRODUCTION OF ORGANISED BEINGS.

1. *General Considerations.*

641. If the changes which living beings undergo during the period of their existence, and the termination of that existence by the separation of their elements at a period more or less remote from their first combination, be regarded as distinguishing them in a striking and evident manner from the masses of inert matter which surround them, still more is their difference manifested in the series of processes, which constitute the function of *Reproduction*. A very unnecessary degree of mystery has been spread around the exercise of this function, not only by general enquirers, but by scientific physiologists. It has been regarded as a process never to be comprehended by man, of which the nature and the laws are alike inscrutable. A fair comparison of it, however, with other functions, will show that it is not in reality less comprehensible or more recondite than any one of them;—that our acquaintance with each depends upon the facility with which it may be submitted to investigation;—and that, if properly enquired into by an extensive survey of the animated world, the real character of the process, its conditions, and its mode of operation, may be understood as completely as those of any other vital phenomenon.

642. It may be considered as a fundamental truth of Physiological Science, that *every living organism has had its origin in a pre-existing organism.* The doctrine of 'spontaneous generation,' or the supposed origination of organised structures *de novo* out of assemblages of inorganic particles, although at different times sustained with a considerable show of argument, based on a specious array of facts, cannot now be said to have any claim whatever to be received as even a possible hypothesis; all the facts on which it claimed to rest having either been themselves disproved, or having been found satisfactorily explicable on the general principle *omne vivum ex ovo.* Thus, the appearance of Animalcules in infusions of decaying organic matter, the springing-up of Fungi in spots to which it would not have been supposed that their germs could have been conveyed, the occurrence of Entozoa in the bodies of various animals into which it seemed almost beyond possibility that their eggs could have been introduced, with other facts of a like nature, may now be accounted for, without any violation of probability, by our increased knowledge of the mode in which these organisms are propagated. Thus it is now well ascertained that the germs of Fungi and of many kinds of Animalcules are diffused through the atmosphere, and are conveyed by its movements in every direction; and that if to decomposing substances, of a kind that would otherwise have been most abundantly peopled by these organisms, such air only be allowed to have access, as has been deprived of its organic germs by filtration (so to speak) through a red-hot

tube or strong sulphuric acid, no living organisms will make their appearance in them; whilst, in a few hours after the exposure of the very same substances to ordinary atmospheric air, it has been found to be crowded with life.* And when it is borne in mind in the case of the Entozoa, that the members of this class are remarkable for the immense number of eggs which most of them produce, for the metamorphoses which many of them are known to undergo, and for the varieties of form under which there is reason to suspect that the same germs may develope themselves (§ 309), it becomes obvious that no adequate proof has yet been afforded that they have been, in any particular case, otherwise than the products of a pre-existing living organism.—This, again, is the conclusion to which all the most general doctrines of Physiology necessarily conduct us. For it is most certain that we know nothing of Vital Force, save as manifested through Organised structures; whilst, on the other hand, the combination of Inorganic matter into Organised structures is one of the most characteristic operations of Vital Force; hence it is scarcely conceivable that any operation of Physical Forces upon Inorganic matter should evolve a living Organism. Nor is such a conception more feasible, if it be admitted that Vital Force stands in such a relation to the Physical Forces, that we may regard the former as a manifestation of the latter, when acting through Organised structures (§ 54); since no Vital force can be manifested (according to this view), and no Organisation can take place, except through a pre-existing organism.

643. It may be further considered as an established Physiological truth, that when placed under circumstances favourable to its complete evolution, every germ will develope itself into the likeness of its parent; drawing into itself, and appropriating by its own assimilative and formative operations, the nutrient materials supplied to it; and repeating the entire series of phases through which its parent may have passed, however multiform these may be.† Now the germs of all tribes of Plants and Animals whatever, bear an extremely close relation to each other in their earliest condition; so that there is no appreciable distinction amongst them, which would enable it to be determined whether a particular molecule is the germ of a Conferva or of an Oak, of a Zoophyte or of a Man. But let each be placed in the conditions it requires; and a gradual evolution of the germ into a complex fabric will take place, the *more general* characters of the new organism preceding the *more special*, as already explained (CHAP. VIII.). These conditions are not different in kind from those which are essential to the process of Nutrition in the adult; for they consist, on the one hand, in a due supply of aliment in the condition in which it can be appropriated; and, on the other hand, in the operation of certain external agencies, especially Heat, which seems to supply the *force* requisite for the developmental process (§ 557). Now although we may not be able to discern any such ostensible differences in the germs of different orders of living beings, as can enable us to discriminate them from each other, yet, seeing so marked a diversity in their operations under circumstances essentially the same, we cannot do otherwise than

* See the experiments of Schulze, in the "Edinb. New Phil. Journ." 1837, p. 165.

† The apparent exceptions to this rule, which have been brought together under the collective term 'Alternation of Generations,' will be presently considered, and will be shown to be only exceptional when misinterpreted (§ 649).

attribute to them distinct properties ; and it will be convenient to adopt the phrase *germinal capacity* as a comprehensive expression of that peculiar endowment, in virtue of which each germ developes itself into a structure of its own specific type, when the requisite forces are brought to bear upon it, and the requisite materials are supplied to it.* Thus, then, every act of Development may be considered as due to the force supplied by *Heat* or some other physical agency, which, operating through the organic germ, exerts itself as formative power ; whilst the mode in which it takes effect is dependent upon the properties or endowments of the substances through which it acts, namely, the germ on the one hand, the alimentary materials on the other,—just as an Electric current transmitted through the different nerves of sense, produces the sensory impressions which are characteristic of each respectively ; or, as the same current, transmitted through one form of Inorganic matter, produces Light and Heat, through another, Chemical change, or through another, Magnetism.

644. In the development of any living being, therefore, from its primordial germ, we have three sets of conditions to study ;—namely, *first*, the physical forces which are in operation ; *second*, the properties of the germ, which these forces call into activity ; and *third*, the properties of the alimentary materials which are incorporated in the organism during its development. There is evidence that each of these may have a considerable influence on the result ; but in the higher organisms it would seem that the second is more dominant than it is in the lower. For among many of the lower tribes, both of Plants and Animals, there is reason to believe that the range of departure from the characters of its parent, which the organism may present, is considerably greater than that of the higher ; and that this is chiefly due to the external conditions under which it has been developed. The forms of a number of species of the lower Fungi, for example, appear to be in great part dependent on the nature of their aliment (§ 272) ; so among the Entozoa, there seems strong reason to believe that those of the *Cystic* order are only *Cestoidea*, that are prevented by the circumstances under which they exist from attaining their full development (§ 309 *b*) ; and the production of a fertile 'queen' or of an imperfect 'worker,' among the Hive-bees, appears to be entirely determined by the food with which the larva is supplied (§ 60). No such variations have been observed among the higher classes ; in which it would seem as if the form attained by each germ is more rigidly determined by its own endowments ; a modification in the other conditions, which in the lower tribes would considerably affect the result, being in them unproductive of any corresponding change. For if such modification be considerable, the organism is unable to adapt itself to it, and consequently either perishes or is imperfectly developed ; whilst, if it be less potent, it produces no obvious effect. Thus, a deficiency of Food in the

* This term is preferred to that of 'germ-power' suggested by Mr. Paget, because the latter seems to imply that the *force* of development exists in the germ itself. Now if this were true, not only must the whole formative power of the adult have been possessed by its first cell-germ, but the whole formative power of all the beings simultaneously belonging to any one race, must have been concentrated in the first cell-germ of their original progenitor. This seems a *reductio ad absurdum* of any such doctrine ; and we are driven back on the assumption (which all observation confirms), that the *force* of development is derived from *external* agencies.

growing state of the higher Animal, will necessarily prevent the attainment of the full size; but it will not exert that influence on the relative development of different parts, that it does among Plants, in which it favours the production of flowers and fruit in place of leaves, or that it seems to exercise in several parallel cases among Animals (§ 359). So, again, a deficiency of Heat may slightly retard the development of the Chick; but if the egg be allowed to remain long without the requisite warmth, the embryo dies, instead of passing into a state of inactivity like that of Reptiles or Insects.—The extent, indeed, to which these external conditions may affect the development of the inferior organisms, must not be in the least judged of by that to which their operation is restricted in the higher; and it is probable that we have yet much to learn on the subject. At present, it may be stated as a problem for determination, whether, from a being of superior organization, *lower* forms of living structure, capable of maintaining an independent existence, and of propagating their kind, can ever originate, by an imperfect action of its formative powers (§ 709). Various morbid growths, such as Cancer-cells, to which the higher organisms are liable, have been looked upon in this light: these have certainly a powerful vitality of their own, which enables them to increase and multiply at the expense of the organism which they infest; and they have also an energetic reproductive power, by which they can propagate their kind, so as to transmit the disease to other organisms, or to remote parts of the same organism: but such growths are not independent; they cannot maintain their own existence, when detached from the organism in which they are developed; and they have not, therefore, the attribute of *a separate individuality*. Various phenomena hereafter to be detailed, however, respecting the 'gemmiparous' production of living beings, when taken in connection with that just cited, seem to render it by no means impossible that the individualization may be more complete in other cases, so that independent beings of a lower type *may possibly* originate in a perverted condition of the formative operations in the higher. But no satisfactory evidence has ever been afforded by experience, that such 'equivocal generation' has actually taken place; and its possibility is here alluded to only as a contingency which it is right to keep in view. That *no higher* type has ever originated through an advance in developmental power, may be safely asserted; for, although various instances have been brought forward to justify the assertion that such is possible, yet these instances entirely fail to establish the analogy that is sought to be drawn from them.*

* Thus the Author of the "Vestiges of the Natural History of Creation" refers to the various modifications which have taken place in our cultivated Plants and Domesticated Animals, in proof that such elevation is possible; quite overlooking the fact that these external influences merely *modify* the development, without *elevating* it, and that these races, if left to themselves, speedily revert to their common specific type. And he adduces the phenomena of metamorphosis—the transformation of the worm-like larva into an Insect, and of a fish-like tadpole into a Frog,—as giving some analogical sanction to the same doctrine; totally overlooking the fact, that these transformations are only part of the ordinary developmental process, by which the complete form of the species is evolved, instead of being transitions from the perfected type of one class to the perfected type of one above it. So, again, he quotes the transformation of the worker-grub of the Hive-Bee, into the fertile Queen, as an example of a similar advance; without regarding the circumstance that the worker is *psychically* higher (according to human ideas, at least,) than the queen, whose instincts appear limited to the performance of her sexual functions; and that the utmost which the fact is

645. The *developmental power* which each germ possesses, under the conditions just now detailed, is manifested, not merely in the first evolution of the germ into its complete specific type, but also in the maintenance of its perfect form, and, within certain limits, by the reproduction of parts that have been destroyed by injury or disease. This reproduction, as Mr. Paget has pointed out,* differs from the ordinary process of nutrition in this,—that "in grave injuries and diseases, the parts that might serve as models for the new materials to be assimilated to, or as tissue-germs to develope new structures, are lost or spoiled; and yet the effects of injury and disease are recovered from, and the right specific form and composition are retained;"—and, again, "that the reproduced parts are formed, not according to any present model, but according to the appropriate specific form, and often with a more strikingly evident design towards that form, as an end or purpose, than we can discern in the natural construction of the body." In the reproduction of the leg of a full-grown Salamander after amputation, which was observed to take place by Spallanzani, it is clear that whilst the process was from the first of a nature essentially similar to that by which its original development took place, it tended to produce, not the leg of a larva, but that of an adult animal. Hence, it is obvious that, through the whole of life, the formative processes are so directed, as to maintain the perfection of the organism, by keeping it up, so far as possible, to the model or archetype that is proper to the epoch of its life which it has attained.—The amount of this regenerating power, however, varies greatly in different classes of organized beings, and at different stages of the existence of the same being; and, as Mr. Paget has pointed out,† it seems to bear an inverse ratio to the degree of development which has previously taken place in each case. Thus in the *Hydra* and other Zoophytes, it would appear (as in Plants) to be almost unlimited; for the developmental process in them is checked at such an early period, that both the form of the organism and the structure of its tissues retain the most simple type (§ 289); and by the subdivision of one individual, no fewer than fifty were produced by Trembley. In this, as probably in all the cases in which new individuals have been obtained by artificial subdivision, there is some natural tendency to their production by the vegetative process of gemmation (§ 647); but this does not always manifest itself. It is a curious fact, that the first attempt at regeneration, in some of these cases, is not always complete; but that successive efforts are made, each of which approximates more and more closely to the perfect type. This was well seen in one of Sir J. G. Dalyell's experiments; for he observed that, having cloven the stem of a *Tubularia* (a Hydroid Zoophyte), after

capable of proving is, that the same germ may be developed into two different forms, according to the circumstances of its early growth. It must always be borne in mind that the character of a species, to be complete, should include *all* its forms, perfect and imperfect, modified and unmodified; since in this mode alone, can that 'capacity for variation' be determined, which is so remarkable a feature in many cases, and is that which specially distinguishes the races of plants and animals that have been subjected to human influence. In no instance has this variation tended to confuse the limits of well-ascertained species; it has merely increased our acquaintance with the number of diversified forms into which the same germ may develope itself.

* "Lectures on Reproduction and Repair."

† Loc. cit.

the natural fall of its head, an imperfect head was at first produced, which soon fell off and was succeeded by another more fully formed ; this in its turn was succeeded by another ; and so on, until the fifth head was produced, which was as complete as the original.

646. As a general statement of the amount of this regenerating power which exists in most of the different classes of animals, has been already given (CHAP. VII.), it is unnecessary here to do more than allude to some of those facts which most strongly bear out the doctrine just laid down.—Next to Zoophytes, there are no animals in which the regenerative power is known to be so strong, as it is in the lower Articulata (as the Cestoid Entozoa, and the inferior Annelida) and in the *Planaria* (§ 309 *d*), which may perhaps be regarded as rather approximating to the Molluscous type ; and here, again, we see that a low grade of general development is favourable to its exercise, and that the spontaneous multiplication which occasionally takes place in these animals by fission or gemmation, is only another form of the same process. In the higher forms of both these sub-kingdoms, as we no longer meet with multiplication by gemmation, so do we find that the reparative power is much more limited ; the only manifestation of it among the fully-formed *Arachnida* and *Crustacea* being the reproduction of limbs, and the power of effecting even this being usually deficient in perfect Insects. The enquiries of Mr. Newport, however, upon the reproductive powers of *Myriapods* and *Insects,* in different stages of their development,* confirm the general principle already stated ; for he has ascertained that in their *larval* condition, Insects can usually reproduce limbs or antennæ ; and that Myriapods, whose highest development scarcely carries them beyond the larvæ of perfect Insects, can regenerate limbs or antennæ, up to the time of their last moult, when, their normal development being completed, the regenerative power seems entirely expended. The *Phasmidæ* and some other Insects of the order *Orthoptera* retain a similar degree of this power in their perfect state ; but these are remarkable for the similarity of their larval and imago states, the latter being attained, as in Arachnida, by a direct course of development, without anything that can be called a 'metamorphosis.'—Little is known of the regenerative power in the higher Mollusca ; but it has been affirmed that the head of the Snail may be reproduced after being cut off, provided the cephalic ganglion be not injured, and an adequate amount of heat be supplied (§ 98).—In Vertebrata, again, it is observable that the greatest reparative power is found among *Batrachian Reptiles,* whose development is altogether lower, and whose life is altogether more vegetative, than that of probably any other group in this sub-kingdom. In Fishes it has been found that portions of the fins which have been lost by disease or accident, are the only parts that are reproduced. But in the Salamander, entire new legs, with perfect bones, nerves, muscles, &c., are reproduced after loss or severe injury of the original members ; and in the Triton a perfect eye has been formed to replace one which had been removed. In the true *Lizards,* an imperfect reproduction of the tail takes place, when a part of it has been broken off ; but the newly-developed portion contains no perfect vertebræ, its centre being occupied by a cartilaginous column, like that of the lowest

* "Philosophical Transactions," 1844.

Fishes. In the warm-blooded Vertebrata generally, as in Man, the power of true reproduction after loss or injury seems limited, as Mr. Paget has pointed out * to three classes of parts :—namely, (1). "Those which are formed entirely by nutritive repetition, like the blood and epethelia, their germs being continually generated *de novo* in the ordinary condition of the body; (2). Those which are of lowest organization, and (which seems of more importance) of lowest chemical character, as the gelatinous tissues, the areolar and tendinous, and the bones; (3). Those which are inserted in other tissues, not as essential to their structure, but as accessories, as connecting or incorporating them with the other structures of vegetative or animal life, such as nerve-fibres and blood-vessels. With these exceptions, injuries or losses are capable of no more than repair, in its more limited sense; *i. e.*, in the place of what is lost, some lowly organized tissue is formed, which fills up the breach, and suffices for the maintenance of a less perfect life."—Yet, restricted as this power is, its operations are frequently most remarkable; and are in no instance, perhaps, more strikingly displayed, than in the re-formation of a whole bone, when the original one has been destroyed by disease. The new bony matter is thrown out, sometimes within, and sometimes around, the dead shaft; and when the latter has been removed, the new structure gradually assumes the regular form, and all the attachments of muscles, ligaments, &c., become as complete as before. A much greater variety and complexity of actions are involved in this process, than in the reproduction of whole organs in the simpler animals; though its effects do not appear so striking. It would seem that in some individuals this regenerating power is retained to a greater degree than it is by the class at large;† and here again we find, that in the early period of development the power is more strongly-exerted than in the adult condition. The most remarkable proof of its persistence even in Man, has been collected by Prof. Simpson; who has brought together numerous cases, in which, after "spontaneous amputation of the limbs of a fœtus in utero," occurring at an early period of gestation, there has obviously been an imperfect effort at the re-formation of the amputated part from the stump.‡

* "Lectures on Reproduction and Repair."

† One of the most curious and well-authenticated instances of this kind is related by Mr. White in his work on the "Regeneration of Animal and Vegetable Substances," 1785, p. 16. "Some years ago, I delivered a lady of rank of a fine boy, who had two thumbs upon one hand, or rather, a thumb double from the first joint, the outer one less than the other, each part having a perfect nail. When he was about three years old, I was desired to take off the lesser one, which I did; but to my great astonishment it grew again, and along with it, the nail. The family afterwards went to reside in London, where his father showed it to that excellent operator, William Bromfield, Esq., surgeon to the Queen's household; who said, he supposed Mr. White, being afraid of damaging the joint, had not taken it wholly out, but he would dissect it out entirely, and then it would not return. He accordingly executed the plan he had described, with great dexterity, and turned the ball fairly out of the socket; notwithstanding this, it grew again, and a fresh nail was formed, and the thumb remained in this state."—The Author has been himself assured by a most intelligent Surgeon, that he was cognizant of a case in which the whole of one ramus of the lower jaw had been lost by disease in a young girl, yet the jaw had been completely regenerated, and teeth were developed and occupied their normal situations in it.

‡ These cases were brought by Prof. Simpson, before the Physiological Section of the British Association, at its Meeting in Edinburgh, Aug., 1850. The Author, having had the opportunity of examining Prof. Simpson's preparations, as well as two living examples, is perfectly satisfied as to the fact.

By the knowledge of these facts and principles, we seem justified in the surmise, that the occurrence of supernumerary or multiple parts is not always due (as usually supposed) to the 'fusion' of two germs, but that it may result from the subdivision of one; for if it be supposed that this subdivision has taken place when the developmental process has advanced no further than in a Hydra or a Planaria, it seems by no means impossible that each part might, as in those creatures, advance in its development up to the attainment of its complete form.

647. There are many tribes, both of Plants and Animals, in which multiplication is effected not only *artificially* but *spontaneously*, by the separation of parts, which, though developed from the same germ in perfect continuity with each other, are capable of maintaining an independent existence, and which, when thus separated, take rank as distinct individuals. This process, which is obviously to be regarded, no less than the preceding, as a peculiar manifestation of the ordinary operations of Nutrition, may take place in either of four different modes—1. In the lowest Cellular Plants, and the simplest Protozoa, every component cell of the aggregate mass that springs from a single germ, being capable of existing independently of the rest, may be regarded as a distinct individual; and thus every act of growth which consists in the multiplication of cells (§ 556), makes a corresponding augmentation in the number of individuals.—2. In many organisms of a somewhat higher type, in which the fabric of each complete individual is made up of several component parts, we find the new growths to be complete repetitions of that from which they are put forth; and thus the composite organism presents the semblance of a collection of individuals united together, so that nothing is needed but the severance of the connection, to resolve it into a number of separate individuals, each perfect in itself. The most characteristic example of this is presented by the *Hydra* (§ 289), which is continually multiplying itself after this fashion; for the buds or 'gemmæ' which it throws off, are not merely structurally but functionally complete (being capable of seizing and digesting their own prey), previously to their detachment from the parent.—3. In by far the larger proportion of cases, on the other hand, the 'gemma' does not possess the complete structure of the parent, at the time of its detachment, but is endowed with the capacity for developing whatever may be deficient. Thus, the bud of a Phanerogamic Plant possesses no roots, and its capacity for independent existence depends upon its power of evolving those organs. On the other hand, the 'zoospore' of an Ulva or a Conferva (§ 267 *a*) is nothing else than a young cell, from which the entire organism is to be evolved after it has been set free; and even in the 'bulbels' of the Marchantia (§ 275), the advance is very little greater. The 'bulbels' of certain Phanerogamic plants, however, bear more resemblance to ordinary buds.—4. In the preceding cases, the organism which is developed by this process, resembles that from which it has been put forth; but there are many cases in which the offset differs in a marked degree from the *stock*, and evolves itself into such a different form, that the two would not be supposed to have any mutual relation, if their affinity were not proved by a knowledge of their history. Sometimes we find that the new individual thus budded off is in every respect as complete as that from which it proceeded, though developed upon a different type; but in other instances it is made up of

little else than a generative apparatus, provided with locomotive instruments to carry it to a distance, its nutritive apparatus being very imperfect. Of the first, we have an example in the development of Medusæ from the Hydroid Polypes (§ 290); and of the second in the peculiar subdivision of certain Annelida, hereafter to be described (§ 715).—Now it is obvious that in this process, no agency is brought into play, that differs in any essential mode from that which is concerned in the ordinary nutritive operation. The multiplication of individuals is performed exactly after the same fashion as the extension of the parent organism; and the very same parts may be regarded as organs belonging to it, or as new individuals, according to their stage of development, and the relation of dependence which they still hold to it. The essence of this operation is *the multiplication of cells by continual subdivision* (§ 556).

648. We have now, on the other hand, to enquire into the nature of the true *Generative* process, by which the original germ is endowed with its developmental capacity; and this we shall find to be of a character precisely the opposite of the preceding. For, under whatever circumstances the generative process is performed, it appears essentially to consist in *the reunion of the contents of two cells,** of which the germ, which is the real commencement of a 'new generation,' is the result. This process is performed under the three following conditions.—1. All the cells of the entire aggregate produced by the previous subdivision, may be capable of thus uniting with each other indiscriminately; there being no indication of any sexual distinction. This is what we see in the simplest Cellular plants (§ 652).—2. All the component cells of each organism may, in like manner, pair with other cells, to produce fertile germs; but there are differences in the shares which they respectively take in the process, which indicate that their endowments are not precisely similar, and that a sexual distinction exists between them, notwithstanding that this is not indicated by any obvious structural character. This condition is seen in the Zygnema and its allies (§ 654).—3. The generative power is restricted to certain cells, which are set apart from the rest of the fabric, and destined to this purpose alone; and the endowments of the two sets are so far different, that the one furnishes the germ, whilst the other supplies the fertilizing influence; whence the one set have been appropriately designated 'germ-cells' and the other 'sperm-cells.' Such is the case in all the higher Plants among which a true generative apparatus has been discovered; and also throughout the Animal kingdom.

649. Thus, then, in the entire process in which a new being originates, possessing like structure and endowments with its parent, two distinct classes of actions participate,—namely, the act of *Generation,* by which the germ is produced,—and the act of *Development,* by which that germ is evolved into the complete organism. The former is an operation altogether *sui generis;* the latter is only a peculiar modification of the Nutritive function; yet it may give origin, as we have seen, to new individuals, by the separation (natural or artificial) of the parts which are capable of existing as such. Now between these two operations there would seem to be a kind of antagonism. Whilst every act of Development tends to

* In very rare instances, it is the reunion of the two parts of the contents of the same cell, which had previously tended to separate from each other, as if in the process of subdivision (§ 653).

diminish the 'germinal capacity,' the act of Generation *renews* it; and thus the tree which has continued to extend itself by budding until its vital energy is well-nigh spent, may develope flowers and mature seeds from which a vigorous progeny shall spring up. But the *multiplication of individuals* does not directly depend upon the act of generation alone; it may be accomplished by the detachment of *gemmæ*, whose production is a simple act of development; and the individuals thus produced are sometimes similar, sometimes dissimilar, to the beings from which they sprang. When they are dissimilar, however, the original type is always reproduced by an intervening act of generation; and *the immediate products of the true generative act always resemble one another.* Hence the phrase "alternation of generations" can only be legitimately employed when the term generation is used to designate a succession of individuals, by whatever process they have originated; an application of it which cannot but lead to a complete obliteration of the essential distinction which the attempt has been here made to draw, between the generative act and the act of gemmation. For when it is said that "generation A produces generation B, which is dissimilar to itself, whilst generation B produces generation C, which is dissimilar to itself but which returns to the form of generation A," * it is entirely left out of consideration that generation A produces (the so-called) generation B by a process of *gemmation;* whilst the process by which generation B produces generation C, is one of *true generation.* So generation C developes D by gemmation, which resembles B; and D, by a true generative act, produces E, which resembles A and C. This distinction, although it may at first sight appear merely verbal, will yet be found of fundamental importance in the appreciation of the true relations of these processes, and of their resulting products. So, in the Author's opinion, the application of the term 'generation' to the *entire product* of the development of any germ originating in a generative act, whether that product consist of a single individual, or of a succession, will be found much more appropriate, and more conducive to the end in view, than the indiscriminate application of it to each succession, whether produced by gemmation or by sexual reunion. It is of great importance to the due comprehension of certain phenomena of Reproduction, which will come under consideration in the Animal kingdom, that the relations of the products of these two processes should be rightly appreciated; and this appreciation of them will, it is believed, be best gained by a careful enquiry into the phenomena of Reproduction in the Vegetable kingdom.

2. *Reproduction in Plants.*

650. Our knowledge of the reproductive process in the Vegetable Kingdom has of late been greatly extended by Microscopic research; and although it is not possible at present to give a complete sketch, much less

* Such is the doctrine propounded in the highly original and ingenious work of Prof. Steenstrup on the "Alternation of Generations," a translation of which has been published by the Ray Society. The Author feels satisfied, from careful and repeated consideration of the phenomena on which this doctrine is founded, that it cannot be accepted in its original form, and that it requires to be modified in the manner indicated in the text. His reasons for this modification, and likewise for his dissent from the hypothesis put forth by his friend Prof. Owen in his treatise on "Parthenogenesis," will be found in the "Brit. and For. Med. Chir. Rev." vol. i. p. 183, and vol. iv. p. 436.

a full account, of the mode in which it is performed in all the principal groups of Plants, still enough has been ascertained in some of the most important, to make it probable that the general doctrines founded upon the phenomena which they present, may be applied to those whose history has been less completely studied. It may be stated as a certainty that the true Generative act is not confined, as was long supposed by most Physiologists, to the *Phanerogamia* or Flowering Plants; but that it is performed by many of the *Cryptogamia;* and as it has been discovered to take place at both extremities of the Cryptogamic series (namely, among the simplest Cellular plants, and in the Ferns), further research may be confidently expected to show, that the intermediate groups form no exception to the general law. So far as our present knowledge extends, this generative act may be performed in three distinct modes, each of which is peculiar to a certain assemblage of Plants.—The *first* of these is the 'conjugation' of two apparently *similar* cells, and the admixture of their contents, whereby a new body, the *sporangium,* is produced; this, so far as we at present know, is peculiar to the lower Algæ; but it may be surmised that a similar process takes place in the lower Fungi and Lichens.—In the *second* case, the generative act is effected by the agency of cells which are manifestly *dissimilar* in their endowments; for the one set develope within themselves certain moving filaments, that convey their influence to the cells of the other set, within which last the germ originates; so that the distinction between 'sperm-cells' and 'germ-cells' is here characteristically shown. This is the mode in which the generative act seems to be performed in the higher Algæ and in Ferns; and probably in all the intermediate tribes of Cryptogamia. In the inferior groups, the fertilized germ appears to be at once thrown upon the world, and is dependent for its supply of food upon its own absorbing and assimilating powers; but in the superior, it is supplied for a time with nutriment prepared and imparted to it by its parent.—In the *third* form of the generative process, which is peculiar to Phanerogamia, there is the same distinction between 'sperm-cells' and 'germ-cells,' but the mode in which the action of the former upon the latter is brought about, is different; for the 'sperm-cell' (pollen-grain) does not evolve self-moving filaments, but puts forth long tubes, which extend themselves until they reach the 'germ-cell,' and thus convey to it the fertilizing influence. Moreover, the germ is supplied with nutriment by its parent, not merely whilst it remains in connection with the organism which evolved it, but for some time subsequently; its continued growth being provided for by the store laid up around it in the *seed;* and it being at the expense of the materials which it thence obtains, that the germ is enabled to develope and put forth the first of those organs, which are subsequently to act as its own instruments of absorption and assimilation.—Each of these processes will now be described in more detail.

651. It is among the lowest tribes of Plants, that we see the closest relation between the functions of Nutrition and Reproduction. For we do not here find, as in the higher Plants, that the development of distinct *organs* is requisite, to enable the descendants of the original germ-cell to exist independently of each other; since this power is possessed by every one of the cells that are produced by its fission (§ 139), so that they have as much title to be accounted distinct 'individuals,' as have the leaf-buds of

the Flowering-plant. Thus, a frond of the *Coccochloris cystifera*, one of the simple Algæ belonging to the family *Palmelleæ*, presents us with a mass of independent cells, held together by a common mucous investment (Fig. 248); and in such a mass we usually find cells in various stages of development. At *a* is seen a simple globular cell, surrounded by a well-defined mucous envelope; at *b* are cells which present an elongated form, and are evidently about to undergo subdivision; at *c* we observe a pair of cells, whose separation has apparently been recent, as they are still surrounded by the same mucous envelope; at *d* we have a group of three cells, each having its own mucous envelope, but all of them still held together by the original investment; the two smaller of these cells are obviously produced by a secondary subdivision of one of the first pair; whilst the other shows by its elongated form that it is about to subdivide; and at *e* is a group of small cells formed by a continuance of the same subdividing process, each of them subsequently increasing in size, acquiring a mucous envelope of its own (as at *a*), and in its turn going through the same series of changes. And as this will take place equally, whether the cell be detached from the general mass, or remain as a component part of it, we cannot refuse to it the designation of a distinct individual. It will be found convenient, however, to distinguish the individuals which are thus related as the products of a single generative act, by the appellation of *co-individuals*.—To what extent this method of multiplication may proceed, has not been ascertained. There is reason to think that there is a limit to its performance, since we find it giving place, at a certain period of the year, to the converse process; namely, the true Generative act. But upon the principle already laid down, we should expect that it may be carried on to a vast extent, under favourable circumstances; since there is here no advance upon the very simplest condition of an organised structure, and consequently but little exhaustion of the 'germinal capacity' (§ 645).

FIG. 248.

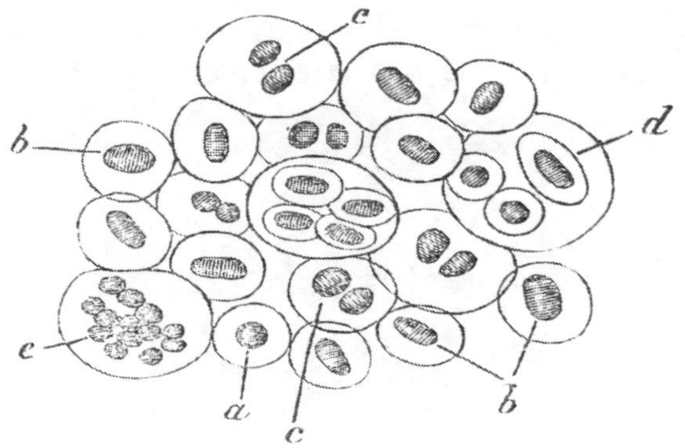

Multiplication of *Coccochloris cystifera* by subdivision of cells; *a*, *b*, *c*, *d*, *e*, various stages of development.

652. As amongst the independent cells of which the simplest Algæ are composed, each seems capable of going through the whole series of Nutritive operations *for* and *by* itself, so there is no perceptible difference in their endowments as regards the Generative act; the process of 'conjugation' appearing to take place indifferently amongst any pair of them. As examples of this process, two instances will be first selected from the families of *Desmideæ* and *Diatomaceæ;* in which its occurrence is a fact of peculiar interest, as tending to show their close alliance to the simpler Algæ, from which they have been separated by Prof. Ehrenberg and other naturalists, who have considered them deserving of a place in the Animal kingdom.—The *Desmideæ* are characterised by the peculiar structure of their cells; for the wall of each consists of two distinct and

symmetrical valves, united to each other by a transverse suture; and the endochrome shows a tendency to separation into two equal halves at the same part. In some instances there is a constriction at the suture, so that the two portions of the cell are connected only by a narrow neck, as in *Euastrum* (Fig. 249, A); but in other cases, as in *Closterium*, the median portion traversed by the suture is the broadest part of the cell. The plants of this family are generally met with in the condition of detached cells or 'fronds;' but in some instances, these are associated into confervoid filaments; the difference simply arising from the continuance of the connection, in the latter case, between the cells which have been multiplied by subdivision, and from the separation of these cells in the former. The first stage in the conjugating process is the approximation of two cells, in whose contained granules an active movement is often seen at the same time. Each of the two cells then gives way at the plane of its transverse division (B), and the four halves discharge their contained 'endochromes' into the central space, so that those of the two cells are thus completely incorporated. At first the new mass is merely surrounded by the mucous envelope, and has no definite shape; but it soon becomes invested by a distinct cell-wall, which presents a regular form, usually globular or spheroidal; and this cell-wall is often furnished with curious prolongations which give it a hispid surface (as seen at B), these being sometimes very long, and furnished with hooks at their extremities. The new body thus formed is termed the *sporangium*. The mode in which it gives origin to the ordinary frond, has not yet been clearly ascertained; but it seems probable that from each sporangium, several new fronds may be produced. Every one of the new cells thus generated may multiply itself by subdivision; and thus each sporangium may develope itself into a large mass of such cells, all having similar forms and endowments. It has been remarked that the sporangia of the Desmideæ (which are for the most part inhabitants of small collections of fresh water, that are liable to be dried up by the heat of summer,) are most abundant in spring; so that it may be surmised that they are not killed by desiccation, like the growing fronds, but are enabled to regenerate the race after the period of drought has passed away.—The mode in which the conjugation and development of the sporangia take place in the entire family of *Desmideæ*, is essentially the same; and even where the cells that have been multiplied by subdivision remain in continuity with each other, so as to form a filament, they dissociate themselves before the conjugation takes place.

FIG. 249.

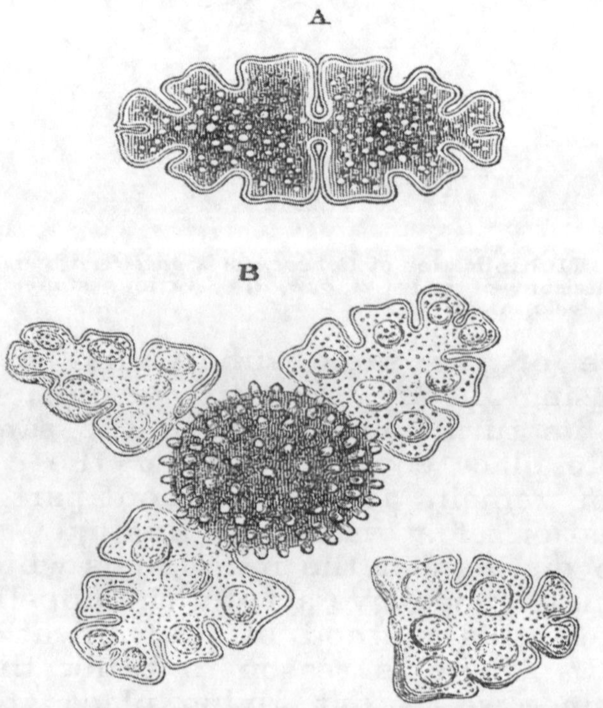

Euastrum oblongum:—A, single frond;—B, two fronds in conjugation, showing the sporangium in the centre between the four emptied valves.

Sometimes it is observable that instead of the complete separation of the two halves of each cell in the act of conjugation, they only partially open for the discharge of the endochrome; and in some of the higher forms of this family, we find the mixture to take place in one of the conjugating cells, within which the sporangium is developed. This happens especially in the genus *Desmidium,* which approaches closely to the conjugating Confervæ.*

653. In the family *Diatomaceæ,* the membranous envelope of each cell is covered by a siliceous incrustation, which usually presents very beautiful and regular markings; and this hard 'shell' is composed of two 'valves,' which are divided longitudinally, and are capable of being separated from each other. Like the plants of the preceding family, the cells may occur either singly (Fig. 250, A, B), or in connection with each other; in the first case they are known as 'frustules;' in the second, they form bands or filaments composed of several cells adherent by their long sides, which cells have all been produced by subdivision from a single

FIG. 250.

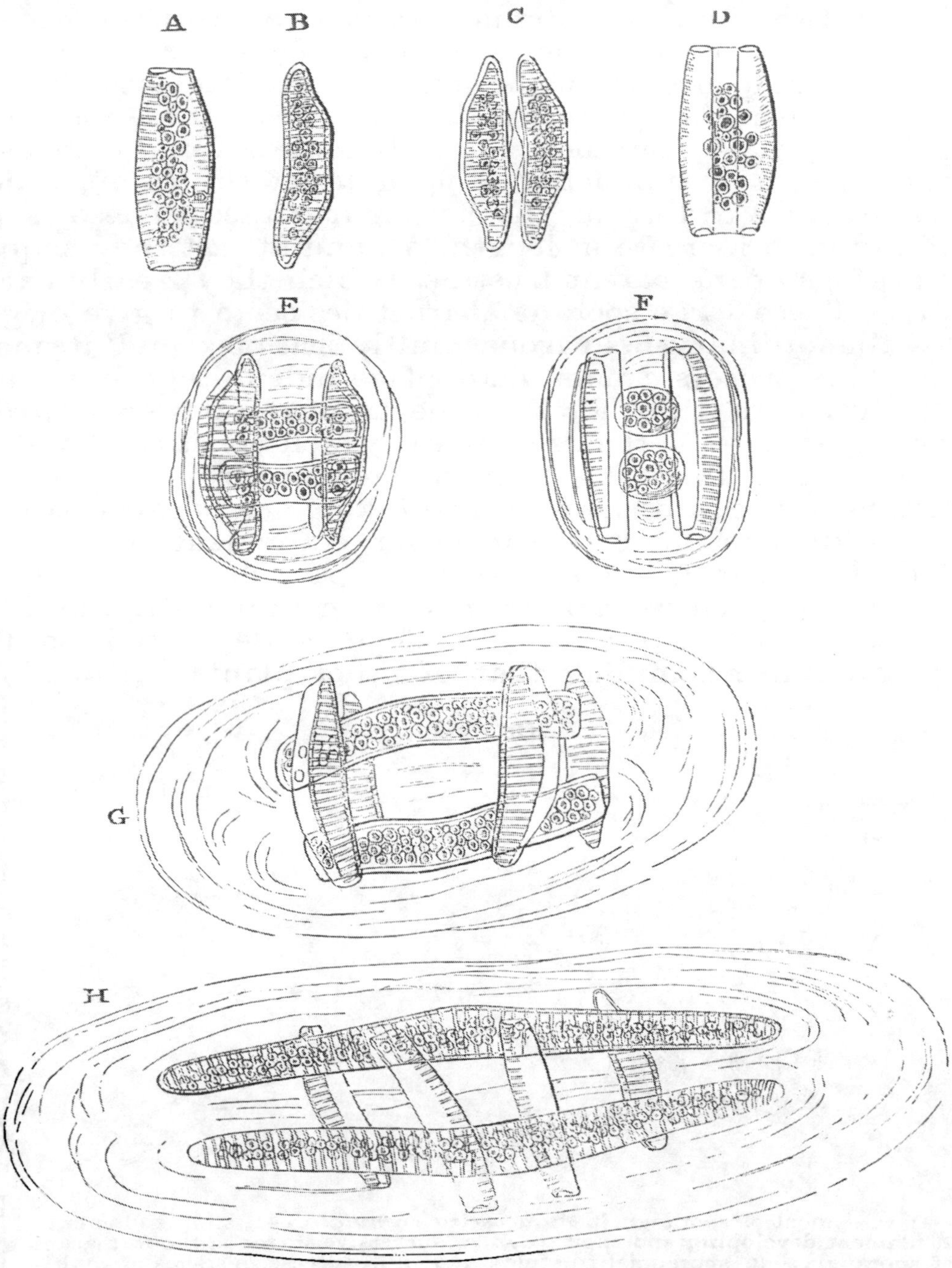

Conjugation of *Eunotia turgida*:—A, single frustule; B, side view of the same; C, two frustules with their concave surfaces in close apposition; D, front view of the same, showing the separation of each frustule longitudinally into two halves; E and F, side and front views after the formation of the sporangia; G, the sporangia more advanced, exceeding the parent frustules in size; H, completion of the sporangia.

* See Mr. Ralfs's admirable Monograph on the "British Desmideæ."

one, and have not become detached from each other. When two frustules (as in *Eunotia turgida*) are about to enter into conjugation, they move towards each other, and come into close approximation by their concave edges (C); the valves of each frustule separate from each other (D), and the inner membrane forms two protuberances, which meet two similar ones of the other frustule. These protuberances soon give way, and the endochromes of the two frustules discharge themselves and mingle, not, however, into *one* mass, but into *two*; these masses are at first irregular in shape; but they shortly become covered with a smooth membrane, and present a symmetrical elongated form (E, F), which presents some resemblance to that of the parent frustule. These 'sporangial frustules' continue to increase in length (G), and at last they acquire the transverse markings of the parent frustules, which they greatly exceed in dimensions (H). These large sporangial frustules seem to give origin to a new brood by the ordinary fissiparous multiplication; and it may be conjectured that this process cannot take place indefinitely, but that the race would die out, were it not for the renewal of the powers of growth by the act of conjugation.—In the *Meloseireæ*, which constitute a filamentous tube of Diatomaceæ, there is a curious departure from the ordinary type; for no conjugation can be here seen to take place between two distinct frustules; but a disturbance of a somewhat equivalent nature takes place in the endochrome of a single frustule, its particles moving from the extremities towards the centre, rapidly increasing in quantity, and aggregating into a mass, round which a new envelope is developed, so that it becomes a sporangium resembling that of other plants of its kind (Fig. 251). At first sight, this might seem an exception to the general rule of the reunion of the contents of *two* cells; but it is more so in appearance than in reality. For in some of the group of *Conjugateæ*, whose conjugation will be presently described (§ 654), the union of endochromes takes place, not between the cells of distinct filaments, but between two contiguous cells of the same filament; and as these two cells formed part of one and the same cell previously to their separation by its subdivision, it is obvious that the reunion of the two halves of the endochrome of the Meloseireæ is only the performance of the same process at an earlier period. The sporangium thus formed is gradually developed into a filament resembling that of the parent, by progressive fissiparous multiplication (Fig. 251, 3, *a*, *b*, *c*).

Fig. 251.

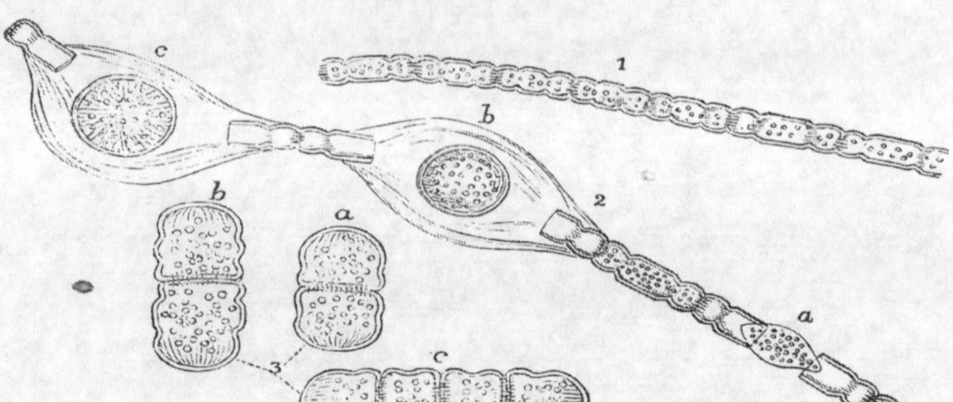

Development of sporangia in *Aulacoseira crenulata*:—1, simple filament; 2, filament developing sporangia; *a*, *b*, *c*, successive stages in the formation of sporangia;—3, sporangial frustules, *a*, *b*, *c*, in successive stages of multiplication.

654. The process of 'conjugation' has now been seen to occur in so many species of the family of *Palmelleæ,* which includes the simplest cellular forms of the ordinary Algæ (such as the *Hæmatococcus,* Figs. 4 and 65, the *Protococcus,* to which the 'red snow' belongs, the *Chlorococcus,* Fig. 248, the *Palmella,* of which one species is known as the 'gory dew,' and many others), that it may be regarded as a part of the regular series of vital actions in every member of that group.—It is, however, in the Zygnema and its allies, constituting the family of *Conjugateæ,* that this process has been longest known and most attentively studied. All these plants are filamentous; that is, the process of subdivision always takes place in the same direction, and the cells thus produced remain in continuity with each other; so that, from a single cell, a long filament is evolved. Still, the component cells of this filament for the most part retain their common relation to the generative as well as to the nutritive function, so that they may all take a share in the conjugating process: the only exception to this yet noticed, being in the case of two species of *Zygnema,* in which the alternate cells alone conjugate.* The conjugation ordinarily takes place between the cells of distinct filaments; these approximate each other, and put forth little protuberances that coalesce and establish a free passage between the cavities of the cells, whose contents then intermingle (Fig. 69). In the true *Zygnema,* the endochrome of one cell passes over entirely into the cavity of the other; and it is within the latter that the sporangia are formed. Further, it may generally be observed that *all* the cells of one filament thus empty themselves, whilst *all* the cells of the other filament become the recipients, the sporangia being developed within them. Here, therefore, we seem to have a foreshadowing of the *sexual* distinction of the reproductive cells, which manifests itself more fully in the higher organisms. In the genus *Mesocarpus,* however, and some other Conjugateæ, the two cells pour their endochromes into a dilatation of the passage that has been formed between them, and it is there that the sporangium is formed.—There are several tribes of the family Conjugateæ, in which the conjugation takes place between the adjacent cells of the same filament; each cell of the conjugating pair putting forth a conical protuberance near the septum, and the two protuberances uniting, so as to permit the mixture of endochromes, as in other instances, the sporangia being formed within one of the cells. Hence, when one of these filaments is 'in fruit,' the alternate cells will be found to contain sporangia, whilst the intermediate ones will be entirely empty.—In all these cases, the development of the filament appears to take place by the subdivision of the sporangium, which is its primordial cell, as in the instances already described.

655. In a considerable proportion of the inferior Algæ, the propagation is commonly effected by a very different process; namely, the separation and dispersion of 'Zoospores.' These bodies are formed without any conjugation, and they originate, in fact, in the *subdivision* of the cells from which they are set free (§ 142). The whole history of these bodies shows, that they are not products of a true generative act, but are merely to be regarded as detached *gemmæ,* formed by a developmental process. For whilst, in the ordinary process of subdivision, the endochrome of each cell

* See Dr. Hassall's "British Fresh-water Algæ," p. 139.

divides into only two parts, and these, when completely separated by a new cell-wall, remain in contiguity within their common envelope, the 'zoospores' are formed by a subdivision of the endochrome into a greater number of parts, each of which acquires a cell-wall of its own; and all the young cells thus produced escape from the general investment, and undergo dispersion by the exercise of the locomotive power with which they are endowed.—Such is the only method of reproduction which has yet been observed in the *Ulvæ* and a considerable number of *Confervæ*. But there is a strong probability from analogy, that a conjugating process takes place even amongst these at *some* stage of their existence; for it does not seem possible that the multiplication by gemmation can take place to an *unlimited* extent in these lower organisms, any more than in the higher. It may perhaps be surmised without any great improbability, that the process of conjugation does not take place among the cells of the *complete* plant, but in some earlier stage of development. In fact, the early stage of the *Confervæ* and *Ulvæ* so strongly resembles the permanent condition of the *Palmelleæ* (Fig. 67), that it is not impossible that some of what have been supposed to be conjugating cells of the latter group, may really belong to the former. For the complete history of the development of the sporangia is as yet so far from having been made out, that it is by no means certain that those which result from the conjugation of a supposed Palmella or Chlorococcus, may not develope themselves into an Ulva or Conferva; and when the phenomena of reproduction in the higher Algæ (§ 656) are compared with those which present themselves in the Ferns (§ 660), it will be seen that there is a strong probability from analogy in favour of such a view.—This point is one peculiarly deserving of further investigation.

656. In passing to the higher Algæ, we encounter a very marked change in the mode in which the Generative act is performed; the essential character of it, however, still remaining the same. In the structure of these plants, as formerly pointed out (§ 268), we no longer meet with the same homogeneousness as in the simpler tribes; the frond has a definite form and structure in each species, instead of being a mere amorphous mass; the several cells composing it no longer repeat each other, but exhibit a great variety of forms and modes of aggregation; and they no longer possess a capacity for independent existence. The Reproductive function is now limited to particular portions of the fabric; and the sexual distinction between the 'sperm-cells' and the 'germ-cells' now becomes apparent.—Although the occurrence of the true generative act has been rendered highly probable in the whole of the sub-classes *Melanospermeæ* (or *Fucales*) and *Rhodospermeæ* (or *Ceramiales*), by the discovery of two distinct sets of organs, analogous to those which are known to concur in this process in the higher plants, yet the precise mode in which their concurrence takes place has not yet been made out. In the Fuci and their allies, the propagation appears to be effected solely by the bodies which are commonly designated 'spores,' but, which should rather be termed *sporangia*. The 'germ-cells' in which these originate, are developed, in the inferior forms of this type, such as *Sporochnus* (Fig. 252), in certain spots upon the general surface of the frond (B); having the form of pear-shaped vesicles (C, *c*), attached at their base, and containing an olive-green endochrome. They are commonly surrounded, however, by

filaments (a, b) whose cells, when examined near the period of maturity, are seen to contain orange - coloured granules of very minute size; and from similar filaments, contained within the conceptacles of the *Fuci*, moving bodies of very peculiar appearance, strongly resembling in their actions the spiral fibres or 'phytozoaires' of the antheridia of Mosses and Ferns (Fig. 255 D), have been observed to issue. These bodies are of an ovoid form; and each of them has a red spot near its middle, whilst from either end proceeds a long thread, which is at first in continual movement. They

FIG. 252.

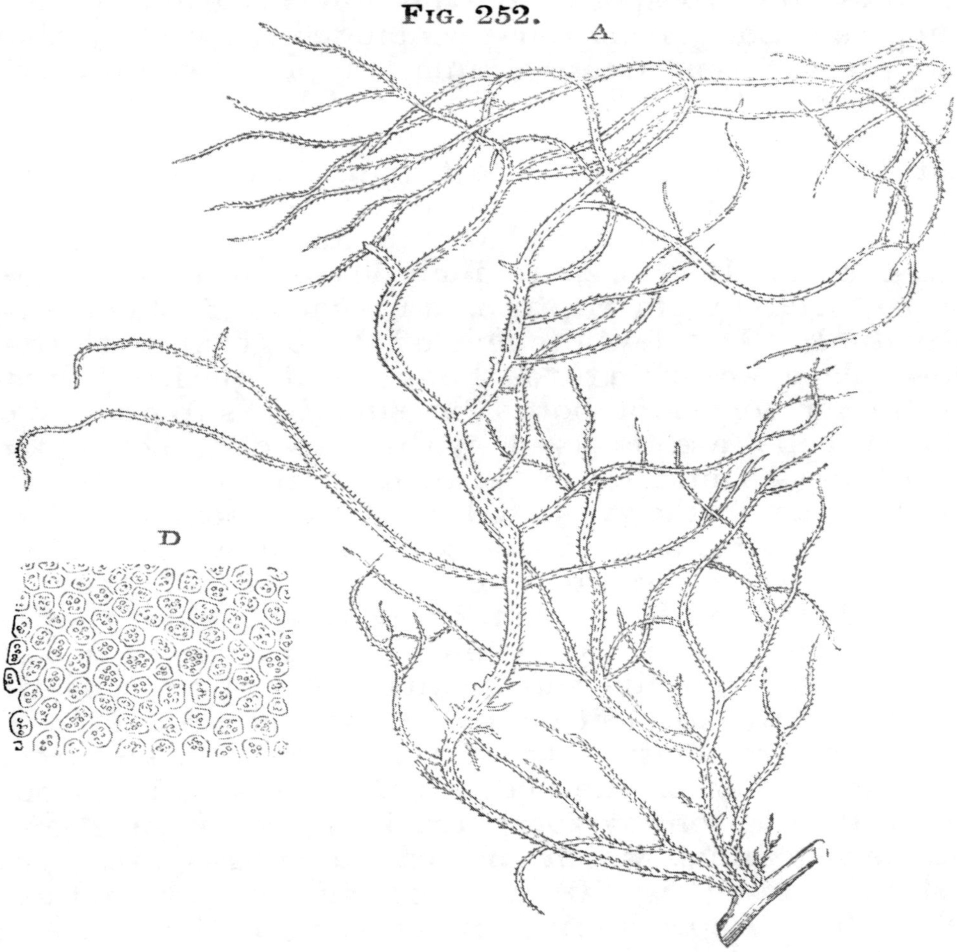

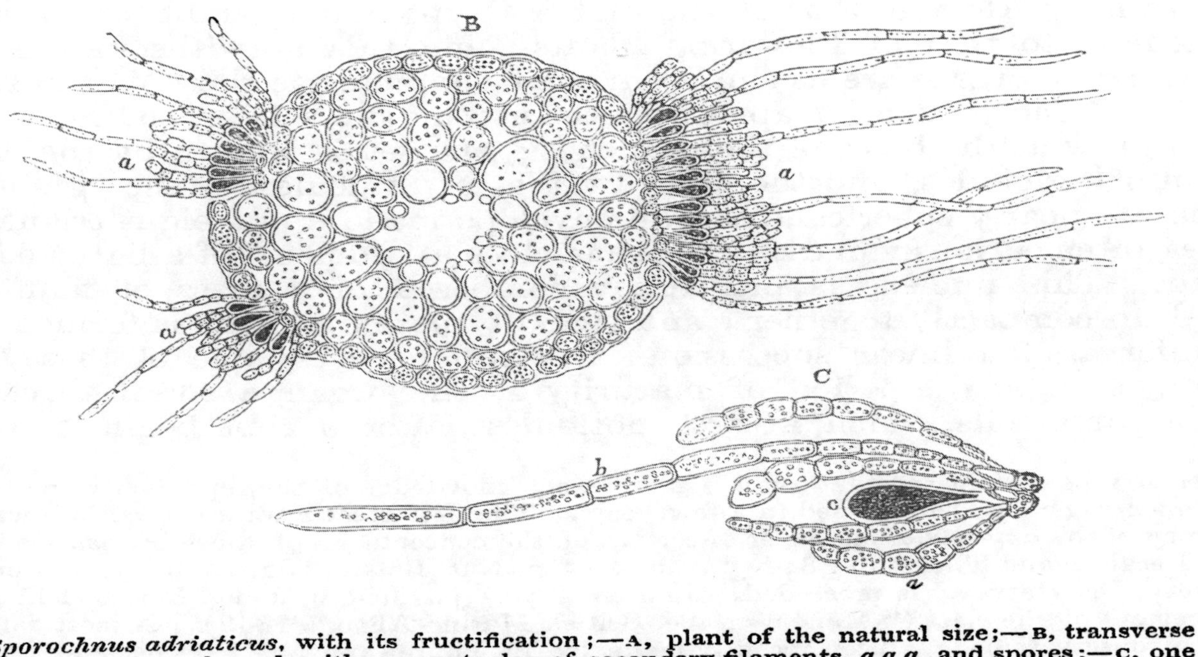

Sporochnus adriaticus, with its fructification;—A, plant of the natural size;—B, transverse section of stem, enlarged, with conceptacles of secondary filaments, *a a a*, and spores;—C, one of the tufts, composed of secondary filaments (antheridia?) *a*, *b*, surrounding the spore *c*, still more enlarged;—D, cellular layer forming the surface of the plant.

become motionless, however, in no long time after their escape, but do not, like the 'zoospores' of the inferior Algæ, give origin to any other structure; and it may be concluded therefore, that if they take any share in the reproductive function (as it would seem probable that they do, from the fact that they are developed in vast numbers, either within conceptacles of their own, or within those which develope the spores), their function is to exert a fertilizing influence upon the germ, which is developed within the sporangium. This last body sometimes contains but a single cell, included within an envelope or 'perispore;' but not unfrequently the perispore includes from two to eight such cells, which are set free by its rupture, and each of them gives origin to a new individual. The development of these germs into the form of the parent takes place according to the general method already described. — In the higher forms of both the superior sub-classes of Algæ, the reproductive cells are evolved within special conceptacles, developed on the principal frond or on branches proceeding from it; and these conceptacles, as formerly stated (§ 268 *a*), sometimes contain the 'sperm-cells' and 'germ-cells' in close approximation, whilst in other instances the 'germ-cells' occupy the upper part of the conceptacle, and the 'sperm-cells' the lower; in other cases, again, they occupy distinct conceptacles, and these may be either on different parts of the same plant, or they may be on different plants. All these variations foreshadow the varieties in the disposition of the sexual organs in Phanerogamia.—Besides this propagation by the true generative process, we find in the *Ceramiales* another method, which seems to be essentially the same with the multiplication by zoospores in the simpler Algæ (§ 268 *a*); but the detached gemmæ, which are called 'tetraspores' from being found in clusters of four (Fig. 70), have no cilia and are not self-moving.*

657. The Reproductive apparatus of the little group of *Characeæ* presents us with a closer approximation than we have as yet seen among the ordinary Algæ, to that of the higher Cryptogamia, and even, in some particulars to that of Flowering Plants; and this notwithstanding that the nutritive organs are as simple in their character as those of Confervæ (§ 269). The generative apparatus consists of two sets of bodies, both of which grow at the bases of the branches; one set is known by the designation of 'globules,' and the other by that of 'nucules.'—The 'globules,' which are nearly spherical, have an envelope made up of eight triangular valves, often curiously marked, which enclose a 'nucleus' of a light reddish colour. This nucleus is principally composed of a mass of filaments, rolled up compactly together; and each of these filaments is formed, like a Conferva, of a linear succession of cells. In every one of these cells there is seen, at the period of maturity of the organ, a spiral thread of two or three coils, which, at first motionless, after a time begins to move

* It may be anticipated that a much more exact knowledge of the reproductive process of the higher Algæ will be obtained in a few years. The first great advance towards it was the discovery of the *antheridial* character of certain of the conceptacles of the Fucaceæ, made by MM. Decaisne and Thuret in 1844 ("Ann. des Sci. Nat. Botan.," Janv. 1845); and on this discovery, Dr. Harvey has proceeded, so far as he was justified in doing, in remodelling the arrangement of the class ("Manual of the British Marine Algæ," 1849). A most valuable series of observations has been recently made by M. Thuret and MM. Derbès and Soulier, of which the general results alone have been yet given to the public ("Ann. des Sci. Nat." Juin, 1850).

and revolve within the cell; and at last the cell-wall gives way, and the spiral thread makes its way out, partially straightens itself, and moves actively through the water for some time, in a tolerably determinate direction. The movements appear partly to depend upon the lashing action of two long and very delicate filaments, with which these bodies, like the 'phytozoaires' of the Fucaceæ, are furnished; and it would seem probable that they are destined to bring some of these curious products of the 'sperm-cells,' into contact with the 'germ-cell,' and thus to originate, by an act essentially similar to that of conjugation, an embryonic body capable of developing itself into a new plant.—The germ-cell is contained within the 'nucule;' the exterior of which body is formed by five spirally-twisted tubes, that give to it a very peculiar aspect. What is the precise nature of its nucleus is as yet uncertain; it would seem probable that this chiefly consists of a mass of starchy matter (resembling that of the 'albumen' of a true seed) in which a single germ-cell is imbedded. For when the nucule is cut across, or ruptured by pressure, a multitude of little grains are forced out, which are proved by the iodine-test to consist of starch; and it has been determined by the observations of Vaucher, that only one embryo is developed from each nucule, and this by a process that much resembles the germination of a Phanerogamic plant.* No *gemmæ* are known to be spontaneously detached from these plants; but they may be multiplied by artificial subdivision, the separated parts continuing to vegetate under favourable circumstances, and developing the entire structure.

658. In regard to the Reproduction of *Lichens* (§ 270), nothing can here be offered in addition to what has been already stated. The 'soredia' of Lichens are certainly to be considered as *gemmæ*, representing the tetraspores of the higher Algæ; whilst the 'asci' may be surmised to be really *sporangia*, the products of a conjugating process; but on the latter point no light has yet been thrown. The development of these plants seems to take place, in all essential particulars, as in Algæ; the simplest forms being first evolved, in every case, by the subdivision of the spore-cells, which are set free from the asci; and the continuance of this process of multiplication, in definite directions, being all that is needed for the evolution of the specific form which gradually manifests itself.—The same ignorance must be confessed respecting the generative process in the *Fungi* (§§ 271–273). In all the cases yet known, the multiplication is accomplished by 'spores;' but these are developed as offsets from the peculiar cells termed *basidia*, and seem to be related to the parent structure rather as gemmæ, than as true generative products. It would seem highly probable from analogy, that something equivalent to the conjugation of the lower Algæ, or to the fertilization of a germ by a spermatic

* It is somewhat remarkable that, although the Characeæ have long been favourite subjects for observation amongst microscopists, much should remain to be determined respecting the structure of the nucule, and the mode in which the generative act is accomplished. The discovery of the spermatic filaments is due to Mr. Varley, who described them in the "Transactions of the Society of Arts" for 1834; they were afterwards independently discovered by M. Thuret, and described by him in the "Ann. des Sci. Nat." for 1840.—It may seem not a little surprising that such a multitude of spermatic filaments should have been provided for the fertilization of a single germ; but it must not be forgotten that precisely the same is seen among the highest Animals; and that if these bodies were less multiplied not one of them might find its way into the nucule.

filament, takes place at some epoch in the life of the Fungi; although the lower forms may repeat the process of multiplication by detached gemmæ many times before the true generative act is again required.

659. In the *Hepaticæ* and *Mosses,* however (§§ 275, 276), we again meet with two distinct sets of sexual organs, besides a provision for the production and detachment of gemmæ or bulbels. The *antheridia* contain sperm-cells, within which are developed spermatic filaments or *phytozoaires,* that closely approximate in appearance and in the nature of their movements to those of the Characeæ; whilst from the *archegonia* (or *pistillidia*) are evolved the 'capsules' containing 'spores,' which are commonly regarded as constituting the proper fructification of these orders. Now it has been observed that when the antheridia and archegonia are developed upon different individuals, as happens in some species of Mosses, the evolution of the latter into capsules does not take place, unless the plants bearing the former are in the neighbourhood; and as it has been shown by M. Valentin's elaborate investigations,* that no impregnation of the contents of the capsule, by the introduction of any external substance into its cavity, can take place *after* the formation of the spores, there is no other mode of accounting for this fact, than by admitting that the 'germ-cell' is contained in the archegonium, that it is fertilized by the products of the 'sperm-cells,' and that the whole mass of spores contained within the capsule is the product of its subsequent subdivision. In this point of view, such a mass would represent the sporangium of the Fuci in every respect, save that the process of subdivision, or 'sporangial gemmation,' has proceeded much further; it would also represent the single embryo of the Flowering-plant, which also is composed of a mass of cells originating in the subdivision of the primitive germ-cell. The 'spores,' however, are capable of existing independently of each other, like the cells of a Palmella; and thus from each of them a distinct individual may arise. In the development of one of these cells into a new plant, the first stage is the rupture of its outer coat, and the protrusion of the delicate cell-wall that lined it. By the subdivision of this primordial cell, an embryonic cluster is soon formed, presenting a very confervoid aspect, so as closely to resemble the mature condition of plants much lower in the scale of organisation. It is by a continuation of the same process, that the 'frond' of the *Hepaticæ,* which presents various gradations from a simple and almost amorphous expansion to an assemblage of parts arranged with some degree of regularity upon an axis, is gradually evolved; and from every part of its under surface proceed radical fibres, which remain in the condition of the simple confervoid filaments first put forth. But in *Mosses,* in which there is a more perfect separation of the axis and its foliaceous appendages, we find the stem gradually developed from one end of the primitive confervoid body or pro-embryo, at first, however, as a mere homogeneous mass of cells; the leaves are then put forth from the axis, one after another; and the structure characteristic of the perfect organs becomes progressively apparent (§ 276). No true root, or descending axis, is ever evolved in Mosses; and the radical filaments, which proceed from the base of the stem, remain in the simple condition of those which are put forth from

* "Linnæan Transactions," vol. xvii.

the pro-embryo. We have seen that in the *Marchantia* a very curious apparatus is developed specially for the production of gemmæ or 'bulbels' (Fig. 68); and this kind of multiplication is very common both among Liverworts and Mosses. It seems to take place, indeed, even in the confervoid state of the pro-embryo, which frequently separates into several parts, from each of which a new plant may be developed. There are certain species of Mosses, which rarely if ever occur with true fructification in certain localities; their propagation being effected almost solely by the spontaneous detachment of gemmæ which are put forth from their stems; and thus among these humble plants, the product of a single sexual operation may attain the age of our largest trees, and may occupy as large a space in the economy of nature.—The foregoing interpretation of the generative process in this group of plants, which had been put forth by Mr. Thwaites* previously to the discovery of the facts to be described in the next paragraph, is so remarkably confirmed by the analogy which they supply, that it may, in the author's opinion, be almost unhesitatingly adopted.

660. It has been by the discoveries recently made in regard to the processes of Generation and development in the class of *Ferns*, that a new light has been thrown upon the nature of these operations, not only in this particular group, but in the Cryptogamia in general. It now appears that what has been hitherto considered the 'fructification' of the Fern,—namely, the collection of *thecæ* containing *spores*, usually found on the under sides or at the edges of the fronds (Fig. 81),—is really an apparatus for gemmiparous production; the 'spore' not being the immediate product of the sexual or true generative operation, but being a free *gemma*, which, when cast off by its parent, is forthwith developed into a structure containing sexual organs, from which the new generation really originates. It will be convenient, however, in describing the process of evolution, to commence with the 'spore,' as the best defined starting-point. This body is a *cell* of irregular form, but usually somewhat pyramidal (Fig. 253, A); its outer wall is formed of a brownish-coloured resisting membrane, in

Fig. 253.

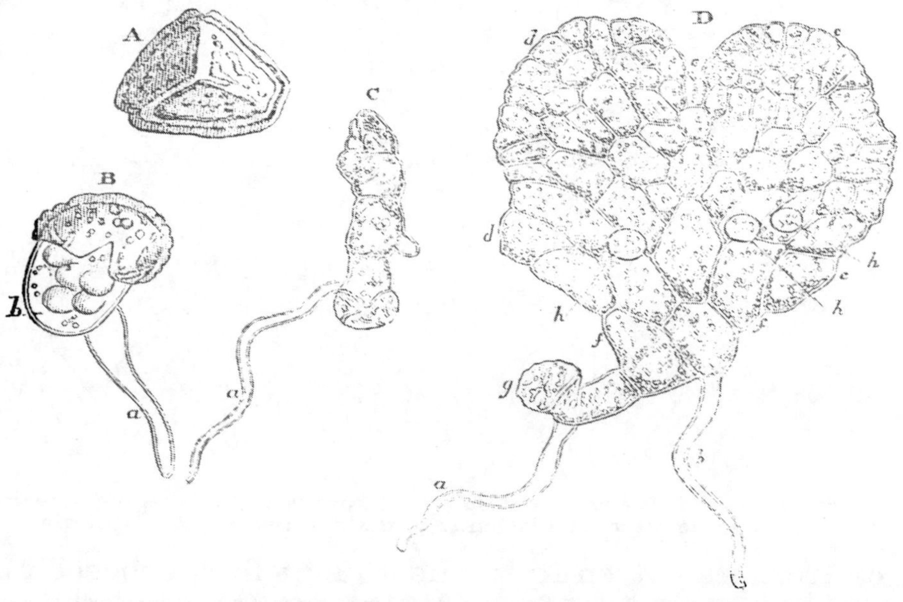

Development of Spore of *Pteris serrulata*;—A, spore as set free from the theca;—B, spore beginning to germinate, putting forth the tubular prolongation *a*, from the principal cell *b*;—C, linear series of cells, formed by the multiplication of the original;—D, pro-embryo taking the form of a leaf-like expansion; *a*, first, and *b*, second radical fibre; *c*, *d*, the two lobes, and *e* the indentation between them; *ff*, under part of the pro-embryo; *g*, external coat of the original spore; *h*, *h*, antheridia.

* "Annals of Natural History," March and November, 1849.

some part of which is a minute aperture; its inner wall is extremely thin and transparent; and the cavity of the cell contains an oleaginous mucilage, in which are usually found three nuclei. Under the influence of warmth, moisture, and light, the spore begins to enlarge, the first indication of its increase being the rounding-off of its angles; then from the orifice in its outer wall is put forth a tubular prolongation of the internal cell-wall (*a*) which serves as a radical fibre, absorbing nourishment from the surface on which the spore is lying; and by this absorption, the inner cell is so distended, that it bursts the external unyielding integument, and is now directly exposed to the influence of light (B). Its contents speedily become green; and the cell itself elongates in a direction opposite to that of the root-fibre, so that it acquires a cylindrical form. A production of new cells then takes

FIG. 254.

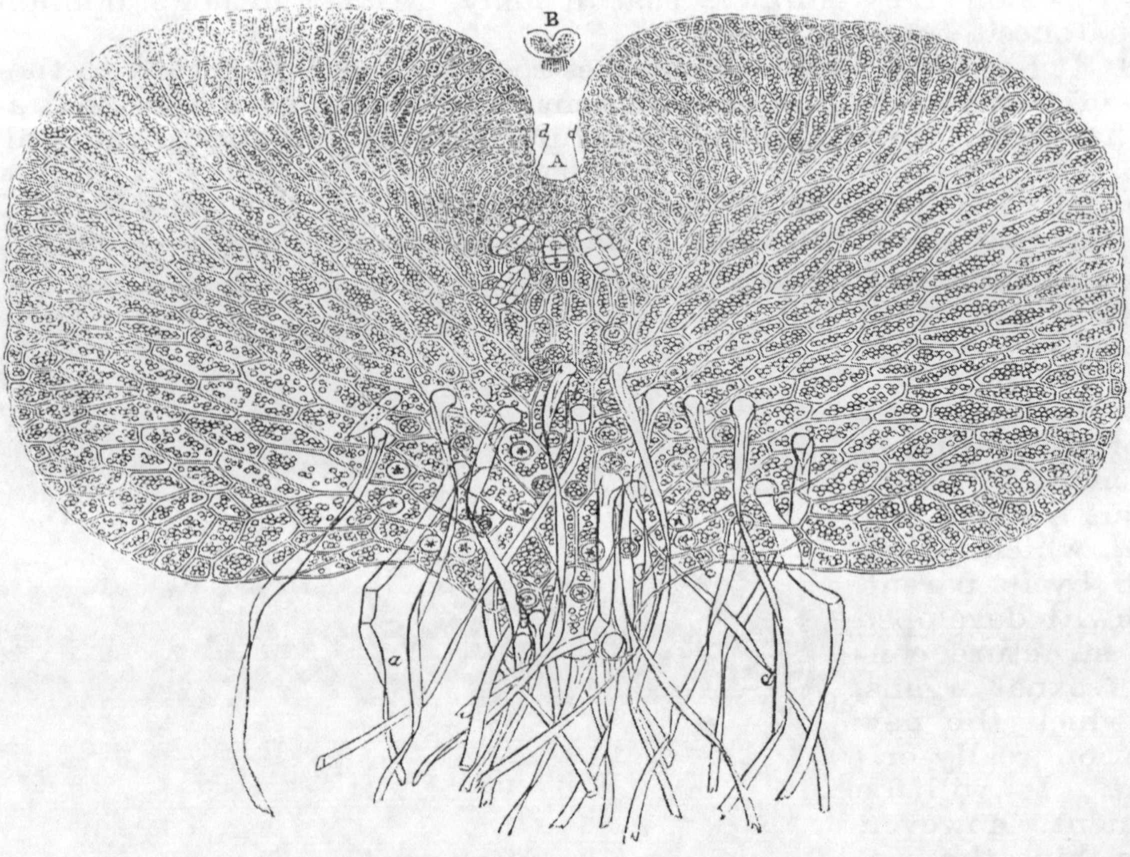

Pro-embryo of *Pteris serrulata*; B, natural size; A, magnified;—*a a*, root-fibres; *b*, antheridia; *c*, the same after the discharge of their contents; *d*, pistillidia.

place from its extremity; and this at first proceeds in a single series, so as to form a kind of confervoid filament (C); but the new growth soon takes place laterally as well as longitudinally, so that a flattened, leaf-like expansion, or "thallus' closely resembling that of a young Marchantia, is soon formed. This thallus varies in its configuration in different species of Ferns; in the species here figured (D), it is bilobed, the two divisions (*c*, *d*) being separated by a kind of notch (*e*); but its essential structure is always the same. From its under surface are developed additional root-fibres, which serve to fix it to the soil, and, at the same time, to supply it with moisture.—To this body, which, in its fully developed form, is represented in Fig. 254, the name of *pro-embryo*, has been appropriately given.*

661. At an early period in the development of the 'pro-embryo,' certain peculiar glandular-looking bodies are seen projecting from its under surface (Fig. 253, D, *h*, *h*); these augment in number with the advance in growth; and at the time of the complete evolution of the 'pro-embryo,' they are seen in considerable numbers (Fig. 254, A, *b*, *c*), especially about its base, near the origins of the radical fibres (*a*, *a*). These bodies owe their origin to a peculiar protrusion which takes place from certain of the cells of the 'pro-embryo' (Fig. 255, A, *a*); this is at first entirely filled with chlorophyll; but soon a peculiar free cell (*b*) is seen in its interior, filled with mucilage and colourless granules. This cell gradually becomes filled with another brood of young cells (B, *e*); it increases considerably

FIG. 255.

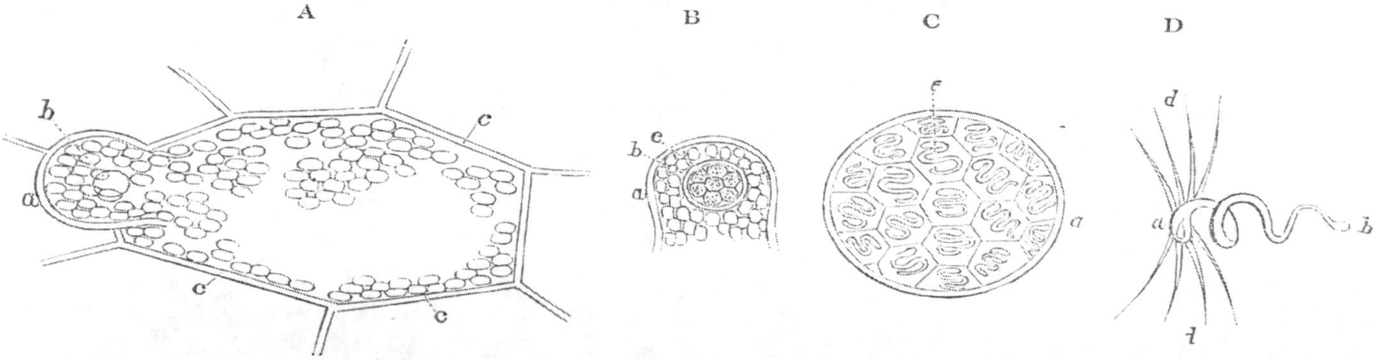

Development of the *antheridia*, and *spermatic filaments*;—A, one of the cells of the pro-embryo, bounded by the cell-wall *c c c*, giving off a projection at *a*, which contains a peculiar secondary cell, *b*;—B, further development of this part; *a*, projection of the parent-cell, *b*. secondary cell or antheridium filled with a new generation of cells *e*;—C, antheridium completely developed and shown on a larger scale; *a*, wall of the secondary cell; *e*, contained cells; each inclosing a spiral filament;—D, one of the spiral filaments (phytozoaires) highly magnified; *a*, large extremity; *b*, other extremity less dilated; *c*, cilia.

in its dimensions, so as to fill the projection which incloses it; this part of the original cavity is now completely cut off from the parent-cell, of which it was an offshoot; and the *antheridium* (as this peculiar cell with its contents is now to be called) henceforth ranks as a distinct and independent organ. Each of the smaller cells contained within the primary cell of the 'antheridium' is seen, towards the period of maturity, to contain a spirally-coiled filament (C); and when the small cells have been set free by the bursting of the antheridium (Fig. 254, A, *c*), they themselves burst and give exit to their spiral-filaments, which have a rapid movement of rotation on their axis, partly dependent upon the long cilia with which they are furnished (Fig. 255, D). Each of these spiral filaments makes from two to three turns; its anterior extremity (*a*) is consi-

* This "Marchantia-like expansion" has been long known as the first production from the spore; and was described in the former editions of this work as a *cotyledon* developed for the elaboration of nourishment for the young Fern which is afterwards to sprout from its centre. It will be seen hereafter, that though it undoubtedly performs an *analogous* function, yet that it cannot be regarded as *homologous* with a cotyledon; its relation to the young fern being that of a parent to its offspring, and not that of a temporary leaf to a more permanent one.

derably enlarged, and seems to contain a minute vesicle; its posterior extremity (*b*) is also somewhat dilated.*

662. Besides these bodies, the 'pro-embryo' bears a small number of others, differing essentially from the preceding in character, and usually occupying a different position; thus in the bilobed frond of *Pteris serrulata*, they are seen near the median indentation (Fig. 254, *d*, *d*). The number of these bodies seems indeterminate; sometimes only three, in other instances as many as eight or more, being seen in 'pro-embryoes' of the same species. Each of them at its origin presents itself only as a slight elevation of the cellular layer of the 'pro-embryo,' within which is a large intercellular space, containing a peculiar cell (Fig. 256, A, *a*), and opening externally by an orifice (*c*) at the summit of the projection; but when fully developed, it is composed of from ten to twelve cells, built up in

Fig. 256.

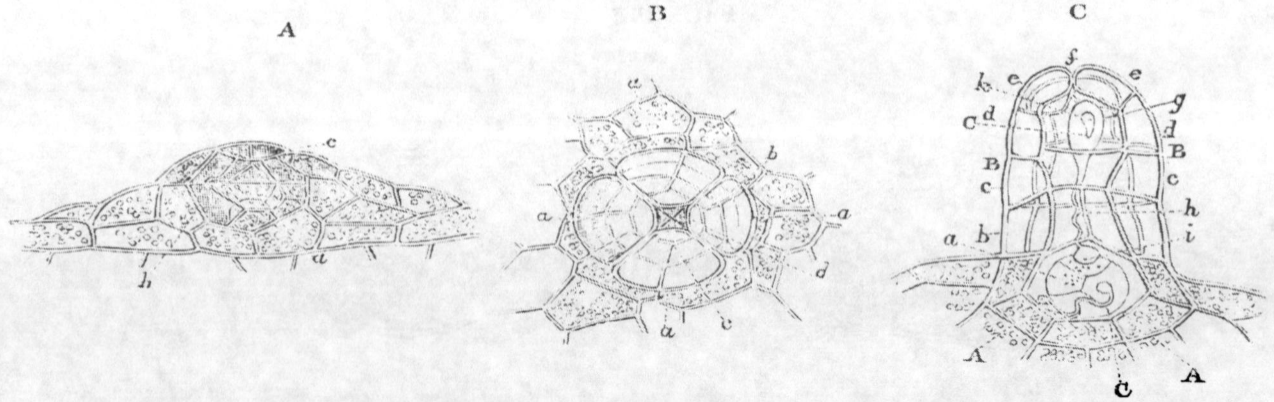

Development of the *pistillidium* of *Pteris serrulata*;— A, side view in its early state; *a*, embryo-sac; *b*, cells of the pistillidium; *c*, opening at the summit;—B, pistillidium in more advanced stage, seen from above; *a a a a*, cells surrounding the base of the cavity; *b*, *c*, *d*, successive layers of cells, the highest enclosing a quadrangular orifice;—C, vertical section at the time of impregnation; A A, cavity containing the embryo-sac *a*; B B, walls of the pistillidium, made up of the four layers of cells, *b*, *c*, *d*, *e*, and having an opening on the summit at *f*; C C, spiral filaments; *g*, large extremity, enclosing a minute body *k*; its thread-like portion *h* lying in the canal of the pistillidium, and its small end *i* dilated into a globular form, and in contact with the embryo-sac.

layers of four cells each, one upon another (B), so as to form a kind of column (C), having a central canal which leads down to the cavity at its base. The subsequent history of this body shows that it is to be considered as a *pistillidium* or female organ; and that the peculiar cell contained in the cavity at its base is to be regarded as a 'germ-cell' or 'embryo-sac.' As the development of the pistillidium is taking place, it appears that certain of the spiral filaments, or *phytozoaires*, set free from the antheridia, penetrate into its cavity; and that one of these comes into peculiar relation with the embryo-sac by its smaller extremity (C, *i*), which dilates into a globular form, and becomes detached from the remainder of the filament. The whole of the filament, as it lies in the canal of the pistillidium, is seen to have enlarged considerably, probably by the absorption of mucus whilst it traverses the under-side of the frond. The contact of the dilated end of the filament, or 'sperm-cell,' with the

* The discovery of these antheridia in the 'pro-embryo' of Ferns was made by M. Naegeli in 1846. He gave an accurate account of the production of the spiral filaments, but entirely overlooked the 'pistillidia' (presently to be described), which he seems to have regarded as antheridia in a different grade of development.

embryo-sac or 'germ-cell,' appears to constitute an act of fecundation, precisely equivalent in all essential particulars to the 'conjugation' of the Algæ; and as the result of this fecundation, a new body, the true 'germ' makes its appearance in the interior of the embryo sac.

663. The germ has at first a globular form, and consists of a homogeneous mass of minute cells (Fig. 257, A, *a*); but as its development proceeds, rudiments of special organs begin to make their appearance; it grows at the expense of the nutriment prepared for it by the 'pro-embryo;' and it soon bursts forth from the cavity of the pistillidium, which organ, in the meantime, is becoming atrophied. In the very beginning of its development, the tendency is seen in the cells of one extremity to grow upwards, to form the stem and leaves; and in those of the other extremity to grow downwards, to form the root. The condition of the germ when these parts are first distinctly evolving themselves, and its relation to the pro-embryo, are shown in Fig. 257, B; in which we see the rudiment of the first leaf at *a*, some of the hairs upon its surface at *e*, the first cells of the leaf-stalk at *h*, those of the terminal bud or 'growing-point' of the stem at *i*, those of the root at *b*, those of the pro-embryo at *l m*, and the fragments of the pistillidium at *c c* and *d*. Already spiral vessels begin to show themselves, as at *f*.—The further progress of this germination may be readily apprehended. When the true root has been sufficiently evolved to serve for the absorption of fluid nutriment, and the first true frond has been expanded to the air, so that the young plant can now elaborate its own alimentary materials, the pro-embryo, whose function is now discharged, decays away. The axis, elongating itself upwards to form the stem, gives off successive leaves; and some or all of these, when fully evolved, bear the *sori*, or clusters of *thecæ*, which produce a a new brood of spores.* Besides this method of multiplication, some

FIG. 257.

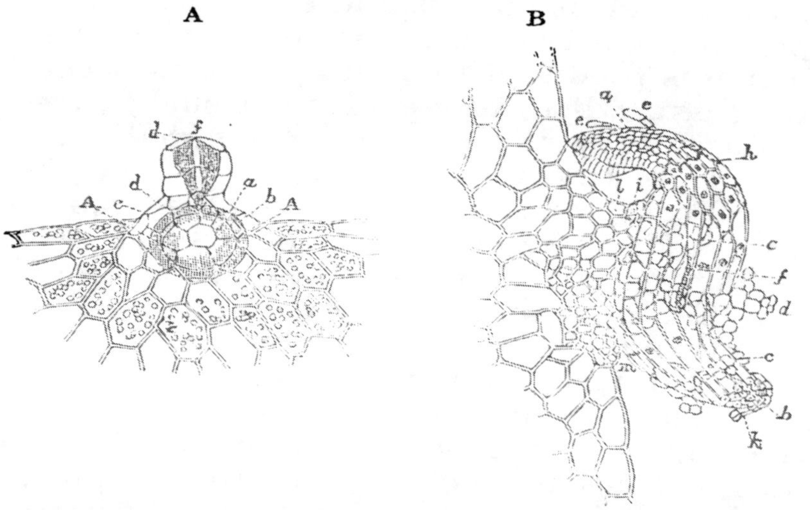

Development of the embryo of *Polypodium aureum*:—A, pistillidium some time after fertilization, containing the embryo-sac A, within which is seen the globular germ-mass, *a*; at *b*, *c*, *d*, are shown the contracted cells of the pistillidium, and at *f* the opening at its summit:—B, embryo much further advanced, and emerging from the pistillidium; *a*, leaf; *b*, root; *c c*, *d*, fragments of the pistillidium; *e*, hairs; *f*, first spiral vessels; *h*, leaf-stalk; *i*, terminal bud or growing-point; *k*, root-sheath; *l*, *m*, portion of pro-embryo supporting the embryo.

* The discovery of the pistillidia and of their contained embryo-sacs, of the fecundation of these by the spermatic filaments, and of the true relation of the young Fern to the pro-embryo, is due to Count Leszczyc-Suminski, whose very beautiful Monograph "Zur Entwickelungs-Geschichte der Farrnkräuter," published at Berlin in 1848, also contains a much more elaborate history of the development of the young Fern, than had been previously given. In the Report upon Count Suminski's observations, presented to the Berlin Academy by Dr. Münter, he states that a part of the fertilizing process above described,—namely, the penetration of the moving filaments into the orifice of the pistillidium,—has been *repeatedly witnessed* by himself and Prof. Link; but it appeared to them that the *phytozoaires* resolved

Ferns possess another, which corresponds with the formation of gemmæ in the Marchantia; bulbels being formed from the tissue of the leaves, either on the surface or in the angles of the lobes; and then dropping off and becoming independent plants.

664. Thus the history of the Fern presents us with a very characteristic example of the so-called "alternation of generations;" and one, too, which remarkably illustrates the general principle which has been already laid down (§ 649) with regard to the true nature of this alternation. Starting from the spore, we must consider the pro-embryo as, in common parlance, a distinct 'individual;' it is completely independent of its parent; and it obtains and elaborates its own nutriment. The life of the pro-embryo terminates, however, with the evolution of the young Fern, which is produced from it by a true process of sexual generation: whilst the Fern, when fully evolved as such, gives origin to a new series of pro-embryoes by the process of gemmation, and dies without having reproduced its kind in any other way. Thus, between each two generations of true Ferns, a pro-embryo intervenes, springing from the first by gemmation, and giving origin to the second by true sexual reproduction; and in like manner, between each two generations of pro-embryoes there is a true Fern, originating in the sexual operations of the one, and producing the other by gemmation. Thus in the case before us, we have a complete exemplification of the general fact, that the so-called 'generations' are not related to one another in the same way; but that whilst their relationship is in the one case that of offspring to parent, it is in the other that of offset to stock. The latter cannot be legitimately held to constitute a distinct 'generation,' if that term be used in the signification in which it has until recently been ordinarily understood in physiology,—namely, the production of a new being by sexual union. And if we carefully examine the case before us, we shall see how little claim the 'pro-embryo' has, to be considered as a generation distinct from that of the Fern which produces the original spore. For let it be supposed that the 'pro-embryo,' instead of being cast off from the Fern in the state of spore, had been developed in continuity with it, we should then have distinctly

themselves into "little heaps of mucus" after their motion had ceased. Whichever statement is the true one (and the latter is in very curious accordance with the observations of Mr. Newport upon the fertilization of the ovum of Frogs, § 684), the essence of the matter remains the same; for in either case, the contents of the sperm-cells and those of the germ-cells are brought into mutual relation.

The Author would add, that having enjoyed the advantage of inspecting Count Suminski's preparations (in company with Mr. Robert Brown and Mr. Thwaites), he was so perfectly satisfied of the correctness of many of his descriptions, as to be the more willing to place confidence in the remainder.

Unfortunately for science, the question of the Sexuality of the Cryptogamia has been made quite a *party* one amongst Botanists; and it may be doubted whether some of them (*e.g.* Prof. Schleiden and his followers) would admit the affirmative upon *any* amount of evidence that the nature of the case admits of. It is not a little curious, that the question of the sexuality of the Phanerogamia was contested in the same spirit, when the doctrine was first propounded by Linnæus; and that many of the Botanists of that day denied the possibility of his assertions, and ridiculed his inferences, with just the same arrogant despotism that is now displayed by certain individuals, towards all who presume to differ from them upon the point under discussion. Belonging himself to no party, and having early espoused the opposite doctrine of the analogy of the spore to the pollen-grain, it may be supposed that the Author would not now abandon this for another, save upon what appears to him preponderating evidence in favour of the latter.

recognised it as the true generative apparatus of the Fern, the equivalent of the flower of the Phanerogamia. In no essential respect, indeed, would it differ; for it is obvious that the antheridia and pistillidia are the equivalents of the anthers and pistils of the flowering-plant; and that the germ is the product of the joint action of the 'sperm-cells' and 'germ-cells' which they respectively contain. That the embryo-sac in the Fern lies naked in the cavity of the pistillidium, from the walls of which is supplied the nourishment which it imbibes,—whilst in the Phanerogamia it is contained within an ovule, and surrounded with a mass of nutriment prepared and stored up there before its impregnation,—seems to constitute the most important difference in the conditions under which the germ is developed in the two groups respectively; and this difference, however valuable in a systematic arrangement of Plants, as serving for the separation and definition of these groups, cannot be considered as having any fundamental importance in a physiological point of view. In every important respect, then, save in the fact of its originating in a detached gemma, and being developed independently, the pro-embryo bears the same relation to the Fern, as a *capitulum* (or head of flowers, such as is borne by the *Compositæ*) bears to the plant which evolves it; the development of the sexual apparatus is necessary to our idea of a perfect 'individual;' and thus we must regard the pro-embryo as really belonging to the preceding Fern-generation, instead of constituting an entirely new one, unless we are prepared to go so far as to maintain that the fact of its detachment gives it a title to the appellation 'new generation,' which would be to extend very widely the usual acceptation of the term.

665. The real import of the Reproductive organs of the *Lycopodiaceæ* and *Equisetaceæ* (§ 277 *b*, *note*), can scarcely yet be regarded as fully ascertained; since the history of their action has not been completely determined. They would seem, however, to hold in some respects an intermediate position between those of Ferns and those of Flowering Plants. For it appears from the late researches of Müller upon the development of two genera of the first of these groups, that the larger reproductive bodies are true *ovules*, each containing an embryo-sac, surrounded by a store of nutriment; and that from within this embryo-sac, the embryo shoots forth, very much as in the Phanerogamia, save that its cotyledons, or first temporary leaves, are developed *during* this process of germination, instead of *previously* to it.* Whether the production of the embryo, however, is the result of fertilization by 'phytozoaires,' as in the Ferns, or by pollen-cells, as in the Flowering Plants, has not yet been determined. In the Equisetaceæ it would seem that the former is the case; antheridia, with their characteristic contents, having recently been discovered in them.† The transition seems to be completed by the curious little group of *Marsileaceæ*, in which the 'phytozoaires' of the Cryptogamia appear to be conveyed to the germ-cell, by a provision closely resembling that which is concerned in the fertilization of the ovule of Phanerogamia. Each 'germ-cell' is here imbedded in a mass of nutritious matter, invested by proper envelopes, forming an ovule; and the 'sperm-cell' seems rather to have the character of a pollen-grain, than that of the

* See "Annals of Natural History," Ser. 1, vol. xix., and Ser. 2, vol. ii.
† "Annales des Sciences Naturelles," Juin, 1850, p. 375.

ordinary antheridial cells of the Cryptogamia. But the ovules and pollen-grains are very commonly developed within the same capsule; and it is not until after they have been set free by the dehiscence of this, that the impregnation of the former by the latter takes place. The first part of this process appears to correspond in all essential particulars with that presently to be described as the mode of fertilization in Phanerogamia; save that the pollen-tube has here direct access to the ovules, instead of being obliged to penetrate the ovarium from without; but after this act has been accomplished, 'phytozoaires' are found in the mucilage enveloping the ovule, which are believed by Naegeli to have issued from the (so called) pollen-tube.* The germination of the Marsileaceæ appears to correspond with that of the Lycopodiaceæ, as regards the development of the cotyledons during the process. In fact, it may be likened to the development of the true embryo of Ferns, in every essential particular, save that the latter derives its nourishmeut from the pro-embryo, the former from the store laid up in the ovule.

666. If the views here set forth respecting the nature of the generative process in the Cryptogamia be correct, it follows that the difference between them and the Phanerogamia does not lie, as has been until lately believed by most Botanists, in the *sexuality* of the latter, and in the *non-sexuality* of the former; but in the mode in which the influence of the 'sperm-cell' is conveyed to the 'germ-cell,' and in the provision made for the subsequent development of the latter. Such differences are consequently much less essential in their character, and admit of a transition being formed by connecting links, between one group and the other, such as appear to be established by the families mentioned in the preceding paragraph.—We have now to enquire into the mode in which the generative function is performed in Phanerogamia; a subject upon which there has been a vast amount of discussion within the last few years, but in regard to the main points of which the greater number of careful and unprejudiced observers seem now agreed. The Generative apparatus is here entirely separated from the Nutritive, and constitutes a distinct organ, the flower; of which, however, the foliaceous portion (constituting the floral envelopes) may remain undeveloped, only the essential parts being evolved. These essential parts usually consist of the *ovarium* containing the *ovules*, and the *anthers* containing the pollen-grains. The ovarium, as formerly explained (§ 278) is commonly made up of several distinct *carpels*, more or less closely united; and each of these carpels may be regarded as a foliaceous organ, whose edges are folded together and have become adherent. So, too, the anthers may be considered as leaf-like organs, in which the development of pollen-grains has taken the place of that of the ordinary parenchyma. A less-developed form of this generative apparatus, however, is seen in the *Coniferœ* and their allies; for the floral envelopes are completely wanting, and the proper sexual organs present a very different aspect from that which characterises the ordinary Phanerogamia. Each 'cone' of a Pine or Fir consists of a mass of 'scales,' very regularly arranged around the axis; and every one of these scales is a carpellary leaf, of which the edges are not folded together, so

* See the memoir of Naegeli on this subject, in the "Annales des Sciences Naturelles," Fevrier, 1848, p. 108, translated from the "Zeitschrift Wissenschaftliche botanik," 1847, p. 188.

that the ovules which they bear are not enclosed. Hence this group of plants is commonly designated as *gymnospermous* or naked-seeded. The 'catkin,' again, is not unlike a smaller cone, each of its scales being a polliniferous leaf. Thus the 'cone' represents a cluster of female or ovuliferous flowers, and the 'catkin' a similar cluster of male or staminiferous; and whilst in the mode of fertilization of their ovules by pollen-grains, and in the history of their embryonic development, the Gymnosperms essentially correspond with Flowering Plants, they bear a very close resemblance to the Lycopodiaceæ in the simplicity of the reproductive organs, as well as in their general aspect and mode of growth.

667. The *pollen-grains* are developed within the parenchyma of the leaf-like organ that is subsequently to become the anther, by a curious process, which seems to indicate the necessity for a special elaboration or preparation of their contents. According to Mr. Henfrey,* the first stage is the development of new cells within those of the ordinary parenchyma; the wall of each of these new cells enclosing the entire contents of that within which it is formed, and the young cell thus entirely filling the original cavity. When the young cell is completely formed, the wall of the cell that enclosed it decays away, leaving the young cell free; and this is now known as one of the 'parent-cells' of the pollen. The protoplasma of each of these parent-cells then divides into two, and then into four portions; so that two septa, generally crossing at right angles, are found, dividing the original cavity into four cells, each of which generally has the form of a quarter of a sphere. Within every one of these cells, termed by Mr. Henfrey the 'special parent-cell,' a new cell-wall is developed around its contents; and this new cell, the 'pollen-grain,' is afterwards left free by the dissolution of the walls of the 'parent-cells' and of the 'special-parent cells' within which it was enclosed.—The pollen-grain, like other cells (§ 136), possesses two envelopes, of which the inner one is extremely delicate, whilst the outer one is firm and resisting, in virtue of the peculiar deposit which takes place upon it while yet within the pollen-cell. This deposit frequently forms prominent ridges, which may cross each other, so as to leave reticulations of a very regular and beautiful aspect; and it always seems to be thinner than usual in one or more spots. The contents of the pollen-grain have at first the character of the ordinary protoplasma; but gradually the fluid becomes more watery, and the granular particles more distinct. Of these particles, some appear to be mucilaginous, others to be oily, and others, again, to consist of starch. In an early stage of the development of the pollen-cell, a regular 'cyclosis' may be seen within it; but this ceases some time before its maturation, and a mere molecular movement is then all that remains.

668. The *ovule* is developed in the midst of the parenchyma of the *placenta* (§ 278), and commences as a single cell, which, by successive subdivisions, forms a projecting mass, that sometimes remains 'sessile' upon the placenta, and sometimes becomes partially detached from it, remaining connected only by a peduncle. The mass of cells that is first formed, is that which afterwards constitutes the *nucleus* of the ovule; as

* "Reports of British Association" for 1848; Part II., p. 84. This account corresponds with that of Naegeli in nearly every essential particular, save that the latter regards the new cells as originating in separate nuclei or cytoblasts, which Mr. Henfrey affirms not to be the case.

its development proceeds, it becomes enveloped in one, two, or even three coats, which are formed by the multiplication of cells that at first constitute merely an annular enlargement at its base. The apex of the nucleus, however, is never entirely covered-in by these investments; a small aperture being always left through them, which is called the *micropyle.* In the interior of the nucleus is a large cavity, which appears to be formed by the enlargement of one of its cells, that grows at the expense of the surrounding tissue (Fig. 258, A). This cavity, which is termed the *embryo-sac,* is at first filled only with protoplasma; but some little time *before* fecundation, there are seen in it a certain number of free cell-nuclei, rarely fewer than three, and frequently more; these are observed especially near the apex or micropylar end of the embryo-sac (C), although, when they are very numerous, some of them are seen near its base. Around these nuclei, free cells of a spheroidal form are developed (D), of which only one is usually destined to be fertilized, and to be the original of the future embryo; all the rest subsequently disappearing, as if their function were to elaborate or prepare the nutriment for the developing germ.—The maturation of the pollen-grains takes place contemporaneously with that of the ovules; and when the former are set free by the rupture of the anther, they fall upon the stigma, and begin to absorb the viscid secretion which bedews its surface. In consequence of this absorption, the inner membrane or proper cell-wall becomes distended, and either breaks through the thinner points of the external envelope, or pushes this before it, so as to form one or more long slender projections, which are known as the *pollen-tubes.* These insinuate themselves between the loosely-aggregated cells of the style, and grow downwards until they reach its base, a distance, in some cases, of several inches. Arrived at the ovarium, they direct themselves towards the micropyles of the ovules; and entering these, they make their way towards the embryo-sac, usually through a channel formed by the diffluence of a sort of cord of peculiar cells that previously passed from the apex of the embryo-sac to that of the mammillary protuberance of the nucleus (B, E). The extremity of the pollen-tube then impinges upon the apex of the embryo-sac itself (F), and sometimes pushes it slightly inwards, so as to have given origin to the idea that it enters its cavity. There is not, however, according to the latest and best observers, any direct communication between the cavity of the pollen-tube and that of the embryo-sac; but whatever admixture may take place between their contents, must occur by transudation through their respective limitary membranes. It is in such an admixture, however, that the act of fertilization appears essentially to consist; for soon after the contact of the pollen-tube with the embryo-sac has taken place, it becomes apparent that *one* of the cells which the latter contained is now undergoing rapid development, and that the remainder are in a state of degeneration. This cell, whose previous existence appears to be now indubitably ascertained, is distinguished as the *germ-vesicle* or *embryonic-vesicle;* and it is from it, when fertilized, that the entire embryonal structure originates. The early processes of development correspond precisely with those which have been already described as taking place throughout the whole of the inferior tribes; for the first cell gives origin, by transverse fission, to a pair; this, again, to four (G); and so on, it being usually in the *terminal* cell of the filament so generated, that the process of multiplication chiefly takes place, just as in the Confervæ. At the same time, the mucilaginous matter which fills up the embryo-sac becomes organised and converted into loose cellular tissue; the cells

FIG. 258.

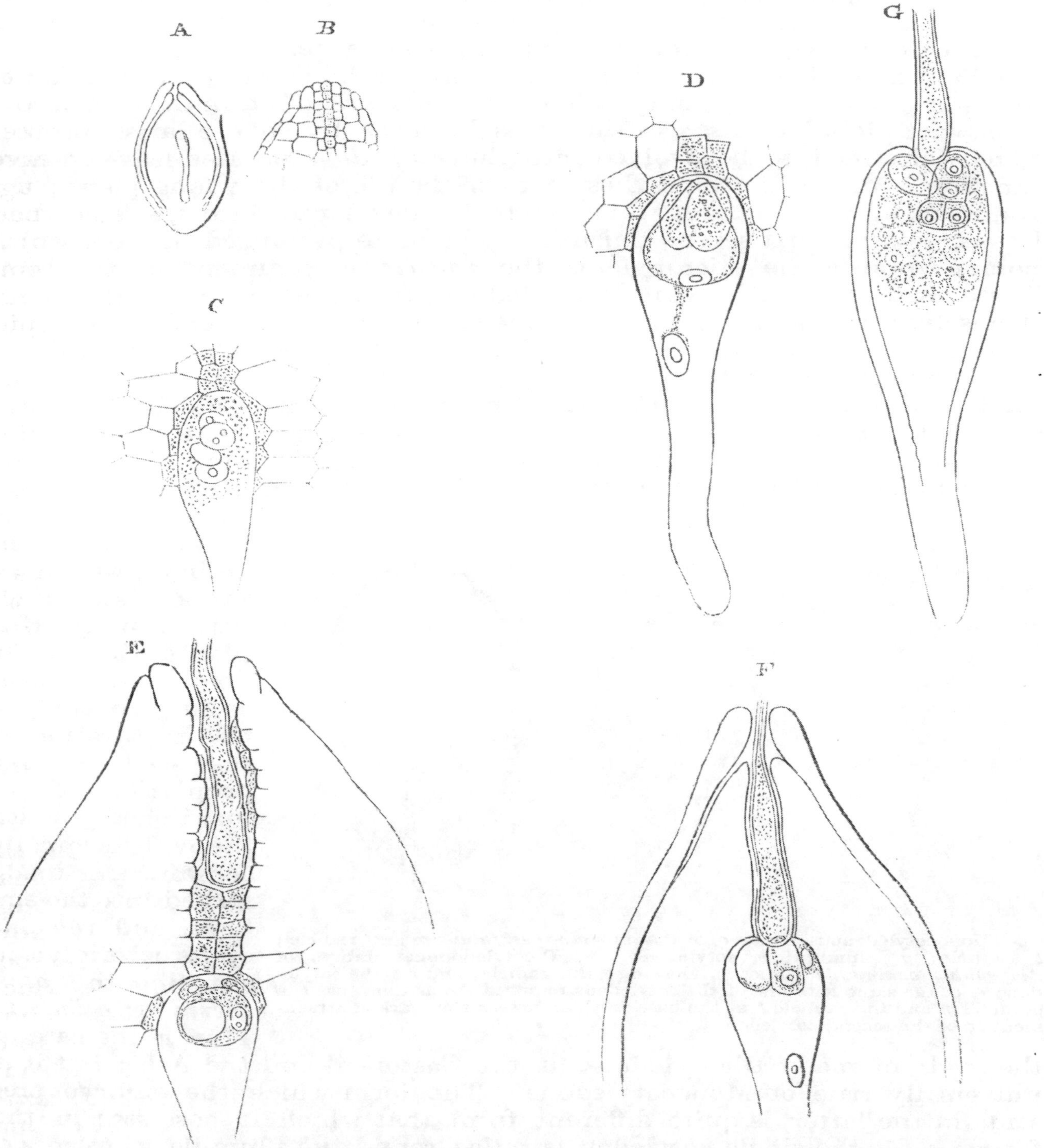

Development of the embryo of *Œnotheraceæ*:—A, longitudinal section of non-fecundated ovulum, showing the embryo-sac in the midst of granular mucilage;—B, longitudinal section of the mammillary projection of the nucleus;—C, upper portion of the embryo-sac, with the surrounding cellular tissue, from a more advanced embryo;—D, the same at a later period;—E, advance of the pollen-tube towards the nucleus;—F, contact of the pollen-tube with the summit of the embryo-sac;—G, embryo beginning to form, within the embryo-sac after fecundation.

being formed around free nuclei (§ 142). As in other cases, however, in which this mode of cell-formation occurs, the tissue thus produced is very transitory in its character; for it deliquesces again as the embryonic mass increases in bulk and presses upon it; and thus the only purpose which it can be imagined to serve, is that of elaborating the nutriment for the growing fabric, which may well be supposed to need such a preparation, in virtue of its superior importance and permanence.

669. The embryo, which is at first a simple filament, usually enlarges most at its lower extremity, where its cells are often multiplied into a somewhat globular mass; of this mass, however, by far the larger proportion is destined to be evolved into the *cotyledons* or seed-leaves, whose function is limited to the earliest part of the life of the young plant; the rudiment of the *plumula,* which is to be developed into the stem and leaves, being at first scarcely visible. The more prolonged portion, which points towards the micropyle, is the *radicle* or rudiment of the root. These parts increase in some Dicotyledons, until they have absorbed unto themselves all the nutriment contained, not only in the embryo-sac, but also in the tissue of the nucleus itself; so that the seed, at its maturity, contains nothing but the embryo, of which the cotyledons are rendered thick and fleshy by the amount of nutritious matter which they have absorbed. This is the case, for example, in the Leguminous family generally, and a good illustration of it is furnished by the Almond (Fig. 259, B, C); in such instances, the remains of the nucleus and of the embryo-sac coalesce to form an envelope to the embryo, within the proper seed-coats. But in many other plants, the cotyledons remain foliaceous; so that, at the maturation of the seed, a large proportion of the nutriment which the ovule originally contained, is unabsorbed into the embryo, and remains on its exterior, constituting the *perisperm* or *albumen.* This is the case in the seeds of many Dicotyledons, as the Castor-oil and the Ash; but it is universally true of Monocotyledons. The form which the embryo presents in the latter, is quite different from that which it possesses in the former; for the single cotyledon is rolled round the plumula, so to speak, in such a manner as completely to enclose it, except where a little fissure is left by the non-adhesion of the edges of the cotyledon (Fig. 259 A, *g*), through which the plumula subsequently escapes. This arrangement is quite conformable to the relation which exists between the stem and all

Fig. 259.

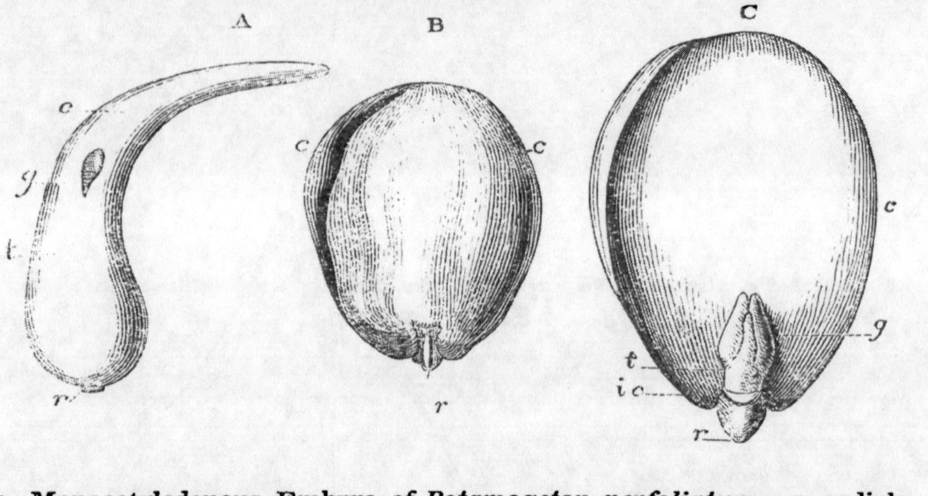

A, Monocotyledonous Embryo of *Potamogeton perfoliatum*; *r*, radicle; *t*, caulicle; *g*, plumula; *c*, cotyledon; — B, Dicotyledonous embryo of *Amygdalus communis* (Almond), showing *r* the radicle, and *c c* the cotyledons; — C, the same with one of the cotyledons removed, so as to expose the plumula *g*, and the caulicle *t*, at the base of which is seen the mark of attachment *i c* of the second cotyledon.

the subsequent leaves of Monocotyledons in general; for they, too, ensheath the axis at their base, as is well seen in the Grasses.

670. The Seed, when completely matured, becomes detached from the placenta, and is set free by the opening of the seed-vessel. In this condition it may remain for a great length of time in a state of complete inaction, without the loss of its vitality, provided that it be secluded from influences which would either force it into growth, or tend to occasion its decomposition; and in fact, this period of 'dormant vitality' may be protracted, in the case of many seeds, to an extent to which there seems no limit (§§ 115–117).—The conditions which are absolutely requisite for the Germination of seeds, are water, oxygen, and a certain amount of warmth; and the process is favoured by the absence of light. When the seed imbibes moisture, the embryonic tissue swells, and the radicle is advanced towards the micropyle, whilst at the same time the seed-coats are commonly ruptured by the internal distension. During these processes, certain chemical changes take place in the starchy matter contained in the perisperm or absorbed into the cotyledons, in virtue of which it is converted into a saccharine compound, probably to be blended with albuminous matter to form a protoplasma for the nutrition of the growing tissue; and these changes involve the production of a certain amount of carbonic acid, by the union of the carbon of the seed with atmospheric oxygen (§ 503). Hence it is that oxygen, as well as water, is required as a condition of this process; and that light is rather injurious than beneficial, since it tends to *fix* the carbon in the vegetable tissues. This conversion is effected by the agency of *diastase*, an albuminous compound stored up in the seed, which seems to act as a 'ferment.' Whilst thus living upon organic compounds previously elaborated by an agency other than its own, thriving best in the dark, and imparting to the air a large quantity of carbonic acid, the germinating plant may be regarded as physiologically very much in the condition of a fungus.—The radicle, however, becoming prolonged by the imbibition of water and by the formation of new tissue at its extremity, tends to grow downwards into the soil; and the plumula then begins to elongate itself in the opposite direction, whilst the minute protuberances which it previously bore, evolve themselves into leaves. The cotyledonary portion is usually the last to quit the seed-coats, and sometimes it remains and withers there; but usually it is carried upwards to the surface, acquires a green colour, and performs the functions of a leaf. In the germination of such Monocotyledons as thus send up their cotyledon (Fig. 260, A), this organ very closely resembles the subsequent leaves in form, and in its relations to the stem; in that of Dicotyledons, on the other hand, the plumula rises up from between the cotyledons (B). In many Gymnosperms, the cotyledons are multiple, forming a sort of verticil, resembling that of their true leaves. —The germinating plant continues to appropriate the nutriment stored up in the cotyledons or remaining in the perisperm, and to employ it as the material for the extension of its permanent fabric; and by the time that this store has been exhausted, the roots and leaves are sufficiently developed to enable them to perform their respective functions, and thus to minister to the further extension of the structure. It is then only that true woody fibre and vessels begin to show themselves in the axis; for during the previous stages of embryonic development, its tissue has

been cellular only; and an almost exclusively cellular organisation is common to young parts at all subsequent periods.

FIG. 260.

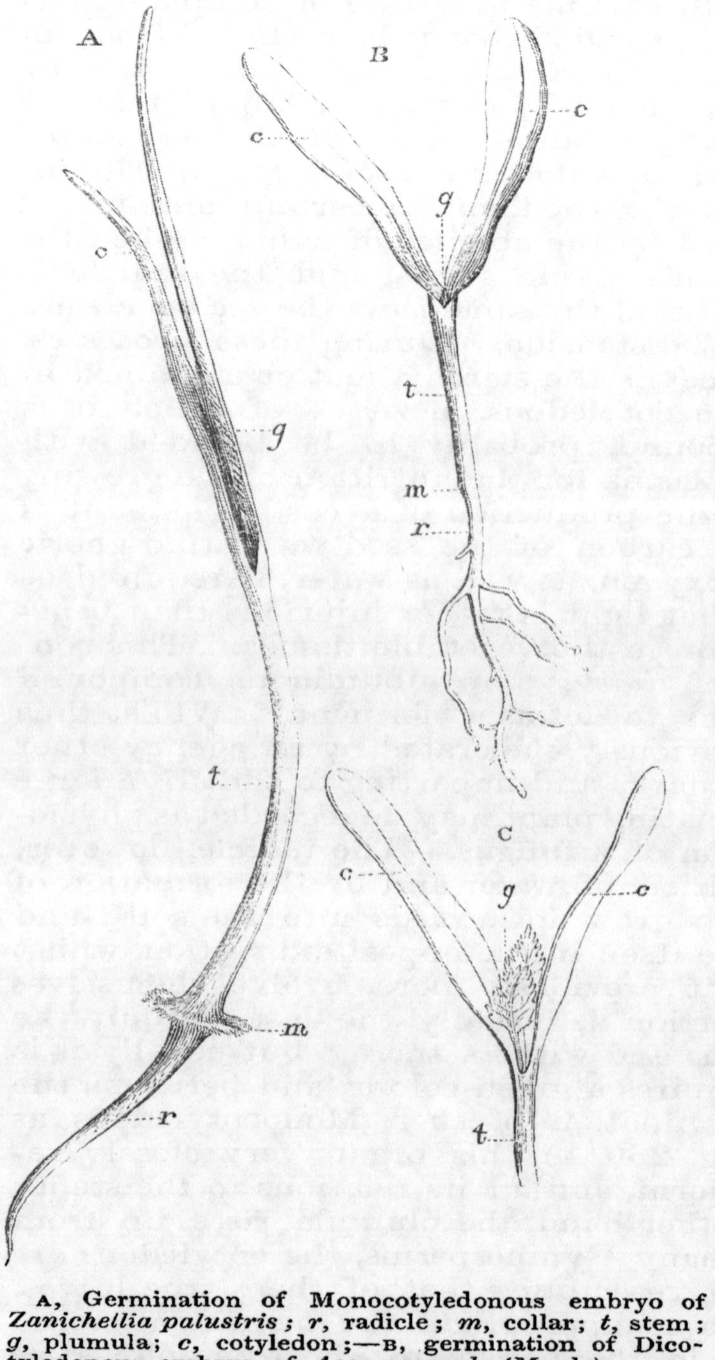

A, Germination of Monocotyledonous embryo of *Zanichellia palustris*; *r*, radicle; *m*, collar; *t*, stem; *g*, plumula; *c*, cotyledon;—B, germination of Dicotyledonous embryo of *Acer negundo* (Maple); *r*, radicle; *m*, collar; *t*, stem; *g*, plumula; *cc*, cotyledons;—C, the upper part of the same in a more advanced state.

671. The plumula may be regarded as the first 'terminal bud' of the growing plant; and with its short stem and radicle forms what is termed a *phyton*. A tree, to whatever dimensions it may attain, is but an aggregation of such phytons; and these are formed in continuous development from the one thus immediately evolved from the seed. The elongation of the stem is provided for by the development of a succession of terminal leaf-buds; each, as the leaves have performed their functions and have died off, being replaced by another. But in nearly all cases (Palms being the principal exception), lateral leaf-buds also are developed, which are, so to speak, the terminal leaf-buds of the branches that are to grow forth from the axis (Fig. 82, A, *d*, *d*); and when these have been developed into branches, they in their turn put forth other lateral buds; and thus an unlimited amount of ramification may take place around the central axis. In all cases, however, the buds originate in the medullary tissue of the stem or branches; being, in fact, produced by the multiplication of its cells at certain points, which usually have some definite relation to each other, so as to give to the whole structure a more or less symmetrical character (§ 335). Thus we have a multitude of parts evolved, which are mere repetitions of each other, and which have scarcely any relation of mutual dependence; and these parts, when detached from the common stock, and placed under circumstances favourable to their growth, will preserve their vitality, and will develope themselves into organisms in all respects similar to that from which they were removed. Such a detachment of leaf-buds sometimes takes place spontaneously (§ 291, *note*), and constitutes the

ordinary mode in which the plant is propagated. In other instances, nothing more is necessary than the artificial separation of the leaf-buds; thus the Sugar-cane is ordinarily propagated by dividing the stem into as many pieces as it possesses lateral buds, and burying these in the ground; and, in like manner, the Potato is multiplied by the division of its 'tuber' (which is really an underground stem) into as many pieces as it has 'eyes.' So, again, there are many trees and plants, of which the branches will take root, when merely cut or broken off, and placed with their ends in the soil; and such, therefore, may be propagated by 'slips' or 'cuttings.' In many other cases, however, more care is necessary; the buds not possessing enough developmental capacity to ensure the production of roots after their complete detachment from the stock; and the evolution of such buds is provided for, either by connecting them with a new stock to which they become adherent, so that it stands in the place of the stem and roots of that from which they have been removed, which operation is termed 'grafting;' or by delaying the final detachment until they have already struck root into the soil, as is practised in the operation of 'layering.' Now in all these cases, the plant, which is developed from the detached *bud*, partakes of the characters of the stock from which it sprang, much more fully than does the plant raised from *seed;* for whilst the latter continues the *species* only, the former reproduces the particular *variety.** Hence, when it is desired to multiply a certain kind of fruit-tree, the buds are employed rather than the seeds. But this method of propagation cannot be carried to an indefinite extent; for although it may not be true (as stated by some), that the life of the 'graft' will only last as long as that of the 'stock' from which it was taken, yet it is almost invariably found that varieties of trees and plants which are thus multiplied, lose their vigour and 'die out' after a certain lapse of time. They all partake, in fact, of the 'germinal capacity,' which was engendered by the original generative act; and every multiplication of parts by the continued subdivision of cells must be regarded (so to speak) as an expenditure of that capacity; so that there is necessarily a limit to this operation, although, as formerly pointed out (§ 645), the limit is much less strict when the extension takes place in structures of a low type, than when *development* of a higher kind is required in addition to mere *growth.*

672. Now from attaching too exclusive consideration to the fact of the independent vitality of leaf-buds, many Botanists have come to the conclusion, that a composite tree is to be regarded as not one individual, but *an aggregate of individuals;* and that each series of buds should rank as a *distinct generation.* Now if such be the acceptation in which the terms 'individual' and 'generation' are to be employed, it will be found that they must be rendered still more comprehensive. In the first place, if we look to what a part may *become,* rather than to what it *is,* as our test of individuality, we must include under this designation, not merely the leaf-buds, but the leaves, of many plants (such as the *Bryophyllum*), which possess the power of developing buds and roots from their margins; and

* It would seem, however, as if, in the case of *annual* plants, the variety is more steadily transmitted by seed; such is certainly true of the Cerealia, in which we find repeated, not merely the general modifications induced by cultivation, but the more particular forms that occasionally arise *de novo.*

this view has been adopted by Prof. Owen, who regards every leaf, and even every modified form of the same fundamental type,—each sepal, petal, stamen, and carpel of a flower,—as entitled to rank as a distinct being.* But if we go *so* far, we must go further still, and admit that each *fragment* of a leaf is entitled to be characterized as a distinct individual; since the leaf of Bryophyllum, even when divided into numerous segments, will continue to vegetate under favourable circumstances, and will evolve itself into a complete plant; and the whole of such a leaf, therefore, must be regarded as an aggregate of individuals. Such a doctrine was actually propounded in regard to the *Hydra*, when its extraordinary power of developing the whole from any single part was first discovered (§ 289); but no Physiologist would at present support such an idea; and as its singleness is not now disputed on account of its repetition of similar parts, and its almost unlimited reproductive capacity, neither (as it seems to the author) should that of the Plant. For it would be just as legitimate to regard as 'distinct individuals' the several tentacula which form a circle round the mouth of the Hydra, as it is to assign this rank to the several petals of a corolla, or the carpels of an ovary. The cells of the simplest Phytozoa *may* be allowed the designation of 'distinct individuals,' if by this be implied no more than that they are complete in themselves, and have the power of separate existence. But although thus functionally independent, they are organically related to each other in the same degree as are the component cells of the most heterogeneous fabric, whether Vegetable or Animal; for the aggregate mass of cells of a Palmella or a Desmidium, produced by the subdivision of the sporangium, is clearly the homologue of the entire embryo of the Phanerogamous plant, which is (like it) composed of a mass of cells, evolved by successive subdivision from the fertilized germ-vesicle; and if every new pair that is produced by the subdivision of a pre-existent cell, is to be regarded as a 'new generation' in the one case, it must in the other also.—The confusion which at present prevails in regard to this subject, and which has given rise to the strangest misapprehensions, has arisen from the same confusion of *functional* and *homological* relationship, which formerly obstructed the progress of Philosophical Anatomy (see CHAP. VIII); and will not be dissipated, until a clear distinction shall have been drawn between the two. The question of 'individuality,' in the usual acceptation of the term, is one entirely of the former kind; for its limits are not established by any other rule than functional capacity; and nothing can be more variable than its degree. Thus, such an individuality exists in the segments of the leaves of one plant, in the entire leaves of a second, in the leaf-buds of a third, in the branches of a fourth, and in the entire axis and appendages of a fifth; whilst in a sixth, the individuality shall entirely depend upon circumstances, its buds not being able to sustain their vitality after their detachment, unless their development be favoured by engrafting them on a living stock.† On the other hand, the question of 'generations' is one entirely of the latter nature; and can only be determined by adhering strictly to homological principles.

* See Prof. Owen's "Parthenogenesis," p. 54, *et seq.*

† If the individuality of leaf-buds be maintained, because they will continue to exist as grafts, the same attribute ought to be allowed to parts of *animals*, *e. g.* teeth, testes, ovaries, &c., which have been removed from one animal and implanted in another, and which have formed new attachments to the latter, and continued to grow.

673. Another view has been suggested, which at first sight appears more worthy of adoption; namely, that a *tree* may be regarded as a collection of *annual plants;* the buds of each year giving origin to those of the next, when their own term of existence is expired.* In a potato, for example, it is argued that each year's growth terminates in the production of tubers or underground stems, which contain the buds that are developed into distinct and independent plants in the ensuing season; these in their turn giving origin to tubers, whose buds are to be developed in a subsequent year. Now what is true of the potato, it is urged, is true of an ordinary tree; the only difference being, that the remains of previous growths are persistent, although dead, and that thus a permanent stem is formed, on which every generation of plants is developed, as it were parasitically, and to which each generation makes an addition that is left behind when the leaves decay.—In this, as in the preceding doctrine, however, it appears to the author that too much account is made of the *leaves,* and too little of the other parts. The leaf is by no means, as some have represented it, the entire plant; it is only the most important of the vegetative organs of the plant. But it cannot maintain an independent existence (save in a few rare cases in which the leaf possesses unusual absorbent powers), unless it is able to develope roots; and it cannot perform the generative act, unless it can evolve the flower. Now other parts of the plant may possess the same independent vitality, if only, like leaf-buds, they can evolve the organs in which they are themselves deficient; thus there are many trees which can be propagated by cuttings of their roots, these having the power, under favourable circumstances, of putting forth leaves; and there are others which can be multiplied in like manner by division of their stems, each cutting being able to put forth both leaves and roots.—Now whilst too much account is made of the leaves as integral components of the plant, too little is made of the *general cellular basis,* from which the leaves originate, and which retains its vitality in every stem, through the whole period of its existence. This cellular basis is the continuous product of that in which the whole fabric has its origin; it is that of which the leaves are offsets, developed for a particular purpose (the elaboration of nutriment for the axis and its other appendages), and ceasing to exist when that purpose is answered; and it retains the power of giving origin to buds from *any part of it* that may be stimulated to increased development. For although it may be quite true that, under *ordinary* circumstances, each year's growth of buds originates in the new tissue formed in the preceding year, yet this tissue is but the extension of the general cellular basis; and, under *extraordinary* circumstances, portions of this at a great distance from the last-formed buds, may develope a new set of foliaceous organs.†—But further,

* See Dr. Harvey 'On the Nature, Longevity, and Size of Trees,' in the "Edinb. Philos. Journ.," Jan. 1847.

† The following, for example, occurred within the Author's own observation. An Elm-tree, which grew to the height of nearly thirty feet before it gave off any branches, had its upper part entirely broken off in a gale of wind, and the stem was left standing, entirely bare of foliage. Its death was considered almost inevitable (and such it was upon Dr. Harvey's theory); but it was thought desirable to give to it a chance of recovery, and nothing else was done than to slope off the top of the stump, so as to prevent the lodgment of rain. The next spring, a great number of buds were developed, along nearly the whole length of the stump, where no buds or branches had grown for many previous years; these, in process of time, became branches; and the topmost branches having gradually changed their direction (in

the doctrine in question is entirely inapplicable to the case of the *leafless* Phanerogamia, such as the *Cactaceæ*. The succulent mass of which their stems are composed, is obviously homologous with that general cellular basis, of which the axis of all the higher plants consists at an early stage of their development, and from which the leaves are developed wherever they exist; whilst its foliaceous surface performs the functions of the leaf, the two organs not being here separated, nor their functions specialised. Now it cannot but be admitted, that it is this cellular mass, which in the Cactaceæ constitutes *the plant;* since here no separate leaves are evolved. And further, we must regard the whole as one integer, unless we are prepared to say that every separate portion of this mass, which can maintain an independent existence, is to be regarded as endowed with a distinct individuality.* Now the duration of this cellular stem of the Cactaceæ is extremely prolonged, its life being very slow; so that there are undoubted instances of plants of this order continuing to exist for 100 years; and their probable term of life is very much longer. There need not, therefore, be the least difficulty in admitting the continued vitality of the general cellular basis of the stem of an ordinary tree, notwithstanding that it may have attained the age of some hundreds or even thousands of years. The parts first formed may have long since decayed away, but a new growth is continually taking place; for the 'cambium-layer' (in the (Exogenous stem) is in a state of continual increment, and the proximity of leaves is not required for the growth of the additional layers of wood and bark into which it developes itself, nothing else being needed than a supply of elaborated sap, which may have been prepared by the leaves of remote parts of the fabric (§ 559).

674. There appears, then, to be no medium between, on the one hand, regarding the entire fabric developed from a single generative act (i. e. the fertilization of a single 'germ-cell' by the contents of a 'sperm-cell') as forming *one organism*, however great may be the multiplication of similar parts, or however independent these parts may be of each other; and the including every product of its own development, whether contemporaneous or successive, as *one generation;*—or, on the other hand, attributing a distinct individuality to every component of the most complex organism, and designating every augmentation of the number of its cells, by the subdivision of those previously existing, as the production of a new generation. In either case, it must be freely admitted, we are forced to do a certain violence to our ordinary conceptions; but if, on the one hand, it seems strange not to admit the proper individuality of a completely independent being, such as a potato-plant which has been developed, in common with several others, from a single tuber, nor to allow that in developing itself from a bud into a complete plant, and in putting forth its leaves and flowers, and in maturing its seed, it passes through a complete generation; on the other hand it seems yet more absurd to regard the human organism, or any other composite fabric, as made up of a congeries of individuals,

accordance with the well-known law) from the horizontal to the perpendicular, now appear like continuations of the stem, and the tree, after an interval of about 27 years, has quite recovered its symmetrical appearance, although its aspect is of course very different from that which it presented before the accident.

* It has been shown that small fragments of the Cactaceæ may thus be made to grow, by grafting them upon other plants of the family, even though the generic alliance be not very close. See Note in p. 902.

and to regard each of these as entering upon a new generation whenever its component cells have been renewed. And it may be the wisest course, perhaps, to invent new terms, rather than to distort the meaning of those in common use.—If we take a retrospect, however, of the whole series of phenomena of Vegetable Reproduction, we see that in all the cases (by far the larger proportion) in which a true generative act is known or believed to take place, the fertilized cell, which is the immediate product of this act, gives origin, by successive subdivisions, to an immense congeries of cells, each having a certain degree of independent vitality ;—that in the simplest Protophytes, each one of these detaches itself from the rest, and lives for and by itself, in a state of *functional* independence of them, except so far as it requires another to conjugate with, whilst there is yet the same *homological* relationship among them all, as exists among the component cells of the adult human organism ;—that at a little higher elevation in the scale, the cells developed from one primitive germ remain attached to each other, and form masses of almost unlimited extent (such as the gigantic fronds of marine Algæ, § 268), all whose parts are in like manner homologically related, whilst they nevertheless continue so similar in their endowments as to possess a great degree of functional independence ;—that as we ascend yet higher we find the evolution of the original germ-cell no longer consisting in the multiplication of homogeneous parts, but involving the development of some of the products of its subdivision into forms very dissimilar to its own and to each other ; and of the organs so developed, a certain combination is requisite for the maintenance of vegetative activity ;—but that even where a considerable degree of mutual dependence is thus established, there is usually, even in the most heterogeneous and highly-specialized Vegetable organism, such a multiplication of similar parts, that many of these can be removed without serious injury to the remainder ; whilst there is frequently also such an amount of reproductive capacity still remaining in each principal organ, that, even when separated from the rest, it can develope whatever parts are deficient, and can thus maintain its existence independently of them. It is only when the whole series of phenomena is thus comprehensively surveyed, that we rightly appreciate the real relationship of the component organs of the most perfect Phanerogamous plant, or of the isolated cells of the simplest Protophyte.*

3. *Reproduction in Animals.*

675. The condition of the Reproductive function in the lower part of the Animal kingdom, may be considered as almost precisely analogous to that which it presents among the higher Plants ; for, on the one hand, we find the act of *Generation* performed under circumstances, which on the whole most resemble those under which it takes place in the Phanerogamia ; whilst, on the other, we see that this is by no means the only—frequently not even the chief—mode in which the multiplication of the

* In the views which he has taken of this subject, the Author finds himself in full accordance with his valued friend Mr. Thwaites ; the acknowledged importance of whose contributions to Cryptogamic Botany gives to his opinions an authority, which is greatly enhanced, in the minds of those who are personally acquainted with him, by the clearness and philosophic acumen of his ordinary habits of thought.

original stock is provided for, repetitions of that stock being produced by a process of self-division or of gemmation, so that a large number of the independent beings ordinarily designated as 'individuals' may result from one act of generation. Such beings, as shown in the preceding paragraph, have a very different relationship to one another, from that which is possessed by the perfect individuals among the higher classes of animals, every one of which contains in itself the whole product of the act of generation in which it originated; being, in fact, as nearly related to each other *by descent*, as are the several parts of the body of the latter, although possessing a functional independence that enables them to maintain a separate existence. As it is peculiarly important that the true nature of this relationship should be kept in view, *the detached portions of the stock originating in a single generative act* will be termed *Zöoids;* whilst by the words 'animal' or 'entire animal, the equivalent of *Zöon*, will be implied, in the lower tribes as in the higher, the *collective product of a single generative act.** Thus, the whole Zoophytic structure produced by continuous gemmation from a single ovum, will be considered as one animal; just as the whole product of a single seed is one plant. And as the *homological* relation of the parts to each other is not disturbed by their detachment, although they are rendered *functionally* independent, the detached gemmæ or zöoids of the *Hydra* are the real equivalents of the connected polypes of the *Laomedea* or other composite Hydroid Zoophyte (§§ 289, 290).

676. Further, as the multiplication of separate zöoids to any extent, is a process of exactly the same order as the growth of the composite structure in which they remain continuous with each other; and as this, again, is obviously of the same nature with the development and growth of the component parts of more heterogeneous organisms higher in the scale, it is obviously correct to include in the former case, as in the latter, under the title of *one generation*, all that intervenes between one generative act and the next. If the phenomena be viewed under this aspect, it will be obvious that the so-called 'alternation of generations' has no real existence; since in every case, the whole series of forms which is evolved by continuous development from one generative act, repeats itself precisely in the products of the next generative act. The alternation which is very frequently presented in the forms of the lower Animals, is between the products of the generative act and the products of gemmation; and the most important difference between them usually consists in this,—that the former do not contain the generative apparatus, which is evolved in the latter alone. Not unfrequently, indeed, it happens that the detached zöoids are little else than combinations of generative organs with a locomotive apparatus adapted for their dispersion; or they may be, in the first instance, nothing else than buds, which will gradually evolve themselves into such a self-moving generative apparatus. Now the relation of these generative zöoids to those which are merely repetitions of the nutri-

* The Author is by no means insensible of the awkwardness of these designations, and will be most glad to take advantage of any better ones that can be suggested. The term *zöoid* was suggested to him by Mr. Huxley, whose researches on the Acalephæ, carried on in the Indian Seas, had led him independently to adopt precisely the same view of the relation of the two processes, as that for which the Author contends. By some of the French Naturalists, the word *Zöonite* has been used in nearly the same sense.

tive organs,—as, for example, the relation between the Medusa-buds and the Polypes of the Coryne or Laomedea (Figs. 95 and 96),—will be seen to be precisely the same as that of the flower-buds to the leaves of a Phanerogamic plant; and whilst, as formerly remarked, we have in the *Vallisneria spiralis* an example of the spontaneous detachment of a bud in which one portion of the reproductive apparatus is already evolved, we have in the separation and diffusion of the spores of the Fern, a still more remarkable example of the detachment of gemmæ in the simplest possible condition, whose whole power of development is directed to the production of a true generative apparatus (§ 664). The generative zöoid may be merely a segment cast off from the body at large, possessing no locomotive power, as in the case of the Tænia; or it may contain a combination of generative and locomotive organs, as in the self-dividing Annelida. But it may possess, not merely locomotive organs, but a complete nutritive apparatus of its own, which is the case in all those instances in which the zöoid is cast off in an early stage of its development, and has to attain an increased size, and frequently also to evolve the generative organs, subsequently to its detachment; of this we have examples in the Medusæ budded-off from Hydraform Zoophytes, and in the solitary Salpæ, as well as in other beings of similar completeness.

677. It has been already pointed out, that what have been called the *fissiparous* and the *gemmiparous* modes of reproduction, are one and the same in their essential nature. Of the former we have the type in the multiplication of cells by *self-division,* which is best seen in the Algæ (§ 140); whilst of the latter we have the type in the multiplication of cells by *out-growth,* which is common in the Fungi (§ 141). The fissiparous method is the one most frequently witnessed among the Protozoa, whose condition most nearly approximates that of the Algæ; and it is occasionally seen, also, among Zoophytes. But in the latter class, as in all the higher tribes in which detached zöoids are produced, gemmation is by far the most usual method of their evolution.

678. The 'act of generation' is accomplished, in Animals, as in the higher Plants, by the union of the contents of a 'sperm-cell' with those of a 'germ-cell;' the latter being that from within which the embryo is evolved, whilst the former supplies some material or influence necessary to its evolution. These 'sperm-cells' and 'germ-cells' are usually developed in special organs, as in all the higher Plants; but in the Hydra, as well, perhaps, as in some others of the lower Animals, they are evolved out of the midst of the ordinary parenchyma, and have no special locality. In many of the lower tribes, again, both sets of generative organs are developed in the same organism, and are capable of conjoint action; the animal is then said to be *hermaphrodite.* There are others in which both sets of organs exist, but they are not capable of conjoint action, so that the concurrence of two individuals is required, each fertilizing the other; such are also termed hermaphrodite, although they differ from the true hermaphrodites in not being self-fertilizing. In the highest members of the Radiated, Articulated, and Molluscous divisions, however, and in all Vertebrata, only one set of sexual organs is possessed by each individual, which is then designated as *monosexual.*—It has been a matter frequently and earnestly discussed, whether the embryo is to be considered as the product of the male or female portion of the generative apparatus; some

having regarded the function of the sperm-cell as limited to the stimulation of the developmental process in an embryonic structure already evolved in the germ-cell; whilst others have considered that the real germ is produced in the sperm-cell, and that it is implanted in the germ-cell in the act of fertilization, to be henceforth dependent upon the female organism for the materials of its development. Now if we cast our eyes back upon the simplest Cryptogamia, we see that there is no proper distinction of male and female amongst them, but that each of the conjugating cells contributes an equal share to the result. And the departures from this equality which are seen in the higher Plants, seem rather to have reference to the part to be taken by the female portion of the apparatus in aiding the subsequent development of the germ, than to its first production; and it seems probable that in the highest animal, as in the lowest plant, the formation of the first embryonic vesicle depends upon the actual intermixture of the contents of two cells, so that neither male nor female can be properly said to supply the germ by itself. Looking, however, to the very equal mode in which the characters of the two parents are mingled in *hybrid* offspring, and to the certainty that the *material* conditions which determine the development of the germ are almost exclusively supplied by the female, it would seem probable that the *dynamical* conditions are in great part furnished by the male.

679. The 'sperm-cells' of Animals usually present a remarkable correspondence with those of the higher Cryptogamia in the nature of their products; which, in the majority of instances, are self-moving filaments, or *Spermatozoa*, closely resembling the *phytozoaires* of Ferns, &c. These were formerly considered in the light of distinct and independent beings, and were termed 'spermatic animalcules;' but it is now universally admitted that they cannot be regarded as having any more independent vitality than blood-corpuscles or epithelium-cells, the latter of which, when ciliated, have at least an equal power of spontaneous movement. The forms of these spermatozoa present a certain degree of variation in the different groups of animals; but these variations cannot be said to have any correspondence with the zoological relations of the animals in which they are seen. Generally speaking, however, each spermatozoon is composed of an ovoid 'body,' with a long filiform 'tail,' strongly resembling the phytozoaire of the Fern (Fig. 255, D); and it seems to be by the undulations of this tail, rather than by cilia attached to the body, that the movements of the corpuscle are effected. An interior organisation was formerly described as traceable in the spermatozoa; but it is now certain that they are homogeneous, or nearly so, throughout; and that the supposed presence of a mouth, anus, &c., has no real foundation. These movements, however, are by no means of the same nature in all instances. Sometimes they consist of lateral undulations, as uniform and moderate in their rate as the vibrations of a pendulum; in other instances, they consist of a series of interrupted jerks, the tail being coiled up into a circle, and then suddenly uncoiled; other spermatozoa, again, advance with a regular screw-like or boring movement, turning rapidly on their axes; lastly, there are certain spermatozoa which have no motor power whatever, but the conditions of these, as will presently be shown, are altogether peculiar. It is obvious that the purpose of these movements is to bring the spermatozoa into contact with the germ-cell; and we shall see

that the circumstances under which that contact takes place, vary greatly in different animals. For in some cases, the fluid containing the spermatozoa is conveyed into the interior of the female generative organs, so that it is ready to fertilize the germ-cells as soon as they are matured ; in other instances it is effused upon the ova at the time of their deposition ; and in other cases, again, it is poured forth into the surrounding liquid, and diffuses its fertilizing power far and wide through this, so as to act upon any ova which may have been deposited in its neighbourhood. The vitality of the spermatozoa is not usually long preserved, after they have been set free from the organism which produced them ; but, like that of muscles, cilia, &c., it is manifested for a longer time in the case of cold-blooded animals, than in that of warm. Thus, the spermatozoa of Birds frequently cease to move within fifteen or twenty minutes after the death of the animal which produced them, or their removal from its body ; in the Mammalia, their motion continues for some time longer, especially if they remain enclosed in their natural organs ; but the spermatozoa of Fishes will continue moving for several days after their expulsion. The most remarkable prolongation of their vitality is seen, however, when, after their expulsion from the male, they are received into the female generative organs ; such is the case with many Mollusca and Articulata, which have a special receptacle or *spermotheca* for storing-up the seminal fluid, in order that it may fertilize the ova as they are successively developed ; the spermatozoa remaining active within this, even for some months. Indeed it appears from the researches of Mr. Goodsir already referred to (§ 316 *d*), that in certain Crustacea the conclusion of the development of the spermatozoa themselves ordinarily takes place in this situation, their formation not being nearly complete at the time when they leave the organs which evolved them. Even in Mammalia, their movements have been found to continue unimpaired for some days after their introduction within the female organs.* The movements of the Spermatozoa are usually rendered irregular, and more speedily brought to a close, by the admixture of fresh water with the spermatic fluid, which diminishes its density and alters their hygroscopic relations to it ; but the admixture of animal fluids of somewhat higher density, such as milk, mucus, serum, or saliva, produces very little effect. The admixture of salt-water, again, scarcely affects the movement ; nor does that of fresh, in the case of those animals whose ova are fertilized by the diffusion of the seminal fluid through it, as happens in many fresh-water Bivalve Mollusks. All such chemical agents as affect the chemical composition of the Spermatozoa, speedily put an end to their movements ; and this result is instantaneously brought about by an electric discharge, although a galvanic current does not seem to have any effect upon them. A higher or a lower temperature than that habitual to the organisms within which they are produced, usually seems to have a retarding influence ; but the motions of the Spermatozoa of Frogs and Fishes have been seen to continue, when

* This fact seems to assist in explaining the phenomena of 'protracted gestation ;' for if at the time when sexual intercourse takes place, no ovum be prepared for fertilization, but the spermatozoa can retain their vitality for one, two, or three weeks, so as to fertilize an ovum matured at the end of that period, the usual term between *intercourse* and *parturition* will be extended by that interval, without any extension of the term between the actual *conception* and *parturition*.

the surrounding medium was beneath the freezing-point; and those of certain fresh-water Gasteropoda have not been arrested by contact with water of 158°—176° Fahr.

680. The mode of development of the Spermatozoa, as first discovered by Wagner, and since more fully demonstrated by Kölliker, is usually as follows.—In the parenchyma of the spermatic organs of some animals, but more commonly in the cavities of tubular vesicular organs, resembling ordinary glands in structure, and designated as *Testes*, there are formed certain large cells, which seem to correspond with the epithelial cells that have been shown to be the agents in the secreting process (§ 185). These parent-cells,—corresponding with the parent-cells within which the real sperm-cells are developed in the Ferns (Fig. 255, c), and with the 'parent-cells of the pollen' in the Phanerogamia (§ 667),—give origin in their interior to a variable number of *vesicles of evolution*, each of which produces a single spermatozoon. The earliest stages of the development of the spermatozoon have not as yet been made out; for on the one hand, the vesicle is filled with granular matter which obscures its other contents; whilst on the other, the spermatozoon itself does not exhibit those sharp distinct contours, dependent upon its high refractive power, which afterwards distinguish it. Gradually, however, the granular matter disappears, and the spermatozoon is distinctly seen lying coiled up within its vesicle. When this is completely matured, it bursts, and gives exit to the contained spermatozoon; and if the parent-cell have previously burst, as usually happens in Mammalia, the spermatozoa henceforth move freely in the spermatic fluid. In Birds, however, it is more common for the parent-cells to continue so long unruptured, that when the spermatozoa are set free from the vesicles of evolution, they are still contained within the enveloping cyst, usually straightening themselves out and lying in bundles; and it is only when the containing cyst at last ruptures, that these bundles are broken up, and the individual spermatozoa dispersed.—The foregoing may be considered as the most complete mode of evolution of the spermatozoa; and it is on this type that the process is most commonly performed. According to Wagner, however, the vesicles of evolution are not always produced within the parent-cells; but the nuclei of these at once resolve themselves into spermatozoa, which then present a solid massive form, as in the Chilopoda, Acaridæ, and Entomostraca; or, as in the Nematoid Entozoa, the spermatozoon is formed by the conversion of the cell-membrane as well as the nucleus of parent-cell. Even where the spermatozoon is formed within its own secondary cell, or vesicle of evolution, it is probably by the metamorphosis of its nucleus that it is really evolved; the process, so far as it can be traced, being extremely analogous to that by which the peculiar stinging organs of the Medusæ and Hydroid Polypes are developed, each of them being a cell whose nucleus is transformed into a fibre, which afterwards projects from one extremity of it by the rupture or solution of the cell-wall.—That the Spermatozoa are the essential constituents of the seminal fluid, and that the latter has not in itself any fertilizing power (which some have attributed to it), may now be regarded as fully proved. There are some cases, as pointed out by Wagner, in which the 'liquor seminis' is altogether absent, so that they constitute the sole element of the semen; whilst on the other hand, they are never wanting in the seminal fluid of animals capable of procreation.

Moreover, there are many animals in which the fecundation of the ovum only takes place after the diffusion of the seminal fluid through water; and it is difficult to imagine that the liquor seminis, in so extremely dilute a condition, can be operative for its fertilization; although, when the vast multitude of spermatozoa discharged at once in such cases is borne in mind, and their power of continued spontaneous movement is taken into account, it seems obvious that a special provision has been made for bringing the spermatozoa and ova into direct contact. Such contact we have every reason to believe to be essential to the act of fecundation. The experiment was long ago tried by Spallanzani, and by Prevost and Dumas, and has been recently repeated by Mr. Newport, of separating the spermatozoa from the liquor seminis by filtration, and of trying their respective effects upon the ovum. Mr. Newport found that when thus separated, and applied to different sets of eggs, those with which either the spermatozoa or the filtering-papers had been placed in contact, were almost universally fertilized, while only a very few of those treated with the liquor seminis were fecundated; and the fertilization of these last is attributed by him, with great probability, to the passage of a few of the spermatozoa through the filtering-paper.*

681. The development of the Spermatozoa is in most cases *periodical;* Man and some of the domesticated races being the only animals in which there is a constant aptitude for procreation. The spermatic organs, which remain for long periods in a state of atrophy, at particular times take on an increased development, and their product is then formed in great abundance. Some of the most remarkable examples of this kind are presented by the class of Fishes; but the contrast is scarcely less notable in the Passerine Birds, whose testes in spring attain to twenty or even thirty times the size and weight which they possess in the winter. It is when the organs are undergoing this rapid increase, that the several stages in the development of the Spermatozoa may be most advantageously studied.

682. The sperm-cell and its contents having been thus described, the 'germ-cell' next presents itself for our consideration. It can scarcely be doubted that this is the real character of the *germinal vesicle* (Fig. 261, *v g*), which is a peculiar cell, with a very well-marked nucleus, termed the *germinal spot* (*t g*), that presents itself in every ovum when matured for fecundation. It is surrounded by a mass of nutrient matter, chiefly composed of albumen and oil-particles, which is known as the *vitellus* or *yolk* (*j*); and the whole is inclosed within an envelope, which is termed the vitelline membrane or yolk-sac (*m v*). This membrane in the Mammal (whose ovum is here represented) is of peculiar thickness and transparency, and is distinguished as the *Zona pellucida.* The size of the ovum depends mainly upon the quantity of yolk which it contains; and this seems proportioned to the grade of development which the embryo is to attain, whilst still dependent upon it. Thus in the Insect, whose larvæ come forth from the egg in a very immature condition, the yolk is very minute, and the bulk of the larva on its emersion bears a marvellously small proportion to that which it soon presents. On the other hand in Birds, whose

* See the "Proceedings of the Royal Society," June 20, 1850.—For the most recent and complete information on the Development and Varieties of the Spermatozoa, see the Article *Semen* in the "Cyclopædia of Anatomy and Physiology."

entire development into the ornithic type is accomplished before quitting the egg, the store of yolk is much larger in proportion, and the young is not nearly so disproportionate to the adult in size. After it has passed forth from the ovarium, and has been fertilized, the proper ovum of many

FIG. 261.

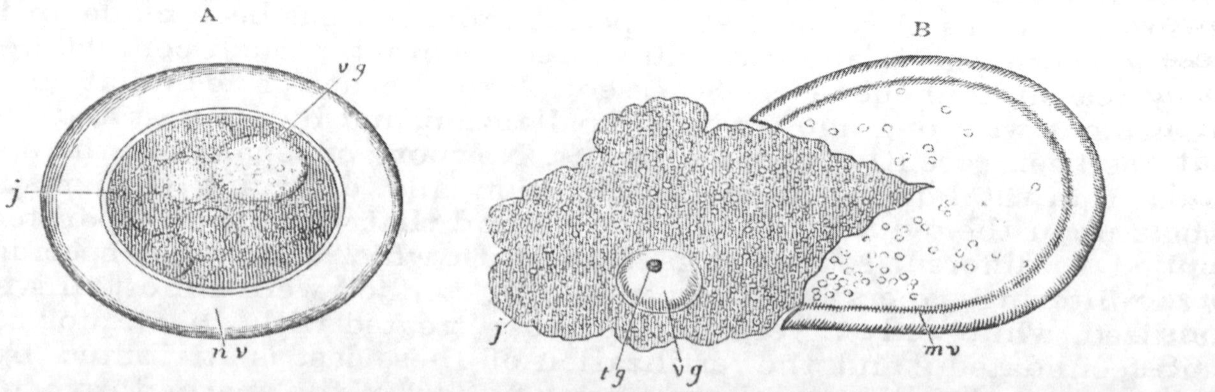

Constituent parts of Mammalian Ovum; A, entire; B, ruptured, with the contents escaping; —*m v*, vitelline membrane; *j*, yolk; *v g*, germinal vesicle; *t g*, germinal spot.

animals receives an additional investment of albumen, which is known as the 'white' of the egg; this is gradually drawn into the yolk as the latter is exhausted, and contributes to the nutrition of the embryo during the latter stages of its development. The ovum of Mammalia, whose yolk is extremely minute, does not (as might be supposed) constitute an exception to the principle just stated; for the embryo makes but a very small advance in development, whilst sustained by the material supplied by the yolk; and is dependent for support, during the whole remainder of its evolution, upon the nutriment which it derives from the parent by the more direct connection subsequently formed (§§ 732, 733).

683. The development of the ovum, like that of the spermatic cells, sometimes takes place in the parenchyma of the germ-preparing organs or ovaries, sometimes within their cavity. In many of the lower animals, the testes and ovaries bear such a close resemblance to one another, as to be quite undistinguishable; and the same is the case in the early condition of the generative apparatus even of Man. In Articulated and Molluscous animals generally, the ovaries, like the testes, have a glandular character; but while the former retain the vesicular type, the latter are often prolonged into convoluted tubes. In the Vertebrata, we have a return to the parenchymatous type of ovarian structure; the ova being evolved in the midst of a very solid fibrous tissue or stroma. Each ovum seems to be developed within a 'parent-cell' of its own, which is called the *ovisac*; and the production of the ovisacs may take place very early in life, for in the ovaries of some animals they can be detected almost as soon as these organs are themselves evolved, and generally present themselves not long afterwards. The germinal vesicle is the part of the ovum which earliest shows itself within the ovisac; and is at first seen in its centre, surrounded by an assemblage of granules which is the commencement of the yolk. This collection gradually augments, and the vitelline membrane is developed around it; and as the ovum advances towards maturity, it draws from the vascular substance around, that amount of albuminous and

oleaginous matters which is appropriate to it. Like the augmented development of the contents of the spermatic organs, that of the ovaries is generally periodical; a large number of ova, in most of the lower tribes of animals, are advancing towards maturity at the same period, and they are discharged either simultaneously or successively; after which, the ovarium relapses into its previous inactivity. In the Human female, however, and in that of many domesticated animals, the difference between these two states is much less marked; and although the complete maturation of the ova, and their escape from the ovary, may only take place at particular intervals, yet there appears to be a continual advance towards that maturation, even during the earlier periods of life. When it has attained its full development, the ovum escapes from the ovisac, which either ruptures or thins away to give it exit; and it is then ready for fecundation. If *not* fecundated, it usually dies within a few days; its continued life being dependent upon the due performance of the operations for which it is destined.

684. The fecundation of the ovum is accomplished by the contact of the spermatozoa; and the place and circumstances of this contact vary considerably, as already pointed out. It does not appear, however, that the essential nature of the fecundating process is in any way influenced by the locality in which it occurs; and what is true of one case is probably true of all. Great difficulties, however, stand in the way of the determination of the changes in the ovum, which immediately precede and follow the act of fecundation; and observers are by no means agreed upon the point. The Germinal Vesicle of the ovum which is approaching maturity, no longer presents its ordinary pellucidity, but becomes obscure; and this obscuration, which has led some to the belief in its entire disappearance, is affirmed by Dr. Barry to be due to the development of a mass of cells in its interior, which sprout, as it were, from the germinal spot, and gradually fill up its cavity. This statement is confirmed by Wagner, and Vogt; whose observations lead them to the belief that, when thus filled with cells, the germinal vesicle bursts and sets them free, so that they become diffused through the yolk. This view is adopted also by Mr. Newport, as the result of his recent observations; and he states that in the Frog, this dissolution of the germinal vesicle and diffusion of its contents takes place as a preparation for fecundation, and not in consequence of it.* The contact of the spermatozoa with the exterior of the ovum appears sufficient to impregnate it, without that actual penetration of the spermatozoon into its interior, which has been believed to take place by Dr. Barry and others; but from the ingenious experiments of Mr. Newport, it appears that the act of impregnation is not instantaneous, and that it may be effected *partially* (so as to occasion *some*, though not all, of the normal changes in the ovum), as well as completely. Availing himself of the agency of caustic potass, which had been ascertained by Frerichs to be a powerful solvent of the spermatozoa, he applied this agent to the ova at determinate periods after the application of these bodies; and he found that when the interval of time between the application of the seminal fluid and that of the solution of potash was only one or two seconds, the 'segmentation' of the yolk took place (§ 685), but no embryoes were

* See the Abstract of his Paper on the 'Impregnation of the Ovum in the Amphibia,' in the "Proceedings of the Royal Society" for June 28, 1850.

produced. When, however, the interval was five seconds, a very few embryoes were formed; but when the interval was fifteen or more seconds, they were produced in greater number. It seemed to Mr. Newport most probable, that the spermatozoon, on coming into contact with the ovum, becomes diffluent; and that sufficient time is required for the passage of the fluid so produced, into the ovum, by endosmosis. It is interesting to compare this with the similar conclusion arrived at by Prof. Link, with regard to the diffluence of the 'phytozoaire' of the Fern, on its entrance into the pistillidium (§ 663, *note*). Should such prove to be the truth, it will harmonize most admirably with the general views already propounded with regard to the nature of the process of fecundation; as the spermatozoon may then be looked upon as a sort of solidification of the contents of the 'sperm-cell,' endowed with a temporary power of spontaneous movement, for the purpose of bringing it into contact with the 'germ-cell,' and resolved by that contact into its original fluid form, in which it is capable of being absorbed, and thus of acting upon the product of the germ-cell within the ovum. It is obvious that the whole process thus comes to bear a very close correspondence to the fertilization of the ovule of the Flowering Plant; the chief difference being, that the pollen-tube carries down the contents of the sperm-cell into more immediate relation with the embryo-sac, and that the cells developed within the latter are not previously set free by its rupture, as are those of the germinal vesicle of Animals. This last, however, may prove not to be a constant occurrence.

685. The first changes consequent upon fecundation are so nearly the same in their character in all Animals, that it is desirable to give such a general account of them, as shall be applicable to the greater number of individual cases; and shall thus serve as a foundation on which to build the special descriptions hereafter to be given, of the principal plans of development which are exhibited during the later stages. In the impregnated ovum of many Entozoa, whose transparency enables their interior to be more clearly discerned than that of most higher animals, a new and peculiar cell is seen in the midst of the yolk; which, from its subsequent history, may obviously be regarded as the equivalent of the 'embryonic vesicle' of Phanerogamia (§ 668). Although the origin of this cell has not been distinctly traced, yet there are strong reasons for the belief that it is one of those which was originally developed within the germinal vesicle, and which, after being set free by its rupture or diffluence, has received the fertilizing influence of the spermatic fluid, and has thus acquired a 'germinal capacity,' the other cells of the same brood remaining undeveloped. In the ova of certain *Entozoa* (as *Cucullanus* and *Ascaris dentata*), and perhaps in those of other Animals belonging to the inferior classes, the development of the embryonic mass commences very much after the same plan as in Plants. The embryonic-vesicle, lying free in the midst of the yolk, subdivides into two, each of these into two others, and so on, according to the regular type of cell-multiplication in a growing part; so that, in place of the single cell, we have first 2, then 4, then 8, then 16, then 32, and so on. As these cells multiply and enlarge, they draw into themselves the nutrient matter which surrounds them; so that, when the whole yolk has been thus absorbed, the yolk-bag is entirely occupied by the embryonic mass, entirely composed of cells descended from the original germ-cell, thus representing, in the manner of its deve-

lopment, the embryo of a Leguminous plant (§ 669). But in other Entozoa (even in some species of *Ascaris*) a different plan is followed; for each of the cells which is produced by the successive fissions of the embryonic vesicle, draws around itself a certain portion of the yolk, which thus successively divides into as many segments as there are embryonic cells; and thus is produced that *segmentation* of the entire yolk, or of a part of it, which is one of the most striking features in the early history of embryonic development in all the higher animals. The several stages of this process, as it takes place in the ovum of *Ascaris acuminata*, are shown in Fig. 262, A, B, C, D, E; and for the sake of comparison, the early

FIG. 262.

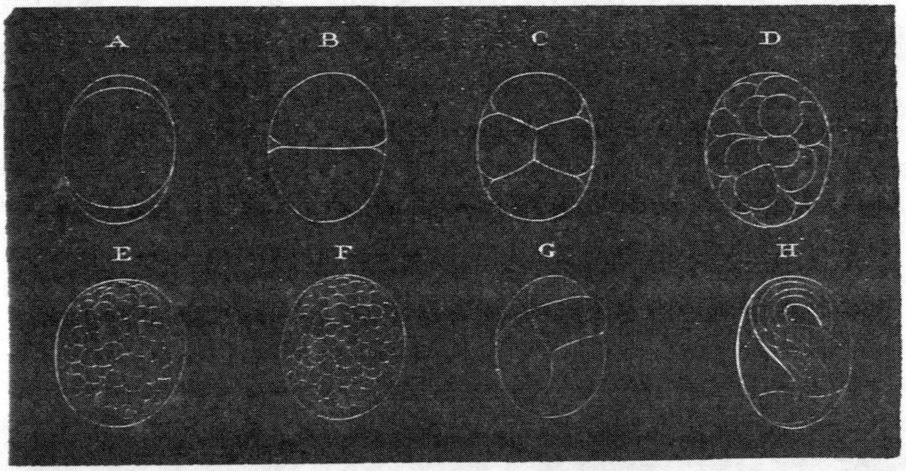

Successive stages of segmentation in the vitellus of the ovum of *Ascaris acuminata*:—A, ovum recently impregnated, the yolk-bag slightly separated from the enveloping membrane; B, first fission into two halves; C, second fission, forming four segments; D, yolk now divided into numerous segments; E, formation of 'mulberry mass' by further segmentation; F, the mass of cells now beginning to show the form of the future worm; G, further progress of its evolution; H, the worm, formed by the conversion of the yolk-cells, now nearly mature.

stages of segmentation in the yolk of the *Mammalian* ovum are shown in Fig. 263, A, B, C. It should be stated, however, in regard to the latter,

FIG. 263.

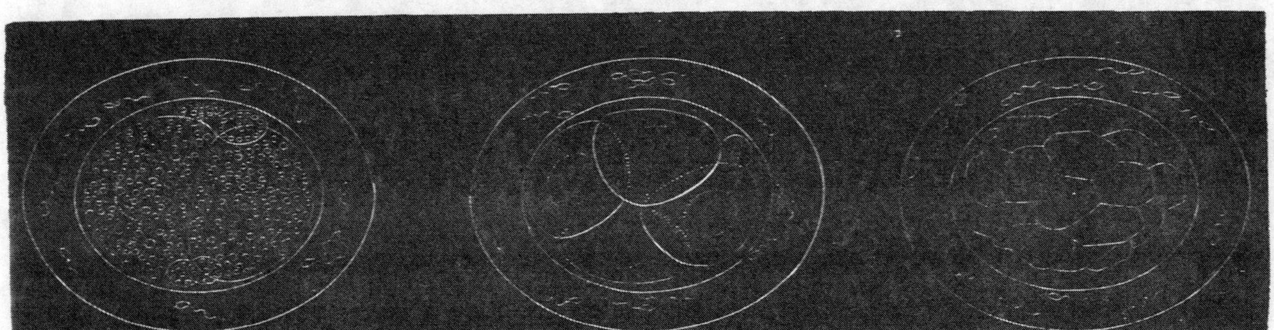

Progressive stages in the segmentation of the vitellus of the *Mammalian* ovum:—A, its first division into two halves; B, subdivision of each half into two; C, further subdivision, producing numerous segments.

that the embryonic vesicle has not yet been clearly made out in its interior, and that the nature of the body which forms the centre of each segment has not been precisely determined; if not actually a cell, however, there can be little doubt that it is a cell-nucleus, and that it is the lineal descendant of the immediate product of the mutual action of the contents of the 'sperm-cell' and 'germ-cell.' When this process has gone on to such an extent as to fill the entire yolk-bag with an aggregation of minute spherical segments of yolk (Fig. 262, E), having each a minute cell or cell-nucleus in its interior, every segment becomes invested with a cell-membrane of its own; so that the result produced is in effect the same, as in the case in which the descendants of the embryonic vesicle drew the yolk into their own cavity; for the yolk-bag is now occupied by a mulberry-like mass of cells, every one of which has a share of the 'germinal capacity' possessed by the original embryonic vesicle; and it is by the transformations of these cells, that the embryonic structures are subsequently evolved.—In a considerable number of the higher animals, however, in whose ova the yolk is large enough to carry on their development, so that they arrive at their characteristic forms while yet within the egg, this segmentation does not take place in the whole yolk, but only in that part of it immediately surrounding the embryonic vesicle, which lies on the surface, instead of in the centre, of the yolk. This may be well observed in Fishes (Fig. 264, A-E); and according to M. Coste, it is by a similar process that

FIG. 264.

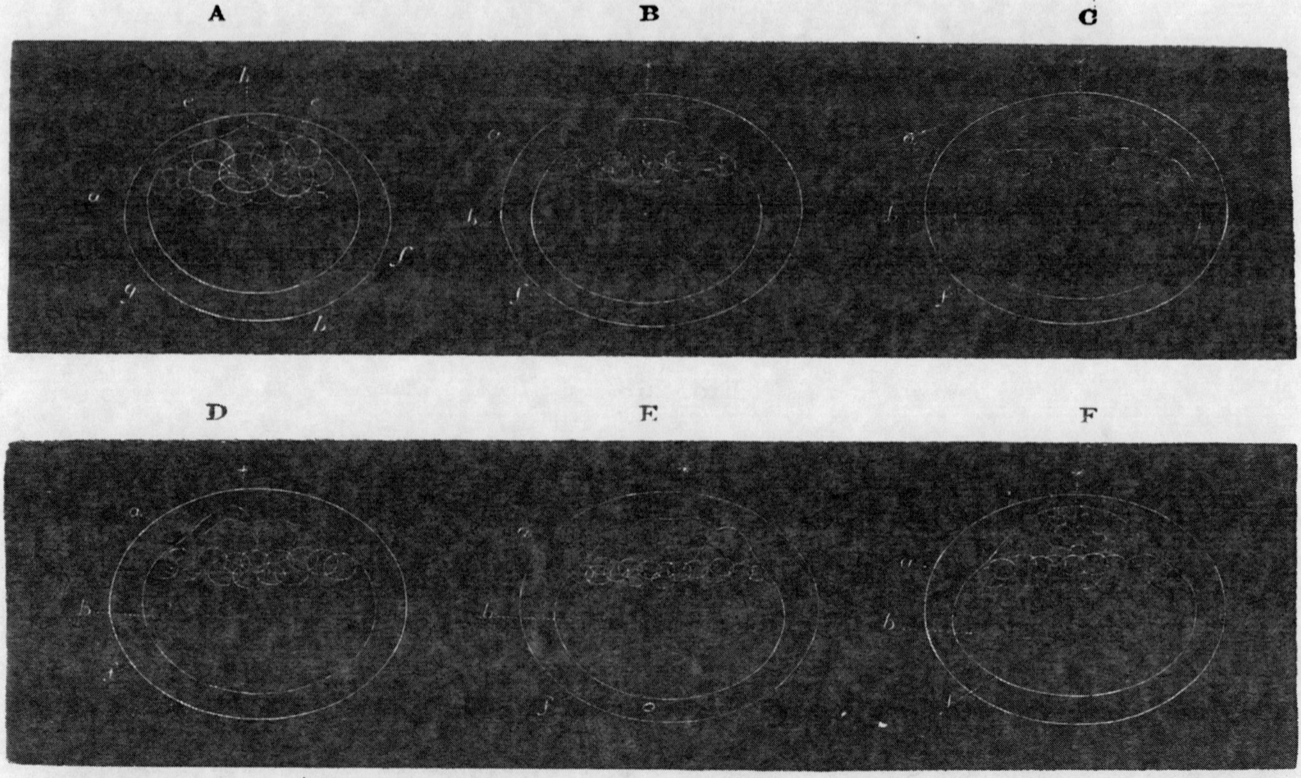

Early stages in the development of the ovum of *Coregonus palæa*. In all the figures, *a* indicates the shell-membrane; *b*, the vitellus; *e*, *e*, oil-globules in the vitellus; *f*, the albumen; *g*, vitelline membrane; *h*, situation of the germinal vesicle; *, germinal mass:—A, ovarian ovum, with a slight elevation in the situation of the germinal vesicle; B, ovum two days after fecundation, showing the germinal elevation, probably produced by the attraction of a portion of the yolk around the embryonic vesicle; C, ovum rather more advanced, showing the furrow indicative of the first cleavage; D, later stage, a second cleavage having taken place, so as to produce four segments; E, further progress of the segmentation; F, development of the 'mulberry-mass' by continued fission.

the *cicatricula* or germ-spot is produced at the surface of the yolk-bag of Birds, whence all subsequent changes emanate. This separation of the yolk into two parts—that which cleaves, and that which does not cleave, —has not yet been observed in any Invertebrata, save the Cephalopoda. It is obvious that the non-cleaving portion of the yolk is to be regarded as a superaddition to the true yolk, which is employed in forming the mulberry mass; being destined to afford the material for the ulterior development of the cells of which this is composed.

686. The most general phenomena of Reproduction having been thus described, we have now to turn our attention to the principal specialities which the process exhibits in the several groups of the Animal kingdom. Of the history of reproduction among the *Protozoa*, our knowledge is at present very limited. The propagation of the greater number of the (so-called) Polygastric *Infusoria* is only known to be effected by the process of subdivision, or *fissiparous* multiplication (Fig. 265), which is analogous, on the one hand, to that which we have seen to prevail in the parallel group of Plants (§ 652), and on the other to that which occurs in the earliest stage of embryonic development in the highest Animal (§ 685). It is curious to observe that the direction of this division is not constant in different individuals of the same species; this being sometimes longitudinal (E, F, G), sometimes transverse (H, I, K). It is possible that

FIG. 265.

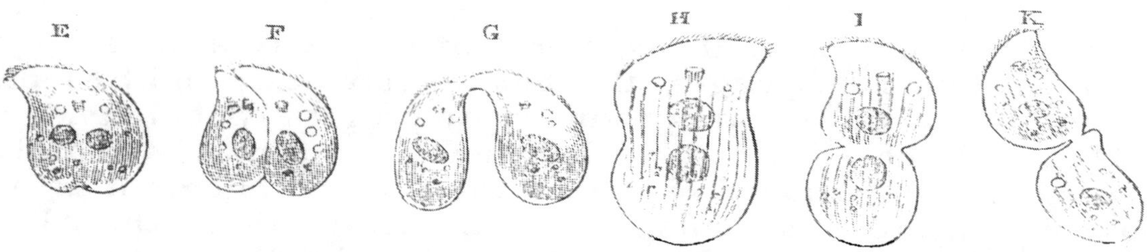

Fissiparous multiplication of *Chilodon cucullulus*; E, F, G, successive stages of longitudinal fission; H, I, K, successive stages of transverse fission.

there may be here some such alternation, as we commonly see in the direction of the subdivision of cells, which remain connected together and are not growing in one direction only (Fig. 14, E, Fig. 36, *a*). Of the extraordinary rapidity with which multiplication may take place after this method, some account has been already given (§ 284 *a*).—Besides the fissiparous mode, the *gemmiparous* is not unfrequently seen, a bud being put forth, which gradually increases in size, acquires the characters of its stock, and at last becomes detached; this is especially observable in the Vorticellinæ (Fig. 88), which, in their stationary habits, strongly remind us of Polypes, whose gemmiparous tendency is so remarkable.—In addition to these methods, it would appear that certain Infusoria, especially the *Kolpodinæ*, propagate by the breaking-up of their own mass into reproductive particles; a method that strongly reminds us of the dispersion of the zoospores of such Algæ as the *Achlya prolifera* (Fig. 6).—Nothing like 'conjugation' has ever been witnessed among undoubted Infusoria; and at present the occurrence of such an act in any tribe whose position is doubtful (as, for example, that of Diatomaceæ, § 653) is considered as an indication of its Vegetable nature. But it can scarcely be doubted that we have yet much to learn with regard to these remarkable beings. Many of the reputed 'Animalcules' are undoubtedly to be transferred to the Vegetable kingdom; many more are only transitory forms of some other kind; and even of the most characteristic forms of the group, it may be very reasonably questioned whether they can go on multiplying by subdivision for ever, and whether a true *generative* act must not occasionally become necessary, in order to reproduce the capacity for continued propagation. It is not an unimportant consideration bearing upon this subject, that the appearance of Animalcules in fluids, into which it seems certain that their germs must have been conveyed by the air, indicates that they must have arrived there in the condition of *ova*, since the Animalcules themselves are killed by desiccation,—a curious fact, when it is borne in mind that the more highly-organized Rotifera may be wafted by the atmosphere, and dispersed over the entire globe, in a perfectly dry state, without the loss of their vitality.

687. Of the higher Protozoa, the *Rhizopoda*, and the *Porifera*, the extent of our knowledge is nearly the same. In these, the new parts which are produced, among the first by the gemmation, and among the second by the subdivision, of the original cells, do not usually become disconnected from each other, but remain in union, so as to form composite structures. This is not always the case, however; for there are many of the (so-called) Foraminifera, in which the buds are soon separated, and maintain a completely independent existence. Here, also, it is yet unknown whether true generative organs exist; and it is quite uncertain, therefore, whether these animals ever present themselves under any other form.—In regard to the Sponges, all that we have any right to affirm is, that these compound masses are produced by progressive subdivision from a single cell; and that this cell is detached from a mass of similar character. As already mentioned (§ 286 *c*), the ciliated 'gemmules' may be pretty certainly regarded as analogous to the zoospores of the Algæ, that is, as *gemmæ*; whilst there seems reason to consider the 'capsules' as the homologues of the *ova* of higher animals. Yet until it shall have been ascertained that they have been produced by something like a sexual process, it cannot be positively affirmed that such is their nature; and when it is borne in mind that each capsule contains *several* reproductive bodies, it will be seen that this *may* be only a peculiar apparatus for enclosing a number of gemmæ in a protective envelope, which shall preserve them from being destroyed by the winter's cold.

688. In the class of *Polypifera*, however, the true generative operation is undoubtedly performed; notwithstanding that the multiplication of independent organisms is in great degree effected by the process of gemmation. The *Hydra* presents us with the simplest possible condition of a sexual apparatus; for no special organs are developed for the evolution of either sperm-cells or ova, these being formed in the substance of the wall of the stomach, the former just beneath the arms, and the latter nearer to the foot (§ 289). It has not yet been ascertained whether the ovum is fertilized before or after its complete extrusion from the body of its parent; and its further development has not been traced. There can be little doubt, however, that it is destined to be developed into the likeness of its parent; since the collections of fresh water inhabited by the Hydra do not contain any organism that could be regarded as an intermediate

form. The process of gemmation seems to continue almost indefinitely, under the influence of warmth and food; and the determining condition of the occurrence of the true generative operation is a diminution of temperature, which, threatening destruction to the parent, calls forth the exercise of the special provision for the perpetuation of the race. In this we trace a marked conformity to the plan, of which we see very striking manifestations in the Vegetable kingdom; thus, it is observable of the fruit-trees of temperate climates, that under the habitual influence of a high temperature, and when copiously supplied with aliment, they will extend themselves by the formation of leaf-buds and branches, and will bear few flowers or even none; whilst, on the other hand, when the temperature is lowered, or a part of their supply of aliment is cut off, the extension of the individual fabric will be checked, but it will bear a much larger quantity of flowers and fruit. Although both sets of generative organs may be developed in a single Hydra, yet it is not uncommon for one to bear sperm-cells only, and for another to produce only ova. Whether the Hydræ be monœcious or diœcious, however, it is probable that the fertilization of their ova is accomplished by the diffusion of the spermatozoa through the water, rather than by any more direct application.—The development of the ova has not yet been studied: but it is probable, from the analogy of other cases, that when the yolk-bag has been filled by the mulberry mass, the cells of the interior liquefy, so as to leave a cavity which becomes the stomach, whilst the cells of the exterior remain to form the walls of that cavity, and absorb its contents;—that a thinning-away takes place in a certain spot of this wall, so as to form the mouth, as happens, in fact, during the development of the gemmæ which bud off from the body of the adult Hydra (Fig. 94);—and that, as in those gemmæ, the tentacula are gradually developed around the mouth, making their first appearance as little knobs, and then progressively elongating themselves until they have attained their normal dimensions.*

689. A more specialized form of the Generative apparatus is seen in the *Helianthoid* and *Asteroid* Zoophytes; in which the ova and spermatozoa are developed in organs that occupy the chambers surrounding the digestive cavity; and these organs already exhibit characters which distinctively mark them. The ovaries of the *Actinia coriacea* (Fig. 266, *e*) are about two hundred in number, and form elongated masses attached along the inner border of the leaflets or vertical partitions (*d*), that radiate inwards from the outer integument (Fig. 98). Each ovary is composed of several horizontal folds or plaits, which, when unfolded, show this structure to be about three times the length it assumes when attached to the leaflet; and each plait is made up of two layers of membrane, between which the ova are developed. Where the ova do not intervene, however, these two layers come into apposition, and form in the first place a kind of mesentery (*f*) by which the ovaries are attached to the leaflet; but they then separate, passing one on each side of the leaflet, so as to line the intervening spaces, and become continuous with the membrane lining the tentacula and digestive cavity, and through this with the external investment. Hence, notwithstanding the size and importance of the ovaries,

* A particular description and figures of the sexual process in the Hydra, are given by Dr. Allen Thomson in the "Edinb. New Phil. Journal" for April, 1847.

their structure is the simplest possible.* The testes (known to be such by the spermatozoa they contain) are developed in the very same chambers with the ovaries ; and have the form of long convoluted tubes (*g*), of very soft consistency, united by a kind of mesentery to the inner border of the ovaries. These 'vermiform filaments,' as they are commonly termed, are very easily lacerated and detached from their situations ; and are frequently seen protruding from the mouths of Actiniæ that have been torn from their natural attachments, or from any wounds that may have been made in the integuments of their base or sides. Their mode of commencement and termination has not been clearly made out ; nor has it been ascertained in what manner their contents are discharged and applied to the ova. These last are probably set free by the thinning-away of the membrane which covers them, and then lie freely in the ovarial thambers. It is certain that chey must be fertilized while yet within the body ; for they are usually retained in its cavities until they have passed through their early stages of development, though the grade which they have attained at the time of their discharge does not seem to be constantly the same.—As far as the history of the embryonic development of the Helianthoid Polypes is yet known, it seems to be essentially as follows. The mulberry-mass, emerging from the egg-covering, acquires cilia upon its surface, by whose action it can be propelled through the water ; and it is known in this condition by the name of 'gemmule.' The cavity of the stomach and the mouth are formed by the process described in the last paragraph ; and the tentacula originate after the same fashion, a single row only being first produced, and new rows subsequently springing up. While this process is going on, the body loses its power of spontaneous movement, and fixes itself to the spot to which it afterwards remains attached. The wall of the stomach is seen to consist of a double membrane, of which the two

FIG. 266

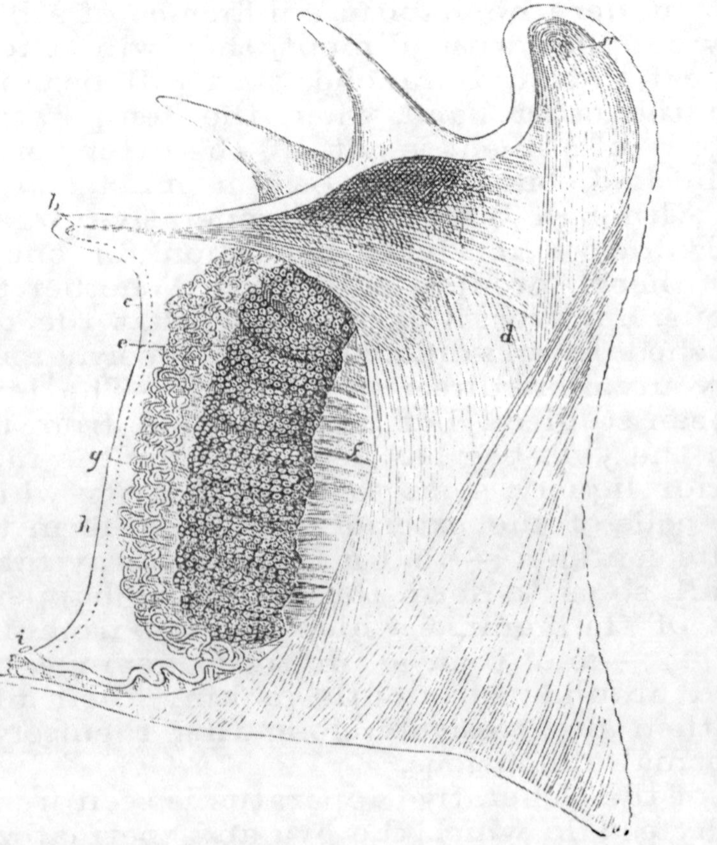

Interior of one of the ovarial chambers of *Actinia coriacea*, showing the generative apparatus: *a*, lip; *b*, border of mouth; *c*, wall of stomach; *d*, muscular partition; *e*, ovary; *f*, mesenteric membrane connecting the ovary to the muscular partition; *g*, spermatic tubulus, or vermiform filament; *h*, membrane connecting it to the ovary; *i*, spermatic duct passing towards the stomach.

* See the 'Anatomy of Actinia Coriacea,' by Mr. T. P. Teale, in the "Transactions of the Leeds Philosophical and Literary Society," vol. i. part I.

layers are at first in contact; but a separation takes place between them, so that the inner, which forms the wall of the stomach, no longer remains connected with the outer, which constitutes the integument, except by the vertical leaflets that divide the intervening spaces.—The mode in which the generative function is performed in the *Asteroid* Polypes, is obviously the same in all essential particulars as that just described; for the ova are developed in a similar situation, that is, between two layers of the folds of membrane projecting into the prolongations of the digestive cavity; and the spermatozoa are evolved from convoluted tubes closely connected with the same plaits (Fig. 103).—The process of extension by gemmation is carried on to an almost indefinite amount in the composite forms of these Zoophytes, as already explained (§§ 291, 292); but it is seldom that the gemmæ spontaneously detach themselves, though they will maintain an independent existence if artificially separated. No instance is known among these orders, in which the gemma developes itself into a form different from that of the stock that put it forth; nor does the embryo ever seem to assume any other type than that of its parent; but, whether produced by gemmation, or by the true generative operation, the new polypes appear to have the same structure and endowments.

690. A very different succession of phenomena is presented, however, by those curious organisms which are usually designated as Compound Hydroida (§ 290). These, looking only to their nutritive apparatus, may be considered as not differing in any essential particular from a Hydra-stock, which has put forth a set of primary and secondary gemmæ that have not yet separated from it (Fig. 94). But their reproduction is carried on upon a type altogether dissimilar: and the curious discoveries which have been made of late years, in regard to the mode in which it is accomplished, reveal a connection previously altogether unsuspected, between this group and the Medusan Acalephæ.—It will be desirable, in the first place, to consider the relation which the polypes of one of these composite structures bear to each other and to the common stock. It will be recollected that the stem and branches consist of a tubular ramification, which is in continuity with the lining membrane of the stomachs of the polypes, whilst the external horny integument is continuous with the bell like polype-cell (Fig. 95). Now it is not here the polype alone, nor the collection of polypes, which constitutes the Zoophyte; for all the polypes may drop off, like the leaves of a tree, and the polypidom may yet retain its vitality, and may put forth new polypes. Moreover, the new polypes formed by gemmation are not evolved from the pre-existing polypes, but from some part of the stem or branches. The external horny coat softens at a certain spot, and the internal membrane protrudes, pushing (as it were) the horny integument before it (Fig. 267, A); this protrusion increases (B, C), and soon comes to present the form of a polype-cell (D), the mouth of which, however, is still closed. After a time the horny integument thins away, and at last opens at the projecting extremity, so that the cell assumes its bell-like aspect (E); but the internal membrane does not as yet undergo the same process, so that no polype-mouth is formed. This operation, however, is effected in the next stage (F); and soon afterwards the tentacula are seen as little wart-like processes around the orifice (G). Up to the time when the mouth is formed, a double current is seen in the stalk of the bud; the particles

ascending into it from the stem, passing round its dilated interior, and then descending in its stalk. The whole history of this development, and the connection of the polypes with the nutritive functions alone, show that they must be considered as standing in the same relation to the stock, as the leaves of a tree to its stem and branches; the lining membrane of the

FIG. 267.

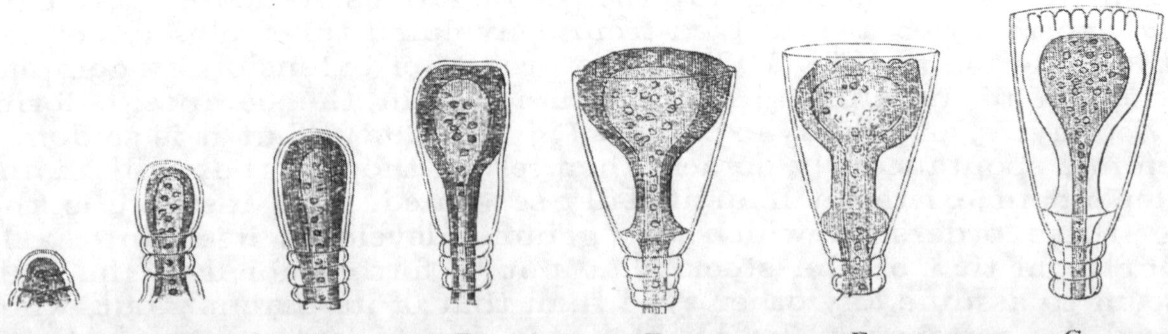

Progressive stages of the development of the polype-bud of *Campanularia gelatinosa.*

polypidom, like the cambium-layer of the tree, being the part in which the developmental capacity chiefly exists. The extension of the original stock, after this fashion, may go on in some species almost indefinitely, the polype-buds, as they are formed, remaining in continuity with the structure of which they are offsets; but there are some zoophytes of this tribe, which are less disposed to ramification, and which, like the Hydra, form detached gemmæ, that henceforth live independently, and develope themselves into new organisms, similar to that from which they were budded-off. Such appears to be the case with the *Tubularidæ* and *Corynidæ.* These buds do not, however, attain the form and condition of the parent previously to their detachment; but rather resemble, in their grade of development, the 'gemmules' or globular bodies first evolved from the ova of many of the lower tribes.* These buds are produced, either singly, or several together, within pear-shaped capsules; and their form and character are not discernible, until they are set free by the rupture of their envelopes. Their expulsion takes place slowly and gradually, and at the time of their emersion their organs are by no means fully developed. In some instances, the body has somewhat the shape of a polype whose tentacula are just beginning to sprout; and the completion of their development soon takes place. But in other cases, the form of the gemma at its first emersion is that of a minute Planaria, being flattened and oblong, and destitute of visible organs; and like that animal, this *planula* (as it has been designated by Sir J. G. Dalyell †) crawls with considerable acti-

* The bodies here alluded to are those which are described by M. Van Beneden in his admirable "Récherches sur les Tubulaires" as *ova.* They do not appear to be evolved, however, as the result of any sexual operation; but rather to be portions of the medullary membrane pinched off, so to speak, at a very early period, though continuing for a time to derive from it their materials for development into polypes. The *free gemmæ* of M. Van Beneden are obviously Medusa-buds.

† "Rare and Remarkable Animals of Scotland," vol. i. — Bodies resembling Sir J. G. Dalyell's *planulæ* are also described by Sars (" Fauna Littoralis Norvegiæ ") as proceeding from capsules developed on the *Podocoryna carnea.*

vity, and changes its direction when it meets with an obstacle. At last, however, it attaches itself, and sends up a stem terminated by a polype-head, which developes itself into the form characteristic of the species. Similar planulæ appear to be evolved from some of the so-called 'ovarian capsules' of *Campanularidæ* and *Sertularidæ*, as will be presently explained.

691. In the provisions hitherto described for the extension of the original fabric, or for the production of new and independent fabrics of the same kind, we have nothing that can be said to represent the true *generative* process; and it is on the arrangement by which this is effected, that recent discoveries have shed such a new and unexpected light. Besides the polype-buds, which, when detached, develope themselves into Zoophytic structures resembling those from which they sprang, the Compound Hydroida develope buds, which have all the essential characters of Medusan *Acalephæ*. The various stages of this process are best seen in the *Tubularidæ* and *Corynidæ;* in which these buds are put forth freely from the stem and branches, or even from the body of the polype itself, instead of being enclosed within a capsule (Fig. 268). In the first instance they are merely, like the polype-buds, little protuberances (*h h*) developed from the soft tissue of the interior, and attached by a footstalk; they soon, however, present a quadrangular form, and four spots are seen, which are the nuclei or centres of growth whence the marginal cirrhi

FIG. 268.

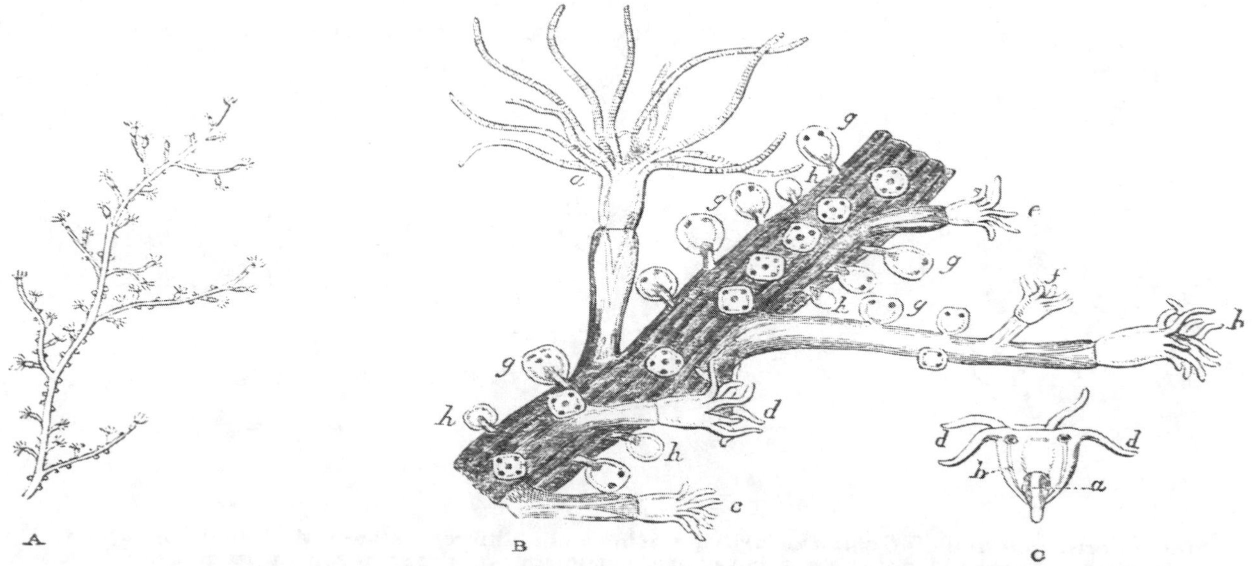

Development of Medusa-buds from *Perigonimus muscoides;*—A, part of a polype-stem of its natural size;—B, portion of the same enlarged, showing *a* a polype with its tentacula expanded, *b c d e f*, other polypes in various states of contraction, *g g*, medusa-buds showing the four nuclei, *h h*, medusa-buds less advanced;—C, medusa-bud more advanced and detached, showing *a* the stomach, *b*, the four radiating canals, *d d*, the marginal cirrhi.

are afterwards to be developed (*g, g*). After a further interval, an aperture forms itself at the most projecting part of the bud, which thenceforth has somewhat the form of a bell (C); and the stomach (*a*) is seen at its deepest portion, with four radiating canals (*b, b*) proceeding to its marginal cirrhi. (See also Fig. 96.) As these buds approach their maturity, they exhibit independent movements, and at last become detached and swim away freely, being propelled, like the ordinary Medusæ, by the rhythmical contractions of the disk. Their further development proceeds, therefore, altogether independently of the organism by which they were evolved; they capture and digest their own food, and thus nourished they increase in size and probably undergo some changes in form; and (although this stage of the process has not yet been determined by actual observation) there can be little doubt that true generative organs are developed in them, and that, by sexual reunion, ova are formed, as in the ordinary Medusæ; these being developed, in their turn, into the Zoophytic structure, whence similar medusa-buds will in due time be put forth.—In the *Campanularidæ* and *Sertularidæ*, similar buds are developed within the capsules which have been usually termed 'ovarian' or 'ovigerous' (Fig. 95, *e*); which capsules, as already mentioned, have been shown by Prof. E. Forbes to be in reality metamorphosed branches (§ 290). Like the polype-buds developed within similar capsules, these medusa-buds spring, not from ova, but from a detached portion of the medullary substance;* and they are gradually developed, even while yet contained within the capsule, into the Medusan type (Fig. 269). Here, too, the marginal cirrhi are seen to spring from the nuclei of elongated cells that form the margin of the primitive disk (A, *b*); and the central stomach (*a*) occupies the place which was at first filled by the projection of the medullary substance. As the time approaches for the opening of the capsule, the cirrhi undergo elongation, and the four canals proceeding from the central stomach to the margin (a disposition of parts especially characteristic of the Medusæ) are now seen (B.) The buds then become detached from their foot-stalks,

FIG. 269.

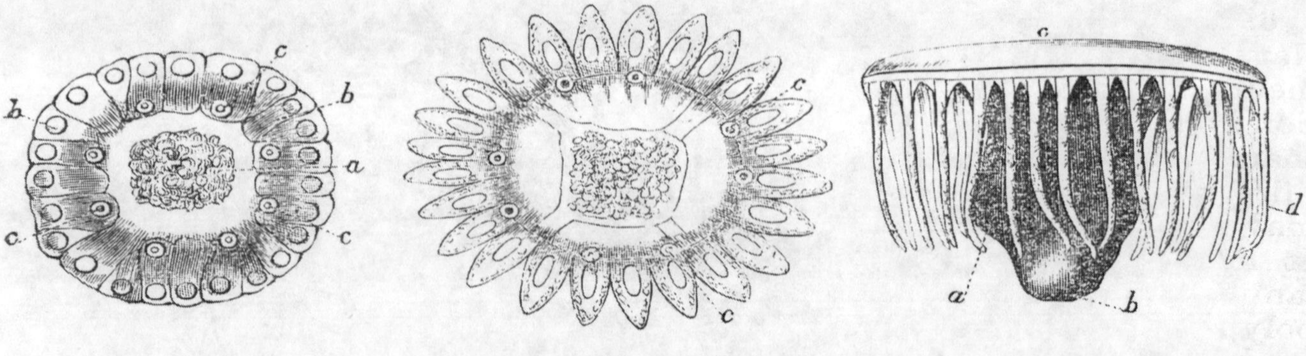

Medusiform gemmæ of *Campanularia gelatinosa* in different stages of development;—A, a gemma not yet escaped from the ovarian capsule, showing *a*, the stomach in progress of formation; *b b*, elongated cells, the rudiments of the cirrhi; *c c*, eye-cells (?);—B, a gemma somewhat more advanced, showing *c c*, the eye cells (?);—C, a medusiform gemma freely swimming through the water after its escape; *a*, the body; *b*, mouth; *c*, disk; *d*, cirrhi.

and begin to exhibit active movements in the interior of their capsules; and at last an opening is formed in the latter, either by the dehiscence of a lid, or by the thinning-away and rupture of the most prominent portion, through which the young Medusæ swim forth, and disperse them-

* Although they are described by Van Beneden as developed from ova, yet it is clear from his own account that such is not the case; and that what he called the vitellus is continuous with the medullary substance of the stem and branches of the zoophyte. See his beautiful "Mémoire sur les Campanulaires," Bruxelles, 1843.

selves actively through the water, their marginal cirrhi being now developed, and the proboscis and stomach being adapted for the reception and digestion of food (c). Here, as in the preceding case, a deficiency must be confessed, as to our knowledge of the subsequent stages from actual observation ; but as this knowledge *has been* completely attained in a case which seems in all essential respects parallel (to be described in the next paragraph), there can be little doubt that, as in the preceding case, the medusa-bud evolves itself into the perfect Medusan type, developes sexual organs, performs the true generative process, and thus produces real ova, from which a zoophytic structure, similar to that which gave off the medusa-buds, is again evolved.—But there are other cases, again, in which the bodies that escape from the 'ovigerous capsules' are not medusa-buds, but ciliated gemmules, which, after a period of active movement, fix themselves, and become the foundations of new polype-structures. Now it is not impossible that some of these may be simple polype-buds or bulbels, like the 'planulæ' set free from the Corynidæ and Tubularidæ ; but in other cases it appears that their origin is different. It has been observed by Lister * and Lovén,† that the medusa-like bodies generated in the interior of the 'ovigerous capsules,' instead of becoming completely detached and freely swimming forth, expand at the summit, generate and set free their ova in the condition of ciliated 'gemmules,' and then, wither like blossoms, to be succeeded by a new expansion. It is not impossible that, in other instances again, the medusa-buds may not even expand themselves at the mouth of the 'ovigerous capsule,' but may perform the generative process in its interior, and may thus set free ciliated gemmules without ever themselves existing in the condition of independent Medusæ.‡ Similar varieties will be shown to exist among the Acalephæ (§ 695).

692. We have thus seen that the animal fabrics which are known to the Naturalist as *Compound Hydroida,* whilst extending and multiplying themselves by ordinary gemmation, perform their true Generative operation through the intermediation of organisms possessing all the essential characters of true *Medusæ ;* and that these may either develope and fertilize their ova whilst still in organic connection with the parent-stock, or may detach themselves from it, and maintain an independent existence. It has now to be shown that the bodies with which the Naturalist is familiar in the condition of true *Medusæ,* really commence life in the polypoid state, and may be regarded as gemmæ budded off from a *Polype-stock.* The very curious history now to be given, has been ascertained by the united labours of many observers ; and the process may be now regarded as having, in all its essential features, been clearly elucidated.—The generative organs of the Medusæ, as already described (§ 294), are contained in four 'ovarial chambers' surrounding the mouth, and opening by orifices of their own (Fig. 106). Within each of these chambers is seen a plaited riband-like membrane, between the folds of which are developed, during the period of fertility, either spermatic cells or ovules. As

* "Philosophical Transactions," 1834. † "Wiegman's Archiv." 1837.

‡ For a fuller account of the observations which have been made on this very interesting subject, and for an attempt to reconcile their apparent discrepancies, see the "Brit. and For. Med. Chir. Rev." for Jan. 1848. Since that time, the production of medusa-buds from *Sertularidæ* has been ascertained to take place by Van Beneden.

these usually exist in separate individuals,* and as no means can be discovered for the conveyance of the seminal fluid of one to the ovarial chambers of the other, it is probable that fecundation is effected in the same manner as in many Mollusca and Fishes,—the fluid being dispersed through the water in the neighbourhood of the ova, with which they thus come into contact. A very curious provision for the protection of the germ during its subsequent development, is found in some Medusæ. Small pear-shaped bags or pouches are appended to the fringes of the tentacula; and into these are conveyed (in what mode, however, is unknown) from four to eight germs, which undergo their development within them; and, when these are perfected, they make their way out by the rupture of the sac. In this curious disposition, we are reminded of the peculiar development of the Marsupial Mammalia.—The embryo emerges from the marsupium in the condition of a ciliated gemmule, of a rather oblong flattened form, very closely resembling an Infusory Animalcule. In this grade of development it has no mouth; but the central cells deliquesce, so as to form a cavity, which is to be the future stomach. The next change consists in the enlargement of one end of the embryo, and the contraction of the other; the latter is developed into a suctorial base or foot, by which the animal attaches itself to some fixed object; whilst in the former a central depression is soon observed, which gradually deepens until it communicates with the internal cavity, thus forming a mouth, whilst four tubercles that are seen around it are gradually elongated into tentacula, which are afterwards augmented in number.—Thus the first product of the ovum is not a Medusa, but a Hydra-like polype; and the resemblance is not limited to external form, but extends to internal structure, and to the mode of existence and multiplication. For this *Strobila* (as the polypoid larva of the Medusa has been termed), attaching itself by its base to a fixed object, spreads its arms (which are furnished with urticating organs) through the water in search of prey (Fig. 270, A); and by their means its food is drawn into the stomach, whence the indigestible residue is ejected through the mouth, just as in the Hydra. Further, the polype multiplies itself by gemmation; and this not only by the development of lateral gemmæ from its body (B), but also in some instances by the evolution of a *stolon* or creeping stem from its base (resembling that of many Zoophytes), from which gemmæ sprout forth, these generally becoming subsequently detached. Thus from a single individual, a large colony (C) may be produced in no very long time, the detached gemmæ most commonly remaining in proximity with each other; and the rate of multiplication, as in the Hydra, is chiefly influenced by the temperature, and by the supply of food. Further, it has been ascertained by Sir J. G. Dalyell, that this creature has the same power of regenerating parts of the body that have been removed, or of reproducing the complete body and tentacula from a portion of it, as that which is possessed by the common Hydra; and that both portions of a bisected Strobila could not merely be developed into new and perfect individuals, but could multiply the stock by gemmation, as before their fission. There appears no definite limit to its continuance in this state; it has been kept by the last-named excellent

* It has recently been asserted by M. Derbès (Ann. des Sci. Nat." Juin, 1850), that the *Cyanæa chrysaora* is hermaphrodite; but it does not appear certain that he has not mistaken the 'thread-cells' or urticating organs for sperm-cells.

observer for several years; and from the entire conformity of its general structure and habits to that of the Hydra, and the absence of any indication of its *larval* nature, it was considered by him as a complete being, "entitled to a distinct position in the *Systema Naturæ*."—A minute examination of its structure, however, shows that it is distinguishable from the ordinary Hydraform polype. The part surrounding the mouth is capable of being protruded as a sort of proboscis; and the aperture of the mouth is then seen to have a quadrangular form. The inner surface of the lips and mouth, and the external surface of the tentacula and body, are covered with cilia; so that currents of water, unless when the mouth is shut, are continually passing in and out from the mouth and along the tentacula. No proper Hydroid polype has its arms or body ciliated. The wall of the stomach consists of two layers, of which the inner one is folded inwards, so as to form four longitudinal projections into the cavity of the stomach, a sort of pouch or blind canal being left between the folds of each plait. These four short canals terminate at the upper end in another canal, that passes between the outer margin of the mouth and the periphery of the disk; and into this circular canal, the tubular cavities of the tentacula also open.* This arrangement is obviously indicative of the Medusan affinities of the Strobila; but it does not render it the less a Polype.

FIG 270.

Strobila (or polypoid state of Medusa) propagating by gemmation:—A, single individual, attached by its base, its tentacula extended in search of food; B, a group of five, the gemmæ remaining in connection with the parent; C, a colony consisting of numerous individuals proceeding from a common stock, but most of them detached from it.

693. The *Strobila* does not always remain, however, in its original condition. Under certain conditions not yet ascertained, it ceases to multiply by ordinary gemmation, although this process may have gone on with regularity and activity for months and even years previously; and enters upon an entirely different series of operations, of which the first is the assumption of a more elongated cylindrical form than it had previously possessed. A constriction or indentation is seen around this cylinder, just below the ring which surrounds the mouth and gives origin to the tentacula; and similar constrictions are soon repeated around the lower parts of the cylinder, so as to give to the whole body somewhat the appearance of a *rouleau* of coins (Fig. 271). Still, however, a sort of fleshy bulb,

* This description is derived from the excellent "Observations on the Development of the Medusæ," by the late Dr. John Reid, contained in the "Annals of Natural History" for Jan. 1848, and in his "Physiological, Pathological, and Anatomical Researches."

somewhat of the form of the original polype, is left at the base or attached extremity (Fig. 272, A). The number of circles is indefinite; and all are

Fig. 271.

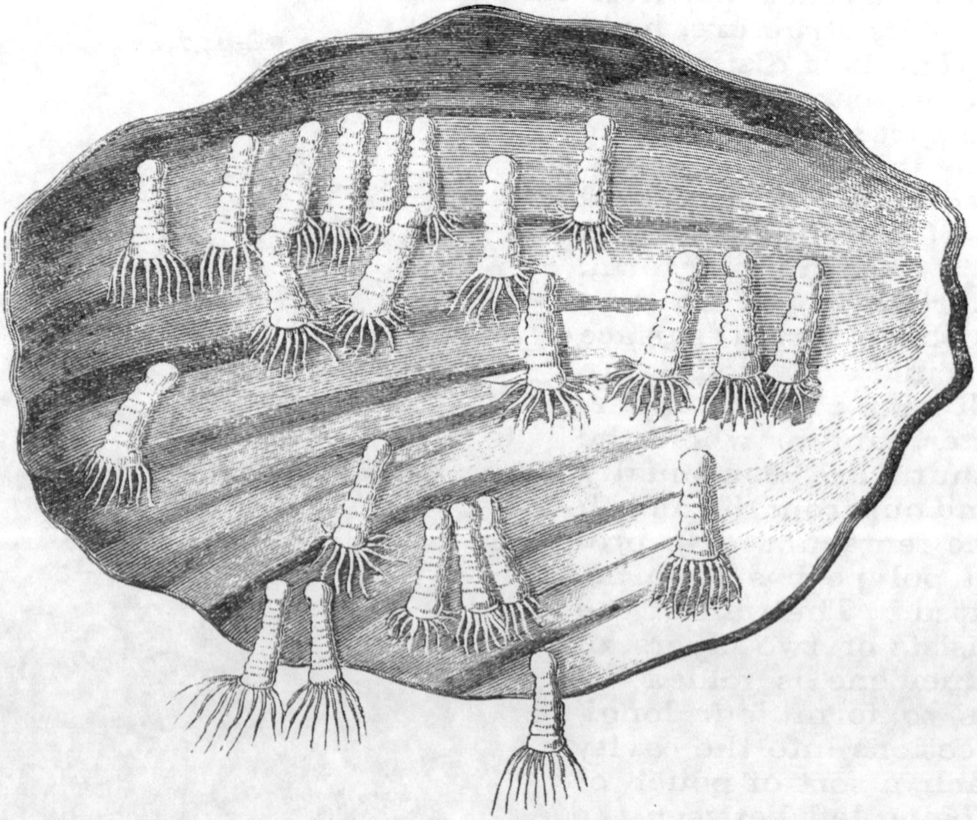

Group of *Strobilæ*, or *Medusan Larvæ*, attached to a shell of Pecten.

not formed at once, new constrictions appearing below, after the upper portions have been detached; as many as 30 or even 40 have thus been produced in one specimen. The constrictions then gradually deepen, so

Fig. 272.

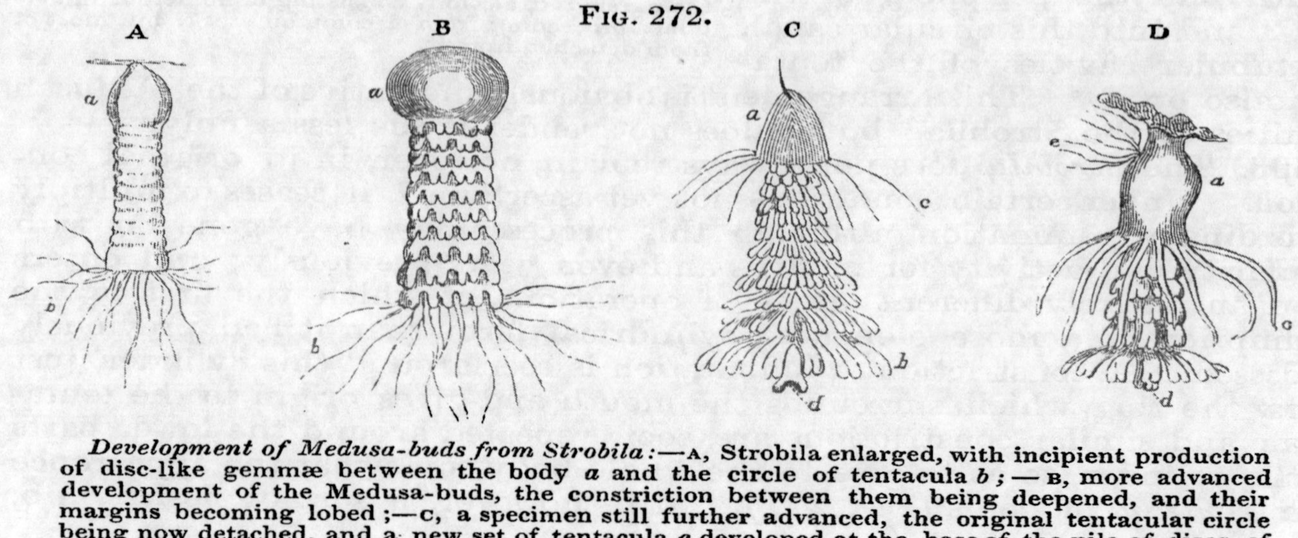

Development of Medusa-buds from Strobila:—A, Strobila enlarged, with incipient production of disc-like gemmæ between the body *a* and the circle of tentacula *b*; — B, more advanced development of the Medusa-buds, the constriction between them being deepened, and their margins becoming lobed;—C, a specimen still further advanced, the original tentacular circle being now detached, and a new set of tentacula *c* developed at the base of the pile of discs, of which those nearest the extremity *b* are most advanced, having acquired the proboscis *d*, and the general form of the Medusa;—D, a Strobila returning to its original polypoid state, several of the Medusa-buds having been detached, the new tentacula *c* being fully developed; and a polype-bud *e* being in process of formation.

as almost to divide the cylinder into a pile of saucer-like bodies; the division being most complete above, and the upper discs usually presenting some increase in their diameter; and whilst this is going on, the edges of the discs become lobed, and the lobes soon present the clefts and ocelli characteristic of the detached Medusæ (B). Up to this period, the tentacula of the original polype surmount the highest of the discs; and a general contraction and relaxation of the whole cylinder, causing the intervals between the discs to be diminished or increased, may be occasionally seen to take place. But before the detachment of the topmost disc, the circle of tentacula around the original mouth disappears; and a new circle is developed upon the summit of the bulb, which remains at the base of the pile of discs (C). At last, the topmost and largest disc begins to exhibit a sort of convulsive action; it becomes detached, and swims freely away; and the same series of changes takes place from above downwards, until the whole pile of discs is detached and converted into free-swimming Medusæ. But the original polypoid body still remains; and may return to its polype-like and original mode of gemmation (D), becoming the progenitor of a new colony of *Strobilæ*, every one of which may develope in its turn a pile of medusa-discs.* The bodies thus detached have all the essential characters of Medusæ. Each consists of an umbrella-like disc, divided at its edge into a variable number of lobes, *eight* being apparently the normal type; and of a stomach, which occupies a considerable proportion of the disc, and projects downwards in the form of a proboscis, in the centre of which is the quadrangular mouth (Fig. 273, A, B). At first there are no indications of genital organs; but these subsequently appear. Each of the lobes of the margin is bifid; and at the bottom of the cleft is a little body which has been supposed to be a rudimentary eye (C); but it is very doubtful if it be really entitled to this designation. As the animal advances towards maturity, the segments or lobes of the border of the disc increase comparatively little in size, whilst the intervals between them gra-

FIG. 273.

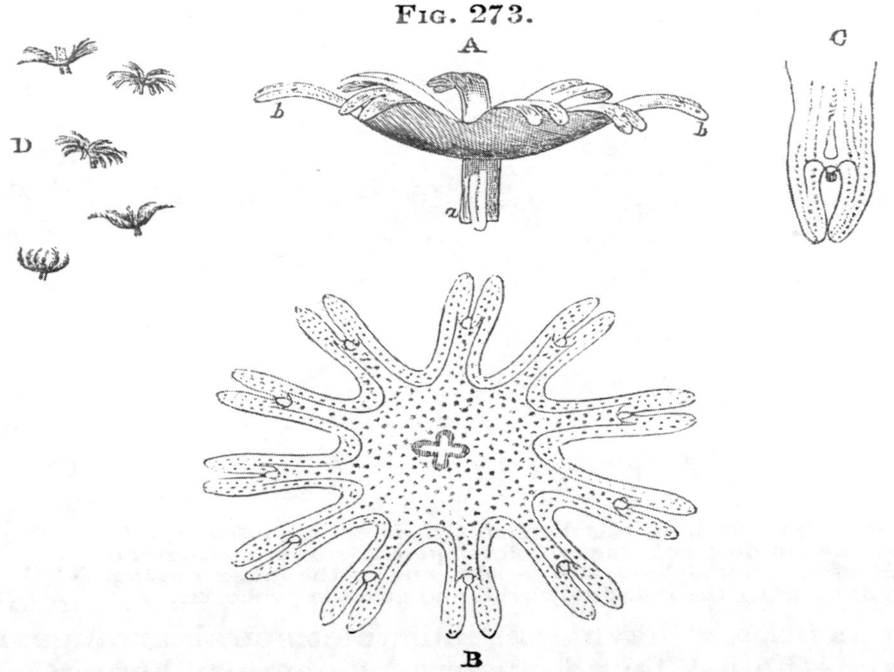

Development of *Medusæ* from detached portions of *Strobila*;—A, individual viewed sideways, and enlarged, showing *a* proboscis, and *b* bifid lobes; B, individual seen from above, showing the bifid lobes of the margin, and the quadrilateral mouth; C, one of the bifid lobes still more enlarged, showing the ocellus at the bottom of the cleft; D, group of young Medusæ, as seen swimming in the water, of the natural size.

* This last fact, which is of fundamental importance in the philosophical interpretation of this wonderful process, was first brought to light by Sir J. G. Dalyell. (See his "Rare and Remarkable Animals of Scotland," vol. i.)

dually fill up; tubular prolongations of the stomach extend themselves over the disc; and its borders become furnished with long, pendent, prehensile cirrhi. The mouth, which even in the youngest detached animal admits of being greatly extended and protruded, is quadrangular, and presents four extensible angles. These angles grow more rapidly than the four-sided oral tube or proboscis; so that, in the more advanced animals, the mouth appears during the growth to have divided or split into four lobes; and the minute serratures which appear on the edges of these, are the commencement of the lobes and fringes that are observed on the tentacula of the adult animal (Fig. 106, B).—The Medusæ which are as yet known to be developed after this mode, belong to the 'hooded-eyed' group of *Pulmograda*; whilst those which are budded-off from the composite Hydroid Zoophytes belong to the 'naked-eyed' subdivision. Even after the complete Medusan form has been attained, propagation occasionally takes place by gemmation; but only, so far as yet observed, in the inferior or 'naked-eyed' group. This curious fact was first observed by Sars in *Cytæis*; from the peduncular stomach of which he observed medusan gemmæ to arise, these being symmetrically arranged around its four sides, but in different stages of advancement (Fig. 274). He also observed in *Thaumantias* a similar gemmation from the ovaries. Both these observations have been verified by Prof. E. Forbes; who has also seen in a species of *Sarsia* an evolution of gemmæ around the whole length of the peduncle, these being in different stages of development, and having an indistinct spiral arrangement; whilst in another *Sarsia* he has found clusters of gemmæ hanging like bunches of grapes from the tubercles at the base of the four marginal cirrhi.*

FIG. 274.

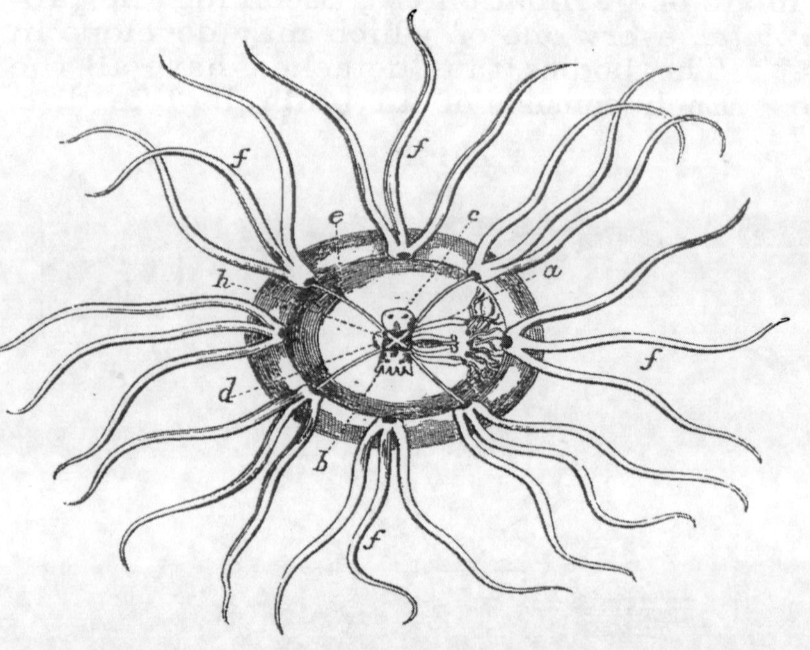

Gemmiparous multiplication of *Cytæis octopunctata*; *a, b, c, d*, buds in different stages of development, sprouting from the wall of the stomach *h*; at *e* is seen one of the four radiating canals; at *f f f* the marginal cirrhi, and at *h* the proboscis.

694. Now if the history of the development and propagation of the Medusæ be compared with that of the corresponding processes in the Hydroid Zoophytes, it will be seen that the two series of phenomena are of precisely the same order. The immediate product of the ovum, in each case, is a Hydraform polype. This lives for an indefinite time in the polypoid condition, and multiplies itself by gemmation; the gemmæ either detaching themselves, so as to form distinct zooids; or remaining in con-

* "Monograph of the British Naked-Eyed Medusæ," published by the Ray Society, 1848, p. 17.

tinuity with the original stock. But to whatever extent this process is carried on, it merely consists in a multiplication of the original Hydra; and all the zooids thus produced have the same *homological* relation to each other, as have the several parts of any single body among the higher Animals. But, under some change either of internal or external conditions, which cannot yet be distinctly stated, the *polypoid* gemmation gives place to the *medusan*,—the multiplication of the products of the original generative act, to the preparation for a repetition of that act. Now although the medusa-buds of the *Strobila* are formed in a different situation from those of the *Coryne* or *Campanularia*, being developed between the body and tentacular circle of the original polype in the one case, and from the sides of the polype-body or from the connecting stem in the other, yet there is otherwise no essential difference between the two operations. In the Strobila, as in the Coryne and Campanularia, the original polype-body remains; so that, after having given off a large number of medusa-buds, it continues its ordinary polypoid-existence, and may repeat the same process at a future time.* It is in the Medusa-buds only, that *generative* organs are evolved; these are subservient, in the one case as in the other, to the performance of the true generative process; and it is from the ova thus produced, that the polypoid larvæ arise.—Now that the whole of this series of processes is the continuation of one and the same developmental operation, and that the polypoid and medusan states should be included in the idea of *one generation,* instead of being regarded as *distinct generations,* will appear from this consideration; that they do, in fact, hold the same relation to each other, both homologically and functionally, as do the nutritive and generative organs in higher animals. The polype is no more a complete organism, than is the larva of an Insect or the tadpole of a Frog; for, although possessed of the capacity to maintain its own life, it has no generative apparatus for the continuance of the species.—So, again, we do not consider a Plant as having attained its complete development, however great may be its extension by leaf-buds, until it has developed flowers and produced fruit. The Medusan zooid is obviously the flower-bud or generative apparatus of the polype; it originates in a process of continuous development, precisely resembling that which evolves the generative organs, both in the annual Plant and in the Animal, *after* the nutritive apparatus has attained nearly its full development; and its special endowments obviously have reference to its peculiar function, that of propagating the race in localities remote from the original stock. In some of the Campanularidæ, as we have seen, the Medusa-buds never become completely detached from the Zoophytic structure, but seem to remain dependent upon it for their nutriment; and here the embryoes seem to have an unusual motor power, which serves for their dispersion.

* The production of Medusæ from the *Strobila* was represented by Sars as the result of *fission;* the fact that the polypoid body remains behind, forms a new set of tentacula, and may again give off polypoid and medusan gemmæ, not being known to him. This view is taken up by Steenstrup, who goes so far as to maintain that the original polypoid animal is really a medusa of the female sex, which acts as a sort of *nurse* to a brood of young ones. How these young ones are developed, whether from ova or gemmæ, he does not clearly explain; but he considers them as a new generation, *alternating* with the polypoid form,—a mode of viewing the subject which tends to confuse the essential distinction between the product of gemmation and that of the true generative act.

In most other cases, however, the Medusa-buds are not only detached, but, being separated when as yet they have attained but a small size and a low grade of development, they are provided with a nutrient apparatus of their own, whereby they can obtain and prepare the needful materials. This is perfectly conformable to what has been shown to take place in the Fern (§ 664). Of the detachment of a self-moving body containing the generative products, we shall meet with several curious examples, among the higher Animals.

695. The Reproductive process among other tribes usually assembled together under the class of *Acalephæ*,* is not less remarkable. In all of them, as it would appear, is there a double method of multiplication, namely, by gemmation, and by true generation; but the results of the former process differ very considerably in the different groups.—In *Beroe*, as probably in other *Ciliograda* (§ 294 *a*), gemmæ are occasionally produced, which seem to evolve themselves into the likeness of their stock; but the ordinary method of reproduction is by true generation, their progeny being apparently developed at once into the parental form. If this prove to be the case (which is not yet certain), it constitutes an important point of agreement betweeen the Ciliograda and the Helianthoid Zoophytes.—The *Cirrhigrada* (§ 294 *b*) are, in Mr. Huxley's view, composite polypoid animals, analogous to the compound Hydroida, although their parts have a very different arrangement. The multiplication of the organs (every one of the so-called tubular cirrhi being according to Mr. H. a distinct polype) takes place by a process of gemmation: but the generative act is provided for in a different mode. The bodies developed from the bases of the cirrhi, and formerly termed ovarian capsules (Fig. 108, D, E), are in his view rather medusiform buds. They become detached in some instances, and swim freely through the water; and in due time produce either spermatozoa or ova. From each of the latter, when fertilized, a single polypoid animal originates; and this gradually evolves the composite structure by gemmation, the central proboscis-like organ being the stock, and the surrounding tubular cirrhi the buds. The cavities of all these have been found by Mr. Huxley to communicate with each other and with the general reservoir, thus forming a polygastric digestive apparatus.—The order *Physograda* includes a great number of forms, whose true nature was long misunderstood, but which are now shown to be, like the preceding, composite organisms, the multiplication of whose parts is effected by gemmation. This is the case with the *Physsophoridæ* and *Diphydæ*, two groups whose relation to the Medusæ seems to consist in this,—that the contractile natatorial organ, which is developed in the latter around the stomach, is developed separately and on one side of it in the former. In each of these, long chains may be formed by the multiplication of similar organs by gemmation; but there is always a separate provision for the generative process, consisting of a

* Since the outline view of this class contained in the earlier portion of this work (§§ 293, 294) was in print, much has been added to our knowledge of it, chiefly by the researches of Mr. Huxley, the Assistant-Surgeon of H.M.S. Rattlesnake, recently engaged in a Surveying Voyage in the Eastern Archipelago. A portion of his results, which will probably tend to modify very greatly the existing arrangements of the classes Acalephæ and Polypifera, are contained in his Paper on 'The Anatomy and Affinities of the Medusæ,' in the "Philosophical Transactions" for 1849, Part II. For further information on some points not there treated of, the Author is indebted to communications made to him by Mr. Huxley himself.

gemma of a peculiar kind, which contains within it either ovisacs or sperm-cells. Both sets of these are formed in each individual of the Physsophoridæ, as in monœcious plants; but only one set in each individual of the Diphydæ, as in diœcious plants. This gemma, in some instances, is not developed into any special form, but is a mere protuberance containing the essential organs within it; but in other cases it has a natatorial disc developed around it; and some of those which are provided with this organ become detached, and swim about after the manner of Medusæ.—The *Physalia*, in Mr. Huxley's view, is simply one of the Physograda, in which the air-vesicle that all possess has undergone an unusual development; and its cirrhi are so many polype-mouths, or portions of a polygastric digestive apparatus. Its reproduction is effected after the same manner; the ova being developed within a medusiform gemma, that possesses a power of independent movement, and the spermatozoa within a gemma of less specialized character; whilst, from the ova which are thus generated and fertilized, a polypoid animal would probably spring, that would gradually evolve itself by gemmation into the likeness of its parent.—Much labour and research will undoubtedly be required to complete our knowledge of this subject; but when it shall have been thoroughly investigated, it will unquestionably be found to be one of the most remarkable pages in the whole volume of Natural History, abounding, though this does, in wonders and beauties throughout.

696. In the class *Echinodermata*, in which the Radiated type is carried to its highest degree of development, the structure of each organism being more complex, and its parts more specialized and mutually dependent, the power of multiplication by gemmation seems to be altogether wanting. Still we find a very extraordinary power of regeneration of lost parts to be retained. The *Ophiurida* can reproduce the arms, and the *Asteriada* can replace the rays, which they may have lost; and although neither arms nor rays can reproduce the body, when separated from it, yet if a small part of the disc (in some species at least) remain attached to the separated arm, the whole may in time be developed from this.* This regenerative power seems to extend, in the *Holothuriada*, to the reproduction of the entire visceral apparatus; and instances of spontaneous fission are said to occur in the same group, especially in those aberrant forms of it which lead towards the Articulated series (§ 296 *e*). In the *Synaptæ*, which exhibit this curious property, there is also another departure from the usual type; the two sets of sexual organs being combined in the same organism, instead of being, as in all other Echinodermata, allotted to distinct individuals. The position of these organs in the different orders has been already noticed in the general account of the class (§§ 296-296 *e*); but it must be remarked that the so-called "ovaries" are as frequently "testes" or spermatic organs; the male and female generative apparatus being scarcely distinguishable from one another in this class, either by their external aspect or their internal structure, except when their respective products are nearly mature.—We have now to trace the history of

* The Author once met with a single ray of a Star-fish in a state of activity, as was evidenced by the energetic movements of its tubular suckers; although it was obvious, from the appearance of the wound, that it must have been detached from the body for some days or even weeks. Might this have continued to live, and have re-formed the body?

embryonic development in this class, so far as it is known; and we shall find it to present some very curious features.

697. Of the orders *Crinoidea* and *Cystidea*, nothing can be said under this head. The fossil remains of the latter group do not afford any clue to the history of their growth from the egg; and little is known in regard to the former, save the fact of the change from the pedunculate to the free state, of which mention has already been made (§ 296).—The accounts of the development of the *Asteriada* given by different trustworthy observers, exhibit so much discrepancy, that there can be little doubt that the process takes place after at least two very diverse plans. The *first* and simplest of these has been witnessed by Sars * in the *Echinaster rubens;* and the observations of Agassiz † are on the whole in accordance with those of that industrious naturalist. In the early stages, the segmentation of the yolk takes place as in other animals (Fig. 275, C, D, E, F); and the embryo comes forth from the egg soon after it has

FIG. 275.

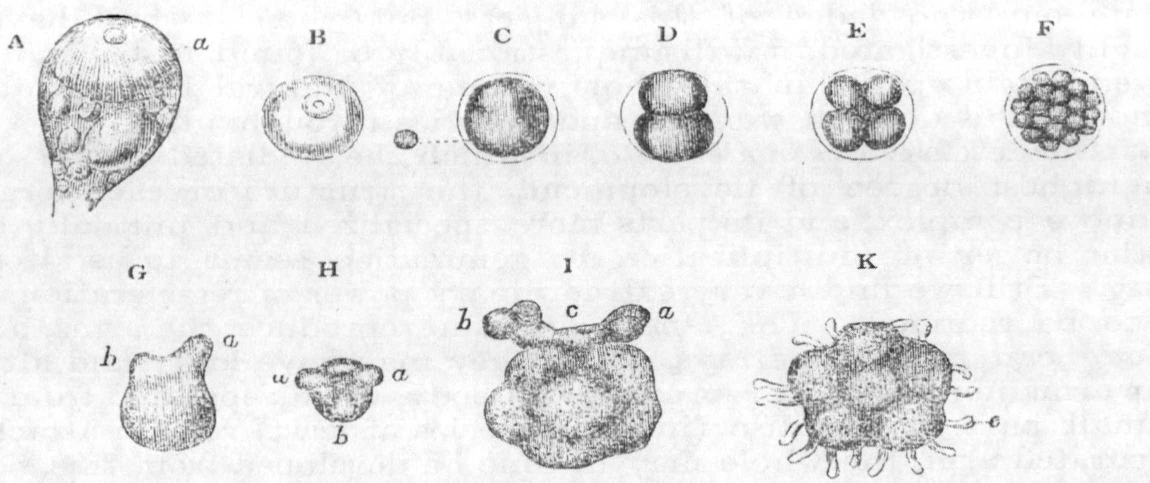

Development of the ovum of *Echinaster rubens;*—A, ovarian cæcum, containing ova in various stages of development, the most mature being seen at *a*; B, ovarian ovum highly magnified, showing the germinal vesicle and spot; C, an ovum after its deposition, the natural size being also shown; D, E, F, successive stages of the segmentation of the yolk; G, embryo soon after its escape from the ovum, showing at *a*, *b*, two of the rudimentary organs of attachment; H, end view of the same, showing all three (*a*, *a*, *b*) of these organs; I, more advanced embryo, attached by four organs, two at *a*, and two at *b*, with a papillary projection *c* between them; K, embryo more advanced, free and furnished with cirrhi *e*.

attained the state of the 'mulberry mass,' and swims freely about, by means of the cilia with which it is covered, in a sort of marsupial chamber which is formed by the drawing-together of the rays of the parent around its mouth. Soon after its emersion, the embryo begins to put forth an organ of attachment, resembling the stem of a Crinoid; this at first possesses two tubercles, then three (G, H), then four, with a fifth smaller one between them (I).‡ At the same time, the principal mass gradually be-

* "Fauna littoralis Norvegiæ," 1846.

† "Lectures on Comparative Embryology," New York, 1849.

‡ From the figures and descriptions of Sars, it would appear as if the organ of adhesion was developed from *one side* of the disc; but it is affirmed by Agassiz that it hangs down as a pedicle from the centre of the under side of the body, and that its apparent *lateral* projection

comes flattened, and shapes itself into five lobes surrounding a central disc; thus sketching out the body and rays. When in this state, it attaches itself to fixed objects by its organ of adhesion; but if detached, it swims through the water by the action of the cilia with which the body and arms are clothed; so that it bears a strong analogy to the *Pentacrinus* in process of conversion into a free-moving *Comatula* (§ 296). At the same time, five double rows of small tubercles may be perceived radiating from the centre of what is to become the ventral surface of the body; these gradually elongate themselves, and become cirrhi (K, *e*), each furnished with a sucker at its extremity. A peculiar tubercle is also seen at the edge of each of the five lobes of the body; and this is the rudiment of the *ocellus*, which is afterwards found at the extremity of each ray. As development proceeds, the primitive organ of adhesion gradually decreases in size, and the animal creeps by means of its cirrhi; and at last the pedicle is drawn (as it were) into the body, the lobes of the body lengthen into rays, the animal loses its ciliograde progression, and the ordinary characters of the Star-fish become apparent.—The progress of the internal organization is thus described by Agassiz. The earliest deposit of calcareous matter takes place around the prominent tubercles of the lower surface; at first in the condition of little isolated crystals, which are formed as nuclei in the cells; and then as a network formed by the coalescence of several of these. Of these networks there are at first ten, symmetrically disposed on the ventral surface, in a manner corresponding to the arrangement of the solid plates in Crinoids; but they gradually increase in number, and more distinctly mark out the rays; new ones being interposed in pairs between those already existing, and small spines projecting from the older ones. The calcareous deposit in the dorsal surface, on the other hand, seems to proceed from a central nucleus above the yolk-mass. The progress of development is obviously from without inwards; the cells on the surface of the yolk-mass being the first to undergo metamorphosis into the permanent structure. Those occupying the central part of the body and pedicle, undergo liquefaction; and a kind of circulation is seen in the latter. Gradually what remains of the yolk-mass is more distinctly circumscribed in the interior of the animal, and forms a central cavity with prolongations extending into the rays; but it is not until the pedicle has contracted itself into a mere vesicle, that the mouth is formed, by the thinning-away of the envelope of the yolk-mass on the lower surface, a little to one side of the base of the pedicle; and it is not until after the formation of the mouth, that the nervous ring can be traced, with its prolongations extending to the ocelli at the extremities of the rays.

698. The *second* plan of development seems much more conformable to what will be presently described as taking place in the Ophiurida and Echinida; for the body first developed from the embryonic mass is a larva of which little remains in the permanent structure, and the Star-fish is budded-off, as it were, from the anterior extremity of this. This larva, which has received the name of *Bipinnaria* from the symmetrical wing-like arrangement of its natatory organs, presents much more resemblance to an Articulated, or to a Vertebrated, than to a Radiated

is simply due to the flexion it undergoes, when the embryo is taken out of the water in which it floats, and is laid upon a plane surface.

animal (Fig. 276). Its body is elongated, and carries at its anterior extremity the portion of the yolk-mass not yet metamorphosed, from which the Star-fish is afterwards to be developed; and into the cavity of this a passage is formed, through what may be termed the mouth of the larva (*a*), which opens in the middle of a transverse furrow; whilst another tube passing forth from it (*b*) seems to answer to an intestine. On either side of the anterior portion of the body are six, or more, narrow fin-like appendages, which are fringed with cilia; and the posterior part of the body is prolonged into a sort of pedicle, bilobed towards its extremity, which also is covered with cilia. The organisation of this larva seems completed, and its movements through the water are very active, before the mass at its anterior extremity presents anything of the aspect of the Star-fish; in this respect corresponding with the movements of the 'pluteus' of the Echinida. The temporary mouth of the larva does not remain as the permanent mouth of the Star-fish; for it is on what is to become the dorsal side of the body; and the true mouth is subsequently formed by the thinning-away of the integument (which has completely enclosed the yolk-mass) on the ventral surface. The young Star-fish is separated from the bipinnarian larva, by the forcible contractions of the connecting pedicle, as soon as the calcareous consolidation of its integument has taken place, and its true mouth has been formed, but long before it has attained the adult condition; and as its ulterior development has not hitherto been observed in any instance, it is not yet known what are the species in which this mode of evolution prevails. The larva continues active for several days after its detachment; but there is no reason to believe that its existence is prolonged for any considerable time; and as the Star-fish is not formed by gemmation from it, but from a portion of the yolk-mass which remained unconsolidated after its completion, it is obvious that the larva does not stand in the same relation to the Star-fish, as the Hydraform polype to its medusa-bud. We shall hereafter see how close is the analogy borne by this temporary apparatus, to the first product of embryonic development in the higher animals.*

FIG. 276.

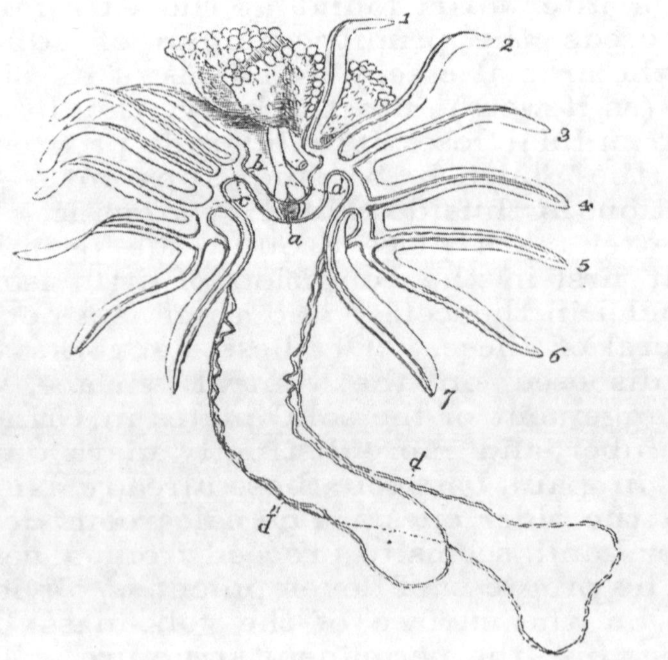

Bipinnaria asterigera, or Larva of Star-fish; *a*, mouth; *a'*, œsophagus; *b*, intestinal tube and anal orifice; *c*, furrow in which the mouth is situated; *d d'*, bilobed peduncle; 1, 2, 3, 4, 5, 6, 7, ciliated arms.

699. For our knowledge of the history of embryonic development

* See the Observations of Koren and Daniellsen (of Bergen) in the "Zoologiske Bidrag," Bergen, 1847 (translated in the "Ann. des Sci. Nat.," Juin, 1847); and the Memoir of

in the orders *Ophiurida* and *Echinida*, we are principally indebted to the researches of Prof. Müller.* The embryo issues forth from the ovum as soon as it has attained, by the repeated segmentation of the yolk, the condition of the 'mulberry-mass;' and the superficial cells of this are covered with cilia, by whose agency it swims freely through the water. So rapid are the early processes of development, that no more than from twelve to twenty-four hours intervene between fecundation and the emersion of the embryo; the division into two, four, or even eight segments taking place within three hours after impregnation. Within a few hours after its emersion, the embryo changes from the spherical into a sub-pyramidal form with a flattened base; and in the centre of this base is a depression, which gradually deepens, so as to form a mouth that communicates with a cavity in the interior of the mass, which is surrounded by a portion of the yolk-mass that has returned to the liquid granular state. The pyramid is at first triangular, but it afterwards becomes quadrangular; and the angles are greatly prolonged round the mouth (or base), whilst the apex of the pyramid is sometimes greatly prolonged in the opposite direction, but is sometimes rounded off into a kind of dome (Fig. 277, A). All parts of this curious body, and especially its most projecting portions, are strengthened by a frame-work of thread-like calcareous rods (*e*). In this condition, the embryo swims freely through the water, being

FIG. 277.

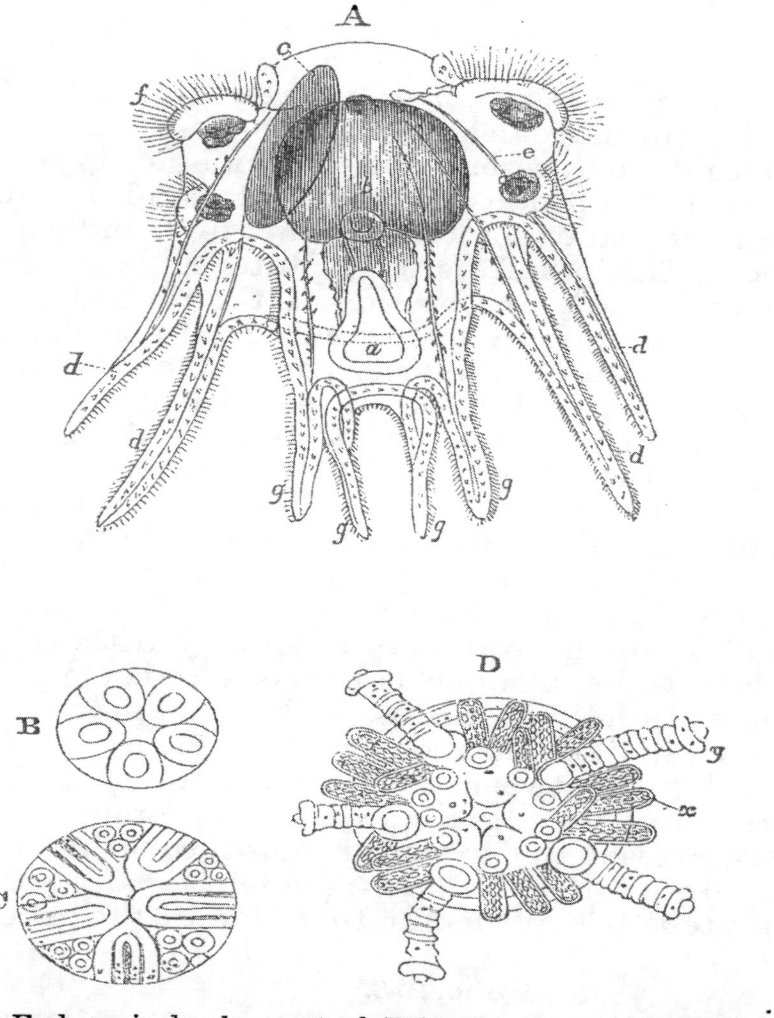

Embryonic development of *Echinus*;—A, *Pluteus-larva* at the time of the first appearance of the disc; *a*, mouth in the midst of the four-pronged proboscis; *b*, granular yolk-mass; *c*, echinoid disc; *d, d, d, d*, four arms of the pluteus-body; *e*, calcareous framework; *f*, ciliated lobes; *g, g, g, g*, ciliated processes of the proboscis:—B, disc, with the first indication of the cirrhi:—C, disc, with the origin of the spines between the cirrhi:—D, more advanced disc, with the cirrhi and spines projecting considerably from the surface. (N.B. In Figs. B, C, and D, the pluteus is not represented, its parts having undergone no change, save in becoming relatively smaller.)

Prof. Müller 'Uber die Larven und die Metamorphose der Echinodermen,' in "Abhaldlungen der Königlichen Akademie der Wissenschaften zu Berlin," 1848.

* 'Ueber die Larven und die Metamorphose der Ophiuren und Seeigel,' in "Abhaldlungen der Königlichen Akademie der Wissenschaften zu Berlin," 1846. See also, for the earlier stages, a Memoir by M. Derbès, in the "Ann. des Sci. Nat.," Août, 1847.

propelled by the action of cilia, which clothe the four angles of the pyramid and its projecting arms, and which are sometimes thickly set upon two or four projecting lobes (*f*); and it has received the designation of *Pluteus*. The 'pluteus' of the *Ophiura* and that of the *Echinus* at first differ very little in their general form and structure; but the 'plutei' of different species vary considerably in the number of their arms, some having as many as thirteen. The mouth is usually surrounded by a sort of proboscis, the angles of which are prolonged into four slender processes (*g, g, g, g*) shorter than the four outer legs, but furnished with a similar calcareous frame-work. In this condition, the 'pluteus' may be said to present the Acalephoid type, corresponding with the Medusæ in its probosciform mouth, and resembling the Beroe in its propulsion by bands of cilia.—The first indication of the production of the young Echinus from its 'pluteus,' is given by the formation of a circular disc (A, *c*), on one side of the central granular yolk-mass (*b*); and this disc soon presents five prominent tubercles (B), which subsequently become elongated into tubular cirrhi. The disc gradually extends itself over the yolk-mass; and between its cirrhi, the rudiments of spines are seen to protrude (C); these, with the cirrhi, increase in length, so as to project against the envelope of the 'pluteus,' and to push themselves through it; whilst, at the same time, the original angular appendages of the 'pluteus' diminish in size, the ciliary movement becomes less active, being superseded by the action of the cirrhi and spines, and the mouth of the 'pluteus' closes up. By the time that the disc has grown over half of the granular sphere, very little of the 'pluteus' remains, except some of the slender calcareous rods; and the number of tentacula and spines rapidly increases. The calcareous frame-work of the shell at first consists, like that of the Star-fishes, of a series of isolated networks developed between the cirrhi; and upon these the first-formed spines rest (D). But they gradually become more consolidated, and extend themselves over the granular mass, so as to form the series of plates. The mouth of the Echinus (which is altogether distinct from that of the 'pluteus') is formed at that side of the granular mass, over which the shell is last extended; and the first indication of it consists in the appearance of five calcareous accretions, which are the summits of the five portions of the frame-work of jaws and teeth that surrounds it. All traces of the original 'pluteus' are now lost; and the larva, which now presents the general aspect of an Echinoid animal, gradually augments in size, multiplies the number of its plates, cirrhi, and spines, evolves itself into its particular generic and specific type, and undergoes various changes of internal structure, tending to the development of the complete organism.—The body of the *Ophiura* takes its origin in a set of little cæca, which extend themselves from the central mass, nearly in the

FIG. 278.

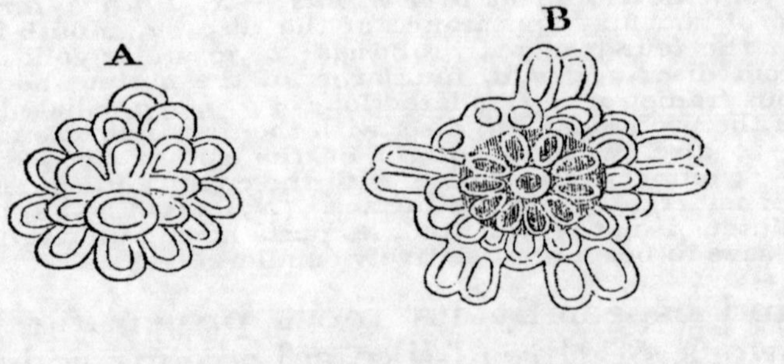

Origin of the *Ophiura* from the side of the vitelline mass in the body of its *Pluteus*:—A, cluster of cæca; B, the form of the body marked out, with the commencement of the arms.

position of the disc of the Echinus (Fig. 278); and these grow into a sort of circular cluster (A), which gradually shapes itself into the form of the body of the Ophiura with incipient arms (B). The other changes take place very much after the same fashion as in the Echinus. —Of the embryonic development of the *Holothuriada*, nothing is yet known.

700. The developmental processes of which a sketch has now been given, undoubtedly constitute a most remarkable series. We here find the portion of the yolk-mass that first undergoes organisation, converted into a structure which is destined only to possess a transient existence, and which disappears entirely by the time that the development of the principal part of the yolk-mass has advanced so far, that it begins to assume the characters of the permanent organism. This, however, is what takes place in the higher Vertebrata; for the structures first developed in the egg of the Bird hold nearly the same relation to the rudimentary chick, that the 'Pluteus' bears to the incipient Echinus or Ophiura, or the 'Bipinnaria' to the incipient Star-fish. The only essential difference consists in this,—that the development of these temporary structures proceeds so much further in the latter case, as to give them more the character of distinct individuals; and they are endowed with self-moving powers, whereby they are dispersed through the water in this stage of their existence, so as to prevent that accumulation in particular localities, which would otherwise result from the comparatively sluggish habits of these animals in their adult condition.

701. Passing on now to the *Molluscous* sub-kingdom, we find in the *Bryozoa* such a close approximation to the Zoophytic type, as regards their multiplication by continuous gemmation, that it is not surprising that their place in the animal series should have been at first mistaken, more especially when the notion prevailed that no true Mollusk ever multiplies itself by this method. So characteristic is this process, indeed, of Bryozoa, that there is no instance known of an animal of this group existing permanently in a solitary condition, like the Hydra or Actinia; since every Bryozoon, on emerging from the egg, tends to evolve itself into a composite structure, the gemmation taking place either from the interior of the tubular stem and branches, which usually connect together the cavities of the several cells, or from the cells themselves, where no stem intervenes. The general plan in which this gemmation commences, is the same as in the Hydraform Zoophytes; for there is first seen a bud-like protuberance of the horny external integument, into which the medullary lining prolongs itself; the cavity thus formed, however, is not to become, as in the group just referred to, the stomach of the new zooid; but it constitutes the chamber surrounding the stomach, which remains completely enclosed by the continuity of the membrane lining the cell with the external coat of the alimentary canal (§ 298). The digestive apparatus takes its origin in a thickening of the medullary membrane (Fig. 277, A, *a*), which projects from one side of the cavity into its interior; and the cells of the central part of this appear to deliquesce and form the digestive cavity, whilst from those which surround it are developed the walls of the alimentary canal, the tentacula, and the muscles which protrude or retract this apparatus. The first appearance of the tentacula is seen at B, *a*; and a more advanced stage in the development, the cell being still closed, is

shown at C. Whilst these organs are being evolved, the foundation is laid for the generative apparatus, in a similar thickening of the medullary membrane in the lower part of the cell (B, *b*); and this in due time acquires its full development, and matures its spermatic cells.—The mode in which the ova of Bryozoa are produced, fertilized, and set free, has been already described (§ 298); of the history of the subsequent development of the embryo, however, very little is known, save that it comes forth from the ovum in the state of a free-swimming ciliated gemmule, from which the first cell, with its highly-organized tenant, is gradually evolved. Of the fresh-water species of this group, some appear to differ from the marine in the separation of the sexual organs, the male and female portions of the apparatus being no longer contained within the same cells. And we further meet, in many species of them, with a very firm horny casing around the eggs, which appears destined to secure them from the effects of the winter's cold, to which the fresh-water Bryozoa are more severely exposed than the marine,—being, for the most part, inhabitants of small collections of water, that are liable to be completely frozen.

FIG. 279.

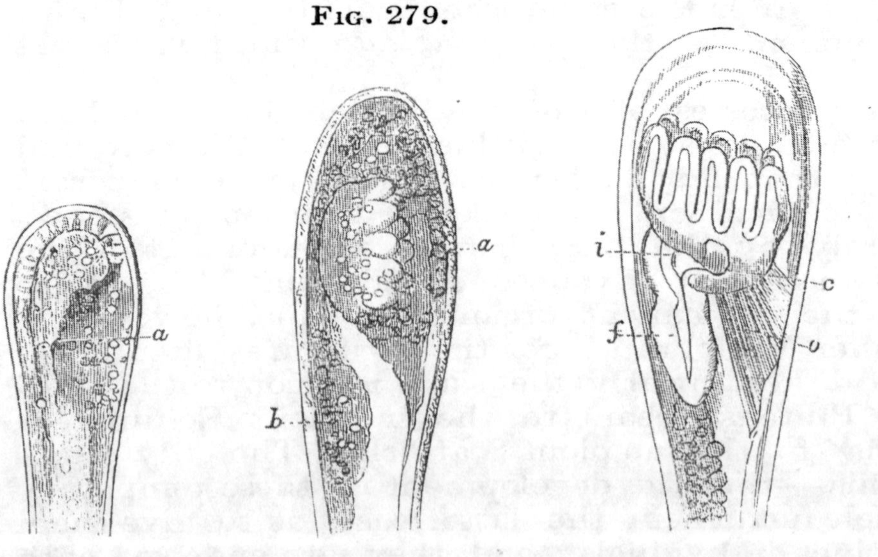

Gemmiparous extension of *Laguncula repens*:—A, a bud in an early stage of development, showing at *a* a thickening of the interior membrane, which is the first rudiment of the digestive apparatus;—B, the same more advanced, the rudimentary tentacula being now in process of formation, and another thickening showing itself at *b*, as the rudiment of the male generative organ;—C, the same, shortly before the opening of the cell; *c*, buccal cavity; *f*, stomach; *i*, intestine; *o*, retractor muscles.

702. The *Tunicata* correspond closely with the Bryozoa in their mode of extension by gemmation; for this is the mode wherein all those clusters are formed, which constitute a large proportion of the entire series of these animals; the solitary species, which do not propagate by this method, being almost as exceptional as they are among Zoophytes. In all that group which corresponds in its general structure with the Ascidian type, the gemmation takes place externally; and it usually proceeds from a sort of tubular stolon, which extends itself from the base of the 'test' (Fig. 121, *t*), and which, like the connecting peduncle of *Perophora* (§ 466), is in connection with the circulating apparatus, so that a double current of blood moves in it. The development of the Ascidian zooid within the buds (t') which arise from this stolon, seems to take place after a mode essentially the same with that of the Bryozoan; but when it has proceeded to an extent that renders the zooid independent of the parent, the circulation through the connecting peduncle usually ceases, so that henceforth the zooid is organically detached from the parent-stock, although remaining included with it in the common enve-

lope. Thus we see that the presence of a general circulation in the clusters of Perophora, is in reality to be regarded as a persistence of an embryonic character originally common to all the Compound Ascidians. In some of these animals, however, as in many Bryozoa, the gemmation of the zooids takes place directly from the exterior of the body, instead of from a prolongation of it into a stolon. In the *Salpidæ*, the production of zooids by gemmation, takes place after a very different fashion; for the stolon from which they are put forth is here *internal;* and large numbers of buds are developed from it at the same time. The process commences at a very early period in the life of the parent-stock which encloses it; for the stolon is distinctly visible before its own embryonic development is completed; and the evolution of the first set of buds commences forthwith, so that it takes place coincidently with the growth of the parent-stock itself. A second group of buds is usually put forth, and even the formation of a third has often commenced, before the first is ready to be detached. All the buds of each group are detached at the same time; they adhere together by their external surfaces, or by special organs of attachment; and they thus form those chains or clusters of 'aggregate' Salpæ, which are frequently found floating on the ocean-surface in the warmer regions of the globe. Thus whilst the several Zooids of the compound Ascidian masses owe their existence the one to the other, and the whole to the parent-stock which laid the foundation of the cluster,—thus strictly resembling the relation of the several leaf-buds of a tree to each other and to the original plumule, the individuals forming the Salpa-chains are related only by their common origin from the same stolon, and the parent-stock forms no part of the chain, but remains as a solitary individual.—According to MM. Löwig and Kölliker, a sort of fissiparous multiplication takes place in some of the *Botryllidæ*, at a very early period of embryonic development; each ovum producing, by its segmental division, not a single individual, but a stellate cluster (Fig. 120). Should such be the case, the phenomenon would be the parallel, in the Animal Kingdom, to the free gemmation of Mosses whilst yet in the confervoid state (§ 659). The fact, however, is denied by Prof. Milne-Edwards, who considers that the cluster is formed by subsequent gemmation from the first individual; and the matter remains open for further investigation.

703. Every Tunicated animal of the *Ascidian* type, whether solitary or composite, possesses two sets of sexual organs; from the position of which, as already pointed out (§ 299 *a*), it may be inferred that the ova are fertilized before they pass forth from the body of the parent. Indeed the changes in which the earlier stages of embryonic development consist, may be usually seen to have taken place previously to the exit of the ovum from the cloaca. In all the members of this group in which the history of the process has been yet observed, the embryo comes forth from the egg in a form quite different from that which it is subsequently to present; and is endowed with the power of free locomotion, so important for the dispersion of animals that remain during the whole remainder of life in a fixed condition. The following is a sketch of the history of the process, as observed by Prof. Milne-Edwards in *Amaroucium proliferum* (Figs. 118, 119). The yolk appears to undergo the usual segmentation within a very short period after fecundation, so that the 'mulberry mass'

(Fig. 280, *b*), is soon produced ; and at the same time there is formed between the yolk and the external membrane of the ovum, a gelatinous, transparent, nearly colourless layer, which apparently becomes the external tunic or 'test' of the young animal. The marginal portion of the yolk then begins to separate from the central mass, so as at first to appear like a sort of ring encircling it (*c*) ; but it is soon observable that the

FIG. 280.

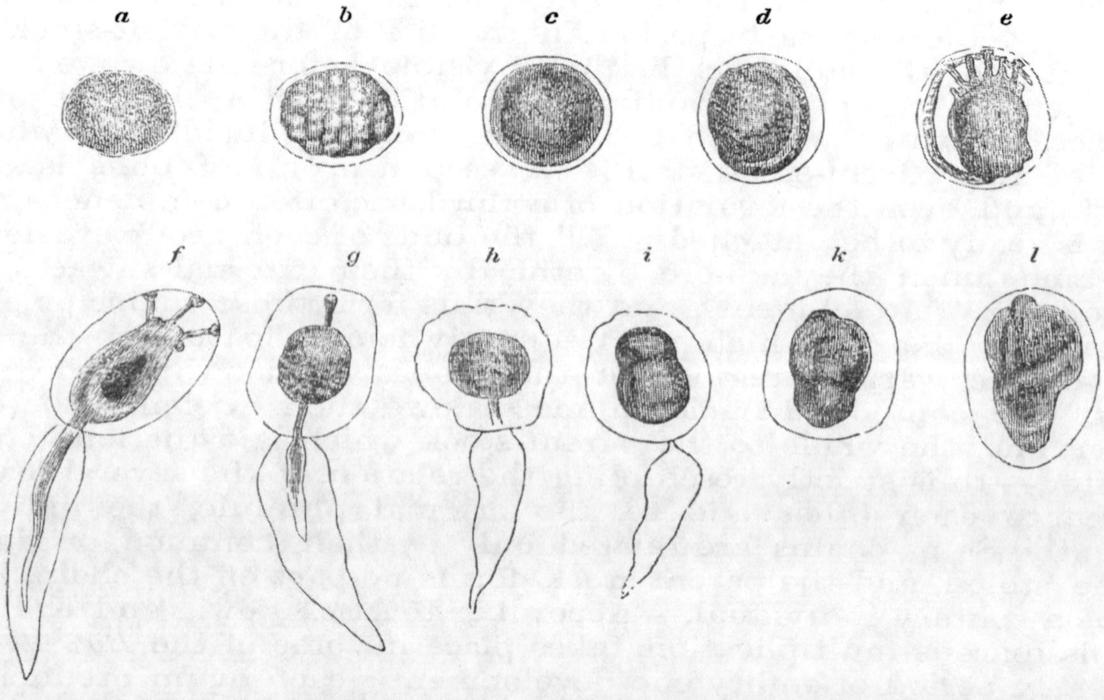

Development of the embryo of *Amaroucium proliferum* ;—*a*, mature ovum, escaped from the ovary; *b*, ovum taken from the cloaca, advanced to the state of 'mulberry mass; *c*, *d*, more advanced states of the ovum, still within the cloaca; *e*, ovum whose embryo is ready to emerge; *f*, larva newly escaped; *g*, larva some hours after fixation; *h*, the same, ten hours after fixation; *i*, the same, twenty hours after fixation; *k*, the same, at the commencement, and *l*, at the conclusion, of the second day after fixation.

apparent ring is really a tapering prolongation, which encircles the central part of the yolk, adhering by its base, and having its extremity free (*d*). A further change takes place previously to the exclusion of the egg ; for the tail-like appendage becomes more and more separated from the central mass ; and the anterior extremity is furnished with five diverging cylindrical processes, which advance towards the border of the egg (*e*), two of these, however, disappearing, whilst the other three increase in size, previously to the exclusion of the egg. Either whilst the egg is still within the cloaca, or after it has escaped from the anal orifice, its envelope bursts, and the larva escapes ; and in this condition (*f*) it presents very much the appearance of a tadpole, the tail being straightened-out, and propelling the body through the water by its lateral undulations. The centre of the body is occupied by a mass of liquid yolk ; and this is continued into the interior of the three anterior processes, each of which has a sort of sucker at its extremity. After swimming about for some hours with an active wriggling movement, the larva attaches itself to some solid body by means of one of these suckers ; if disturbed from its position, it at first swims about as before; but it soon completely loses its activity, and becomes permanently attached; and important changes speedily manifest themselves in its interior. The prolongations of the central yolk-substance into the anterior processes and tail, are gradually retracted (*g*, *h*), so that the whole of it is once more concentrated into one mass; and the tail, now consisting only of the gelatinous envelope, is either detached entire from the body, by the contraction of the connecting portion (*i*), or withers and is thrown off gradually in shreds. At this stage, a division of the central mass into two unequal parts is indicated; and the anterior part exhibits a deep yellow annular patch, circumscribing a paler yellow spot, which marks the position of the future mouth; whilst in the posterior portion there is a clear spot, in which the development of the heart commences. The formation of the internal organs takes place very rapidly; so that by the end of the second day of the sedentary state, the outlines of the branchial sac, and of the stomach and intestine, may be traced, no external orifices, however, being as yet formed. The pulsation of the heart commences on the third day, and the formation of the branchial and anal orifices takes place on the fourth; and the ciliary currents are immediately established through the branchial sac and alimentary canal. It is interesting to remark, that for some time after the organs of nutrition have come into full action, no trace whatever can be seen of those of reproduction; and that the entire form of the young Amaroucium (Fig. 281) corresponds rather with that of the *Didemnian* group, in which the post-abdomen is not developed (Fig. 121), than with that of the *Polyclinians*, which it is ultimately to possess (Fig. 119). The elongation of the posterior portion so as to form the post-abdomen (enclosing the heart and generative apparatus), does not commence until towards the end of the second week, when the formation of the genital organs, from a mass of granular matter between the heart and the intestine, begins to show itself.—The embryonic development of other Ascidians, solitary as well as composite, takes place in a manner essentially the same with the foregoing; a free-moving, tadpole-like larva being first produced in all the instances in which the process has been yet observed; and this attaching itself by some part of its surface, before the development of the future organism has made much progress.

FIG. 281.

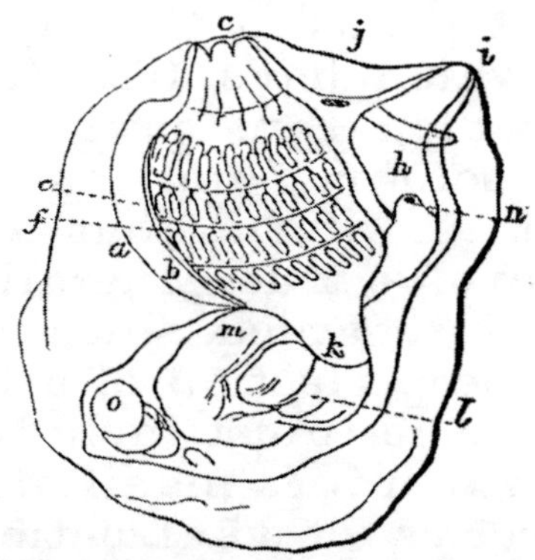

Anatomy of more advanced embryo of *Amaroucium proliferum*, showing its progress towards the adult condition (Fig. 119);—*a*, general tegumentary envelope; *b*, proper tunic of the individual; *c*, mouth; *e*, branchial sac; *f*, thoracic sinus; *h*, cloaca; *i*, anus; *j*, ganglion; *k*, œsophagus; *l*, stomach; *m*, intestine; *n*, its termination in the cloaca; *o*, heart.

704. The embryonic development of the *Salpidæ*, however, presents some very curious distinctive features. As already mentioned (§ 299 *b*), these animals exist under two distinct forms, the 'solitary,' and the 'aggregate;' every solitary Salpa having an aggregate form, which answers to it and alternates with it. It is in the aggregate Salpæ alone, that sexual organs exist; and each individual contains both ovaries and testes. Nevertheless it seems probable that they are not self-fertilizing; since the development of the spermatic organ has been found in several instances to be more tardy than that of the ovary; so that it is probable that the ovaries of all the zooids of one chain are fertilized by the spermatozoa developed

by another. Each zooid usually produces but a single ovum, and propagates but once in life; and this ovum is fertilized whilst yet within the body of the parent, doubtless by spermatozoa drawn in with the branchial current. Instead of being expelled, however, in an early stage of its embryonic development, the ovum is retained within the body of the parent; and it there forms an adhesion to a peculiar organ, that resembles, in all essential particulars, the *placenta* of the Mammal (§ 733), the fœtal and maternal vessels interpenetrating as it were, without communicating, in such a manner as to allow the transudation of fluids from one to the other. The embryo does not become detached, until the greater number of its organs have nearly attained their full development; the alternating pulsations of the heart so characteristic of the class, together with the respiratory movements, having been established for some time; and the internal stolon from which the gemmæ are to be put forth, being already present in the form of a short delicate filament, whose edges possess serrations that indicate the points of origin of the future buds.— Thus, starting, as before, from the completion of the generative act, we find a 'solitary' Salpa produced, which, like the first-formed Compound Ascidian, has the power of multiplying its kind by gemmation; and the chief difference between the two cases consists in this, that the original Salpa-stock never evolves true generative organs in its own body, but that these, as in the Hydraform Zoophytes, are developed in distinct zooids, which have, like the solitary Salpa, the power of free locomotion, and which thus propagate the race in distant localities. To speak of the 'solitary' and 'aggregate' Salpæ as constituting two distinct generations, is to affirm that the former are complete organisms; but as they never evolve true generative organs, it is obvious that they have no title to be so regarded, since the evolution of the genital apparatus is essential to the perfection of every fully-organised individual. The character of the species must include, as is now generally admitted, *both* the forms under which it is manifested; and it is necessary to bring together the entire series of vital actions performed by both, in order to possess that complete physiological history of any species of Salpa, which is afforded, in higher animals, by the life of any single individual, if hermaphrodite, or by that of any pair, if diœcious. The production of the 'aggregate' Salpæ must, it is evident, be regarded as a process of *development*, whilst that of the 'solitary' is the only true *generation*; consequently, this example does not, any more than those already cited, afford support to the doctrine of "Alternation of Generations," although usually cited as one of its most characteristic instances. — It is interesting to remark in conclusion, that through the whole of this very interesting group, we find a provision for the *dispersion* of each species in one state or another of its existence. In the *Ascidians*, it is the embryo which is self-moving, and the adult which is fixed; whilst in the *Salpians*, it is the embryo which is fixed, and the adult which is self-moving. In the one case, as in the other, the motions appear to be purely automatic; but in the embryo they are performed by those simple rhythmical contractions in cellular tissues, which correspond with those of the heart in the early stage of its development; whilst in the adult they seem to be reflex, depending on the instrumentality of nerves and muscles.*

* For the most recent information on the Reproduction of the Tunicata, the following

705. In the *Conchiferous Acephala* we meet with no indications whatever of the power of gemmiparous multiplication; the true generative process being the only means by which their Reproduction is accomplished. This is true alike of the *Brachiopoda* (§ 300), and of the *Lamellibranchiata* (§ 301); and in both of these, moreover, we meet with this further departure from the type of the Tunicata, that the male and female generative organs are in most instances disposed on distinct individuals. These organs, however, so closely resemble one another in their position and external aspect, that as they have been known in some instances to be ovaria, the eggs having been found in them when mature, it has been presumed that they were so in all; and this notwithstanding the observation of Leeuwenhöek, that in some individuals this organ contains spermatozoa instead of ovules. There is now no doubt, however, that the sexes are really as distinct as in Fishes; and that the fertilizing materials diffused through the water by a Conchiferous Mollusk of one sex, are drawn into the cavity of the shell by the respiratory currents of the other, and there come in contact with the ova. Our surprise at this mode of propagation is diminished by the recollection, that the Conchifera are very gregarious in their habits, few or none habitually leading solitary lives; and it has been ascertained by experiment, that single individuals are *not* capable of continuing the race,—the proximity of one, at least, of each sex being required for the purpose. The genus *Cyclas*, however, constitutes an exception to the general law, male and female organs having been detected in each of its individuals; and more extended observation will perhaps discover the same combination in other instances. The ova, when fertilized and discharged from the ovarium, do not, in general, at once quit the cavity of the shell, but are conveyed to the branchiæ, where they are retained until the included embryoes are ready to come forth; so that the aeration of their contained fluids (a change as necessary to them as to the adult animal) is effectually provided for. In many Conchifera, a distinct layer of albumen is found between the yolk-membrane and the general envelope; and this is sometimes so considerable in amount, that the yolk-bag forms but a small proportion of the whole ovum.—The history of embryonic development has been less studied in this class than in most others; and but little is known respecting it. It is pretty certain, however, that the young issue forth from the egg in the condition of ciliated gemmules, like those of Bryozoa; and that they swim about in the fluid included between the valves of the parent, in which state they have been mistaken for Infusory Animalcules. At a later period, the embryoes may sometimes be found to be provided with shells, whilst yet swimming freely by ciliary action in the interior of the shell of the parent; and it can scarcely be doubted that

Treatises and Memoirs should be especially consulted:—" Observations sur les Ascidies Composées des Cotes de La Manche," by Prof. Milne-Edwards, Paris, 1839;" "On the Alternation of Generations," by Prof. Steenstrup (translation published by the Ray Society, 1845); 'De la Composition et de la Structure des Envelloppes des Tuniciers,' by MM. Löwig and Kölliker, in "Ann. des Sci. Nat.," Avril, 1846; 'Observations sur le Génération et la Développement des Biphores,' by M. Krohn, in "Ann. des Sci. Nat.," Juillet, 1846; "Fauna Littoralis Norvegiæ," by M. Sars, 1846; and "Récherches sur l'Embryogenie, &c. des Ascidies simples," by M. Van Beneden, Brussels, 1847. A paper confirming and extending M. Krohn's observations on the Salpidæ, and tracing the history of the Reproduction of the Pyrosomidæ, has just been presented to the Royal Society by Mr. Huxley.

they are dispersed in this condition, and that they do not lose the power of free locomotion, until they have attained an advanced grade of development. The foot is one of the last organs to be evolved, as might be expected, when it is borne in mind that it is an organ belonging to only a part of the group; thus presenting another example of the principle, that the more special condition arises out of the more general.

706. In the *aquatic* orders of the class *Gasteropoda,* we still find the sexes distinct. Among the least active forms of this group, such as the *Patella* and *Chiton,* it is probable that fertilization is effected, as in the Conchifera, without the actual congress of two individuals; but in the higher, the spermatic duct of the male terminates in a projecting organ, adapted to convey its fluid within the oviduct of the female. In the aquatic Gasteropoda possessing spiral shells, the ovary in the female and the testis in the male occupy a corresponding position,—the higher part of the cavity of the shell. In the *Paludina vivipara,* the ova are delayed in a dilatation of the oviduct near its extremity, until the young are so completely matured, that they are hatched there, so as to pass out alive; but in most, if not in all, other cases, they are deposited by the parent, before the development of the embryo has proceeded far. Frequently, however, they are provided with an additional protective covering or *nidamentum,* which is formed by large glands situated near the termination of the oviduct. This nidamentum has different forms in the several species which produce it. In some instances it is a sort of gelatinous mass, in which the ova are imbedded with greater or less regularity. But in general it is composed of a large number of distinct sacs, each containing a few eggs; and these are connected together by a sort of footstalk. In the common *Buccinum undatum* (whelk), these sacs are flattened spheres, and are united together in the manner of bunches of fruit; very large masses of them are often to be picked up on our shores. In the *Pyrula,* they are flattened disc-like cases, united into a single string by a pedicle connecting the centre of each disc with that of the next. In the *Doris* they are spherical, and are arranged side by side in rows; a large number of which, being united, form a riband-like mass. Mr. Darwin speaks of one of these, produced by a Doris of the Falkland Islands about $3\frac{1}{2}$ inches long, which measured nearly twenty inches in length by half an inch in breadth; and which, on a moderate computation, must have contained 600,000 eggs, a number which is far from being without parallel in these most fertile animals.—The *pulmoniferous* Gasteropoda, on the other hand, are hermaphrodite, each individual possessing male and female organs; but they are not capable of self-impregnation, the congress of two being necessary, and each fertilizing the ova of the other. The eggs of these species are deposited singly in the earth, and are hatched by the warmth of the sun. They appear capable of undergoing very severe treatment without the loss of their fertility (§§ 65, 103).—The reproductive apparatus in the active little *Pteropoda* nearly resembles that of the pulmonary Gasteropods; the male and female organs are united in the same individual, but the congress of two is required. In the *Clio,* which alone has been minutely examined in this respect, the male organs are of very large size; the testis occupying a great part of the cavity of the body, and the penis being of extraordinary length.

707. The history of the Embryonic development of Gasteropoda presents a series of facts of great interest; and as this class may be considered as the type of the Molluscous group, we may probably regard the course of its development as that which is most characteristic of the sub-kingdom as a whole. Numerous and valuable observations have been made upon the evolution of the ova of various members of the class, both 'testaceous' and 'naked;' and it may be stated as their uniform result, that the young of the latter, as well as of the former, are provided with a simple spiral shell at their exit from the egg, although they may subsequently cast this off; and further, that the young of the aquatic species are provided with instruments for locomotion, which they do not possess in their adult condition; the obvious purpose being here, as in other cases, to disperse them over a wider area than that through which the sluggish movements they perform in their perfect state would enable them to extend.—The ova of Gasteropoda generally contain a considerable amount of albumen, surrounding the vitelline sac; and this is subsequently absorbed into the embryonic mass, in proportion as the material directly supplied by the yolk is exhausted. The most minute account yet given of the embryonic development of any Gasteropod, being that of M. Vogt, whose observations were made upon *Acteon viridis*, one of the Nudibranchiate order, it will be from this source that the following sketch will be chiefly derived. The early changes which the yolk undergoes, are conformable in all essential particulars to the type already described; but with this difference, that the process of segmentation divides it into unequal instead of equal parts; and that in this mode, a distinction is very early manifested between a peripheral layer of small cells, and the included mass of which the cells are much larger. It is not certain, however, that this peculiarity is common to Gasteropods generally. When the subdivided yolk has attained the condition of the 'mulberry mass,' a curious alternating revolution begins to take place in this body, within the egg; two or three turns being made in one direction, and the same number in a reverse direction. This movement, which seems due to ciliary action, may be well observed in the ova of the common *Lymnæus stagnalis*, in which it continues during a large portion of the period that elapses before the escape of the embryo from the ovum. The embryonic mass soon begins to show a bi-lateral symmetry, and before very long a subdivision is indicated into the anterior or cephalic portion, and the posterior or visceral. The parts which are first developed, are by no means those which are most characteristic of the adult; for the cephalic portion is soon extended on each side into a lobe very much resembling the fin-like expanse out of the Pteropoda (Fig. 130), which is furnished with long cilia (Fig. 282, *h*); it is also observed to contain the auditory vesicle, or rather its 'otolithe' (*k*), which, although so early developed, remains in the same rudimentary state during the whole of life; and it is from this part, also, that the prominence (*i*) is put forth, which is afterwards to be evolved into the foot or muscular disc of the animal. The formation of these parts has made considerable progress, by the end of the fourth day after the deposit of the fertilized ovum, when the ventral portion of the embryo merely consists of a mass of cells, in which not even the outline of its future organs can be seen, though a breaking-up of the mass into groups of cells begins to

show itself. On the next day, the shell first makes its appearance, as a very thin layer over the lower part of the ventral mass; and this extends itself on subsequent days, until, by the eighth, it becomes large enough to enclose the embryo completely, when the latter contracts itself (B, *m m′*). During this period, the formation of the internal organs

FIG. 282.

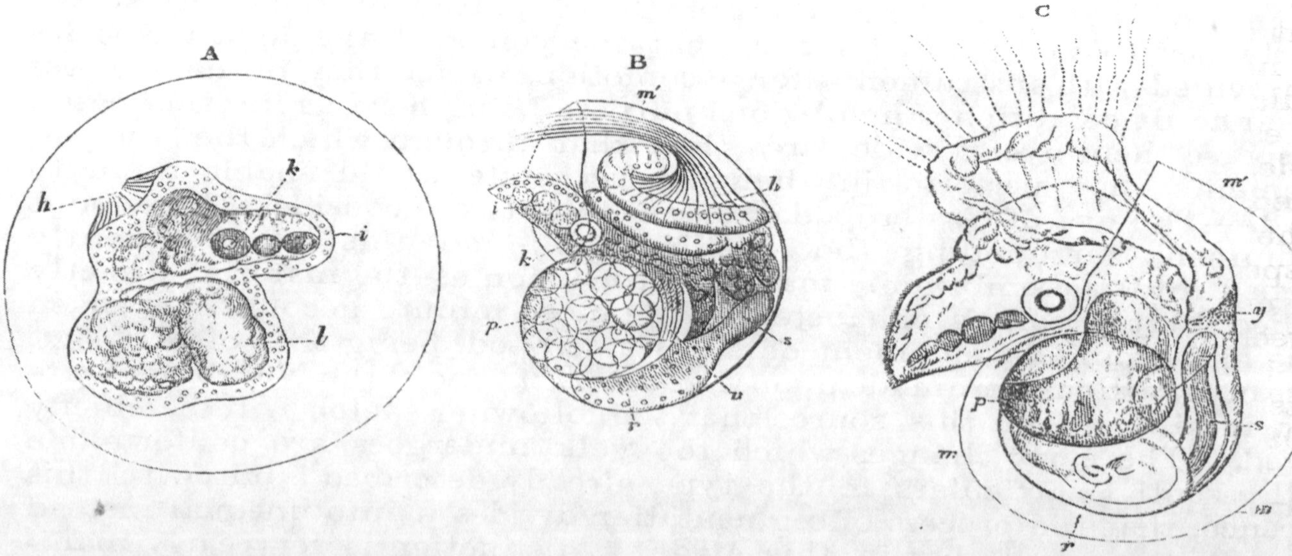

Development of embryo of *Acteon viridis*, as seen in the lateral aspect:—A, at the end of the fourth day; B, seventh day; C, twenty-fifth day; *h*, rotatory organs; *i*, foot; *k*, auditory vesicle; *l*, ventral portion of the embryo; *m m′*, shell; *p*, liver; *r*, stomach; *s*, intestine; *u*, suspensor muscle; *y*, black pigment of the hood of the mantle.

is rapidly taking place; for at its termination, the stomach, *r*, and intestine, *s*, are clearly distinguishable, as is also the suspensor muscle, *u*; but the liver, which is so important and characteristic an organ in this group, exists only as a mass of untransformed cells (*p*). The movements of the embryo now change their character; it projects itself from the shell, expands its ciliated lobes, and sets the cilia in vibration, and after a while draws itself into its shell again, very much after the manner of a Rotifer or a Bryozoon; and when it is completely retracted, the foot closes the orifice of the shell, like an operculum. It is curious to trace the regular rhythmical movement of *rotation* gradually giving place to these less constant and apparently more spontaneous actions. At the period when the embryo is ready for emersion, its movements have become as active as the narrowness of its prison permits; and when it is set free by the rupture of its envelope, it swims forth by the action of its ciliated lobes, and these also serve to bring food into its mouth, which does not possess at this period any trace of the reducing apparatus subsequently to be developed in it. Its condition at that time is seen at C; the principal change from the state shown in the preceding figure, being in the condition of the liver, which is now in progress of transformation into the perfect type. The 'mantle,' also, is now very distinct from the subjacent parts.

708. The subsequent history of the process has not been fully traced

out; but it is doubtless in this earlier portion that its chief peculiarity consists. The transformation of the entire yolk into the substance of the embryo, and the origination of all the organs of the latter in the cells that are formed by the subdivision of the former, is doubtless a very important feature in the process. The order in which the organs are evolved,—the ciliated lobes and foot, the otolithes and auditory vesicles, the shell, the mantle and operculum, the liver and intestine,—is also extremely remarkable. But the most curious fact of all, and the one which is most significant of the predominance of the organs of *vegetative* over those of *animal* life, in this group, is the *entire absence* of all trace either of nervous or circulating systems, at a time when the general structure of the embryo, and especially the visceral apparatus, has made such great progress, as to enable it to lead an active life, and to digest and assimilate its own food immediately on its emersion from the egg. Further it may be observed in the course of this development, that the progress from the general to the special is on the whole extremely well marked; the difference being first manifested between the containing and the contained parts, then between the cephalic and visceral, and then between the several groups of cells into which the component mass of the latter breaks up, each taking on its own distinct method of evolution.

709. A very curious phenomenon has been noticed by several observers, which seems like an imperfect gemmiparous production in the early embryo. It is not unfrequently seen that some of the cells of the vitelline mass detach themselves from the principal cluster, become clothed with long cilia, and continue to move about actively within the egg, until the escape of the embryo. It is even affirmed by Nordnann, that they increase by partial subdivision, and that thus from a single detached cell may be produced a cluster having a very definite form, and furnished with long cilia, so as very strongly to resemble a parasitic animal. It has not been shown, however, that these bodies ever advance to a higher condition, or are capable of generating their kind; and the correct view is probably to regard them (with Vogt) simply as portions of the embryonic mass, exactly resembling those that form the ciliated lobes, which, being detached from the rest, preserve their vitality for an unusually long time,—such vitality, however, not being different in kind from that of an ordinary ciliated epithelium-cell, though greater in degree.—It is affirmed by Agassiz, however, that the vitelline mass sometimes divides itself spontaneously into two portions, and that each of these may become a perfect animal; which statement, if correct, lends confirmation to the hypothesis already suggested (§ 646) respecting the origin of double monsters.*

710. All the species of the class *Cephalopoda*, so far as is at present known, are diœcious, the male and female organs being disposed on separate individuals. There is, nevertheless, a remarkable similarity between them, both in their general aspect, and in certain peculiarities which they present. The testis of the male consists of a capacious membranous sac, which, when opened, is found to contain a mass of short branching cæca,

* On the Embryology of the Gasteropoda, see especially the admirable Memoir of M. Vogt, in the "Ann. des Sci. Nat.," Juillet, 1846, and the various preceding Memoirs there referred to; the Memoir of M. Nordnann on *Tergipes*, translated into the same Journal for March, 1846; and the Lectures of Prof. Agassiz on "Comparative Embryology," already referred to.

attached to a small portion of its inner surface; these cæca, however, have not any orifice for the discharge of their contents, which appear to escape into the general cavity by the rupture of their walls. From this cavity it is conveyed by a duct, which, after passing through other accessory glandular structures, enters a wide muscular sac, where a remarkable change is effected in the condition of the spermatozoa. A number of them are clustered together, and enclosed in peculiar investments, which are known under the name of *spermatophora* or *the moving filaments of Needham,* and which are obviously analogous to the nidamental investments of the ova in the female. These are from half to two-thirds of an inch in length; and each consists of an external, transparent, and cylindrical case, in which is contained the proper sperm-sac. A little more than the anterior third of this is spirally disposed, and to this part Needham applied the term *screw;* next follows a short portion, which he calls the *sucker;* then a still smaller and cup-shaped part; and lastly an oblong and spongy bag, in which are contained the minute spermatozoa. When moistened with water, these bodies commence a series of alternate contractions and relaxations, by which the filament within is moved forwards, and the screw with its compressed spire is thrust forcibly against the anterior part of the capsule. This capsule in a short time becomes ruptured; by degrees the sucker and cup advance; and, as soon as they have escaped from the end of the cylinder, the spongy tail is forcibly driven out, and generally with so much violence as to break it into several pieces, thus giving exit to its contained spermatozoa. These movements are certainly not caused by the exercise of any distinctly-animal powers residing in the spermatophora; they are partly dependent upon the peculiar properties possessed by the membranes and filaments in relation to water; and they will be exhibited long after the death of the Cephalopod, if the filaments be taken out of the sac and placed in that fluid. Their function is thus evidently to diffuse the spermatozoa through the surrounding medium, in such a manner that they may find their way into the midst of the large clusters of ova deposited by the female; these are probably fertilized after their extrusion from her body (as in Fishes and Batrachia), since, in most species at least, the intromittent organ does not seem long enough to convey the fecundating fluid within it. As to the degree of actual congress between the sexes, there are various accounts, probably relating to different species.—The ovarium of the female, like the testis of the male, consists of a large sac with thickened walls; and, if this be opened at a time when the ovules are in an advanced stage of development, it will be found to contain a cluster of little egg-shaped bodies, attached to a small part of the inner wall of the sac, by short pedicles which principally consist of blood-vessels. These bodies consist of a portion of the substance of the sac, with its lining membrane, raised up by the development of the ovisacs; and each of them contains an ovule, which, when ready for extrusion, escapes, by the gradual thinning and final rupture of its envelope, into the general cavity. The ruptured membrane remains in the form of a cup; and we thus witness in this class the first appearance of the *calyx,* which will be described as developed from the external surface of the ovarium in the oviparous Vertebrata. From the cavity of the ovarium proceeds the oviduct, which conveys the ova into a glandular body, where they receive a nidamental investment, the nature and form of which differs in the various species. It may be specially mentioned, however, that the shell of the *Argonaut* (Paper Nautilus) is very probably to be regarded in the light of a nidamental receptacle for the ova; as there is strong reason to believe it to be peculiar to the female of that animal, and the eggs are attached to its involuted portion by long filamentary stalks. The vitelline sac is not surrounded (as in Gasteropoda) by a layer of fluid albumen, but is immediately invested by the chorion or general envelope.

711. The history of Embryonic development in the Cephalopoda, which has been very carefully studied by Prof. Kölliker,* differs in many important particulars from that of the lower Mollusca; and these are, for the most part, characters of approximation to the Vertebrata. For, in the first place, the process of segmentation does not take place in the whole yolk, but only in the portion of it nearest to the first germ-cell; and from this arises a cluster of cells, which lies upon the surface of the yolk, and sends out an extension all around. This extension, the representative of the 'germinal membrane' of the Fowl's egg, constitutes the yolk-sac; whilst from the original cluster, or at least from the cells of its outer layer, are produced the several organs of the embryo. Thus, when the process of development commences, the first structures are seen rising off (so to speak) one end of the yolk-sac; but as they gradually draw to themselves and appropriate the substance of the yolk, the embryonic mass increases, and the yolk-sac diminishes; until at last, at the time of the emersion of the embryo from the egg, the contents of the yolk-sac being nearly exhausted, it presents itself as a mere appendage to the embryo.—The mode and order of appearance of the principal organs, are very different from what we might expect. The first part of the germ-mass that becomes distinct, is a slight central elevation that is afterwards to become the visceral portion, covered with its mantle; around and beneath this, on either side, is an elevation that subsequently forms half of the funnel, the two halves being at first widely separated from each other; and around this, again, is a bilobed expansion which forms the cephalic portion, each lobe bearing one of the eyes, and sending-off from its under side (that nearest the yolk) as many projections as the species is to possess arms. The position of the mouth is indicated at the junction of these two lobes. Gradually the visceral portion becomes more elevated from the surface of the yolk-sac, by the augmentation of its own substance; and in so doing, the two halves of the funnel approximate each other until they join. The mantle becomes very distinct from the included parts, and is extended posteriorly into a fin-like organ on either side. The cephalic lobes are still (like those of Gasteropoda) very large in proportion to the rest of the body; and the eyes which they bear are very early developed in these animals, as if for the purpose of guiding their active movements on their emersion from the egg. The arms increase in length, extending themselves over the yolk-sac, which they seem (as it were) to embrace; and thus the peduncle by which it is connected with the embryo comes to be nearly in their centre. The development of the various organs in the visceral mass takes place in an order much nearer to that of Vertebrata than to that of Gasteropoda; for the evolution of the circu-

* "Entwickelungsgeschichte der Cephalopoden;" Zurich, 1844.

lating and permanent respiratory apparatus takes place *pari passu* with that of the digestive; and that of the nervous system is not far behind. At the time of the escape of the common *Sepia* from the egg, the first layers of its shell are found between the folds of the mantle in the dorsal region; the ink-bag is charged with its characteristic secretion; nearly all the organs characteristic of the adult are distinguishable, though not as yet in their relative proportions; and the young animal is capable of swimming actively through the water, both by means of its lateral fins, and by its cephalic arms, which are furnished with a connecting web that remains permanent in some species. The yolk-bag is not yet completely emptied, and it is found in the midst of the circle of arms, communicating with the stomach by a duct that passes down parallel to the œsophagus; it is gradually emptied and drawn in, however, and its original connection is only indicated by a cæcal protrusion from the anterior part of the 'crop.'—This is the first example we have yet encountered, in which the embryonic development is so far completed within the egg, that the young animal comes forth in the general condition of its parent; and it seems obviously connected with the fact, that the yolk is here formed in much larger proportional amount; so that it serves, not merely to supply the materials for the production of the first embryonic mass, but also for its continued growth, and for its evolution into the several organs characteristic of the adult form.

712. We now return to a much simpler order of phenomena, that is presented to us in the inferior part of the *Articulated* series; in which we find an almost complete reversion to the Zoophytic type, as regards the repetition of similar parts in the segments of the body, the formation of these by successive gemmation, and their power of independent existence. This is especially seen in the lower *Entozoa,* the simplest of whose forms carry us back to the type of the Protozoa, in which multiplication takes place by mere cell-division, every product of such division repeating the original form, and being able to live detached from the rest (§ 309 *a*); and here, too, we observe a deficiency of the true generative apparatus, which circumstance, taken in connection with the peculiarity of the products of their internal gemmation, indicates that the phase of complete evolution of these curious beings is not yet known, but that they are the early forms of some higher type of Entozoon, probably of the Trematode order.—The propagation of the *Cystic* Entozoa (§ 309 *b*), so far as at present understood, seems entirely gemmiparous; and here, too, it seems probable that true generative organs are evolved in another phase of their development, in which, in fact, they would belong to another order.—In the *Cestoid* Entozoa, we find the generative apparatus most extraordinarily developed; the so-called 'body' being little else than a longitudinal repetition of generative buds, each of which (like an ordinary flower) contains both sets of sexual organs, and is self-fertilizing (§ 309 *c*); and these may be detached from the anterior portion, or 'head,' and may retain their vitality for some time, giving exit to the eggs, when they have arrived at maturity.—In the *Trematoda,* also, the generative apparatus attains a high development, and occupies a large part of the body, as shown in Fig. 283. Both kinds of sexual organs are present in each individual; but they are not self-fertilizing, and the congress of two is consequently necessary, each impregnating the ova of the other, as in the pulmoniferous Gasteropods.

The male organs consist of a set of extremely long and convoluted seminiferous tubules, occupying the central part of the body; and these discharge their secretion by two trunks into a common canal, which terminates in the penis or intromittent organ, situated just behind the anterior sucker. The ovaries, situated along the margin of the body, consist of the ramifications of the two oviducts, within which the ova are produced; and these oviducts terminate in a single large canal doubled-up into numerous convolutions, which, being appropriated to the retention of the ova for a time, whilst they undergo a further maturation, has been designated as a 'uterus.' The outlet of the female generative apparatus, or 'vulva,' is in the immediate neighbourhood of the male genital organ.—In the *Nematoid* Entozoa, the generative apparatus is arranged upon the diœcious type, the male and female organs being peculiar to distinct individuals; and the ova are fertilized by sexual congress. A sufficient account of the mode in which the function is performed in this group, has already been given in the general survey of it (§ 309 *e*).

FIG. 283.

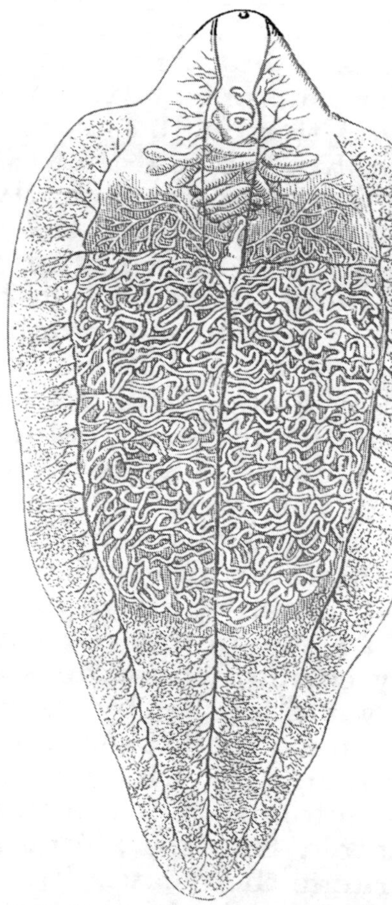

Generative apparatus of *Fasciola hepatica* (Fluke): – in the central portion are seen the testes, composed of convoluted tubuli; in the lateral and posterior parts, the ovaries; in the median and anterior portion, the uterus containing ova arrived at maturity; a pair of vascular trunks is seen, passing over the uterus, and ramifying on each side of the orifices of the organs of generation.

713. Of the history of Embryonic Development in the Entozoa, much still remains to be known. It appears that in the *Nematoidea*, which constitute the highest order of the class, the entire mass of cells formed by the repeated segmentation of the yolk, goes to be transformed into the embryonic structure, as shown in Fig. 262, F, G, H; and it is probable, therefore, that the same takes place in the lower orders. —The embryo of the *Cestoidea* is developed, not (like that of the Nematoidea) into the likeness of the adult worm, but merely into that of the portion termed its 'head,' from which all the generative segments subsequently spring by gemmation; and it appears to be to the non-development of this portion of the organism, occasioned by the unfavourable circumstances under which it is existing, that the *Cystic* modification is due, in which the reproduction takes place by gemmation only, just as a plant that is prevented from flowering will throw out additional leaf-buds.—The *Trematoda* present the very curious phenomenon of larval gemmation, that reminds us of that of the *Strobila* of Medusæ (§ 693); for each single ovum is developed, by the subdivision of its germinal mass, not into one, but into numerous individuals, each of which becomes complete in itself; and it appears that the same process may be repeated even in a later stage, the complete Distoma being the product of a second gemmation (§ 309 *d*). This, too, has been described as a case of 'alternation of generations;' but it is obvious that it has no title to be so regarded, if each generation be

held to consist of a complete organism; since the bodies into which the ova are first developed, have scarcely a higher animality than the ciliated 'gemmules' of Zoophytes, and are not more unlike the adult, than is the germinal membrane proceeding from the cicatricula of the Fowl's egg (§ 728), to the chick into which it will develope itself.

714. In the reproduction of the *Rotifera*, many points remain to be elucidated. It is not certain whether or not gemmation takes place in this class; but, as already pointed out (§ 310 *b*), there seems a probability of the occurrence of this mode of increase in its lower forms. It is usually believed that in the great bulk of this group, the generative apparatus is hermaphrodite, and that each individual is capable of self-fertilization; but this view has been recently brought into question; and it is certain that in one genus, at least, the male and female are distinct, and that an actual congress is necessary for the fertilization of the ova (§ 310 *b*). In this Animalcule, moreover, we have the remarkable phenomenon of a being possessed of a complete male generative apparatus, yet altogether destitute of organs of nutrition. Such a being is in itself not unlike the male generative apparatus which is budded-off by many Acalephæ (§ 695); but it differs in its origin, being developed at once from the egg, and not thrown off as a bud.—The ova of Rotifera are sometimes incubated within the body of the female parent, so that the young comes forth alive; are sometimes carried about by her, attached to the neighbourhood of the tail, and are sometimes freely deposited and left to themselves. The same species, indeed, which is at one time viviparous, bringing its embryoes to maturity within a few hours, may at another time deposit ova, enveloped in a thick horny casing, whose development shall not take place for a long period. This variation is by no means uncommon elsewhere, especially among Zoophytes; and it seems destined to preserve the race through a depression of temperature, which would be fatal to the adult animal.—The history of the development of the embryo is in all essential respects the same as in the Nematoid Entozoa; for the cells of the 'mulberry mass' may be observed to arrange themselves into subordinate groups, each of which evolves some one of the principal organs; and thus the whole animal is completed, or nearly so, at the time of its emersion from the egg, nothing like a metamorphosis being here observable. The whole process of development is completed, in some cases, within twenty-four hours; and thus, notwithstanding the small number of eggs which each individual produces at once, a most rapid multiplication takes place under favourable circumstances (§ 310 *b*).

714 *a*. In the class of *Annelida*, we still find that gemmation performs a very important part in the act of Reproduction: the multiplication of similar segments, which is so remarkable in many members of this group, being almost entirely due to it; and a spontaneous division sometimes taking place, by which the parts thus produced are detached from one another, sometimes in such a condition that they must be regarded as perfect individuals, whilst in other cases they seem but little more elevated in the scale of animality, than are the detached oviferous segments of the Tænia. The complete reproduction by gemmation succeeded by spontaneous fission, may be seen to take place in *Nais;* a worm which, though aquatic in its habits, belongs to the order Terricolæ. After the number of segments in the body has been greatly multiplied by gemma-

tion, a separation of those of the posterior portion begins to take place; a constriction forms itself about the beginning of the posterior third of the body, in front of which the alimentary canal undergoes a dilatation, whilst on the segment behind it, a proboscis and eyes are developed, so as to form the head of the young animal which is to be budded-off; and in due time, by the narrowing of the constriction, a complete separation is effected, and the young animal thenceforth leads an independent life. Not unfrequently, however, before its detachment, a new set of segments is developed in front of it, which is in like manner provided with a head, and separated from the main body by a partial constriction; and the same process may be repeated a second and even a third time; so that we may have in this animal the extraordinary phenomenon of four worms that are afterwards to exist as separate individuals, united end to end, receiving nourishment by one mouth, and possessing but one anal orifice. — A similar phenomenon has been observed in several genera of the Dorsibranchiate order; but the gemmæ thus detached are not complete individuals, for each consists of little else than a generative apparatus, with the addition of locomotive organs; thus bearing a similar relation to the parent-stock, which does not form generative organs of its own, with that which is borne by the Medusa-buds to the Polype-stock. In the one case, as in the other, it must be improper to reckon the generative segments as a new generation, since they are merely the 'complement' of the organism that would be incomplete without them. As many as six of these generative offsets have been seen in continuity with each other, and with the parent stock, by Prof. Milne-Edwards; the most posterior being evidently the oldest, and the one in direct connection with the parent consisting as yet but of a few segments, and being obviously the youngest. A similar detachment of the generative segments has been observed among the *Tubicolæ*. — There are several Annelida which may be multiplied by artificial subdivision, each part being able to grow up into the likeness of the perfect animal; though they do not spontaneously reproduce themselves in this mode.

715. Of the *Dorsibranchiate* Annelida, some appear to be hermaphrodite and self-impregnating, but in many others the sexes are distinct; and the generative organs, whether of one or both kinds, are, as in the Cestoid Entozoa, repeated in every segment of the body. In the *Tubicolæ* it appears that the sexes are nearly always distinct; and as their peculiar mode of life does not allow the congress of two individuals, it is probable that the fertilization of the ova is effected by the diffusion of the seminal fluid through the surrounding water. This is, of course, much more likely to be effectual, owing to the gregarious habits of these animals. Among the *Terricolæ*, whose organisation is altogether higher than that of the proper Annelida, the generative organs, although both sexes are frequently combined in the same individual, are not multiplied to the same extent; for they are usually restricted to a small number of segments, and each set opens externally by a single orifice; as is seen, for example, in the *Nais* (Fig. 284). They are not, however, self-impregnating; but a double congress takes place, as in the Planariæ, &c. — In the *Earth-worm*, towards the end of the summer, there is developed around the body a thick and broad belt; this is an apparatus for suction, by which the worms are held together during the congress. It is remarkable

that the ova do not escape through the ducts which serve to convey the spermatic fluid to the ovaria ; but the ovaria burst, when distended with mature ova, and allow their contents to be dispersed through the interior of the animal. In this respect, the process of reproduction in the Earth-worm bears a striking analogy to that which we have witnessed in Flowering-plants ; for in the latter, the fertilizing influence is transmitted down the minute canals of the style, and the seeds escape, when ripe, by the dehiscence of the walls of their envelope. The ova of the Earth-worm pass backwards between the integument and the intestine, to the anal extremity ; and in their progress, they gradually undergo their development, and are expelled from the parent, either as completely-formed worms, or surrounded by a dense and tough case, which gives them the character of pupæ. Whether they are produced in the perfect or in the pupal form, depends on the nature of the soil which the worms are inhabiting; in a light and loose soil, the young quit the parent prepared to act for themselves ; but in a tough clayey soil, they continue the pupal form for some time, so as to arrive at a still higher degree of development, before commencing to maintain an independent existence. A similar mode of expelling the ova has been observed in some species of Dorsibranchiata, some of which are provided with marsupial sacs, in which the ova are matured.—A very different one, however, is witnessed in the *Leech.* The ova are developed in a ring-like mass encircling the body; and this is thrown off altogether by violent efforts on the part of the animal, the body being withdrawn from within it. A sort of cocoon, open at both extremities, is thus produced, which contains from twelve to fourteen ova, enclosed in a protecting substance furnished by the mucous glands of the parent ; and the young, after their escape from the ova, quit the cocoon through the openings left by its body.

Fig. 284.

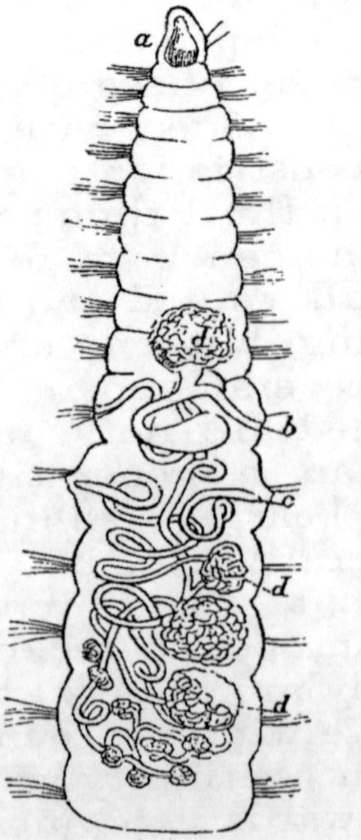

Generative organs of *Nais filiformis; a*, head; *b*, aperture of male organs; *c*, aperture of female organs ; *d*, *d*, ovaria.

716. In the history of the Development of the several orders of Annelida, there exists, as already pointed out (§ 311), a very marked diversity; for whilst the young of the Terricolæ and Suctoria do not usually come forth from the egg, until they have acquired the characteristic form of the parent (although the number of segments may be subsequently augmented), the embryoes of the Dorsibranchiata and Tubicolæ come forth in a state of far less advancement, and only acquire their perfect form by such a series of changes, as deserves the designation of a metamorphosis. So far as has been yet observed, there is a very close conformity in the earliest states of all these embryoes. They come forth from the egg in a condition very little more advanced than the ciliated gemmules of the polypes; consisting of a nearly globular mass of untransformed cells, certain parts of the surface of which are covered with cilia, symmetrically disposed. In the course of a few hours, however, this embryonic mass elongates, and indications of a segmental division become apparent ; so that by the end of the first day after its emersion, there may be distin-

guished in the embryo of *Terebella* (Fig. 285, A) the cephalic segment, *a*, the first segment of the body, *b*, which is thickly covered with cilia, a second narrower and non-ciliated segment, *c*, and the caudal ciliated segment ×. A little later (B), a new segment, *d*, is seen, interposed between the penultimate and the caudal segments; and the dark internal granular mass is observed to have extended itself to the extremity of the body, forming the first outline of the intestinal tube. The number of segments continues to increase by the process of gemmation, each new one being budded-off from the caudal extremity of the penultimate segment, so as to be interposed betwixt it and the original caudal segment, which is thus progressively removed from the cephalic, with which it was at first in close proximity. Thus in the larva in the more advanced stage represented at C, we find that the segments *e*, *f*, *g*, *h*, *i*, *j*, have been interposed between the caudal segment × and the segment *d* of the previous growth (B); at the same time, the cephalic segment has become more developed, and the eye-spots are now very distinct; and each segment of the body is furnished with a pair of setigerous appendages, closely resembling those of the adult. The evolution of the internal organs, also, has made considerable progress; so that we clearly distinguish the pharynx, *p*, the œsophagus, *q*, the stomach, *r*, (the walls of which are still tinged with the yolk-substance), and the intestine, *s*, as in process of formation. During this early period of their development, the embryoes remain in the midst of the gelatinous substance in which the ova were at first imbedded, and which appears to serve as the common 'albumen' for the whole collection; and from this substance they would seem to derive nutriment by imbibition, since they increase considerably in size before they are capable of receiving food through the mouth. It is only when the digestive apparatus and locomotive organs have attained a grade of development which enables the larvæ to obtain food for themselves, that they emerge from their gelatinous bed; and whatever may be their ultimate destination, they lead for a time a life of activity. When the communication is first established between the pharyngeal cavity and the external surface, so as to form the mouth, the interior of the passage, as well as the commencement of the intestinal canal, are lined with cilia; and it is by their agency that the young animal obtains its food, until other organs are developed. At this period, neither circulating apparatus nor distinct blood can be detected; nor can the presence of a nervous system be affirmed, although the existence of ocelli, and the activity of the movements of the larva, seem to justify the presumption that it is in process

FIG. 285.

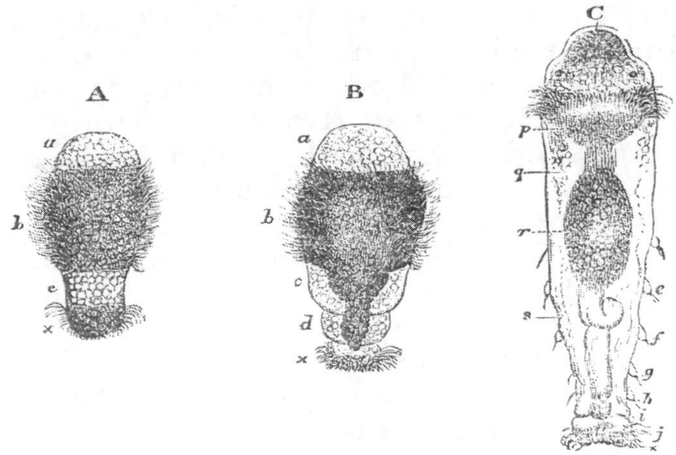

Early stages of development of *Terebella nebulosa*; A, larva one day after its emersion from the egg; B, larva more advanced, but still apodal; C, further progress of the same, with first appearance of appendages; *a*, cephalic segment; *b*, ciliated segment; *c*, third ring; *d*, *e*, *f*, *g*, *h*, *i*, *j*, successively formed segments; × anal segment; *p*, pharynx: *q*, œsophagus; *r*, stomach; *s*, intestine.

of formation.—The further development of the embryonic into the perfect form need not be traced in detail. It consists partly in the successive multiplication of segments, each new one being budded-off from the caudal extremity of the one that was formed last before; and partly in the successive development of new organs, especially those constituting the circulatory, respiratory, nervo-muscular, and generative apparatus. In the *Tubicolæ,* the tubular envelope is usually formed after the larvæ have passed a few days in the condition of *Errantia;* and from that time the development of their locomotive apparatus takes a retrograde rather than an advancing direction.*

717. In the class *Myriapoda,* there is no known instance of multiplication by fission, either natural or artificial; and the act of reproduction is solely accomplished, therefore, through the generative apparatus. The sexual organs are always diœcious, and the fertilization of the ova is accomplished by actual congress. As already mentioned (§ 312 *a*), the bodies of Myriapods present a considerable repetition of the generative as well as of other organs; but still there is by no means the same degree of segmental independence amongst them, as may be seen in the typical Annelida.—The history of the development of the embryo within the egg, seems to correspond in its main features with that of Insects, which will be presently described; whilst in the later changes which are seen after its emersion, we are reminded of the Annelida; for the length of the body is greatly augmented by the successive addition of new segments, and these are formed by gemmation from the penultimate segment. It would seem as if the 'germinal capacity' were still expended (as in the lower classes) in the act of *growth* rather than of *development;* and as if the continued production of similar parts of a lower grade, were incompatible with the evolution of the parts previously formed into a higher type. We see, however, in the Myriapoda, that the advance of development is occasionally marked by the *aggregation* of segments that were originally distinct. This is especially the case in the *Scolopendridæ,* in which the multiplication of segments never takes place to anything like the same extent that it does in the *Iulidæ,* and in whose organisation there is obviously a greater approach to the concentration manifested in the higher Articulata. Thus the head, according to Mr. Newport, is composed of *eight* segments, which are often consolidated into one piece like the head of Insects; and each moveable division of the body is in reality composed of two distinct segments, originally separate, but anchylosed together at an early period of their formation; their original distinctness being usually marked by the persistence of the separate ganglia and pairs of legs. The ganglia however sometimes coalesce, especially in the anterior part of the body of the *Polydesmidæ.*†

718. It is very remarkable that, in animals so highly organised as *Insects,* we should find a very marked example of gemmiparous multiplication, occurring as part of the regular history of the race. Such, however, is only the case in the genus *Aphis;* the curious history of whose double method of reproduction has been already detailed (§ 315 *d*). Now when

* See the elaborate Memoir of Prof. Milne-Edwards 'Sur le Développement des Annélides,' in the "Ann. des Sci. Nat.," Mars, 1845.

† See Mr. Newport's valuable Papers on the Myriapoda in the "Linnæan Transactions," vol. xix., and in the "Philosophical Transactions" for 1841, 1843, and 1844.

we compare this series of phenomena with those of a similar kind which have already occupied our attention, it becomes obvious that this is a case of *larval gemmation*, and that, notwithstanding the long succession of broods of 'zooids' which may be thus produced, the 'generation' cannot be said to be completed, until a perfect pair of male and female Aphides shall have been evolved, capable of continuing their race by the process of true sexual generation. This evolution may be postponed for a much longer period than usual, by preventing the animals from being subjected to that depression of temperature, which, at the end of the warm season, ordinarily seems to check the gemmiparous multiplication, and to call into exercise the true sexual operation.* Thus, notwithstanding the great increase in the number of independent beings, not one of the viviparous larvæ becomes a complete organism; for so long as this method of multiplication is continuing, the type of the perfect insect is never evolved, and no true sexual characters make their appearance.†

719. All *perfect* Insects possess true Generative organs; and it is by them alone, that their reproduction is accomplished. Throughout the whole of this immense group, the sexual organs are constructed upon a plan essentially the same. They are invariably diœcious; and the outlet, both of the male and female organs, is situated at the posterior extremity of the body. The testes are always formed of cæcal tubuli, resembling those of glands in general; but there is such a variety in the mode in which these are disposed, that as many as twenty-four different types of these organs have been enumerated. The spermatic duct is often furnished with a vesicular dilatation, which serves as a reservoir for that fluid; and it terminates in a penis or intromittent organ, which is usually enclosed in a pair of valves prolonged from the last segment of the abdomen. The penis, or some neighbouring part, is frequently furnished with recurved hooks, that take a firm hold of the female during the act of coition. The ovaria of the female are formed upon very much the same plan with the testes; and present a similar variety in the arrangements of their parts. The number and the degree of prolongation of their cæca, are usually in relation with the fertility of the species; and this is greatest in those Social Insects, in which a large proportion of the individuals never attain their full sexual perfection, the number of fertile females being very small. The development of the several parts of the ovum may be advantageously studied in this class; since the same ovarian cæcum very commonly contains ova in different stages of development. Here, as elsewhere (§ 683), the ovisac appears to be the part first formed, as the parent-cell of the ovum; this at first closely embraces the germinal vesicle, but soon a granular matter, the vitellus, is interposed between them; and when this has accumulated, a proper envelope, the vitelline membrane, is developed around it. The females of many Insects are provided with a *spermotheca*, or receptacle for the seminal fluid of the male; into this the fluid is received during coition; and there it is retained, without the loss of the vitality of the spermatozoa, for many

* See "Burmeister's Entomology" (translated by Shuckard), p. 310.

† The viviparous larvæ are usually spoken of as 'females;' but they have no more real title to this designation, than is possessed by a Hydra which is budding-off new polypes from its body. For a fuller and more controversial exposition of the Author's views on this subject, see the "Brit. and For. Med.-Chir. Rev." vol. iv. p. 443.

weeks, so that each ovum is fertilized as it approaches the outlet. This outlet is frequently prolonged into an ovipositor; an instrument produced from the last segment of the abdomen (and apparently homologous with the sheath of the penis in the male), by which the ova may be conveyed into situations peculiarly appropriate for their reception. Special instruments are frequently developed in connection with the ovipositor, by which the parent is enabled to bore, or even to saw-out, a fitting receptacle; and the various provisions which are made in these and other ways, for the protection of the eggs, and for the supply of food to the larvæ when hatched, are among the most curious manifestations of the *instincts* of this wonderful class.

720. The Development of the Insect-larva within the egg, takes place upon a plan not very dissimilar to that, which has been already described in the inferior Articulata. The entire yolk is divided by segmentation in the usual manner; but this subdivision takes place most minutely, and the formation of cells occurs most completely, at the peripheral portion; so that the first condition of the embryo within the egg, like that of the free-swimming embryo of Annelida, is a somewhat elongated body, composed of the vitelline mass included in a cellular envelope. This envelope, however, is first completed on that which is to become the *ventral* surface of the body; and on the dorsal aspect a broad open space is for some time left, through which the yolk may be clearly seen. This simple elongated body, however, soon exhibits a constriction, which marks out the cephalic portion; and other segmental divisions early begin to show themselves, thus indicating the Articulated character of the animal, almost before its internal organisation can be said to have commenced. The dorsal opening is gradually closed; and the cellular membrane formed around the yolk presents a considerable increase in thickness, owing to the formation of new layers of cells. At the same time, the cells upon the surface of the included vitellus are undergoing subdivision and metamorphosis; and these form the walls of the alimentary canal, which thus includes the residue of the yolk. Various collections of cells are shortly seen, which are the respective foundations of the different organs that are to be developed in the larva; but the parts that usually first present an approach to the characters they present after the emersion of the embryo from the ovum, are those which, being concerned in mastication, are most directly subservient to that function which the larvæ exercise with such extraordinary energy, immediately on their entrance into the world. The mouth and anus are formed, as in other cases, by the thinning-away of the envelope which at first included the entire cavity; and not unfrequently the second of these orifices has not made its appearance, at the time of the escape of the larva from the egg. Of the different grades of development which the larva may present on its emersion, and of the phenomena of the subsequent metamorphosis, as full an account has already been given, as the limits of this work permit (§§ 315 *e*—*g*).* It may be added, however, that the number of segments formed in the egg is never increased by subsequent gemmation.

721. Although in the higher *Crustacea*, as in Insects generally, the

* The most complete history yet published, of the Development of the ovum of Insects, is contained in Prof. Kölliker's "Observationes de primâ Insectorum genesi," Turici, 1842.

power of multiplication by gemmation seems entirely wanting, yet in the *Entomostracous* group we have several exemplifications of it, which seem to correspond in their essential features with the case of the Aphis (§ 317 *a*).—The conformation of the sexual organs, which constitute the only instruments of propagation in the higher members of the class, has been already described ; and a sketch has also been given of the present state of our knowledge of the metamorphoses undergone by Crustaceans, as well in the lower group as in the higher (§§ 316 *d*, 317 *a*, *c*). It may here be added, however, that the male often possesses a very different size and conformation from the female, not merely among the Entomostraca and Suctoria, but even in the higher orders of Crustacea ; so that distinct species and even genera have been made out of what were really the opposite sexes of one and the same.—The early stages of the development of the embryo within the ovum, have not yet been clearly made out. It would seem as if, in the higher Crustacea, as in the Arachnida, the germinal membrane springs from a spot on one side, instead of arising, as in Insects, out of the peripheral portion of the 'mulberry mass,' by the more minute subdivision of its cells. However, it gradually extends itself around the yolk, the dorsal region, as in Insects, being the last to close-in ; and whilst it is yet but a disc on one side of the yolk, it exhibits a distinct separation between the cephalic portion and the trunk, other indications of segmental division making themselves apparent as the development proceeds. Here, too, as in Insects, the parts about the mouth are the first to attain some degree of completeness.—In the Entomostraca, on the other hand, it would seem as if the whole mass of the subdivided yolk went at once to transform itself into the embryonic structures, as in the lower Articulata.* This part of the subject, however, has been as yet but very imperfectly studied. —In most of those cases in which the young come forth from the egg in a condition very different from that which they are ultimately to assume, an increase takes place in the number of segments ; and this increase is sometimes very considerable. It does not take place, however, as in the Annelida and Myriapoda, at the caudal extremity only ; for the last segment of each of the three great divisions of the body, the head, the thorax, and the abdomen, seems to possess the gemmiparous power ; and thus the number of segments in each of these regions becomes separately augmented.—The physiology of Reproduction in the curious group of *Cirrhipoda*, so far as it is yet understood, has been already explained (§ 318 *a*).

722. In the class of *Arachnida*, there does not exist, so far as is yet known, any example of gemmiparous reproduction ; but looking to the very low grade of development of some of the *Acaridæ*, and the embryonic condition in which some of their organs and tissues remain, it would not seem improbable that the same mode of multiplication may present itself among them, as we have seen to exist in the Aphides and in certain Entomostracous Crustacea. This expectation may seem to be justified by the large amount of regenerative power (another manifestation of the same 'germinal capacity),' which exists even in the

* See the Memoir of M. Van Beneden 'Sur le Developpement des Nicothoés,' in "Ann. des Sci. Nat.," Juin, 1850.

highest Arachnida; for this extends, as in the Crustacea, to the reproduction of entire limbs.—The sexes are universally separated in this class; and the ova are fertilized by a complete congress. The very curious mode in which the seminal fluid of the male Spider is introduced into the female passages, has been already described; and the provisions which exist in the Araneidæ for the protection of the ova, have also been noticed (§ 319 *e*). In the Scorpionidæ, the ova are retained within the body until they are hatched; so that the young are born alive.—The development of the ovum has been hitherto chiefly studied in the Spider; and the observations of Herold upon this point, valuable as they are, leave much to be elucidated in the early history of the process. The vitelline sac is invested in an outer tegument, or chorion; and between the two is a layer of albumen, as in the Fowl's egg. The vitellus, as in the ovum of Cephalopods, Fishes, and Birds, consists of two parts, the 'germ-yolk,' and the 'food-yolk;' and it is the former alone, which undergoes the process of cleavage. A *cicatricula*, or 'germ-spot,' is thus formed on one side of the principal mass of the yolk; and from this a cellulo-membranous expansion takes place, which gradually invests the whole vitellus; the part which is to become the dorsal region of the animal, being, as in other Articulata, the last to close-in. A thickening of this 'germinal membrane' takes place in the seat of the original germ-spot; and it is here that the foundation is laid for the development of the principal organs. In the first place, the part which is to become the cephalic segment is separated by a constriction from the principal mass; the four segments of the thorax are indicated by parallel fissures on either side; and the abdominal portion is marked-off by a deeper constriction. The rudiments of the principal appendages to the head soon begin to bud forth from the cephalic segment; and the characteristic group of simple eyes early shows itself. The rudiments of the thoracic members soon appear, budding-forth from their respective segments; and the simple alimentary canal is formed around the vitelline mass; at first wide, but gradually contracting, as the material of the yolk is appropriated to the formation of the tissues. The mouth and anus, as in other cases, do not make their appearance until a later period; but even up to this time, although the posterior part of the dorsal vessel is seen along the upper curvature of the abdomen, the integument has not completely closed-in over the dorsal portion of the thorax. This closure, however, gradually takes place; the cephalic and thoracic members are perfected; the development of the internal organs advances in a corresponding degree; and the young Spider comes forth from the egg in a form and condition which differ very little from that of its parent, whose size it gradually attains by simple *growth*.—Thus the whole *nisus* of development in the Spider appears to be directed, even from the earliest period, to the evolution of the complete organism; and no part can be said to be evolved for a mere temporary purpose, the entire germinal membrane remaining persistent as the integument of the Animal.—For the development of the *Acaridæ*, see § 319 *d*.

723. In no animal belonging to the *Vertebrated* series, do we ever witness the least approach, at least after its characteristic form has been attained, to the production of a distinct individual by gemmation; although, as already pointed out (§ 646), it scarcely seems improbable

that a partial separation may take place in the germinal mass at an early period of its development, which may give rise to the subsequent evolution of supernumerary parts, or even to the duplication of the entire structure. In the one case, the additional part derives its nutriment from the organism of which it is an outgrowth, but possesses a 'germinal capacity' of its own ; in the other case, the product of each portion of the germ may become completely independent, and may develope itself into the likeness of the parent. Such a method of accounting for 'monstrosities by excess' will be found (the Author believes) in strict accordance with sound physiological principles, and is at least as legitimate as the doctrine usually received, which regards them as consequent upon the partial fusion of germs originally distinct ; whilst it is much more readily applicable to cases of 'monstrosity by inclusion.' In these last, organs and tissues that seem like fragments of a fœtal body, are found imbedded in the midst of another organism ; and this not merely in the bodies of females, in which their presence might be attributed to the abnormal development of the normal generative products, but in those of males, in which no such origin could be assigned to them, and in which gemmation at an early period of embryonic life seems the only admissible alternative. —The only indication of the persistence of this 'germinal capacity' in the organism which has once attained the Vertebrated type, is that which is exhibited in the restoration or reparation of lost or injured parts ; and on this point there is nothing to add to what has been already stated (§ 646).

724. The type of the true Generative apparatus in the class of *Fishes*, presents little elevation above that which it possesses in the higher Invertebrata. For both the testes and ovaria of the greater number of Osseous Fishes seem to be constructed upon the same plan with those of the Cephalopoda ; each of these organs having a common cavity into which their products are received, and from which they are directly conveyed to the outlet by an efferent duct. In all the Cartilaginous Fishes, however, and in a few of the Osseous, the arrangement resembles that which prevail among the higher Vertebrata ; for the ovary is unprovided with an excretory duct, and the ova, when set free from it, fall into the cavity of the abdomen. In the Cyclostome fishes, they find their way out of this cavity by a simple orifice situated behind the anus ; an arrangement that reminds us of that which exists in the Earth-worm. In the Sharks and Rays, on the other hand, the margin of the orifice is prolonged inwards, so as to form a sort of funnel, into which the ova are received when they escape from the ovary ; an arrangement which obviously foreshadows the genital canal and the Fallopian tubes of Mammalia (§ 322 *k*).—In the Osseous Fishes generally, there is no sexual congress ; but the ova discharged by the female are fertilized by the seminal fluid diffused through the water of the neighbourhood by the male. In the case of the Stickleback and a few other fishes, however, which construct nests, it appears that when the ova have been deposited in these, the male discharges his seminal fluid upon them. Of the ova for whose fertilization no such special provision exists, a large proportion are likely to be unproductive ; and of the young actually hatched, the greater number soon perish for want of protection by their parents, falling a prey to the voracity of other fishes. Hence arises the necessity for the enormous fecundity of these

animals, and for the size of their generative organs. It has been calculated that above a million of eggs are produced at once by a single Cod; and in other species the amount may be still greater.—A very curious fact in the history of the generative system of Fishes, is that the male is frequently capable of the reproductive act, when far from his own full growth. Several mistakes have arisen from this source; the young with fully-developed milts having been mistaken for adults; and having been described as distinct species, on account of the difference of their size and markings, both of which subsequently undergo a change. Thus the fish commonly known as the Parr, is actually the young of the Salmon in the second year.—Among the Cartilaginous Fishes, the fertilization of the ova is usually effected whilst they are yet within the oviducts, by an actual congress of the sexes; and there are several of these, in which the embryo undergoes nearly its entire development within the body of the parent, so that the young are produced alive. In some species of Ray and Shark, the oviduct is found to have a sort of uterine enlargement near its termination, in which the ova are delayed for some time, and in which they receive a degree of additional assistance from the parent, which foreshadows the more complex provision for this object in Mammalia. There is no *direct* communication between the blood-vessels of the parent and those of the fœtus; but the membranes of the latter absorb nutriment from the extremely vascular lining of the oviduct of the former, to such an extent that, in the *Torpedo*, the weight of the mature fœtus is between two and three times that of the egg.* The ova of the higher Cartilaginous fishes, in which their fertilization is secured by sexual congress, and in which a provision of some kind exists for their protection during the period of embryonic development, are much fewer in number than those of the Osseous.—One of the most curious provisions for the early protection of the offspring in this class, is that which we find in the *Syngnathidæ* or Pipe-fishes. The *male*, in most of these, has a pouch on the under side of its body, formed by the meeting of two folds of skin; into this, the eggs deposited by the female are conveyed; and here they remain, until the young attain a considerable degree of development. Even when able to swim about by themselves, they seek the protection afforded by this curious contrivance; which closely resembles that well known to exist in the Marsupial Mammalia,—differing from it, nevertheless, in the absence of any special means of affording nutrition to the embryo. In some species of this group, however, the pouch is absent; but the ova are received into a set of hemispherical depressions on the under side of the abdomen, to which they attach themselves by their gelatinous envelope. The *Gobius niger* constructs a regular nest among sea-weeds, in which it deposits its ova; and these it watches with maternal care until they are hatched. A very elaborate nest is constructed by the male of the fresh-water *Stickleback;* and it is carefully watched by him during the period of embryonic development, in order to guard the eggs from the voracity of the females, who are continually endeavouring to make a prey of its contents.

725. In *Reptiles*, we find the generative apparatus exhibiting a manifest

* In these animals, the temporary branchial filaments are extremely long, and appear to serve as special absorbing organs. In Dr. Davy's opinion, their function is particularly subservient to the development of the electrical apparatus. See his "Anatomical and Physiological Researches," vol. i. p. 67.

advance, in degree of organisation, from the highest form in which it exists in Fishes (§ 324 *u*); and the provision for the fertilization of the ova is such as more effectually to secure it. In the *Batrachia*, as in most Fishes, the eggs are fertilized after they are extruded from the genital orifice of the female; but that of the male is in close contact with it at this period, so that the seminal fluid is most advantageously applied to them. In the higher Reptiles, however, there is an actual introduction of the male fluid into the oviducts of the female; this is accomplished in *Ophidia* by a temporary eversion of the orifices of the spermatic ducts; but in the higher *Sauria* and *Chelonia*, there is a regular intromittent organ, which is, however, sometimes completely bifid, and always grooved. The groove is completed, by the swelling of the organ at the time of excitement, into a canal, which receives the spermatic ducts. The ovaria are constructed upon the same general plan as in the higher Fishes; but are usually more compact in their form. The ova escaping from them are received into the funnel of the oviducts, and are conveyed by them into the *cloaca*, which receives both their orifices, as well as those of the ureters and rectum. Of the *Batrachia*, one species, the Land Salamander, is viviparous; and each oviduct is dilated, near its termination, into a sort of uterine cavity, in which the young are developed. Among the higher Reptiles of this country, moreover, the Viper, Slow-worm, and common Lizard, frequently produce their young alive; the egg being retained until the embryo has attained its full development, and the enveloping membrane being usually burst (as Mr. Bell* considers) in the act of parturition. With these animals, as indeed with all Reptiles, a high degree of heat is necessary for the maturation of the eggs; a portion of this heat seems to be developed within the body of the parent herself, by some peculiar excitement of the calorifying power which these animals possess (§ 620); but the gravid female may often be seen basking in the sun, acquiring from its rays the caloric which she is not herself capable of generating. In the oviparous Reptiles, it is not common to find the parent affording any protection to the eggs when once deposited, or to the young produced from them. The *Pipa Americana* (Surinam Toad), however, is an exception; in this species, the female has on her back a number of cells hollowed in the integument; in these the eggs are placed by the male, when she has deposited them; and here they remain for about 80 days, during which time the embryoes undergo their metamorphosis, so as to come forth as perfect Frogs. It has been satisfactorily determined that the gelatinous mass in which the ova of Frogs are imbedded, serves for the first nutriment of the young Tadpoles on their emergence from the egg; and the same is probably the case in many other instances.

726. In *Birds*, the generative apparatus does not manifest any great advance beyond the form it presents in Reptiles (§ 325 *l*); but there is always a provision for the fertilization of the ova, by sexual congress, soon after they have been discharged from the ovaria. The ovarium, which is usually developed on one side only, consists of a thin and condensed layer of cellular parenchyma, constituting a *stroma*, in which the ovules are evolved; its size is much smaller in proportion to the whole body, except at the period of the development of the ova, than in most of the

* "History of British Reptiles," pp. 35, 63.

classes we have yet considered. It is covered by an envelope of its own, and then by the peritoneum; its surface is usually smooth; but, when the ova are being developed, their size causes them to project more or less, so as to form a number of convexities upon it (Fig. 188). As the ovisacs enlarge, they gradually project from the ovarium, carrying before them its envelopes; and at last, the ovarium presents almost the appearance of a bunch of grapes,—the ovisacs hanging from it only by a short peduncle, which contains vessels. Each of these projecting bodies, therefore, contains an ovum, which is enveloped in its ovisac; around this is a thin layer of the stroma; this, again, is enclosed in the proper envelope of the ovarium, consisting of two layers, of which the inner one is vascular; and the peritoneum envelopes the whole. At last the ovum escapes by the rupture of its coverings; and these remain as a sort of cup, which is termed the *calyx*, and which subsequently disappears by absorption.—The oviduct commences by a wide slit, which receives the ovum at its escape from the ovary; at its lower part it dilates into a thick glandular sac, which secretes the shell. The ovum, during its passage along the oviduct, receives, as in most of the higher animals, an additional layer of albumen; and this is surrounded by a membrane, which has been commonly regarded as composed simply of the hardened external layer of the albumen, but which has in reality a regular fibrous texture (§ 160); and is properly analogous to the *chorion* of Mammalia (§§ 325 *l*, 727). This membrane is composed of two layers, which separate at the blunt end of the egg, to include a bubble of air; its outer surface is somewhat shaggy, and sends little processes into the shell, which is formed by the calcification of a similar membrane. The outer part of the albumen is extremely watery; but the inner part is viscid, and clings closely to the yolk. The yolk-bag is held in its place within the egg, by two twisted cords, termed *chalazæ*, which appear composed of coagulated albumen; these arise, by a funnel-shaped expansion, from the sides of the yolk-bag opposite the poles of the egg, towards which they pass. In this manner, the yolk-bag is prevented from quitting the central part of the egg; but it is permitted to rise towards the side; and, as it is rather lighter than the albumen, it will always approach the shell, whichever side the egg may be resting on. Further, it is permitted to turn on its own axis: and in this manner, the *cicatricula* or germ-spot (§ 728), being specifically lighter than the rest, will always be turned uppermost. By this very simple contrivance, two necessary conditions are provided for; the embryo is placed, during its development, in the most advantageous position for receiving the influence of the maternal heat; and is also brought into the nearest possible relation with the external air, by which the aeration of its fluids is effected. The former condition is indispensable throughout the class, except in a few instances, in which the solar heat is sufficiently intense, or in which artificial heat is designedly substituted (§ 625). The latter, here as elsewhere, is absolutely necessary to the development of the embryo: and the shell, being porous, does not interpose any obstacle to the aeration of the fluids it contains.* Of the various provisions so remarkable in this class, as in

* An attempt was made some time since, to show that no respiratory process takes place in the egg through the medium of the shell-membrane; and that the development of the embryo takes place with equal perfection, when the air is completely excluded from the interior. The result, however, was completely fallacious, in consequence of the imperfect

that of Insects, for the protection and maintenance of the young, some notice has already been taken (§ 325 *l*). In general it is to be remarked, that the attention which the young receive after they break the shell, is prolonged in proportion as the plumage, and especially the feathers of flight, are to be of a more perfect character; so that, in this comparatively trifling variation, we have an illustration of the general law,—that the higher the grade of development which the being is ultimately to attain, the more is it assisted in the early stages by its parent.

727. This law is remarkably exemplified in the class *Mammalia*, which unquestionably ranks at the head of the Animal kingdom, in respect to degree of intelligence and general elevation of structure. It is the universal and most prominent characteristic of this class, that the young are retained within the body of the parent, until they have made considerable progress in their development; and that they are afterwards nourished by a secretion from the blood of the female parent. In regard to the degree of development attained by the young, however, at the period of parturition, there is much variation in the different orders; and the lower tribes of the class are so truly intermediate, as to the mode in which this part of the generative function is performed, between the oviparous Birds and Reptiles, and the true viviparous Mammalia, as to be appropriately separated as a distinct sub-class, that of *Ovo-viviparous*, or more properly *Implacental*, Mammalia (§ 326). And, as already pointed out, this inferior grade manifests itself also in the general development of these animals; especially in their osteology, which presents many oviparous characters, and in the low type of conformation of their brain. At the other extremity in the scale of development, we find Man; in whom the dependence of the offspring on the parent is extended through a much longer period than in any other animal; and this has an obvious relation to the high amount of intelligence which he is ultimately to attain.—The transition from the generative apparatus of Birds, to that of the lowest Mammalia, is by no means abrupt; and the passage from the latter to the highest forms which the class presents, is very gradual (§ 326 *z*).—The ovum itself corresponds, in all essential characters, with that of Oviparous animals;—chiefly differing in its minute size, in proportion to that of the ovisac (§ 683). Its external envelope is thicker than the ordinary yolk-bag of oviparous animals; and has received the name of *Zona pellucida*, from its appearing under the microscope (when the ovum is flattened by compression) as a broad transparent ring* (Fig. 261 *m v*). In the immature ovisac, the space between its inner layer and the ovum is for the most part filled up with cells; these, however, gradually dissolve away, especially on the side nearest the surface of the ovary, whilst an albuminous fluid is effused from the deeper part of the ovisac, which pushes before it the residual layer that immediately surrounds the ovum, forming the *discus proligerus*, and thus carries it against the opposite wall. At this period, the vascular tunic enveloping the ovisac, which is

nature of the means of exclusion employed; and the very accurate experiments of Schwann leave no doubt, that the access of oxygen is as necessary to the embryo, in all but the very earliest period of development, as it is to the adult. See "Brit. and For. Med. Rev." vol. x. p. 229.

* This *Zona pellucida* has been described by some writers under the name of *chorion*; but the true chorion is afterwards formed around it.

derived (as in Birds) from the parenchyma of the ovary, is acquiring great thickness and consistency; and the two membranes together form the structure known as the *Graafian Follicle*, which was formerly regarded as peculiar to the Mammalia, but which is really analogous to the inner part of the *calyx* of Birds and other Ovipara. When the ovum has attained its full development, it is carried to the side of the Graafian follicle nearest the surface of the ovary, by the effusion of fluid on the side more deeply imbedded; and there it remains, until the rupture of the envelopes allows of its exit from the cavity. This rupture is preceded by a gradual thinning of these membranes at that point; but on the opposite side, the outer or vascular tunic becomes much thickened; and a fleshy substance of a reddish-yellow colour (entirely composed of cells) sprouts from the interior of the ovisac, which it gradually fills after the discharge of the ovum, forming the *corpus luteum*.—When set free from the ovarium, the ovum is received into the Fallopian tube; and its fertilization appears to take place almost immediately on its entrance into that canal, if it have not been previously accomplished. During its passage towards the uterus, it acquires an additional envelope, the true *chorion*, which has subsequently to perform very important functions in the nutrition of the embryo. This is at first seen as a layer of cells in contact with the outer surface of the Zona pellucida; it usually, however, imbibes fluid, which separates it from the Zona pellucida; and a change takes place in its own structure, by the alteration in the form of its cells, which extend themselves and interlace in various directions, so as to give the fabric a fibrous texture, and to produce asperities on its outer surface. In its situation and mode of production, the chorion is evidently homologous with the *membrana testæ* of Birds; and the fluid which it absorbs is comparable to the albumen of the egg of the latter. The connection between the two is clearly established by the ovum of the *Monotremata*, in which there is a distinct albumen around the yolk-bag, contained in a membrane, which,—as these ova are not to be incubated, but are destined to receive their early development within the uterus,—must subsequently become the chorion. In all Mammiferous animals, the oviduct, instead of immediately conveying the ovum out of the body, deposits it in the receptacle provided for its further development; within which it forms a new connection with the parent, and is supplied with nutriment from the fluids of the latter, until it has arrived at a state of completeness usually corresponding with that which the Chick presents, when it emerges from its shelly covering.

728. It now remains for us to consider the peculiarities which distinguish the development of the embryo of the *Vertebrata*, from the process already described in the other Sub-Kingdoms. These peculiarities chiefly consist in this;—that the permanent structure is developed from a very limited portion only of the 'mulberry mass' (§ 685); and that the remainder of this, with that extension which it forms over the large mass of 'food-yolk' that forms the bulk of the vitelline body in Birds and Fishes, and probably in the higher Reptiles, has but a very subordinate and temporary office,—being destined (like the cotyledonous expansion of the higher Plants) solely for the imbibition and assimilation of nutriment, during the early period of evolution. In the ova of Batrachia and Mammalia, there is no 'food-yolk' in addition to the

'germ-yolk;' and as the entire vitellus consists of the latter, the whole of it is involved in the process of segmentation. This arrangement, which is common to them with Invertebrata generally (save the highest of the Articulated and Molluscous series), appears to be related to the fact, that in neither of these two cases is the development of the embryo carried on far at the expense of the vitellus; for that of the Frog proceeds no further within the ovum, than to the production of a fish-like larva; and the continued development of the Mammal is very early provided for by a different arrangement.—The evolution of the fractions of the vitellus (Fig. 286, A) into true cells, by the formation of a cell-wall around them (§ 685), takes place soonest, as well as most completely, at its peripheral portion; and its cells arrange themselves, in the act of formation, into a kind of membrane lining the yolk-bag, at the same

FIG. 286.

A B

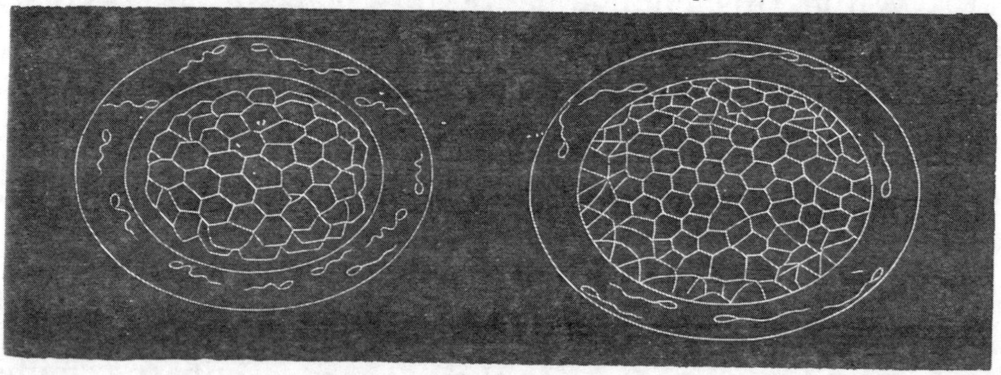

Latter stage in the segmentation of the yolk of the *Mammalian Ovum*; at A is shown the "mulberry mass" formed by the minute subdivision of the vitelline spheres; at B, a further increase has brought its surface into contact with the vitelline membrane, against which the spherules are flattened.

time assuming a pentagonal or hexagonal shape from mutual pressure, so as to resemble pavement-epithelium (B). Of the globular masses of the interior, those nearest the surface seem to be developed into true cells, and to increase the thickness of the membrane already formed by the more superficial layer; but the spherules of the interior appear for the most part to liquefy again; so that the embryonic mass is now in the condition of a cellular stratum, known as the 'germinal membrane,' inclosing a liquid yolk,—a condition which is closely paralleled by the embryo of Zoophytes, previously to the formation of the oral aperture.—In the ova of Fishes and Birds, however, and probably in those of true Reptiles, there is a large food-yolk, in which the segmentation does not take place; and the mode in which this becomes enclosed by the 'germinal membrane,' is by the extension of the peripheral portion of the *cicatricula*, which seems to be formed by the segmentation of the 'germ-yolk.'—The 'germinal membrane' is sometimes termed the *blastoderma*; and the new envelope which is formed by it, between the vitellus and its original sac, has been termed by Bischoff the *blastodermic vesicle*. This vesicle, very soon after its formation, presents at one point an opaque, roundish spot, which is produced by an accumulation of cells and nuclei of less transparency than elsewhere; within this, which is termed the *area germinativa*, all the structures of the permanent organism originate. The germinal membrane increases in extent and thickness, by the formation of new cells (whose mode of production has not been clearly made out); and it subdivides into two layers, which, although both at first composed of cells, soon present distinctive characters, and are concerned in very different ulterior operations. The outer one of these is commonly known as the *serous* layer; but being the one in whose substance the foundation is laid for the vertebral column and the nervous system, it is sometimes called the *animal* layer. The inner one is usually known as the *mucous* layer; and being the one chiefly concerned in the formation of the nutritive apparatus, it is sometimes called the *vegetative* layer. This division is at first most evident in the neighbourhood of the area germinativa; but it soon extends from this point, and implicates nearly the whole of the germinal membrane.

729. The 'area germinativa' (Fig. 287 *ag*), at its first appearance, has a rounded form; but it soon loses this, becoming first oval, and then pear-shaped. While this change is taking place in it, there gradually appears in its centre a clear space, termed the *area pellucida;* and this is bounded externally by a more opaque circle (whose opacity is due to the greater accumulation of cells and nuclei in that part than in the area pellucida), which subsequently becomes the *area vasculosa.* In the formation of these two spaces, both the serous and mucous layers of the germinal membrane seem to take their share; but the foundation of the embryonic structure, known as the *primitive trace,* is laid in the serous lamina only. This consists in a shallow groove, lying between two oval masses (*v*), known as the *laminæ dorsales.* The form of these changes with that of the area pellucida; at first they are oval, then pyriform, and at last become of a guitar shape. At the same time, they rise more and more from the surface of the area pellucida, so as to form two ridges of higher elevation, with a deeper groove between them; and the summits of these ridges tend to approach each other, and gradually unite, so as to convert the groove into a tube. At the same time, the anterior portion of the groove dilates into three recesses or vesicles, which indicate the position of the three principal divisions of the Encephalon, afterwards to be developed as the *prosencephalon,* the *mesencephalon,* and the *epencephalon* (Fig. 290 B, *x, y, z*). The most internal parts of these laminæ, bounding the bottom and sides of the groove, appear to furnish the rudiments of the nervous centres which this cranio-vertebral canal is to contain; whilst the outer parts are developed into the rudiments of the vertebral column and cranium. Even before the laminæ dorsales have closed over the primitive groove, a few square-shaped, at first indistinct, plates, which are the rudiments of vertebræ (Fig. 288, A, *q, q*), begin to appear at about the middle of each. The position of the bodies of the

FIG. 287.

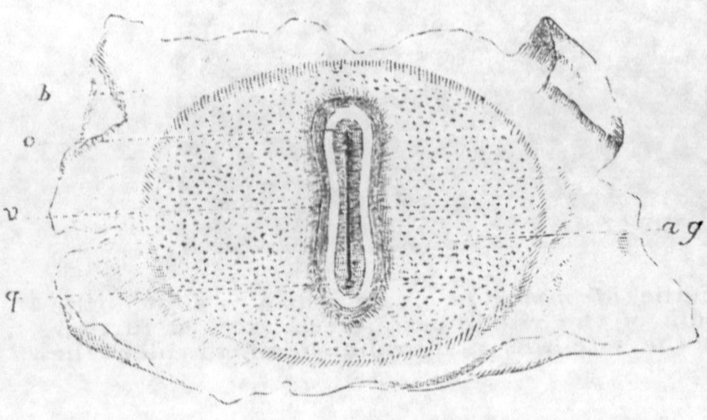

The germ, and surrounding parts, from a more advanced Uterine Ovum; *b*, blastoderma, or germinal membrane; *a g*, area germinativa; *c*, cephalic extremity of the germ; *v*, first indications of vertebræ; *q*, caudal extremity.

vertebræ is indicated at this period, by a distinct cylindrical rod of nucleated cells, termed the *chorda dorsalis ;* and this retains its embryonic type in the Myxinoid Fishes (§ 322). While this is going on, an accumulation of cells takes place between the two laminæ of the germinal membrane at the 'area vasculosa ;' and these cells speedily form them-

FIG. 288.

A B

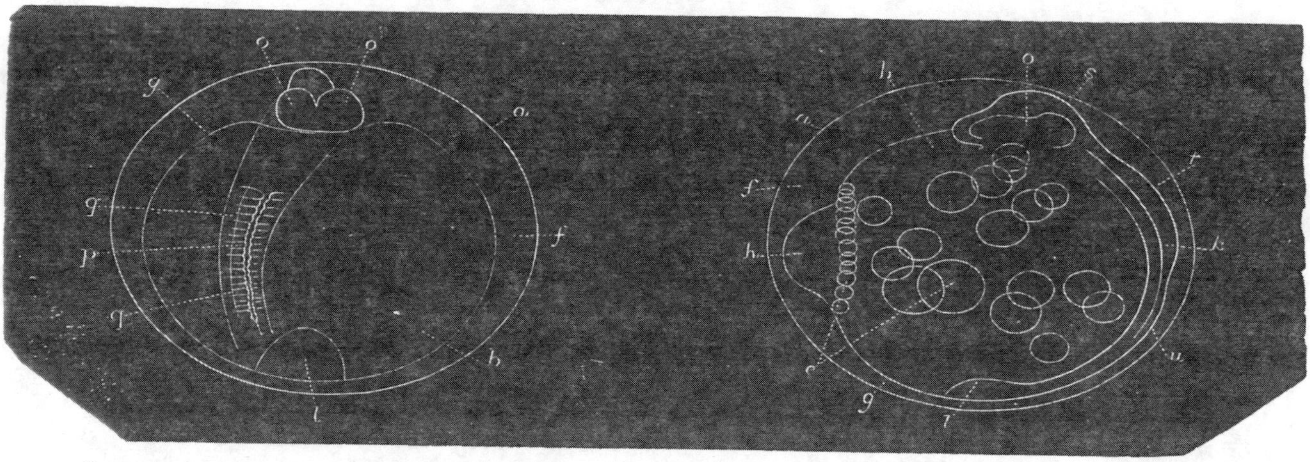

Ovum of *Coregonus palæa*, eleven days after fecundation ; A, as seen in front through the transparent ovum ; B, as seen sideways ; — *a*, shell-membrane ; *b*, yolk ; *e e*, oil globules ; *f*, albumen ; *g g*, vitelline membrane ; *h*, yolk-vesicle ; *k*, trunk of the embryo ; *l*, tail ; *o o*, optic lobes ; *p*, chorda dorsalis ; *q q*, vertebral divisions ; *u*, curve of the trunk.

selves into a distinct layer, the *vascular* lamina, in which the first blood-vessels of the embryo are developed, as will be presently described (§ 730). From the dorsal laminæ on either side, a prolongation passes outwards and then downwards, forming what is known as the *ventral lamina ;* in this are developed the ribs and the tranverse processes of the vertebræ ; and the two have the same tendency to meet on the median line, and thus to close-in the abdominal cavity, which the dorsal laminæ have to inclose the spinal cord. At the same time, the layers of the 'germinal membrane,' which lie beyond the extremities of the embryo, are folded in, so as to make a depression on the yolk ; and their folded margins gradually approach one another under the abdomen. The first rudiment of the intestinal canal presents itself as a channel along the under surface of the embryonic mass, formed by the rising-up of the inner layer of the germinal membrane into a ridge on either side. The two ridges gradually arch over and meet, so as to form a tube, which is thus (so to speak) pinched off the general vitelline sac; but it remains in connection with this by means of an unclosed portion, which constitutes the 'vitelline duct.' In oviparous animals generally, the yolk-bag, as it is emptied of its contents, is gradually drawn into the abdominal cavity ; but in Mammalia, and also in the higher Cartilaginous Fishes, it is cut off from the intestine by the obliteration of the vitelline duct, and the complete closure of the abdominal parietes around the peduncle. The minute yolk-bag of Mammalia is known under the name of the 'umbilical vesicle.'

730. Whilst these new structures are being produced, a very remarkable change is taking place in that part of the serous lamina which surrounds

the 'area pellucida.' This rises up on either side in two folds; and these gradually approach one another, at last meeting in the space between the general envelope and the embryo, and thus forming an additional investment to the latter. As each fold contains two layers of membrane, a double envelope is thus formed; of this, the outer lamina adheres to the general envelope; whilst the inner remains as a distinct sac, to which the name of *Amnion* is given. (See Figs. 289 and 292.) This takes place during the third day in the Chick; and at about the same period, a very important provision for the future support of the embryo begins to be made, by the development of blood-vessels and the formation of blood. Hitherto, the embryonic structure has been nourished by direct absorption of the alimentary materials supplied to it by the yolk; in the same manner as the simplest Cellular Plant is developed at the expense of the carbonic acid, moisture, &c., which it obtains for itself from the surrounding elements. But its increasing size, and the necessity for a more free communication between its parts than any structure consisting of cells alone can permit, call for the development of vessels, through which the nutritious fluid may be conveyed. These vessels are first seen in that part of the Vascular lamina of the germinal membrane, which immediately surrounds the embryo; and they form a network, bounded by a circular channel, which is known under the name of the *Vascular Area* (§ 484, Fig. 214). This gradually extends itself, until the vessels spread over the whole of the membrane containing the yolk. The first blood-discs appear to be formed from the nuclei of the cells, whose cavities have become continuous with each other to form the vessels (§ 175); and from these, the subsequent blood-discs of the first series are probably generated. This network of blood-vessels serves the purpose of absorbing the nutritious matter of the Yolk, and of conveying it towards the embryonic structures, which are now in process of rapid development. The first movement of the fluid is *towards* the embryo; and this can be witnessed before any distinct heart is evolved. The same process of absorption from the yolk, and of conversion into blood, probably continues as long as there is any alimentary material left in the sac. The mode in which the heart is formed, and the subsequent phases of its development, as well as that of the principal blood-vessels, have been already described (§§ 485-489); it is only necessary here to add, that the mass of cells in which it originates may be considered as a thickened portion of the vascular layer. The trunks which connect the circulating system of the embryo with that of the 'vascular area,' are called *Omphalo-Mesenteric, Meseraic,* or *Vitelline* vessels. It was formerly believed, that the nutrient matter of the yolk passes directly through the vitelline duct, into the (future) digestive cavity of the embryo, and is from it absorbed into its structure; but there can now be little doubt, that the vitelline vessels are the real agents of its absorption, and that they convey it to the tissues in process of formation.

FIG. 289.

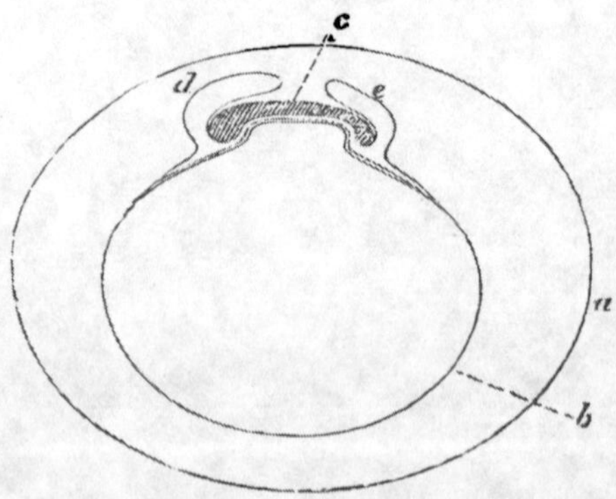

Diagram of Ovum at the commencement of the formation of the Amnion;—*a*, chorion; *b*, yolk-sac; *c*, embryo; *d*, and *e*, folds of the serous layer rising up to form the amnion.

They do, in fact, correspond to the Mesenteric veins of Invertebrated animals, which are the sole agents in the absorption of nutriment from their digestive cavity (§ 426); and the blastodermic vesicle may be regarded as the temporary stomach of the embryo,—remaining as the permanent stomach in the Radiated tribes. Previously to the ninth day of incubation (in the Fowl's egg), a series of folds are formed by the lining membrane of the vesicle, which project into its cavity; these become gradually deeper and more crowded, as the bag diminishes in size by the absorption of its contents. The vitelline vessels that ramify upon the vesicle, send into these folds (or valvulæ conniventes) a series of inosculating loops, which immensely increase the extent of this absorbing apparatus. But these minute vessels are not in immediate contact with the yolk; for there intervenes between them a layer of nucleated cells, which is easily washed away. It was from the colour of these, communicated to the vessels beneath, that Haller termed the latter *vasa lutea*; when the cellular layer is removed, the vessels present their usual aspect. There seems good reason to believe that these cells are the real agents in the process of absorbing and assimilating the nutritive matter of the yolk; and that they deliver this up to the vessels, by themselves undergoing rupture or dissolution, being replaced by newly-formed layers.

731. The development of the embryo of *Fishes* takes place without any further change of plan; the nutrient material supplied by the yolk being gradually taken into the circulation, through the medium of the germinal membrane, and being applied to the nutrition of the various tissues and

FIG. 290.

A B

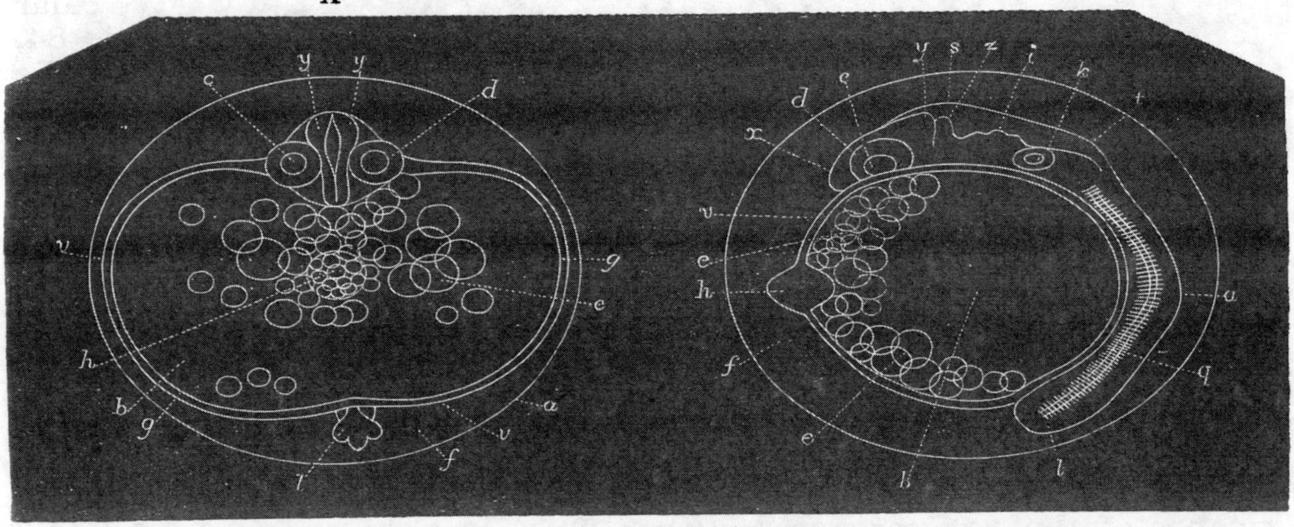

Ovum of *Coregonus palœa*, sixteen days after impregnation;—A, seen in front;—B, seen sideways;—*a*, shell-membrane; *b*, yolk; *c*, crystalline lens; *d*, choroidal system; *e e*, oil-globules; *f*, albumen; *g*, vitelline membrane; *h*, yolk-vesicle; *i*, keel of the encephalon; *k*, ear; *l*, tail; *q*, vertebral divisions; *s*, cephalic bend; *t*, nuchal bend; *v*, epidermoidal stratum; *x*, prosencephalon; *y*, mesencephalon; *z*, epencephalon.

organs which successively make their appearance. Up to the time of the emersion of the fœtal fish from the egg, the aeration of the blood is provided for in no other way, than by its distribution upon the surface of the yolk-bag; and even after it has come forth, whilst the yolk-bag is still far from being emptied, and hangs down from its abdomen, and the gills have not yet come into play, the dispersion of the blood over the vascular area still serves to aerate it, as well as to enable it to absorb the nutritive contents yet remaining to be appropriated. In most of the Osseous Fishes, the blood-vessels which ramify upon the yolk-bag may be regarded as part of the portal system; for they branch off from the intestinal veins, and return into the vena cava. As the yolk diminishes in size, and the permanent respiration is established, the blood is transmitted more directly through the liver to the heart; by the enlargement of vessels that were at first capillary, into regular trunks. In the higher Cartilaginous fishes, however, the blood sent to the respiratory surface is derived from an arterial trunk, as in all the higher Vertebrata.—The general course of the development of the principal organs in Fish, may be understood from the accom-

FIG. 291.

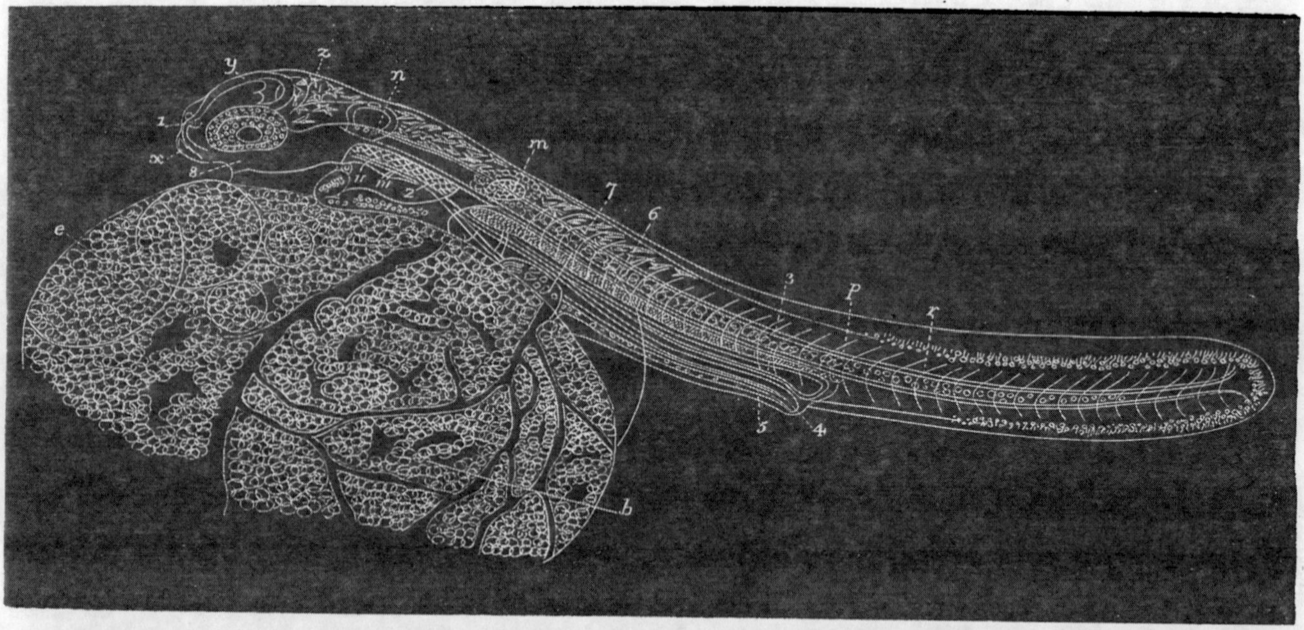

Embryo of *Coregonus palæa,* thirty three days after fecundation;—*b*, yolk; *e*, oil-globules; *m*, pectoral fin; *n*, ear; *p*, chorda dorsalis; *r*, sheath of the dorsal cord; *x*, prosencephalon; *y*, mesencephalon; *z*, epencephalon; 1, pineal gland; 2, buccal intestine; 3, ureter; 4, anus; 5, ventral intestine; 6, kidneys; 7, liver; 8, mouth; II and III, second and third branchial fissures.

panying figures; the history of the evolution of each principal organ has been already given under its own head; with the exception of that of the Nervous system, which will be found in its proper place (§§ 782–786).

732. In all the higher Vertebrata, we find that at an early period of embryonic development, provision is made for a more effectual performance of the respiratory function. This consists in the development of a peculiar membrane, the *Allantois;* which sprouts forth from the caudal extremity of the embryo, at first as a little mass of cells, whose interior is soon observed to be hollow; and which continues to enlarge, until it surrounds the entire embryo (Figs. 292–295). In *Reptiles* and *Birds,* it extends itself around the yolk-sac, intervening between it and the membrane of the shell; and the porosity of the latter allows the air to have free access to the blood, which is copiously distributed upon it. It is a remarkable circumstance that this membrane should not merely serve

as a respiratory organ, but that it should also stand to the Corpora Wolffiana, or temporary kidneys, in the light of a receptacle for their secretion,

FIG. 292.*

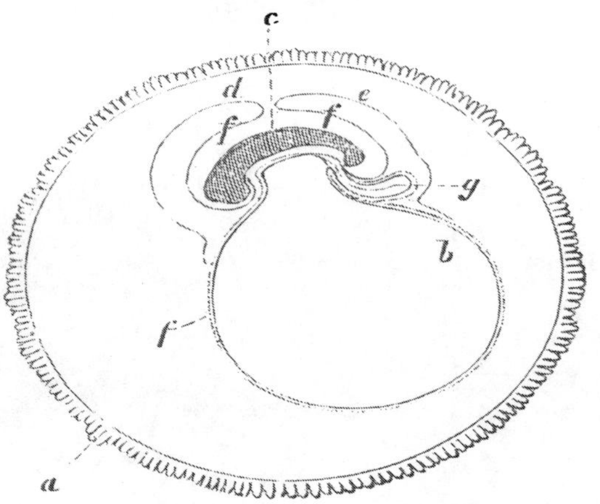

FIG. 293.†

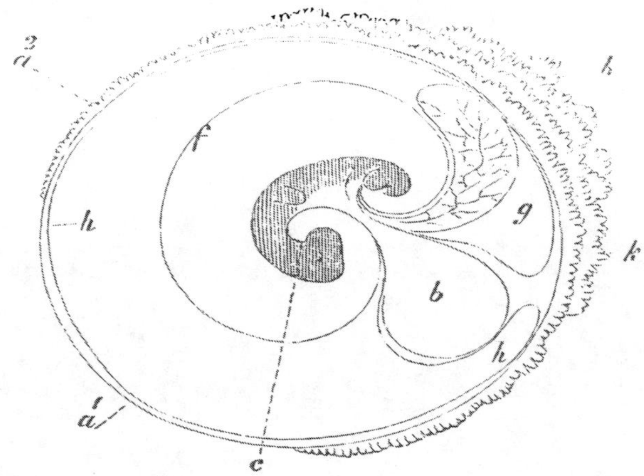

* Diagram of *Human Ovum*, with the Amnion in process of formation:—*a*, the chorion; *b*, the yolk-bag, surrounded by the serous and vascular laminæ; *c*, the embryo; *d*, *e*, and *f*, external and internal folds of the serous layer, forming the amnion; *g*, incipient allantois.

† Diagram representing a *Human Ovum* in second month:—*a* 1, smooth portion of chorion; *a* 2, villous portion of chorion; *k*, *k*, elongated villi, beginning to collect into Placenta; *b*, yolk-sac or umbilical vesicle; *c*, embryo; *f*, amnion (inner layer); *g*, allantois; *h*, outer layer of amnion, coalescing with chorion.

or urinary bladder. The greater portion of this allantois is afterwards cast off by the contraction of its pedicle; but a part of its root is usually retained, to be converted into the permanent urinary bladder.—In *Mammalia*, however, the chief office of the allantois is to convey the vessels of the embryo to the internal surface of the Chorion (Fig. 294); the tufts of which are rendered vascular by the penetration of these vessels into them (Fig. 295); and in this mode the embryo is nourished for a time, the absorbent tufts drawing their supplies by endosmose from the very vascular lining of the uterus.

733. Whilst the foregoing changes have been taking place in the ovum, the mucous membrane lining the uterus has undergone an important change; for it has become extremely thick and vascular, and its follicles are greatly enlarged, and appear to secrete a nutritive material from the blood. When the tufts of the chorion come into apposition with it, they insinuate themselves into these glandular follicles; and absorb the material which has been elaborated by them for the support of the embryo. Each villus of the chorion contains a capillary loop; this is inclosed in a layer of cells, and this again in a lamina of basement-membrane; the whole constituting the *fœtal* tuft. On the other hand, this tuft is in apposition with a layer of cells covering the orifices of the uterine glandulæ, and these lie upon the basement-membrane of its mucous lining; forming the *maternal* portion of the apparatus. Such is the highest form which it assumes in the *Implacental* Mammalia (§ 326); the ovum being ejected from the uterus, before any further changes have taken place. In the higher orders, however, we find such a concentration of this arrangement in one or more spots, as is adapted to bring the fœtal and maternal blood-vessels into closer approximation. This is accomplished, in the Ruminants, by an enlargement and multiplication of the

vascular tufts at several points of the surface of the chorion, so as to form what are termed the *fœtal cotyledons;* and these come into relation with a corresponding series of *maternal cotyledons,* formed by the enlargement

FIG. 294.*

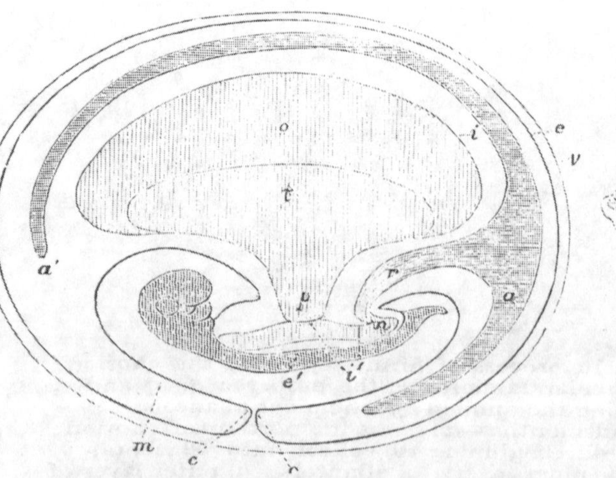

FIG. 295.†

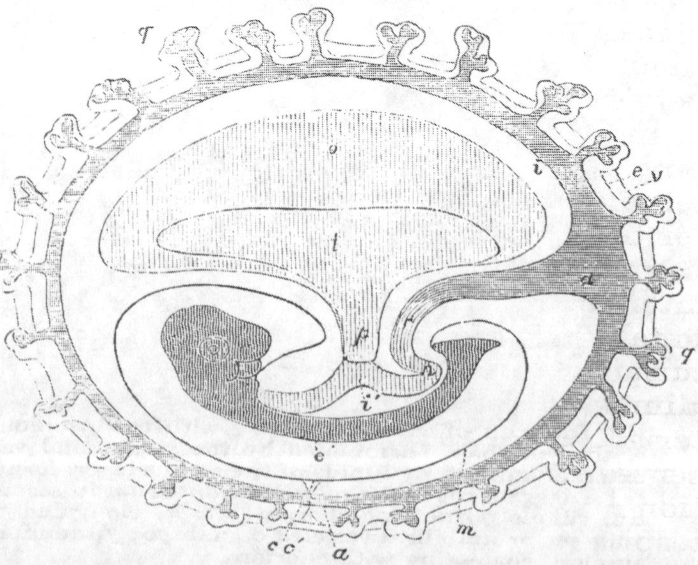

* Diagram of a Human Ovum in a more advanced stage;—*a a'*, the allantois developing itself around the embryo; *c c'*, the cephalic and caudal prolongations of the amniotic sac, nearly united; *e*, serous layer of the germinal membrane; *e'*, vertebral structure of the fœtus; *i*, mucous layer of the germinal membrane, partly converted into the rudimentary intestine *i'*, and partly forming the umbilical vesicle *o*; *m*, portion of the serous layer of the germinal membrane reflected to form the amnion; *n*, rudimentary portion of the intestine, representing the rectum; *p*, pedicle of the umbilical vesicle; *r*, pedicle of the allantois; *t*, limit of the area vasculosa; *v*, vitelline membrane.

† Diagram of an Ovum still more advanced, the two prolongations of the allantois having now met and united at *a*, and the amniotic sac being now complete; the external layer of the germinal membrane, *e*, is now disappearing; vascular villosities, *q q*, are being developed from the allantois; and the intestine *i'* is forming its first loop.—The other references as in the preceding figure.

and multiplication of the blood-vessels of the lining membrane of the uterus. No inosculation takes place between these two sets of vessels; but the loopings of each are received into the interspaces between those of the other; and each tuft is covered with its own layer of basement-membrane and of cells. In Man, however, and in the Quadrumana, Carnivora, and other orders, a still more concentrated arrangement is found. The vascular tufts are enormously developed from one portion only of the chorion; and they dip down, as it were, into a cavity, whose inner wall is formed by an extension of the lining membrane of the large veins or sinuses of the uterus. The former, which are the ultimate ramifications of the umbilical arteries, constitute the *fœtal* portion of the *Placenta* (as this compound organ is termed); whilst the latter, which constitutes its *maternal* portion, is irregularly subdivided by partitions, formed by reflexions of its wall over the fœtal tufts; and each of these reflexions carries with it an extension of the cellular layer that covers the mucous surface, so that the extremity of every loop of fœtal vessels forming a placental villus is invested first by a fœtal, and then by a maternal, layer of cells, a space being usually observed to intervene between them (Fig. 296). The blood is conveyed into the placental cavity by the 'curling arteries' of the uterus; and after it has been brought into relation with that of the fœtus, it is returned to the uterus by large apertures communicating with its sinuses. Thus the placental tufts, in which the fœtal blood is circulating, are bathed (as it were) in the maternal blood, just as the branchiæ of aquatic animals are bathed in the medium they inhabit; and an interchange can freely take place by endosmose between the two fluids, although no more direct communication exists between the two systems of vessels. This interchange is subservient to two most important purposes. The fœtal tufts draw from the maternal blood the materials which are required for the nutrition of the embryo, these materials having been first elaborated by the two sets of intervening cells; and in this character, the fœtal tufts resemble the villi of the intestinal surface, which dip down into the fluids of the alimentary canal, and absorb the nutritive materials which they furnish. But the placenta serves also as the instrument for the depuration of the blood of the fœtus, especially by the respiratory process; for the aerated blood of the mother will necessarily impart oxygen to that of the fœtus, and will carry off its carbonic acid; so that in this respect the fœtal tufts may be likened functionally, as well as structurally, to the gills of aquatic animals. But it is probable that during the early stages of embryonic life, before the proper depurating organs of the fœtus have come into full activity, other excretory matter may be got rid of through the same channel; and there even seems a strong probability, that the fœtal blood may so react upon the maternal, as to communicate to it some of the properties which it has derived from its male parent. Such, at least, seems the most feasible explanation of the fact, which has now been observed to take place in a great number of instances, that if a mare, cow, bitch, or other mammiferous female (not even excepting the human species), be impregnated in the first instance by a male possessing some marked peculiarity, and on subsequent occasions by a different male, the offspring of the second parentage will bear in greater or less degree the characters of the first; these peculiarities becoming less and less obvious on each successive occasion.*

Fig. 296.

Extremity of a Placental villus:—*a*, external membrane of the villus, continuous with the lining membrane of the vascular system of the mother; *b*, external cells of the villus, belonging to the placental decidua; *c*, *c*, germinal centres of the external cells; *d*, the space between the maternal and fœtal portions of the villus; *e*, the internal membrane of the villus, continuous with the external membrane of the chorion; *f*, the internal cells of the villus, belonging to the chorion; *g*, the loop of umbilical vessels.

734. Notwithstanding the peculiarity of the provision which is made in Mammalia for the nutrition of the embryo, the earlier part of its course of development presents little to distinguish it from that of the inferior classes. It may, in fact, be said, that all the most important parts of the apparatus of organic life, and even the fundamental portions of that of animal life, are developed upon the same general plan in all Vertebrata; and that the special peculiarities of each class only gradually evolve themselves. The conditions under which the alimentary canal (§§ 409, 729), the heart and blood-vessels (§§ 485, 730), the liver (§ 589), the corpora

* See the very ingenious papers on this subject by Dr. Harvey, in the "Edinb. Monthly Journal" for 1849 and 1851; and his pamphlet "On a remarkable effect of Cross-breeding," Edinburgh, 1851.

wolffiana (§ 599), the vertebral column (§ 729), the nervous centres (§ 782), and the eye and ear (§§ 826, 830), first present themselves, exhibit no essential differences in the Fish, Reptile, Bird, or Mammal; and the first considerable divergence from the common type is shown in that alteration in the structure of the Heart, and in the arrangement of the great vessels, which distinguishes the air-breathing Vertebrata from the Fish; whilst in the latter, the original conformation is retained, and is made subservient to the distribution of the blood in the branchial filaments. So, again, at a later period, the Corpora Wolffiana give place in the air-breathing Vertebrata to the permanent Kidneys, but remain persistent in the Fish. In like manner, the course of development of the vertebral elements very early shows those peculiarities, which are characteristic of the type of the Fish; whilst it is not until a subsequent period, that in these or any other parts (save in the provisions already described for the nutrition of the fœtus and the aeration of its blood) is there anything that can be accounted distinctive of either of the higher classes of Vertebrata.—The accompanying figures exhibit an early phase of the evolution of the Mammalian embryo, in which most of the organs are in a very rudimentary condition. It is in a state not much further advanced than this, that the embryo of *Monotremata* and *Marsupialia* is expelled from the uterus; in the higher orders, however, it is retained in that cavity, and continues to derive its support

FIG. 297.

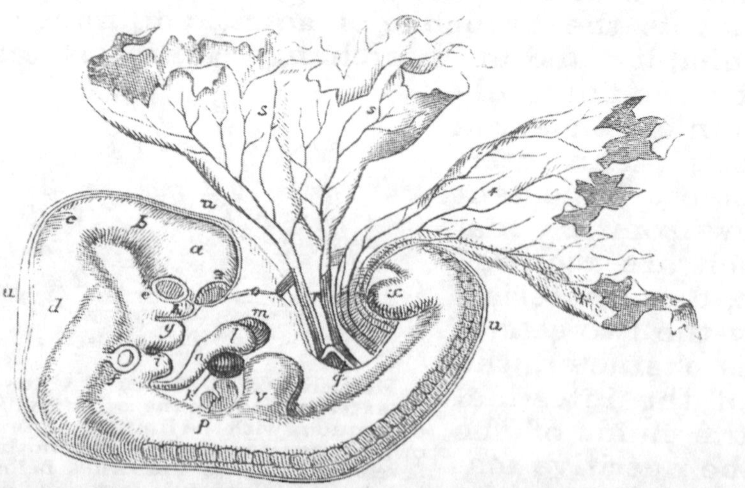

Embryo of *Dog*, twenty-five days after the last copulation, enlarged three times;—*a*, prosencephalon; *b*, deutencephalon; *c*, mesencephalon; *d*, epencephalon; *e*, eye; *f*, auditory vesicle; *g*, *h*, *i*, visceral arches; *k*, right auricle; *l*, right ventricle; *m*, left ventricle; *n*, bulbus aorticus; *o*, pericardium; *p*, liver; *q*, loop of intestine, communicating through *r*, the ductus omphalo-mesentericus, with *s*, *s*, the umbilical vesicle; *t*, allantois; *u*, amnion; *v*, anterior extremity; *x*, posterior extremity; *z*, nose.

FIG. 298.

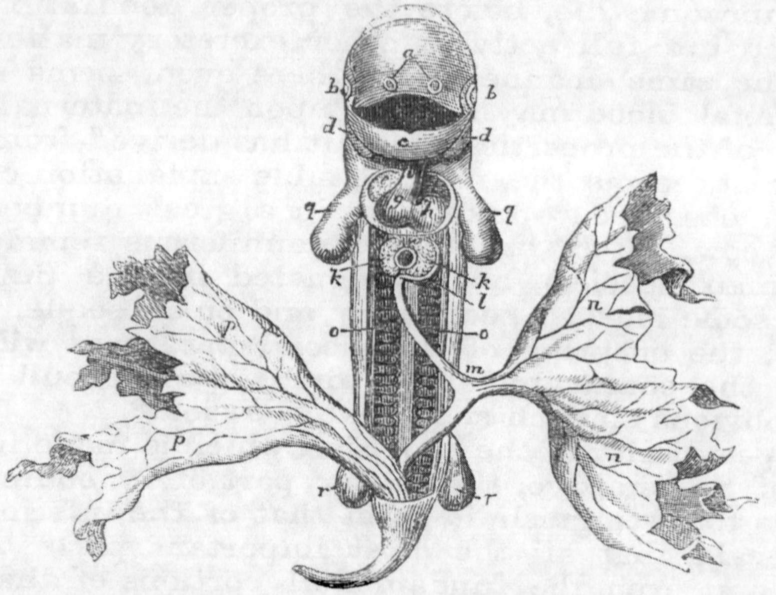

The same Embryo, straightened, and seen in front;—*a*, *a*, nostrils; *b*, *b*, eyes; *c*, *c*, first visceral arch, forming the lower jaw; *d*, *d*, second visceral arches; *e*, right auricle; *f*, left auricle; *g*, right ventricle; *h*, left ventricle; *i*, aortic bulb; *k*, *k*, liver, between the two lobes of which is seen the divided orifice of the omphalo-mesenteric vein; *l*, stomach; *m*, intestine, communicating with the umbilical vesicle *n*, *n*; *o*, *o*, corpora Wolffiana; *p*, allantois; *q*, *q*, anterior extremities; *r*, *r*, posterior extremities.

through the placenta, until nearly all the principal organs in the body, save the generative, are prepared for the active performance of their respective functions. Still there are considerable differences in the degree of independence manifested by the young Mammalia of different orders at the epoch of their birth; and to the most important of these allusion has been already made (§ 326 *bb*).

735. Although nothing is certainly known in regard to the causes which influence the Sex of the offspring, some facts which have been observed on this curious subject are worthy of being here adverted to, with the purpose of stimulating further enquiry. It is stated by Mr. Knight, that several kinds of monœcious Plants can be made to produce solely-male or solely-female flowers, by regulating the quantity of light and heat under which they are grown. If the heat be excessive, compared with the quantity of light which the plant receives, male flowers only appear;—but if light be in excess, female flowers alone will be produced. In this case it is evident that both sets of sexual organs exist in a rudimentary state; and that one or the other kind is developed according to external circumstances. To the same industrious experimenter we owe some interesting facts in regard to Animals, to which this explanation is not applicable. He remarks that, in flocks or herds of domesticated quadrupeds, it is no uncommon thing to meet with *females*, whose offspring is almost invariably of the same sex, although it may have resulted from intercourse with several different males; whilst, on the other hand, he has never met with *males* that exhibited any such uniformity in the sex of their offspring with different females. Hence he concludes that the female parent exercises the chief influence in determining the sex. An experiment upon the fecundation of Birds, which he states to have been frequently repeated, gave the following curious result. When the female was kept without intercourse with the male, up to nearly the time of laying, so that the eggs had advanced very far in their development at the time of fertilization, the proportion of males among the offspring was very large,—commonly about six out of seven.* Some observations which have been made on the Human species tend to show, that there is a probability of a majority in the number of one sex over the other, according to the relative ages of the parents;—the male children predominating in those families in which the father's age is considerably above the mother's, and *vice versâ*.†

736. In no tribe of animals save Mammalia, do we find that the young are nourished exclusively, for some time after their birth, by a fluid secreted from the blood of the maternal parent; the nearest approach to this is presented by the Pigeon (§ 325 *l*). Of the variety in the condition which the Mammary glands present, in the different orders of the class, some account has been already given (§§ 326 *aa*, 578); and it is only requisite here to describe the most important characters of the fluid they secrete. The *Milk* of most Mammalia consists of Water, holding in solution a peculiar albuminous substance termed Caseine, and various Saline ingredients, together with (in most instances) a certain form of Sugar; and having Oleaginous globules suspended in it.—The most important difference between Caseine and Albumen consists in the fact, that the

* "Selection from Mr. Knight's Physiological Papers," pp. 347, 357.

† Quetelet "Sur L'Homme," tom. i. pp. 52, 53.

former *is not* coagulated by heat, which precipitates the latter; and that it *is* coagulated by acetic acid, which has no effect upon the latter. Caseine is very remarkable for the facility with which it may be coagulated by the contact of certain animal membranes; thus a piece of 'rennet,' which is nothing else than the dried stomach of a calf, will coagulate the caseine of 1800 times its weight of milk. Further, Caseine appears to surpass albumen in its power of combining with the phosphates of lime and magnesia, and rendering them soluble; and it seems to be in this mode, that the earthy phosphates, which are so important for the consolidation of the bones of the suckling animal, are introduced into its system. —Save in the larger proportion of these substances, the Saline matter of Milk is nearly the same as that of blood.—The Sugar of Milk is peculiar as containing nearly 12 per cent. of water; so that it may really be considered as a hydrate of sugar. It is nearly identical in its composition with starch; and it appears to be directly formed at the expense of the farinaceous elements of the food. Sugar of milk is chiefly remarkable for its proneness to metamorphosis into lactic acid, under the influence of a decomposing animal membrane which acts as a 'ferment;' and it is thought by some, that the agency of such a membrane in occasioning the coagulation of caseine, is first exerted in producing lactic acid, which in its turn acts upon the caseine.—The Oleaginous globules, which consist of the substance called butyrine, are surrounded by a thin pellicle, that keeps them from coalescing whilst the milk is at rest; but when it is agitated, the envelopes of the oil-globules are broken, and they coalesce, so as to form butter. It has been ascertained that butyric acid is one of the products of the change in sugar produced by the contact of putrescent animal membranes; and it can scarcely be doubted that it is directly producible by the transformation of that substance in the living body.—Thus ordinary Milk contains the three classes of organic principles, which form the chief part of the food of animals, namely, the albuminous, the saccharine, and the oleaginous; together with those mineral compounds, which are required for the development and consolidation of the fabric of the infant. It would appear, however, that the combination of all these is not necessary; but rather has reference to the composition of the food, on which the animal is habitually supported. Thus, in the Carnivora, so long as they live upon a purely animal diet, the milk contains no sugar, which principle is altogether wanting in the food of the adult; but if they be fed upon a mixed diet, sugar soon appears in their milk. Amongst the different species of Herbivorous animals, the proportion of the several ingredients varies considerably; and it is also liable to variation in the same individual, according to the nature of the food, the amount of exercise taken by the animal, and other circumstances. Thus in the milk of the Cow, Goat, and Sheep, the average proportions of Caseine, Butter, and Sugar are nearly the same one with another, each amounting to from 3 to 5 per cent. In the milk of the Ass and Mare, on the other hand, the proportion of Caseine is under 2 per cent., the oleaginous constituents are scarcely traceable, whilst the sugar and allied substances rise to nearly 9 per cent. In the Human Female, the saccharine and oleaginous elements are both present in large amount; whilst the Caseine forms a moderate proportion.—The proportion of the saccharine and oleaginous elements appears to be specially affected by the amount in which these are present in the food, and by the degree in which the quantity ingested is consumed by the respiratory process. Thus, a low external temperature, and out-door exercise, by increasing the production of carbonic acid from the lungs, occasion a consumption of the oleaginous and saccharine matters which might otherwise pass into the milk, and thus diminish the amount of cream. On the other hand, exercise favours the secretion of caseine ; which would seem to show, that this ingredient is derived from the disintegration of the azotised tissues. Thus in Switzerland, the cattle which pasture in exposed situations, and which are obliged to use a great deal of muscular exertion, yield a very small quantity of butter, but an unusually large proportion of cheese ; yet the same cattle, when stall-fed, give a large quantity of butter, and very little cheese.

4. *On the Laws of the Exercise of the Reproductive Function.*

737. When we contemplate the immense number of diversified forms, which the study of the Organised creation brings under our notice, and witness these distinct forms perpetuated, as it would seem, by the process of Reproduction, so as to constitute separate races, the question naturally arises, whether all these had a different origin ; or whether the characters of any of them have been so modified in the course of time, as to lead to the belief in a diversity of origin among those, which were at first really identical. When it can be *shown* that two races have had a separate origin, they are regarded as of *different species;* and, in the absence of proof, this is *inferred,* when we see some peculiarity of organisation, characteristic of each, so constantly transmitted from parent to offspring, that the one cannot be supposed to have lost, or the other to have acquired it, through any known operation of physical causes. It cannot be regarded as an unimportant question to the Naturalist, to ascertain what these constant distinctions are ; whilst it is an investigation of high interest, in a Physiological point of view, to trace the modifying influence of external circumstances upon the structure and functions of living beings, and to enquire how far the results of such influences may be transmitted hereditarily, so that the differences produced by them may be perpetuated. Where races which have originally sprung from a common stock present marked differences, they are spoken of as *varieties ;* and the variety may be *transient,* from its peculiarity manifesting a tendency to disappear, or *permanent,* where it continues to be transmitted without change. The uncertainty of the limits of species is daily becoming more and more evident ; and every Naturalist is aware, that a very large number of races are usually considered as having a distinct origin, when they are nothing more than permanent varieties of a common stock. Whilst the exertions of the enterprising discoverer are adding to our already enormous list of species, from the unexplored resources of foreign lands, the skill of the horticulturist and of the breeder is exerted to produce new varieties of species already in our catalogues ; and it has unfortunately too often happened, that a new specific name has been invented for the latter as well as for the former ; and that a mere hybrid or transient variety has thus taken the rank of a species, to the confusion of all true principles of arrangement. The philosophic naturalist, on the other hand, aims to reduce the number of species, by investigating the degree of variation which each is

liable to undergo, the forms it assumes at different periods of its existence, the permanent characters by which it may be distinguished during its whole life, the habits which are natural to it, the degree in which these may be changed by the influence of circumstances; and, in fine, he endeavours to become acquainted with the *whole* natural history of a reputed species, before separating it from another to which it may be closely allied.

738. Many examples may be given, of the success with which this mode of investigation is now being prosecuted. The belief which is gaining ground, that many diversified forms of the simpler Cryptogamia may arise from similar germs developed under different circumstances, has already been noticed (§ 272); and among the higher Plants, the experiments of Mr. Herbert on the primrose, cowslip, oxslip, and polyanthus (which he proves to be all varieties of one species), are sufficient evidence of the important results which would probably accrue from a similar investigation in other quarters. The uncertainty of all principles of arrangement founded upon arbitrary characters, has been demonstrated by the fact recently published,* that the flowers and pseudo-bulbs of three reputed *genera* of Orchideous plants have been produced by the same individual.† —In Zoology, again, it is only necessary to recal the phenomena which have been arranged under the head of 'alternation of generations,' to see how diverse may be the forms presented by the very same being at different epochs of its existence. The great influence of external circumstances in modifying the form of shells, has been pointed out by Mr. J. E. Gray;‡ who has shown, among other instances, that what have been regarded as *six* distinct species of *Murex* are in reality but different states of *one;* and Mr. S. Stutchbury has been equally successful in reducing the number of species of *Patella, Cypræa,* and *Oliva,* by attending to the changes of form which each individual undergoes in the progress of its development. Many instances might be related, in proof of the uncertainty of reputed specific distinctions among higher classes; thus, Insects have been seen presenting the characters of different species on the two sides of the body; and it is now certain that an erroneous multiplication of species among Birds, especially in the migrating tribes, has been occasioned by their change of plumage at different seasons.

739. The Naturalist endeavours to simplify the pursuit of his science, by the adoption of easily-recognised external characters, as the basis of his classification of the multitudinous forms which he brings together; but such can only be safely employed, when indicative of peculiarities in internal structure, which are found to be little subject to variation, and which are not liable to be affected by the influence of physical causes. The colour of flowers, for example, is liable to so much alteration from the influence of soil and climate, that it is seldom regarded as of itself any test of the unity or diversity of species. In certain moths and butterflies, on the other hand, the uniform appearance of particular spots on the wings is held sufficient to constitute a specific character, because it is never

* "Linn. Trans." vol. xvii.

† This fact has also come under the Author's own notice in the Durdham Down Nursery, near Bristol, two of the genera being the same as in the instance just quoted, but the third a different one, so that *four* may thus be regarded as of the same species.

‡ "Philos. Transactions," 1833.

known to vary in those kinds; and it would probably be found associated, if the examination were pushed far enough, with some unequivocal differences in the configuration of internal organs: in other cases, however, there is considerable tendency to change; but, as in the colouring of plants, there are usually certain limits within which the varieties of shade are restrained. Sometimes one sex varies extremely little, while the form and colour of the other present much diversity. Amidst all these difficulties attending the discrimination of species from structural characters alone, it is not unreasonable to enquire, if there be any other means of effecting the object with greater certainty. This subject has been fully considered by Dr. Prichard in his elaborate work on the Physical History of Man; all that can be here considered, are the laws according to which the intermixture of species, and the transmission of hereditary or acquired peculiarities from parent to offspring, appear to take place.—The conclusion which has now been attained on the first of these points, and which (if stated in a sufficiently general form) is equally applicable to both the Animal and Vegetable kingdoms, may be regarded as one of the most valuable tests which the naturalist possesses. In Plants, the stigma of the flower of one species may be fertilized with the pollen of an allied species; and, from the seeds produced, plants of an intermediate character may be raised. But these *hybrid* plants will not perpetuate the race; for, although they may ripen their seed for one or two generations, they will not continue to reproduce themselves beyond the third or fourth. But, if the intervention of one of the parent species be used, its stigma being fertilized by the pollen of the hybrid, or *vice versâ*, a mixed race may be kept up for some time longer; but it will then have a manifest tendency to return to the form of the parent whose intervention has been employed. Where, on the other hand, the parents were themselves only varieties, the hybrid is only another variety, and its powers of reproduction are rather increased than diminished; so that it may continue to propagate its own race, or may be used for the production of other varieties, almost *ad infinitum*. In this way many beautiful new varieties of garden flowers have been obtained, especially among such species as have a natural tendency to change their aspect.* Amongst animals, the limits of hybridity are more narrow, since the hybrid is totally unable to continue its race with one of its own kind; and although it may be fertile with one of its parent species, the progeny will of course be nearer in character to the pure blood, and the race will ultimately merge into it.† In Animals, as among Plants, the mixed offsprings originating from different races within the limits of the same species, generally exceed in vigour, and in the ten-

* There are many instances in which foreign plants, that have been introduced into this country under different specific names, have been found capable of producing fertile hybrids; in these cases a more accurate examination of the original locality has generally shown, that the parents were nothing more than permanent varieties, or even hybrids naturally occurring between other varieties. This is particularly the case with many of the South American genera, such as that elegant garden flower, the *Calceolaria;* and this is probably the explanation of the almost indefinite number of splendid varieties, well known to horticulturists, which may be obtained from the South American *Amaryllis*.

† One or two instances have been mentioned, in which a mule has, from union with a similar animal, produced offspring; but this is certainly the extreme limit, since no one has ever maintained that the race can be continued further than one generation, without admixture with one of the parent species.

dency to multiply, the parent races from which they are produced, so as often to gain ground upon the older varieties, and gradually to supersede them. Thus, the mixture of the European races with the Hindoo and South American, has produced tribes of such superior characters of body, and of such rapid tendency to multiplication, that there is reason to believe that they will ultimately become the dominant powers in the community.* The general principle, then, is that beings of distinct species, or descendants from stocks originally different, cannot produce a mixed race which shall possess the capability of perpetuating itself; whilst the union of varieties has a tendency to produce a race superior in energy and fertility to its parents.

740. In examining into the characters of the different species of Plants and Animals with which different regions on the earth's surface are peopled, the Naturalist soon becomes aware, that there are many kinds which are restricted to particular localities, whilst others are diffused extensively or even universally over the globe;—that there are some spots (especially insular ones), of which the aboriginal inhabitants are almost entirely different from those elsewhere found;—and yet that amongst these, there will always be found species holding the same rank with regard to the remainder, and thus representing each other in different countries.† Thus, the species of Plants and Animals, originally inhabiting the eastern and western hemispheres, were probably almost entirely different, until the agency of man changed their geographical distribution; and almost the same may be said of the species north and south of the equator. On the other hand, Man, and his constant attendants the dog and the house-fly, exist in every quarter of the globe. Again, we find in New Holland no quadrupeds which do not belong to the order *Marsupialia* or *Monotremata*, with the exception of a dog which is believed to have been introduced by man, and to have run wild; and none of these species are found elsewhere. The greater part of the plants also belong to distinct genera; and those included in the genera occurring elsewhere constitute distinct species,—with scarcely any exception but among the Cryptogamia, the distribution of which seems more extended than that of Flowering-plants. The Flora of insular situations, if at a great distance from land, contains very few species which occur elsewhere. Thus, among the Flowering-plants of St. Helena, which is so far removed even from the western shores of Africa, there have been found, out of 61 native species, only *two* or *three* which exist in any other part of the globe. From these and many similar facts it appears fair to conclude, that every species of plant and animal had originally a distinct locality, from which it has been dispersed, according to the capabilities possessed by its structure of adapting itself to changes in its external conditions, its own locomotive powers, and the degree in which it is subject to external agencies. "What is a rare plant," says Decandolle, "but one which is so organised that it can only live in a

* Several additional instances of this kind are related in Dr. Prichard's work, vol. i. p. 147, and in Mr. G. Combe's "Constitution of Man," chapter v.

† "We see in two distant countries a similar relation between the Plants and Insects of the same families, though the species of both are different. When Man is the agent in introducing into a country a new species, this relation is often broken; as one instance of this I may mention that the leaves of the cabbages and lettuces, which in England afford food to such a multitude of slugs and caterpillars, in the gardens near Rio are untouched." Darwin, "Journal of the Voyage of the Beagle."

particular locality, and which perishes in all others; such a plant is incapable of assuming different forms. What, on the other hand, is a common plant? It is one robust enough to exist in very different localities, and under very different circumstances, and which will therefore put on many different forms." Plants, then, are liable to run into varieties, in proportion as they are more robust, more common, or more cultivated; and some native species are, from this cause, domesticated with greater difficulty than many exotics. Precisely the same may be said of Animals; those which have the power of adaptation to differences of temperature, food, &c. are most universally diffused; while those that can only exist within narrower limits of variation, are restricted to the neighbourhood of their original locality.*

741. It becomes a most interesting question, then, to determine what are the changes which may be produced by the influence of external circumstances, and how far these are hereditarily transmissible. On this subject, a few facts may be stated, which will give an insight into the nature of the enquiry; but it is one which deserves more attention than it has yet received, since it is not only essential to the correctness of all Natural-history classifications, but is connected with some of the highest questions in Physiological science.—One of the most obvious distinctions, where it is well marked, is that of *size;* and yet, as we have already seen (§§ 378, 379), it is one peculiarly open to fallacy. Not only the size of the entire organism, but the relative development of individual parts, may be greatly modified by the supply of food; this is especially the case in Plants, whose tissues are simple, and whose different organs closely resemble each other in elementary constitution. Thus, cultivation often converts a single flower into a double one, by the metamorphosis of its stamens into petals, or by the development of a row of petals previously abortive, or by the change of the small tubular florets of a 'composite' flower (like those composing the 'disc' or 'eye' of the Dahlia) into the flat expanded florets which constitute the 'ray.' Cultivation has a similar effect in obliterating the spines, prickles, and thorns, from the surface of many plants; a change which was fancifully, but not improperly, termed by Linnæus "the taming of wild fruits." The instances of such alterations effected by external agency in the Vegetable kingdom, are almost innumerable; and they are not confined to structure, being observed in habit also. Thus, many plants, which are *annuals* in a cold climate, become *perennial* if transported to the torrid zone; and plants which are usually *biennial,* forming their organs of vegetation one year, and those of fructification in the second, and then perishing, may be converted into *annuals* by heat, or into *triennials* by cold. It is very difficult, however, to say how far the varieties thus created may become permanent by their hereditary transmission. The usual principle is, that propagation by seeds will only reproduce the *species,* the *race* not being continued with any certainty. In most Plants which have been much altered by cultivation—such as the Apple, the Cabbage, or the Dahlia—the seeds, if dropped on a poor soil, will produce offspring which approximates to the original type of

* The geographical distribution of organised beings is a question of the highest interest to the Physiologist as well as to the Naturalist; and it is one of those which requires the utmost elucidation it can obtain from the combined researches of both. It is a department of enquiry which is at present being most successfully prosecuted by Prof. E. Forbes.

the species; whilst from the seeds of the *Cerealia* (corn-grains), which are believed to have been originally grasses of some very different aspect, no other forms are ever produced, which might assist in the solution of the curious problem of their origin. It is not improbable that, as among animals, varieties which arise from some peculiarity in the constitution of the being itself, are more liable to be reproduced in the offspring, than those which are simply the result of external agencies. It is evident, at least, that here also the capability of undergoing such modifications, is that which renders the species most truly valuable to man.

742. Amongst Animals, the various breeds of domestic cattle, of the horse, dog, &c. afford abundant evidence of the modifying influence of external conditions; since there is little doubt that they have respectively originated from single stocks, and that their peculiarities have been engrafted, as it were, upon their specific characters. Between the Shetland pony and the Arabian racer, for example, or between the Newfoundland dog and the Italian greyhound, there would seem much greater difference than between the Lion and Tiger (the skulls of which are so much alike that even Cuvier was not always able to distinguish them), or between various other species of the Feline tribe, which, from the incapability of domestication, have not been exposed to such influences. But that these domesticated races, however different their external characters, have a common origin, is indicated by the perfect freedom with which they breed together; and by the fact that, whenever they return to a state of nature,—as is the case with the dogs introduced by the Spaniards into Cuba, and the horses and wild cattle which now overspread the plains of South America,—the differences of breed disappear, and a common form is possessed by all the individuals. It is not a little curious, too, that instincts which must have remained dormant for many generations during the domesticated condition of the race, should re-appear when this change takes place in its habits; thus, among the wild horses of South America, there is the same tendency to associate in herds, under the protection of a leader, as among those of Asia, whose ancestors are not known to have been ever reduced to subjection. "It seems reasonable to conclude," as Sir C. Lyell has justly remarked, "that the power bestowed on the horse, the dog, the ox, the sheep, the cat, and many species of domestic fowls, of supporting almost every climate, was given expressly to enable them to follow man throughout all parts of the globe, in order that he may obtain their services, and they our protection." "Unless some animals had manifested in a wild state an aptitude to second the efforts of man, their domestication would never have been attempted. If they had all resembled the wolf, the fox, and the hyæna, the patience of the experimentalist would have been exhausted by innumerable failures, before he at last succeeded in obtaining some imperfect results; so, if the first advantages derived from the cultivation of plants, had been elicited by as tedious and costly a process as that by which we now make some slight additional improvement in certain races, we should have remained to this day in ignorance of the greater number of their useful qualities."

743. How all the varieties have been produced, which are now so numerous and striking, is a question much more easily asked than replied to satisfactorily. That very important changes may be induced by the direct influence of climate, food, and habits of life, upon successive gene-

rations, cannot be doubted; since experience shows that particular races of animals, placed under new conditions, gradually undergo modifications which adapt them to those conditions. Thus, of the number of new varieties which have sprung up amongst the domesticated races first introduced into South America by the Spaniards, there are many which are clearly traceable to climatic influences; and even within a recent period, numerous examples of a similar modification have occurred. For example, Sir C. Lyell mentions that some Englishmen engaged in conducting the operations of the Real del Monte Company in Mexico, carried out with them some greyhounds of the best breed, to hunt the hares which abound in that country. The great platform which is the scene of sport, is at an elevation of about 9000 feet above the level of the sea, and the mercury in the barometer stands habitually at the height of about 19 inches. It was found that the greyhounds could not support the fatigues of a long chase in this attenuated atmosphere; and before they could come up with their prey, they lay down gasping for breath; but these same animals have produced whelps, which have grown up, and are not in the least degree incommoded by the want of density in the air, but run down the hares with as much ease as do the fleetest of their race in this country.—But peculiarities sometimes arise, to all appearance *de novo*, which cannot be attributed with the same degree of probability to external agencies; thus, it is by no means uncommon to find individuals of the human species with six fingers and six toes; and such peculiarities are more likely to be continued hereditarily, than are those which have been acquired. Sometimes advantage has been taken by man, of spontaneous variations of this kind, for some purpose useful to him; and he has exerted his skill to perpetuate them. The following example is of comparatively recent occurrence. In the year 1791, one of the ewes on the farm of Seth Wright in the state of Massachusets, produced a male lamb, which, from the singular length of its body and the shortness of its legs, received the name of the *otter* breed. This physical conformation, incapacitating the animal from leaping fences, appeared to the farmers around so desirable, that they wished it continued. Wright determined on breeding from this ram, and the first year obtained only two with the same peculiarities; in the following years, he obtained greater numbers; and, when the offspring became capable of breeding with one another, a new and strongly-marked variety, before unknown to the world, was established.*—This history shows the influence which the circumstance of a scanty population may have formerly had on the production of varieties, both in the human and other species. At the present time, any peculiarity which may occasionally arise, speedily merges by intermixture, and returns to the common standard; but it may be surmised that, in the older ages of the world, some individuals in which a peculiarity existed, may have been so far separated from the rest, as to necessitate frequent union among themselves, so that the character would be rendered still more marked, instead of disappearing; and, being propagated for a few generations, would be rendered permanent.

744. Acquired peculiarities, on the other hand, are seldom reproduced in the offspring, unless they have a relation with the natural habits and

* "Philosophical Transactions," 1813.

physical wants of the species; but, when this relation exists, they may be transmitted as regularly as the specific character. Thus, in dogs, the relative perfection of the organs of sight and smell, perhaps also of hearing, varies much in different breeds, and their mode of hunting their prey undergoes a corresponding change; but in these cases, no new instinct is developed, the difference merely consisting in the relative proportion of those already existing; and the new peculiarities have an intimate relation to the habits of the animal in a wild state. For example, in a mongrel race of dogs employed by the inhabitants of the banks of the Magdalena almost exclusively in hunting the white-lipped Pecari, a peculiar instinct appears to have become hereditary, like that of the pointers and other dogs of this country. The address of these dogs consists in restraining their ardour, and attaching themselves to no animal in particular, but keeping the whole herd in check. Now among these dogs some are found, which, the very first time they are taken to the woods, are acquainted with this mode of attack; whereas, a dog of another breed starts forward at once, is surrounded by the Pecari, and, whatever may be its strength, is destroyed in a moment.* It is impossible not to recognise in many acquired habits, however, something more than a relation to the instincts necessary for the preservation of the species; they evidently arise, in part at least, from the connection of the race with Man. This is more particularly exemplified in the instance of the breed of Shepherds' Dogs, which often display an extraordinary hereditary sagacity respecting their peculiar vocation; as well as in cases which have been frequently mentioned, where the descendants of dogs to which peculiar tricks have been taught, have displayed an unusual aptitude for learning the same. It may then be considered, that the capability of undergoing such modifications is a part of the psychical as well as structural character of the Dog, even in a wild state; and that his relation to man may have as important an influence on *his* hereditary propensities, as the supply of their physical wants has on animals of other species. The same may, perhaps, be said of the Horse, in the races of which we find peculiar habits transmitted from parent to offspring, which are the pure results of human instruction.—It is from the want of this relation towards either the natural habits of the species or their subserviency to man, that habits acquired by other animals do not become hereditary. Thus, Pigs have been taught to hunt and point game with great activity and steadiness, and other learned individuals of the same species have been taught to spell; but these acquirements have in no instance been transmitted to the offspring, not being the result of the development or modification of any instinctive propensity naturally existing. In like manner, however artificially the forms of domesticated animals may have been altered in all the individuals of successive generations, the usual character of the species and variety is maintained in each one of the offspring; unless, as sometimes happens, this alteration happens to coincide with natural varieties of the species. Thus instances are on record, in which Dogs, that have been deprived of their tails by accident or design, have produced puppies with a similar deficiency; but as breeds

* Some curious instances of a similar propagation of acquired peculiarities connected with the natural habits of the race, are given by Mr. Knight, "Phil. Trans." 1837; the most remarkable, perhaps, are the facts related of the Retriever.

of tail-less dogs have spontaneously arisen, there would be a stronger tendency to the perpetuation of this acquired peculiarity, than when no such peculiarity naturally occurred. There can be no doubt that much has yet to be learned, of the influence of the state of the parent upon the development of the offspring; and that, though credulity and the love of the marvellous have been the occasion of many strange fictions being transmitted to us, we are by no means justified in rejecting the doctrine without further enquiry.* And when it is borne in mind, that the races of animals among which the so-called *spontaneous* variations are most apt to spring up, are also those which are most susceptible of the modifying influence of external conditions, it seems highly probable that these spontaneous variations in the offspring are really attributable to the influence of external agencies in modifying the constitution of the parent.

CHAPTER XIX.

SENSIBLE MOTIONS OF LIVING BEINGS.

745. ALTHOUGH we are ordinarily accustomed to think of Animals as self-moving, and to regard Plants as altogether destitute of the power of spontaneous motion, yet, as already pointed out on several occasions, this distinction between the two kingdoms is one of which the validity cannot be sustained. For the 'zoospores' of many of the inferior Algæ are as active in their movements, as are the Animalcules which they so much resemble; and are there some Protophytes, which seem to be in a state of more or less active motion during the whole period of their lives, at least until the occurrence of 'conjugation.' Thus the *Oscillatoriæ*, which are long filamentous cells, have a movement of alternate flexion and extension, writhing like worms in pain; sometimes they appear to twist spirally, and then to project themselves forwards by straightening again. These movements are greatly influenced by temperature and light, being more active in heat and sunshine than at a low temperature and in shade; and they are checked by any strong chemical agents, which also put a stop to the motions of Animalcules inhabiting the same water. And most of the *Diatomaceæ* move slowly through the water, as if by the action of cilia; although these organs are not distinctly discernible.† Among the higher Algæ, and probably in the whole Cryptogamic series, the 'phytozoaires' possess an activity in no degree inferior to that of the spermatozoa of Animals. And although the different mode in which the reproductive function is performed in the Phanerogamia, renders it unnecessary that the contents of the sperm-cells should possess such a power of spontaneous movement, and the entire organisms are firmly rooted to the ground

* Montgomery, "On the Signs of Pregnancy," p. 16. See also Walker "On Intermarriage," pp. 275—8, and Dr. Harvey "On a remarkable Case of Cross-breeding," for several examples of the influence of the mental condition of the mother at the time of conception, upon the offspring, in various domesticated animals.

† The *Bacillaria paradoxa* has a very remarkable movement; for the long staff-like segments of which the plant is composed, slide edgeways over one another until they become almost completely detached, and then return and slide in the opposite direction; repeating this movement with a rhythmical regularity.

during the whole of life, yet we shall hereafter find that they exhibit movements of one part upon another, which are scarcely inferior in character to those of Zoophytes, these also being fixed during the greater part of their existence, although its earliest period has been passed by them in a condition resembling that of the zoospores of the inferior Algæ.

746. Now with regard to *ciliary* movements, it may be asserted without hesitation that they do not in the least indicate consciousness or self-determining power on the part of the beings which exhibit them ; and there is no reason whatever, why the fact of their performance by any organism should be regarded as entitling it to a place in the Animal kingdom. And it cannot but induce us to look at those movements of Animalcules, which are due to ciliary action, as of a purely automatic character, when we see that movements of a precisely similar nature are performed by beings whose place is indubitably in the Vegetable kingdom. In fact, as already pointed out (CHAP. V.), there is no other line of demarcation to be drawn between the Protophyta and Protozoa, than that which is based upon the nature of their respective aliments, and the mode in which these are obtained and appropriated.—And the same may be said of those rhythmical motions which are exhibited by Oscillatoriæ, and by 'phytozoaires' and 'spermatozoa ;' their very uniformity and constancy being indications that they do not proceed from a self-determining power, but are entirely automatic. Such movements, as already pointed out, appear to result from the expenditure of a certain amount of Vital force, which becomes re-converted into the Physical, through the peculiarity of the material substratum which is their instrument (§§ 50–54).

747. The sensible motions of the higher Plants, however, which are usually performed under the influence of stimuli applied to their tissues, or to certain parts of them, appear referable to a somewhat different category ; being, in fact, manifestations of the same property of *contractility*, as that which exists in certain Animal tissues, but especially in the muscular. This property of *contractility* on the application of a stimulus, may be readily distinguished from the *elasticity* which is simply due to the mechanical relation of the particles composing the tissue ; the latter being retained as long as there is no evident decomposition, whilst the former is an essentially *vital* endowment. An elastic ligament, when stretched, tends to contract only in virtue of the disturbance of its molecular arrangement, that has been produced by the force applied to it ; but a muscle which contracts powerfully upon the stimulus of a simple touch, or upon one of a still less mechanical nature, can do so only by a property of its own, which is connected with its attributes as a living being. In the lowest and simplest Animals, whatever degree of contractility is possessed, appears to be almost equally diffused through the system (§§ 283 *a*, 289) ; and we can neither discover in them any structure specially endowed with this property, nor anything resembling a nervous system fitted to call it into exercise. In proportion as we ascend the scale, however, we find a distinct muscular structure evolved, in which the general contractility of the body becomes, as it were, concentrated ; and, in proportion to its development and complexity, it supersedes the corresponding but more feeble powers of the remainder of the tissues. It is now almost entirely subjected to the nervous system ; and all those parts of it, which are not connected with the functions of organic life merely, are rendered

subservient to the will, and thus become the instruments of its operation upon the place and condition of the body.

748. Of the evident movements observable in the higher Plants, there are some which take place as a part of the regular series of phenomena of growth and reproduction, and which must be regarded as the ordinary operations of the vital forces which they have derived from the light, heat, &c. that have sustained their organic activity; but there are others which are performed in respondence to excitement of a mechanical kind. The immediate connection of these movements with the organic functions, in the first class of instances, and the indication they would seem to give of consciousness and sensibility in the second, have led many persons to seek for an explanation of them in the hypothesis of the existence of a nervous system in these beings. But it will be seen, if the question be fairly investigated, that, whilst no evidence of its presence is furnished by the minutest anatomical research, no argument for its operation can be deduced from the phenomena observed. In the simplest and most intelligible instances of sensible motions in Plants, the change is the result of the contraction of the part to which the stimulus is applied; this contraction being consequent upon the change of form of its cells, which must be regarded as a manifestation of their peculiar vital endowments (§ 138). Thus, the leaf of the wild Lettuce exudes, when the plant is in flower, the milky juice contained in its vesicles, if these be irritated by the touch; and the contraction of the poison-gland of the Nettle, when the tubular hair which surmounts it is pressed, forces out its secretion, and produces urtication. So, again, if the base of the filament of the Berberry be touched with the point of a pin, the stamen immediately bends over and touches the style. In this case, the movement is produced by the peculiar contractility of the tissue on the interior side of the filament, which, when called into operation by the application of a stimulus, necessarily occasions the flexion of the stalk. This peculiar irritability has a relation with the functions of the flower; since, when called into play (as it frequently is) by the contact of insects, the fertilization of the stigma will be assisted. Many similar instances might be adduced, in which a corresponding operation is connected with the process of reproduction in Plants.—There are cases of more complexity, however, in which an irritation of one part produces motion in a distant and apparently-unconnected organ. Thus, in the *Dionœa muscipula* (Venus' fly-trap), the contact of any subtance with one of the three prickles which stand upon each lobe of the leaf, will occasion the closure of the lobes together, by a change taking place in their leaf-stalk. And in the *Mimosa pudica* (Sensitive Plant), any irritation applied to one of the leaflets will occasion, not only its own movement towards its fellow, but the depression of the rib from which it springs; and, if the plant be healthy, a similar depression will be produced in the principal leaf-stalk, and even in the petioles of other leaves. This propagation of the effect of the stimulus, from one part to another more or less remote, is usually accomplished in Animals through the nervous system; but there is little doubt that in Plants the transmission is made through an entirely different channel. The experiments of Dutrochet upon the Mimosa have shown it to be through the vascular system, that an irritation in one part is made to produce movement in a distant organ; and there can be little

doubt that the same is true of the Dionæa also. Where each leaflet of the Mimosa is implanted upon its rib, there is a little swelling or intumescence ; this is more evident where the lateral ribs join the central one ; and it is of considerable size at the base of the petiole, where it is articulated with the stem. The experiments which have been made upon its properties, have been performed, therefore, in the latter situation ; but the description of their results will apply equally well to the rest. The intumescence consists of a succulent tissue, which, on the upper side, appears very distensible, and on the lower very irritable. In the usual position of the leaf or leaflet, the distension of the two sides seems equally balanced ; but anything which causes an increase of fluid on the upper side, or a contraction of the vesicles on the lower, will obviously give rise to flexion of the stalk. The latter effect may be readily produced by touching that part of the intumescence itself ; and then the leaf or leaflet will be depressed by the contraction of the part *immediately* irritated, just as in the case of the stamen of the Berberry. The same result follows the stimulation of this part by an electric spark, by the concentration of the sun's rays upon it with a burning glass, or by chemical agents ; and if, instead of applying a temporary stimulus, whose effect is speedily recovered from, a notch be made in the lower side of the intumescence, the balance between its resistance and the expansive tendency of the upper side is then permanently destroyed, and the stalk remains depressed. Now, supposing the lower side to be in its usual condition, flexion of the stalk may result also from whatever distends the vesicles of the upper part of the intumescence ; and this is the mode in which the movement is usually effected. For a stimulus applied to any part of the leaf, will cause a contraction of its vesicles ; and the fluid expelled from them is carried by the circulating system to the distensible portion of the intumescence belonging to each leaflet, and to that of the petiole itself.—It appears, then, that these evident motions are readily explicable on the supposition, that *contractility* is a property of various tissues of Plants, and that this may be excited by stimuli of a physical nature. To suppose more, would be unphilosophical because unnecessary.

749. There are other movements, however, arising from causes which originate in the system itself, of which some notice should be taken. Such are, the folding of the flowers and drooping of the leaves, known as the *sleep* of plants. These phenomena seem due to a diminution in the activity of those vital processes, by which the turgescence of the soft parts of the structure is maintained ; and this diminution appears partly to result from the withdrawal of the usual stimuli, especially light, and to be in part of a periodical character. For it is found that artificial light and warmth will cause many flowers and leaves to erect themselves for a time ; and that, by proper management, the usual periods may be completely reversed. But the phenomenon cannot be altogether explained on this principle, since there are many plants of which the flowers only expand in the night, and which must be kept in darkness to prevent them from closing ; and in the present state of our knowledge, we can only consider this *periodicity*, like that periodical cessation or augmentation of particular functions which are common to nearly all organised beings, as a part of that regular train of operations, which is characteristic of each living being, and which proceeds from the original endowments of its organism, called

into exercise by forces external to itself.—One other spontaneous Vegetable motion may be instanced, as of a very inexplicable character; that of the *Hedysarum gyrans*, a Bengalese plant, each of whose petioles supports three leaflets, of which the central one is large and broad, and the two lateral ones, which are situated opposite to each other, small and narrow. The position of the central leaflet appears peculiarly influenced by light: for in the day-time it is usually horizontal; by the action of strong solar light it is raised towards the stalk, whilst in the evening it bends downwards; and it is manifestly depressed, if placed in the shade for a few minutes only. The small lateral leaves are in incessant motion; they describe an arch forwards towards the middle leaflet, and then another backwards towards the footstalk; and this by revolving on their articulation with the petiole. They pass over the space in 30 or 40 seconds, and then remain quiet for nearly a minute; the leaflets do not move together, but in opposite directions, one usually rising while the other is sinking; the inflexion downwards is generally performed more rapidly and uniformly than that upwards, which occasionally takes place by starts. These movements continue night and day; being slower, however, in cold nights, and more rapid in warm and moist weather. They seem less affected by mechanical or chemical stimuli, than do those of any other plant; and continue for a longer time in separated parts.

750. One class of spontaneous Vegetable movements has been shown by Dutrochet to be due to the action of Endosmose (§ 415) in the organs which execute them. This is particularly the case in various seed-vessels, which burst when ripe, in such a manner as to eject their contents with force,—as in the instance of the *Momordica elaterium* (common Squirting-Cucumber). His experiments upon the capsule of the Balsam termed *Impatiens noli-me-tangere* are particularly interesting. The valves of this capsule, when the fruit is ripe, suddenly spring from each other and curl inwards, scattering the seeds to some distance. Now an examination of the tissue of the valves shows, that the outer part consists of much larger vesicles than the inner; and that the fluids contained in it are the densest. According to the law of Endosmose, the fluids contained in the tissue of the interior will have a tendency to pass into the vesicles of the exterior; and it will distend them in such a manner, as to produce a disposition in that side to expand, when permitted to do so, whilst the inner side has an equal disposition to contract. This at last occurs from the separation of their edges consequent upon their ripening; and then each valve rolls inwards. If, however, the valves be placed in a fluid more dense than that contained in the exterior vesicles, such as syrup or gum-water, these will be emptied on the same principle, and the valves will become straight, or even curl outwards.—A very curious movement, which probably depends upon a similar cause, may be observed in a little Fungus, which is not uncommon in some parts of Britain, named *Carpobolus*, from its peculiar manner of scattering its fruit. The sporules are collected into one mass, and inclosed in a globular bag, which is called a *sporangium*. This lies in a cavity, of which the inner wall is capable of separating itself from the outer, and of suddenly everting itself, so as to project in a globular form from the mouth of the cavity which it previously lined. This sudden eversion ejects the sporangium (with a degree of violence which for so minute a plant is very remarkable) from the cavity in which it was formed;

the mouth of this, which was at first nearly closed, spontaneously dilating itself as the sporules are mature. The eversion of the membrane is probably due to a change having taken place in the relative distension of the cells forming its inner and outer layer, which would operate much as in the capsule of the Balsam.

751. Among the simplest *Protozoa,* it seems as if the change of form of the single cell of which each individual is composed, were the sole means of movement which it possesses (§ 283 *a*) ; and this change of form seems rather to be due to operations taking place *within* the cell, than to occur in respondence to mechanical irritation applied to its exterior,—a circumstance which throws some light upon the phenomena just now described as occurring in Plants (§ 749). In these movements it is impossible to imagine with any probability that consciousness can participate ; nor can it be supposed that a nervous system can be their instrument. Now although the movements of the *Hydra* (§ 289), and of other Zoophytes of its class, may *appear* to indicate consciousness and self-determining power, yet it is very doubtful whether such endowments can justly be assigned to these animals. For their contractile tissue is of the simplest possible character, resembling that which is found in a very early state of newly-forming parts of higher Animals ; and the most careful scrutiny has failed to discover the faintest vestiges of a Nervous System in them, notwithstanding that the extreme transparency of their bodies renders them peculiarly favourable subjects for the most minute examination. And further, when their movements are fairly compared with those of higher animals, it becomes obvious that they resemble those by which food is immediately introduced into the stomach, the tentacula of the Hydra being homologous with the œsophageal muscles of Man ; and as we know that the latter act without our will, and even without our consciousness, it would be inconsistent with sound philosophy to regard them as dependent upon such springs of action in the Hydra, without some very cogent evidence that they are so.—Again, the rhythmical movements of the disc of the *Pulmograde Acalephæ* (§ 294), whereby they are propelled through the water, bear a much closer resemblance to the rhythmical contractions of the heart of higher animals, than they do to any other of their actions ; and as the latter are performed without any exercise of will, and even without the guidance of consciousness, it seems reasonable to suppose that the former are so likewise. And such an interpretation is confirmed by what is known of the nervous system of these animals ; for its extent of diffusion is so limited, that it is impossible to imagine the whole contractile tissue of the disc to be influenced by it ; and, moreover, portions of the disc entirely separated from the rest, and not containing any portion of the nervous centres, will continue their alternating contractions and relaxations, just like the heart of a cold-blooded animal taken out of its body.—In the young of the *Compound Ascidians* (§ 703), again, we find an active movement of the embryonic mass through the water, effected by the lateral undulations of the tadpole-like tail ; and this at a time when this organ consists of nothing but cells, and when it is certain that the formation of the nervous system has not even commenced. Such movements are strictly comparable, on the one hand, with the rhythmical movements of the Oscillatoriæ or the Hedysarum among Plants, and on the other, with the rhythmical contractions of the heart in the embryo of

the higher Animals, when as yet its walls consist of nothing but cells (§ 485). They cannot be attributed to any external stimulation, and must be regarded as a part of the regular series of vital operations of the cells which exhibit them, as the ciliary vibrations are of the ciliated epithelium-cells, or as the acts of secretion or reproduction are of the cells of glands or ovaries (§ 51).

752. There are many movements in higher Animals, which, though performed by the instrumentality of the Muscular tissue, are perfectly involuntary, take place without exciting any consciousness of their occurrence, and appear to be essentially independent of Nervous agency, although capable of being affected by it. All the actions of this class are *immediately* subservient to the maintenance of the Organic functions; thus we find the propulsion of food along the alimentary canal effected by the peristaltic contractions, alternating with relaxations, of its proper muscular coat; and the circulation of the blood in great measure dependent upon the alternating contractions and relaxations of the muscular walls of the heart. Between these two sets of actions, however, there is this important difference;—that whilst the former, like the closure of the fly-trap of the Dionæa, or the bending-down of the filaments of the Berberry, are chiefly dependent upon the application of a stimulus, which calls into activity the contractile power of the tissues,—the latter, like the rhythmical movements of Plants previously referred to, take place without any external stimulation, the regular alternation of contraction and relaxation being apparently *their* peculiar manifestation of vital activity. Thus we find that the peristaltic movements of the alimentary canal are for the most part excited by the contact of solid or liquid substances with its lining membrane; and that, when they are not otherwise taking place, they may be called into activity by a slight mechanical irritation applied to its external surface, or, in a warm-blooded animal at least, by the admission of cold air into the abdominal cavity. On the other hand, the alternate contractions and relaxations of the heart will continue with extraordinary regularity (in a cold-blooded animal), after the organ has been removed from the body, and has been completely drained of its blood;* and it seems impossible to refer them to the agency of any stimulation derived from an external source. But, on the other hand, the peristaltic movements of the alimentary canal may be often seen to take place without any ostensible stimulation; and the alternating movements of the heart, after they have ceased, may often be re-excited by a slight mechanical irritation. Hence it is obvious, that both kinds of action are referable to the same endowment, namely, a *motility* which is inherent in the tissues, and is a part of their vital endowments, and which may manifest itself either *spontaneously*, or *in respondence to a stimulus*: but that the former method is the rule in the case of the heart, the exception in that of the alimentary canal; whilst the latter method is the rule in the case of the alimentary canal, the exception in that of the heart.—The endowment in question, however, is not limited in the higher animals to the heart and alimentary canal; for it pervades, in a greater or less degree, the walls of their blood-vessels and absorbents, those of the principal gland-ducts, those of the

* It has been stated by Dr. Mitchell of Philadelphia, that the heart of a Sturgeon which had been cut out and hung up to dry, continued its rhythmical movements, until its tissues had lost so much of their fluid, that a crackling sound was heard with each contraction.

urinary bladder, uterus, &c., all of which parts are furnished with non-striated muscular fibres (§ 225), or with those elongated contractile cells, which have been recently shown by Prof. Kölliker to take a considerable share in the movements of the contractile tissues of even the highest animals.*

752 *a.* By far the larger proportion of the sensible movements of higher Animals, however, are called forth by the instrumentality of the Nervous System; through which the contractility of the Muscular apparatus is called into exercise. This is the case, not only with all those actions that are designated as *voluntary,* being the result of an individual self-determining power, but also with many that are as *involuntary* or *automatic* as those which we have been just considering; these last being connected, more or less directly, with the maintenance of the organic functions, or, in some mode or other, with the conservation of the bodily fabric. We shall find that these are performed with no more dependence on consciousness, than the rhythmical contractions of the heart, or the peristaltic movements of the alimentary canal; and that the nervous system is made to participate in them simply as a *conductor* of stimuli, from the parts which receive the impressions, to those which are to respond to them by contraction; the conditions of Animal existence requiring that this conduction shall be performed with greater rapidity and energy, than we see to take place in the parallel cases that present themselves among Plants (§ 759).—It is very interesting, however, to observe, that the form of Muscular tissue which most readily obeys the stimulus of innervation, and which is the instrument of the most purely Animal functions, is that which most perfectly exhibits that *cellular* type of structure, which seems common to all the parts most actively engaged in the vital operations; and that the act of muscular contraction in the highest Animal is due to the same kind of change in the form of the cells of the ultimate fibrillæ, as that which produces the sensible motions of Plants (§ 224). The only essential difference, then, between the contractile tissues of Plants and those of the higher Animals, consists in this; that the latter are capable of being called into activity by a kind of stimulus—the Nerve-force—which does not operate in the former. But this speciality of endowment does not supersede the more general susceptibility which exists among the inferior grades; for we find that the 'striated muscular fibre,' although called into action more readily and powerfully by the stimulus of Innervation than by any other, is yet induced to change its state, like the contractile tissues of Plants, by electrical, chemical, or mechanical irritation; whilst the 'non-striated fibre' responds so much more readily to this kind of irritation, than it does to the stimulus of Innervation, that it is even difficult to demonstrate the action of the latter upon it (§ 231).—A general account of the Movements which are characteristic of the different classes of Animals has already been given (CHAP. VIII.); and to enter into any detail as to the various mechanical adaptations, by which the force generated by Muscular contraction is applied to their production, would carry us far beyond the limits of the present Treatise.

* See Siebold and Kölliker's "Zeitschrift für Physiologie," Band i., 1849.

CHAPTER XX.

FUNCTIONS OF THE NERVOUS SYSTEM.

1. *General Considerations.*

753. We have now completed our survey of the entire series of operations which are common to the Vegetable and Animal kingdoms; and it only remains for us to consider those which are peculiar to the latter. Such are the various actions of the Nervous System,—an apparatus to which we find among Plants nothing in the least analogous; and the possession of which, therefore, must be considered as the peculiar and characteristic endowment of the Animal. Still, as already pointed out on several occasions, we cannot justly regard it as universally present throughout those classes, which possess on other grounds a title to be ranked in the Animal kingdom; for there are many such beings in which we seem entitled to affirm that it is absent; and there are many others, in which, if present at all, it exists in a condition quite rudimentary, and can take little share in the actions of the organism. But the life of all such beings is chiefly *vegetative* in its nature; their movements are not dissimilar in kind to those which we see in Plants; and their title to a place in the Animal kingdom chiefly rests upon the nature of their food, the mode in which they appropriate it, and their resemblance to the embryonic conditions of beings undoubtedly belonging to that group. In proportion, however, as we ascend the Animal series, do we find the Nervous System presenting more and more of structural complication, and obviously acquiring more and more of functional importance; so that in Vertebrated animals, and more especially in Man, it obviously becomes that part of the organism, to whose welfare everything else is rendered subordinate (§ 320). And we observe that this is the case, not merely in virtue of its direct instrumentality as the organ of Mind; but also as the medium through which the conditions are provided for carrying on many of the Organic functions, which we find to be brought into more and more intimate dependence upon it, in proportion as we rise in the scale. Thus the Ingestion of food, and the Aeration of the circulating fluid, which are provided for in many of the lower tribes of animals by ciliary action, in which Nervous agency has no participation whatever, are made to depend in the higher upon the Nervo-muscular apparatus, which, without interfering with those organic processes that take place (so to speak) in the *penetralia* of the system, guards the *portals* for entrance and exit. But we find that, when thus brought into intimate relation with the Organic functions, the Nervous system is employed in its very simplest mode of operation; that which does not involve Intelligence, Will, or even Instinct (in the proper sense of that term), but which may take place independently of all Consciousness, by the simple *reflexion* of an impression, conveyed *to* a ganglionic centre by one set of fibres proceeding towards it from the periphery, along another set which passes *from* it to the muscles, and calls them into operation (§ 244). This

'reflex function,' therefore, by which 'automatic' movement is effected without the necessary participation even of consciousness, is the simplest application of the Nervous System in the Animal body. We shall presently see reason to believe, that a very large proportion of the movements of many of the lower animals are of this reflex character; and that they are not necessarily accompanied by sensation, although this may usually be aroused by the same cause which produces them. As we rise, however, in the scale of Animal existence, we find the *reflex* movements forming a smaller and smaller proportion of the whole; until, in Man, they constitute so limited a part of that entire series of movements of which the Nervous system is the agent, that their very existence has been overlooked.

754. But the main purpose of the Nervous System is to serve as the instrument of those *Psychical** powers, which are altogether peculiar to the Animal. Now when we attempt to analyse these attributes, we may resolve them, like the properties of the material body, into different groups. We find that the *first* excitement of all mental changes, whether these involve the action of the *feelings* or of the *reason*, depends upon *sensations*; which are produced by impressions made upon the nerves of certain parts of the body, and are conveyed by these to a particular ganglionic centre, which is termed the *sensorium*, being the part in which Sensation, or the capability of *feeling* external impressions, especially resides. Now there are numerous actions, especially among the lower Animals, which seem to be as far removed from the influence of the Will, as little directed by Intelligence, and hence as truly 'automatic' in their nature, as the 'reflex' movements themselves; but which, nevertheless, seem to depend upon sensation for their excitement. The movement seems, in some instances, to be immediately consequent upon the production of the sensation; of this kind are the act of sneezing, and the sudden start produced by a loud and unexpected sound, in ourselves. But in other instances it appears rather to proceed from an internal feeling excited by the sensation, more akin to the Emotions of Man; which, without any calculation of consequences, any intentional adaptation of means to ends, any exertion of the reason, or any employment of a discriminating Will, may produce an action, or train of actions, as directly and obviously adapted to the well-being of the individual, as we have seen those of the reflex character to be. It is impossible to say, in regard to many of the actions of the lower animals, to what extent they involve feelings or emotions at all analogous to those which *we* experience; and it would seem better to apply the generic term *Consensual* to those, in which the Sensation excites the motor action, either immediately, or through the agency of an indiscriminating impulse excited by the sensation. This class will include all the purely 'instinctive' actions of the lower Animals; which make up, with the reflex, nearly the whole of the Animal functions in many tribes, and which are peculiarly elaborate in their character, and wonderful in their results, in the class of Insects. It is obvious that such actions, not being the result of any intelligent choice or voluntary direction on the part of the animals which perform

* This term is used to designate the sensorial and mental endowments of Animals, in the most comprehensive acceptation of those terms.

them, are not less truly 'automatic' than those belonging to the 'reflex' group; the only essential difference being, that the latter are performed without the necessary participation of sensation, the excitement of which appears to be essential to the performance of the latter.—The Automatic movements are found to be gradually brought under the control of the Intelligence and Will, as we rise towards Man, in whom those faculties are most strongly developed, so as to keep the consensual as well as the reflex actions in subordination to the more elevated purposes of his existence.

755. There are many sensations, however, which do not thus immediately give rise to muscular movements; their operation being rather that of stimulating to action the Intellectual powers. There can be little doubt that *all* Mental processes are dependent, in the first instance, upon Sensations; which serve to the Mind the same kind of purpose, that food and air fulfil in the economy of the body. If we could imagine a being to come into the world with its mental faculties fully prepared for action, but destitute of any power of receiving sensations, these faculties would never be aroused from the condition in which they are in profound sleep; and the being must remain in a state of complete unconsciousness, because there is nothing of which it can be made sensible, no kind of Idea which can be aroused within it. But after the mind has once been in active operation, the destruction of all *future* power of receiving sensations would not reduce it again to the inactive condition; for sensations and ideas are so stored up in the mind, by the power of Memory, that it can feed (as it were) upon the *past*. Now the ideas which are excited by sensations, if associated with feelings of pleasure or pain, constitute those Emotions, or 'moving powers of the mind,' which are, either directly or indirectly, the springs of the greater part of our actions. When strongly excited, the emotions may produce movements which are purely involuntary, and which the Will may not be able to restrain; but in their normal exercise, they act in subordination to the reasoning processes, and their operation upon the movements of the body is through the medium of volition. It is, indeed, the peculiar character of voluntary movement, that it is the result of the exercise of the Reasoning powers, being the expression of a definite *purpose*, of a *designed* adaptation of means to ends, on the part of the individual performing it; instead of proceeding from a mere blind indiscriminating impulse, such as that which is the mainspring of the instinctive and emotional actions. It is in Man, that we find the highest development of the reasoning faculties; but it is quite absurd to limit them to him, as some have done; since no impartial observer can doubt, that many of the lower animals can execute reasoning processes, as complete in their way as those of Man, though much more limited in their scope.

756. Thus, then, we have to consider the Nervous system under three heads;—*first*, as the instrument of the Reflex actions;—*second*, as the instrument of the Consensual and Instinctive actions;—*third*, as the instrument of the Intellectual processes, and of Voluntary movements. Now there is good reason to believe, that to each of these groups of actions a particular portion of the Nervous Centres, with its afferent and efferent nerves, may be assigned; one ganglion, or series of ganglia, being limited in its function to the reception of simple *impressions*, and to the

origination of *reflex* movements consequent thereon; whilst another ganglion, or series of ganglia, is set apart for the reception of impressions of which the mind becomes conscious as *sensations*, and for the origination of the *consensual* movements which these excite; whilst a third, which is superadded to the foregoing in Vertebrated animals, receiving from the proper sensorium the sensations which have been there called up, is the instrument through which these sensations call *ideas* into existence, by which these ideas are employed in the various processes of *intellection*, and by which the *voluntary* determination is brought to bear upon the muscular apparatus. But we shall find reason to believe, that just as this Cerebral ganglion is called into action through the sensory portion of the automatic apparatus, so does it react through the motor portion of the same apparatus; and that it has no more direct connection with the muscular system, than it has with the organs of sense.

757. What has been hitherto said refers exclusively to that division of the Nervous system, which is concerned in the reception of impressions, the production of sensations, and the stimulation of muscles to contraction; and as these are all purely animal functions, it has been called the *nervous system of animal life*. There is another set of ganglia and nerves, however, which constitute what is termed the *Sympathetic* or visceral system; this is distributed to the various parts of the nutritive and secretory apparatus, its fibres forming a plexus upon the walls of the blood-vessels, and being distributed with them to all parts of the body; and hence it has been designated the *nervous system of organic life*. We have seen reason to believe, however, that the functions of Nutrition and Secretion are not themselves dependent upon nervous agency (§ 364); and the relation of the Sympathetic system to them, therefore, can scarcely be so intimate as that designation would seem to imply. Probably it is the vehicle for the involuntary action of the emotions upon those functions; and it may perhaps serve also to harmonize them with each other. —We shall now take a brief survey of the comparative structure of the Nervous system in the principal groups of Animals, and enquire what actions may be justly attributed to its several parts in each instance; commencing with those in which the structure is the simplest, and the variety of actions the smallest; and passing on gradually to those, in which the structure is increased in complexity by the addition of new and distinct parts, and in which the actions present a corresponding variety.

2. *Comparative View of the Structure and Actions of the Nervous System, in the principal groups of the Animal Kingdom.*

758. In the beings which have been associated under the designation of *Protozoa*, no definite traces of a Nervous System can be discovered; and it is maintained by many Physiologists, that, in these animals, the nervous matter is present in a "diffused form,"—that is to say, incorporated with the tissues; but it would be difficult to assign a valid reason for a supposition so gratuitous. An arrangement of this kind cannot be required to confer on the individual parts of the organism their vital properties; since these exist, to as great an extent, in beings which are allowed to be entirely destitute of it, namely the entire Vegetable king-

dom. The simplest office of a nervous system is, as we have seen (§ 752), to establish a communication between parts specially modified to receive impressions, and others particularly adapted to respond to them. Where every portion of the body has similar endowments, there can be no object in such a communication; just as, where every part of the surface is equally capable of absorption, and every part of the tissue equally permeated by nutrient fluid, there is no necessity for a circulating system. The motions exhibited by Animals of these lowest classes, would seem to be scarcely less directly dependent upon external stimuli, than are those of Plants; being, in fact, the result of the general diffusion of that exalted degree of irritability, which is restricted in most plants to particular parts of the structure (§ 751).

759. No valid ground can be shown for assigning a higher character to the movements even of *Zoophytes*, such as the *Hydra* and *Actinia*, or their composite allies; since all those actions which are concerned in the prehension and ingestion of food, are, as already remarked, really analogous rather to the peristaltic movements of the œsophageal muscles and stomach of higher animals, than to the motions of their limbs; and the few which they exhibit, that are not referable to this category, appear to be the result of the direct stimulation of light or other physical agencies, which may operate by producing organic changes in their tissues, as in those of Plants. At any rate it may safely be affirmed, that if these Zoophytes do possess any rudiment of a Nervous System, and if any of their actions are to be attributed to its instrumentality, its participation in the general course of their vital actions must be very trifling. We have at present, it must be acknowledged, no certain means of appreciating the degree of sensibility possessed by the lowest members of the Animal kingdom. The motions which follow the impressions of external agents, are our only means of judging of its possession by a particular being; and in the interpretation of these phenomena, the greatest care is requisite to avoid being misled by false analogies. Much error has probably arisen, from comparing the manifestations of life exhibited by creatures of this doubtful character, with those of the highest Animals; and thence inferring that, because motions are witnessed in the former which bear some analogy to those of the latter, they must be equally dependent on a nervous system. But, when it is considered how completely vegetative is the life of such beings, and how closely all their motions are connected with the performance of their organic functions, it would seem obvious that the general comparison should be made with Plants, rather than with Animals; and that we should seek the assistance of principles of a higher character, only when those we already possess are insufficient to explain the phenomena. A nervous system would seem to be required only in a being possessed of a number of distinct organs, whose actions are of such a character that they cannot be brought into mutual relation, without a more immediate and direct communication than that afforded by the circulating system, which, as we have seen, is the only bond of union possessed by Plants between distant parts. In the lowest and simplest Animals, whatever degree of contractility exists, appears to be almost equally diffused through the system; and we neither find any special sensory organs, adapting one part more than another to the reception of impressions, nor do we observe any portion of the structure peculiarly endowed

with the power of motion; neither can we discover anything like a nervous system fitted to receive such impressions, and to excite respondence to them in distant parts. To use the forcible expression of Sir Gilbert Blane—"Mr. Hunter, by a happy turn of expression, calls the function of the nervous system *internuncial.* It is evident that some such principle must exist in the complicated system of the superior Animals, in order to establish that connection which constitutes each individual a WHOLE." But where all the parts act for themselves, there is, as we have seen, no necessity for such an internuncial communication; and consequently, although, when united, their functions all tend towards the maintenance of the system to which they belong, they are capable of being separated from it and from each other, without these functions being necessarily abolished.

760. It is in the higher *Radiata*, that we find the first definite indications of the existence of a connected Nervous system. It is probable that such exists in all the *Acalephæ*, although the softness of their tissues renders it difficult of detection. According to Ehrenberg, two nervous circles may be detected in the *Medusa*;—one running along the margin of the mantle, and furnished with eight ganglia, from which filaments proceed to the eight red spots which he supposes to be eyes,—whilst the other is disposed around the entrance to the stomach, and is furnished with four ganglia, from which filaments proceed to the tentacula. A nervous ring has also been detected by Prof. Agassiz in *Sarsia*, one of the 'naked-eyed' Pulmogrades; and he states that it is entirely composed of ganglionic cells.* In *Beroe*, it is affirmed by Dr. Grant that a nervous ring exists round the mouth, furnished with eight ganglia, from each of which a filament passes towards the other extremity of the body, while others are sent to the lips and tentacula.—In the *Echinodermata*, however, its manifestations are much less equivocal. In the *Asterias*, for instance, a ring of nervous matter surrounds the mouth, and sends three filaments to each of the rays; of these, one seems to traverse its length, and the two others to be distributed on the cæcal prolongations of the stomach. In the species examined and figured by Tiedemann, no ganglionic enlargements of this ring seem to exist, and it seems not improbable that, as in the Medusæ, the entire ring is composed of vesicular nervous matter; but this element is usually collected into distinct ganglia, which are found at those points of the ring whence the branches diverge, the number of these ganglia being always equal to that of the rays. In those species which possess ocelli at the extremities of the rays, the nervous cord proceeding towards each swells into a minute ganglion in its neighbourhood.—In the *Echinus*, the arrangement of the nervous system follows the same general plan; the filaments which diverge from the oral ring being distributed (in the absence of rays) to the complicated dental apparatus, whilst others pass along the course of the vessels to the digestive organs. The transition between the Radiata and Articulata, presented by the *Holothuria* and *Sipunculus*, is peculiarly well marked in the

* "Contributions to the Natural History of the Acalephæ of North America," Part I., p. 232.—It is considered by Prof. Agassiz that the movements of these Acalephæ are voluntary and exhibit a purpose; but his statements on this subject exhibit a very strange confusion of ideas; and the Author is assured by Mr. Huxley that he has never seen the slightest evidence of anything beyond automatic action in these animals.

nervous system of these animals; for the ring which encircles the mouth is here comparatively small, but two filaments traverse the length of their prolonged bodies, running near the abdominal surface, which is their situation in the Articulated classes.—When we compare the character of the nervous system of these Radiated classes with that of higher Animals of more heterogeneous structure, we find that every segment of the body which is similar to the rest, is connected with a ganglionic centre, that seems subservient to the functions of its own division alone, and to have little communication with or dependence upon the remainder; these centres being all apparently similar to each other in their endowments. This is the case, indeed, not merely with the tribes we are now considering, but with the lower Articulata; the chief difference being, that the individual portions are here disposed in a radiate manner round a common centre, whilst in the latter they are longitudinally arranged. But when the different organs are so far specialized as to be confined to distinct portions of the system, and each part consequently becomes possessed of a different structure, and is appropriated to a separate function, this repetition of parts in the nervous system no longer exists; its individual portions assume special and distinct offices; and they are brought into much closer relation to one another, by means of the *commissures* or connecting fibres, which form a large part of the nervous masses in the higher Animals. It is evident that, between the most simple and the most complex forms of this system, there must be a number of intermediate gradations,—each of them having a relation with the general form of the body, its structure and economy, and the specialization of its distinct functions. This will be found, on careful examination, to be strictly the case; and yet, with a diversity of its parts, as great as exists in the conformation of any other organs, its essential character will be found to be the same throughout.

761. Among the *Molluscous* classes, there is no repetition of parts like that just described; and the nervous system partakes of that general want of symmetry in the body, which seems so characteristic of the predominance of the vegetative organs in these animals (§ 297). Its ganglionic centres are for the most part disposed near the mouth, their actions being apparently destined to little else than the supply of the digestive organs; and their size is usually proportional to the development of the organs of special sensation which are connected with them, and to the energy of the masticatory movements required for the reduction of the food. Where, however, unusually active powers of locomotion are possessed, we commonly find ganglia situated in the neighbourhood of the organs destined to serve this purpose; but the most constant of all the ganglia in the Mollusca is that which is specially connected with the gills, and is hence termed the *branchial* ganglion. This, indeed, is the only one which distinctly presents itself in the lowest animals of this type; those, namely, belonging to the groups of *Bryozoa* and *Tunicata*. In many of the former group, but most distinctly in the fresh-water species, a small body may be seen at the base of the tentacular circle (Fig. 114, A, *a*), between the oral and anal orifices; which, from its appearance, and its correspondence in position with the undoubted ganglion of the Tunicata, may be regarded almost with certainty as a nervous centre. No branches, however, have been seen to proceed from it, either to the tentacula, or to

the muscles; but the activity of the movements of these animals, with the perfection of their muscular structure (the fibres being striated in many species), would seem to indicate that nervous instrumentality is concerned in them.—In the Tunicata there is no difficulty in discovering the nervous ganglion, which is always situated between the oral and anal orifices (Fig. 117, B, *f*, Fig. 119, *j*, Fig. 121, *j*), being sometimes single, and sometimes bilobed or even double; this ganglion is connected by branches with the sensory tentacula which guard the oral orifice, and branches extend from it also over the muscular coat of the branchial sac, filaments being specially transmitted to the oral and anal sphincters. Other ganglia of smaller size have been described by Meckel as situated between the stomach and the branchial sac; but it is doubtful whether the bodies in question belong to the nervous system at all.—No beings possessed of a complex internal structure, a distinct stomach and alimentary tube, a pulsating heart and ramifying vascular apparatus, with branchial appendages for aerating the blood, and highly-developed secretory and reproductive organs, can be imagined to spend the period of their existence in a mode more completely vegetative than these. The continuous and equable current of water which enters the cavity of the mantle, and which serves at the same time to convey food to the stomach and to aerate the blood, is maintained, without any nervous agency, by the vibrations of the cilia with which the walls of the cavity are clothed; and it is only when substances come in contact with the sensitive tentacula, which ought not to enter the orifice, or when they have been introduced and are to be expelled, or when some external irritation is applied, that we see anything like a definite muscular movement, indicating the agency of a nervous system. The ganglion appears to control the circular sphincters which surround the orifices of the mantle, as well as the muscular fibres which are spread in a network over the whole sac; by the contraction of the former, the ingress of improper substance is prevented; and by that of the latter, the fluid contained in the cavity is violently ejected, as occurs when the animal is alarmed by any external contact, or is excited to the same action by any internal irritation. This ganglion, then, must be regarded as chiefly *branchial*; and, from the analogy of higher animals, we should be led to regard its actions as altogether 'automatic,' the movements just described being analogous to those of coughing and sneezing in Man. Rudimentary organs of special sense, however, are found in some Tunicata; and we must regard this ganglion as the seat of whatever consciousness these animals may possess.

762. In the *Conchifera*, there is a greater separation of the nervous centres, in accordance with the higher development of the general organisation; but the branchial ganglion is still the principal one, and this is found near the posterior extremity of the body. At the sides of the œsophagus, however, there are seen two small ganglia (Fig. 299, *a*, *a*), united by a band which arches over it, and usually, also, by another pair of filaments which meet beneath it; in some of the higher Conchifera (such as the *Mactra*) these two ganglia nearly coalesce into a single mass above the œsophagus, and are then seen to be the evident analogues of the *cephalic* ganglia of the Articulata and Vertebrata. These ganglia are connected by nervous filaments (probably of an *afferent* character) with the sensory tentacula that guard the oral orifice; and they send backwards a

pair of trunks, *e*, which enter the *branchial* ganglion, whose branches are extensively distributed to all parts of the respiratory apparatus, siphons, &c., and also to the adductor muscle. In the *Solen*, and other Conchifera which possess a foot, a third principal nervous centre, the *pedal* ganglion (*b*), is found at its base, and seems peculiarly connected with its actions; but where the foot is absent, as in the *Oyster* tribe, this does not exist. The pedal ganglion has always an immediate connection with the cephalic; and in general, this connection is quite distinct from that of the branchial ganglion, so that the pedal and branchial ganglia have not even an apparent connection with each other. Sometimes, however, the pedal ganglion has no such obviously distinct communication with the cephalic, being situated on the nervous cords which pass backwards from the cephalic to the branchial; and the pedal and branchial ganglia thus seem to be connected together. The normal relations of the parts, however, are not altered by such an arrangement; for the trunks which connect the branchial ganglia with the cephalic pass *over*, not *through*, the former, which has its own band of union as structurally distinct as in the Solen.—Besides these principal centres, we meet with numerous smaller ones (*f*, *f*,) upon the nervous cords which proceed from them to the different parts of the general muscular envelope.

Fig. 299.

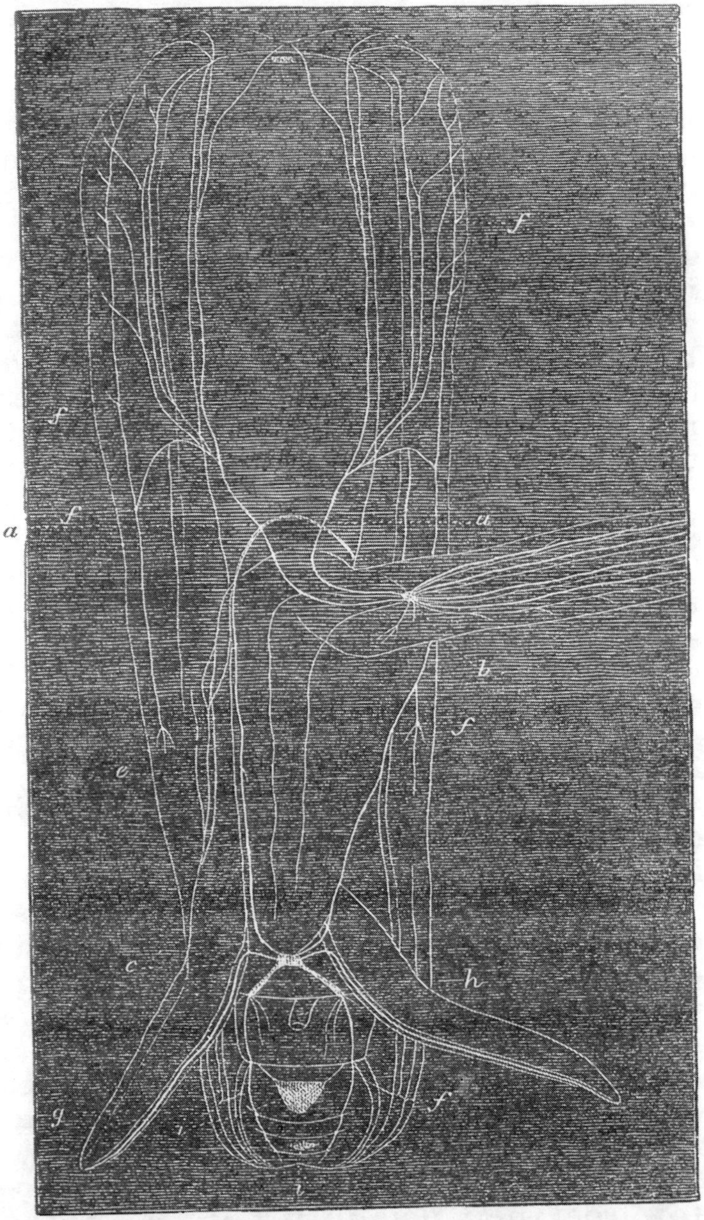

Nervous system of *Solen*;—*a*, *a*, cephalic ganglia, connected by a transverse band arching over the œsophagus; *b*, pedal ganglion, the branches of which are distributed to the powerful muscular foot; *c*, branchial ganglion, the branches of which proceed to the gills *g*, the siphons, *i*, *i*, and other parts; *e*, trunks connecting the cephalic and branchial ganglia; *f*, *f*, *f*, *f*, minute ganglia on the branches distributed to the mantle; *h*, anus.

763. A good deal of variety exists in this class, in regard to the actions to which the nervous system is subservient. Many of those which have no foot (such as the *Oyster*) are attached for life to the place where they originally fix themselves; and no evident motion is exhibited by these, save the opening and closure of the shell, which corresponds with the dilatation and contraction of the sac of the mantle in the Tunicata. This is principally accomplished by the posterior ganglion, which supplies the

adductor muscle. Other species, being unattached, are enabled to swim by the flapping of the valves in the water. Among those which have a foot, this organ is employed for many different purposes, such as burrowing, leaping, &c.; which are generally executed in such a manner, as to imply that they are in some degree under the direction of consciousness. And it is in the most free and active of these Bivalves, that we have the most distinct indications of the existence of organs of special sensation; these, however, where they exist, are not restricted to any particular part of the body, but are disposed along the free margins of the lobes of the mantle. For reasons which will be given hereafter, it appears probable that, whilst the *general* movements of the foot are directed by the cephalic ganglia, under the guidance of sensations, the *particular* actions by which it fixes itself on a given surface, and adapts its disc to the inequalities which it encounters, may be produced simply by impressions conveyed to this ganglion through its afferent nerves, and reflected through its motor fibres, in which sensations are not necessarily concerned. The same may be inferred respecting the actions of the branchial ganglion; which, so far as they participate in the respiratory function, may probably be considered as simply reflex; but when the adductor muscle is caused by it to draw together the valves, either for protection from threatened injury, or for the propulsion of the body through the water, the movement being prompted or directed by consciousness, we must regard the cephalic ganglia as its original source.—The whole course of the lives of these animals shows them to be so little elevated in the scale of psychical endowment, that we can scarcely regard the motions executed by them as possessing a voluntary character. The greater part of them are concerned in protecting the animal from danger, and in the prehension of its food; and may be compared in the higher animals to the closing of the glottis against irritating matters, and to the contraction of the pharynx in swallowing.

764. As the head is not otherwise indicated, in the two preceding classes of Mollusca, than by the position of the mouth, and as the organs of special sensation are but very imperfectly evolved, it is not to be wondered at that the cephalic ganglia should be inferior in size to those more connected with powerful and active muscles. But in the higher classes it is very different; and in proportion as we meet with evidence of the possession of the senses of sight, hearing, &c., and find these organs no longer scattered over the periphery of the body, but collected upon a 'head,' do we observe a greater concentration of the ganglionic system towards their neighbourhood.—In the lower species of *Gasteropoda*, the general distribution of the nervous system is not very dissimilar to that which has been last described. Thus, in *Carinaria* (Fig. 129), the cephalic ganglia are still found separate, lying at the sides of the œsophagus, and connected by a transverse band; and these receive the optic nerves and tentacular filaments. Besides other branches transmitted to the neighbouring organs, two principal trunks are sent backwards, which unite in a large ganglion situated among the viscera; and from this, nerves proceed to the foot, respiratory organs, and posterior part of the trunk; so that it may be regarded as combining the functions of the pedal and branchial ganglia, as is further indicated by its quadrilobate form. Generally, however, the cephalic ganglia are much larger in

proportion to the rest, than they are in Conchifera ; and they show a tendency to approximate each other, and to gain a position *above* the œsophagus, as we see in *Paludina* (Fig. 128, *u*, *u*) ; or even to meet upon it in a single mass, as in *Aplysia* (Fig. 127, *d*). The *branchial* ganglion is constantly to be met with, but its position is extremely variable, like that of the organs to whose action it is related ; it is generally found in the immediate neighbourhood of the respiratory organs ; and, as in the preceding group, it is always connected with the cephalic centres by a distinct commissural trunk. It is not unfrequently conjoined into one mass with the pedal or palleal centres ; but its presence may always be recognised by the distribution of its nerves, as well as by its separate connection with the cephalic ganglia. The position of the *pedal* ganglion, which is constantly found in the Gasteropoda, and which is generally *double* though the foot is single, also varies, but in a less degree ; since it is usually near the ventral surface, at no great distance behind the head. In addition to the foregoing centres, which are common to the Gasteropoda with the higher Conchifera, we find that the muscular mantle usually has a pair of ganglia of its own, which are termed the *palleal ;* these are sometimes united with the branchial, sometimes with the pedal, according, perhaps, as the movements of the mantle most participate in the respiratory acts, or in the general locomotion of the body ; but they are recognized by the distribution of their nerves, and by their separate connection with the cephalic ganglia. Thus in Aplysia, the cord of communication which unites the single supra-œsophageal ganglion (Fig. 127, *d*) with the two infra-œsophageal ganglia (*e*, *e*), is triple ; the latter are known, by the distribution of their nerves, to combine the functions of pedal and palleal ganglia, whence we can account for their double connection with the cephalic ; and the third cord passes through them to the branchial ganglion (*m*), which is single, and lies at the posterior part of the body. The pedal, branchial, and palleal ganglia are all found separate from one another in some instances ; but in other cases they are all brought together in the neighbourhood of the œsophagus. This is the usual arrangement in the 'naked' Gasteropods, whether pulmonated or branchiferous. It is in the former that the concentration is carried to its greatest extent ; thus in the *Slug*, we find but a single nervous mass beneath the œsophagus, in which the three pairs of ganglia have become fused. In the Nudibranchiata, on the other hand, these ganglionic centres for the most part remain distinct, although they are aggregated together.—Besides the foregoing system of ganglia and nerves, we find, in many of the Gasteropoda, a separate system connected with the complicated apparatus of manducation and deglutition, which this class usually possesses. The nerves which supply these organs do not proceed from the cephalic ganglia, but from a distinct centre, which is usually placed in advance of them upon the mouth or pharynx ; and their ramifications proceed also along the œsophagus to the stomach. This set of nerves and ganglia, which is even more important from its relative development in other classes, may be called from its distribution the *stomato-gastric* system. We shall see that analogous ganglionic centres, and nerves of very similar distribution, exist in Vertebrata ; but that they no longer constitute a distinct apparatus, being fused, as it were, into the general Cerebro-spinal system.

765. In our examination of the Nervous System of the Mollusca, we have hitherto met with a progressive multiplication of its centres, in accordance with the progressive complication of the organism, and especially with the increase of its sensori-motor powers. To the purely 'reflex' actions, immediately connected with the maintenance of organic life, we have found that general locomotive movements are superadded, which, as they involve the guidance of sensations, must be regarded as belonging to the 'consensual' class; and as a distinct foot is developed, the muscular coat of the mantle thickened, a complicated apparatus for the reduction of the food provided, and more perfect organs of special sense evolved, do we find a corresponding development of ganglia which are their centres of action. In no instance, however, are these ganglia repetitions of each other, except on the two sides of the median line; and there is no evidence that the cephalic ganglia have any other function, than that of receiving sensory impressions, communicating them (so to speak) to the consciousness, and calling forth respondent muscular movements. Nor is there anything in the habits of life of these animals, which indicates that their actions are of a nature above the automatic.

FIG. 300.

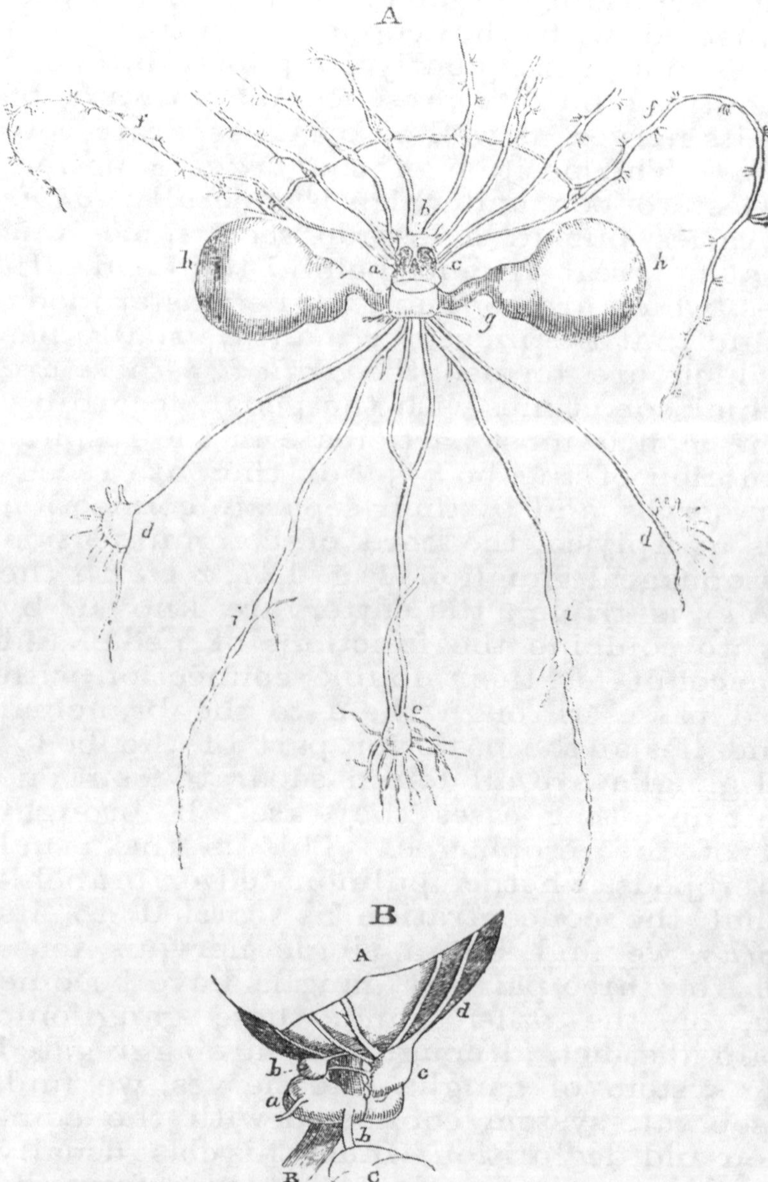

Nervous system of *Argonauta argo*; A, as seen in front; B, as viewed in profile, showing the relations of the nervous centres to the buccal mass A, the œsophagus B, and the eye C;—*a*, cephalic ganglion; *b*, buccal ganglion; *c*, sub-œsophageal ganglion; *d*, *d*, stellate ganglia of the mantle; *e*, visceral ganglion; *f*, nerves of the arms, with ganglionic enlargements; *g*, optic nerves; *h*, *h*, eyes; *i*, *i*, branchial nerves with their ganglia.

766. The Nervous System of the *Cephalopoda* exhibits an obvious approach towards that of Vertebrated animals, in the concentration of the cephalic ganglia into one mass, which, though still perforated by the œsophagus, lies almost entirely above it, and is sometimes protected by plates of cartilage that constitute the rudiment of a neuro-skeleton. In the *Nautilus*, however, which is the type of the inferior or *Tetrabranchiate* order of this class, the general distribution of this system corresponds pretty closely with that seen in the higher Gas-

teropoda. The œsophagus is still encircled with a ganglionic ring, of which the upper part gives off the optic nerves, whilst the lower supplies the mouth and tentacula, and sends trunks backwards into the shell. The trunk which supplies the internal tentacula, and what is regarded as the olfactory organ, has a small ganglion situated upon it; and other ganglia, which probably belong to the stomato-gastric and sympathetic systems, are found on the nerves distributed to the viscera. In the *Dibranchiate* order, whose habits are more active and whose general organisation is higher, we find a somewhat different arrangement (Figs. 300–302). The organ of vision here attains an increased development and importance; an organ of hearing evidently exists; and the whole surface of the body is possessed of sensibility. In *Argonauta*, the cephalic ganglia are united into a single large mass on the median line (Fig. 300, *a*), from the lateral portions of which the optic nerves are given off; but in *Sepia* and *Octopus*, the optic ganglia (Figs. 301, 302, *b*, *b*) are distinct from the median portion, which is considered by Prof. Owen as an olfactive ganglion. The nerves which supply the buccal apparatus are usually connected with a distinct ganglion in advance of the cephalic (Fig. 300, *b*, Fig. 302, *e*); but sometimes this ganglion is fused into the general mass, as happens in Octopus (Fig. 301). Besides this *buccal* ganglion, we find in Sepia a *labial* ganglion (Fig. 302, *f*), which supplies the parts immediately surrounding the entrance to the mouth. The nerves which supply the arms, proceed, in all Cephalopods, from the anterior part of the sub-œsophageal mass; and in the Dibranchiata, whose arms are furnished with suckers, the nerve-trunks (Fig. 300, *f*, *f*) are beset with a series of ganglia in connection with those organs, every sucker possessing a ganglion of its own. It has been shown by Dr. Sharpey, that the principal trunk of each arm may be divided into two distinct tracts; in one of which there is nothing but *fibrous* structure, forming a direct communication between the suckers and the cephalic ganglia; whilst in the other are contained the *ganglia* which peculiarly appertain to the suckers, and which are connected with them by distinct filaments: so that each sucker has a separate relation with a ganglion of its own, whilst all are alike connected with the cephalic ganglia, and are placed under their control. We see the results of this arrangement, in the modes in which the contractile power of the suckers may be called into operation. When the animal embraces any substance with its arm (being directed to this

FIG. 301.

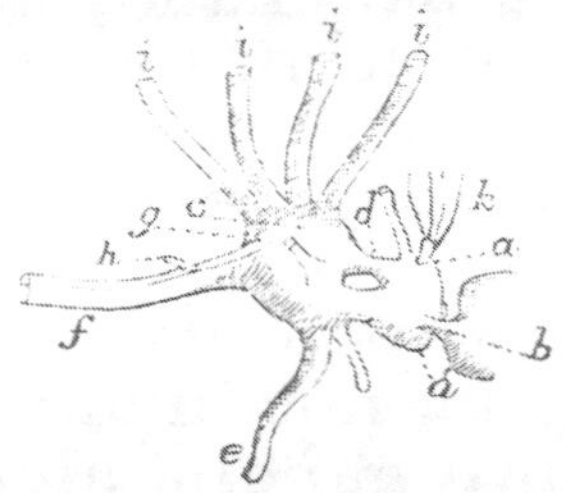

Nervous centres of *Octopus*; — *a*, pharyngeal ganglion; *a'*, proper cephalic ganglion, *b*, optic ganglion; *c*, sub-œsophageal ganglion; *d*, aperture giving passage to the œsophagus; *e*, one of the great palleal nerves; *f*, principal visceral nerve; *g*, nerve of the funnel; *h*, auditory nerves; *i*, *i*, *i*, *i*, nerves of the arms; *k*, buccal nerves.

FIG. 302.

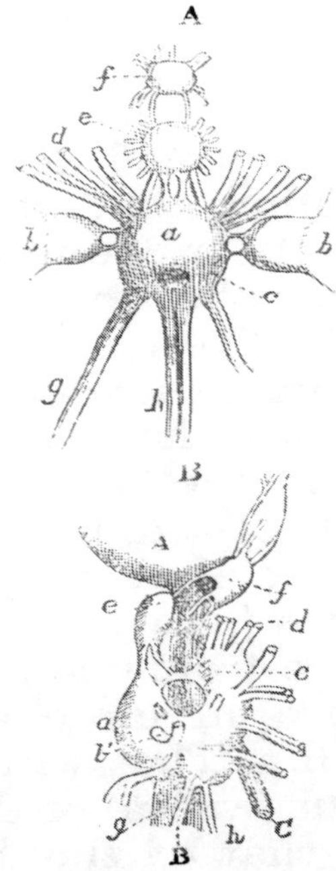

Nervous centres of *Sepia officinalis*;—A, viewed from above; B, the same as seen from the side, with its relations to A, the buccal mass, B, the œsophagus, and C the aorta;—*a*, cephalic ganglion; *b*, *b*, optic ganglion; *c*, sub-œsophageal ganglion; *d*, *d*, nerves of the arms; *e*, buccal ganglion; *f*, labial ganglion; *g*, nerves of the mantle; *h*, visceral nerve.

action by its sight or other sensation) it can bring all the suckers simultaneously to bear upon it; evidently by the transmission of an impulse along the motor cords, that proceed from the cephalic ganglia to the suckers. On the other hand, any individual sucker may be made to contract and attach itself, by placing a substance in contact with it alone; and this action will take place equally well, when the arm is separated from the body, or even in a small piece of the arm when recently severed from the rest,—thus proving that, when it is directly excited by an impression made upon itself, it is a *reflex* act, quite independent of the cephalic ganglia, not involving sensation, and taking place through the medium of its own ganglia alone. In the *Octopus,* which swims by a kind of circular fin (Fig. 131), formed by a membrane connecting the tentacula, a curious connection exists between the nerves radiating to these from the cephalic ganglia; for, at the base of the arms, a nervous ring is found (like that of the Asterias § 760), which unites them all, and probably contributes to harmonise their actions.—From the posterior part of the sub-œsophageal mass, proceed the nerve trunks which supply the general surface of the mantle; and these enter two large stellate ganglia (Fig. 300, *d, d*), before separating into their ramifications. Other trunks, also furnished with ganglia of their own (*i, i*), supply the respiratory apparatus; and in the centre we find a single or double trunk, which passes towards the digestive apparatus, and enters the visceral ganglion (*e*), which may probably be considered as belonging to the sympathetic system.—Thus we have traced in the Cephalopoda, the highest development of a nervous system formed to minister chiefly to the nutritive functions; we shall now follow that of the Articulata, in which the locomotive powers are especially predominant; and we shall afterwards find that the Vertebrata combine the types characteristic of both.

767. The plan on which the Nervous system is distributed in the subkingdom *Articulata,* exhibits a remarkable uniformity throughout the whole series; whilst its character gradually becomes more elevated, as we trace it from the lowest to the highest divisions of the group. It usually consists of a double nervous cord, studded with ganglia at intervals; and the more alike the different segments, the more equal are these ganglia. The two filaments of the nervous cord are sometimes at a considerable distance from one another, and their ganglia distinct; but more frequently they are in close apposition, and the ganglia appear single and common to both. That which may be regarded as the typical conformation of the nervous system of this group, is seen in the ganglionic cord of *Scolopendra,* or in that of the larvæ of most Insects, such as that of *Sphinx ligustri* (Fig. 303). Here we see the nervous cord nearly uniform throughout, its two halves being separated, however, in the anterior portion of the body; the ganglia are disposed at tolerably regular intervals, are similar to each other in size (with the exception of the last, which is formed by the coalescence of two), and every one supplies its own segment, and has little connection with any other. The two filaments of the cord diverge behind the head, to enclose the œsophagus; above which we find a pair of ganglia, that receive the nerves of the eyes and antennæ. We shall find that, in the higher classes, the inequality in the formation and office of the different segments, and the increased powers of special sensation, involve a considerable change in the nervous system, which is concentrated

about the head and thorax. In the simplest Vermiform tribes, on the other hand, we lose all trace of separate ganglia, the nervous cord passing without evident enlargement from one extremity to the other. Whatever may be the degree of multiplication of the ganglia of the trunk, they seem but repetitions of one another; the functions of each segment being the same with those of the rest. The *cephalic* ganglia, however, are always larger and more important; they are connected with the organs of special sense; and they evidently possess a power of directing and controlling the movements of the entire body, whilst the power of each ganglion of the trunk is confined to its own segment. —The longitudinal ganglionic cord of Articulata occupies a position, which seems at first sight altogether different from that of the nervous system of Vertebrated animals; being found in the neighbourhood of the *ventral* or inferior surface of their bodies, instead of lying just beneath their *dorsal* or upper surface. From the history of their development (§ 720), however, and from some other considerations, it has been suggested that the *whole* body of these animals may be considered as in an inverted position; the part in which the segmentation is first distinguished in Insects being the equivalent of the dorsal region in Vertebrata, and that over which the germinal membrane is the last to close-in, being homologous with the ventral region. This view applies also to the position of the 'dorsal vessel,' which would then be on the ventral side of the axis, as in Vertebrata. Regarded under this aspect, the longitudinal nervous tract of Articulata corresponds with the Spinal cord of Vertebrated animals in *position*, as we shall find that it does in *function*.

768. When the structure of the chain of ganglia is more particularly

Fig. 303.

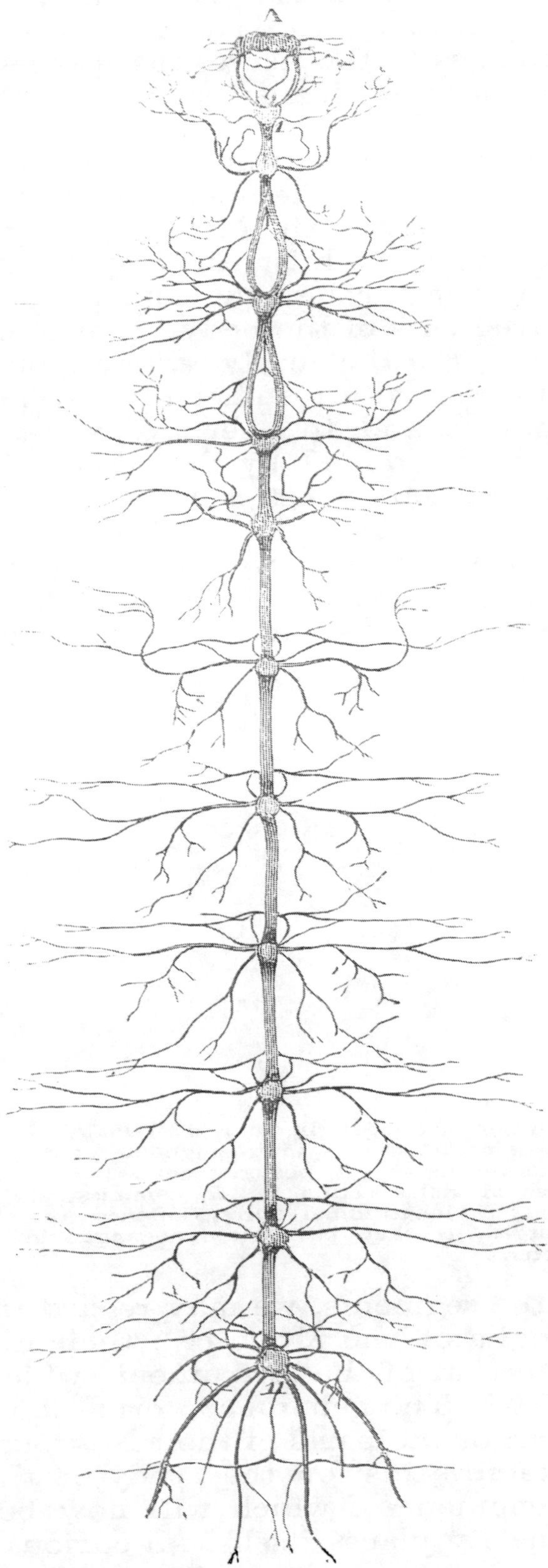

Nervous System of Larva of *Sphinx ligustri*:—A, cephalic ganglia; 1-11, ganglia of the trunk.

inquired into, it is found to consist of two distinct tracts; one of which is composed of nerve *fibres* only, and passes backwards from the cephalic ganglia, over the surface of all the ganglia of the trunk, giving off branches to the nerves that proceed from them; whilst the other includes the *ganglia* themselves (Fig. 305, A). Hence, as in the Mollusca, every part of the body has two sets of nervous connections; one with the cephalic ganglia; and the other with the ganglion of its own segment. Impressions made upon the afferent fibres, which proceed from any part of the body to the cephalic ganglia, become *sensations* when conveyed to the latter; whilst, in respondence to these, the *consensual* impulses, operating through the cephalic ganglia, harmonize and direct the general movements of the body, by means of the efferent nerves proceeding from them. For the purely *reflex* operations, on the other hand, the ganglia of the ventral cord are sufficient; each one ministering to the actions of its own segment, and, to a certain extent also, to those of other segments. It has been ascertained by the careful dissections of Mr. Newport, to whom we owe all our most accurate knowledge of the structure of the Nervous system in Articulated animals,* that of the fibres constituting the roots, by which the nerves are implanted in the ganglia, some pass into the vesicular matter of the ganglion, and, after coming into relation with its vesicular substance, pass out again on the same side (Fig. 304, *f, k*); whilst a second set, after traversing the vesicular matter, pass out by the trunks proceeding from the opposite side of the same ganglion; and a third set run along the portion of the cord which connects the ganglia of different segments, and enter the nervous trunks that issue from them, at a distance of one or more ganglia above or below. Thus it appears, that an impression conveyed by an afferent fibre to any ganglion, may excite a motion in the muscles of the same side of its own segment; or in those of the opposite side; or in those of segments at a greater or less distance, according to the point at which the efferent fibres leave the cord. And as the function of these ganglia is altogether related to the locomotive actions of the segments, we may regard them as so many repetitions of the *pedal* ganglia of the Mollusca; their multiplication being in precise accordance with that of the instruments which they supply.

FIG 304.

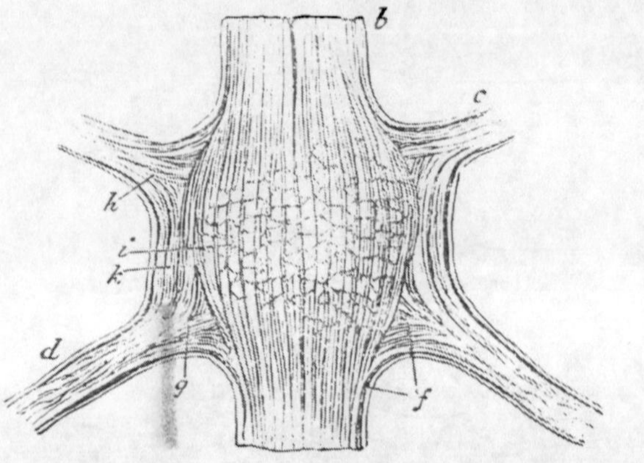

Portion of the ganglionic tract of *Polydesmus maculatus*: — *b*, inter-ganglionic cord; *c*, anterior nerves; *d*, posterior nerves; *f, k*, fibres of reflex action; *g, h*, commissural fibres; *i*, longitudinal fibres, softened and enlarged, as they pass through ganglionic matter.

769. The general conformation of Articulated animals, and the arrangement of the parts of their Nervous system, render them peculiarly favourable subjects for the study of the *reflex* actions; some of the principal phenomena of which will now be described.—The *Mantis religiosa* customarily places itself in a curious position, especially when threatened or

* See his successive papers in the "Philosophical Transactions" from 1832 to 1843, and particularly his Memoir on the Nervous System of Myriapoda in the last-named year.

attacked, resting upon its two posterior pairs of legs, and elevating its thorax with the anterior pair, which are armed with powerful claws: now if the anterior segment of the thorax, with its attached members, be removed, the posterior part of the body will still remain balanced upon the four legs which belong to it, resisting any attempts to overthrow it, recovering its position when disturbed, and performing the same agitated movements of the wings and elytra as when the unmutilated insect is irritated: on the other hand, the detached portion of the thorax, which contains a ganglion, will, when separated from the head, set in motion its long arms, and impress their hooks on the fingers which hold it.—If the head of a *Centipede* be cut off, whilst it is in motion, the body will continue to move onwards by the action of the legs; and the same will take place in the separate parts, if the body be divided into several distinct portions. After these actions have come to an end, they may be excited again, by irritating any part of the nervous centres, or the cut extremity of the nervous cord. The body is moved forwards by the regular and successive action of the legs, as in the natural state; but its movements are always forwards, never backwards, and are only directed to one side, when the forward movement is checked by an interposed obstacle. Hence although they might *seem* to indicate consciousness and a guiding will, they do not so in reality; for they are carried on, as it were, mechanically; and show no direction of object, no avoidance of danger. If the body be opposed in its progress by an obstacle of not more than half of its own height, it mounts over it, and moves directly onwards, as in its natural state; but if the obstacle be equal to its own height, its progress is arrested, and the cut extremity of the body remains forced up against the opposing substance,—*the legs still continuing to move.*—If, again, the nervous cord of a Centipede be divided in the middle of the trunk, so that the hinder legs are cut off from connection with the cephalic ganglia, they will continue to move, but not in harmony with those of the fore part of the body; being completely paralysed, so far as the animal's controlling power is concerned; though still capable of performing reflex movements, by the influence of their own ganglia, which may thus continue to propel the body, in opposition to the determinations of the animal itself.—The case is still more remarkable, when the nervous cord is not merely divided, but a portion of it is entirely removed from the middle of the trunk; for the anterior legs still remain obedient to the animal's control; the legs of the segments, from which the nervous cord has been removed, are altogether motionless; whilst those of the posterior segments continue to act, through the reflex powers of their own ganglia, in a manner which shows that the animal has no power of checking or directing them.*

770. The stimulus to the reflex movements of the legs, in the foregoing cases, appears to be given by the contact of the extremities with the solid surface on which they rest. In other instances, the appropriate impression can only be made by the contact of liquid; thus a *Dytiscus* (a kind of water-beetle) having had its cephalic ganglia removed, remained motionless, so long as it rested upon a dry surface; but when cast into water, it executed the usual swimming motions with great energy and rapidity, striking all its comrades to one side by its violence, and persisting in these

* See Newport in the "Philosophical Transactions" for 1843.

for more than half an hour.—Other movements, again, may be excited through the respiratory surface. Thus, if the head of a *Centipede* be cut off, and, while it remains at rest, some irritating vapour (such as that of ammonia or muriatic acid) be caused to enter the air tubes on one side of the trunk, the body will be immediately bent in the opposite direction, so as to withdraw itself as much as possible from the influence of the vapour: if the same irritation be then applied on the other side, the reverse movement will take place; and the body may be caused to bend in two or three different curves, by bringing the irritating vapour into the neighbourhood of different parts of either side. This movement is evidently a reflex one, and serves to withdraw the entrances of the air-tubes from the source of irritation; in the same manner as the acts of coughing and sneezing in the higher animals cause the expulsion, from the air-passages, of solid, liquid, or gaseous irritating matters, which may have found their way into them.

771. From these and similar facts it appears, that the ordinary movements of the legs and wings of Articulated animals are of a reflex nature, and may be effected solely through the ganglia with which these organs are severally connected; whilst in the perfect being, they are harmonised, controlled, and directed by impulses which act through the *cephalic* ganglia and the nerves proceeding from them. There is strong reason to believe, that the operations to which these ganglia are subservient, are almost entirely of a *consensual* nature; being immediately prompted by sensations, chiefly those of sight, and seldom or never by any processes of a truly rational character. When we attentively consider the habits of these animals, we find that their actions, though evidently directed to the attainment of certain ends, are very far from being of the same spontaneous nature, or from possessing the same *designed* adaptation of means to ends, as those performed by ourselves, or by the more intelligent Vertebrata, under like circumstances. We judge of this by their unvarying character,—the different individuals of the same species executing precisely the same movements, when the circumstances are the same; and by the very elaborate nature of the mental operations, which would be required, in many instances, to arrive at the same results by an effort of reason. Of such we cannot have a more remarkable example, than is to be found in the operations of Bees, Wasps, and other social Insects; which construct habitations for themselves, upon a plan which the most enlightened human intelligence, working according to the most refined geometrical principles, could not surpass; but which yet do so without education communicated by their parents, or progressive attempts of their own, and with no trace of hesitation, confusion, or interruption,—the different individuals of a community all labouring effectively to one purpose, because their automatic impulses (producing what are usually termed *instinctive* actions) are all of the same nature. (See §§ 308, 799).

772. Not only are the locomotive ganglia multiplied in accordance with the repetition of segments and members; but the *respiratory* ganglia are multiplied in like manner, in accordance with the repetition of the respiratory organs. The respiratory division of the nervous system consists of a chain of minute ganglia, lying upon the larger cord, and sending off its delicate nerves between those that proceed from the ganglia of the latter, as seen in Fig. 305, C. These respiratory ganglia and their nerves

are best seen in the thoracic portion of the cord, where the cords of communication between the pedal ganglia diverge or separate from one another. And this is particularly the case in the Pupa state, when the whole cord is being shortened, and their divergence is increased. The thoracic portion of the cord, in the Pupa of the Sphinx ligustri, is shown in Fig. 305, B; which represents the 2nd, 3rd, and 4th double ganglia of the ventral cord, the cords of connection between them, here widely diverging laterally, and the small respiratory ganglia, which are connected with each other by delicate filaments that pass over the ganglia of the ventral cord, and which send off lateral branches, that are distributed to the air-tubes and other parts of the respiratory apparatus, and communicate with those of the other system.

Fig. 305.

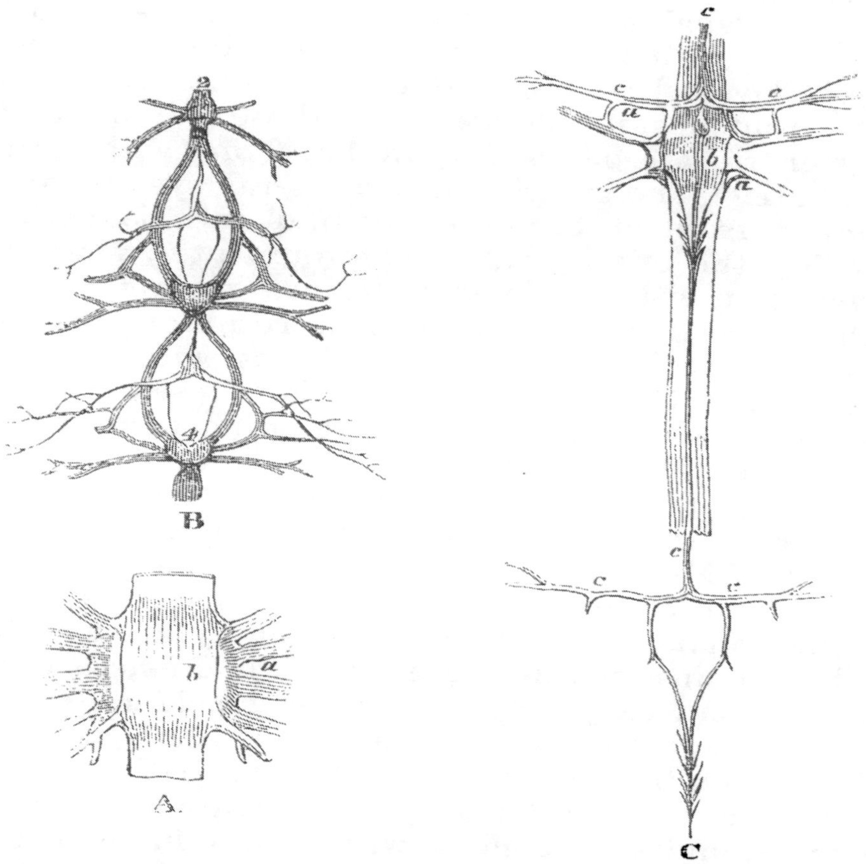

Parts of the Nervous System of Articulata.—A, single ganglion of *Centipede*, much enlarged, showing the distinctness of the purely fibrous tract, *b*, from the ganglionic column, *a*:—B, portion of the double cord from the thorax of the pupa of *Sphinx ligustri*, showing the respiratory ganglia and nerves between the ganglia (2, 3, 4), and the separated cords of the locomotive system:—C, view of the two systems combined, showing their arrangement in the larva; *a*, ganglion of ventral column; *b*, fibrous tract passing over it; *c*, *c*, respiratory system of nerves distinct from both.

773. The apparatus for the ingestion and preparation of food has its own system of ganglionic centres and nerves in nearly all Articulata. Thus in the *Leech*, we find a minute ganglion existing at the base of each of the three teeth (or rather jaws) by which incisions are made; these ganglia are connected with each other, and with the cephalic, by slender filaments; and they also seem to be in connection with other filaments, which may be traced along the alimentary canal. In *Iulus*, and in the larva of *Sphinx ligustri*, Mr. Newport has detected a central and two lateral ganglia, situated in close proximity with the cephalic, and communicating with them; from the former of these a 'recurrent' trunk (Fig. 306, *l*) passes backwards along the œsophagus to the stomach, apparently corresponding in its distribution with the gastric division of the par vagum in Vertebrata. The lateral ganglia (*h*, *m*), however, seem rather to belong to the *Sympathetic* or visceral system; their filaments being chiefly distributed upon the dorsal vessel, and upon the intestinal canal.—A more complex stomato-gastric system is found in Insects which are remarkable

for their masticating powers. Thus in the *Gryllotalpa vulgaris* (mole-cricket), we find it consisting of two divisions, the *median*, and the *lateral*. The former seems to originate in a small ganglion, situated (as in the Sphinx) anteriorly and inferiorly to the cephalic mass, with which it communicates by a connecting branch on each side; from this ganglion, nerves proceed to the walls of the buccal cavity, the mandibles, &c.; but its principal trunk is sent backwards beneath the pharynx. The ramifications of this 'recurrent' are distributed along the œsophageal tube and dorsal vessel; whilst the trunk passes downwards to the stomach, where its branches inosculate with those supplied by the lateral system, and enter a pair of small ganglia, from which most of the visceral nerves radiate. The ganglia of the lateral system are two on either side, lying, one in front of the other, behind and beneath the cephalic masses, with which the anterior pair communicate; from these, two cords pass backwards on either side, one derived from the anterior, and the other from the posterior; and these cords run along the sides of the œsophagus and dorsal vessel, and after inosculating with the branches of the median system, enter the two cœliac ganglia, whose branches radiate to the abdominal viscera.—In this, as in the preceding case, it would seem as if the *median* portion of this system partly resembles the gastric portion of the Par Vagum of Vertebrata, with the portion of the medulla oblongata which serves as its ganglionic centre; corresponding also, like the buccal ganglion of Mollusca, with the third division of the Fifth pair in Vertebrata, by which the muscles of mastication are especially supplied, and with the glosso-pharyngeal, which is especially concerned in the act of deglutition. The union between these nerves in Fishes, near their origin, is extremely close; and they may almost be considered as proceding from the same ganglionic centre. On the other hand, the *lateral* ganglia seem more analogous to the centres of the Sympathetic system in Vertebrata; especially in the connection of their branches with all the other systems of nerves, and in the share which they have in the formation of the cœliac ganglia. And this view of the relative functions of the two divisions included under the general term of 'stomato-gastric system,' is strengthened by the fact, that the connection between the Sympathetic system and the Par Vagum is peculiarly intimate in Fishes, so that the two sets of nerves can scarcely be isolated from each other.

774. Although the *cephalic* ganglia are usually larger than those of the ventral trunk, yet their relative size varies considerably. They receive nerve-trunks from the eyes, antennæ, and other sensory organs; and the history of their development shows that they are to be regarded as composed of several distinct pairs, which are fused (as it were) into one mass on either side. According to Mr. Newport, the cephalic ganglia of the Centipede are formed by the coalescence of the ganglia of the four segments of which the anterior portion of the head is composed; whilst the first sub-œsophageal ganglion is in like manner composed of those of four segments which have coalesced to form the posterior part of the head. The relative development of the cephalic ganglia, however, is so closely accordant with that of the visual organs,—as we see, not only in comparing different species with each other, but in comparing the *larva* and *imago* states of the same Insects,—that there can be no doubt that they are to be regarded as chiefly *optic* ganglia, corresponding with the

optic lobes of Fishes. There is not the least trace, in Articulata generally, of anything that can be fairly considered homologous with the cerebrum and cerebellum of Vertebrata ; the first subœsophageal ganglion, which has been likened to the latter, being really homologous (as the distribution of its nerves abundantly proves) with the Medulla Oblongata.

775. We shall now briefly notice the principal varieties of this general plan, which are presented to us in the principal subdivisions of the Articulated series.—In the *Entozoa*, no other than *cephalic* ganglia (Fig. 141, *f*) have yet been detected ; these give off a band which encircles the œsophagus ; and send back a pair of longitudinal filaments that run towards the caudal extremity of the body, diverging from one another more or less widely, and giving off branches at intervals, that encircle the body. No ganglionic enlargements can be detected in these ; but it does not seem improbable that they may contain nerve-cells. The arrangement of the supposed nervous system in the *Rotifera*, has been already described (§ 310 *a*).—In the *Annelida* we are led to that condition of the nervous system, which has been spoken of as typical of the group of Articulata ; for, whilst the soft-skinned species, in which there are neither organs of special sensation, nor distinct members for propulsion, have scarcely any ganglionic enlargements on the nervous cord, the higher tribes, in which the division into segments becomes distinct, and in which the animal relies for locomotion more upon the action of its members than upon that of its trunk, have ganglia regularly disposed at intervals corresponding with the division into segments. These ganglia, however, as in the inferior *Myriapoda*, in which the segments of the body are very numerous, are often so small and so closely set together, that they seem almost to form one continuous tract (Fig. 306). The cephalic ganglia are usually of small bulk in the Annelida, in accordance with the very imperfect development of their organs of special sense ; but as we pass from them to the lower Myriapoda, and thence through the higher forms of that class to the Insects which are most distinguished for their visual powers, we find a very remarkable increase in the relative size of these organs.—In harmony with the increased development of the sensori-motor portion of the nervous system, do we find the respiratory, stomato-gastric, and visceral centres becoming more distinct, and obviously rising in functional importance.

FIG. 306.

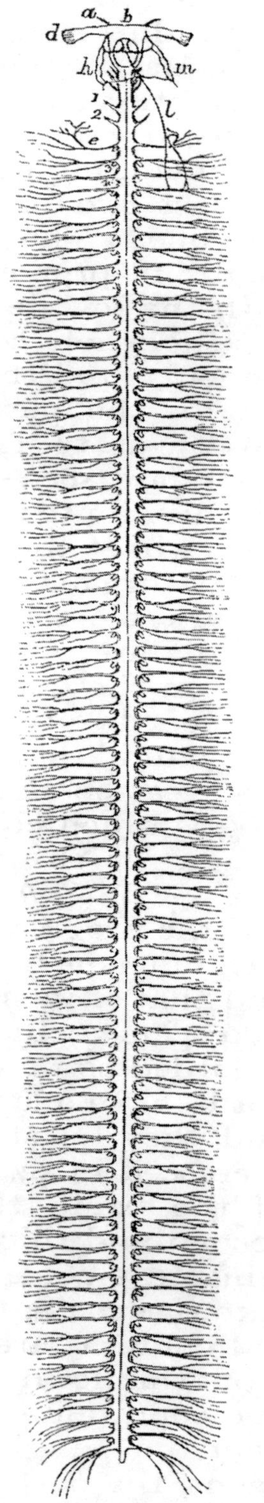

Nervous System of *Iulus terrestris* ; — *a*, antennal nerves ; *b*, cephalic ganglia ; *d*, optic nerves ; *h*, *m*, lateral ganglia ; *l*, recurrent nerve ; 1, 2, 3, 4, ganglia of the ventral cord.

776. The nervous system of *Insects*, like the rest of their organs, presents very different aspects at the different stages of their metamorphosis ; and these have a

peculiarly interesting relation with the general characters and habits of the animals. The *Larva*, as formerly stated (§ 315 *e*), may be regarded as, in almost every respect, on a level with the Annelida; all its segments are equal, or nearly so; all are nearly alike concerned in the function of locomotion; and its nervous cords, with their ganglia, are consequently disposed with great uniformity. The number of segments being always 13 (including the head as one), that of the ganglia is usually the same; the last two or even three ganglia, however, are frequently consolidated into one. The cephalic ganglia, placed in front of the œsophagus, are small in proportion to the size they subsequently attain, in conformity with the low development of the organs of special sensation. Throughout the whole column of the larva, the separation of its lateral halves is evident; and this is a character peculiar to the lower Articulated tribes; for, in the perfect Insects, Crustacea, &c., its divisions approximate so closely as to leave no space between them. The small respiratory filaments are seen to come off a little above the ganglionic nerves, and these are distributed to the stigmata, and to the muscles concerned in respiration, whilst the others ramify on the general surface and supply the locomotive organs. When the larva is about to assume the Pupa state, a very remarkable series of changes takes place in the nervous system; for the ganglia are rapidly approximated, in accordance with the sudden diminution in the length of the body; but the cords themselves are not yet shortened, so that they assume a sinuous form, and, in the thoracic region the lateral halves are more widely separated than before (Fig. 305, B). No great change is yet seen in the ganglia themselves; but the œsophageal ring is much contracted; and the filaments proceeding to the rudimentary wings, which now make their appearance, begin to attain a considerable size.—The *Sphinx ligustri* remains for several months in the *Pupa* state; and the progressive alterations in its nervous system may, therefore, be very advantageously watched. It appears that, between the time of the first and that of its second metamorphosis, very considerable changes gradually take place, which all tend towards its final development. In what may be regarded as its characteristic form in the pupa state, the inter-ganglionic cords have adapted themselves to the shortened dimensions of the body, and they lie straight as in the larva; the cephalic ganglia have greatly increased in size, and are in such close proximity with the first ganglion of the trunk, that the œsophageal aperture is now much contracted; the second and third ganglia of the trunk, from which the nerves pass to the wings, are considerably enlarged, whilst the fourth and fifth have coalesced into one mass, to which the sixth also closely approximates; the abdominal columns are but little altered, their ganglia, however, are now somewhat smaller in proportion to the rest.—These changes conduct us towards the condition of the nervous system in the *Imago* or perfect Insect. The cephalic ganglia have now undergone an enormous increase in development, the part connected with the eyes being particularly enlarged; and they extend over the œsophageal canal so much as to conceal it, uniting themselves closely with the first ganglion of the trunk. The second ganglion has entirely shifted its position and receded towards the middle of the thorax; the third has quite disappeared, seeming to have coalesced in part with the second, and in part with the one below it, as well as with their connecting cords. The next ganglion appears to contain the nervous matter,—not only of the fourth and fifth, which have evidently coalesced to form it,—but of the sixth and seventh, which have become obliterated, though their nerves are still given off from the cord. The remaining ganglia have undergone but little change, but are much smaller in proportion to the rest. This alteration is evidently conformable to that which has taken place in the condition of the locomotive apparatus,—the number of legs having now diminished to six, which proceed from the three segments of the thorax, with which the wings also are connected.

777. We see, then, that the tendency of the metamorphosis is to concentrate the ganglionic portion of the nervous system in the head and thorax; the former being the position of the organs of special sensation, the latter the situation of the locomotive apparatus. A lateral concentration may be frequently observed, as well as a longitudinal one; for in some larvæ the two cords are quite distinct, and are separated by a considerable interval; and these approximate in the Imago into a single column. There are many Insects in which the concentration is carried much farther than in the instance now described; the abdominal ganglia being almost entirely obliterated, and the nervous centres restricted to the head and thorax. This is partly the case in the *Melolontha* (Cock-chafer), whose thorax contains three contiguous ganglionic masses, from which nerves radiate to the wings and legs, and others pass backwards into the abdomen, where no ganglia exist. The greatest concentration exists, however, in the orders *Homoptera* and *Hemiptera.*—It is interesting to observe that in many *Lepidoptera* and *Hymenoptera,* which are remarkable for rapid and powerful flight, the nerves supplying both pairs of wings are united at their origins. On the other hand, in many Insects which are not remarkable for velocity or equability of motion, the nerves supplying each wing originate separately, and have little communication, just as in the larva of the Sphinx; and in the *Coleoptera,* in which the upper pair, or *elytra* (§ 314), are motionless during flight, the nerves frequently remain entirely separate. Hence it is not unfairly argued by Mr. Newport, that this common origin of the nerves is subservient to the uniformity and equability of the actions of the wings, required in Insects of rapid and powerful flight. This arrangement reminds us, on the one hand, of the circular trunk that connects the nerves of the arms of Octopus (§ 766); and on the other, of the plexiform arrangement of the nerves of the extremities in the higher Vertebrata.

778. The *Crustacea* present us with nearly as great a variety in the condition of the nervous apparatus, as do the Insect tribes (§ 316 *c*). In some of the least-developed species of this class, the nervous filaments are scarcely perceptible. In many more, in which the equality of the segments of the body indicates an affinity with the class Myriapoda, the nervous system almost exactly resembles that of the Centipede or higher Annelida. In higher orders, however, we perceive, as in Insects, a tendency to the concentration of the ganglia in the thorax, and to the increase in the size of the cephalic centres. This is seen in the *Lobster,* in which, although none of the ganglia are obliterated, the last seven are small in comparison with the first five. But it is in the short-bodied *Crabs,* that this concentration becomes most apparent. We here very commonly find but one thoracic mass, from which the whole of the trunk is supplied with nerves; and this conformation evidently leads us towards the Mollusca, in which

there is a similar tendency to the concentration of the nervous matter around the œsophagus.—The arrangement of the nervous system in the *Cirrhipoda* is essentially the same; its chief peculiarity consisting in the extremely low development of the cephalic ganglia, in conformity with that absence of organs of special sense, which is characteristic of the adult state of these animals (§ 318 *a*).—The distribution of the nervous system in the *Arachnida*, which corresponds pretty closely with that which prevails in the higher Crustacea, has been already sufficiently described (§ 319 *b*).*

779. Proceeding now to the *Vertebrated* series, we find, as heretofore pointed out, that their Nervous System constitutes a far more important portion of the entire organism, than it does in any Invertebrated animal; and that, in its most characteristic forms, it combines the locomotive centres of the Articulata with the sensorial centres of the Mollusca; possessing in addition two organs, the *Cerebrum* and *Cerebellum*, to which nothing analogous can be detected in any of the inferior classes (§§ 320, 320 *p*).—That which may be regarded as the fundamental portion of the nervous system in Vertebrata, is the *Cranio-spinal axis;* which consists of the *Medulla Spinalis* or *Spinal Cord*, of its anterior prolongation termed the *Medulla Oblongata*, and of the chain of *Sensory Ganglia* which forms the superior continuation of the latter. The whole of this axis lies above the alimentary canal; and there is consequently no œsophageal ring, like that of Articulated and Molluscous animals; but the two lateral strands of the cranio-spinal axis diverge from each other as they enter the cranium, so as to leave the space which is termed the *fourth ventricle* (Fig. 307). This cavity communicates anteriorly with the *third* ventricle, which separates the lateral halves of the anterior portion of the sensorial apparatus; and posteriorly with the spinal canal, which intervenes between the two lateral halves of the spinal cord. This last, however, like the space between the lateral halves of the ventral cord in the higher Articulata, is nearly obliterated in Man and the Mammalia, although sufficiently distinguishable in Fishes.—The Spinal Cord consists of a continuous tract of grey matter, enclosed within strands of longitudinal fibres; and it may thus be regarded as analogous to the ganglionic chain of the Articulata. Below the medulla oblongata, its endowments appear nearly similar throughout; for all the nerves which proceed from

* The view here given of the Physiology of the Nervous System of Invertebrated Animals, was first developed by the Author, in his "Prize Thesis" on that subject, published in 1839. The prevalent doctrine at that period regarding the actions of the ventral cord of Articulata, was that of Mr. Newport; who maintained that the fibrous tract was *motor*, and the ganglionic *sensory*, likening them respectively to the anterior and posterior columns of the spinal cord. In opposition to this doctrine, the Author adduced a number of facts and arguments, which appeared to him to prove unequivocally that the ganglia of the ventral cord are centres of reflex action; and his views were very early adopted by Professors Owen, Sharpey, and other distinguished Physiologists in this country. Mr. Newport, however, having remained unconvinced, determined to re-investigate the subject for himself; and in his valuable paper in the "Philosophical Transactions" for 1843, avowed his adoption of the 'reflex' doctrine, which he strengthened by additional facts of great importance, drawn both from anatomical and from experimental enquiry.—As the Author has reason to believe that this doctrine is now generally accepted by Physiologists, both at home and abroad, as the correct interpretation of the phenomena presented by the Nervous System of Invertebrata, he has not thought it necessary to do more in this place than explain and illustrate it.

it are distributed to the sensory surfaces and to the locomotive organs. In some Vertebrata, whose form resembles that of the Articulata (such as the Eel and Serpent), there is no difference in the size or distribution of the several pairs of nerves, as no extremities are developed; but in other cases, the size of the trunks proceeding to the anterior and posterior extremities is much greater than that of the nerves given off from the other segments of the cord; and the quantity of grey matter at their roots is correspondingly increased. In these trunks, both afferent and efferent fibres are bound up; but they separate at their roots, or junction with the spinal cord,—the afferent being connected with the side of the cord nearest the surface of the back,—and the motor with that next the viscera. Both these roots have two sets of connections; some of each enter the grey substance of the cord, in which they seem lost; whilst others are continuous with the fibrous portion of the cord, and are thus put in connection either with other segments or with the encephalic centres. In this respect, then, there is a precise correspondence between the spinal column of Vertebrata and the ventral cord of Insects; and in the former, as in the latter, does experiment indicate, that each segment of the cord has a certain degree of independence; reflex actions being excitable through it, so long as the circle of afferent and motor nerves, and their ganglionic centre, are in an active and uninjured state, even though it be completely separated from all the rest.—At the upper portion of the spinal cord, however, there is a series of ganglionic enlargements, having several distinct functions. From the Medulla Oblongata proceed the chief nerves which are subservient to the respiratory actions, and also those concerned in mastication and deglutition; so that this may be regarded as combining the respiratory and the stomato-gastric ganglia.—Above or in front of this, again, we find Auditory, Optic, and Olfactive ganglia, corresponding to the various subdivisions of the cephalic ganglia in the Invertebrata; these receive trunks from their respective organs of sensation, and may probably be regarded as *sensorial* centres, or seats of consciousness for the impressions which they severally transmit. The 'cranio-spinal axis' constitutes the whole nervous system of *Amphioxus*, in which there seems nothing that in the least represents a Cerebrum or Cerebellum (§ 321 *a*); and among the *Cyclostome* fishes generally, the condition of this apparatus is but little higher, save as regards the larger development of the sensory ganglia.

FIG. 307.

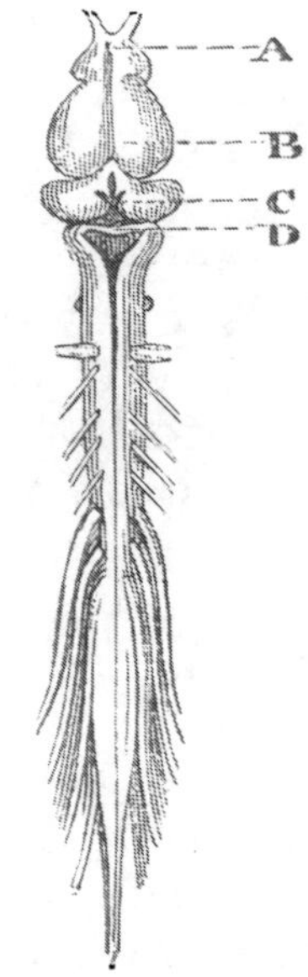

Nervous Centres in *Frog*:—A, olfactive ganglia; B, cerebral hemispheres; C, optic ganglia; D, cerebellum, so small as not to cover the fourth ventricle.

780. But in all higher Vertebrata, we find superimposed (as it were) upon the Sensory ganglia, the bodies which are known as the *Cerebral Hemispheres* or *ganglia*; whilst superimposed upon the Medulla Oblongata, we find the *Cerebellum*. The former constitute the mass of the Brain in the Mammalia; covering-in and obscuring the sensory ganglia so completely, that the fundamental importance of these is by no means generally recognised. In Fishes, however, the proportion between the two sets of centres is entirely reversed, the rudiments of the cerebral hemispheres being usually inferior in size to the optic ganglia alone. The

intermediate classes present us with a succession of gradations from the one type to the other, as regards not merely the size of the Cerebrum, but also its complexity of structure; and a very close relation may be seen between the degree of development which it exhibits, and the degree of *Intelligence* of the species. It is a point which is especially worthy of note, that no sensory nerves terminate directly in the Cerebrum, nor do any motor nerves issue directly from it; and there seems a strong probability, that there is not, as was formerly supposed, a direct continuity between any of the nerve-fibres distributed to the body, and the medullary substance of the Cerebrum. For whilst the nerves of 'special' sense have their own ganglionic centres, it cannot be shown that the nervous fibres of 'general' sense, which either enter the cranium as part of the cephalic nerves, or which pass up from the cranio-spinal axis, have any higher destination than the ganglionic masses termed *Thalami Optici*, which undoubtedly form part of the group of sensorial centres. So, the motor fibres which pass forth from the cranium, either into the cephalic nerve-trunks, or into the motor columns of the spinal cord, cannot be certainly said to have an origin higher than the *Corpora Striata;* which, like the Thalami, are most assuredly to be regarded as ganglionic centres, possessing considerable independence of the Cerebrum, though formerly regarded as mere appendages to it. And we shall find strong physiological ground for the belief, that the Cerebrum has no communication with the external world, otherwise than by the sensori-motor apparatus which ministers to the automatic actions; receiving through the sensory ganglia that consciousness of external objects and events, which is the spring of its intellectual or emotional operations; and communicating its voluntary determinations to the motor part of the same system, to be worked-out (so to speak) by it, through the instrumentality of the muscles upon which it plays.—The *Cerebellum*, in like manner, presents a great difference in relative development in the several classes of Vertebrata; being in the lowest a mere thin lamina of nervous matter on the median line, only partially covering-in the 'fourth ventricle'; whilst in the highest it is a mass of considerable size, having two lateral lobes or hemispheres, in addition to its central portion. It is connected with both columns of the spinal cord; and experiment leads to the belief, that its chief office is to combine the individual actions of different members, into the complex and nicely-balanced movements required for progression of various kinds, and, in Man, for the execution of the various operations which his intelligence prompts him to undertake. — We shall now briefly glance at the relative development and position of these parts, in the different classes of Vertebrata.

781. Commencing with *Fishes*, we find a series of four distinct ganglionic masses, arranged in a line which is nearly continuous, from behind forwards, with that of the Spinal Cord; of these, the posterior is usually single, and on the median plane, whilst the others are in pairs.— 1. The posterior, from its position and connections, is evidently to be regarded in the light of a *Cerebellum;* and it bears a much larger proportion to the rest, in this class, than in any other. — 2. The pair in front of this are not the hemispheres of the Cerebrum, as their large size in some instances (the Cod, for instance) might lead us to suppose; but they are immediately connected with the Optic nerve, which, in fact, termi-

nates in them, and are therefore to be considered (like the chief part of the cephalic masses of Invertebrated animals) as *Optic lobes* or *ganglia.* They seem, however, in some degree to represent also the *Thalami Optici* of higher animals, as will be seen in the next paragraph. — 3. In front of these are the bodies usually considered as representing the *Cerebral Hemispheres;* which are small, generally destitute of convolutions, and possessing no ventricle in their interior, — except in the Sharks and Rays, in which they are much more highly developed than in the Osseous fishes. In the latter, in fact, these bodies seem to be the homologues of the portion of the mass lying *beneath* the ventricle in the higher Cartilaginous fishes, which is obviously the representative of the *Corpus Striatum;* so that, among ordinary Fishes, there is little or no trace of the true *Cerebrum* or

FIG. 308.

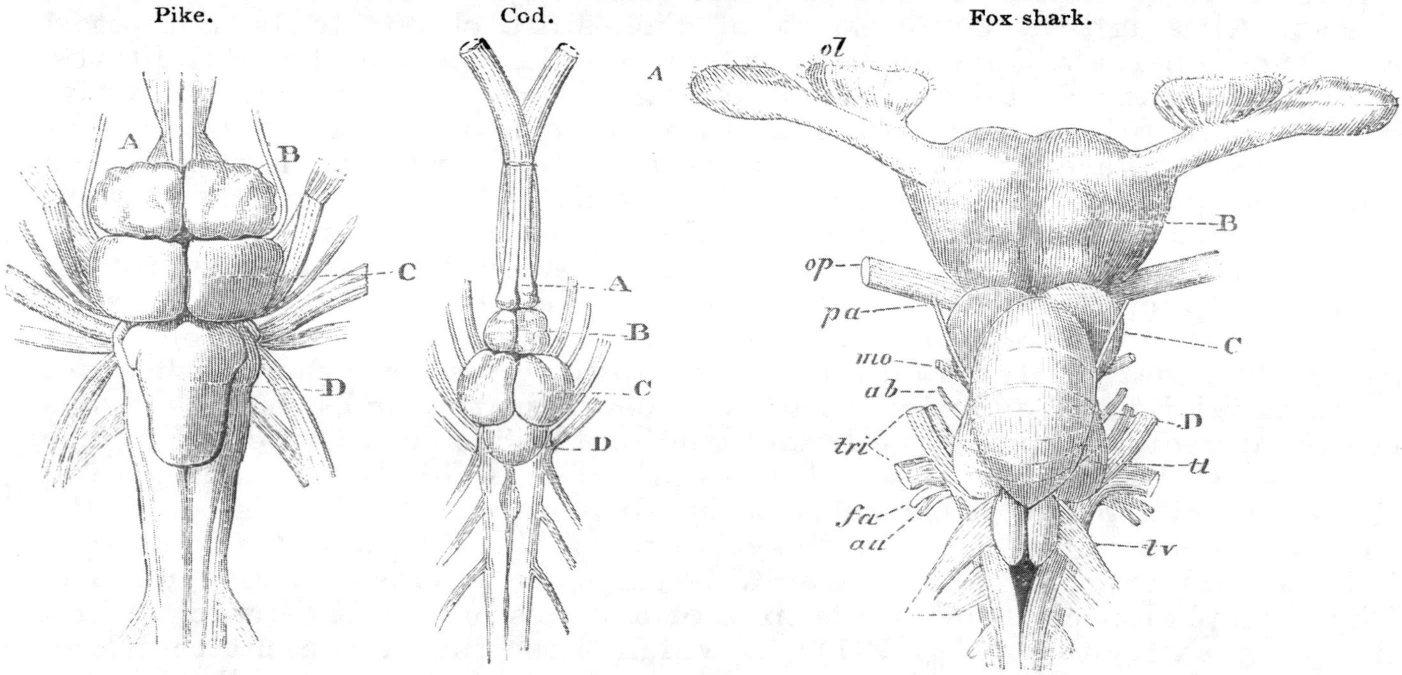

Brains of *Fishes:* — A, olfactive lobes or ganglia; B, cerebral hemispheres; C, optic lobes; D, cerebellum;—*ol*, olfactory nerve; *op*, optic nerve; *pa*, patheticus; *mo*, motor oculi; *ab*, abducens; *tri*, trifacial; *fa*, facial; *au*, auditory; *vag*, vagus; *tt*, tubercles or ganglia of the trifacial; *tv*, tubercles of the vagus.

Hemispheric ganglion, which makes its first appearance in the tribe most distinguished by the elevation of its general structure.—4. Anterior to these is another pair of ganglionic enlargements, from which the Olfactory nerves arise; and these are, therefore, correctly designated as the *Olfactive ganglia.* In some instances, these ganglia are not immediately seated upon the prolonged spinal cord, but are connected with it by long peduncles; this is the case in the Sharks; and we are thus led to perceive the real nature of the portion of the trunk of the Olfactory nerve in Man, which lies within the cranium, and of its bulbous expansion on the Ethmoid bone. — Besides these principal ganglionic enlargements, there are often smaller ones, with which other nerves are connected. Thus, in the Shark, we find a pair of tubercles of considerable size, at the origin of the Trifacial nerves; and another pair, in most Fishes, at the roots of the Vagi. In some instances, too, distinct *Auditory ganglia* present themselves; as in the Carp.—The Spinal Cord differs much in its proportions in different tribes of this class. In the Eel and other Vermiform fishes, it is of nearly uniform size throughout; and, in the lowest of these, the cerebral ganglia are scarcely more prominent than are those of the leech or caterpillar. In proportion as distinct locomotive members are developed, do we find enlargements of the spinal cord corresponding with the origins of their nerves, just as in the ganglionic column of Insects; and where the anterior members are very powerful, as in the *Trigla* (gurnard), these enlargements have an evidently ganglionic character. In such species as the *Lophius* (frog-fish), in which the nutritive system is enormously developed at the expense of activity of locomotion, and the animal is thus constructed more upon the Molluscous type, the nervous centres are confined to the neighbourhood of the head; for the true spinal cord soon separates into a bundle of nerves, which act only as conductors.

782. Although the Optic Lobes of Fishes are chiefly to be compared with the Tubercula Quadrigemina, which are the real ganglia of the Optic nerve in higher Vertebrata, their analogy is not so complete to these bodies in the fully-formed Brain of Man, as it is to certain parts which occupy their place at an earlier period. The third ventricle, which is quite distinct from the Corpora Quadrigemina, is hollowed out, as it were, from the floor of the Optic Lobes of Fishes; and the 'anterior commissure' bounds its front; hence these must be considered as analogous to the Thalami Optici and parts surrounding the third ventricle, as well as to the Corpora Quadrigemina. This is made evident by the fact, observed by Müller, that, in the Lamprey, the Optic Lobes of other Fishes are represented by two pairs of ganglionic centres; the one, which encloses the third ventricle, being the homologue of the Thalami Optici of higher animals; and the other, in which the optic nerves chiefly terminate, being the representative of their Corpora Quadrigemina. With this condition, the early state of the Brain in the embryo of the Bird and Mammiferous animal, and even in Man himself, bears a very close correspondence. The Encephalon consists at this time of a series of vesicles, arranged in a line with each other (Fig. 297); of which those that represent the Cerebrum are the smallest, whilst that which represents the Cerebellum is the largest. The latter (or *Epencephalon*), as in Fishes, is single, covering the fourth ventricle on the dorsal surface of the Medulla Oblongata. Anterior to this, is the single vesicle of the Corpora Quadrigemina (or *Mesencephalon*), from which the Optic Nerve chiefly arises; this has in its interior a cavity, the ventricle of Sylvius, which exists even in the adult Bird, where the Corpora Quadrigemina are pushed from each other, as it were, by the increased development of the Cerebral hemispheres. In front of this is the vesicle of the Third Ventricle (or *Deutencephalon*), which contains also the Thalami; as development proceeds, this, like the preceding, is covered by the enlarged hemispheres; whilst its roof becomes cleft anteriorly on the median line, so as to form the anterior entrance to the cavity. Still more anteriorly is the double vesicle (or *Prosencephalon*), which represents the hemispheres of the Cerebrum; this has a cavity on each side, the floor of which is formed by the Corpora Striata. The cavity of the cerebral vesicles has at first no opening, except into that of the third ventricle; at a later period is formed that fissure on the inferior and posterior side, which (under the name of the fissure of Sylvius) enables the

membranes enveloping the brain to be reflected into the lateral ventricles. —Thus it will be seen that the real analogy between the brain of the Human fœtus, and that of the adult Fish, is not so close as, from the resemblance in their external form, might have been supposed. In the small proportion which the Cerebral Hemispheres bear to the other parts, there is evidently a very close correspondence; and this extends also to the general simplicity of their structure, the absence of convolutions, and the deficiency of commissures. But there is a much nearer analogy between the *fœtal* brain of the Fish, and the *fœtal* brain of the Mammal; indeed, at the earliest period of their formation, they could scarcely be distinguished; during their advance to the permanent condition, however, each undergoes changes, which are so much more decided in the higher animals than in the lower, that in the latter there seems comparatively little departure from the fœtal condition, whilst in the former the condition appears entirely changed.

783. We have, then, in Fishes, and in the early Human embryo, this remarkable condition of the Encephalic mass,—that it is evidently made up of a series of distinct ganglionic centres, of which the portions representing the Cerebral Hemispheres are usually the smallest, being obviously an addition to the remainder, whose existence is independent of them. Thus, in passing from before backwards, we meet, 1st, with the Olfactive ganglia; 2d, with the Corpora Striata, overlaid with the mere rudiment of a Cerebrum; 3d, with the Thalami Optici, inclosing the third ventricle; 4th, with the Corpora Quadrigemina, or proper Optic Ganglia; and 5th, with the Cerebellum. Besides these, we have centres for the Auditory and Gustative nerves, or proper Auditory and Gustative ganglia, lodged in the Medulla Oblongata. All these ganglionic centres have their own distinct connections with the Medulla Oblongata; except the Hemispheres, which do not appear to communicate with it, except through the medium of the bodies on which they are superposed. We shall probably form the most correct view of their relations, if, excluding the Cerebrum and Cerebellum, we regard them as collectively homologous with the Cephalic ganglia of Invertebrated animals, which, as we have seen, are the immediate centres of the nerves of sensation, and are connected with the ganglia in the trunk by fibrous cords which represent the Medulla Oblongata. The size of the Cephalic ganglia, in the higher Invertebrata, is chiefly dependent upon the development of the visual organs, which are the principal guides in the movements of these animals; but, as Mr. Newport's researches on their embryonic development have shown, they are really composed of several pairs of distinct ganglionic centres (§ 774); and it is interesting, also, to remark, that the situation of the rudimentary organ of hearing in the Gasteropodous Mollusca is precisely analogous to that of the Auditory ganglion in the Vertebrata, the auditory sacculi being lodged in the posterior lobes of their cephalic ganglia. The Optic and Olfactive ganglia of Vertebrated animals receive nerves of sensation from the organs situated in their neighbourhood, and seem to give off motor nerves in the fibrous peduncles which connect them with the motor tract of the Medulla Oblongata. The Thalami Optici and Corpora Striata, on the other hand, appear to be the ganglionic centres of fibres entirely transmitted through the Spinal Cord, as they do not directly receive or give off any nerve-trunks, saving that the former receive some of the roots of the Optic

nerve; and the special connection of the former with the Sensory tract, and of the latter with the Motor, with other reasons hereafter to be given, lead to the belief that these are the ganglionic centres of 'common' or tactile sensations, and of the movements immediately excited by them; whilst the passage of some of the filaments of the Optic Nerve into the Thalami, may not improbably be interpreted as ministering to that peculiarly-intimate connection, which exists between the senses of Sight and Touch.—Thus, we may consider this series of ganglionic centres as forming, with the Spinal Cord (of which they constitute the encephalic representation), an *automatic* apparatus, exactly comparable with that of the Insect; and on this the Cerebrum is superposed, in such a manner as to be obviously an independent organ, receiving its stimulus to action from the sensorial centres, and transmitting its motor impulses through the same channel.

784. The Brain of *Reptiles* does not show any considerable advance in its general structure, above that of Fishes; but the Cerebral Hemispheres are usually much larger in proportion to the Optic lobes; whilst the Cerebellum is smaller. The very low development of the Cerebellum is especially seen in the Frog, (Fig. 307), in which it is so small as not even to cover-in the 'fourth ventricle;' but it is common to nearly the whole group. The deficiency in commissures still exists to a great extent. The 'anterior commissure' in front of the 'third ventricle,' is the only uniting band which can be distinctly traced in Fishes; and Reptiles have, in addition to this, a layer of uniting fibres which may be compared to the 'fornix;' but as yet, there is no vestige of a true 'corpus callosum,' or great transverse commissure of the Hemispheres. The distinction between the tubercula quadrigemina, and the parts inclosing the third ventricle, is more obvious than in Fishes; in fact, the optic ganglia of Reptiles correspond pretty closely with the vesicle of the tubercula quadrigemina, or mesencephalon, in the brain of the fœtal Mammal.—The Nervous Centres of *Batrachia*, like all their other organs, present, in the Tadpole state, the characters of those of Fishes; and these are partly retained by the 'perennibranchiate' species during the whole of their existence. In the Frog and its allies, however, the Encephalon acquires the true Reptilian type; and a marked change takes place in the condition of the Spinal Cord. For in the Tadpole-condition, this organ is elongated, and of nearly uniform size throughout; but in proportion as the tail is atrophied and the extremities are developed, do we find the spinal cord relatively shortened, and presenting enlargements at the parts in which the nerves of the limbs originate, especially those of the posterior extremities. Similar enlargements are seen in Turtles, in connection with the nerves of both the anterior and posterior extremities, which take a nearly equal share in the general movements of the body. On the other hand, in Serpents, as in Eels and other snake-like Fishes, the Spinal Cord is of nearly uniform size throughout; and as it gives off, in some instances, more than 300 pairs of nerves, it is obvious that only a

Fig. 309.

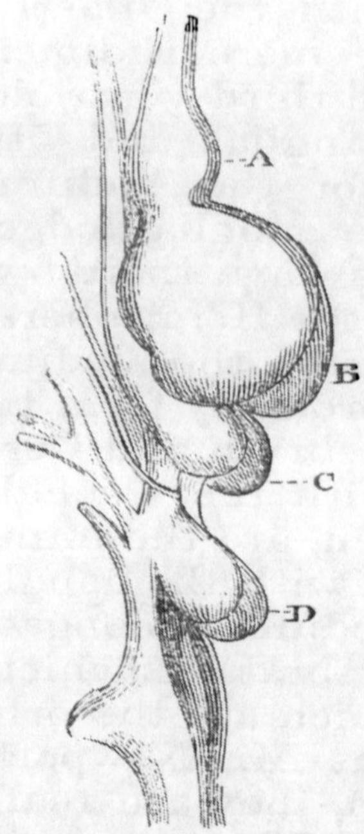

Brain of *Turtle*:—A, olfactive ganglia; B, cerebral hemispheres; C, optic ganglia; D, cerebellum.

very small proportion of their fibres can have any direct connection with the Encephalon, through the longitudinal strands of the Medulla Oblongata.

785. In the Brain of *Birds,* our attention is at once attracted by the increased development of the Cerebral Hemispheres, which extend forwards so as to conceal the Olfactive ganglia, and arch backwards so as partly to cover the Optic ganglia (here called the Corpora Bigemina), which are separated from each other and thrown to either side, being now quite distinct from the Thalami Optici. The Cerebellum also is much increased in size, proportionably to the Medulla Oblongata and its ganglia; and it is sometimes marked with transverse lines, which indicate the intermixture of grey and white matter in its substance; there is as yet, however, no appearance of a division into hemispheres. On drawing apart the hemispheres of the Cerebrum, the Corpora Striata, Optic Thalami, and Corpora Bigemina or Optic Ganglia, are seen beneath them; and their collective size still bears a considerable proportion to that of the whole Encephalon. The Optic Ganglia are still hollow, as they are in the embryo condition of Man. Indeed the Brain of the Human fœtus, about the twelfth week, will bear comparison, in many respects, with that of the Bird. The Cerebral hemispheres, much increased in size, and arching back over the Thalami and Optic ganglia, but destitute of convolutions, and imperfectly con-

FIG. 310.

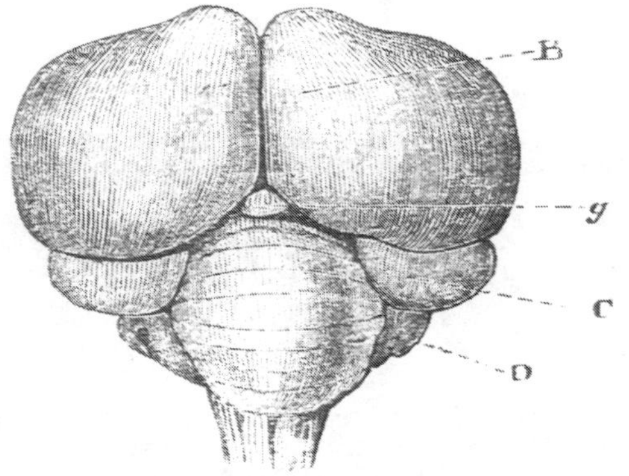

Brain of *Buzzard;* the olfactive ganglia are concealed beneath B, the hemispheres; C, optic ganglia; D, cerebellum; *g*, pineal gland.

FIG. 311.

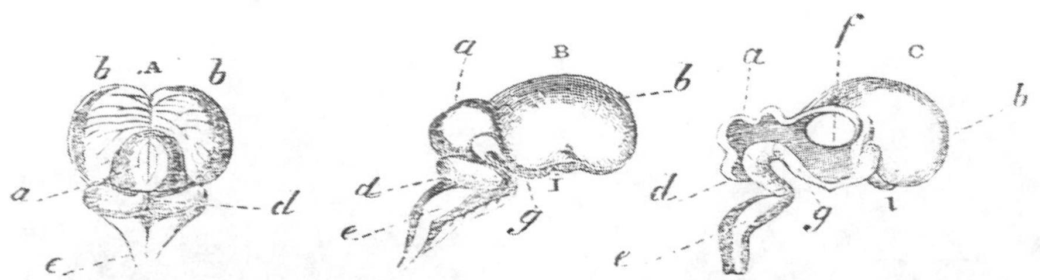

Brain of *Human Embryo* at twelfth week. A, seen from behind; B, side view; C, sectional view;—*a*, corpora quadrigemina; *bb*, hemispheres; *d*, cerebellum; *e*, medulla oblongata; *f*, optic thalamus; *g*, floor of third ventricle; I, olfactory nerve.

nected by commissures,—the large cavity still existing in the Optic ganglia, and freely communicating with the third ventricle,—and the imperfect evolution of the Cerebellum,—make the correspondence in the general condition of the two very considerable.

786. The Brain of the lowest *Mammalia* presents but a slight advance upon that of Birds, in regard both to the relative proportions of its parts, and to their degree of development. Thus, in the *Marsupialia,* the Hemispheres exhibit scarcely any convolutions; the great transverse commissure, or 'corpus callosum,' is deficient; and, as in all the Oviparous Vertebrata, the rudimentary Cerebrum represents, not the entire cerebrum of Man, but its *anterior* lobe only. There is gradually to be noticed, however, in ascending the scale, a backward prolongation of the Cerebral hemispheres, so that first the Optic ganglia, and then the Cerebellum, are

covered by them; and this extension corresponds with the development of the *middle* lobe and its great commissure. The Cerebellum partly shows itself, however, in all but the Quadrumana, when we look at the brain from above downwards; in the *Rodentia*, which are in this respect among the lowest of the true Viviparous Mammalia, nearly the whole of the Cerebellum is uncovered. In proportion to the increase of the Cerebral hemispheres, there is a diminution in the size of the ganglia immediately connected with the organs of sense; and this in comparison, not only with the rest of the Encephalon, but even with the Spinal Cord; so that in Man the Tubercula Quadrigemina are absolutely smaller than they are in many animals of far inferior size. The internal structure of the Hemispheres becomes more complex, in the same proportion as their size and the depth of the convolutions increase; and in Man all these conditions present themselves in a far higher degree than in any other animal. In fact, it is only among the Ruminantia, Pachydermata, Carnivora, and Quadrumana, that regular convolutions can be said to exist; and it is only in the

FIG. 312.

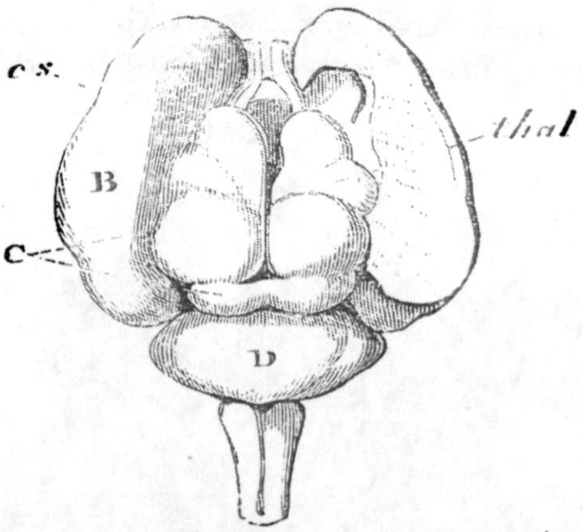

Brain of *Squirrel*, laid open; the hemispheres, B, being drawn to either side, to show the subjacent parts; C, the optic lobes; D, cerebellum; *thal*, thalamus opticus; *c s*, corpus striatum.

FIG. 313.

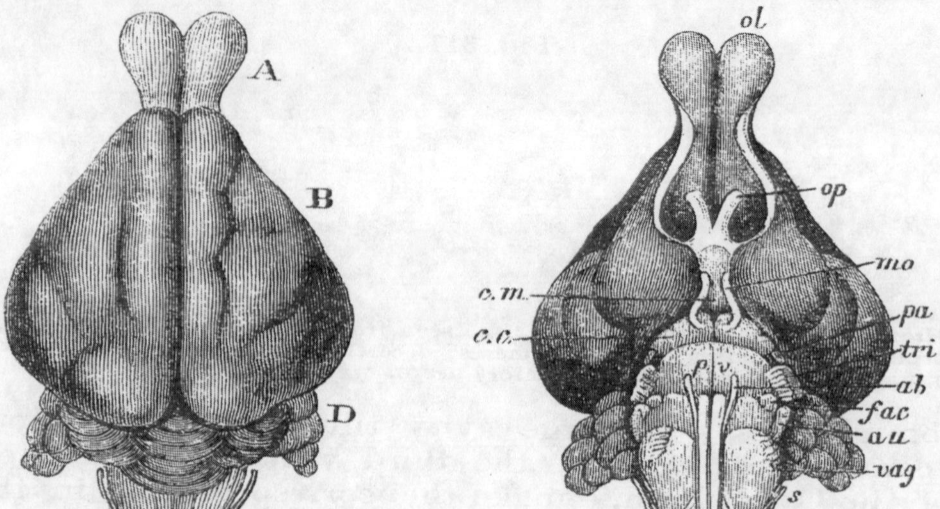

Upper and under surface of Brain of *Rabbit*, A, B, D, as before; *ol*, olfactive lobes; *op*, optic nerve; *mo*, motor oculi; *cm*, corpora mamillaria; *cc*, crus cerebri; *pv*, pons varolii; *pa*, patheticus; *tri*, trifacial; *ab*, abducens; *fac*, facial; *au*, auditory; *vag*, vagus; *s*, spinal accessory; *hyp*, hypoglossal.

higher Carnivora and Quadrumana, that there is any indication of the existence of *posterior* lobes; the presence of which is marked by the development of the posterior cornua of the lateral ventricles, and by the position of the hippocampus major. All these phases are distinguishable in the development of the brain of the Human embryo; for up to the end of the third month, the hemispheres present only the rudiments of anterior lobes, and do not even cover-in the thalami; during the fourth and part of the fifth months, the middle lobes are developed on their posterior aspect, and cover the tubercula quadrigemina; and the posterior lobes, of which there was no previous rudiment, subsequently begin to sprout from the back of the middle lobes, remaining separated from them by a distinct furrow, however, even in the brain of the mature fœtus, and sometimes in that of older persons.—The correspondence between the bulbous expansion of the Olfactive nerves in Mammalia, and the Olfactive lobes of the lower Vertebrata, is made evident by the presence, in both instances, of a cavity which communicates with the lateral ventricle on each side; it is in Man only, that this cavity is wanting. The external form of the Corpora Quadrigemina of Mammalia, differs from that of the Optic ganglia of Birds, owing to the division of the former into anterior and posterior eminences (the nates and testes); and there is also an internal difference, occasioned by the contraction of the cavity or ventricle, which now only remains as the 'aqueduct of Sylvius.' The Auditory ganglia are lodged in the substance of the Medulla Oblongata, forming the 'grey nuclei' of the 'posterior pyramids;' and similar nuclei in the 'restiform bodies' are the ganglionic centres of the Glosso-pharyngeal nerves, and probably minister to the sense of Taste.—The Cerebellum is chiefly remarkable for the development of its lateral parts or hemispheres; the central portion, sometimes called the 'vermiform process,' is relatively less developed than in the lower Vertebrata, in which it forms the whole of the organ.

787. Thus, when we analyse the entire *Cerebro-Spinal* system of Vertebrata, we find that it may be resolved into the following fundamentally-distinct parts:—1. A system of ganglia subservient to the reflex actions of the organs of *locomotion,* and corresponding with the chain of pedal or locomotive ganglia that makes up the chief part of the ventral cord of the Articulata; in this system, the grey or vesicular matter forms one continuous tract, which occupies the interior of the *Spinal Cord.*—2. A ganglionic centre for the movements of *respiration,* and another for those of *mastication* and *deglutition;* these, with part of the preceding, make up the proper substance of the *Medulla Oblongata.*—3. A series of ganglia, in immediate connection with the organs of *Special Sense;* these are situated within the cranium, at the anterior extremity of the Medulla Oblongata; and in the lowest Vertebrata, they constitute by far the largest portion of the entire Encephalon.—4. The *Cerebellum,* which is a sort of off-shoot from the upper extremity of the Medulla Oblongata, lying behind the preceding.—5. The *Cerebral Hemispheres,* a pair of ganglionic masses, which lie upon the ganglia of special sense, capping them over more or less completely, according to their relative development.—Of these, the first three may be considered as constituting the *automatic* portion of the nervous centres; whilst the Cerebrum is certainly the original source of all *voluntary* movements; and the Cerebellum seems to contribute to the adjustment and combination of the individual acts, by which the directions of the Will are worked out, through the instrumentality of the automatic apparatus, in the manner to be presently explained.

788. The development of the *Sympathetic* or Visceral system of nerves, in the Vertebrated classes, advances *pari passu* with that of the Cerebro-Spinal; from which it gradually becomes more distinct. In many Fishes, as in the Invertebrata, the two are so blended that it is difficult to separate them, the visceral nerves appearing to be derived exclusively from the cerebro-spinal system; but, as we ascend the scale, the former is seen to possess centres of its own; and in Mammalia it becomes a system of great complexity, having two large ganglia (the semilunar) in the abdomen, from which filaments are distributed to all the digestive organs, and a regular series along the spine. It communicates with each of the spinal nerves near their roots, as well as with most of the cerebral; and interchanges filaments with them. It forms a plexus which is minutely distributed upon the large vascular trunks, and which probably accompanies their ramifications into every part of the system.

789. It is in Vertebrated animals, that we meet with the greatest complexity in the actions of the Nervous system, and that we experience the greatest difficulty in determining the attributes of each of its parts. This is due, on the one hand, to the increased variety in its *modi operandi;* for whilst the actions of the lower tribes belong for the most part to the Automatic group, many of those of Vertebrata are dependent upon Voluntary determinations, and some upon Emotional impulses; and it is not always easy to assign their true source. On the other hand, a difficulty arises out of the peculiarity in the arrangement of their several ganglionic centres, which are for the most part combined together in one continuous mass, so that they cannot be isolated from one another without the infliction of such injuries, as must considerably interfere with the performance of their appropriate actions. Still, taking Nature as our guide, and availing ourselves of the experiments which she has prepared for us in the different natural combinations of these ganglionic centres, it seems possible to attain to very definite conclusions in regard to all the most general questions involved in the inquiry; and to these alone will it be desirable for us here to restrict ourselves.

790. The *Cranio-Spinal Axis,*—including the Spinal Cord, Medulla Oblongata, and Sensory Ganglia,—may be considered as the representative of the entire nervous system of Articulated animals (save of that portion which is homologous with the Sympathetic); and, like it, seems to be the centre of all the actions which may be designated as *automatic.* Although it was long held as a physiological truth, that the principal part of the Sensory fibres passes up to the grey matter which forms the surface of the Cerebral hemispheres, and that the fibres which are subservient to voluntary Motion originate in the same situation, and pass downwards through the spinal cord and nerve-trunks to the muscles; yet anatomical inquiry fails to sanction such a conclusion, and tends, concurrently with the results of physiological investigation, to the conclusion, that no sensory fibres pass upwards beyond the chain of sensory ganglia (including the thalami optici), and that no motor fibres really originate from any higher point than the corpora striata. With this chain of ganglionic centres, which constitute the real *sensorium,* the vesicular matter of the Cerebral or Hemispheric ganglia is connected, by that vast collection of fibres which radiate from their central portion to their surface; but these fibres may be regarded as simply *commissural,* connecting the grey matter

of the cerebral surface with the thalami optici, corpora striata, and other ganglionic centres, whose endowments are altogether distinct. It is, then, in the Cranio-Spinal Axis, that all the afferent and sensory nerves terminate; and it must be through it, therefore, that the Cerebrum is acted upon by external impressions. On the other hand, it is in the same Axis that all the motor nerves originate; and it must be through it, therefore, that the Cerebrum is brought into communication with the Muscular apparatus.

791. That this Cranio-Spinal Axis is a distinct centre of automatic action, and does not derive its power (as formerly supposed) from the Cerebrum, is made evident by a variety of considerations. Thus, Infants are sometimes born without any Cerebrum or Cerebellum; and such have existed for several hours or even days, breathing, crying, sucking, and performing various other movements. The Cerebrum and Cerebellum have been experimentally removed from Birds and young Mammalia, thus reducing these beings to a similar condition; and all their vital operations have, nevertheless, been so regularly performed, as to enable them to live for weeks, or even months. In the *Amphioxus*, as already stated (§ 321 *a*), we have an example of a completely-formed adult animal, in which no rudiment of a Cerebrum or Cerebellum can be detected. And in ordinary profound sleep, or in apoplexy, the functions of these organs are so completely suspended, that the animal is, in all essential particulars, in the same condition for a time as if destitute of them. It is possible, indeed, to reduce a Vertebrated animal to the condition (so far as its nervous system is concerned) of an Ascidian Mollusk (§ 761); for it may continue to exist for some time, when not merely the Cerebrum and Cerebellum have been removed from above, but when nearly the whole Spinal Cord has been removed from below,—that part only of the latter being left (namely, the Medulla Oblongata), which, being the centre of the respiratory actions, corresponds to the single ganglion of the Tunicata. On the other hand, no Vertebrated animal can exist by its Encephalon alone, the Spinal Axis being destroyed or removed; for the reflex actions of the latter are so essential to the continuance of its respiration, and consequently of its circulation, that if they be suspended (by the destruction of the portion of the Cord which is concerned in them), all the organic functions must soon cease.

792. That the actions performed by the Spinal Cord are of a purely *reflex* nature,—consisting in the excitement of muscular movements, in respondence to external impressions, without the necessary intervention of sensation,—appears to be a necessary inference from the facts that have been brought to light by experiment and observation. Experiments on the nature of this function are best made upon cold-blooded animals; as their general functions are less disturbed by the effects of severe injuries of the nervous system, than are those of Birds and Mammals. When the Cerebrum has been removed, or its functions have been suspended by a severe blow upon the head, a variety of motions may be excited by their appropriate stimuli. Thus, if the edge of the eye-lid be touched with a straw, the lid immediately closes. If liquid be poured into the mouth, or a solid substance be pushed within the grasp of the muscles of deglutition, it is swallowed. If the foot be pinched, or burned with a lighted taper, it is withdrawn; and (if the creature experimented on be a Frog) the

animal will leap away, as if to escape from the source of irritation. If the cloaca be irritated with a probe, the hind-legs will endeavour to push it away.—Now the performance of these as well as of other movements, many of them most remarkably adapted to an evident purpose, might be supposed to indicate, that *sensations* are called up by the impressions; and that the animal can not only *feel*, but can *voluntarily* direct its movements, so as to get rid of the irritation which annoys it. But such an inference would be inconsistent with other facts.—In the first place, the motions performed by an animal under such circumstances, are never spontaneous, but are always excited by a *stimulus* of some kind. Thus, a decapitated Frog, after the first violent convulsive movements occasioned by the operation have passed away, remains at rest until it is touched; and then the leg, or its whole body, may be thrown into sudden action, which immediately subsides again. In the same manner, the act of swallowing is not performed, except when it is excited by the contact of food or liquid; and even the respiratory movements, spontaneous as they seem to be, would not continue, unless they were continually re-excited by the presence of venous blood in the vessels. These movements are all *necessarily* linked with the stimulus that excites them;—that is, the same stimulus will always produce the same movement, when the condition of the body is the same. Hence it is evident, that the judgment and will are not concerned in producing them; and that the *adaptiveness* of the movements is no proof of the existence of consciousness and discrimination in the being which executes them,—the adaptation being made *for* that being, by the peculiar structure of its nervous apparatus, which causes a certain movement to be executed in respondence to a given impression,—not *by* it. An animal thus circumstanced may be not unaptly compared to an automaton; in which particular movements, adapted to produce a given effect, are produced by touching certain springs. Here the adaptation was in the mind of the maker or designer of the automaton; and so it evidently is, in regard to the reflex or consensual movements of animals, as well as with respect to the various operations of their nutritive system, over which they have no control, yet which concur most admirably to a common end.

793. Again, we find that such movements may be performed, not only when the Encephalon has been removed, the spinal cord remaining entire, but also when the Spinal Cord has been itself cut across, so as to be divided into two or more portions,—each of them completely isolated from each other, and from other parts of the nervous centres. Thus, if the head of a Frog be cut off, and its spinal cord be divided in the middle of the back, so that its fore-legs remain connected with the upper part, and its hind legs with the lower, each pair of members may be excited to movement by a stimulus applied to itself; but the two pairs will not exhibit any consentaneous motions, as they will do when the spinal cord is undivided. Or, if the Spinal cord be cut across, without the removal of the Brain, the lower limbs may be *excited* to movement, by an appropriate stimulus, though they are completely paralysed to the *will;* whilst the upper remain under the control of the animal, as completely as before. And when this separation happens to be made in the Human subject, by accidental injury or by disease, it is found that if it be complete, there is not only a total want of voluntary control over the lower extremities,

but a complete absence of sensation also,—the individual not being in the least conscious of any impression made upon them. When the lower segment of the Cord remains sound, and its nervous connections with the limbs are unimpaired, distinct reflex movements may be excited in the limbs, by stimuli directly applied to them, as, for instance, by pinching the skin, tickling the sole of the foot, or applying a hot plate to its surface; and this without the least sensation, on the part of the patient, either of the cause of the movement, or of the movement itself.

794. This fact, taken in connection with the preceding experiments, both upon Vertebrated and Articulated animals, distinctly proves that Sensation is *not* a necessary link in the chain of 'reflex' actions; but that all which is required is an *afferent* fibre, capable of receiving the impression made upon the surface, and of conveying it to the centre; a *ganglionic centre*, composed of vesicular nervous substance, into which the afferent fibre passes; and an *efferent* fibre, capable of transmitting the motor impulse, from the ganglionic centre, to the muscle which is to be thrown into contraction.—These conditions are realized in the Spinal Cord. We may have reflex actions excited through any one isolated segment of it, as through a single ganglion of the ventral cord of Articulata; but they are then confined to the parts supplied by the nerves of that segment. Thus, if the spinal cord of a Frog be divided just above the origin of the crural nerves, the hind-legs may be thrown into reflex contraction by various stimuli applied to themselves; but the fore-legs will exhibit no movement of this kind. But when the brain has been removed, and the Spinal Cord is left entire, movements may be excited in distant parts,—as, for example, in the fore-legs, by any powerful irritation of the posterior extremities,—and *vice versâ*. This is particularly well seen in the convulsive movements, which take place in certain disordered states of the nervous system; a slight local irritation being sufficient to throw almost any muscles of the body into a state of energetic action. And a similar state may be artificially induced, by applying strychnine (in solution) to the Spinal Cord of a decapitated Frog.

795. The particular reflex actions to which the Spinal Cord (using that term in its limited sense, as excluding the Medulla Oblongata) is subservient, are mostly connected with the organic functions. They are chiefly of an *expulsive* kind; being destined to force-out the contents of various cavities of the body. Thus the ordinary acts of defecation and urination, the ejaculatio seminis, and parturition, are all reflex movements, over which the will has but very little control, when once the stimulus by which they are excited has come into full action.—But the movements of the posterior extremities are occasionally due to the purely automatic action of the Spinal Cord. It has been already noticed, that these may be excited, even in Man, when the spinal cord has been severed in the middle without injury to its lower segment; and it is remarkable that gentle stimuli, applied to the skin of the sole of the foot, appear the most capable of producing them. We have seen how completely, in the lower animals, the acts of progression may be sustained, by the repeated stimulus of the contact of the ground, or of fluid, without any influence from the Cephalic ganglia; the power of these being limited, it would seem, to the control and direction of them. And there is strong reason to believe, that, so far as the ordinary acts of locomotion are concerned,

the movements of the inferior extremities in Man may be performed on the same plan, being continued by reflex power, when once set in action by the Will, whilst we are walking steadily onwards,—the mind being at the same time occupied by some train of thought, which engrosses its whole attention (§ 801). Even when the mind is sufficiently on the alert to guide, direct, and control the motions of the limbs, their separate actions appear to be performed without any direct agency of the will. It is certain that, in Birds, the movements of flight may be performed after the removal of the Cerebrum.

796. The *Medulla Oblongata* does not differ in any other essential particular from the Spinal Cord (of which it may be considered as the cranial prolongation), than this; that whilst the ganglionic portion of the latter is made up of the centres which minister to general locomotion, being homologous with the repetition of the *pedal* ganglia in the ventral cord of Articulated animals, the former contains the centres of the automatic movements of Deglutition and Respiration, and may therefore be regarded as representing their *stomato-gastric* and *respiratory* ganglia.—Both these actions are purely automatic in their character; and the Will can only restrain them, when the stimulus which calls them forth is not acting with any great degree of potency. The act of *Swallowing* is excited by the contact of solid or fluid matters with the membrane lining the fauces; and the impression, conveyed to the Medulla Oblongata by an afferent nerve (the 'glosso-pharyngeal'), excites there a reflex impulse, which, being transmitted along a motor nerve (chiefly the pharyngeal portion of the 'par vagum') to the muscles, calls them into those combined and consecutive movements, which are requisite for the reception of the food from the buccal cavity, and for its propulsion down the œsophagus. These movements may be excited in a state of complete unconsciousness; or after the removal of the Cerebrum from above, and of the Spinal Cord from below. And when we suppose that we are swallowing *voluntarily*, the action of the will is limited to the mere excitation of the reflex movement, by the carrying-back of the solid or liquid to be swallowed, upon the tongue, into contact with the lining membrane of the fauces. When this contact has once been effected, scarcely any power of the will could prevent the consecutive movement. The acts of *Prehension* of food with the lips, though usually effected by voluntary power in the adult, seem to be capable of taking place as a part of the reflex operation of the Medulla Oblongata, in the Infant, as in the lower animals. This is particularly evident in the prehension of the nipple by the lips of the infant, and in the act of suction which the contact of that body (or of any resembling it) seems to excite. The experiments provided for us by nature, in the production of anencephalous monstrosities, fully prove that the integrity of the nervous connection of the lips and respiratory organs with the Medulla Oblongata, is alone sufficient for the performance of this action; and experiments upon young animals, from which the brain has been removed, establish the same fact. Thus Mr. Grainger found that, upon introducing his finger, moistened with milk, or with sugar and water, between the lips of a puppy thus mutilated, the act of suction was excited; and not merely the act of suction itself, but other movements having a relation to it; for as the puppy lay on its side, sucking the finger, it pushed out its feet, in the same manner as young pigs exert

theirs in compressing the sow's dugs. The act of *Mastication*, again, although usually considered a voluntary one, comes by habit to be performed without any exertion of the will; and may then be referred to that class denominated 'secondarily automatic' (§ 801).—The movements of *Respiration* are purely reflex in their character, although capable of being imitated, or to a certain extent controlled, by an exertion of the will. The purpose of this subjection, which seems peculiar to the higher animals, is evidently to render them subservient to the production of Vocal Sounds; to which they obviously could not minister, if they were as independent of voluntary control and direction, as they seem to be in the lower animals. But this subjection only exists, when the stimulus does not act with more than moderate force, that is, when the blood is undergoing its normal aeration; for if the process be checked for a few seconds, the demand for aeration becomes so strong, that the will can no longer restrain the movement to which it prompts. The stimulus to this action chiefly originates in the lungs, and is either the presence of venous blood in their capillaries, or of carbonic acid in their air-cells. This impression (which need not be strong enough to arouse the consciousness) is conveyed to the Medulla Oblongata by the pulmonic portion of the 'par vagum'; and a reflex impulse is then transmitted through the 'phrenic' and 'intercostal' nerves, to the diaphragm and the muscles of the ribs, which produces the act of inspiration. It is not through the 'par vagum' alone, however, that excitor impressions are conveyed; for the nerves of the general surface, and especially of the face, are subservient to this function; and it is through them that the *first* inspiration is excited, by the contact of cold air with the skin of the new-born Mammal. It appears, too, that the Cerebrum is concerned in the maintenance of the respiratory movements, not volitionally, however, but automatically; the circulation of imperfectly arterialized blood through its vessels, occasioning an impulse to be transmitted from it to the respiratory nerves, through the medium of their ganglionic centre. So when the respiratory efforts are very powerfully called forth, various other parts of the nervo-muscular apparatus are put in action, besides those just mentioned.

797. The chain of *Sensory Ganglia*, which forms nearly the entire Encephalon of Fishes (§ 781), but which is overlaid and obscured in Man and the higher Vertebrata by the relatively enormous development of the Cerebrum, may be regarded as constituting the true *Sensorium;* that is, as the seat of *consciousness*, to which impressions made upon the nerves of sense are carried, and through which the individual is rendered cognizant of them. There is abundant evidence that this endowment does not exist in the locomotive, stomato-gastric, or respiratory ganglia, of which the Spinal Cord and the principal part of the Medulla Oblongata are made up; whilst on the other hand, there is adequate proof that the presence of a Cerebrum is not necessary to its possession. For these ganglia are obviously homologous with the Cephalic ganglia of Invertebrated animals; and if the latter be the instruments of consciousness and the seat of sensibility (to deny which, would be to refuse these endowments to Invertebrated animals altogether), there is no reason to doubt that the former are so likewise, their relations to the nerves of sense being precisely the same. There is no adequate reason for the belief, that the

addition of the Cerebral Hemispheres, in the Vertebrated series, *alters* the endowments of the Sensory Ganglia on which they are superimposed; on the contrary, we have everywhere seen that the addition of ganglionic centres, as instruments of new functions, leaves those which were previously existing, in the discharge of their original duties. — So far as the results of experiments can be relied on, they afford a corroboration of these views, by showing that sensory impressions can be felt, and automatic movements excited or directed, through the medium of these ganglia, after the complete removal of the Cerebrum. Thus if a bird be thus mutilated, it maintains its equilibrium, and recovers it when it has been disturbed; if pushed, it walks; if thrown into the air, it flies. A pigeon deprived of its cerebrum has been observed to seek out the light parts of a partially illuminated room in which it was confined, and to avoid objects that lay in its way; and at night, when sleeping with closed eyes and its head under its wing, it raised its head and opened its eyes upon the slightest noise. It is scarcely possible to believe that these movements were purely reflex; since they seem obviously to indicate the guiding influence of sensation. — The results of other experiments, which have been made upon the sensory centres themselves, and upon the organs from which they derive their impressions, manifest the remarkable influence of these impressions in guiding the ordinary movements of the lower animals; these being completely disturbed, and a new set of purely automatic movements being substituted, when the ordinary relations of the organs were interfered with. Thus it has been ascertained by Flourens, that a vertiginous movement may be induced in pigeons, by simply blindfolding one eye; and Longet has produced the same effect, by evacuating the humours of one eye. These vertiginous movements are more decided and prolonged, when, instead of one eye being blinded, one of the optic ganglia is removed; the animal continuing to turn itself towards the injured side, as if rotating on an axis. — The results of the experiments of M. Flourens upon the portion of the Auditory nerve proceeding to the 'semi-circular canals' (§ 825), are still more extraordinary. Section of the horizontal semi-circular canal in Pigeons, on both sides, induces a rapid jerking horizontal movement of the head, from side to side; and a tendency to turn to one side, which manifests itself whenever the animal attempts to walk forwards. Section of a vertical canal, whether the superior or inferior, of both sides, is followed by a violent vertical movement of the head. And section of the horizontal and vertical canals, at the same time, causes horizontal and vertical movements. Section of either canal on one side only, is followed by the same effect as when the canal is divided on both sides; but this is inferior in intensity. The movements continue to be performed during several months. In Rabbits, section of the horizontal canal is followed by the same movements as those exhibited by Pigeons; and they are even more constant, though less violent. Section of the anterior vertical canal causes the animal to make continued forward 'somersets;' whilst section of the posterior vertical canal occasions continual backward 'somersets.' The movements cease when the animal is in repose; and they recommence when it begins to move, increasing in violence as its motion is more rapid. These curious results are supposed by M. Flourens to indicate, that the nerve supplying the semi-circular canals does not minister to the sense of hearing, but to the direction of the movements of the animal; but they are fully explicable, upon the supposition, that the normal function of the semi-circular canals is to indicate to the animal the *direction* of sounds, and that its movements are partly determined by these; so that a destruction of one or other of them will produce an irregularity of movement (resulting, as it would seem, from a sort of giddiness on the part of the animal), just as when one of the eyes of a bird is covered or destroyed, as in the experiments previously cited.

798. Notwithstanding that, in Man, the high development of *Intelligence* supersedes in great degree the operations of *Instinct*, we still find that there are in ourselves certain movements, which can be distinguished as neither voluntary nor reflex, and which are examples of that method of operation, which seems to be the chief source of the actions of the lower Vertebrata, as of the Invertebrated classes in general. These movements are as automatic and involuntary as are the proper 'reflex;' but they differ from them in this, that they are only excited by impressions of which we are conscious, that is, by sensations; and hence they are conveniently designated as *consensual.*—As examples of this group, we may advert to the act of Vomiting, produced by various causes which act through the organs of sense; such as the sight of a loathsome object, a disagreeable smell, or a nauseous taste. The excitement of the act of Sneezing by a dazzling light, is another example of the same kind; for even if it be granted, that the act of sneezing is ordinarily excited through the reflex system alone (which is by no means certain), there can be no doubt that in this instance it cannot be brought into play without a sensation actually felt. The same may be said of the Laughter which sometimes involuntarily bursts forth, at the provocation of some sight or sound, to which no distinct ludicrous idea or emotion can be attached; and of that resulting from the act of tickling, in which case it is most certainly occasioned by the sensation, and by that alone. The start produced by a loud and unexpected sound, and the closure of the eyes to a dazzling light, or on the sudden approach of a body that might injure them (which has been observed to take place in cases in which the eyelids could not be *voluntarily* closed), are additional examples of the same kind. It is in certain morbid states, however, that the direct influence of Sensations in occasioning and governing movements, in a manner not to be accounted for either by reflex or voluntary action, is most remarkably manifested. Thus in cases of excessive irritation of the retina, which renders the eye most painfully sensitive to even a feeble amount of light,—the state designated as *photophobia,*—the eyelids are drawn together spasmodically, with such force as to resist very powerful efforts to open them; and if they be forcibly drawn apart, the pupil is frequently rolled beneath the upper lid, much further than it could be carried by a voluntary effort. And in painful affections of the walls of the chest, we may observe the usual movements of the ribs to be very much abridged; the dependence of this abridgment upon the painful sensation which they occasion, being most evident in those instances in which the affection is confined to one side,—for there is then a marked curtailment in *its* movements, whilst those of the other side may take place as usual; a difference which cannot be reflex, and which the Will cannot imitate. Again, in some Convulsive disorders, we may notice that the paroxysms are excited by causes, which

act through the organs of special sense; thus in Hydrophobia, we observe the immediate influence of the sight or the sound of liquids, and of the slightest currents of air; and in many Hysteric subjects, the sight of a paroxysm in another individual is the most certain means of inducing it in themselves.

799. When we contrast the actions of Man and of the higher Vertebrata, with those of the lower, we cannot but perceive that we gradually lose the indications of *Intelligence* and *Will*, as the sources of the movements of the animal; whilst we see a corresponding predominance of those, which are commonly denominated *Instinctive*, and which are performed (as it would appear) in immediate respondence to certain sensations —without any *intentional* adaptation of means to ends on the part of the individual, although such adaptiveness doubtless exists in the actions themselves, being a consequence of the original constitution of the nervous system of each animal performing them. It cannot be doubted by any person who has attentively studied the characters of the lower animals, that many of them possess psychical endowments, corresponding with those which we term the Intellectual powers and Moral feelings in Man; but in proportion as these are undeveloped, in that proportion is the animal under the dominion of those Instinctive impulses, which, so far as its own consciousness is concerned, may be designated as blind and aimless, but which are ordained by the Creator for its protection from danger, and for the supply of its natural wants. The same may be said of the Human infant, or of the Idiot, in whom the reasoning powers are undeveloped. Instinctive actions may in general be distinguished from those which are the result of voluntary power guided by reason, chiefly by the two following characters:—1. Although, in many cases, experience is required to give the Will command over the muscles concerned in its operations, no experience or education is required, in order that the different actions, which result from an Instinctive impulse, may follow one another with unerring precision.—2. These actions are always performed by the same species of animal, nearly, if not exactly, in the same manner; presenting no such variation in the means adapted to the object in view, and admitting of no such improvement in the progress of life, or in the succession of ages, as we observe in the habits of individual men, or in the manners and customs of nations, that are adapted to the attainment of any particular ends, by those voluntary efforts which are guided by reason. —The fact, too, that these Instinctive actions are often seen to be performed under circumstances rendering them nugatory, as reason informs us, for the ends which they are to accomplish (as when the domesticated Beaver builds a dam across its apartment, or when the Bee tries to produce a queen from drone-larvæ), is an additional proof, that such actions are prompted, like the reflex and consensual movements we have been just enquiring into, by an impulse which immediately results from a particular impression or sensation, and not by anticipation of the effect which the action will produce. And so far as can be determined by the foregoing tests, we find the comparative predominance of the Instinctive actions and the Intelligent, to be in close accordance with the relative development of the Cranio-Spinal Axis and the Cerebrum. Moreover, some of the actions which have been designated as Instinctive, such as that of sucking in the Infant (§ 796), have been shown to be purely 'reflex,' not even

involving consciousness. On the whole, then, we seem fully justified in the conclusion, that the Instinctive actions of animals are truly 'automatic' in their character, and that they are performed by the instrumentality of the Cranio-Spinal Axis in Vertebrata, and of the Ganglia which correspond to them in Invertebrated animals; and that they originate in impressions made upon the afferent nerves, either by external objects, or by changes taking place in the internal organisation (such, for example, as the periodic development of the sexual organs), which impressions excite respondent movements, either through the ganglia of 'consensual' or those of 'reflex' action, according as they are or are not of a nature to require the *consciousness* of the animal to be acted on, as a link in the chain of their operation upon the muscular system.*

800. But we may trace the agency of the Sensory ganglia, in Man and the higher Vertebrata, not merely in their direct and independent operation on the muscular system, but also in the manner in which they participate in all Voluntary actions. There can be no doubt that, in every exertion of the *will* upon the muscular system, we are guided by the sensations communicated through the afferent nerves, which indicate to the Sensorium the state of the muscle. Many interesting cases are on record, which show the necessity of this Muscular Sense, for determining voluntary contraction of the muscle. Thus, Sir C. Bell (who first prominently directed attention to this class of facts, under the designation of the *Nervous Circle*), mentions an instance of a woman, who was deprived of it in her arms, without losing the motor power; and who stated, that she could not sustain anything in her hands (not even her child), by the strongest effort of her will, unless she kept her eyes constantly fixed upon it; the muscles losing their power, and the hands dropping the object, as soon as the eyes were withdrawn from it. Here the employment of the *visual* sense supplied the deficiency of the *muscular;* but instead of being inseparably connected, as the latter is in the state of health, with the action of the muscle, the former could be only brought to bear on it by an effort of the will; and the sustaining power was therefore dependent, not upon the immediate influence of the will upon the muscle, but upon the voluntary direction of the sight towards the object to be supported.—Again, in the production of vocal sounds, the nice adjustment of the muscles of the larynx, which is requisite to pro-

* The highest development of the purely Instinctive tendencies, with the least possible interference of Intelligence, is to be found in the class of Insects; and above all in the order *Hymenoptera*, and in that of *Neuroptera*, which is nearly allied to it. It is, of course, impossible to draw the line between the two sources of action, with complete precision; but we observe, in the habits of Bees and other social Insects, every indication of the absence of a power of choice, and of the entire domination of instinctive propensities called into action by sensations. Thus, although Bees display the greatest art in the construction of their habitations, and execute a variety of curious contrivances, beautifully adapted to variations in their circumstances, the constancy with which individuals and communities will act alike under the same conditions, appears to preclude the idea of their possessing any inherent power of spontaneously departing from the line of action, to which they are tied down by the constitution of their Nervous system. We do not find one individual or one community clever, and another stupid; nor do we ever witness a disagreement, or any appearance of indecision, as to the course of action to be pursued by the several members of any republic. The actions of all tend to one common end, simply because they are performed in respondence to impulses, which all alike share. For a Bee to be destitute of its peculiar tendency to build at certain angles, would be as remarkable as for a Human being to be destitute of the desire to eat, when his system should require food.

duce determinate tones, can only be learned in the first instance under the guidance of the sensation of the sounds produced, and can only be effected by an act of the will, in obedience to a mental conception (a sort of inward sensation) of the tone to be uttered; which conception cannot be formed, unless the sense of hearing has previously brought similar tones to the mind. Hence it is, that persons who are born *deaf*, are also *dumb*. They may have no malformation of the organs of speech; but they are incapable of uttering distinct vocal sounds or musical tones, because they have not the guiding conception, or recalled sensation, of the nature of these. By long training, and by efforts directed by the muscular sense of the larynx itself, some persons thus circumstanced have acquired the power of speech; but the want of a sufficiently definite control over the vocal muscles, is always very evident in their use of the organ.—The conjoint movement of the two eyes, which concur to direct their axes towards the same object, are among the most interesting of these actions, in which Volition and Consensual action are alike concerned; and they afford an excellent illustration of the necessity for guiding sensations, to determine the actions of muscles. The sensations, however, are not so much those of the muscles themselves, as those received through the visual organ; but the former appear capable of continuing to guide the harmonious movements of the eyeballs, when the sense of sight has been lost. It is a striking peculiarity of these movements, that, in the majority of them, two muscles or combinations of muscles of opposite action are in operation at once; thus, when the eyes are made to rotate in a horizontal plane, the internal rectus of one side acts with the external rectus of the other. In most other cases, there is a difficulty in performing two opposite movements, on the two sides, at the same time. Thus, if we move the right hand as if winding on a reel, and *afterwards* make the left hand revolve in a contrary direction, no difficulty is experienced; but if we attempt to move the two *at the same time* in contrary directions, we shall find it almost impossible.

801. It is not difficult to account, on the foregoing principles, for the fact which has been a source of great perplexity to Metaphysicians and Physiologists,—that movements which were at first performed voluntarily, and which even required a distinct effort of the will for each, may become, by habitual repetition, so far independent of the will, that they are performed when the whole attention of the mind is bestowed upon some other train of action. Thus we all know that, in walking along an accustomed road, we frequently occupy our minds with some perfectly continuous chain of reasoning; and yet our limbs continue to move under us with regularity, until we are surprised by finding ourselves at the place of our destination, or perhaps at some other which we had not intended to visit, but to which habit has conducted us. Or we may read aloud for a long time, without having in the least degree comprehended the meaning of the words we have uttered; our attention having been closely engaged by some engrossing thoughts or feelings within. Or a musician may play a well-known piece of music, whilst carrying on an animated conversation.—Some Metaphysicians have explained these facts by supposing that (as the mind cannot *will* two different things at the same time) the Volition is in a sort of vibratory condition between the two sets of actions, now prompting one and now the other. But it would seem much more

conformable to the analogy afforded by other physiological phenomena, to regard these, with Hartley, as "secondarily automatic;" that is, as taking the place, in Man, of those actions which are primarily and purely automatic in many of the lower animals. We shall see that, even when most purely Voluntary, these actions are performed by the instrumentality of the automatic apparatus (§ 806); and the influence of habit gradually links-on the movements to the sensations which at first guided them, in such a manner that the latter at last come to be themselves adequate excitors of the movement, when the series has been once commenced by an exertion of the will. It has been thought by some to be a sufficient proof of the voluntary nature of these movements, that we can check them at any time by an effort of the will; but this we do only when the attention has been recalled to them, so that the Cerebrum, liberated as it were from its previous self-occupation, resumes its usual play upon the automatic centres. In the performance of such habitual actions, it would seem as if, the first *start* having been given by the will, the sensation involved in each movement becomes the stimulus to the next, and so on, until the habitual series is concluded, or the attention is called back to them.—This view is confirmed by the fact, that in cases of severe injury of the Brain, in which Intelligence and Will seem completely in abeyance, habitual actions may often be excited.

802. That the *Cerebellum* is in some way connected with the powers of *motion*, might be inferred from its connection with the antero-lateral columns of the Spinal Cord, as well as with the posterior; and the comparative size of the organ, in different orders of Vertebrated animals, gives us some indication of what the nature of its function may be. For we find its degree of development corresponding pretty closely with the variety and complexity of the Muscular movements, which are habitually executed by the species; the organ being the largest in those animals, which require the *combined* effort of a great variety of muscles to maintain their usual position, or to execute their ordinary movements; whilst it is the smallest in those, which require no muscular exertion for the one purpose, and little combination of different actions for the other. Thus in animals that habitually rest and move upon four legs, which is the case with most Reptiles, there is comparatively little occasion for any organ to combine and harmonize the actions of their several muscles; and in these, the Cerebellum is usually small. But among the more active predaceous Fishes (as the Shark), Birds of the most powerful and varied flight (as the Swallow), and such Mammals as can maintain the erect position, and can use their extremities for other purposes than support and motion, we find the Cerebellum of much greater size, relatively to the remainder of the Encephalon. There is a marked advance in this respect, as we ascend through the series of Quadrumanous Animals, from the Baboons, which usually walk on all-fours, to the semi-erect Apes, which often stand and move on their hind-legs only. The greatest development of the Cerebellum is found in Man, who surpasses all other animals in the number and variety of the combinations of muscular movement, which his ordinary actions involve, as well as of those which he is capable, by practice, of learning to execute.—From experiments upon all classes of Vertebrated animals, it has been found that, when the Cerebellum is removed, the power of walking, springing, flying, standing, or maintaining the equili-

brium of the body, is destroyed. It does not seem that the animal has in any degree lost the *voluntary* power over its individual muscles; but it cannot *combine* their actions for any general movements of the body. The *reflex* movements, such as those of respiration, remain unimpaired. When an animal thus mutilated is laid on its back, it cannot recover its former posture; but it moves its limbs, or flutters its wings, and evidently is not in a state of stupor. When placed in the erect position, it staggers and falls like a drunken man,—not, however, without making efforts to maintain its balance. Phrenologists, who attribute a different function to the Cerebellum, have attempted to put aside these results, on the ground that the severity of the operation is alone sufficient to produce them; but, as we have already seen (§ 797), many animals may be subjected to a much more severe operation,—the removal of the Cerebral hemispheres,—without the loss of the power of combining and harmonizing the muscular actions, provided the Cerebellum be left uninjured.—Thus, then, the idea of the functions of the Cerebellum, which we derive from Comparative Anatomy, seems fully borne out by the results of experiment; and it is also consistent with the conclusions to which observation of Pathological phenomena seems to lead. Some of these phenomena, however, appear to indicate that either the median lobe of the Cerebellum, or some collection of ganglionic matter in its vicinity, is the centre of *sexual* sensibility, which seems to be distinct from 'common' or 'tactile' sensibility.

803. It appears to be through the instrumentality of the *Cerebrum* or *Hemispheric Ganglia*, that the sensations awakened in the Sensorium give rise to *Ideas;* which then become the material (so to speak) of all the higher psychical operations. It is not alone, however, by the exercise of the reasoning faculties, leading to voluntary determinations, that mental states react upon the motor apparatus; for we find that *emotional* excitement will give rise to movements, which are as involuntary as the automatic, and which the reason and will may in vain endeavour to keep in check. And the distinctness of their character is further obvious from this, that cases of paralysis not unfrequently occur (especially in the facial nerve, through which the muscles of 'expression' are for the most part excited to action), in which the muscles are obedient to an emotional impulse, though the will exerts no power over them; whilst in other instances, the will may exert its due influence, and yet the emotional state cannot manifest itself. There are, moreover, several disordered states of the nervous system (such as Chorea and Hysteria), in which irregular or convulsive movements, totally unrestrainable by the will, are directly consequent upon emotional excitement.—It has been supposed by some, that the Emotional movements of Man and the higher animals, might be ranked in the same category with the Instinctive actions of the lower; and that the Desires of the former are comparable to the Instinctive propensities of the latter. But this comparison is erroneous; for such propensities, as already shown, are nothing else than tendencies to perform given movements in respondence to particular sensations, without any *idea* of the purpose of the movement, or of the object which has excited it; whereas an Emotion involves an idea of the object which has excited it, and a Desire involves a conception of the object to be attained. The *imitative* actions will afford a good example of the difference between a propensity and a desire. The former is manifested in such imitative

actions as are purely consensual, the sensation in each case exciting the movement automatically; as when we yawn involuntarily, from seeing or hearing the action performed by another; or as when infants learn to perform many of the movements which they witness in adults. But in other instances, imitative actions are voluntary; being the result of a *desire* to perform them, which involves a distinct idea of the object; and being at the same time a source of pleasure to the performer, which is the spring of the desire. Thus we find the two sources of action to be so distinct, that the tendency to involuntary or automatic imitation may be very strong in an individual, who is utterly unable to mimic or imitate voluntarily, and who has no conscious inclination to do so.

804. Now when the Emotions and Moral Feelings are analyzed, we find them to be in like manner complex in their nature; being made up by the association of *ideas*, which are Cerebral in their seat, with the simple *feelings* of pleasure and pain, which are probably localized in the Sensorium. Thus, Benevolence may be defined to be the pleasurable idea of the happiness of others. The whole class of Selfish emotions, on the other hand, is nothing else than the pleasurable contemplation of objects of supposed value to self. Combativeness, again, is the pleasurable idea of antagonism to others; Veneration, the pleasurable contemplation of rank or perfections superior to our own; Hope, the pleasurable anticipation of future enjoyment; Cautiousness, a combination of the painful contemplation of future evil, with the pleasurable idea of the precautions taken to prevent it.—Now when emotions are excited by external sensations, these emotions may act downwards through the automatic system, producing movements which may be in direct antagonism to the Will. Thus we may see or hear something ludicrous, which involuntarily provokes laughter, although we may have the strongest possible motives for desiring to restrain it. This downward action of the emotions appears, then, to have its immediate seat in the sensory ganglia, in which pleasurable and painful feelings are excited by the ideas formed in the cerebrum; and it is by the strong excitement of these feelings, that the emotional movements are called into play. No ideas purely intellectual, that is, not associated with *feelings*, will give rise to movements resembling the emotional.—The purely Emotional actions are not always directly excited, however, by *external* sensations; for they may result from the operations of the Mind itself. Thus, involuntary laughter may result from a ludicrous *idea*, called up by some train of association, and having no obvious connection with the sensation which first set this process in operation; and the various movements of the face and person, by which Actors endeavour to express strong Emotions, are only effectual in conveying their meaning, when they result from the actual working of the emotions in the mind of the performer, who has, by an effort of the will, identified himself (so to speak) with the character he personates. A still more remarkable case is that, in which paroxysms of Hysterical convulsion, in themselves beyond the power of the Will to excite or to control, are brought on by a voluntary effort; which seems to act by "getting up," so to speak, the state of feeling, which is the immediate cause of the disordered movements. In all these instances, and others of like nature, it would seem as if the agency of the Cerebrum produced the same condition in the Sensory ganglia and their motor fibres, as that which is more directly

excited by sensations received through their own afferent nerves. It may be reasonably surmised, then, that the Sensory ganglia, like the Cephalic ganglia which are the instruments of the Instinctive actions of the lower animals, can only be excited to action by stimuli *immediately* operating upon them; but that these stimuli may be either Sensations directly originating in external objects, or Conceptions resulting from the impressions left by those objects, of which there is strong reason to believe that the Cerebrum is the storehouse. — The Emotions are concerned in Man, however, in many actions which are in themselves strictly voluntary. Unless they be so strongly excited as to get the better of the Will, they do not operate downwards upon the motor system, but upwards upon the Cerebral; supplying the chief *motives* by which the voluntary determinations are guided (§ 805).

805. The faculty of *Memory* appears to depend upon the instrumentality of the Cerebral hemispheres; impressions made upon the Organs of Sense not being remembered, unless they are at once registered (as it were) in this part of the nervous centres. This faculty is one of those first awakened in the opening mind of the Infant; and it is one of which we find traces in animals, that seem to be otherwise governed by pure Instinct. It obviously affords the first step towards the exercise of the Reasoning powers; since no *experience* can be obtained without it; and the foundation of all intelligent adaptation of means to ends, lies in the application of the knowledge which has been acquired, and stored up in the mind. There is strong reason to believe, that no impression of this kind, once made upon the Cerebrum, is ever entirely lost, except through disease or accident, which will frequently destroy the memory altogether, or will annihilate the recollection of some particular class of objects or of words. All memory, however, seems to depend upon the principle of Suggestion; one idea being linked with another, or with a particular sensation, in such a manner as to be called up by its recurrence; and a period of many years frequently intervening, without that combination of circumstances presenting itself, which is requisite to arouse the dormant impression of some early event. Sometimes this combination occurs in dreaming, delirium, or insanity; and ideas are recalled, of which the mind, in a state of healthy activity, has no remembrance.—It is upon the ideas aroused in the mind by Sensorial changes, or recalled by *Conception,* or evolved by the process of *Reflection* (in which the mind perceives its own operations, and traces relations amongst its objects of thought), or generated by the *Imagination* (which really acts, however, rather by combining into new forms, than by creating altogether *de novo*), that all acts of *Reasoning* are based. These consist, for the most part, in the aggregation and collocation of ideas; the decomposition of complex ideas into more simple ones, and the combination of simple ideas into general expressions; in which are exercised the faculty of *Comparison,* by which the relations and connections of ideas are perceived,—that of *Abstraction,* by which we fix our attention on any particular qualities of the object of our thought, and isolate it from the rest,—and that of *Generalisation,* by which we grasp in our minds some definite notions in regard to the general relations of those objects. These are the processes chiefly concerned in the simple acquirement of Knowledge; with which class of operations, the Emotional part of our nature has very little participation. But in

those modes of exercise of our Reasoning powers, which are chiefly concerned in the determination of our actions, the Emotions, &c., are largely concerned. They chiefly (if not solely) act upon the reasoning powers, by modifying the form in which the ideas are presented to the mind; whether these ideas are directly excited by external sensations, or whether they are called up by an act of the Memory, or result from the exercise of the Imagination. For as they essentially consist of feelings of pleasure and pain, connected with certain classes of ideas, the former produce a *desire* of the objects to which they relate, the latter a *repugnance* to them. They thus have a most important influence upon the *Judgment,* which is formed by the comparison of certain kinds of ideas; and they may consequently modify the Volitional determination, or act of the Will, which is consequent upon this, and which may either be directed towards the further operations of the mind itself, or may exert an immediate influence on the bodily frame, by the agency of the Nervous System. In either case, it is the characteristic distinction of a Volitional operation, that *means are intentionally adapted to ends,* in accordance with the belief of the mind as to their mutual relations. Upon the correctness of that decision, will depend the power of the action to accomplish what the mind had in view.

806. Although Physiologists have been accustomed to regard the Will as *directly determining* all those muscular movements which are usually distinguished as Voluntary, yet a careful analysis of the process fully bears out the inferences which might be erected upon the considerations already advanced,—that the influence of the Will is *not* directly conveyed to the muscles by fibres beginning in the cerebral convolutions and proceeding to the muscles, but that it is exerted through the Automatic centres. For it has been shown that these Automatic centres (the Sensory Ganglia, Medulla Oblongata, and Spinal Cord) receive all the sensory nerves, and give origin to all the motor; and that the fibres which pass between the cerebral convolutions and the sensory ganglia, probably serve merely to bring these centres into mutual relation, and are not continuous with those of any nerves, either sensory or motor.—Now every one who has attentively considered the nature of what we are accustomed to call *voluntary* action, has been struck with the fact, that the Will simply determines the *result,* not the special movements by which that result is brought about. If it were otherwise, we should be dependent upon our anatomical knowledge, for our power of performing even the simplest movements of the body. Again, there are very few cases in which we can single-out any individual muscle, and put it in action independently of others; and the cases in which we *can* do so, are those in which a single muscle is concerned in producing the result,—as in the elevation of the eyelid; and we then really single-out the muscle, by 'willing' the result. Thus, then, however startling the position may at first appear, we have a right to affirm that the Will cannot exert any direct or immediate power over the muscles; but that its determinations are carried into effect through an intermediate mechanism, which, without any further guidance on our own parts, selects and combines the particular muscles whose contractions are requisite to produce the desired movement. We have seen that the Sensorium (or collection of sensory ganglia) *plays,* so to speak, upon the Cerebrum; sending to it sensations, whereby its

peculiar activity as an instrument of purely mental operations is called forth; and, in return, the Cerebrum appears to *play* downwards upon the motor portion of the automatic apparatus, sending it volitional impulses, which excite its motorial activity. And, hence, it follows that *all* the movements which are performed by the instrumentality of the cerebro-spinal nervous system, are in themselves *automatic;* and that the peculiarity in their character,—whether Reflex, Instinctive, Emotional, or Voluntary,—is due to the speciality of the source and seat of the impulses which respectively originate them.

807. The nerves of the *Sympathetic* System possess a certain degree of power of exciting Muscular contractions, in the various parts to which they are distributed. Thus by irritating them, immediately after the death of an animal, contractions may be excited in any part of the alimentary canal, from the pharynx to the rectum, according to the trunks which are irritated, — in the heart, after its ordinary movements have ceased,—in the aorta, vena cava, and thoracic duct,—in the ductus choledochus, uterus, fallopian tubes, vas deferens, and vesiculæ seminales. But the very same contractions may be excited, by irritating the roots of the Spinal nerves, from which the Sympathetic trunks receive their white fibres; and there is, consequently, strong reason to believe, that the *motor* power of the latter is entirely dependent upon the Cerebro-spinal system. Whatever *sensory* endowments the Sympathetic trunks possess, are probably to be referred to the same connection. In the ordinary condition of the body, these are not manifested. The parts exclusively supplied by Sympathetic trunks do not appear to be in the least degree sensible; and no sign of pain is given, when the Sympathetic trunks themselves are irritated. But in certain diseased conditions of those organs, violent pains are felt in them; and these pains can only be produced, through the medium of fibres communicating with the sensorium through the spinal nerves.—It is difficult to speak with any precision, as to the functions of the Sympathetic system. There is much reason to believe, however, that it constitutes the channel, through which the passions and emotions of the mind affect the Organic functions; and this especially through its power of regulating the calibre of the arteries. We have examples of the influence of these states upon the Circulation, in the palpitation of the heart which is produced by an agitated state of feeling; in the Syncope, or suspension of the heart's action, which sometimes comes on from a sudden shock; in the acts of blushing and turning pale, which consists in the dilatation or contraction of the small arteries; in the sudden increase of the salivary, lachrymal, and mammary secretions, under the influence of particular states of mind, which increase is probably due to the temporary dilatation of the arteries that supply the glands, as in the act of blushing; and in many other phenomena. It is probable that the Sympathetic system not only thus brings the Organic functions into relation with the Animal: but that it also tends to harmonize the former with each other, so as to bring the various acts of Secretion, Nutrition, &c., into mutual conformity. Of the distinctive function of the grey or organic fibres, we have no certain knowledge; but they not improbably may have some direct influence upon the chemical processes, which are involved in these changes, and may thus affect the *quality* of the secretions; whilst the office of the white fibres is rather to

regulate the diameter of the blood-vessels supplying the glands, and thus to determine the *quantity* of their products.

3. *General Summary.*

808. A retrospective view of the ground over which we have now passed, will lead us to some interesting conclusions.

I. It has been shown that the movements occasionally exhibited in the Vegetable kingdom, do not imply the existence, in its members, of consciousness or of a guiding will; being merely dependent upon that property of *contractility* upon the application of a stimulus, which may be regarded as due to the peculiar manner in which the elements of the contractile tissues are combined and arranged. Hence Vegetables may be considered as peculiarly, though not entirely, constituting the *Kingdom of Organic Life.*

II. In immediate connection with the lowest of the Vegetable kingdom, are the lowest of the Animal tribes, the *Protozoa* and lower *Radiata.* Here we see indications of the same simple contractility which Plants enjoy, but this has a more important relation to the well-being of the individual, and forms a more prominent part of its vital actions.

III. In addition to the movements thus produced, we witness others, in the higher *Radiata,* and in the lower *Articulata* and *Mollusca,* which are consequent upon impressions made on distant parts and transmitted through the nervous system; and some, even, which seem to be performed under the guidance of sensation; although none which clearly evince intelligence and design on the part of the Animals themselves. Proceeding still further, we observe these 'instinctive' movements becoming more complex in their character, and more refined and special in their objects; until we arrive at the class of *Insects,* which seem to possess the highest development of the *Instinctive* faculties, of any known animals. But we do not find *Intelligence* by any means increasing in the same ratio. On the contrary, it remains very low; and its power of modifying the dictates of Instinct, when these happen (from particular causes) to be erroneous, is very slight. If we further inquire, in what *orders* of Insects this power is most strikingly manifested, we shall have little hesitation in fixing upon the *Hymenoptera* and *Neuroptera*;—which include the Bee, Wasp, Ant, White-Ant, and other social Insects. Now it is not a little remarkable, that Insects should, of all classes of Animals, be most distinguished for *locomotive* power (as compared with their size); and that, of all Insects, the Hymenoptera and Neuroptera possess this power in the highest degree. It is evident that the higher kinds of instinctive actions, including the peculiar endowments of the nervous and muscular systems just referred to, have for their object the maintenance of *animal* life, as distinguished on the one hand from the mere *organic* life of Vegetables, and from the *mental* or *psychical* life of higher beings, on the other. Hence we should regard Insects, and especially the Hymenoptera and Neuroptera, as typical of the *Kingdom of Animal Life.* In this, we find the movements, which are produced by the direct contractility of the tissues stimulated, bearing a smaller and still smaller proportion to the whole; and at last restricted merely to the parts immediately concerned in the maintenance of the organic functions, with which they always remain associated (§ 752).

IV. Ascending from the Articulated through the *Vertebrated* series, we observe a gradually-increasing development of the reasoning powers or Intelligence; and a gradual fading-away of the instincts, which become subordinate to the higher psychical faculties. A comparison between the habits of Birds and those of Insects, will put this in a striking light. It has been shown that several points of correspondence exist between these two classes, indicating that they hold a corresponding rank in their respective sub-kingdoms. But whilst all the actions of Insects appear to be under the guidance of pure unvarying instinct, those of Birds, whilst evidently prompted by similar impulses, are yet capable of great modification in each individual (according to the peculiar circumstances in which it may happen to be placed) by the influence of its reasoning faculties. Ascending to Man, we find the pure instincts so completely subordinated to the higher psychical faculties, that it is only when the latter are yet dormant, as in infancy, or undeveloped, as in complete idiocy, that their uncontrolled operation is witnessed. It is easy to perceive the final cause for this change. If the organisation of the Human system had been adapted to perform all the actions necessary for the continued maintenance of its existence, with the same certainty and freedom from voluntary effort as we perceive where pure instinct is the governing principle, and if all his sensations had given rise to intuitive perceptions, instead of those perceptions being acquired by the exercise of his mind,—it is evident that external circumstances would have created no stimulus to the improvement of his intellectual powers, and that the strength of his instinctive propensities would have diminished the freedom of his moral agency. Although, therefore, to all the actions *immediately* necessary for the maintenance of his own existence, and for the continuance of his race, a powerful instinct strongly impels him, these propensities could not be gratified, if the means were not provided by the exercise of the mental powers, which he enjoys in a degree far exceeding those of any other terrestrial being.— Hence we should be led to regard his place in the Animal Kingdom, as being not at its head or in its centre, but at the extreme most remote from its point of contact with the kingdom of Organic life,— in fact, at the point at which we may believe it to touch another Kingdom, that of *pure Intelligence.* Such a view tends to show the true nobility of Man's rational and moral nature; and the mode in which he may most effectually fulfil the ends for which his Creator designed him. He may learn from it the evil of yielding to those merely animal tendencies,— those "fleshly lusts which war against the soul,"— that are characteristic of beings so far below him in the scale of existence; and the dignity of those pursuits which exercise the intellect, and which expand and strengthen those lofty moral feelings, which he alone of terrestrial beings is capable of entertaining.

809. In tracing the progressive complication of the Psychical manifestations, during the early life of the Human being, a remarkable correspondence may be observed with that gradual increase in mental endowments, which is to be remarked in ascending the Animal scale. The first movements of an Infant are evidently of a purely automatic character, and are directed solely to the supply of its physical wants; they are thus analogous to the instinctive actions of the lowest animals possessed of a nervous system. The new sensations which are constantly

being excited by surrounding objects, call into exercise the dormant powers of mind; notions are acquired of the character and position of external objects; and the simple processes of association, with its concomitant—memory, are actively engaged during the first months of an infant's life. At the same time an attachment to persons and places begins to manifest itself. All these are the characteristics of the great majority of the lower Vertebrata, so far, at least, as our knowledge of their springs of action enables us to form a judgment. As the child advances in age, the powers of observation are strengthened; the perceptions become more complete; those powers of reflection are called out, which prompt him to reason upon the causes of what he observes, and to perform actions resulting from more complicated mental processes than those which guide the infant; and, at the same time, we observe the development of the moral feelings, which are at first manifested only towards beings who are the objects of sense. Among the more sagacious quadrupeds, it is easy to discover instances of reasoning as close and prolonged as that which usually takes place in early childhood; and the attachment of the dog to man is evidently influenced by moral feelings, of which the latter is the object. "Man," it was expressively said by Burns, "is the God of the Dog." Up to this point, then, we observe nothing peculiar in the character of man; and it is only when his higher intellectual and moral endowments begin to manifest themselves, especially those relating to an invisible Being, that we can point to any obvious distinction between the immortal ψυχη of man, and the transitory πνευμα of the brutes that perish. May we not regard *these* as *here* existing but as the germs or rudiments of those higher and more exalted faculties, which the human mind shall possess, when, purified from the dross of earthly passions, and enlarged into the comprehension of the whole scheme of Creation, the soul of man shall reflect, without shade or diminution, the full effulgence of the Love and Power of its Maker?

[In the foregoing outline of the Structure and Actions of the Nervous System, the Author has simply aimed to express his own present convictions, and not to trace the history of the enquiry. He deems it but just, however, to state, that, although he has been led to abandon some parts of Dr. Marshall Hall's system of doctrines, which he formerly embraced, he still regards the results of that gentleman's enquiries as worthy of a place among the most important physiological discoveries of any age. In his progress towards what he now believes to be the true view of the relation of the Spinal Cord and its nerves to the Cerebrum, he has been partly guided by the views of Messrs. Todd and Bowman; from whom, however, he dissents, in regarding the Sensory Ganglia as parts of the Automatic apparatus, and in assigning to them functions altogether independent of the Cerebrum. His views on this subject were formed as far back as the year 1837; and were expressed in a paper, "On the Voluntary and Instinctive Actions of Living Beings," published in the "Edinb. Med. and Surg. Journ.," No. CXXXII. In his general ideas of the relations of the different modes of nervous activity, and of their predominance in different tribes of animals respectively, he finds himself in close and unexpected accordance with Unzer, whose work entitled "Erste Gründe einer Physiologie der eigentlichen thierischen Natur thierischer Körper," published in 1771, displays an insight into the physiology of the Nervous System, which is the more wonderful, when the imperfect state of Anatomical knowledge (and especially of Comparative Anatomy) at that period is borne in mind.]

CHAPTER XXI.

OF SENSATION, AND THE ORGANS OF SENSE.

1. *Of Sensation in General.*

810. It seems probable that all Animals which possess a definite Nervous System, have a greater or less degree of *consciousness* of the impressions made upon it, whether by external objects, or by changes taking place in their own organism; and to this consciousness, we give the name of *sensibility.* It is very important to bear in mind, that although we commonly refer our sensations to the parts on which the impressions are made, and speak of these parts as possessing sensibility, we really use incorrect language; since that of which we are actually conscious, is the change in the central *sensorium,* resulting from the excitement of its nervous polarity by the force transmitted from the periphery; and the difference between what are called 'sensible' and 'insensible' parts of the body consists in this, that the former can receive and transmit the impressions which thus arouse the consciousness, whilst the latter are unable to do so. This is evident from two facts; *first,* that if the nervous communication of any 'sensible' part with the sensorium be interrupted, no impressions, however violent, can make themselves felt; and *second,* that if the trunk of the nerve be irritated or pinched, any where in its course, the pain which is felt is referred, not to the point injured, but to the surface to which these nerves are distributed. Hence the well-known fact, that, for some time after the amputation of a limb, the patient feels pains, which he refers to the fingers or toes that have been removed; this continues, until the irritation of the cut extremities of the nervous trunks has subsided.

811. It would seem probable that, among the lower tribes of Animals, there exists no other kind of sensibility, than that termed 'general' or 'common;' which exists, in a greater or less degree, in almost every part of the bodies of the higher. It is by this, that we feel those impressions, made upon our bodies by the objects around us, which produce the various modifications of pain, the sense of contact or resistance, and others of a similar character. From what was formerly stated (§ 249) of the dependence of the impressibility of the sensory nerves, upon the activity of the circulation in the neighbourhood of their extremities, it is obvious that no parts destitute of blood-vessels can receive such impressions, or (in common language) can possess sensibility. Accordingly we find that the Hair, Nails, Teeth, Cartilages, and other parts that are altogether extra-vascular, are themselves destitute of sensibility; although certain parts connected with them, such as the bulb of the hair, or the vascular membrane lining the pulp-cavity of the tooth, may be acutely sensitive. Again, in Tendons, Ligaments, Fibro-cartilages, Bones, &c., whose substance contains very few vessels, there is but a very low amount of sensibility. On the other hand, the Skin and other parts, which are peculiarly adapted to receive such impressions, are extremely vascular; and it is interesting to

observe, that some of the tissues just mentioned become acutely sensible, when new vessels form in them in consequence of diseased action. It does not necessarily follow, however, that parts should be sensible in a degree proportional to the amount of blood they may contain; for this blood may be sent to them for other purposes, and they may contain but a small number of sensory nerves. Thus, it is a condition necessary to the action of Muscles, that they should be copiously supplied with blood; but they are by no means acutely sensible; and, in like manner, Glands, which receive a large amount of blood for their peculiar purposes, are far from possessing a high degree of sensibility.

812. But besides the 'general' or 'common' sensibility, which is diffused over the greater part of the body, in most animals, there are certain parts, which are endowed with the property of receiving impressions of a peculiar or 'special' kind, such as sounds or odours, that would have no influence on the rest; and the sensations which these excite, being of a kind very different from those already mentioned, arouse ideas in our minds, which we should never have gained without them. Thus, although we can acquire a knowledge of the shape and position of objects by the touch, we could form no notion of their colour without sight, of their sounds without hearing, or of their odours without smell. The nerves which convey these 'special' impressions are not able to receive those of a 'common' kind; thus the eye, however well fitted for seeing, would not feel the touch of the finger, if it were not supplied by branches from the Fifth pair, as well as by the Optic. Nor can the different nerves of special sensation be affected by impressions, that are adapted to operate on others; thus the ear cannot distinguish the slightest difference between a luminous and a dark object; nor could the eye distinguish a sounding body from a silent one, except when the vibrations can be *seen*. But Electricity and other Physical Forces, when applied to the several nerves of special sense, may excite the sensations peculiar to each (§§ 246, 247).

813. It is important to keep in mind the distinction between the *sensations* themselves, and the *ideas* which are the immediate results of those sensations, when they are perceived by the mind. These ideas relate to the *cause* of the sensation, or the object by which the impression is made. Thus, the formation of the picture of an object upon the retina, produces a certain impression upon the optic nerve; which, being conveyed to the sensorium, excites a corresponding sensation; and with this, in all ordinary cases, we immediately connect an idea of the nature of the object. So closely, indeed, is this idea usually related to the sensation, that we are not in the habit of making a distinction between them. We find that some of these *perceptions* or elementary notions are *intuitive*; that is, they are prior to all experience, and are *necessarily* connected with the sensation which produces them, as reflex movements are with the impression that excites them. This seems to be the case, for example, with regard to *erect vision*. There is no reason whatever to think, that either infants or any of the lower animals see objects in an inverted position, until they have corrected their notion by the touch; for there is no reason why the inverted picture on the retina should give rise to the idea of the inversion of the object. The picture is so received by the mind, as to convey to us an idea of the position of external objects, which harmonizes with the ideas we derive through the touch; and whilst we are

in such complete ignorance of the manner in which the mind becomes conscious of the sensation at all, we need not feel any difficulty about the mode in which this conformity is effected. But in Man, the attaching definite ideas to certain groups of lines, colours, &c., with respect to the objects they represent, is a subsequent process, in which experience and memory are essentially concerned; as we see particularly well, in cases of no unfrequent occurrence, in which the sense of sight has been acquired comparatively late in life, and in which the mode of using it, and of connecting the sensations received through it with those received through the touch, has had to be learned by a long-continued training. The elementary notions thus formed, which may, by long habit, present themselves as immediately and unquestionably, as if they were intuitive, are termed *acquired* perceptions.—It is probable that, among the lower animals, the proportion of *intuitive* perceptions is much greater than in Man; whilst, on the other hand, his power of acquiring perceptions is much greater than theirs. So that, whilst the young of the lower animals very soon become possessed of all the knowledge, which is necessary for the acquirement of their food, the construction of their habitations, &c., their range is very limited, and they are incapable of attaching any ideas to a great variety of objects, of which the human mind takes cognizance. This correspondence between the acquired perceptions of Man, and the intuitive perceptions of many of the lower animals, is strikingly evident in regard to the power of measuring distance; which is acquired very gradually by the Human infant, or by a person who has first obtained the faculty of sight later in life; but which is obviously possessed by many of the lower animals, to whose maintenance it is essential, immediately upon their entrance into the world. Thus a Flycatcher, immediately after its exit from the egg, has been known to peck-at and capture an insect,—an action which requires a very exact appreciation of distance, as well as a power of precisely regulating the muscular movements in accordance with it.

2. *Of the Sense of Touch.*

814. By the sense of Touch is usually understood that modification of the common sensibility of the body, of which the surface of the Skin is the especial seat, but which exists also in some of its internal reflexions. In some animals, as in Man, nearly the whole exterior of the body is endowed with it, in no inconsiderable degree; whilst in others, as the greater number of Mammalia, most Birds, Reptiles, and Fishes, and a large proportion of the Invertebrata, the greater part of the body is so covered with hairs, scales, bony or horny plates, shells of various kinds, complete horny envelopes, &c., as to be nearly insensible; and the faculty is restricted to particular portions of the surface, or to organs projecting from it, which often possess a peculiarly high degree of this endowment. Even in Man, the acuteness of the sensibility of the cutaneous surface varies greatly in different parts; being greatest at the extremities of the fingers, and in the lips; and least in the skin of the trunk, arm, and thigh.

815. The impressions that produce the sense of Touch, are ordinarily received through the sensory *papillæ,* which are minute elevations of the surface, enclosing loops of capillary vessels (Fig. 23), and branches of the

sensory nerves (§ 242).—True papillary organs have not yet been discovered in any Invertebrata; or even in Fishes, Serpents, or Chelonians. —In the soft-skinned *Batrachia*, an imperfect papillary structure is extensively diffused over the surface; but on the thumb of the male Frog, and probably on that of other Batrachia, large papillæ are developed at the season of sexual excitement. In many Lizards, a papillary structure is found on the under surface of the toes; and in the Chameleon it exists also in the integuments of its prehensile tail.—In Birds, the only parts of the skin on which tactile papillæ seem to exist, are on the under surface of the toes, and on the web of the Palmipedes, on which sensory impressions are made that guide the movements of the feet; and on the bill of the Duck tribe, which is plunged into the mud, &c., in search of food.—Among Mammalia, we find that the papillary structure is especially developed on those parts of the tegumentary surface, which are of the most important use in appreciating the qualities of the food, or in guiding the movements of the instruments of locomotion. Thus in the Quadrumana generally, both hands and feet are thickly set with papillæ; and in those which have a prehensile tail, the surface of this organ possesses them in abundance. In the Carnivorous and Herbivorous Mammals, whose extremities are furnished with claws or encased in hoofs, we find the lips and the parts surrounding the nostrils to be the chief seat of tactile sensibility, and to be copiously furnished with papillæ; and this is especially the case with those, which have the lips or nostrils prolonged into a snout or proboscis, as the Pig, the Rhinoceros, the Tapir, and the Elephant. In the Mole, too, the papillary structure is remarkably developed at the end of the snout.—The papillary apparatus, and the sense to which it ministers, are not confined, however, to the tegumentary surface of the exterior of the body; for we find that the tongue of many animals is copiously furnished with sensory papillæ; and it is probable, from the experience of Man, that only a part of these minister to the sense of Taste, but that the remainder are tactile (§ 818).

816. A very different set of instruments is developed for this sense, however, in certain animals, either in place of, or in addition to, the true papillary apparatus. Thus, in Articulated animals generally, the jointed appendages to the head, known as *antennæ* and *palpi*, are undoubtedly instruments of touch; and this function they may execute most efficiently, notwithstanding the density of their covering. For just as a blind man judges of the proximity and characters of objects, by the impressions communicated to his hand by the contact of the stick with which he examines them, so may an Insect or a Crustacean receive sensory impressions upon the nerves distributed to the basal joints of their long antennæ, although the organs themselves may be as insensible (or rather, as unimpressible) as the stick.* The antennæ, when prolonged, seem to guide the movements of the animal; whilst, on the other hand, the palpi rather

* The Author is acquainted with a blind gentleman who exhibits a remarkable dexterity in the use of his stick in guiding his movements; and has been informed by him, that much of his power of discrimination depends upon the flexibility, elasticity, &c. of this instrument; so that, when he has chanced to lose or break the one to which he has been accustomed, it is often long before he can obtain another that shall suit him as well.—This circumstance seems to throw some light upon the remarkable variety which is seen in the conformation of the *antennæ* of Insects; as it may be imagined that each is adapted to receive and to communicate impressions of a particular class, adapted to the wants of the species.

appear to minister to the cognizance of objects brought into the neighbourhood of the mouth, and to have for their chief office to guide in the selection of food.—In many of the higher animals, the *hairs* are most delicate instruments of touch; for although themselves insensible, their bulbs are seated upon cutaneous papillæ, copiously provided with nerves and blood-vessels, in such a manner that any motion or vibration communicated to the hair must produce an impression upon the papilla at its base. Such an organisation is found in those long stiff hairs, which are known as the *vibrissæ* or 'whiskers' of the Feline tribe, and which are particularly large in the Seal. These sensitive hairs are also highly developed in many of the Rodentia, such as the Hare and Rabbit; and it has been proved by experiment, that, if they be cut off, the animal loses in great degree its power of guiding its movements in the dark.

817. The only idea communicated to our minds, when this sense is exercised in its simplest form, is that of *resistance;* and we cannot acquire a notion of the size or shape of an object, nor can we judge correctly as to the nature of its surface, through this sense alone, unless we *move* the object over our own sensory organ, or pass the latter over the former. By the various degrees of resistance which we then encounter, we form our estimate of the hardness or softness of the body. By the impressions made upon our sensory papillæ, when they are passed over its surface, we form our idea of its smoothness or roughness. But it is through the *muscular* sense, which renders us cognizant of the relative position of the fingers, of the amount of movement the hand has performed in passing over the object, and of other impressions of like nature, that we acquire our notions of the size and figure of the object; and hence we perceive that the sense of Touch, without the power of giving motion to the tactile organ, would have been of comparatively little use. It is chiefly in the *variety* of movements, of which the hand of Man is capable,—thus conducive as they are, not merely to his prehensile powers, but to the exercise of his sensory endowments,—that it is superior to that of every other animal; and it cannot be doubted, that this affords us a very important means of acquiring information in regard to the external world, and especially of correcting many vague and fallacious notions, which we should derive from the sense of Sight, if used alone. The power of *tactile discrimination* does not by any means bear a constant relation to the degree of 'common sensibility' in a part, that is, to its susceptibility to impressions which produce pain; for the latter may depend simply upon the amount of nerves supplied to the surface; and this may be great in parts which have few papillæ, and which are supplied by nerves whose central terminations are so closely blended, that impressions made on points which are in near approximation to each other cannot be distinguished. The sensory apparatus contained in the integuments would seem to be necessary for the exercise of the *sense of temperature;* for it appears from the recent experiments of Prof. Weber, that if the integuments be removed, the application of hot or cold bodies only causes *pain,* their elevation or depression of temperature not being perceived; and the same is the case, when hot or cold bodies are applied to the nerve-trunks. It is worthy of note that there are many cases on record, in which the sense of Temperature has been lost, whilst the ordinary Tactile sense remains; and it is sometimes preserved, when there is a complete loss of

every other kind of sensibility. So again we find that the *subjective* sensations of temperature,—that is, sensations which originate from changes in the body itself, not from external impressions,—are frequently excited quite independently of those of contact or resistance; a person being sensible of heat or of chilliness in some part of his body, without any real alteration of its temperature, and without any corresponding affection of the tactile sensations. And further, it is to be remarked, that whilst, for the exercise of the Tactile sense, absolute *contact* between the impressible surface and the solid body is required, the influence of temperature may be communicated by radiations from a distance.

3. *Of the Sense of Taste.*

818. The sense of Taste, like that of Touch, is excited by the direct contact of particular substances with certain parts of the body: but it is of a much more refined nature than touch; inasmuch as it communicates to us a knowledge of properties, which that sense would not reveal to us. All substances, however, do not make an impression on the organ of Taste. Some have a strong savour, others a slight one, and others are altogether insipid. The cause of these differences is not altogether understood; but it may be remarked that, in general, bodies which cannot be dissolved in water, alcohol, &c., and which thus cannot be presented to the gustative papillæ in a state of solution, have no taste. A considerable part of the impression produced by many substances taken into the mouth, is received through the sense of Smell, rather than through that of Taste. There are many substances, however, which have no aromatic or volatile character; and whose taste, though not in the least dependent upon the action of the nose, is nevertheless of a powerful character.—The sense of Taste has for its chief purpose, to direct animals in their choice of food; hence its organ is always placed at the entrance to the digestive canal. In higher animals, the Tongue is the principal seat of it; but other parts of the mouth are also capable of receiving the impression of certain savours. The mucous membrane which covers the tongue, is copiously supplied with papillæ, of various forms and sizes. Those of simplest structure closely resemble the cutaneous papillæ; but there are others, which resemble clusters of such papillæ, each being composed of a fasciculus of looped capillaries with a bundle of nerve-fibres, whose precise mode of termination it has not yet been found possible to ascertain. These *fungiform* papillæ, which are covered with a very thin epithelium, are probably the special instruments of the sense of taste; for the exercise of which it seems probable that the sapid substance should penetrate (in solution) to the interior of the papilla.—The *conical* papillæ, on the other hand, being furnished with thick epithelial investments, which are sometimes prolonged into horny spines, would seem destined chiefly to mechanical purposes; thus in the Felines, in which tribe the spines are most remarkably developed, they form a most efficient rasp, by which the bones

FIG. 314.

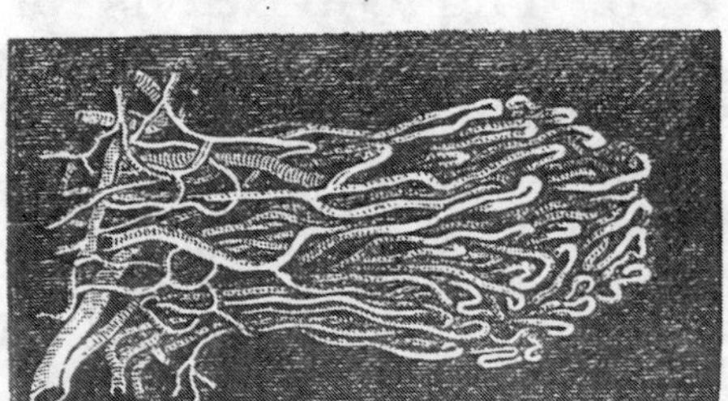

Capillary plexus of *fungiform papilla* of Human Tongue.

of their prey may be stripped of the smallest particles of flesh that may adhere to them; and in other cases, in which the investments are rather of a brush-like nature, they probably serve to cleanse the teeth from adhering particles, whilst the tactile sensibility of the papillæ directs the muscular movements of the tongue.—Of the degree of Taste possessed by different animals, it is impossible to form an accurate judgment, without a more accurate means of discrimination than we possess, between the Gustative and Tactile papillæ of the tongue. And we have not any certain knowledge, how far the sense of Taste may be exercised without a papillary structure.

4. *Of the Sense of Smell.*

819. Certain bodies possess the property of exciting sensations of a peculiar nature, which cannot be perceived by the organs of taste or touch, but which seem to depend upon the diffusion of the particles of the substance through the surrounding air, in a state of extreme minuteness. As the solubility of a substance in liquid seems a necessary condition of its exciting the sense of taste, so does its volatility, or tendency to a vaporous state, appear requisite for its possession of Odorous properties. Most volatile substances are more or less odorous; whilst those which do not readily transform themselves into vapour, usually possess little or no fragrance in the liquid or solid state, but acquire strong odorous properties, as soon as they are converted into vapour, — by the aid of heat, for example. There are some solid substances, which possess very strong odorous properties, without losing weight in any appreciable degree by the diffusion of their particles through the air. This is the case, for example, with Musk; a grain of which has been kept freely exposed to the air of a room, whose door and windows were constantly open, for a period of ten years; during which time the air, thus continually changed, was completely impregnated with the odour of musk; and yet, at the end of that time, the particle was not found to have perceptibly diminished in weight. We can only attribute this result to the extreme minuteness of the division of the odorous particles of this substance. There are other odorous solids, such as Camphor, which rapidly lose weight by the loss of particles from their surface, when freely exposed to the air.

820. The conditions of the sense of Smell are best studied in the higher animals. The membrane over which the olfactive nerve is distributed, is extremely vascular, and is covered with a thick pulpy epithelium; and the nerves lose themselves in its substance, apparently becoming divested of the 'white substance of Schwann,' and presenting a close resemblance to the 'gelatinous' fibres (§ 238). The odoriferous medium must be brought into contact with this surface; and the surface itself must be neither dry, nor clogged with too large an amount of fluid secretion. — How far any sense of smell exists in the lower Invertebrata, cannot be satisfactorily determined; but it would seem not improbable that even where no special organ is apparent, some part of the general surface may be endowed with Olfactive sensibility. Thus, the *Pulmoniferous Gasteropods* seem guided to their food by its scent; and either the entrance to the respiratory sac, or some part of the soft spongy mantle, or of the integuments of the head, may be adapted to the cognizance of odours, by the distribution of an olfactive nerve to its surface. In the

Nudibranchiata, there is strong reason to believe that the dorsal tentacula, which frequently have a peculiar laminated structure, are the organs of this sense.* In the *Nautilus*, a peculiar laminated body, strongly resembling the olfactive organ of fish, has been considered by Prof. Owen as the special organ of Smell; and although no such organ has been discovered in the *Dibranchiate* Cephalopods, yet there is reason to believe that they enjoy the sense; and it may not improbably be located in the external lips.—Among the higher Articulata, there is ample reason to believe that the sense of Smell exists, although there is considerable uncertainty regarding its special instrument. That many *Insects* are guided to their food, to the proper nidus for their eggs, and to the opposite sex of their own species, and are even informed of the proximity of their natural enemies, by odorous emanations, can scarcely be doubted by any one who has watched their habits and experimented upon their actions. Some Entomologists have supposed the seat of the olfactory sense in Insects to be in their antennæ, others in the palpi, and others in the entrances to their tracheæ. The latter supposition must be considered as very improbable; and it derives no real support from experiment, since the movements which may be produced in a decapitated insect, by bringing acrid vapours into proximity with the stigmata, are evidently analogous to those of coughing and sneezing, which are excited in Man by similar acrid vapours, not through the sense of Smell, but through the 'common' sensibility of the membrane lining the respiratory passages. It is very difficult to determine by experiment, whether the antennæ or the palpi are the most probable instruments of the olfactive sense; and it seems not unlikely that, as the antennæ and palpi are appendages of a similar class, the sense of Smell may not be constantly localized in either of them, but that it may be assigned to one or to the other, according to the modifications they respectively require for the performance of their other offices.—The same may be stated in regard to the olfactive organ of the *Crustacea*. The manner in which Crabs and Lobsters are attracted by odorous bait placed in closed traps, makes it almost certain that they must possess some sense of Smell; and there can be little doubt that its instrument will be found in the basal joint of either the first or the second pair of antennæ.†—The Olfactive organ of Vertebrata usually corresponds with that of Man in its general characters; but in *Fishes* it is not connected with the respiratory passages, and has no posterior nares, its sole orifice being in front. In all air-breathing Vertebrata, it may be considered as a diverticulum from the commencement of the respiratory tube; and is obviously placed there, not only in order that the respiratory current may effectually introduce the odoriferous medium, but also that it may serve to warn its possessor of the presence of such odoriferous emanations, as cannot be breathed with impunity. In many instances, the chief purpose of the sense of Smell appears to be, to direct the animals to their food; in other cases, to warn them of the proximity of their enemies. The former is the case, more or less, in most Carnivora; the latter in Herbivora. The

* See Embleton, in "Ann. of Nat. Hist." March, 1849.

† It has been customary to regard, with Rosenthal, the sacculus in the first pair of antennæ as the olfactive organ, and that in the second as the auditory organ. Dr. A. Farre, however, has adduced strong reasons for reversing this opinion ("Philosophical Transactions," 1843); and Mr. Huxley has adopted and strengthened Dr. Farre's views (§ 823).

development of the olfactory organ, as measured by the size of the olfactive ganglia and nerves, and by the extent of the surface over which their branches are distributed, varies greatly in different tribes. In *Reptiles* there is but little provision for the extension of the olfactive surface; and there is no evidence that the sense of Smell is more than very feebly developed in them. In *Birds,* the nasal cavity is of considerable size, and its lining membrane is spread over the 'turbinated bones,' which project into its cavity; still there is reason to believe that much of what has been set down to the account of Smell in Birds, is really attributable to their acute sight, and that in no Birds is there any approach in this respect, to those Mammals which are most distinguished for the acuteness of their scent. Of the organ of Smell in *Mammalia* generally, and of its absence in the Cetacea, sufficient mention has already been made (§ 326 *w*).

5. *Of the Sense of Hearing.*

821. By this sense we become acquainted with the sounds produced by bodies in a certain state of vibration; the vibrations being propagated through the surrounding medium, by the corresponding waves or undulations which they produce in it. Although air is the usual medium through which sound is propagated, yet liquids or solids may answer the same purpose. On the other hand, no sound can be propagated through a perfect vacuum.—It is a fact of much importance, in regard to the action of the Organ of Hearing, that sonorous vibrations which have been excited, and are being transmitted, in a medium of one kind, are not imparted with the same readiness to others. The following conclusions have been drawn from experimental inquiries on this subject.

I. Vibrations excited in solid bodies, may be transmitted to water without much loss of their intensity; although not with the same readiness that they would be communicated to another solid.

II. On the other hand, vibrations excited in water lose something of their intensity in being propagated to solids; but they are returned, as it were, by these solids to the liquid, so that the sound is more loudly heard in the neighbourhood of these bodies, than it would otherwise have been.

III. The sonorous vibrations are much more weakened in the transmission of solids to air; and those of air make but little impression on solids.

IV. Sonorous vibrations in water are transmitted but feebly to air; and those which are taking place in air are with difficulty communicated to water; but the communication is rendered more easy, by the intervention of a membrane extended between them.

The application of these conclusions, in the Physiology of Hearing, will be presently apparent.

822. The essential part of an Organ of Hearing is a nerve endowed with the peculiar property of receiving and transmitting sonorous undulations; and it is by no means indispensable that any other special provision should be made for this purpose, since the Auditory nerve may be spread out over any surface which will be affected by the undulations of the surrounding medium. Hence we must not imagine the sense to be absent, wherever we cannot discover a definite organ for the purpose. On

the other hand, we are not to suppose that animals possess a distinct auditory sense, merely because we find them possessed of organs homologous with those which are certainly the instruments of that sense in higher animals; for they may be so rudimentary in their degree of development, that we can scarcely imagine them to possess any considerable functional capacity. Such is the case among the *Acalephæ*, the lowest animals in which any such organs have been detected. Along the margin of the disc, in many species, both 'naked-eyed' and 'hooded-eyed,' are to be found peculiar sacculi, containing either calcareous granules, or a single large highly-refracting spheroidal body, so closely resembling the auditory vesicles in the embryo of Gasteropoda (Fig. 282, *k*), that there can scarcely be a doubt of their homologous nature. No similar bodies have yet been discovered among the Echinodermata or the Tunicated Acephala; but traces of them have been detected in some Conchifera. —It is in *Gasteropoda*, that the Organ of Hearing first presents itself in a condition sufficiently elevated to put an end to all question of its character. It is as yet nothing more, however, than a sac containing one or more calcareous concretions, termed *otolithes*, which are observed to be in continual vibration; and this sac is imbedded in the posterior part of the œsophageal nervous collar. In the *Cephalopoda*, we find the auditory sac detached from the ganglionic mass, with which it remains connected by the Auditory nerve; and the sac itself is lodged in a cavity excavated in the cartilage that supports the cephalic nervous centres, an arrangement that evidently foreshadows the proper 'sense-capsule,' which it acquires in Vertebrata. No greater degree of complexity, however, is attained by the instrument itself, notwithstanding the close approximation which Cephalopods present to Vertebrata in the general elaborateness of their organisation.

823. There can be no doubt that the higher *Articulata* possess the sense of Hearing, although its special organ cannot always be detected. Thus it is obvious that the movements of *Insects* are guided by sounds, the sexes being frequently attracted to each other by the sounds which one of them has the power of producing; and it is probable that the organ of hearing is enclosed in the basal joints of the antennæ. In *Crustacea* its presence seems much less equivocally indicated; and yet it appears probable from recent enquiries, that what has been usually considered the organ of Hearing (§ 316 *c*) is in reality an organ of Smell, and that the true organ of Hearing is in the basal joint of the *first* pair of antennæ. For in *Lucifer*, as described by Mr. Huxley,* we find in this situation a sacculus containing a single spherical strongly-refracting otolithe, so closely resembling the organ of Hearing in Gasteropoda, that there can scarcely be a doubt of its analogous character; although it is completely enclosed in the crustaceous envelope of the member in which it is situated. In *Palæmon* there is a departure from this type, which leads us towards the ordinary form of the organ in Crustacea; for in the envelope of the basal joint of the antennæ is a narrow fissure, which opens into a pyriform cavity, contained within a membranous sac that lies within the substance of the joint. The anterior extremity of the sac is enveloped in a mass of pigment-granules; whilst on the side which is opposite to the fissure, a

* "Annals of Natural History," April, 1851.

series of hairs with bulbous bases are attached, on which seems to rest a large ovoidal strongly-refracting otolithe. The antennal nerve gives off branches which terminate at the bases of the hairs; and it can scarcely be doubted that it is through the vibrations of the otolithe transmitted to them, that the sonorous undulations are perceived. The ordinary structure of the organ of hearing in the Macrourous Decapods, as described by Dr. A. Farre,* presents a very curious modification of this type. The auditory sac, enclosed in the basal joint of the smaller antennæ, communicates with the external surface by a small valvular aperture; and instead of calcareous 'otolithes,' particles of siliceous sand are found in its interior, which appear to have entered by this orifice. These, in the ordinary position of the animal, will rest upon the extremities of a row of hair-like processes, which project into the interior of the sac; and each of these processes contains a row of cells, which are probably nerve-vesicles. This sacculus is connected with the cephalic ganglia, by a nerve-trunk distinct from that which supplies the antenna itself; and this trunk forms a plexus which surrounds the sac, but is peculiarly abundant beneath the row of hair-like processes. There can scarcely be a reasonable doubt of the Auditory nature of this organ, the connection between this and the more usual forms of the organ of hearing being supplied by Palæmon and Lucifer; and it is interesting to find the place of otolithes here taken by particles of sand introduced from without,—a provision which reminds us (as Dr. A. Farre has justly remarked) of the introduction of stones into the stomach of Granivorous Birds, in which they answer the purpose of gastric teeth. A portion of the shell of the joint is unconsolidated in the Lobster; so that the mouth of the auditory sac is only covered-in by a membrane, which may be considered as the representative of that which closes the 'fenestra ovalis' of the vestibule of Vertebrata; but in many other Decapods, the shelly covering is complete over the whole joint, save at the valvular orifice. The fluids of the auditory sac will be thrown into vibration by undulations of the surrounding medium, communicated either through the membrane covering the 'fenestra ovalis,' or through the shelly investment; and the vibrations will be strengthened by the presence of the sandy particles, which also, there is reason to believe, will, by their own vibration, make stronger impressions upon the hair-like processes, than the liquid itself would do.—No special organ of hearing has yet been discovered in *Arachnida.*

824. A special organ of Hearing exists in all *Vertebrated* animals save the *Amphioxus;* but in the lowest Cyclostome Fishes, it presents little or no advance in its development, upon the type on which it is constructed in the Cephalopods. For it consists of a simple sac, lodged in the cranial cartilage, and having no direct external communication; this sac is filled with fluid, and contains otolithes; and the auditory nerve is distributed upon its walls. In ascending through the series of *Cyclostomi,* however, we find the 'semicircular canals' successively developed; only one of these passages being present in *Myxine,* two in the *Lamprey,* and three in all the higher members of the class. At the same time, provision is made for the more direct action of the sonorous undulations in the surrounding

* "Philosophical Transactions," 1843.

medium, upon the fluid contained in the auditory sac or labyrinth; for a portion of its investing cartilage or bone is deficient at one part of the external surface, the aperture being only closed by a membrane, through which the communication can readily take place. Some rudiments of a tympanic cavity, interposed between this membrane and the external surface, may be found in certain Osseous Fishes; and in several tribes, the organ of hearing possesses a peculiar connection with the air-bladder, which appears to be a foreshadowing of the 'Eustachian tube' of higher classes.—In the true *Reptiles,* a considerable advance is constantly to be found in the character of the Auditory organ; for a tympanic cavity is added, with a membranous 'drum' and a chain of bones; and a rudiment of a 'cochlea' is generally discoverable, which has a separate opening (fenestra rotunda) into the tympanic cavity. The 'membrana tympani' is usually visible externally, but it is sometimes covered by the skin; the cavity of the tympanum communicates with the fauces by an 'Eustachian tribe.' Among the *Amphibia* there is a considerable variation in the structure of the organ of Hearing; for whilst some possess the tympanic apparatus, others are entirely destitute of it.—In *Birds* the structure of the Ear is essentially the same as in the higher Reptiles. A distinct 'cochlea' always exists, though its form is not spiral, but nearly straight; and its cavity is divided into two passages by a membranous partition, on which the ramifications of the auditory nerve are spread out. There is no external ear, save in a few species of nocturnal Birds; but the tympanic cavity communicates with cavities in the cranial bones, which are thus filled with air; and these, by increasing the extent of surface, produce a more powerful resonance.

825. In the ordinary *Mammalia,* the organ of Hearing is formed upon the same general plan as it presents in Man; in the Monotremata, however, it more approaches that of Birds. All the Mammalia, save the aquatic tribes, have an external ear, and this is sometimes of an enormous size in proportion to the dimensions of the body, as is seen especially in the Bats. Moreover in several tribes it can be turned in any direction by muscular movement, so as most advantageously to receive the faintest sounds from any quarter. The canal (Fig. 315, 8) into which the external ear reflects the sonorous vibrations, passes inwards until it is closed by the *membrana tympani* or 'drum of the ear' (9), which forms the external wall of the tympanic cavity. Within this cavity, which communicates with the throat by the 'Eustachian tube' (13), there is a series of small bones (10), which serve to establish a connection between the membrana tympani and that which covers-in the 'fenestra ovalis.' The long handle of the 'malleus' is attached to the membrana tympani, and near its base it gives attachment to the 'tensor tympani' muscle; its head is received into a hollow on the body of the 'incus,' which also has a long process that is connected with the 'stapes;' whilst the oval extremity of the 'stapes' is attached to the membrane covering the 'fenestra ovalis,' or entrance to the labyrinth, this little bone being also connected with the minute 'stapedius' muscle, which regulates its movements.—The purpose of this Tympanic apparatus is evidently to receive the sonorous vibrations from the air, and to transmit them to the membranous wall of the labyrinth; in such a manner that the vibrations thus excited in the latter may be much more powerful,

than they would be if the air acted immediately upon it, as in the lower Vertebrata. The usual condition of the membrana tympani appears to be rather lax; and, when in this condition, it vibrates in accordance with grave or deep tones. By the action of the 'tensor tympani' (a little muscle lodged in the Eustachian tube, 13) it may be tightened, so as to vibrate in accordance with sharper or higher tones; but it will then be less able to receive the impressions of deeper sounds. This state we may easily induce artificially, by holding the breath, and forcing air from the throat into the Eustachian tube, so as to make the membrane bulge out by pressure from within; or by exhausting the cavity by an effort at inspiration, with the mouth and nostrils closed, which will cause the membrane to be pressed inwards by the external air. In either case, the hearing is immediately found to be imperfect; but the deficiency relates only to grave sounds, acute ones being heard even more plainly than before. There is a different limit to the acuteness of the sounds, of which the ear can naturally take cognizance, in different persons. If the sound be so high in pitch, that the membrana tympani cannot vibrate in unison with it, the individual will not hear it, although it may be loud; and it has been noticed, that certain individuals cannot hear the very shrill tones produced by particular Insects, or even Birds, which are distinctly audible to others.

FIG. 315.

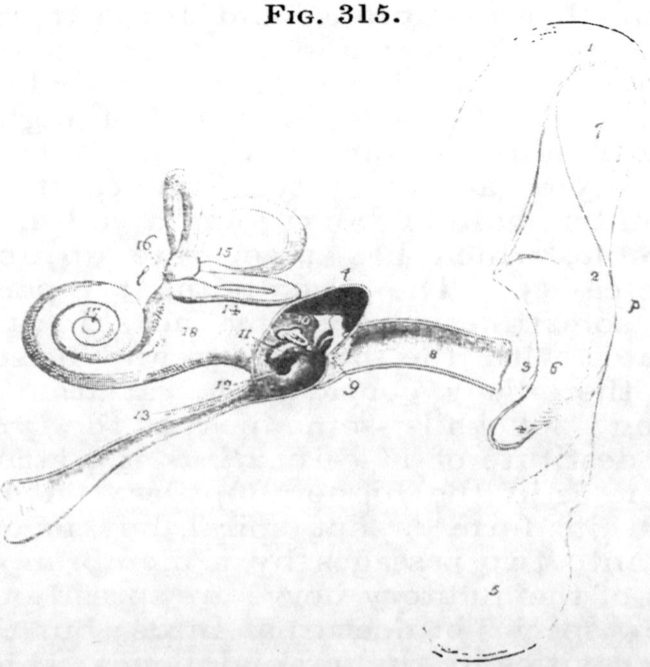

A diagram of the Ear. *p.* The pinna. *t.* The tympanum. *l.* The labyrinth. 1. The upper part of the helix. 2. The antihelix. 3. The tragus. 4. The antitragus. 5. The lobulus. 6. The concha. 7. The upper part of the fossa scaphoidea. 8. The meatus. 9. The membrana tympani, divided by the section. 10. The three little bones, malleus, incus, and stapes, crossing the area of the tympanum; the foot of the stapes blocks up the fenestra ovalis upon the inner wall of the tympanum. 11. The promontory. 12. The fenestra rotunda; the dark opening above the ossicula leads into the mastoid cells. 13. The Eustachian tube; the little canal upon this tube contains the tensor tympani muscle in its passage to the tympanum. 14. The vestibule. 15. The three semicircular canals, horizontal, perpendicular, and oblique. 16. The ampullæ upon the perpendicular and horizontal canals. 17. The cochlea. 18. A depression between the convexities of the two tubuli which communicate with the tympanum and vestibule; the one is the scala tympani, terminating at 12; the other is the scala vestibuli.

825 *a.* The vibrations thus transmitted to the membrane covering the fenestra ovalis, will act upon the fluid contained within the 'labyrinth,' which consists of the 'vestibule' (Fig. 315, 14), the 'semicircular canals' (15), and the 'cochlea' (17). The vestibule is evidently the part that corresponds with the simple cavity, which constitutes the entire organ of hearing in the lower animals; and the others may be regarded as extensions of it for particular purposes.—In all Vertebrata save the lowest Fishes, *three* semicircular canals exist; and they uniformly lie in three different planes, at right angles to each other, corresponding to the bottom and two adjoining sides of a cube; hence it has been supposed, and with

much apparent probability, that they assist in producing the idea of the *direction* of sounds.—The form of the cochlea is nearly that of a snail-shell, being a spiral canal, excavated in the solid bone, and making about two turns and a-half round a central pillar; this canal, however, is divided into two by a partition which runs along its entire length, and which is partly formed by a thin lamina of bone, and partly by a delicate membrane. The two passages do not communicate with each other, except at the summit of the helix; and at their lower end they terminate differently, one opening freely into the vestibule, and the other communicating with the cavity of the tympanum by an aperture termed the 'fenestra rotunda'(12), which is closed by a membrane.—These cavities are lined by a membrane, on which the ramifications of the auditory nerve are minutely distributed, and in which may be found cells that appear to be nerve-vesicles; and the distribution of this nerve is peculiarly close and abundant on the 'lamina spiralis' of the cochlea. Hence the nerve will be affected by the sonorous undulations into which the included fluid is thrown; and these will probably be rendered more free than they are in those forms of the auditory cavity in which it is completely enclosed within bony walls, by the existence of the second orifice leading from the cochlea to the tympanic cavity. The cochlea has been supposed to be the organ which enables us to judge of the *pitch* of sounds; an idea that derives some confirmation from the correspondence between the development of the cochlea in different animals, and the variety in the pitch (or length of the scale) of the sounds which it is important that they should hear distinctly, especially the voices of their own kind.—A pair of 'otolithes,' formed of particles of carbonate and phosphate of lime cemented together by animal mucus, is found in the vestibular sac.

826. The history of the development of the Auditory organ in the higher Vertebrata, presents a series of facts of great interest, of which the following is an outline.—The apparatus takes its origin in a portion of the Epencephalic vesicle, or 'vesicle of the medulla oblongata' (Fig. 290, B, *k*), which protrudes on either side, its cavity at first communicating with that of the vesicle, which remains permanent as the 'fourth ventricle.' As its protrusion increases, it becomes elongated and pear-shaped, and is only connected with the central mass by a pedicle whose canal gradually closes up; the sac thus cut off becomes the vestibular cavity, and the pedicle the auditory nerve. At first there is no vestige either of cochlea, semicircular canals, or tympanic apparatus; but the sac presents the simple character which it permanently retains in the Cephalopoda and the lower Fishes. Gradually, however, the semicircular canals are developed, by a contraction and folding-in of the walls of the vestibular sac; and the cochlea is probably formed as an offset from it. At the same time, the formation of cartilage, and subsequently of bone, takes place around the auditory sac and its prolongations, forming the 'sense-capsule,' which, in the higher Vertebrata, coalesces with the vertebral elements to form the temporal bone (§ 326 *k*).—It is very interesting to remark, that the membranous labyrinth, between the eighth and thirteenth days in the Chick, has a structure almost precisely similar to that of the retinal expansion of the same period; consisting, like it, of a distinct but very delicate fibrous mesh, in the spaces between which are deposited a quantity of granular matter and numerous nucleated cells, whilst its exterior is composed of a dense mass of nuclei, almost precisely

analogous to the granular particles which form a large part of the entire substance of the retina (§ 829).*

6. *Of the Sense of Sight.*

827. By the faculty of Sight, we are enabled to take cognizance of *luminous* impressions; and through these, we become acquainted with the form, size, colour, position, &c., of the objects that transmit or reflect light. But such knowledge can only be acquired (so far, at least, as we have the means of judging) through the medium of an optical instrument, which shall form upon the expanded surface of the Optic nerve an exact picture of surrounding objects, resembling that which is formed by the Camera Obscura; and it is from the conveyance of the impression produced by this picture to the Sensorium, that we derive the consciousness of that impression, on which we base our notions of the objects which have produced it. This optical instrument, or Eye, may exist, as we shall presently see, under a great variety of conditions; but it will be only when it is sufficiently perfect to form a distinct picture, that any definite knowledge of surrounding objects can be obtained through its means. There are, however, in many of the lower animals, certain coloured spots, which, from their position, nervous connections, and resemblance to the lowest forms of undoubted visual organs, must be regarded as rudimentary eyes; but all the information to which we can suppose these to be subservient, will be the reception of vague impressions of light and darkness. The lowest animals in which any such organs have been distinguished,† are the *Pulmograde Acalephæ;* in which every one of the ocelliform spots that are situated at regular intervals along the margin of the disc (Fig. 106, A, *ff*), is composed of a collection of pigmentary granules, superposed upon a ganglionic enlargement of a nervous thread, which radiates towards it from the circular collar. In the sub-order *Steganophthalmata,* these ocelli are protected by membranous coverings or hoods.—Similar ocelliform spots have been observed by Prof. E. Forbes at the extremities of the rays of certain *Asteriada,* where also they are connected with the extremities of nerve-filaments, and are protected by a peculiar arrangement of minute spines around each of them; but in no other Echinodermata have they been yet observed.—Among the lower *Articulata* the development of the visual apparatus does not seem to pass this rudimentary grade. In the animals belonging to the class of *Entozoa,* no trace of visual organs has been detected, except among the *Planariæ* and such of their allies of the Trematode order as do not inhabit the bodies of other animals, but range freely in search of their food; these, as also *Leeches,* have numerous eye-spots disposed about the head, by which it may be supposed that their movements are in some degree guided. Similar eye-spots may be seen upon the cephalic projection of the *Rotifera* (Fig. 144, *b*); and they become much more distinct in the *Dorsibranchiate Annelida,* which are active in their habits, although it does not appear that even in them does any proper optical instrument

* See Mr. H. Gray, in "Philosophical Transactions," 1850.

† No account is here taken of the red spots seen in many Polygastric Animalcules, since there is not the least reason for believing them to be ocelli, similar spots having been seen in many undoubtedly Vegetable cells (§ 284).

exist, adapted to form a representation of neighbouring objects.—The same may be said of the lower Mollusca. Ocelliform spots have been observed in the neighbourhood of the oral orifice of several *Tunicata;* and similar spots are very obvious upon the margin of the mantle of Pectens and other free-swimming *Conchifera.* As a general rule, however, this latter class would seem to be more destitute of even a rudiment of visual organs, than is the former.

828. The Eyes of most of the higher *Articulated* animals, are constructed upon the *composite* type; each of the masses that is situated upon either side of the head, being made up of an aggregation of single eyes, every one of which is in itself a complete visual instrument, but is adapted to receive and to bring to a focus only those rays which come to it in one particular direction. In most *Insects,* each composite eye forms a large hemispherical protuberance, which occupies a considerable part of the side of the head (Fig. 316); and when examined with a microscope, its surface is seen to be divided into a vast number of 'facets,' which are usually hexagonal. The number of these facets, every one of which is the cornea of a distinct eye, is usually very great (§ 315 *b*). Behind the cornea is a layer of dark pigment, which takes the place and serves the purpose of the 'iris' in the eyes of Vertebrata; and this is perforated by a central aperture or 'pupil,' through which the rays of light which have traversed the cornea, gain access to the interior of the eye. When a vertical section is made of one of these composite eyes, it is seen that each separate eye is the frustrum of a pyramid, of which the cornea forms the large end, or base (Fig. 317, *a*), whilst the small end abuts upon a bulbous expansion of the optic nerve; the interior of this pyramid is occupied by a transparent substance (*b*), which represents the 'vitreous humour;' and the pyramids are separated from each other by a layer of dark pigment, which completely encloses them, save at the pupillary apertures, and also at a corresponding set of apertures at their smaller ends, where the pigment is perforated by the fibres of the optic nerve (*c*), of which one proceeds to each separate eye. Each facet of the common cornea, or 'corneule,' is usually convex on both its surfaces; and thus acts as a lens, the focus of which has been ascertained by experiment to be equivalent to the length of the transparent pyramid behind it; so that the image produced by the lens will fall upon the extremity of the filament of the optic nerve which passes to its truncated end. The rays which have passed through the several 'corneules' are prevented from mixing with each other, by means of the layer of black pigment which surrounds each cone; and thus no rays, except those which correspond with the axis of the cone, can reach the fibres of the optic nerve. Hence it is evident that each separate eye must have an extremely limited range of vision; being adapted to

FIG. 316.

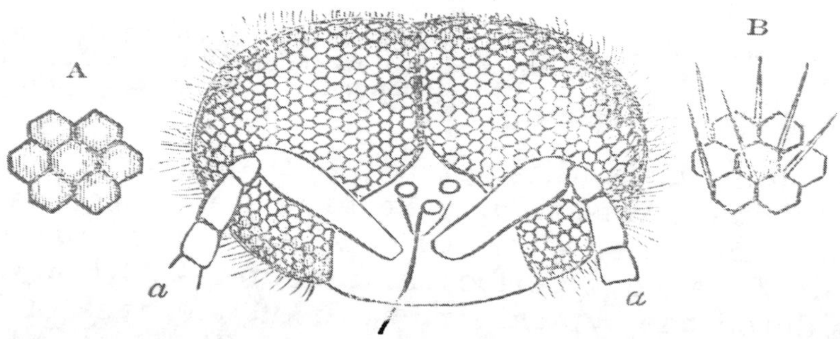

Head and Eyes of the *Bee,* showing the division into facets; *a, a,* antennæ;—A, facets enlarged;—B, the same with hairs growing between them.

receive but a very small pencil of rays proceeding from a single point in any object; and as these eyes are usually immoveable, they would afford but very imperfect information of the position of surrounding objects, were it not for their enormous multiplication, by which a separate eye is provided (so to speak) for each point to be viewed. No two of the separate eyes, save those upon the opposite sides of the head which are directed exactly forwards, can form an image of the same point at the same time; but the combined action of all of them may give to the Insect, it may be imagined, as distinct a picture as that which *we* obtain by a very different organisation. At any rate, it seems certain, from observation of the movements of Insects, that the vision by which they are guided must be very perfect and acute.—Although the foregoing may be considered the typical structure of the eyes of Insects, yet there are various slight departures from it in the different subdivisions of the class. Thus in some cases, the posterior surface of each 'corneule' is concave, and a space is left between it and the iris, which seems to be occupied by a watery fluid or 'aqueous humour;' in other instances, again, this space is occupied by a double-convex body, which seems to represent the 'crystalline lens;' and there are cases in which this 'crystalline lens' is found behind the iris, the number of eyes being reduced, and each individual eye being larger, so that the entire aggregate approaches, both in its structure and its mode of action, to that of Arachnida and certain Crustacea. Besides their composite eyes, Insects usually possess a small number of rudimentary single eyes, resembling those of the Arachnida; these are seated upon the top of the head, and are called *stemmata*. Their precise use is unknown; but that they have considerable influence on the direction of the movements, appears from the fact, that if the stemmata of a Bee be covered with paint, on being let go it will fly continually upwards,—a fact which seems related to those already mentioned, in regard to the influence of visual sensations upon automatic movements (§ 797). It is remarkable that the larvæ of Insects which undergo a 'complete' metamorphosis, only possess simple eyes; the composite eyes being developed, at the same time with the wings and other parts which are characteristic of the Imago state, during the latter part of the Pupa condition.—In the higher *Crustacea*, the structure of the Eyes is nearly the same as in Insects; but the compound masses are not so large relatively to the bulk of the body, and the number of distinct eyes is not nearly so great. In the lower Crustacea, however, as in Myriapoda, the visual organs much more

FIG. 317.

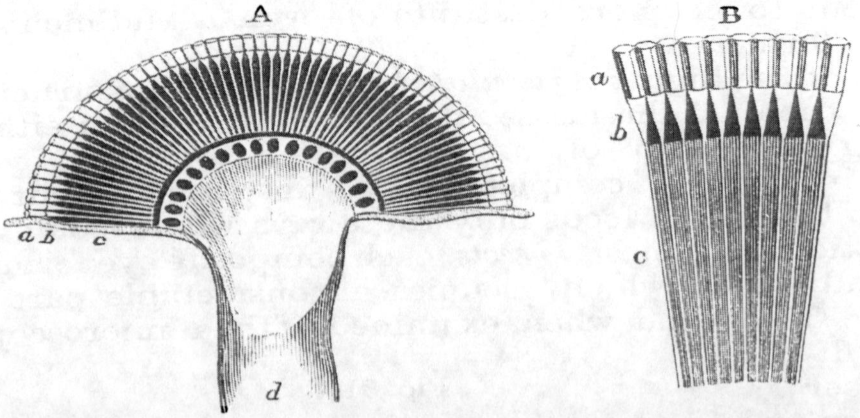

A, Section of the eye of *Melolontha vulgaris* (Cockchaffer): B, a portion more highly magnified:—*a*, facets of the cornea; *b*, transparent pyramids surrounded with pigment; *c*, fibres of the optic nerve; *d*, trunk of the optic nerve.

closely approximate the type of structure which they present in the Arachnida; each aggregate mass being composed of a small number of 'simple' eyes, of which every one has its own separate cornea, as well as its own crystalline lens and vitreous humour; and these in some instances being altogether detached from each other. Among some of the *Suctorial Crustacea*, the visual organs are altogether wanting in their state of full development, although they are uniformly present in their early condition; and the same may be said of the *Cirrhipoda.*—Among *Arachnida,* which in this as in many other respects present an approximation to Vertebrata, we find a great reduction in the number of eyes, which are never more than eight in number (sometimes being only two), and are to be compared with the 'stemmata' of Insects, rather than with their compound eyes. These eyes are sometimes collected into one mass on the summit of the cephalo-thorax (Fig. 165, *c*), and are sometimes disposed symmetrically and separately on the two sides of the median line. In the Scorpions, we find two large eyes placed on the dorsal aspect of the cephalo-thorax, near the median line; and three pairs of smaller ones, which are placed on the outer margins of the same division of the body. The larger eyes are described by Müller as each possessing a 'cornea,' which is convex anteriorly and concave posteriorly; and a nearly globular 'crystalline lens,' resembling that of Fishes, whose anterior surface lies in the hollow of the cornea, while its posterior rests upon the 'vitreous humour' without being imbedded in it. The 'vitreous humour' is a nearly hemispherical mass of soft granular matter, being almost flat in front, and very convex behind; over its posterior surface is spread the 'retina,' or expansion of the optic nerve; and this is covered by a thick layer of pigment, which passes inwards in front of the vitreous humour, so as to form a sort of iris, the pupillary aperture of which, however, exceeds the diameter of the crystalline lens.

829. Among those classes which constitute the higher division of the Molluscous series, in virtue of the possession of a distinct 'head,' the presence of visual organs is by no means constant; many Gasteropoda and Pteropoda being destitute of them altogether, and others possessing ocelliform spots, which may be concluded to be rudimentary eyes, from their similarity in position to the eyes of those which undoubtedly possess visual powers. The eyes are always very minute, however, in proportion to the bulk of the body, and in no instance do they present a high type of structure; their general organisation, indeed, bears a close resemblance to that which has been described in the eye of the Scorpion. In the *Cephalopoda,* we find the visual organs presenting a much larger size, and attaining a much higher grade of development, in accordance with their greater functional activity in directing the rapid and energetic movements practised by a large proportion of these animals. We here find nearly all the principal parts, which are characteristic of the eye of higher animals; namely, a cornea, an anterior chamber filled with an aqueous fluid enclosed in a distinct capsule, a crystalline lens of globular form (as in Fishes), a large posterior chamber filled with vitreous humour, a tough fibrous or 'sclerotic' coat, a vascular 'choroid' coat within this, covered by black pigment upon its inner surface, and a retinal expansion. The relations of this last to the optic ganglion, however, are very peculiar. This ganglion is situated almost close to the back of the eye; and instead of transmitting a single optic nerve, as in

higher animals, it gives off a multitude of filaments, which separately pierce the sclerotic coat, and then form a plexus between this and the choroid, which has been mistaken for the retina. The true retina, however, is a very thin lamella, apparently composed of vesicular nerve-substance, which is found between the pigment and the membrane inclosing the vitreous humour; but the connection of this with the network of nerve-tubes on the outside of the pigmentary layer, has not yet been made out. No proper 'iris' exists in the eyes of Cephalopoda; but its place is supplied by a partial prolongation of the sclerotic coat over the front of the crystalline, a central pupillary aperture being left. The cornea is not, like the true cornea of higher animals, a transparent continuation of the sclerotic coat, but is a modification of the general integument, analogous rather to the external or conjunctival layer of the cornea of Vertebrata; it is remarkable that in some Cephalopoda, it should be perforated by an orifice of considerable size, through which the capsule of the crystalline lens projects into the external medium.—There are but few exceptions to the general fact of the presence of distinct visual organs in the *Vertebrata;* and the type upon which these are formed, presents scarcely any essential diversity in the different classes (Fig. 318). In all we find the Eye of nearly globular shape, with a projection in front caused by the greater curvature of the cornea. It is enclosed in a tough fibrous envelope, or 'sclerotic' coat (1), which supports the transparent cornea (2); and this is lined by the vascular 'choroid' (3), of which the muscular 'iris' (6), that is interposed between the anterior and posterior chambers, may be regarded as in some sort a continuation. The choroid is lined by a layer of black pigment; and between this and the hyaloid membrane (or capsule of the vitreous humour) is interposed the 'retina' (8). The optic nerve, in all instances, is a single trunk, which perforates the sclerotic and choroid coats at the back of the eye, and then spreads out into a fibrous network, which is covered on its free surface by a layer of vesicular or ganglionic

FIG. 317.

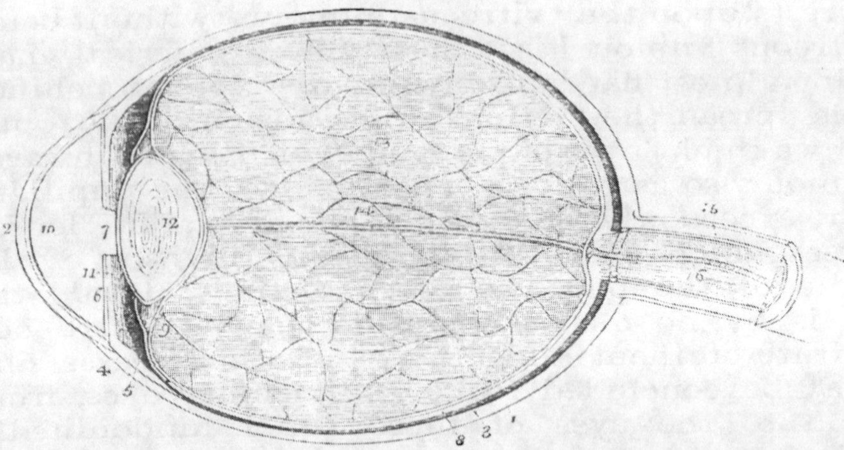

A longitudinal section of the globe of the *Human Eye*. 1. The sclerotic, thicker behind than in front. 2. The cornea, received within the anterior margin of the sclerotic, and connected with it by means of a bevelled edge. 3. The choroid, connected anteriorly with (4) the ciliary muscle, and (5) the ciliary processes. 6. The iris. 7. The pupil. 8. The third layer of the eye, the retina, terminating anteriorly by an abrupt border at the commencement of the ciliary processes. 9. The canal of Petit, which encircles the lens (12); the thin layer in front of this canal is the zonula ciliaris, a prolongation of the vascular layer of the retina to the lens. 10. The anterior chamber of the eye, containing the aqueous humour: the lining membrane by which the humour is secreted is represented in the diagram. 11. The posterior chamber. 12. The lens, more convex behind than before, and enclosed in its proper capsule. 13. The vitreous humour enclosed in the hyaloid membrane, and in cells formed in its interior by that membrane. 14. A tubular sheath of the hyaloid membrane, which serves for the passage of the artery of the capsule of the lens. 15. The neurilemma of the optic nerve. 16. The arteria centralis retinæ, embedded in the centre of the optic nerve.

matter. It is of this, indeed, that the nervous portion of the retina is chiefly composed; and the nerve-vesicles here found can scarcely be distinguished from those of the grey matter of the brain. This is, perhaps, the best example we possess, of the general doctrine formerly propounded (§ 245), that the changes commencing at the peripheral extremities of the afferent nerves, are probably effected by the agency of cells, like those which originate at the central origins of the efferent. The refractive media contained in the interior of the eye are always three in number, and of different densities; the anterior chamber (10) being occupied by the 'aqueous humour,' which is a limpid watery fluid; the 'crystalline lens' (12), a substance of considerable firmness, whose convexity is much less in the terrestrial than in the aquatic Vertebrata, being placed immediately behind the iris; and the great bulk of the posterior chamber being occupied by the 'vitreous humour' (13), which is of the consistence of thin jelly.—The chief peculiarities which the Visual organs present in the classes of Fishes (§ 322 *i*), Reptiles (§ 324 *s*), Birds (§ 325 *i*), and Mammals (§ 326 *w*), have already been sufficiently noticed.

830. The development of the Eye in the higher Vertebrata commences by a protrusion from a part of the anterior cerebral vesicle, representing the 'vesicle of the thalami optici' (Fig. 297), which is at that time hollow; and the cavity of the protrusion is continuous with that of the vesicle itself, which remains as the 'third ventricle.' The protrusion is lined, like the cerebral vesicle, with granular matter, which gradually becomes distinctly cellular, forming a layer of truly ganglionic character; and whilst this change is taking place, the protrusion increases, becomes pear-shaped, and is at last connected only by a narrow pedicle with the vesicle from which it sprang. This pedicle closes up, so as completely to separate the two cavities; and the one which has been thus budded-forth constitutes the rudiment of the eye, whilst the other goes on to form the ganglionic bodies at the base of the cerebrum, and the connecting pedicle becomes the optic nerve, which connects the retina with its ganglionic centre. The spherical extremity of the protrusion is absorbed, and the retina, or vesicular lining, becomes attached to the margin of the lens, which is in the mean time developed in the interior of the cavity, and is at first completely surrounded by the retina. The formation of the coats of the Eye takes place subsequently; the development even of the 'fibrous lamina' and of the 'membrana Jacobi' of the retina itself, not taking place until after its cellular layer has been very distinctly formed.* —It is a curious circumstance, and one not very easy to account for, that the development of the Eye should take place from the Prosencephalic and not from the Mesencephalic vesicle; as it is in the latter that the proper 'optic ganglia' originate, with which the optic nerves come at last to have their principal connection, their connection with the 'thalami optici' being much less close.

831. In the most perfect form of the Eye, such as that presented by Vertebrata generally, the luminous rays which diverge from the several points of any object, and fall upon the front of the cornea, are refracted by its convex surface, whilst passing through it into the eye, and are made to converge slightly. They are brought more closely together by the crystalline lens, which they reach after passing through the pupil;

* See Mr. H. Gray in "Philosophical Transactions," 1850.

and its refracting influence, together with that produced by the vitreous humor, is such as to cause the rays, that issued from each point, to meet in a focus on the retina. In this manner, a complete inverted image is formed, as shown in Fig. 319, which represents a vertical section of the eye, and the general course of the rays in its interior; those which issue from the point A being brought to a focus at D, whilst those diverging from B are made to converge upon the retina at C. The Retina, which is itself so thin as to be nearly transparent, is spread over the layer of black pigment, which lines the choroid coat. The purpose of this is evidently to *absorb* the rays of light that form the picture, immediately after they have passed through the retina; in this manner, they are prevented from being reflected from one part of the interior of the globe to another; which would cause great confusion and indistinctness in the picture. Hence it is that, in those *albino* individuals (both of the Human race, and among the lower animals), in whose eyes this pigment is deficient, vision is extremely imperfect, except in a very feeble light; for the vascularity of the choroid and iris is such, as to give to these membranes a bright red hue, which enables them powerfully to reflect the light that reaches the interior of the eye, when they are not prevented from doing so by the interposition of the pigmentary layer. — The eye is so constructed, as to avoid certain errors and defects, to which all ordinary optical instruments are liable. One of these imperfections, termed *spherical aberration*, results from the fact, that the rays of light, passing through a convex lens whose curvature is circular, are *not all* brought to their proper foci; those which have passed through the exterior of the lens, being made to converge sooner than those which have traversed its central portion. The result of this imperfection is, that the image is deficient in clearness, unless only the central part of the lens be employed.—The other source of imperfection is what is termed *chromatic aberration;* and it results from the unequal degree, in which the differently-coloured rays are refracted, so that they are brought to a focus at different points. The violet rays, being the most refrangible, are soonest brought to a focus; and the red, being the least refrangible, have their focus at the greatest distance from the lens. Hence it is impossible to obtain an image by an ordinary lens, in which the colours of the object are accurately represented; for the foci of its differently-coloured portions will be different; and its white rays will be decomposed, so that the outlines will be surrounded by coloured fringes.—The Optician is enabled to correct the effects of these aberrations, by combining lenses of different densities and curvatures; so arranged as to correct each other's errors, without neutralizing the refractive power. This is precisely the plan adopted in the construction of the Eye; which, when perfectly formed, and in a healthy state, forms an accurate picture of the object upon the retina, free from either spherical or chromatic aberration. This is effected by the combination of *humours* of different densities, having curvatures precisely adapted to the required purpose.

FIG. 319.

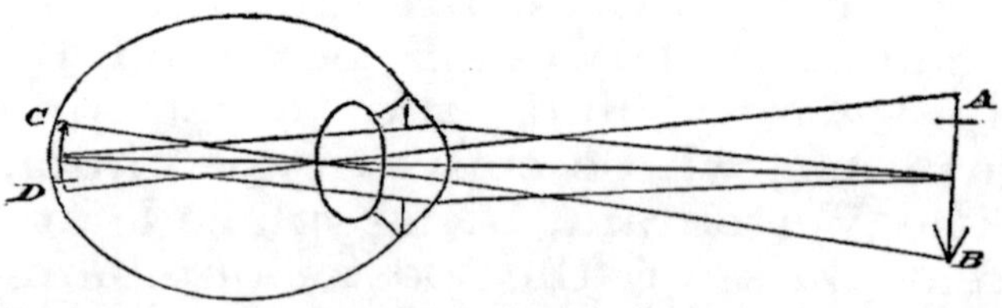

832. The power, by which a healthy well-formed eye can accommodate itself to the distinct vision of objects at varying distances, is a very remarkable one; and its rationale is not yet perfectly understood. According to the laws of Optics, the picture of a near object can only be distinct, when formed more remotely from the lens than the picture of a distant object. Consequently when the eye, that has been looking at a distant object, and has seen it clearly, is turned to a near object, a distinct picture of the latter cannot be formed without some alteration, either in the distance of the cornea or of the lens from the retina, or in the curvature of their refracting surfaces. It seems most probable that the adjustment is chiefly effected by the 'ciliary muscle,' which is a collection of muscular fibres, radiating from the junction of the cornea and sclerotic, to the 'ciliary processes' of the choroid, which hold the lens in its place; for by the contraction of this muscle, the lens will be drawn forwards, and the eye thus adapted for the vision of near objects. In the human eye, the ciliary muscle consists of 'non-striated' fibres; but in the eyes of Birds, in which it is much more highly developed, it is formed of 'striated' fibres; whilst, by the partial ossification of the anterior portion of the sclerotic coat, its origin is more securely fixed.* It does not seem improbable that the 'pecten' or peculiar erectile organ in the vitreous humour of the Bird's eye (§ 325 *i*), may be subservient to this adjustment; for, when the lens is carried forwards, a vacuity must exist behind it, which the entrance of blood into this plexus of vessels will supply. It seems quite certain, from observation of the actions of Birds, that they must possess the power of adapting their eyes to distinct vision at different distances, in a higher degree than any other animals. The 'choroid gland' of the Fish's eye (§ 322 *i*) may not improbably answer a similar purpose.—The adjustment is probably in all cases, as in Man, a purely 'automatic' action, taking place without any voluntary effort, whenever the visual sense is directed by the will to a special object; and it is a very good example of that class of actions, in which *sensation* is a necessary link in the chain of reflexion (§ 798).

833. Another automatic action, which adapts the eye for distinct vision under varying degrees of light, is the alteration in the diameter of the 'pupil.' This is effected by the muscular structure of the 'iris,' which is made up in many animals (though not distinctly so in Man) of two sets of fibres, a circular and a radiating; the aperture of the pupil being diminished by the contraction of the former, when the light is powerful, so as to exclude its excess from the interior of the eye; and being augmented by the contraction of the latter, when the light is faint, so as to admit the greatest possible number of rays. The contraction of the pupil also takes place when vision is directed to any very near object; and its purpose appears then to be, to prevent the rays from entering the eye at such a wide angle, as would render it impossible for them to be all brought to their proper foci, and would thus produce an indistinct image. In either case, the regulation of the diameter of the pupil is an action with which the will has nothing to do, and which it cannot effect by any direct effort.—It is worthy of note, that in Birds the fibres of the iris are very strongly marked, and that they are of the

* See Messrs. Todd and Bowman's "Physiological Anatomy and Physiology of Man," vol. ii. p. 27.

'striated' kind. The diameter of the pupil is often seen to vary in them, without any change in the amount of light, or any alteration in the position of the eyes, whence it has been supposed that they possess a voluntary power over this movement; but the fact is probably rather, that the alteration takes place in virtue of a change in the direction of the sight from a near to a distant object, or *vice versâ*, which may occur without any obvious difference in the position of the eyes, when the two objects are in the same line, and the eyes are placed at the sides of the head, and not in front, — as in the Parrot, in which this change has been most frequently observed. When the eyes are so situated that both of them can be directed to the same object, their axes are made to converge in it; and the angle at which they meet, which will be very acute when the object is distant, increases rapidly as the object is approximated to the eyes. It is from the 'muscular sense,' which informs us of the condition of the muscles thus brought into conjoint action, that we derive our chief information as to the *distance* of objects, by an intuitive interpretation of the impressions thus made upon our consciousness. It is quite certain that in Man, this intuitive apprehension is *acquired;* for the infant, or a person who has newly become possessed of vision, is only able to form it after a long course of experimental training. It is equally certain, on the other hand, that in many of the lower animals it must be *congenital;* since they perform actions, which manifest a power of accurately estimating distances, and of regulating their muscular movements accordingly, immediately on their entrance into the world.

834. A considerable variety exists among Vertebrated animals, in regard to the position of the Eyes, and the degree in which they possess the same range of vision. It would seem as if a very extensive range of vision posteriorly, is necessary in such timorous animals as the Ruminants, which are almost always on the watch for enemies, and seek their safety in flight; that perfection of the visual sense, which (as will be presently shown) can only be gained by the combined use of the two eyes, being sacrificed in them to the power of seeing round the whole horizon at once, which they possess in virtue of the lateral position of their eyes. Where the position of the eyes is such that their spheres of vision are entirely distinct (as happens in many Fishes), the optic nerve proceeding from each eye passes direct to the ganglion on the opposite side; but where, as most commonly happens, the range of one eye overlaps (so to speak) that of the other, so that both see the same object at once, if it be within that overlapping portion, a different arrangement of the fibres of the optic nerves prevails; for those of such parts of the retinæ of both eyes, as look towards the same side (*i. e.* the inner portion of the retina of the right eye, and the outer portion of the retina of the left, and *vice versâ*), pass to the ganglion of the opposite side; so that each eye is connected with both ganglia. The effect, however, will still be the same; namely, to make each ganglion the seat of the visual impressions originating from sources on the opposite side of the body; and the purpose of this is probably, to bring the guiding sensations of sight into relation with the muscular movements they regulate. For, by a decussation of fibres in the Medulla Oblongata, the Sensory Ganglia above it are connected with *opposite* sides of the Spinal Cord below it; and thus, the sensations of visual objects on the *right* side of the body being formed in the *left* optic

ganglion, their influence will be directly conveyed to the right side of the spinal cord, and to the muscles whose nerves originate from it.—Although the forward position of the eyes, by greatly diminishing the range of vision, might seem to detract from the advantage which is conferred by the possession of *two* eyes, yet it confers a great advantage of a different kind; for it is by the intuitive combination of the two *dissimilar* pictures, which are formed of any near object upon the retinæ of the two eyes, that we gain the impression of its projection or solidity. That the pictures *are* dissimilar, is easily shown by holding up a thin book in such a manner that its back shall be in a line with the nose, and at a moderate distance from it; and the experimenter who looks at the book, first with one eye, and then with the other, will find that he gains a different view of the object with each eye, when used separately; so that if he were to represent it, as he actually sees it under these circumstances, he would have two perspective delineations differing from one another, because drawn from different points. But on looking at the object with the two eyes conjointly, there is no confusion between these pictures; nor does the mind dwell upon either of them singly; but the union of the two intuitively gives us the idea of a solid *projecting* body,—such an idea as we could only have otherwise acquired by the exercise of the sense of touch. That this is really the case, has been proved by experiments with a very ingenious instrument, the Stereoscope, invented by Prof. Wheatstone; which is so contrived, as to bring to the two eyes, by reflection from mirrors, two different pictures, such as would be accurate representations of a solid object, as seen by the two eyes respectively. When the arrangement is such, as to bring the images of these pictures to those parts of the retinæ, which would have been occupied by the images of the solid (supposing *that* to have been before the eyes), the mind will perceive, not one or other of the single representations of the object, nor a confused union of the two, but a body projecting in *relief*, the exact counterpart of that from which the drawings were made.* It is doubtless by a similar intuitive interpretation, that we recognise the erect position of objects, notwithstanding the inversion of their images upon our retinæ. This is certainly not a matter of experience; nor is it capable of explanation (as some have supposed) by a reference to the direction in which the rays fall upon the retina. It is the *Mind* which rectifies the inversion; and it is just as difficult to understand how the inverted image upon the retina should be taken cognizance of by the mind at all, as it is to comprehend how it should thus be rectified. In fact, there is no real connection whatever, between the inversion of the image upon the retina, and that wrong perception of external objects, which some have thought would be its necessary consequence.

* The most wonderful reproduction, to the *mind's eye*, of the solid body, is effected when the two pictures employed are *photographic* representations (either 'daguerreotypes' or 'talbotypes') taken at the proper angular distance from each other.

CHAPTER XXII.

OF THE SOUNDS PRODUCED BY ANIMALS.

835. The purposes of Animal existence frequently involve the necessity of such a communication between one individual and another, as can only be made by the production of Sounds on the one part, and by the Hearing of them on the other. The most general requirement of this kind, is probably that arising out of the Generative function, in those tribes in which the congress of two individuals is requisite; and we find that one or both of the sexes are often provided with the means of producing sounds, which indicate their presence to such of the opposite sex as may be within hearing of them. The sounds produced by Invertebrated animals are not vocal or articulate in their character, although they frequently serve to intimate the state of tranquillity or excitement of the individuals that utter them; and it is only in the higher Vertebrata, in which the apparatus of Voice is so constructed as to be capable of a great variety of actions, and is brought into relation with the respiratory organs, that it acquires its most expressive character. The restriction of articulate language to Man, is rather a result of the superiority of his *mental* than of his *vocal* endowments; since many Birds can execute a perfect imitation of the sounds which he utters, although incapable of attaining to more than a general comprehension of their import, and this rather in the 'concrete,' than by the formation of any 'abstract' notions of the meaning of the separate words which they imitate.

836. Although certain of the *Nudibranchiate* Gasteropods have been occasionally heard to produce a clear bell-like sound, yet their mode of generating it is unknown; and they are the only animals of the Molluscous series, in which such an endowment has ever been observed.—It is among *Insects*, more than in any other class of Invertebrata, that the power of generating sounds is met with. Some of these seem necessarily to arise during the ordinary movements of these animals; such is the 'hum' which is produced in flight, and which varies in its character from the dull droning sound of the common 'shard-borne beetle,' to the shrill trumpet of the gnat and mosquito, that serves to give warning of the proximity of these blood-thirsty insects. Generally speaking, the Insects that fly with the greatest force and rapidity, and with wings seemingly motionless (owing to the extreme rapidity of their vibrations), make the most noise; whilst those that fly gently and leisurely, and can be seen to fan the air with their wings (as is the case with most Lepidoptera), yield little or no sound. It appears, however, from the experiments of Burmeister, that, in Bees and Flies, the sound is not so much produced by the simple motion of the wings, to which it is commonly attributed, as by the vibrations of a little membranous plate situated in each of the posterior spiracles of the thorax; for if the apertures of these be stopped, no sound is heard, even though the wings remain in movement. Other sounds are produced by the act of mastication; thus the noise occasioned by the armies of Locusts, when incalculable millions of powerful jaws are

in action at the same time, has been compared to the crackling of a flame of fire driven by the wind.—The sounds which are produced by special means, with direct reference to mutual communication, are generated in a great variety of modes. Thus, the neuters or 'soldiers' among the *Termites* make a vibratory sound, rather shriller and quicker than the ticking of a watch, by striking hard substances with their mandibles; and this seems to answer the purpose of keeping the 'labourers,' who answer it with a kind of hiss, alert and at their work. The well-known sound termed the 'death-watch,' is produced in a similar manner by the *Anobium*, a small beetle that burrows in old timber; if the signal be answered, it is continually repeated; whilst if no answer be returned, the animal changes its situation before again making its presence known. The noise exactly resembles that produced by tapping moderately with the nail upon the table; and the insect may often be brought to answer this imitation, as well as the real sound of its own kind. A very curious sound, the mode of whose production has not been certainly ascertained, is given out by the *Sphinx atropos* (death's-head-moth), when confined or taken into the hand; this sound has been likened to the cry of a mouse, but is more plaintive and even lamentable. The peculiar 'hum' given off by Bees during their ordinary labours in the hive, is frequently so modified as to be (to all appearance) a means of communication between them; thus it assumes a sharp angry tone, when the hive has been disturbed, especially if some of the bees have been killed; it is changed to a low and plaintive sound, when the queen has been taken away; and this is exchanged for a cheerful humming, which is speedily diffused through the entire community, when she is restored.—Of the sounds that seem especially to have reference to sexual communication, the most remarkable are those of the *Grasshopper*, and of the *Cicada* tribes. In the former the sounds are produced by the attrition of the anterior pairs of wings against each other, one of the nervures being furnished with a rough file-like edge, which is made to pass over the nervures of the opposite wing; and the sound is augmented by the resonance of a certain part of the wing, that is surrounded by peculiarly strong nervures, between which the thin membrane is tightly stretched, so that it acts as a 'tympanum.' The sound-producing organs of the latter, however, are situated internally, and are somewhat complex in their structure. Their essential part seems to be a tense membrane, stretched across a cavity in the last segment of the thorax on either side, which is drawn-in or forced-out by the action of two opposing bundles of muscular fibres; and it is found that even in a dead specimen the sound is produced, when these muscles are pulled and suddenly let go. Externally to this apparatus are other membranous plates, whose office appears to be to increase the sound by resonance; and so effectually do they act, that a certain Cicada of Brazil is said to be audible at the distance of a mile, which is as if a Man of ordinary stature possessed a voice that could be heard all over the world. The sound is often kept up for some hours; resembling the 'hum' of Bees in its continuity, rather than the interrupted 'chirp' of the Crickets.

837. Among *Vertebrata*, the production of Sounds appears to be confined to the air-breathing classes; no Fishes being known to possess the means of generating them. In *Reptiles*, it is at the point where the trachea opens

into the front of the pharynx, that the vibratory apparatus is situated, which gives out sounds when air is forced through it from the lungs. The sounds produced by animals of this class, however, are of a very simple and inexpressive kind. Thus from Turtles, Serpents, and ordinary Lizards, we hear nothing else than a 'hiss,' occasioned by the passage of the air through the narrow fissure of the glottis; this sound being often much prolonged, owing to the great capacity of their lungs. In Frogs, a 'croak' is produced by the vibration of the lips of the glottis itself; and in the larger Crocodiles, Alligators, &c., this croak is augmented in its volume, and becomes a 'roar.' In these orders we find the rudiments of the proper 'larynx' of Mammals; which, however, are chiefly developed in the male sex.

838. In *Birds*, the situation of the vocal organ is very different. The summit of the trachea is furnished with a 'larynx,' provided with cartilages and muscles, in which many of the parts of the larynx of Man can be recognised; but the use of this seems to have reference entirely to the regulation of the ingress and egress of air. The vocal sounds for which Birds, as a class, are so remarkable, are formed by an organ which is altogether peculiar to them, and which is situated at the *lower* extremity of the trachea, just at its bifurcation into the bronchial tubes. The structure of this 'inferior larynx' varies greatly in different species, being most complex in those which are able to produce the greatest varieties; and in some Birds which are entirely voiceless (such as the Storks), the organ is entirely wanting. The two or three lowest rings of the trachea are usually consolidated into one, and in the interior of this a cross-bone runs from front to back, which has a vertical semilunar membrane prolonged upwards from its upper edge. This seems analogous in its action to the vibrating tongues or plates of 'reed-instruments' of music; and it will be put in motion by currents of air passing on either side of it. Each of the bronchial tubes has its own glottis for the regulation of the passage of air; and we also find a part of the walls of both the bronchial tubes and of the trachea to be composed of a thin membrane, stretched tensely between cartilages, which serves as a 'tympanum,' to increase the loudness of the sounds by resonance. Sometimes we find special dilatations of the trachea, which are apparently subservient to the same purpose. The actions of the vocal apparatus are regulated in the Singing-birds by at least five pairs of muscles, besides those which increase or diminish the length of the trachea, a change which is of considerable importance in modifying the *pitch* of the notes. In those Birds which can imitate the articulate sounds of the Human voice, such as the Parrot, the Raven, and the Mocking-bird, the tongue is employed in their production.

839. In *Mammals* the same 'larynx' is made to answer the double purpose of regulating the ingress and egress of air, and of producing vocal tones. There are few, if any, of this class, which have not the power of producing some vocal sound; and the structure of their larynx generally corresponds pretty closely with that of Man, which will be briefly described as an example of the most complete form of vocal apparatus.—The Larynx is built-up, as it were, upon the *Cricoid* cartilage (Fig. 320, x *w r u*), which surmounts the trachea, and which might be considered as its highest ring, modified in form, its depth from above downwards being much greater

posteriorly than anteriorly. This is embraced, as it were, by the Thyroid cartilage (G E H); which is articulated to the sides of the Cricoid by its lower horns, round the extremities of which it may be considered to rotate, as on a pivot. In this manner, the front of the *Thyroid* cartilage may be lifted up, or depressed, by the muscles which act upon it; whilst the position of its posterior part is but little changed. Upon the upper surface of the back of the Cricoid cartilage, are seated the two small *Arytenoid* cartilages (N F); these are so tied to the cricoid by a bundle of strong ligaments (B B), as to have a sort of rotation upon an articulating surface, which enables them to be approximated or separated from each other,—their inner edges being nearly parallel in the first case, but slanting away from each other in the second. To the summit of these cartilages are attached the *Chordæ Vocales*, or vocal ligaments (T U), composed of yellow fibrous or elastic tissue. These stretch across to the front of the Thyroid cartilage; and it is upon their condition and relative situation, that the absence or the production of vocal tones, and all their modifications of pitch, depend. They are rendered tense by the depression of the front of the Thyroid cartilage, and relaxed by its elevation; by which action the *pitch* of the tones is regulated. But for the production of any vocal tones whatever, they must be brought into a nearly parallel condition, by the mutual approximation of the points of the arytenoid cartilages to which they are attached; whilst in the intervals of vocalization, these are separated, so that the *rima glottidis*, or fissure between the chordæ vocales, assumes the form of a narrow V, with its point directed backwards. Thus there are two sets of movements concerned in the act of vocalization;—the regulation of the relative position of the Vocal Cords, which is effected by the movements of the Arytenoid cartilages;—and the regulation of their tension, which is determined by the movements of the Thyroid cartilage. The Arytenoid cartilages are made to diverge from one another by means of the *Crico-arytenoidei postici* of the two sides (N *l*, N *l*), which proceed from their outer corners, and turn somewhat round the edge of the Cricoid, to be attached to the lower part of its back; their action is to draw the outer corners of the Arytenoid cartilages outwards and downwards, so that the points to which the vocal ligaments are attached are separated from one another, and the rima glottidis is thrown open. The action of these muscles is antagonized by that of the *Arytenoideus transversus*, which draws together the Arytenoid cartilages; and by that of the

FIG. 320.

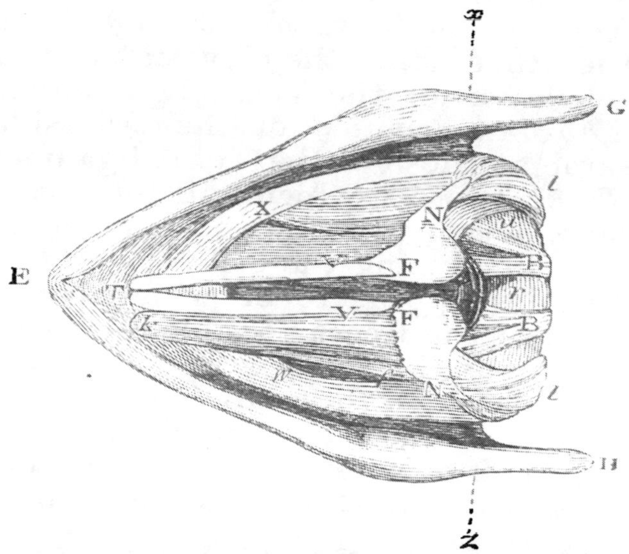

Bird's-eye view of *Human Larynx* from above:—G E H, the thyroid cartilage, embracing the ring of the cricoid *r u* X *w*, and turning upon the axis *x z*, which passes through the lower horns; N F, N F, the arytenoid cartilages, connected by the arytenoideus transversus; T V, T V, the vocal ligaments; N X, the right crico-arytenoideus lateralis (the left being removed); V *k f*, the left thyro-arytenoideus (the right being removed); N *l*, N *l*, the crico-arytenoidei postici; B B, the crico-arytenoid ligaments.

Crico-arytenoidei laterales of the two sides (N X), which run forwards and downwards from the outer corners of the Arytenoid cartilages, and tend by their contraction to bring together their anterior points, to which the Vocal ligaments are attached.—The depression of the front of the Thyroid cartilage, and the consequent tension of the Vocal ligaments, is occasioned by the conjoint action of the *Crico-thyroidei* of the two sides, which occasions the Thyroid and Cricoid cartilages to rotate, the one upon the other, at the articulation formed by the inferior cornua of the former; and this action will be assisted by the *Sterno-thyroidei*, which tend to depress the front of the Thyroid cartilage, by pulling from a fixed point below. On the other hand, the elevation of the front of the Thyroid cartilage, and the relaxation of the Vocal ligaments, are effected by the contraction of the *Thyro-arytenoidei* of the two sides (V *k f*), whose attachments are the same as those of the Vocal ligaments themselves; and this is aided by the *Thyro-hyoidei*, which will tend to draw up the front of the Thyroid cartilage, acting from a fixed point above.

840. During the ordinary acts of inspiration and expiration, the Chordæ Vocales appear to be widely separated from each other, and to be in a state of the freest possible relaxation. In order to produce a vocal sound, they must be made to approach one another, and their inner faces must be brought into parallelism; both of which ends are accomplished by the rotation of the Arytenoid cartilages: whilst, at the same time, they must be put into a certain degree of tension, by the depression of the Thyroid cartilage. Both of these movements take place consentaneously, and are mutually adapted to each other; the vocal ligaments being approximated, and the rima glottidis consequently narrowed, at the same time that their tension is increased.—It has been fully proved by the researches of Willis, Müller, and others, that the action of the Vocal ligaments, in the production of sound, bears no resemblance to that of vibrating strings; and that it is not comparable to that of the mouth-piece of the flute-pipes of the Organ; but that it is, in all essential particulars, the same with that of the 'reeds' of the Hautboy or Clarionet, or the 'tongues' of the Accordion or Concertina. An 'artificial larynx' has been constructed on this principle, which may be made to produce sounds very similar to the vocal tones of Man. Its general arrangement may be understood from Fig. 321; in which C is the pipe for the passage of air, D a ring at its summit for the attachment of the flexible vibrating plates, B its long and narrow orifice, A a pin that serves as a fixed point from which the tension may take place, whilst E, F are two bits of cork glued to the corners of the vibrating plates, by which they may be more conveniently moved and strained, so as to bring the edges of the slit G H near together and into parallelism, and to regulate their tension.—The loudness of the voice is

FIG. 321.

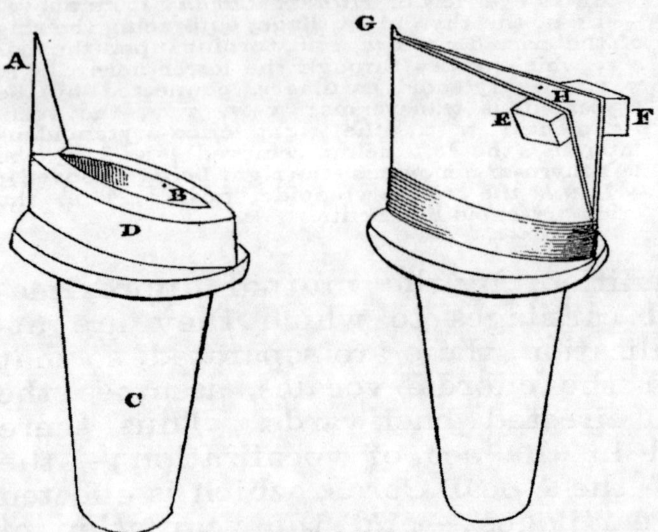

Artificial Larynx.

often increased by some special apparatus of resonance. This is particularly the case with the 'Howling Monkeys' of America, whose larynx possesses several pouches opening from it, one of which is excavated in the substance of the hyoid bone itself. Although these Monkeys are of inconsiderable size, yet their voices are louder than the roaring of lions; that of a single individual is distinctly audible at a distance of two miles; and when a number of them are congregated together, the effect is terrific.

841. The actions of the Larynx are among the most interesting examples of the 'automatic' operation of Volition (§ 806). For the will cannot influence the state of contraction of any of the vocalizing muscles, *except in the act of vocalization;* and it is requisite for the performance of this act, that the tone to be produced should have been previously conceived (however momentarily) in the mind, so that this conception takes the place of the guiding sensation, in regulating the actions of the muscles which produce it. When this cannot be formed, in consequence of congenital absence of the sense of hearing, the power of producing vocal tones can only be acquired by attention to the muscular sensations; and the result of this is very imperfect.—The vocal sounds produced by the action of the larynx, are of very different characters; and may be distinguished into the *cry,* the *song,* and the ordinary or acquired *voice.* The *cry* is generally a sharp sound, having little modulation or accuracy of pitch, and being usually disagreeable in its timbre or quality. It is that by which animals express their unpleasing emotions, especially pain or terror; and the Human infant, like many of the lower animals, can utter no other sound.—In *song,* by the regulation of the vocal cords, definite and sustained musical tones are produced, which can be changed or modulated at the will of the individual. Different species of Birds have their respective songs; which are partly instinctive, and partly acquired by education. In Man, the power of song is entirely acquired; but some individuals possess a much greater facility in acquiring it than others; this superiority appearing to depend on their more precise conception of the tones to be sounded, and on their power of more ready imitation, as well as on differences in the construction of the larynx itself. The larynx of an accomplished vocalist, obedient to the expression of the emotions, as well as to the dictates of the will, may be said to be the most perfect musical instrument ever constructed.—The *voice* is a sound more resembling the cry, in regard to the absence of any sustained musical tone; but it differs from the cry, both in the quality of its tone, and in the modulation of which it is capable by the will. The power of producing *articulate* sounds, from the combination of which *Speech* results, is altogether independent of the Larynx; being due to the action of the muscles of the mouth, tongue, and palate. Distinctly articulate sounds may be produced without any vocal or laryngeal tone, as when we *whisper;* and it has been experimentally shown, that the only condition necessary for this mode of speech, is the propulsion of a current of air through the mouth, from back to front. On the other hand, we may have the most perfect laryngeal tone, without any articulation; as in the production of musical sounds, not connected with words.

CONCLUSION.

It scarcely appears fitting to bring to a close this general survey of the Organised Creation, without the remark, that if little has been said, in the course of it, upon the Evidences of Design presented by the structure of Living Beings, it is because it has been thought, that, when the perfect adaptation which exists between all their minute details, and the harmony of the parts they have to perform in the grand system of the Universe, were being explained and demonstrated, it might be safely left to the mind of the reader to draw those inferences, which it is perhaps impossible for any soundly-judging person to avoid making, who is unwarped by the pride of human reason, or by that tendency to practical disregard of them, which, in so many instances, is mistaken for a valid argument on the side of disbelief. When we consider the universality of this adaptation, so constant that it cannot be the effect of chance,—and the consummate harmony of the whole result, so immeasurably transcending the highest efforts of human genius,—it seems scarcely possible to arrive at any other rational conclusion, than that the Universe, with all that it contains, is the work of One Almighty and Benevolent Mind.

All our Science, then, is but an investigation of the mode in which the Creator acts; its highest 'laws' are but expressions of the mode in which He manifests His agency to us. And when the Physiologist is inclined to dwell unduly upon his capacity for penetrating the secrets of Nature, it may be salutary for him to reflect, that, even should he succeed in placing his department of study upon a level with those Physical Sciences, in which the most complete knowledge of 'causation' (using that term in the sense of 'unconditional sequence') has been acquired, and in which the highest generalizations have been attained, he is still as far as ever from being able to comprehend that Power, which is the 'efficient cause' alike of the simplest and most minute, and of the most complicated and most majestic phenomena of the Universe. But when Man shall have passed through this embryo state, and shall have undergone that metamorphosis, in which everything whose purpose was temporary shall be thrown aside, and his permanent or immortal essence shall alone remain, then, we are encouraged to believe, his finite mind will be brought into nearer connection with the Infinite, and his highest aspirations after Truth, Beauty, and Goodness will be gratified by the disclosure of their Source, and by the increase of his power of approach towards it. The Philosopher who has attained the highest summit of mortal wisdom, is he who, if he use his faculties aright, has the clearest perception of the limits of human knowledge, and the most earnest desires for the lifting of that veil which separates him from the Unseen. He, then, has the strongest motives for that humility of spirit and purity of heart, without which, we are assured, none shall see God.

INDEX.

The Numbers refer to the Paragraphs.

The Numbers refer to the Paragraphs.

The Numbers refer to the Paragraphs.

The Numbers refer to the Paragraphs.

The Numbers refer to the Paragraphs.

D.

The Numbers refer to the Paragraphs.

E.

The Numbers refer to the Paragraphs.

F.

The Numbers refer to the Paragraphs.

The Numbers refer to the Paragraphs.

The Numbers refer to the Paragraphs.

M.

The Numbers refer to the Paragraphs.

N.

The Numbers refer to the Paragraphs.

The Numbers refer to the Paragraphs.

The Numbers refer to the Paragraphs.

The Numbers refer to the Paragraphs.

S.

The Numbers refer to the Paragraphs.

The Numbers refer to the Paragraphs.

The Numbers refer to the Paragraphs.

THE END.